Errata and Corrigenda number 3 : January 1975 have been entered
number 4 : January 1976
number 5 : January 1977
number 6 : January 1978
number 7 : January 1979
number 8 : January 1980
number 9 : January 1981

SPECIFIC HEAT

Nonmetallic Solids

THERMOPHYSICAL PROPERTIES OF MATTER
The TPRC Data Series

A Comprehensive Compilation of Data by the
Thermophysical Properties Research Center (TPRC), Purdue University

Y. S. Touloukian, Series Editor
C. Y. Ho, Series Technical Editor

Volume 1. Thermal Conductivity–Metallic Elements and Alloys
Volume 2. Thermal Conductivity–Nonmetallic Solids
Volume 3. Thermal Conductivity–Nonmetallic Liquids and Gases
Volume 4. Specific Heat–Metallic Elements and Alloys
Volume 5. Specific Heat–Nonmetallic Solids
Volume 6. Specific Heat–Nonmetallic Liquids and Gases
Volume 7. Thermal Radiative Properties–Metallic Elements and Alloys
Volume 8. Thermal Radiative Properties–Nonmetallic Solids
Volume 9. Thermal Radiative Properties–Coatings
Volume 10. Thermal Diffusivity
Volume 11. Viscosity
Volume 12. Thermal Expansion–Metallic Elements and Alloys
Volume 13. Thermal Expansion–Nonmetallic Solids

New data on thermophysical properties are being constantly accumulated at TPRC. Contact TPRC and use its interim updating services for the most current information.

THERMOPHYSICAL PROPERTIES OF MATTER
VOLUME 5

SPECIFIC HEAT

Nonmetallic Solids

Y. S. Touloukian
Director
Thermophysical Properties Research Center
and
Distinguished Atkins Professor of Engineering
School of Mechanical Engineering
Purdue University
and
Visiting Professor of Mechanical Engineering
Auburn University

E. H. Buyco
Assistant Professor of Engineering
Purdue University
Calumet Campus
Formerly
Assistant Senior Researcher
Thermophysical Properties Research Center
Purdue University

IFI/PLENUM • NEW YORK-WASHINGTON • 1970

Library of Congress Catalog Card Number 73-129616

SBN (13-Volume Set) 306-67020-8

SBN (Volume 5) 306-67025-9

IFI/Plenum Data Corporation is a subsidiary of
Plenum Publishing Corporation
227 West 17th Street, New York, N.Y. 10011

Distributed in Europe by Heyden & Son, Ltd.
Spectrum House, Alderton Crescent
London N.W. 4, England

Printed in the United States of America

"In this work, when it shall be found that much is omitted, let it not be forgotten that much likewise is performed..."

SAMUEL JOHNSON, A.M.

From last paragraph of Preface to his two-volume *Dictionary of the English Language,* Vol. I, page 5, 1755, London, Printed by Strahan.

Foreword

In 1957, the Thermophysical Properties Research Center (TPRC) of Purdue University, under the leadership of its founder, Professor Y. S. Touloukian, began to develop a coordinated experimental, theoretical, and literature review program covering a set of properties of great importance to science and technology. Over the years, this program has grown steadily, producing bibliographies, data compilations and recommendations, experimental measurements, and other output. The series of volumes for which these remarks constitute a foreword is one of these many important products. These volumes are a monumental accomplishment in themselves, requiring for their production the combined knowledge and skills of dozens of dedicated specialists. The Thermophysical Properties Research Center deserves the gratitude of every scientist and engineer who uses these compiled data.

The individual nontechnical citizen of the United States has a stake in this work also, for much of the science and technology that contributes to his well-being relies on the use of these data. Indeed, recognition of this importance is indicated by a mere reading of the list of the financial sponsors of the Thermophysical Properties Research Center; leaders of the technical industry of the United States and agencies of the Federal Government are well represented.

Experimental measurements made in a laboratory have many potential applications. They might be used, for example, to check a theory, or to help design a chemical manufacturing plant, or to compute the characteristics of a heat exchanger in a nuclear power plant. The progress of science and technology demands that results be published in the open literature so that others may use them. Fortunately for progress, the useful data in any single field are not scattered throughout the tens of thousands of technical journals published throughout the world. In most fields, fifty percent of the useful work appears in no more than thirty or forty journals. However, in the case of TPRC, its field is so broad that about 100 journals are required to yield fifty percent. But that other fifty percent! It is scattered through more than 3500 journals and other documents, often items not readily identifiable or obtainable. Nearly 50,000 references are now in the files.

Thus, the man who wants to use existing data, rather than make new measurements himself, faces a long and costly task if he wants to assure himself that he has found all the relevant results. More often than not, a search for data stops after one or two results are found—or after the searcher decides he has spent enough time looking. Now with the appearance of these volumes, the scientist or engineer who needs these kinds of data can consider himself very fortunate. He has a single source to turn to; thousands of hours of search time will be saved, innumerable repetitions of measurements will be avoided, and several billions of dollars of investment in research work will have been preserved.

However, the task is not ended with the generation of these volumes. A critical evaluation of much of the data is still needed. Why are discrepant results obtained by different experimentalists? What undetected sources of systematic error may affect some or even all measurements? What value can be derived as a "recommended" figure from the various conflicting values that may be reported? These questions are difficult to answer, requiring the most sophisticated judgment of a specialist in the field. While a number of the volumes in this Series do contain critically evaluated and recommended data, these are still in the minority. The data are now being more intensively evaluated by the staff of TPRC as an integral part of the effort of the National Standard Reference Data System (NSRDS). The task of the National Standard Reference Data System is to organize and operate a comprehensive program to prepare compilations of critically evaluated data on the properties of substances. The NSRDS is administered by the National Bureau of Standards under a directive from the Federal Council for Science and Technology, augmented by special legislation of the Congress of the United States. TPRC is one of the national resources participating in the National Standard Reference Data System in a united effort to satisfy the needs of the technical community for readily accessible, critically evaluated data.

As a representative of the NBS Office of Standard Reference Data, I want to congratulate Professor Touloukian and his colleagues on the accomplishments represented by this Series of reference data books. Scientists and engineers the world over are indebted to them. The task ahead is still an awesome one and I urge the nation's private industries and all concerned Federal agencies to participate in fulfilling this national need of assuring the availability of standard numerical reference data for science and technology.

EDWARD L. BRADY
Associate Director for Information Programs
National Bureau of Standards

Preface

Thermophysical Properties of Matter, the TPRC Data Series, is the culmination of twelve years of pioneering effort in the generation of tables of numerical data for science and technology. It constitutes the restructuring, accompanied by extensive revision and expansion of coverage, of the original *TPRC Data Book*, first released in 1960 in loose-leaf format, 11″ × 17″ in size, and issued in June and December annually in the form of supplements. The original loose-leaf *Data Book* was organized in three volumes: (1) metallic elements and alloys, (2) nonmetallic elements, compounds, and mixtures which are solid at N.T.P., and (3) nonmetallic elements, compounds, and mixtures which are liquid or gaseous at N.T.P. Within each volume, each property constituted a chapter.

Because of the vast proportions the *Data Book* began to assume over the years of its growth and the greatly increased effort necessary in its maintenance by the user, it was decided in 1967 to change from the loose-leaf format to a conventional publication. Thus, the December 1966 supplement of the original *Data Book* was the last supplement disseminated by TPRC.

While the manifold physical, logistic, and economic advantages of the bound volume over the loose-leaf oversize format are obvious and welcome to all who have used the unwieldy original volumes, the assumption that this work will no longer be kept on a current basis because of its bound format would not be correct. Fully recognizing the need of many important research and development programs which require the latest available information, TPRC has instituted a *Data Update Plan* enabling the subscriber to inquire, by telephone if necessary, for specific information and receive, in many instances, same-day response on any new data processed or revision of published data since the latest edition. In this context, the TPRC Data Series departs drastically from the conventional handbook and giant multivolume classical works, which are no longer adequate media for the dissemination of numerical data of science and technology without a continuing activity on contemporary coverage. The loose-leaf arrangements of many works fully recognize this fact and attempt to develop a combination of bound volumes and loose-leaf supplement arrangements as the work becomes increasingly large. TPRC's *Data Update Plan* is indeed unique in this sense since it maintains the contents of the TPRC Data Series current and live on a day-to-day basis between editions. In this spirit, I strongly urge all purchasers of these volumes to complete in detail and return the *Volume Registration Certificate* which accompanies each volume in order to assure themselves of the continuous receipt of annual listing of corrigenda during the life of the edition.

The TPRC Data Series consists initially of 13 independent volumes. The initial ten volumes will be published in 1970, and the remaining three by 1972. It is also contemplated that subsequent to the first edition, each volume will be revised, updated, and reissued in a new edition approximately every fifth year. The organization of the TPRC Data Series makes each volume a self-contained entity available individually without the need to purchase the entire Series.

The coverage of the specific thermophysical properties represented by this Series constitutes the most comprehensive and authoritative collection of numerical data of its kind for science and technology.

Whenever possible, a uniform format has been used in all volumes, except when variations in presentation were necessitated by the nature of the property or the physical state concerned. In spite of the wealth of data reported in these volumes, it should be recognized that all volumes are not of the same degree of completeness. However, as additional data are processed at TPRC on a continuing basis, subsequent editions will become increasingly more complete and up to date. Each volume in the Series basically comprises three sections, consisting of a text, the body of numerical data with source references, and a material index.

The aim of the textual material is to provide a complementary or supporting role to the body of numerical data rather than to present a treatise on the subject of the property. The user will find a basic theoretical treatment, a comprehensive presentation of selected works which constitute reviews, or compendia of empirical relations useful in estimation of the property when there exists a paucity of data or when data are completely lacking. Established major experimental techniques are also briefly reviewed.

The body of data is the core of each volume and is presented in both graphical and tabular format for convenience of the user. Every single point of numerical data is fully referenced as to its original source and no secondary sources of information are used in data extraction. In general, it has not been possible to critically scrutinize all the original data presented in these volumes, except to eliminate perpetuation of gross errors. However, in a significant number of cases, such as for the properties of liquids and gases and the thermal conductivity of all the elements, the task of full evaluation, synthesis, and correlation has been completed. It is hoped that in subsequent editions of this continuing work, not only new information will be reported but the critical evaluation will be extended to increasingly broader classes of materials and properties.

The third and final major section of each volume is the material index. This is the key to the volume, enabling the user to exercise full freedom of access to its contents by any choice of substance name or detailed alloy and mixture composition, trade name, synonym, etc. Of particular interest here is the fact that in the case of those properties which are reported in separate companion volumes, the material index in each of the volumes also reports the contents of the other companion volumes.* The sets of companion volumes are as follows:

Thermal conductivity:	Volumes 1, 2, 3
Specific heat:	Volumes 4, 5, 6
Radiative properties:	Volumes 7, 8, 9
Thermal expansion:	Volumes 12, 13

The ultimate aims and functions of TPRC's Data Tables Division are to extract, evaluate, reconcile, correlate, and synthesize all available data for the thermophysical properties of materials with the result of obtaining internally consistent sets of property values, termed the "recommended reference values." In such work, gaps in the data often occur, for ranges of temperature, composition, etc. Whenever feasible, various techniques are used to fill in such missing information, ranging from empirical procedures to detailed theoretical calculations. Such studies are resulting in valuable new estimation methods being developed which have made it possible to estimate values for substances and/or physical conditions presently unmeasured or not amenable to laboratory investigation. Depending on the available information for a particular property and substance, the end product may vary from simple tabulations of isolated values to detailed tabulations with generating equations, plots showing the concordance of the different values, and, in some cases, over a range of parameters presently unexplored in the laboratory.

The TPRC Data Series constitutes a permanent and valuable contribution to science and technology. These constantly growing volumes are invaluable sources of data to engineers and scientists, sources in which a wealth of information heretofore unknown or not readily available has been made accessible. We look forward to continued improvement of both format and contents so that TPRC may serve the scientific and technological community with ever-increasing excellence in the years to come. In this connection, the staff of TPRC is most anxious to receive comments, suggestions, and criticisms from all users of these volumes. An increasing number of colleagues are making available at the earliest possible moment reprints of their papers and reports as well as pertinent information on the more obscure publications. I wish to renew my earnest request that this procedure become a universal practice since it will prove to be most helpful in making TPRC's continuing effort more complete and up to date.

It is indeed a pleasure to acknowledge with gratitude the multisource financial assistance received from over fifty of TPRC's sponsors which has made the continued generation of these tables possible. In particular, I wish to single out the sustained major support being received from the Air Force Materials Laboratory–Air Force Systems Command, the Office of Standard Reference Data–National Bureau of Standards, and the Office of Advanced Research and Technology–National Aeronautics and Space Administration. TPRC is indeed proud to have been designated as a National Information Analysis Center for the Department of Defense as well as a component of the National

*For the first edition of the Series, this arrangement was not feasible for Volume 7 due to the sequence and the schedule of its publication. This situation will be resolved in subsequent editions.

Standard Reference Data System under the cognizance of the National Bureau of Standards.

While the preparation and continued maintenance of this work is the responsibility of TPRC's Data Tables Division, it would not have been possible without the direct input of TPRC's Scientific Documentation Division and, to a lesser degree, the Theoretical and Experimental Research Divisions. The authors of the various volumes are the senior staff members in responsible charge of the work. It should be clearly understood, however, that many have contributed over the years and their contributions are specifically acknowledged in each volume. I wish to take this opportunity to personally thank those members of the staff, research assistants, graduate research assistants, and supporting graphics and technical typing personnel without whose diligent and painstaking efforts this work could not have materialized.

Y. S. TOULOUKIAN

Director
Thermophysical Properties Research Center
Distinguished Atkins Professor of Engineering

Purdue University
Lafayette, Indiana
July 1969

Introduction to Volume 5

This volume of *Thermophysical Properties of Matter*, the TPRC Data Series, was initiated in recent years and follows the general format of the Center's work on thermal conductivity.

The volume comprises three major sections: the front text material together with its bibliography, the main body of numerical data and its references, and the material index.

The text material is intended to assume a role complementary to the main body of numerical data, the presentation of which is the primary purpose of this volume. It is felt that a concise discussion of the theoretical nature of the property under consideration together with a review of predictive procedures and recognized experimental techniques will be appropriate in a major reference work of this kind. The extensive reference citations given in the text should lead the interested reader to a highly comprehensive literature for a detailed study. It is hoped, however, that enough detail is presented for this volume to be self-contained for the practical user.

The main body of the volume consists of the presentation of numerical data compiled over the years in a most comprehensive and meticulous manner. The scope of coverage includes most nonmetallic materials of engineering importance which are in the solid state at normal temperature and pressure. The extraction of all data directly from their original sources ensures freedom from errors of transcription. Furthermore, some gross errors appearing in the original source documents have been corrected. The organization and presentation of the data together with other pertinent information in the use of the tables and figures are discussed in detail in the text of the section entitled *Numerical Data*.

It is regrettable that the authors have not yet had the time to review and evaluate critically the extensive data compiled in this volume. However, it is hoped that the user will be able to exercise proper selectivity and discretion among conflicting sets of data based on the extensive information reported for each set in the accompanying specification tables.

As stated earlier, all data have been obtained from their original sources and each data set is so referenced. TPRC has in its files all documents cited in this volume. Those that cannot be readily obtained elsewhere are available from TPRC in microfiche form.

The material index at the end of this volume covers the contents of all three companion volumes (Volumes 4, 5, and 6) on specific heat. It is hoped that the user will find these comprehensive indices helpful.

This work has grown out of activities made possible principally through the support of the Air Force Materials Laboratory–Air Force Systems Command, under the monitorship of Mr. John H. Charlesworth. In the preparation of this volume we have drawn most heavily upon the scientific literature and hence we feel a debt of gratitude to the authors of the referenced articles.

While this volume is primarily intended as a reference work for the designer, researcher, experimentalist, and theoretician, the teacher at the graduate level may also use it as a teaching tool to point out to his students the topography of the state of knowledge on the specific heat of nonmetals. We believe there is also much food for reflection by the specialist and the academician concerning the meaning of "original" investigation and its "information content."

The authors are keenly aware of the possibility of many weaknesses in a work of this scope. We hope that we will not be judged too harshly and that we will receive suggestions regarding references omitted, additional material groups needing more detailed treatment, improvements in presentation, and, most important, any inadvertent errors. If the *Volume Registration Certificate* accompanying this volume is returned, the reader will assure himself of receiving annually a list of corrigenda as possible errors come to our attention.

Lafayette, Indiana
July 1969

Y. S. Touloukian
E. H. Buyco

Contents

Material Index

GROUPING OF MATERIALS AND LIST OF FIGURES AND TABLES

1. ELEMENTS

2. OXIDES

2. OXIDES (continued)

2. OXIDES (continued)

3. ANTIMONIDES

4. ARSENIDES

5. BERYLLIDES

6. BORIDES

6. BORIDES (continued)

7. CARBIDES

8. GERMANIDES

9. IODIDES

10. PHOSPHIDES

11. SELENIDES

12. SILICIDES

12. SILICIDES (continued)

13. SULFIDES

13. SULFIDES (continued)

14. TELLURIDES

15. BROMIDES

16. CHLORIDES

16. CHLORIDES (continued)

17. FLUORIDES

17. FLUORIDES (continued)

18. HYDRIDES

19. NITRIDES

20. CARBONATES

21. NITRATES and NITRITES

22. SULFATES

22. SULFATES (continued)

23. GLASSES and CERMETS

24. OXYGEN COMPOUNDS

24. OXYGEN COMPOUNDS (continued)

24. OXYGEN COMPOUNDS (continued)

Theory, Estimation, and Measurement

Notation

A	Grüneisen constant; Cross-sectional area
a	Lattice constant: Empirical constant
b	Empirical constant
c, C	Heat capacity of mass m, specific heat per unit mass
C_a, C_f	Constant which depends on particular type of lattice and on crystal structure, respectively
C_e	Electronic specific heat
C_p, C_v	Specific heat at constant pressure and constant volume, respectively
d	Density
e	Base of natural logarithm, 2.71828
E	Total energy of an oscillator, particle, or system; Internal energy; Voltage
H	Enthalpy
$(\Delta H)_f$	Heat of fusion
h	Planck constant, 6.6262×10^{-27} erg sec
I	Electrical current
J, J'	Quantum mechanical exchange constants
K	Calibration factor in ice drop calorimeter
k	Boltzmann constant, 1.3806×10^{-16} erg K^{-1}
L	Linear dimension
m	Mass of a particle, system, or specimen
m_e	Mass of an electron
n	Integer, 0, 1, 2, 3, . . .
N_A	Avogadro's number, 6.0222×10^{23} g-mol^{-1}
N_e	Number of electrons per gram atom
p	Momentum of a particle; Pressure of a gas
q	Direction coordinate from equilibrium position
Q	Amount of heat absorbed or removed from the system
R	Gas constant, 8.3143 J K^{-1} g-mol^{-1}
s	Spin vector
T	Temperature, K
t	Time
V	Volume
v	Specific volume
W	Work done on or by the system
x, x_m	$h\nu/kT$ and $h\nu_D/kT$, respectively, as used in equation (17)
X_i	Atomic mole or mass fraction of ith component in an alloy or mixture
α, α_f	Coefficient of thermal linear expansion, and a constant which depends on crystal structure, respectively
β	Coefficient of isobaric volumetric expansion; Constant in Debye cube law
γ	Constant in the electronic specific heat relation (26)
θ_D, θ_E	Characteristic Debye temperature and Einstein temperature, $h\nu_D/k$ and $h\nu/k$, respectively
ν	Frequency of oscillation of a particle
ν_D	Debye frequency
ω	Natural angular frequency
ρ	Electrical resistivity
ρ_e	Number of free electrons per unit volume
ϵ	Energy of an oscillation
π	Mathematical constant, 3.14159 . . .
κ_T	Isothermal compressibility, as used in equation (36)

Theory of Specific Heat of Solids

1. INTRODUCTION

Rapid advances in the frontiers of science and technology have brought about a general realization of the fact that the present limitations in many technical developments are a direct result of inadequate knowledge of the thermophysical properties of materials. In the high-temperature range ($T > 1000$ K), interest in the determination of specific heats of materials has been hastened because of the requirements in space programs as well as industrial applications. The need for data at high temperatures has advanced our knowledge in many areas of solid state studies such as lattice vibrations, energy levels in magnetic solids, electronic distributions, and many other atomic and molecular phenomena.

The measurement of specific heat at cryogenic temperatures ($C_p \cong C_v$ for $T \leq 4$ K) provides us with a direct means to test theoretical models of a system. For instance, precise specific heat measurements were needed to test the validity of Debye's and Einstein's theory for specific heat of solids at low temperatures. Finally, knowledge of accurate specific heat data at low temperature is very useful in studies of cryogenic techniques.

2. DEFINITIONS

When a quantity of heat Q is added to a system so that there is a change in temperature, $T_2 - T_1$, then the mean heat capacity of the mass m of the substance is defined by

$$\bar{c} = \frac{Q}{T_2 - T_1} \tag{1}$$

The limiting value of the above ratio as the temperature changes by dT is defined as the true heat capacity, i.e.,

$$c = \frac{đQ^*}{dT} \tag{2}$$

*$đQ$ is used instead of dQ to indicate that it is not an exact differential.

In order to obtain a quantity that is independent of the mass, m, of a substance, equation (2) is divided by m; i.e.,

$$C = \frac{c}{m} = \frac{đQ}{m\,dT} \tag{3}$$

The quantity q represents the amount of heat per unit mass, so that equation (3) may also be written as

$$C = \frac{đq}{dT} \tag{4}$$

Raising the temperature of a unit mass of a substance by an amount dT, however, does not define the process in a thermodynamic sense; for instance, it will take a different amount of heat $đq$ if the process is at constant pressure than when the process is at constant volume. As a matter of fact there are an infinite number of different processes for a system at temperature T to change to a temperature $T + dT$. It is clear, therefore, that an infinite number of specific heats could also be defined for a substance. The two processes that are most commonly used in thermodynamics are those at constant volume and constant pressure. For these two processes equation (4) may be written

$$C_p = \left(\frac{đq}{dT}\right)_p \tag{5}$$

and

$$C_v = \left(\frac{đq}{dT}\right)_v \tag{6}$$

Experimentally, the values of the specific heat measured are either at constant pressure, C_p, or at constant volume, C_v. The units most commonly used for specific heat are cal g^{-1} K^{-1}, Btu lb^{-1} F^{-1}, joules kg^{-1} K^{-1}. The units for molar or atomic specific heat are cal g-mol^{-1} K^{-1}, Btu lb-mol^{-1} F^{-1}, joules kg-mol^{-1} K^{-1}, cal g-atom^{-1} K^{-1}, joules kg-atom^{-1} K^{-1}, etc.

3. DULONG AND PETIT'S LAW

In 1819 Dulong and Petit [9] published the results of their measurements on the specific heat at constant pressure of thirteen solid elements at room temperature. From these measurements, they observed that the product of the specific heat at constant pressure and the atomic weight was approximately a constant, about 6 cal g-atom^{-1} K^{-1}. Subsequent researches, extending from 1840 to 1862, revealed the general applicability of the Dulong and Petit's law to several metallic elements, when the specific heat at constant pressure was determined at temperatures sufficiently below their melting point but not far below room temperature. During the same period an important extension of Dulong and Petit's law was applied to chemical compounds, i.e., the molar specific heat of a compound is equal to the sum of the atomic specific heats of its constituent elements. This law which is generally referred to as the Kopp–Neumann law [32] has also been applied to predict the atomic specific heat of alloys. For alloys, the atomic specific heat is equal to the sum of the product of the atomic specific heat of each constituent element and its atomic fraction. If an alloy consists of elements 1, 2, 3, . . . , n, with atomic fraction $X_1, X_2, X_3, \ldots, X_n$ and atomic specific heat $C_{p1}, C_{p2}, C_{p3}, \ldots, C_{pn}$, then the atomic specific heat of the alloy is

$$C_p = \sum_{i=1}^{n} X_i C_{p_i} \tag{7}$$

Equation (7) should be applied with caution for alloys especially near magnetic and phase transitions. Bottema and Jaeger [5] have applied the Kopp–Neumann law to the alloy Ag_3Au and they found that the experimental data on the specific heat at constant pressure of this alloy agree closely with the calculated values between 0 C to 400 C. Between 400 C and 800 C, the values obtained from the Kopp–Neumann law were 0.5 percent to 1.8 percent higher than the experimental results. Buyco [46] calculated the specific heat of the alloys of aluminum, beryllium, nickel, and iron between 300 K to 1000 K and found the calculated values agree with the experimental data to within 5 percent.

The theoretical justification of the law of Dulong and Petit was demonstrated by Boltzmann in 1871. The results obtained previously by Dulong and Petit also follow from Boltzmann's equipartition of energy theorem. Complete and detailed derivation of this theorem is discussed elsewhere [15, 20, 21, 33, and 43].

The following is a brief exposition. The energy of a linear harmonic oscillator consists of kinetic and potential energies, i.e.,

$$E = \frac{p^2}{2m} + \frac{m\omega^2 q^2}{2} \tag{8}$$

where p is the momentum, m is the mass, ω is the natural angular frequency, q is the distance from equilibrium position, and E is the total energy of an oscillator. From the theorem of equipartition of energy [15, 20, 21, 31], each degree of freedom contributes $(kT/2)$ to the energy of a particle in equilibrium. A three-dimensional oscillator which has six degrees of freedom will therefore have an internal energy of $3\,kT$ at thermal equilibrium. A gram-atom of an element has N_A atoms; hence, the internal energy is $3N_AkT$. The specific heat at constant volume is obtained by differentiating the internal energy with respect to temperature at constant volume, i.e.,

$$\left(\frac{\partial E}{\partial T}\right)_v = C_v = 3N_A k \tag{9}$$

where N_A is the Avogadro constant and k is the Boltzmann constant. The product of Avogadro constant and Boltzmann constant is equal to the gas constant R. Therefore:

$$C_v = 3R \cong 5.96 \text{ cal mol}^{-1}\text{ K}^{-1}$$

Hence, the Dulong and Petit value of about 6 cal mol^{-1} deg^{-1} for the specific heat of metallic solids can be accounted for on the basis of classical statistical mechanics. However, the observation of Dulong and Petit was short lived. In 1875 Weber [48] showed that the atomic specific heat of silicon, boron, and carbon are considerably lower than the values predicted by Dulong and Petit. For example, the atomic specific heat of crystalline silicon, boron, and diamond were found to be 4.8, 2.7, and 1.8 cal mol^{-1} deg^{-1}, respectively, at room temperature. Subsequent specific heat measurements at low temperatures ($T < 300$ K) revealed that the specific heat of solids increased rapidly with temperature and almost leveled off about their Debye temperature. Classical theory does not explain this behavior for solids. It should also be noted that classical theory encounters the same difficulty in the behavior of molar specific heats.

4. EINSTEIN'S SPECIFIC HEAT THEORY

Einstein [10] proposed a simple model to account

for the decrease in the specific heat at low temperatures below the value $3R$ per mole which was obtained at elevated temperatures. His oversimplified physical model considers the thermal properties of the vibrations of a lattice of N_A atoms as a set of $3N_A$ independent harmonic oscillators in one dimension, each with the same frequency, ν. He then quantized the energy of the oscillators in accordance with the results obtained by Planck. According to Planck, a harmonic oscillator does not have a continuous energy spectrum but can accept energy values equal to an integer times $h\nu$, where ν is the frequency of oscillations and h is the Planck constant. Hence the possible energy levels of an oscillator may be given by

$$\epsilon = nh\nu \qquad n = 0, 1, 2, 3, \ldots$$

The average energy of an oscillator at temperature T, according to the well known Planck formula [7, 20, 21, 32], is

$$\bar{\epsilon} = \frac{h\nu}{\exp(h\nu/kT) - 1} \tag{10}$$

In Einstein's model the vibrational energy of a solid element containing N_A atoms is $3N_A$ times the average energy of an oscillator, i.e.,

$$\bar{E} = 3N_A \frac{h\nu}{\exp(h\nu/kT) - 1} \tag{11}$$

The results obtained from quantum mechanics however showed that the average energy of an oscillator [7, 15] should be written as

$$\bar{\epsilon} = \frac{h\nu}{2} + \frac{h\nu}{\exp(h\nu/kT) - 1} \tag{12}$$

instead of as in equation (10).

The result obtained for the specific heat by differentiating equation (10) is the same as that obtained from equation (12). In any case the specific heat for one atom of an element is

$$\left(\frac{\partial E}{\partial T}\right)_v = C_v = \frac{3N_A k(h\nu/kT)^2 \exp(h\nu/kT)}{[\exp(h\nu/kT) - 1]^2} \tag{13}$$

For convenience, the characteristic Einstein temperature defined by $\theta_E = h\nu/k$ may be introduced in equation (13) to obtain

$$C_v = \frac{3R(\theta_E/T)^2 \exp(\theta_E/T)}{[\exp(\theta_E/T) - 1]^2} \tag{14}$$

In the high-temperature range with $T \gg \theta_E$ [15, 20, 21, 32], equation (14) upon expansion in power series becomes

$$C_v \simeq 3R\left[1 - \frac{1}{12}\left(\frac{\theta_E}{T}\right)^2\right] \tag{15}$$

When the value of $[(\theta_E/T)^2/12]$ is such that it is very much smaller than 1, then Einstein's theory yields the classical Dulong and Petit value of 6 cal mol^{-1} deg^{-1}.

In the low-temperature region $T \ll \theta_E$, equation (14) may be written approximately as

$$C_v \simeq 3R\left(\frac{\theta_E}{T}\right)^2 \exp(-\theta_E/T) \tag{16}$$

According to equation (16), the low-temperature specific heat of solids should approach zero exponentially. Experimental evidence indicates that C_v approaches zero more slowly than this. The reason for the discrepancy between Einstein's theoretical prediction and the experimental results may be explained on the basis of the assumption made in the theory that each atom in a solid vibrates independently of the others but with precisely the same frequency. However, in spite of the weakness in Einstein's theory, his pioneering work opened the way for the application of quantum theory to the specific heat of solids.

5. DEBYE'S SPECIFIC HEAT THEORY

From the point of view of the wave whose wavelength is large compared with the interatomic distances, a crystal may appear like a continuum. The fundamental assumption of Debye [6] is that the continuum model may be employed for all possible vibrational modes of the crystal. Debye has given a limit to the total number of vibrational modes equal to $3N_A$, where N_A is the number of atoms in a gram atom of an element. In this case, the frequency spectrum which corresponds to an ideal continuum is cut off in order to comply with a total of $3N_A$ modes. This procedure should provide a maximum frequency ν_D (Debye frequency) which is common to both the longitudinal and transverse modes. By associating with each vibrational mode a harmonic oscillator of the same frequency, Debye obtained the following expression [7, 15, 20, 21, 32] for the vibrational energy:

$$\bar{E} = 9N_A h\nu_D \left(\frac{kT}{h\nu_D}\right)^4 \int_0^{x_m} \frac{x^3\,dx}{e^x - 1} \tag{17}$$

where

$$x = h\nu/kT \qquad x_m = h\nu_D/kT$$

Clearly, when $T \gg \theta_D$, x_m is small compared with unity for the whole integration range. In this case $e^x - 1 \cong x$ so that equation (17) could easily be integrated to obtain the expression

$$\bar{E} \cong 3N_A kT \tag{18}$$

Then

$$\left(\frac{\partial \bar{E}}{\partial T}\right)_v = C_v = 3N_A k = 3R \cong 6 \text{ cal mol}^{-1} \text{ deg}^{-1}$$

a result agreeing with classical theory.

At very low temperatures, $T \ll \theta_D$, the upper limit of integration in equation (17) may be replaced by infinity since $h\nu/kT \to \infty$ as $T \to 0$. It is now possible to integrate equation (17) as follows [51]

$$\int_0^\infty \frac{x^3\,dx}{e^x - 1} = 6\sum_1^\infty \frac{1}{n^4} = \frac{\pi^4}{15} \tag{19}$$

Hence

$$\bar{E} = \frac{3}{5}\pi^4 N_A kT\left(\frac{T}{\theta_D}\right)^3 \tag{20}$$

and

$$C_v = \left(\frac{\partial \bar{E}}{\partial T}\right)_v = \frac{12}{5}\pi^4 N_A k\left(\frac{T}{\theta_D}\right)^3 \tag{21}$$

or

$$C_v = \frac{12}{5}\pi^4 R\left(\frac{T}{\theta_D}\right)^3 \tag{22}$$

For one atom or one mole of a substance, $R = 1.987$ cal mol^{-1} deg^{-1} so that equation (22) may be written as

$$C_v = 464.5\left(\frac{T}{\theta_D}\right)^3 \text{ cal mol}^{-1} \text{ deg}^{-1} \qquad T < \left(\frac{\theta_D}{50}\right) \tag{23}$$

Debye's theory predicts a cube law dependence of the specific heat of the elements for temperatures $T < (\theta_D/10)$. The range of validity of this law [15] has now been restricted to $T < (\theta_D/50)$ as a result of more recent theoretical work on specific heat studies. The predictions of Debye's theory agree quite well with experimental values of the specific heat of solids and is a definite improvement over Einstein's work.

Due to improved calorimetric measurements at low temperatures ($T < 5$ K), in recent years accurate specific heat values revealed that Debye's equation for C_v does not fit the experimental results precisely. Furthermore, it was observed that θ_D, which according to Debye's theory is a constant, did in fact vary with temperature. The deficiency of the Debye theory may be explained on the basis of the approximation made in treating solids as a continuous elastic media and neglecting the discreteness of the atoms.

Further improvements on Debye's theory was developed by Born and Karman [4]. They calculated the frequency spectrum by considering the lattice modes of vibration for a particular crystal structure under investigation. The method is involved so that one is referred to the original work [4] for detailed discussion.

6. ELECTRONIC SPECIFIC HEAT

In 1900, Drude [8] suggested a model for a free-electron theory of metals. He assumed that metals contain free electrons in thermal equilibrium with the atoms of the solid. He further assumed that the potential energy of the free electrons is equal to the product of the number of electrons per unit volume and the average energy of an electron. The essential feature in the problem is the determination of the number of electrons with energy between E and $E + dE$. Classical theory using Maxwell–Boltzmann statistics [2, 8, 15, 20, 21, 32, 43], would give an expression for the electronic specific heat as

$$C_e = \tfrac{3}{2}N_e k \tag{24}$$

Using Fermi–Dirac statistics [7, 15, 19, 20, 21, 31, 32], the following expression for the electronic specific heat may be obtained at low temperatures:

$$C_e = \pi^2 R(2m_e k/h^2)\left(\frac{\pi}{3\rho_e}\right)^{2/3} T \tag{25}$$

or simply

$$C_e = \gamma T \tag{26}$$

where ρ_e is the number of free electrons per unit volume, γ is the proportionality constant, T is the absolute temperature, N_e is the number of electrons per gram atom, m_e is the mass of an electron, k is the Boltzmann constant, h is the Planck constant, R is the gas constant, and C_e is the electronic specific heat.

The specific heat of metals below the Debye temperature and "very much" below the Fermi temperature [15, 19, 20, 21, 32] may be expressed as the sum of the electronic specific heat and the lattice specific heat, i.e.,

$$C_v = \gamma T + \beta T^3 \quad (27)$$

Indeed, this relationship has been verified by accurate low temperature specific heat measurements. At sufficiently low temperature ($T < 1$ K) the electronic specific heat is dominant, while at high temperatures the lattice contribution is predominant.

7. MAGNETIC SPECIFIC HEAT

There are two types of materials that exhibit a magnetic contribution to the total specific heat: namely, the ferromagnetic and the ferrimagnetic materials.

A ferromagnet is a material [7, 15, 20, 21, 32] that contains a spontaneous magnetic moment. This means that this material possesses a magnetic moment even in the absence of an external magnetic field. This type of material exhibits a magnetic ordering with parallel alignment of adjacent spins. A ferromagnetic material has a Curie temperature, T_c, which is defined as the temperature above which magnetization disappears, and the material becomes paramagnetic. The Curie temperature separates the ordered ferromagnetic phase from the disordered paramagnetic phase.

An antiferromagnet is a material [7, 15, 20, 21, 32], that has spins which are ordered in an antiparallel arrangement. There is no net magnetic moment at temperatures below the Néel temperature. Hysteresis is usually observed and a sharp maximum in the susceptibility curve is exhibited. Above the Néel temperature, the spins are said to be free, and the material becomes paramagnetic. In some ways ferrimagnetic materials are similar to the ferromagnetic materials except that in the former the adjacent spins are unequal and antiparallel. The Néel temperature may be defined for ferrimagnetic material as the temperature separating the ordered ferrimagnetic phase from the disordered paramagnetic phase.

For ferri- and ferromagnets, the internal energy [7, 15, 20, 21, 32], is given by the expression

$$\bar{E} = 4\pi V(2\alpha_f J s a^2)\left(\frac{kT}{2\alpha_f J s a^2}\right)^{5/2} \int_0^x \frac{x^4\, dx}{e^{x^2} - 1} \quad (28)$$

At low temperatures the upper limit for x may be taken equal to infinity and hence the integral may be easily determined. Differentiating equation (28) gives the magnetic specific heat [15]

$$C_M = \frac{d\bar{E}}{dT} = C_f N_A k \left(\frac{kT}{2Js}\right)^{3/2} \quad (29)$$

where α_f and C_f are constants which depend upon crystal structure, a is the lattice constant, J is the quantum mechanical exchange constant, k is the Boltzmann constant, N_A is the Avogadro number, s is the magnitude of the spin vector, and V is the volume of the material.

Equation (29) shows that at low temperatures the ferromagnetic contribution to the specific heat is proportional to the three-halves power of the absolute temperature. For metals which are ferromagnetic [15], the total specific heat is equal to the sum of the electronic, lattice, and magnetic terms, i.e.,

$$C_v = \gamma T + \beta T^3 + \delta T^{3/2} \quad (30)$$

For ferrimagnets, which are electrical insulators, [15], the electronic term is negligible compared with the other terms, so that the total specific heat may be given by the expression

$$C_v = \beta T^3 + \delta T^{3/2} \quad (31)$$

Both sides of equation (31) may be divided by $T^{3/2}$ to give

$$C_v / T^{3/2} = \beta T^{3/2} + \delta \quad (32)$$

A plot of $C_v/T^{3/2}$ versus $T^{3/2}$ should give a straight line with slope β and intercept δ.

For the case of antiferromagnetic materials [15], the expressions for the mean internal energy is

$$\bar{E} = 4\pi V(2\alpha_a J' s a^2)\left(\frac{kT}{2\alpha_a J' s a^2}\right)^4 \int_0^x \frac{x^3\, dx}{e^x - 1} \quad (33)$$

The upper limit for integration may be taken as equal to infinity at low temperatures so that differentiation of equation (33) gives the magnetic specific heat [15, 28]

$$C_M = C_a N_A k \left(\frac{kT}{2J's}\right)^3 \quad (34)$$

where C_a is a constant which depends upon the type of lattice and J' is the magnitude of the exchange constant.

The striking difference between the contributions to the specific heat exhibited by ferromagnets and ferrimagnets is the $T^{3/2}$ dependence in the former and T^3 dependence in the latter. Hence for antiferromagnetic materials, the temperature dependence is of the same form as the Debye's T^3 formula. The separation of the spin wave contribution from the lattice specific heat in antiferromagnetic materials is indeed very difficult.

8. LOW-TEMPERATURE SPECIFIC HEAT

The specific heat of solids is ordinarily measured at constant pressure. The specific heat at constant volume is that which is obtained if the interatomic distance is kept constant as the temperature changes. The specific heat at constant volume, C_v, may be assumed to be approximately equal to the specific heat at constant pressure, C_p, at cryogenic temperatures. At high temperatures, $C_p > C_v$. This difference is obtained from the classical thermodynamic relations

$$C_p - C_v = -T\left(\frac{\partial V}{\partial T}\right)_p^2 \Big/ \left(\frac{\partial V}{\partial p}\right)_T \quad (35)$$

From the definition of the isothermal compressibility

$$\kappa_T = -\left(\frac{\partial V}{\partial p}\right)_T \Big/ V \quad (36)$$

and the isobaric coefficient of volumetric expansion

$$\beta = \left(\frac{\partial V}{\partial T}\right)_p \Big/ V \quad (37)$$

Using equations (36) and (37), equation (35) may be written as

$$C_p - C_v = \frac{TV\beta^2}{\kappa_T} \quad (38)$$

By rearranging equation (38), this may also be written as

$$C_p - C_v = \left(\frac{V\beta^2}{\kappa_T C_p^2}\right) C_p^2 T = AC_p^2 T \quad (39)$$

where

$$A = \frac{V\beta^2}{\kappa_T C_p^2}$$

The parameter A is called the Grüneisen constant, which is actually only approximately constant [15] over a wide range of temperature. If A is calculated at any one temperature from values of V, β, and κ_T, it may be used [15, 20, 21, 32] to calculate $C_p - C_v$ over a wide range of temperature without introducting a serious error.

For isotropic substances, the isothermal coefficient of volumetric expansion may be written in terms of the coefficient of linear expansion

$$\beta = \left(\frac{\partial V}{\partial T}\right)_p \Big/ V = 3\left[\left(\frac{\partial L}{\partial T}\right)_p \Big/ L\right] = 3\alpha \quad (40)$$

Hence, from equation (38)

$$C_p - C_v = \frac{9\alpha^2}{\kappa_T} TV = \left(\frac{9V\alpha^2}{\kappa_T C_p^2}\right) C_p^2 T \quad (41)$$

where

$$A = \frac{9V\alpha^2}{\kappa_T C_p^2}$$

In the absence of contributions from magnetic and nuclear specific heat, the expression for C_v for most metals has been shown [15, 20, 21, 32] to be

$$C_v = \gamma T + \beta T^3 \quad (27)$$

where γT is the electronic contribution and βT^3 is the lattice contribution. For nonmetals, the electronic contribution may be very small compared with the lattice term so that

$$C_v = \beta T^3 \quad (42)$$

When the nuclear quadrupole moment interacts with the electronic field gradient of the lattice and the electron, then the total specific heat of the substance is given as

$$C_v = \gamma T + \beta T^3 + \alpha T^{-2} \quad (43)$$

where αT^{-2} is the nuclear contribution to the total specific heat.

9. NORMAL AND SUPERCONDUCTING MATERIALS

At a certain critical temperature (superconducting temperature), several materials exhibit superconducting behavior [15, 20, 21, 32]. Below this temperature, the specific heat of a superconducting material is found to depart significantly from the values obtained for a normally behaving material. It is also found that if an external magnetic field of sufficient strength is applied while the specific heat of the material is being measured, the values obtained correspond to what the normal values would be. Hence, the specific heat values obtained experimentally in the presence of sufficient external magnetic field below the superconducting critical temperature are referred to as the normal specific heat (C_N) while the values obtained in the absence of a magnetic field are referred to as superconducting specific heat (C_S). For example, the critical superconducting temperatures of aluminum and niobium are approximately 1.196 K and 9.22 K, respectively.

Other Major Sources of Data

There exists in the literature a number of reference sources which, while less extensive in scope, may nevertheless prove valuable to the reader. While it is not the intent here to cite every available review, it is felt that the following works, listed in chronological order, are of particular significance. One should note that most of the citations do not present critical evaluation of the data they report.

Furukawa, Saba, and Reilly [12] report on the critical analysis of the thermodynamic properties of copper, silver, and gold between 0 and 300 K. A tabulation is given for the values of specific heat C_p, enthalpy $H - H_0^0$, entropy S^0, Gibbs energy $G - H_0^0$, enthalpy function $(H - H_0^0)/T$ and Gibbs energy function $(G - H_0^0)/T$. The report also contains a comparison of the values of the electronic coefficient of the specific heat and the 0 K limiting Debye characteristic temperature with their selected values. An appraisal of low-temperature calorimetry is also given.

Touloukian [44] edited a handbook entitled *Thermophysical Properties of High Temperature Solid Materials* consisting of nine books totaling more than 8500 pages. The properties covered in the handbook are density, melting point, heat of fusion, heat of vaporization, heat of sublimation, electrical resistivity, specific heat at constant pressure, thermal conductivity, thermal diffusivity, thermal linear expansion, thermal radiative properties (absorptance, emittance, reflectance, and transmittance), and vapor pressure. Generally, only materials with melting points above 800 K are included, except for materials within the categories of polymers, plastics, and composites.

Touloukian, Gerritsen, and Moore [45], *Thermophysical Properties Research Literature Retrieval Guide*, consisting of a set of three books, contains references for 33,700 research documents on thermophysical properties of matter. The properties covered are thermal conductivity, specific heat at constant pressure, viscosity, thermal radiative properties (emissivity, absorptivity, reflectivity, transmissivity), optical constants (total and spectral), diffusion coefficient, thermal diffusivity, and Prandtl number. This publication supersedes the earlier works of this series (Volume I, 1960 and Volume II, 1963), and constitutes an enlarged and consolidated definitive work reporting the total literature through June 1964.

Schick [29] edited a comprehensive work entitled *Thermodynamics of Certain Refractory Compounds*. Volume 2 of this work includes thermodynamic properties of borides, carbides, nitrides, and oxides of 31 elements in the temperature range from 0 to 6000 K. Over 160 thermodynamic tables, together with comprehensive discussions, are presented.

Moeller et al.'s [24] compilation on *Thermophysical Properties of Thermal Insulating Materials* should prove useful in cryogenic and high temperature applications. The properties included in this compilation are thermal conductivity, thermal linear expansion, specific heat, total normal emittance, thermal diffusivity, compressive strength, density, melting point, and modulus of elasticity. Various experimental methods for determining thermal properties are described and their accuracies are indicated.

Wood and Deem [52] report on the compilation of specific heat, thermal linear expansion, and thermal conductivity data for materials of possible structural usefulness above 1500 K. Data are presented graphically with notations as to measurement methods and test conditions.

Hultgren, Orr, Anderson, and Kelley [16] published their book on the *Selected Values of Thermodynamic Properties of Metals and Alloys* in 1963. This book presents in tabular form heat capacity, enthalpy, entropy, free energy function, and vapor pressure. In some cases the heat of fusion, melting point, and other transition temperatures are also given. For the binary alloys, phase diagrams are included.

Eldridge and Deem [11] issued a report under the auspices of the Data and Publication Panel of

ASTM–ASME joint committee on effects of temperature on the properties of metals. The metals covered are Al, Co, Fe, Mg, Mo, Ni, and their alloys. The properties included are thermal conductivity, thermal linear expansion, specific heat, electrical resistivity, density, emissivity, diffusivity, and magnetic permeability. Emphasis is given to data over a range from cryogenic (2 K) to elevated temperatures (2800 K).

Johnson [17] edited a compendium of the properties of materials at low temperatures. The first phase of the compendium covers properties of ten fluids (Part I), properties of solids (Part II), and an extensive bibliography of references (Part III). The properties covered are density, expansivity, thermal conductivity, specific heat, enthalpy, heats of transition, phase equilibria, dielectric constants, adsorption, surface tension, and viscosity for solid, liquid, and gas phases of He, H_2, Ne, N_2, O_2, air, CO, F_2, A, and NH_3. Data sheets, primarily in graphic form, are presented for "best values" of data collected. The sources of the materials used, other references, and tables of selected values with appropriate comments are furnished with each data sheet.

Kelley's [18] bulletin contains the then-available high-temperature specific heat data for the elements and inorganic compounds. The thermodynamic properties are listed in tables and algebraic expressions for their representations are also given.

Stull and Sinke [40] published their well-known reference work on the *Thermodynamic Properties of the Elements* in 1956. This book reports specific heat as well as thermodynamic property values for the elements in their condensed and gaseous state. A search of the literature was made by the authors through 1955. Whenever experimental data were not available, reasonable estimates were made in order to fill the gaps in information. A tabulation of thermodynamic values from 298.15 K to 3000 K is given for the elements.

Methods for the Measurement of the Specific Heat of Solids

1. INTRODUCTION

There are few methods for the practical and precise determination of the specific heat of solids. Although many variants and minor modifications or improvements are reported in the various references cited in this section, the most important ones are described in detail in reference [54]. References [55] to [61] also constitute major works on calorimetry including various specialized applications.

The primary methods for the measurement of the specific heat of solids which are commonly used are the method of mixtures or drop method, adiabatic method, comparative method, pulse-heating method, and modifications of these. A number of specific calorimetric techniques are briefly described in this section.

The method of mixtures [14, 37, 50] is widely employed for measuring specific heats of solids above room temperature. This method frequently gives accurate results in a temperature range where no phase transition exists. The usual method consists of dropping the substance under investigation from a furnace temperature into a calorimeter (at room or ice temperature) and the quantity that is obtained directly is the change in enthalpy. Heat capacities are obtained from these values by differentiation, i.e., $C_p = (\partial H/\partial T)_p$. This method is inherently not suitable for use with substances which undergo phase transitions over the temperature range of interest or whose specific heat is highly temperature sensitive.

Various methods of obtaining directly the true specific heat based on the Nernst calorimeter [38, 42, 47, 49] have been used successfully in obtaining precise data in the temperature range below room temperature. Attempts to use this method at moderately high temperatures have not produced accurate results because of heat exchange with the surroundings. This method involves the measurement of energy required to raise the temperature of the substance over small temperature intervals from a fraction of a degree to a few degrees.

2. NERNST-TYPE ADIABATIC VACUUM CALORIMETER

A typical adiabatic vacuum calorimeter consists of a block over which an insulated coil of platinum wire is wound. The block may be either a solid sample under investigation or a container for the solid sample. The block is suspended by leads in a vacuum-tight container. The container is cooled in a dewar containing liquid air, hydrogen, or helium, depending on the temperature range involved. At the start of the operation, the vacuum-tight container is filled with helium gas at very low pressure while the block is cooled to the bath temperature by heat transfer through the helium gas. After the block has been cooled, the gas is removed by pumping and a known amount of heat is applied to the platinum coil by means of electric current for a given time interval. The temperature rise of the block is measured by means of a suitable resistance thermometer. The specific heat is then determined from the measured heat input and temperature change of the sample. Improved versions of the Nernst-type adiabatic calorimeter are described by Taylor and Smith [42], Wallace *et al.* [47], and Westrum [49].

The calorimeter assembly which is discussed by Wallace *et al.* [47] consisted of the sample container, the thermal shields, the outer jacket with associated radiation shields, and the vacuum system. Figure 1 presents a schematic diagram of the calorimeter.

3. MODIFIED ADIABATIC CALORIMETER

A modification of the direct method has been applied successfully by Schmidt and Leidenfrost [30] to obtain the specific heats of powders and granular materials from 273 K to 773 K. The determination of specific heats was carried out for Mond Nickel (99.85% Ni) with an accuracy of 0.6 percent.

The theory of the method as employed for a continuously heated adiabatic calorimeter for measuring powders and granular materials is discussed in detail in reference [30].

Consider a calorimeter and sample system with negligible heat loss to the surroundings, then the heat input may be expressed as

$$\frac{dQ}{dt} = mC_p\frac{dT}{dt} + W_c\frac{dT}{dt} \tag{44}$$

where dQ/dt is the heat input per unit time, T is the temperature, t is the time, m is the mass of the specimen, W_c is the thermal constant of calorimeter body and heater element, energy per degree, and C_p is the specific heat of specimen. From equation (44),

$$C_p = \frac{1}{m}\left[\frac{dQ/dt}{dT/dt} - W_c\right] \tag{45}$$

It is desirable to achieve as small a temperature variation as possible if the specific heat is assumed

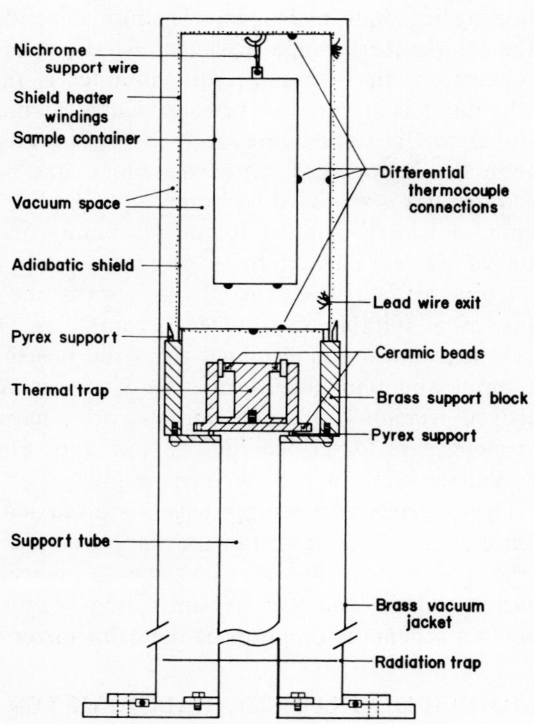

Fig. 1. Schematic diagram for adiabatic specific heat calorimeter [47].

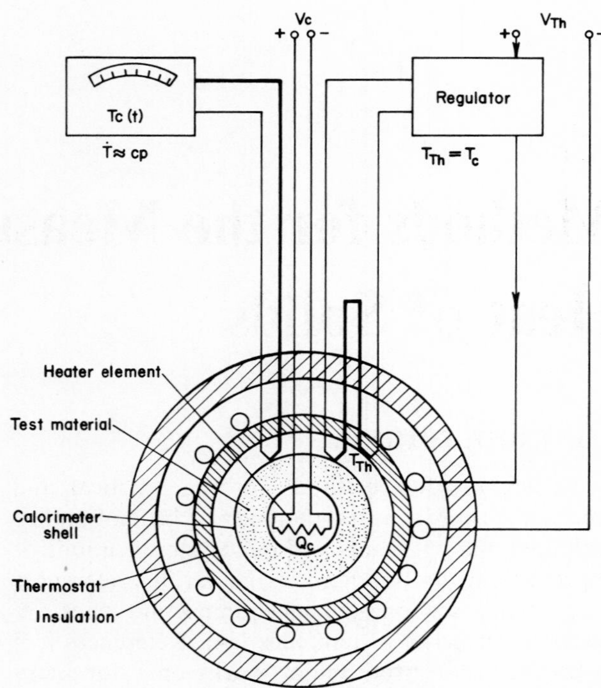

Fig. 2. Schematic diagram for spherical adiabatic calorimeter [30].

constant during each measurement interval. On the other hand, this temperature variation must be large enough to lend itself to precision measurement. The heating must be such that steady-state condition is reached within a reasonable length of time. Schmidt and Leidenfrost [30] have shown that for powders or granular materials of low thermal diffusivity, the following assumptions can be satisfied well enough to yield accurate measurements:

1. The temperature field is dependent only on time and the radial coordinate.
2. The sample is uniformly homogeneous, and its properties are constant over small temperature differences.
3. The sum of the heat capacities of the calorimeter body and its inside heater is small compared with the heat capacity of the sample mass.

The experimental arrangement of the apparatus is shown in schematic form in Fig. 2.

4. DROP ICE CALORIMETER

In this method [13] the heat given off by the sample is used to melt a portion of the ice in an equilibrium ice-water bath and the resulting change

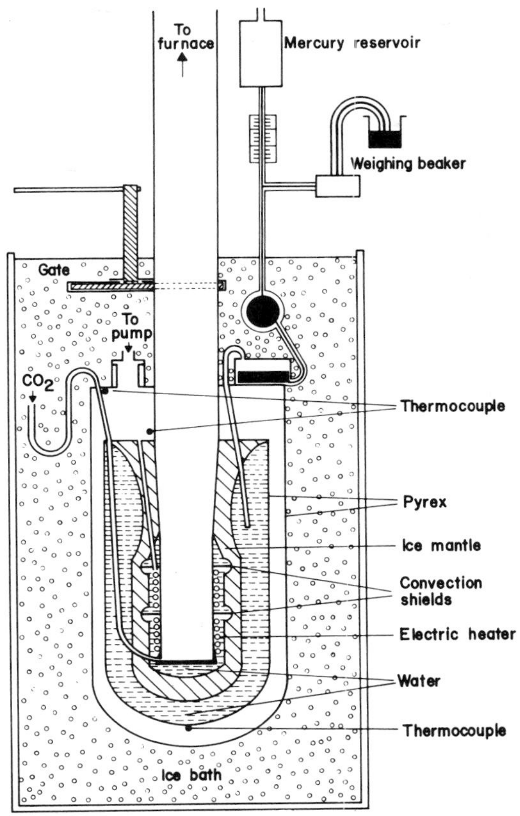

Fig. 3. Schematic diagram for drop ice calorimeter [13].

in volume of the bath is measured by the change in height of a mercury column. The calibration factor for a particular calorimeter (ratio of heat input to mass of mercury displaced by melted ice) is determined from the following expression:

$$K = \Delta H_f/(v_i - v_w)d_m \tag{46}$$

where K is the calibration factor, ΔH_f is the heat of fusion of ice, v_i is the specific volume of ice, v_w is the specific volume of water, and d_m is the density of mercury.

The calibration factor K relates the enthalpy change of the specimen to the height of the mercury column. Values of $(H_T - H_{273.15})$ are then determined for various initial specimen temperatures. These data are either represented graphically or by a suitable empirical relation. The specific heat curve is either derived from the graphically smoothed enthalpy data or from the equation

$$C_p = \frac{d}{dT}(H_T - H_{273.15})_p$$

A schematic drawing of the ice calorimeter is shown in Fig. 3. A central well is provided to receive the specimen whose enthalpy is to be determined. An electric heater, sheathed in a metal tube, is soldered on the outside of the well in order to introduce known amounts of heat for calibration purposes. The lower portion of the well is surrounded by two coaxial glass vessels which provide an insulating space between the inner ice-water system and the surrounding ice bath. Any volume change resulting from the melting of ice in the inner vessel displaces an equivalent volume of mercury and is collected in a beaker and weighed to account for the change in mercury in the calorimeter. A special gate prevents heat transfer from above to the calorimeter along the central well.

5. DROP ISOTHERMAL WATER CALORIMETER

In the drop water calorimeter a sample is heated in the furnace and dropped into the calorimeter

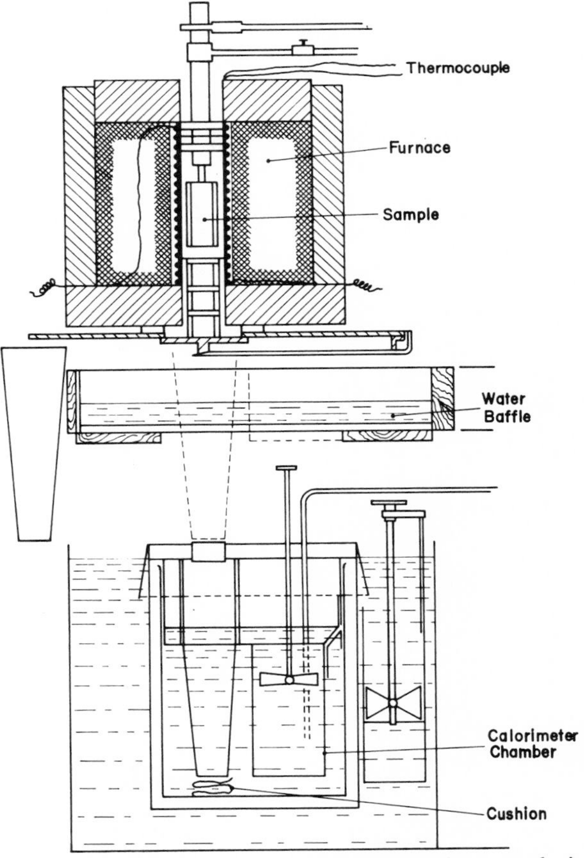

Fig. 4. Schematic diagram for drop isothermal water calorimeter [50].

proper, which consists of a water bath with free air space above. The water in the bath is stirred to assure uniform temperature. The calorimeter is enclosed by an isothermal jacket and the top is covered with copper plates which have a constant temperature because of their high thermal conductivity. The rise in the temperature of the calorimeter is measured with great accuracy by using a Beckmann thermometer or a sensitive thermopile. The enthalpy change of the specimen is determined from the known heat capacity of the calorimeter and its temperature rise. The enthalpy change may be referred to either 273.15 K or 298.15 K. In either case the specific heat is obtained from the smoothed enthalpy data by either graphical or analytical differentiation, i.e.,

$$C_p = \frac{d(H_T - H_{298.15})_p}{dT}$$

A schematic drawing [50] is shown in Fig. 4 to illustrate the details of the apparatus.

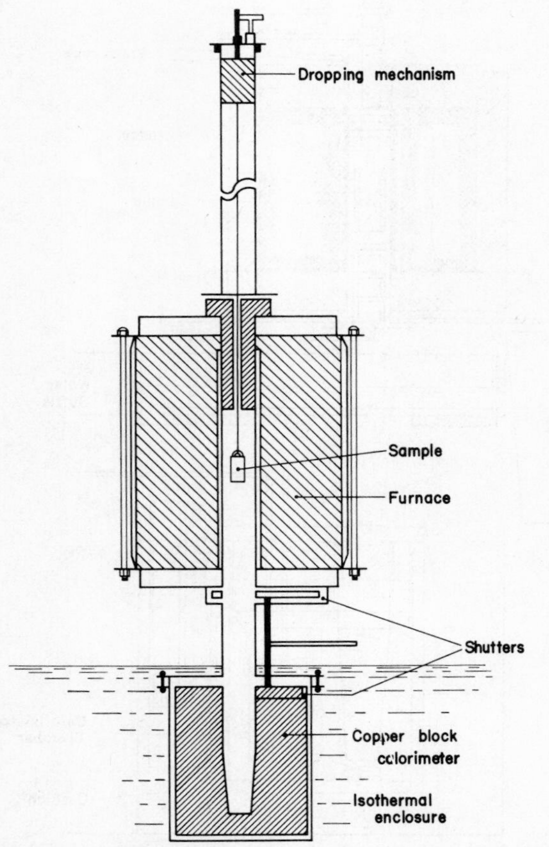

Fig. 5. Schematic diagram for drop isothermal copper block calorimeter [37].

6. DROP COPPER BLOCK CALORIMETER

This drop calorimeter employs a copper block which is submerged in an isothermal oil bath. The temperature of the calorimeter is measured using a special bridge network of copper and manganin resistances. The heat released from the sample is distributed to the copper block because of its high thermal conductivity. Generally it takes some time to achieve uniform heat distribution. The change in enthalpy of the specimen is measured in terms of the amount of heat absorbed by the copper block in changing from its initial temperature to its final temperature. This value is then corrected to 298.15 K so that the tabulated enthalpy values of the specimen are referred to 298.15 K, that is, $H_T - H_{298.15}$. The specific heat as a function of temperature may then be derived from the smoothed enthalpy data obtained either graphically or from the equation

$$C_p = \frac{d}{dT}(H_T - H_{298.15})_p$$

A schematic diagram according to Southard [37] is shown in Fig. 5.

7. PULSE-HEATING METHOD

The pulse-heating method of measuring specific heat is very attractive, particularly for materials that are electrical conductors. This method was first discussed by Avramescu [1] and later modified by other investigators [2, 25, 39, 41]. The method involves the rapid heating of small samples in vacuum. Voltage probes are attached across the central portion of the sample wire which is then mounted in a high-vacuum system. The sample is connected to an electrical circuit consisting of a large storage battery, a variable resistor, a fixed resistor, and a high-current relay controlled by a timing circuit which determines the duration of the pulse. A schematic diagram of a typical circuit [41] for the measurement of specific heat is shown in Fig. 6. The current flowing through the specimen and the voltage drop across the central portion are measured simultaneously as a function of time. The specific resistance at each time interval is calculated from the relationship $\rho = AE/LI$, where A is the cross-sectional area of sample, E is the voltage, I is the current, and L is the distance between voltage probes. This specific electrical resistance is then plotted as a function of time. The specific heat at any temperature T is given by the equation

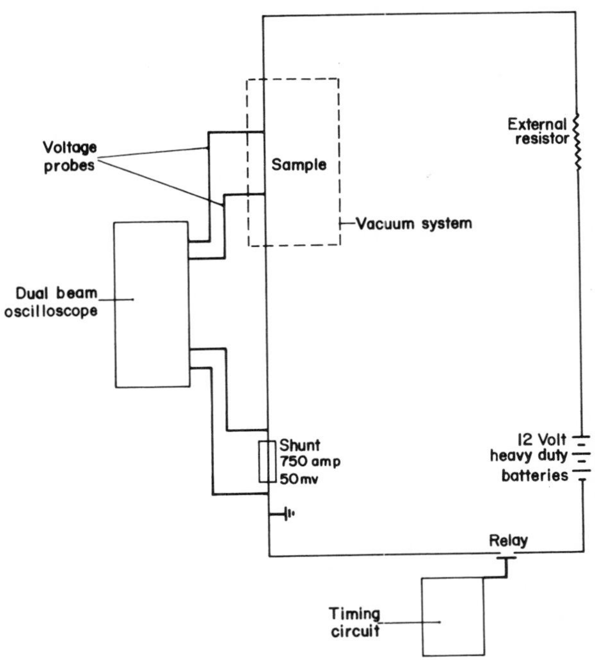

Fig. 6. Schematic diagram of circuit for specific heat measurement using pulse-heating method [41].

$$C_p = \frac{EI(d\rho/dT)}{Jm(d\rho/dt)} \tag{47}$$

where C_p is the specific heat, cal g^{-1} K^{-1}, J is the conversion factor, 4.184 joules cal^{-1}, m is the mass of sample between voltage probes, grams, $d\rho/dT$ is the temperature coefficient of the resistance at temperature T, $d\rho/dt$ is the time rate of change of resistivity at temperature T, and ρ is the electrical resistivity of sample.

8. COMPARATIVE METHOD

The method consists of placing a specimen with its temperature-monitoring thermocouple in a refractory container of low thermal conductivity and in turn placing this in a furnace whose temperature is maintained constant above or below the specimen temperature. The container is calibrated by determining its heating rate when empty and then with a reference sample of known specific heat. Separate electrical heating circuits are usually provided for the specimen and the shield so that their temperature will rise equally and simultaneously in order to reduce heat losses. The specific heat C_{p2} of the unknown specimen is calculated from the following relation:

$$\frac{C_{p2}W_2}{C_{p1}W_1} = \frac{\Delta t_2/\Delta T_2 - \Delta t_r/\Delta T_r}{\Delta t_1/\Delta T_1 - \Delta t_r/\Delta T_r} \tag{48}$$

where $(\Delta t/\Delta T)$ is the slope of a time–temperature curve, and the subscripts r, 1, and 2 represent the empty container, the container with specimen 1, and the container with specimen 2, respectively. The papers by Boggs and Wiebelt [3] and Smith [34] give excellent accounts in the use of this method.

Irreproducible heating or cooling conditions and differences in thermal conductivity between the unknown and reference specimen usually account for the inaccuracies encountered in this method.

References to Text

1. Avramescu, A., "Temperature Variation of the True Specific Heat of Conductivity Copper and Conductivity Aluminum up to the Melting Point," *Z. Tech. Physik* **20**, 213–17, 1939.
2. Baxter, H., "Determination of Specific Heat of Metals," *Nature* **153**, 316, 1944.
3. Boggs, J. H. and Wiebelt, J. A., "An Investigation of a Particular Comparative Method of Specific Heat Determination in the Temperature Range of 1500 F to 2600 F," USAEC TID–5734, 1–91, 1960.
4. Born, M. and Karman, T., "Vibrations in Space Lattices," *Physik Z.* **13**, 297–309, 1912.
5. Bottema, J. A. and Jaeger, F. M., "The Law of Additive Atomic Heats in Intermetallic Compounds," *Proc. Acad. Sci. Amsterdam* **35**, 928–31, 1932.
6. Debye, P., "The Theory of Specific Heat," *Ann. Physik* **39** (4), 789–839, 1912.
7. Dekker, A. J., *Solid State Physics*, Prentice-Hall, Inc., 1–525, 1961.
8. Drude, P., "The Electronic Theory of Metal," *Ann. Physik* **1**, 566–613, 1900.
9. Dulong, P. L. and Petit, A. T., *Ann. Chim.* **10**, 395–413, 1819.
10. Einstein, A., "The Planck's Theory of Radiation and the Theory of Specific Heat," *Ann. Physik* **22** (4), 180–90, 1907.
11. Eldridge, E. A. and Deem, H. W., "Report on Physical Properties of Metals and Alloys from Cryogenic to Elevated Temperatures," ASTM–STP–296, 1–206, 1961.
12. Furukawa, G. T., Saba, W. G., and Reilly, M. L., "Critical Analysis of the Heat-Capacity Data of the Literature and Evaluation of Thermodynamic Properties of Copper, Silver, and Gold from 0 to 300 K," NSRDS–NBS 18, 1–49, 1968.
13. Ginnings, D. C. and Corruccini, R. J., "An Improved Ice Calorimeter—The Determination of its Calibration Factor and the Density of Ice at 0 C," *J. Res. Natl. Bur. Std.* **38**, 583–91, 1947.
14. Ginnings, D. C. and Furukawa, G. T., "Heat Capacity Standards for the Range 14 to 1200 K," *J. Am. Chem. Soc.* **75**, 522–7, 1953.
15. Gopal, E. S. R., *Specific Heats at Low Temperatures*, Plenum Press, 1–111, 1966.
16. Hultgren, R., Orr, R. L., Anderson, P. D., and Kelley, K. K., *Selected Values of Thermodynamic Properties of Metals and Alloys*, John Wiley and Sons, Inc., 1–963, 1963.
17. Johnson, V. J. (Editor), "A Compendium of the Properties of Materials at Low Temperature," WADD–TR–60–56, Pt. 2, 1–333, 1960. [AD 249 786]
18. Kelley, K. K., "Data on Theoretical Metallurgy. XIII. High-Temperature Heat Capacity and Entropy Data for the Elements and Inorganic Compounds," U.S. Bur. Mines Bull. 584, 1–232, 1960.
19. King, A. L., *Thermophysics*, W. H. Freeman and Company, 1–369, 1962.
20. Kittel, C., *Introduction to Solid State Physics*, John Wiley and Sons, Inc., 122–5, 1963.
21. Kittel, C., *Elementary Solid State Physics*, John Wiley and Sons, Inc., 49–52, 1962.
22. Lehman, G. W., "Thermal Properties of Refractory Materials," WADD–TR–60–581, 1–19, 1960. [AD 247 411], [PB 160 804]
23. Levinson, L. S., "High Temperature Drop Calorimeter," *Rev. Sci. Instr.* **36** (6), 639–42, 1962.
24. Moeller, C. E., Loser, J. B., Thompson, M. B., Snyder, W. E., and Hopkins, V., "Thermophysical Properties of Thermal Insulating Materials," ASD–TDR–64–5, 1–362, 1964. [AD 601 535], [N64 22689]
25. Nathan, A. M., "A Dynamic Method for Measuring the Specific Heat of Metals," *J. Appl. Phys.* **22**, 234–5, 1951.
26. Parker, W. J., Jenkins, R. J., Butler, C. P., and Abbott, G. L., "Flash Method of Determining Thermal Diffusivity, Heat Capacity and Thermal Conductivity," *J. Appl. Phys.* **32**, 1679–84, 1961.
27. Reif, F., *Fundamentals of Statistical and Thermal Physics*, McGraw-Hill Book Co., Inc., 1–651, 1965.
28. Sachs, M., *Solid State Theory*, McGraw-Hill Book Co., Inc., 143–68, 1963.
29. Schick, H. L. (Editor), *Thermodynamics of Certain Refractory Compounds, Vol.* 2, Academic Press, 1–775, 1966.
30. Schmidt, E. O. and Leidenfrost, W., "Adiabatic Calorimeter for Measurements of Specific Heats of Powder and Granular Materials at 0 C to 500 C," *ASME 2nd Symp. Thermophysical Properties*, Princeton, N.J., 178–84, 1962.
31. Sears, F. W., *An Introduction to Thermodynamics, The Kinetic Theory of Gases, and Statistical Mechanics*, Addison-Wesley Publishing Co., Inc., 1–373, 1964.
32. Seitz, F., *The Modern Theory of Solids*, McGraw-Hill Book Co., Inc., 38–9, 1940.
33. Slater, J. C., *Introduction to Chemical Physics*, McGraw-Hill Book Co., Inc., 1939.
34. Smith, C. S., "A Simple Method for Thermal Analysis Permitting Quantitative Measurements of Specific and Latent Heats," *Trans. AIME* **137**, 1936, 1940.
35. Smith, D. F., Kaylor, C. E., Walden, G. E., Taylor, A. R., and Gayle, J. B., "Construction, Calibration and Operation of Ice Calorimeter," U.S. Bur. Mines Rept. Invest. 5832, 1–20, 1961.

36. Sommerfeld, A., "The Electronic Theory of Metals," *Naturwiss* **15**, 825–32, 1927.
37. Southard, J. C., "A Modified Calorimeter for High Temperatures. The Heat Content of Silica, Wollastonite, and Thorium Dioxide above 25°," *J. Am. Chem. Soc.* **63**, 3142–6, 1941.
38. Sterrett, K. F., Blackburn, D. A., Bestul, A. B., Chang, S. S., and Horman, J., "An Adiabatic Calorimeter for the Range 10 K to 360 K," *J. Res. Natl. Bur. Std.* **69C**, 19–26, 1965.
39. Strittmater, R. C., Pearson, G. J., and Danielson, G. C., "Measurements of Specific Heats by a Pulse Method," *Proc. Iowa Acad. Sci.* **64**, 466–70, 1957.
40. Stull, D. R. and Sinke, G. C., *Thermodynamic Properties of the Elements*, Am. Chem. Soc., 1–234, 1956.
41. Taylor, R. E. and Finch, R. A., "The Specific Heats and Resistivities of Molybdenum, Tantalum, and Rhenium," *J. Less-Common Metals* **6**, 283–94, 1964.
42. Taylor, A. R. and Smith, D. F., "Construction, Calibration, and Operation of a Low-Temperature Adiabatic Calorimeter," U.S. Bur. Mines Rept. Invest. 5974, 1–17, 1962.
43. Tolman, R. C., *Principles of Statistical Mechanics*, Oxford Univ. Press, London, 1938.
44. Touloukian, Y. S. (Editor), *Thermophysical Properties of High Temperature Solid Materials*, MacMillan Co., Vols. 1, 2, 3, 4, 5, and 6, 1–8500, 1967.
45. Touloukian, Y. S., Gerritsen, J. K., and Moore, N. Y., *Thermophysical Properties Research Literature Retrieval Guide*, Plenum Press, 2nd Ed., Books 1, 2, and 3, 1967.
46. Touloukian, Y. S. (Editor), "Recommended Values of the Thermophysical Properties of Eight Alloys, Major Constituents and Their Oxides," TPRC Rept. 16, 323–46, 1966.
47. Wallace, W. E., Craig, R. S., and Johnston, W. V., "An Adiabatic Calorimeter for the Range 15 C to 290 C," U.S. At. Energy Comm., NYD–6328, 1–16, 1966.
48. Weber, H. F., "The Specific Heat of Elements Carbon, Boron, and Silicon," *Phil. Mag.* **49**, 161–301, 1875.
49. Westrum, E. F., Jr., "Cryogenic Calorimetric Contributions to Chemical Thermodynamics," *J. Chem. Educ.* **39** (9), 443–54, 1962.
50. White, W. P., "Specific Heat Determination at Higher Temperatures," *Am. J. Sci.* **47** (4), 1–59, 1919.
51. Whittaker, E. T. and Watson, G. N., *Modern Analysis*, Cambridge Univ. Press, 4th edition, 1938.
52. Wood, W. D. and Deem, H. W., "Thermal Properties of High-Temperature Materials," RSIC–202, 1–399, 1964. [AD 455 069]
53. Zemansky, M. W., *Heat and Thermodynamics*, McGraw-Hill Book Co., Inc., 1–484, 1957.
54. McCullough, J. P. and Scott, D. W. (Editors), *Experimental Thermodynamics, Volume I, Calorimetry of Non-Reacting Systems*, Plenum Press (New York)/Butterworths (London), 1968.
55. White, W. P., *The Modern Calorimeter*, Chemical Catalog Co., New York, 1928.
56. Swietoslawski, W., *Microcalorimetry*, Reinhold, NewYork, 1964.
57. Calvet, E. and Prat, H., *Microcalorimétrie*, Masson et Cie, Paris, 1956.
58. Roth, W. A. and Becker, F., *Kalorimetrische Methoden zur Bestimmung chemischer Reaktionswärmen*, F. Vieweg, Braunschweig, 1956.
59. Rossini, F. D. (Editor), *Experimental Thermochemistry*, Vol. I, Interscience, New York, 1956.
60. Weissberger, A. (Editor), "*Calorimetry*" *in Technique of Organic Chemistry Vol. I. Physical Methods of Organic Chemistry*, Chap. X, Interscience, New York, 1959.
61. Skinner, H. A. (Editor), *Experimental Thermochemistry*, Vol. II, Interscience, London, 1962.

Numerical Data

Data Presentation and Related General Information

1. SCOPE OF COVERAGE

The materials studied in this volume consist of nonmetallic elements, oxides, and other nonmetallic compounds and mixtures. The nonmetallic elements and compounds are listed in the table of contents in alphabetical order according to chemical name. The data presented are original experimental data on the specific heat of these materials as reported by various investigators. These data were extracted from the world's technical and scientific literature, United States Government Publications, Doctoral and Masters dissertations, data supplied by private companies, and special reports of major research centers throughout the world. The range of temperatures covered is from zero degree Kelvin to the melting point and beyond. For most high-temperature materials, no information is found in the liquid range.

2. PRESENTATION OF DATA

The data for all substances are presented in graphical and tabular form together with a specification table for each substance. The specification table gives the temperature range, the original reference number, the curve number, reported estimates of error, year of publication of the original document, specimen designation, and such other pertinent information as composition or purity of sample, test environment, mechanical, chemical, and thermal history of the test specimen, etc., to the extent provided in the original source document. The data for the specific heat of the materials are plotted on a log–log scale for comparative evaluation. When several sets of data are coincident, the graphical plotting of all of them would lead to confusion. For this reason, some of the sets of data points are omitted from the figures. They are, however, reported in the data tables and specification tables.

The numerical data are presented in double columns. The temperature T is in degrees Kelvin, and the specific heat C_p in calories per gram per degree Kelvin. A unique curve number is assigned to each set of data. This corresponds exactly to the number which also appears in the specification table and on the figure.

The two general types of data that are obtainable from the literature are the true specific heat data obtained directly from the results of measurements using, for instance, the Nernst-type calorimeter and the derived true specific heat data, deduced from direct enthalpy measurements using the drop technique. In the latter type an empirical equation has been fitted by the authors to the enthalpy data by least squares technique and specific heat obtained by differentiation. The results are usually tabulated at rounded temperature intervals.

3. SYMBOLS AND ABBREVIATIONS USED IN THE FIGURES AND TABLES

Symbol	*Definition*	*Units*
T	Temperature	degree Kelvin, K
C_p	Constant pressure specific heat	cal g^{-1} K^{-1}
C_v	Constant volume specific heat	cal g^{-1} K^{-1}
M. P.	Melting point	degree Kelvin, K
T. P.	Transition point	degree Kelvin, K
s. c.	Superconducting	
N	Normal	
c	Cubic	
f.c.c.	Face-centered cubic	
b.c.c.	Body-centered cubic	
h	Hexagonal	
c.p.h.	Close-packed hexagonal	

CONVERSION FACTORS FOR UNITS OF SPECIFIC HEAT

MULTIPLY ↓ by appropriate factor to OBTAIN →	cal$_{th}$ g-mol^{-1} C^{-1}	cal$_{th}$ g^{-1} C^{-1}	cal$_{IT}$ g-mol^{-1} C^{-1}	cal$_{IT}$ g^{-1} C^{-1}	J g-mol^{-1} K^{-1}	J g^{-1} K^{-1}	J kg-mol^{-1} K^{-1}	J kg^{-1} K^{-1}	Btu$_{th}$ lb^{-1} F^{-1}	Btu$_{IT}$ lb^{-1} F^{-1}
cal$_{th}$ g-mol^{-1} C^{-1}	1	1/M	0.999331	0.999331/M	4.184	4.184/M	4.184 x 10^3	4.184/M x 10^3	1/M	0.999331/M
cal$_{th}$ g^{-1} C^{-1}	M	1	0.999331M	0.999331	4.184M	4.184	4.184M x 10^3	4.184 x 10^3	1	0.999331
cal$_{IT}$ g-mol^{-1} C^{-1}	1.00067	1.00067/M	1	1/M	4.1868	4.1868/M	4.1868 x 10^3	(4.1868/M) x 10^3	1.00067/M	1/M
cal$_{IT}$ g^{-1} C^{-1}	1.00067M	1.00067	M	1	4.1868M	4.1868	4.1868M x 10^3	4.1868 x 10^3	1.00067	1
J g-mol^{-1} K^{-1}	0.239006	0.239006/M	0.238846	0.238846/M	1	1/M	1 x 10^3	1 x 10^3/M	0.239006/M	0.238846/M
J g^{-1} K^{-1}	0.239006M	0.239006	0.238846M	0.238846	M	1	M x 10^3	10^3	0.239006	0.238846
J kg-mol^{-1} K^{-1}	2.39006 x 10^{-4}	(2.39006/M) x 10^{-4}	2.38846 x 10^{-4}	(2.38846/M) x 10^{-4}	10^{-3}	10^{-3}/M	1	1/M	(2.39006/M) x 10^{-4}	(2.38846/M) x 10^{-4}
J kg^{-1} K^{-1}	2.39006M x 10^{-4}	2.39006 x 10^{-4}	2.38846M x 10^{-4}	2.38846 x 10^{-4}	M x 10^{-3}	10^{-3}	M	1	2.39006 x 10^{-4}	2.38846 x 10^{-4}
Btu$_{th}$ lb^{-1} F^{-1}	M	1	0.999331M	0.999331	4.184M	4.184	4.184M x 10^3	4.184 x 10^3	1	0.999331
Btu$_{IT}$ lb^{-1} F^{-1}	1.00067M	1.00067	M	1	4.1868M	4.1868	4.1868M x 10^3	4.1868 x 10^3	1.00067	1

Classification of Materials

Classification	Limits of composition (weight percent)*			
	X_1	$X_1 + X_2$	X_2	X_3
1. Nonmetallic elements	>99.5	—	<0.2	<0.2
2. Compounds	>95.0	—	<2.0	<2.0
3. Binary mixtures (or solutions)	—	⩾95.0	⩾2.0	⩽2.0
4. Multiple mixtures (or solutions)	—	⩾95.0	>2.0	>2.0
	—	<95.0	⩾2.0	⩽2.0
	—	<95.0	>2.0	>2.0
	⩽95.0	—	<2.0	<2.0

$^*X_1 \geq X_2 \geq X_3 \geq X_4 \geq \ldots$

4. CONVERSION FACTORS FOR UNITS OF SPECIFIC HEAT

The conversion factors given in the table on page 20a are based upon the following basic definitions:

$1\ \text{lb} = 0.45359237\ \text{kg}$*
$1\ \text{cal}_{\text{th}} = 4.184\ (\text{exactly})\ \text{J}$*
$1\ \text{cal}_{\text{IT}} = 4.1868\ (\text{exactly})\ \text{J}$*
$1\ \text{Btu}_{\text{th}}\ \text{lb}^{-1}\ \text{F}^{-1} = 1\ \text{cal}_{\text{th}}\ \text{g}^{-1}\ \text{C}^{-1}$†
$1\ \text{Btu}_{\text{IT}}\ \text{lb}^{-1}\ \text{F}^{-1} = 1\ \text{cal}_{\text{IT}}\ \text{g}^{-1}\ \text{C}^{-1}$†

The subscripts "th" and "IT" designate "thermochemical" and "International Steam Table," respectively.

5. CLASSIFICATION OF MATERIALS

The classification scheme as shown in the table for nonmetallic solids contained in this volume is based upon the chemical composition of the material. This scheme is mainly for the convenience of material grouping and data organization, and is not intended to be used as definitions for the various material groups.

6. CONVENTION FOR BIBLIOGRAPHIC CITATION

For the following types of documents the bibliographic information is cited in the sequences given below.

Journal Article:

a. Author(s)—The names and initials of all authors are given. The last name is written first, followed by initials.
b. Title of article—In this volume, the titles of the journal articles listed in the *References to Text* are given, but not of those listed in the *References to Data Sources*.
c. Journal title—The abbreviated title of the journal as in *Chemical Abstracts* is given.
d. Series, volume, and number—If the series is designated by a letter, no comma is used between the letter for series and the numeral for volume, and they are underlined together. In case series is also designated by a numeral, a comma is used between the numeral for series and the numeral for volume, and only the numeral representing volume is underlined. No comma is used between the numerals representing volume and number. The numeral for number is enclosed in parentheses.
e. Pages—The inclusive page numbers of the article.
f. Year—The year of publication.

Report:

a. Author(s).
b. Title of report—In this volume, the titles of the reports listed in the *References to Text* are given, but not of those listed in the *References to Data Sources*.
c. Name of the responsible organization.
d. Report, or bulletin, circular, technical note, etc.
e. Number

*National Bureau of Standards, "New Values for the Physical Constants Recommended by NAS–NRC," *NBS Tech. News Bull.* **47**(10), 175–7, 1963.

†Mueller, E. F. and Rossini, F. D., "The Calory and the Joule in Thermodynamics and Thermochemistry," *Am. J. Phys.* **12**(1), 1–7, 1944.

f. Part
g. Pages
h. Year
i. ASTIA's AD number—This is given in square brackets whenever available.

Book:

a. Author(s)
b. Title
c. Volume
d. Edition
e. Publisher
f. Place of publication
g. Pages
h. Year.

7. CRYSTAL STRUCTURES, TRANSITION TEMPERATURES, AND OTHER PERTINENT PHYSICAL CONSTANTS OF THE ELEMENTS

The table on the following pages contains information on the crystal structure, transition temperatures, and certain other pertinent physical constants of each element. This information is very useful in data analysis and synthesis. However, no attempt has been made to critically evaluate the temperatures/constants given in the table and they should not be considered recommended values. This table has an independent series of numbered references which immediately follow the table.

CRYSTAL STRUCTURES, TRANSITION TEMPERATURES, AND OTHER PERTINENT PHYSICAL CONSTANTS OF THE ELEMENTS

Name	Atomic Number	Atomic Weight[a]	Density,[b] kg m$^{-3}\cdot 10^{-3}$	Crystal Structure	Phase Transition Temp., K	Superconducting Transition Temp., K	Curie Temp., K	Néel Temp., K	Debye Temperature at 0 K, K	Debye Temperature at 298 K, K	Melting Point, K	Boiling Point, K	Critical Temp., K
Actinium	89	(227)	10.07[1c]	f.c.c.[2]					124[3]	100[4] (at~50 K)	1323[5]	3200±300[6]	
Aluminum	13	26.9815	2.702[5]	f.c.c.[7]		1.196[5] 1.17[8] 1.18[9]			423±5[3]	390[3]	933.2[3,10]	2723[29]	8650[11] 7740[109]
Americium	95	(243)	11.7[5]	Double c.p.h.[2]							1473[29]	2880[108]	
Antimony	51	121.75	6.684[29]	r.[2] (?) ? (?) ? (?)	367.8[13] (?-?) 690[13] (?-?)	2.6[8] (Sb II, high-pressure modification)			150[3]	200[14]	903.7[13] 903.65[23]	1907±10[3]	2989[15]
Argon	18	39.948	0.0017824[29] (at 273.2 K and 1 atm)	f.c.c.[16]						90[4] (at~45 K)	83.8[17]	87.29[13]	151[15]
Arsenic	33	74.9216	5.73[29] (gray, at 287.2 K) 4.7[29] (black) 2.0[29] (yellow)	r.[7] (gray) c.[5] (yellow)					236[3]	275[18]	1090[13] (35.8 atm) subl. 886[5]	1090[13] (35.8 atm)	
Astatine	85	(210)									573.2[19]	650[20]	
Barium	56	137.34	3.5[29]	b.c.c.[2] (α) ? (β)[13]	648[13,21] (α-β)				110.5±1.8[22]	116[23]	998.2[5]	1910[3]	3663[15] 3920[109]
Berkelium	97	(249)											
Beryllium	4	9.0122	1.85[29]	c.p.h.[2] (α) b.c.c.[2] (β)	1533[24] (α-β)	~6[108] ~8.4[108]			1160[25]	1031[3]	1550[26]	3142±100[3]	6153[15]
Bismuth	83	208.980	9.78[29]	r.[2]		3.9[8] (Bi II, at 25 kbar) 7.2[8] (Bi III, at 27 kbar)			119±2[3]	116±5[3]	544.525[3,111]	1824±8[3]	4620[27]
Boron	5	10.811	2.50[42]	Simple r.[2] (α) r.[2] (β)	1473[2] (α-β)				1315[53]	1362[3]	2573[5]	4050±100[30]	
Bromine	35	79.909	3.119[29]	orthorh.[16]							266.0[17]	331.93[29]	~~589~~[15] 584

[a] Atomic weights are based on $^{12}C = 12$ as adopted by the International Union of Pure and Applied Chemistry in 1961; those in parentheses are the mass numbers of the isotopes of longest known half-life.

[b] Density values are given at 293.2 K unless otherwise noted.

[c] Superscript numbers designate references listed at the end of the table.

Name	Atomic Number	Atomic Weight[a]	Density,[b] $kg\,m^{-3}\cdot 10^{-3}$	Crystal Structure	Phase Transition Temp., K	Superconducting Transition Temp., K	Curie Temp., K	Néel Temp., K	Debye Temperature at 0 K, K	Debye Temperature at 298 K, K	Melting Point, K	Boiling Point, K	Critical Temp., K
Cadmium	48	112.40	8.65[29]	c.p.h.[2] b.c.c.[4] (?)		0.56[5] 0.52[9]			252±48[3]	221[3] 170[4] (b.c.c., at ~85 K)	594.18[3,10] Subl. 594.1[13] (at 0.11 mm Hg)	1038[3]	1903[15] 3560[109]
Calcium	20	40.08	1.55[29]	f.c.c.[7] (α) b.c.c.[7] (β)	737[62] (α–β)				234±5[3]	230[3]	1123[19] Subl. 1123[13] (at 0.35 mm Hg)	1765[3]	3267[15]
Californium	98	(251)											
Carbon (amorphous)	6	12.01115	1.8~2.1[29]								Subl. 3925–3970[5]	4473[5]	
Carbon (diamond)	6	12.01115	3.51[29]	d.[16]					2240±5[31]	1874[3]	>3823[5]	5100[5]	
Carbon (graphite)	6	12.01115	2.26[29] (α)	h.[2] (α) r.[2] (β)					402±11[3]	1550[3]	Subl. 3925–3970[5]	4473[5]	
Cerium	58	140.12	6.90[29]	f.c.c. (α)[32] Double c.p.h.?[8] (β) f.c.c. (γ)[32] b.c.c. (δ)[32]	103±5[33] (α–β) 263±5[33] (β–γ) 1003[32] (γ–δ)			13[32]	146[3]	138[34]	1077[26]	3972[3]	10400[109]
Cesium	55	132.905	1.873[29]	b.c.c.[2]					40±5[3]	43[23]	301.9[29] Subl. 301.9[13] (at 1.2 μHg)	939[35]	2060[113,114,115] 1900[109]
Chlorine	17	35.453	0.003214[29] (at 273.2 K)	t.[16]						115[4,36] (at ~58 K)	172.2[26]	239.10[13]	417[15]
Chromium	24	51.996	7.16[42]	c.p.h.[17,d] (α) b.c.c.[7] (β)	~299[17] (α–β)[d]			311[37]	598±32[3]	424[3]	2118[38]	2918±35[3]	
Cobalt	27	58.9332	8.862[42]	c.p.h.[7] (α) f.c.c.[17] (β)	690[39] (α–β)		1400[40]		452±17[3]	386[3]	1765[3,10]	3229[3]	
Copper	29	63.54	8.933[29]	f.c.c.[2]					342±2[3]	310[3]	1356[3,10]	2811±20[41]	8500[11] 8280[109]
Curium	96	(247)	7[42]	Double c.p.h.[8]									
Dysprosium	66	162.50	8.556[42]	c.p.h.[2] (α) b.c.c.[2] (β)	Near m.p.[2] (α–β)			174[43] 83.5[43] (ferro–antiferromag.)	172±35[3]	158[44]	1773[12]	3011[44]	7640[109]

[d]Close-packed hexagonal crystalline modification of chromium may be formed by electrodeposition below 293 K under special conditions of deposition process. This c.p.h. form is unstable and will irreversibly transform into b.c.c. form on heating.

Name	Atomic Number	Atomic Weight[a]	Density,[b] kg m^{-3} · 10^{-3}	Crystal Structure	Phase Transition Temp., K	Superconducting Transition Temp., K	Curie Temp., K	Neel Temp., K	Debye Temperature at 0 K, K	Debye Temperature at 298 K, K	Melting Point, K	Boiling Point, K	Critical Temp., K
Einsteinium	99	(254)											
Erbium	68	167.26	9.06[42]	c.p.h.[2] (α) b.c.c.[2] (β)	1643[2] (α–β)		19[4]	80[4]	134±10[45]	163[44]	1770[26]	3000[3]	7250[109]
Europium	63	151.96	5.245[28]	b.c.c.[7]				~90[4]	127[3]		1099[5]	1971[46]	4600[109]
Fermium	100	(253)											
Fluorine	9	18.9984	0.001695[29] (at 273.2 K and 1 atm)	c.[108] (β–F_2)							53.58[5]	85.24[13]	144[15]
Francium	87	(223)							39[3]		300.2[19]	879[108]	
Gadolinium	64	157.25	7.87[42]	c.p.h.[2] (α) b.c.c.[2] (β)	1535[32] (α–β)		292[40]		170[3]	155±3[3]	1579[19]	3540[3]	8670[109]
Gallium	31	69.72	5.91[29]	orthorh.[4] (α) t.[4] (β)	275.6[13] (α–β) (at 8.86 x 10^6 mm Hg)	1.091[5] 7.2[38] (Ga II, high-pressure modification)			317[3]	240[14] 125[4] (tetra at ~63 K)	302.93[5] 275.6[13] (at 8.86 x 10^6 mm Hg)	2510[3]	7620[27]
Germanium	32	72.59	5.36[29]	d.[7]		5.5[47] (at ~118 kbar) 8.4[108]			378±22[3]	403[3]	1210.6[5]	3100[3]	5642[15]
Gold	79	196.967	19.3[42]	f.c.c.[7]					165±1[3]	178±8[3]	1336.2[3, 10] 1336.15[23]	3240[3]	9500[11] 8060[109]
Hafnium	72	178.49	13.28[42]	c.p.h.[48] (α) b.c.c.[48] (β)	2023±20[48] (α–β)	0.16[9] 0.35[108]			256±5[3]	213[23]	2495[19]	4575±150[49]	
Helium	2	4.0026	0.0001785[29] (at 273.2 K and 1 atm)	c.p.h.[16]						30[4] (at~15 K)	3.45[29] 1.8±0.2[17] (at 30 atm)	4.216[13] 4.22[23]	5.3[15]
Holmium	67	164.930	8.80[29]	c.p.h.[2] (α) b.c.c.[2] (β)	Near m.p.[50] (α–β)		20[4]	132[4]	114±7[45]	161[44]	1734[19]	3228[51]	
Hydrogen	1	1.00797	0.00008987[29] (at 273.2 K and 1 atm)	c.p.h.[16]						116[36] (para., at~58 K) 105[36] (ortho., at~53 K)	13.8±0.1[17]	20.39[13] 20.37[23]	33.3[15]
Indium	49	114.82	7.3[29]	f.c.t.[7]		3.4035[5]			108.8±0.3[3]	129[14]	429.76[3, 110]	2279±6[3]	4377[15] 7050[109]
Iodine	53	126.9044	4.93[29]	orthorh.[16]						105[4] (at~53 K)	386.8[29] subl. 298.16[13] (at 0.31 mm Hg)	457.50[29]	785[15]
Iridium	77	192.2	22.5[42]	f.c.c.[7]		0.14[5, 9]			425±5[3]	228[3]	2716[3, 10]	4820±30[3]	

Name	Atomic Number	Atomic Weight[a]	Density,[b] kg m^{-3} · 10^{-3}	Crystal Structure	Phase Transition Temp., K	Superconducting Transition Temp., K	Curie Temp., K	Néel Temp., K	Debye Temperature at 0 K, K	Debye Temperature at 298 K, K	Melting Point, K	Boiling Point, K	Critical Temp., K
Iron	26	55.847	7.87[28]	b.c.c.-ferromag.[7] (α) b.c.c.-paramag.[7] (β) f.c.c.[7] (γ) b.c.c.[7] (δ)	1183[2] (β–γ) 1673[13] (γ–δ)		1043[40]		457 ±12[3]	373[3]	1810[19]	3160[20]	~~10350~~[27] 6750[123] 9400[109]
Krypton	36	83.80	0.003708[29] (at 273.2 K and 1 atm)	f.c.c.[16]						60[4] (at∼30 K)	116.6[5]	119.93[13]	209.4[15]
Lanthanum	57	138.91	6.18[42]	Double c.p.h.[8] (α) f.c.c.[2] (β) b.c.c.[2] (γ)	583[32] (α–β) 1141[32] (β–γ)	4.9[8] (α) 6.3[8] (β)			142 ±3[52]	135 ±5[44]	1193[5]	3713 ±70[3]	10500[109]
Lawrencium	103	(257)											
Lead	82	207.19	11.34[29]	f.c.c.[2]		7.193[5]			102 ±5[3]	87 ±1[3]	600.576[3,111]	2022 ±10[41]	5400[27] 4760[109]
Lithium	3	6.939	0.534[29]	b.c.c.[7]	Martensitic transformation at low temp.[56]				352 ±17[3]	448[3]	453.7[19]	1599[13]	4150[11] 3720[109]
Lutetium	71	174.97	9.85[29]	c.p.h.[2] (α) b.c.c.[2] (β)	Near m.p.[50] (α–β)				210[54]	116[3]	1923[19]	4140[3]	
Magnesium	12	24.312	1.74[29]	c.p.h.[7]					396 ±54[3]	330[3]	923[55]	1385[3]	3530[109]
Manganese	25	54.9380	7.43(α)[28] 7.29(β)[28] 7.18(γ)[28]	~~c.[7] (α)~~ b.c.c.[93] (α) c.[7] (β) ~~f.c.t.[7] (γ)~~ f.c.c.[93] (γ) b.c.c.[7] (δ)	1000[13] (α–β) 1374[13] (β–γ) 1410[13] (γ–δ)			95[5]	418 ±32[3]	363[3]	1517 ±3[5]	2360[13]	6050[109]
Mendelevium	101	(256)											
Mercury	80	200.59	13.546[29] 14.19[29] (at 234.25 K)	r.[7] (α) b.c.t.-pressure induced structure (β)	Martensitic transformation at low temp.[56]	4.153[5] (α) 3.949[5] (β)			∼ 75[58]	92 ±8[3]	234.28[3,10]	629.73[3,10]	1733[27] 1705[109]
Molybdenum	42	95.94	10.24[42]	b.c.c.[2]		0.92[5,9]			459 ±11[3]	377[3]	2883[13]	5785 ±175[3]	17000[11] 16800[109]
Neodymium	60	144.24	7.007[29]	Double c.p.h.[8] (α) b.c.c.[32] (β)	1135[32] (α–β)			8[4] (ordinary) 19[4] (special)	159[3]	148 ±8[3]	1292[19]	2956[60]	7900[109]
Neon	10	20.183	0.0009002[29] (at 273.2 K and 1 atm)	f.c.c.[16]						60[4] (at∼30 K)	24.48[5]	27.23[5] 27.06[23]	44.5[15]

Name	Atomic Number	Atomic Weight[a]	Density,[b] kg m^{-3} · 10^{-3}	Crystal Structure	Phase Transition Temp., K	Superconducting Transition Temp., K	Curie Temp., K	Néel Temp., K	Debye Temperature at 0 K, K	Debye Temperature at 298 K, K	Melting Point, K	Boiling Point, K	Critical Temp., K
Neptunium	93	(237)	20.46[42]	orthorh.[2] (α) t.[2] (β) b.c.c.[2] (γ)	551[2] (α–β) 813[2] (β–γ)				121[3]	163[3]	913.2[5]	4150[3]	
Nickel	28	58.71	8.90[42]	f.c.c.[7]			631[40]		427 ±14[3]	345[3]	1726[3,10] 1726 ±4[61]	3055[63]	6294[15] 11750[109]
Niobium	41	92.906	8.57[42]	b.c.c.[7]		9.13[5] 9.09[8] 9.1[9]			241 ±13[3]	260[64]	2741 ±27[3] 2688[65]	4813[66]	19000[109]
Nitrogen	7	14.0067	0.0012506[29]	c.[16] (α) h.[107] (β)	35.62[13] (α–β)					70[4] (at∼35 K)	63.29[5]	77.34[13,23]	126.2[15]
Nobelium	102	(254)											
Osmium	76	190.2	22.48[29]	c.p.h.[2]		0.655[5] 0.65[8]			500[67]	400[68]	3283 ±10[69]	5300 ±100[70]	
Oxygen	8	15.9994	0.001429[29] (at 273.2 K and 1 atm)	b.c.orthorh.[7] (α) r.[7] (β) c.[7] (γ)	23.876 ± 0.01[112] (α–β) 43.818 ± 0.01[112] (β–γ)					250[4] (at∼125 K) 500[36] (at∼250 K)	54.8[5]	90.19[13] 90.18[23]	154.8[15]
Palladium	46	106.4	12.02[28]	f.c.c.[2]					283 ±16[3]	275[14]	1825[3,10]	3200[3]	
Phosphorus	15	30.9738	1.82[29] (β) 2.22[29] (γ) 2.69[29] (δ)	h. ?[7] (α) b.c.c.[7] (β) c.[7] (γ) f.c.orthorh.[17] (δ)	196[71] (α–β) 298.16[13] (β–γ) 298.16[13] (β–δ)				193[3] (white) 325[3] (red)	576[3] (white) 800[3] (red)	317.3[5] (white) 1300[72] (black)	553[13]	993.8[15]
Platinum	78	195.09	21.45[29]	f.c.c.[2]					234 ±1[3]	225 ±5[3]	2042[3,10]	4100[3]	8280[15]
Plutonium	94	(242)	19.737[29] (at 298.2 K)	Simple monocl.[2] (α) b.c. monocl.[2] (β) f.c. orthorh.[2] (γ) f.c.c.[2] (δ) b.c.t.[2] (δ') b.c.c.[2] (ε)	396.7[73] (α–β) 475[73] (β–γ) 591.4[73] (γ–δ) 729[73] (δ–δ') 757 ±3[73] (δ'–ε)				171[74]	176[74]	912.7[5]	3727[75]	
Polonium	84	(210)	9.3[29] (α) 9.5[29] (β)	Simple c.[7] (α) r.[7] (β)	327 ±1.5[76] (α–β)				81[3]		527.2[5]	1235[20]	2281[15]
Potassium	19	39.102	0.86[29]	b.c.c.[7]					89.4 ±0.5[3]	100[3]	336.8[5]	1027[35]	2450[11] 2140[109]
Praseodymium	59	140.907	6.769[29]	Double c.p.h.[8] (α) b.c.c.[2] (β)	1071[32] (α–β)			25[77]	85 ±1[45]	138[78]	1192 ±2[79]	3616[80]	8900[109]

Name	Atomic Number	Atomic Weight[a]	Density,[b] kg m^{-3} · 10^{-3}	Crystal Structure	Phase Transition Temp., K	Superconducting Transition Temp., K	Curie Temp., K	Néel Temp., K	Debye Temperature at 0 K, K	Debye Temperature at 298 K, K	Melting Point, K	Boiling Point, K	Critical Temp., K
Promethium	61	(145)		h.[7] (α) b.c.c.[120] (β)	1185[120] (α–β)			6[120]			1353 ± 10[81]	2730[3]	
Protactinium	91	(231)	15.37[42]	b.c.t.[2]		1.4[9]			159[3]	262[3]	1503[5]	4680[3]	
Radium	88	(226)	5[29]						89[3]		973.2[5]	1900[3]	
Radon	86	(222)	0.00973[29] (at 273.2 K and 1 atm)	f.c.c.[7]						400[4] (at ~ 200 K)	202.2[5]	211[13]	377.16[15]
Rhenium	75	186.2	21.1[42]	c.p.h.[2]		1.698[26]			429 ± 22[3]	275[23]	3453[5]	6035 ± 135[3]	20000[11]
Rhodium	45	102.905	12.45[42]	f.c.c.[7]	possible transformation at 1373–1473 K[57]				480 ± 32[3]	350[3]	2233[3,10,82]	3960 ± 60[3]	
Rubidium	37	85.47	1.53[29]	b.c.c.[2]					54 ± 4[3]	59[23]	312.04[5]	959[35]	2100[113,115,116] 2030[109]
Ruthenium	44	101.07	12.2[29]	c.p.h.[7] (α) ? (β) ? (γ) ? (δ)	1308[13,121] (α–β) 1473[13,121] (β–γ) 1773[13,121] (γ–δ)	0.49[5,9]			600[67]	415[3]	2523 ± 10[69]	4325 ± 25[3]	
Samarium	62	150.35	7.54[29]	r.[32] (α) b.c.c.[32] (β)	1190[32] (α–β)		14[8]	106[8]	116[45]	184 ± 4[3]	1345.2[83]	2140[3]	5400[109]
Scandium	21	44.956	3.00[42]	c.p.h.[2] (α) b.c.c.[2] (β)	1607[2] (α–β)				470 ± 80[52]	476[3]	1812[5]	3537 ± 30[3]	
Selenium	34	78.96	4.50[29] (α) 4.80[29] (β)	monocl.[7] (α) h.[7] (β) amorphous[7]	304[84,117] (vitrification) 398[13] (vit.–β) 423[13] (α–β)	7.3[85] (at ~ 118 kbar)			151.7 ± 0.4[86]	89[36] (at ~ 45 K) 150[4] (at ~ 75 K)	490.2[5]	1009[13] (Se_6) 958.0[13] ($Se_{4.37}$) 1027[13] (Se_2)	1757[15]
Silicon	14	28.086	2.33[42]	d.[7]		7.5[47] (at 118–128 kbar)			647 ± 11[3]	692[3]	1685 ± 2[87]	2753[28]	5159[15]
Silver	47	107.870	10.5[29]	f.c.c.[2]					228 ± 3[3]	221[3]	1234.0[3,13]	2468 ± 15[41]	7460[11]
Sodium	11	22.9898	0.9712[29]	b.c.c.[2]	Martensitic transformation at low temp.[56]				157 ± 1[3]	155 ± 5[3]	371.0[13]	1154[35]	2800[11] 2400[109]
Strontium	38	87.62	2.60[28]	f.c.c.[88] (α) c.p.h.[7] (β) b.c.c.[7] (γ)	488[88] (α–β) 878[88] (β–γ)				147 ± 1[22]	148[23]	1042[5]	1645[3]	3059[15] 3810[109]
Sulfur	16	32.064	2.07[29] (α) 1.96[29] (β)	r.[7] (α) monocl.[7] (β)	368.6[13] (α–β)				200[3] (β)	527[89] (α) 250[89] (α, at 40 K)	386.0[5] (α) 392.2[5] (β) Subl. 368.6[13] (at 0.0047 mm Hg)	717.75[3,10]	1313[15]
Tantalum	73	180.948	16.6[42]	b.c.c.[2]		4.483[5] 4.48[9]			247 ± 13[3]	225[14]	3269[5]	5760 ± 60[3]	22000[11]

Name	Atomic Number	Atomic Weight[a]	Density,[b] kg m^{-3} · 10^{-3}	Crystal Structure	Phase Transition Temp., K	Superconducting Transition Temp., K	Curie Temp., K	Neel Temp., K	Debye Temperature at 0 K, K	Debye Temperature at 298 K, K	Melting Point, K	Boiling Point, K	Critical Temp., K
Technetium	43	(99)	11.50[29]	c.p.h.[2]		8.22[5] 11.2[9]			351[3]	422[3]	2473 ± 50[5]	5300[3]	
Tellurium	52	127.60	6.24[29] (α) 6.00[5] (amorph.)	h.[7] (α) ? (β)[7] amorph.[5]	621[13] (α–β)	3.3[8] (Te II, at 56 kbar)			141 ± 12[3]		722.7[5]	1163 ± 1[3]	2329[15]
Terbium	65	158.924	8.25[29]	c.p.h.[2,32] (α) b.c.c.[2] (β)	Near m.p.[2] (α–β)		219[90]	230[90]	150[91]	158[44]	1629[19]	3810[3]	
Thallium	81	204.37	11.85[29]	c.p.h.[2] (α) b.c.c.[2] (β)	508.3[5] (α–β)	2.39[5] 2.38[8] 2.37[9]			88 ± 1[3]	96[14]	576.2[19]	1939[92]	3219[15]
Thorium	90	232.038	11.7[42]	f.c.c.[2] (α) b.c.c.[2] (β)	1673 ± 25[93] (α–β)	1.368[5] 1.37[9]			170[94]	100[14]	2023[19]	4500[20]	14550[109]
Thulium	69	168.934	9.32[29]	c.p.h.[2] (α) b.c.c.[2] (β)	Near m.p.[50] (α–β)		22[95] (ferro.–antiferro.)	53[96]	127 ± 1[45]	167[44]	1818[5]	2266[97]	6430[109]
Tin	50	118.69	5.750[29] (α) 7.31[29] (β)	f.c.c.[7] (α) b.c.t.[7] (β) r.[29] (?)	286.2 ± 3[98] (α–β)	3.722[5] (β)			236 ± 24[3] (gray) 196 ± 9[3] (white)	254[3] (gray) 170[14] (white)	505.06[3,10]	2766 ± 14[3]	8000[11] 9300[109]
Titanium	22	47.90	4.5[29]	c.p.h.[7] (α) b.c.c.[7] (β)	1155[13] (α–β)	0.39[5,9]			426 ± 5[3]	380[14]	1953[99]	3586[100]	
Tungsten	74	183.85	19.3[29]	b.c.c.[2]		0.011[122]			388 ± 17[3]	312 ± 3[3]	3653[3,10,13]	6000 ± 200[3]	23000[11]
Uranium	92	238.03	19.07[28]	orthorh.[7] (α) t.[7] (β) b.c.c.[7] (γ)	37±2[118] (α_0–α) 938[13] (α–β) 1049[13] (β–γ)	0.68[5] (α) 1.80[5] (γ)			200[94]	300[3]	1405.6 ± 0.6[101]	3950 ± 250[102]	12500[27] 12000[109]
Vanadium	23	50.942	6.1[28]	b.c.c.[2]		5.3[5] 5.03[9]			326 ± 54[3]	390[14]	2192 ± 2[61]	3582 ± 42[3]	11200[109]
Xenon	54	131.30	0.005851[29] (at 273.2 K and 1 atm)	f.c.c.[16]							161.2[26]	165.1[13]	289.75[15]
Ytterbium	70	173.04	7.02[42]	f.c.c.[32] (α) b.c.c.[32] (β)	1071[2,5] (α–β)				118[103]		1097[12]	1970[3]	4420[109]
Yttrium	39	88.905	4.47[29]	c.p.h.[32] (α) b.c.c.[32] (β)	1753[119] (α–β)				268 ± 32[3]	214[104]	1798[119]	3670[105]	8950[109]
Zinc	30	65.37	7.140[29]	c.p.h.[2]		0.875[5] 0.85[9]			316 ± 20[3]	237 ± 3[3]	692.655[3,110]	1175[106]	2169[15] 2910[109]
Zirconium	40	91.22	6.57[59]	c.p.h.[7] (α) b.c.c.[7] (β)	1135[13] (α–β)	0.546[5] 0.55[9]			289 ± 24[3]	250[14]	2125[19]	4650[20]	12300[109]

REFERENCES

(Crystal Structures, Transition Temperatures, and Other Pertinent Physical Constants of the Elements)

1. Farr, J.D., Giorgi, A.L., and Bowman, M.G., USAEC Rept. LA-1545, 1-13, 1953.
2. Elliott, R.P., Constitution of Binary Alloys, 1st Suppl., McGraw-Hill, 1965.
3. Gschneider, K.A, Jr., Solid State Physics (Sietz, F. and Turnbull, D., Editors), 16, 275-426, 1964.
4. Gopal, E.S.R., Specific Heat at Low Temperatures, Plenum Press, 1966.
5. Weast, R.C. (Editor), Handbook of Chemistry and Physics, 47th Ed., The Chemical Rubber Co., 1966-67.
6. Foster, K.W. and Fauble, L.G., J. Phys. Chem., 64, 958-60, 1960.
7. The Institution of Metallurgists, Annual Yearbook, pp. 68-73, 1960-61.
8. Meaden, G.T., Electrical Resistance of Metals, Plenum Press, 1965.
9. Matthias, B.T., Geballe, T.H., and Compton, V.B., Rev. Mod. Phys., 35, 1-22, 1963.
10. Stimson, H.F., J. Res. NBS, 42, 209, 1949.
11. Grosse, A.V., Rev. Hautes Tempér. et Réfract., 3, 115-46, 1966.
12. Spedding, F.H. and Daane, A.H., J. Metals, 6 (5), 504-10, 1954.
13. Rossini, F.D., Wagman, D.D., Evans, W.H., Levine, S., and Jaffe, I., NBS Circ. 500, 537-822, 1952.
14. deLaunay, J., Solid State Physics, 2, 219-303, 1956.
15. Gates, D.S. and Thodos, G., AIChE J., 6 (1), 50-4, 1960.
16. Gray, D.E. (Coordinating Editor), American Institute of Physics Handbook, McGraw-Hill, 1957.
17. Sasaki, K. and Sekito, S., Trans. Electrochem. Soc., 59, 437-60, 1931.
18. Anderson, C.T., J. Am. Chem. Soc., 52, 2296-300, 1930.
19. Trombe, F., Bull. Soc. Chim. (France), 20, 1010-2, 1953.
20. Stull, D.R. and Sinke, G.C., Thermodynamic Properties of the Elements in Their Standard State, American Chemical Soc., 1956.
21. Rinck, E., Ann. Chim. (Paris), 18 (10), 455-531, 1932.
22. Roberts, L.M., Proc. Phys. Soc. (London), B70, 738-43, 1957.
23. Zemansky, M.W., Heat and Thermodynamics, 4th Ed., McGraw-Hill, 1957.
24. Martin, A.J. and Moore, A., J. Less-Common Metals, 1, 85, 1959.
25. Hill, R.W. and Smith, P.L., Phil. Mag., 44 (7), 636-44, 1953.
26. Moffatt, W.G., Pearsall, G.W., and Wulff, J., The Structure and Properties of Materials, Vol. I, pp. 205-7, 1964.
27. Grosse, A.V., Temple Univ. Research Institute Rept., 1-40, 1960.
28. Lyman, T. (Editor), Metals Handbook, Vol. 1, 8th Ed., American Soc. for Metals, 1961.
29. Lange, N.A. (Editor), Handbook of Chemistry, Revised 10th Edition, McGraw-Hill, 1967.
30. Paule, R.C., Dissertation Abstr., 22, 4200, 1962.
31. Burk, D.L. and Friedberg, S.A., Phys. Rev., 111 (5), 1275-82, 1958.
32. Spedding, F.H. and Daane, A.H. (Editors), The Rare Earths, John Wiley, 1961.
33. McHargue, C.J., Yakel, H.L., and Letter, C.K., ACTA Cryst., 10, 832-33, 1957.
34. Arajs, S. and Colvin, R.V., J. Less-Common Metals, 4, 159-68, 1962.
35. Bonilla, C.F., Sawhney, D.L., and Makansi, M.M., Trans. Am. Soc. Metals, 55, 877, 1962.
36. Rosenberg, H.M., Low Temperature Solid State Physics, Oxford at Clarendon Press, 1965.
37. Arajs, S., J. Less-Common Metals, 4, 46-51, 1962.
38. Edwards, A.R. and Johnstone, S.T.M., J. Inst. Metals, 84 (8), 313-7, 1956.
39. Lagneborg, R. and Kaplow, R., ACTA Metallurgica, 15 (1), 13-24, 1967.
40. Kittel, C., Introduction to Solid State Physics, 3rd Ed., John Wiley, 1967.
41. Kirshenbaum, A.D. and Cahill, J.A., J. Inorg. and Nucl. Chem., 25 (2), 232-34, 1963.
42. Touloukian, Y.S. (Ed.), Thermophysical Properties of High Temperature Solid Materials, MacMillan, Vol. 1, 1967.
43. Griffel, M., Skochdopole, R.E., and Spedding, F.H., J. Chem. Phys., 25 (1), 75-9, 1956.
44. Gschneidner, K.A., Jr., Rare Earth Alloys, Van Nostrand, 1961.
45. Dreyfus, B., Goodman, B.B., Lacaze, A., and Trolliet, G., Compt. Rend., 253, 1764-6, 1961.

46. Spedding, F.H., Hanak, J.J., and Daane, A.H., Trans. AIME, 212, 379, 1958.
47. Buckel, W. and Wittig, J., Phys. Lett. (Netherland), 17 (3), 187-8, 1965.
48. Deardorff, D.K. and Kata, H., Trans. AIME, 215, 876-7, 1959.
49. Panish, M.B. and Reif, L., J. Chem. Phys., 38 (1), 253-6, 1963.
50. Miller, A.E. and Daane, A.H., Trans. AIME, 230, 568-72, 1964.
51. Spedding, F.H. and Daane, A.H., USAEC Rept. IS-350, 22-4, 1961.
52. Montgomery, H. and Pells, G.P., Proc. Phys. Soc. (London), 78, 622-5, 1961.
53. Kaufman, L. and Clougherty, E.V., ManLabs, Inc., Semi-Annual Rept. No. 2, 1963.
54. Lounasmaa, O.V., Proc. 3rd Rare Earth Conf., 1963, Gordon and Breach, New York, 1964.
55. Baker, H., WADC TR 57-194, 1-24, 1957.
56. Reed, R.P. and Breedis, J.F., ASTM STP 387, pp. 60-132, 1966.
57. Hansen, M., Constitution of Binary Alloys, 2nd Edition, McGraw-Hill, p. 1268, 1958.
58. Smith, P.L., Conf. Phys. Basses Temp., Inst. Intern. du Froid, Paris, 281, 1956.
59. Powell, R.W. and Tye, R.P., J. Less-Common Metals, 3, 202-15, 1961.
60. Yamamoto, A.S., Lundin, C.E., and Nachman, J.F., Denver Res. Inst. Rept., NP-11023, 1961.
61. Oriena, R.A. and Jones, T.S., Rev. Sci. Instr., 25, 248-51, 1954.
62. Smith, J.F., Carlson, O.N., and Vest, R.W., J. Electrochem. Soc., 103, 409-13, 1956.
63. Edwards, J.W. and Marshal, A.L., J. Am. Chem. Soc., 62, 1382, 1940.
64. Morin, F.J. and Maita, J.P., Phys. Rev., 129 (3), 1115-20, 1963.
65. Pendleton, W.N., ASD-TDR-63-164, 1963.
66. Woerner, P.F. and Wakefield, G.F., Rev. Sci. Instr., 33 (12), 1456-7, 1962.
67. Walcott, N.M., Conf. Phys. Basses Temp., Inst. Intern. du Froid, Paris, 286, 1956.
68. White, G.K. and Woods, S.B., Phil. Trans. Roy. Soc. (London), A251 (995), 273-302, 1959.
69. Douglass, R.W. and Adkins, E.F., Trans. AIME, 221, 248-9, 1961.
70. Panish, M.B. and Reif, L., J. Chem. Phys., 37 (1), 128-31, 1962.
71. Bridgman, P.W., J. Am. Chem. Soc., 36 (7), 1344-63, 1914.
72. Slack, G.A., Phys. Rev., A139 (2), 507-15, 1965.
73. Sandenaw, T.A. and Gibney, R.B., J. Phys. Chem. Solids, 6 (1), 81-8, 1958.
74. Sandenaw, T.A., Olsen, C.E., and Gibney, R.B., Plutonium 1960, Proc. 2nd Intern. Conf. (Grison, E., Lord, W.B.H., and Fowler, R.D., Editors), 66-79, 1961.
75. Mulford, R.N.R., USAEC Rept. LA-2813, 1-11, 1963.
76. Goode, J.M., J. Chem. Phys., 26 (5), 1269-71, 1957.
77. Cable, J.W., Moon, R.M., Koehler, W.C., and Wollan, E.O., Phys. Rev. Letters, 12 (20), 553-5, 1964.
78. Murao, T., Progr. Theoret. Phys. (Kyoto), 20 (3), 277-86, 1958.
79. Grigor'ev, A.T., Sokolovskaya, E.M., Budennaya, L.D., Iyutina, I.A., and Maksimona, M.V., Zhur. Neorg. Khim., 1, 1052-63, 1956.
80. Daane, A.H., USAEC AECD-3209, 1950.
81. Weigel, F., Angew. Chem., 75, 451, 1963.
82. Nassau, K. and Broyer, A.M., J. Am. Ceram. Soc., 45 (10), 474-8, 1962.
83. McKeown, J.J., State Univ. of Iowa, Ph.D. Dissertation, 1-113, 1958.
84. Abdullaev, G.B., Mekhtiyeva, S.I., Abdinov, D.Sh., and Aliev, G.M., Phys. Letters, 23 (3), 215-6, 1966.
85. Wittig, J., Phys. Rev. Letters, 15 (4), 159, 1965.
86. Fukuroi, T. and Muto, Y., Tohoku Univ. Res. Inst. Sci. Rept., A8, 213-22, 1956.
87. Olette, M., Compt. Rend., 244, 1033-6, 1957.
88. Sheldon, E.A., and King, A.J., ACTA Cryst., 6, 100, 1953.
89. Eastman, E.D. and McGavock, W.C., J. Am. Chem. Soc., 59, 145-51, 1937.
90. Arajs, S. and Colvin, R.V., Phys. Rev., A136 (2), 439-41, 1964.
91. Roach, P.R. and Lounasmaa, O.V., Bull. Am. Phys. Soc., 7, 408, 1962.
92. Shchukarev, S.A., Semenov, G.A., and Rat'kovskii, I.A., Zh. Neorgan. Khim., 7, 469, 1962.
93. Pearson, W.B., A Handbook of Lattice Spacings and Structures of Metals and Alloys, Pergamon Press, 1958.

94. Smith, P. L. and Walcott, N. M., Conf. Phys. Basses Temp., Inst. Intern. du Froid, 283, 1956.

95. Davis, D. D. and Bozorth, R. M., Phys. Rev., 118 (6), 1543-5, 1960.

96. Aliev, N. G. and Volkenstein, N. V., Soviet Physics - JETP, 22 (5), 997-8, 1966.

97. Spedding, F. H., Barton, R. J., and Daane, A. H., J. Am. Chem. Soc., 79, 5160, 1957.

98. Raynor, G. V. and Smith, R. W., Proc. Roy. Soc. (London), A244, 101-9, 1958.

99. Savitskii, E. M. and Burhkanov, G. S., Zhur. Neorg. Khim., 2, 2609-16, 1957.

100. Argent, B. B. and Milne, J. G. C., Niobium, Tantalum, Molybdenum and Tungsten, Elsevier Publ. Co. (Quarrell, A. G., Editor), pp. 160-8, 1961.

101. Argonne National Laboratory, USAEC Rept. ANL-5717, 1-67, 1957.

102. Holden, A. N., Physical Metallurgy of Uranium, Addison-Wesley, 1958.

103. Lounasmaa, O. V., Phys. Rev., 129, 2460-4, 1963.

104. Jennings, L. D., Miller, R. E., and Spedding, F. H., J. Chem. Phys., 33 (6), 1849-52, 1960.

105. Ackerman, R. J. and Rauh, E. G., J. Chem. Phys., 36 (2), 448-52, 1962.

106. Rosenblatt, G. M. and Birchenall, C. E., J. Chem. Phys., 35 (3), 788-94, 1961.

107. Streib, W. E., Jordan, T. H., and Lipscomb, W. N., J. Chem. Phys., 37 (12), 2962-5, 1962.

108. Samsonov, G. V. (Editor), Handbook of the Physicochemical Properties of the Elements, Plenum Press, 1968.

109. Kopp, I. Z., Russ. J. Phys. Chem., 41 (6), 782-3, 1967.

110. Stimson, H. F., in Temperature, Its Measurement and Control in Science and Industry (Herzfeld, C. M., Ed.), Vol. 3, Part 1, Reinhold, New York, pp. 59-66, 1962.

111. McLaren, E. H., in Temperature, Its Measurement and Control in Science and Industry (Herzfeld, C. M., Ed.), Vol. 3, Part 1, Reinhold, New York, pp. 185-98, 1962.

112. Orlova, M. P., in Temperature, Its Measurement and Control in Science and Industry (Herzfeld, C. M., Ed.), Vol. 3, Part 1, Reinhold, New York, pp. 179-83, 1962.

113. Grosse, A. V., J. Inorg. Nucl. Chem., 28, 2125-9, 1966.

114. Hochman, J. M. and Bonilla, C. F., in Advances in Thermophysical Properties at Extreme Temperatures and Pressures (Gratch, S., Ed.), ASME 3rd Symposium on Thermophysical Properties, Purdue University, March 22-25, 1965, ASME, pp. 122-30, 1965.

115. Dillon, I. G., Illinois Institute of Technology, Ph. D. Thesis, June 1965.

116. Hochman, J. M., Silver, I. L., and Bonilla, C. F., USAEC Rept. CU-2660-13, 1964.

117. Abdullaev, G. B., Mekhtieva, S. I., Abdinov, D. Sh., Aliev, G. M., and Alieva, S. G., Phys. Status Solidi, 13 (2), 315-23, 1966.

118. Fisher, E. S. and Dever, D., Phys. Rev., 2, 170 (3), 607-13, 1968.

119. Beaudry, B. J., J. Less-Common Metals, 14 (3), 370-2, 1968.

120. Williams, R. K. and McElroy, D. L., USAEC Rept. ORNL-TM 1424, 1-32, 1966.

121. Jaeger, F. M. and Rosenbaum, E., Proc. Nederland Akademie van Wetenschappen, 44, 144-52, 1941.

122. Gibson, J. W. and Hein, R. A., Phys. Letters, 12 (25), 688-90, 1964.

123. Grosse, A.V., Research Institute of Temple Univ., Report on USAEC Contract No. AT (30-1)-2082, 1-71, 1965.

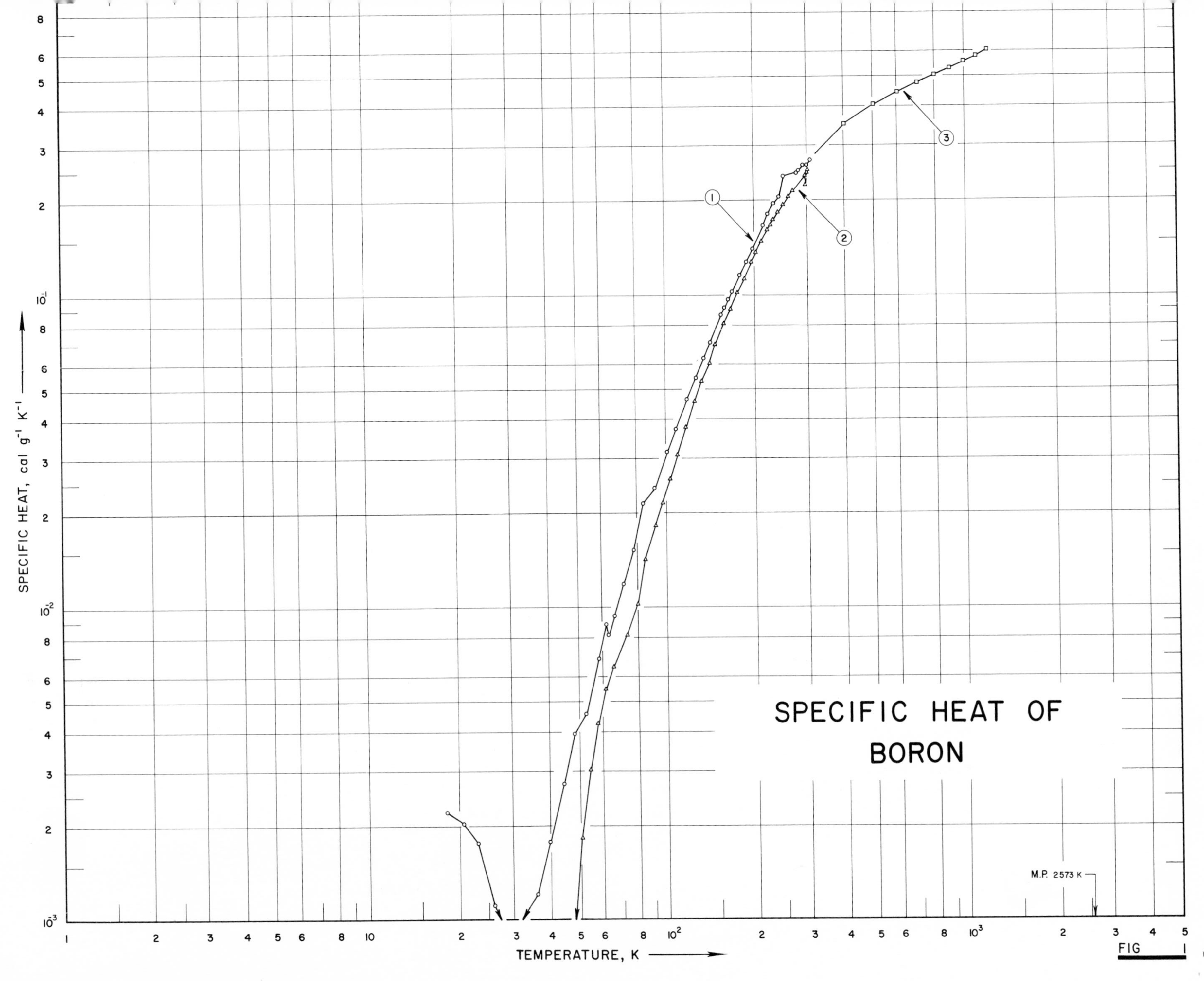
SPECIFIC HEAT OF
BORON
SPECIFIC HEAT, cal g^{-1} K^{-1}
TEMPERATURE, K
M.P. 2573 K
1
2
3

FIG 1

SPECIFICATION TABLE NO. 1 SPECIFIC HEAT OF BORON

(Impurity $<2.00\%$ each; total impurities $<5.00\%$)

[For Data Reported in Figure and Table No. 1]

Curve No.	Ref. No.	Year	Temp. Range, K	Reported Error, %	Name and Specimen Designation	Composition (weight percent), Specifications and Remarks
1	92	1951	18-308			Extremely pure; amorphous.
2	92	1951	17-304			Extremely pure; crystalline; heated under vacuum to 1700-1900 C.
3	162	1960	298-1200		Boron III	0.08 Si, 0.06 Na, 0.04 Fe and 0.02 Ni; amorphous; sample supplied by the Fairmount Chemical Company; sealed in gold ampules.

DATA TABLE NO. 1 SPECIFIC HEAT OF BORON

[Temperature, T, K; Specific Heat, C_p, Cal g^{-1} K^{-1}]

T	C_p
CURVE 1	
18.25	2.201 x 10^{-3}
20.55	2.044
23.04	1.757
25.89	1.110
29.08	8.602 x 10^{-4}*
35.98	1.202 x 10^{-3}
39.70	1.776
44.43	2.710
48.52	3.940
52.97	4.551
58.71	6.873
62.25	8.806
63.10	8.121
66.10	9.379
71.20	1.193 x 10^{-2}
77.03	1.539
83.79	2.154
91.78	2.405
100.83	3.154
108.93	3.737
118.06	4.625
127.04	5.457
135.43	6.299
142.40	7.067
155.10	8.676
159.49	9.120
163.88	9.694
168.71	1.027 x 10^{-1}
178.62	1.156
187.96	1.281
197.89	1.405
215.29	1.680
223.73	1.833
233.59	1.973
243.69	2.083
252.69	2.415
277.62	2.477*
279.62	2.471*
283.18	2.476*
283.85	2.530*
288.97	2.547*
291.10	2.622
296.58	2.610*
300.26	2.629
CURVE 1 (cont.)	
303.26	2.629 x 10^{-1}*
308.29	2.725
CURVE 2	
16.90	2.312 x 10^{-4}*
19.47	4.625*
21.89	6.937*
24.90	8.232*
27.84	5.920*
30.48	5.550*
32.74	4.782*
35.47	3.811*
40.48	3.691*
43.87	5.938*
48.12	1.508 x 10^{-3}
50.96	1.831
54.51	3.006
57.77	4.246
61.46	5.485
65.23	6.484
72.71	8.158
79.58	1.023 x 10^{-2}
84.74	1.434
91.66	1.828
97.02	2.167
103.11	2.585
109.72	3.080
116.81	3.765
125.43	4.588
133.00	5.309
140.54	6.086
147.98	6.928
157.86	8.103
166.08	9.000
175.54	1.020 x 10^{-1}
185.96	1.141
195.77	1.280
202.71	1.375
211.43	1.490
220.70	1.617
227.43	1.690
232.75	1.752
241.07	1.850
CURVE 2 (cont.)	
251.28	1.954 x 10^{-1}
261.67	2.078
270.29	2.172
297.74	2.263
296.44	2.434
301.79	2.487
303.71	2.516
CURVE 3	
298	2.643 x 10^{-1}*
400	3.562
500	4.099
600	4.490
700	4.814
800	5.102
900	5.369
1000	5.623
1100	5.868
1200	6.108

*Not shown on plot

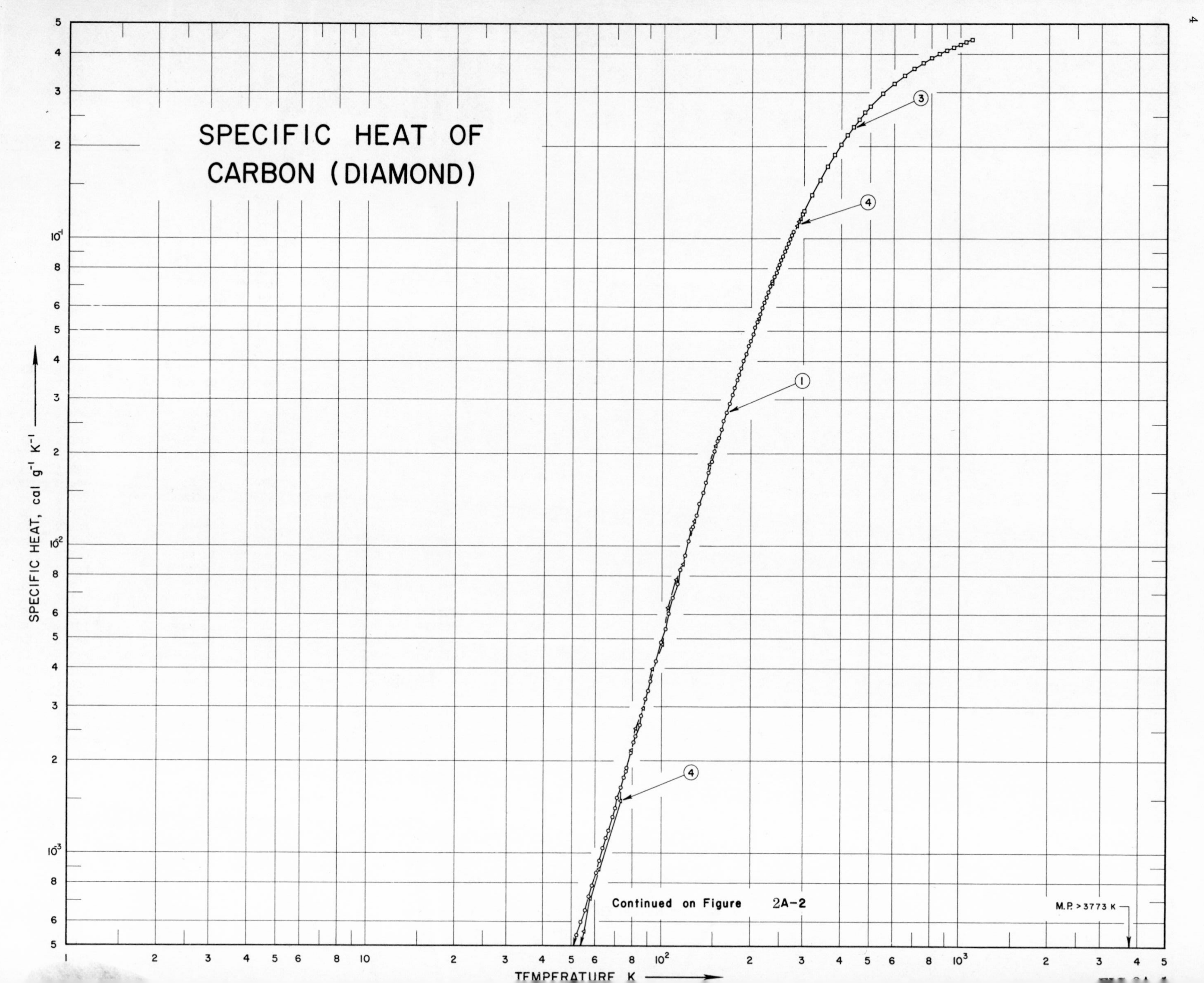
SPECIFIC HEAT OF
CARBON (DIAMOND)
SPECIFIC HEAT, cal g^{-1} K^{-1}
TEMPERATURE K
Continued on Figure 2A-2
M.P. > 3773 K
3
4
1
4

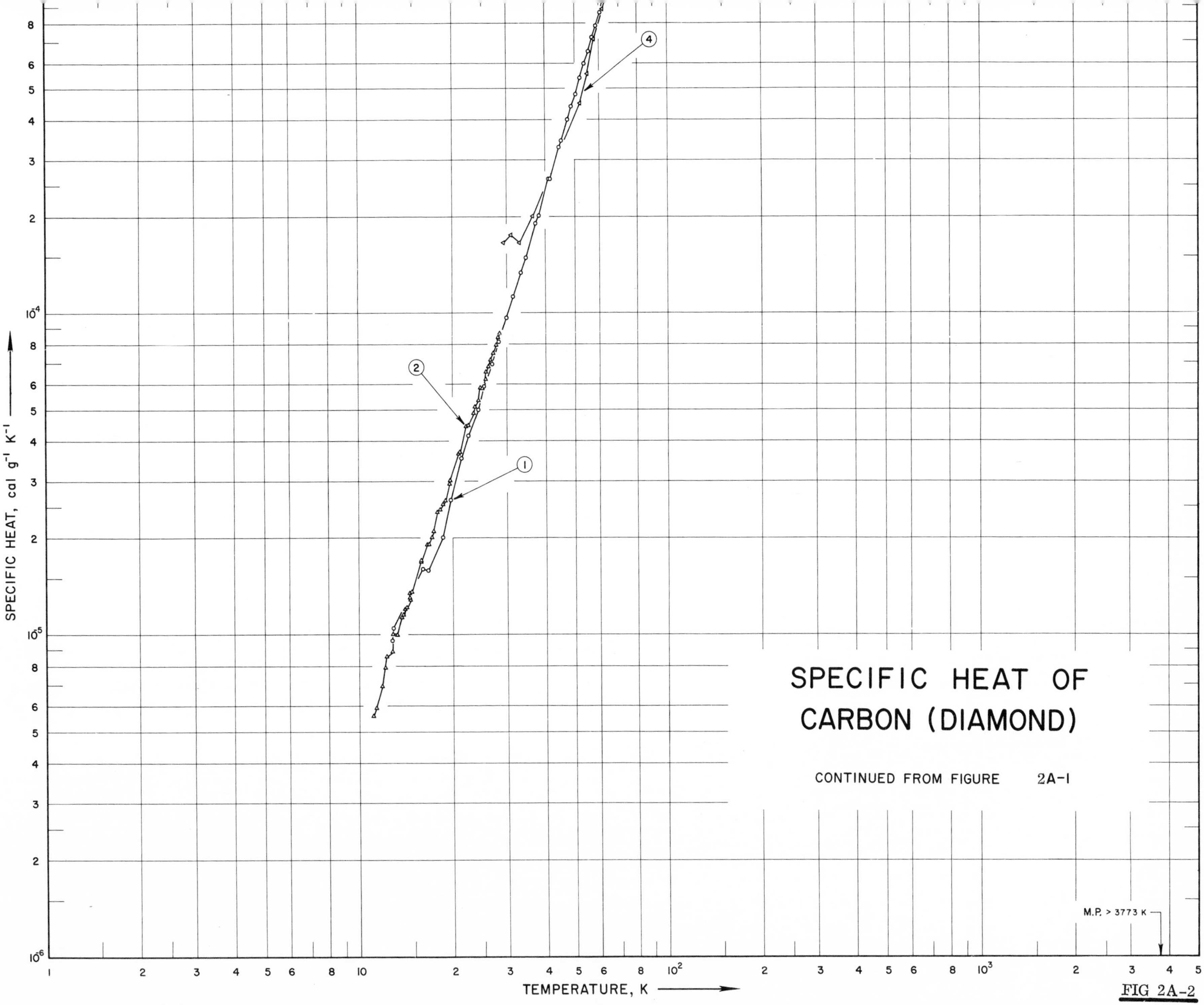

SPECIFIC HEAT OF CARBON (DIAMOND)
CONTINUED FROM FIGURE 2A-1
SPECIFIC HEAT, cal g⁻¹ K⁻¹
TEMPERATURE, K
M.P. > 3773 K
1
2
4

FIG 2A-2

SPECIFICATION TABLE NO. 2-A SPECIFIC HEAT OF CARBON (DIAMOND)

(Impurity $<$ 2.00% each; total impurities $<$ 5.00%)

[For Data Reported in Figure and Table No. 2-A]

Curve No.	Ref. No.	Year	Temp. Range, K	Reported Error, %	Name and Specimen Designation	Composition (weight percent), Specifications and Remarks
1	1	1958	12- 272	< 6.0	Commercial diamond	Commercial grade; under helium atmosphere.
2	2	1958	11- 200	< 6.0	Diamond chips	High purity.
3	3	1962	298-1100	0.4	Diamond	
4	4	1953	30- 300		Fragmented bort	Traces of Al, Mg; low concentrations Fe, Si; region of misalignment; 20% crystals were fluorescent.

DATA TABLE NO. 2-A SPECIFIC HEAT OF CARBON (DIAMOND)

[Temperature, T, K; Specific Heat, C_p, Cal $g^{-1} K^{-1}$]

T	C_p
CURVE 1	
12.833	9.58 x 10^{-6}
12.968	1.05 x 10^{-5}
16.015	1.60
16.745	1.58
18.631	2.01
19.757	2.72
21.304	3.53
22.464	4.18
24.100	5.00
25.276	5.90
26.994	6.92
28.299	8.12
29.992	9.75
31.332	1.12 x 10^{-4}
33.407	1.34
34.606	1.48
37.316	1.90
38.046	2.02
41 319	2.61
41.325	2.61
44.491	3.28
45.228	3.44
47.570	4.01
48.985	4.41
50.450	4.82
52.045	5.40
53.980	5.98
55.597	6.54
57.333	7.21
58.957	7.87
60.540	8.62
62.189	9.42
63.648	1.031 x 10^{-3}
65.392	1.124
66.750	1.194
68.591	1.311
70.070	1.401
71.787	1.512
73.386	1.637
74.995	1.763
CURVE 1 (cont.)	
76.409	1.850 x 10^{-3}
76.685	1.899
79.165	2.122*
79.452	2.129
80.868	2.302
81.972	2.412*
82.410	2.462
84.808	2.715
85.550	2.803*
87.112	2.978
88.700	3.198
90.190	3.382
92.201	3.633
96.025	4.231
99.904	4.843
103.085	5.385
106.381	6.092
109.756	6.791
113.212	7.561
116.719	8.408
120.283	9.367
123.697	1.035 x 10^{-2}
126.975	1.133
127.870	1.154
131.154	1.262
134.485	1.378
137.805	1.495
141.116	1.620
144.495	1.747
147.945	1.892
151.444	2.040
154.989	2.201
156.186	2.252
159.510	2.405
162.867	2.569
166.298	2.727
169.777	2.919
173.316	3.107
176.730	3.289
180.042	3.479
CURVE 1 (cont.)	
182.415	3.623 x 10^{-2}
185.742	3.808
189.157	4.010
192.641	4.227
196.182	4.449
199.611	4.660
203.157	4.901
206.809	5.139
210.341	5.387
215.014	5.701
218.511	5.955
222.107	6.207
225.595	6.467
229.264	6.723
232.816	7.003
236.364	7.259
239.816	7.527
243.176	7.784
246.452	8.037
249.159	8.239
252.541	8.517
256.039	8.794
259.648	9.105
263.168	9.360
266.789	9.678
270.507	9.972
274.134	1.028 x 10^{-1}
277.675	1.057
CURVE 2	
11.138	5.6 x 10^{-6}
11.382	5.9
11.812	6.9
12.172	7.9
12.320	8.58
12.846	1.00 x 10^{-5}
12.880	8.83 x 10^{-6}
13.308	9.92*
13.380	9.92
CURVE 2 (cont.)	
13.774	1.13 x 10^{-5}
13.844	1.16
14.202	1.22
14.232	1.22
14.540	1.31
14.587	1.35
14.676	1.29
14.756	1.37
15.744	1.72
15.800	1.69
16.549	1.90
16.720	1.91
17.271	2.01
17.400	2.09
17.811	2.40
18.236	2.45
18.655	2.55
18.996	2.61
19.546	2.95
19.686	3.07
21.023	3.67
21.183	3.69
22.209	4.47
22.429	4.48
23.319	4.89
23.577	5.12
24.238	5.37
24.648	5.87
24.934	5.82
25.517	6.27
25.619	6.58
26.188	6.82
26.559	7.15
27.066	7.50
27.626	7.98
28.070	8.38
28.418	8.63
77.880	1.993* x 10^{-3}*
79.061	2.102*
80.178	2.199*
CURVE 2 (cont.)	
81.236	2.291* x 10^{-3}*
82.239	2.406*
79.278	2.025*
80.160	2.209*
81.907	2.359*
83.533	2.546*
195.490	4.357* x 10^{-2}*
197.400	4.540*
199.000	4.636*
CURVE 3	
298.15	1.218 x 10^{-1}
300	1.233
320	1.399
340	1.565
360	1.727
380	1.886
400	2.038
420	2.184
440	2.323
460	2.456
480	2.582
500	2.702
550	2.973
600	3.210
650	3.415
700	3.593
750	3.747
800	3.883
850	4.002
900	4.110
950	4.208
1000	4.302
1050	4.392
1100	4.483
CURVE 4	
29.47	1.7 x 10^{-4}
31.00	1.8
33.08	1.7
36.53	2.0*
41.63	2.7*
44.07	3.2*
52.18	4.5
55.20	5.7
58.31	7.2
62.54	8.89
66.40	1.06* x 10^{-3}*
69.36	1.30*
73.20	1.48
74.91	1.70*
76.16	1.85*
79.57	2.17
82.36	2.56
87.34	2.97
93.70	3.98
100.33	4.80
106.71	6.32
112.68	7.77
118.02	8.75
123.42	1.03 x 10^{-2}*
129.11	1.21*
134.78	1.40*
141.38	1.64
146.55	1.86
152.64	2.12*
158.46	2.40*
164.10	2.70*
169.53	2.91*
175.06	3.14*
179.78	3.49*
185.29	3.78*
191.11	4.14*
196.00	4.43*
200.29	4.68*
205.77	5.05*
210.85	5.36*

*Not shown on plot

DATA TABLE NO. 2-A (continued)

T	C_p
CURVE 4 (cont.)	
213.09	5.50×10^{-2}
216.67	5.78*
226.77	6.54*
231.79	6.77*
236.78	7.20
241.93	7.62*
246.39	7.88*
250.76	8.24*
257.11	8.73*
265.99	9.52*
271.55	1.000×10^{-1}*
277.48	1.043*
281.26	1.079*
285.22	1.118
290.80	1.149
293.51	1.172
299.07	1.224*
300.57	1.234

*Not shown on plot

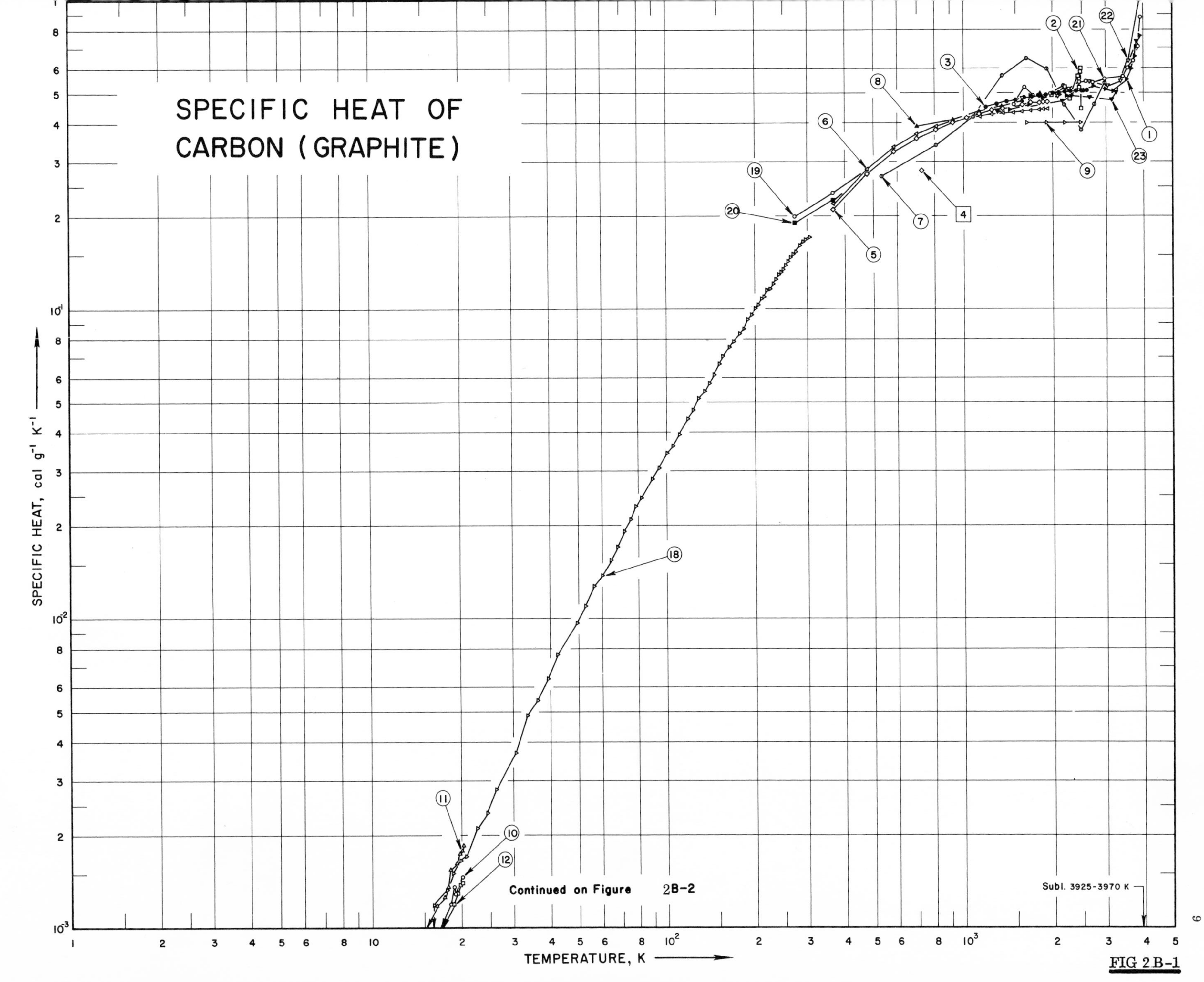
SPECIFIC HEAT OF
CARBON (GRAPHITE)
SPECIFIC HEAT, cal g⁻¹ K⁻¹
TEMPERATURE, K
Subl. 3925-3970 K
Continued on Figure 2B-2
1
2
3
4
5
6
7
8
9
10
11
12
18
19
20
21
22
23

FIG 2B-1

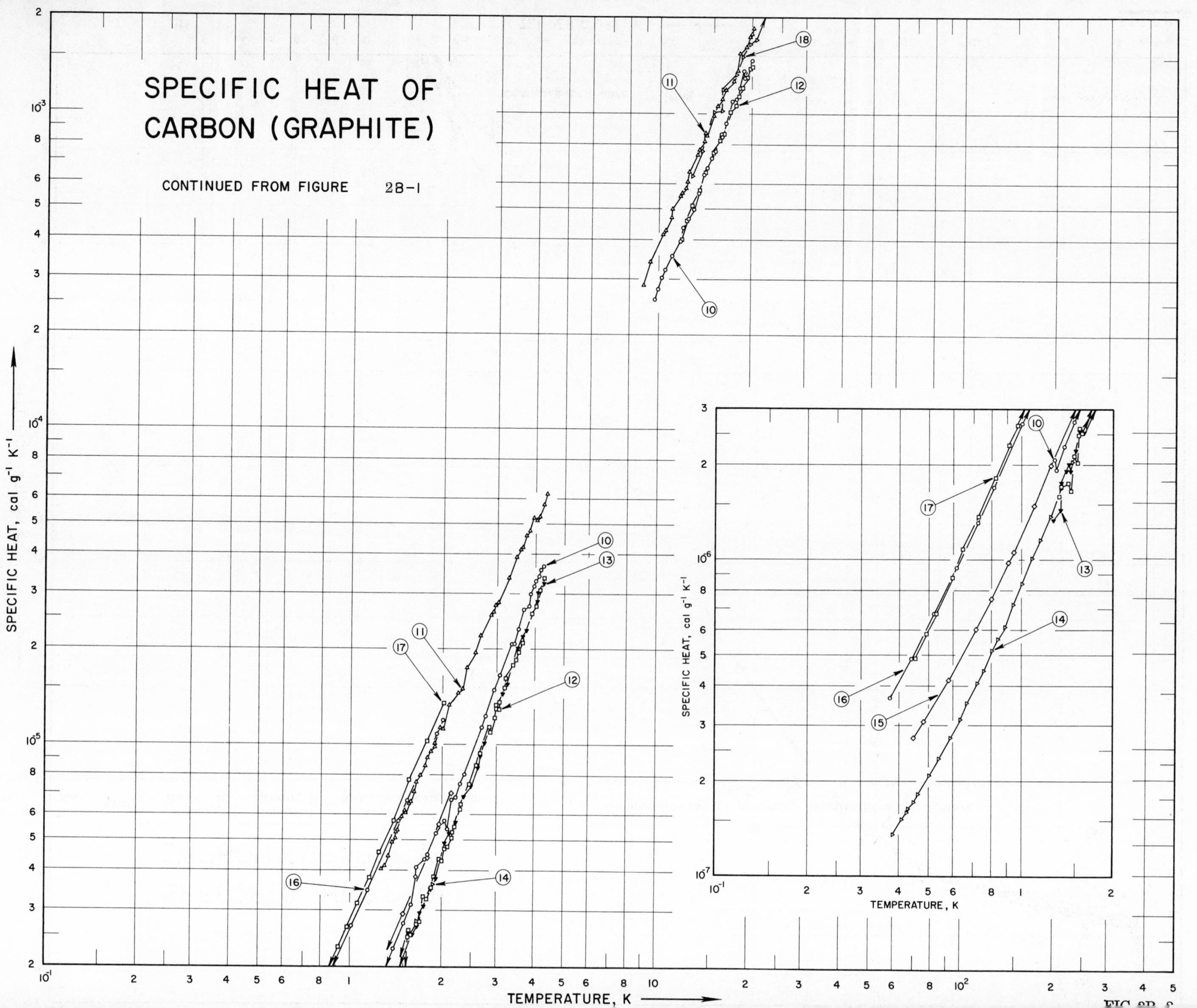
SPECIFIC HEAT OF
CARBON (GRAPHITE)
CONTINUED FROM FIGURE 2B-1
SPECIFIC HEAT, cal g^{-1} K^{-1}
TEMPERATURE, K
10
11
12
13
14
15
16
17
18

SPECIFICATION TABLE NO. 2-B SPECIFIC HEAT OF CARBON (GRAPHITE)

(Impurity < 2.00% each; total impurities < 5.00%)

[For Data Reported in Figure and Table No. 2-B]

Curve No.	Ref. No.	Year	Temp. Range, K	Reported Error, %	Name and Specimen Designation	Composition (weight percent), Specifications and Remarks
1	5	1960	1447-1676	± 5.0	3474 D	Very fine-grained and uniform; extruded; density = 65.0 lb ft^{-3}.
2	6	1962	1993-2483	± 5.0	Carbon graphite	
3	7	1965	1200-2600	0.14		< 0.001 Al, < 0.001 Si, 0.0001 B, Ca, and Mg.
4	8	1962	729		Graphite brick	Density = 131 lb ft^{-3}.
5	9	1960	366-1922	± 1.0	7087	Grade 7087 graphite; extruded; sealed in 95% argon - 5% hydrogen; density = 106 5 lb ft^{-3}(75 F).
6	9	1960	366-1922	± 1.0	GBH	Grade GBH graphite; molded; sealed in 95% argon - 5% hydrogen; density = 109.0 lb ft^{-3}(75 F).
7	10	1962	533-3033	≤ 5.0	ATJ	99.5 C, 0.2 Si, 0.1 Fe, trace Ca, Mg; molded and fired; density = 110.3 10 ft^{-3}.
8	11	1960	699-1811	< 2.9	ATJ	Sealed in helium.
9	10	1962	1644-2477	≤ 5.0	CS	
10	13	1958	1-20	≤ 0.8	H-CS-II pile graphite	0.4 ash, 0.005 gas, trace Fe, Mg, Ni.
11	13	1958	1-20	≤ 0.8	SA-25	< 0.05 ash, 0.01 gas, slight trace B, Fe, Si.
12	13	1958	1-20	≤ 0.8	CNG	< 0.05 ash, 0.001 gas, trace Mg, Ni, slight trace B; Canadian Natural Graphite.
13	13	1958	1-20	≤ 0.8	CNG-B	0.0004 B, Boronated Canadian Natural Graphite.
14	14	1963	0.4-2	± 2.0	NMG	Natural Madagascar Graphite; pumped for at least 3 days at room temperature; > 100 μ average crystallite dimension; very low degree of stacking faults.
15	14	1963	0.4-2	± 2.0	Pile graphite	Pumped for at least 3 days at room temperature; 240 Å average crystallite dimensions; low degree of stacking faults.
16	14	1963	0.4-2	± 2.0	Graphitized lamp-black SA 25	Pumped for at least 3 days at room temperature; average crystallite dimensions L_e = 125 Å, L_a = 90 Å; high degree of stacking faults.
17	14	1963	0.4-2	± 2.0	Pyrographite	Pumped for at least 3 days at room temperature; average crystallite dimensions L_e = 265 Å, L_a = 200 Å; very high degree of stacking faults.
18	15	1953	13-300		CS	High purity.
19	16	1956	273-1922		GBH	Under He atmosphere.
20	16	1956	273-1922		7087	Under He atmosphere.
21	17	1957	13-300		Acheson	0.27 Al; 0.07 Fe, 0.07 Si, 0.06 Pb, trace, Ca, Mg, Cu, Cr, V, Be, Ti and B; 1.5 x 10^{21} cm^{-1} total integrated neutron flux; exposed in Hanford reactor for several years at 30 C; kept in a vacuum for 24 hours.
22	12	1955	17-300	≤ 5.0	Ceylon natural graphite	0.06 ash, large foliated crystals.

DATA TABLE NO. 2-B SPECIFIC HEAT OF CARBON (GRAPHITE)

[Temperature, T, K; Specific Heat, C_p, Cal g^{-1} K^{-1}]

CURVE 1

T	C_p
1447	4.41×10^{-1}
1676	4.57
1694	4.90
1839	4.90
1868	4.81
2103	4.91
2232	4.91
2278	4.54*
2555	5.09
2722	5.25
2778	5.23*
3000	5.14
3077	4.30*
3198	5.07
3221	5.05*
3525	5.52
3656	5.97
3711	6.00*
3719	5.59*
3739	5.84*
3767	6.58
3818	7.02
3905	7.65

CURVE 2

T	C_p
1993	4.93×10^{-1}
2188	5.26
2298	4.80
2298	5.26
2429	5.70
2452	5.24
2461	5.88
2471	6.01
2483	4.48

CURVE 3

T	C_p
1200	4.54×10^{-1}
1300	4.63
1400	4.70
1500	4.76
1600	4.83
1700	4.86
1800	4.91
1900	4.94×10^{-1}
2000	4.96
2100	5.00
2200	5.03
2300	5.05
2400	5.08
2500	5.09
2600	5.11

CURVE 4

T	C_p
729	2.80×10^{-1}
729	2.79*
729	2.80*

CURVE 5

T	C_p
366	2.11×10^{-1}
478	2.75
589	3.24
700	3.56
811	3.80
922	4.00
1033	4.17
1144	4.29
1255	4.40
1366	4.48
1478	4.55*
1589	4.60
1700	4.64
1811	4.68
1866	4.69

CURVE 6

T	C_p
366	2.20×10^{-1}
478	2.84
589	3.35
700	3.68
811	3.90
922	4.05*
1033	4.15*
1144	4.22
1255	4.26
1366	4.30
1478	4.33×10^{-1}
1589	4.36
1700	4.39
1811	4.41
1866	4.42

CURVE 7

T	C_p
533	2.7×10^{-1}
811	3.4
1089	4.2
1366	5.7
1644	6.5
1922	6.0
2200	4.6
2478	3.8
2755	4.6
3033	5.4

CURVE 8

T	C_p
700	3.90×10^{-1}
922	4.11
1144	4.32
1366	4.53
1589	4.73
1811	4.94

CURVE 9

T	C_p
1644	4.0×10^{-1}
1922	4.0
2200	4.0
2478	4.0

CURVE 10

T	C_p
1.288	2.071×10^{-6}
1.308	1.927
1.389	2.277
1.495	2.748
1.580	3.133
1.649	3.830
1.658	4.115
1.753	4.372
1.799	4.504×10^{-6}
1.915	5.292
2.037	5.783
2.074	5.427
2.149	6.727
2.200	6.850*
2.202	6.833
2.285	7.504
2.344	8.040*
2.369	8.016
2.688	1.137×10^{-5}
2.760	1.231
2.937	1.478
3.066	1.658
3.341	2.083
3.420	2.097
3.519	2.317
3.590	2.429
3.596	2.479*
3.660	2.652
3.802	2.748
3.863	2.993
3.901	3.022*
3.960	3.170
4.000	3.289
4.050	3.357
4.097	3.395*
4.127	3.401
4.177	3.561
4.189	3.479*
4.217	3.518*
4.279	3.736*
4.285	3.757
9.719	2.581×10^{-4}
9.959	2.765
10.233	3.004
10.577	3.189
10.902	3.608*
11.211	3.529
11.477	3.933*
11.766	3.927
12.085	4.325
12.595	4.622
13.147	4.965
13.602	5.548
14.163	6.380×10^{-4}
14.429	6.697
14.774	7.067*
14.951	7.174
15.355	7.539
16.130	8.533
16.186	8.392
16.571	9.250
17.443	1.083×10^{-3}
17.924	1.117
18.347	1.199
18.922	1.365
19.052	1.282
20.114	1.469

CURVE 11

T	C_p
1.267	4.076×10^{-6}
1.296	4.197
1.336	4.493
1.371	4.945
1.405	5.152
1.415	5.334
1.427	5.397
1.473	5.968
1.518	6.237
1.520	6.131
1.544	6.563*
1.562	6.550
1.573	6.629
1.620	7.112
1.641	7.602
1.696	8.027
1.717	8.135*
1.765	8.608
1.788	9.100
1.826	9.325*
1.836	9.593
1.856	9.542*
1.883	9.892
1.890	1.017×10^{-5}
1.957	1.132
1.976	1.095*
2.010	1.123
2.025	1.152*
2.066	1.212×10^{-5}*
2.092	1.316
2.230	1.454
2.325	1.517
2.394	1.751
2.534	1.801*
2.538	1.954
2.695	2.205
2.857	2.535*
2.869	2.577
2.907	2.632
2.955	2.768
3.006	2.812
3.051	2.785*
3.120	2.997*
3.265	3.383
3.455	3.923
3.532	3.877*
3.562	4.076
3.619	4.228
3.716	4.597
3.820	4.753
3.944	5.232
4.032	5.110
4.130	5.265
4.284	5.732
4.364	6.218
8.922	2.847×10^{-4}
9.375	3.378
10.369	4.122
10.515	4.271
11.013	4.692
11.029	4.700*
11.129	4.940
11.744	5.412*
11.882	5.489
11.960	5.584
12.364	5.773
12.452	6.069
12.637	6.522
13.434	7.382
13.643	7.617
14.170	8.433*
14.179	8.147
14.254	8.567

*Not shown on plot

DATA TABLE NO. 2-B (continued)

CURVE 11 (contd)

T	C_p
14.378	8.483 x 10^{-4}
15.119	9.842
15.524	1.054 x 10^{-3}
16.034	1.117
16.589	1.187
17.557	1.267
18.101	1.357
18.476	1.547
19.352	1.622
19.825	1.741
20.034	1.737
20.138	1.774
20.337	1.851

CURVE 12

T	C_p
1.347	1.592 x 10^{-6}
1.365	1.718
1.406	1.666
1.421	1.883*
1.427	1.762
1.480	2.019*
1.481	2.256*
1.493	2.145
1.500	2.027*
1.537	2.028
1.545	2.477
1.549	2.307*
1.553	2.432*
1.583	2.519
1.595	2.617
1.603	2.589*
1.620	2.397*
1.652	2.722
1.677	3.089
1.683	2.770
1.687	2.889*
1.689	2.899*
1.735	3.315
1.735	3.132*
1.750	3.256
1.788	3.252*
1.794	3.497
1.827	3.544*
1.831	3.540
1.847	3.725*
1.849	3.522*
1.889	3.844

CURVE 12 (contd)

T	C_p
1.950	4.363 x 10^{-6}
1.964	4.357*
1.982	4.311
2.008	4.412*
2.016	4.398*
2.030	4.707
2.056	4.942
2.061	4.892*
2.070	4.773*
2.111	5.078*
2.142	5.073
2.151	5.303
2.198	5.517
2.254	6.105*
2.296	6.259
2.299	6.500
2.454	7.510
2.567	8.625*
2.575	8.617
2.657	9.400
2.832	1.117 x 10^{-5}*
2.843	1.104*
2.847	1.136
2.860	1.090
2.954	1.222
2.992	1.292
2.997	1.337
3.052	1.291
3.060	1.369
3.082	1.424*
3.170	1.515
3.205	1.622
3.226	1.587*
3.364	1.775*
3.372	1.784
3.457	1.854
3.497	1.987*
3.521	2.017*
3.534	1.952
3.649	2.094
3.738	2.336*
3.904	2.578
4.043	2.739
4.152	3.077
4.294	3.354
11.922	3.984 x 10^{-4}
12.384	4.563
12.866	5.104

CURVE 12 (contd)

T	C_p
13.125	5.217 x 10^{-4}*
13.664	5.648
14.119	6.283*
14.427	6.467
14.885	7.121
15.180	7.476
15.393	7.408*
15.894	8.157
16.209	8.633*
16.311	8.567
17.113	9.825*
17.137	1.007
17.237	1.027*
17.846	1.053
18.223	1.126
18.741	1.197
19.124	1.226
19.477	1.295*
19.846	1.383
19.987	1.394*
20.110	1.397

CURVE 13

T	C_p
1.275	1.327 x 10^{-6}
1.348	1.439
1.366	1.746
1.418	1.904
1.438	2.008
1.458	1.900
1.499	2.212*
1.515	2.226
1.547	2.353*
1.555	2.556
1.600	2.532
1.653	2.691
1.690	2.904
1.703	3.060*
1.741	3.087
1.779	3.285*
1.835	3.509
1.900	3.799
1.977	4.315*
2.042	4.894
2.128	5.243
2.214	5.676
2.287	6.174*
2.350	6.871

CURVE 13 (contd)

T	C_p
2.488	7.402 x 10^{-6}
2.601	8.458
2.666	9.208
2.745	1.011 x 10^{-5}
2.945	1.219*
3.059	1.362*
3.131	1.430
3.238	1.549
3.327	1.658*
3.439	1.843*
3.530	2.005
3.643	2.188
3.738	2.281
3.844	2.473*
3.964	2.571*
4.051	2.813
4.178	3.085
4.276	3.241*
4.291	3.232

CURVE 14

T	C_p
0.381_6	1.36 x 10^{-7}
0.409_0	1.51
0.425_0	1.59
0.427_5	1.63
0.447_5	1.70
0.461_3	1.81
0.501_1	2.07
0.540_4	2.35
0.590_5	2.73
0.635_2	3.12
0.668_6	3.50
0.721_4	4.07
0.757_1	4.46
0.808_9	5.16
0.841_9	5.62
0.887_9	6.14
0.942_9	7.20
1.009_5	8.44
1.089_9	1.001 x 10^{-6}
1.156_3	1.158
1.250_7	1.372
1.351_6	1.680
1.464_4	2.039
1.663_6	2.769
1.872_2	3.632

CURVE 15

T	C_p
0.447_7	2.74 x 10^{-7}
0.482_9	3.10
0.582_1	4.19
0.717_5	6.02
0.802_5	7.52
0.912_0	9.83
0.946_9	1.067 x 10^{-6}
1.106_6	1.484
1.268_9	1.998
1.490_6	2.948
1.772_9	4.425
1.953_2	5.623
2.139_5	7.047

CURVE 16

T	C_p
0.374_7	3.65 x 10^{-7}
0.440_1	4.87
0.521_7	6.74
0.619_4	9.43
0.725_2	1.313 x 10^{-6}
0.818_5	1.690
1.013_8	2.697
1.142_5	3.487
1.441_4	5.772
1.544_5	6.696*
1.780_9	9.156*
1.915_1	1.079 x 10^{-5}
2.006_8	1.191

CURVE 17

T	C_p
0.454_6	4.88 x 10^{-7}
0.491_8	5.81
0.528_7	6.77
0.598_1	5.98
0.647_9	1.071 x 10^{-6}
0.729_8	1.371
0.823_7	1.819
0.911_6	2.301
0.976_6	2.669
1.057_8	3.159
1.150_0	3.814
1.246_7	4.568
1.378_6	5.774
1.562_0	7.758
1.771_7	1.024 x 10^{-5}
2.013_4	1.355

CURVE 18

T	C_p
12.92	6.3 x 10^{-4}
13.77	7.6
15.15	1.01 x 10^{-3}
16.06	1.02
16.20	1.18
17.88	1.32
18.73	1.51
19.93	1.65
20.77	1.70
22.67	2.11
24.47	2.36
26.29	2.82
30.56	3.69
33.57	4.86
36.33	5.46
39.36	6.39
42.48	7.67
49.44	9.67
52.95	1.10 x 10^{-2}
56.48	1.27
60.20	1.38
64.16	1.55
67.90	1.71
71.49	1.92
75.17	2.09
78.80	2.32
82.39	2.46
89.52	2.83
94.23	3.07
100.39	3.43
106.59	3.61
112.49	3.94
118.14	4.40
123.88	4.72
129.43	5.06
135.19	5.45
140.57	5.75
146.37	6.18
152.42	6.68
158.23	7.01
163.93	7.55
169.40	7.88
177.55	8.37
183.68	8.62
189.63	9.25
195.24	9.64
200.81	1.006 x 10^{-1}
206.11	1.036
211.10	1.087

*Not shown on plot

DATA TABLE NO. 2-B (continued)

T	C_p
CURVE 18 (contd)	
215.80	1.099×10^{-1}
220.40	1.157
225.30	1.165
230.44	1.213
235.55	1.252
240.48	1.291
245.20	1.322
249.33	1.348
254.34	1.388
259.69	1.421
264.92	1.473
269.89	1.507
271.49	1.513*
275.53	1.539
279.94	1.572*
282.37	1.589*
284.07	1.602
286.99	1.625*
291.21	1.659
293.22	1.652*
294.71	1.653*
296.95	1.684
299.02	1.702*
300.57	1.714
CURVE 19	
273	1.98×10^{-1}
366	2.38
478	2.81
589	3.18
700	3.51
811	3.78
922	4.00
1033	4.17
1144	4.28
1255	4.35
1366	4.36
1478	4.32
1589	4.23
1700	4.08
1811	3.88
1866	3.77
1922	3.64

T	C_p
CURVE 20	
373	1.80×10^{-1}
366	2.26
478	2.75
589	3.18
700	3.54
811	3.85
922	4.09
1033	4.28
1144	4.40
1255	4.46
1366	4.46
1478	4.40
1589	4.28
1700	4.10
1811	3.86
1866	3.71
1922	3.55
CURVE 21*	
12.82	1.14×10^{-3}
13.99	1.42
14.68	1.51
15.22	1.61
16.02	1.82
16.44	1.86
17.86	2.21
18.07	2.20
20.05	2.39
20.23	2.49
22.48	2.76
27.08	4.18
29.38	4.95
31.26	5.25
34.37	7.09
37.04	7.37
39.74	8.83
42.48	9.33
45.60	1.06×10^{-2}
49.42	1.13
53.55	1.41
57.45	1.51
61.02	1.74
68.15	1.96
72.20	2.24
76.46	2.51
80.69	2.69

T	C_p
CURVE 21 (contd)	
80.91	2.71×10^{-2}
85.56	2.94
85.68	2.95
90.28	3.17
91.15	3.24
94.90	3.42
95.39	3.46
101.21	3.77
106.36	4.06
111.19	4.33
116.14	4.68
118.22	4.79
121.08	4.94
127.74	5.32
132.64	5.68
137.56	5.98
141.65	6.28
147.05	6.61
152.61	7.15
158.01	7.56
163.40	7.98
169.29	8.46
175.14	8.70
181.23	9.20
187.63	9.68
194.03	1.013×10^{-1}
200.40	1.058
206.87	1.108
222.77	1.221
230.87	1.303
236.57	1.312
242.73	1.378
248.60	1.410
259.51	1.501
263.89	1.527
268.23	1.563
275.71	1.610
279.25	1.623
282.65	1.638
286.22	1.674
289.90	1.727
293.64	1.721
296.59	1.772
300.25	1.792
303.68	1.827

T	C_p
CURVE 22*	
17.70	1.12×10^{-3}
18.46	1.45
21.65	1.67
23.45	1.94
25.47	2.22
27.92	2.62
31.25	3.62
33.31	4.48
35.76	5.02
38.64	6.39
41.63	7.19
44.75	8.12
47.87	8.83
51.66	1.17×10^{-2}
55.67	1.20
59.78	1.42
63.94	1.61
68.06	1.70
72.14	1.93
76.11	2.22
83.14	2.53
88.05	2.72
93.06	3.02
97.88	3.15
102.48	3.41
106.86	3.68
111.16	3.90
115.93	4.12
120.92	4.47
125.96	4.67
130.37	5.00
134.89	5.19
139.32	5.42
143.90	5.66
148.51	6.01
156.61	6.58
169.88	7.19
178.12	7.95
189.64	8.62
195.16	8.72
202.28	9.32
208.75	9.77
216.79	1.032×10^{-1}
223.28	1.070
229.27	1.097
233.61	1.150
235.64	1.147

T	C_p
CURVE 22 (contd)	
241.62	1.177×10^{-1}
242.02	1.176
247.78	1.231
249.61	1.243
253.36	1.253
260.20	1.308
261.54	1.322
265.40	1.311
265.42	1.364
270.77	1.363
271.08	1.367
276.41	1.427
279.62	1.412
281.47	1.430
287.75	1.481
290.87	1.520
293.29	1.518
294.69	1.537
295.78	1.524
300.45	1.587

* Not shown on plot

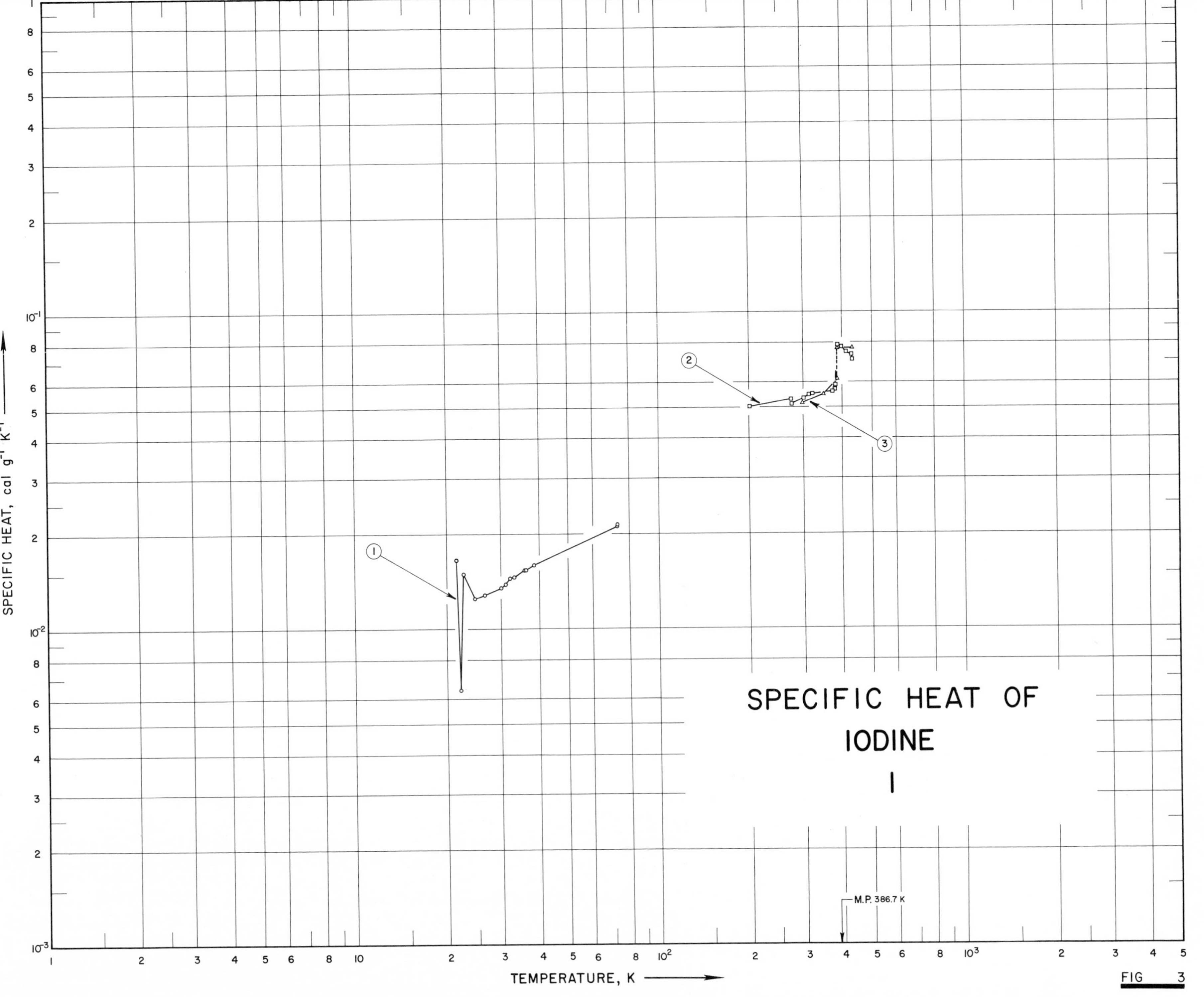

SPECIFIC HEAT OF
IODINE
I
SPECIFIC HEAT, cal g⁻¹ K⁻¹
TEMPERATURE, K
M.P. 386.7 K
1
2
3

FIG 3

SPECIFICATION TABLE NO. 3 SPECIFIC HEAT OF IODINE

(Impurity <2.00% each; total impurities <5.00%)

[For Data Reported in Figure and Table No. 3]

Curve No.	Ref. No.	Year	Temp. Range, K	Reported Error, %	Name and Specimen Designation	Composition (weight percent), Specifications and Remarks
1	453	1916	22-73			
2	454	1936	202-433			
3	455	1938	298-433			Merck reagent; purified.

DATA TABLE NO. 3 SPECIFIC HEAT OF IODINE

[Temperature, T, K; Specific Heat, C_p, Cal g^{-1} K^{-1}]

T	C_p	T	C_p
CURVE 1		**CURVE 3**	
Series 1		298.15	5.153×10^{-2}
		300	5.153*
22.0	6.422×10^{-3}	350	5.495
30.0	1.351×10^{-2}	(s) 386.75	6.150
32.2	1.458	(ℓ) 386.75	7.683
36.4	1.548	400	7.683*
38.5	1.592	433	7.683
Series 2			
21.5	1.663×10^{-2}		
22.7	1.489		
24.6	1.257		
26.5	1.288		
28.6	1.422		
31.0	1.391		
33.2	1.466		
35.9	1.540		
72.8	2.120		
72.9	2.147		
CURVE 2			
202.2	5.00×10^{-2}		
202.9	5.05*		
274.6	5.31		
275.1	5.12		
275.3	5.23*		
275.5	5.20*		
301.5	5.34		
301.9	5.40*		
314.8	5.48		
315.0	5.49*		
322.8	5.52		
348.7	5.60*		
375.7	5.61		
380.3	5.70		
383.2	5.92		
389.4	7.88		
398.8	7.77		
413.2	7.48		
413.9	7.37		
432.7	7.10		

*Not shown on plot

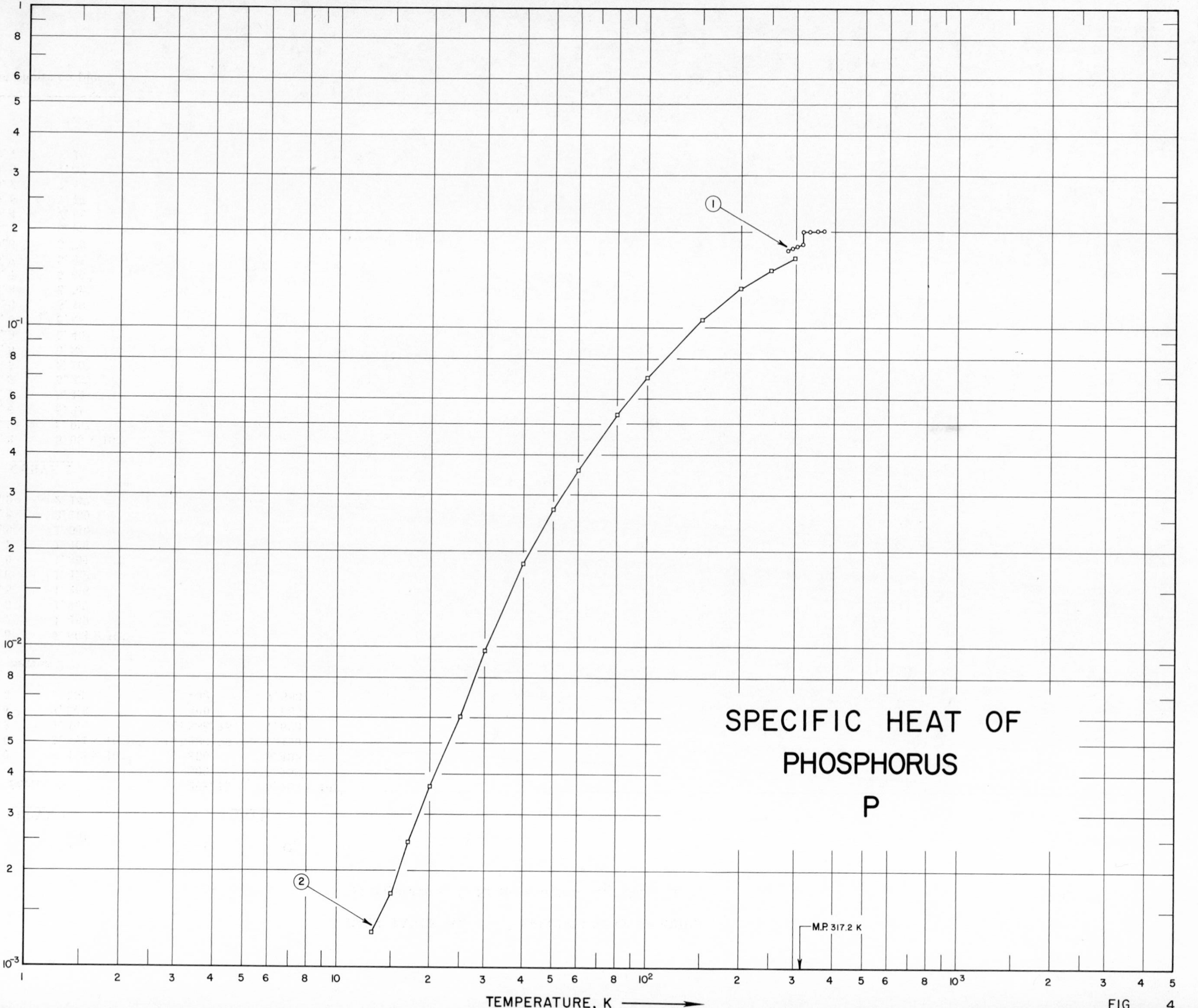
SPECIFIC HEAT OF
PHOSPHORUS
P
SPECIFIC HEAT, cal g^{-1} K^{-1}
TEMPERATURE, K
M.P. 317.2 K
1
2

FIG 4

SPECIFICATION TABLE NO. 4 SPECIFIC HEAT OF PHOSPHORUS

(Impurity < 2.00% each; total impurities < 5.00)

[For Data Reported in Figure and Table No. 4]

Curve No.	Ref. No.	Year	Temp. Range, K	Reported Error, %	Name and Specimen Designation	Composition (weight percent), Specifications and Remarks
1	456	1942	283-370			
2	457	1965	13-298		Black Phosphorus	99.12 P, ~0.3 C and ~0.3 Pb.

DATA TABLE NO. 4 SPECIFIC HEAT OF PHOSPHORUS

[Temperature, T, K; Specific Heat, C_p, Cal g^{-1} K^{-1}]

T	C_p
CURVE 1	
283.15	1.754 x 10^{-1}
293.15	1.777
303.15	1.800
(s) 317.35	1.833
(ℓ) 317.35	2.002
323.15	2.004*
333.15	2.009
343.15	2.012*
353.15	2.015
363.15	2.017*
370.15	2.019
CURVE 2	
13	1.28 x 10^{-3}
15	1.69
17	2.45
20	3.664
25	6.028
30	9.715
40	1.825 x 10^{-2}
50	2.700
60	3.577
80	5.353
100	6.993
150	1.064 x 10^{-1}
200	1.336
250	1.513
298.15	1.665

*Not shown on plot

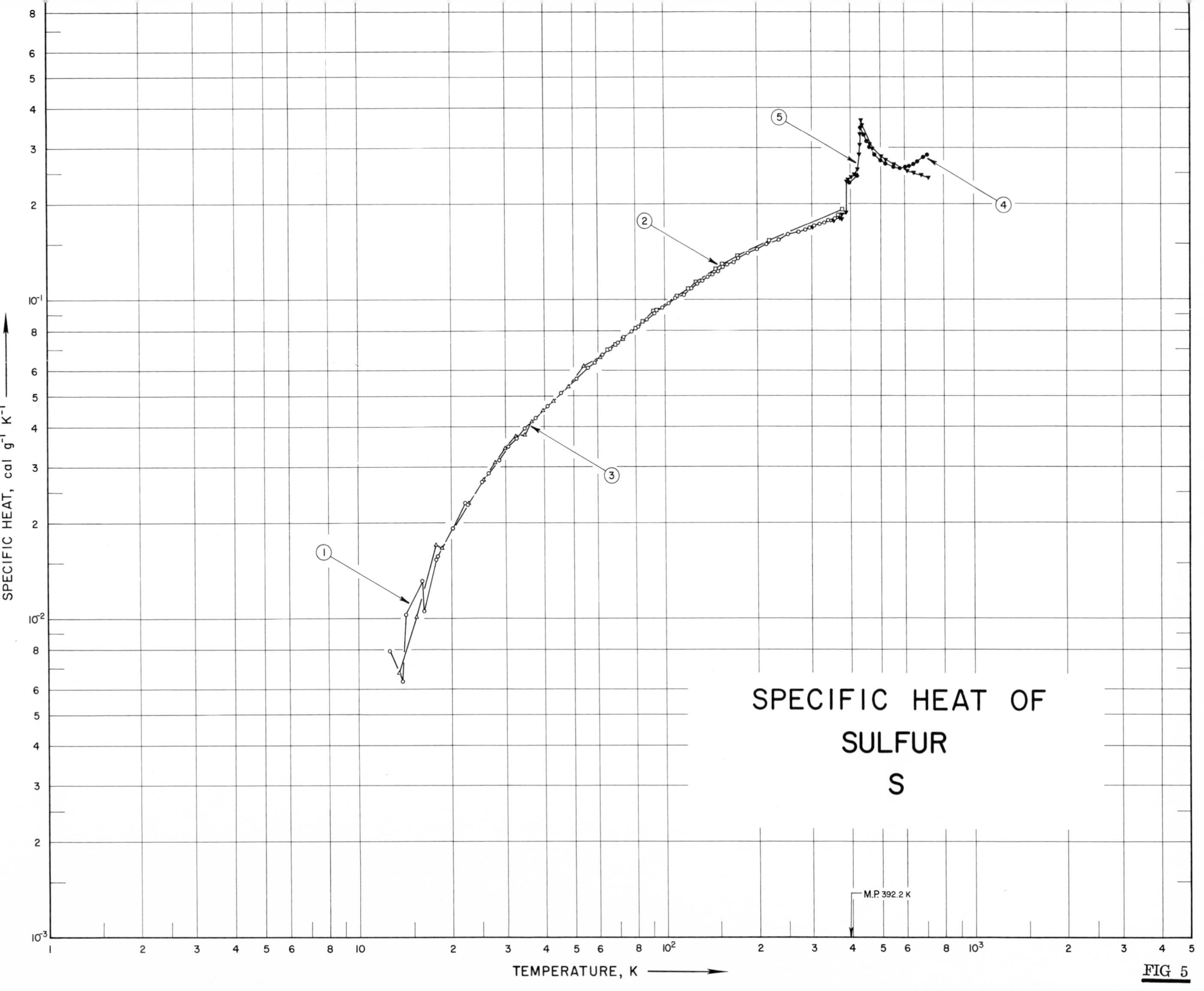

SPECIFIC HEAT OF
SULFUR
S
M.P. 392.2 K
TEMPERATURE, K
SPECIFIC HEAT, cal g⁻¹ K⁻¹
1
2
3
4
5

FIG 5

SPECIFICATION TABLE NO. 5 SPECIFIC HEAT OF SULFUR

(Impurity <2.00% each; total impurities <5.00)

[For Data Reported in Figure and Table No. 5]

Curve No.	Ref. No.	Year	Temp. Range, K	Reported Error, %	Name and Specimen Designation	Composition (weight percent), Specifications and Remarks
1	450	1937	13-366	0.5	Rhombic	0.001 CS_2; pure crystallized commercial product melted and maintained at 120 C for several days; distilled in Pyrex Glass and recrystallized; freed of CS_2 by heating at 85 ±5 C at 1 mm pressure for 8 days.
2	450	1937	65-376	0.5	Monoclinic	Same as above.
3	450	1937	14-61	0.5		A mixture of 60% monoclinic form and 40% rhombic form.
4	451	1954	393-713			
5	452	1959	303-718	0.1-0.2	Rhombic	99.999 S; sample supplied by NBS.

DATA TABLE NO. 5 SPECIFIC HEAT OF SULFUR

[Temperature, T, K; Specific Heat, C_p, Cal $g^{-1} K^{-1}$]

T	C_p
CURVE 1	
12.68	7.923 x 10^{-3}
13.85	6.363
14.26	1.032 x 10^{-2}
16.13	1.323
16.33	1.064
17.99	1.541
18.16	1.575
20.26	1.925
20.27	1.968*
22.23	2.311
22.61	2.283
25.18	2.692
26.38	2.860
28.66	3.157
30.71	3.475
32.62	3.687
34.68	3.974
37.70	4.286
41.10	4.660
45.79	5.122
51.39	5.696
55.90	6.132
58.95	6.379
59.82	6.482*
62.54	6.759
63.99	6.828*
66.33	7.018
68.84	7.286*
70.04	7.352
70.34	7.383*
73.22	7.651
74.04	7.658*
74.61	7.648*
78.12	7.966
82.44	8.278
87.66	8.706
92.81	9.092
96.29	9.323*
98.83	9.498
100.83	9.626*
103.92	9.782
105.25	9.838*
109.80	1.013 x 10^{-1}
117.13	1.044
122.76	1.094

T	C_p
CURVE 1 (cont.)	
128.63	1.125 x 10^{-1}
132.02	1.149*
133.29	1.152
134.02	1.163*
137.45	1.172*
139.32	1.182
139.63	1.177*
144.50	1.205
145.01	1.207*
150.39	1.233
150.40	1.234*
150.40	1.234*
152.21	1.252*
155.44	1.272
157.92	1.277*
160.16	1.296
163.26	1.303*
165.08	1.313*
165.31	1.316*
167.75	1.319
169.16	1.333*
169.50	1.315*
174.26	1.331*
174.76	1.354
180.25	1.366*
181.33	1.360*
187.56	1.401
188.86	1.417*
194.00	1.423*
196.67	1.434*
200.83	1.448
202.25	1.462*
207.38	1.466*
208.25	1.479*
214.59	1.475*
215.69	1.495
220.99	1.513*
222.50	1.518*
226.56	1.528*
229.53	1.536*
236.46	1.547
236.51	1.555*
244.79	1.577*
249.02	1.585*
253.63	1.605

T	C_p
CURVE 1 (cont.)	
257.12	1.600 x 10^{-1}*
258.85	1.604*
263.79	1.623*
265.98	1.619*
272.91	1.634
280.15	1.648*
280.47	1.652*
287.32	1.663
289.12	1.661*
289.26	1.659*
290.17	1.679*
295.48	1.672*
296.46	1.660*
296.55	1.685
297.71	1.658*
298.28	1.673*
301.40	1.690*
302.32	1.689*
303.97	1.689*
306.13	1.705
311.46	1.718*
311.95	1.742*
312.42	1.720*
319.98	1.740
320.92	1.745*
323.17	1.744*
324.37	1.762*
325.87	1.763*
330.34	1.749
331.32	1.775*
335.76	1.734*
338.11	1.770*
340.22	1.777
344.16	1.765*
346.87	1.787*
347.86	1.801*
353.74	1.798*
357.23	1.806
361.21	1.772*
364.02	1.851*
364.59	1.769*
365.60	1.854

T	C_p
CURVE 2	
64.83	6.993 x 10^{-2}
68.82	7.268
72.82	7.583
80.12	8.178
84.76	8.559
91.98	8.238
94.31	9.283
98.25	9.573*
102.55	9.825*
110.08	1.027 x 10^{-1}
119.59	1.086
127.24	1.136
147.42	1.256
154.91	1.297
172.41	1.374
219.80	1.538
375.14	1.924*
376.16	1.926
CURVE 3	
13.51	6.769 x 10^{-3}
15.41	1.017
17.75	1.712
18.65	1.678
20.27	1.937*
22.72	2.305
25.38	2.726
27.73	3.097
29.99	3.450
32.33	3.740
34.66	3.784
36.55	4.173
39.86	4.510
43.04	4.835
48.17	5.356
54.19	6.220
61.43	6.634
CURVE 4	
393	2.34 x 10^{-1}
403	2.36*
413	2.39*
423	2.46

T	C_p
CURVE 4 (cont.)	
433	3.47 x 10^{-1}
443	3.28
453	3.14
463	3.02
473	2.92*
483	2.85
493	2.79*
503	2.74
513	2.70*
523	2.67
533	2.64*
543	2.62*
553	2.61
563	2.60*
573	2.59*
583	2.59
593	2.59*
603	2.60
613	2.61*
623	2.62
633	2.64*
643	2.66
653	2.68*
663	2.70
673	2.73*
683	2.76*
693	2.79
703	2.82*
713	2.85
CURVE 5	
303	1.699 x 10^{-1}
313	1.716*
323	1.733*
333	1.749*
343	1.765*
353	1.780
363	1.795*
368.54	1.802
368.54	1.844*
373	1.803*
374	1.789
374	1.858
383	1.877*

T	C_p
CURVE 5 (cont.)	
388.357	1.888 x 10^{-1}
388.357	2.364
393	2.384
403	2.425
413	2.475
423	2.569
429	2.854
430	3.054
431	3.314
432	3.616*
433	3.671
434	3.629*
435	3.581
436	3.529*
437	3.490*
443	3.325*
453	3.165*
463	3.064
473	2.984
483	2.918*
493	2.864*
503	2.818
513	2.779*
523	2.742
533	2.710*
543	2.681*
553	2.654
563	2.629*
573	2.607*
583	2.587*
593	2.569*
603	2.553*
613	2.538
623	2.525*
633	2.513*
643	2.502
653	2.491*
663	2.480*
673	2.469*
683	2.457
693	2.443*
703	2.428*
713	2.409*
717.75	2.400

* Not shown on plot

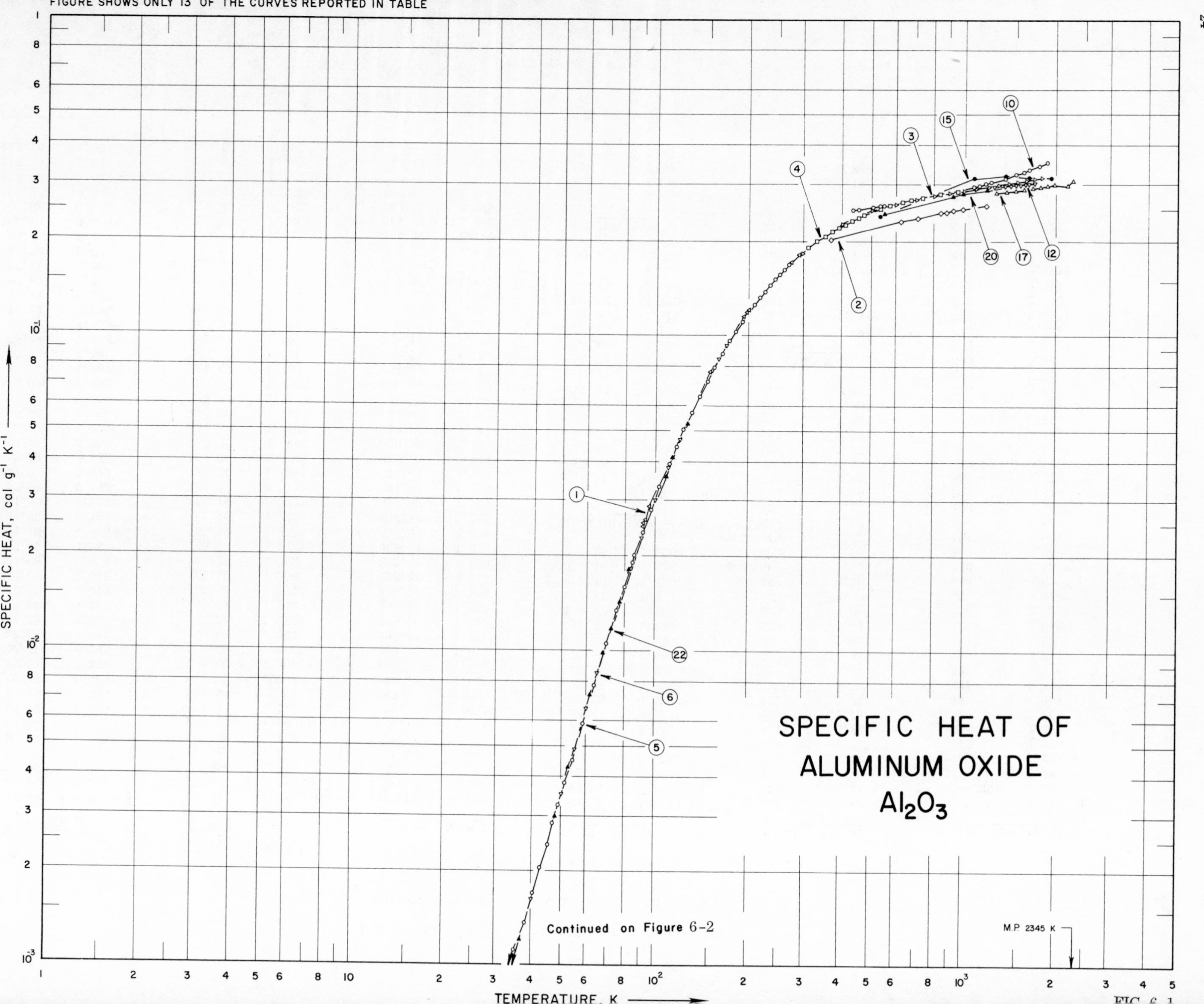

FIGURE SHOWS ONLY 13 OF THE CURVES REPORTED IN TABLE
SPECIFIC HEAT OF
ALUMINUM OXIDE
Al2O3
SPECIFIC HEAT, cal g-1 K-1
TEMPERATURE, K
Continued on Figure 6-2
M.P. 2345 K

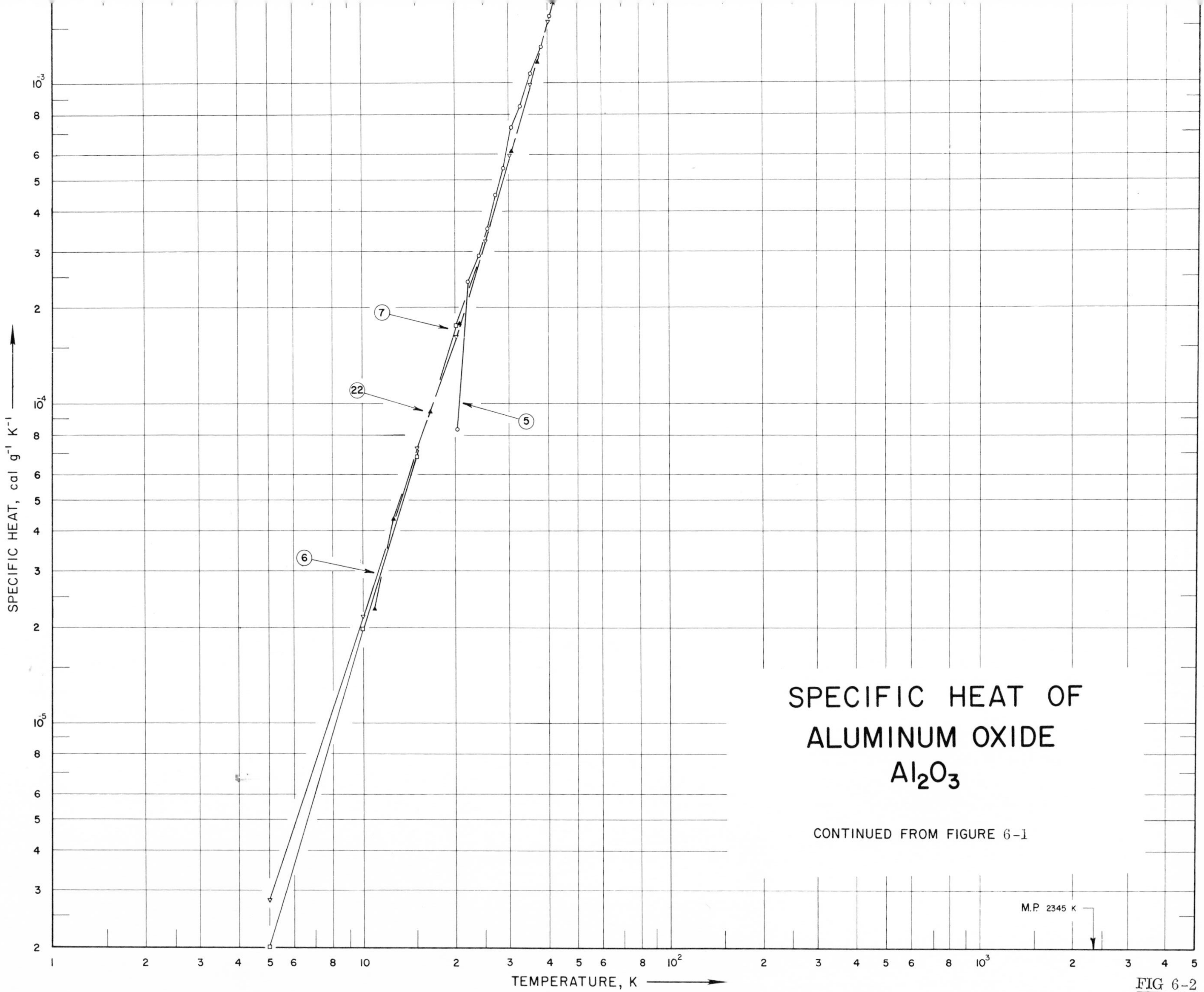

SPECIFIC HEAT OF
ALUMINUM OXIDE
Al_2O_3
CONTINUED FROM FIGURE 6-1
M.P. 2345 K
SPECIFIC HEAT, cal g^{-1} K^{-1}
TEMPERATURE, K
5
6
7
22

FIG 6-2

SPECIFICATION TABLE NO. 6 SPECIFIC HEAT OF ALUMINUM OXIDE Al_2O_3

[For Data Reported in Figure and Table No. 6]

Curve No.	Ref. No.	Year	Temp. Range, K	Reported Error, %	Name and Specimen Designation	Composition (weight percent), Specifications and Remarks
1	18	1926	91-291			99. 3 Al_2O_3.
2	19	1929	369-1187			Sapphire.
3	20	1945	298-1800	± 0. 4		100 Al_2O_3; natural almost colorless sapphire.
4	21	1947	273-1173	± 0. 2		Corundum, synthetic sapphire; 0. 02 - 0. 03 impurities mostly SiO_2.
5	22	1950	20-295			Synthetic sapphire; 0. 02 SiO_2.
6	23	1953	5-1200	0. 2		Corundum, synthetic sapphire; 0. 01 - 0. 02 impurities.
7	24	1956	5-1200	0. 2		Corundum, synthetic sapphire; 99. 98 Al_2O_3, 0. 005 each Fe, Si, and 0. 002 Cr.
8	25	1956	337-923			Synthetic sapphire; calorimetric standard.
9	26	1958	312-689			Pure Al_2O_3.
10	27	1958	435-1884	2. 9		Al_2O_3; polycrystalline.
11	28	1960	65-300			α-Al_2O_3.
12	29	1960	1089-1700			Synthetic sapphire.
13	30	1960	325-986	0. 5		99. 997 Al_2O_3, 0. 0013 Cr, 0. 001 Fe, 0. 001 Mo, and 0. 0004 Cu.
14	31	1960	533-1228	2. 9		Synthetic sapphire.
15	32	1962	533-1922	≤5. 0		100 Al_2O_3; density 233 lb ft^{-3}.
16	33	1961	342-764			Natl. Bur. Std. standard sample; screened on 0. 065 and treated with hot HCl to remove traces of Fe.
17	34	1961	1273-2273	1. 3		Synthetic sapphire.
18	35	1962	53-291	± 0. 10		Calorimetry conference synthetic sapphire.
19	36	1962	283-303	0. 10		Calorimetry conference standard.
20	37	1963	552-1385			Synthetic sapphire.
21	38	1963	1300-2000	≤3. 0		98. 7 Al_2O_3 and 1. 0 SiO_2; sintered; under argon atmosphere.
22	39	1965	10-353	0. 1		Synthetic sapphire.

DATA TABLE NO. 6 SPECIFIC HEAT OF ALUMINUM OXIDE Al_2O_3

[Temperature, T, K; Specific Heat, C_p, Cal g^{-1} K^{-1}]

T	C_p
CURVE 1	
91.1	2.49 x 10^{-2}
91.7	2.55
93.0	2.60
95.6	2.86
150.6	7.68
193.3	1.155 x 10^{-1}
197.2	1.190
200.1	1.218
275.1	1.726
276.4	1.743*
288.6	1.800*
291.3	1.813
CURVE 2	
369	2.030 x 10^{-1}
625	2.315
711	2.385
841	2.484
876	2.482
926	2.524
998	2.545
1009	2.502*
1187	2.606
CURVE 3	
298.15	1.845 x 10^{-1}
300	1.857*
400	2.291
500	2.508
600	2.640
700	2.731
800	2.801
900	2.858
1000	2.907
1100	2.952
1200	2.993
1300	3.031
1400	3.068
1500	3.104
1600	3.138
1700	3.172
1800	3.205

T	C_p
CURVE 4	
273	1.731 x 10^{-1}*
293	1.830*
313	1.922
333	2.007
353	2.085
373	2.157
393	2.224
413	2.285
433	2.341
453	2.392
473	2.438
493	2.480*
513	2.518*
533	2.552
553	2.583*
573	2.611
593	2.637*
613	2.660*
633	2.681
653	2.701*
673	2.719
693	2.736*
713	2.753*
733	2.769
753	2.784*
773	2.799*
793	2.813*
813	2.827
833	2.840*
853	2.853*
873	2.865*
893	2.877*
913	2.888*
933	2.899*
953	2.909
973	2.919*
993	2.928*
1013	2.937*
1033	2.945*
1053	2.953*
1073	2.960*
1093	2.967*
1113	2.974*
1123	2.981*
1153	2.988*
1173	2.995*

T	C_p
CURVE 5	
20.19	8.5 x 10^{-5}
21.91	2.48 x 10^{-4}
23.85	2.99
25.40	3.65
26.95	4.65
28.54	5.61
30.39	7.57
32.29	8.80
34.95	1.114 x 10^{-3}
38.04	1.355
40.60	1.696
42.97	2.046
45.05	2.413
46.96	2.832
48.92	3.230
51.06	3.780
54.13	4.472
58.74	5.857
63.72	7.730
69.60	1.051 x 10^{-2}
75.01	1.340
79.48	1.594
84.58	1.901
85.76	2.003
91.34	2.388
97.32	2.803
103.59	3.312
111.03	3.919
117.48	4.469
123.70	5.033
130.82	5.700
138.85	6.475
146.94	7.193
154.99	7.969
164.11	8.805
173.28	9.674
181.42	1.044 x 10^{-1}
189.85	1.116
199.02	1.193*
208.20	1.266
216.93	1.334
225.44	1.396
234.46	1.459
243.08	1.520
251.59	1.578
260.24	1.636
CURVE 5 (cont.)	
269.44	1.700 x 10^{-1}
278.05	1.758*
285.61	1.800*
294.85	1.854*
CURVE 6	
5	2.8 x 10^{-6}
10	2.2 x 10^{-5}
15	7.41
20	1.66 x 10^{-4}
25	3.322
30	6.158
35	1.026 x 10^{-3}
40	1.619
45	2.436*
50	3.497
55	4.850
60	6.517
65	8.486
70	1.074 x 10^{-2}
75	1.329*
80	1.616*
85	1.933*
90	2.272
95	2.631*
100	3.010
110	3.823
120	4.700
130	5.617*
140	6.544*
150	7.499*
160	8.436
170	9.362
180	1.026 x 10^{-1}*
190	1.114*
200	1.199*
210	1.280*
220	1.358*
230	1.432*
240	1.503*
250	1.570*
260	1.635*
270	1.696*
280	1.754*

T	C_p
CURVE 6 (cont.)	
290	1.809 x 10^{-1}*
298.16	1.852*
300	1.861*
310	1.911*
320	1.957*
330	2.002*
340	2.044*
350	2.083*
360	2.122*
370	2.158*
380	2.192*
390	2.224*
400	2.255*
410	2.283*
420	2.310*
430	2.336*
440	2.360*
450	2.384*
460	2.406*
470	2.427*
480	2.448*
490	2.467*
500	2.486*
510	2.503*
520	2.521*
530	2.538*
540	2.553*
550	2.569*
560	2.583*
570	2.598*
580	2.611*
590	2.625*
600	2.637*
610	2.650*
620	2.662*
630	2.673*
640	2.684*
650	2.695*
660	2.705*
670	2.715*
680	2.725*
690	2.734*
700	2.743*
720	2.761*
740	2.777*
CURVE 6 (cont.)	
760	2.792 x 10^{-1}*
780	2.807*
800	2.821*
820	2.834*
840	2.846*
860	2.857*
880	2.868*
900	2.879*
920	2.889*
940	2.899*
960	2.907*
980	2.916*
1000	2.924*
1020	2.932*
1040	2.939*
1060	2.946*
1080	2.953*
1100	2.960*
1120	2.966*
1140	2.972*
1160	2.977*
1180	2.983*
1200	2.988*
CURVE 7	
5	2.0 x 10^{-6}
10	2.0 x 10^{-5}
15	7.0
20	1.8 x 10^{-4}
25	3.33*
30	6.17*
35	1.03 x 10^{-3}*
40	1.62*
45	2.438*
50	3.497*
55	4.852*
60	6.514*
65	8.486*
70	1.074 x 10^{-2}*
75	1.328*
80	1.616*
85	1.933*
90	2.272*
95	2.630*

*Not shown on plot

DATA TABLE NO. 6 (continued)

T	C_p
CURVE 7 (cont.)*	
100	3.010×10^{-2}
105	3.408
110	3.826
115	4.257
120	4.702
125	5.155
130	5.617
135	6.083
140	6.554
145	7.025
150	7.496
155	7.968
160	8.436
165	8.901
170	9.362
175	9.817
180	1.026×10^{-1}
185	1.071
190	1.114
195	1.156
200	1.199
205	1.240
210	1.280
215	1.319
220	1.358
225	1.395
230	1.432
235	1.468
240	1.503
245	1.538
250	1.571
255	1.603
260	1.635
265	1.666
270	1.696
273.16	1.715
275	1.726
280	1.754
285	1.782
290	1.809
295	1.836
298.16	1.852
300	1.861
305	1.886
310	1.911
315	1.934

T	C_p
CURVE 7 (cont.)*	
320	1.957×10^{-1}
325	1.980
330	2.002
335	2.022
340	2.044
345	2.064
350	2.083
360	2.122
370	2.158
380	2.191
390	2.224
400	2.254
410	2.283
420	2.310
430	2.336
440	2.360
450	2.384
460	2.406
470	2.427
480	2.447
490	2.467
500	2.486
510	2.503
520	2.521
530	2.538
540	2.553
550	2.568
560	2.583
570	2.598
580	2.611
590	2.624
600	2.637
610	2.650
620	2.661
630	2.672
640	2.684
650	2.695
660	2.705
670	2.715
680	2.725
690	2.734
700	2.743
720	2.760
740	2.777
760	2.769
780	2.807

T	C_p
CURVE 7 (cont.)*	
800	2.820×10^{-1}
820	2.834
840	2.846
860	2.857
880	2.868
900	2.879
920	2.889
940	2.898
960	2.907
980	2.916
1000	2.924
1020	2.932
1040	2.939
1060	2.946
1080	2.953
1100	2.959
1120	2.966
1140	2.971
1160	2.977
1180	2.983
1200	2.988
CURVE 8*	
337.35	2.020×10^{-1}
423.55	2.315
523.15	2.530
627.25	2.669
732.05	2.771
825.05	2.848
923.35	2.899
CURVE 9*	
311.92	1.920×10^{-1}
321.65	1.964
331.35	2.008
341.05	2.048
350.65	2.086
360.35	2.123
515.15	2.516
524.95	2.532
534.55	2.548
544.15	2.562
553.85	2.577
563.55	2.591

T	C_p
CURVE 9 (cont.)*	
660.05	2.702×10^{-1}
669.75	2.711
679.35	2.721
689.05	2.731
CURVE 10	
434.82	2.519×10^{-1}
457.04	2.535
507.59	2.572
533.15	2.591
550.37	2.603
616.48	2.651*
627.59	2.659*
696.48	2.709*
750.38	2.748*
810.93	2.792*
837.04	2.810*
872.04	2.836*
892.04	2.851*
938.71	2.884*
1012.59	2.938*
1079.82	2.987
1088.71	2.992*
1123.71	3.018
1179.82	3.059
1239.26	3.102
1273.71	3.127*
1366.48	3.193
1413.15	3.228*
1486.48	3.281
1492.59	3.285*
1574.26	3.344
1644.26	3.395
1662.04	3.408*
1716.48	3.447
1763.71	3.481
1820.93	3.523*
1883.15	3.568

T	C_p
CURVE 11*	
65	8.493×10^{-3}
70	1.079×10^{-2}
80	1.611
90	2.266
100	3.027
110	3.835
120	4.708
130	5.649
140	6.591
150	7.513
160	8.444
170	9.395
180	1.033×10^{-1}
190	1.123
200	1.208
210	1.289
220	1.365
230	1.439
240	1.512
250	1.582
260	1.646
270	1.704
280	1.755
290	1.817
300	1.868
CURVE 12	
1088.9	2.956×10^{-1}*
1144.4	2.973*
1200.0	2.987*
1255.5	3.000
1311.1	3.012*
1366.6	3.023
1422.2	3.032
1477.8	3.040
1533.3	3.047
1588.9	3.054
1644.4	3.060
1700.0	3.065
CURVE 13*	
325.54	1.980×10^{-1}
327.54	1.993
357.71	2.104
377.89	2.186

T	C_p
CURVE 13 (cont.)*	
379.91	2.186×10^{-1}
425.59	2.330
425.63	2.334
476.52	2.445
476.86	2.445
478.52	2.457
522.47	2.522
524.49	2.522
473.98	2.618
575.89	2.623
622.49	2.665
624.52	2.665
653.85	2.700
673.16	2.733
675.18	2.733
726.61	2.780
728.66	2.768
748.94	2.794
750.93	2.789
780.04	2.818
782.03	2.806
816.16	2.848
818.02	2.846
850.45	2.853
885.62	2.890
887.59	2.890
935.61	2.904
986.36	2.930
986.82	2.937
CURVE 14*	
533.15	2.63×10^{-1}
672.04	2.72
810.93	2.80
949.82	2.89
1088.71	2.97
1227.59	3.06
CURVE 15	
533.15	2.40×10^{-1}
810.93	2.80*
1088.70	3.20
1366.48	3.25
1644.26	3.20
1922.04	3.20

*Not shown on plot

DATA TABLE NO. 6 (continued)

T	C_p
CURVE 16*	
Series 1	
344	2. 08 x 10^{-1}
380	2. 27
406	2. 30
436	2. 37
492	2. 47
515	2. 52
544	2. 57
566	2. 60
591	2. 63
614	2. 65
638	2. 66
662	2. 68
685	2. 71
703	2. 73
721	2. 75
740	2. 78
762	2. 81
Series 2	
342	2. 04
378	2. 19
405	2. 28
433	2. 33
461	2. 41
490	2. 47
513	2. 52
538	2. 56
561	2. 60
585	2. 64
609	2. 66
630	2. 69
653	2. 71
680	2. 75
701	2. 78
722	2. 80
743	2. 84
764	2. 86
CURVE 17	
1273. 15	2. 859 x 10^{-1}
1373. 15	2. 886
1473. 15	2. 913
1573. 15	2. 941
1673. 15	2. 969
1773. 15	2. 996
1873. 15	3. 024
1973. 15	3. 051
2173. 15	3. 016
2273. 15	3. 134
CURVE 18*	
53. 377	4. 375 x 10^{-3}
56. 621	5. 353
59. 604	6. 342
62. 273	7. 392
66. 033	8. 915
68. 349	9. 998
85. 703	1. 989 x 10^{-2}
89. 379	2. 242
91. 860	2. 417
94. 004	2. 566
95. 559	2. 681
98. 257	2. 885
102. 140	3. 187
197. 890	1. 183 x 10^{-1}
201. 027	1. 212
205. 199	1. 246
206. 616	1. 259
209. 860	1. 284
213. 329	1. 312
213. 339	1. 310
215. 275	1. 304
279. 172	1. 760
281. 572	1. 772
287. 205	1. 805
290. 935	1. 832
CURVE 19*	
Series 1	
286. 582	1. 8348 x 10^{-2}
288. 313	1. 8384
289. 798	1. 8502
CURVE 19 (cont.)*	
Series 1 (cont.)	
291. 532	1. 8602 x 10^{-2}
293. 265	1. 8673
294. 752	1. 8735
296. 241	1. 8831
297. 731	1. 8940
299. 219	1. 8986
300. 701	1. 9044
Series 2	
283. 368	1. 8140
284. 851	1. 8230
286. 334	1. 8321
288. 313	1. 8370
289. 798	1. 8479
291. 284	1. 8573
292. 771	1. 8651
294. 258	1. 8754
295. 745	1. 8774
297. 234	1. 8475
298. 723	1. 9050
300. 212	1. 9001
301. 703	1. 9104
303. 194	1. 9198
Series 3	
284. 356	1. 8178
287. 324	1. 3827
288. 808	1. 8463
290. 541	1. 8533
291. 779	1. 8555
293. 266	1. 8638
295. 001	1. 8711
296. 241	1. 8833
297. 730	1. 8886
299. 219	1. 8990
300. 709	1. 9063
302. 200	1. 9131
303. 691	1. 9238
CURVE 20	
551. 95	2. 46 x 10^{-1}
920. 65	2. 79
996. 55	2. 85*
1097. 75	2. 92*
1190. 25	2. 92
1282. 55	3. 04*
1385. 55	3. 03*
CURVE 21*	
1300	3. 014 x 10^{-1}
1350	3. 012
1400	3. 007
1450	3. 034
1500	3. 047
1550	3. 046
1600	3. 059
1650	3. 067
1700	3. 071
1750	3. 074
1800	3. 078
1850	3. 085
1900	3. 081
1950	3. 081
2000	3. 076
CURVE 22	
Series 1	
10. 908	2. 344 x 10^{-5}
12. 554	4. 454
16. 613	9. 611
20. 606	1. 828 x 10^{-4}
25. 456	3. 586*
30. 449	6. 376
36. 793	1. 207 x 10^{-3}
42. 775	2. 025*
47. 765	2. 970
52. 831	4. 236
Series 2	
61. 852	7. 199
63. 953	8. 038*
67. 952	9. 782
CURVE 22 (cont.)	
Series 2 (cont.)	
72. 016	1. 176 x 10^{-2}
79. 084	1. 566*
82. 068	1. 818
85. 474	1. 963*
90. 695	2. 326*
Series 3	
76. 972	1. 436
81. 864	1. 730*
84. 823	1. 929*
90. 311	2. 302*
116. 260	4. 359*
118. 022	4. 527*
Series 4	
96. 546	2. 745*
102. 199	3. 181*
108. 197	3. 664*
113. 743	4. 149
119. 886	4. 688*
126. 580	5. 295
133. 229	5. 900*
139. 889	6. 535*
146. 889	7. 196
154. 230	7. 909
Series 5*	
154. 837	7. 961
162. 007	8. 624
169. 569	9. 337
177. 166	1. 002 x 10^{-1}
184. 408	1. 067
191. 346	1. 128
198. 666	1. 188
206. 352	1. 252
213. 751	1. 313
221. 771	1. 372
238. 598	1. 496
CURVE 22 (cont.)	
Series 6*	
215. 015	1. 243 x 10^{-1}
211. 700	1. 295
220. 632	1. 362
229. 248	1. 426
237. 578	1. 488
245. 660	1. 544
266. 072	1. 673
274. 230	1. 723
282. 280	1. 770
289. 906	1. 810
297. 546	1. 849
Series 7*	
242. 519	1. 525
249. 961	1. 567
259. 270	1. 630
266. 884	1. 675
260. 867	1. 639
269. 134	1. 691
277. 198	1. 739
286. 847	1. 793
Series 8*	
308. 929	1. 902
315. 828	1. 942
322. 649	1. 970
336. 049	2. 030
340. 584	2. 048
347. 161	2. 070
353. 623	2. 101

*Not shown on plot

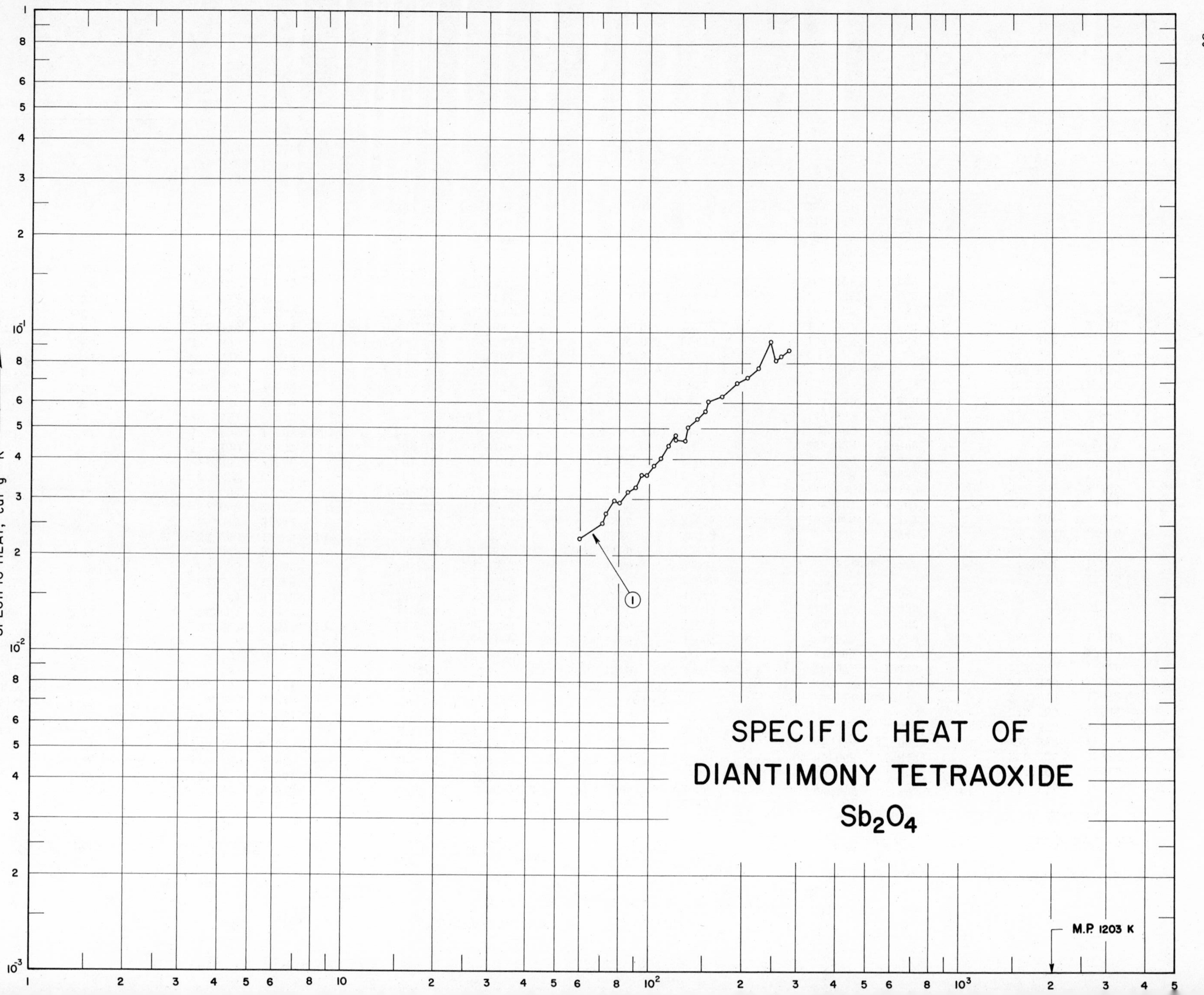
SPECIFIC HEAT OF
DIANTIMONY TETRAOXIDE
Sb_2O_4
M.P. 1203 K
1
SPECIFIC HEAT, cal g⁻¹ K⁻¹

SPECIFICATION TABLE NO. 7 SPECIFIC HEAT OF DIANTIMONY TETRAOXIDE Sb_2O_4

[For Data Reported in Figure and Table No. 7]

Curve No.	Ref. No.	Year	Temp. Range, K	Reported Error, %	Name and Specimen Designation	Composition (weight percent), Specifications and Remarks
1	40	1930	73-285	1		Correct ratio for Sb_2O_4, prepared by boiling mixture of Sb_2O_3 and HNO_3; washed free of HNO_3, dried, and heated at 850 C under vacuum; density 6.47 g cm^{-1} at 23.8 C.

DATA TABLE NO. 7 SPECIFIC HEAT OF DIANTIMONY TETRAOXIDE Sb_2O_4

[Temperature, T, K; Specific Heat, C_p, Cal g^{-1} K^{-1}]

T	C_p
CURVE 1	
72.8	2.705 x 10^{-2}
77.3	2.978
80.1	2.947
85.6	3.162
90.6	3.259
94.4	3.584
98.5	3.577
104.3	3.847
115.7	4.436
122.6	4.787
59.8	2.258
70.8	2.522
134.5	5.041
143.8	5.389
165.6	6.094
193.7	6.950
209.7	7.220
226.1	7.785
100.1	3.577*
103.7	3.743*
109.0	4.036
114.8	4.309*
118.8	4.472*
123.1	4.608
131.3	4.589
153.1	5.672
172.7	6.299
200.9	7.070*
229.3	7.707*
247.8	9.379
256.4	8.204
266.9	8.459
271.9	8.488*
278.8	8.618*
284.9	8.836

*Not shown on plot

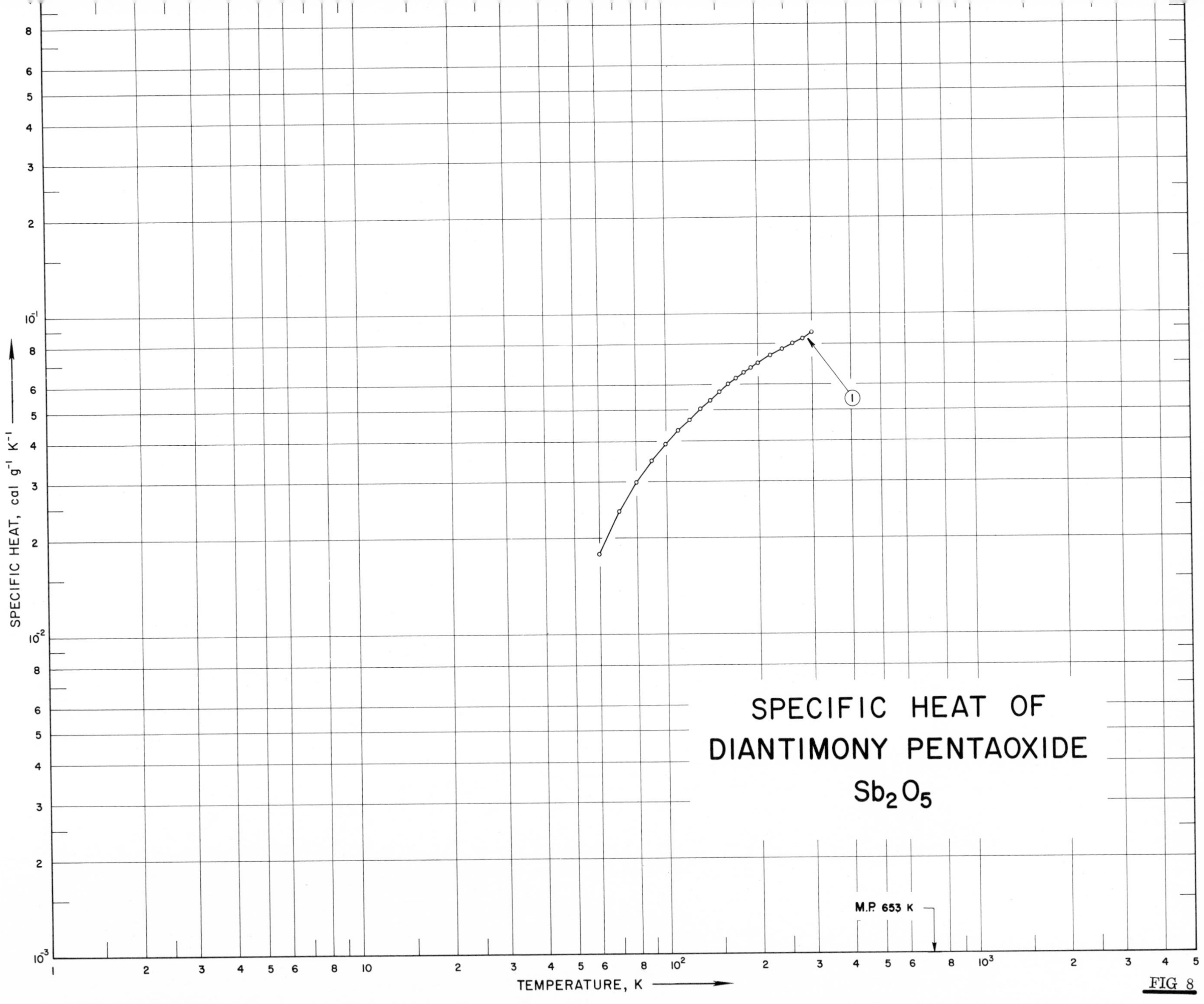
SPECIFIC HEAT OF
DIANTIMONY PENTAOXIDE
Sb_2O_5
M.P. 653 K
TEMPERATURE, K
SPECIFIC HEAT, cal g^{-1} K^{-1}

FIG 8

SPECIFICATION TABLE NO. 8 SPECIFIC HEAT OF DIANTIMONY PENTAOXIDE Sb_2O_5

[For Data Reported in Figure and Table No. 8]

Curve No.	Ref. No.	Year	Temp. Range, K	Reported Error, %	Name and Specimen Designation	Composition (weight percent), Specifications and Remarks
1	40	1930	60-300	1		Obtained from measurement of two hydrated samples.

DATA TABLE NO. 8 SPECIFIC HEAT OF DIANTIMONY PENTAOXIDE Sb_2O_5

[Temperature, T, K; Specific Heat, C_p, Cal g^{-1} K^{-1}]

T	C_p
	CURVE 1
60	1.777 x 10^{-2}
70	2.420
80	2.986
90	3.487
100	3.913
110	4.328
120	4.689
130	5.036
140	5.376
150	5.706
160	6.006
170	6.281
180	6.547
190	6.785
200	7.008
210	7.212*
220	7.416
230	7.601*
240	7.756
250	7.901*
260	8.046
270	8.204*
280	8.368
290	8.538*
300	8.726

*Not shown on plot

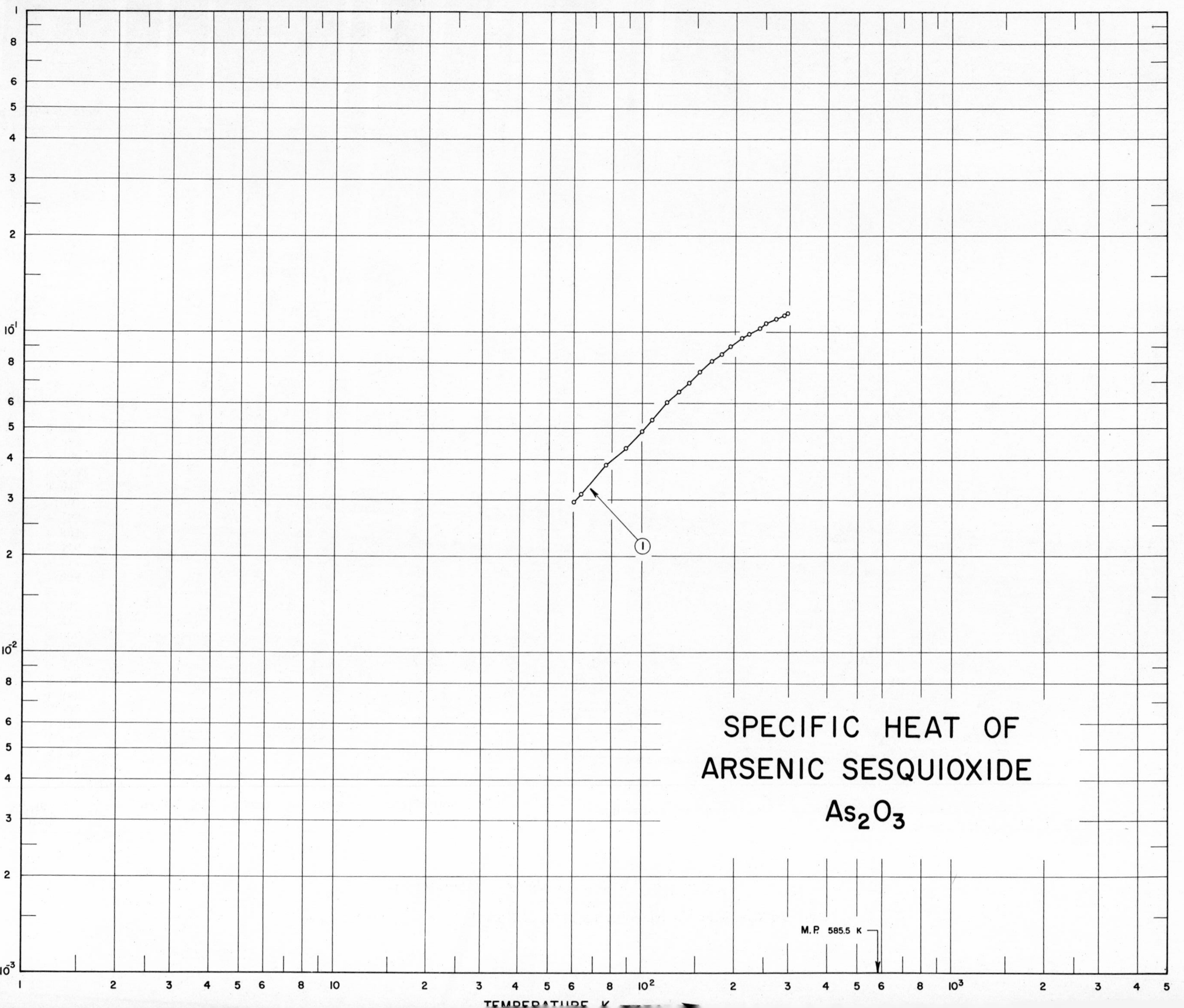

SPECIFIC HEAT OF
ARSENIC SESQUIOXIDE
As_2O_3
M.P. 585.5 K
SPECIFIC HEAT, cal g^{-1} K^{-1}

SPECIFICATION TABLE NO. 9 SPECIFIC HEAT OF ARSENIC SESQUIOXIDE As_2O_3

[For Data Reported in Figure and Table No. 9]

Curve No.	Ref. No.	Year	Temp. Range, K	Reported Error, %	Name and Specimen Designation	Composition (weight percent), Specifications and Remarks
1	104	1930	60-296			>99. 8 As_2O_3; crystalline of octahedral structure; density 3. 85 g cm^{-3}.

DATA TABLE NO. 9 SPECIFIC HEAT OF ARSENIC SESQUIOXIDE As_2O_3

[Temperature, T, K; Specific Heat, C_p, Cal g^{-1} K^{-1}]

T	C_p
CURVE 1	
60.2	2.950 x 10^{-2}
63.9	3.133
76.9	3.863
89.0	4.369
100.0	4.902
108.1	5.330
121.1	6.053
132.0	6.543
143.0	6.973
154.9	7.528
168.9	8.155
181.0	8.549
194.7	9.059
210.4	9.600
223.0	9.893
241.9	1.035 x 10^{-1}
253.7	1.073
272.2	1.103
288.7	1.135
291.6	1.142*
296.6	1.151

*Not shown on plot

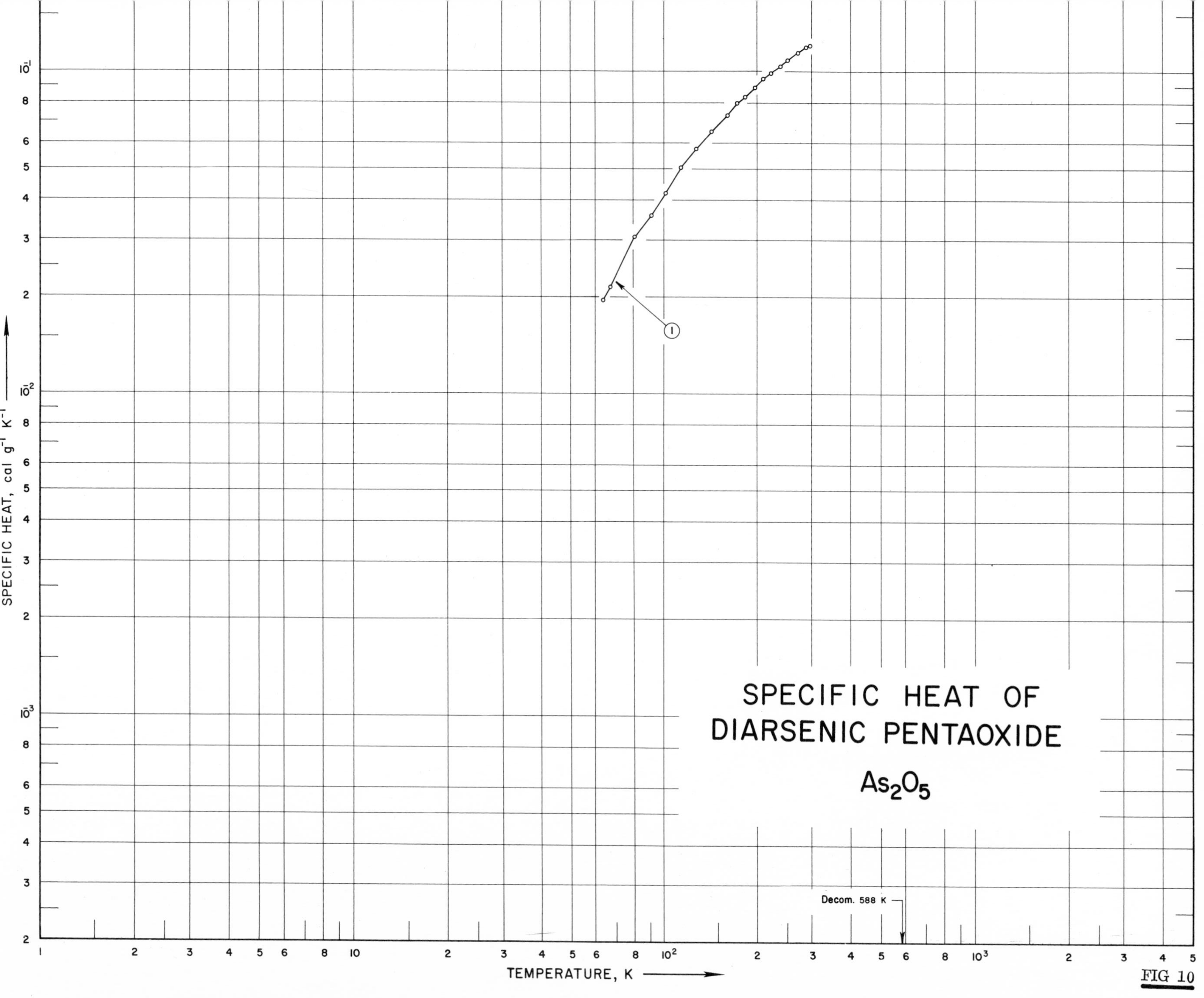
SPECIFIC HEAT OF
DIARSENIC PENTAOXIDE
As2O5
Decom. 588 K
1
SPECIFIC HEAT, cal g-1 K-1
TEMPERATURE, K
FIG 10

SPECIFICATION TABLE NO. 10 SPECIFIC HEAT OF DIARSENIC PENTAOXIDE As_2O_5

[For Data Reported in Figure and Table No. 10]

Curve No.	Ref. No.	Year	Temp. Range, K	Reported Error, %	Name and Specimen Designation	Composition (weight percent), Specifications and Remarks
1	309	1930	63-296			>99.7 As_2O_5 and 0.2 As_2O_3; crystalline aggregates; density 4.32 g cm^{-3}.

DATA TABLE NO. 10 SPECIFIC HEAT OF DIARSENIC PENTAOXIDE As_2O_5

[Temperature, T, K; Specific Heat, C_p, Cal $g^{-1} K^{-1}$]

T	C_p
CURVE 1	
63.5	1.962 x 10^{-2}
67.1	2.152
80.8	3.088
91.6	3.595
101.9	4.224
114.0	5.032
127.8	5.789
143.5	6.568
160.9	7.398
168.5	7.790*
173.2	8.007
181.5	8.386*
184.1	8.390
196.9	9.003
209.8	9.521
222.0	9.947
239.1	1.048 x 10^{-1}
251.1	1.091
270.3	1.153
287.1	1.194
291.7	1.204*
296.2	1.204

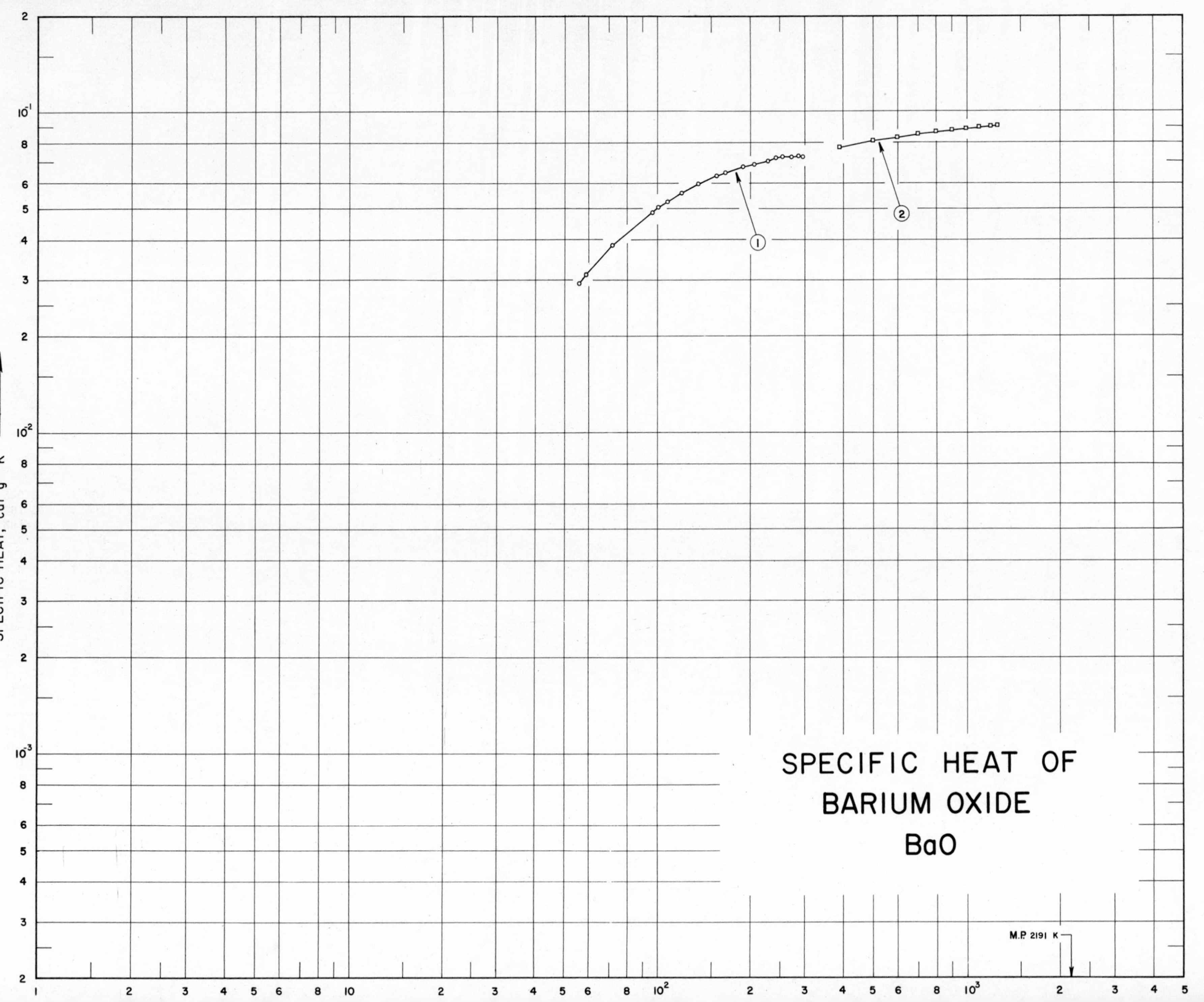
SPECIFIC HEAT OF
BARIUM OXIDE
BaO
SPECIFIC HEAT, cal g⁻¹ K⁻¹
M.P. 2191 K
1
2

SPECIFICATION TABLE NO. 11 SPECIFIC HEAT OF BARIUM OXIDE BaO

[For Data Reported in Figure and Table No. 11]

Curve No.	Ref. No.	Year	Temp. Range, K	Reported Error, %	Name and Specimen Designation	Composition (weight percent), Specifications and Remarks
1	41	1935	56-299			Kahlbaum best grade.
2	42	1951	390-1262			99 BaO; and ~1 SiO_2.

DATA TABLE NO. 11 SPECIFIC HEAT OF BARIUM OXIDE BaO

[Temperature, T, K; Specific Heat, C_p, Cal g^{-1} K^{-1}]

T	C_p
CURVE 1	
56.1	2.912 x 10^{-2}
59.2	3.104
72.1	3.845
97.0	4.870
101.5	5.027
102.8	5.075*
109.1	5.267
120.6	5.602
136.6	5.956
156.2	6.322
166.8	6.471
190.2	6.742
207.1	6.879
230.1	7.010
230.5	7.107*
244.2	7.205
250.1	7.199*
255.7	7.225
259.4	7.244*
274.3	7.218
287.6	7.296
298.6	7.244
CURVE 2	
390	7.721 x 10^{-2}
400	7.770*
500	8.129
600	8.355
700	8.518
800	8.648
900	8.758
1000	8.856
1100	8.946
1200	9.031
1262	9.082

*Not shown on plot

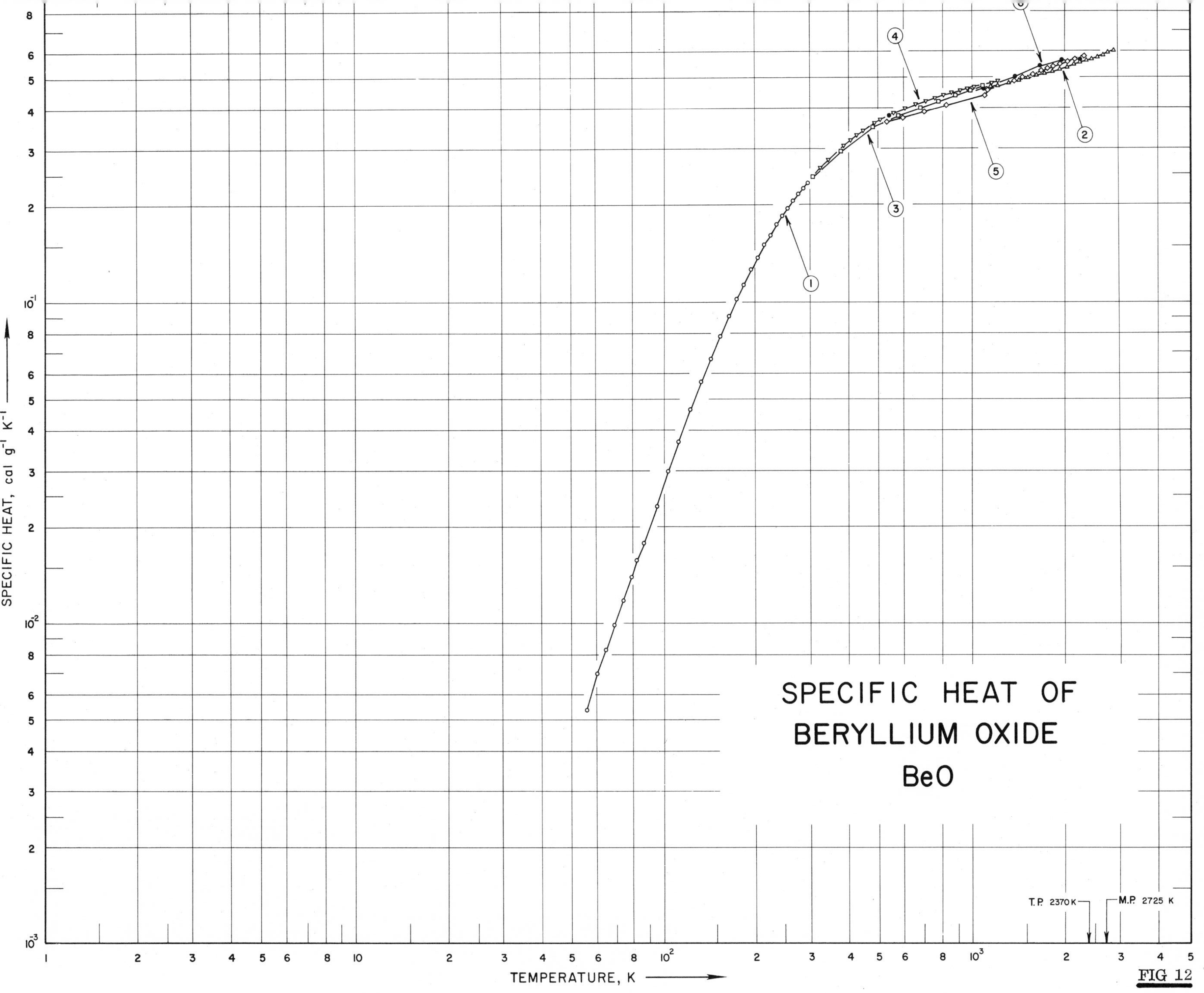
SPECIFIC HEAT OF
BERYLLIUM OXIDE
BeO
T.P. 2370 K
M.P. 2725 K
TEMPERATURE, K
SPECIFIC HEAT, cal g⁻¹ K⁻¹
1
2
3
4
5

FIG 12

SPECIFICATION TABLE NO. 12 SPECIFIC HEAT OF BERYLLIUM OXIDE BeO

[For Data Reported in Figure and Table No. 12]

Curve No.	Ref. No.	Year	Temp. Range, K	Reported Error, %	Name and Specimen Designation	Composition (weight percent), Specifications and Remarks
1	43	1939	55-292			99. 6 BeO.
2	44	1960	1200-2820	0. 25-0. 5		99. 9 BeO, with impurities of Al, Ni, Cu, Zn, Ag, Fe, and Ti; pressed and sintered with 0. 1 glucose as cementing substance at 1400 - 1800 C.
3	45	1962	303-1073	± 3		
4	46	1963	298-1200	0. 40		99. 96 BeO, 0. 01 Si, 0. 007 Al, 0. 002 K, 0. 002 Na, 0. 001 Cs, 0. 001 Fe, < 0. 001 Ca, < 0. 001 Cu, < 0. 00005 Li, < 0. 00005 Mg; supplied by Norton Co. ; pressed, fired at 1800 C and sintered.
5	47	1963	526-2280	± 5		99. 5 BeO, 0. 0090 Si, 0. 0050 Al, 0. 0020 Mo, 0. 0010 Ca, 0. 0010 Cr, 0. 0010 Fe, 0. 0010 Na, 0. 0010 Ni, 0. 0003 Mn, ≤ 0. 0001 B, Cd, Li, ≤ 0. 0001 Co, Cu; supplied by Brush Beryllium Co. ; cold pressed; density 179 lb ft^{-3}.
6	48	1963	533-2200	≤ 5		Sample supplied by Zirconium Corp. of America; crushed in hardened steel mortar to pass 100-mesh screen; pressed and sintered; density at 25 C before exposure: apparent density (AST M method B311-58) 183 lb ft^{-3}, true density (by immersion in xylene) 187 lb ft^{-3}.

DATA TABLE NO 12 SPECIFIC HEAT OF BERYLLIUM OXIDE BeO

[Temperature, T, K; Specific Heat, C_p, Cal $g^{-1}K^{-1}$]

T	C_p
CURVE 1	
55.5	5.356×10^{-3}
59.9	6.954
64.0	8.233
68.4	9.832
73.2	1.175×10^{-2}
78.0	1.399
81.3	1.578
81.5	1.583
85.7	1.778
94.5	2.306
103.4	2.966
111.9	3.681
122.1	4.636
132.1	5.683
142.0	6.695
152.2	7.846
162.8	9.097
172.5	1.026×10^{-1}
182.0	1.1443
192.4	1.2730
202.3	1.3933
212.5	1.5172
222.3	1.6351
232.6	1.7606
242.4	1.8685
252.3	1.9772
262.2	2.0895
272.4	2.1938
282.2	2.2846
288.4	2.3273*
292.4	2.3773
CURVE 2	
1142	4.741×10^{-1}
1200	4.789
1300	4.873
1400	4.957
1500	5.040
1600	5.124
1700	5.207
1800	5.291
1900	5.374
2000	5.458
CURVE 2 (cont.)	
2100	5.541×10^{-1}
2200	5.625
2300	5.709
2400	5.792
2500	5.876
2600	5.959
2700	6.043
2800	6.126*
2820	6.143
CURVE 3	
303.15	2.48×10^{-1}
373.15	2.98
473.15	3.54
573.15	3.84
673.15	4.08
773.15	4.26
873.15	4.43
973.15	4.60
1073.15	4.76
CURVE 4	
298.15	2.440×10^{-1}*
300	2.457*
320	2.637
340	2.803
360	2.956*
380	3.097
400	3.230
420	3.348
440	3.454
460	3.551*
480	3.640
500	3.721
550	3.899
600	4.046
650	4.171
700	4.278
750	4.372
800	4.455
850	4.529
900	4.597
CURVE 4 (cont.)	
950	4.659×10^{-1}*
1000	4.717
1050	4.771
1100	4.822*
1150	4.870
1200	4.916
CURVE 5	
527.04	3.668×10^{-1}
589.82	3.777
693.71	3.954
818.15	4.157
1009.26	4.453
1112.04	4.604
1202.59	4.732*
1358.70	4.943
1440.37	5.047
1555.93	5.189
1653.71	5.303
1744.82	5.405
1812.04	5.478
1899.82	5.568
2005.37	5.672
2138.71	5.794
2247.04	5.886*
2277.59	5.911
CURVE 6	
533.15	3.85×10^{-1}
810.93	4.17*
1088.71	4.70
1366.48	5.10
1644.26	5.50
1922.04	5.70
2199.82	5.70

*Not shown on plot

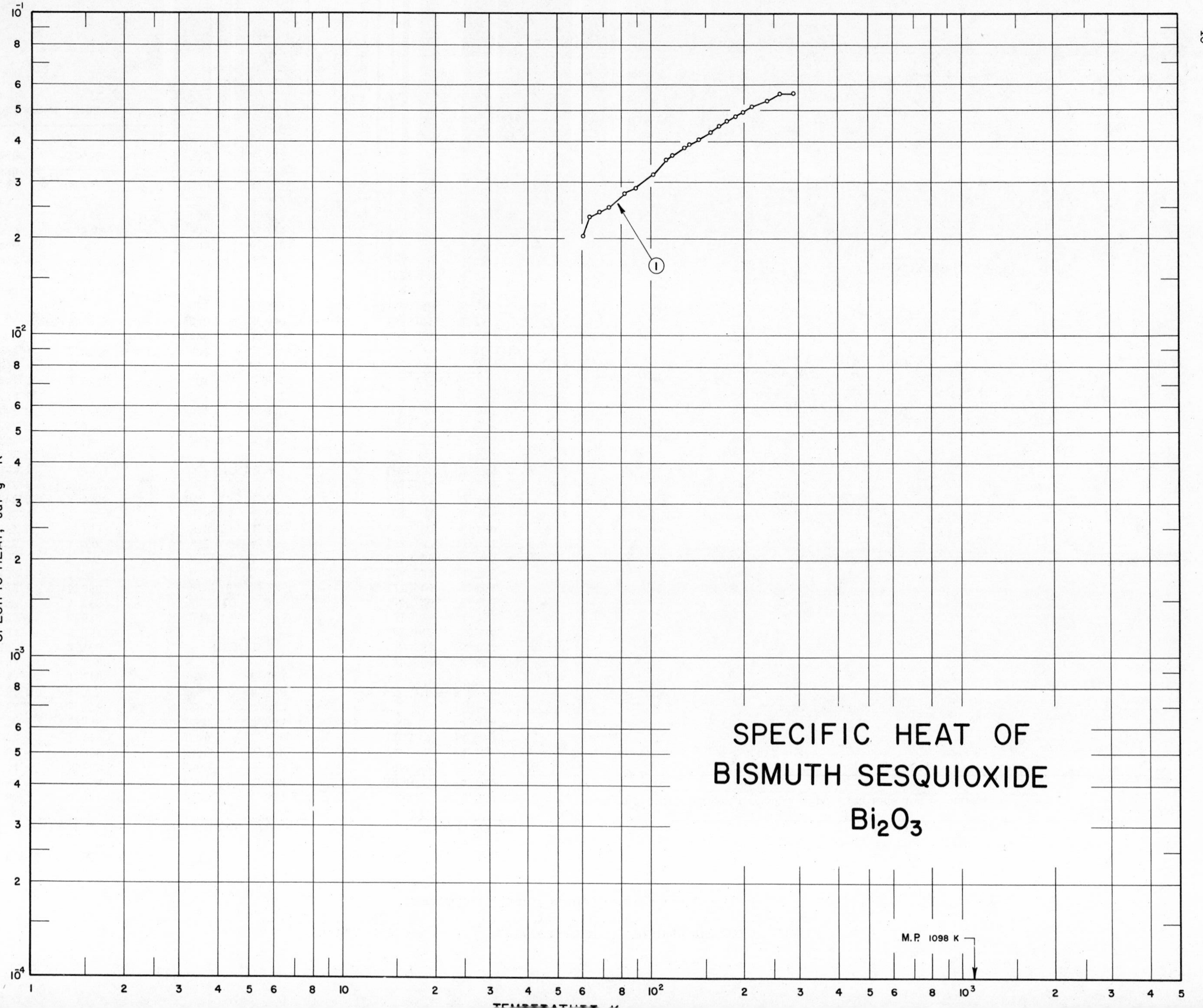

SPECIFIC HEAT OF
BISMUTH SESQUIOXIDE
Bi_2O_3
M.P. 1098 K
SPECIFIC HEAT, cal g^{-1} K^{-1}
1

SPECIFICATION TABLE NO. 13 SPECIFIC HEAT OF BISMUTH SESQUIOXIDE Bi_2O_3

[For Data Reported in Figure and Table No. 13]

Curve No.	Ref. No.	Year	Temp. Range, K	Reported Error, %	Name and Specimen Designation	Composition (weight percent), Specifications and Remarks
1	49	1930	60-289			99. 6 B_2O_3; density 9. 33 g cm^{-3} at 23. 3 C.

DATA TABLE NO. 13 SPECIFIC HEAT OF BISMUTH SESQUIOXIDE Bi_2O_3

[Temperature, T, K; Specific Heat, C_p, Cal g^{-1} K^{-1}]

T	C_p
CURVE 1	
60. 6	2. 039 x 10^{-2}
63. 7	2. 337
68. 7	2. 412
73. 6	2. 498
82. 6	2. 758
89. 3	2. 858
101. 4	3. 159
113. 0	3. 500
118. 6	3. 618
129. 2	3. 820
143. 1	4. 062
155. 7	4. 283
166. 4	4. 474
175. 8	4. 612
188. 5	4. 796
198. 2	4. 925
134. 7	3. 914
213. 3	5. 122
238. 4	5. 339
289. 3	5. 633
262. 1	5. 616
271. 4	5. 642*
279. 7	5. 695*

*Not shown on plot

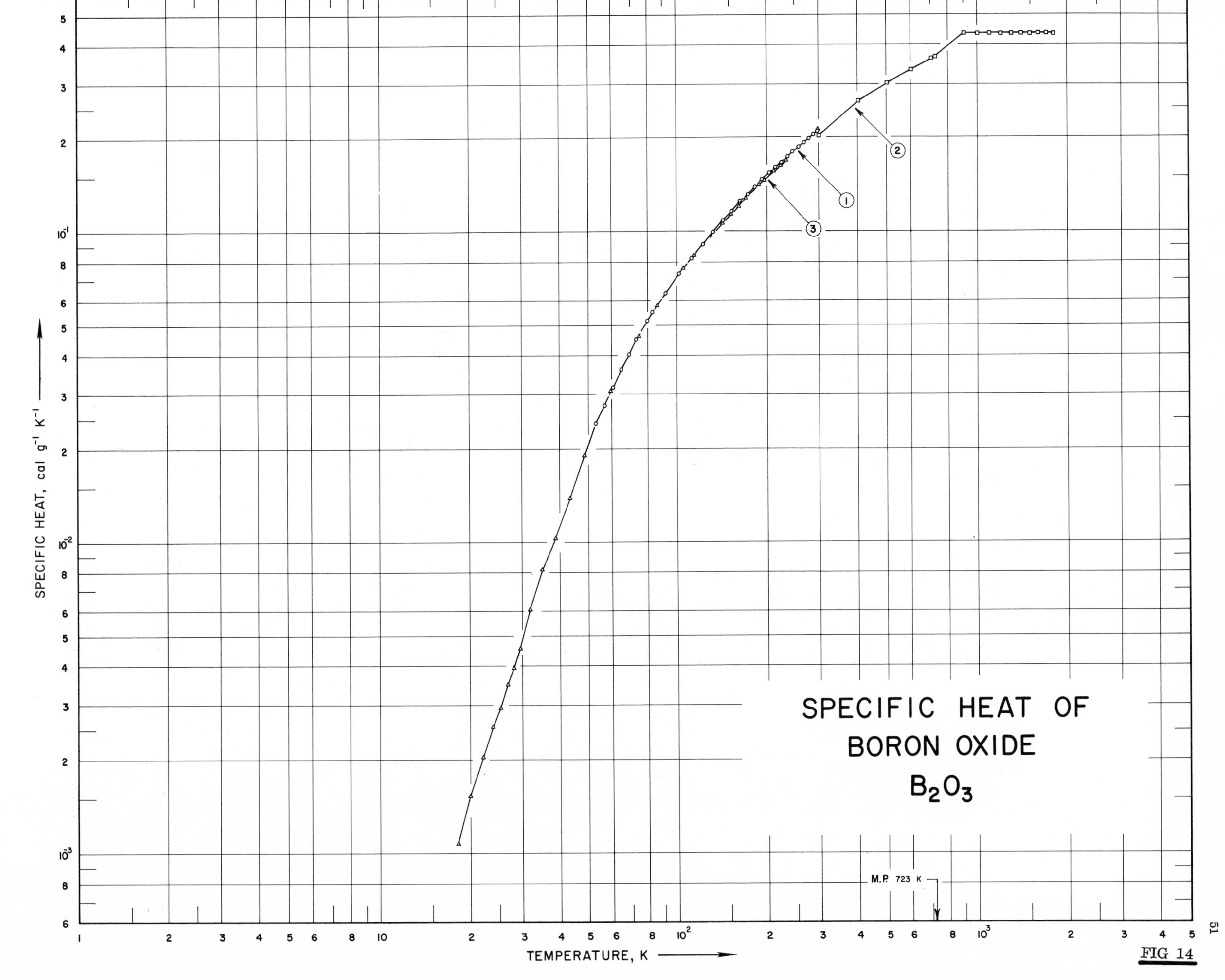
SPECIFIC HEAT OF
BORON OXIDE
B_2O_3
SPECIFIC HEAT, cal g⁻¹ K⁻¹
TEMPERATURE, K
M.P. 723 K
1
2
3

FIG 14

SPECIFICATION TABLE NO. 14 SPECIFIC HEAT OF BORON OXIDE B_2O_3

[For Data Reported in Figure and Table No. 14]

Curve No.	Ref. No.	Year	Temp. Range, K	Reported Error, %	Name and Specimen Designation	Composition (weight percent), Specifications and Remarks
1	50	1941	52-295			99. 7 B_2O_3, 0. 10 H_2O, 0. 10 accounted for in impurities of original boric acid, 0. 10 unaccounted; prepared from boric acid by heating 1 week at 120 C after which temperature raised 10 C daily up to 200 C for one day, resulting crystals heated 400 C for two days.
2	51	1941	289-1800			Two samples of boron sesquioxide glass; 99. 30 - 99. 79 B_2O_3, 0. 06 - . 055 A_2O.
3	52	1950	18-295			B_2O_3; moisture free to about 0. 1%; prepared by heating for one week boric acid (0. 05 impurity) at 120 - 130 C after which temperature increased 10 C per day until it remained at 200 C for one day; resulting mixture crystallized at 200 C over three day period; temperature raised to 400 C for two days under vacuum; resulting material crushed, screened and heated three more days at 400 C under vacuum.

DATA TABLE NO. 14 SPECIFIC HEAT OF BORON OXIDE B_2O_3

[Temperature, T, K; Specific Heat, C_p, Cal g^{-1} K^{-1}]

T	C_p
CURVE 1	
52.9	2.434×10^{-2}
56.3	2.760
60.1	3.165
64.3	3.620
68.4	4.057
72.7	4.512
79.4	5.188
82.7	5.511
91.5	6.340
101.8	7.345
112.0	8.258
122.2	9.137
132.6	1.002×10^{-1}
143.2	1.092
153.2	1.167
163.7	1.257
174.0	1.332
183.8	1.404
194.5	1.482
204.3	1.551
214.5	1.617
224.3	1.689
234.9	1.759
244.7	1.818
255.1	1.887
265.6	1.950
275.7	2.005
285.6	2.066
295.1	2.114
CURVE 2	
298.15	2.046×10^{-1}
300	2.060*
400	2.646
500	3.032
600	3.341
700	3.612
(s) 723	3.671
(l) 900	4.372
1000	4.372
1100	4.372
1200	4.372
1300	4.372
CURVE 2 (cont.)	
1400	4.372×10^{-1}
1500	4.372
1600	4.372
1700	4.372
1800	4.372
CURVE 3	
18.08	1.073×10^{-3}
19.94	1.535
21.98	2.035
23.64	2.549
25.19	2.921
26.55	3.489
27.95	3.937
29.30	4.585
31.57	6.081
34.81	8.171
38.65	1.035×10^{-2}
43.33	1.390
48.48	1.903
53.53	2.488*
59.10	3.069
63.20	3.488*
69.72	4.056*
74.23	4.632
79.58	5.095*
85.90	5.800
89.71	6.200
104.82	7.634
114.39	8.420
123.32	9.182
132.19	9.917
141.77	1.071×10^{-1}
150.98	1.147
160.31	1.216
169.98	1.290
187.01	1.416
195.35	1.475
204.19	1.539
210.32	1.579
220.39	1.644
230.55	1.707
235.87	1.743
CURVE 3 (cont.)	
247.38	1.829×10^{-1}
257.32	1.882
273.10	1.982
284.00	2.069
296.60	2.151

*Not shown on plot

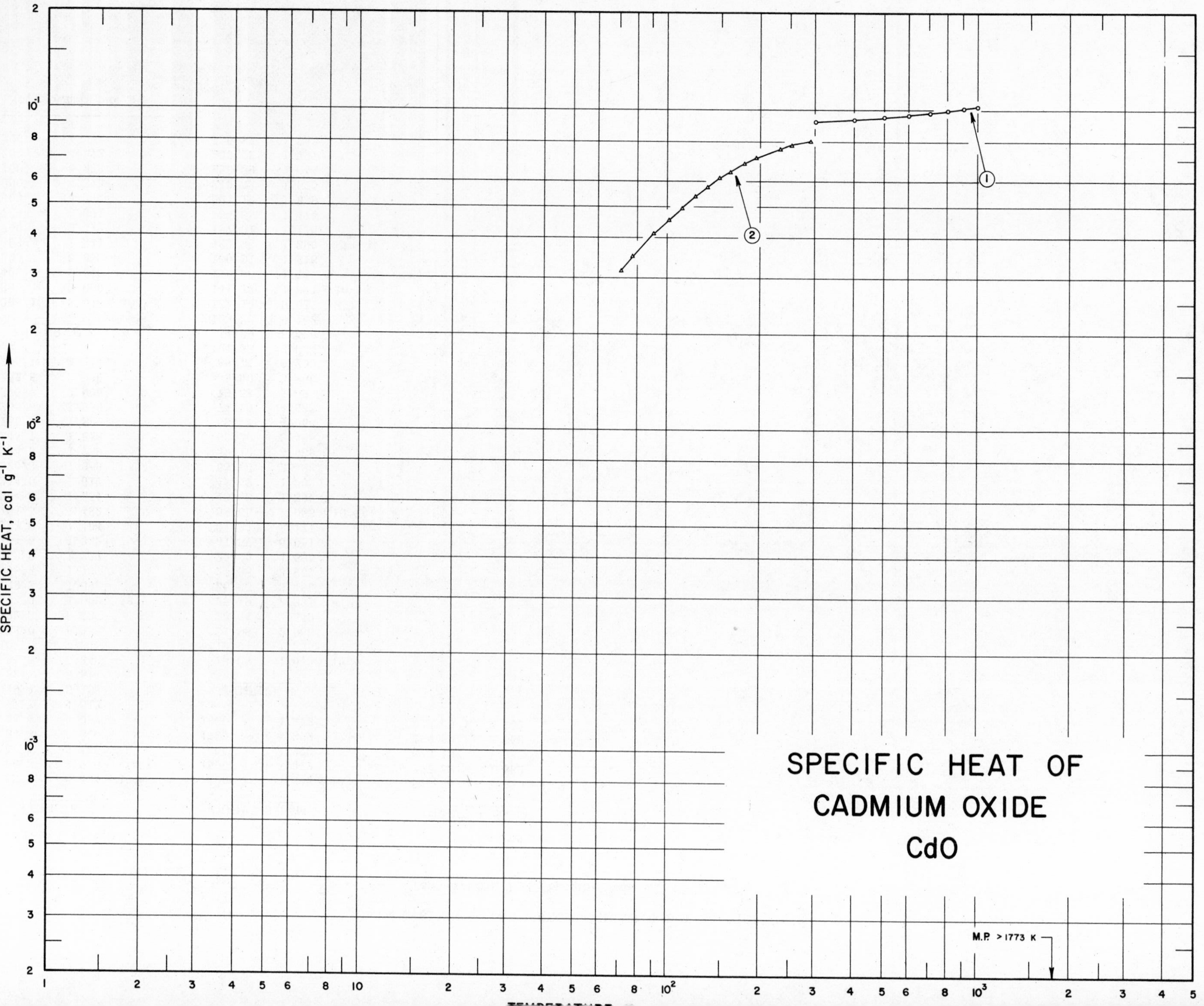
SPECIFIC HEAT OF
CADMIUM OXIDE
CdO
M.P. > 1773 K
1
2
SPECIFIC HEAT, cal g⁻¹ K⁻¹

SPECIFICATION TABLE NO. 15 SPECIFIC HEAT OF CADMIUM OXIDE CdO

[For Data Reported in Figure and Table No. 15]

Curve No.	Ref. No.	Year	Temp. Range, K	Reported Error, %	Name and Specimen Designation	Composition (weight percent), Specifications and Remarks
1	54	1965	300-1000	1.3		99.9525 CdO; supplied by J. T. Baker Chemical Co..
2	152	1928	71-253			Crystalline; prepared by heating pure cadmium oxide in an open platinum dish at 1100 C for 3 days.

DATA TABLE NO. 15 SPECIFIC HEAT OF CADMIUM OXIDE CdO

[Temperature, T, K; Specific Heat, C_p, Cal g^{-1} K^{-1}]

T	C_p
CURVE 1	
300	9.174×10^{-2}
400	9.330
500	9.486
600	9.642
700	9.798
800	9.953
900	1.011×10^{-1}
1000	1.026
CURVE 2	
71.3	3.176×10^{-2}
77.9	3.501
90.5	4.117
101.1	4.551
112.2	4.986
124.2	5.407
136.1	5.810
148.0	6.188
160.9	6.471
288.0	8.061*
289.2	8.061*
290.4	8.022
151.9	6.266*
177.1	6.833
194.9	7.138
232.5	7.694
253.5	7.835

*Not shown on plot

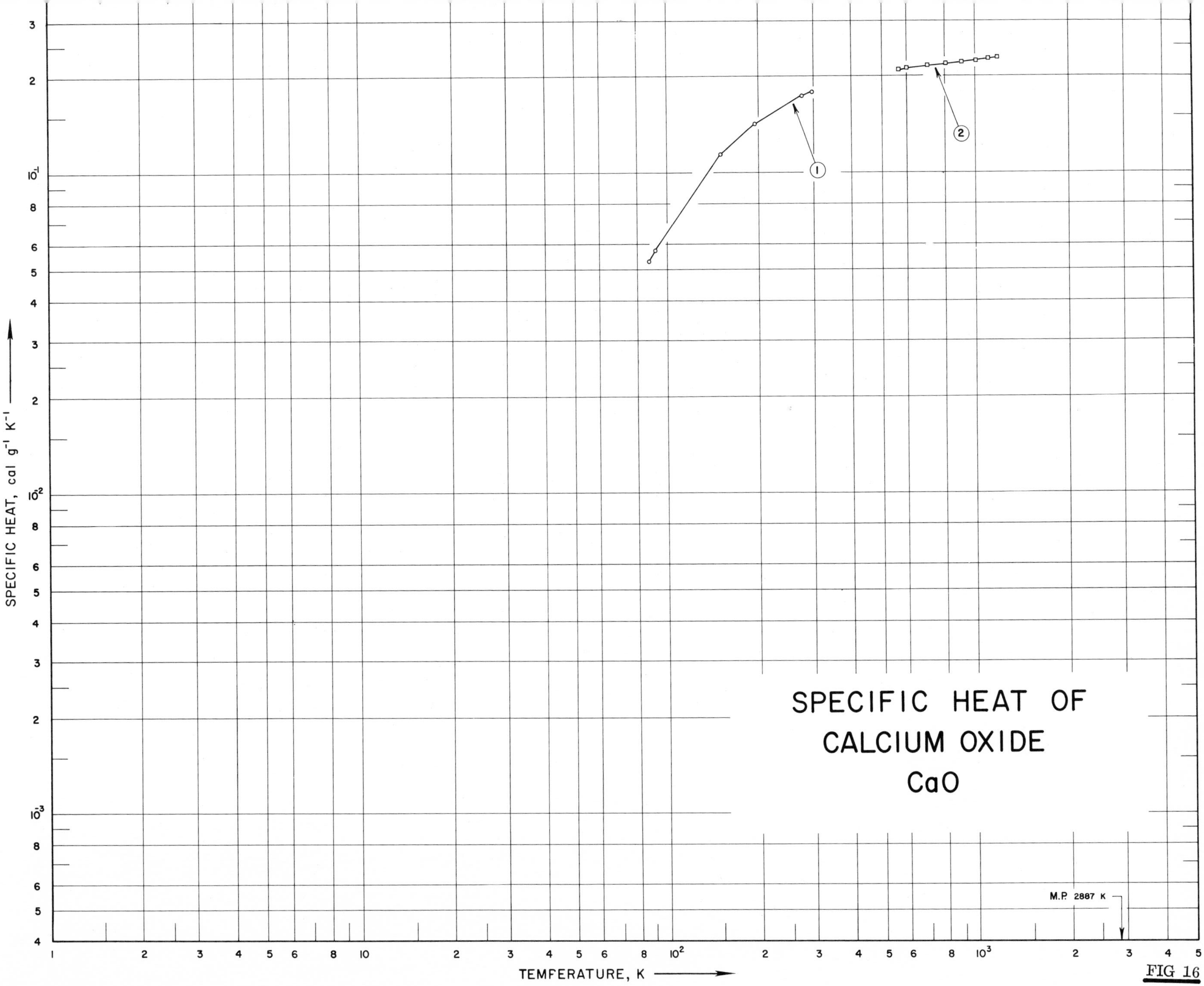

SPECIFIC HEAT OF
CALCIUM OXIDE
CaO
SPECIFIC HEAT, cal g⁻¹ K⁻¹
TEMFERATURE, K
M.P. 2887 K
1
2

FIG 16

SPECIFICATION TABLE NO. 16 SPECIFIC HEAT OF CALCIUM OXIDE CaO

[For Data Reported in Figure and Table No. 16]

Curve No.	Ref. No.	Year	Temp. Range, K	Reported Error, %	Name and Specimen Designation	Composition (weight percent), Specifications and Remarks
1	53	1926	87-293	1. 0		98. 8 CaO, 0. 4 H_2O.
2	42	1951	563-1176			Material obtained by calcinating CaO_3 at 800 C in vacuum.

DATA TABLE NO. 16 SPECIFIC HEAT OF CALCIUM OXIDE CaO

[Temperature, T, K; Specific Heat, C_p, Cal g^{-1} K^{-1}]

T	C_p
CURVE 1	
87.2	5.35×10^{-2}
87.7	5.36*
91.1	5.79
92.2	5.91*
150.3	1.158×10^{-1}
194.0	1.446
197.1	1.474*
275.4	1.762
277.8	1.772*
282.1	1.794*
292.7	1.808
CURVE 2	
563	2.130×10^{-1}
600	2.148
700	2.189
800	2.223
900	2.252
1000	2.278
1100	2.302
1176	2.320

*Not shown on plot

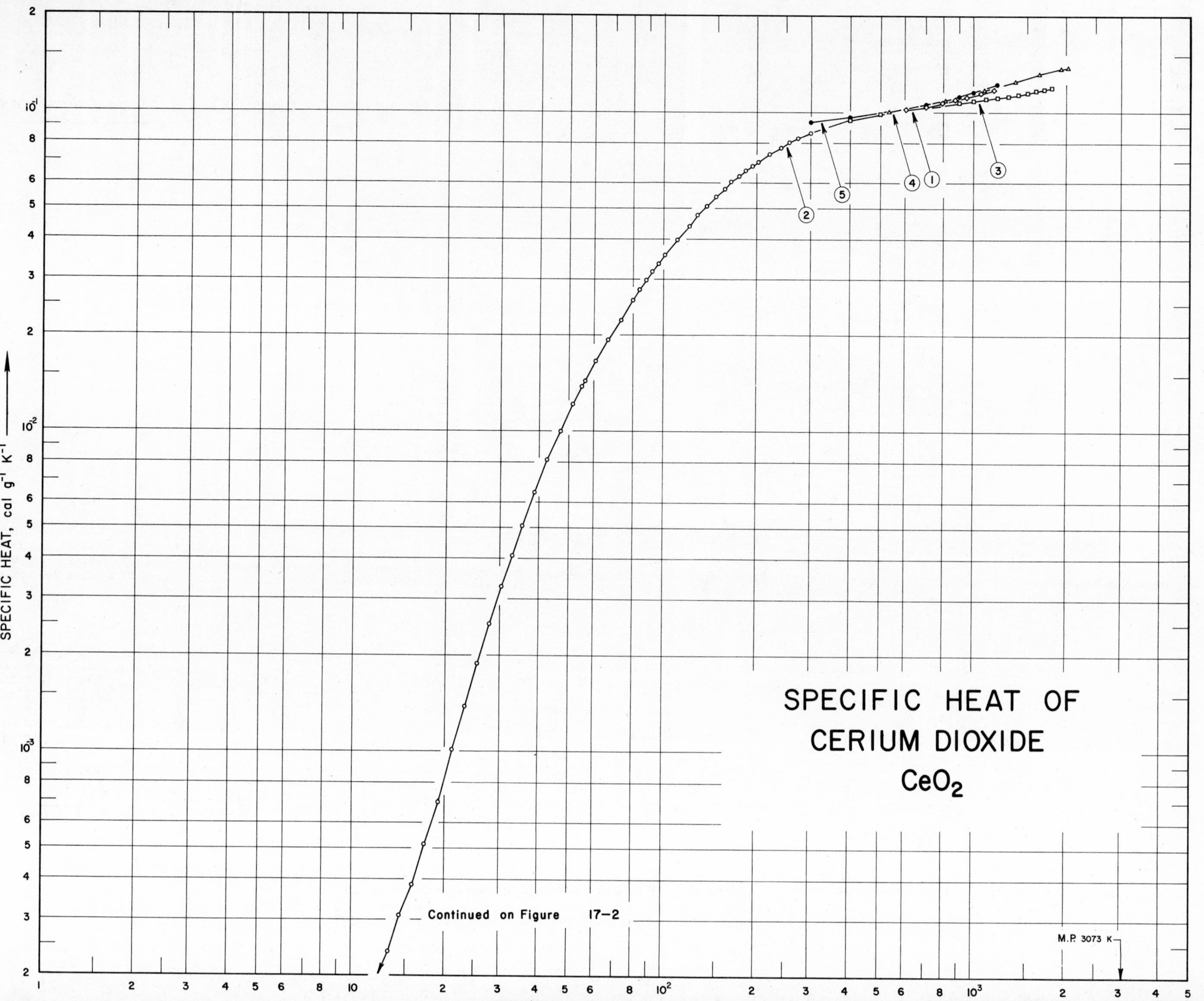
SPECIFIC HEAT OF
CERIUM DIOXIDE
CeO_2
M.P. 3073 K
Continued on Figure 17-2
SPECIFIC HEAT, cal g^{-1} K^{-1}
1
2
3
4
5

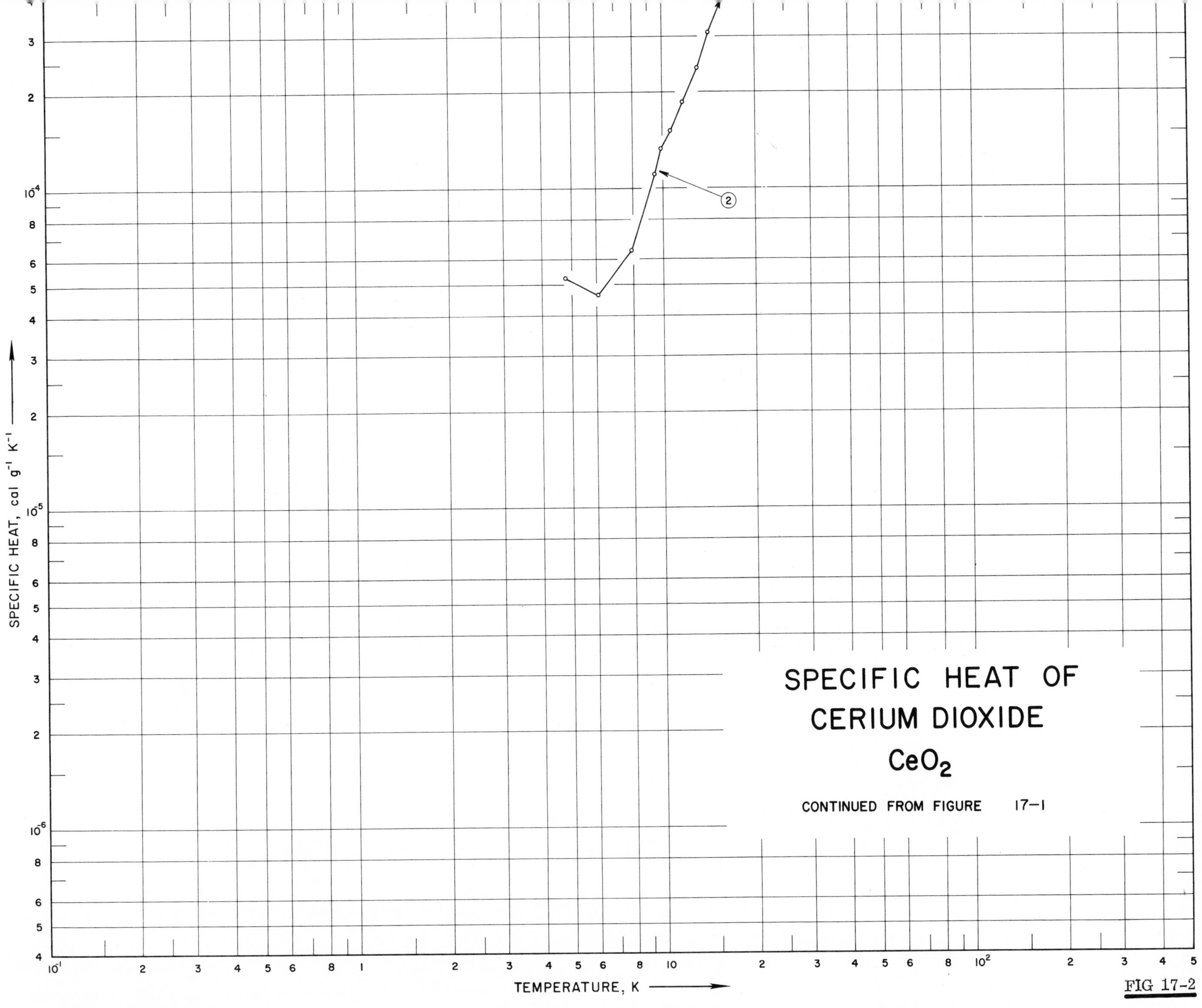

SPECIFIC HEAT OF
CERIUM DIOXIDE
CeO2
CONTINUED FROM FIGURE 17-1
SPECIFIC HEAT, cal g-1 K-1
TEMPERATURE, K
2

FIG 17-2

SPECIFICATION TABLE NO. 17 SPECIFIC HEAT OF CERIUM DIOXIDE CeO_2

[For Data Reported in Figure and Table No. 17]

Curve No.	Ref. No.	Year	Temp. Range, K	Reported Error, %	Name and Specimen Designation	Composition (weight percent), Specifications and Remarks
1	55	1960	608-1172	0. 2		99. 9 CeO_2.
2	285	1961	5-302	0. 1-10		~99. 98 CeO_2; prepared by precipitating cerium hydroxide using gaseous ammonia and igniting in air to dioxide at 1000 C for 72 hrs.
3	57	1961	298-1800	0. 2		99. 9 CeO_2; supplied by Lindsay Chemical Co. , heated at 1050 C for 1 hr.
4	48	1962	533-2044	≤5		Composition before exposure: 82. 2 Ce, 0. 2 Zr, 0. 1 Ca; after exposure: 81. 5 Ce, 0. 10 C; supplied by Zirconium Corp. of America; crushed in hardened steel mortar to pass 100-mesh screen, pressed and sintered; density at 25 C before exposure: apparent density (ASTM method B311-58) 412 lb ft^{-3}, true density (by immersion in xylene) 429 lb ft^{-3}; after exposure: apparent density 410 lb ft^{-3}, true density 422 lb ft^{-3}.
5	54	1965	300-1200	0. 5		Spectroscopically pure; supplied by Johnson, Mathey and Co. Ltd. , London.

DATA TABLE NO. 17 SPECIFIC HEAT OF CERIUM DIOXIDE CeO_2

[Temperature, T, K; Specific Heat, C_p, Cal $g^{-1}K^{-1}$]

T	C_p	T	C_p	T	C_p
CURVE 1		CURVE 2 (cont.)		CURVE 3 (cont.)	
608.1	1.020×10^{-1}	Series 2		1600	1.168×10^{-1}
702.3	1.046	84.27	2.770×10^{-2}	1700	1.180
797.3	1.073	92.55	3.155	1800	1.193
874.1	1.095	101.64	3.561		
951.9	1.117	110.92	3.963	CURVE 4	
1066.2	1.150	120.31	4.356*	533.15	1.00×10^{-1}
1171.7	1.180	129.59	4.726	810.93	1.09
		138.75	5.073	1088.71	1.17
CURVE 2		147.95	5.404	1366.48	1.25
Series 1		157.15	5.717	1644.26	1.32
4.81	5.228×10^{-5}	166.34	6.007	1922.04	1.37
6.14	4.647	175.63	6.281	2044.26	1.38
7.97	6.391	185.01	6.536		
9.40	1.104×10^{-4}	194.26	6.774	CURVE 5	
9.94	1.336	203.32	6.989	300	9.255×10^{-2}
10.70	1.511	212.20	7.187*	400	9.581
11.74	1.859	220.94	7.367	500	9.912*
13.15	2.382	229.62	7.535*	600	1.024×10^{-1}*
14.38	3.079	239.51	7.704	700	1.056
15.66	3.335	247.55	7.855*	800	1.089*
17.16	5.113	256.58	8.000	900	1.122
19.00	6.914	265.54	8.134*	1000	1.154
21.10	1.011×10^{-3}	274.58	8.267	1100	1.187*
23.26	1.383	283.63	8.384*	1200	1.220
25.49	1.877	292.64	8.500*		
27.84	2.492	301.69	8.599		
30.37	3.259				
32.82	4.084	CURVE 3			
35.37	5.026	298.15	8.549×10^{-2}*		
38.85	6.426	300	8.572*		
42.85	8.105	400	9.407		
47.02	9.976	500	9.856		
51.65	1.210×10^{-2}	600	1.015×10^{-1}*		
56.66	1.444	700	1.038*		
55.32	1.380	800	1.057*		
60.96	1.655	900	1.073		
66.86	1.938	1000	1.089		
73.12	2.235	1100	1.103		
80.15	2.573	1200	1.117		
88.36	2.963	1300	1.130		
96.76	3.345	1400	1.143		
121.27	4.398	1500	1.155		

*Not shown on plot

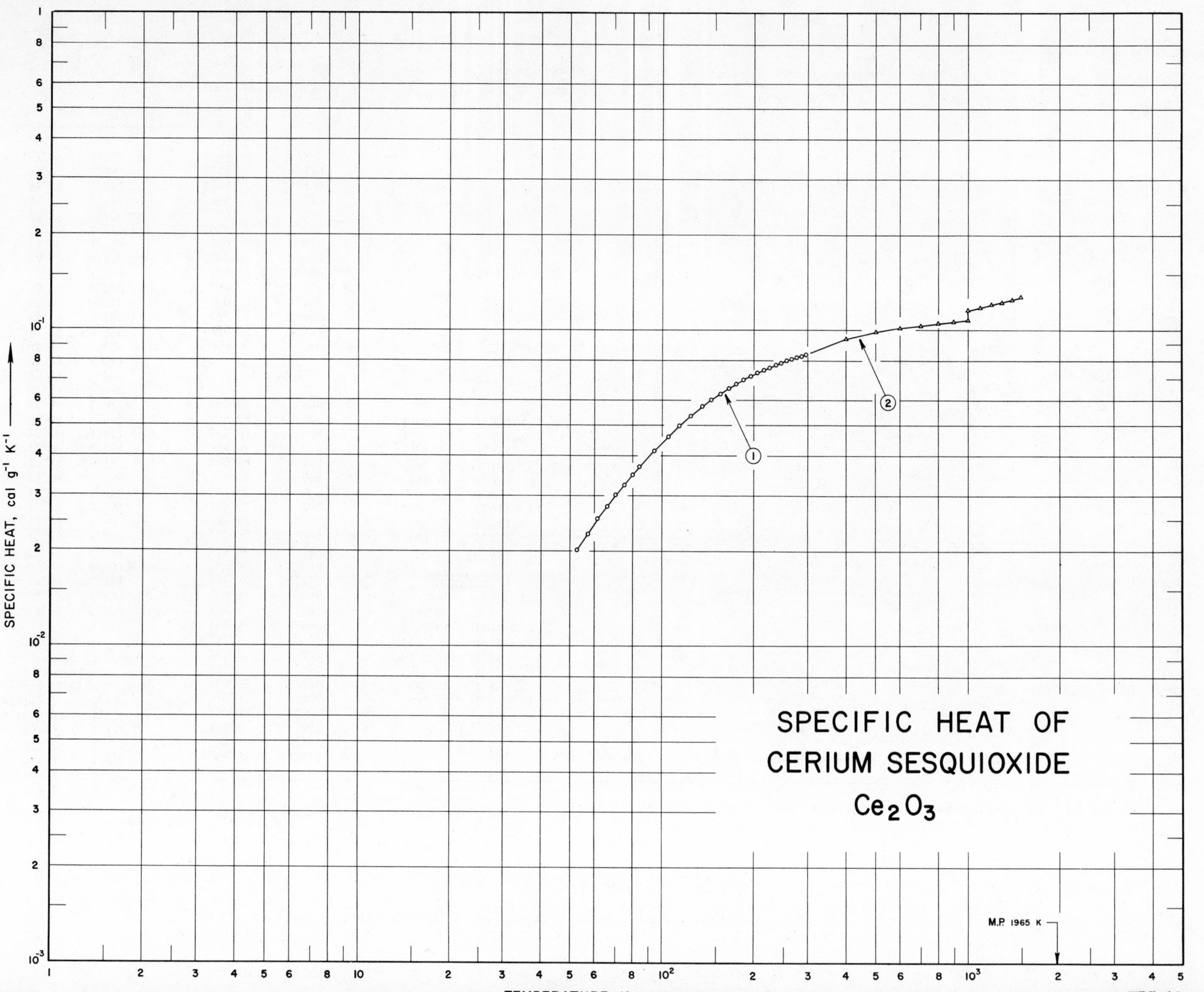
SPECIFIC HEAT OF
CERIUM SESQUIOXIDE
Ce_2O_3
M.P. 1965 K
SPECIFIC HEAT, cal g^{-1} K^{-1}

SPECIFICATION TABLE NO. 18 SPECIFIC HEAT OF CERIUM SESQUIOXIDE Ce_2O_3

[For Data Reported in Figure and Table No. 18]

Curve No.	Ref. No.	Year	Temp. Range, K	Reported Error, %	Name and Specimen Designation	Composition (weight percent), Specifications and Remarks
1	58	1963	50-298	0.10		$Ce_2O_{3.33}$, 0.02 Al_2O_3, $<$0.002 C; measured under nitrogen atmosphere.
2	59	1963	298-1500			$Ce_2O_{3.33}$; 99.9 CeO_2, 0.02 Al_2O_3, 0.001 C.

DATA TABLE NO. 18 SPECIFIC HEAT OF CERIUM SESQUIOXIDE Ce_2O_3

[Temperature, T, K; Specific Heat, C_p, Cal $g^{-1} K^{-1}$]

T	C_p
CURVE 1	
52.96	2.007×10^{-2}
57.27	2.250
61.73	2.522
66.17	2.761
70.66	2.995
75.18	3.235
80.02	3.482
84.61	3.689
94.47	4.140
105.03	4.594
114.78	4.981
124.76	5.350
135.97	5.728
145.67	6.005
155.74	6.276
165.85	6.520
176.10	6.763
186.13	6.955
196.02	7.136
206.40	7.312
216.39	7.461
226.39	7.604
236.19	7.741
245.62	7.866
256.84	7.990
266.35	8.083
276.97	8.186
286.81	8.274
296.61	8.351
CURVE 2	
(α) 298.15	8.3611×10^{-2}*
300	8.3874*
400	9.3399
500	9.8246
600	1.0125×10^{-1}
700	1.0339
800	1.0506
900	1.0647
(α) 1000	1.0770
(β) 1000	1.1515
1100	1.1762
1200	1.2006
CURVE 2 (cont.)	
(β) 1300	1.2249×10^{-1}
1400	1.2489
(β) 1500	1.2729

*Not shown on plot

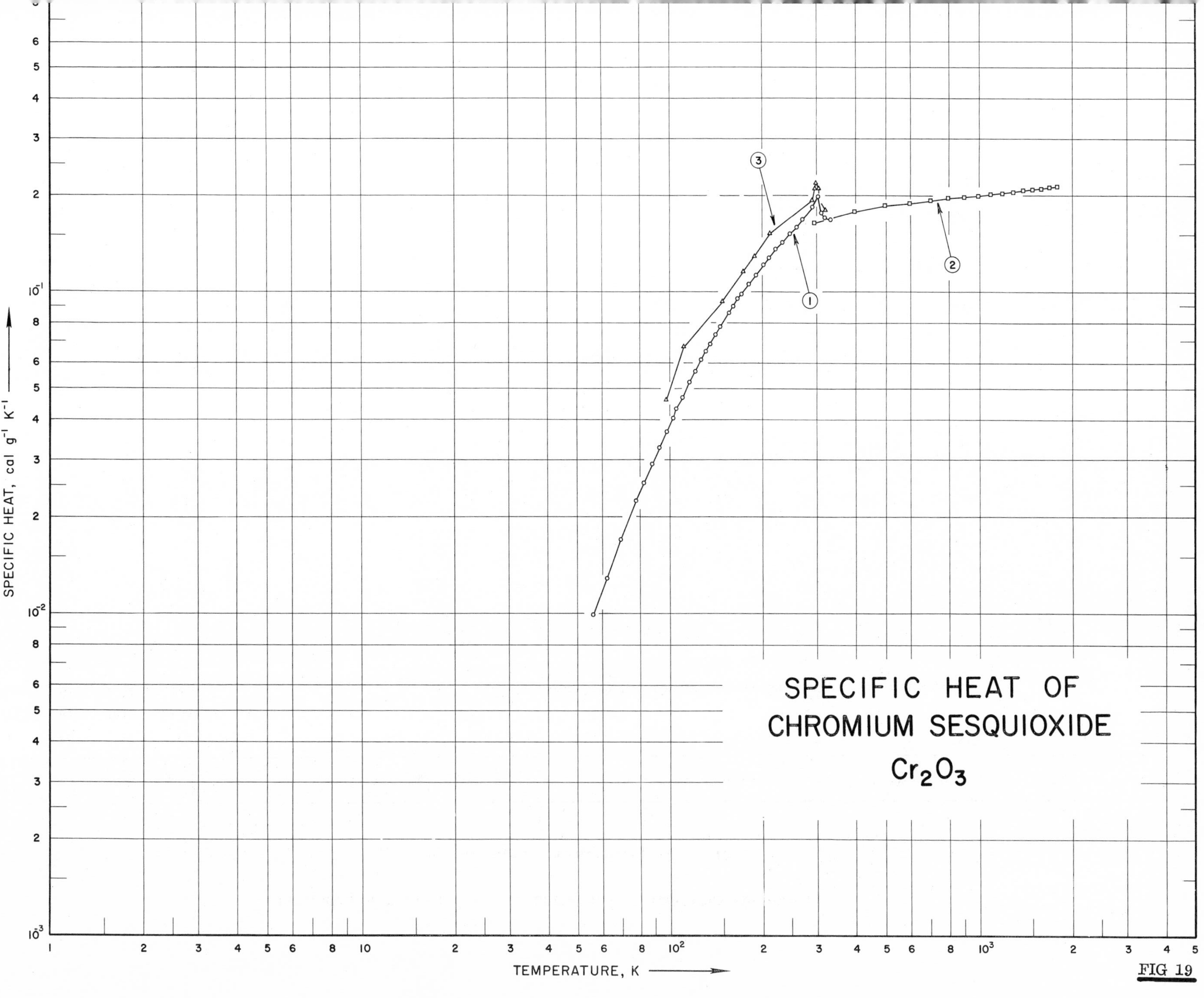
SPECIFIC HEAT OF
CHROMIUM SESQUIOXIDE
Cr_2O_3
TEMPERATURE, K
SPECIFIC HEAT, cal g^{-1} K^{-1}
1
2
3

FIG 19

SPECIFICATION TABLE NO. 19 SPECIFIC HEAT OF CHROMIUM SESQUIOXIDE Cr_2O_3

[For Data Reported in Figure and Table No. 19]

Curve No.	Ref. No.	Year	Temp. Range, K	Reported Error, %	Name and Specimen Designation	Composition (weight percent), Specifications and Remarks
1	60	1937	56-335			100 Cr_2O_3.
2	61	1944	298-1800			
3	62	1952	98-322			Sample prepared by firing ammonium chromate in air at 1000 C.

DATA TABLE NO. 19 SPECIFIC HEAT OF CHROMIUM SESQUIOXIDE Cr_2O_3

[Temperature, T, K; Specific Heat, C_p, Cal g^{-1} K^{-1}]

T	C_p
CURVE 1	
56.3	9.893 x 10^{-3}
62.5	1.288 x 10^{-2}
69.4	1.695
78.5	2.248
83.0	2.543
88.1	2.901
93.0	3.264
98.1	3.659
98.6	3.703*
102.9	4.058
105.9	4.333
107.4	4.426*
111.1	4.691
117.2	5.230
122.2	5.674
127.4	6.126
132.1	6.514
136.4	6.874
141.7	7.341
147.0	7.782
157.4	8.598
162.2	9.005
168.5	9.505
172.7	9.821
176.6	1.0137 x 10^{-1}*
182.6	1.0584
187.4	1.0933*
192.1	1.1314
199.6	1.1827*
203.6	1.2123
207.9	1.2459*
212.6	1.2781
217.9	1.3117*
223.7	1.3557
229.0	1.3906*
234.9	1.4307
239.7	1.4570*
247.6	1.5130
260.4	1.5919
273.3	1.6800
276.7	1.7037*
281.8	1.7432*
293.4	1.8438
CURVE 1 (cont.)	
294.9	1.8524 x 10^{-1}*
296.2	1.8708*
299.6	1.8998*
299.7	1.9024*
301.5	1.9320*
302.4	1.9432*
302.6	1.9445*
303.7	1.9622*
304.4	1.9728*
305.5	1.9787
306.6	1.9491*
306.7	1.9280*
308.3	1.9294*
309.7	1.8090*
311.1	1.7912*
311.6	1.7623*
313.9	1.7550
316.8	1.7096*
320.2	1.6899
324.8	1.6754*
329.9	1.6761*
335.6	1.6669
CURVE 2	
298.15	1.643 x 10^{-1}
300	1.647*
400	1.781
500	1.851
600	1.896
700	1.928
800	1.954
900	1.977
1000	1.987
1100	2.016
1200	2.033
1300	2.050
1400	2.067
1500	2.083
1600	2.099
1700	2.115
1800	2.130
CURVE 3	
98	4.605 x 10^{-2}
112	6.742
149	9.308
174	1.1577 x 10^{-1}
190	1.2959
214	1.5130
291	1.9340
294	1.9734*
298	2.1050
300	2.1938
307	2.1017
310	1.8748*
320	1.8156*
330	1.8156*
329	1.7892*
322	1.7991

*Not shown on plot

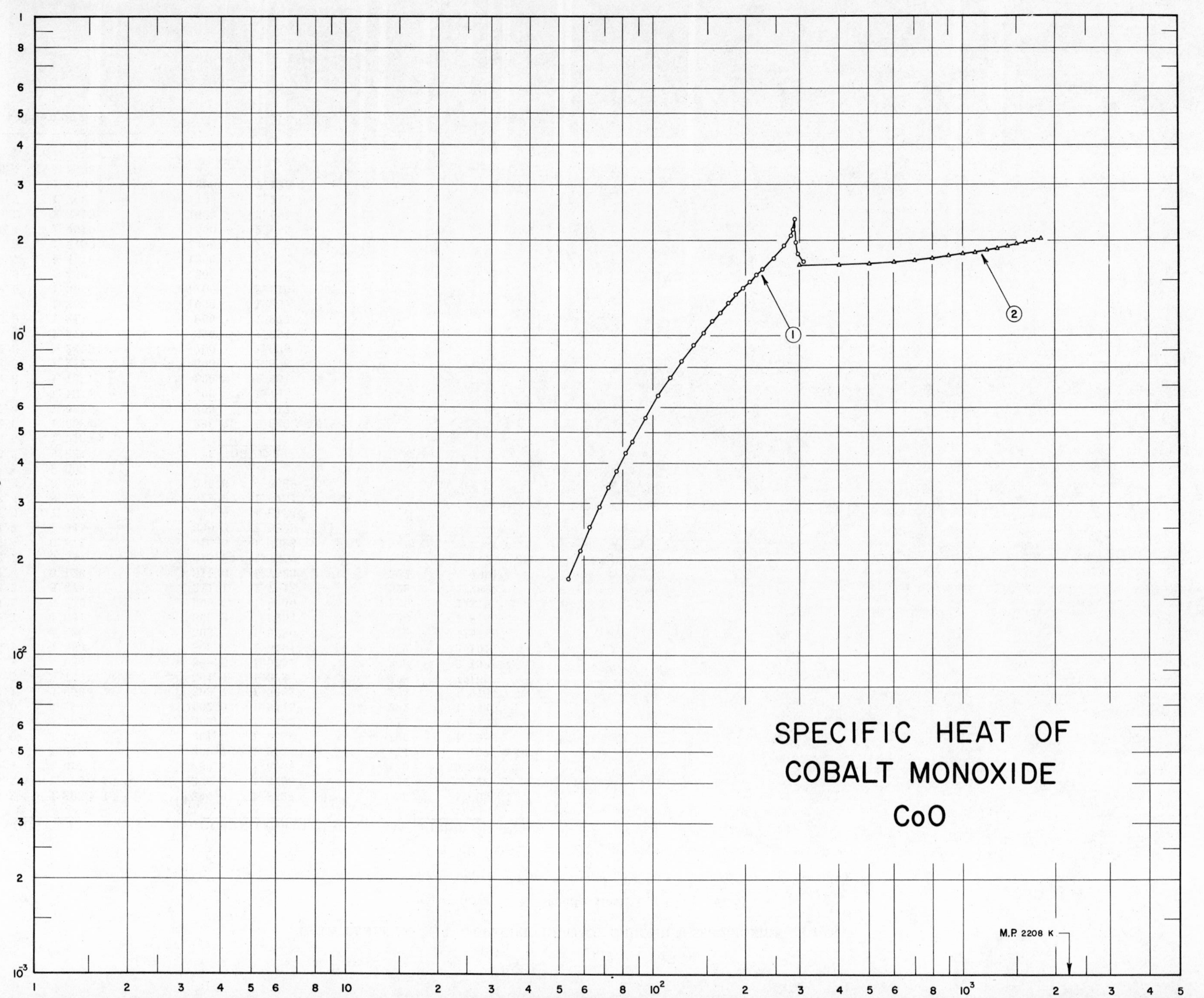
SPECIFIC HEAT OF
COBALT MONOXIDE
CoO
M.P. 2208 K
SPECIFIC HEAT, cal g^{-1} K^{-1}
1
2

SPECIFICATION TABLE NO. 20 SPECIFIC HEAT OF COBALT MONOXIDE CoO

[For Data Reported in Figure and Table No. 20]

Curve No.	Ref. No.	Year	Temp. Range, K	Reported Error, %	Name and Specimen Designation	Composition (weight percent), Specifications and Remarks
1	63	1957	53-308			CoO, 78.61 Co; prepared from recrystallized reagent grade cobaltous sulfate heptahydrate.
2	64	1958	298-1800	0.5		78.61 Co, 21.36 O_2, 0.02 SiO_2, 0.01 S; reheated 72 hrs in air at 1180 - 1230 C, then 28 hrs in helium at 1150 - 1160 C.

DATA TABLE NO. 20 SPECIFIC HEAT OF COBALT MONOXIDE CoO

[Temperature, T, K; Specific Heat, C_p, Cal g^{-1} K^{-1}]

T	C_p	T	C_p
CURVE 1		CURVE 2	
53.61	1.747 x 10^{-2}	298.15	1.681 x 10^{-1}
58.27	2.133	300	1.681*
62.80	2.530	400	1.682
67.20	2.929	500	1.697
71.72	3.352	600	1.718
76.31	3.783	700	1.741
81.76	4.308	800	1.766
85.71	4.687	900	1.791
94.37	5.524	1000	1.817
104.79	6.536	1100	1.844
114.22	7.426	1200	1.870
124.28	8.367	1300	1.897
135.71	9.390	1400	1.924
145.57	1.027 x 10^{-1}	1500	1.951
155.66	1.113	1600	1.978
165.67	1.193	1700	2.005
175.82	1.272	1800	2.032
185.97	1.351		
195.82	1.419		
206.10	1.490		
215.94	1.558		
225.96	1.627		
235.85	1.694*		
245.65	1.761		
256.13	1.841*		
266.10	1.926		
276.03	2.028*		
279.74	2.083		
282.15	2.121*		
284.06	2.173		
285.48	2.228*		
286.36	2.252*		
286.48	2.287*		
287.37	2.345		
288.27	2.255*		
289.27	2.147*		
290.33	1.969		
291.44	1.889*		
292.59	1.841*		
294.25	1.803		
296.82	1.775*		
297.34	1.771*		
299.87	1.750*		
303.40	1.734*		
307.61	1.722		

*Not shown on plot

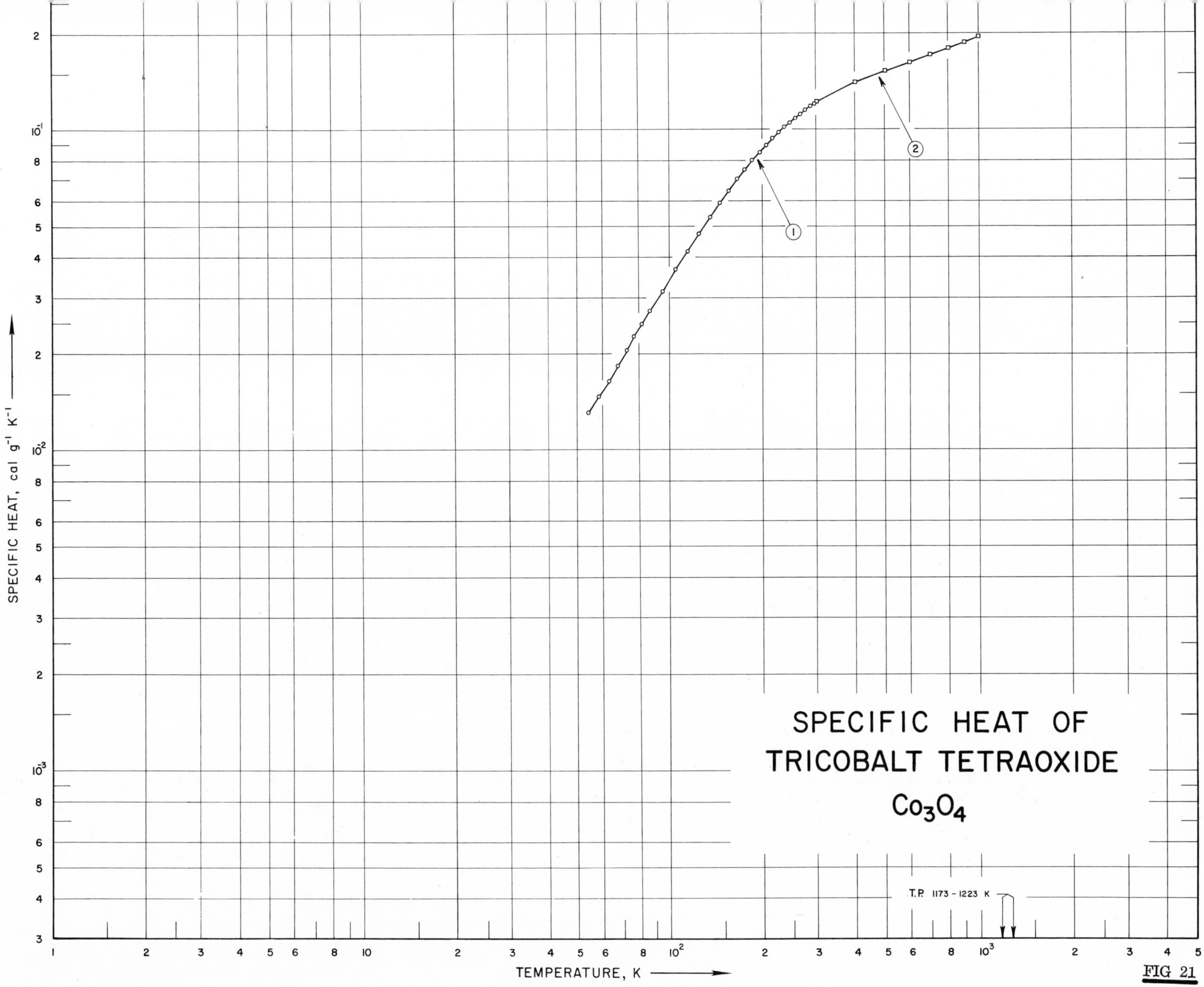

SPECIFIC HEAT OF TRICOBALT TETRAOXIDE Co_3O_4

SPECIFIC HEAT, cal g^{-1} K^{-1}

TEMPERATURE, K

T.P. 1173 - 1223 K

(1)

(2)

FIG 21

SPECIFICATION TABLE NO. 21 SPECIFIC HEAT OF TRICOBALT TETRAOXIDE Co_3O_4

[For Data Reported in Figure and Table No. 21]

Curve No.	Ref. No.	Year	Temp. Range, K	Reported Error, %	Name and Specimen Designation	Composition (weight percent), Specifications and Remarks
1	63	1957	54-296			Cobalt spinel Co_3O_4; prepared from recrystallized reagent grade cobalt sulfate heptahydrate.
2	64	1958	298-1000	0. 5		73. 40 Co; prepared from reagent grade cobalt sulfate heptahydrate by heating in air 15 days at 850 C and 16 hrs at 900 C; quenched to room temperature.

DATA TABLE NO. 21 SPECIFIC HEAT OF TRICOBALT TETRAOXIDE Co_3O_4

[Temperature, T, K; Specific Heat, C_p, Cal g^{-1} K^{-1}]

T	C_p
CURVE 1	
54.02	1.309×10^{-2}
58.55	1.472
63.05	1.641
67.68	1.843
72.32	2.050
76.78	2.261
81.40	2.473
86.53	2.722
94.93	3.143
104.98	3.675
114.67	4.190
124.81	4.755
136.03	5.369
145.76	5.921
155.91	6.482
165.95	7.014
176.02	7.524
186.06	8.039
196.02	8.500
206.38	8.990
216.33	9.418
226.40	9.833
236.23	1.022×10^{-1}
245.77	1.057
256.47	1.095
266.42	1.129
276.33	1.162
286.92	1.192
296.34	1.216
CURVE 2	
298.15	1.225×10^{-1}*
300	1.229
400	1.416
500	1.540
600	1.640
700	1.729
800	1.811
900	1.890
1000	1.966

*Not shown on plot

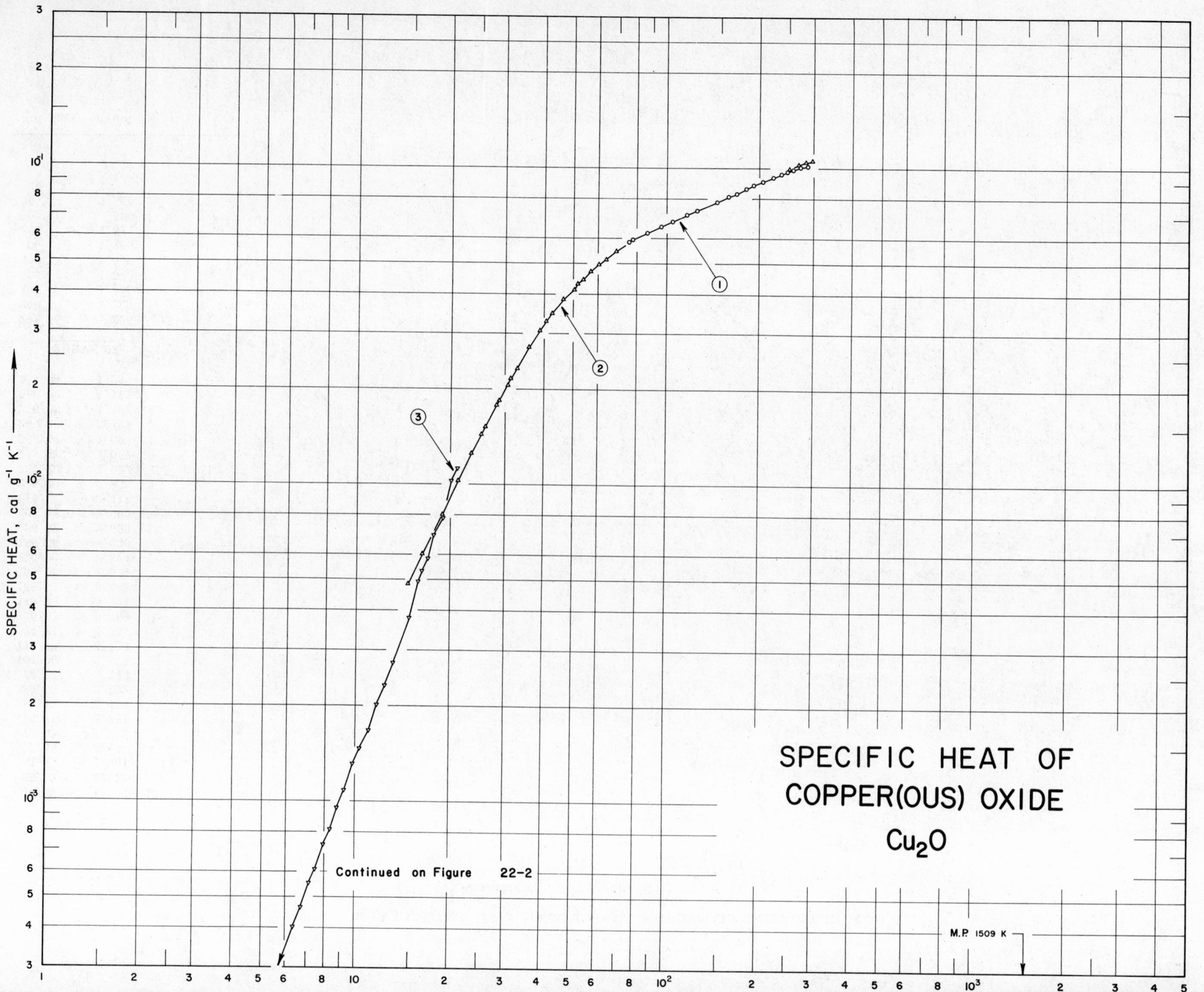

SPECIFIC HEAT OF
COPPER(OUS) OXIDE
Cu_2O
Continued on Figure 22-2
M.P. 1509 K
1
2
3
SPECIFIC HEAT, cal g^{-1} K^{-1}

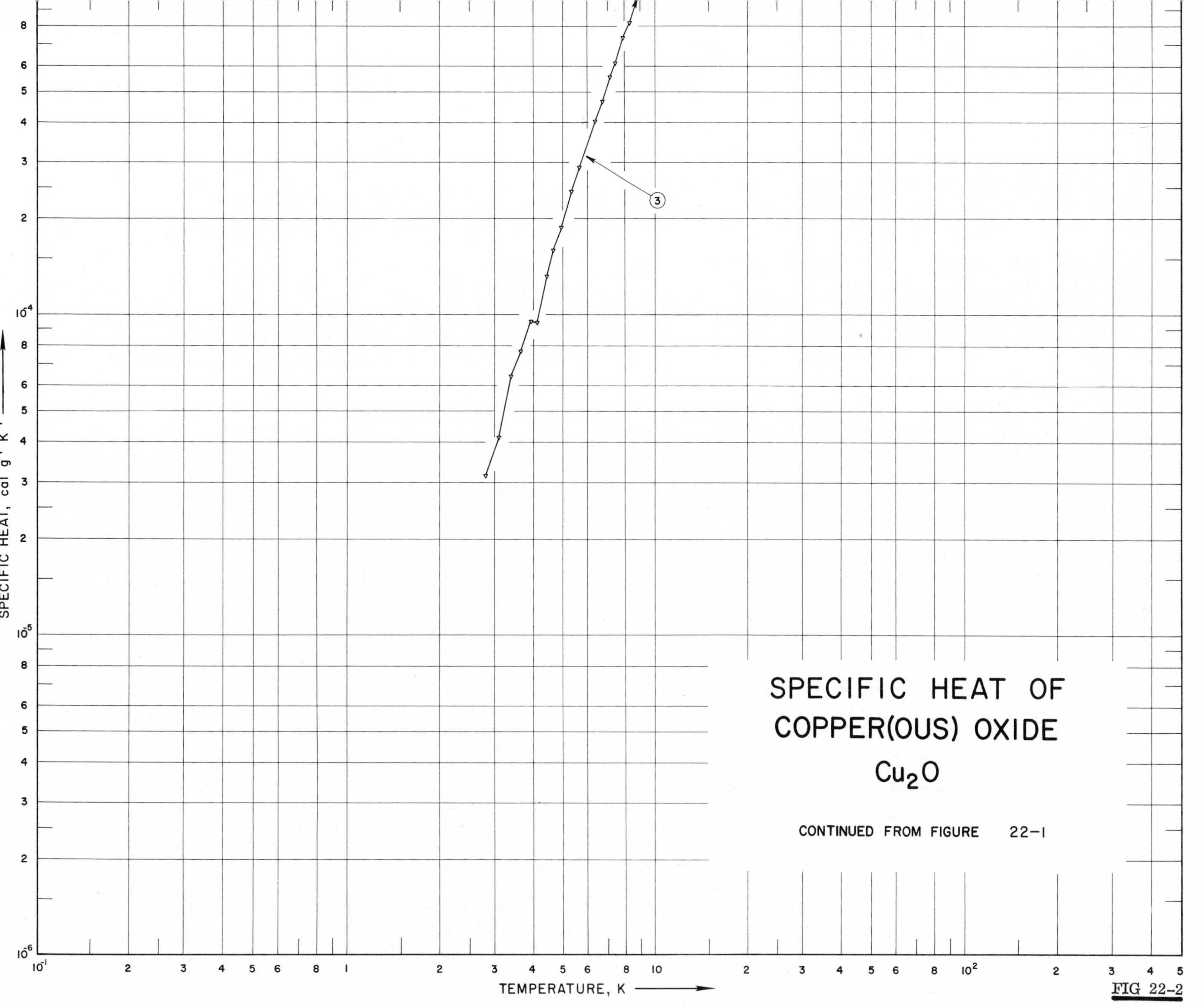
SPECIFIC HEAT OF
COPPER(OUS) OXIDE
Cu_2O
CONTINUED FROM FIGURE 22-1
SPECIFIC HEAT, cal g^{-1} K^{-1}
TEMPERATURE, K
3

FIG 22-2

SPECIFICATION TABLE NO. 22 SPECIFIC HEAT OF COPPER(OUS) OXIDE Cu_2O

[For Data Reported in Figure and Table No. 22]

Curve No.	Ref. No.	Year	Temp. Range, K	Reported Error, %	Name and Specimen Designation	Composition (weight percent), Specifications and Remarks
1	65	1929	76-291			100 Cu_2O.
2	66	1951	14-300			99. 8 Cu_2O; prepared by precipitation of a warm Fehling solution with dextrose.
3	67	1962	2-19			99. 4 Cu_2O; finely divided red powder; prepared by reduction of warm Fehling solution with dextrose.

DATA TABLE NO. 22 SPECIFIC HEAT OF COPPER(OUS) OXIDE Cu_2O

[Temperature, T, K; Specific Heat, C_p, Cal $g^{-1} K^{-1}$]

T	C_p	T	C_p	T	C_p
CURVE 1		CURVE 2 (cont.)		CURVE 3	
75.9	5.847 x 10^{-2}	47.04	3.914 x 10^{-2}	2.794	3.151 x 10^{-5}
78.4	5.954	50.34	4.164	3.087	4.157
87.0	6.245	51.84	4.322	3.375	6.441
96.9	6.538	54.00	4.478	3.635	7.699
105.3	6.794	57.01	4.716	3.937	9.522
117.0	7.115	57.82	4.760*	4.107	9.445
125.6	7.346	60.65	4.983	4.406	1.328 x 10^{-4}
147.0	7.842	61.65	5.013*	4.662	1.596
159.9	8.149	64.01	5.172	4.943	1.882
170.0	8.310	69.44	5.470	5.313	2.449
182.9	8.611	76.39	5.801*	5.671	2.898
192.9	8.855	82.80	6.052*	6.384	4.015
206.1	9.072	88.31	6.254*	6.760	4.649
223.5	9.365	94.28	6.465*	7.138	5.518
236.1	9.575	101.73	6.689*	7.496	6.125
247.6	9.736	109.75	6.886*	7.910	7.328
259.0	9.869	116.62	7.070*	8.335	8.195
252.1	9.771*	124.00	7.259*	8.778	9.529
273.5	1.004 x 10^{-1}	131.66	7.475*	9.269	1.086 x 10^{-3}
289.1	1.013	138.81	7.664*	9.811	1.313
291.0	1.025*	145.42	7.832*	10.385	1.471
		153.63	8.020*	11.019	1.688
CURVE 2		160.73	8.195*	11.723	2.033
		166.88	8.328*	12.462	2.326
14.72	4.869 x 10^{-3}	174.06	8.488*	13.196	2.743
16.47	6.036	181.20	8.642*	14.951	3.795
18.99	8.076	187.89	8.782*	15.975	4.939
21.31	1.028 x 10^{-2}	194.87	8.928*	16.423	5.343
23.49	1.253	203.79	9.075*	17.076	5.837
25.22	1.448	212.62	9.264*	17.793	6.917
26.00	1.530	220.84	9.410*	19.081	7.814
28.22	1.793	229.00	9.578*	20.207	1.031 x 10^{-2}
28.75	1.846	236.16	9.690*	21.090	1.126
30.53	2.068	243.54	9.802*		
31.31	2.157	252.12	9.934		
32.84	2.347	259.67	1.008 x 10^{-1}*		
35.70	2.710	269.67	1.024		
38.89	3.087	275.47	1.033*		
40.87	3.280	283.67	1.044		
42.72	3.498	291.67	1.054*		
43.39	3.543*	299.64	1.063		
46.69	3.861				

*Not shown on Plot

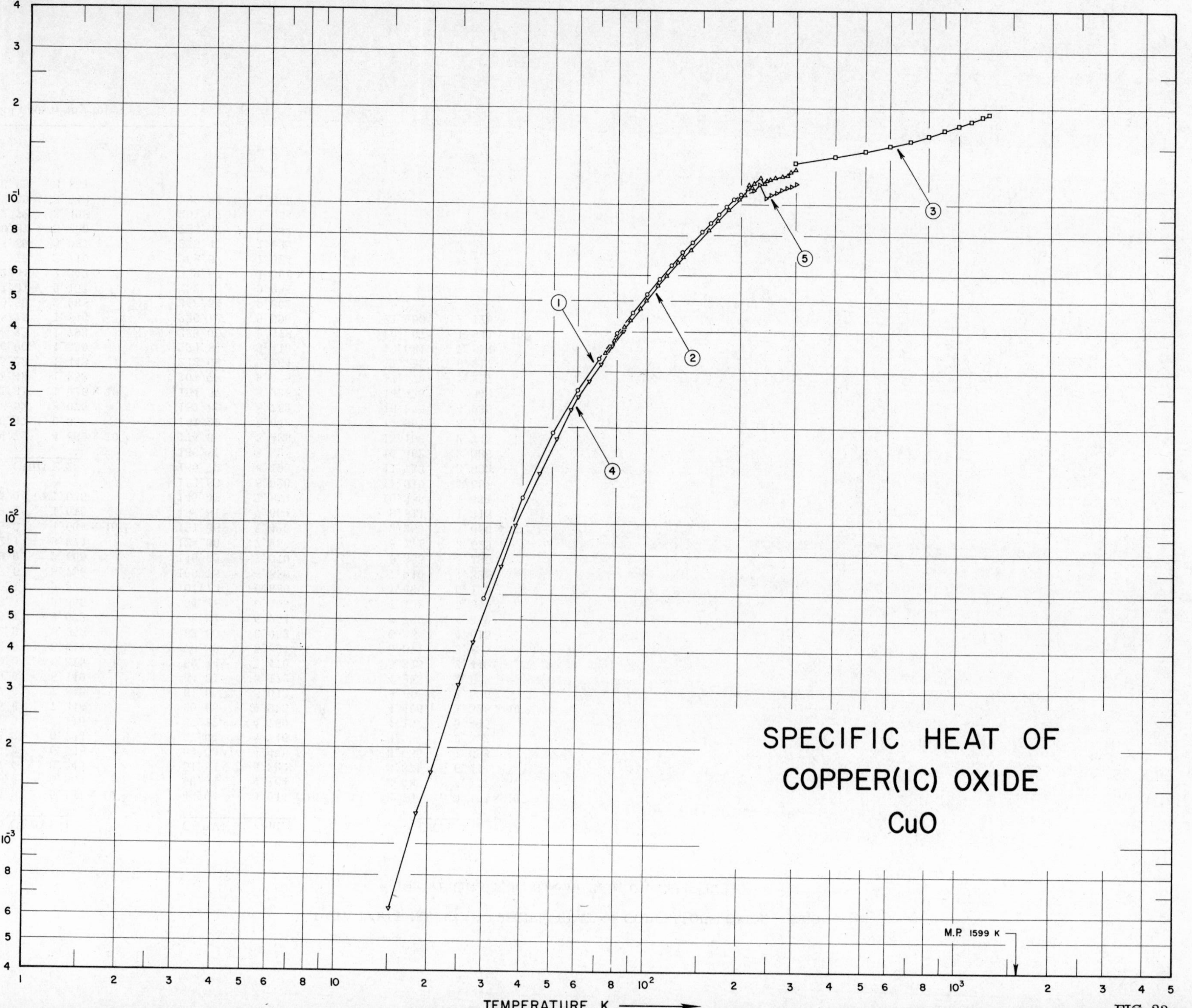
SPECIFIC HEAT OF
COPPER(IC) OXIDE
CuO
M.P. 1599 K
TEMPERATURE, K
SPECIFIC HEAT, cal g⁻¹ K⁻¹
1
2
3
4
5

SPECIFICATION TABLE NO. 23 SPECIFIC HEAT OF COPPER (IC) OXIDE CuO

[For Data Reported in Figure and Table No. 23]

Curve No.	Ref. No.	Year	Temp. Range, K	Reported Error, %	Name and Specimen Designation	Composition (weight percent), Specifications and Remarks
1	68	1928	30-200			Kahlbaum purity.
2	65	1929	71-302			99. 4 CuO and 0. 6 Cu_2O.
3	69	1933	298-1253			
4	70	1953	15-297			99. 95 CuO; prepared by heating electrolytic copper (99. 95 Cu) at 800 C in furnace for several days, repeatedly ground and oxidized.
5	71	1954	200-300			

DATA TABLE NO. 23 SPECIFIC HEAT OF COPPER (IC) OXIDE, CuO

[Temperature, T, K; Specific Heat, C_p, Cal $g^{-1}K^{-1}$]

CURVE 1

T	C_p
30	5.8 x 10^{-3}*
40	1.2 x 10^{-2}*
50	1.91*
60	2.595*
70	3.256
80	3.917
90	4.55
100	5.20*
110	5.81*
120	6.41*
130	7.02*
140	7.57*
150	8.15
160	8.70
170	9.25
180	9.76
190	1.03 x 10^{-1}
200	1.07

CURVE 2

T	C_p
71.3	3.244 x 10^{-2}*
73.7	3.407
75.8	3.550*
78.7	3.739
81.8	3.926*
84.8	4.114
88.2	4.321
91.7	4.493*
95.2	4.701
99.5	4.994
108.5	5.552
111.6	5.733*
115.2	5.940
119.2	6.203*
122.2	6.428*
125.2	6.556
142.1	7.469*
199.5	1.054 x 10^{-1}*
203.5	1.078*
210.1	1.116

CURVE 2 (cont.)

T	C_p
212.3	1.149 x 10^{-1}
214.4	1.128*
219.9	1.166
221.3	1.159*
224.4	1.179*
237.9	1.158
241.1	1.182
248.7	1.197
258.3	1.204
130.7	6.812 x 10^{-2}
133.7	6.954*
138.1	7.171
153.8	8.032*
157.2	8.252
163.0	8.552*
169.2	8.868
183.5	9.601
211.1	1.121 x 10^{-1}*
217.0	1.126
230.5	1.203
240.7	1.179*
281.2	1.238
284.9	1.255*
287.5	1.258
299.4	1.292
301.6	1.291*
193.8	1.025*
197.6	1.036
201.1	1.049*
204.5	1.069
208.1	1.105*
211.6	1.134*
216.0	1.121*
219.2	1.135*
228.8	1.184*
246.0	1.174*
261.2	1.206*
280.6	1.239*
284.9	1.246*
289.1	1.261*
293.6	1.265*

CURVE 2 (cont.)

T	C_p
204.7	1.073 x 10^{-1}
209.7	1.106*
212.6	1.146*
229.4	1.184*
233.5	1.155*
238.2	1.155*
261.5	1.211*
265.7	1.209*
271.4	1.218

CURVE 3

T	C_p
298.15	1.345 x 10^{-1}
300	1.346*
400	1.407
500	1.467
600	1.528
700	1.588
800	1.648
900	1.709
1000	1.769
1100	1.829
1200	1.890
1253	1.921

CURVE 4

T	C_p
218.60	1.144 x 10^{-1}
227.52	1.168*
235.81	1.161*
243.96	1.166*
253.11	1.184*
262.92	1.195*
272.57	1.210*
282.06	1.233*
289.47	1.255*
297.23	1.271*
196.60	1.024*
204.39	1.071*
213.70	1.126*
222.93	1.162*

CURVE 4 (cont.)

T	C_p
231.68	1.158 x 10^{-1}*
240.26	1.160*
248.52	1.173*
187.12	9.780 x 10^{-2}*
196.44	1.025 x 10^{-1}*
205.38	1.075*
214.58	1.130*
223.76	1.165
232.53	1.159
240.96	1.162
249.12	1.173
15.14	6.286 x 10^{-4}
18.46	1.245 x 10^{-3}
20.50	1.660
25.08	3.118
27.84	4.249
34.01	7.317
37.96	9.819
45.48	1.432 x 10^{-2}
51.56	1.8254
57.17	2.254
60.30	2.482
65.42	2.790
71.17	3.147
77.42	3.570
84.02	3.987
90.70	4.423*
97.97	4.823*
105.40	5.307*
112.95	5.752*
121.73	6.250*
130.62	6.745*
138.50	7.177*
146.57	7.661*
155.84	8.186*
165.06	8.643*
174.47	9.130*
183.73	9.626*
192.32	1.006 x 10^{-1}*
200.99	1.052*
209.82	1.109*

CURVE 5

T	C_p
200	1.075 x 10^{-1}*
210	1.088*
220	1.100
230	1.157
240	1.050
250	1.069
260	1.088
270	1.113
280	1.125
290	1.138
300	1.157

* Not shown on Plot

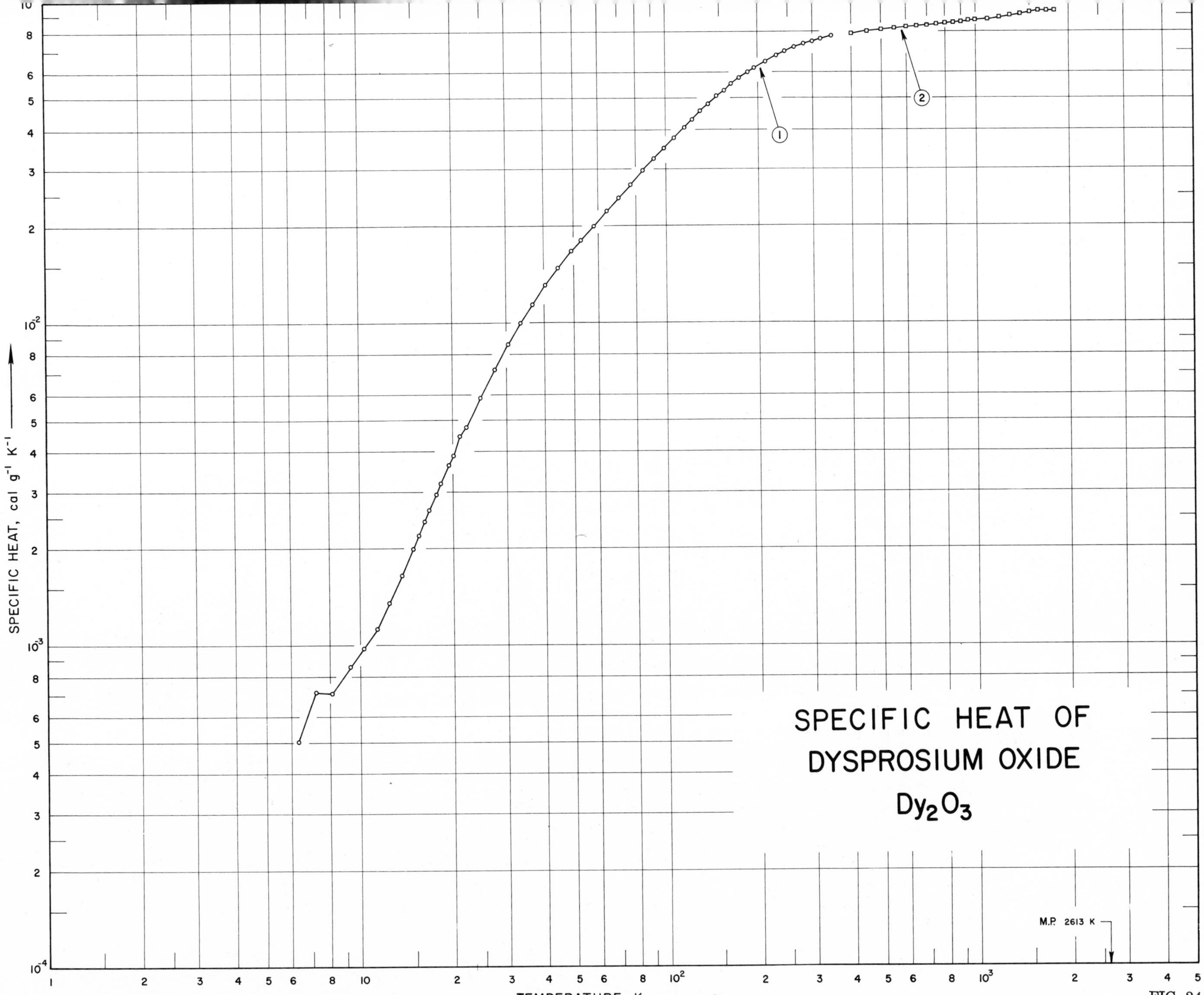

SPECIFIC HEAT OF DYSPROSIUM OXIDE Dy_2O_3

SPECIFIC HEAT, cal g^{-1} K^{-1}

TEMPERATURE, K

M.P. 2613 K

(1)

(2)

FIG 24

SPECIFICATION TABLE NO. 24 SPECIFIC HEAT OF DYSPROSIUM OXIDE Dy_2O_3

[For Data Reported in Figure and Table No. 24]

Curve No.	Ref. No.	Year	Temp. Range, K	Reported Error, %	Name and Specimen Designation	Composition (weight percent), Specifications and Remarks
1	72	1962	6-346	0. 1		99. 9 Dy_2O_3, 0. 015 Y, 0. 010 C, and 0. 010 Si; powder specimen; supplied by Michigan Chemical Co; helium atmosphere.
2	59	1963	400-1800			99. 9 Dy_2O_3; dried at 1100 - 1200 C.

DATA TABLE NO. 24 SPECIFIC HEAT OF DYSPROSIUM OXIDE, Dy_2O_3

[Temperature, T, K; Specific Heat, C_p, Cal $g^{-1}K^{-1}$]

T	C_p
CURVE 1	
Series 1	
6.36	5.013×10^{-4}
7.24	7.131
8.16	7.105
9.31	8.552
10.35	9.785
11.42	1.126×10^{-3}
12.49	1.354
13.64	1.649
14.90	1.995
16.27	2.421
17.79	2.949
19.52	3.603
21.20	4.426
Series 2	
15.54	2.190
16.87	2.627
18.42	3.174
20.17	3.866
22.33	4.735
24.72	5.858
27.55	7.169
30.52	8.555
33.55	9.963
36.89	1.146×10^{-2}
40.66	1.308
44.92	1.483
49.57	1.668
Series 3	
53.15	1.803
58.44	1.999
64.37	2.221
70.77	2.446
77.39	2.684
84.82	2.966
CURVE 1 (cont.)	
Series 4	
84.33	2.948×10^{-2}
91.92	3.217
99.39	3.475
107.09	3.742
115.16	4.020
122.74	4.273
Series 5	
130.43	4.523
138.70	4.775
147.25	5.030
155.99	5.268
165.06	5.501
174.35	5.727
Series 6	
177.48	5.791*
186.19	5.979
194.78	6.153
203.32	6.311*
212.31	6.467
221.54	6.611*
230.64	6.740
Series 7	
239.84	6.877*
245.45	6.936
254.80	7.046*
263.29	7.139
272.50	7.239*
281.86	7.319
291.18	7.397*
300.40	7.464
309.76	7.534
319.15	7.590
CURVE 1 (cont.)	
Series 7 (cont.)	
328.56	7.641×10^{-2}*
337.99	7.700
346.58	7.735
CURVE 2	
400	7.869×10^{-2}
450	7.983
500	8.076
550	8.156
600	8.227
650	8.291
700	8.351
750	8.408
800	8.462
850	8.514
900	8.564
950	8.614
1000	8.662
1050	8.709*
1100	8.756
1150	8.802*
1200	8.848
1250	8.893*
1300	8.938
1350	8.983*
1400	9.027
1450	9.071*
1500	9.115
1550	9.159*
1600	9.249
1650	9.249*
1700	9.249
1750	9.249*
1800	9.249

*Not shown on Plot

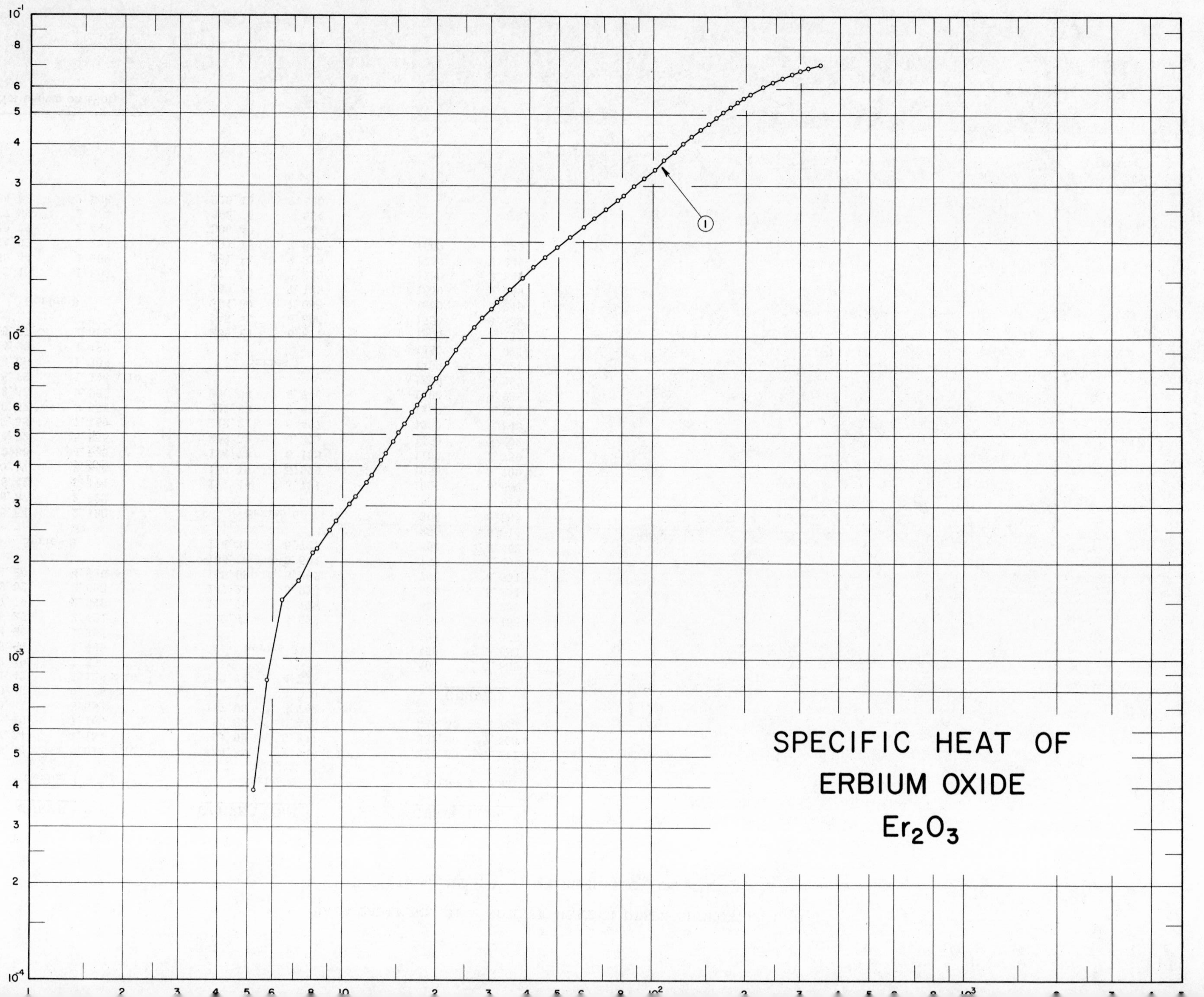
SPECIFIC HEAT OF
ERBIUM OXIDE
Er_2O_3
SPECIFIC HEAT, cal g^{-1} K^{-1}
1

SPECIFICATION TABLE NO. 25 SPECIFIC HEAT OF ERBIUM OXIDE Er_2O_3

[For Data Reported in Figure and Table No. 25]

Curve No.	Ref. No.	Year	Temp. Range, K	Reported Error, %	Name and Specimen Designation	Composition (weight percent), Specifications and Remarks
1	72	1962	5-346	0. 1		99. 9 Er_2O_3, 0. 035 Tm, 0. 010 Ca, 0. 010 Dy, 0. 010 Si, and 0. 003 Ho; powder specimen; sample supplied by the Michigan Chemical Co; measured in helium atmosphere.

DATA TABLE NO. 25 SPECIFIC HEAT OF ERBIUM OXIDE, Er_2O_3

[Temperature, T, K; Specific Heat, C_p, Cal $g^{-1}K^{-1}$]

T	C_p
CURVE 1	
Series 1	
80.89	2.776×10^{-2}
87.29	2.959
94.21	3.145
101.76	3.348
109.79	3.571
117.87	3.791
126.06	4.013
134.50	4.235
143.25	4.455
151.94	4.664
160.54	4.857
169.21	5.046
178.22	5.226
Series 2	
5.26	3.895×10^{-4}
5.78	8.549
6.46	1.514×10^{-3}
7.28	1.749
8.06	2.136
9.12	2.502
10.53	3.033
11.98	3.542
13.32	4.151
14.58	4.766
15.89	5.398
17.43	6.164
19.07	6.991
Series 3	
8.33	2.191×10^{-3}
9.57	2.685
11.00	3.195
12.41	3.728
13.76	4.371
15.19	5.048
16.76	5.840
CURVE 1 (cont.)	
Series 3 (cont.)	
18.38	6.625×10^{-3}
20.02	7.485
21.67	8.313
23.31	9.155
24.96	9.971
26.64	1.076×10^{-2}
28.39	1.154
30.23	1.237
32.51	1.333
Series 4	
28.46	1.157*
31.56	1.290
34.66	1.417
38.06	1.542
41.65	1.661
45.47	1.783
49.75	1.912
54.76	2.056
60.14	2.207
65.47	2.355
71.16	2.502
77.79	2.680
84.94	2.893*
179.34	5.249*
188.84	5.422
197.78	5.584
207.09	5.736
Series 5	
208.99	5.764*
217.95	5.900*
227.01	6.026
236.11	6.143*
244.97	6.264
253.90	6.360*
262.98	6.457
CURVE 1 (cont.)	
Series 5 (cont.)	
272.02	6.551×10^{-2}*
281.19	6.638
290.33	6.716*
299.53	6.789
308.44	6.855*
317.78	6.920
327.16	6.980*
336.58	7.038*
345.64	7.085

* Not shown on Plot

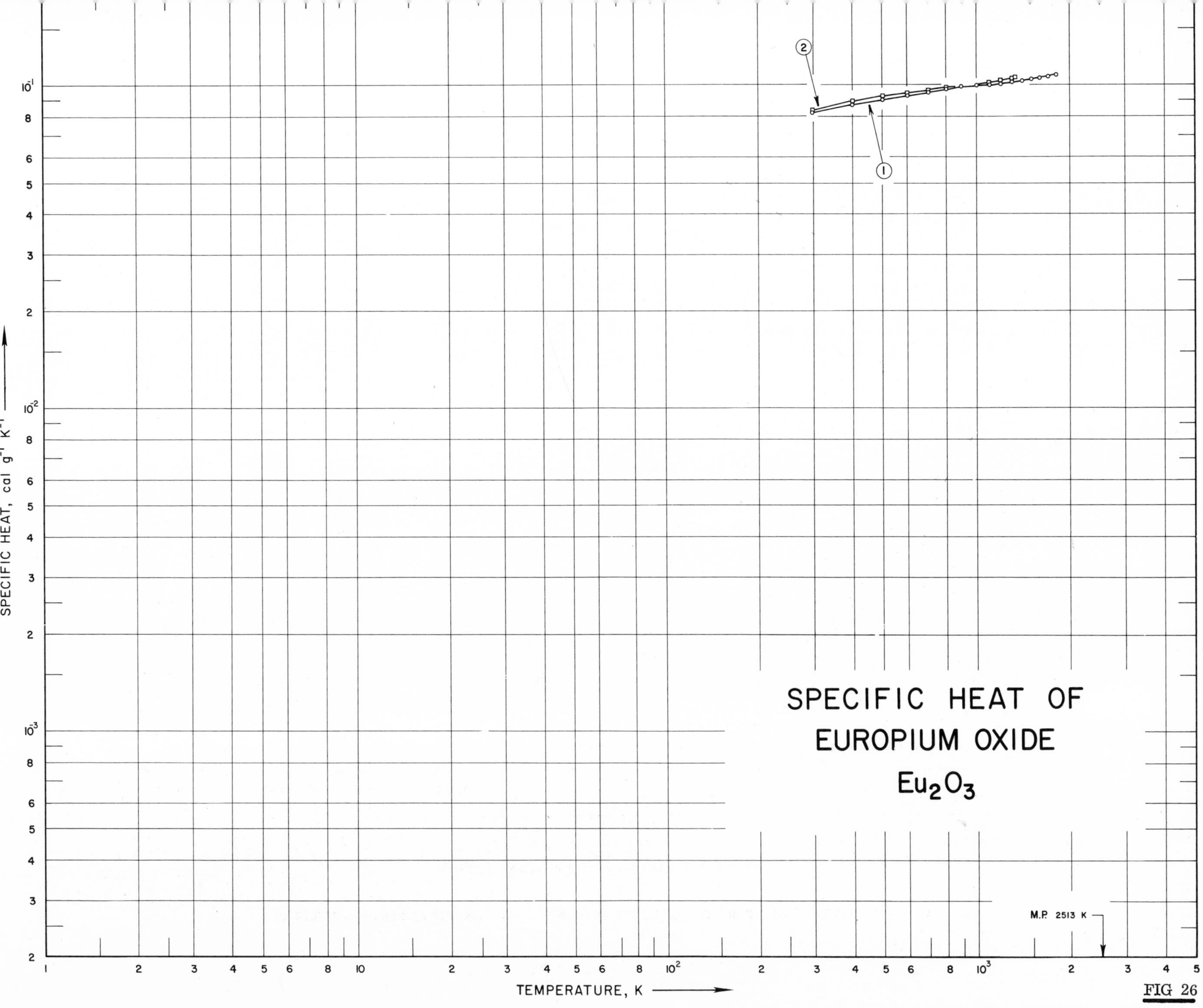
SPECIFIC HEAT OF
EUROPIUM OXIDE
Eu_2O_3
SPECIFIC HEAT, cal g^{-1} K^{-1}
TEMPERATURE, K
M.P. 2513 K
1
2
10^{-1}
10^{-2}
10^{-3}
10
10^2
10^3

FIG 26

SPECIFICATION TABLE NO. 26 SPECIFIC HEAT OF EUROPIUM OXIDE Eu_2O_3

[For Data Reported in Figure and Table No. 26]

Curve No.	Ref. No.	Year	Temp. Range, K	Reported Error, %	Name and Specimen Designation	Composition (weight percent), Specifications and Remarks
1	73	1962	298-1800	0. 2		99. 9 Eu_2O_3; monoclinic; measured in helium atmosphere.
2	73	1962	298-1350	0. 2		99. 9 Eu_2O_3; cubic; measured in helium atmosphere.

DATA TABLE NO. 26 SPECIFIC HEAT OF EUROPIUM OXIDE, Eu_2O_3

[Temperature, T, K; Specific Heat, C_p, Cal $g^{-1}K^{-1}$]

T	C_p
CURVE 1	
298	8.295×10^{-2}
300	8.307*
400	8.779
500	9.096
600	9.352
700	9.580
800	9.792
895	9.986
895	9.884*
900	9.890*
1000	1.001×10^{-1}
1100	1.013
1200	1.025
1300	1.036
1400	1.048
1500	1.060
1600	1.072
1700	1.084
1800	1.095
CURVE 2	
298	8.463×10^{-2}
300	8.479*
400	9.023
500	9.342
600	9.572
700	9.760
800	9.926
900	1.008×10^{-1}*
1000	1.022*
1100	1.036
1200	1.050
1300	1.063
1350	1.070

*Not shown on Plot

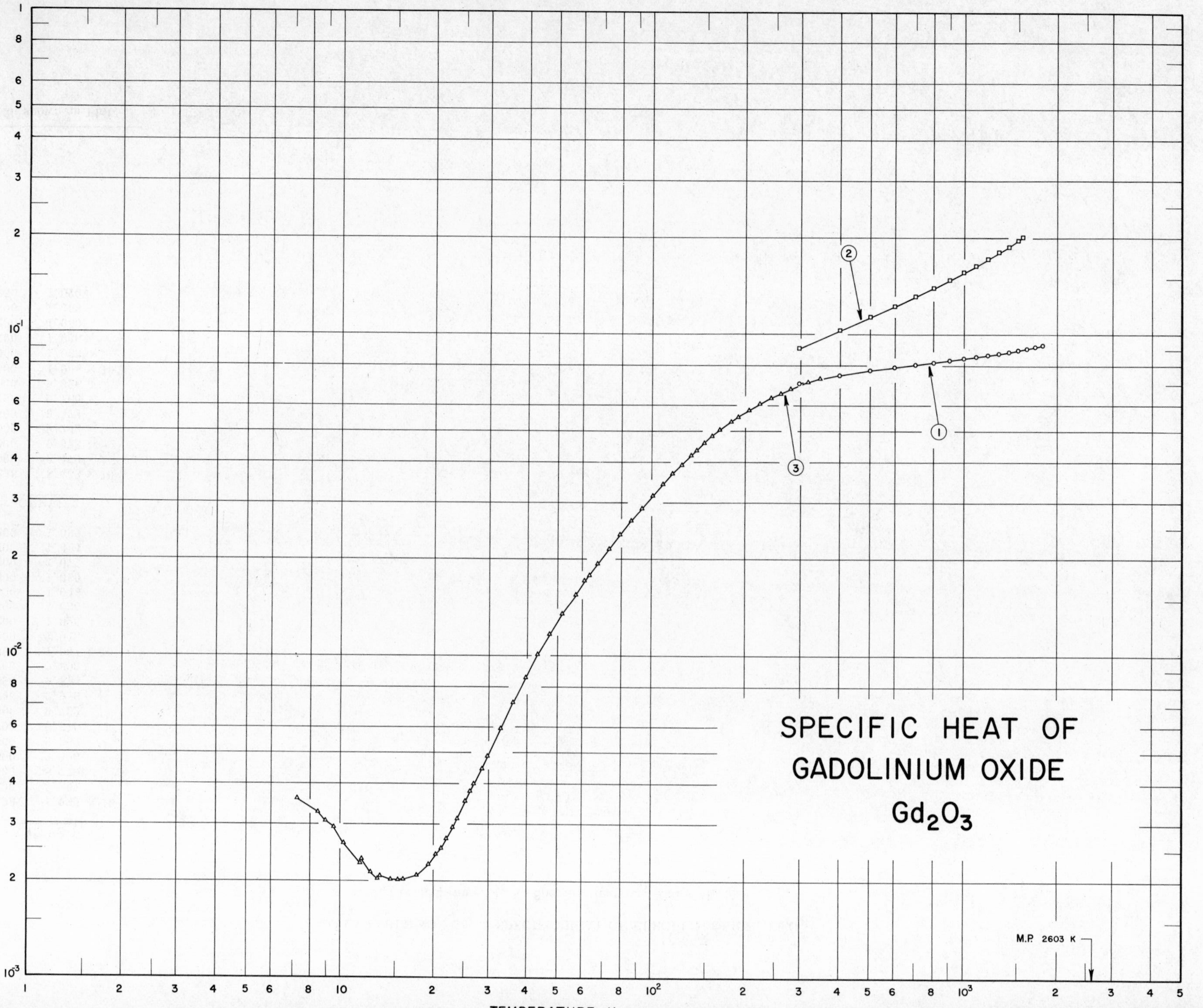

SPECIFIC HEAT OF
GADOLINIUM OXIDE
Gd_2O_3
M.P. 2603 K
SPECIFIC HEAT, cal g⁻¹ K⁻¹
TEMPERATURE, K
1
2
3

SPECIFICATION TABLE NO. 27 SPECIFIC HEAT OF GADOLINIUM OXIDE Gd_2O_3

[For Data Reported in Figure and Table No. 27]

Curve No.	Ref. No.	Year	Temp. Range, K	Reported Error, %	Name and Specimen Designation	Composition (weight percent), Specifications and Remarks
1	73	1962	298-1800	0. 2		99. 9 Gd_2O_3; monoclinic; measured in helium atmosphere.
2	73	1962	298-1550	0. 2		99. 9 Gd_2O_3; cubic; measured in helium atmosphere.
3	72	1962	7-346	0. 1		99. 90 Gd_2O_3, 0. 045 Y, 0. 020 Si, 0. 010 Ca, and 0. 0075 Eu; sample supplied by Michigan Chemical Co; pelleted under pressure of 2000 - 4000 psi; fired at 1170 K; measured in helium atmosphere.

DATA TABLE NO. 27 SPECIFIC HEAT OF GADOLINIUM OXIDE, Gd_2O_3

[Temperature, T, K; Specific Heat C_p, Cal $g^{-1}K^{-1}$]

T	C_p
CURVE 1	
298	7.028×10^{-2}
300	7.040*
400	7.478
500	7.734
600	7.917
700	8.066
800	8.197
900	8.318
1000	8.432
1100	8.542
1200	8.649
1300	8.754
1400	8.857
1500	8.959
1600	9.061
1700	9.161
1800	9.262
CURVE 2	
298	9.052×10^{-2}
300	9.084*
400	1.039×10^{-1}
500	1.141
600	1.233
700	1.319
800	1.402
900	1.484
1000	1.565
1100	1.645
1200	1.725
1300	1.804
1400	1.884
1500	1.963
1550	2.002
CURVE 3	
Series 1	
139.94	4.375×10^{-2}
148.25	4.599
156.97	4.822×10^{-2}
165.89	5.035
Series 2	
7.28	3.600×10^{-3}
8.48	3.274
8.98	3.062
10.21	2.620
11.71	2.342
13.18	2.030
14.53	2.028
16.00	2.022
17.63	2.088
19.32	2.251
21.14	2.527
23.11	2.935
25.48	3.542
28.58	4.488
Series 3	
8.45	3.230*
9.52	2.941
10.56	2.557*
11.56	2.273
12.52	2.138
13.50	2.066
14.48	2.028*
15.45	2.014
16.56	2.036*
17.77	2.099*
19.03	2.218*
20.45	2.414
22.04	2.701
23.92	3.128
26.34	3.790
Series 4	
27.01	3.995
29.80	4.894
32.71	5.939×10^{-3}
35.95	7.170
39.52	8.560
43.30	1.005×10^{-2}
47.34	1.167
51.97	1.351
57.25	1.555
63.11	1.784
Series 5	
61.16	1.710
67.25	1.940
73.10	2.155
Series 6	
79.21	2.380
86.08	2.640
93.33	2.886
101.12	3.150
109.02	3.420
116.95	3.679
125.30	3.943
134.35	4.212
143.89	4.480*
153.66	4.737*
163.29	4.974*
172.50	5.183*
181.55	5.377
190.55	5.553
Series 7	
196.38	5.658*
205.30	5.815
214.55	5.967*
223.70	6.105
232.62	6.232*
241.46	6.359
250.54	6.469*
259.86	6.579
269.20	6.681×10^{-2}*
278.61	6.781
288.05	6.866*
297.72	6.955*
307.61	7.037*
317.42	7.115
327.16	7.183*
336.80	7.250*
346.08	7.305

* Not shown on Plot

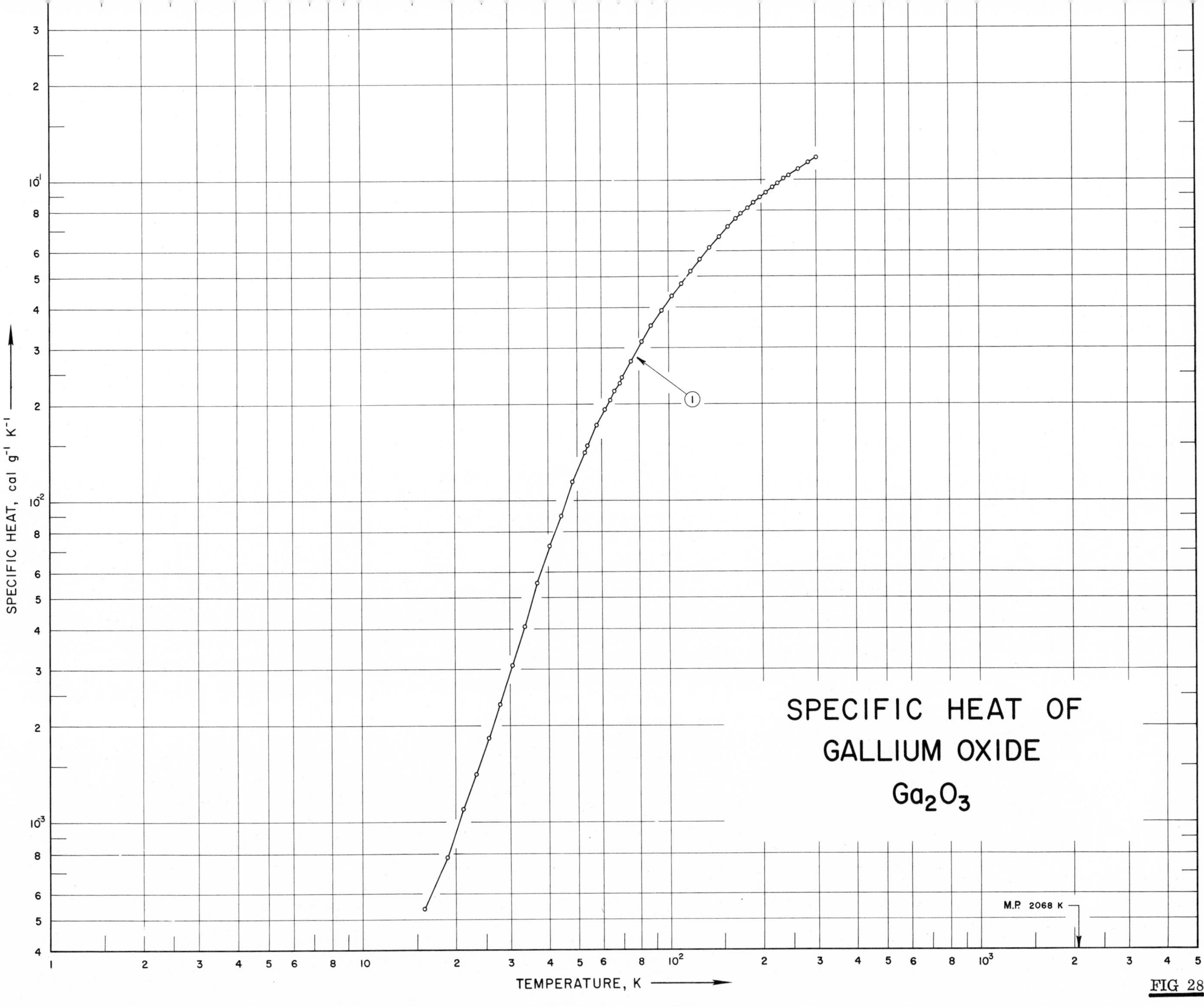

SPECIFIC HEAT OF
GALLIUM OXIDE
Ga_2O_3
M.P 2068 K
TEMPERATURE, K
SPECIFIC HEAT, cal g^{-1} K^{-1}
1

FIG 28

SPECIFICATION TABLE NO. 28 SPECIFIC HEAT OF GALLIUM OXIDE Ga_2O_3

[For Data Reported in Figure and Table No. 28]

Curve No.	Ref. No.	Year	Temp. Range, K	Reported Error, %	Name and Specimen Designation	Composition (weight percent), Specifications and Remarks
1	74	1952	15-300			98.67 βGa_2O_3, 1.16 SiO_2, 0.1 ZnO, 0.05 each Fe_2O_3, Al_2O_3, 0.02 SnO_2, 0.01 MgO, 0.008 CuO, 0.001 each V_2O_5, MoO_3, PbO, and MnO; corrected for impurities.
2	75	1958	53-298			99.9 Ga_2O_3, 0.05 ZnO, $<$0.01 other impurities.

DATA TABLE NO. 28 SPECIFIC HEAT OF GALLIUM OXIDE, Ga_2O_3

[Temperature, T, K; Specific Heat, C_p, Cal $g^{-1}K^{-1}$]

T	C_p	T	C_p
CURVE 1		CURVE 1 (cont.)	
15.82	5.39×10^{-4}	226.86	9.742×10^{-2}
18.89	7.79	235.78	1.001×10^{-1}
21.22	1.10×10^{-3}	247.01	1.035
23.43	1.41	254.44	1.056*
25.64	1.84	264.03	1.085
27.94	2.34	272.90	1.111*
30.60	3.08	273.47	1.111*
33.56	4.07	283.04	1.135
36.91	5.559	292.26	1.159*
40.54	7.218	300.79	1.174
44.29	8.979		
48.61	1.150×10^{-2}	CURVE 2*	
53.02	1.417		
54.23	1.485	53.81	1.486×10^{-2}
58.08	1.721	58.36	1.747
58.60	1.762*	62.79	2.002
61.78	1.922	67.37	2.271
64.15	2.061	71.99	2.547
66.61	2.200	76.45	2.807
69.27	2.331	80.29	3.047
69.65	2.343*	84.16	3.266
70.54	2.414	94.92	3.898
70.59	3.396*	105.09	4.489
75.54	2.721	114.63	5.016
76.45	2.774*	124.55	5.554
82.03	3.127	135.99	6.146
83.34	3.206*	145.48	6.626
88.46	3.500	155.76	7.112
89.07	3.537*	165.77	7.560
95.37	3.900	175.70	7.992
103.09	4.341	186.26	8.424
110.41	4.751	196.26	8.803
118.32	5.192	206.43	9.182
127.15	5.666	216.26	9.512
136.41	6.119	226.28	9.854
146.70	6.631	236.27	1.016×10^{-1}
156.80	7.122	246.10	1.044
165.76	7.517	256.50	1.076
173.88	7.859	266.30	1.110
182.01	8.184	276.63	1.126
189.83	8.493	286.83	1.149
198.92	8.813	296.34	1.173
207.91	9.107	298.15	1.175
217.62	9.464		

* Not shown on Plot

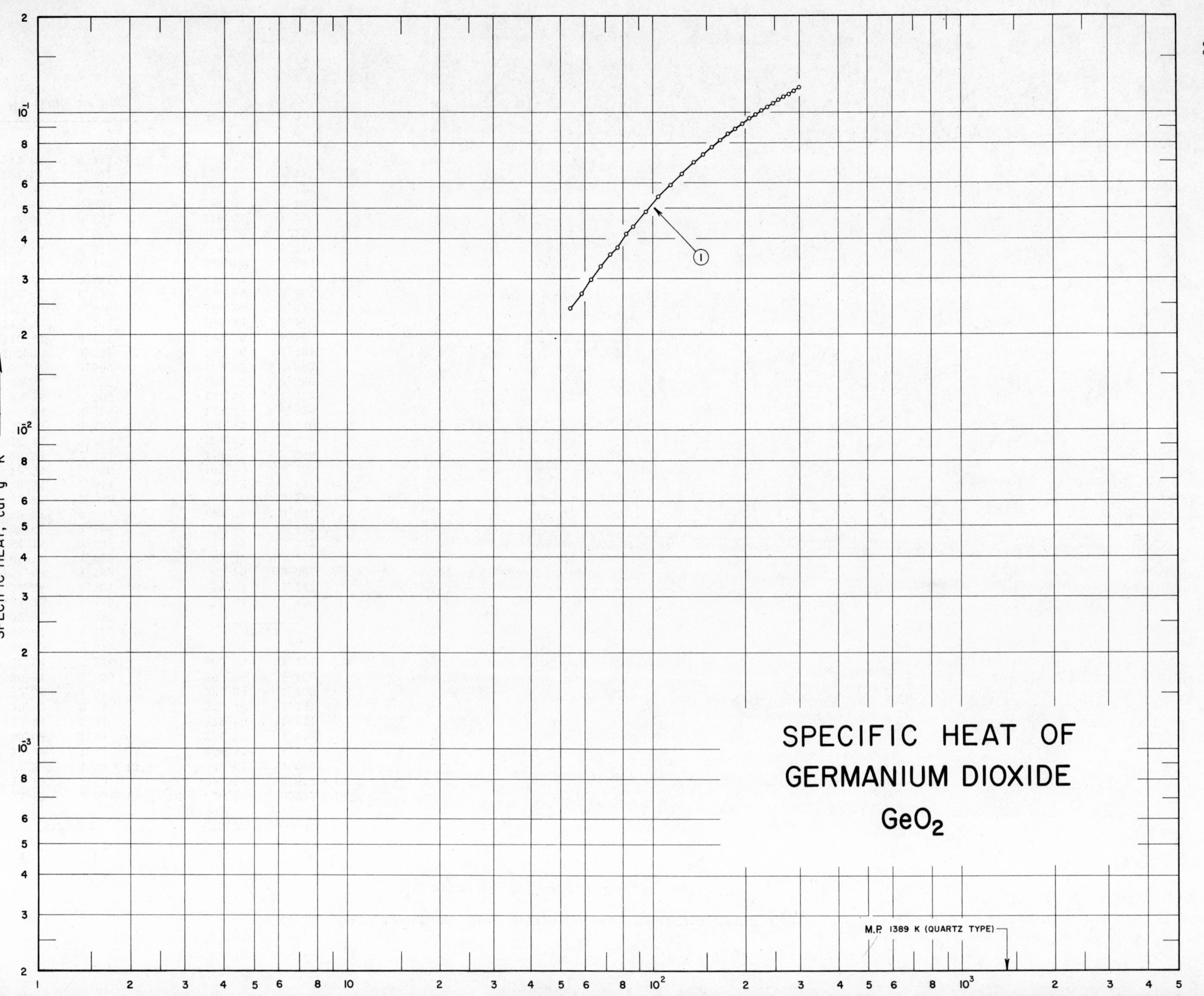

SPECIFIC HEAT OF
GERMANIUM DIOXIDE
GeO_2
M.P. 1389 K (QUARTZ TYPE)
1
SPECIFIC HEAT, cal g^{-1} K^{-1}

SPECIFICATION TABLE NO. 29 SPECIFIC HEAT OF GERMANIUM DIOXIDE GeO_2

[For Data Reported in Figure and Table No. 29]

Curve No.	Ref. No.	Year	Temp. Range, K	Reported Error, %	Name and Specimen Designation	Composition (weight percent), Specifications and Remarks
1	75	1958	53-298			99. 99 GeO_2.

DATA TABLE NO. 29 SPECIFIC HEAT OF GERMANIUM DIOXIDE, GeO_2

[Temperature, T, K; Specific Heat, C_p, Cal $g^{-1} K^{-1}$]

T	C_p
CURVE 1	
53.97	2.413×10^{-2}
58.36	2.682
62.88	2.969
67.42	3.260
72.13	3.553
76.51	3.744
81.73	4.126
86.03	4.370
94.90	4.872
104.91	5.421
114.50	5.903
124.61	6.396
136.30	6.929
145.73	7.348
155.95	7.758
166.05	8.147
176.03	8.513
186.14	8.867
196.08	9.185
206.48	9.535
216.39	9.837
226.11	1.013×10^{-1}
236.21	1.040
245.77	1.065
256.32	1.095
266.51	1.119
276.41	1.142
286.79	1.165*
296.24	1.188
298.15	1.190

*Not shown on Plot

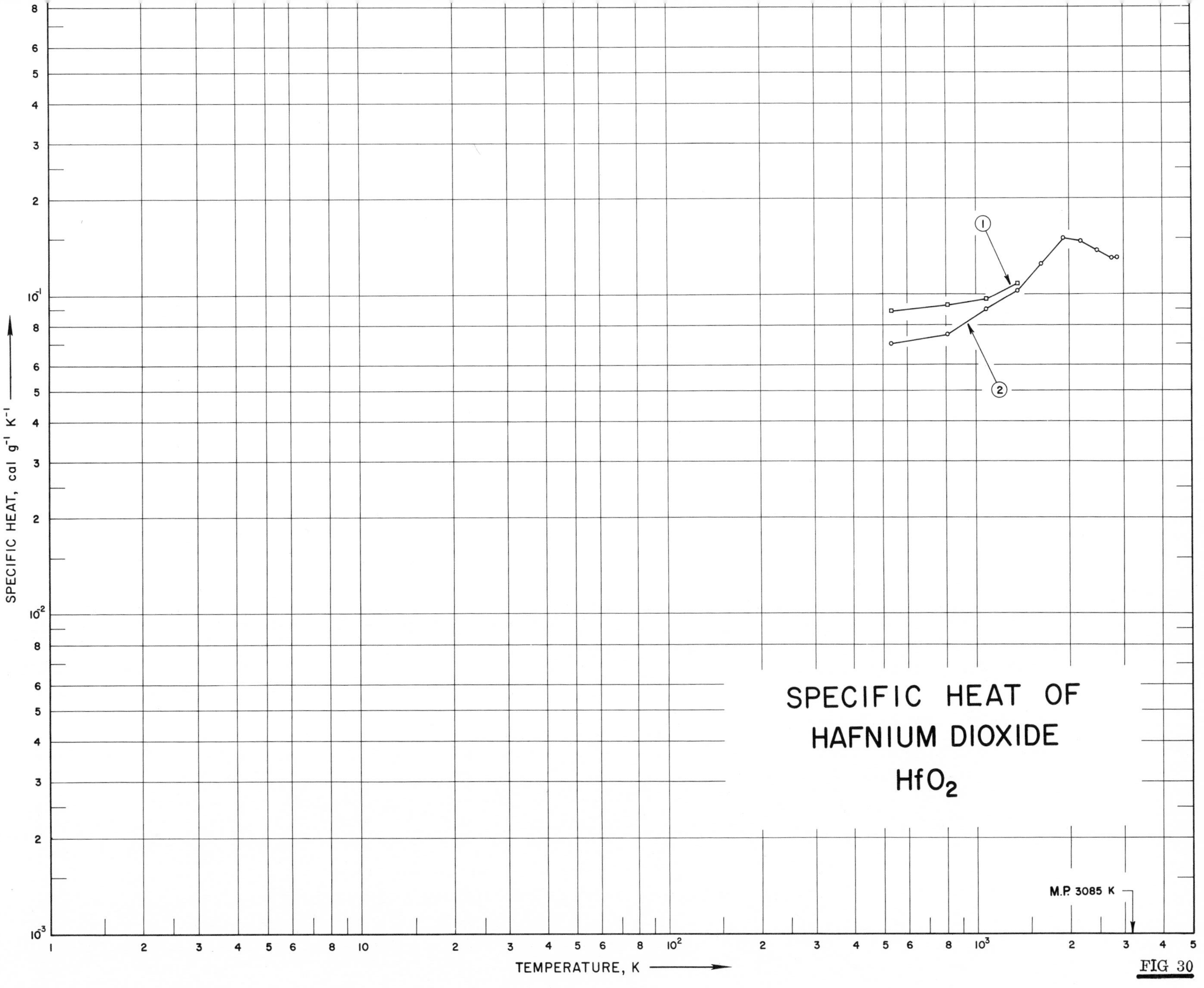
SPECIFIC HEAT OF
HAFNIUM DIOXIDE
HfO_2
M.P. 3085 K
TEMPERATURE, K
SPECIFIC HEAT, cal g^{-1} K^{-1}
1
2

FIG 30

SPECIFICATION TABLE NO. 30 SPECIFIC HEAT OF HAFNIUM DIOXIDE HfO_2

[For Data Reported in Figure and Table No. 30]

Curve No.	Ref. No.	Year	Temp. Range, K	Reported Error, %	Name and Specimen Designation	Composition (weight percent), Specifications and Remarks
1	76	1961	533-1366	5.0		97.0 HfO_2, 2.3 Zr; sample made by spraying powdered ZrO_2 using powder gun with 90 $ft^3\ hr^{-1}$ N_2, 10 $ft^3\ hr^{-1}$ H_2 plasma gas and 10 $ft^3\ hr^{-1}$ carrier gas; density 524 lb ft^{-3}.
2	48	1962	533-2894			Before exposure: 82.0 Hf, 2.5 Fe, 0.3 Mg, 0.10 Ca, 0.10 Ti; after exposure: 84.0 Hf, 0.5a C; sample supplied by Zirconium Corp. of America; crushed in a hardened steel mortar to pass 100-mesh screen; pressed and sintered; density at 25 C, before exposure: apparent density (ASMT method B311-58) 561 lb ft^{-3}, true density (by immersion in xylene) 595 lb ft^{-3}; after exposure: apparent density 557 lb ft^{-3}, true density 601 lb ft^{-3}.

DATA TABLE NO. 30 SPECIFIC HEAT OF HAFNIUM DIOXIDE HfO_2

[Temperature, T, K; Specific Heat, C_p, Cal $g^{-1}K^{-1}$]

T	C_p
CURVE 1	
533.15	8.85×10^{-2}
810.93	9.25
1088.71	9.66
1366.48	1.08×10^{-1}
CURVE 2	
533.15	7.0×10^{-2}
810.93	7.5
1088.71	9.0
1366.48	1.03×10^{-1}
1644.26	1.25
1922.04	1.50
2199.82	1.47
2477.59	1.38
2755.37	1.30
2894.26	1.31

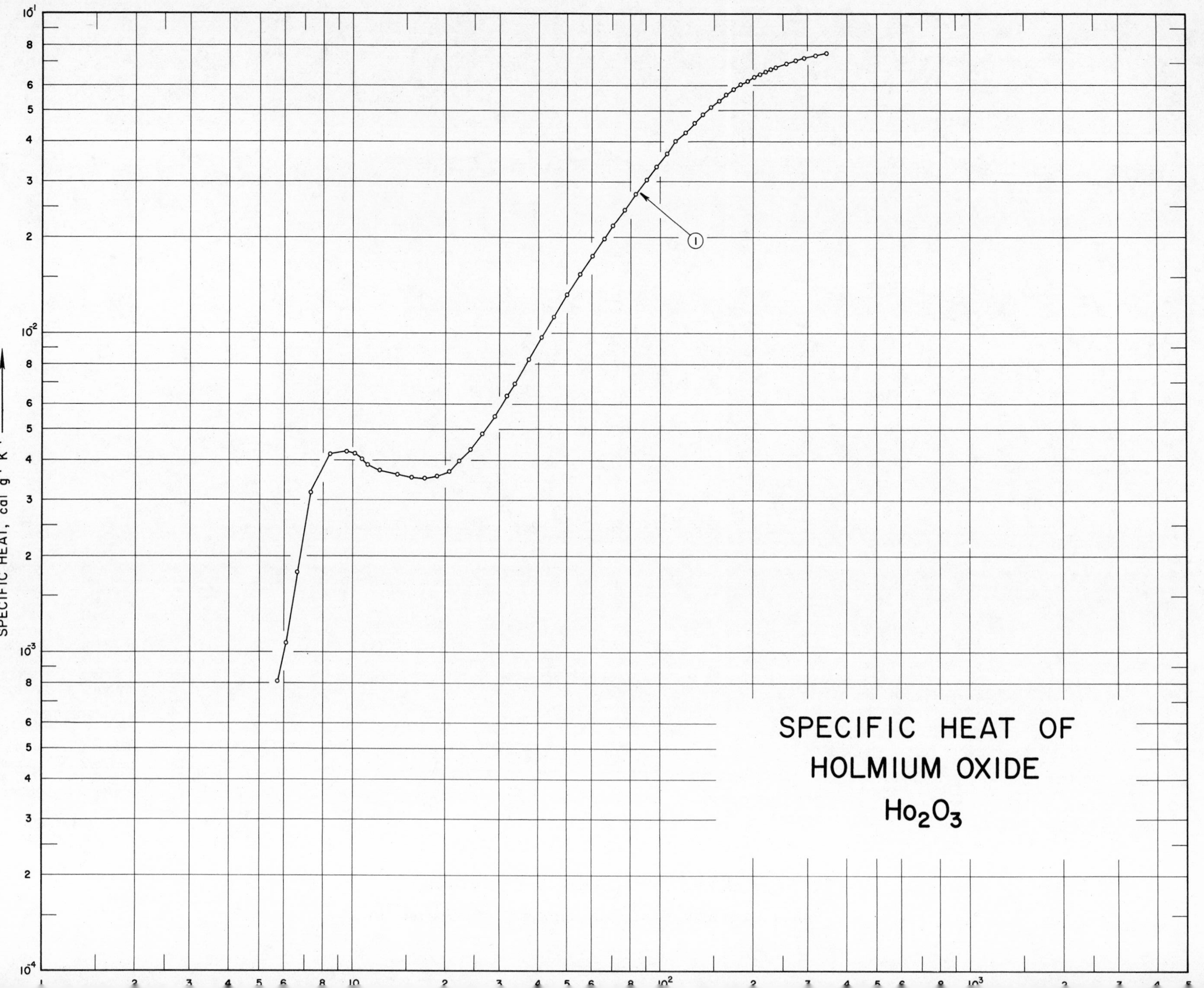
SPECIFIC HEAT OF
HOLMIUM OXIDE
Ho2O3
SPECIFIC HEAT, cal g-1 K-1
1

SPECIFICATION TABLE NO. 31 SPECIFIC HEAT OF HOLMIUM OXIDE Ho_2O_3

[For Data Reported in Figure and Table No. 31]

Curve No.	Ref. No.	Year	Temp. Range, K	Reported Error, %	Name and Specimen Designation	Composition (weight percent), Specifications and Remarks
1	72	1963	6-346	0. 1		99. 9 Ho_2O_3, 0. 010 Ca, 0. 010 Er, and 0. 010 Si; powder specimen; sample supplied by the Michigan Chemical Co; measured in helium atmosphere.

DATA TABLE NO. 31 SPECIFIC HEAT OF HOLMIUM OXIDE, Ho_2O_3

[Temperature, T, K; Specific Heat, C_p, Cal $g^{-1}K^{-1}$]

T	C_p	T	C_p
CURVE 1		**CURVE 1 (cont.)**	
Series 1		Series 3	
70.63	2.159×10^{-2}	5.75	8.098×10^{-4}
77.08	2.430	6.11	1.063×10^{-3}
83.72	2.722	6.64	1.773
90.54	3.075	7.32	3.157
97.87	3.314	8.49	4.184
105.81	3.637	9.61	4.248
113.89	3.960	10.84	4.020
121.35	4.248	12.36	3.713
129.99	4.560	14.00	3.607
138.60	4.846	15.62	3.522
147.16	5.113	17.20	3.501
155.96	5.362	18.81	3.552
165.02	5.595	20.58	3.687
174.24	5.809	22.39	3.962
183.42	6.000	24.35	4.303
192.85	6.177	26.58	4.811
202.33	6.330	29.09	5.486
211.49	6.465	31.85	6.338
220.62	6.590		
Series 2		Series 4	
229.08	6.696	10.20	4.195
238.25	6.796	11.26	3.853
247.41	6.886*	12.49	3.697*
256.66	6.974	13.86	3.618*
267.87	7.053*	15.35	3.530*
275.00	7.122	33.61	6.910
284.06	7.185*	37.25	8.223
293.10	7.243	41.08	9.644
302.21	7.296*	45.31	1.123×10^{-2}
311.55	7.344*	50.10	1.319
320.91	7.400	55.32	1.528
329.92	7.439*	60.63	1.747
338.42	7.484*	66.18	1.977
346.71	7.503	72.31	2.227*
		79.44	2.533*

* Not shown on Plot

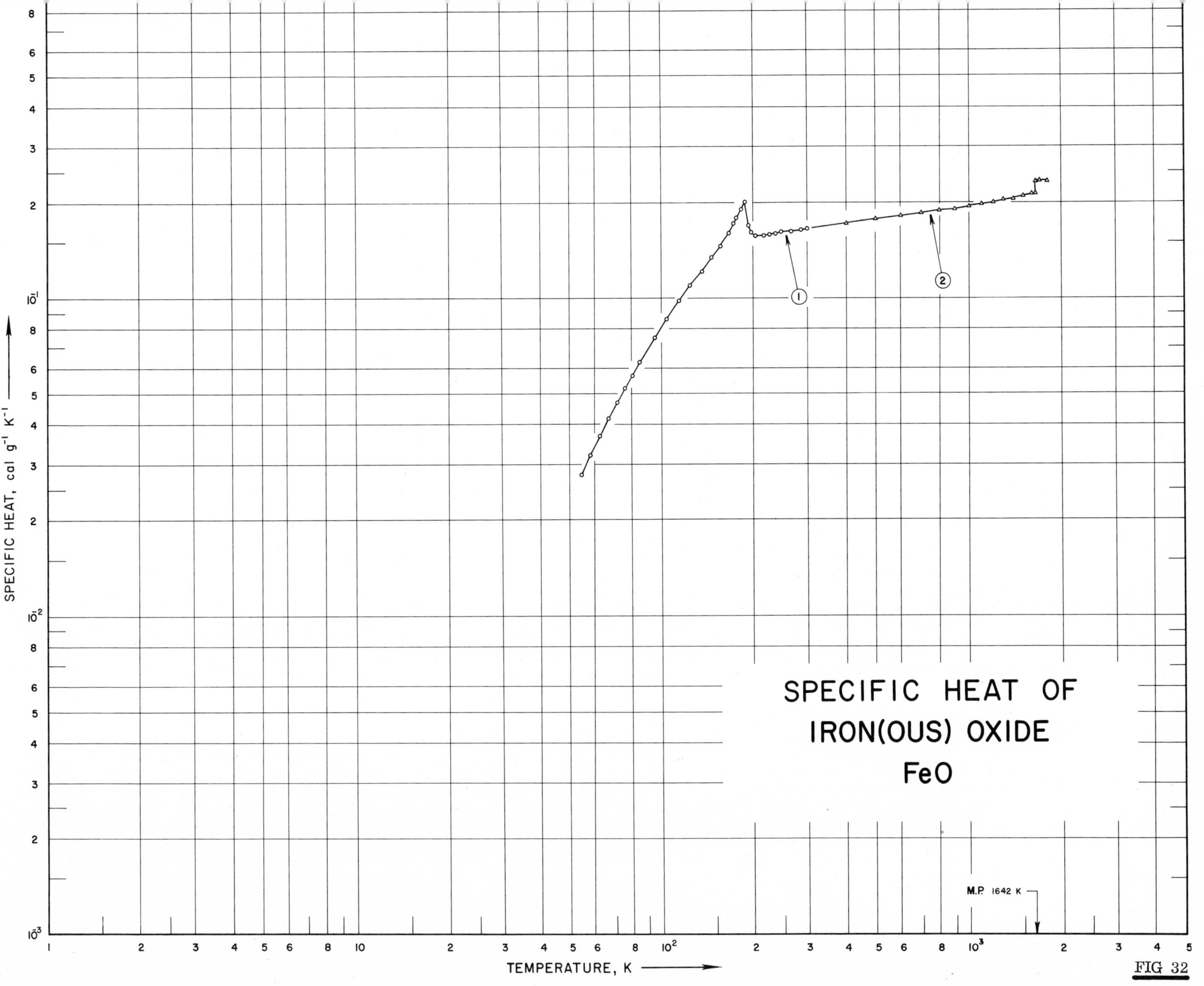
SPECIFIC HEAT OF
IRON(OUS) OXIDE
FeO
M.P. 1642 K
TEMPERATURE, K
SPECIFIC HEAT, cal g⁻¹ K⁻¹
1
2

FIG 32

SPECIFICATION TABLE NO. 32 SPECIFIC HEAT OF IRON (OUS) OXIDE $Fe_{0.947}O$

[For Data Reported in Figure and Table No. 32]

Curve No.	Ref. No.	Year	Temp. Range, K	Reported Error, %	Name and Specimen Designation	Composition (weight percent), Specifications and Remarks
1	77	1951	54-298			76. 60 Fe, 23. 18 O_2, 0. 17 SiO_2.
2	78	1951	298-1800			76. 60 Fe, 23. 18 O, 0. 17 Si; heated 4. 9 to 9. 3 days at 1150 K and quenched.

DATA TABLE NO. 32 SPECIFIC HEAT OF IRON(OUS) OXIDE, $Fe_{0.947}O$

[Temperature, T, K; Specific Heat, C_p, Cal $g^{-1}K^{-1}$]

T	C_p	T	C_p
CURVE 1		CURVE 2	
54.37	2.803×10^{-2}	298.55	1.670×10^{-1}*
58.24	3.212	300	1.672*
62.44	3.693	400	1.747
66.91	4.210	500	1.789
71.42	4.725	600	1.840
75.91	5.247	700	1.876
80.22	5.760	800	1.910
84.96	6.322	900	1.942
94.95	7.511	1000	1.973
104.72	8.656	1100	2.004
114.66	9.874	1200	2.034
124.81	1.104×10^{-1}	1300	2.064
135.97	1.235	1400	2.094
146.31	1.354	1500	2.123
155.93	1.473	1600	2.153
165.84	1.616	1650	2.168
167.52	1.646*	1650	2.366
172.56	1.736	1700	2.366
175.89	1.800	1800	2.366
177.03	1.817*		
180.58	1.877*		
183.33	1.916		
185.54	1.944*		
185.92	1.961*		
187.67	2.019		
190.10	1.983*		
193.50	1.704		
196.32	1.633		
197.97	1.608*		
202.87	1.582		
206.65	1.578*		
216.52	1.584		
226.21	1.595		
236.06	1.605		
245.84	1.614		
257.20	1.624*		
266.08	1.639		
276.05	1.648*		
286.29	1.659		
296.46	1.665*		
298.16	1.669		

*Not shown on Plot

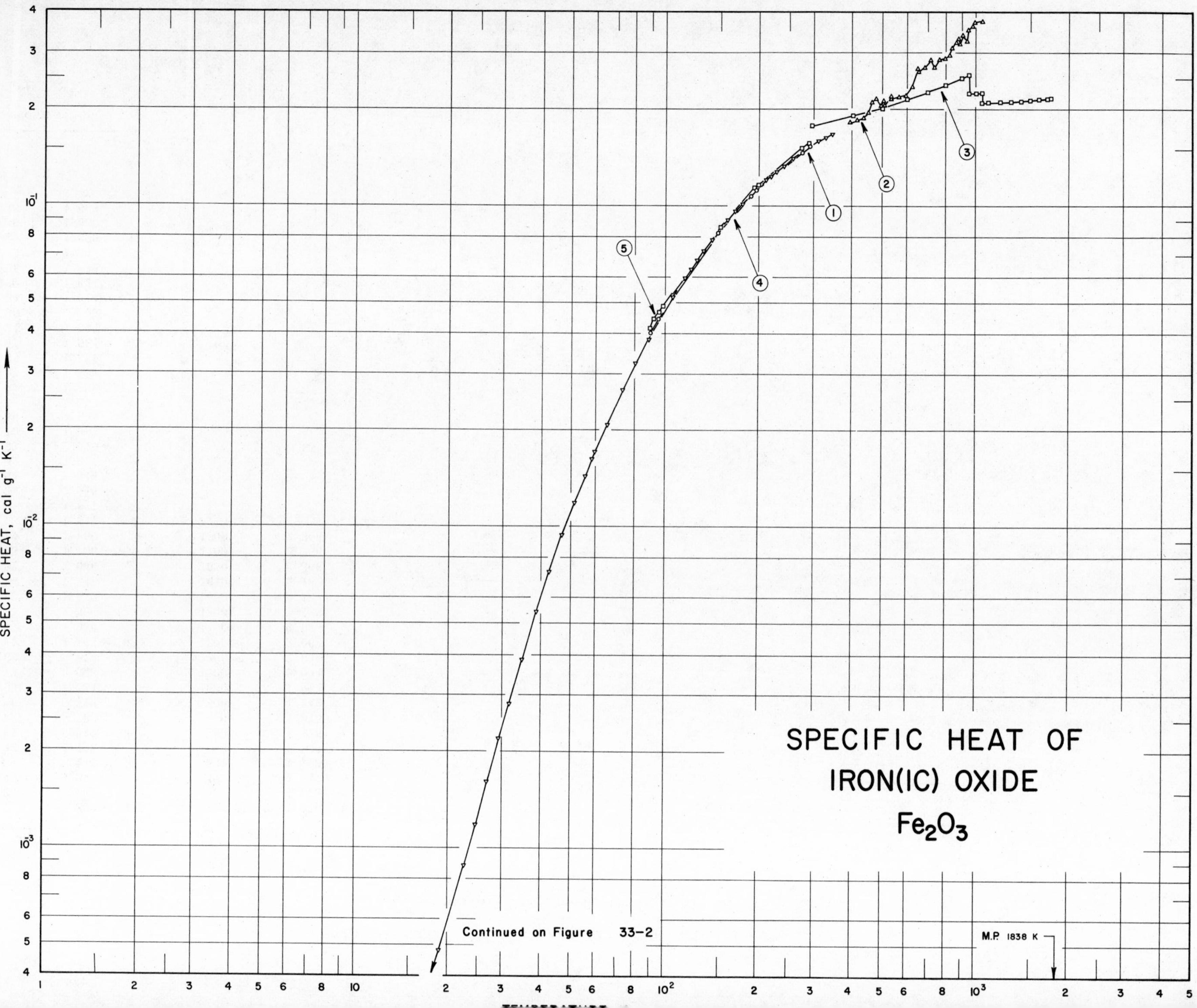
SPECIFIC HEAT OF
IRON(IC) OXIDE
Fe_2O_3
SPECIFIC HEAT, cal g^{-1} K^{-1}
Continued on Figure 33-2
M.P. 1838 K
1
2
3
4
5

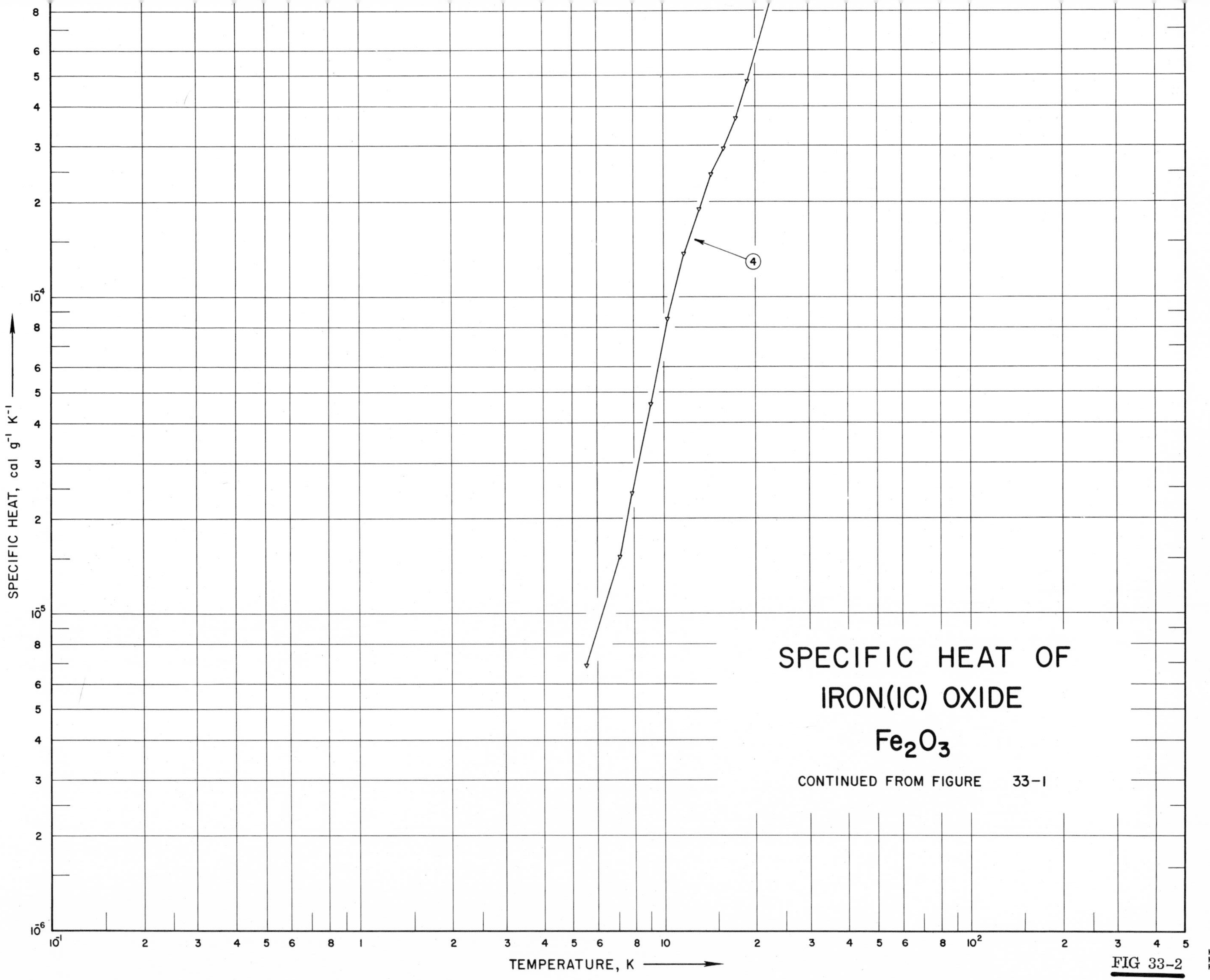
SPECIFIC HEAT OF
IRON(IC) OXIDE
Fe2O3
CONTINUED FROM FIGURE 33-1
SPECIFIC HEAT, cal g⁻¹ K⁻¹
TEMPERATURE, K
4

FIG 33-2

SPECIFICATION TABLE NO. 33 SPECIFIC HEAT OF IRON (IC) OXIDE Fe_2O_3

[For Data Reported in Figure and Table No. 33]

Curve No.	Ref. No.	Year	Temp. Range, K	Reported Error, %	Name and Specimen Designation	Composition (weight percent), Specifications and Remarks
1	18	1926	89-292			Specular: 99. 2 Fe_2O_3, 0. 5 H_2O, and 0. 5 SiO_2.
2	79	1926	391-1051	≤0. 5		98. 73 Fe_2O_3, 0. 620 SiO_2, 0. 345 Mn, 0. 117 Fe_3O_4, 0. 028 C, 0. 027 S, and 0. 02 P; finely powdered crystalline Fe_2O_3.
3	78	1951	298-1750			69. 86 Fe; prepared from reagent grade $FeCl_2$.
4	80	1959	5-345			0. 01 Mn, <0. 01 Al, Co, Mg, Ni, and S, and <0. 001 Ca, Cu, and Sn.
5	18	1926	88-289			Kahlbaum's purity; 99. 5 Fe_2O_3 and 0. 5 H_2O.

DATA TABLE NO. 33 SPECIFIC HEAT OF IRON(IC) OXIDE, Fe_2O_3

[Temperature, T, K; Specific Heat, C_p, Cal $g^{-1}K^{-1}$]

T	C_p
CURVE 1	
89.8	4.04 x 10^{-2}
90.3	4.14*
90.5	4.10*
95.5	4.45
148.4	8.27
186.9	1.082 x 10^{-1}
192.8	1.109*
196.3	1.128
275.9	1.477
278.1	1.488*
289.1	1.534*
291.9	1.539
CURVE 2	
Series 2	
391.0	1.82 x 10^{-1}*
393.5	1.84
414.0	1.86
450.5	1.97
490.5	2.04*
493.0	2.07
508.0	2.07
533.0	2.17
Series 3	
419.2	1.87*
435	1.89
463	2.11
479.5	2.17
483.7	2.06*
505.5	2.14
535	2.22
567	2.20
592.5	2.23
626.5	2.38
654.5	2.64
682	2.72*
685.5	2.73
701.5	2.70
CURVE 2 (cont.)	
Series 3 (cont.)	
715.5	2.87 x 10^{-1}
737.5	2.71
763	2.88
799	2.91
823	2.98
840	3.14
880	3.35
904	3.42
Series 4	
864	3.26
870.5	3.20
889	3.22
936	3.28
941	3.58
973	3.67
991.5	3.76
1051	3.80
CURVE 3	
α298.15	1.793 x 10^{-1}
300	1.796*
400	1.922
500	2.044
600	2.163
700	2.281
800	2.399
900	2.516
α950	2.575
β950	2.254
1000	2.254
β1050	2.254
γ1050	2.101
1100	2.106
1200	2.118
1300	2.128
1400	2.140
1500	2.151
CURVE 3 (cont.)	
1600	2.162 x 10^{-1}
1700	2.172
γ1750	2.178
CURVE 4	
Series 1	
58.61	1.636 x 10^{-2}
65.30	2.083
73.20	2.683
80.93	3.242
88.57	3.837
96.83	4.470*
106.36	5.202
116.07	5.951
121.39	6.356
127.04	6.781
133.82	7.282
141.85	7.852
150.70	8.466*
159.58	9.079
168.38	9.631
177.24	1.017 x 10^{-1}
185.88	1.068*
194.32	1.115*
202.65	1.160*
210.85	1.202
219.61	1.245
228.19	1.285
Series 2	
223.58	1.263*
232.31	1.304*
241.06	1.343
294.68	1.380*
258.22	1.413
266.75	1.446
275.38	1.478*
283.78	1.507*
292.35	1.534*
CURVE 4 (cont.)	
Series 2 (cont.)	
301.23	1.563 x 10^{-1}
310.20	1.592
319.04	1.616*
327.77	1.640
336.53	1.664*
345.42	1.687
Series 3	
5.58	6.887 x 10^{-6}
7.10	1.503 x 10^{-5}
7.87	2.379
9.02	4.571
10.31	8.453
11.65	1.365 x 10^{-4}
13.02	1.885
14.39	2.442
15.72	2.937
17.25	3.651
18.97	4.803
22.70	8.804
24.65	1.189 x 10^{-3}
26.89	1.607
29.24	2.197
31.56	2.801
34.80	3.859
38.75	5.424
42.76	7.239
46.99	9.411
51.51	1.194 x 10^{-2}
55.81	1.447
59.71	1.709
CURVE 5	
88.3	4.12 x 10^{-2}
91.8	4.42
95.8	4.65
98.3	4.85
150.3	8.58
CURVE 5 (cont.)	
194.3	1.141 x 10^{-1}
198.5	1.162
274.9	1.521
276.7	1.529*
286.9	1.551*
289.4	1.561

*Not shown on Plot

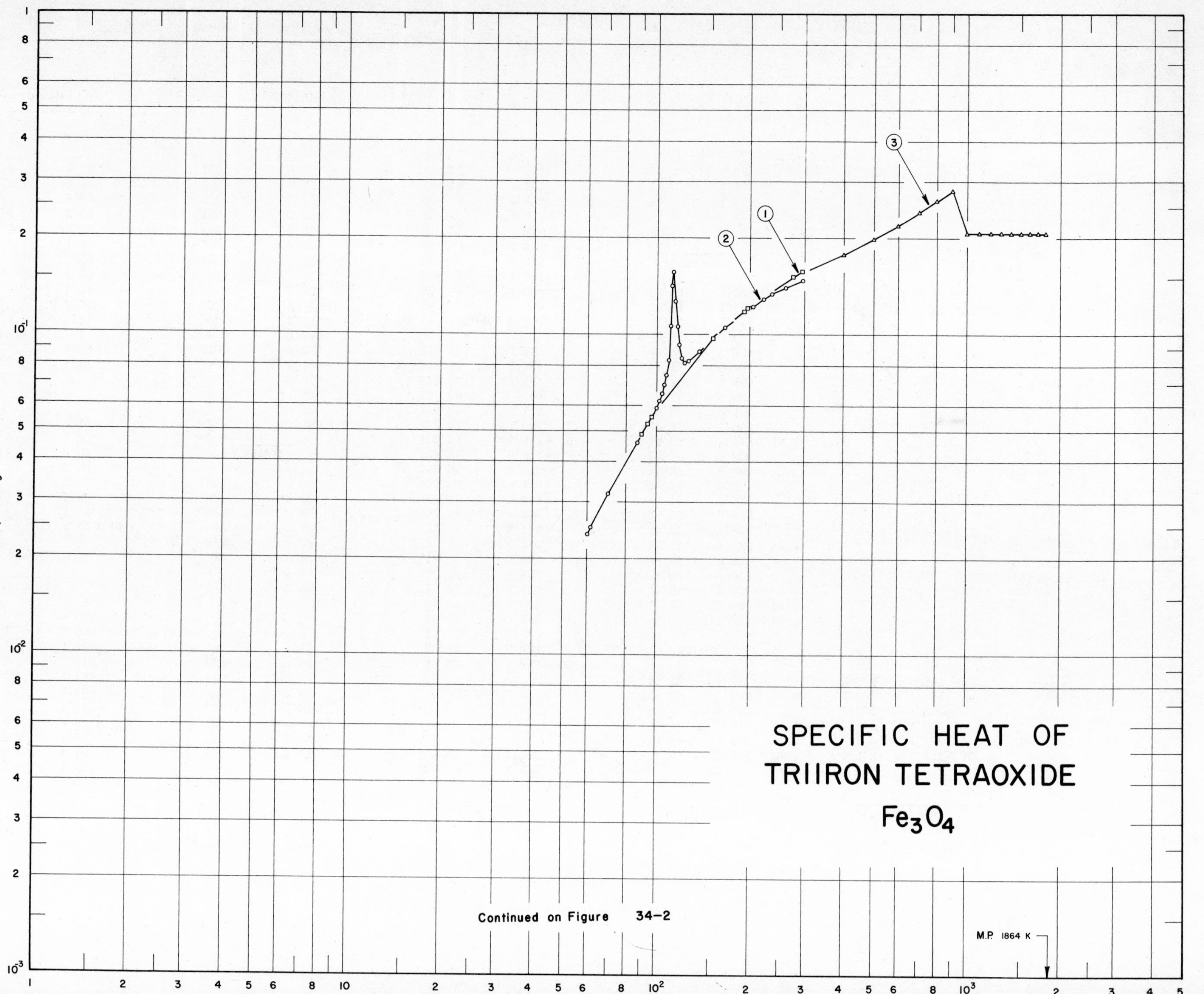
SPECIFIC HEAT OF
TRIIRON TETRAOXIDE
Fe_3O_4
SPECIFIC HEAT, cal g^{-1} K^{-1}
Continued on Figure 34-2
M.P. 1864 K
1
2
3

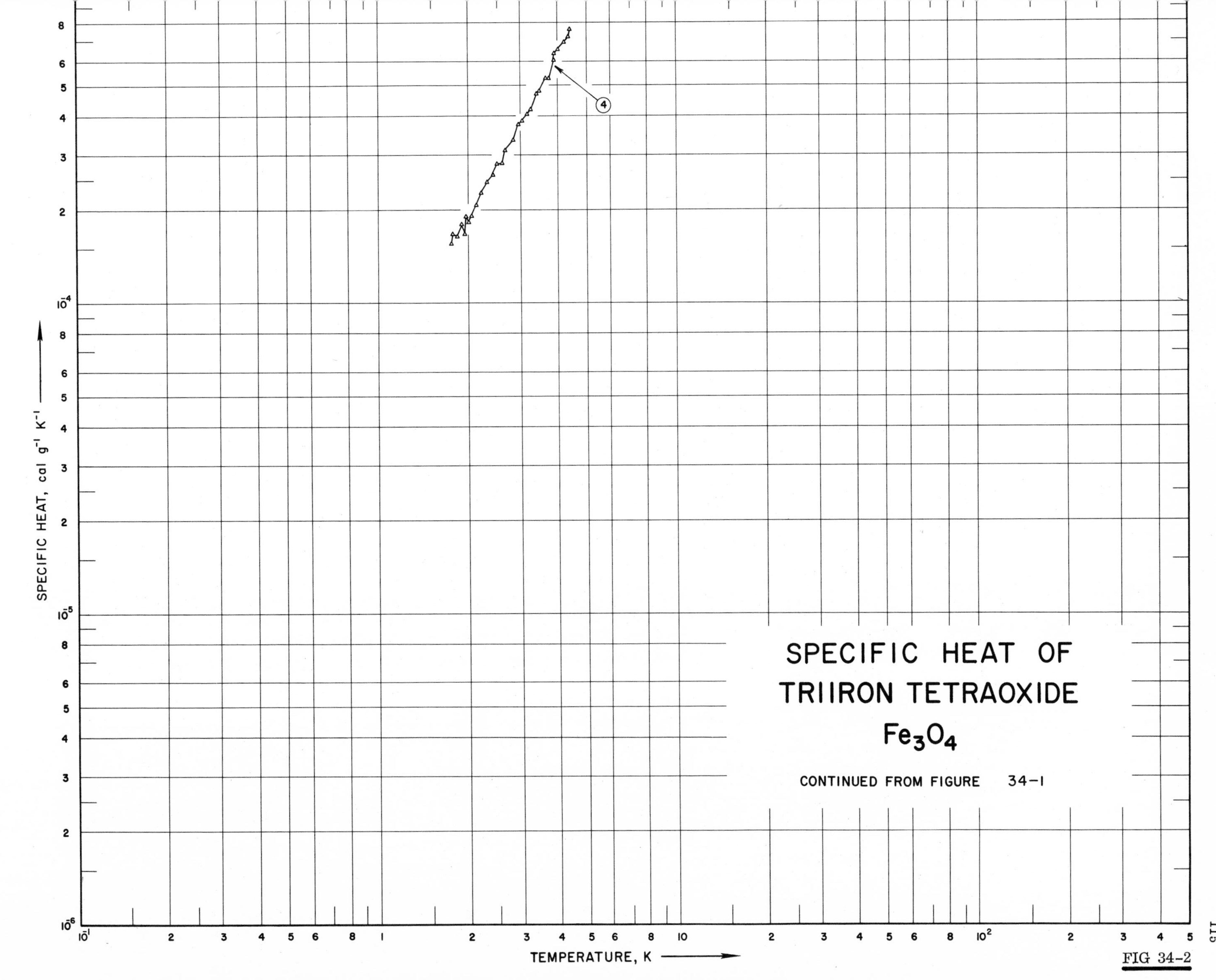

SPECIFIC HEAT OF
TRIIRON TETRAOXIDE
Fe3O4
CONTINUED FROM FIGURE 34-1
SPECIFIC HEAT, cal g-1 K-1
TEMPERATURE, K
4

FIG 34-2

SPECIFICATION TABLE NO. 34 SPECIFIC HEAT OF TRIIRON TETRAOXIDE Fe_3O_4

[For Data Reported in Figure and Table No. 34]

Curve No.	Ref. No.	Year	Temp. Range, K	Reported Error, %	Name and Specimen Designation	Composition (weight percent), Specifications and Remarks
1	18	1926	90-295			99. 00 Fe_3O_4, 0. 63 Fe_2O_3, and 0. 37 others.
2	65	1929	60-300			99. 00 Fe_3O_4, 0. 63 Fe_2O_3, and 0. 37 others.
3	78	1951	298-1800			72. 16 Fe, 27. 54 O_3, 0. 22 SiO_2; prepared from ferric oxide by heating 8 hrs at 1630 K under vacuum.
4	81	1956	1. 8-4. 2			Natural magnetite crystal.

DATA TABLE NO. 34 SPECIFIC HEAT OF TRIIRON TETRAOXIDE Fe_3O_4

[Temperature, T, K; Specific Heat, C_p, Cal g^{-1} K^{-1}]

T	C_p
CURVE 1	
90.0	4.88 x 10^{-2}
90.2	4.93*
94.2	5.24
96.9	5.50
153.2	9.68
193.5	1.186 x 10^{-1}
197.2	1.205
276.3	1.513
278.7	1.525*
292.1	1.558*
295.0	1.570
CURVE 2	
297.3	1.480 x 10^{-1}
299.7	1.485*
60.5	2.373 x 10^{-2}
62.0	2.492
70.5	3.166
87.6	4.591
100.7	5.904
103.0	6.210
105.1	6.564
107.0	6.950
108.8	7.458
110.4	8.314
112.0	1.063 x 10^{-1}
113.6	1.422
114.7	1.566
115.9	1.274
117.6	1.062
119.4	9.294 x 10^{-2}
121.8	8.413
124.8	8.184
127.5	8.275
138.9	8.866
153.7	9.773*
167.5	1.052 x 10^{-1}
191.4	1.165*
206.2	1.235
222.1	1.296
CURVE 2 (cont.)	
235.7	1.347 x 10^{-1}
239.2	1.358*
231.3	1.325*
246.0	1.374*
261.1	1.411
CURVE 3	
α298.15	1.565 x 10^{-1}*
300	1.569*
400	1.778
500	1.986
600	2.193
700	2.402
800	2.610
900	2.818
α1000	2.073
β1100	2.073
1200	2.073
1300	2.073
1400	2.073
1500	2.073
1600	2.073
1700	2.073
β1800	2.073
CURVE 4	
1.780	1.67 x 10^{-4}
1.845	1.65
1.953	1.68
2.011	1.83
2.058	1.91
2.142	2.08
2.211	2.29
2.327	2.461
2.437	2.609
2.511	2.815
2.603	2.832
2.687	3.104
2.851	3.364
CURVE 4 (cont.)	
2.966	3.778 x 10^{-4}
3.068	3.871
3.167	4.081
3.269	4.241
3.410	4.746
3.483	4.839
3.567	4.963*
3.667	5.360
3.776	5.305
3.906	6.086
4.059	6.609
4.231	6.955
4.398	7.264
4.416	7.288
4.427	7.677
1.768	1.56
1.832	1.63*
1.907	1.80
1.970	1.90
2.050	1.96*
2.181	2.27*
2.273	2.471*
2.697	3.035*
2.732	3.200*
3.933	6.399
4.150	6.906*
4.363	7.247

*Not shown on Plot

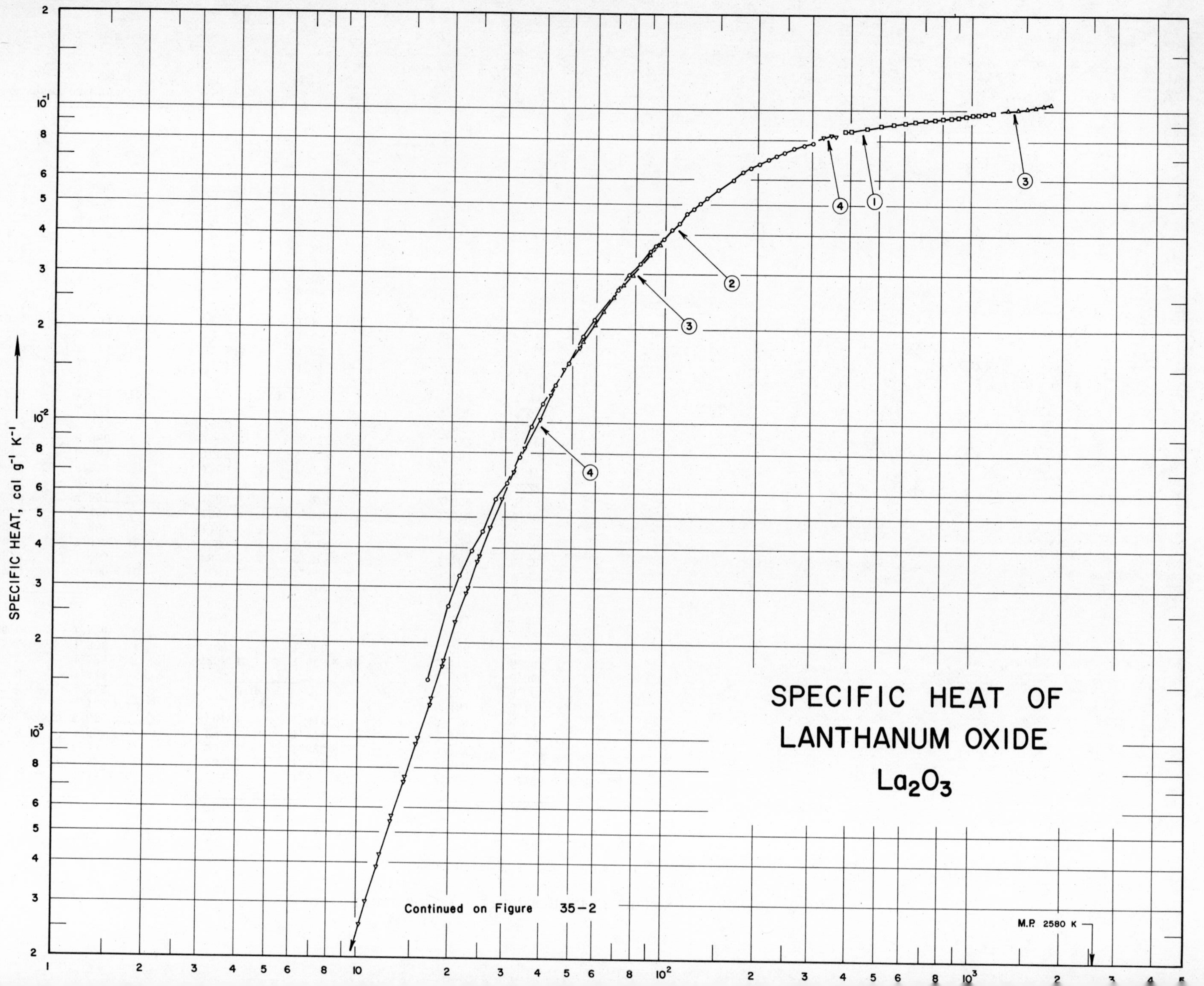
SPECIFIC HEAT OF
LANTHANUM OXIDE
La_2O_3
Continued on Figure 35-2
M.P. 2580 K
SPECIFIC HEAT, cal g^{-1} K^{-1}
1
2
3
3
4
4

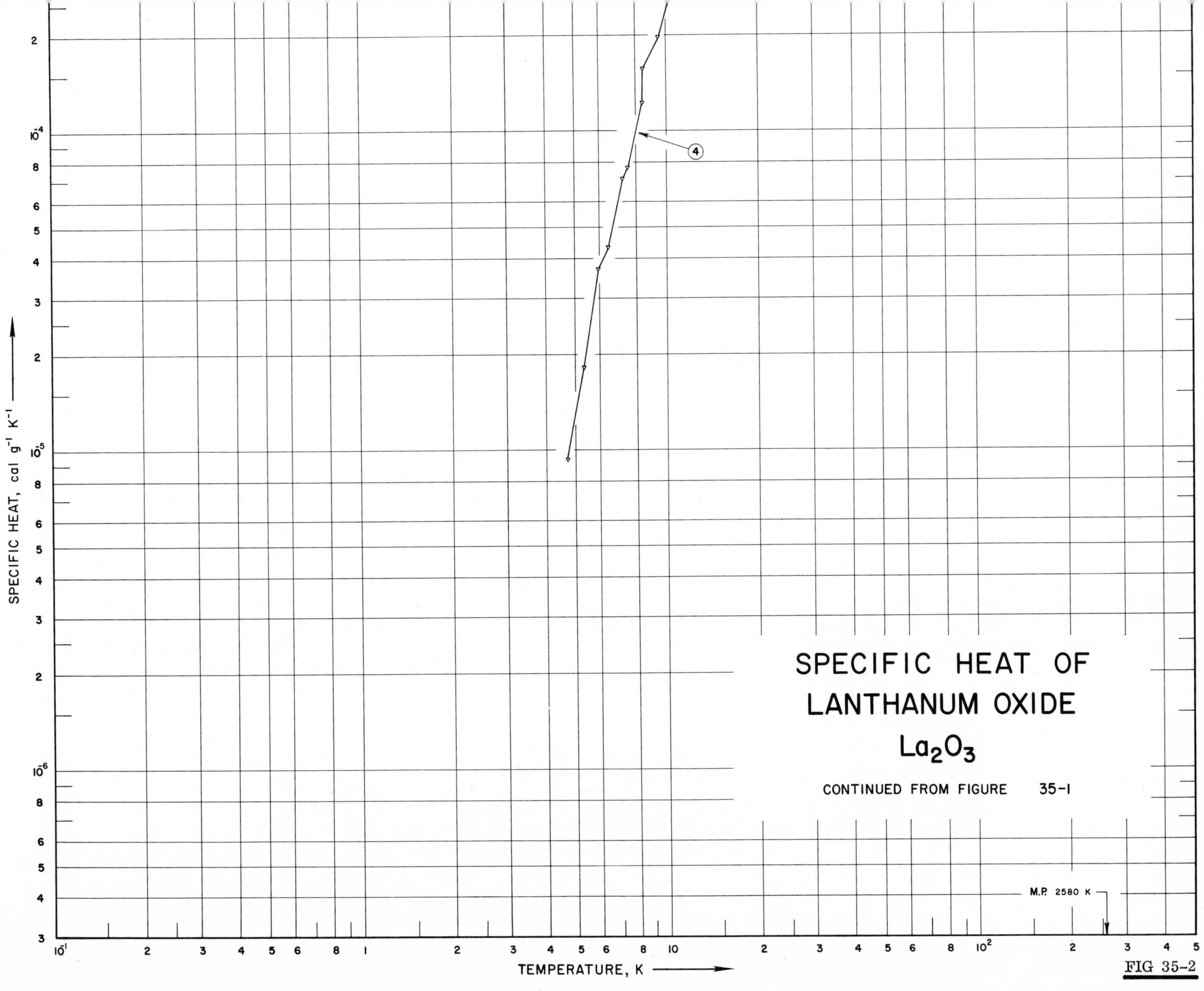

SPECIFIC HEAT OF
LANTHANUM OXIDE
La_2O_3
CONTINUED FROM FIGURE 35-1
M.P. 2580 K
4
TEMPERATURE, K
SPECIFIC HEAT, cal g^{-1} K^{-1}
10^{-1}
1
10
10^2
10^{-4}
10^{-5}
10^{-6}

FIG 35-2

SPECIFICATION TABLE NO. 35 SPECIFIC HEAT OF LANTHANUM OXIDE La_2O_3

[For Data Reported in Figure and Table No. 35]

Curve No.	Ref. No.	Year	Temp. Range, K	Reported Error, %	Name and Specimen Designation	Composition (weight percent), Specifications and Remarks
1	82	1951	383-1171			
2	83	1959	16-300			99.997 La_2O_3, 0.00015 Fe_2O_3; sample supplied by the Lindsay Chemical Co.; heated to constant weight at 950 C for 2 hrs in air to decompose hydroxide or carbonate.
3	84	1961	53-1800	0.1		99.997 La_2O_3; measured under vacuum.
4	72	1962	5-355	0.1		99.997 La_2O_3; sample supplied by Lindsay Chemical Co.; pelleted under pressure 2000-4000 psi; fired at 1170 K; measured in helium atmosphere.

DATA TABLE NO. 35 SPECIFIC HEAT OF LANTHANUM OXIDE, La_2O_3

[Temperature, T, K; Specific Heat, C_p, Cal $g^{-1}K^{-1}$]

CURVE 1

T	C_p
383	8.533 x 10^{-2}
400	8.607
450	8.786
500	8.927
550	9.044
600	9.144
650	9.233
700	9.313
750	9.386
800	9.455
850	9.520
900	9.583
950	9.643
1000	9.701
1050	9.757
1100	9.812
1150	9.867*
1171	9.889

CURVE 2

T	C_p
16.91	1.538 x 10^{-3}
19.55	2.609
21.27	3.275
23.35	3.922
25.36	4.509
27.78	5.739
30.39	6.479
33.12	7.833
36.13	9.772
39.73	1.155 x 10^{-2}
43.95	1.337
48.52	1.556
53.73	1.899
58.49	2.146
69.81	2.665
75.77	2.971
82.12	3.215
88.71	3.514
92.13	3.662
98.17	3.888

CURVE 2 (cont.)

T	C_p
104.46	4.140 x 10^{-2}
110.89	4.370
117.22	4.598
123.45	4.836
129.89	5.026
136.43	5.251
142.67	5.416*
148.65	5.586
155.04	5.727*
166.03	5.995
171.57	6.160*
177.35	6.333
183.07	6.347*
188.81	6.518
201.66	6.742
215.03	6.959
221.90	7.097*
228.54	7.160
235.16	7.253*
241.61	7.338
254.11	7.528*
260.07	7.577
272.36	7.693*
279.79	7.758
287.18	7.794*
294.27	7.876*
300.30	7.872

CURVE 3

T	C_p
53.62	1.831 x 10^{-2}
58.39	2.064
62.92	2.294
67.74	2.527
72.52	2.763
77.27	2.975
83.17	3.238*
88.13	3.456
95.16	3.744
105.16	4.143*
114.79	4.502*

CURVE 3 (cont.)

T	C_p
124.92	4.861 x 10^{-2}
136.13	5.220*
146.13	5.521*
156.10	5.785*
166.13	6.043*
176.23	6.267*
187.09	6.494*
196.39	6.660*
206.99	6.853*
215.84	6.997*
226.23	7.151*
236.12	7.286*
245.72	7.399*
256.29	7.522*
266.07	7.626*
275.99	7.725*
286.69	7.829*
296.34	7.924*
398	8.556
400	8.565
500	8.912
600	9.146
700	9.327
800	9.480
900	9.616
1000	9.742
1100	9.861
1200	9.973
1300	1.009 x 10^{-1}
1400	1.020
1500	1.030
1600	1.041
1700	1.051
1800	1.062

CURVE 4

T	C_p
Series 1	
5.30	1.8 x 10^{-5}
6.38	4.30
7.45	7.67
8.39	1.57 x 10^{-4}
10.14	2.55
11.51	3.898
12.82	5.432
14.07	7.213
15.49	9.515
17.04	1.271 x 10^{-3}
18.74	1.691
20.54	2.231
22.42	2.876
24.41	3.637
Series 2	
4.69	9.2 x 10^{-6}
5.90	3.68 x 10^{-5}
7.17	7.06
8.34	1.23 x 10^{-4}
9.44	1.96
10.65	3.04
11.82	4.266
12.95	5.617
14.26	7.458
15.70	9.944
17.29	1.329 x 10^{-3}
18.99	1.756
20.82	2.232*
22.74	2.999
24.73	3.763
26.82	4.650
29.19	5.730
31.84	7.001
34.79	8.480*
38.27	1.025 x 10^{-2}*

CURVE 4 (cont.)

T	C_p
Series 3	
34.48	8.318 x 10^{-3}
38.81	1.028 x 10^{-2}
42.33	1.235
46.93	1.474
52.20	1.748
57.54	2.020*
62.94	2.294*
68.40	2.558*
74.05	2.821*
80.14	3.108*
87.07	3.431*
Series 4	
83.51	3.267 x 10^{-2}*
90.51	3.577*
97.82	3.869*
105.38	4.164*
113.35	4.465*
121.18	4.748*
128.88	5.009*
137.29	5.279*
146.06	5.540*
154.72	5.776*
160.14	5.911*
168.82	6.123*
177.41	6.316*
186.16	6.501*
195.26	6.679*
204.56	6.841*
213.89	6.995*
223.06	7.133*
230.45	7.240*
240.01	7.375*

CURVE 4 (cont.)

T	C_p
Series 5	
235.71	7.311 x 10^{-2}*
245.33	7.434*
254.60	7.547*
263.62	7.648*
272.54	7.744*
281.47	7.830*
290.35	7.912*
299.17	7.989*
308.12	8.063*
317.20	8.127*
326.30	8.192
355.51	8.250
345.12	8.314

*Not shown on Plot

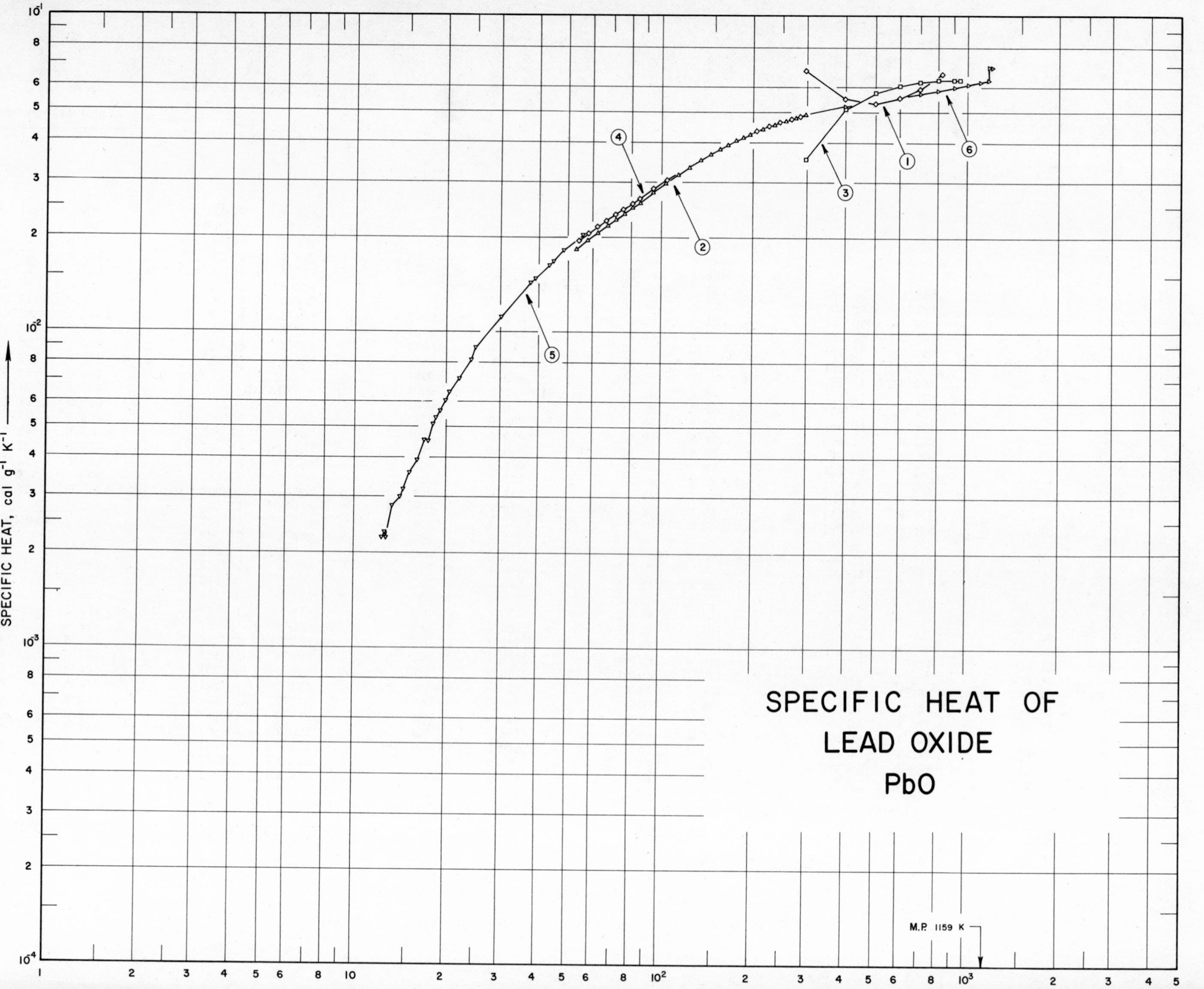
SPECIFIC HEAT OF
LEAD OXIDE
PbO
M.P. 1159 K
SPECIFIC HEAT, cal g^{-1} K^{-1}

SPECIFICATION TABLE NO. 36 SPECIFIC HEAT OF LEAD OXIDE PbO

[For Data Reported in Figure and Table No. 36]

Curve No.	Ref. No.	Year	Temp. Range, K	Reported Error, %	Name and Specimen Designation	Composition (weight percent), Specifications and Remarks
1	85	1942	298-823		Red PbO	Red lead monoxide; dried in vacuum desiccator with potassium hydroxide and later for 2 wks with anhydrous magnesium perchlorate; heated at 140 C and later at 400 C.
2	86	1958	53-298		Red PbO	92. 69 Pb; prepared by heating electrolytic lead dioxide in a vacuo at 430 - 480 C for 8 wks.
3	85	1942	298-943		Yellow PbO	Yellow lead monoxide; prepared by heating red monoxide at 600 C for 3 hrs.
4	86	1958	54-302		Yellow PbO	92. 84 Pb; prepared by heating lead carbonate 560 - 580 C for 80 hrs, 725 C for 10 hrs; quench to room temperature.
5	87	1960	12-303		Yellow PbO	99. 5 PbO; crystalline; dried at 150 C and 0. 05 mm Hg; measured under 20 mm Hg helium.
6	88	1961	300-1200	0. 01	Yellow PbO	

DATA TABLE NO. 36 SPECIFIC HEAT OF LEAD OXIDE, PbO

[Temperature, T, K; Specific Heat, C_p, Cal $g^{-1}K^{-1}$]

CURVE 1

T	C_p
298	6.767 x 10^{-2}
300	6.723*
400	5.512
500	5.329
600	5.550
700	5.961
800	6.475*
823	6.602

CURVE 2

T	C_p
53.40	1.835 x 10^{-2}
58.07	1.951
62.71	2.064
67.26	2.173
71.67	2.279
76.13	2.379
81.30	2.490
85.97	2.588
94.64	2.768
104.71	2.969
114.78	3.154
124.89	3.333
135.91	3.513
145.79	3.666
155.84	3.805
165.83	3.930
176.34	4.050
186.35	4.159
195.99	4.247
206.38	4.347*
216.27	4.425
226.09	4.502*
236.73	4.574
245.91	4.623*
256.39	4.691
266.62	4.749*
276.58	4.798
286.79	4.847*
296.27	4.901*
298.15	4.906

CURVE 3

T	C_p
298	3.541 x 10^{-2}
300	3.589
400	5.104
500	5.759
600	6.076
700	6.233
800	6.304
900	6.326
943	6.326

CURVE 4

T	C_p
54.13	1.950 x 10^{-2}
58.39	2.056
62.59	2.153
66.82	2.258
71.12	2.357
75.57	2.452
81.03	2.569
85.21	2.654
94.91	2.851
105.21	3.043
114.74	3.218*
124.80	3.391*
135.95	3.565*
145.71	3.715*
155.95	3.852*
165.87	3.973*
176.28	4.085*
185.83	4.183*
196.21	4.277*
206.23	4.367*
216.48	4.442*
226.22	4.516
236.04	4.588*
245.94	4.646
256.39	4.704*
266.26	4.753
276.22	4.803*
286.42	4.847
296.11	4.901*
298.15	4.906*

CURVE 5

T	C_p
12.47	2.24 x 10^{-3}
12.70	2.33
12.74	2.24
13.42	2.82
14.23	3.00
14.55	3.18
15.35	3.59
16.25	3.94
17.03	4.570
17.60	4.525
18.31	5.108
18.54	5.377
18.92	5.466*
19.25	5.601
19.48	5.645*
20.05	6.049
20.08	6.093*
20.61	6.228*
20.85	6.497
21.09	6.676*
22.15	7.124
24.12	8.155
25.04	8.916
30.42	1.116 x 10^{-2}
37.83	1.443
39.11	1.488
43.86	1.640
45.00	1.689
48.84	1.828
49.99	1.855*
56.63	2.034
58.73	2.083*
61.70	2.151*
64.57	2.209*
71.03	2.366*
74.63	2.446*
77.76	2.505*
81.09	2.594*
84.00	2.657*
86.83	2.715*
89.56	2.782*
97.44	2.917*

CURVE 5 (cont.)

T	C_p
100.53	2.971 x 10^{-2}*
105.88	3.074*
107.99	3.109*
111.01	3.154*
114.00	3.213*
117.37	3.257*
119.05	3.289*
123.55	3.365*
125.17	3.387*
127.78	3.432*
130.36	3.463*
132.07	3.495*
135.02	3.540*
138.00	3.584*
139.87	3.625*
143.83	3.670*
145.99	3.701*
149.45	3.750*
152.88	3.795*
154.97	3.822*
161.68	3.925*
167.23	3.965*
170.73	4.019*
172.95	4.050*
178.20	4.073*
186.42	4.171*
189.09	4.180*
193.05	4.225*
193.43	4.248*
197.26	4.265*
200.14	4.283*
201.15	4.301*
204.28	4.328*
204.59	4.333*
208.62	4.364*
209.14	4.368*
213.94	4.386*
217.86	4.422*
218.99	4.422*
223.69	4.463*
225.77	4.480*
229.76	4.539*

CURVE 5 (cont.)

T	C_p
229.98	4.503 x 10^{-2}*
233.05	4.534*
237.21	4.552*
240.65	4.610*
241.36	4.570*
244.41	4.637*
250.30	4.651*
256.16	4.700*
266.17	4.740*
271.91	4.785*
274.93	4.790*
289.95	4.861*
294.47	4.884*
295.38	4.888*
298.39	4.893*
299.18	4.906*
299.48	4.902*
301.37	4.920*
302.74	4.924*

CURVE 6

T	C_p
300	4.903 x 10^{-2}*
400	5.210
500	5.418*
600	5.586*
700	5.736
800	5.876
900	6.010
1000	6.140
1100	6.268
1165.7	6.352
1165.7	6.960
1200	6.960

*Not shown on Plot

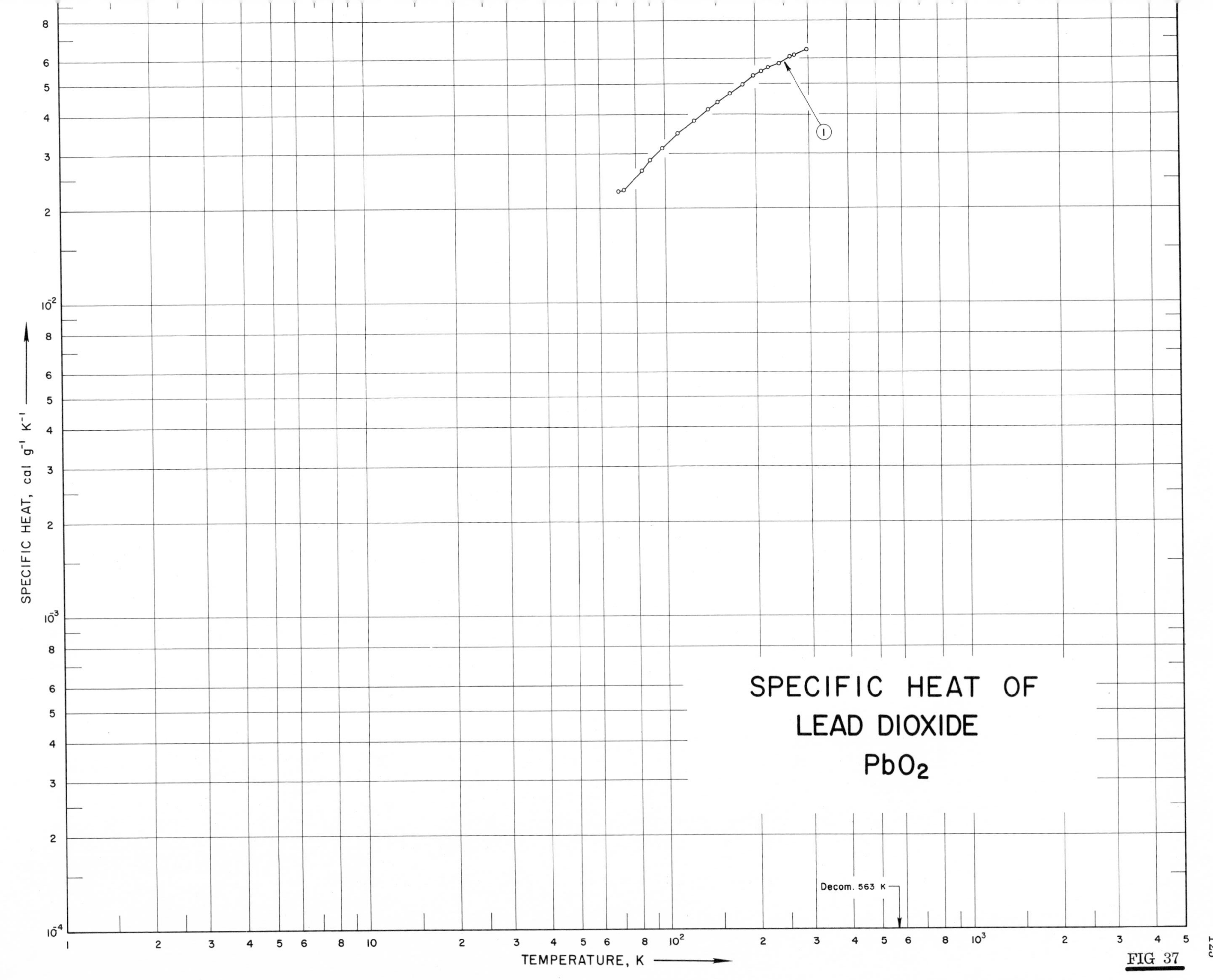
SPECIFIC HEAT OF
LEAD DIOXIDE
PbO_2
Decom. 563 K
1
TEMPERATURE, K
SPECIFIC HEAT, cal g^{-1} K^{-1}

FIG 37

SPECIFICATION TABLE NO. 37 SPECIFIC HEAT OF LEAD DIOXIDE PbO_2

[For Data Reported in Figure and Table No. 37]

Curve No.	Ref. No.	Year	Temp. Range, K	Reported Error, %	Name and Specimen Designation	Composition (weight percent), Specifications and Remarks
1	311	1929	70-297			99.5 theoretical amount of active oxygen; prepared by electrolysis of acid solution of lead nitrate.

DATA TABLE NO. 37 SPECIFIC HEAT OF LEAD DIOXIDE PbO_2

[Temperature, T, K; Specific Heat, C_p, Cal g^{-1} K^{-1}]

T	C_p
CURVE 1	
69.9	2.270 x 10^{-2}
73.1	2.378
84.5	2.653
89.8	2.870
98.9	3.129
111.6	3.485
126.0	3.829
140.7	4.168
151.8	4.373
166.9	4.695
183.4	4.992
198.4	5.318
212.9	5.481
227.9	5.644
242.0	5.832
261.6	6.104
270.5	6.200
298.1	6.446*
297.2	6.451

*Not shown on plot

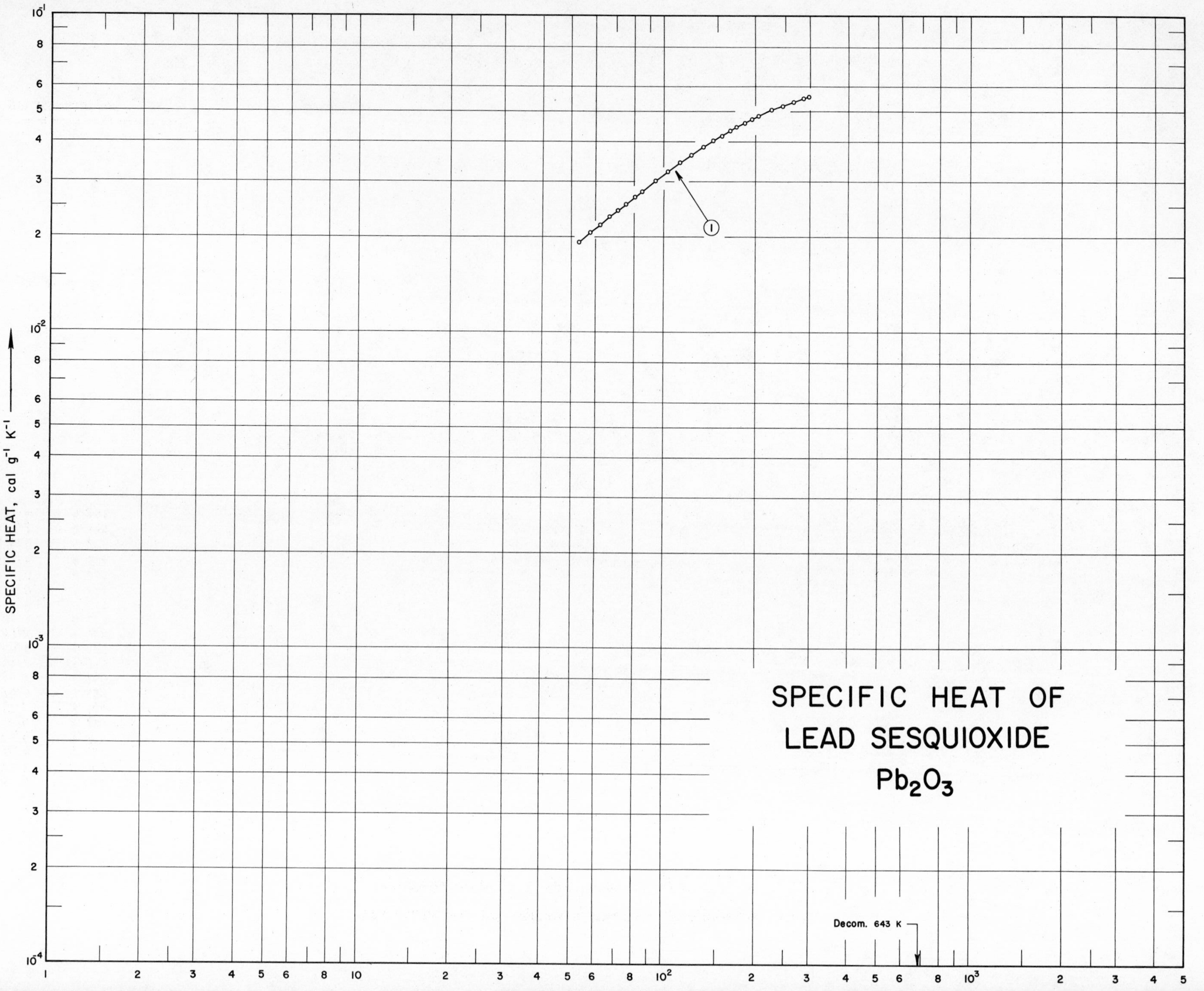
SPECIFIC HEAT OF
LEAD SESQUIOXIDE
Pb2O3
Decom. 643 K
1
SPECIFIC HEAT, cal g-1 K-1

SPECIFICATION TABLE NO. 38 SPECIFIC HEAT OF LEAD SESQUIOXIDE Pb_2O_3

[For Data Reported in Figure and Table No. 38]

Curve No.	Ref. No.	Year	Temp. Range, K	Reported Error, %	Name and Specimen Designation	Composition (weight percent), Specifications and Remarks
1	86	1958	53-298			89.64 Pb.

DATA TABLE NO. 38 SPECIFIC HEAT OF LEAD SESQUIOXIDE, Pb_2O_3

[Temperature, T, K; Specific Heat, C_p, Cal $g^{-1}K^{-1}$]

T	C_p
CURVE 1	
53.36	1.923 x 10^{-2}
58.06	2.060
62.66	2.188
67.03	2.314
71.50	2.431
75.95	2.545
81.00	2.677
85.48	2.783
94.96	3.008
104.75	3.227
114.88	3.438
124.78	3.635
136.53	3.854
145.90	4.018
155.87	4.180
166.02	4.329
175.85	4.459
186.19	4.595
196.05	4.710
206.42	4.829
216.45	4.931*
226.13	5.028
235.97	5.112*
245.91	5.199
256.50	5.285*
266.45	5.365
276.38	5.426*
287.09	5.493
296.56	5.558*
298.15	5.566

*Not shown on Plot

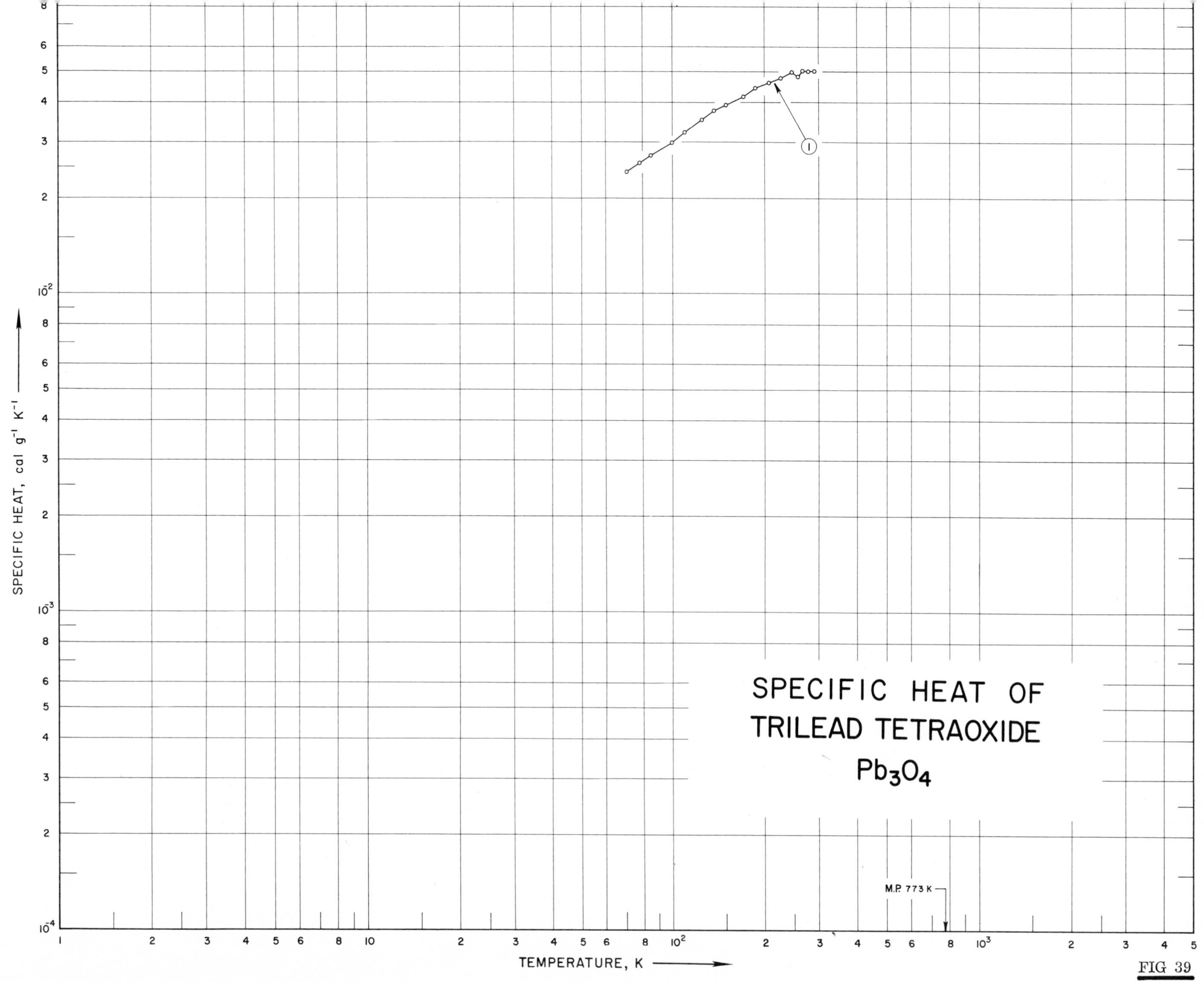
SPECIFIC HEAT OF
TRILEAD TETRAOXIDE
Pb3O4
M.P. 773 K
1
SPECIFIC HEAT, cal g-1 K-1
TEMPERATURE, K

FIG 39

SPECIFICATION TABLE NO. 39 SPECIFIC HEAT OF TRILEAD TETRAOXIDE Pb_3O_4

[For Data Reported in Figure and Table No. 39]

Curve No.	Ref. No.	Year	Temp. Range, K	Reported Error, %	Name and Specimen Designation	Composition (weight percent), Specifications and Remarks
1	311	1929	71-292			97. 0% of theoretical active oxygen, reduction by hydrogen showed 90. 62 Pb (90. 66 Pb theo.); prepared by decomposition of electrolytic lead dioxide in a stirred bath of molten potassium nitrate at 460 C, then washed out and dried at 120 C.

DATA TABLE NO. 39 SPECIFIC HEAT OF TRILEAD TETRAOXIDE Pb_3O_4

[Temperature, T, K; Specific Heat, C_p, Cal g^{-1} K^{-1}]

T	C_p
CURVE 1	
71.5	2.429 x 10^{-2}
78.5	2.599
85.6	2.744
100.1	2.997
111.7	3.257
127.1	3.543
138.6	3.766
151.3	3.950
172.1	4.196
187.4	4.425
208.8	4.618
227.1	4.774
247.1	4.994
259.8	4.813
266.2	5.023
278.2	5.013
292.6	5.026

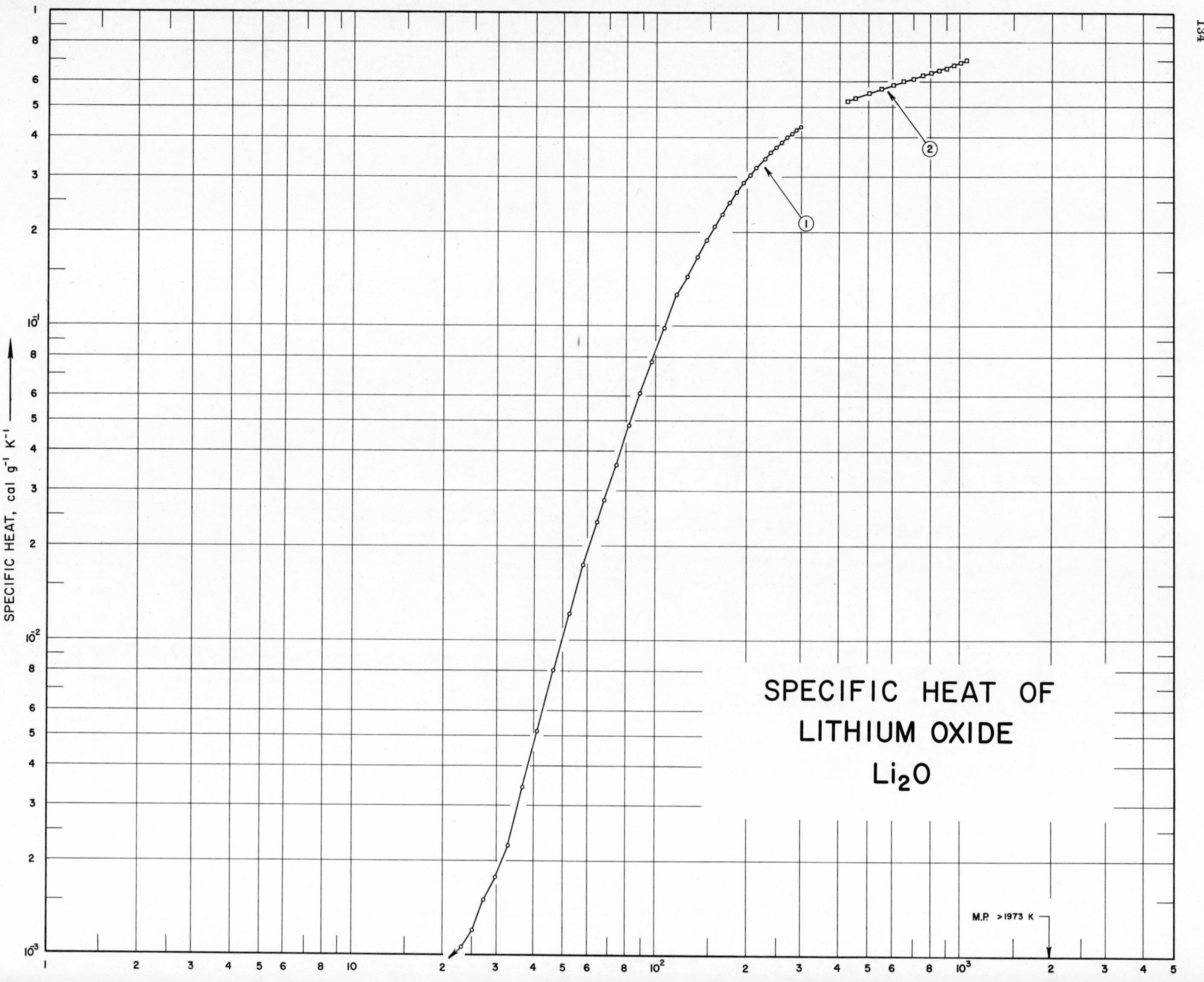
SPECIFIC HEAT OF
LITHIUM OXIDE
Li2O
M.P. >1973 K
SPECIFIC HEAT, cal g-1 K-1
1
2

SPECIFICATION TABLE NO. 40 SPECIFIC HEAT OF LITHIUM OXIDE Li_2O

[For Data Reported in Figure and Table No. 40]

Curve No.	Ref. No.	Year	Temp. Range, K	Reported Error, %	Name and Specimen Designation	Composition (weight percent), Specifications and Remarks
1	89	1951	17-298			99.74 Li_2O and 0.26 CaO; heat treated in nickel crucible at 1000 to 1300 C for 3 to 5 hrs; corrected for impurities.
2	90	1955	424-1050			99.2 Li_2O, 0.8 LiOH; prepared from 99.9 Li_2O_2.

DATA TABLE NO. 40 SPECIFIC HEAT OF LITHIUM OXIDE, Li_2O

[Temperature, T, K; Specific Heat, C_p, Cal $g^{-1}K^{-1}$]

T	C_p	T	C_p
CURVE 1		CURVE 2	
17.06	6.62×10^{-4} *	425	5.238×10^{-1}
20.68	9.64*	450	5.357
23.03	1.06×10^{-3}	500	5.565
25.01	1.20	550	5.745
27.23	1.50	600	5.906
29.87	1.77	650	6.055
32.81	2.24	700	6.193
36.59	3.41	750	6.325
41.04	5.15	800	6.451
46.44	8.03	850	6.573
52.12	1.22×10^{-2}	900	6.691
58.05	1.75	950	6.807
64.51	2.39	1000	6.921
68.01	2.80	1050	7.034
74.26	3.63		
81.75	4.840		
88.90	6.124		
97.03	7.751		
107.30	9.913		
117.64	1.227×10^{-1}		
127.38	1.448		
137.49	1.675		
146.97	1.891		
155.73	2.082		
164.96	2.277		
174.90	2.482		
184.52	2.687		
194.17	2.877		
204.23	3.049		
208.11	3.113*		
214.43	3.207		
227.44	3.432		
238.53	3.590		
248.71	3.731		
258.38	3.863		
268.63	4.001		
279.01	4.121		
288.74	4.235		
298.89	4.330		

* Not shown on Plot

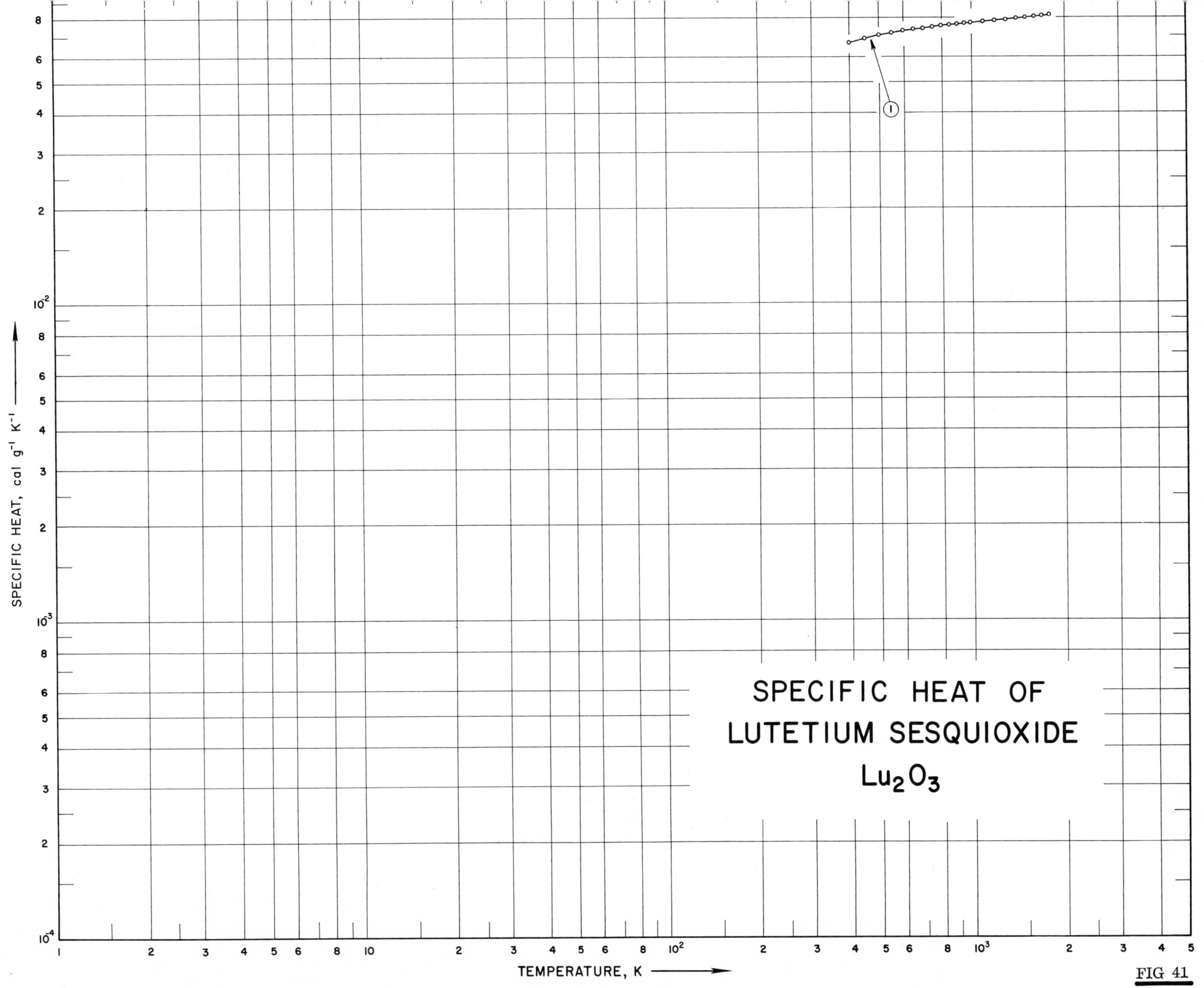
SPECIFIC HEAT OF
LUTETIUM SESQUIOXIDE
Lu_2O_3
1
TEMPERATURE, K
SPECIFIC HEAT, cal g^{-1} K^{-1}

FIG 41

SPECIFICATION TABLE NO. 41 SPECIFIC HEAT OF LUTETIUM SESQUIOXIDE Lu_2O_3

[For Data Reported in Figure and Table No. 41]

Curve No.	Ref. No.	Year	Temp. Range, K	Reported Error, %	Name and Specimen Designation	Composition (weight percent), Specifications and Remarks
1	59	1963	400-1800			99.9 Lu_2O_3; dried at 1100 - 1200 C.

DATA TABLE NO. 41 SPECIFIC HEAT OF LUTETIUM SESQUIOXIDE, Lu_2O_3

[Temperature, T, K; Specific Heat, C_p, Cal $g^{-1}K^{-1}$]

T	C_p
CURVE 1	
400	6.642 x 10^{-2}
450	6.857
500	7.018
550	7.142
600	7.242
650	7.325
700	7.395
750	7.456
800	7.509
850	7.558
900	7.602
950	7.643
1000	7.681
1100	7.750
1200	7.814
1300	7.873
1400	7.930
1500	7.984
1600	8.036
1700	8.087
1800	8.136

*Not shown on Plot

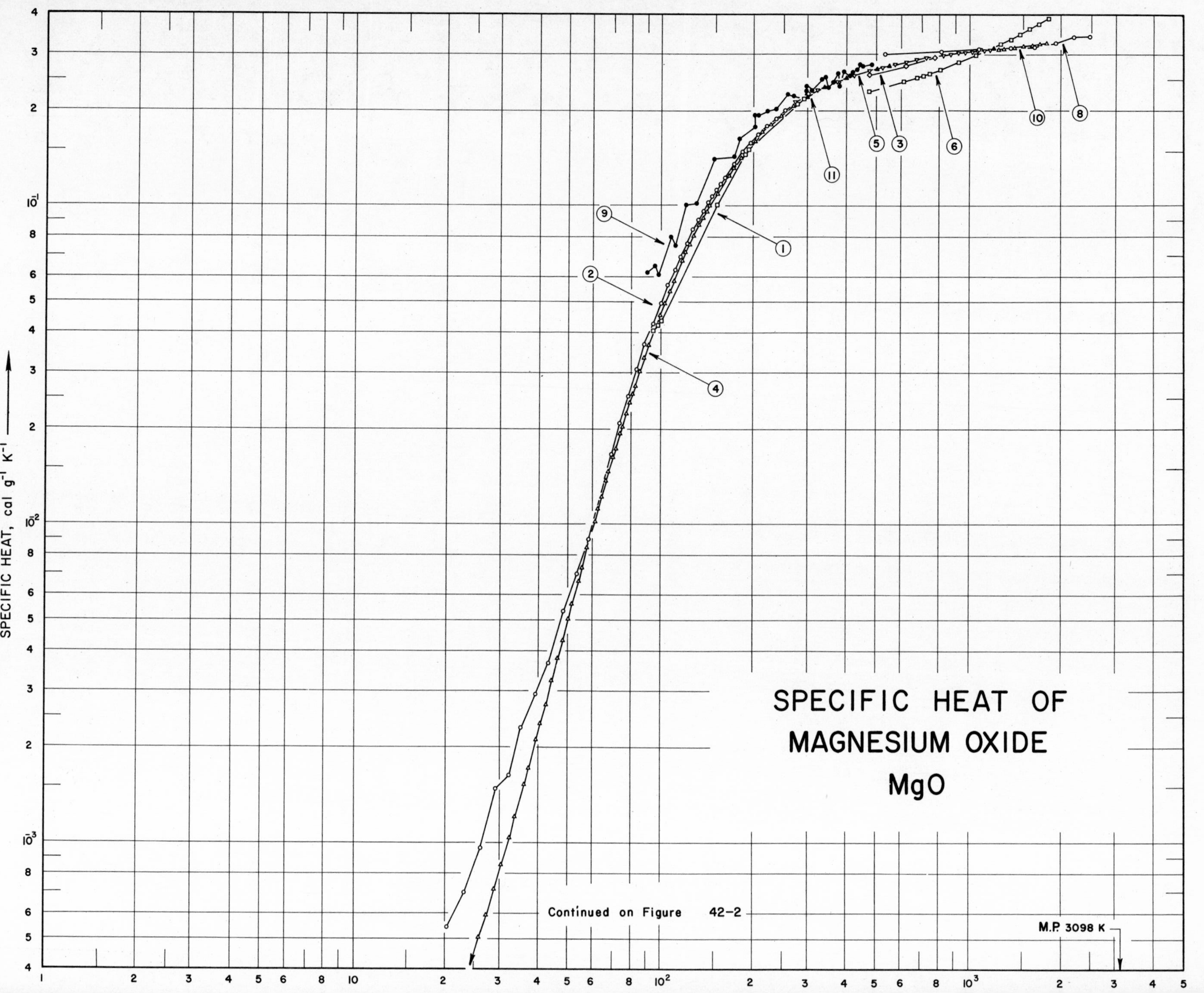
SPECIFIC HEAT OF
MAGNESIUM OXIDE
MgO
Continued on Figure 42-2
M.P. 3098 K
SPECIFIC HEAT, cal g^{-1} K^{-1}

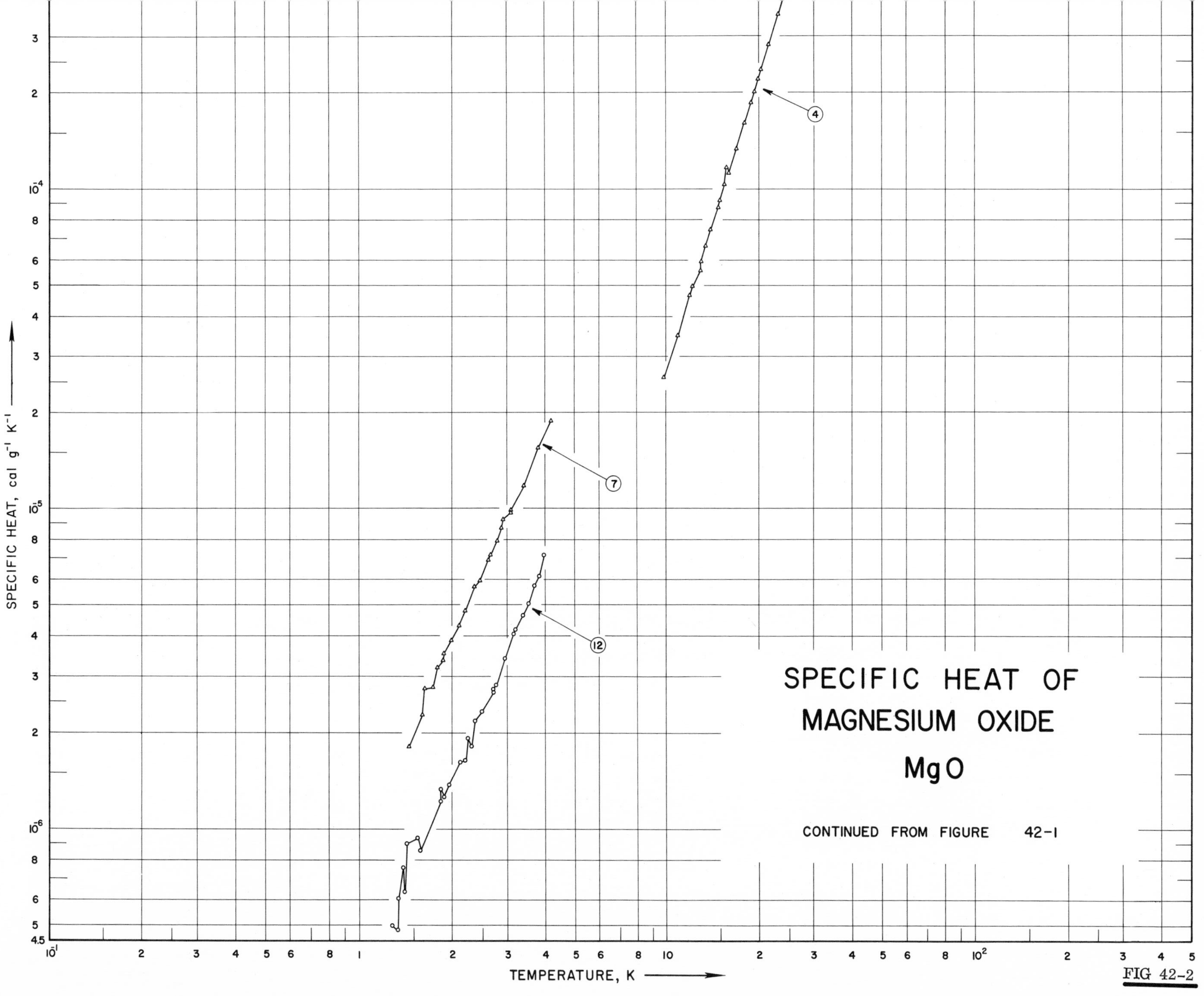
SPECIFIC HEAT OF
MAGNESIUM OXIDE
MgO
CONTINUED FROM FIGURE 42-1
4
7
12
SPECIFIC HEAT, cal g^{-1} K^{-1}
TEMPERATURE, K
3
2
10^{-4}
8
6
5
4
3
2
10^{-5}
8
6
5
4
3
2
10^{-6}
8
6
5
4.5
10^{-1}
2
3
4
5
6
8
1
2
3
4
5
6
8
10
2
3
4
5
6
8
10^2
2
3
4
5

FIG 42-2

SPECIFICATION TABLE NO. 42 SPECIFIC HEAT OF MAGNESIUM OXIDE MgO

[For Data Reported in Figure and Table No. 42]

Curve No.	Ref. No.	Year	Temp. Range, K	Reported Error, %	Name and Specimen Designation	Composition (weight percent), Specifications and Remarks
1	18	1926	94-291			High grade sample, fused magnesium oxide; supplied by Norton Co.
2	91	1937	20-300			95 MgO and ~5 $Mg(OH)_2$; sample prepared by decomposing $Mg(OH)_2$ under vacuum at 300 C and raised to 350 C at the end of decomposition; corrected for $Mg(OH)_2$.
3	92	1950	473-773		Magnesia	MgO.
4	93	1959	9-270	≤0. 5		Single crystals; supplied by Norton Co. ; heated in a vacuum at 150 C for 18 hrs; 10^{-6} mm Hg vacuum; helium atmosphere.
5	94	1960	273-1173	0. 25		Bal. MgO, 0. 025 Ca, 0. 020 Fe, 0. 009 Si, 0. 008 Mn, 0. 004 Al, <0. 002 Na, <0. 001 Ag, <0. 001 Cr, and <0. 001 Cu; single crystal; supplied by Norton Co.
6	95	1961	475-1811	3. 0		>99. 0 MgO, <0. 5 Si, and <0. 3 Mn; density 186 lb ft^{-3}; measured under helium atmosphere.
7	96	1962	1. 3-4. 2			Surface area of 166 m^2 g^{-1}.
8	48	1962	533-2478	≤5. 0		Before exposure: 59. 4 Mg, 0. 3 Fe, 0. 3 Si, 0. 2 Ca, and 0. 1 Al; after exposure: 60. 1 Mg and <0. 05 C; sample supplied by Zirconium Corp. of America; crushed in hardened steel mortar to pass 100-mesh screen; pressed and sintered; before exposure: apparent density (ASTM method B311-58) 206 lb ft^{-3}, true density (by immersion in xylene) 219 lb ft^{-3}; after exposure: apparent density 218 lb ft^{-3}, true density 224 lb ft^{-3}.
9	97	1962	90-481	≤5. 0		MgO crystal, probably pure; Hanova liquid platinum was applied on the specimen's front surface for opaqueness and then painted with Parson's black for constant absorptivity; Hanova liquid platinum coatings were applied also on specimen's rear surface to obtain good conductive surface.
10	98	1963	298-1800	0. 1		99. 93 MgO, 0. 04 Al_2O_3, and 0. 01 SiO_2; macro crystalline.
11	46	1963	298-1200	0. 40		99. 9 MgO, 0. 025 Ca, 0. 02 Fe, 0. 009 Si, 0. 008 Mn, 0. 004 Al, <0. 002 Na, <0. 001 Ag, and <0. 001 Cr, <0. 001 Cu; fused.
12	96	1962	1. 3-4			Surface area of 13. 1 m^2 g^{-1}.

DATA TABLE NO. 42 SPECIFIC HEAT OF MAGNESIUM OXIDE MgO

[Temperature, T, K; Specific Heat, C_p, Cal $g^{-1}K^{-1}$]

T	C_p
CURVE 1	
94.1	4.02 x 10^{-2}
97.9	4.42
98.5	4.50*
100.0	4.70
151.8	1.101 x 10^{-1}
188.4	1.452
193.4	1.502
275.3	2.090
278.0	2.137*
288.3	2.170*
291.0	2.184
CURVE 2	
20.34	5.46 x 10^{-4}
23.05	6.99
26.08	9.65
29.16	1.48 x 10^{-3}
32.07	1.63
35.17	2.30
39.15	2.93
43.40	3.67
48.30	5.33
53.52	6.97
58.26	8.95
69.01	1.66 x 10^{-2}
73.62	2.07
78.46	2.535
83.53	3.073
88.77	3.673
94.26	4.291
99.94	4.950
105.51	5.623
111.23	6.305
116.04	6.910
121.57	7.617
127.56	8.452
132.60	9.065
137.25	9.638
142.03	1.031 x 10^{-1}
146.78	1.072
151.39	1.127
156.19	1.170
161.39	1.233
CURVE 2 (cont.)	
166.76	1.287 x 10^{-1}*
172.75	1.353
178.83	1.419*
184.81	1.481
190.51	1.534*
195.81	1.582
200.72	1.629*
205.76	1.664*
210.20	1.708
215.36	1.744*
220.34	1.789
225.06	1.819*
230.50	1.853*
236.51	1.887
242.55	1.932*
248.29	1.966*
253.74	1.999
258.99	2.033*
264.18	2.062*
269.15	2.080*
274.10	2.105*
278.78	2.142*
283.90	2.164*
289.19	2.196*
294.25	2.232*
300.68	2.254
CURVE 3	
473	2.58 x 10^{-1}
623	2.76
773	2.92
CURVE 4	
9.759	2.585 x 10^{-5}
9.799	2.617*
9.844	2.659*
10.849	3.463
10.878	3.575*
10.914	3.587*
11.892	4.617
11.911	4.651*
12.138	4.966
CURVE 4 (cont.)	
12.898	5.569 x 10^{-5}
12.906	5.907
13.450	6.614
13.905	7.442
14.686	8.762
14.898	9.196
15.471	1.042 x 10^{-4}
15.696	1.174
15.896	1.126
16.019	1.124*
16.873	1.340
16.909	1.343*
17.921	1.601
18.101	1.650*
18.123	1.669*
18.851	1.861
18.932	1.887*
19.194	2.003
19.937	2.210
20.263	2.354
20.548	2.424*
21.500	2.833
23.104	3.513
24.216	4.086
25.867	5.046
27.247	5.954
28.823	7.142
30.419	8.511
32.262	1.034 x 10^{-3}
33.692	1.201
36.005	1.521
37.264	1.718
39.501	2.103
40.904	2.377
42.940	2.835
44.504	3.232
46.547	3.810
48.208	4.329
50.287	5.053
51.740	5.614
54.183	6.628
55.670	7.313
57.902	8.429
59.175	9.107*
CURVE 4 (cont.)	
61.092	1.019 x 10^{-2}
62.701	1.117
64.358	1.224
66.494	1.368
67.915	1.468
70.232	1.639
71.463	1.734
73.910	1.933
74.978	2.022
77.430	2.239
79.223	2.403
80.845	2.563
82.406	2.716
85.494	3.029
88.488	3.344
91.417	3.675
92.052	3.724*
95.624	4.123*
99.306	4.542
102.902	4.969
106.484	5.403
110.069	5.837
113.639	6.254*
117.191	6.713
120.730	7.157
124.263	7.596
127.625	8.013
129.700	8.253*
133.280	8.717
136.904	9.159
140.559	9.608
144.252	1.005 x 10^{-1}
147.803	1.047*
148.840	1.059*
152.216	1.098
155.549	1.137*
158.847	1.173*
162.244	1.212*
165.739	1.249
169.320	1.288*
172.989	1.325
175.409	1.351*
178.719	1.384*
182.142	1.420*
CURVE 4 (cont.)	
185.672	1.451 x 10^{-1}*
189.309	1.486*
192.864	1.519*
196.348	1.550*
199.765	1.579*
201.580	1.596
204.901	1.623*
208.165	1.651*
211.376	1.677
214.540	1.703
217.813	1.726*
221.192	1.753*
224.522	1.778*
226.412	1.791
229.672	1.815*
232.890	1.838*
236.068	1.859*
239.351	1.881
242.736	1.903*
246.081	1.925
249.388	1.938*
250.598	1.952*
253.750	1.972*
256.967	1.994
260.152	2.009*
263.308	2.028*
266.436	2.045*
269.793	2.063
CURVE 5	
273.15	2.110 x 10^{-1}
323.15	2.309
373.15	2.454
423.15	2.565
473.15	2.652
523.15	2.723
573.15	2.781
623.15	2.830
673.15	2.871
723.15	2.907
773.15	2.938*
823.15	2.965
873.15	2.989
CURVE 5 (cont.)	
923.15	3.010 x 10^{-1}
973.15	3.029
1023.15	3.046
1073.15	3.061
1123.15	3.076
1173.15	3.088
CURVE 6	
474.82	2.297 x 10^{-1}
612.04	2.457
678.15	2.535
709.82	2.572
741.48	2.609
802.04	2.680*
819.82	2.700
924.82	2.823
1052.59	2.973*
1135.37	3.070
1270.93	3.228
1369.82	3.344
1473.71	3.465
1579.82	3.590
1689.82	3.718
1703.71	3.735
1811.48	3.861
CURVE 7	
1.612	2.743 x 10^{-6}
1.717	2.771
1.850	3.379
2.430	5.975
2.628	7.190
2.860	8.708
3.084	9.721
3.388	1.187 x 10^{-5}
3.785	1.557
4.188	1.893

* Not shown on plot

DATA TABLE NO. 42 (continued)

T	C_p
CURVE 7 (cont.)	
Series 2	
1.431	1.810×10^{-6}
1.587	2.276
1.707	2.849*
1.774	3.191
1.858	3.544
1.968	3.888
2.091	4.319
2.180	4.800
2.336	5.702
3.586	6.924
2.753	7.943
2.895	9.277
3.084	9.982
CURVE 8	
533.15	3.00×10^{-1}
810.93	3.05
1088.71	3.10
1366.48	3.13
1644.26	3.15
1922.04	3.25
2199.82	3.40
2477.59	3.40
CURVE 9	
90.15	6.2×10^{-2}
95.15	6.5
98.15	6.1
108.15	8.0
111.15	7.5
120.15	1.01×10^{-1}
131.15	1.02
149.15	1.40
173.15	1.43
180.15	1.64
201.15	1.78
201.15	1.93
208.15	1.93
221.15	1.98
237.15	2.02
245.15	2.03*

T	C_p
CURVE 9 (cont.)	
258.15	2.24×10^{-1}
269.15	2.22
296.15	2.30
296.15	2.35*
297.15	2.19*
297.15	2.39
301.15	2.40*
302.15	2.22*
309.15	2.31
332.15	2.50
341.15	2.54
349.15	2.38
361.15	2.46
375.15	2.62
379.15	2.40
393.15	2.65
410.15	2.56
419.15	2.64
431.15	2.68
437.15	2.65*
441.15	2.78
451.15	2.75
481.15	2.78
CURVE 10	
298	2.2133×10^{-1}*
300	2.2229*
350	2.4149*
400	2.5438*
450	2.6361*
500	2.7055*
550	2.7600*
600	2.8044*
650	2.8416*
700	2.8736*
750	2.9019*
800	2.9271*
850	2.9502*
900	2.9715*
950	2.9913*
1000	3.0101*
1050	3.0279*
1100	3.0450*
1150	3.0614*
1200	3.0774*

T	C_p
CURVE 10 (cont.)	
1250	3.0928×10^{-1}
1300	3.1079
1350	3.1227*
1400	3.1372
1450	3.1515*
1500	3.1656
1550	3.1794*
1600	3.1931
1650	3.2067*
1700	3.2202
1750	3.2335*
1800	3.2468
CURVE 11	
298.15	2.209×10^{-1}
300	2.217*
320	2.297*
340	2.365
360	2.427*
380	2.481*
400	2.528
420	2.570*
440	2.607*
460	2.641*
480	2.671*
500	2.699
550	2.759
600	2.809*
650	2.852*
700	2.888*
750	2.921*
800	2.950*
850	2.976*
900	2.999*
950	3.020*
1000	3.040*
1050	3.058*
1100	3.074*
1150	3.089*
1200	3.102

T	C_p
CURVE 12	
Series 1	
1.315	4.825×10^{-7}
1.371	7.587
1.410	8.992
1.819	1.330×10^{-6}
1.932	1.374
Series 2	
1.267	4.985×10^{-7}
1.335	6.058
1.396	6.342
1.525	9.354
1.559	8.542
1.819	1.222×10^{-6}
1.853	1.254
2.099	1.617
2.177	1.646
2.161	1.665*
2.282	1.809
2.664	2.736
2.724	2.813
2.684	2.686*
2.226	1.919
2.341	2.170
2.458	2.324
2.674	2.666
2.916	3.398
3.135	4.046
3.355	4.635
3.504	5.077
3.794	6.159
2.971	3.420*
3.176	4.198
3.411	4.683*
3.666	5.736
3.923	7.155

* Not shown on plot

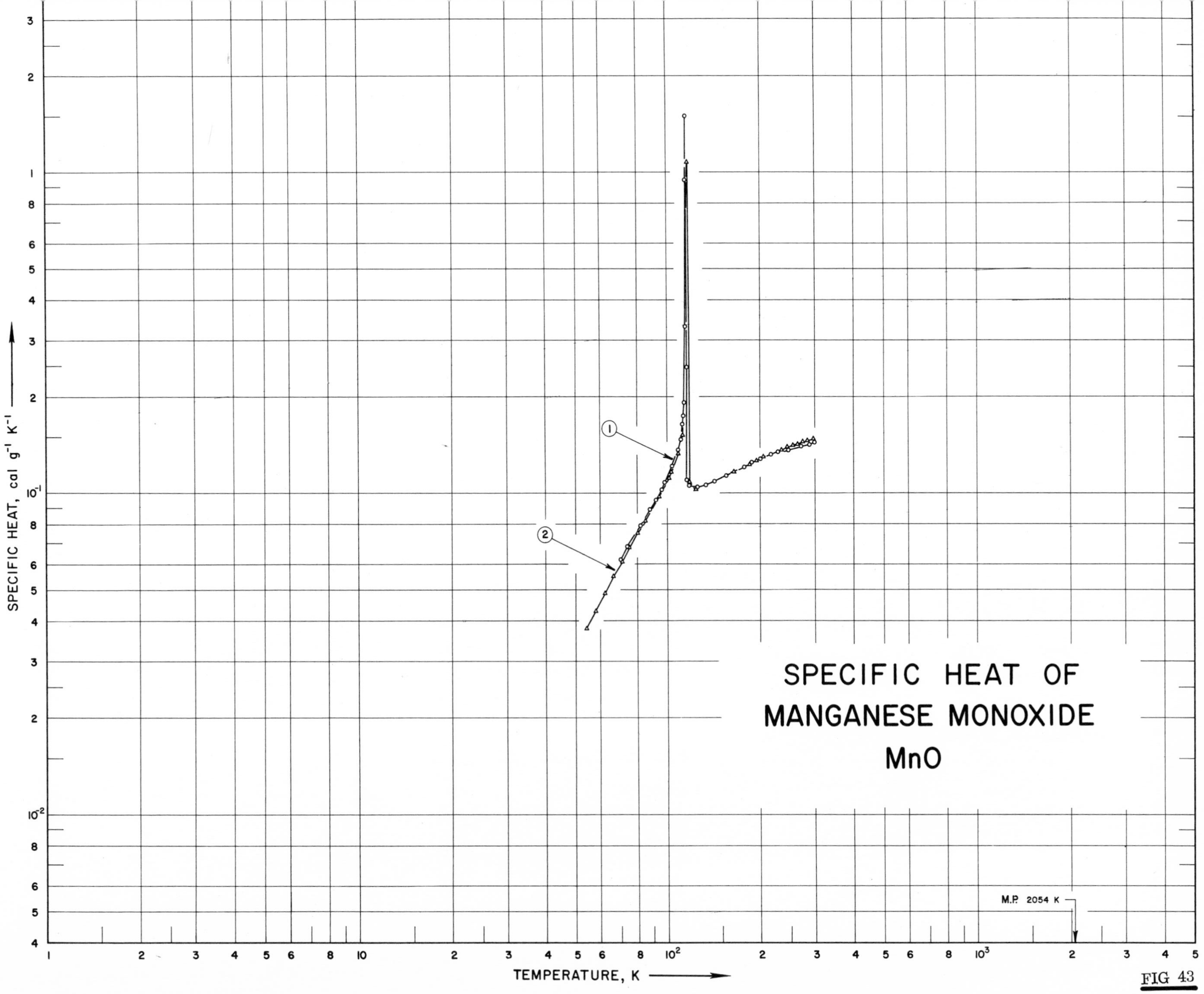
SPECIFIC HEAT OF
MANGANESE MONOXIDE
MnO
M.P. 2054 K
TEMPERATURE, K
SPECIFIC HEAT, cal g^{-1} K^{-1}
1
2

FIG 43

SPECIFICATION TABLE NO. 43 SPECIFIC HEAT OF MANGANESE MONOXIDE MnO

[For Data Reported in Figure and Table No. 43]

Curve No.	Ref. No.	Year	Temp. Range, K	Reported Error, %	Name and Specimen Designation	Composition (weight percent), Specifications and Remarks
1	99	1928	70-300			99. 0 MnO; finely crystalline bright green product; prepared by reduction of amorphous Mn_3O_4.
2	77	1951	54-298			99. 85 MnO, 0. 030 available oxygen, and 0. 005 S; prepared from electrolytic manganese.

DATA TABLE NO. 43 SPECIFIC HEAT OF MANGANESE MONOXIDE MnO

[Temperature, T, K; Specific Heat, C_p, Cal g^{-1} K^{-1}]

T	C_p	T	C_p
CURVE 1		CURVE 2	
70.4	6.240 x 10^{-2}	54.69	3.818 x 10^{-2}
74.3	6.864	58.42	4.320
82.1	7.965	62.45	4.903
88.5	8.949	66.80	5.534
92.1	9.582	71.27	6.174
96.5	1.038 x 10^{-1}	75.73	6.814
98.9	1.093	80.55	7.538
104.8	1.236	85.21	8.250
108.8	1.362*	94.71	9.832
109.1	1.386	102.10	1.126 x 10^{-1}
109.9	1.428*	103.99	1.172
111.0	1.485	106.04	1.225*
112.0	1.548*	109.76	1.346
113.0	1.651	113.32	1.525
113.9	1.762	114.94	1.737*
114.8	1.943	116.25	1.906*
115.5	3.341	117.78	1.081 x 10^{0}
115.7	9.541	119.19	1.094 x 10^{-1}
115.9	1.5164 x 10^{0}	120.69	1.060*
116.2	3.378 x 10^{-1}*	121.41	1.053*
116.5	2.488	124.63	1.044
117.1	1.105	135.74	1.065*
118.5	1.066	145.90	1.102*
118.9	1.061*	155.66	1.135*
126.8	1.053	165.74	1.174
128.6	1.055*	175.92	1.210*
134.5	1.070	185.99	1.244
143.3	1.099	196.02	1.275
155.6	1.145	206.26	1.302
178.0	1.216	216.20	1.331*
189.7	1.259	226.16	1.356*
203.3	1.293	236.00	1.378
217.5	1.334	245.69	1.400
229.7	1.352	256.05	1.418
247.1	1.371	265.94	1.437
289.1	1.432	276.07	1.456
293.7	1.446*	286.28	1.472
300.2	1.452	296.40	1.485*
272.9	1.409	298.16	1.486

*Not shown on plot

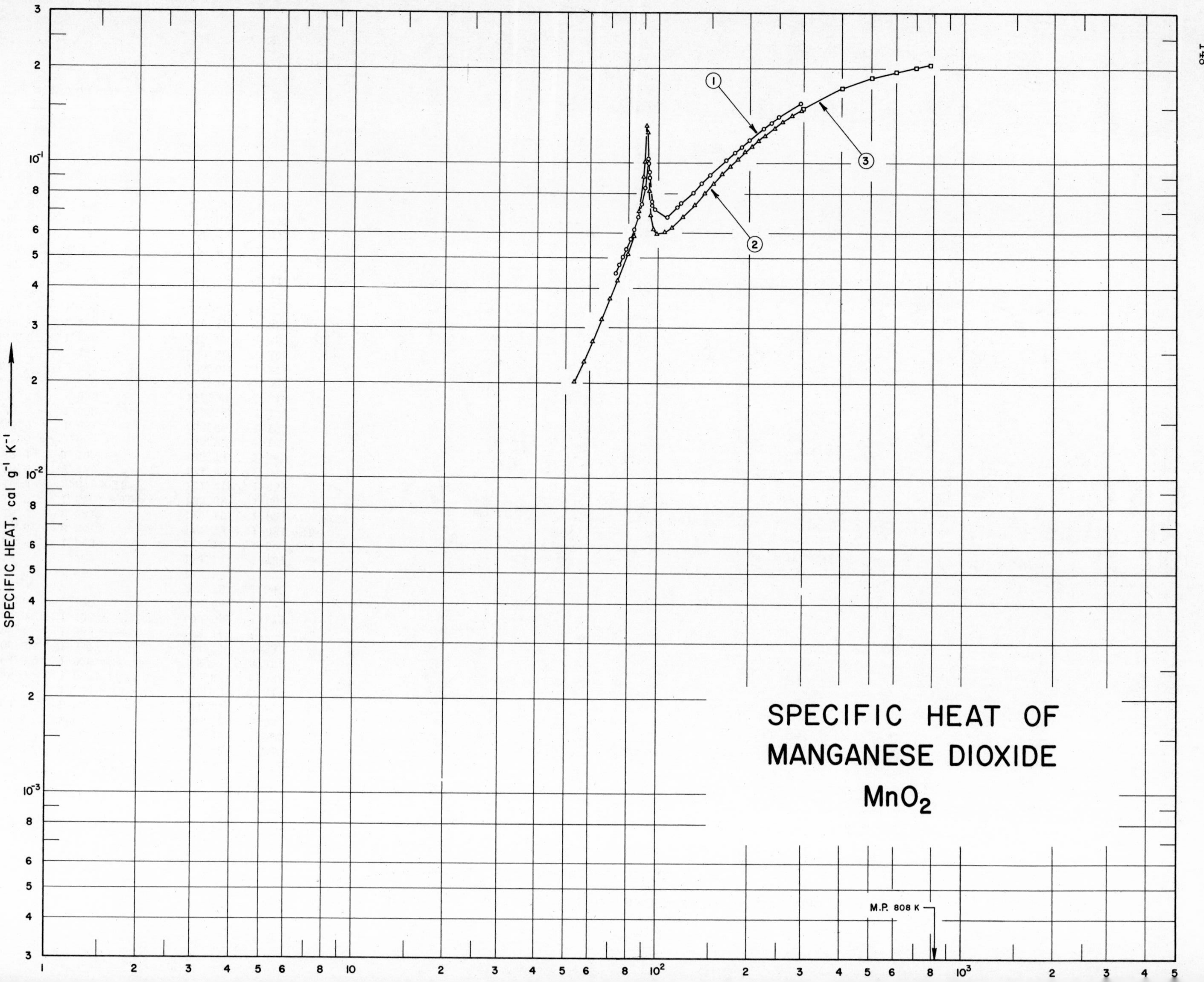
SPECIFIC HEAT OF
MANGANESE DIOXIDE
MnO_2
M.P. 808 K
SPECIFIC HEAT, cal g⁻¹ K⁻¹
1
2
3

SPECIFICATION TABLE NO. 44 SPECIFIC HEAT OF MANGANESE DIOXIDE MnO_2

[For Data Reported in Figure and Table No. 44]

Curve No.	Ref. No.	Year	Temp. Range, K	Reported Error, %	Name and Specimen Designation	Composition (weight percent), Specifications and Remarks
1	99	1928	72-293			99.6 MnO_2; prepared by heating the nitrate at 170 K.
2	100	1943	53-294			99.88 MnO_2; pulverized and heated at 500 C in a stream of pure oxygen.
3	101	1943	298-780			100.00 MnO_2; density 318.5 lb ft^{-3}.

DATA TABLE NO. 44 SPECIFIC HEAT OF MANGANESE DIOXIDE MnO_2

[Temperature, T, K; Specific Heat, C_p, Cal g^{-1} K^{-1}]

CURVE 1

T	C_p
72.5	4.457×10^{-2}*
72.8	4.487
74.6	4.732*
74.8	4.774
76.7	5.029
78.8	5.313
81.1	5.744
83.4	6.177
86.0	6.727
88.3	7.419
90.2	8.350
90.3	8.446*
91.8	9.788*
92.5	1.039×10^{-1}
92.9	9.958×10^{-2}
93.2	9.969*
93.7	9.363
93.9	8.983
95.4	7.533
95.6	7.368
97.4	7.12
107.2	6.73
115.4	7.229
119.7	7.492
130.2	8.058
139.6	8.619
148.8	9.197
166.9	1.020×10^{-1}
179.3	1.089
188.7	1.135
199.5	1.190
221.9	1.293
236.6	1.345
249.8	1.402
292.6	1.564*
293.8	1.556

CURVE 2

T	C_p
53.5	2.017×10^{-2}
57.3	2.349
61.1	2.715
65.5	3.199
69.8	3.704
73.9	4.257×10^{-2}
79.8	5.157
80.4	5.286*
83.3	5.891
84.3	6.174*
86.7	7.003
87.6	7.470*
89.4	9.028
90.1	1.014×10^{-1}
91.4	1.313
92.1	1.256
93.5	8.187×10^{-2}
94.4	6.827
96.2	6.183
97.4	6.058*
99.2	5.977*
100.4	5.957*
102.0	5.965*
103.3	5.981*
105.4	6.031
106.7	6.071*
111.1	6.254
112.5	6.318*
120.7	6.747
122.2	6.833*
132.0	7.399
132.9	7.447*
142.6	8.020
143.4	8.077*
152.5	8.605
153.4	8.653*
163.3	9.219
164.1	9.270*
173.6	9.779
174.3	9.824*
183.9	1.034
184.5	1.037*
194.4	1.089
195.0	1.093*
204.6	1.137
205.2	1.141*
214.7	1.184
215.3	1.187*
224.8	1.227
225.4	1.232×10^{-2}*
235.9	1.275*
241.1	1.296
245.7	1.316*
252.5	1.341*
256.4	1.354
266.3	1.391*
276.1	1.423
285.8	1.450*
294.7	1.479

CURVE 3

T	C_p
298.15	1.491×10^{-1}
300	1.504*
400	1.743
500	1.872
600	1.954
700	2.015
780	2.056

*Not shown on plot

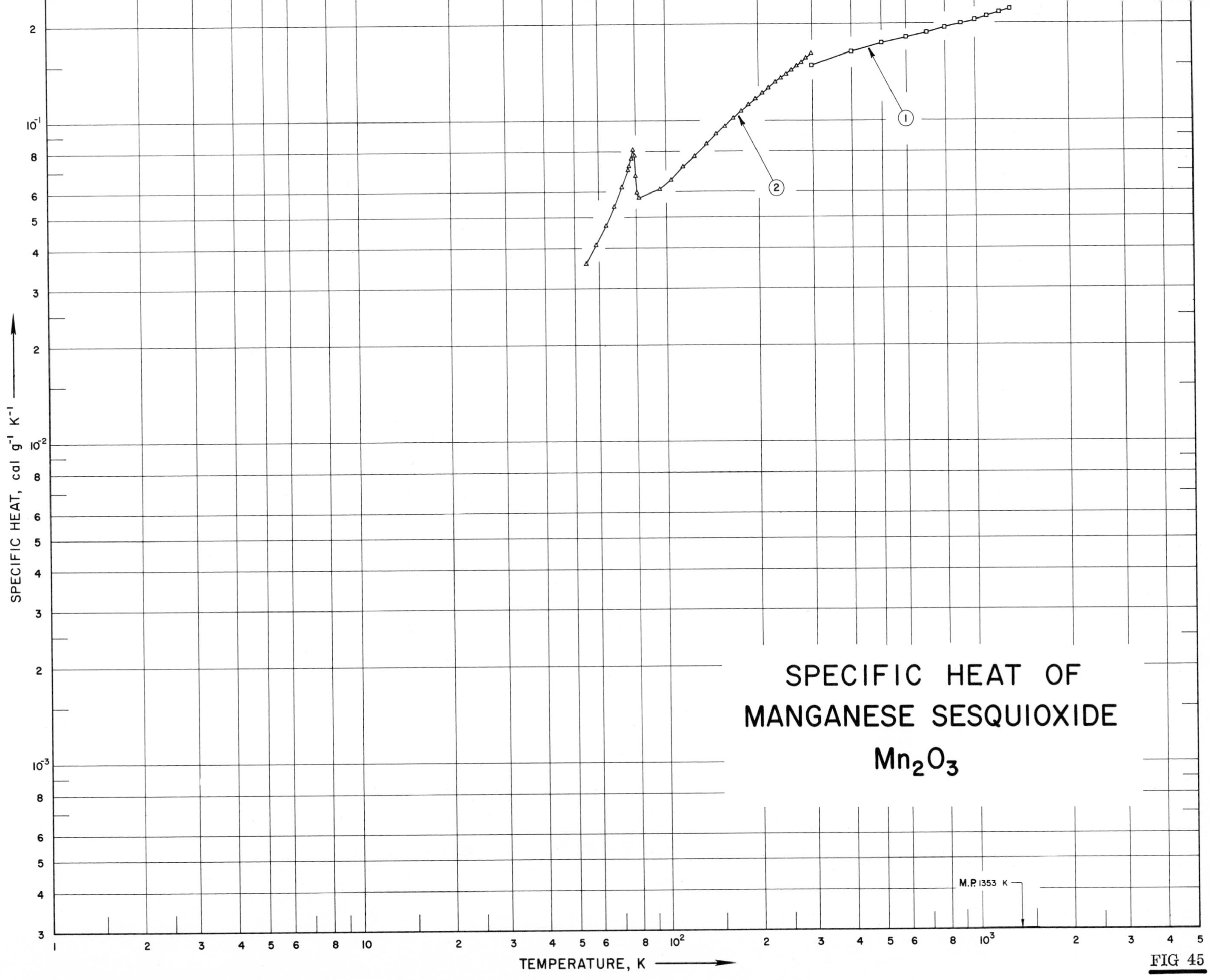
SPECIFIC HEAT OF
MANGANESE SESQUIOXIDE
Mn_2O_3
M.P. 1353 K
TEMPERATURE, K
SPECIFIC HEAT, cal g^{-1} K^{-1}
1
2

FIG 45

SPECIFICATION TABLE NO. 45 SPECIFIC HEAT OF MANGANESE SESQUIOXIDE Mn_2O_3

[For Data Reported in Figure and Table No. 45]

Curve No.	Ref. No.	Year	Temp. Range, K	Reported Error, %	Name and Specimen Designation	Composition (weight percent), Specifications and Remarks
1	102	1954	298-1300			Mn_2O_3, 69. 64 Mn, and 10. 13 O_2.
2	103	1954	54-298			69. 64 Mn, 10. 13 O_2 (theoretical 69. 59 and 10. 14).

DATA TABLE NO. 45 SPECIFIC HEAT OF MANGANESE SESQUIOXIDE Mn_2O_3

[Temperature, T, K; Specific Heat, C_p, Cal g^{-1} K^{-1}]

T	C_p
CURVE 1	
298.15	1.495 x 10^{-1}
300	1.498*
400	1.651
500	1.750
600	1.828
700	1.896
800	1.959
900	2.019
1000	2.077
1100	2.134
1200	2.189
1300	2.245
CURVE 2	
54.39	3.596 x 10^{-2}
58.81	4.114
63.23	4.736
67.51	5.417
71.77	6.218
75.32	7.051
75.94	7.272
77.28	7.652
78.28	8.121
79.33	7.804
79.98	6.759
80.39	5.997
81.92	5.784
83.68	5.737*
84.27	5.734*
95.17	6.136
104.60	6.582
114.70	7.203
124.51	7.798
135.88	8.489
146.36	9.109
155.68	9.648
166.07	1.023 x 10^{-1}
176.08	1.078
185.97	1.129
196.09	1.177
206.31	1.225
216.43	1.270
CURVE 2 (cont.)	
228.60	1.323 x 10^{-1}
237.35	1.359
246.09	1.397
256.10	1.442
266.35	1.491
276.01	1.530
286.36	1.581
296.25	1.621*
298.16	1.630

*Not shown on plot

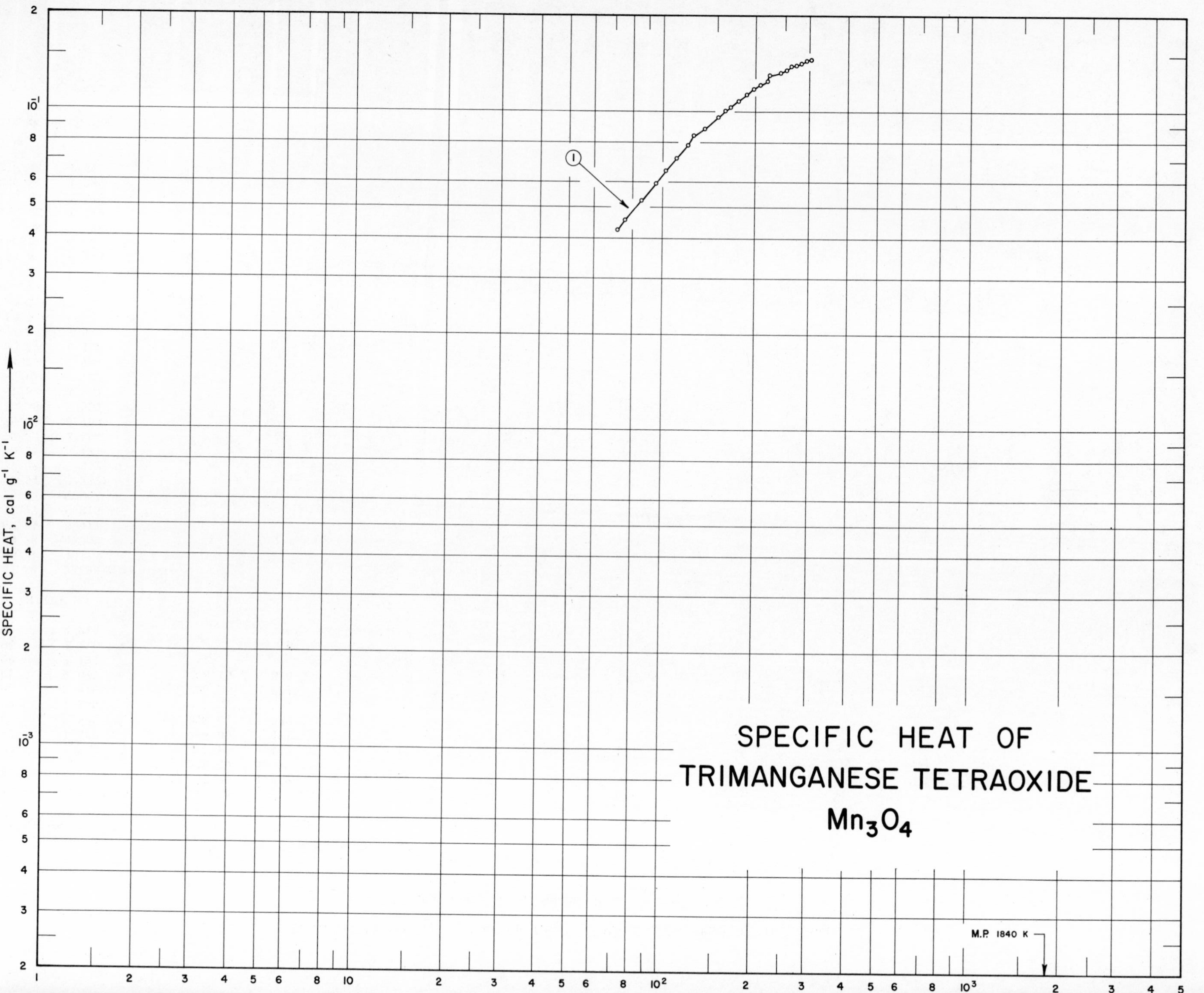
SPECIFIC HEAT OF
TRIMANGANESE TETRAOXIDE
Mn_3O_4
M.P. 1840 K
SPECIFIC HEAT, cal g^{-1} K^{-1}
1

SPECIFICATION TABLE NO. 46 SPECIFIC HEAT OF TRIMANGANESE TETRAOXIDE Mn_3O_4

[For Data Reported in Figure and Table No. 46]

Curve No.	Ref. No.	Year	Temp. Range, K	Reported Error, %	Name and Specimen Designation	Composition (weight percent), Specifications and Remarks
1	99	1928	72-305			Finely crystalline; perpared from C. P. grade MnS.

DATA TABLE NO. 46 SPECIFIC HEAT OF TRIMANGANESE TETRAOXIDE Mn_3O_4

[Temperature, T, K; Specific Heat, C_p, Cal g^{-1} K^{-1}]

T	C_p
CURVE 1	
72.2	4.250 x 10^{-2}
76.0	4.571
85.5	5.292
95.6	5.978
103.6	6.564
111.9	7.189
121.5	7.875
127.8	8.413
138.2	8.867
151.9	9.675
160.0	1.009 x 10^{-1}
167.4	1.047
177.6	1.097
188.0	1.148
199.1	1.195
200.4	1.196
209.4	1.234
220.0	1.256
223.2	1.318
241.4	1.341
252.5	1.367
263.1	1.401
271.5	1.418
283.4	1.432
295.3	1.457
305.2	1.462

*Not shown on plot

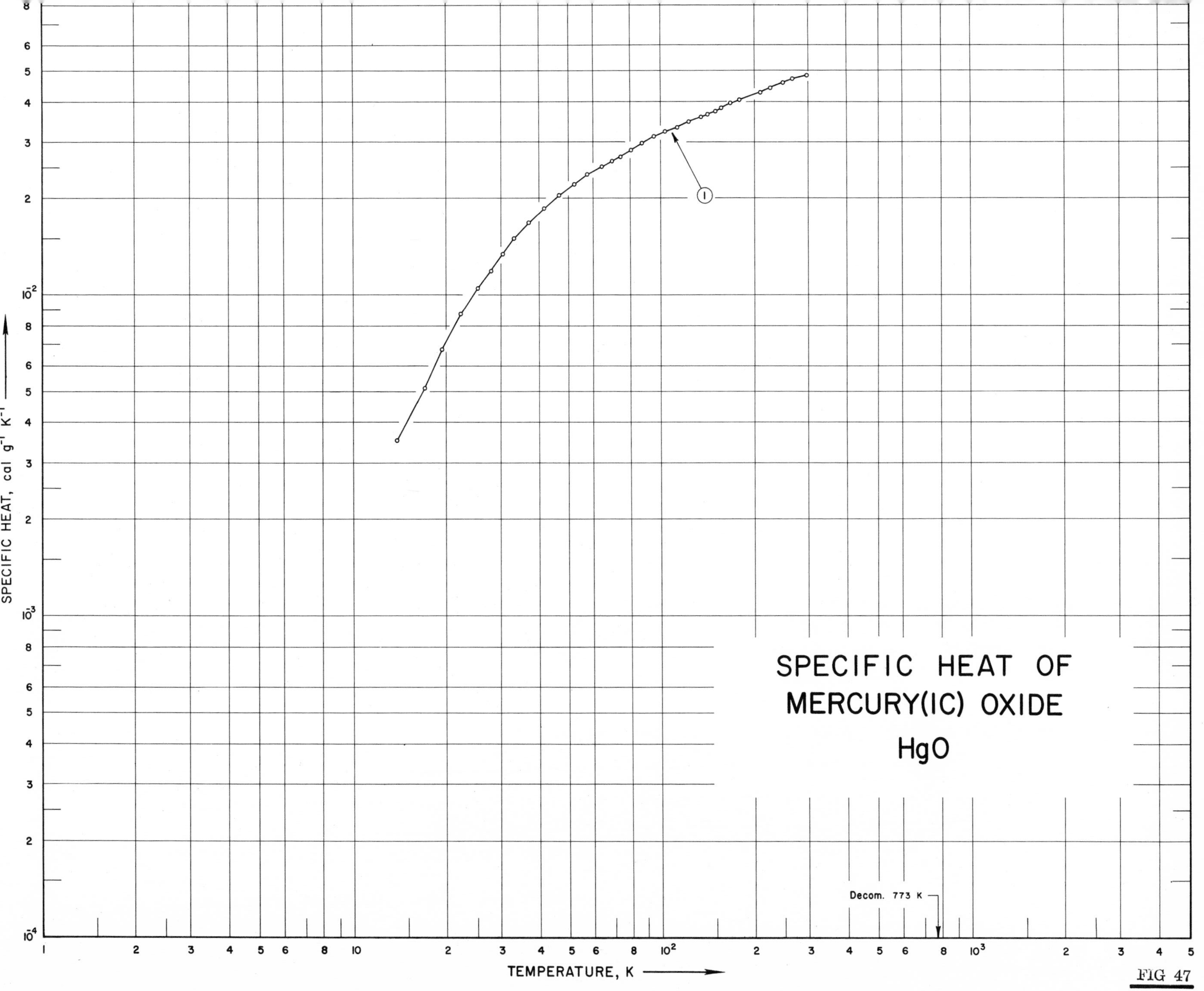
SPECIFIC HEAT OF
MERCURY(IC) OXIDE
HgO
1
Decom. 773 K
TEMPERATURE, K
SPECIFIC HEAT, cal g^{-1} K^{-1}

FIG 47

SPECIFICATION TABLE NO. 47 SPECIFIC HEAT OF MERCURY (IC) OXIDE HgO

[For Data Reported in Figure and Table No. 47]

Curve No.	Ref. No.	Year	Temp. Range, K	Reported Error, %	Name and Specimen Designation	Composition (weight percent), Specifications and Remarks
1	105	1953	15-298	≤0.5		Red modification, 0.020 insol. HCl, 0.010 nonvolatile matter, 0.001 Cl_2, 0.004 Fe, 0.005 SO_4, and 0.003 total N_2.

DATA TABLE NO. 47 SPECIFIC HEAT OF MERCURY (IC) OXIDE HgO

[Temperature, T, K; Specific Heat, C_p, Cal g^{-1} K^{-1}]

T	C_p
CURVE 1	
14.89	3.509 x 10^{-3}
17.06	5.106
19.45	6.759
22.33	8.702
25.41	1.051 x 10^{-2}
28.10	1.192
30.69	1.346
33.49	1.500
37.24	1.676
41.78	1.867
46.96	2.047
52.24	2.210
57.58	2.380
58.36	2.374*
64.59	2.513
69.17	2.610
74.56	2.700
80.84	2.840
87.52	2.962
95.56	3.111
104.34	3.229
114.42	3.331
124.94	1.464
136.03	3.594
142.70	3.667
147.34	3.721*
150.84	3.734
158.15	3.850
169.51	3.958
180.87	4.055
210.58	4.295
216.19	4.330*
227.55	4.418
240.44	4.493
240.62	4.498
250.73	4.599
250.86	4.579*
262.10	4.680*
268.37	4.703
278.18	4.758*
291.23	4.824*
298.24	4.840

*Not shown on plot

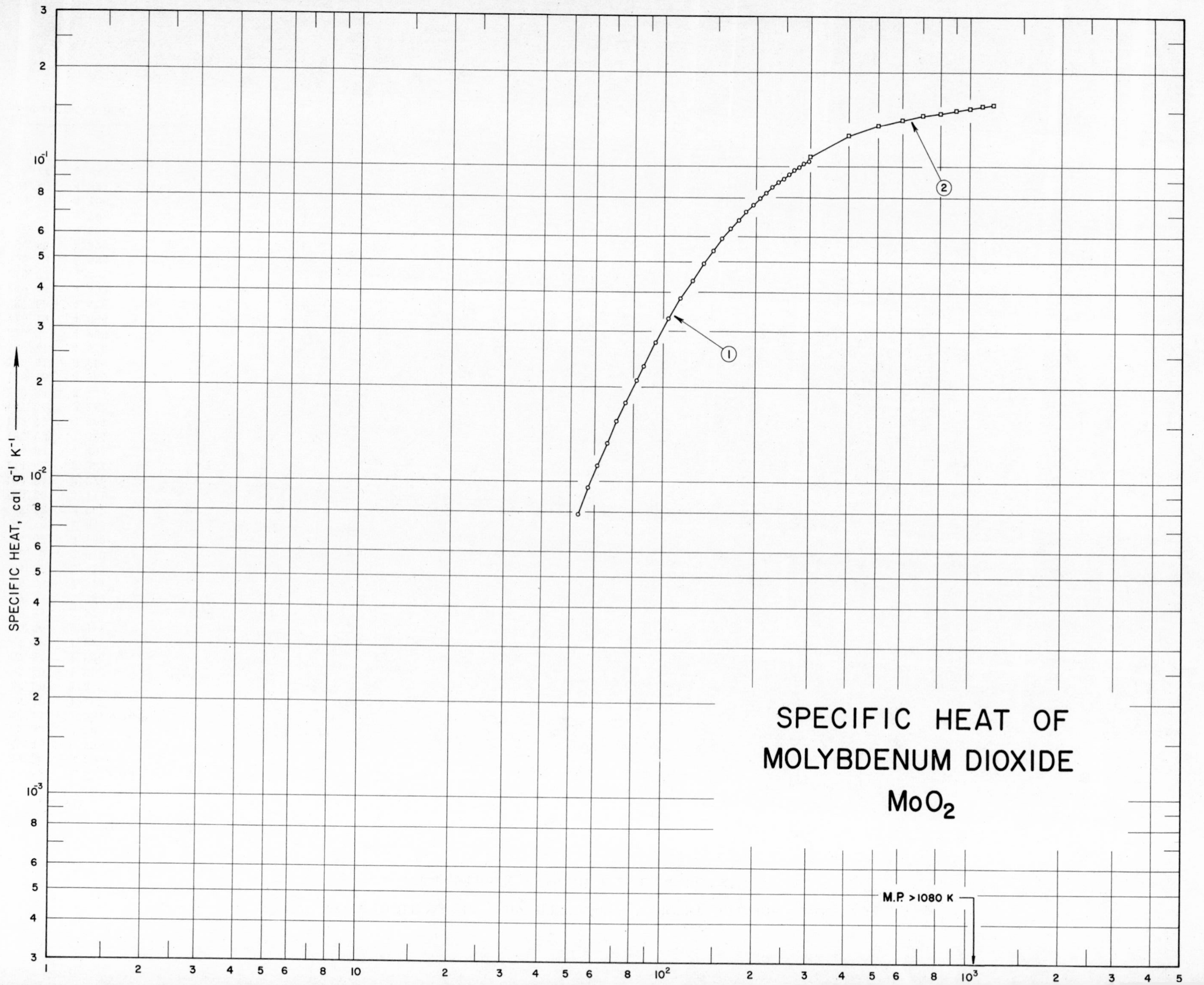
SPECIFIC HEAT OF
MOLYBDENUM DIOXIDE
MoO_2
M.P. >1080 K
SPECIFIC HEAT, cal g^{-1} K^{-1}
1
2

SPECIFICATION TABLE NO. 48 SPECIFIC HEAT OF MOLYBDENUM DIOXIDE MoO_2

[For Data Reported in Figure and Table No. 48]

Curve No.	Ref. No.	Year	Temp. Range, K	Reported Error, %	Name and Specimen Designation	Composition (weight percent), Specifications and Remarks
1	75	1958	53-296			74. 99 Mo.
2	54	1965	300-1200	1. 8		Supplied by Climax Molybdenum Co.

DATA TABLE NO. 48 SPECIFIC HEAT OF MOLYBDENUM DIOXIDE MoO_2

[Temperature, T, K; Specific Heat, C_p, Cal g^{-1} K^{-1}]

T	C_p
CURVE 1	
53.31	7.871×10^{-3}
57.41	9.544
61.61	1.126×10^{-2}
66.05	1.321
70.92	1.552
75.82	1.778
82.14	2.096
86.86	2.328
94.94	2.757
104.96	3.291
114.84	3.810
125.00	4.349
135.77	4.906
145.61	5.403
155.99	5.898
165.87	6.353
175.90	6.776
186.08	7.202
196.06	7.570
206.29	7.949
215.51	8.270
226.01	8.613
236.26	8.918
245.67	9.168
256.32	9.458
266.21	9.723
276.10	9.958
286.48	1.021×10^{-1}
296.00	1.043*
CURVE 2	
300	1.072×10^{-1}
400	1.257
500	1.352
600	1.412
700	1.455
800	1.488
900	1.517
1000	1.542
1100	1.566
1200	1.587

*Not shown on plot

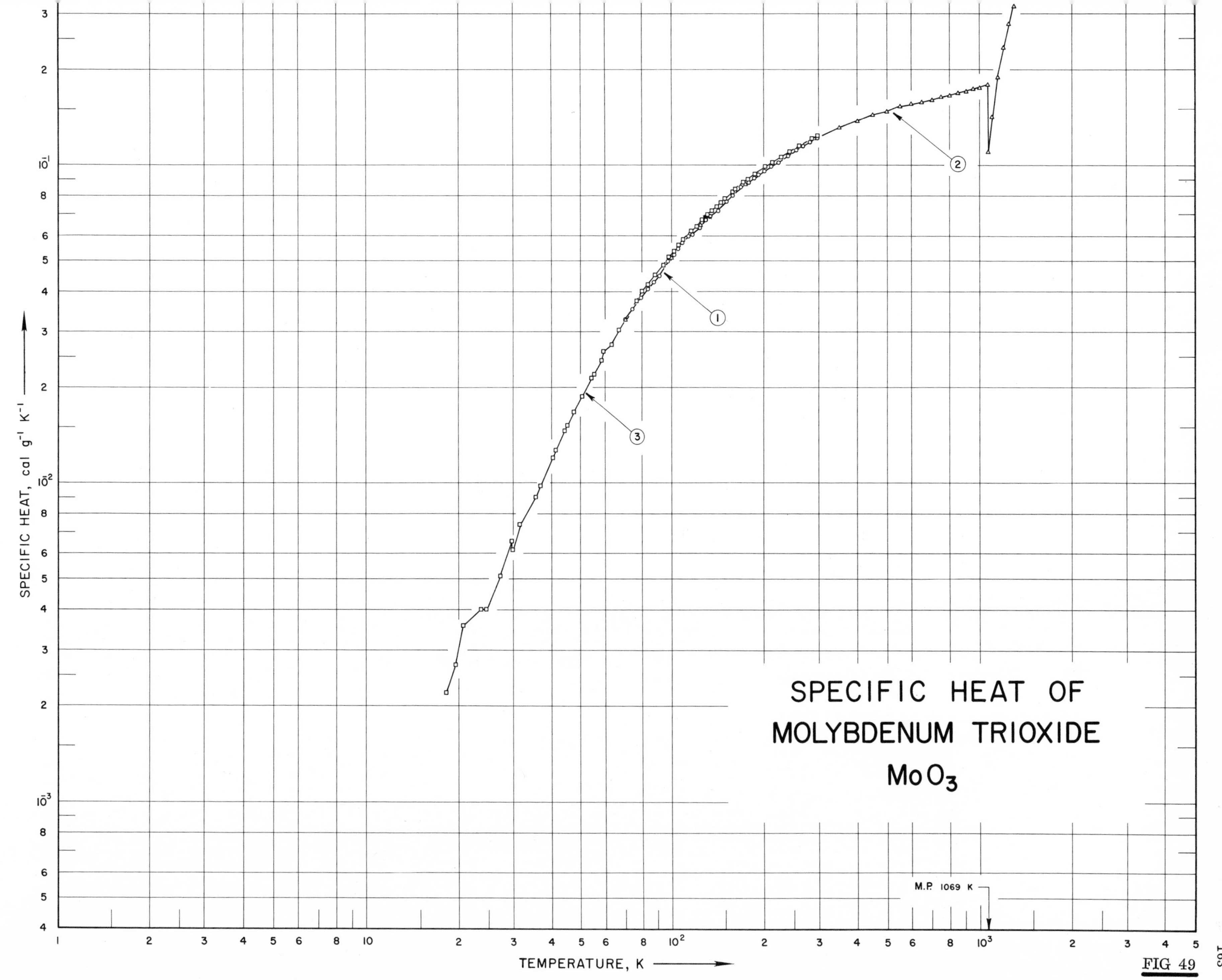

SPECIFIC HEAT OF
MOLYBDENUM TRIOXIDE
MoO_3
M.P. 1069 K
SPECIFIC HEAT, cal g⁻¹ K⁻¹
TEMPERATURE, K
1
2
3

FIG 49

SPECIFICATION TABLE NO. 49 SPECIFIC HEAT OF MOLYBDENUM TRIOXIDE MoO_3

[For Data Reported in Figure and Table No. 49]

Curve No.	Ref. No.	Year	Temp. Range, K	Reported Error, %	Name and Specimen Designation	Composition (weight percent), Specifications and Remarks
1	106	1943	70-299			99.9 MoO_3; small transparent rhombic crystal.
2	107	1953	70-1300			MoO_3; 66.8 ± 0.05 Mo, 0.005 non-volatile with HCl at 450 C, 0.001 insoluble in NH_3, and trace of heavy metals and alkaline metals.
3	108	1956	20-300			MoO_3, C.P. grade.

DATA TABLE NO. 49 SPECIFIC HEAT OF MOLYBDENUM TRIOXIDE MoO_3

[Temperature, T, K; Specific Heat, C_p, Cal g^{-1} K^{-1}]

T	C_p
CURVE 1	
70.05	3.286 x 10^{-2}
74.48	3.54
79.75	3.84
83.95	4.10
87.70	4.30
91.15	4.50
97.55	5.00
101.79	5.27
105.58	5.49
109.22	5.72
114.00	5.98
117.65	6.08
124.72	6.38
129.36	6.75
134.07	6.947
142.25	7.235
144.65	7.420*
151.50	7.725
154.87	7.864*
159.51	8.101
163.47	8.267*
169.45	8.545
179.40	8.879
182.50	9.087*
186.37	9.143
192.46	9.358
201.05	9.622
208.92	9.872*
212.87	9.969
220.05	1.018 x 10^{-1}*
224.12	1.027
230.39	1.048*
234.69	1.059*
240.15	1.079
250.67	1.109*
254.86	1.120
260.30	1.133*
268.20	1.153
280.32	1.181
294.72	1.216*
298.73	1.223

T	C_p
CURVE 2	
70	3.283 x 10^{-2}
75	3.565*
100	5.143
125	6.520
150	7.677*
175	8.740
225	1.035 x 10^{-1}*
250	1.104*
273.16	1.165*
275	1.170*
298.16	1.227*
300	1.226*
350	1.329
400	1.398
450	1.453
500	1.497
550	1.556
600	1.569
650	1.600
700	1.629
750	1.656
800	1.683
850	1.708
900	1.732
950	1.756
1000	1.779
1050	1.801*
(s) 1068.36	1.810
(l) 1068.36	1.113
1100	1.440
1150	1.920
1200	2.375
1250	2.806
1300	3.212

T	C_p
CURVE 3	
Series 1	
94.01	4.875 x 10^{-2}
98.19	5.133
102.18	5.387
106.02	5.619
109.73	5.837
113.31	6.043*
116.79	6.232
121.10	6.469
126.20	6.750
131.10	7.006
135.85	7.257
140.49	7.477
145.03	7.695
149.46	7.898
153.79	8.087*
158.04	8.280
162.22	8.452
167.21	8.649*
172.98	8.882
178.65	9.089
176.10	9.007*
181.69	9.215*
187.18	9.408
192.58	9.606*
197.89	9.779*
203.13	9.940
208.28	1.013 x 10^{-1}*
213.38	1.027
218.34	1.043*
223.30	1.058*
228.21	1.069
233.06	1.083*
237.86	1.097*
242.62	1.109
Series 2	
241.77	1.107 x 10^{-1}*
246.45	1.120*
251.10	1.131*
255.70	1.143*

T	C_p
CURVE 3 (cont.)	
Series 2 (cont.)	
260.27	1.154 x 10^{-1}
264.79	1.165*
269.28	1.178*
273.74	1.188*
278.15	1.200*
282.54	1.208*
286.89	1.219
291.20	1.230*
295.49	1.239*
299.76	1.246
Series 3	
55.25	2.204 x 10^{-2}
58.72	2.448
63.10	2.775
67.02	3.050
70.62	3.303*
73.97	3.525*
77.12	3.750
80.11	4.025
83.83	4.216
88.54	4.535
92.96	4.825*
97.17	5.071*
101.19	5.332*
Series 4	
19.60	2.69 x 10^{-3}
24.65	4.02
27.47	5.11
29.89	6.59
31.86	7.406
35.64	9.039
40.69	1.202 x 10^{-2}
44.48	1.459
47.72	1.685
50.56	1.881
54.53	2.152
59.43	2.605

T	C_p
CURVE 3 (cont.)	
Series 5	
18.27	2.20 x 10^{-3}
20.86	3.58
23.68	4.03
30.02	6.18
36.99	9.851
41.58	1.269 x 10^{-2}
45.27	1.520
48.36	1.731*
51.12	1.917*
55.02	2.184*

*Not shown on plot

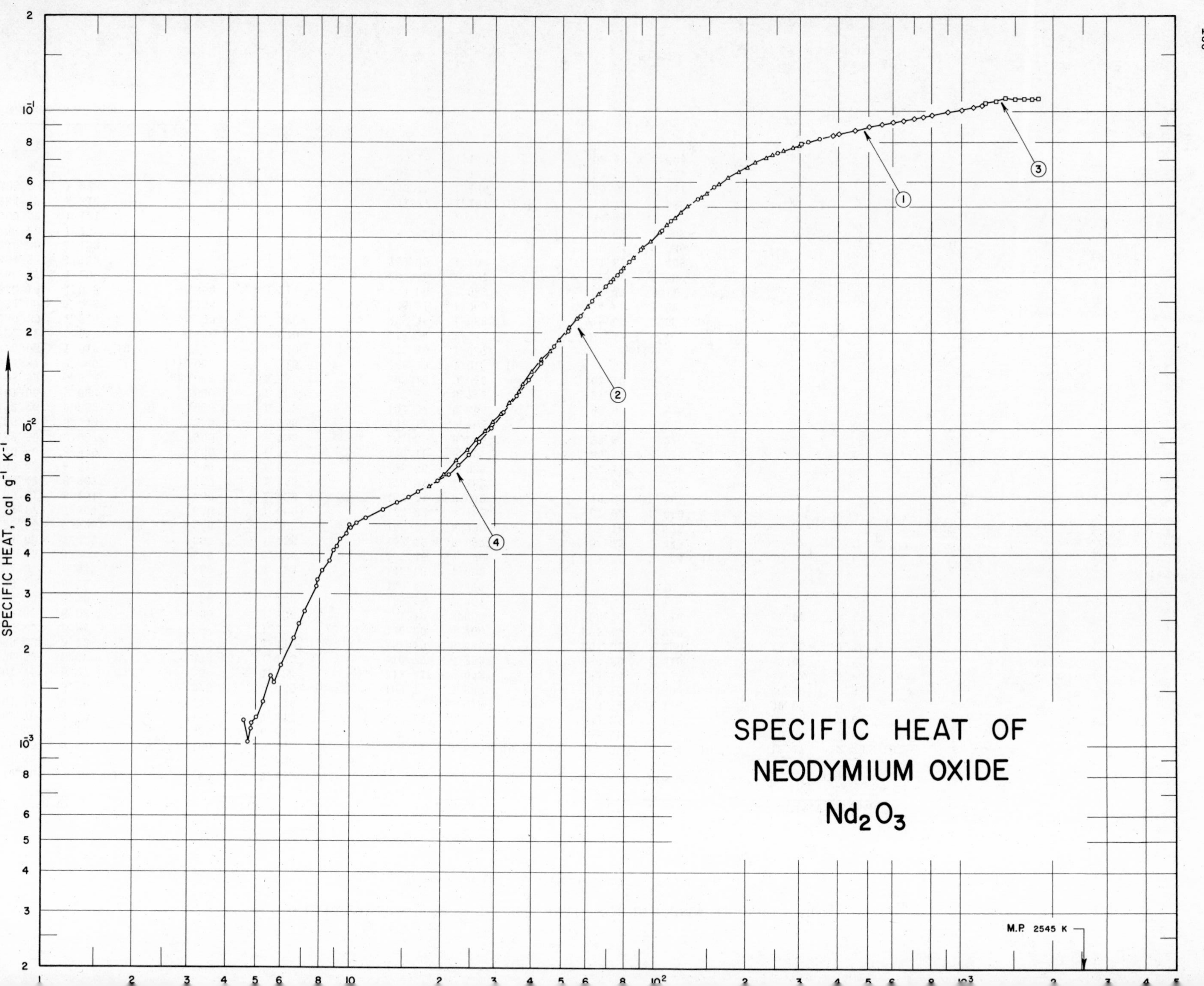
SPECIFIC HEAT OF
NEODYMIUM OXIDE
Nd_2O_3
M.P. 2545 K
SPECIFIC HEAT, cal g^{-1} K^{-1}
1
2
3
4
10^{-1}
10^{-2}
10^{-3}
10
10^2
10^3

SPECIFICATION TABLE NO. 50 SPECIFIC HEAT OF NEODYMIUM OXIDE Nd_2O_3

[For Data Reported in Figure and Table No. 50]

Curve No.	Ref. No.	Year	Temp. Range, K	Reported Error, %	Name and Specimen Designation	Composition (weight percent), Specifications and Remarks
1	82	1951	383-1171			
2	83	1958	18-298			99. 9 Nd_2O_3, <0. 1 Pr_5O_{11}, and <0. 1 Sm_2O_3; sample supplied by the Lindsay Chemical Co. ; heated to constant weight at 950 C for 24 hrs in air to decompose hydroxides or carbonates.
3	73	1962	298-1795	0. 2		99. 9 Nd_2O_3; hexagonal; measured in helium atmosphere.
4	72	1962	5-346	0. 1		99. 9 Nd_2O_3; sample supplied by the Lindsay Chemical Co; pelleted under 2000 - 4000 psi; fired at 1170 K; measured in helium atmosphere.

DATA TABLE NO. 50 SPECIFIC HEAT OF NEODYMIUM OXIDE, Nd_2O_3

[Temperature, T, K; Specific Heat, C_p, Cal $g^{-1}K^{-1}$]

T	C_p
CURVE 1	
383	8.427×10^{-2}
400	8.526
450	8.774
500	8.976
550	9.147
600	9.298
650	9.434
700	9.560
750	9.678
800	9.790
850	9.898*
900	1.000×10^{-1}
950	1.010*
1000	1.020
1050	1.030*
1100	1.039
1150	1.049*
1171	1.053
CURVE 2	
18.26	6.565×10^{-3}
20.28	7.198
22.34	7.944
24.27	8.586
26.05	9.231
27.78	9.831
29.43	1.053×10^{-2}
31.37	1.116
33.43	1.208
35.71	1.303
36.94	1.394
37.70	1.422*
39.43	1.517
40.25	1.557*
42.46	1.660
42.86	1.677*
45.58	1.762
48.91	1.908
52.63	2.091
CURVE 2 (cont.)	
55.47	2.241×10^{-2}
56.72	2.303*
60.47	2.443
60.97	2.482*
65.67	2.676
72.70	2.997
78.53	3.210
84.78	3.481
90.89	3.752
96.86	3.924
103.38	4.198
109.57	4.423
115.67	4.639
121.88	4.872
128.14	5.058
134.41	5.227*
141.00	5.413
147.30	5.593
161.32	5.949
167.68	6.075*
173.97	6.222
180.27	6.346*
186.97	6.499
193.54	6.632*
200.02	6.714
206.45	6.835*
212.45	6.925
222.88	7.056*
229.88	7.191
236.50	7.272*
243.29	7.349
250.12	7.381*
263.43	7.549
277.60	7.700*
284.76	7.788
291.49	7.843*
298.13	7.846
CURVE 3	
α 298	7.904×10^{-2}*
300	7.920
400	8.542*
500	8.944*
600	9.259*
700	9.532*
800	9.784*
900	1.002×10^{-1}*
1000	1.026*
1100	1.048*
1200	1.070
1300	1.092
α1395	1.113
β1395	1.106*
1400	1.106*
1500	1.106
1500	1.106
1600	1.106
1700	1.106
β1795	1.106
CURVE 4	
Series 1	
71.90	2.928×10^{-2}
77.46	3.167
Series 2	
5.29	1.376×10^{-3}
Series 3	
4.84	1.183
5.59	1.661
6.60	2.187
7.87	3.320
9.10	4.265
CURVE 4 (cont.)	
Series 4	
4.83	1.129×10^{-3}
5.72	1.584
6.89	2.422
7.94	3.308*
8.89	4.122
9.98	4.963
11.39	5.207
12.88	5.561
14.27	5.858
15.56	6.075
16.78	6.330
18.05	6.494*
19.47	6.836
21.03	7.195
22.69	7.688
24.54	8.295
26.66	9.085
29.07	1.008×10^{-2}
31.95	1.131
35.14	1.277
38.59	1.436
42.45	1.613
47.02	1.824
Series 5	
4.72	1.022×10^{-3}
5.65	1.608*
6.83	2.386*
7.82	3.177
8.61	3.825
9.30	4.488
9.93	4.841*
10.51	5.014
11.08	5.186*
11.63	5.237*
12.17	5.391*
12.69	5.510*
13.50	5.751*
CURVE 4 (cont.)	
Series 6	
5.03	1.221×10^{-3}
6.06	1.789
7.19	2.648
8.16	3.554
9.14	4.413*
9.47	4.663*
9.79	4.666
10.10	4.856
10.39	4.984*
10.69	4.966*
10.98	5.082*
12.20	5.406*
13.83	5.769*
15.03	5.994*
16.41	6.244*
Series 7	
4.59	1.207
5.58	1.551*
6.76	2.294*
7.81	3.183*
8.66	3.908*
9.37	4.562*
10.99	5.141*
12.11	5.326*
13.39	5.694*
14.71	5.947*
16.11	6.149*
17.69	6.452*
19.39	6.794*
21.21	7.272*
Series 8	
42.54	1.617×10^{-2}*
46.67	1.808*
51.11	2.014*
56.13	2.240*
61.43	2.480*
CURVE 4 (cont.)	
Series 9	
52.04	2.054×10^{-2}
57.15	2.286
62.66	2.538
69.03	2.812
75.60	3.085
82.55	3.388
89.96	3.694
97.40	3.965*
105.32	4.256
113.45	4.544
121.56	4.823*
129.91	5.082*
139.08	5.367*
138.21	5.344
146.85	5.587*
155.66	5.813
164.47	6.030*
173.55	6.235*
182.78	6.428*
191.79	6.604*
200.81	6.767*
209.92	6.907*
218.98	7.044*
227.75	7.165*
226.66	7.154*
235.82	7.275*
244.88	7.388*
253.83	7.486
263.04	7.590*
272.47	7.685*
281.77	7.769*
291.02	7.852*
300.13	2.929*
309.16	7.995
318.44	8.063
327.93	8.131*
337.34	8.194*
346.43	8.256

*Not shown on Plot

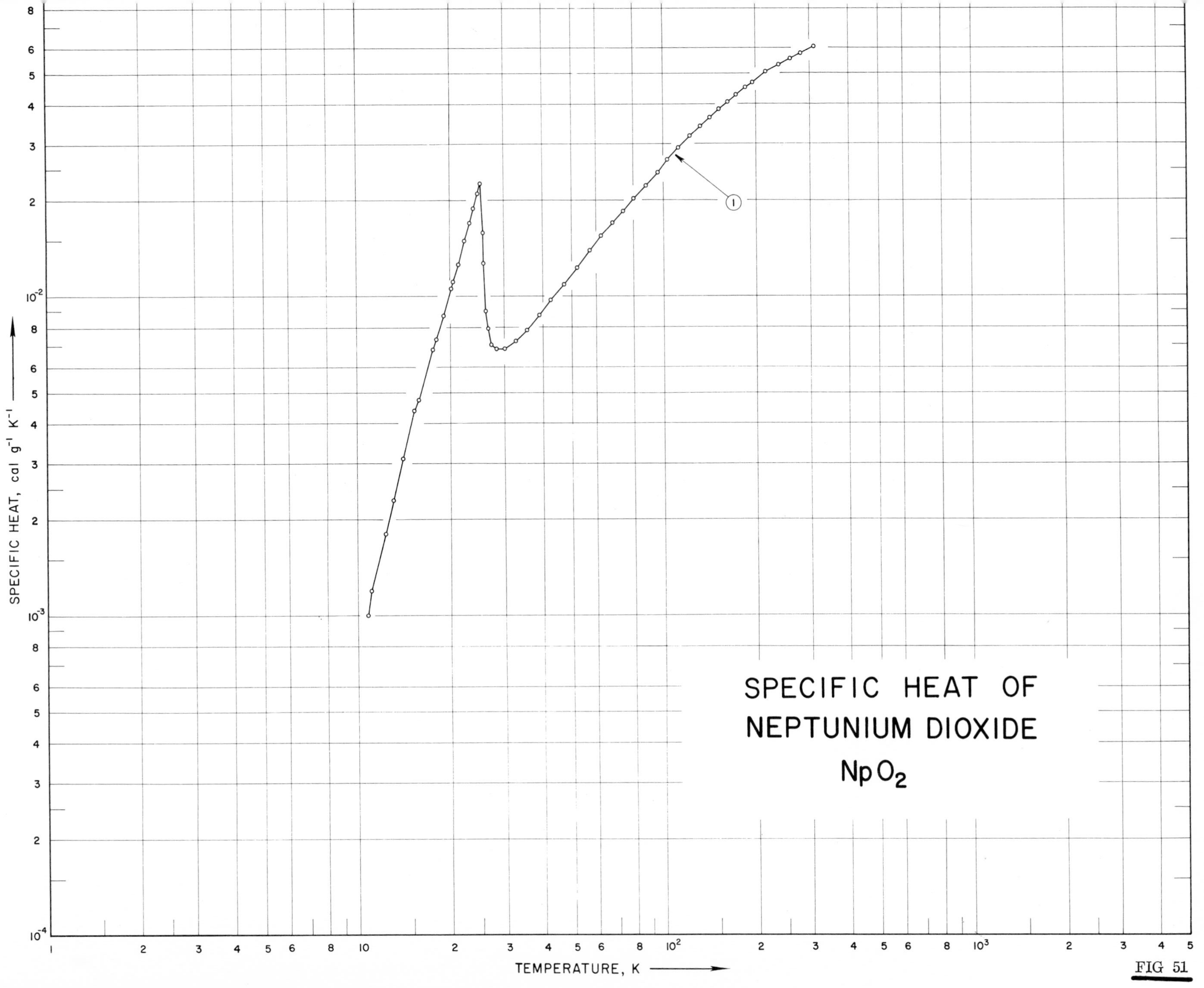
SPECIFIC HEAT OF
NEPTUNIUM DIOXIDE
NpO_2
SPECIFIC HEAT, cal g^{-1} K^{-1}
TEMPERATURE, K
1

FIG 51

SPECIFICATION TABLE NO. 51 SPECIFIC HEAT OF NEPTUNIUM DIOXIDE NpO_2

[For Data Reported in Figure and Table No. 51]

Curve No.	Ref. No.	Year	Temp. Range, K	Reported Error, %	Name and Specimen Designation	Composition (weight percent), Specifications and Remarks
1	109	1953	10-312			99. 9^+ NpO_2, < 0. 1 total Cr, Fe, and Ca; Np^{237} prepared by U^{238} (n, 2n) U^{237} β^{-1} Np^{237}, hydroxide precipitated from acid solution and ignited to constant weight in air in Pt boat at 700 C.

DATA TABLE NO. 51 SPECIFIC HEAT OF NEPTUNIUM DIOXIDE, NpO_2

[Temperature, T, K; Specific Heat, C_p, Cal $g^{-1}K^{-1}$]

T	C_p	T	C_p
CURVE 1		CURVE 1 (cont.)	
Series 1		Series 2 (cont.)	
10.82	1.0×10^{-3}	25.71	1.59×10^{-2}
12.44	1.8	26.08	8.99×10^{-3}
14.09	3.1	26.52	7.95*
15.90	4.75	27.13	7.25*
17.67	6.80	27.99	6.95*
19.17	8.70	52.43	1.24×10^{-2}*
20.39	1.06×10^{-2}	57.52	1.39
21.42	1.26	62.78	1.54
22.44	1.49	68.40	1.70
23.29	1.70	74.18	1.85
23.98	1.88	80.73	2.02
24.63	2.10	88.10	2.22
25.21	2.26	96.52	2.44
25.84	1.27	104.79	2.67
26.60	7.91×10^{-3}	113.97	2.91
27.38	7.06	123.87	3.17
28.41	6.84	133.28	3.40
30.13	6.84	143.35	3.62
32.71	7.24	153.87	3.854
35.62	7.84	164.38	4.077
39.07	8.77	174.89	4.289
42.99	9.74	185.78	4.504
47.44	1.09×10^{-2}	196.70	4.683
52.27	1.23	207.36	4.869*
		217.73	5.010
Series 2		228.55	5.155*
		239.12	5.300
11.08	1.2×10^{-3}	249.80	5.374*
13.12	2.3	260.69	5.526
15.40	4.39	271.48	5.682*
18.10	7.32	282.04	5.779
20.54	1.11×10^{-2}	292.42	5.846*
22.47	1.52*	302.64	5.868*
23.88	1.87*	312.69	6.002
24.76	2.15*		
25.27	2.26*		

*Not shown on Plot

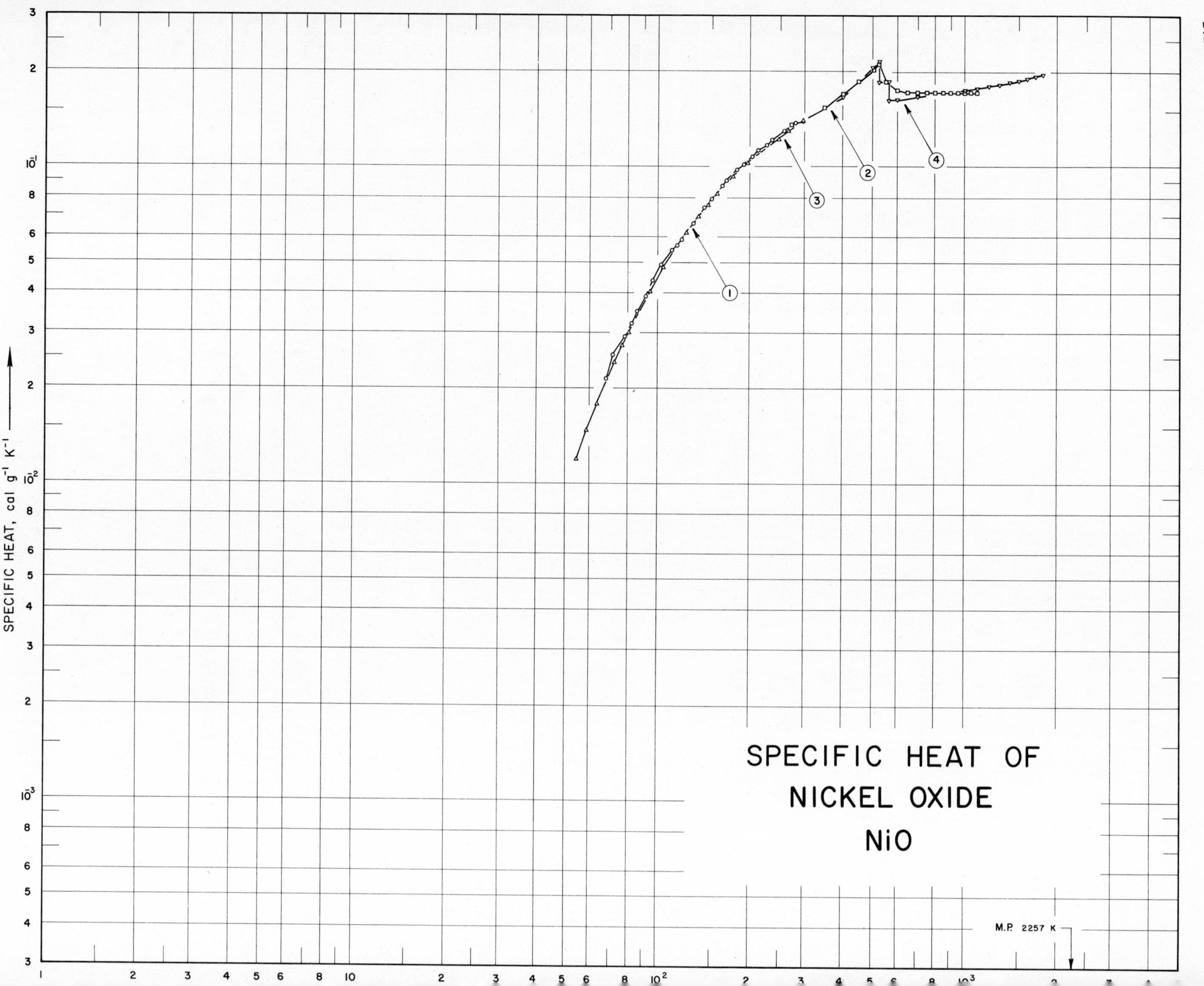
SPECIFIC HEAT OF
NICKEL OXIDE
NiO
M.P. 2257 K
SPECIFIC HEAT, cal g⁻¹ K⁻¹
1
2
3
4

SPECIFICATION TABLE NO. 52 SPECIFIC HEAT OF NICKEL OXIDE NiO

[For Data Reported in Figure and Table No. 52]

Curve No.	Ref. No.	Year	Temp. Range, K	Reported Error, %	Name and Specimen Designation	Composition (weight percent), Specifications and Remarks
1	110	1940	68-297			<0. 2 impurities; transparent cubic crystals.
2	111	1955	273-1100			78. 51 - 78. 54 Ni and 0. 01 - 0. 1 Si; prepared by decomposing $Ni(NO_3)_2 \cdot 6\ H_2O$ and heating 8 hrs at 1000 C.
3	63	1957	54-296			99. 96 NiO, 0. 05 CoO, 0. 02 acid insoluble, and 0. 01 Na_2O; prepared from reagent grade hexahydrate of nickelous nitrate and nickelous sulfate.
4	64	1958	298-1800	0. 4		99. 96 NiO, 0. 05 CoO, 0. 01 No_2O, and 0. 02 acid insoluble; prepared from reagent grade nickelous nitrate hexahydrate and nickelous sulfate hexahydrate.

DATA TABLE NO. 52 SPECIFIC HEAT OF NICKEL OXIDE, NiO

[Temperature, T, K; Specific Heat, C_p, Cal $g^{-1}K^{-1}$]

T	C_p
CURVE 1	
68.05	2.151 x 10^{-2}
71.96	2.555
78.19	2.925
82.07	3.216
85.89	3.514
86.70	3.539*
91.08	3.918
94.42	4.141*
96.32	4.414
103.13	4.943
111.70	5.480
115.71	5.678
120.33	5.975
131.10	6.69
141.72	7.47
150.08	7.94
151.86	8.06*
162.37	8.75
168.09	9.10
172.61	9.40*
181.32	9.85
183.01	1.00 x 10^{-1}*
190.75	1.02
204.20	1.09
213.13	1.14
225.34	1.18
235.37	1.22
247.81	1.27*
257.49	1.31
277.20	1.374*
280.32	1.396
296.68	1.413
CURVE 2	
273.16	1.357 x 10^{-1}
300	1.414*
350	1.547
400	1.700
450	1.865
500	2.036
523.16	2.117
550	1.86
600	1.75
650	1.728
700	1.728
750	1.728
800	1.728
850	1.728
900	1.728
950	1.728
1000	1.728
1050	1.728
1100	1.728
CURVE 3	
54.28	1.20 x 10^{-2}
58.93	1.488
63.49	1.795
68.10	2.117*
72.47	2.433
76.74	2.745
80.22	3.009
83.81	3.271*
94.66	4.090
104.97	4.871
114.57	5.566*
CURVE 3 (cont.)	
124.60	6.279 x 10^{-2}
135.88	7.053
145.55	7.679
155.86	8.327
165.78	8.886*
175.84	9.424
185.78	9.941*
195.94	1.042 x 10^{-1}
206.13	1.089*
216.19	1.132*
225.85	1.173*
236.07	1.211*
245.63	1.246
256.29	1.285*
266.15	1.318
276.01	1.349*
286.43	1.381*
295.94	1.412
CURVE 4	
α298.15	1.418 x 10^{-1}*
300	1.420*
400	1.670
500	2.055
α525	2.162
β525	1.858
550	1.858*
β565	1.858
γ565	1.639
600	1.648
700	1.675
800	1.702*
CURVE 4 (cont.)	
900	1.729 x 10^{-1}*
1000	1.756
1100	1.783
1200	1.810
1300	1.837
1400	1.864
1500	1.891
1600	1.918
1700	1.945
γ1800	1.972

*Not shown on Plot

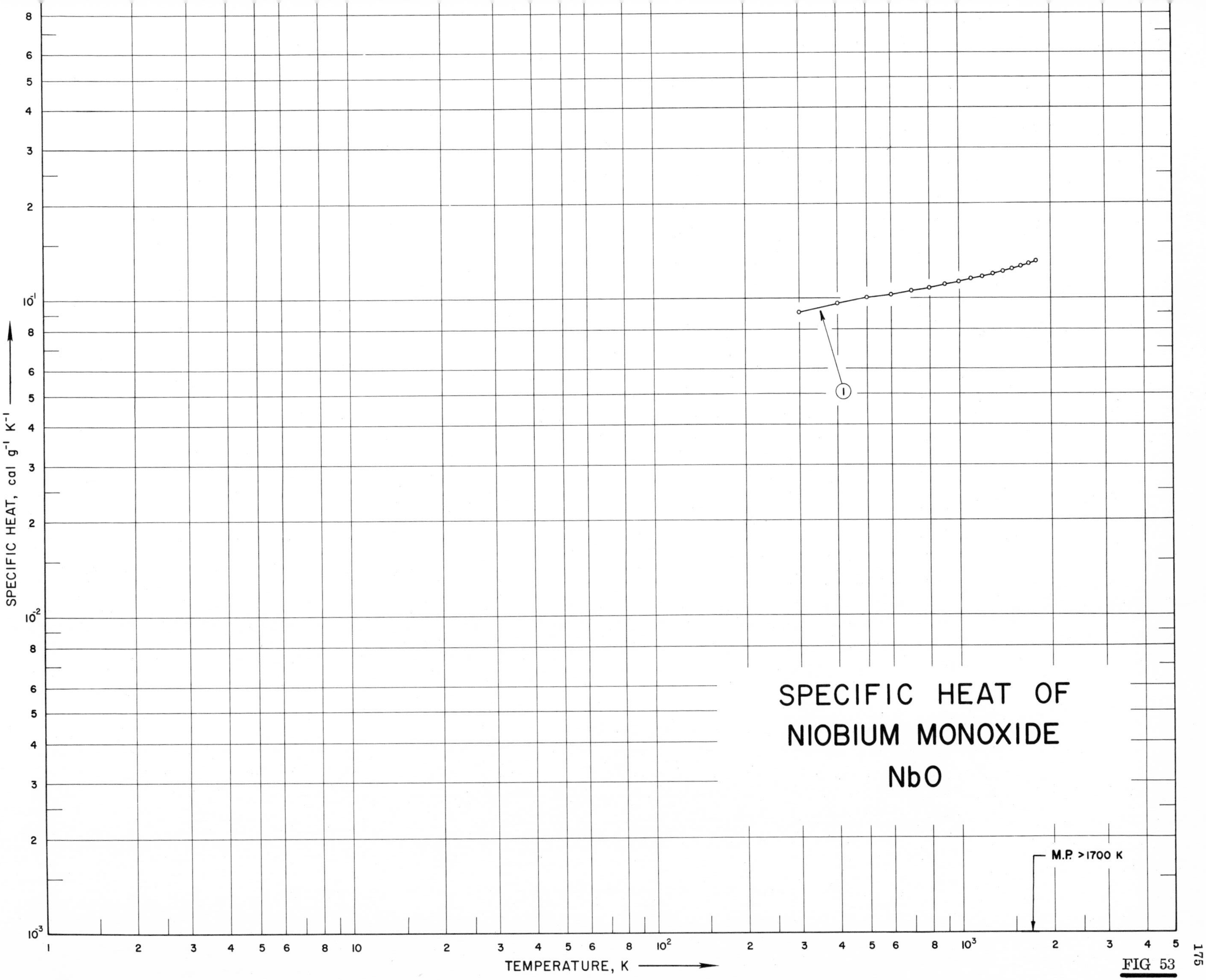

SPECIFIC HEAT OF
NIOBIUM MONOXIDE
NbO
SPECIFIC HEAT, cal g^{-1} K^{-1}
TEMPERATURE, K
M.P. >1700 K
1

FIG 53

SPECIFICATION TABLE NO. 53 SPECIFIC HEAT OF NIOBIUM MONOXIDE NbO

[For Data Reported in Figure and Table No. 53]

Curve No.	Ref. No.	Year	Temp. Range, K	Reported Error, %	Name and Specimen Designation	Composition (weight percent), Specifications and Remarks
1	112	1960	300-1810			Prepared synthetically from high purity Nb_2O_5, carbothermic Nb and carbon black.

DATA TABLE NO. 53 SPECIFIC HEAT OF NIOBIUM MONOXIDE, NbO

[Temperature, T, K; Specific Heat, C_p, Cal $g^{-1}K^{-1}$]

T	C_p
CURVE 1	
300	9.063×10^{-2}
400	9.632
500	1.001×10^{-1}
600	1.031
700	1.059
800	1.084
900	1.107
1000	1.130
1100	1.153
1200	1.176
1300	1.198
1400	1.220
1500	1.242
1600	1.264
1700	1.286
1800	1.308

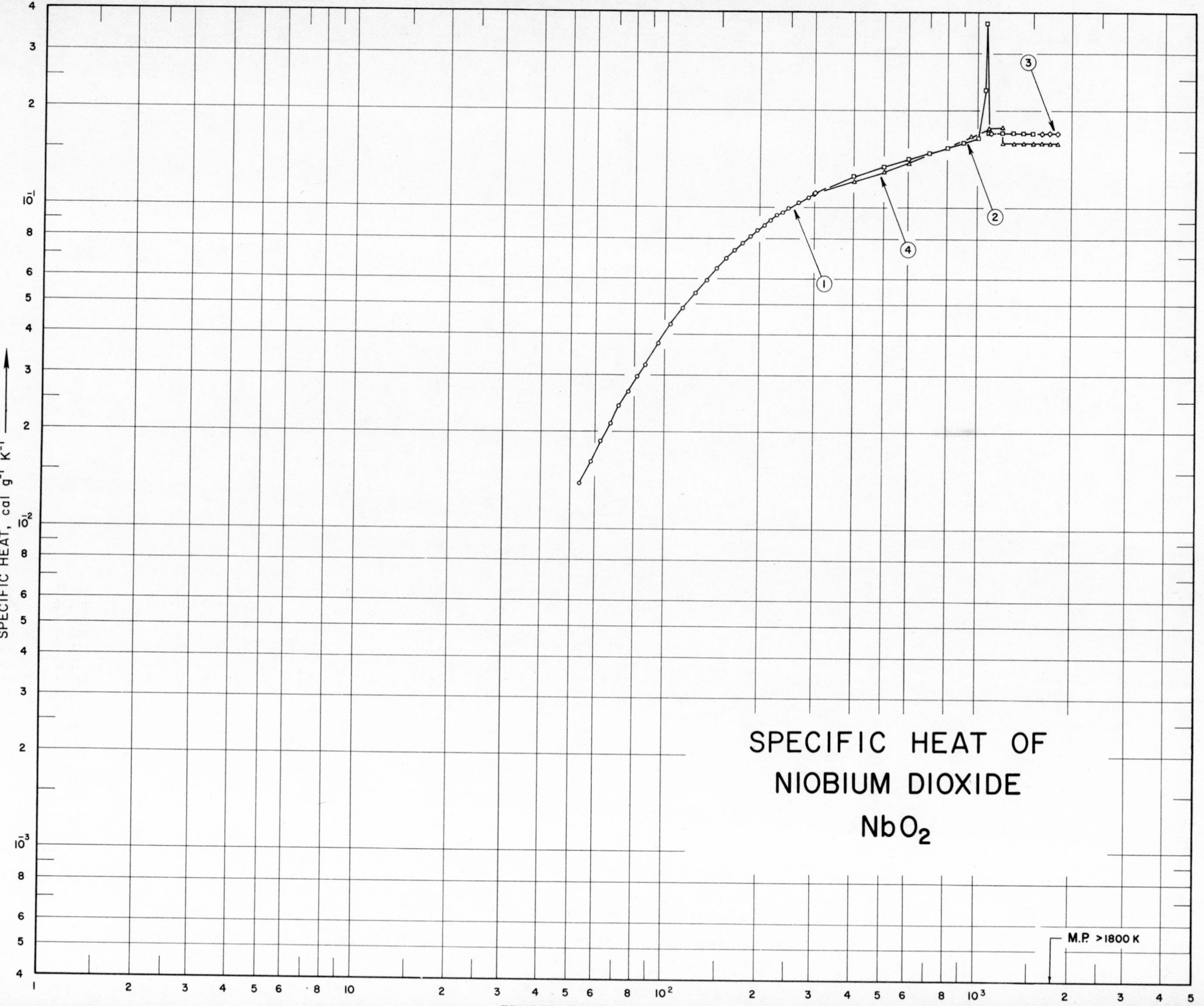
SPECIFIC HEAT OF
NIOBIUM DIOXIDE
NbO_2
M.P. >1800 K
SPECIFIC HEAT, cal g^{-1} K^{-1}
1
2
3
4

SPECIFICATION TABLE NO. 54 SPECIFIC HEAT OF NIOBIUM DIOXIDE NbO_2

[For Data Reported in Figure and Table No. 54]

Curve No.	Ref. No.	Year	Temp. Range, K	Reported Error, %	Name and Specimen Designation	Composition (weight percent), Specifications and Remarks
1	75	1958	53-296			99. 9 NbO_2.
2	113	1960	298-1500	1		74. 42 Nb.
3	112	1960	300-1800			
4	57	1961	298-1800	0. 2		99. 90 NbO_2; prepared by reduction of niobium pentoxide with hydrogen for 4 hrs at 950 - 1000 C; given 4 hrs more treatment in hydrogen at 950-1000 C.

DATA TABLE NO. 54 SPECIFIC HEAT OF NIOBIUM DIOXIDE, NbO_2

[Temperature, T, K; Specific Heat, C_p, Cal $g^{-1}K^{-1}$]

T	C_p	T	C_p	T	C_p
CURVE 1		**CURVE 2 (cont.)**		**CURVE 4 (cont.)**	
53.45	1.387 x 10^{-2}	1000	1.649 x 10^{-1}	β1090	1.777 x 10^{-1}
58.11	1.622	1010	1.654*	1100	1.777*
62.58	1.874	1059	2.330	β1200	1.777
67.17	2.135	1069	3.716	γ1200	1.589
71.77	2.405	1080	1.704	1300	1.589
76.56	2.676	1100	1.704*	1400	1.589
81.56	2.977	1200	1.704	1500	1.589
86.26	3.249	1300	1.704	1600	1.589
95.12	3.763	1400	1.704	1700	1.589
104.85	4.320	1500	1.704	γ1800	1.589
114.69	4.856				
124.96	5.415	**CURVE 3**			
136.02	5.966				
145.71	6.448	300	1.108 x 10^{-1}		
156.03	6.913	400	1.251*		
165.95	7.330	500	1.344*		
176.26	7.735	600	1.417*		
185.96	8.118	700	1.481*		
196.07	8.462	800	1.540*		
206.31	8.799	900	1.595*		
216.49	9.103	1000	1.649*		
226.25	9.407	1100	1.704		
236.23	9.647	1200	1.704*		
245.71	9.904	1300	1.704*		
256.53	1.017 x 10^{-1}*	1400	1.704*		
266.63	1.038	1500	1.704*		
276.19	1.059*	1600	1.704		
286.82	1.078	1700	1.704		
296.30	1.100*	1800	1.704		
CURVE 2		**CURVE 4**			
298.15	1.104 x 10^{-1}*	α298.15	1.100 x 10^{-1}*		
300	1.108*	300	1.102*		
400	1.251	400	1.207		
500	1.344	500	1.296		
600	1.417	600	1.380		
700	1.481	700	1.461*		
800	1.539	800	1.540*		
900	1.595	900	1.618*		
		α950	1.657		

*Not shown on Plot

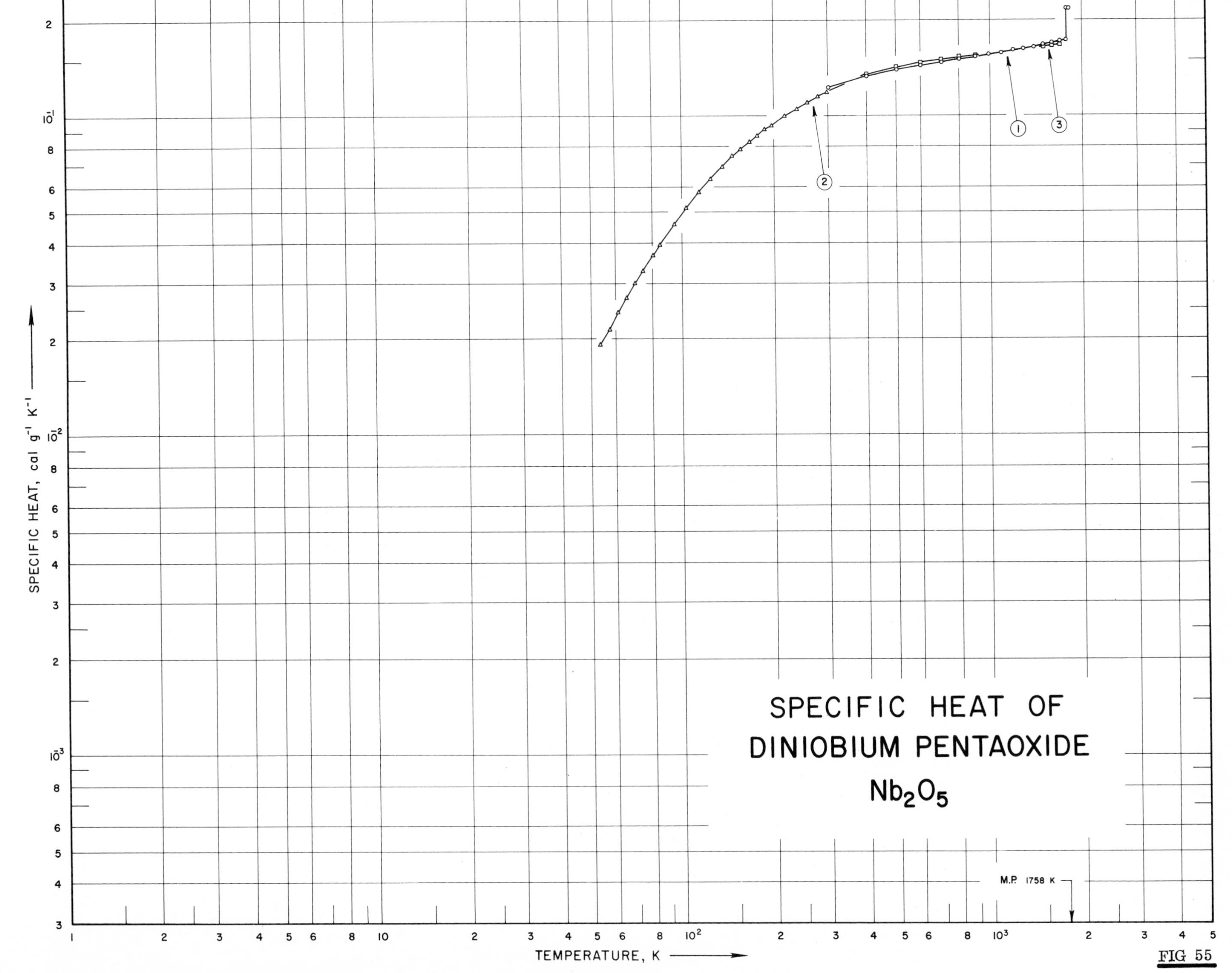
SPECIFIC HEAT OF
DINIOBIUM PENTAOXIDE
Nb_2O_5
M.P. 1758 K
TEMPERATURE, K
SPECIFIC HEAT, cal g⁻¹ K⁻¹
1
2
3

FIG 55

SPECIFICATION TABLE NO. 55 SPECIFIC HEAT OF DINIOBIUM PENTAOXIDE Nb_2O_5

[For Data Reported in Figure and Table No. 55]

Curve No.	Ref. No.	Year	Temp. Range, K	Reported Error, %	Name and Specimen Designation	Composition (weight percent), Specifications and Remarks
1	114	1953	300-1810	≤0. 5		0. 03 Si, <0. 05 Mg, and <0. 01 Ti; crystalline; heated to 1050 C before measurements.
2	103	1954	53-298			<0. 10 impurities.
3	112	1960	300-1700			

DATA TABLE NO. 55 SPECIFIC HEAT OF DINIOBIUM PENTAOXIDE Nb_2O_5

[Temperature, T, K; Specific Heat, C_p, Cal $g^{-1}K^{-1}$]

CURVE 1

T	C_p
300	1.223×10^{-1}
400	1.333
500	1.395
600	1.438
700	1.473
800	1.503
900	1.529
1000	1.555
1100	1.579
1200	1.602
1300	1.625
1400	1.647
1500	1.669
1600	1.691
1700	1.712
(s) 1785	1.731
(l) 1785	2.178
1800	2.178*
1810	2.178

CURVE 2

T	C_p
53.24	1.93×10^{-2}
57.12	2.15
61.25	2.42
65.58	2.71
70.06	3.00
74.52	3.28
80.17	3.66
84.45	3.94
94.75	4.60
104.36	5.18
114.54	5.79
124.62	6.35
136.03	6.94
146.39	7.44
155.98	7.86
166.28	8.30
176.36	8.67
186.11	9.05
196.13	9.32
206.15	9.68*

CURVE 2 (cont.)

T	C_p
216.60	1.00×10^{-1}
228.17	1.03*
236.41	1.05
245.82	1.074*
256.19	1.100
266.49	1.120*
276.76	1.145
286.50	1.165*
296.64	1.185*
298.16	1.187

CURVE 3

T	C_p
300	1.192×10^{-1}*
400	1.340
500	1.415
600	1.462
700	1.495
800	1.522
900	1.544
1000	1.564*
1100	1.582*
1200	1.599*
1300	1.615*
1400	1.630*
1500	1.646
1600	1.661
1700	1.676

*Not shown on Plot

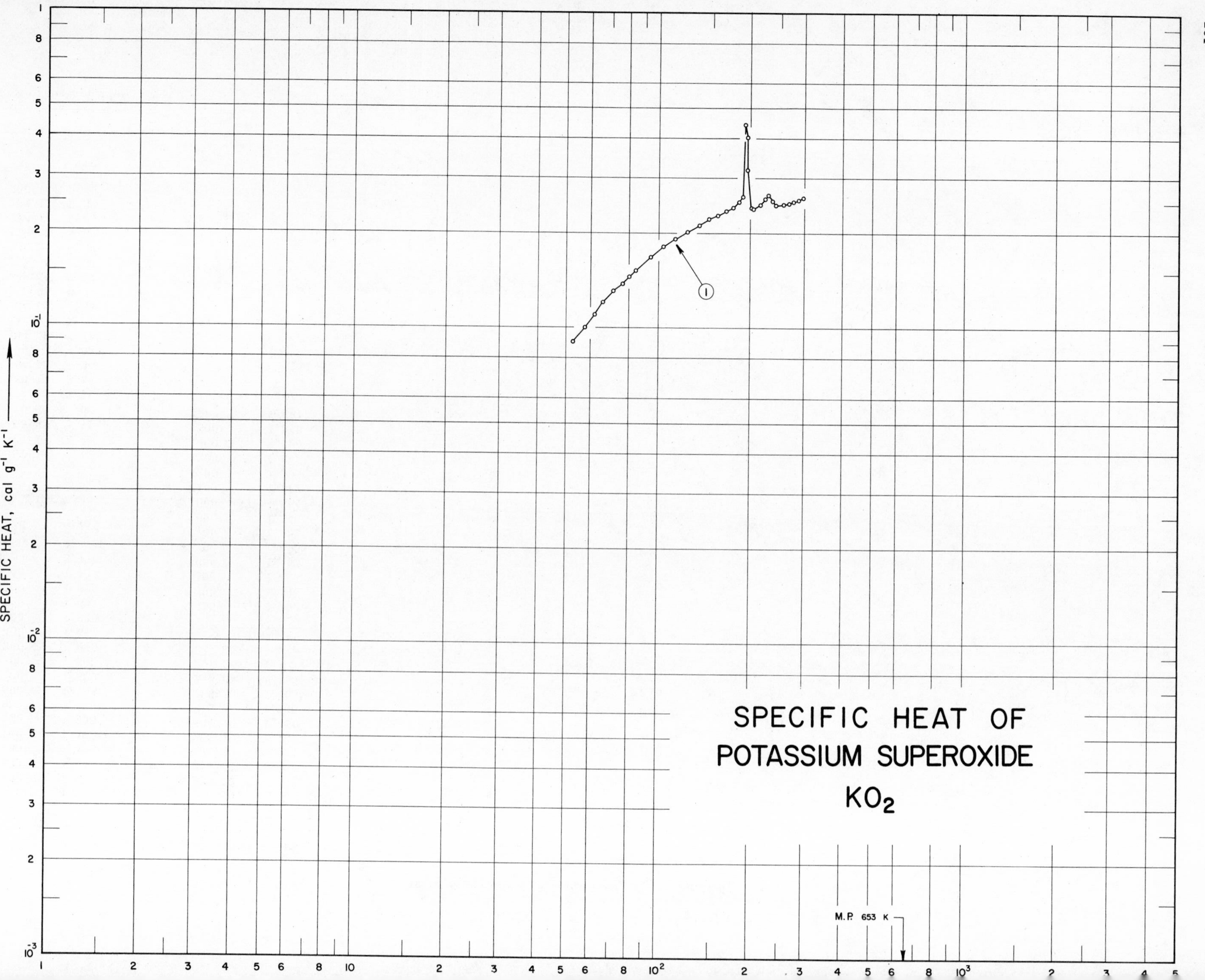
SPECIFIC HEAT OF
POTASSIUM SUPEROXIDE
KO2
M.P. 653 K
1
SPECIFIC HEAT, cal g-1 K-1

SPECIFICATION TABLE NO. 56 SPECIFIC HEAT OF POTASSIUM SUPEROXIDE KO_2

[For Data Reported in Figure and Table No. 56]

Curve No.	Ref. No.	Year	Temp. Range, K	Reported Error, %	Name and Specimen Designation	Composition (weight percent), Specifications and Remarks
1	240	1953	52–296			92. 4 KO_2, 3. 5 Na_2O_3, and 4. 1 KCO_3; corrected for impurities.

DATA TABLE NO. 56 SPECIFIC HEAT OF POTASSIUM SUPEROXIDE KO_2

[Temperature, T, K; Specific Heat, C_p, Cal g^{-1} K^{-1}]

T	C_p
CURVE 1	
52.59	9.053×10^{-2}
57.29	1.006×10^{-1}
62.05	1.116
66.72	1.218
71.49	1.312
76.19	1.398
80.00	1.463
84.07	1.533
94.83	1.695
104.70	1.820
114.88	1.940
125.06	2.045
136.09	2.145
146.17	2.229
156.04	2.300
166.36	2.374
176.21	2.446
183.54	2.525
186.22	2.567*
188.63	2.647
192.51	4.474
195.31	4.076
195.86	3.206
200.25	2.435
205.02	2.415
206.44	2.426*
216.68	2.494
219.56	2.522*
224.64	2.581
226.25	2.605*
228.96	2.651
232.87	2.633*
236.37	2.550
236.93	2.537*
241.47	2.478
246.14	2.476*
256.71	2.494
266.24	2.518*
276.41	2.547
286.62	2.575
296.72	2.602

*Not shown on plot

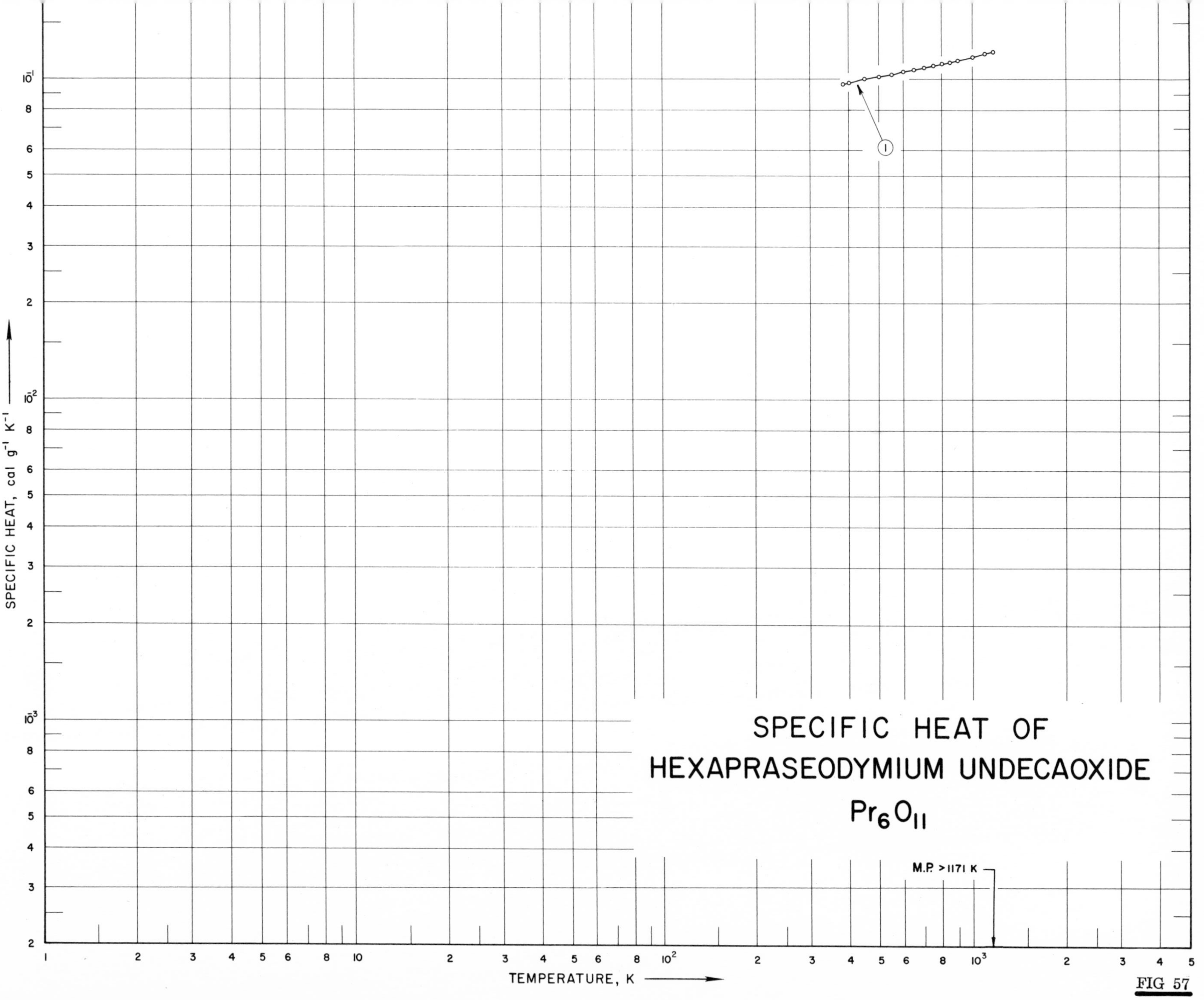
SPECIFIC HEAT OF
HEXAPRASEODYMIUM UNDECAOXIDE
Pr_6O_{11}
M.P. > 1171 K
1
TEMPERATURE, K
SPECIFIC HEAT, cal g^{-1} K^{-1}

FIG 57

SPECIFICATION TABLE NO. 57 SPECIFIC HEAT OF HEXAPRASEODYMIUM UNDECAOXIDE Pr_6O_{11}

[For Data Reported in Figure and Table No. 57]

Curve No.	Ref. No.	Year	Temp. Range, K	Reported Error, %	Name and Specimen Designation	Composition (weight percent), Specifications and Remarks
1	82	1951	383-1171			99.5 Pr_6O_{11}.

DATA TABLE NO. 57 SPECIFIC HEAT OF HEXAPRASEODYMIUM UNDECAOXIDE Pr_6O_{11}

[Temperature, T, K; Specific Heat, C_p, Cal $g^{-1}K^{-1}$]

T	C_p
CURVE 1	
383	9.689×10^{-2}
400	9.784
450	1.003×10^{-1}
500	1.025
550	1.044
600	1.061
650	1.078
700	1.094
750	1.109
800	1.124
850	1.138
900	1.152
950	1.166*
1000	1.180
1050	1.194*
1100	1.207
1150	1.221*
1171	1.226

*Not shown on Plot

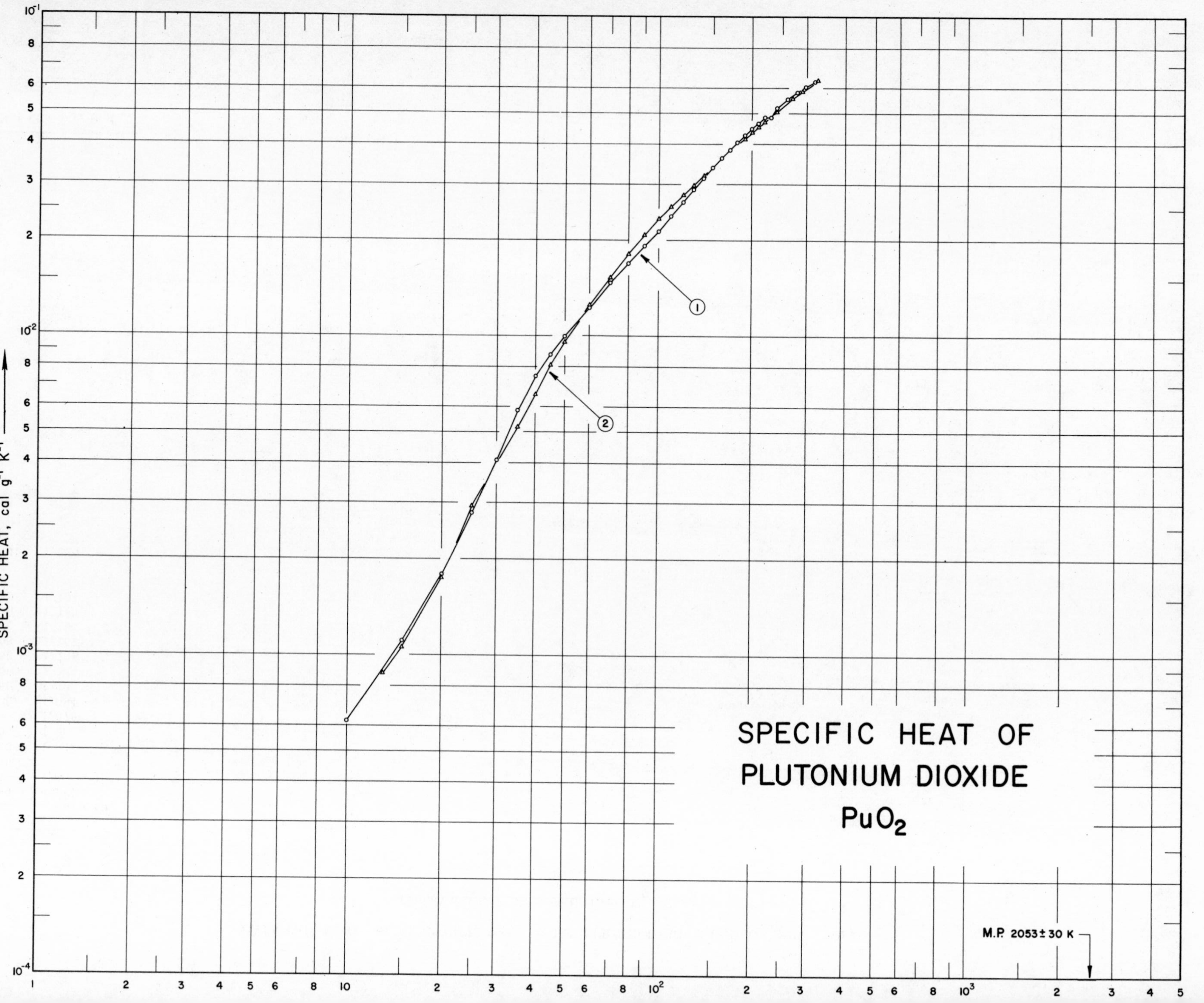
SPECIFIC HEAT OF
PLUTONIUM DIOXIDE
PuO_2
SPECIFIC HEAT, cal g^{-1} K^{-1}
M.P. 2053±30 K
1
2

SPECIFICATION TABLE NO. 58 SPECIFIC HEAT OF PLUTONIUM DIOXIDE PuO_2

[For Data Reported in Figure and Table No. 58]

Curve No.	Ref. No.	Year	Temp. Range, K	Reported Error, %	Name and Specimen Designation	Composition (weight percent), Specifications and Remarks
1	115	1962	10-320			87. 30 Pu, after initial preparation: 0. 010 Ni, 0. 008 Si, 0. 004 C, 0. 004 Fe, <0. 002 Cr, 0. 0004 Cu, 0. 0004 Mn, and 0. 0043 ppm other elements, after being compressed twice in steel die: 0. 320 Fe, 0. 240 Ni, 0. 055 Am, 0. 030 Si, 0. 030 Mn, 0. 0025 Cu, 0. 0196 other elements; pressed at 50, 000 psi and fired in air atmosphere at 1650 - 1700 C for 5 hrs.
2	116	1963	13-325			Same as above.

DATA TABLE NO. 58 SPECIFIC HEAT OF PLUTONIUM DIOXIDE PuO_2

[Temperature, T, K; Specific Heat, C_p, Cal g^{-1} K^{-1}]

T	C_p	T	C_p
CURVE 1		CURVE 2 (cont.)	
10	6.204×10^{-4}	25	2.92×10^{-3}
15	1.102×10^{-3}	30	4.12*
20	1.796	35	5.18
25	2.766	40	6.57
30	4.088	45	8.10
35	5.825	50	9.56
40	7.482	60	1.25×10^{-2}
45	8.723	70	1.53
50	9.964	80	1.80
60	1.237×10^{-2}	90	2.07
70	1.467	100	2.34
80	1.693	110	2.55
90	1.912	120	2.77
100	2.124	130	2.96
110	2.369	140	3.18
120	2.628	150	3.39*
130	2.883	160	3.58*
140	3.131	170	3.80*
150	3.369	180	3.98*
160	3.602	190	4.16
170	3.825	200	4.38
180	4.044	210	4.53
190	4.252	220	4.70
200	4.456	230	4.89*
210	4.650	240	5.07
220	4.836	250	5.25*
230	4.836	260	5.43*
240	5.190	270	5.58
250	5.354*	280	5.73*
260	5.511	290	5.88
270	5.661*	298.15	5.99*
280	5.803	300	6.02*
290	5.937*	310	6.17*
298.15	6.040	320	6.28*
300	6.066*	325	6.35
310	6.186*		
320	6.299		
CURVE 2			
13	8.76×10^{-4}		
15	1.06×10^{-3}		
20	1.75		

* Not shown on plot

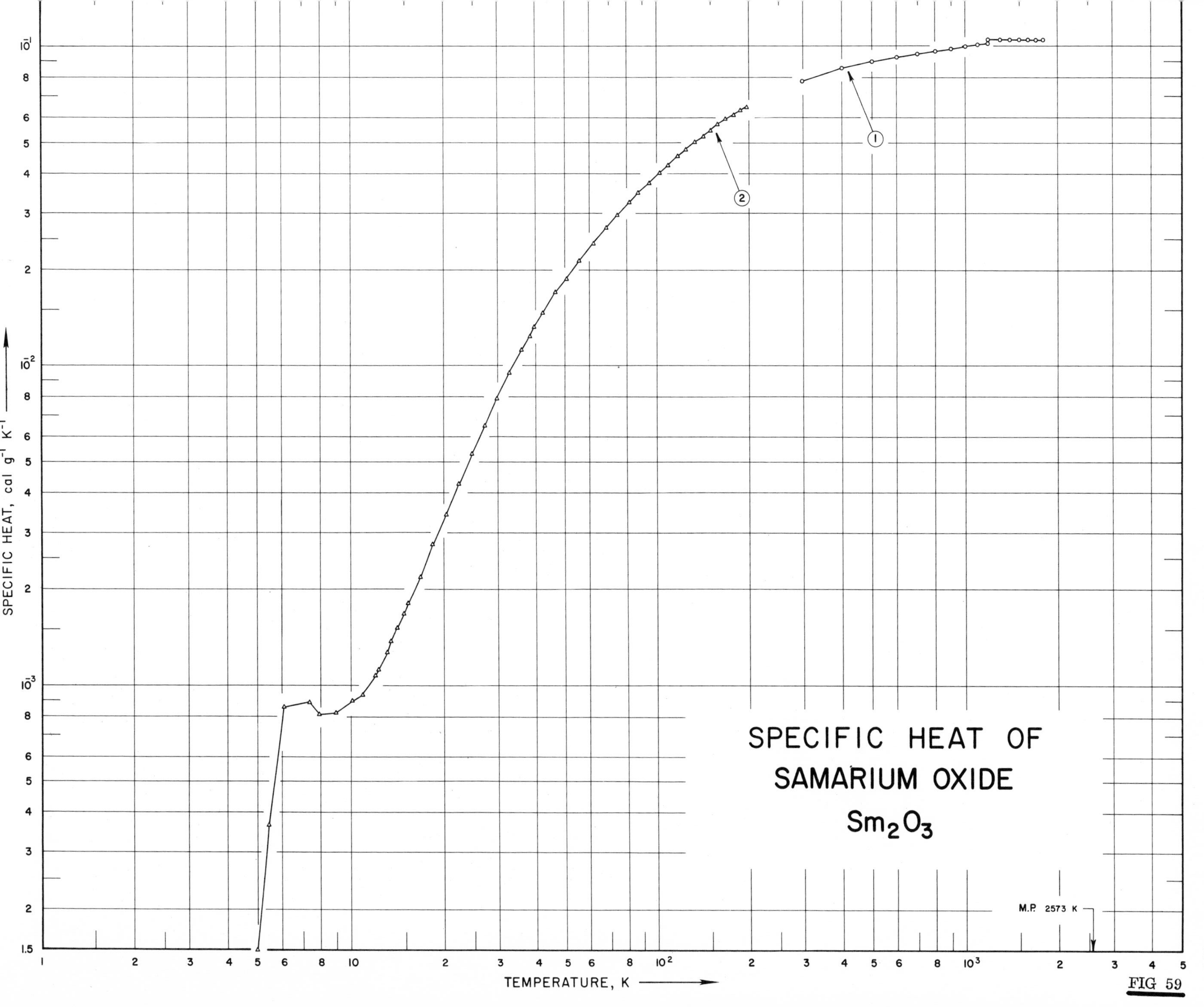
SPECIFIC HEAT OF
SAMARIUM OXIDE
Sm_2O_3
M.P. 2573 K
1
2
TEMPERATURE, K
SPECIFIC HEAT, cal g^{-1} K^{-1}

FIG 59

SPECIFICATION TABLE NO. 59 SPECIFIC HEAT OF SAMARIUM OXIDE Sm_2O_3

[For Data Reported in Figure and Table No. 59]

Curve No.	Ref. No.	Year	Temp. Range, K	Reported Error, %	Name and Specimen Designation	Composition (weight percent), Specifications and Remarks
1	73	1962	298-1798	0. 2		99. 9 Sm_2O_3; monoclinic; measured in helium atmosphere.
2	72	1962	5-346	0. 1		99. 9 Sm_2O_3, 0. 035 Ca, 0. 020 Si, and 0. 010 Eu; supplied by Michigan Chemical Co; pelleted under 2000 - 4000 psi; fired at 1170 K; measured under helium atmosphere.
3	73	1962	298-1149	0. 2		99. 9 Sm_2O_3; cubic, measured under helium atmosphere.

DATA TABLE NO. 59 SPECIFIC HEAT OF SAMARIUM OXIDE, Sm_2O_3

[Temperature, T, K; Specific Heat, C_p, Cal $g^{-1}K^{-1}$]

T	C_p
CURVE 1	
α 298	7.826×10^{-2}
300	7.848
400	8.580
500	8.991
600	9.274
700	9.498
800	9.690
900	9.864
1000	1.003×10^{-1}
1100	1.018
α1195	1.032
β1195	1.058
1200	1.058*
1300	1.058
1400	1.058
1500	1.058
1600	1.058
1700	1.058
β1798	1.058
CURVE 2	
Series 1	
4.99	1.491×10^{-4}
5.42	3.671
6.09	8.575
7.34	8.890
8.91	8.231
10.04	8.948
10.94	9.320
11.99	1.078×10^{-3}
13.03	1.270
14.05	1.508
15.27	1.807
16.73	2.188
18.44	2.753
20.34	3.430
22.35	4.293
24.56	5.311
27.03	6.527
CURVE 2 (cont.)	
Series 1 (cont.)	
29.73	7.961×10^{-3}
32.62	9.510
35.85	1.130×10^{-2}
39.49	1.326
Series 2	
7.90	8.145×10^{-4}*
8.90	8.030*
9.88	8.862*
11.01	9.492*
12.23	1.121×10^{-3}
13.47	1.385
14.84	1.683
38.03	1.249×10^{-2}
42.28	1.475
46.67	1.703
50.31	1.886
55.74	2.147
61.72	2.425
68.27	2.714
74.75	2.981
81.63	3.272
Series 1A	
73.12	2.918×10^{-2}*
80.20	3.215*
86.96	3.493
94.11	3.756
101.74	4.024
109.47	4.291
116.82	4.537
124.56	4.786
132.79	5.033
141.07	5.268
149.55	5.495
158.36	5.716
167.07	5.925
CURVE 2 (cont.)	
Series 1A (cont.)	
176.42	6.126×10^{-2}
186.08	6.321
195.45	6.496
Series 2A	
201.42	6.599
210.67	6.751
219.84	6.897
228.80	7.029
237.73	7.155
246.47	7.279
255.24	7.385
264.07	7.494
272.87	7.591
282.01	7.686
291.40	7.783
300.76	7.875
310.13	7.961
319.56	8.044
329.07	8.122
338.52	8.188
346.78	8.251
CURVE 3*	
298	7.9422×10^{-2}
300	7.9621
400	8.6599
500	9.0611
600	9.3456
700	9.5749
800	9.7749
900	9.9579
1000	1.0130×10^{-1}
1100	1.0296
1149	1.0375

* Not shown on Plot

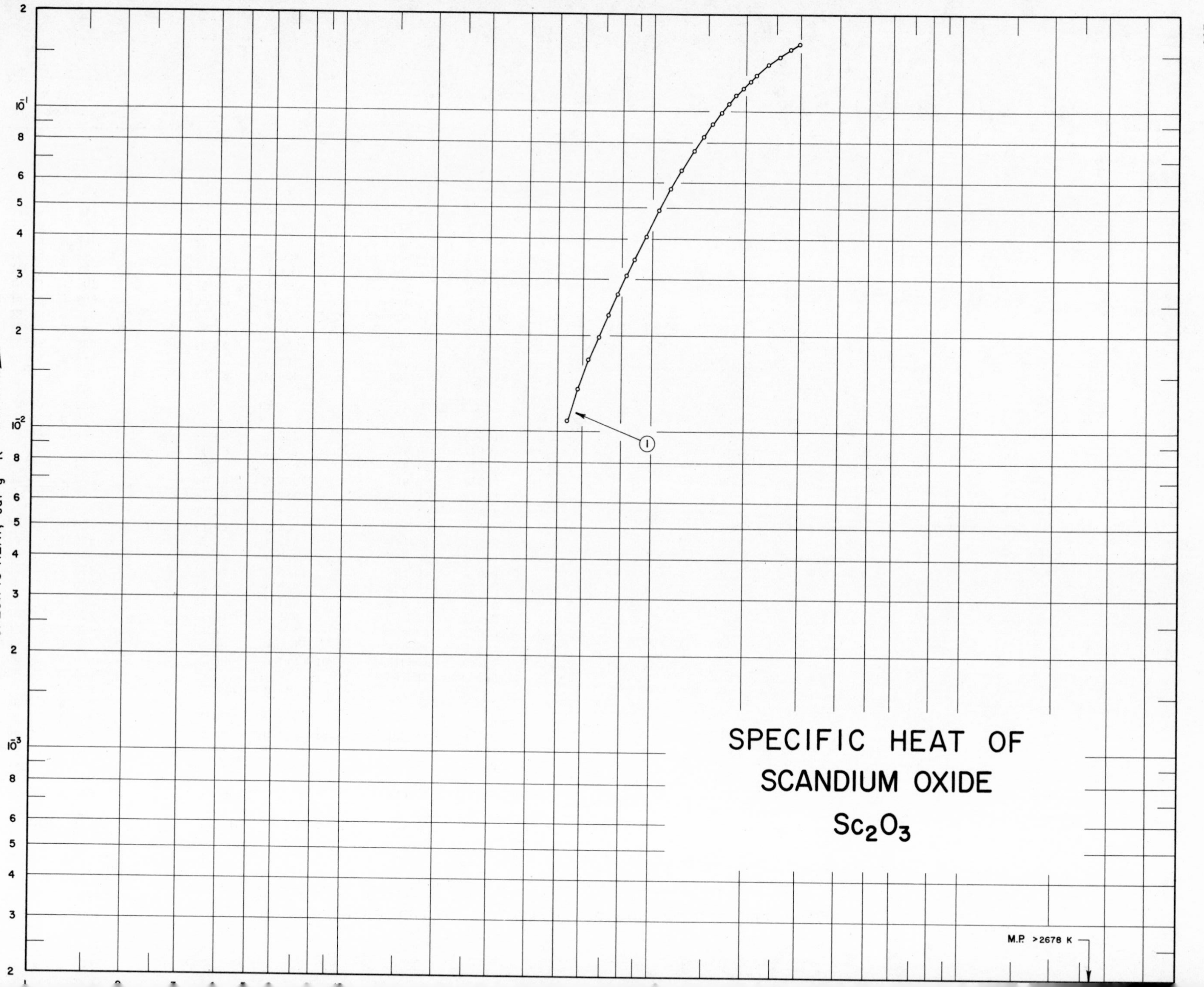

SPECIFIC HEAT OF
SCANDIUM OXIDE
Sc2O3
M.P. >2678 K
1
SPECIFIC HEAT, cal g-1 K-1
2
10-1
8
6
5
4
3
2
10-2
8
6
5
4
3
2
10-3
8
6
5
4
3
2

SPECIFICATION TABLE NO. 60 SPECIFIC HEAT OF SCANDIUM OXIDE Sc_2O_3

[For Data Reported in Figure and Table No. 60]

Curve No.	Ref. No.	Year	Temp. Range, K	Reported Error, %	Name and Specimen Designation	Composition (weight percent), Specifications and Remarks
1	58	1963	53-296	0.10		99.9 Sc_2O_3; measured in nitrogen atmosphere.

DATA TABLE NO. 60 SPECIFIC HEAT OF SCANDIUM OXIDE, Sc_2O_3

[Temperature, T, K; Specific Heat, C_p, Cal $g^{-1}K^{-1}$]

T	C_p
CURVE 1	
53.33	1.083×10^{-2}
57.88	1.361
62.66	1.684
67.40	1.978
72.37	2.312
77.45	2.692
82.61	3.088
87.45	3.454
95.25	4.086
105.01	4.914
114.82	5.768
124.49	6.587
135.77	7.541
145.51	8.361
156.08	9.180
166.13	9.919
176.21	1.064×10^{-1}
186.12	1.129
196.38	1.191
206.47	1.249
216.17	1.302
226.23	1.355*
236.46	1.405
246.33	1.447*
256.42	1.490
266.74	1.527*
276.64	1.564
286.92	1.595*
296.27	1.629

*Not shown on Plot

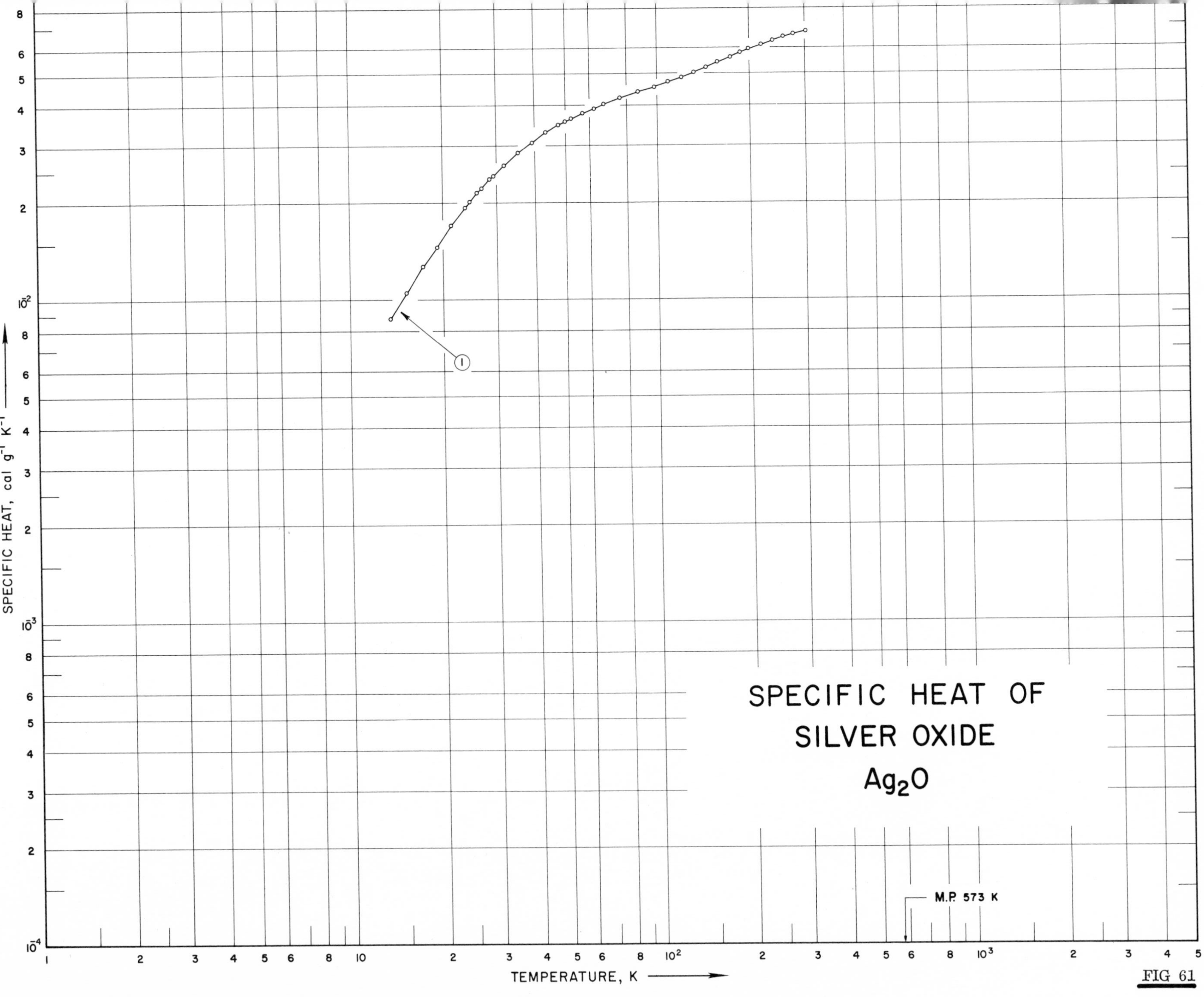
SPECIFIC HEAT OF
SILVER OXIDE
Ag_2O
M.P. 573 K
TEMPERATURE, K
SPECIFIC HEAT, cal g^{-1} K^{-1}
1

FIG 61

SPECIFICATION TABLE NO. 61 SPECIFIC HEAT OF SILVER OXIDE Ag_2O

[For Data Reported in Figure and Table No. 61]

Curve No.	Ref. No.	Year	Temp. Range, K	Reported Error, %	Name and Specimen Designation	Composition (weight percent), Specifications and Remarks
1	117	1962	14-302			~99. 9 Ag_2O, 0. 05 Cu, 0. 04 H_2O,and 0. 01 other impurities; prepared by long heating in contact with water at 325 C under 200 atmosphere of O_2; annealed; measured in helium atmosphere.

DATA TABLE NO. 61 SPECIFIC HEAT OF SILVER OXIDE Ag_2O

[Temperature, T, K; Specific Heat, C_p, Cal $g^{-1}K^{-1}$]

CURVE 1

Series 1

T	C_p
13.5	8.760×10^{-3}
15.28	1.053×10^{-2}
17.22	1.277
19.12	1.463
21.25	1.705
23.58	1.946
25.93	2.153
28.44	2.386
31.51	2.619
35.05	2.870
39.10	3.098
43.44	3.314
47.74	3.491
52.22	3.651
57.03	3.797
62.09	3.923
67.01	4.048
76.12	4.225
81.40	4.320*
87.16	4.419
92.87	4.496*
98.27	4.578
103.66	4.656*
109.25	4.729
114.86	4.820*
120.55	4.898
126.47	4.997*
132.62	5.079
138.80	5.183*
144.95	5.260
150.98	5.338*
157.12	5.433
163.07	5.511*
168.82	5.580*
174.90	5.665
180.95	5.735*
186.92	5.808
193.02	5.882*

CURVE 1 (cont.)

Series 1 (cont.)

T	C_p
198.85	5.955×10^{-2}
204.77	6.011*
211.08	6.080*
217.57	6.153
223.79	6.210*
230.21	6.274*
237.15	6.330
243.70	6.408*
250.06	6.464*
256.28	6.503
262.56	6.568*
268.81	6.624*
275.47	6.671
281.52	6.714*
288.07	6.740*
294.66	6.775*
301.67	6.796

Series 2

T	C_p
24.50	2.020
26.63	2.218
29.17	2.434
32.43	2.680*
36.25	2.943*
40.49	3.167*
45.38	3.383*
50.52	3.590
55.35	3.750*
60.11	3.884*
64.78	3.996*
69.64	4.099*
74.36	4.181*
78.97	4.272*

*Not shown on Plot

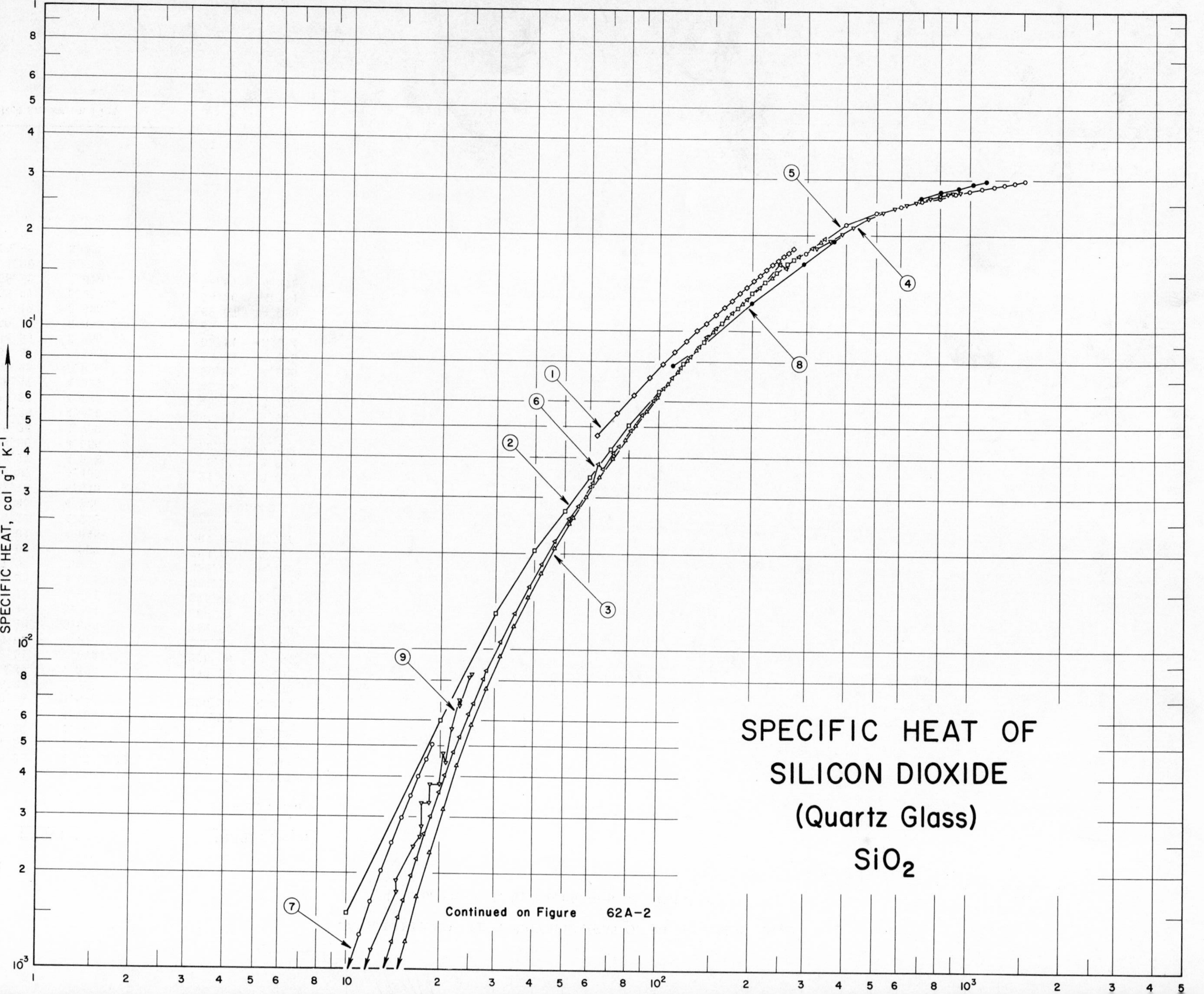
SPECIFIC HEAT OF
SILICON DIOXIDE
(Quartz Glass)
SiO_2
Continued on Figure 62A-2
SPECIFIC HEAT, cal g^{-1} K^{-1}
1
2
3
4
5
6
7
8
9

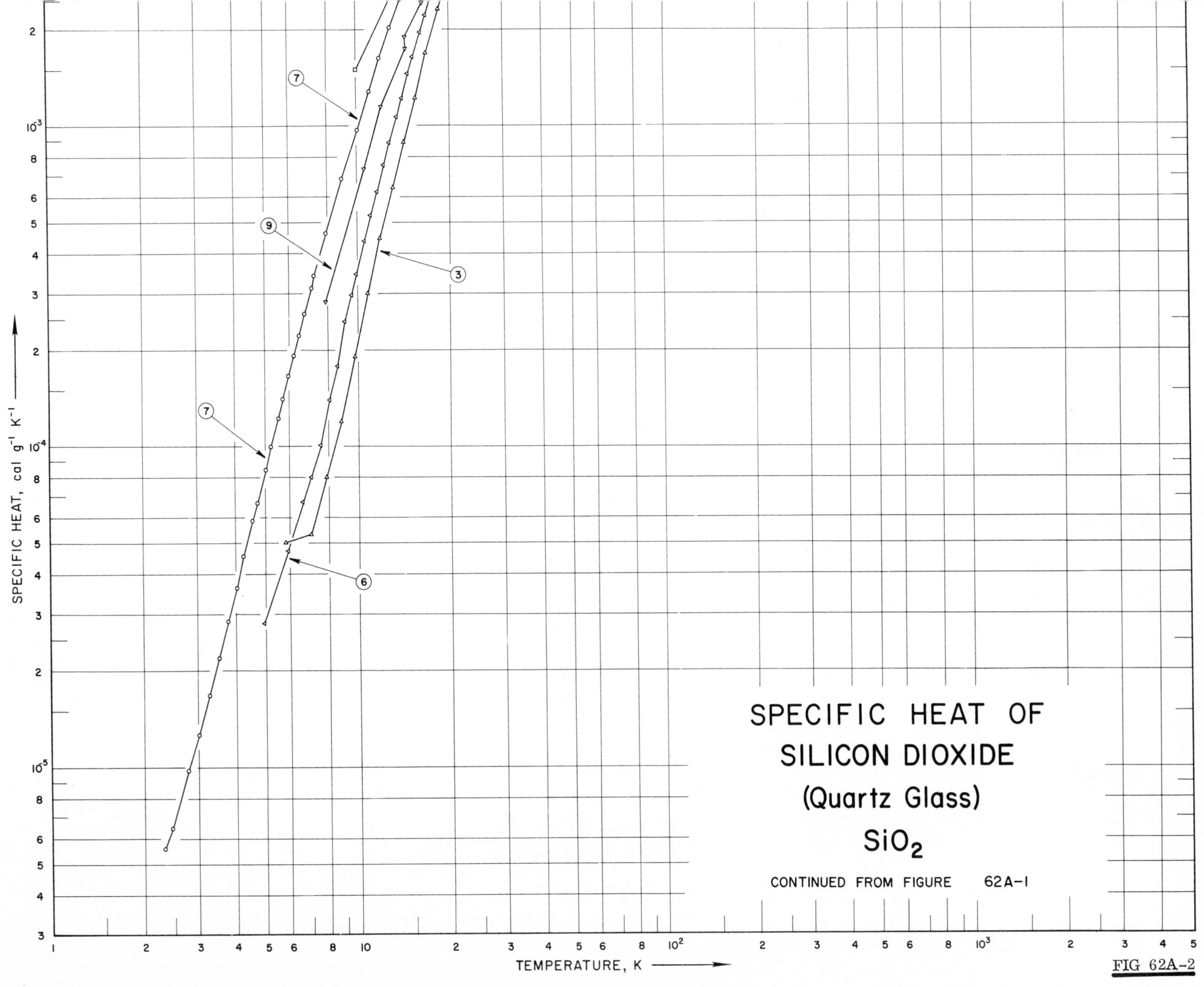
SPECIFIC HEAT OF
SILICON DIOXIDE
(Quartz Glass)
SiO_2
CONTINUED FROM FIGURE 62A-1
SPECIFIC HEAT, cal g^{-1} K^{-1}
TEMPERATURE, K
7
9
3
7
6

FIG 62A-2

SPECIFICATION TABLE NO. 62-A SPECIFIC HEAT OF SILICON DIOXIDE (Quartz Glass) SiO_2

[For Data Reported in Figure and Table No. 62-A]

Curve No.	Ref. No.	Year	Temp. Range, K	Reported Error, %	Name and Specimen Designation	Composition (weight percent), Specifications and Remarks
1	118	1911	63-273			Glass.
2	119	1921	10-273			Glass; amorphous.
3	120	1956	5-344	5 at 5K 1 at 10K <0. 1 above 25K		99. 97 SiO_2; specimen from Brazil; vitreous silica; irradiated with neutrons for 30 days at 35 C with damaging flux of 2. 5 x 10^{19} nvt.
4	121	1936	323-935			Transparent without blemish; from Brazil.
5	122	1941	298-1520			99. 95 SiO_2 and 0. 05 residue; silica glass; prepared from transparent vitreosil glass tubing by crushing and screening out fines.
6	120	1956	5-343	5 at 5K 1 at 10K <0. 1 above 25K		99. 97 SiO_2; specimen from Brazil; vitreous silica; irradiated with neutrons for 30 days at 35 C with damaging flux of 7. 7 x 10^9 nvt.
7	123	1959	2-18			Optical quality; annealed at 1100 C after crushing.
8	9	1960	111-1144			Sample obtained from Hanovia Chemical Co; fused silica.
9	124	1962	8-25			Silica; irradiated with 7. 7 x 10^{19} n cm^{-2} fast neutrons.

DATA TABLE NO. 62-A SPECIFIC HEAT OF SILICON DIOXIDE (Quartz Glass) SiO_2

[Temperature, T, K; Specific Heat, C_p, Cal $g^{-1}K^{-1}$]

CURVE 1

T	C_p
63.15	4.7×10^{-2}
73.15	5.5
83.15	6.3
93.15	7.15
103.15	7.9
113.15	8.6
123.15	9.3
133.15	1.00×10^{-1}
143.15	1.06
153.15	1.13
163.15	1.19
173.15	1.25
183.15	1.32
193.15	1.38
203.15	1.44
213.15	1.50
223.15	1.56
233.15	1.62
243.15	1.67
253.15	1.72
263.15	1.77
273.15	1.82

CURVE 2

T	C_p
10	1.5×10^{-3}
20	6.0
30	1.3×10^{-2}
40	2.05
50	2.72
60	3.48
70	4.26
80	5.06
100	6.43
120	7.90
140	9.23
160	1.06×10^{-1}
180	1.19
200	1.32
220	1.43
240	1.53
260	1.63
273	1.682

CURVE 3

T	C_p
5.81	5.0×10^{-5}
7.00	5.3
7.90	8.0
8.81	1.2×10^{-4}
9.78	1.91
10.83	3.01
11.95	4.49
13.13	6.46
14.28	8.94
15.58	1.235×10^{-3}
16.99	1.696
18.61	2.325
20.51	3.182
22.79	4.352
25.38	5.833
28.15	7.573
31.14	9.57
34.48	1.19×10^{-2}
38.11	1.46
42.01	1.746
46.60	2.094
51.85	2.496
53.24	2.601
58.85	3.034
64.61	3.490
71.13	3.998
78.01	4.544
84.30	5.051
91.32	5.604
99.05	6.201
107.16	6.824
115.54	7.463
124.07	8.105
132.77	8.743
141.62	9.368*
150.25	9.966
159.11	1.056×10^{-1}
166.51	1.103
176.33	1.164*
185.91	1.222
195.25	1.276*
204.63	1.329*
214.07	1.380*
223.64	1.431*

CURVE 3 (cont.)

T	C_p
233.65	1.482×10^{-1}
243.67	1.532*
253.60	1.580*
263.50	1.626*
273.44	1.670*
283.43	1.716*
285.09	1.720*
295.05	1.762*
305.04	1.803*
314.84	1.841*
324.48	1.879*
334.16	1.916
344.04	1.953*

CURVE 4

T	C_p
323.95	1.830×10^{-1}
356.55	1.935
388.85	2.028
420.95	2.128
472.65	2.262
528.65	2.362
576.25	2.441
624.25	2.497
673.35	2.556
747.85	2.636
801.35	2.676
851.85	2.706
887.85	2.706
935.85	2.748

CURVE 5

T	C_p
298.15	1.765×10^{-1}
300	1.775*
400	2.161
500	2.359
600	2.484
700	2.574
800	2.645
900	2.705
1000	2.759
1100	2.809
1200	2.855

CURVE 5 (cont.)

T	C_p
1300	2.899×10^{-1}
1400	2.942
1500	2.984*
1520	2.992

CURVE 6

Series 1

T	C_p
192.60	1.265×10^{-1}
211.70	1.371
221.46	1.422*

Series 2

T	C_p
61.96	3.357×10^{-2}
64.00	3.826
74.65	4.337
81.43	4.883
88.58	5.449
96.26	6.041
104.08	6.646
110.31	7.120
118.38	7.712
126.67	8.347
135.13	8.961
143.93	9.580
153.22	1.021×10^{-1}
163.09	1.085*
173.12	1.149
182.74	1.207*
192.11	1.261*
226.45	1.448*
236.64	1.499*
246.69	1.631
256.62	1.598
266.49	1.641*
276.39	1.684*
282.66	1.712
292.21	1.753*
302.13	1.794*
312.34	1.834
322.28	1.870*
332.78	1.912*
343.45	1.952

CURVE 6 (cont.)

Series 3

T	C_p
4.95	2.8×10^{-5}
5.91	4.7
6.60	6.7
7.55	1.0×10^{-4}
8.59	1.78
9.59	2.96
10.58	4.38
11.66	6.26
12.84	8.89
14.08	1.22×10^{-3}
15.43	1.65
16.94	2.224
18.73	3.014
20.84	4.078
23.17	5.342
25.65	6.800
28.32	8.60
31.25	1.06×10^{-2}
34.72	1.30
38.53	1.58
42.46	1.856
46.81	2.199
51.84	2.573

Series 4

T	C_p
7.0	8.0×10^{-5}
8.08	1.4×10^{-4}
9.07	2.46
9.96	3.46
11.12	5.29
12.29	7.54
13.51	1.07×10^{-3}
14.84	1.46
16.29	1.969
17.95	2.661
19.91	3.583
22.15	4.792
24.69	6.283
27.60	8.077
30.85	1.03×10^{-2}
34.40	1.28*
38.15	1.55*

CURVE 6 (cont.)

Series 4 (cont.)

T	C_p
42.18	1.840×10^{-2}*
46.67	2.192*
55.15	2.826
60.58	3.249
66.16	3.688
71.98	4.131
190.96	1.255×10^{-1}*
200.48	1.309*
210.10	1.361*

CURVE 7

T	C_p
2.344	5.58×10^{-6}
2.465	6.457
2.790	9.802
3.018	1.2531×10^{-5}
3.278	1.6692
3.522	2.1884
3.784	2.8441
4.031	3.6080
4.267	4.5316
4.565	5.8663
4.719	6.6485
5.017	8.4724
5.225	9.536
5.504	1.2179×10^{-4}
5.724	1.4072
5.989	1.6617
6.226	1.9222
6.488	2.2284
6.769	2.6028
7.129	3.1337
7.293	3.4249
7.976	4.6348
8.982	6.8831
10.039	9.7439
11.015	1.2884×10^{-3}
11.991	1.6436
12.993	2.047
14.006	2.4980
15.048	2.9839
16.062	3.4898

*Not shown on plot

DATA TABLE NO. 62-A (continued)

T	C_p
CURVE 7 (cont.)	
17.051	4.004×10^{-3}
18.015	4.540
18.980	5.077
CURVE 8	
111.5	7.8×10^{-2}
144	9.3*
199	1.23×10^{-1}
293	1.64
366	1.92
477	2.26*
588	2.48*
699	2.63
810	2.75
922	2.84
1033	2.92
1144	2.98
CURVE 9	
Series 1	
12.1	1.15×10^{-3}
14.6	1.75
16.5	2.43
18.5	3.31
20.9	4.43
Series 2	
17.5	2.80×10^{-3}
18.6	3.78
20.0	3.78
21.9	5.61
23.2	6.69
24.9	8.16
Series 3	
7.9	2.83×10^{-4}
10.6	7.32
14.4	1.75×10^{-3}*
17.5	3.31

T	C_p
CURVE 9 (cont.)	
Series 4	
14.6	1.90×10^{-3}
17.4	2.60
20.5	4.76
23.1	6.97
25.4	8.37

* Not shown on plot

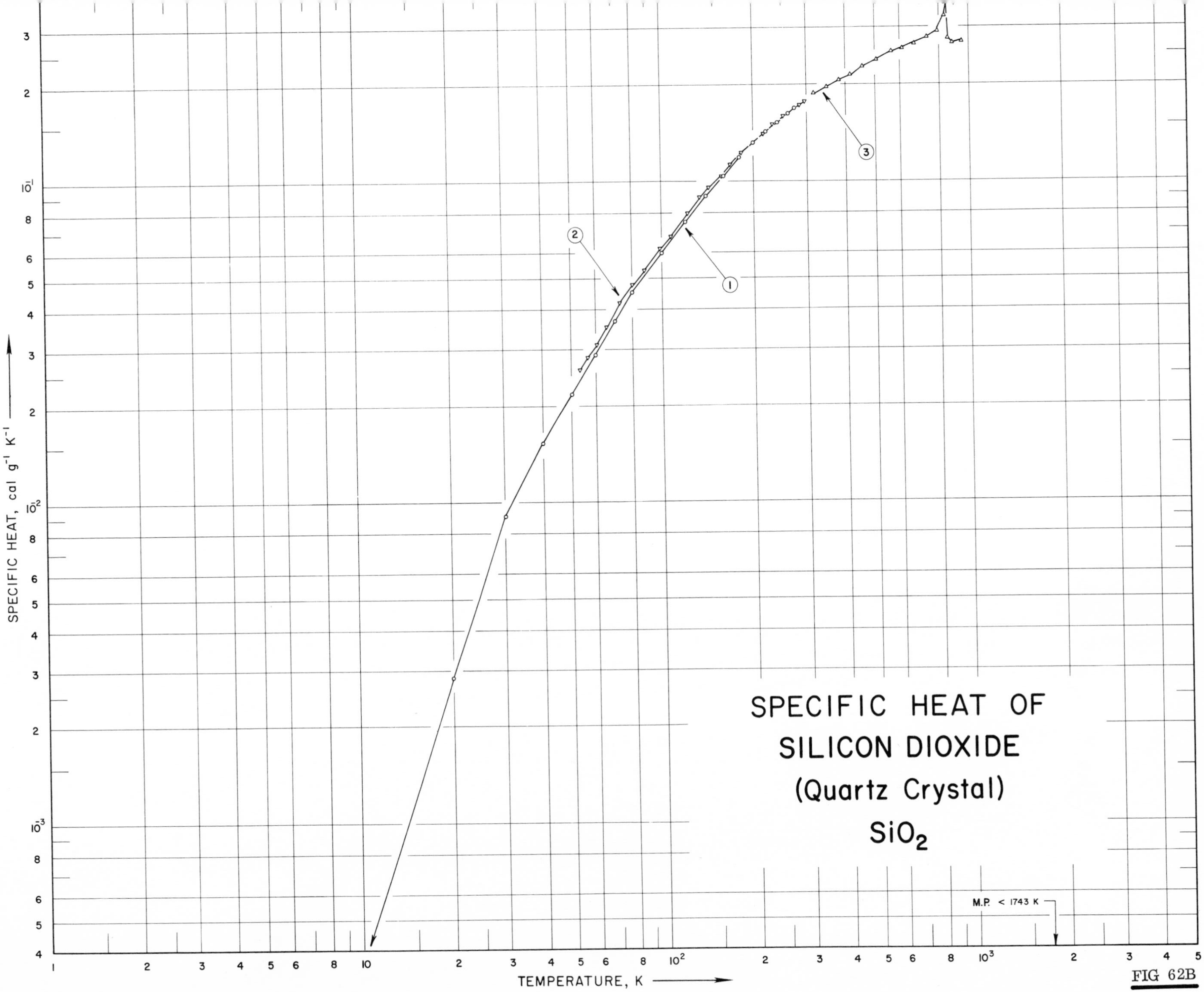

SPECIFIC HEAT OF
SILICON DIOXIDE
(Quartz Crystal)
SiO_2
SPECIFIC HEAT, cal g^{-1} K^{-1}
TEMPERATURE, K
M.P. < 1743 K
1
2
3

FIG 62B

SPECIFICATION TABLE NO. 62-B SPECIFIC HEAT OF SILICON DIOXIDE (Quartz Crystal) SiO_2

[For Data Reported in Figure and Table No. 62-B]

Curve No.	Ref. No.	Year	Temp. Range, K	Reported Error, %	Name and Specimen Designation	Composition (weight percent), Specifications and Remarks
1	119	1921	10-273			
2	125	1936	53-296			99.93 SiO_2, and 0.07 impurities; α-quartz; density 2.6378 g cm^{-2} at 22.2 C.
3	121	1936	317-949			

DATA TABLE NO. 62-B SPECIFIC HEAT OF SILICON DIOXIDE (Quartz Crystal) SiO_2

[Temperature, T, K; Specific Heat, C_p, Cal g^{-1} K^{-1}]

T	C_p
CURVE 1	
10	3.66 x 10^{-4}*
20	2.83 x 10^{-3}
30	8.99
40	1.53 x 10^{-2}
50	2.17
60	2.87
70	3.66
80	4.50
100	5.96
120	7.46
140	8.96
160	1.04 x 10^{-1}
180	1.18
200	1.31
220	1.42
240	1.52
260	1.62
273	1.689
CURVE 2	
53.4	2.57 x 10^{-2}
56.4	2.80
60.3	3.08
65.4	3.50
72.8	4.17
80.0	4.76
87.5	5.27
98.9	6.12
107.9	6.70
122.1	7.91
134.1	8.85
143.8	9.54
156.2	1.035 x 10^{-1}
169.1	1.12
184.8	1.22
197.7	1.29*
216.1	1.39
234.7	1.49
252.1	1.58
272.0	1.70*
285.0	1.72
296.1	1.76

T	C_p
CURVE 3	
316.65	1.872 x 10^{-1}
334.05	1.925*
348.85	1.960
367.55	2.011
381.25	2.057
390.65	2.092*
403.35	2.129*
414.65	2.145
428.95	2.192*
441.15	2.222*
456.35	2.271
475.95	2.314*
506.65	2.377
534.15	2.443*
565.45	2.516
585.25	2.541*
610.35	2.589
640.65	2.634*
670.15	2.683
702.95	2.723*
732.25	2.791
758.55	2.851*
777.45	2.912*
792.55	2.949
808.75	3.036*
820.65	3.100*
827.25	3.183*
834.65	3.271
840.65	3.460*
844.35	3.610
855.05	2.783
866.55	2.689*
883.45	2.698
899.95	2.703*
921.05	2.714*
949.45	2.721

*Not shown on plot

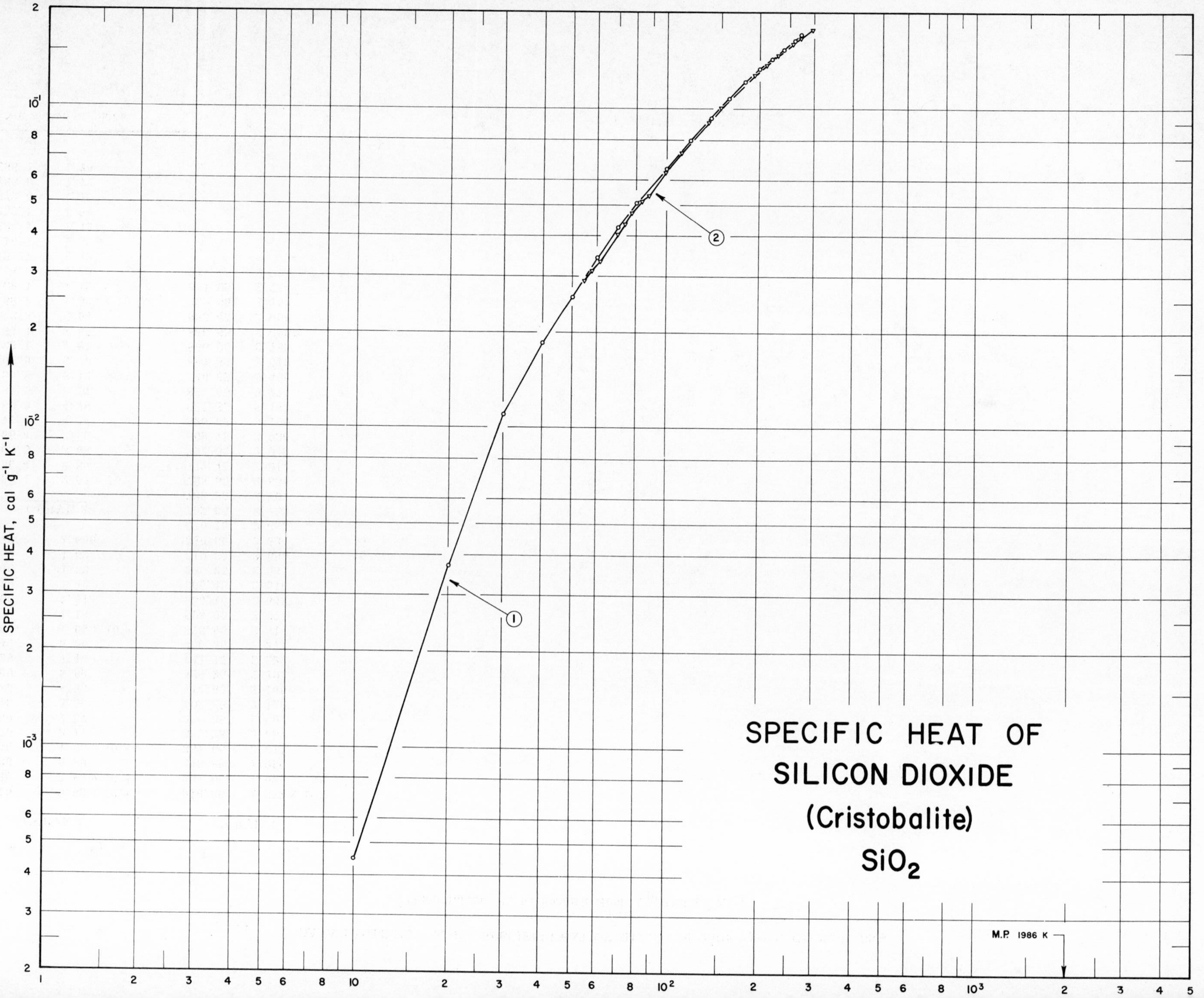
SPECIFIC HEAT OF
SILICON DIOXIDE
(Cristobalite)
SiO_2
M.P. 1986 K
SPECIFIC HEAT, cal g^{-1} K^{-1}
1
2

SPECIFICATION TABLE NO. 62-C SPECIFIC HEAT OF SILICON DIOXIDE (CRISTOBALITE) SiO_2

[For Data Reported in Figure and Table No. 62-C]

Curve No.	Ref. No.	Year	Temp. Range, K	Reported Error, %	Name and Specimen Designation	Composition (weight percent), Specifications and Remarks
1	119	1921	10-273			
2	125	1936	54-297			99.99 SiO_2; crystobalite; density 2.3201 g cm^{-3} at 23.3 C.

DATA TABLE NO. 62-C SPECIFIC HEAT OF SILICON DIOXIDE (CRISTOBALITE) SiO_2

[Temperature, T, K; Specific Heat, C_p, Cal g^{-1} K^{-1}]

T	C_p
CURVE 1	
10	4.5 x 10^{-4}
20	3.7 x 10^{-3}
30	1.1 x 10^{-2}
40	1.85
50	2.57
60	3.40
70	4.26
80	5.06
100	6.43
120	7.93
140	9.36
160	1.08 x 10^{-1}
180	1.21
200	1.34
220	1.44
240	1.54
260	1.64
273	1.709
CURVE 2	
54.8	2.89 x 10^{-2}
57.8	3.10
61.3	3.33
69.9	4.03
73.7	4.36
77.7	4.71
83.9	5.11
88.0	5.37
99.8	6.31
112.0	7.25
120.4	6.78*
136.8	9.05
150.7	1.005 x 10^{-1}
164.0	1.09*
178.2	1.18*
193.7	1.27
210.9	1.365
229.5	1.47
241.4	1.525*
257.1	1.595
272.2	1.66
297.2	1.755
297.3	1.77*

*Not shown on plot

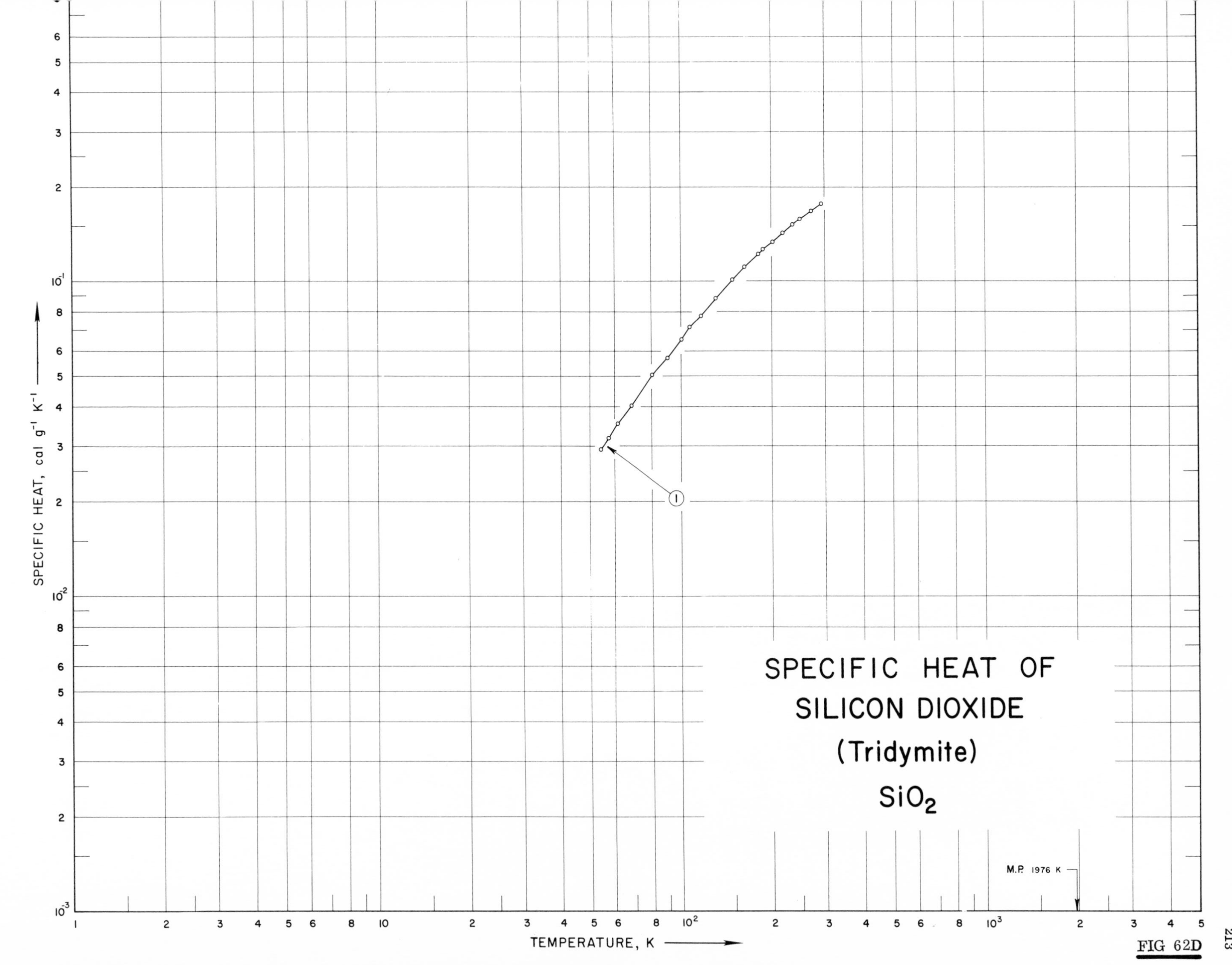

SPECIFIC HEAT OF
SILICON DIOXIDE
(Tridymite)
SiO_2
M.P. 1976 K
1
TEMPERATURE, K
SPECIFIC HEAT, cal g^{-1} K^{-1}

FIG 62D

SPECIFICATION TABLE NO. 62-D SPECIFIC HEAT OF SILICON DIOXIDE (TRIDYMITE) SiO_2

[For Data Reported in Figure and Table No. 62-D]

Curve No.	Ref. No.	Year	Temp. Range, K	Reported Error, %	Name and Specimen Designation	Composition (weight percent), Specifications and Remarks
1	125	1936	54-295			99.46 SiO_2; tridymite; density 2.2777 g cm^{-3} at 23.7 C.

DATA TABLE NO. 62-D SPECIFIC HEAT OF SILICON DIOXIDE (TRIDYMITE) SiO_2

[Temperature, T, K; Specific Heat, C_p, Cal g^{-1} K^{-1}]

T	C_p
CURVE 1	
54.2	2.931 x 10^{-2}
57.6	3.170
61.5	3.517
68.6	4.003
80.5	5.026
90.8	5.710
100.4	6.563
108.8	7.148
117.7	7.796
131.4	8.836
149.3	1.010 x 10^{-1}
164.3	1.113
181.6	1.220
187.9	1.258
202.6	1.339
219.5	1.429
235.4	1.515
249.1	1.579
271.8	1.679
278.1	1.704*
290.8	1.757*
294.9	1.769

*Not shown on plot

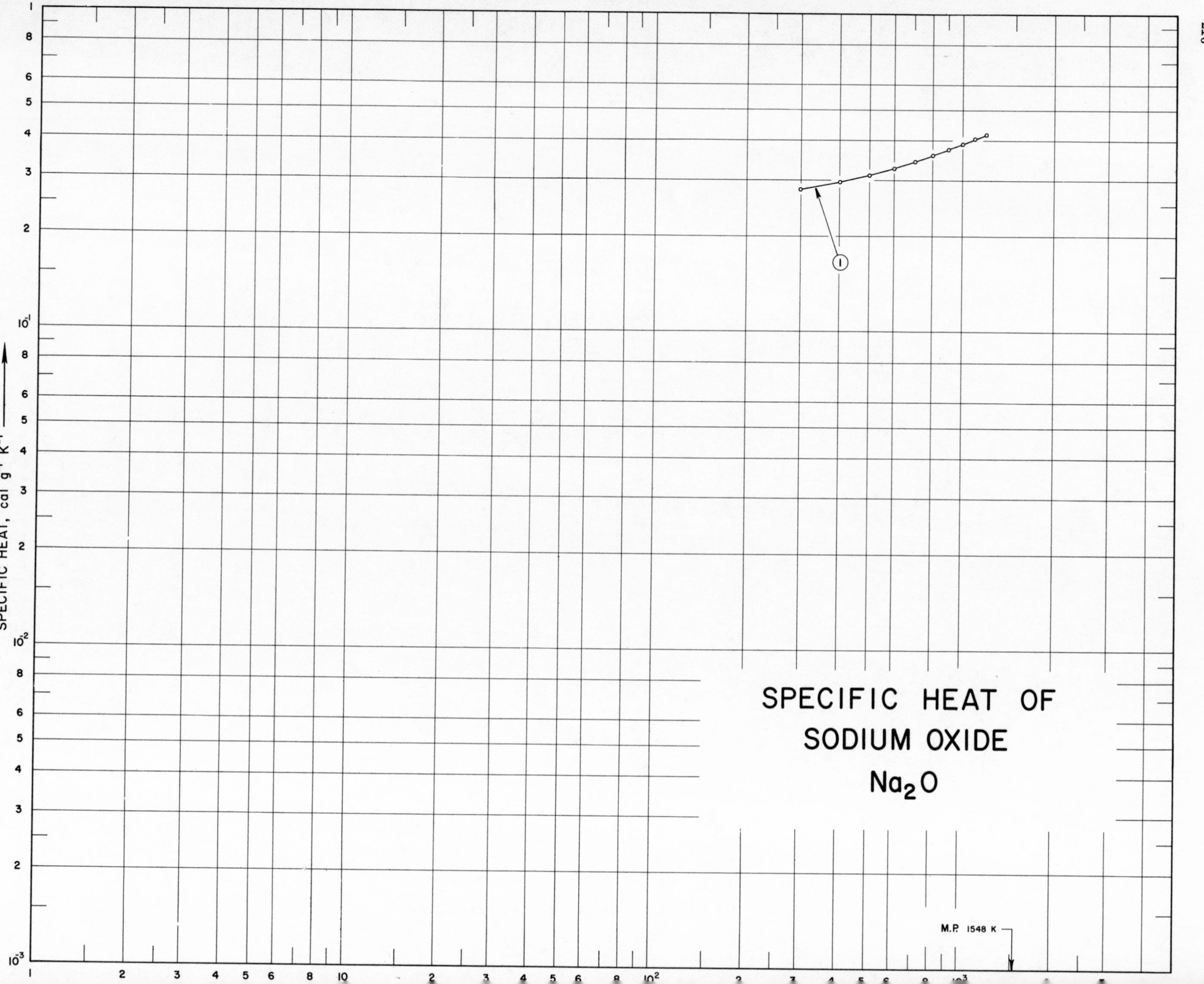
SPECIFIC HEAT OF
SODIUM OXIDE
Na_2O
M.P. 1548 K
SPECIFIC HEAT, cal g^{-1} K^{-1}

SPECIFICATION TABLE NO. 63 SPECIFIC HEAT OF SODIUM OXIDE Na_2O

[For Data Reported in Figure and Table No. 63]

Curve No.	Ref. No.	Year	Temp. Range, K	Reported Error, %	Name and Specimen Designation	Composition (weight percent), Specifications and Remarks
1	126	1960	298-1170	±2		96.76 Na_2O, ± 0.33 Na_2CO_3 and 0.91 Na_2O_2.

DATA TABLE NO. 63 SPECIFIC HEAT OF SODIUM OXIDE, Na_2O

[Temperature, T, K; Specific Heat, C_p, Cal $g^{-1}K^{-1}$]

T	C_p
	CURVE 1
298	2.813 x 10^{-1}
300	2.816*
400	2.976
500	3.135
600	3.294
700	3.454
800	3.613
900	3.773
1000	3.932
1100	4.091
1170	4.203

*Not shown on Plot

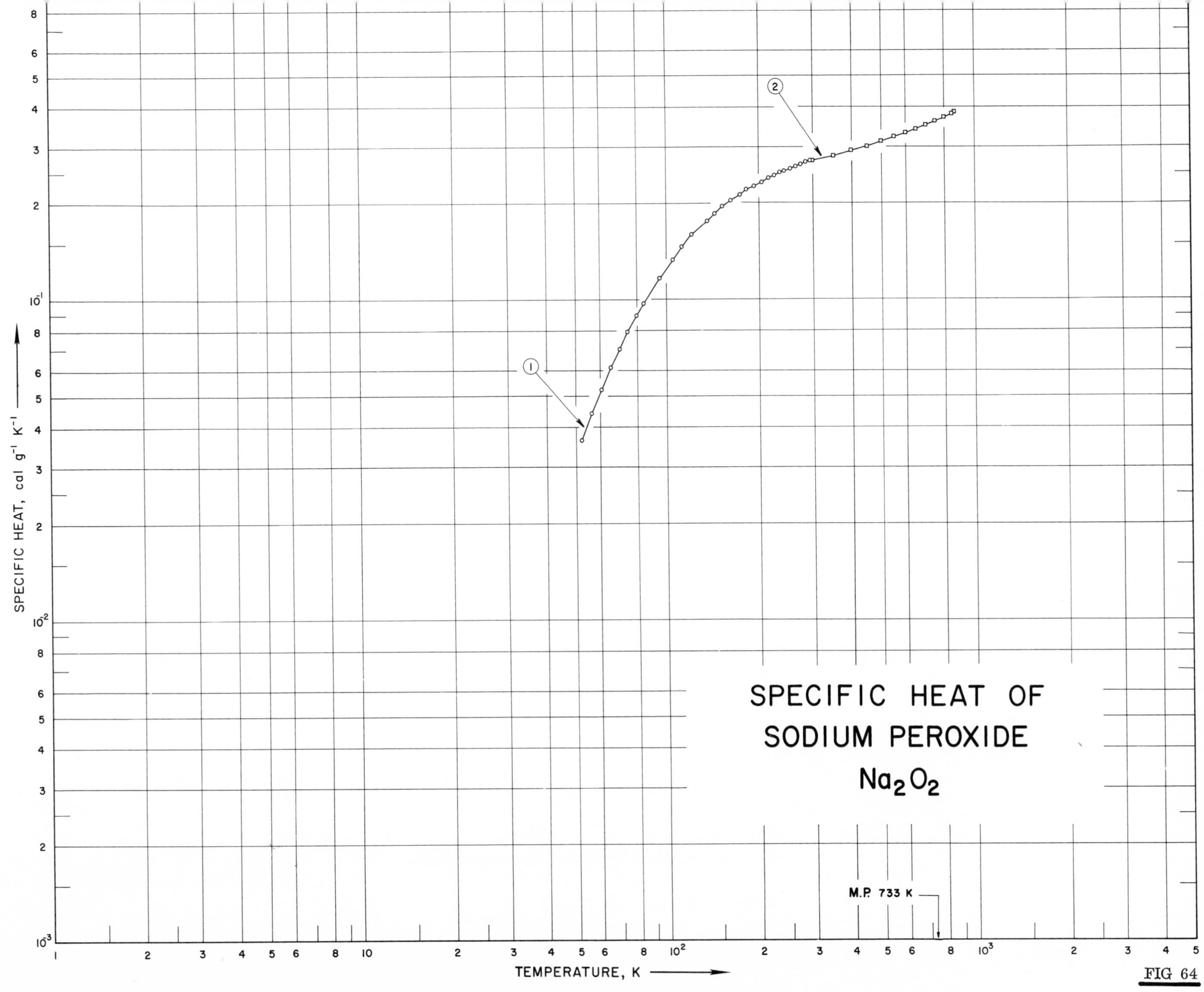
SPECIFIC HEAT OF
SODIUM PEROXIDE
Na_2O_2
M.P. 733 K
TEMPERATURE, K
SPECIFIC HEAT, cal g^{-1} K^{-1}

FIG 64

SPECIFICATION TABLE NO. 64 SPECIFIC HEAT OF SODIUM PEROXIDE Na_2O_2

[For Data Reported in Figure and Table No. 64]

Curve No.	Ref. No.	Year	Temp. Range, K	Reported Error, %	Name and Specimen Designation	Composition (weight percent), Specifications and Remarks
1	240	1953	52-298			94.0 ~~Na_2O_3~~ Na_2O_2, 3.6 ~~NaO~~ Na_2O, and 2.4 ~~$NaCO_3$~~ Na_2CO_3; corrected estimates for NaO small.
2	241	1959	298-869			98.3 Na_2O_2, 1.5 Na_2CO_3, and 0.2 Na_2O.

DATA TABLE NO. 64 SPECIFIC HEAT OF SODIUM PEROXIDE, Na_2O_2

[Temperature, T, K; Specific Heat, C_p, Cal $g^{-1}K^{-1}$]

T	C_p
CURVE 1	
52.31	3.654 x 10^{-2}
56.81	4.414
61.35	5.277
65.80	6.163
70.33	7.039
74.93	7.917
80.22	8.934
84.44	9.731
95.78	1.177 x 10^{-1}
106.35	1.349
114.49	1.476
123.72	1.607
137.21	1.776
145.08	1.864
154.62	1.960
164.79	2.054
175.81	2.149
184.78	2.212
195.21	2.269
205.61	2.344
215.99	2.405
225.87	2.458
235.68	2.501
244.94	2.532
255.27	2.586
265.47	2.626
275.32	2.662
285.55	2.703
296.32	2.726*
298.16	2.737
CURVE 2	
298	2.738 x 10^{-1}*
300	2.742
350	2.842
400	2.942
450	3.042
500	3.142
550	3.242
600	3.342
650	3.442
CURVE 2 (cont.)	
700	3.542 x 10^{-1}
750	3.642
800	3.742
850	3.842
869	3.880

* Not shown on Plot

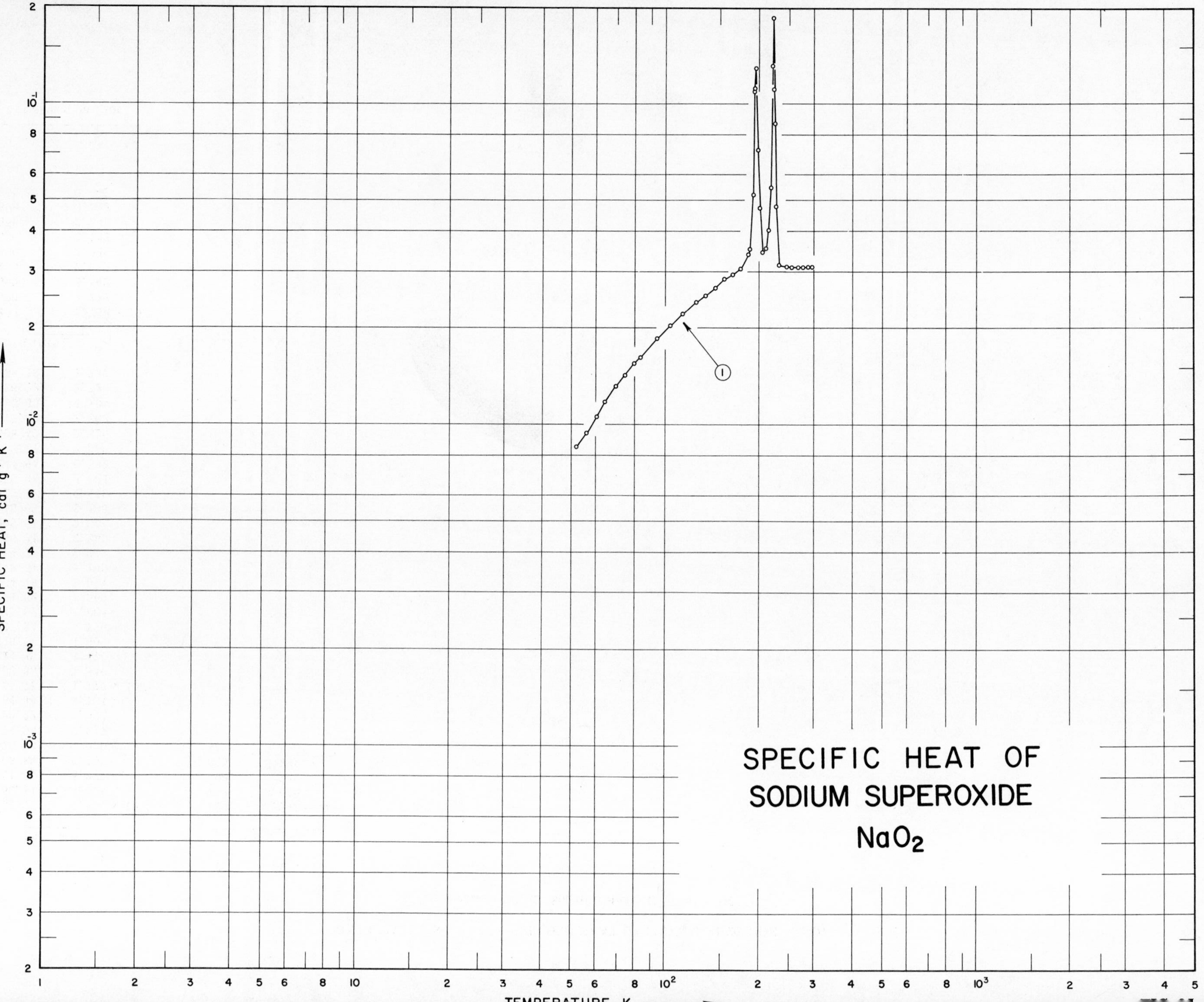
SPECIFIC HEAT OF
SODIUM SUPEROXIDE
NaO2
SPECIFIC HEAT, cal g-1 K-1
TEMPERATURE, K
1

SPECIFICATION TABLE NO. 65 SPECIFIC HEAT OF SODIUM SUPEROXIDE NaO_2

[For Data Reported in Figure and Table No. 65]

Curve No.	Ref. No.	Year	Temp. Range, K	Reported Error, %	Name and Specimen Designation	Composition (weight percent), Specifications and Remarks
1	240	1953	52-296			92.5 NaO_2, 6.0 Na_2O_2, and 1.5 Na_2CO_3; corrected for impurities.

DATA TABLE NO. 65 SPECIFIC HEAT OF SODIUM SUPEROXIDE NaO_2

[Temperature, T, K; Specific Heat, C_p, Cal g^{-1} K^{-1}]

T	C_p
CURVE 1	
52.13	8.524×10^{-2}
56.02	9.471
60.41	1.067×10^{-1}
64.97	1.196
69.61	1.323
74.36	1.442
79.71	1.563
83.56	1.640
94.46	1.871
104.34	2.056
114.45	2.236
126.97	2.429
135.91	2.558
145.86	2.695
155.74	2.855
165.83	2.957
175.79	3.098
185.70	3.402
187.15	3.540
191.51	5.268
194.23	1.105×10^{0}
194.38	1.133
196.23	1.316
198.49	7.264×10^{-1}
201.82	4.798
206.02	3.478
206.03	3.458*
210.30	3.584
214.99	3.947*
215.87	4.082
219.11	5.548
221.79	1.341×10^{0}
223.27	1.882
224.55	1.123
225.02	8.700×10^{-1}
227.98	4.860
232.14	3.184
235.94	3.158*
245.75	3.138
255.89	3.127
266.19	3.129
276.01	3.127
286.26	3.133
296.46	3.133

*Not shown on plot

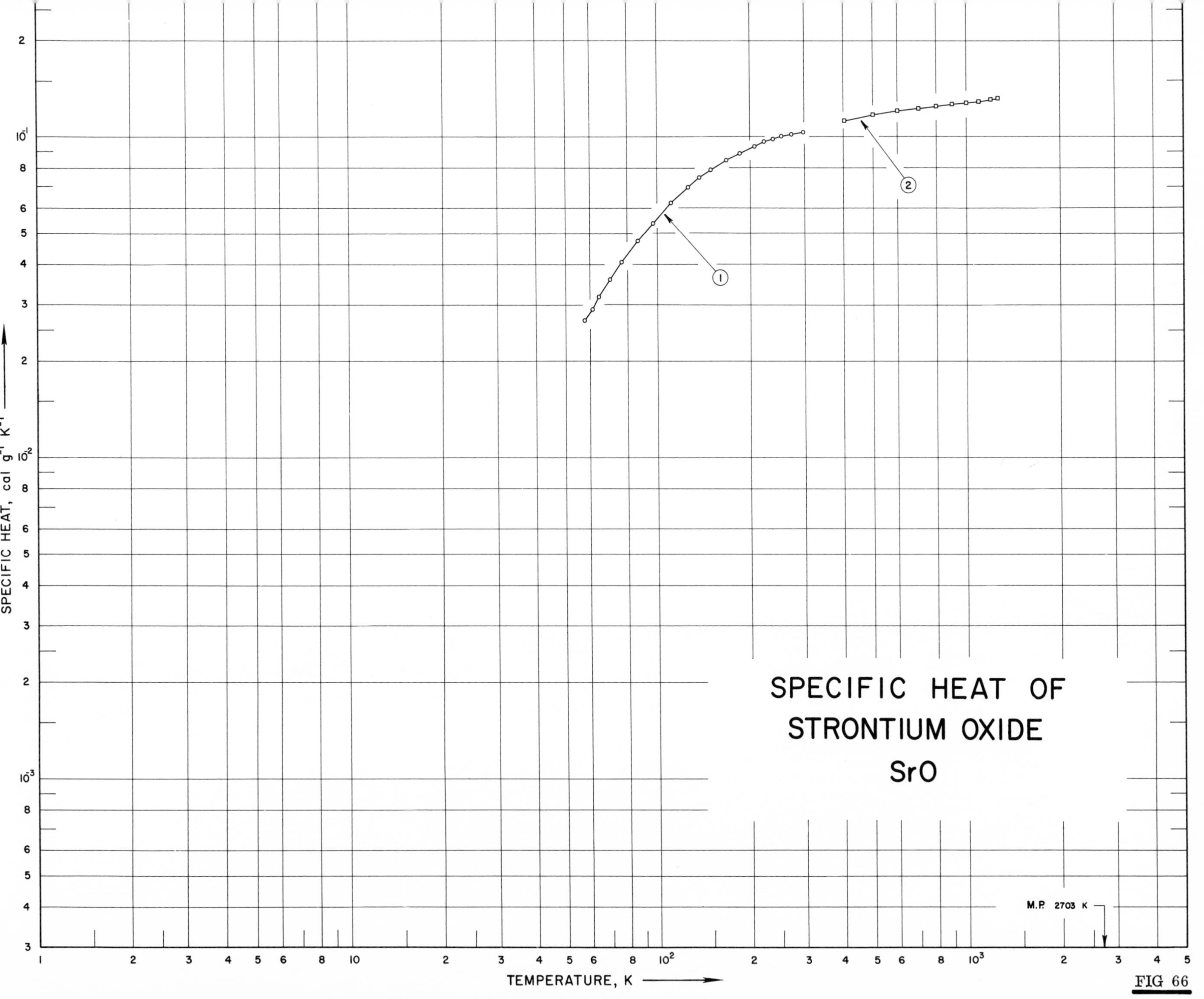
SPECIFIC HEAT OF
STRONTIUM OXIDE
SrO
M.P. 2703 K
TEMPERATURE, K
SPECIFIC HEAT, cal g^{-1} K^{-1}
1
2

FIG 66

SPECIFICATION TABLE NO. 66 SPECIFIC HEAT OF STRONTIUM OXIDE SrO

[For Data Reported in Figure and Table No. 66]

Curve No.	Ref. No.	Year	Temp. Range, K	Reported Error, %	Name and Specimen Designation	Composition (weight percent), Specifications and Remarks
1	41	1935	58-298			Kahlbaum best grade; impurities mainly carbonate; -14 + 35 mesh size; measured under vacuum.
2	42	1951	405-1265			SrO obtained by thermal decomposition of $SrCO_3$ at 1000 C in a vacuum.

DATA TABLE NO. 66 SPECIFIC HEAT OF STRONTIUM OXIDE, SrO

[Temperature, T, K; Specific Heat, C_p, Cal $g^{-1}K^{-1}$]

T	C_p
CURVE 1	
57.9	2.680×10^{-2}
61.2	2.897
64.3	3.163
70.0	3.589
76.8	4.071
86.8	4.725
97.1	5.378
111.2	6.219
126.1	6.948
138.4	7.455
149.9	7.890
168.6	8.451
185.3	8.873
207.7	9.307
222.6	9.630
238.3	9.823
254.4	1.002×10^{-1}
273.3	1.016
279.9	1.025*
290.4	1.030*
298.4	1.039
CURVE 2	
405	1.128×10^{-1}
500	1.175
600	1.207
700	1.231
800	1.250
900	1.267
1000	1.282
1100	1.295
1200	1.308
1265	1.317

*Not shown on Plot

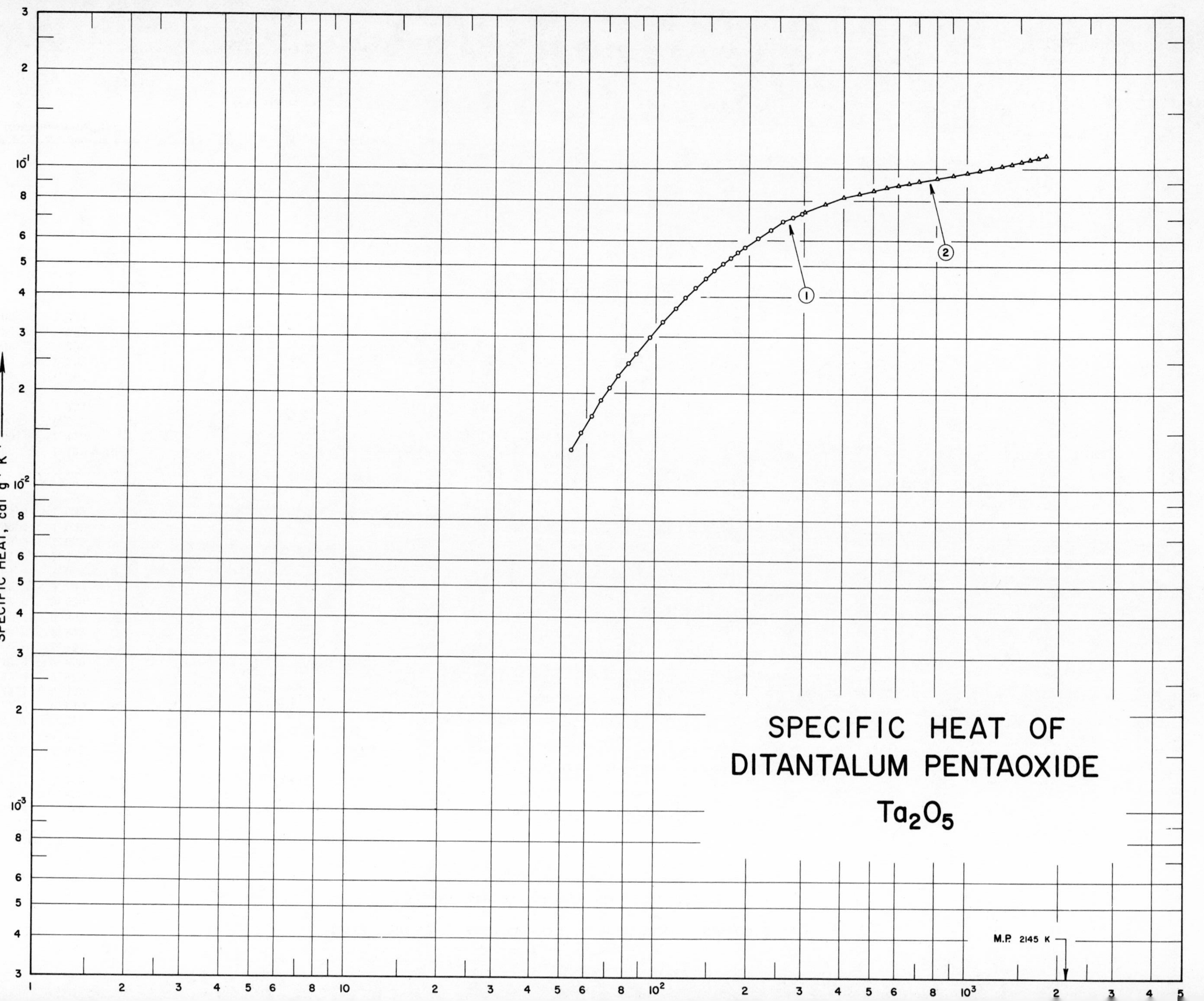
SPECIFIC HEAT OF
DITANTALUM PENTAOXIDE
Ta2O5
SPECIFIC HEAT, cal g-1 K-1
M.P. 2145 K
1
2

SPECIFICATION TABLE NO. 67 SPECIFIC HEAT OF DITANTALUM PENTAOXIDE Ta_2O_5

[For Data Reported in Figure and Table No. 67]

Curve No.	Ref. No.	Year	Temp. Range, K	Reported Error, %	Name and Specimen Designation	Composition (weight percent), Specifications and Remarks
1	127	1940	53-294			Virtually atomic weight purity; compressed into pellets.
2	114	1953	298-1800	≤ 0.4		Heated to 1200 C before measurement.

DATA TABLE NO. 67 SPECIFIC HEAT OF DITANTALUM PENTAOXIDE Ta_2O_5

[Temperature, T, K; Specific Heat, C_p, Cal $g^{-1}K^{-1}$]

T	C_p	T	C_p
CURVE 1		CURVE 2 (cont.)	
53.4	1.329 x 10^{-2}	800	9.351 x 10^{-2}
57.6	1.499	850	9.449*
62.1	1.693	900	9.544
66.6	1.891	950	9.635*
70.9	2.069	1000	9.724
75.4	2.257	1050	9.810*
81.5	2.474	1100	9.895
86.2	2.648	1150	9.979*
95.6	2.970	1200	1.006 x 10^{-1}
105.3	3.328	1250	1.014*
115.7	3.665	1300	1.022
124.9	3.959	1350	1.030*
134.9	4.269	1400	1.038
145.0	4.550	1450	1.046*
154.7	4.806	1500	1.054
165.0	5.055	1550	1.062*
174.6	5.283	1600	1.070
184.4	5.498	1650	1.077*
194.5	5.704	1700	1.085
203.4	5.895*	1750	1.093*
214.2	6.085	1800	1.100
224.5	6.273*		
234.7	6.461		
245.2	6.642*		
255.8	6.825		
265.8	6.934*		
275.5	7.049		
285.0	7.189*		
294.2	7.266		
CURVE 2			
298	7.307 x 10^{-2}*		
300	7.330		
350	7.799		
400	8.130		
450	8.380		
500	8.579		
550	8.747		
600	8.892		
650	9.021		
700	9.139		
750	9.248		

*Not shown on Plot

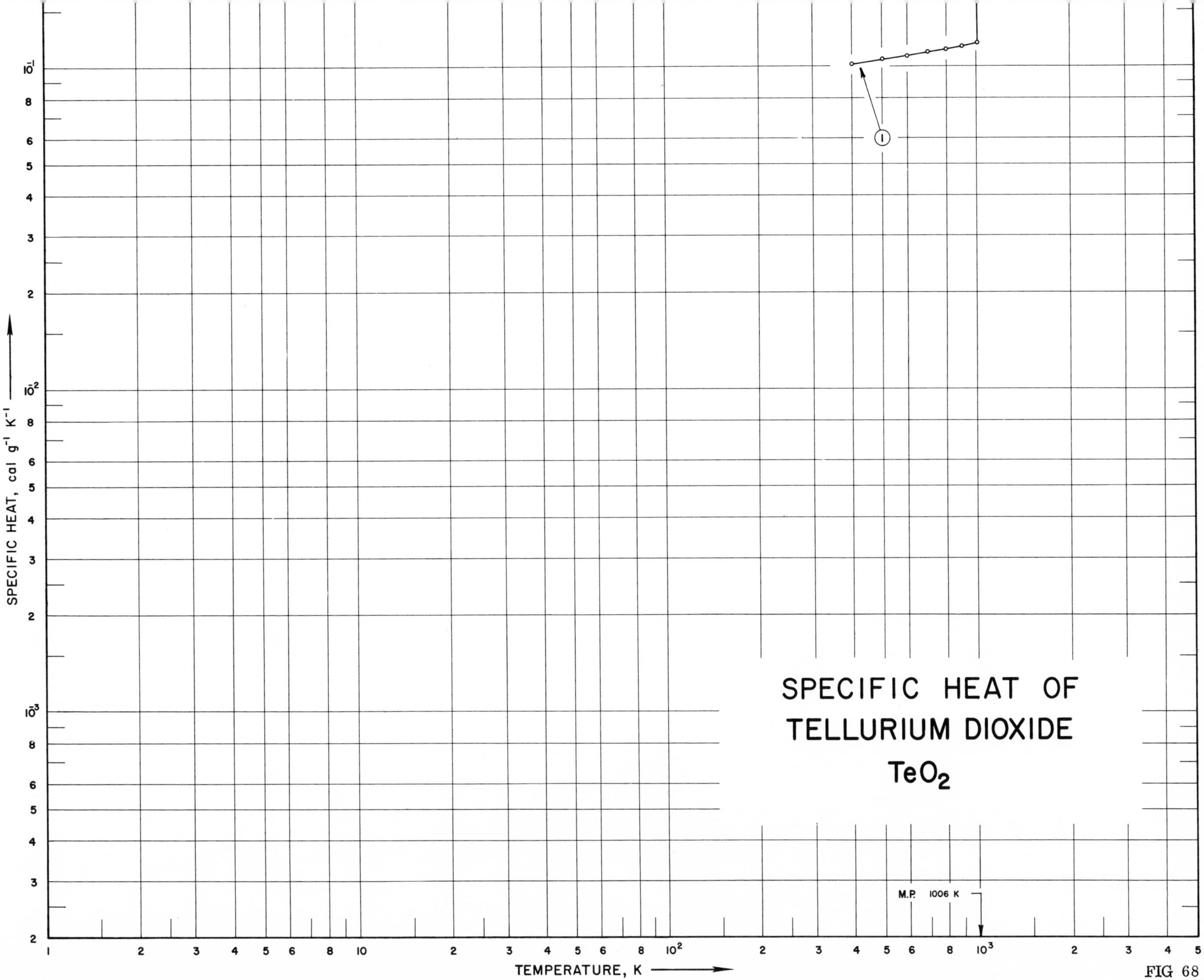

SPECIFIC HEAT OF TELLURIUM DIOXIDE

TeO_2

FIG 68

SPECIFICATION TABLE NO. 68 SPECIFIC HEAT OF TELLURIUM DIOXIDE TeO_2

[For Data Reported in Figure and Table No. 68]

Curve No.	Ref. No.	Year	Temp. Range, K	Reported Error, %	Name and Specimen Designation	Composition (weight percent), Specifications and Remarks
1	128	1962	400-1200	0.5		Spectroscopically pure with only traces of Ag, Ca, Na, Si, Mn; supplied by the Johnson, Matthey Co., Ltd.; sealed under argon atmosphere.

DATA TABLE NO. 68 SPECIFIC HEAT OF TELLURIUM DIOXIDE, TeO_2

[Temperature, T, K; Specific Heat, C_p, Cal $g^{-1}K^{-1}$]

T	C_p
CURVE 1	
400	1.016×10^{-1}
500	1.055
600	1.086
700	1.113
800	1.138
900	1.163
1000	1.187*
1006.16	1.188
1006.16	1.719
1100	1.722
1200	1.726

*Not shown on Plot

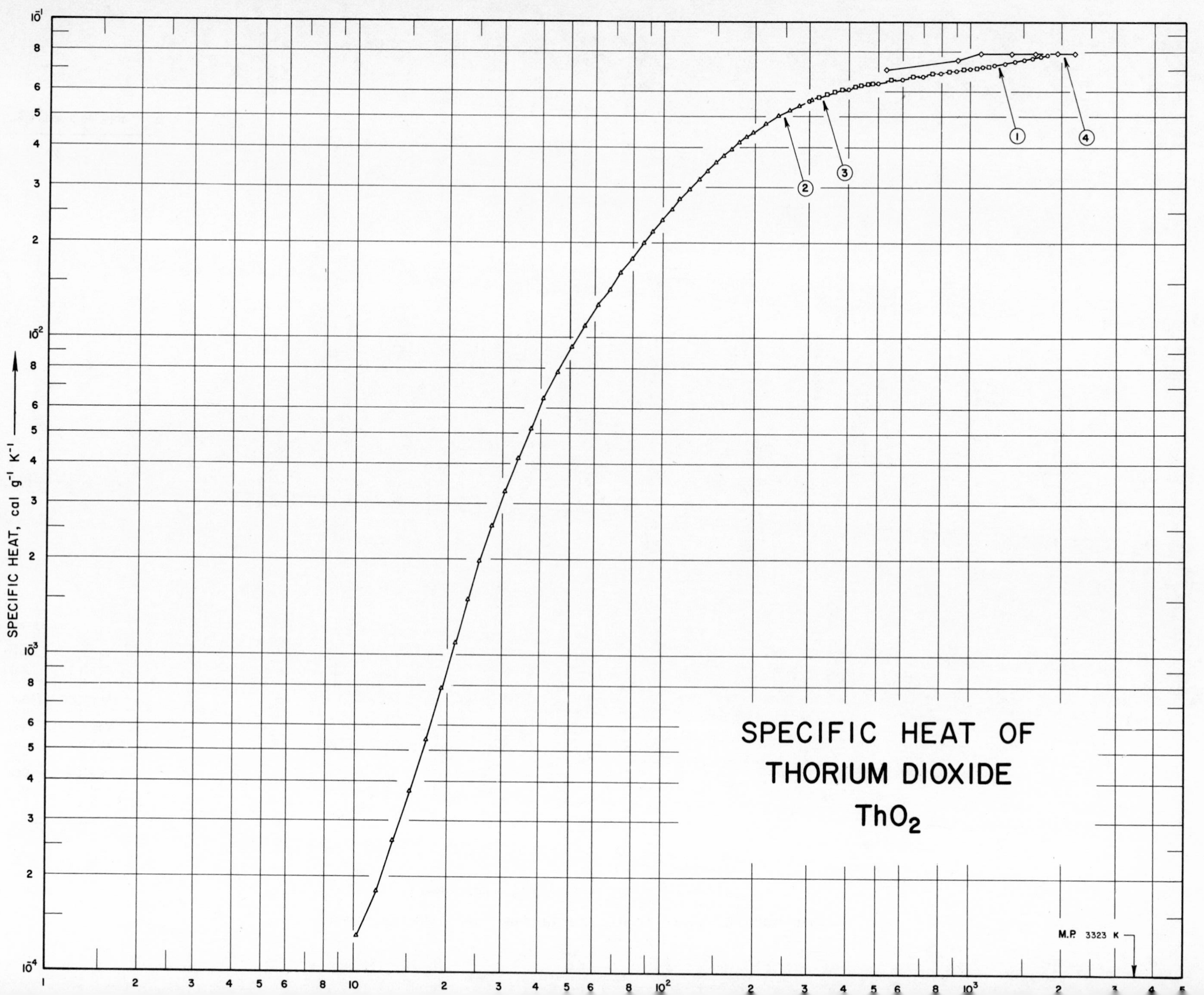
SPECIFIC HEAT OF
THORIUM DIOXIDE
ThO_2
SPECIFIC HEAT, cal g^{-1} K^{-1}
M.P. 3323 K
1
2
3
4

SPECIFICATION TABLE NO. 69 SPECIFIC HEAT OF THORIUM DIOXIDE ThO_2

[For Data Reported in Figure and Table No. 69]

Curve No.	Ref. No.	Year	Temp. Range, K	Reported Error, %	Name and Specimen Designation	Composition (weight percent), Specifications and Remarks
1	122	1941	299-1790	0. 5		99. 28 ThO_2, 0. 26 common metals, 0. 46 rare earth.
2	129, 130	1953	10-305			Thoria ThO_2, 0. 015 max.rare earth, 0. 005 each Al, Si, 0. 004 La, and < 0. 005 others.
3	131	1961	298-1200	0. 3-0. 5	Sample 1	99. 95 ThO_2, 0. 01 Al, 0. 005 Ca, 0. 005 Cu, 0. 004 Fe, <0. 001 B, and <0. 0005 Cr; supplied by the Lindsay Chemical Co; pressed; fired, sintered; density 605 lb ft^{-3}.
4	48	1962	533-2200	≤5. 0		Sample supplied by Zirconium Corp. of America; crushed in hardened steel mortar to pass 100-mesh screen, pressed and sintered; density at 25 C, before exposure: apparent density (ASTM method B311-58) 568 lb ft^{-3}, true density (by immersion in xylene) 604 lb ft^{-3}.

DATA TABLE NO. 69 SPECIFIC HEAT OF THORIUM DIOXIDE, ThO_2

[Temperature, T, K; Specific Heat, C_p, Cal $g^{-1}K^{-1}$]

CURVE 1

T	C_p
298.15	5.588 x 10^{-2}
300	5.601*
400	6.081
500	6.351
600	6.538
700	6.686
800	6.813
900	6.928
1000	7.036
1100	7.139
1200	7.238
1300	7.335
1400	7.430
1500	7.525
1600	7.618
1700	7.710
1790	7.793

CURVE 2

T	C_p
10.19	1.3 x 10^{-4}
11.85	1.8
13.44	2.6
15.15	3.7
17.09	5.38
19.14	7.84
21.18	1.09 x 10^{-3}
23.27	1.49
25.46	1.97
27.83	2.55
30.52	3.28
33.59	4.161
37.00	5.183
40.93	6.410
45.48	7.803
50.49	9.352
55.86	1.097 x 10^{-2}
61.63	1.271
61.43	1.265*
67.15	1.433
72.93	1.602
79.41	1.788
86.35	1.990

CURVE 2 (cont.)

T	C_p
61.39	1.263 x 10^{-2}*
67.11	1.453*
74.95	1.658*
84.37	1.932
92.34	2.157
99.61	2.355
106.97	2.554
114.43	2.752
122.30	2.955
130.77	3.165
139.55	3.378
148.37	3.577
157.30	3.770
166.27	3.949
175.73	4.127
185.75	4.301
195.89	4.468
185.66	4.297*
195.64	4.460*
205.69	4.615*
215.83	4.755
226.06	4.888*
236.26	5.013
246.35	4.748*
256.45	5.229
266.58	5.327*
276.62	5.418
286.58	5.501*
296.52	5.577*
305.40	5.645

CURVE 3

T	C_p
298.15	5.590 x 10^{-2}*
300	5.609*
320	5.749
340	5.870
360	5.973
380	6.060
400	6.139*
420	6.207
440	6.268
460	6.325

CURVE 3 (cont.)

T	C_p
480	6.370 x 10^{-2}
500	6.423*
550	6.522
600	6.605*
650	6.681
700	6.745*
750	6.806
800	6.859*
850	6.912
900	6.957*
950	7.007
1000	7.048*
1050	7.094
1100	7.135*
1150	7.177
1200	7.215*

CURVE 4

T	C_p
533.15	7.0 x 10^{-2}
810.93	7.5
1088.71	7.9
1366.48	7.9
1644.26	7.9
1922.04	7.9
2199.82	7.9

* Not shown on Plot

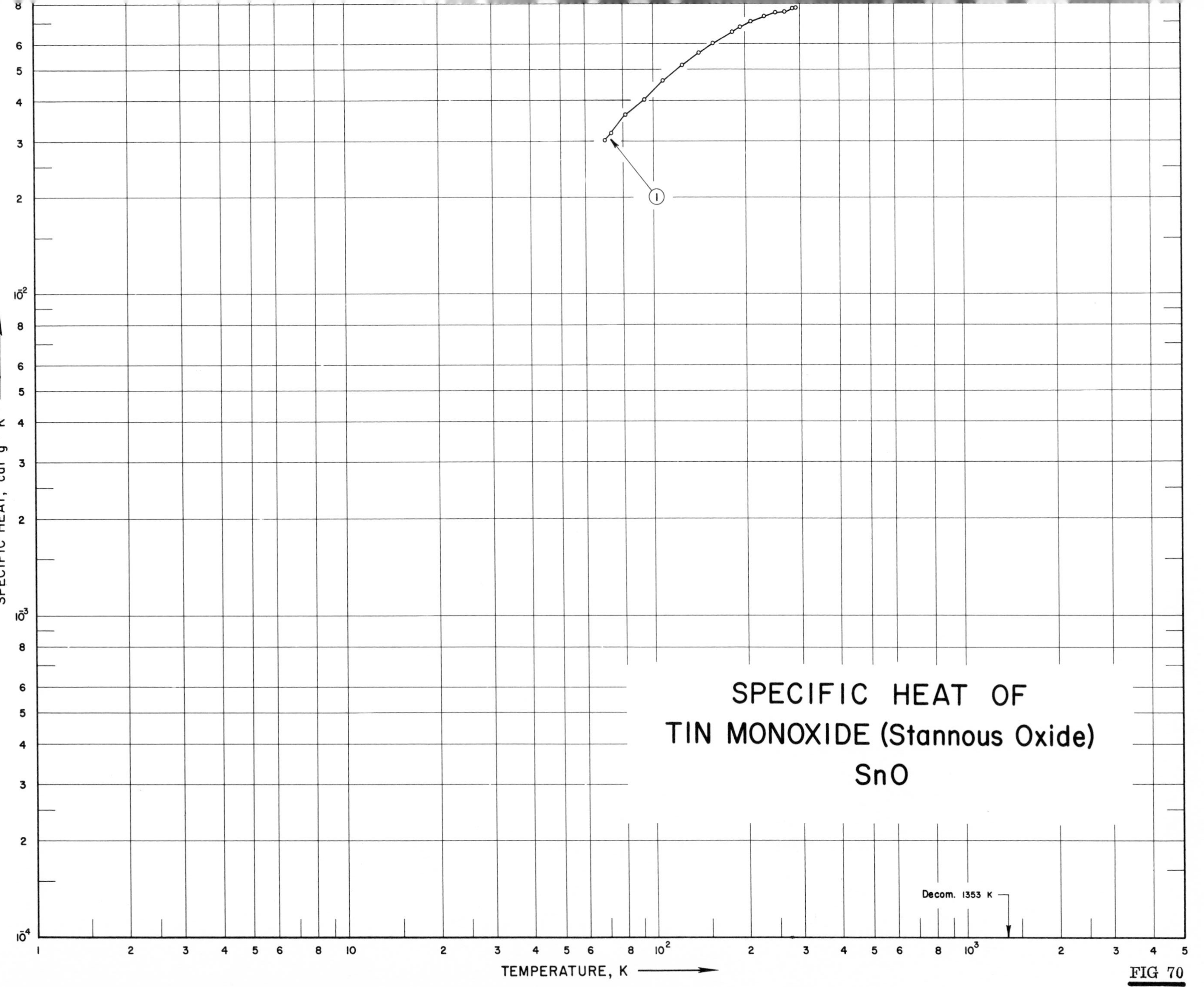
SPECIFIC HEAT OF
TIN MONOXIDE (Stannous Oxide)
SnO
Decom. 1353 K
1
SPECIFIC HEAT, cal g^{-1} K^{-1}
TEMPERATURE, K

FIG 70

SPECIFICATION TABLE NO. 70 SPECIFIC HEAT OF TIN MONOXIDE (STANNOUS OXIDE) SnO

[For Data Reported in Figure and Table No. 70]

Curve No.	Ref. No.	Year	Temp. Range, K	Reported Error, %	Name and Specimen Designation	Composition (weight percent), Specifications and Remarks
1	132	1929	69-292			98.0 SnO; prepared by precipitation of $Sn(OH)_2$ with ammonia from a boiling solution of pure $SnCl_2$.

DATA TABLE NO. 70 SPECIFIC HEAT OF TIN MONOXIDE (STANNOUS OXIDE), SnO

[Temperature, T, K; Specific Heat, C_p, Cal $g^{-1}K^{-1}$]

T	C_p
CURVE 1	
69.6	3.019 x 10^{-2}
73.0	3.192
81.7	3.645
93.3	4.033
108.1	4.641
125.0	5.178
140.8	5.638
156.9	6.047
181.7	6.575
193.5	6.809
209.7	7.079
230.2	7.362
251.3	7.550
268.3	7.602
284.3	7.803
292.5	7.810

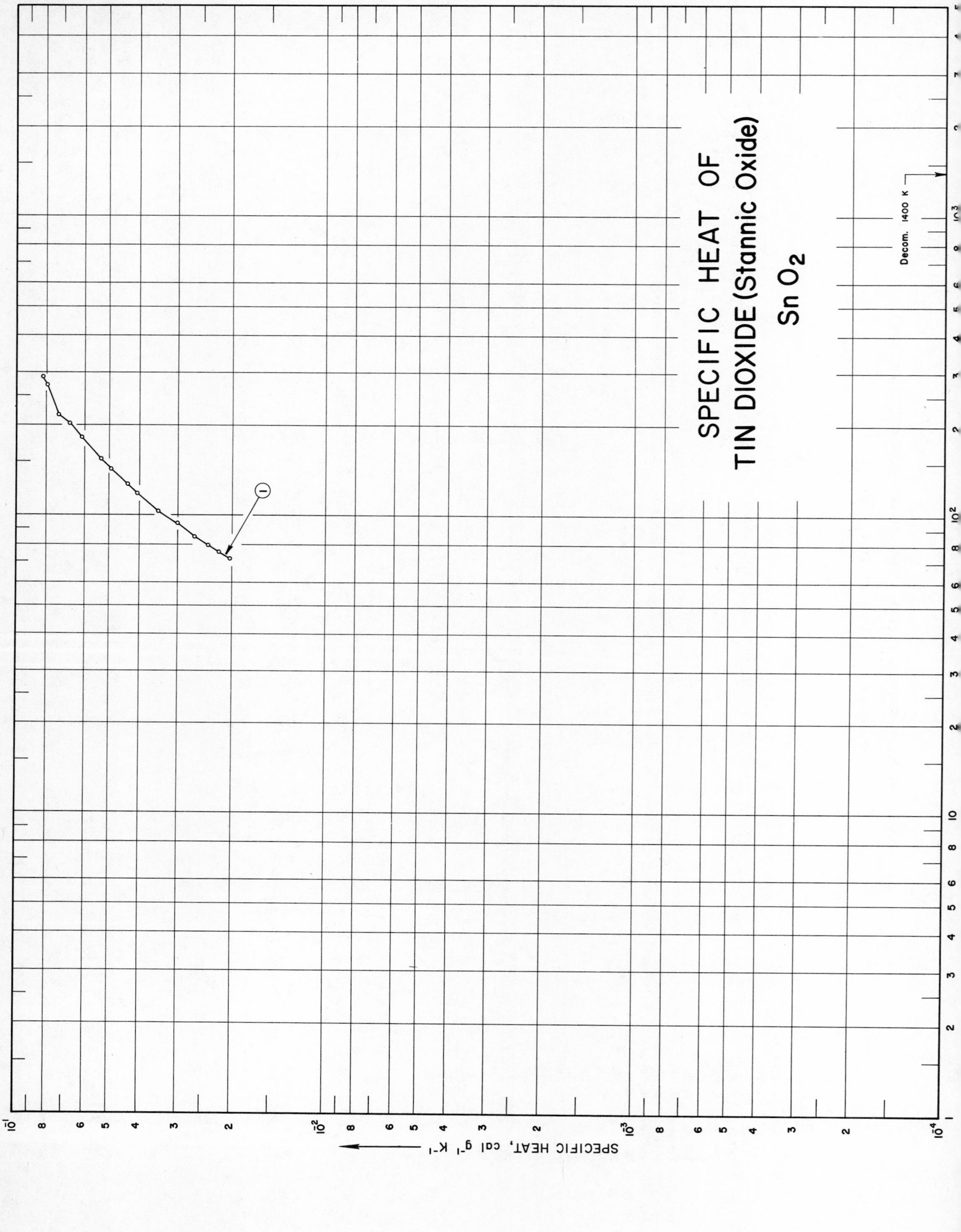

SPECIFIC HEAT OF TIN DIOXIDE (Stannic Oxide) SnO_2

SPECIFICATION TABLE NO. 71 SPECIFIC HEAT OF TIN DIOXIDE (STANNIC OXIDE) SnO_2

[For Data Reported in Figure and Table No. 71]

Curve No.	Ref. No.	Year	Temp. Range, K	Reported Error, %	Name and Specimen Designation	Composition (weight percent), Specifications and Remarks
1	132	1929	72-289			>99. 0 SnO_2, prepared by action of dilute HNO_3 on pure electrolytic tin.

DATA TABLE NO. 71 SPECIFIC HEAT OF TIN DIOXIDE (STANNIC OXIDE), SnO_2

[Temperature, T, K; Specific Heat, C_p, Cal $g^{-1}K^{-1}$]

T	C_p
CURVE 1	
71.8	2.048 x 10^{-2}
75.2	2.205
79.3	2.401
84.8	2.651
93.8	3.007
103.6	3.478
118.2	4.091
126.0	4.393
124.2	4.938
154.6	5.377
181.8	6.184
202.7	6.755
215.5	7.332
271.6	7.976
273.4	8.062*
287.2	8.162*
289.4	8.235

* Not shown on Plot

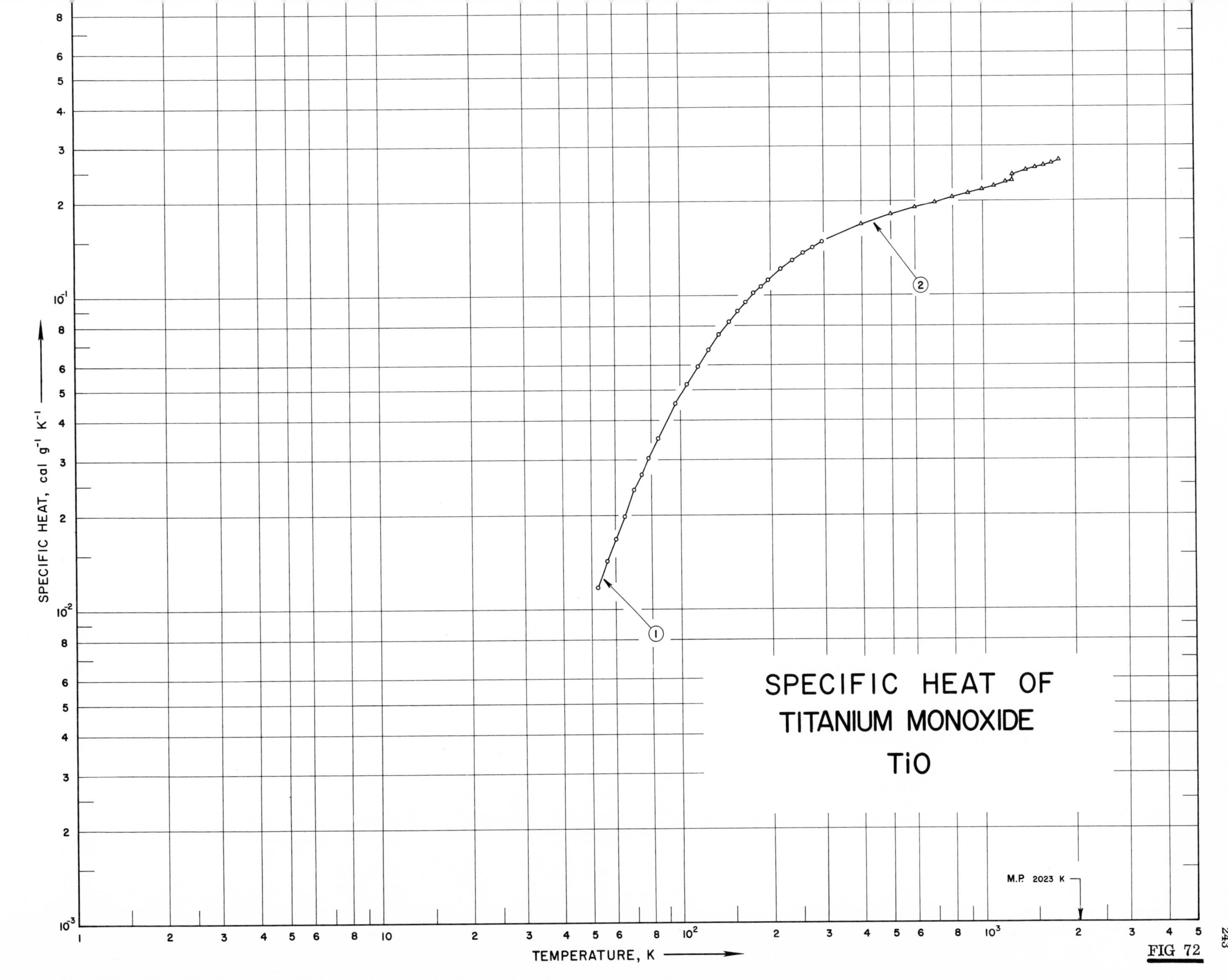
SPECIFIC HEAT OF
TITANIUM MONOXIDE
TiO
M.P. 2023 K
SPECIFIC HEAT, cal g^{-1} K^{-1}
TEMPERATURE, K
1
2

FIG 72

SPECIFICATION TABLE NO. 72 SPECIFIC HEAT OF TITANIUM MONOXIDE TiO

[For Data Reported in Figure and Table No. 72]

Curve No.	Ref. No.	Year	Temp. Range, K	Reported Error, %	Name and Specimen Designation	Composition (weight percent), Specifications and Remarks
1	133	1946	53-296			99.2 TiO, 0.7 Si, and 0.1 Ti.
2	134	1946	298-1800	1		99.2 TiO, 0.1 TiC, and 0.7 Si; measured in helium atmosphere.

DATA TABLE NO. 72 SPECIFIC HEAT OF TITANIUM MONOXIDE, TiO

[Temperature, T, K; Specific Heat, C_p, Cal $g^{-1}K^{-1}$]

T	C_p
CURVE 1	
52.6	1.18×10^{-2}
56.7	1.43
60.6	1.690
64.8	1.989
69.7	2.410
74.2	2.704
78.6	3.049
84.5	3.518
96.9	4.516
105.7	5.228
115.1	5.977
125.1	6.759
135.1	7.510
145.8	8.268
155.4	8.925
165.5	9.565
176.2	1.021×10^{-1}
185.9	1.075
196.3	1.128
205.8	1.177*
215.8	1.222
226.1	1.264*
235.8	1.300
246.2	1.337*
256.2	1.375
266.2	1.406*
276.2	1.434
286.1	1.462*
296.3	1.490*
CURVE 2	
α298.15	1.495×10^{-1}*
300	1.499*
400	1.698
500	1.820
600	1.912
700	1.989
800	2.059
900	2.125
1000	2.188
1100	2.250
1200	2.309
α1264	2.347
CURVE 2 (cont.)	
β1264	2.448×10^{-1}
1300	2.465*
1400	2.512
1500	2.559
1600	2.606
1700	2.652
β1800	2.700

*Not shown on Plot

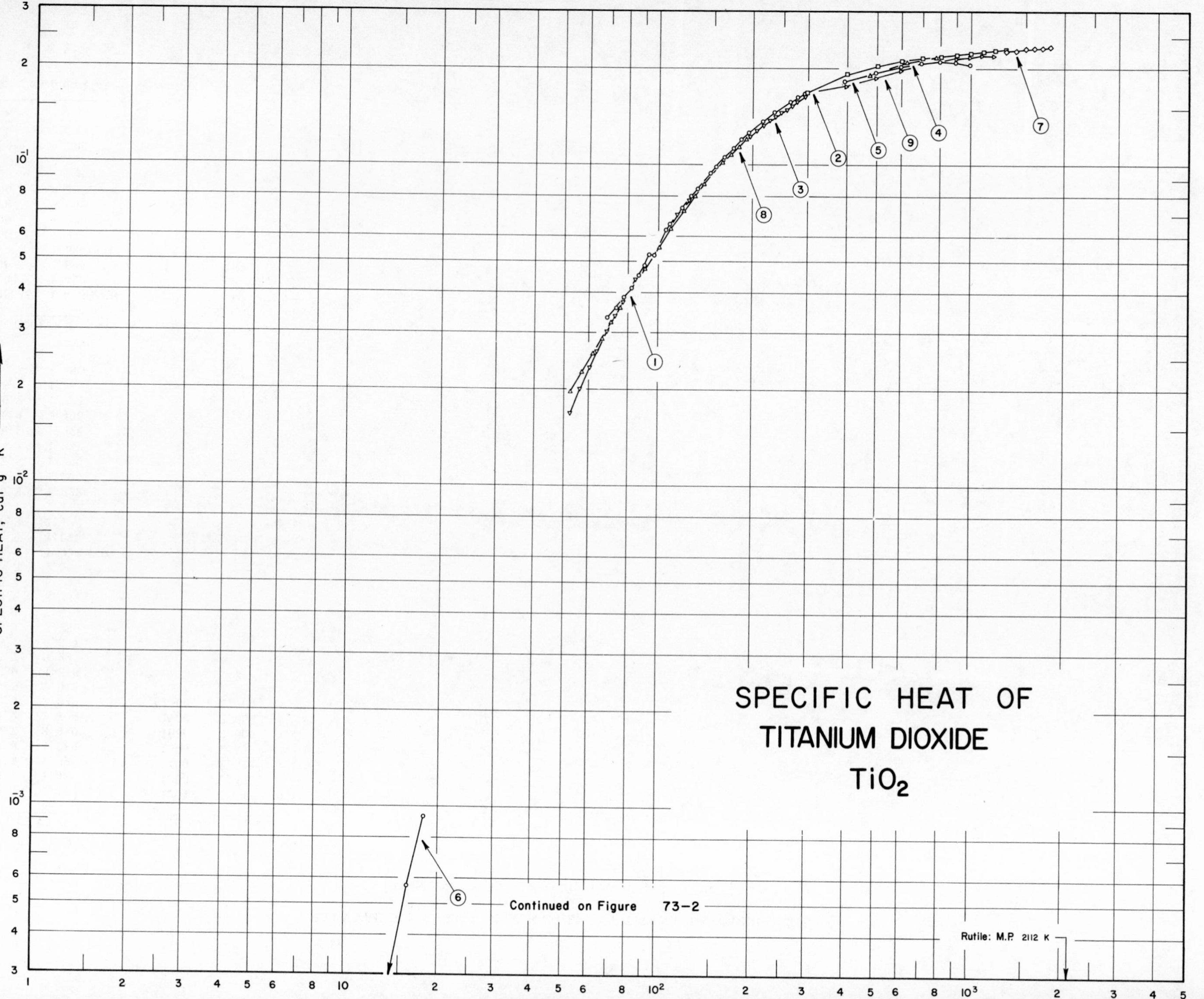
SPECIFIC HEAT OF
TITANIUM DIOXIDE
TiO_2
SPECIFIC HEAT, cal g^{-1} K^{-1}
Continued on Figure 73-2
Rutile: M.P. 2112 K
1
2
3
4
5
6
7
8
9

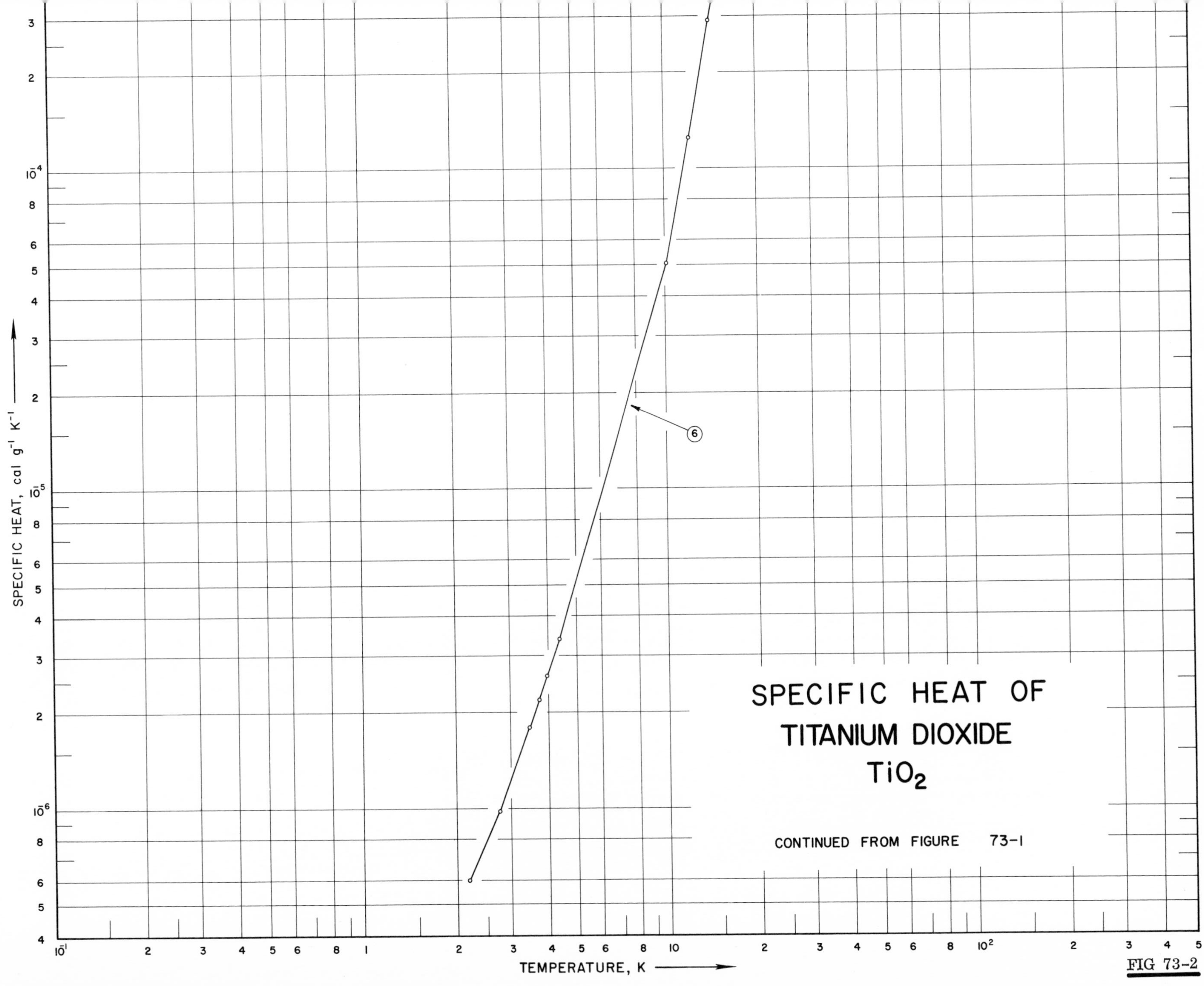
SPECIFIC HEAT OF
TITANIUM DIOXIDE
TiO2
CONTINUED FROM FIGURE 73-1
6
TEMPERATURE, K
SPECIFIC HEAT, cal g-1 K-1

FIG 73-2

SPECIFICATION TABLE NO. 73 SPECIFIC HEAT OF TITANIUM DIOXIDE TiO_2

[For Data Reported in Figure and Table No. 73]

Curve No.	Ref. No.	Year	Temp. Range, K	Reported Error, %	Name and Specimen Designation	Composition (weight percent), Specifications and Remarks
1	135	1939	69-295			< 0.4 SiO_2; powdered, pressed into pellets.
2	134	1946	298-1300	0.5	Anatase	99.07 TiO_2, 0.30 SiO_2, 0.15 CaO, and 0.07 others; dried 4 hrs at 1050 C.
3	136	1947	53-298	± 0.3	Anatase	99.3 TiO_2 and 0.3 H_2O; density 242 lb ft^{-3}; corrected for H_2O.
4	92	1950	473-773			Doubtful accuracy.
5	137	1956	393-993		Anatase	X-ray showed only lines of anatase; synthetically prepared from doubly distilled $TiCl_4$, heated 4 hrs at 565 C; density 243.7 lb ft^{-3}.
6	138	1958	2-18		Rutile	Transparent, but slightly yellow.
7	134	1946	298-1800	0.2	Rutile	97.90 TiO_2, 0.55 ZrO_2, 0.5 SiO_2, 0.27 V_2O_5, 0.15 CaO, 0.15 Fe_2O_3, 0.12 Al_2O_3, and 0.10 others.
8	136	1947	53-298	± 0.3	Rutile	99.7 TiO_2; density 265 lb ft^{-3}.
9	137	1956	293-1193		Rutile	X-ray showed no lines of anatase; white with slight yellow cast; synthetically prepared from doubly distilled $TiCl_4$; heated 1.5 hrs at 930 C; density 259.1 lb ft^{-3}.

DATA TABLE NO. 73 SPECIFIC HEAT OF TITANIUM DIOXIDE, TiO_2

[Temperature, T, K; Specific Heat, C_p, Cal $g^{-1}K^{-1}$]

T	C_p
CURVE 1	
Series 1	
85.40	4.48 x 10^{-2}*
89.15	4.84*
93.23	5.26*
106.02	6.31*
112.09	6.67*
119.50	7.35
128.35	8.01*
137.52	8.76*
153.80	9.87
163.53	1.06 x 10^{-1}
174.28	1.13
184.34	1.20
194.77	1.26
215.08	1.37*
224.70	1.42*
234.22	1.46*
244.20	1.50*
264.22	1.58*
274.23	1.62*
295.07	1.68*
Series 2	
68.78	3.34 x 10^{-2}
73.06	3.57
77.80	3.87
82.10	4.12
86.13	4.51
90.05	4.87
94.32	5.24
105.60	6.27
109.69	6.52
134.65	8.42
146.68	9.42
277.09	1.621 x 10^{-1}
286.40	1.656*
291.90	1.672
CURVE 2	
298.15	1.688 x 10^{-1}
300	1.695*
CURVE 2 (cont.)	
400	1.926 x 10^{-1}
500	2.042
600	2.110
700	2.156
800	2.191
900	2.219
1000	2.244
1100	2.265
1200	2.285
1300	2.303
CURVE 3	
52.5	1.683 x 10^{-2}
56.2	2.000
60.1	2.340
63.9	2.638
68.7	3.024
72.9	3.383
77.2	3.748
84.8	4.402
95.2	5.330*
104.5	6.125
115.2	6.992
125.3	7.772*
134.9	8.479*
145.7	9.229*
155.5	9.905
165.6	1.056 x 10^{-1}*
175.6	1.119*
185.6	1.174
195.9	1.230
205.8	1.283
216.3	1.337
226.3	1.377
235.8	1.421
246.2	1.458
256.1	1.498
266.3	1.539
276.5	1.579*
286.0	1.612
295.8	1.648*
298.16	1.654
CURVE 4	
473	1.91 x 10^{-1}
623	2.10
773	2.18
CURVE 5	
393	1.837 x 10^{-1}
493	1.949
593	2.034
693	2.086
793	2.108
893	2.099
993	2.060
CURVE 6	
2.16	5.982 x 10^{-7}
2.72	9.870
3.42	1.795 x 10^{-6}
3.69	2.183
3.92	2.602
4.31	3.380
10.0	5.085 x 10^{-5}
12.0	1.256 x 10^{-4}
14.0	2.931
16.0	5.683
18.0	9.332
CURVE 7	
298.15	1.688 x 10^{-1}*
300	1.695*
400	1.920*
500	2.031*
600	2.098*
700	2.142*
800	2.174*
900	2.201*
1000	2.224*
1100	2.244*
1200	2.261*
1300	2.279
1400	2.294
1500	2.309
CURVE 7 (cont.)	
1600	2.324 x 10^{-1}
1700	2.338
1800	2.352
CURVE 8	
52.5	1.967 x 10^{-2}
57.0	2.260
61.9	2.594
66.0	2.887
70.8	3.247
75.4	3.596
79.5	3.914*
90.5	4.775
100.1	5.554
109.7	6.340
120.2	7.184
130.5	8.005
140.4	8.766
150.7	9.516*
160.7	1.022 x 10^{-1}
170.6	1.086
180.7	1.150
190.9	1.208
201.1	1.263*
211.2	1.317*
221.2	1.365*
231.1	1.406*
240.9	1.447*
251.1	1.491*
261.3	1.529*
271.1	1.564*
281.0	1.397*
290.7	1.627*
297.7	1.645*
298.16	1.647*
CURVE 9	
293	1.681 x 10^{-1}*
393	1.778
493	1.886
593	1.979*
693	2.056
CURVE 9 (cont.)	
793	2.116 x 10^{-1}*
893	2.160
993	2.190
1093	2.204
1193	2.203

* Not shown on Plot

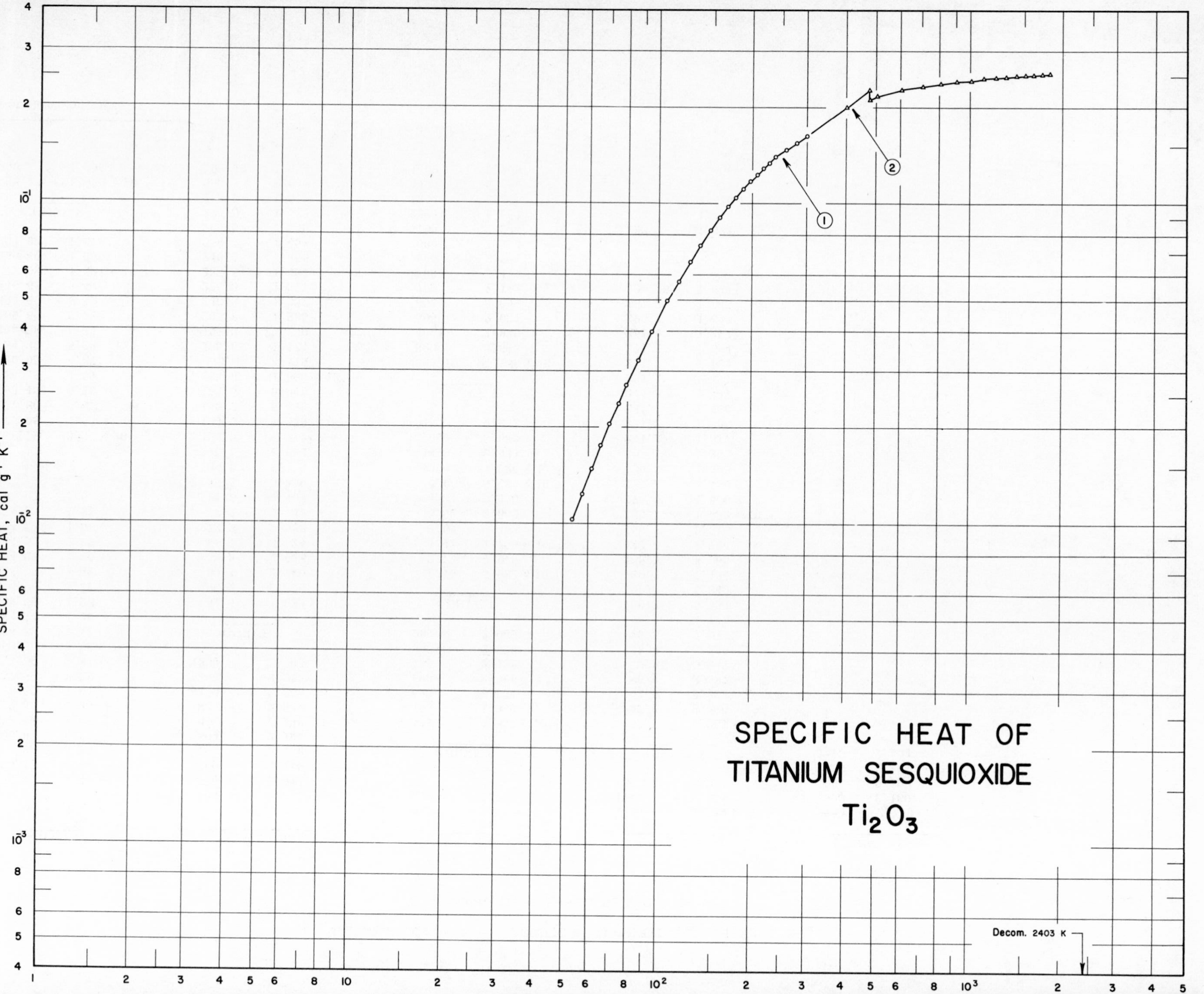
SPECIFIC HEAT OF
TITANIUM SESQUIOXIDE
Ti_2O_3
Decom. 2403 K
SPECIFIC HEAT, cal g^{-1} K^{-1}
1
2

SPECIFICATION TABLE NO. 74 SPECIFIC HEAT OF TITANIUM SESQUIOXIDE Ti_2O_3

[For Data Reported in Figure and Table No. 74]

Curve No.	Ref. No.	Year	Temp. Range, K	Reported Error, %	Name and Specimen Designation	Composition (weight percent), Specifications and Remarks
1	133	1946	53-298			99.4 Ti_2O_3, 0.3 SiO_2, and 0.3 TiC.
2	134	1946	298-1800	0.2-1.5		99.4 Ti_2O_3, 0.3 TiC, and 0.3 SiO_2; prepared from finely ground reaction mixture of C and TiO_2 by heating 20 hrs at 1400 C.

DATA TABLE NO. 74 SPECIFIC HEAT OF TITANIUM SESQUIOXIDE, Ti_2O_3

[Temperature, T, K; Specific Heat, C_p, Cal $g^{-1}K^{-1}$]

T	C_p	T	C_p
CURVE 1		CURVE 2 (cont.)	
53.0	1.023 x 10^{-2}	800	2.373 x 10^{-1}
57.0	1.234	900	2.405
61.1	1.481	1000	2.431
65.4	1.750	1100	2.453
69.7	2.039	1200	2.471
74.3	2.361	1300	2.487
78.8	2.693	1400	2.502
85.6	3.215	1500	2.516
94.8	3.951	1600	2.529
106.4	4.924	1700	2.541
115.2	5.680	β1800	2.553
125.2	6.524		
135.2	7.357		
145.6	8.198		
155.5	8.985		
165.5	9.722		
175.6	1.045 x 10^{-1}		
185.6	1.109		
195.9	1.173		
205.9	1.234		
216.2	1.292		
226.2	1.344		
235.7	1.399		
235.8	1.401*		
240.3	1.437*		
244.7	1.430*		
246.2	1.433*		
256.0	1.470		
266.1	1.507*		
276.2	1.545		
286.1	1.579*		
296.4	1.612*		
298.16	1.618		
CURVE 2			
α298.15	1.618 x 10^{-1}*		
300	1.624*		
400	1.997		
α473	2.268		
β473	2.136		
500	2.173		
600	2.269		
700	2.330		

* Not shown on Plot

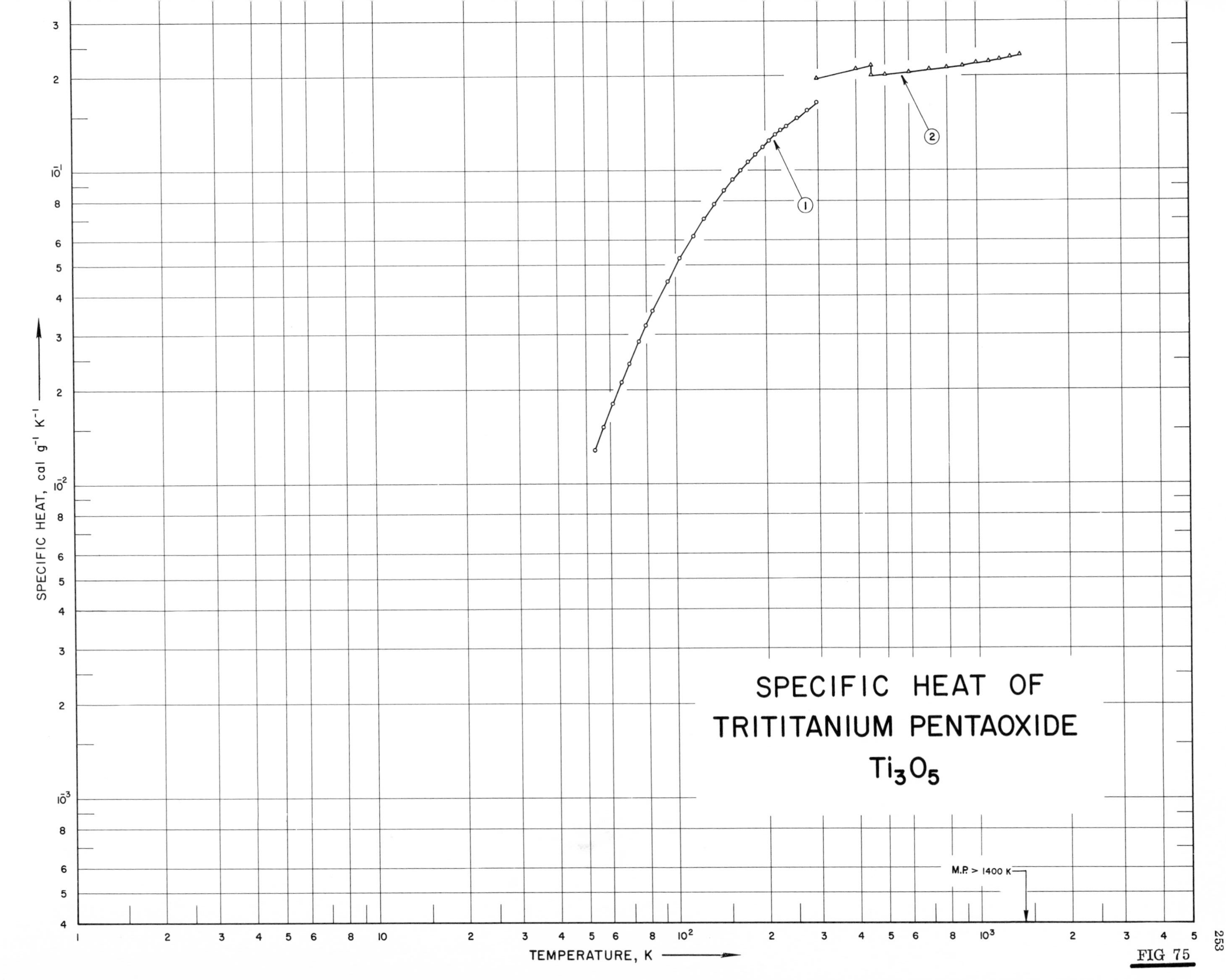
SPECIFIC HEAT OF
TRITITANIUM PENTAOXIDE
Ti_3O_5
M.P. > 1400 K
TEMPERATURE, K
SPECIFIC HEAT, cal g^{-1} K^{-1}
1
2

FIG 75

SPECIFICATION TABLE NO. 75 SPECIFIC HEAT OF TRITITANIUM PENTAOXIDE Ti_3O_5

[For Data Reported in Figure and Table No. 75]

Curve No.	Ref. No.	Year	Temp. Range, K	Reported Error, %	Name and Specimen Designation	Composition (weight percent), Specifications and Remarks
1	133	1946	53-298			99.1 Ti_3O_5, 0.7 SiO_2, and 0.2 TiC.
2	134	1946	298-1400	0.2-4		99.1 Ti_3O_5, 0.7 SiO_2, and 0.2 TiC; prepared by reduction of TiO_2 with C under vacuum for 8 hrs at 1300 C.

DATA TABLE NO. 75 SPECIFIC HEAT OF TRITITANIUM PENTAOXIDE Ti_3O_5

[Temperature, T, K; Specific Heat, C_p, Cal $g^{-1}K^{-1}$]

T	C_p
CURVE 1	
53.1	1.286×10^{-2}
56.9	1.523
61.0	1.803
65.4	2.111
69.6	2.434
75.0	2.854
79.4	3.208
83.9	3.576
94.1	4.420
104.3	5.284
115.1	6.200
125.2	7.045
135.1	7.845
145.8	8.672
155.7	9.383
165.7	1.009×10^{-1}
175.6	1.078
185.7	1.139
195.9	1.198
205.5	1.254
216.1	1.309
226.1	1.356
235.8	1.399
246.0	1.445*
256.1	1.489
266.2	1.529*
276.3	1.568
286.3	1.609*
296.6	1.649*
298.16	1.654
CURVE 2	
α298.15	1.978×10^{-1}
300	1.981*
400	2.113
α450	2.179
β450	2.02
500	2.04
600	2.07
700	2.11
CURVE 2 (cont.)	
800	2.15×10^{-1}
900	2.18
1000	2.22
1100	2.25
1200	2.29
1300	2.32
β1400	2.36

*Not shown on Plot

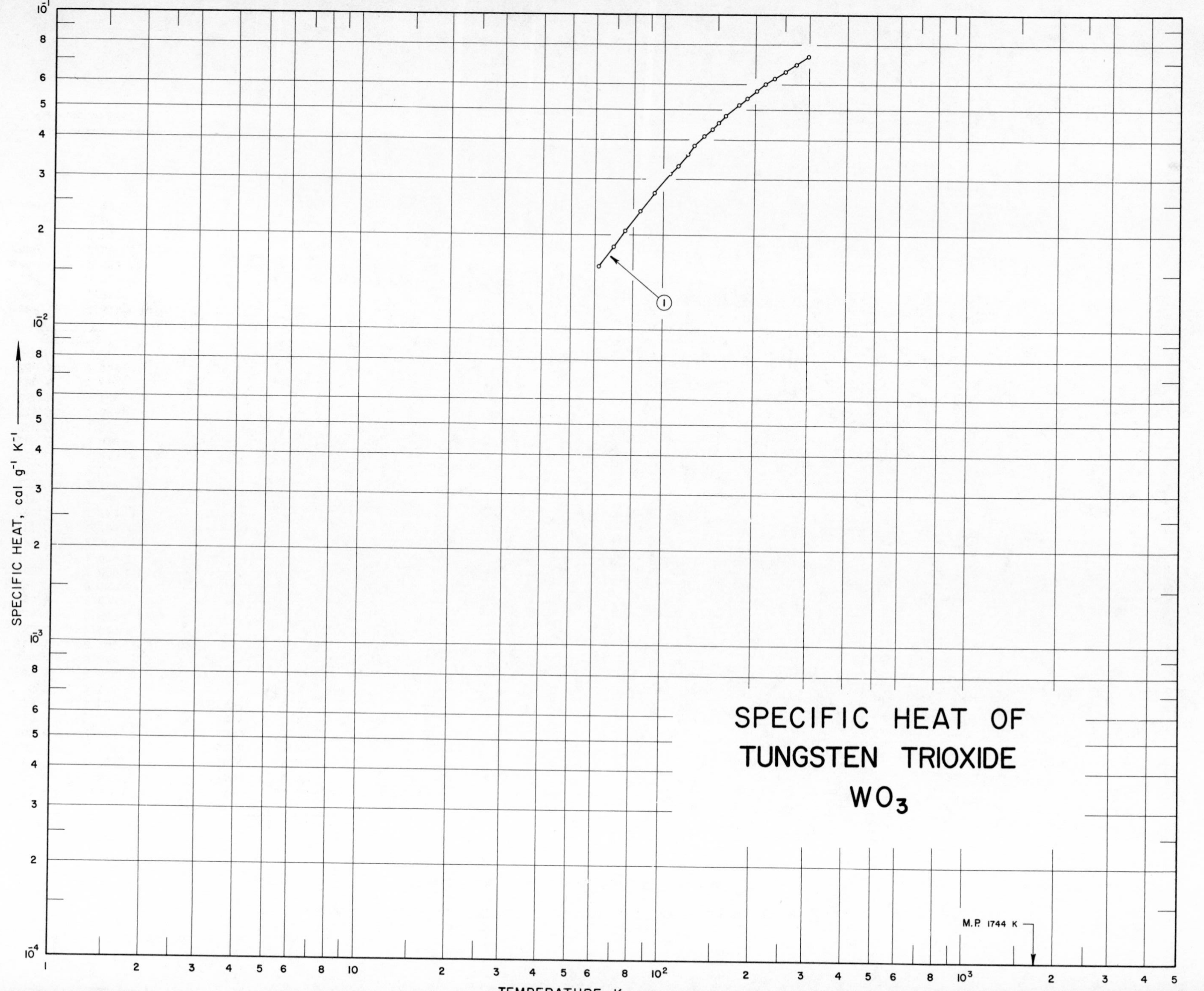

SPECIFIC HEAT OF
TUNGSTEN TRIOXIDE
WO3
M.P. 1744 K
1
SPECIFIC HEAT, cal g⁻¹ K⁻¹

SPECIFICATION TABLE NO. 76 SPECIFIC HEAT OF TUNGSTEN TRIOXIDE WO_3

[For Data Reported in Figure and Table No. 76]

Curve No.	Ref. No.	Year	Temp. Range, K	Reported Error, %	Name and Specimen Designation	Composition (weight percent), Specifications and Remarks
1	106	1945	63-299	± 0.3		Rhombohedral crystal.

DATA TABLE NO. 76 SPECIFIC HEAT OF TUNGSTEN TRIOXIDE, WO_3

[Temperature, T, K; Specific Heat, C_p, Cal $g^{-1}K^{-1}$]

T	C_p
CURVE 1	
62.90	1.59 x 10^{-2}
63.37	1.64*
68.98	1.84
71.55	1.97*
75.25	2.06
79.68	2.25*
84.20	2.37
88.90	2.53*
93.80	2.72
105.22	3.14
112.44	3.31
120.52	3.63
126.86	3.854
136.30	4.126
144.96	4.327
152.88	4.566
160.16	4.801
167.68	4.979*
177.86	5.206
183.46	5.335*
189.08	5.452
203.95	5.778
212.20	5.963*
216.13	6.054
230.95	6.323
236.40	6.399*
242.85	6.531*
251.79	6.626
267.50	6.887*
273.68	6.989
287.03	7.232*
299.20	7.402

*Not shown on Plot

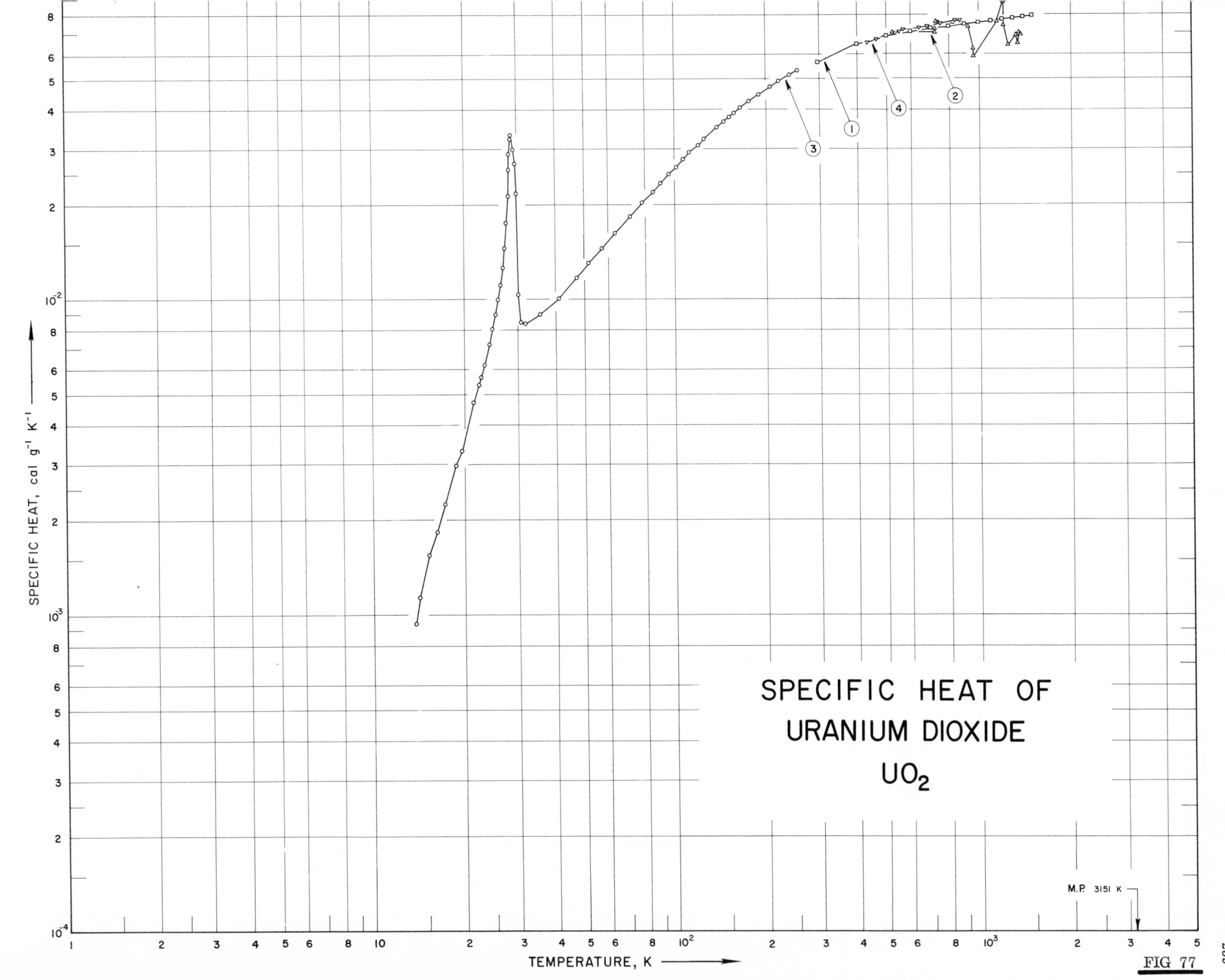
SPECIFIC HEAT OF
URANIUM DIOXIDE
UO_2
M.P. 3151 K
TEMPERATURE, K
SPECIFIC HEAT, cal g^{-1} K^{-1}
1
2
3
4

FIG 77

SPECIFICATION TABLE NO. 77 SPECIFIC HEAT OF URANIUM DIOXIDE UO_2

[For Data Reported in Figure and Table No. 77]

Curve No.	Ref. No.	Year	Temp. Range, K	Reported Error, %	Name and Specimen Designation	Composition (weight percent), Specifications and Remarks
1	139	1947	298-1500	0. 1		UO_2, 88. 26 U (theoretically 88. 15).
2	140	1949	525-1378	≤15		99. 7 UO_2; average values of Cp from 50 C to T C.
3	141, 142	1952	14-255	0. 2		99. 3 UO_2 and 0. 7 UO_3, traces of other metal oxides; 88. 6 U, powder of well crystallized particles.
4	143	1958	433-876	0. 8		88. 0 U and < 0. 01 Fe, Si; brown powder; prepared by reduction of U_3O_8 with hydrogen at 800 C.

DATA TABLE NO. 77 SPECIFIC HEAT OF URANIUM DIOXIDE UO_2

[Temperature, T, K; Specific Heat, C_p, Cal g^{-1} K^{-1}]

T	C_p
CURVE 1	
298	5.638×10^{-2}
300	5.661*
400	6.433
500	6.823
600	7.062
700	7.230
800	7.360
900	7.468
1000	7.563
1100	7.648
1200	7.727
1300	7.802
1400	7.874
1500	7.944
CURVE 2	
525.15	7.0×10^{-2}
526.15	7.0*
528.15	6.9*
722.15	7.2
728.15	7.6
728.15	7.0
927.15	7.3
939.15	7.3*
961.15	6.2
964.15	5.9
1153.15	7.6
1155.15	7.7*
1161.15	7.7*
1208.15	8.8
1212.15	7.4
1252.15	6.4
1336.15	6.9
1350.15	6.5
1359.15	7.0
1378.15	6.9
CURVE 3	
Series 1	
14.15	1.14×10^{-3}
16.23	1.84
18.77	2.99×10^{-3}
21.49	4.724
41.14	1.003×10^{-2}
47.20	1.169
51.99	1.301
57.23	1.452
Series 2	
13.68	9.40×10^{-4}
15.27	1.55×10^{-3}
17.29	2.25
19.61	3.32
22.69	5.691
28.37	2.890×10^{-2}
28.57	3.235
28.83	3.184*
29.07	3.165*
29.32	2.985
29.60	2.694
29.89	2.165
30.28	1.037
30.88	8.449×10^{-3}
31.35	8.368*
31.94	8.402
32.72	8.468*
35.67	8.975
Series 3	
22.36	5.368×10^{-3}
23.32	6.217
24.12	7.187
24.06	7.057*
24.78	8.013
25.40	8.961
25.96	9.960
26.45	1.115×10^{-2}
26.91	1.260
27.30	1.451
27.64	1.749
28.05	2.123
28.16	2.572×10^{-2}
28.36	2.933*
28.54	3.226*
28.71	3.325
28.89	3.258*
29.06	3.196*
Series 4	
63.62	1.617×10^{-2}
71.60	1.837
78.86	2.024
85.08	2.191
90.52	2.344
96.22	2.496
102.04	2.624
108.09	2.774
113.37	2.924
120.06	3.067
126.17	3.214
139.71	3.519
146.20	3.654
152.45	3.777
158.61	3.899
165.21	4.025
171.86	4.140*
176.57	4.247
183.81	4.354*
190.95	4.454
193.57	4.536*
200.83	4.643*
207.85	4.714
214.56	4.806*
221.36	4.928
228.91	4.991*
234.97	5.080*
241.70	5.154
248.45	5.228*
255.18	5.302
CURVE 4	
433.15	6.50×10^{-2}
467.15	6.67
494.15	6.80*
501.15	6.87*
502.15	6.87*
512.15	6.96*
553.15	7.04
554.15	7.02*
571.15	7.15
606.15	7.18*
644.15	7.24
649.15	7.32*
682.15	7.34
692.15	7.36*
702.15	7.34*
714.15	7.45*
730.15	7.39*
743.15	7.42*
757.15	7.48
766.15	7.42*
781.15	7.62*
798.15	7.60*
819.15	7.62*
830.15	7.60*
840.15	7.64
850.15	7.68*
865.15	7.68*
876.15	7.68

* Not shown on plot

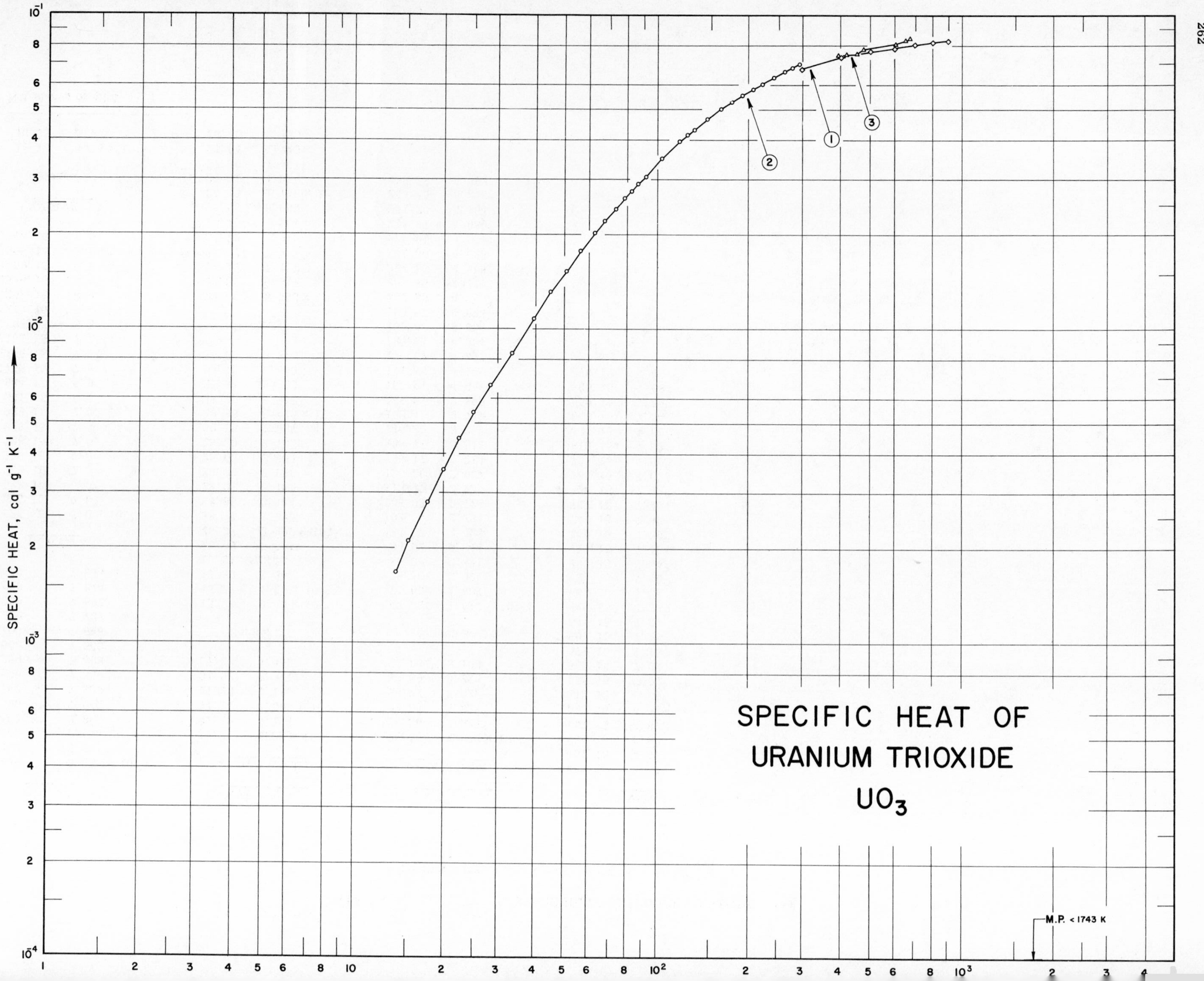
SPECIFIC HEAT OF
URANIUM TRIOXIDE
UO_3
SPECIFIC HEAT, cal g^{-1} K^{-1}
M.P. < 1743 K
1
2
3

SPECIFICATION TABLE NO. 78 SPECIFIC HEAT OF URANIUM TRIOXIDE UO_3

[For Data Reported in Figure and Table No. 78]

Curve No.	Ref. No.	Year	Temp. Range, K	Reported Error, %	Name and Specimen Designation	Composition (weight percent), Specifications and Remarks
1	139	1947	298-900			UO_3, 83. 02 U (theoretically 83. 22 U).
2	141, 142	1952	13-294	0. 2 above 35K		UO_3, 0. 003 H_2O; prepared by decomposing uranyl nitrate 8 hrs at 300 C, ground dried 3 hrs at 100 C.
3	143	1958	392-673	1. 3		83. 00 U and < 0. 003 Fe; amorphous orange powder; prepared by ignition of $UO_4 \cdot 2H_2O$ at 280 - 300 C.

DATA TABLE NO. 78 SPECIFIC HEAT OF URANIUM TRIOXIDE UO_3

[Temperature, T, K; Specific Heat, C_p, Cal g^{-1} K^{-1}]

T	C_p	T	C_p
CURVE 1		**CURVE 2 (cont.)**	
298	6.816 x 10^{-2}	199.74	5.761 x 10^{-2}
300	6.834*	207.09	5.866
400	7.428	214.29	5.995*
500	7.750	221.08	6.093
600	7.966	227.84	6.184*
700	8.131	234.84	6.292*
800	8.270	241.83	6.397
900	8.393	248.74	6.523*
		256.04	6.596*
CURVE 2		263.45	6.680
		270.89	6.827*
13.81	1.69 x 10^{-3}	278.46	6.882
15.24	2.13	285.84	6.890*
17.52	2.81	294.51	7.047
19.96	3.580		
22.32	4.506	**CURVE 3**	
24.92	5.460		
28.31	6.638	392.15	7.50 x 10^{-2}
33.20	8.414	419.15	7.56
39.03	1.090 x 10^{-2}	451.15	7.60
44.73	1.326	473.15	7.85
50.48	1.540	485.15	7.87*
56.59	1.793	601.15	8.13
62.98	2.044	651.15	8.37
67.60	2.222	673.15	8.50
73.60	2.449		
78.77	2.642		
82.51	2.779		
86.83	2.933		
92.09	3.083		
104.49	3.538		
119.30	3.992		
126.32	4.202		
132.86	4.366		
139.55	4.537*		
146.17	4.698		
152.71	4.848*		
156.57	4.939*		
163.16	5.083		
169.56	5.205*		
176.25	5.348		
183.55	5.478*		
190.46	5.603		

* Not shown on plot

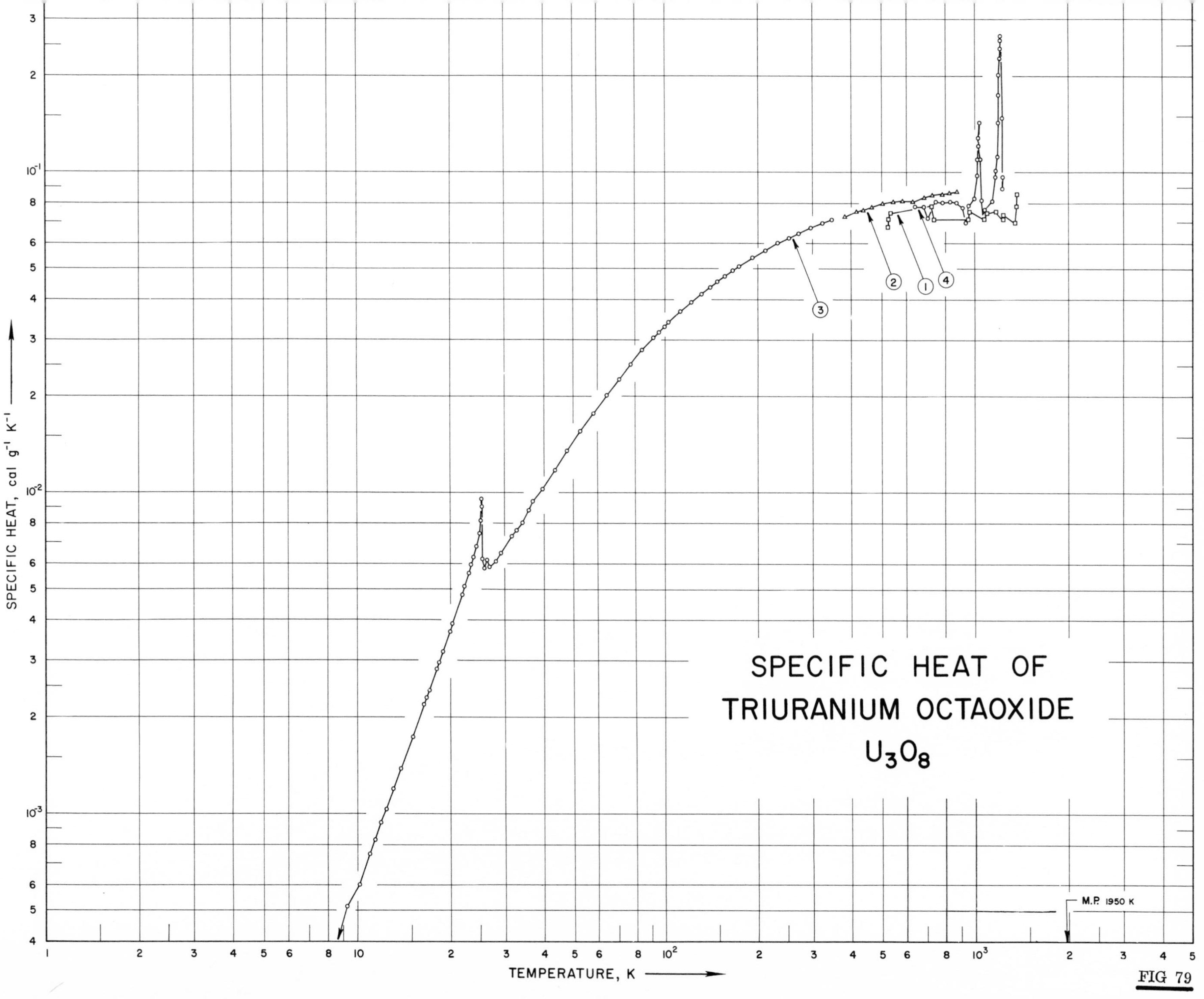
SPECIFIC HEAT OF
TRIURANIUM OCTAOXIDE
U_3O_8
SPECIFIC HEAT, cal g^{-1} K^{-1}
TEMPERATURE, K
M.P. 1950 K
1
2
3
4
5
6
8
10
2
3
4
5
6
8
10^2
2
3
4
5
6
8
10^3
2
3
4
5
4
5
6
8
10^{-3}
2
3
4
5
6
8
10^{-2}
2
3
4
5
6
8
10^{-1}
2
3
1
2
3
4

FIG 79

SPECIFICATION TABLE NO. 79 SPECIFIC HEAT OF TRIURANIUM OCTOXIDE U_3O_8

[For Data Reported in Figure and Table No. 79]

Curve No.	Ref. No.	Year	Temp. Range, K	Reported Error, %	Name and Specimen Designation	Composition (weight percent), Specifications and Remarks
1	144	1949	526-1365	≤15		Nearly 100 U_3O_8; average values of Cp from 50 C to TC.
2	143	1958	380-875	0. 6		84. 79 U, 0. 002 Fe, Si; deep olive green powder; prepared by ignition of $UO_2(NO_3)_2 \cdot 6H_2O$ at 850 C.
3	145	1959	5-347	0. 1-1		0. 020 Si, 0. 006 Al, 0. 003 Mg, 0. 002 Ni, 0. 001 Fe, 0. 0003 Cu, and 0. 000008 > B; prepared from uranyl nitrate hexahydrate, the U_3O_8 produced was reduced to UO_2 by heating in dry purified hydrogen gas at 500 C until H_2O formation ceased, temperature raised to 1200 C and sample kept at this temperature for 4 hrs before cooling to room temperature, oxidized in air at 800 C to constant weight and then heated 7 days at 800 C under vacuum, cooled to room temperature over 2 months.
4	146	1961	641-1233	1-3		84. 78 U.

DATA TABLE NO. 79 SPECIFIC HEAT OF TRIURANIUM OCTOXIDE U_3O_8

[Temperature, T, K; Specific Heat, C_p, Cal g^{-1} K^{-1}]

T	C_p
CURVE 1	
526.15	6.8 x 10^{-2}
528.15	7.2
533.15	7.5
725.15	7.9
726.15	8.0*
739.15	7.2
951.15	7.2
961.15	7.6
1086.15	7.2
1095.15	7.5
1097.15	7.5*
1174.15	7.6
1242.15	7.2
1242.15	7.4
1247.15	7.5*
1350.15	7.0
1357.15	7.9
1364.15	8.5*
1365.15	8.6
CURVE 2	
380.15	7.300 x 10^{-2}
392.15	7.401*
415.15	7.565
422.15	7.576*
437.15	7.666
459.15	7.778*
463.15	7.828*
464.15	7.831
485.15	7.945*
504.15	8.030
508.15	8.03
513.15	8.009*
526.15	8.093*
528.15	8.080*
530.15	8.047*
543.15	8.131
555.15	8.143*
558.15	8.169*
572.15	8.181*
573.15	8.186*
575.15	8.206*
586.15	8.206
CURVE 2 (cont.)	
590.15	8.237 x 10^{-2}*
597.15	8.237*
600.15	8.169*
607.15	8.186*
615.15	8.250*
623.15	8.260*
624.15	8.143*
631.15	8.174
635.15	8.219*
637.15	8.247*
648.15	8.307*
652.15	8.313*
653.15	8.237*
661.15	8.345*
667.15	8.300*
676.15	8.402*
688.15	8.388
707.15	8.440*
715.15	8.458*
718.15	8.464*
729.15	8.484*
730.15	8.503
741.15	8.529*
753.15	8.541*
755.15	8.554*
761.15	8.557*
766.15	8.554*
772.15	8.582*
788.15	8.597
792.15	8.630*
804.15	8.643*
805.15	8.633*
821.15	8.621*
827.15	8.694
829.15	8.696*
841.15	8.659*
849.15	8.681*
850.15	8.659*
875.15	8.757
CURVE 3	
Series 1	
95.80	3.180 x 10^{-2}
103.64	3.426
112.78	3.696
122.11	3.953
131.23	4.189
140.23	4.403
148.76	4.593
157.14	4.768
165.91	4.936
175.01	5.109
184.09	5.272*
193.22	5.424
202.34	5.570*
211.63	5.714
221.06	5.852*
Series 2	
214.07	5.748 x 10^{-2}*
223.37	5.869*
232.88	6.010
242.51	6.146*
251.93	6.260
261.09	6.369*
270.15	6.458
279.15	6.564*
288.06	6.661*
296.92	6.741
305.72	6.820*
314.45	6.895*
323.32	6.965
332.11	7.029*
340.03	7.085*
346.86	7.132
Series 3	
24.80	7.505 x 10^{-3}
26.78	5.909
29.16	6.510
31.62	7.335
34.24	8.098
CURVE 3 (cont.)	
Series 3 (cont.)	
36.93	9.451
39.63	1.031 x 10^{-2}
Series 4	
13.80	1.394 x 10^{-3}
15.07	1.749
16.76	2.314
18.43	2.987
20.36	3.931
22.33	5.134
24.21	6.812
26.45	6.207
29.37	6.580*
32.59	7.671
35.90	8.887
39.52	1.023 x 10^{-2}*
43.56	1.183
47.96	1.359
52.85	1.557
58.24	1.775
64.29	2.022
70.92	2.280
77.69	2.539
84.49	2.790
91.87	3.053
99.98	3.313
Series 5	
4.77	9.261 x 10^{-5}*
5.36	1.354 x 10^{-4}*
6.25	1.840*
7.32	2.624*
8.37	3.728*
9.22	5.189
10.11	6.056
10.99	7.516
11.95	9.428
13.06	1.206 x 10^{-3}
11.46	8.347 x 10^{-4}
12.44	1.041 x 10^{-3}
CURVE 3 (cont.)	
Series 5 (cont.)	
13.59	1.337 x 10^{-3}*
14.96	1.735*
16.47	2.205
18.09	2.844
19.96	3.708
21.92	4.849
23.07	5.664
23.46	5.996
23.83	6.330
24.18	6.695*
24.51	7.093*
24.83	7.554*
25.13	8.219
25.42	7.594*
25.76	5.860
26.12	5.786*
26.50	5.822*
26.87	5.868*
27.23	5.938*
27.58	6.022*
28.01	6.152
Series 6	
17.11	2.445 x 10^{-3}
18.94	3.214
22.17	5.015*
23.22	5.800*
24.17	6.680*
24.73	7.388*
24.96	7.709*
25.11	8.252*
25.24	9.602
25.35	9.113
25.41	8.173*
25.49	6.247
25.58	6.145*
25.65	5.972*
25.73	5.930*
25.81	5.810*
25.90	5.774*
26.00	5.761*
CURVE 3 (cont.)	
Series 6 (cont.)	
26.15	5.779 x 10^{-3}*
26.34	5.747*
26.48	5.919*
26.68	5.794*
27.97	6.160*
CURVE 4	
641.15	7.85 x 10^{-2}
662.15	7.86*
683.15	7.85
705.15	7.21
727.15	7.97*
748.15	8.12
769.15	8.15*
789.15	8.08
809.15	8.14*
831.15	8.12
853.15	8.08*
873.15	8.09
893.15	7.98*
914.15	7.80
935.15	7.0
956.15	7.93
977.15	8.13*
998.15	8.37
1019.15	9.80
1023.35	1.109 x 10^{-1}
1027.55	1.295
1031.75	1.217
1035.95	1.425*
1042.15	1.445
1047.35	1.110
1052.55	8.23 x 10^{-2}
1063.15	7.24*
1083.15	7.70
1145.15	8.20
1167.15	9.76
1171.55	1.022 x 10^{-1}
1175.95	1.044*
1180.35	1.067*
1184.75	1.128

*Not shown on plot

DATA TABLE NO. 79 (continued)

T	C_p
CURVE 4 (cont.)	
1189.15	1.432×10^{-1}
1193.55	1.750
1197.95	2.012
1202.35	2.269
1206.75	2.447
1211.15	2.585
1215.15	2.663
1220.15	2.457*
1224.55	1.481
1228.95	8.90×10^{-2}
1233.35	9.70

* Not shown on plot

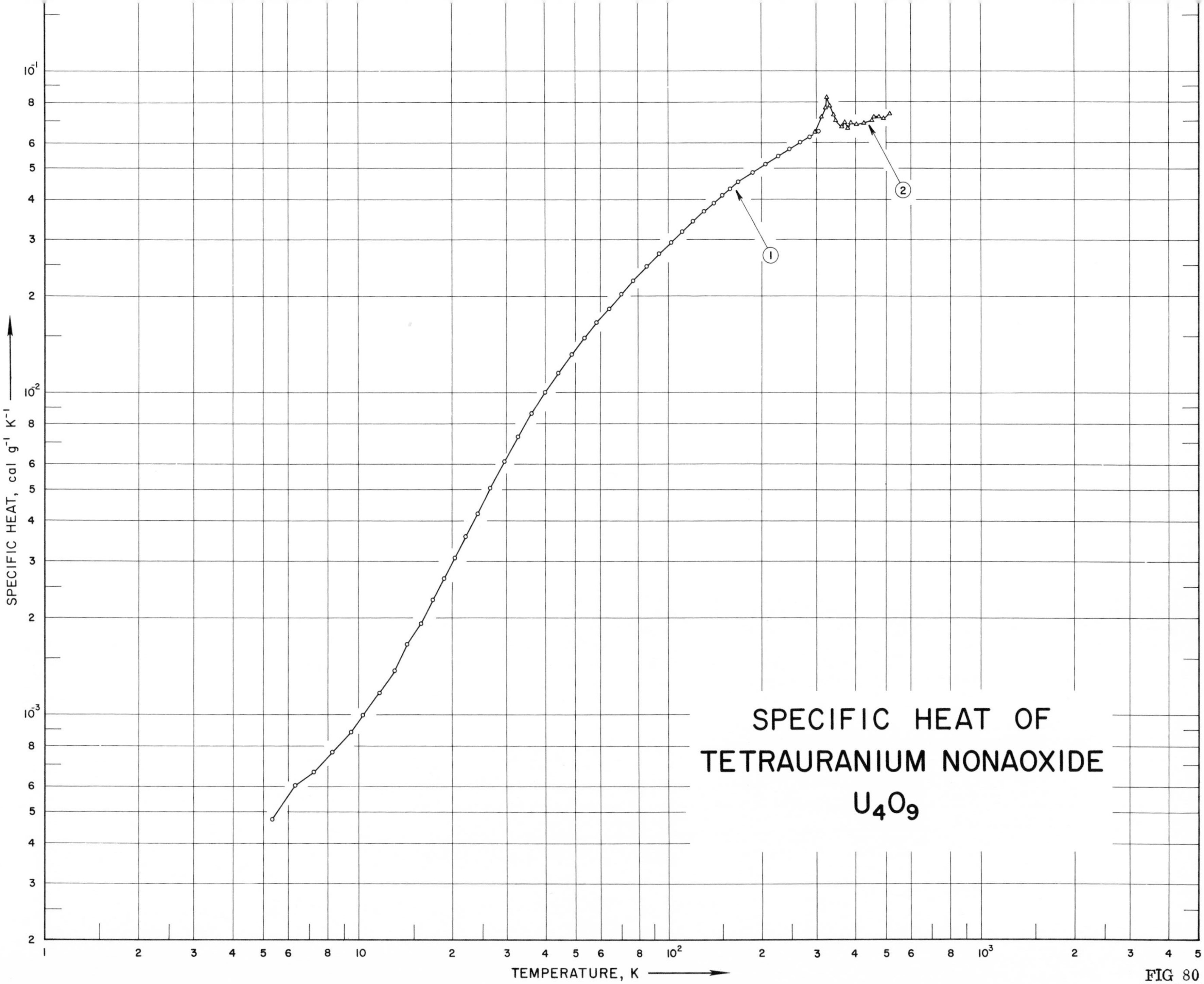

SPECIFIC HEAT OF
TETRAURANIUM NONAOXIDE
U_4O_9
SPECIFIC HEAT, cal g^{-1} K^{-1}
TEMPERATURE, K
1
2
10^{-1}
10^{-2}
10^{-3}
10
10^{2}
10^{3}

FIG 80

SPECIFICATION TABLE NO. 80 SPECIFIC HEAT OF TETRAURANIUM NONAOXIDE U_4O_9

[For Data Reported in Figure and Table No. 80]

Curve No.	Ref. No.	Year	Temp. Range, K	Reported Error, %	Name and Specimen Designation	Composition (weight percent), Specifications and Remarks
1	147	1957	5-310	0. 1 above 5K 1 at 14 K and 5 at 5 K		U_4O_9, 86. 81 ± 0. 08 U (86. 86 theoretical), 13. 14 ± 0. 02 O_2 (13. 14 theoretical), 0. 007 Si, 0. 001 Al, 0. 001 Cu, 0. 0007 Fe, 0. 0005 Mg, 0. 0001 Pb, and 0. 00002 B; preparation: U_2O_8 reduced to UO_2 by heating in alumina boats in a stream of anhydrous hydrogen at 500 C until evolution of H_2O ceased, then heating for additional 4 hrs at 1200 C, cooled in H_2 atmosphere, stoichiometric UO_2 and U_3O_8 were mixed in quartz tubes and evaluated; mixture heated 7 days at 800 C and gradually cooled to 20 C over 2 months period.
2	148	1965	297-515			Prepared from 99. 9 powdered UO_2; UO_2 was reduced in dry hydrogen gas 12 - 20 hrs, part of UO_2 was oxidized in air at 650 C for 5 hrs to U_3O_8. Stoichiometric mixture of UO_2 and U_3O_8 were ground and mixed carefully to homogeneous, heated at 950 C for 180 hrs and then gradually cooled down to room temperature for 100 hrs.

DATA TABLE NO. 80 SPECIFIC HEAT OF TETRAURANIUM NONAOXIDE U_4O_9

[Temperature, T, K; Specific Heat, C_p, Cal g^{-1} K^{-1}]

T	C_p
CURVE 1	
Series 1	
5.37	4.743 x 10^{-4}
6.32	6.020
7.23	6.659
8.28	7.662
9.46	8.848
10.42	9.943
11.71	1.168 x 10^{-3}
13.05	1.377
14.44	1.654
15.91	1.926
17.46	2.270
18.98	2.654
20.49	3.083
22.18	3.582
24.13	4.220
26.62	5.070
29.57	6.111
32.76	7.299
36.29	8.617
40.00	1.003 x 10^{-2}
44.28	1.154
48.90	1.318
53.72	1.484
58.72	1.651
64.17	1.830
70.46	2.024
77.61	2.240
77.65	2.243*
85.46	2.477
93.72	2.708
102.64	2.947
Series 2	
63.05	1.793 x 10^{-2}*
69.39	1.992*
76.58	2.209*
84.39	2.444*
92.59	2.675*
101.49	2.916*
111.14	3.173
120.91	3.423
CURVE 1 (cont.)	
Series 2 (cont.)	
130.54	3.663 x 10^{-2}
140.24	3.894
150.01	4.113
159.70	4.320
169.27	4.515
169.14	4.511*
178.59	4.692*
187.98	4.864
197.33	5.019*
206.79	5.177
216.47	5.321*
226.29	5.467
236.22	5.604*
246.27	5.747
256.46	5.877*
266.75	6.015
276.88	6.143*
286.92	6.265
296.88	6.381*
305.82	6.518
Series 3	
9.28	8.483 x 10^{-4}*
10.65	1.003 x 10^{-3}*
11.98	1.795*
13.26	1.423*
14.56	1.669*
15.99	1.945*
17.53	2.285*
CURVE 2	
297	6.49 x 10^{-2}
301	6.44*
305	7.09*
309	7.04*
313	7.22
317	7.47*
321	7.72
325	8.31
329	8.36
CURVE 2 (cont.)	
333	7.82 x 10^{-2}
337	7.64*
341	7.36
347	7.04
355	6.86*
363	6.72
371	6.96
379	6.68
387	6.91
395	6.89*
403	6.85
411	6.91*
419	6.86*
427	6.91
435	6.82*
443	7.00*
451	7.05
459	7.22
467	7.08*
475	7.23
483	7.08*
491	7.15
499	7.05*
507	7.00*
515	7.40

*Not shown on plot

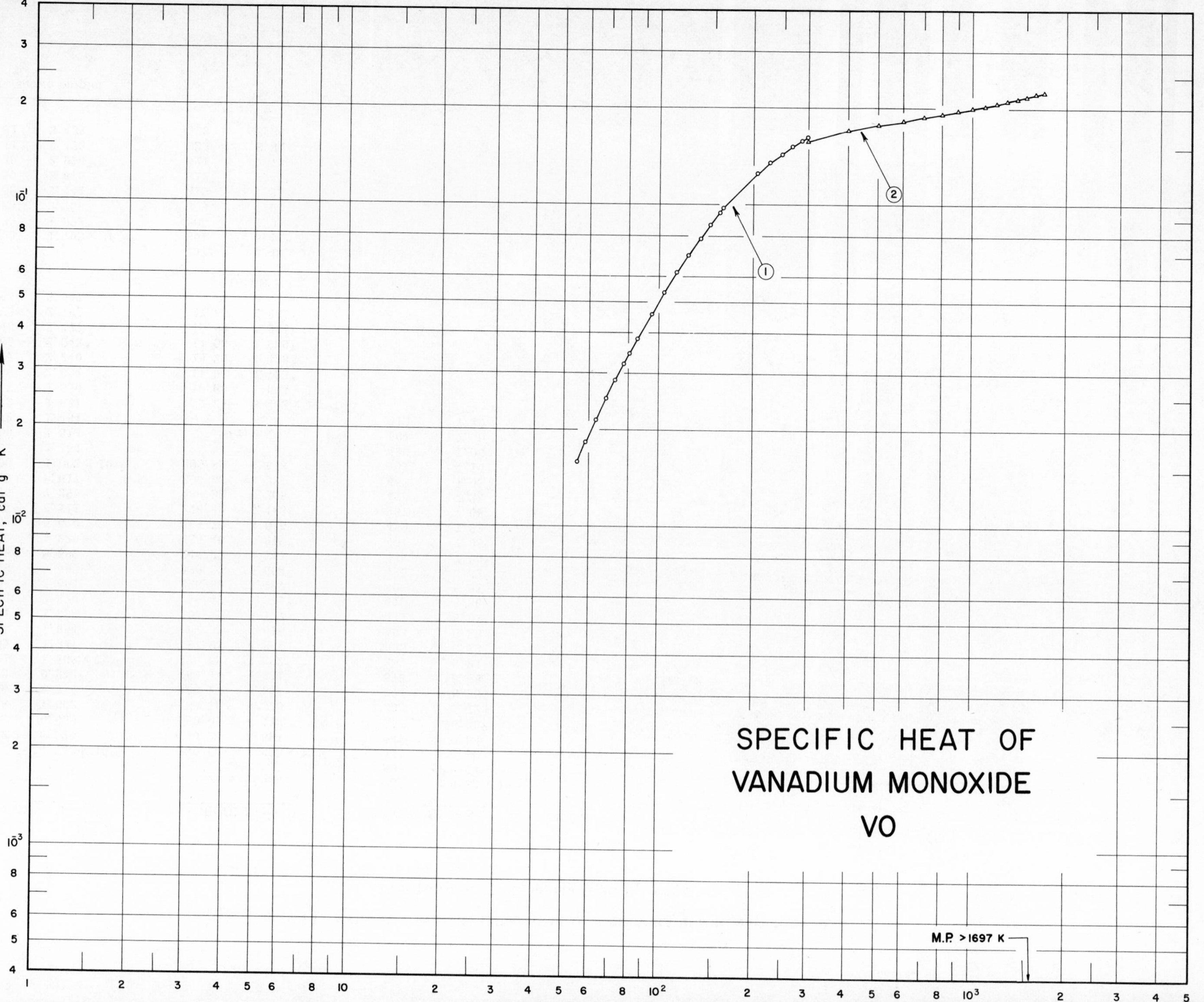

SPECIFIC HEAT OF
VANADIUM MONOXIDE
VO
M.P. >1697 K
SPECIFIC HEAT, cal g^{-1} K^{-1}
1
2

SPECIFICATION TABLE NO. 81 SPECIFIC HEAT OF VANADIUM MONOXIDE VO

[For Data Reported in Figure and Table No. 81]

Curve No.	Ref. No.	Year	Temp. Range, K	Reported Error, %	Name and Specimen Designation	Composition (weight percent), Specifications and Remarks
1	77	1951	54-298			98.2 VO; 74.72 V, 0.92 Si, 0.23 NaO, and 0.10 Fe.
2	102	1954	298-1700			98.2 VO, small amounts of V_2O_3.

DATA TABLE NO. 81 SPECIFIC HEAT OF VANADIUM MONOXIDE VO

[Temperature, T, K; Specific Heat, C_p, Cal g^{-1} K^{-1}]

T	C_p
CURVE 1	
54.97	1.585×10^{-2}
58.61	1.818
63.11	2.134
68.07	2.496
72.87	2.844
77.61	3.198
80.49	3.432
85.40	3.806
94.98	4.563
104.53	5.323
114.72	6.157
124.62	6.951
135.96	7.860
146.22	8.680
155.95	9.419
160.46	9.788
206.39	1.251×10^{-1}
216.23	1.303*
226.13	1.352
237.24	1.403*
245.79	1.439
256.10	1.479*
266.20	1.519
275.58	1.552*
286.34	1.589
296.36	1.619*
298.16	1.622
CURVE 2	
298.15	1.578×10^{-1}
300	1.581*
400	1.706
500	1.781
600	1.837
700	1.885
800	1.927
900	1.966
1000	2.004
1100	2.040
1200	2.076
1300	2.111
1400	2.145
1500	2.180
CURVE 2 (cont.)	
1600	2.214×10^{-1}
1700	2.248

*Not shown on plot

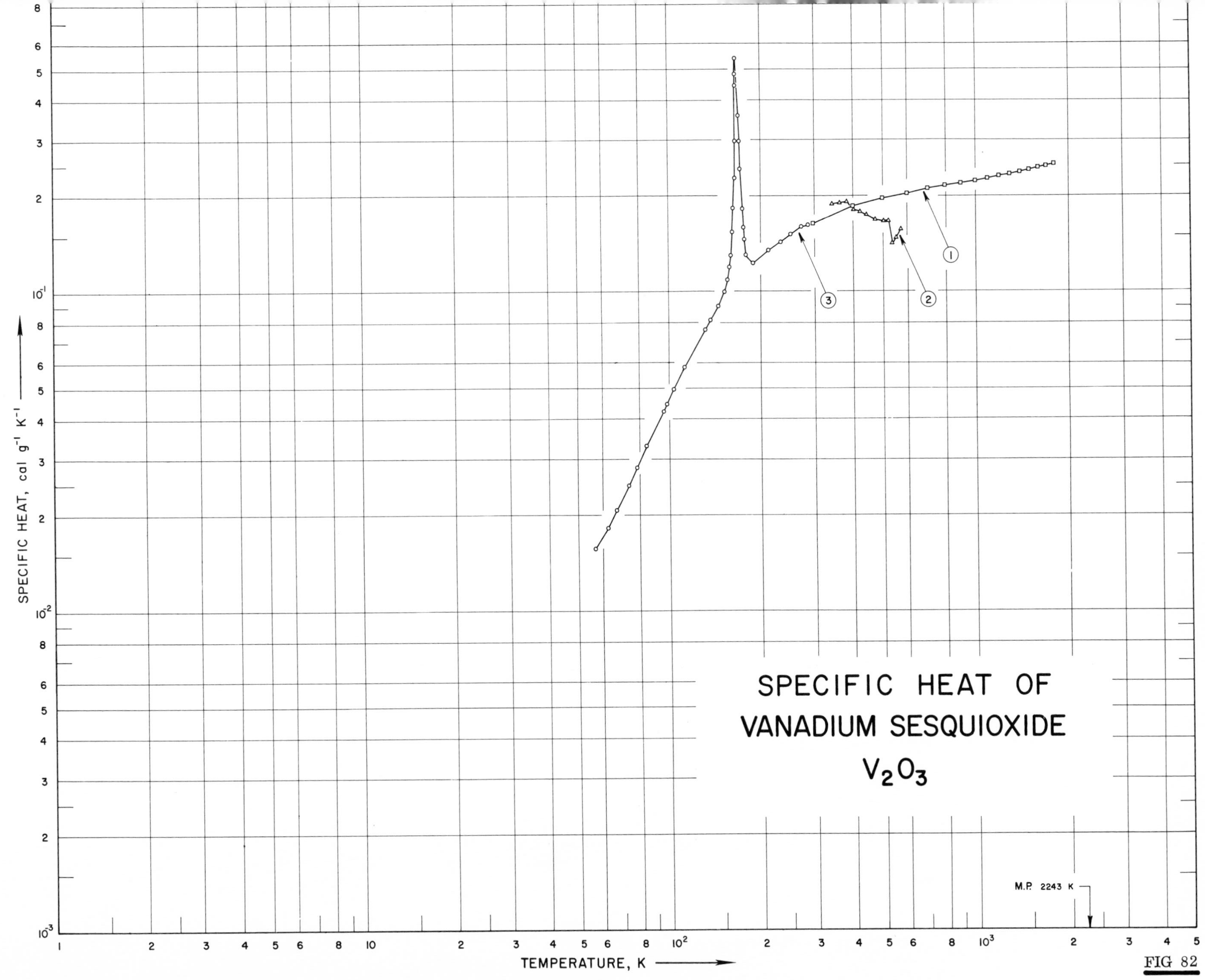

SPECIFIC HEAT OF
VANADIUM SESQUIOXIDE
V_2O_3
SPECIFIC HEAT, cal g^{-1} K^{-1}
TEMPERATURE, K
M.P. 2243 K
1
2
3

FIG 82

SPECIFICATION TABLE NO. 82 SPECIFIC HEAT OF VANADIUM SESQUIOXIDE V_2O_3

[For Data Reported in Figure and Table No. 82]

Curve No.	Ref. No.	Year	Temp. Range, K	Reported Error, %	Name and Specimen Designation	Composition (weight percent), Specifications and Remarks
1	149	1947	298-1800			67. 89 V; prepared by heating pure V_2O_5 in a silica flask at 800 C in a stream of pure hydrogen until no further water vapor has evolved.
2	150	1951	343-573	2. 0		
3	151	1936	57-287			Powder compressed into pellets at pressure of 2 tons per in.^{2}; density 4. 83 g cm^{-3} at 22. 0 C.

DATA TABLE NO. 82 SPECIFIC HEAT OF VANADIUM SESQUIOXIDE V_2O_3

[Temperature, T, K; Specific Heat, C_p, Cal g^{-1} K^{-1}]

CURVE 1

T	C_p
298.15	1.646 x 10^{-1}
300	1.651*
400	1.859
500	1.972
600	2.048
700	2.106
800	2.155
900	2.199
1000	2.239
1100	2.277
1200	2.313
1300	2.349
1400	2.384
1500	2.418
1600	2.451
1700	2.485
1800	2.518

CURVE 2

T	C_p
343.15	1.89 x 10^{-1}
353.15	1.89*
363.15	1.90
373.15	1.91*
383.15	1.92
393.15	1.89*
403.15	1.83
413.15	1.81*
423.15	1.79
433.15	1.76*
443.15	1.75
453.15	1.73*
463.15	1.71*
473.15	1.69
483.15	1.68*
493.15	1.67*
503.15	1.67
513.15	1.67*
523.15	1.67
538.15	1.43
543.15	1.45*
553.15	1.48
563.15	1.52*
573.15	1.56

CURVE 3

Series 1

T	C_p
56.9	1.571 x 10^{-2}
62.5	1.829
66.9	2.077
73.4	2.488
78.4	2.825
84.6	3.308
96.2	4.225
113.5	5.803
132.4	7.618
145.7	9.033
213.8	1.357x 10^{-1}
234.7	1.448
251.3	1.524
273.0	1.604
287.2	1.628
98.9	4.461x 10^{-2}
104.4	4.974
137.7	8.139

Series 2

T	C_p
165.7	2.276 x 10^{-1}
189.9	1.241
153.2	1.008
156.9	1.097
159.2	1.202
161.0	1.303
163.0	1.547
164.7	1.868
166.6	2.971
167.7	4.448
168.4	4.856
169.2	5.406
170.9	3.582
171.9	2.976
173.0	2.437
175.2	1.835
176.7	1.598
178.2	1.460
180.4	1.313
182.0	1.288*
185.2	1.242*

CURVE 3 (cont.)

Series 2 (cont.)

T	C_p
153.8	1.020 x 10^{-1}*
161.6	1.394*
168.4	4.506*
169.4	5.094*
178.8	1.406*
187.9	1.238*
192.3	1.250*
157.6	1.129*
186.3	1.241*
153.1	1.004*
193.9	1.257*
152.9	1.002*
157.4	1.115*

*Not shown on plot

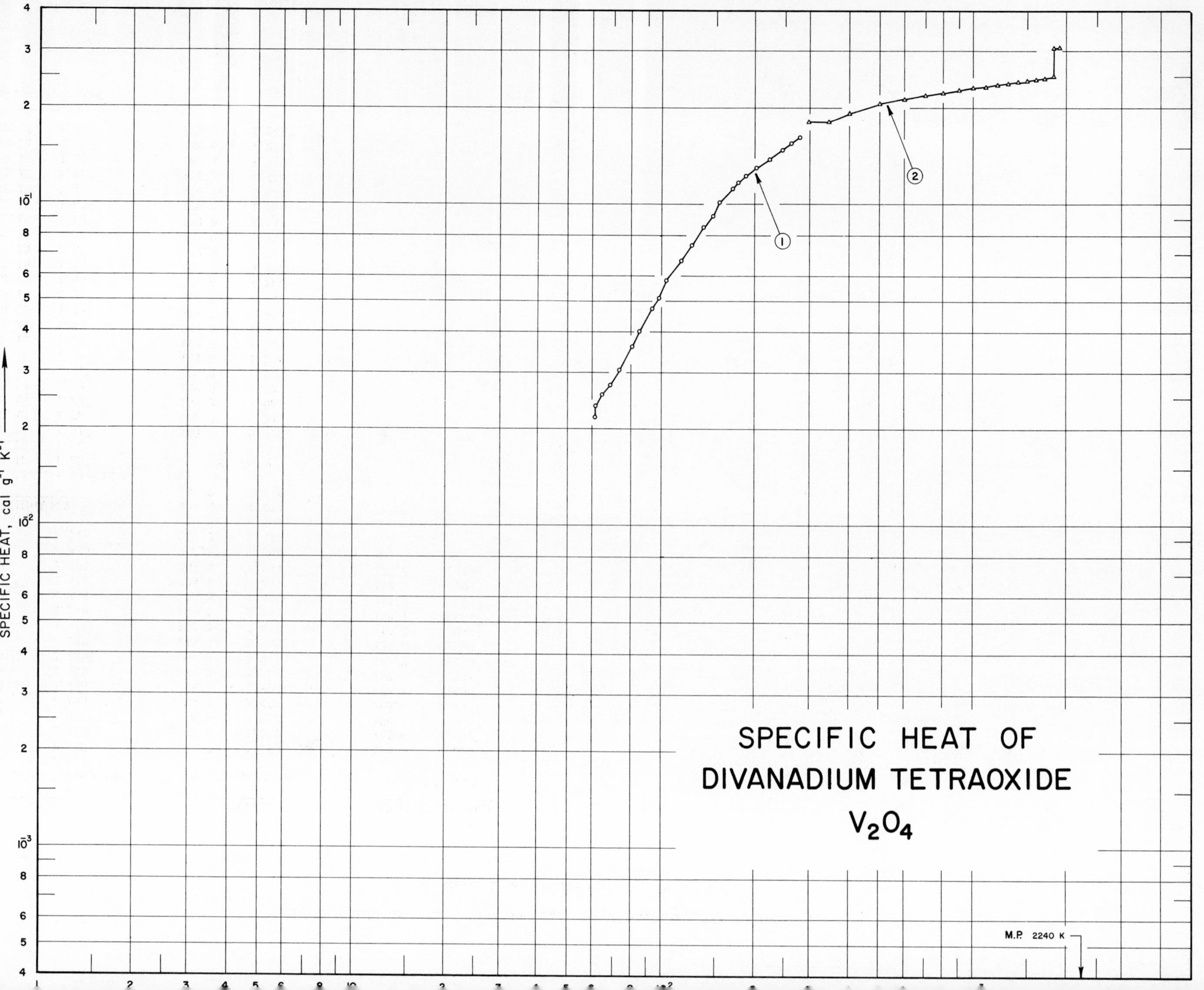
SPECIFIC HEAT OF
DIVANADIUM TETRAOXIDE
V_2O_4
SPECIFIC HEAT, cal g^{-1} K^{-1}
M.P. 2240 K
1
2

SPECIFICATION TABLE NO. 83 SPECIFIC HEAT OF DIVANADIUM TETRAOXIDE V_2O_4

[For Data Reported in Figure and Table No. 83]

Curve No.	Ref. No.	Year	Temp. Range, K	Reported Error, %	Name and Specimen Designation	Composition (weight percent), Specifications and Remarks
1	151	1936	61-279			Powder compressed into pellets at 2 tons per in.^{2} pressure; density 4.260 at 21.4 C.
2	149	1947	298-1900			61.45 V; prepared from pure vanadium trioxide by controlled oxidation with air in a platinum vessel at 300 C; placed in silica flask, evacuated at room temperature and given prolonged heat treatment below 600 C.

DATA TABLE NO. 83 SPECIFIC HEAT OF DIVANADIUM TETRAOXIDE V_2O_4

[Temperature, T, K; Specific Heat, C_p, Cal $g^{-1} K^{-1}$]

T	C_p
CURVE 1	
Series 1	
61.4	2.188 x 10^{-2}
68.8	2.751
73.5	3.061
80.5	3.612
84.9	4.020
93.6	4.744
98.1	5.136
104.7	5.816
115.7	6.703
125.6	7.492
137.4	8.505
147.0	9.228
154.7	1.014 x 10^{-1}
170.1	1.222
176.9	1.177
188.0	1.234
203.7	1.310
212.3	1.347*
224.2	1.394
234.0	1.436*
245.8	1.493
262.6	1.555
270.9	1.591*
279.4	1.625
Series 2	
61.4	2.363 x 10^{-2}
64.9	2.587
100.3	5.375*
144.7	9.138*
157.1	1.056*
187.6	1.228*
165.2	1.136*
157.3	1.045*
162.2	1.091*

T	C_p
CURVE 2	
(α) 298.15	1.803 x 10^{-1}
300	1.803*
(α) 345	1.803
(β) 345	1.823*
400	1.936
500	2.064
600	2.142
700	2.198
800	2.241
900	2.277
1000	2.309
1100	2.338
1200	2.364
1300	2.390
1400	2.414
1500	2.438
1600	2.461
1700	2.483
1800	2.506*
(β) 1818	2.510
(l) 1818	3.074
1900	3.074

*Not shown on plot

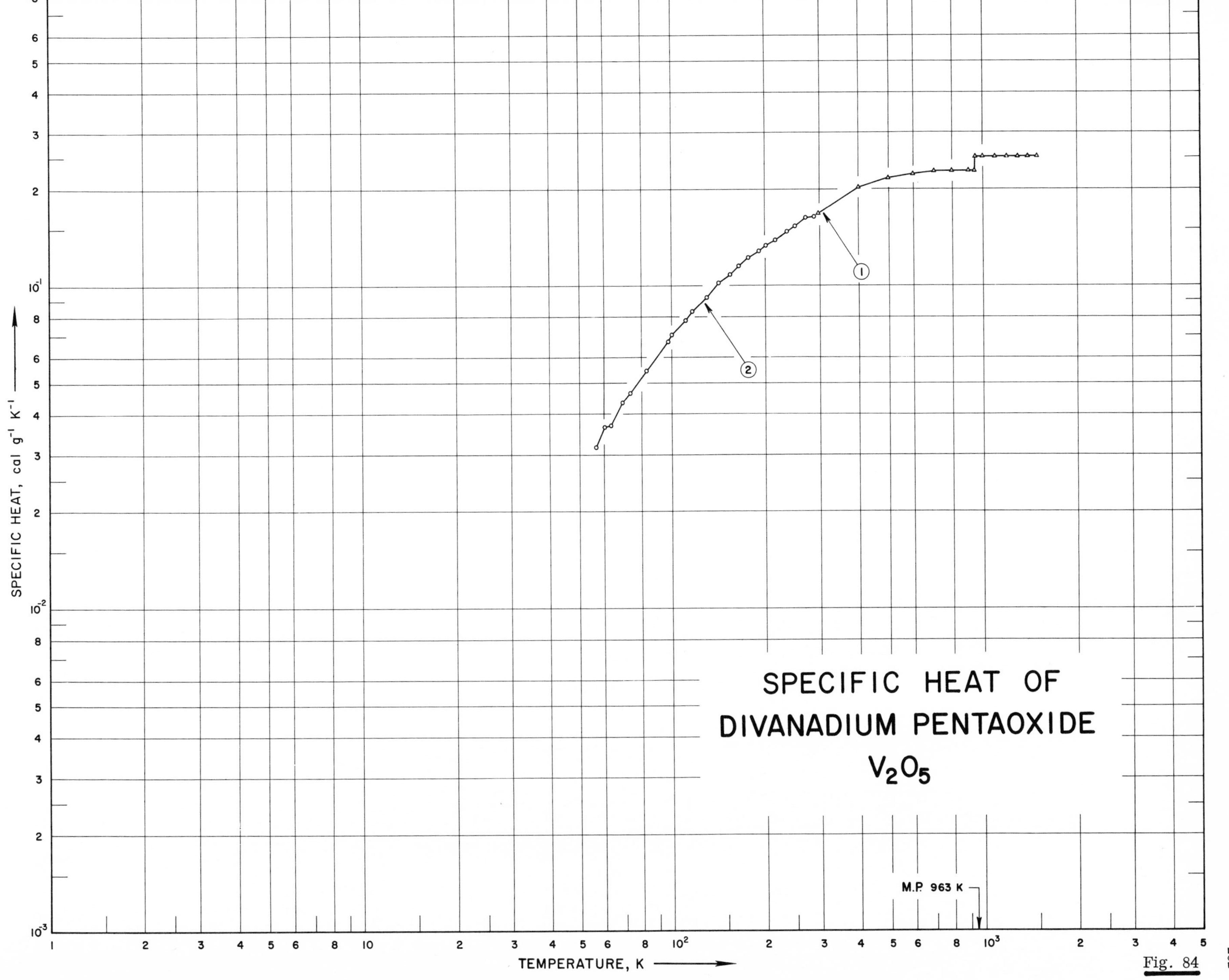
SPECIFIC HEAT OF
DIVANADIUM PENTAOXIDE
V_2O_5
SPECIFIC HEAT, cal g^{-1} K^{-1}
TEMPERATURE, K
M.P. 963 K
1
2

Fig. 84

SPECIFICATION TABLE NO. 84 SPECIFIC HEAT OF DIVANADIUM PENTAOXIDE V_2O_5

[For Data Reported in Figure and Table No. 84]

Curve No.	Ref. No.	Year	Temp. Range, K	Reported Error, %	Name and Specimen Designation	Composition (weight percent), Specifications and Remarks
1	149	1947	298-1500			55.96 V; prepared from purified ammonium vanadate by heating in platinum vessel in a stream of pure hydrogen at 440-460 C for 7 days.
2	151	1936	57-290			Powder compressed into pellets at pressure of 2 tons per $in.^2$.

DATA TABLE NO. 84 SPECIFIC HEAT OF DIVANADIUM PENTAOXIDE V_2O_5

[Temperature, T, K; Specific Heat, C_p, Cal g^{-1} K^{-1}]

T	C_p
CURVE 1	
298.15	1.677 x 10^{-1}
300	1.686*
400	2.018
500	2.160
600	2.228
700	2.260
800	2.273
900	2.276
(s) 943	2.275
(l) 943	2.507
1000	2.507
1100	2.507
1200	2.507
1300	2.507
1400	2.507
1500	2.507
CURVE 2	
56.8	3.174 x 10^{-2}
60.4	3.659
63.7	3.689
69.6	4.352
73.9	4.675
83.4	5.454
98.0	6.701
100.9	7.053
111.9	7.801
117.7	8.340
130.4	9.203
143.9	1.020 x 10^{-1}
155.3	1.089
165.2	1.151
178.0	1.222
191.4	1.286
202.6	1.343
217.8	1.390
223.8	1.434*
235.7	1.476
251.0	1.548
271.3	1.631
279.9	1.647*
289.5	1.644

*Not shown on plot

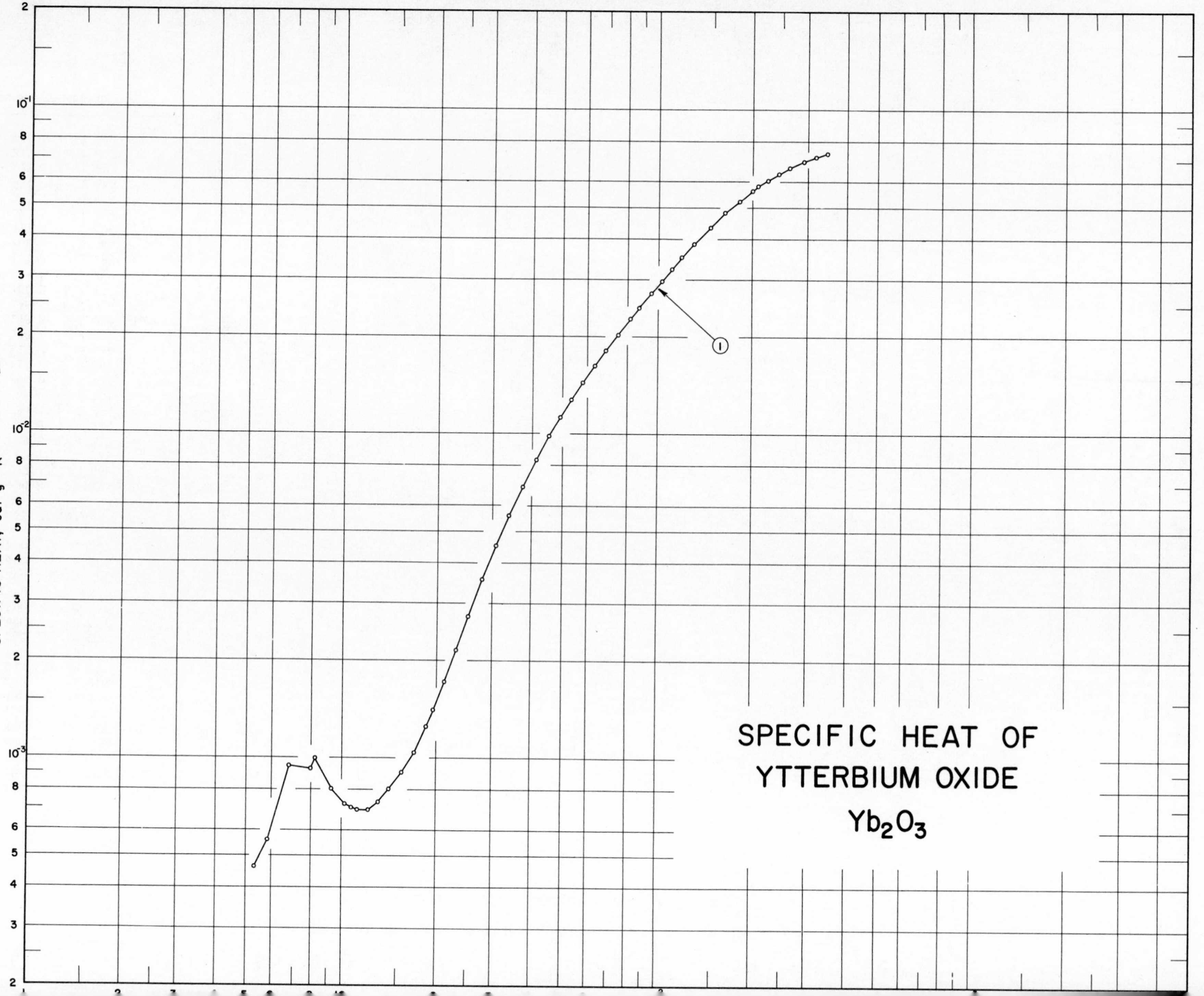
SPECIFIC HEAT OF
YTTERBIUM OXIDE
Yb2O3
1
SPECIFIC HEAT, cal g-1 K-1
10-1
10-2
10-3

SPECIFICATION TABLE NO. 85 SPECIFIC HEAT OF YTTERBIUM OXIDE Yb_2O_3

[For Data Reported in Figure and Table No. 85]

Curve No.	Ref. No.	Year	Temp. Range, K	Reported Error, %	Name and Specimen Designation	Composition (weight percent), Specifications and Remarks
1	72	1962	5-346	0.1		99.9 Yb_2O_3, 0.050 Lu, 0.010 Ca, and 0.010 Si; powder specimen; supplied by Michigan Chemical Company; measured in helium atmosphere.

DATA TABLE NO. 85 SPECIFIC HEAT OF YTTERBIUM OXIDE Yb_2O_3

[Temperature, T, K; Specific Heat, C_p, Cal g^{-1} K^{-1}]

T	C_p
CURVE 1	
Series 1	
87.12	2.455 x 10^{-2}
95.08	2.715
103.01	2.975
111.08	3.242
119.48	3.516
Series 2	
5.32	4.643 x 10^{-4}
5.85	5.608
6.80	9.439
7.97	9.287
9.27	8.036
10.76	7.054
Series 3	
8.21	9.973 x 10^{-4}
9.22	8.095*
10.29	7.232
11.29	6.952
12.15	6.952
13.07	7.358
14.17	8.019
15.55	9.034
17.03	1.048 x 10^{-3}
18.53	1.254
Series 4	
19.55	1.411 x 10^{-3}
21.18	1.726
23.05	2.154
25.27	2.746
27.94	3.553
30.88	4.532
33.89	5.598
37.32	6.867
41.28	8.323
45.37	9.874
49.26	1.132 x 10^{-2}
53.34	1.279
CURVE 1 (cont.)	
Series 4 (cont.)	
57.98	1.446 x 10^{-2}
63.08	1.627
68.45	1.811
74.68	2.020
81.67	2.263
Series 5	
123.56	3.643 x 10^{-2}*
130.85	3.874
138.66	4.113*
147.00	4.360
155.54	4.598*
164.30	4.829
173.23	5.052*
182.45	5.271
191.73	5.479*
200.86	5.663
209.88	5.836
Series 6	
214.95	5.933 x 10^{-2}*
224.22	6.088
233.32	6.237*
242.54	6.379
251.72	6.494*
255.06	6.544*
263.89	6.648
272.94	6.752*
281.98	6.846*
291.15	6.930
300.41	7.016*
309.84	7.093*
319.34	7.161
328.76	7.224*
338.09	7.280*
346.72	7.336

*Not shown on plot

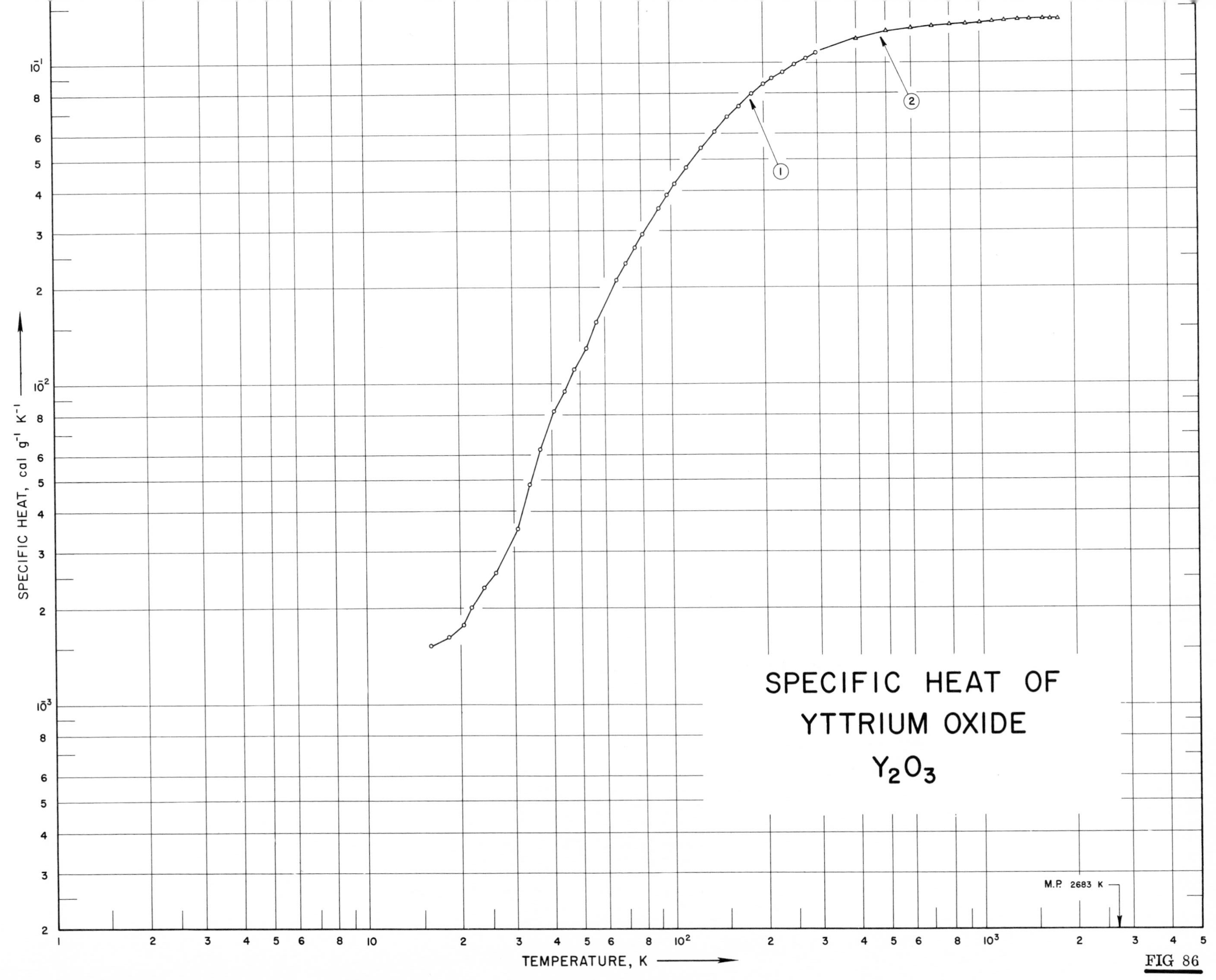

SPECIFIC HEAT OF
YTTRIUM OXIDE
Y_2O_3
M.P. 2683 K
TEMPERATURE, K
SPECIFIC HEAT, cal g^{-1} K^{-1}

FIG 86

SPECIFICATION TABLE NO. 86 SPECIFIC HEAT OF YTTRIUM OXIDE Y_2O_3

[For Data Reported in Figure and Table No. 86]

Curve No.	Ref. No.	Year	Temp. Range, K	Reported Error, %	Name and Specimen Designation	Composition (weight percent), Specifications and Remarks
1	83	1959	16-298			>99.9 Y_2O_3, < 0.01 Gd_2O_3, <0.01 Dy_2O_3, and <0.02 Ho_2O_3, supplied by Lindsay Chemical Company; heated to constant weight at 950 C for 24 hours in air to decompose hydroxides or carbonates.
2	73	1962	298-1799	0.2		99.99 Y_2O_3; measured in helium atmosphere.

DATA TABLE NO. 86 SPECIFIC HEAT OF YTTRIUM OXIDE Y_2O_3

[Temperature, T, K; Specific Heat, C_p, Cal $g^{-1}K^{-1}$]

T	C_p
CURVE 1	
15.96	1.528×10^{-3}
18.37	1.625
20.45	1.776
21.63	2.006
23.87	2.334
26.01	2.582
30.85	3.547
33.86	4.863
36.67	6.275
40.93	8.215
44.34	9.499
47.86	1.112×10^{-2}
52.15	1.298
56.33	1.566
65.78	2.105
70.75	2.387
76.13	2.662
80.83	2.946
81.81	3.003*
91.08	3.549
93.81	3.676*
97.02	3.896
100.06	4.065*
102.83	4.228
106.21	4.390*
108.42	4.530*
112.44	4.756
118.75	5.124*
125.23	5.460
131.89	5.766*
138.96	6.134
146.21	6.483*
153.27	6.816
160.92	7.143*
166.50	7.369
173.02	7.653*
176.95	7.759
184.37	8.064
186.58	8.153*
193.37	8.374*
200.19	8.645
207.07	8.844*
212.86	8.977*
CURVE 1 (cont.)	
214.27	8.990×10^{-2}
219.22	9.136*
225.62	9.269*
232.05	9.428
238.41	9.597*
240.33	9.654*
246.77	9.787*
253.61	9.955
263.59	1.021×10^{-1}*
266.48	1.020*
276.56	1.044
279.94	1.055*
283.31	1.057*
292.36	1.081*
298.26	1.084
CURVE 2	
(α) 298	1.0883×10^{-1}*
300	1.0916*
400	1.1998
500	1.2527
600	1.2839
700	1.3048
800	1.3203
900	1.3325
1000	1.3428
1100	1.3518
1200	1.3599
1300	1.3674*
(α) 1330	1.3695
(β) 1330	1.3728*
1400	1.3728
1500	1.3728
1600	1.3728
1700	1.3728
(β) 1799	1.3728

*Not shown on plot

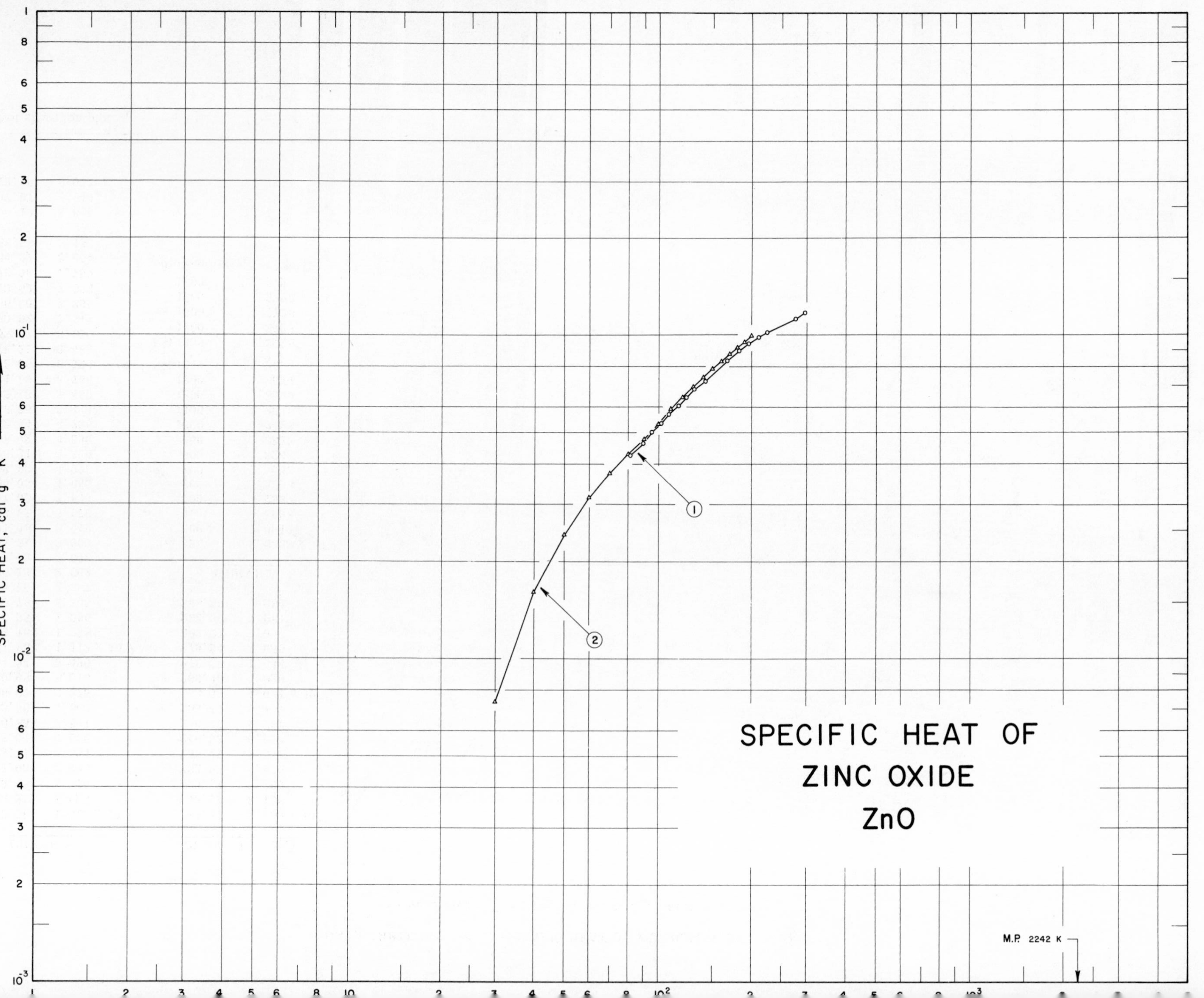

SPECIFIC HEAT OF
ZINC OXIDE
ZnO
M.P. 2242 K
SPECIFIC HEAT, cal g⁻¹ K⁻¹
1
2

SPECIFICATION TABLE NO. 87 SPECIFIC HEAT OF ZINC OXIDE ZnO

[For Data Reported in Figure and Table No. 87]

Curve No.	Ref. No.	Year	Temp. Range, K	Reported Error, %	Name and Specimen Designation	Composition (weight percent), Specifications and Remarks
1	152	1928	81-298			99.9 ZnO; microcrystalline.
2	68	1928	30-200			

DATA TABLE NO. 87 SPECIFIC HEAT OF ZINC OXIDE ZnO

[Temperature, T, K; Specific Heat, C_p, Cal g^{-1} K^{-1}]

T	C_p
CURVE 1	
81.8	4.286×10^{-2}
84.2	4.376*
89.7	4.662
95.8	5.014
102.5	5.376
109.0	5.713
116.6	6.072
120.1	6.240*
124.2	6.435
128.7	6.697*
131.2	6.852
134.9	6.943*
139.2	7.122*
142.9	7.244
167.0	8.378
174.7	8.674*
183.2	8.981
189.7	9.275*
196.8	9.458
204.7	9.825*
210.8	9.894
220.1	1.0101×10^{-1}*
225.0	1.0221
228.4	1.0375*
231.8	1.0338*
277.3	1.1387
282.5	1.1551*
284.8	1.1590*
286.9	1.1707*
289.1	1.1649*
295.3	1.1680*
297.9	1.1847
CURVE 2	
30	7.4×10^{-3}
40	1.62×10^{-2}
50	2.43
60	3.15
70	3.76
80	4.30
90	4.79
100	5.33
CURVE 2 (cont.)	
110	5.92×10^{-2}
120	6.46
130	6.96
140	7.45
150	7.91
160	8.36
170	8.77
180	9.19
190	9.56
200	9.95

* Not shown on plot

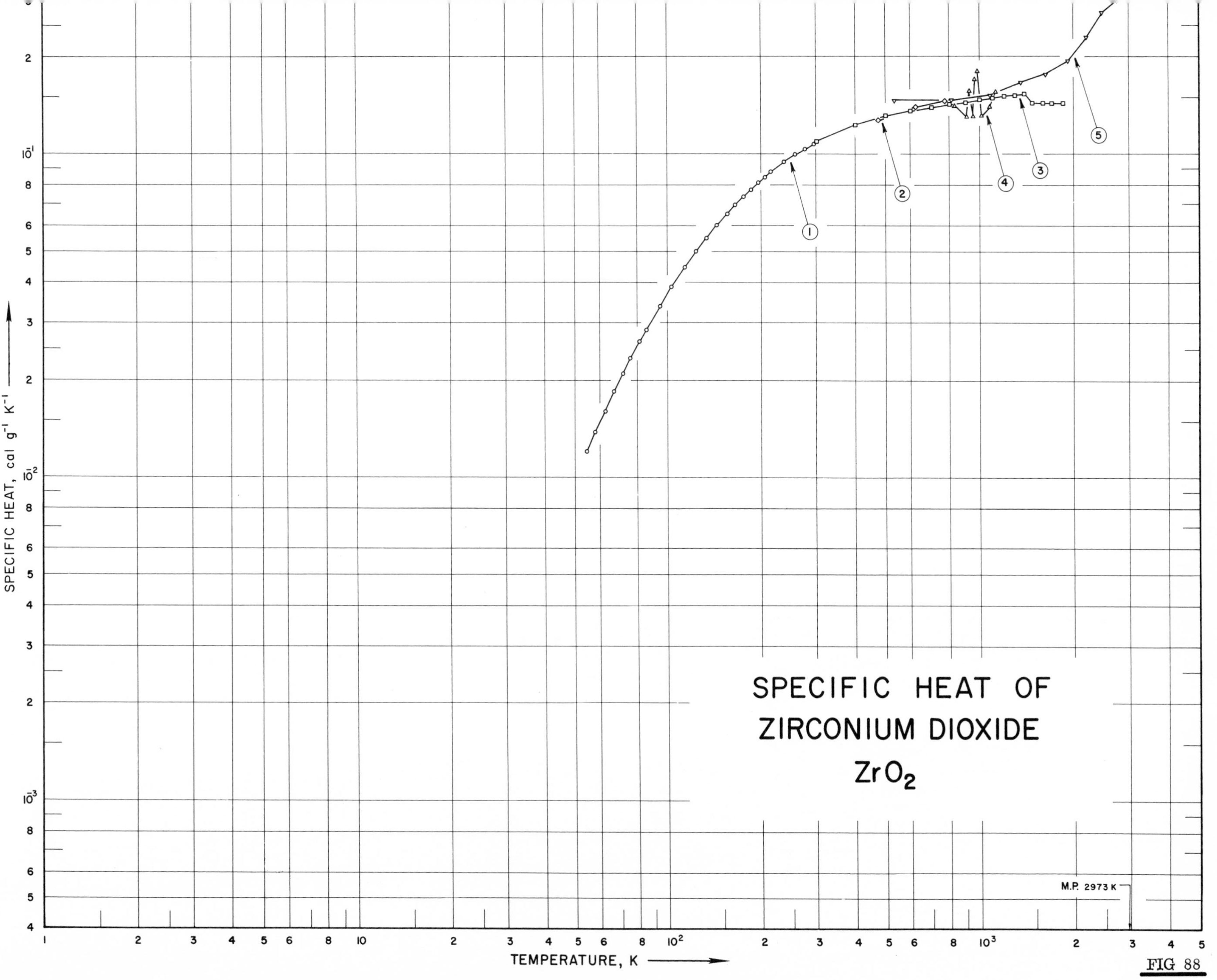

SPECIFIC HEAT OF
ZIRCONIUM DIOXIDE
ZrO_2
M.P. 2973 K
TEMPERATURE, K
SPECIFIC HEAT, cal g⁻¹ K⁻¹

FIG 88

SPECIFICATION TABLE NO. 88 SPECIFIC HEAT OF ZIRCONIUM DIOXIDE ZrO_2

[For Data Reported in Figure and Table No. 88]

Curve No.	Ref. No.	Year	Temp. Range, K	Reported Error, %	Name and Specimen Designation	Composition (weight percent), Specifications and Remarks
1	153	1944	54-295	0.3		99.14 ZrO_2, 0.30 SiO_2, 0.20 TiO_2, 0.07 CaO, and ≤0.05 other oxides; corrected for impurities.
2	92	1950	473-773			
3	154	1950	298-1850	0.2		ZrO_2 and 1.25 H_f; x-ray diffraction showed only nonoclinic oxide.
4	155	1961	836-1127			Sample A, B, and C one inch diameter; density 271 lb ft^{-3}, sample D, density 342 lb ft^{-3}.
5	48	1962	533-2755	≤ 5		Before exposure: 70.7 Zr, 1.0 Ca, 0.1 Al, 0.1 Mg, 0.1 Si, and 0.1 Ti, after exposure: 71.2 Zr, and 0.15 C; sample supplied by Zirconium Corp. of America; crushed in hardened steel mortar to pass 100-mesh screen; pressed and sintered; density at 25 C, apparent density (ASTM method B311-58) 334 lb ft^{-3}, true density (by immersion in xylene) 351 lb ft^{-3}, after exposure: apparent density 334 lb ft^{-3}, true density 357 lb ft^{-3}.

DATA TABLE NO. 88 SPECIFIC HEAT OF ZIRCONIUM DIOXIDE ZrO_2

[Temperature, T, K; Specific Heat, C_p, Cal g^{-1} K^{-1}]

T	C_p
CURVE 1	
54.3	1.195 x 10^{-2}
57.9	1.373
62.2	1.589
66.8	1.834
71.4	2.088
75.7	2.319
81.0	2.606
85.3	2.844
94.6	3.369
103.6	3.871
114.0	4.440
124.3	4.994
134.3	5.487
144.9	6.005
155.4	6.505
165.0	6.932
175.3	7.350
185.1	7.740
195.4	8.139
205.1	8.472
215.0	8.814
225.2	9.106*
235.7	9.406
245.5	9.681*
255.6	9.933
265.9	1.019 x 10^{-1}*
276.1	1.042
285.8	1.061*
295.0	1.077
CURVE 2	
473	1.27 x 10^{-1}
623	1.40
773	1.46
CURVE 3	
(α) 298	1.087 x 10^{-1}*
300	1.091
400	1.238
500	1.314
600	1.362
CURVE 3 (cont.)	
(α) 700	1.397 x 10^{-1}
800	1.425
900	1.448
1000	1.469
1100	1.489
1200	1.507
1300	1.524
1400	1.541
(α) 1478	1.445
(β) 1478	1.445
1500	1.445*
1600	1.445
1700	1.445
1800	1.445*
1850	1.445
CURVE 4	
Series 1	
836.16	1.41 x 10^{-1}
867.16	
Series 2	
924.16	1.57 x 10^{-1}
963.16	1.70
986.16	1.80
Series 3	
907.16	1.31 x 10^{-1}
907.16	1.33*
907.16	1.33*
953.16	1.30
1016.66	1.31
1085.16	1.40
1127.16	1.56
CURVE 5	
533.15	1.47 x 10^{-1}
810.93	1.47
1088.71	1.52
1366.48	1.67
1644.26	1.77
1922.04	1.95
2199.82	2.30
2477.59	2.75
2755.37	3.00

* Not shown on plot

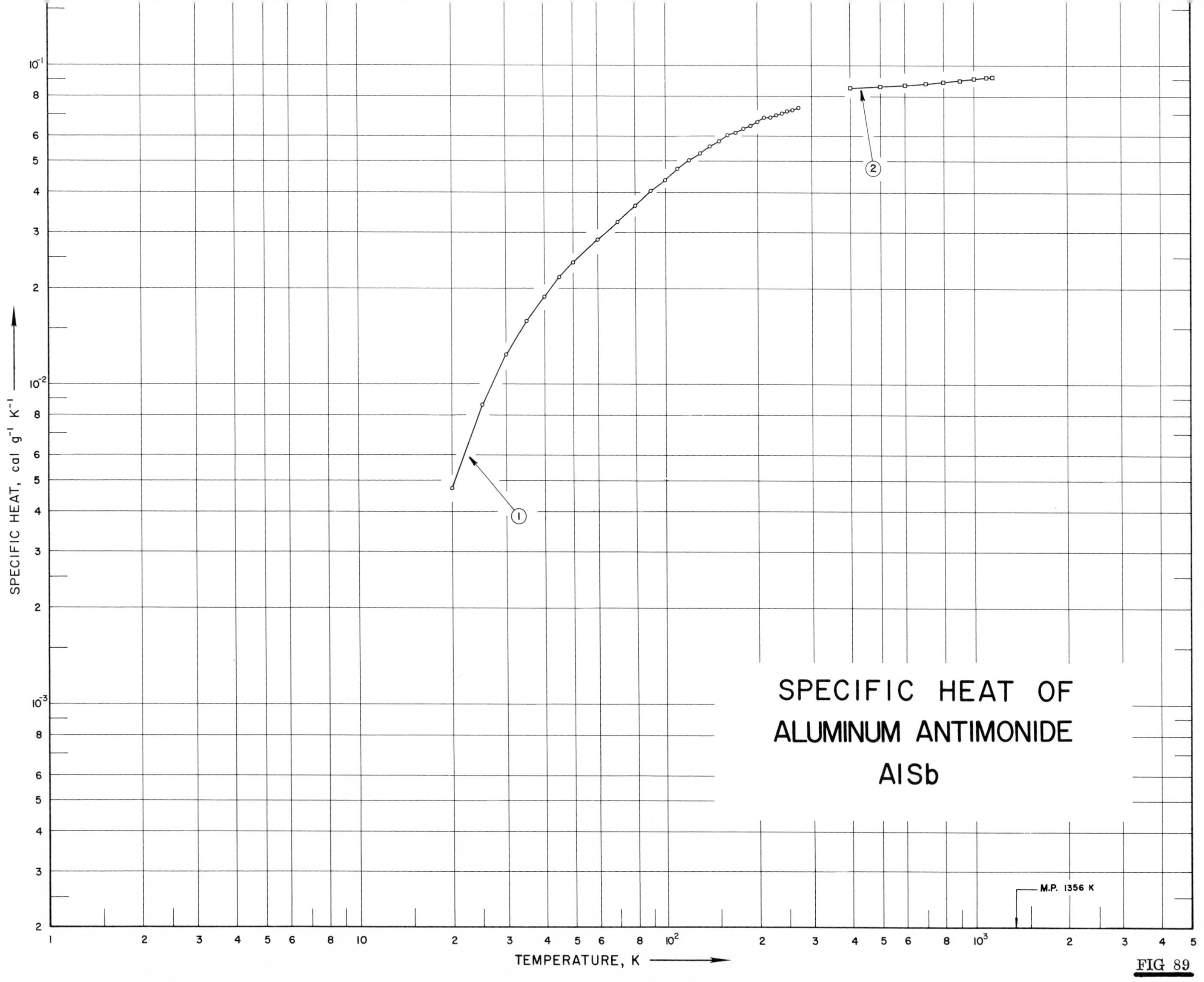
SPECIFIC HEAT OF
ALUMINUM ANTIMONIDE
AlSb
M.P. 1356 K
SPECIFIC HEAT, cal g⁻¹ K⁻¹
TEMPERATURE, K
FIG 89

SPECIFICATION TABLE NO. 89 SPECIFIC HEAT OF ALUMINUM ANTIMONIDE AlSb

[For Data Reported in Figure and Table No. 89]

Curve No.	Ref. No.	Year	Temp. Range, K	Reported Error, %	Name and Specimen Designation	Composition (weight percent), Specifications and Remarks
1	156	1963	20-273	≤2.0		
2	157	1963	400-1150			~99.99 AlSb.

DATA TABLE NO. 89 SPECIFIC HEAT OF ALUMINUM ANTIMONIDE AlSb

[Temperature, T, K; Specific Heat, C_p, Cal g^{-1} K^{-1}]

T	C_p
CURVE 1	
20	4.71×10^{-3}
25	8.60
30	1.24×10^{-2}
35	1.587
40	1.895
45	2.165
50	2.404
60	2.844
70	3.247
80	3.644
90	4.027
100	4.390
110	4.719
120	5.015
130	5.303
140	5.573
150	5.798
160	6.001
170	6.176
180	6.332
190	6.476
200	6.616
210	6.860
220	6.858
230	6.969
240	7.075
250	7.180
260	7.272
270	7.328*
273.2	7.342
CURVE 2	
400	8.458×10^{-2}
500	8.5590
600	8.66
700	8.7608
800	8.8616
900	8.9624
1000	9.0634
1100	9.1642
1150	9.2146

*Not shown on plot

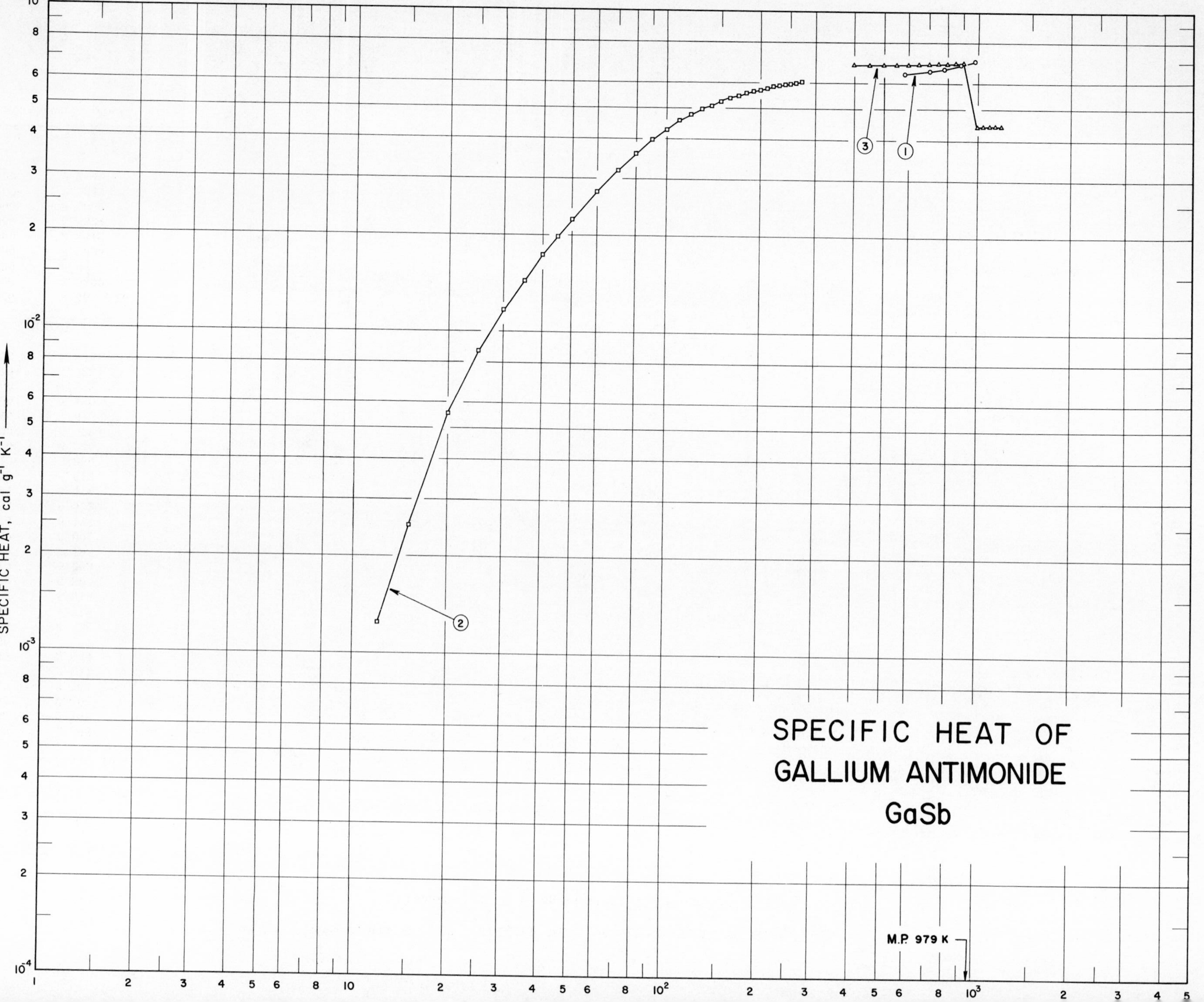

SPECIFIC HEAT OF
GALLIUM ANTIMONIDE
GaSb
M.P. 979 K
SPECIFIC HEAT, cal g^{-1} K^{-1}
1
2
3

SPECIFICATION TABLE NO. 90 SPECIFIC HEAT OF GALLIUM ANTIMONIDE GaSb

[For Data Reported in Figure and Table No. 90]

Curve No.	Ref. No.	Year	Temp. Range, K	Reported Error, %	Name and Specimen Designation	Composition (weight percent), Specifications and Remarks
1	158	1961	586-980	≤0. 2		99. 99 GaSb; obtained by melting stoichiometric amounts of pure metals together in evacuated quartz ampules.
2	156	1963	12-273	≤2. 0		
3	157	1963	400-1200			~99. 99 GaSb.

DATA TABLE NO. 90 SPECIFIC HEAT OF GALLIUM ANTIMONIDE GaSb

[Temperature, T, K; Specific Heat, C_p, Cal g^{-1} K^{-1}]

T	C_p	T	C_p
CURVE 1		CURVE 3	
585.85	6.405×10^{-2}	(s)400	6.8302×10^{-2}
704.35	6.593	450	6.8416
785.05	6.721	500	6.8532
884.95	6.880	550	6.8648
980.35	7.032	600	6.8764
		650	6.8880
CURVE 2		700	6.8996
		750	6.9112
12	1.25×10^{-3}	800	6.9228
15	2.50	850	6.9344
20	5.54	(s)900	6.9460
25	8.64		
30	1.167×10^{-2}	(l) 1000	4.4184×10^{-2}
35	1.441	1050	4.4184
40	1.713	1100	4.4184
45	1.974	1150	4.4184
50	2.225	(l) 1200	4.4184
60	2.716		
70	3.173		
80	3.591		
90	3.953		
100	4.257		
110	4.506		
120	4.726		
130	4.920		
140	5.087		
150	5.228		
160	5.343		
170	5.440		
180	5.529		
190	5.611		
200	5.688		
210	5.759		
220	5.818		
230	5.870		
240	5.909		
250	5.943		
260	5.975		
270	6.004*		
273.2	6.011		

*Not shown on plot

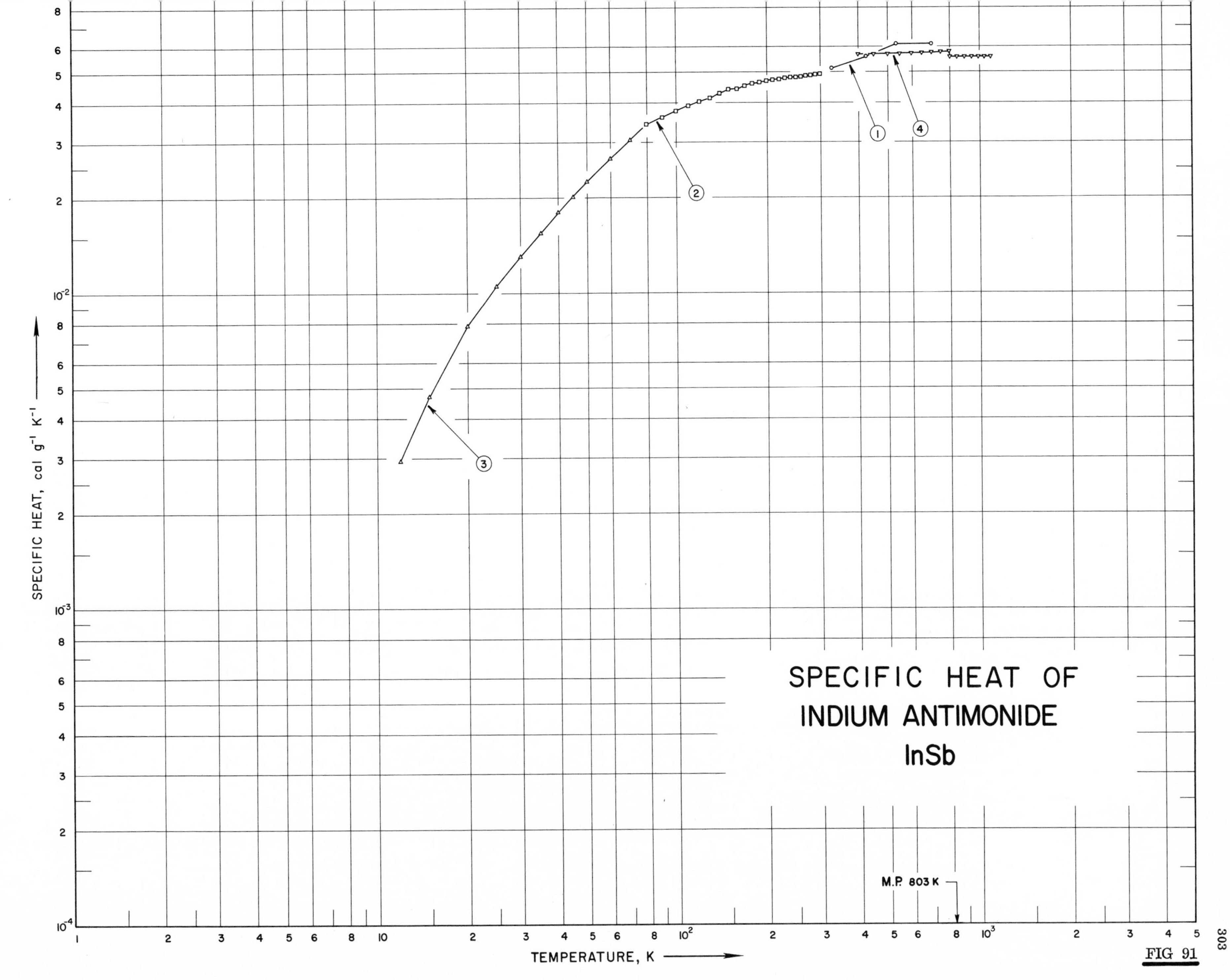
SPECIFIC HEAT OF
INDIUM ANTIMONIDE
InSb
M.P. 803 K
TEMPERATURE, K
SPECIFIC HEAT, cal g^{-1} K^{-1}
1
2
3
4

FIG 91

SPECIFICATION TABLE NO. 91 SPECIFIC HEAT OF INDIUM ANTIMONIDE InSb

[For Data Reported in Figure and Table No. 91]

Curve No.	Ref. No.	Year	Temp. Range, K	Reported Error, %	Name and Specimen Designation	Composition (weight percent), Specifications and Remarks
1	159	1958	328-698			51.44 Sb, 0.001 Fe, 0.001 Mg, 0.001 Pb, 0.001 Sn, 0.0001 Cu, and 0.0001 Si.
2	160	1959	80-300	3-7		Polycrystalline.
3	156	1963	12-273	≤2		
4	157	1963	400-1100			~99.99 InSb.

DATA TABLE NO. 91 SPECIFIC HEAT OF INDIUM ANTIMONIDE InSb

[Temperature, T, K; Specific Heat, C_p, Cal g^{-1} K^{-1}]

T	C_p
CURVE 1	
328	5.2 x 10^{-2}
423	5.6
533	6.2
698	6.2
CURVE 2	
80	3.45 x 10^{-2}
90	3.63
100	3.8
110	3.95
120	4.06
130	4.19
140	4.31
150	4.4
160	4.48
170	4.56
180	4.63
190	4.68
200	4.72
210	4.76
220	4.78
230	4.82
240	4.85
250	4.86
260	4.89
270	4.91
280	4.93
290	4.95
300	4.97
CURVE 3	
12	2.96 x 10^{-3}
15	4.77
20	7.90
25	1.060 x 10^{-2}
30	1.319
35	1.568
40	1.810
45	2.047
50	2.276
60	2.699
70	3.070

T	C_p
CURVE 3 (cont.)	
80	3.382*
90	3.641*
100	3.851*
110	4.014*
120	4.153*
130	4.275*
140	4.382*
150	4.472*
160	4.546*
170	4.607*
180	4.658*
190	4.703*
200	4.746*
210	4.789*
220	4.825*
230	4.858*
240	4.886*
250	4.912*
260	4.935*
270	4.954*
273.2	4.960*
CURVE 4	
(s) 400	5.72 x 10^{-2}
450	5.74
500	5.75
550	5.76
600	5.78
650	5.80
700	5.81
750	5.84
(s) 800	5.86
(l) 803	5.62 x 10^{-2}
850	5.62
900	5.62
950	5.62
1000	5.62
1050	5.62
(l) 1100	5.62

*Not shown on plot

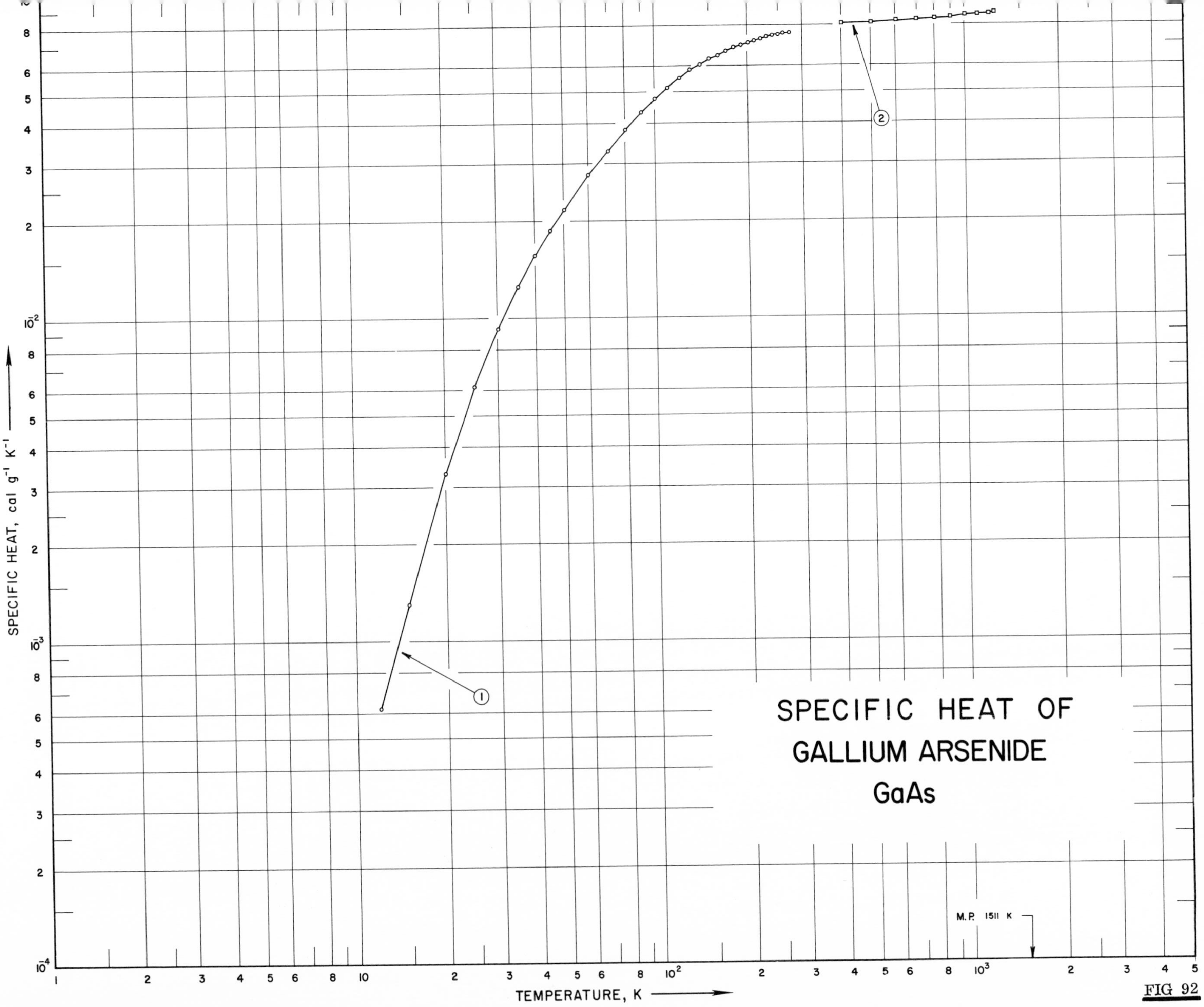

SPECIFIC HEAT OF
GALLIUM ARSENIDE
GaAs

FIG 92

SPECIFICATION TABLE NO. 92 SPECIFIC HEAT OF GALLIUM ARSENIDE GaAs

[For Data Reported in Figure and Table No. 92]

Curve No.	Ref. No.	Year	Temp. Range, K	Reported Error, %	Name and Specimen Designation	Composition (weight percent), Specifications and Remarks
1	156	1963	12-273	≤2		
2	157	1963	400-1250			~99. 99 GaAs.

DATA TABLE NO. 92 SPECIFIC HEAT OF GALLIUM ARSENIDE GaAs

[Temperature, T, K; Specific Heat, C_p, Cal g^{-1} K^{-1}]

T	C_p
CURVE 1	
12	6.2×10^{-4}
15	1.3×10^{-3}
20	3.30
25	6.15
30	9.26
35	1.25×10^{-2}
40	1.563
45	1.866
50	2.160
60	2.731
70	3.291
80	3.821
90	4.317
100	4.768
110	5.158
120	5.501
130	5.801
140	6.080
150	6.303
160	6.495
170	6.658
180	6.804
190	6.942
200	7.069
210	7.181
220	7.283
230	7.366
240	7.436
250	7.495
260	7.536
270	7.575*
273.2	7.579
CURVE 2	
400	8.1012×10^{-2}
500	8.1734
600	8.2454
700	8.3176
800	8.3898
900	8.4620
1000	8.5342
1100	8.6064
1200	8.6786
1250	8.7146

*Not shown on plot

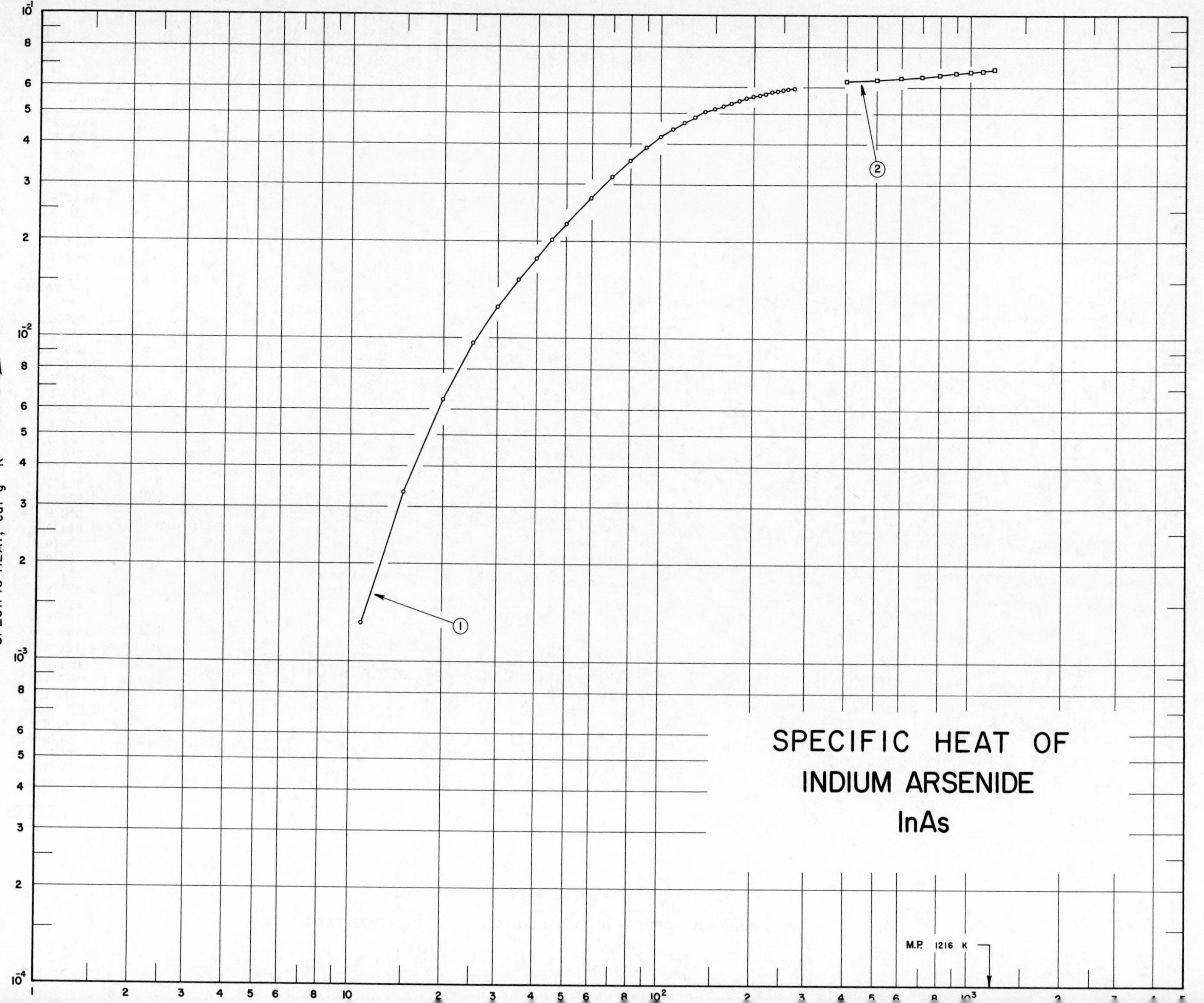
SPECIFIC HEAT OF
INDIUM ARSENIDE
InAs
SPECIFIC HEAT, cal g^{-1} K^{-1}
M.P. 1216 K
1
2

SPECIFICATION TABLE NO. 93 SPECIFIC HEAT OF INDIUM ARSENIDE InAs

[For Data Reported in Figure and Table No. 93]

Curve No.	Ref. No.	Year	Temp. Range, K	Reported Error, %	Name and Specimen Designation	Composition (weight percent), Specifications and Remarks
1	156	1963	11-273	≤2.0		
2	157	1963	400-1200			99.99 InAs.

DATA TABLE NO. 93 SPECIFIC HEAT OF INDIUM ARSENIDE InAs

[Temperature, T, K; Specific Heat, C_p, Cal g^{-1} K^{-1}]

T	C_p
CURVE 1	
11	1.318 x 10^{-3}
15	3.331
20	6.472
25	9.614
30	1.252 x 10^{-2}
35	1.522
40	1.775
45	2.022
50	2.266
60	2.730
70	3.162
80	3.552
90	3.906
100	4.212
110	4.470
120	4.684
130	4.874
140	5.027
150	5.174
160	5.292
170	5.395
180	5.481
190	5.556
200	5.627
210	5.692
220	5.750
230	5.808
240	5.861
250	5.905
260	5.942
270	5.970*
273.2	5.977
CURVE 2	
400	6.2810 x 10^{-2}
500	6.3518
600	6.4226
700	6.4934
800	6.5642
900	6.6352
1000	6.7060
1100	6.7768
1200	6.8476

*Not shown on plot

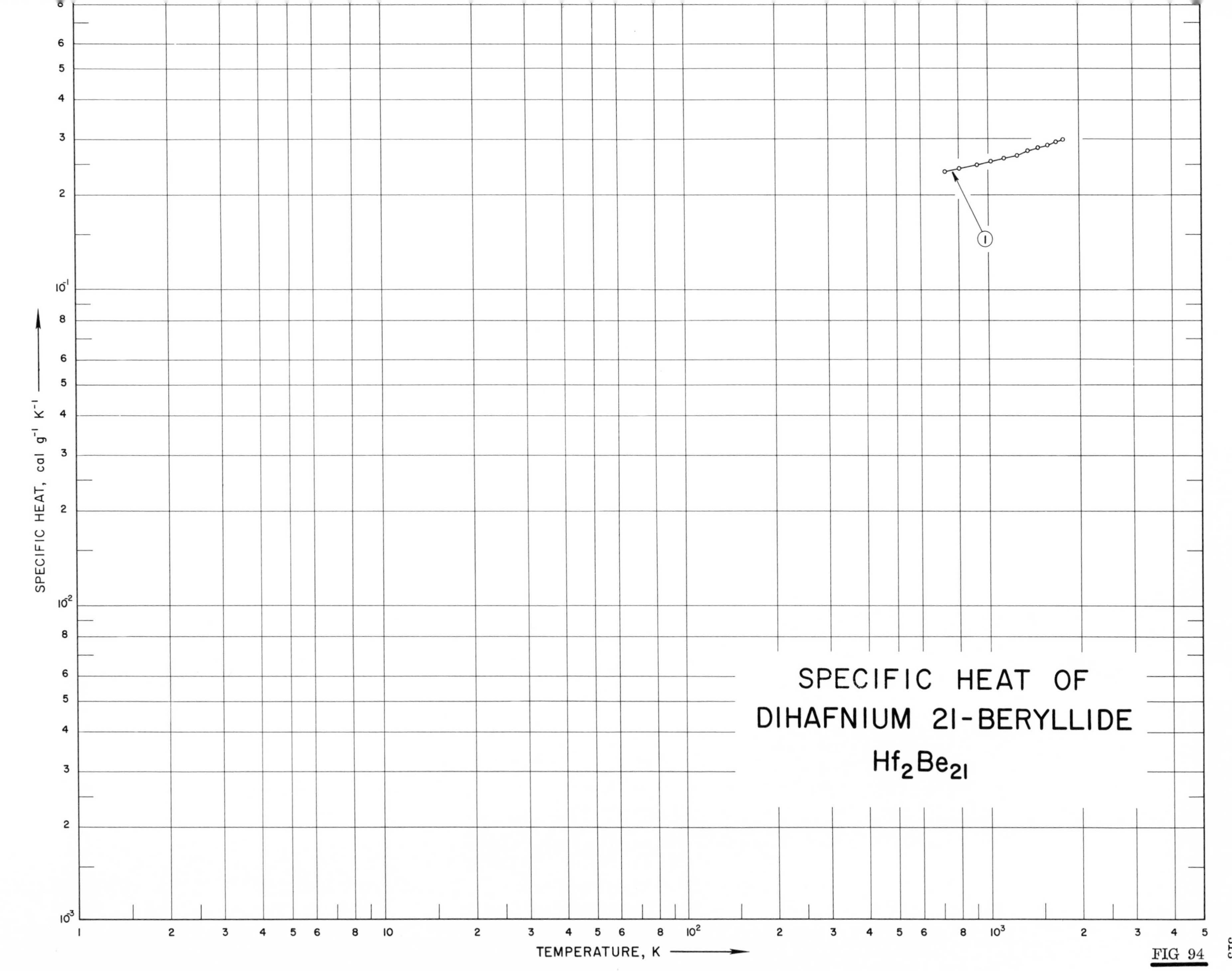
SPECIFIC HEAT OF
DIHAFNIUM 21-BERYLLIDE
Hf2Be21
TEMPERATURE, K
SPECIFIC HEAT, cal g-1 K-1
1

FIG 94

SPECIFICATION TABLE NO. 94 SPECIFIC HEAT OF DIHAFNIUM 21-BERYLLIDE Hf_2Be_{21}

[For Data Reported in Figure and Table No. 94]

Curve No.	Ref. No.	Year	Temp. Range, K	Reported Error, %	Name and Specimen Designation	Composition (weight percent), Specifications and Remarks
1	162	1961	728-1783			34.1 Be; hot pressed.

DATA TABLE NO. 94 SPECIFIC HEAT OF DIHAFNIUM 21-BERYLLIDE Hf_2Be_{21}

[Temperature, T, K; Specific Heat, C_p, Cal $g^{-1}K^{-1}$]

T	C_p
CURVE 1	
727.594	2.341 x 10^{-1}
810.928	2.390
922.039	2.455
1033.150	2.520
1144.261	2.585
1255.372	2.649
1366.483	2.714
1477.594	2.779
1588.706	2.844
1699.817	2.909
1783.150	2.957

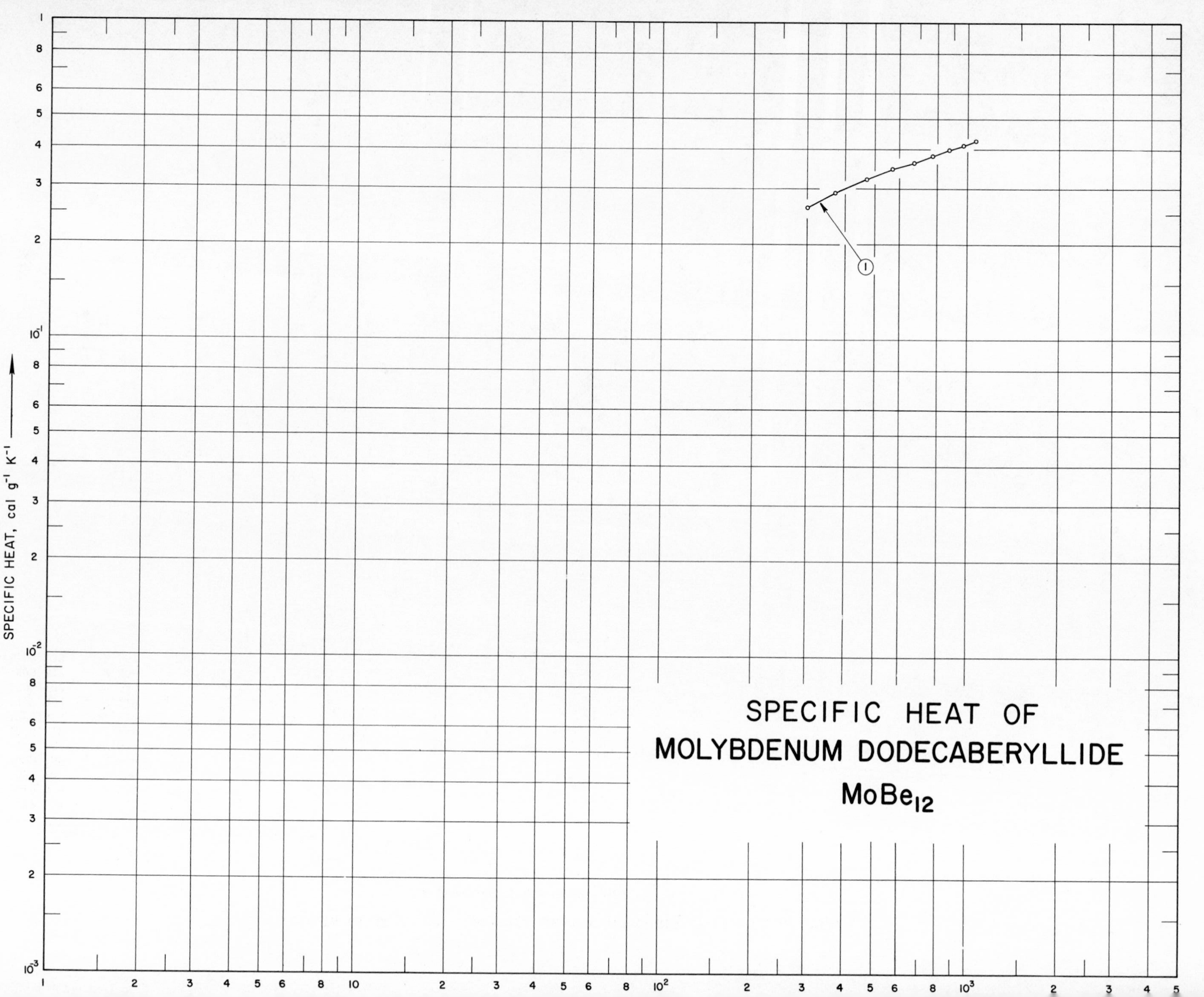
SPECIFIC HEAT OF
MOLYBDENUM DODECABERYLLIDE
MoBe12
SPECIFIC HEAT, cal g-1 K-1
1

SPECIFICATION TABLE NO. 95 SPECIFIC HEAT OF MOLYBDENUM DODECABERYLLIDE $MoBe_{12}$

[For Data Reported in Figure and Table No. 95]

Curve No.	Ref. No.	Year	Temp. Range, K	Reported Error, %	Name and Specimen Designation	Composition (weight percent), Specifications and Remarks
1	45	1963	303-1073	<3		

DATA TABLE NO. 95 SPECIFIC HEAT OF MOLYBDENUM DODECABERYLLIDE $MoBe_{12}$

[Temperature, T, K; Specific Heat, C_p, Cal $g^{-1}K^{-1}$]

T	C_p
CURVE 1	
303.15	2.61×10^{-1}
373.15	2.93
473.15	3.22
573.15	3.46
673.15	3.64
773.15	3.81
873.15	3.98
973.15	4.13
1073.15	4.28

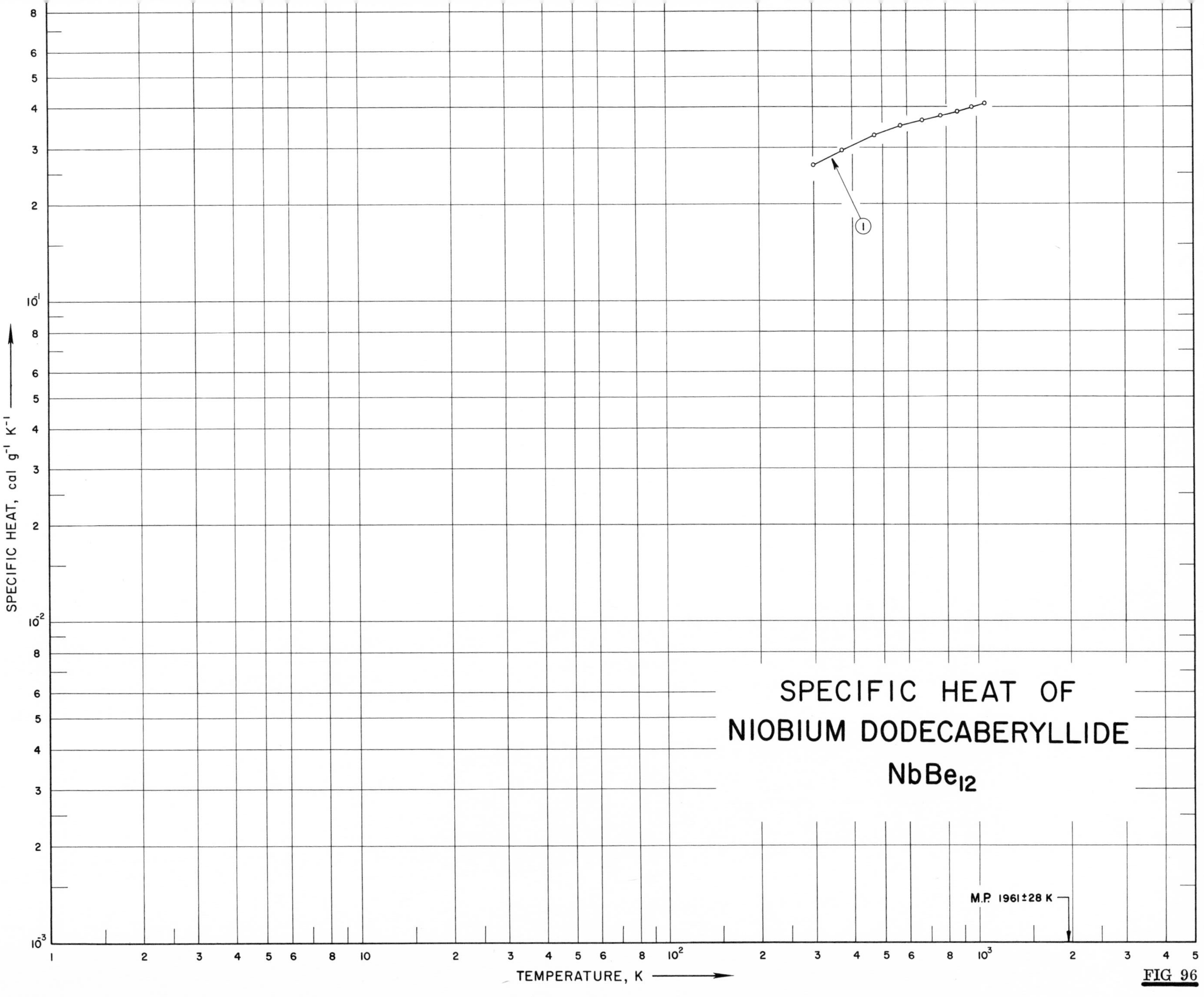
SPECIFIC HEAT OF
NIOBIUM DODECABERYLLIDE
$NbBe_{12}$
M.P. 1961±28 K
1
SPECIFIC HEAT, cal g^{-1} K^{-1}
TEMPERATURE, K
FIG 96

SPECIFICATION TABLE NO. 96 SPECIFIC HEAT OF NIOBIUM DODECABERYLLIDE $NbBe_{12}$

[For Data Reported in Figure and Table No. 96]

Curve No.	Ref. No.	Year	Temp. Range, K	Reported Error, %	Name and Specimen Designation	Composition (weight percent), Specifications and Remarks
1	45	1962	303-1073	<3		

DATA TABLE NO. 96 SPECIFIC HEAT OF NIOBIUM DODECABERYLLIDE $NbBe_{12}$

[Temperature, T, K; Specific Heat, C_p, Cal $g^{-1}K^{-1}$]

T	C_p
CURVE 1	
303.15	2.65 x 10^{-1}
373.15	2.94
473.15	3.27
573.15	3.50
673.15	3.65
773.15	3.76
873.15	3.87
973.15	4.00
1073.15	4.11

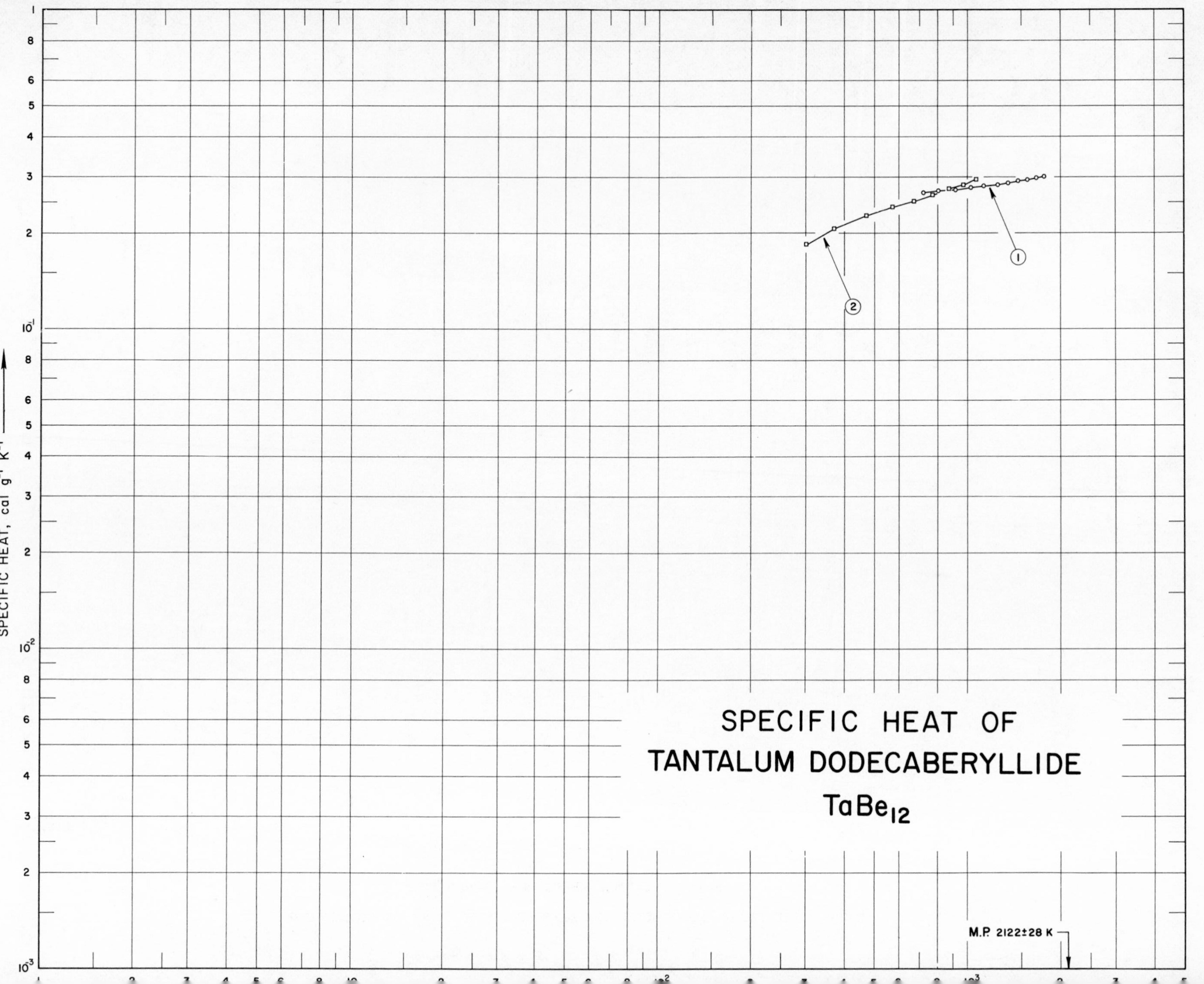
SPECIFIC HEAT OF
TANTALUM DODECABERYLLIDE
TaBe12
SPECIFIC HEAT, cal g-1 K-1
M.P. 2122±28 K
1
2

SPECIFICATION TABLE NO. 97 SPECIFIC HEAT OF TANTALUM DODECABERYLLIDE $TaBe_{12}$

[For Data Reported in Figure and Table No. 97]

Curve No.	Ref. No.	Year	Temp. Range, K	Reported Error, %	Name and Specimen Designation	Composition (weight percent), Specifications and Remarks
1	162	1961	728-1783			Single phase composition; hot pressed.
2	45	1962	303-1073	<3		

DATA TABLE NO. 97 SPECIFIC HEAT OF TANTALUM DODECABERYLLIDE $TaBe_{12}$

[Temperature, T, K; Specific Heat, C_p, Cal $g^{-1}K^{-1}$]

T	C_p
CURVE 1	
727.60	2.6806 x 10^{-1}
810.90	2.7060
922.02	2.7398
1033.16	2.7735
1144.27	2.8073
1255.38	2.8411
1366.49	2.8749
1477.60	2.9087
1588.72	2.9425
1699.83	2.9763
1783.16	3.0016
CURVE 2	
303.15	1.85 x 10^{-1}
373.15	2.07
473.15	2.27
573.15	2.41
673.15	2.53
773.15	2.64
873.15	2.75
973.15	2.84
1073.15	2.95

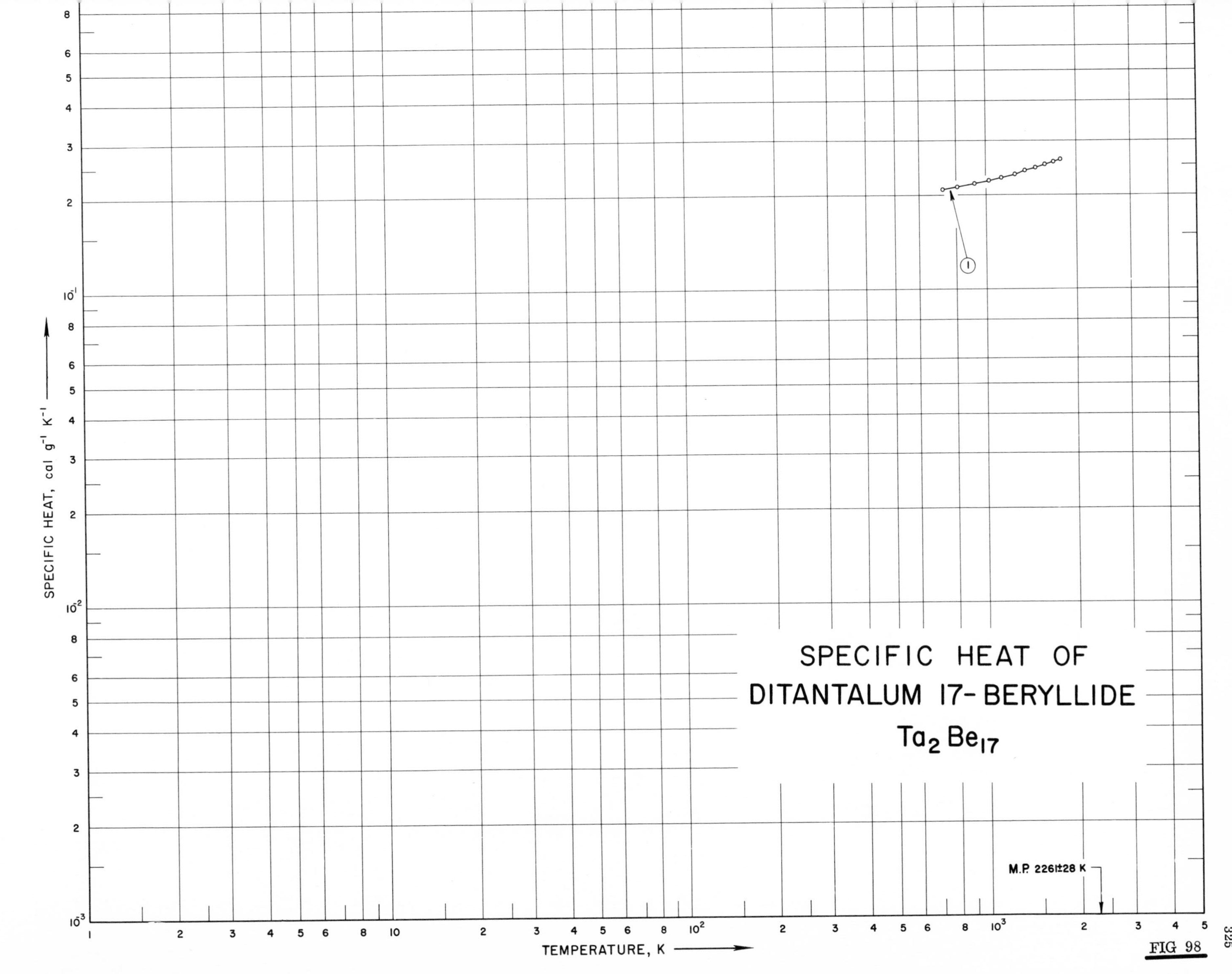
SPECIFIC HEAT OF
DITANTALUM 17-BERYLLIDE
Ta_2Be_{17}
SPECIFIC HEAT, cal g⁻¹ K⁻¹
TEMPERATURE, K
M.P. 2261±28 K
1

FIG 98

SPECIFICATION TABLE NO. 98 SPECIFIC HEAT OF DITANTALUM 17-BERYLLIDE Ta_2Be_{17}

[For Data Reported in Figure and Table No. 98]

Curve No.	Ref. No.	Year	Temp. Range, K	Reported Error, %	Name and Specimen Designation	Composition (weight percent), Specifications and Remarks
1	162	1961	728–1783			Single phase composition; hot pressed.

DATA TABLE NO. 98 SPECIFIC HEAT OF DITANTALUM 17-BERYLLIDE Ta_2Be_{17}

[Temperature, T, K; Specific Heat, C_p, Cal $g^{-1}K^{-1}$]

T	C_p
CURVE 1	
727.60	2.0930×10^{-1}
810.90	2.1330
922.02	2.1864
1033.16	2.2398
1144.27	2.2932
1255.38	2.3466
1366.49	2.4000
1477.60	2.4534
1588.72	2.5068
1699.82	2.5603
1783.16	2.6003

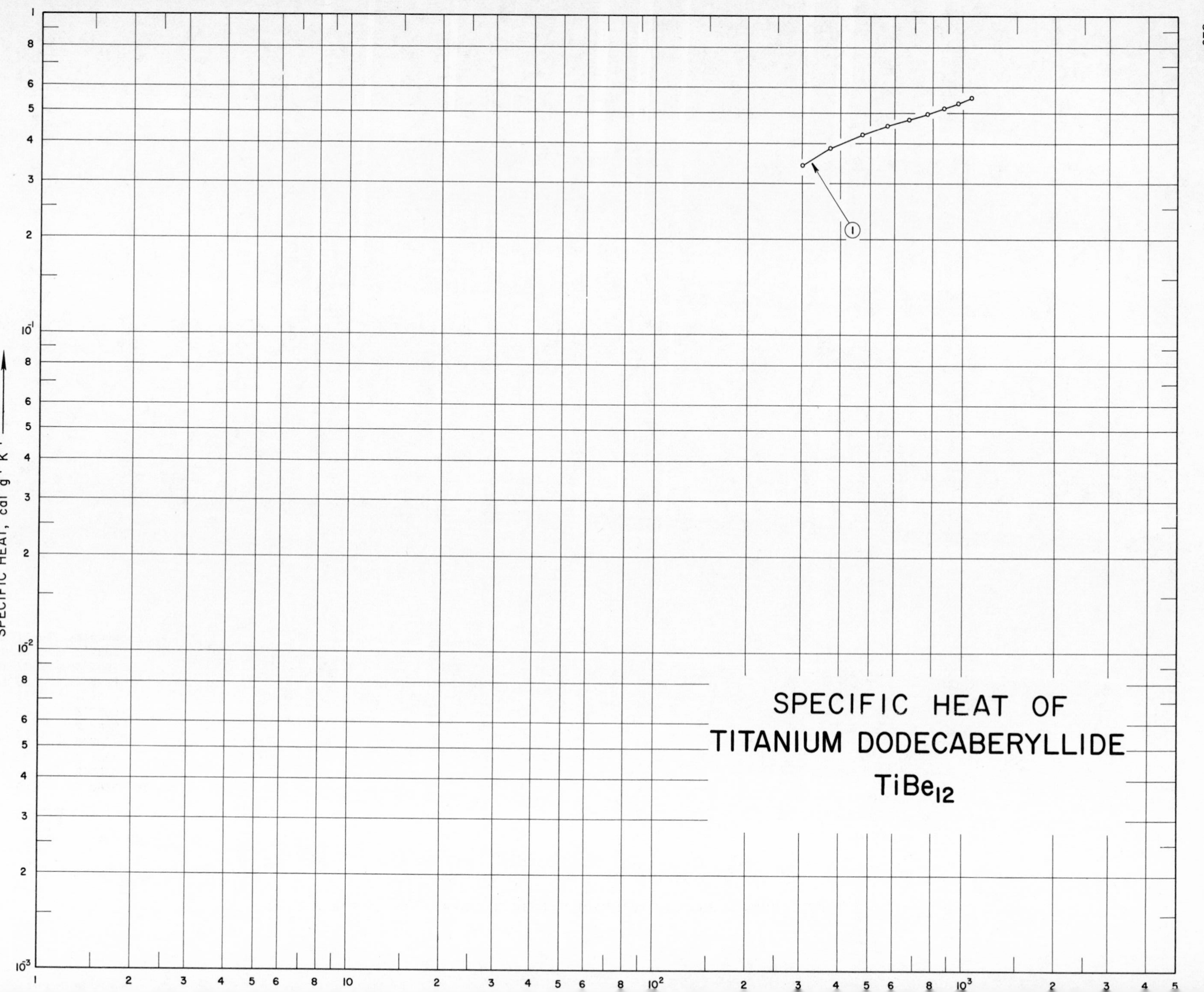
SPECIFIC HEAT OF
TITANIUM DODECABERYLLIDE
TiBe12
SPECIFIC HEAT, cal g-1 K-1
1

SPECIFICATION TABLE NO. 99 SPECIFIC HEAT OF TITANIUM DODECABERYLLIDE $TiBe_{12}$

[For Data Reported in Figure and Table No. 99]

Curve No.	Ref. No.	Year	Temp. Range, K	Reported Error, %	Name and Specimen Designation	Composition (weight percent), Specifications and Remarks
1	45	1963	303-1073	<3		

DATA TABLE NO. 99 SPECIFIC HEAT OF TITANIUM DODECABERYLLIDE $TiBe_{12}$

[Temperature, T, K; Specific Heat, C_p, Cal $g^{-1}K^{-1}$]

T	C_p
CURVE 1	
303.15	3.41 x 10^{-1}
373.15	3.86
473.15	4.28
573.15	4.54
673.15	4.76
773.15	4.97
873.15	5.17
973.15	5.37
1073.15	5.57

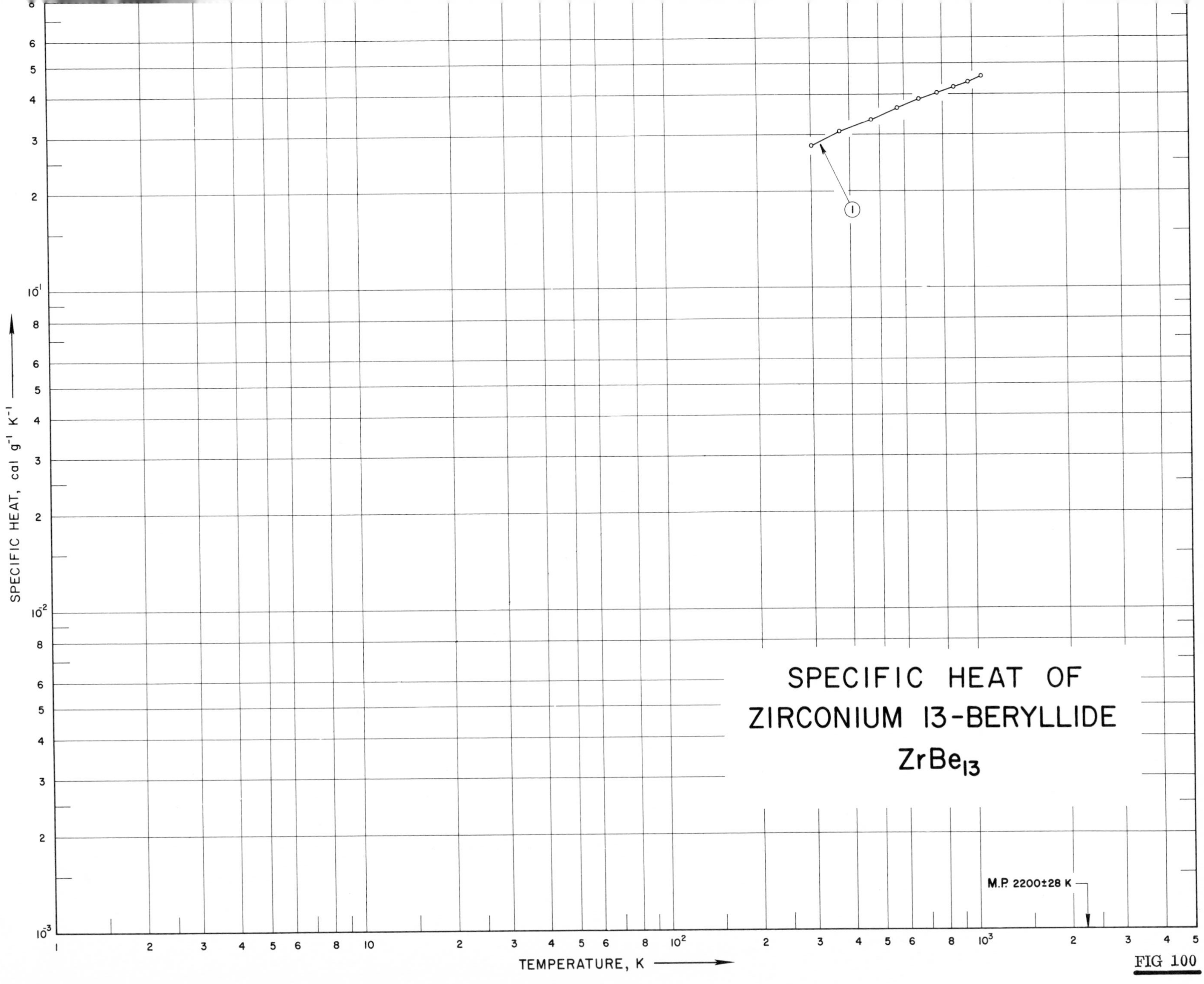
SPECIFIC HEAT OF
ZIRCONIUM 13-BERYLLIDE
$ZrBe_{13}$
M.P. 2200±28 K
1
SPECIFIC HEAT, cal g⁻¹ K⁻¹
TEMPERATURE, K

FIG 100

SPECIFICATION TABLE NO. 100 SPECIFIC HEAT OF ZIRCONIUM 13-BERYLLIDE $ZrBe_{13}$

[For Data Reported in Figure and Table No. 100]

Curve No.	Ref. No.	Year	Temp. Range, K	Reported Error, %	Name and Specimen Designation	Composition (weight percent), Specifications and Remarks
1	45	1962	303-1073	<3		

DATA TABLE NO. 100 SPECIFIC HEAT OF ZIRCONIUM 13-BERYLLIDE $ZrBe_{13}$

[Temperature, T, K; Specific Heat, C_p, Cal $g^{-1}K^{-1}$]

T	C_p
CURVE 1	
303.15	2.76 x 10^{-1}
373.15	3.07
473.15	3.38
573.15	3.63
673.15	3.85
773.15	4.03
873.15	4.20
973.15	4.37
1073.15	4.54

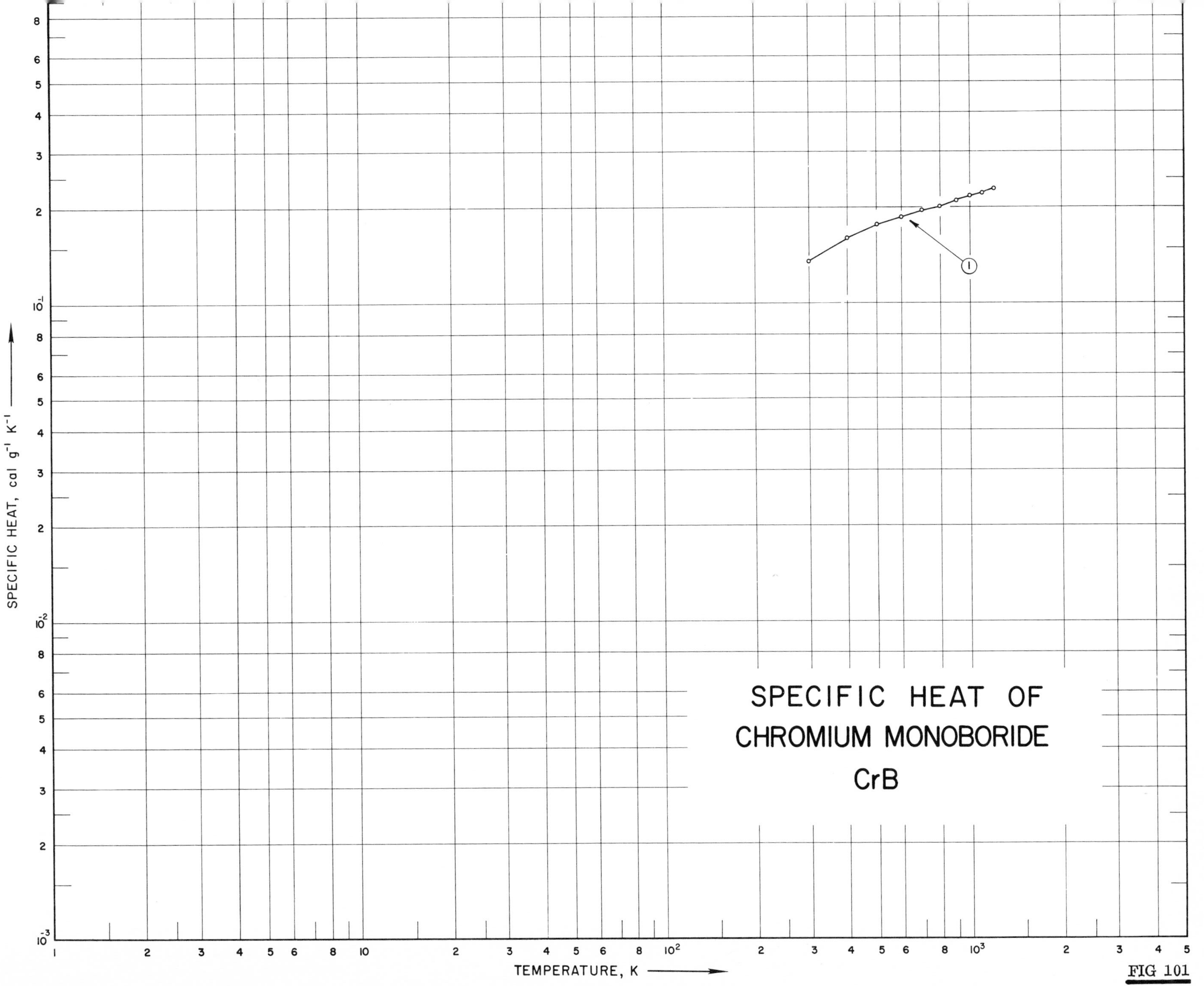

SPECIFIC HEAT OF
CHROMIUM MONOBORIDE
CrB
TEMPERATURE, K
SPECIFIC HEAT, cal g⁻¹ K⁻¹

FIG 101

SPECIFICATION TABLE NO. 101 SPECIFIC HEAT OF CHROMIUM MONOBORIDE CrB

[For Data Reported in Figure and Table No. 101]

Curve No.	Ref. No.	Year	Temp. Range, K	Reported Error, %	Name and Specimen Designation	Composition (weight percent), Specifications and Remarks
1	163	1962	298-1200	0.5		Traces of impurities.

DATA TABLE NO. 101 SPECIFIC HEAT OF CHROMIUM MONOBORIDE CrB

[Temperature, T, K; Specific Heat, C_p, Cal g^{-1} K^{-1}]

T	C_p
CURVE 1	
298.16	1.364 x 10^{-1}
300	1.371*
400	1.618
500	1.764
600	1.872
700	1.962
800	2.040
900	2.113
1000	2.183
1100	2.250
1200	2.315

*Not shown on plot

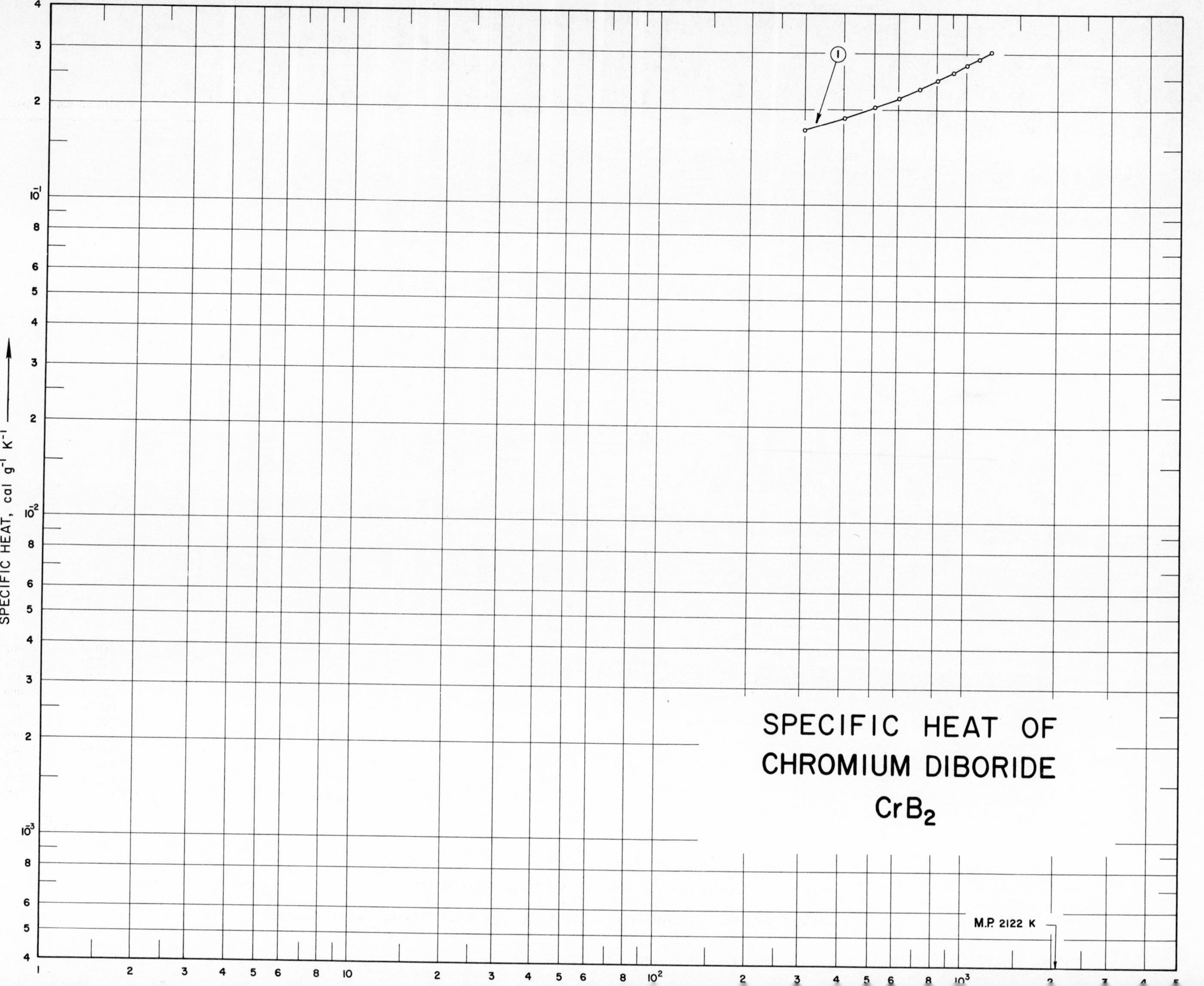
SPECIFIC HEAT OF
CHROMIUM DIBORIDE
CrB2
M.P. 2122 K
SPECIFIC HEAT, cal g⁻¹ K⁻¹
1

SPECIFICATION TABLE NO. 102 SPECIFIC HEAT OF CHROMIUM DIBORIDE CrB_2

[For Data Reported in Figure and Table No. 102]

Curve No.	Ref. No.	Year	Temp. Range, K	Reported Error, %	Name and Specimen Designation	Composition (weight percent), Specifications and Remarks
1	163	1962	298-1200			Traces of impurities.

DATA TABLE NO. 102 SPECIFIC HEAT OF CHROMIUM DIBORIDE CrB_2

[Temperature, T, K; Specific Heat, C_p, Cal $g^{-1} K^{-1}$]

T	C_p
CURVE 1	
298.16	1.738 x 10^{-1}
300	1.743*
400	1.888
500	2.033
600	2.179
700	2.326
800	2.471
900	2.616
1000	2.762
1100	2.907
1200	3.054

*Not shown on plot

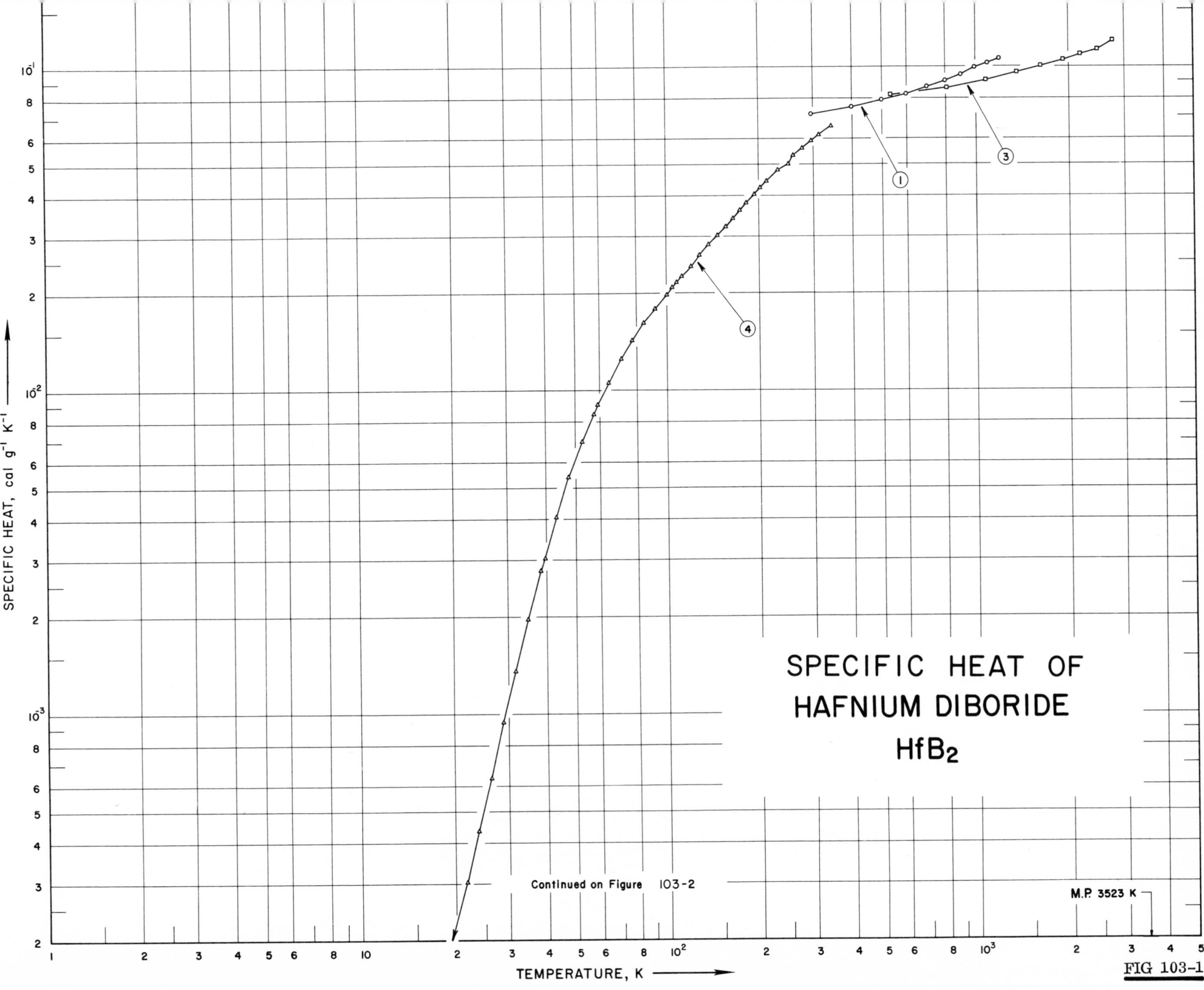
SPECIFIC HEAT OF
HAFNIUM DIBORIDE
HfB_2
Continued on Figure 103-2
M.P. 3523 K
TEMPERATURE, K
SPECIFIC HEAT, cal g^{-1} K^{-1}
1
3
4

FIG 103-1

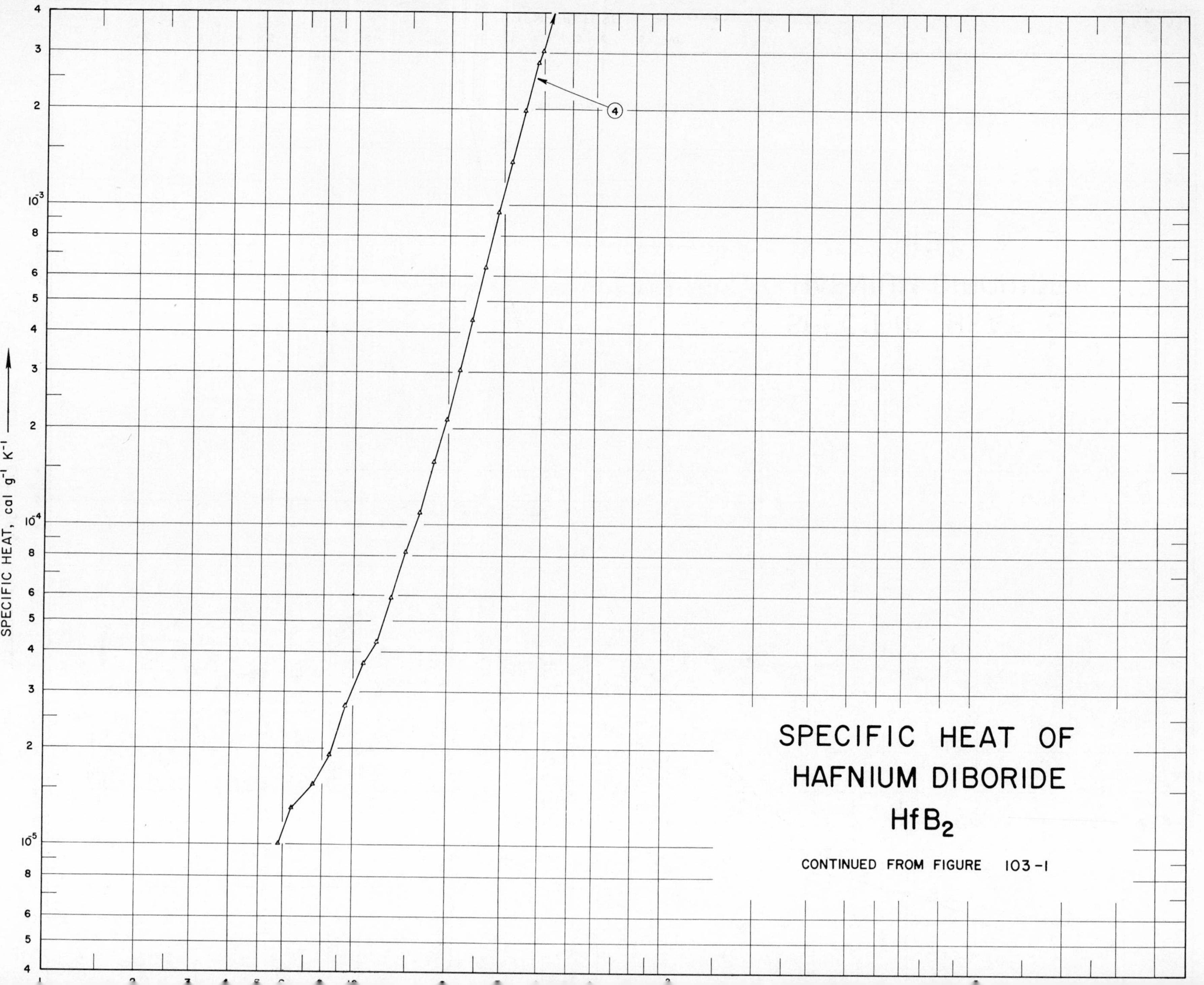
SPECIFIC HEAT OF
HAFNIUM DIBORIDE
HfB_2
CONTINUED FROM FIGURE 103-1
4
SPECIFIC HEAT, cal g^{-1} K^{-1}
4
3
2
10^{-3}
8
6
5
4
3
2
10^{-4}
8
6
5
4
3
2
10^{-5}
8
6
5
4

SPECIFICATION TABLE NO. 103 SPECIFIC HEAT OF HAFNIUM DIBORIDE HfB_2

[For Data Reported in Figure and Table No. 103]

Curve No.	Ref. No.	Year	Temp. Range, K	Reported Error, %	Name and Specimen Designation	Composition (weight percent), Specifications and Remarks
1	163	1962	298-1200	0.5		Traces of impurities.
2	164	1962	500-1200			Single phase composition; sample supplied by the Carborundum Co.
3	48	1962	533-2755	≤5		Before exposure: 89.5 Hf, 10.0 B, 3.5 Fe, 1.5 Zr, 0.1 Mg, 0.1 Ti, and 0.01 C; after exposure: 89.4 Hf, 10.5 B, and 0.77 C; sample supplied by the Carborundum Co; crushed in a hardened steel mortar to pass 100-mesh screen; hot pressed; density at 25 C, before exposure: apparent density (ASTM method B311-58) 666 lb ft^{-3}, true density (by immersion in xylene) 674 lb ft^{-3}, after exposure: apparent density 629 lb ft^{-3}, true density, 641 lb ft^{-3}.
4	165	1963	5-345			88.98 Hf, 10.97 B, 0.16 C, 0.01 - 0.1 Zr, 0.001 Cr, 0.001 Cu, 0.001 Mg, 0.0042 N, 0.0030 Fe, 0.0030 Ti, 0.0026 O, and 0.0010 Si; zone-refined.

DATA TABLE NO. 103 SPECIFIC HEAT OF HAFNIUM DIBORIDE HfB_2

[Temperature, T, K; Specific Heat, C_p, Cal $g^{-1} K^{-1}$]

T	C_p
CURVE 1	
298.17	7.111×10^{-2}
300	7.121*
400	7.510
500	7.905
600	8.295
700	8.690
800	9.080
900	9.474
1000	9.864
1100	1.026×10^{-1}
1200	1.065
CURVE 2*	
500	7.902×10^{-2}
600	8.294
700	8.686
800	9.078
900	9.470
1000	9.863
1100	1.026×10^{-1}
1200	1.065
CURVE 3	
533.15	8.20×10^{-2}
810.93	8.60
1088.71	9.10
1366.48	9.60
1644.26	1.01×10^{-1}
1922.04	1.05
2199.82	1.09
2477.59	1.13
2755.37	1.20
CURVE 4	
5.80	9.994×10^{-6}
6.39	1.299×10^{-5}
7.42	1.549
8.41	1.899
9.49	2.698
10.78	3.698
11.90	4.297
13.20	5.896
14.66	8.195
CURVE 4 (cont.)	
16.34	1.084×10^{-4}
18.07	1.569
19.86	2.129
21.83	3.048
23.88	4.352
26.23	6.371
28.90	9.424
31.69	1.361×10^{-3}
34.85	1.969
38.46	2.774
39.51	3.048
43.38	4.099
47.74	5.402
52.69	6.941
57.51	8.465
59.28	9.020
64.53	1.065×10^{-2}
71.10	1.255
77.69	1.434
84.62	1.620
92.06	1.797
100.35	1.981
108.86	2.167
98.88	1.947*
105.62	2.097
113.15	2.260
121.33	2.434
129.80	2.624
138.61	2.820
147.43	3.015
156.37	3.211
165.45	3.410
174.37	3.608
183.57	3.809
187.92	3.892*
194.89	4.057
203.93	4.243
212.95	4.430
222.14	4.617*
231.51	4.801
240.90	4.982*
250.20	5.012
259.35	5.330
268.38	5.469*
277.33	5.617
286.45	5.767*
CURVE 4 (cont.)	
295.76	5.906
305.10	6.045*
314.43	6.179
323.70	6.305*
334.41	6.438*
344.93	6.572

* Not shown on plot

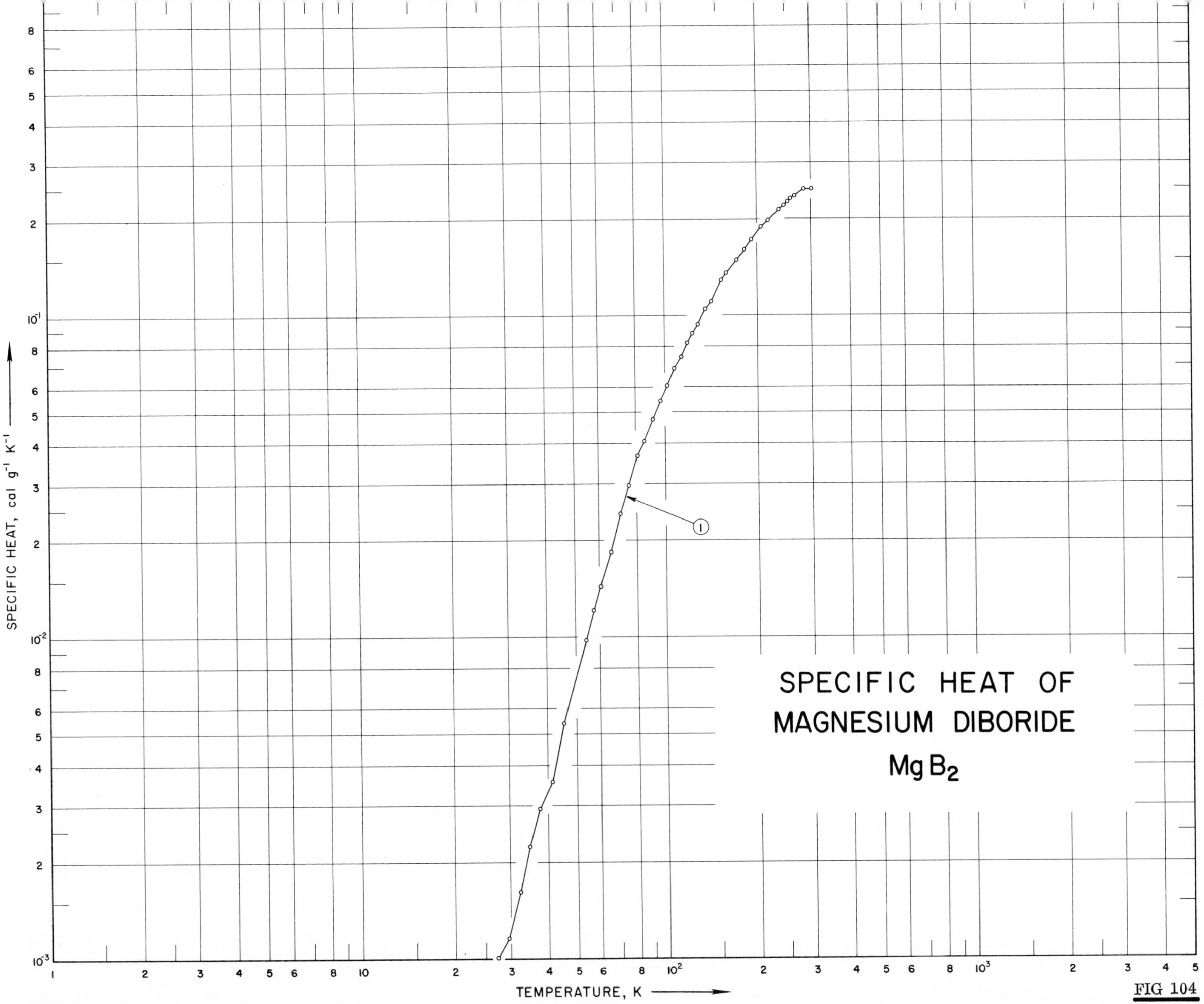

SPECIFIC HEAT OF MAGNESIUM DIBORIDE MgB_2

FIG 104

SPECIFICATION TABLE NO. 104 SPECIFIC HEAT OF MAGNESIUM DIBORIDE MgB_2

[For Data Reported in Figure and Table No. 104]

Curve No.	Ref. No.	Year	Temp. Range, K	Reported Error, %	Name and Specimen Designation	Composition (weight percent), Specifications and Remarks
1	166, 167	1956	21-304			93.90 MgB_2, 3.69 B, 1.08 MgB_4, 0.73 MgO, 0.46 Mg and 0.14 other impurities; prepared by heating stoichiometric amounts of Mg and B 3 hrs at 900 ± 25 C in helium atmosphere; corrected for impurities.

DATA TABLE NO. 104 SPECIFIC HEAT OF MAGNESIUM DIBORIDE MgB_2

[Temperature, T, K; Specific Heat, C_p, Cal $g^{-1} K^{-1}$]

T	C_p	T	C_p
CURVE 1		CURVE 1 (cont.)	
21.12	2.177×10^{-4}*	183.37	1.614×10^{-1}
23.05	4.789*	208.21	1.871
25.06	7.402*	248.39	2.220*
27.19	1.001×10^{-3}	279.19	2.449*
29.66	1.154	300.14	2.484*
32.41	1.611	237.71	2.184*
34.92	2.242	238.69	2.151*
37.75	2.917	254.87	2.266
41.37	3.549	287.69	2.475*
45.43	5.399	304.22	2.489*
54.12	9.797	304.06	2.487
57.42	1.206×10^{-2}		
60.99	1.439		
65.42	1.848		
70.26	2.412		
75.60	2.950		
81.29	3.666		
85.52	4.073		
91.29	4.763		
97.04	5.410		
102.83	6.065		
108.58	6.869		
114.13	7.448		
119.20	8.201		
124.40	8.802		
129.79	9.403		
293.42	2.486×10^{-1}*		
136.05	1.050		
140.35	1.104		
154.92	1.287		
173.53	1.494		
194.73	1.736		
219.27	1.980		
236.77	2.146		
238.63	2.118*		
246.14	2.201		
253.63	2.304*		
259.61	2.328		
266.88	2.367		
274.04	2.424*		
286.69	2.485		
298.81	2.502*		
26.44	1.263×10^{-3}*		
128.41	9.575×10^{-2}*		
145.03	1.168×10^{-1}*		
160.29	1.364		

*Not shown on plot

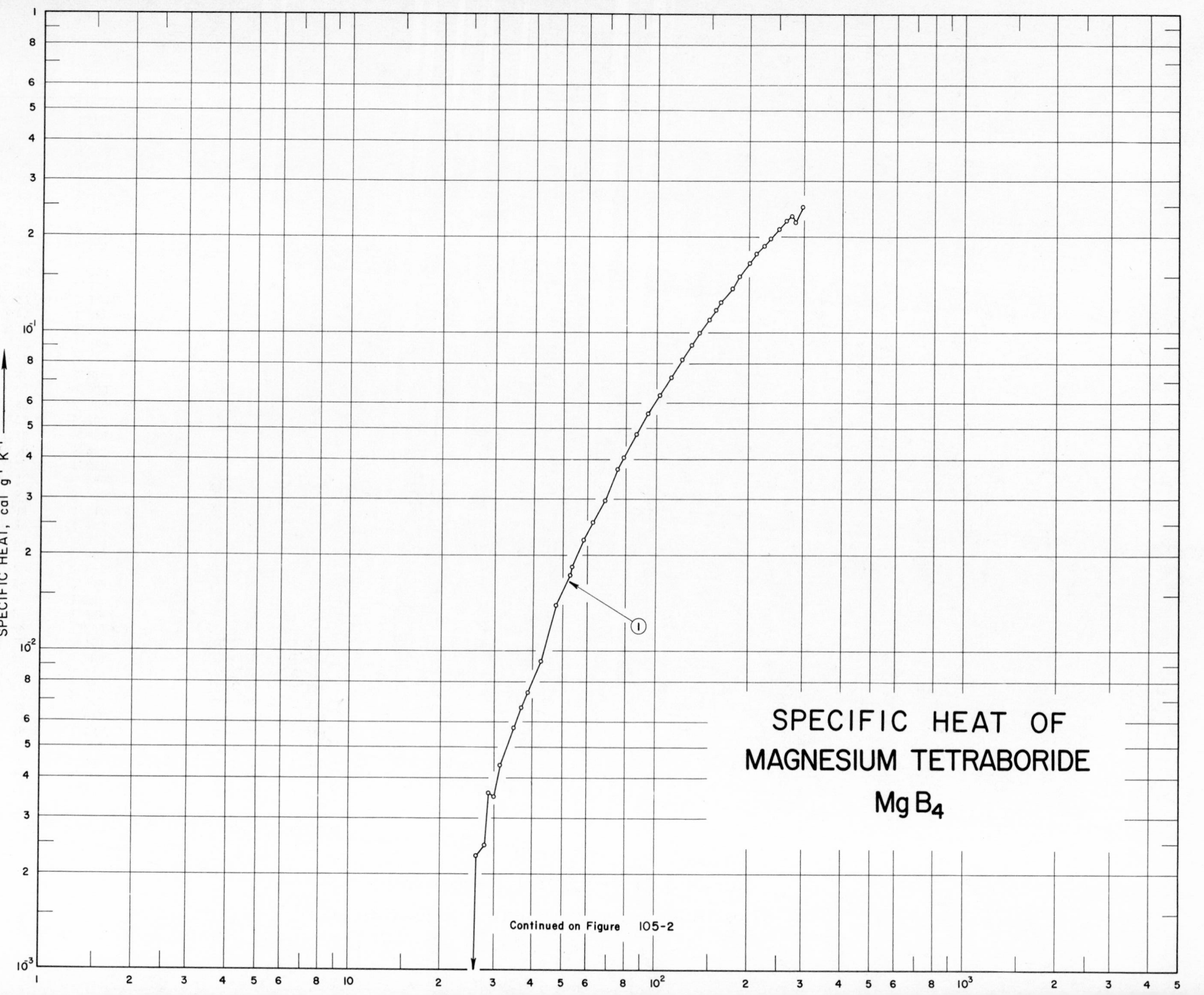
SPECIFIC HEAT OF
MAGNESIUM TETRABORIDE
MgB_4
1
Continued on Figure 105-2
SPECIFIC HEAT, cal g^{-1} K^{-1}

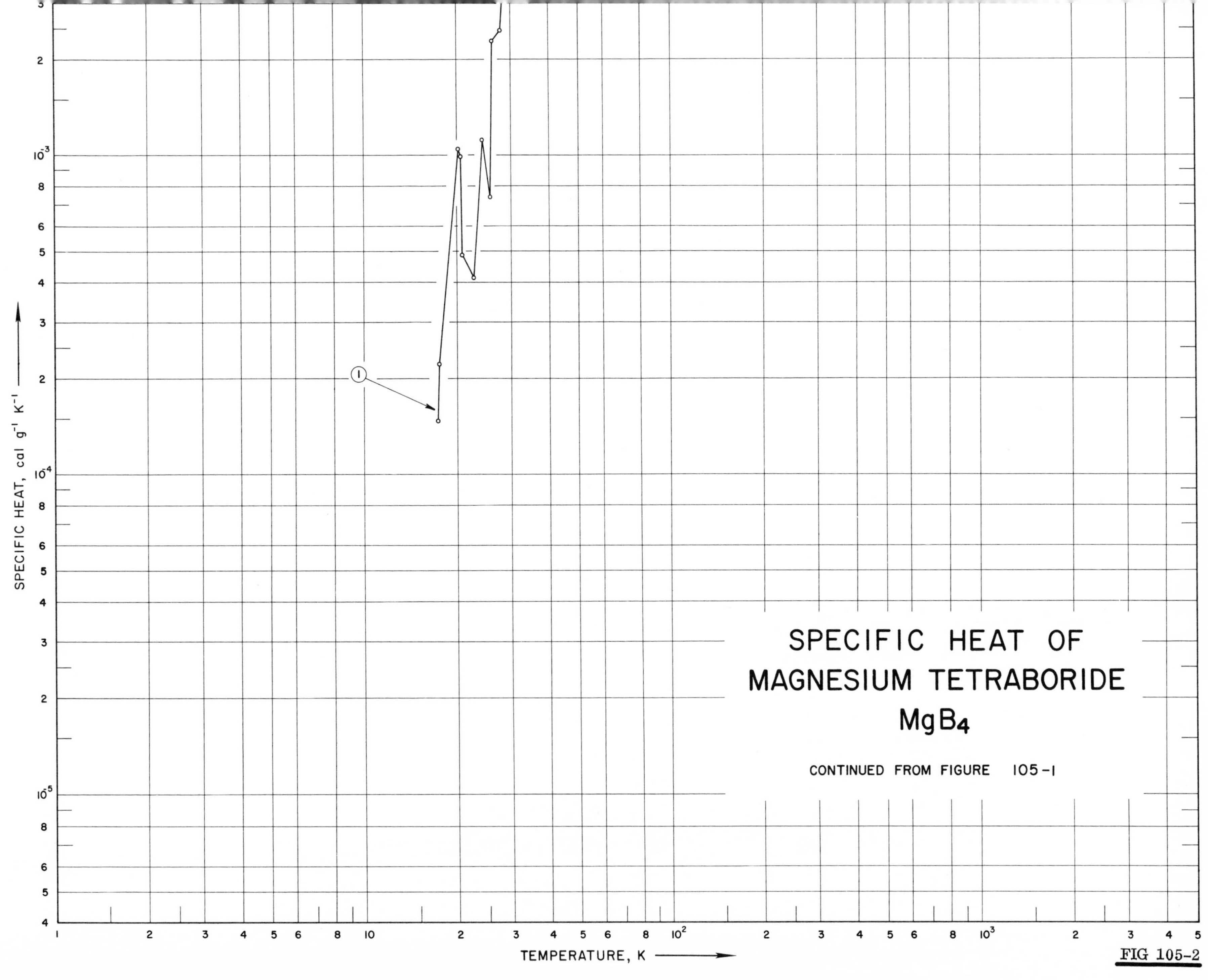

SPECIFIC HEAT OF
MAGNESIUM TETRABORIDE
MgB4
CONTINUED FROM FIGURE 105-I
SPECIFIC HEAT, cal g-1 K-1
TEMPERATURE, K

FIG 105-2

SPECIFICATION TABLE NO. 105 SPECIFIC HEAT OF MAGNESIUM TETRABORIDE MgB_4

[For Data Reported in Figure and Table No. 105]

Curve No.	Ref. No.	Year	Temp. Range, K	Reported Error, %	Name and Specimen Designation	Composition (weight percent), Specifications and Remarks
1	166, 167	1956	17-300			89.42 MgB_4, 10.32 B, and 0.25 other impurities; prepared by heating stoichiometric amounts of Mg and B 3 hrs at 900 ± 25 C in helium atmosphere; corrected for impurities.

DATA TABLE NO. 105 SPECIFIC HEAT OF MAGNESIUM TETRABORIDE MgB_4

[Temperature, T, K; Specific Heat, C_p, Cal g^{-1} K^{-1}]

T	C_p	T	C_p
CURVE 1		CURVE 1 (cont.)	
17.34	1.480×10^{-4}	201.24	1.652×10^{-1}
17.52	2.220	212.85	1.776
20.29	1.051×10^{-3}	220.17	1.762*
20.56	9.918×10^{-4}	225.01	1.872
20.73	4.885	225.11	1.931*
22.66	4.145	235.27	1.971
24.38	1.125×10^{-3}	242.98	2.038*
25.83	7.401×10^{-4}	251.01	2.116
26.26	2.294×10^{-3}	257.37	2.163*
27.94	2.472	265.74	2.247
28.72	3.597	267.17	2.252*
29.82	3.508	271.52	2.283*
31.45	4.382	276.60	2.326
34.55	5.729	280.31	2.359*
36.82	6.646	281.27	2.448*
38.53	7.431	283.39	2.223
42.77	9.296	284.62	2.419*
47.81	1.397×10^{-2}	285.26	2.410*
52.96	1.739	290.41	2.433*
53.71	1.843	295.49	2.469*
58.32	2.234	299.53	2.490
62.81	2.542		
68.65	2.977		
69.03	3.008*		
75.11	3.733		
75.69	3.684*		
78.84	4.062		
86.28	4.790		
94.01	5.551		
103.08	6.347		
112.87	7.237		
121.84	8.200		
130.84	9.132		
139.20	9.977		
149.05	1.100×10^{-1}		
154.62	1.154*		
155.81	1.180		
162.66	1.249		
169.74	1.313*		
177.27	1.380		
179.64	1.376*		
187.17	1.499		
188.99	1.512*		

*Not shown on plot

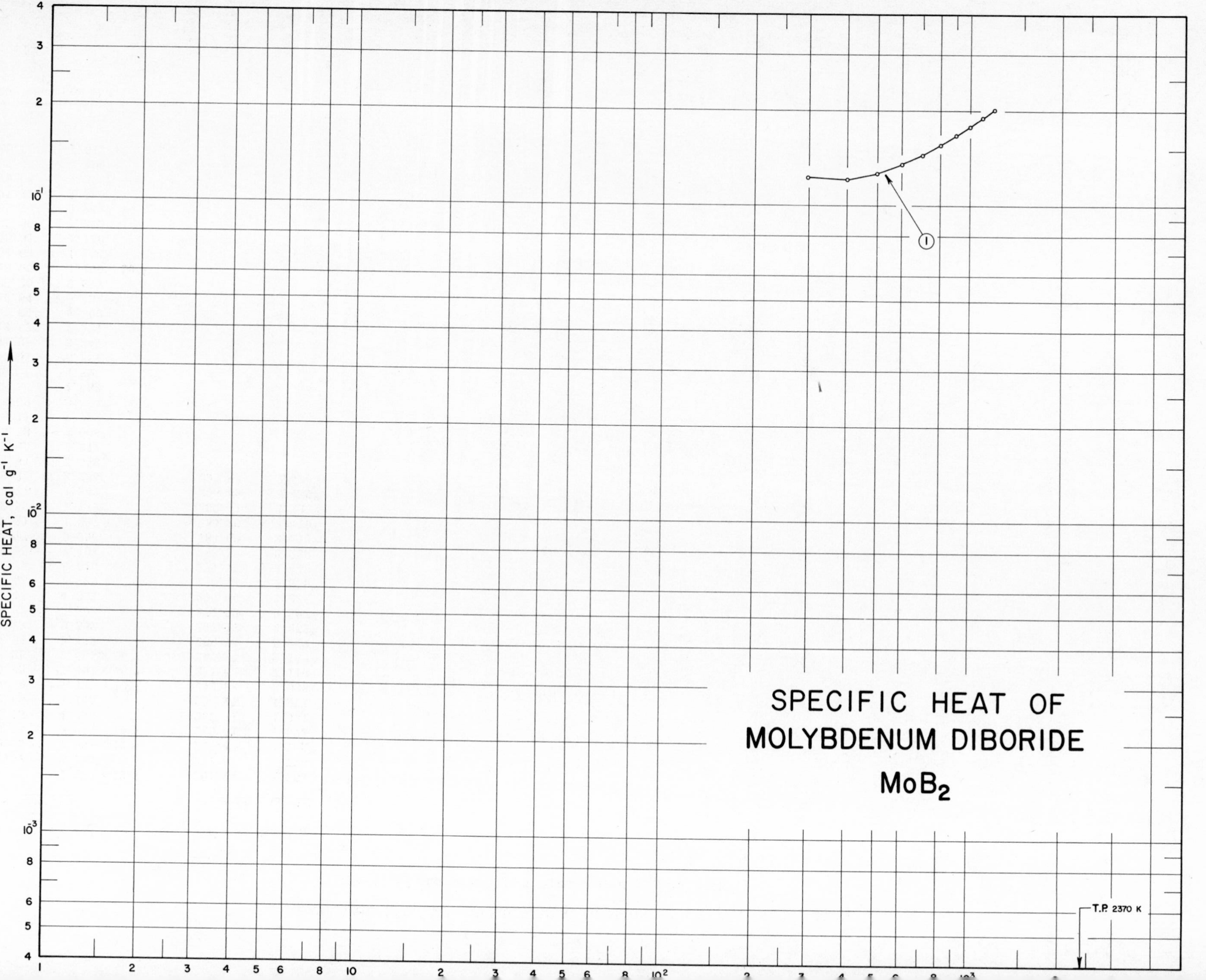
SPECIFIC HEAT OF
MOLYBDENUM DIBORIDE
MoB2
T.P. 2370 K
1
SPECIFIC HEAT, cal g⁻¹ K⁻¹

SPECIFICATION TABLE NO. 106 SPECIFIC HEAT OF MOLYBDENUM DIBORIDE MoB_2

[For Data Reported in Figure and Table No. 106]

Curve No.	Ref. No.	Year	Temp. Range, K	Reported Error, %	Name and Specimen Designation	Composition (weight percent), Specifications and Remarks
1	163	1962	298-1200	0.5		Traces of impurities.

DATA TABLE NO. 106 SPECIFIC HEAT OF MOLYBDENUM DIBORIDE MoB_2

[Temperature, T, K; Specific Heat, C_p, Cal $g^{-1} K^{-1}$]

T	C_p
CURVE 1	
298.16	1.229 x 10^{-1}
300	1.231*
400	1.211
500	1.266
600	1.350
700	1.448
800	1.553
900	1.664
1000	1.776
1100	1.890
1200	2.006

* Not shown on plot

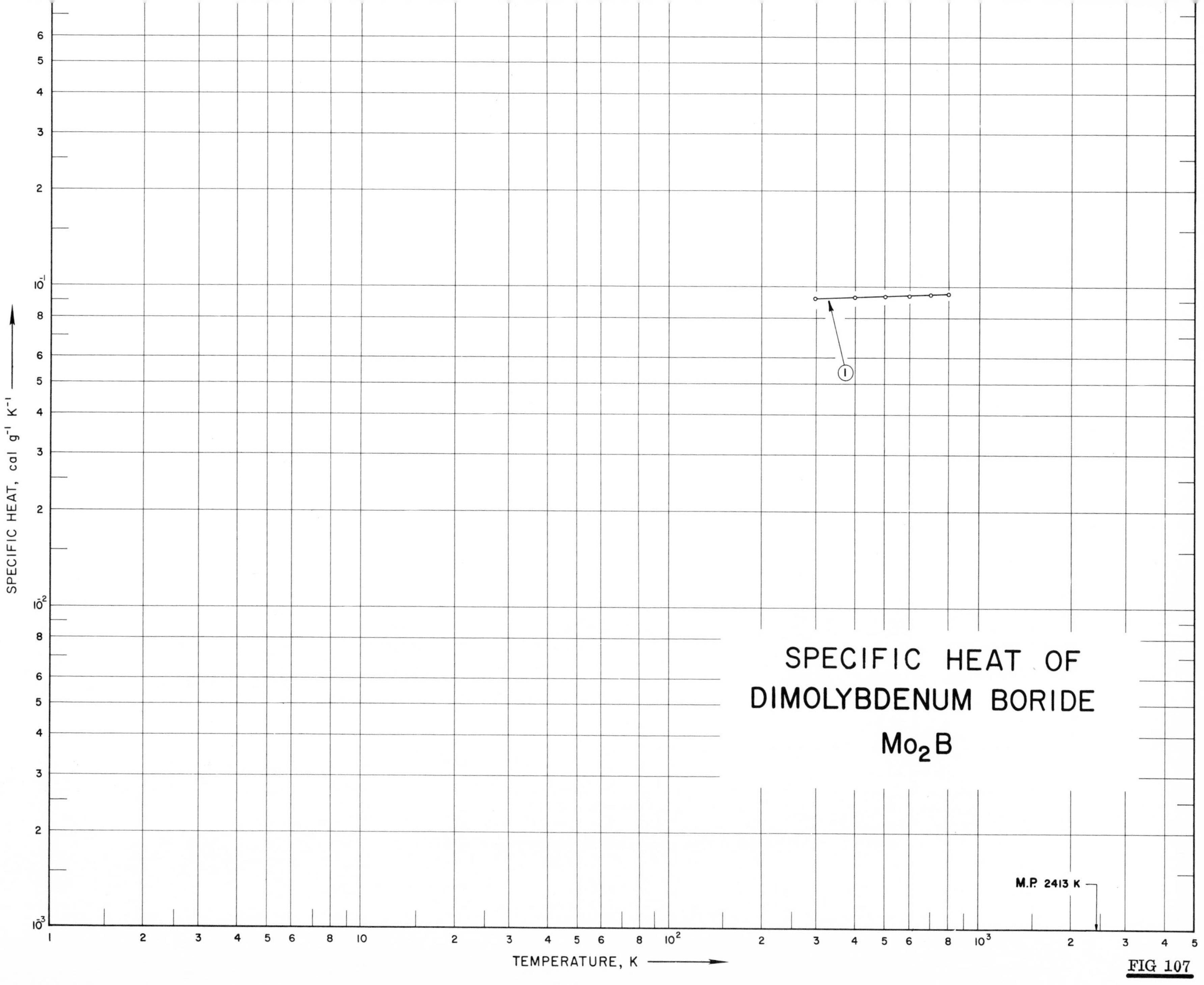
SPECIFIC HEAT OF
DIMOLYBDENUM BORIDE
Mo_2B
M.P. 2413 K
SPECIFIC HEAT, cal g^{-1} K^{-1}
TEMPERATURE, K
1

FIG 107

SPECIFICATION TABLE NO. 107 SPECIFIC HEAT OF DIMOLYBDENUM BORIDE Mo_2B

[For Data Reported in Figure and Table No. 107]

Curve No.	Ref. No.	Year	Temp. Range, K	Reported Error, %	Name and Specimen Designation	Composition (weight percent), Specifications and Remarks
1	163	1962	298-800	0.5		Traces of impurities.

DATA TABLE NO. 107 SPECIFIC HEAT OF DIMOLYBDENUM BORIDE Mo_2B

[Temperature, T, K; Specific Heat, C_p, Cal $g^{-1} K^{-1}$]

T	C_p
CURVE 1	
298.16	9.270 x 10^{-2}
300	9.280*
400	9.334
500	9.408
600	9.468
700	9.532
800	9.596

*Not shown on plot

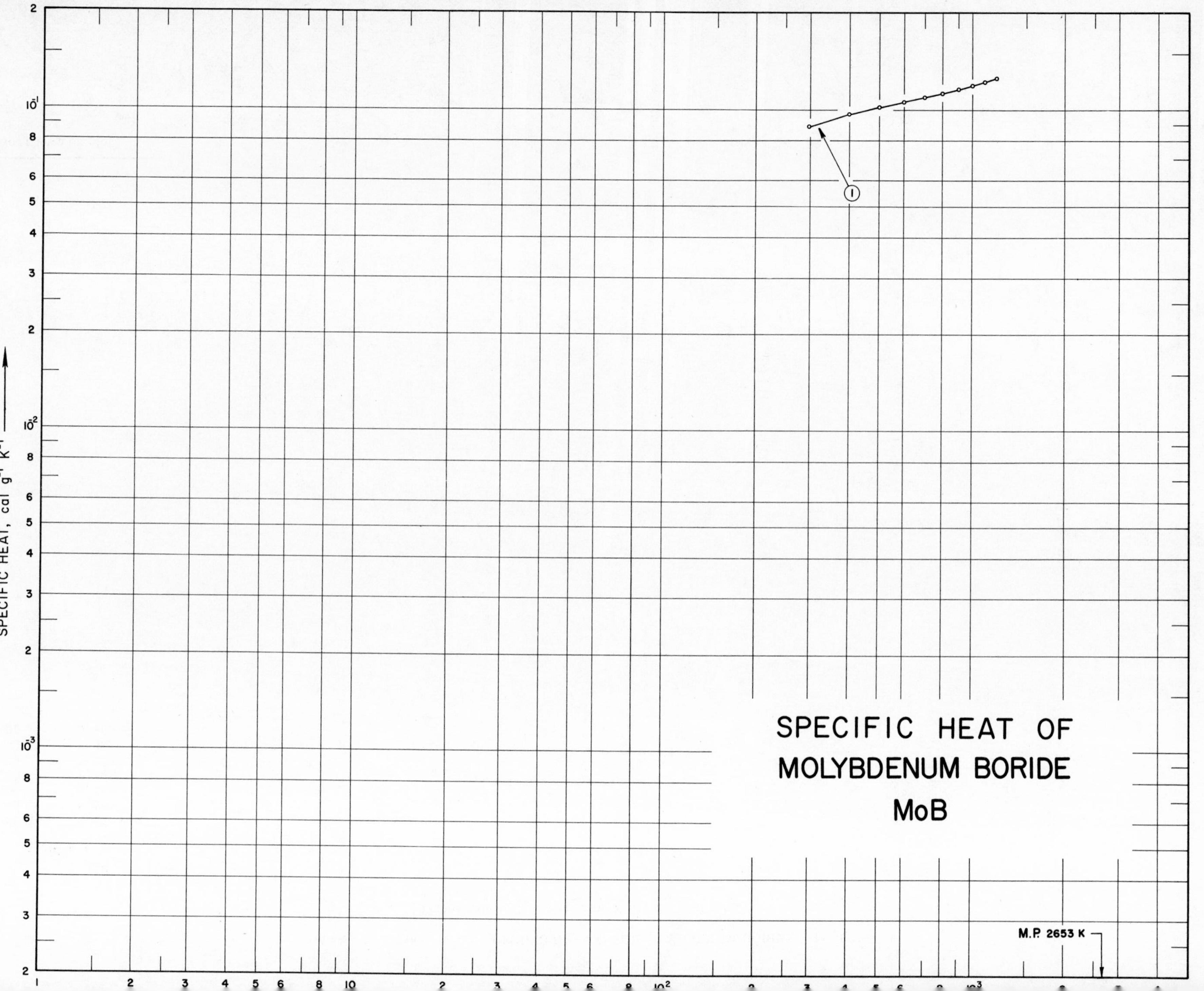
SPECIFIC HEAT OF
MOLYBDENUM BORIDE
MoB
M.P. 2653 K
1
SPECIFIC HEAT, cal g⁻¹ K⁻¹

SPECIFICATION TABLE NO. 108 SPECIFIC HEAT OF MOLYBDENUM BORIDE MoB

[For Data Reported in Figure and Table No. 108]

Curve No.	Ref. No.	Year	Temp. Range, K	Reported Error, %	Name and Specimen Designation	Composition (weight percent), Specifications and Remarks
1	163	1962	298-1200	0.50		Traces of impurities.

DATA TABLE NO. 108 SPECIFIC HEAT OF MOLYBDENUM BORIDE MoB

[Temperature, T, K; Specific Heat, C_p, Cal $g^{-1} K^{-1}$]

T	C_p
CURVE 1	
298.16	8.824 x 10^{-2}
300	8.843*
400	9.649
500	1.017 x 10^{-1}
600	1.059
700	1.095
800	1.129
900	1.162
1000	1.192
1100	1.222
1200	1.252

*Not shown on plot

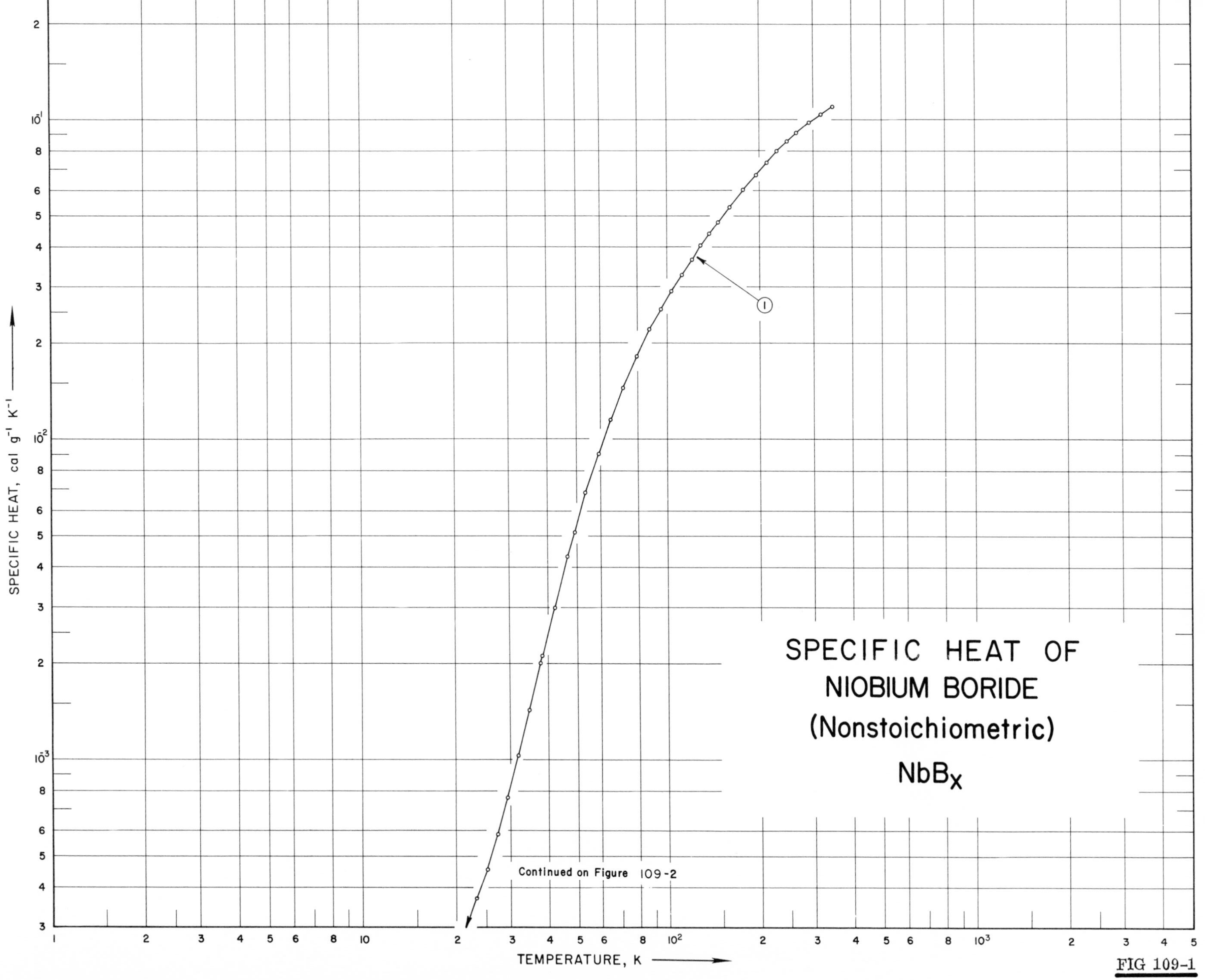

SPECIFIC HEAT OF
NIOBIUM BORIDE
(Nonstoichiometric)
NbB_x
1
Continued on Figure 109-2
TEMPERATURE, K
SPECIFIC HEAT, cal g^{-1} K^{-1}

FIG 109-1

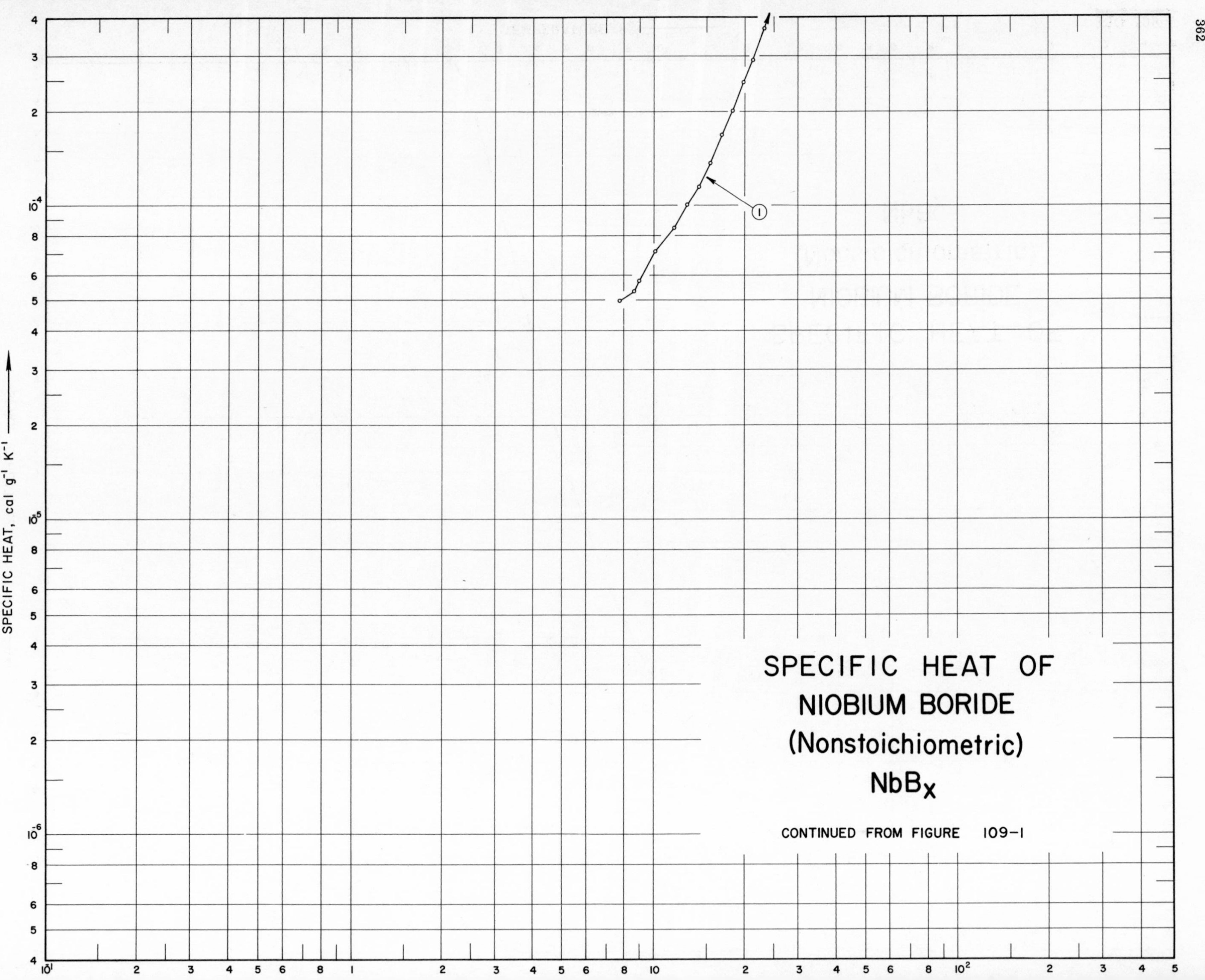
SPECIFIC HEAT OF
NIOBIUM BORIDE
(Nonstoichiometric)
NbB_x
CONTINUED FROM FIGURE 109-1
SPECIFIC HEAT, cal g^{-1} K^{-1}
1

SPECIFICATION TABLE NO. 109 SPECIFIC HEAT OF NIOBIUM BORIDE (Nonstoichiometric) NbB_x

[For Data Reported in Figure and Table No. 109]

Curve No.	Ref. No.	Year	Temp. Range, K	Reported Error, %	Name and Specimen Designation	Composition (weight percent), Specifications and Remarks
1	168	1963	8-346	0.1-5	$NbB_{1.963}$	~99.868, impurities, 0.13 Ti, 0.001 Fe, 0.001 Si, 0.0142 C, 0.0066 O, and 0.0055 N; zone melted.

DATA TABLE NO. 109 SPECIFIC HEAT OF NIOBIUM BORIDE (Nonstoichiometric) NbB_x

[Temperature, T, K; Specific Heat, C_p, Cal $g^{-1} K^{-1}$]

T	C_p	T	C_p
CURVE 1		CURVE 1 (cont.)	
Series 1		Series 3 (cont.)	
7.74	4.994×10^{-5}	138.05	4.418
8.94	5.783	147.04	4.802
		151.78	5.007*
Series 2		160.23	5.363
		169.10	5.730*
8.64	5.345×10^{-5}	178.07	6.093
10.09	7.185	186.83	6.437*
11.74	8.499	195.44	6.770
12.96	1.008×10^{-4}	203.03	7.060*
14.14	1.157	211.65	7.380
15.46	1.376	220.48	7.694*
16.85	1.691	229.49	8.005
18.28	2.006	238.23	8.297*
19.76	2.488	247.12	8.583
21.39	2.935	256.05	8.852*
23.19	3.680	265.06	9.118
25.05	4.539	273.19	9.348*
27.04	5.835	282.10	9.593*
29.25	7.605	290.82	9.823
31.85	1.030×10^{-3}	299.72	1.005×10^{-1}*
34.75	1.434	309.17	1.028*
37.92	1.999	318.83	1.050
		328.42	1.071*
Series 3		337.71	1.090*
		346.05	1.107*
38.46	2.110×10^{-3}		
42.23	2.989		
46.81	4.314	Series 4	
49.22	5.115		
53.94	6.839	313.40	1.037*
59.15	8.972	322.59	1.057*
64.98	1.155×10^{-2}	331.91	1.078*
71.67	1.455	341.21	1.097*
79.63	1.824	347.74	1.109
87.98	2.217		
95.92	2.566		
104.09	2.926		
112.42	3.294		
120.72	3.660		
129.29	4.037		

*Not shown on plot

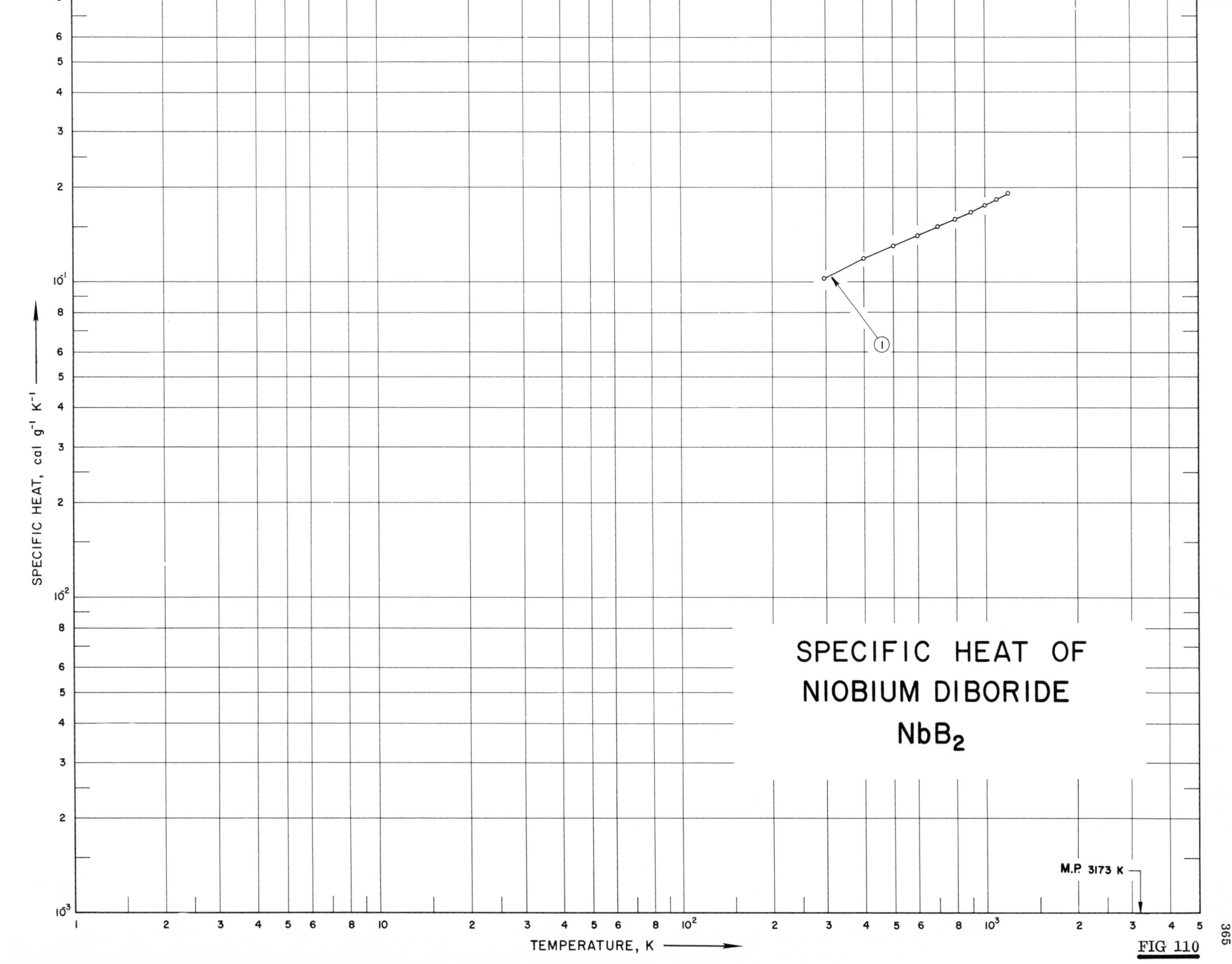
SPECIFIC HEAT OF
NIOBIUM DIBORIDE
NbB_2
M.P. 3173 K
TEMPERATURE, K
SPECIFIC HEAT, cal g^{-1} K^{-1}
1
FIG 110

SPECIFICATION TABLE NO. 110 SPECIFIC HEAT OF NIOBIUM DIBORIDE NbB_2

[For Data Reported in Figure and Table No. 110]

Curve No.	Ref. No.	Year	Temp. Range, K	Reported Error, %	Name and Specimen Designation	Composition (weight percent), Specifications and Remarks
1	163	1962	298-1200	0.5		Traces of impurities.

DATA TABLE NO. 110 SPECIFIC HEAT OF NIOBIUM DIBORIDE NbB_2

[Temperature, T, K; Specific Heat, C_p, Cal $g^{-1} K^{-1}$]

T	C_p
CURVE 1	
298.16	1.031×10^{-1}
300	1.035*
400	1.192
500	1.309
600	1.409
700	1.503
800	1.592
900	1.679
1000	1.765
1100	1.848
1200	1.933

*Not shown on plot

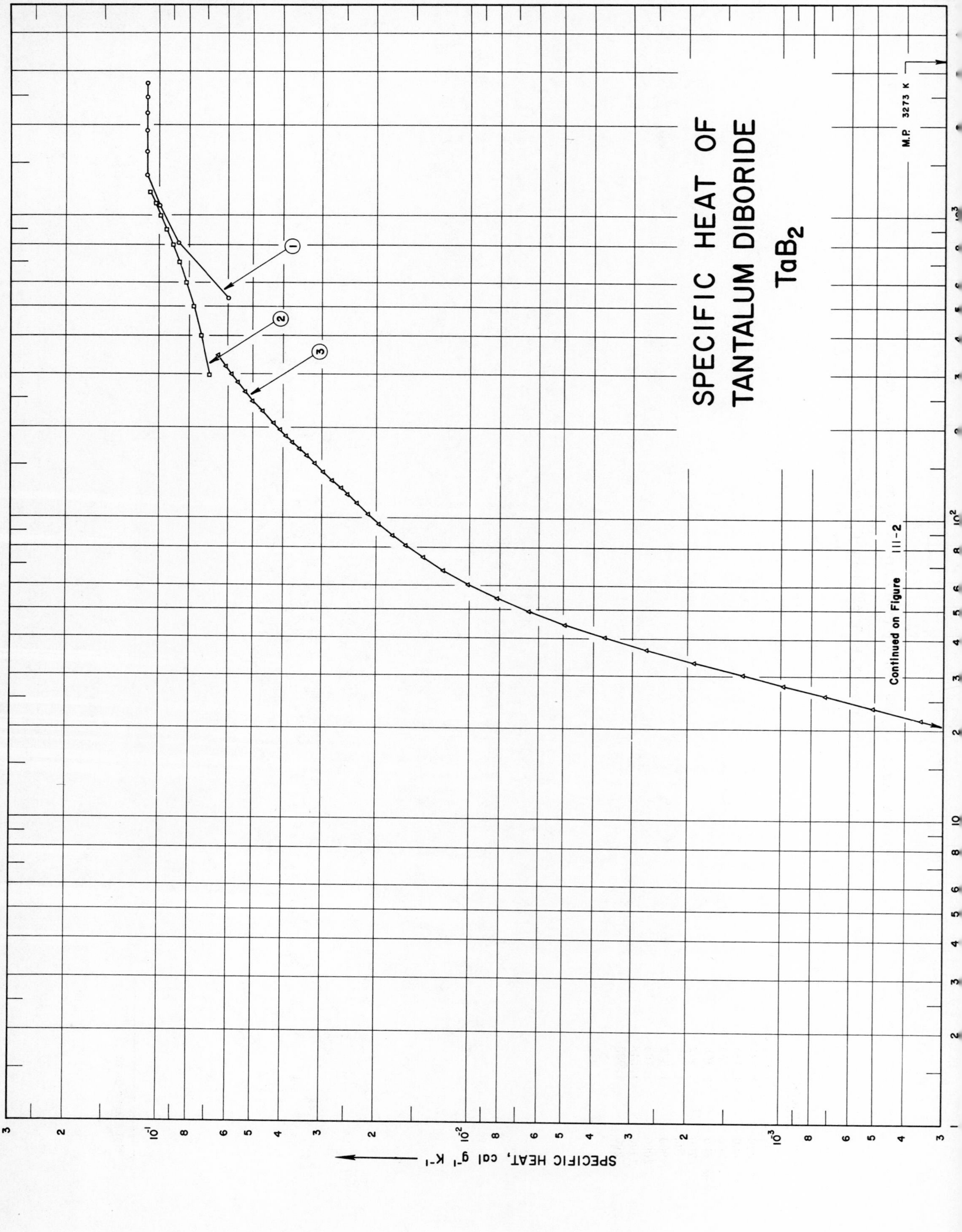

SPECIFIC HEAT OF TANTALUM DIBORIDE TaB_2

Continued on Figure III-2

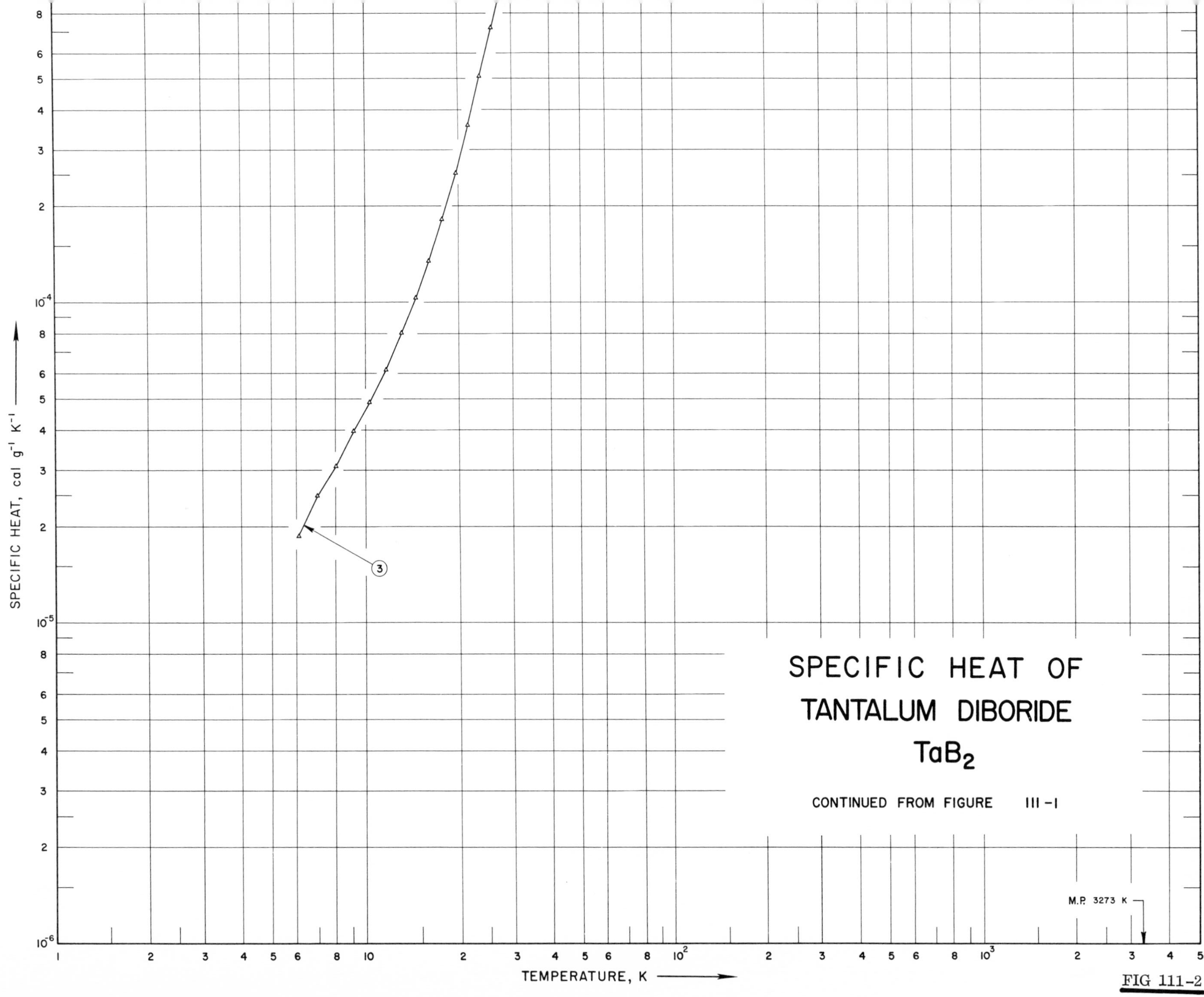
SPECIFIC HEAT OF
TANTALUM DIBORIDE
TaB_2
CONTINUED FROM FIGURE III-I
M.P. 3273 K
3
TEMPERATURE, K
SPECIFIC HEAT, cal g^{-1} K^{-1}

FIG 111-2

SPECIFICATION TABLE NO. 111 SPECIFIC HEAT OF TANTALUM DIBORIDE TaB_2

[For Data Reported in Figure and Table No. 111]

Curve No.	Ref. No.	Year	Temp. Range, K	Reported Error, %	Name and Specimen Designation	Composition (weight percent), Specifications and Remarks
1	32	1962	533-2755	≤5		Sample supplied by General Electric Co; pressed and sintered; density 756 lb ft^{-3}.
2	163	1962	298-1200	0.50		Traces of impurities.
3	169	1963	6-345		$TaB_{2.11}$	88. 80 Ta, 11. 21 B, 0. 001-0. 01 Ti, Si, and Cr, 0. 0200 C, 0. 0029 O, 0. 0022 N, polycrystalline, single phase; zone-refined.

DATA TABLE NO. 111 SPECIFIC HEAT OF TANTALUM DIBORIDE TaB_2

[Temperature, T, K; Specific Heat, C_p, Cal $g^{-1} K^{-1}$]

CURVE 1

T	C_p
533.15	6.0×10^{-2}
810.93	8.7
1088.71	1.0×10^{-1}
1366.48	1.1
1644.26	1.1
1922.04	1.1
2199.82	1.1
2477.59	1.1
2755.37	1.1

CURVE 2

T	C_p
298.16	6.901×10^{-2}
300	6.916*
400	7.336
500	7.780
600	8.214
700	8.644
800	9.078
900	9.513
1000	9.942
1100	1.038×10^{-1}
1200	1.081

CURVE 3

Series 1

T	C_p
126.62	2.604×10^{-2}
134.73	2.778
143.67	2.969
152.53	3.158
161.50	3.350
170.52	3.537
179.67	3.725
188.71	3.912
197.77	4.097
207.30	4.287
216.98	4.478*
226.38	4.658
235.74	4.830*
245.06	5.001
254.17	5.152*
263.08	5.296×10^{-2}
271.94	5.440*
281.02	5.581
290.40	5.720*
299.73	5.852
309.07	5.980*
318.35	6.101
327.49	6.220*
336.58	6.330*
345.64	6.431

Series 2

T	C_p
6.10	1.865×10^{-5}
7.01	2.503
8.06	3.092
9.15	3.975
10.31	4.908
11.67	6.184
13.10	8.049
14.54	1.040×10^{-4}
16.08	1.355
17.80	1.835
19.65	2.547
21.60	3.583
23.67	5.104
25.91	7.254
27.97	9.845
30.24	1.331×10^{-3}
33.24	1.911
36.68	2.712
40.35	3.698
44.54	4.954
49.26	6.464
54.43	8.137
60.47	1.010×10^{-2}
67.25	1.217
74.37	1.414
81.55	1.608
88.20	1.776
96.16	1.955
104.24	2.126×10^{-2}
112.58	2.305
120.96	2.485

*Not shown on plot

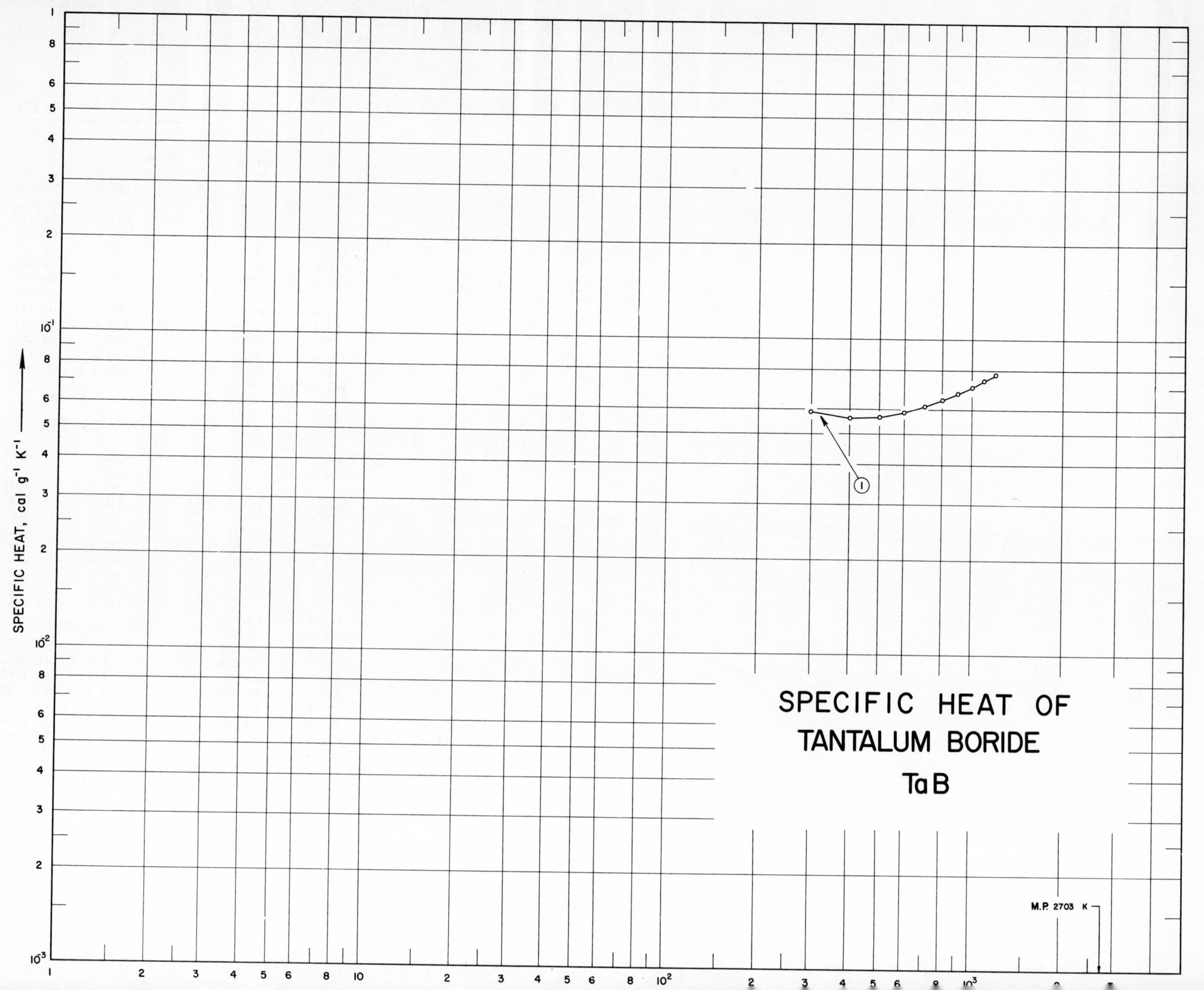
SPECIFIC HEAT OF
TANTALUM BORIDE
TaB
M.P. 2703 K
SPECIFIC HEAT, cal g^{-1} K^{-1}

SPECIFICATION TABLE NO. 112 SPECIFIC HEAT OF TANTALUM BORIDE TaB

[For Data Reported in Figure and Table No. 112]

Curve No.	Ref. No.	Year	Temp. Range, K	Reported Error, %	Name and Specimen Designation	Composition (weight percent), Specifications and Remarks
1	163	1962	298-1200	0.5		Traces of impurities.

DATA TABLE NO. 112 SPECIFIC HEAT OF TANTALUM BORIDE TaB

[Temperature, T, K; Specific Heat, C_p, Cal $g^{-1} K^{-1}$]

T	C_p
CURVE 1	
298.16	5.893 x 10^{-2}
300	5.898*
400	5.627
500	5.695
600	5.893
700	6.154
800	6.446
900	6.758
1000	7.081
1100	7.415
1200	7.755

*Not shown on plot

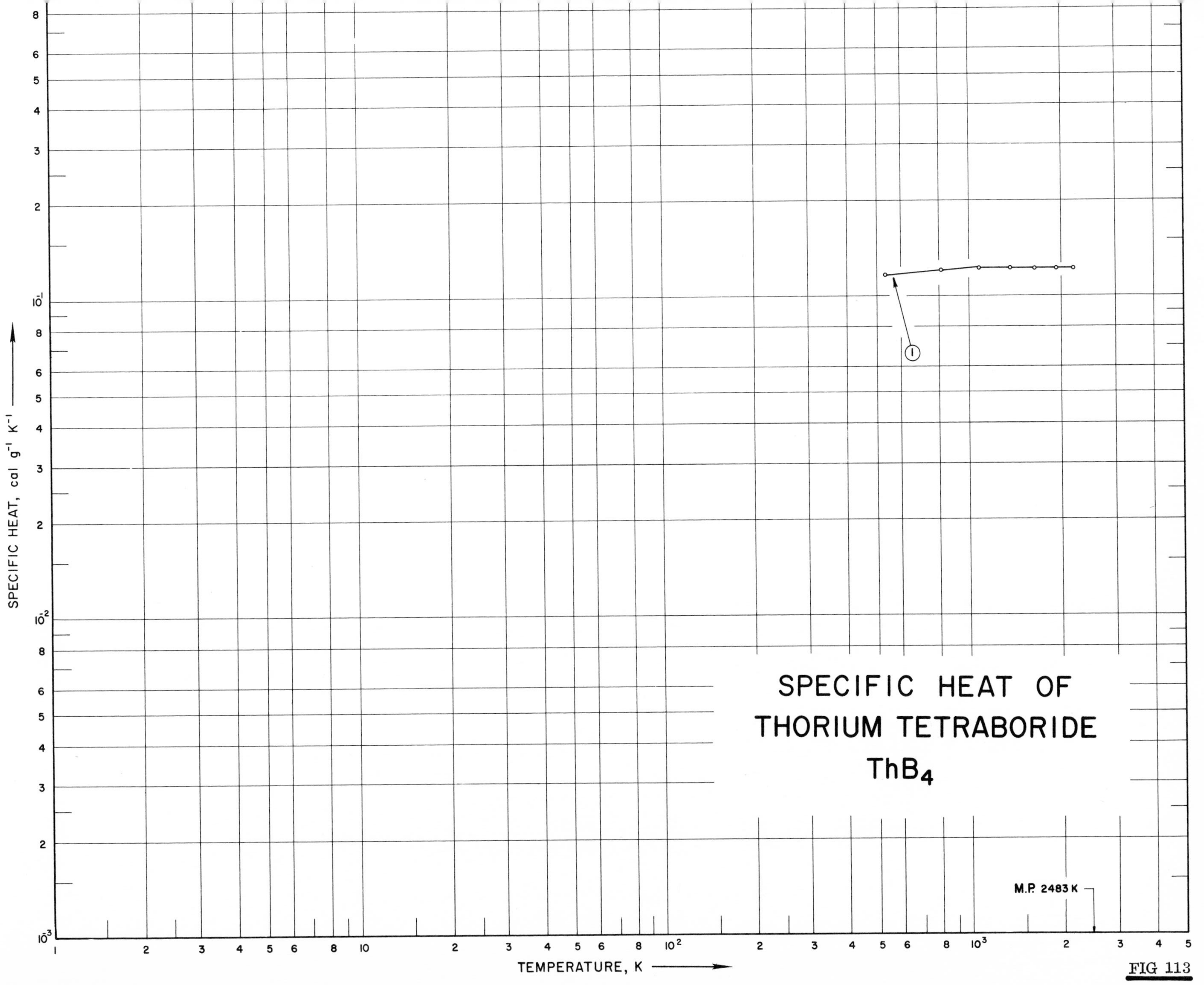
SPECIFIC HEAT OF
THORIUM TETRABORIDE
ThB4
SPECIFIC HEAT, cal g⁻¹ K⁻¹
TEMPERATURE, K
M.P. 2483 K
1

FIG 113

SPECIFICATION TABLE NO. 113 SPECIFIC HEAT OF THORIUM TETRABORIDE ThB_4

[For Data Reported in Figure and Table No. 113]

Curve No.	Ref. No.	Year	Temp. Range, K	Reported Error, %	Name and Specimen Designation	Composition (weight percent), Specifications and Remarks
1	48	1962	533-2200	≤5		Sample supplied by The Carborundum Co.; crushed in hardened steel mortar to pass 100-mesh screen; hot pressed; density at 25 C, before exposure: apparent density (ASTM method B311-58) 485 lb ft^{-3}, true density (by immersion in xylene) 510 lb ft^{-3}.

DATA TABLE NO. 113 SPECIFIC HEAT OF THORIUM TETRABORIDE ThB_4

[Temperature, T, K; Specific Heat, C_p, Cal g^{-1} K^{-1}]

T	C_p
CURVE 1	
533.15	1.18×10^{-1}
810.93	1.22
1088.71	1.24
1366.48	1.24
1644.26	1.24
1922.04	1.24
2199.82	1.24

*Not shown on plot

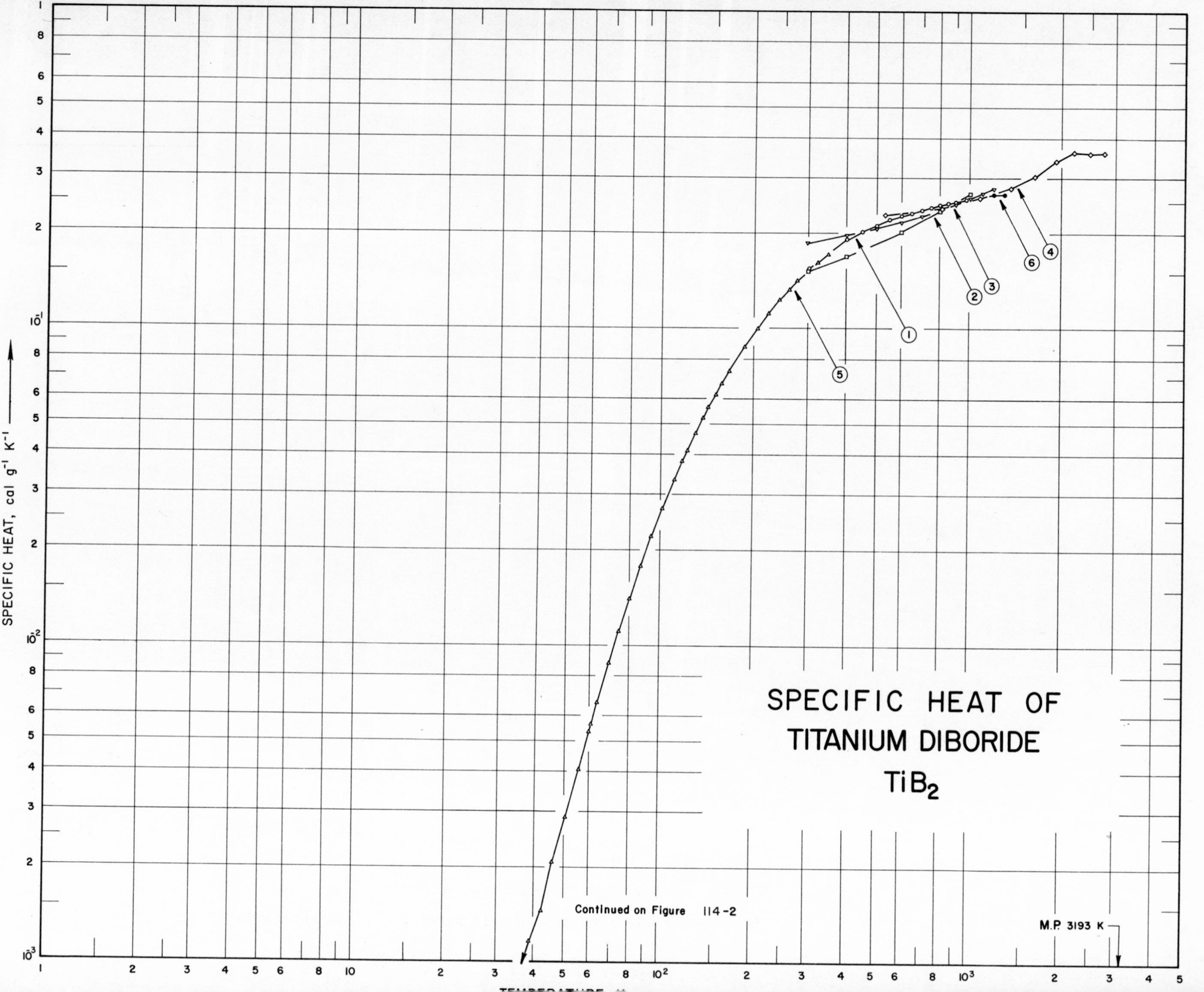
SPECIFIC HEAT OF
TITANIUM DIBORIDE
TiB_2
M.P. 3193 K
Continued on Figure 114-2
SPECIFIC HEAT, cal g^{-1} K^{-1}
1
2
3
4
5
6

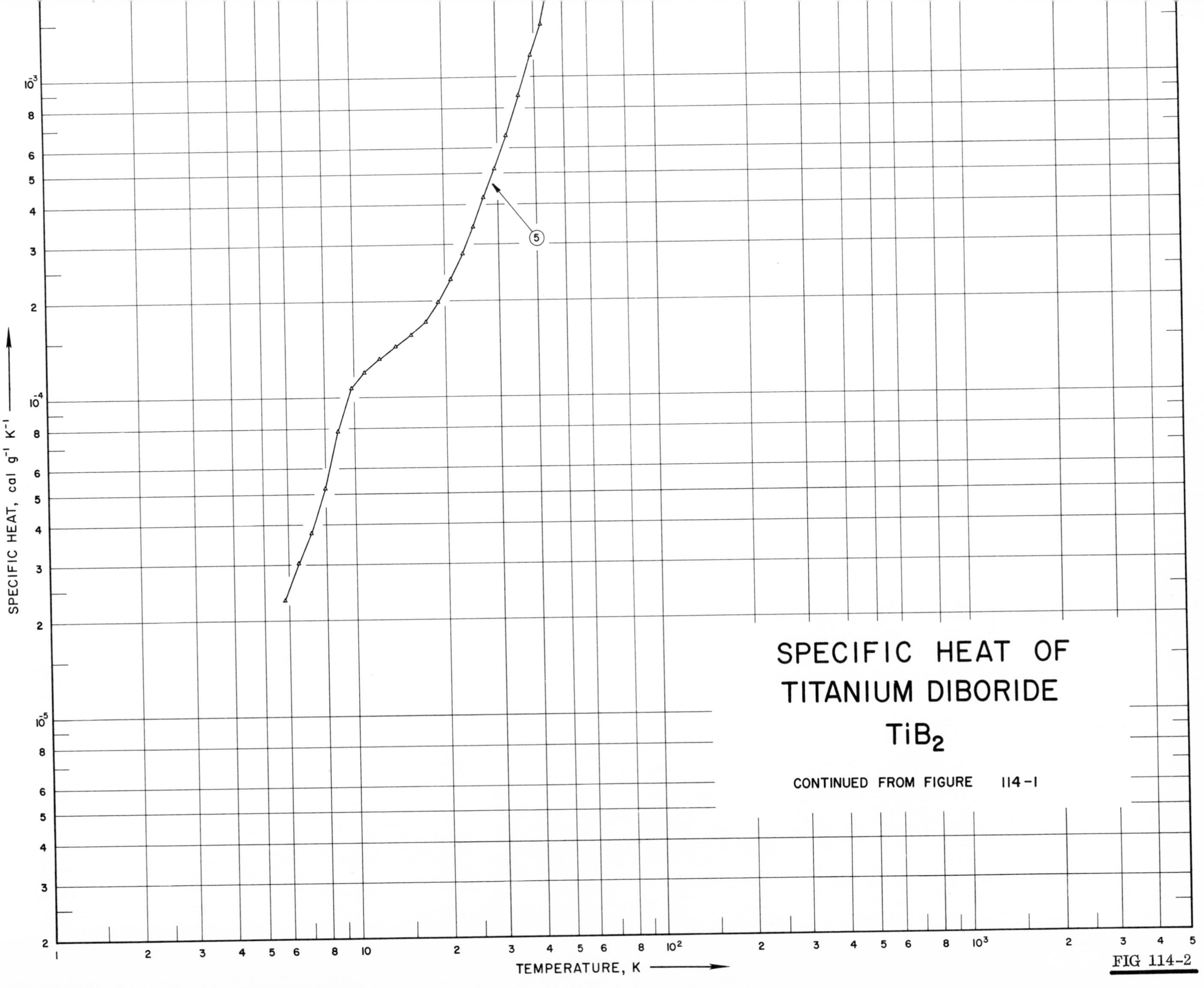

SPECIFIC HEAT OF
TITANIUM DIBORIDE
TiB_2
CONTINUED FROM FIGURE 114-1
5
TEMPERATURE, K
SPECIFIC HEAT, cal g^{-1} K^{-1}

FIG 114-2

SPECIFICATION TABLE NO. 114 SPECIFIC HEAT OF TITANIUM DIBORIDE TiB_2

[For Data Reported in Figure and Table No. 114]

Curve No.	Ref. No.	Year	Temp. Range, K	Reported Error, %	Name and Specimen Designation	Composition (weight percent), Specifications and Remarks
1	310	1957	303-973			99.7 TiB_2, 0.2 B, and 0.1 Fe.
2	171	1959	300-1000			
3	163	1962	298-1200			Traces of impurities.
4	48	1962	533-2755	≤5		Before exposure: 69.8 Ti, 29.5 B, 0.6 N_2, 0.4 Fe, 0.3 V, and 0.2 C, after exposure: 69.3 Ti, 28.7 B, 0.32 C, and 0.3 N_2; sample supplied by The Carborundum Co; crushed in hardened steel mortar to pass 100-mesh screen; hot pressed; density at 25 C before exposure: apparent density (ASTM method B311-58) 281 lb ft^{-3}, true density (by immersion in xylene) 285 lb ft^{-3}, after exposure: apparent density 260 lb ft^{-3}, true density 264 lb ft^{-3}.
5	169	1963	6-347			68.85 Ti, 30.48 B, 0.12 C, 0.11 O, 0.10 N, and 0.06 Fe, spectrographic analyses: 0.10 Co and Cr, 0.001 Ni and Si, 0.001 - 0.01 Mg, Al, and Mo, and 0.001 others; sample supplied by the Millmaster Chemical Co; zone-refined.
6	172	1964	273-1300			69.6 Ti, 28.0 B, and 0.97 C, monolithnic titanium diboride; hot pressed; measured in an argon atmosphere.

DATA TABLE NO. 114 SPECIFIC HEAT OF TITANIUM DIBORIDE TiB_2

[Temperature, T, K; Specific Heat, C_p, Cal $g^{-1} K^{-1}$]

CURVE 1

T	C_p
303.15	1.556 x 10^{-1}
400	1.910
450	2.033
500	2.131
550	2.213
600	2.280
650	2.337
700	2.386
750	2.428
800	2.465
850	2.498
900	2.527
973.15	2.564

CURVE 2

T	C_p
300	1.533 x 10^{-1}
400	1.698
600	2.028
800	2.357
1000	2.688

CURVE 3

T	C_p
298.16	1.872 x 10^{-1}
300	1.877*
400	1.979
500	2.081
600	2.183
700	2.286
800	2.388*
900	2.488
1000	2.591
1100	2.693
1200	2.795

CURVE 4

T	C_p
533.15	2.30 x 10^{-1}
810.93	2.40
1088.71	2.60
1366.48	2.80
1644.26	3.05
1922.04	3.40
2199.82	3.62 x 10^{-1}
2477.59	3.60
2755.37	3.60

CURVE 5

Series 1

T	C_p
138.02	5.256 x 10^{-2}
143.63	5.678
150.87	6.207
158.58	6.765
166.94	7.373

Series 2

T	C_p
111.37	3.343 x 10^{-2}
118.25	3.828
122.47	4.128
130.28	4.694
137.86	5.250*

Series 3

T	C_p
5.79	2.302 x 10^{-5}
6.44	3.021
7.10	3.740
7.93	5.178
8.78	7.768
9.75	1.064 x 10^{-4}
10.83	1.194
12.12	1.309
13.68	1.438
15.40	1.554
17.17	1.712
18.96	1.971
20.76	2.316
22.60	2.791
24.68	3.380
26.88	4.186
29.26	5.150
32.02	6.559
35.27	8.717
38.83	1.169 x 10^{-3}
42.32	1.551 x 10^{-3}
45.95	2.071
50.37	2.874
55.50	4.087
60.74	5.673

Series 4

T	C_p
59.74	5.354 x 10^{-3}
63.37	6.605
69.02	8.802
74.23	1.109 x 10^{-2}
80.02	1.404
86.75	1.785
93.87	2.205
101.92	2.710
110.68	3.300*
121.48	4.055*
128.42	4.551*
135.88	5.102*
143.78	5.683*
152.39	6.312*
161.33	6.963*
169.96	7.588*
178.39	8.185*
187.32	8.806
196.74	9.452*
206.02	1.006 x 10^{-1}
215.15	1.066*
224.41	1.125
233.90	1.182*
243.40	1.238
252.62	1.291*
261.69	1.340
259.06	1.326*
268.23	1.374*
277.24	1.421
286.27	1.466*
295.40	1.509*
304.48	1.551*
313.54	1.591*
322.91	1.630
332.18	1.667 x 10^{-1}*
340.53	1.701*
347.25	1.726

CURVE 6

T	C_p
273.15	1.404 x 10^{-1}*
298.15	1.539*
300	1.548*
400	1.912*
500	2.131*
600	2.276*
700	2.381*
800	2.460*
900	2.523*
1000	2.614*
1100	2.641*
1200	2.667
1300	2.673

*Not shown on plot

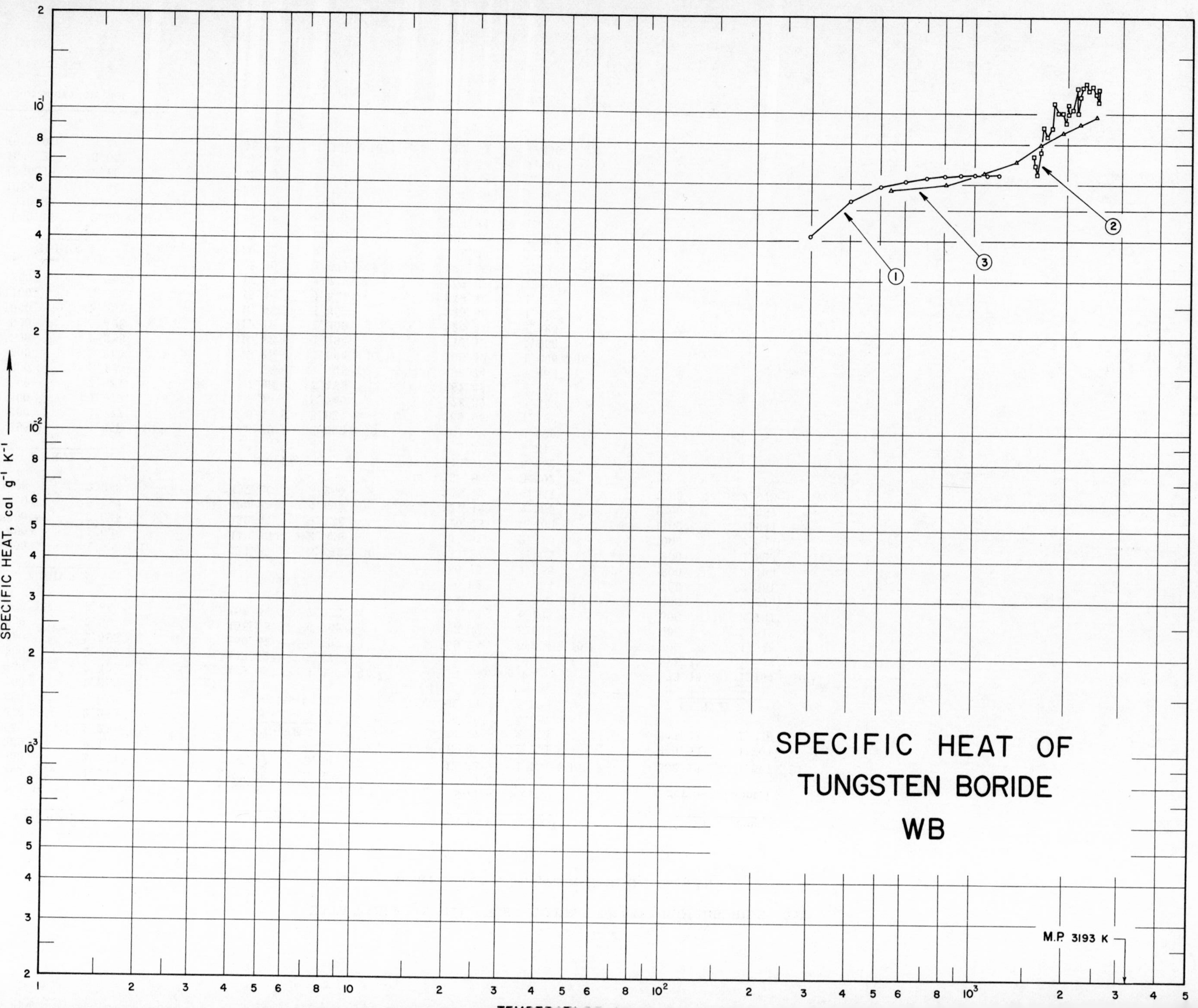

SPECIFIC HEAT OF
TUNGSTEN BORIDE
WB
M.P. 3193 K
SPECIFIC HEAT, cal g⁻¹ K⁻¹
1
2
3

SPECIFICATION TABLE NO. 115 SPECIFIC HEAT OF TUNGSTEN BORIDE WB

[For Data Reported in Figure and Table No. 115]

Curve No.	Ref. No.	Year	Temp. Range, K	Reported Error, %	Name and Specimen Designation	Composition (weight percent), Specifications and Remarks
1	163	1962	298-1200	0.5		Traces of impurities.
2	173	1962	1556-2518	±5		93.7 W, 5.16 B, 0.16 O, 0.09 C, 0.01 N and ~0.88 impurities; hot pressed.
3	48	1962	533-2478	≤5		Before exposure: 95.3 W, 4.7 B, 0.2 Nb, Fe, Si, and V, and 0.1 Zr, after exposure: 94.8 W, 5.2 B, 0.09 C; sample supplied by The Carborundum Co; crushed in hardened steel mortar to pass 100-mesh screen; hot pressed; density at 25 C, before exposure: apparent density (ASTM B311-58) 950 lb ft^{-3}, true density (by immersion in xylene) 955 lb ft^{-3}, after exposure: apparent density 918 lb ft^{-3}, true density 924 lb ft^{-3}.

DATA TABLE NO. 115 SPECIFIC HEAT OF TUNGSTEN BORIDE WB

[Temperature, T, K; Specific Heat, C_p, Cal g^{-1} K^{-1}]

T	C_p
CURVE 1	
298.16	4.115×10^{-2}
300	4.151*
400	5.363
500	5.902
600	6.180
700	6.329
800	6.416
900	6.457
1000	6.478
1100	6.488
1200	6.483
CURVE 2	
1556	7.4×10^{-2}
1574	6.9
1597	6.5
1630	7.6
1674	9.1
1688	9.1*
1718	8.5
1775	9.1
1819	1.08×10^{-1}
1868	1.02
1920	1.01
1961	9.4×10^{-2}
2000	1.07×10^{-1}
2000	1.0
2057	1.04
2074	1.07*
2086	1.02*
2118	1.21
2129	1.01*
2169	1.23*
2192	1.14
2232	1.22
2247	1.17*
2281	1.10*
2286	1.25
2321	1.18
2345	1.29*
2379	1.18
2393	1.23*
CURVE 2 (cont.)	
2399	1.25×10^{-1}*
2458	1.16
2518	1.09
2518	1.20
CURVE 3	
533.15	5.8×10^{-2}
810.93	6.0
1088.71	6.5
1366.48	7.1
1644.26	8.0
1922.04	8.8
2199.82	9.3
2477.59	9.8

* Not shown on plot

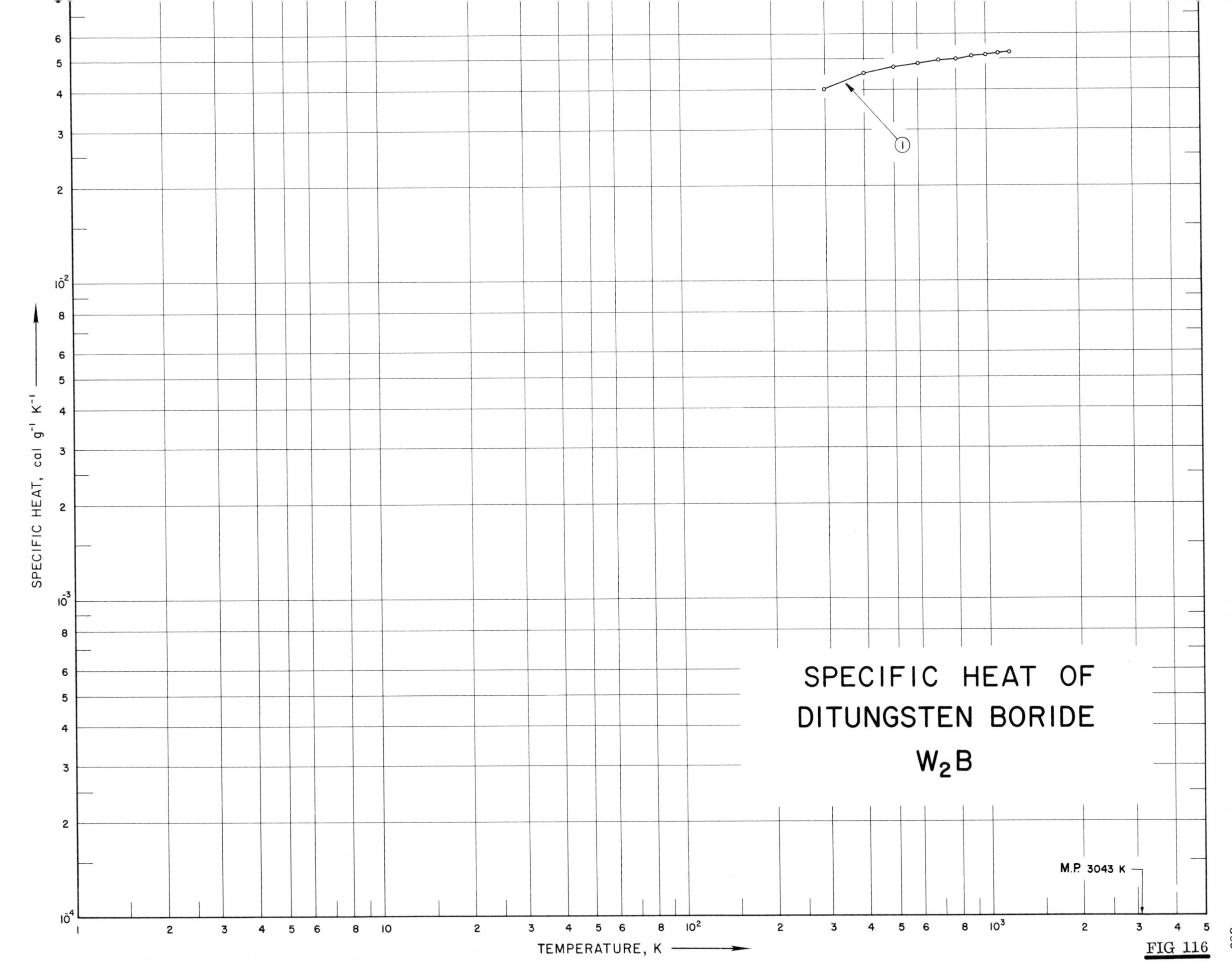
SPECIFIC HEAT OF
DITUNGSTEN BORIDE
W_2B
M.P. 3043 K
1
SPECIFIC HEAT, cal g⁻¹ K⁻¹
TEMPERATURE, K

FIG 116

SPECIFICATION TABLE NO. 116 SPECIFIC HEAT OF DITUNGSTEN BORIDE W_2B

[For Data Reported in Figure and Table No. 116]

Curve No.	Ref. No.	Year	Temp. Range, K	Reported Error, %	Name and Specimen Designation	Composition (weight percent), Specifications and Remarks
1	163	1962	298-1200	0.5		Traces of impurities.

DATA TABLE NO. 116 SPECIFIC HEAT OF DITUNGSTEN BORIDE W_2B

[Temperature, T, K; Specific Heat, C_p, Cal $g^{-1} K^{-1}$]

T	C_p
CURVE 1	
298.16	4.058×10^{-2}
300	4.071*
400	4.512
500	4.737
600	4.874
700	4.975
800	5.051
900	5.117
1000	5.176
1100	5.228
1200	5.279

* Not shown on plot

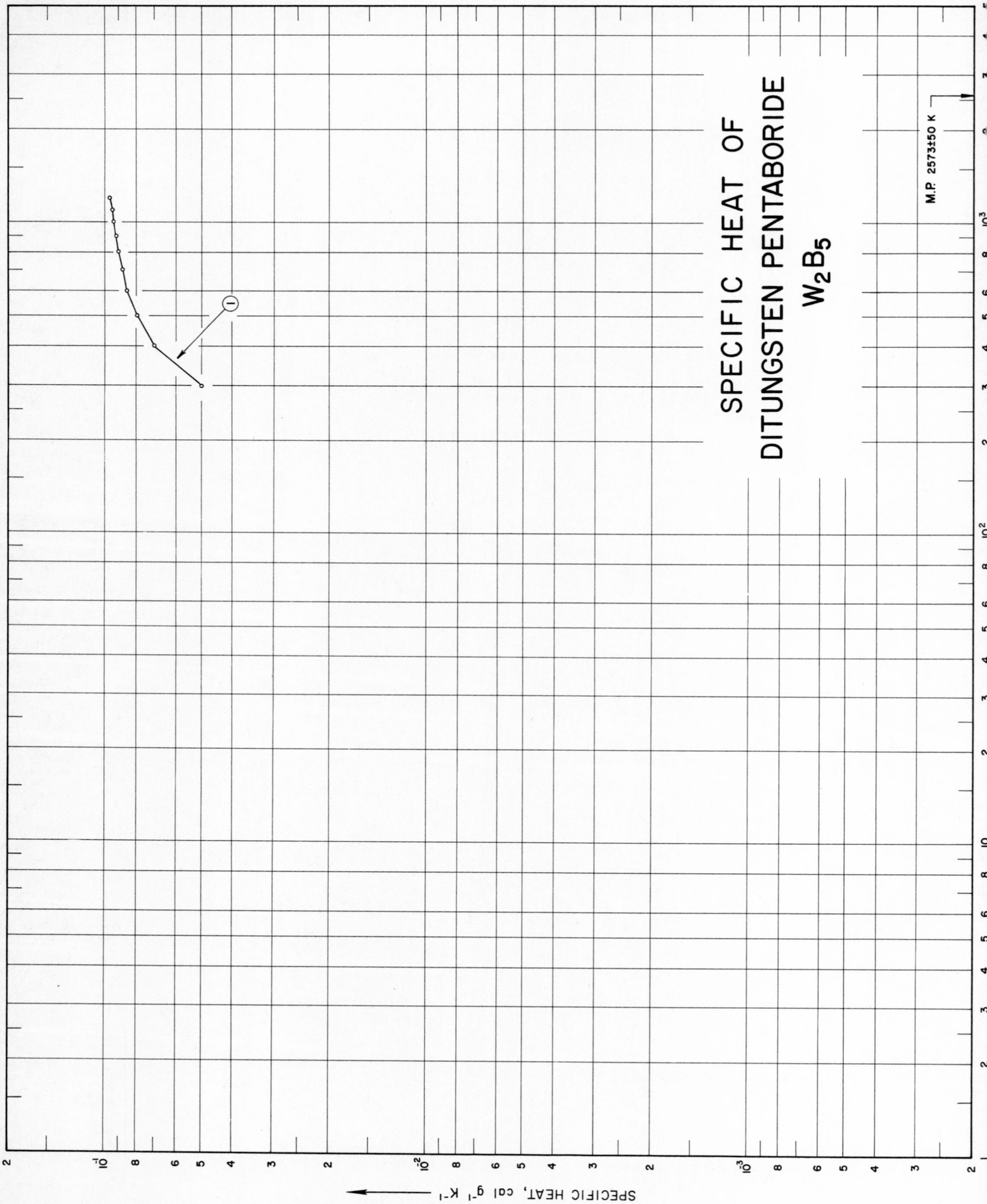

SPECIFIC HEAT OF DITUNGSTEN PENTABORIDE W_2B_5

SPECIFICATION TABLE NO. 117 SPECIFIC HEAT OF DITUNGSTEN PENTABORIDE W_2B_5

[For Data Reported in Figure and Table No. 117]

Curve No.	Ref. No.	Year	Temp. Range, K	Reported Error, %	Name and Specimen Designation	Composition (weight percent), Specifications and Remarks
1	163	1962	298-1200	0.50		Traces of impurities.

DATA TABLE NO. 117 SPECIFIC HEAT OF DITUNGSTEN PENTABORIDE W_2B_5

[Temperature, T, K; Specific Heat, C_p, Cal g^{-1} K^{-1}]

T	C_p
CURVE 1	
298.16	5.001 x 10^{-2}
300	5.055*
400	7.030
500	7.971
600	8.505
700	8.846
800	9.086
900	9.266
1000	9.408
1100	9.527
1200	9.629

*Not shown on plot

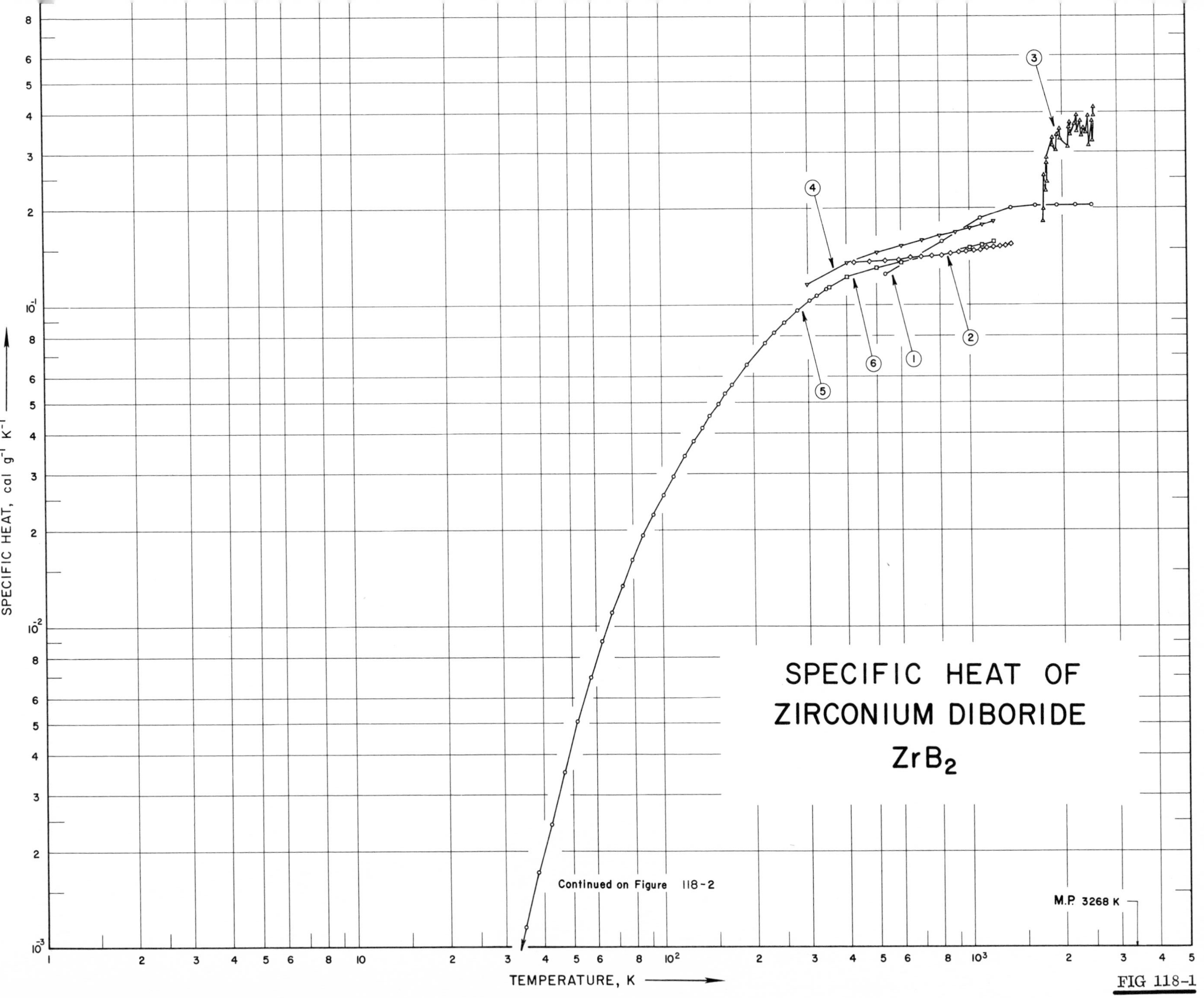

SPECIFIC HEAT OF
ZIRCONIUM DIBORIDE
ZrB2
Continued on Figure 118-2
M.P. 3268 K
TEMPERATURE, K
SPECIFIC HEAT, cal g-1 K-1
1
2
3
4
5
6

FIG 118-1

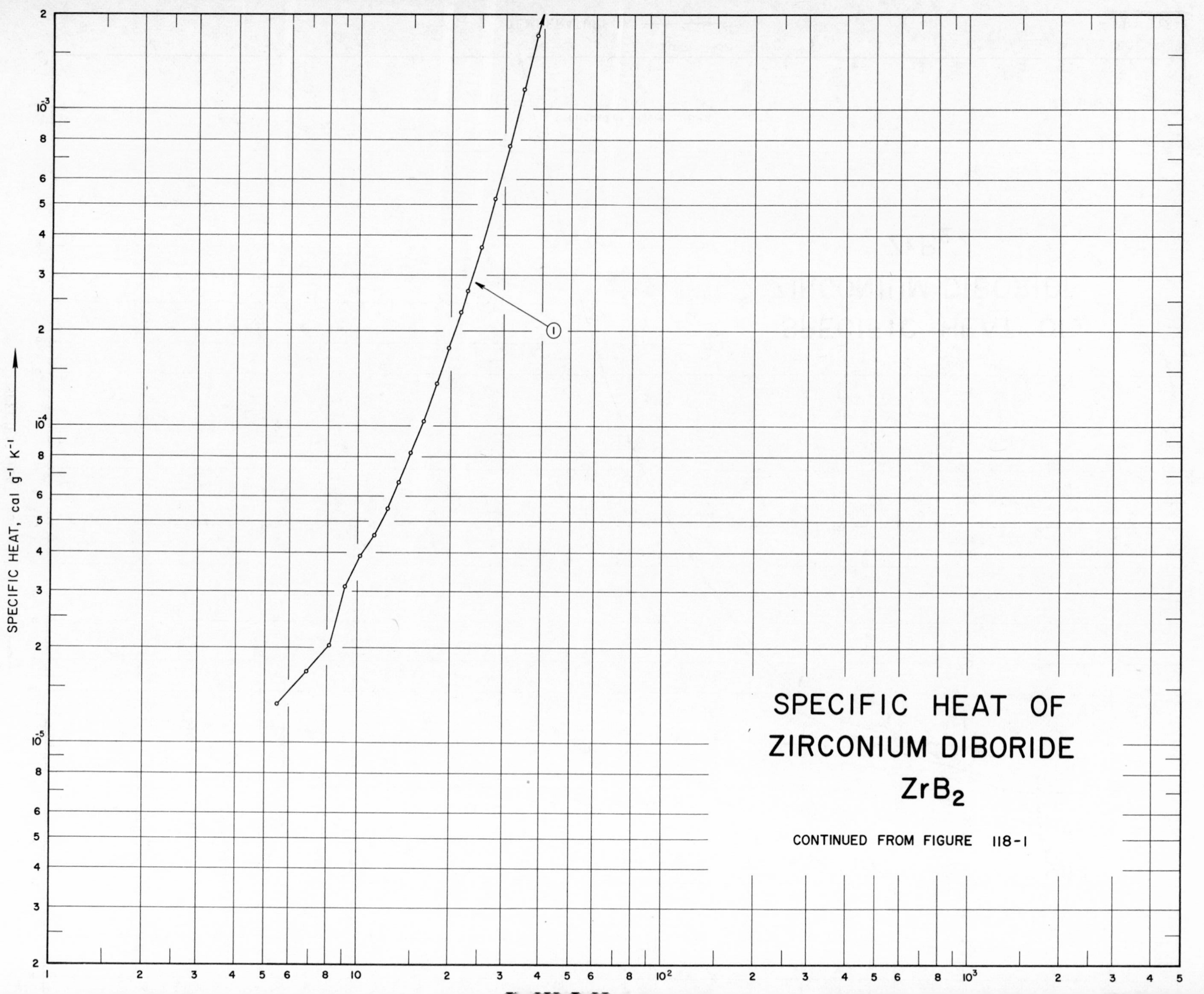
SPECIFIC HEAT OF
ZIRCONIUM DIBORIDE
ZrB_2
CONTINUED FROM FIGURE 118-1
1
SPECIFIC HEAT, cal g^{-1} K^{-1}

SPECIFICATION TABLE NO. 118 SPECIFIC HEAT OF ZIRCONIUM DIBORIDE ZrB_2

[For Data Reported in Figure and Table No. 118]

Curve No.	Ref. No.	Year	Temp. Range, K	Reported Error, %	Name and Specimen Designation	Composition (weight percent), Specifications and Remarks
1	32	1962	533-2478	≤5		78.7 Zr, 17.6 B, 0.36 C, remainder Al, Ca, Fe, Ni, Si, and Ti; sample supplied by Norton Co; hot pressed; density 258 lb ft^{-3}.
2	76	1961	422-1366	5		Major constituent ZrB_2, 1.60 O_2, 1.4 N_2; sample made by spraying powder HfC using powder gun with 80 ft^3 hr^{-1} N_2-plasma gas and 10 ft^3 hr^{-1} N_2-carrier gas.
3	173	1962	1739-2521	±5		78.94 Zr, 16.86 B, 1.36 O, <0.5 Ni, 0.37 total C, <0.3 Al, <0.25 Hf, <0.02 V, 0.18 Fe, 0.14 N, 0.14 Ti, 0.1 Cr, <0.05 Cu, <0.01 Mg, Mn, and Nb.
4	163	1962	298-1200	0.5		Traces of impurities.
5	164	1962	5-345	0.1-5		Sample A: 98.62 ZrB_2; Sample B: 99.61 ZrB_2; spectrochemical analysis: Ag, Cu, Ti, V, Cr, and Mn, 0.0010, Al and Mg 0.0010 - 0.0100, Hf and Fe 0.0010 - 0.1000, and Si 0.0100 - 0.1000; zone-refined.
6	174	1964	400-1200			99.3 ZrB_2, 0.01 - 0.1 Si, 0.001 - 0.1 Hf and Fe, 0.0215 C, 0.0134 N, 0.001 - 0.01 Mg, 0.0052 O, 0.001 Ag, Ca, Cu, Ti, V, Cr, and Mn; zone refined.

DATA TABLE NO. 118 SPECIFIC HEAT OF ZIRCONIUM DIBORIDE ZrB_2

[Temperature, T, K; Specific Heat, C_p, Cal $g^{-1} K^{-1}$]

CURVE 1

T	C_p
533.15	1.25×10^{-1}
810.93	1.58
1088.71	1.86
1366.48	2.01
1644.26	2.05
1922.04	2.05
2199.82	2.05
2477.59	2.05

CURVE 2

T	C_p
422.04	1.362×10^{-1}
477.59	1.374
533.15	1.385
588.71	1.396
644.26	1.407
699.82	1.418
755.37	1.429
810.93	1.440
866.48	1.452
922.04	1.463
977.59	1.474
1033.15	1.485
1088.71	1.496
1144.26	1.507
1199.82	1.518
1255.37	1.530
1310.93	1.541
1366.48	1.552

CURVE 3

T	C_p
1739	1.83×10^{-1}
1742	2.00
1745	2.55
1749	2.05*
1765	2.28
1776	2.79
1786	2.88
1786	2.42
1852	3.33
1856	3.15
1909	3.04
1914	2.98*
1929	3.39×10^{-1}
1961	3.55
1966	3.30
1974	3.26*
1983	3.59*
2097	3.11
2100	3.58
2115	3.72
2121	3.39
2127	3.44*
2169	3.72*
2183	3.67
2224	3.92
2232	3.47
2253	3.45*
2265	3.52*
2295	3.75
2298	3.49*
2306	3.37
2327	3.61*
2345	3.57
2382	3.65*
2384	3.46
2408	3.46*
2414	3.46*
2437	3.89
2440	3.89*
2443	3.14
2498	3.76
2500	3.26
2504	3.32*
2507	3.17*
2507	3.71*
2518	4.15
2521	3.89

CURVE 4

T	C_p
298.16	1.163×10^{-1}
300	1.166*
400	1.355
500	1.462
600	1.538
700	1.598
800	1.650×10^{-1}
900	1.697
1000	1.742
1100	1.785
1200	1.826

CURVE 5

T	C_p
Series 1	
67.62	1.107×10^{-2}
73.24	1.345
79.29	1.615
86.15	1.928
93.23	2.241
100.79	2.574
108.87	2.935
Series 2	
5.50	1.329×10^{-5}
6.86	1.684
8.12	2.038
9.16	3.102
10.21	3.899
11.42	4.520
12.55	5.494
13.67	6.646
14.93	8.242
16.41	1.046×10^{-4}
Series 3	
18.05	1.365×10^{-4}
19.76	1.772
21.62	2.259
22.80	2.694
25.11	3.695
27.90	5.211
31.13	7.692
34.70	1.160×10^{-3}
38.37	1.707
42.29	2.450
46.80	3.520
51.88	5.052
57.41	6.975×10^{-3}
62.44	8.936
68.91	1.160×10^{-2}*
Series 4	
119.42	3.412×10^{-2}
127.64	3.784
135.78	4.153
144.07	4.531
152.81	4.922
161.47	5.310
169.96	5.681
172.64	5.805*
181.64	6.181*
190.24	6.545
198.74	6.895*
207.14	7.232*
217.19	7.622
224.60	7.905*
233.49	8.216
242.55	8.532*
251.46	8.843
260.17	9.116*
268.66	9.384*
277.25	9.639
286.43	9.902*
295.78	1.016×10^{-1}*
304.56	1.039
312.70	1.059*
320.44	1.078
328.67	1.097*
336.97	1.114*
345.25	1.131

CURVE 6

T	C_p
400	1.220×10^{-1}
500	1.303
600	1.359
700	1.405*
800	1.444*
900	1.479×10^{-1}*
1000	1.513
1100	1.545
1200	1.575

*Not shown on plot

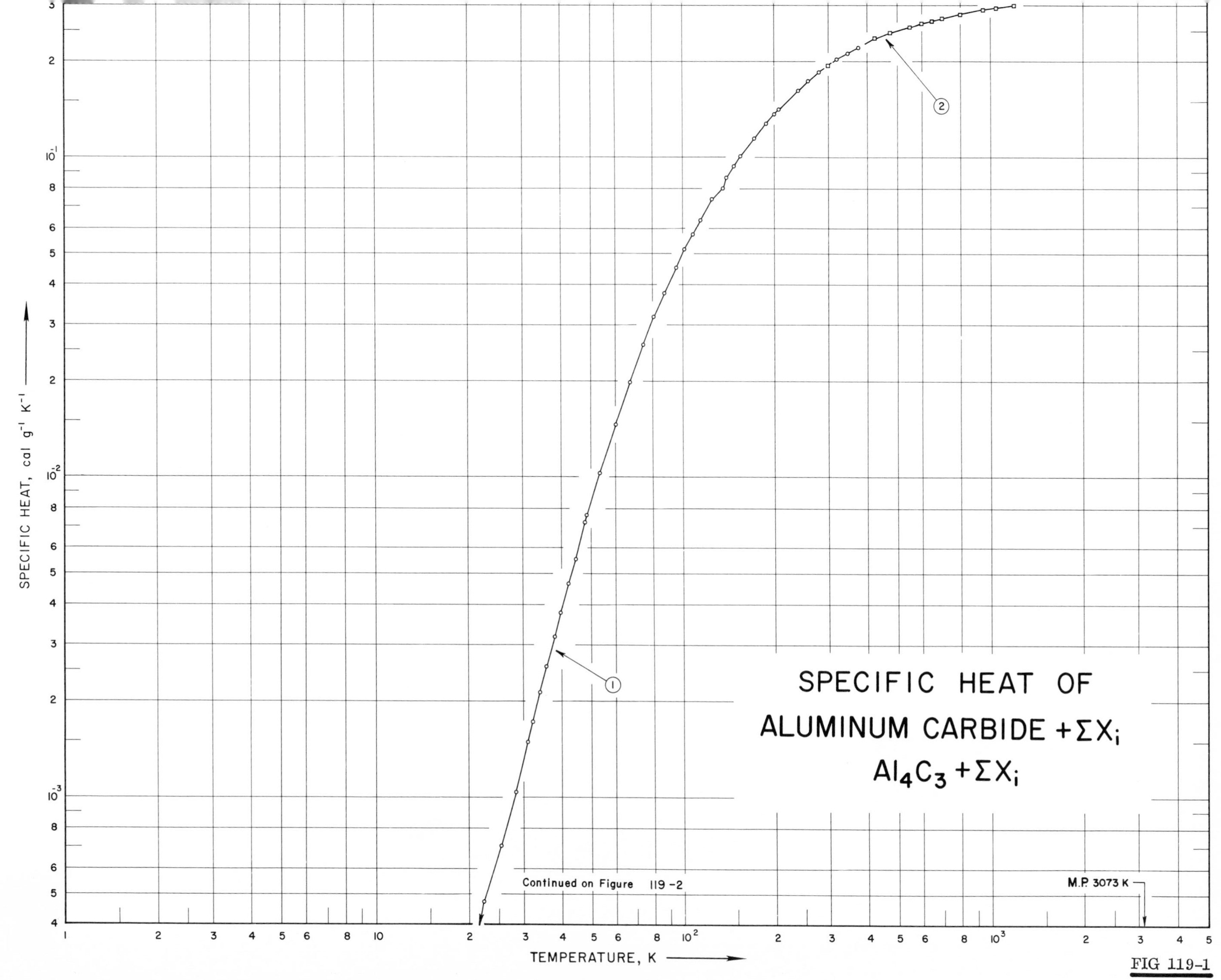
SPECIFIC HEAT OF
ALUMINUM CARBIDE +ΣXi
Al4C3 +ΣXi
Continued on Figure 119-2
M.P. 3073 K
TEMPERATURE, K
SPECIFIC HEAT, cal g-1 K-1
1
2

FIG 119-1

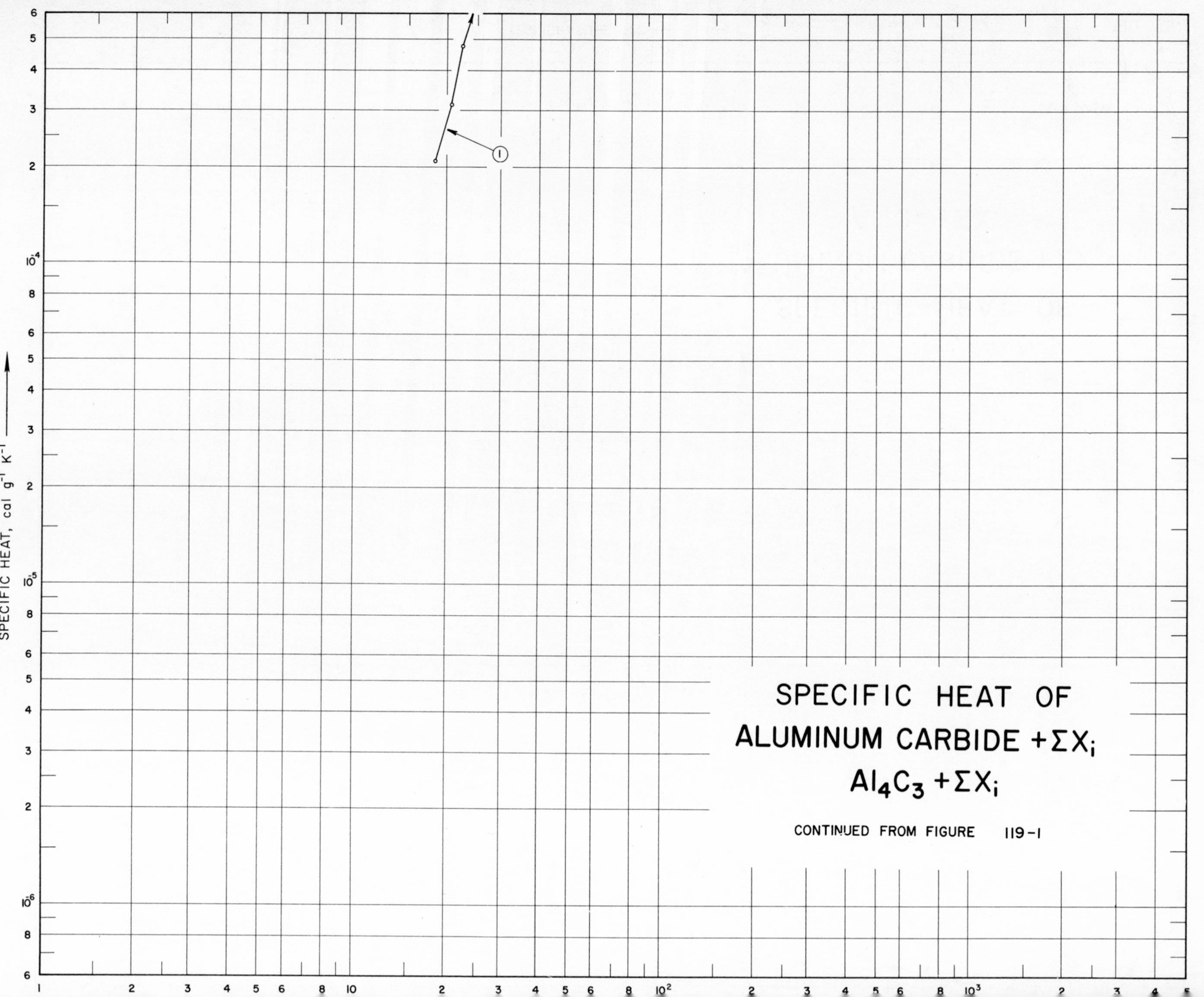
SPECIFIC HEAT OF
ALUMINUM CARBIDE +ΣX_i
Al_4C_3 +ΣX_i
CONTINUED FROM FIGURE 119-1
SPECIFIC HEAT, cal g^-1 K^-1
1
10^-4
10^-5
10^-6
10
10^2
10^3

SPECIFICATION TABLE NO. 119 SPECIFIC HEAT OF ALUMINUM CARBIDE + ΣX_i $Al_4C_3 + \Sigma X_i$

[For Data Reported in Figure and Table No. 119]

Curve No.	Ref. No.	Year	Temp. Range, K	Reported Error, %	Name and Specimen Designation	Composition (weight percent), Specifications and Remarks
1	170	1964	18-375			94.8 Al_4C_3, 2.1 Al_2O_3, 1.2 free Al, 0.8 free C, and 1.3 AlN; spectrochemical analysis: 0.10 - 1.0 Fe, 0.01 - 0.1 Si, 0.01 - 0.1 V, 0.001 - 0.01 Cr, 0.001 - 0.01 Cu, 0.001 - 0.01 Mg; 0.001 - 0.01 N, 0.001 - 0.01 Ti, 0.001 - 0.01 Zr, 0.0001 - 0.001 Ca, 0.0001 - 0.001 Mn.
2	170	1964	300-1200			Same as above.

DATA TABLE NO. 119 SPECIFIC HEAT OF ALUMINUM CARBIDE + ΣX_i $Al_4C_3 + \Sigma X_i$

[Temperature, T, K; Specific Heat, C_p, Cal g^{-1} K^{-1}]

T	C_p
CURVE 1	
Series 1	
80.93	3.161×10^{-2}
87.53	3.777
95.52	4.507
102.19	5.174
109.19	5.795
115.26	6.375
120.96	6.913
126.36	7.421*
Series 2	
83.37	3.394×10^{-2}*
83.78	3.433*
92.00	4.186*
99.30	4.858*
105.92	5.486*
112.10	6.073*
118.39	6.671*
125.29	7.322*
137.59	8.005
140.09	8.693
148.27	9.429
156.82	1.018×10^{-1}
165.07	1.088*
173.09	1.155
181.15	1.219*
189.92	1.287
199.00	1.355*
Series 3	
208.26	1.422×10^{-2}
194.84	1.324*
204.78	1.397*
214.36	1.464*
223.62	1.526*
232.60	1.584*
241.34	1.638
250.22	1.691*
259.24	1.742
CURVE 1 (cont.)	
Series 4	
19.44	2.557×10^{-4}
21.28	3.553
23.19	4.947
26.12	7.770
28.95	1.146×10^{-3}
31.83	1.624
34.70	2.251
37.94	3.149
Series 5	
18.36	2.075×10^{-4}
20.59	3.221
22.31	4.748
25.39	7.056
28.22	1.043×10^{-3}
31.07	1.494
34.22	2.145
38.06	3.196
42.30	4.687
Series 6	
32.27	1.722×10^{-3}
35.97	2.580
39.90	3.790
44.38	5.527
48.77	7.601
53.73	1.038×10^{-2}
60.19	1.467
67.19	1.993
Series 7	
47.95	7.195×10^{-3}
53.80	1.042×10^{-2}*
60.27	1.471*
67.12	1.989*
74.44	2.590
82.58	3.321*
CURVE 1 (cont.)	
Series 8*	
53.67	1.033×10^{-2}
59.91	1.444
67.04	1.984
73.82	2.534
80.93	3.169
88.80	3.897
96.70	4.618
104.96	5.394
113.34	6.192
Series 9	
202.70	1.382×10^{-1}*
212.73	1.453*
222.81	1.521*
233.19	1.587*
244.99	1.660*
257.76	1.773*
270.12	1.801*
282.14	1.862
Series 10*	
228.26	1.556×10^{-1}
238.86	1.623
249.41	1.686
260.04	1.746
282.92	1.866
Series 11	
287.59	1.888×10^{-1}*
299.20	1.943*
310.67	1.994*
322.19	2.042
CURVE 1 (cont.)	
Series 12*	
280.88	1.856×10^{-1}
291.76	1.909
303.40	1.962
318.92	2.013
326.34	2.060
337.69	2.103
348.85	2.143
Series 13*	
332.84	2.084×10^{-1}
344.32	2.127
355.74	2.167
367.15	2.205
Series 14	
332.39	2.082×10^{-1}*
338.24	2.104*
349.39	2.145*
354.70	2.165*
365.61	2.202*
376.38	2.235
CURVE 2	
300	1.947×10^{-1}*
310	1.993*
320	2.036*
330	2.077*
340	2.114*
350	2.150*
360	2.184*
370	2.215*
373.15	2.225*
380	2.245*
390	2.273*
400	2.300*
425	2.360
450	2.414*
CURVE 2 (cont.)	
475	2.462×10^{-1}
500	2.505*
550	2.580
600	2.643
650	2.697
700	2.742
750	2.782*
800	2.818
850	2.849*
900	2.877*
950	2.902
1000	2.924*
1050	2.945
1100	2.964*
1150	2.981*
1200	2.997

*Not shown on plot

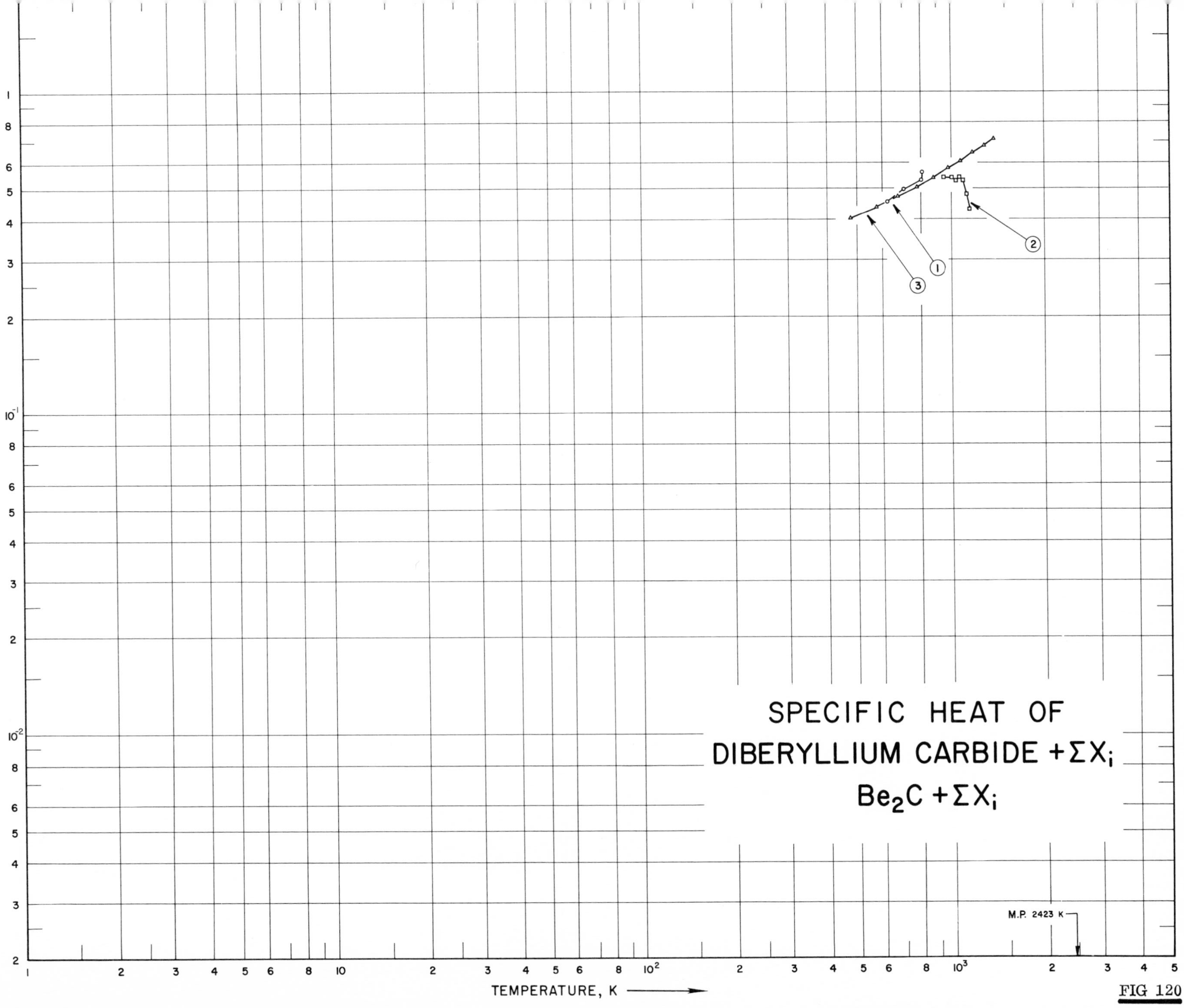
SPECIFIC HEAT OF
DIBERYLLIUM CARBIDE $+\Sigma X_i$
$Be_2C + \Sigma X_i$
SPECIFIC HEAT, cal g^{-1} K^{-1}
TEMPERATURE, K
M.P. 2423 K
1
2
3

FIG 120

SPECIFICATION TABLE NO. 120 SPECIFIC HEAT OF DIBERYLLIUM CARBIDE + ΣX_i $Be_2C + \Sigma X_i$

[For Data Reported in Figure and Table No. 120]

Curve No.	Ref. No.	Year	Temp. Range, K	Reported Error, %	Name and Specimen Designation	Composition (weight percent), Specifications and Remarks
1	175	1948	625-804	±15		56. 5 Be, 3. 14 free C and 0. 005 free Fe, powdered sample.
2	176	1948	946-1150	±25		Left end: 52. 45 Be and 1. 94 free C, middle: 48. 65 Be and 3. 39 free C, right end: 50. 24 Be and 1. 96 free C (theor. 60. 05 Be); sample supplied by the Norton Co; hot pressed.
3	177	1950	473-1373	±15		Before test: 80 Be_2C, most impurities were oxides and nitrides, after test: 74 Be_2C.

DATA TABLE NO. 120 SPECIFIC HEAT OF DIBERYLLIUM CARBIDE + ΣX_i $Be_2C + \Sigma X_i$

[Temperature, T, K; Specific Heat, C_p, Cal g^{-1} K^{-1}]

T	C_p
CURVE 1	
625.15	4.56 x 10^{-1}
622.15	4.46*
704.15	4.95
702.15	5.67*
699.65	5.40*
807.15	5.60
804.15	5.30
CURVE 2	
946.15	5.4 x 10^{-1}
970.15	5.4
1008.15	5.4
1042.15	5.3
1066.15	5.4
1091.15	5.3
1127.15	4.8
1150.15	4.3
CURVE 3	
473.15	4.04 x 10^{-1}
573.15	4.38
673.15	4.72
773.15	5.06
873.15	5.40
973.15	5.74
1073.15	6.08
1173.15	6.42
1273.15	6.76
1373.15	7.10

*Not shown on plot

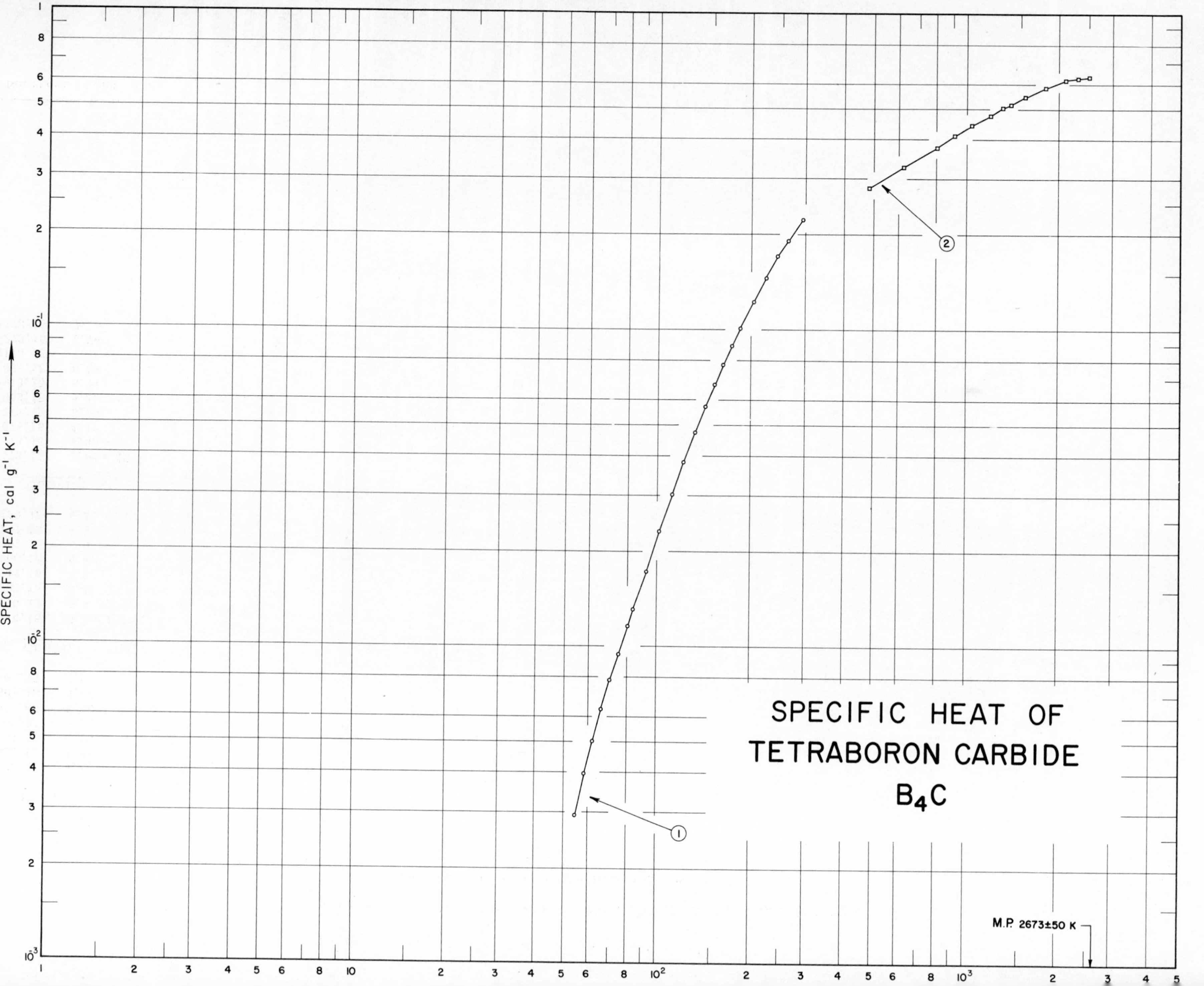
SPECIFIC HEAT OF
TETRABORON CARBIDE
B4C
SPECIFIC HEAT, cal g-1 K-1
M.P. 2673±50 K
1
2

SPECIFICATION TABLE NO. 121 SPECIFIC HEAT OF TETRABORON CARBIDE B_4C

[For Data Reported in Figure and Table No. 121]

Curve No.	Ref. No.	Year	Temp. Range, K	Reported Error, %	Name and Specimen Designation	Composition (weight percent), Specifications and Remarks
1	50	1941	54-294	0.5		96 B_4C and 4 free and included graphite; corrected for graphite.
2	47	1963	482-2510	±5		75.97 B, 21.18 C, 0.07 B_2O_7, 0.27 Fe, 0.40 Si, and 0.015 Al_2O_3; sample supplied by the Carborundum Co; hot pressed at 3940F; density 156 lb ft^{-3}.

DATA TABLE NO. 121 SPECIFIC HEAT OF TETRABORON CARBIDE B_4C

[Temperature, T, K; Specific Heat, C_p, Cal g^{-1} K^{-1}]

T	C_p	T	C_p
CURVE 1		CURVE 2 (cont.)	
54.5	2.930 x 10^{-3}	1316.48	5.040 x 10^{-1}
58.5	3.961	1398.71	5.199
62.3	5.028	1481.48	5.349*
66.2	6.348	1567.59	5.493
70.6	7.813	1634.26	5.596*
75.0	9.459	1803.71	5.828
80.1	1.154 x 10^{-2}	1933.15	5.975*
83.6	1.311	2133.15	6.149
92.1	1.744	2272.04	6.233*
102.0	2.317	2383.15	6.279
112.7	3.062	2510.93	6.307
122.7	3.843		
133.0	4.759		
143.4	5.733		
153.5	6.723		
163.6	7.790		
174.1	8.933		
184.3	1.007 x 10^{-1}		
194.0	1.115		
204.1	1.234		
214.0	1.351		
224.3	1.469		
234.5	1.589		
244.7	1.704		
254.8	1.818		
264.9	1.928		
275.3	2.049		
284.5	2.143		
294.3	2.235		
CURVE 2			
481.48	2.820 x 10^{-1}		
624.82	3.279		
800.37	3.797		
840.93	3.910*		
915.93	4.112		
1049.26	4.449		
1092.04	4.551*		
1200.93	4.798		
1205.93	4.809*		
1247.04	4.897*		
1258.15	4.920*		

*Not shown on plot

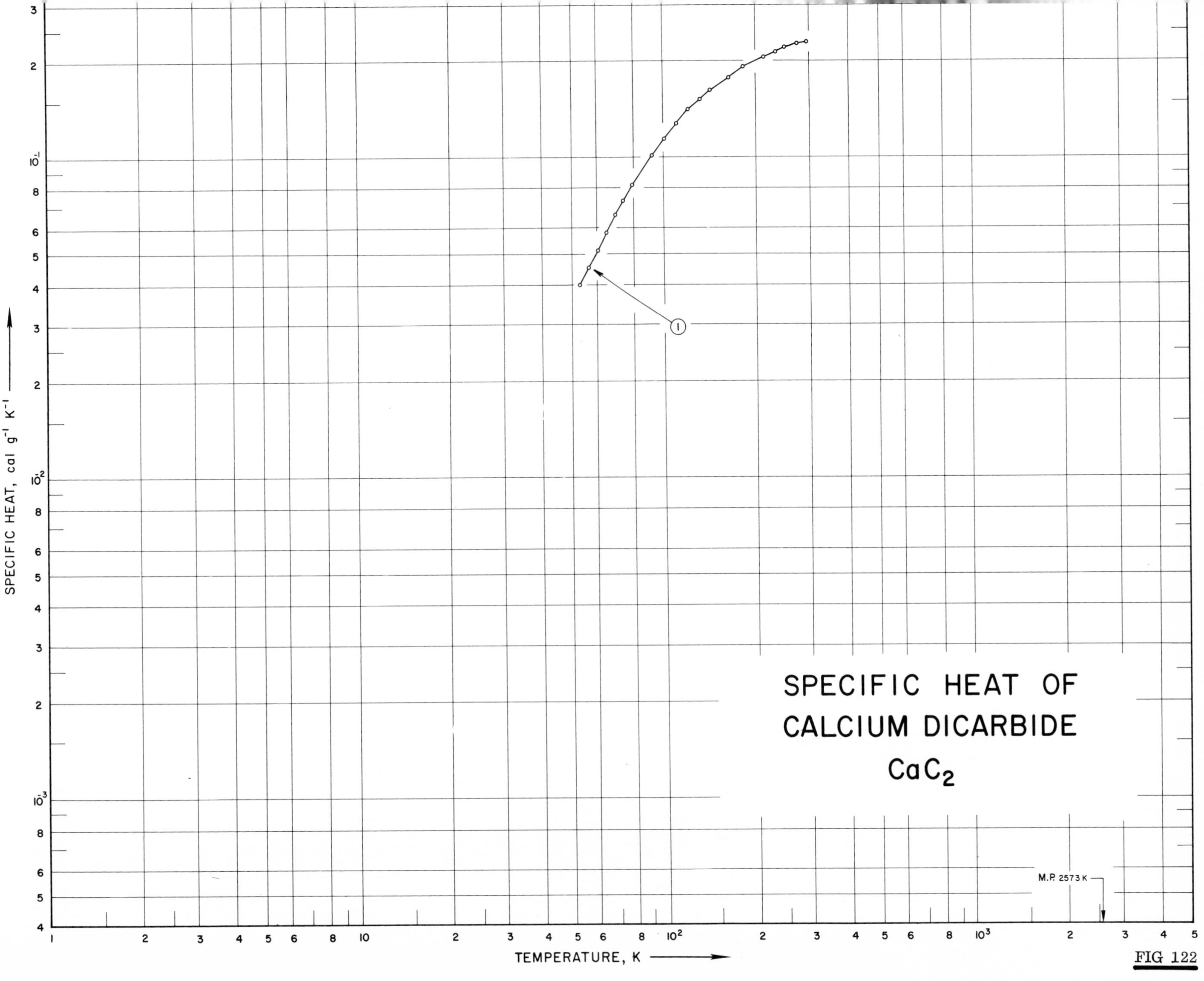

SPECIFIC HEAT OF
CALCIUM DICARBIDE
CaC_2
M.P. 2573 K
1
TEMPERATURE, K
SPECIFIC HEAT, cal g^{-1} K^{-1}

SPECIFICATION TABLE NO. 122 SPECIFIC HEAT OF CALCIUM DICARBIDE CaC_2

[For Data Reported in Figure and Table No. 122]

Curve No.	Ref. No.	Year	Temp. Range, K	Reported Error, %	Name and Specimen Designation	Composition (weight percent), Specifications and Remarks
1	178	1941	53-295			91.0 CaC_2, 6.47 CaO, 1.15 SiO_2, 0.77 Al_2O_3, 0.29 Fe, 0.2 C, and 0.08 MgO; sample supplied by the National Carbide Corp; corrected for impurities.

DATA TABLE NO. 122 SPECIFIC HEAT OF CALCIUM DICARBIDE CaC_2

[Temperature, T, K; Specific Heat, C_p, Cal g^{-1} K^{-1}]

T	C_p
CURVE 1	
53.0	4.02 x 10^{-2}
56.8	4.55
60.7	5.16
65.0	5.88
69.7	6.65
74.0	7.35
79.6	8.28
83.2	8.79*
92.0	1.016 x 10^{-1}
101.9	1.156
112.5	1.296
122.7	1.422
133.0	1.534
143.7	1.640
153.6	1.715*
164.1	1.797
174.0	1.867*
184.3	1.933
194.6	1.991
204.9	2.036*
214.7	2.078
224.5	2.117*
235.1	2.154
245.0	2.192*
254.9	2.220
265.2	2.243*
275.8	2.285
285.7	2.309*
295.0	2.318

*Not shown on plot

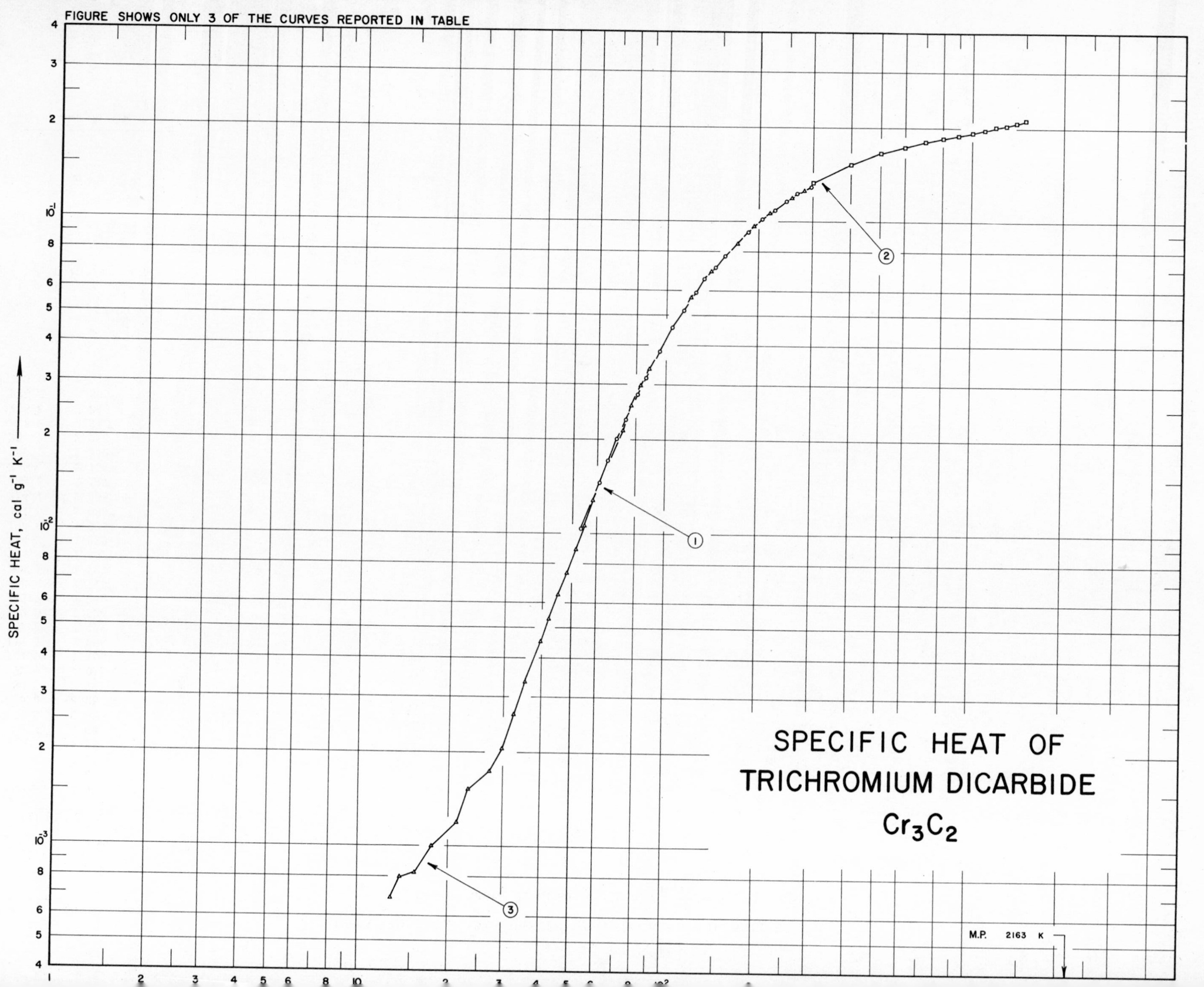

FIGURE SHOWS ONLY 3 OF THE CURVES REPORTED IN TABLE
SPECIFIC HEAT, cal g^{-1} K^{-1}
SPECIFIC HEAT OF
TRICHROMIUM DICARBIDE
Cr_3C_2
M.P. 2163 K
1
2
3

SPECIFICATION TABLE NO. 123 SPECIFIC HEAT OF TRICHROMIUM DICARBIDE Cr_3C_2

[For Data Reported in Figure and Table No. 123]

Curve No.	Ref. No.	Year	Temp. Range, K	Reported Error, %	Name and Specimen Designation	Composition (weight percent), Specifications and Remarks
1	61	1944	53-295			98.67 Cr_3C_2, 0.98 Cr_2O_3, and 0.25 uncombined C.
2	61	1944	298-1500			98.77 Cr_3C_2, 0.98 Cr_2O_3, and 0.25 uncombined C.
3	179	1953	12-300			86.2 Cr, 13.2 ± 4 C (theor. 86.67 Cr and 13.3 C), small amounts of metallic Cr, trace of Al, Cu, Fe, and Mg; prepared from Cr_2O_3 and lampblack in H_2 furnace at 1525 C.
4	180	1954	273-1200			86.2 Cr, 13.2 ± 4 C, small amounts of Cr, traces of Al, Cu, Fe, and Mg.

DATA TABLE NO. 123 SPECIFIC HEAT OF TRICHROMIUM DICARBIDE Cr_3C_2

[Temperature, T, K; Specific Heat, C_p, Cal g^{-1} K^{-1}]

T	C_p
CURVE 1	
53.4	1.041 x 10^{-2}
57.2	1.235*
61.1	1.463
65.1	1.719
69.4	2.001
74.2	2.329
81.0	2.499
86.1	3.159
95.6	3.847
105.3	4.530
114.9	5.196
125.5	5.909
135.3	6.543
145.4	7.148
156.0	7.765
165.7	8.298*
175.5	8.792*
185.9	9.286
196.0	9.753*
206.2	1.017 x 10^{-1}
216.0	1.058*
226.1	1.096
236.3	1.130*
276.5	1.169
256.4	1.198*
266.2	1.230
276.8	1.260*
286.2	1.281*
295.2	1.295
CURVE 2	
298.15	1.298 x 10^{-1}*
300	1.304
400	1.535
500	1.658
600	1.740
700	1.801
800	1.852
900	1.896
CURVE 2 (cont.)	
1000	1.937 x 10^{-1}
1100	1.975
1200	2.011
1300	2.046
1400	2.081
1500	2.114
CURVE 3	
12.90	6.720 x 10^{-4}
13.82	7.998
15.51	8.220
17.66	9.997
21.25	1.189 x 10^{-3}
23.16	1.522
27.63	1.727
29.82	2.055
32.44	2.644
35.22	3.355
39.54	4.510
42.18	5.332
45.16	6.365
48.28	7.492
51.48	8.909
54.57	1.060 x 10^{-2}
58.12	1.289
62.28	1.483*
66.82	1.721*
71.99	2.177
77.15	2.577
79.91	2.769*
82.91	2.988
84.42	2.969*
88.17	3.366
93.52	3.749*
99.41	4.152*
105.89	4.612*
113.13	5.126*
121.83	5.704
128.70	6.115*
CURVE 3 (cont.)	
135.20	6.532 x 10^{-2}*
141.84	6.948
148.84	7.403*
156.16	7.881*
162.84	8.270*
170.12	8.459
177.87	8.936*
185.97	9.347*
193.60	9.653
201.01	1.004 x 10^{-1}*
208.14	1.030*
212.17	1.042*
218.88	1.068
225.22	1.088*
231.21	1.106*
236.86	1.126*
242.12	1.143*
247.33	1.166*
251.75	1.182*
256.06	1.192
260.64	1.210*
265.06	1.224*
269.40	1.237*
272.49	1.246*
276.42	1.260*
280.48	1.265
282.61	1.275*
284.40	1.284*
286.23	1.281*
288.70	1.286*
290.07	1.289*
293.70	1.297*
297.30	1.300*
300.71	1.311*
CURVE 4*	
273.15	1.247 x 10^{-1}
300.00	1.321
400.00	1.501
CURVE 4 (cont.)*	
500.00	1.613 x 10^{-1}
600.00	1.698
700.00	1.770
800.00	1.835
900.00	1.896
1000.00	1.955
1100.00	2.012
1200.00	2.068

*Not shown on plot

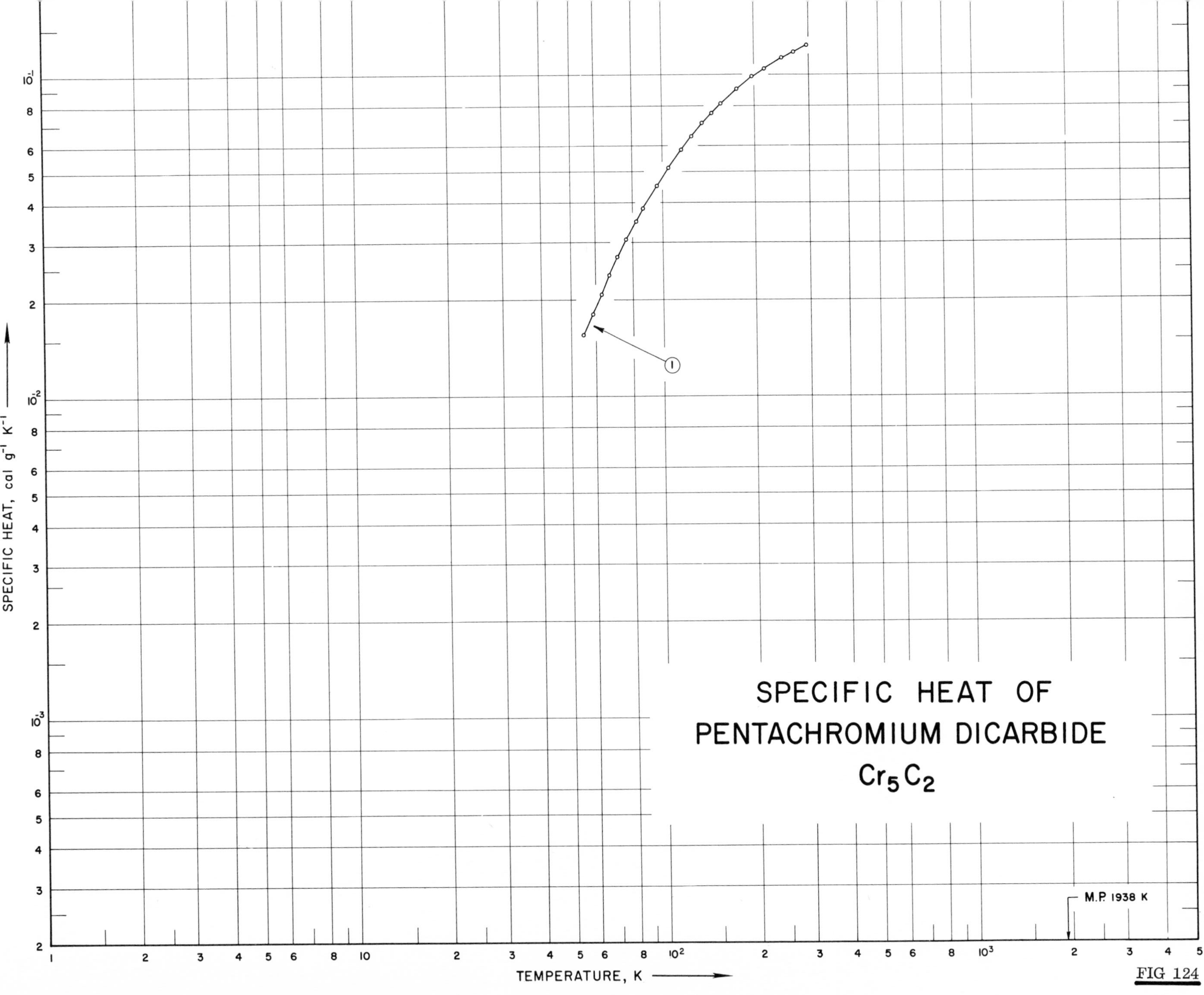
SPECIFIC HEAT OF
PENTACHROMIUM DICARBIDE
Cr_5C_2
SPECIFIC HEAT, cal g⁻¹ K⁻¹
TEMPERATURE, K
M.P. 1938 K
1

FIG 124

SPECIFICATION TABLE NO. 124 SPECIFIC HEAT OF PENTACHROMIUM DICARBIDE Cr_5C_2

[For Data Reported in Figure and Table No. 124]

Curve No.	Ref. No.	Year	Temp. Range, K	Reported Error, %	Name and Specimen Designation	Composition (weight percent), Specifications and Remarks
1	61	1944	53-295			99.6 Cr_5C_2, 0.4 Cr_2O_3.

DATA TABLE NO. 124 SPECIFIC HEAT OF PENTACHROMIUM DICARBIDE Cr_5C_2

[Temperature, T, K; Specific Heat, C_p, Cal g^{-1} K^{-1}]

T	C_p
CURVE 1	
54.8	1.553 x 10^{-2}
58.7	1.802
62.8	2.082
66.9	2.391
71.4	2.727
76.3	3.084
79.3	3.298*
82.2	3.516
86.7	3.851
96.3	4.538
105.3	5.164
116.0	5.868
125.5	6.477
136.4	7.111
146.2	7.635
156.0	8.121
166.6	8.628*
176.2	9.033
186.8	9.445*
197.8	9.867
206.6	1.016 x 10^{-1}*
216.3	1.045
226.7	1.078*
236.8	1.103*
246.7	1.132
256.5	1.153*
266.9	1.180
276.8	1.203*
286.0	1.216*
295.3	1.230

* Not shown on plot

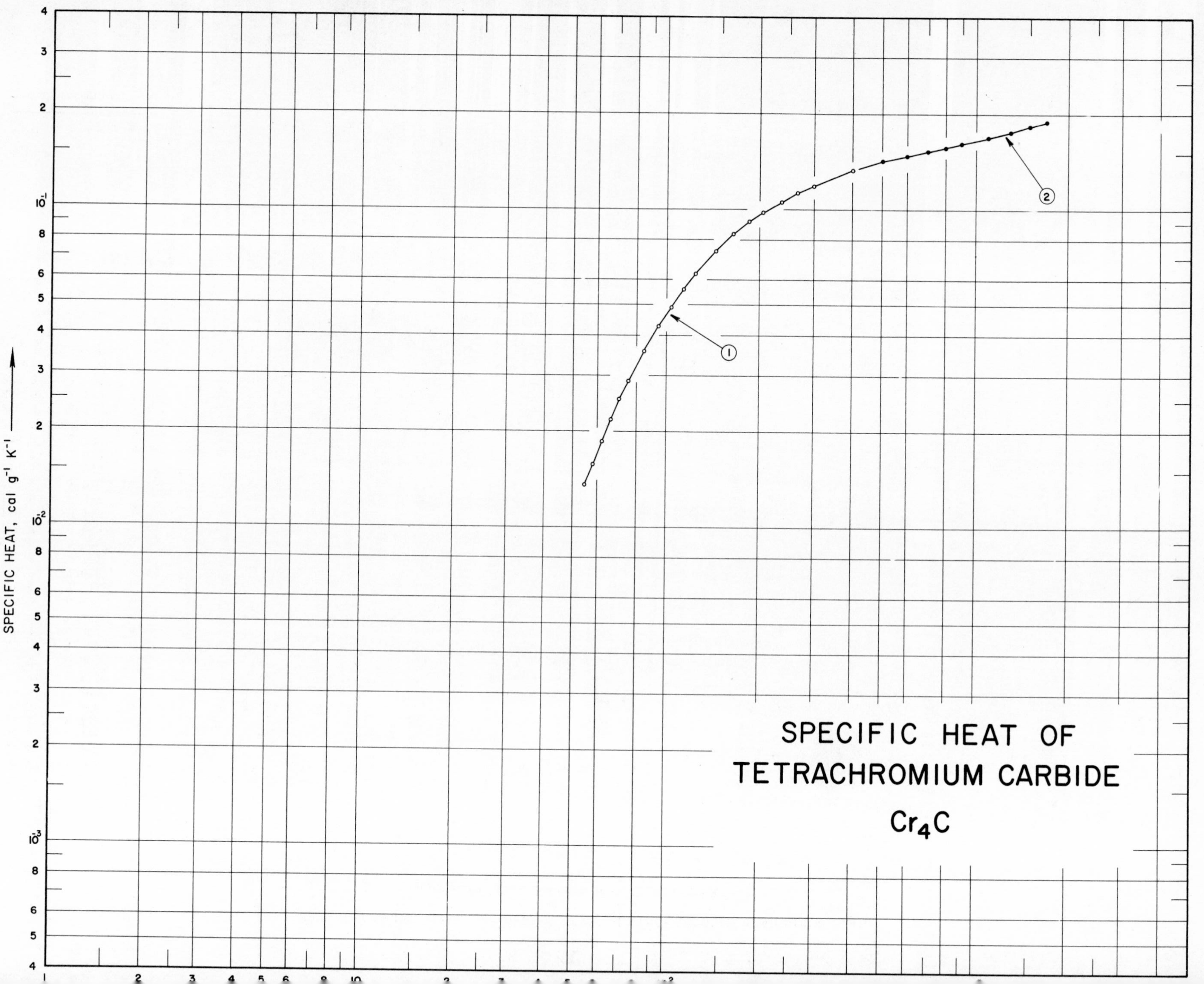
SPECIFIC HEAT OF
TETRACHROMIUM CARBIDE
Cr_4C
SPECIFIC HEAT, cal g^{-1} K^{-1}
1
2

SPECIFICATION TABLE NO. 125 SPECIFIC HEAT OF TETRACHROMIUM CARBIDE Cr_4C

[For Data Reported in Figure and Table No. 125]

Curve No.	Ref. No.	Year	Temp. Range, K	Reported Error, %	Name and Specimen Designation	Composition (weight percent), Specifications and Remarks
1	61	1944	53-295			~100 Cr_4C.
2	61	1944	298-1700			~100 Cr_4C.

DATA TABLE NO. 125 SPECIFIC HEAT OF TETRACHROMIUM CARBIDE Cr_4C

[Temperature, T, K; Specific Heat, C_p, Cal g^{-1} K^{-1}]

T	C_p
CURVE 1	
54.6	1.351 x 10^{-2}
58.1	1.583
62.2	1.866
66.5	2.192
70.7	2.521
75.5	2.885
80.3	3.259*
84.5	3.575
93.9	4.273
103.1	4.926
113.6	5.626
124.0	6.276
134.6	6.889*
144.6	7.398
154.4	7.871*
164.9	8.334
174.7	8.712*
185.0	9.121
195.1	9.461*
205.4	9.761
215.2	1.003 x 10^{-1}*
225.5	1.031*
236.1	1.057
245.6	1.082*
255.3	1.103*
266.1	1.126
276.5	1.150*
285.8	1.162*
295.0	1.170*
CURVE 2	
298.15	1.178 x 10^{-1}*
300	1.181
400	1.326
500	1.411
600	1.472
700	1.523
800	1.567
900	1.608
1000	1.647*
1100	1.685
1200	1.721*
CURVE 2 (cont.)	
1300	1.757 x 10^{-1}
1400	1.793*
1500	1.828
1600	1.863*
1700	1.897

*Not shown on plot

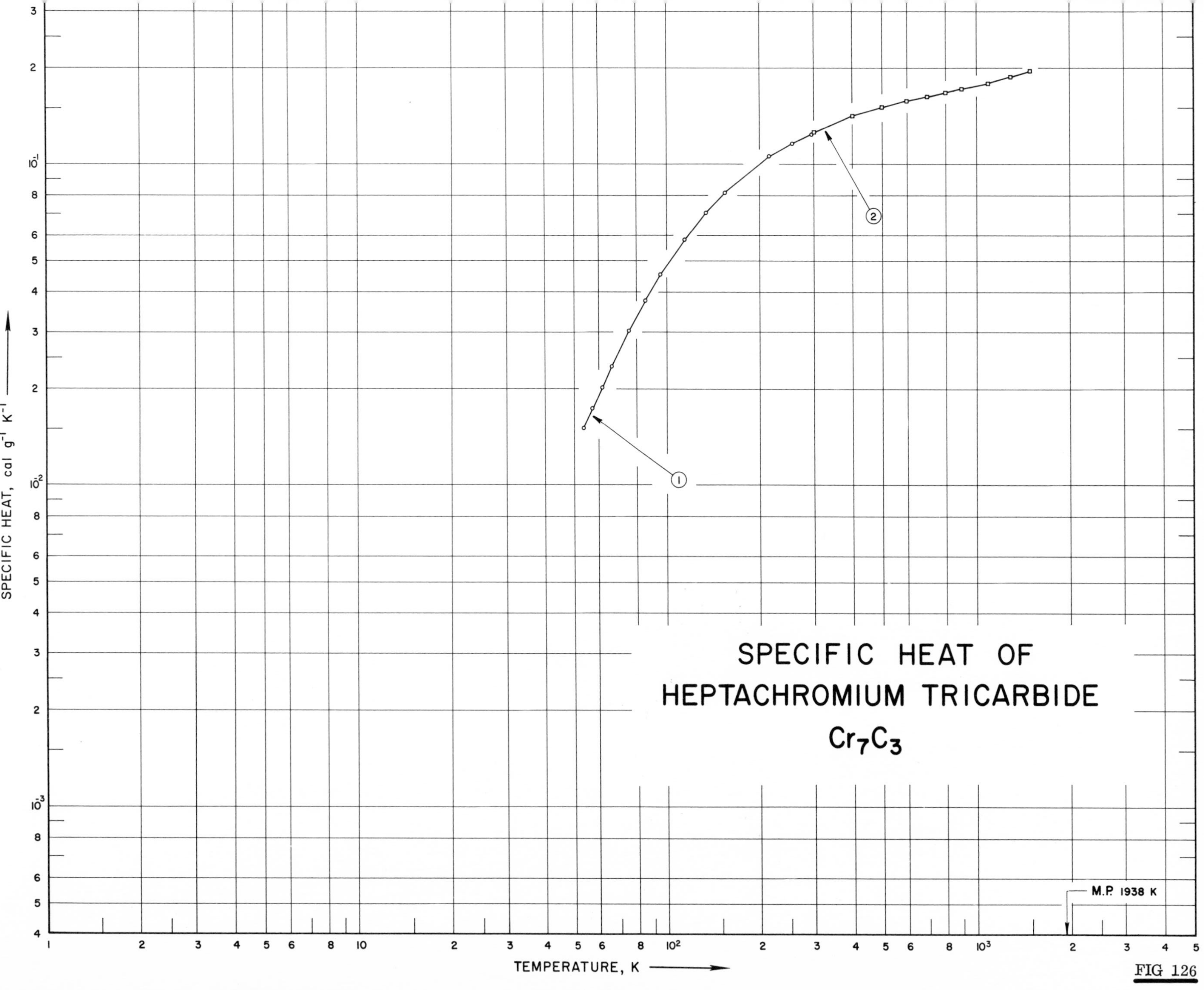
SPECIFIC HEAT OF
HEPTACHROMIUM TRICARBIDE
Cr_7C_3
M.P. 1938 K
SPECIFIC HEAT, cal g⁻¹ K⁻¹
TEMPERATURE, K
1
2

FIG 126

SPECIFICATION TABLE NO. 126 SPECIFIC HEAT OF HEPTACHROMIUM TRICARBIDE Cr_7C_3

[For Data Reported in Figure and Table No. 126]

Curve No.	Ref. No.	Year	Temp. Range, K	Reported Error, %	Name and Specimen Designation	Composition (weight percent), Specifications and Remarks
1	61	1944	53-295			~100 Cr_7C_3.
2	61	1944	298-1500	max. dev. 0.6 av. dev. 0.2		~100 Cr_7C_3.

DATA TABLE NO. 126 SPECIFIC HEAT OF HEPTACHROMIUM TRICARBIDE Cr_7C_3

[Temperature, T, K; Specific Heat, C_p, Cal g^{-1} K^{-1}]

T	C_p
CURVE 1	
53.8	1.509 x 10^{-2}
57.4	1.740
61.7	2.023
66.4	2.369
71.0	2.702
75.6	3.032
81.1	3.439*
85.8	3.789
95.6	4.501
105.4	5.171*
115.2	5.829
125.2	6.468*
135.2	7.061
145.7	7.631*
156.1	8.166
165.9	8.638*
176.3	9.080
185.9	9.490*
196.5	9.888*
206.3	1.021 x 10^{-1}*
216.5	1.053
226.2	1.084*
236.7	1.112*
246.5	1.140*
256.4	1.164
266.0	1.188*
276.6	1.212*
286.0	1.229*
295.3	1.243
CURVE 2	
298	1.247 x 10^{-1}*
300	1.252
400	1.411
500	1.503
600	1.570
700	1.625
800	1.673
900	1.717
1000	1.759*
1100	1.799
CURVE 2 (cont.)	
1200	1.838 x 10^{-1}*
1300	1.877
1400	1.915*
1500	1.953

*Not shown on plot

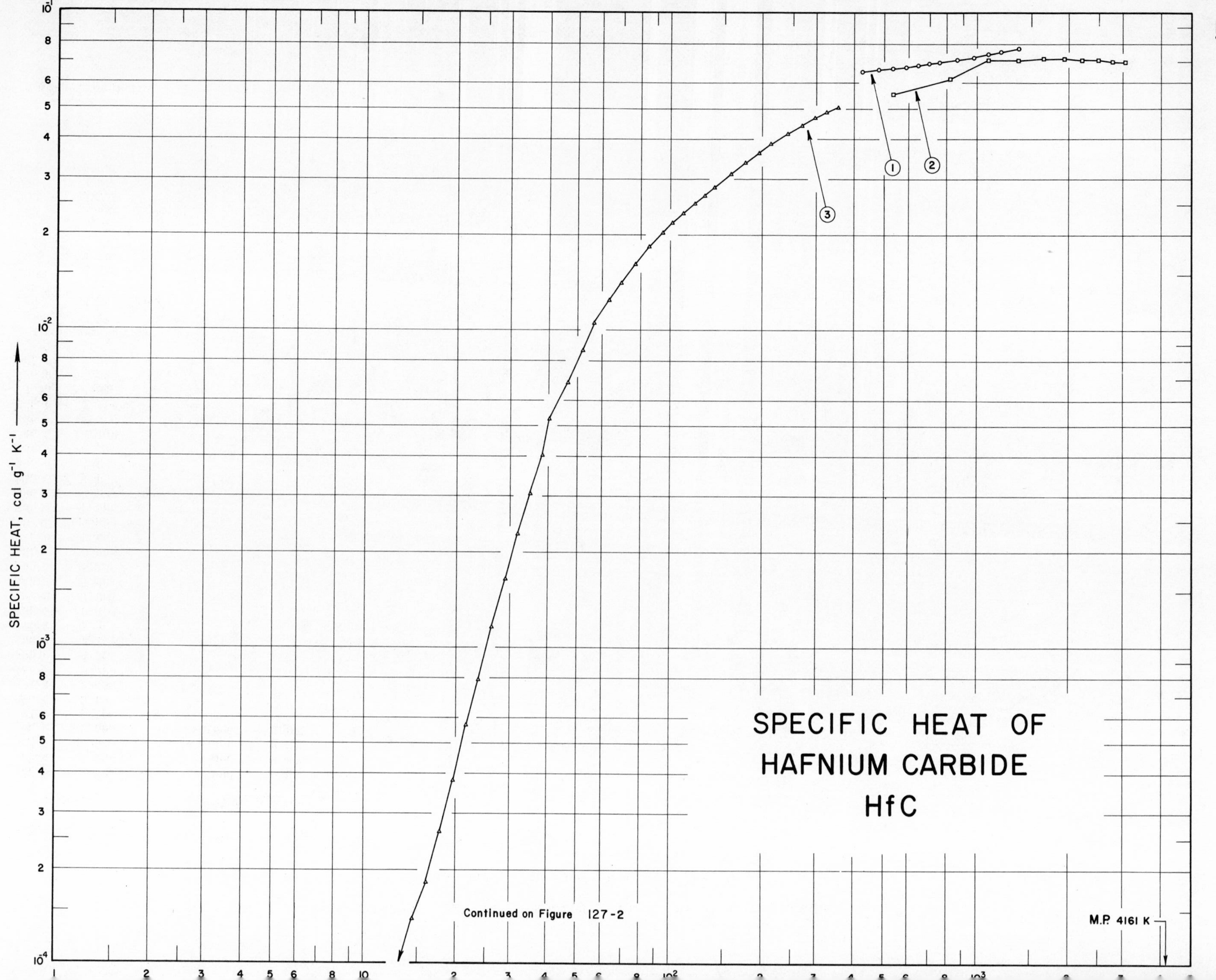

SPECIFIC HEAT OF
HAFNIUM CARBIDE
HfC
SPECIFIC HEAT, cal g⁻¹ K⁻¹
Continued on Figure 127-2
M.P. 4161 K
1
2
3

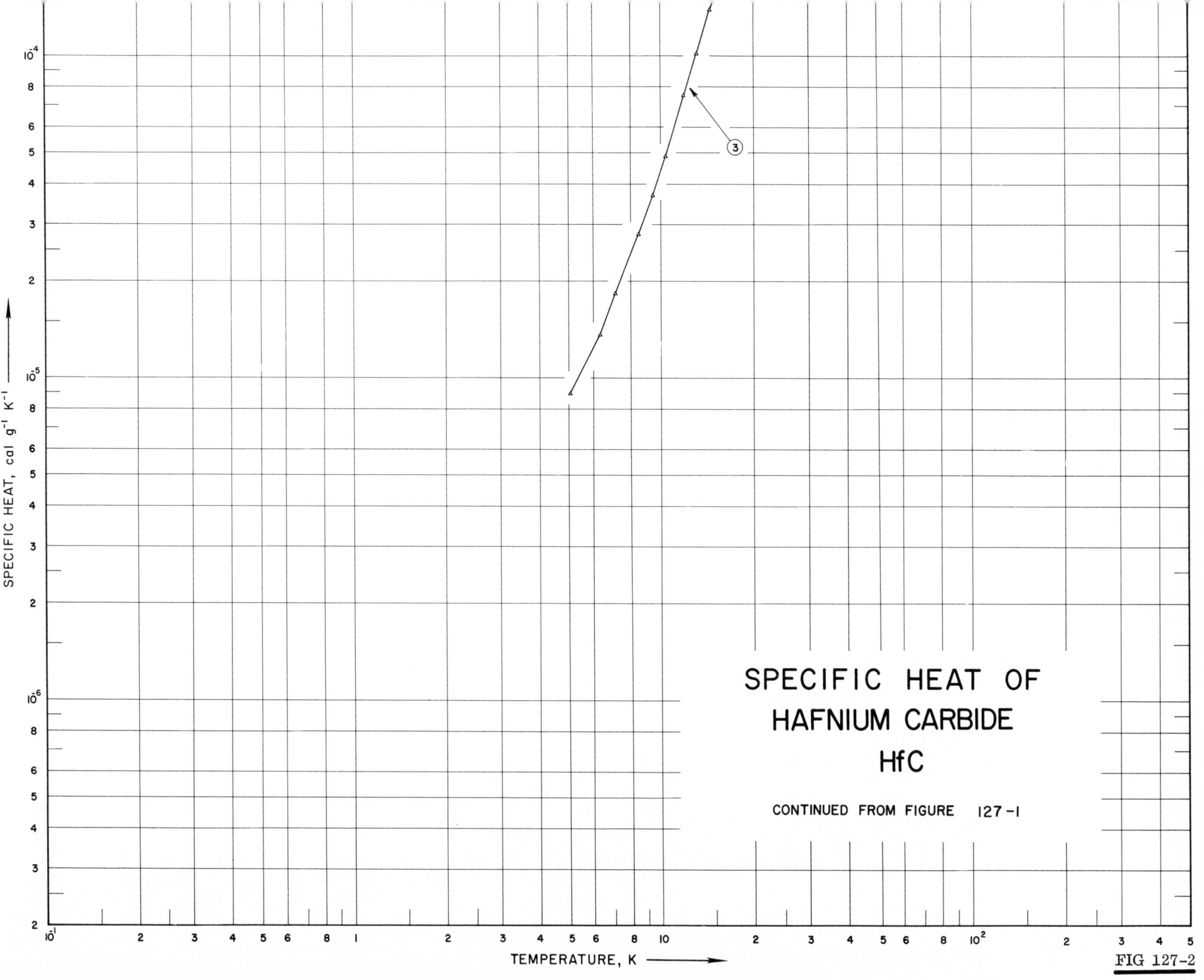

SPECIFIC HEAT OF
HAFNIUM CARBIDE
HfC
CONTINUED FROM FIGURE 127-1
SPECIFIC HEAT, cal g⁻¹ K⁻¹
TEMPERATURE, K
3

FIG 127-2

SPECIFICATION TABLE NO. 127 SPECIFIC HEAT OF HAFNIUM CARBIDE HfC

[For Data Reported in Figure and Table No. 127]

Curve No.	Ref. No.	Year	Temp. Range, K	Reported Error, %	Name and Specimen Designation	Composition (weight percent), Specifications and Remarks
1	76	1961	422-1366	5. 0		Before spraying 79. 8 Hf, 5. 38 C; sample made from powder HfC by spraying with powder gun using nitrogen-hydrogen plasma gas and 8 ft^3 hr^{-1} nitrogen carrier gas.
2	32	1962	533-3033	≤5		Before exposure: 93. 8 Hf, 5. 85 C, 1. 0 N, 0. 3 Ti, and 0. 2 Al, after exposure 93. 9 Hf, 5. 73 C, 0. 9 N; sample supplied by the Carborundum Co; hot pressed; density at 25 C, before exposure: apparent density (ASTM method B311-58) 700 lb ft^{-3}, true density (by immersion in xylene) 750 lb ft^{-3}, after exposure: apparent density 700 lb ft^{-3}, true density 736 lb ft^{-3}.
3	165	1964	5-350			Bal. Hf, 6. 12 C, 0. 035 Zr, 0. 031 N, 0. 005 Fe, 0. 003 O, 0. 002 Ti, Si, 0. 001 H, Cu, and Mg; zone-refined.

DATA TABLE NO. 127 SPECIFIC HEAT OF HAFNIUM CARBIDE HfC

[Temperature, T, K; Specific Heat, C_p, Cal g^{-1} K^{-1}

T	C_p
CURVE 1	
422.03	6.540×10^{-2}
477.59	6.620
533.15	6.690
588.70	6.760
644.26	6.830
699.81	6.900
755.37	6.970
810.92	7.040*
866.48	7.110
922.03	7.190*
977.59	7.260
1033.15	7.330*
1088.70	7.400
1144.26	7.470*
1199.81	7.540
1255.37	7.610*
1310.92	7.680*
1366.78	7.760
CURVE 2	
533.15	5.500×10^{-2}
810.92	6.700
1088.70	7.100
1366.48	7.100
1644.26	7.200
1922.03	7.200
2199.81	7.100
2477.59	7.100
2755.37	7.000
3033.15	7.000
CURVE 3	
5.09	8.924×10^{-6}
6.39	1.365×10^{-5}
7.16	1.837
8.50	2.782
9.50	3.675
10.50	4.882
12.09	7.507
13.25	1.018×10^{-4}
14.59	1.391
CURVE 3 (cont.)	
16.15	1.801×10^{-4}
17.89	2.662
19.75	3.774
21.75	5.664
23.81	7.880
26.36	1.152×10^{-3}
29.03	1.636
31.88	2.266
34.91	3.041
38.29	4.009
42.24	5.256
46.93	6.823
52.22	8.615
57.97	1.052×10^{-2}
63.77	1.270
69.85	1.418
77.17	1.616
85.67	1.831
94.77	2.024
95.28	2.034*
102.52	2.175
111.39	2.341
120.68	2.506
129.96	2.664
139.20	2.819
148.56	2.967*
157.91	3.112
167.14	3.249*
176.27	3.382
185.46	3.509*
194.79	3.633
203.96	3.751*
212.90	3.863
221.97	3.970*
231.05	4.074*
240.11	7.172
249.29	4.265*
258.53	4.359*
267.88	4.441
277.07	4.525*
286.20	4.606*
295.35	4.678
304.47	4.748*
CURVE 3 (cont.)	
313.74	4.815×10^{-2}*
322.37	4.873
331.16	4.931*
339.68	4.982*
346.90	5.030*
350.00	5.046

*Not shown on plot

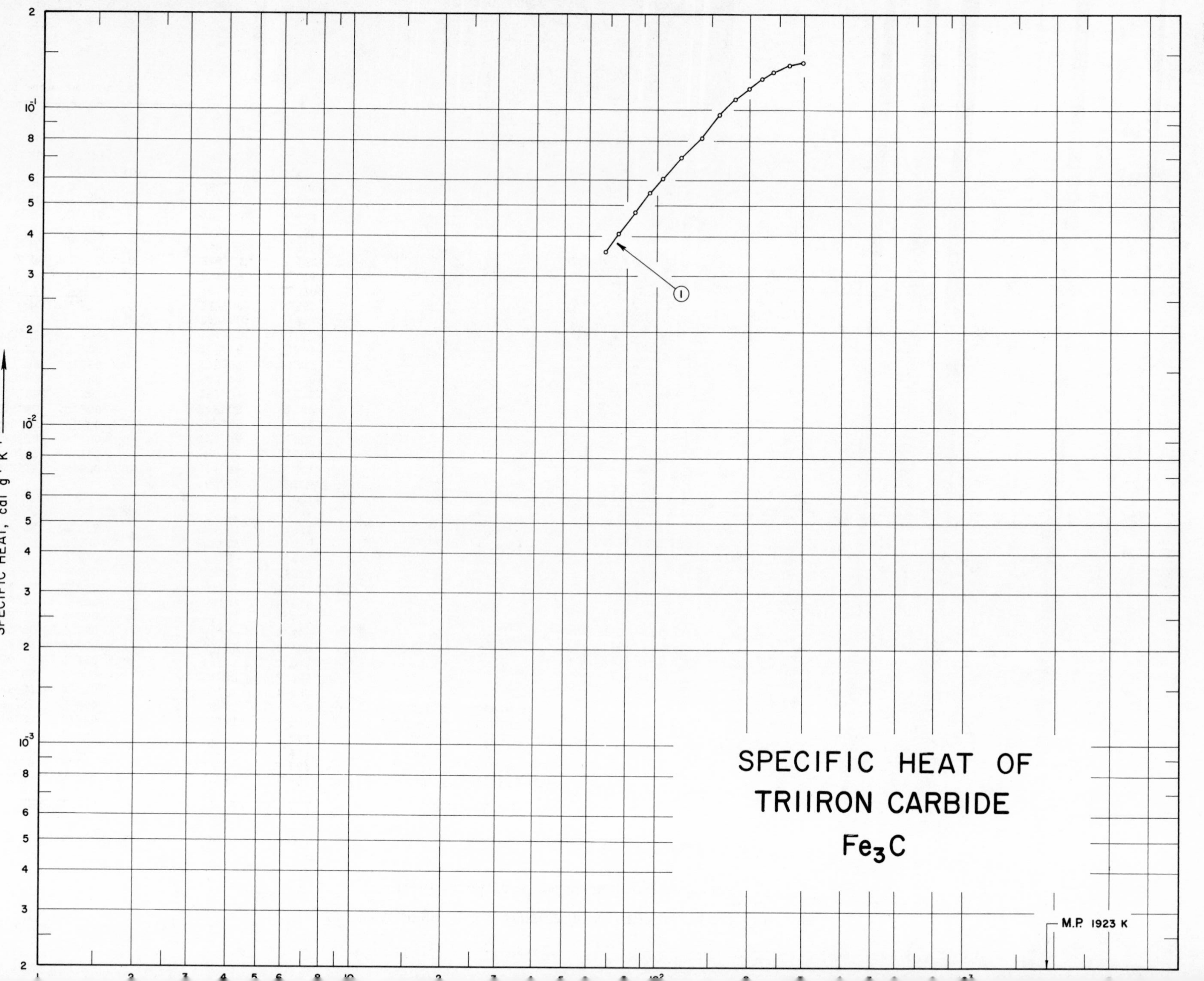
SPECIFIC HEAT, cal g⁻¹ K⁻¹
1
SPECIFIC HEAT OF
TRIIRON CARBIDE
Fe3C
M.P. 1923 K

SPECIFICATION TABLE NO. 128 SPECIFIC HEAT OF TRIIRON CARBIDE Fe_3C

[For Data Reported in Figure and Table No. 128]

Curve No.	Ref. No.	Year	Temp. Range, K	Reported Error, %	Name and Specimen Designation	Composition (weight percent), Specifications and Remarks
1	181	1939	68-298	2		

DATA TABLE NO. 128 SPECIFIC HEAT OF TRIIRON CARBIDE Fe_3C

[Temperature, T, K; Specific Heat, C_p, Cal g^{-1} K^{-1}]

T	C_p
CURVE 1	
68.00	3.570 x 10^{-2}
75.00	4.060
80.00	4.450*
85.00	4.751
90.00	5.135*
95.00	5.492
100.00	5.848*
105.00	6.099
110.00	6.450*
120.00	7.007
130.00	7.491
140.00	8.149
150.00	8.867*
160.00	9.608
170.00	1.029 x 10^{-1}*
180.00	1.087
190.00	1.133*
200.00	1.172
210.00	1.208*
220.00	1.252
230.00	1.282*
240.00	1.317
250.00	1.345*
260.00	1.372*
270.00	1.386
280.00	1.400*
290.00	1.405*
298.00	1.411

*Not shown on plot

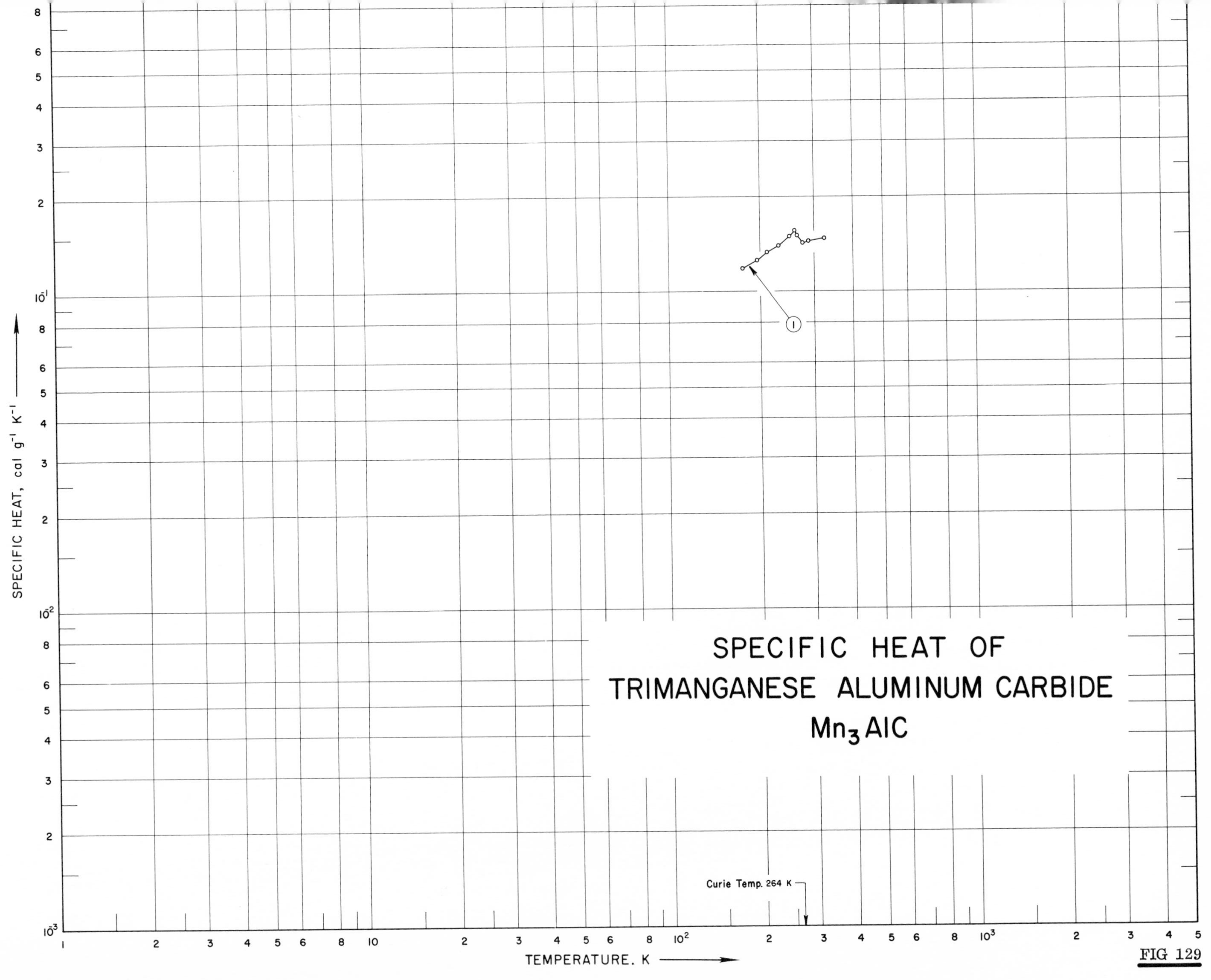
SPECIFIC HEAT OF
TRIMANGANESE ALUMINUM CARBIDE
Mn_3AlC
Curie Temp. 264 K
1
SPECIFIC HEAT, cal g⁻¹ K⁻¹
TEMPERATURE, K

FIG 129

SPECIFICATION TABLE NO. 129 SPECIFIC HEAT OF TRIMANGANESE ALUMINUM CARBIDE Mn_3AlC

[For Data Reported in Figure and Table No. 129]

Curve No.	Ref. No.	Year	Temp. Range, K	Reported Error, %	Name and Specimen Designation	Composition (weight percent), Specifications and Remarks
1	182	1962	176-325	0. 5		>99. 9 Mn_3AlC; prepared from >99. 9% purity materials.

DATA TABLE NO. 129 SPECIFIC HEAT OF TRIMANGANESE ALUMINUM CARBIDE Mn_3AlC

[Temperature, T, K; Specific Heat, C_p, Cal g^{-1} K^{-1}]

T	C_p
CURVE 1	
176.15	1.187 x 10^{-1}
186.15	1.222*
197.15	1.257
211.15	1.335
216.15	1.338*
230.15	1.391
242.15	1.448*
250.65	1.496
254.15	1.526*
258.15	1.548*
260.15	1.548*
262.15	1.555
266.15	1.511
271.15	1.448*
272.15	1.430*
277.15	1.421
279.15	1.426*
289.15	1.440
305.15	1.458*
315.15	1.461*
325.15	1.475

*Not shown on plot

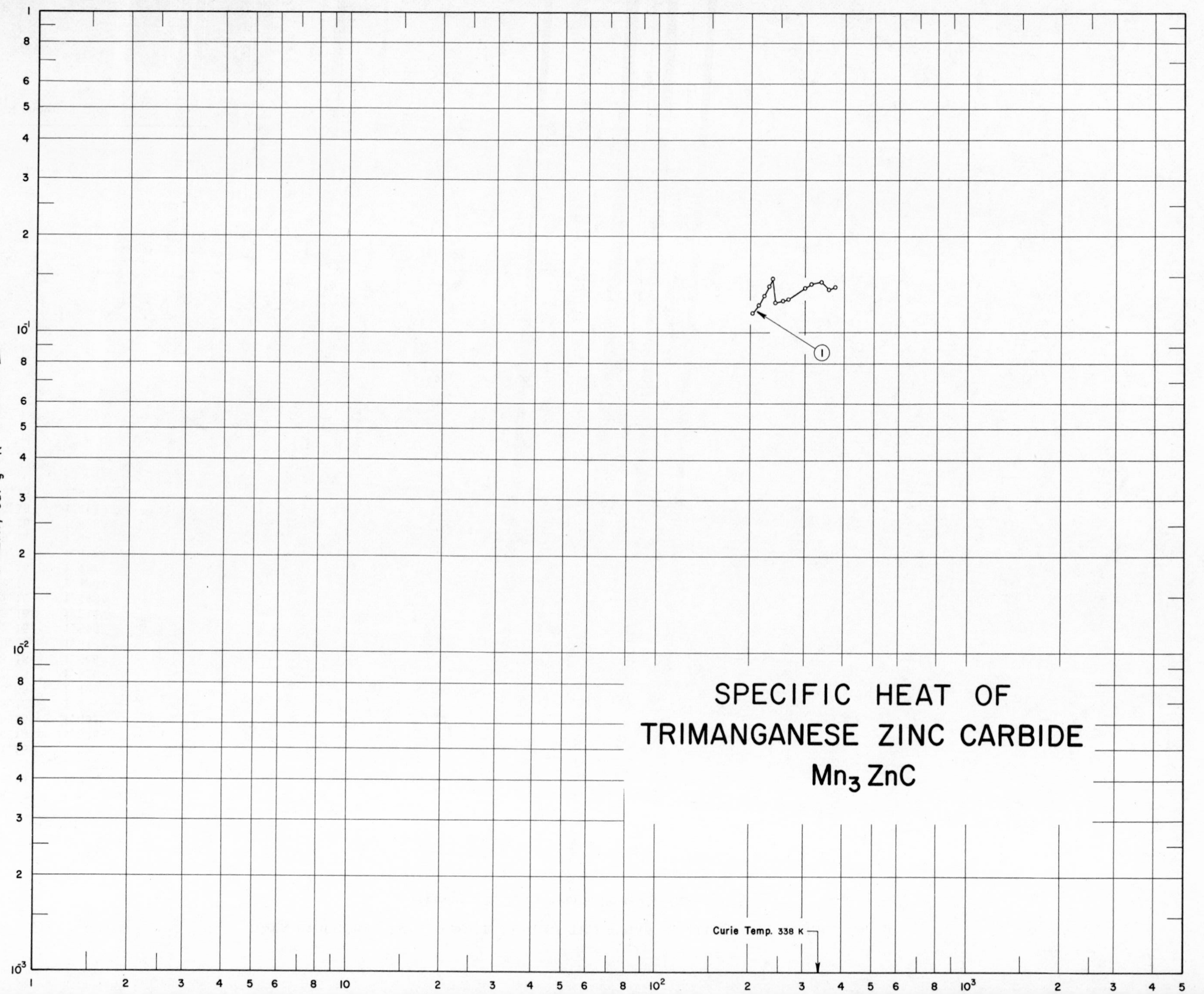
SPECIFIC HEAT OF
TRIMANGANESE ZINC CARBIDE
Mn3 ZnC
Curie Temp. 338 K
1
SPECIFIC HEAT, cal g-1 K-1

SPECIFICATION TABLE NO. 130 SPECIFIC HEAT OF TRIMANGANESE ZINC CARBIDE Mn_3ZnC

[For Data Reported in Figure and Table No. 130]

Curve No.	Ref. No.	Year	Temp. Range, K	Reported Error, %	Name and Specimen Designation	Composition (weight percent), Specifications and Remarks
1	182	1962	204-375	0.5		>99.9 Mn_3ZnC; prepared from >99.9% purity materials.

DATA TABLE NO. 130 SPECIFIC HEAT OF TRIMANGANESE ZINC CARBIDE Mn_3ZrC

[Temperature, T, K; Specific Heat, C_p, Cal g^{-1} K^{-1}]

T	C_p
CURVE 1	
204	1.160 x 10^{-1}
214	1.218
218	1.243*
223	1.301
225	1.313*
225	1.325*
231	1.400
234	1.425*
234	1.445*
242	1.247
246	1.247*
255	1.264
266	1.276
302	1.383
305	1.379*
315	1.420
323	1.420*
330	1.445*
333	1.449*
340	1.449
343	1.445*
347	1.408*
347	1.396*
353	1.387*
355	1.375*
358	1.375
359	1.387*
367	1.375*
367	1.400*
375	1.396

*Not shown on plot

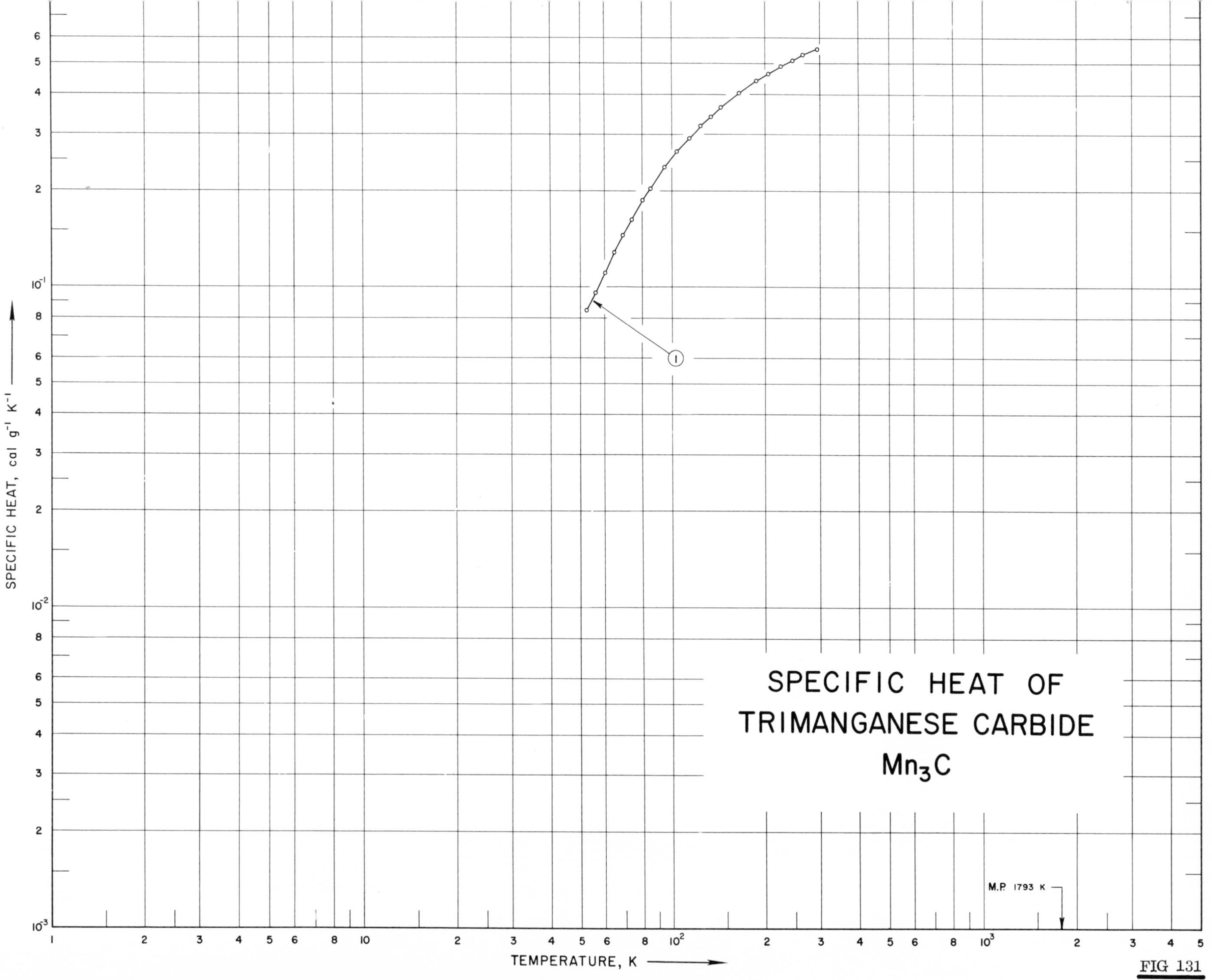
SPECIFIC HEAT OF
TRIMANGANESE CARBIDE
Mn_3C
M.P. 1793 K
1
SPECIFIC HEAT, cal g^{-1} K^{-1}
TEMPERATURE, K

FIG 131

SPECIFICATION TABLE NO. 131 SPECIFIC HEAT OF TRIMANGANESE CARBIDE Mn_3C

[For Data Reported in Figure and Table No. 131]

Curve No.	Ref. No.	Year	Temp. Range, K	Reported Error, %	Name and Specimen Designation	Composition (weight percent), Specifications and Remarks
1	183	1943	52-295			98. 8 Mn_3C; 1. 2 Mn not in separate phase, 93. 15 Mn, 6. 71 C (theor. 93. 21 Mn and 6. 79 C), <0. 02 H_2 and 0. 015 inorganic residue; prepared by heating electrolytic Mn and high purity C for 72 hrs at 850 K in vacuum, corrected for excess Mn.

DATA TABLE NO. 131 SPECIFIC HEAT OF TRIMANGANESE CARBIDE Mn_3C

[Temperature, T, K; Specific Heat, C_p, Cal g^{-1} K^{-1}]

T	C_p
CURVE 1	
52.60	8.464×10^{-2}
56.00	9.594
60.30	1.111×10^{-1}
64.80	1.283
69.20	1.454
74.10	1.639
80.60	1.874
85.10	2.035
94.70	2.372
104.30	2.652
114.20	2.921
124.70	3.198
134.40	3.415
144.80	3.650
155.20	3.857*
165.10	4.041
175.40	4.211*
188.60	4.423
195.80	4.528*
205.50	4.657
215.70	4.782*
225.70	4.909
235.50	5.004*
245.30	5.119
255.50	5.219*
265.50	5.316
275.60	5.406*
285.30	5.473*
295.20	5.551

*Not shown on plot

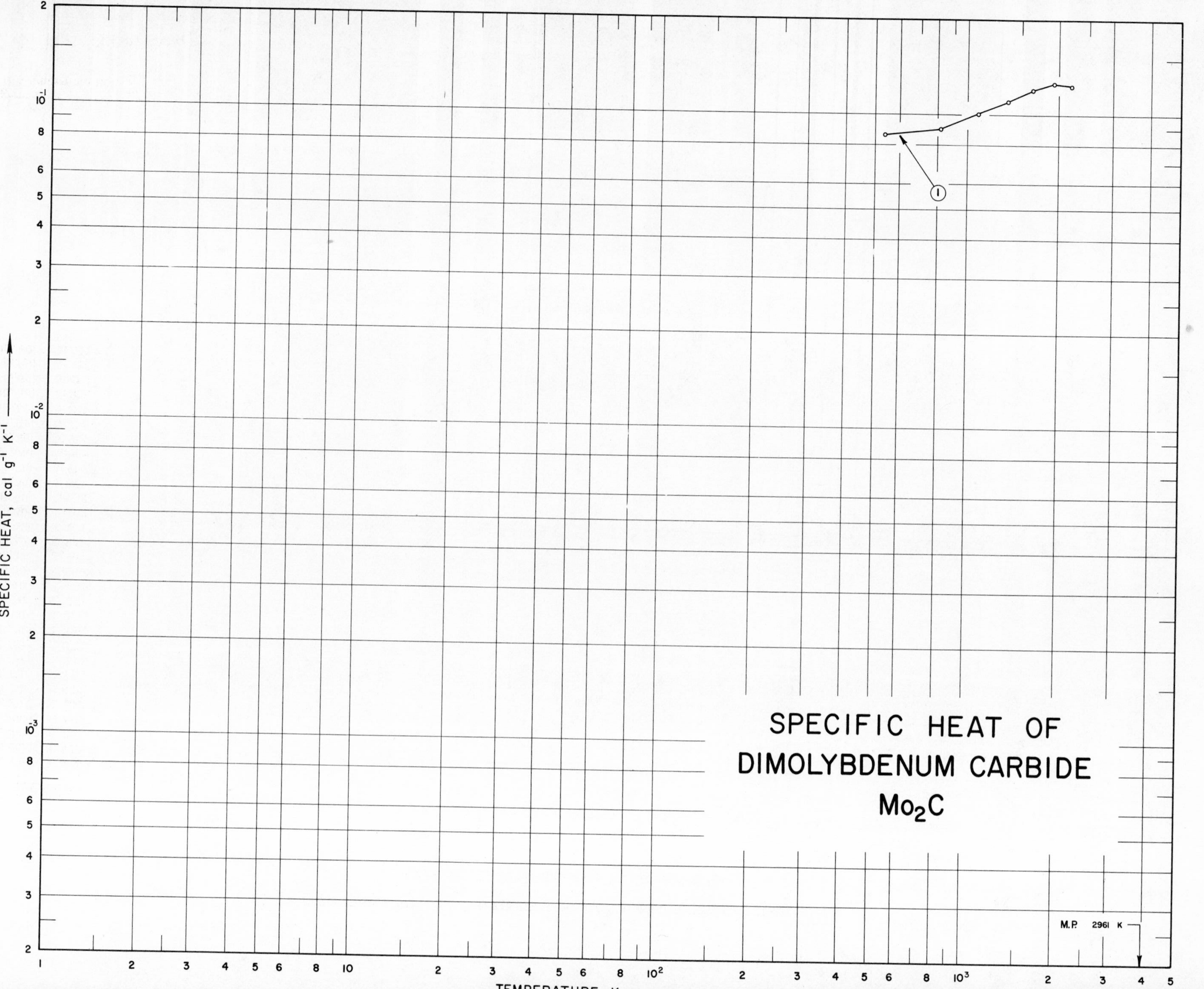

SPECIFIC HEAT OF
DIMOLYBDENUM CARBIDE
Mo_2C
M.P. 2961 K
1
SPECIFIC HEAT, cal g⁻¹ K⁻¹
TEMPERATURE, K

SPECIFICATION TABLE NO. 132 SPECIFIC HEAT OF DIMOLYBDENUM CARBIDE Mo_2C

[For Data Reported in Figure and Table No. 132]

Curve No.	Ref. No.	Year	Temp. Range, K	Reported Error, %	Name and Specimen Designation	Composition (weight percent), Specifications and Remarks
1	48	1962	533-2200	≤5		Before exposure: 92. 1 Mo, 5. 51 C, 2. 0 Si, 0. 6 Fe, 0. 3 Ti, 0. 2 Al, and <0. 1 N, after exposure: 92. 4 Mo, 5. 39 C, <0. 1 N; sample supplied by the Carborundum Co. ; crushed in hardened steel mortar to pass 100-mesh screen; hot pressed; density at 25 C, before exposure: apparent density (ASTM method B311-58) 554 lb ft^{-3}, true density (by immersion in xylene) 560 lb ft^{-3}, after exposure: apparent density 535 lb ft^{-3}, true density 542 lb ft^{-3}.

DATA TABLE NO. 132 SPECIFIC HEAT OF DIMOLYBDENUM CARBIDE Mo_2C

[Temperature, T, K; Specific Heat, C_p, Cal g^{-1} K^{-1}]

T	C_p
CURVE 1	
533.15	8.600×10^{-2}
810.93	9.000
1088.71	1.010×10^{-1}
1366.48	1.100
1644.26	1.200
1922.04	1.260
2199.82	1.240

*Not shown on plot

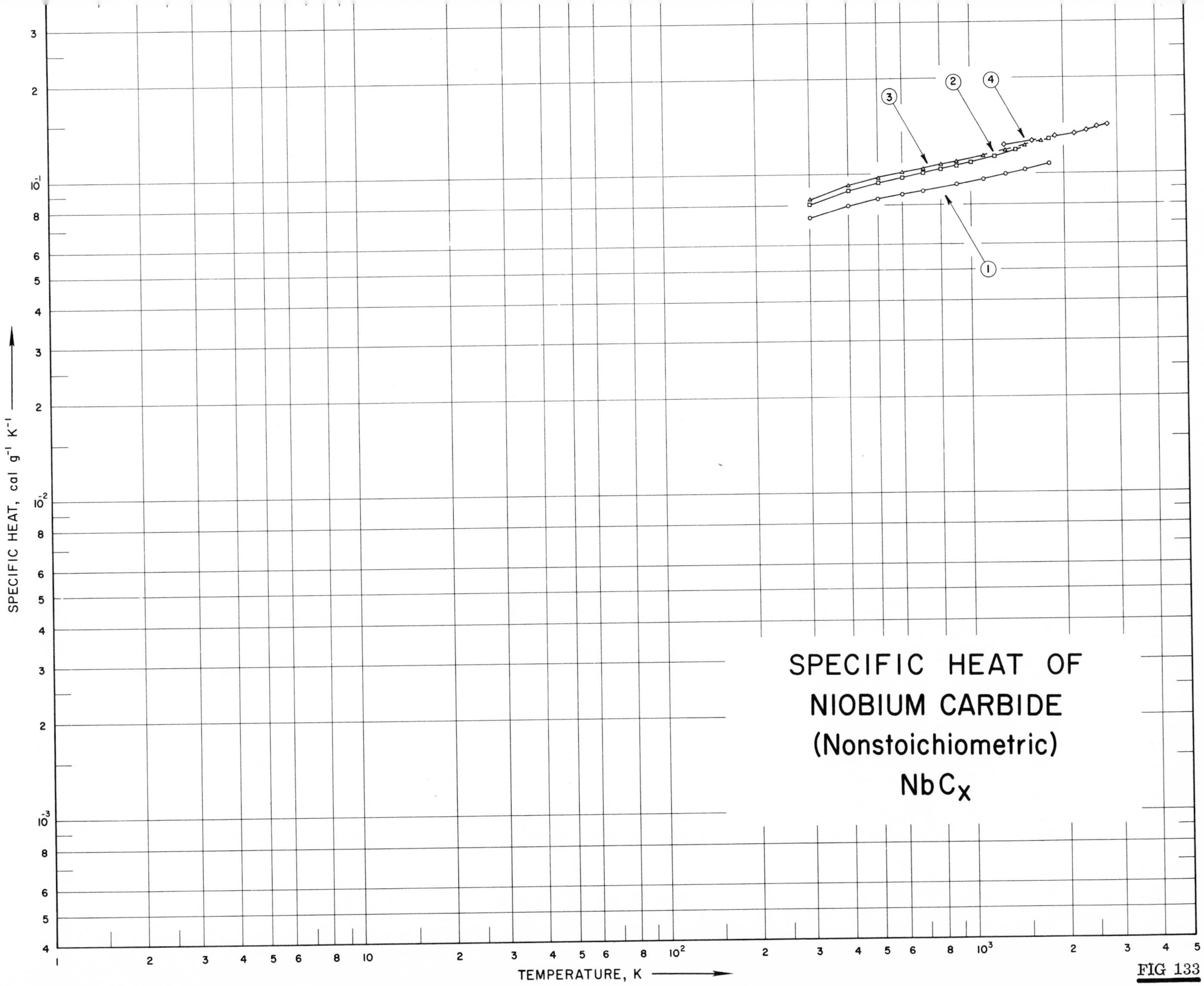

SPECIFIC HEAT OF NIOBIUM CARBIDE (Nonstoichiometric) NbC_x

SPECIFIC HEAT, cal g^{-1} K^{-1}

TEMPERATURE, K

1 2 3 4

FIG 133

SPECIFICATION TABLE NO. 133 SPECIFIC HEAT OF NIOBIUM CARBIDE (Nonstoichiometric) NbC_x

[For Data Reported in Figure and Table No. 133]

Curve No.	Ref. No.	Year	Temp. Range, K	Reported Error, %	Name and Specimen Designation	Composition (weight percent), Specifications and Remarks
1	112	1960	300-1800		$NbC_{0.50}$	
2	112	1960	300-1800		$NbC_{0.769}$	
3	112	1960	300-1800		$NbC_{0.847}$	
4	184	1963	1289-2778	1.3	$NbC_{0.97}$	88.78 Nb, 11.10 C, <0.05 free C, 0.10 W, 0.06 Fe, and <0.05 Ta; hot pressed.

DATA TABLE NO. 133 SPECIFIC HEAT OF NIOBIUM CARBIDE (Nonstoichiometric) NbC_x

[Temperature, T, K; Specific Heat, C_p, Cal g^{-1} K^{-1}]

T	C_p
CURVE 1	
300	7.330 x 10^{-2}
400	7.987
500	8.372
600	8.655
700	8.877
800	9.079*
900	9.261
1000	9.443*
1100	9.605
1200	9.777*
1300	9.939
1400	1.010 x 10^{-1}*
1500	1.025
1600	1.041*
1700	1.057*
1800	1.073
CURVE 2	
300	8.077 x 10^{-2}
400	8.892
500	9.392
600	9.765
700	1.008 x 10^{-1}
800	1.036
900	1.062
1000	1.087
1100	1.111*
1200	1.135
1300	1.159*
1400	1.182
1500	1.204*
1600	1.228*
1700	1.250*
1800	1.273
CURVE 3	
300	8.363 x 10^{-2}
400	9.266
500	9.789
600	1.016 x 10^{-1}
700	1.046
CURVE 3 (cont.)	
800	1.073 x 10^{-1}
900	1.096
1000	1.120*
1100	1.141
1200	1.162*
1300	1.184
1400	1.204*
1500	1.224
1600	1.245*
1700	1.264
1800	1.284*
CURVE 4	
1289	1.230 x 10^{-1}
1383	1.240*
1487	1.250*
1582	1.260
1675	1.270*
1789	1.290*
1887	1.300
1975	1.310*
2178	1.330
2283	1.350*
2374	1.360
2482	1.370*
2586	1.390
2682	1.400*
2778	1.410

*Not shown on plot

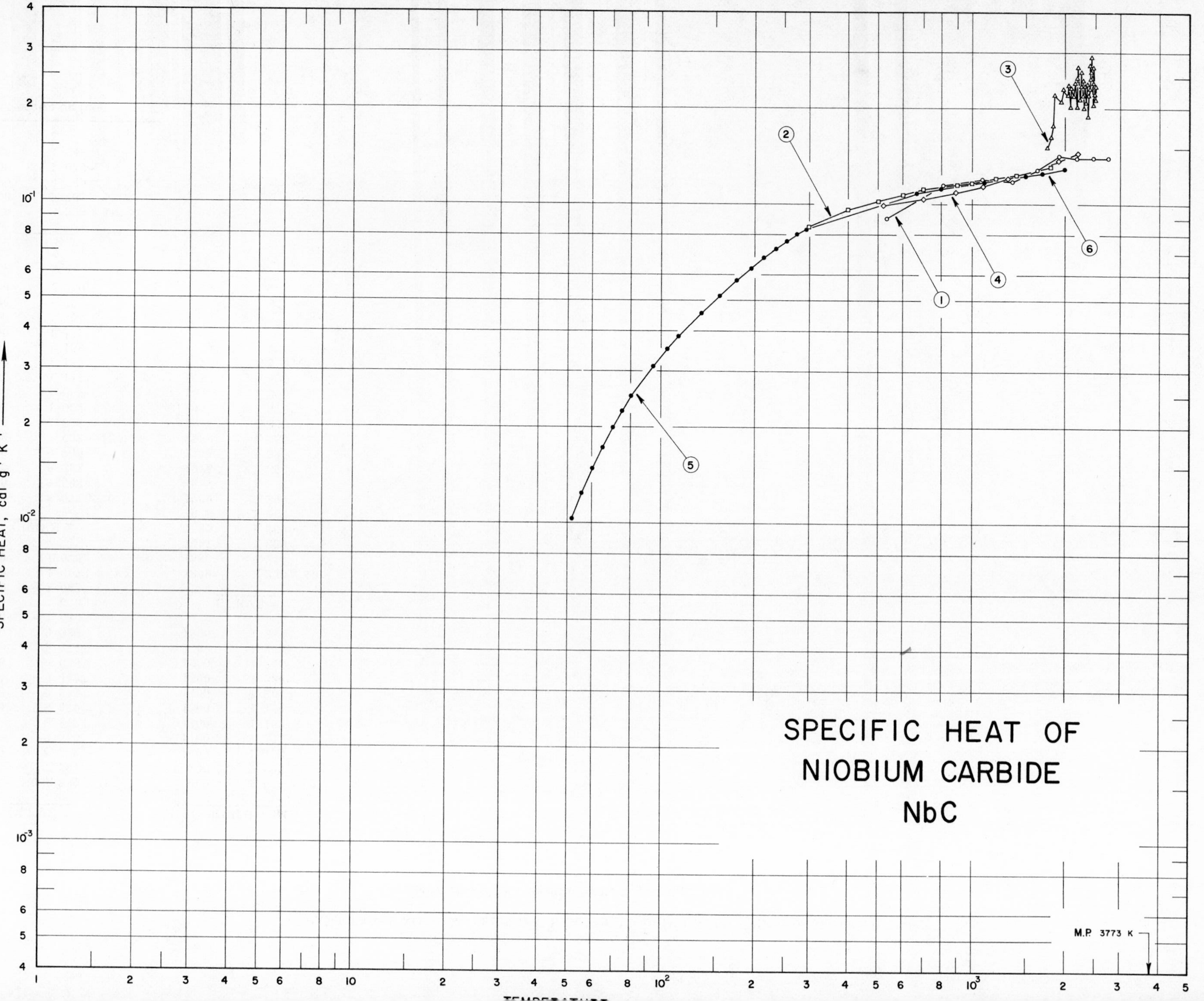

SPECIFIC HEAT OF
NIOBIUM CARBIDE
NbC
M.P. 3773 K
SPECIFIC HEAT, cal g⁻¹ K⁻¹

SPECIFICATION TABLE NO. 134 SPECIFIC HEAT OF NIOBIUM CARBIDE NbC

[For Data Reported in Figure and Table No. 134]

Curve No.	Ref. No.	Year	Temp. Range, K	Reported Error, %	Name and Specimen Designation	Composition (weight percent), Specifications and Remarks
1	32	1962	533-2755	≤5		>88.42 Nb, 11.3 C, 0.1 Fe, 0.10 W, 0.07 N, <0.01 Cr, Mg, Mn, Ni, Si, Sn, Ti, and Zr; sample supplied by Kennametal, Inc.; hot pressed; density 476 lb ft^{-3}.
2	112	1960	300-1800			
3	173	1962	1763-2529	5		86.66 Nb, 10.81 total C.
4	185	1962	522-2208			11.79 total C, 10.84 combine C, 0.94 free C; sprayed sample.
5	186	1964	51-296	0.1		88.17 Nb, 11.74 total C, 11.35 combined C, <0.05 N_2, 0.03 O_2, 0.02 Ti, 0.006 Ag, and 0.002 Mn.
6	186	1964	298-2000	0.1		Same as above.

DATA TABLE NO. 134 SPECIFIC HEAT OF NIOBIUM CARBIDE NbC

[Temperature, T, K; Specific Heat, C_p, Cal g^{-1} K^{-1}]

T	C_p
CURVE 1	
533.15	9.000 x 10^{-2}
810.93	1.150 x 10^{-1}
1088.71	1.200
1366.48	1.180
1644.26	1.280
1922.04	1.420
2199.82	1.400
2477.59	1.400
2755.37	1.400
CURVE 2	
300	8.503 x 10^{-2}
400	9.666
500	1.029 x 10^{-1}
600	1.070
700	1.102
800	1.128*
900	1.152
1000	1.172
1100	1.192*
1200	1.212
1300	1.231*
1400	1.249
1500	1.266*
1600	1.284*
1700	1.301*
1800	1.318
CURVE 3	
1763	1.510 x 10^{-1}
1817	1.630
1840	1.780
1868	2.210
1952	2.100
1961	2.080*
1970	2.300
1971	2.250*
1983	2.250*
2051	2.200
2053	2.370
CURVE 3 (cont.)	
2068	2.340 x 10^{-1}*
2077	2.320*
2091	2.020
2095	2.340
2115	2.360*
2141	2.190
2146	2.310*
2154	2.100*
2154	2.160*
2158	2.410
2193	2.030
2199	2.450
2206	2.700
2237	2.140
2237	2.240*
2263	2.360
2265	2.100*
2265	2.690
2278	2.090*
2286	2.590*
2293	2.410*
2298	2.450
2301	2.000
2308	1.990*
2339	2.360
2342	2.330*
2345	2.080
2346	2.040*
2351	1.990*
2360	2.380
2367	2.360*
2376	1.880
2376	2.200
2409	2.300
2414	2.220*
2419	2.480
2421	2.660*
2421	2.730
2425	2.590
2440	2.900
2451	2.580*
2453	2.680
CURVE 3 (cont.)	
2457	2.860 x 10^{-1}*
2464	2.400
2471	2.060
2475	2.280*
2475	2.060*
2477	2.090*
2488	2.230
2494	2.330*
2506	2.360*
2507	2.210*
2515	2.130
2518	2.350
2529	2.240*
CURVE 4	
522.59	9.927 x 10^{-2}
704.82	1.040 x 10^{-1}
890.37	1.091
1090.93	1.146
1315.93	1.207
1595.37	1.285*
1923.15	1.375
2208.15	1.454
CURVE 5	
51.96	1.043 x 10^{-2}
55.88	1.248
60.33	1.487
65.09	1.738
70.17	1.996
75.27	2.251
80.75	2.505*
84.80	2.686
94.71	3.104
105.01	3.505
114.55	3.860
124.45	4.201*
135.76	4.576
145.81	4.901*
155.67	5.199
CURVE 5 (cont.)	
166.05	5.505 x 10^{-2}*
176.03	5.793
185.95	6.059*
196.60	6.337
206.09	6.581*
216.35	6.828
226.10	7.067*
236.14	7.278
245.85	7.478*
256.45	7.685
266.20	7.858*
276.33	8.039
286.60	8.211*
296.11	8.368
CURVE 6	
298.15	8.371 x 10^{-2}*
300	8.404*
400	9.610*
500	1.024 x 10^{-1}*
600	1.064*
670	1.086
670	1.097*
700	1.105*
800	1.128
900	1.147*
1000	1.164*
1100	1.179
1200	1.194*
1300	1.207*
1400	1.221*
1500	1.233
1600	1.246*
1700	1.258
1800	1.270*
1900	1.282*
2000	1.294

* Not shown on plot

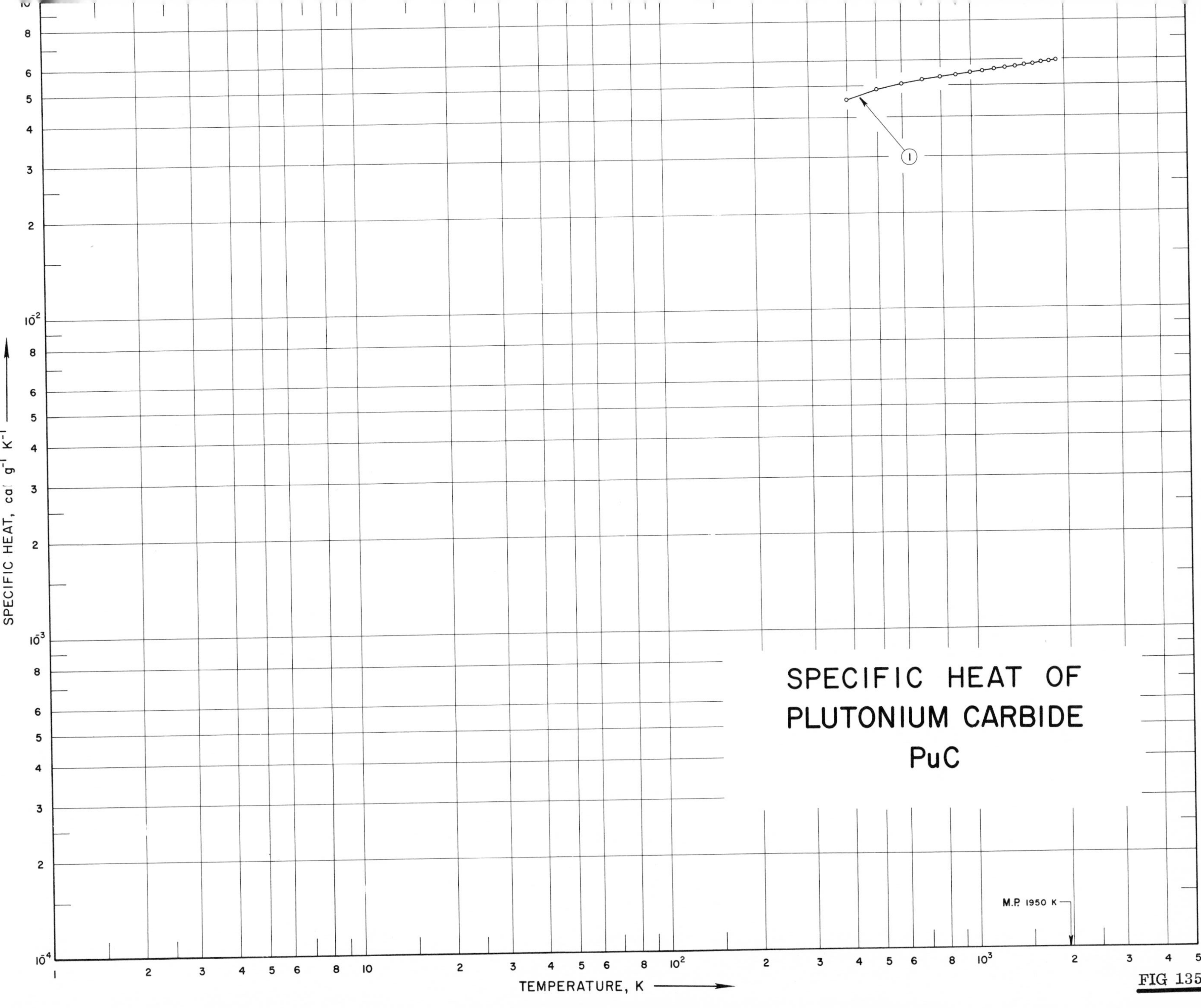
SPECIFIC HEAT OF
PLUTONIUM CARBIDE
PuC
M.P. 1950 K
TEMPERATURE, K
SPECIFIC HEAT, cal g^{-1} K^{-1}
1

FIG 135

SPECIFICATION TABLE NO. 135 SPECIFIC HEAT OF PLUTONIUM CARBIDE PuC

[For Data Reported in Figure and Table No. 135]

Curve No.	Ref. No.	Year	Temp. Range, K	Reported Error, %	Name and Specimen Designation	Composition (weight percent), Specifications and Remarks
1	187	1963	400-1900			Single phase stoichiometric PuC.

DATA TABLE NO. 135 SPECIFIC HEAT OF PLUTONIUM CARBIDE PuC

[Temperature, T, K; Specific Heat, C_p, Cal g^{-1} K^{-1}]

T	C_p
CURVE 1	
400	4.535 x 10^{-2}
500	4.866
600	5.059
700	5.205
800	5.311
900	5.398
1000	5.472
1100	5.539
1200	5.602
1300	5.661
1400	5.713
1500	5.772
1600	5.819
1700	5.874
1800	5.921
1900	5.969

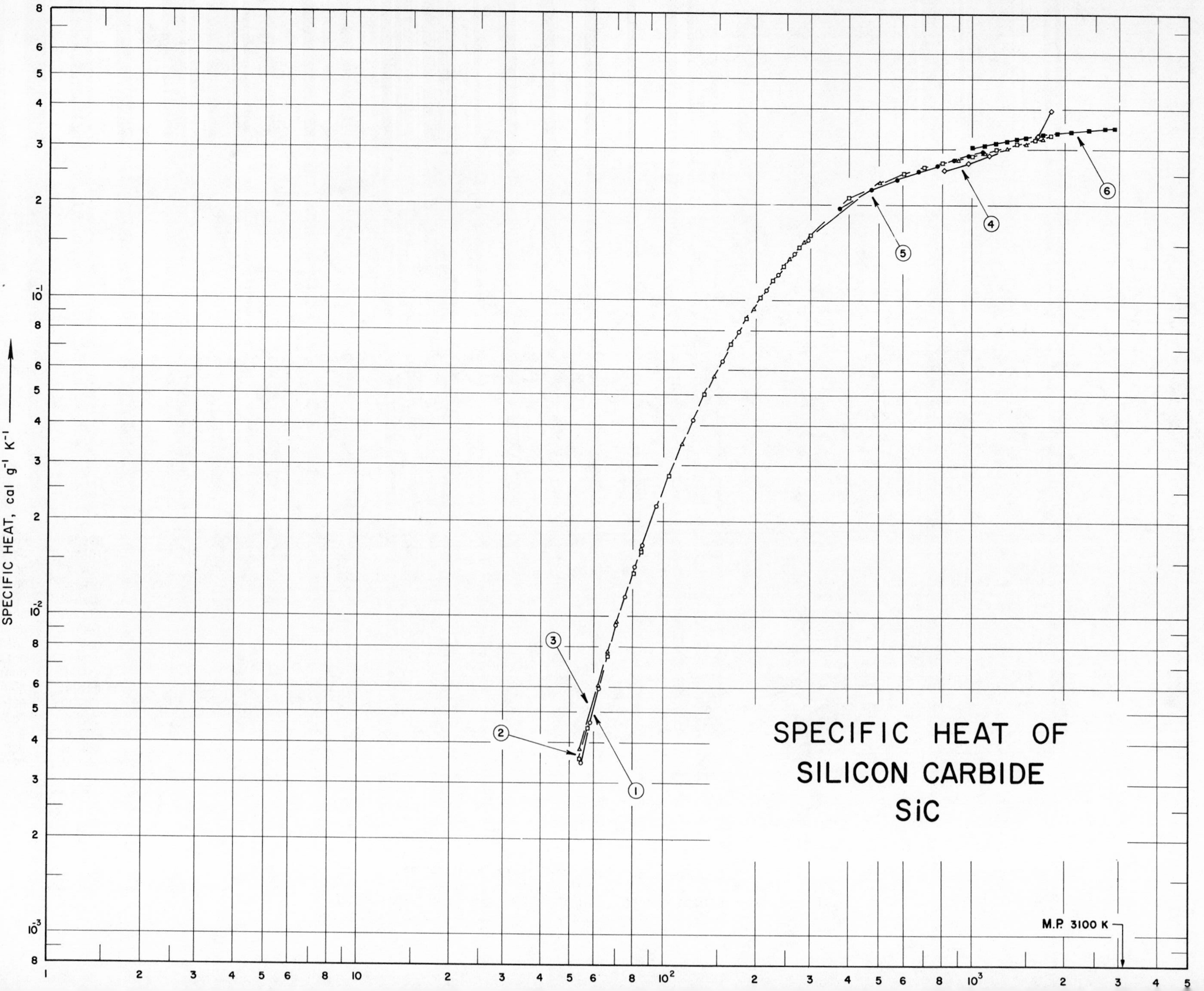
SPECIFIC HEAT OF
SILICON CARBIDE
SiC
M.P. 3100 K
SPECIFIC HEAT, cal g^{-1} K^{-1}
1
2
3
4
5
6

SPECIFICATION TABLE NO. 136 SPECIFIC HEAT OF SILICON CARBIDE SiC

[For Data Reported in Figure and Table No. 136]

Curve No.	Ref. No.	Year	Temp. Range, K	Reported Error, %	Name and Specimen Designation	Composition (weight percent), Specifications and Remarks
1	50	1941	54-294	0.5		99.0 SiC and 0.6 SiO_2; corrected for SiO_2 impurity.
2	188	1952	53-1800		hexagonal type II	99.73 SiC, 69.84 Si, 29.89 C, 0.18 Fe, 0.08 Al, and <0.01 Ca.
3	188	1952	53-1700		cubic	<1.0 hexagonal SiC, 0.34 free C, 0.17 SiO_2, 0.06 Al, 0.013 free Si, and 0.004 Fe.
4	189	1958	810-1810	3		As received: 67.46 Si, 28.58 C, 0.73 Al, 0.58 Fe, and 0.48 CaO, after test: 68.12 Si, 27.29 C, 1.47 Al, 0.44 CaO, and 0.32 Fe; density 193.4 lb ft^{-3}.
5	45	1962	303-1073	<3		96.5 SiC, 2.5 Si, 0.4 C, and 0.4 Al.
6	190	1964	1000-2900	~2		87.07 SiC, 12.0 free C, 0.73 Fe; measured in an argon atmosphere; density 27.9 g cm^{-3}.

DATA TABLE NO. 136 SPECIFIC HEAT OF SILICON CARBIDE SiC

[Temperature, T, K; Specific Heat, C_p, Cal g^{-1} K^{-1}]

CURVE 1

T	C_p
54.30	3.469 x 10^{-3}
58.20	4.642
62.10	5.940
66.20	7.562*
70.60	9.384
75.20	1.153 x 10^{-2}
80.70	1.430
85.40	1.682
94.90	2.231
104.60	2.830*
114.70	3.519*
124.70	4.220
135.10	4.974*
145.20	5.700*
155.40	6.431
165.70	7.195*
176.00	7.969
185.60	8.652*
195.80	9.379*
205.40	1.008 x 10^{-1}*
215.60	1.081
225.90	1.153*
235.80	1.218
245.60	1.283*
255.70	1.346*
265.70	1.402
276.10	1.474*
284.80	1.518*
294.60	1.568

CURVE 2

T	C_p
53.57	3.564 x 10^{-3}
57.39	4.462
61.72	5.835*
66.25	7.477
70.85	9.326*
75.55	1.140 x 10^{-2}*
80.46	1.378
84.98	1.614
95.07	2.197*
104.54	2.785
114.73	3.481*
124.76	4.170*

CURVE 2 (cont.)

T	C_p
136.16	5.006 x 10^{-2}
146.11	5.740*
155.86	6.454*
166.15	7.230
176.27	7.969*
186.21	8.707
196.13	9.426*
206.29	1.012 x 10^{-1}
216.27	1.085*
226.18	1.153
236.01	1.217*
245.80	1.283
256.06	1.347*
266.22	1.409*
275.99	1.467
286.20	1.529*
296.36	1.580*
298.15	1.594*
300.00	1.607
400.00	2.099
500.00	2.352*
600.00	2.512
700.00	2.627
800.00	2.719
900.00	2.797*
1000.00	2.866
1100.00	2.930*
1200.00	2.990
1300.00	3.047*
1400.00	3.102
1500.00	3.156*
1600.00	3.209
1700.00	3.261*
1800.00	3.312

CURVE 3

T	C_p
53.91	3.838 x 10^{-3}
57.58	4.684*
61.88	6.079
66.41	7.791
70.93	9.618
75.51	1.160 x 10^{-2}*
80.35	1.407

CURVE 3 (cont.)

T	C_p
84.72	1.641 x 10^{-2}*
95.13	2.237*
104.60	2.823*
114.73	3.521
124.64	4.208*
135.82	5.011*
146.09	5.762
155.89	6.481*
166.01	7.240*
175.89	7.974*
186.03	8.717*
195.90	9.433
206.08	1.015 x 10^{-1}*
216.20	1.087*
226.08	1.156*
236.12	1.223*
245.86	1.290*
256.27	1.353*
266.27	1.418*
276.14	1.477*
286.40	1.536
296.47	1.590*
298.17	1.602*
300.00	1.615*
400.00	2.102*
500.00	2.352
600.00	2.508*
700.00	2.621*
800.00	2.710*
900.00	2.785
1000.00	2.852*
1100.00	2.913
1200.00	2.970*
1300.00	3.025
1400.00	3.078*
1500.00	3.129
1600.00	3.179*
1700.00	3.229

CURVE 4

T	C_p
810.93	2.560 x 10^{-1}
977.59	2.710
1144.26	2.860

CURVE 4 (cont.)

T	C_p
1310.93	3.010 x 10^{-1}*
1477.59	3.160*
1644.26	3.310
1810.93	3.970

CURVE 5

T	C_p
303.15	1.63 x 10^{-1}*
373.15	1.96
473.15	2.24
573.15	2.42
673.15	2.55
773.15	2.67
873.15	2.77
973.15	2.87
1073.15	2.95

CURVE 6

T	C_p
1000	3.033 x 10^{-1}
1100	3.098
1200	3.152
1300	3.200
1400	3.241
1500	3.274
1600	3.305*
1700	3.332
1800	3.356*
1900	3.377
2000	3.396*
2100	3.413
2200	3.430*
2300	3.444*
2400	3.456
2500	3.468*
2600	3.480*
2700	3.489
2800	3.499*
2900	3.509

*Not shown on plot

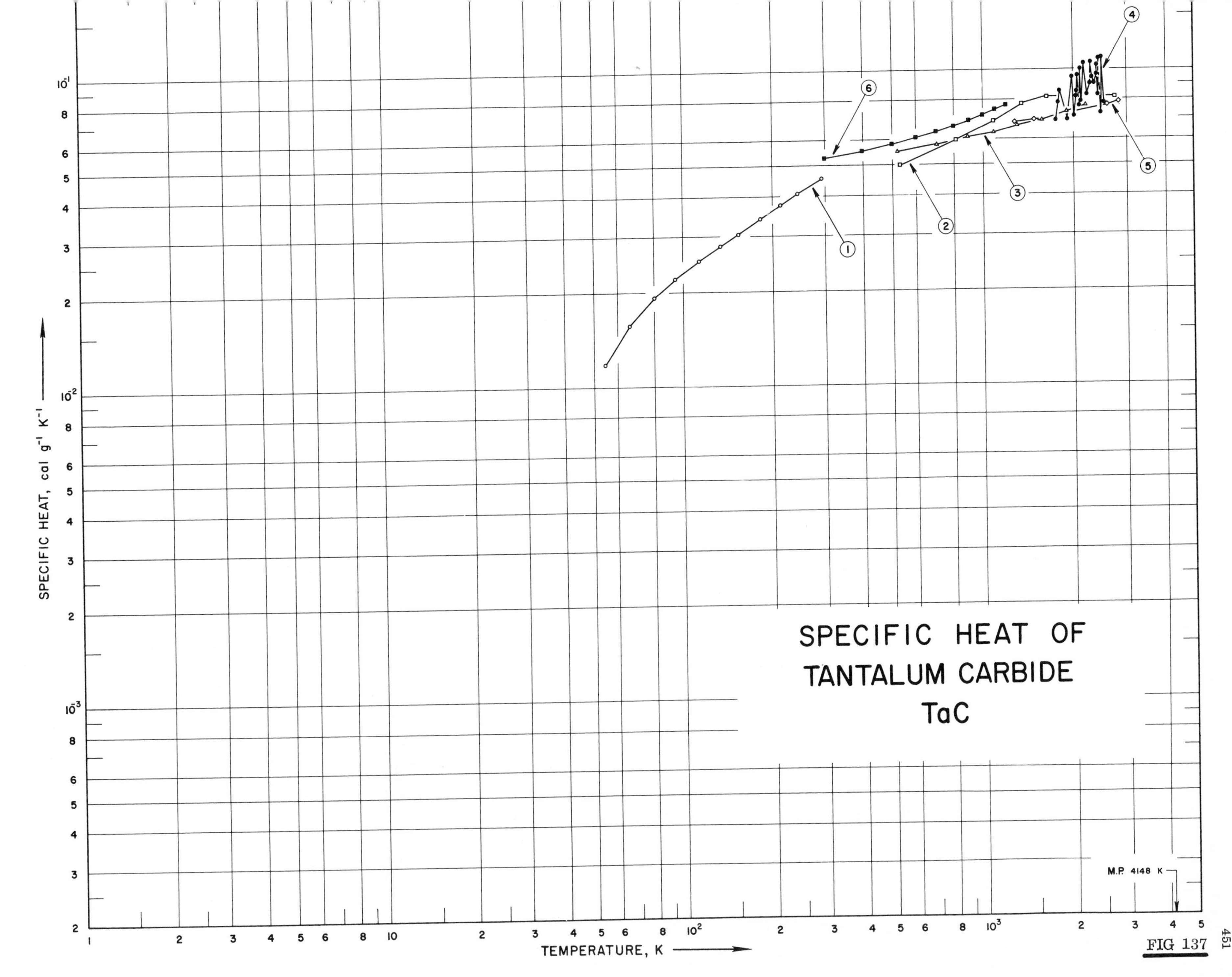
SPECIFIC HEAT OF
TANTALUM CARBIDE
TaC
SPECIFIC HEAT, cal g^{-1} K^{-1}
TEMPERATURE, K
M.P. 4148 K
1
2
3
4
5
6

FIG 137

SPECIFICATION TABLE NO. 137 SPECIFIC HEAT OF TANTALUM CARBIDE TaC

[For Data Reported in Figure and Table No. 137]

Curve No.	Ref. No.	Year	Temp. Range, K	Reported Error, %	Name and Specimen Designation	Composition (weight percent), Specifications and Remarks
1	127	1940	54-294			99.95 TaC, 0.03 excess C, 0.02 other impurities.
2	32	1962	533-2755	≤5		>93.75 Ta, 6.14 C, 0.10 W, <0.01 Al, Ca, Cb, Fe, Mg, Na, Ni, Si, Sn, Ti, and Zr; sample supplied by Kennametal, Inc.; hot pressed; density 476 lb ft^{-3}.
3	185	1962	523-2214			Bal. Ta, 6.29 total C, 6.24 combined C, and 0.019 free C; sprayed sample.
4	173	1962	1763-2544	5		90.48 Ta, 6.27 total C, 0.04 free C, 1.62 Nb, 0.9 Ti, 0.22 Hf, 0.20 Fe, 0.18 O, 0.17 Zr, 0.09 N, 0.07 Mn, <0.01 Cr, Cu, and Ni; hot pressed at 1350 C and 2000 psi for 1 hr; density 82.2 lb ft^{-3} (theoretically: density 90.4 lb ft^{-3}).
5	184	1963	1296-2843	1.8		92.14 Ta, 6.21 C, 0.80 W, 0.50 Nb, 0.20 Fe, and <0.05 free C.
6	54	1965	300-1200	0.7		Traces of Ca, Cu, and Si; sample supplied by the Carborundum Co.

DATA TABLE NO. 137 SPECIFIC HEAT OF TANTALUM CARBIDE TaC

[Temperature, T, K; Specific Heat, C_p, Cal g^{-1} K^{-1}]

T	C_p
CURVE 1	
54.60	1.179 x 10^{-2}
58.10	1.299*
62.10	1.430*
66.50	1.572
71.10	1.710*
75.60	1.827*
80.30	1.936
80.60	1.945*
84.40	2.023*
84.90	2.034*
94.50	2.214
104.30	2.376*
114.70	2.528
124.80	2.669*
134.80	2.807
144.80	2.935*
154.50	3.059
164.60	3.184*
174.20	3.306*
184.50	3.429
194.30	3.548*
204.50	3.669*
214.60	3.780
224.50	3.895*
234.50	3.999*
244.70	4.102
255.10	4.204*
265.30	4.293*
275.10	4.367*
284.60	4.437*
293.90	4.532*
294.50	4.544
CURVE 2	
533.15	5.00 x 10^{-2}
810.93	6.00
1088.71	6.90
1366.48	7.80
1644.26	8.20
1922.04	8.20*
2199.82	8.20*
2477.59	8.20*
2755.37	8.20
CURVE 3	
523.15	5.526 x 10^{-2}
706.48	5.855
893.71	6.126
1090.37	6.383
1315.93	6.663
1585.37	6.987
1904.26	7.365
2214.26	7.730
CURVE 4	
1763	6.910 x 10^{-2}
1780	7.860
1789	8.000*
1804	8.550
1939	6.930
1997	9.450
2023	7.100
2052	8.220
2058	7.930*
2068	8.570
2076	8.320*
2078	9.530
2115	7.680
2125	1.011 x 10^{-1}
2132	7.910 x 10^{-2}
2188	1.054 x 10^{-1}
2234	8.330 x 10^{-2}
2258	8.240*
2291	9.050
2297	1.065 x 10^{-1}
2309	1.070*
2316	9.460 x 10^{-2}
2321	9.220*
2349	8.880*
2360	9.080
2379	8.920*
2396	9.350*
2401	9.660
2406	8.450*
2408	1.027 x 10^{-1}
2426	8.310 x 10^{-2}
2433	9.170*
2448	8.370*
CURVE 4 (cont.)	
2448	1.086 x 10^{-1}
2455	1.080*
2460	8.580 x 10^{-2}*
2460	9.720*
2465	1.065 x 10^{-1}*
2472	7.280 x 10^{-2}
2474	1.090 x 10^{-1}
2474	9.580 x 10^{-2}*
2507	8.780*
2544	7.820
CURVE 5	
1296	6.830 x 10^{-2}
1401	6.910*
1488	6.970*
1491	6.970
1606	7.050*
1703	7.110*
1809	7.190*
1919	7.260*
1930	7.270*
2065	7.360*
2209	7.460*
2318	7.540*
2426	7.610*
2535	7.690*
2628	7.750
2642	7.760*
2746	7.830
2843	7.900
CURVE 6	
300	5.312 x 10^{-2}
400	5.582
500	5.851
600	6.126
700	6.396
800	6.665
900	6.935
1000	7.204
1100	7.499
1200	7.743

*Not shown on plot

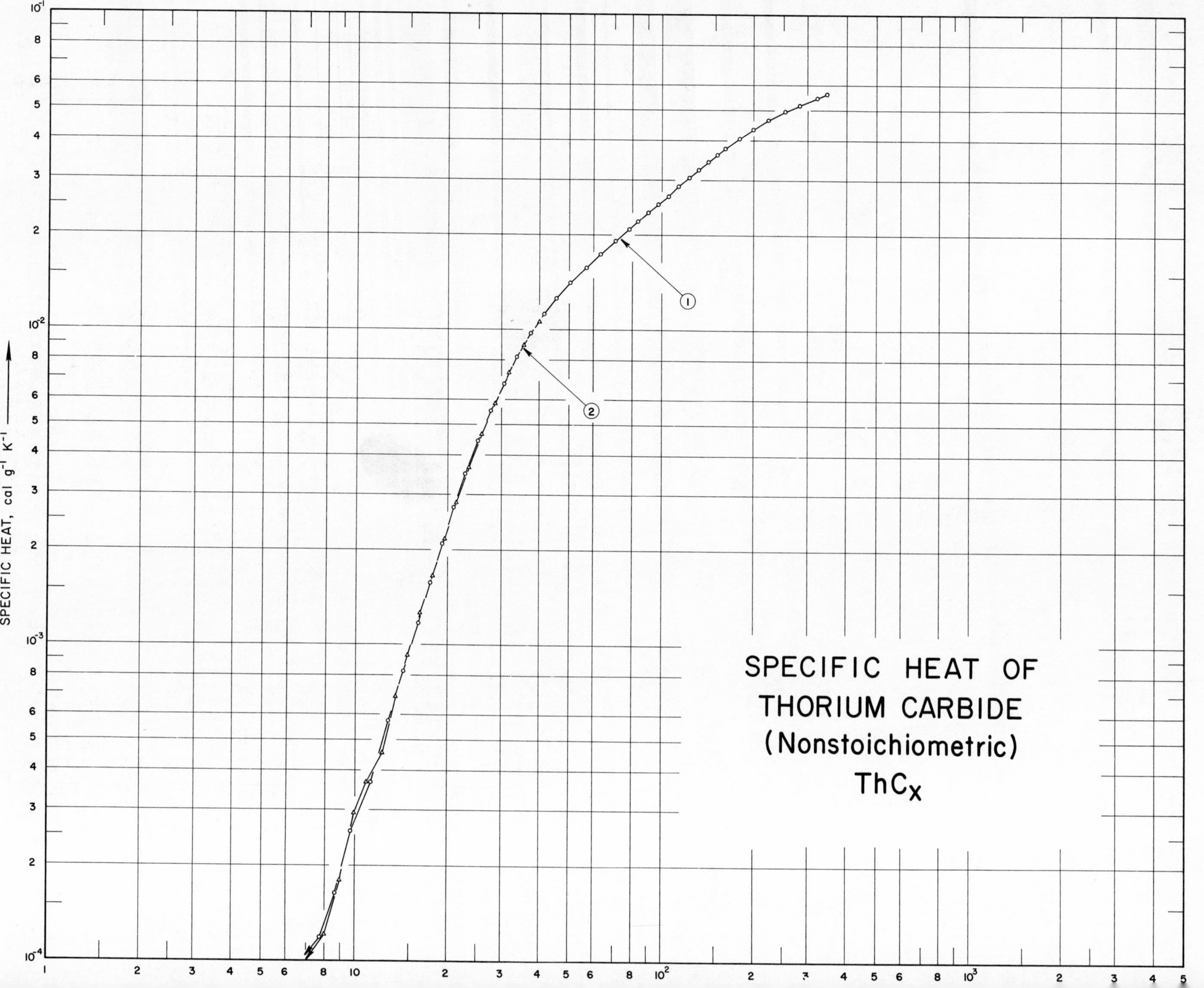
SPECIFIC HEAT OF
THORIUM CARBIDE
(Nonstoichiometric)
ThC$_x$
1
2
SPECIFIC HEAT, cal g^{-1} K^{-1}

SPECIFICATION TABLE NO. 138 SPECIFIC HEAT OF THORIUM CARBIDE (Nonstoichiometric) ThC_x

[For Data Reported in Figure and Table No. 138]

Curve No.	Ref. No.	Year	Temp. Range, K	Reported Error, %	Name and Specimen Designation	Composition (weight percent), Specifications and Remarks
1	191	1965	6-346	0.1	$ThC_{1.93}$, hypostoichiometric Thorium Dicarbide	8.99 C, 1.758 W, 0.0575 O, 0.00268 N, 0.00228 H; turned and melted several times, crushed to powder, and remelted several more times.
2	192	1965	6-345		ThC_2	99.21 ThC_2, 0.79 free C; prepared from thorium powder and graphite; pressed at 700 Kg cm^{-2} into 0.8 cm dia pellets and heated for 30 min under vacuum at 2100 C; pellets were crushed and ground, heated for 5 hrs at 2000 C, then reground and reheated for 5 hrs at 2000 C to homogenize the sample.

DATA TABLE NO. 138 SPECIFIC HEAT OF THORIUM CARBIDE (Nonstoichiometric) ThC_x

[Temperature, T, K; Specific Heat, C_p, Cal g^{-1} K^{-1}]

T	C_p
CURVE 1	
Series 1	
83.82	2.203 x 10^{-2}
90.61	2.352
97.88	2.504
105.62	2.668
114.54	2.858
123.70	3.052
132.92	3.241
142.48	3.430
152.03	3.609
161.56	3.778
170.92	3.933*
180.32	4.079
189.96	4.218*
199.53	4.347
Series 2	
6.09	5.871 x 10^{-5}*
6.87	1.018 x 10^{-4}*
7.72	1.096
8.65	1.644
9.73	2.583
11.28	3.679
12.87	5.754
14.40	8.270
16.02	1.178 x 10^{-3}
17.56	1.581
19.19	2.090
20.99	2.724
22.97	3.492
25.18	4.415
27.72	5.511
30.60	6.772
33.84	8.193
37.73	9.786
41.77	1.123 x 10^{-2}
45.78	1.257
50.71	1.407
57.00	1.575
63.61	1.744
70.86	1.909
78.76	2.086
CURVE 1 (cont.)	
Series 3*	
61.78	1.699 x 10^{-2}*
68.00	1.848*
74.52	1.990*
81.68	2.158*
89.80	2.338
Series 4*	
74.30	1.986 x 10^{-2}
81.54	2.153
89.74	2.338
Series 5	
195.71	4.303 x 10^{-2}*
205.11	4.424*
214.69	4.537*
224.40	4.650
234.14	4.754*
243.88	4.853*
253.52	4.938
263.25	5.025*
273.07	5.109*
282.82	5.186
292.57	5.265*
302.30	5.347*
311.90	5.399*
321.38	5.461
330.76	5.515*
339.13	5.569*
346.38	5.604
CURVE 2	
Series 1*	
75.30	2.003 x 10^{-2}
81.73	2.160
Series 2*	
83.21	2.198 x 10^{-2}
90.97	2.378
99.82	2.552
109.60	2.760
119.31	2.967
128.96	3.168
Series 3	
5.98	7.420 x 10^{-5}*
6.87	9.763*
7.99	1.211 x 10^{-4}
8.96	1.796
9.92	2.929
10.96	3.671
12.21	4.567
13.51	6.834
14.84	9.295
16.34	1.261 x 10^{-3}
17.94	1.660
19.60	2.160
21.42	2.827
23.52	3.659
25.94	4.679
28.64	5.889
31.81	7.315
35.70	8.928
40.18	1.060 x 10^{-2}
45.34	1.229*
Series 4*	
40.63	1.081 x 10^{-2}
45.40	1.233
50.27	1.376
55.91	1.527
62.42	1.702
69.70	1.873
77.63	2.055
Series 5*	
134.25	3.276 x 10^{-2}
143.16	3.457
152.77	3.632
162.86	3.815
166.97	3.896
176.77	4.044
186.89	4.196
197.06	4.335
207.30	4.473
217.76	4.593
228.44	4.713
239.29	4.827
250.09	4.926
260.59	5.014
270.79	5.098
279.87	5.166
289.65	5.251
299.21	5.318
308.55	5.374
317.82	5.427
327.01	5.480
336.12	5.558
345.16	5.613

* Not shown on plot

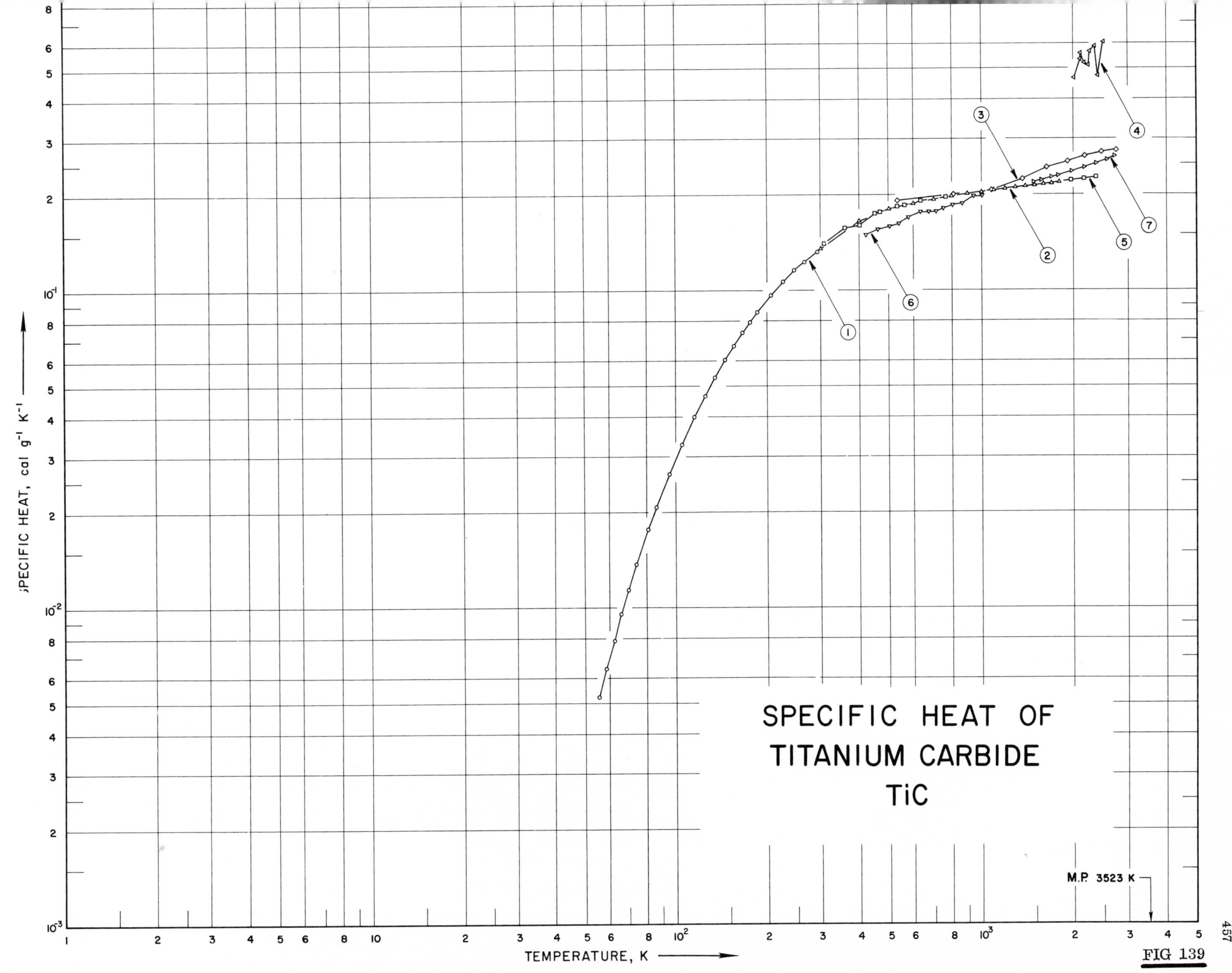
SPECIFIC HEAT OF
TITANIUM CARBIDE
TiC
SPECIFIC HEAT, cal g⁻¹ K⁻¹
TEMPERATURE, K
M.P. 3523 K
1
2
3
4
5
6
7

FIG 139

SPECIFICATION TABLE NO. 139 SPECIFIC HEAT OF TITANIUM CARBIDE TiC

[For Data Reported in Figure and Table No. 139]

Curve No.	Ref. No.	Year	Temp. Range, K	Reported Error, %	Name and Specimen Designation	Composition (weight percent), Specifications and Remarks
1	193	1944	55-294			96.08 TiC, 1.82 TiO_2, and 0.06 unaccounted; data corrected for impurities.
2	194	1946	298-1800			99.0 TiC, 0.4 unreacted Ti (theo. 29.56 Ti and 19.85 C); prepared by heating powdered Ti with 99.7 pure C in a vacuum at 1300 C; density 300 lb ft^{-3}.
3	48	1962	533-2755	≤5		Before exposure: 79.8 Ti, 19.2 C, 0.9 N, 0.6 Fe, and 0.5 Zr, after exposure: 79.2 Ti, 19.0 C, 0.6 N; sample supplied by the Carborundum Co; hot pressed; crushed in hardened steel mortar to pass 100-mesh screen; density at 25 C, before exposure: apparent density (ASTM method B311-58) 295 lb ft^{-3}, true density (by immersion in xylene) 298 lb ft^{-3}, after exposure: apparent density 292 lb ft^{-3}, true density 300 lb ft^{-3}.
4	173	1962	2023-2121	5		77.41 Ti, 17.78 total C, 0.43 free C, 0.50 free Ti, 0.56 Zr, 0.12 Fe, <0.1 Al, Hf, 0.09 Nb, 0.004 Cu, and <0.001 Mg; solid pieces machined to specification using diamond tool and electric discharge technique.
5	195	1963	298-2372			80.3 ± 0.3 Ti, 19.3 C, <0.2 metallic impurities; sample supplied by the Carborundum Co; density 298 lb ft^{-3}.
6	195	1963	423-1005	±5		79.2 Ti, 20.2 C, <0.2 metallic impurities; sample supplied by the Norton Co; density 296 lb ft^{-3}.
7	196	1965	1274-2722	1.4		79.42 Ti, 18.82 combined C, 0.71 free C, <0.1 W, and 0.04 O_2; lattice parameter a_0 = 4.327 ± 0.0002 Å; hot pressed in graphite dies and then electrospark machined.

DATA TABLE NO. 139 SPECIFIC HEAT OF TITANIUM CARBIDE TiC

[Temperature, T, K; Specific Heat, C_p, Cal g^{-1} K^{-1}]

T	C_p
CURVE 1	
55.1	5.208 x 10^{-3}
58.6	6.410
62.3	7.895
65.9	9.531
69.8	1.145 x 10^{-2}
74.2	1.370
81.2	1.761
86.5	2.073
95.7	2.649
105.3	3.272
115.6	3.988
125.7	4.655
135.6	5.325
146.0	6.019
156.2	6.693
166.5	7.348
176.2	7.929
186.3	8.534
196.8	9.137*
206.5	9.644
216.8	1.018 x 10^{-1}*
226.3	1.065
236.0	1.108*
246.2	1.156
256.0	1.196*
266.4	1.239
276.3	1.277*
285.7	1.307*
294.9	1.333
CURVE 2	
298.15	1.342 x 10^{-1}*
300	1.351
400	1.655
500	1.802
600	1.889
700	1.946
800	1.988
900	2.021
1000	2.048
1100	2.072
1200	2.093
1300	2.113
CURVE 2 (cont.)	
1400	2.131
1500	2.148
1600	2.165
1700	2.181
1800	2.197
CURVE 3	
533.15	1.930 x 10^{-1}
810.93	2.000
1088.71	2.080
1366.48	2.250
1644.26	2.440
1922.04	2.550
2199.82	2.640
2477.59	2.720
2755.37	2.760
CURVE 4	
2023	4.670 x 10^{-1}
2113	5.350
2186	5.200
2279	5.640
2353	5.890
2449	6.040*
2506	6.020
2412	4760
2252	5.100
2121	5.590
CURVE 5	
Series 1	
298.15	1.340 x 10^{-1}*
565.15	1.860
637.15	1.910
768.15	1.970
470.15	1.770
403.15	1.600
CURVE 5 (cont.)	
Series 2	
309.15	1.400 x 10^{-1}
361.15	1.570
451.15	1.750
481.15	1.780*
535.15	1.840
565.15	1.860*
587.15	1.880*
698.15	1.940*
691.15	1.940*
801.15	1.990*
816.15	1.990*
Series 3	
1573.15	2.160 x 10^{-1}*
1673.15	2.180*
1773.15	2.190*
1873.15	2.210*
1973.15	2.220
2073.15	2.240*
2173.15	2.250
2273.15	2.270*
2373.15	2.280
CURVE 6	
423.15	1.490 x 10^{-1}
442.15	1.530*
462.15	1.560
482.15	1.580*
501.15	1.590
521.15	1.610*
540.15	1.630
559.15	1.660*
579.15	1.700
598.15	1.730*
617.15	1.760*
635.15	1.770
654.15	1.770*
673.15	1.770
693.15	1.770*
712.15	1.770
CURVE 6 (cont.)	
732.15	1.790 x 10^{-1}*
751.15	1.810
771.15	1.830*
790.15	1.850*
809.15	1.860
829.15	1.870*
849.15	1.870*
868.15	1.880
888.15	1.900*
908.15	1.930*
927.15	1.970*
946.15	1.990
965.15	2.000*
985.15	2.000*
1005.15	1.990
CURVE 7	
1274	2.11 x 10^{-1}*
1389	2.16*
1481	2.19
1576	2.22
1696	2.27
1775	2.29*
1780	2.30
1890	2.34*
1987	2.37
2075	2.40*
2184	2.44
2274	2.48*
2377	2.51
2473	2.55*
2569	2.58
2674	2.62*
2722	2.64

*Not shown on plot

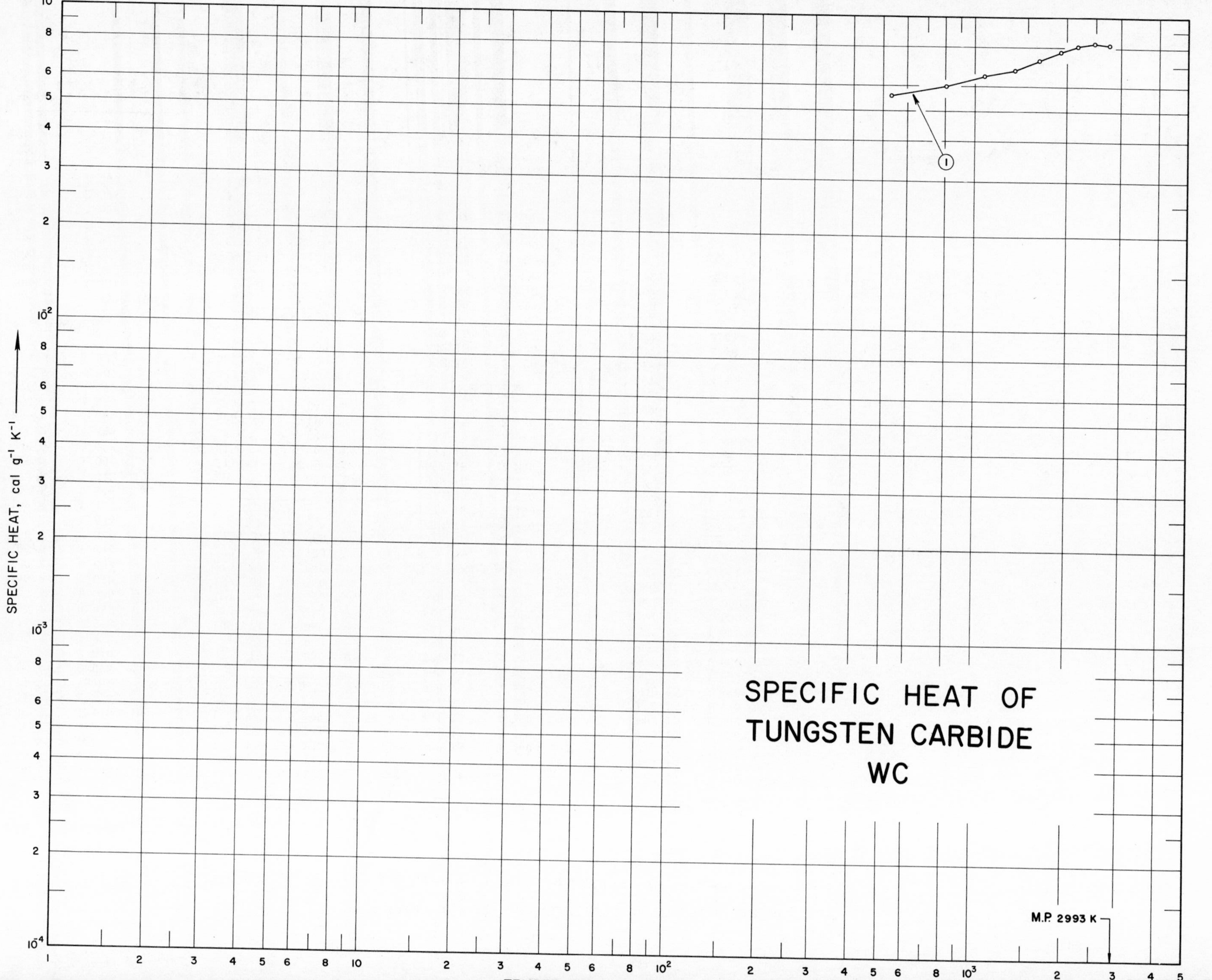

SPECIFIC HEAT OF
TUNGSTEN CARBIDE
WC
M.P. 2993 K
1
SPECIFIC HEAT, cal g⁻¹ K⁻¹

SPECIFICATION TABLE NO. 140 SPECIFIC HEAT OF TUNGSTEN CARBIDE WC

[For Data Reported in Figure and Table No. 140]

Curve No.	Ref. No.	Year	Temp. Range, K	Reported Error, %	Name and Specimen Designation	Composition (weight percent), Specifications and Remarks
1	48	1962	533-2755	≤5		Before exposure: 93.9 W, 6.15 C, 0.4 Fe, 0.3 Si, and <0.1 N, after exposure: 94.1 W, 5.95 C, and <0.1 N; sample supplied by the Carborundum Co; hot pressed; crushed in a hardened steel mortar to pass 100-mesh screen; density 25 C, before exposure: apparent density (ASTM method B311-58) 935 lb ft^{-3}, true density (by immersion in xylene) 942 lb ft^{-3}, after exposure: apparent density 818 lb ft^{-3}, true density 918 lb ft^{-3}.

DATA TABLE NO. 140 SPECIFIC HEAT OF TUNGSTEN CARBIDE WC

[Temperature, T, K; Specific Heat, C_p, Cal g^{-1} K^{-1}]

T	C_p
CURVE 1	
533.15	5.600 x 10^{-2}
810.93	6.000
1088.71	6.500
1366.48	6.800
1644.26	7.300
1922.04	7.800
2199.82	8.100
2477.59	8.300
2755.37	8.200

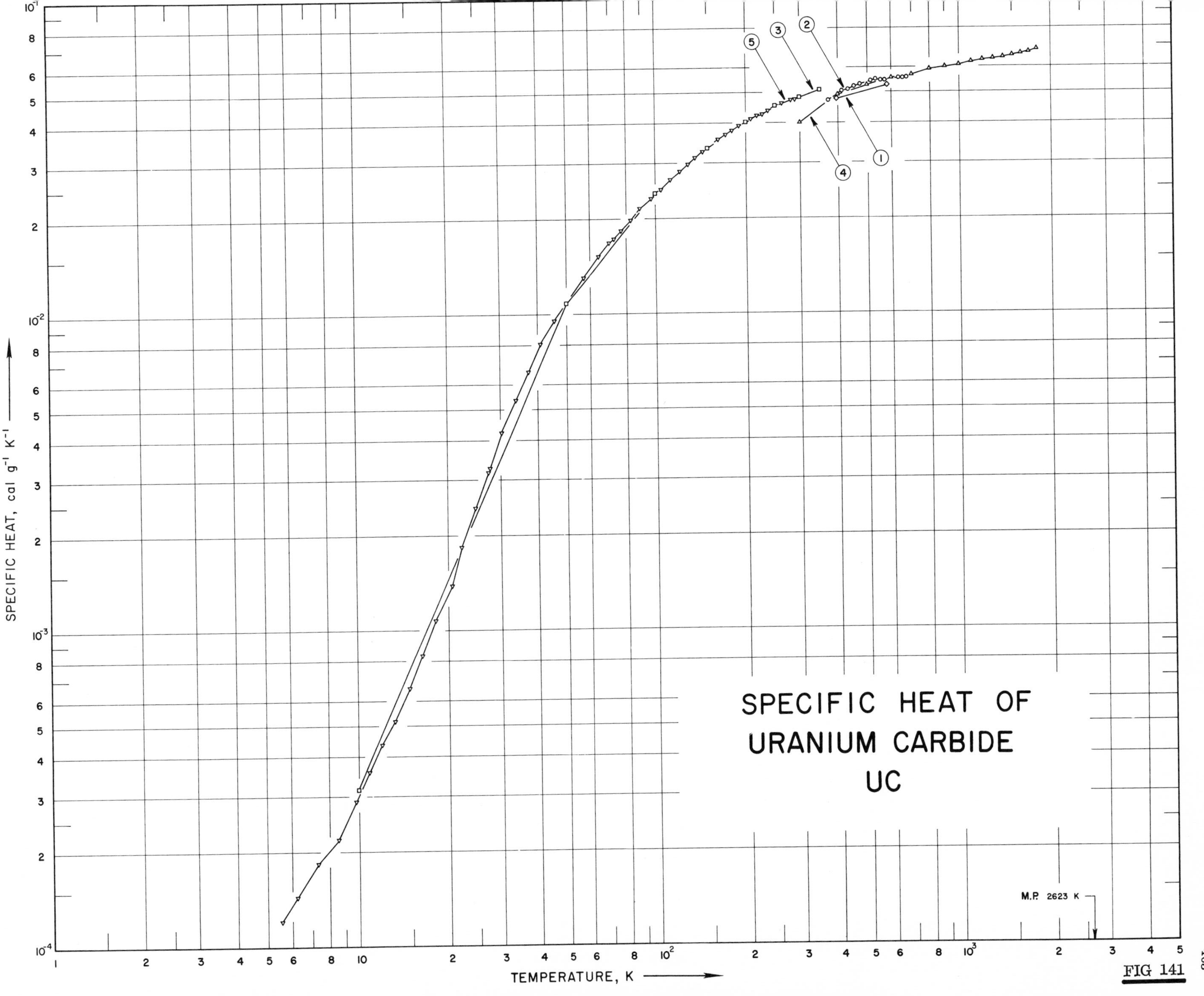

SPECIFIC HEAT OF
URANIUM CARBIDE
UC
SPECIFIC HEAT, cal g⁻¹ K⁻¹
TEMPERATURE, K
M.P. 2623 K
1
2
3
4
5

FIG 141

SPECIFICATION TABLE NO. 141 SPECIFIC HEAT OF URANIUM CARBIDE UC

[For Data Reported in Figure and Table No. 141]

Curve No.	Ref. No.	Year	Temp. Range, K	Reported Error, %	Name and Specimen Designation	Composition (weight percent), Specifications and Remarks
1	197	1958	398, 523			~95. 185 U, 4. 815 total C, 0. 054 free C.
2	198	1962	373-673	3. 0		99. 93 U, 0. 044 C, 0. 00601 Fe, 0. 004 Si, 0. 0032 N, 0. 0018 Mg, and 0. 00172 Al; this is the composition of U from which UH_3 was prepared; prepared by solid-phase reaction of uranium hydride and carbon at 1100 C for 2 hrs in 10^{-5} mm Hg vacuum.
3	199	1962	10-350			Average composition, 94. 18 U, 5. 134 C; measured in a helium atmosphere.
4	200	1964	300-1800	≤4		95. 12 U, 4. 88 C, 0. 0135 O, and 0. 006 N.
5	201	1965	5-345	0. 1-5		94. 16 U, 5. 01 total C, 0. 07 free C.

DATA TABLE NO. 141 SPECIFIC HEAT OF URANIUM CARBIDE UC

[Temperature, T, K; Specific Heat, C_p, Cal g^{-1} K^{-1}]

T	C_p
CURVE 1	
398	4.800 x 10^{-2}
523	5.300
CURVE 2	
373.15	4.780 x 10^{-2}
393.15	4.900*
413.15	5.070
433.15	5.110
453.15	5.220
473.15	5.300
493.15	5.410*
513.15	5.420
533.15	5.500
553.15	5.490
573.15	5.490
593.15	5.520*
613.15	5.530*
633.15	5.560
653.15	5.560
673.15	5.590
CURVE 3	
10	3.159 x 10^{-4}
50	1.094 x 10^{-2}
100	2.413
150	3.353
200	4.031
250	4.507
300	4.858
350	5.102
298.15	4.383*
CURVE 4	
300	4.0478 x 10^{-2}
400	4.91
500	5.344
600	5.6096
700	5.7936
800	5.9355
900	6.0525

T	C_p
CURVE 4 (cont.)	
1000	6.1537 x 10^{-2}
1100	6.2453
1200	6.3301
1300	6.4097
1400	6.486
1500	6.5594
1600	6.632
1700	6.702
1800	6.822
CURVE 5	
Series 1	
70.77	1.683 x 10^{-2}
77.22	1.845
83.05	1.999
89.76	2.158
97.21	2.323
105.51	2.499
113.99	2.677
121.79	2.831
129.42	2.989
136.94	3.125
145.20	3.259
154.43	3.418*
163.09	3.554
171.92	3.685
180.90	3.801
190.31	3.919
199.76	4.033*
209.05	4.124
218.35	4.219
227.82	4.286
237.23	4.401
246.47	4.478*
255.52	4.549*
264.39	4.610
273.24	4.686*
282.25	4.742
291.11	4.796
300.00	4.872*

T	C_p
CURVE 5 (cont.)	
Series 2	
5.59	1.200 x 10^{-4}
6.29	1.440
7.34	1.839
8.58	2.199
9.80	2.879
10.96	3.559
12.01	4.359
13.33	5.198
14.81	6.598
16.34	8.317
18.09	1.076 x 10^{-3}
20.06	1.396
22.23	1.843
24.66	2.451
27.26	3.199
Series 3	
27.51	3.279 x 10^{-3}
30.45	4.219
33.82	5.382
37.43	6.662
41.57	8.089
45.76	9.505
50.67	1.107 x 10^{-2}*
57.26	1.308
64.66	1.520
73.02	1.739
297.50	4.815*
307.09	4.892*
316.82	4.950*
326.76	5.003*
336.63	5.052*
345.99	5.074*

*Not shown on plot

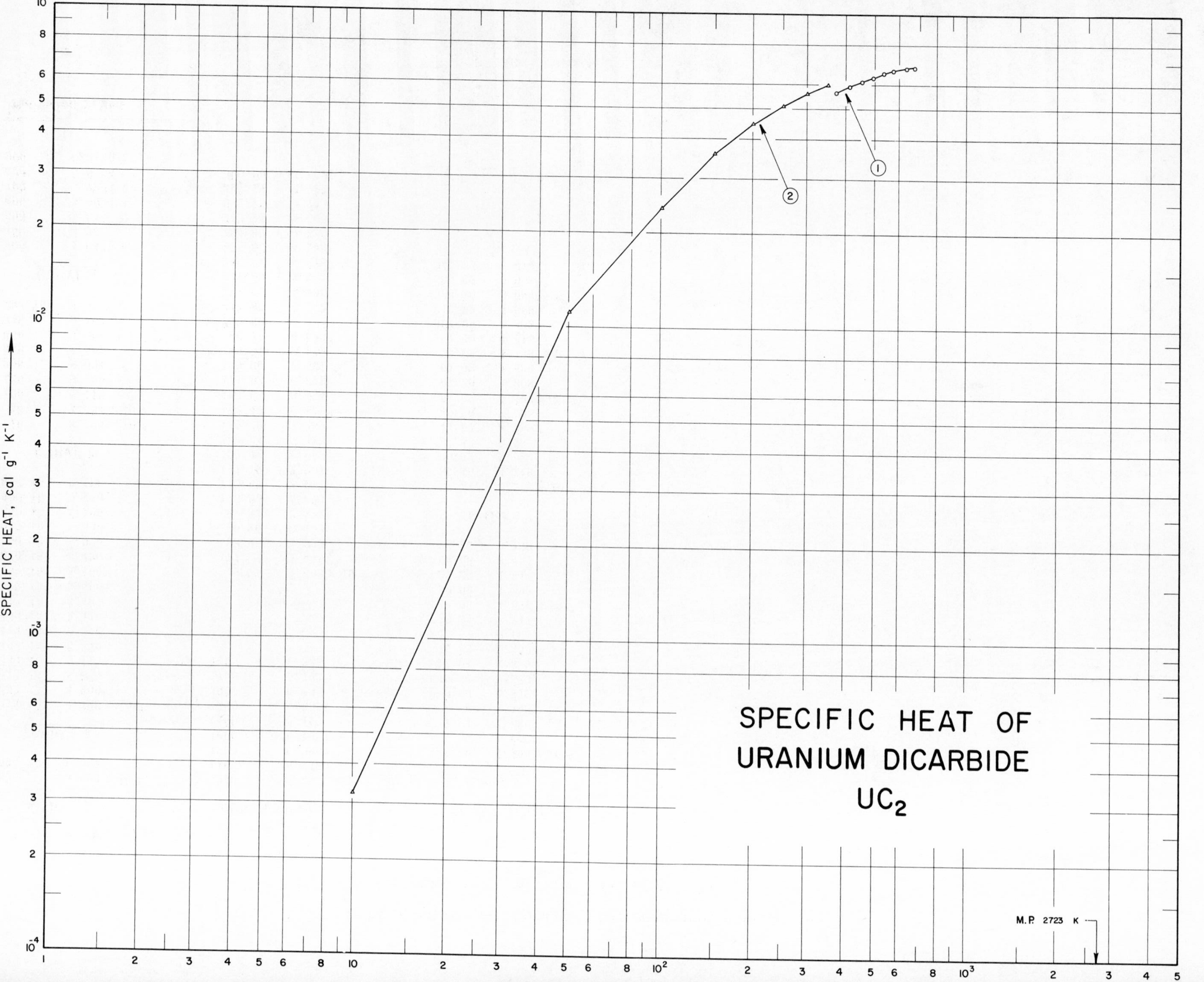
SPECIFIC HEAT OF
URANIUM DICARBIDE
UC2
SPECIFIC HEAT, cal g-1 K-1
M.P. 2723 K
1
2

SPECIFICATION TABLE NO. 142 SPECIFIC HEAT OF URANIUM DICARBIDE UC_2

[For Data Reported in Figure and Table No. 142]

Curve No.	Ref. No.	Year	Temp. Range, K	Reported Error, %	Name and Specimen Designation	Composition (weight percent), Specifications and Remarks
1	198	1962	373-673	3.0		99.93 U, 0.044 C, 0.00601 Fe, 0.004 Si, 0.0032 N, 0.0018 Mg, and 0.00172 Al; this is the composition of U from which UH_3 was prepared; prepared by two-step solid-phase reaction of stoichiometric mixture of UH_3 and C at 1100 C for 2 hrs under vacuum; reheated at 1700 C for 2 hrs.
2	199	1962	10-350			90.79 U, 9.20 C; measured in a helium atmosphere.

DATA TABLE NO. 142 SPECIFIC HEAT OF URANIUM DICARBIDE UC_2

[Temperature, T, K; Specific Heat, C_p, Cal g^{-1} K^{-1}]

T	C_p
CURVE 1	
373.15	5.650 x 10^{-2}
393.15	5.750*
413.15	5.910
433.15	5.990*
453.15	6.110
473.15	6.180*
493.15	6.310
513.15	6.420*
533.15	6.500
553.15	6.550*
573.15	6.610
593.15	6.690*
613.15	6.720*
633.15	6.730
653.15	6.740*
673.15	6.740
CURVE 2	
10	3.243 x 10^{-4}
50	1.119 x 10^{-2}
100	2.430
150	3.594
200	4.487
250	5.113
300	5.613
350	5.964
298.15	5.590*

*Not shown on plot

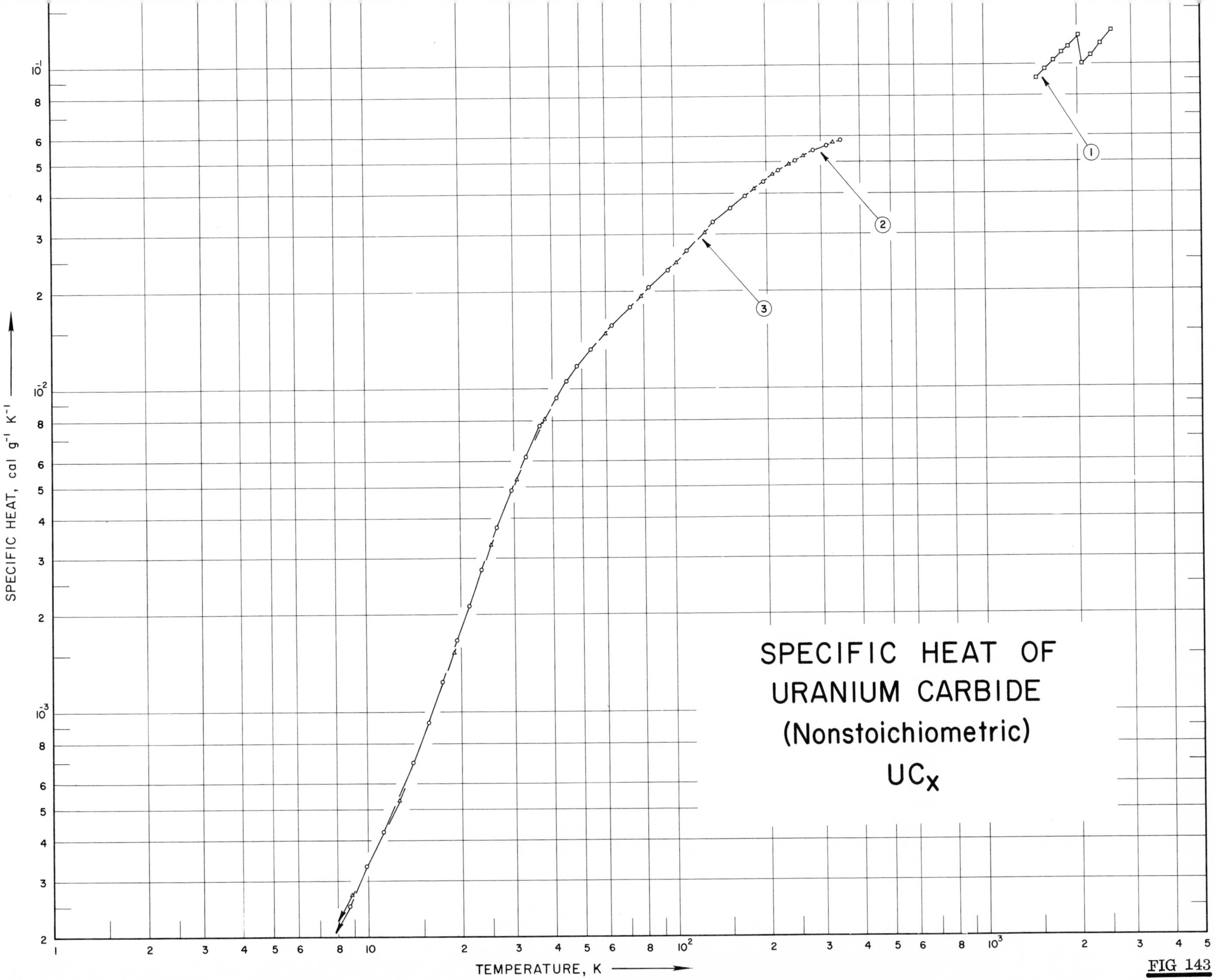

SPECIFIC HEAT OF URANIUM CARBIDE (Nonstoichiometric) UC_x

SPECIFIC HEAT, cal g^{-1} K^{-1}

TEMPERATURE, K

1 2 3

FIG 143

SPECIFICATION TABLE NO. 143 SPECIFIC HEAT OF URANIUM CARBIDE (Nonstoichiometric) UC_x

[For Data Reported in Figure and Table No. 143]

Curve No.	Ref. No.	Year	Temp. Range, K	Reported Error, %	Name and Specimen Designation	Composition (weight percent), Specifications and Remarks
1	202	1963	1484-2581	2	$UC_{1.93}$	90.88 U, 8.9 total C, 0.13 free C, and 0.022 O_2.
2	201	1965	5-346	0.1-5	$UC_{1.9}$	90.79 U, 9.20 total C, and 1.0 free C.
3	203	1965	6-345	0.1-5	$UC_{1.94}$	91.18 ± 0.07 U, 8.91 ± 0.03 C, 0.0035 ± 0.0005 O_2, and <0.05 free C.

DATA TABLE NO. 143 SPECIFIC HEAT OF URANIUM CARBIDE (Nonstoichiometric) UC_x

[Temperature, T, K; Specific Heat, C_p, Cal $g^{-1} K^{-1}$]

T	C_p
CURVE 1	
1484	9.1 x 10^{-2}
1490	9.1*
1587	9.7
1689	1.03 x 10^{-1}
1794	1.09
1878	1.14
2016	1.23*
2021	1.23
2081	1.00
2086	1.01*
2213	1.07
2277	1.11*
2393	1.17
2483	1.21*
2581	1.27
CURVE 2	
Series 1	
95.20	2.334 x 10^{-2}
102.59	2.491*
110.31	2.668
118.20	2.859*
126.37	3.032*
134.78	3.254
143.53	3.423*
152.39	3.607
161.17	3.806*
169.96	3.944
178.50	4.105*
187.05	4.255*
195.56	4.385
204.22	4.515*
210.47	4.588*
219.01	4.722
228.22	4.841*
237.63	4.937*
247.03	5.083
256.23	5.159*
265.33	5.244*
273.56	5.320*
282.86	5.420
CURVE 2 (cont.)	
Series 2	
265.62	5.244 x 10^{-2}*
275.10	5.347*
284.07	5.431*
293.79	5.489*
303.63	5.585*
312.07	5.638
321.16	5.707*
329.81	5.769*
338.07	5.826*
346.01	5.865
Series 3	
5.36	1.188 x 10^{-4}*
5.90	1.303*
6.59	1.342*
7.52	1.955*
8.70	2.491
9.89	3.335
11.17	4.255
12.52	5.328*
13.98	6.976
15.63	9.276
17.43	1.242 x 10^{-3}
19.38	1.667
21.22	2.135
23.37	2.764
26.16	3.726
29.24	4.868
32.66	6.171
36.34	7.628
40.51	9.123*
44.29	1.042 x 10^{-2}*
48.97	1.192*
54.58	1.353*
Series 4	
41.18	9.391 x 10^{-3}
44.76	1.058 x 10^{-2}
48.38	1.173
CURVE 2 (cont.)	
Series 4 (cont.)	
49.37	1.200 x 10^{-2}*
53.54	1.326
58.20	1.457*
62.97	1.572
67.49	1.713*
72.06	1.798
Series 5	
41.88	9.621 x 10^{-3}*
46.53	1.119 x 10^{-2}*
51.91	1.280*
58.18	1.460*
64.14	1.618*
69.39	1.740*
75.68	1.882*
82.91	2.058
90.80	2.242*
Series 6*	
80.93	2.009 x 10^{-2}
89.04	2.200
98.01	2.396
106.97	2.591
115.56	2.787
123.87	2.986
131.96	3.170
140.19	3.354
CURVE 3	
6.08	1.370 x 10^{-4}
6.86	1.664*
7.78	2.101*
8.87	2.721
10.01	3.482*
11.25	4.194*
12.58	5.319
13.94	6.803*
15.43	8.812*
CURVE 3 (cont.)	
17.10	1.162 x 10^{-3}*
18.90	1.534
20.76	1.991*
22.79	2.564*
25.09	3.298
27.59	4.173*
30.52	5.259
34.10	6.646*
37.82	8.066
41.86	9.528*
46.27	1.102 x 10^{-2}*
50.51	1.237*
54.86	1.364*
59.65	1.496
65.00	1.633*
71.33	1.781*
78.21	1.942
85.54	2.116*
93.32	2.290*
101.26	2.466
109.55	2.657*
117.74	2.849*
125.93	3.037
134.64	3.236*
144.14	3.444*
153.80	3.645*
162.99	3.826*
172.34	4.002*
181.73	4.167
191.02	4.321*
200.24	4.466*
209.21	4.596
218.21	4.721*
227.36	4.842*
236.41	4.952
245.35	5.058*
254.48	5.154*
263.82	5.247
273.12	5.337*
282.27	5.421*
291.43	5.505*
300.61	5.577*
309.67	5.650*
CURVE 3 (cont.)	
318.69	5.715 x 10^{-2}*
327.67	5.776
336.58	5.837*
345.48	5.894*

*Not shown on plot

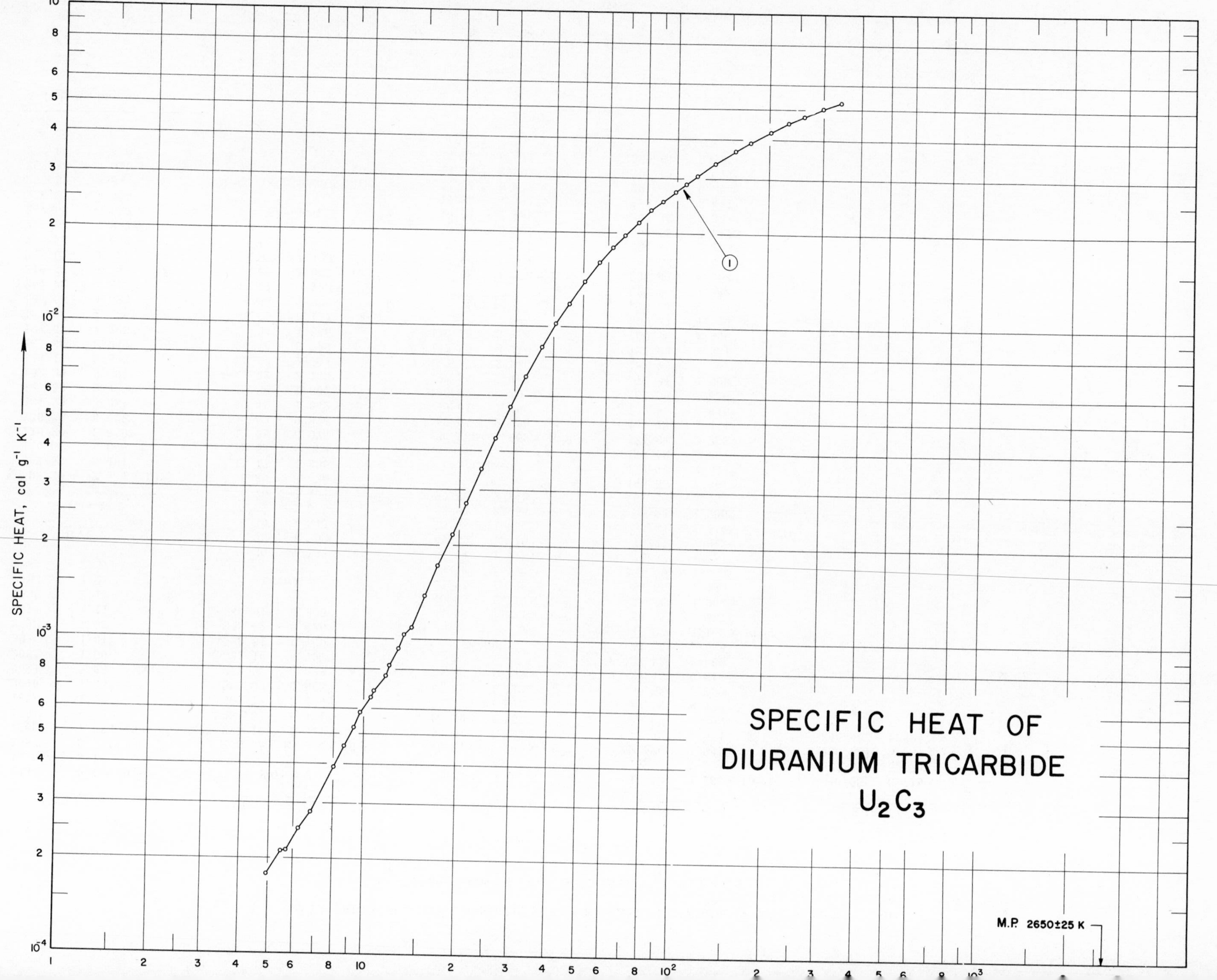
SPECIFIC HEAT OF
DIURANIUM TRICARBIDE
U_2C_3
M.P. 2650±25 K
SPECIFIC HEAT, cal g⁻¹ K⁻¹
1

SPECIFICATION TABLE NO. 144 SPECIFIC HEAT OF DIURANIUM TRICARBIDE U_2C_3

[For Data Reported in Figure and Table No. 144]

Curve No.	Ref. No.	Year	Temp. Range, K	Reported Error, %	Name and Specimen Designation	Composition (weight percent), Specifications and Remarks
1	203	1965	5-345	0.1-5		92.9 ± 0.1 U, 6.94 ± 0.03 C, 0.002 ± 0.0005 O_2, 0.003 N_2, and <0.05 free C.

DATA TABLE NO. 144 SPECIFIC HEAT OF DIURANIUM TRICARBIDE U_2C_3

[Temperature, T, K; Specific Heat, C_p, Cal g^{-1} K^{-1}]

T	C_p
CURVE 1	
Series 1	
4.92	1.796×10^{-4}
5.48	2.109
6.23	2.499
8.69	4.589
9.74	5.838
10.77	6.834
12.04	8.279
13.41	1.041×10^{-3}
Series 2	
5.70	2.128×10^{-4}
6.81	2.812
8.06	3.905
9.32	5.253
10.52	6.522
11.71	7.654
12.89	9.373
14.14	1.097×10^{-3}
15.53	1.386
17.16	1.722
Series 3	
19.02	2.160×10^{-3}
21.10	2.724
23.60	3.501
26.19	4.401
29.15	5.520
32.69	6.946
36.79	8.631
40.85	1.027×10^{-2}
45.18	1.196
50.80	1.406
56.60	1.607
62.45	1.799
Series 4	
54.54	1.537×10^{-2}*
61.66	1.774*
68.17	1.969

T	C_p
CURVE 1 (cont.)	
Series 4 (cont.)	
75.29	2.159×10^{-2}
82.94	2.367
90.65	2.546
99.20	2.717
108.26	2.893
117.09	3.058
125.85	3.212*
134.62	3.359
Series 5	
129.60	3.275×10^{-2}*
138.02	3.413*
146.60	3.540*
155.50	3.673
164.71	3.796*
174.03	3.917
183.31	4.026*
192.57	4.130*
202.08	4.235
211.71	4.327*
221.35	4.427*
231.29	4.511
241.30	4.604*
251.12	4.682*
260.89	4.753
270.62	4.825*
280.45	4.895*
290.41	4.964*
300.52	5.026
310.77	5.091*
320.91	5.141*
330.92	5.210*
338.87	5.247*
344.78	5.284

*Not shown on plot

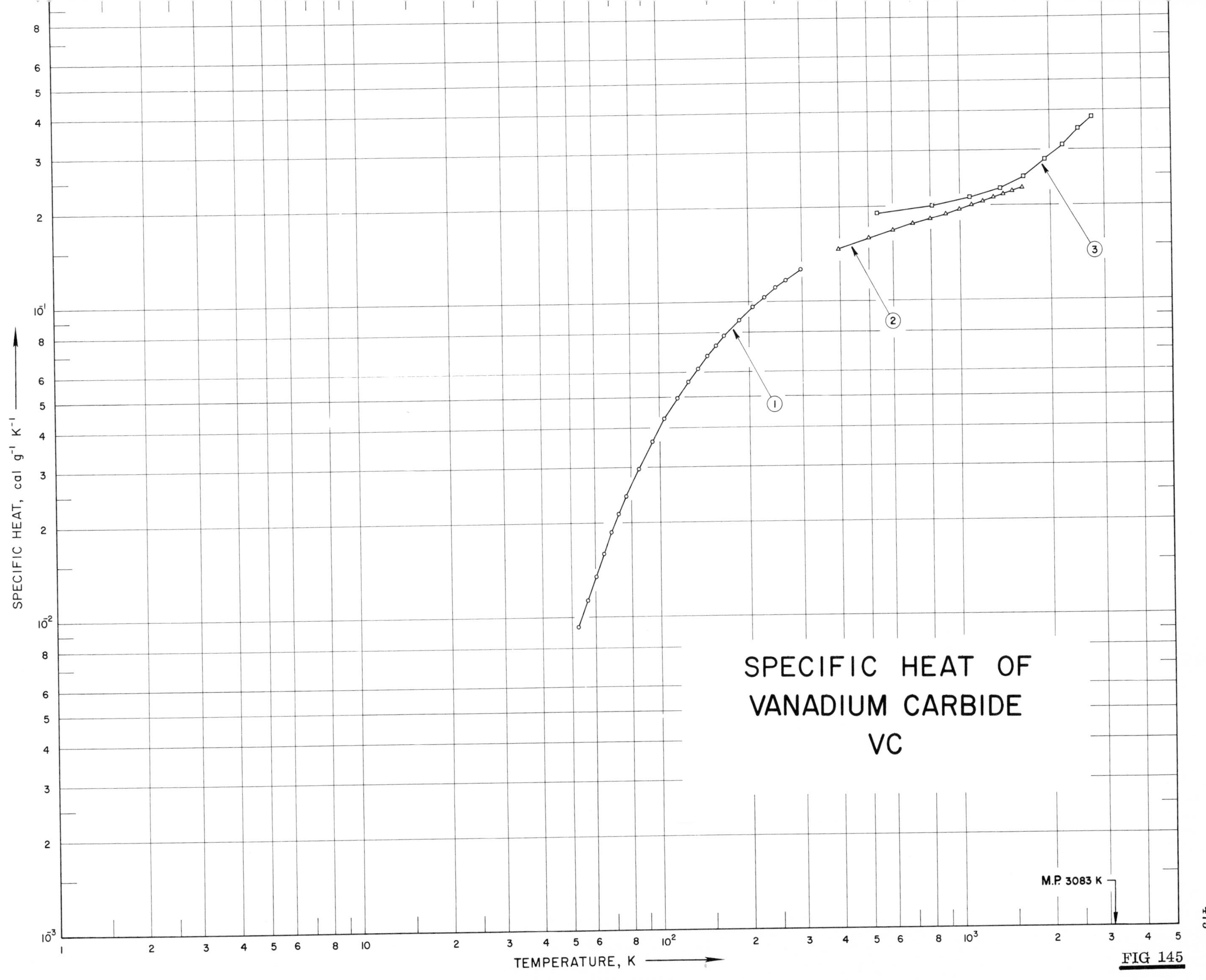
SPECIFIC HEAT OF
VANADIUM CARBIDE
VC
SPECIFIC HEAT, cal g⁻¹ K⁻¹
TEMPERATURE, K
M.P. 3083 K
1
2
3

FIG 145

SPECIFICATION TABLE NO. 145 SPECIFIC HEAT OF VANADIUM CARBIDE VC

[For Data Reported in Figure and Table No. 145]

Curve No.	Ref. No.	Year	Temp. Range, K	Reported Error, %	Name and Specimen Designation	Composition (weight percent), Specifications and Remarks
1	204	1949	52-298			80.90 V and 19.04 C; heated in a vacuum 26 hrs at 1300 - 1350 C, product analyzed and adjusted in composition after 12 and 22 hrs of heating.
2	205	1949	397-1611			Same as above.
3	48	1962	533-2755	≤5		Before exposure: 81.0 V, 18.6 C, 0.7 Fe, and 0.5 N, after exposure: 80.8 V, 18.4 C, and 0.3 N; sample supplied by the Carborundum Co; hot pressed; crushed in a hardened steel mortar to pass 100-mesh screen; density at 25 C, before exposure: apparent density (ASTM method B311-58) 338 lb ft^{-3}, true density (by immersion in xylene) 342 lb ft^{-3}, after exposure: apparent density 312 lb ft^{-3}, true density 318 lb ft^{-3}.

DATA TABLE NO. 145 SPECIFIC HEAT OF VANADIUM CARBIDE VC

[Temperature, T, K; Specific Heat, C_p, Cal g^{-1} K^{-1}]

T	C_p
CURVE 1	
52.50	9.363×10^{-3}
56.50	1.140×10^{-2}
60.20	1.351
64.30	1.599
68.50	1.866
72.70	2.146
77.00	2.433
85.00	2.972
94.70	3.632
104.40	4.282
115.10	4.975
125.20	5.599
135.10	6.185
145.60	6.774
155.30	7.298
165.50	7.816
175.50	8.318*
185.60	8.779
195.90	9.226*
205.80	9.628
216.20	1.005×10^{-1}*
226.00	1.041
235.80	1.076*
246.10	1.110
256.20	1.144*
266.10	1.172
276.20	1.203*
286.30	1.238*
297.00	1.264*
298.16	1.266
CURVE 2	
397.00	1.470×10^{-1}
400	1.474*
500	1.596
600	1.687
700	1.762
800	1.829
900	1.892
1000	1.951
1100	2.009
1200	2.066
CURVE 2 (cont.)	
1300	2.121×10^{-1}
1400	2.176
1500	2.231
1600	2.285*
1611	2.291
CURVE 3	
533.15	1.900×10^{-1}
810.93	2.000
1088.71	2.130
1366.48	2.270
1644.26	2.460
1922.04	2.800
2199.82	3.100
2477.59	3.500
2755.37	3.800

*Not shown on plot

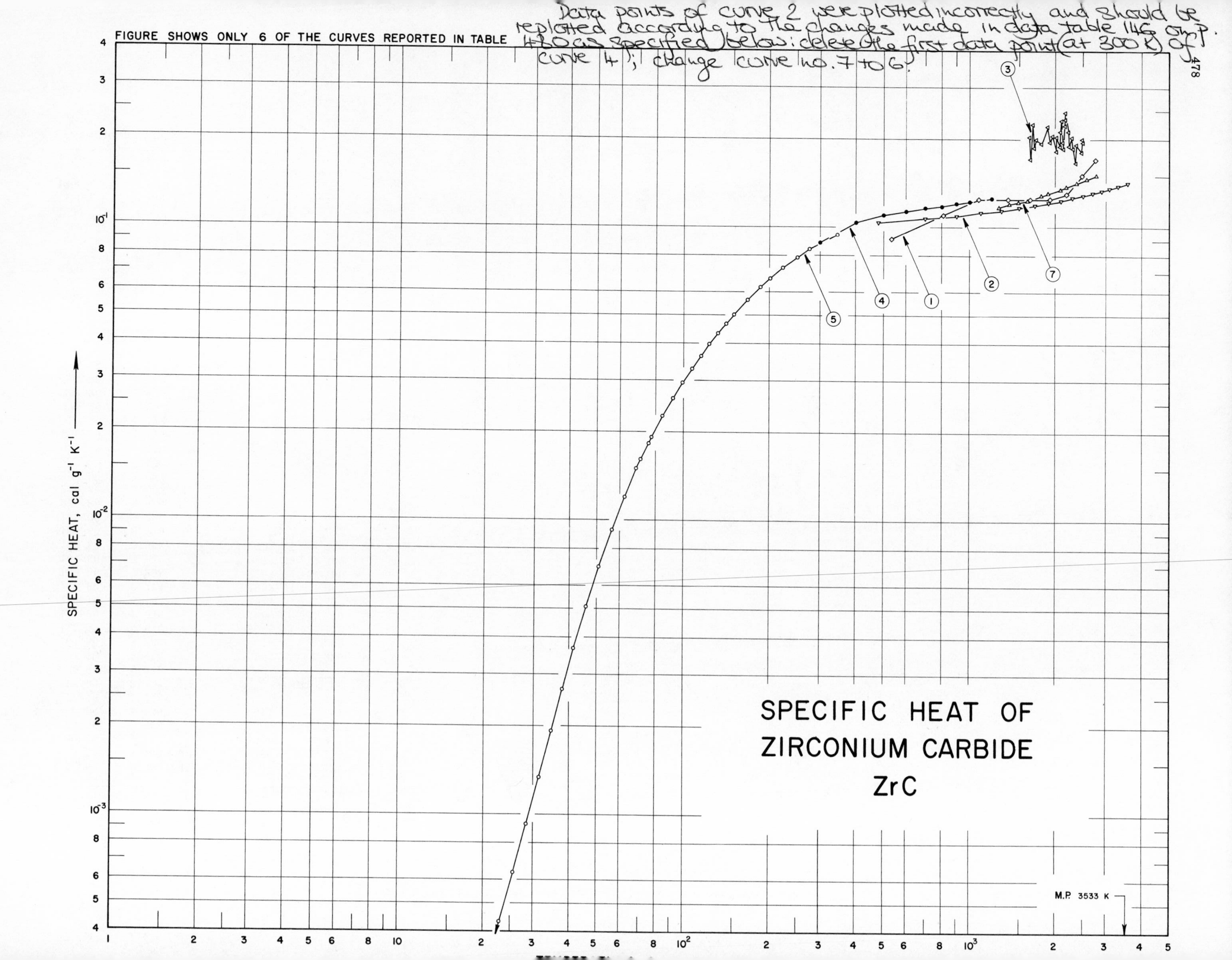
FIGURE SHOWS ONLY 6 OF THE CURVES REPORTED IN TABLE
Data points of curve 2 were plotted incorrectly and should be replotted according to the changes made in data table 146 on p. 480 as specified below: delete the first data point (at 300 K) of curve 4; change curve no. 7 to 6.
SPECIFIC HEAT, cal g^{-1} K^{-1}
SPECIFIC HEAT OF
ZIRCONIUM CARBIDE
ZrC
M.P. 3533 K
1
2
3
4
5
7
10
10^2
10^3
10^{-1}
10^{-2}
10^{-3}

SPECIFICATION TABLE NO. 146 SPECIFIC HEAT OF ZIRCONIUM CARBIDE ZrC

[For Data Reported in Figure and Table No. 146]

Curve No.	Ref. No.	Year	Temp. Range, K	Reported Error, %	Name and Specimen Designation	Composition (weight percent), Specifications and Remarks
1	32	1962	533-2755	≤5		Before exposure: 88.5 Zr, 11.0 C, 0.5 N, and 0.1 total impurities, after exposure: 89.3 Zr, 10.8 C, and 0.5 N; sample supplied by the General Electric Co; pressed and sintered; density at 25 C, before exposure: apparent density (ASTM method B311-58) 392 lb ft^{-3}, true density (by immersion in xylene) 395 lb ft^{-3}, after exposure: apparent density 289 lb ft^{-3}, true density 292 lb ft^{-3}.
2	185	1962	480-3573			11.76 total C and 11.09 combined C; sprayed sample.
3	173	1962	1639-2499	5		84.23 Zr, 10.46 total C, 1.21 O_2, 1.03 N, 0.78 Hf, <0.3 Nb, 0.25 free C, <0.2 Ni, 0.19 Fe, 0.15 Cu, 0.14 Ti, 0.12 Al, 0.07 V, 0.02 Mn, and <0.01 Mg; solid pieces machined to specifications by using diamond tools and electric discharge techniques.
4	206	1963	300-1200			89.27 Zr, 11.20 C, 0.005 O_2 and 0.067 N_2, impurities: quantitative spectrographic analysis, 0.12 Ti, 0.07 B, 0.07 Fe, 0.001 Si, semi-quantitative spectrographic analysis, 0.01 Al and Sn, 0.001 - 0.01 Mg, 0.001 Ca, Cu, Hf, Mn, Mo, and Pb; zone-refined.
5	206	1963	5-345	0.1-5		89.27 Zr, 11.20 C, 0.005 O_2, and 0.067 N_2, impurities quantitative spectrographic analysis, 0.12 Ti, 0.07 B, 0.07 Fe, and 0.001 Si, semiquantitative analysis, 0.01 Al and Sn, 0.001 - 0.01 Mg, 0.0001 Ca, Cu, Hf, Mn, Mo, and Pb; zone-refined.
6	54	1965	300-1200	±0.70		Traces of Ca, Si, Al, Mn, and Fe; sample supplied by the Carborundum Co.
7	196	1965	1275-2788	1.4		88.24 Zr, 11.13 combined C, 0.24 free C, <0.1 Hf, and 0.05 O_2; lattice parameter $a_0 = 4.6953 \pm 0.0002$ Å; hot pressed; in graphite dies and then electrospark machined.

DATA TABLE NO. 146 SPECIFIC HEAT OF ZIRCONIUM CARBIDE ZrC

[Temperature, T, K; Specific Heat, C_p, Cal g^{-1} K^{-1}]

T	C_p
CURVE 1	
533.15	9.100 x 10^{-2}
810.93	1.100 x 10^{-1}
1088.71	1.250
1366.48	1.250
1644.26	1.250
1922.04	1.250
2199.82	1.300
2477.59	1.500
2755.37	1.700
CURVE 2	
400	1.038 x 10^{-1}
500	1.041*
700	1.069
900	1.095
1100	1.120
1300	1.145
1500	1.169
1700	1.194
1900	1.217
2100	1.241
2300	1.265
2500	1.285
2700	1.312
2900	1.336
3100	1.360
3300	1.383
3500	1.407*
3573	1.415
CURVE 3	
1639	2.040 x 10^{-1}
1648	1.710
1675	2.250
1696	1.870
1721	2.000
1799	1.930
1814	1.950*
1897	2.210
1915	1.960
1955	2.060
2017	1.820
CURVE 3 (cont.)	
2023	2.040 x 10^{-1}
2057	2.040*
2071	2.080*
2080	2.020*
2080	2.080*
2085	2.080*
2094	1.890
2094	2.130
2100	2.210*
2107	2.320
2113	2.240*
2113	2.240*
2118	2.100*
2132	1.960
2138	2.130*
2146	2.050*
2147	1.860
2164	2.230
2166	2.150*
2169	2.340
2172	2.100*
2175	2.470
2189	2.310*
2198	2.250*
2206	2.080*
2232	2.110
2238	2.170*
2243	2.110*
2249	1.890
2252	2.010*
2286	2.030
2298	2.130*
2306	2.000*
2310	1.860
2364	1.670
2364	1.910
2367	1.820*
2403	1.780*
2411	1.760*
2420	1.810*
2471	1.800
2474	1.990
2483	1.870*
2486	1.810*
CURVE 3 (cont.)	
2492	1.960 x 10^{-1}*
2495	1.910*
2499	1.960
CURVE 4	
300	8.806 x 10^{-2}
400	1.046 x 10^{-1}
500	1.101
600	1.136
700	1.164
800	1.186
900	1.205
1000	1.224
1100	1.240*
1200	1.257
CURVE 5	
Series 1	
71.02	1.607 x 10^{-2}
77.48	1.916
84.53	2.260
92.01	2.604
99.71	2.944
107.56	3.286
115.66	3.631
123.93	3.978
132.58	4.335
141.47	4.680
150.35	5.020
159.30	5.341*
168.31	5.656
177.30	5.954*
186.29	6.240
195.30	6.511*
222.32	7.242
231.69	7.471*
240.94	7.689*
250.03	7.883
258.94	8.069*
267.76	8.236*
276.48	8.406
292.81	8.692*
CURVE 5 (cont.)	
Series 1 (cont.)	
301.41	8.820 x 10^{-2}*
336.31	9.303*
345.16	9.417
*Series 2**	
313.69	9.007 x 10^{-2}
322.55	9.133
331.36	9.258
Series 3	
5.59	1.066 x 10^{-5}*
6.75	1.550*
7.85	2.325*
9.11	3.294*
10.43	4.650*
11.81	6.200*
13.30	8.429*
14.83	1.143 x 10^{-4}*
16.52	1.541*
18.44	2.121*
20.59	3.013*
22.95	4.331
25.57	6.336
28.44	9.252
31.45	1.340 x 10^{-3}
34.62	1.913
37.91	2.661
41.60	3.665
45.97	5.064
50.94	6.942
56.46	9.259
62.43	1.200 x 10^{-2}
68.86	1.501
75.78	1.831
201.94	6.698
205.90	6.814*
209.81	6.913*
213.67	7.014*
217.49	7.117*
CURVE 6*	
300	9.446 x 10^{-2}
400	1.046 x 10^{-1}
500	1.101
600	1.136
700	1.164
800	1.186
900	1.205
1000	1.224
1100	1.240
1200	1.257
CURVE 7	
1275	1.17 x 10^{-1}
1383	1.20
1489	1.22
1601	1.24
1692	1.26*
1794	1.29
1892	1.31
2089	1.35
2191	1.38
2282	1.40*
2384	1.42
2493	1.44*
2591	1.47
2686	1.49*
2788	1.51

* Not shown on plot

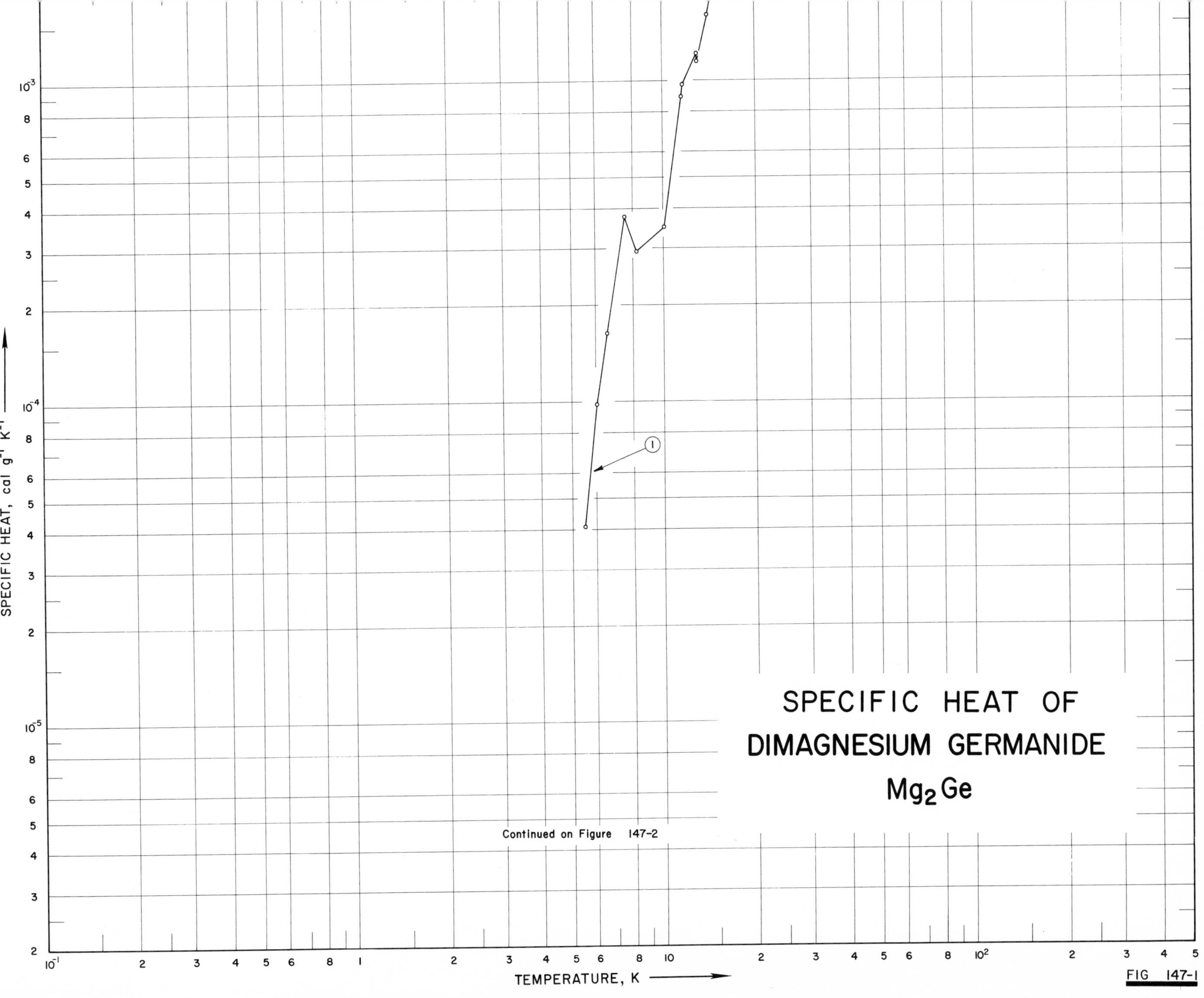

SPECIFIC HEAT OF
DIMAGNESIUM GERMANIDE
Mg_2Ge
Continued on Figure 147-2
1
TEMPERATURE, K
SPECIFIC HEAT, cal g^{-1} K^{-1}
FIG 147-1

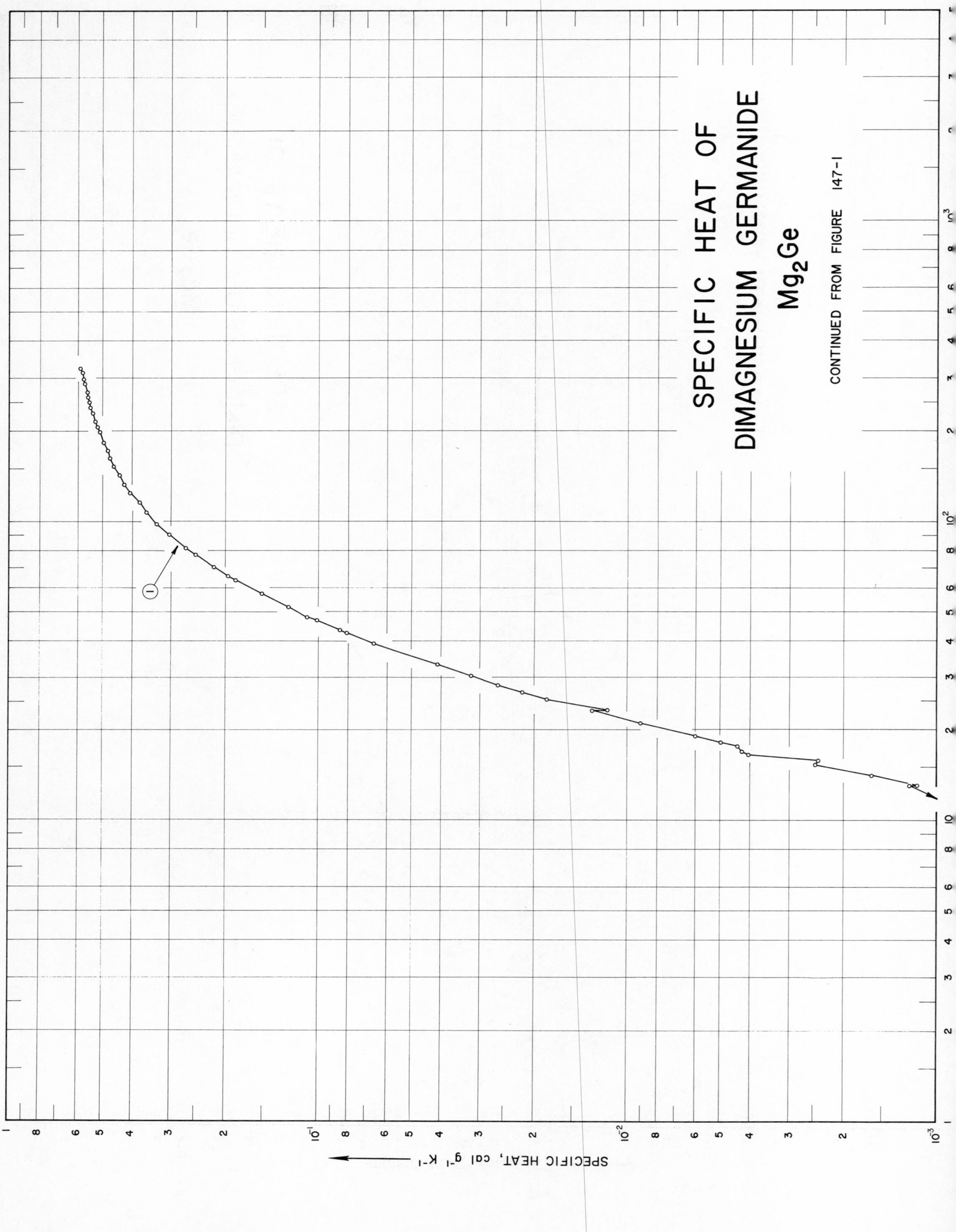

SPECIFIC HEAT OF
DIMAGNESIUM GERMANIDE
Mg_2Ge
CONTINUED FROM FIGURE 147-1
SPECIFIC HEAT, cal g^{-1} K^{-1}
1

SPECIFICATION TABLE NO. 147 SPECIFIC HEAT OF DIMAGNESIUM GERMANIDE Mg_2Ge

[For Data Reported in Figure and Table No. 147]

Curve No.	Ref. No.	Year	Temp. Range, K	Reported Error, %	Name and Specimen Designation	Composition (weight percent), Specifications and Remarks
1	407	1966	6-325			Prepared from stoichiometric proportions of Mg (99.99) and Ge; 40 ohm cm resistivity.

DATA TABLE NO. 147 SPECIFIC HEAT OF DIMAGNESIUM GERMANIDE Mg_2Ge

[Temperature, T, K; Specific Heat, C_p, Cal $g^{-1}K^{-1}$]

T	C_p
CURVE 1	
Series 1	
6.07	9.900×10^{-5}
7.53	3.795×10^{-4}
11.69	9.817
13.04	1.163×10^{-3}
14.16	1.625
15.34	2.475
16.58	4.067
17.69	4.422
19.06	6.039
23.35	1.165×10^{-2}
25.38	1.823
28.34	2.623
Series 2	
39.39	6.616×10^{-2}
43.77	8.514
48.31	1.090×10^{-1}
66.32	1.968
Series 3	
82.0	2.692×10^{-1}
90.8	3.057
99.1	3.344
107.6	3.617
116.4	3.798
125.2	4.082
133.1	4.254
143.0	4.417
152.7	4.606
162.9	4.741
172.9	4.826
183.8	4.988
199.7	5.112
206.4	5.206
215.8	5.300
Series 4	
219.9	5.322×10^{-1}*
230.1	5.395
240.3	5.485

T	C_p
CURVE 1 (cont.)	
Series 4 (cont.)	
250.4	5.510×10^{-1}
260.4	5.574
270.3	5.628
277.0	5.639*
289.4	5.725
298.7	5.760
308.4	5.833*
315.7	5.823
325.0	5.918
Series 5	
5.51	4.125×10^{-5}
6.58	1.650×10^{-4}
8.27	2.970
10.14	3.547
11.62	8.992
13.01	1.229×10^{-3}
14.35	1.633*
15.72	2.425
16.92	4.282
18.12	4.991
19.48	6.146*
21.02	9.075
23.24	1.312×10^{-2}
26.78	2.186
30.46	3.209
33.26	4.125
Series 6	
39.1	6.559×10^{-2}*
42.7	8.110*
47.0	1.011×10^{-1}
52.0	1.256
57.8	1.527
64.1	1.863
70.7	2.179
77.8	2.512

T	C_p
CURVE 1 (cont.)	
Series 7*	
82.4	2.696×10^{-1}
91.6	3.074
100.5	3.387
109.7	3.656
119.5	4.155
140.8	4.363
152.2	4.565
163.8	4.767
175.1	4.879
186.1	5.020
196.9	5.121
207.7	5.230
218.3	5.369
Series 8*	
228.7	5.392×10^{-1}
238.9	5.475
249.0	5.527
259.0	5.567
268.8	5.602
278.7	5.658
288.5	5.724
298.2	5.718
307.9	5.780

* Not shown on plot

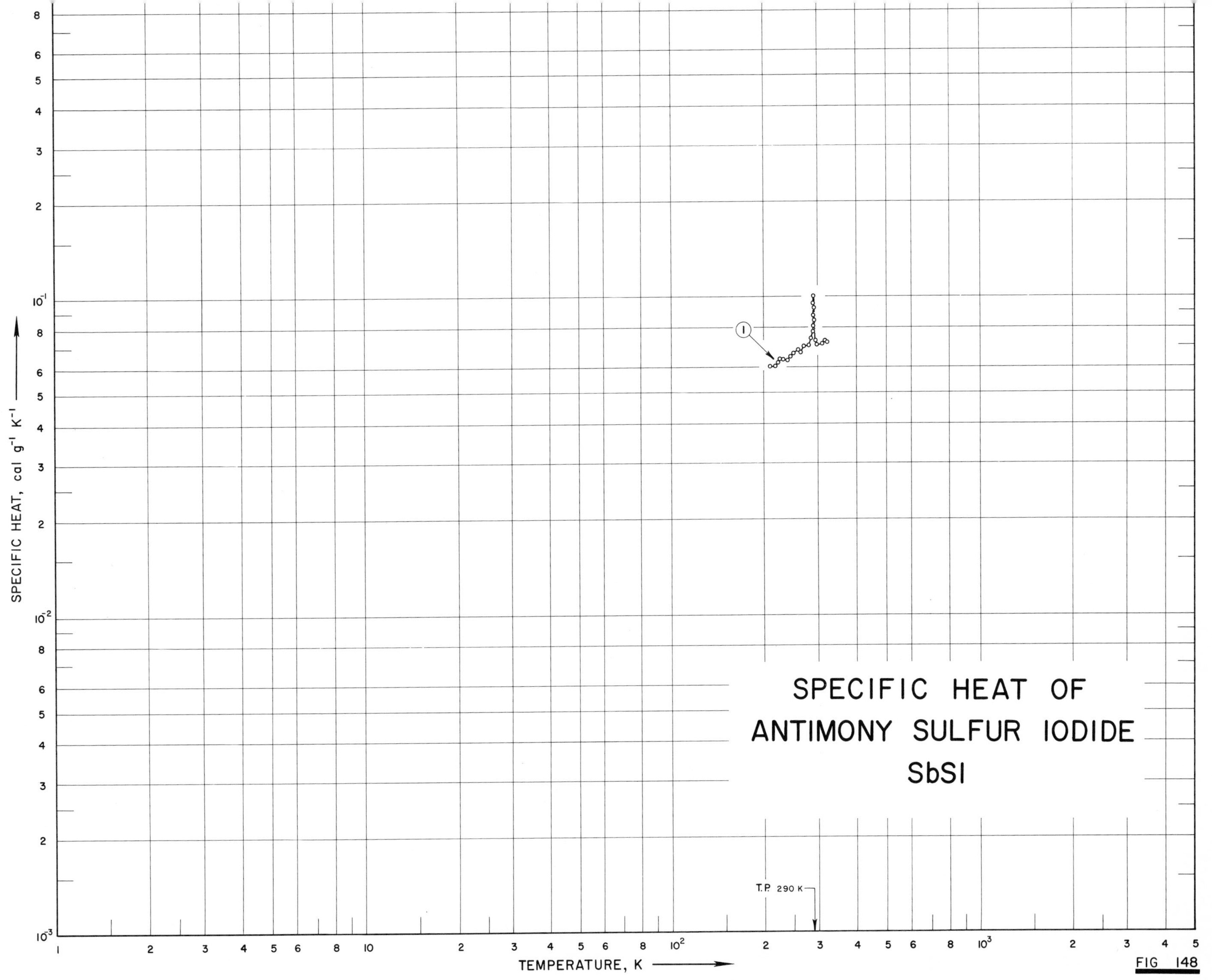

SPECIFIC HEAT OF
ANTIMONY SULFUR IODIDE
SbSI
T.P. 290 K
1
SPECIFIC HEAT, cal g⁻¹ K⁻¹
TEMPERATURE, K

FIG 148

SPECIFICATION TABLE NO. 148 SPECIFIC HEAT OF ANTIMONY SULFUR IODIDE SbSI

[For Data Reported in Figure and Table No. 148]

Curve No.	Ref. No.	Year	Temp. Range, K	Reported Error, %	Name and Specimen Designation	Composition (weight percent), Specifications and Remarks
1	336	1965	210-325			99.9999 SbSI; polycrystalline; prepared from SbI_3 and Sb_2S_3; sealed in evacuated pyrex ampule; heated to 500 C; melt cooled to room temperature in two days.

DATA TABLE NO. 148 SPECIFIC HEAT OF ANTIMONY SULFUR IODIDE SbSI

[Temperature, T, K; Specific Heat, C_p, Cal $g^{-1}K^{-1}$]

T	C_p
CURVE 1	
Series I	
210.25	6.02 x 10^{-2}
211.25	5.95*
212.05	5.98*
214.25	5.98*
215.15	6.13*
215.85	6.06*
217.35	6.02*
218.25	5.98*
219.15	6.02
220.05	6.13*
Series II	
217.65	6.13 x 10^{-2}*
218.85	6.16*
219.75	6.16*
220.65	6.16*
221.85	6.06*
222.75	6.20*
223.65	6.23
224.85	6.30*
225.75	6.27*
227.45	6.34
228.35	6.27*
229.55	6.27*
230.65	6.20*
231.55	6.27*
232.45	6.38
233.55	6.23*
234.45	6.34
235.35	6.31*
238.15	6.37*
239.35	6.37*
240.15	6.34
241.05	6.34*
242.15	6.37*
243.15	6.34*
243.85	6.45*
244.95	6.45*
245.75	6.55
246.75	6.52*
247.75	6.52*
248.55	6.45*
249.65	6.63*

T	C_p
CURVE 1 (cont.)	
250.75	6.63 x 10^{-2}
251.55	6.63*
254.55	6.63*
255.75	6.77*
256.45	6.59*
257.55	6.77*
258.35	6.77*
259.25	6.80*
260.55	6.73
261.55	6.73*
262.35	6.73*
263.35	6.77*
264.25	6.84*
265.25	6.77
266.05	6.84*
267.05	6.95*
267.95	6.98*
268.95	7.02*
269.95	7.09*
270.85	6.98
271.55	6.91*
272.35	7.09*
273.41	6.91*
274.19	6.95*
274.97	7.05*
275.75	6.98*
276.52	6.91*
277.29	6.91*
278.06	6.88*
278.83	7.05*
281.90	7.05
282.67	7.37*
283.45	7.41*
284.15	7.30*
285.25	7.30*
285.95	7.45
286.75	7.37*
287.45	7.34*
288.25	7.62*
288.75	7.80*
288.95	7.91*
289.45	8.34*
289.75	7.87*
290.45	7.77
290.65	8.12*

T	C_p
CURVE 1 (cont.)	
291.25	1.01 x 10^{-1}*
291.45	8.87 x 10^{-2}*
291.45	9.58
291.65	1.01 x 10^{-1}*
291.85	1.05*
291.95	9.94 x 10^{-2}*
292.15	9.26
292.25	9.30*
292.45	9.62*
293.15	8.48
293.85	7.84*
294.55	7.69*
295.55	7.45*
296.45	7.30
297.15	7.20*
297.95	7.27*
299.05	7.12
299.75	7.12*
300.45	7.16*
301.15	6.91*
301.95	7.05*
302.75	7.02*
303.55	7.09*
312.15	7.16
312.95	7.09*
314.15	7.09*
317.65	7.30
318.35	7.16*
319.05	7.09*
320.05	7.26*
320.75	7.23*
321.45	7.20*
322.15	7.16*
323.15	7.20*
323.95	7.12*
324.65	7.20

* Not shown on plot

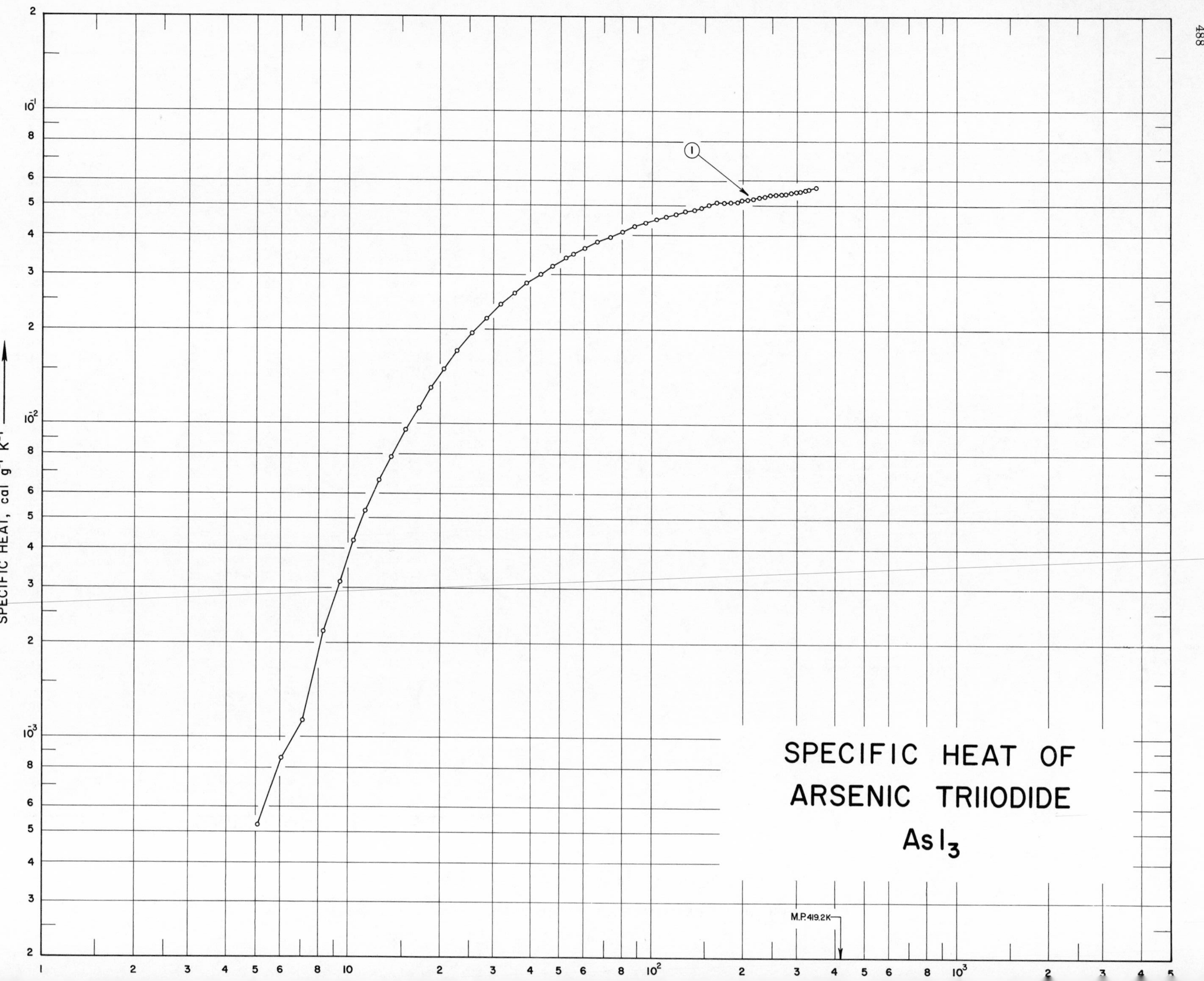

SPECIFIC HEAT OF
ARSENIC TRIIODIDE
AsI_3
M.P. 419.2K
SPECIFIC HEAT, cal g^{-1} K^{-1}

SPECIFICATION TABLE NO. 149 SPECIFIC HEAT OF ARSENIC TRIIODIDE AsI_3

[For Data Reported in Figure and Table No. 149]

Curve No.	Ref. No.	Year	Temp. Range, K	Reported Error, %	Name and Specimen Designation	Composition (weight percent), Specifications and Remarks
1	327	1955	5-349	±0.15-6.0		83.30 I and 16.45 As, (83.60 and 16.44 theo).

DATA TABLE NO. 149 SPECIFIC HEAT OF ARSENIC TRIIODIDE AsI_3

[Temperature, T, K; Specific Heat, C_p, Cal $g^{-1}K^{-1}$]

T	C_p	T	C_p	T	C_p
CURVE 1		**CURVE 1 (cont.)**		**CURVE 1 (cont.)**	
Series I		Series III		Series VII*	
5.11	5.27 x 10^{-4}	197.77	5.217 x 10^{-2}	256.05	5.456 x 10^{-2}
6.08	8.56	207.28	5.230	266.16	5.504
7.17	1.40 x 10^{-3}	216.95	5.294	276.28	5.511
8.33	2.19	226.99	5.335	286.49	5.520
9.43	3.14	237.22	5.377	296.83	5.548
10.46	4.28	247.54	5.417	305.07	5.570
11.48	5.331	257.83	5.452	315.20	5.605
12.64	6.540	268.14	5.454	325.33	5.643
13.92	7.886	278.56	5.476	335.59	5.678
15.44	9.591	289.04	5.520	345.85	5.715
17.02	1.134 x 10^{-2}	299.47	5.550		
18.69	1.308	309.85	5.586	Series VIII	
20.60	1.504	320.20	5.625		
22.89	1.725	330.52	5.669	238.00	5.381 x 10^{-2}
25.50	1.957	340.84	5.711*	248.52	5.419
28.45	2.189	348.61	5.744	258.86	5.480
31.70	2.421			266.63	5.504
35.19	2.638	Series IV*		271.88	5.509
38.95	2.840			277.13	5.520
43.07	3.033	142.59	4.923 x 10^{-2}	282.47	5.528
47.66	3.228	151.76	4.986		
52.70	3.415	161.01	5.054		
		170.79	5.094		
Series II		180.74	5.147		
55.21	3.501 x 10^{-2}	Series V*			
60.44	3.665				
66.63	3.839	151.13	4.991 x 10^{-2}		
73.47	3.999	158.04	5.019		
80.79	4.161	163.22	5.057		
88.21	4.306	166.23	5.072		
95.86	4.422	169.42	5.105		
104.09	4.532				
112.53	4.635	Series VI*			
121.01	4.727				
129.39	4.811	196.72	5.212 x 10^{-2}		
137.85	4.881	207.35	5.256		
146.45	4.945	218.09	5.298		
155.21	5.011	228.67	5.340		
163.96	5.153	239.08	5.384		
172.85	5.107	249.34	5.432		
182.33	5.153				
192.02	5.195				

* Not shown on plot

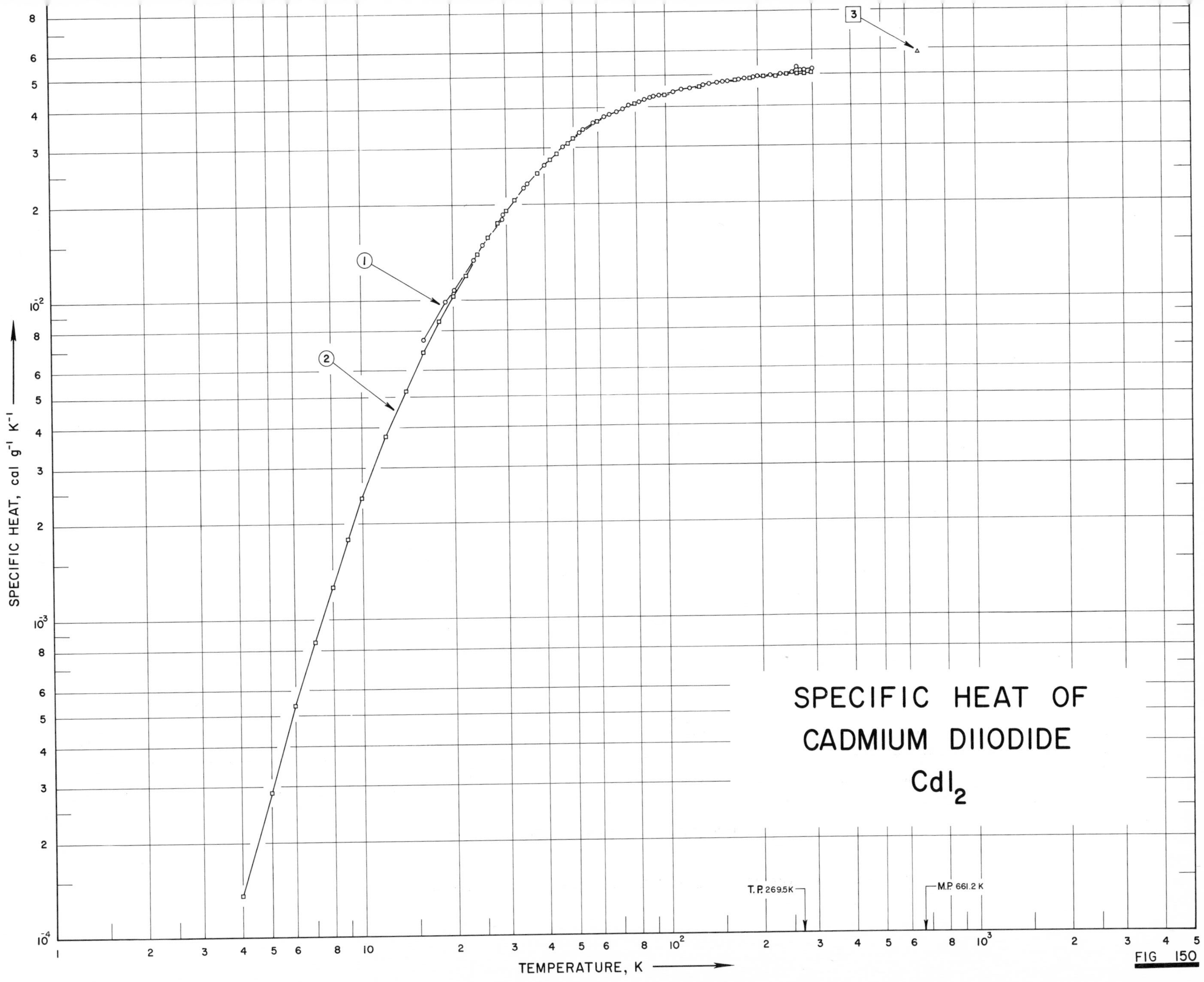

SPECIFIC HEAT OF
CADMIUM DIIODIDE
CdI_2
T.P. 269.5K
M.P. 661.2 K
TEMPERATURE, K
SPECIFIC HEAT, cal g^{-1} K^{-1}
1
2
3
FIG 150

SPECIFICATION TABLE NO. 150 SPECIFIC HEAT OF CADMIUM DIIODIDE CdI_2

[For Data Reported in Figure and Table No. 150]

Curve No.	Ref. No.	Year	Temp. Range, K	Reported Error, %	Name and Specimen Designation	Composition (weight percent), Specifications and Remarks
1	303	1955	16-299	< 0.2-0.5		C.P. analyzed CdI_2; sample supplied by the J.T. Baker Chem. Co; dried in oven for 20 hrs at 115 C; cooled in dessicator.
2	248	1959	2.0-300	1.0		
3	242	1961	661.2			

DATA TABLE NO. 150 SPECIFIC HEAT OF CADMIUM DIIODIDE CdI_2

[Temperature, T, K; Specific Heat, C_p, Cal $g^{-1}K^{-1}$]

T	C_p
CURVE 1	
Series I	
16.08	7.531 x 10^{-3}
20.24	1.085 x 10^{-2}
25.05	1.493
29.33	1.865
34.37	2.264
40.08	2.666
46.40	3.034
Series II	
18.97	9.936 x 10^{-2}
23.31	1.345 x 10^{-2}
29.10	1.795
35.32	2.328
40.60	2.678*
46.54	3.044*
52.77	3.358
54.13	3.425
58.35	3.597
63.43	3.763
69.46	3.892
76.38	4.048
82.89	4.184
Series III	
54.20	3.418 x 10^{-2}*
59.61	3.648*
65.91	3.819
73.10	3.975
79.82	4.118*
86.18	4.241
92.27	4.331
Series IV	
89.73	4.308 x 10^{-2}
95.72	4.355
101.53	4.424*
107.18	4.485
113.84	4.544
121.57	4.599
129.12	4.653*
CURVE 1 (cont.)	
Series V	
109.69	4.504 x 10^{-2}*
115.17	4.559*
120.54	4.599*
126.95	4.642*
133.55	4.686
140.87	4.730
148.06	4.766
155.15	4.800
161.30	4.828
168.23	4.853*
Series VI	
133.60	4.681 x 10^{-2}*
140.92	4.737*
148.11	4.765*
155.20	4.794*
161.55	4.823*
168.48	4.846*
175.33	4.878
182.12	4.897
188.85	4.925*
195.52	4.942
202.15	4.964
208.73	4.973*
215.28	4.976*
221.79	4.993
Series VII	
177.10	4.882 x 10^{-2}*
183.88	4.909*
190.61	4.931*
197.29	4.946*
202.52	4.953*
211.49	4.976*
220.38	4.998*
229.21	5.018*
237.97	5.037
246.67	5.062
252.51	5.073*
261.13	5.100
269.66	5.194*
CURVE 1 (cont.)	
Series VIII	
259.11	5.087 x 10^{-2}*
265.43	5.140*
271.71	5.155
277.96	5.167*
285.30	5.190
293.73	5.190*
302.12	5.237
Series IX	
247.39	5.067 x 10^{-2}*
256.01	5.092*
264.58	5.149*
273.51	5.171*
281.92	5.179*
290.30	5.198*
298.65	5.209*
Series X	
209.26	4.981 x 10^{-2}*
215.78	4.991*
222.27	5.006*
228.73	5.012*
235.15	5.033*
241.54	5.049*
247.90	5.057*
254.23	5.058*
260.52	5.087*
261.10	5.103*
267.34	5.220*
273.54	5.155*
279.75	5.180*
285.94	5.192*
292.11	5.205*
298.24	5.205*
Series XI	
260.21	5.093 x 10^{-2}*
266.52	5.155*
258.77	5.073*
264.08	5.086*
CURVE 1 (cont.)	
268.40	5.178 x 10^{-2}*
272.69	5.151*
264.82	5.122*
267.13	5.209*
269.41	5.297
271.70	5.139*
274.01	5.171*
276.32	5.154*
CURVE 2	
2	1.67 x 10^{-5}*
3	5.61*
4	1.37 x 10^{-4}
5	2.88
6	5.39
7	8.54
8	1.27 x 10^{-3}
9	1.79
10	2.41
12	3.76
14	5.23
16	6.90
18	8.61
20	1.04 x 10^{-2}
22	1.21
24	1.40
26	1.58
28	1.75
30	1.92
32	2.08
34	2.23*
36	2.37*
38	2.51
40	2.64*
42	2.766
44	2.881
46	2.990*
48	3.100
50	3.204
55	3.431*
60	3.625
65	3.787*
70	3.918*
CURVE 2 (cont.)	
75	4.036 x 10^{-2}*
80	4.129
85	4.209*
90	4.272*
95	4.329*
100	4.378
105	4.425*
110	4.471*
115	4.512*
120	4.554*
125	4.592*
130	4.630
135	4.663*
140	4.696*
145	4.726*
150	4.756*
155	4.781*
160	4.803*
165	4.825*
170	4.844
175	4.860*
180	4.879*
185	4.893*
190	4.910
195	4.923*
200	4.934*
205	4.948*
210	4.959
215	4.967*
220	4.978*
225	4.986*
230	4.994
235	5.000*
240	5.005*
245	5.011*
250	5.016
255	5.022*
260	5.027*
265	5.033*
270	5.038
275	5.044*
280	5.049*
285	5.057
290	5.063*
CURVE 2 (cont.)	
295	5.068 x 10^{-2}*
298.16	5.074*
300	5.077
CURVE 3	
661.2	5.87 x 10^{-2}

* Not shown on plot

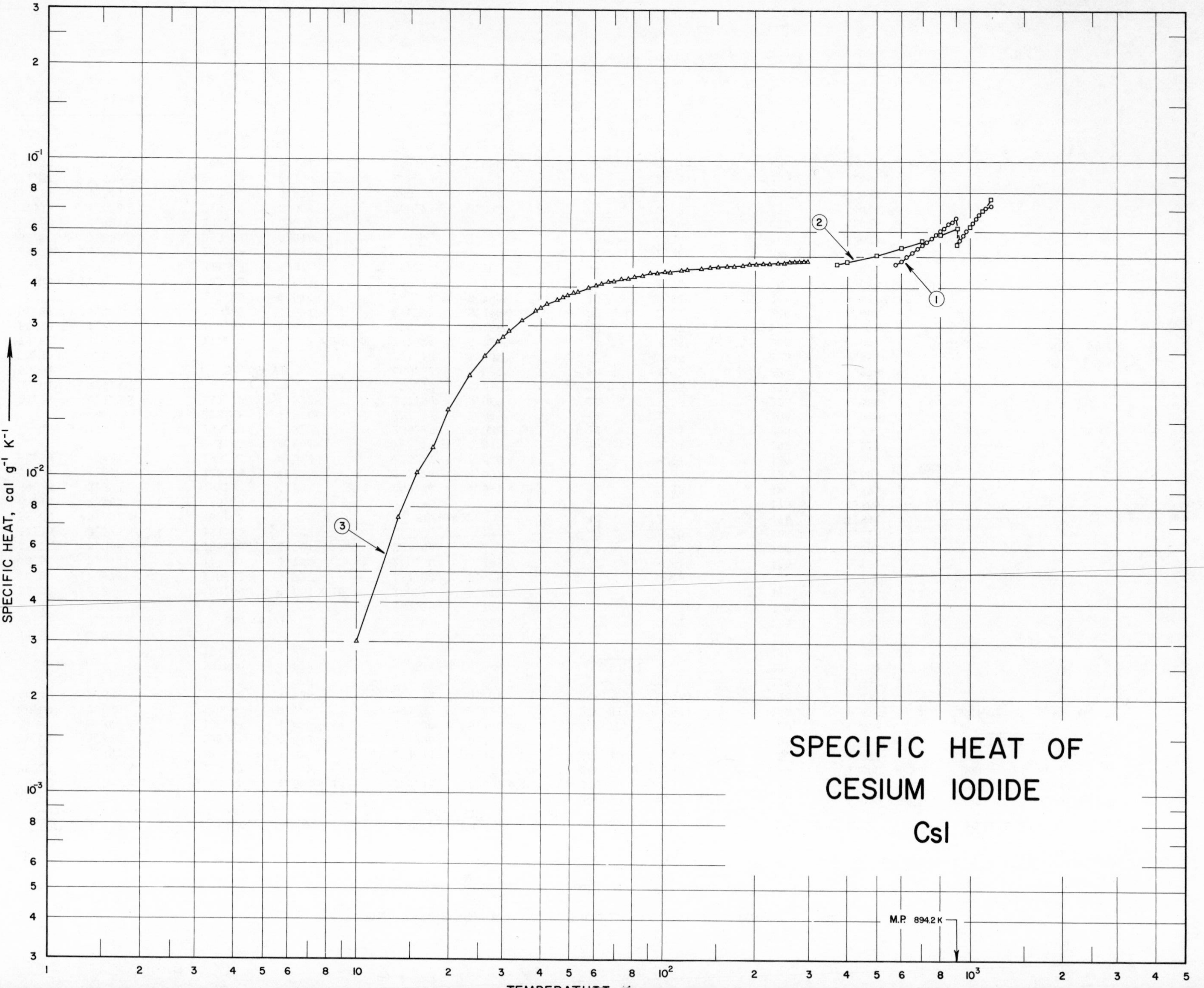
SPECIFIC HEAT OF
CESIUM IODIDE
CsI
M.P. 894.2 K
SPECIFIC HEAT, cal g⁻¹ K⁻¹
1
2
3

SPECIFICATION TABLE NO. 151 SPECIFIC HEAT OF CESIUM IODIDE CsI

[For Data Reported in Figure and Table No. 151]

Curve No.	Ref. No.	Year	Temp. Range, K	Reported Error, %	Name and Specimen Designation	Composition (weight percent), Specifications and Remarks
1	251	1958	575-1175			
2	252	1961	370-1172			99.80 CsI.
3	253	1963	14-298	0.3		> 99.989 CsI, 0.001-0.01 Ca, Na, 0.0001-0.001 Al and K; chemically pure grade sample was recrystallized several times from water; dried at 260 to 300 C in a vacuum oven for several hrs.

DATA TABLE NO. 151 SPECIFIC HEAT OF CESIUM IODIDE CsI

[Temperature, T, K; Specific Heat, C_p, Cal $g^{-1}K^{-1}$]

T	C_p
CURVE 1	
575	4.747 x 10^{-2}
600	4.886
625	5.026
650	5.167
675	5.309
700	5.453
725	5.597
750	5.742
775	5.888
800	6.034
825	6.181
850	6.329
875	6.477
900	6.624
(s) 907	6.666*
(l) 907	5.491*
925	5.643
950	5.850
975	6.051
1000	6.246
1025	6.436
1050	6.622
1075	6.803
1100	6.982
1125	7.156
1150	7.328
1175	7.497
CURVE 2	
370	4.754 x 10^{-2}
400	4.830
500	5.090
600	5.357
700	5.625
800	5.895
(s) 904	6.177
(l) 908	5.489*
1000	6.238*
1100	7.051*
1172	7.636

T	C_p
CURVE 3	
10.00	3.002 x 10^{-3}
13.51	7.417
15.52	1.028 x 10^{-2}
17.06	1.242
19.80	1.638
23.39	2.100
26.24	2.413
28.92	2.676
30.00	2.775
31.53	2.898
34.83	3.143
38.73	3.366
40.00	3.425
42.27	3.528
45.53	3.654
47.45	3.721
48.59	3.751*
49.12	3.771
51.48	3.848
52.00	3.859
54.74	3.931*
57.62	3.992
58.50	4.010*
60.64	4.054
63.31	4.104
63.99	4.116*
66.40	4.155
69.62	4.190
73.59	4.234
77.43	4.285
81.14	4.326
85.71	4.374
91.11	4.405
96.33	4.433
101.42	4.460
106.39	4.478
115.66	4.518
124.97	4.561
133.98	4.591
142.77	4.617
152.63	4.641
162.78	4.662
172.45	4.682
181.69	4.697
191.87	4.718
CURVE 3 (cont.)	
201.94	4.736 x 10^{-2}
207.13	4.744*
213.72	4.754
224.08	4.761
230.28	4.770*
239.24	4.782
250.06	4.794
259.14	4.809
269.17	4.819
278.63	4.826
288.16	4.849
298.39	4.857

* Not shown on plot

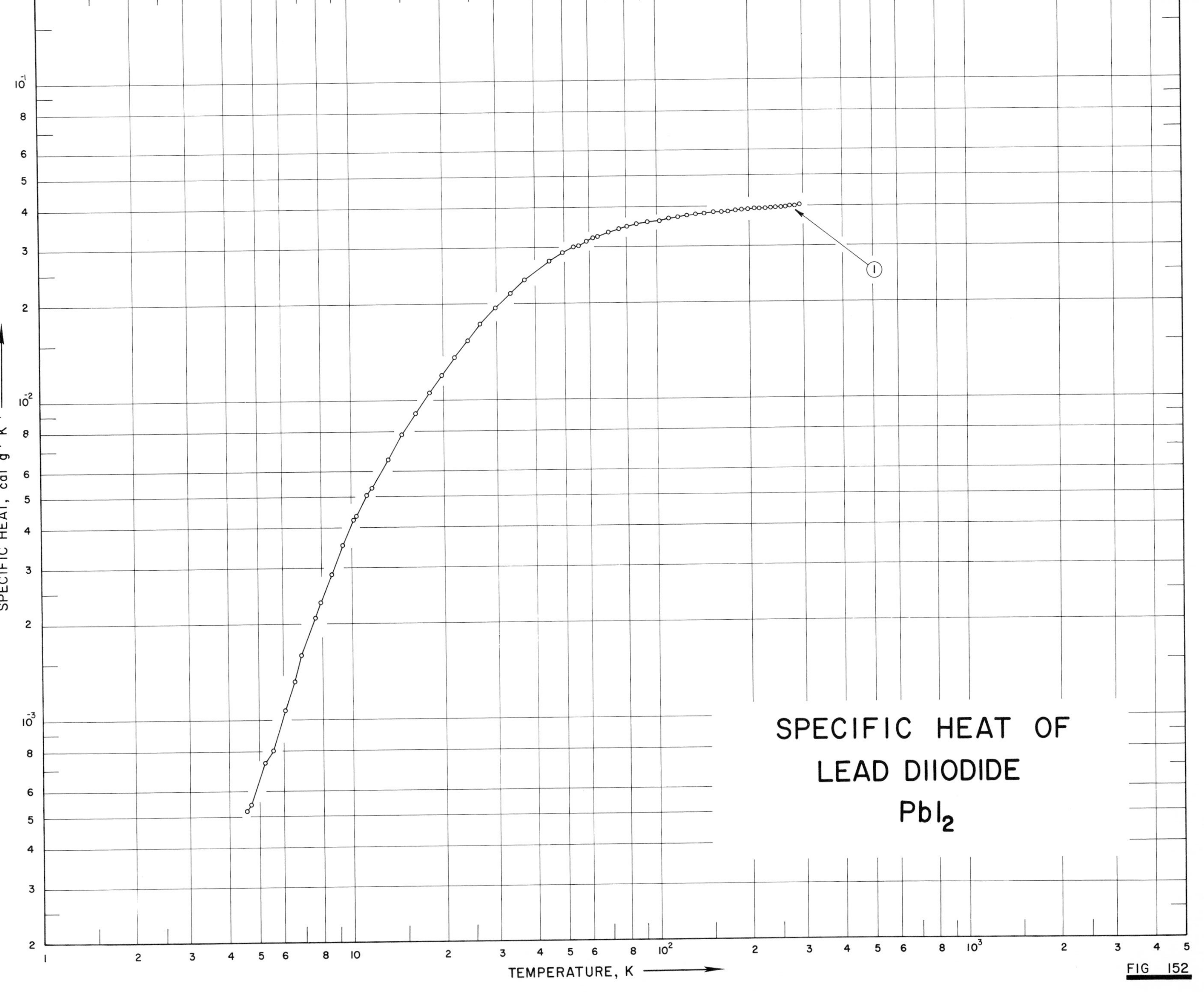
SPECIFIC HEAT OF
LEAD DIIODIDE
PbI_2
SPECIFIC HEAT, cal g⁻¹ K⁻¹
TEMPERATURE, K
1

FIG 152

SPECIFICATION TABLE NO. 152 SPECIFIC HEAT OF LEAD DIIODIDE PbI_2

[For Data Reported in Figure and Table No. 152]

Curve No.	Ref. No.	Year	Temp. Range, K	Reported Error, %	Name and Specimen Designation	Composition (weight percent), Specifications and Remarks
1	327	1955	4.7-291	0.15-6.0		55.10 I and 44.82 Pb, (55.05 and 44.95 theo); hemispherical pelletizing sample.

DATA TABLE NO. 152 SPECIFIC HEAT OF LEAD DIIODIDE PbI_2

[Temperature, T, K; Specific Heat, C_p, Cal $g^{-1}K^{-1}$]

T	C_p	T	C_p
CURVE 1		CURVE 1 (cont.)	
Series I		102.00	3.629×10^{-2}
4.70	5.423×10^{-4}	109.99	3.672
5.52	8.026	118.20	3.713
6.53	1.323×10^{-3}	126.73	3.750
7.97	2.343	135.28	3.779
8.67	2.885	143.90	3.802
10.24	4.230	154.11	3.831
11.29	5.054	162.76	3.848
		171.74	3.863
Series II		172.09	3.870*
		181.19	3.878
4.52	5.206×10^{-4}	190.28	3.893
5.21	7.375	199.37	3.906
6.07	1.085×10^{-3}	208.51	3.922
6.89	1.605	217.72	3.928
7.68	2.082	227.02	3.939
8.49	2.820*	236.24	3.948
9.37	3.514	245.43	3.959
10.41	4.360	254.60	3.969
11.69	5.338	263.72	3.978
13.17	6.522	272.77	4.000
14.79	7.850	281.86	4.002
16.34	9.130	291.00	4.011
18.07	1.056×10^{-2}		
19.89	1.205	Series IV	
21.85	1.366		
24.08	1.542	80.69	3.468×10^{-2}
26.72	1.739		
29.83	1.954	Series V	
33.31	2.173		
37.11	2.386	159.09	3.837×10^{-2}*
44.92	2.724	167.55	3.854*
49.36	2.876		
55.37	3.045		
62.25	3.199		
Series III			
53.82	3.004×10^{-2}		
59.13	3.132		
64.24	3.238		
69.79	3.325		
75.89	3.408		
86.13	3.529		
93.74	3.583		

* Not shown on plot

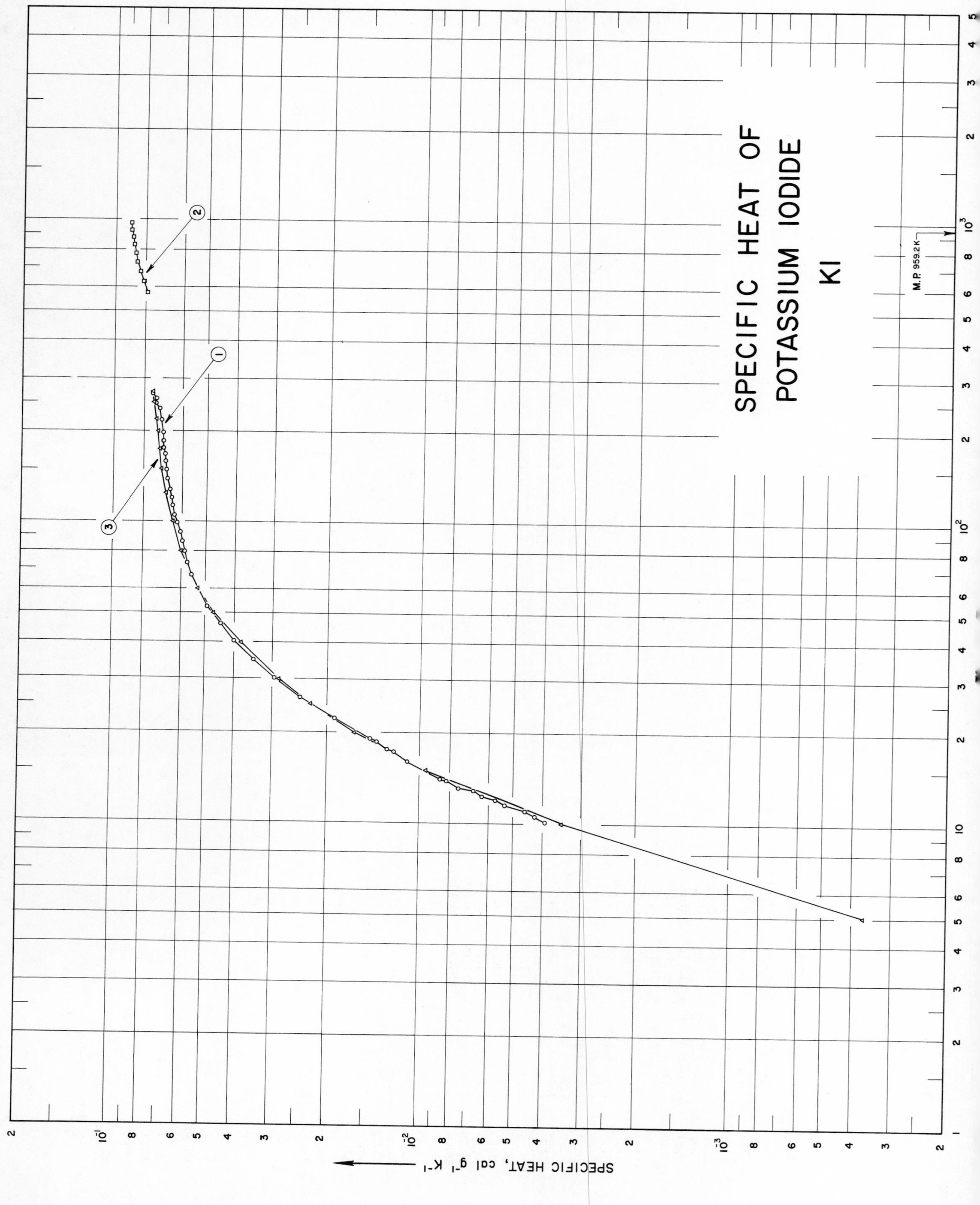

SPECIFIC HEAT OF POTASSIUM IODIDE KI

SPECIFIC HEAT, cal g^{-1} K^{-1}

M.P. 959.2 K

1

2

3

SPECIFICATION TABLE NO. 153 SPECIFIC HEAT OF POTASSIUM IODIDE KI

[For Data Reported in Figure and Table No. 153]

Curve No.	Ref. No.	Year	Temp. Range, K	Reported Error, %	Name and Specimen Designation	Composition (weight percent), Specifications and Remarks
1	243	1949	10-288			
2	244	1953	573-973			Sample supplied by the Harshaw Chem. Co; measured under dry helium gas.
3	245	1957	2.5-270	0.2-2		High purity, optical quality; sample supplied by the Harshaw Chem. Co.

DATA TABLE NO. 153 SPECIFIC HEAT OF POTASSIUM IODIDE KI

[Temperature, T, K; Specific Heat, C_p, Cal $g^{-1}K^{-1}$]

CURVE 1

T	C_p
10.2	3.9 x 10^{-3}
10.6	4.2
11.0	4.6
11.5	5.3
12.0	5.7
12.3	6.3
12.9	6.7
13.2	7.5
13.9	8.2
14.1	8.6
14.9	9.4*
15.1	9.8*
16.1	1.1 x 10^{-2}
16.5	1.1*
17.3	1.22
17.7	1.28
18.7	1.39
19.1	1.45
20.4	1.64
22.4	1.89
26.2	2.46
30.5	2.98
35.0	3.49
40.4	4.01
46.2	4.48
52.5	4.93
59.9	5.30*
66.8	5.58
73.6	5.78
79.7	5.88
86.0	5.96
87.1	5.98*
92.6	6.10
99.1	6.22
106.4	6.36
113.8	6.46
120.8	6.53
127.5	6.58
132.4	6.71*
139.2	6.72
149.4	6.77
159.9	6.81
167.9	6.87
176.2	6.90
186.3	6.93

CURVE 1 (cont.)

T	C_p
198.6	6.98 x 10^{-2}
219.1	7.07
238.0	7.18
248.7	7.40
258.2	7.36
269.9	7.57

CURVE 2

T	C_p
573	7.918 x 10^{-2}
623	8.157
673	8.360
723	8.545
773	8.657
823	8.770
873	8.845
923	8.910
973	8.960

CURVE 3

T	C_p
2.5	3.928 x 10^{-5}*
3	6.717*
5	3.675 x 10^{-4}
10	3.479 x 10^{-3}
15	9.542
20	1.635 x 10^{-2}
25	2.285
30	2.891
40	3.830
50	4.731
60	5.310
80	6.048
100	6.479
125	6.804
150	7.018
175	7.159
200	7.271
220	7.370
250	7.451
270	7.518

* Not shown on plot

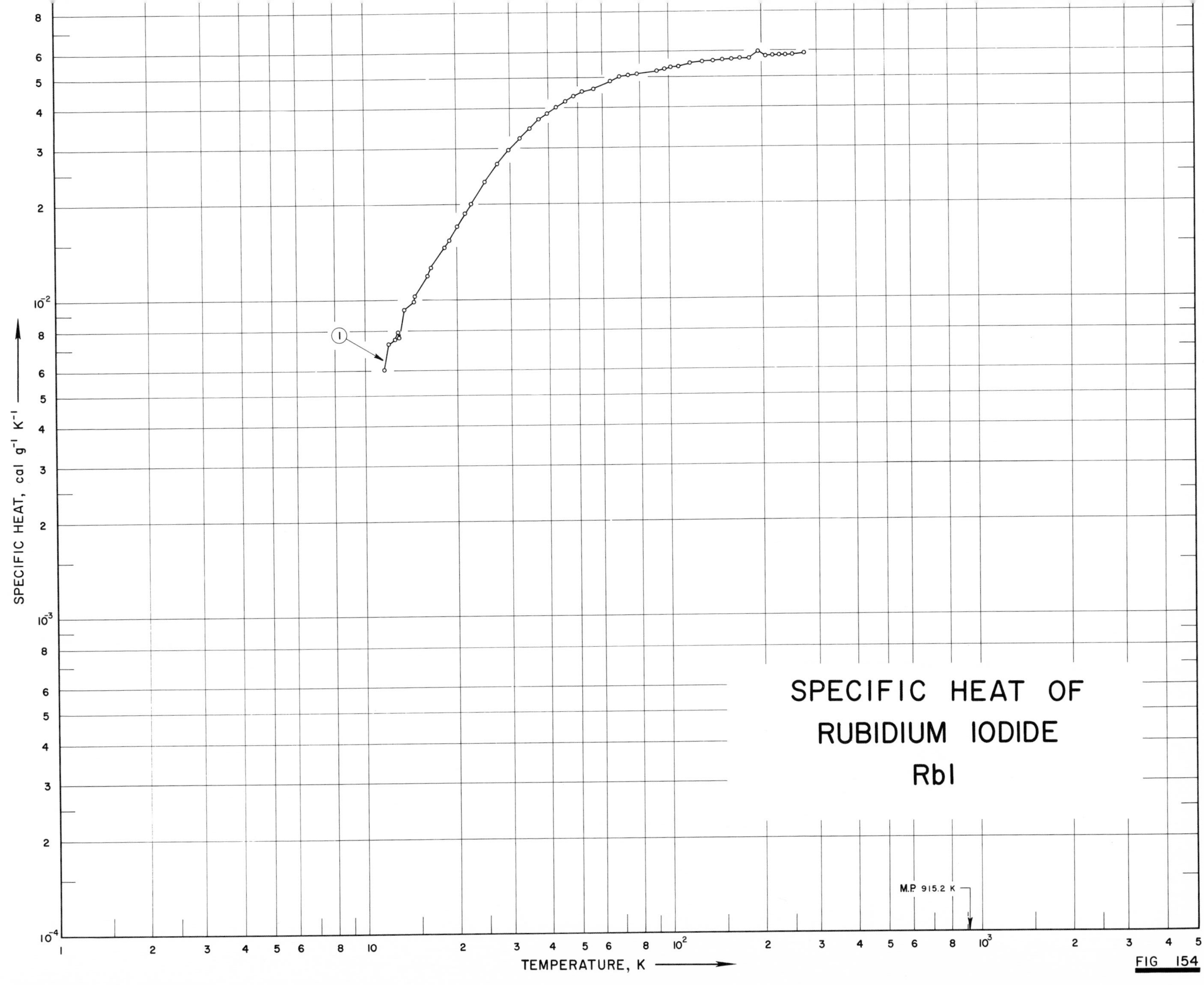

SPECIFIC HEAT OF
RUBIDIUM IODIDE
RbI
SPECIFIC HEAT, cal g⁻¹ K⁻¹
TEMPERATURE, K
M.P. 915.2 K
1
FIG 154

SPECIFICATION TABLE NO. 154 SPECIFIC HEAT OF RUBIDIUM IODIDE RbI

[For Data Reported in Figure and Table No. 154]

Curve No.	Ref. No.	Year	Temp. Range, K	Reported Error, %	Name and Specimen Designation	Composition (weight percent), Specifications and Remarks
1	243	1949	12-277			

DATA TABLE NO. 154 SPECIFIC HEAT OF RUBIDIUM IODIDE RbI

[Temperature, T,K; Specific Heat, C_p, Cal $g^{-1}K^{-1}$]

T	C_p
CURVE 1	
11.6	6.0 x 10^{-3}
12.0	7.3
12.6	7.5
12.8	7.9
12.9	7.6
13.5	9.3
14.5	9.89
14.6	1.03 x 10^{-2}
16.2	1.20
16.6	1.27
16.7	1.25*
18.4	1.47
19.0	1.54
20.2	1.71
20.3	1.74*
21.4	1.87
22.3	2.01
22.8	2.04*
24.8	2.36
24.9	2.35
27.3	2.68
29.8	2.97
32.3	3.23
34.8	3.46
37.3	3.71
39.8	3.85
42.5	4.05
45.6	4.22
48.9	4.39
51.9	4.53
56.5	4.61
64.3	4.89
68.6	5.01
73.5	5.06
78.7	5.12
91.9	5.25
97.0	5.31
102.2	5.37
107.6	5.42
112.7	5.47*
117.9	5.52
123.3	5.54*
129.0	5.58
134.4	5.58*
140.0	5.61

T	C_p
CURVE 1 (cont.)	
144.9	5.62 x 10^{-2}*
150.3	5.65
155.6	5.67*
161.2	5.69
166.7	5.72*
172.2	5.73
177.9	5.73*
184.6	5.73
189.5	5.75*
197.8	6.00
208.4	5.81
208.8	5.79*
219.3	5.84
225.1	5.83*
230.3	5.84
235.7	5.85*
241.7	5.84
247.9	5.84*
254.8	5.85
260.5	5.91*
276.9	5.91

* Not shown on plot

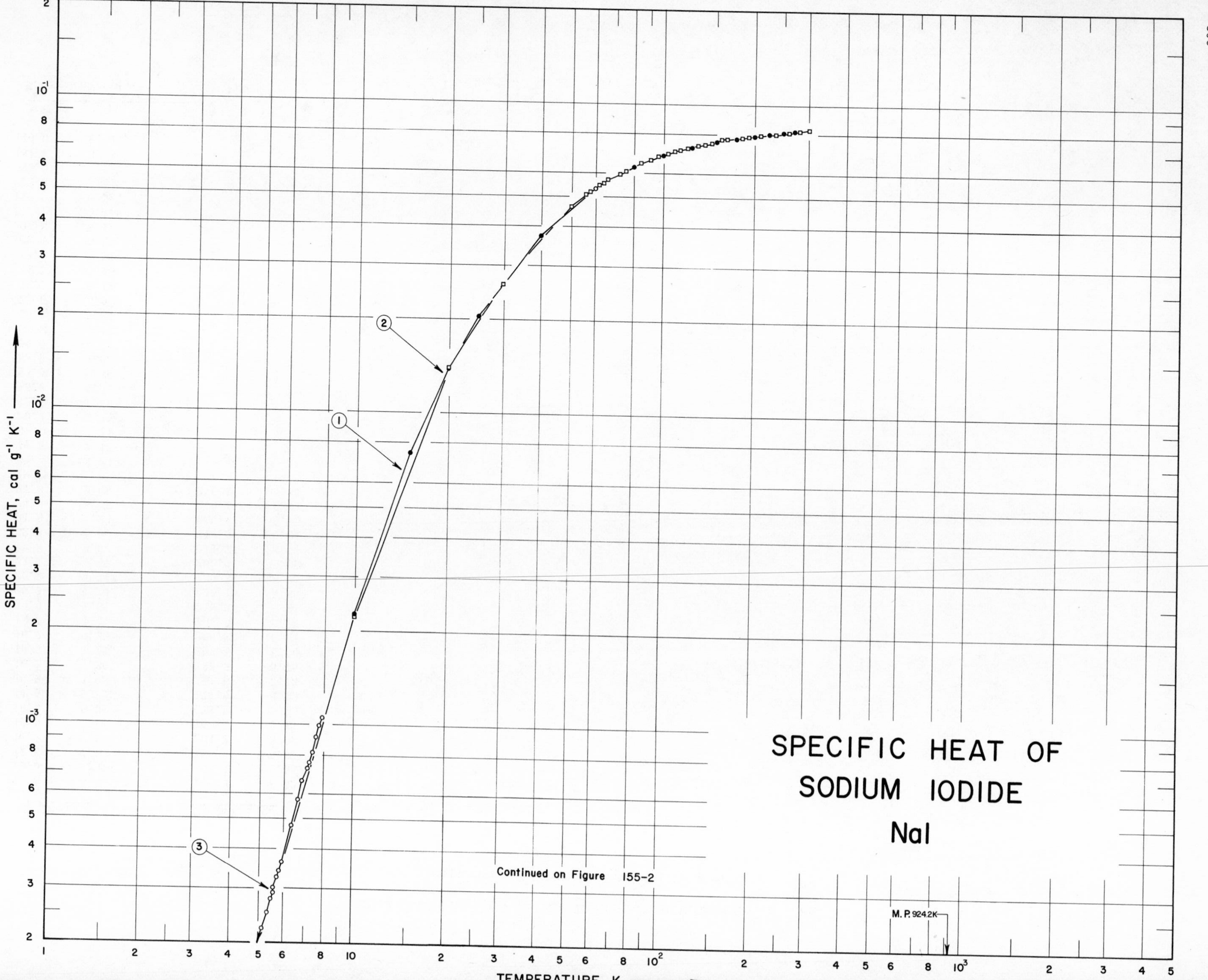

Continued on Figure 155-2

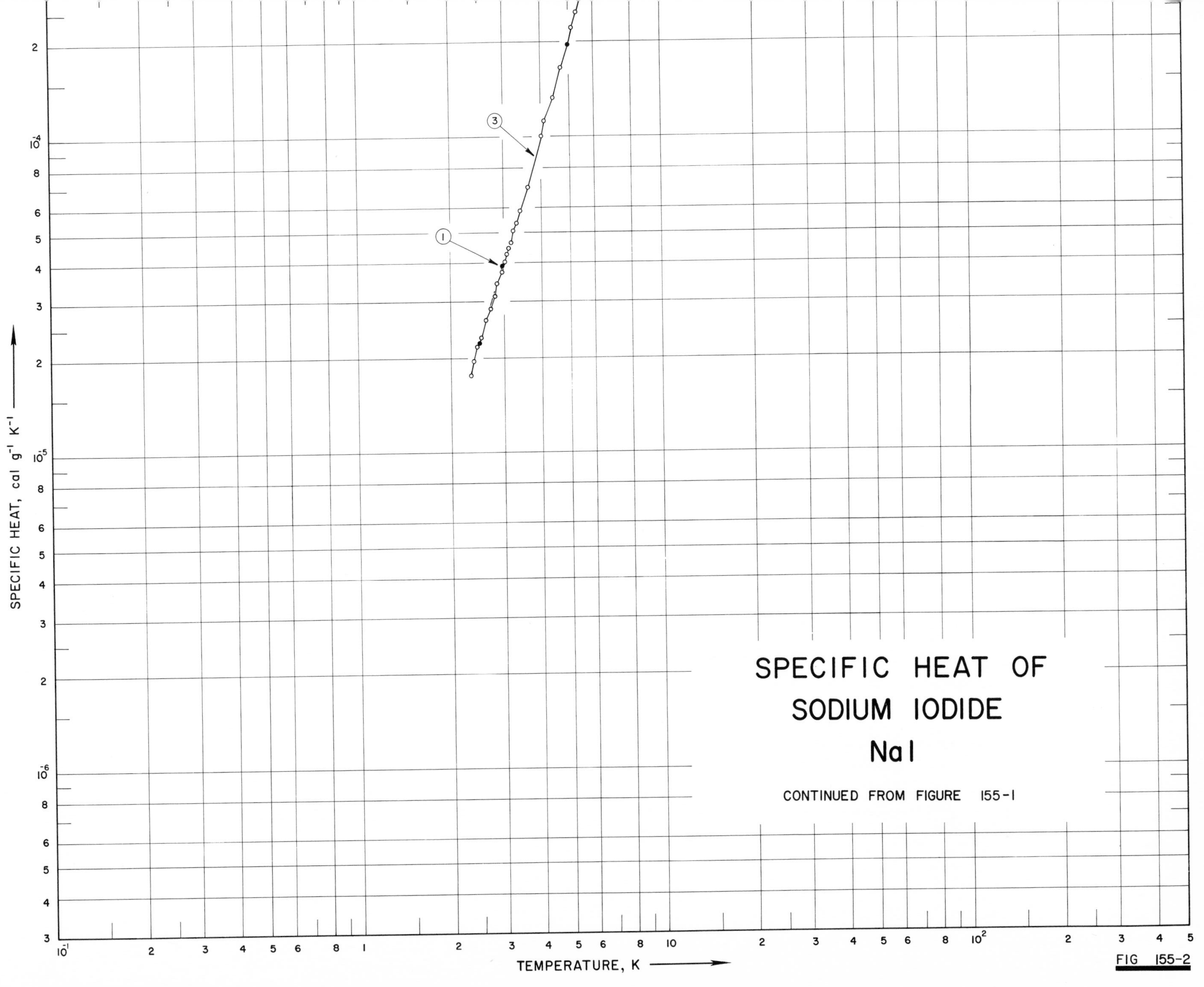

SPECIFIC HEAT OF
SODIUM IODIDE
NaI
CONTINUED FROM FIGURE 155-1
SPECIFIC HEAT, cal g⁻¹ K⁻¹
TEMPERATURE, K
3
1

FIG 155-2

SPECIFICATION TABLE NO. 155 SPECIFIC HEAT OF SODIUM IODIDE NaI

[For Data Reported in Figure and Table No. 155]

Curve No.	Ref. No.	Year	Temp. Range, K	Reported Error, %	Name and Specimen Designation	Composition (weight percent), Specifications and Remarks
1	304	1957	2.5–270	0.2–2		High purity, optical quality; sample supplied by the Harshaw Chem. Co; measured under a helium atmosphere.
2	246	1964	56–301	0.3		99.9889 NaI, 0.001–0.01 K, 0.0001–0.001 Ca, and 0.001 Mg; melted under an atmosphere of dry hydrogen.
3	337	1960	2.3–7.9			High purity, optically clear single crystal; sample supplied by the Harshaw Chemical Co; measured in high vacuum.

DATA TABLE NO. 155 SPECIFIC HEAT OF SODIUM IODIDE NaI

[Temperature, T,K; Specific Heat, Cp, Cal $g^{-1}K^{-1}$]

T	C_p
CURVE 1	
2.5	2.242×10^{-5}
3	3.916
5	1.961×10^{-4}
10	2.236×10^{-3}
15	7.472
20	1.409×10^{-2}
25	2.060
30	2.664*
40	3.753
50	4.638*
60	5.336
80	6.289
100	6.861
125	7.305
150	7.605
175	7.802
200	7.949
225	8.069
250	8.169
270	8.246
CURVE 2	
56.32	5.116×10^{-2}
57.98	5.221
59.88	5.342*
61.87	5.468
64.05	5.591
66.65	5.727
69.40	5.856
72.21	5.982
75.15	6.105
78.25	6.[illegible]32*
81.41	6.356*
84.50	6.467
87.66	6.570*
90.62	6.645
93.63	6.727*
96.87	6.805
100.91	6.891*
104.00	6.956
107.18	7.017*
109.85	7.072
110.76	7.084*
113.81	7.138
CURVE 2 (cont.)	
117.12	7.189×10^{-2}*
120.67	7.248
124.19	7.299*
127.66	7.352*
131.09	7.400
134.50	7.442*
137.88	7.449
140.99	7.509*
144.47	7.552
148.03	7.585*
151.61	7.620*
155.17	7.649*
158.69	7.682*
163.12	7.716
165.67	7.734
167.51	7.746
169.11	7.760*
170.94	7.771
174.40	7.794
177.89	7.819*
181.35	7.844
184.85	7.860*
188.42	7.881*
192.03	7.901
195.61	7.926*
196.81	7.928
200.34	7.947
203.86	7.961*
207.36	7.976*
210.85	7.992
214.31	8.007*
217.36	8.018*
220.78	8.030*
224.19	8.047*
228.02	8.060*
228.34	8.062*
231.40	8.071*
235.02	8.085
238.33	8.101*
238.96	8.105*
242.28	8.112*
246.89	8.126*
250.98	8.145*
252.79	8.149*
255.20	8.160*
CURVE 2 (cont.)	
256.82	8.166×10^{-2}*
259.14	8.174
262.12	8.187*
264.79	8.200*
267.79	8.207*
270.80	8.216*
274.31	8.228*
278.04	8.240*
281.67	8.256
285.28	8.272*
288.86	8.280*
292.44	8.291*
295.05	8.297*
297.70	8.306*
301.16	8.309
CURVE 3	
2.321	1.775×10^{-5}
2.387	1.957
2.462	2.177
2.538	2.316
2.643	2.653
2.721	2.871
2.806	3.140
2.890	3.460
2.979	3.749*
3.054	4.001
3.059	4.113*
3.103	4.263
3.137	4.386*
3.152	4.439
3.202	4.620
3.211	4.707*
3.274	5.012
3.304	5.109*
3.343	5.321
3.358	5.406*
3.439	5.829
3.477	5.961*
3.650	6.985
4.061	1.000×10^{-4}
4.142	1.030
4.441	1.345
4.483	1.361*
CURVE 3 (cont.)	
4.728	1.665×10^{-4}
4.752	1.660
4.958	1.955*
4.980	1.934*
5.149	2.229
5.181	2.231*
5.318	2.492
5.466	2.768
5.581	2.909
5.585	3.003
5.707	3.250
5.738	3.209*
5.819	3.400
5.910	3.634
6.390	4.748
6.678	5.746
6.834	6.057
6.980	6.634*
7.114	7.252
7.246	7.532
7.396	8.186
7.584	9.106
7.752	9.960
7.909	1.056×10^{-3}

* Not shown on plot

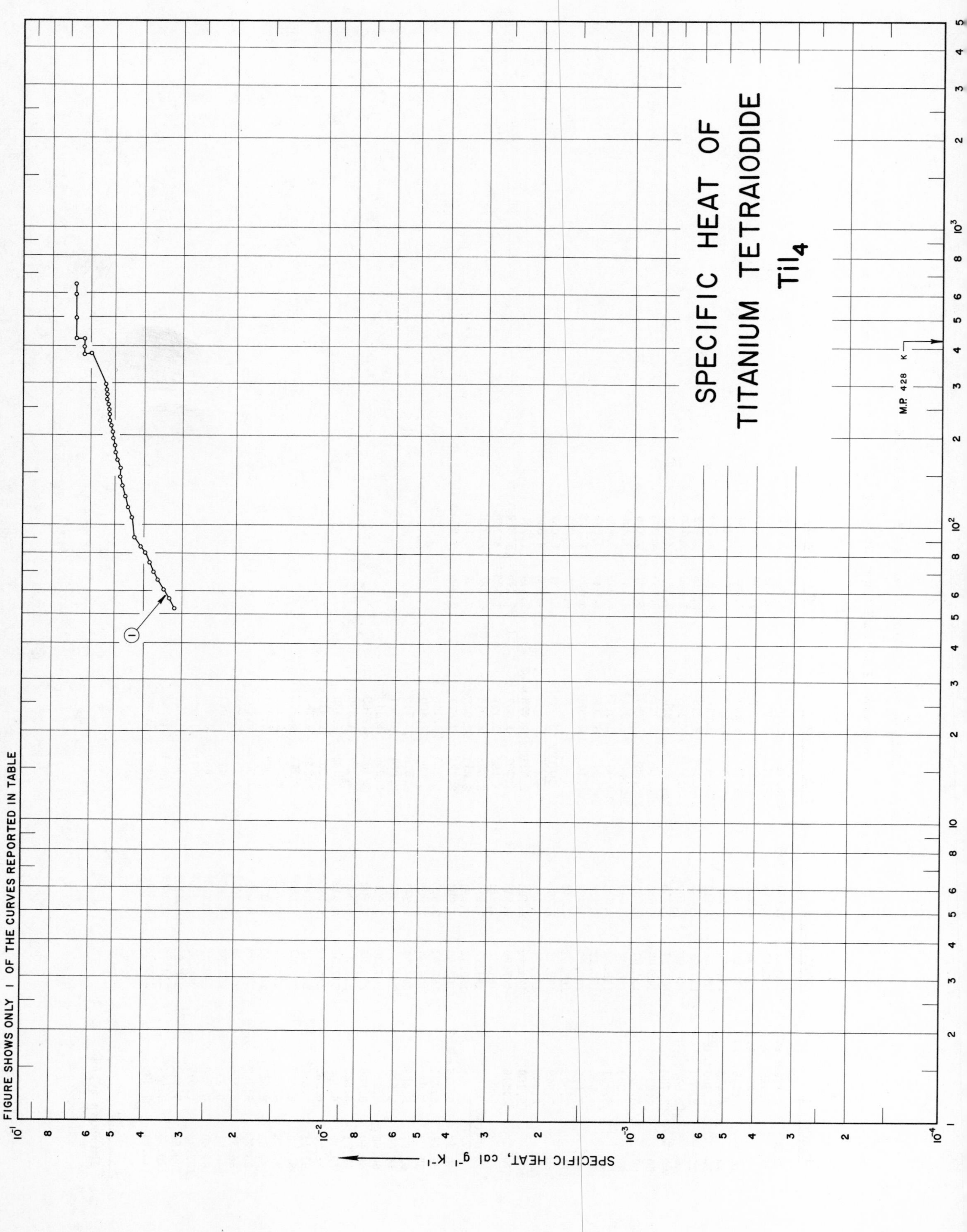

SPECIFIC HEAT OF TITANIUM TETRAIODIDE TiI_4

FIGURE SHOWS ONLY 1 OF THE CURVES REPORTED IN TABLE

SPECIFIC HEAT, cal g^{-1} K^{-1}

10^{-1} 8 6 5 4 3 2 10^{-2} 8 6 5 4 3 2 10^{-3} 8 6 5 4 3 2 10^{-4}

1 2 3 4 5 6 8 10 2 3 4 5 6 8 10^2 2 3 4 5 6 8 10^3 2 3 4 5

M.P. 428 K

1

SPECIFICATION TABLE NO. 156 SPECIFIC HEAT OF TITANIUM TETRAIODIDE TiI_4

[For Data Reported in Figure and Table No. 156]

Curve No.	Ref. No.	Year	Temp. Range, K	Reported Error, %	Name and Specimen Designation	Composition (weight percent), Specifications and Remarks
1	247	1961	298-650			90.87 I, 8.68 Ti and 0.18 Cl, (91.38 and 8.62 theo).
2	247	1961	52-297			

DATA TABLE NO. 156 SPECIFIC HEAT OF TITANIUM TETRAIODIDE TiI_4

[Temperature, T,K; Specific Heat, C_p, Cal $g^{-1}K^{-1}$]

T	C_p
CURVE 1	
298.15	5.404 x 10^{-2}
300	5.418*
α 379	6.001
β 379	6.372
400	6.372
β 428	6.372
(l) 428	6.732
500	6.732
600	6.732
650	6.732
CURVE 2	
52.40	3.197 x 10^{-2}
56.47	3.319
60.74	3.465
65.37	3.618
69.93	3.746
74.77	3.861
80.79	3.993
84.87	4.074
90.07	4.174
105.79	4.417
115.15	4.549
125.08	4.660
135.96	4.763
145.69	4.849
155.82	4.921
165.97	4.981
175.93	5.038
186.01	5.089
195.90	5.127
206.18	5.154
216.40	5.202
226.26	5.231
236.47	5.254
246.15	5.280
256.72	5.305
266.60	5.334
276.68	5.359
286.90	5.388
296.89	5.400*

* Not shown on plot

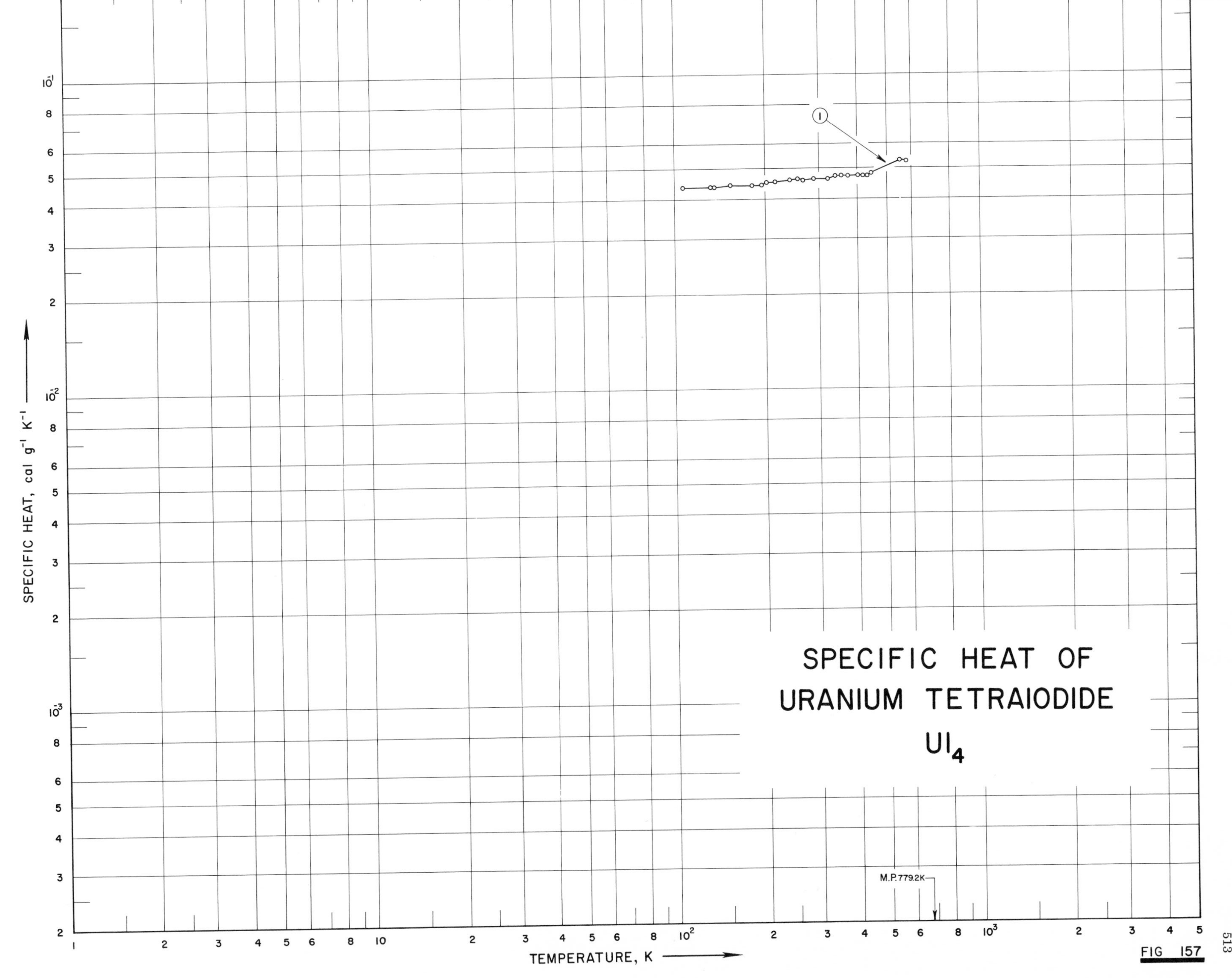
SPECIFIC HEAT OF
URANIUM TETRAIODIDE
UI_4
SPECIFIC HEAT, cal g⁻¹ K⁻¹
TEMPERATURE, K
M.P. 779.2K
1
FIG 157

SPECIFICATION TABLE NO. 157 SPECIFIC HEAT OF URANIUM TETRAIODIDE UI_4

[For Data Reported in Figure and Table No. 157]

Curve No.	Ref. No.	Year	Temp. Range, K	Reported Error, %	Name and Specimen Designation	Composition (weight percent), Specifications and Remarks
1	325	1959	107-597			0.07 impurities, mainly Fe and Si; heat treated.

DATA TABLE NO. 157 SPECIFIC HEAT OF URANIUM TETRAIODIDE UI_4

[Temperature, T, K; Specific Heat, C_p, Cal $g^{-1}K^{-1}$]

T	C_p
CURVE 1	
107	4.391 x 10^{-2}
133	4.404
138	4.391
154	4.419
181	4.477
196	4.535
203	4.520
216	4.563
241	4.606
258	4.635
268	4.606
290	4.649
299	4.678*
323	4.678
333	4.691*
341	4.721
357	4.663
357	4.734*
375	4.706
393	4.734*
405	4.706
412	4.777*
419	4.753*
420	4.721
426	4.706*
432	4.749*
435	4.736
438	4.792*
439	4.807*
444	4.764*
(s)449	4.826
(l)551	5.308
552	5.308*
562	5.280*
563	5.323*
570	5.381*
577	5.323*
583	5.280
589	5.327*
593	5.280*
597	5.308*

*Not shown on plot

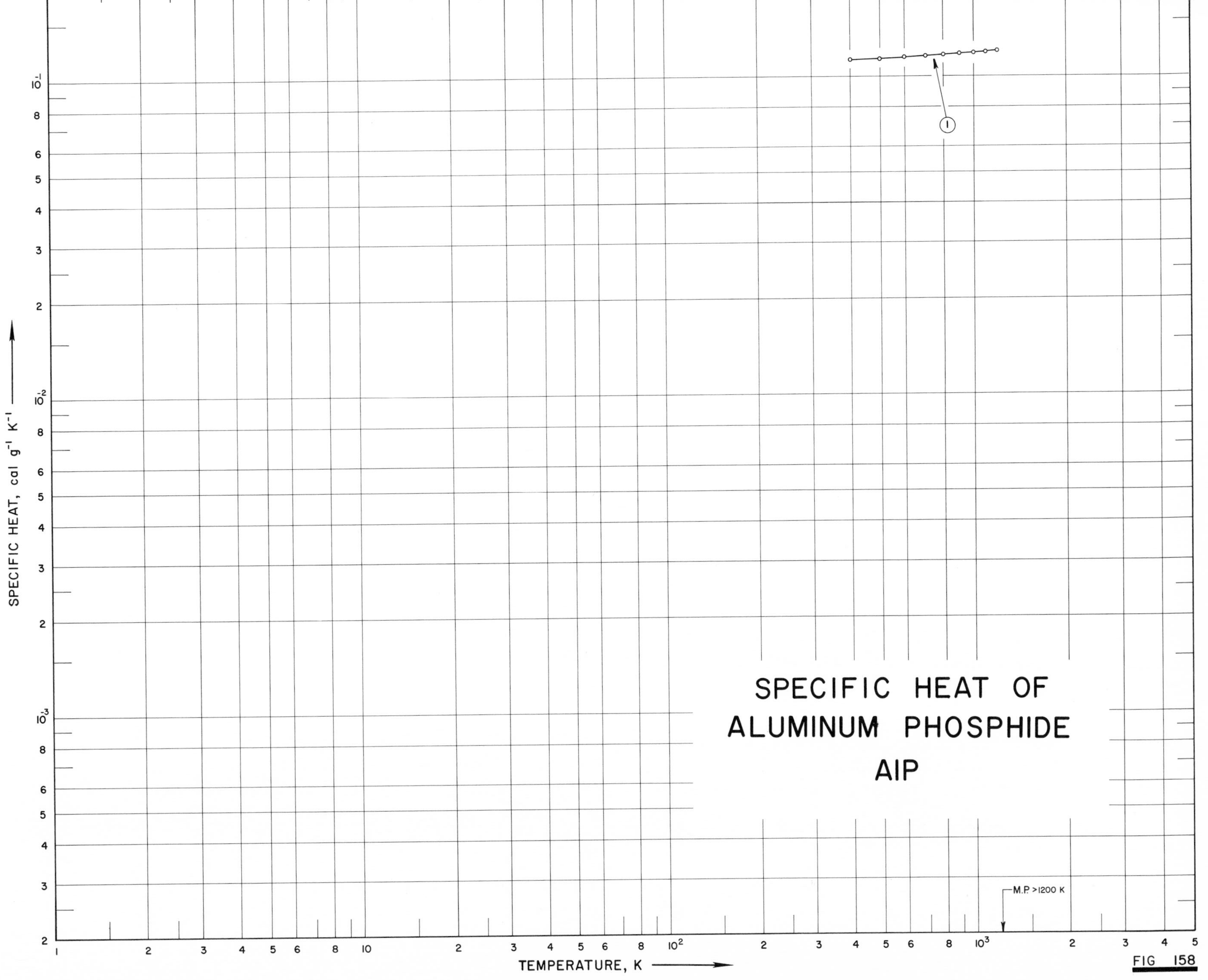
SPECIFIC HEAT OF
ALUMINUM PHOSPHIDE
AlP
1
M.P. >1200 K
SPECIFIC HEAT, cal g^{-1} K^{-1}
TEMPERATURE, K
FIG 158

SPECIFICATION TABLE NO. 158 SPECIFIC HEAT OF ALUMINUM PHOSPHIDE AlP

[For Data Reported in Figure and Table No. 158]

Curve No.	Ref. No.	Year	Temp. Range, K	Reported Error, %	Name and Specimen Designation	Composition (weight percent), Specifications and Remarks
1	157	1963	400-1200			~99.99 AlP.

DATA TABLE NO. 158 SPECIFIC HEAT OF ALUMINUM PHOSPHIDE AlP

[Temperature, T, K; Specific Heat, C_p, Cal $g^{-1}K^{-1}$]

T	C_p
CURVE 1	
400	1.141×10^{-1}
500	1.149
600	1.157
700	1.166
800	1.174
900	1.182
1000	1.190
1100	1.198
1200	1.21

* Not shown on plot

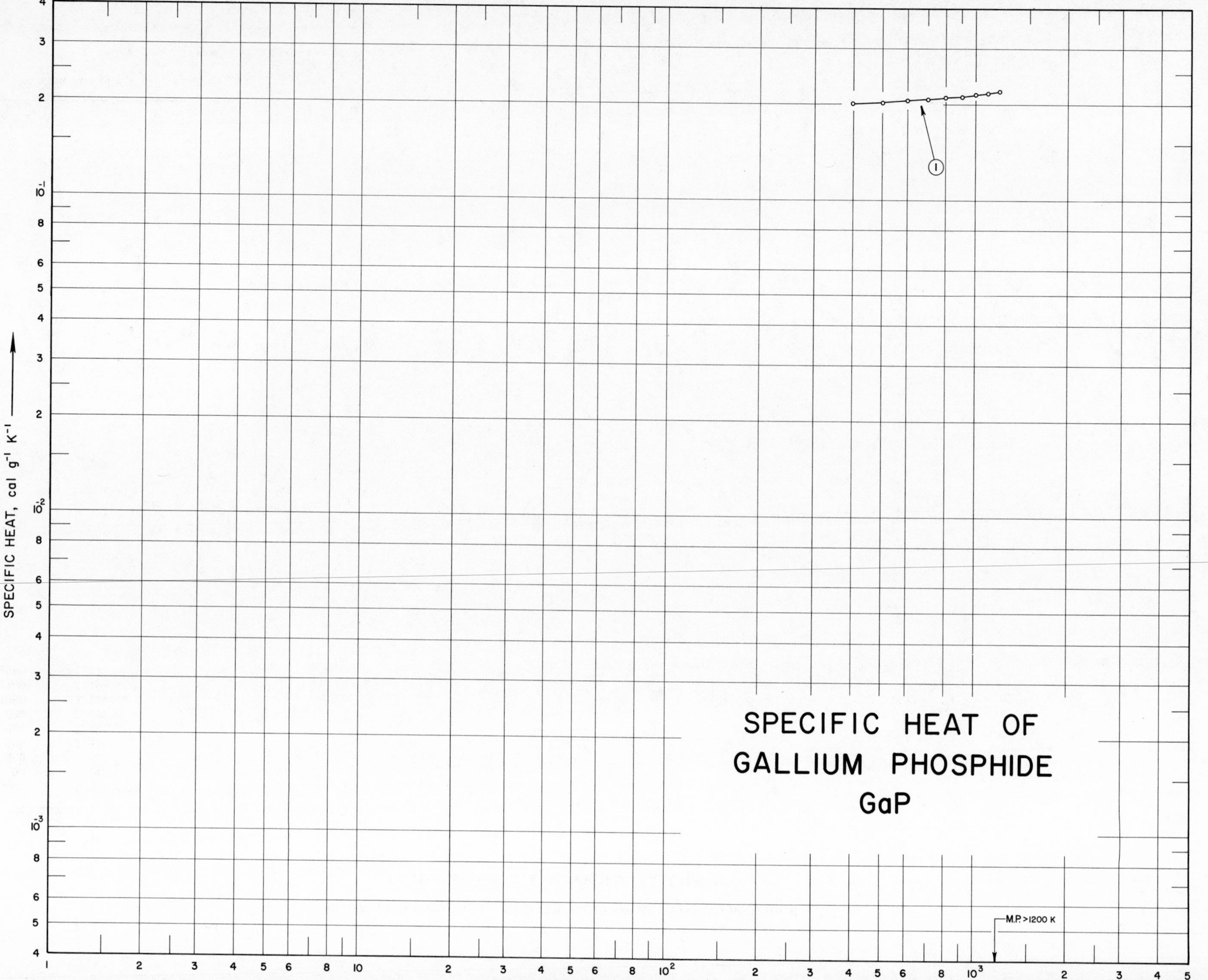
SPECIFIC HEAT OF
GALLIUM PHOSPHIDE
GaP
M.P.>1200 K
1
SPECIFIC HEAT, cal g⁻¹ K⁻¹

SPECIFICATION TABLE NO. 159 SPECIFIC HEAT OF GALLIUM PHOSPHIDE GaP

[For Data Reported in Figure and Table No. 159]

Curve No.	Ref. No.	Year	Temp. Range, K	Reported Error, %	Name and Specimen Designation	Composition (weight percent), Specifications and Remarks
1	157	1963	400-1200			~99.99 GaP.

DATA TABLE NO. 159 SPECIFIC HEAT OF GALLIUM PHOSPHIDE GaP

[Temperature, T,K; Specific Heat, C_p, Cal $g^{-1}K^{-1}$]

T	C_p
CURVE 1	
400	2.020×10^{-1}
500	2.044
600	2.068
700	2.091
800	2.115
900	2.139
1000	2.162
1100	2.186
1200	2.210

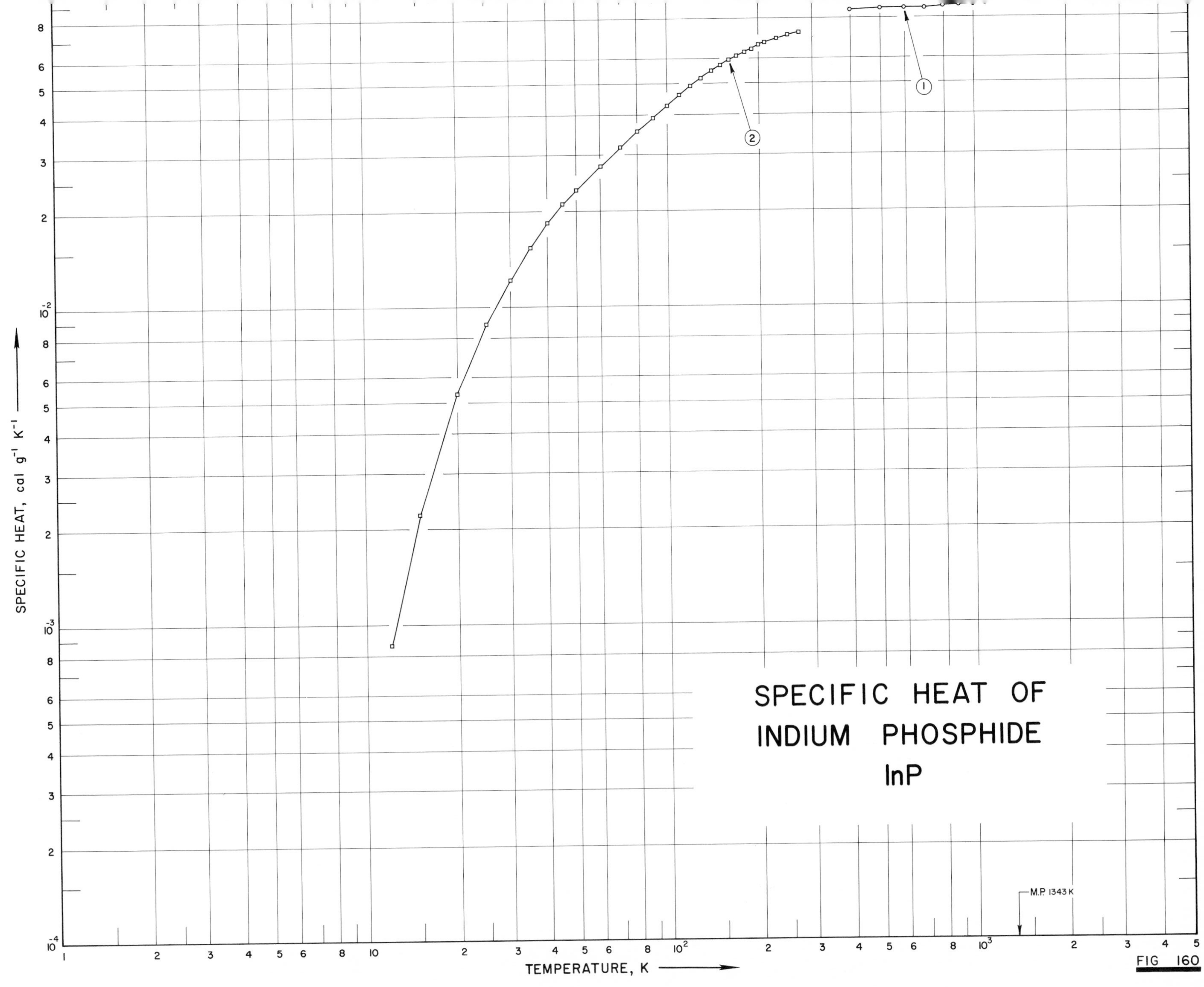
SPECIFIC HEAT OF
INDIUM PHOSPHIDE
InP
SPECIFIC HEAT, cal g⁻¹ K⁻¹
TEMPERATURE, K
M.P. 1343 K
1
2
FIG 160

SPECIFICATION TABLE NO. 160 SPECIFIC HEAT OF INDIUM PHOSPHIDE InP

[For Data Reported in Figure and Table No. 160]

Curve No.	Ref. No.	Year	Temp. Range, K	Reported Error, %	Name and Specimen Designation	Composition (weight percent), Specifications and Remarks
1	157	1963	400-1200			~99.99 InP.
2	156	1963	12-273.2	≤2.0		

DATA TABLE NO. 160 SPECIFIC HEAT OF INDIUM PHOSPHIDE InP

[Temperature, T, K; Specific Heat, C_p, Cal $g^{-1}K^{-1}$]

T	C_p
CURVE 1	
400	8.522×10^{-2}
500	8.578
600	8.634
700	8.689
800	8.745
900	8.801
1000	8.856
1100	8.912
1200	8.968
CURVE 2	
12	8.6×10^{-4}
15	2.22×10^{-3}
20	5.32
25	8.89
30	1.22×10^{-2}
35	1.547
40	1.845
45	2.113
50	2.346
60	2.775
70	3.180
80	3.567
90	3.940
100	4.296
110	4.623
120	4.932
130	5.226
140	5.504
150	5.748
160	5.956
170	6.137
180	6.303
190	6.462
200	6.609
210	6.745
220	6.859*
230	6.955
240	7.044*
250	7.124
260	7.202*
270	7.280*
273.2	7.298

* Not shown on plot

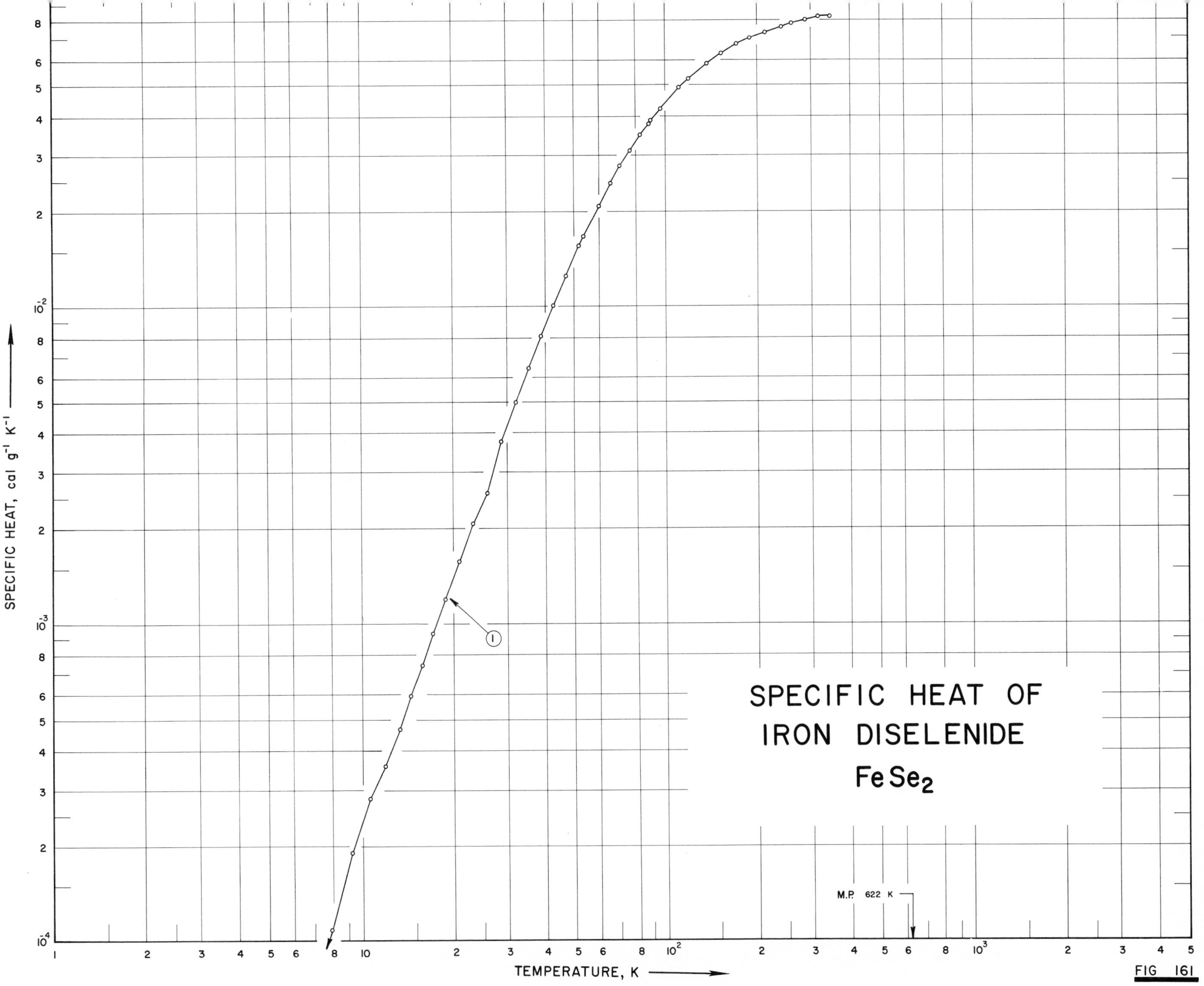

SPECIFIC HEAT OF IRON DISELENIDE $FeSe_2$

FIG 161

SPECIFICATION TABLE NO. 161 SPECIFIC HEAT OF IRON DISELENIDE $FeSe_2$

[For Data Reported in Figure and Table No. 161]

Curve No.	Ref. No.	Year	Temp. Range, K	Reported Error, %	Name and Specimen Designation	Composition (weight percent), Specifications and Remarks
1	214	1962	5-347	0.1-5		Impurities: ~0.01 Ni, Si, and ~0.001 Mn, high purity selenide impurities: 0.0002 Cl, 0.0008 Fe, 0.0004 Na, 0.0003 K, and 0.0012 non-volatile matter; orthorhombic marcasite-type structure; made from high purity iron and selenium; heated slowly to 1000 C in an evacuated and sealed silica tube for 2 hrs and cooled to room temp; heated at 340 C for one month and then cooled slowly to room temperature during a period of 1 month.

DATA TABLE NO. 161 SPECIFIC HEAT OF IRON DISELENIDE $FeSe_2$

[Temperature, T, K; Specific Heat, C_p, Cal g^{-1} K^{-1}]

T	C_p
CURVE 1	
Series 1	
53.66	1.669 x 10^{-2}
60.13	2.082
65.85	2.455
71.07	2.781
76.51	3.115
82.66	3.492
89.61	3.885
Series 2	
5.36	3.742 x 10^{-5}*
6.81	7.110*
7.91	1.095 x 10^{-4}
9.24	1.913
10.59	2.821
11.96	3.593
13.23	4.697
14.47	5.932
15.74	7.447
17.15	9.370
18.81	1.205 x 10^{-3}
20.80	1.582
23.12	2.091
25.80	2.607
28.81	3.787
32.05	5.005
35.43	6.451
38.97	8.168
42.73	1.013 x 10^{-2}
47.10	1.264
51.96	1.560
Series 3	
87.69	3.776 x 10^{-2}
96.27	4.220
104.45	4.609*
112.32	4.950
120.26	5.266
128.67	5.570*
137.28	5.872
145.74	6.094*
154.32	6.318
CURVE 1 (cont.)	
Series 3 (cont.)	
163.26	6.527 x 10^{-2}*
172.22	6.710
181.34	6.882*
190.93	7.042
200.38	7.184*
203.92	7.230*
213.37	7.354
222.97	7.470*
232.35	7.574*
241.81	7.681
251.35	7.765*
260.79	7.845
270.31	7.929*
279.94	8.009*
289.43	8.074
298.68	8.158*
308.07	8.214*
317.67	8.233
327.45	8.224*
337.48	8.257*
347.46	8.299

* Not shown on plot

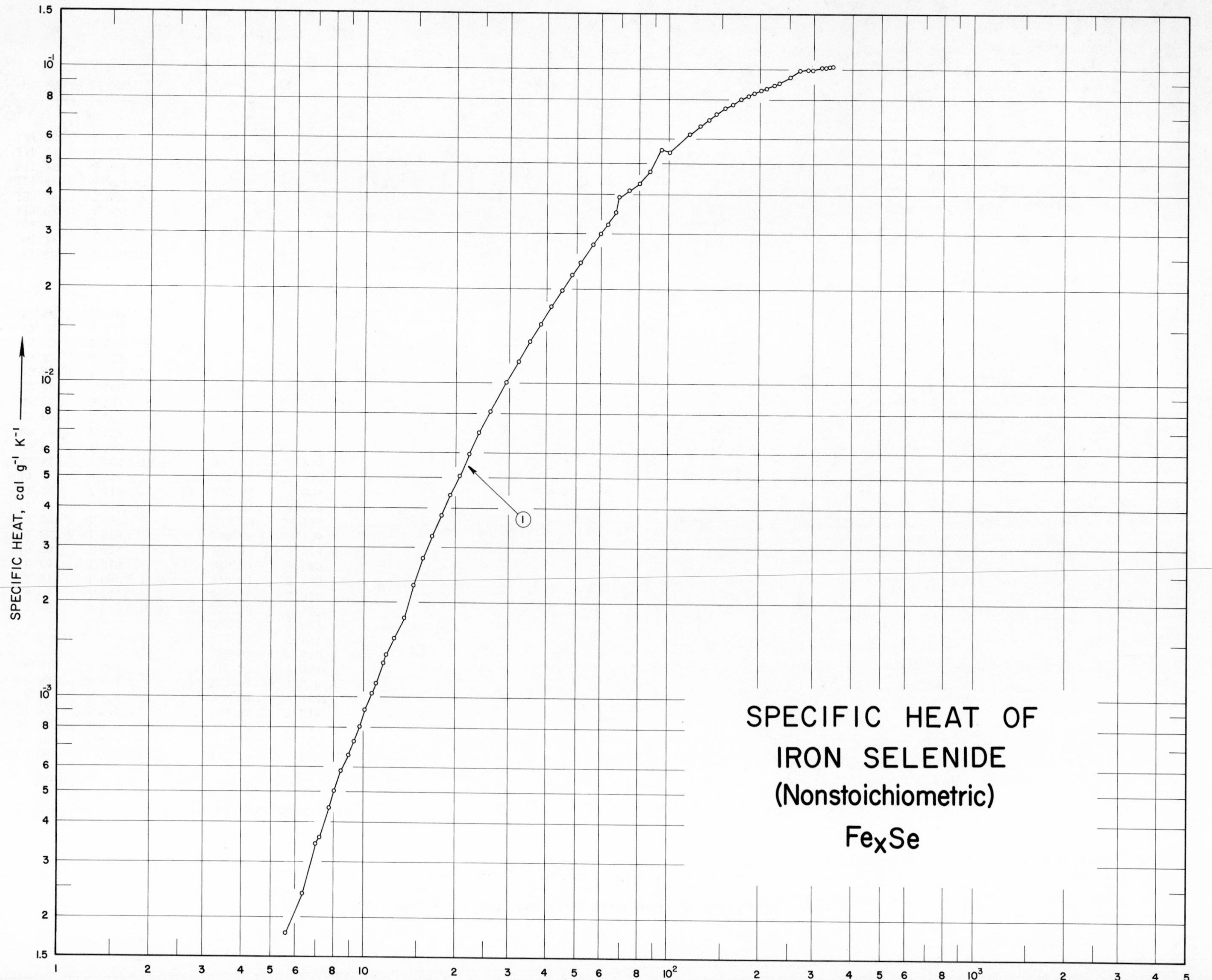

SPECIFIC HEAT OF
IRON SELENIDE
(Nonstoichiometric)
Fe_xSe
SPECIFIC HEAT, cal g^{-1} K^{-1}
1

SPECIFICATION TABLE NO. 162 SPECIFIC HEAT OF IRON SELENIDE (nonstoichiometric) Fe_xSe

[For Data Reported in Figure and Table No. 162]

Curve No.	Ref. No.	Year	Temp. Range, K	Reported Error, %	Name and Specimen Designation	Composition (weight percent), Specifications and Remarks
1	226	1959	6-347	0.1-1.0		99.979 $Fe_{1.042}Se$, 0.01 Ni, 0.01 Si, and 0.001 Mn; fused for 4 hrs at 1050 C; cooled to room temperature; fragmented under dry nitrogen; homogenized at 350 C for 30 days and cooled to room temperature over 30 days.

DATA TABLE NO. 162 SPECIFIC HEAT OF IRON SELENIDE Fe_xSe (nonstoichiometric)

[Temperature, T, K; Specific Heat, C_p, Cal $g^{-1} K^{-1}$]

T	C_p
CURVE 1	
Series 1	
51.58	2.440 x 10^{-2}
56.71	2.787
63.13	3.226
69.08	3.954
74.43	4.176
80.50	4.346
87.17	4.742
94.34	5.577
101.47	5.488
109.61	5.882*
118.34	6.268
127.34	6.628
136.29	6.950
144.93	7.250
154.15	7.526
163.86	7.798
173.52	8.042
182.90	8.249
192.20	8.467
201.68	8.641
211.17	8.799
223.65	9.014
232.96	9.126
242.36	9.300*
251.68	9.427
261.19	9.540*
270.80	9.658
280.33	9.765*
289.71	9.868
299.16	9.963
308.73	1.006 x 10^{-1}
318.32	1.015
327.94	1.023
337.51	1.031
347.23	1.039
Series 2	
7.02	3.424 x 10^{-4}
7.77	4.466
8.48	5.806
9.31	7.295
10.15	9.081
11.03	1.102 x 10^{-3}

T	C_p
CURVE 1 (cont.)	
Series 2 (cont.)	
11.97	1.370 x 10^{-3}
Series 3	
5.60	1.787 x 10^{-4}
6.38	2.382
7.25	3.573
8.06	5.062
8.91	6.551
9.76	8.188
10.67	1.027 x 10^{-3}
11.62	1.280
12.63	1.548
13.65	1.897
14.67	2.297
15.77	2.775
16.89	3.257
18.07	3.799
19.29	4.402
20.65	5.080
22.30	5.952
24.17	6.972
26.26	8.146
29.47	1.006 x 10^{-2}
32.30	1.179
35.32	1.368
38.30	1.559
41.49	1.764
45.04	1.998
48.40	2.226
54.52	2.651*
60.40	3.003
64.63	3.324*
67.59	3.522
70.56	3.706*
73.53	3.908*
76.75	4.107*
80.36	4.338*
83.94	4.551*
87.50	4.761*
91.11	4.955*
94.92	5.154*
99.11	5.365*
103.51	5.584*

* Not shown on plot

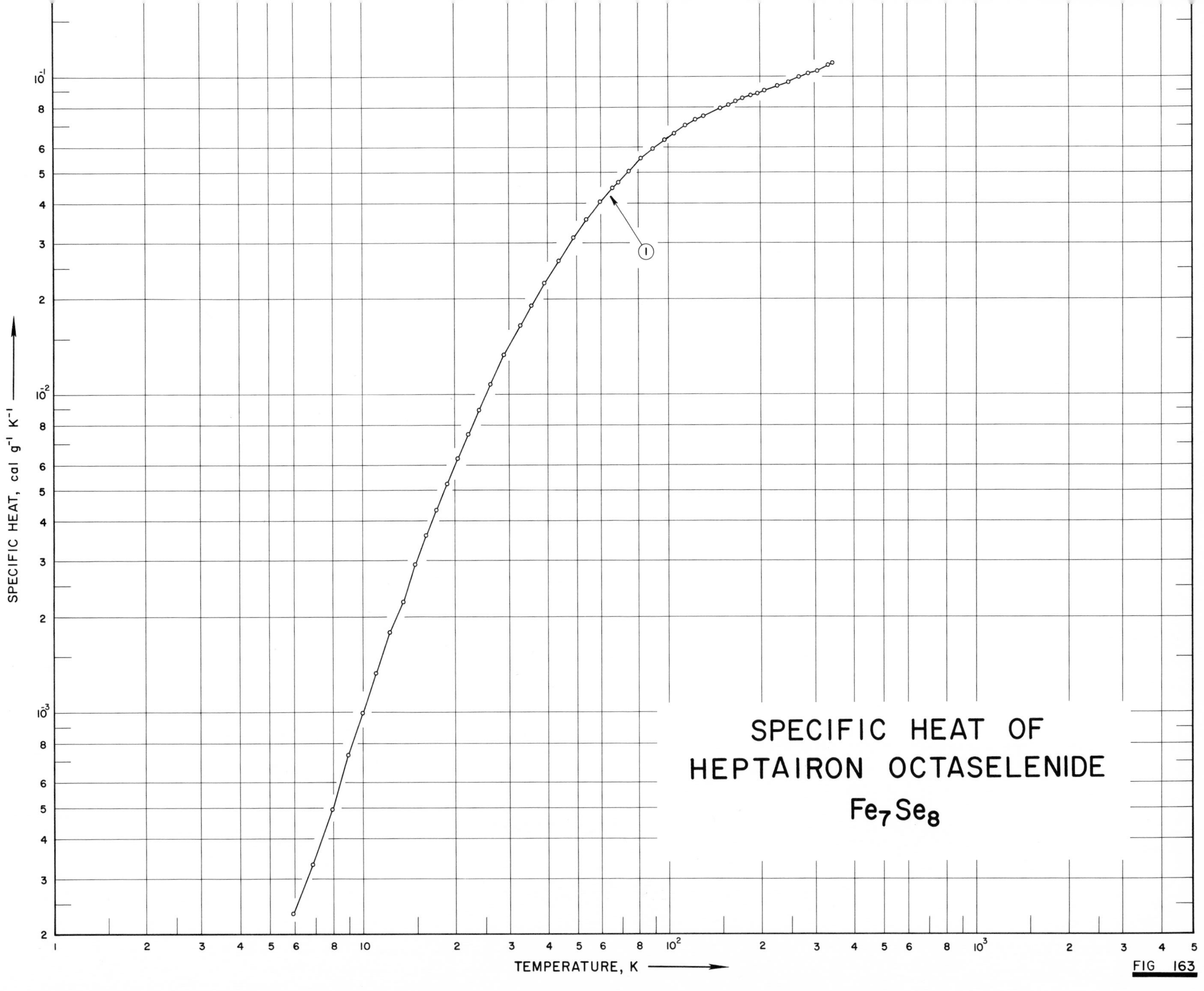
SPECIFIC HEAT OF
HEPTAIRON OCTASELENIDE
Fe7Se8
SPECIFIC HEAT, cal g-1 K-1
TEMPERATURE, K
1
FIG 163

SPECIFICATION TABLE NO. 163 SPECIFIC HEAT OF HEPTAIRON OCTASELENIDE Fe_7Se_8

[For Data Reported in Figure and Table No. 163]

Curve No.	Ref. No.	Year	Temp. Range, K	Reported Error, %	Name and Specimen Designation	Composition (weight percent), Specifications and Remarks
1	226	1959	6-345	0.1-1.0		99.979 Fe_7Se_8, 0.01 Ni, 0.01 Si, and 0.001 Mn; fused for 4 hrs at 1050 C; cooled to room temperature; fragmented under dry nitrogen; homogenized at 350 C for 30 days and cooled to room temperature over 30 days.

DATA TABLE NO. 163 SPECIFIC HEAT OF HEPTAIRON OCTASELENIDE Fe_7Se_8

[Temperature, T, K; Specific Heat, C_p, Cal g^{-1} K^{-1}]

T	C_p
CURVE 1	
5.91	2.347×10^{-4}
6.86	3.373
7.92	4.987
8.90	7.334
9.95	9.974
11.11	1.335×10^{-3}
12.35	1.789
13.58	2.325
14.84	2.925
16.13	3.607
17.45	4.385
18.84	5.254
20.34	6.301
22.02	7.505
23.91	8.959
26.14	1.078×10^{-2}
28.86	1.312
32.66	1.656
35.41	1.914
39.01	2.253
43.23	2.644
48.46	3.121
53.75	3.576
59.40	4.031
65.61	4.493
68.25	4.673
74.25	5.057
81.67	5.524
89.51	5.952
97.40	6.317
105.92	6.675
114.67	7.001
123.20	7.386
131.46	7.529
139.76	7.754*
148.64	7.972
152.28	8.060*
158.23	8.184
166.24	8.341
176.54	8.533
187.24	8.724
197.70	8.891
207.95	9.047
218.00	9.199*
228.16	9.341
238.43	9.495*
CURVE 1 (cont.)	
248.51	9.636×10^{-2}
258.43	9.770*
268.37	9.911
278.34	1.005×10^{-1}*
288.43	1.020
298.72	1.034*
308.98	1.048
319.12	1.064*
331.24	1.084
344.70	1.106

* Not shown on plot

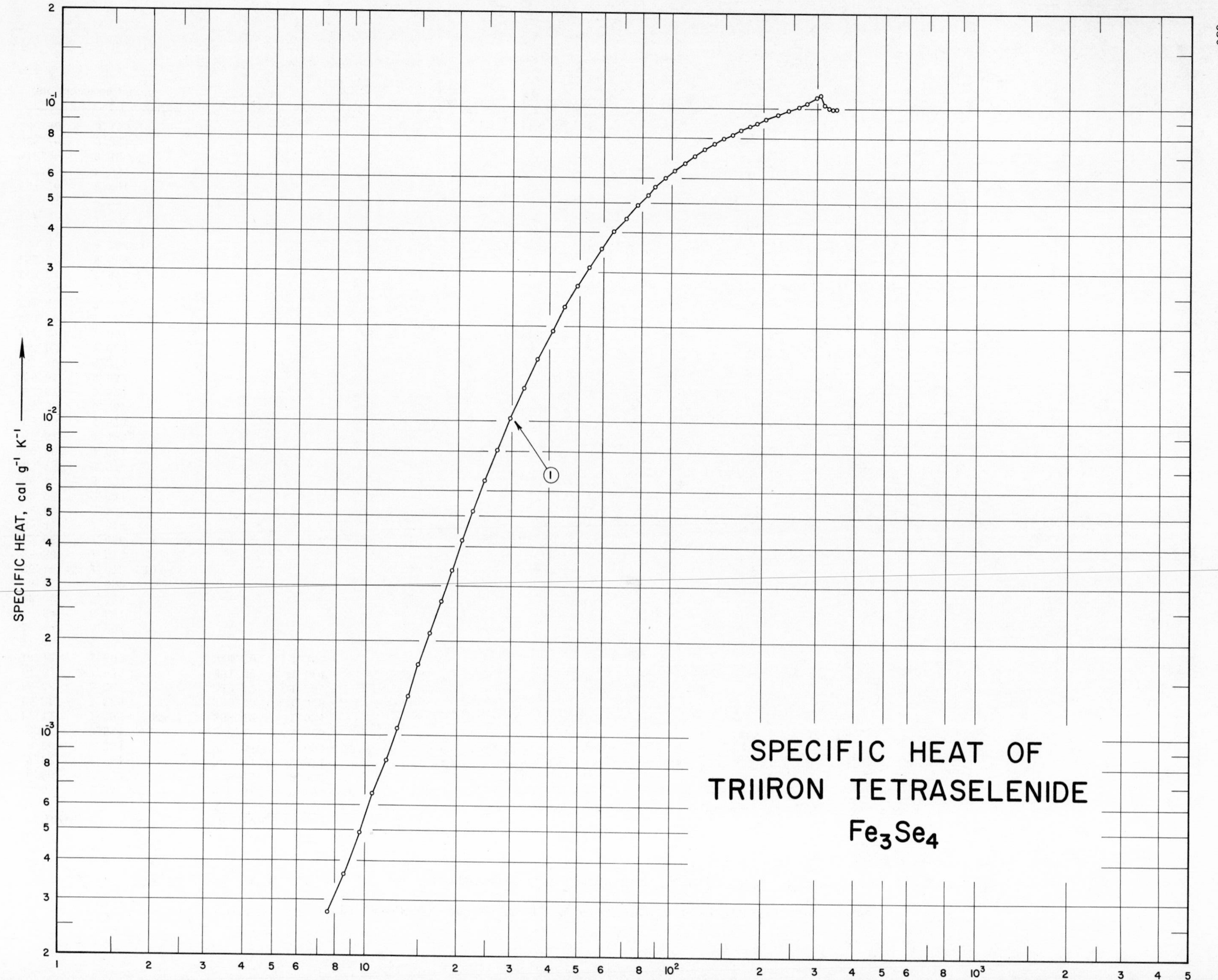
SPECIFIC HEAT OF
TRIIRON TETRASELENIDE
Fe_3Se_4
SPECIFIC HEAT, cal g⁻¹ K⁻¹
1

SPECIFICATION TABLE NO. 164 SPECIFIC HEAT OF TRIIRON TETRASELENIDE Fe_3Se_4

[For Data Reported in Figure and Table No. 164]

Curve No.	Ref. No.	Year	Temp. Range, K	Reported Error, %	Name and Specimen Designation	Composition (weight percent), Specifications and Remarks
1	226	1959	6-348			99.979 Fe_3Se_4, 0.01 Ni, 0.01 Si, and 0.001 Mn; fused for 4 hrs at 1050 C; cooled to room temperature; fragmented under dry nitrogen atmosphere, homogenized at 350 C for 30 days and cooled to room temperature over 30 days.

DATA TABLE NO. 164 SPECIFIC HEAT OF TRIIRON TETRASELENIDE Fe_3Se_4

[Temperature, T, K; Specific Heat, C_p, Cal g^{-1} K^{-1}]

T	C_p
CURVE 1	
Series 1	
81.54	5.243 x 10^{-2}*
88.16	5.646
95.16	6.012
102.57	6.367
110.73	6.735
119.66	7.084
129.18	7.425
138.85	7.719
148.51	8.029
159.09	8.299
169.51	8.553
180.21	8.781
190.88	8.995
194.78	9.073*
204.34	9.228
213.76	9.414*
223.01	9.557
232.23	9.700*
241.53	9.852
250.85	9.996*
260.04	1.013 x 10^{-1}
269.14	1.028*
278.38	1.044
287.76	1.064*
297.22	1.089
306.81	1.102
316.74	1.026
327.01	1.001
337.29	9.997 x 10^{-2}
347.55	9.988
Series 2	
5.57	1.303 x 10^{-4}*
6.70	2.027*
7.58	2.751
8.53	3.620
9.62	4.923
10.68	6.516
11.72	8.398
12.77	1.067 x 10^{-3}
13.85	1.335
14.99	1.696
16.23	2.124

T	C_p
CURVE 1 (cont.)	
Series 2 (cont.)	
17.62	2.688 x 10^{-3}
19.11	3.378
20.69	4.208
22.44	5.223
24.45	6.516
26.82	8.164
29.55	1.025 x 10^{-2}
32.71	1.281
36.30	1.588
40.51	1.959
44.84	2.346
49.04	2.724
53.74	3.135
59.00	3.579
64.70	4.047
70.75	4.493
77.06	4.930
83.85	5.376
274.20	1.035 x 10^{-1}*
283.64	1.053*
291.43	1.070*
297.63	1.088*
301.85	1.099*
304.07	1.105*
305.72	1.110*
306.83	1.112*
307.93	1.107*
309.03	1.102*
310.14	1.094*
311.24	1.078*
312.33	1.063*
313.42	1.047*
314.52	1.035*
316.19	1.022*
318.42	1.012*
320.66	1.007*
322.91	1.004*
325.16	1.002*
329.10	9.996 x 10^{-2}*
336.77	9.981*
346.44	9.986*

* Not shown on plot

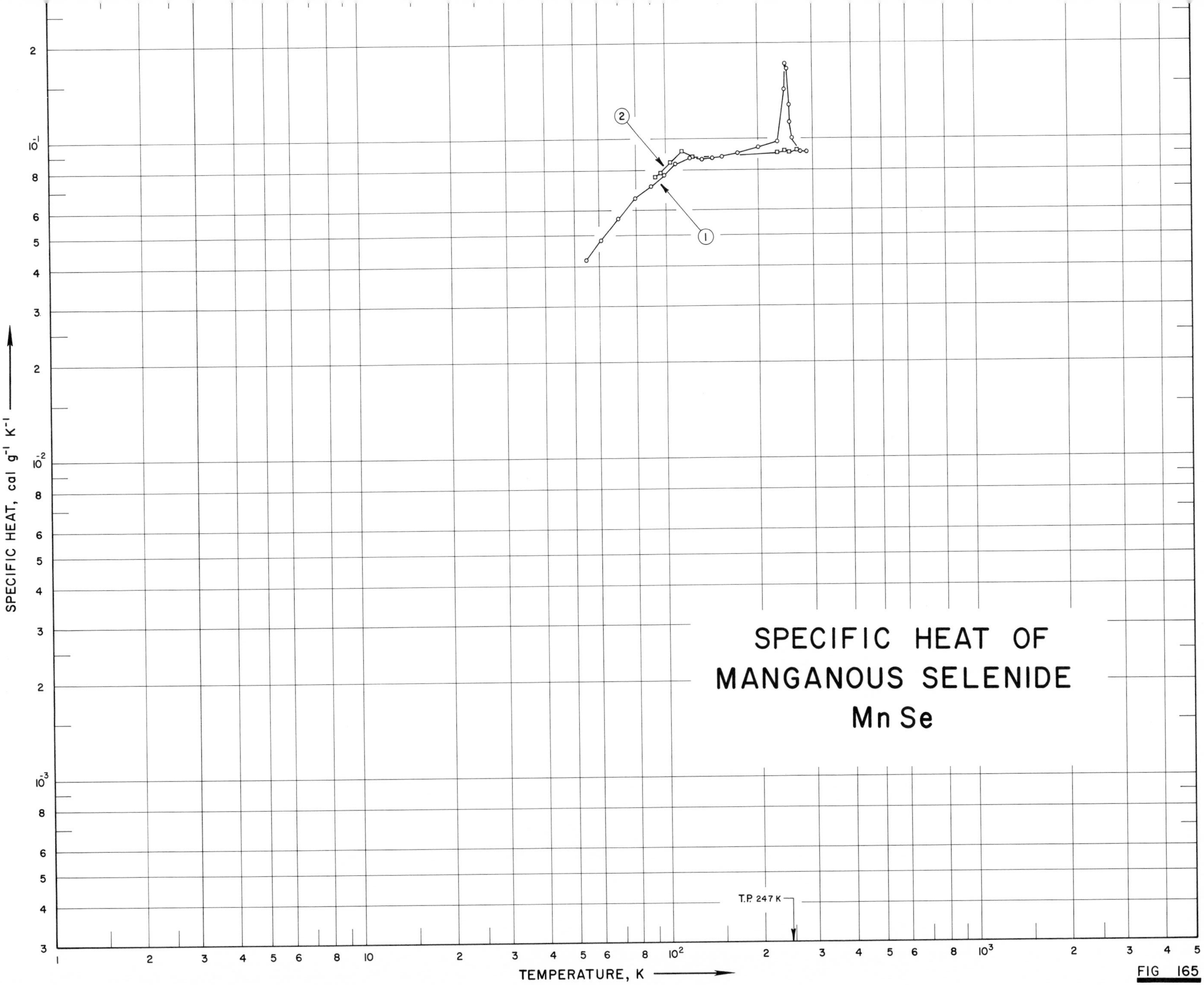

SPECIFIC HEAT OF
MANGANOUS SELENIDE
Mn Se
SPECIFIC HEAT, cal g⁻¹ K⁻¹
TEMPERATURE, K
T.P. 247 K
1
2

FIG 165

SPECIFICATION TABLE NO. 165 SPECIFIC HEAT OF MANGANOUS SELENIDE MnSe

[For Data Reported in Figure and Table No. 165]

Curve No.	Ref. No.	Year	Temp. Range, K	Reported Error, %	Name and Specimen Designation	Composition (weight percent), Specifications and Remarks
1	221	1939	54-287			99.33 MnSe.
2	221	1939	93-266			Same as above.

DATA TABLE NO. 165 SPECIFIC HEAT OF MANGANOUS SELENIDE MnSe

[Temperature, T, K; Specific Heat, C_p, Cal $g^{-1} K^{-1}$]

T	C_p
CURVE 1	
54.3	4.20 x 10^{-2}
57.4	4.50*
61.0	4.85
69.6	5.68
79.2	6.54
81.3	6.59*
82.5	6.67*
84.8	6.87*
86.3	6.94*
89.7	7.16
92.1	7.34*
93.4	7.37*
96.4	7.611*
99.0	7.789
100.3	7.865*
101.5	7.909*
102.1	7.909*
104.3	8.156*
104.4	8.126*
107.4	8.290*
108.5	8.395
110.3	8.485*
111.0	8.462*
112.4	8.656*
113.1	8.679*
114.7	8.813*
115.8	8.880*
118.1	8.806*
118.4	8.828*
120.3	8.737
120.9	8.664*
121.8	8.624*
123.5	8.641*
123.8	8.612*
124.6	8.634*
127.5	8.612*
128.8	8.641*
132.7	8.675
133.3	8.701*
142.4	8.753
153.1	8.865
162.0	8.985*
171.1	9.089
180.7	9.194*
190.6	9.336*
200.0	9.426
CURVE 1 (cont.)	
209.3	9.545 x 10^{-2}*
218.4	9.642*
222.3	9.635*
231.0	9.829*
237.0	1.019 x 10^{-1}*
239.5	1.096*
242.7	1.432
245.2	1.727
249.3	1.672
251.7	1.291
254.4	1.137
257.7	1.003
261.1	9.590 x 10^{-2}*
264.7	9.231*
268.5	9.171*
273.4	9.112
277.1	9.127*
280.4	9.097*
287.0	9.112
CURVE 2	
92.5	7.603 x 10^{-2}
95.9	7.835
104.8	8.432
113.7	9.119
122.9	8.783
230.3	9.075
233.9	9.067*
237.9	9.067*
243.6	9.104
251.4	9.097
259.0	9.157*
266.4	9.209

* Not shown on plot

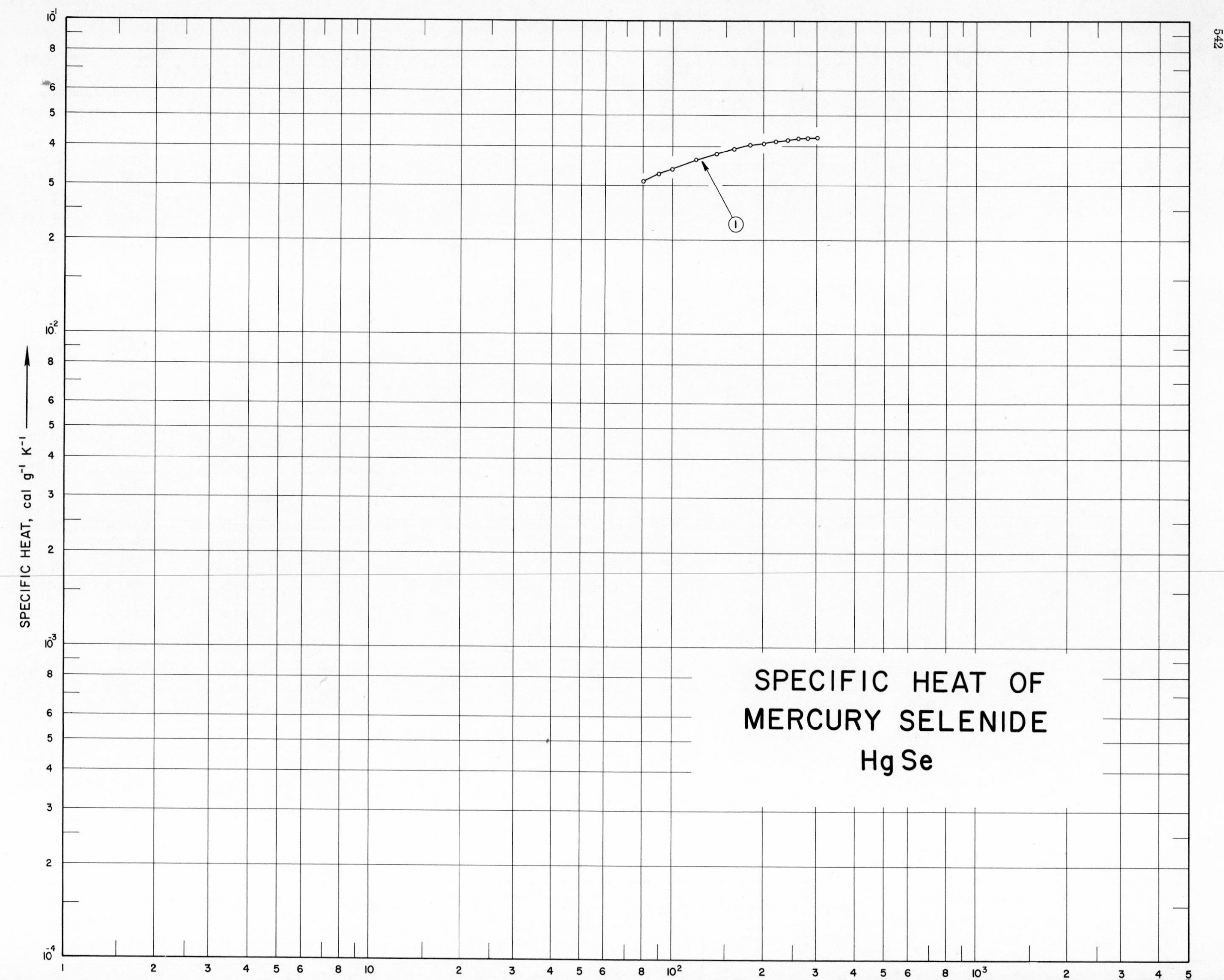
SPECIFIC HEAT OF
MERCURY SELENIDE
Hg Se
SPECIFIC HEAT, cal g^{-1} K^{-1}
1

SPECIFICATION TABLE NO. 166 SPECIFIC HEAT OF MERCURY SELENIDE HgSe

[For Data Reported in Figure and Table No. 166]

Curve No.	Ref. No.	Year	Temp. Range, K	Reported Error, %	Name and Specimen Designation	Composition (weight percent), Specifications and Remarks
1	160	1959	80-300	3-7		Polycrystalline.

DATA TABLE NO. 166 SPECIFIC HEAT OF MERCURY SELENIDE HgSe

[Temperature, T, K; Specific Heat, C_p, Cal $g^{-1} K^{-1}$]

T	C_p
CURVE 1	
80	3.09×10^{-2}
90	3.25
100	3.39
120	3.61
140	3.77
160	3.91
180	4.01
200	4.08
220	4.14
240	4.18
260	4.22
280	4.25
300	4.26

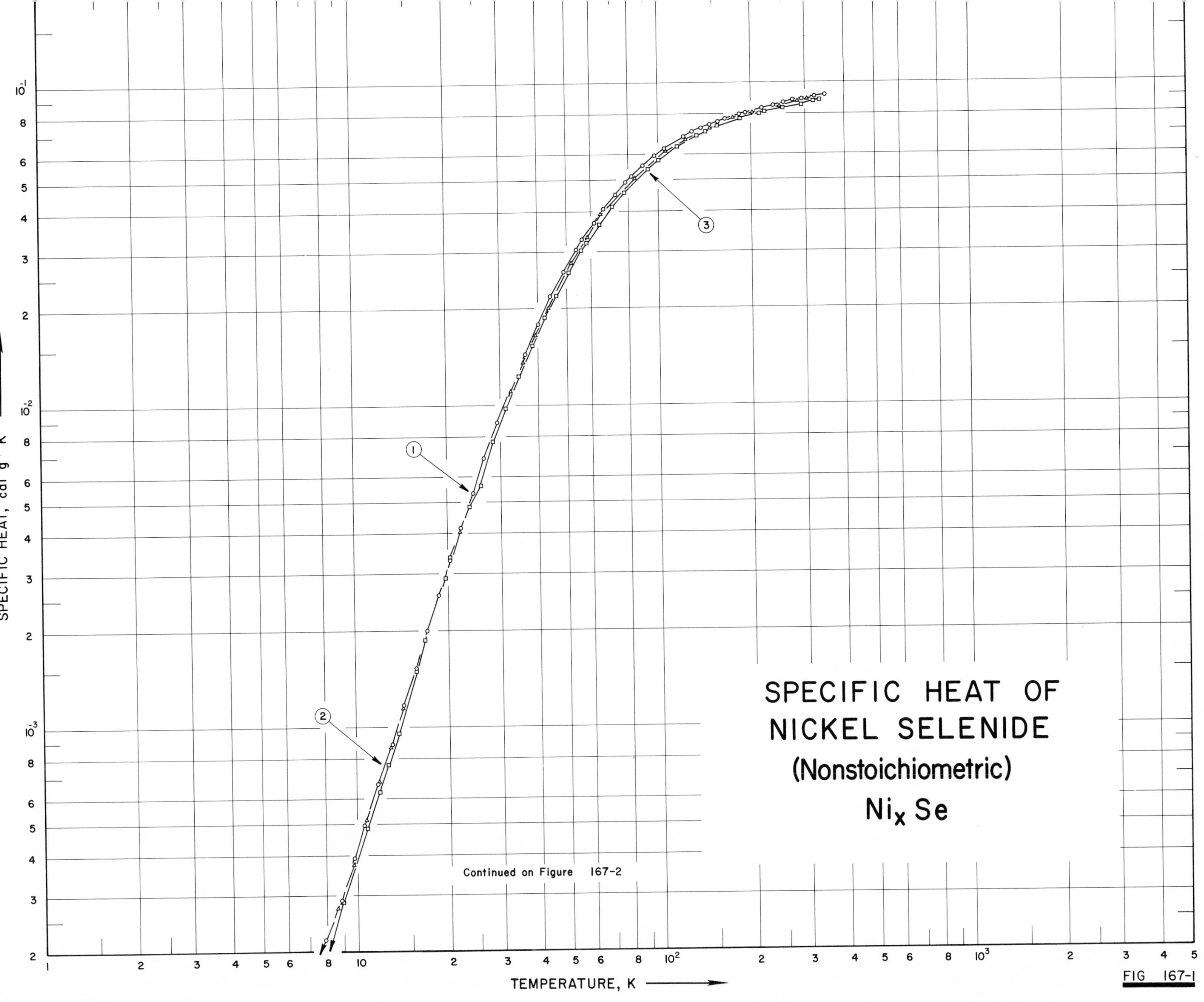

SPECIFIC HEAT OF
NICKEL SELENIDE
(Nonstoichiometric)
$Ni_x Se$
Continued on Figure 167-2
TEMPERATURE, K
SPECIFIC HEAT, cal g⁻¹ K⁻¹
1
2
3
FIG 167-1

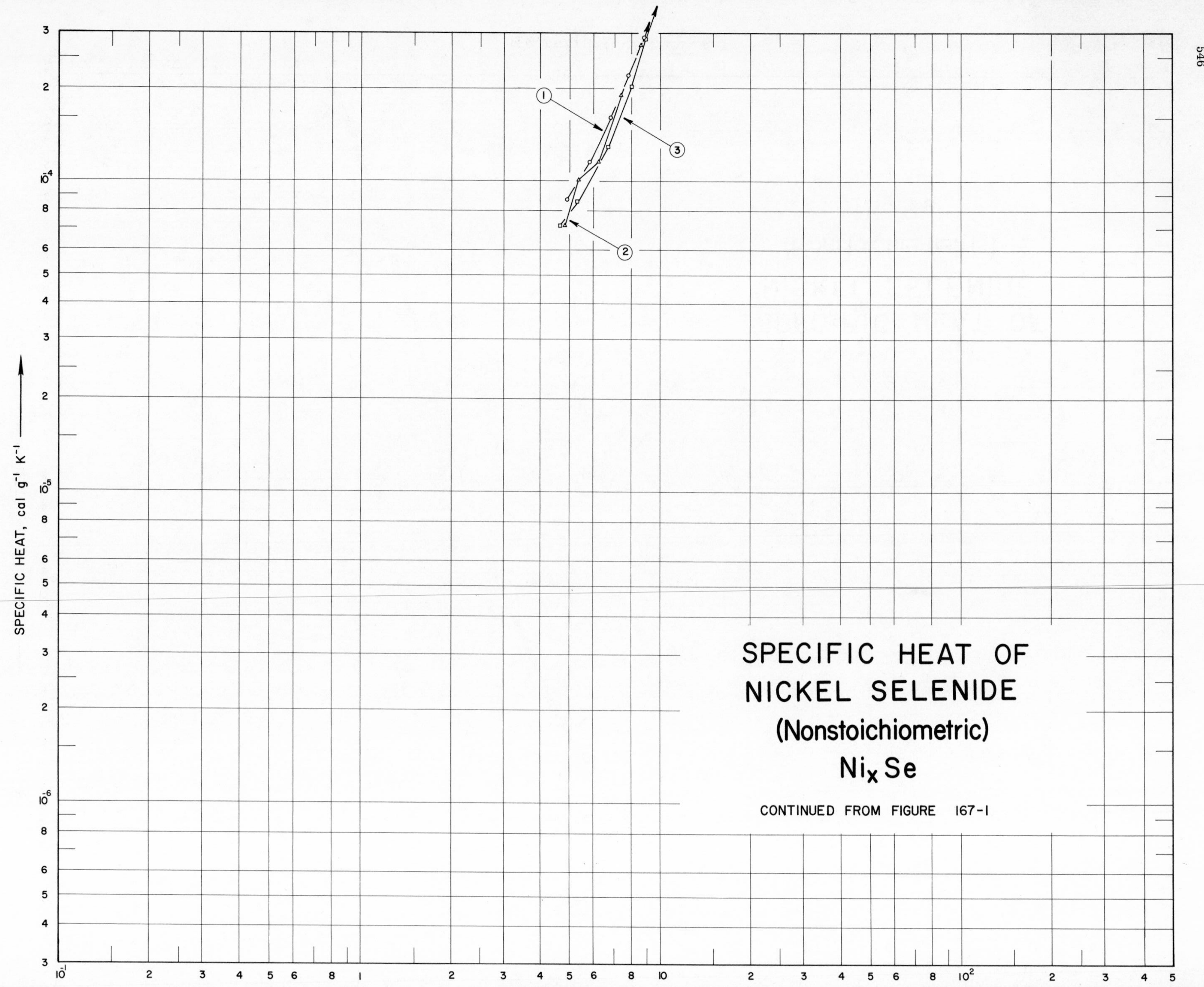
SPECIFIC HEAT, cal g⁻¹ K⁻¹
SPECIFIC HEAT OF
NICKEL SELENIDE
(Nonstoichiometric)
$Ni_x Se$
CONTINUED FROM FIGURE 167-1

SPECIFICATION TABLE NO. 167 SPECIFIC HEAT OF NICKEL SELENIDE (nonstoichiometric) Ni_xSe

[For Data Reported in Figure and Table No. 167]

Curve No.	Ref. No.	Year	Temp. Range, K	Reported Error, %	Name and Specimen Designation	Composition (weight percent), Specifications and Remarks
1	227	1960	5-347	0.1-1.0		$Ni_{0.95}Se$; Ni impurities: 0.01 Al, 0.005 Mg, 0.005 Si, 0.001 each Ca, Co, and Fe, and 0.0001 each Ba, Cu, Cr, and Mn, Se impurities: 0.0002 Cl, 0.00008 Fe, 0.00004 Na and 0.00003 K; prepared by fusion of high purity nickel and selenium; fused for 2 hrs at 1050, 1000, 950 C, respectively, and cooled; fragmented and annealed at 550 C for 7 days and cooled over a period of 2 days; measured in a helium atmosphere.
2	227	1960	5-346	0.1-1.0		$Ni_{0.875}Se$; same as above.
3	227	1960	5-347	0.1-1.0		$Ni_{0.80}Se$; same as above.

DATA TABLE NO. 167 SPECIFIC HEAT OF NICKEL SELENIDE Ni_xSe (nonstoichiometric)

[Temperature, T, K; Specific Heat, C_p, Cal $g^{-1} K^{-1}$]

T	C_p
CURVE 1	
Series 1	
56.71	3.286 x 10^{-2}
61.59	3.692
66.77	4.083
72.47	4.507
78.89	4.934
Series 2	
4.91	8.684 x 10^{-5}
5.88	1.158 x 10^{-4}
6.88	1.592
7.89	2.171
8.84	2.895
9.76	3.908
10.66	4.921
11.78	6.658
13.03	8.829
14.32	1.171 x 10^{-3}
15.65	1.520
17.08	1.994
18.62	2.558
20.27	3.280
22.09	4.197
24.11	5.326
26.48	6.830
29.29	8.822
32.58	1.141 x 10^{-2}*
36.14	1.442
40.01	1.785
44.32	2.175
48.96	2.600
53.97	3.051
Series 3	
82.09	5.125 x 10^{-2}
89.30	5.535
97.37	5.910
105.89	6.250
113.91	6.538
121.59	6.782
129.88	7.014
138.58	7.227
147.33	7.409
CURVE 1 (cont.)	
Series 3 (cont.)	
156.19	7.576 x 10^{-2}
165.03	7.709
174.05	7.843*
183.04	7.958
182.04	7.948*
191.02	8.055
200.06	8.146*
209.17	8.244*
218.35	8.331
227.56	8.399*
237.01	8.489
246.47	8.587*
255.78	8.652
264.93	8.703*
274.18	8.770
283.53	8.825*
292.84	8.886
302.13	8.922*
311.55	8.959*
321.06	9.000
330.48	9.042*
339.86	9.087*
347.33	9.113
CURVE 2	
4.82	7.193 x 10^{-5}
5.39	1.007 x 10^{-4}
6.28	1.151
7.42	1.870
8.60	2.733
9.70	3.740
10.76	5.179
11.84	6.762
12.98	8.689
14.27	1.147 x 10^{-3}
15.66	1.519*
17.11	1.984*
18.57	2.528*
20.13	3.179*
21.94	4.057
24.12	5.242*
26.66	6.811*
29.47	8.750*
CURVE 2 (cont.)	
32.51	1.103 x 10^{-2}
35.80	1.365
39.40	1.662
43.27	2.005
47.41	2.367*
51.88	2.758
52.95	2.850*
58.25	3.302
64.59	3.801
71.20	4.267*
77.54	4.683*
84.02	5.086
90.82	5.448*
98.27	5.788*
106.39	6.117
114.65	6.412*
123.00	6.668
131.36	6.898*
139.77	7.098*
148.46	7.291
157.25	7.459*
166.19	7.600*
175.18	7.727
184.20	7.843*
193.28	7.956*
202.40	8.054
211.59	8.141*
220.65	8.223*
229.79	8.297*
239.01	8.363*
248.11	8.435
257.09	8.492*
265.97	8.558*
274.98	8.635*
283.88	8.707
292.94	8.773*
302.05	8.855
311.05	8.920*
320.15	8.973*
329.40	9.023*
338.66	9.088*
345.99	9.102*
CURVE 3	
Series 1	
156.88	7.361 x 10^{-2}
165.84	7.504*
174.93	7.632*
176.40	7.649*
185.38	7.767
194.29	7.872*
203.10	7.960*
212.01	8.046
221.09	8.130
230.22	8.206*
239.42	8.278*
248.70	8.335*
254.12	8.369
263.27	8.432*
272.43	8.486*
281.60	8.539*
290.82	8.591
300.14	8.639*
309.57	8.695*
319.10	8.758
328.61	8.818*
337.72	8.882
346.76	8.942*
Series 2	
4.68	7.147 x 10^{-5}
5.36	8.576
5.48	8.576*
6.73	1.286 x 10^{-4}
8.03	2.001
8.97	2.859
9.87	3.716*
10.87	4.860
11.80	6.275
12.67	7.690
13.60	9.563
14.38	1.129 x 10^{-3}*
15.68	1.489
16.89	1.862
18.15	2.308*
19.63	2.902
21.49	3.759
23.56	4.858
CURVE 3 (cont.)	
Series 2 (cont.)	
25.77	6.164 x 10^{-3}
28.30	7.784
31.10	9.763
34.58	1.244 x 10^{-2}
38.33	1.544
42.13	1.861
46.18	2.187
50.76	2.590
55.95	3.019
58.08	3.182
64.04	3.649
70.82	4.134
77.82	4.591
84.76	5.017*
92.43	5.400
100.64	5.768
108.85	6.091*
116.87	6.368
125.13	6.619*
134.31	6.865
142.84	7.061
151.64	7.241*

* Not shown on plot

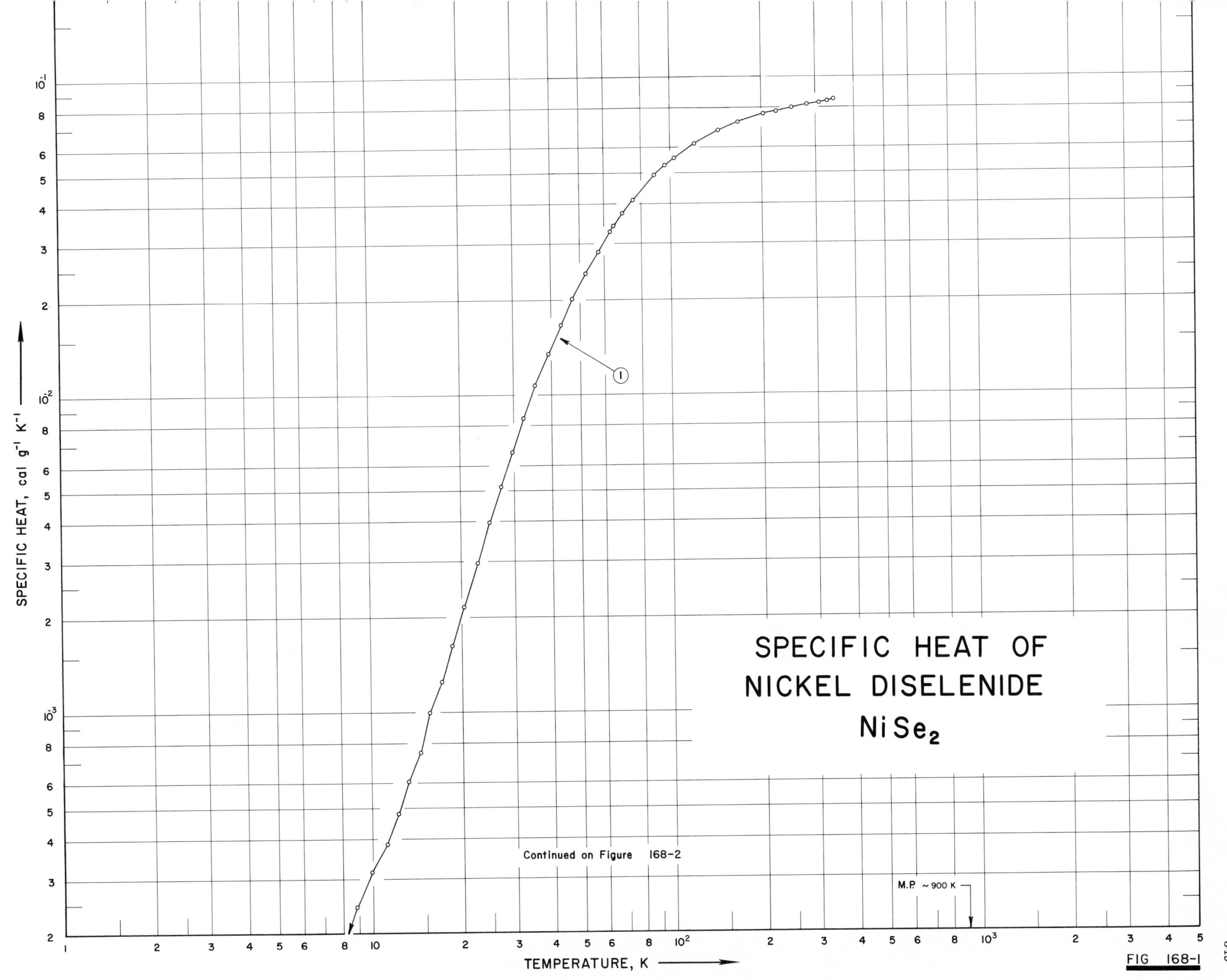
SPECIFIC HEAT OF
NICKEL DISELENIDE
NiSe2
SPECIFIC HEAT, cal g⁻¹ K⁻¹
TEMPERATURE, K
Continued on Figure 168-2
M.P. ~ 900 K
1
FIG 168-1

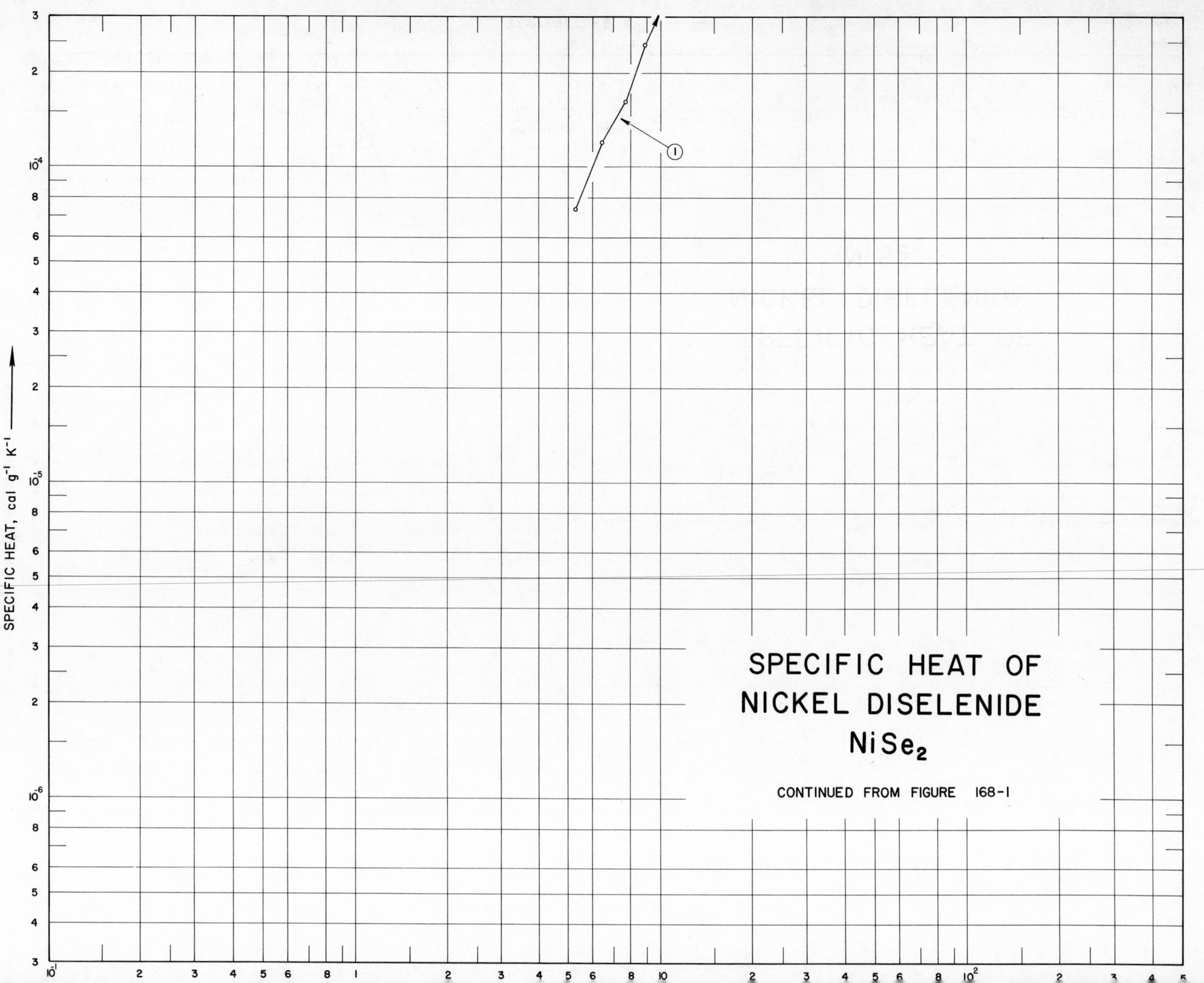
SPECIFIC HEAT OF
NICKEL DISELENIDE
NiSe$_2$
CONTINUED FROM FIGURE 168-1
SPECIFIC HEAT, cal g^{-1} K^{-1}
1

SPECIFICATION TABLE NO. 168 SPECIFIC HEAT OF NICKEL DISELENIDE $NiSe_2$

[For Data Reported in Figure and Table No. 168]

Curve No.	Ref. No.	Year	Temp. Range, K	Reported Error, %	Name and Specimen Designation	Composition (weight percent), Specifications and Remarks
1	214	1962	5-345	0.1-5		Impurities, 0.01 Al, 0.005 Mg, Si, 0.001 Ca, Co, Fe, 0.0001 Cr, Cu, and Mn; prepared from nickel oxide by reduction with H_2 at 500 C for 5 hrs; after cooling was fragmented and heated with H_2 at 1000 C for 4 hrs; mixture of Ni and Se was heated in an evacuated and sealed silica tube at 800 C for 1 day; then temperature lowered to 400 C for 3 days; product crushed and heated at 400 C for 1 wk; heated to 500 C for 1 wk; annealed at 300 C for 1 wk.

DATA TABLE NO. 168 SPECIFIC HEAT OF NICKEL DISELENIDE $NiSe_2$

[Temperature, T, K; Specific Heat, C_p, Cal g^{-1} K^{-1}]

T	C_p	T	C_p
CURVE 1		CURVE 1 (cont.)	
Series 1		Series 2 (cont.)	
64.88	3.428×10^{-2}	11.20	3.831×10^{-4}
69.16	3.731*	12.27	4.805
76.04	4.122	13.35	6.052
81.54	4.531*	14.54	7.455
88.85	4.940	15.77	9.832
96.10	5.284	17.10	1.251×10^{-3}
103.76	5.604	18.61	1.621
112.23	5.924*	20.40	2.157
120.73	6.202	22.57	2.955
129.02	6.443*	24.89	3.982
136.89	6.643*	27.20	5.165
144.63	6.800	29.67	6.624
152.32	6.970*	32.48	8.480
151.60	6.952*	35.67	1.076×10^{-2}
159.52	7.090*	39.26	1.354
167.82	7.234	43.29	1.677
176.80	7.363*	47.44	2.023
186.18	7.483*	52.26	2.425
195.62	7.594*	57.17	2.827
204.99	7.691	62.87	3.277
214.17	7.778*	69.18	3.733
223.30	7.852		
232.28	7.926*		
241.27	8.000*		
250.48	8.060		
262.05	8.124*		
271.20	8.184*		
280.33	8.231		
289.44	8.286*		
298.50	8.328*		
307.57	8.364		
316.73	8.406*		
325.99	8.448		
335.39	8.485*		
344.95	8.526		
Series 2			
5.27	7.385×10^{-5}		
6.43	1.200×10^{-4}		
7.69	1.616		
8.92	2.447		
10.06	3.139		

* Not shown on plot

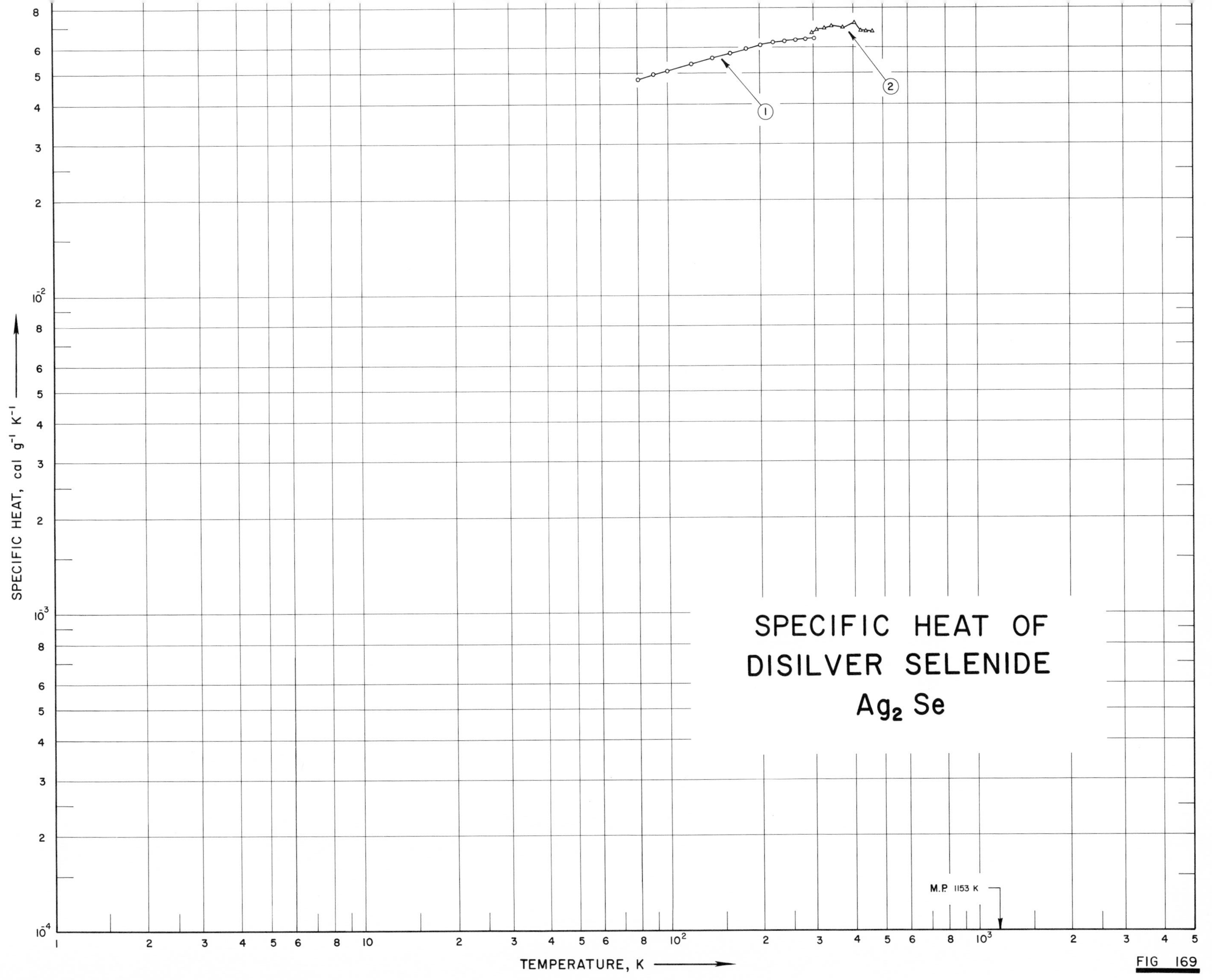
SPECIFIC HEAT OF
DISILVER SELENIDE
Ag_2Se
SPECIFIC HEAT, cal g⁻¹ K⁻¹
TEMPERATURE, K
M.P. 1153 K
1
2

FIG 169

SPECIFICATION TABLE NO. 169 SPECIFIC HEAT OF DISILVER SELENIDE Ag_2Se

[For Data Reported in Figure and Table No. 169]

Curve No.	Ref. No.	Year	Temp. Range, K	Reported Error, %	Name and Specimen Designation	Composition (weight percent), Specifications and Remarks
1	160	1959	80-300	3-7		Polycrystalline
2	228	1962	308-444	5.0		Spectroscopically pure; single crystal; zone melted under controlled vapor pressure; cooled to 150 C at 5-10 C per hr; annealed for several hrs below transition.

DATA TABLE NO. 169 SPECIFIC HEAT OF DISILVER SELENIDE Ag_2Se

[Temperature, T, K; Specific Heat, C_p, Cal g^{-1} K^{-1}]

T	C_p
CURVE 1	
80	4.77 x 10^{-2}
90	4.95
100	5.1
120	5.36
140	5.6
160	5.8
180	5.99
200	6.14
220	6.24
240	6.31
260	6.36
280	6.4
300	6.42
CURVE 2	
Series 1	
308.0	6.880 x 10^{-2}
318.4	6.940*
333.4	6.970*
341.4	7.070
350.2	7.090*
358.2	7.220*
367.4	6.860*
374.4	6.920
392.4	7.080*
399.2	7.190*
405.4	7.270
410.7	6.780*
419.2	6.830*
430.0	6.850
436.9	6.800*
443.9	6.800
Series 2	
295.3	6.710 x 10^{-2}
303.4	6.800*
314.4	6.940*
326.0	6.900
333.6	7.050*
342.4	7.080*
350.0	7.110*
363.5	7.150*
366.2	7.000*
372.4	6.850*
CURVE 2 (cont.)	
Series 2 (cont.)	
379.2	6.880 x 10^{-2}*
388.4	7.000*
398.2	7.050*
414.2	6.900*
420.2	6.700*
436.3	6.930*
442.9	6.750*
452.5	6.790*
465.3	6.740

* Not shown on plot

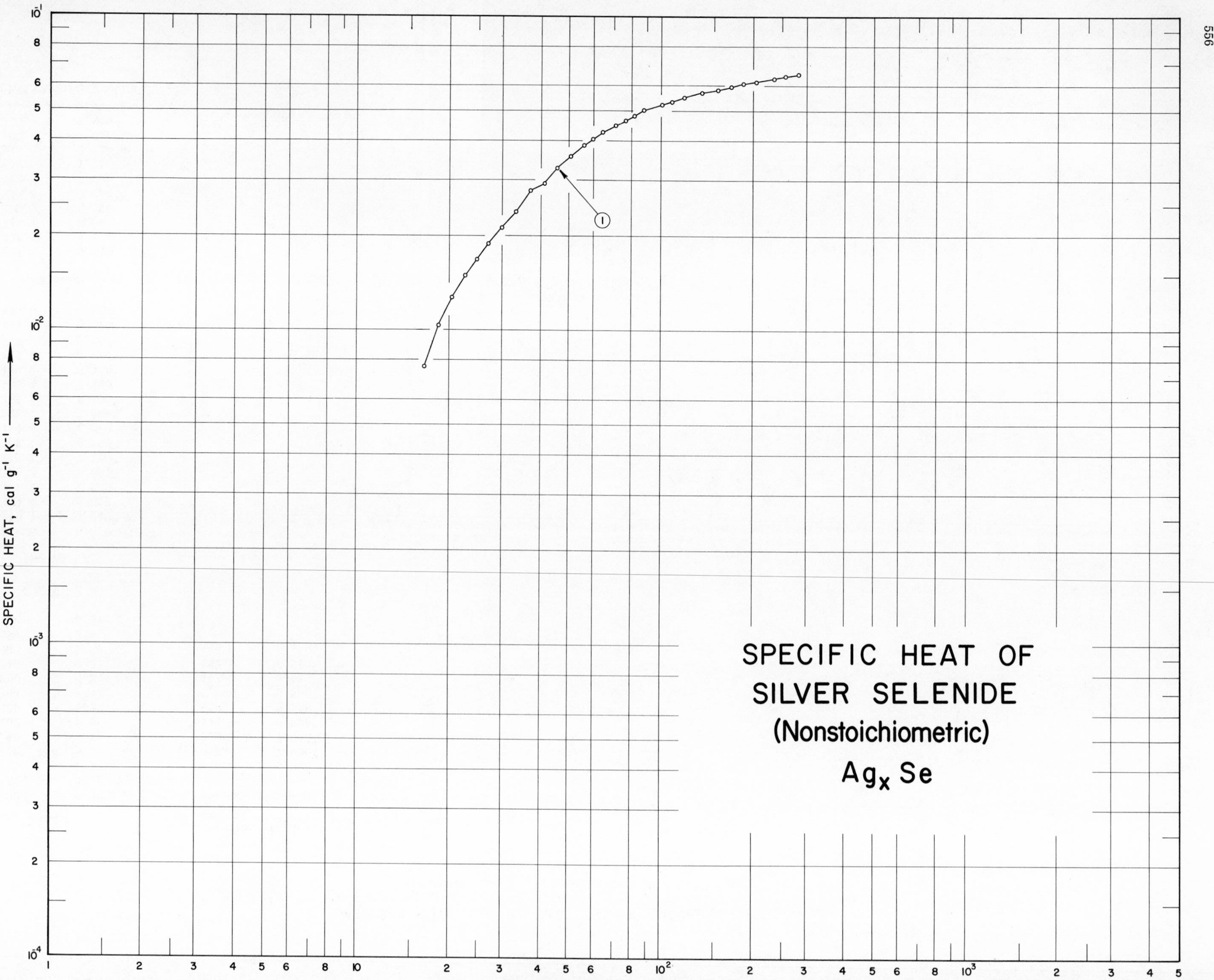

SPECIFIC HEAT OF
SILVER SELENIDE
(Nonstoichiometric)
$Ag_x Se$
SPECIFIC HEAT, cal g^{-1} K^{-1}
1

SPECIFICATION TABLE NO. 170 SPECIFIC HEAT OF SILVER SELENIDE (nonstoichiometric) Ag_xSe

[For Data Reported in Figure and Table No. 170]

Curve No.	Ref. No.	Year	Temp. Range, K	Reported Error, %	Name and Specimen Designation	Composition (weight percent), Specifications and Remarks
1	225	1962	17-283		$Ag_{1.99}Se$	99.99 $Ag_{1.99}Se$; crushed under argon atmosphere.

DATA TABLE NO. 170 SPECIFIC HEAT OF SILVER SELENIDE Ag_xSe (nonstoichiometric)

[Temperature, T, K; Specific Heat, C_p, Cal $g^{-1} K^{-1}$]

T	C_p
CURVE 1	
16.66	7.696×10^{-3}
18.53	1.039×10^{-2}
20.57	1.277
22.63	1.492
24.74	1.686
27.00	1.887
29.87	2.111
33.20	2.384
37.00	2.789
41.33	2.936
45.93	3.286
50.73	3.572
56.03	3.872
60.06	4.032
64.82	4.274
71.24	4.495
77.13	4.672
82.85	4.829
88.23	5.016
91.96	5.040*
101.25	5.234
109.88	5.371
120.90	5.534
129.13	5.629*
137.34	5.708
145.99	5.789*
155.32	5.888
159.87	5.926*
170.60	5.987
178.89	6.062*
187.75	6.127
197.36	6.174*
207.49	6.236
207.83	6.225
216.34	6.270*
217.27	6.283*
235.28	6.331
240.89	6.365*
246.28	6.396*
258.04	6.460
264.47	6.511*
271.07	6.545*
277.34	6.552*
283.48	6.583

* Not shown on plot

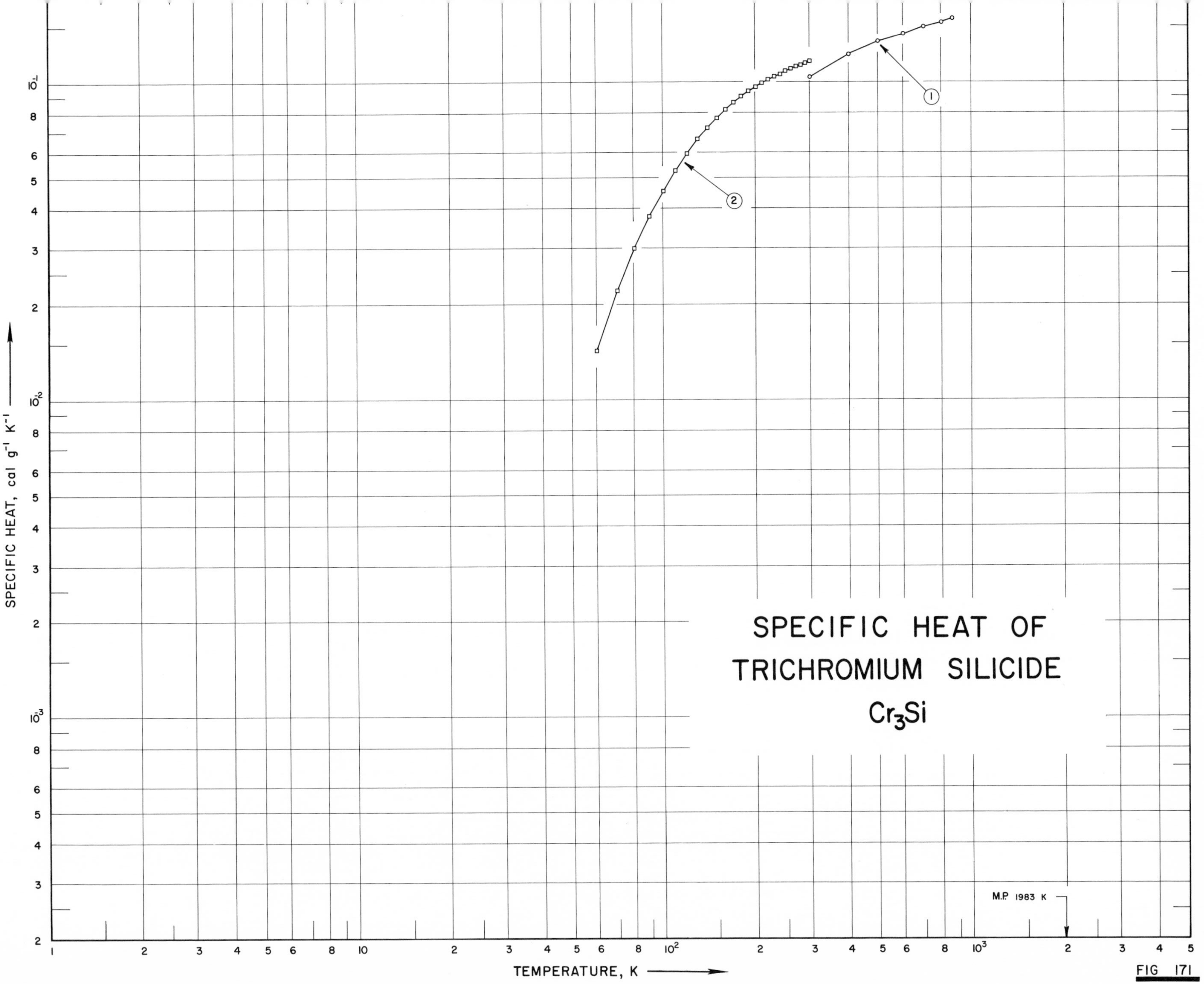

SPECIFIC HEAT OF
TRICHROMIUM SILICIDE
Cr_3Si
SPECIFIC HEAT, cal g^{-1} K^{-1}
TEMPERATURE, K
M.P. 1983 K
1
2
10^{-1}
10^{-2}
10^{-3}
10
10^2
10^3

FIG 171

SPECIFICATION TABLE NO. 171 SPECIFIC HEAT OF TRICHROMIUM SILICIDE Cr_3Si

[For Data Reported in Figure and Table No. 171]

Curve No.	Ref. No.	Year	Temp. Range, K	Reported Error, %	Name and Specimen Designation	Composition (weight percent), Specifications and Remarks
1	229	1961	298-873	±2.0		Stoichiometric Cr_3Si; measured in an argon atmosphere.
2	418	1965	60-300			Prepared from single crystal silicon (>99.999 Si), and electrolytic chromium (99.98 Cr).

DATA TABLE NO. 171 SPECIFIC HEAT OF TRICHROMIUM SILICIDE Cr_3Si

[Temperature, T, K; Specific Heat, C_p, Cal g^{-1} K^{-1}]

T	C_p
CURVE 1	
298	1.046 x 10^{-1}*
300	1.051
400	1.240
500	1.352
600	1.436
700	1.505
800	1.566
873	1.608
CURVE 2	
60	1.4387 x 10^{-2}
70	2.2198
80	3.0071
90	3.7914
100	4.5549
110	5.2976
120	5.9727
130	6.6479
140	7.2192
150	7.7386
160	8.2060
170	8.6475
180	9.0370
190	9.4006
200	9.6603
210	9.9719
220	1.0232 x 10^{-1}
230	1.0491
240	1.0699
250	1.0907
260	1.1089
270	1.1270
273.15	1.1322*
280	1.1426
290	1.1581
298.15	1.1738*
300	1.1738

*Not shown on plot

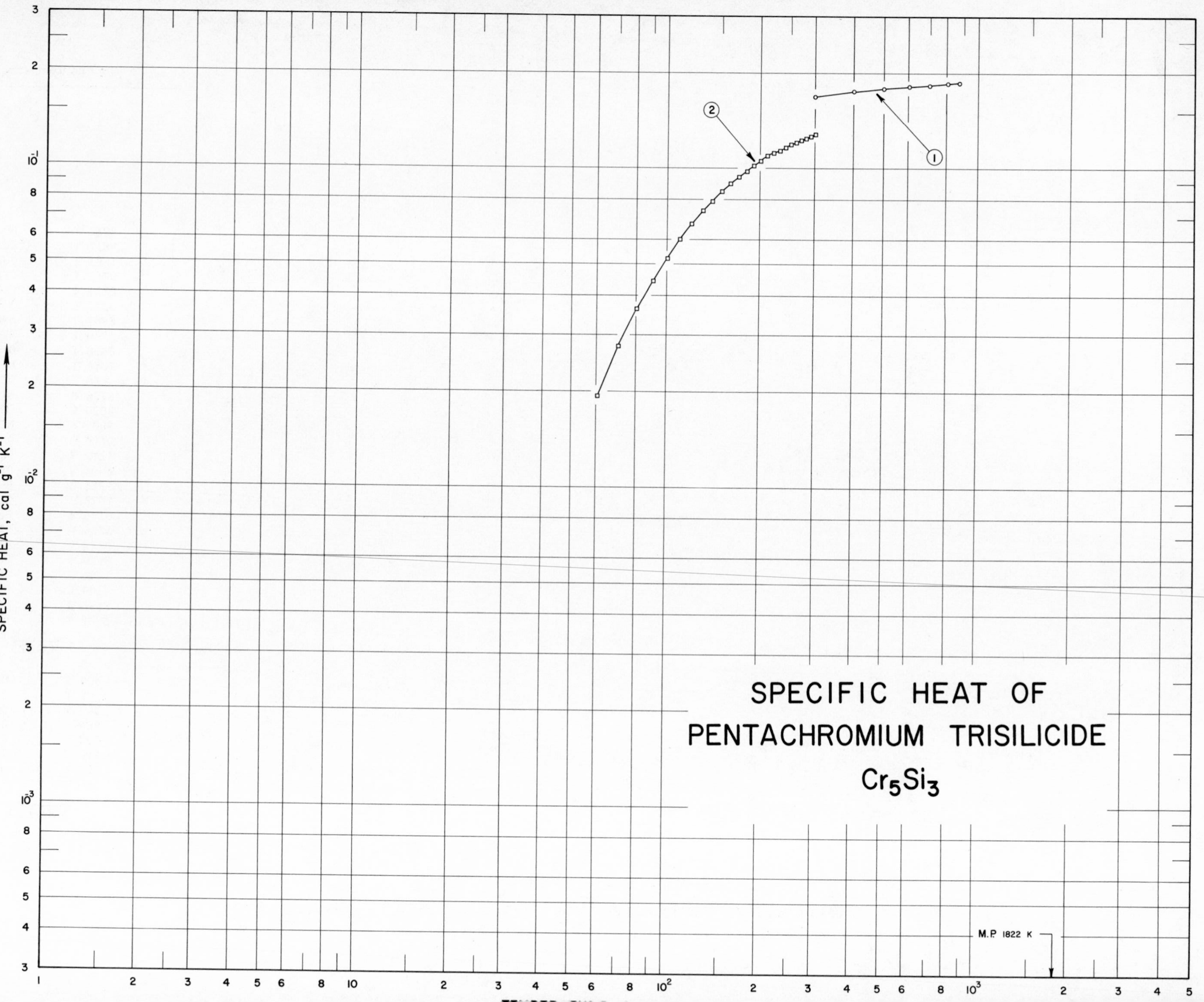

SPECIFIC HEAT OF
PENTACHROMIUM TRISILICIDE
Cr_5Si_3
M.P. 1822 K
SPECIFIC HEAT, cal g^{-1} K^{-1}
1
2

SPECIFICATION TABLE NO. 172 SPECIFIC HEAT OF PENTACHROMIUM TRISILICIDE Cr_5Si_3

[For Data Reported in Figure and Table No. 172]

Curve No.	Ref. No.	Year	Temp. Range, K	Reported Error, %	Name and Specimen Designation	Composition (weight percent), Specifications and Remarks
1	229	1961	298-873	±2.0		Stoichiometric Cr_5Si_3; measured in an argon atmosphere.
2	418	1965	60-300			Prepared from single crystal silicon (>99.999 Si) and electrolytic chromium (99.98 Cr).

DATA TABLE NO. 172 SPECIFIC HEAT OF PENTACHROMIUM TRISILICIDE Cr_5Si_3

[Temperature, T, K; Specific Heat, C_p, Cal g^{-1} K^{-1}]

T	C_p
CURVE 1	
298	1.6976 x 10^{-1}*
300	1.6990
400	1.7505
500	1.7843
600	1.8112
700	1.8349
800	1.8568
873	1.8721
CURVE 2	
60	1.9329 x 10^{-2}
70	2.7772
80	3.6271
90	4.4547
100	5.2489
110	5.9988
120	6.7209
130	7.3874
140	7.9429
150	8.4983
160	8.9982
170	9.4425
180	9.8314
190	1.0220 x 10^{-1}
200	1.0553
210	1.1081
220	1.1164
230	1.1442
240	1.1692
250	1.1942
260	1.2136
270	1.2331
273.15	1.2386*
280	1.2497
290	1.2664
298.15	1.2802*
300	1.2830

* Not shown on plot

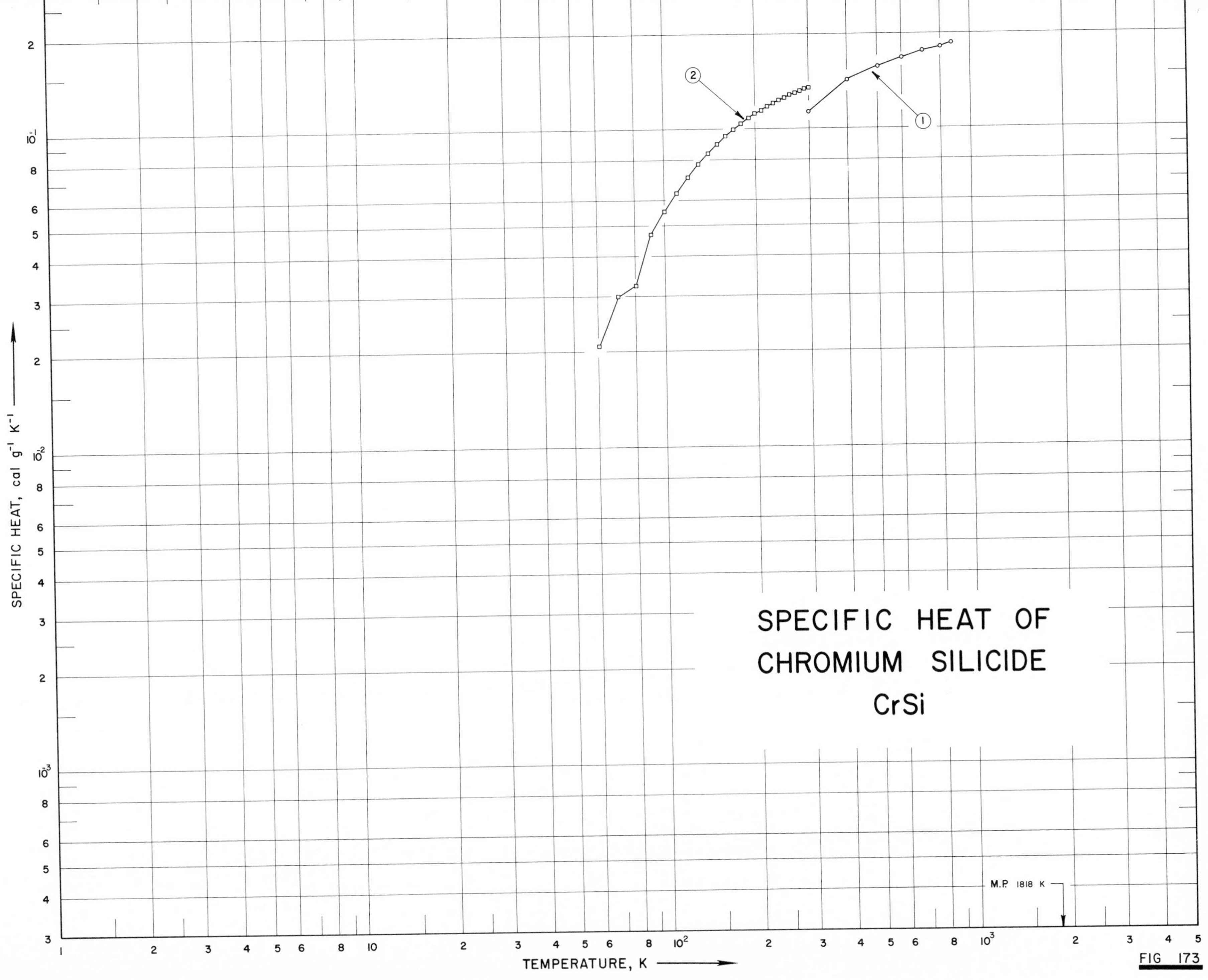
SPECIFIC HEAT OF
CHROMIUM SILICIDE
CrSi
M.P. 1818 K
1
2
TEMPERATURE, K
SPECIFIC HEAT, cal g⁻¹ K⁻¹
FIG 173

SPECIFICATION TABLE NO. 173 SPECIFIC HEAT OF CHROMIUM SILICIDE CrSi

[For Data Reported in Figure and Table No. 173]

Curve No.	Ref. No.	Year	Temp. Range, K	Reported Error, %	Name and Specimen Designation	Composition (weight percent), Specifications and Remarks
1	229	1961	298-873	±2.0		Stoichiometric CrSi; measured in an argon atmosphere.
2	418	1965	60-300			Prepared from single crystal silicon (>99.999 Si) and electrolytic chromium (99.98 Cr).

DATA TABLE NO. 173 SPECIFIC HEAT OF CHROMIUM SILICIDE CrSi

[Temperature, T, K; Specific Heat, C_p, Cal $g^{-1} K^{-1}$]

T	C_p
CURVE 1	
298	1.1485 x 10^{-1}
300	1.1566*
400	1.4328
500	1.5835
600	1.6850
700	1.7631
800	1.8287
873	1.8719
CURVE 2	
60	2.0772 x 10^{-2}
70	2.9845
80	3.2293
90	4.6678
100	5.5034
110	6.2973
120	7.0435
130	7.7597
140	8.3865
150	8.9535
160	9.4908
170	9.9683
180	1.0386 x 10^{-1}
190	1.0774
200	1.1132
210	1.1461
220	1.1759
230	1.2057
240	1.2296
250	1.2535
260	1.2744
270	1.2953
273.15	1.3012*
280	1.3132
290	1.3340
298.15	1.3460*
300	1.3490

* Not shown on plot

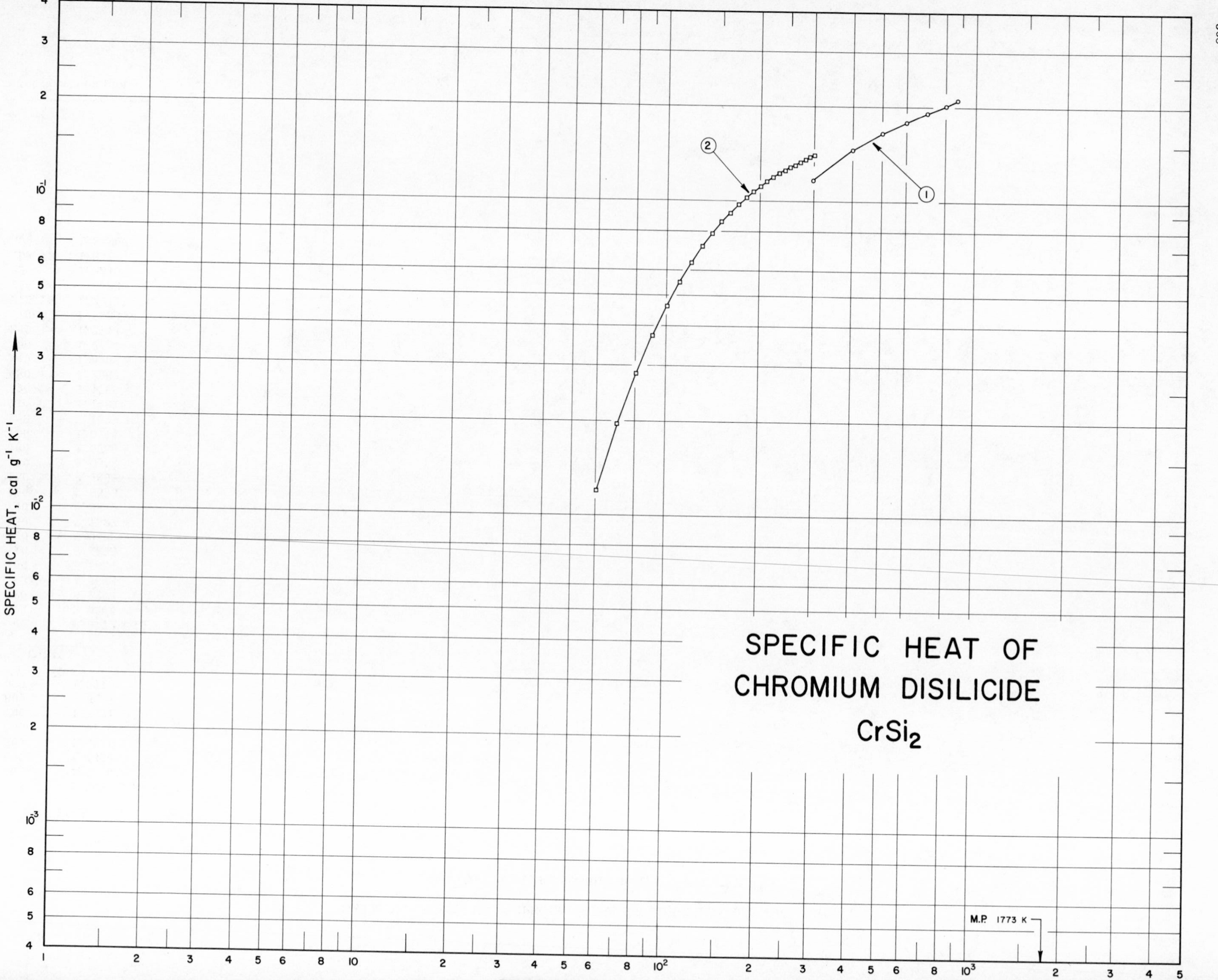

SPECIFIC HEAT OF
CHROMIUM DISILICIDE
$CrSi_2$
M.P. 1773 K
SPECIFIC HEAT, cal g^{-1} K^{-1}
1
2

SPECIFICATION TABLE NO. 174 SPECIFIC HEAT OF CHROMIUM DISILICIDE $CrSi_2$

[For Data Reported in Figure and Table No. 174]

Curve No.	Ref. No.	Year	Temp. Range, K	Reported Error, %	Name and Specimen Designation	Composition (weight percent), Specifications and Remarks
1	229	1961	298-873	±2.0		Stoichiometric $CrSi_2$; measured in an argon atmosphere.
2	418	1965	60-300			Prepared from single crystal silicon (>99.999 Si) and electrolytic chromium (99.98 Cr).

DATA TABLE NO. 174 SPECIFIC HEAT OF CHROMIUM DISILICIDE $CrSi_2$

[Temperature, T, K; Specific Heat, C_p, Cal g^{-1} K^{-1}]

T	C_p
CURVE 1	
298	1.1772×10^{-1}
300	1.1849*
400	1.4699
500	1.6541
600	1.7987
700	1.9245
800	2.0403
873	2.1210
CURVE 2	
60	1.1998×10^{-2}
70	1.9555
80	2.8106
90	3.7254
100	4.6401
110	5.5151
120	6.3503
130	7.1590
140	7.8882
150	8.5511
160	9.1477
170	9.7442
180	1.0241×10^{-1}
190	1.0739
200	1.1169
210	1.1600
220	1.1965
230	1.2296
240	1.2595
250	1.2893
260	1.3125
270	1.3390
273.15	1.3456*
280	1.3655
290	1.3854
298.15	1.4053*
300	1.4086

* Not shown on plot

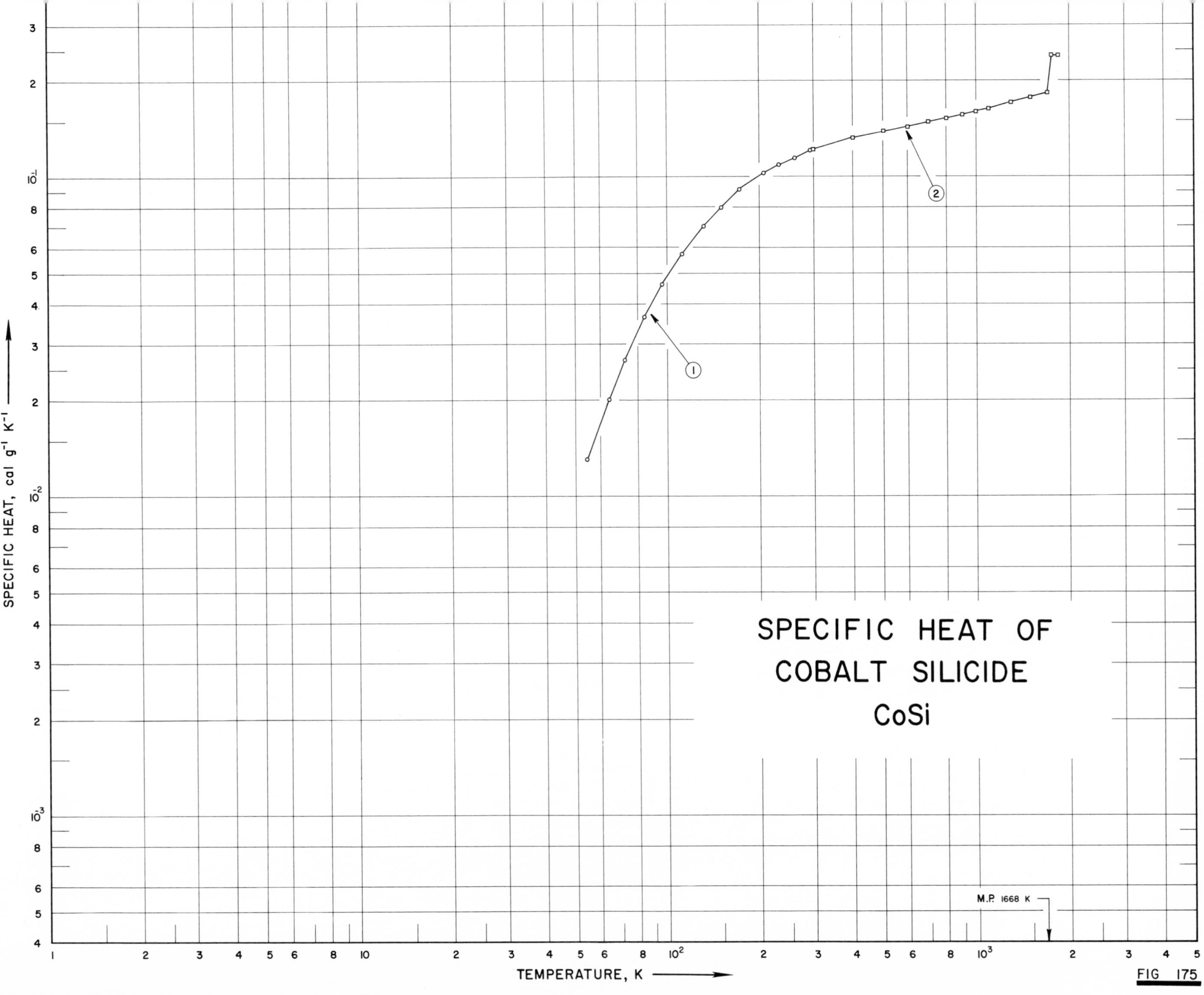
SPECIFIC HEAT OF
COBALT SILICIDE
CoSi
SPECIFIC HEAT, cal g⁻¹ K⁻¹
TEMPERATURE, K
M.P. 1668 K
1
2
FIG 175

SPECIFICATION TABLE NO. 175 SPECIFIC HEAT OF COBALT SILICIDE CoSi

[For Data Reported in Figure and Table No. 175]

Curve No.	Ref. No.	Year	Temp. Range, K	Reported Error, %	Name and Specimen Designation	Composition (weight percent), Specifications and Remarks
1	230	1964	54-294			Stoichiometric CoSi; (99.98 Co) and (>99.997 Si) melted together in purified argon.
2	230	1964	298-1850			Same as above.

DATA TABLE NO. 175 SPECIFIC HEAT OF COBALT SILICIDE CoSi

[Temperature, T, K; Specific Heat, C_p, Cal $g^{-1} K^{-1}$]

T	C_p
CURVE 1	
54.28	1.305 x 10^{-2}
56.92	1.514*
58.93	1.675*
64.13	2.013
65.75	2.103*
70.77	2.532*
72.57	2.679
78.46	3.131*
79.99	3.285*
84.28	3.650
85.65	3.856*
90.38	4.156*
91.94	4.254*
96.53	4.628
98.74	4.787*
105.98	5.257*
112.22	5.776
118.69	6.320
124.44	6.647*
131.95	7.075
138.58	7.487*
145.91	7.855*
151.82	8.056
155.73	8.336*
162.82	8.589*
166.21	8.715*
173.89	9.174
181.87	9.424*
184.06	9.632*
196.73	9.978*
207.01	1.028 x 10^{-1}
218.29	1.060*
232.30	1.098
242.21	1.113*
253.54	1.133*
260.77	1.150
268.67	1.151*
270.54	1.170*
284.37	1.180*
286.05	1.199*
290.42	1.199*
292.09	1.209*
293.86	1.215

T	C_p
CURVE 2	
298.15	1.218 x 10^{-1}
300	1.219*
400	1.329
500	1.395
600	1.447
700	1.491
800	1.533
900	1.571
1000	1.607
1100	1.642
1200	1.675*
1300	1.711
1400	1.744*
1500	1.777
1600	1.813*
1700	1.846
1750	2.401
1800	2.401*
1850	2.401

* Not shown on plot

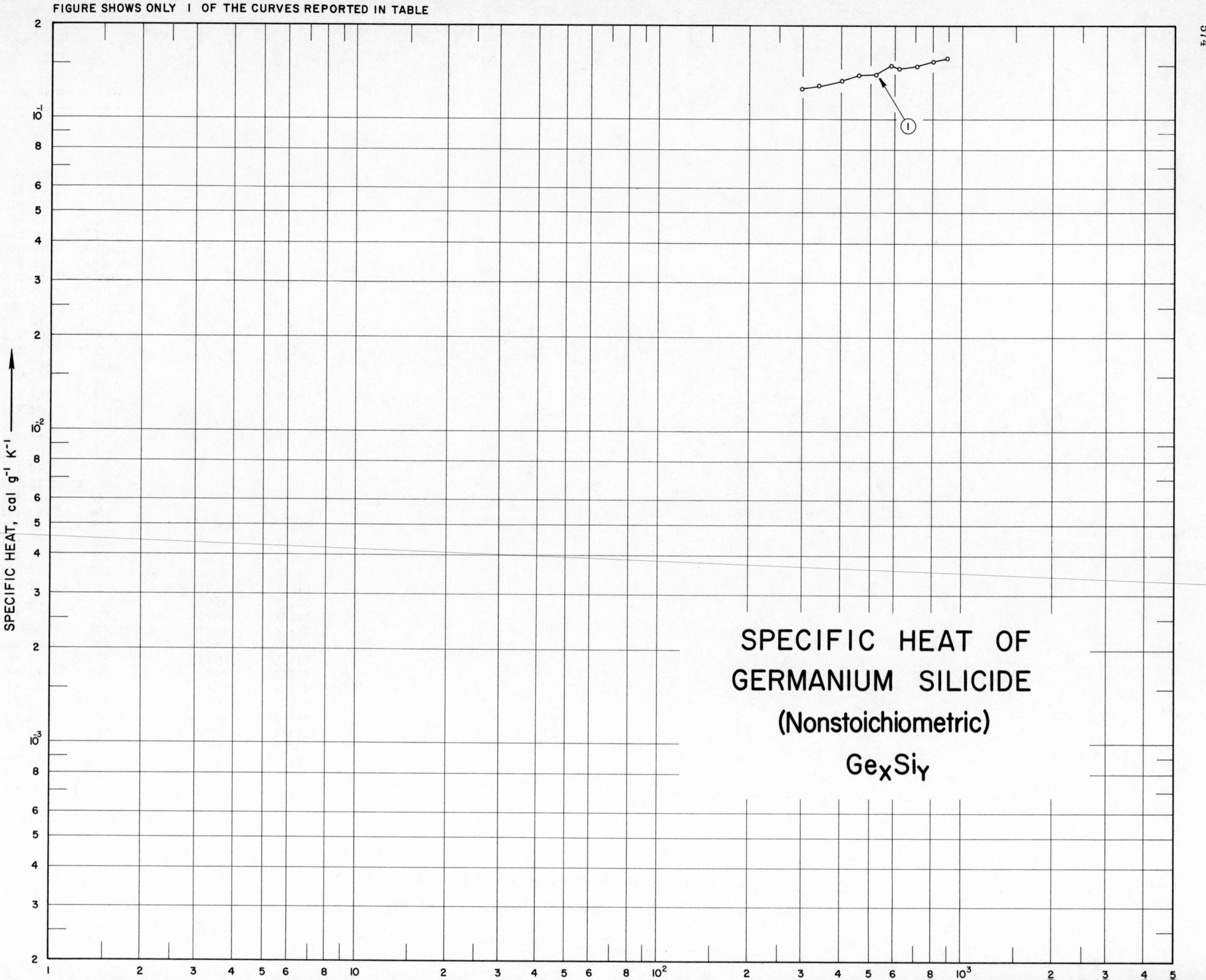
FIGURE SHOWS ONLY 1 OF THE CURVES REPORTED IN TABLE
SPECIFIC HEAT OF
GERMANIUM SILICIDE
(Nonstoichiometric)
Ge_XSi_Y
SPECIFIC HEAT, cal g^{-1} K^{-1}
1

SPECIFICATION TABLE NO. 176 SPECIFIC HEAT OF GERMANIUM SILICIDE Ge_xSi_y (nonstoichiometric)

[For Data Reported in Figure and Table No. 176]

Curve No.	Ref. No.	Year	Temp. Range, K	Reported Error, %	Name and Specimen Designation	Composition (weight percent), Specifications and Remarks
1	231	1964	296-892	2.0-3.0	Si-Ge 46; n-type	52.6 Ge and 47.4 Si; 1.7 x 10^{-3} ohm cm resistivity.
2	231	1964	295-799	2.0-3.0	Si-Ge 75; p-type	Same as above; 2.0 x 10^{-3} ohm cm resistivity.

DATA TABLE NO. 176 SPECIFIC HEAT OF GERMANIUM SILICIDE Ge_xSi_y (nonstoichiometric)

[Temperature, T, K; Specific Heat, C_p, Cal g^{-1} K^{-1}]

T	C_p
CURVE 1	
296.8	1.257 x 10^{-1}
296.0	1.229*
338.0	1.270
401.2	1.326
432.5	1.351*
457.0	1.380
522.0	1.397
551.0	1.408*
582.7	1.484
587.4	1.408*
620.9	1.459
681.0	1.505*
712.3	1.473
743.2	1.545*
779.2	1.530*
806.4	1.534
846.1	1.534*
869.7	1.502*
891.5	1.564
CURVE 2*	
295.4	1.247 x 10^{-1}
356.1	1.305
383.7	1.369
413.6	1.415
451.0	1.411
329.1	1.308
473.7	1.402
509.5	1.412
548.1	1.415
590.8	1.466
623.6	1.428
653.9	1.444
685.8	1.451
735.1	1.480
764.5	1.486
799.3	1.556

*Not shown on plot

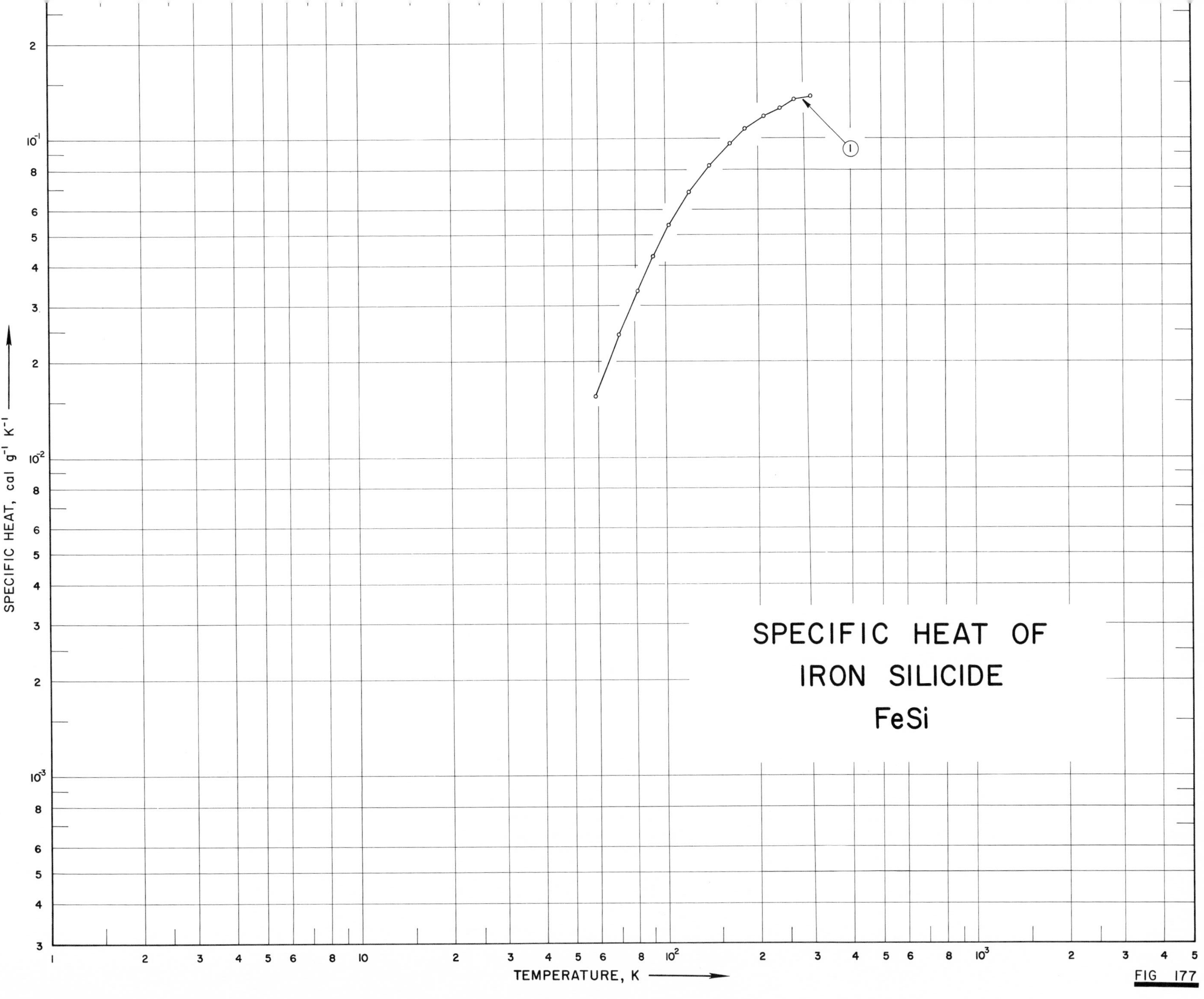

SPECIFIC HEAT OF
IRON SILICIDE
FeSi
1
SPECIFIC HEAT, cal g⁻¹ K⁻¹
TEMPERATURE, K
FIG 177

SPECIFICATION TABLE NO. 177 SPECIFIC HEAT OF IRON SILICIDE FeSi

[For Data Reported in Figure and Table No. 177]

Curve No.	Ref. No.	Year	Temp. Range, K	Reported Error, %	Name and Specimen Designation	Composition (weight percent), Specifications and Remarks
1	419	1963	59-298			34.0 Si.

DATA TABLE NO. 177 SPECIFIC HEAT OF IRON SILICIDE $FeSi$

[Temperature, T, K; Specific Heat, C_p, Cal $g^{-1} K^{-1}$]

T	C_p
CURVE 1	
58.97	1.56×10^{-2}
61.11	1.75*
62.53	1.83*
64.30	1.97*
70.13	2.43
77.07	3.02*
81.23	3.36
84.16	3.66*
85.28	3.74*
91.54	4.29
97.65	4.86*
103.60	5.37
110.30	5.95*
112.68	6.22*
120.70	6.80
125.66	7.19*
128.56	7.52*
133.85	7.86*
140.26	8.26
146.53	8.73*
154.76	9.18*
163.60	9.68
172.54	1.01×10^{-1}*
178.80	1.06*
182.46	1.07
189.41	1.10*
192.34	1.09*
197.91	1.12*
210.29	1.17
217.66	1.20*
226.24	1.22*
237.02	1.25
245.45	1.27*
255.47	1.30*
264.00	1.32
298.16	1.36

* Not shown on plot

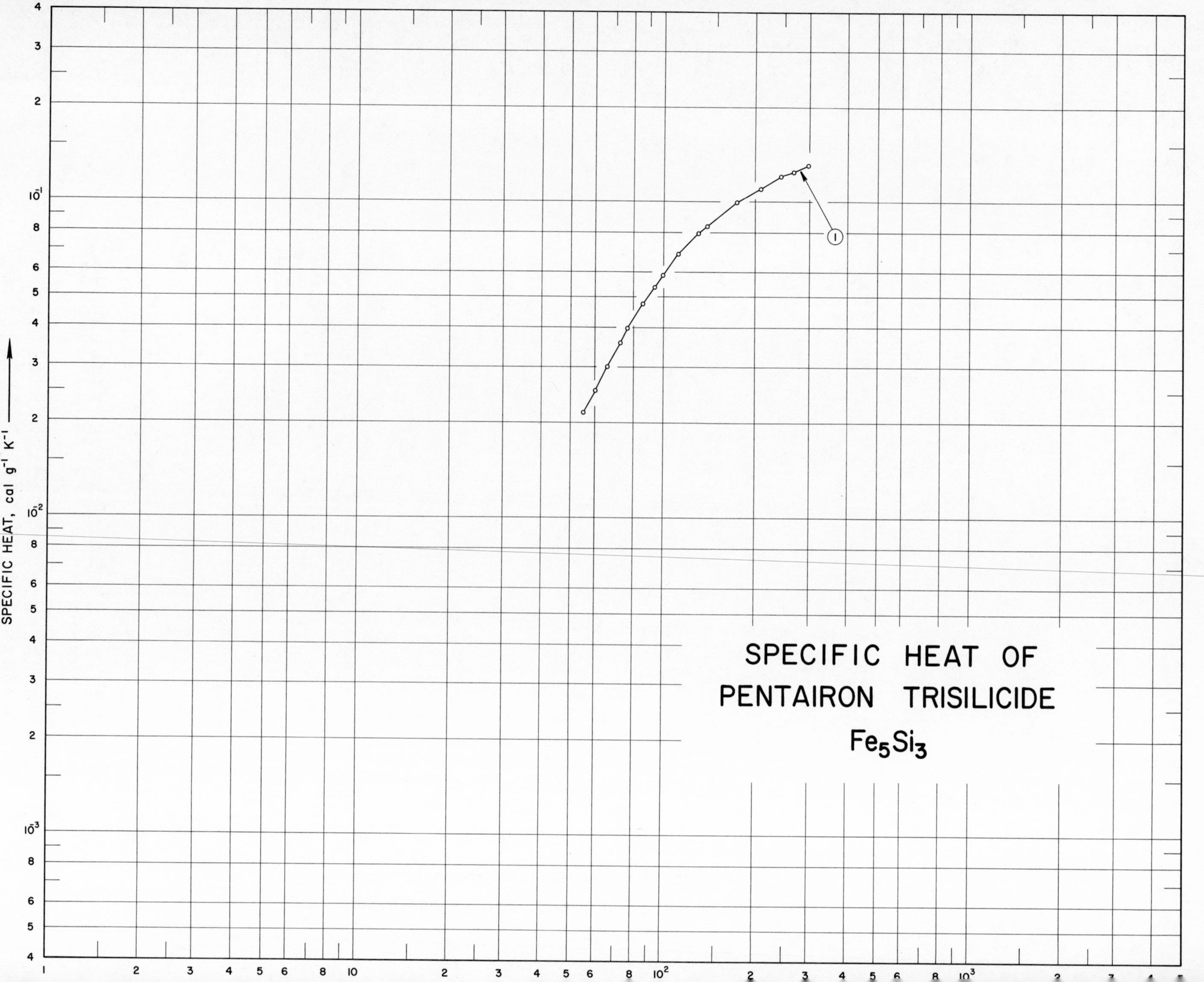
SPECIFIC HEAT OF
PENTAIRON TRISILICIDE
Fe_5Si_3
SPECIFIC HEAT, cal g^{-1} K^{-1}
1

SPECIFICATION TABLE NO. 178 SPECIFIC HEAT OF PENTAIRON TRISILICIDE Fe_5Si_3

[For Data Reported in Figure and Table No. 178]

Curve No.	Ref. No.	Year	Temp. Range, K	Reported Error, %	Name and Specimen Designation	Composition (weight percent), Specifications and Remarks
1	419	1963	55-298			24.02 Si.

DATA TABLE NO. 178 SPECIFIC HEAT OF PENTAIRON TRISILICIDE Fe_5Si_3

[Temperature, T, K; Specific Heat, C_p, Cal g^{-1} K^{-1}]

T	C_p
CURVE 1	
55.25	2.16 x 10^{-2}
60.74	2.54
66.38	3.02
73.05	3.59
77.13	3.97
78.60	4.06*
86.58	4.78
94.66	5.38
100.49	5.87
113.30	6.85
122.99	7.51*
131.14	7.98
140.33	8.39
152.35	9.10*
164.06	9.55*
175.14	1.00 x 10^{-1}
185.64	1.04*
198.66	1.07*
210.16	1.10
222.02	1.14*
231.97	1.18*
233.45	1.17*
238.20	1.18*
244.20	1.21
246.30	1.21*
254.02	1.22*
258.65	1.23*
260.90	1.23
266.29	1.26*
268.88	1.26*
273.30	1.27*
277.29	1.27*
286.34	1.29*
298.16	1.31
CURVE 2	
300	1.32 x 10^{-1}
400	1.37
500	1.44
600	1.50
700	1.56
800	1.62
900	1.68
1000	1.74
CURVE 2 (cont.)	
1100	1.80 x 10^{-1}
1200	1.86
1300	1.92

*Not shown on plot

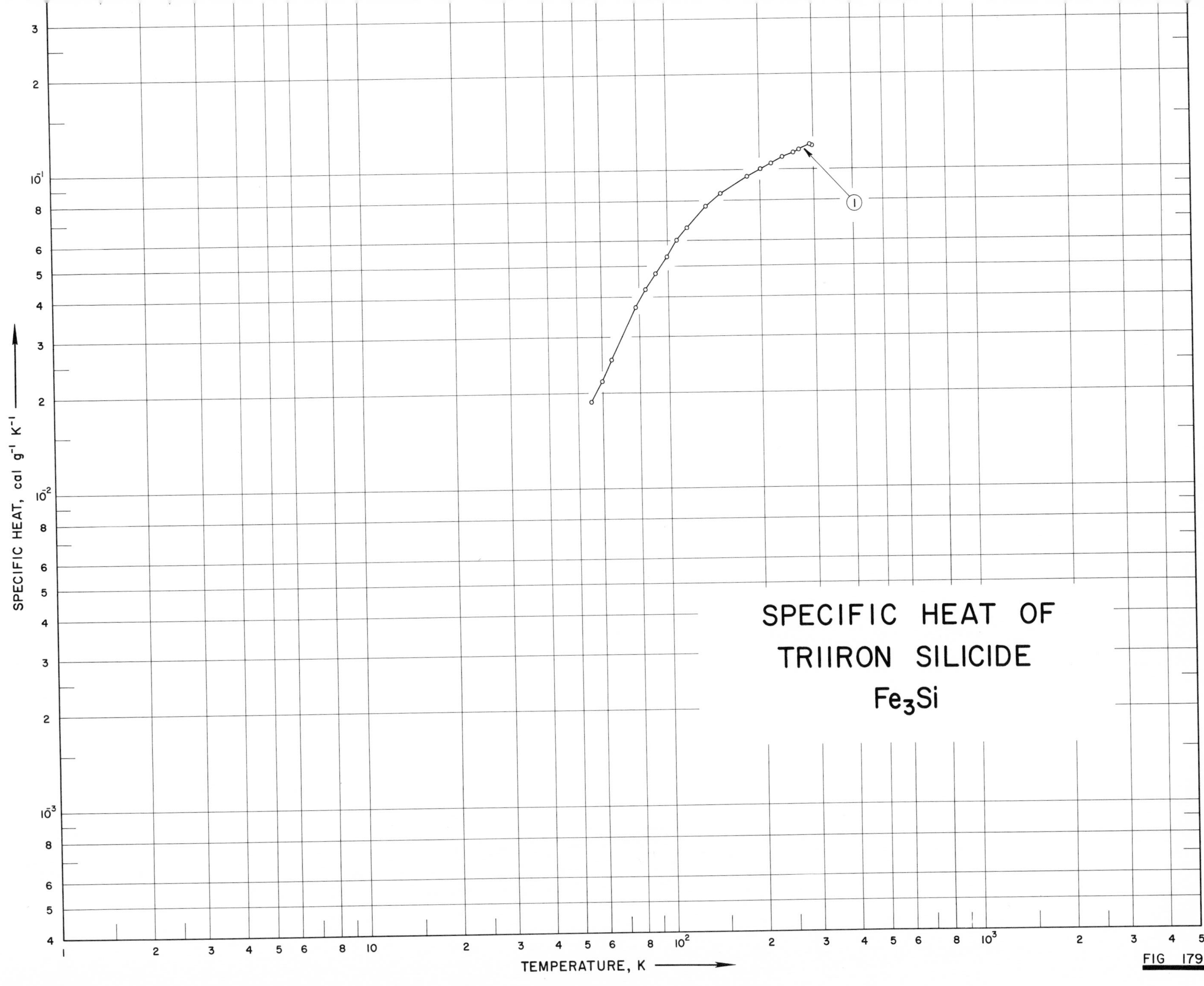
SPECIFIC HEAT OF
TRIIRON SILICIDE
Fe_3Si
SPECIFIC HEAT, cal g⁻¹ K⁻¹
TEMPERATURE, K
1

FIG 179

SPECIFICATION TABLE NO. 179 SPECIFIC HEAT OF TRIIRON SILICIDE Fe_3Si

[For Data Reported in Figure and Table No. 179]

Curve No.	Ref. No.	Year	Temp. Range, K	Reported Error, %	Name and Specimen Designation	Composition (weight percent), Specifications and Remarks
1	420	1962	55-299	1.0		14.25 Si; stoichiometrically close to Fe_3Si.

DATA TABLE NO. 179 SPECIFIC HEAT OF TRIIRON SILICIDE Fe_3Si

[Temperature, T, K; Specific Heat, C_p, Cal $g^{-1} K^{-1}$]

T	C_p
CURVE 1	
55.25	1.88 x 10^{-2}
60.44	2.19
64.95	2.55
78.49	3.73
84.80	4.25
91.63	4.76
92.79	4.86*
99.08	5.38
107.90	6.02
116.58	6.61
124.86	7.15*
134.39	7.73
150.12	8.49
157.58	8.73*
166.89	9.16*
184.20	9.59
193.40	9.92*
201.29	1.01 x 10^{-1}
210.62	1.03*
219.96	1.05
229.89	1.07*
239.29	1.10
249.27	1.12*
259.35	1.13
259.55	1.13*
270.14	1.16
281.03	1.17*
294.05	1.20
297.83	1.20*
299.45	1.19

* Not shown on plot

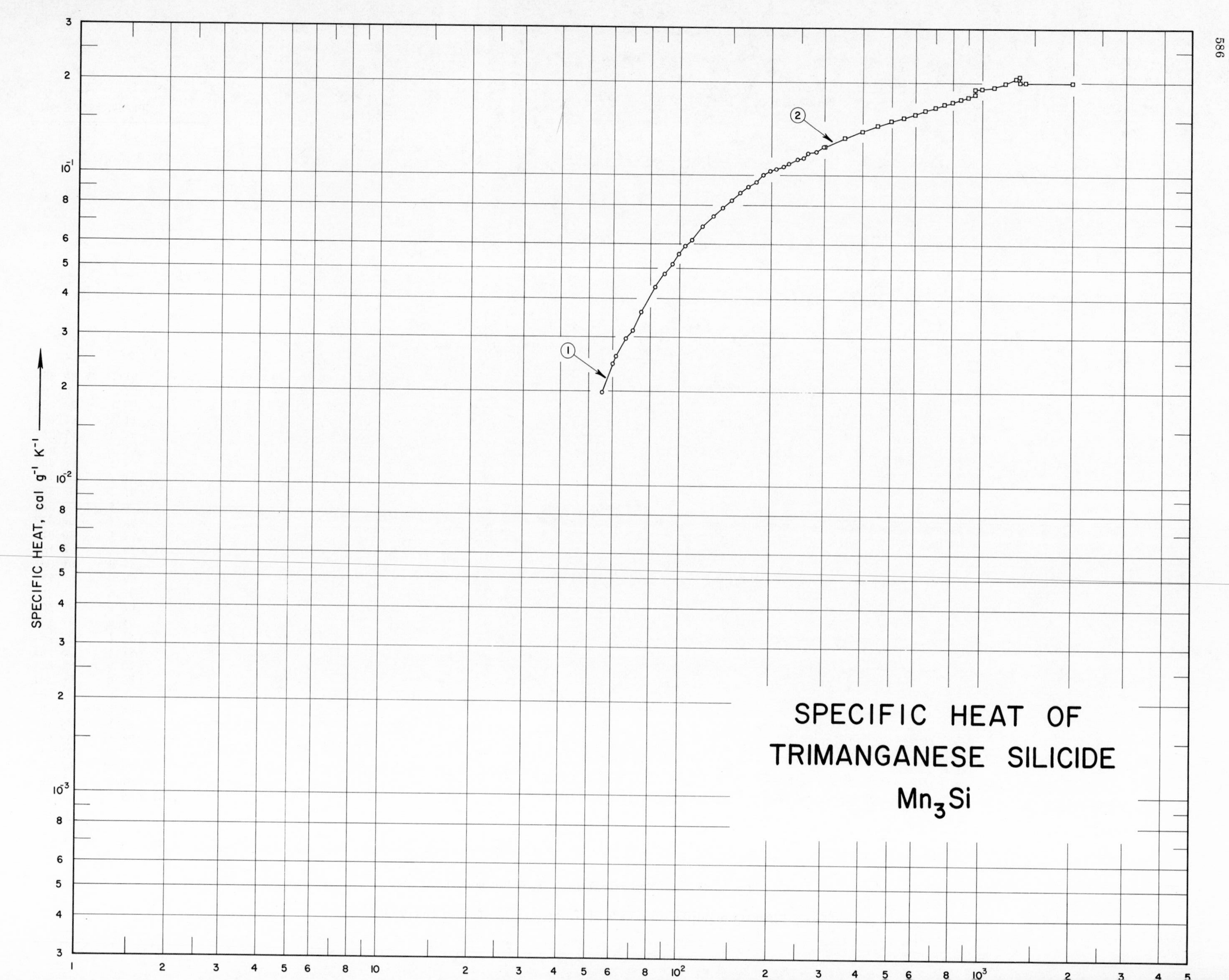

SPECIFIC HEAT OF
TRIMANGANESE SILICIDE
Mn_3Si
SPECIFIC HEAT, cal g⁻¹ K⁻¹
1
2

SPECIFICATION TABLE NO. 180 SPECIFIC HEAT OF TRIMANGANESE SILICIDE Mn_3Si

[For Data Reported in Figure and Table No. 180]

Curve No.	Ref. No.	Year	Temp. Range, K	Reported Error, %	Name and Specimen Designation	Composition (weight percent), Specifications and Remarks
1	421	1965	56-300			Prepared from double refined electrolytic manganese (99.98 Mn) and single crystal silicon (99.997 Si, resistivity = 3 ohm cm); single phase; vacuum annealed for 24 hrs at 900 C.
2	421	1965	300-2000			Same as above.

DATA TABLE NO. 180 SPECIFIC HEAT OF TRIMANGANESE SILICIDE Mn_3Si

[Temperature, T, K; Specific Heat, C_p, Cal g^{-1} K^{-1}]

T	C_p	T	C_p
CURVE 1		CURVE 2	
55.53	1.990 x 10^{-2}	300	1.240 x 10^{-1}*
60.33	2.467	350	1.323
61.70	2.591	400	1.394
66.67	2.964	450	1.450
70.11	3.130	500	1.500
74.83	3.607	550	1.545
82.89	4.333	600	1.587
89.32	4.809	650	1.627
94.66	5.162	700	1.665
99.32	5.535	750	1.702
104.78	5.908	800	1.738
109.04	6.157	850	1.774
113.55	6.509*	900	1.809
118.00	6.820	950	1.844
122.56	7.048*	950	1.911
129.85	7.380	1000	1.919
138.15	7.857	1100	1.946
142.98	8.064*	1200	1.996
148.64	8.353	1300	2.062
152.23	8.479*	(s) 1343	2.095
158.06	8.790	(ℓ) 1343	2.007
163.13	8.975*	1400	2.007
168.32	9.163	1500	2.007
173.34	9.286*	1600	2.007
179.02	9.535	1700	2.007
183.39	9.866*	1800	2.007
188.01	1.007 x 10^{-1}	1900	2.007
193.26	1.010*	2000	2.007
194.04	1.024*		
198.17	1.032		
203.15	1.045*		
208.14	1.055		
214.92	1.072*		
219.15	1.072		
223.79	1.078*		
230.76	1.094		
239.01	1.111*		
246.59	1.130		
257.15	1.146		
265.61	1.179		
283.61	1.192		
299.53	1.240		

*Not shown on plot

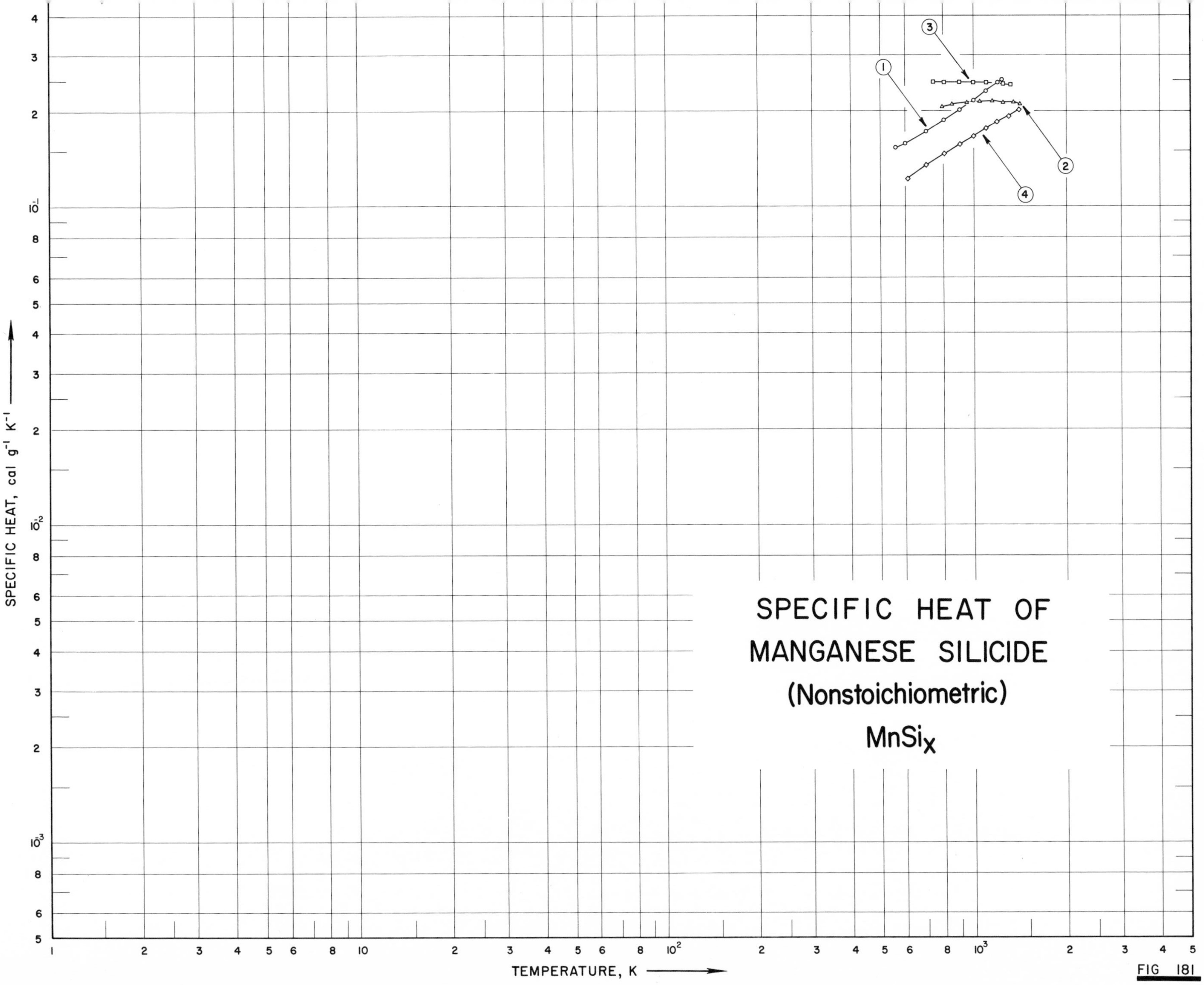

SPECIFIC HEAT OF MANGANESE SILICIDE (Nonstoichiometric) $MnSi_x$

SPECIFIC HEAT, cal g^{-1} K^{-1}

TEMPERATURE, K

FIG 181

SPECIFICATION TABLE NO. 181 SPECIFIC HEAT OF MANGANESE SILICIDE (nonstoichiometric) $MnSi_x$

[For Data Reported in Figure and Table No. 181]

Curve No.	Ref. No.	Year	Temp. Range, K	Reported Error, %	Name and Specimen Designation	Composition (weight percent), Specifications and Remarks
1	232	1963	558-1241	1.5		$MnSi_{0.3223}$.
2	232	1963	792-1418	1.1		$MnSi_{1.0}$.
3	232	1963	743-1332	1.6		$MnSi_{0.5458}$.
4	232	1963	610-1402	2.8		$MnSi_{2.234}$.

DATA TABLE NO. 181 SPECIFIC HEAT OF MANGANESE SILICIDE $MnSi_x$ (nonstoichiometric)

[Temperature, T, K; Specific Heat, C_p, Cal g^{-1} K^{-1}]

T	C_p
CURVE 1	
558	1.535 x 10^{-1}
560	1.538*
562	1.540*
564	1.543*
566	1.546*
568	1.548*
570	1.551*
572	1.553*
574	1.556*
576	1.559*
578	1.562*
580	1.564*
582	1.567*
584	1.570*
586	1.572*
588	1.575*
590	1.578*
592	1.581*
594	1.584*
596	1.587*
598	1.589*
600	1.592
650	1.661*
700	1.731
750	1.802*
800	1.874
850	1.947*
900	2.019
950	2.092*
1000	2.165
1050	2.239*
1100	2.312
1150	2.385*
1200	2.459
1241	2.520
CURVE 2	
792	2.075 x 10^{-1}
800	2.079*
850	2.103
900	2.123*
950	2.135
1000	2.145*
1050	2.149
1100	2.151*
CURVE 2 (cont.)	
1150	2.150 x 10^{-1}
1200	2.147*
1250	2.143
1300	2.137*
1350	2.131
1400	2.123*
1418	2.119
CURVE 3	
743	2.479 x 10^{-1}
750	2.479*
800	2.478
850	2.475*
900	2.472
950	2.467*
1000	2.464
1050	2.459*
1100	2.454
1150	2.449*
1200	2.445*
1250	2.440
1300	2.434*
1332	2.431
CURVE 4	
610	1.228 x 10^{-1}
700	1.354
800	1.472
900	1.577
1000	1.673
1100	1.764
1200	1.852
1300	1.936
1402	2.021

* Not shown on plot

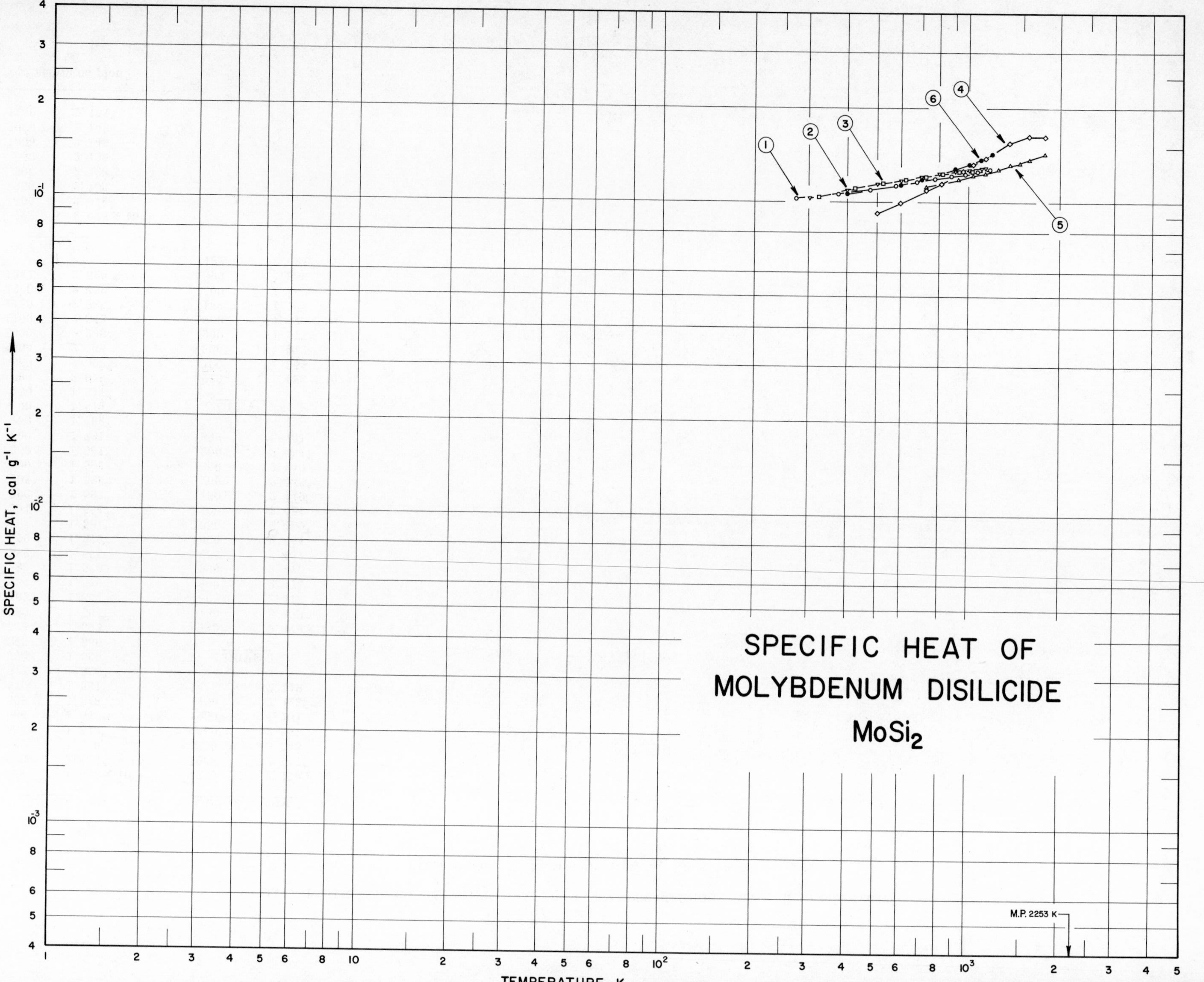
SPECIFIC HEAT OF
MOLYBDENUM DISILICIDE
$MoSi_2$
SPECIFIC HEAT, cal g⁻¹ K⁻¹
TEMPERATURE, K
M.P. 2253 K
1
2
3
4
5
6

SPECIFICATION TABLE NO. 182 SPECIFIC HEAT OF MOLYBDENUM DISILICIDE $MoSi_2$

[For Data Reported in Figure and Table No. 182]

Curve No.	Ref. No.	Year	Temp. Range, K	Reported Error, %	Name and Specimen Designation	Composition (weight percent), Specifications and Remarks
1	233	1953	273-1173	3.0		0.8 Fe, 0.50 O_2, 0.34 N_2 and 0.17 C.
2	234	1954	300-1148			97.8 $MoSi_2$, 1.4 Fe_2O_3, and 0.89 SiO_2; density =371 lb ft^{-3}.
3	235	1956	303-1148			97.8 $MoSi_2$ and 1.4 Fe_2O_3.
4	95	1961	501-1797	3.0		61.5-63.5 Mo, 35-37 Si; slip cast; measured in a helium atmosphere; density = 362 lb ft^{-3}.
5	162	1961	728-1783			Single-phase composition; hot pressed.
6	54	1965	300-1200	0.7		Sample supplied by The Carborundum Co.

DATA TABLE NO. 182 SPECIFIC HEAT OF MOLYBDENUM DISILICIDE $MoSi_2$

[Temperature, T, K; Specific Heat, C_p, Cal g^{-1} K^{-1}]

T	C_p
CURVE 1	
273	1.019×10^{-1}
373	1.060
473	1.098
573	1.134
673	1.165
773	1.193
873	1.217
973	1.238
1073	1.256
1173	1.270
CURVE 2	
303	1.018×10^{-1}
400	1.095
500	1.142
598	1.174
598	1.172*
620	1.172*
700	1.206
800	1.230
900	1.250
1000	1.265
1100	1.278
1148	1.283
CURVE 3	
303	1.019×10^{-1}*
323	1.039
373	1.078*
423	1.108
473	1.132*
523	1.151
573	1.167*
598	1.174*
598	1.172*
623	1.182
673	1.198*
723	1.212
773	1.225*
823	1.236
873	1.245*
923	1.254
973	1.262*
CURVE 3 (cont.)	
1023	1.268×10^{-1}
1073	1.275*
1148	1.283*
CURVE 4	
501	9.25×10^{-2}
596	9.94
729	1.091×10^{-1}
811	1.150
955	1.255
1047	1.322
1143	1.392
1249	1.469*
1364	1.552
1373	1.558*
1471	1.63*
1578	1.62
1690	1.62*
1797	1.62
CURVE 5	
727	1.128×10^{-1}
811	1.152*
922	1.184
1033	1.216
1144	1.248
1255	1.280
1366	1.312
1478	1.344
1589	1.376
1700	1.408*
1783	1.432
CURVE 6	
300	1.019×10^{-1}*
400	1.072
500	1.131
600	1.149
700	1.183
800	1.225
900	1.273
1000	1.323
CURVE 6 (cont.)	
1100	1.376×10^{-1}
1200	1.430

* Not shown on plot

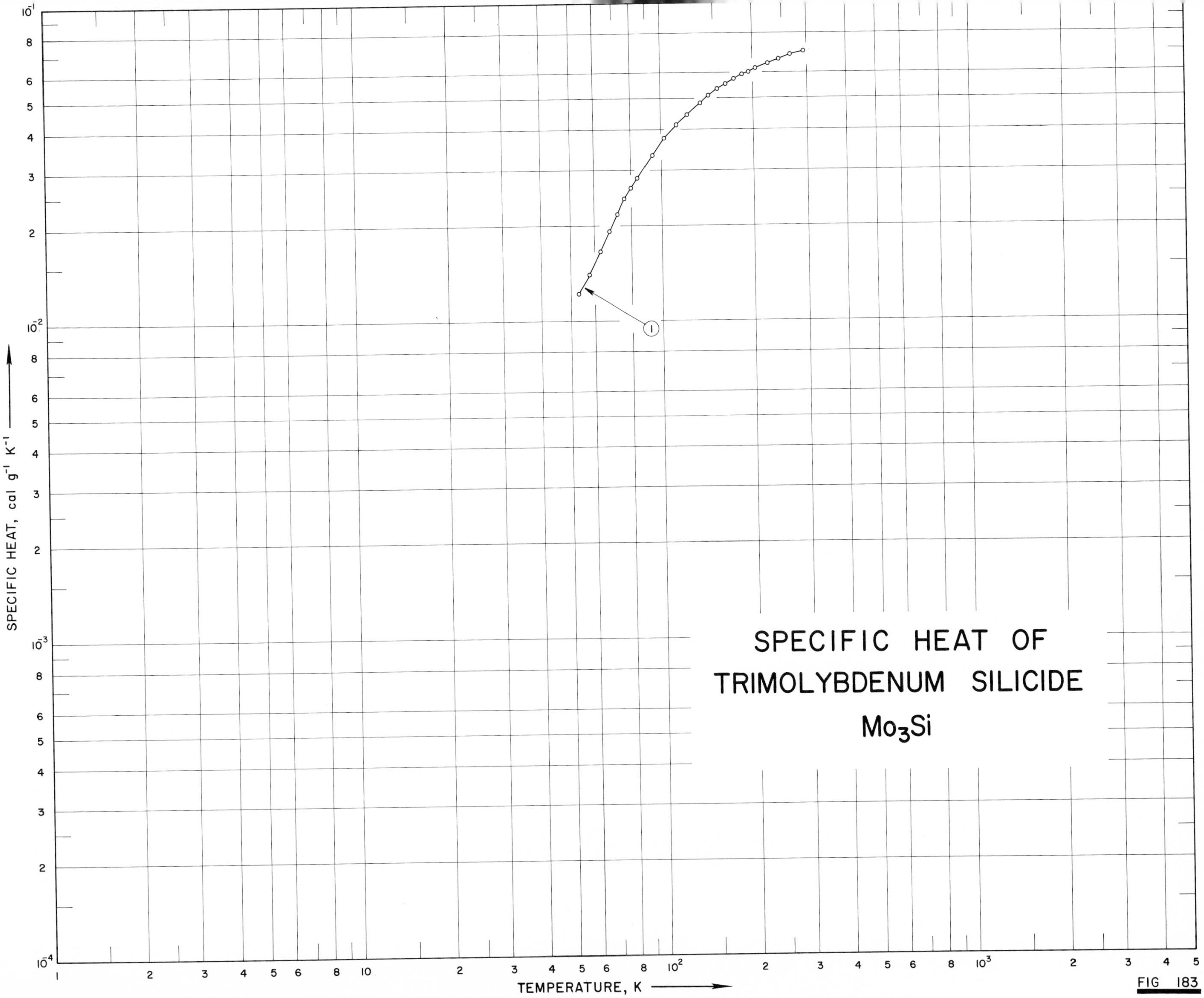

SPECIFIC HEAT OF
TRIMOLYBDENUM SILICIDE
Mo_3Si
SPECIFIC HEAT, cal g^{-1} K^{-1}
TEMPERATURE, K
1

FIG 183

SPECIFICATION TABLE NO. 183 SPECIFIC HEAT OF TRIMOLYBDENUM SILICIDE Mo_3Si

[For Data Reported in Figure and Table No. 183]

Curve No.	Ref. No.	Year	Temp. Range, K	Reported Error, %	Name and Specimen Designation	Composition (weight percent), Specifications and Remarks
1	422	1958	54-296			91.25 Mo and 8.24 Si (91.11 and 8.89 theor.).

DATA TABLE NO. 183 SPECIFIC HEAT OF TRIMOLYBDENUM SILICIDE Mo_3Si

[Temperature, T, K; Specific Heat, C_p, Cal $g^{-1} K^{-1}$]

T	C_p
CURVE 1	
53.67	1.220 x 10^{-2}
57.86	1.409
62.71	1.667
67.50	1.931
72.18	2.185
76.97	2.437
80.38	2.617
84.57	2.826
94.56	3.308
104.93	3.762
114.88	4.146
124.50	4.485
136.08	4.850
145.57	5.120
155.87	5.371
165.64	5.577
175.81	5.768
186.45	5.955
195.83	6.085
206.06	6.241
216.19	6.358*
226.32	6.457
236.15	6.571*
245.62	6.657
256.35	6.752*
266.64	6.837
276.14	6.898*
286.38	6.968*
295.99	7.044

* Not shown on plot

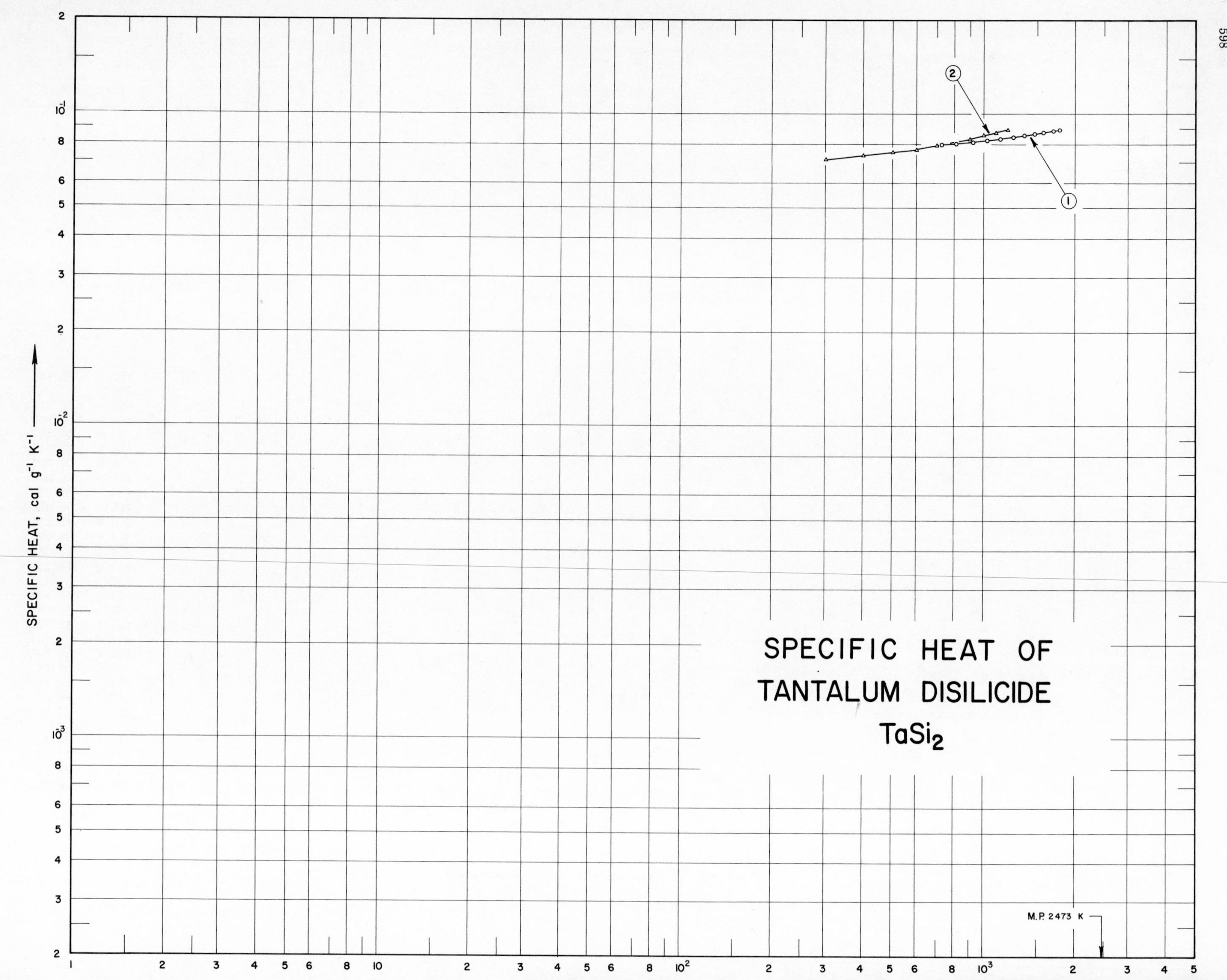
SPECIFIC HEAT OF
TANTALUM DISILICIDE
$TaSi_2$
M.P. 2473 K
SPECIFIC HEAT, cal g^{-1} K^{-1}
1
2

SPECIFICATION TABLE NO. 184 SPECIFIC HEAT OF TANTALUM DISILICIDE $TaSi_2$

[For Data Reported in Figure and Table No. 184]

Curve No.	Ref. No.	Year	Temp. Range, K	Reported Error, %	Name and Specimen Designation	Composition (weight percent), Specifications and Remarks
1	162	1961	728-1783			Single-phase; prepared by solid-state reaction of constituent elements at 2370 F; hot pressed.
2	54	1965	300-1200	2.0		73.40 Ta, 24.35 Si and 1.30 C; sample supplied by The Carborundum Co.

DATA TABLE NO. 184 SPECIFIC HEAT OF TANTALUM DISILICIDE $TaSi_2$

[Temperature, T, K; Specific Heat, C_p, Cal g^{-1} K^{-1}]

T	C_p
CURVE 1	
728	8.015 x 10^{-2}
811	8.090
922	8.189
1033	8.289
1144	8.388
1255	8.488
1366	8.587
1478	8.687
1589	8.786
1700	8.886
1783	8.960
CURVE 2	
300	7.152 x 10^{-2}
400	7.351
500	7.553
600	7.751
700	7.954
800	8.152*
900	8.350
1000	8.553
1100	8.751
1200	8.953

* Not shown on plot

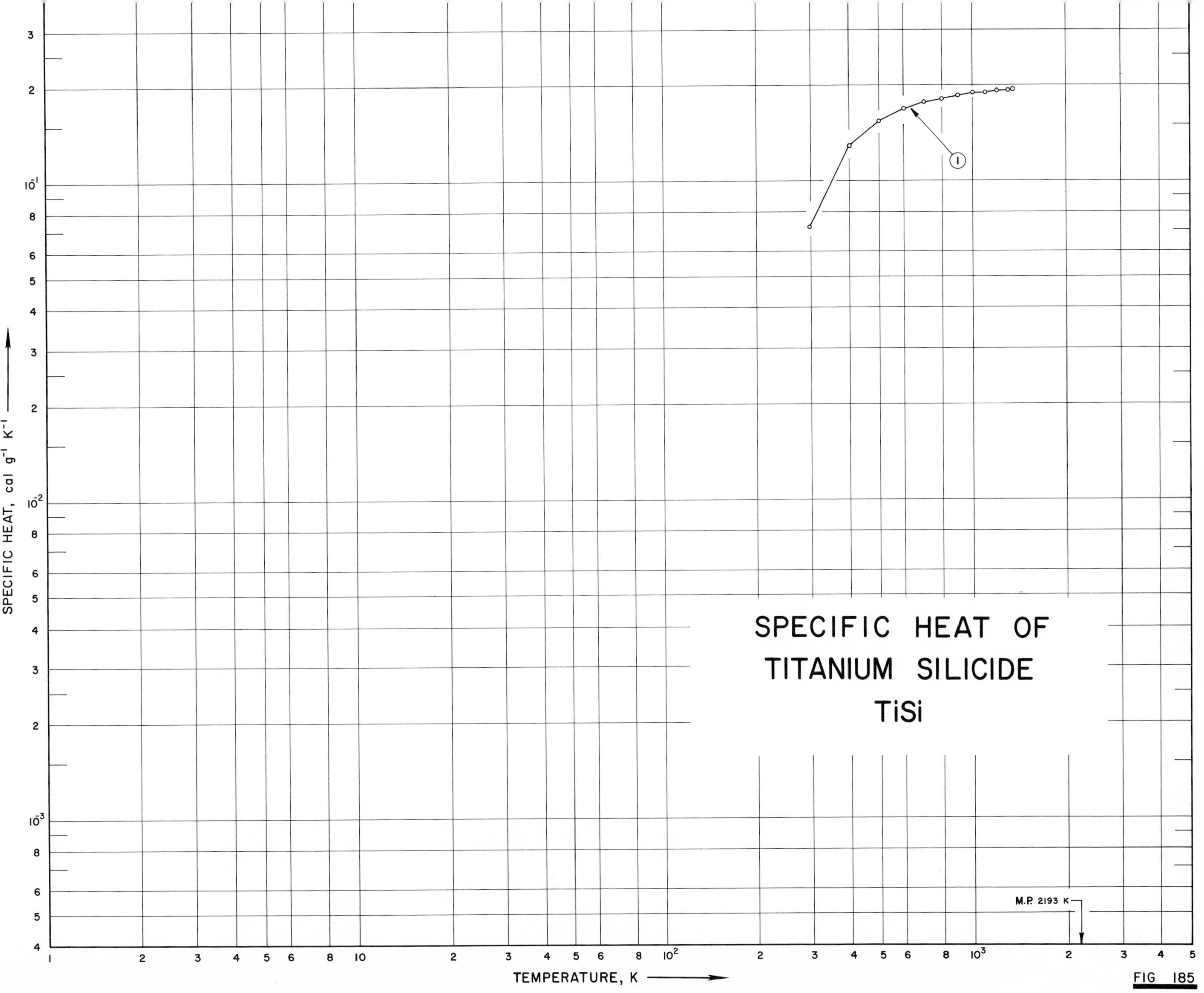
SPECIFIC HEAT OF
TITANIUM SILICIDE
TiSi
1
M.P. 2193 K
SPECIFIC HEAT, cal g^{-1} K^{-1}
TEMPERATURE, K
FIG 185

SPECIFICATION TABLE NO. 185 SPECIFIC HEAT OF TITANIUM SILICIDE TiSi

[For Data Reported in Figure and Table No. 185]

Curve No.	Ref. No.	Year	Temp. Range, K	Reported Error, %	Name and Specimen Designation	Composition (weight percent), Specifications and Remarks
1	236	1959	298-1350	±1.8		

DATA TABLE NO. 185 SPECIFIC HEAT OF TITANIUM SILICIDE TiSi

[Temperature, T, K; Specific Heat, C_p, Cal g^{-1} K^{-1}]

T	C_p
CURVE 1	
298.15	7.230×10^{-2}
300	7.392*
400	1.304×10^{-1}
500	1.566
600	1.707
700	1.794
800	1.849
900	1.887
1000	1.915
1100	1.934
1200	1.950
1300	1.962
1350	1.966

*Not shown on plot

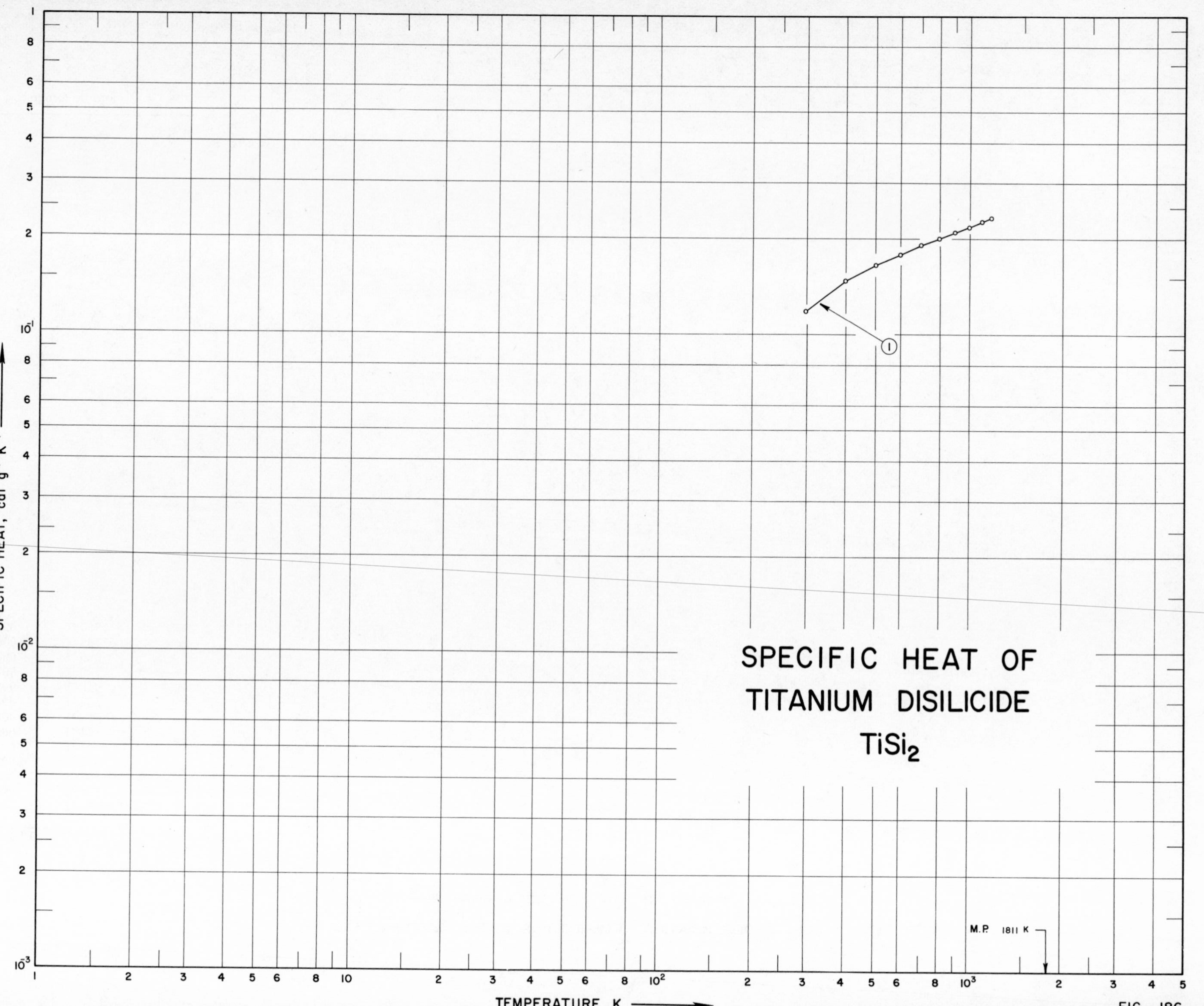
SPECIFIC HEAT OF
TITANIUM DISILICIDE
$TiSi_2$
1
M.P. 1811 K
SPECIFIC HEAT, cal g^{-1} K^{-1}
TEMPERATURE, K
FIG. 186

SPECIFICATION TABLE NO. 186 SPECIFIC HEAT OF TITANIUM DISILICIDE $TiSi_2$

[For Data Reported in Figure and Table No. 186]

Curve No.	Ref. No.	Year	Temp. Range, K	Reported Error, %	Name and Specimen Designation	Composition (weight percent), Specifications and Remarks
1	236	1958	298-1180	±1.9		

DATA TABLE NO. 186 SPECIFIC HEAT OF TITANIUM DISILICIDE $TiSi_2$

[Temperature, T, K; Specific Heat, C_p, Cal g^{-1} K^{-1}]

T	C_p
CURVE 1	
298.15	1.182 x 10^{-1}
300	1.189*
400	1.482
500	1.660
600	1.793
700	1.906
800	2.006
900	2.101
1000	2.191
1100	2.279
1180	2.347

*Not shown on plot

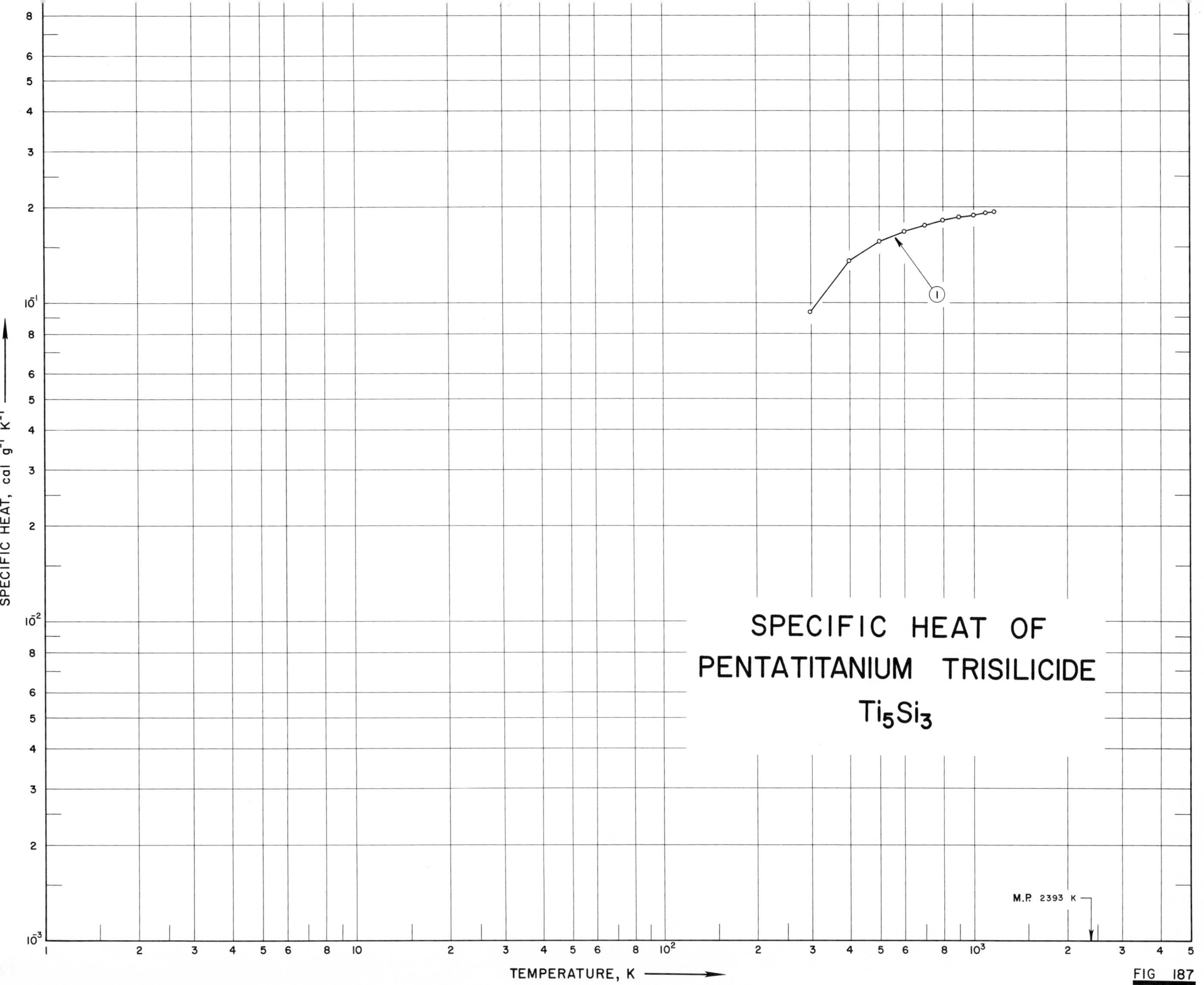

SPECIFIC HEAT OF
PENTATITANIUM TRISILICIDE
Ti_5Si_3
M.P. 2393 K
SPECIFIC HEAT, cal g^{-1} K^{-1}
TEMPERATURE, K
FIG 187

SPECIFICATION TABLE NO. 187 SPECIFIC HEAT OF PENTATITANIUM TRISILICIDE Ti_5Si_3

[For Data Reported in Figure and Table No. 187]

Curve No.	Ref. No.	Year	Temp. Range, K	Reported Error, %	Name and Specimen Designation	Composition (weight percent), Specifications and Remarks
1	236	1958	298-1170	±3.0		

DATA TABLE NO. 187 SPECIFIC HEAT OF PENTATITANIUM TRISILICIDE Ti_5Si_3

[Temperature, T, K; Specific Heat, C_p, Cal g^{-1} K^{-1}]

T	C_p
CURVE 1	
298.15	9.318×10^{-2}
300	9.433*
400	1.358×10^{-1}
500	1.560
600	1.678
700	1.756
800	1.812
900	1.857
1000	1.894
1100	1.926
1170	1.946

* Not shown on plot

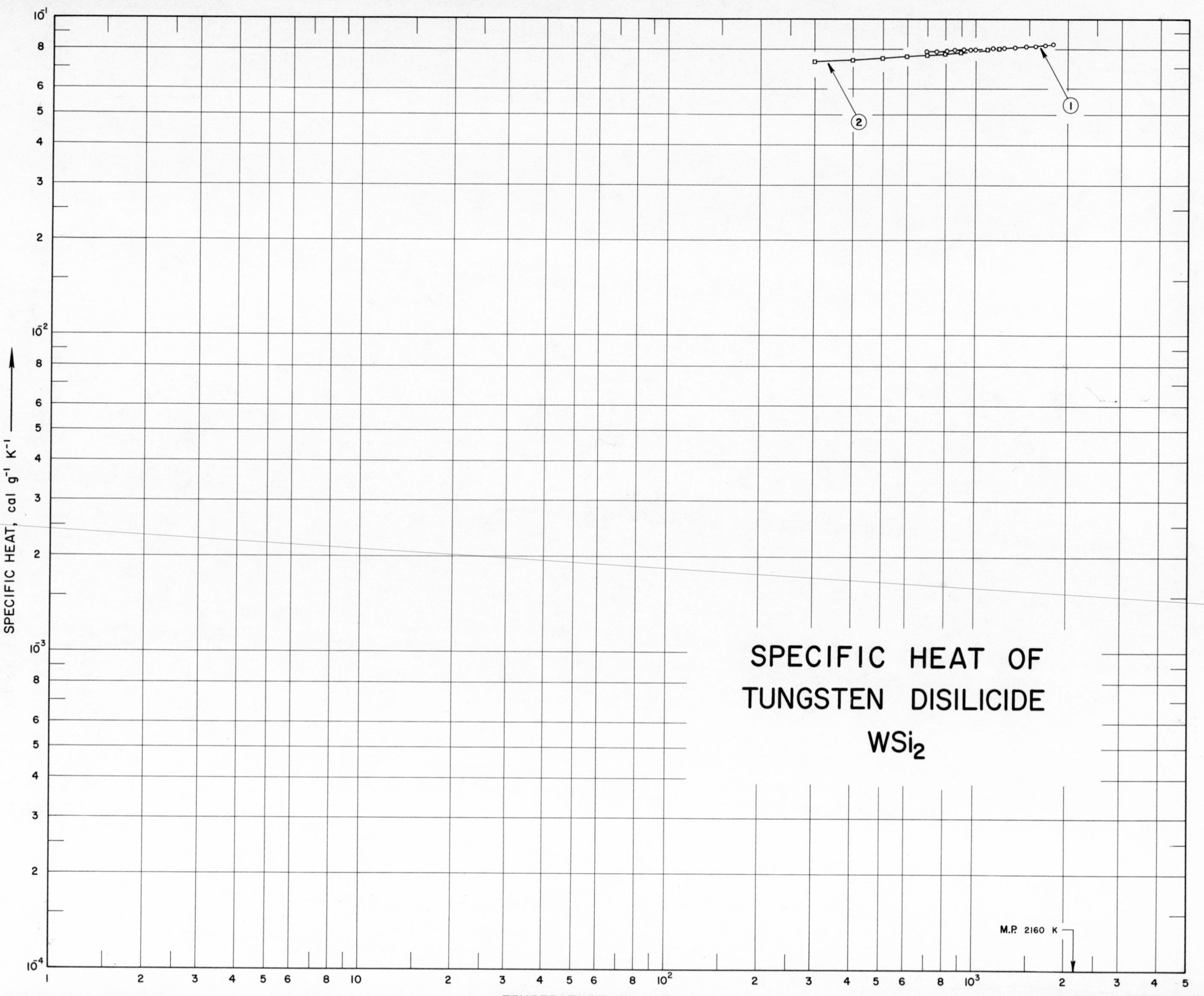
SPECIFIC HEAT OF
TUNGSTEN DISILICIDE
WSi_2
M.P. 2160 K
1
2
SPECIFIC HEAT, cal g^{-1} K^{-1}

SPECIFICATION TABLE NO. 188 SPECIFIC HEAT OF TUNGSTEN DISILICIDE WSi_2

[For Data Reported in Figure and Table No. 188]

Curve No.	Ref. No.	Year	Temp. Range, K	Reported Error, %	Name and Specimen Designation	Composition (weight percent), Specifications and Remarks
1	162	1961	700-1794			Single-phase; prepared by solid-state reaction of constituent elements at 2370 F; hot pressed.
2	54	1965	300-1200	1.7		75.97 W and 23.32 Si; sample supplied by The Carborundum Co.

DATA TABLE NO. 188 SPECIFIC HEAT OF TUNGSTEN DISILICIDE WSi_2

[Temperature, T, K; Specific Heat, C_p, Cal g^{-1} K^{-1}]

T	C_p
CURVE 1	
700	7.959 x 10^{-2}
755	7.978
811	7.997
866	8.016
922	8.035
978	8.054
1033	8.073
1089	8.091*
1144	8.110
1200	8.129*
1255	8.148
1311	8.167*
1366	8.186
1422	8.205*
1478	8.224
1533	8.243*
1589	8.262
1644	8.281*
1700	8.300
1755	8.319*
1794	8.332
CURVE 2	
300	7.366 x 10^{-2}
400	7.453
500	7.541
600	7.628
700	7.716
800	7.803
900	7.891
1000	7.978*
1100	8.066
1200	8.158

* Not shown on plot

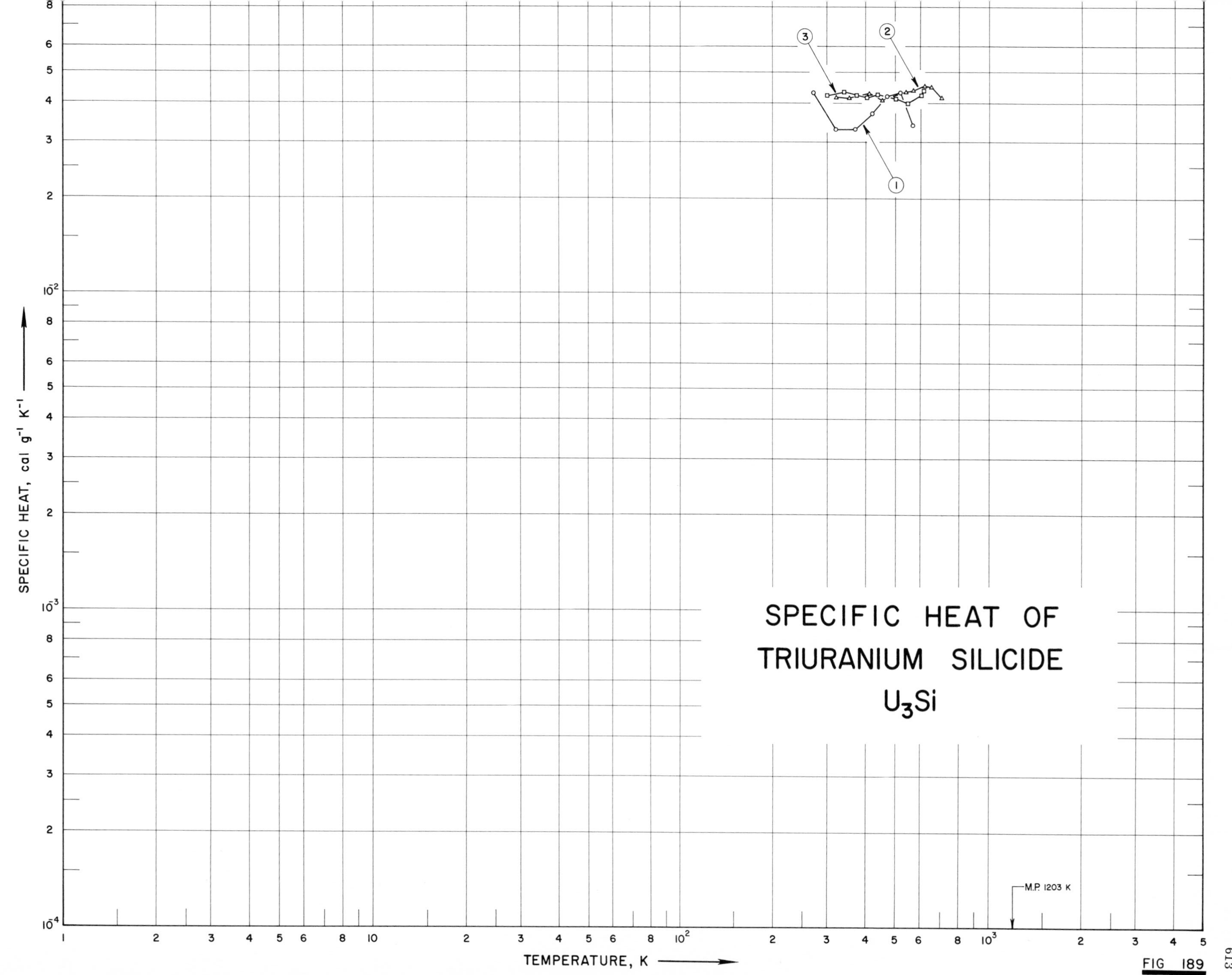
SPECIFIC HEAT OF
TRIURANIUM SILICIDE
U_3Si
SPECIFIC HEAT, cal g^{-1} K^{-1}
TEMPERATURE, K
M.P. 1203 K
1
2
3

FIG 189

SPECIFICATION TABLE NO. 189 SPECIFIC HEAT OF TRIURANIUM SILICIDE U_3Si

[For Data Reported in Figure and Table No. 189]

Curve No.	Ref. No.	Year	Temp. Range, K	Reported Error, %	Name and Specimen Designation	Composition (weight percent), Specifications and Remarks
1	237	1957	273-573			0.05 Fe, 0.01 Al, Mn, and $<$0.01 each of others; sintered.
2	195	1963	324-708	5.0	Sample 535	96.1 U, 3.9 Si, prepared by arc melting and casting uranium and silicon; annealed for 24 hrs at 800 C; density = 894 lb ft^{-3}.
3	195	1963	307-621	5.0	Sample 533	96.9 U, 3.9 Si; prepared by arc melting and casting uranium and silicon; annealed for 24 hrs at 800 C; density = 905 lb ft^{-3}.

DATA TABLE NO. 189 SPECIFIC HEAT OF TRIURANIUM SILICIDE U_3Si

[Temperature, T, K; Specific Heat, C_p, Cal g^{-1} K^{-1}]

T	C_p
CURVE 1	
273	4.3 x 10^{-2}
323	3.3
373	3.3
423	3.7
473	4.2
523	4.3
573	3.4
CURVE 2	
324	4.150 x 10^{-2}
334	4.180*
345	4.190*
357	4.120
368	4.120*
380	4.160*
391	4.230*
402	4.280*
414	4.280
425	4.220*
437	4.150*
448	4.100*
458	4.090
472	4.140*
484	4.230*
496	4.220*
513	4.240*
519	4.140*
531	4.260*
543	4.320
554	4.340*
566	4.350*
577	4.370
589	4.400*
601	4.410*
612	4.490*
624	4.530
635	4.550*
646	4.540*
658	4.480
669	4.340*
682	4.140*
708	4.140
CURVE 3	
Series 1	
376	4.250 x 10^{-2}
382	4.240*
390	4.180*
399	4.230*
406	4.140
416	4.180*
421	4.250*
429	4.300*
440	4.250
448	4.240*
457	4.190*
466	4.170*
472	4.130*
483	4.070*
489	4.040*
462	4.300*
472	4.260*
483	4.180*
495	4.100*
504	4.110
516	4.060*
528	3.980*
539	3.940*
554	3.970
560	3.990*
573	4.050*
589	4.090*
597	4.210*
610	4.210
621	4.380
Series 2	
307	4.240 x 10^{-2}
318	4.240*
323	4.250*
328	4.270*
333	4.290*
338	4.310*
344	4.320
349	4.320*
CURVE 3 (cont.)	
Series 2 (cont.)	
354	4.390 x 10^{-2}
360	4.290*
366	4.290*
370	4.390*
377	4.360*
383	4.320*
389	4.330*
436	4.290*
441	4.320*
448	4.210*
454	4.190*
459	4.020*
464	4.060*
468	4.130*
474	4.150*
479	4.170*
483	4.180*
488	4.200*
492	4.210*
497	4.210*
507	4.240*
512	4.240*
516	4.230*
521	4.240*
Series 3*	
424	4.150 x 10^{-2}
429	4.160
436	4.160
444	4.170
450	4.180
458	4.180
465	4.190
473	4.200
480	4.190
488	4.180
496	4.170
503	4.170
511	4.190
519	4.220
CURVE 3 (cont.)	
Series 3 (cont.)	
527	4.250 x 10^{-2}
536	4.280
544	4.280
552	4.270
561	4.270
568	4.260
574	4.250
579	4.210

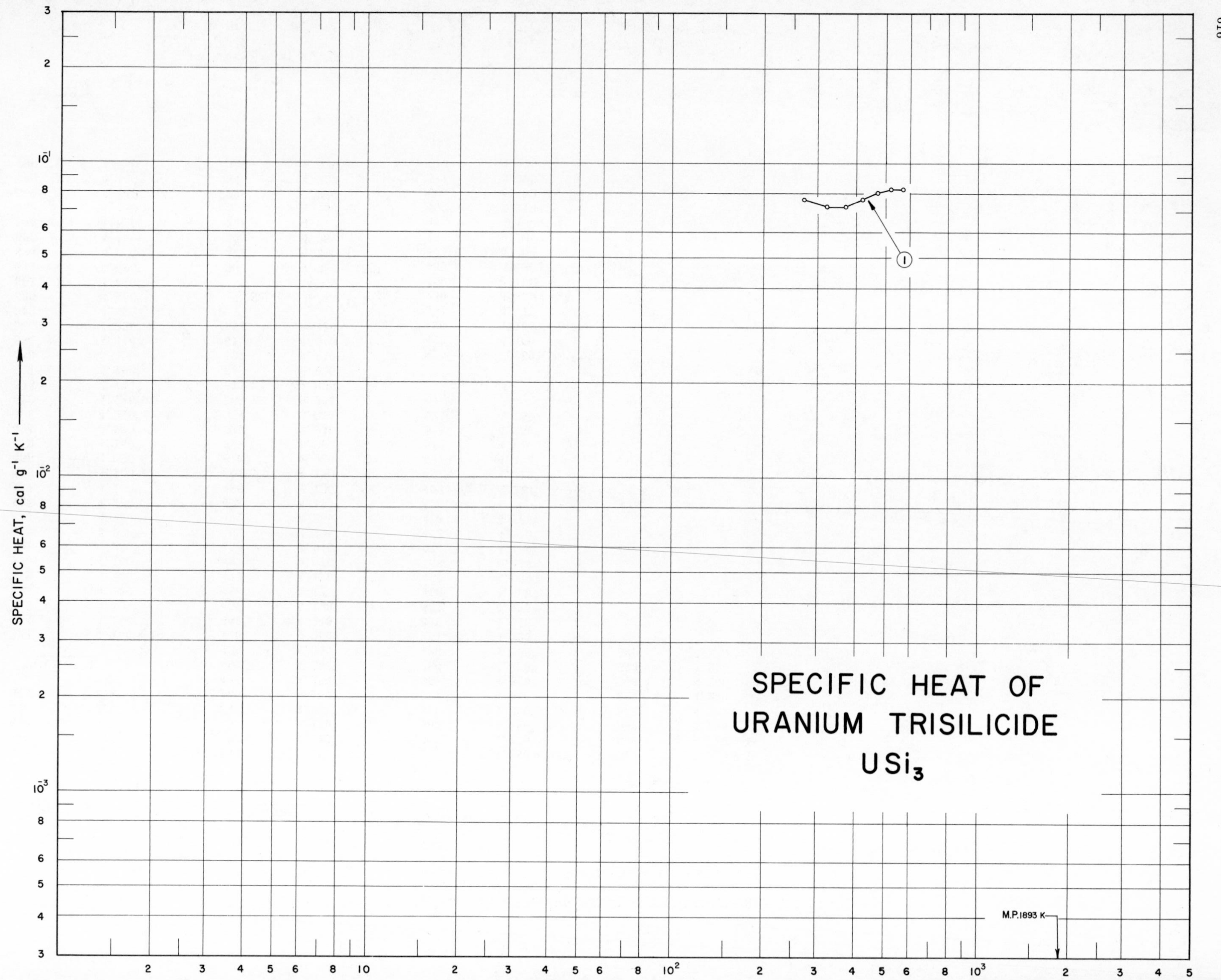
SPECIFIC HEAT OF
URANIUM TRISILICIDE
USi_3
M.P. 1893 K
1
SPECIFIC HEAT, cal g^{-1} K^{-1}
10^{-1}
10^{-2}
10^{-3}
10
10^2
10^3

SPECIFICATION TABLE NO. 190 SPECIFIC HEAT OF URANIUM TRISILICIDE USi_3

[For Data Reported in Figure and Table No. 190]

Curve No.	Ref. No.	Year	Temp. Range, K	Reported Error, %	Name and Specimen Designation	Composition (weight percent), Specifications and Remarks
1	237	1957	273-573			1.0 W, 0.3 Fe, 0.09 Al, 0.05 Cu, and <0.01 others; sintered.

DATA TABLE NO. 190 SPECIFIC HEAT OF URANIUM TRISILICIDE USi_3

[Temperature, T, K; Specific Heat, C_p, Cal g^{-1} K^{-1}]

T	C_p
CURVE 1	
273	7.7×10^{-2}
323	7.3
373	7.3
423	7.7
473	8.1
523	8.3
573	8.3

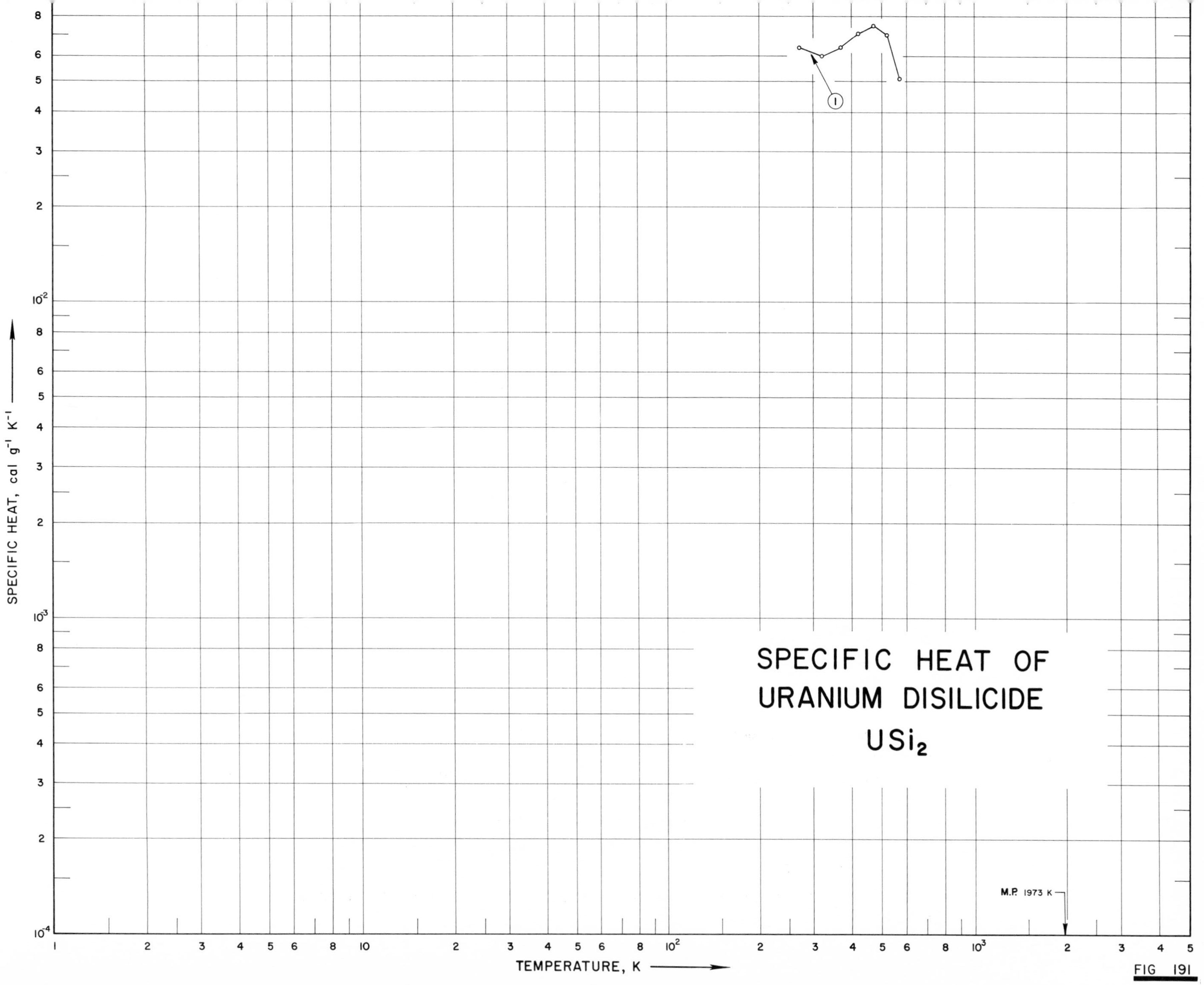
SPECIFIC HEAT OF
URANIUM DISILICIDE
USi$_2$
M.P. 1973 K
TEMPERATURE, K
SPECIFIC HEAT, cal g^{-1} K^{-1}
1
FIG 191

SPECIFICATION TABLE NO. 191 SPECIFIC HEAT OF URANIUM DISILICIDE USi_2

[For Data Reported in Figure and Table No. 191]

Curve No.	Ref. No.	Year	Temp. Range, K	Reported Error, %	Name and Specimen Designation	Composition (weight percent), Specifications and Remarks
1	237	1957	273-573			0.08 Fe, 0.07 Al, 0.03 Cu, and < 0.01 each of others; sintered.

DATA TABLE NO. 191 SPECIFIC HEAT OF URANIUM DISILICIDE USi_2

[Temperature, T, K; Specific Heat, C_p, Cal g^{-1} K^{-1}]

T	C_p
CURVE 1	
273	6.4 x 10^{-2}
323	6.0
373	6.4
423	7.1
473	7.5
523	7.0
573	5.1

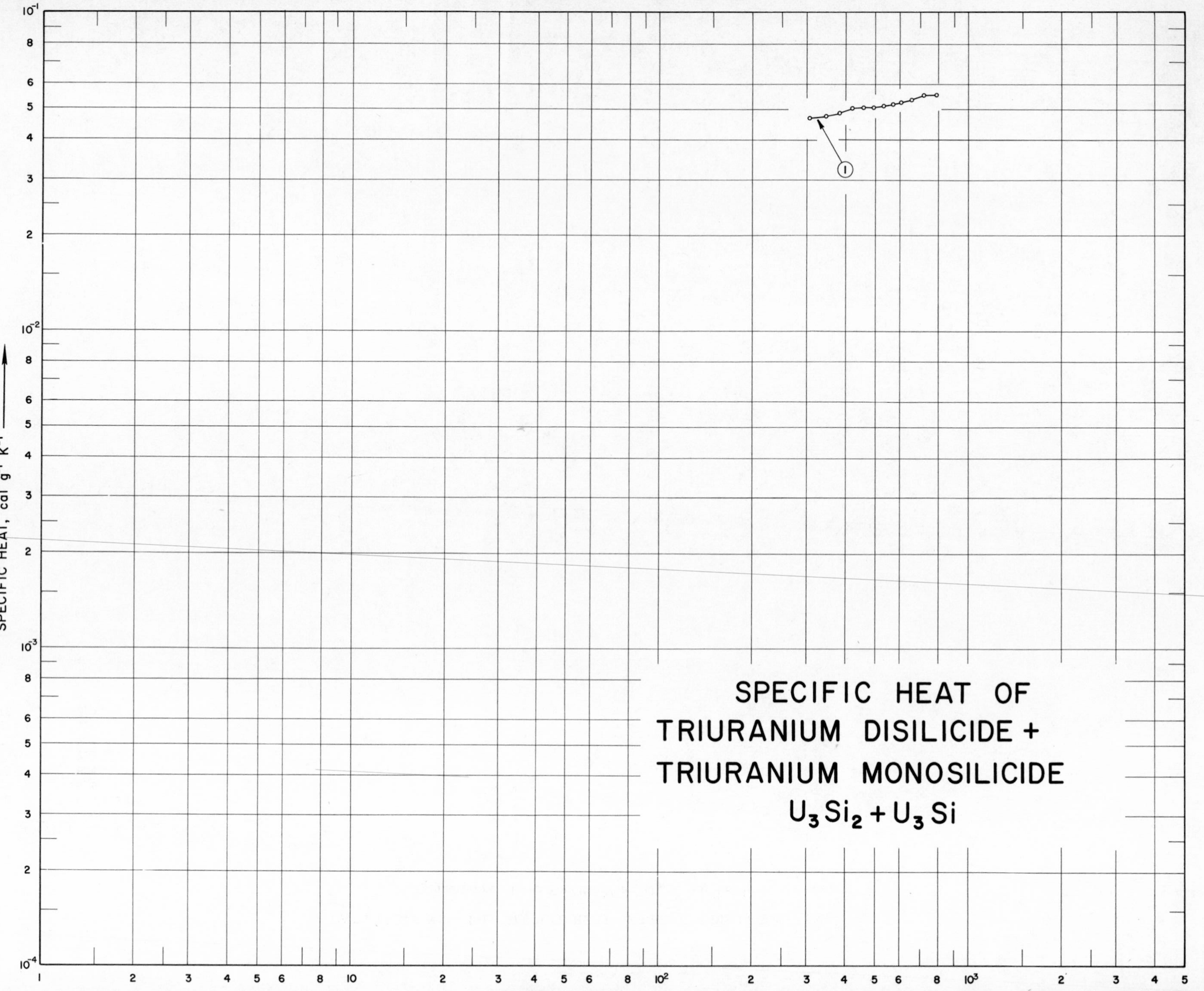
SPECIFIC HEAT OF
TRIURANIUM DISILICIDE +
TRIURANIUM MONOSILICIDE
$U_3Si_2 + U_3Si$
SPECIFIC HEAT, cal g^{-1} K^{-1}
1
10^{-1}
10^{-2}
10^{-3}
10^{-4}
1
10
10^2
10^3

SPECIFICATION TABLE NO. 192 SPECIFIC HEAT OF TRIURANIUM DISILICIDE + TRIURANIUM MONOSILICIDE $U_3Si_2 + U_3Si$

[For Data Reported in Figure and Table No. 192]

Curve No.	Ref. No.	Year	Temp. Range, K	Reported Error, %	Name and Specimen Designation	Composition (weight percent), Specifications and Remarks
1	195	1963	308-795	±5.0	Sample 539	93.9 U and 6.1 Si; prepared by arc melting and casting uranium and silicon; annealed for 24 hrs at 800 C; density = 805 lb ft^{-3}.

DATA TABLE NO. 192 SPECIFIC HEAT OF TRIURANIUM DISILICIDE + TRIURANIUM MONOSILICIDE $U_3Si_2 + U_3Si$

[Temperature, T, K; Specific Heat, C_p, Cal $g^{-1} K^{-1}$]

T	C_p
CURVE 1	
Series 1	
308	4.690 x 10^{-2}
328	4.720*
347	4.750
367	4.810*
385	4.890
404	4.970*
423	5.030
441	5.050*
460	5.050
479	5.050*
498	5.050
517	5.080*
535	5.110
554	5.160*
573	5.190
591	5.210*
610	5.230
629	5.250*
648	5.290*
666	5.350
685	5.490*
703	5.510*
721	5.520
739	5.510*
758	5.500*
776	5.500*
795	5.510
Series 2*	
337	4.500 x 10^{-2}
357	4.510
377	4.520
397	4.560
417	4.630
436	4.700
456	4.750
476	4.770
495	4.770
515	4.760
535	4.760
555	4.770
575	4.800
595	4.840
CURVE 1 (cont.)	
Series 2 (cont.)*	
615	4.900 x 10^{-2}
635	4.950
655	4.990
674	5.010
694	5.020
714	5.020
734	5.040
754	5.120
773	5.280
792	5.510
Series 3*	
349	4.710 x 10^{-2}
369	4.730
389	4.780
408	4.800
428	4.840
447	4.880
467	4.900
487	4.930
506	4.960
526	4.970
545	4.980
565	5.000
586	5.020
606	5.040
625	5.070
645	5.100
664	5.150
684	5.200
693	5.260

* Not shown on plot

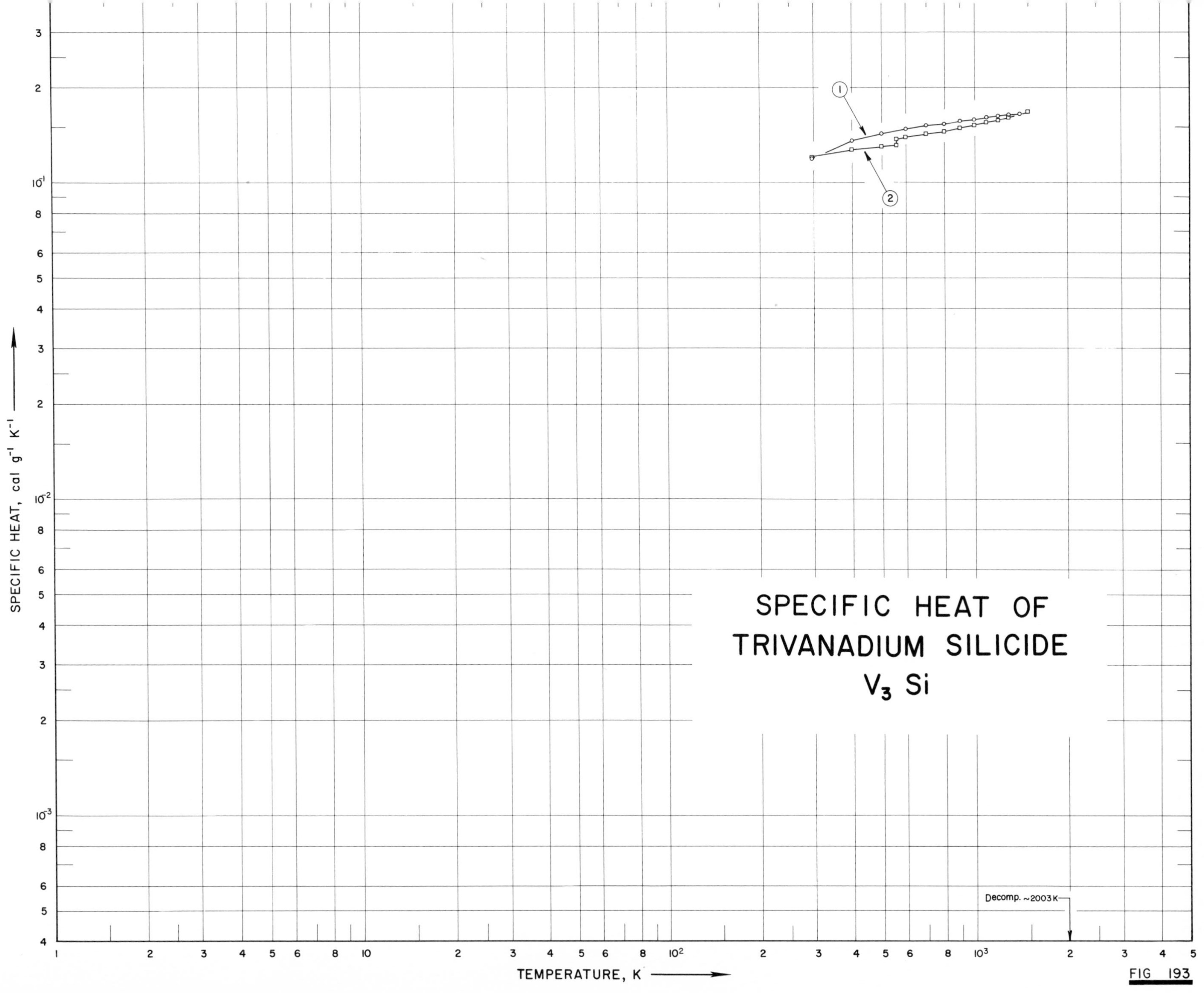
SPECIFIC HEAT OF
TRIVANADIUM SILICIDE
V_3 Si
SPECIFIC HEAT, cal g^{-1} K^{-1}
TEMPERATURE, K
Decomp. ~2003K
1
2
FIG 193

SPECIFICATION TABLE NO. 193 SPECIFIC HEAT OF TRIVANADIUM SILICIDE V_3Si

[For Data Reported in Figure and Table No. 193]

Curve No.	Ref. No.	Year	Temp. Range, K	Reported Error, %	Name and Specimen Designation	Composition (weight percent), Specifications and Remarks
1	238	1962	298-1310	±2.2		~98.0 V_3Si.
2	239	1963	298-1500	0.10		~99.0 V_3Si, impurities 1.0 V_5Si_3; crystalline.

DATA TABLE NO. 193 SPECIFIC HEAT OF TRIVANADIUM SILICIDE V_3Si

[Temperature, T, K; Specific Heat, C_p, Cal $g^{-1} K^{-1}$]

T	C_p
CURVE 1	
298	1.196 x 10^{-1}
300	1.200*
400	1.354
500	1.432
600	1.482
700	1.518
800	1.547
900	1.571
1000	1.593
1100	1.613
1200	1.632
1300	1.651
1310	1.653
CURVE 2	
298.15	1.2072 x 10^{-1}
300	1.2091*
400	1.2758
500	1.3066
560	1.3177
560	1.3823
600	1.3951
700	1.4270
800	1.4590
900	1.4909
1000	1.5299
1100	1.5548
1200	1.5868
1300	1.6187
1400	1.6507*
1500	1.6826

* Not shown on plot

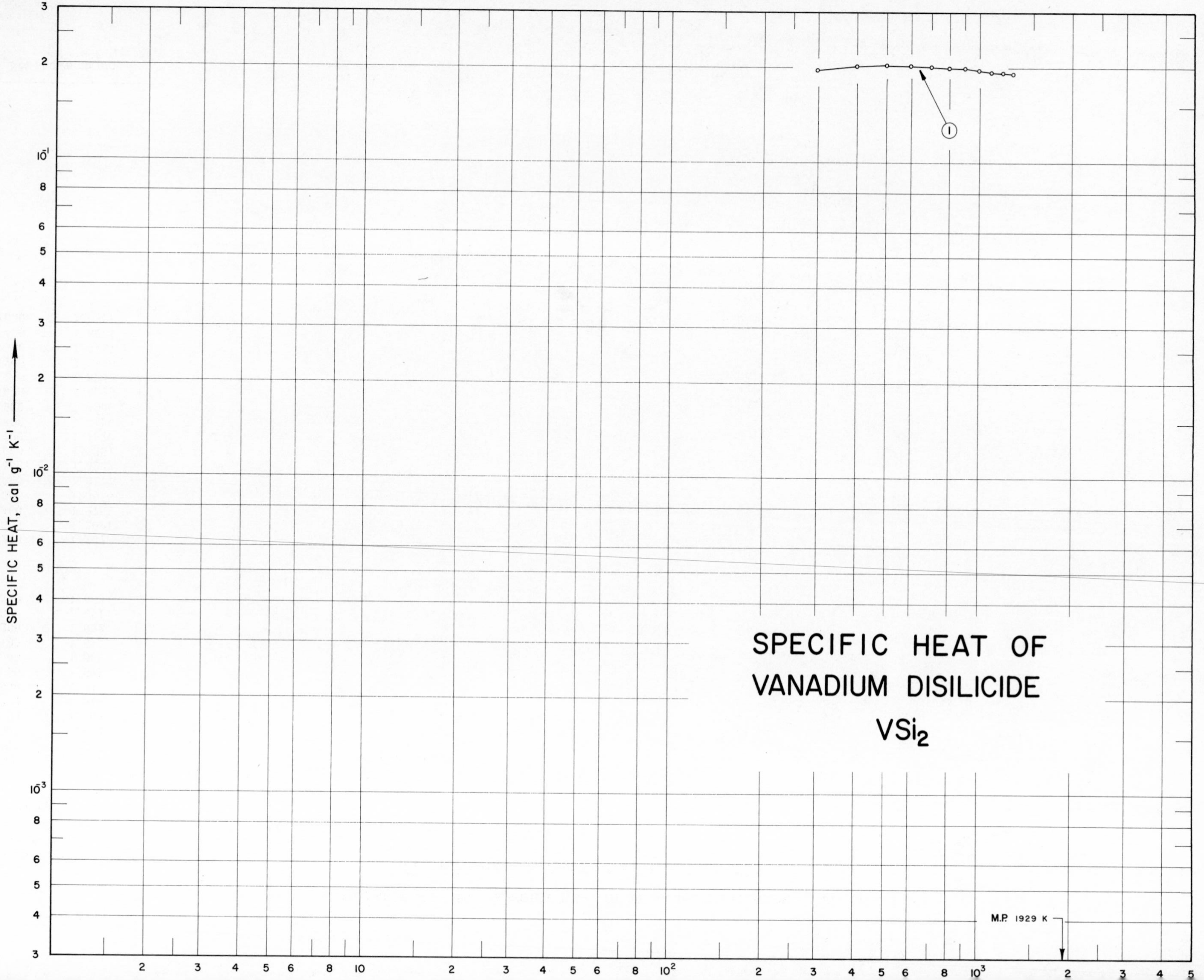

SPECIFIC HEAT OF VANADIUM DISILICIDE VSi_2

SPECIFIC HEAT, cal g^{-1} K^{-1}

M.P. 1929 K

1

SPECIFICATION TABLE NO. 194 SPECIFIC HEAT OF VANADIUM DISILICIDE VSi_2

[For Data Reported in Figure and Table No. 194]

Curve No.	Ref. No.	Year	Temp. Range, K	Reported Error, %	Name and Specimen Designation	Composition (weight percent), Specifications and Remarks
1	238	1962	290-1290	±1.2		~98.0 VSi_2.

DATA TABLE NO. 194 SPECIFIC HEAT OF VANADIUM DISILICIDE VSi_2

[Temperature, T, K; Specific Heat, C_p, Cal g^{-1} K^{-1}]

T	C_p
CURVE 1	
298	1.957×10^{-1}
300	1.958*
400	2.006
500	2.017
600	2.014
700	2.004
800	1.990
900	1.975
1000	1.957
1100	1.939
1200	1.921
1290	1.904

* Not shown on plot

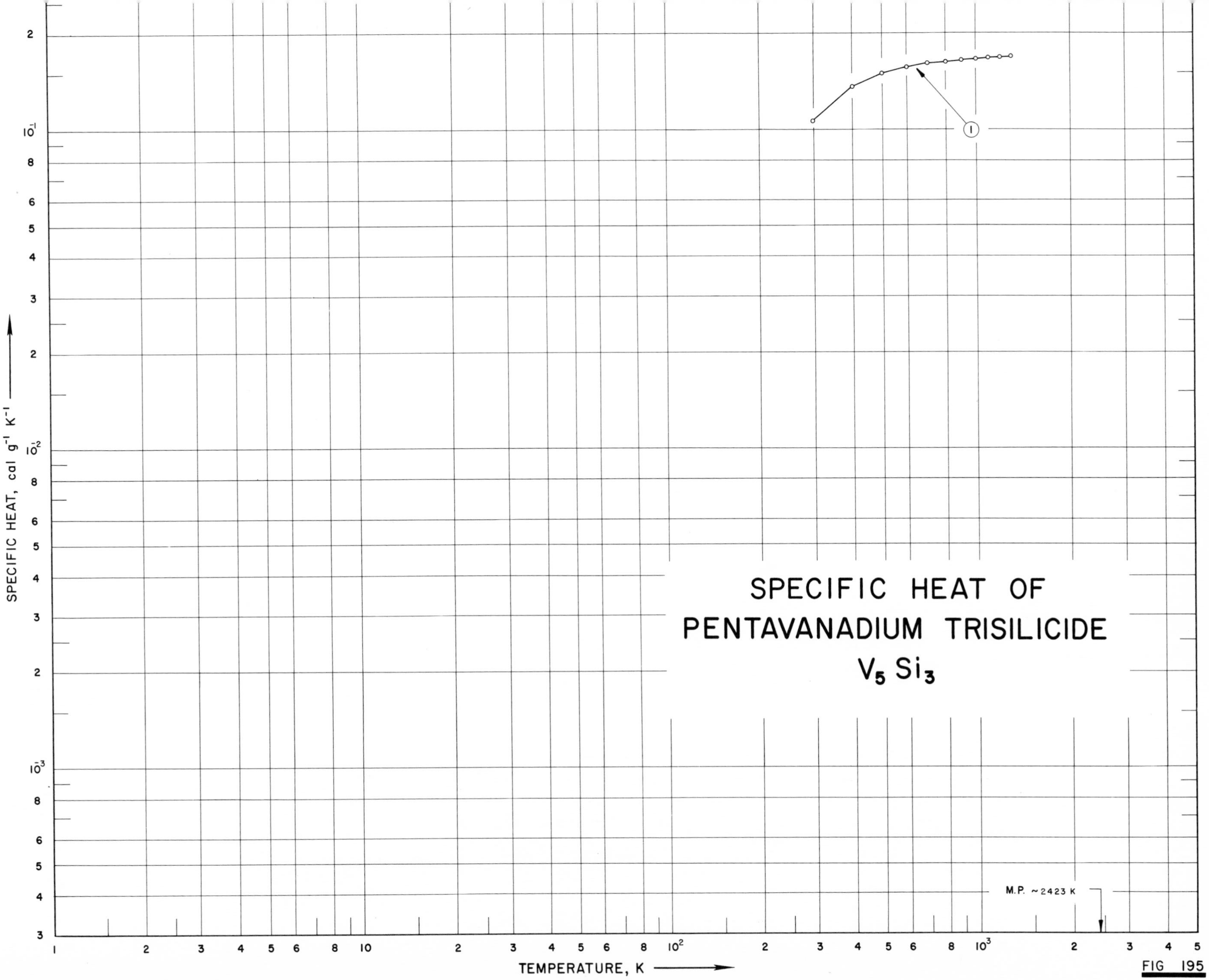

SPECIFIC HEAT OF
PENTAVANADIUM TRISILICIDE
V_5Si_3

FIG 195

SPECIFICATION TABLE NO. 195 SPECIFIC HEAT OF PENTAVANADIUM TRISILICIDE V_5Si_3

[For Data Reported in Figure and Table No. 195]

Curve No.	Ref. No.	Year	Temp. Range, K	Reported Error, %	Name and Specimen Designation	Composition (weight percent), Specifications and Remarks
1	238	1962	298-1290	±1.9		~98.0 V_5Si_3.

DATA TABLE NO. 195 SPECIFIC HEAT OF PENTAVANADIUM TRISILICIDE V_5Si_3

[Temperature, T, K; Specific Heat, C_p, Cal g^{-1} K^{-1}]

T	C_p
CURVE 1	
298	1.061 x 10^{-1}
300	1.070*
400	1.356
500	1.490
600	1.562
700	1.605
800	1.635
900	1.654
1000	1.668
1100	1.680
1200	1.688
1290	1.694

* Not shown on plot

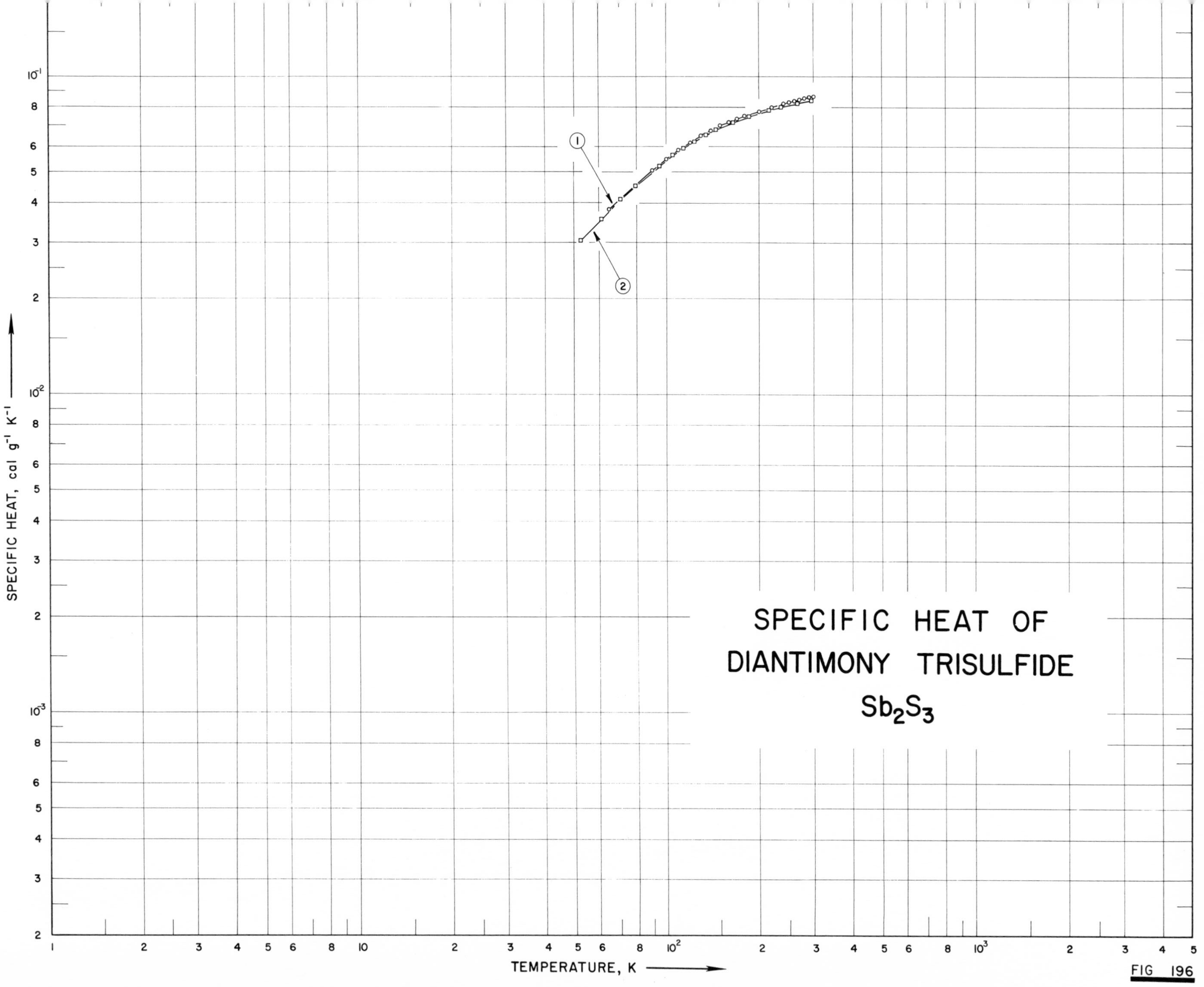
SPECIFIC HEAT OF
DIANTIMONY TRISULFIDE
Sb_2S_3
SPECIFIC HEAT, cal g⁻¹ K⁻¹
TEMPERATURE, K
1
2

FIG 196

SPECIFICATION TABLE NO. 196 SPECIFIC HEAT OF DIANTIMONY TRISULFIDE Sb_2S_3

[For Data Reported in Figure and Table No. 196]

Curve No.	Ref. No.	Year	Temp. Range, K	Reported Error, %	Name and Specimen Designation	Composition (weight percent), Specifications and Remarks
1	28	1960	65-300			Probably pure.
2	338	1962	53-296	0.3		71.73 Sb and 28.33S; mixture of pure antimony and sulfur was heated slowly to 450 C; held at 450 C for 7 days; cooled to room temperature.

DATA TABLE NO. 196 SPECIFIC HEAT OF DIANTIMONY TRISULFIDE Sb_2S_3

[Temperature, T, K; Specific Heat, C_p, Cal $g^{-1}K^{-1}$]

T	C_p
CURVE 1	
65	3.83 x 10^{-2}
70	4.09*
80	4.59*
90	5.08
100	5.49
110	5.86
120	6.20
130	6.52
140	6.78
150	7.02
160	7.20
170	7.36
180	7.51
190	7.64*
200	7.76
210	7.89*
220	8.01
230	8.11*
240	8.20
250	8.32
260	8.39
270	8.46
280	8.54
290	8.62
300	8.68
CURVE 2	
52.55	3.059 x 10^{-2}
56.77	3.297*
61.37	3.562
66.22	3.836*
71.19	4.110
76.05	4.357*
79.85	4.548
84.29	4.751*
95.00	5.234
105.30	5.646
114.52	5.952
124.58	6.267
135.85	6.565
145.56	6.803
155.73	7.021*
165.91	7.198
176.06	7.360*
CURVE 2 (cont.)	
186.84	7.498 x 10^{-2}
195.99	7.622*
216.28	7.845
266.25	7.951*
236.28	8.028
245.84	8.090*
256.43	8.178*
266.23	8.246
276.50	8.293*
286.56	8.364*
296.47	8.425

* Not shown on plot

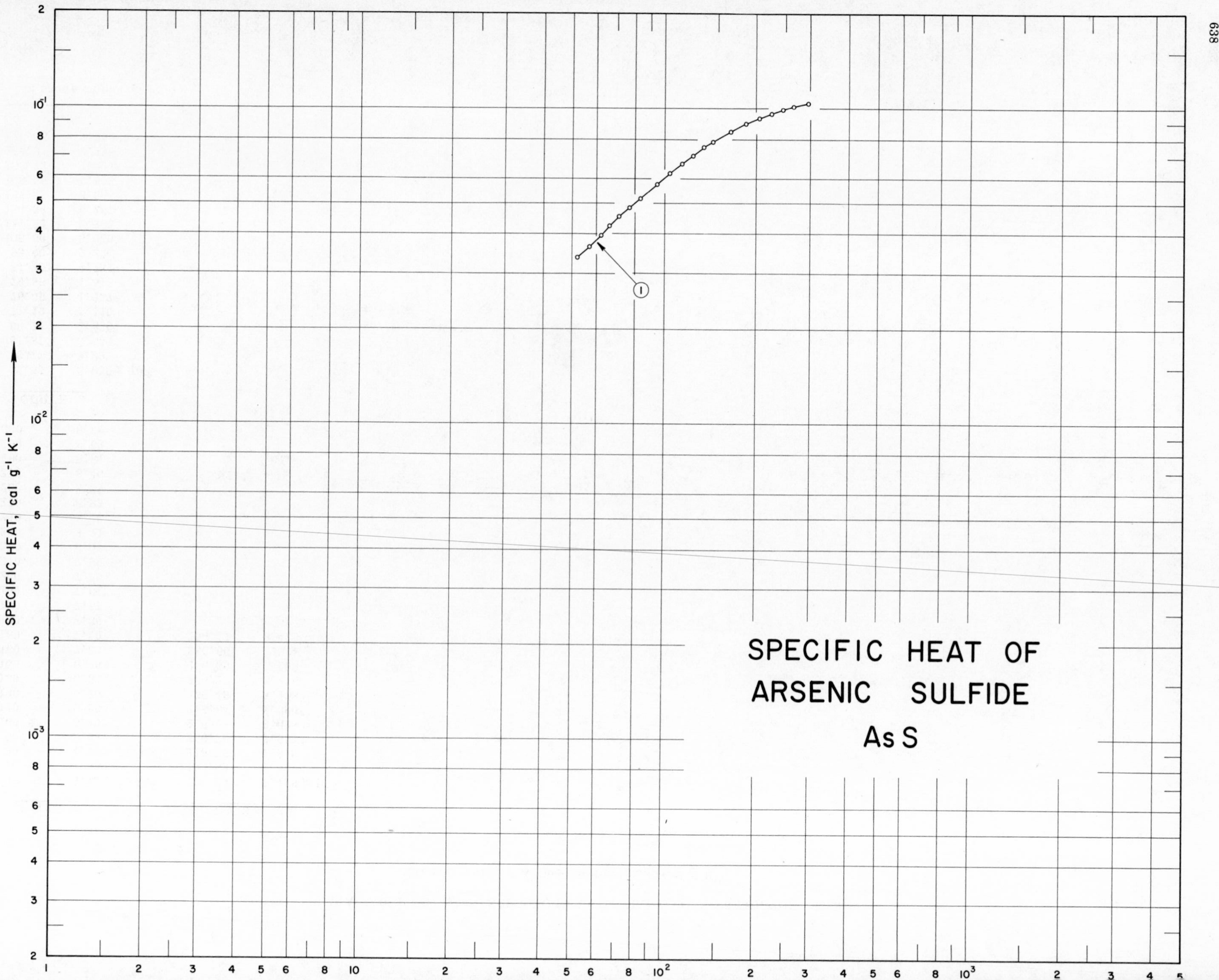
SPECIFIC HEAT OF
ARSENIC SULFIDE
As S
SPECIFIC HEAT, cal g⁻¹ K⁻¹
1

SPECIFICATION TABLE NO. 197 SPECIFIC HEAT OF ARSENIC SULFIDE AsS

[For Data Reported in Figure and Table No. 197]

Curve No.	Ref. No.	Year	Temp. Range, K	Reported Error, %	Name and Specimen Designation	Composition (weight percent), Specifications and Remarks
1	339	1964	52-296	≤ 0.3		70.12 As and 30.01 S, (70.03 and 29.97 theo); prepared from pure arsenic and sulfur by reacting them at 310 C.

DATA TABLE NO. 197 SPECIFIC HEAT OF ARSENIC SULFIDE AsS

[Temperature, T, K; Specific Heat, C_p, Cal $g^{-1}K^{-1}$]

T	C_p
CURVE 1	
52.34	3.383 x 10^{-2}
57.01	3.665
62.15	3.992
66.93	4.267
71.94	4.557
77.05	4.853
79.88	5.009*
83.62	5.205
94.91	5.791
104.85	6.279
114.53	6.714
124.57	7.137
135.74	7.562
145.46	7.892
155.77	8.216*
166.03	8.496
175.93	8.747*
186.19	8.985
195.91	9.166*
206.18	9.365
217.27	9.552*
226.33	9.692
236.00	9.833*
245.82	9.964
256.46	1.009 x 10^{-1}*
266.42	1.019
276.58	1.030*
286.79	1.040*
296.30	1.049

* Not shown on plot

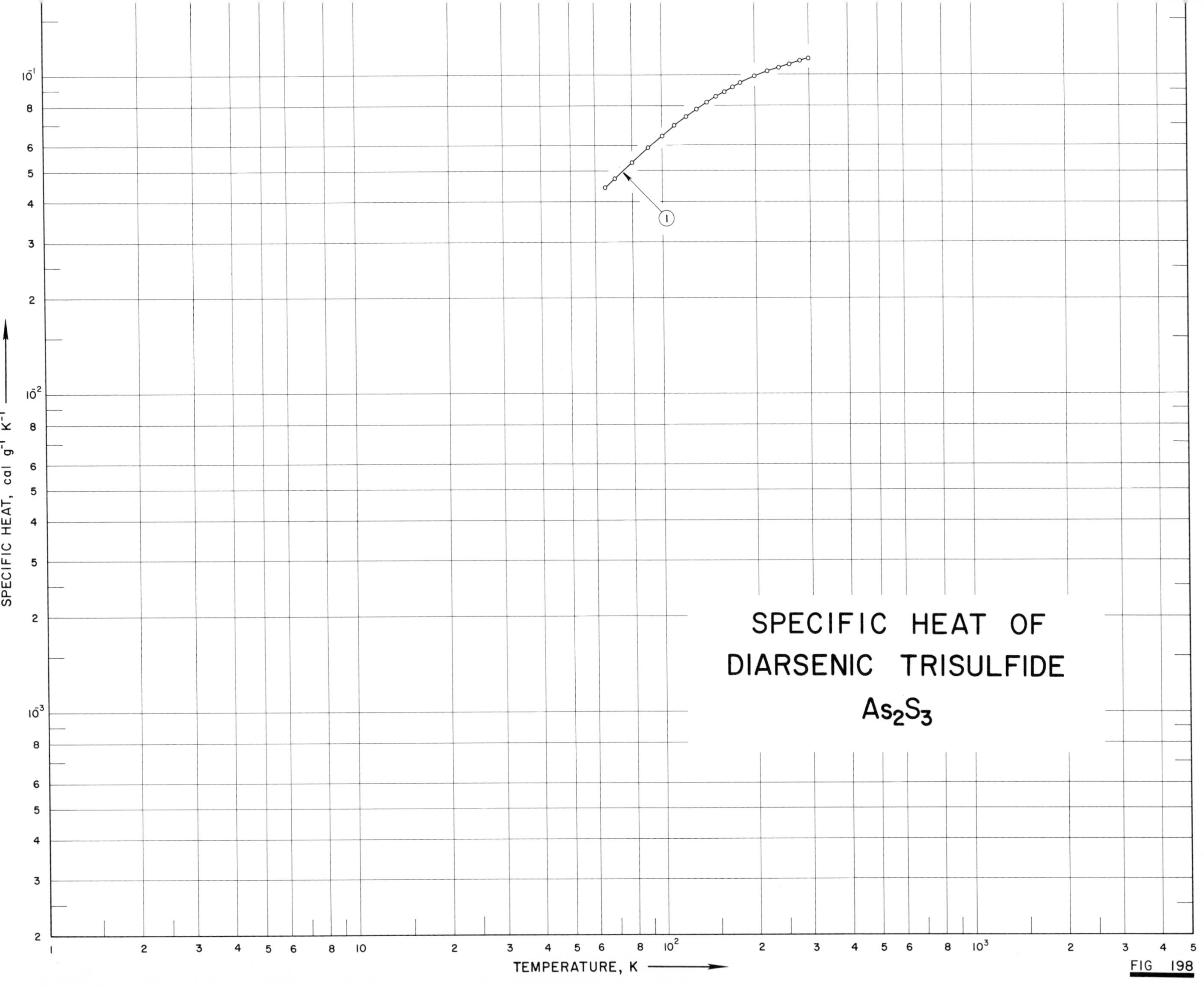
SPECIFIC HEAT OF
DIARSENIC TRISULFIDE
As_2S_3
SPECIFIC HEAT, cal g⁻¹ K⁻¹
TEMPERATURE, K
1

FIG 198

SPECIFICATION TABLE NO. 198 SPECIFIC HEAT OF DIARSENIC TRISULFIDE As_2S_3

[For Data Reported in Figure and Table No. 198]

Curve No.	Ref. No.	Year	Temp. Range, K	Reported Error, %	Name and Specimen Designation	Composition (weight percent), Specifications and Remarks
1	28	1960	65-300			Probably pure.

DATA TABLE NO. 198 SPECIFIC HEAT OF DIARSENIC TRISULFIDE As_2S_3

[Temperature, T, K; Specific Heat, C_p, Cal $g^{-1}K^{-1}$]

T	C_p
CURVE 1	
65	4.43×10^{-2}
70	4.74
80	5.32
90	5.93
100	6.48
110	6.97
120	7.44
130	7.84
140	8.23
150	8.58
160	8.88
170	9.19
180	9.47
190	9.71*
200	9.92
210	1.01×10^{-1}*
220	1.03
230	1.04*
240	1.06
250	1.07*
260	1.08
270	1.09*
280	1.11
290	1.12*
300	1.13

* Not shown on plot

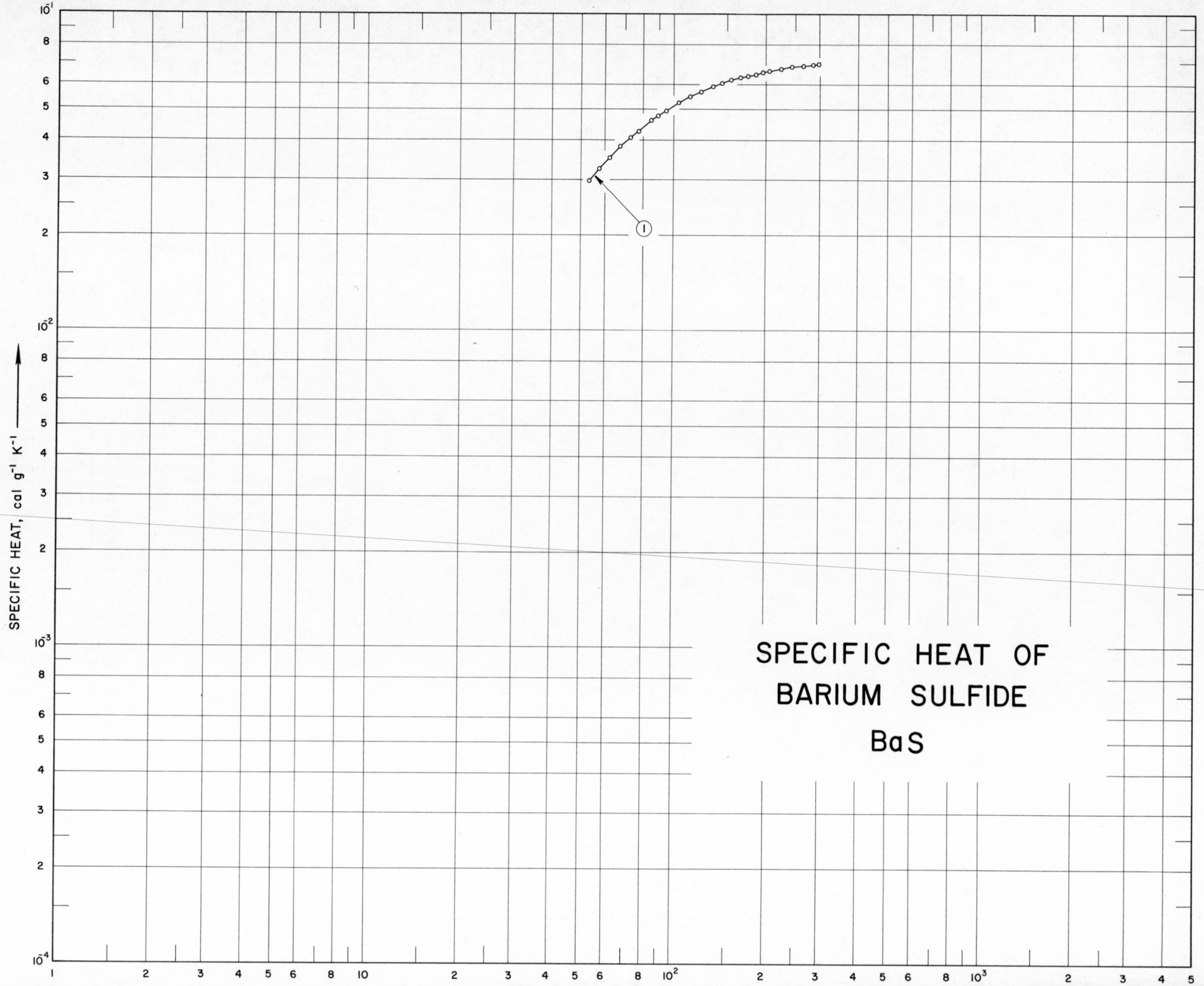
SPECIFIC HEAT OF
BARIUM SULFIDE
BaS
SPECIFIC HEAT, cal g^{-1} K^{-1}
1

SPECIFICATION TABLE NO. 199 SPECIFIC HEAT OF BARIUM SULFIDE BaS

[For Data Reported in Figure and Table No. 199]

Curve No.	Ref. No.	Year	Temp. Range, K	Reported Error, %	Name and Specimen Designation	Composition (weight percent), Specifications and Remarks
1	305	1960	54-298			99.53 BaS, 0.22 $BaSO_4$, 0.04 SiO_2; prepared from reagent grade barium sulfate; ignited at 850 C and reduced by hydrogen at 1000 C.

DATA TABLE NO. 199 SPECIFIC HEAT OF BARIUM SULFIDE BaS

[Temperature, T,K; Specific Heat, C_p, Cal $g^{-1}K^{-1}$]

T	C_p
CURVE 1	
53.71	2.989 x 10^{-2}
57.93	3.252
62.63	3.533
67.78	3.833
73.04	4.093
77.26	4.283
85.15	4.610
89.44	4.776
95.01	4.966
105.00	5.268
114.51	5.501
124.78	5.718
135.91	5.914
145.70	6.079
155.68	6.203
166.17	6.310
176.30	6.398
186.40	6.481
196.60	6.540
206.51	6.622
216.59	6.675*
226.26	6.723
236.07	6.764*
245.99	6.811
256.55	6.847*
266.90	6.876
276.56	6.912*
286.71	6.935
296.15	6.965*
298.15	6.965

* Not shown on plot

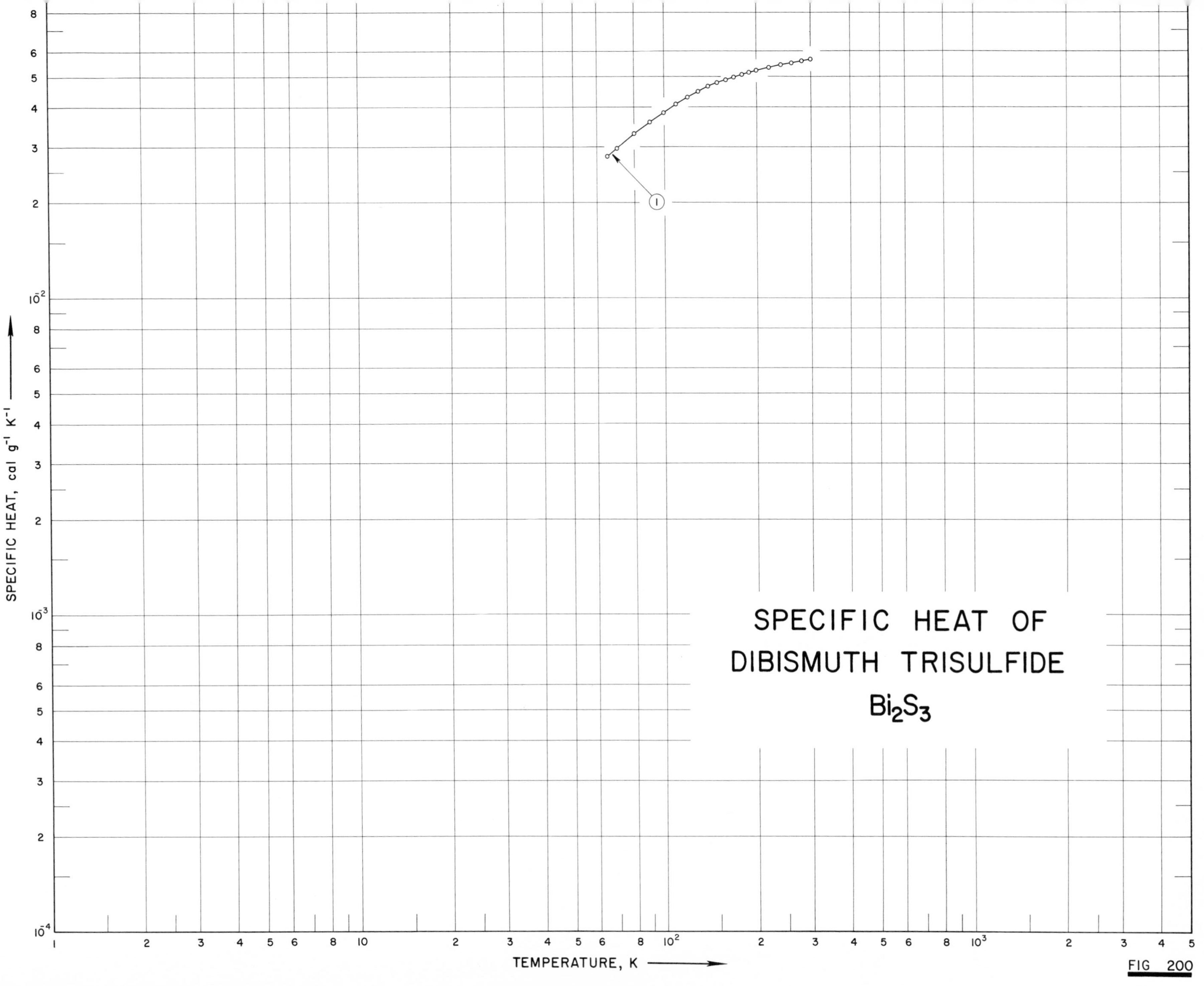
SPECIFIC HEAT OF
DIBISMUTH TRISULFIDE
Bi_2S_3
SPECIFIC HEAT, cal g⁻¹ K⁻¹
TEMPERATURE, K
1

FIG 200

SPECIFICATION TABLE NO. 200 SPECIFIC HEAT OF DIBISMUTH TRISULFIDE Bi_2S_3

[For Data Reported in Figure and Table No. 200]

Curve No.	Ref. No.	Year	Temp. Range, K	Reported Error, %	Name and Specimen Designation	Composition (weight percent), Specifications and Remarks
1	28	1960	65-300			

DATA TABLE NO. 200 SPECIFIC HEAT OF DIBISMUTH TRISULFIDE Bi_2S_3

[Temperature, T, K; Specific Heat, C_p, Cal $g^{-1}K^{-1}$]

T	C_p
CURVE 1	
65	2.80 x 10^{-2}
70	2.97
80	3.31
90	3.60
100	3.86
110	4.10
120	4.31
130	4.49
140	4.67
150	4.79
160	4.89
170	4.99
180	5.08
190	5.16
200	5.23
210	5.28*
220	5.35
230	5.39*
240	5.44
250	5.46*
260	5.51
270	5.57*
280	5.61
290	5.64*
300	5.68

* Not shown on plot

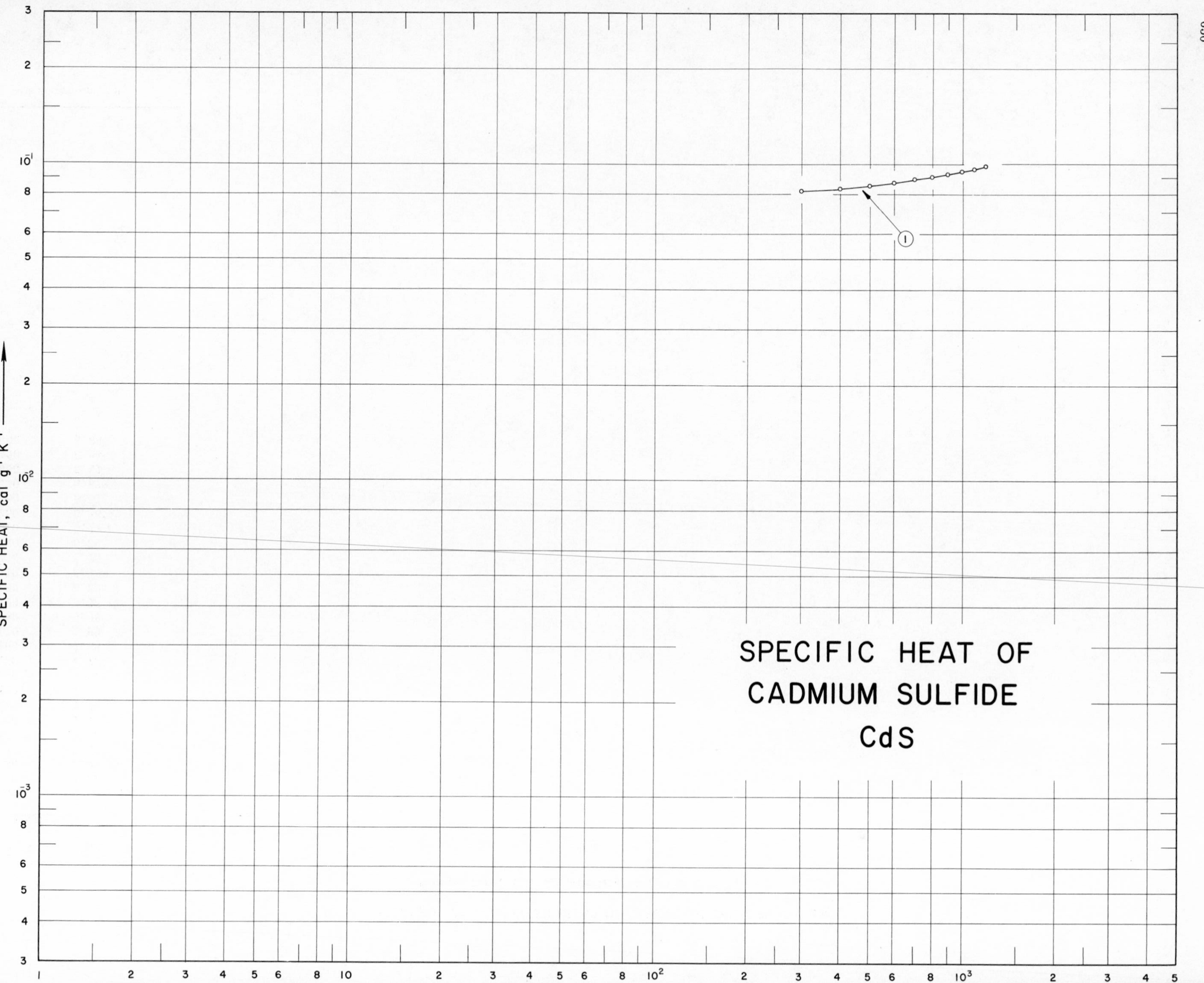
SPECIFIC HEAT OF
CADMIUM SULFIDE
CdS
SPECIFIC HEAT, cal g^{-1} K^{-1}
1

SPECIFICATION TABLE NO. 201 SPECIFIC HEAT OF CADMIUM SULFIDE CdS

[For Data Reported in Figure and Table No. 201]

Curve No.	Ref. No.	Year	Temp. Range, K	Reported Error, %	Name and Specimen Designation	Composition (weight percent), Specifications and Remarks
1	54	1965	300-1200	1.5		~100 Cd S, 0.0005-0.0010 Si, and 0.00005 Mg; sample supplied by the Eagle-Picher Co.

DATA TABLE NO. 201 SPECIFIC HEAT OF CADMIUM SULFIDE CdS

[Temperature, T, K; Specific Heat, C_p, Cal $g^{-1}K^{-1}$]

T	C_p
CURVE 1	
300	8.210×10^{-2}
400	8.397
500	8.577
600	8.763
700	8.943
800	9.130
900	9.310
1000	9.497
1100	9.677
1200	9.864

* Not shown on plot

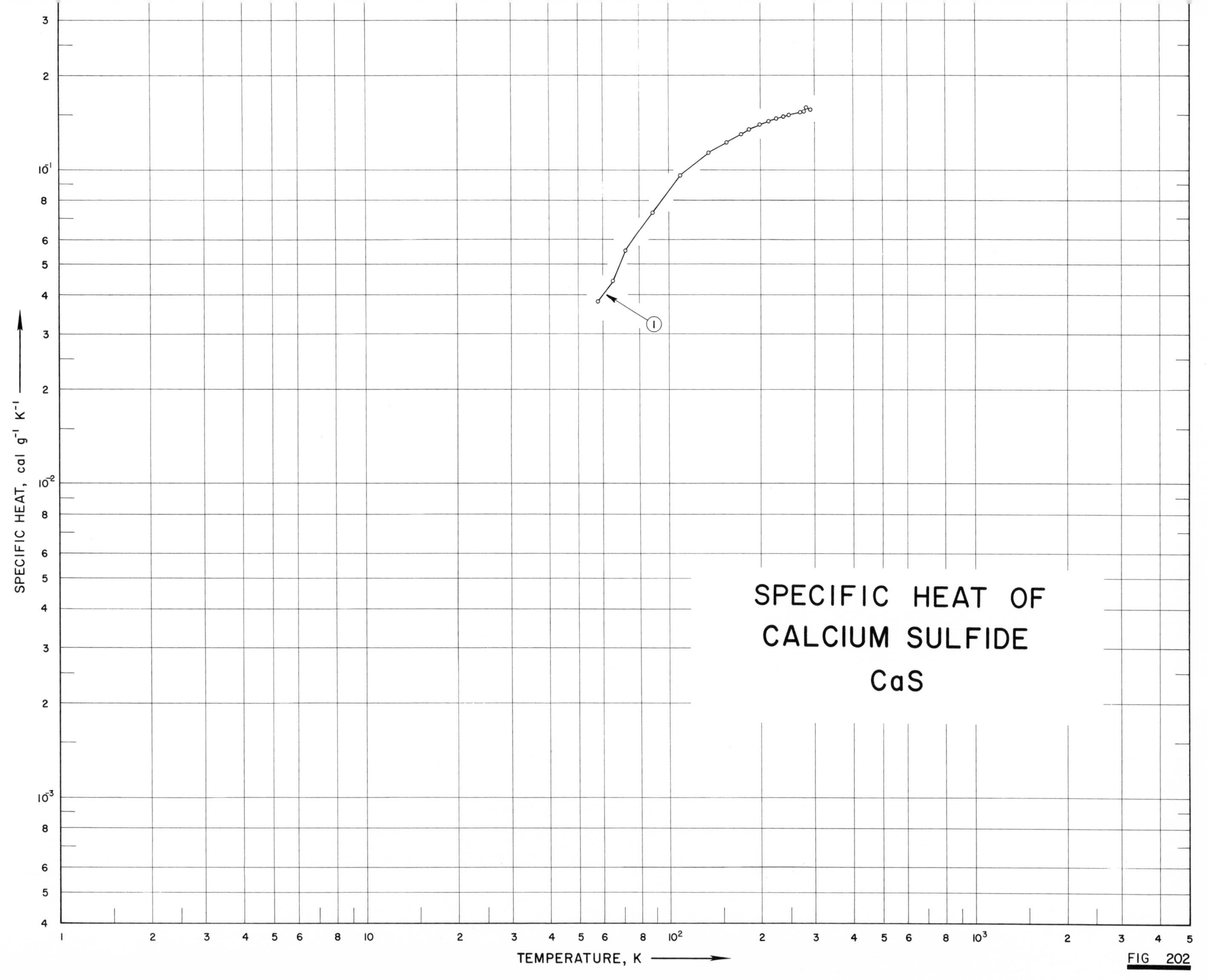
SPECIFIC HEAT OF
CALCIUM SULFIDE
CaS
SPECIFIC HEAT, cal g⁻¹ K⁻¹
TEMPERATURE, K
1
FIG 202

SPECIFICATION TABLE NO. 202 SPECIFIC HEAT OF CALCIUM SULFIDE CaS

[For Data Reported in Figure and Table No. 202]

Curve No.	Ref. No.	Year	Temp. Range, K	Reported Error, %	Name and Specimen Designation	Composition (weight percent), Specifications and Remarks
1	340	1931	58-295			Prepared from calcium sulfate; density = 2.56 g cm^{-3} at 23.6 C.

DATA TABLE NO. 202 SPECIFIC HEAT OF CALCIUM SULFIDE CaS

[Temperature, T, K; Specific Heat, C_p, Cal $g^{-1}K^{-1}$]

T	C_p
CURVE 1	
58.1	3.843×10^{-2}
65.1	4.471
72.0	5.569
89.5	7.358
110.8	9.683
136.4	1.140×10^{-1}
155.9	1.233
174.7	1.306
185.3	1.358
200.5	1.406
214.6	1.445
226.9	1.464
239.4	1.485
249.6	1.507
271.2	1.538
279.5	1.547
285.5	1.592
290.1	1.564*
294.9	1.569

* Not shown on plot

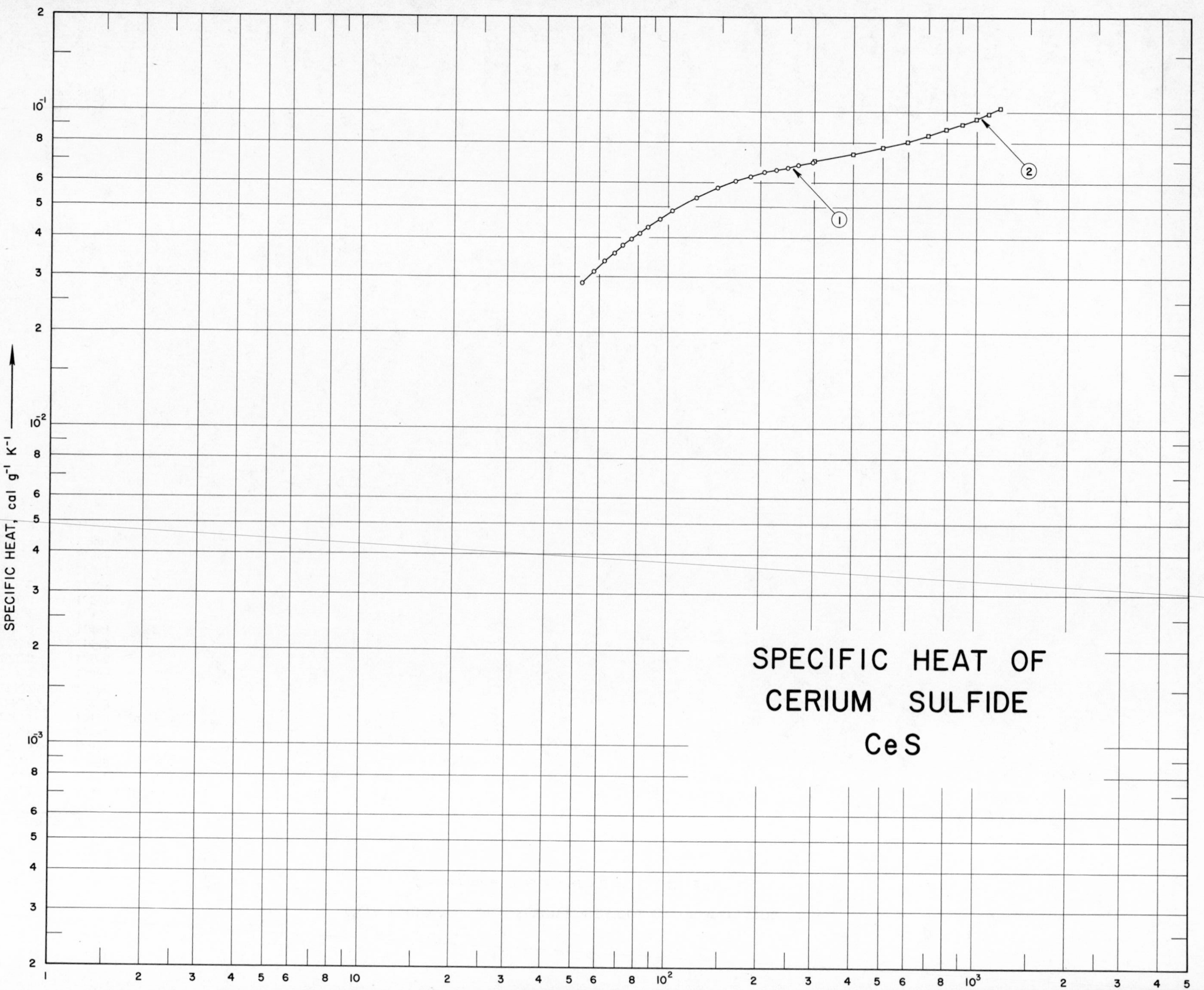

SPECIFIC HEAT OF
CERIUM SULFIDE
CeS
SPECIFIC HEAT, cal g⁻¹ K⁻¹
1
2

SPECIFICATION TABLE NO. 203 SPECIFIC HEAT OF CERIUM SULFIDE CeS

[For Data Reported in Figure and Table No. 203]

Curve No.	Ref. No.	Year	Temp. Range, K	Reported Error, %	Name and Specimen Designation	Composition (weight percent), Specifications and Remarks
1	306	1959	53-297	0.3		81.40 Ce (81.38 theo). 18.56 S (18.62 theo).
2	54	1965	300-1200	3.4		~100 CeS; traces of Cd and Mg.

DATA TABLE NO. 203 SPECIFIC HEAT OF CERIUM SULFIDE CeS

[Temperature, T,K; Specific Heat, C_p, Cal $g^{-1}K^{-1}$]

T	C_p
CURVE 1	
53.13	2.879 x 10^{-2}
57.88	3.128
62.88	3.369
67.44	3.576
71.97	3.771
76.67	3.958
81.60	4.134
86.30	4.301
94.53	4.572
104.82	4.882
114.66	5.126*
124.84	5.362
135.94	5.587*
145.86	5.762
156.19	5.923*
166.24	6.051
176.18	6.162*
186.33	6.266
196.35	6.353*
206.50	6.440
216.40	6.510*
226.37	6.580
236.26	6.638*
245.98	6.696
256.30	6.748*
266.30	6.800
276.39	6.835*
286.65	6.882*
296.72	6.934
CURVE 2	
300	6.987 x 10^{-2}
400	7.347
500	7.713
600	8.073
700	8.439
800	8.805
900	9.165
1000	9.531
1100	9.891
1200	1.026 x 10^{-1}

* Not shown on plot

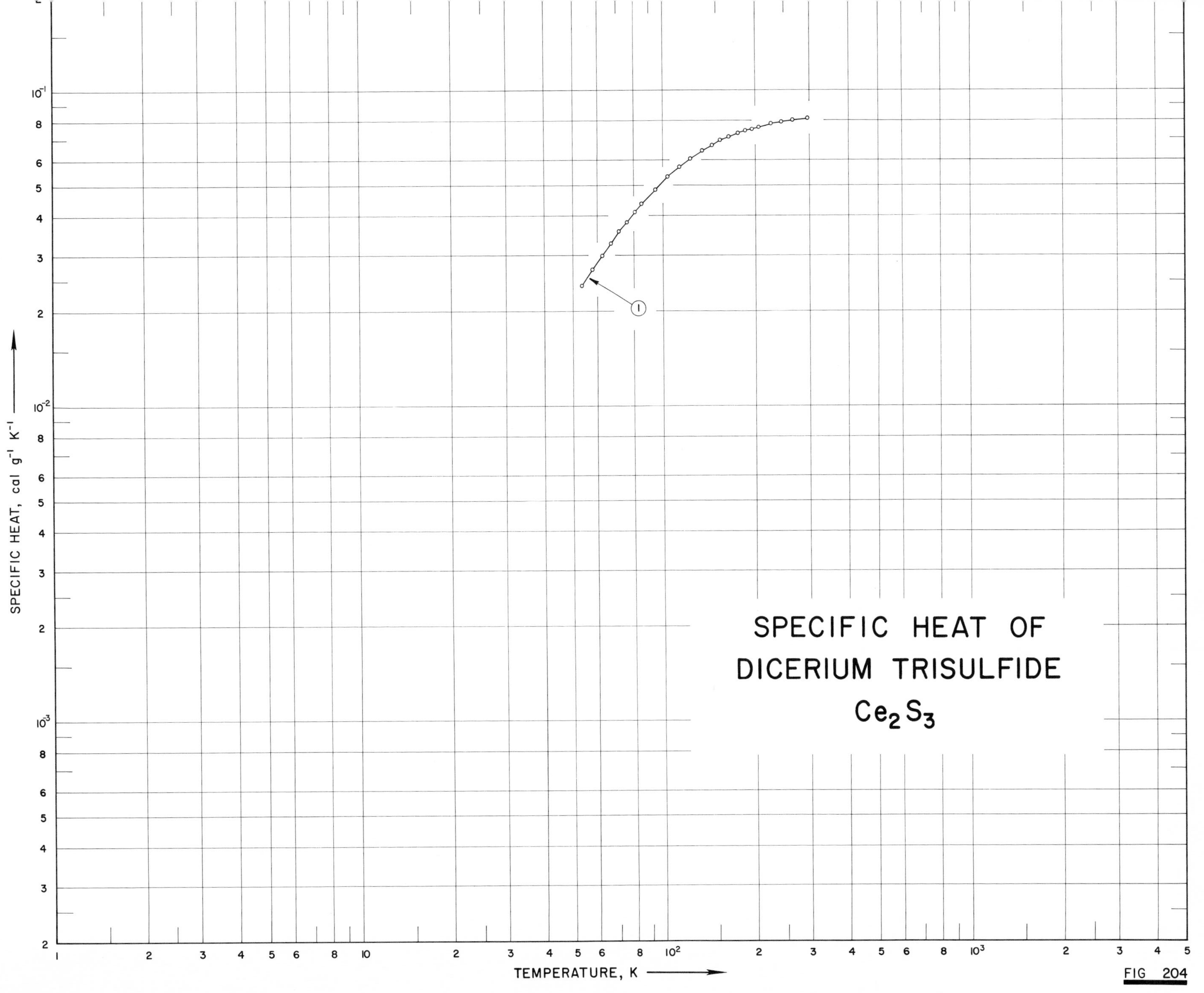
SPECIFIC HEAT OF
DICERIUM TRISULFIDE
Ce_2S_3
SPECIFIC HEAT, cal g^{-1} K^{-1}
TEMPERATURE, K
1

FIG 204

SPECIFICATION TABLE NO. 204 SPECIFIC HEAT OF DICERIUM TRISULFIDE Ce_2S_3

[For Data Reported in Figure and Table No. 204]

Curve No.	Ref. No.	Year	Temp. Range, K	Reported Error, %	Name and Specimen Designation	Composition (weight percent), Specifications and Remarks
1	306	1959	54-296	0.3		74.27 Ce (74.45 theo) and 25.3 S (25.55 theo).

DATA TABLE NO. 204 SPECIFIC HEAT OF DICERIUM TRISULFIDE Ce_2S_3

[Temperature, T, K; Specific Heat, C_p, Cal $g^{-1}K^{-1}$]

T	C_p
CURVE 1	
53.63	2.402×10^{-2}
58.23	2.701
62.72	2.996
67.16	3.294
71.52	3.565
76.07	3.838
81.34	4.141
85.92	4.396
94.88	4.864
104.97	5.345
114.63	5.735
124.65	6.094
135.80	6.442
145.55	6.710
155.87	6.938
165.81	7.124
177.62	7.326
186.21	7.443
196.05	7.547
206.50	7.666
216.37	7.749*
226.30	7.844
236.44	7.903*
245.95	7.950
256.43	8.017*
266.29	8.049
276.31	8.091*
286.55	8.128*
296.11	8.174

* Not shown on plot

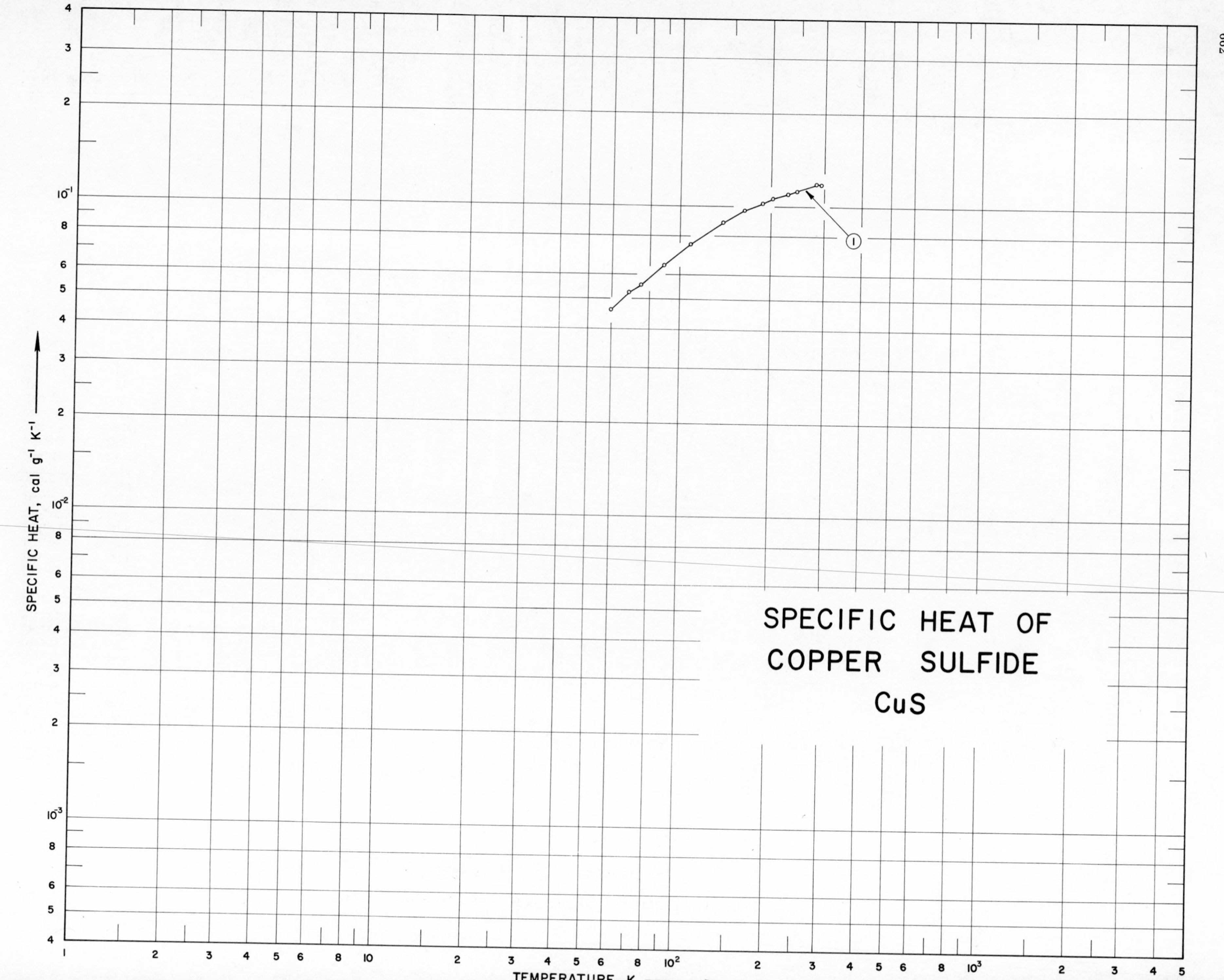
SPECIFIC HEAT OF
COPPER SULFIDE
CuS
SPECIFIC HEAT, cal g^{-1} K^{-1}
TEMPERATURE, K
1

SPECIFICATION TABLE NO. 205 SPECIFIC HEAT OF COPPER SULFIDE CuS

[For Data Reported in Figure and Table No. 205]

Curve No.	Ref. No.	Year	Temp. Range, K	Reported Error, %	Name and Specimen Designation	Composition (weight percent), Specifications and Remarks
1	341	1932	60–295		Covellite	Pure; natural mineral; density = 4.64 g cm^{-3} at 21.2 C.

DATA TABLE NO. 205 SPECIFIC HEAT OF COPPER SULFIDE CuS

[Temperature, T, K; Specific Heat, C_p, Cal $g^{-1}K^{-1}$]

T	C_p
CURVE 1	
59.8	4.615×10^{-2}
61.9	4.756*
68.6	5.271
75.1	5.588
89.0	6.461
109.9	7.572
139.0	8.898
164.4	9.703
177.6	1.006×10^{-1}*
188.0	1.027
203.2	1.062
215.4	1.086*
229.7	1.104
245.4	1.130
284.1	1.189
289.9	1.161*
294.6	1.186

* Not shown on plot

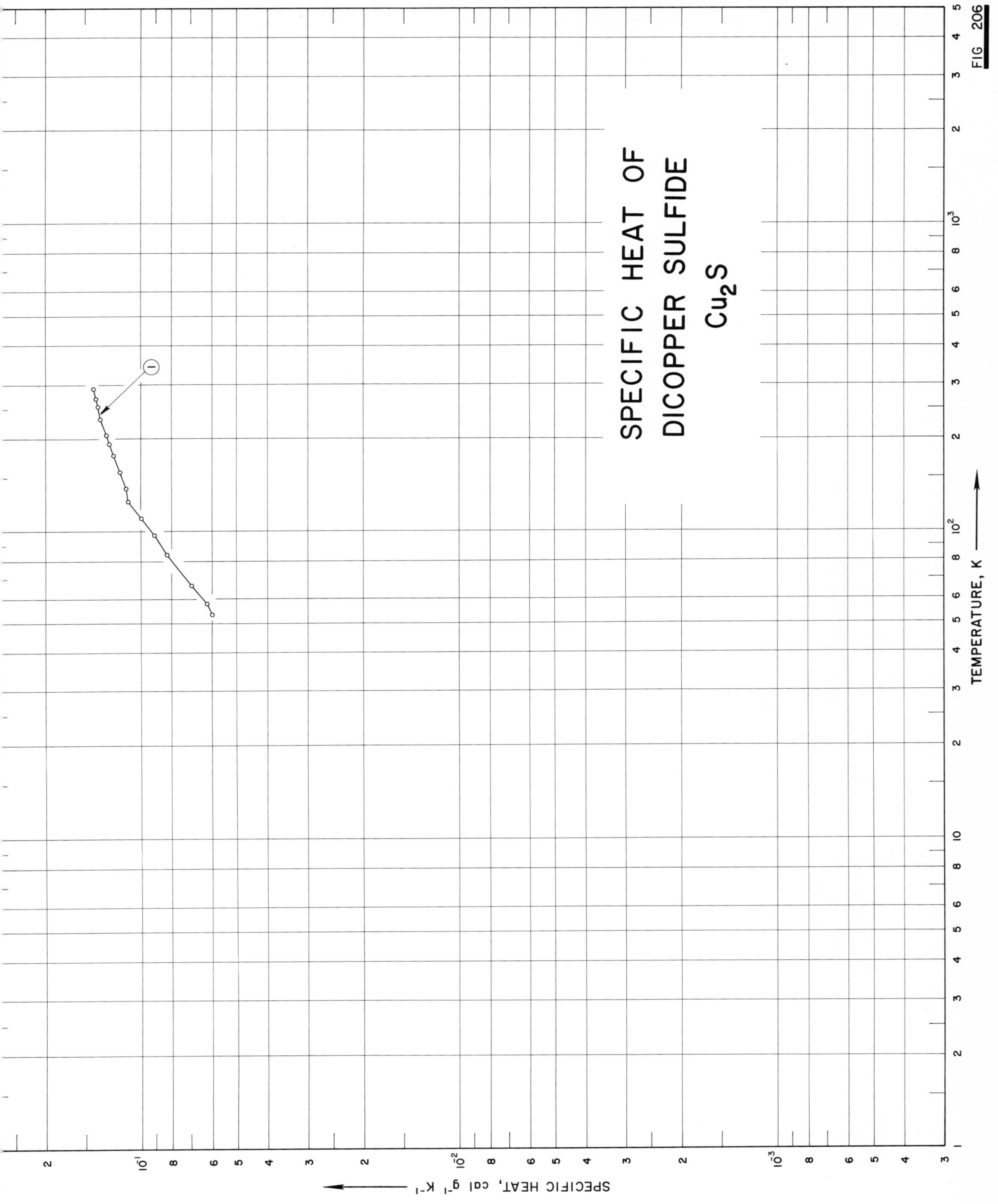

SPECIFIC HEAT OF DICOPPER SULFIDE Cu_2S

SPECIFIC HEAT, cal g^{-1} K^{-1}

TEMPERATURE, K

FIG 206

SPECIFICATION TABLE NO. 206 SPECIFIC HEAT OF DICOPPER SULFIDE Cu_2S

[For Data Reported in Figure and Table No. 206]

Curve No.	Ref. No.	Year	Temp. Range, K	Reported Error, %	Name and Specimen Designation	Composition (weight percent), Specifications and Remarks
1	341	1932	54-292			99.8 Cu_2S; crystalline; density = 5.76 g cm^{-3} at 22.4 C.

DATA TABLE NO. 206 SPECIFIC HEAT OF DICOPPER SULFIDE Cu_2S

[Temperature, T, K; Specific Heat, C_p, Cal $g^{-1}K^{-1}$]

T	C_p
CURVE 1	
53.7	6.010 x 10^{-2}
56.1	6.012*
58.0	6.256
62.6	6.709*
66.1	6.994
71.3	7.459*
84.4	8.364
97.6	9.171
111.6	1.011 x 10^{-1}
126.0	1.100
139.0	1.122
157.0	1.174
177.2	1.237
193.3	1.263
206.8	1.299
210.4	1.306*
232.5	1.352
243.0	1.362*
255.6	1.375
271.4	1.399
285.4	1.416*
292.2	1.424

*Not shown on plot

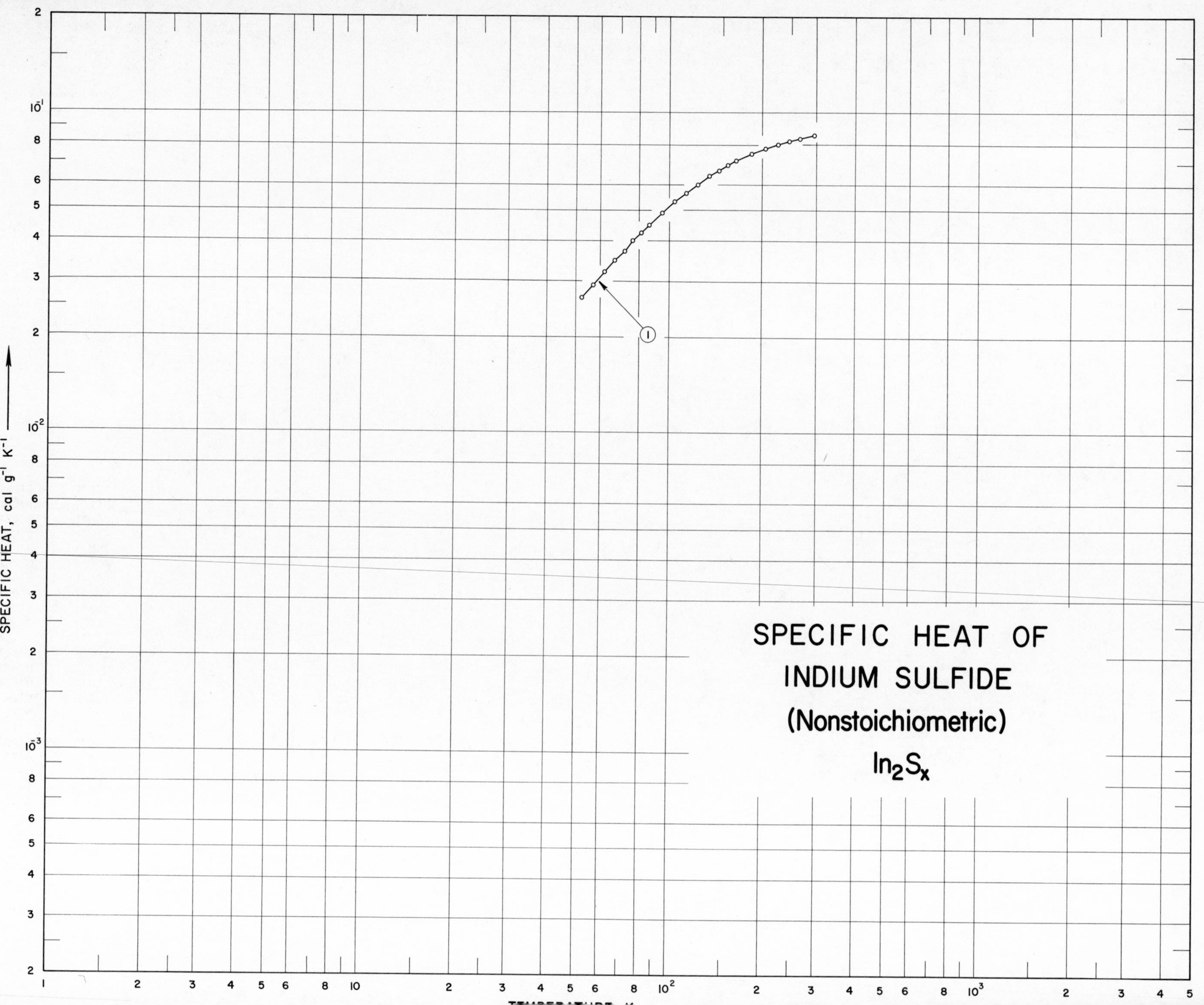

SPECIFIC HEAT OF
INDIUM SULFIDE
(Nonstoichiometric)
In2Sx
SPECIFIC HEAT, cal g-1 K-1
1

SPECIFICATION TABLE NO. 207 SPECIFIC HEAT OF DIINDIUM SULFIDE (nonstoichiometric) In_2S_x

[For Data Reported in Figure and Table No. 207]

Curve No.	Ref. No.	Year	Temp. Range, K	Reported Error, %	Name and Specimen Designation	Composition (weight percent), Specifications and Remarks
1	338	1962	53-297			$In_2S_{2.93}$; 70.49 In, 28.84 S, 0.56 H_2O and 0.1 SiO_2; data corrected for impurities.

DATA TABLE NO. 207 SPECIFIC HEAT OF DIINDIUM SULFIDE In_2S_x (nonstoichiometric)

[Temperature, T, K; Specific Heat, C_p, Cal $g^{-1}K^{-1}$]

T	C_p
CURVE 1	
52.66	2.66 x 10^{-2}
57.26	2.92
62.39	3.208
67.32	3.492
72.02	3.739
76.70	4.002
81.74	4.243
86.42	4.496
95.14	4.904
104.85	5.322
114.28	5.683
124.64	6.051
135.81	6.409
145.69	6.687
156.10	6.929
165.88	7.145
175.99	7.340*
185.91	7.519
196.08	7.664*
206.23	7.800
216.07	7.917*
226.16	8.038
236.05	8.140*
245.88	8.217
256.48	8.310*
266.42	8.393
277.33	8.449*
286.84	8.526*
296.75	8.600

* Not shown on plot

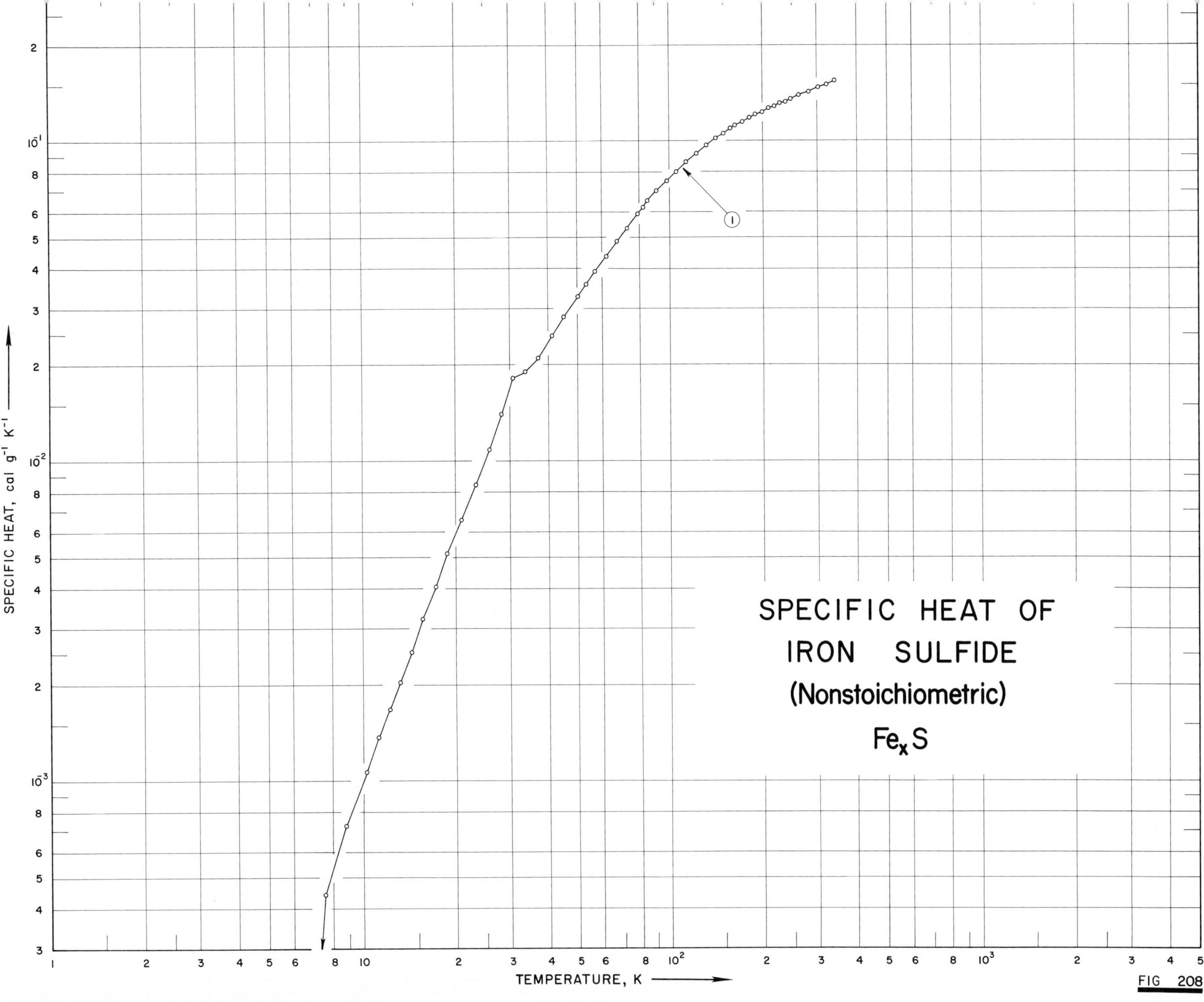
SPECIFIC HEAT OF
IRON SULFIDE
(Nonstoichiometric)
Fe_xS
SPECIFIC HEAT, cal g^{-1} K^{-1}
TEMPERATURE, K
1
FIG 208

SPECIFICATION TABLE NO. 208 SPECIFIC HEAT OF IRON SULFIDE (nonstoichiometric) Fe_xS

[For Data Reported in Figure and Table No. 208]

Curve No.	Ref. No.	Year	Temp. Range, K	Reported Error, %	Name and Specimen Designation	Composition (weight percent), Specifications and Remarks
1	307	1959	6-347	0.1	sulfur-rich pyrrhotite	$Fe_{0.877}S$; iron impurities, ~0.01 Ni, S and ~0.001 Mn, sulfur purified by double distillation; prepared by reacting stoichiometric amounts of iron and sulfur in electric furnace at 800 C; after reaction has gone almost to completion, it was cooled; heated at 800 C for 7 days; cooled to 100 C per day to room temperature; fragmented; homogenized at 290 C and after 30 days cooled to room temperature over a period of 6 days.

DATA TABLE NO. 208 SPECIFIC HEAT OF IRON SULFIDE Fe_xS (nonstoichiometric)

[Temperature, T,K; Specific Heat, C_p, Cal $g^{-1}K^{-1}$]

T	C_p
CURVE 1	
Series I	
82.50	6.233 x 10^{-2}
85.92	6.535
91.76	7.015
98.77	7.537
106.21	8.076
114.28	8.619
123.44	9.199
132.83	9.736
141.63	1.020 x 10^{-1}
150.19	1.061
158.69	1.100
164.39	1.123
173.39	1.156
182.49	1.190
190.68	1.219
200.89	1.247
210.12	1.275
219.45	1.298
228.87	1.323
238.50	1.346
248.25	1.368
Series II	
53.65	3.599 x 10^{-2}
57.16	3.921
62.08	4.390
67.34	4.881
73.04	5.390
79.29	5.953
86.06	6.548*
93.74	7.160*
101.84	7.764*
108.69	8.250*
117.20	8.811*
126.00	9.357*
134.89	9.848*
149.76	1.060 x 10^{-1}*
169.95	1.145*
194.10	1.228*
223.47	1.309*
240.61	1.351*
252.46	1.378*
262.52	1.400
CURVE 1 (cont.)	
272.68	1.421 x 10^{-1}*
282.83	1.441
293.09	1.462*
303.38	1.481
313.63	1.499*
323.84	1.518
334.05	1.536*
344.27	1.553
Series III	
246.87	1.366 x 10^{-1}*
256.33	1.387*
266.32	1.408*
276.40	1.430*
286.51	1.448*
296.68	1.468*
306.86	1.487*
317.01	1.506*
327.12	1.524*
337.27	1.542*
347.48	1.561*
Series IV	
7.55	4.405 x 10^{-4}
8.88	7.255
10.21	1.073 x 10^{-3}
11.22	1.376
12.18	1.687
13.16	2.044
14.30	2.534
15.67	3.213
17.23	4.081
18.76	5.181
20.90	6.579
23.19	8.453
25.72	1.088 x 10^{-2}
28.34	1.402
30.88	1.825
33.70	1.904
37.26	2.104
41.41	2.459
45.69	2.848
50.18	3.268
55.22	3.738*
60.62	4.248*
CURVE 1 (cont.)	
Series V	
5.82	2.714 x 10^{-5}*
6.52	9.254*
7.20	2.394 x 10^{-4}*
8.48	7.020*
Series VI	
5.97	3.578 x 10^{-5}*
6.36	7.279*
7.88	6.391 x 10^{-4}*
10.17	1.055 x 10^{-3}*
Series VII*	
6.07	4.318 x 10^{-5}
7.90	6.305 x 10^{-4}
7.92	6.428
9.79	9.377
29.06	1.498 x 10^{-2}
30.57	1.783
32.21	1.843
33.97	1.900
35.87	1.999

* Not shown on plot

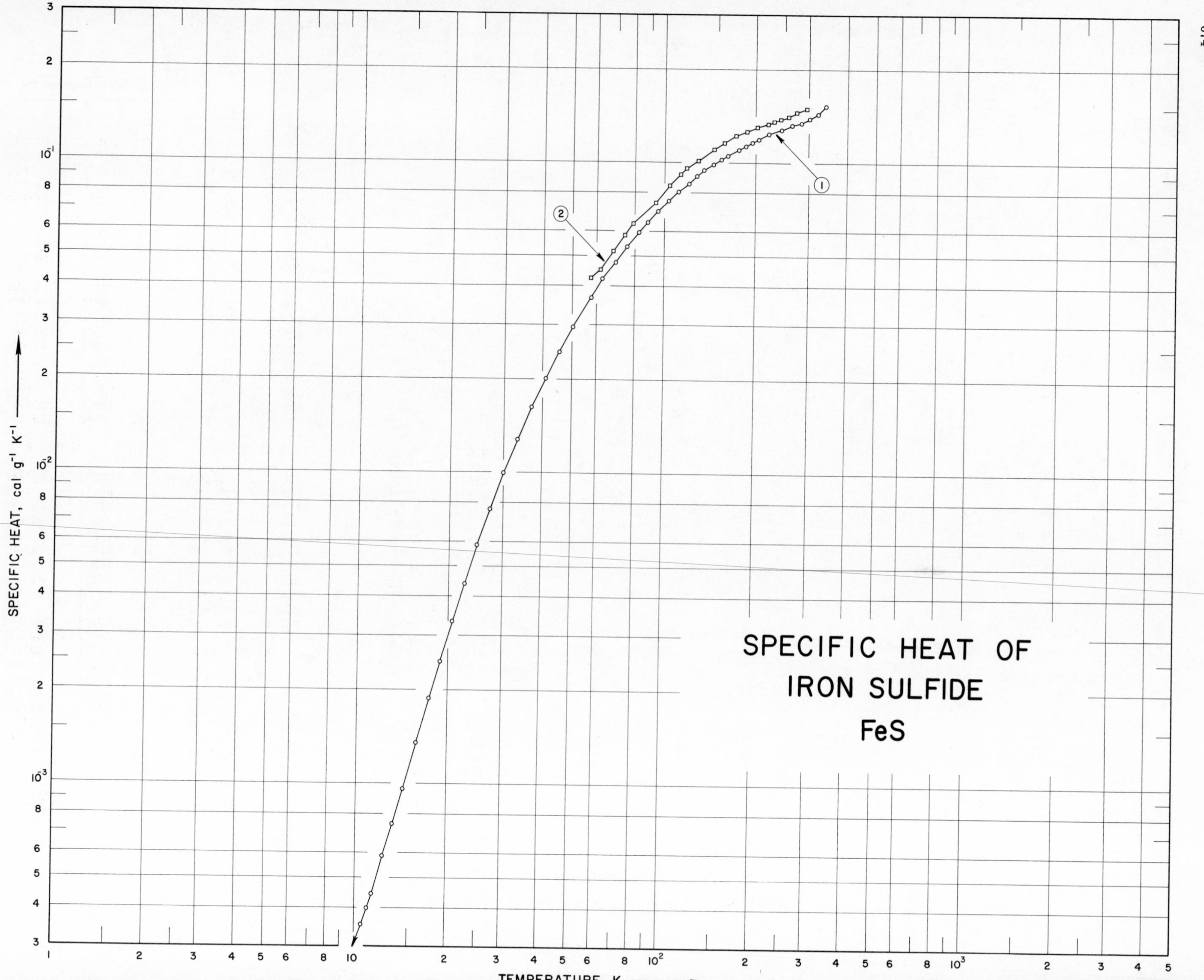

SPECIFIC HEAT OF
IRON SULFIDE
FeS
TEMPERATURE, K
SPECIFIC HEAT, cal g⁻¹ K⁻¹
1
2

SPECIFICATION TABLE NO. 209 SPECIFIC HEAT OF IRON SULFIDE FeS

[For Data Reported in Figure and Table No. 209]

Curve No.	Ref. No.	Year	Temp. Range, K	Reported Error, %	Name and Specimen Designation	Composition (weight percent), Specifications and Remarks
1	307	1959	7-345	0.1	Iron-rich pyrrhotite	Iron impurities: ~0.01 Ni, S, and ~0.001 Mn, sulfur purified by double distillation; prepared by reacting Fe and S in electric furnace at 800 C; after reaction has gone almost to completion it was cooled heated at 800 C for 7 days; cooled to 100 C per day to room temperature; fragmented; homogenized at 290 C and after 30 days cooled to room temperature over a period of 6 days.
2	340	1931	58-296			Prepared from mixture of iron oxides made from strips of pure ingot iron by heating in air several days at 900 C; density = 4.65 g cm^{-3} at 23.9 C.

DATA TABLE NO. 209 SPECIFIC HEAT OF IRON SULFIDE FeS

[Temperature, T, K; Specific Heat, C_p, Cal $g^{-1}K^{-1}$]

T	C_p
CURVE 1	
Series I	
88.72	6.412 x 10^{-2}
95.96	6.951
104.18	7.527
112.76	8.090
121.00	8.593
129.31	9.061
137.68	9.476
146.16	9.868
154.88	1.024 x 10^{-1}
163.96	1.059
Series II	
6.97	9.782 x 10^{-5}*
8.32	1.820 x 10^{-4}*
9.05	2.252*
9.86	2.969*
10.56	3.594
11.42	4.459
12.39	5.892
13.32	7.473
14.38	9.691
15.84	1.366 x 10^{-3}
17.42	1.882
18.90	2.476
20.63	3.303
22.63	4.403
24.85	5.826
27.29	7.623
30.16	9.961
33.46	1.283 x 10^{-2}
37.06	1.620
41.11	2.009
45.71	2.457
50.67	2.950
Series III	
7.05	1.012 x 10^{-4}*
7.80	1.285*
8.48	1.808*
9.11	2.320*
9.75	2.855*

T	C_p
CURVE 1 (cont.)	
47.93	2.680 x 10^{-2}*
52.81	3.157*
58.03	3.666
63.80	4.224
69.89	4.779
76.19	5.341
83.16	5.952
90.82	6.572*
159.12	1.040 x 10^{-1}*
168.14	1.073*
177.57	1.105
187.20	1.134
196.73	1.161
206.16	1.185
215.88	1.208*
225.75	1.231
235.53	1.252*
245.35	1.272
255.25	1.291*
265.27	1.310
275.34	1.330*
285.33	1.349
294.91	1.366*
304.92	1.390
314.90	1.416*
324.74	1.442
334.71	1.487*
344.70	1.529
CURVE 2	
57.9	4.242 x 10^{-2}
60.1	4.344*
62.0	4.509
68.4	5.182
71.1	5.479*
74.5	5.823
79.4	6.328
93.7	7.408
105.7	8.432
113.5	9.113
119.9	9.551
130.4	1.017 x 10^{-1}
146.4	1.104

T	C_p
CURVE 2 (cont.)	
159.0	1.166 x 10^{-1}
173.1	1.222
188.2	1.259
204.6	1.301
221.2	1.340
231.6	1.356
244.6	1.373
258.0	1.408
275.8	1.455
296.0	1.493

* Not shown on plot

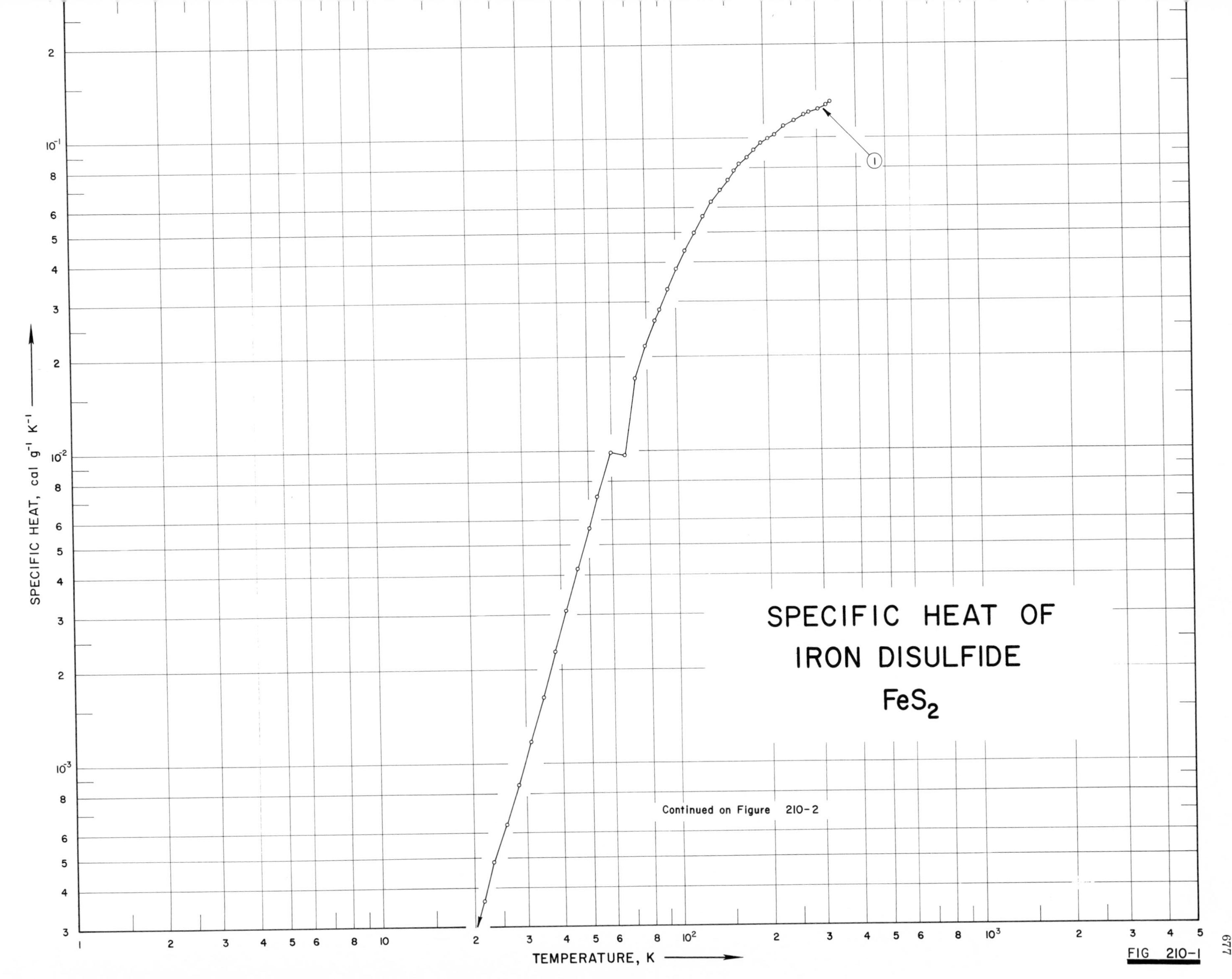
SPECIFIC HEAT OF
IRON DISULFIDE
FeS_2
Continued on Figure 210-2
1
SPECIFIC HEAT, cal g⁻¹ K⁻¹
TEMPERATURE, K
FIG 210-1

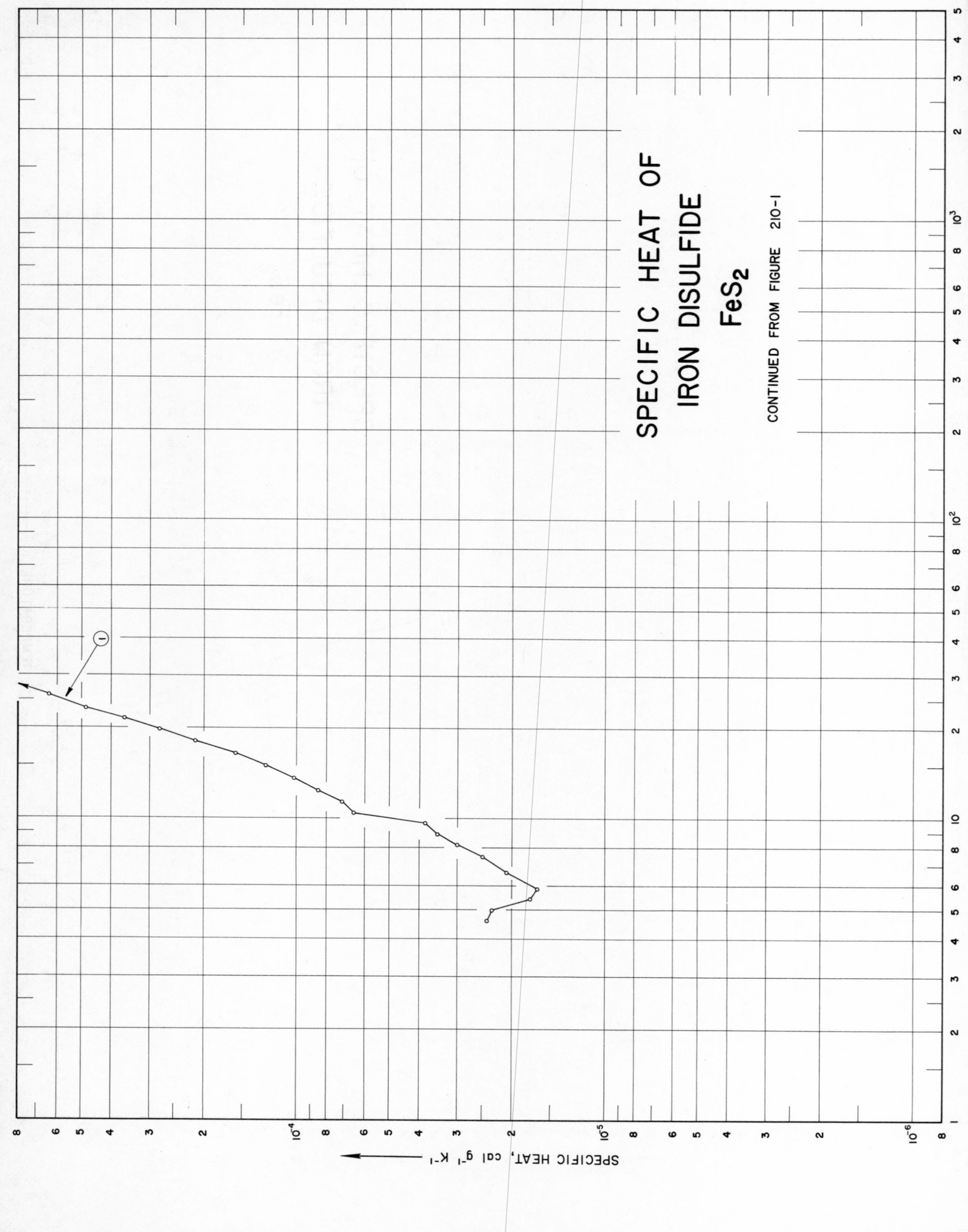

SPECIFIC HEAT OF
IRON DISULFIDE
FeS_2

CONTINUED FROM FIGURE 210-1

SPECIFIC HEAT, cal g^{-1} K^{-1} →

①

SPECIFICATION TABLE NO. 210 SPECIFIC HEAT OF IRON DISULFIDE FeS_2

[For Data Reported in Figure and Table No. 210]

Curve No.	Ref. No.	Year	Temp. Range, K	Reported Error, %	Name and Specimen Designation	Composition (weight percent), Specifications and Remarks
1	214	1962	5-346	0.1-5.0	Pyrite	53.45 ± 0.04 S, 46.53 ± 0.03 Fe, 0.008 each Mn, Si and 0.0075 Ni; sample supplied by Bosmo grube, Nordland, Norway; crushed to 30-80 mesh powder.

DATA TABLE NO. 210 SPECIFIC HEAT OF IRON DISULFIDE FeS_2

[Temperature, T, K; Specific Heat, C_p, Cal $g^{-1}K^{-1}$]

T	C_p
CURVE 1	
Series I	
88.33	2.841×10^{-2}
94.32	3.291
101.09	3.811
108.46	4.375
116.55	4.993
125.07	5.618
133.89	6.240
142.51	6.812
151.07	7.350
159.84	7.860
Series II	
4.60	2.417×10^{-5}
5.06	2.334
5.43	1.750
5.89	1.667
6.65	2.084
7.49	2.500
8.16	3.001
8.87	3.501
9.62	3.834
10.49	6.584
11.37	6.668
12.45	8.501
13.65	1.017×10^{-4}
14.98	1.267
16.46	1.584
18.06	2.142
19.77	2.792
21.56	3.642
23.56	4.842
25.78	6.376
28.28	8.501
31.23	1.174×10^{-3}
34.63	1.624
38.18	2.283
41.86	3.094
45.88	4.209
50.15	5.644
53.99	7.180
59.86	9.893
66.09	9.660
CURVE 1 (cont.)	
72.30	1.704×10^{-2}
79.05	2.160
85.37	2.618
Series III	
165.68	8.296×10^{-2}
175.23	8.667
185.37	9.147
195.90	9.602
206.03	9.995
215.79	1.035×10^{-1}
225.20	1.066*
234.29	1.092
243.43	1.118
252.56	1.142
261.95	1.166*
271.37	1.184
275.57	1.193*
284.53	1.214
293.71	1.230*
302.98	1.247
312.19	1.263
321.44	1.276
330.09	1.289*
338.12	1.299*
346.11	1.306

* Not shown on plot

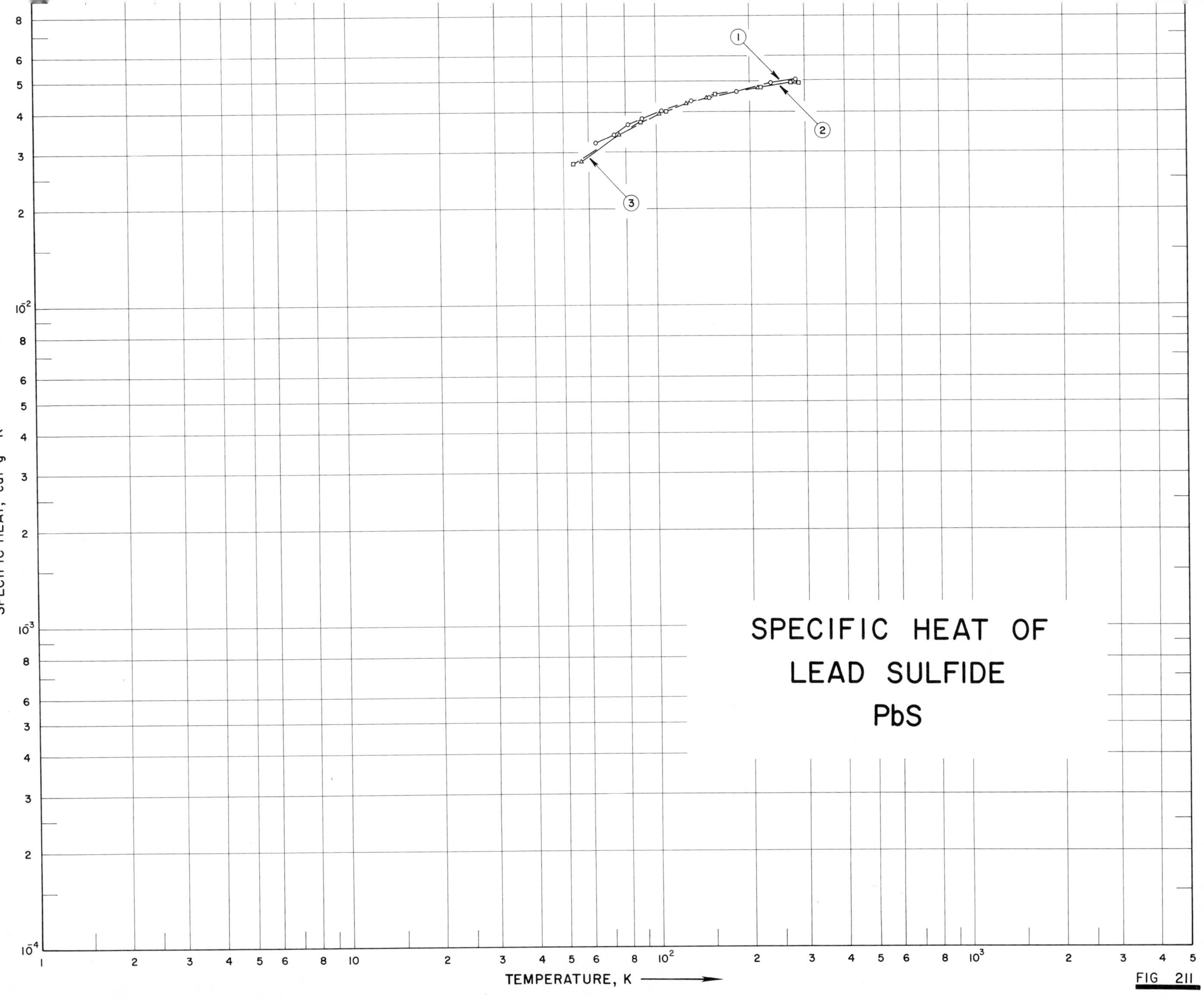
SPECIFIC HEAT OF
LEAD SULFIDE
PbS
SPECIFIC HEAT, cal g⁻¹ K⁻¹
TEMPERATURE, K
1
2
3
FIG 211

SPECIFICATION TABLE NO. 211 SPECIFIC HEAT OF LEAD SULFIDE PbS

[For Data Reported in Figure and Table No. 211]

Curve No.	Ref. No.	Year	Temp. Range, K	Reported Error, %	Name and Specimen Designation	Composition (weight percent), Specifications and Remarks
1	308	1918	64-283	< 1.0	Galena	
2	341	1932	54-289		Galena	Density = 7.57 g cm^{-3} at 22.4 C.
3	341	1932	54-282		Synthetic lead sulfide	Prepared by the carbon dioxide disulfide method; density = 7.57 g cm^{-3} at 22.4 C.

DATA TABLE NO. 211 SPECIFIC HEAT OF LEAD SULFIDE PbS

[Temperature, T, K; Specific Heat, C_p, Cal $g^{-1}K^{-1}$]

T	C_p
CURVE 1	
63.8	3.22 x 10^{-2}
66.6	3.33*
70.0	3.38*
73.1	3.49
75.9	3.51*
78.8	3.61*
81.7	3.68
84.6	3.71*
87.4	3.79*
90.1	3.81
96.9	3.95*
100.0	4.01*
105.3	4.05
109.2	4.05*
111.8	4.11*
114.3	4.13*
121.9	4.17*
124.2	4.221*
131.3	4.318
133.6	4.322*
142.3	4.447*
149.1	4.435
155.0	4.556*
161.7	4.531*
168.4	4.577*
175.3	4.698*
180.0	4.652*
182.3	4.669
196.6	4.765*
196.7	4.807*
197.6	4.786*
198.7	4.798*
235.0	4.932
237.0	4.957*
280.7	5.020*
282.7	5.082
CURVE 2	
53.7	2.769 x 10^{-2}
56.5	2.800*
63.0	3.067*
75.6	3.414*
89.2	3.741
108.1	4.027
CURVE 2 (cont.)	
138.2	4.393 x 10^{-2}*
156.5	4.522
172.1	4.698*
201.9	4.748*
218.3	4.798
232.7	4.819*
247.5	4.828*
271.3	4.915
280.4	4.924*
282.6	4.907*
289.4	4.911
CURVE 3	
54.0	2.729 x 10^{-2}*
57.3	2.826
61.1	2.958*
64.2	3.128*
76.2	3.430
103.8	3.975
126.4	4.280
147.4	4.464
215.7	4.794
233.7	4.807*
281.7	4.961

* Not shown on plot

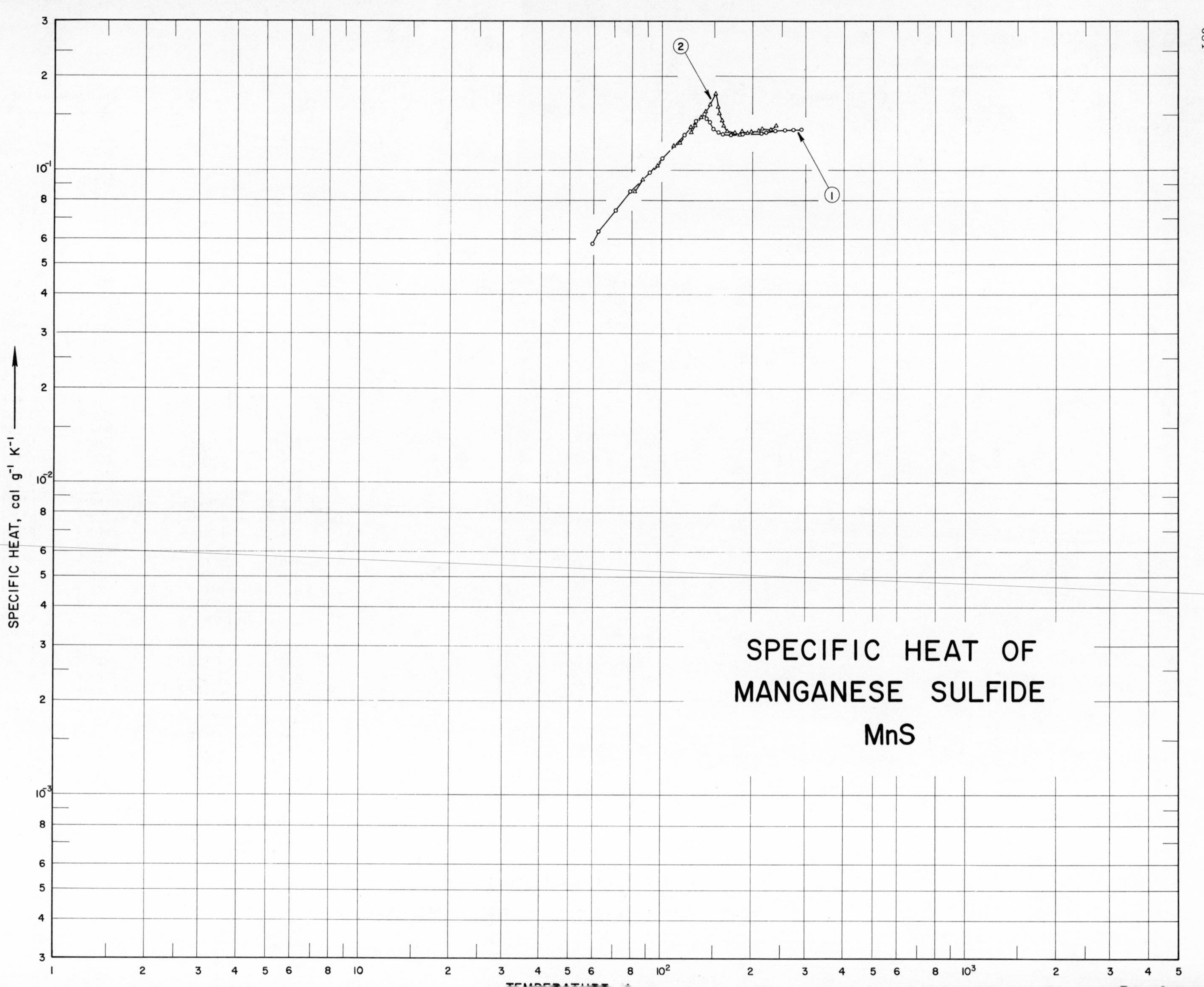
SPECIFIC HEAT OF
MANGANESE SULFIDE
MnS
SPECIFIC HEAT, cal g⁻¹ K⁻¹
1
2

SPECIFICATION TABLE NO. 212 SPECIFIC HEAT OF MANGANESE SULFIDE MnS

[For Data Reported in Figure and Table No. 212]

Curve No.	Ref. No.	Year	Temp. Range, K	Reported Error, %	Name and Specimen Designation	Composition (weight percent), Specifications and Remarks
1	340	1931	60-297			Prepared from C.P. grade $MnSO_4$; density = 3.93 g cm^{-3} at 21.9 C.
2	342	1966	82-240			0.07 O_2; powder sample pressed into 1/2 in.dia pellet using 5K bar pressure; density = 87% of single crystal density.

DATA TABLE NO. 212 SPECIFIC HEAT OF MANGANESE SULFIDE MnS

[Temperature, T,K; Specific Heat, C_p, Cal $g^{-1}K^{-1}$]

T	C_p
CURVE 1	
Series I	
59.8	5.826×10^{-2}
62.3	6.388
71.0	7.455
79.1	8.584
91.8	9.805
102.0	1.094×10^{-1}
109.5	1.190*
119.1	1.300
131.3	1.443
145.7	1.435
291.4	1.359
Series II	
124.6	1.352×10^{-1}
128.5	1.385*
131.2	1.435*
132.8	1.451*
134.3	1.476*
135.7	1.493*
137.1	1.509*
138.4	1.524
139.8	1.527*
141.1	1.490*
141.6	1.474
142.9	1.454*
144.2	1.444*
145.5	1.436*
146.8	1.433
148.8	1.368
150.7	1.340*
151.9	1.340*
153.2	1.323*
155.0	1.317*
157.0	1.314*
158.2	1.304
215.3	1.314
Series III	
60.7	5.936×10^{-2}*
69.8	7.238*
82.8	9.032*
109.3	1.194×10^{-1}*
CURVE 1 (cont.)	
121.3	1.323×10^{-1}*
129.3	1.412*
138.0	1.521*
139.4	1.520*
140.7	1.500*
143.0	1.463*
144.2	1.451*
145.4	1.442*
146.7	1.440*
147.9	1.389*
149.2	1.362*
150.4	1.346*
151.7	1.336*
Series IV	
133.2	1.463×10^{-1}*
138.2	1.529*
139.4	1.531*
139.4	1.521*
147.2	1.432*
151.5	1.337*
152.7	1.331*
153.9	1.323*
155.1	1.320
156.3	1.317*
157.3	1.313*
158.4	1.315*
159.6	1.315*
160.7	1.306*
161.8	1.310*
163.0	1.312*
165.2	1.309*
166.3	1.308*
167.4	1.310*
168.5	1.316*
169.2	1.308*
170.1	1.306
174.3	1.302*
182.4	1.301*
187.4	1.304
193.1	1.308*
199.3	1.308*
212.9	1.322*
216.8	1.324*
220.5	1.323*
CURVE 1 (cont.)	
Series V	
218.9	1.325×10^{-1}*
222.7	1.328
226.2	1.325*
230.3	1.327*
232.5	1.329*
236.0	1.331*
239.5	1.333
246.0	1.338*
257.5	1.353
262.8	1.355*
268.0	1.354*
273.0	1.356
278.1	1.360*
283.4	1.366*
289.6	1.370*
295.3	1.370*
Series VI	
175.7	1.302×10^{-1}*
195.9	1.309*
204.4	1.314*
296.9	1.371*
CURVE 2	
81.6	8.549×10^{-2}
86.9	9.369
92.2	9.868*
97.0	1.047×10^{-1}
110.6	1.204
110.7	1.175*
113.3	1.233*
114.0	1.215*
114.6	1.222*
116.5	1.234
118.5	1.278*
118.6	1.266*
119.3	1.293*
120.6	1.293*
122.1	1.313*
122.2	1.301*
122.8	1.324*
125.4	1.376
CURVE 2 (cont.)	
125.5	1.334×10^{-1}
125.6	1.347*
127.7	1.382*
129.0	1.401
131.0	1.432*
131.8	1.441*
132.2	1.430*
133.6	1.461*
133.6	1.467*
135.4	1.483
135.5	1.497*
136.2	1.487
136.6	1.513*
137.4	1.511*
138.5	1.517*
139.1	1.537*
139.8	1.523*
140.9	1.558
142.1	1.560*
142.8	1.595*
143.1	1.556*
144.7	1.623*
145.0	1.630
146.4	1.620*
146.6	1.611*
147.2	1.651*
148.3	1.697*
148.9	1.739*
149.5	1.686*
150.0	1.737*
150.1	1.745*
151.0	1.753*
151.4	1.768
151.8	1.669*
152.5	1.714*
152.5	1.717*
152.8	1.667*
153.6	1.643*
153.8	1.625*
154.1	1.608
154.7	1.615*
154.7	1.576*
155.9	1.529
156.0	1.545*
156.2	1.507*
CURVE 2 (cont.)	
156.6	1.500×10^{-1}*
157.1	1.522*
157.2	1.462*
158.2	1.450
158.4	1.458*
158.5	1.438*
159.0	1.412*
160.4	1.460*
160.4	1.407*
161.4	1.391
162.9	1.384*
163.9	1.361*
166.4	1.344
167.6	1.362*
169.0	1.333*
171.5	1.345*
171.6	1.349*
175.5	1.328*
179.1	1.317*
182.2	1.305
185.2	1.334
193.3	1.316
198.0	1.330
211.3	1.341
214.2	1.340*
216.7	1.362
228.0	1.358*
228.8	1.366*
230.3	1.350
232.3	1.354*
232.8	1.358*
236.6	1.354*
240.1	1.398

* Not shown on plot

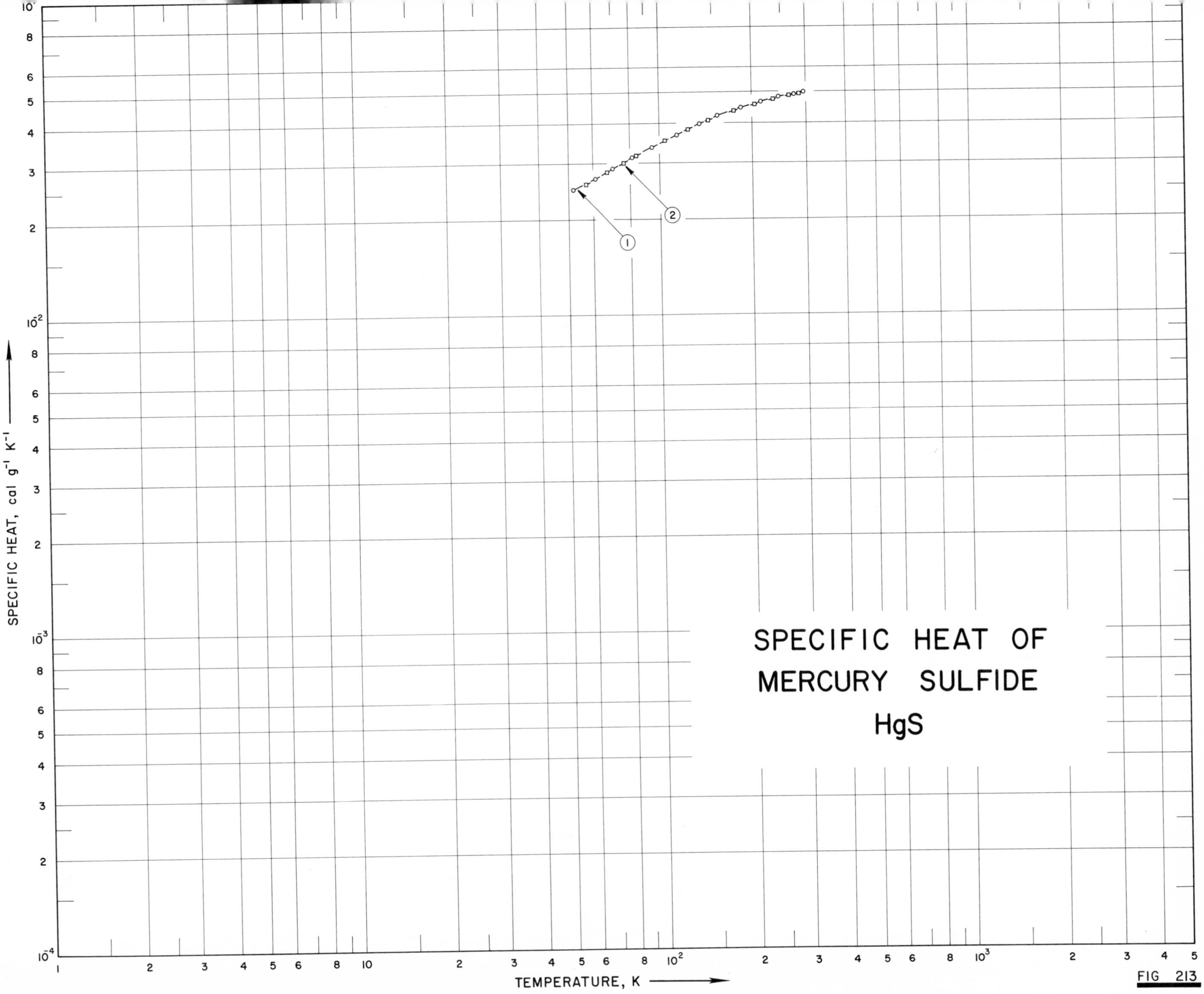

SPECIFIC HEAT OF MERCURY SULFIDE HgS

FIG 213

SPECIFICATION TABLE NO. 213 SPECIFIC HEAT OF MERCURY SULFIDE HgS

[For Data Reported in Figure and Table No. 213]

Curve No.	Ref. No.	Year	Temp. Range, K	Reported Error, %	Name and Specimen Designation	Composition (weight percent), Specifications and Remarks
1	343	1962	52-297		Sample A; Cinnabar	86.16 Hg and 13.82 S (86.22 and 13.78 theo).
2	343	1962	53-297		Sample B	99.9 Hg S.

DATA TABLE NO. 213 SPECIFIC HEAT OF MERCURY SULFIDE HgS

[Temperature, T, K; Specific Heat, C_p, Cal $g^{-1}K^{-1}$]

T	C_p
CURVE 1	
52.16	2.473 x 10^{-2}
56.76	2.576*
61.27	2.681
65.51	2.779*
70.27	2.886
75.31	2.988*
81.60	3.114
86.19	3.198*
94.98	3.358
105.69	3.544*
114.95	3.684
124.96	3.828*
136.22	3.974
146.03	4.097*
155.92	4.202
165.99	4.293*
176.23	4.379*
186.13	4.461
196.24	4.521*
206.38	4.590*
216.25	4.654
226.16	4.710*
236.21	4.766*
245.87	4.835
256.74	4.899*
266.71	4.904*
276.52	4.908
286.83	4.951*
296.51	4.994
CURVE 2	
52.61	2.474 x 10^{-2}*
57.21	2.576
62.16	2.695*
67.03	2.806
71.85	2.913*
76.55	3.008
79.90	3.074*
84.08	3.155
94.55	3.348*
104.83	3.525
114.52	3.674*
124.93	3.823
135.84	3.963*
CURVE 2 (cont.)	
145.79	4.087 x 10^{-2}
155.96	4.192*
166.07	4.286*
176.14	4.371
186.03	4.448*
196.06	4.508*
206.15	4.581
216.25	4.637*
226.42	4.697*
236.20	4.745
245.89	4.788*
256.63	4.835*
266.69	4.878
276.54	4.899*
286.87	4.934
296.75	4.977*

* Not shown on plot

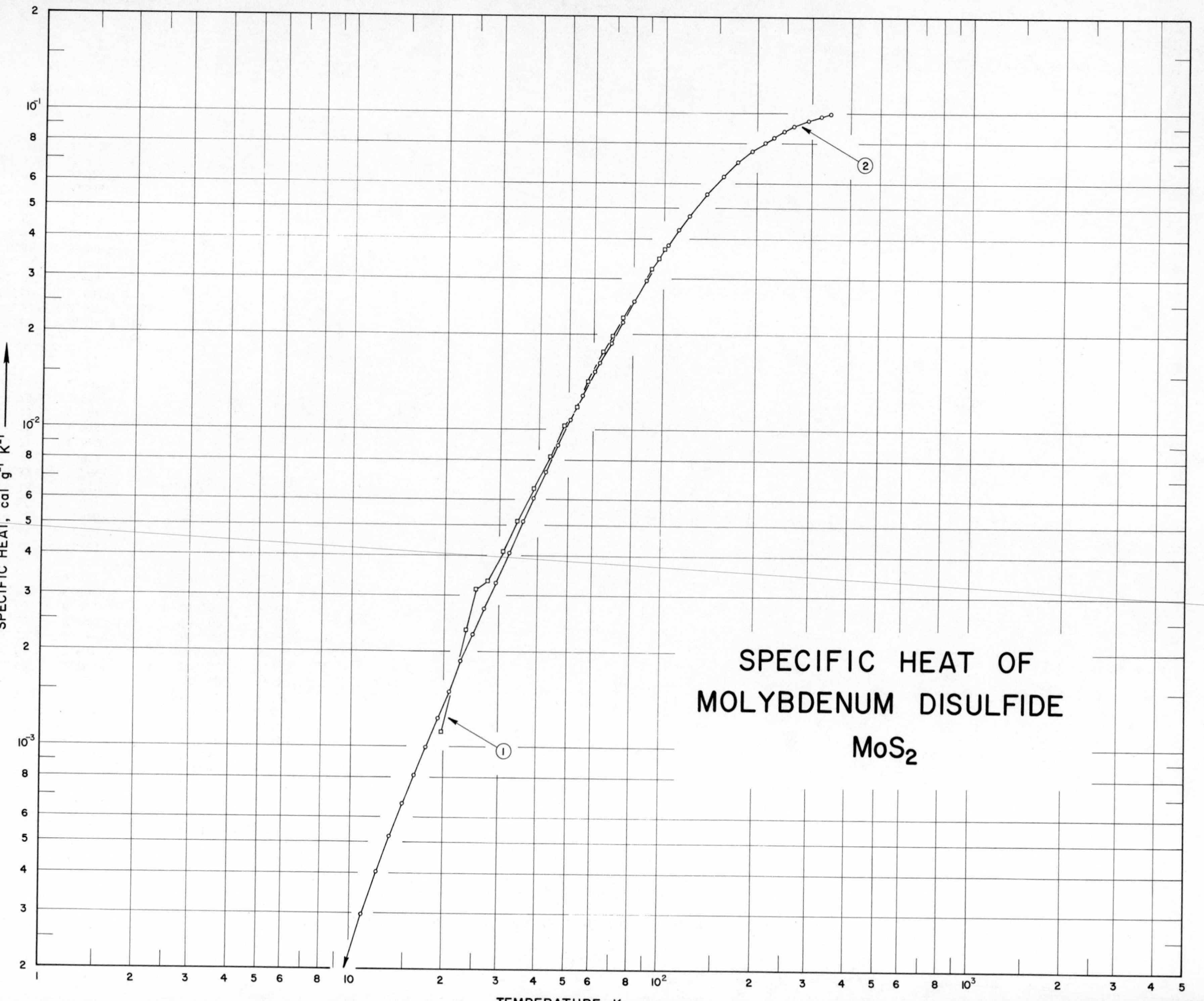
SPECIFIC HEAT OF
MOLYBDENUM DISULFIDE
MoS2
SPECIFIC HEAT, cal g-1 K-1
TEMPERATURE, K
1
2

SPECIFICATION TABLE NO. 214 SPECIFIC HEAT OF MOLYBDENUM DISULFIDE MoS_2

[For Data Reported in Figure and Table No. 214]

Curve No.	Ref. No.	Year	Temp. Range, K	Reported Error, %	Name and Specimen Designation	Composition (weight percent), Specifications and Remarks
1	108	1956	20-102			Contains some free carbon and oil.
2	327	1955	6-346	0.15-6.0		59.8 Mo, 40.3 S (59.94 and 40.06 theo), 0.1 gangue and 0.0 ± 0.1 C; sample obtained by purifying natural molybdenite from Lyndock Township, Ontario, Canada.

DATA TABLE NO. 214 SPECIFIC HEAT OF MOLYBDENUM DISULFIDE MoS_2

[Temperature, T, K; Specific Heat, C_p, Cal $g^{-1}K^{-1}$]

T	C_p	T	C_p
CURVE 1		**CURVE 2 (cont.)**	
19.69	1.118 x 10^{-3}	216.29	8.140 x 10^{-2}*
23.55	2.343	226.04	8.358
25.37	3.124	235.80	8.552*
27.53	3.330	245.46	8.739
30.88	4.117	255.14	8.902*
34.18	5.142	264.98	9.058
38.63	6.510	274.92	9.195*
43.71	8.228	284.89	9.339*
48.58	1.028 x 10^{-2}	294.94	9.439
53.43	1.175	305.13	9.564*
55.32	1.313*	315.29	9.683*
57.66	1.424	325.43	9.782
59.44	1.505*	335.58	9.876*
64.56	1.760	345.67	9.976
69.06	1.975		
74.58	2.253	Series II	
80.96	2.594*		
86.68	2.917*	5.74	3.623 x 10^{-5}*
91.99	3.212	6.92	6.184*
96.94	3.474	7.87	9.745*
101.54	3.727	8.88	1.487 x 10^{-4}*
		9.89	2.149*
CURVE 2		10.98	2.967
Series I		12.20	4.029
		13.51	5.254
62.80	1.622 x 10^{-2}	14.80	6.622
68.23	1.887	16.12	8.121
74.32	2.194	17.55	9.932
80.93	2.548	19.13	1.224 x 10^{-3}
88.29	2.954	20.82	1.493
96.01	3.376*	22.79	1.862
104.35	3.826	24.86	2.261
112.93	4.285	26.93	2.717
121.24	4.718	29.24	3.286
129.59	5.135*	32.21	4.092
138.38	5.538	35.64	5.122
146.77	5.923*	38.52	6.072
155.99	6.303	42.21	7.371
165.33	6.659*	46.30	8.952
174.75	6.990	50.77	1.078 x 10^{-2}
184.21	7.290*	55.51	1.284
193.47	7.571	60.86	1.535
202.89	7.815*	60.84	1.524*
212.34	8.046		

* Not shown on plot

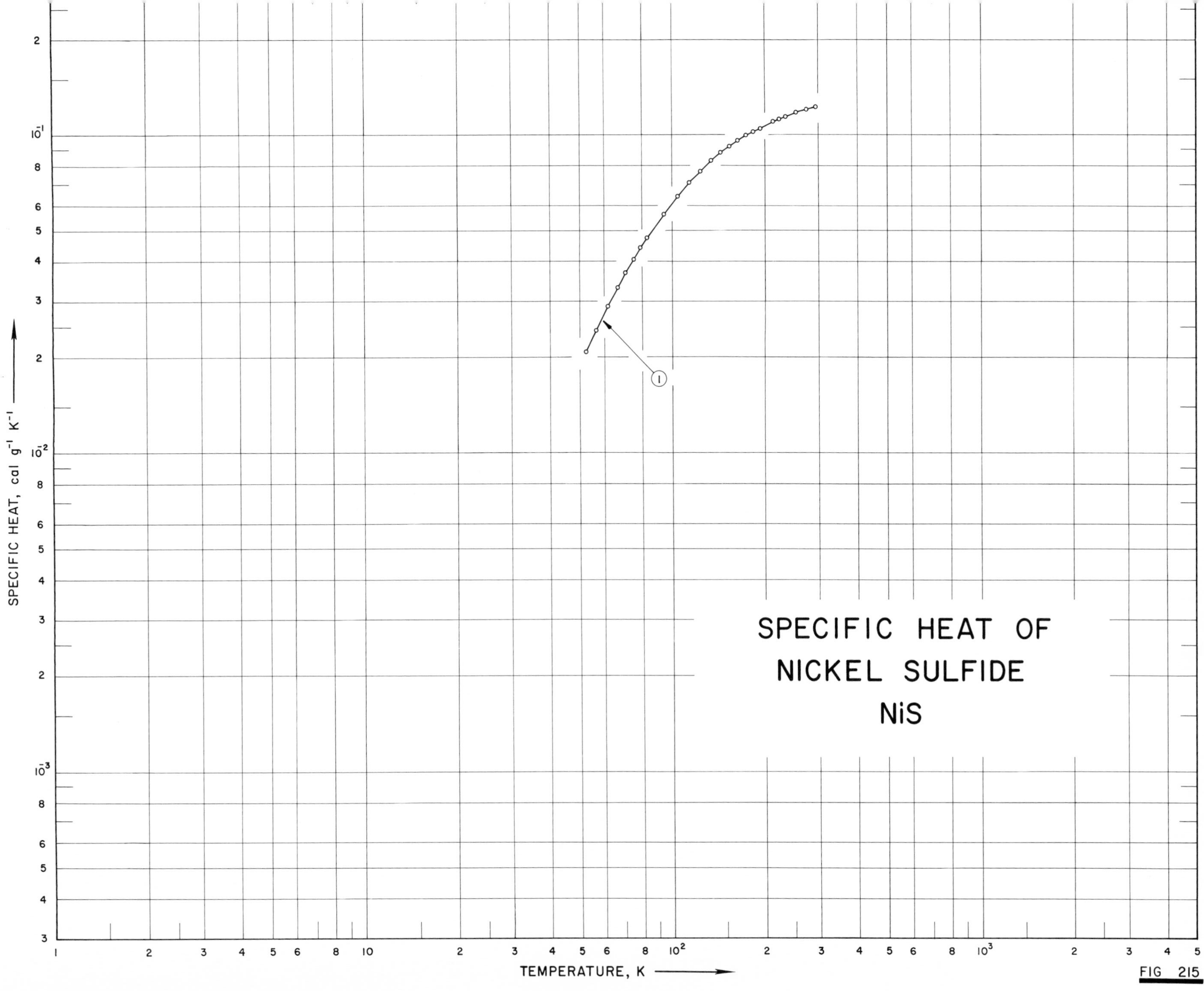
SPECIFIC HEAT OF
NICKEL SULFIDE
NiS
SPECIFIC HEAT, cal g⁻¹ K⁻¹
TEMPERATURE, K
1
FIG 215

SPECIFICATION TABLE NO. 215 SPECIFIC HEAT OF NICKEL SULFIDE NiS

[For Data Reported in Figure and Table No. 215]

Curve No.	Ref. No.	Year	Temp. Range, K	Reported Error, %	Name and Specimen Designation	Composition (weight percent), Specifications and Remarks
1	339	1964	52-296	≤ 0.3		64.56 Ni (64.68 theo) and 0.12 acid insoluble material; prepared from pure nickel oxide, and sulfur at temperatures between 530-580 C for 3.5 weeks; heated in vacuum for 3 hrs at 400 C and then slowly cooled to room temperature.

DATA TABLE NO. 215 SPECIFIC HEAT OF NICKEL SULFIDE NiS

[Temperature, T, K; Specific Heat, C_p, Cal $g^{-1}K^{-1}$]

T	C_p
CURVE 1	
52.44	2.099 x 10^{-2}
56.51	2.451
61.68	2.918
66.56	3.348
70.76	3.715
75.07	4.097
79.20	4.456
83.19	4.788
94.51	5.687
105.04	6.471
114.62	7.119
124.90	7.745
136.02	8.358
145.96	8.850
155.71	9.266
165.95	9.667
175.82	1.001 x 10^{-1}
185.95	1.032
196.15	1.061
216.40	1.110
226.22	1.131
236.33	1.150
245.76	1.169*
256.70	1.187
266.52	1.203*
276.51	1.215
286.74	1.225*
296.27	1.239

* Not shown on plot

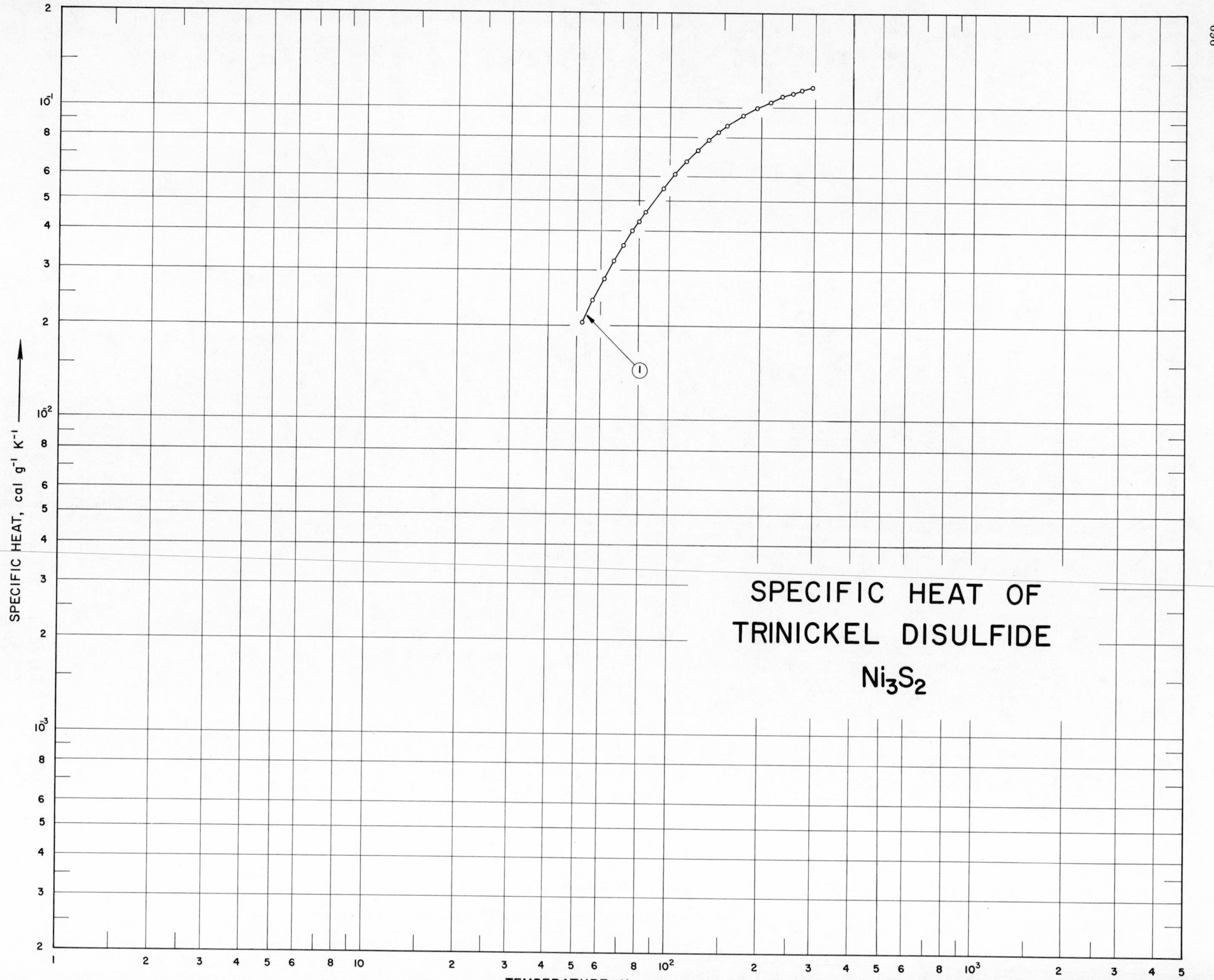

SPECIFIC HEAT OF
TRINICKEL DISULFIDE
Ni_3S_2
SPECIFIC HEAT, cal g^{-1} K^{-1}
1

SPECIFICATION TABLE NO. 216 SPECIFIC HEAT OF TRINICKEL DISULFIDE Ni_3S_2

[For Data Reported in Figure and Table No. 216]

Curve No.	Ref. No.	Year	Temp. Range, K	Reported Error, %	Name and Specimen Designation	Composition (weight percent), Specifications and Remarks
1	339	1964	53-297	≤ 0.3		73.14 Ni (73.31 theo); prepared from pure nickel oxide and sulfur by repeatedly heating the mixture at 200-525 C until sulfur content of product is slightly lower than theoretical value.

DATA TABLE NO. 216 SPECIFIC HEAT OF TRINICKEL DISULFIDE Ni_3S_2

[Temperature, T, K; Specific Heat, C_p, Cal $g^{-1}K^{-1}$]

T	C_p
CURVE 1	
52.52	2.069 x 10^{-2}
56.95	2.424
61.88	2.839
66.60	3.229
71.48	3.619
76.24	4.010
80.04	4.308
84.39	4.637
96.08	5.507
105.38	6.143
114.92	6.743
124.79	7.309
136.03	7.891
146.22	8.353
155.91	8.757
165.94	9.123*
176.06	9.452
185.88	9.739*
195.79	9.993
216.46	1.048 x 10^{-1}
226.26	1.066*
236.31	1.085
245.79	1.101*
256.29	1.116
266.92	1.132*
273.31	1.144
286.55	1.157*
296.54	1.169

* Not shown on plot

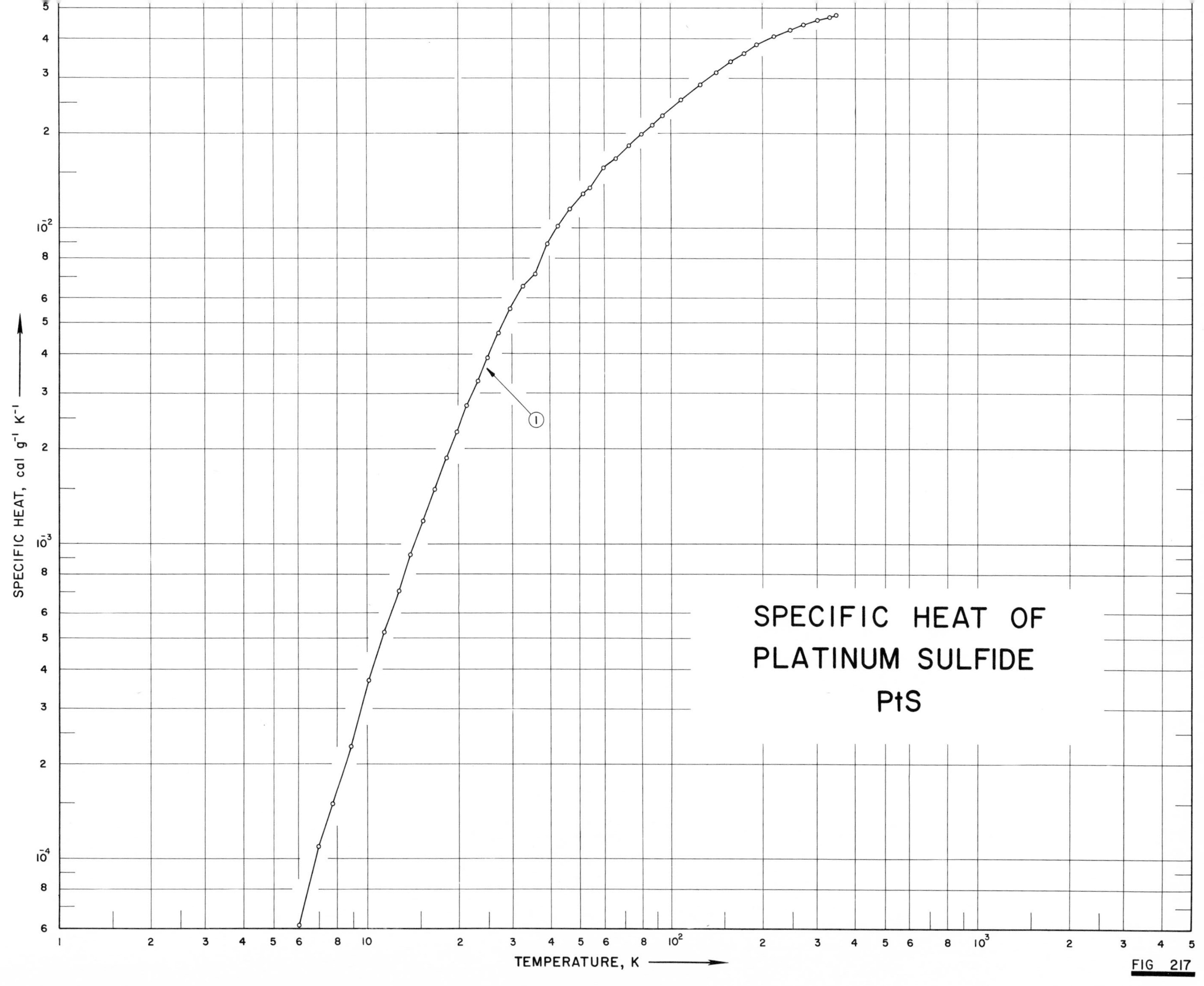

SPECIFIC HEAT OF
PLATINUM SULFIDE
PtS
SPECIFIC HEAT, cal g^{-1} K^{-1}
TEMPERATURE, K
1
FIG 217

SPECIFICATION TABLE NO. 217 SPECIFIC HEAT OF PLATINUM SULFIDE PtS

[For Data Reported in Figure and Table No. 217]

Curve No.	Ref. No.	Year	Temp. Range, K	Reported Error, %	Name and Specimen Designation	Composition (weight percent), Specifications and Remarks
1	223	1961	6-348	~0.1		~100 PtS, 0.009 volatile material, < 0.001 Fe, Pb, 0.0007 Pd, and 0.0001 Au; Pts synthesized in 2 steps; prepared by heating appropriate amount of elements in evacuated and sealed silica tube at 750 C for one day; sintered product crushed and heated with stoichiometric finely divided platinum at 900 C for 2 days; resulting dark grey powder was annealed at 500 C for 2 days and cooled to room temperature over a period of 7 days; density = 624 16 ft^{-3}.

DATA TABLE NO. 217 SPECIFIC HEAT OF PLATINUM SULFIDE PtS

[Temperature, T, K; Specific Heat, C_p, Cal $g^{-1}K^{-1}$]

T	C_p
CURVE 1	
Series I	
54.00	1.349×10^{-2}
59.79	1.558
65.98	1.662
73.12	1.824
80.35	1.981
87.24	2.127
94.52	2.274
Series II	
6.01	6.163×10^{-5}
6.99	1.096×10^{-4}
7.77	1.492
8.80	2.285
10.11	3.693
11.47	5.230
12.74	7.087
13.98	9.201
15.32	1.177×10^{-3}
16.74	1.497
18.22	1.861
19.70	2.259
21.32	2.736
23.11	3.285
24.99	3.896
27.16	4.644
29.67	5.542
37.54	6.577
35.75	7.198
39.16	8.919
42.88	1.012×10^{-2}
46.97	1.150
51.33	1.277
Series III	
94.19	2.267×10^{-2}*
101.10	2.402*
108.84	2.551
117.40	2.710*
125.65	2.857
132.91	2.985*
141.06	3.118
149.44	3.249*

T	C_p
CURVE 1 (cont.)	
157.89	3.374×10^{-2}
166.37	3.491*
174.93	3.596
183.61	3.688*
192.20	3.801
201.20	3.892*
210.37	3.972*
219.30	4.054
228.33	4.127*
237.28	4.200*
246.33	4.271
255.30	4.325*
263.94	4.385*
272.49	4.431
260.99	4.352*
269.64	4.412*
278.25	4.462*
286.71	4.511*
295.57	4.606*
304.40	4.597
313.45	4.642*
322.60	4.671*
331.86	4.699
341.21	4.734*
347.63	4.763

* Not shown on plot

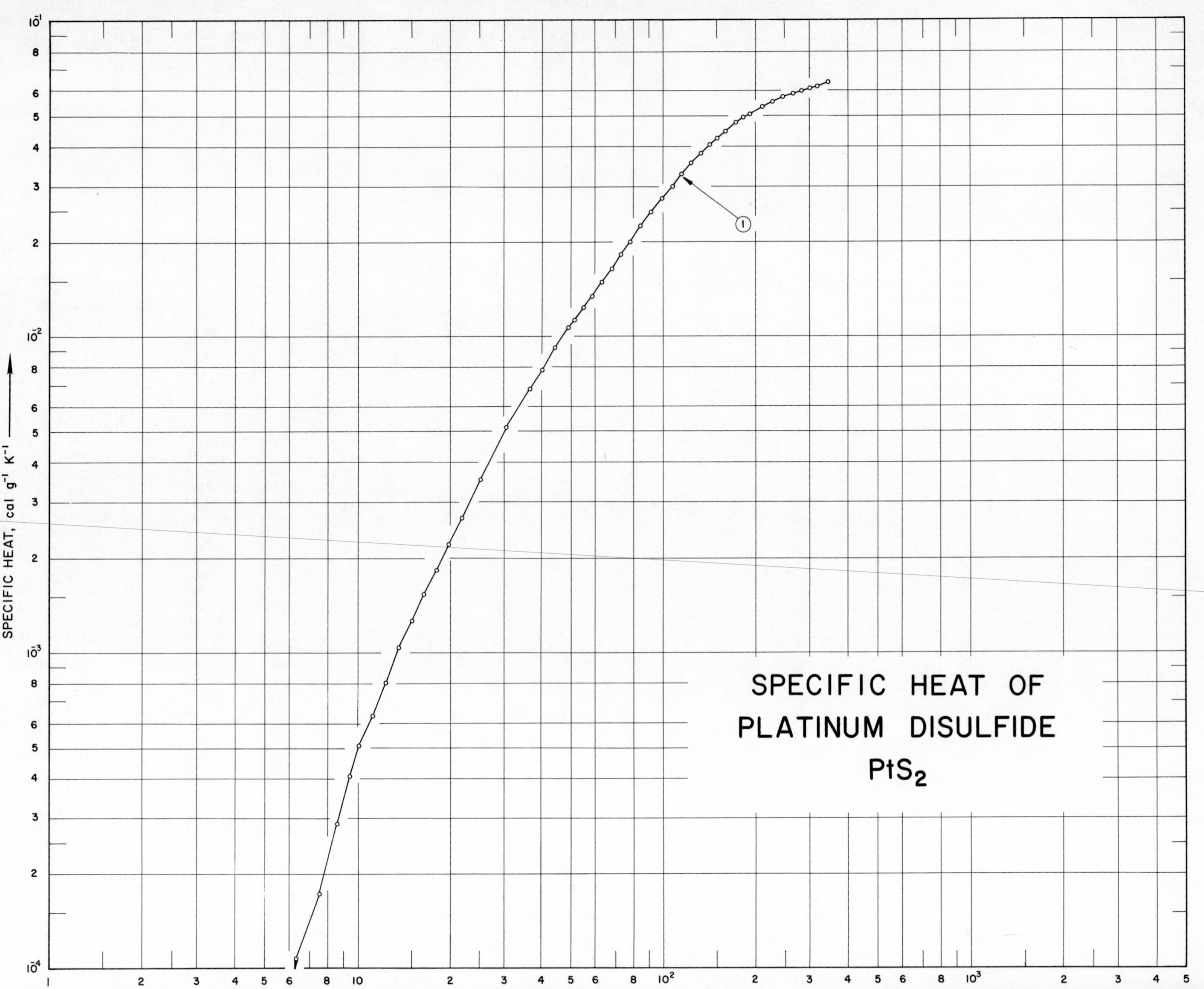
SPECIFIC HEAT OF
PLATINUM DISULFIDE
PtS2
SPECIFIC HEAT, cal g-1 K-1
1

SPECIFICATION TABLE NO. 218 SPECIFIC HEAT OF PLATINUM DISULFIDE PtS_2

[For Data Reported in Figure and Table No. 218]

Curve No.	Ref. No.	Year	Temp. Range, K	Reported Error, %	Name and Specimen Designation	Composition (weight percent), Specifications and Remarks
1	224	1961	5-347	~0.1		~99.0 PtS_2, high purity, < 0.090 volatile material. < 0.001 Fe, Pb, 0.0007 Pd, and 0.0001 Au; synthesized from stoichiometric amounts of the elements in an evacuated and sealed silica tube by heating to 750 C for one day; annealed at 500 C for several days; cooled slowly to room temperature over another 7 days; measured in a helium atmosphere.

DATA TABLE NO. 218 SPECIFIC HEAT OF PLATINUM DISULFIDE PtS_2

[Temperature, T, K; Specific Heat, C_p, Cal $g^{-1}K^{-1}$]

CURVE 1

Series I

T	C_p
78.16	1.982×10^{-2}
84.65	2.221
91.96	2.476
99.50	2.735
107.27	2.990
115.89	3.269
124.78	3.549
133.27	3.795
142.09	4.030
151.26	4.254
160.30	4.454
165.57	4.558*
174.98	4.744
184.41	4.911
193.23	5.057
202.13	5.176*
211.12	5.304
220.12	5.412*
229.28	5.516
238.66	5.612*
247.99	5.703
257.14	5.787*
266.19	5.844
275.17	5.898*
284.07	5.979
292.96	6.041*
301.95	6.099
311.11	6.157*
320.57	6.196
330.17	6.238*
339.37	6.292*
347.31	6.327

Series II

T	C_p
51.46	1.123×10^{-2}
55.22	1.236
58.99	1.342
63.28	1.489
68.26	1.649
73.24	1.812

CURVE 1 (cont.)

Series III

T	C_p
10.17	5.092×10^{-4}
11.21	6.327
12.38	8.063
13.68	1.038×10^{-3}
15.06	1.265
16.52	1.520
18.11	1.832
19.89	2.207
21.90	2.673

Series IV

T	C_p
5.42	3.472×10^{-5}
6.33	1.080×10^{-4}
7.53	1.736
8.61	2.893
9.42	4.089
10.30	5.401*
11.26	6.365*
12.37	7.947*
13.65	1.030×10^{-3}*
19.12	2.037*
20.79	2.411*
22.97	2.955*
25.28	3.549
27.84	4.251*
30.54	5.019
33.45	5.871*
36.68	6.824
40.14	7.881
44.80	9.262
49.56	1.069×10^{-2}
54.33	1.213*
54.28	1.209*
59.20	1.362*
64.36	1.528*
70.93	1.741*
78.52	2.001*
86.06	2.271*
98.86	2.557*

CURVE 1 (cont.)

Series V

T	C_p
262.55	5.833×10^{-2}*
262.07	5.783*
271.08	5.852*
280.06	5.960*

* Not shown on plot

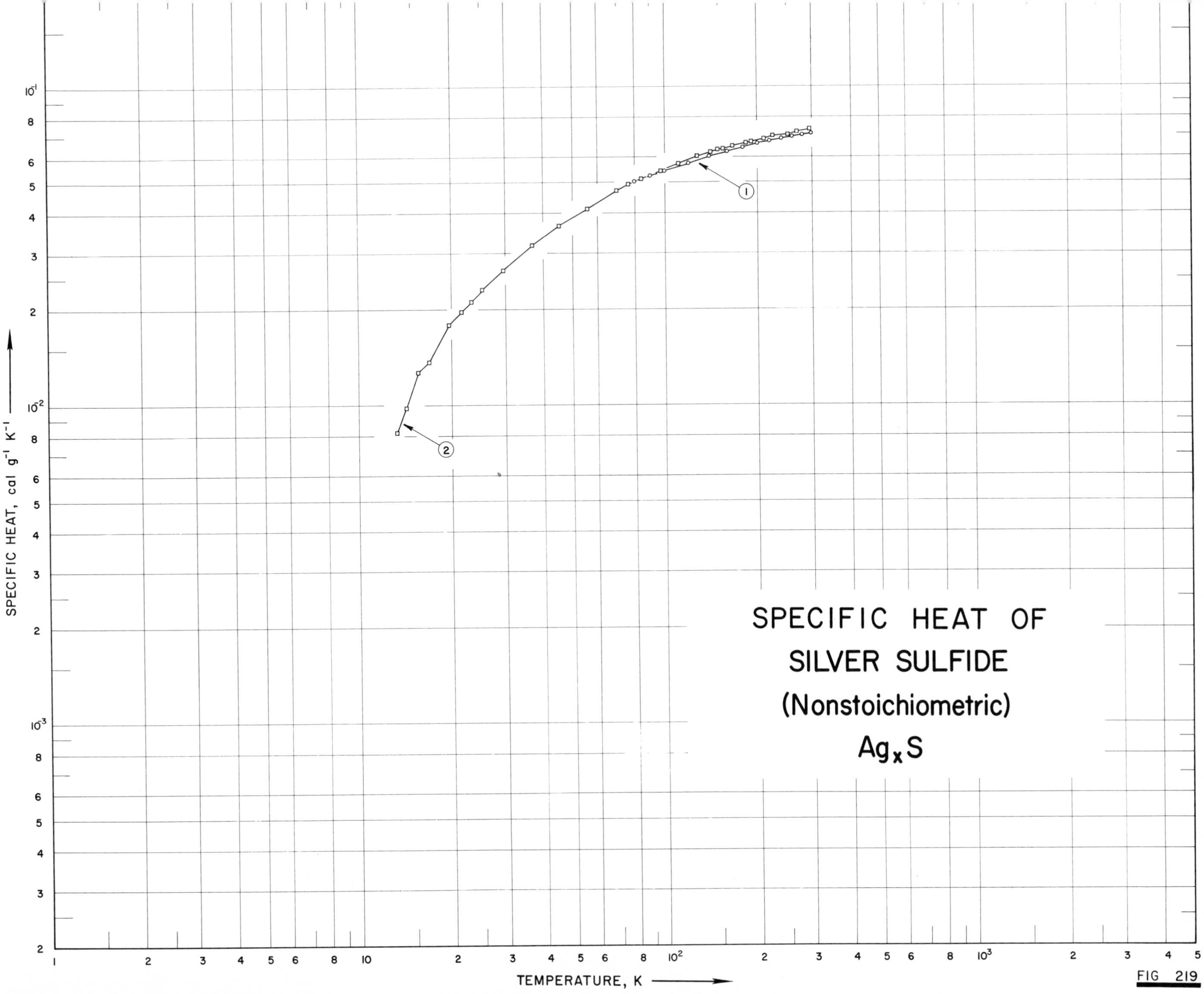

SPECIFIC HEAT OF SILVER SULFIDE (Nonstoichiometric) Ag_xS

SPECIFIC HEAT, cal g^{-1} K^{-1}

TEMPERATURE, K

FIG 219

SPECIFICATION TABLE NO. 219 SPECIFIC HEAT OF SILVER SULFIDE (nonstoichiometric) Ag_xS

[For Data Reported in Figure and Table No. 219]

Curve No.	Ref. No.	Year	Temp. Range, K	Reported Error, %	Name and Specimen Designation	Composition (weight percent), Specifications and Remarks
1	160	1959	80-300	3.0-7.0		Ag_2S; polycrystalline.
2	225	1962	13-296			99.99 $Ag_{1.99}S$; crushed under argon atmosphere.

DATA TABLE NO. 219 SPECIFIC HEAT OF SILVER SULFIDE (nonistoichiometric) Ag_xS

[Temperature, T, K; Specific Heat, C_p, Cal $g^{-1}K^{-1}$]

T	C_p
CURVE 1	
80	5.049 x 10^{-2}
90	5.277
100	5.460
120	5.799
140	6.054
160	6.270
180	6.465
200	6.633
220	6.756
240	6.876
260	6.972
280	7.068
300	7.155
CURVE 2	
13.23	8.267 x 10^{-3}
14.14	9.848
15.56	1.285 x 10^{-2}
16.95	1.378
19.60	1.807
21.58	1.974
23.17	2.120
23.91	2.225*
25.14	2.310
27.26	2.488*
28.22	2.590*
29.67	2.663
32.87	2.910*
36.62	3.194
40.85	3.457*
45.56	3.668
55.91	4.138
61.58	4.409*
69.85	4.738
72.16	4.819*
76.86	4.952
80.78	5.042*
84.38	5.147
86.74	5.236*
91.04	5.398*
97.29	5.455
97.41	5.568*
112.46	5.751
123.49	5.994*

T	C_p
CURVE 2 (cont.)	
128.81	6.075 x 10^{-2}
135.47	6.132*
141.10	6.213
148.96	6.229*
149.41	6.338
155.47	6.387
161.71	6.464*
167.65	6.517
177.17	6.565*
184.62	6.650
192.46	6.736
200.15	6.768*
210.18	6.885
217.08	6.946*
224.87	7.003
251.00	7.088
245.33	7.124*
254.35	7.157*
260.17	7.181*
268.66	7.234
282.40	7.299*
296.15	7.372

* Not shown on plot

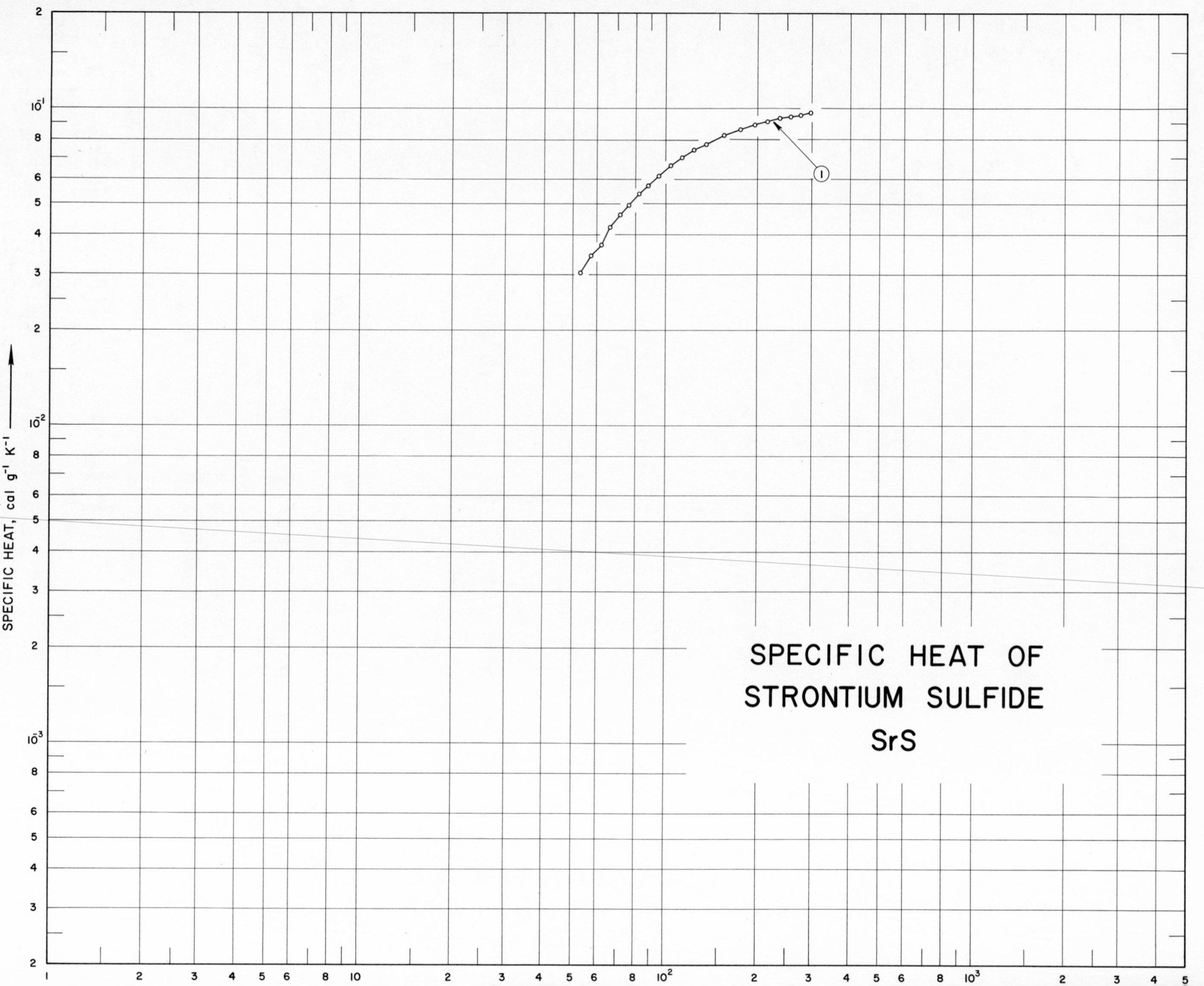
SPECIFIC HEAT OF
STRONTIUM SULFIDE
SrS
SPECIFIC HEAT, cal g⁻¹ K⁻¹
1

SPECIFICATION TABLE NO. 220 SPECIFIC HEAT OF STRONTIUM SULFIDE SrS

[For Data Reported in Figure and Table No. 220]

Curve No.	Ref. No.	Year	Temp. Range, K	Reported Error, %	Name and Specimen Designation	Composition (weight percent), Specifications and Remarks
1	305	1960	53-298			99.16 SrS, 0.83 $SrSO_4$, 0.01 SiO_2; prepared from reagent grade strontium carbonate and HCl; heated in a stream of pure hydrogen at 1000 C for 4-5 hrs.

DATA TABLE NO. 220 SPECIFIC HEAT OF STRONTIUM SULFIDE SrS

[Temperature, T, K; Specific Heat, C_p, Cal $g^{-1}K^{-1}$]

T	C_p
CURVE 1	
53.25	3.054 x 10^{-2}
57.72	3.450
62.24	3.741
66.86	4.230
71.74	4.617
76.67	4.988
82.82	5.404
88.05	5.733
95.77	6.171
105.19	6.649
114.78	7.045
124.97	7.425
136.74	7.784
146.15	8.049*
156.51	8.289
166.41	8.480*
176.53	8.655
186.23	8.789*
196.40	8.906
206.37	9.040*
216.54	9.148
226.20	9.240*
236.18	9.324
246.33	9.390*
256.83	9.466
266.40	9.541*
276.90	9.599
286.79	9.658*
296.27	9.725*
298.15	9.725

* Not shown on plot

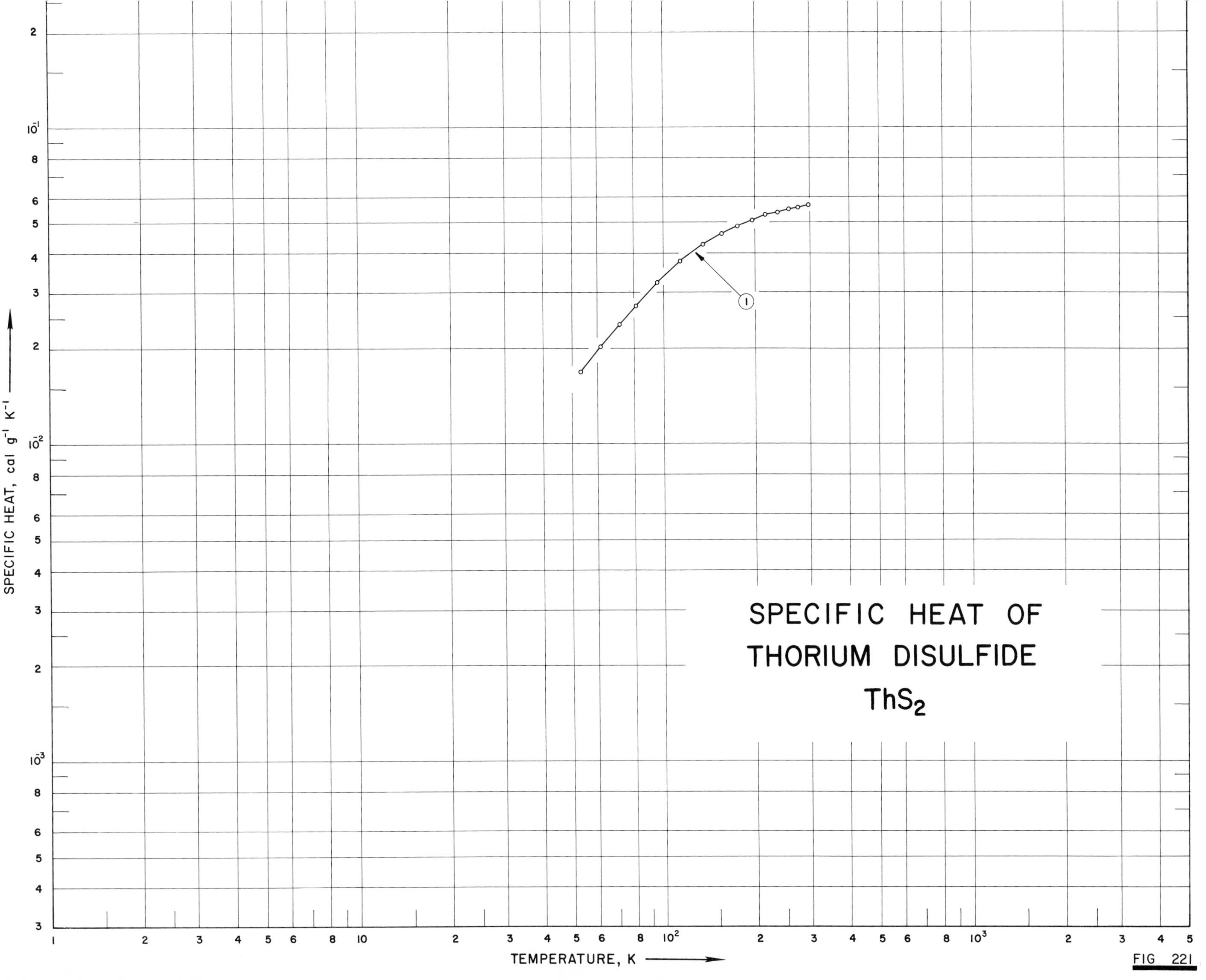
SPECIFIC HEAT OF
THORIUM DISULFIDE
ThS_2
SPECIFIC HEAT, cal g^{-1} K^{-1}
TEMPERATURE, K
1
FIG 221

SPECIFICATION TABLE NO. 221 SPECIFIC HEAT OF THORIUM DISULFIDE ThS_2

[For Data Reported in Figure and Table No. 221]

Curve No.	Ref. No.	Year	Temp. Range, K	Reported Error, %	Name and Specimen Designation	Composition (weight percent), Specifications and Remarks
1	306	1959	53-297	0.3		78.54 Th and 21.43 S (78.35 and 21.65 theo).

DATA TABLE NO. 221 SPECIFIC HEAT OF THORIUM DISULFIDE ThS_2

[Temperature, T, K; Specific Heat, C_p, Cal $g^{-1}K^{-1}$]

T	C_p
CURVE 1	
53.24	1.678 x 10^{-2}
57.43	1.839*
61.96	2.019
66.49	2.195*
71.07	2.375
75.83	2.549*
81.16	2.701
85.94	2.915*
95.05	3.205
105.06	3.508*
114.65	3.758
124.76	3.998*
135.97	4.237
145.72	4.430*
155.49	4.588
166.02	4.740*
175.92	4.855
186.03	4.973*
195.93	5.061
206.36	5.152*
216.33	5.247
225.91	5.314*
236.56	5.368
245.74	5.429*
256.21	5.490
266.05	5.534*
275.98	5.581
286.66	5.625*
296.18	5.672

* Not shown on plot

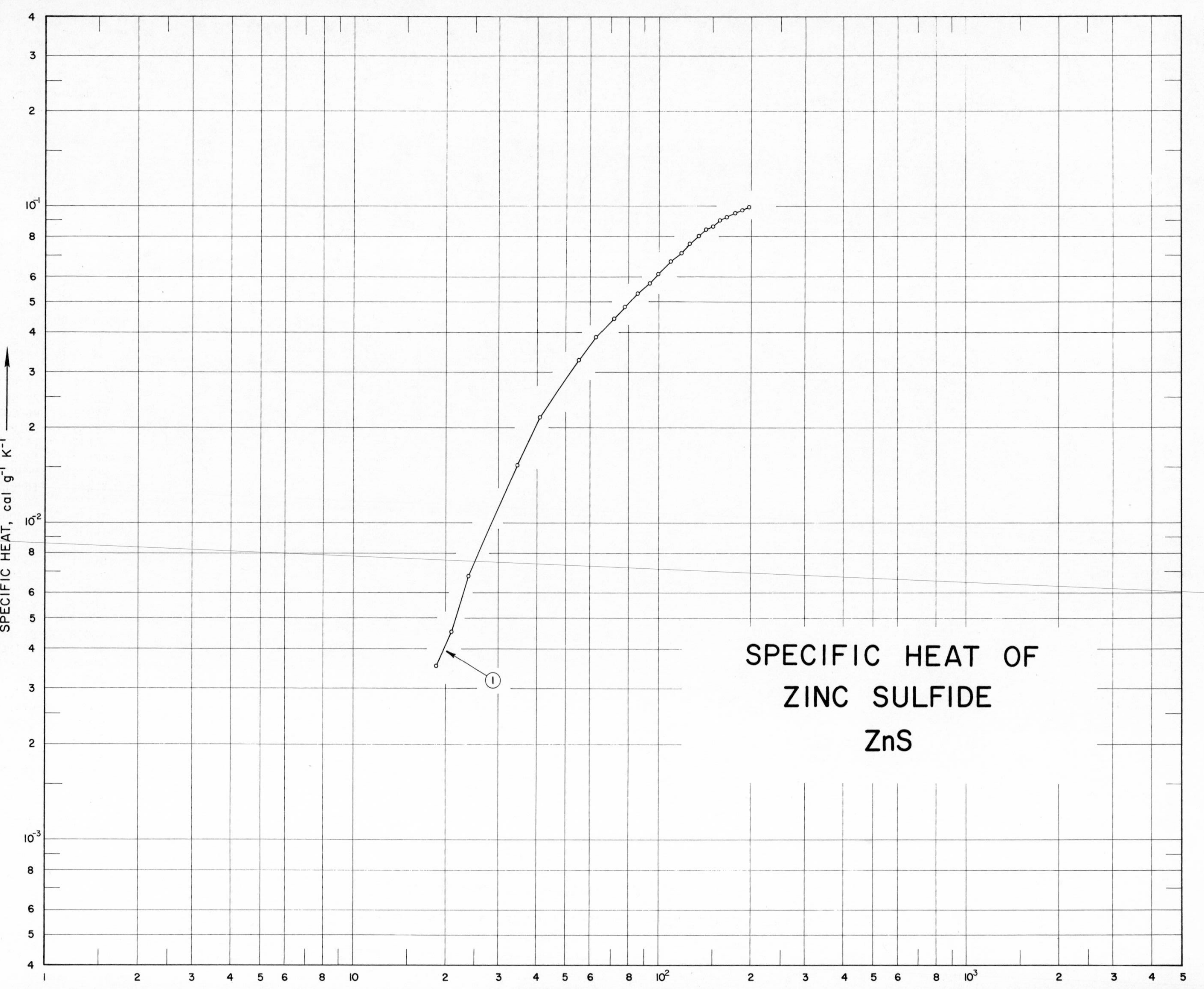

SPECIFIC HEAT OF
ZINC SULFIDE
ZnS
SPECIFIC HEAT, cal g^{-1} K^{-1}
1

SPECIFICATION TABLE NO. 222 SPECIFIC HEAT OF ZINC SULFIDE ZnS

[For Data Reported in Figure and Table No. 222]

Curve No.	Ref. No.	Year	Temp. Range, K	Reported Error, %	Name and Specimen Designation	Composition (weight percent), Specifications and Remarks
1	68	1928	19-196			

DATA TABLE NO. 222 SPECIFIC HEAT OF ZINC SULFIDE ZnS

[Temperature, T,K; Specific Heat, C_p, Cal $g^{-1}K^{-1}$]

T	C_p
CURVE 1	
18.65	3.542×10^{-3}
20.95	4.533
23.75	6.809
34.20	1.546×10^{-2}
40.90	2.167
55.00	3.289
62.90	3.864
71.50	4.427
77.20	4.845
85.80	5.346
93.10	5.749
99.80	6.194
109.40	6.736
117.90	7.178
126.40	7.617
135.10	8.102
142.70	8.462
150.30	8.612
158.40	9.006
167.90	9.284
177.90	9.514
187.00	9.731
196.30	9.900

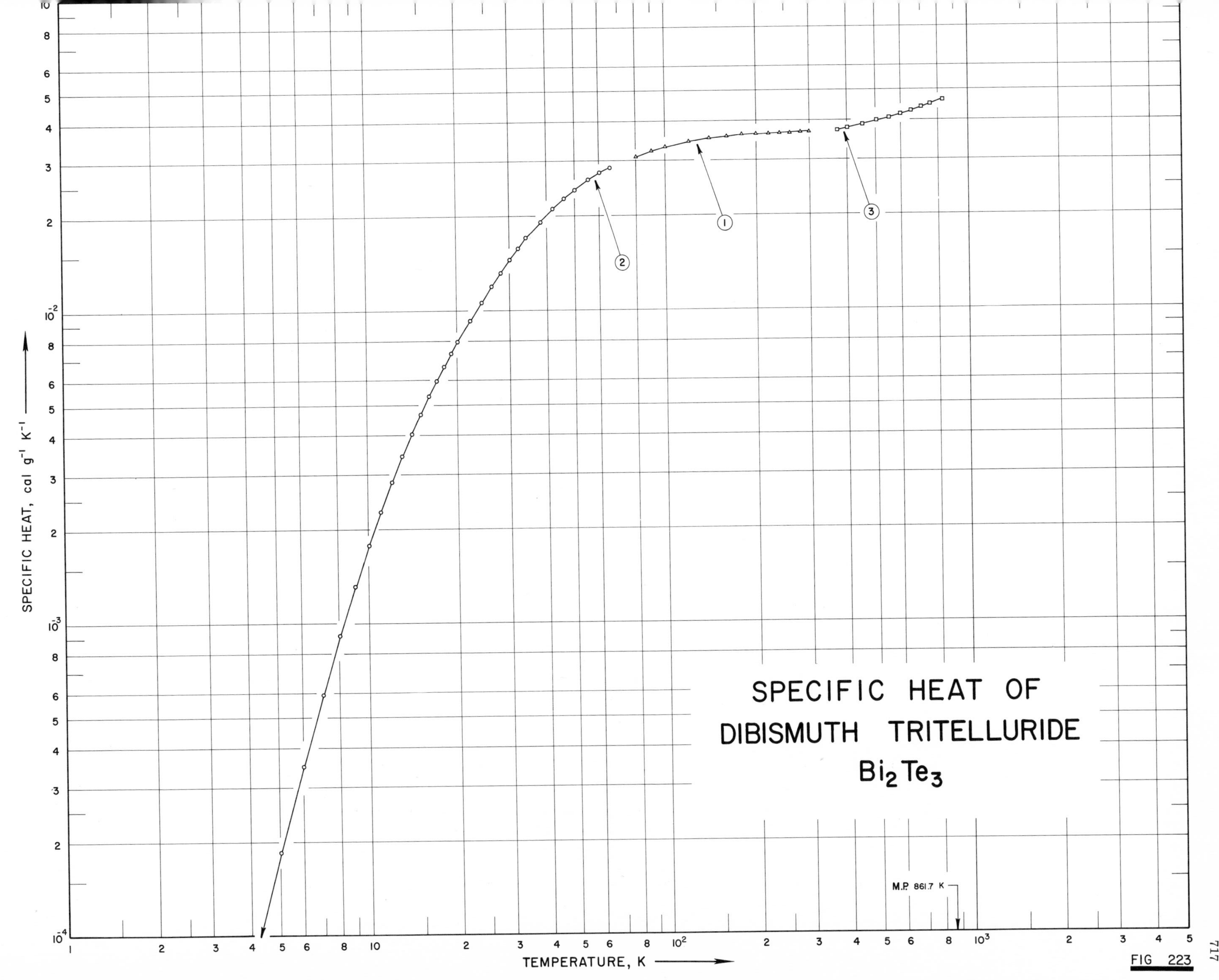
SPECIFIC HEAT OF
DIBISMUTH TRITELLURIDE
Bi_2Te_3
SPECIFIC HEAT, cal g^{-1} K^{-1}
TEMPERATURE, K
M.P. 861.7 K
1
2
3
10^{-4}
10^{-3}
10^{-2}
10^{2}
10^{3}

FIG 223

SPECIFICATION TABLE NO. 223 SPECIFIC HEAT OF DIBISMUTH TRITELLURIDE Bi_2Te_3

[For Data Reported in Figure and Table No. 223]

Curve No.	Ref. No.	Year	Temp. Range, K	Reported Error, %	Name and Specimen Designation	Composition (weight percent), Specifications and Remarks
1	160	1959	80-300	3-7		Polycrystalline.
2	218	1961	1. 5-65	1. 0-3. 0		Recrystallized twice to increase the purity; coated with polymer varnish BF-2.
3	219	1962	373-823	2. 0		Zone-refined.

DATA TABLE NO. 223 SPECIFIC HEAT OF DIBISMUTH TRITELLURIDE Bi_2Te_3

[Temperature, T, K; Specific Heat, C_p, Cal $g^{-1}K^{-1}$]

T	C_p
CURVE 1	
80	3.09×10^{-2}
90	3.23
100	3.33
120	3.45
140	3.54
160	3.59
180	3.62
200	3.64
220	3.66
240	3.67
260	3.68
280	3.70
300	3.71
CURVE 2	
1.5	3.259×10^{-6}
2	6.518*
3	2.785×10^{-5}
4	7.821*
5	1.843×10^{-4}
6	3.496
7	5.925
8	9.154
9	1.324×10^{-3}
10	1.786
11	2.290
12	2.850
13	3.442
14	4.0587
15	4.681
16	5.332
17	5.984
18	6.666
19	7.317
20	7.969
22	9.302
24	1.066×10^{-2}
26	1.200
28	1.330
30	1.460
32	1.588
34	1.706
36	1.816*
38	1.920
40	2.014*
CURVE 2 (cont.)	
42	2.115×10^{-2}
44	2.207*
46	2.293
48	2.379*
50	2.447
55	2.616
60	2.755
65	2.859
CURVE 3	
373	3.7460×10^{-2}
400	3.8000
450	3.9000
500	4.0000
550	4.1000
600	4.2000
650	4.3000
700	4.4000
750	4.5000
800	4.6000
823	4.6460

* Not shown on plot

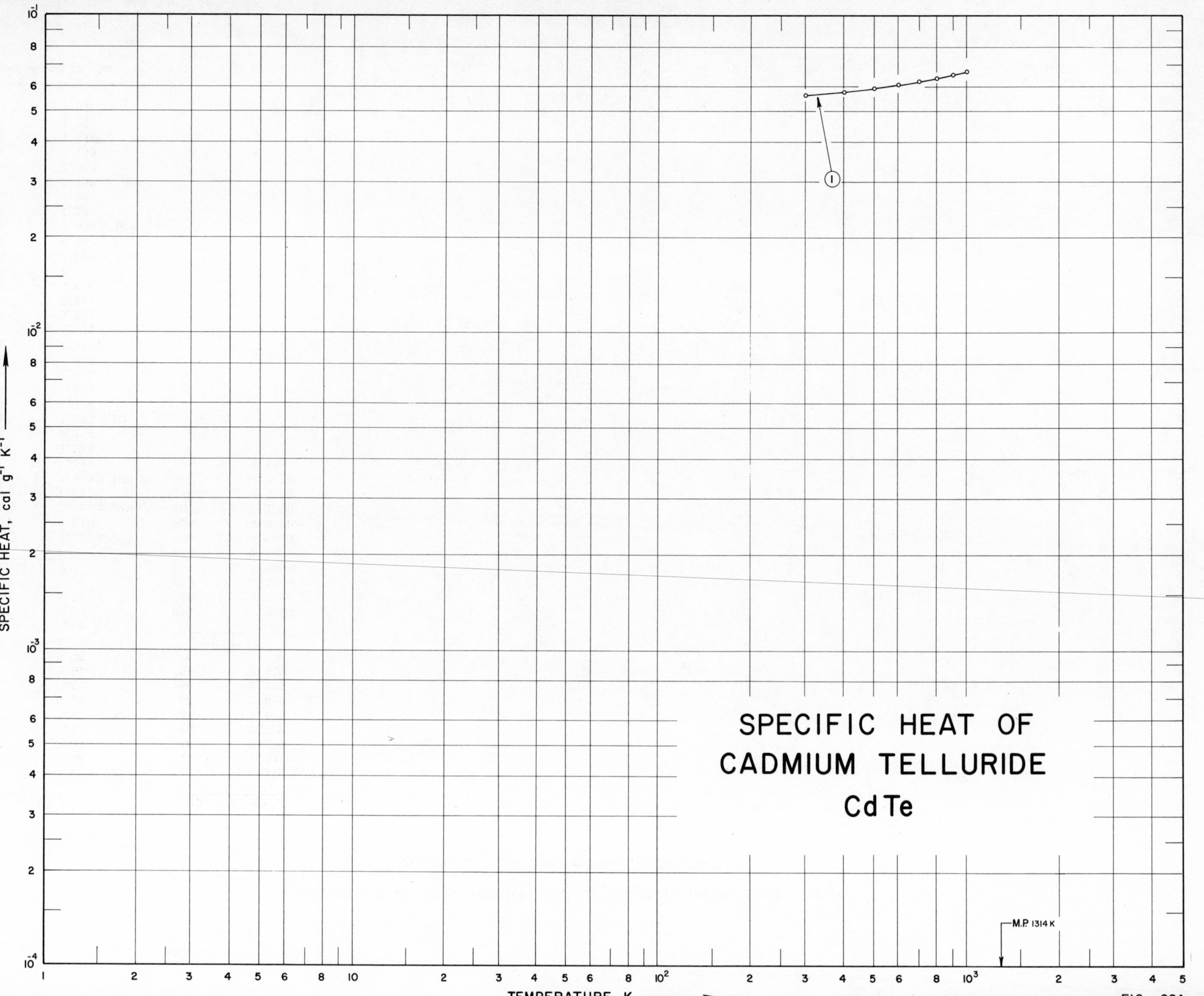
SPECIFIC HEAT OF
CADMIUM TELLURIDE
CdTe
M.P. 1314 K
SPECIFIC HEAT, cal g^{-1} K^{-1}
1

SPECIFICATION TABLE NO. 224 SPECIFIC HEAT OF CADMIUM TELLURIDE CdTe

[For Data Reported in Figure and Table No. 224]

Curve No.	Ref. No.	Year	Temp. Range, K	Reported Error, %	Name and Specimen Designation	Composition (weight percent), Specifications and Remarks
1	54	1965	300-1000	4.5		~100 CdTe; traces of Sn.

DATA TABLE NO. 224 SPECIFIC HEAT OF CADMIUM TELLURIDE CdTe

[Temperature, T, K; Specific Heat, C_p, Cal $g^{-1}K^{-1}$]

T	C_p
CURVE 1	
300	5.612×10^{-2}
400	5.762
500	5.912
600	6.062
700	6.208
800	6.358
900	6.508
1000	6.658

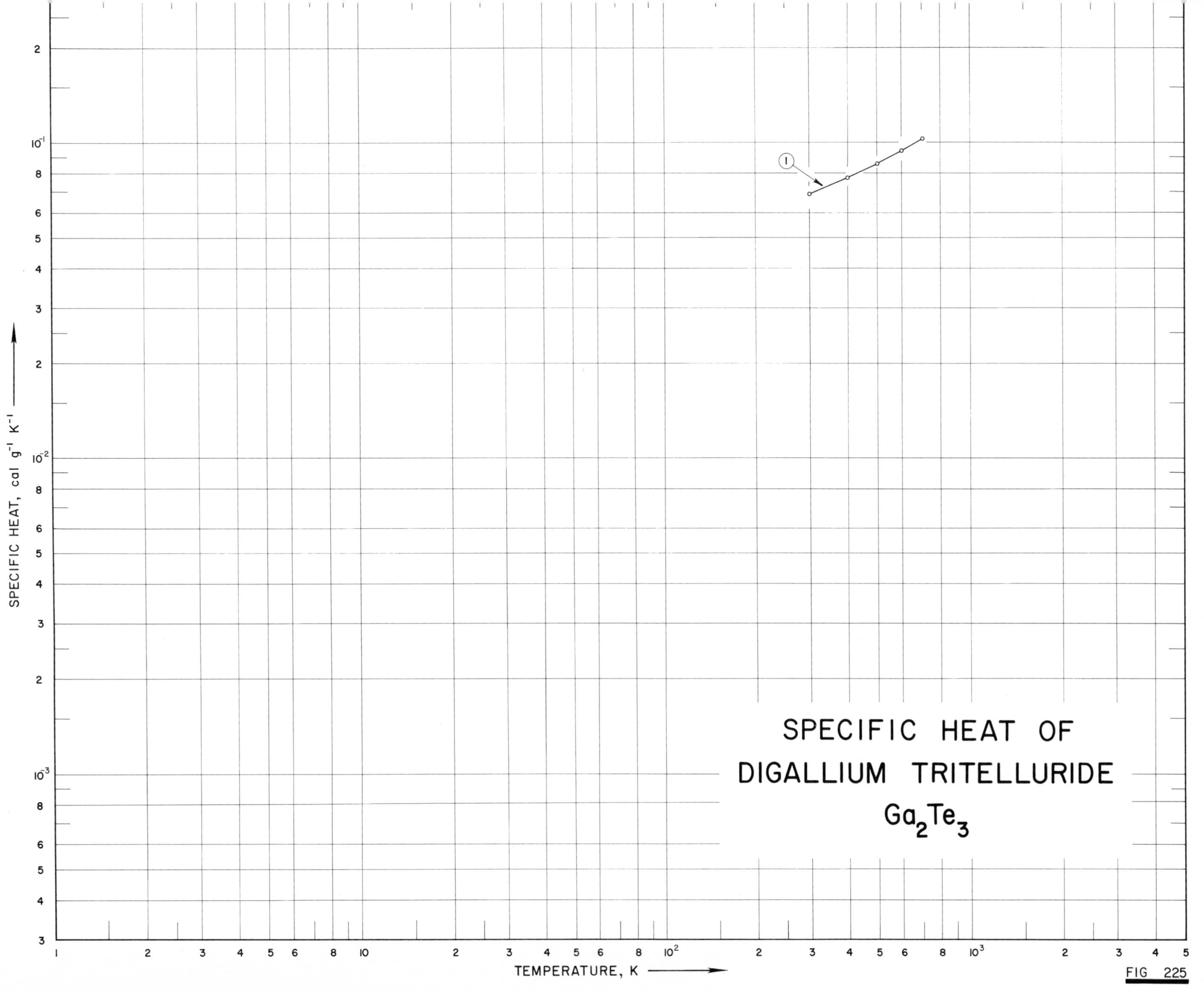
SPECIFIC HEAT OF
DIGALLIUM TRITELLURIDE
Ga_2Te_3
SPECIFIC HEAT, cal g^{-1} K^{-1}
TEMPERATURE, K
1

FIG 225

SPECIFICATION TABLE NO. 225 SPECIFIC HEAT OF DIGALLIUM TRITELLURIDE Ge_2Te_3

[For Data Reported in Figure and Table No. 225]

Curve No.	Ref. No.	Year	Temp. Range, K	Reported Error, %	Name and Specimen Designation	Composition (weight percent), Specifications and Remarks
1	423	1963	300-700			

DATA TABLE NO. 225 SPECIFIC HEAT OF DIGALLIUM TRITELLURIDE Ga_2Te_3

[Temperature, T, K; Specific Heat, C_p, Cal $g^{-1}K^{-1}$]

T	C_p
CURVE 1	
300	6.907×10^{-2}
400	7.759
500	8.611
600	9.446
700	1.031×10^{-1}

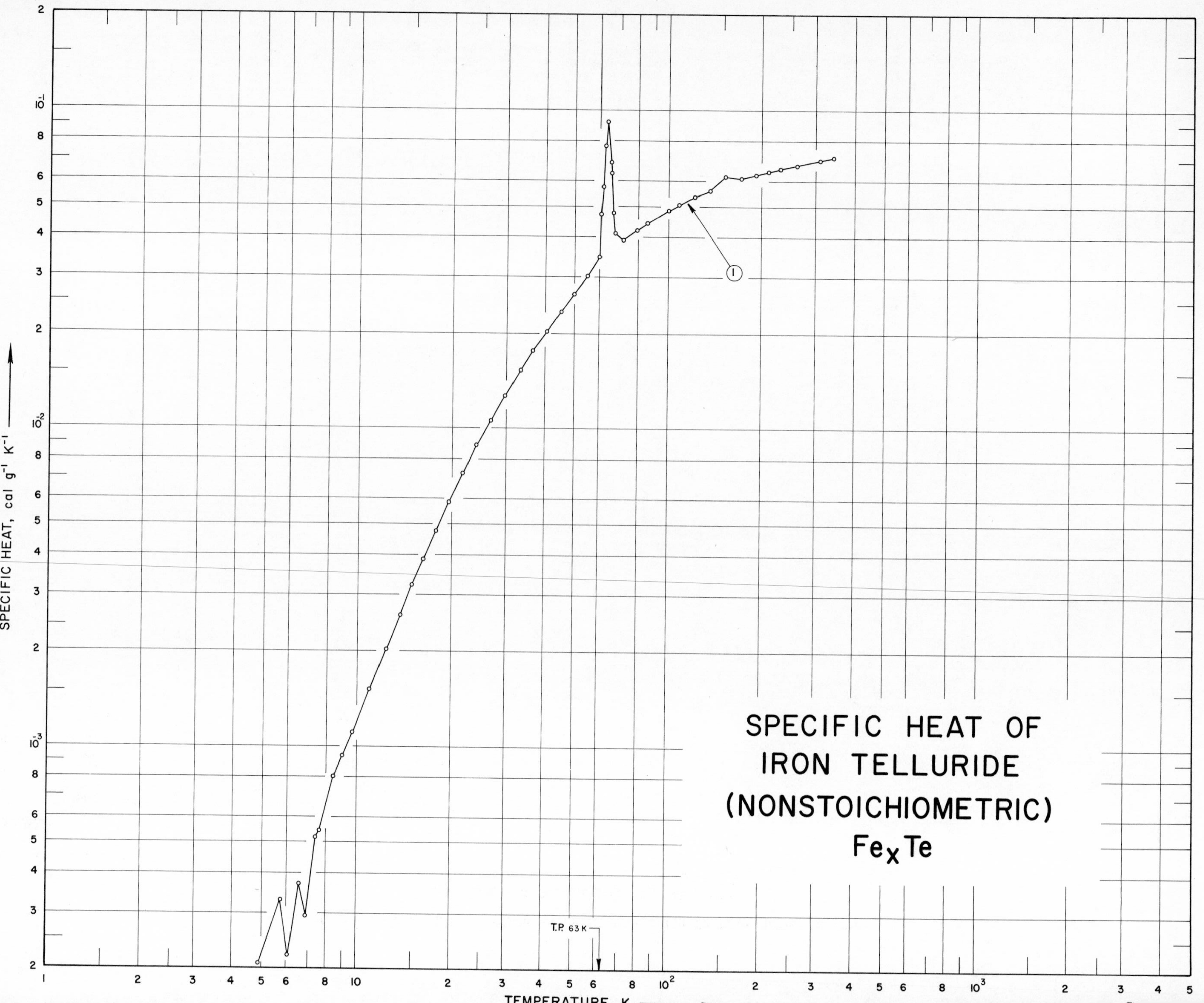

SPECIFIC HEAT OF
IRON TELLURIDE
(NONSTOICHIOMETRIC)
Fe$_x$Te
1
T.P. 63 K
TEMPERATURE, K
SPECIFIC HEAT, cal g^{-1} K^{-1}

SPECIFICATION TABLE NO. 226 SPECIFIC HEAT OF IRON TELLURIDE (nonstoichiometric) Fe_xTe

[For Data Reported in Figure and Table No. 226]

Curve No.	Ref. No.	Year	Temp. Range, K	Reported Error, %	Name and Specimen Designation	Composition (weight percent), Specifications and Remarks
1	32	1962	6-341	0.1-5.0		~100 $Fe_{1.111}Te$; iron impurities: 0.01 each Ni and Si, and 0.001 Mn, tellurium impurities: 0.01 Fe and traces of Al, Pb, and Mg; heated for 1 wk at 700 C and cooled; fragmented; heated at 400 C for 2 wks and cooled to room temperature at the rate of 50 C per day.

DATA TABLE NO. 226 SPECIFIC HEAT OF IRON TELLURIDE Fe_xTe (nonstoichiometric)

Temperature, T, K; Specific Heat, C_p, Cal $g^{-1}K^{-1}$

T	C_p
CURVE 1	
Series I	
121.50	5.378 x 10^{-2}
126.46	5.468*
136.37	5.615
143.21	5.763*
152.48	6.216
162.00	6.011*
171.70	6.116
181.40	6.216*
191.11	6.306
200.75	6.380*
210.35	6.448
219.86	6.512*
229.35	6.575
Series II	
6.09	2.204 x 10^{-4}
6.89	2.947
7.60	5.431
8.41	8.014
8.97	9.332
9.65	1.107 x 10^{-3}
10.93	1.513
12.33	2.025
13.62	2.589
14.84	3.211
16.15	3.875
17.66	4.761
19.45	5.868
21.54	7.239
23.93	8.863
26.61	1.064 x 10^{-2}
29.55	1.276
Series III	
4.87	2.083 x 10^{-4}
5.70	3.290
6.56	3.707
7.37	5.188
8.38	8.014*
9.43	1.044 x 10^{-3}*
10.64	1.418*
11.89	1.845*
CURVE 1 (cont.)	
30.08	1.313 x 10^{-2}*
33.11	1.526
36.41	1.753
40.39	2.021
45.06	2.332
49.92	2.662
54.98	3.015
55.26	3.034*
60.56	4.756
65.62	6.380
71.92	3.948
79.24	4.203
85.93	4.449
92.70	4.663*
100.20	4.878
108.36	5.086
116.65	5.278*
124.93	5.452*
Series IV	
51.66	2.781 x 10^{-2}*
56.78	3.150*
60.08	3.470
61.85	5.784
63.61	9.248
65.41	6.538*
67.30	4.109
69.32	3.891*
71.87	3.920*
75.08	4.059*
Series V	
52.06	2.808 x 10^{-2}*
57.35	3.198*
60.49	3.543*
61.60	4.718*
62.44	7.730
63.08	9.364*
63.67	9.248*
64.25	8.948*
64.77	8.341*
65.28	6.907
66.24	4.790
CURVE 1 (cont.)	
67.49	4.040 x 10^{-2}*
68.60	3.908*
70.58	3.895*
226.85	6.554*
236.34	6.617*
245.87	6.680*
242.07	6.628*
250.49	6.680*
260.03	6.738
269.74	6.791*
279.49	6.839*
289.33	6.886*
299.46	6.928*
309.83	7.002
320.23	7.049*
330.82	7.097*
341.65	7.144

* Not shown on plot

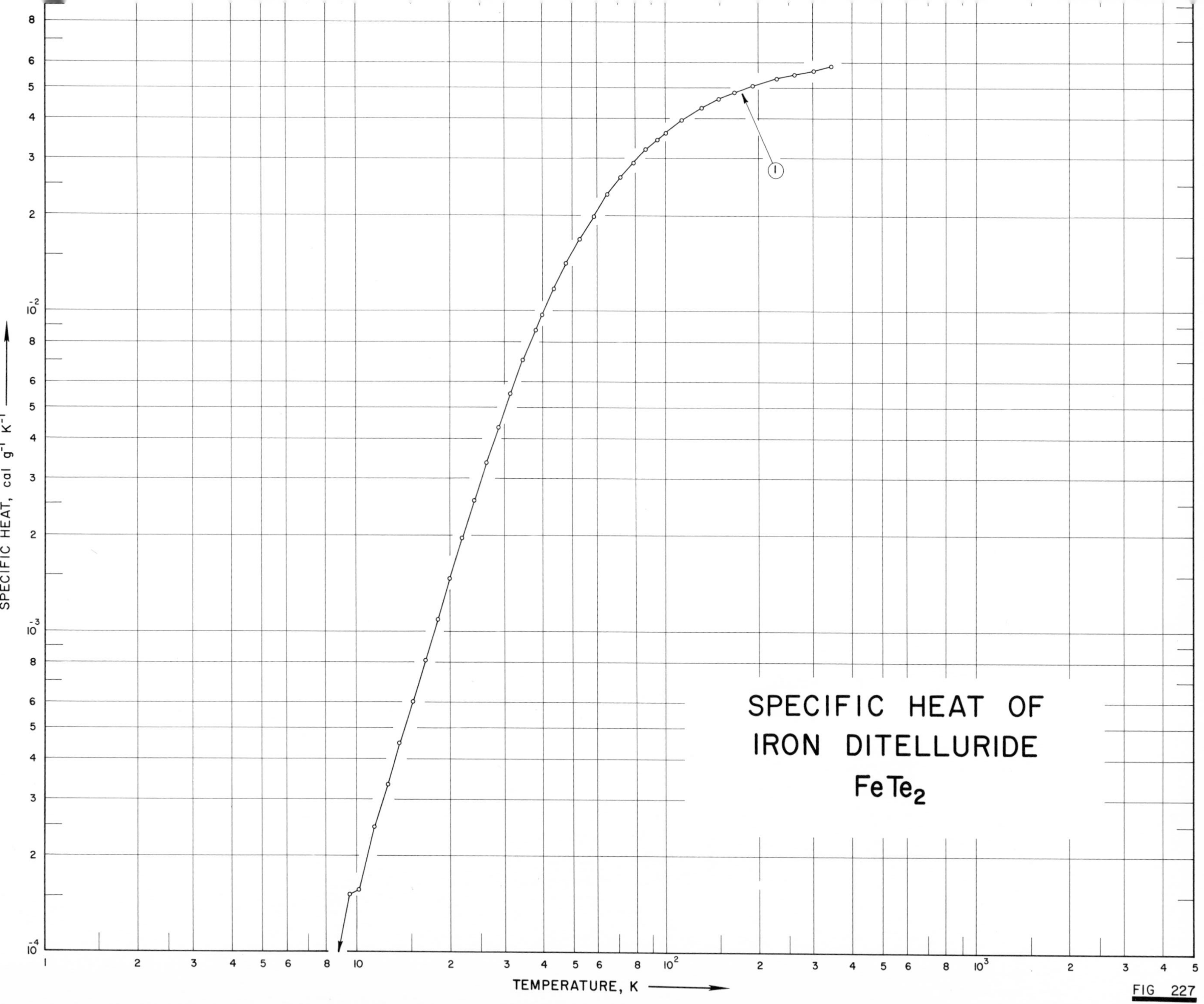
SPECIFIC HEAT OF
IRON DITELLURIDE
$FeTe_2$
1
SPECIFIC HEAT, cal g^{-1} K^{-1}
TEMPERATURE, K

FIG 227

SPECIFICATION TABLE NO. 227 SPECIFIC HEAT OF IRON DITELLURIDE $FeTe_2$

[For Data Reported in Figure and Table No. 227]

Curve No.	Ref. No.	Year	Temp. Range, K	Reported Error, %	Name and Specimen Designation	Composition (weight percent), Specifications and Remarks
1	220	1959	7-345	0.1-5.0		~100 $FeTe_2$; iron impurities: 0.01 each Ni and Si, and 0.001 Mn, tellurium impurities: 0.01 Fe, and traces of Al, Mg and Pb; heated for 1 wk at 700 C and cooled; fragmented; heated at 400 C for 2 wks and cooled to room temperature at the rate of 50 C per day.

DATA TABLE NO. 227 SPECIFIC HEAT OF IRON DITELLURIDE $FeTe_2$

[Temperature, T, K; Specific Heat, C_p, Cal $g^{-1}K^{-1}$]

T	C_p	T	C_p
CURVE 1		CURVE 1 (cont.)	
Series I			
100.24	3.633 x 10^{-2}	181.90	4.999 x 10^{-2}*
105.54	3.771*	191.57	5.082
113.73	3.970	201.28	5.156*
122.79	4.176*	210.72	5.221*
131.28	4.340	219.96	5.279*
139.58	4.485*	229.40	5.340
148.20	4.616	238.90	5.394*
157.46	4.742*	248.62	5.443*
166.97	4.854	251.27	5.449*
		261.34	5.500
Series II		271.24	5.542*
		281.38	5.581*
6.83	4.629 x 10^{-5}*	291.66	5.635*
7.71	6.333*	302.20	5.677
8.67	9.419*	312.84	5.716*
9.52	1.517 x 10^{-4}	323.49	5.754*
10.24	1.569	334.18	5.806*
11.45	2.469	344.88	5.838
12.65	3.343		
13.86	4.501		
15.15	6.044		
16.61	8.133		
18.20	1.099 x 10^{-3}		
19.90	1.472		
21.77	1.961		
23.83	2.581		
26.09	3.388		
28.52	4.366		
31.22	5.574		
34.32	7.047		
37.73	8.763		
39.60	9.792		
43.43	1.184 x 10^{-2}		
47.81	1.421		
52.96	1.694		
58.71	1.994		
64.89	2.320		
71.66	2.636		
79.28	2.919		
86.65	3.201		
94.02	3.446		
168.89	4.864*		
174.75	4.928*		

* Not shown on plot

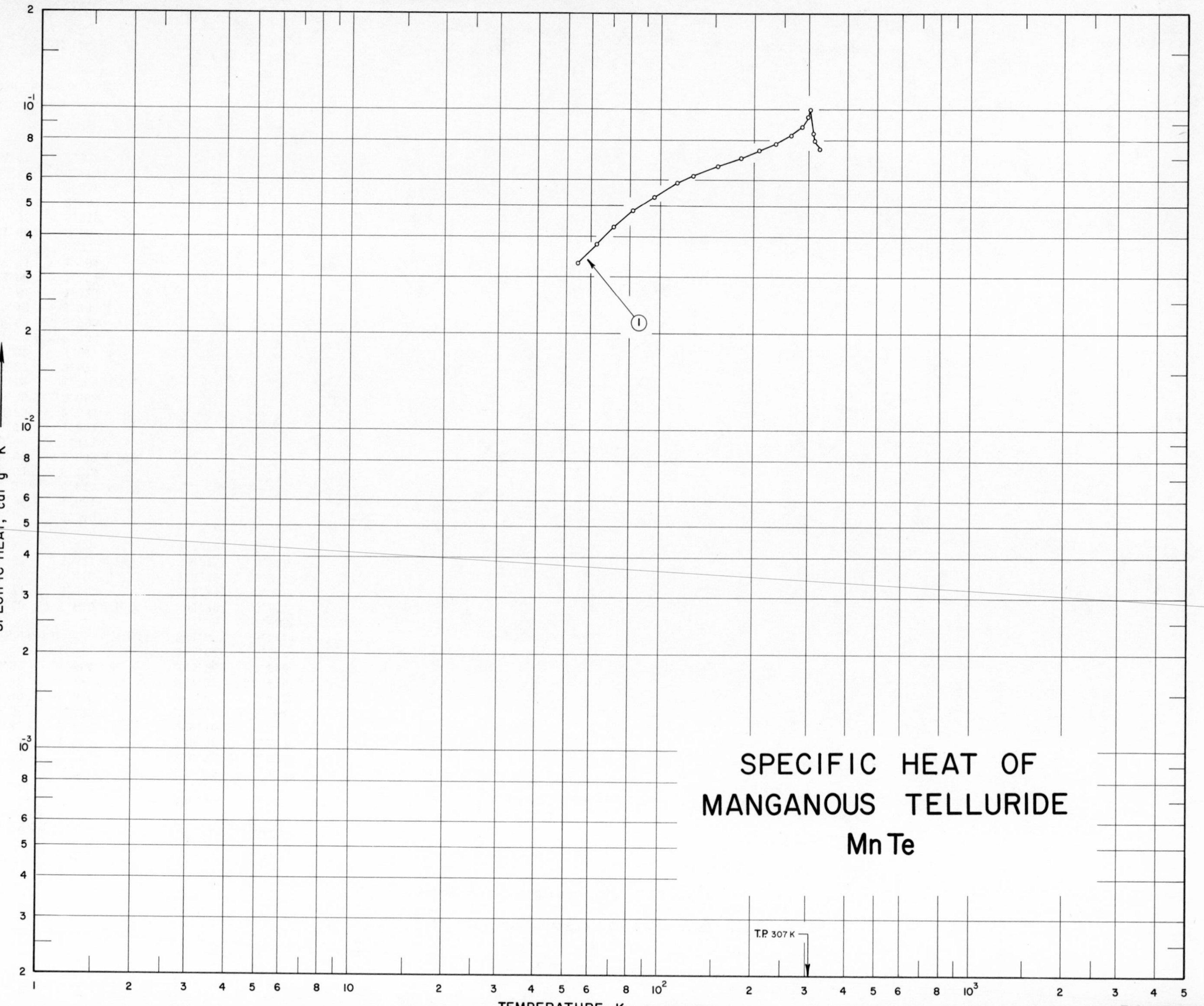
SPECIFIC HEAT OF
MANGANOUS TELLURIDE
MnTe
T.P. 307 K
1
SPECIFIC HEAT, cal g^{-1} K^{-1}

SPECIFICATION TABLE NO. 228 SPECIFIC HEAT OF MANGANOUS TELLURIDE MnTe

[For Data Reported in Figure and Table No. 228]

Curve No.	Ref. No.	Year	Temp. Range, K	Reported Error, %	Name and Specimen Designation	Composition (weight percent), Specifications and Remarks
1	221	1939	55-327			99.57 MnTe.

DATA TABLE NO. 228 SPECIFIC HEAT OF MANGANOUS TELLURIDE MnTe

[Temperature, T, K; Specific Heat, C_p, Cal $g^{-1}K^{-1}$]

T	C_p	T	C_p
CURVE 1		**CURVE 1 (cont.)**	
54.5	3.31 x 10^{-2}	304.4	1.007 x 10^{-1}
58.3	3.56*	307.4	1.004*
62.5	3.81	310.5	8.404 x 10^{-2}
66.4	4.05*	313.7	7.992
70.7	4.31	316.9	7.730*
75.3	4.56*	320.4	7.637*
81.5	4.82	323.7	7.533*
82.0	4.84*	327.0	7.549
85.6	4.99*		
86.6	5.03*		
95.2	5.35		
100.2	5.505*		
104.9	5.654*		
109.4	5.780*		
113.6	5.911		
117.6	6.037*		
119.9	6.092*		
121.4	6.114*		
127.6	6.223		
136.2	6.371*		
144.8	6.502*		
153.8	6.651		
162.8	6.788*		
172.3	6.946*		
181.6	7.051		
190.7	7.182*		
199.8	7.330*		
208.3	7.445		
217.5	7.565*		
227.0	7.691*		
236.2	7.850		
245.7	8.015*		
255.1	8.152*		
264.4	8.338		
273.3	8.541*		
277.2	8.634*		
281.0	8.754*		
286.0	8.891		
289.8	9.028*		
293.3	9.171*		
295.2	9.395*		
297.8	9.499		
298.1	9.466*		
301.5	9.751*		

* Not shown on plot

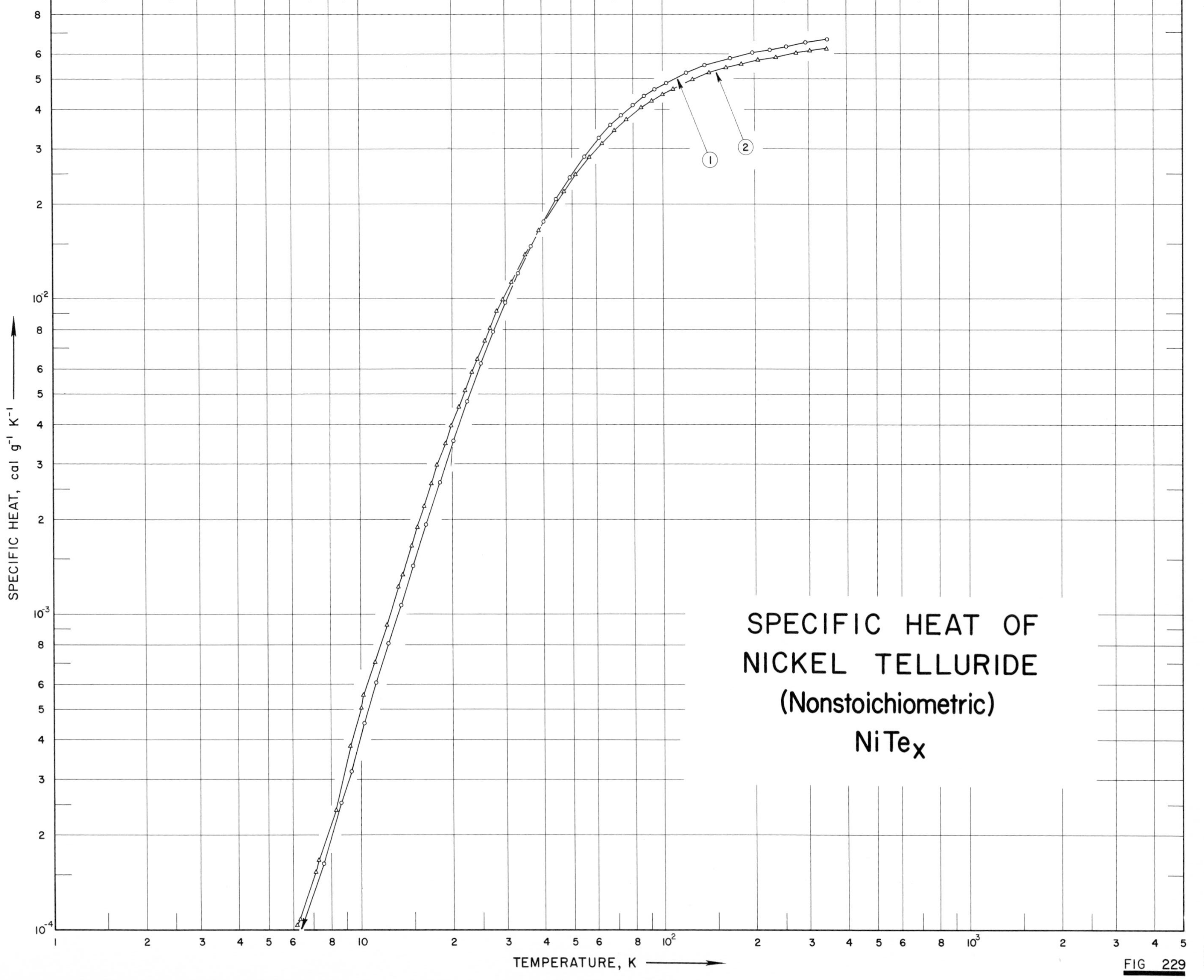
SPECIFIC HEAT OF
NICKEL TELLURIDE
(Nonstoichiometric)
NiTe$_x$
SPECIFIC HEAT, cal g^{-1} K^{-1}
TEMPERATURE, K
1
2
FIG 229

SPECIFICATION TABLE NO. 229 SPECIFIC HEAT OF NICKEL TELLURIDE (nonstoichiometric) $NiTe_x$

[For Data Reported in Figure and Table No. 229]

Curve No.	Ref. No.	Year	Temp. Range, K	Reported Error, %	Name and Specimen Designation	Composition (weight percent), Specifications and Remarks
1	222	1958	5-347	0.1		>99.99 $NiTe_{1.1}$; fused for 2 hrs at 1000 C, cooled, fragmented, annealed in vacuo at 500 C for 30 days; cooled to room temperature over a period of 2 days, additional heating for 2 wks at temperature gradually decreasing from 500 to 300 C.
2	222	1958	6-346	0.1		>99.99 $NiTe_{1.5}$; same as above.

DATA TABLE NO. 229 SPECIFIC HEAT OF NICKEL TELLURIDE NiTe (nonstoichiometric)

[Temperature, T, K; Specific Heat, C_p, Cal $g^{-1}K^{-1}$]

T	C_p
CURVE 1	
Series I	
5.44	6.857×10^{-5}*
6.44	1.044×10^{-4}*
7.56	1.625
8.65	2.532
Series II	
5.41	6.752×10^{-5}*
6.38	9.811*
7.33	1.477×10^{-4}*
8.34	2.205*
9.36	3.186
10.34	4.505
11.34	6.077
12.44	8.070
13.63	1.074×10^{-3}
14.93	1.431
16.44	1.930
18.23	2.630
20.20	3.554
22.43	4.728
24.95	6.235
27.44	7.881
30.08	9.742
33.18	1.206×10^{-2}
36.60	1.470
40.37	1.762
44.61	2.082
49.47	2.435
55.05	2.821
51.25	2.567*
56.11	2.891*
61.85	3.248
67.67	3.567
73.54	3.847
79.97	4.131
87.07	4.404
94.38	4.632
87.69	4.427*
95.21	4.657*
103.23	4.867
111.74	5.060*
120.27	5.229
123.47	5.281*
CURVE 1 (cont.)	
129.99	5.391×10^{-2}
138.62	5.515
147.64	5.627*
157.35	5.734*
167.52	5.827
177.63	5.906*
187.57	5.976*
197.10	6.035
206.52	6.088*
216.13	6.137*
225.83	6.182
235.58	6.227*
245.31	6.268*
255.16	6.311
265.11	6.357*
275.09	6.415*
285.08	6.469*
295.18	6.513
298.82	6.521*
308.92	6.559*
318.95	6.591*
328.91	6.619*
338.80	6.640*
347.16	6.668
CURVE 2	
Series I	
6.33	1.080×10^{-4}
7.29	1.669
8.34	2.399
9.28	3.818
10.24	5.558
11.20	7.017
12.27	9.226
13.44	1.231×10^{-3}
14.72	1.653
16.22	2.204
17.97	2.967
19.95	3.965
22.06	5.141
24.30	6.477
26.75	8.081
29.50	9.939
CURVE 2 (cont.)	
Series II	
6.18	1.040×10^{-4}
7.12	1.529
8.14	2.329*
9.11	3.549*
10.07	5.038
11.13	6.897*
12.43	9.626*
13.82	1.342×10^{-3}
15.41	1.891
17.18	2.604
19.03	3.478
21.08	4.562
23.32	5.878
25.71	7.391
28.36	9.160
31.53	1.135×10^{-2}
35.12	1.385
38.88	1.648
42.83	1.914*
47.10	2.194
51.87	2.493
57.19	2.802
63.15	3.124
69.84	3.439
76.60	3.722
73.70	3.602*
78.72	3.805*
85.34	4.051
92.54	4.270
100.77	4.486
109.14	4.674
117.37	4.834*
126.07	4.988
134.78	5.119*
143.56	5.233
152.66	5.335*
162.16	5.430
171.91	5.512*
181.61	5.584
191.29	5.650*
200.90	5.706*
206.43	5.733
216.26	5.783*
226.08	5.826*
CURVE 2 (cont.)	
235.89	5.869×10^{-2}
245.75	5.910*
255.65	5.947*
265.60	5.984*
275.68	6.019
285.61	6.053*
295.57	6.082*
305.68	6.112
315.65	6.137*
325.63	6.158*
335.65	6.186*
345.72	6.215

*Not shown on plot

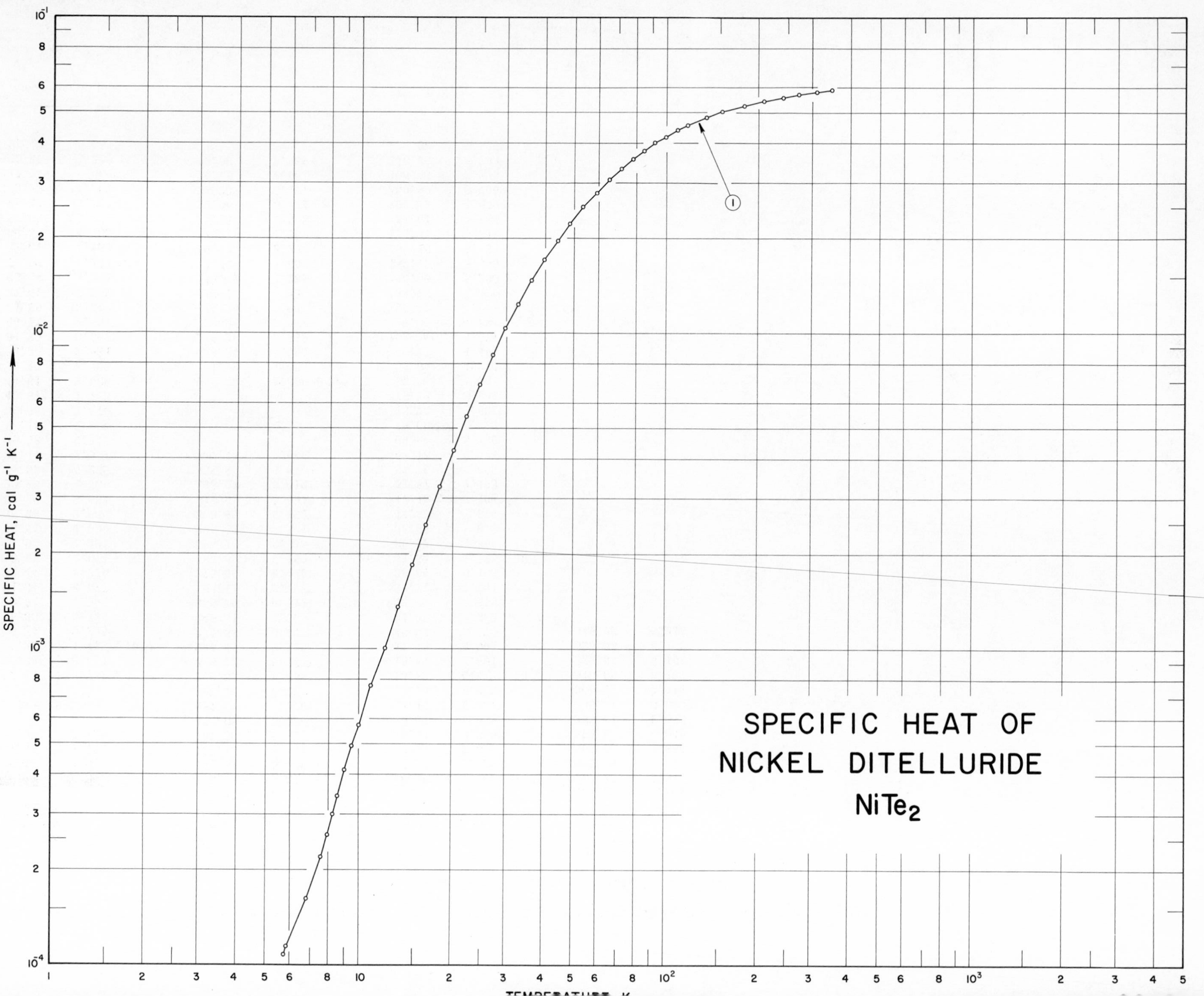
SPECIFIC HEAT OF
NICKEL DITELLURIDE
NiTe$_2$
SPECIFIC HEAT, cal g^{-1} K^{-1}
1

SPECIFICATION TABLE NO. 230 SPECIFIC HEAT OF NICKEL DITELLURIDE $NiTe_2$

[For Data Reported in Figure and Table No. 230]

Curve No.	Ref. No.	Year	Temp. Range, K	Reported Error, %	Name and Specimen Designation	Composition (weight percent), Specifications and Remarks
1	222	1958	6-348	0.1		>99.99 $NiTe_2$; fused for 2 hrs at 1000 C, cooled, fragmented, annealed in vacuo at 500 C for 30 days, cooled to room temperature over a period of 2 days; additional heating for 2 wks at temperature gradually decreasing from 500 to 300 C.

DATA TABLE NO. 230 SPECIFIC HEAT OF NICKEL DITELLURIDE $NiTe_2$

[Temperature, T, K; Specific Heat, C_p, Cal $g^{-1}K^{-1}$]

T	C_p	T	C_p
CURVE 1		CURVE 1 (cont.)	
Series I			
		44.06	1.963×10^{-2}
5.86	1.147×10^{-4}	48.63	2.232
6.79	1.615	53.72	2.507
7.56	2.198	59.32	2.789
8.53	3.412	65.36	3.070
9.48	4.922	71.74	3.327
		78.15	3.567
Series II		84.81	3.796
		92.01	4.010
5.75	1.080×10^{-4}	99.89	4.196
6.86	1.644*	108.39	4.380*
7.91	2.580	109.78	4.409
8.98	4.148	118.18	4.563
9.94	5.744	126.70	4.707*
10.94	7.645	135.20	4.830
12.08	1.008×10^{-3}	143.73	4.940*
13.36	1.358	152.58	5.033
14.82	1.843	161.71	5.125*
16.40	2.467	171.18	5.203*
18.15	3.263	180.91	5.274
20.05	4.263	190.65	5.341*
22.14	5.441	200.27	5.398*
		209.80	5.447
Series III		219.36	5.494*
		212.33	5.458*
5.90	1.175×10^{-4}*	221.75	5.501*
6.76	1.567*	231.53	5.550*
7.40	2.064*	241.59	5.593
8.26	2.991	251.78	5.632*
9.23	4.539*	262.05	5.665*
10.26	6.193*	275.32	5.702
11.38	8.477*	282.57	5.735*
12.66	1.153×10^{-3}*	292.75	5.767*
14.11	1.597*	302.97	5.793*
15.76	2.210*	301.47	5.793*
17.68	3.038*	311.64	5.817
19.89	4.145*	321.43	5.843*
22.17	5.457*	330.87	5.868*
24.51	6.892	340.47	5.885*
24.37	6.803*	348.23	5.903
26.99	8.509		
29.82	1.043×10^{-2}		
32.83	1.244		
36.19	1.470		
39.95	1.713		

* Not shown on plot

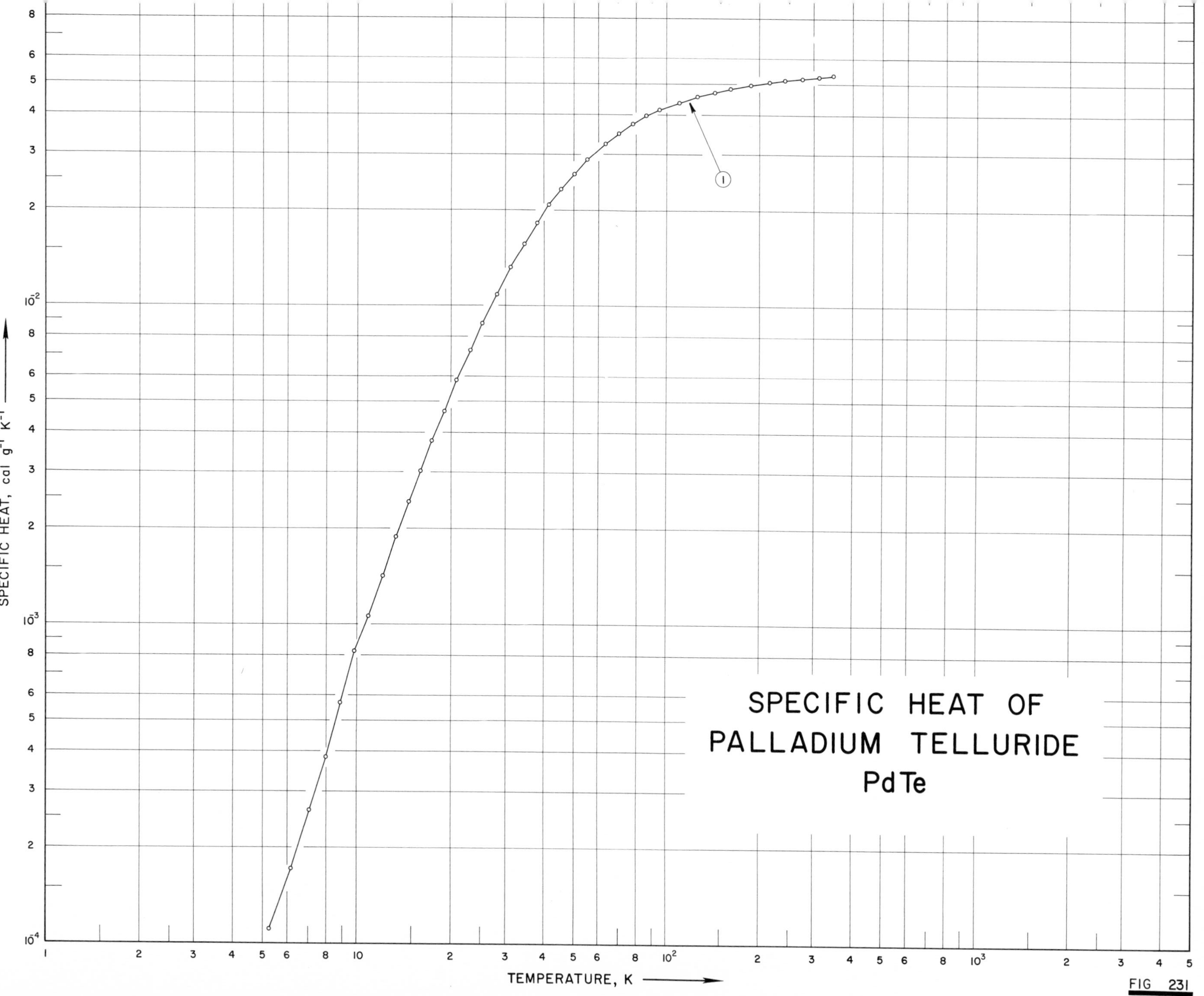
SPECIFIC HEAT OF
PALLADIUM TELLURIDE
PdTe
SPECIFIC HEAT, cal g^{-1} K^{-1}
TEMPERATURE, K
1
FIG 231

SPECIFICATION TABLE NO. 231 SPECIFIC HEAT OF PALLADIUM TELLURIDE PdTe

[For Data Reported in Figure and Table No. 231]

Curve No.	Ref. No.	Year	Temp. Range, K	Reported Error, %	Name and Specimen Designation	Composition (weight percent), Specifications and Remarks
1	223	1961	5-347	~0.1		Impurities: 0.006 volatile matter, 0.007 Au, 0.005 Pt, 0.003 Ag, 0.002 Fe, 0.001 Rh, and 0.0002 Pb; prepared from stoichiometric amounts of palladium and tellurium heated in evacuated and sealed silica tube to 800 C; kept in molten state for 2 hrs, cooled, annealed for 7 days at 500 C and cooled to room temperature over a two-day period.
2	424	1965	2-7			$PdTe_{1.04}$; prepared from 99.8 Pd supplied by Johnson, Matthey and Co., Ltd., and 99.999 Te supplied by the American Smelting and Refining Co.; annealed 1 wk at 450 C.
3	424	1965	1.6-7			PdTe; same as above.

DATA TABLE NO. 231 SPECIFIC HEAT OF PALLADIUM TELLURIDE PdTe

[Temperature, T,K; Specific Heat, C_p, Cal $g^{-1}K^{-1}$]

CURVE 1

T	C_p
5.21	1.111 x 10^{-4}
6.17	1.709
7.06	2.607
7.97	3.846
8.89	5.696
9.82	8.218
10.89	1.065 x 10^{-3}
12.14	1.420
13.41	1.893
14.68	2.425
15.99	3.039
17.42	3.777
19.03	4.679
20.91	5.816
23.09	7.239
25.37	8.824
28.25	1.092 x 10^{-2}
31.28	1.318
34.51	1.558
38.04	1.816
41.85	2.071
45.91	2.330
50.42	2.596
55.63	2.869
57.71	2.975*
63.61	3.237
70.56	3.488
78.54	3.735
86.71	3.953
95.05	4.120
103.27	4.258*
110.97	4.365
118.60	4.460*
126.80	4.550
135.61	4.629*
144.51	4.698
153.32	4.761*
161.82	4.815
160.22	4.812*
161.29	4.812*
170.34	4.860*
179.39	4.898*
188.52	4.932
197.80	4.977*
206.63	5.003*

CURVE 1 (cont.)

T	C_p
215.90	5.033 x 10^{-2}
225.32	5.061*
234.65	5.086*
243.73	5.117
252.83	5.136*
262.19	5.158*
258.23	5.140*
267.35	5.158*
276.52	5.181
285.76	5.206*
295.02	5.228*
304.21	5.240*
313.44	5.254
322.78	5.272*
332.13	5.285*
340.56	5.293*
347.12	5.305

CURVE 2*

T	C_p
2.028	3.51 x 10^{-5}
2.028	4.93
2.059	5.13
2.059	3.53
2.118	3.50
2.118	5.37
2.207	5.64
2.324	5.80
2.351	5.63
2.399	5.55
2.407	5.28
2.407	4.08
2.440	5.25
2.556	5.43
2.556	4.55
2.603	5.50
2.659	5.62
2.804	5.93
2.804	5.51
3.010	6.37
3.237	7.47
3.237	7.51
3.400	8.21
3.551	9.11
3.851	1.09 x 10^{-4}

CURVE 2 (cont.)*

T	C_p
4.037	1.23 x 10^{-4}
4.315	1.40
4.519	1.62
4.853	1.98
5.058	2.21
5.675	3.00
6.634	4.61
7.031	5.45
7.283	6.06

CURVE 3*

T	C_p
1.647	1.69 x 10^{-5}
1.647	3.42
1.738	1.97
1.856	2.51
1.856	4.20
1.966	4.92
1.966	3.11
2.118	5.71
2.118	3.93
2.209	4.51
2.324	5.33
2.324	6.71
2.407	5.81
2.440	6.81
2.556	6.46
2.556	6.85
2.804	8.35
2.804	6.93
3.010	1.04 x 10^{-4}
3.010	7.89 x 10^{-5}
3.237	1.28 x 10^{-4}
3.237	8.91 x 10^{-5}
3.400	1.01 x 10^{-4}
3.400	1.44
3.551	1.58
3.703	1.24
3.703	1.84
3.851	2.00
3.851	1.35
3.943	2.16
4.037	2.29
4.037	1.50
4.065	2.32

CURVE 3 (cont.)*

T	C_p
4.093	2.30 x 10^{-4}
4.150	2.17
4.191	2.10
4.251	1.97
4.315	1.74
4.315	1.88
4.401	1.90
4.539	1.95
4.853	2.30
4.853	2.29
5.321	2.89
5.321	2.89
5.675	3.43
5.929	3.92
6.314	4.58
6.634	5.29
7.031	6.18
7.283	6.84

* Not shown on plot

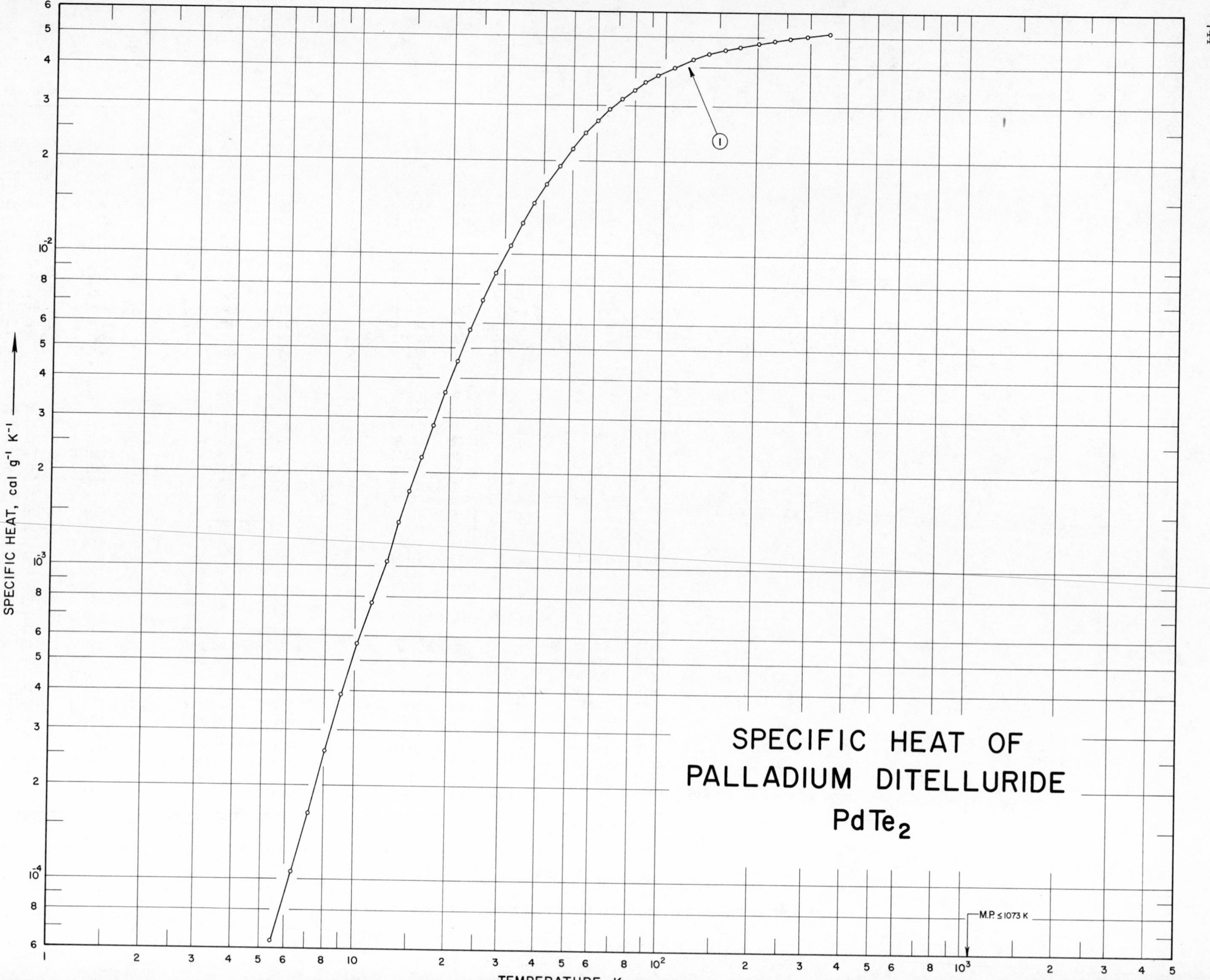
SPECIFIC HEAT OF
PALLADIUM DITELLURIDE
PdTe$_2$
M.P. ≤ 1073 K
TEMPERATURE, K
SPECIFIC HEAT, cal g^{-1} K^{-1}
1

SPECIFICATION TABLE NO. 232 SPECIFIC HEAT OF PALLADIUM DITELLURIDE $PdTe_2$

[For Data Reported in Figure and Table No. 232]

Curve No.	Ref. No.	Year	Temp. Range, K	Reported Error, %	Name and Specimen Designation	Composition (weight percent), Specifications and Remarks
1	224	1961	5-346	~0.1		99.999 tellurium composition, palladium impurities: 0.007 Au, 0.006 SiO_2, 0.005 Pt, 0.003 Ag, 0.002 Fe, 0.0002 Pb, 0.001 Rh, and 0.006 volatile matter; prepared by heating mixture of elements at 800 C, kept molten for 2 hrs, after cooling, the sample was fragmented, sealed in a silica tube and annealed at 500 C for 7 days and cooled to room temperature over a period of two days.

DATA TABLE NO. 232 SPECIFIC HEAT OF PALLADIUM DITELLURIDE $PdTe_2$

[Temperature, T, K; Specific Heat, C_p, Cal $g^{-1}K^{-1}$]

T	C_p	T	C_p
CURVE 1 Series I		CURVE 1 (cont.)	
		16.13	2.244×10^{-3}
106.97	3.996×10^{-2}	17.55	2.837
115.44	4.126*	19.15	3.612
123.62	4.242	21.00	4.557
131.64	4.331*	23.06	5.721
139.67	4.411	25.40	7.107
147.95	4.485*	28.00	8.725
156.84	4.549	28.64	9.123*
166.07	4.615*	31.23	1.078×10^{-2}
175.21	4.668	34.08	1.260
173.55	4.651*	37.24	1.462
182.89	4.709*	40.93	1.687
192.27	4.745*	45.18	1.936
201.40	4.787	49.96	2.198
210.33	4.823*	54.99	2.453
210.19	4.823*	60.26	2.694
219.26	4.859*	65.95	2.936
228.28	4.886	72.16	3.157
237.29	4.914*	79.21	3.384
247.17	4.944*	86.75	3.594
255.19	4.967	94.35	3.765
264.33	4.983*	102.14	3.911*
273.44	4.997*		
282.61	5.033*		
291.69	5.055		
300.74	5.072*		
309.84	5.088*		
391.07	5.094*		
328.32	5.121*		
337.48	5.141*		
346.37	5.155		
Series II			
5.42	6.360×10^{-5}		
6.30	1.161×10^{-4}		
7.13	1.632		
8.07	2.572		
9.07	3.880		
10.14	5.677		
11.32	7.657		
12.53	1.039×10^{-3}		
13.73	1.393		
13.59	1.349*		
14.81	1.741		

* Not shown on plot

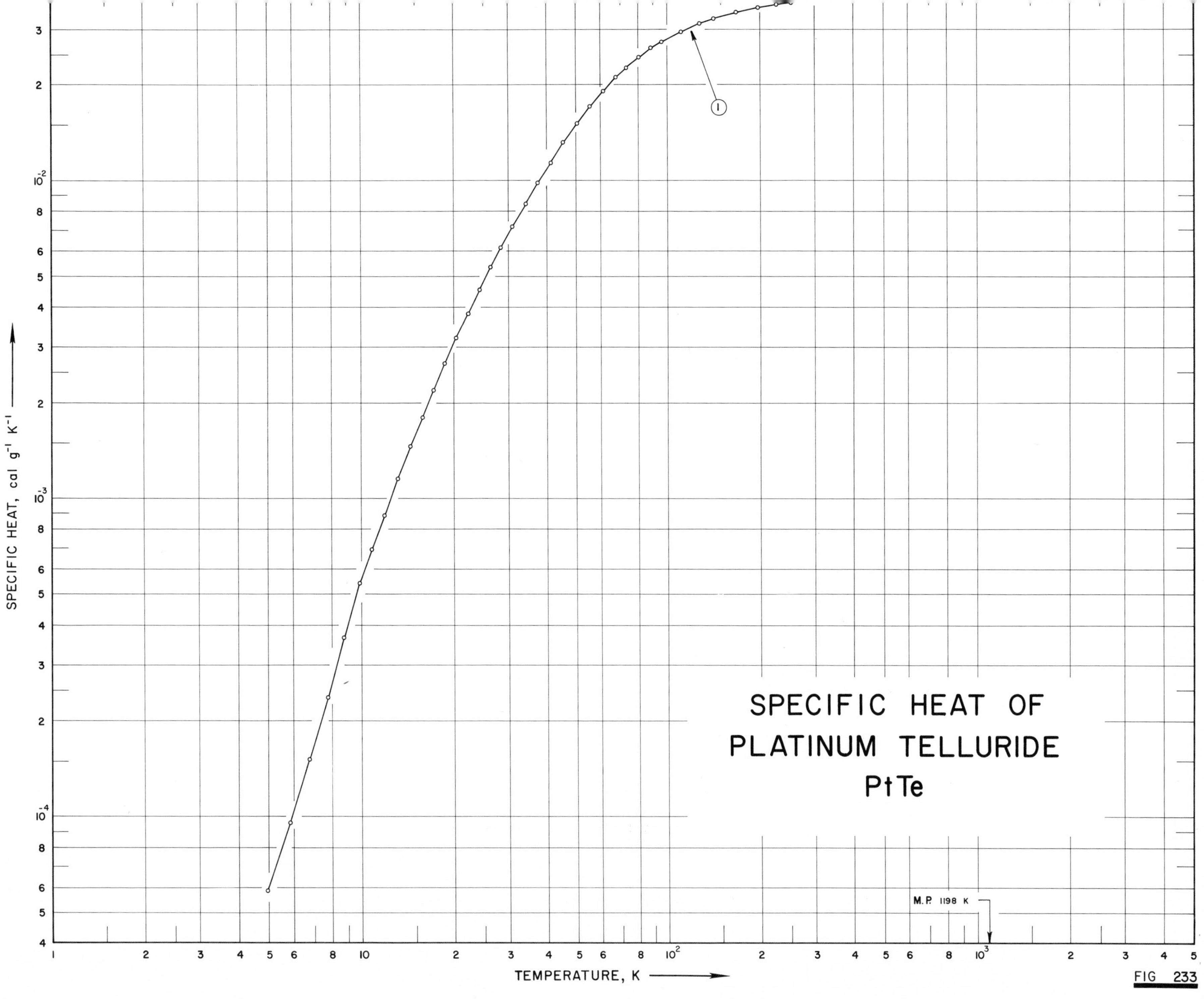

SPECIFIC HEAT OF
PLATINUM TELLURIDE
PtTe
1
M.P. 1198 K
SPECIFIC HEAT, cal g⁻¹ K⁻¹
TEMPERATURE, K

FIG 233

SPECIFICATION TABLE NO. 233 SPECIFIC HEAT OF PLATINUM TELLURIDE PtTe

[For Data Reported in Figure and Table No. 233]

Curve No.	Ref. No.	Year	Temp. Range, K	Reported Error, %	Name and Specimen Designation	Composition (weight percent), Specifications and Remarks
1	223	1961	5-347	~0.10		99.999 Te, platinum impurities: 0.009 volatile material, < 0.001 Fe, Pb, 0.0017 Pd, and 0.0001 Au; prepared by allowing elements to react at 1000 C for 6 hrs, then raising temperature to 1200 C for 1 hr to melt sample; cool to room temperature overnight, melted, broken into fragments, annealed at 500 C for 7 days and slowly cooled to room temperature for 7 days; density = 750 lb ft^{-3}.

DATA TABLE NO. 233 SPECIFIC HEAT OF PLATINUM TELLURIDE PtTe

[Temperature, T, K; Specific Heat, C_p, Cal $g^{-1}K^{-1}$]

T	C_p
CURVE 1	
Series I	
67.93	2.107×10^{-2}
73.71	2.262
81.17	2.445
88.74	2.606
96.25	2.731
103.91	2.843*
111.67	2.946
119.76	3.034*
128.22	3.116
136.77	3.188*
142.66	3.233
151.34	3.284*
160.22	3.339*
169.23	3.379
178.35	3.419*
187.73	3.458*
197.16	3.490
206.61	3.518*
216.02	3.541*
225.33	3.563
234.50	3.587*
243.63	3.610*
252.67	3.621
261.16	3.645*
Series II	
4.92	5.888×10^{-5}
5.83	9.606
6.78	1.518×10^{-4}
7.76	2.386
7.67	2.355*
8.73	3.650
9.84	5.426
10.83	6.938
11.95	8.853
13.19	1.155×10^{-3}
14.46	1.467
15.76	1.810
17.12	2.197
18.64	2.665
20.32	3.201
22.16	3.827
24.14	4.534
CURVE 1 (cont.)	
26.30	5.355×10^{-3}
28.40	6.170
30.92	7.186
34.06	8.482
37.40	9.864
41.20	1.145×10^{-2}
45.52	1.322
50.34	1.511
55.60	1.705
61.37	1.903
67.61	2.094*
Series III	
261.12	3.633×10^{-2}*
270.18	3.651*
279.15	3.665
Series IV	
259.71	3.638×10^{-2}*
268.84	3.651*
277.99	3.665*
287.14	3.680*
296.08	3.696*
304.91	3.710*
313.80	3.714
322.95	3.721*
332.36	3.735*
340.92	3.749*
347.07	3.755

*Not shown on plot

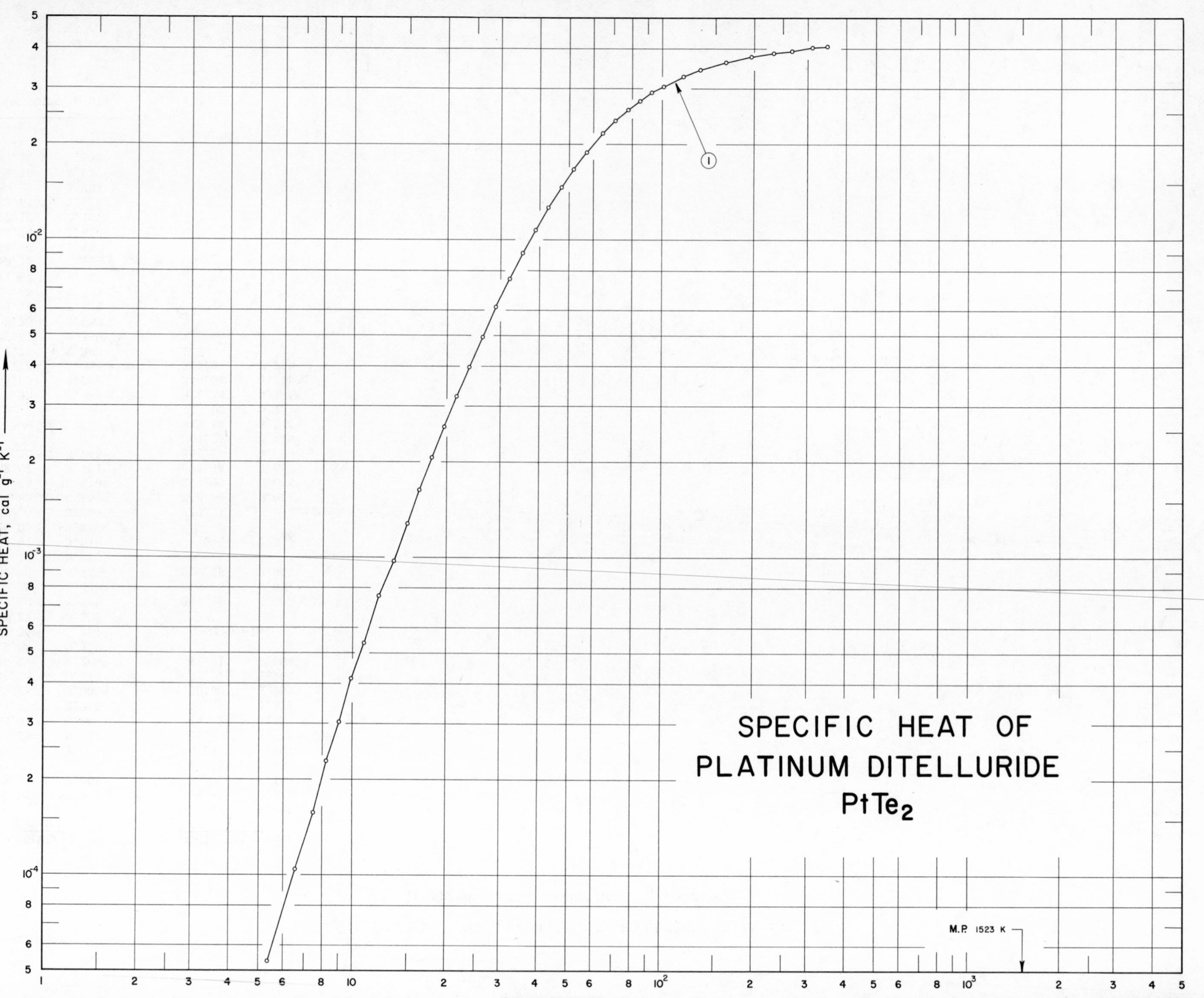
SPECIFIC HEAT OF
PLATINUM DITELLURIDE
PtTe2
M.P. 1523 K
1
SPECIFIC HEAT, cal g-1 K-1

SPECIFICATION TABLE NO. 234 SPECIFIC HEAT OF PLATINUM DITELLURIDE $PtTe_2$

[For Data Reported in Figure and Table No. 234]

Curve No.	Ref. No.	Year	Temp. Range, K	Reported Error, %	Name and Specimen Designation	Composition (weight percent), Specifications and Remarks
1	224	1961	5-347	~0.1		High purity; platinum impurities: <0.001 Fe, Pb, 0.0007 Pd, 0.0001 Au, 0.009 volatile materials; prepared by reacting the elements at 1000 C for 5 hrs, annealed at 500 C for 7 days and cooled to room temperature for 7 days.

DATA TABLE NO. 234 SPECIFIC HEAT OF PLATINUM DITELLURIDE $PtTe_2$

[Temperature, T, K; Specific Heat, C_p, Cal $g^{-1}K^{-1}$]

T	C_p	T	C_p
CURVE 1 Series I		CURVE 1 (cont.) Series III	
64.49	2.164 x 10^{-2}	150.33	3.558 x 10^{-2}*
70.99	2.374	159.11	3.615*
77.99	2.569	167.91	3.662*
85.00	2.747	161.51	3.629*
92.84	2.907	170.18	3.669*
101.23	3.048	178.89	3.709*
109.34	3.166*	187.73	3.744*
117.58	3.269	196.44	3.775
126.02	3.362*	205.12	3.800*
134.67	3.440	213.88	3.831*
143.70	3.513*	222.94	3.855*
161.41	3.622	232.31	3.875
		241.69	3.908*
Series II		250.92	3.922*
		260.07	3.940*
5.37	5.330 x 10^{-5}	266.80	3.948
6.57	1.044 x 10^{-4}	275.73	3.966*
7.48	1.577	283.06	3.979*
8.22	2.287	291.97	3.997*
9.02	3.042	301.00	4.008*
9.89	4.153	310.14	4.024
10.89	5.396	319.43	4.035*
12.13	7.550	328.74	4.051*
13.50	9.749	337.87	4.059*
14.92	1.281 x 10^{-3}	346.72	4.068
16.36	1.639		
17.90	2.059		
19.62	2.578		
21.51	3.209		
23.64	3.971		
26.14	4.948		
28.96	6.138		
32.08	7.541		
35.35	9.058		
38.92	1.079 x 10^{-2}		
42.90	1.266		
47.14	1.463		
51.80	1.668		
57.10	1.889		
63.45	2.131*		
70.39	2.356*		

* Not shown on plot

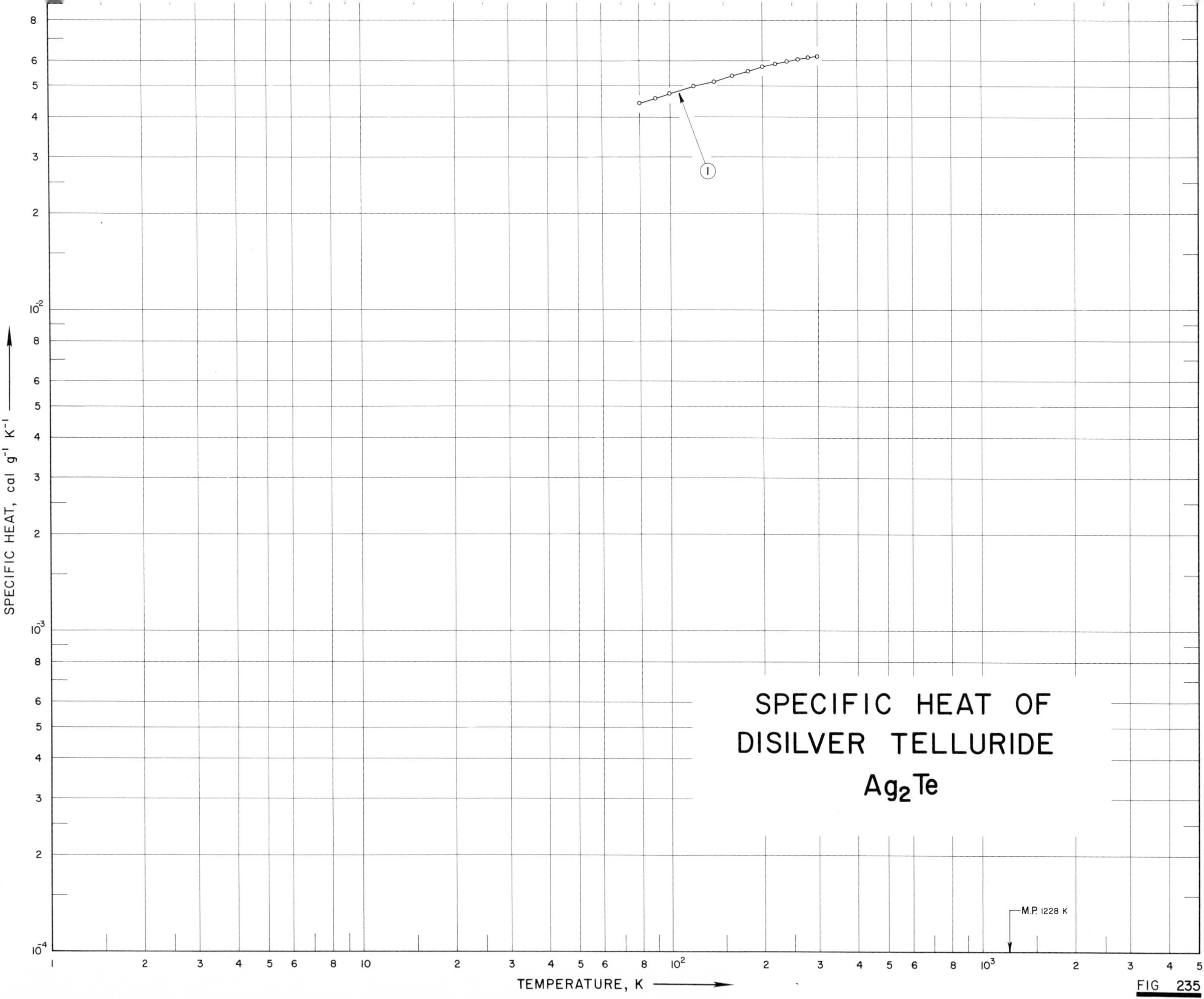

SPECIFIC HEAT OF
DISILVER TELLURIDE
Ag_2Te
1
M.P. 1228 K
SPECIFIC HEAT, cal g^{-1} K^{-1}
TEMPERATURE, K
FIG 235

SPECIFICATION TABLE NO. 235 SPECIFIC HEAT OF DISILVER TELLURIDE Ag_2Te

[For Data Reported in Figure and Table No. 235]

Curve No.	Ref. No.	Year	Temp. Range, K	Reported Error, %	Name and Specimen Designation	Composition (weight percent), Specifications and Remarks
1	160	1959	80-300	3-7		Polycrystalline.

DATA TABLE NO. 235 SPECIFIC HEAT OF DISILVER TELLURIDE Ag_2Te

[Temperature, T, K; Specific Heat, C_p, Cal $g^{-1}K^{-1}$]

T	C_p
CURVE 1	
80	4.41×10^{-2}
90	4.59
100	4.71
120	4.95
140	5.17
160	5.38
180	5.56
200	5.73
220	5.87
240	5.96
260	6.05
280	6.12
300	6.18

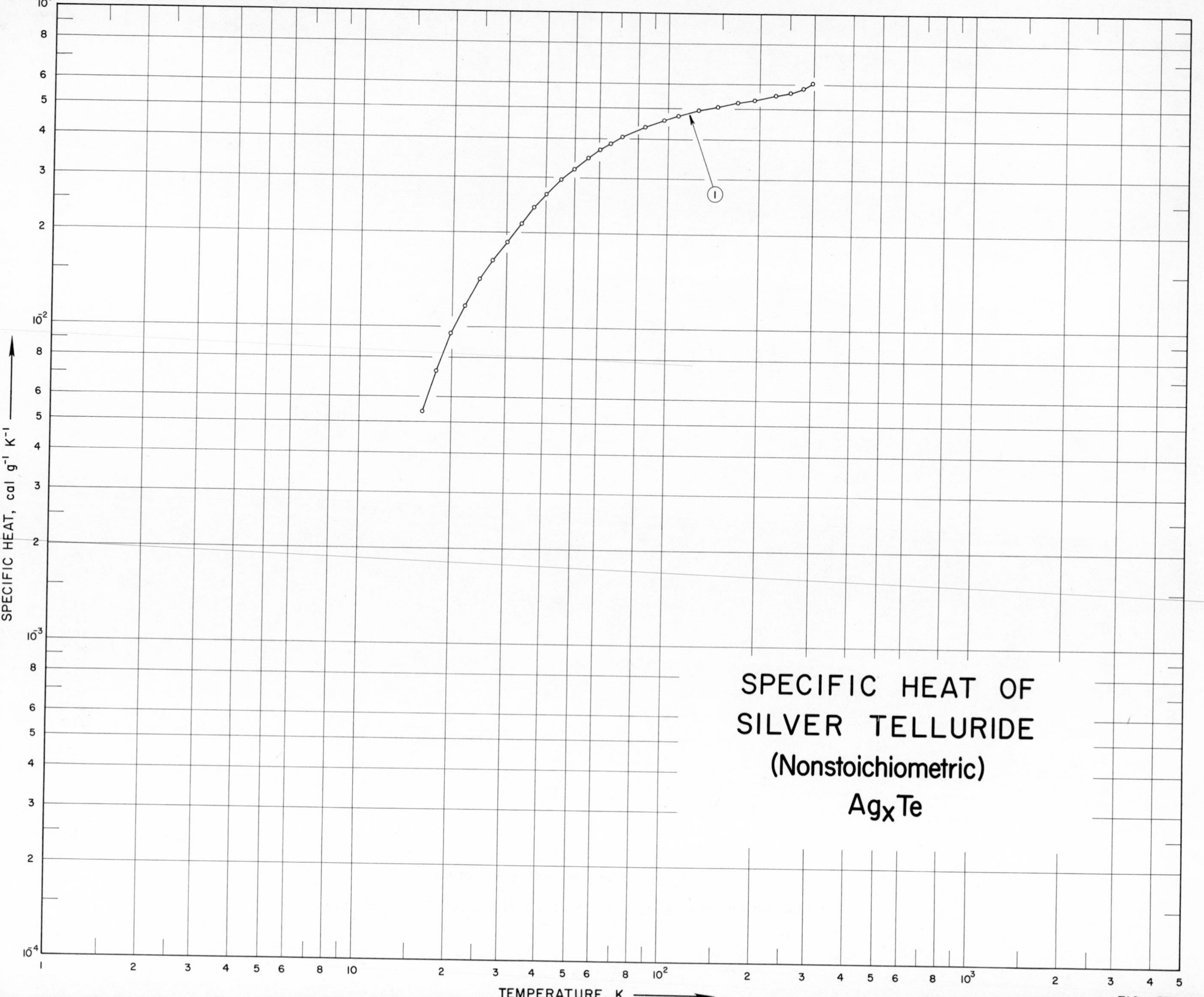

SPECIFIC HEAT OF
SILVER TELLURIDE
(Nonstoichiometric)
Ag_xTe
1
TEMPERATURE, K
SPECIFIC HEAT, cal g^{-1} K^{-1}

SPECIFICATION TABLE NO. 236 SPECIFIC HEAT OF SILVER TELLURIDE (nonstoichiometric) Ag_xTe

[For Data Reported in Figure and Table No. 236]

Curve No.	Ref. No.	Year	Temp. Range, K	Reported Error, %	Name and Specimen Designation	Composition (weight percent), Specifications and Remarks
1	225	1962	16-296			99.99 $Ag_{1.88}Te$; crushed under argon pressure.

DATA TABLE NO. 236 SPECIFIC HEAT OF SILVER TELLURIDE Ag_xTe (nonstoichiometric)

[Temperature, T, K; Specific Heat, C_p, Cal $g^{-1}K^{-1}$]

T	C_p
CURVE 1	
16.16	5.391×10^{-3}
17.91	7.268
19.90	9.539
22.10	1.169×10^{-2}
24.65	1.423
27.37	1.632
30.25	1.856
33.50	2.135
36.82	2.408
40.44	2.653
45.36	2.950
50.09	3.192
55.49	3.467
60.59	3.683
65.62	3.843
71.65	4.040
84.83	4.367
97.60	4.591
109.38	4.733
126.51	4.945
145.70	5.094
154.12	5.166*
161.87	5.197*
170.15	5.269
185.28	5.348*
192.59	5.369
200.58	5.418*
207.64	5.448*
214.12	5.490*
220.67	5.518
226.05	5.542*
233.15	5.569*
244.47	5.624*
245.72	5.606*
251.79	5.645
257.70	5.690*
269.38	5.802*
277.27	5.839
283.34	5.908*
289.54	5.966*
296.16	6.033

* Not shown on plot

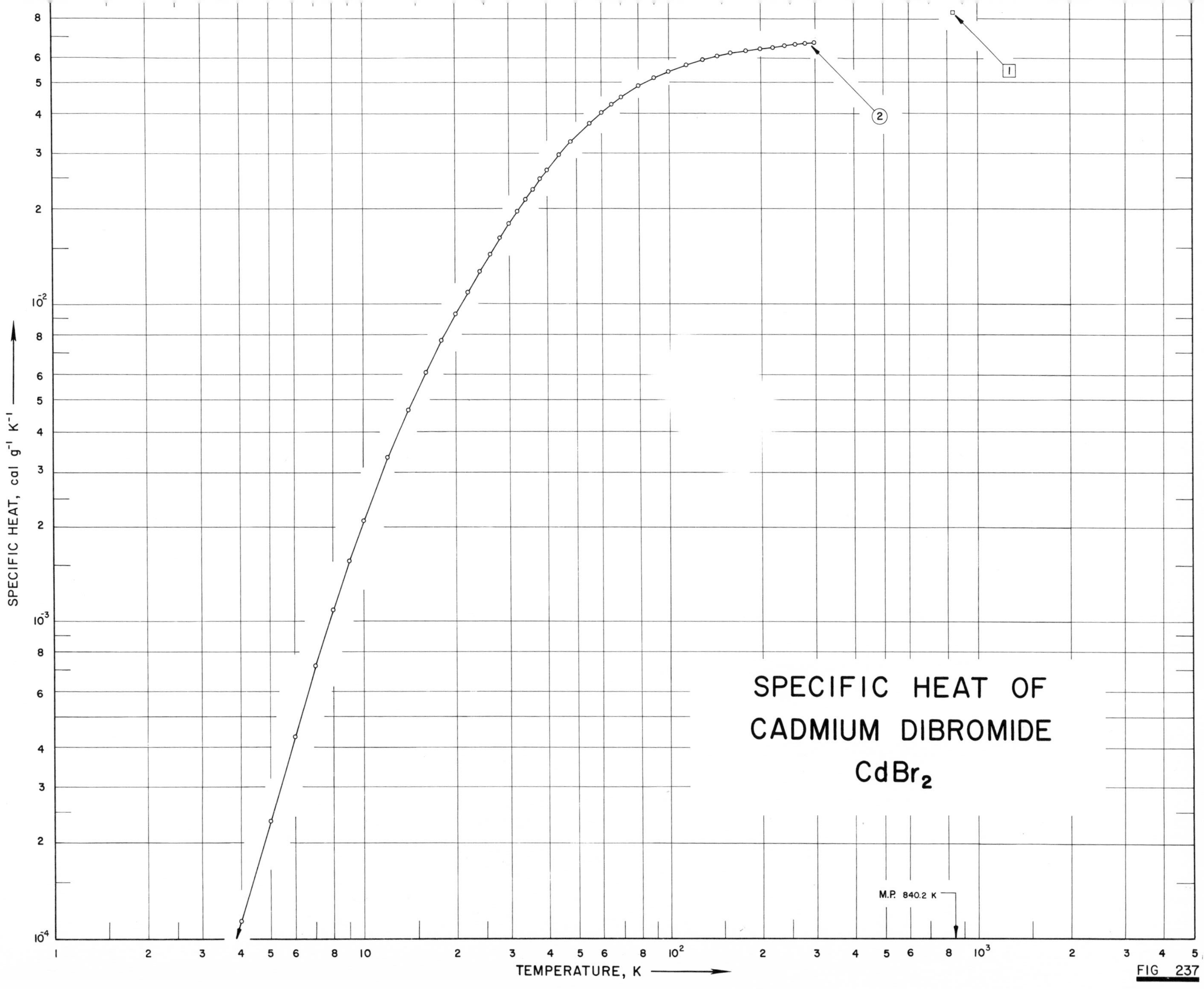
SPECIFIC HEAT OF
CADMIUM DIBROMIDE
$CdBr_2$
M.P. 840.2 K
TEMPERATURE, K
SPECIFIC HEAT, cal g^{-1} K^{-1}
1
2
FIG 237

SPECIFICATION TABLE NO. 237 SPECIFIC HEAT OF CADMIUM DIBROMIDE $CdBr_2$

[For Data Reported in Figure and Table No. 237]

Curve No.	Ref. No.	Year	Temp. Range, K	Reported Error, %	Name and Specimen Designation	Composition (weight percent), Specifications and Remarks
1	242	1961	841.2			
2	312	1960	2-300	< 1.0		

DATA TABLE NO. 237 SPECIFIC HEAT OF CADMIUM DIBROMIDE $CdBr_2$

[Temperature, T, K; Specific Heat, C_p, Cal $g^{-1}K^{-1}$]

T	C_p	T	C_p
CURVE 1		CURVE 2 (cont.)	
841.2	8.38×10^{-2}	110.00	5.632×10^{-2}
		115.00	5.713
CURVE 2		120.00	5.790*
		125.00	5.864*
2.00	1.397×10^{-5}*	130.00	5.930
3.00	4.706*	135.00	5.992*
4.00	1.147×10^{-4}	140.00	6.047*
5.00	2.353	145.00	6.099
6.00	4.338	150.00	6.147*
7.00	7.279	155.00	6.191*
8.00	1.096×10^{-3}	160.00	6.228
9.00	1.551	165.00	6.261*
10.00	2.081	170.00	6.290*
12.00	3.309	175.00	6.312*
14.00	4.669	180.00	6.334
16.00	6.139	185.00	6.360*
18.00	7.720	190.00	6.378*
20.00	9.338	195.00	6.397*
22.00	1.099×10^{-2}	200.00	6.415
24.00	1.272	205.00	6.433*
26.00	1.448	210.00	6.455*
28.00	1.625	215.00	6.470*
30.00	1.801	220.00	6.489
32.00	1.978	225.00	6.507*
34.00	2.151	230.00	6.525*
36.00	2.331	235.00	6.540*
38.00	2.504	240.00	6.558
40.00	2.665	245.00	6.573*
42.00	2.823*	250.00	6.588*
44.00	2.974	255.00	6.606*
46.00	3.128*	260.00	6.621
48.00	3.276	265.00	6.636*
50.00	3.419*	270.00	6.654*
55.00	3.746	275.00	6.669*
60.00	4.036	280.00	6.683
65.00	4.290	285.00	6.702*
70.00	4.514	290.00	6.713*
75.00	4.720*	295.00	6.727*
80.00	4.900	298.15	6.735*
85.00	5.062*	300.00	6.739
90.00	5.202		
95.00	5.327*		
100.00	5.437		
105.00	5.540*		

* Not shown on plot

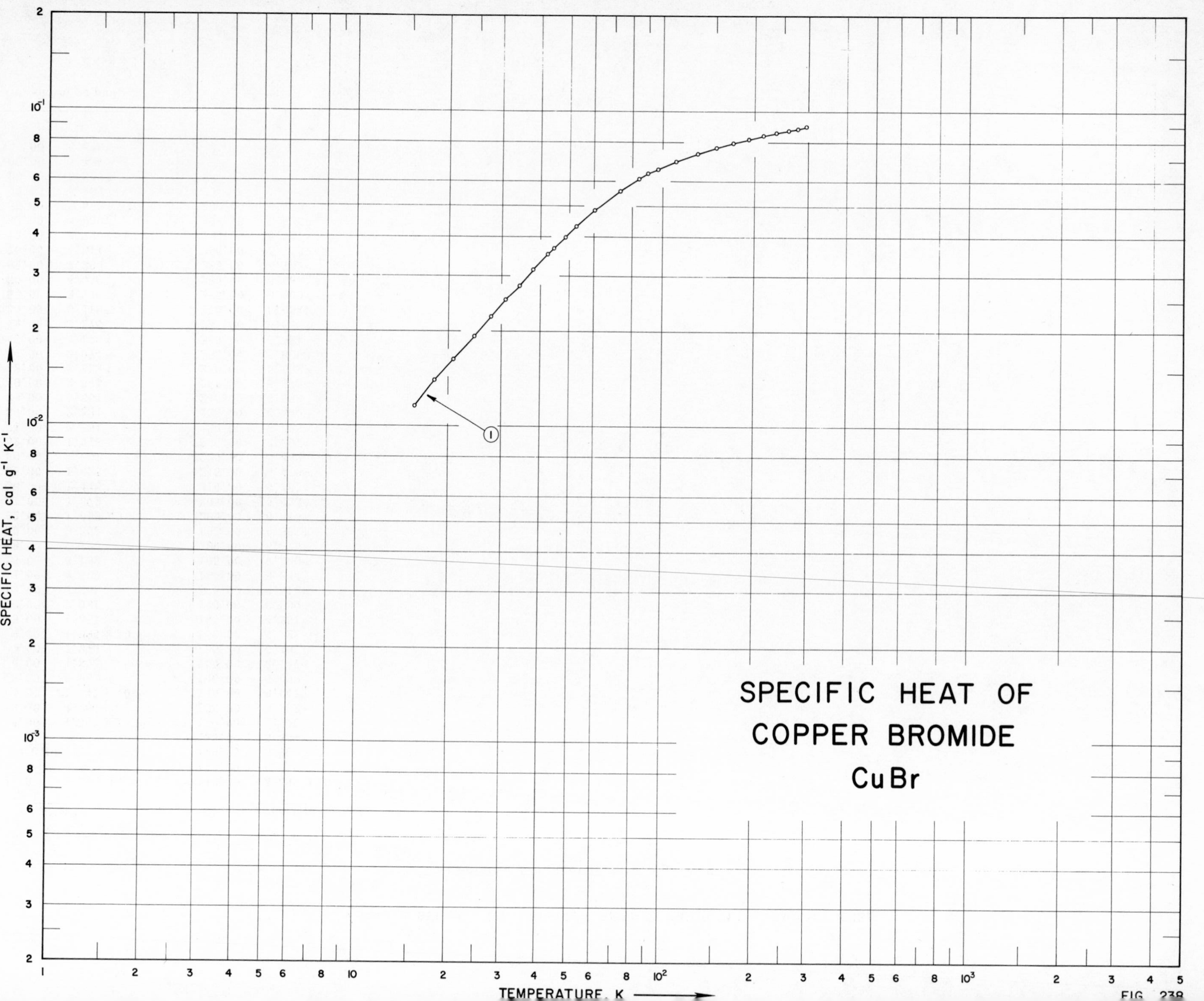
SPECIFIC HEAT OF
COPPER BROMIDE
CuBr
SPECIFIC HEAT, cal g⁻¹ K⁻¹
TEMPERATURE, K
1
FIG. 238

SPECIFICATION TABLE NO. 238 SPECIFIC HEAT OF COPPER BROMIDE CuBr

[For Data Reported in Figure and Table No. 238]

Curve No.	Ref. No.	Year	Temp. Range, K	Reported Error, %	Name and Specimen Designation	Composition (weight percent), Specifications and Remarks
1	313	1952	16-296			> 99.92 CuBr.

DATA TABLE NO. 238 SPECIFIC HEAT OF COPPER BROMIDE $CuBr$

[Temperature, T, K; Specific Heat, Cp, Cal $g^{-1}K^{-1}$]

T	Cp
CURVE 1	
15.51	1.175×10^{-2}
18.09	1.413
20.78	1.654
24.25	1.958
27.71	2.262
30.89	2.555
34.29	2.836
37.90	3.187
42.47	3.570
44.45	3.725
48.54	4.046
52.96	4.377
54.34	4.525*
60.35	4.949
66.11	5.327*
73.09	5.685
78.58	5.988*
84.00	6.211
84.73	6.278*
89.87	6.453
96.94	6.663
104.44	6.869*
112.43	7.067
131.67	7.297*
131.27	7.485
141.41	7.660
150.84	7.820
189.93	7.939*
170.11	8.071
180.57	8.176*
192.01	8.322
203.50	8.399*
214.73	8.518
225.10	8.594*
236.25	8.713
247.46	8.769*
258.27	8.866
268.22	8.894*
277.77	8.971
286.91	9.076*
295.83	9.146

* Not shown on plot

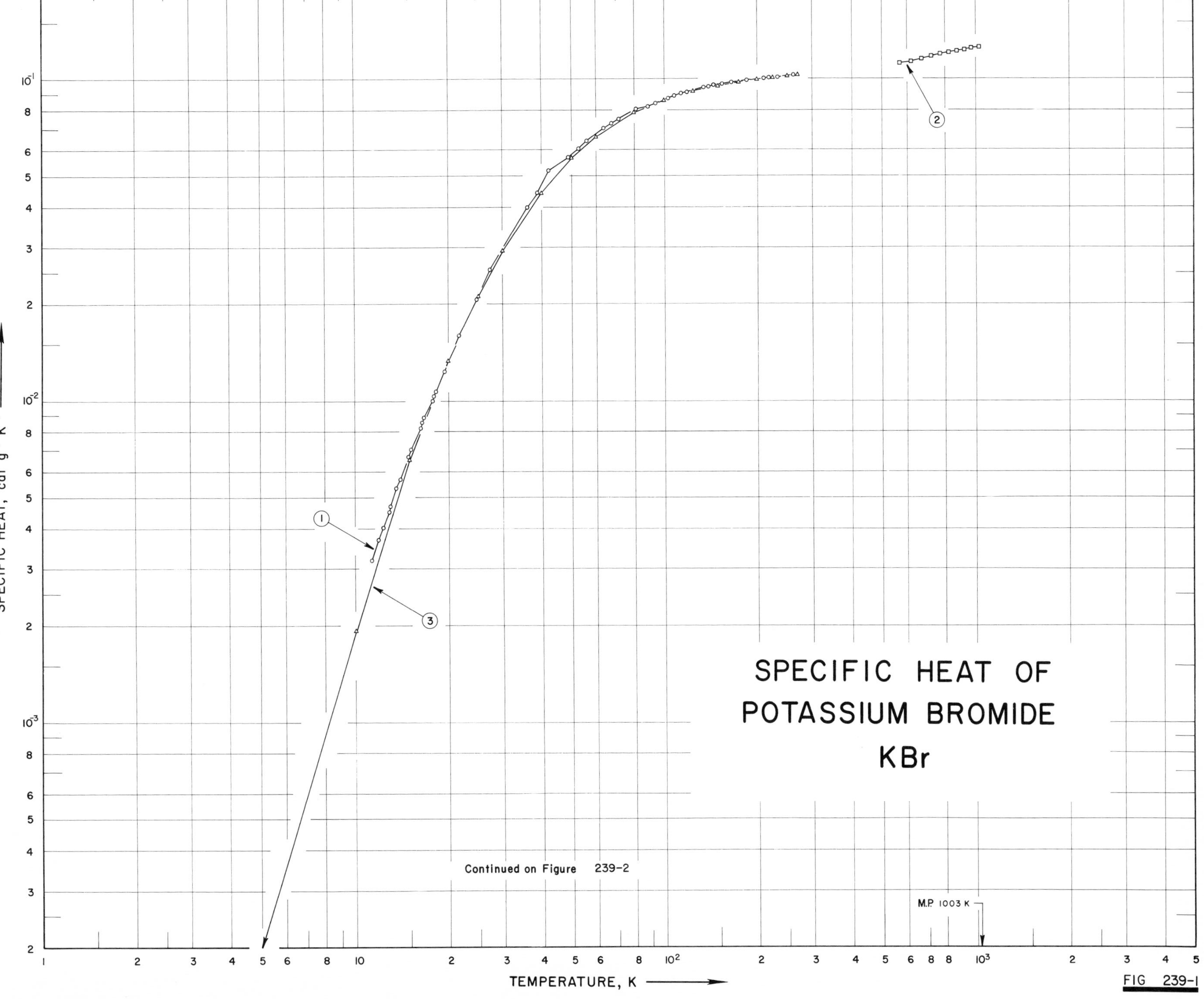
SPECIFIC HEAT OF
POTASSIUM BROMIDE
KBr
SPECIFIC HEAT, cal g⁻¹ K⁻¹
TEMPERATURE, K
M.P. 1003 K
Continued on Figure 239-2
1
2
3
FIG 239-1

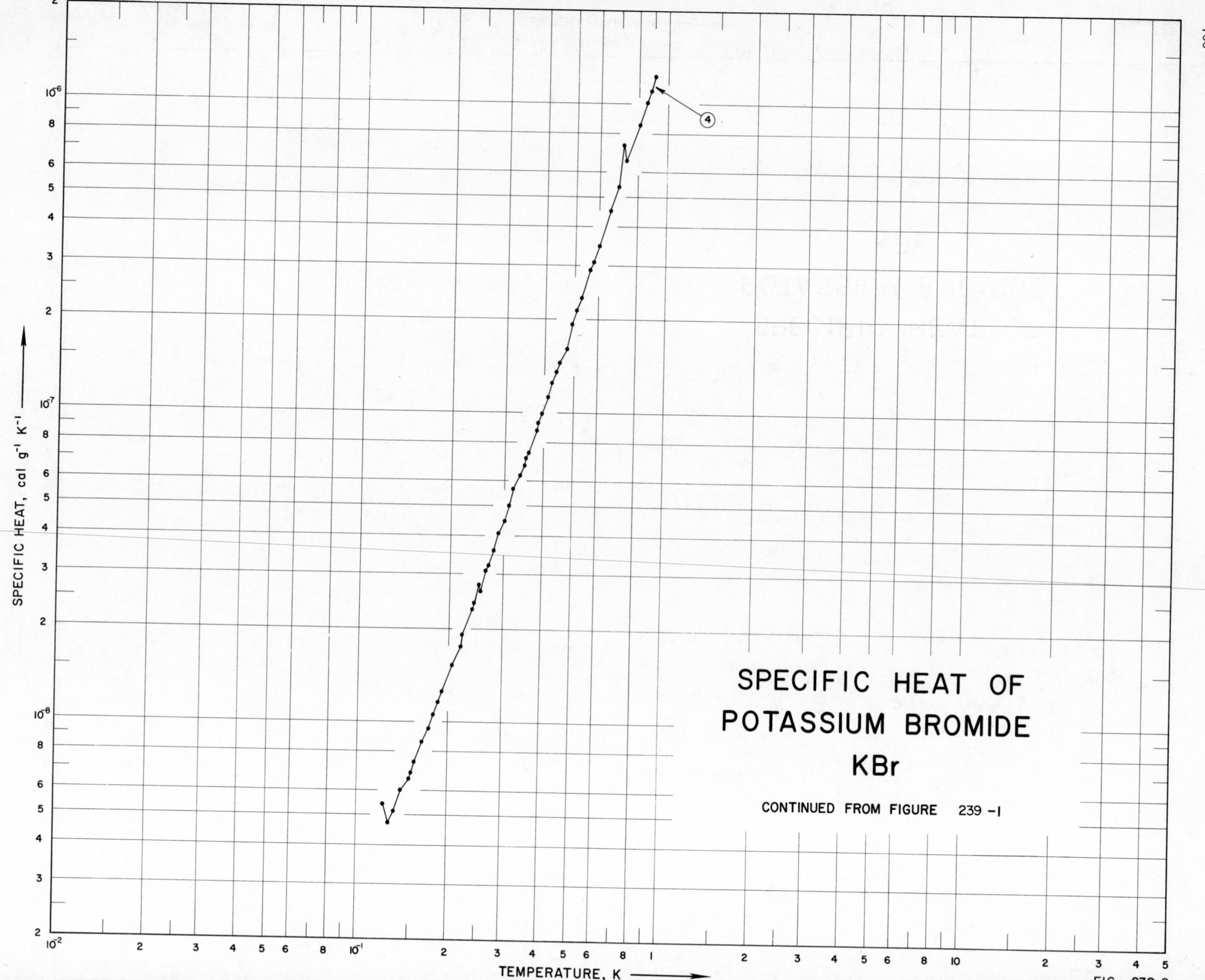
SPECIFIC HEAT OF
POTASSIUM BROMIDE
KBr
CONTINUED FROM FIGURE 239 -I
TEMPERATURE, K
SPECIFIC HEAT, cal g^{-1} K^{-1}
4

SPECIFICATION TABLE NO. 239 SPECIFIC HEAT OF POTASSIUM BROMIDE KBr

[For Data Reported in Figure and Table No. 239]

Curve No.	Ref. No.	Year	Temp. Range, K	Reported Error, %	Name and Specimen Designation	Composition (weight percent), Specifications and Remarks
1	243	1949	11-270			Sample supplied by the Harshaw Chem. Co.; measured under dry helium gas.
2	244	1953	573-1003	0.1-0.2		High purity, optical quality; sample supplied by the Harshaw Chem. Co.; measured under helium atmosphere.
3	245	1957	3-270	0.2-2		Very small traces of Co, Cu, and Mn; single crystal; sample supplied by the Harshaw Chem. Co.
4	314	1963	0.1-0.9			

DATA TABLE NO. 239 SPECIFIC HEAT OF POTASSIUM BROMIDE KBr

[Temperature, T, K; Specific Heat, C_p, Cal $g^{-1}K^{-1}$]

T	C_p
CURVE 1	
11.30	3.193×10^{-3}
11.90	3.697
12.10	3.865*
12.40	4.033
12.90	4.537
13.00	4.705
13.60	4.378
14.00	5.714
14.10	5.882*
14.90	6.722
15.20	7.058
15.40	7.226*
16.30	8.235
16.50	8.571
16.70	8.907
17.80	1.008×10^{-2}
18.00	1.042
18.30	1.076
19.50	1.244
19.60	1.260*
21.70	1.613
24.80	2.084
27.40	2.571
29.90	2.975*
36.10	4.000
38.90	4.453
42.30	5.243
45.50	5.344*
49.00	5.747
52.80	6.117
56.20	6.470
59.80	6.789*
63.80	7.075
67.80	7.344
71.70	7.579
80.80	7.966*
81.30	8.100
84.70	8.151*
89.40	8.285
94.40	8.487
99.00	8.655*
103.70	8.789
108.80	8.924
114.00	9.075
CURVE 1 (cont.)	
119.50	9.176×10^{-2}
124.80	9.293*
130.00	9.394
135.30	9.461
140.60	9.529
146.10	9.613
155.90	9.730
161.40	9.764*
167.00	9.831
173.10	9.882*
179.70	9.915*
186.40	9.982
192.80	1.003×10^{-1}*
198.50	1.005*
205.10	1.008*
211.90	1.010
219.20	1.013
225.90	1.017*
232.80	1.018
246.80	1.025*
254.00	1.027*
262.20	1.035
269.90	1.039*
CURVE 2	
573	1.126×10^{-1}
623	1.147
673	1.168
723	1.185
773	1.201
823	1.216
873	1.230
923	1.244
973	1.258
1003	1.265
CURVE 3	
3	4.067×10^{-5}*
5	1.991×10^{-4}*
10	1.948×10^{-3}
15	6.554
20	1.347×10^{-2}
CURVE 3 (cont.)	
25	2.138×10^{-2}
30	2.949
40	4.467
50	5.704
60	6.650
80	7.916
100	8.659
125	9.222
150	9.587
175	9.839
200	1.001×10^{-1}
225	1.015
250	1.029
270	1.039
CURVE 4	
Series 1	
0.2188	1.816×10^{-8}*
0.2189	1.768
0.2199	1.839*
0.2205	1.925
0.2248	1.925*
0.2253	1.962*
0.2397	2.323
0.2428	2.421
0.2502	2.789
0.2534	2.658
0.2542	2.797*
0.2552	2.800*
0.2628	3.084
0.2675	3.132*
0.2690	3.217
0.2786	3.552*
0.2787	3.592
0.2887	4.091
0.2898	3.998*
0.2973	4.324*
0.3005	4.4703*
0.3013	4.466
0.3101	4.987*
0.3120	5.009
0.3209	5.698
CURVE 4 (cont.)	
0.3241	5.527×10^{-8}*
0.3291	5.849*
0.3373	6.285
0.3400	6.413*
0.3492	6.793
0.3520	7.156
0.3589	7.466
0.3802	8.788
0.3851	9.272
0.3931	1.004×10^{-7}*
0.3963	9.949×10^{-8}
0.4129	1.127×10^{-7}*
0.4157	1.144*
0.4159	1.1345
0.4268	1.257
0.4322	1.285*
0.4427	1.364
0.4520	1.459
0.4796	1.629
0.4962	1.946
0.5145	2.158
0.5303	2.361
0.5663	2.911
0.5814	3.096
0.6033	3.490
0.6588	4.512
0.7221	7.422
0.8123	8.531
0.8585	1.012×10^{-6}
0.8885	1.103
0.9108	1.204
Series 2	
0.1236	5.431×10^{-9}
0.1272	4.742
0.1339	5.165
0.1412	6.011
0.1414	5.957*
0.1504	6.562
0.1508	6.755*
0.1515	6.853*
0.1521	6.867
0.1554	7.434
CURVE 4 (cont.)	
0.1658	8.627×10^{-9}
0.1732	9.527
0.1742	9.684*
0.1744	9.515*
0.1783	1.027×10^{-8}*
0.1792	1.060
0.1793	1.044*
0.1851	1.155*
0.1853	1.157
0.1906	1.245*
0.1916	1.252
0.1919	1.262*
0.2051	1.534
0.2056	1.516*
0.2299	2.019*
0.2439	2.423*
0.2445	2.451*
0.2620	2.935*
0.2902	3.958*
0.3221	5.354*
0.3393	6.146*
0.3594	7.371*
0.3841	8.943*
0.4059	1.069×10^{-7}*
0.4467	1.412*
0.4576	1.524*
0.5309	2.377*
0.5983	3.403*
0.6009	3.482*
0.6529	4.462*
0.6961	5.401
0.7398	6.590
0.8614	1.024×10^{-6}*

* Not shown on plot

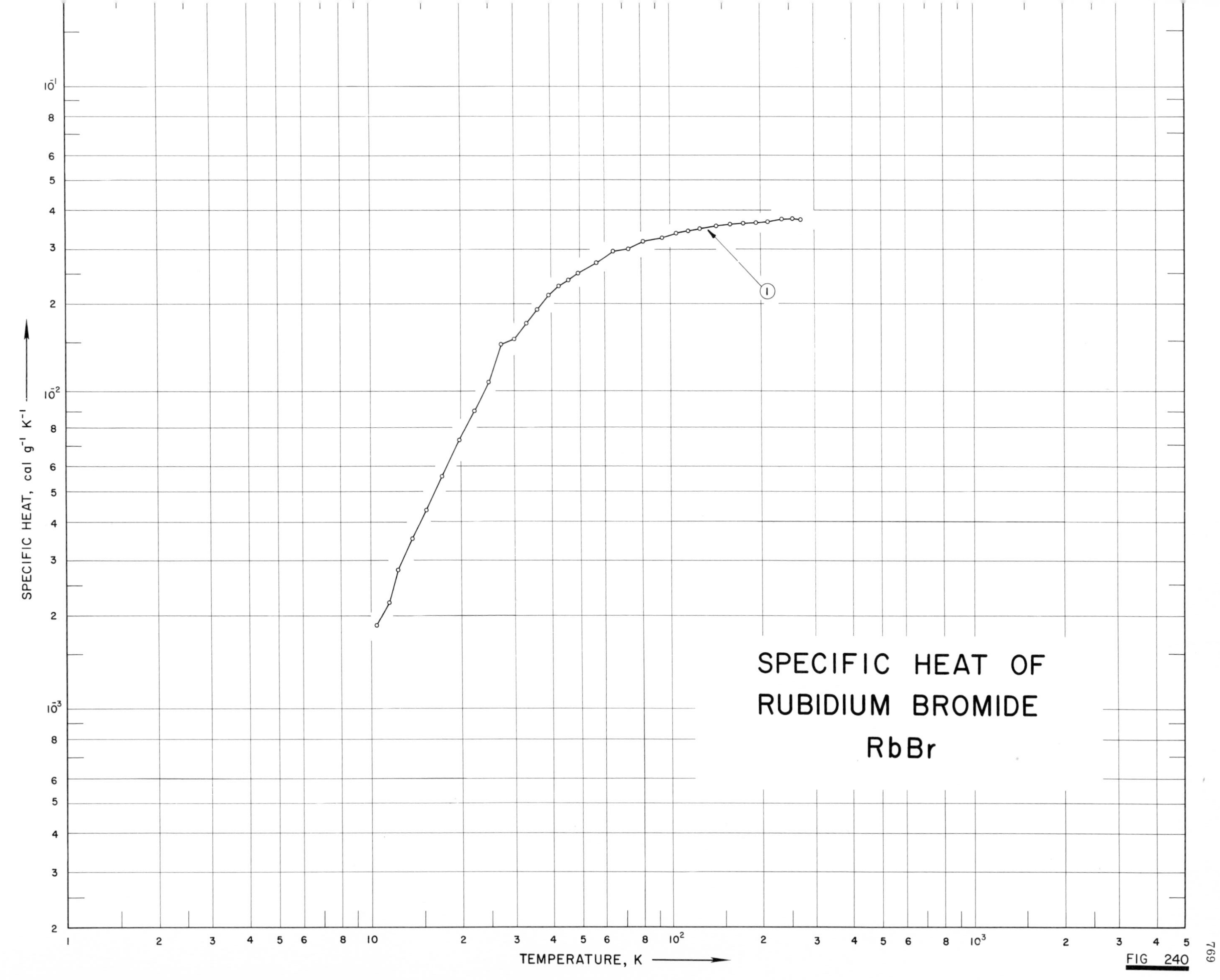
SPECIFIC HEAT OF
RUBIDIUM BROMIDE
RbBr
1
TEMPERATURE, K
SPECIFIC HEAT, cal g⁻¹ K⁻¹
FIG 240

SPECIFICATION TABLE NO. 240 SPECIFIC HEAT OF RUBIDIUM BROMIDE RbBr

[For Data Reported in Figure and Table No. 240]

Curve No.	Ref. No.	Year	Temp. Range, K	Reported Error, %	Name and Specimen Designation	Composition (weight percent), Specifications and Remarks
1	243	1949	10-273			

DATA TABLE NO. 240 SPECIFIC HEAT OF RUBIDIUM BROMIDE RbBr

[Temperature, T, K; Specific Heat, Cp, Cal $g^{-1}K^{-1}$]

T	Cp	T	Cp
CURVE 1		CURVE 1 (cont.)	
10.5	1.88×10^{-3}	187.5	3.70×10^{-2}*
11.6	2.24	193.3	3.67
12.4	2.84	199.2	3.73*
13.8	3.57	200.1	3.73*
15.4	4.42	206.1	3.75*
17.4	5.69	212.2	3.75
19.8	7.38	217.9	3.76*
22.2	9.20	223.7	3.76*
24.8	1.13×10^{-2}	229.6	3.76*
27.3	1.52	235.9	3.78
30.3	1.55	242.4	3.78*
33.2	1.75	249.2	3.79*
36.2	1.94	256.5	3.81
39.4	2.17	264.6	3.79*
42.8	2.31	272.7	3.78
46.2	2.44		
49.6	2.58		
52.8	2.70*		
56.7	2.79		
60.5	2.89*		
64.5	2.96		
68.5	3.03*		
72.7	3.09		
76.9	3.15*		
81.5	3.21		
86.5	3.24*		
86.8	3.24*		
94.5	3.31		
99.9	3.38*		
105.8	3.41		
110.6	3.45*		
116.0	3.48		
121.4	3.52*		
126.8	3.54		
132.2	3.56*		
137.6	3.58*		
143.1	3.61		
148.4	3.62*		
153.8	3.63*		
159.3	3.65		
164.8	3.67*		
170.5	3.67*		
176.0	3.68		
181.7	3.69*		

* Not shown on plot

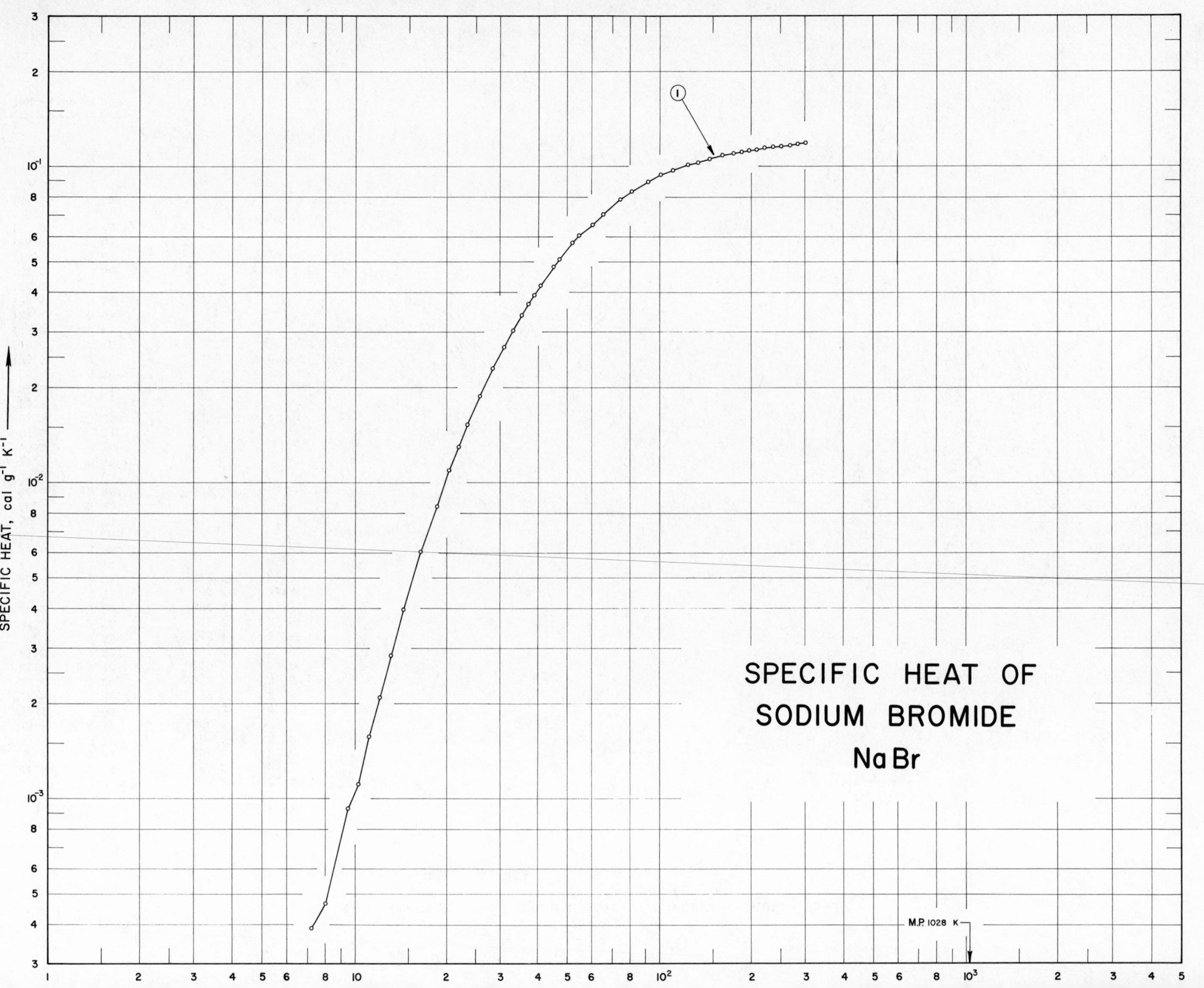

SPECIFIC HEAT OF
SODIUM BROMIDE
NaBr
M.P. 1028 K
SPECIFIC HEAT, cal g⁻¹ K⁻¹
1

SPECIFICATION TABLE NO. 241 SPECIFIC HEAT OF SODIUM BROMIDE NaBr

[For Data Reported in Figure and Table No. 241]

Curve No.	Ref. No.	Year	Temp. Range, K	Reported Error, %	Name and Specimen Designation	Composition (weight percent), Specifications and Remarks
1	246	1964	7-302	0.3		> 99.979 NaB, < 0.02 K, 0.0001-0.001 Ca and Mg; crystalline structure; reagent grade product recrystallized from water and dried overnight at 600 C.

DATA TABLE NO. 241 SPECIFIC HEAT OF SODIUM BROMIDE NaBr

[Temperature, T, K; Specific Heat, C_p, Cal $g^{-1}K^{-1}$]

T	C_p	T	C_p	T	C_p
CURVE 1		CURVE 1 (cont.)		CURVE 1 (cont.)	
7.21	3.800 x 10^{-4}	102.00	9.418 x 10^{-2}	246.35	1.163 x 10^{-1}*
8.00	4.675	104.71	9.512*	250.10	1.165
9.45	9.320	107.95	9.631*	253.83	1.169*
10.25	1.118 x 10^{-3}	111.50	9.751	257.55	1.171*
11.11	1.575	114.73	9.861*	261.26	1.173*
12.03	2.099	117.91	9.959*	264.94	1.175*
13.09	2.848	121.05	1.005 x 10^{-1}*	268.61	1.177
14.45	3.994	124.14	1.014	271.95	1.179*
16.48	6.016	127.20	1.022*	275.63	1.180*
18.62	8.436	130.47	1.030*	279.40	1.183*
20.34	1.092 x 10^{-2}	134.01	1.038	284.23	1.186
21.86	1.307	137.13	1.045*	287.29	1.187*
23.36	1.523	140.59	1.053*	292.89	1.190*
25.70	1.877	144.01	1.059*	298.22	1.194*
28.48	2.309	147.41	1.066	301.63	1.196
30.82	2.687	150.77	1.071*		
33.03	3.048	154.12	1.076*		
35.25	3.393	157.60	1.083*		
37.07	3.682	161.21	1.087		
38.80	3.938	164.80	1.093*		
40.86	4.223	168.41	1.097*		
42.93	4.535*	171.18	1.101*		
45.06	4.824	171.99	1.103*		
47.27	5.115	175.50	1.106		
49.63	5.431*	178.98	1.111*		
52.15	5.742	179.13	1.111*		
54.88	6.059	182.54	1.114*		
57.75	6.378*	186.23	1.119		
58.67	6.476*	189.93	1.122*		
60.50	6.558	193.60	1.127*		
60.55	6.559*	197.23	1.130		
62.87	6.914*	200.88	1.133*		
63.30	6.953*	204.51	1.137*		
65.72	7.175	208.07	1.139		
68.69	7.431*	211.61	1.142*		
71.68	7.675*	215.14	1.145*		
74.55	7.893	218.65	1.148*		
77.74	8.008*	222.14	1.150		
81.23	8.366	225.61	1.151*		
84.76	8.596*	229.07	1.154*		
88.39	8.806*	232.52	1.156*		
91.92	8.980	235.95	1.158		
95.32	9.140*	239.37	1.159*		
98.64	9.281*	242.77	1.161*		

* Not shown on plot

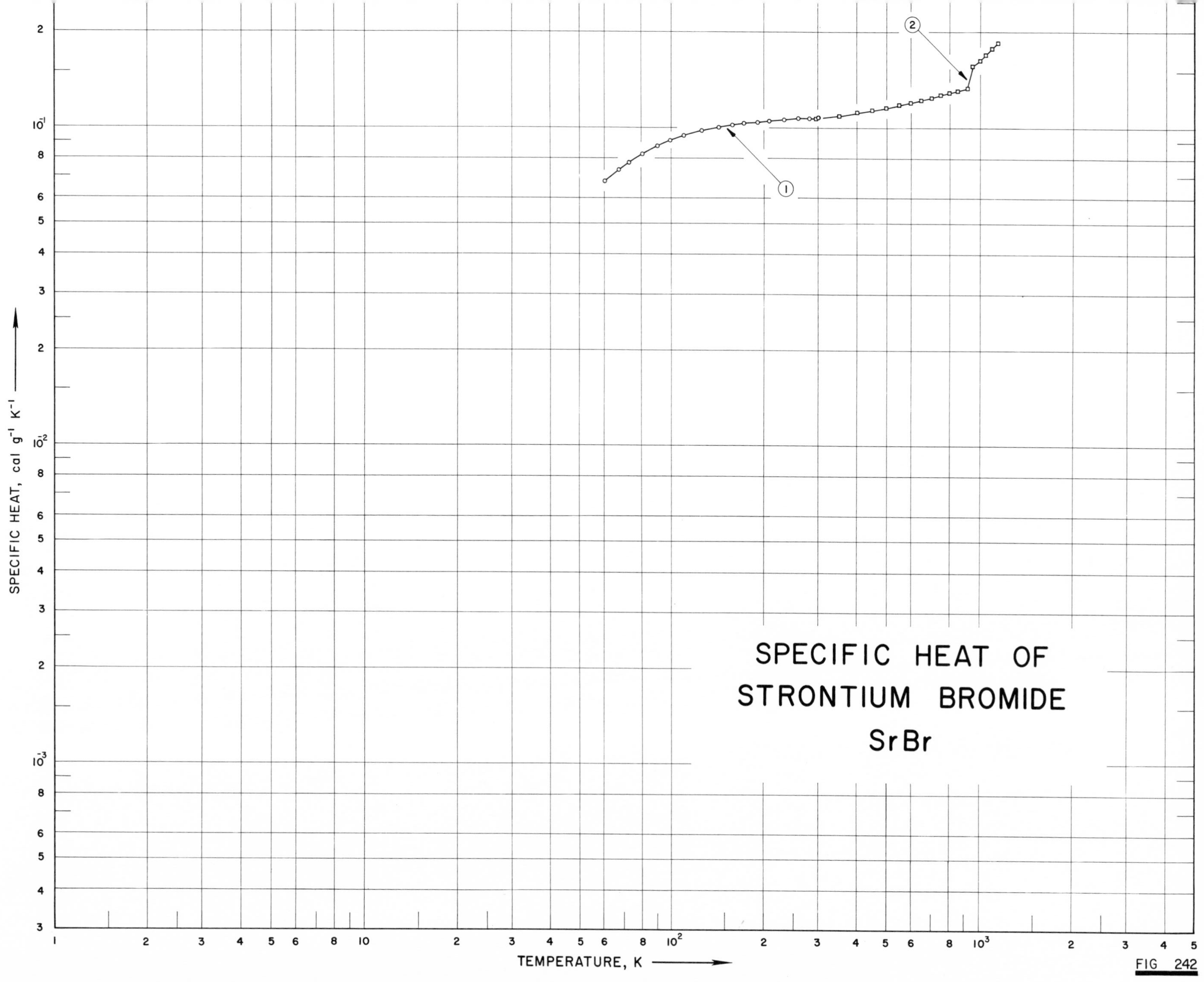

SPECIFIC HEAT OF
STRONTIUM BROMIDE
SrBr
SPECIFIC HEAT, cal g⁻¹ K⁻¹
TEMPERATURE, K
1
2
FIG 242

SPECIFICATION TABLE NO. 242 SPECIFIC HEAT OF STRONTIUM BROMIDE SrBr

[For Data Reported in Figure and Table No. 242]

Curve No.	Ref. No.	Year	Temp. Range, K	Reported Error, %	Name and Specimen Designation	Composition (weight percent), Specifications and Remarks
1	315	1962	60-302			0.1-0.01 Ca; heated 4 hrs to 400-450 C.
2	315	1962	298-1150			Same as above.

DATA TABLE NO. 242 SPECIFIC HEAT OF STRONTIUM BROMIDE $SrBr$

[Temperature, T, K; Specific Heat, C_p, Cal $g^{-1}K^{-1}$]

T	C_p
CURVE 1	
60.30	6.815 x 10^{-2}
63.60	7.149*
67.55	7.394
70.73	7.633*
73.26	7.788
77.57	8.087*
80.92	8.266
84.15	8.445*
87.25	8.607*
90.28	8.798
93.28	8.893*
96.21	9.025*
99.52	9.162
103.18	9.299*
106.74	9.365*
110.56	9.467
112.93	9.556*
117.82	9.646*
127.28	9.807
132.29	9.903*
137.73	9.992*
143.05	1.009 x 10^{-1}
148.24	1.016*
153.38	1.020*
158.44	1.024
163.43	1.029*
168.37	1.033*
173.28	1.035
178.47	1.035*
183.38	1.020*
183.91	1.039*
187.18	1.040*
192.48	1.047
197.75	1.050*
199.91	1.051*
202.96	1.054*
208.13	1.057
213.25	1.059*
218.34	1.061*
223.38	1.063*
228.38	1.066*
233.34	1.068
236.36	1.069*
242.12	1.070*
CURVE 1 (cont.)	
247.83	1.075 x 10^{-1}*
253.50	1.075*
259.12	1.075
264.70	1.075*
270.22	1.075*
270.88	1.075*
275.70	1.074*
281.12	1.074
286.48	1.076*
297.07	1.076*
301.86	1.076
CURVE 2	
298.15	1.074 x 10^{-1}
300.00	1.074*
350.00	1.098
400.00	1.121
450.00	1.144
500.00	1.166
550.00	1.188
600.00	1.210
650.00	1.232
700.00	1.254
750.00	1.275
800.00	1.297
850.00	1.319
900.00	1.341*
(s)916.00	1.348
(ℓ)916.00	1.341*
950.00	1.570
1000.00	1.642
1050.00	1.714
1100.00	1.786
1150.00	1.859

*Not shown on plot

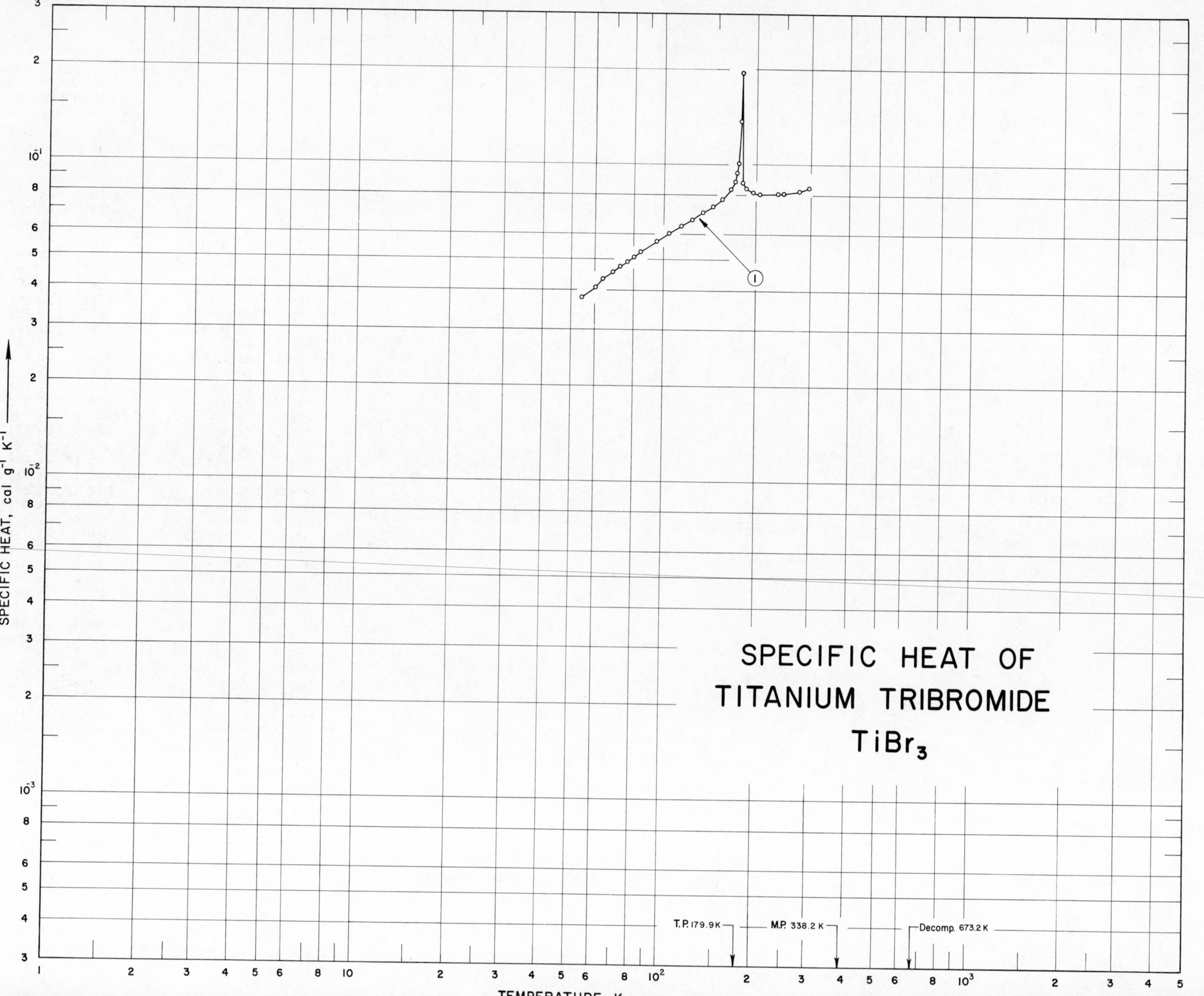
SPECIFIC HEAT OF
TITANIUM TRIBROMIDE
$TiBr_3$
T.P. 179.9 K
M.P. 338.2 K
Decomp. 673.2 K
TEMPERATURE, K
SPECIFIC HEAT, cal g⁻¹ K⁻¹
1

SPECIFICATION TABLE NO. 243 SPECIFIC HEAT OF TITANIUM TRIBROMIDE $TiBr_3$

[For Data Reported in Figure and Table No. 243]

Curve No.	Ref. No.	Year	Temp. Range, K	Reported Error, %	Name and Specimen Designation	Composition (weight percent), Specifications and Remarks
1	247	1961	55-298			83.18 Br (83.35 theo.) and 16.74 Ti (16.65 theo.).

DATA TABLE NO. 243 SPECIFIC HEAT OF TITANIUM TRIBROMIDE TiB_3

[Temperature, T, K; Specific Heat, Cp, Cal $g^{-1}K^{-1}$]

T	Cp
CURVE 1	
54.63	3.793×10^{-2}
60.27	4.099
64.37	4.332
68.71	4.565
72.75	4.756
76.55	4.930
80.32	5.096
84.42	5.295
95.12	5.701
105.19	6.052
114.90	6.397
124.85	6.692
135.92	7.050
146.03	7.374
155.92	7.763
165.94	8.354
166.93	8.448*
170.68	8.861
173.82	9.400
175.82	1.018×10^{-1}
176.43	1.039*
178.47	1.373
180.02	1.952
181.77	8.771×10^{-2}
183.85	8.559*
186.20	8.427
187.62	8.392*
196.31	8.184
206.54	8.065
216.85	8.062*
226.44	8.058*
236.61	8.090
246.46	8.118
257.13	8.149*
266.85	8.187*
276.82	8.260
286.96	8.364*
297.88	8.462

*Not shown on plot

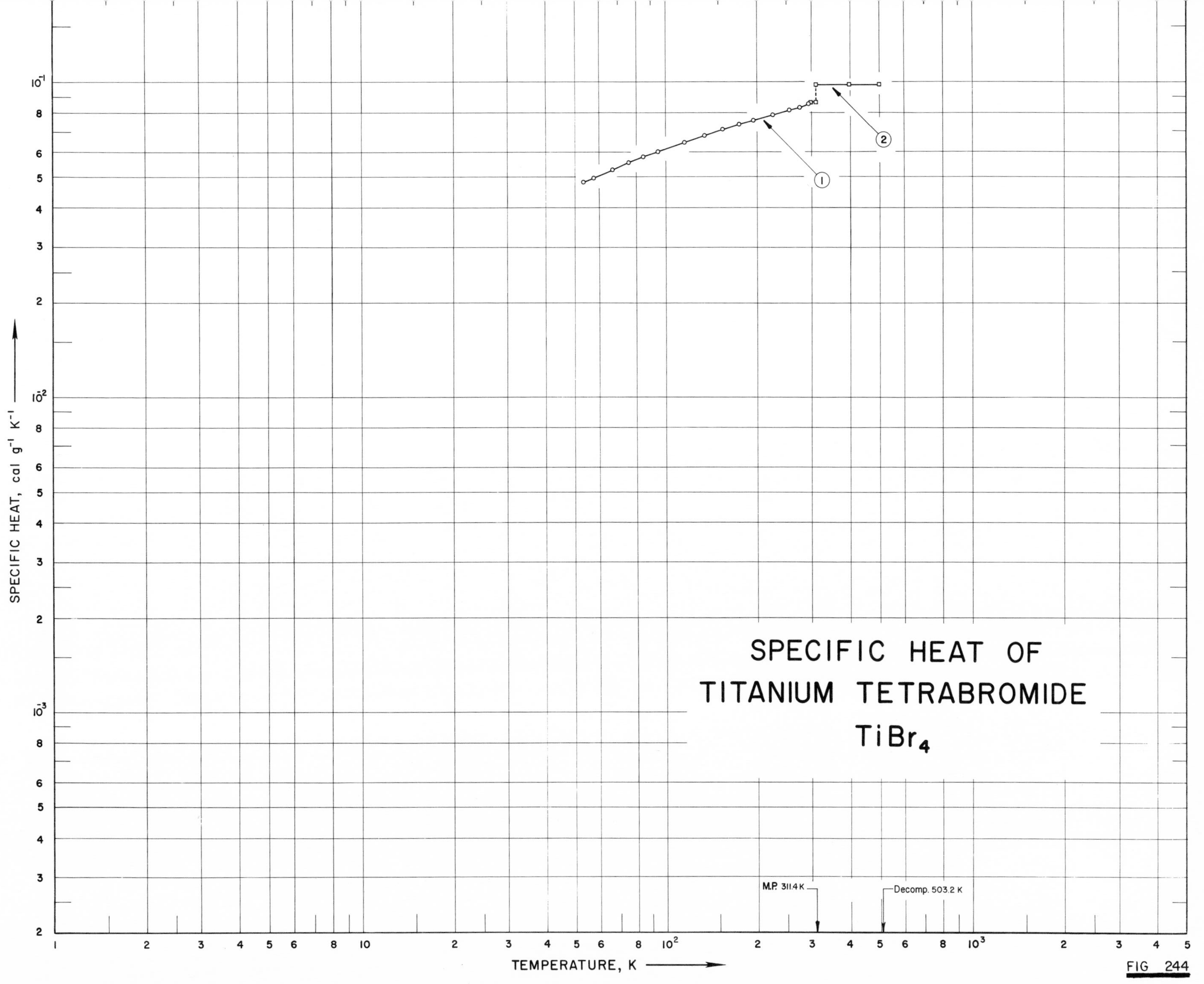
SPECIFIC HEAT OF
TITANIUM TETRABROMIDE
$TiBr_4$
M.P. 311.4 K
Decomp. 503.2 K
1
2
SPECIFIC HEAT, cal g^{-1} K^{-1}
TEMPERATURE, K
FIG 244

SPECIFICATION TABLE NO. 244 SPECIFIC HEAT OF TITANIUM TETRABROMIDE $TiBr_4$

[For Data Reported in Figure and Table No. 244]

Curve No.	Ref. No.	Year	Temp. Range, K	Reported Error, %	Name and Specimen Designation	Composition (weight percent), Specifications and Remarks
1		1961	54-296			99.998 $TiBr_4$; 86.88 Br (86.88 theo.) and 13.04 Ti (13.03 theo.).
2	247	1961	298-500			

DATA TABLE NO. 244 SPECIFIC HEAT OF TITANIUM TETRABROMIDE $TiBr_4$

[Temperature, T K; Specific Heat, Cp, Cal $g^{-1}K^{-1}$]

	T	Cp
	CURVE 1	
	53.71	4.826×10^{-2}
	58.09	4.971
	62.69	5.134*
	67.16	5.297
	71.64	5.436*
	76.27	5.575
	81.04	5.719*
	85.22	5.806
	95.07	6.015
	105.36	6.228*
	117.34	6.456
	125.54	6.603*
	136.13	6.780
	145.94	6.949*
	155.86	7.101
	165.96	7.223*
	176.03	7.351
	186.03	7.452*
	195.84	7.574
	206.30	7.694*
	216.20	7.778*
	226.91	7.882
	236.13	7.974*
	245.80	8.059*
	256.26	8.156
	266.49	8.246*
	276.43	8.328
	289.71	8.445*
	296.34	8.551
	CURVE 2	
	298.15	8.624×10^{-2}*
	300.00	8.624
(s)	311.40	8.624
(l)	311.40	9.875
	400.00	9.875
	500.00	9.875

* Not shown on plot

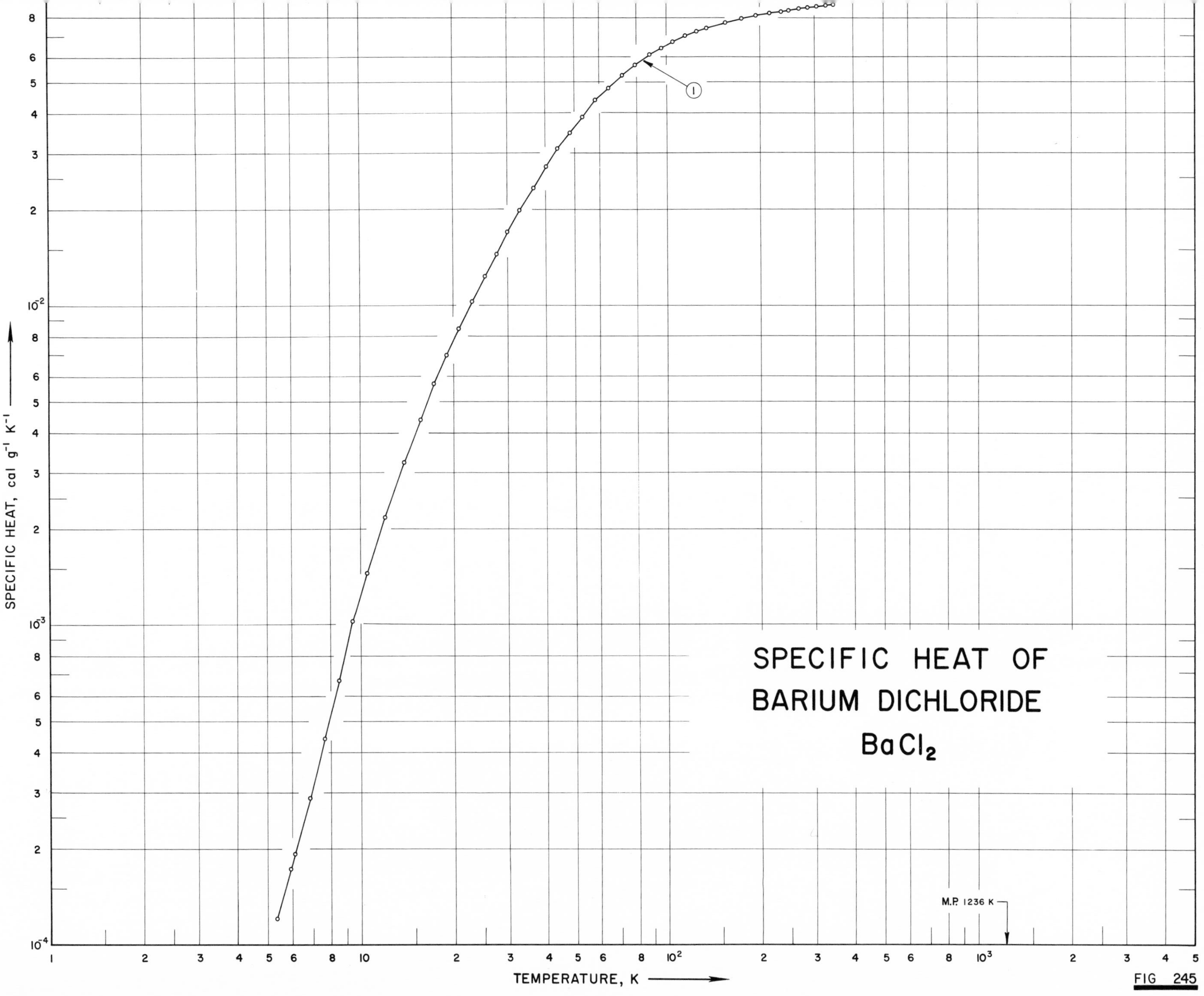

SPECIFIC HEAT OF
BARIUM DICHLORIDE
$BaCl_2$
M.P. 1236 K
1
TEMPERATURE, K
SPECIFIC HEAT, cal g^{-1} K^{-1}
FIG 245

SPECIFICATION TABLE NO. 245 SPECIFIC HEAT OF BARIUM DICHLORIDE $BaCl_2$

[For Data Reported in Figure and Table No. 245]

Curve No.	Ref. No.	Year	Temp. Range, K	Reported Error, %	Name and Specimen Designation	Composition (weight percent), Specifications and Remarks
1	317	1966	5-346	≤ 5.0		99.9 $BaCl_2$; prepared by dehydrating reagent grade $BaCl_2 \cdot 2H_2O$ in HCl atmosphere.

DATA TABLE NO. 245 SPECIFIC HEAT OF BARIUM DICHLORIDE $BaCl_2$

[Temperature, T, K; Specific Heat, Cp, Cal $g^{-1}K^{-1}$]

T	Cp	T	Cp
CURVE 1		CURVE 1 (cont.)	
Series 1		10.59	1.469×10^{-3}
		12.15	2.180
249.74	8.447×10^{-2}	13.94	3.246
258.29	8.471*	15.71	4.437
268.31	8.519	17.45	5.710
277.74	8.543*	19.09	7.001
287.12	8.586	20.87	8.490
297.11	8.620*	23.03	1.037×10^{-2}
307.69	8.663	25.30	1.244
318.23	8.687*	27.59	1.460
328.71	8.720	30.13	1.706
338.12	8.744*	33.03	1.993
346.44	8.759	36.56	2.346
		40.44	2.727
Series 2		44.36	3.101
		48.54	3.488
75.18	5.431×10^{-2}*	53.29	3.900
81.82	5.792*	58.61	4.334
88.85	6.124	64.90	4.801
96.98	6.431	72.12	5.254
106.40	6.739	79.66	5.681
116.41	7.012		
126.24	7.241		
135.97	7.434		
145.80	7.597*		
155.76	7.736		
165.86	7.856*		
176.12	7.957		
186.23	8.053		
196.21	8.125		
206.08	8.192		
215.85	8.259		
225.64	8.312*		
235.46	8.370		
245.20	8.413		
Series 3			
5.92	1.729×10^{-4}		
5.36	1.201		
6.11	1.921		
6.87	2.881		
7.68	4.418		
8.50	6.723		
9.44	1.023×10^{-3}		

* Not shown on plot

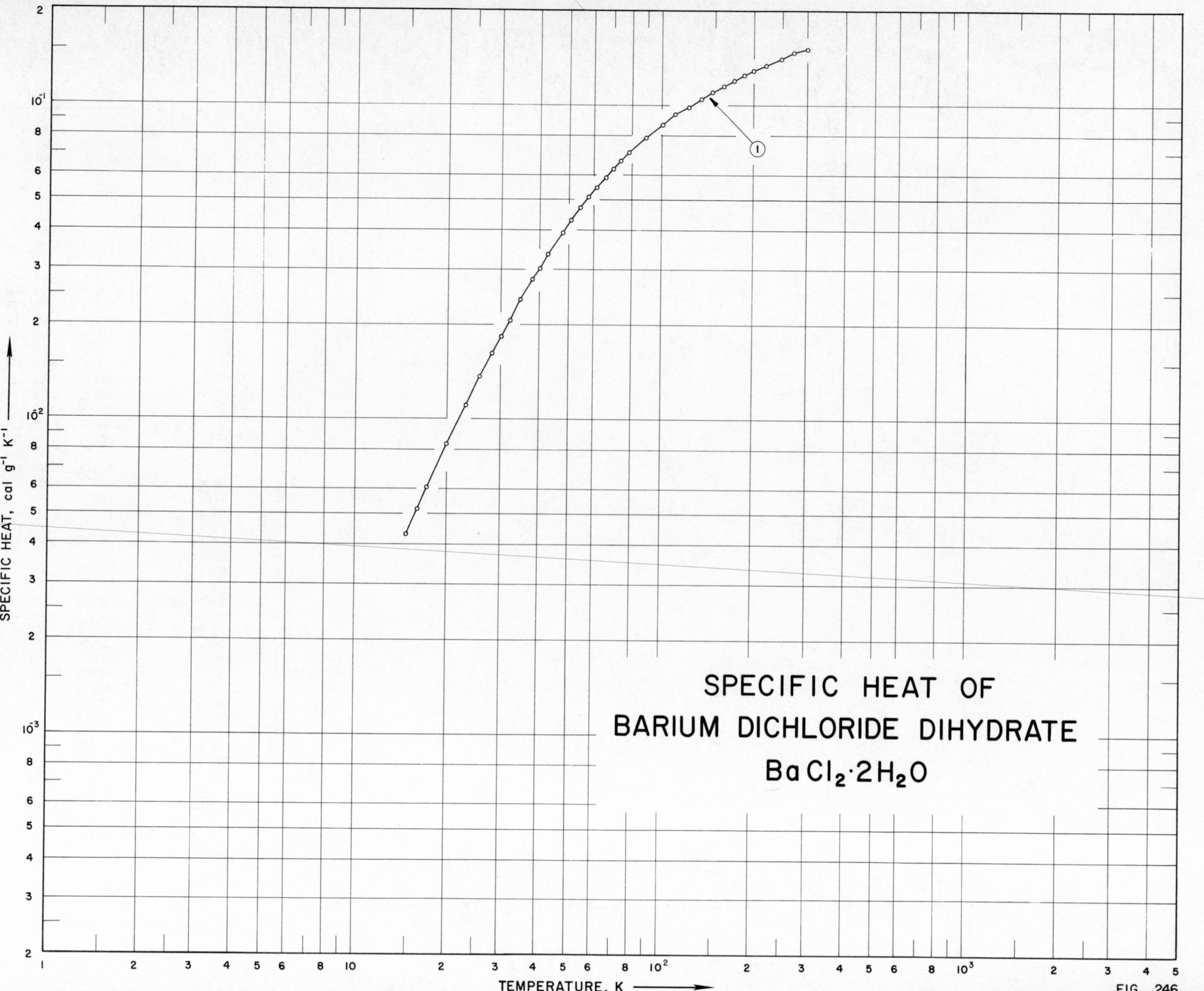
SPECIFIC HEAT OF
BARIUM DICHLORIDE DIHYDRATE
$BaCl_2 \cdot 2H_2O$
SPECIFIC HEAT, cal g⁻¹ K⁻¹
TEMPERATURE, K
1

FIG 246

SPECIFICATION TABLE NO. 246 SPECIFIC HEAT OF BARIUM DICHLORIDE DIHYDRATE $BaCl_2 \cdot 2H_2O$

[For Data Reported in Figure and Table No. 246]

Curve No.	Ref. No.	Year	Temp. Range, K	Reported Error, %	Name and Specimen Designation	Composition (weight percent), Specifications and Remarks
1	318	1936	15–301			C.P. grade $BaCl_2$ was crystallized three times from redistilled water; dried at 105 K and rehydrated in a dessicator containing water to obtain the theoretical amount of water.

DATA TABLE NO. 246 SPECIFIC HEAT OF BARIUM DICHLORIDE DIHYDRATE $BaCl_2 \cdot 2H_2O$

[Temperature, T, K; Specific Heat, Cp, Cal $g^{-1}K^{-1}$]

T	Cp
CURVE 1	
14.87	4.339×10^{-3}
16.18	5.198
17.24	6.099
20.00	8.350
23.16	1.117×10^{-2}
25.69	1.375
28.14	1.625
30.21	1.846
32.13	2.067
34.87	2.411
37.98	2.791
40.45	3.025
42.91	3.381
48.02	3.942
51.15	4.326
54.60	4.723
58.06	5.120
61.78	5.476
66.12	5.914
69.82	6.283
70.78	6.377*
73.92	6.676
78.67	7.093
83.68	7.494*
89.08	7.900
94.72	8.293
100.61	8.681
106.37	9.005*
112.10	9.361
117.92	9.631
123.62	9.889
129.49	1.021×10^{-1}*
135.64	1.056
141.68	1.083*
147.84	1.109
153.76	1.132*
160.01	1.155
166.86	1.178*
173.45	1.201
180.25	1.226*
187.15	1.254
193.41	1.271*
200.27	1.295
CURVE 1 (cont.)	
207.32	1.314×10^{-1}*
214.06	1.333*
221.02	1.349
229.72	1.369*
236.41	1.383*
244.76	1.415
254.54	1.433*
263.58	1.446*
272.18	1.474
281.17	1.493*
287.97	1.500*
297.07	1.518*
301.28	1.520

*Not shown on plot

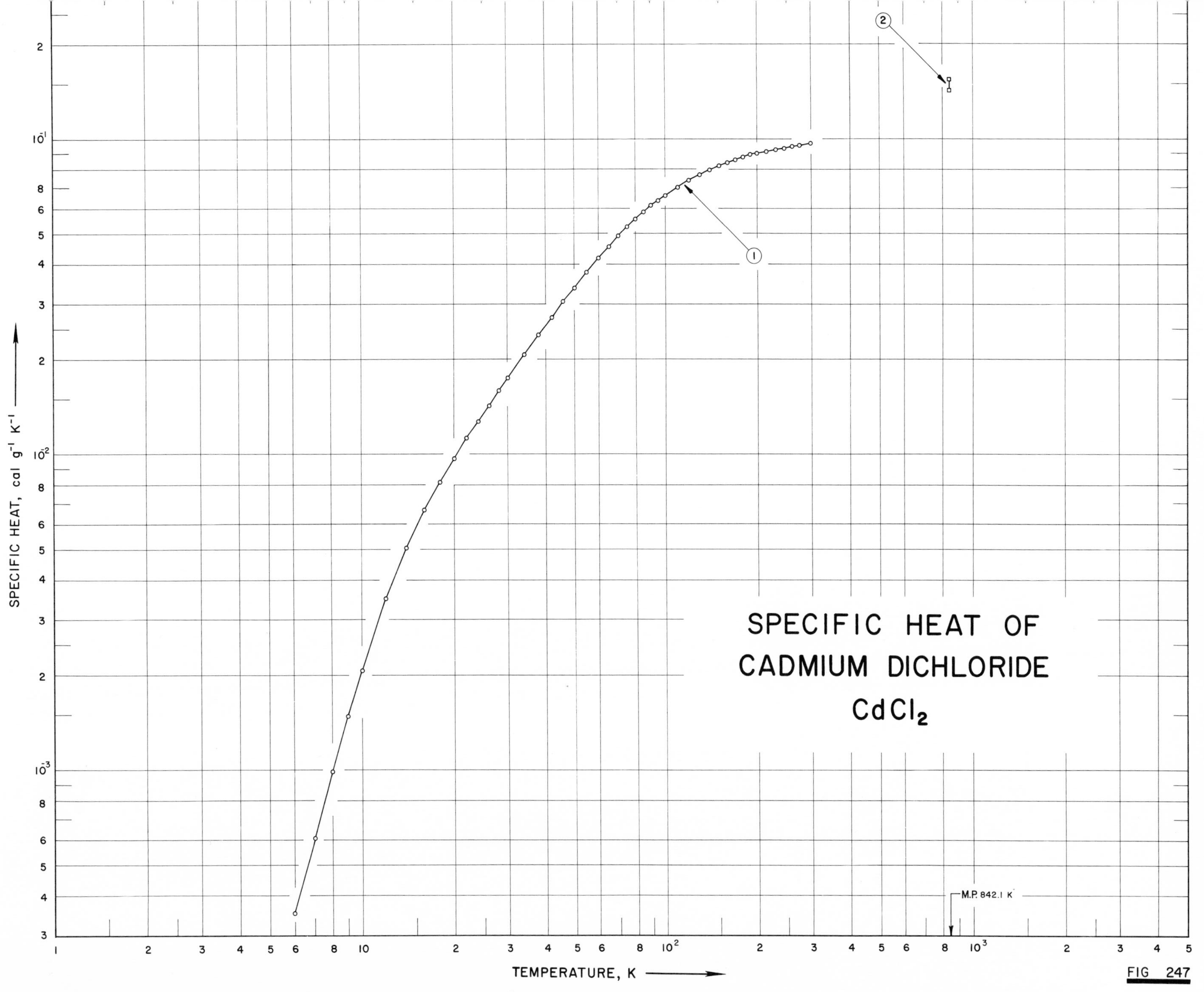
SPECIFIC HEAT OF
CADMIUM DICHLORIDE
$CdCl_2$
SPECIFIC HEAT, cal g^{-1} K^{-1}
TEMPERATURE, K
M.P. 842.1 K
1
2

FIG 247

SPECIFICATION TABLE NO. 247 SPECIFIC HEAT OF CADMIUM DICHLORIDE $CdCl_2$

[For Data Reported in Figure and Table No. 247]

Curve No.	Ref. No.	Year	Temp. Range, K	Reported Error, %	Name and Specimen Designation	Composition (weight percent), Specifications and Remarks
1	248	1959	2-300	1.0		
2	249	1960	842.1			Purified

DATA TABLE NO. 247 SPECIFIC HEAT OF CADMIUM DICHLORIDE $CdCl_2$

[Temperature, T, K; Specific Heat, Cp, Cal $g^{-1}K^{-1}$]

T	Cp	T	Cp
CURVE 1		CURVE 1 (cont.)	
2	1.20×10^{-5}	135	7.95×10^{-2}*
3	4.09	140	8.03
4	9.66	145	8.15*
5	1.94×10^{-4}	150	8.27
6	3.55	155	8.37*
7	6.11	160	8.47
8	9.93	165	8.56*
9	1.49×10^{-3}	170	8.65
10	2.08	175	8.72*
12	3.51	180	8.79
14	5.06	185	8.86*
16	6.72	190	8.92
18	8.24	195	8.97*
20	9.77	200	9.02
22	1.13×10^{-2}	205	9.07*
24	1.28	210	9.12*
26	1.44	215	9.17
28	1.60	220	9.21*
30	1.76	225	9.25*
32	1.93*	230	9.29
34	2.09	235	9.32*
36	2.25*	240	9.36*
38	2.41	245	9.39
40	2.57*	250	9.43*
42	2.74	255	9.46*
44	2.90*	260	9.49
46	3.07	265	9.53*
48	3.24*	270	9.55*
50	3.40	275	9.59
55	3.80	280	9.61*
60	4.21	285	9.65*
65	4.59	290	9.68*
70	4.95	295	9.71*
75	5.31	298.16	9.73*
80	5.63	300	9.74
85	5.93		
90	6.20		
95	6.45	CURVE 2	
100	6.67		
105	6.87*	(s) 842.1	1.55×10^{-1}
110	7.08	(l) 842.1	1.43
115	7.26*		
120	7.44		
125	7.60*		
130	7.75		

* Not shown on plot

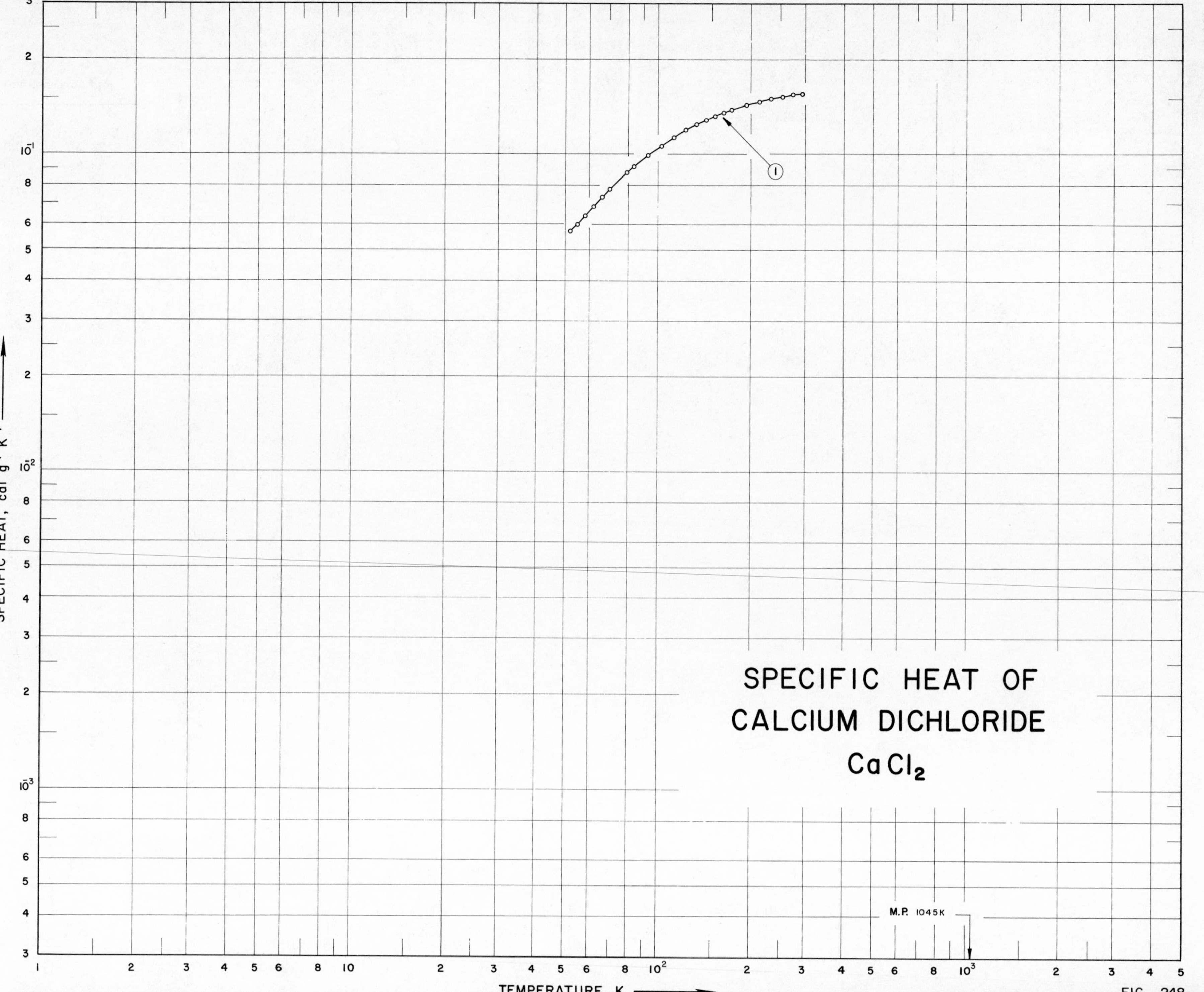
SPECIFIC HEAT OF
CALCIUM DICHLORIDE
$CaCl_2$
1
M.P. 1045 K
SPECIFIC HEAT, cal g^{-1} K^{-1}
TEMPERATURE, K

SPECIFICATION TABLE NO. 248 SPECIFIC HEAT OF CALCIUM DICHLORIDE $CaCl_2$

[For Data Reported in Figure and Table No. 248]

Curve No.	Ref. No.	Year	Temp. Range, K	Reported Error, %	Name and Specimen Designation	Composition (weight percent), Specifications and Remarks
1	250	1943	53-295			35.89 Ca (36.11 theor.), 63.82 Cl (63.89 theor.) and 0.59 $MgCl_2$; prepared from pure calcite by dissolving in C.P. HCl followed by evaporation to obtain the hydrated crystals; crushed and slowly heated in a vacuum over several days to 130 C; dehydration completed by passing steam of dry hydrogen chloride over several days gradually raising the temperature to 740 C. (corrected for $MgCl_2$)

DATA TABLE NO. 248 SPECIFIC HEAT OF CALCIUM DICHLORIDE CaCl

[Temperature, T, K; Specific Heat, Cp, Cal $g^{-1}K^{-1}$]

T	Cp
CURVE 1	
52.6	5.732 x 10^{-2}
55.3	6.042
58.6	6.442
62.5	6.900
66.4	7.385
70.3	7.825
80.0	8.814
84.2	9.208
93.6	1.002 x 10^{-1}
103.3	1.075
113.4	1.141
123.7	1.201
134.3	1.254
144.3	1.296
154.0	1.332
164.4	1.366
174.3	1.391
184.6	1.415*
194.5	1.440
204.8	1.459*
214.6	1.472
224.7	1.488*
234.9	1.502
244.8	1.516*
255.0	1.530
265.7	1.538*
275.8	1.551
285.8	1.555*
295.1	1.561

* Not shown on plot

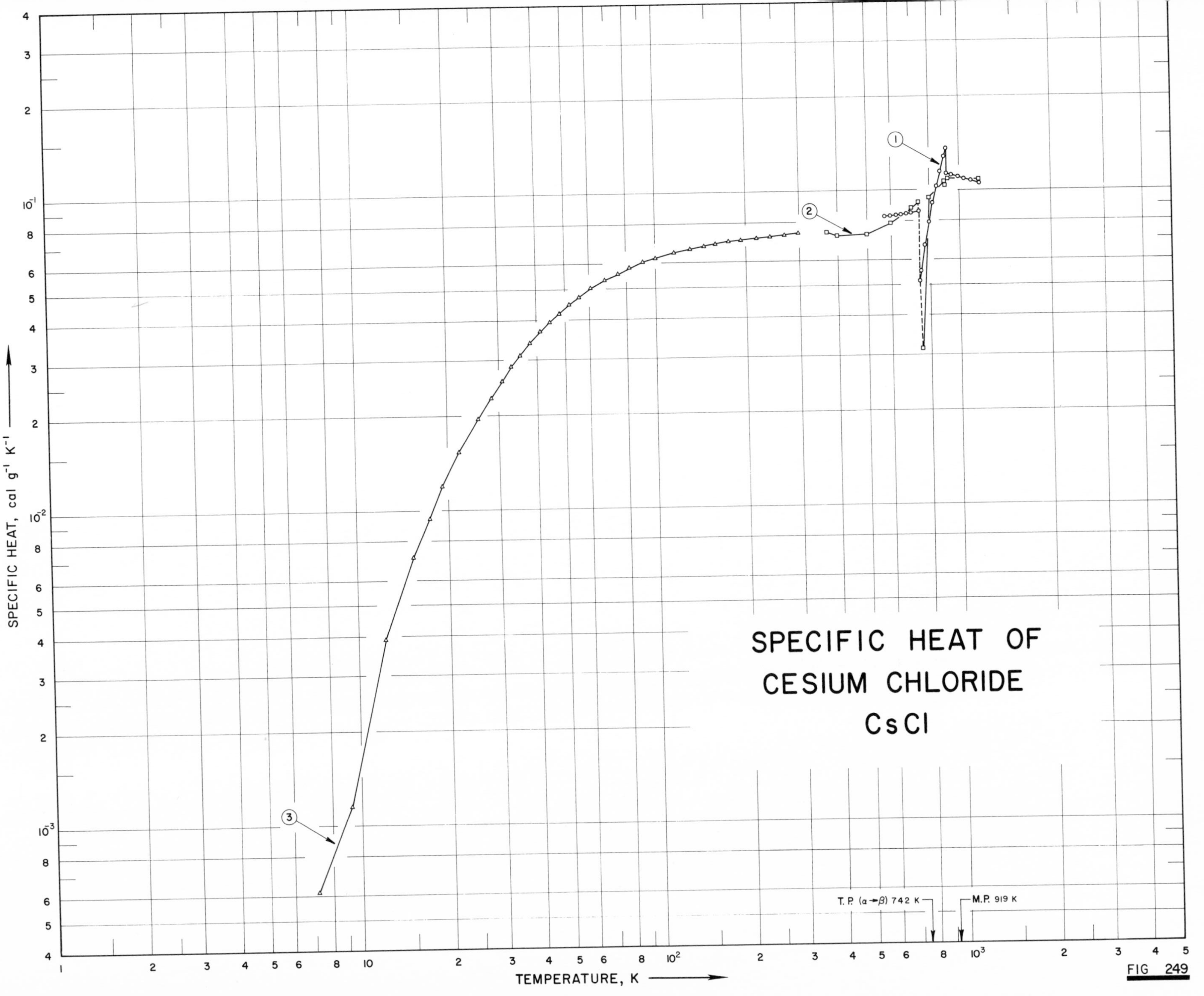
SPECIFIC HEAT OF
CESIUM CHLORIDE
CsCl
SPECIFIC HEAT, cal g⁻¹ K⁻¹
TEMPERATURE, K
T.P. (α→β) 742 K
M.P. 919 K
1
2
3
FIG 249

SPECIFICATION TABLE NO. 249 SPECIFIC HEAT OF CESIUM CHLORIDE CsCl

[For Data Reported in Figure and Table No. 249]

Curve No.	Ref. No.	Year	Temp. Range, K	Reported Error, %	Name and Specimen Designation	Composition (weight percent), Specifications and Remarks
1	251	1958	575–1175			< 0.1 each Li, K, Na and 0.01 Ca, total impurities <0.2.
2	252	1961	370–1170			99.80 CsCl, < 0.1 each Li, Na, K, and < 0.01 Ca; resublimed; dried in a vacuum oven for several hrs.
3	253	1963	7–299	0.3		99.99 $CsCl_2$, 0.001–0.01 Al, Ca, K, Na, and Rb; chemically pure sample recrystallized several times from water and dried at temp. between 260–300 C in a vacuum for several hrs.

DATA TABLE NO. 249 SPECIFIC HEAT OF CESIUM CHLORIDE CsCl

[Temperature, T, K; Specific Heat, C_p, Cal $g^{-1}K^{-1}$]

T	C_p
CURVE 1	
575	8.361 x 10^{-2}
600	8.375
625	8.403
650	8.442
675	8.491
700	8.548
725	8.613*
α742	8.661
β742	5.174
750	5.560
775	6.768
800	7.976
825	9.183
850	1.039 x 10^{-1}
875	1.160
900	1.281
β918	1.368
ℓ918	1.144
925	1.141*
950	1.131
975	1.121*
1000	1.112
1025	1.103*
1050	1.095
1075	1.087
1100	1.079
1125	1.072*
1150	1.065*
1175	1.058
CURVE 2	
370	7.441 x 10^{-2}
400	7.246
500	7.304
600	7.934
700	8.836
α740	9.242
β753	3.015
800	9.541
900	1.075 x 10^{-1}
β905	1.046
ℓ924	1.098
1000	1.093*
1100	1.085*
1170	1.080

T	C_p
CURVE 3	
7.19	6.177 x 10^{-4}
9.29	1.556 x 10^{-3}
12.24	3.938
15.24	7.115
17.31	9.497
19.18	1.202 x 10^{-2}
21.99	1.545
25.36	1.962
28.05	2.288
30.55	2.579
32.96	2.870
35.28	3.112
38.18	3.405
41.43	3.705
44.64	3.969
48.10	4.239
51.96	4.514
55.95	4.758
58.47	4.903*
61.29	5.059
64.75	5.221*
68.36	5.375
72.14	5.518*
75.78	5.644
79.28	5.772*
82.68	5.875
85.98	5.968*
88.79	6.039*
91.69	6.101
94.80	6.163*
97.85	6.221*
100.85	6.274
106.47	6.358*
110.08	6.414*
117.13	6.512
124.32	6.607*
132.32	6.691
137.05	6.734*
146.38	6.812
150.96	6.845
159.21	6.903
168.07	6.958*
176.79	7.008
185.38	7.051*
193.61	7.088
202.22	7.119*
CURVE 3 (cont.)	
209.43	7.147 x 10^{-2}*
217.04	7.178
224.81	7.195*
230.22	7.213*
240.63	7.251
250.42	7.287*
259.26	7.317*
269.44	7.352
279.54	7.394*
285.39	7.412*
293.64	7.430*
299.38	7.447

* Not shown on plot

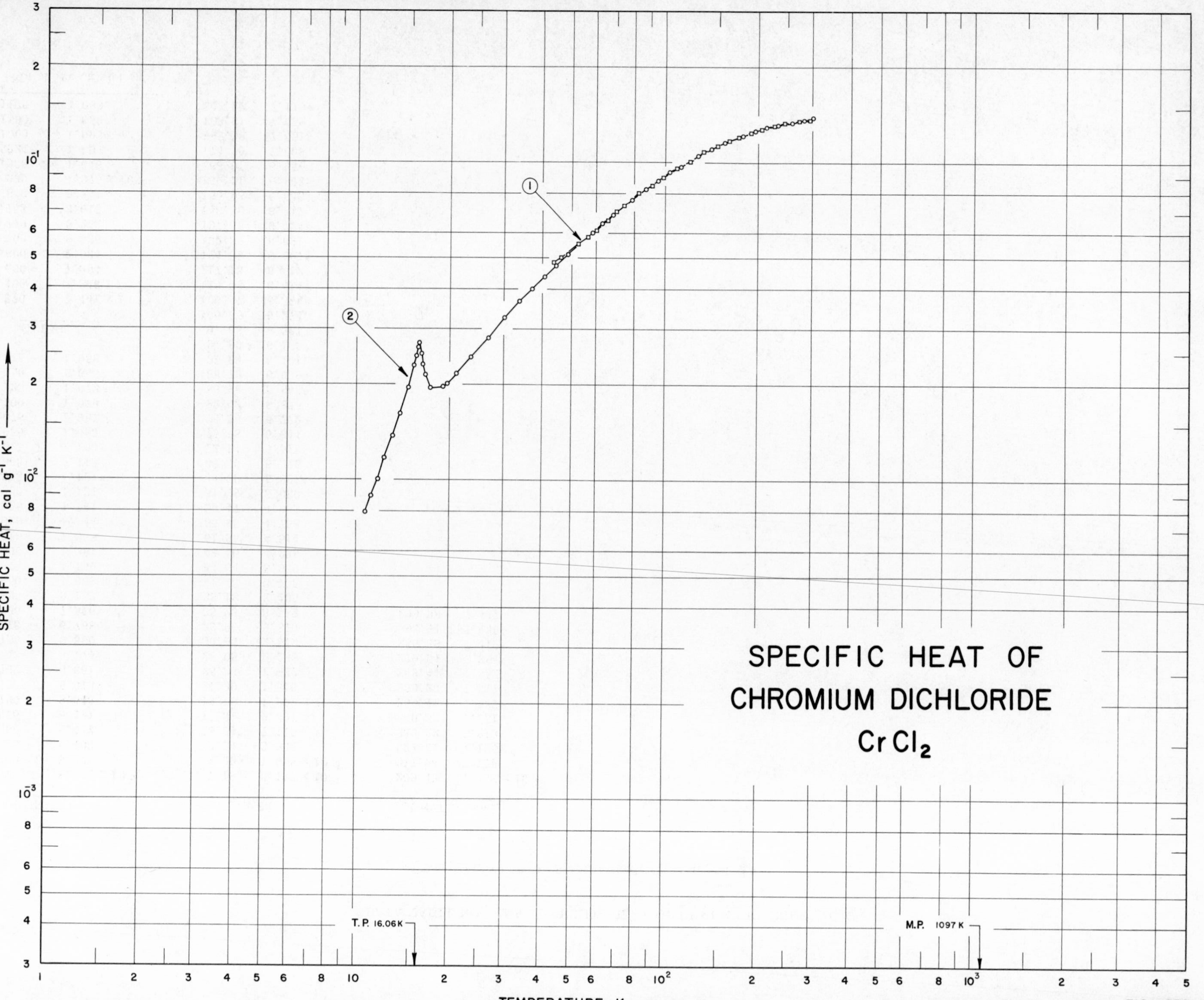

SPECIFIC HEAT OF
CHROMIUM DICHLORIDE
Cr Cl2
T.P. 16.06 K
M.P. 1097 K
1
2
SPECIFIC HEAT, cal g-1 K-1

SPECIFICATION TABLE NO. 250 SPECIFIC HEAT OF CHROMIUM DICHLORIDE $CrCl_2$

[For Data Reported in Figure and Table No. 250]

Curve No.	Ref. No.	Year	Temp. Range, K	Reported Error, %	Name and Specimen Designation	Composition (weight percent), Specifications and Remarks
1	60	1937	43-296			Grayish white substance; prepared from $CrCl_3$ by reduction of H_2 containing small amount of HCl.
2	254	1961	11-299			57.59 Cl (57.69 theo.), 42.21 Cr (42.31 theo.), 0.14 insol. in H_2O, and 0.001 each Fe and Ni; prepared from violet powdered $CrCl_3$ by passing HCl and H_2 gases in a furnace at 780 C.

DATA TABLE NO. 250 SPECIFIC HEAT OF CHROMIUM DICHLORIDE $CrCl_2$

[Temperature, T, K; Specific Heat, C_p, Cal $g^{-1}K^{-1}$]

T	C_p
CURVE 1	
43.6	4.870×10^{-2}
46.4	5.049
49.5	5.231*
52.9	5.570
54.7	5.660*
56.7	5.842
58.2	5.955*
60.2	6.116
65.9	6.588
70.0	7.012
74.3	7.330
82.5	8.019
91.1	8.477
99.0	8.981
109.8	9.640
121.8	1.017×10^{-1}
134.9	1.082
148.3	1.130
162.4	1.172
174.3	1.211
191.2	1.244
206.8	1.273
227.3	1.308
245.3	1.332
273.5	1.356
295.1	1.361*
296.1	1.363
CURVE 2	
Series 1	
53.60	5.617×10^{-2}*
58.69	6.044
63.37	6.455
68.46	6.861
73.89	7.273*
78.87	7.651
83.13	7.985*
87.27	8.274
91.49	8.550*
95.94	8.819
100.30	9.079*
104.58	9.323
CURVE 2 (cont.)	
108.49	9.543×10^{-2}*
113.23	9.795
118.43	1.005×10^{-1}*
123.67	1.031*
128.96	1.055
133.94	1.077*
137.85	1.093*
142.68	1.110
147.52	1.130*
152.45	1.145*
157.78	1.163
163.21	1.180*
168.45	1.193*
174.26	1.210*
179.84	1.223
185.15	1.235*
190.67	1.246*
196.66	1.257
201.71	1.269*
208.56	1.277*
214.92	1.290
221.10	1.299*
227.45	1.307*
233.36	1.315
239.32	1.324*
245.71	1.329*
251.72	1.334*
258.15	1.339
264.93	1.348*
271.62	1.360*
277.92	1.362*
284.69	1.368
291.70	1.374×10^{-1}*
298.50	1.381*
301.83	1.389
Series 2	
10.73	7.997×10^{-3}
11.24	8.990
11.79	1.015×10^{-2}
12.43	1.185
13.14	1.385
13.86	1.630
CURVE 2 (cont.)	
14.63	1.956×10^{-2}
15.31	2.307
16.22	2.520
17.40	1.960
19.02	1.975
21.03	2.171
23.50	2.451
26.70	2.824
30.47	3.266
33.90	3.657
37.13	4.017
40.87	4.398
44.73	4.775
48.80	5.170
52.64	5.530*
Series 3	
14.68	1.974×10^{-2}*
15.33	2.317*
15.80	2.616
16.21	2.595*
16.68	2.166
17.16	1.959*
17.85	1.900*
18.48	1.929*
19.05	1.964
19.60	2.020
Series 4	
15.60	2.465×10^{-2}
15.92	2.693*
16.11	2.472*
16.20	2.636*
16.49	2.327
16.77	2.099*
17.03	1.985*
Series 5	
15.63	2.506×10^{-2}*
15.96	2.709
16.26	2.546*
CURVE 2 (cont.)	
16.56	2.270×10^{-2}*
16.87	2.050*
17.20	1.961*
Series 6	
271.89	1.355*
279.19	1.361*

*Not shown on plot

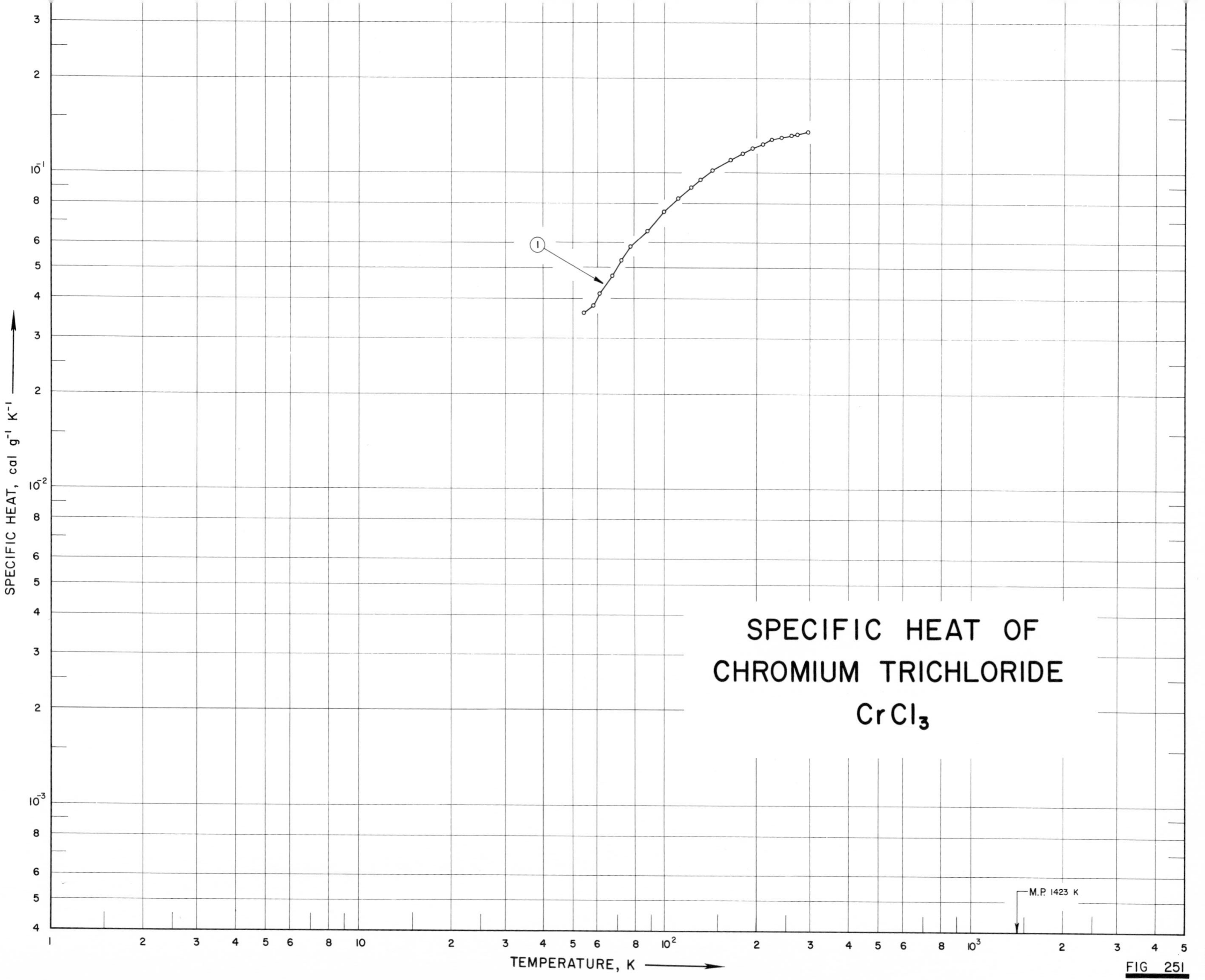
SPECIFIC HEAT OF
CHROMIUM TRICHLORIDE
$CrCl_3$
SPECIFIC HEAT, cal g^{-1} K^{-1}
TEMPERATURE, K
M.P. 1423 K
1
FIG 251

SPECIFICATION TABLE NO. 251 SPECIFIC HEAT OF CHROMIUM TRICHLORIDE $CrCl_3$

[For Data Reported in Figure and Table No. 251]

Curve No.	Ref. No.	Year	Temp. Range, K	Reported Error, %	Name and Specimen Designation	Composition (weight percent), Specifications and Remarks
1	60	1937	54-297			Brilliant violet substance; prepared by passing Cl_2 gas through mixture of Cr_2O_3 and carbon at 800 C; compressed into pellets and dried by passing dry chloride containing a small amount of CCl_4 at 300 C; evacuated to remove excess chlorine.

DATA TABLE NO. 251 SPECIFIC HEAT OF CHROMIUM TRICHLORIDE $CrCl_3$

[Temperature, T, K; Specific Heat, Cp, Cal $g^{-1}K^{-1}$]

T	Cp
CURVE 1	
54.4	3.602×10^{-2}
58.4	3.801
61.2	4.165
67.4	4.749
72.4	5.302
77.7	5.872
88.4	6.573
99.8	7.570
112.0	8.341
123.3	9.035
132.8	9.559
144.9	1.025×10^{-1}
165.9	1.110
181.6	1.162
195.7	1.203
211.2	1.240
225.6	1.289
243.0	1.301
261.9	1.320
272.6	1.328*
273.5	1.341
295.4	1.355*
296.9	1.363

* Not shown on plot

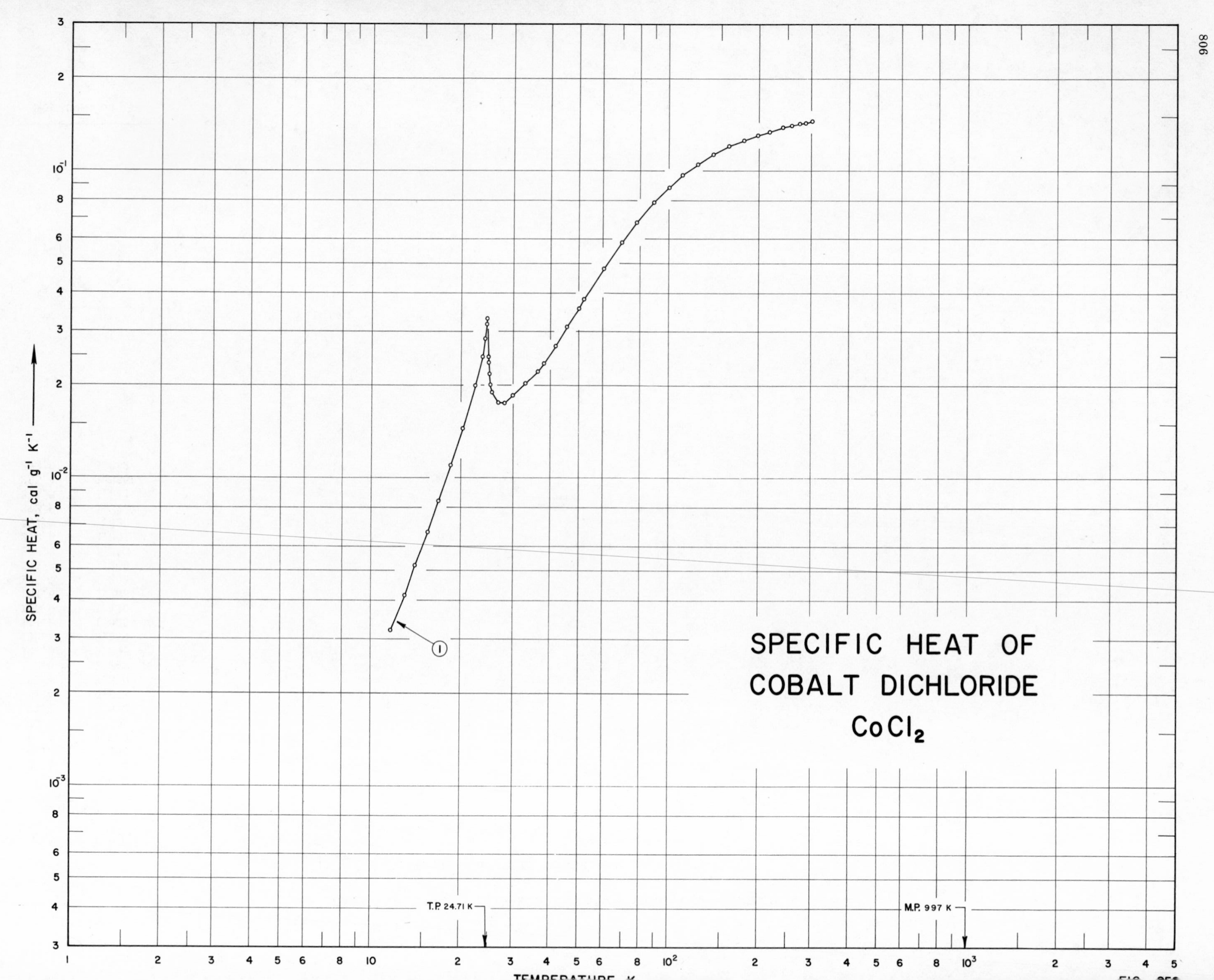
SPECIFIC HEAT OF
COBALT DICHLORIDE
$CoCl_2$
SPECIFIC HEAT, cal g^{-1} K^{-1}
T.P. 24.71 K
M.P. 997 K
1

SPECIFICATION TABLE NO. 252 SPECIFIC HEAT OF COBALT DICHLORIDE $CoCl_2$

[For Data Reported in Figure and Table No. 252]

Curve No.	Ref. No.	Year	Temp. Range, K	Reported Error, %	Name and Specimen Designation	Composition (weight percent), Specifications and Remarks
1	255	1961	12–304			45.27 Co (45.39 theor.), 54.65 Cl (54.61 theor.), 0.001 each Ca, Si, and Fe; prepared from $Na_3Co(No_2)_6$ by dehydrating in a stream of HCl at 800 C.

DATA TABLE NO. 252 SPECIFIC HEAT OF COBALT DICHLORIDE $CoCl_2$

[Temperature, T,K; Specific Heat, Cp, Cal $g^{-1}K^{-1}$]

T	Cp
CURVE 1	
Series 1	
52.39	3.841 x 10^{-2}
56.74	4.346*
60.78	4.825
65.28	5.355*
69.72	5.884
74.23	6.386*
78.35	6.822
81.62	7.188*
85.44	7.577*
89.22	7.948
92.87	8.271*
96.53	8.587*
100.34	8.887
104.16	9.172*
108.27	9.480*
111.99	9.734
116.38	1.003 x 10^{-1}
120.77	1.030*
125.24	1.057
129.59	1.080*
133.84	1.101*
137.56	1.119*
141.94	1.137
145.96	1.156*
150.27	1.172*
154.55	1.189*
159.48	1.206
164.48	1.222*
169.64	1.238*
174.13	1.251*
178.29	1.261
183.48	1.275*
188.75	1.285*
193.84	1.298*
198.98	1.310
204.24	1.322*
209.86	1.333*
215.25	1.343*
218.98	1.348
224.76	1.357*
230.66	1.362*
236.40	1.373*
241.52	1.382

T	Cp
CURVE 1 (cont.)	
246.68	1.384 x 10^{-1}*
252.62	1.397*
258.44	1.405
264.90	1.409*
269.72	1.417*
275.47	1.425
282.17	1.430*
288.16	1.432
294.09	1.438*
298.09	1.446*
303.57	1.451
Series 2	
11.69	3.20 x 10^{-3}
13.02	4.17
14.18	5.24
15.54	6.71
16.88	8.46
18.64	1.11 x 10^{-2}
20.41	1.46
22.42	2.00
24.85	2.39
27.50	1.78*
30.98	1.87
33.89	2.04
36.47	2.24
38.15	2.38
41.82	2.71
45.87	3.11
50.17	3.57
Series 3	
22.72	2.08 x 10^{-2}*
23.68	2.49
24.25	2.86
24.36	2.95*
24.51	3.18
24.69	3.19*
24.86	2.35*
25.01	2.19
25.13	2.09*
25.26	2.01
25.41	1.96*

T	Cp
CURVE 1 (cont.)	
25.59	1.91 x 10^{-2}
25.80	1.87*
26.14	1.82*
26.51	1.80*
26.92	1.78
27.42	1.77*
28.13	1.77
Series 4	
24.45	3.08 x 10^{-2}*
24.58	3.32
24.71	3.15*
24.84	2.50
25.00	2.21*
25.14	2.09*
Series 5	
291.70	1.455 x 10^{-1}*
297.91	1.429*

* Not shown on plot

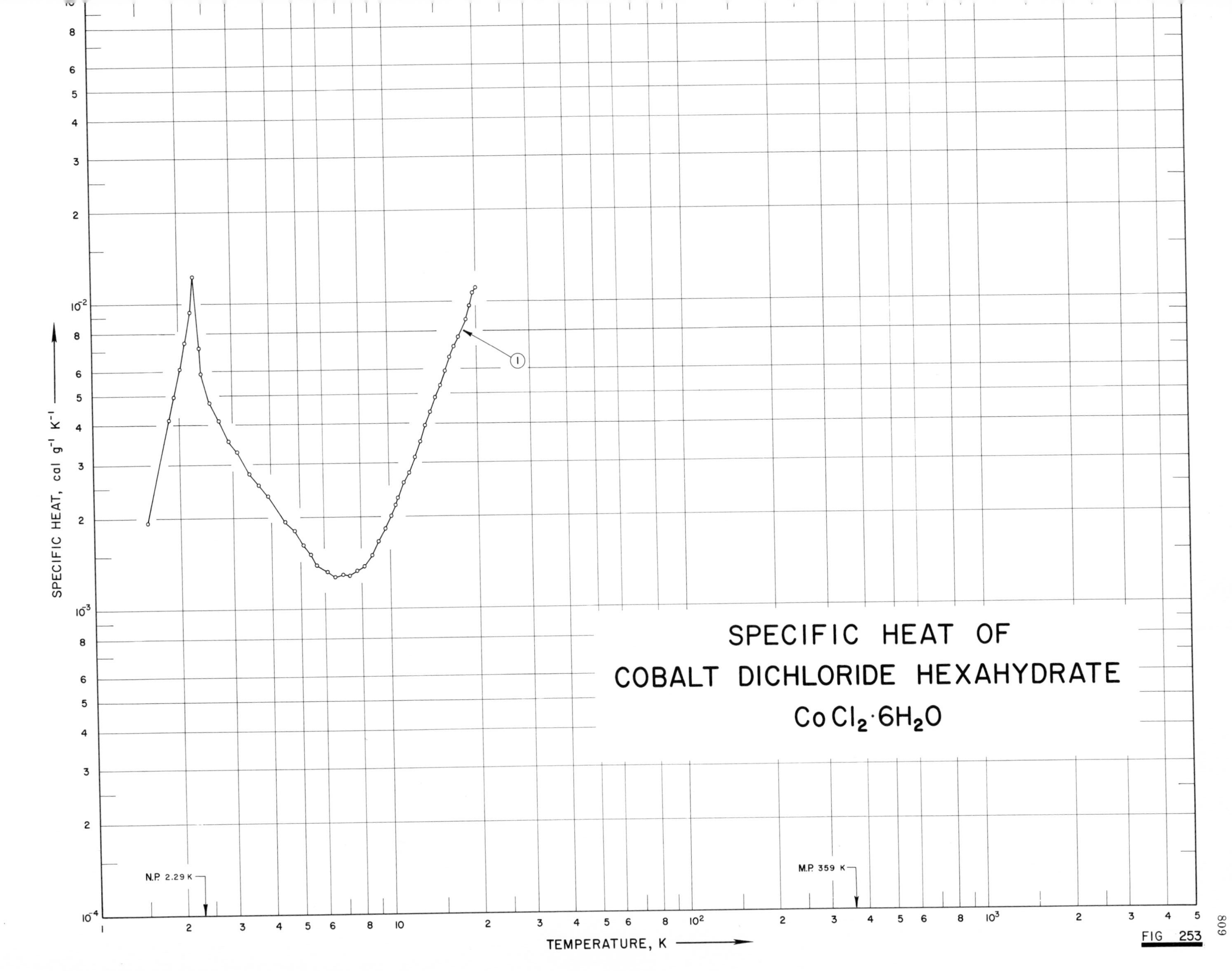

SPECIFIC HEAT OF
COBALT DICHLORIDE HEXAHYDRATE
$CoCl_2 \cdot 6H_2O$
SPECIFIC HEAT, cal g^{-1} K^{-1}
TEMPERATURE, K
N.P. 2.29 K
M.P. 359 K
1

FIG 253

SPECIFICATION TABLE NO. 253 SPECIFIC HEAT OF COBALT DICHLORIDE HEXAHYDRATE $CoCl_2 \cdot 6H_2O$

[For Data Reported in Figure and Table No. 253]

Curve No.	Ref. No.	Year	Temp. Range, K	Reported Error, %	Name and Specimen Designation	Composition (weight percent), Specifications and Remarks
1	256	1959	1.5-20			Impurities: 0.06 substances not precipitated by $(NH_4)_2S$ as (SO_4), 0.05 Ni, 0.02 Zn, 0.004 insol., 0.004 sulphide, 0.002 each Fe and ammonium; reagent grade; sample supplied by Fischer Scheinlifre Supply Co.

DATA TABLE NO. 253 SPECIFIC HEAT OF COBALT DICHLORIDE HEXAHYDRATE $CoCl_2 \cdot 6H_2O$

[Temperature, T, K; Specific Heat, Cp, Cal $g^{-1}K^{-1}$]

T	Cp	T	Cp
CURVE 1		CURVE 1 (cont.)	
1.5269	1.946×10^{-3}	16.0910	6.641×10^{-3}
1.8316	4.216	16.6930	7.229
1.9084	5.002	17.3090	7.734
2.0014	6.170	17.8750	8.658*
2.0870	7.532	18.4090	8.784
2.1700	9.474	18.9080	9.751
2.2494	1.244×10^{-2}	19.3660	1.072×10^{-2}
2.3448	7.254×10^{-3}	19.7950	1.122
2.3681	5.977		
2.5112	4.800		
2.6952	4.172		
2.8913	3.575		
3.0891	3.312		
3.3666	2.787		
3.6250	2.568		
3.8760	2.360		
4.4122	1.942		
4.7498	1.806		
5.0361	1.619		
5.3512	1.511		
5.6220	1.388		
6.0752	1.321		
6.4688	1.271		
6.8450	1.291		
7.2534	1.281		
7.6630	1.339		
8.1258	1.381		
8.6108	1.496		
9.0914	1.665		
9.5798	1.822		
10.0280	2.017		
10.4360	2.190		
10.1600	2.064*		
10.6630	2.307		
11.1650	2.589		
11.6780	2.791		
12.1990	3.140		
12.7380	3.522		
13.2930	3.912		
13.8720	4.413		
14.4380	4.918		
14.9960	5.380		
15.5470	5.968		

* Not shown on plot

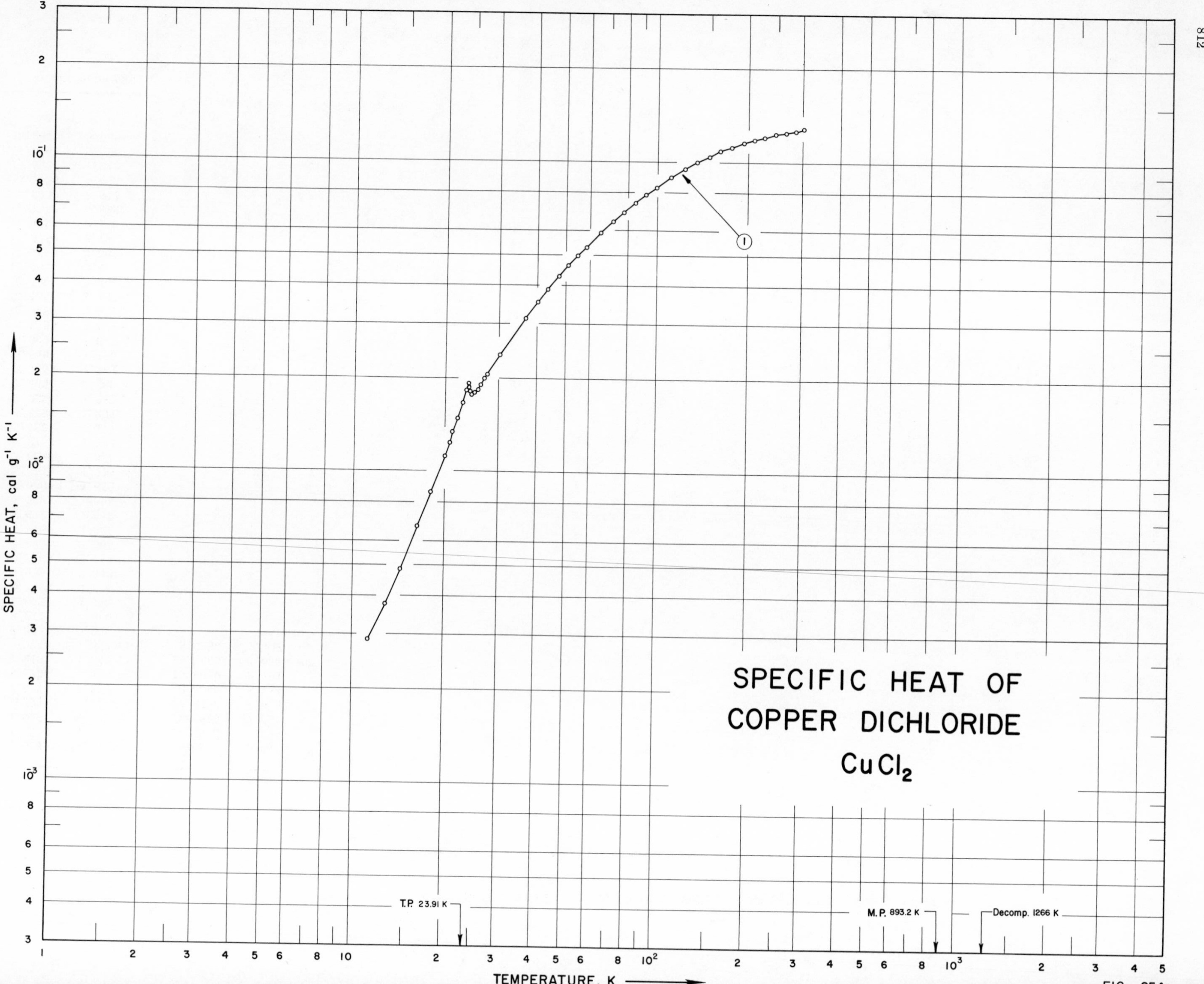
SPECIFIC HEAT OF
COPPER DICHLORIDE
$CuCl_2$
SPECIFIC HEAT, cal g^{-1} K^{-1}
TEMPERATURE, K
T.P. 23.91 K
M.P. 893.2 K
Decomp. 1266 K
1

SPECIFICATION TABLE NO. 254 SPECIFIC HEAT OF COPPER DICHLORIDE $CuCl_2$

[For Data Reported in Figure and Table No. 254]

Curve No.	Ref. No.	Year	Temp. Range, K	Reported Error, %	Name and Specimen Designation	Composition (weight percent), Specifications and Remarks
1	254	1961	11-303			52.85 Cl_2 (52.74 theo.), 47.16 Cu (47.26 theo.), 0.001 each Al, Mg, Si, and Ag; dark brown powder; prepared from $CuCl_2 \cdot 2H_2O$ by dehydration.

DATA TABLE NO. 254 SPECIFIC HEAT OF COPPER DICHLORIDE $CuCl_2$

[Temperature, T, K; Specific Heat, Cp, Cal $g^{-1}K^{-1}$]

T	Cp
CURVE 1	
Series 1	
300.55	1.283×10^{-1}
302.36	1.283
303.04	1.282
Series 2	
11.47	2.864×10^{-3}
12.98	3.741
14.49	4.835
16.40	6.627
18.10	8.546
20.02	1.115×10^{-2}
22.13	1.489
24.24	1.805
27.09	1.984
30.42	2.372
33.59	2.751
36.77	3.124
40.39	3.532
43.76	3.887
47.49	4.273
51.19	4.641
Series 3	
54.94	4.992×10^{-2}
58.63	5.326
62.05	5.625*
65.33	5.904
68.73	6.172*
71.97	6.422
75.05	6.652*
77.88	6.862
81.28	7.131*
84.81	7.382
88.27	7.609*
91.91	7.839
95.69	8.070*
99.41	8.278
103.29	8.494*
107.36	8.710*
111.50	8.918
115.70	9.126*
119.62	9.312*
CURVE 1 (cont.)	
123.64	9.498×10^{-2}
127.63	9.662*
131.76	9.833*
135.04	9.967
139.29	1.012×10^{-1}*
143.62	1.028*
148.04	1.043
152.73	1.058*
157.54	1.072*
162.18	1.084
166.93	1.098*
171.77	1.110*
176.77	1.122
181.97	1.133*
187.36	1.145*
193.06	1.155
198.80	1.165*
204.33	1.174*
209.69	1.180
215.07	1.192*
220.58	1.201*
226.36	1.209
232.59	1.217*
239.22	1.224*
245.93	1.230
252.58	1.238*
259.08	1.244*
265.46	1.250
272.29	1.258*
279.29	1.262*
285.72	1.268
292.36	1.273*
298.09	1.276*
Series 4	
20.86	1.243×10^{-2}
21.42	1.343
21.93	1.439*
22.41	1.537*
22.97	1.670
23.51	1.822
24.03	1.870
24.58	1.779*
28.16	1.790
CURVE 1 (cont.)	
25.75	1.839×10^{-2}
26.35	1.900
27.04	1.975*
27.70	2.051
Series 5	
23.54	1.822×10^{-2}*
23.70	1.882*
23.85	1.897*
23.99	1.897*
24.13	1.852*
24.28	1.807*
24.42	1.778*
24.58	1.778
Series 6	
23.93	1.919×10^{-2}*
24.07	1.874*
24.22	1.822*
24.36	1.793*
Series 7	
282.83	1.267×10^{-1}*
290.04	1.270*

* Not shown on plot

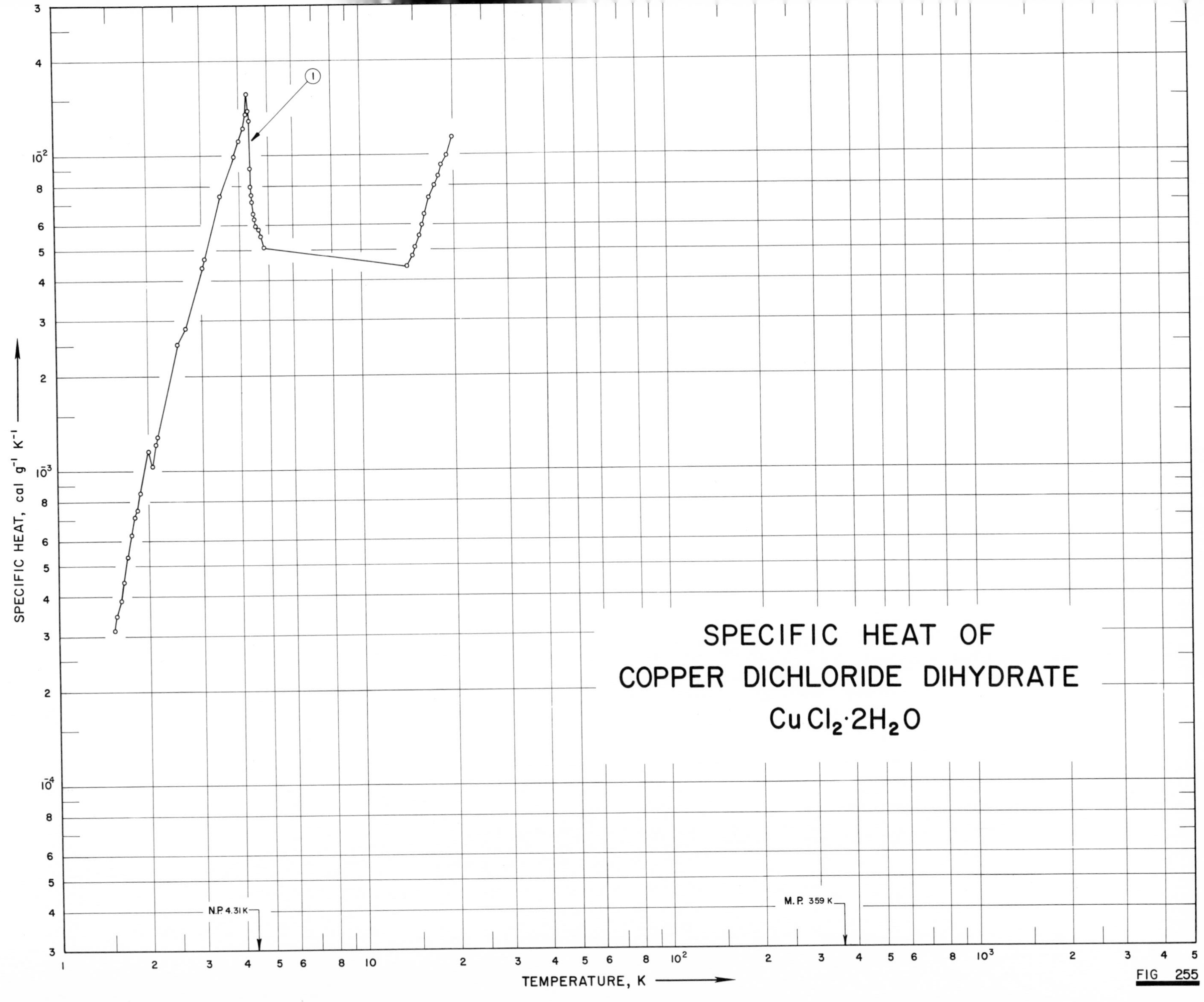

SPECIFIC HEAT OF
COPPER DICHLORIDE DIHYDRATE
CuCl₂·2H₂O
SPECIFIC HEAT, cal g⁻¹ K⁻¹
TEMPERATURE, K
N.P. 4.31 K
M.P. 359 K
1

FIG 255

SPECIFICATION TABLE NO. 255 SPECIFIC HEAT OF COPPER DICHLORIDE DIHYDRATE $CuCl_2 \cdot 2H_2O$

[For Data Reported in Figure and Table No. 255]

Curve No.	Ref. No.	Year	Temp. Range, K	Reported Error, %	Name and Specimen Designation	Composition (weight percent), Specifications and Remarks
1	257	1952	1.5-20			Purissimium grade; sample supplied by Brocades; specimen recrystallized three times from crude salt; measured under helium atmosphere.

DATA TABLE NO. 255 SPECIFIC HEAT OF COPPER DICHLORIDE DIHYDRATE $CuCl_2 \cdot 2H_2O$

[Temperature, T, K; Specific Heat, C_p, Cal $g^{-1}K^{-1}$]

T	C_p
CURVE 1	
Series 1	
2.492	2.529×10^{-3}*
2.500	2.529
2.662	2.840
2.676	2.876*
2.719	2.876*
3.028	4.430
3.041	4.537*
3.061	4.601*
3.079	4.719
3.470	7.403
3.485	7.501*
3.500	7.501*
3.514	7.641*
3.535	7.739*
3.867	9.912
3.882	1.021×10^{-2}*
3.901	1.023*
3.923	1.053*
4.009	1.117
4.035	1.140*
4.059	1.150*
4.084	1.182*
Series 2	
4.188	1.227×10^{-2}
4.213	1.326*
4.243	1.431*
4.264	1.579
4.308	1.590*
4.313	1.385
4.402	7.459×10^{-3}
4.450	6.533
4.477	6.744*
4.496	6.281
4.536	6.169*
4.566	5.987
4.596	5.930*
4.625	5.832
4.654	5.664*
4.709	5.510
4.777	5.384*
CURVE 1 (cont.)	
4.848	5.089×10^{-3}
4.921	5.005*
Series 3	
1.534	3.126×10^{-4}
1.556	3.477
1.585	3.561*
1.612	3.898
1.655	4.430
1.700	5.356
1.764	6.309
1.806	7.164
1.845	7.515
1.887	8.510
2.020	1.162×10^{-3}
2.067	1.044
2.114	1.227
2.150	1.280
4.091	1.227×10^{-2}*
4.136	1.286*
4.156	1.308*
4.175	1.338*
4.210	1.374*
4.232	1.409*
4.249	1.350
4.263	1.468*
4.265	1.490*
4.285	1.560*
4.310	1.396*
4.341	1.286
4.365	9.099×10^{-3}
4.386	7.921
4.407	7.332*
4.431	7.108
Series 4	
14.093	4.458×10^{-3}
14.183	4.402*
14.256	4.514*
14.476	4.585*
14.698	4.795
14.910	5.117
CURVE 1 (cont.)	
15.113	5.272×10^{-3}*
15.215	5.314*
15.408	5.552
15.605	5.790*
15.781	6.071
15.954	6.351*
16.121	6.505
16.281	6.772*
16.426	6.688*
16.624	7.304
16.825	7.206*
17.017	7.627*
17.205	7.781*
17.387	8.033
17.514	8.230*
17.692	8.496*
17.851	8.580
18.028	8.622*
18.219	9.001*
18.315	9.295
18.691	9.478*
18.880	9.772*
19.069	9.996
19.226	1.032×10^{-2}*
19.428	1.046*
19.621	1.108*
19.813	1.145

* Not shown on plot

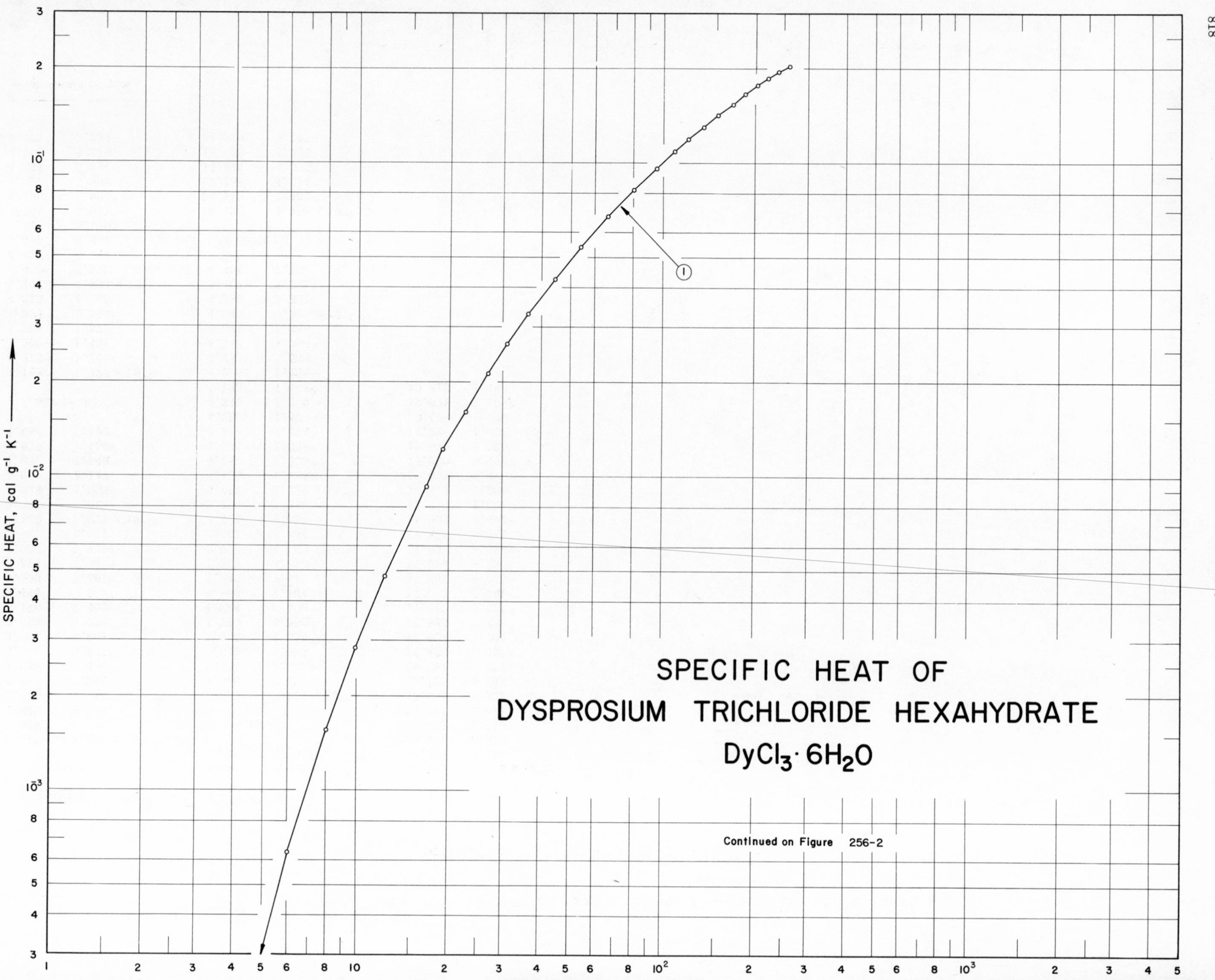
SPECIFIC HEAT OF
DYSPROSIUM TRICHLORIDE HEXAHYDRATE
$DyCl_3 \cdot 6H_2O$
Continued on Figure 256-2
SPECIFIC HEAT, cal g^{-1} K^{-1}
1

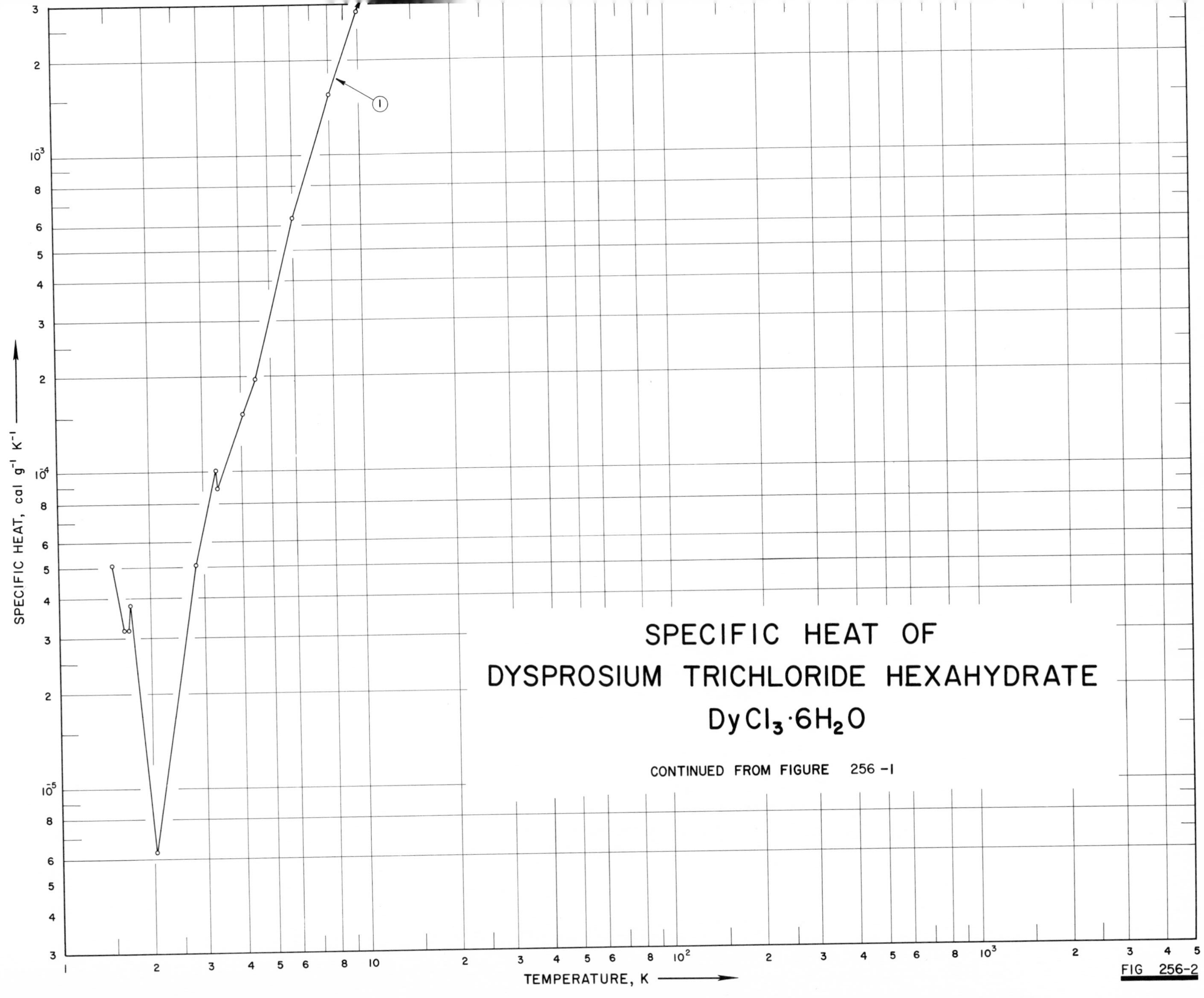
SPECIFIC HEAT OF
DYSPROSIUM TRICHLORIDE HEXAHYDRATE
$DyCl_3 \cdot 6H_2O$
CONTINUED FROM FIGURE 256-1
SPECIFIC HEAT, cal g^{-1} K^{-1}
TEMPERATURE, K
1
FIG 256-2

SPECIFICATION TABLE NO. 256 SPECIFIC HEAT OF DYSPROSIUM TRICHLORIDE HEXAHYDRATE $DyCl_3 \cdot 6H_2O$

[For Data Reported in Figure and Table No. 256]

Curve No.	Ref. No.	Year	Temp. Range, K	Reported Error, %	Name and Specimen Designation	Composition (weight percent), Specifications and Remarks
1	319	1961	1.5-260			

DATA TABLE NO. 256 SPECIFIC HEAT OF DYSPROSIUM TRICHLORIDE HEXAHYDRATE $DyCl_3 \cdot 6H_2O$

[Temperature, T, K; Specific Heat, C_p, Cal $g^{-1}K^{-1}$]

T	C_p
CURVE 1	
1.51	5.07×10^{-5}
1.63	3.17
1.69	3.17
1.72	3.80
2.04	6.34×10^{-6}
2.83	5.07×10^{-5}
3.32	1.01×10^{-4}
3.35	8.88×10^{-5}
4.07	1.52×10^{-4}
4.49	1.97
6.16	6.34
8.06	1.56×10^{-3}
9.95	2.85
12.37	4.81
14.46	6.77
16.76	9.28
19.15	1.23×10^{-2}
22.62	1.61
26.97	2.15
31.03	2.66
36.45	3.31
44.54	4.28
54.44	5.43
66.57	6.765
80.96	8.217
95.88	9.612
109.66	1.091×10^{-1}
122.39	1.192
136.47	1.302
152.39	1.422
170.02	1.545
187.69	1.662
205.56	1.773
223.52	1.869
240.99	1.958
260.13	2.047

* Not shown on plot

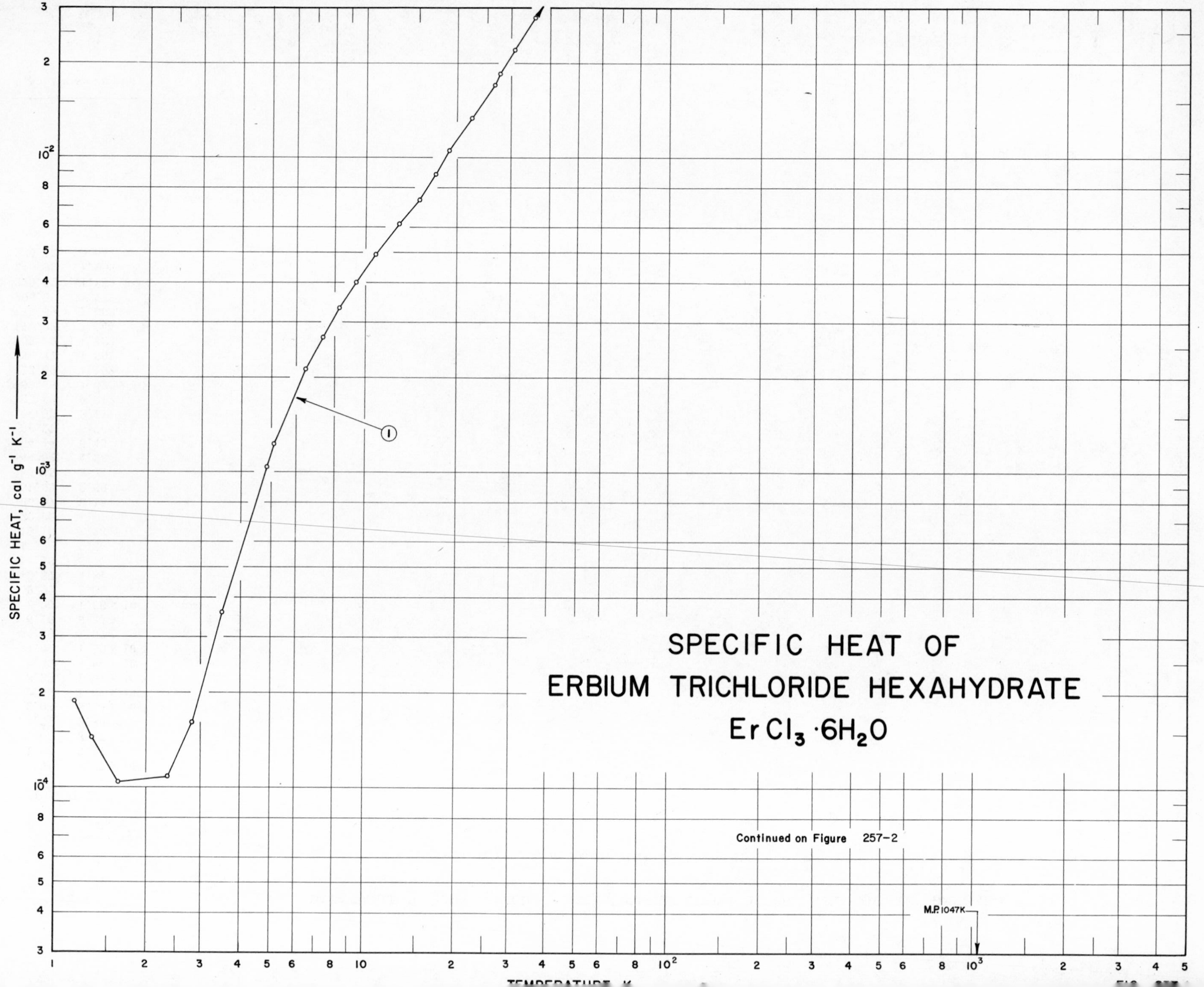
SPECIFIC HEAT OF
ERBIUM TRICHLORIDE HEXAHYDRATE
$ErCl_3 \cdot 6H_2O$
Continued on Figure 257-2
M.P. 1047K
SPECIFIC HEAT, cal g^{-1} K^{-1}
1

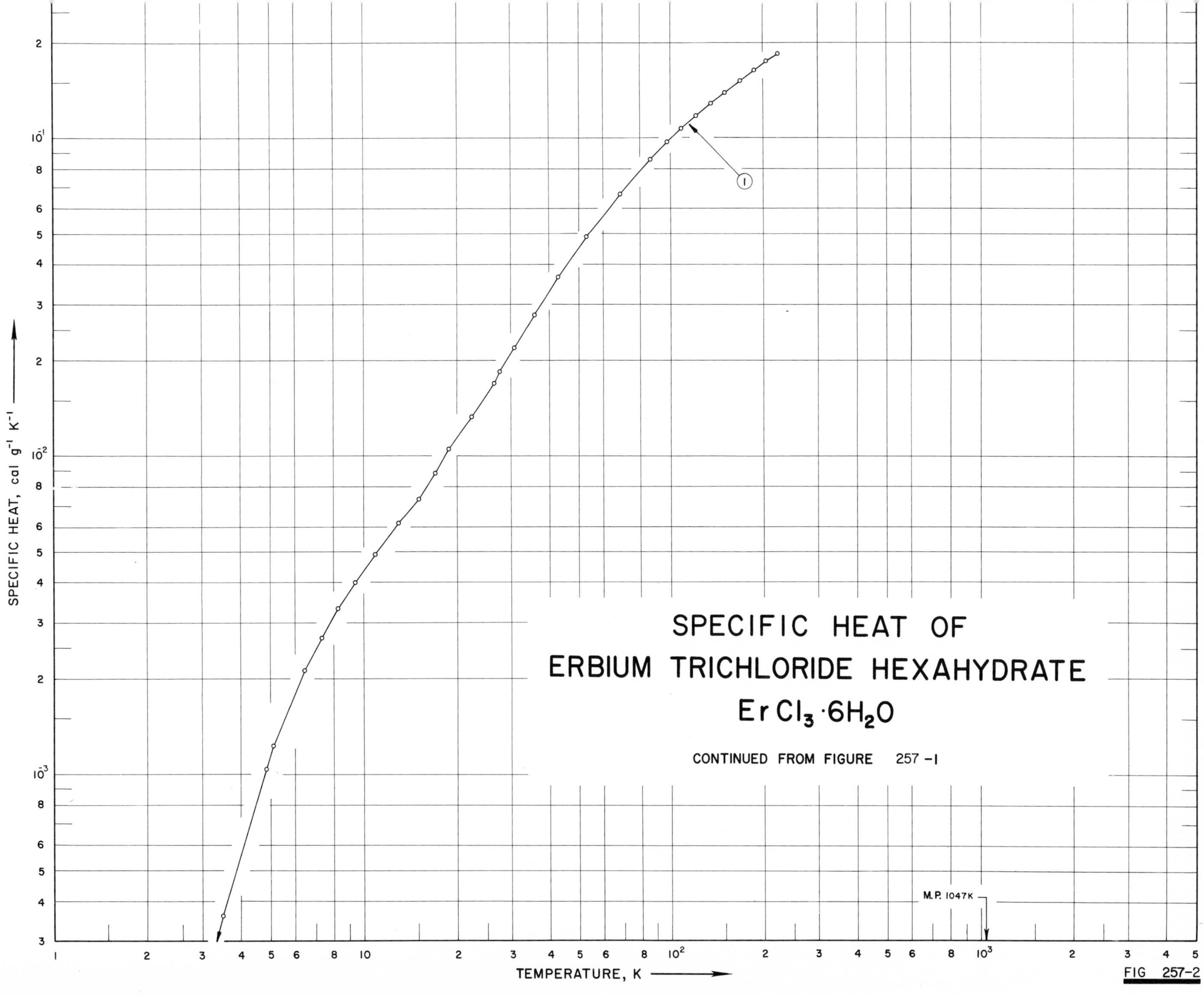
SPECIFIC HEAT OF
ERBIUM TRICHLORIDE HEXAHYDRATE
$ErCl_3 \cdot 6H_2O$
CONTINUED FROM FIGURE 257-1
M.P. 1047 K
TEMPERATURE, K
SPECIFIC HEAT, cal g^{-1} K^{-1}
1
FIG 257-2

SPECIFICATION TABLE NO. 257 SPECIFIC HEAT OF ERBIUM TRICHLORIDE HEXAHYDRATE $ErCl_3 \cdot 6H_2O$

[For Data Reported in Figure and Table No. 257]

Curve No.	Ref. No.	Year	Temp. Range, K	Reported Error, %	Name and Specimen Designation	Composition (weight percent), Specifications and Remarks
1	320	1961	1.2-224			Prepared from 99.96 Er_2O_3.

DATA TABLE NO. 257 SPECIFIC HEAT OF ERBIUM TRICHLORIDE HEXAHYDRATE $ErCl_3 \cdot 6H_2O$

[Temperature, T, K; Specific Heat, C_p, Cal $g^{-1}K^{-1}$]

T	C_p
CURVE 1	
1.17	1.878×10^{-4}
1.34	1.446
1.63	1.046
2.36	1.089
2.81	1.615
3.50	3.613
4.86	1.048×10^{-3}
5.12	1.231
6.48	2.125
7.36	2.691
8.30	3.348
9.42	4.014
10.98	4.940
13.06	6.174
15.14	7.382
17.02	8.879
18.86	1.061×10^{-2}
22.33	1.332
22.41	1.354*
26.54	1.705
27.65	1.855
30.87	2.213
35.98	2.799
43.10	3.687
53.35	4.940
68.71	6.732
86.65	8.616
98.12	9.768
108.84	1.073
109.49	1.082*
122.39	1.187
136.29	1.297
151.89	1.408
169.34	1.528
187.29	1.657
205.57	1.763
224.05	1.856

* Not shown on plot

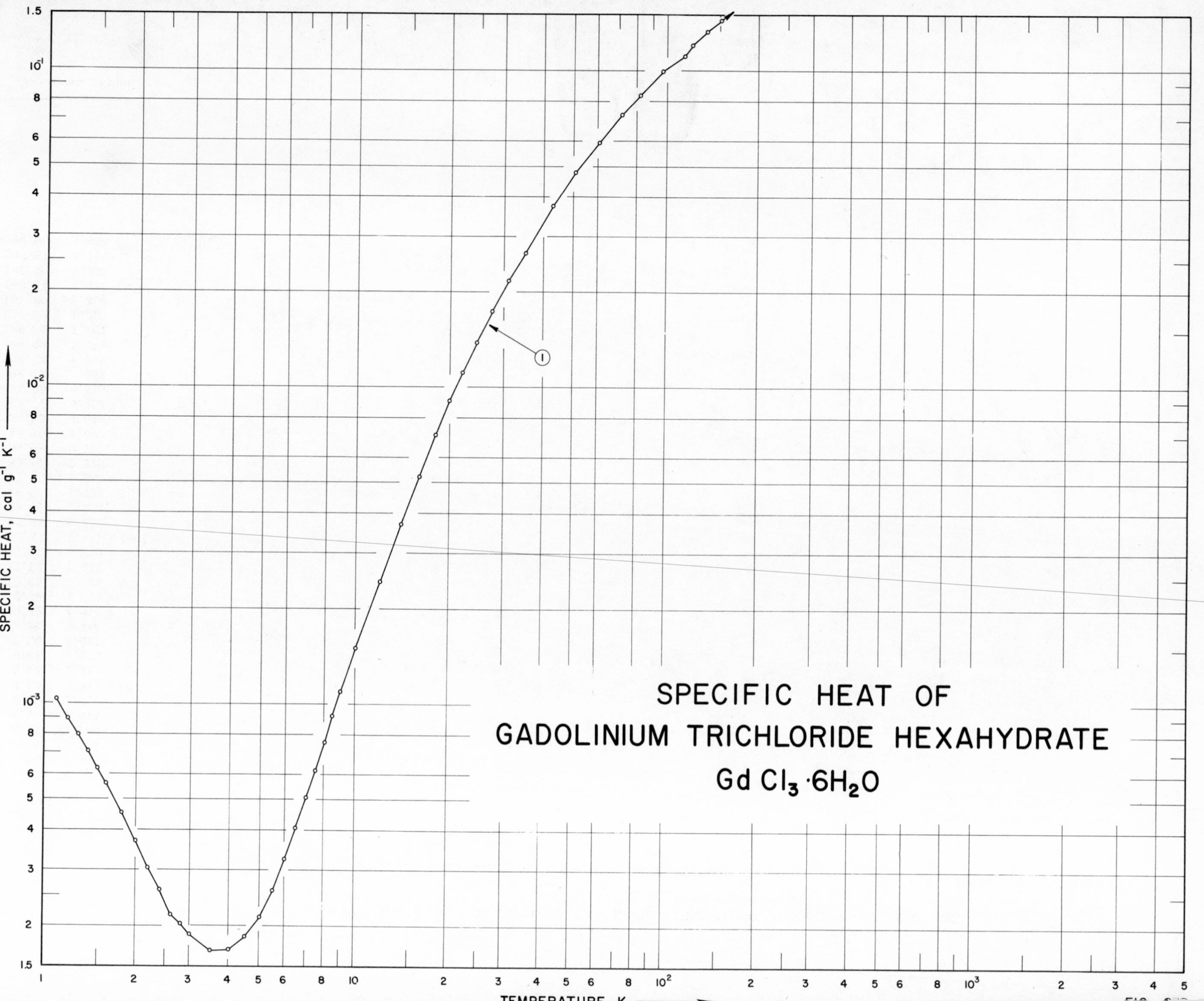
SPECIFIC HEAT OF
GADOLINIUM TRICHLORIDE HEXAHYDRATE
Gd Cl$_3$·6H$_2$O
SPECIFIC HEAT, cal g^{-1} K^{-1}
1

SPECIFICATION TABLE NO. 258 SPECIFIC HEAT OF GADOLINIUM TRICHLORIDE HEXAHYDRATE $GdCl_3 \cdot 6H_2O$

[For Data Reported in Figure and Table No. 258]

Curve No.	Ref. No.	Year	Temp. Range, K	Reported Error, %	Name and Specimen Designation	Composition (weight percent), Specifications and Remarks
1	321	1961	1.1-259			Prepared from 99.9 Gd_2O_3 and HCl.

DATA TABLE NO. 258 SPECIFIC HEAT OF GADOLINIUM TRICHLORIDE HEXAHYDRATE $GdCl_3 \cdot 6H_2O$

[Temperature, T, K; Specific Heat, C_p, Cal $g^{-1}K^{-1}$]

T	C_p	T	C_p
CURVE 1		CURVE 1 (cont.)	
1.1	1.039×10^{-3}	137.0	1.330×10^{-1}
1.2	8.966×10^{-4}	152.0	1.448
1.3	7.982	170.0	1.585*
1.4	7.068	187.0	1.707*
1.5	6.258	205.0	1.820*
1.6	5.615	224.0	1.925*
1.8	4.534	242.0	2.012*
2.0	3.692	259.0	2.088*
2.2	3.055		
2.4	2.605		
2.6	2.161		
2.8	2.039		
3.0	1.878		
3.5	1.679		
4.0	1.685		
4.5	1.852		
5.0	2.135		
5.5	2.592		
6.0	3.261		
6.5	4.071		
7.0	5.081		
7.5	6.181		
8.0	7.564		
8.5	9.197		
9.0	1.097×10^{-3}		
10.0	1.502		
12.0	2.444		
14.0	3.692		
16.0	5.223		
18.0	7.094		
20.0	9.062		
22.0	1.118×10^{-2}		
24.5	1.382		
27.5	1.744		
31.0	2.169		
35.0	2.663		
43.0	3.732		
51.0	4.758		
61.0	5.946		
72.0	7.256		
82.5	8.362		
97.5	9.921		
110.5	1.114×10^{-1}		
122.5	1.215		

* Not shown on plot

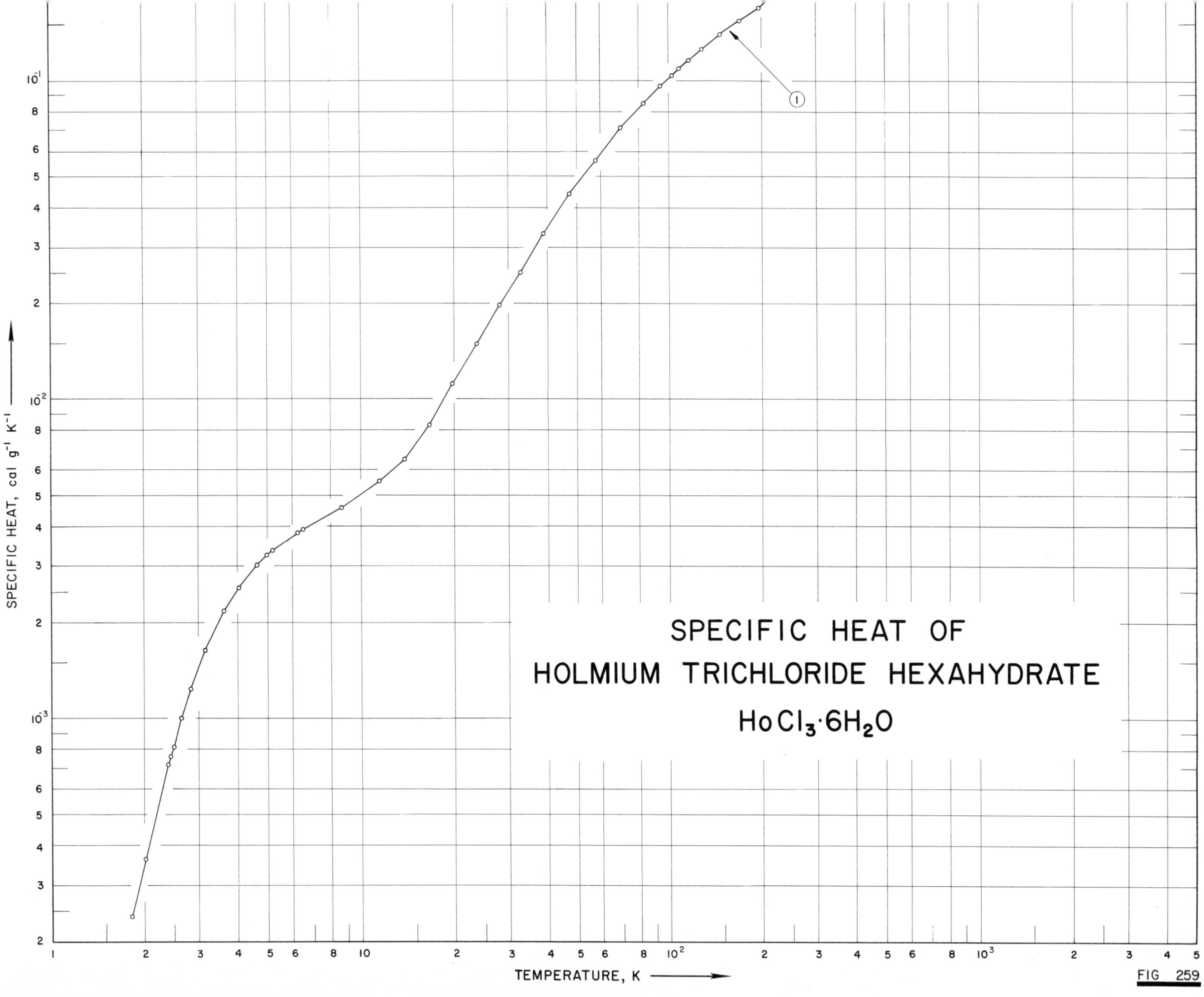
SPECIFIC HEAT OF
HOLMIUM TRICHLORIDE HEXAHYDRATE
$HoCl_3 \cdot 6H_2O$
SPECIFIC HEAT, cal g^{-1} K^{-1}
TEMPERATURE, K
1

FIG 259

SPECIFICATION TABLE NO. 259 SPECIFIC HEAT OF HOLMIUM TRICHLORIDE HEXAHYDRATE $HoCl_3 \cdot 6H_2O$

[For Data Reported in Figure and Table No. 259]

Curve No.	Ref. No.	Year	Temp. Range, K	Reported Error, %	Name and Specimen Designation	Composition (weight percent), Specifications and Remarks
1	320	1961	1.4-231			Prepared from 99.9 Ho_2O_3.

DATA TABLE NO. 259 SPECIFIC HEAT OF HOLMIUM TRICHLORIDE HEXAHYDRATE $HoCl_3 \cdot 6H_3O$

[Temperature, T, K; Specific Heat, C_p, Cal $g^{-1}K^{-1}$]

T	C_p
CURVE 1	
1.39	1.266 x 10^{-4}*
1.82	2.400
2.01	3.635
2.38	7.169
2.44	7.617
2.50	8.146
2.64	1.007 x 10^{-3}
2.83	1.244
3.15	1.642
3.63	2.180
4.04	2.570
4.63	3.043
4.97	3.263
5.18	3.371
6.23	3.849
6.51	3.944
8.69	4.631
11.52	5.582
13.93	6.508
16.73	8.385
19.80	1.126 x 10^{-2}
23.62	1.499
28.09	1.987
32.72	2.512
39.01	3.326
47.76	4.425
58.01	5.649
70.37	7.132
83.60	8.499
94.98	9.614
103.09	1.029 x 10^{-1}*
103.83	1.044
109.31	1.093
117.71	1.160
118.17	1.168*
129.98	1.261
130.27	1.259*
148.00	1.394
149.21	1.401*
170.12	1.545
190.74	1.687
211.28	1.809
230.76	1.913

* Not shown on plot

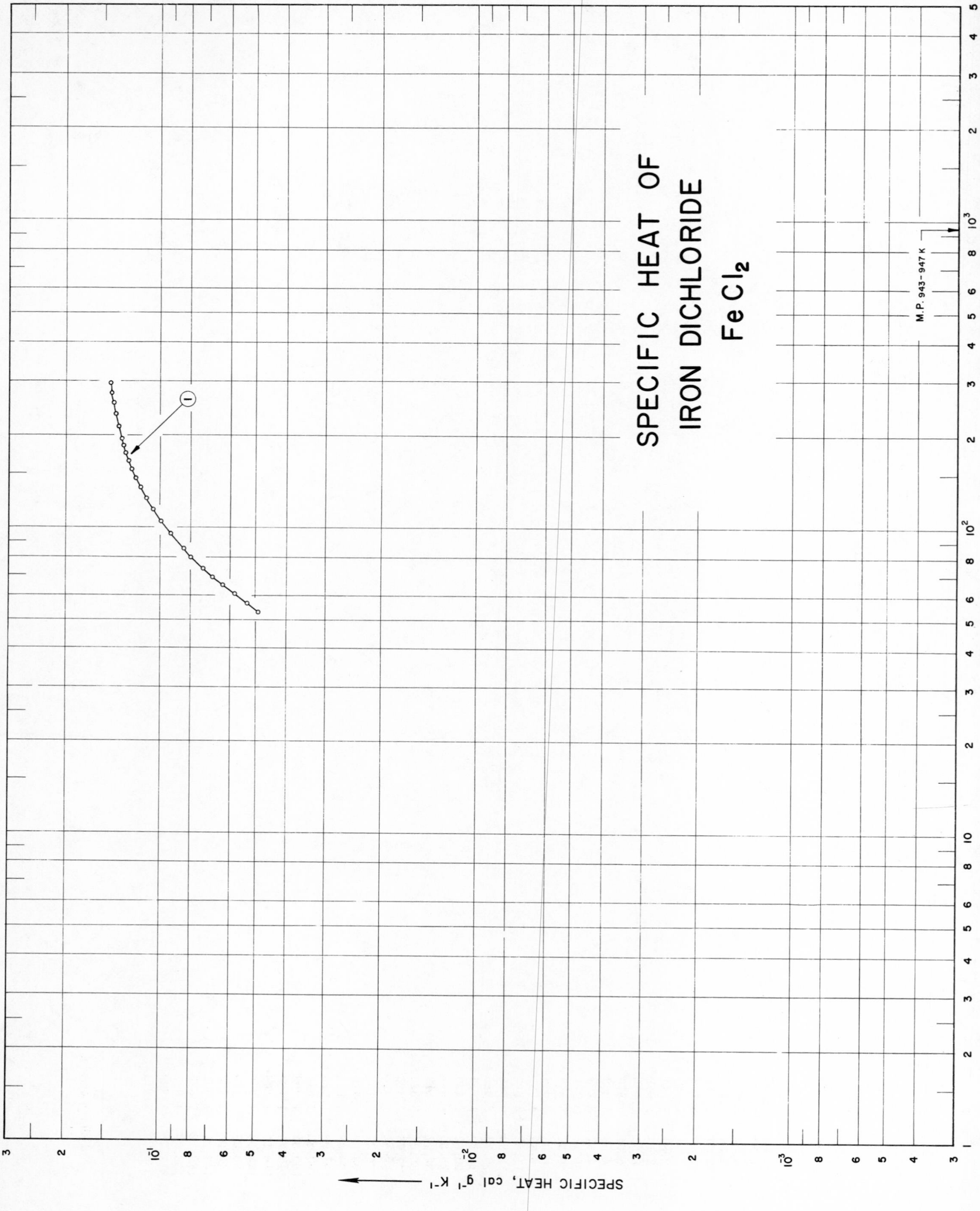

SPECIFIC HEAT OF IRON DICHLORIDE $FeCl_2$

SPECIFICATION TABLE NO. 260 SPECIFIC HEAT OF IRON DICHLORIDE $FeCl_2$

[For Data Reported in Figure and Table No. 260]

Curve No.	Ref. No.	Year	Temp. Range, K	Reported Error, %	Name and Specimen Designation	Composition (weight percent), Specifications and Remarks
1	250	1943	53-295			44.13 Fe (44.06 theo.), 55.85 Cl (55.94 theo.); prepared from $FeCl_2 \cdot 4H_2O$ by heating slowly in vacuum to 200 and then in steam of dry HCl, temp gradually raised to 550 C.

DATA TABLE NO. 260 SPECIFIC HEAT OF IRON DICHLORIDE

[Temperature, T, K; Specific Heat, C_p, Cal $g^{-1}K^{-1}$]

T	C_p
CURVE 1	
53.2	4.885 x 10^{-2}
56.7	5.289
60.8	5.790
65.2	6.337
69.3	6.827
73.6	7.297
80.6	7.991
85.5	8.433
95.4	9.269
105.1	9.940
115.2	1.056 x 10^{-1}
125.2	1.110
135.6	1.158
145.8	1.197
155.5	1.230
165.9	1.259
175.3	1.284
185.3	1.306
195.2	1.327
205.0	1.341*
214.9	1.357
225.1	1.373*
235.2	1.385
245.6	1.398*
255.4	1.407
265.6	1.415*
275.9	1.427
285.7	1.430*
295.0	1.437

*Not shown on plot

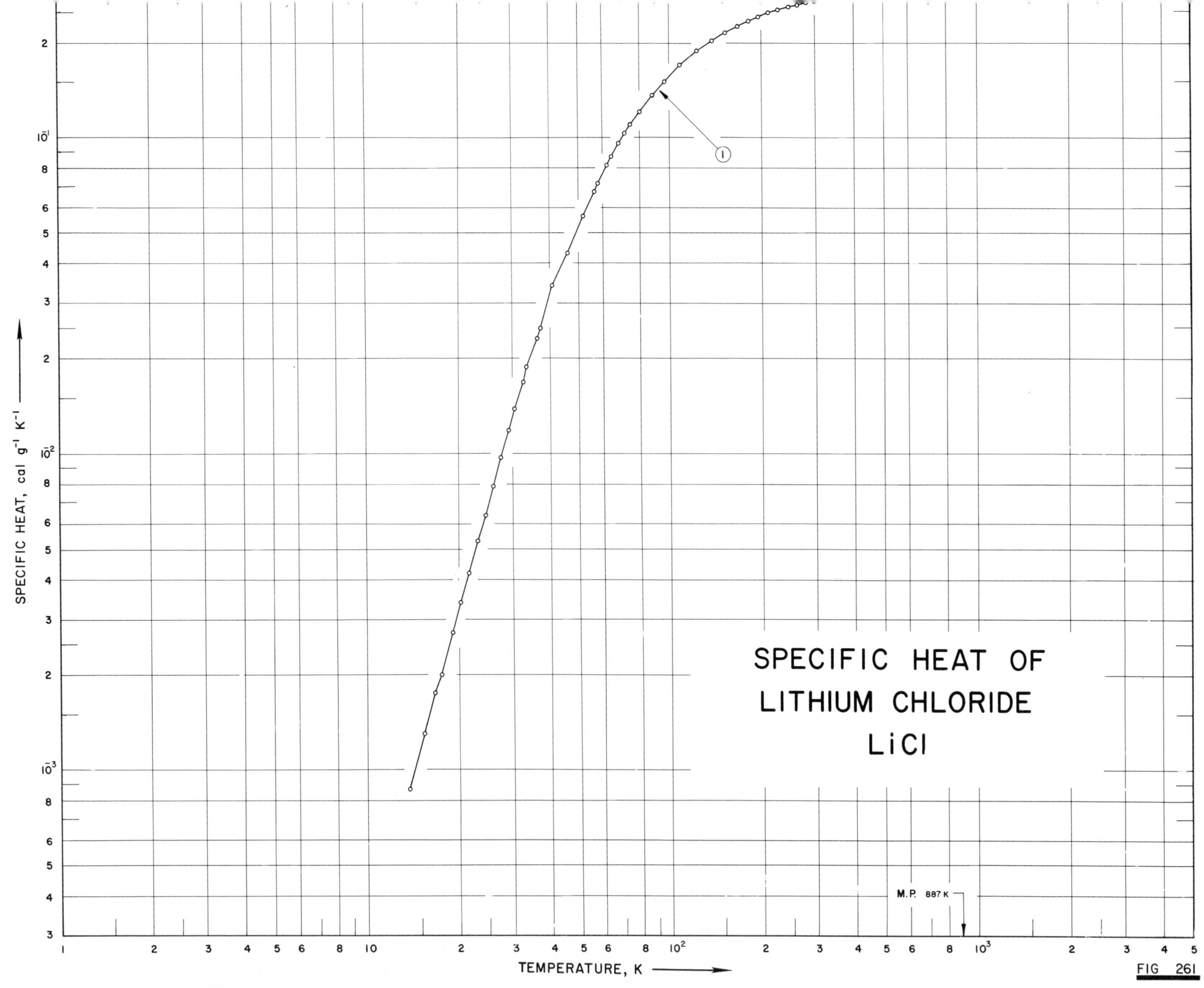
SPECIFIC HEAT OF
LITHIUM CHLORIDE
LiCl
SPECIFIC HEAT, cal g⁻¹ K⁻¹
TEMPERATURE, K
M.P. 887 K
1
FIG 261

SPECIFICATION TABLE NO. 261 SPECIFIC HEAT OF LITHIUM CHLORIDE LiCl

[For Data Reported in Figure and Table No. 261]

Curve No.	Ref. No.	Year	Temp. Range, K	Reported Error, %	Name and Specimen Designation	Composition (weight percent), Specifications and Remarks
1	322	1960	14-320	0.1-1		99.1 LiCl, 0.06 Na, 0.03 K, 0.006 alkalinity, 0.005 SO_4, 0.005 insol. matter, 0.002 Ca and 0.0005 NO_3; purified by recrystallization and fused.

DATA TABLE NO. 261 SPECIFIC HEAT OF LITHIUM CHLORIDE LiCl

[Temperature, T, K; Specific Heat, C_p, Cal $g^{-1}K^{-1}$]

T	C_p
CURVE 1	
Series 1	
16.65	1.761 x 10^{-3}
19.07	2.734
21.64	4.244
24.40	6.419
27.49	9.778
30.59	1.397 x 10^{-2}
33.81	1.890
37.37	2.517
41.70	3.383*
46.58	4.449*
51.39	5.561*
56.60	6.794
62.36	8.192
68.53	9.644
74.98	1.106 x 10^{-1}
81.43	1.239*
88.01	1.362*
95.08	1.486*
102.03	1.597*
109.30	1.701
116.72	1.798*
124.12	1.890
131.52	1.965*
138.83	2.038
146.06	2.104*
153.34	2.161
160.92	2.219*
168.45	2.265
175.72	2.306*
182.99	2.349
190.21	2.384*
197.56	2.421
204.91	2.454*
212.60	2.484
219.84	2.512*
227.52	2.547
235.07	2.563*
242.55	9.657*
246.90	2.594
249.94	2.367*
255.11	2.608*
263.92	2.638
CURVE 1 (cont.)	
272.31	2.656 x 10^{-1}
280.78	2.677
290.11	2.689*
299.64	2.709
309.77	2.723*
320.42	2.742
Series 2	
13.77	8.707 x 10^{-4}
15.42	1.310 x 10^{-3}
17.54	2.004
20.38	3.406
23.03	5.301
25.84	7.905
29.27	1.194 x 10^{-2}
32.67	1.692
36.36	2.325
40.83	3.430
46.05	4.337
51.94	5.665
58.29	7.187
64.71	8.737
71.80	1.038 x 10^{-1}
80.03	1.208
88.52	1.370
97.03	1.517

* Not shown on plot

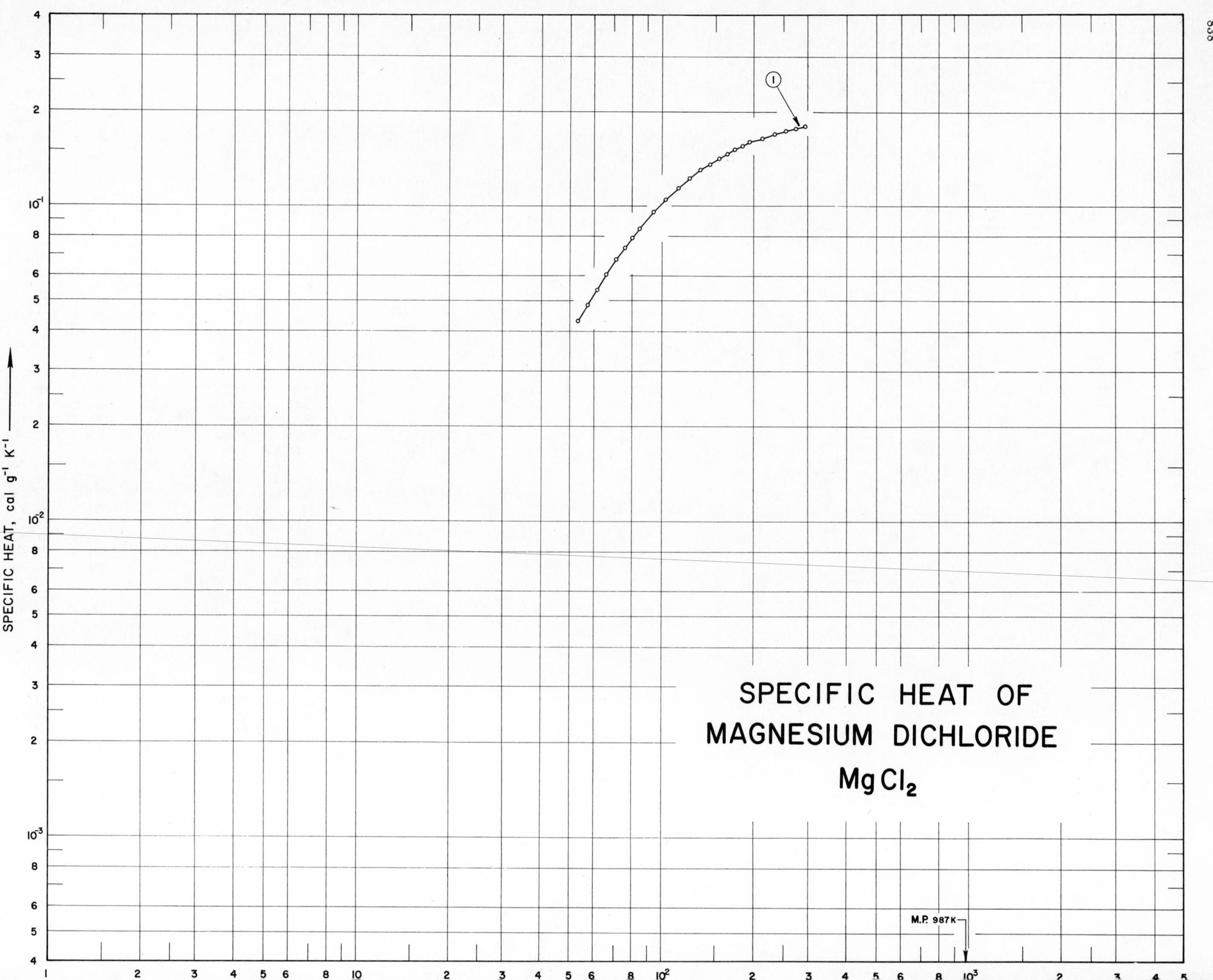
SPECIFIC HEAT OF
MAGNESIUM DICHLORIDE
$MgCl_2$
M.P. 987 K
1
SPECIFIC HEAT, cal g^{-1} K^{-1}

SPECIFICATION TABLE NO. 262 SPECIFIC HEAT OF MAGNESIUM DICHLORIDE $MgCl_2$

[For Data Reported in Figure and Table No. 262]

Curve No.	Ref. No.	Year	Temp. Range, K	Reported Error, %	Name and Specimen Designation	Composition (weight percent), Specifications and Remarks
1	250	1943	54-295			74.25 Cl (74.46 theo), 25.74 Mg (25.54 theo), 0.2 MgO; prepared from magnesium ammonium chloride hexahydrate by heating slowly to 200 C and then by a stream of dry HCl to 400 C; heated several more days in stream of dry HCl at 500 to 600 C; evacuated and cooled. (corrected for MgO impurities)

DATA TABLE NO. 262 SPECIFIC HEAT OF MAGNESIUM DICHLORIDE $MgCl_2$

[Temperature, T, K; Specific Heat, Cp, Cal $g^{-1}K^{-1}$]

T	Cp
CURVE 1	
53.6	4.320×10^{-2}
57.6	4.851
61.8	5.418
66.3	6.064
71.4	6.765
76.1	7.370
80.4	7.925
84.8	8.453
94.2	9.522
103.9	1.049×10^{-1}
114.2	1.144
123.9	1.223
134.9	1.302
144.6	1.362
155.3	1.420
165.1	1.467
174.6	1.506
185.1	1.547
195.0	1.584
204.8	1.608*
215.3	1.636
225.1	1.660*
235.2	1.682
245.0	1.702*
255.3	1.722
266.1	1.739*
276.5	1.758
286.1	1.768*
295.4	1.783

* Not shown on plot

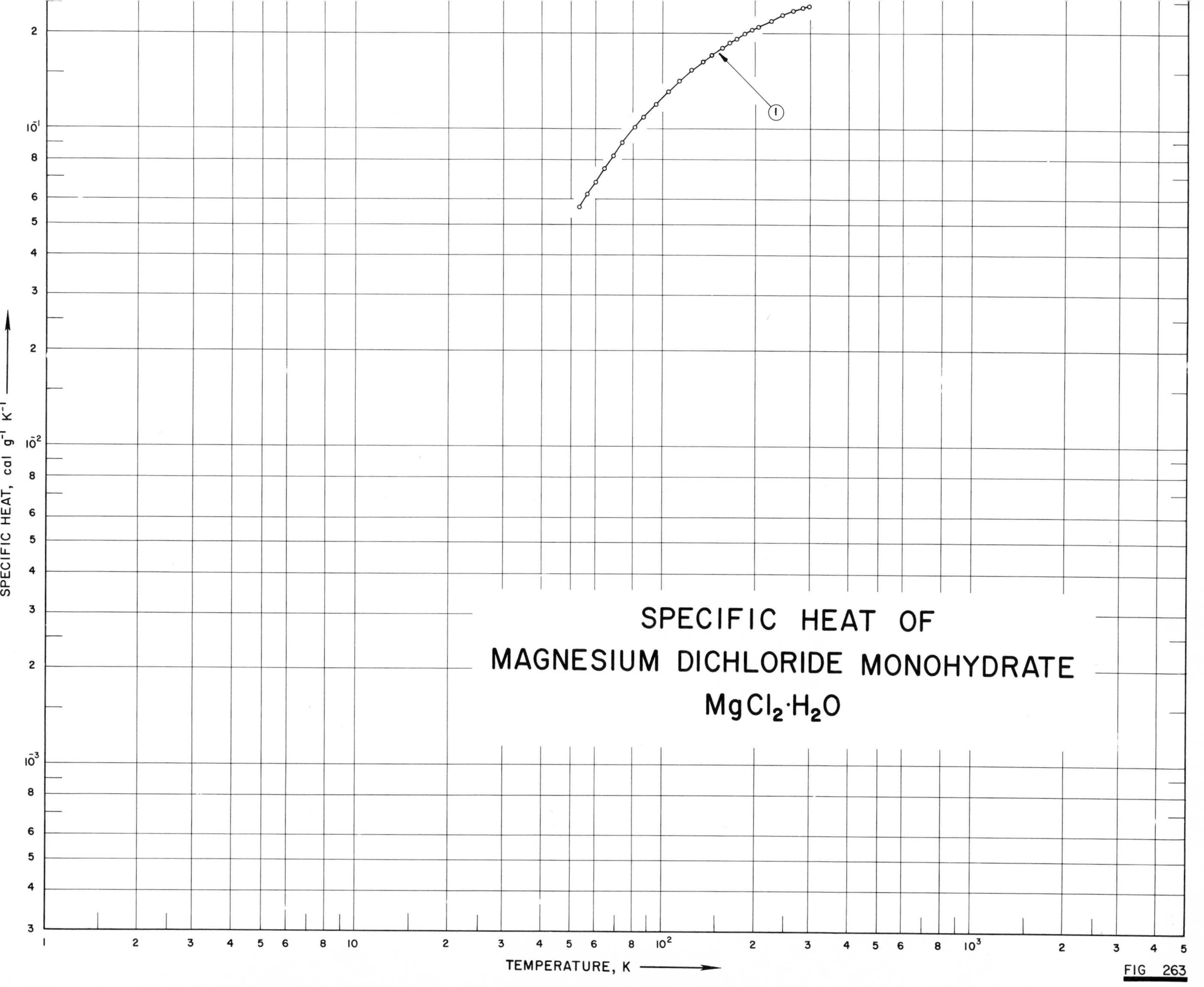
SPECIFIC HEAT OF
MAGNESIUM DICHLORIDE MONOHYDRATE
$MgCl_2 \cdot H_2O$
SPECIFIC HEAT, cal g^{-1} K^{-1}
TEMPERATURE, K
1
FIG 263

SPECIFICATION TABLE NO. 263 SPECIFIC HEAT OF MAGNESIUM DICHLORIDE MONOHYDRATE $MgCl_2 \cdot H_2O$

[For Data Reported in Figure and Table No. 263]

Curve No.	Ref. No.	Year	Temp. Range, K	Reported Error, %	Name and Specimen Designation	Composition (weight percent), Specifications and Remarks
1	323	1943	54-298			62.44 Cl (62.62 theo), 21.57 Mg (21.47 theo), and 0.14 MgO; prepared from stoichiometric quantities of dehydrated and anhydrous magnesium chloride; heated 16 hrs at 120-140 C under vacuum. (corrected for MgO inpurities)

DATA TABLE NO. 263 SPECIFIC HEAT OF MAGNESIUM DICHLORIDE MONOHYDRATE $MgCl_2 \cdot H_2O$

[Temperature, T,K; Specific Heat, Cp, Cal $g^{-1}K^{-1}$]

T	Cp
CURVE 1	
53.5	5.697×10^{-2}
56.9	6.241
60.5	6.808
64.6	7.497
69.1	8.227
74.1	9.024
81.7	1.009×10^{-1}*
81.9	1.011
86.7	1.079*
87.0	1.082
95.7	1.198
105.3	1.310
114.8	1.418
125.6	1.531
135.7	1.626
145.7	1.708
157.2	1.800
166.0	1.863
175.8	1.923
185.6	1.986
196.3	2.050
206.0	2.094
216.4	2.144*
226.3	2.192
236.3	2.234*
246.2	2.274
256.3	2.312*
266.4	2.352
275.9	2.388*
285.9	2.408
295.7	2.429*
298.2	2.426

*Not shown on plot

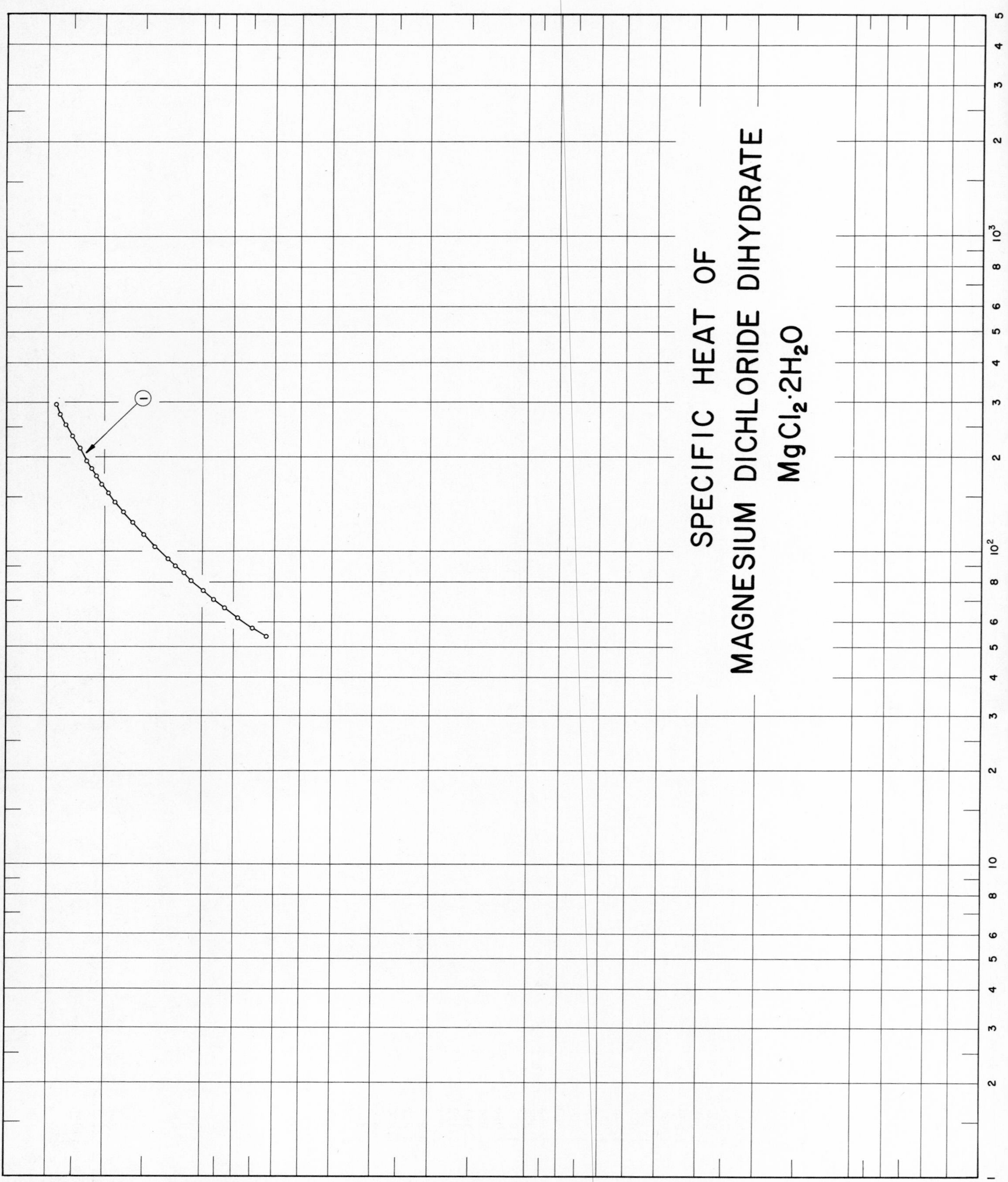

SPECIFIC HEAT OF MAGNESIUM DICHLORIDE DIHYDRATE

$MgCl_2 \cdot 2H_2O$

SPECIFICATION TABLE NO. 264 SPECIFIC HEAT OF MAGNESIUM DICHLORIDE DIHYDRATE $MgCl_2 \cdot 2H_2O$

[For Data Reported in Figure and Table No. 264]

Curve No.	Ref. No.	Year	Temp. Range, K	Reported Error, %	Name and Specimen Designation	Composition (weight percent), Specifications and Remarks
1	323	1943	54-295			54.04 Cl (54.03 theo), 18.62 Mg (18.53 theo), and 0.01 MgO; prepared from tetrahydrate by heating in a stream of dry HCl at temperatures gradually increasing from 170-220 C until the required amount of water was removed; aged for 7 hrs at 103 C.

DATA TABLE NO. 264 SPECIFIC HEAT OF MAGNESIUM DICHLORIDE DIHYDRATE $MgCl_2 \cdot 2H_2O$

[Temperature, T, K; Specific Heat, Cp, Cal $g^{-1}K^{-1}$]

T	Cp
CURVE 1	
54.1	6.491 x 10^{-2}
57.6	7.117
62.1	7.907
66.8	8.783
70.8	9.423
75.1	1.011 x 10^{-1}
81.2	1.100
85.6	1.166
90.2	1.236
94.9	1.301
104.7	1.430
114.4	1.550
124.7	1.676
134.8	1.787
145.3	1.892
155.2	1.988
165.6	2.084
175.4	2.163
185.7	2.244
195.9	2.324
206.1	2.390*
215.9	2.449
225.8	2.515*
236.2	2.576
246.2	2.643*
256.2	2.696
266.3	2.752*
276.3	2.806
285.7	2.849*
295.0	2.887

* Not shown on plot

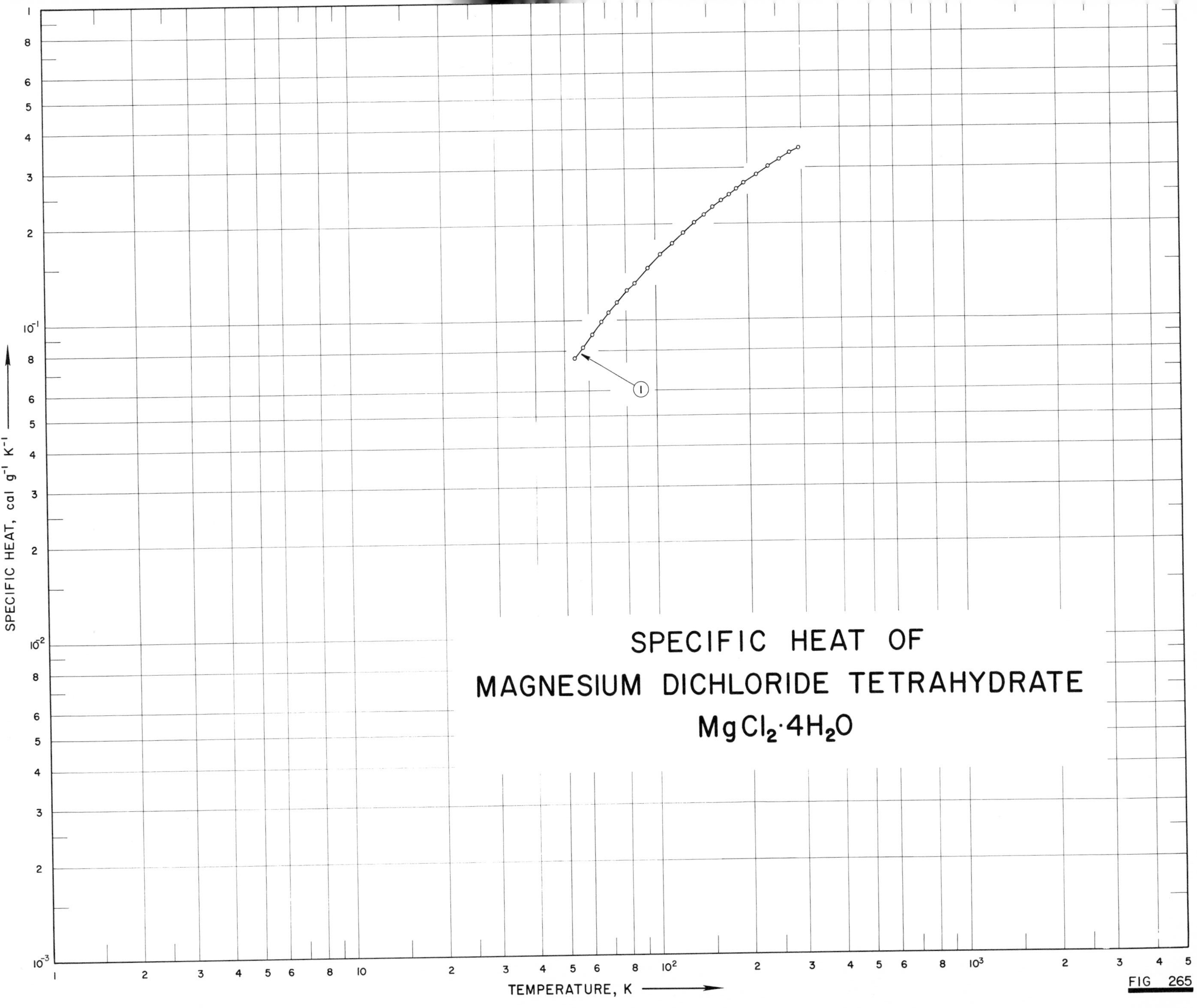
SPECIFIC HEAT OF
MAGNESIUM DICHLORIDE TETRAHYDRATE
$MgCl_2 \cdot 4H_2O$
1
SPECIFIC HEAT, cal g^{-1} K^{-1}
TEMPERATURE, K
FIG 265

SPECIFICATION TABLE NO. 265 SPECIFIC HEAT OF MAGNESIUM DICHLORIDE TETRAHYDRATE $MgCl_2 \cdot 4H_2O$

[For Data Reported in Figure and Table No. 265]

Curve No.	Ref. No.	Year	Temp. Range, K	Reported Error, %	Name and Specimen Designation	Composition (weight percent), Specifications and Remarks
1	323	1943	54-296			42.39 Cl (42.39 theo), 14.61 Mg (14.54 theo), and 0.012 MgO; prepared from hexahydrate by heating in air at 100-103 C for 6 days.

DATA TABLE NO. 265 SPECIFIC HEAT OF MAGNESIUM DICHLORIDE TETRAHYDRATE $MgCl_2 \cdot 4H_2O$

[Temperature, T, K; Specific Heat, Cp, Cal $g^{-1}K^{-1}$]

T	Cp
CURVE 1	
54.1	7.585×10^{-2}
57.5	8.195
61.8	8.995
66.4	9.892
70.3	1.060×10^{-1}
74.9	1.139
81.0	1.237
85.7	1.303
94.5	1.450
104.5	1.595
114.2	1.733
124.6	1.874
135.2	2.012
145.1	2.132
155.6	2.252
165.3	2.363
175.6	2.467
185.4	2.567
196.0	2.674
205.8	2.757*
216.0	2.846
225.9	2.930*
236.4	3.013
246.1	3.093*
255.9	3.167
265.9	3.245*
276.0	3.320
385.7	3.381*
295.5	3.433

*Not shown on plot

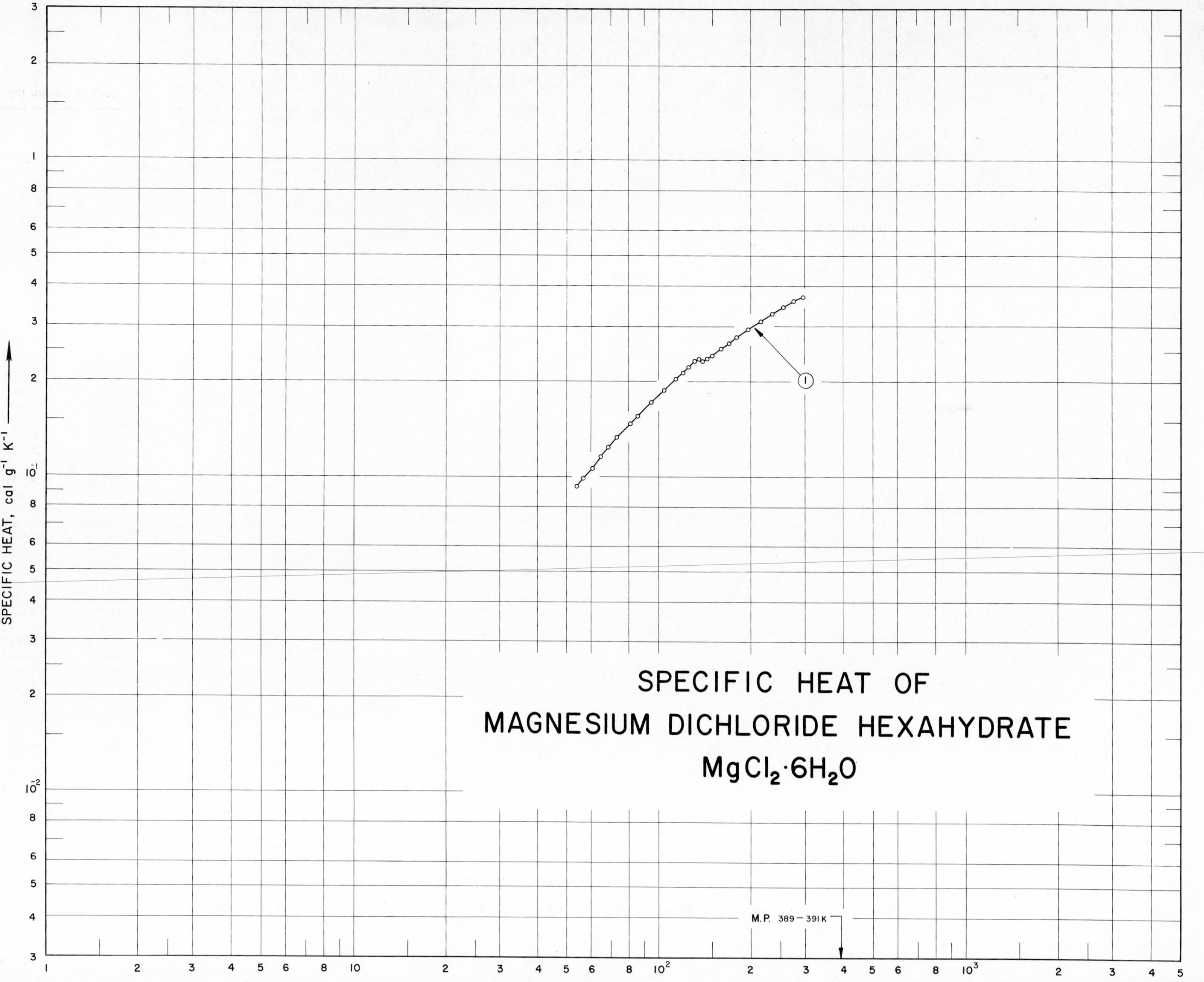
SPECIFIC HEAT OF
MAGNESIUM DICHLORIDE HEXAHYDRATE
$MgCl_2 \cdot 6H_2O$
M.P. 389 – 391 K
SPECIFIC HEAT, cal g^{-1} K^{-1}

SPECIFICATION TABLE NO. 266 SPECIFIC HEAT OF MAGNESIUM DICHLORIDE HEXAHYDRATE $MgCl_2 \cdot 6H_2O$

[For Data Reported in Figure and Table No. 266]

Curve No.	Ref. No.	Year	Temp. Range, K	Reported Error, %	Name and Specimen Designation	Composition (weight percent), Specifications and Remarks
1	323	1943	54-296			Meck's reagent grade; 34.81 Cl (34.88 theo), and 12.05 Mg (11.96 theo); stored over 80% H_2SO_4 for seven days at room temperature.

DATA TABLE NO. 266 SPECIFIC HEAT OF MAGNESIUM DICHLORIDE HEXAHYDRATE $MgCl_2 \cdot 6H_2O$

[Temperature, T,K; Specific Heat, Cp, Cal $g^{-1}K^{-1}$]

T	Cp
CURVE 1	
54.10	9.295×10^{-2}
56.90	9.880
60.60	1.065×10^{-1}
64.70	1.156
68.60	1.237
73.10	1.327
80.90	1.470
80.90	1.471*
85.20	1.549
94.70	1.718
104.30	1.876
104.40	1.876*
114.10	2.032
115.00	2.046*
117.40	2.087*
120.70	2.138
123.80	2.189*
124.40	2.196*
126.40	2.227
128.30	2.256*
130.00	2.285
132.20	2.311
134.00	2.338*
134.90	2.341*
135.90	2.354
137.60	2.354*
139.40	2.324
141.20	2.327*
142.90	2.343*
144.60	2.361
145.50	2.373*
146.20	2.380*
147.80	2.400*
149.40	2.416
151.00	2.437*
155.30	2.484*
159.00	2.533
165.70	2.607*
169.60	2.648
175.50	2.708*
179.70	2.766
185.80	2.822*
195.90	2.921
205.70	3.005*
CURVE 1 (cont.)	
216.10	3.098×10^{-1}
226.10	3.179*
235.90	3.265
245.80	3.352*
255.50	3.439
265.90	3.502*
276.20	3.582
285.80	3.635*
295.80	3.696

*Not shown on plot

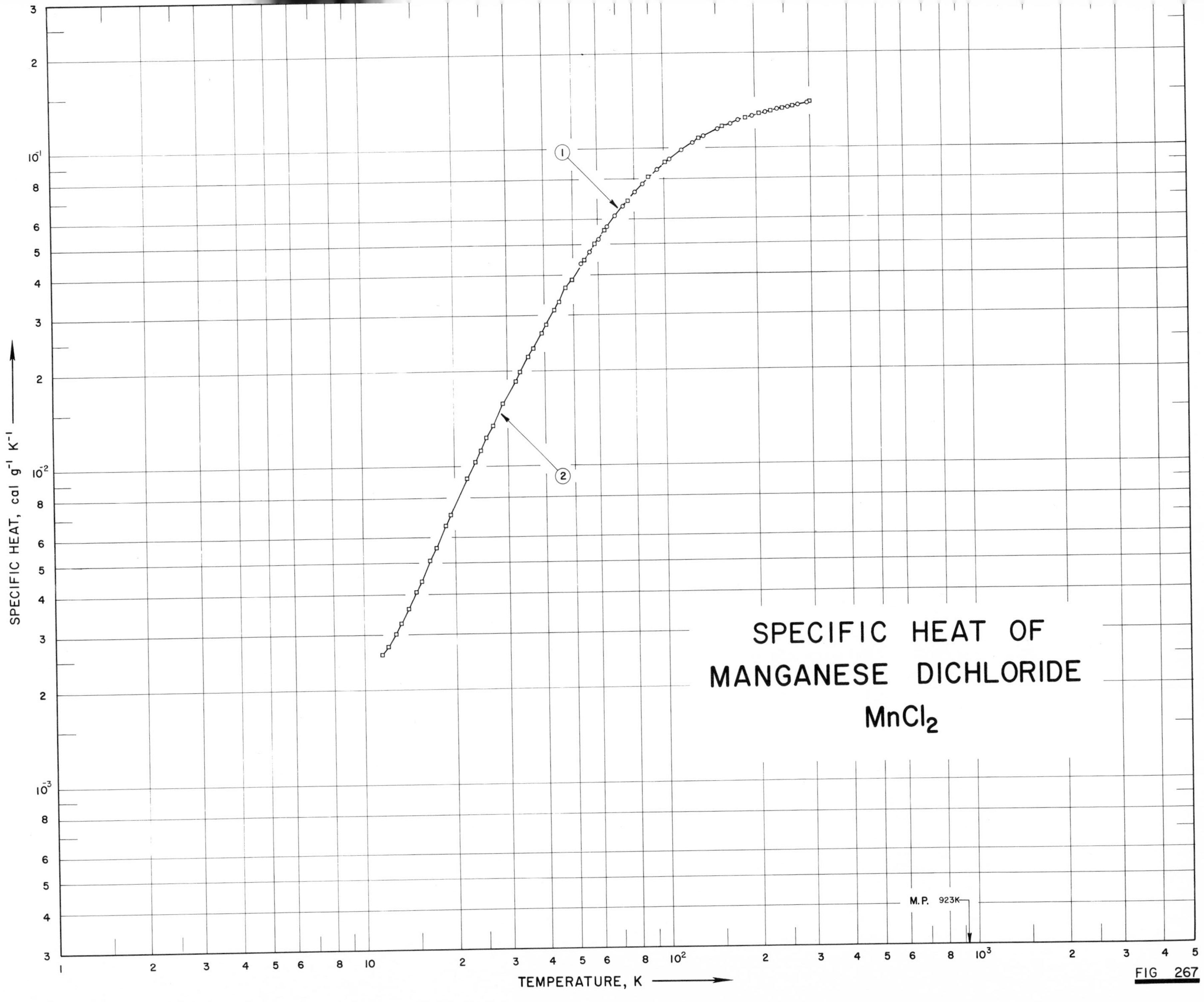
SPECIFIC HEAT OF
MANGANESE DICHLORIDE
MnCl2
SPECIFIC HEAT, cal g-1 K-1
TEMPERATURE, K
M.P. 923K
1
2
FIG 267

SPECIFICATION TABLE NO. 267 SPECIFIC HEAT OF MANGANESE DICHLORIDE $MnCl_2$

[For Data Reported in Figure and Table No. 267]

Curve No.	Ref. No.	Year	Temp. Range, K	Reported Error, %	Name and Specimen Designation	Composition (weight percent), Specifications and Remarks
1	250	1943	53-295			56.30 Cl (56.35 theo), and 43.42 Mn (43.65 theo); prepared from $MnCl_2 \cdot 4H_2O$ by heating in vacuum to 200 C; temp raised slowly to avoid fusion; dehydration completed by passing dry HCl at 620.
2	324	1962	11-300			56.35 Cl (56.35 theo), 43.71 Mn (43.65 theo), 0.001 Si, Fe, Cu, and Ag; prepared from analytical reagent $MnCl_2 \cdot 4H_2O$; heated with HCl stream at 780-800 C.

DATA TABLE NO. 267 SPECIFIC HEAT OF MANGANESE DICHLORIDE $MnCl_2$

[Temperature, T, K; Specific Heat, Cp, Cal $g^{-1}K^{-1}$]

T	Cp
CURVE 1	
53.1	4.388×10^{-2}
56.6	4.758
60.5	5.211
64.8	5.721
68.8	6.186
73.0	6.635
80.4	7.350
85.2	7.784
95.2	8.622
104.8	9.298
115.2	9.941
124.9	1.048×10^{-1}
135.5	1.097
150.6	1.152
156.0	1.171*
165.9	1.200
176.1	1.223
186.1	1.247*
195.6	1.268
205.8	1.283
215.5	1.299
225.8	1.314*
235.9	1.325
245.8	1.338*
256.1	1.349
266.0	1.358*
276.0	1.370
285.7	1.378*
295.0	1.381
CURVE 2	
Series I	
54.28	4.477×10^{-2}
58.99	5.047
63.37	5.561
67.83	6.063*
71.96	6.501*
75.77	6.899
79.68	7.297*
84.69	7.785*
88.91	8.153
92.68	8.463*
96.13	8.725*
100.99	9.075

T	Cp
CURVE 2 (cont.)	
105.77	9.409×10^{-2}*
111.39	9.750*
116.49	1.004×10^{-1}*
121.33	1.031*
125.88	1.055*
130.35	1.078
134.41	1.095*
138.92	1.115*
142.68	1.127*
147.53	1.146*
151.57	1.161*
156.02	1.173
159.33	1.185*
163.34	1.197*
167.40	1.208*
171.07	1.216*
174.87	1.227*
178.63	1.236*
181.73	1.244*
186.02	1.252
190.18	1.263*
193.96	1.267*
197.92	1.276*
201.78	1.281*
205.27	1.289
208.99	1.291*
212.26	1.297*
216.28	1.305*
220.51	1.309*
224.66	1.317
228.57	1.320*
232.59	1.325*
236.51	1.331*
241.36	1.333*
246.10	1.341
252.22	1.346*
258.28	1.353*
264.86	1.359
271.25	1.367*
277.49	1.371*
282.23	1.372*
288.38	1.376*
294.39	1.385*
300.04	1.384*
300.07	1.393

T	Cp
CURVE 2 (cont.)	
Series II	
11.97	2.749×10^{-3}
13.19	3.234
14.77	4.069
16.49	5.102
18.57	6.564
21.87	9.242
24.28	1.135×10^{-2}
26.72	1.360
28.97	1.581
31.82	1.869
32.98	1.991
36.50	2.367
40.45	2.815
44.74	3.327
49.34	3.880
53.78	4.417*
Series III	
11.48	2.599×10^{-3}
12.66	3.004
13.96	3.608
15.41	4.394
17.27	5.618
19.30	7.120
Series IV	
21.38	8.813×10^{-3}*
23.29	1.047×10^{-2}
25.44	1.242
31.76	1.861*
35.11	2.223
38.91	2.649
43.10	3.131
47.67	3.677
Series V	
299.03	1.388×10^{-1}*

*Not shown on plot

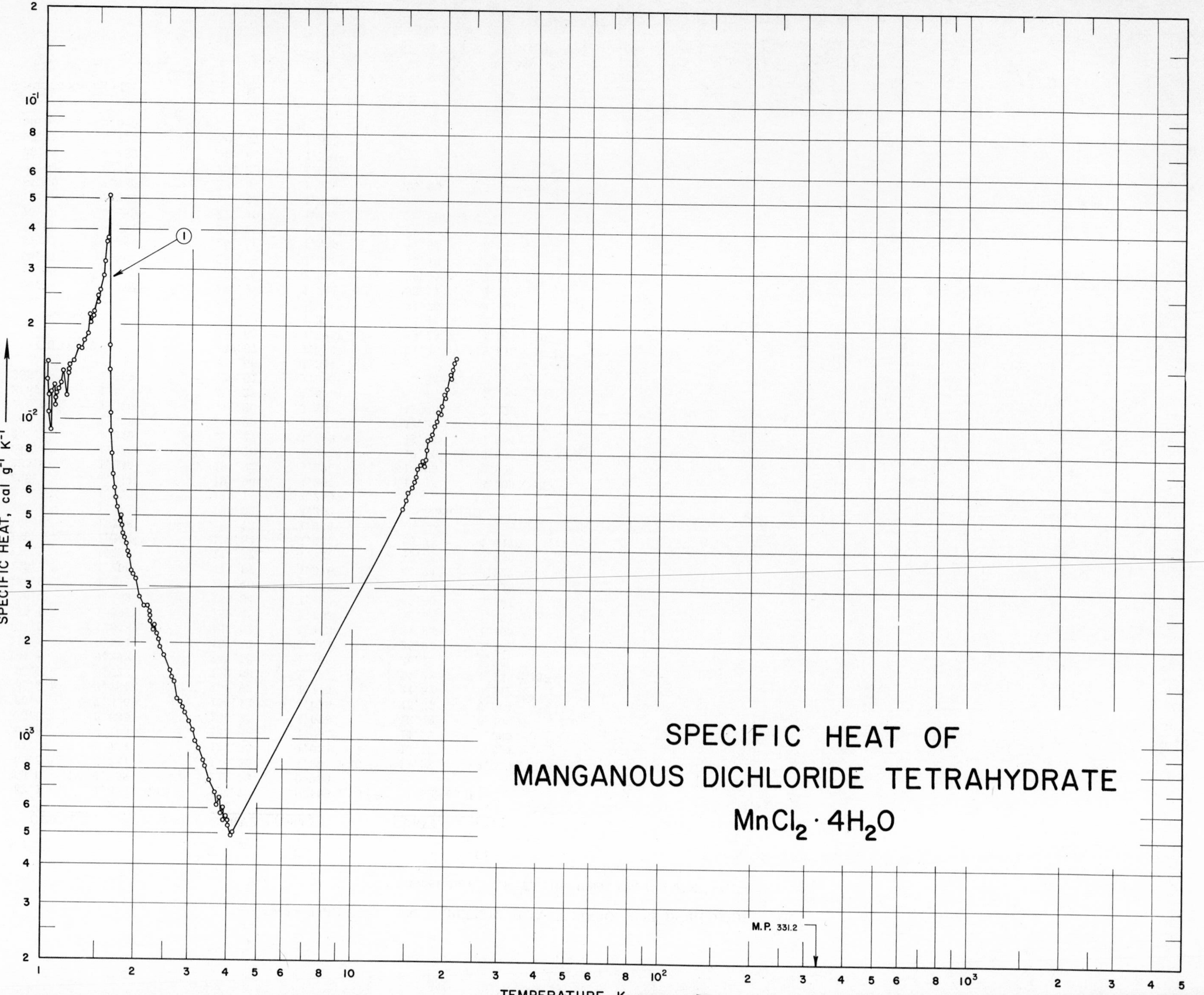
SPECIFIC HEAT OF
MANGANOUS DICHLORIDE TETRAHYDRATE
$MnCl_2 \cdot 4H_2O$
M.P. 331.2
1
SPECIFIC HEAT, cal g^{-1} K^{-1}
TEMPERATURE K

SPECIFICATION TABLE NO. 268 SPECIFIC HEAT OF MANGANOUS DICHLORIDE TETRAHYDRATE $MnCl_2 \cdot 4H_2O$

[For Data Reported in Figure and Table No. 268]

Curve No.	Ref. No.	Year	Temp. Range, K	Reported Error, %	Name and Specimen Designation	Composition (weight percent), Specifications and Remarks
1	262	1953	1–22			99.8 $MnCl_2 \cdot 4H_2O$; sample supplied by Hopkins and William Ltd.; measured under 10^{-4} mole of He gas.

DATA TABLE NO. 268 SPECIFIC HEAT OF MANGANESE DICHLORIDE TETRAHYDRATE $MnCl_2 \cdot 4H_2O$

[Temperature, T,K; Specific Heat, Cp, Cal $g^{-1}K^{-1}$]

T	Cp
CURVE 1	
Series I	
1.026	1.53 x 10^{-2}
1.033	1.34
1.044	1.06
1.056	9.29 x 10^{-3}
1.059	1.23 x 10^{-2}
1.069	1.22*
1.079	1.29
1.089	1.17
1.103	1.22
1.121	1.23*
1.138	1.32
1.178	1.21
1.219	1.50
1.254	1.55
1.289	1.70
1.323	1.73*
1.351	1.79
1.379	1.88
1.408	2.03
1.439	2.14
1.478	2.37
1.786	5.00 x 10^{-3}
1.838	4.28
1.869	4.08
1.896	3.73
1.928	3.68*
1.964	3.27
2.003	3.18
2.127	2.61
2.179	2.61
2.231	2.43
2.278	2.19
Series II	
1.043	1.22 x 10^{-2}
1.052	1.38*
1.061	1.29*
1.066	1.17*
1.091	1.26*
1.100	1.29*
1.127	1.34*
1.164	1.44

T	Cp
CURVE 1 (cont.)	
1.203	1.45 x 10^{-2}
1.243	1.55*
1.286	1.62*
1.321	1.68
1.353	1.80*
1.390	1.87*
1.421	2.07*
1.454	2.21
1.482	2.48
1.523	2.57
1.553	2.86
1.573	3.18
1.608	3.39*
1.631	1.45
1.650	9.20 x 10^{-3}
1.666	7.87
1.684	6.75
1.702	6.11
1.724	5.72
1.742	5.30
1.780	4.81
1.817	4.40
1.876	3.82
1.937	3.37
2.004	3.15*
2.063	2.78
2.230	2.32
2.408	1.93
2.484	1.82
Series III	
1.042	1.09 x 10^{-2}*
1.079	1.11
1.113	1.26
1.155	1.29*
1.199	1.41
1.241	1.51*
1.404	2.15
1.436	2.23*
1.471	2.40*
1.508	2.62*
1.545	2.81*
1.568	3.14*
1.575	3.39*

T	Cp
CURVE 1 (cont.)	
1.581	3.23 x 10^{-2}*
1.587	3.66
1.594	3.43*
1.600	3.76
1.605	4.32*
1.610	4.37*
1.615	4.84*
1.620	5.13
1.629	1.73
1.646	1.05
1.654	9.18 x 10^{-3}*
1.687	6.94*
1.743	5.47*
1.797	4.66
1.836	4.13*
1.887	3.77*
1.938	3.33*
1.983	3.07*
2.047	2.78*
2.101	2.65*
2.161	2.42*
2.225	2.25*
2.262	2.25*
Series IV	
2.232	2.50 x 10^{-3}
2.262	2.33*
2.285	2.26*
2.305	2.27
2.317	2.19*
2.326	2.22*
2.327	2.20*
2.353	2.13
2.356	2.14*
2.390	2.04
2.421	2.03*
2.598	1.63
2.644	1.56
2.681	1.51
2.688	1.50*
2.708	1.47*
2.719	1.39*
2.747	1.33
2.785	1.40*

T	Cp
CURVE 1 (cont.)	
2.820	1.30 x 10^{-3}
2.871	1.25
2.915	1.20
2.946	1.20*
2.997	1.13
3.033	1.11*
3.065	1.06*
3.085	1.06
3.120	1.04*
3.152	9.84 x 10^{-4}
3.191	9.57*
3.214	9.60*
3.230	9.30
3.269	8.85*
3.316	8.72*
3.354	8.51
3.385	8.19
3.513	7.36
3.545	7.05*
3.653	6.76
3.700	6.16
3.744	6.40
3.790	6.52
3.816	5.82
3.840	5.91*
3.866	5.91*
3.890	6.05
3.914	5.53
3.960	5.70
4.005	5.68*
4.010	5.52
4.040	5.31
4.060	5.30*
4.119	4.98
4.120	5.15*
4.180	5.08
Series V*	
1.765	4.81 x 10^{-3}
1.785	4.34
1.810	4.24
1.815	3.96
2.225	2.25
2.340	2.15

T	Cp
CURVE 1 (cont.)	
2.540	1.69 x 10^{-3}
2.560	1.67
3.430	8.15 x 10^{-4}
3.440	8.21
3.440	8.21
4.025	5.86
4.070	5.25
Series VI	
14.54	5.37 x 10^{-3}
14.85	5.71
15.16	6.01
15.41	6.11*
15.56	6.28
15.70	6.38*
15.86	6.53
16.04	6.78
16.23	7.17
16.42	7.22*
16.62	7.43
16.83	7.60
17.02	7.32
17.27	7.67
17.40	8.22
17.58	8.83
17.87	8.94*
18.02	9.20
18.16	9.18*
18.31	9.29*
18.45	9.78
18.59	9.73*
18.75	1.01 x 10^{-2}
18.94	1.08
19.13	1.08*
19.32	1.07
19.49	1.14
19.66	1.20*
19.83	1.23
19.99	1.21
20.15	1.28
20.73	1.43
20.87	1.39
21.01	1.45*
21.14	1.49

T	Cp
CURVE 1 (cont.)	
21.26	1.50 x 10^{-2}*
21.39	1.56
21.54	1.58*
21.67	1.59
Series VII*	
14.53	5.63 x 10^{-3}
15.91	6.04
15.39	6.38
15.90	6.75
16.38	7.60
16.87	8.16
17.28	8.33
17.46	9.01
17.58	8.86
17.82	9.14
18.02	9.92
18.27	9.79
18.51	1.02 x 10^{-2}
18.71	1.09
18.98	1.09
19.24	1.11
19.46	1.20
19.53	1.26
19.72	1.24
19.97	1.24
20.20	1.34
20.40	1.30
20.66	1.46
20.85	1.47
21.08	1.47
21.40	1.59
21.69	1.47
21.93	1.80

*Not shown on plot

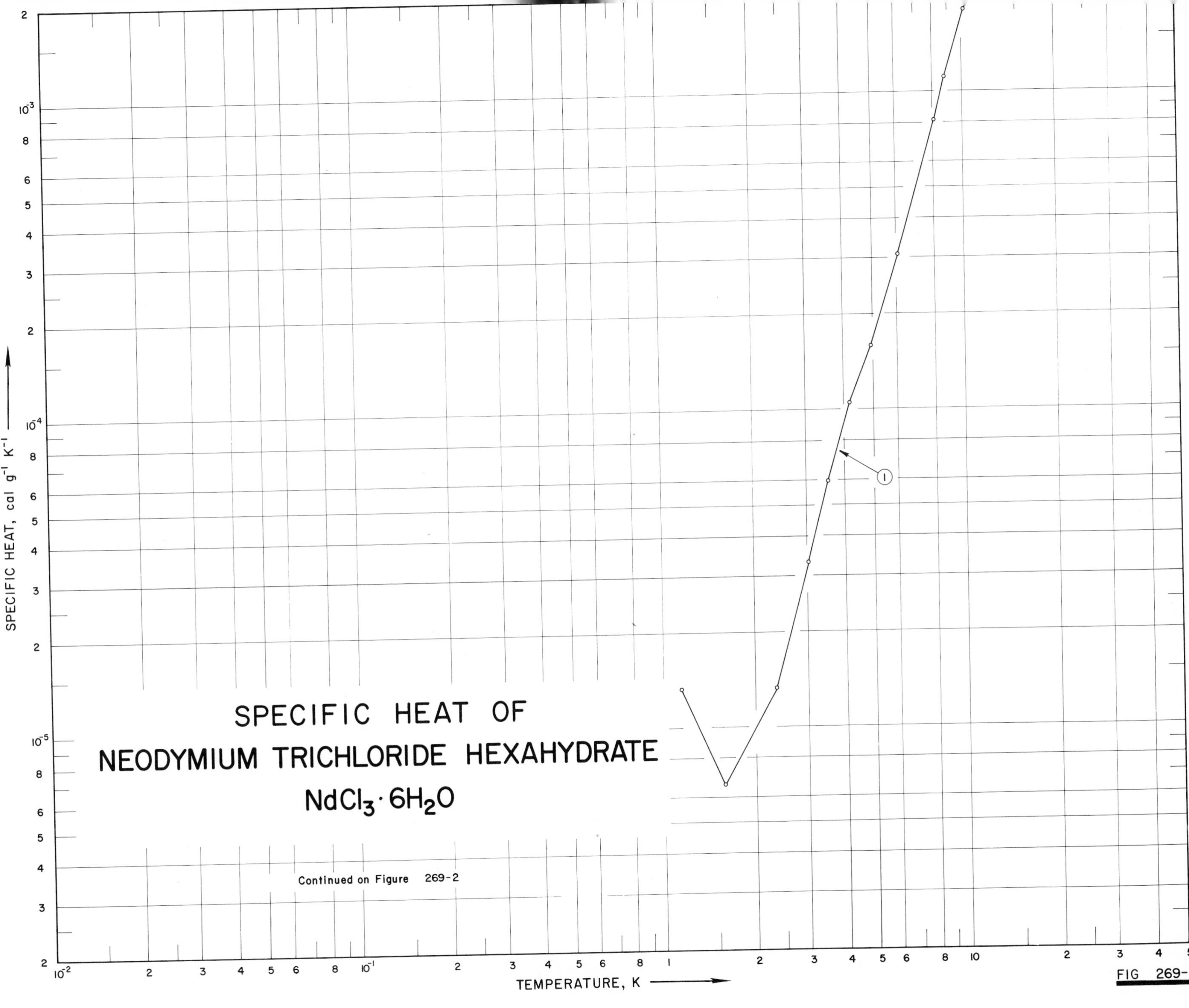
SPECIFIC HEAT OF
NEODYMIUM TRICHLORIDE HEXAHYDRATE
$NdCl_3 \cdot 6H_2O$
Continued on Figure 269-2
SPECIFIC HEAT, cal g^{-1} K^{-1}
TEMPERATURE, K
1
FIG 269-1

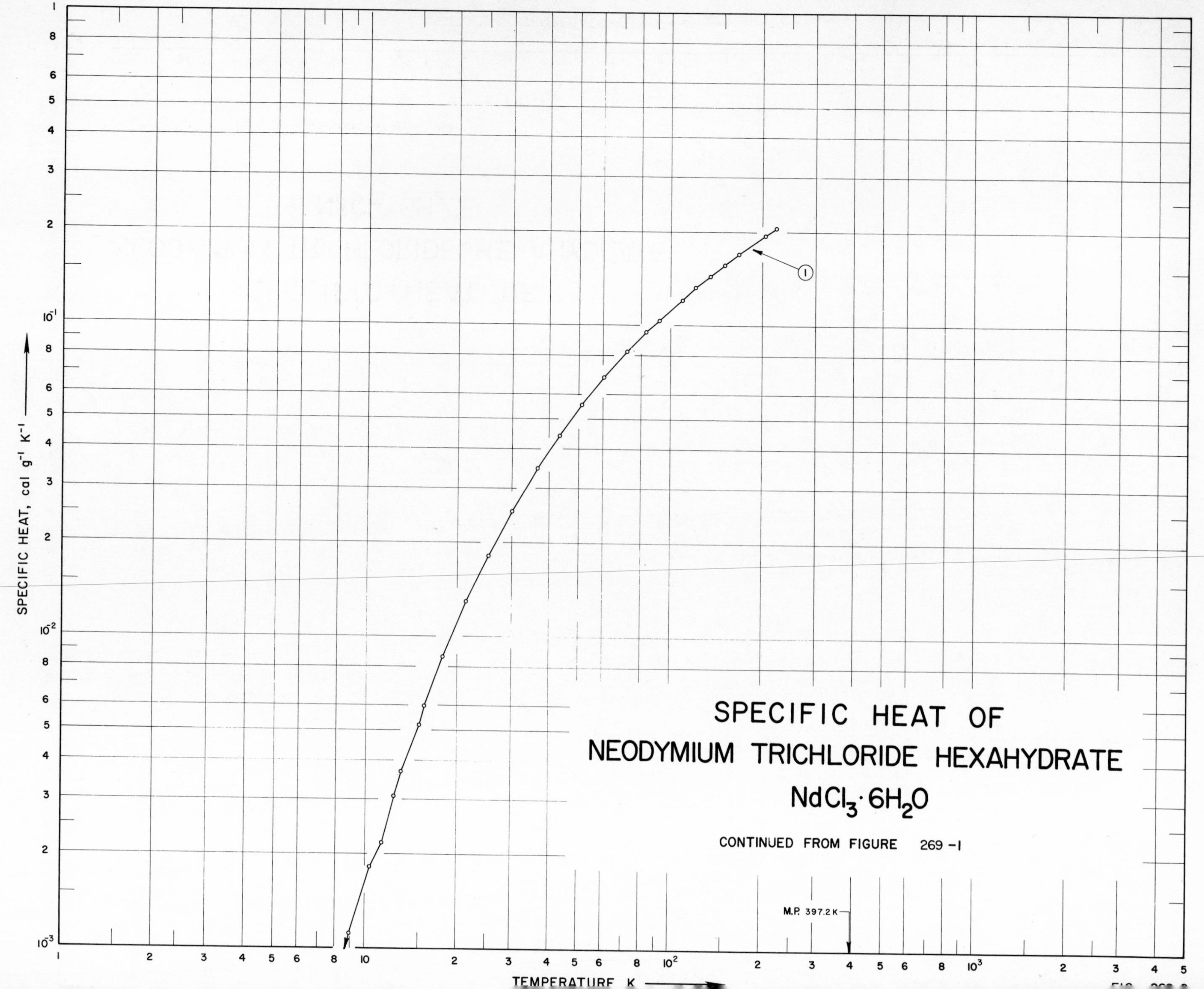
SPECIFIC HEAT OF
NEODYMIUM TRICHLORIDE HEXAHYDRATE
NdCl3·6H2O
CONTINUED FROM FIGURE 269-1
M.P. 397.2 K
1
SPECIFIC HEAT, cal g⁻¹ K⁻¹
TEMPERATURE, K

SPECIFICATION TABLE NO. 269 SPECIFIC HEAT OF NEODYMIUM TRICHLORIDE HEXAHYDRATE $NdCl_3 \cdot 6H_2O$

[For Data Reported in Figure and Table No. 269]

Curve No.	Ref. No.	Year	Temp. Range, K	Reported Error, %	Name and Specimen Designation	Composition (weight percent), Specifications and Remarks
1	319	1961	1.2-223			

DATA TABLE NO. 269 SPECIFIC HEAT OF NEODYMIUM TRICHLORIDE HEXAHYDRATE $NdCl_3 \cdot 6H_2O$

[Temperature, T,K; Specific Heat, Cp, Cal $g^{-1}K^{-1}$]

T	Cp
CURVE 1	
1.16	1.333×10^{-5}
1.58	6.663×10^{-6}
2.37	1.333×10^{-5}
3.05	3.332
3.56	5.997
4.24	1.066×10^{-4}
4.64	1.266
4.98	1.599
6.18	3.198
8.21	8.196
8.94	1.119×10^{-3}
10.42	1.826
11.14	2.186
12.45	3.078
13.17	3.685
15.02	5.211
15.68	5.977
17.95	8.609
18.25	9.015*
21.37	1.299×10^{-2}
21.55	1.317*
25.34	1.818
25.35	1.821*
30.18	2.523
30.36	2.557*
36.59	3.448
43.40	4.452
51.35	5.578
60.93	6.837
72.00	8.269
83.15	9.515
92.36	1.043×10^{-1}
109.73	1.221
122.47	1.332
136.37	1.450
151.72	1.572
168.84	1.703
205.12	1.957
205.13	1.952*
221.52	2.053*
223.37	2.065

*Not shown on plot

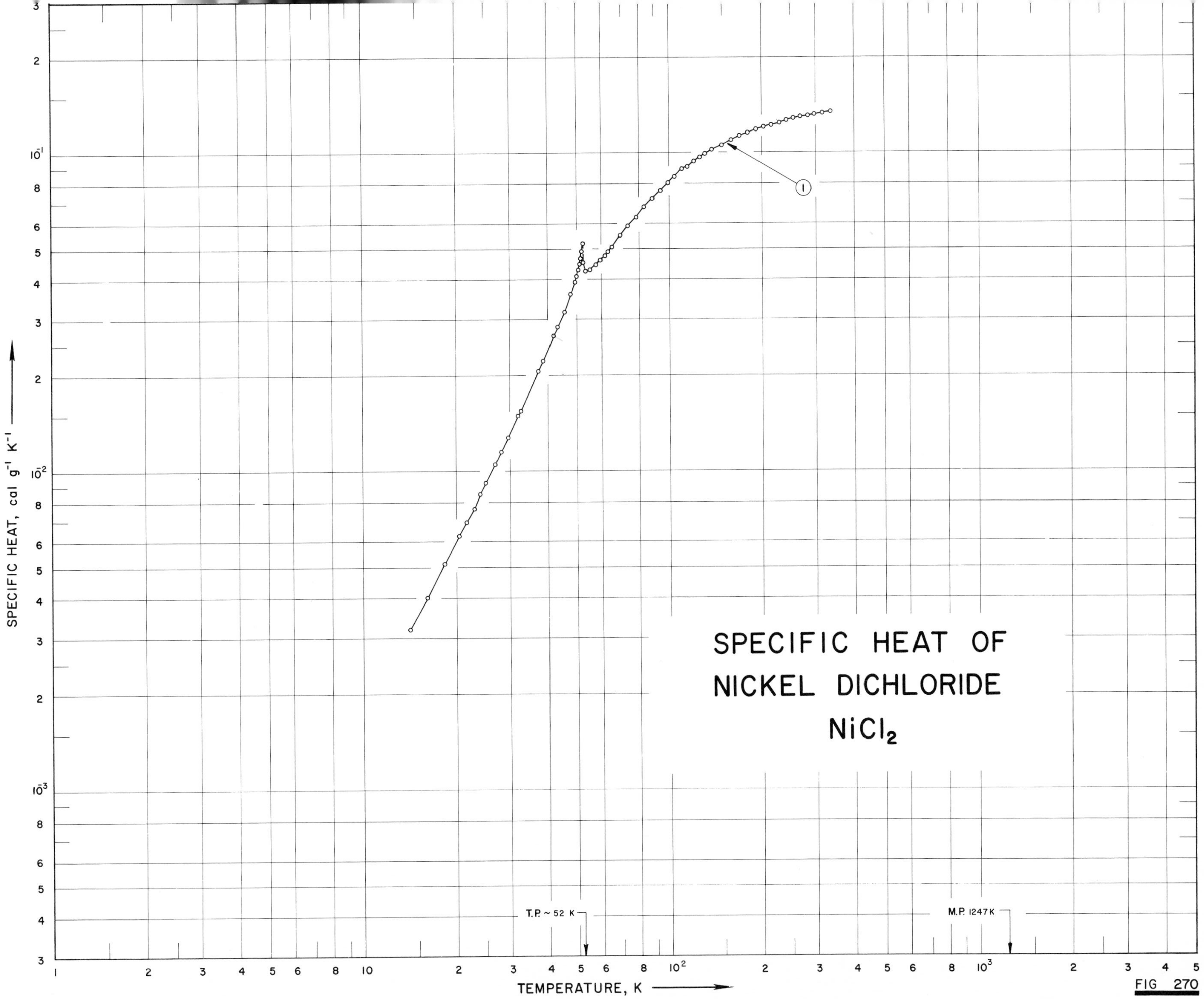

SPECIFIC HEAT OF NICKEL DICHLORIDE $NiCl_2$

FIG 270

SPECIFICATION TABLE NO. 270 SPECIFIC HEAT OF NICKEL DICHLORIDE $NiCl_2$

[For Data Reported in Figure and Table No. 270]

Curve No.	Ref. No.	Year	Temp. Range, K	Reported Error, %	Name and Specimen Designation	Composition (weight percent), Specifications and Remarks
1	264	1952	15-300	1.0-5		54.715 Cl_2 (54.716 theo), 45.29 Ni (45.284 theo); density = 3.54 g ml^{-1}.

DATA TABLE NO. 270 SPECIFIC HEAT OF NICKEL DICHLORIDE $NiCl_2$

[Temperature, T, K; Specific Heat Cp, Cal $g^{-1}K^{-1}$]

T	Cp
CURVE 1	
Series A	
272.60	1.298×10^{-1}*
280.41	1.306*
288.40	1.312*
296.84	1.321
305.90	1.328*
315.87	1.333
325.62	1.347*
336.36	1.349
Series B	
79.91	6.558×10^{-2}*
84.56	6.882*
Series C	
61.45	4.784×10^{-2}*
65.03	5.079
69.21	5.501
73.56	5.909
78.27	6.328
83.48	6.789
88.75	7.227
94.07	7.636
99.49	8.040
104.95	8.402
110.45	8.904
115.84	9.074
121.35	9.382
126.77	9.660
132.28	9.922
138.02	1.016×10^{-1}
143.38	1.039*
148.64	1.059
154.08	1.078*
159.62	1.096
165.05	1.113*
170.36	1.128
176.03	1.143*
181.61	1.157
187.07	1.170*
192.85	1.182
198.52	1.194*
CURVE 1 (cont.)	
204.05	1.205×10^{-1}
210.05	1.215*
216.40	1.227
222.94	1.237*
229.28	1.245
235.42	1.255*
241.76	1.262
248.13	1.272*
254.41	1.278
260.89	1.286*
268.00	1.293
275.23	1.300*
283.07	1.307
Series D	
14.14	3.217×10^{-3}
16.03	4.019
18.34	5.154
20.47	6.343
22.72	7.715
23.90	8.525
26.80	1.062×10^{-2}
29.56	1.287
32.57	1.562
38.51	2.225
41.92	2.678
45.35	3.183
47.85	3.639
49.27	3.941
49.93	4.109
50.58	4.282
51.20	4.482
55.36	4.304
56.22	4.394*
57.28	4.414*
57.83	4.479
58.35	4.510*
58.90	4.556*
59.42	4.593*
59.93	4.623
60.44	4.658*
60.94	4.718*
61.43	4.766*
63.15	4.901
CURVE 1 (cont.)	
Series E	
57.02	4.405×10^{-2}*
59.60	4.600*
62.16	4.816*
Series F	
20.57	6.473×10^{-3}*
22.35	7.530*
24.93	9.282
28.07	1.169×10^{-2}
32.19	1.526*
38.24	2.184*
42.45	2.727*
46.61	3.402*
46.70	4.384*
59.31	4.576*
62.05	4.800*
Series G	
21.56	6.975×10^{-3}
24.16	8.726*
27.61	1.126×10^{-2}*
31.89	1.503
37.22	2.070
43.15	2.854
47.54	3.577*
50.04	4.161*
51.45	4.610*
51.62	4.670
51.77	4.740*
51.94	4.815*
52.04	4.913
52.19	5.013*
52.32	5.198
52.45	4.955*
52.59	4.535
52.96	4.333*
53.56	4.264
54.15	4.268*
54.73	4.286*
55.79	4.330*
58.22	4.497*
61.62	4.778

* Not shown on plot

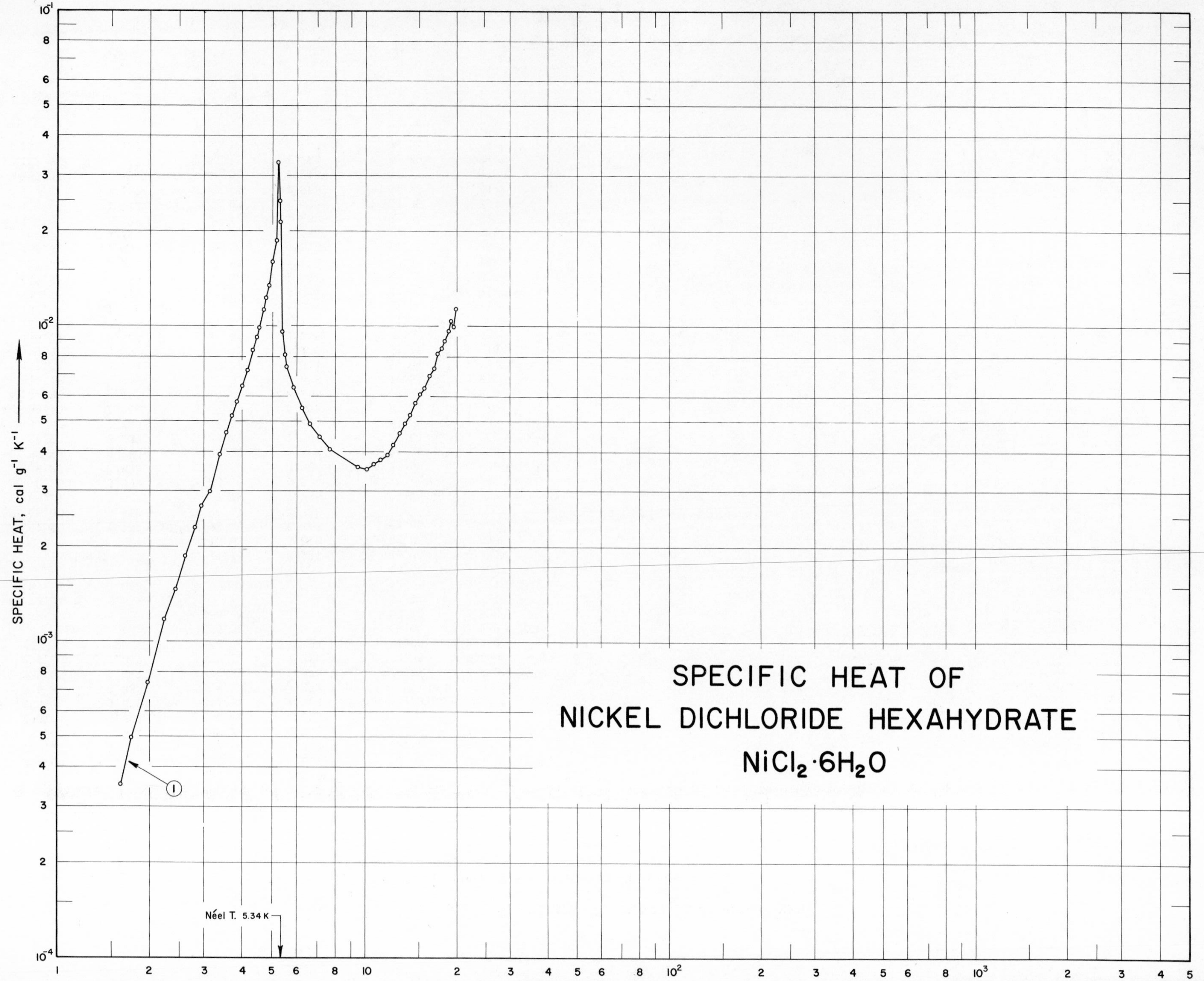

SPECIFIC HEAT OF NICKEL DICHLORIDE HEXAHYDRATE

$NiCl_2 \cdot 6H_2O$

SPECIFICATION TABLE NO. 271 SPECIFIC HEAT OF NICKEL DICHLORIDE HEXAHYDRATE $NiCl_2 \cdot 6H_2O$

[For Data Reported in Figure and Table No. 271]

Curve No.	Ref. No.	Year	Temp. Range, K	Reported Error, %	Name and Specimen Designation	Composition (weight percent), Specifications and Remarks
1	256	1959	1.6-20			0.08 alkalies and earth, 0.01 Zn, 0.009 Co, and insoluable, 0.006 nitrogen compounds, 0.005 Pb, 0.001 Cu, 0.0007 Fe and 0.0004 sulfate.

DATA TABLE NO. 271 SPECIFIC HEAT OF NICKEL DICHLORIDE HEXAHYDRATE $NiCl_2 \cdot 6H_2O$

[Temperature, T, K; Specific Heat, Cp, Cal $g^{-1}K^{-1}$]

T	Cp
CURVE 1	
Series I	
1.6104	3.542×10^{-4}
1.7412	4.981
1.9740	7.434
2.2363	1.186×10^{-3}
2.4239	1.472
2.6182	1.879
2.8012	2.311
2.9495	2.683
3.1500	2.998
3.3805	3.924
3.5624	4.619
3.7186	5.221
3.8587	5.772
4.0222	6.491
4.1842	7.286
4.3488	8.448
4.4688	9.222
4.5520	9.895
4.6108	1.032×10^{-2}*
4.7348	1.138
4.7550	1.179*
4.7740	1.184*
4.7998	1.244
4.8472	1.274*
4.9088	1.354
4.9848	1.437*
5.0743	1.603
5.1828	1.873
5.3095	2.154
5.5198	8.145×10^{-3}
5.8648	6.411
6.2788	5.503
6.6208	4.914
Series II	
4.6444	1.062×10^{-2}*
4.7041	1.126*
4.7918	1.203*
4.9148	1.565*
5.0452	1.525*
5.1764	1.836*
5.2652	3.300
5.3122	2.504
5.4142	9.647×10^{-3}

T	Cp
CURVE 1 (cont.)	
5.5826	7.476×10^{-3}
5.8208	6.500*
6.1765	5.675*
6.6360	5.074*
7.1344	4.472
7.6764	4.085
9.4747	3.584
10.1090	3.542
10.6880	3.668
11.2590	3.799
11.8050	3.912
12.3580	4.249
12.9580	4.586
13.5640	4.922
14.1260	5.259
14.6440	5.721
15.1720	6.100
15.7120	6.395
16.2690	6.984
16.8420	7.362
17.3720	8.204
17.8640	8.540
18.3330	9.003
18.7740	9.718
19.1840	1.048×10^{-2}
19.5880	9.970×10^{-3}
19.9760	1.140×10^{-2}

* Not shown on plot

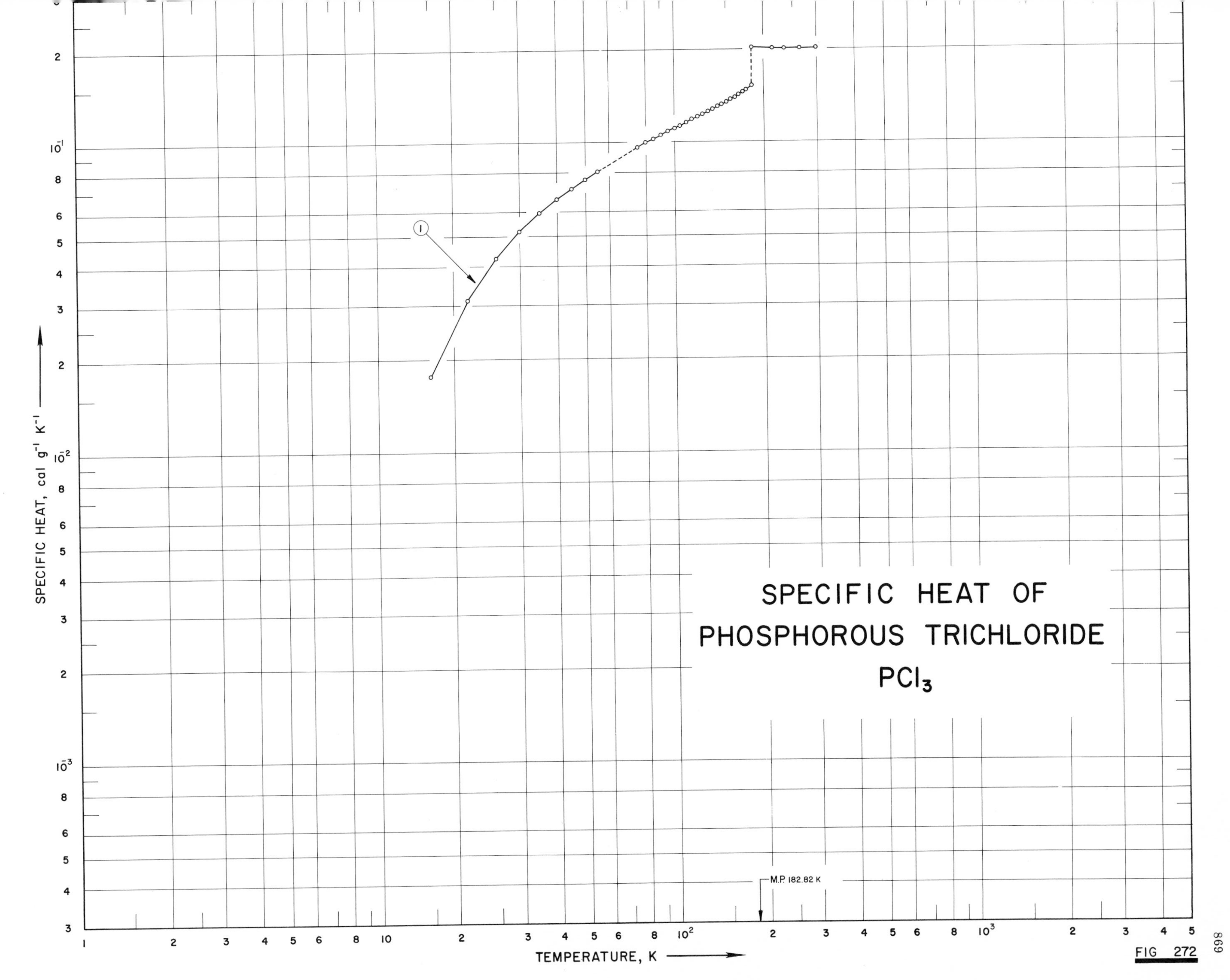
SPECIFIC HEAT OF
PHOSPHOROUS TRICHLORIDE
PCl_3
M.P. 182.82 K
SPECIFIC HEAT, cal g^{-1} K^{-1}
TEMPERATURE, K
FIG 272

SPECIFICATION TABLE NO. 272 SPECIFIC HEAT OF PHOSPHORUS TRICHLORIDE PCl_3

[For Data Reported in Figure and Table No. 272]

Curve No.	Ref. No.	Year	Temp. Range, K	Reported Error, %	Name and Specimen Designation	Composition (weight percent), Specifications and Remarks
1	265	1960	15-300			99.71 PCl_3; reagent grade.

DATA TABLE NO. 272 SPECIFIC HEAT OF PHOSPHOROUS TRICHLORIDE PCl_3

[Temperature, T, K; Specific Heat, Cp, Cal $g^{-1}K^{-1}$]

T	Cp	T	Cp
CURVE 1		**CURVE 1 (cont.)**	
Solid		210	2.011 x 10^{-1}*
		215	2.009
15	1.794 x 10^{-2}	220	2.008*
20	3.118	225	2.007*
25	4.274	230	2.006*
30	5.201	235	2.005
35	5.975	240	2.004*
40	6.617	245	2.004*
45	7.167	250	2.004*
50	7.659	255	2.005*
55	8.117	260	2.005*
		265	2.006
Transition		270	2.007*
		273.16	2.008*
75	9.698 x 10^{-2}	275	2.009*
80	1.003 x 10^{-1}	280	2.010*
85	1.035	285	2.011*
90	1.064	290	2.013*
95	1.090	295	2.014*
100	1.115	298.16	2.015*
105	1.141	300	2.016
110	1.168		
115	1.193		
120	1.217		
125	1.241		
130	1.265		
135	1.288		
140	1.311		
145	1.334		
150	1.357		
155	1.381		
160	1.406		
165	1.431		
170	1.456		
175	1.482		
180	1.508*		
182.82	1.524		
Liquid			
182.82	2.029 x 10^{-1}		
185	2.027*		
190	2.023*		
195	2.019*		
200	2.016*		
205	2.013*		

*Not shown on plot

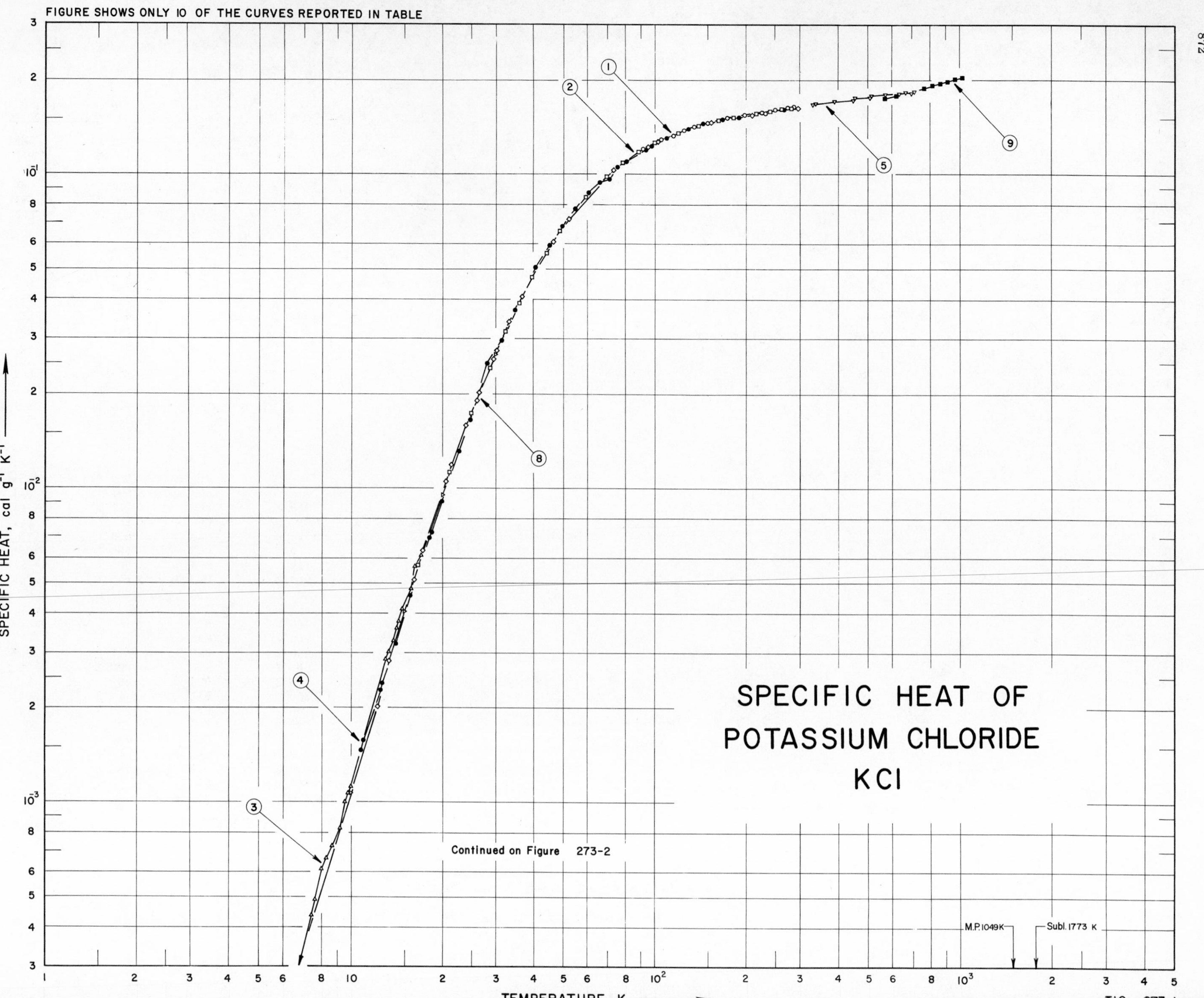
FIGURE SHOWS ONLY 10 OF THE CURVES REPORTED IN TABLE
SPECIFIC HEAT OF
POTASSIUM CHLORIDE
KCl
Continued on Figure 273-2
M.P. 1049K
Subl. 1773 K
SPECIFIC HEAT, cal g⁻¹ K⁻¹
1
2
3
4
5
8
9

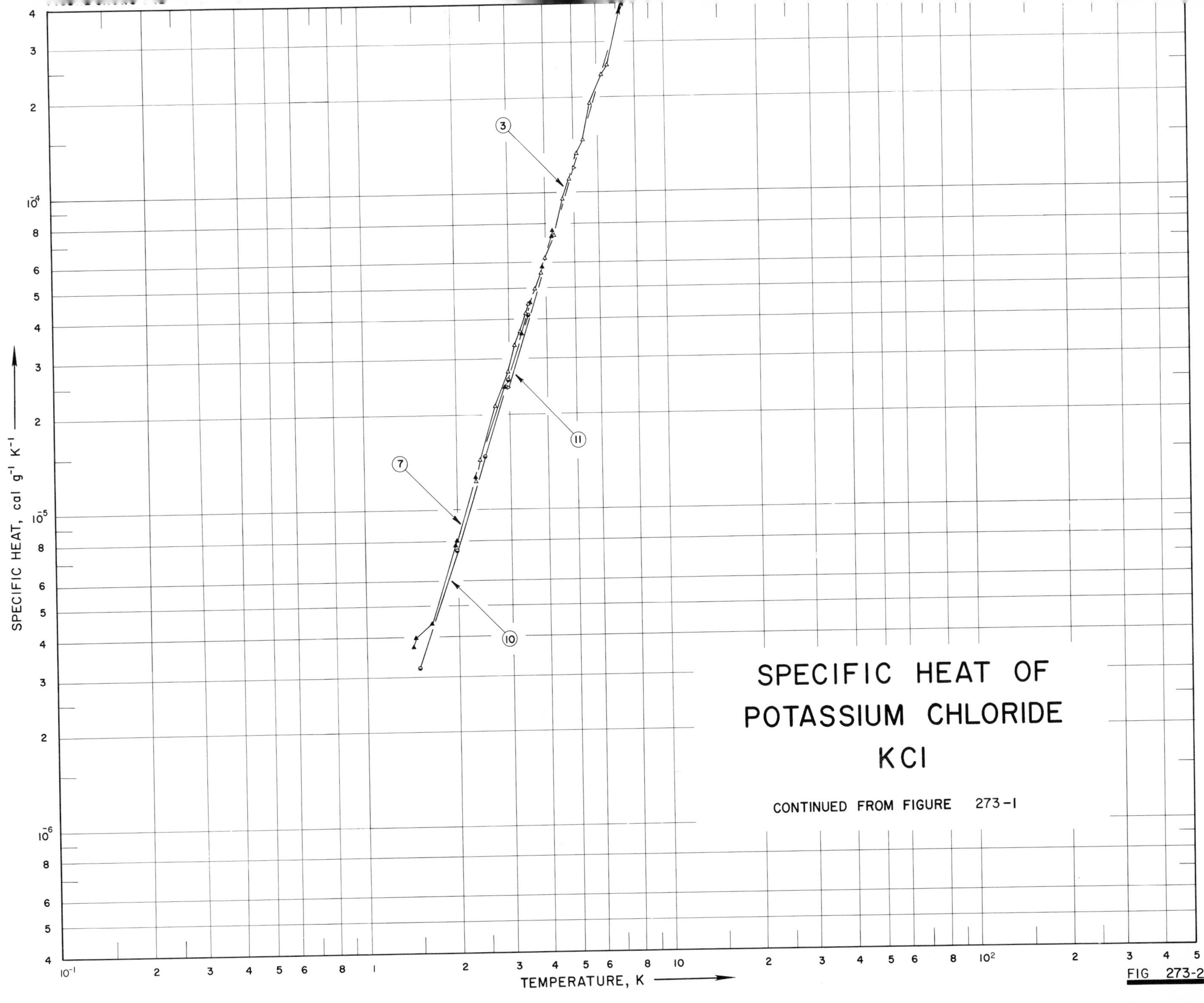
SPECIFIC HEAT OF
POTASSIUM CHLORIDE
KCl
CONTINUED FROM FIGURE 273-1
SPECIFIC HEAT, cal g⁻¹ K⁻¹
TEMPERATURE, K
3
11
7
10

FIG 273-2

SPECIFICATION TABLE NO. 273 SPECIFIC HEAT OF POTASSIUM CHLORIDE KCl

[For Data Reported in Figure and Table No. 273]

Curve No.	Ref. No.	Year	Temp. Range, K	Reported Error, %	Name and Specimen Designation	Composition (weight percent), Specifications and Remarks
1	266	1933	92-287			C.P. pure; recrystallized 3 or 4 times; dried under high vacuum 2 days; heated to 700 C.
2	266	1933	17-285			Same as above; dried under high vacuum 5 days; fused with subsequent grinding.
3	267	1935	2.3-17			
4	243	1949	11-268			
5	268	1951	335-718			
6	268	1951	335-721			
7	269	1953	1.5-4.3			Single crystal; 37 mm dia, 88 mm length (18.9 g); sample supplied by the Harshaw Chemical Co.
8	270	1954	12-298			
9	244	1955	573-1023			
10	271	1955	1.5-4.0			Single crystal; 3 cm dia, 8 cm length, (160 g).
11	245	1957	2.5-270	0.2-2		High purity, optical quality; sample supplied by the Harshaw Chemical Co; measured in a helium atmosphere.
12	272	1961	313-388	1.0		
13	272	1961	313-388	1.0		
14	273	1962	60-350	0.2		Measured in 10^{-5}mm Hg vacuum atmosphere.

DATA TABLE NO. 273 SPECIFIC HEAT OF POTASSIUM CHLORIDE KCl

[Temperature, T,K; Specific Heat, Cp, Cal $g^{-1}K^{-1}$]

T	Cp
CURVE 1	
92.26	1.210 x 10^{-1}
96.84	1.240*
101.70	1.269*
106.86	1.298
111.94	1.321*
116.96	1.348
121.93	1.367*
126.86	1.390
136.58	1.424
141.44	1.439*
146.29	1.455*
151.13	1.466
152.99	1.474*
155.97	1.481*
158.65	1.482*
164.31	1.501*
169.88	1.508*
175.64	1.524
181.29	1.529*
187.01	1.540*
192.72	1.547*
198.44	1.560
204.18	1.560*
209.95	1.568*
215.74	1.571*
221.59	1.577*
227.42	1.587
233.30	1.589*
239.20	1.600*
245.11	1.612*
251.05	1.618
257.02	1.623*
263.03	1.628*
269.06	1.632*
275.12	1.640
281.20	1.651*
287.30	1.655
CURVE 2	
16.69	5.73 x 10^{-3}
21.21	1.13 x 10^{-2}
25.06	1.746
28.81	2.428
32.41	3.165
36.00	3.913
CURVE 2 (cont.)	
39.86	4.739 x 10^{-2}
44.25	5.619
49.27	6.607
54.44	7.606*
59.61	8.485
64.94	9.246*
69.92	9.879
74.62	1.041 x 10^{-1}*
79.11	1.094
84.10	1.147*
89.28	1.186
90.80	1.194*
94.32	1.224*
96.10	1.235*
101.31	1.266
106.43	1.295*
111.48	1.320*
116.47	1.345*
121.41	1.364
126.32	1.387*
131.19	1.403*
136.04	1.423*
141.17	1.443
146.58	1.455*
152.30	1.469*
158.02	1.483*
163.45	1.494
168.88	1.504*
172.60	1.511*
174.31	1.513*
177.99	1.520*
179.39	1.524*
179.76	1.524*
183.41	1.528
184.85	1.530*
185.22	1.536
188.83	1.538*
190.32	1.540*
194.28	1.544*
195.81	1.547*
199.74	1.551*
201.32	1.555*
205.23	1.561*
206.85	1.568*
210.73	1.568*
212.40	1.572*
CURVE 2 (cont.)	
217.26	1.573 x 10^{-1}*
217.97	1.576
219.04	1.577*
223.61	1.583*
224.65	1.584*
229.29	1.589*
230.28	1.591*
235.01	1.591*
235.93	1.597*
240.70	1.599
241.62	1.603*
246.41	1.607*
247.32	1.607*
252.15	1.610
252.68	1.612*
257.80	1.614*
257.91	1.618*
262.94	1.611*
263.70	1.623
268.15	1.616*
269.52	1.630*
273.34	1.630*
278.54	1.636*
284.68	1.643
CURVE 3	
Series I	
2.99	2.76 x 10^{-5}
3.16	3.35
3.28	3.70
3.37	3.89*
3.43	4.21
3.52	4.53
3.62	4.80*
3.71	5.04
3.81	5.37*
3.86	5.69
3.92	5.93*
4.01	6.30
4.10	6.60*
4.29	7.43
CURVE 3 (cont.)	
Series II	
2.35	1.26 x 10^{-5}
2.42	1.45
2.44	1.90*
2.62	2.61*
2.71	2.16
Series III	
4.60	9.71 x 10^{-5}
4.87	1.12 x 10^{-4}
5.14	1.35
5.40	1.50
5.72	1.95
6.14	2.12
6.24	2.38
6.52	2.57
6.73	2.55*
7.45	4.40
7.65	4.94
7.89	5.23*
8.00	6.17
8.22	6.28*
8.33	6.68
8.46	6.76*
8.70	7.30
8.94	7.73*
9.23	8.32
Series IV	
9.53	1.01 x 10^{-3}
9.81	1.08
10.06	1.12
13.01	2.87
13.22	3.03
13.55	3.19*
13.85	3.30
14.20	3.62
14.47	3.81
14.73	4.16
15.84	4.83
16.34	5.69
16.67	5.87*
17.09	6.14
CURVE 4	
10.8	6.71 x 10^{-4}
11.1	4.02
12.6	1.07 x 10^{-3}
12.7	1.21
14.1	1.61
14.2	1.61
15.7	2.41
15.7	2.28
17.0	2.82
18.2	3.49
18.5	3.62
19.9	4.56
22.7	6.57
24.8	8.32
28.2	1.26 x 10^{-2}
31.5	1.49
34.9	2.52
41.2	2.54
45.5	2.99
50.1	3.43
55.3	3.92
61.3	4.39
66.5	1.14 x 10^{-1}
71.8	1.15
76.8	1.20
81.0	5.53 x 10^{-2}
90.4	5.97
94.2	6.02
98.0	6.20
101.7	6.26
105.8	6.41
110.7	6.56
116.0	6.65
121.1	6.83
125.8	6.88
130.7	7.00
136.1	7.11
142.3	7.16
148.9	7.28
155.8	7.31
161.7	7.39
168.2	7.51
174.5	7.56
176.9	7.63
178.8	7.56
181.5	7.61
CURVE 4 (cont.)	
191.6	7.66 x 10^{-2}
198.4	7.70
201.8	7.71
204.3	7.74
207.8	7.73
213.4	7.86
220.4	7.83
227.1	7.85
233.2	7.89
239.6	7.93
244.9	7.91
251.6	7.97
259.2	8.01
267.6	8.07
CURVE 5	
334.55	1.667 x 10^{-1}
340.15	1.676
393.75	1.712
450.75	1.732
458.75	1.755
513.75	1.763
519.15	1.781
577.75	1.800
639.55	1.809
671.35	1.838
700.65	1.827
717.65	1.849
CURVE 6	
335.05	1.666 x 10^{-1}
343.85	1.679
366.65	1.701
377.25	1.688
389.05	1.708*
427.95	1.724
433.95	1.726
480.55	1.745*
521.85	1.765*
527.25	1.774
574.35	1.794*
579.45	1.800
608.25	1.790

*Not shown on plot

DATA TABLE NO. 273 (continued)

T	Cp
CURVE 6 (cont.)	
635.75	1.818×10^{-1}
681.65	1.834*
692.05	1.825
701.55	1.840
714.65	1.830*
721.35	1.840
CURVE 7	
Series I	
4.231	7.386×10^{-5}
4.230	7.386*
3.912	5.861*
3.929	5.944
3.567	4.548*
3.567	4.567
3.312	3.531*
3.316	3.660
2.924	2.472*
2.925	2.472
Series II	
1.466	2.999×10^{-6}
1.434	2.742
1.441	2.727*
1.656	4.426
1.667	4.395*
1.670	4.399*
1.985	7.829
2.012	8.072
2.324	1.286×10^{-5}
2.322	1.241*
2.335	1.258*
4.247	7.739*
4.293	7.707
CURVE 8	
12.21	2.01×10^{-3}
13.37	2.84
14.03	3.23*
15.90	4.79*
16.26	5.15
17.36	6.37

T	Cp
CURVE 8 (cont.)	
20.68	1.06×10^{-2}
20.82	1.09*
20.88	1.09*
21.32	1.13*
21.56	1.20
21.75	1.24*
24.06	1.580
24.18	1.600*
26.15	1.918
26.31	1.972*
26.48	1.977*
26.59	2.024
28.68	2.424*
28.77	2.445*
29.64	2.587
30.00	2.679*
30.27	2.735*
30.43	2.764
33.43	3.408
33.51	3.426*
33.56	3.396*
36.13	3.996*
36.19	4.002*
36.89	4.104
36.41	4.688*
39.46	4.709*
39.56	4.764*
45.76	5.974*
46.73	6.134
47.08	6.173*
52.00	5.757*
52.71	7.247
60.12	8.394*
60.61	8.473*
66.57	9.420*
67.30	9.522*
73.98	1.034×10^{-1}
74.54	1.041*
80.71	1.113*
80.99	1.108*
89.09	1.185*
89.32	1.185*
94.50	1.216*
95.92	1.225
98.33	1.239*
99.53	1.247*

T	Cp
CURVE 8 (cont.)	
103.36	1.273×10^{-1}*
104.21	1.290
104.63	1.280*
110.84	1.309*
111.66	1.315*
119.00	1.3630*
119.30	1.3566*
120.34	1.3665*
121.23	1.3582*
139.56	1.4306*
140.22	1.4270*
144.41	1.4355*
145.74	1.4408*
154.48	1.4664*
155.38	1.4644
156.64	1.4686*
157.39	1.4705*
165.00	1.4930*
165.85	1.4867*
167.62	1.4926*
168.67	1.4868*
186.79	1.5242*
187.85	1.5269*
197.40	1.5356*
199.37	1.5483*
200.85	1.5448*
203.14	1.5515*
204.67	1.5563*
212.92	1.5574
214.09	1.5640*
223.27	1.5708*
224.57	1.5776*
227.68	1.5779*
229.81	1.5833*
234.50	1.5790*
245.90	1.5948*
248.22	1.5961*
252.41	1.6027*
253.96	1.6015*
261.94	1.6163*
263.32	1.6090*
274.91	1.6176*
274.94	1.6202*
275.15	1.6204*
275.53	1.6212*
276.47	1.6230*

T	Cp
CURVE 8 (cont.)	
276.76	1.6177×10^{-1}*
284.06	1.6223*
284.47	1.6259*
285.35	1.6290*
292.06	1.6306*
292.83	1.6338*
297.68	1.6362
CURVE 9	
573.15	1.758×10^{-1}
623.15	1.794
673.15	1.828*
723.15	1.862*
773.15	1.894
823.15	1.926
873.15	1.956
923.15	1.986
973.15	2.014
1023.15	2.042
CURVE 10	
1.5	3.21×10^{-6}
2.0	7.50
2.5	1.49×10^{-5}
3.0	2.61
3.5	4.20
4.0	6.32*
CURVE 11	
2.5	1.502×10^{-5}*
3	2.468
5	1.212×10^{-4}
10	1.078×10^{-3}
15	4.051
20	9.563
25	1.740×10^{-2}*
30	2.671
40	4.778*
50	6.783*
60	8.490*
80	1.099×10^{-1}*
100	1.258*
125	1.382*

T	Cp
CURVE 11 (cont.)	
150	1.463×10^{-1}*
175	1.516*
200	1.553*
225	1.583*
250	1.606*
270	1.622*
CURVE 12*	
313	1.69×10^{-1}
318	1.69
323	1.69
328	1.70
333	1.70
335	1.70
343	1.70
348	1.70
353	1.71
358	1.71
363	1.71
368	1.72
373	1.72
378	1.72
383	1.73
388	1.73
CURVE 13*	
313	1.70
318	1.70
323	1.69
328	1.69
333	1.70
335	1.70
343	1.71
348	1.71
353	1.71
358	1.72
363	1.72
368	1.72
373	1.73
378	1.74
383	1.73
388	1.74

T	Cp
CURVE 14*	
60	8.417×10^{-2}
65	9.205
70	9.856
75	1.042×10^{-1}
80	1.094
85	1.141
90	1.182
95	1.219
100	1.251
105	1.279
110	1.306
115	1.331
120	1.353
125	1.373
130	1.392
135	1.410
140	1.426
145	1.439
150	1.453
155	1.465
160	1.477
165	1.488
170	1.497
175	1.506
180	1.514
185	1.522
190	1.529
195	1.539
200	1.545
205	1.552
210	1.559
215	1.565
220	1.571
225	1.577
230	1.583
235	1.587
240	1.592
245	1.596
250	1.600
255	1.606
260	1.610
265	1.615
270	1.619
275	1.623
280	1.627
285	1.631

* Not shown on plot

DATA TABLE NO. 273 (continued)

T	Cp
CURVE 14 (cont.)*	
290	1.635×10^{-1}
295	1.639
298.16	1.640
300	1.642
305	1.646
310	1.650
315	1.654
320	1.658
325	1.662
330	1.666
335	1.670
340	1.674
345	1.679
350	1.685

* Not shown on plot

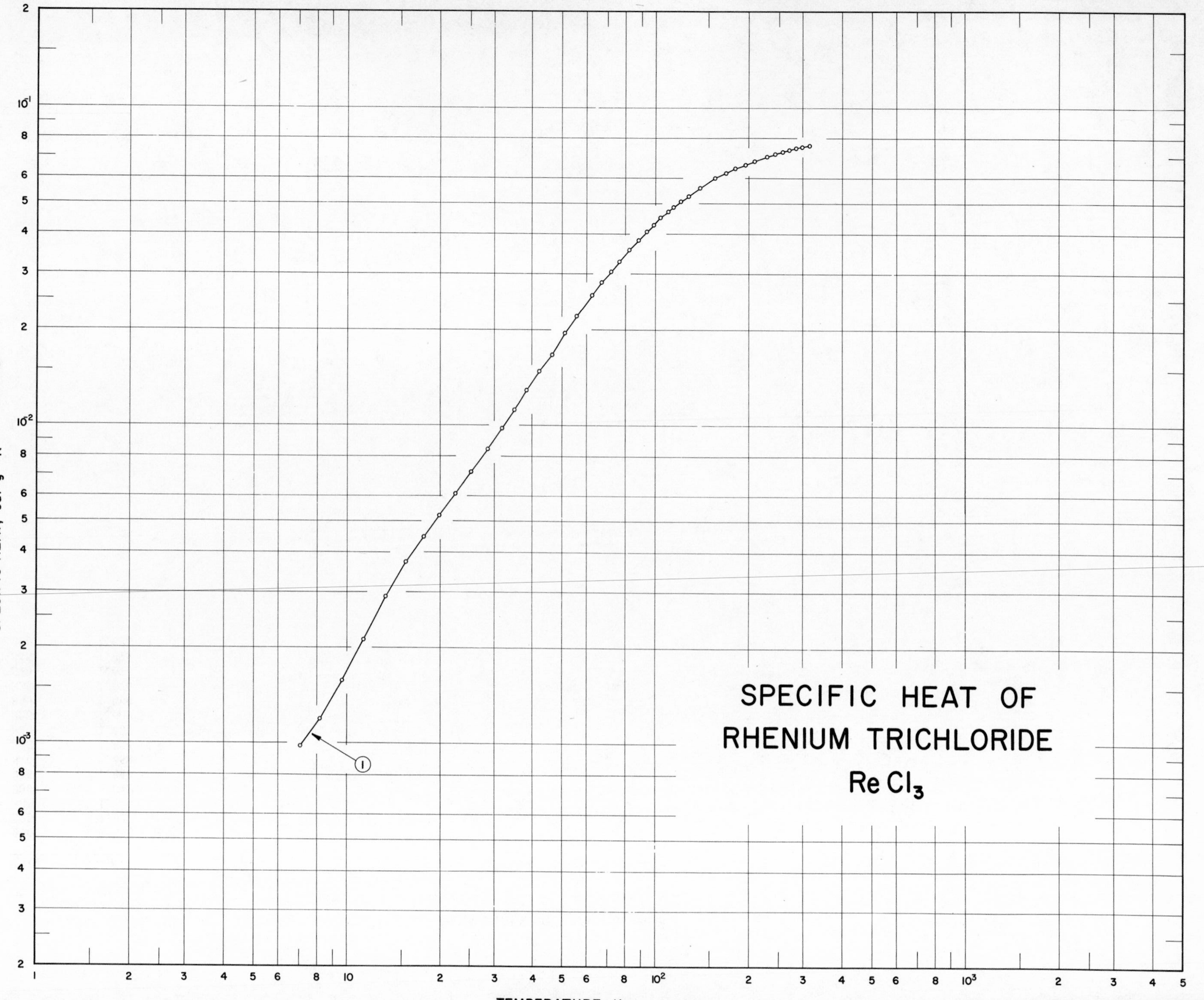

SPECIFIC HEAT OF
RHENIUM TRICHLORIDE
Re Cl$_3$
SPECIFIC HEAT, cal g^{-1} K^{-1}
1

SPECIFICATION TABLE NO. 274 SPECIFIC HEAT OF RHENIUM TRICHLORIDE $ReCl_3$

[For Data Reported in Figure and Table No. 274]

Curve No.	Ref. No.	Year	Temp. Range, K	Reported Error, %	Name and Specimen Designation	Composition (weight percent), Specifications and Remarks
1	274	1966	7-312	0.1-5		36.36 ± 0.03 Cl (36.355 theo.); density = 4.66 g cm^{-3}.

DATA TABLE NO. 274 SPECIFIC HEAT OF RHENIUM TRICHLORIDE $ReCl_3$

[Temperature, T,K; Specific Heat, Cp, Cal $g^{-1}K^{-1}$]

T	Cp
CURVE 1	
Series I	
298.14	7.554×10^{-2}*
305.47	7.585*
Series II	
115.61	4.888×10^{-2}
121.99	5.090
128.16	5.284
134.13	5.452*
139.94	5.602
145.62	5.749*
156.78	6.002
163.06	6.118*
169.22	6.241
175.29	6.351*
181.83	6.453
188.84	6.563*
195.74	6.648
202.57	6.744*
209.31	6.823
215.41	6.891*
222.66	6.973*
229.82	7.041
236.92	7.103*
243.96	7.168
250.93	7.226*
257.85	7.277
264.72	7.325*
271.54	7.376
278.30	7.424*
285.02	7.475
291.69	7.513*
298.33	7.547
304.94	7.575*
311.54	7.609
Series III	
58.69	2.354×10^{-2}*
62.85	2.580
67.49	2.820
72.16	3.053
76.98	3.295

T	Cp
CURVE 1 (cont.)	
82.61	3.572×10^{-2}
88.51	3.849
94.02	4.081
99.28	4.290
104.86	4.505
110.75	4.720
Series IV	
7.08	9.844×10^{-4}
8.17	1.196×10^{-3}
9.66	1.583
11.38	2.133
13.31	2.912
15.54	3.743
17.69	4.495
19.86	5.240
22.40	6.136
25.26	7.199
28.55	8.453
31.76	9.837
34.92	1.127×10^{-2}
38.44	1.291
42.36	1.486
46.74	1.674
51.42	1.964
55.93	2.206

* Not shown on plot

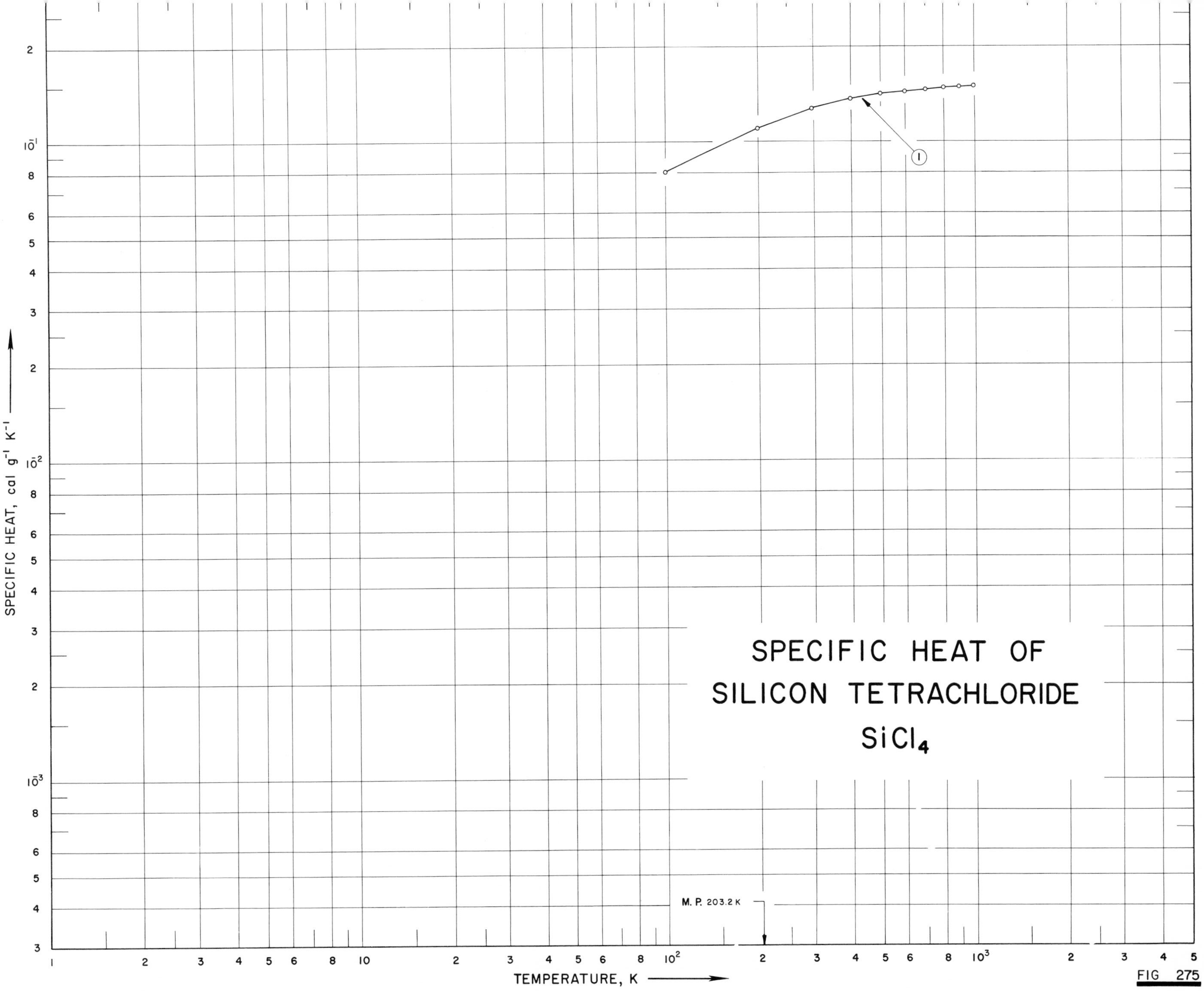

SPECIFIC HEAT OF SILICON TETRACHLORIDE $SiCl_4$

SPECIFIC HEAT, cal g^{-1} K^{-1}

TEMPERATURE, K

M. P. 203.2 K

FIG 275

SPECIFICATION TABLE NO. 275 SPECIFIC HEAT OF SILICON TETRACHLORIDE $SiCl_4$

[For Data Reported in Figure and Table No. 275]

Curve No.	Ref. No.	Year	Temp. Range, K	Reported Error, %	Name and Specimen Designation	Composition (weight percent), Specifications and Remarks
1	275	1953	100-1000			

DATA TABLE NO. 275 SPECIFIC HEAT OF SILICON TETRACHLORIDE $SiCl_4$

[Temperature, T,K; Specific Heat, Cp, Cal $g^{-1}K^{-1}$]

T	Cp
CURVE 1	
100	8.034×10^{-2}
200	1.109×10^{-1}
298.16	1.273*
300	1.275
400	1.364
500	1.414
600	1.444
700	1.463
800	1.476
900	1.485
1000	1.491

* Not shown on plot

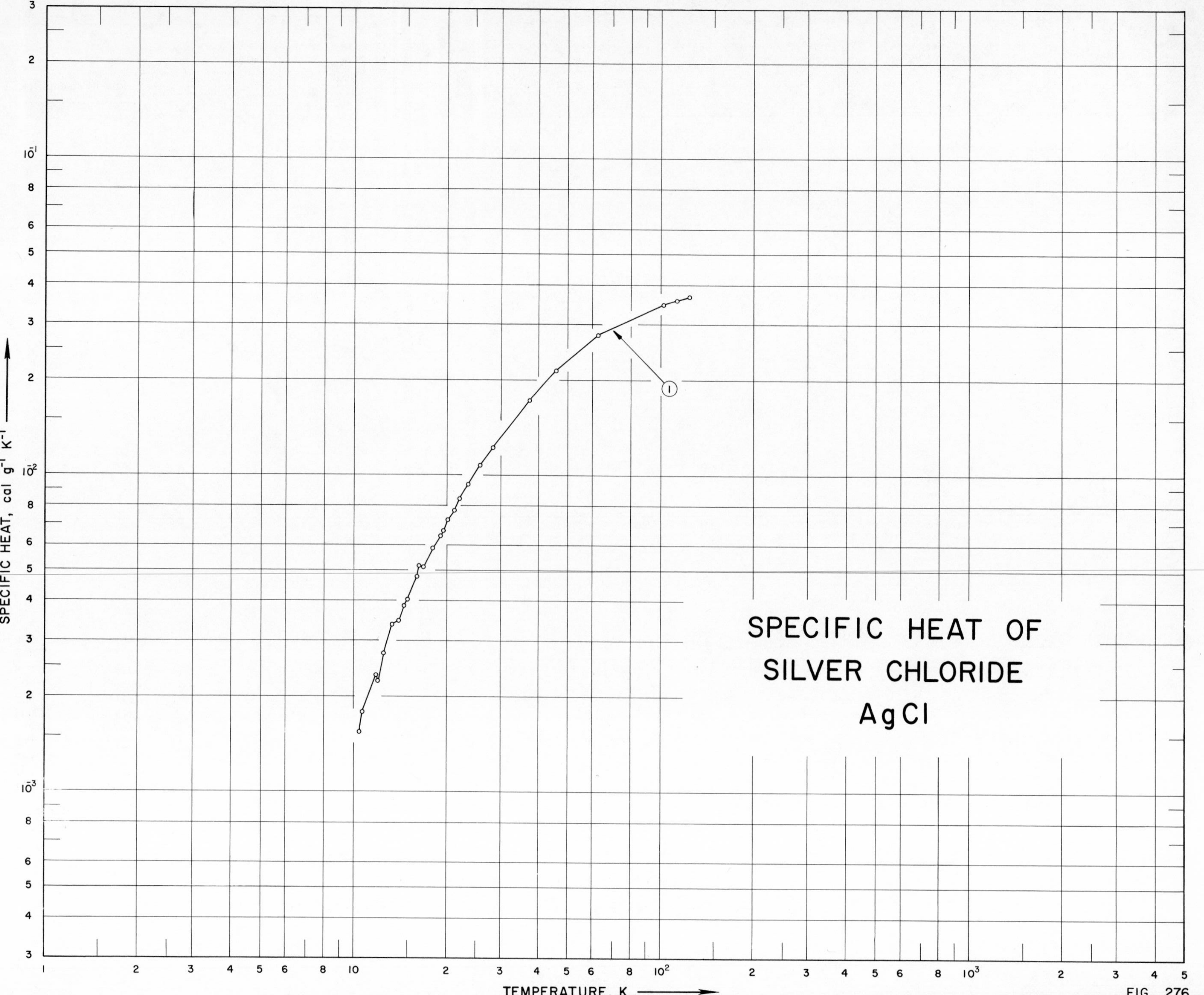
SPECIFIC HEAT OF
SILVER CHLORIDE
AgCl
SPECIFIC HEAT, cal g⁻¹ K⁻¹
TEMPERATURE, K
1

FIG 276

SPECIFICATION TABLE NO. 276 SPECIFIC HEAT OF SILVER CHLORIDE AgCl

[For Data Reported in Figure and Table No. 276]

Curve No.	Ref. No.	Year	Temp. Range, K	Reported Error, %	Name and Specimen Designation	Composition (weight percent), Specifications and Remarks
1	68	1928	11-126			

DATA TABLE NO. 276 SPECIFIC HEAT OF SILVER CHLORIDE AgCl

[Temperature, T, K; Specific Heat, Cp, Cal $g^{-1}K^{-1}$]

T	Cp
CURVE 1	
10.50	1.556×10^{-3}
10.73	1.807
11.83	2.365
12.04	2.254
12.64	2.763
13.40	3.412
14.04	3.510
14.67	3.914
15.10	4.096
16.27	4.849
16.49	5.205
16.90	5.191
18.13	5.924
18.32	6.126*
19.22	6.482
19.65	6.719
20.25	7.284
21.30	7.780
22.20	8.491
23.60	9.419
24.00	9.670
25.90	1.085×10^{-2}
28.5	1.235
37.5	1.741
46.3	2.153
63.3	2.785
103.1	3.493
114.0	3.592
125.6	3.698

* Not shown on plot

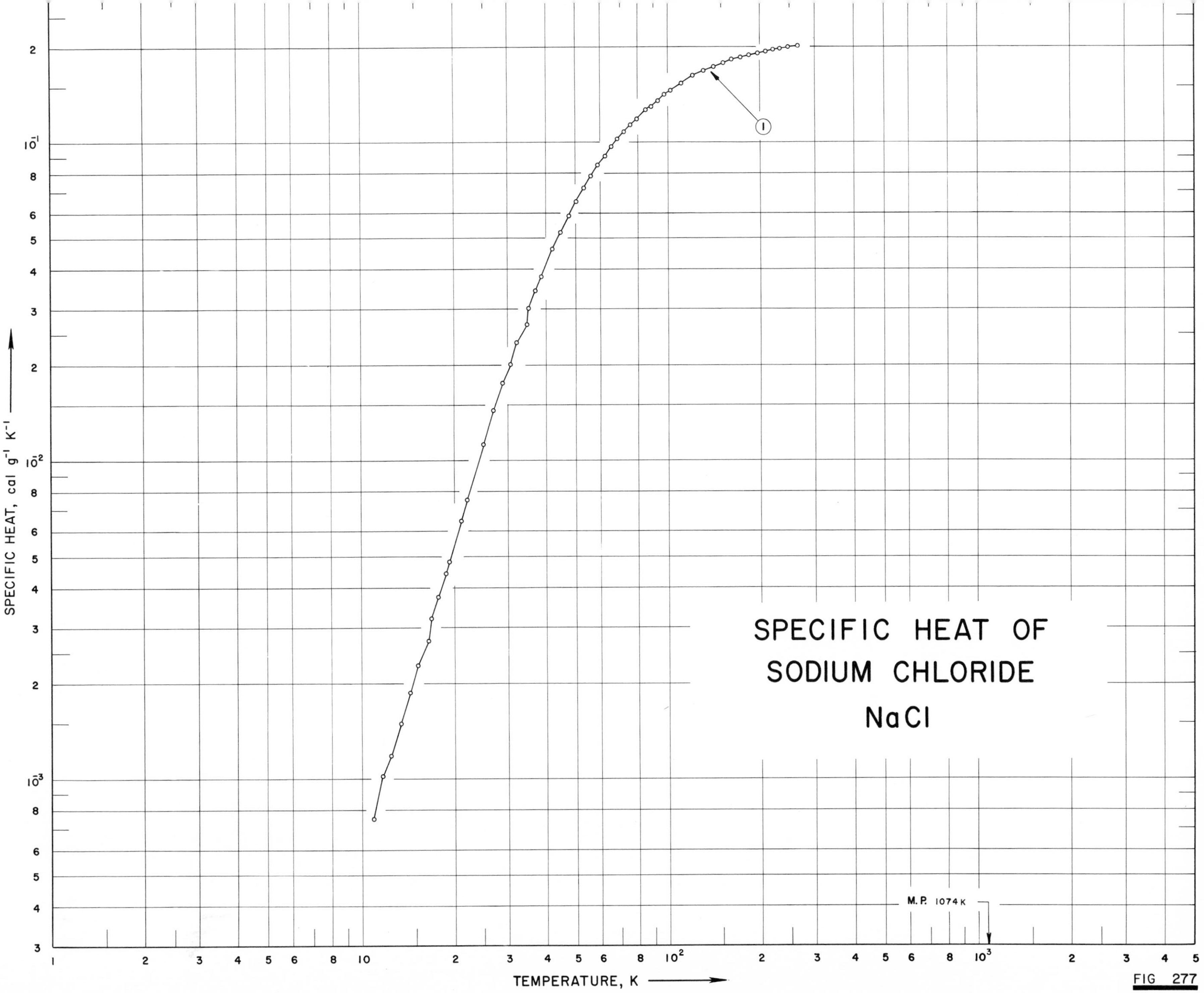

SPECIFIC HEAT OF SODIUM CHLORIDE NaCl

FIG 277

SPECIFICATION TABLE NO. 277 SPECIFIC HEAT OF SODIUM CHLORIDE $NaCl$

[For Data Reported in Figure and Table No. 277]

Curve No.	Ref. No.	Year	Temp. Range, K	Reported Error, %	Name and Specimen Designation	Composition (weight percent), Specifications and Remarks
1	243	1949	11-268			

DATA TABLE NO. 277 SPECIFIC HEAT OF SODIUM CHLORIDE NaCl

[Temperature, T, K; Specific Heat, Cp, Cal $g^{-1}K^{-1}$]

T	Cp	T	Cp
CURVE 1		CURVE 1 (cont.)	
10.9	7.53 x 10^{-4}	132.6	1.687
11.7	1.03 x 10^{-3}	137.7	1.714*
12.5	1.20	142.9	1.740
13.5	1.51	148.0	1.766*
14.4	1.88	153.2	1.786
15.3	2.29	158.6	1.806*
16.6	2.74	163.9	1.825
16.9	3.22	169.4	1.842*
17.8	3.76	174.9	1.859
18.9	4.45	180.6	1.874*
19.4	4.86	186.3	1.888
21.2	6.50	192.2	1.903*
22.2	7.56	198.1	1.916
25.1	1.13 x 10^{-2}	198.1	1.916*
27.2	1.46	204.2	1.929*
29.0	1.76	210.3	1.940
30.7	2.02	216.4	1.949*
32.2	2.36	222.5	1.966
34.9	2.70	228.5	1.966*
35.4	3.04	235.7	1.975
37.2	3.450	241.2	1.980*
38.4	3.809	247.4	1.990
42.4	4.658	253.9	1.998*
45.1	5.256	260.6	2.006*
48.0	5.920	267.5	2.007
50.7	6.557		
53.6	7.214		
56.6	7.871		
59.4	8.518		
62.6	9.086		
65.8	9.705		
69.2	1.028 x 10^{-1}		
72.7	1.088		
76.4	1.143		
80.2	1.198		
85.7	1.271		
89.4	1.308		
93.9	1.363		
98.5	1.4161		
103.1	1.463		
107.7	1.507*		
112.3	1.547		
117.2	1.585*		
122.4	1.639		
127.7	1.657*		

* Not shown on plot

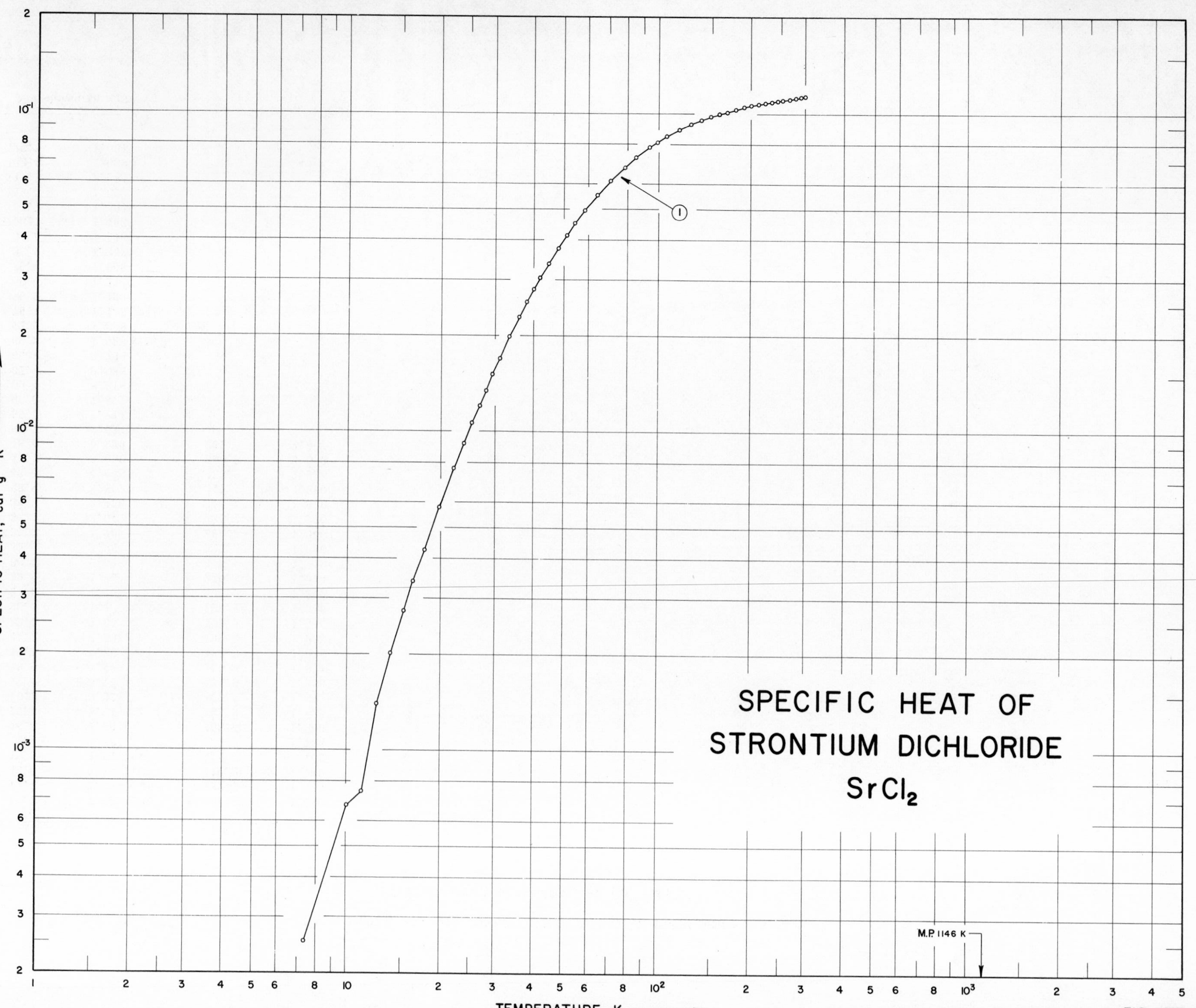

SPECIFIC HEAT OF
STRONTIUM DICHLORIDE
$SrCl_2$
M.P. 1146 K
1
SPECIFIC HEAT, cal g^{-1} K^{-1}

SPECIFICATION TABLE NO. 278 SPECIFIC HEAT OF STRONTIUM DICHLORIDE $SrCl_2$

[For Data Reported in Figure and Table No. 278]

Curve No.	Ref. No.	Year	Temp. Range, K	Reported Error, %	Name and Specimen Designation	Composition (weight percent), Specifications and Remarks
1	276	1963	7-300			> 99.99 $SrCl_2$, impurities: 0.005-0.0005 Na, 0.002-0.0002 Ca, Al, and Ba, and traces of Cu, Mg, and Mn; recrystallized and dried at 600 C for 3 hrs.

DATA TABLE NO. 278 SPECIFIC HEAT OF STRONTIUM DICHLORIDE $SrCl_2$

[Temperature, T, K; Specific Heat, Cp, Cal $g^{-1}K^{-1}$]

T	Cp	T	Cp	T	Cp
CURVE 1		CURVE 1 (cont.)		CURVE 1 (cont.)	
7.38	2.530 x 10^{-4}	105.93	8.469 x 10^{-2}*	224.12	1.089 x 10^{-1}
10.02	6.750	107.44	8.532	226.40	1.091*
11.22	7.444	109.92	8.644*	229.33	1.093*
12.49	1.413 x 10^{-3}	110.79	8.682*	231.72	1.095*
13.87	2.031	114.08	8.813*	235.22	1.097
15.26	2.750	117.87	8.959	238.10	1.100*
16.47	3.406	121.58	9.095*	242.01	1.103*
17.80	4.277	125.24	9.214*	245.46	1.106
19.85	5.829	128.83	9.335	247.78	1.107*
22.16	7.746	131.27	9.412*	250.67	1.109*
23.92	9.229	132.74	9.453*	254.13	1.112*
25.39	1.071 x 10^{-2}	135.30	9.530*	254.96	1.112
26.79	1.211	138.53	9.615	257.29	1.114*
28.13	1.359	142.28	9.709*	259.62	1.115*
29.57	1.520	144.83	9.772*	262.18	1.117*
31.20	1.719	146.57	9.808*	264.50	1.118*
33.42	1.989	148.84	9.861	268.24	1.120
35.99	2.300	151.54	9.919*	271.99	1.123*
38.22	2.576	152.87	9.951*	276.51	1.126*
40.22	2.820	155.48	9.999*	277.14	1.127*
42.20	3.063	158.53	1.006 x 10^{-1}	280.67	1.130
44.95	3.398	160.96	1.010*	286.05	1.140*
48.18	3.799	163.09	1.015*	289.54	1.142*
51.34	4.179	165.52	1.018*	292.36	1.144
54.48	4.541	168.40	1.023	294.20	1.145*
56.10	4.702*	170.03	1.026*	298.46	1.147*
57.35	4.858*	174.44	1.033*	300.43	1.149
58.65	4.996	174.49	1.033*		
60.63	5.205*	177.28	1.037*		
64.27	5.589	179.97	1.041		
64.69	5.625*	182.90	1.045*		
67.87	5.925*	186.66	1.051*		
70.85	6.196	191.50	1.057		
71.58	6.257*	194.19	1.059*		
75.21	6.558*	196.52	1.062*		
78.65	6.838	198.46	1.065*		
81.94	7.093*	201.47	1.068		
85.10	7.328	203.86	1.070*		
88.16	7.533*	206.64	1.073*		
91.31	7.717*	208.54	1.074*		
94.04	7.873	210.74	1.077*		
97.22	8.043*	213.26	1.079		
99.94	8.171	215.47	1.082*		
101.85	8.279*	217.22	1.083*		
103.74	8.366*	219.84	1.086*		

* Not shown on plot

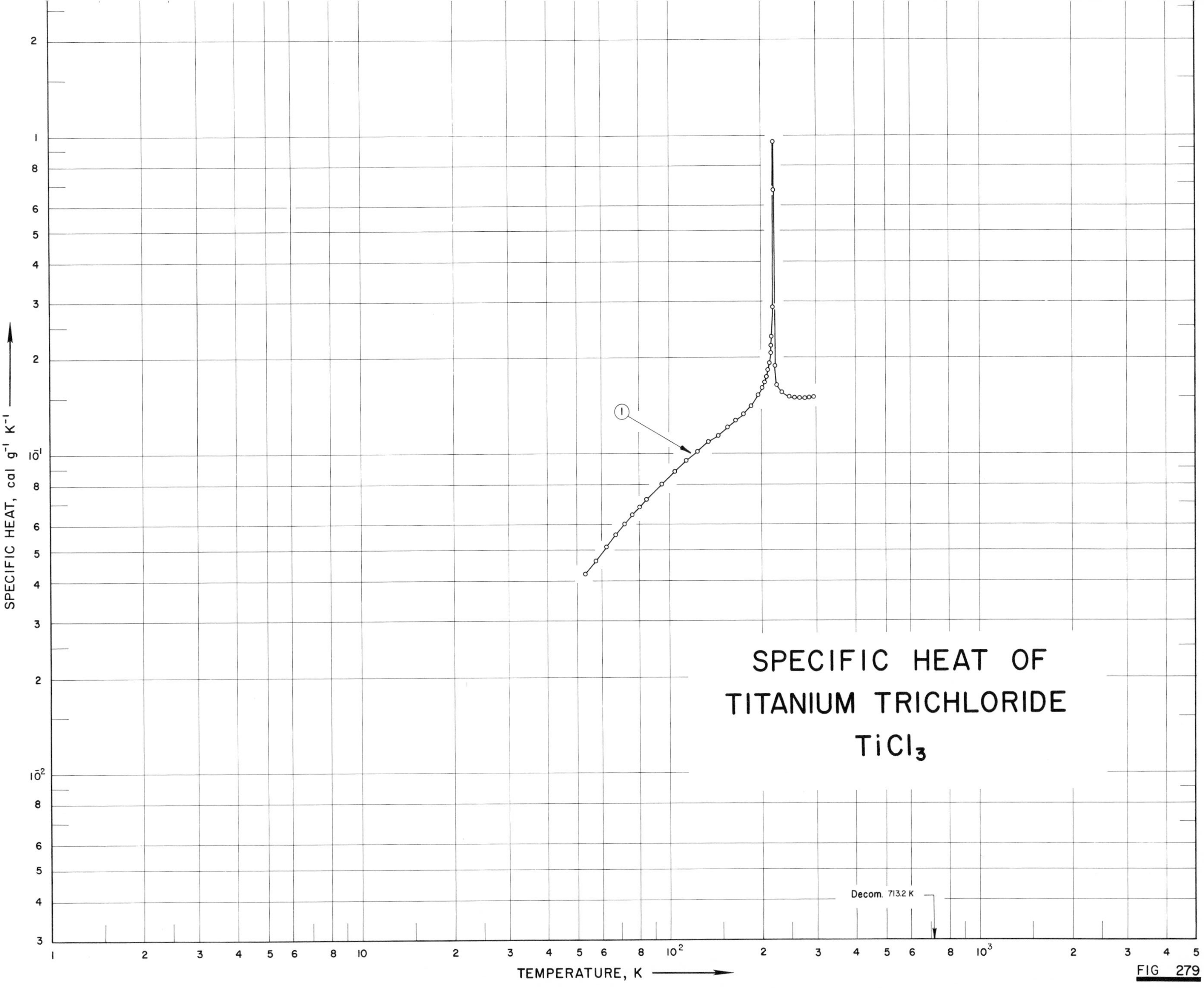
SPECIFIC HEAT OF
TITANIUM TRICHLORIDE
TiCl3
1
Decom. 713.2 K
SPECIFIC HEAT, cal g-1 K-1
TEMPERATURE, K
FIG 279

SPECIFICATION TABLE NO. 279 SPECIFIC HEAT OF TITANIUM TRICHLORIDE $TiCl_3$

[For Data Reported in Figure and Table No. 279]

Curve No.	Ref. No.	Year	Temp. Range, K	Reported Error, %	Name and Specimen Designation	Composition (weight percent), Specifications and Remarks
1	247	1961	53-296			68.98 Cl (68.95 theo.), 31.10 Ti (31.05 theo.), and <0.05 O_2.

DATA TABLE NO. 279 SPECIFIC HEAT OF TITANIUM TRICHLORIDE $TiCl_3$

[Temperature, T,K; Specific Heat, Cp, Cal $g^{-1}K^{-1}$]

T	Cp
CURVE 1	
53.27	4.219×10^{-2}
57.71	4.641
62.30	5.120
67.00	5.600
71.72	6.044
76.26	6.452
80.56	6.839
84.91	7.221
94.91	8.044
105.26	8.816
114.74	9.516
124.73	1.020×10^{-1}
136.24	1.091
145.98	1.147
156.00	1.210
165.90	1.272
175.94	1.339
186.12	1.420
196.15	1.525
202.17	1.613
205.61	1.678
206.23	1.693*
208.52	1.755
210.95	1.844
212.96	1.935
214.86	2.072
215.74	2.199
216.65	2.323
218.23	2.893
219.26	9.516
219.95	6.709
221.76	1.886
225.13	1.650
226.14	1.632*
229.16	1.596*
233.26	1.563
236.45	1.546*
237.69	1.540*
246.09	1.517
256.30	1.503
266.09	1.498
276.09	1.498
286.46	1.502
296.09	1.503

* Not shown on plot

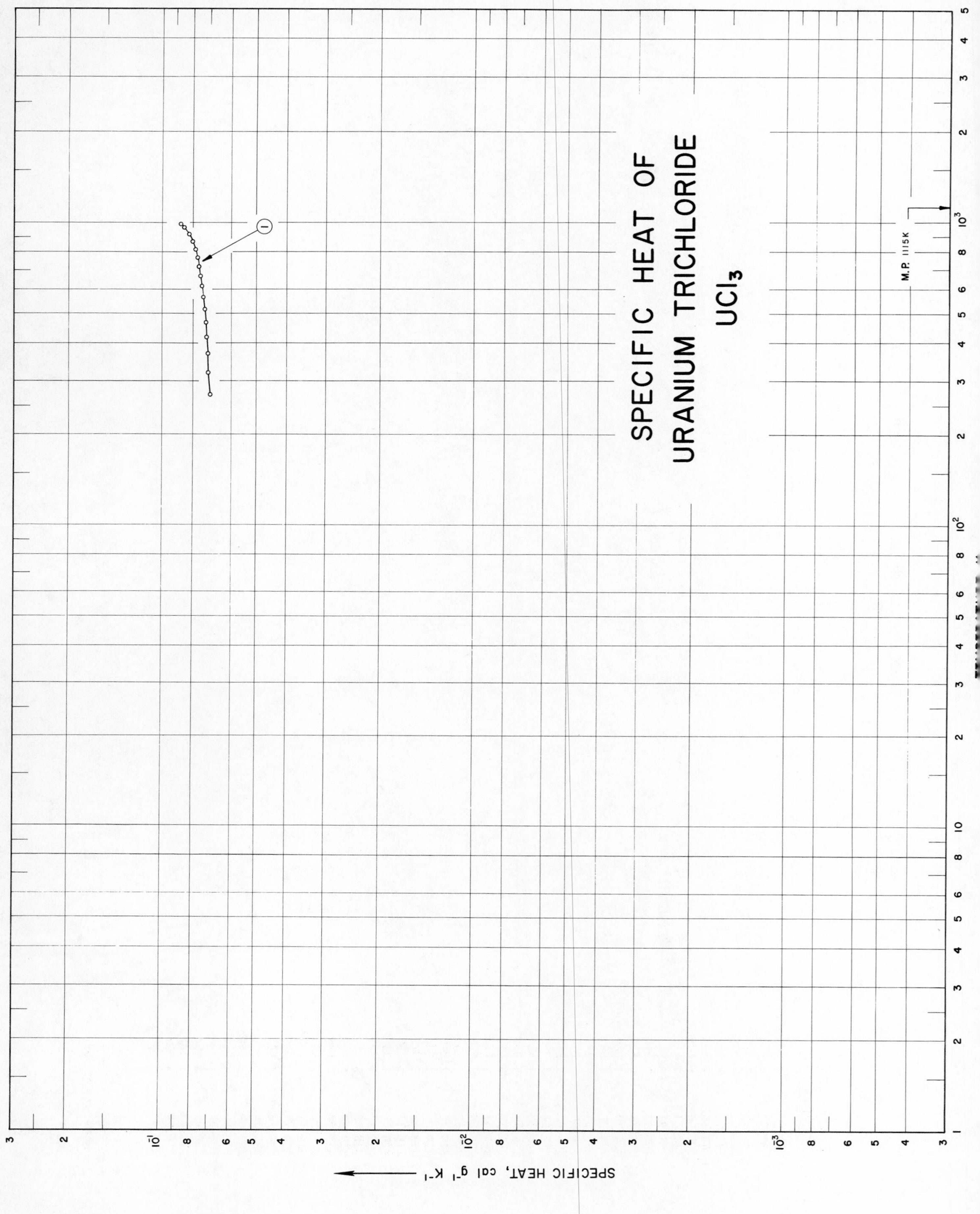

SPECIFIC HEAT OF URANIUM TRICHLORIDE UCl_3

SPECIFIC HEAT, cal g^{-1} K^{-1} →

M.P. 1115K

①

3, 2, 10^{-1}, 8, 6, 5, 4, 3, 2, 10^{-2}, 8, 6, 5, 4, 3, 2, 10^{-3}, 8, 6, 5, 4, 3

1, 2, 3, 4, 5, 6, 8, 10, 2, 3, 4, 5, 6, 8, 10^2, 2, 3, 4, 5, 6, 8, 10^3, 2, 3, 4, 5

SPECIFICATION TABLE NO. 280 SPECIFIC HEAT OF URANIUM TRICHLORIDE UCl_3

[For Data Reported in Figure and Table No. 280]

Curve No.	Ref. No.	Year	Temp. Range, K	Reported Error, %	Name and Specimen Designation	Composition (weight percent), Specifications and Remarks
1	277	1947	273-998			99.97 UCl_3, 0.13 insoluble, 0.08 misc. metals, 0.02 Na, 0.013 Fe, 0.01 Mg, and 0.006 Si; sample supplied by E.C. Evers of Brown University; prepared from sublimed UCl_4 by reduction with H_2.

DATA TABLE NO. 280 SPECIFIC HEAT OF URANIUM TRICHLORIDE UCl_3

[Temperature, T, K; Specific Heat, Cp, Cal $g^{-1}K^{-1}$]

T	Cp
CURVE 1	
273.15	7.06 x 10^{-2}
323.15	7.112
373.15	7.167
423.15	7.224
473.15	7.284
523.15	7.348
573.15	7.416
623.15	7.489
673.15	7.568
723.15	7.655
773.15	7.755
823.15	7.877
875.15	8.036
923.15	8.256
973.15	8.573
998.15	8.774

* Not shown on plot

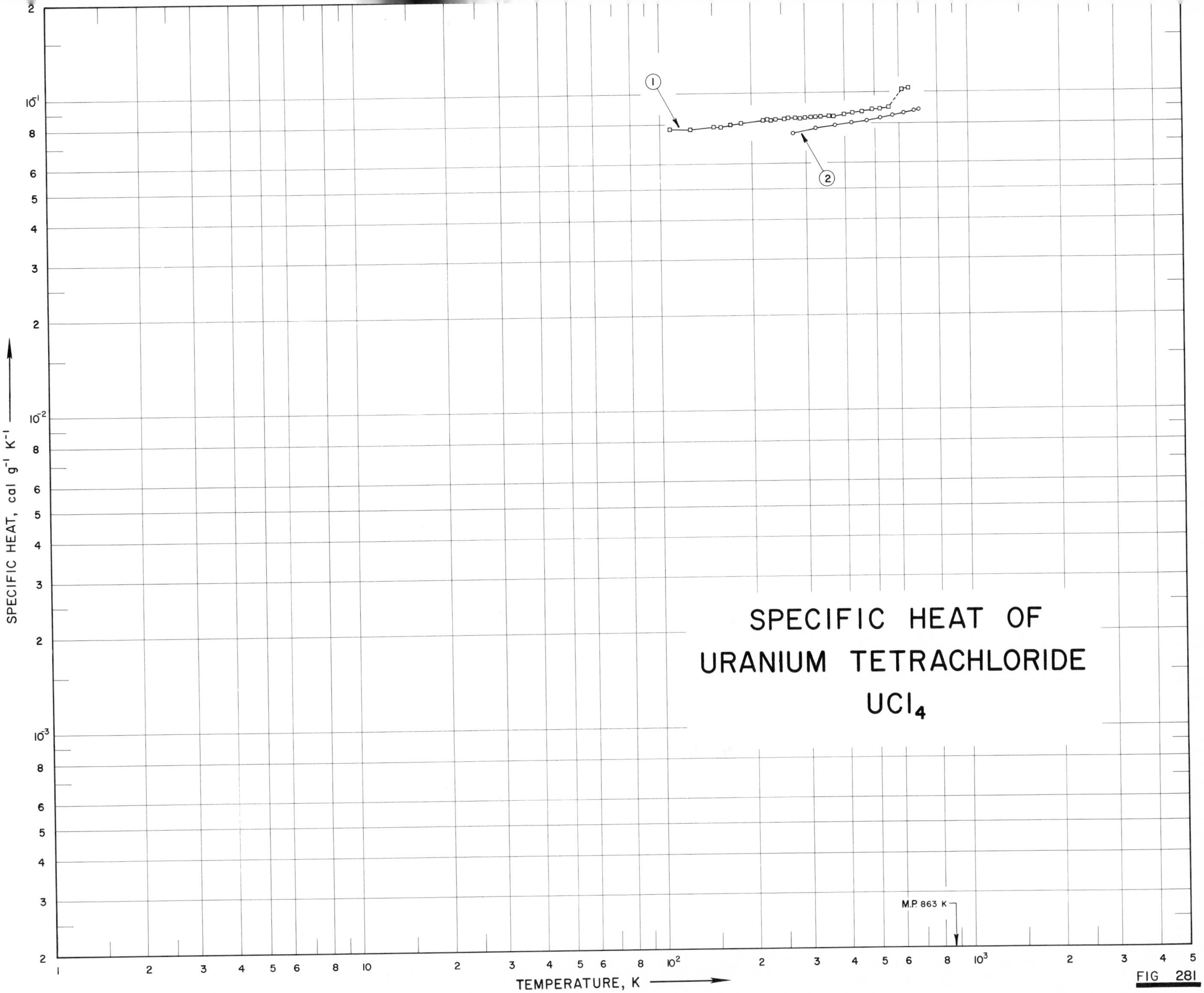

SPECIFIC HEAT OF URANIUM TETRACHLORIDE UCl_4

FIG 281

SPECIFIC HEAT, cal g^{-1} K^{-1}

TEMPERATURE, K

M.P. 863 K

1

2

SPECIFICATION TABLE NO. 281 SPECIFIC HEAT OF URANIUM TETRACHLORIDE UCl_4

[For Data Reported in Figure and Table No. 281]

Curve No.	Ref. No.	Year	Temp. Range, K	Reported Error, %	Name and Specimen Designation	Composition (weight percent), Specifications and Remarks
1	277	1947	273–698			99.97 UCl_4, 1.0–0.1 Ni, 0.1 misc metals (mainly Fe, 0.05), 0.02 Na, 0.01 each Ca, Mg, and 0.005 Al; prepared by reduction with H_2.
2	325	1959	108–647	± 0.5		Impurities: 0.1 mainly Fe and Si.

DATA TABLE NO. 281 SPECIFIC HEAT OF URANIUM TETRACHLORIDE UCl_4

[Temperature, T,K; Specific Heat, Cp, Cal $g^{-1}K^{-1}$]

T	Cp	T	Cp
CURVE 1		CURVE 1 (cont.)	
273.15	7.48 x 10^{-2}	373	8.551 x 10^{-2}*
323.15	7.745	379	8.459*
373.15	7.923	398	8.551
423.15	8.056	399	8.575*
473.15	8.189	410	8.596*
523.15	8.329	413	8.619*
573.15	8.473	425	8.643
623.15	8.617	427	8.619*
673.15	8.762	441	8.643*
698.15	8.834	445	8.643*
		457	8.735
CURVE 2		459	8.667*
		475	8.735*
108	7.756 x 10^{-2}	492	8.759*
127	7.743	494	8.851
151	7.880	495	8.759*
158	7.859	500	8.804*
171	7.972	502	8.759*
185	8.067	522	8.898
192	8.088*	523	8.875*
194	8.043*	528	8.875*
218	8.227	539	8.920*
223	8.159*	545	8.920*
225	8.275	546	8.920*
232	8.203	547	8.898*
240	8.275	551	8.898*
257	8.227	556	9.012*
264	8.367	558	8.991
270	8.367*		
274	8.367*		
278	8.367		
280	8.411*		
288	8.343		
295	8.367*		
299	8.367		
312	8.388		
312	8.367*		
324	8.388		
336	8.411		
340	8.411*		
343	8.388*		
358	8.435		
359	8.435*		
362	8.459*		
370	8.435		

* Not shown on plot

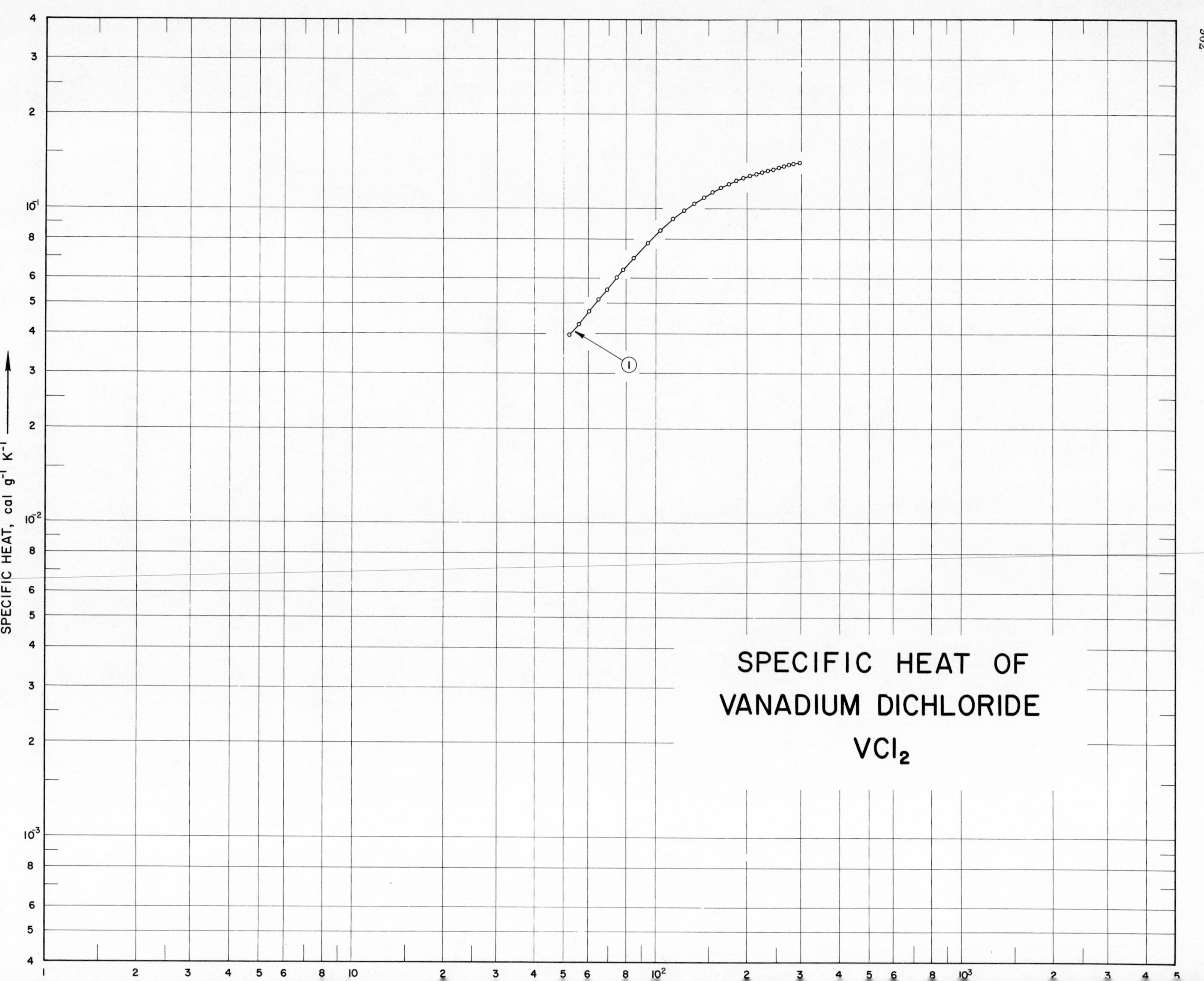
SPECIFIC HEAT OF
VANADIUM DICHLORIDE
VCl_2
1
SPECIFIC HEAT, cal g^{-1} K^{-1}

SPECIFICATION TABLE NO. 282 SPECIFIC HEAT OF VANADIUM DICHLORIDE VCl_2

[For Data Reported in Figure and Table No. 282]

Curve No.	Ref. No.	Year	Temp. Range, K	Reported Error, %	Name and Specimen Designation	Composition (weight percent), Specifications and Remarks
1	326	1947	53-298			99.8 VCl_2; 57.95 Cl_2 (58.19 theo.), 42.09 V (41.81 theo.), and 0.2 H_2O.

DATA TABLE NO. 282 SPECIFIC HEAT OF VANADIUM DICHLORIDE VCl_2

[Temperature, T,K; Specific Heat, Cp, Cal $g^{-1}K^{-1}$]

T	Cp
CURVE 1	
52.5	3.995×10^{-2}
56.3	4.313
60.8	4.742
65.3	5.161
69.7	5.575
74.6	6.055
78.8	6.454
85.0	7.001
94.8	7.831
104.4	8.583
115.0	9.322
125.1	9.946
135.1	1.050×10^{-1}
145.7	1.099
155.5	1.141
165.5	1.177
175.5	1.213
185.6	1.242
196.0	1.262
205.9	1.285
216.2	1.305
226.2	1.322
235.8	1.336
246.3	1.350
256.1	1.370
266.3	1.379
276.2	1.392
286.4	1.404
296.5	1.414*
298.2	1.416

*Not shown on plot

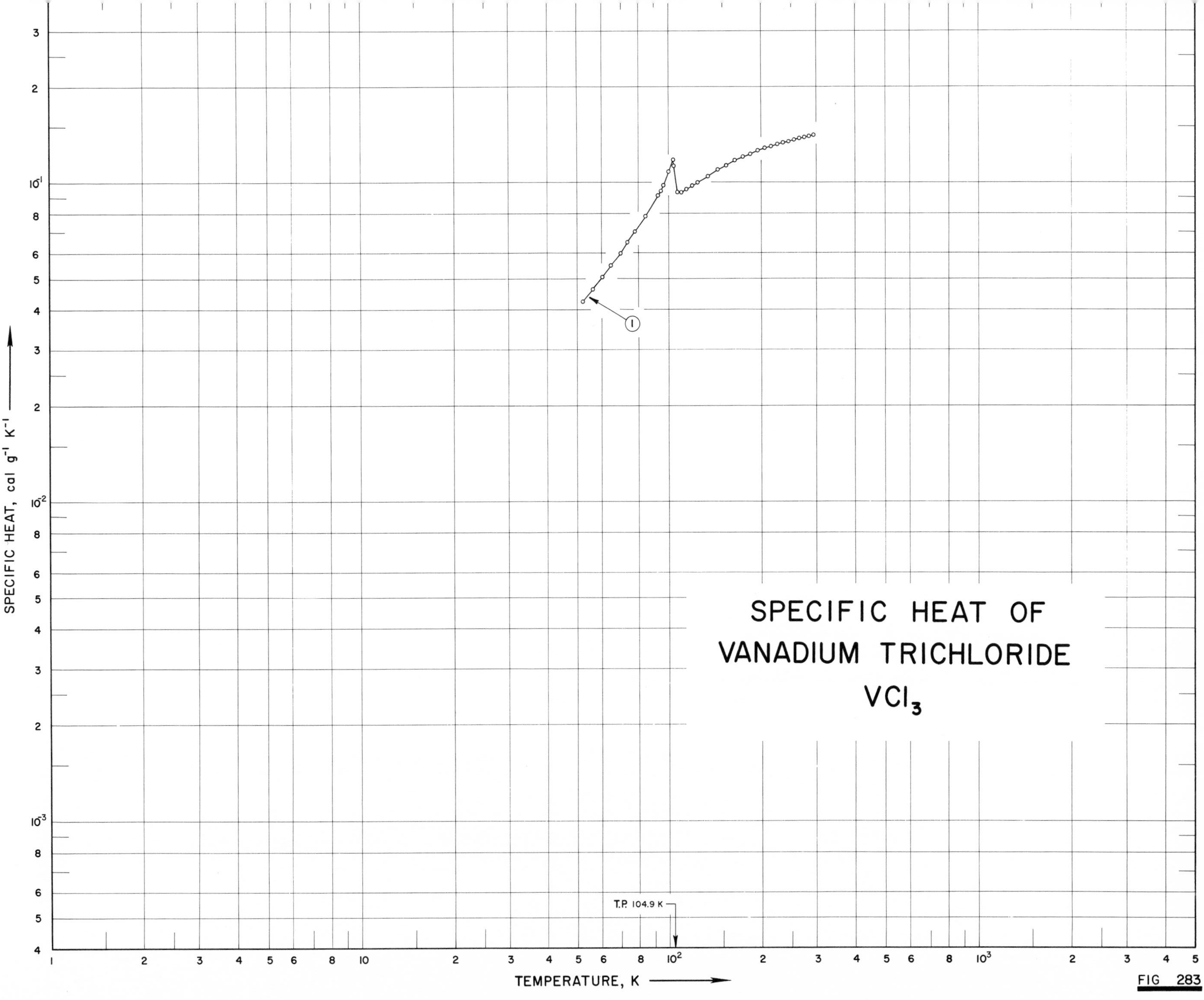
SPECIFIC HEAT OF
VANADIUM TRICHLORIDE
VCl_3
SPECIFIC HEAT, cal g^{-1} K^{-1}
TEMPERATURE, K
T.P. 104.9 K
1
FIG 283

SPECIFICATION TABLE NO. 283 SPECIFIC HEAT OF VANADIUM TRICHLORIDE VCl_3

[For Data Reported in Figure and Table No. 283]

Curve No.	Ref. No.	Year	Temp. Range, K	Reported Error, %	Name and Specimen Designation	Composition (weight percent), Specifications and Remarks
1	326	1947	53-298			99.6 VCl_3; 67.45 Cl (67.61 theo.), 32.41 V (32.39 theo.), and 0.4 H_2O; prepared from freshly distilled VCl_4 which had been obtained by passing chlorine gas over ferrovanadium at 250 C; heated at 160 C in stream of dry CO_2 for 24 hrs.

DATA TABLE NO. 283 SPECIFIC HEAT OF VANADIUM TRICHLORIDE VCl_3

[Temperature, T, K; Specific Heat, Cp, Cal $g^{-1}K^{-1}$]

T	Cp
CURVE 1	
52.5	4.265×10^{-2}
56.6	4.658
60.6	5.078
64.9	5.521
69.2	6.002
73.6	6.515
78.0	7.049
84.6	7.888
92.8	9.102
94.7	9.433
96.8	9.833
100.6	1.086×10^{-1}
104.1	1.178
104.4	1.134
107.5	9.338
110.9	9.312
115.6	9.541
120.6	9.795
125.2	1.003
135.0	1.053
145.7	1.100
155.4	1.139
165.4	1.176
175.7	1.210
185.6	1.237
195.9	1.261
206.0	1.285
216.3	1.307
226.3	1.323
236.0	1.341
246.6	1.352
256.5	1.369
266.5	1.378
276.2	1.391
286.3	1.402
296.5	1.417*
298.15	1.416

* Not shown on plot

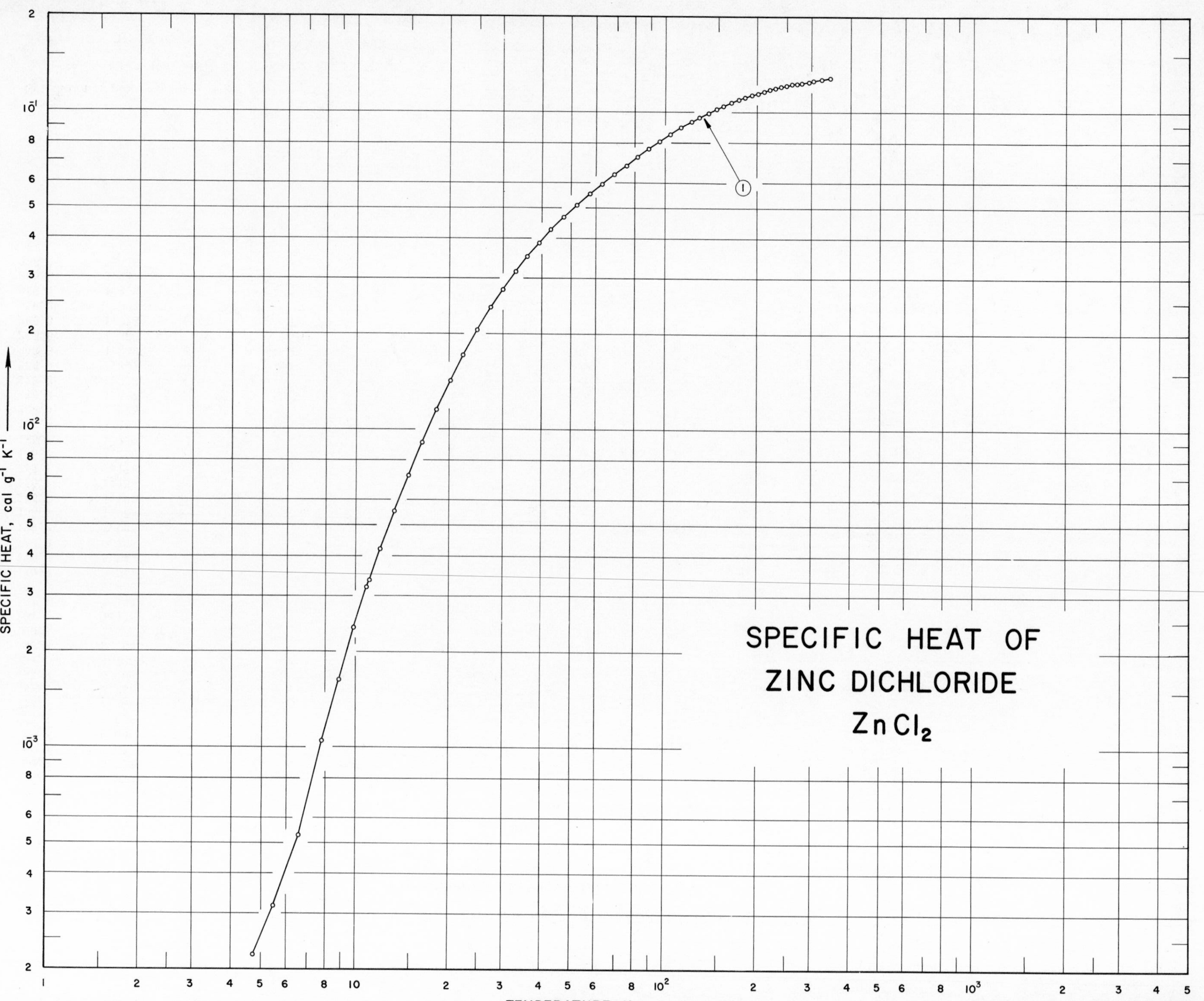
SPECIFIC HEAT OF
ZINC DICHLORIDE
$ZnCl_2$
SPECIFIC HEAT, cal g^{-1} K^{-1}
1

SPECIFICATION TABLE NO. 284 SPECIFIC HEAT OF ZINC DICHLORIDE $ZnCl_2$

[For Data Reported in Figure and Table No. 284]

Curve No.	Ref. No.	Year	Temp. Range, K	Reported Error, %	Name and Specimen Designation	Composition (weight percent), Specifications and Remarks
1	327	1956	5-346	0.15-6.0		51.90 Cl_2 (52.03 theo.), and 47.97 Zn (47.97 theo.); prepared by passing HCl gas dried with $CaSO_4$ over molten zinc (99.998 Zn) at 700 C.

DATA TABLE NO. 284 SPECIFIC HEAT OF ZINC DICHLORIDE $ZnCl_2$

[Temperature, T,K; Specific Heat, Cp, Cal $g^{-1}K^{-1}$]

T	Cp
CURVE 1	
Series I	
11.12	3.360×10^{-3}
12.15	4.248*
13.39	5.547*
14.88	7.212*
16.54	9.245*
18.36	1.160×10^{-2}
20.33	1.427
22.46	1.724
24.87	2.065
27.52	2.429
30.32	2.804*
Series II	
4.73	2.201×10^{-4}
5.48	3.155
6.61	5.283
7.81	1.049×10^{-3}
8.86	1.636
9.88	2.377
10.91	3.192
12.07	4.211
13.38	5.525
14.83	7.176
16.45	9.135
30.08	2.765
33.05	3.153
36.09	3.518
39.39	3.884
43.13	4.261
47.52	4.672
52.42	5.093
57.68	5.525
63.24	5.942
69.20	6.361
75.61	6.791
82.33	7.240
89.00	7.654
96.84	8.085
105.17	8.518
113.99	8.944
123.00	9.333
130.75	9.641
CURVE 1 (cont.)	
139.21	9.949×10^{-2}
147.89	1.024×10^{-1}
156.83	1.050
166.16	1.075
175.55	1.097
184.75	1.116
193.81	1.133
202.95	1.148
212.27	1.163
221.69	1.176
231.26	1.189
241.07	1.201
251.01	1.211
261.00	1.222
271.05	1.228
281.11	1.239
Series III	
276.54	1.232×10^{-1}*
286.30	1.242*
296.44	1.250
306.78	1.258
317.04	1.264*
327.21	1.271
337.34	1.277*
346.44	1.283

* Not shown on plot

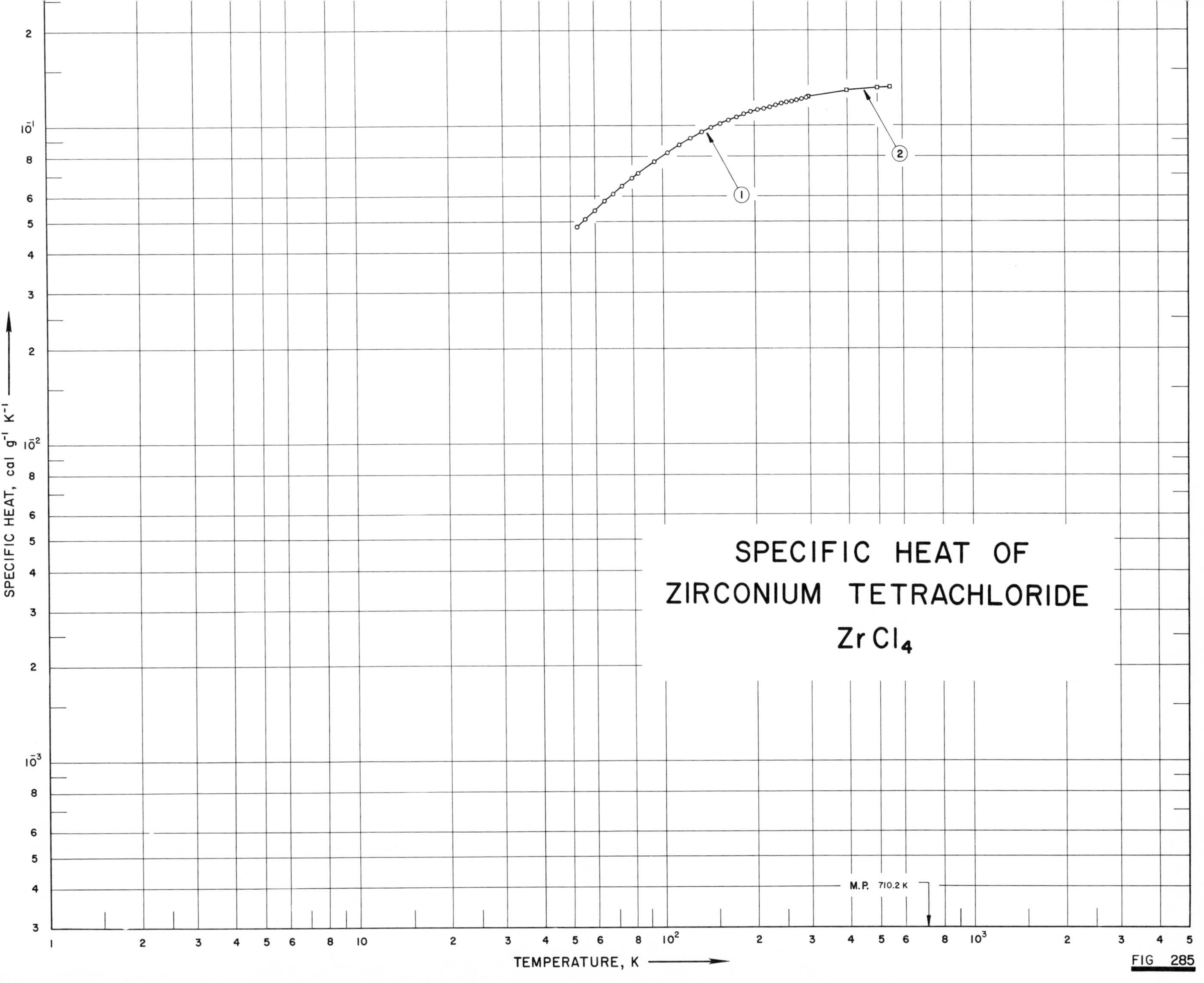
SPECIFIC HEAT OF
ZIRCONIUM TETRACHLORIDE
$ZrCl_4$
M.P. 710.2 K
TEMPERATURE, K
SPECIFIC HEAT, cal g^{-1} K^{-1}
FIG 285

SPECIFICATION TABLE NO. 285 SPECIFIC HEAT OF ZIRCONIUM TETRACHLORIDE $ZrCl_4$

[For Data Reported in Figure and Table No. 285]

Curve No.	Ref. No.	Year	Temp. Range, K	Reported Error, %	Name and Specimen Designation	Composition (weight percent), Specifications and Remarks
1	217	1950	53-296			39.21 Zr, 0.75 Hf; corrected for impurities.
2	278	1950	298-550	0.3		39.21 Zr, 0.75 Hf.

DATA TABLE NO. 285 SPECIFIC HEAT OF ZIRCONIUM TETRACHLORIDE $ZrCl_4$

[Temperature, T, K; Specific Heat, Cp, Cal $g^{-1}K^{-1}$]

T	Cp
CURVE 1	
52.6	4.823×10^{-2}
55.9	5.089
60.0	5.419
64.9	5.806
69.3	6.136
74.0	6.462
80.0	6.857
83.9	7.093
94.8	7.707
104.6	8.217
114.6	8.685
124.7	9.101
136.1	9.539
146.3	9.856
156.2	1.015×10^{-1}
166.3	1.041
176.3	1.067
186.4	1.087
196.4	1.105
206.7	1.122
216.6	1.138
226.7	1.150
236.4	1.163
246.3	1.174
256.5	1.187
266.4	1.199
276.5	1.209
286.8	1.217
296.7	1.228
CURVE 2	
298.15	1.229×10^{-1}*
300	1.231
400	1.292
500	1.320
550	1.325

* Not shown on plot

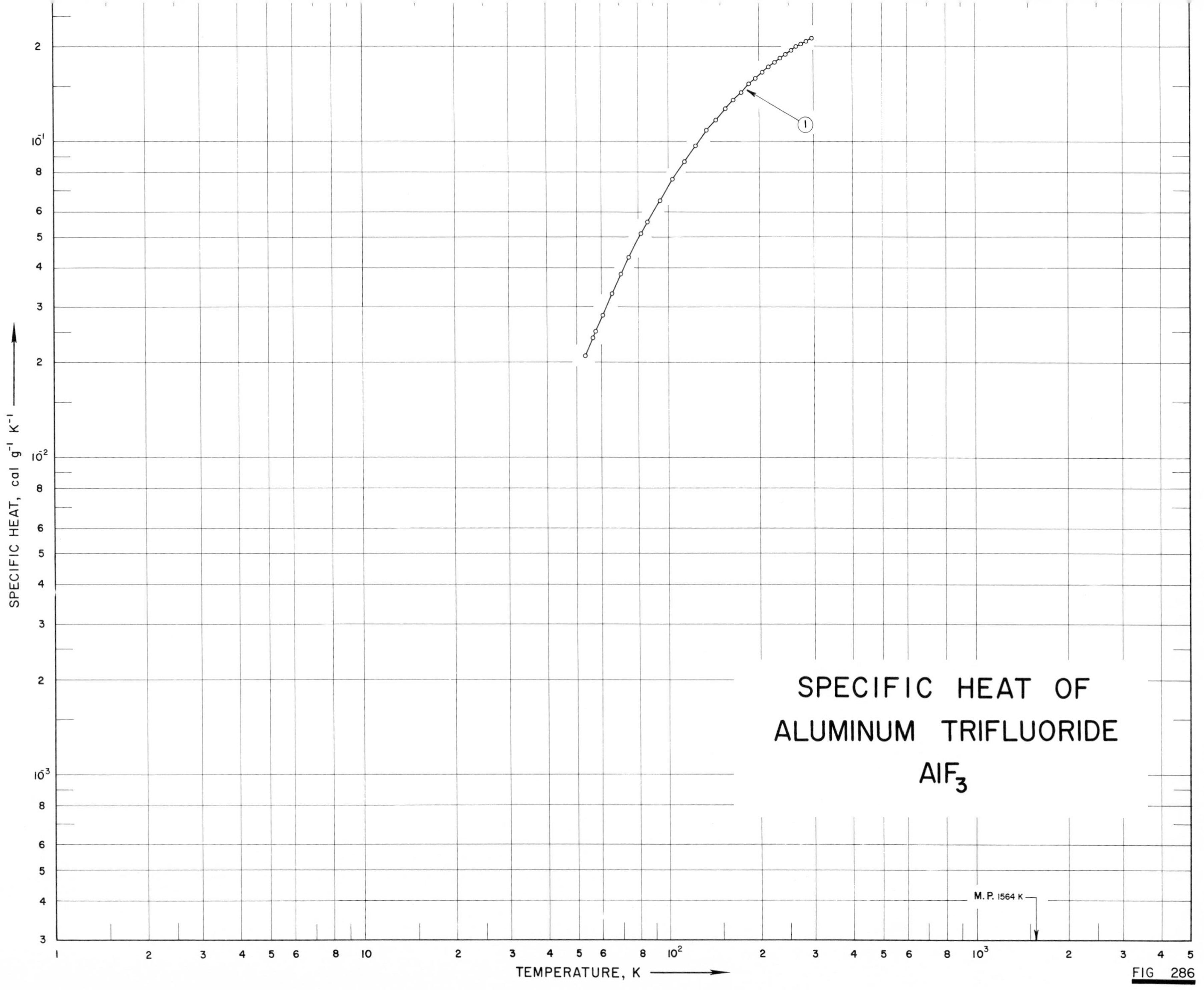
SPECIFIC HEAT OF
ALUMINUM TRIFLUORIDE
AlF3
SPECIFIC HEAT, cal g-1 K-1
TEMPERATURE, K
M.P. 1564 K
1

FIG 286

SPECIFICATION TABLE NO. 286 SPECIFIC HEAT OF ALUMINUM TRIFLUORIDE AlF_3

[For Data Reported in Figure and Table No. 286]

Curve No.	Ref. No.	Year	Temp. Range, K	Reported Error, %	Name and Specimen Designation	Composition (weight percent), Specifications and Remarks
1	279	1956	54-298			32.12 Al (32.13 theo.), 0.06 Ca, 0.008 Si, Fe, 0.005 Mg, Na, 0.003 Ti and others; sample supplied by the Reduction Research Laboratory, Kaiser Aluminum and Chem. Corp.

DATA TABLE NO. 286 SPECIFIC HEAT OF ALUMINUM TRIFLUORIDE AlF_3

[Temperature, T, K; Specific Heat, Cp, Cal $g^{-1}K^{-1}$]

T	Cp
CURVE 1	
53.65	2.112×10^{-2}
56.69	2.397
57.89	2.513
61.01	2.834
65.62	3.313
70.32	3.818
74.88	4.314
82.19	5.125
86.30	5.579
94.60	6.505
104.82	7.626
114.53	8.654
124.39	9.682
135.56	1.081×10^{-1}
145.35	1.176
155.75	1.272
165.58	1.355
175.81	1.438
185.93	1.518
195.84	1.590
206.14	1.659
215.91	1.724
226.20	1.789
235.93	1.846
245.55	1.896
256.18	1.952
266.31	2.000
276.22	2.047
286.59	2.089
296.10	2.129*
298.15	2.137

* Not shown on plot

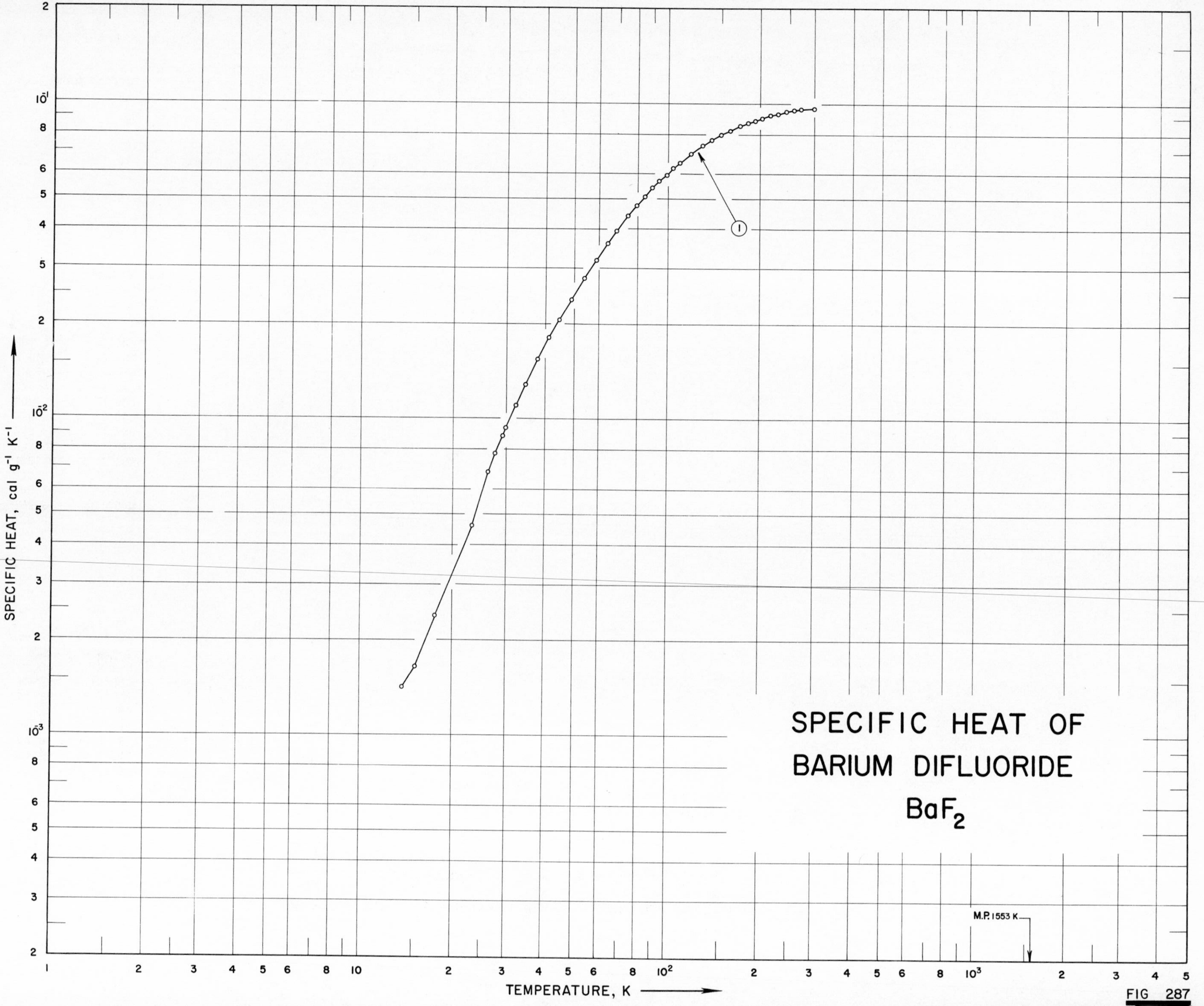

SPECIFIC HEAT OF
BARIUM DIFLUORIDE
BaF_2
M.P. 1553 K
TEMPERATURE, K
SPECIFIC HEAT, cal g⁻¹ K⁻¹
FIG 287

SPECIFICATION TABLE NO. 287 SPECIFIC HEAT OF BARIUM DIFLUORIDE BaF_2

[For Data Reported in Figure and Table No. 287]

Curve No.	Ref. No.	Year	Temp. Range, K	Reported Error, %	Name and Specimen Designation	Composition (weight percent), Specifications and Remarks
1	280	1938	14-301			99.7 ± 0.5 Ba; crystalline; prepared by melting C.P. barium nitrate and potassium fluoride in a platinum crucible.

DATA TABLE NO. 287 SPECIFIC HEAT OF BARIUM DIFLUORIDE BaF_2

[Temperature, T, K; Specific Heat, Cp, Cal $g^{-1}K^{-1}$]

T	Cp	T	Cp
CURVE 1		CURVE 1 (cont.)	
13.79	1.426 x 10^{-3}	203.93	9.040 x 10^{-2}
15.25	1.654	210.14	9.103*
17.72	2.395	216.91	9.205
23.23	5.133	223.27	9.251*
26.20	6.844	229.73	9.308
27.70	7.814	237.32	9.405*
29.35	8.897	244.95	9.479
30.00	9.410	252.69	9.513*
32.43	1.118 x 10^{-2}	259.32	9.559
34.73	1.295	266.35	9.587*
38.08	1.551	272.87	9.633
41.45	1.814	273.94	9.644*
44.96	2.065	279.44	9.639*
49.20	2.395	293.08	9.639*
54.24	2.795	300.69	9.684
59.34	3.200		
64.36	3.616		
64.64	3.639*		
68.65	3.958		
74.76	4.443		
79.83	4.768		
84.72	5.099		
89.48	5.412		
94.41	5.698		
99.52	5.954		
104.78	6.262		
110.11	6.542		
115.03	6.753*		
119.76	6.952		
124.79	7.163*		
130.01	7.363		
134.94	7.540*		
139.63	7.694		
144.10	7.785*		
149.75	7.990		
155.28	8.104*		
160.66	8.218		
166.14	8.338*		
173.17	8.509		
177.59	8.566*		
178.32	8.629*		
183.31	8.709		
188.17	8.777*		
193.05	8.863		
198.09	8.926*		

* Not shown on plot

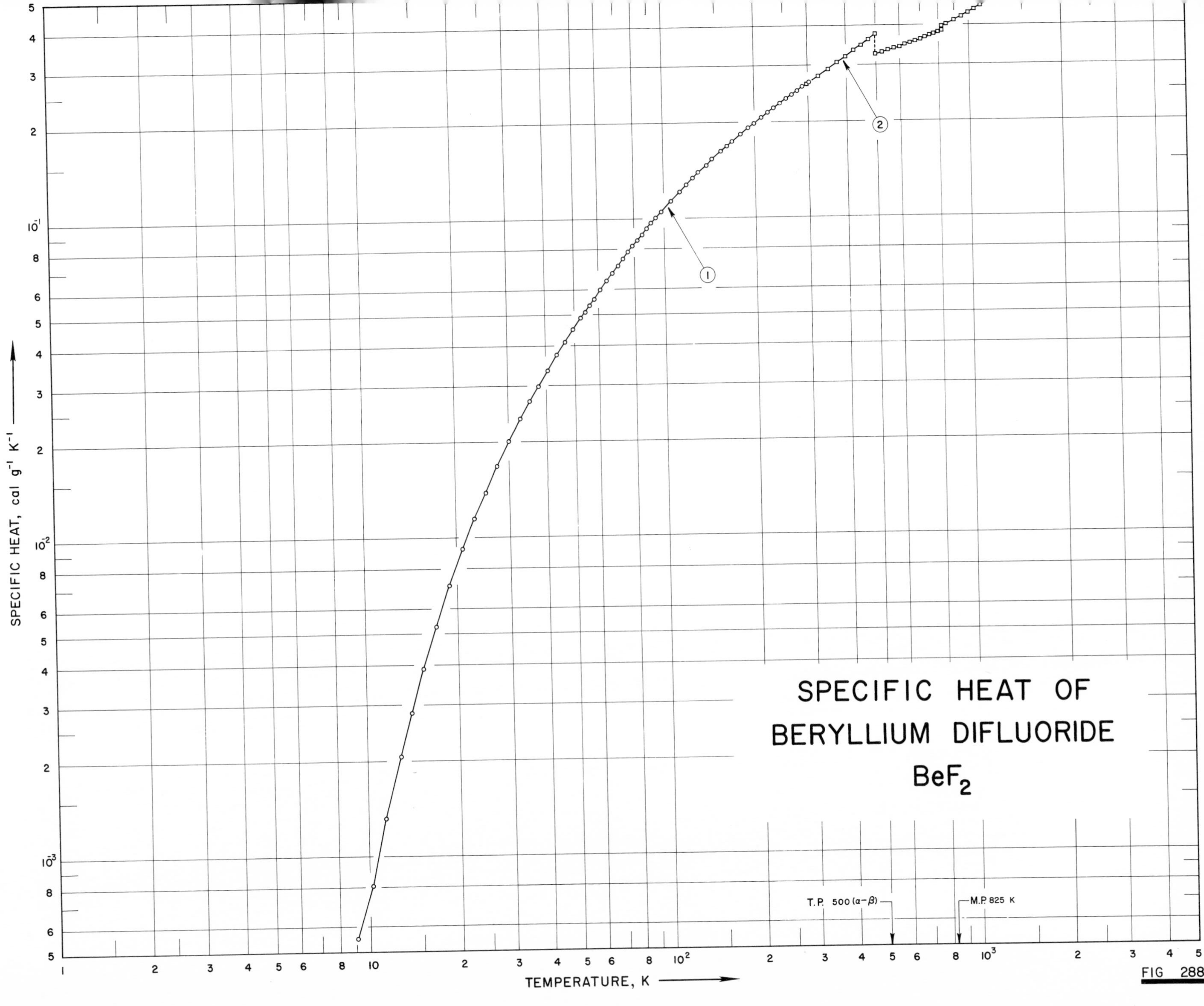
SPECIFIC HEAT OF
BERYLLIUM DIFLUORIDE
BeF_2
SPECIFIC HEAT, cal g^{-1} K^{-1}
TEMPERATURE, K
T.P. 500 ($\alpha-\beta$)
M.P. 825 K
FIG 288

SPECIFICATION TABLE NO. 288 SPECIFIC HEAT OF BERYLLIUM DIFLUORIDE BeF_2

[For Data Reported in Figure and Table No. 288]

Curve No.	Ref. No.	Year	Temp. Range, K	Reported Error, %	Name and Specimen Designation	Composition (weight percent), Specifications and Remarks
1	281	1965	8-304	< 0.3		78.8 ± 2.0 F and 19.9 ± 1.0 Be (80.83 and 19.17 theo.), 0.3 O_2, and 0.06 BeO; crystalline.
2	281	1965	298-1200	0.2		0.35 BeO; glassy form; prepared by heating crystalline BeF_2 to 600 C.

DATA TABLE NO. 288 SPECIFIC HEAT OF BERYLLIUM DIFLUORIDE BeF_2

[Temperature, T, K; Specific Heat, Cp, Cal $g^{-1}K^{-1}$]

T	Cp	T	Cp	T	Cp	T	Cp
CURVE 1		CURVE 1 (cont.)		CURVE 1 (cont.)		CURVE 2 (cont.)	
7.90	4.0 x 10^{-4}	106.83	1.143 x 10^{-1}	243.54	2.2934 x 10^{-1}*	850	4.010 x 10^{-1}
9.10	5.5	109.96	1.177*	247.31	2.3183*	900	4.127
10.23	8.1	113.18	1.211	250.95	2.3419*	950	4.244
11.48	1.3 x 10^{-3}	116.39	1.245*	255.03	2.3700	1000	4.361
12.82	2.1	119.59	1.279	259.04	2.3949*	1050	4.478
14.00	2.83	122.84	1.311*	261.61	2.4104*	1100	4.595
15.36	3.89	126.05	1.344	265.55	2.4368	1150	4.712
17.01	5.28	129.19	1.376*	269.44	2.4600*	1200	4.829
18.92	7.11	131.56	1.399	273.30	2.4859*		
20.91	9.27	132.26	1.406*	277.11	2.5080		
22.95	1.16 x 10^{-2}	135.41	1.437*	280.88	2.5312*		
25.06	1.39	138.70	1.469	284.71	2.5553*		
27.45	1.68	140.18	1.482*	288.60	2.5795		
30.16	2.02	142.05	1.501*	292.44	2.6018*		
32.89	2.372	144.59	1.523*	296.25	2.6231*		
35.44	2.697	145.44	1.531	300.04	2.6450*		
37.83	2.999	148.24	1.557*	303.79	2.6674		
40.75	3.365	151.82	1.589*				
43.95	3.765	155.39	1.621	CURVE 2			
46.89	4.135	158.95	1.652*				
49.79	4.512	162.51	1.684	298.15	2.623 x 10^{-1}		
52.87	4.905	166.17	1.715*	300	2.634*		
54.56	5.114	169.82	1.746	325	2.772		
55.97	5.293*	173.36	1.776*	350	2.910		
56.64	5.376	176.85	1.805*	375	3.050		
58.64	5.629	180.34	1.833	400	3.189		
58.83	5.661*	183.68	1.860*	425	3.329		
61.44	5.997	187.38	1.889*	450	3.467		
61.51	6.007*	191.44	1.922	475	3.608		
64.51	6.390	195.61	1.953*	α500	3.746		
67.53	6.767	199.09	1.981	β500	3.240		
70.52	7.135	202.52	2.006*	525	3.287		
73.48	7.501	205.91	2.030*	550	3.333		
76.42	7.873	207.71	2.045*	575	3.380		
79.26	8.226	211.16	2.069	600	3.427		
82.02	8.575	214.74	2.096*	625	3.474		
85.02	8.943	218.28	2.122*	650	3.521		
88.23	9.337	221.77	2.1453	675	3.567		
91.37	9.700	225.30	2.1707*	700	3.614		
94.45	1.005 x 10^{-1}	228.88	2.1943*	725	3.661		
96.52	1.028*	232.29	2.2175	750	3.706		
97.51	1.039*	235.87	2.2411*	775	3.752		
98.68	1.053	239.63	2.2670*	800	3.799		
100.58	1.073*	239.73	2.2674*	β825	3.846		
103.68	1.108*	243.10	2.2911	(1)825	3.952		

* Not shown on plot

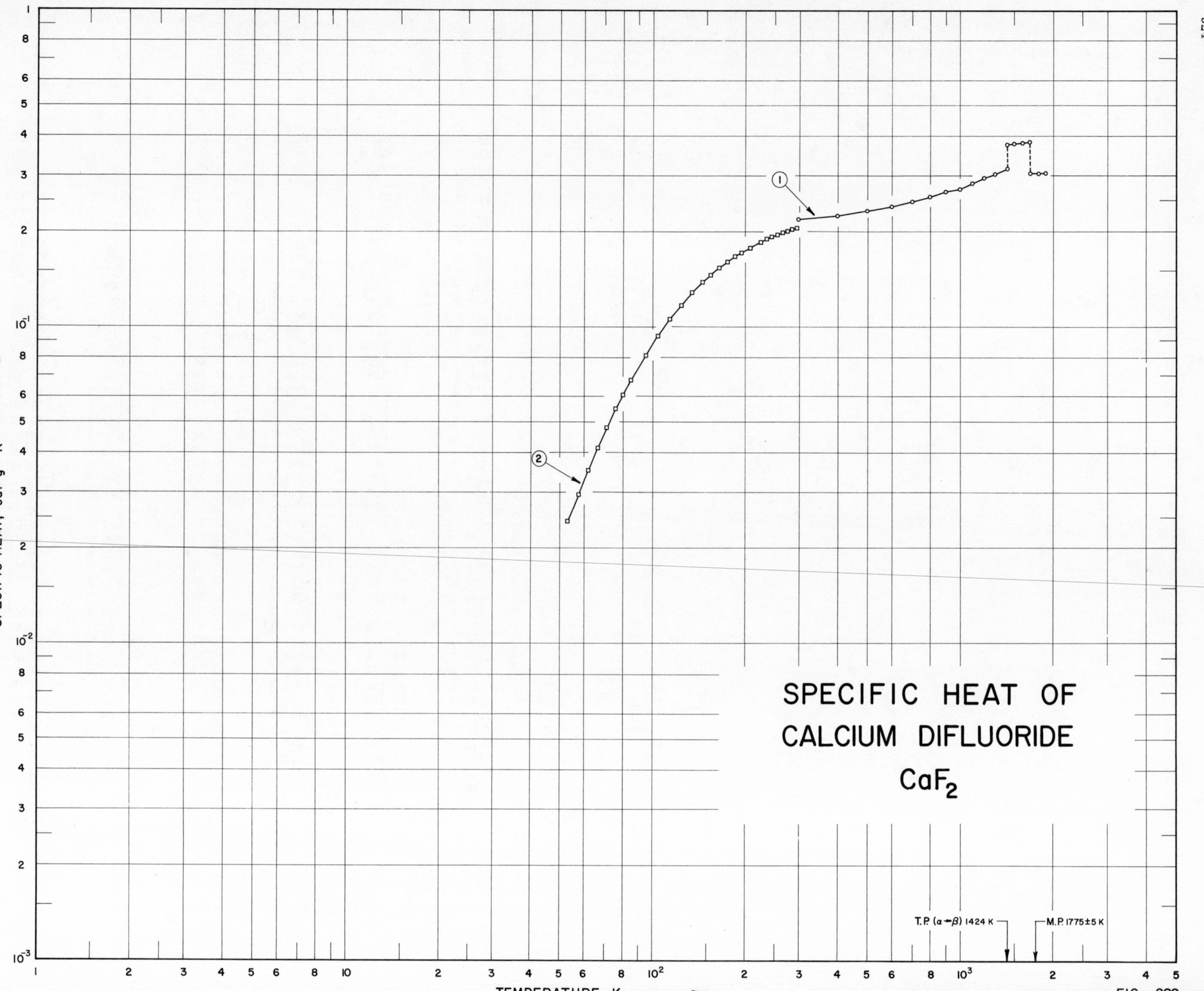

SPECIFIC HEAT OF
CALCIUM DIFLUORIDE
CaF_2
T.P. (α→β) 1424 K
M.P. 1775±5 K
1
2
SPECIFIC HEAT, cal g⁻¹ K⁻¹

SPECIFICATION TABLE NO. 289 SPECIFIC HEAT OF CALCIUM DIFLUORIDE CaF_2

[For Data Reported in Figure and Table No. 289]

Curve No.	Ref. No.	Year	Temp. Range, K	Reported Error, %	Name and Specimen Designation	Composition (weight percent), Specifications and Remarks
1	282	1945	298-1800			51.27 Ca.
2	283	1949	54-297	0.1-0.3		51.27 Ca (51.33 theo.); large natural fluorite crystals.

DATA TABLE NO. 289 SPECIFIC HEAT OF CALCIUM DIFLUORIDE CaF_2

[Temperature, T, K; Specific Heat, Cp, Cal $g^{-1}K^{-1}$]

T	Cp	T	Cp
CURVE 1		**CURVE 2 (cont.)**	
298.15	2.177 x 10^{-1}	226.2	1.856 x 10^{-1}
300	2.178*	236.4	1.895
400	2.242	245.8	1.926
500	2.322	256.3	1.958
600	2.408	266.0	1.985
700	2.496	276.0	2.008
800	2.587	286.4	2.029
900	2.678	296.5	2.049
1000	2.770		
1100	2.862		
1200	2.954		
1300	3.047		
1400	3.140*		
α 1424	3.162		
β 1424	3.762		
1500	3.786		
1600	3.818		
β 1691	3.847		
(l) 1691	3.058		
1700	3.058		
1800	3.058		
CURVE 2			
53.51	2.444 x 10^{-2}		
57.55	2.957		
62.04	3.532		
66.74	4.166		
71.40	4.819		
76.25	5.507		
80.43	6.094		
85.32	6.782		
95.04	8.115		
104.51	9.366		
114.48	1.059 x 10^{-1}		
124.37	1.171		
135.5	1.288		
146.0	1.383		
155.6	1.463		
165.9	1.541		
175.7	1.609		
186.0	1.670		
195.9	1.719		
208.2	1.778		
216.4	1.817*		

* Not shown on plot

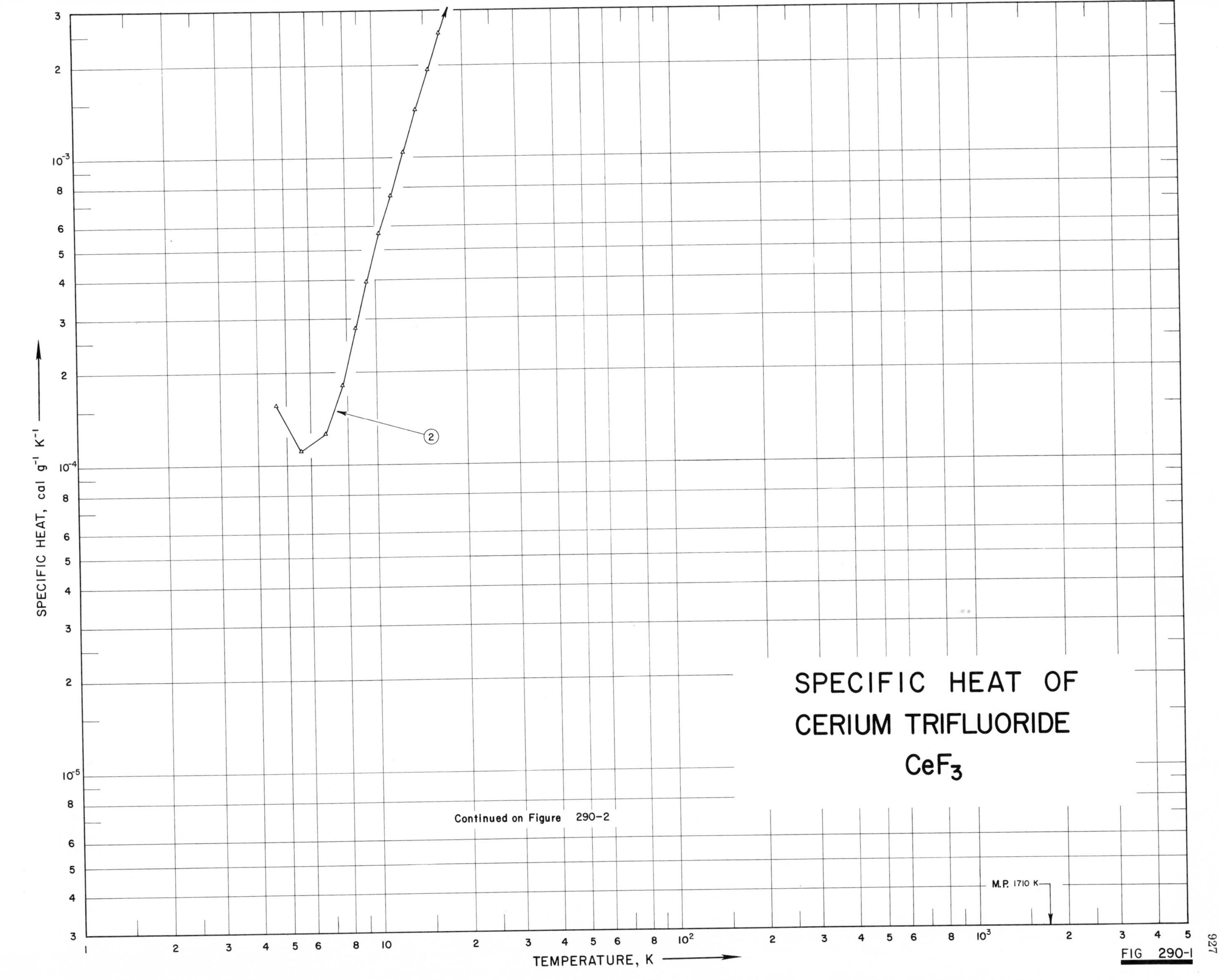
SPECIFIC HEAT OF
CERIUM TRIFLUORIDE
CeF_3
SPECIFIC HEAT, cal g⁻¹ K⁻¹
TEMPERATURE, K
M.P. 1710 K
2
Continued on Figure 290-2

FIG 290-1

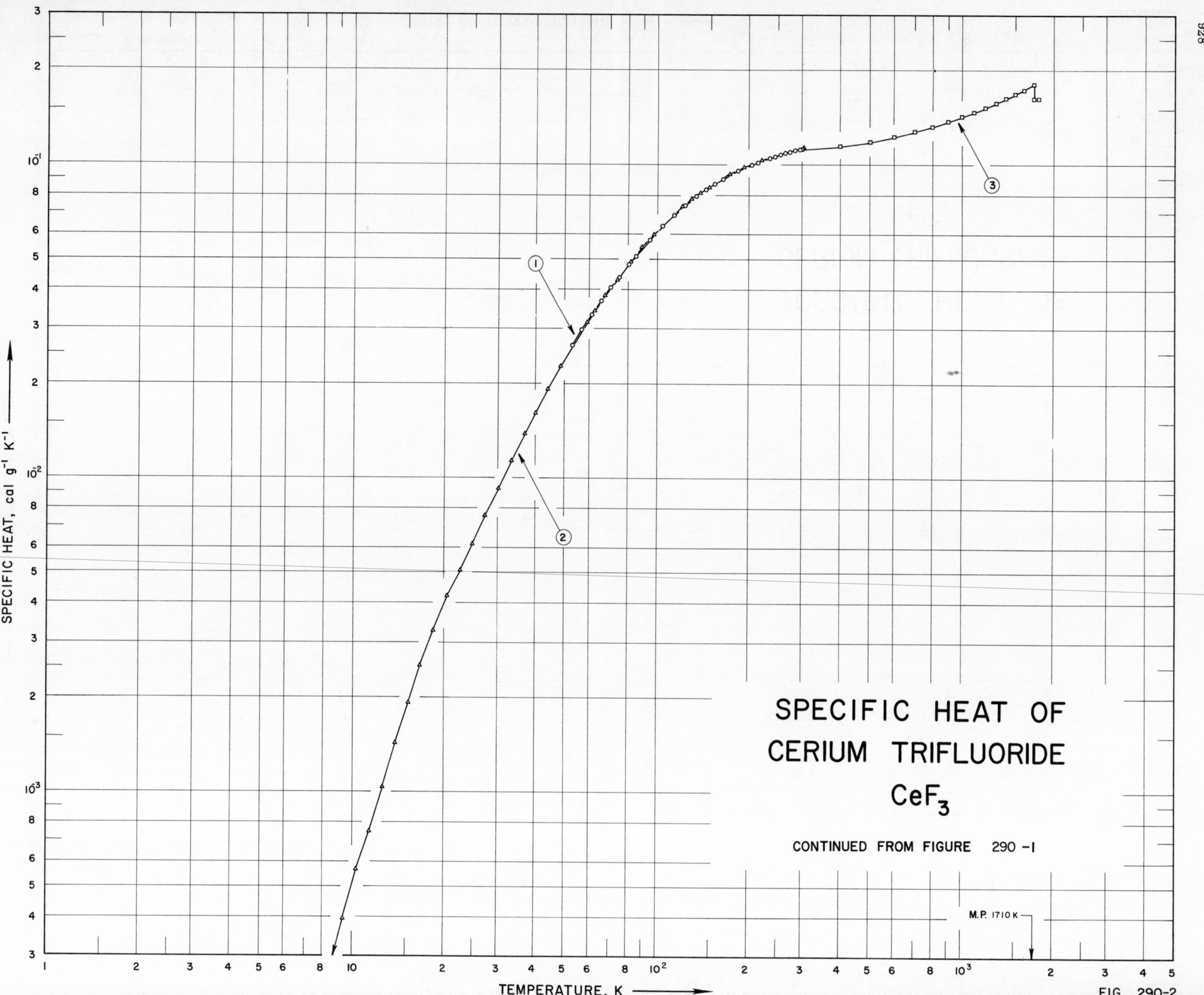
SPECIFIC HEAT OF
CERIUM TRIFLUORIDE
CeF_3
CONTINUED FROM FIGURE 290 -1
M.P. 1710 K
TEMPERATURE, K
SPECIFIC HEAT, cal g^{-1} K^{-1}
1
2
3
FIG 290-2

SPECIFICATION TABLE NO. 290 SPECIFIC HEAT FOR CERIUM TRIFLUORIDE CeF_3

[For Data Reported in Figure and Table No. 290]

Curve No.	Ref. No.	Year	Temp. Range, K	Reported Error, %	Name and Specimen Designation	Composition (weight percent), Specifications and Remarks
1	284	1959	53–296	0.1–0.5		71.14 Ce, 28.87 F, < 0.1 Na, 0.01 Ti, < 0.01 Ca, < 0.01 K, and 0.001 Mg; heated for 25 hrs at 700 C.
2	285	1961	5–304	0.1–10		71.12 Ce, and 28.88 F; prepared by addition of HF to ceric ammonium nitrate solution; precipitate dried at 110 C and ignited; heated at 500 C in the presence of HF gas.
3	284	1959	298–1800	0.1–0.5		71.14 Ce, 28.87 F, < 0.1 Na, 0.01 Ti, < 0.01 each Ca and K, and 0.001 Mg; heated for 25 hrs at 700 C.

DATA TABLE NO. 290 SPECIFIC HEAT OF CERIUM TRIFLUORIDE CeF_3

[Temperature, T, K; Specific Heat, Cp, Cal $g^{-1}K^{-1}$]

T	Cp	T	Cp	T	Cp
CURVE 1		CURVE 2 (cont.)		CURVE 3	
53.21	2.661 x 10^{-2}	157.19	8.806 x 10^{-2}*	298.15	1.124 x 10^{-1}*
57.27	2.987	166.02	9.101*	300	1.124
61.63	3.328	174.91	9.359	400	1.149
66.20	3.685	184.09	9.577*	500	1.188
70.94	4.054	193.54	9.811	600	1.232
75.64	4.396	202.97	1.001 x 10^{-1}*	700	1.279
81.18	4.803	212.41	1.020*	800	1.328
85.93	5.134	221.99	1.035	900	1.378
94.12	5.758	231.60	1.051*	1000	1.428
105.14	6.387	241.08	1.066*	1100	1.478
114.59	6.914	250.55	1.078*	1200	1.529
124.76	7.432	260.06	1.091*	1300	1.580
136.31	7.959	269.45	1.102*	1400	1.631
145.79	8.355	278.95	1.113*	1500	1.682
156.28	8.720	288.55	1.123*	1600	1.733
166.19	9.045	296.88	1.132*	1700	1.784*
176.33	9.324*	303.99	1.138	(s) 1732	1.801
186.28	9.583			(l) 1732	1.623
196.02	9.796*	Series II		1800	1.623
206.50	9.998				
216.36	1.019 x 10^{-1}	4.65	1.6 x 10^{-4}		
226.52	1.037*	5.60	1.1		
236.42	1.053	6.75	1.3		
246.00	1.065	7.74	1.8		
256.40	1.079	8.56	2.8		
266.63	1.091	9.39	4.0		
276.60	1.101	10.30	5.68		
286.74	1.112	11.40	7.51		
296.24	1.123	12.60	1.04 x 10^{-3}		
		13.82	1.43		
CURVE 2		15.27	1.93		
		16.70	2.55		
Series I		18.50	3.29		
		20.52	4.22		
62.96	3.440 x 10^{-2}	22.65	5.118		
67.81	3.830	24.95	6.204		
74.47	4.340	27.44	7.614		
82.00	4.914*	30.22	9.278		
89.77	5.468	33.46	1.140 x 10^{-2}		
97.82	5.991	36.89	1.383		
105.97	6.488*	40.07	1.615		
114.18	6.960*	44.19	1.929		
122.67	7.386	48.65	2.285		
131.47	7.817	53.65	2.687*		
139.38	8.152	59.46	3.158		
148.33	8.497				

* Not shown on plot

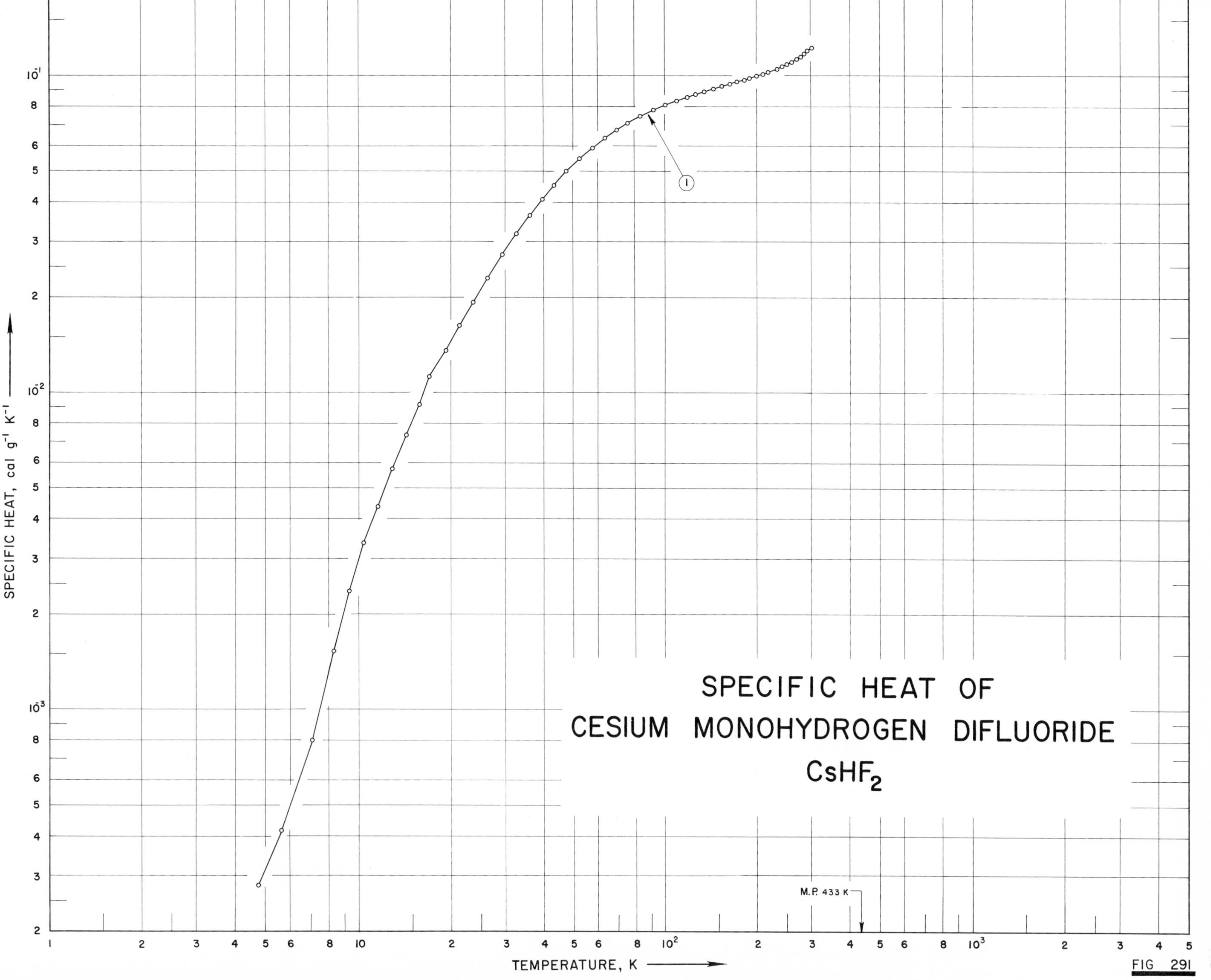
SPECIFIC HEAT OF
CESIUM MONOHYDROGEN DIFLUORIDE
$CsHF_2$
M.P. 433 K
TEMPERATURE, K
SPECIFIC HEAT, cal g^{-1} K^{-1}

FIG 291

SPECIFICATION TABLE NO. 291 SPECIFIC HEAT OF CESIUM MONOHYDROGEN DIFLUORIDE $CsHF_2$

[For Data Reported in Figure and Table No. 291]

Curve No.	Ref. No.	Year	Temp. Range, K	Reported Error, %	Name and Specimen Designation	Composition (weight percent), Specifications and Remarks
1	286	1961	5-303			< 0.1 Ca, Mg, and Na; prepared from > 99.5 Cs_2CO_3; heated 24 hrs at 300 C.

DATA TABLE NO. 291 SPECIFIC HEAT OF CESIUM MONOHYDROGEN DIFLUORIDE $CsHF_2$

[Temperature, T, K; Specific Heat, Cp, Cal $g^{-1}K^{-1}$]

T	Cp
CURVE 1	
Series 1	
58.01	5.945×10^{-2}
57.93	5.939*
63.79	6.393
69.83	6.771
76.16	7.137
83.91	7.498
92.31	7.829
101.20	8.108
110.08	8.353
119.07	8.574
118.34	8.556*
127.28	8.760
136.21	8.940
145.54	9.121
155.09	9.289
164.38	9.435
173.53	9.586
182.71	9.725
191.83	9.865
201.06	1.001×10^{-1}
210.18	1.016
218.96	1.029
198.58	9.970×10^{-2}*
207.46	1.011×10^{-1}*
216.36	1.025*
225.29	1.040*
234.28	1.056
243.30	1.075
252.31	1.090
261.25	1.109
270.05	1.129
278.70	1.151
287.00	1.175
291.91	1.201
302.74	1.231
Series 2	
4.76	2.792×10^{-4}
5.65	4.188
7.10	8.027
8.32	1.536×10^{-3}
9.36	2.379
CURVE 1 (cont.)	
10.45	3.362×10^{-3}
11.62	4.386
12.91	5.753
14.31	7.364
15.82	9.190
17.18	1.125×10^{-2}
19.33	1.362
21.39	1.638
23.69	1.931
26.33	2.314
29.35	2.730
32.61	3.178
36.01	3.631
39.65	4.091
43.51	4.540
47.84	5.013
52.78	5.494

* Not shown on plot

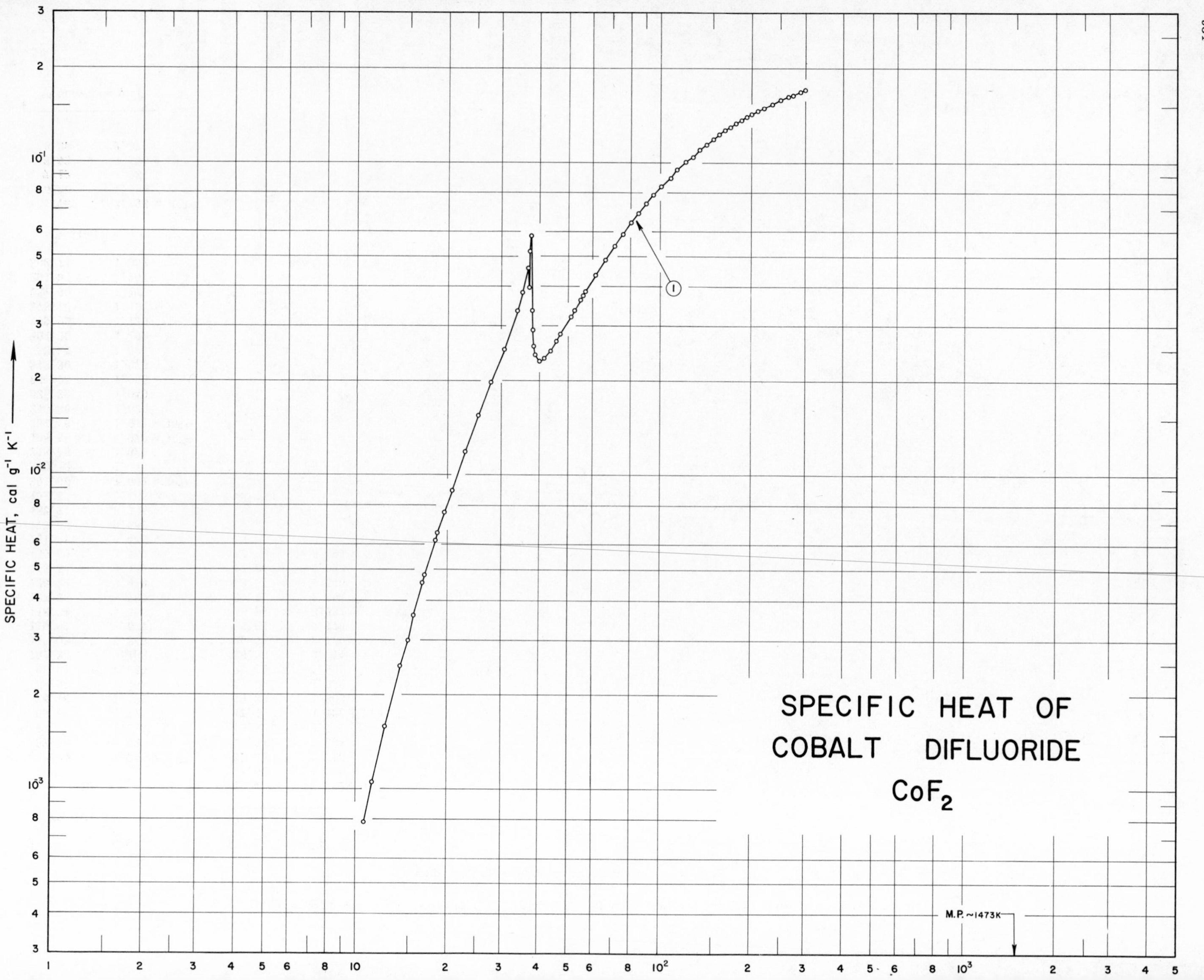
SPECIFIC HEAT OF
COBALT DIFLUORIDE
CoF2
SPECIFIC HEAT, cal g-1 K-1
M.P. ~1473K
1

SPECIFICATION TABLE NO. 292 SPECIFIC HEAT OF COBALT DIFLUORIDE CoF_2

[For Data Reported in Figure and Table No. 292]

Curve No.	Ref. No.	Year	Temp. Range, K	Reported Error, %	Name and Specimen Designation	Composition (weight percent), Specifications and Remarks
1	287	1955	11-302	0.5		60.805 Co (60.80 theo.), 0.04 Fe, 0.005 Si, 0.004 Mg, 0.003 Ni, 0.002 Mn, 0.001 Al, and Cu; prepared from commercial sodium cobaltinitrite; CoF_2 heated at 750 C in the presence of anhydrous HF.

DATA TABLE NO. 292 SPECIFIC HEAT OF COBALT DIFLUORIDE CoF_2

[Temperature, T, K; Specific Heat, Cp, Cal $g^{-1}K^{-1}$]

T	Cp
CURVE 1	
Series I	
51.92	3.277 x 10^{-2}*
56.38	3.769
61.78	4.374
66.41	4.893
Series II	
52.71	3.364 x 10^{-2}
57.34	3.871
62.80	4.489*
67.09	4.970*
71.24	5.418
75.69	5.898
80.50	6.403
85.19	6.888
90.19	7.375
95.61	7.866
101.57	8.394
107.97	8.930
114.61	9.458
128.45	1.048 x 10^{-1}
135.88	1.097
143.27	1.144
150.40	1.186
157.44	1.226
164.66	1.264
171.34	1.296
178.75	1.331
186.40	1.365
194.09	1.396
201.52	1.426
211.02	1.460
218.92	1.486
226.57	1.511*
235.08	1.537
242.70	1.558*
250.90	1.581
258.63	1.602*
265.97	1.623

T	Cp
CURVE 1 (cont.)	
Series III	
10.73	7.840 x 10^{-4}
11.41	1.052 x 10^{-3}
12.58	1.589
14.80	2.992
16.64	4.549
18.65	6.581
20.71	8.954
22.97	1.197 x 10^{-2}
25.36	1.559
27.92	1.991
30.73	2.538
33.99	3.368
37.01	3.986
42.27	2.435*
47.48	2.826
51.32	3.217
55.28	3.646
Series IV	
14.06	2.476 x 10^{-3}
15.54	3.580
16.98	4.828
18.33	6.220
19.61	7.644
Series V	
34.19	3.399 x 10^{-2}*
35.42	3.837
36.19	4.187*
36.79	4.580
37.35	5.215
37.99	3.887*
38.88	2.438
40.10	2.329
41.72	2.371
43.77	2.502
45.92	2.682

T	Cp
CURVE 1 (cont.)	
Series VI	
37.442	5.354 x 10^{-2}*
37.517	5.529*
37.590	5.725*
37.661	5.828
37.730	5.787*
37.802	5.694*
37.869	4.353*
37.969	3.363
38.086	2.930
38.217	2.723*
38.373	2.600
38.534	2.507*
38.697	2.455*
38.863	2.424*
39.029	2.393*
39.257	2.363*
Series VII	
37.694	5.808 x 10^{-2}*
37.768	5.746*
37.850	4.642*
Series VIII	
275.45	1.642 x 10^{-1}
283.33	1.661*
291.79	1.678
299.89	1.702*
122.34	1.004
273.64	1.643*
301.64	1.704

* Not shown on plot

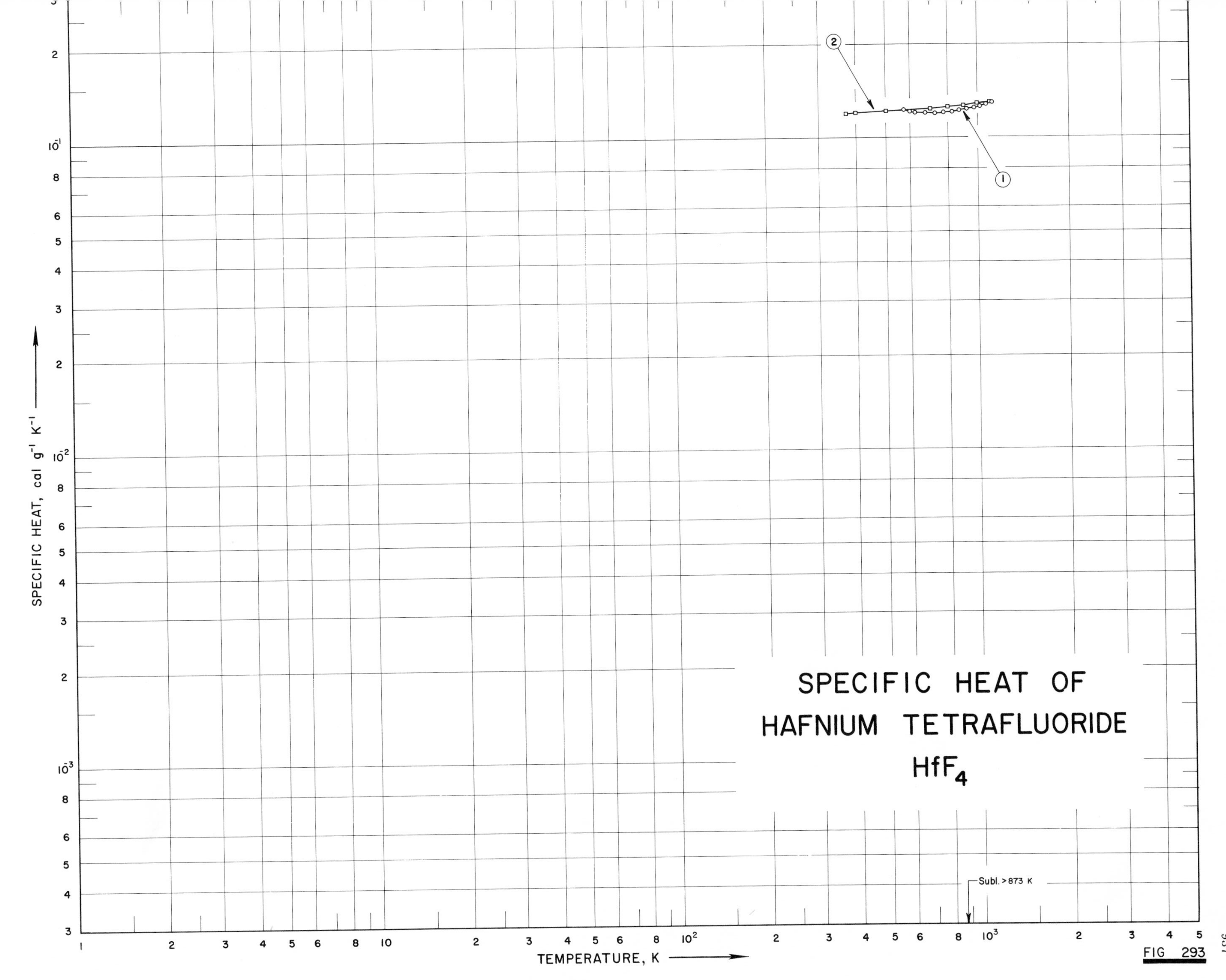
SPECIFIC HEAT OF
HAFNIUM TETRAFLUORIDE
HfF4
Subl. > 873 K
1
2
TEMPERATURE, K
SPECIFIC HEAT, cal g⁻¹ K⁻¹
FIG 293

SPECIFICATION TABLE NO. 293 SPECIFIC HEAT OF HAFNIUM TETRAFLUORIDE HfF_4

[For Data Reported in Figure and Table No. 293]

Curve No.	Ref. No.	Year	Temp. Range, K	Reported Error, %	Name and Specimen Designation	Composition (weight percent), Specifications and Remarks
1	251	1958	575-1125			1.0 Zr.
2	252	1961	370-1105			99.0 HfF_4, 1.0 Zr, and traces of Mg, Al, V, and Fe.

DATA TABLE NO. 293 SPECIFIC HEAT OF HAFNIUM TETRAFLUORIDE HfF_4

[Temperature, T, K; Specific Heat, Cp, Cal $g^{-1}K^{-1}$]

T	Cp
CURVE 1	
575	1.224×10^{-1}
600	1.215
625	1.208
650	1.203*
675	1.201
700	1.199*
725	1.199
750	1.201*
775	1.203
800	1.206*
825	1.210
850	1.215*
875	1.220
900	1.226*
925	1.233
950	1.240*
975	1.247
1000	1.254*
1025	1.262
1050	1.271*
1075	1.279
1100	1.288*
1125	1.297
CURVE 2	
370	1.198×10^{-1}
400	1.202
500	1.215
600	1.228*
700	1.241
800	1.254
900	1.267
1000	1.280
1105	1.294

* Not shown on plot

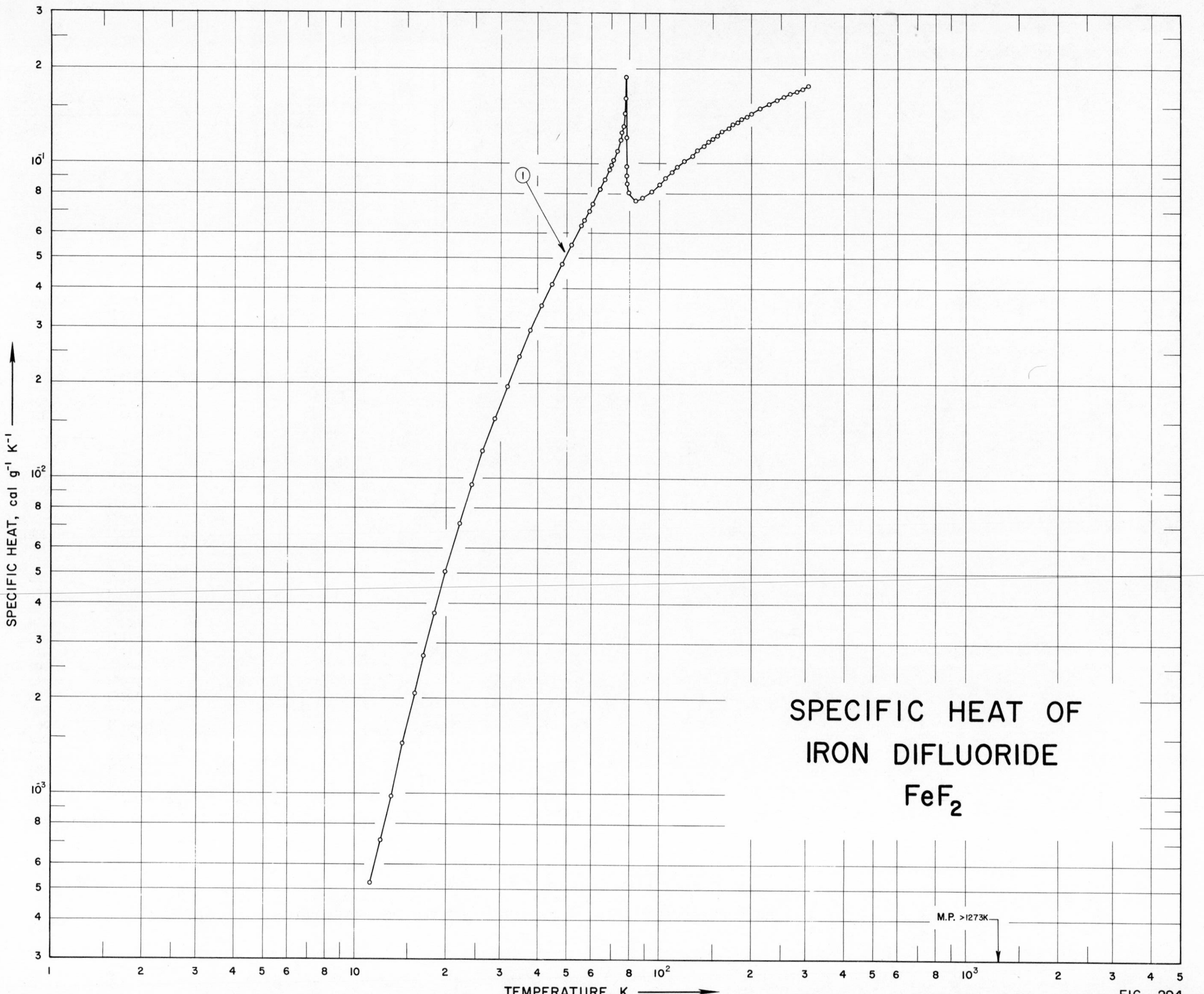
SPECIFIC HEAT OF
IRON DIFLUORIDE
FeF_2
1
M.P. >1273K
SPECIFIC HEAT, cal g^{-1} K^{-1}
TEMPERATURE, K

FIG 294

SPECIFICATION TABLE NO. 294 SPECIFIC HEAT OF IRON DIFLUORIDE FeF_2

[For Data Reported in Figure and Table No. 294]

Curve No.	Ref. No.	Year	Temp. Range, K	Reported Error, %	Name and Specimen Designation	Composition (weight percent), Specifications and Remarks
1	287	1955	11-307	0.5		59.47 Fe (59.51 theo.), 0.005 V, 0.004 Al, and 0.001 each Mn, Mg, and Ni; prepared from $FeCl_2$; $FeCl_2$ heated in HF atmosphere at 1050 C.

DATA TABLE NO. 294 SPECIFIC HEAT OF IRON DIFLUORIDE FeF_2

[Temperature, T, K; Specific Heat, Cp, Cal $g^{-1}K^{-1}$]

T	Cp
CURVE 1	
Series I	
51.83	5.485 x 10^{-2}*
54.60	6.045*
59.34	7.061
64.51	8.285
70.15	9.854
75.62	1.255 x 10^{-1}
81.71	8.090 x 10^{-2}*
88.44	7.763
94.63	8.124
100.46	8.547
106.20	8.975
111.83	9.393
116.90	9.770
122.40	1.016 x 10^{-1}
128.14	1.057
134.15	1.099
140.19	1.138
146.29	1.175
Series II	
11.33	5.219 x 10^{-4}
12.28	7.137
13.25	9.800
14.47	1.449 x 10^{-3}
15.80	2.077
16.95	2.738
18.34	3.739
19.94	5.060
22.17	7.190
27.29	9.512
26.44	1.220 x 10^{-2}
28.88	1.545
31.73	1.956
34.77	2.427
37.96	2.948
41.35	3.526
44.74	4.134
48.22	4.784
51.91	5.495
55.95	6.328
CURVE 1 (cont.)	
Series III	
57.15	6.579 x 10^{-2}
60.89	7.410
64.03	8.159*
66.76	8.860
69.17	9.531
71.34	1.022 x 10^{-1}
73.30	1.096
75.08	1.185
76.67	1.308*
78.12	1.375*
79.89	8.058 x 10^{-2}
82.01	7.653*
84.08	7.611
86.29	7.660*
88.78	7.781*
Series IV	
76.392	1.277 x 10^{-1}*
Series V	
76.566	1.296 x 10^{-1}*
77.143	1.362*
77.599	1.442
77.897	1.522*
78.062	1.607
78.151	1.668*
78.235	1.754*
78.315	1.876
78.411	1.209
78.536	9.718 x 10^{-2}
78.677	9.132
78.823	8.801*
78.973	8.578
79.252	8.305*
79.754	8.044*
80.362	7.860*
Series VI	
78.202	1.727 x 10^{-1}*
78.279	1.826*
78.358	1.593*
CURVE 1 (cont.)	
Series VII	
149.26	1.193 x 10^{-1}
155.58	1.231
161.81	1.264
168.27	1.299
174.68	1.330
180.99	1.359
187.47	1.389
194.19	1.418
200.91	1.445
207.79	1.472*
214.50	1.496
221.48	1.522*
228.50	1.546
235.52	1.567*
242.49	1.590
249.11	1.611*
255.27	1.626
261.88	1.643*
268.58	1.661
275.22	1.681*
282.22	1.696
289.24	1.712*
295.91	1.727
307.30	1.760

* Not shown on plot

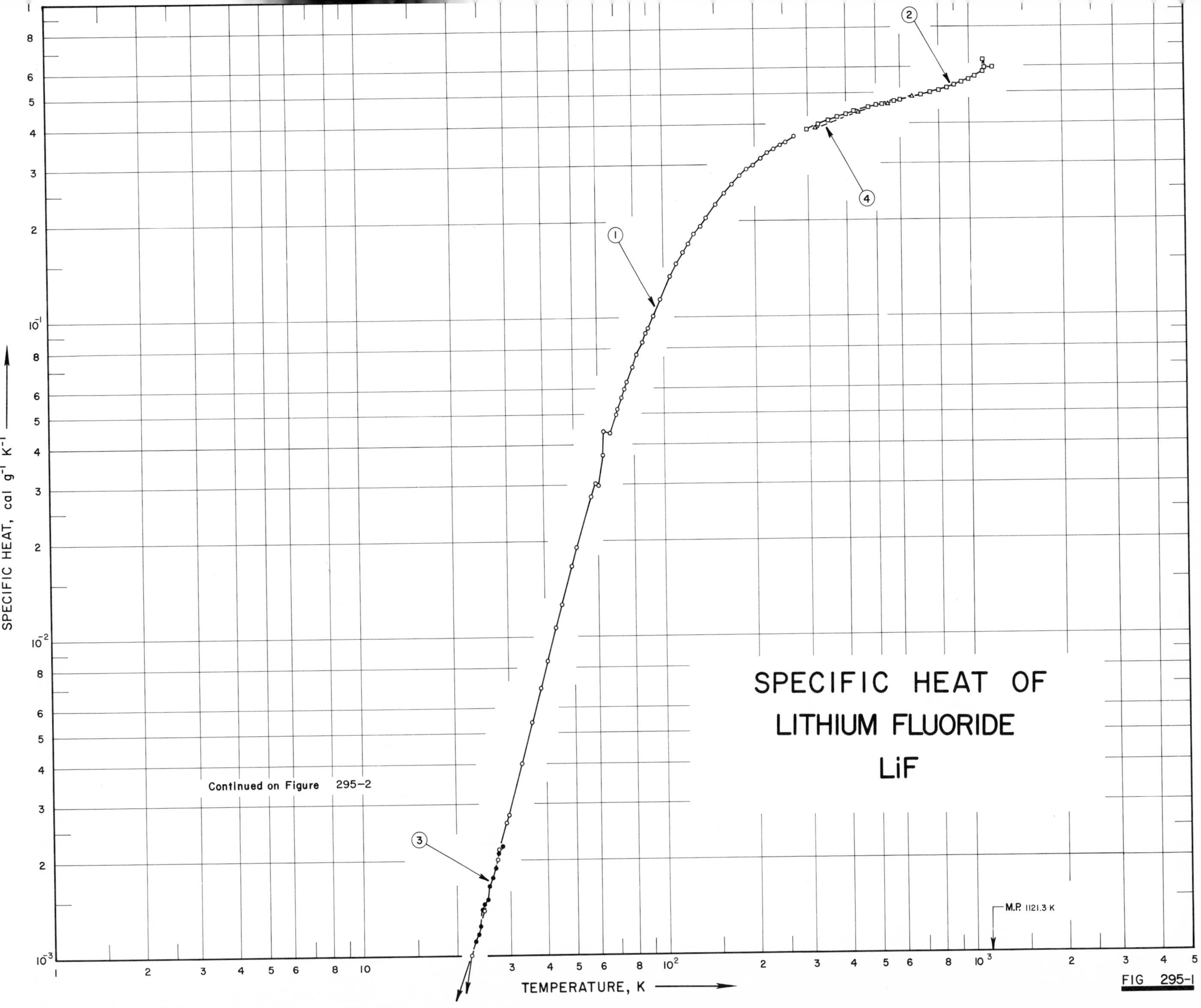

SPECIFIC HEAT OF
LITHIUM FLUORIDE
LiF
Continued on Figure 295-2
M.P. 1121.3 K
SPECIFIC HEAT, cal g⁻¹ K⁻¹
TEMPERATURE, K
FIG 295-1

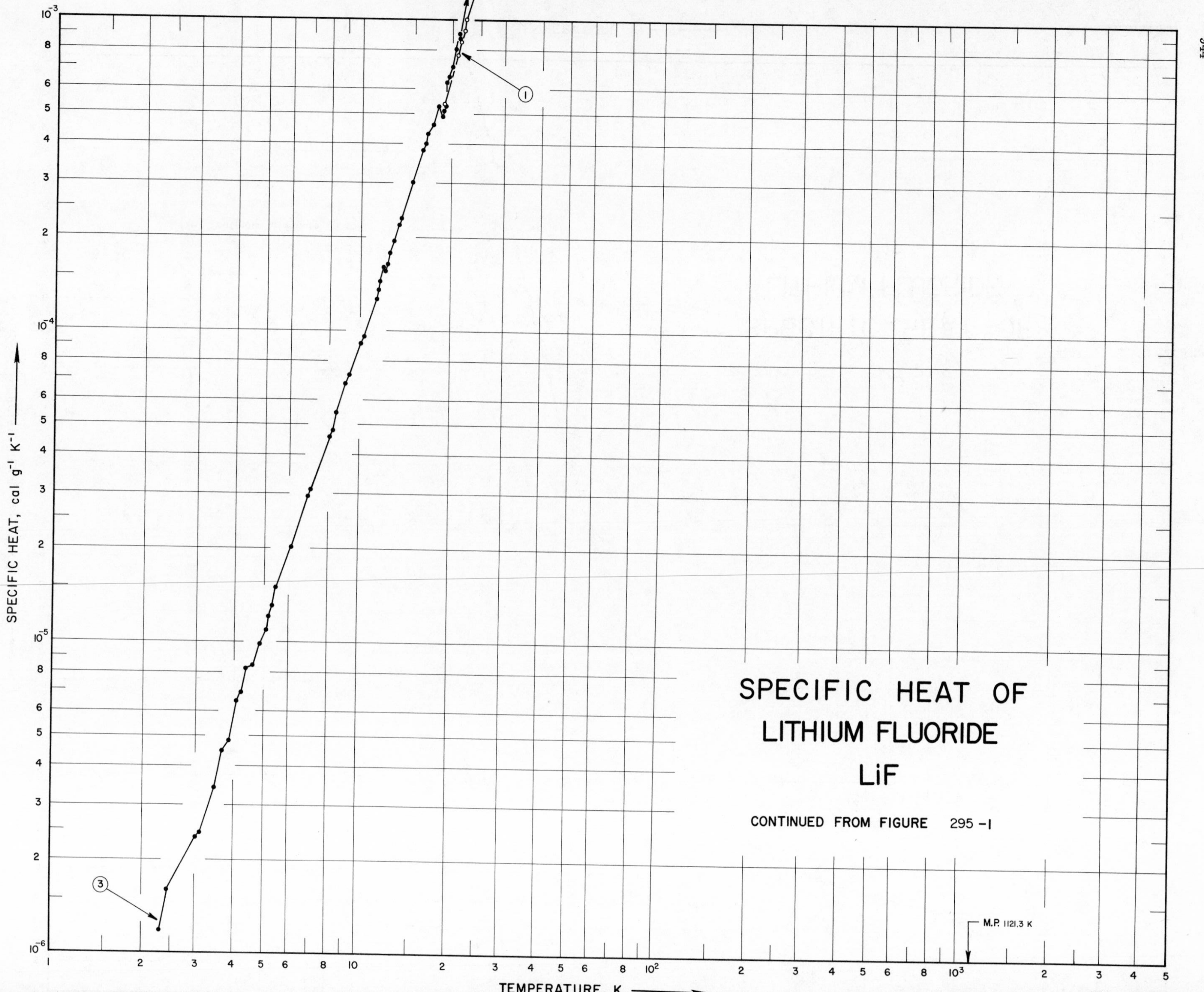

SPECIFIC HEAT OF
LITHIUM FLUORIDE
LiF
CONTINUED FROM FIGURE 295-I
M.P. 1121.3 K
TEMPERATURE, K
SPECIFIC HEAT, cal g⁻¹ K⁻¹
1
3

SPECIFICATION TABLE NO. 295 SPECIFIC HEAT OF LITHIUM FLUORIDE LiF

[For Data Reported in Figure and Table No. 295]

Curve No.	Ref. No.	Year	Temp. Range, K	Reported Error, %	Name and Specimen Designation	Composition (weight percent), Specifications and Remarks
1	243	1949	19-272			
2	288	1953	298-1200	0.5		Before test: a few thousandths of 1% each Ca, Mg, Na, and Ni; traces of Al, Cr, Fe, and Si, after test: addition of a few hundredths of 1% Cr and a few thousandths of 1% Al, Mn, and Si; single crystals; sample supplied by the Harshaw Chemical Co; slowly crystallized.
3	289	1955	2-28			Artificial crystal (197 gm).
4	328	1956	318-658	± 0.5		

DATA TABLE NO. 295 SPECIFIC HEAT OF LITHIUM FLUORIDE LiF

[Temperature, T, K; Specific Heat, Cp, Cal $g^{-1}K^{-1}$]

T	Cp
CURVE 1	
18.8	5.4 x 10^{-4}
19.0	5.4*
20.8	7.710
21.4	8.481
22.0	9.252
22.1	1.002 x 10^{-3}
24.6	1.388
24.8	1.388
27.3	2.005
27.6	2.159
29.3	2.621
29.9	2.776
33.1	4.009
35.9	5.397
38.5	6.939
40.7	8.404
43.4	1.072 x 10^{-2}
45.5	1.264
49.2	1.665
51.1	1.912
57.2	2.753
59.3	3.030
60.4	2.984
62.7	3.716
63.3	4.410
66.4	4.364
66.5	4.410*
69.6	4.988
70.4	5.181
72.7	5.628
74.3	5.991
75.6	6.299
79.0	6.985
81.9	7.656
85.5	8.389
87.8	8.905
89.1	9.229
93.2	1.017 x 10^{-1}
98.3	1.140
106.9	1.348
112.1	1.471
117.2	1.590
122.4	1.709
127.6	1.823
132.9	1.937
CURVE 1 (cont.)	
138.4	2.050 x 10^{-1}
143.8	2.154*
149.1	2.256
153.4	2.352*
159.5	2.440
164.5	2.525*
169.6	2.607
174.7	2.658*
179.7	2.760
184.6	2.828*
188.9	2.904
193.3	2.941*
198.6	2.995
200.5	3.029*
202.7	3.091*
210.9	3.145
216.3	3.201*
221.7	3.261
226.9	3.291*
232.0	3.359
237.4	3.402*
242.9	3.452
248.5	3.498*
254.0	3.536
259.9	3.581*
265.9	3.638*
271.7	3.676
CURVE 2	
298.16	3.861 x 10^{-1}
300	3.871*
325	4.000
350	4.112
375	4.209
400	4.295
425	4.372
450	4.441*
475	4.503
500	4.561
525	4.614
550	4.664*
575	4.711
600	4.756
650	4.841*
CURVE 2 (cont.)	
700	4.923 x 10^{-1}
750	5.005
800	5.089
850	5.177
900	5.273
950	5.376
1000	5.489
1050	5.614
1100	5.752*
(s) 1121.3	5.815
(l) 1121.3	6.365
1125	5.979
1150	5.979*
1175	5.979*
1200	5.979
CURVE 3	
Series I	
2.44	1.596 x 10^{-6}
4.84	9.872
5.14	1.095 x 10^{-5}
6.14	2.013
7.03	3.085
8.08	4.550
8.28	4.782
8.49	5.445
9.03	6.764
9.30	7.219
10.10	9.101
10.32	9.563
11.31	1.273 x 10^{-4}
11.53	1.365
11.66	1.458
11.91	1.604
12.14	1.566
12.38	1.643
12.59	1.758*
14.83	3.008
15.01	3.008*
17.32	4.620
18.05	5.283
18.22	5.622
CURVE 3 (cont.)	
Series II	
2.31	1.188 x 10^{-6}
3.10	2.437
3.67	4.489
4.09	6.463
4.37	8.252
4.59	8.407
5.20	1.203 x 10^{-5}
5.23	1.234*
5.34	1.311
5.46	1.481
6.90	2.938
12.52	1.782 x 10^{-4}
12.90	1.959
13.34	2.190
13.73	2.314
16.02	3.802
16.44	3.995
16.73	4.327
18.62	5.923
19.05	6.286
21.18	9.024
24.36	1.396 x 10^{-3}
24.65	1.450
Series III	
3.02	2.360 x 10^{-6}
3.46	3.424
3.86	4.843
4.23	6.972
18.81	5.746 x 10^{-4}*
19.12	6.309
19.51	6.579
20.05	7.057
20.67	8.021
21.24	8.715
23.08	1.118 x 10^{-3}
23.50	1.172
23.89	1.249
24.24	1.404*
24.59	1.404*
24.93	1.458*
25.25	1.519
CURVE 3 (cont.)	
25.69	1.650 x 10^{-3}
26.28	1.766
26.95	1.897
27.69	2.113
28.33	2.213
CURVE 4	
317.69	3.93 x 10^{-1}
319.39	3.96*
379.70	4.18*
439.56	4.36
441.38	4.37*
549.58	4.64
551.90	4.65*
656.01	4.82*
658.02	4.86

* Not shown on plot

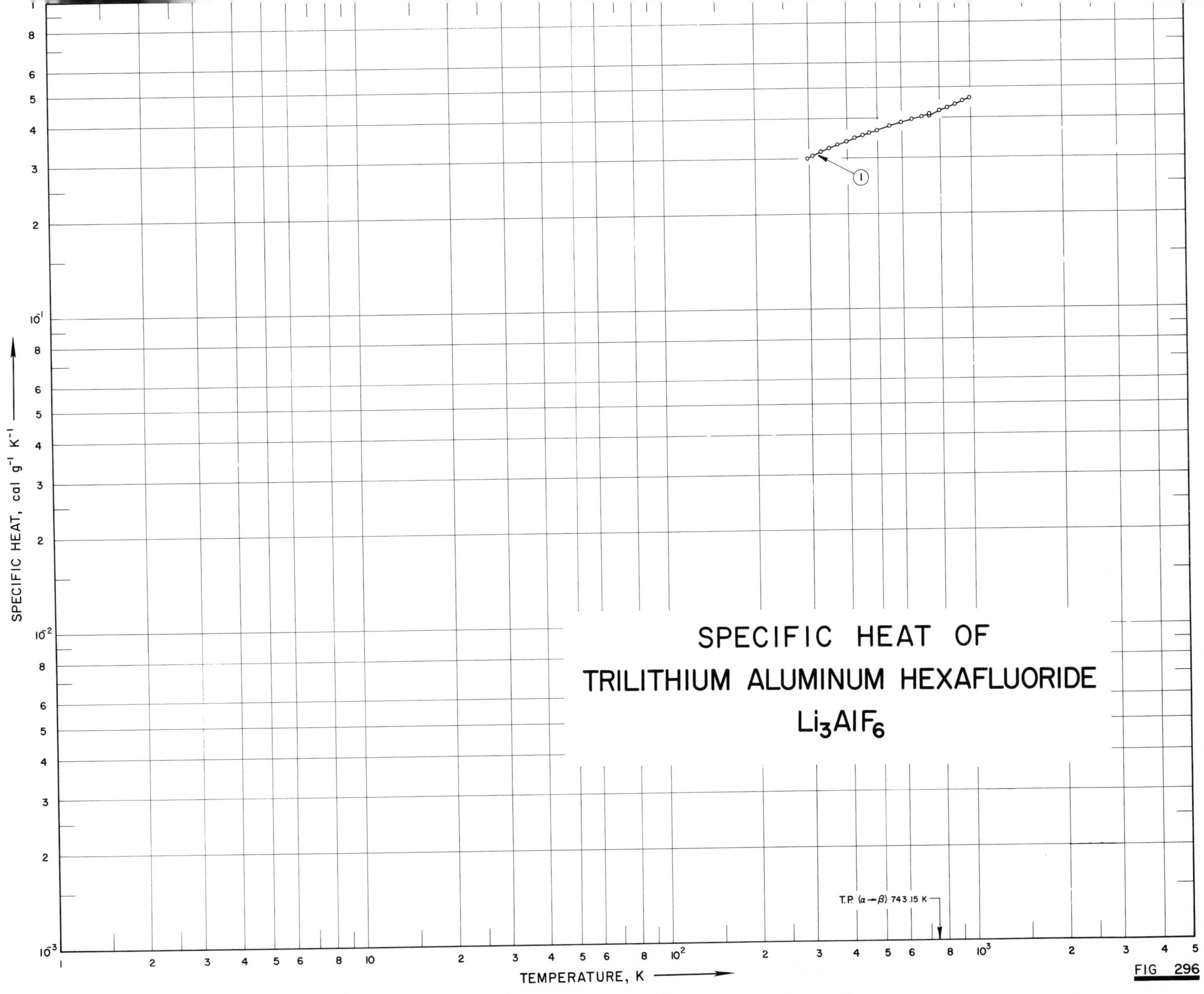
SPECIFIC HEAT OF
TRILITHIUM ALUMINUM HEXAFLUORIDE
Li_3AlF_6
1
T.P. (α→β) 743.15 K
TEMPERATURE, K
SPECIFIC HEAT, cal g⁻¹ K⁻¹
FIG 296

SPECIFICATION TABLE NO. 296 SPECIFIC HEAT OF TRILITHIUM ALUMINUM HEXAFLUORIDE Li_3AlF_6

[For Data Reported in Figure and Table No. 296]

Curve No.	Ref. No.	Year	Temp. Range, K	Reported Error, %	Name and Specimen Designation	Composition (weight percent), Specifications and Remarks
1	170	1964	298-1000			70.43 F, 16.74 Al and 12.92 Li.

DATA TABLE NO. 296 SPECIFIC HEAT OF TRILITHIUM ALUMINUM HEXAFLUORIDE Li_3AlF_6

[Temperature, T, K; Specific Heat, Cp, Cal $g^{-1}K^{-1}$]

T	Cp
CURVE 1	
298.15	2.994 x 10^{-1}
300	3.003*
310	3.051
320	3.097*
330	3.140
340	3.182*
350	3.221
360	3.259*
370	3.295*
373.15	3.306
380	3.330*
390	3.363
400	3.395
425	3.471
450	3.541
475	3.605
500	3.666
550	3.776
600	3.875
650	3.965
700	4.049
α 743.15	4.117
β 743.15	4.099*
750	4.112*
800	4.212
850	4.314
900	4.417
950	4.523
1000	4.630

* Not shown on plot

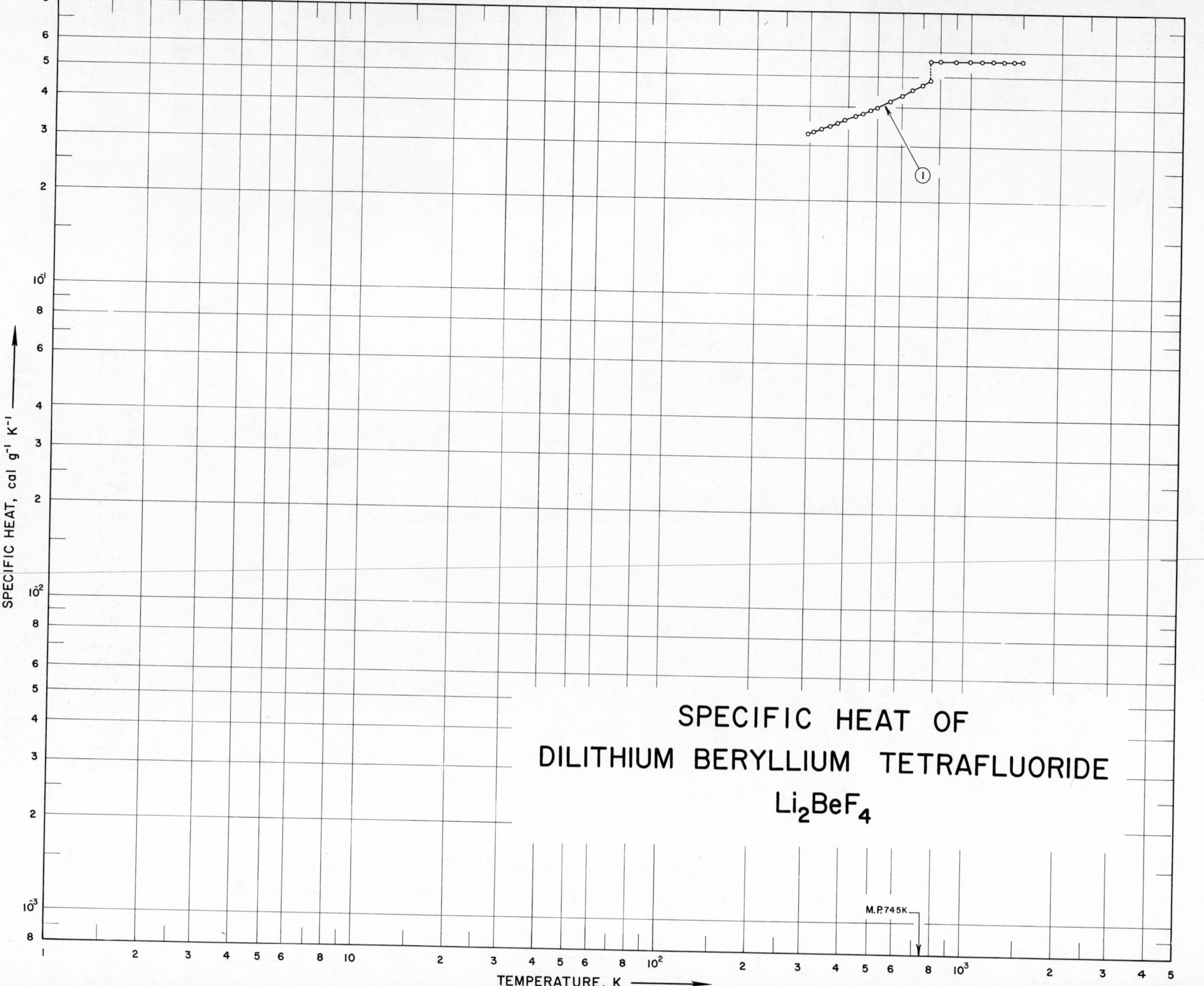

SPECIFIC HEAT OF
DILITHIUM BERYLLIUM TETRAFLUORIDE
Li_2BeF_4
M.P. 745K
1
TEMPERATURE, K
SPECIFIC HEAT, cal g^{-1} K^{-1}

SPECIFICATION TABLE NO. 297 SPECIFIC HEAT OF DILITHIUM BERYLLIUM TETRAFLUORIDE Li_2BeF_4

[For Data Reported in Figure and Table No. 297]

Curve No.	Ref. No.	Year	Temp. Range, K	Reported Error, %	Name and Specimen Designation	Composition (weight percent), Specifications and Remarks
1	170	1964	298-1500			76.86 F, 14.03 Li, and 9.11 Be, 0.001-0.01 Ag, Al, Cu, Fe, K, Mn, Na, Si, Ti, and Zr, 0.0001-0.001 Ba, Ni, and Pb.

DATA TABLE NO. 297 SPECIFIC HEAT OF DILITHIUM BERYLLIUM TETRAFLUORIDE Li_2BeF_4

[Temperature, T, K; Specific Heat, Cp, Cal $g^{-1}K^{-1}$]

T	Cp
CURVE 1	
298.15	3.269 x 10^{-1}
300	3.276*
310	3.312
320	3.348*
330	3.384
340	3.420*
350	3.456
360	3.492*
370	3.529
373.15	3.540*
380	3.565*
390	3.601
400	3.637*
425	3.727
450	3.818
475	3.908
500	3.998
550	4.199
600	4.359
650	4.540
700	4.721
(s)745	4.883
(l)745	5.613
750	5.613*
800	5.613
850	5.613*
900	5.613
950	5.613*
1000	5.613
1050	5.613*
1100	5.613
1150	5.613*
1200	5.613
1250	5.613*
1300	5.613
1350	5.613*
1400	5.613
1450	5.613*
1500	5.613

* Not shown on plot

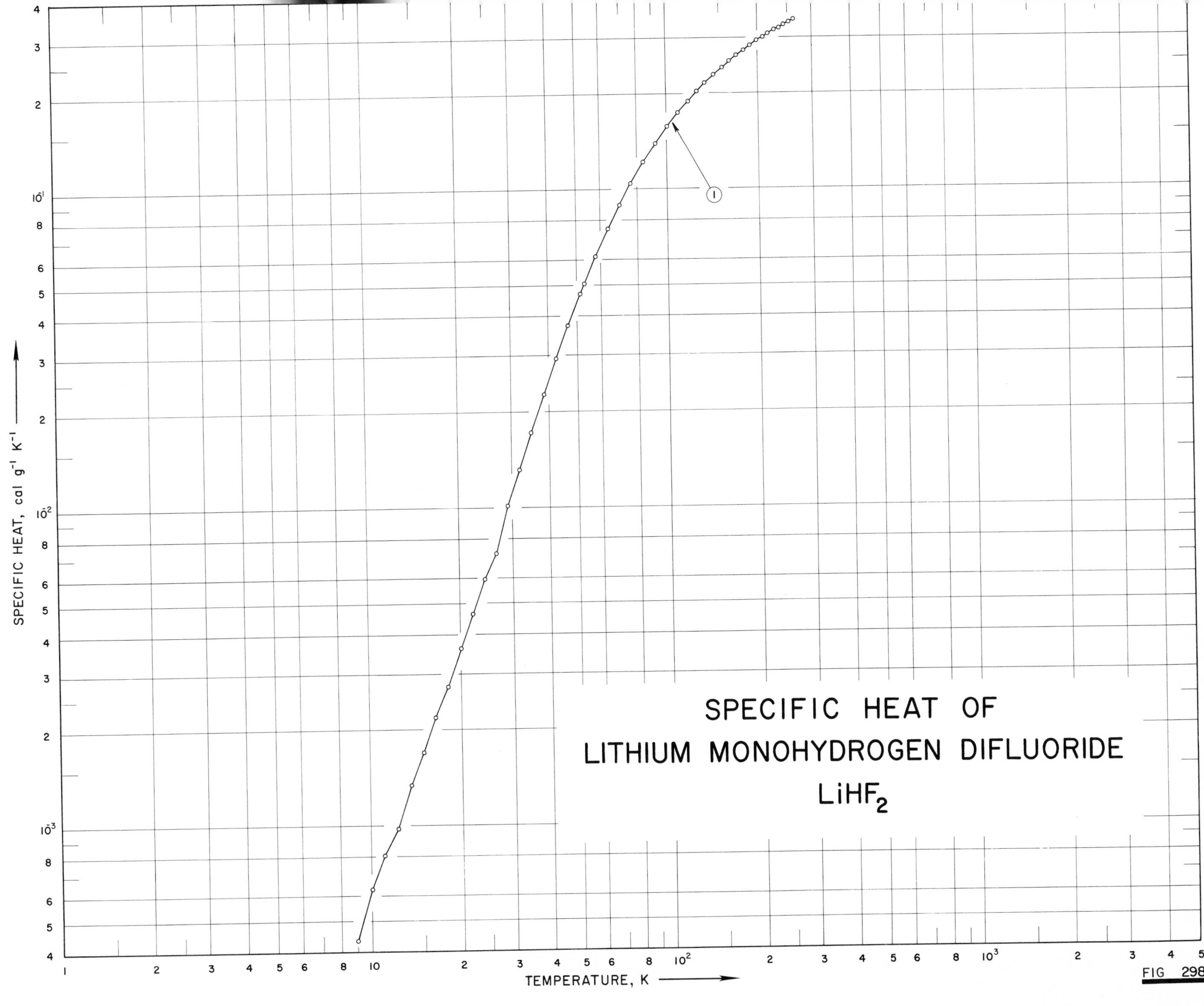
SPECIFIC HEAT OF
LITHIUM MONOHYDROGEN DIFLUORIDE
LiHF2
SPECIFIC HEAT, cal g-1 K-1
TEMPERATURE, K
FIG 298

SPECIFICATION TABLE NO. 298 SPECIFIC HEAT OF LITHIUM MONOHYDROGEN DIFLUORIDE $LiHF_2$

[For Data Reported in Figure and Table No. 298]

Curve No.	Ref. No.	Year	Temp. Range, K	Reported Error, %	Name and Specimen Designation	Composition (weight percent), Specifications and Remarks
1	290	1961	7-301			99.48 $LiHF_2$ and 0.52 LiF; prepared by addition of metal carbonate to boiling reagent aqueous 48% HF solution; refrigerated at -10 C.

DATA TABLE NO. 298 SPECIFIC HEAT OF LITHIUM MONOHYDROGEN DIFLUORIDE $LiHF_2$

[Temperature, T, K; Specific Heat, Cp, Cal $g^{-1}K^{-1}$]

T	Cp
CURVE 1	
Series I	
109.21	1.730×10^{-1}*
118.45	1.898
126.20	2.027
134.89	2.163
143.98	2.294
152.89	2.416
161.77	2.527
170.86	2.631
180.33	2.734
189.93	2.831
199.35	2.940
208.59	3.010
217.69	3.090
226.72	3.171
235.72	3.230
244.00	3.293
253.55	3.356
262.45	3.419
271.53	3.480
281.10	3.543
291.07	3.608
301.36	3.674
Series II	
6.57	1.741×10^{-4}
7.84	2.829
9.01	4.353
10.11	6.311
11.19	8.053
12.39	9.794
13.71	1.349×10^{-3}
15.10	1.719
16.53	2.198
18.18	2.742
20.07	3.613
22.01	4.636
24.13	5.963
26.54	7.182
29.08	1.012×10^{-2}
31.86	1.321
34.99	1.722
38.60	2.272
42.58	2.949
CURVE 1 (cont.)	
46.81	3.754×10^{-2}
51.47	4.716
53.04	5.056
57.98	6.150
63.82	7.513
69.90	8.925
76.55	1.045×10^{-1}
84.12	1.219
92.36	1.399
101.02	1.574
109.74	1.741

* Not shown on plot

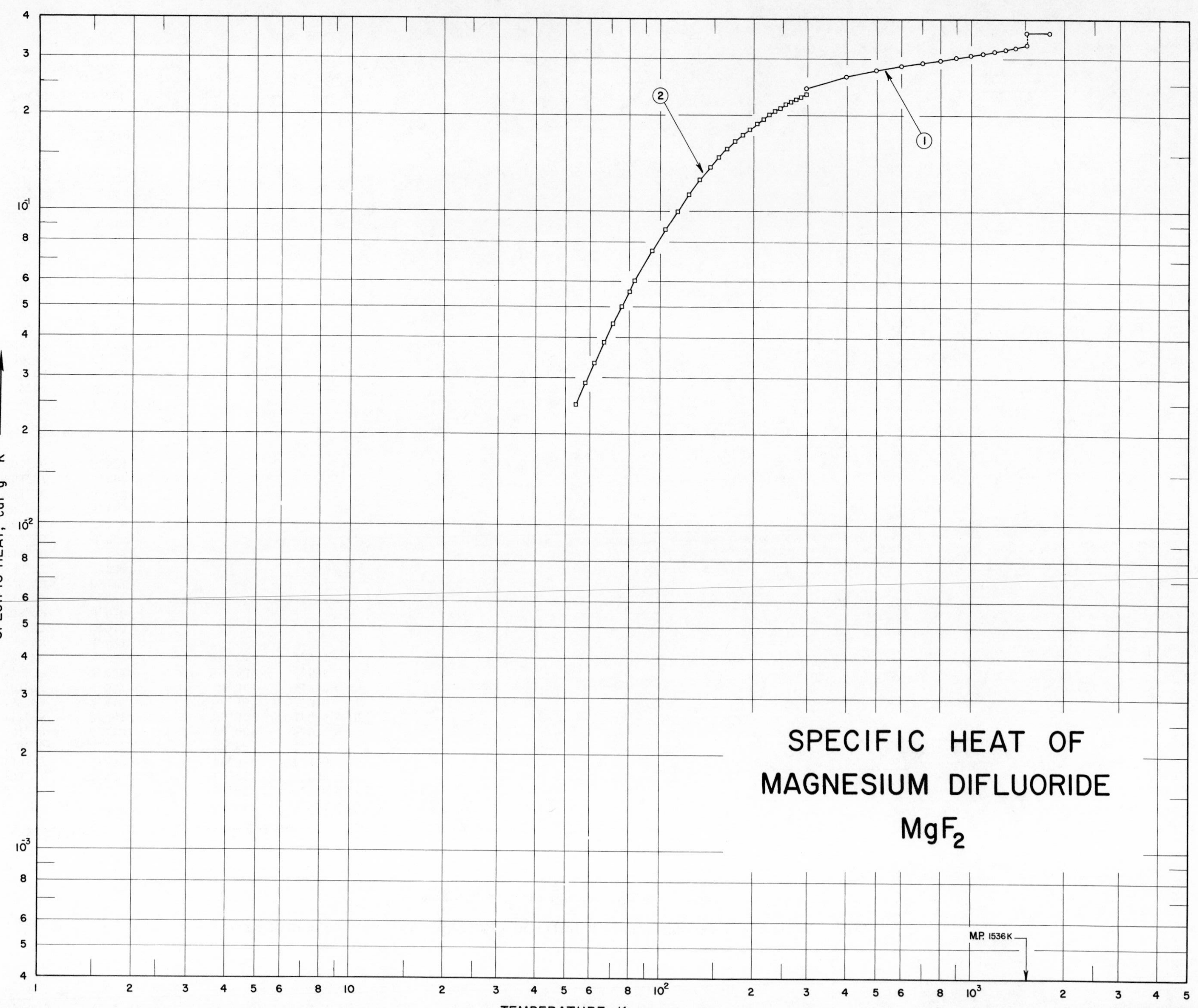
SPECIFIC HEAT OF
MAGNESIUM DIFLUORIDE
MgF_2
M.P. 1536 K
SPECIFIC HEAT, cal g^{-1} K^{-1}
1
2

SPECIFICATION TABLE NO. 299 SPECIFIC HEAT OF MAGNESIUM DIFLUORIDE MgF_2

[For Data Reported in Figure and Table No. 299]

Curve No.	Ref. No.	Year	Temp. Range, K	Reported Error, %	Name and Specimen Designation	Composition (weight percent), Specifications and Remarks
1	282	1945	298-800			38.9 Mg; prepared from Baker C.P. MgO; treated with hot 48% HF for 16 hrs and then drying at 400 C.
2	283	1949	54-298	0.3		38.97 Mg (39.02 theo.); purified from Baker, analyzed reagent of 0.5 SO_4 and < 0.3 Ca; treated with hot 48% HF for 16 hrs and dried at 450 C.

DATA TABLE NO. 299 SPECIFIC HEAT OF MAGNESIUM DIFLUORIDE MgF_2

[Temperature, T, K; Specific Heat, Cp, Cal $g^{-1}K^{-1}$]

T	Cp	T	Cp
CURVE 1		**CURVE 2 (cont.)**	
298.5	2.440×10^{-1}	256.20	2.169×10^{-1}
300	2.446*	266.30	2.208
400	2.658	276.20	2.248
500	2.777	286.50	2.286
600	2.861	296.50	2.317*
700	2.928	298.16	2.325
800	2.985		
900	3.037		
1000	3.086		
1100	3.132		
1200	3.177		
1300	3.221		
1400	3.265		
1500	3.307*		
(s) 1536	3.323		
(l) 1536	3.622		
1600	3.622*		
1700	3.622*		
1800	3.622		
CURVE 2			
54.22	2.491×10^{-2}		
58.05	2.900		
62.12	3.366		
66.64	3.897		
71.12	4.451		
75.72	5.036		
80.20	5.611		
83.62	6.085		
94.70	7.509		
104.30	8.762		
114.54	1.008×10^{-1}		
124.76	1.132		
135.83	1.265		
146.10	1.375		
155.72	1.475		
166.02	1.572		
176.00	1.660		
186.00	1.741		
196.00	1.815		
206.30	1.886		
216.70	1.954		
226.40	2.012		
236.20	2.063		
246.10	2.113		

* Not shown on plot

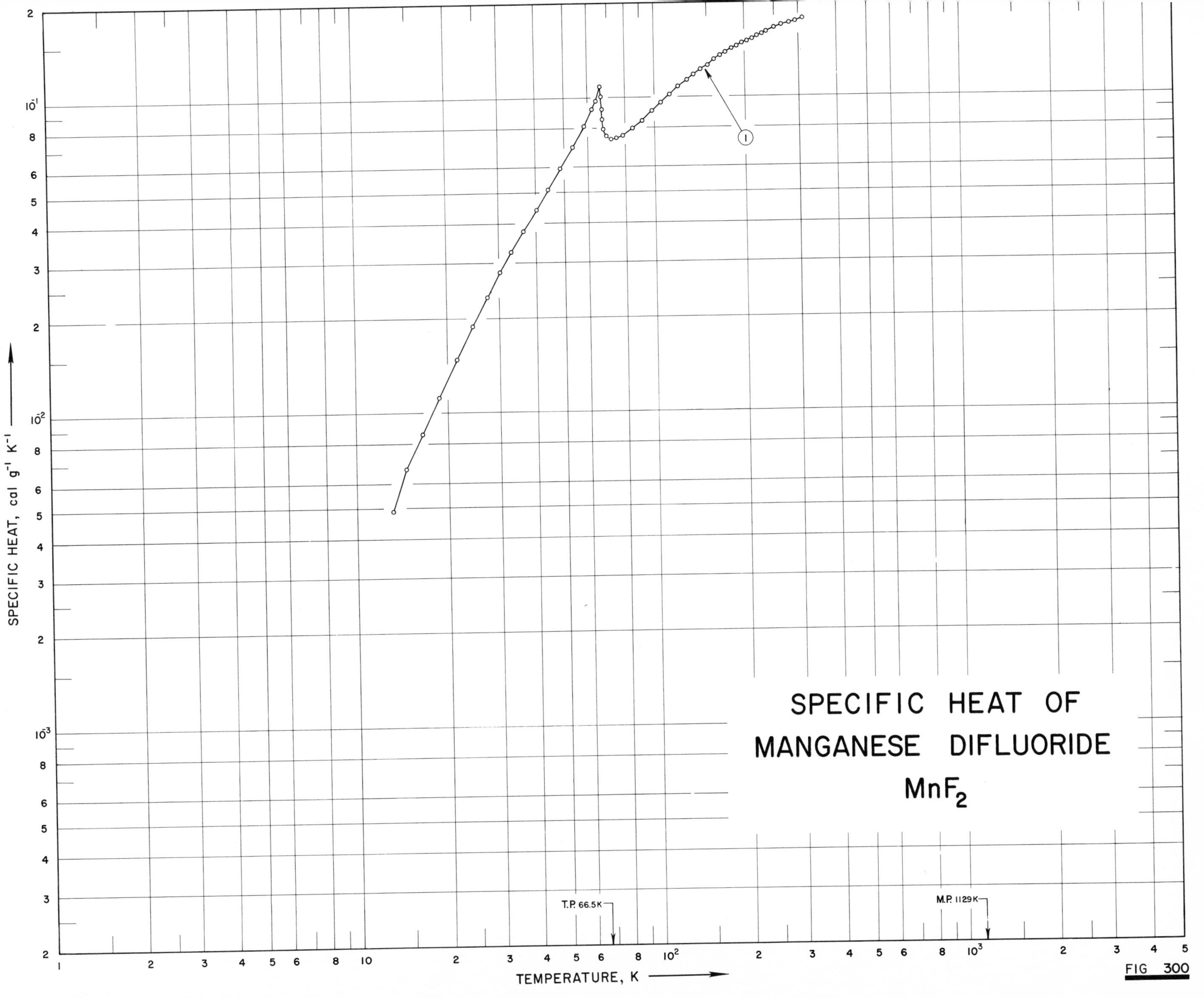
SPECIFIC HEAT OF
MANGANESE DIFLUORIDE
MnF2
1
T.P. 66.5K
M.P. 1129K
SPECIFIC HEAT, cal g-1 K-1
TEMPERATURE, K
FIG 300

SPECIFICATION TABLE NO. 300 SPECIFIC HEAT OF MANGANESE DIFLUORIDE MnF_2

[For Data Reported in Figure and Table No. 300]

Curve No.	Ref. No.	Year	Temp. Range, K	Reported Error, %	Name and Specimen Designation	Composition (weight percent), Specifications and Remarks
1	291	1942	13-310	0.2-5		58.9 Mn (59.11 theo.); prepared by precipitating $MnCO_3$ from solution of analytical reagent manganous sulfate by means of a solution of sodium carbonate containing sufficient sodium bicarbonate; heated 5 hrs at 250 C to remove volatile impurities.

DATA TABLE NO. 300 SPECIFIC HEAT OF MANGANESE DIFLUORIDE MnF_2

[Temperature, T, K; Specific Heat, Cp, Cal $g^{-1}K^{-1}$]

T	Cp
CURVE 1	
Series I	
264.62	1.680×10^{-1}*
271.67	1.698*
278.61	1.708*
Series II	
61.94	9.060×10^{-2}*
65.53	1.004×10^{-1}*
70.04	7.590*
74.81	7.427*
79.40	7.594
Series III	
13.18	4.928×10^{-3}
14.68	6.693
16.60	8.630
18.82	1.130×10^{-2}
21.56	1.489
24.49	1.896
27.49	2.348
30.36	2.803
33.16	3.254
36.44	3.781
40.32	4.435
44.33	5.144
48.68	5.973
53.47	6.951
58.51	8.109
62.29	9.168
64.08	9.766
65.23	1.028×10^{-1}*
66.05	1.064*
66.55	1.075*
66.99	1.004
67.47	8.504×10^{-2}
67.98	7.947
68.59	7.741*
69.48	7.571
70.72	7.453*
72.67	7.398
75.77	7.452
CURVE 1 (cont.)	
Series IV	
66.11	1.069×10^{-1}*
66.50	1.076
66.83	1.055*
67.25	9.181×10^{-2}
67.74	8.092*
68.27	7.865*
68.79	7.709*
Series V	
80.12	7.630×10^{-2}*
85.60	7.991
91.85	8.463
98.98	9.026
106.35	9.619
113.78	1.019×10^{-1}
121.47	1.077
129.06	1.130
136.29	1.177
143.72	1.224
151.23	1.269
158.77	1.313
166.35	1.352
173.67	1.387
181.11	1.419
188.77	1.451
196.56	1.483
204.57	1.510
212.28	1.538
220.14	1.564
Series VI	
221.52	1.570×10^{-1}*
228.09	1.587
235.84	1.613
243.44	1.629*
250.96	1.652
258.26	1.673*
265.77	1.691
273.38	1.700*
280.98	1.715
CURVE 1 (cont.)	
288.46	1.729×10^{-1}*
295.60	1.743
303.11	1.756*
310.22	1.770
Series VII	
78.02	7.531×10^{-2}*

* Not shown on plot

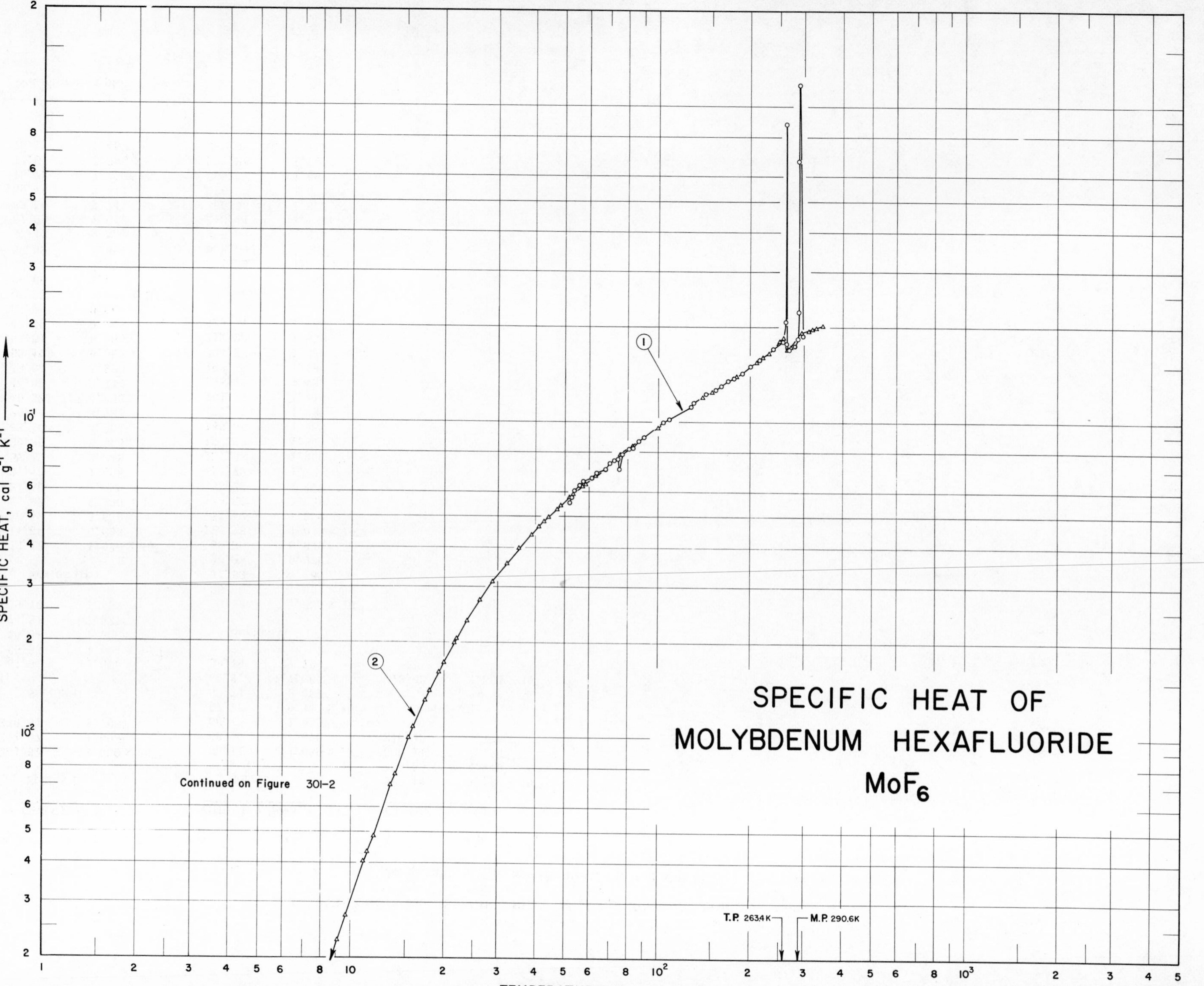

SPECIFIC HEAT OF
MOLYBDENUM HEXAFLUORIDE
MoF_6
T.P. 263.4K
M.P. 290.6K
1
2
Continued on Figure 301-2
SPECIFIC HEAT, cal g^{-1} K^{-1}
TEMPERATURE, K

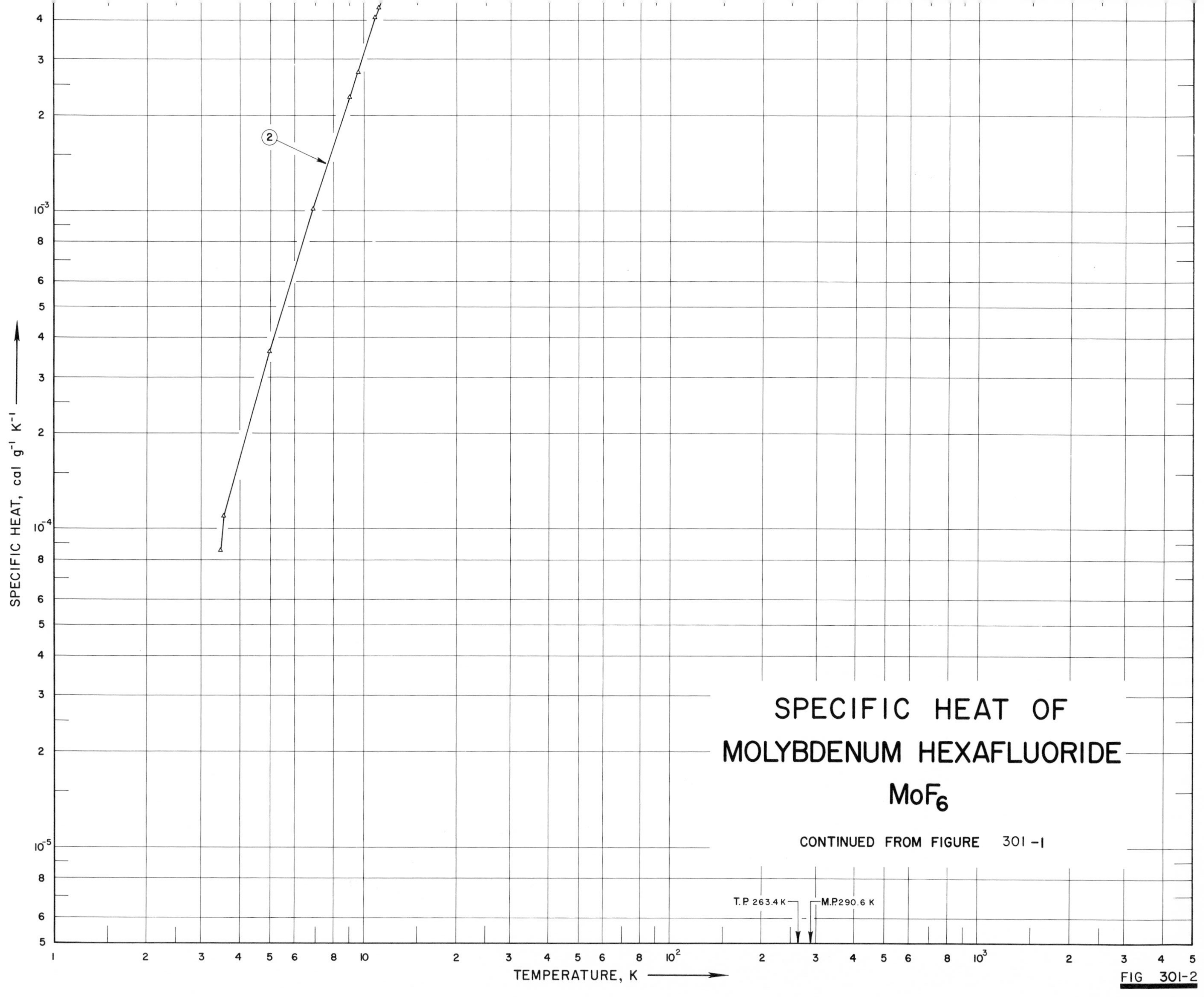
SPECIFIC HEAT OF
MOLYBDENUM HEXAFLUORIDE
MoF_6
CONTINUED FROM FIGURE 301-1
T.P. 263.4 K
M.P. 290.6 K
SPECIFIC HEAT, cal g^{-1} K^{-1}
TEMPERATURE, K
2
FIG 301-2

SPECIFICATION TABLE NO. 301 SPECIFIC HEAT OF MOLYBDENUM HEXAFLUORIDE MoF_6

[For Data Reported in Figure and Table No. 301]

Curve No.	Ref. No.	Year	Temp. Range, K	Reported Error, %	Name and Specimen Designation	Composition (weight percent), Specifications and Remarks
1	292	1957	53-298			Commercial molybdenum hexafluoride was purified by multiple trap-to-trap distillation.
2	329	1966	4-347	0.1-5.0		MoF_6 prepared by direct reaction of fluorine at 400 C and molybdenum.

DATA TABLE NO. 301 SPECIFIC HEAT OF MOLYBDENUM HEXAFLUORIDE MoF_6

[Temperature, T,K; Specific Heat, Cp, Cal $g^{-1}K^{-1}$]

T	Cp
CURVE 1	
52.10	5.578 x 10^{-2}
53.09	5.892
54.10	6.088
55.20	6.240*
56.21	6.345
57.82	6.502
59.77	6.516*
61.74	6.678
63.77	6.936
75.87	7.098
68.06	7.102
70.26	7.407
72.50	7.579
74.77	7.669
77.05	7.912
78.91	7.950*
78.97	8.055*
79.07	8.093*
80.99	8.203
81.12	8.141*
81.20	8.336*
83.92	8.284
87.37	8.722
90.48	8.965
92.80	8.898*
104.90	9.994
109.66	1.024 x 10^{-1}
128.21	1.125
131.71	1.160
134.72	1.168*
144.27	1.232
155.12	1.269
157.55	1.305*
158.56	1.300*
160.21	1.298*
161.46	1.306
170.57	1.356
172.81	1.375*
175.14	1.370*
177.98	1.385
180.78	1.386*
189.36	1.438
192.09	1.462*
200.75	1.508
207.53	1.540*
216.35	1.589
225.24	1.621*
CURVE 1 (cont.)	
239.44	1.711 x 10^{-1}
240.76	1.728*
243.02	1.737*
242.42	1.760*
247.94	1.757*
253.80	1.808*
257.57	1.816
259.75	1.891*
261.90	1.902*
261.98	1.898*
262.87	2.094
263.37	8.837
264.95	1.789
265.08	1.757*
268.55	1.705
270.79	1.711*
273.33	1.723*
274.20	1.740*
280.12	1.752
282.25	1.746*
284.78	1.779*
286.33	1.791*
287.33	1.820*
287.75	1.850
288.98	2.242
289.02	2.265*
289.78	4.848*
289.91	5.264*
289.93	5.481*
290.09	6.738
290.19	1.173 x 10^{0}
292.92	2.102 x 10^{-1}*
294.48	1.930*
298.09	1.886
CURVE 2	
Series I	
9.67	2.734 x 10^{-3}
11.81	4.897
13.90	7.669
15.98	1.086 x 10^{-2}
18.03	1.417
20.04	1.743
22.12	2.078
24.30	2.418*
CURVE 2 (cont.)	
26.69	2.784 x 10^{-2}*
29.33	3.170*
32.25	3.572
35.55	3.994
39.09	4.414
Series II	
41.38	4.671 x 10^{-2}
44.59	5.016
48.89	5.458
53.52	5.907
58.72	6.390
64.30	6.885*
Series III	
3.55	8.574 x 10^{-5}
5.08	3.620 x 10^{-4}
6.97	1.015 x 10^{-3}
9.10	2.282
11.35	4.382
13.49	7.088
15.45	1.001 x 10^{-2}
17.46	1.325
21.68	2.007
23.95	2.364
26.41	2.742
29.08	3.134
31.99	3.537*
Series IV	
32.72	3.635 x 10^{-2}*
35.81	4.026*
39.27	4.433*
43.16	4.864
47.62	5.326
52.41	5.798
57.53	6.280
63.26	6.795
69.53	7.330
76.61	7.906
CURVE 2 (cont.)	
Series V	
83.49	8.440 x 10^{-2}*
91.82	9.050*
100.88	9.646
110.70	1.027 x 10^{-1}*
121.11	1.090*
131.23	1.148*
141.13	1.202
151.22	1.256
Series VI	
161.19	1.306 x 10^{-1}*
171.18	1.356*
181.29	1.407
191.36	1.456
201.39	1.506
211.43	1.556
221.35	1.605
231.22	1.657
241.22	1.726*
251.30	1.817
Series VII	
258.24	1.850 x 10^{-1}
Series VIII	
267.22	1.710 x 10^{-1}
272.54	1.730
277.88	1.754
283.11	1.776
287.77	1.834
Series IX	
296.47	1.930 x 10^{-1}
301.48	1.941*
306.53	1.948*
311.62	1.958
316.73	1.966*
321.87	1.976
327.03	1.985*
332.21	1.995
337.40	2.005*
CURVE 2 (cont.)	
342.59	2.015 x 10^{-1}*
347.52	2.030
Series X	
3.64	1.096 x 10^{-4}
5.12	3.763*
7.00	1.038 x 10^{-3}*
8.99	2.210*
11.09	4.082
13.37	6.912*
15.40	9.913*
17.29	1.296 x 10^{-2}
19.30	1.622
21.40	1.962*
23.63	2.313*
26.03	2.685*
28.63	3.071*
31.48	3.468*
34.65	3.883*
Series XI*	
296.33	1.930 x 10^{-1}
301.39	1.941
306.48	1.948
311.61	1.955
316.78	1.966
321.96	1.975
327.13	1.986
332.23	1.994
337.30	2.004
342.35	2.014
347.42	2.030
Series XII*	
238.93	1.708 x 10^{-1}
243.93	1.739
248.94	1.784
253.93	1.834
258.92	1.875

* Not shown on plot

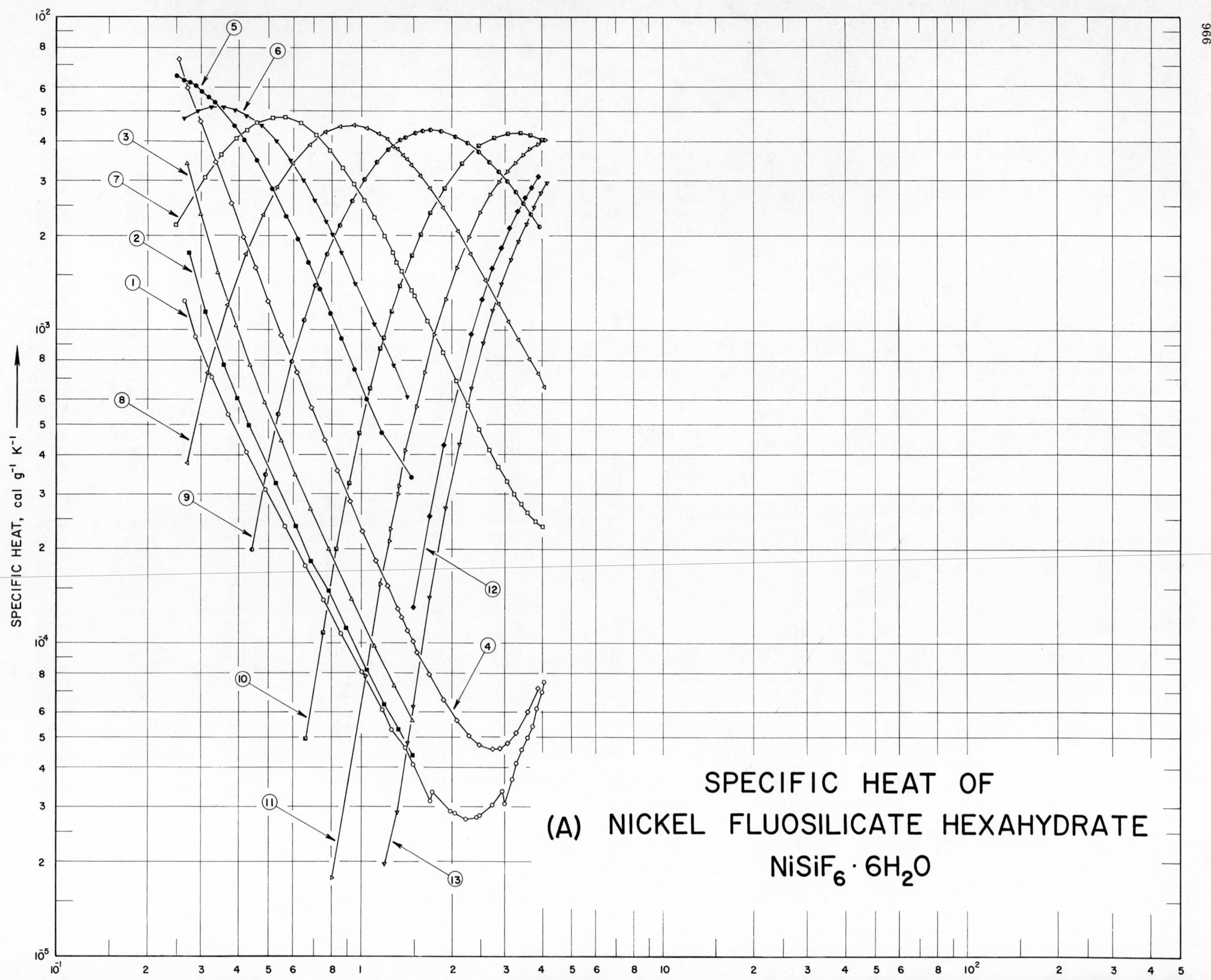
SPECIFIC HEAT OF
(A) NICKEL FLUOSILICATE HEXAHYDRATE
$NiSiF_6 \cdot 6H_2O$
SPECIFIC HEAT, cal g^{-1} K^{-1}

SPECIFICATION TABLE NO. 302 SPECIFIC HEAT OF NICKEL FLUOSILICATE HEXAHYDRATE $NiSiF_6 \cdot 6H_2O$ A

[For Data Reported in Figure and Table No. 302]

Curve No.	Ref. No.	Year	Temp. Range, K	Reported Error, %	Name and Specimen Designation	Composition (weight percent), Specifications and Remarks
1	330	1966	0.26–4.0		H = 0G	High purity sample, $2.4 \pm 0.2 \times 10^{-6}$ Na; spherical single crystal; magnetic field parallel to c-axis.
2	330	1966	0.28–1.5		H = 248G	Same as above.
3	330	1966	0.27–1.5		H = 497G	Same as above.
4	330	1966	0.26–3.9		H = 993G	Same as above.
5	330	1966	0.25–1.5		H = 1987G	Same as above.
6	330	1966	0.27–1.4		H = 2980G	Same as above.
7	330	1966	0.25–4.0		H = 4968G	Same as above.
8	330	1966	0.27–4.1		H = 9935G	Same as above.
9	330	1966	0.44–3.95		H = 19 872G	Same as above.
10	330	1966	0.66–4.0		H = 39 752G	Same as above.
11	330	1966	0.8 –4.15		H = 59 638G	Same as above.
12	330	1966	1.5 –3.9		H = 79 528G	Same as above.
13	330	1966	1.2 –4.2		H = 89 471G	Same as above.

DATA TABLE NO. 302 SPECIFIC HEAT OF NICKEL FLUOSILICATE HEXAHYDRATE, $NiSiF_6 \cdot 6H_2O$ A

[Temperature, T, K; Specific Heat, C_p, Cal $g^{-1} K^{-1}$]

T	C_p
CURVE 1	
Series 1	
1.266	5.331×10^{-5}
1.491	4.113
1.728	3.357
1.988	2.929
2.228	2.764
2.495	2.830
2.746	3.060
2.951	3.390
3.175	3.686
3.426	4.607
3.720	5.463
4.001	7.042
Series 2	
0.266	1.250×10^{-3}
0.291	9.576×10^{-4}
0.327	7.108
0.370	5.426
0.425	4.100
0.491	3.106
0.569	2.363
0.660	1.787
0.755	1.389
0.865	1.079
1.011	8.128×10^{-5}
1.179	6.154
1.406	4.673
1.705	3.159
2.050	2.896
2.438	2.797
2.729	2.995
3.000	3.093
3.279	4.081
3.774	5.002
3.844	6.220
4.083	7.569
CURVE 2	
0.276	1.777×10^{-3}
0.312	1.152
0.359	7.799×10^{-4}
0.396	6.078
CURVE 2 (cont.)	
0.434	4.989×10^{-4}
0.530	3.258
0.615	2.376
0.688	1.843
0.785	1.474
0.899	1.122
1.050	8.260×10^{-5}
1.199	6.384
1.336	5.331
1.485	4.410
CURVE 3	
0.273	3.422×10^{-3}
0.302	2.353
0.344	1.530
0.393	1.040
0.437	7.799×10^{-4}
0.488	5.937
0.551	4.466
0.614	3.475
0.688	2.695
0.789	2.011
0.937	1.399
1.111	9.807×10^{-5}
1.286	7.338
1.471	5.693
CURVE 4	
Series 1	
1.367	1.218×10^{-4}
1.496	1.017
1.698	7.997×10^{-5}
1.883	6.647
2.085	5.693
2.290	5.068
2.494	4.772
2.757	4.607
2.912	4.640
3.070	4.837
3.291	5.199
3.591	6.055
3.896	7.207
CURVE 4 (cont.)	
Series 2	
0.256	7.338×10^{-3}
0.274	5.989
0.302	4.676
0.338	3.459
0.380	2.550
0.417	1.991
0.455	1.589
0.501	1.247
0.554	9.675×10^{-4}
0.620	7.371
0.692	5.663
0.765	4.482
0.841	3.584
0.929	2.950
1.022	2.287
1.125	1.849
1.225	1.534
1.326	1.293
1.431	1.102
1.545	9.379×10^{-5}
CURVE 5	
0.252	6.549×10^{-3}
0.266	6.318
0.278	6.220
0.292	6.055
0.305	5.825
0.320	5.561
0.336	5.364
0.359	5.035*
0.389	4.531
0.421	4.058
0.462	3.495
0.515	2.847
0.574	2.333
0.624	1.965
0.675	1.659
0.737	1.362
0.797	1.142
0.869	9.412×10^{-4}
0.957	7.536
1.055	6.048
1.176	4.742
1.475	3.409
CURVE 6	
Series 1	
0.267	4.772×10^{-3}
0.296	5.035
0.326	5.167
0.359	5.167
0.392	5.068
0.428	4.864
0.476	4.505
0.533	4.011
0.593	3.498
0.652	3.024
Series 2	
0.645	3.110×10^{-3}*
0.709	2.606
0.772	2.231
0.811	2.024
0.867	1.774
0.962	1.418
1.116	1.053
1.284	7.766×10^{-4}
1.430	6.131
CURVE 7	
Series 1	
1.328	1.652×10^{-3}
1.484	1.349
1.673	1.073
1.874	8.523×10^{-4}
2.085	6.911
2.284	5.759
2.482	4.857
2.692	4.176
2.876	3.686
3.063	3.301
3.247	3.008
3.416	2.797
3.589	2.633
3.821	2.465
4.053	2.376
CURVE 7 (cont.)	
Series 2	
0.254	2.172×10^{-3}
0.312	3.093
0.353	3.659
0.395	4.113
0.429	4.387
0.464	4.601
0.520	4.781
0.572	4.785
0.640	4.601
0.717	4.209
0.797	3.765
0.877	3.327
0.954	2.945
1.031	2.613
1.116	2.294
1.204	2.004
1.286	1.774
1.372	1.570
1.513	1.300
CURVE 8	
Series 1	
1.284	3.886×10^{-3}
1.380	3.646
1.486	3.396
1.597	3.133
1.710	2.866
1.897	2.475
2.121	2.080
2.347	1.764
2.618	1.461
2.887	1.221
3.106	1.073
3.352	9.379×10^{-4}
3.652	8.128
3.908	7.306
4.138	6.647
Series 2	
0.272	3.784×10^{-4}
0.319	7.371
CURVE 8 (cont.)	
Series 2 (cont.)	
0.371	1.211×10^{-3}
0.425	1.767
0.483	2.359
0.539	2.886
0.604	3.442
0.685	3.946
0.774	4.321
0.863	4.498
0.959	4.535
1.059	4.446
1.157	4.265
1.247	4.041
1.333	3.807
1.436	3.528
CURVE 9	
Series 1	
0.442	2.067×10^{-4}
0.490	3.478
0.540	5.440
0.597	7.997
0.655	1.089×10^{-3}
0.712	1.399
0.778	1.764
0.855	2.178
0.940	2.613
1.035	3.047
1.137	3.452
1.237	3.781
1.364	4.064
Series 2	
1.269	3.860×10^{-3}*
1.328	3.985*
1.423	4.150
1.527	4.271
1.622	4.337
1.718	4.383
1.867	4.331
2.071	4.186
2.277	3.975

* Not shown on plot

DATA TABLE NO. 302 (continued)

T	C_p
CURVE 9 (cont.)	
Series 2 (cont.)	
2.481	3.725×10^{-3}
2.692	3.452
2.882	3.225
3.073	3.001
3.286	2.774
3.483	2.564
3.695	2.363
3.951	2.152
CURVE 10	
Series 1	
0.660	4.969×10^{-5}
0.751	1.089×10^{-4}
0.834	2.004
0.917	3.268
0.992	4.719
1.074	6.552
1.162	8.786
1.264	1.152×10^{-3}
1.402	1.537*
Series 2	
1.191	9.477×10^{-4}
1.356	1.399×10^{-3}
1.480	1.744
1.582	2.027
1.710	2.379
1.904	2.860
2.183	3.422
2.496	3.896
2.779	4.133
3.097	4.252
3.416	4.265
3.693	4.215
4.014	4.071
CURVE 11	
Series 1	
1.244	2.132×10^{-4}
1.329	2.998×10^{-4}
1.429	4.225*
1.534	5.746
1.637	7.371
1.771	9.708
1.928	1.260×10^{-3}
2.095	1.583
2.300	1.984
2.513	2.386
2.708	2.718
2.897	2.991
3.089	3.232
3.283	3.449
3.475	3.633
3.676	3.791
3.905	3.946
4.151	4.071
Series 2	
0.801	1.777×10^{-5}
1.042	7.898
1.169	1.557×10^{-4}
1.259	2.336
1.339	3.182
1.418	4.160
CURVE 12	
1.498	1.310×10^{-4}
1.693	2.564
1.892	4.317
2.108	6.812*
2.332	9.741
2.533	1.260×10^{-3}
2.740	1.583
2.932	1.836
3.136	2.126
3.341	2.406
3.528	2.652
3.707	2.863
3.922	3.103
CURVE 13	
Series 1	
1.195	1.974×10^{-5}
1.422	4.837
Series 2	
1.311	2.896×10^{-5}
1.485	6.285
1.696	1.408×10^{-4}
1.918	2.708
2.131	4.341
2.341	6.559
2.564	9.115
2.768	1.165×10^{-3}
2.962	1.418
3.164	1.682
3.354	1.925
3.560	2.188
3.789	2.475
4.022	2.751
4.220	2.962

* Not shown on plot

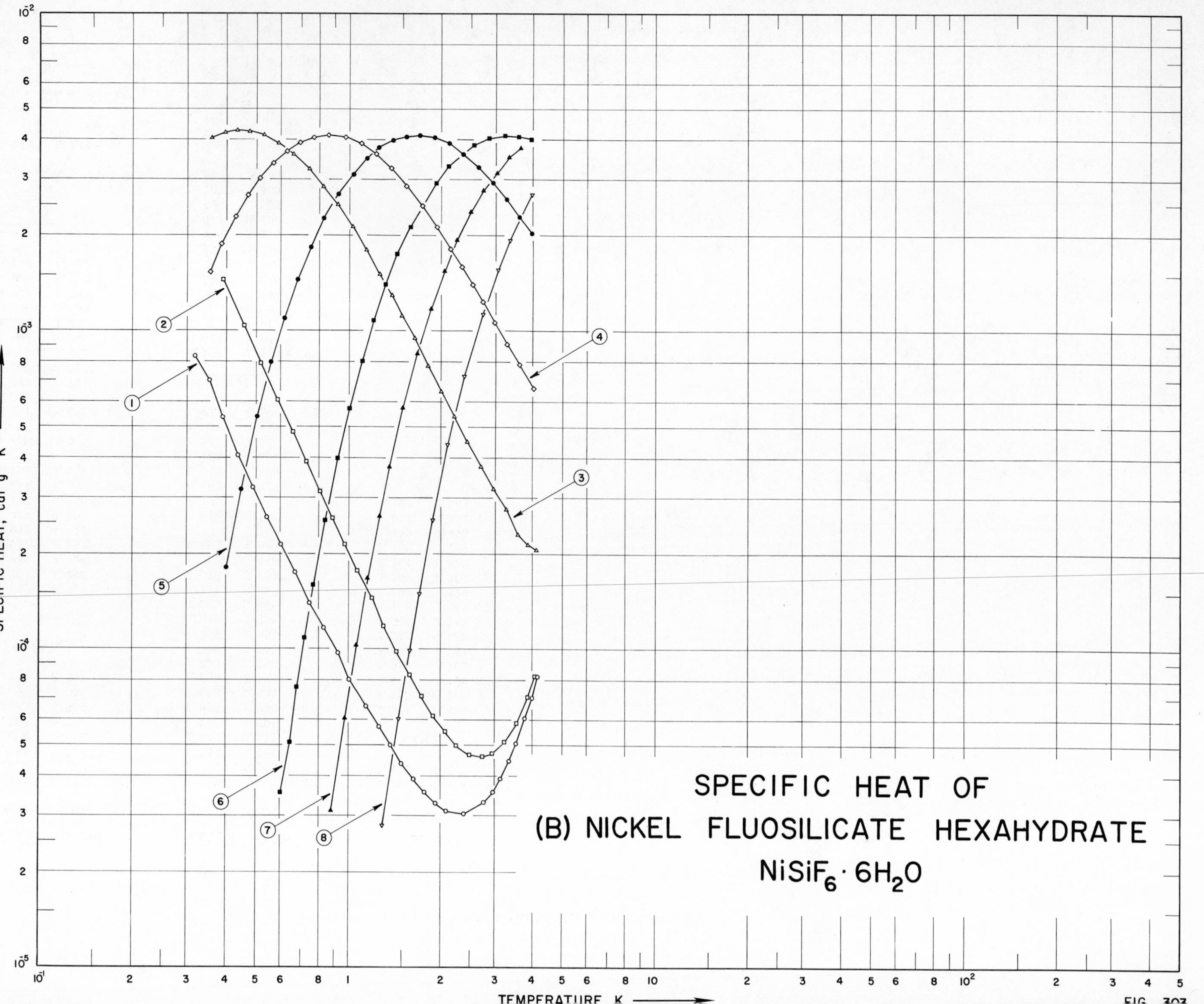

SPECIFIC HEAT OF
(B) NICKEL FLUOSILICATE HEXAHYDRATE
$NiSiF_6 \cdot 6H_2O$
SPECIFIC HEAT, cal g^{-1} K^{-1}
TEMPERATURE, K
1
2
3
4
5
6
7
8

SPECIFICATION TABLE NO. 303 SPECIFIC HEAT OF NICKEL FLUOSILICATE HEXAHYDRATE $NiSiF_6 \cdot 6H_2O$ B

[For Data Reported in Figure and Table No. 303]

Curve No.	Ref. No.	Year	Temp. Range, K	Reported Error, %	Name and Specimen Designation	Composition (weight percent), Specifications and Remarks
1	331	1967	0.3 -4.2		H = 0G	3.5 cm spherical single crystal, magnetic field perpendicular to the c-axis.
2	331	1967	0.4 -4.0		H = 993G	Same as above.
3	331	1967	0.36-4.1		H = 4968G	Same as above.
4	331	1967	0.35-4.0		H = 9935G	Same as above.
5	331	1967	0.4 -3.9		H = 19 872G	Same as above.
6	331	1967	0.6 -3.9		H = 39 752G	Same as above.
7	331	1967	0.8 -4.0		H = 59 638G	Same as above.
8	331	1967	1.3 -3.9		H = 89 471G	Same as above.

DATA TABLE NO. 303 SPECIFIC HEAT OF NICKEL FLUOSILICATE HEXAHYDRATE $NiSiF_6 \cdot 6H_2O$ B

[Temperature, T, K; Specific Heat, C_p, Cal g^{-1} K^{-1}]

T	C_p
CURVE 1	
0.318	8.376 x 10^{-4}
0.354	7.003
0.392	5.361
0.437	4.089
0.488	3.244
0.544	2.609
0.604	2.143
0.671	1.752
0.748	1.428
0.832	1.172
0.925	9.745 x 10^{-5}
1.031	8.029
1.144	6.702
1.260	5.730
1.374	4.986
1.496	4.371
1.630	3.917
1.770	3.561
1.926	3.270
2.082	3.108
2.384	3.043
2.641	3.140
2.762	3.302
2.955	3.561
3.118	3.917
3.331	4.468
3.522	5.051
3.758	6.054
3.985	7.025
4.167	8.223
CURVE 2	
0.391	1.459 x 10^{-3}
0.456	1.042
0.518	7.903 x 10^{-4}
0.586	6.093
0.656	4.827
0.728	3.885
0.806	3.137
0.888	2.580
0.974	2.137
1.070	1.771
1.180	1.450
1.304	1.188
CURVE 2 (cont.)	
1.437	9.842 x 10^{-5}
1.579	8.320
1.731	7.123
1.894	6.184
2.068	5.504
2.257	4.986
2.499	4.662
2.732	4.630
2.941	4.727
3.223	5.148
3.540	5.892
3.851	7.058
4.077	8.256
CURVE 3	
0.362	4.089 x 10^{-3}
0.397	4.222
0.435	4.283
0.478	4.267
0.531	4.131
0.592	3.911
0.662	3.607
0.744	3.241
0.829	2.853
0.920	2.503
1.024	2.138
1.143	1.800
1.265	1.517
1.378	1.306
1.496	1.124
1.641	9.505 x 10^{-4}
1.817	7.835
2.005	6.537
2.211	5.423
2.444	4.510
2.699	3.756
2.969	3.215
3.264	2.762
3.562	2.308
3.855	2.140
4.136	2.069
CURVE 4	
0.355	1.529 x 10^{-3}
0.387	1.878
0.430	2.298
0.471	2.675
0.516	3.028
0.570	3.372
0.630	3.685
0.695	3.926
0.770	4.063
0.863	4.138
0.975	4.078
1.097	3.891
1.227	3.613
1.375	3.261
1.548	2.869
1.739	2.485
1.946	2.123
2.157	1.817
2.346	1.594
2.532	1.409
2.732	1.240
2.980	1.070
3.271	9.120 x 10^{-4}
3.610	7.890
4.022	6.627
CURVE 5	
0.401	1.800 x 10^{-4}
0.448	3.186
0.504	5.397
0.559	7.971
0.618	1.107 x 10^{-3}
0.682	1.459
0.751	1.841
0.833	2.273
0.925	2.708
1.028	3.119
1.146	3.498
1.263	3.775
1.391	3.973
1.541	4.094
1.709	4.120
1.906	4.065
2.124	3.872
CURVE 5 (cont.)	
2.364	3.595 x 10^{-3}
2.642	3.275
2.947	2.931
3.262	2.601
3.600	2.296
3.978	2.025
CURVE 6	
0.601	3.529 x 10^{-5}
0.645	5.083
0.683	7.576
0.721	1.088 x 10^{-4}
0.770	1.593
0.840	2.548
0.923	3.956
1.007	5.747
1.101	8.026
1.208	1.086 x 10^{-3}
1.322	1.404
1.446	1.753
1.587	2.133
1.748	2.534*
1.927	2.932
2.119	3.290
2.325	3.591*
2.555	3.844
2.851	4.039
3.220	4.124
3.594	4.094
3.962	4.011
CURVE 7	
0.876	3.108 x 10^{-5}
0.975	6.087
1.064	1.033 x 10^{-4}
1.158	1.684
1.259	2.626
1.359	3.768
1.501	5.763
1.673	8.541
1.854	1.179 x 10^{-3}
2.051	1.556
2.253	1.948
CURVE 7 (cont.)	
2.491	2.376 x 10^{-3}
2.747	2.784
3.025	3.155
3.326	3.535
3.643	3.786
4.001	4.019*
CURVE 8	
1.285	2.817 x 10^{-5}
1.457	6.054
1.583	9.939
1.700	1.502 x 10^{-4}
1.877	2.558
2.107	4.413
2.381	7.213
2.726	1.136 x 10^{-3}
3.063	1.561
3.351	1.938
3.670	2.321*
3.984	2.696

* Not shown on plot

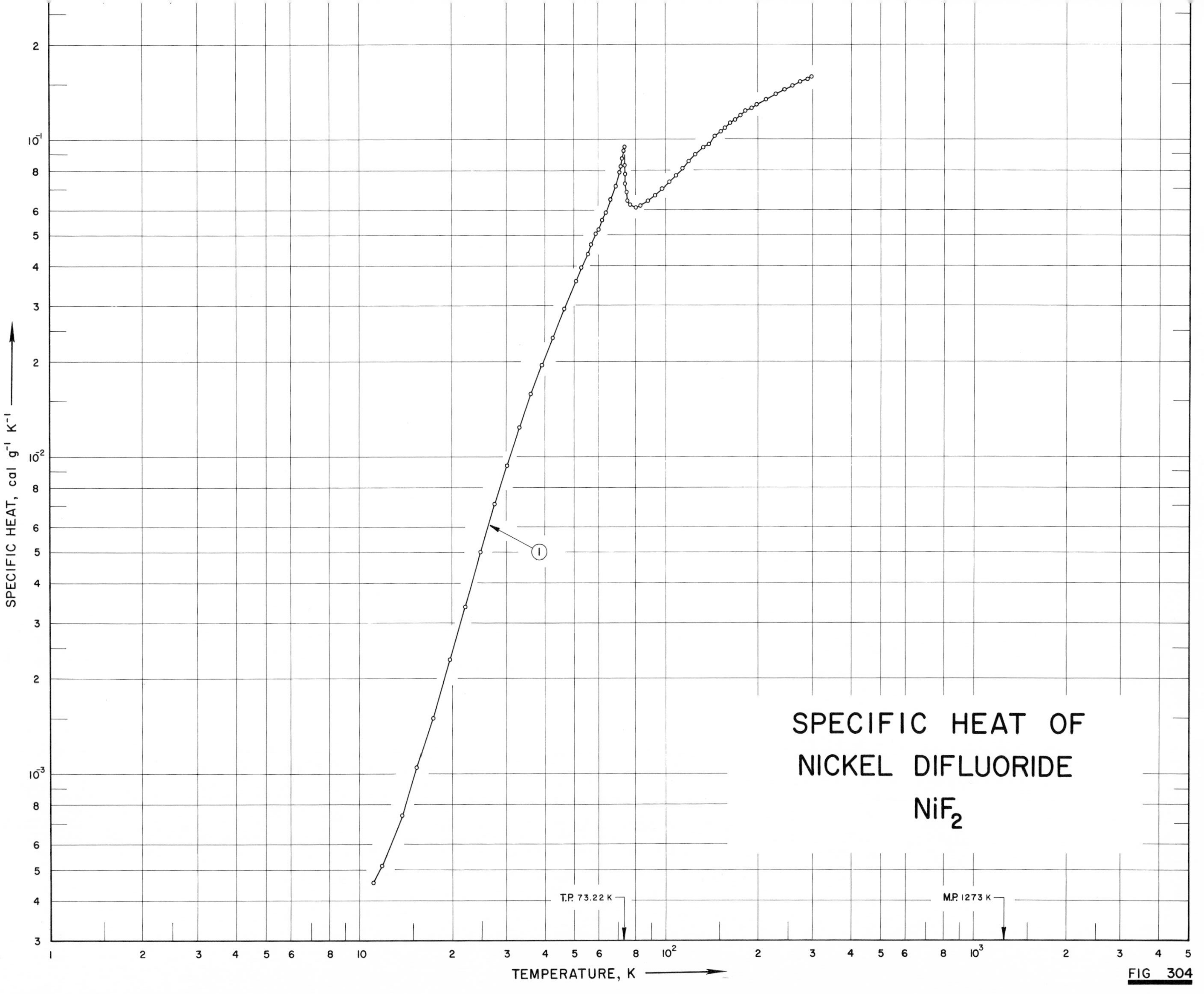
SPECIFIC HEAT OF
NICKEL DIFLUORIDE
NiF_2
SPECIFIC HEAT, cal g⁻¹ K⁻¹
TEMPERATURE, K
T.P. 73.22 K
M.P. 1273 K
1
FIG 304

SPECIFICATION TABLE NO. 304 SPECIFIC HEAT OF NICKEL DIFLUORIDE NiF_2

[For Data Reported in Figure and Table No. 304]

Curve No.	Ref. No.	Year	Temp. Range, K	Reported Error, %	Name and Specimen Designation	Composition (weight percent), Specifications and Remarks
1	293	1955	11-298	0.2-3		0.005 Cu, 0.004 Co, 0.003 Fe and 0.001 Mn; single crystals of optically anistropic material.

DATA TABLE NO. 304 SPECIFIC HEAT OF NICKEL DIFLUORIDE NiF_2

[Temperature, T, K; Specific Heat, C_p, Cal g^{-1} K^{-1}]

T	C_p
CURVE 1	
Series 1	
81.52	6.163 x 10^{-2}*
84.63	6.265*
88.35	6.442
Series 2	
53.13	3.966 x 10^{-2}
57.39	4.699
62.27	5.638
66.72	6.650*
73.33	7.917*
77.68	6.218*
82.32	6.192*
Series 3	
88.78	6.471 x 10^{-2}*
93.09	6.706
98.05	7.029
103.01	7.371
108.30	7.744
114.03	8.144
119.90	8.563
126.22	8.986
132.78	9.439
139.07	9.673
145.54	1.028 x 10^{-1}
152.20	1.067
156.92	1.094
163.35	1.128
169.87	1.162
176.70	1.197
183.79	1.231
190.98	1.265
198.08	1.293
205.90	1.325*
213.03	1.346
220.58	1.367*
227.40	1.397
236.39	1.426*
243.24	1.449
251.28	1.468*
258.75	1.484

T	C_p
CURVE 1 (cont.)	
Series 3 (cont.)	
266.19	1.503 x 10^{-1}*
274.30	1.527
282.40	1.546*
290.31	1.565*
298.25	1.590*
Series 4	
11.14	4.551 x 10^{-4}
11.88	5.171
13.76	7.446
15.42	1.055 x 10^{-3}
17.40	1.510
19.73	2.296
22.09	3.382
24.74	5.006
27.54	7.105
30.23	9.432
33.21	1.245 x 10^{-2}
36.25	1.585
39.37	1.960
42.67	2.388
46.53	2.944
50.85	3.599
59.27	5.061
74.95	6.881
Series 5	
55.69	4.389 x 10^{-2}
60.27	5.236
63.73	5.946
66.27	6.531
68.72	7.189
70.79	7.920
73.05	9.294
73.84	7.282
74.51	6.681*
75.16	6.460
75.88	6.330*
76.94	6.255
78.29	6.184*
80.10	6.147
83.25	6.210

T	C_p
CURVE 1 (cont.)	
Series 6	
72.64	9.019 x 10^{-2}*
72.85	9.256*
73.00	9.380*
73.13	9.494*
73.23	9.536
73.34	9.050*
73.45	8.346
73.57	7.829
73.69	7.467*
73.81	7.229*
73.93	7.095*
74.06	6.981*
74.19	6.857*
74.32	6.785*
74.44	6.743*
74.58	6.660*
74.71	6.609*
74.84	6.567*
74.97	6.516*
75.10	6.474*
75.23	6.454*
Series 7	
71.53	8.295 x 10^{-2}
71.71	8.346*
71.90	8.460*
72.09	8.615*
72.24	8.770
72.35	8.822*
72.46	8.884*
72.57	8.987*
72.68	9.060*
72.79	9.205*
72.89	9.287*
72.99	9.380*
73.10	9.494*
73.20	9.546*
73.30	9.298*
73.41	8.481*

T	C_p
CURVE 1 (cont.)	
Series 8	
81.25	6.160 x 10^{-2}*
83.84	6.235*
216.10	1.363 x 10^{-1}*
233.52	1.388*
230.55	1.410*
237.50	1.433*
299.99	1.582

* Not shown on plot

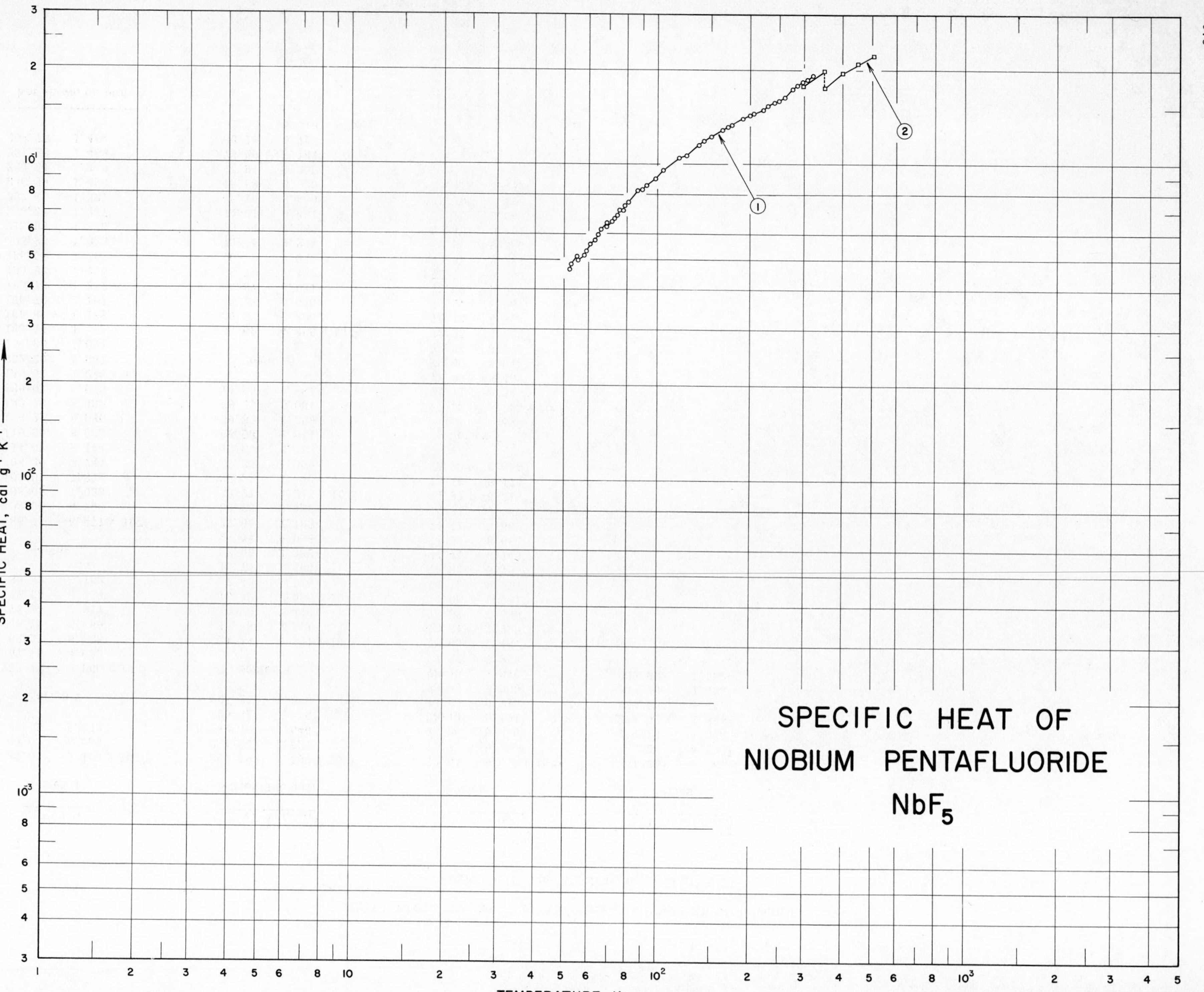
SPECIFIC HEAT OF
NIOBIUM PENTAFLUORIDE
NbF_5
SPECIFIC HEAT, cal g⁻¹ K⁻¹
1
2

SPECIFICATION TABLE NO. 305 SPECIFIC HEAT OF NIOBIUM PENTAFLUORIDE NbF_5

[For Data Reported in Figure and Table No. 305]

Curve No.	Ref. No.	Year	Temp. Range, K	Reported Error, %	Name and Specimen Designation	Composition (weight percent), Specifications and Remarks
1	332	1957	52-321			0.9 impurity of metallic Nb. (corrected for impurity)
2	322	1957	298-506			Same as above.

DATA TABLE NO. 305 SPECIFIC HEAT OF NIOBIUM PENTAFLUORIDE NbF_5

[Temperature, T, K; Specific Heat, C_p, Cal g^{-1} K^{-1}]

CURVE 1

T	C_p
52.40	4.66 x 10^{-2}
52.80	4.85
55.10	5.14
55.39	4.98
58.23	5.15
59.12	5.34
60.79	5.59
63.01	5.76
64.68	5.99
66.10	6.24
66.66	6.20*
68.94	6.55
68.95	6.36
71.57	6.62
73.08	6.75
73.23	6.89*
74.51	6.90
75.87	7.18
78.10	7.16
78.60	7.36*
78.76	7.43
81.21	7.60
82.09	7.58*
86.79	8.26
89.63	8.35
92.56	8.56
98.92	9.05
105.01	9.57
107.04	9.64*
117.78	1.05 x 10^{-1}
120.04	1.05*
125.79	1.06
137.01	1.16
139.39	1.16*
142.14	1.19
145.01	1.19*
147.50	1.21*
149.92	1.22
152.91	1.24*
163.40	1.28
165.93	1.29*
170.52	1.31
175.53	1.34
187.71	1.38*
189.74	1.40

CURVE 1 (cont.)

T	C_p
194.39	1.40 x 10^{-1}*
200.24	1.43
203.41	1.45*
206.56	1.45
221.52	1.50
223.19	1.50*
226.15	1.54*
229.03	1.55
237.22	1.57*
240.85	1.58
249.21	1.59
255.53	1.62*
258.25	1.63*
260.86	1.63
274.20	1.69*
274.57	1.73*
277.64	1.74
287.48	1.78
291.35	1.81*
293.98	1.80*
298.60	1.82
304.97	1.85*
308.02	1.86
311.07	1.87*
313.50	1.87*
320.72	1.92

CURVE 2

T	C_p
298.2	1.775 x 10^{-1}
350.7	1.984
350.7	1.754
400.0	1.956
450.0	2.097
506.5	2.209

* Not shown on plot

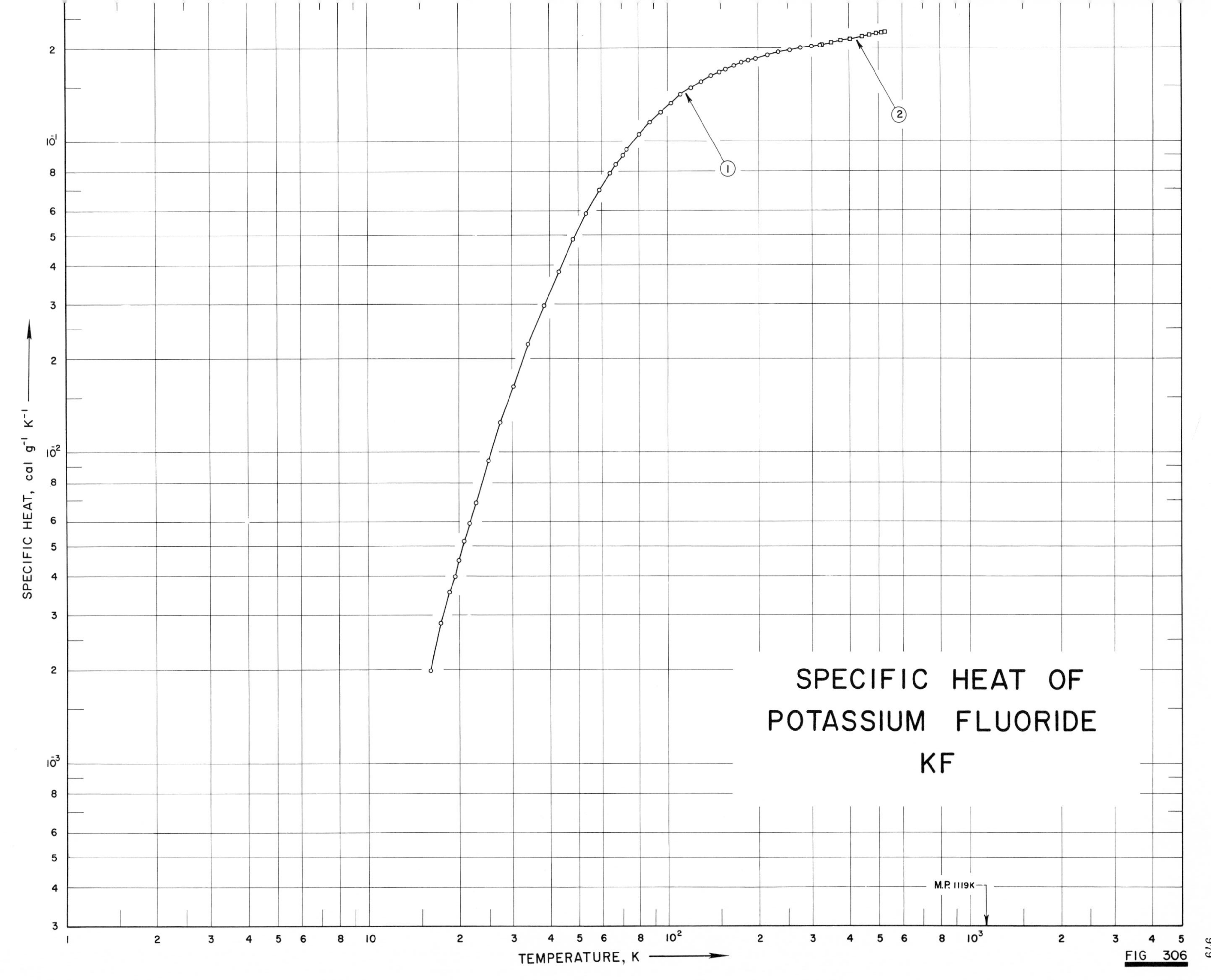

SPECIFIC HEAT OF
POTASSIUM FLUORIDE
KF
M.P. 1119K
SPECIFIC HEAT, cal g⁻¹ K⁻¹
TEMPERATURE, K
FIG 306

SPECIFICATION TABLE NO. 306 SPECIFIC HEAT OF POTASSIUM FLUORIDE KF

[For Data Reported in Figure and Table No. 306]

Curve No.	Ref. No.	Year	Temp. Range, K	Reported Error, %	Name and Specimen Designation	Composition (weight percent), Specifications and Remarks
1	333	1949	16-323	0.2-1.0		
2	333	1949	325-530	0.2-1.0		

DATA TABLE NO. 306 SPECIFIC HEAT OF POTASSIUM FLUORIDE KF

[Temperature, T, K; Specific Heat, C_p, Cal g^{-1} K^{-1}]

T	C_p
CURVE 1	
16.05	1.992×10^{-3}
17.46	2.835
18.55	3.556
19.43	3.995
19.95	4.491
20.70	5.195
21.57	5.906
22.63	6.877
24.91	9.402
27.42	1.254×10^{-2}
30.33	1.628
33.97	2.233
38.37	2.969
43.08	3.820
48.28	4.845
53.43	5.890
58.99	6.999
58.50	6.932*
64.22	7.927
67.04	8.422
70.51	9.006
73.31	9.436
80.57	1.057×10^{-1}
87.81	1.156
95.06	1.244
102.76	1.326
111.41	1.412
120.39	1.486
129.84	1.557
140.10	1.625
148.20	1.671
157.41	1.710
167.19	1.751
176.98	1.790
186.40	1.819
196.10	1.849
206.05	1.874*
215.18	1.899
223.28	1.911*
233.33	1.936
244.33	1.950*
255.45	1.967
266.69	1.983*
277.66	1.998
288.86	2.007*
CURVE 1 (cont.)	
300.35	2.014×10^{-1}
311.35	2.029*
322.63	2.035
CURVE 2	
324.6	2.041×10^{-1}
350.2	2.074
376.2	2.105
403.7	2.134
443.7	2.174
468.3	2.196
492.2	2.217
512.8	2.234
529.9	2.248

* Not shown on plot

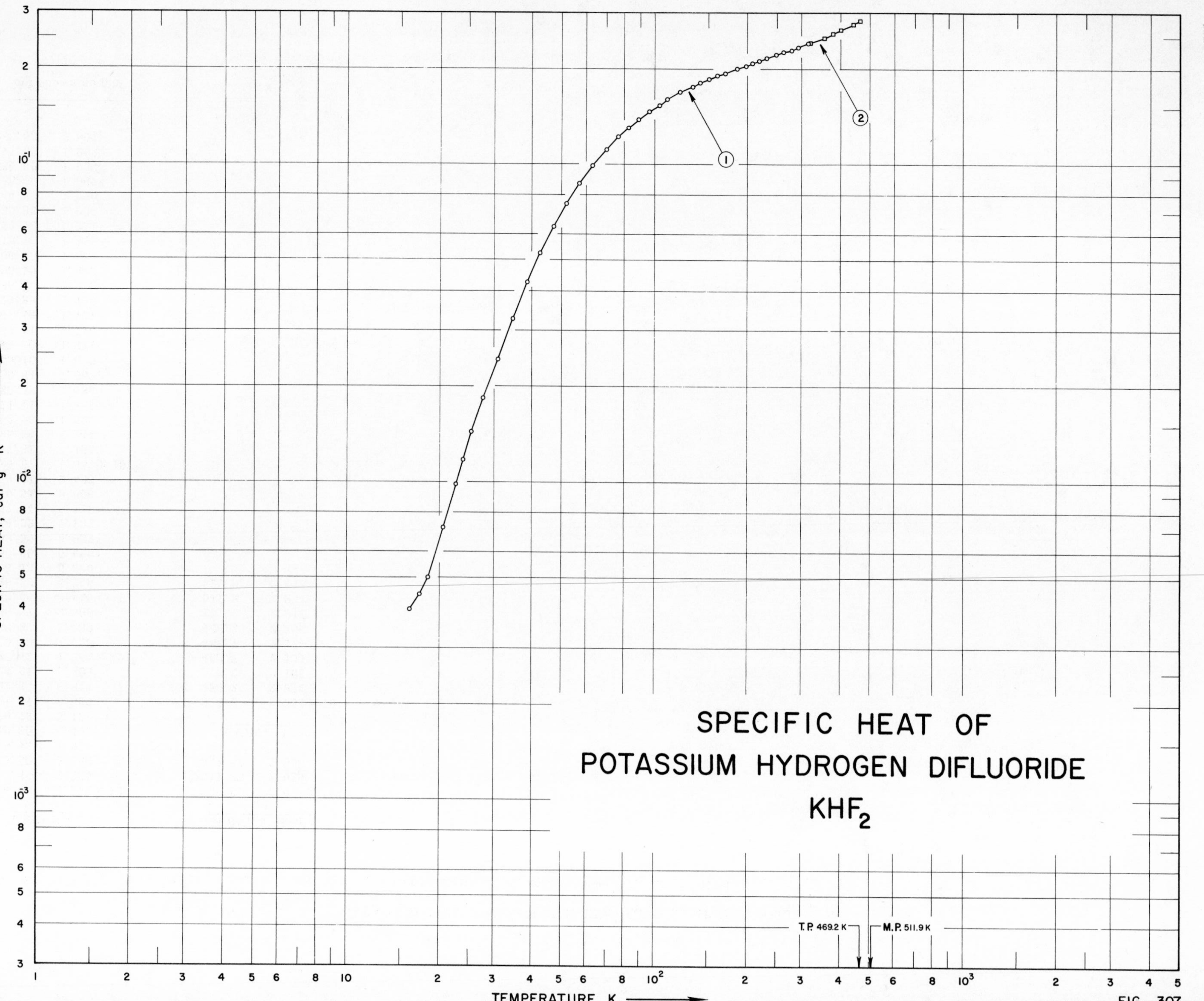

SPECIFIC HEAT OF
POTASSIUM HYDROGEN DIFLUORIDE
KHF_2
T.P. 469.2 K
M.P. 511.9 K
1
2
SPECIFIC HEAT, cal g⁻¹ K⁻¹
TEMPERATURE, K

FIG. 307

SPECIFICATION TABLE NO. 307 SPECIFIC HEAT OF POTASSIUM HYDROGEN DIFLUORIDE KHF_2

[For Data Reported in Figure and Table No. 307]

Curve No.	Ref. No.	Year	Temp. Range, K	Reported Error, %	Name and Specimen Designation	Composition (weight percent), Specifications and Remarks
1	333	1949	16–316	0.2		0.02 Na and 0.001 H_2SiF_6.
2	333	1949	322–465	0.2		Same as above.

DATA TABLE NO. 307 SPECIFIC HEAT OF POTASSIUM HYDROGEN DIFLUORIDE KHF_2

[Temperature, T, K; Specific Heat, C_p, Cal $g^{-1} K^{-1}$]

T	C_p
CURVE 1	
16.02	3.956 x 10^{-3}
17.34	4.417
18.45	4.993
20.55	7.196
22.64	9.846
23.90	1.175 x 10^{-2}
25.49	1.435
27.73	1.836
30.95	2.425
34.71	3.267
38.71	4.269
42.73	5.260
47.28	6.351
52.03	7.518
57.42	8.703
63.12	9.918
63.33	9.898*
70.17	1.111
71.85	1.137*
76.83	1.223 x 10^{-1}
78.03	1.235*
78.99	1.248*
83.17	1.307
85.67	1.341*
89.44	1.384
92.38	1.417*
96.58	1.461
99.32	1.488*
104.10	1.536
106.90	1.561*
111.62	1.600
113.91	1.617*
115.18	1.630*
122.63	1.682
124.21	1.695*
131.71	1.744*
133.18	1.755
140.73	1.801*
142.10	1.808
149.73	1.846*
151.82	1.855
158.84	1.890*
161.48	1.900
168.08	1.929*
170.46	1.938
177.38	1.965*
CURVE 1 (cont.)	
179.28	1.977 x 10^{-1}*
180.58	1.970*
186.89	2.002
189.03	2.015*
196.34	2.038*
198.44	2.045
206.19	2.074*
208.52	2.078
216.25	2.110*
219.17	2.120
221.57	2.124*
229.65	2.147*
232.30	2.164
239.39	2.179*
242.91	2.192
248.89	2.214
254.28	2.228*
258.25	2.238*
263.38	2.257
268.56	2.267*
274.89	2.288*
279.83	2.285
284.18	2.319*
294.33	2.340
304.09	2.379*
315.79	2.402
CURVE 2	
321.6	2.406 x 10^{-1}
356.4	2.508
380.8	2.581
404.3	2.650
443.7	2.767
463.7	2.826*
465.3	2.831

* Not shown on plot

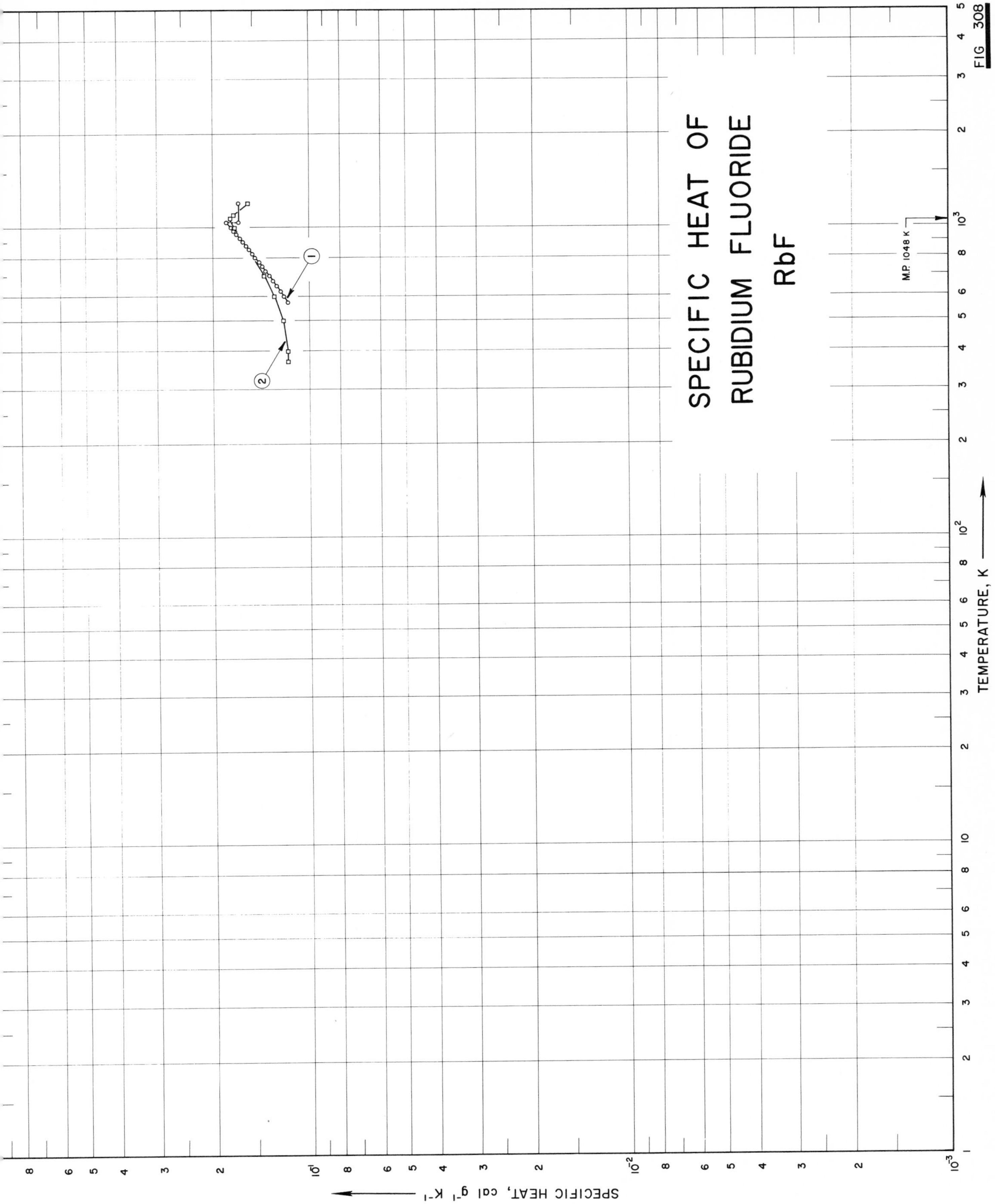

SPECIFIC HEAT OF RUBIDIUM FLUORIDE

RbF

SPECIFIC HEAT, cal g^{-1} K^{-1}

TEMPERATURE, K

M.P. 1048 K

① ②

FIG 308

SPECIFICATION TABLE NO. 308 SPECIFIC HEAT OF RUBIDIUM FLUORIDE RbF

[For Data Reported in Figure and Table No. 308]

Curve No.	Ref. No.	Year	Temp. Range, K	Reported Error, %	Name and Specimen Designation	Composition (weight percent), Specifications and Remarks
1	251	1958	575-1200			0.2 KF and a trace of NaF.
2	252	1961	370-1200			99.8 RbF.

DATA TABLE NO. 308 SPECIFIC HEAT OF RUBIDIUM FLUORIDE RbF

[Temperature, T, K; Specific Heat, C_p, Cal $g^{-1} K^{-1}$]

T	C_p
CURVE 1	
575	1.174×10^{-1}
600	1.206
625	1.240
650	1.273
675	1.307
700	1.342
725	1.376
750	1.411
775	1.446
800	1.481
825	1.516
850	1.552
875	1.587
900	1.623
925	1.659
950	1.694
975	1.730
1000	1.766
1025	1.802*
(s) 1048	1.836
(l) 1048	1.669
1050	1.669*
1075	1.669*
1100	1.669*
1125	1.669*
1150	1.669*
1175	1.669*
1200	1.669
CURVE 2	
Solid	
370	1.174×10^{-1}
400	1.173
500	1.215
600	1.295
700	1.391
800	1.497*
900	1.607*
1000	1.722
1035	1.763*
CURVE 2 (cont.)	
Liquid	
1075	1.773×10^{-1}
1100	1.731
1200	1.564

* Not shown on plot

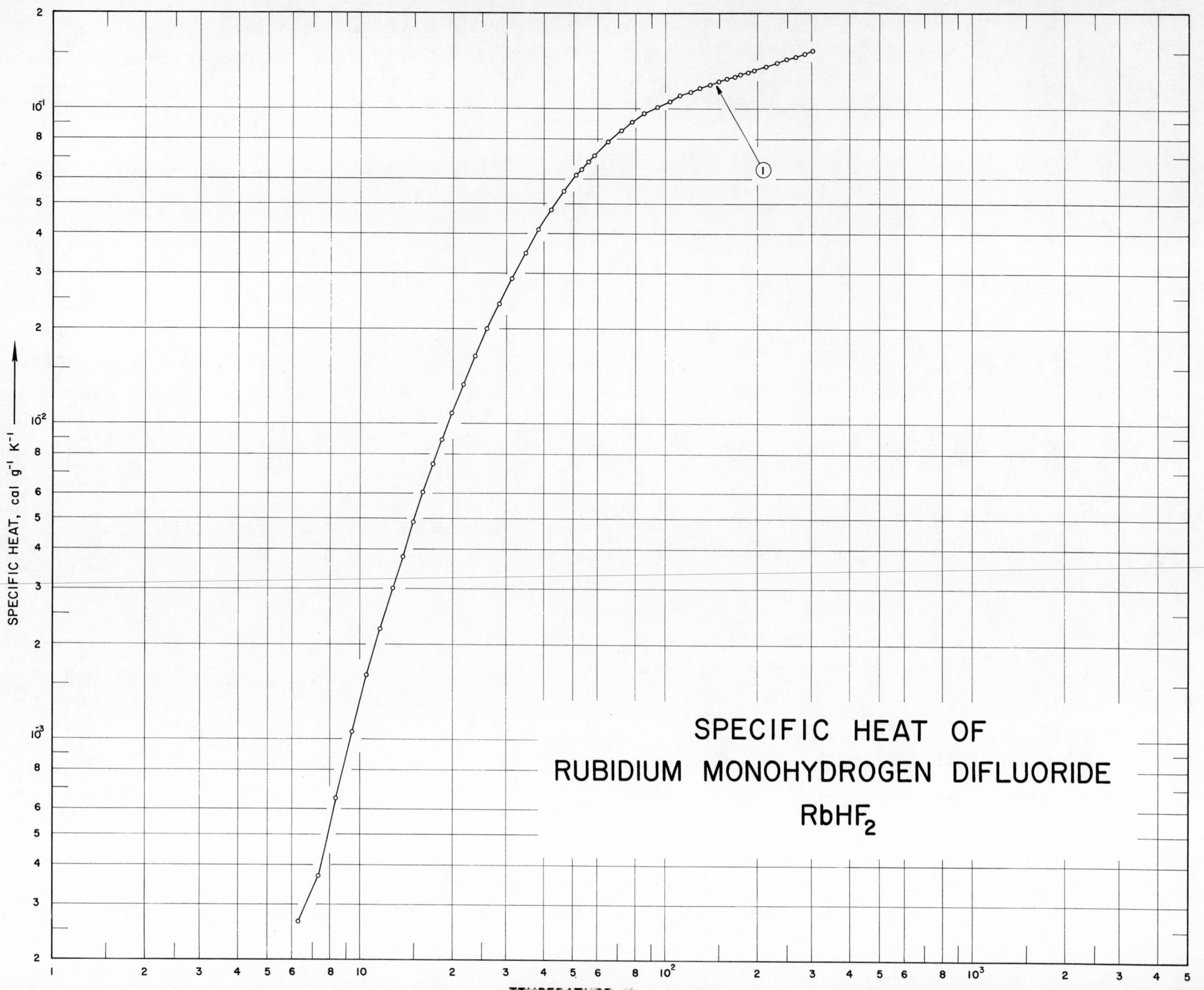
SPECIFIC HEAT OF
RUBIDIUM MONOHYDROGEN DIFLUORIDE
RbHF2
SPECIFIC HEAT, cal g⁻¹ K⁻¹
1

SPECIFICATION TABLE NO. 309 SPECIFIC HEAT OF RUBIDIUM MONOHYDROGEN DIFLUORIDE $RbHF_2$

[For Data Reported in Figure and Table No. 309]

Curve No.	Ref. No.	Year	Temp. Range, K	Reported Error, %	Name and Specimen Designation	Composition (weight percent), Specifications and Remarks
1	286	1961	5-303			Fluoride 99.97 ± 0.15% of theoretical.

DATA TABLE NO. 309 SPECIFIC HEAT OF RUBIDIUM MONOHYDROGEN DIFLUORIDE $RbHF_2$

[Temperature, T, K; Specific Heat, C_p, Cal g^{-1} K^{-1}]

T	C_p
CURVE 1	
Series 1	
53.29	6.425×10^{-2}
58.62	7.123
65.29	7.890
71.70	8.507
78.39	9.085
85.99	9.655
94.42	1.016×10^{-1}
103.38	1.062
112.37	1.100
121.54	1.137
130.93	1.170
140.40	1.198
149.82	1.225
159.10	1.248
168.20	1.270
177.15	1.289
186.07	1.308
195.05	1.326
204.02	1.344*
213.02	1.362
222.11	1.378*
231.24	1.395
240.21	1.414*
249.12	1.432
258.04	1.446*
267.02	1.463
276.08	1.480
285.14	1.497
294.21	1.516*
303.24	1.534
Series 2	
5.34	1.607×10^{-4}
6.34	2.651
7.31	3.695
8.31	6.507
9.41	1.068×10^{-3}
10.50	1.607
11.67	2.241
12.76	3.004
13.76	3.800
14.87	4.900
CURVE 1 (cont.)	
Series 2 (cont.)	
15.99	6.065×10^{-3}
17.22	7.462
18.43	8.916
19.86	1.084×10^{-2}
21.64	1.334
23.70	1.647
25.95	2.005
28.44	2.417
31.29	2.909
34.64	3.493
38.41	4.146
42.34	4.794
46.61	5.471
51.37	6.176
56.40	6.844

* Not shown on plot

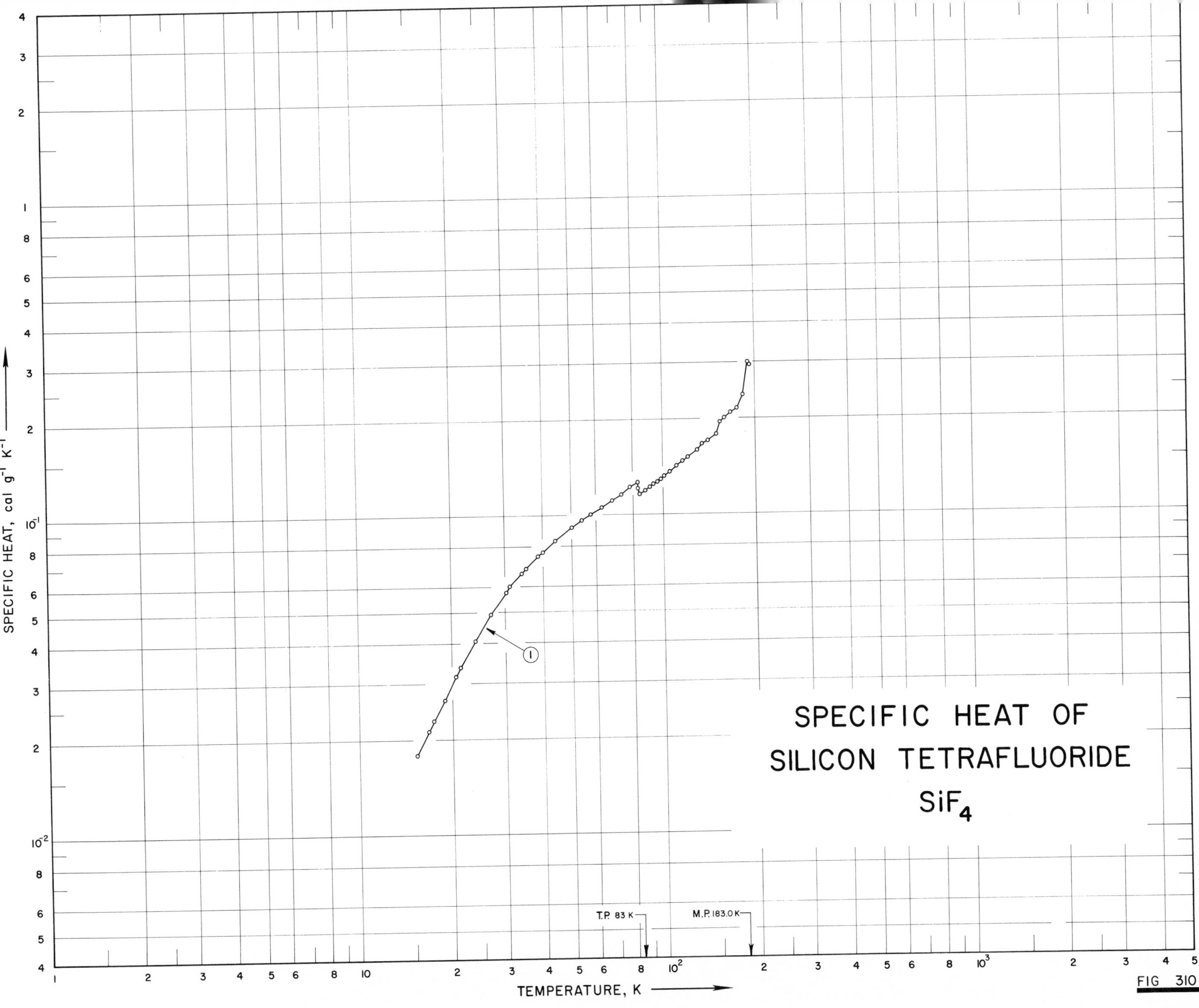
SPECIFIC HEAT OF
SILICON TETRAFLUORIDE
SiF_4
T.P. 83 K
M.P. 183.0 K
TEMPERATURE, K
SPECIFIC HEAT, cal g⁻¹ K⁻¹

FIG 310

SPECIFICATION TABLE NO. 310 SPECIFIC HEAT OF SILICON TETRAFLUORIDE SiF_4

[For Data Reported in Figure and Table No. 310]

Curve No.	Ref. No.	Year	Temp. Range, K	Reported Error, %	Name and Specimen Designation	Composition (weight percent), Specifications and Remarks
1	295	1963	15-194	~2		99.97 SiF_4.

DATA TABLE NO. 310 SPECIFIC HEAT OF SILICON TETRAFLUORIDE SiF_4

[Temperature, T, K; Specific Heat, C_p, Cal g^{-1} K^{-1}]

T	C_p	T	C_p
CURVE 1		CURVE 1 (cont.)	
15.34	1.772 x 10^{-2}	103.75	1.334 x 10^{-1}*
16.75	2.090	106.35	1.359
17.49	2.261	108.90	1.384*
18.80	2.633	111.45	1.412
20.61	3.137	114.10	1.437*
21.31	3.326	116.60	1.460
23.77	4.015	119.14	1.484*
24.02	4.101*	121.59	1.504
26.93	4.872	124.22	1.529*
27.27	5.001*	126.62	1.551*
30.42	5.743	129.34	1.578
31.14	5.955	131.72	1.600*
34.07	6.551	132.42	1.611*
35.20	6.792	133.46	1.609*
38.62	7.419	134.52	1.626*
40.00	7.627	136.20	1.652
44.14	8.282	136.87	1.661*
45.65	8.473*	138.59	1.661*
50.09	9.071	141.21	1.697
54.02	9.580	143.95	1.717*
56.05	9.762*	146.46	1.744*
57.67	9.992	149.35	1.772
59.22	1.011 x 10^{-1}*	151.75	1.798*
62.90	1.056	154.61	1.940
64.32	1.068*	156.91	1.971*
68.02	1.108	159.75	1.993
69.35	1.128*	164.76	2.040*
73.10	1.156	167.35	2.070
74.64	1.171*	169.81	2.092*
78.14	1.217	171.14	2.096*
79.90	1.233*	172.43	2.131*
80.87	1.242*	174.88	2.135*
83.17	1.258	175.24	2.143
83.25	1.260	177.39	2.165*
84.84	1.159	179.94	2.200*
85.82	1.170*	180.67	2.205*
87.94	1.188	180.89	2.272*
88.21	1.196*	182.25	2.355
89.79	1.207*	190.07	2.984
90.79	1.214	191.91	2.946*
92.74	1.234*	192.44	2.965*
93.43	1.241	193.88	2.929
96.03	1.266		
98.66	1.288		
101.29	1.311		

* Not shown on plot

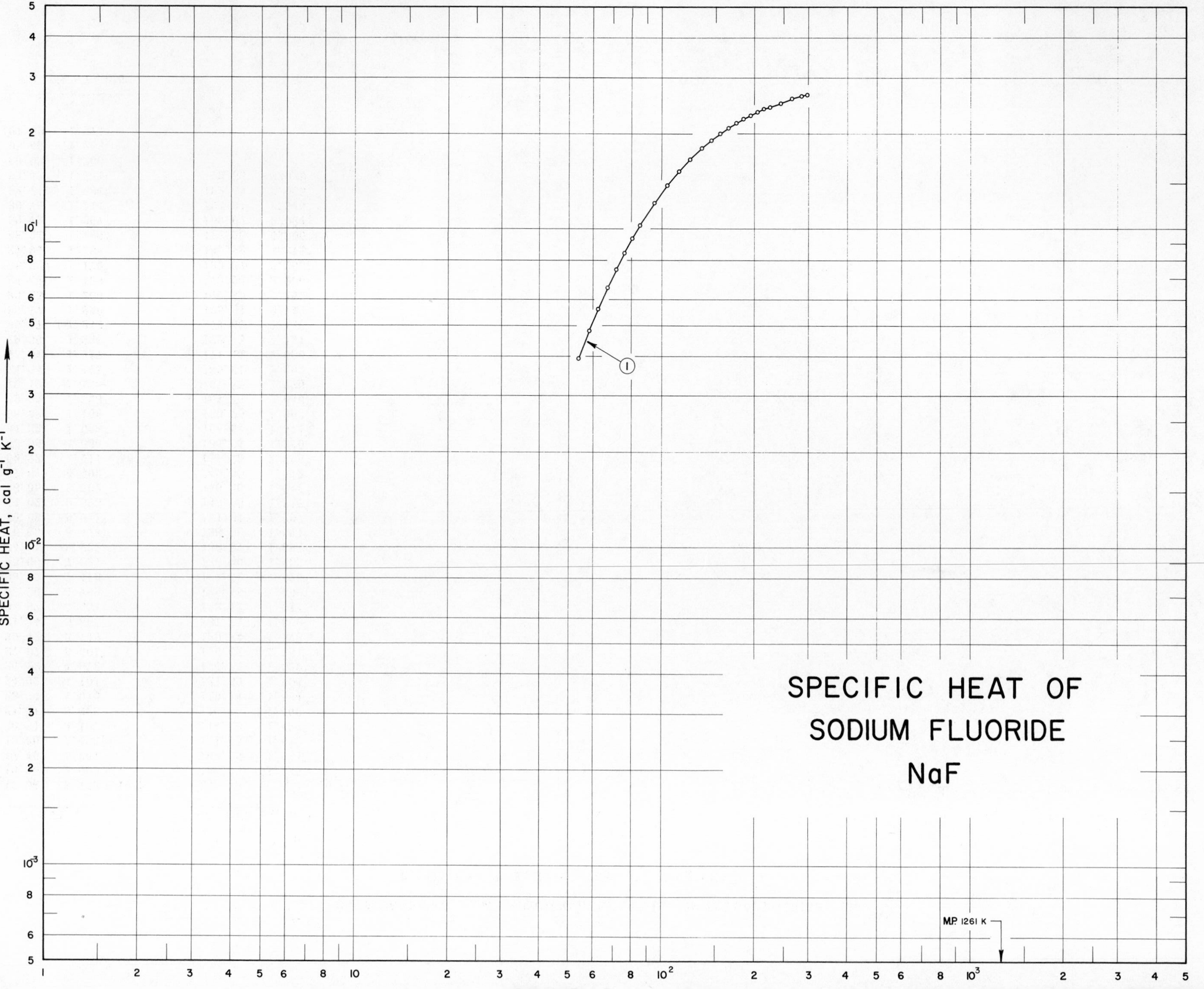

SPECIFIC HEAT OF
SODIUM FLUORIDE
NaF
M.P. 1261 K
1
SPECIFIC HEAT, cal g⁻¹ K⁻¹

SPECIFICATION TABLE NO. 311 SPECIFIC HEAT OF SODIUM FLUORIDE NaF

[For Data Reported in Figure and Table No. 311]

Curve No.	Ref. No.	Year	Temp. Range, K	Reported Error, %	Name and Specimen Designation	Composition (weight percent), Specifications and Remarks
1	279	1956	54-298			Analytical reagent powder; sample supplied by the Mallinckrodt Chem. Co; heated at 700 C immediately before specific heat measurements.

DATA TABLE NO. 311 SPECIFIC HEAT OF SODIUM FLUORIDE NaF

[Temperature, T, K; Specific Heat, C_p, Cal g^{-1} K^{-1}]

T	C_p
CURVE 1	
54.01	3.951×10^{-2}
58.57	4.822
62.80	5.668
67.29	6.597
71.81	7.537
76.33	8.466
80.98	9.419
85.79	1.037×10^{-1}
95.40	1.217
105.04	1.383
114.61	1.532
124.46	1.669
135.98	1.811
145.74	1.919
155.77	2.015
165.84	2.099
175.84	2.173
186.05	2.242
195.79	2.298
205.89	2.353
216.14	2.401
225.80	2.446
235.96	2.486*
245.66	2.517
256.40	2.555*
266.20	2.598
276.20	2.612*
287.15	2.641
295.86	2.660*
298.95	2.665

* Not shown on plot

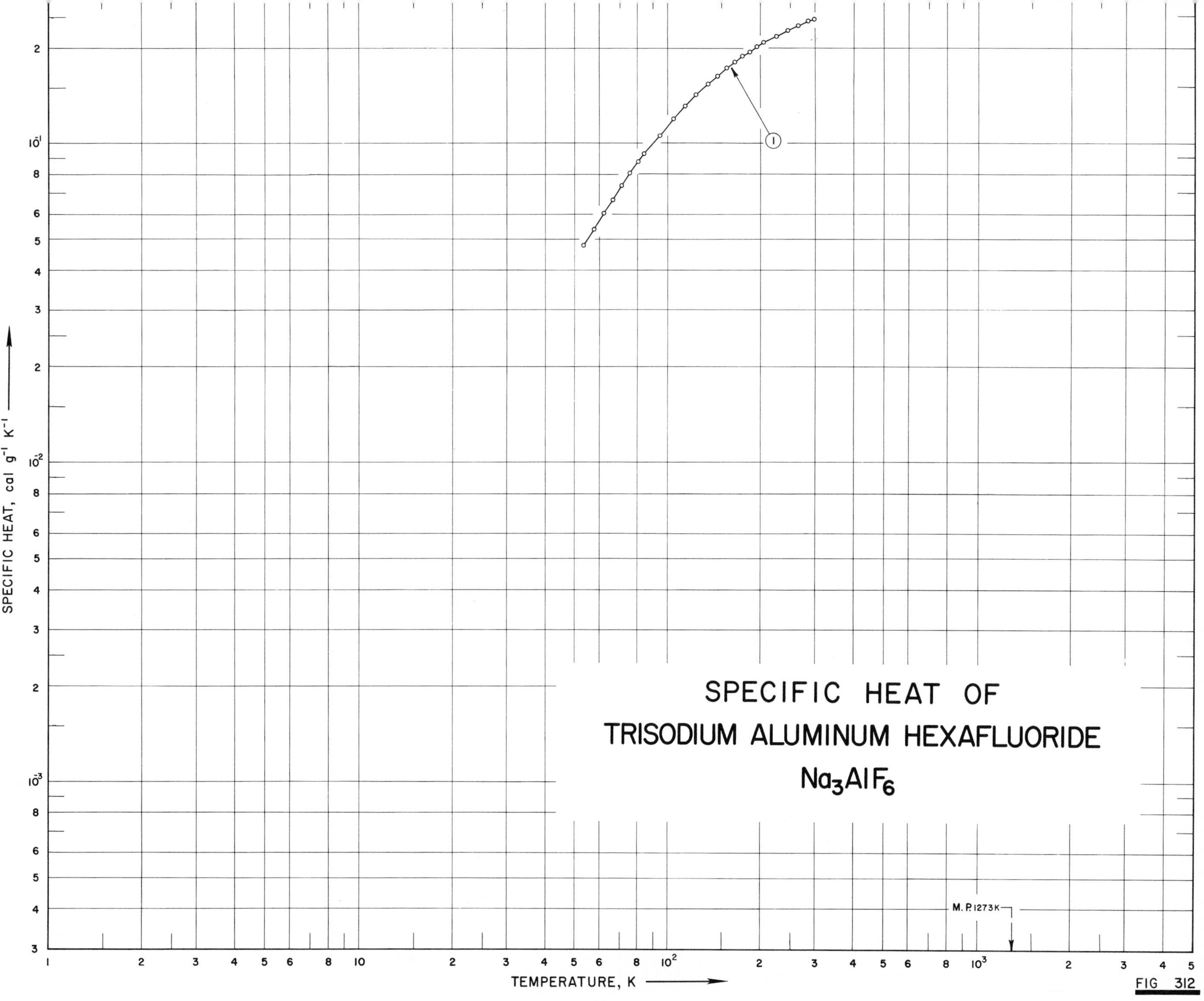
SPECIFIC HEAT OF
TRISODIUM ALUMINUM HEXAFLUORIDE
Na_3AlF_6
SPECIFIC HEAT, cal g^{-1} K^{-1}
TEMPERATURE, K
M.P. 1273K
1
FIG 312

SPECIFICATION TABLE NO. 312 SPECIFIC HEAT OF TRISODIUM ALUMINUM HEXAFLUORIDE Na_3AlF_6

[For Data Reported in Figure and Table No. 312]

Curve No.	Ref. No.	Year	Temp. Range, K	Reported Error, %	Name and Specimen Designation	Composition (weight percent), Specifications and Remarks
1	279	1956	54-298			32.76 Na (32.85 theo), 13.01 Al (12.85 theo), 0.036 K, and 0.007 Li; sample supplied by the Aluminum Co. of America, Research Laboratories; corrected to stoichiometric composition.

DATA TABLE NO. 312 SPECIFIC HEAT OF TRISODIUM ALUMINUM HEXAFLUORIDE Na_3AlF_6

[Temperature, T, K; Specific Heat, C_p, Cal g^{-1} K^{-1}]

T	C_p
CURVE 1	
53.66	4.811 x 10^{-2}
57.94	5.425
62.32	6.078
66.53	6.730
70.94	7.402
75.55	8.097
80.79	8.835
84.45	9.340
94.77	1.070 x 10^{-1}
105.20	1.202
114.69	1.316
124.69	1.427
136.00	1.542
145.51	1.636
155.82	1.728
165.66	1.806
175.91	1.881
185.85	1.948
195.95	2.012
206.18	2.076
216.05	2.128*
226.06	2.177
236.20	2.224*
245.81	2.269
256.33	2.311*
266.23	2.348
276.24	2.382*
286.47	2.418
296.00	2.452*
298.15	2.458

* Not shown on plot

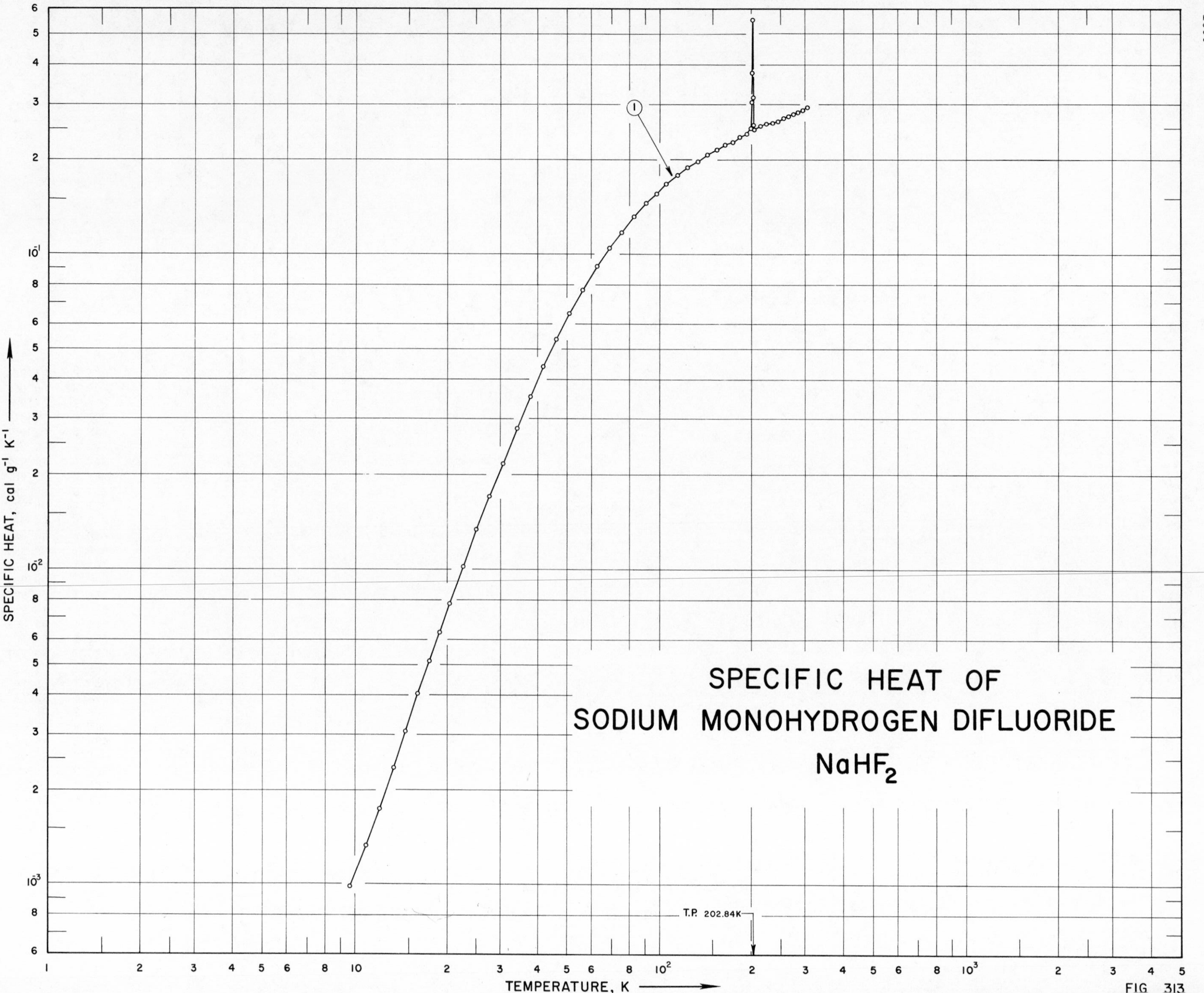
SPECIFIC HEAT OF
SODIUM MONOHYDROGEN DIFLUORIDE
NaHF$_2$
1
T.P. 202.84K
SPECIFIC HEAT, cal g^{-1} K^{-1}
TEMPERATURE, K
FIG 313

SPECIFICATION TABLE NO. 313 SPECIFIC HEAT OF SODIUM MONOHYDROGEN DIFLUORIDE $NaHF_2$

[For Data Reported in Figure and Table No. 313]

Curve No.	Ref. No.	Year	Temp. Range, K	Reported Error, %	Name and Specimen Designation	Composition (weight percent), Specifications and Remarks
1	290	1961	6-305			Obtained from Na_2CO_3.

DATA TABLE NO. 313 SPECIFIC HEAT OF SODIUM MONOHYDROGEN DIFLUORIDE $NaHF_2$

[Temperature, T, K; Specific Heat, C_p, Cal g^{-1} K^{-1}]

CURVE 1

Series 1

T	C_p
5.73	2.097 x 10^{-4}*
8.15	5.000*
9.67	9.839
10.84	1.339 x 10^{-3}
12.07	1.742
13.32	2.355
14.61	3.065
16.02	4.000
17.53	5.129
18.92	6.323
20.43	7.774
22.63	1.023 x 10^{-2}
25.17	1.344
27.71	1.710
30.64	2.168
34.06	2.802
37.82	3.550
41.98	4.413
46.44	5.394
51.28	6.500
56.56	7.708
57.23	7.855*
63.14	9.197
69.40	1.054 x 10^{-1}
75.58	1.179
83.07	1.319
90.92	1.456
98.57	1.570
106.90	1.684
115.85	1.794
124.79	1.892
134.42	1.989
144.75	2.076
154.92	2.160
164.96	2.229
174.96	2.279
184.93	2.353
194.91	2.411
204.83	2.581
215.04	2.561
225.59	2.611
236.03	2.615
246.38	2.661

CURVE 1 (cont.)

Series 1 (cont.)

T	C_p
256.62	2.703 x 10^{-1}
266.70	2.752
276.65	2.790
286.45	2.836
296.11	2.881
305.09	2.926

Series 2

T	C_p
172.99	2.277 x 10^{-1}*
184.09	2.336*
196.04	2.436*
202.24	3.048
204.99	2.497
212.28	2.544*
222.90	2.616*
233.39	2.606*

Series 3

T	C_p
186.08	2.356 x 10^{-1}*
193.96	2.405*
196.78	2.424*
197.90	2.419*
199.02	2.445*
200.13	2.461*
200.96	2.482*
201.51	2.531*
202.06	2.587*
202.54	3.616*
203.00	3.161
203.79	2.481*
204.88	2.495*
209.67	2.429*
214.95	2.568*
217.03	2.573*
218.60	2.582*
220.16	2.597*
222.22	2.621*
224.20	2.629*
226.24	2.629*
228.29	2.584*
233.41	2.606*
241.54	2.647*

CURVE 1 (cont.)

Series 4

T	C_p
200.04	2.563 x 10^{-1}*
208.81	2.521
218.03	2.584
223.34	2.632
224.61	2.648
225.30	2.653
225.91	2.645
226.44	2.603
226.96	2.598
227.51	2.587
228.32	2.581*

Series 5*

T	C_p
199.34	2.445 x 10^{-1}
208.17	2.490
216.65	2.537
225.00	2.574
233.24	2.610

Series 6

T	C_p
196.40	2.419 x 10^{-1}*
198.57	2.437*
199.60	2.453*
200.56	2.471*
201.27	2.513
201.72	2.592*
202.06	2.629*
202.28	2.645*
202.49	2.789*
202.68	3.774
202.84	5.553
203.02	2.521*
203.68	2.486*
204.76	2.490*
209.90	2.531*
217.63	2.576*
222.64	2.618*
224.14	2.640*
224.90	2.637*
225.40	2.644*
225.90	2.661*
226.40	2.619*

CURVE 1 (cont.)

Series 6 (cont.)

T	C_p
226.90	2.589 x 10^{-1}*
227.66	2.586*

Series 7*

T	C_p
183.22	2.344 x 10^{-1}
191.83	2.392
198.12	2.427
201.86	2.616
206.11	2.486
211.05	2.518
216.22	2.545
222.92	2.577
231.16	2.594
239.49	2.631

* Not shown on plot

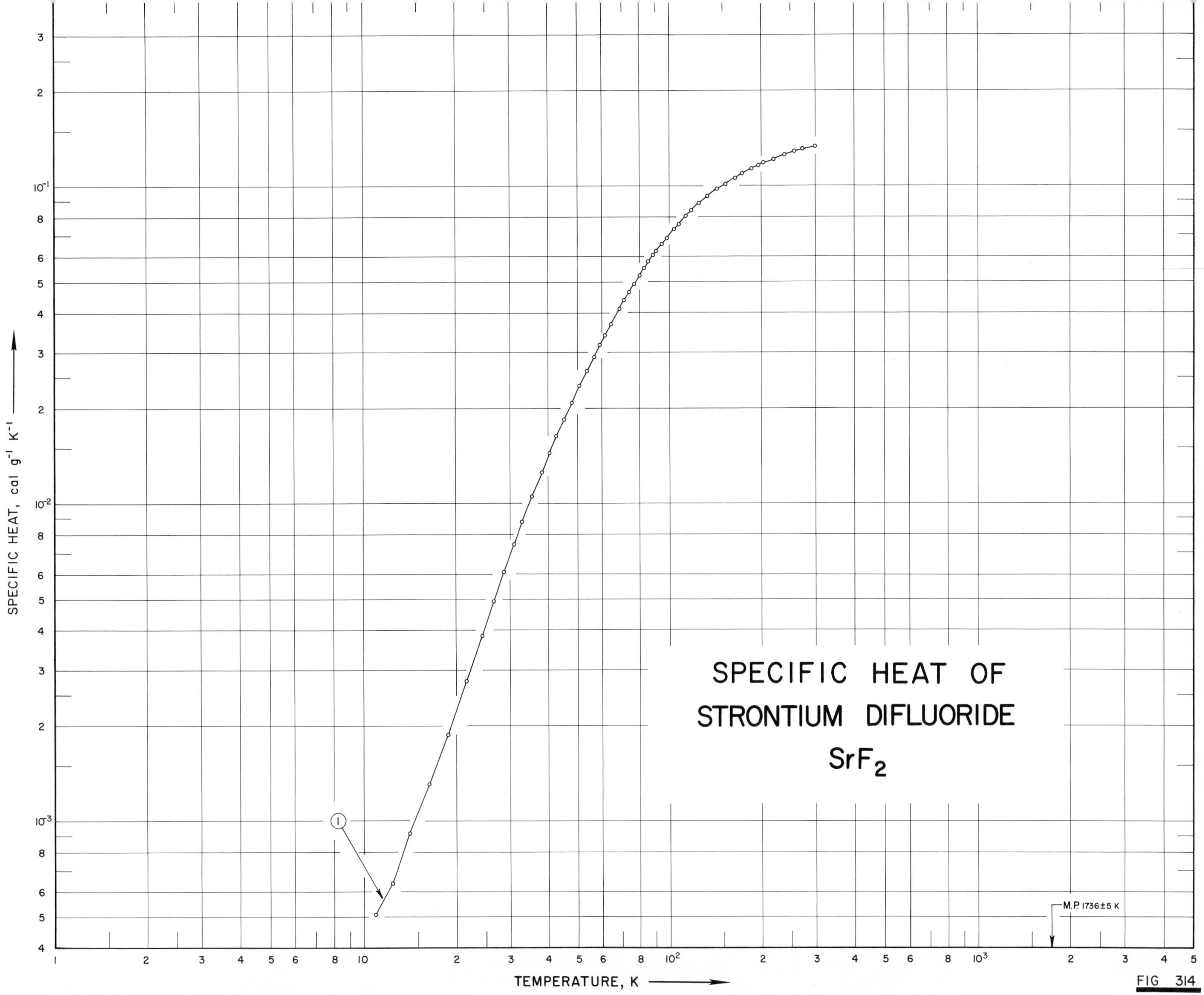

SPECIFIC HEAT OF STRONTIUM DIFLUORIDE SrF_2

FIG 314

SPECIFICATION TABLE NO. 314 SPECIFIC HEAT OF STRONTIUM DIFLUORIDE SrF_2

[For Data Reported in Figure and Table No. 314]

Curve No.	Ref. No.	Year	Temp. Range, K	Reported Error, %	Name and Specimen Designation	Composition (weight percent), Specifications and Remarks
1	276	1963	11-300			Impurities: 0.001-0.01 Ca, K, traces of Cu, Fe, and Mg; washed and dried for several hrs at 600 C.

DATA TABLE NO. 314 SPECIFIC HEAT OF STRONTIUM DIFLUORIDE SrF_2

[Temperature, T, K; Specific Heat, C_p, Cal g^{-1} K^{-1}]

T	C_p
CURVE 1	
10.94	5.095 x 10^{-4}
12.38	6.369
14.08	9.155
16.47	1.314 x 10^{-3}
18.93	1.879
21.56	2.754
24.36	3.845
26.63	4.928
28.73	6.090
30.94	7.419
32.96	8.773
35.42	1.051 x 10^{-2}
38.22	1.258
40.59	1.443
42.86	1.630
45.34	1.835
48.04	2.076
50.94	2.349
53.81	2.621
56.64	2.893
57.05	2.934
59.25	3.153
59.64	3.191
61.45	3.375
62.02	3.431
64.45	3.678
66.74	3.905
68.91	4.118
71.43	4.368
74.30	4.649
77.26	4.937
80.22	5.236
81.79	5.378
83.22	5.519
84.29	5.617
85.99	5.781
86.91	5.864
88.65	6.023
89.44	6.091
91.21	6.244
92.24	6.365
93.69	6.453
95.29	6.583
95.99	6.645
98.24	6.827
CURVE 1 (cont.)	
98.67	6.863 x 10^{-2}
101.38	7.079
101.63	7.096
104.49	7.318
104.68	7.333
106.27	7.450
107.88	7.571
109.20	7.667
110.98	7.798
112.21	7.881
114.00	8.001
115.26	8.092
116.97	8.201
119.45	8.375
121.14	8.486
123.99	8.661
125.51	8.747
126.82	8.835
128.51	8.932
131.78	9.116
134.99	9.298
138.09	9.459
141.13	9.617
144.13	9.760
147.23	9.911
150.14	1.004 x 10^{-1}
153.01	1.017
155.85	1.029
161.20	1.050
162.26	1.054
165.57	1.067
168.84	1.080
172.04	1.091
174.58	1.099
177.10	1.107
180.24	1.117
183.34	1.127
186.42	1.136
189.48	1.144
192.48	1.144
192.51	1.153
193.73	1.157
197.39	1.166
198.50	1.169
201.18	1.176
CURVE 1 (cont.)	
204.40	1.183 x 10^{-1}
208.88	1.193
212.68	1.202
216.55	1.210
220.40	1.218
224.11	1.225
227.23	1.230
231.21	1.238
235.78	1.245
239.88	1.253
244.94	1.261
249.15	1.267
253.34	1.274
254.81	1.276
257.50	1.280
257.54	1.280
258.99	1.282
261.63	1.286
263.14	1.288
265.74	1.292
270.86	1.299
273.87	1.303
273.88	1.303
278.66	1.310
282.09	1.314
285.13	1.318
289.96	1.323
295.29	1.328
300.05	1.334

* Not shown on plot

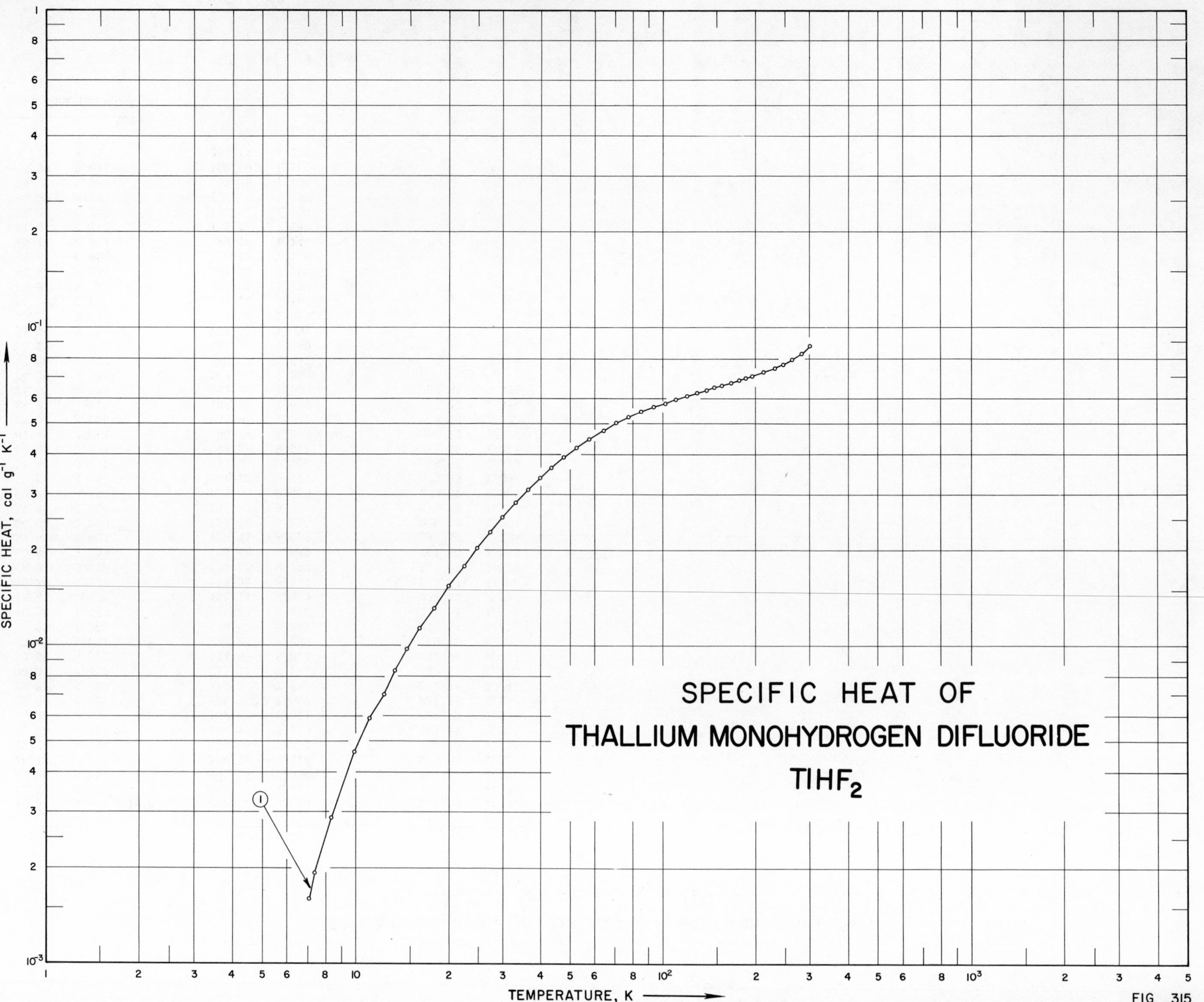
SPECIFIC HEAT OF
THALLIUM MONOHYDROGEN DIFLUORIDE
TlHF₂
TEMPERATURE, K
SPECIFIC HEAT, cal g⁻¹ K⁻¹
1
FIG 315

SPECIFICATION TABLE NO. 315 SPECIFIC HEAT OF THALLIUM MONOHYDROGEN DIFLUORIDE $TlHF_2$

[For Data Reported in Figure and Table No. 315]

Curve No.	Ref. No.	Year	Temp. Range, K	Reported Error, %	Name and Specimen Designation	Composition (weight percent), Specifications and Remarks
1	286	1961	7-301			

DATA TABLE NO. 315 SPECIFIC HEAT OF THALLIUM MONOHYDROGEN DIFLUORIDE $TlHF_2$

[Temperature, T, K; Specific Heat, C_p, Cal g^{-1} K^{-1}]

T	C_p	T	C_p
CURVE 1		CURVE 1 (cont.)	
Series 1		Series 2 (cont.)	
158.10	6.643×10^{-2}	57.63	4.495×10^{-2}
167.53	6.742	52.78	4.232
177.00	6.857	58.52	4.532
186.18	6.964	64.34	4.799
195.73	7.076	70.59	5.029
204.83	7.169	77.45	5.255
213.85	7.272	85.06	5.481
222.80	7.379	93.50	5.674
231.86	7.490	102.15	5.838
220.22	7.338	110.95	5.994
229.13	7.453	120.01	6.138
238.03	7.568	129.01	6.270
246.94	7.695	138.04	6.393
255.88	7.827	147.11	6.508
264.91	7.975	156.26	6.619
273.98	8.139		
283.05	8.320		
292.07	8.537		
300.97	8.804		
Series 2			
7.08	1.606×10^{-3}		
7.36	1.939		
8.34	2.884		
9.84	4.651		
11.14	5.920		
12.28	7.009		
13.45	8.377		
11.69	9.778		
16.06	1.130×10^{-2}		
17.82	1.313		
19.92	1.542		
22.28	1.789		
24.72	2.039		
27.24	2.287		
29.98	2.548		
33.02	2.827		
36.29	3.108		
39.74	3.383		
43.43	3.650		
47.65	3.932		
52.50	4.224		

* Not shown on plot

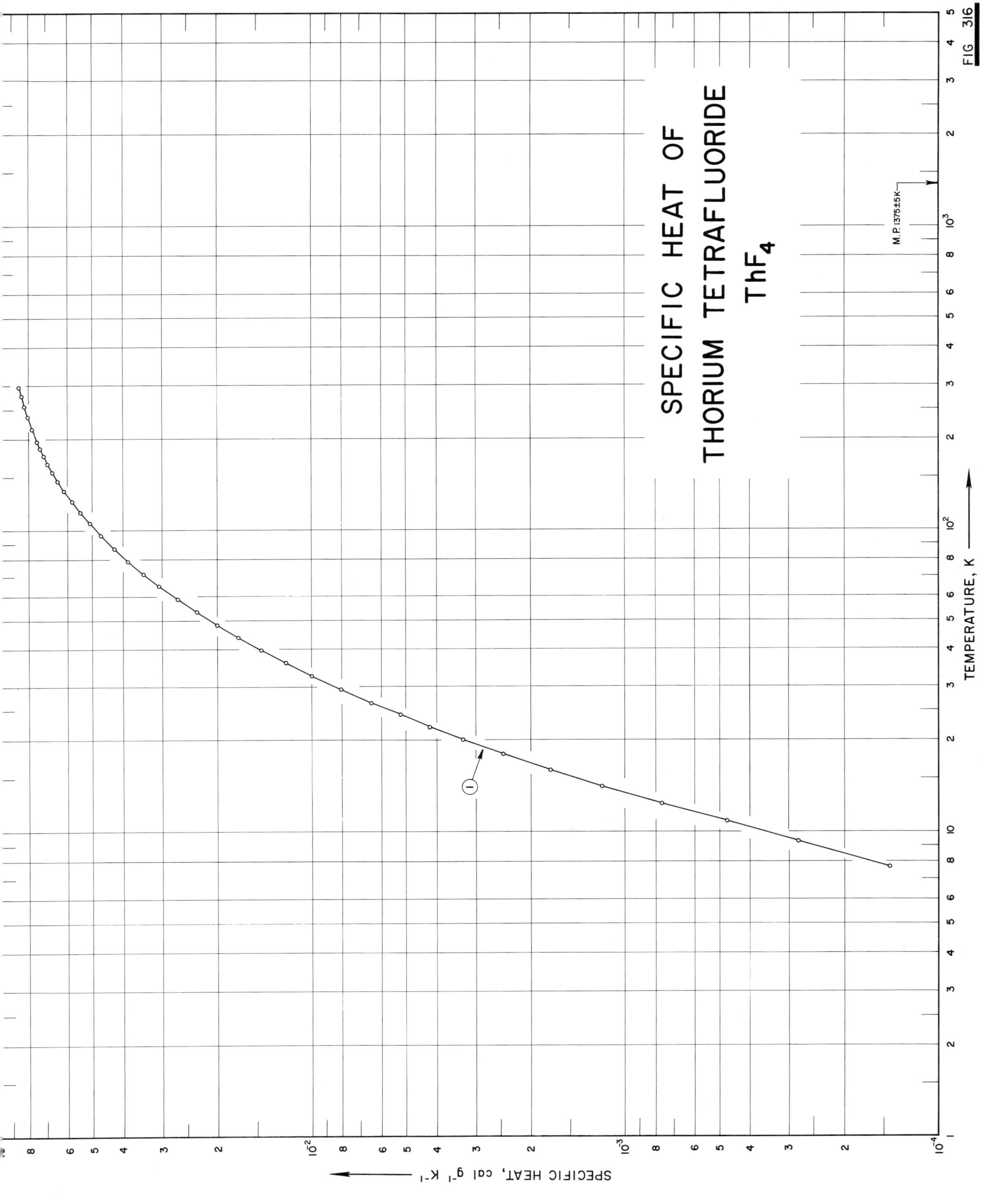

SPECIFIC HEAT OF
THORIUM TETRAFLUORIDE
ThF_4
SPECIFIC HEAT, cal g⁻¹ K⁻¹
TEMPERATURE, K
M.P. 1375±5K
1

FIG 316

SPECIFICATION TABLE NO. 316 SPECIFIC HEAT OF THORIUM TETRAFLUORIDE ThF_4

[For Data Reported in Figure and Table No. 316]

Curve No.	Ref. No.	Year	Temp. Range, K	Reported Error, %	Name and Specimen Designation	Composition (weight percent), Specifications and Remarks
1	296	1954	6-298	0.1-5		75.33 Th (75.33 theo), 24.6 F (24.67 theo); 0.13 unconverted ThO_2, 0.007 Ca, 0.001 Mg, <0.001 K, 0.0005 each Fe, Na, Ni, and <0.0005 each Ag, Al, Be, Bi, Co, Cr, Cu, Li, Mn, Pb, and Sn; prepared by hydrofluorination of pure sample of electrically fused THO_2.

DATA TABLE NO. 316 SPECIFIC HEAT OF THORIUM TETRAFLUORIDE ThF_4

[Temperature, T, K; Specific Heat, C_p, Cal $g^{-1} K^{-1}$]

T	C_p
CURVE 1	
5.54	5.193 x 10^{-5}*
7.67	1.428 x 10^{-4}
9.31	2.791
10.86	4.738
12.47	7.659
14.12	1.185 x 10^{-3}
15.98	1.733
18.02	2.454
20.07	3.291
22.14	4.229
24.21	5.238
26.59	6.494
29.43	8.078
32.62	9.990
36.06	1.212 x 10^{-2}
39.79	1.451
44.02	1.721
48.54	2.015
53.48	2.334
59.02	2.684
59.42	2.708*
65.42	3.077
71.92	3.450
79.31	3.859
87.48	4.284
96.65	4.706
106.08	5.105
115.58	5.475
125.48	5.829
135.92	6.163
146.33	6.465
156.48	6.721
166.57	6.958
176.73	7.166
186.98	7.358
197.04	7.523
206.90	7.676*
216.84	7.809
227.16	7.942*
237.60	8.059
247.88	8.172*
258.06	8.270
268.18	8.364*
278.25	8.445
288.25	8.519*
298.17	8.588

* Not shown on plot

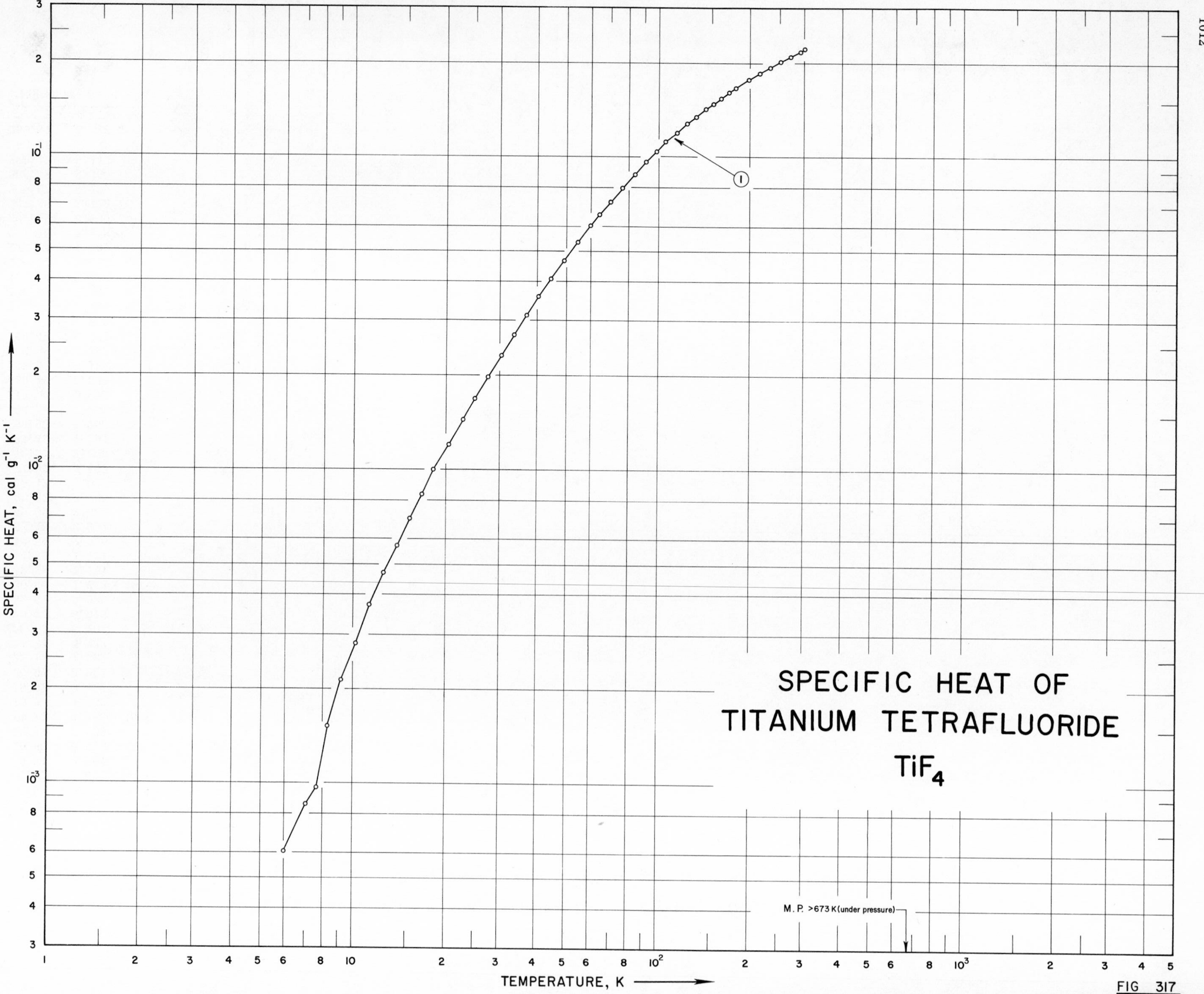
SPECIFIC HEAT OF
TITANIUM TETRAFLUORIDE
TiF4
1
M. P. >673 K (under pressure)
TEMPERATURE, K
SPECIFIC HEAT, cal g-1 K-1
FIG 317

SPECIFICATION TABLE NO. 317 SPECIFIC HEAT OF TITANIUM TETRAFLUORIDE TiF_4

[For Data Reported in Figure and Table No. 317]

Curve No.	Ref. No.	Year	Temp. Range, K	Reported Error, %	Name and Specimen Designation	Composition (weight percent), Specifications and Remarks
1	297	1961	6-302	0.1-5		99.9 TiF_4; sample supplied by the General Chemical Division of Allied Chemical and Dye Corp; finely divided white sample; prepared by reaction of element F_2 on pure TiO_2 and purified by sublimation in nickel reactors; measured in a nitrogen atmosphere.

DATA TABLE NO. 317 SPECIFIC HEAT OF TITANIUM TETRAFLUORIDE TiF_4

[Temperature, T, K; Specific Heat, C_p, Cal $g^{-1} K^{-1}$]

T	C_p
CURVE 1	
Series 1	
59.43	5.944×10^{-2}*
64.57	6.584
70.02	7.224
76.58	7.994
84.02	8.854
91.68	9.701
99.40	1.045×10^{-1}
107.44	1.123
116.12	1.203
125.04	1.283
133.96	1.356
142.87	1.424
151.90	1.487
161.26	1.553
170.85	1.615
180.15	1.672
189.34	1.723*
198.31	1.773
207.09	1.818*
215.89	1.862
224.78	1.906*
233.81	1.947
243.02	1.989*
252.21	2.023
261.69	2.063*
271.80	2.103
282.34	2.146*
292.74	2.184
301.54	2.224
Series 2	
6.32	6.053×10^{-4}
7.06	8.555
7.68	9.685
8.35	1.525×10^{-3}
9.26	2.147
10.21	2.809
11.35	3.721
12.68	4.705
13.94	5.755
15.28	6.990

T	C_p
CURVE 1 (cont.)	
Series 2 (cont.)	
16.76	8.370×10^{-3}
18.42	1.000×10^{-2}
20.52	1.211
22.80	1.457
25.01	1.700
27.60	1.991
30.53	2.338
33.59	2.713
36.93	3.129
40.63	3.585
44.75	4.096
49.43	4.689
54.70	5.350
60.38	6.057
Series 3	
246.39	1.998×10^{-1}*
255.26	2.036*
264.83	2.077*
274.54	2.110*
284.09	2.153*
292.34	2.183*
300.45	2.213*

* Not shown on plot

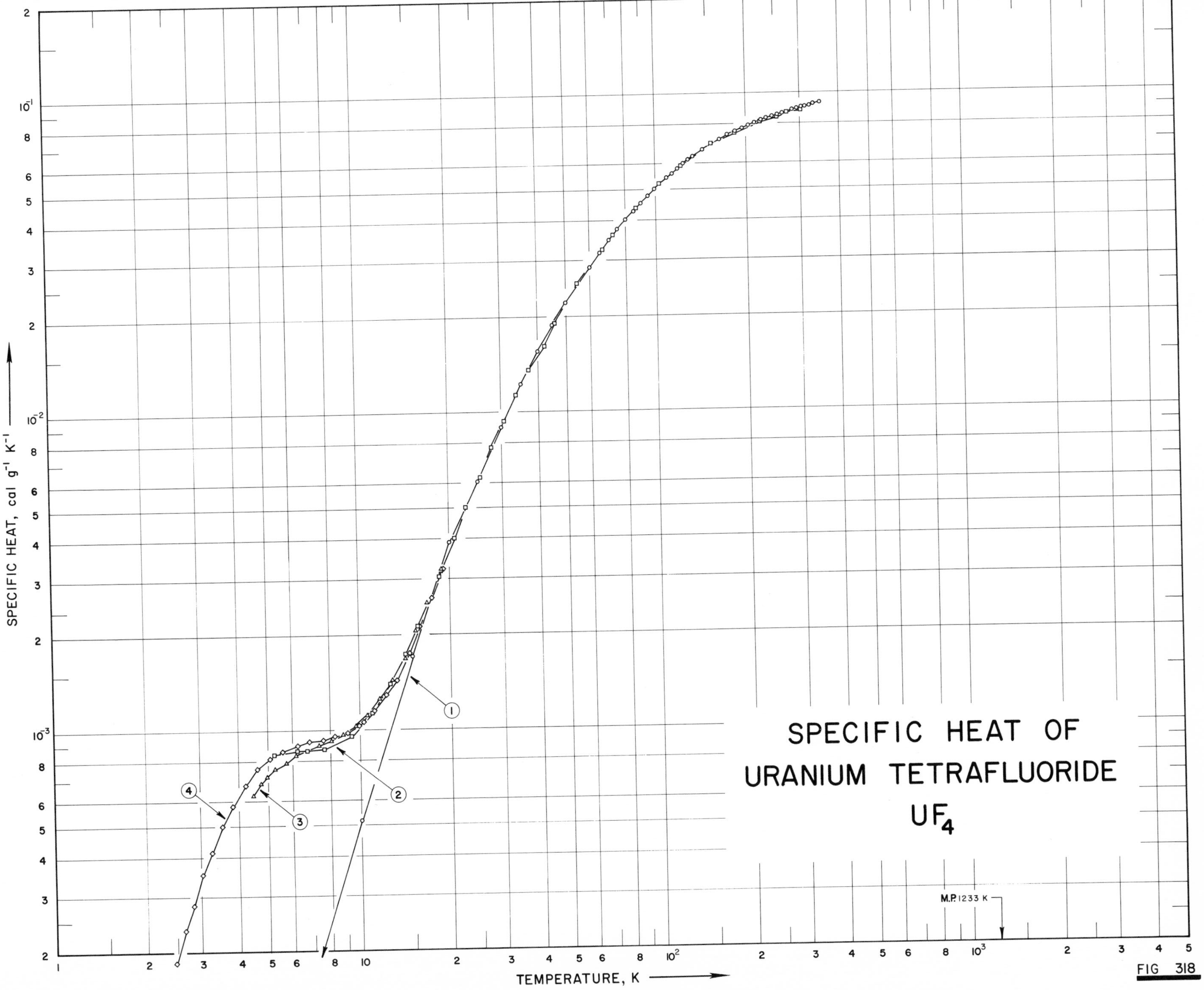

SPECIFIC HEAT OF URANIUM TETRAFLUORIDE UF_4

SPECIFIC HEAT, cal g^{-1} K^{-1}

TEMPERATURE, K

M.P. 1233 K

1 2 3 4

FIG 318

SPECIFICATION TABLE NO. 318 SPECIFIC HEAT OF URANIUM TETRAFLUORIDE UF_4

[For Data Reported in Figure and Table No. 318]

Curve No.	Ref. No.	Year	Temp. Range, K	Reported Error, %	Name and Specimen Designation	Composition (weight percent), Specifications and Remarks
1	298	1948	5-350			2.0 UO_2F_2.
2	299	1955	5-304			75.83 ± 0.07 U and 24.18 ± 0.06 F (75.80 and 24.20 theo), 0.006 ± 0.003 oxygen corrected to 0.05 UO_2F_2, < 0.01 Pt, < 0.005 Ti, and < 0.002 each Ag, Al, As, Be, Bi, Ca, Co, Cr, Cu, Fe, K, Li, Mg, Mn, Mo, Na, Ni, P, Pb, Sb, Sn, Ta, Zr, and 0.0015 rare earth; sample supplied by the Malinckrodt Chem. Co; prepared from sample by sublimating twice under vacuum at 1100 C for several hrs.
3	300	1960	4.4-18	0.4-2.0		75.83 ± 0.07 U, 24.18 ± 0.06 F, < 0.01 Pt, <0.005 Ti, and < 0.002 each Ag, Al, As, Be, Bi, Ca, Co, Cr, Cu, Fe, K, Li, Mg, Mn, Mo, Na, Ni, P, Pb, Sb, Sn, Ta and Zr; powdered form.
4	300	1960	1.6-19	0.4-2.0		75.8 U, 24.18 F, 0.004 Mo, 0.0015 Si, 0.001 Cu, 0.0006 Mg, 0.0005 Al, 0.0003 Ni, 0.0002 Zr, 0.00008 Cr, 0.00002 Ba, and 0.00001 each Li and Mn; granular; melted by induction heating at about 1000 C under helium atmosphere.

DATA TABLE NO. 318 SPECIFIC HEAT OF URANIUM TETRAFLUORIDE UF_4

[Temperature, T, K; Specific Heat, C_p, Cal g^{-1} K^{-1}]

T	C_p
CURVE 1	
5.00	6.089×10^{-5}*
10.00	5.176×10^{-4}
15.00	1.705×10^{-3}
20.00	3.905
25.00	6.036
30.00	8.981
35.00	1.229×10^{-2}
40.00	1.560
45.00	1.891
50.00	2.224
55.00	2.558
60.00	2.871
65.00	3.185
70.00	3.487
75.00	3.771
80.00	4.045
85.00	4.307
90.00	4.558
95.00	4.796
100.00	5.020
105.00	5.240
110.00	5.450
115.00	5.652
120.00	5.842
125.00	6.027
130.00	6.203
135.00	6.368
140.00	6.520*
145.00	6.665
150.00	6.804*
155.00	6.937
160.00	7.066*
165.00	7.187
170.00	7.301*
175.00	7.409
180.00	7.509
185.00	7.606
190.00	7.696*
195.00	7.782
200.00	7.864*
205.00	7.943
210.00	8.018*
215.00	8.092
220.00	8.162*
225.00	8.228
CURVE 1 (cont.)	
230.00	8.290×10^{-2}*
235.00	8.350
240.00	8.408*
245.00	8.464
250.00	8.516*
255.00	8.567
260.00	8.616*
265.00	8.662
270.00	8.708*
273.16	8.738*
275.00	8.755
280.00	8.801*
285.00	8.844
290.00	8.886*
295.00	8.926
298.16	8.952*
300.00	8.967*
305.00	9.006
310.00	9.045*
315.00	9.084
320.00	9.122*
325.00	9.159
330.00	9.195*
335.00	9.231
340.00	9.266*
345.00	9.301*
350.00	9.334
CURVE 2	
5.26	8.406×10^{-4}
6.26	8.660
7.69	8.724
9.38	9.552
11.08	1.130×10^{-3}
12.71	1.404
14.23	1.748
15.73	2.136
17.30	2.617
18.95	3.229
20.80	4.005
22.94	5.012
25.36	6.266
27.97	7.737
30.69	9.380
CURVE 2 (cont.)	
33.73	1.133×10^{-2}
37.22	1.363
41.13	1.622
45.39	1.906
49.92	2.209
54.80	2.528
60.11	2.868
60.38	2.885
66.31	3.254
72.63	3.620
79.57	4.009
86.96	4.407
95.22	4.801
104.44	5.209
114.06	5.601
123.95	5.970
134.43	6.327
145.14	6.651
155.50	6.932
165.59	7.174
175.31	7.384
184.84	7.565
194.39	7.734
204.20	7.890
214.29	8.024
224.48	8.164
234.69	8.288
244.92	8.396
255.00	8.492
264.94	8.584
274.75	8.664
284.43	8.734
294.11	8.801
303.79	8.864
CURVE 3	
Series 1	
4.442	6.244×10^{-4}
4.726	6.817
4.972	7.158
5.293	7.556
5.713	7.966
6.142	8.317
CURVE 3 (cont.)	
Series 1 (cont.)	
6.712	8.660×10^{-4}
7.350	8.966
8.050	9.291
8.814	9.676
8.820	9.660*
9.737	1.031×10^{-3}
10.680	1.120
11.748	1.257
12.888	1.449
14.072	1.697
15.456	2.057
16.976	2.522
18.497	3.053
Series 2	
4.177	6.674×10^{-4}*
Series 3	
2.486	1.856×10^{-4}
2.652	2.350
2.808	2.799
3.014	3.528
3.253	4.184
3.524	4.989
3.830	5.836
4.197	6.715
4.606	7.514
5.062	8.154
5.595	8.651
6.226	8.998
6.875	9.211
7.544	9.374
8.280	9.552
9.094	9.835
9.982	1.036×10^{-3}
Series 4	
10.304	1.061×10^{-3}
11.231	1.156
12.248	1.292
CURVE 3 (cont.)	
Series 4 (cont.)	
13.354	1.433×10^{-3}
14.603	1.764
15.998	2.136
17.481	2.614
19.158	3.238
Series 5	
1.630	3.439×10^{-5}
1.826	4.916
1.993	7.323
2.188	1.111×10^{-4}
2.370	1.544
2.554	2.044
2.753	2.636
2.957	3.264
3.191	3.996
3.476	4.856
Series 6	
1.303	1.305×10^{-5}
1.486	2.133
1.645	3.534
1.810	4.712
2.036	7.174
2.231	1.207×10^{-4}
2.418	1.668
Series 7	
13.753	1.567×10^{-3}
15.063	1.879
16.350	2.261
17.878	2.773
Series 8	
14.951	1.850×10^{-3}
16.247	3.222
17.626	2.678
19.200	3.276

* Not shown on plot

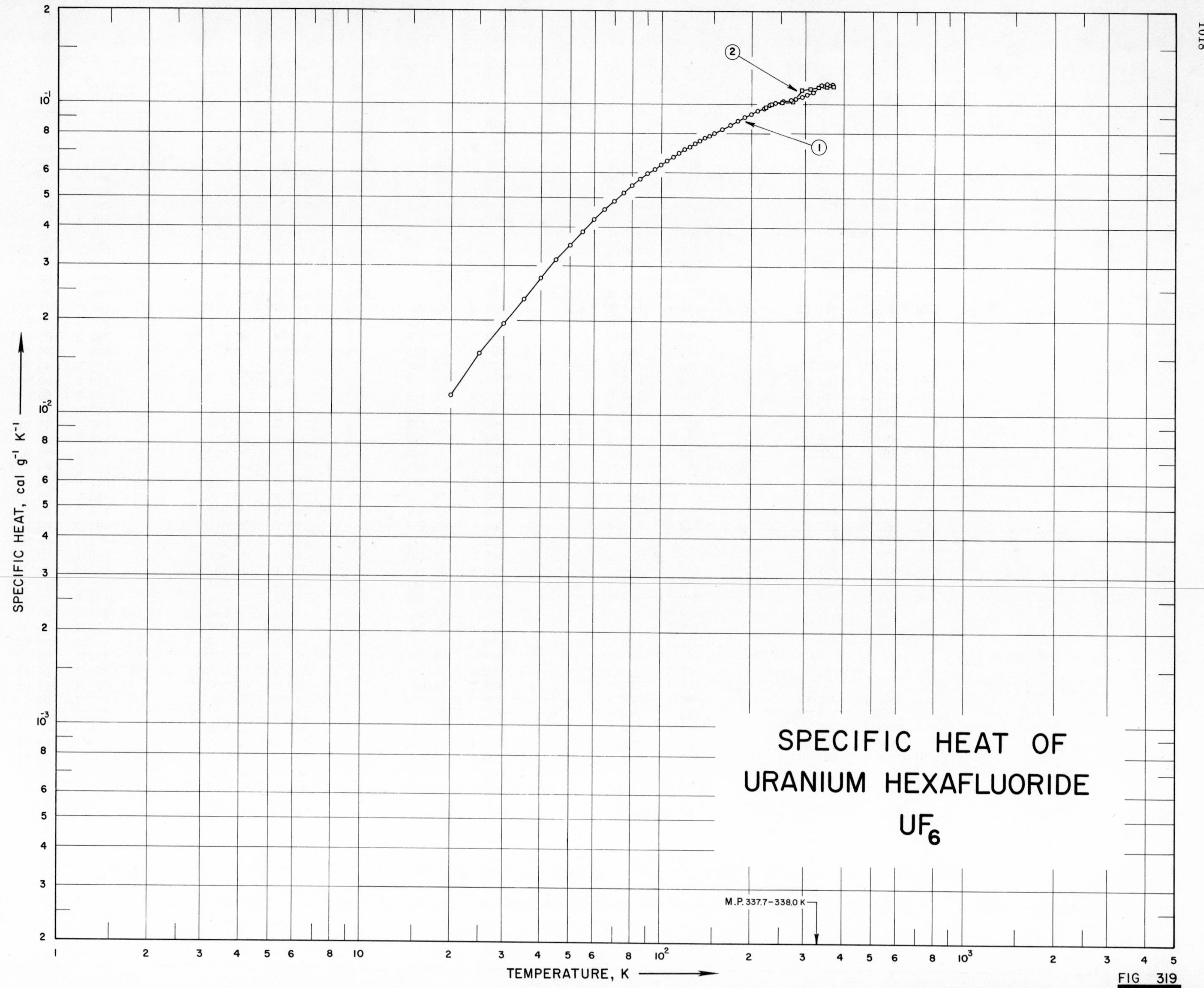
SPECIFIC HEAT OF
URANIUM HEXAFLUORIDE
UF6
M.P. 337.7–338.0 K
SPECIFIC HEAT, cal g⁻¹ K⁻¹
TEMPERATURE, K
1
2

FIG 319

SPECIFICATION TABLE NO. 319 SPECIFIC HEAT OF URANIUM HEXAFLUORIDE UF_6

[For Data Reported in Figure and Table No. 319]

Curve No.	Ref. No.	Year	Temp. Range, K	Reported Error, %	Name and Specimen Designation	Composition (weight percent), Specifications and Remarks
1	298	1948	20–370			2 x 10^{-1} mole fraction impurities.
2	301	1953	223–373	1.0		SiF_4 chief impurity; purified by repeated distillation; under 10^{-5} mm vacuum.

DATA TABLE NO. 319 SPECIFIC HEAT OF URANIUM HEXAFLUORIDE UF_6

[Temperature, T, K; Specific Heat, C_p, Cal g^{-1} K^{-1}]

T	C_p	T	C_p
CURVE 1		CURVE 1 (cont.)	
20	1.154×10^{-2}	245	1.025×10^{-1}*
25	1.571	250	1.035*
30	1.966	255	1.045
35	2.349	260	1.055*
40	2.737	265	1.065*
45	3.124	270	1.075
50	3.505	273.16	1.082*
55	3.879	275	1.085*
60	4.232	280	1.095
65	4.565	283	1.105*
70	4.875	290	1.115*
75	5.176	295	1.125
80	5.454	298.16	1.132*
85	5.720	300	1.136*
90	5.961	305	1.148
95	6.181	310	1.161*
100	6.387	315	1.174*
105	6.581	320	1.187
110	6.766	325	1.201*
115	6.949	330	1.215*
120	7.123	335	1.229
125	7.289	(s) 337.21	1.232*
130	7.450	(l) 337.21	1.295*
135	7.604	340	1.297
140	7.754	345	1.302*
145	7.895	350	1.307*
150	8.038	355	1.312
155	8.173*	360	1.316*
160	8.306	365	1.320*
165	8.438*	370	1.324
170	8.565		
175	8.691*	CURVE 2	
180	8.813		
185	8.925*	233.15	9.71×10^{-2}
190	9.051	243.15	9.83
195	9.165*	253.15	1.01×10^{-1}
200	9.275	273.15	1.05
205	9.386*	293.15	1.11
210	9.503	313.15	1.19
215	9.595*	333.15	1.26
220	9.721	343.15	1.35*
225	9.827*	353.15	1.37
230	9.931	363.15	1.39*
235	1.004×10^{-1}*	373.15	1.41
240	1.014		

* Not shown on plot

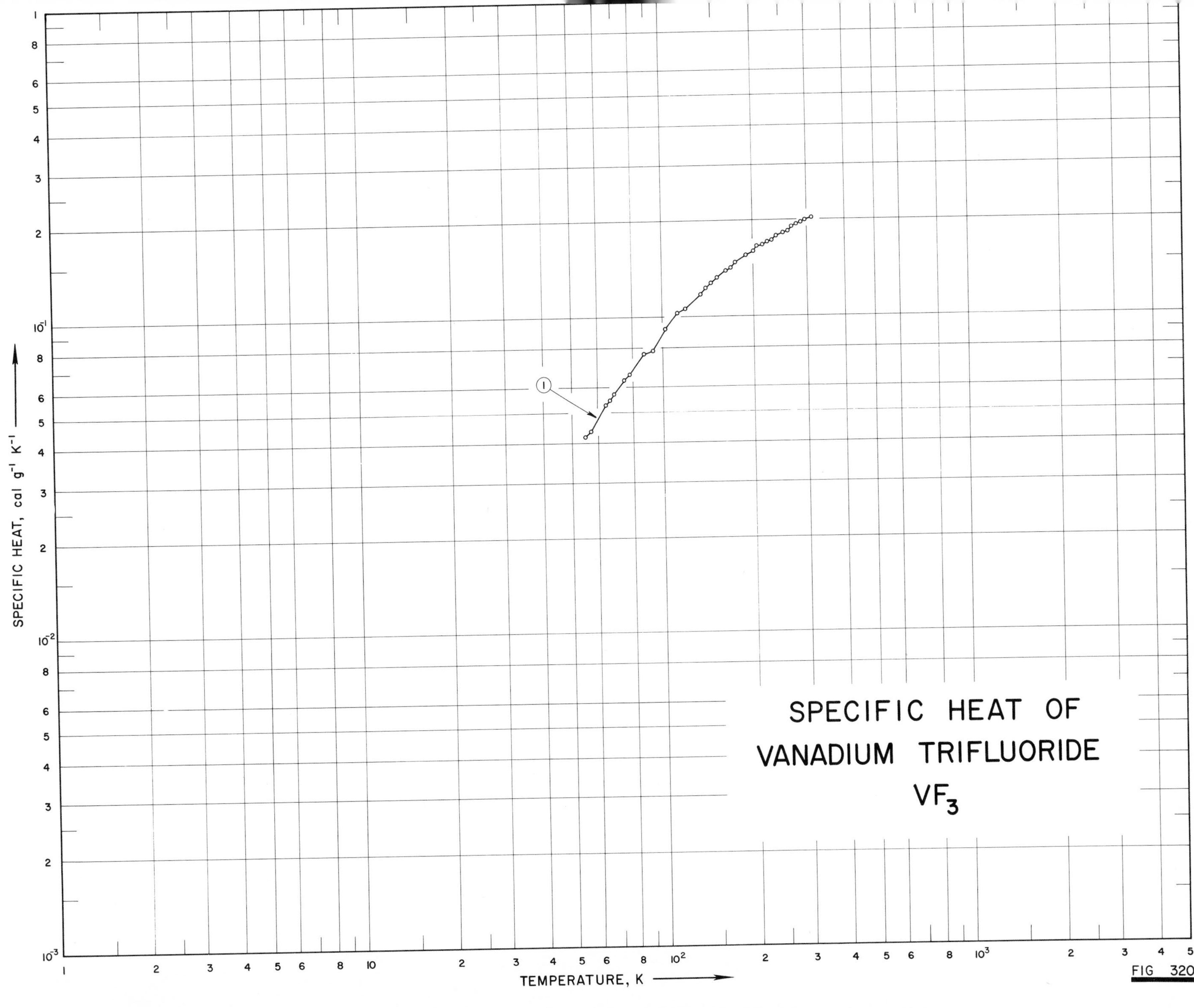
SPECIFIC HEAT OF
VANADIUM TRIFLUORIDE
VF_3
SPECIFIC HEAT, cal g⁻¹ K⁻¹
TEMPERATURE, K
1

FIG 320

SPECIFICATION TABLE NO. 320 SPECIFIC HEAT OF VANADIUM TRIFLUORIDE VF_3

[For Data Reported in Figure and Table No. 320]

Curve No.	Ref. No.	Year	Temp. Range, K	Reported Error, %	Name and Specimen Designation	Composition (weight percent), Specifications and Remarks
1	332	1957	55-315			52.8 F_2 and 46.5 V.

DATA TABLE NO. 320 SPECIFIC HEAT OF VANADIUM TRIFLUORIDE VF_3

[Temperature, T, K; Specific Heat, C_p, Cal g^{-1} K^{-1}]

T	C_p	T	C_p
CURVE 1		CURVE 1 (cont.)	
54.89	4.225 x 10^{-2}	269.61	1.928 x 10^{-1}
57.77	4.391	273.24	1.921*
64.19	5.318	275.52	1.924*
66.53	5.466	279.78	1.960
68.79	5.744	284.23	1.977*
74.64	6.318	287.16	1.973*
78.11	6.578	289.98	1.982
87.40	7.643	292.87	1.975*
90.31	7.819	295.61	1.992
102.87	9.153	298.33	2.003
105.41	9.265*	305.89	2.035*
113.83	1.028 x 10^{-1}	308.46	2.035*
116.71	1.047*	311.87	2.038
118.11	1.059*	315.04	2.051
120.30	1.070		
128.45	1.153*		
135.39	1.190		
140.57	1.241		
143.65	1.231*		
146.38	1.298		
153.77	1.334		
155.98	1.334*		
163.57	1.400		
169.50	1.429		
173.07	1.482*		
177.43	1.487		
181.41	1.501*		
188.03	1.546*		
190.76	1.572*		
191.61	1.557		
194.03	1.581*		
197.67	1.599*		
200.82	1.614		
206.62	1.678		
209.83	1.673*		
213.76	1.684		
216.34	1.694		
218.93	1.706*		
225.66	1.735		
228.94	1.748*		
232.03	1.755		
239.88	1.800		
248.05	1.832*		
254.74	1.836		
263.90	1.871		
266.72	1.924*		

* Not shown on plot

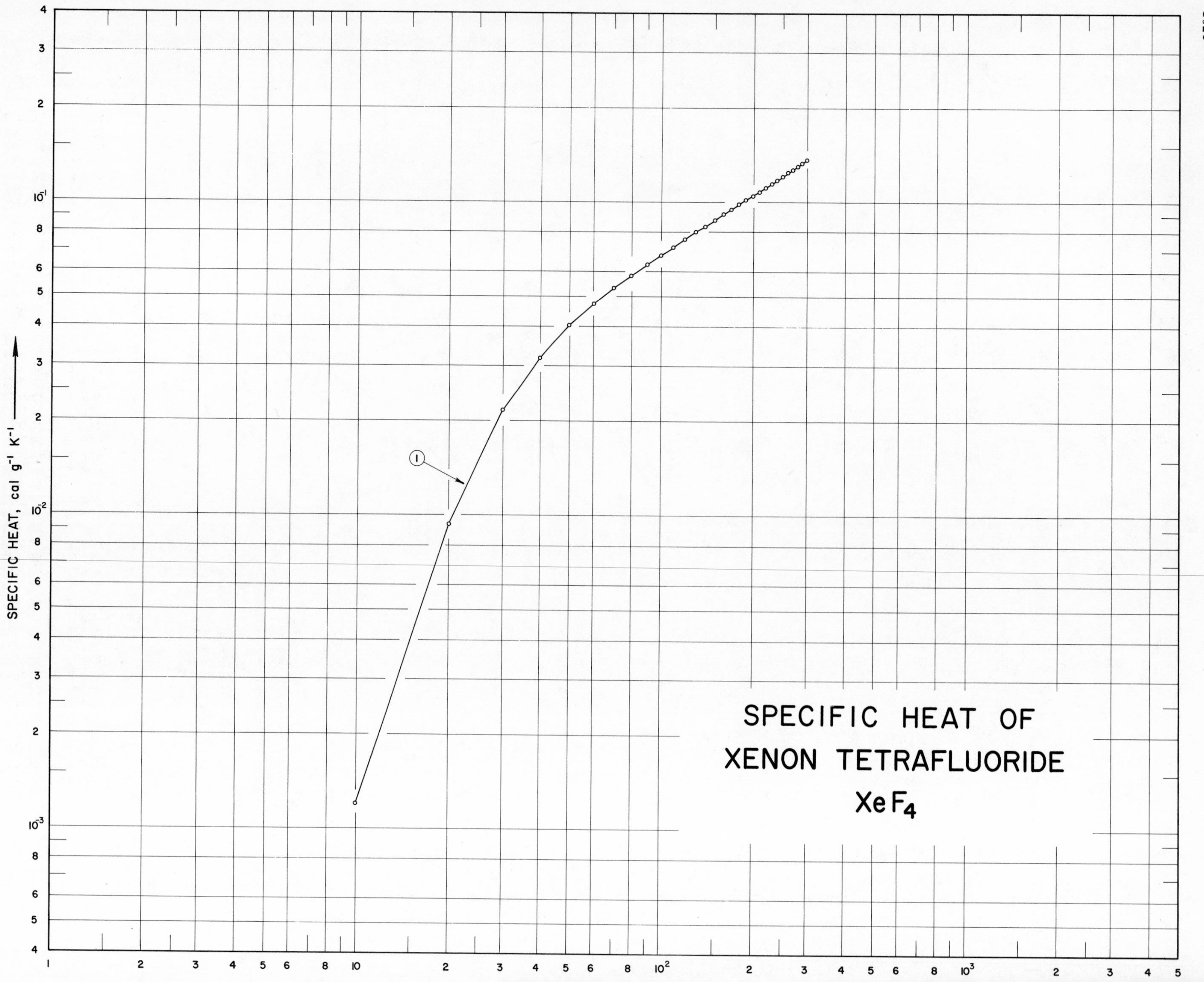
SPECIFIC HEAT OF
XENON TETRAFLUORIDE
XeF_4
SPECIFIC HEAT, cal g^{-1} K^{-1}
1

SPECIFICATION TABLE NO. 321 SPECIFIC HEAT OF XENON TETRAFLUORIDE XeF_4

[For Data Reported in Figure and Table No. 321]

Curve No.	Ref. No.	Year	Temp. Range, K	Reported Error, %	Name and Specimen Designation	Composition (weight percent), Specifications and Remarks
1	334	1963	10-300			Prepared from high purity xenon and fluorine by heating to 450 C for 2.5 hrs; rapidly quenched in water.

DATA TABLE NO. 321 SPECIFIC HEAT OF XENON TETRAFLUORIDE XeF_4

[Temperature, T, K; Specific Heat, C_p, Cal g^{-1} K^{-1}]

T	C_p
CURVE 1	
10	1.200×10^{-3}
20	9.300
30	2.170×10^{-2}
40	3.180
50	4.030
60	4.740
70	5.320
80	5.830
90	6.310
100	6.760
110	7.190
120	7.600
130	8.000
140	8.380
150	8.740
160	9.120
170	9.470
180	9.820
190	1.016×10^{-1}
200	1.051
210	1.082
220	1.116
230	1.147
240	1.181
250	1.213
260	1.247
270	1.279
280	1.312
290	1.342
298.16	1.367*
300	1.373

* Not shown on plot

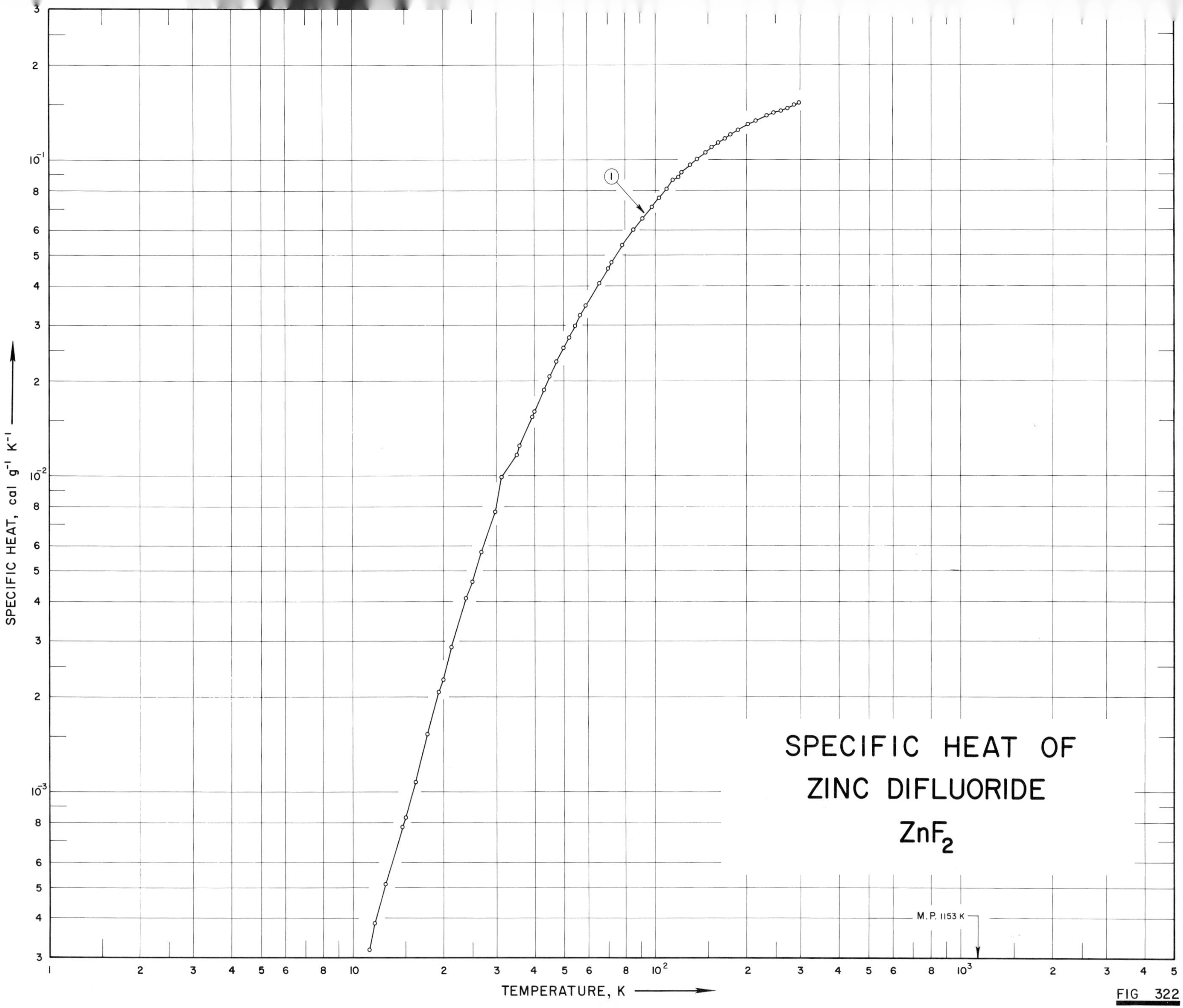

SPECIFIC HEAT OF ZINC DIFLUORIDE ZnF_2

SPECIFIC HEAT, cal g^{-1} K^{-1} →

TEMPERATURE, K →

(1)

M.P. 1153 K

FIG 322

SPECIFICATION TABLE NO. 322 SPECIFIC HEAT OF ZINC DIFLUORIDE ZnF_2

[For Data Reported in Figure and Table No. 322]

Curve No.	Ref. No.	Year	Temp. Range, K	Reported Error, %	Name and Specimen Designation	Composition (weight percent), Specifications and Remarks
1	302	1955	11-299	0.5-3		63.22 Zr (63.24 theo), 0.006 Al, 0.005 Fe, 0.003 Na, 0.002 Ni, Cu, and 0.001 Mg and K; prepared from $ZnCO_3$; sintered; heated in HF atmosphere to 980 C and slowly cooled.

DATA TABLE NO. 322 SPECIFIC HEAT OF ZINC DIFLUORIDE ZnF_2

[Temperature, T, K; Specific Heat, C_p, Cal g^{-1} K^{-1}]

T	C_p
CURVE 1	
Series 1	
59.18	3.480×10^{-2}
54.53	3.008
58.80	3.438*
65.25	4.088
72.14	4.755
78.81	5.400
85.27	6.018
91.62	6.583
98.14	7.147
104.09	7.633
110.31	8.119
117.03	8.638
124.09	9.150
131.60	9.643
139.40	1.013×10^{-1}
147.63	1.060
155.00	1.102
162.90	1.140
170.80	1.177
178.91	1.212
Series 2	
11.03	3.192×10^{-4}
11.89	3.869
12.96	5.127
14.68	7.738
16.19	1.083×10^{-3}
17.87	1.538
19.48	2.089
21.47	2.883
23.94	4.063
26.82	5.746
29.72	7.700
32.72	9.934
35.93	1.251×10^{-2}
39.25	1.541
43.10	1.890
47.56	2.315
52.10	2.756
56.98	3.254
CURVE 1 (cont.)	
Series 3	
188.59	1.251×10^{-1}
195.71	1.278*
203.00	1.302
210.03	1.324*
217.10	1.345
224.10	1.365*
233.54	1.390
240.41	1.405*
247.27	1.421
253.92	1.435*
260.57	1.449
267.49	1.462*
272.66	1.474
279.50	1.485*
286.36	1.499
292.97	1.508*
299.42	1.522

* Not shown on plot

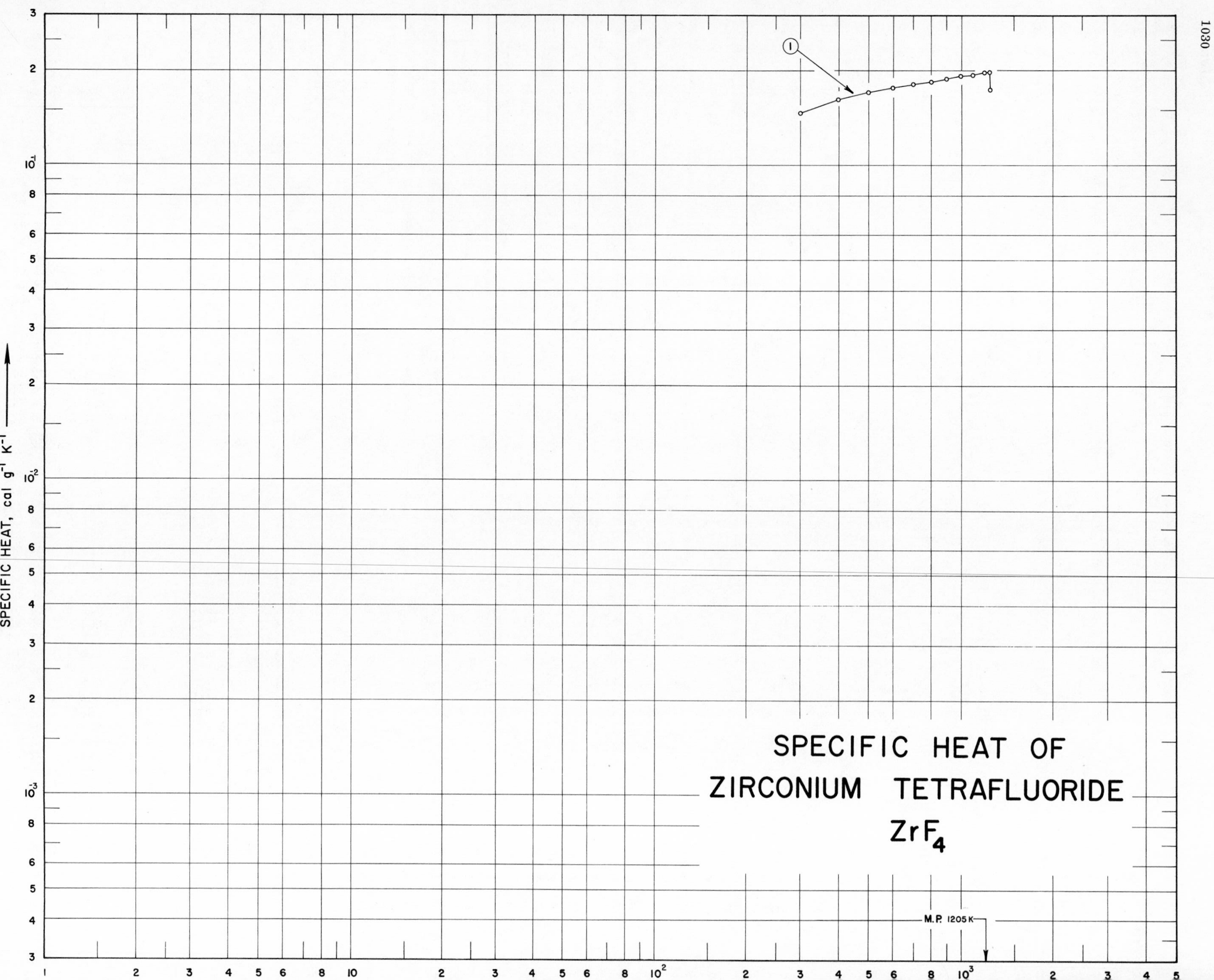
SPECIFIC HEAT OF
ZIRCONIUM TETRAFLUORIDE
ZrF_4
M.P. 1205 K
SPECIFIC HEAT, cal g⁻¹ K⁻¹
1

SPECIFICATION TABLE NO. 323 SPECIFIC HEAT OF ZIRCONIUM TETRAFLUORIDE ZrF_4

[For Data Reported in Figure and Table No. 323]

Curve No.	Ref. No.	Year	Temp. Range, K	Reported Error, %	Name and Specimen Designation	Composition (weight percent), Specifications and Remarks
1	335	1962	300-1205			54.6 Zr and 44.9 F (54.55 and 45.45 theo); sealed under 8 mm Hg helium.

DATA TABLE NO. 323 SPECIFIC HEAT OF ZIRCONIUM TETRAFLUORIDE ZrF_4

[Temperature, T, K; Specific Heat, C_p, Cal g^{-1} K^{-1}]

T	C_p
CURVE 1	
300	1.484 x 10^{-1}
400	1.621
500	1.709
600	1.771
700	1.817
800	1.852
900	1.884
1000	1.915
1100	1.946
1200	1.976
(s) 1205	1.978
(l) 1205	1.734

* Not shown on plot

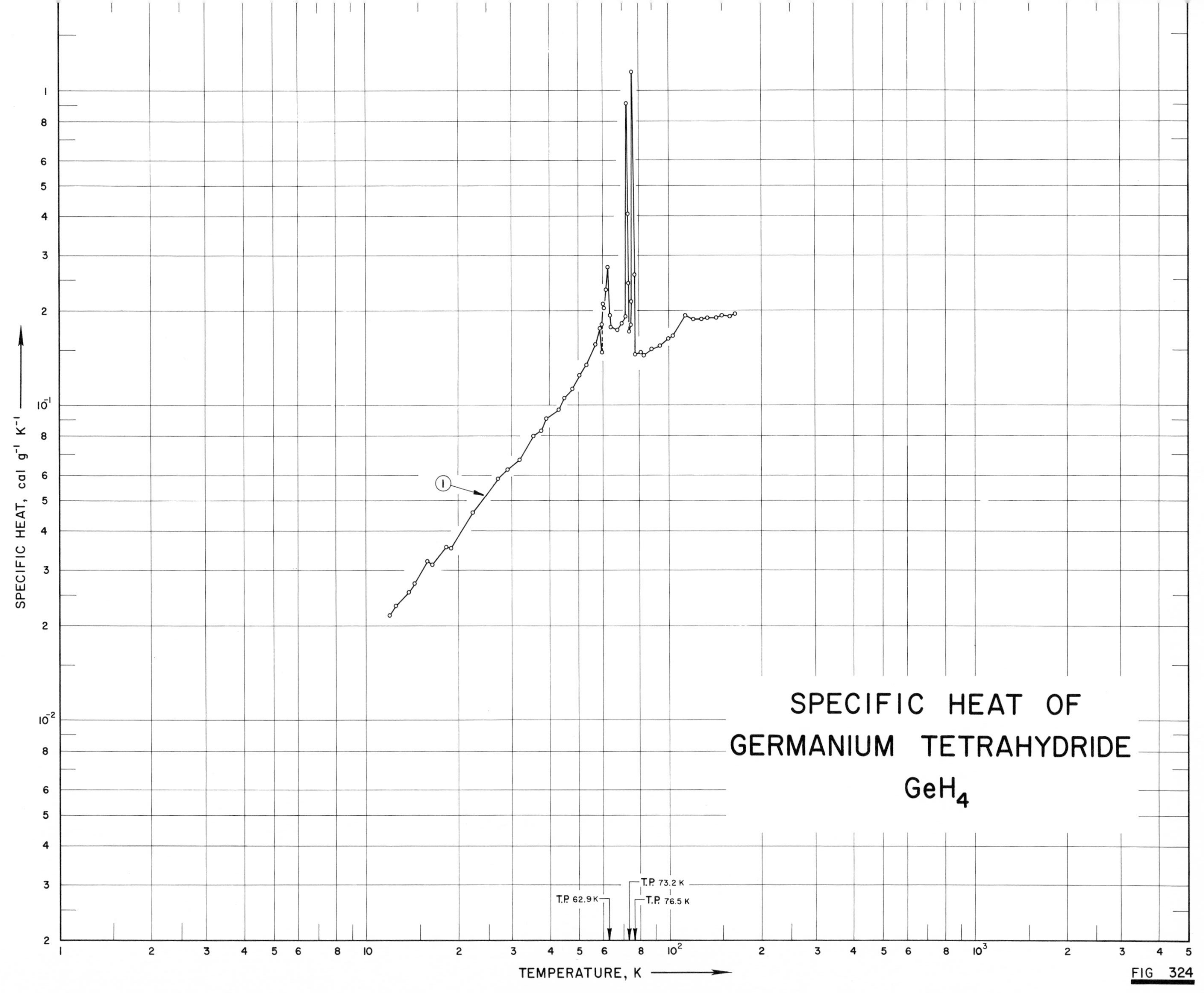
SPECIFIC HEAT OF
GERMANIUM TETRAHYDRIDE
GeH_4
SPECIFIC HEAT, cal g^{-1} K^{-1}
TEMPERATURE, K
T.P. 62.9 K
T.P. 73.2 K
T.P. 76.5 K
1
FIG 324

SPECIFICATION TABLE NO. 324 SPECIFIC HEAT OF GERMANIUM TETRAHYDRIDE GeH_4

[For Data Reported in Figure and Table No. 324]

Curve No.	Ref. No.	Year	Temp. Range, K	Reported Error, %	Name and Specimen Designation	Composition (weight percent), Specifications and Remarks
1	408	1942	12-165			

DATA TABLE NO. 324 SPECIFIC HEAT OF GERMANIUM TETRAHYDRIDE GeH_4

[Temperature, T, K; Specific Heat, C_p, Cal $g^{-1}K^{-1}$]

T	C_p	T	C_p
CURVE 1		CURVE 1 (cont.)	
12.0	2.153 x 10^{-2}	72.2	1.919 x 10^{-1}
12.6	2.310	72.8	9.136
13.8	2.558	73.5	4.059
14.5	2.715	73.9	2.454
16.0	3.211	74.1	1.853*
16.6	3.119	74.4	1.723
18.4	3.550	75.1	1.801
19.0	3.511	75.3	1.801*
22.3	4.581	75.4	2.140
27.1	5.860	76.3	1.146
29.2	6.265	77.5	2.597
31.9	6.748	77.6	1.462
35.5	8.001	78.2	1.475*
37.7	8.327	81.0	1.475
39.1	9.123	82.6	1.462*
40.3	9.227*	83.1	1.449
43.2	9.710	88.1	1.514
45.2	1.069 x 10^{-1}	89.3	1.514*
48.1	1.139	93.7	1.553
50.7	1.250	93.8	1.553*
53.4	1.357	94.6	1.553*
57.1	1.566	99.9	1.631
57.6	1.605*	101.9	1.671*
58.1	1.644*	103.9	1.671
59.0	1.723*	112.7	1.932
59.1	1.762	121.1	1.879
59.5	1.723*	127.9	1.879
59.9	1.475	134.9	1.892
59.9	1.814	143.5	1.892
60.5	2.101	150.6	1.932
60.8	2.049	158.4	1.919
61.8	2.336	165.4	1.958
62.1	2.441*		
62.6	2.662*		
62.7	2.767		
63.6	1.932		
63.7	1.853*		
64.3	1.775		
65.1	1.788*		
65.6	1.762*		
66.4	1.762*		
67.7	1.749		
69.7	1.788*		
69.9	1.827		
71.3	1.866*		

* Not shown on plot

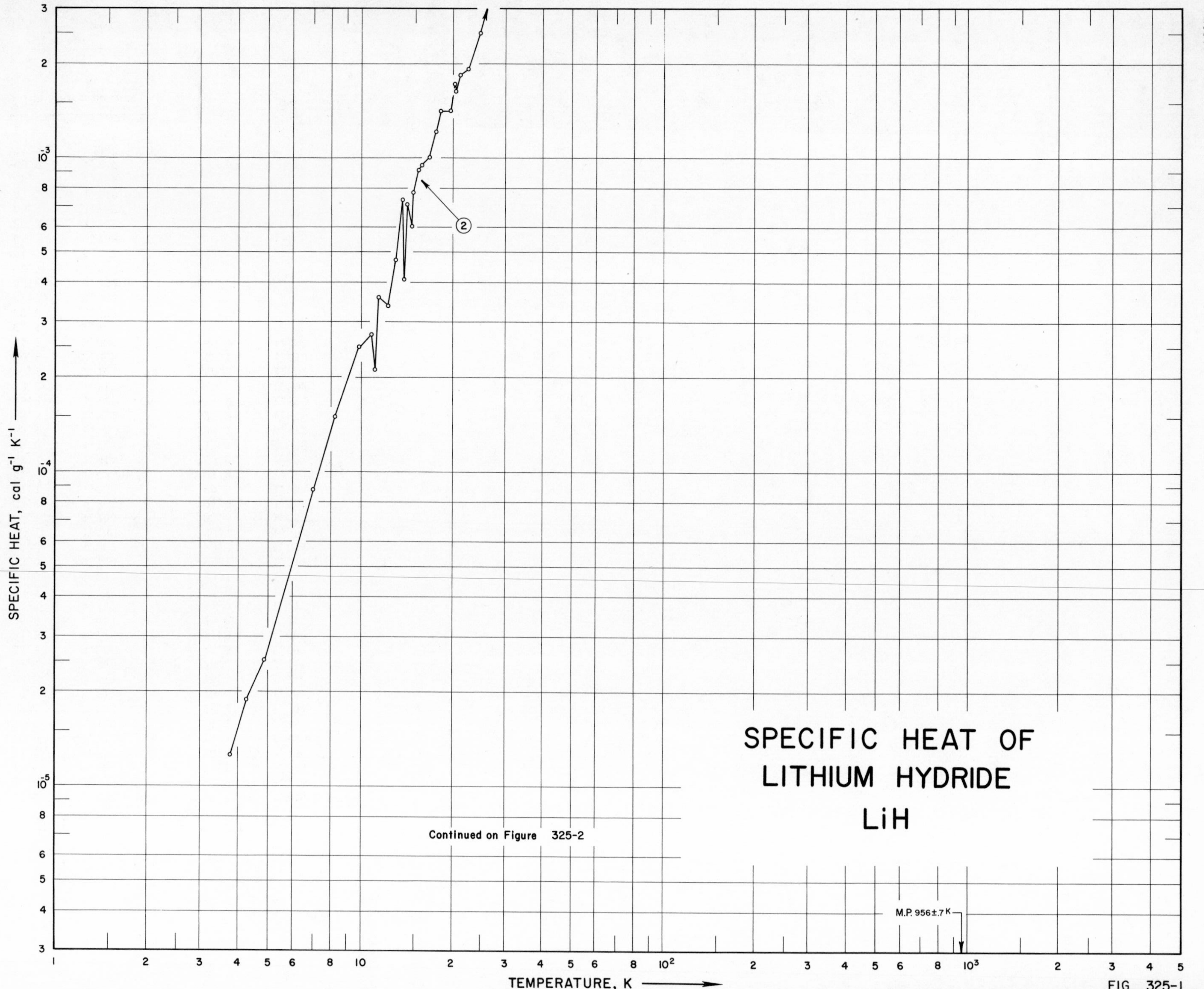
SPECIFIC HEAT OF
LITHIUM HYDRIDE
LiH
SPECIFIC HEAT, cal g⁻¹ K⁻¹
TEMPERATURE, K
M.P. 956±7 K
Continued on Figure 325-2
2

FIG 325-1

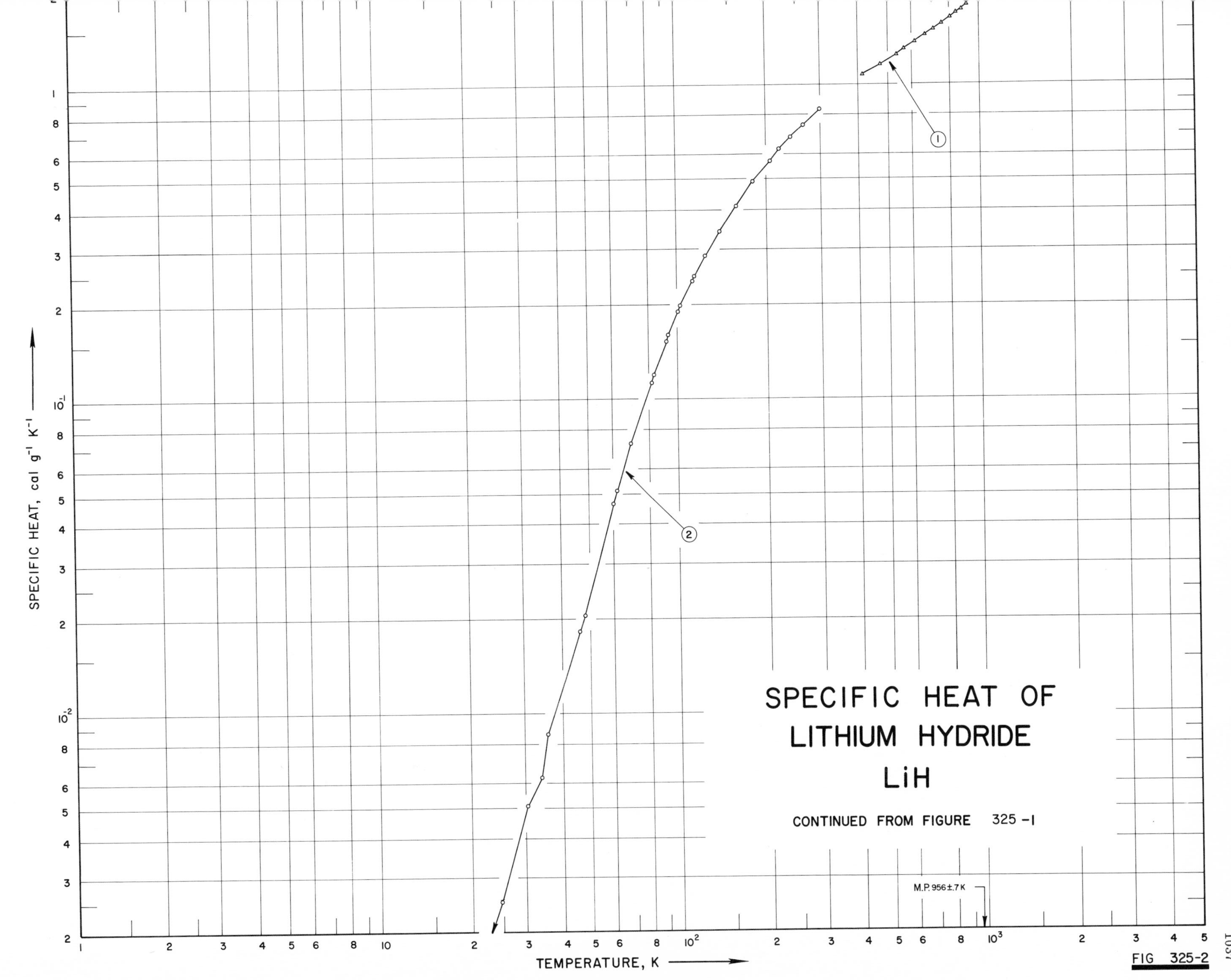
SPECIFIC HEAT OF
LITHIUM HYDRIDE
LiH
CONTINUED FROM FIGURE 325 -I
M.P. 956±7 K
TEMPERATURE, K
SPECIFIC HEAT, cal g^{-1} K^{-1}
1
2

FIG 325-2

SPECIFICATION TABLE NO. 325 SPECIFIC HEAT OF LITHIUM HYDRIDE LiH

[For Data Reported in Figure and Table No. 325]

Curve No.	Ref. No.	Year	Temp. Range, K	Reported Error, %	Name and Specimen Designation	Composition (weight percent), Specifications and Remarks
1	27	1958	413-914	0.66-2.9		Measured in helium atmosphere.
2	207	1961	3.7-296			99.8 LiH, fairly course, slightly colored crystals; dry nitrogen atmosphere at temperature above 12 K; helium cryostat at temperature below 12 K; results follow Debye law from 0-40 K.

DATA TABLE NO. 325 SPECIFIC HEAT OF LITHIUM HYDRIDE LiH

[Temperature, T,K; Specific Heat, C_p, Cal $g^{-1}K^{-1}$]

T	C_p	T	C_p
CURVE 1		CURVE 2 (cont.)	
413	1.074×10^{-2}	24.55	2.516×10^{-3}
476	1.164	30.10	5.096
533	1.246	33.52	6.228
567	1.295	35.30	8.593
615	1.364	45.54	1.824×10^{-2}
664	1.434	47.22	2.051
704	1.491	59.76	4.680
753	1.562	61.53	5.133
775	1.594*	68.77	7.285
805	1.637	80.17	1.132×10^{-1}
811	1.645*	82.29	1.208
840	1.688	90.14	1.527
869	1.729*	92.27	1.609
876	1.739	99.83	1.916
899	1.771*	101.75	1.993
914	1.794	111.25	2.386
		113.39	2.460
CURVE 2		123.17	2.865
		125.61	2.925*
3.72	1.258×10^{-5}	137.92	3.439
4.28	1.887	140.00	3.529*
4.89	2.516	155.90	4.116
6.98	8.807	158.64	4.217*
8.20	1.510×10^{-4}	178.51	4.936
9.82	2.516	181.71	5.039*
10.85	2.768	203.89	5.759
11.03	2.139	207.35	5.878*
11.42	3.649	208.08	5.897*
12.20	3.397	210.16	5.975*
12.90	4.781	217.83	6.234
13.60	7.423	220.54	6.292*
13.71	4.152	228.32	6.551*
14.01	7.172	229.60	6.580*
14.53	6.039	236.42	6.813
14.71	7.801	237.92	6.847*
15.33	9.185	245.03	7.028*
15.70	9.562	247.53	7.116*
16.65	1.019×10^{-3}	260.97	7.472
17.50	1.220	264.18	7.570*
18.17	1.422	292.65	8.278*
19.65	1.422	295.50	8.353
20.30	1.724		
20.39	1.648		
21.02	1.850		
22.48	1.925		

* Not shown on plot

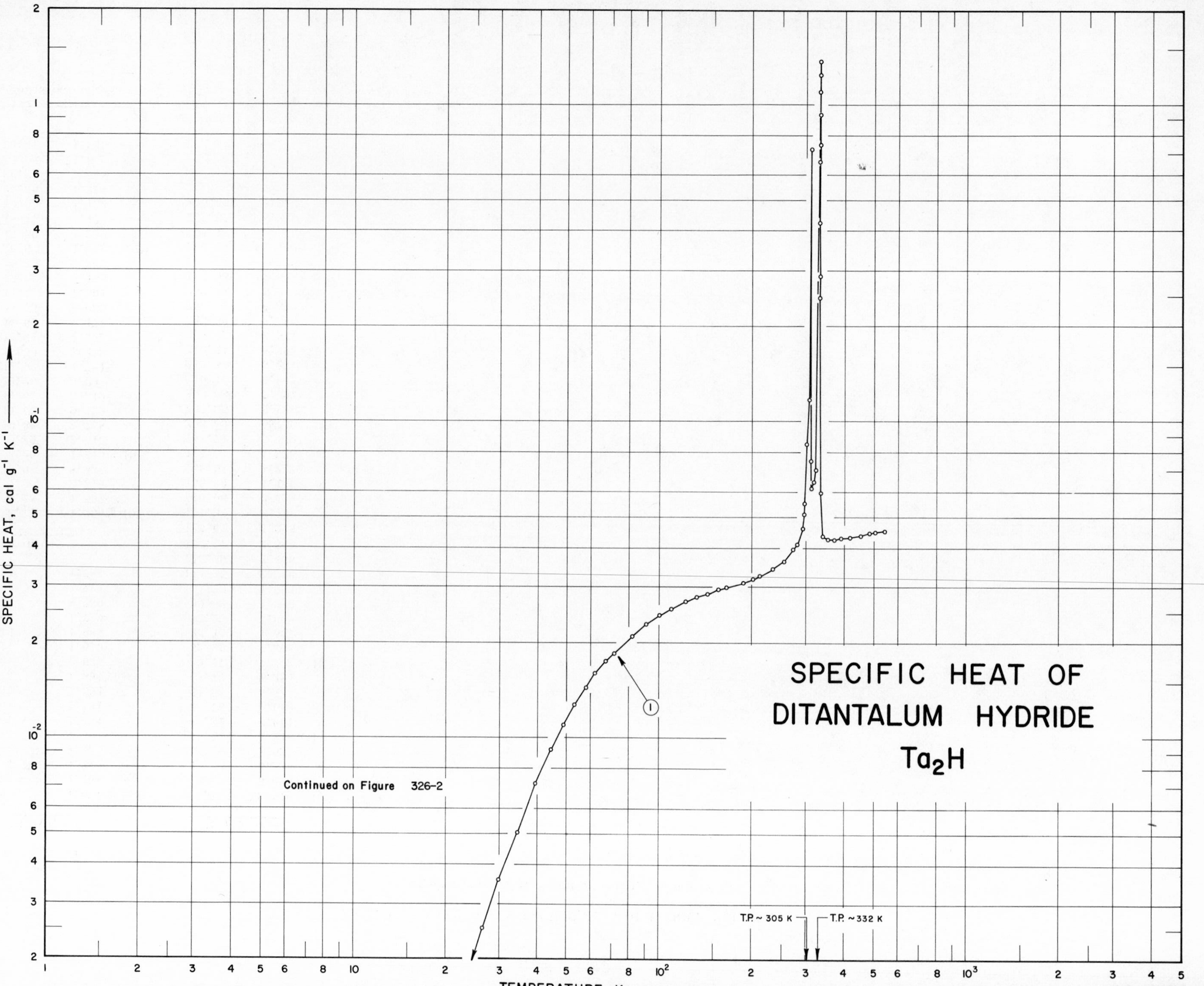

SPECIFIC HEAT OF
DITANTALUM HYDRIDE
Ta_2H
T.P. ~ 305 K
T.P. ~ 332 K
1
Continued on Figure 326-2
SPECIFIC HEAT, cal g^{-1} K^{-1}

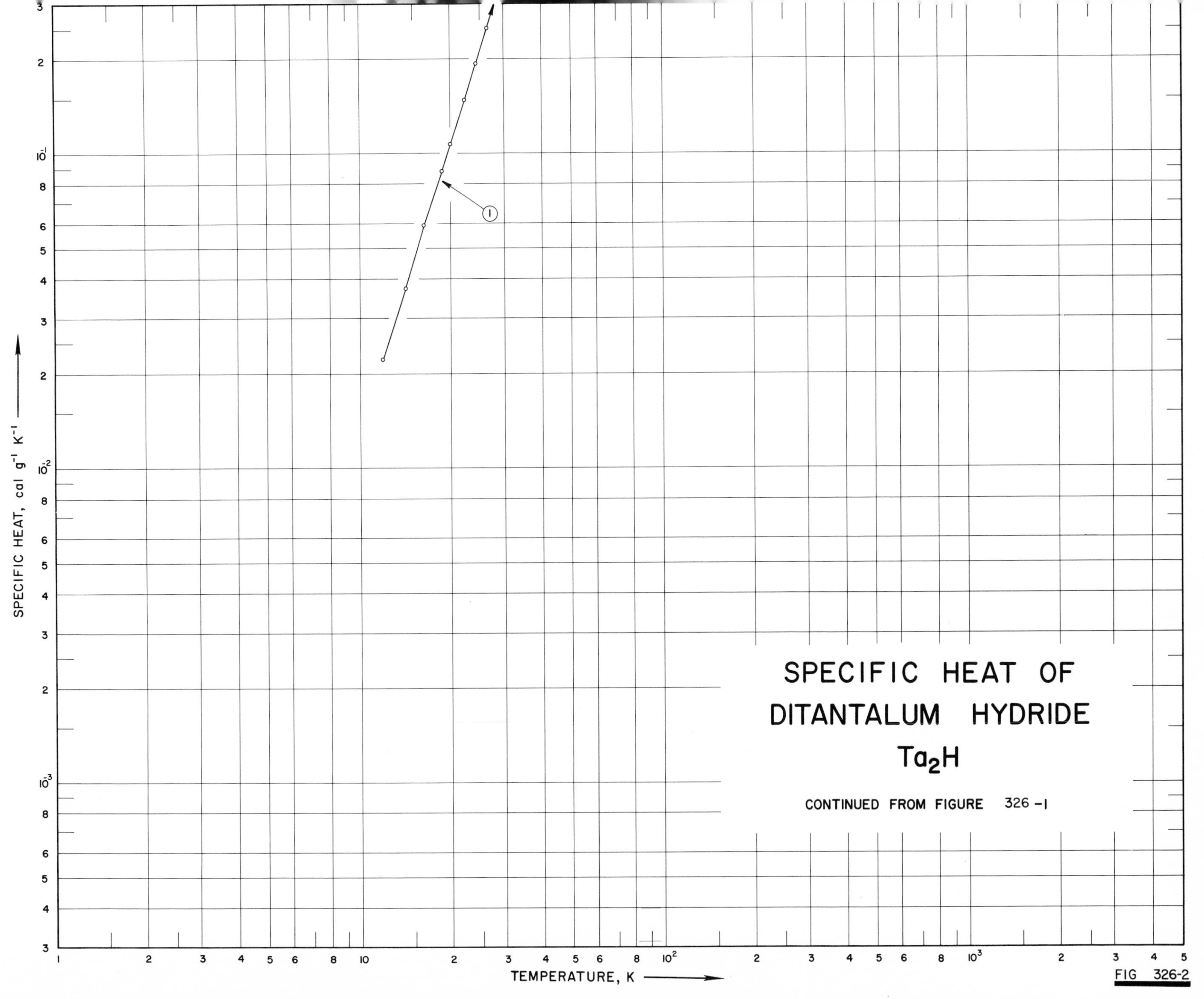

SPECIFIC HEAT OF
DITANTALUM HYDRIDE
Ta_2H
CONTINUED FROM FIGURE 326-1
SPECIFIC HEAT, cal g^{-1} K^{-1}
TEMPERATURE, K
1

FIG 326-2

SPECIFICATION TABLE NO. 326 SPECIFIC HEAT OF DITANTALUM HYDRIDE Ta_2H

[For Data Reported in Figure and Table No. 326]

Curve No.	Ref. No.	Year	Temp. Range, K	Reported Error, %	Name and Specimen Designation	Composition (weight percent), Specifications and Remarks
1	409	1961	12-552			Ta and H_2 were supplied by the Fansteel Metallurgical Corp. and the Air Reduction Corp., respectively; prepared from annealed 99.9^+ Ta and 99.0^+ H_2.

DATA TABLE NO. 326 SPECIFIC HEAT OF DITANTALUM HYDRIDE Ta_2H

[Temperature, T, K; Specific Heat, C_p, Cal $g^{-1}K^{-1}$]

T	C_p
CURVE 1	
Series I	
71.91	1.878 x 10^{-2}
76.75	2.007*
82.00	2.130
87.67	2.250*
91.45	2.314
96.23	7.390*
101.65	2.472
106.98	2.543*
110.90	2.593
116.71	2.659*
122.37	2.716
127.91	2.766*
133.36	2.813
138.72	2.856*
144.42	2.895
150.46	2.935*
156.41	2.972
162.76	3.007*
169.53	3.043
176.23	3.078*
182.82	3.115*
189.35	3.150
195.81	3.185*
202.20	3.221
205.52	3.260*
Series II	
281.25	4.178 x 10^{-2}
286.75	4.347*
292.14	4.637
297.12	5.591
301.14	8.591
304.08	1.210 x 10^{-1}
306.67	1.196
310.07	7.652 x 10^{-2}
314.52	6.254
319.21	6.641*
323.30	7.132
CURVE 1 (cont.)	
Series III	
296.86	5.142 x 10^{-2}
300.29	7.452*
304.00	1.186 x 10^{-1}*
307.03	1.165*
310.90	6.925 x 10^{-2}*
315.76	6.326*
319.00	6.581
Series IV	
200.48	3.211 x 10^{-2}*
207.60	3.254*
214.92	3.301
222.06	3.352*
228.83	3.404*
235.54	3.464
242.19	3.532*
248.73	3.605*
255.18	3.688
261.52	3.777*
267.76	3.879*
273.95	3.998
280.13	4.144*
286.30	4.326*
Series V	
11.99	2.204 x 10^{-4}
14.18	3.748
16.48	5.952
18.76	8.818
20.02	1.080 x 10^{-3}
22.19	1.499
24.21	1.956
26.44	2.546
29.78	3.593
34.28	5.219
39.42	7.242
44.41	9.270
49.08	1.113 x 10^{-2}
53.58	1.287
58.00	1.451
62.66	1.611
67.82	1.768
73.13	1.911*
CURVE 1 (cont.)	
Series VI*	
284.20	4.252 x 10^{-2}
288.67	4.408
293.01	4.665
297.08	5.485
300.58	7.799
303.27	1.137 x 10^{-1}
305.50	1.266
307.75	1.109
310.62	6.856 x 10^{-2}
314.10	6.202
317.64	6.463
321.07	6.799
324.37	7.259
Series VII	
304.63	1.219 x 10^{-1}*
305.68	1.239*
306.75	1.193*
307.92	1.105*
308.91	9.066 x 10^{-2}*
309.83	7.361*
310.87	6.414*
311.89	6.143*
312.84	6.136*
313.80	6.171*
314.74	6.215*
326.04	7.604*
329.16	9.905*
330.95	4.308 x 10^{-1}
331.51	1.124 x 10^{0}
331.81	1.260
332.14	9.506 x 10^{-1}
332.56	7.604
332.96	1.149 x 10^{0}*
333.24	1.396
333.62	7.420 x 10^{-1}*
335.55	7.688 x 10^{-2}*
339.64	4.394
344.47	4.312*
349.31	4.301*
354.14	4.298
CURVE 1 (cont.)	
Series VIII	
334.36	1.271 x 10^{-1}*
335.75	5.821 x 10^{-2}*
347.44	4.315*
353.05	4.305*
359.09	3.905*
265.55	4.305*
372.23	4.307
379.07	4.313*
385.74	4.316*
392.39	4.324
399.04	4.335*
405.67	4.344*
412.30	4.350*
418.91	4.357
425.53	4.371*
432.14	4.386*
438.74	4.386*
445.33	4.397*
451.90	4.406
458.47	4.423*
465.00	4.454*
471.57	4.444*
478.10	4.456*
484.62	4.473
491.15	4.474*
Series IX	
330.37	2.494 x 10^{-1}
331.03	6.722
331.34	1.003 x 10^{0}*
331.58	1.105*
331.82	1.045*
332.10	8.041 x 10^{-1}*
332.44	7.115*
332.76	9.399*
333.09	1.205 x 10^{0}*
333.34	9.776 x 10^{-1}*
333.91	2.930
335.84	6.008 x 10^{-2}
CURVE 1 (cont.)	
Series X	
496.10	4.487 x 10^{-2}*
504.09	4.510
512.08	4.530*
520.05	4.537*
528.08	4.569*
536.09	4.571*
544.06	4.594*
552.05	4.618

* Not shown on plot

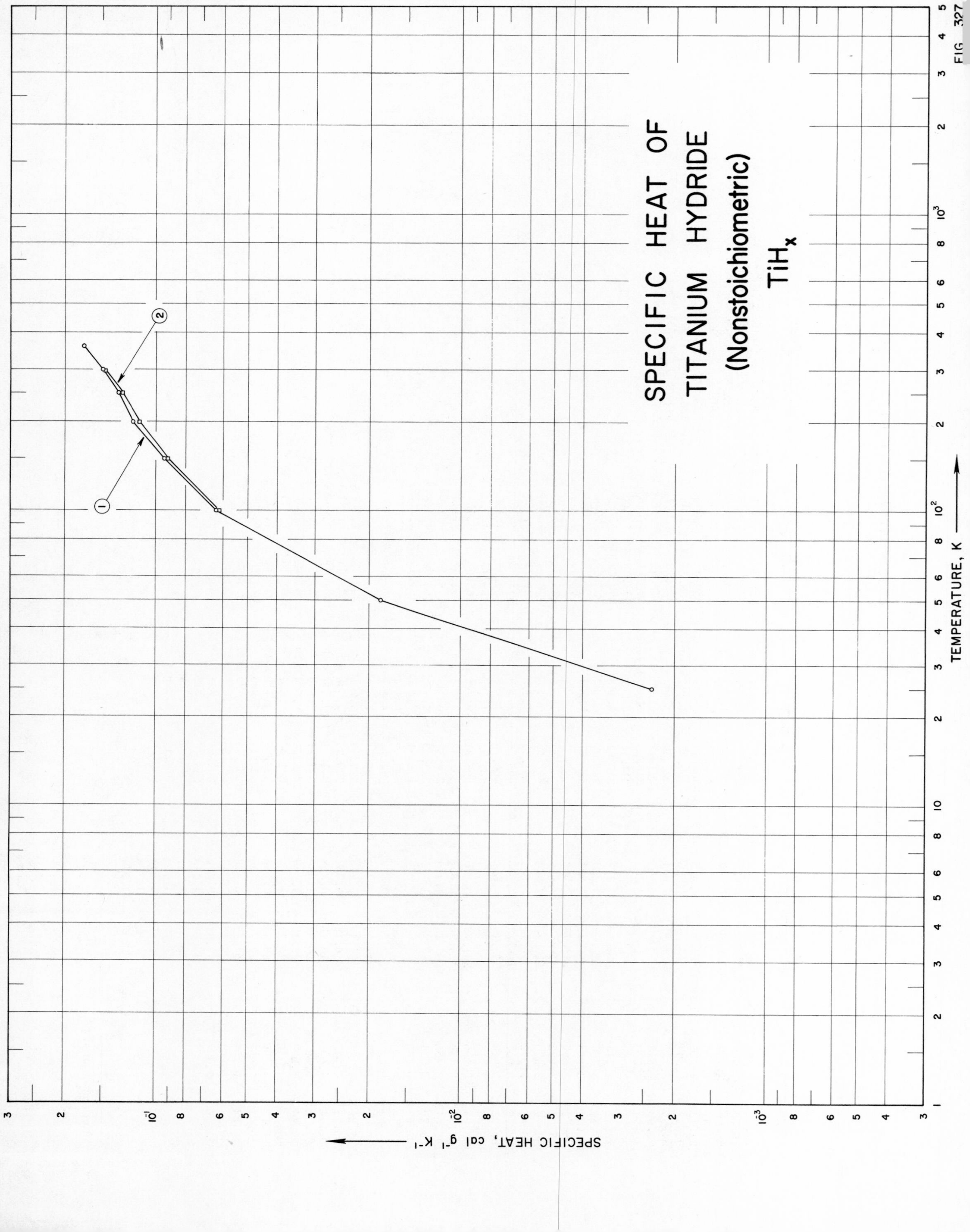

SPECIFIC HEAT OF TITANIUM HYDRIDE (Nonstoichiometric) TiH_x

SPECIFIC HEAT, cal g^{-1} K^{-1}

TEMPERATURE, K

1

2

FIG. 327

SPECIFICATION TABLE NO. 327 SPECIFIC HEAT OF TITANIUM HYDRIDE (nonstoichiometric) TiH_x

[For Data Reported in Figure and Table No. 327]

Curve No.	Ref. No.	Year	Temp. Range, K	Reported Error, %	Name and Specimen Designation	Composition (weight percent), Specifications and Remarks
1	410	1962	25-360			$TiH_{1.607}$.
2	410	1962	100-360			$TiH_{1.718}$.

DATA TABLE NO. 327 SPECIFIC HEAT OF TITANIUM HYDRIDE TiH_x (nonstoichiometric)

[Temperature, T, K; Specific Heat, C_p, Cal $g^{-1}K^{-1}$]

T	C_p
CURVE 1	
25	2.37×10^{-3}
50	1.84×10^{-2}
100	6.39
150	9.42
200	1.19×10^{-1}
250	1.34
298.15	1.50*
300	1.51
360	1.74
CURVE 2	
100	6.25×10^{-2}
150	9.19
200	1.13×10^{-1}
250	1.30
298.15	1.47
300	1.48*
360	1.72*

* Not shown on plot

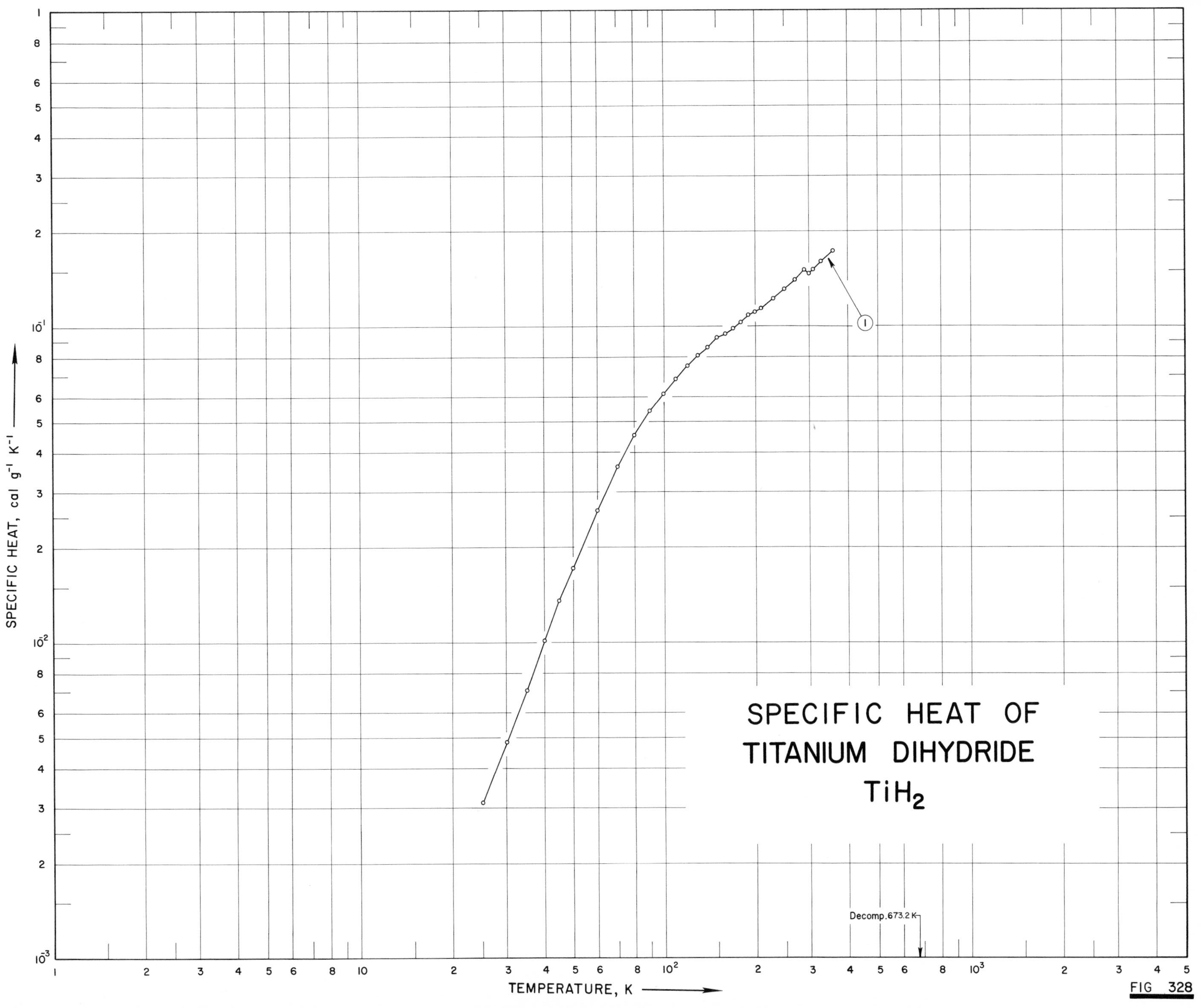
SPECIFIC HEAT OF
TITANIUM DIHYDRIDE
TiH2
1
Decomp. 673.2 K
TEMPERATURE, K
SPECIFIC HEAT, cal g⁻¹ K⁻¹
FIG 328

SPECIFICATION TABLE NO. 328 SPECIFIC HEAT OF TITANIUM DIHYDRIDE TiH_2

[For Data Reported in Figure and Table No. 328]

Curve No.	Ref. No.	Year	Temp. Range, K	Reported Error, %	Name and Specimen Designation	Composition (weight percent), Specifications and Remarks
1	208	1960	25-360	1.0		>99.5 Ti; prepared by direct absorption of known volumes of hydrogen with >99.5 Ti.

DATA TABLE NO. 328 SPECIFIC HEAT OF TITANIUM DIHYDRIDE TiH_2

[Temperature, T, K; Specific Heat, C_p, Cal $g^{-1}K^{-1}$]

T	C_p
CURVE 1	
25	3.13×10^{-3}
30	4.85
35	7.05
40	1.01×10^{-2}
45	1.36
50	1.74
60	2.62
70	3.639
80	4.570
90	5.449
100	6.195
110	6.890
120	7.535
130	8.117
140	8.643
150	9.289
160	9.487
170	9.896
180	1.030×10^{-1}
190	1.090
200	1.110
210	1.151
220	1.191*
230	1.233
240	1.276*
250	1.319
260	1.364*
270	1.414
273.16	1.430*
280	1.470*
288.5	1.560*
290	1.522
298.16	1.470*
300	1.479
310	1.523
320	1.567*
330	1.611
340	1.656*
350	1.701*
360	1.745

* Not shown on plot

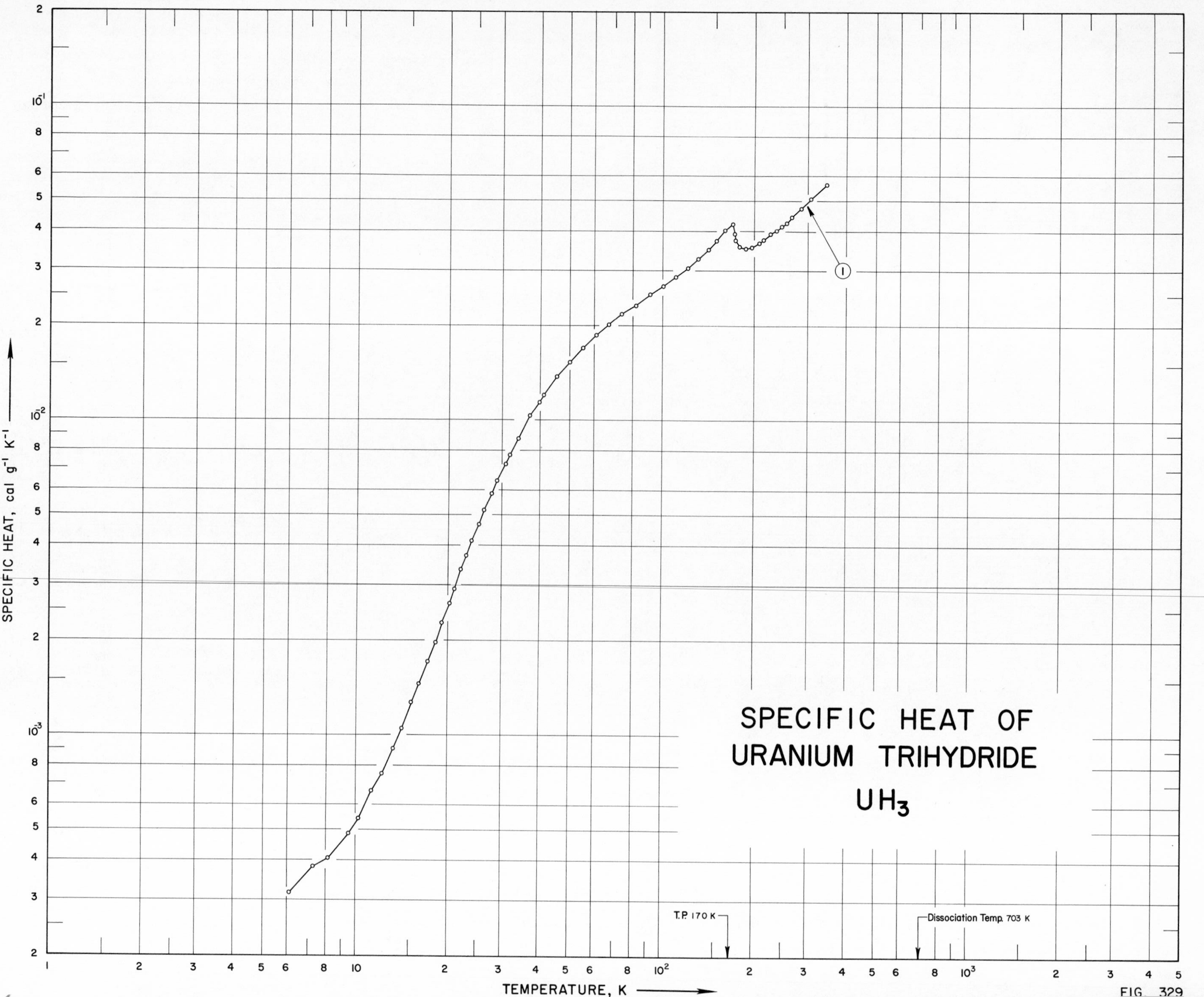
SPECIFIC HEAT OF
URANIUM TRIHYDRIDE
UH_3
T.P. 170 K
Dissociation Temp. 703 K
TEMPERATURE, K
SPECIFIC HEAT, cal g^{-1} K^{-1}
1
FIG. 329

SPECIFICATION TABLE NO. 329 SPECIFIC HEAT OF URANIUM TRIHYDRIDE UH_3

[For Data Reported in Figure and Table No. 329]

Curve No.	Ref. No.	Year	Temp. Range, K	Reported Error, %	Name and Specimen Designation	Composition (weight percent), Specifications and Remarks
1	209	1959	6-347	<5		99.6 UH_3 and 0.06 O_2; prepared by direct reaction of high purity uranium metal and hydrogen above 200 C.

DATA TABLE NO. 329 SPECIFIC HEAT OF URANIUM TRIHYDRIDE UH_3

[Temperature, T,K; Specific Heat, C_p, Cal $g^{-1}K^{-1}$]

T	C_p
CURVE 1	
Series I	
7.30	3.816 x 10^{-4}
9.54	4.894
11.38	6.637
13.26	9.001
15.24	1.265 x 10^{-3}
17.22	1.717
19.13	2.277
21.04	2.937
23.05	3.725
25.32	4.691
27.95	5.890
30.86	7.250
34.03	8.748
37.61	1.037 x 10^{-2}
41.57	1.206
45.84	1.373
50.56	1.542
55.62	1.701
61.30	1.865
67.49	2.022
74.65	2.180
82.65	2.349
91.26	2.514
100.80	2.677
111.19	2.863
121.54	3.057
131.94	3.266
142.20	3.498
151.18	3.729
160.37	4.011
170.13	4.222
180.00	3.555
189.80	3.524
199.31	3.578
209.17	3.663
218.64	3.766
228.00	3.875
237.91	4.003
248.03	4.141
257.70	4.276
267.36	4.417
277.20	4.563*
287.09	4.712
297.06	4.865*

T	C_p
CURVE 1 (cont.)	
307.28	5.027 x 10^{-2}
317.40	5.185*
327.14	5.334*
336.88	5.488*
346.94	5.637
Series II	
6.13	3.2 x 10^{-4}
8.15	4.1
10.17	5.43
12.16	7.55
14.15	1.05 x 10^{-3}
16.17	1.45
18.18	1.98
20.17	2.63
22.15	3.36
24.12	4.169
26.41	5.185
29.08	6.404
31.99	7.794
Series III	
160.14	4.001 x 10^{-2}*
162.09	4.072*
164.01	4.146*
165.91	4.222*
167.79	4.301*
169.64	4.405*
171.48	4.376*
173.37	3.941
175.34	3.710
177.35	3.607*
179.37	3.560*
181.40	3.534*
183.43	3.520*
185.45	3.515*
Series IV*	
168.95	4.376 x 10^{-2}
169.45	4.393
169.96	4.426
170.49	4.442

T	C_p
CURVE 1 (cont.)	
171.03	4.434 x 10^{-2}
171.57	4.359
172.12	4.206
Series V*	
187.86	3.525 x 10^{-2}
189.92	3.526
191.98	3.536
Series VI*	
168.23	4.326 x 10^{-2}
168.77	4.351
169.31	4.376
169.85	4.417
170.39	4.438
170.92	4.451
171.46	4.359
172.00	4.227
172.55	4.101

* Not shown on plot

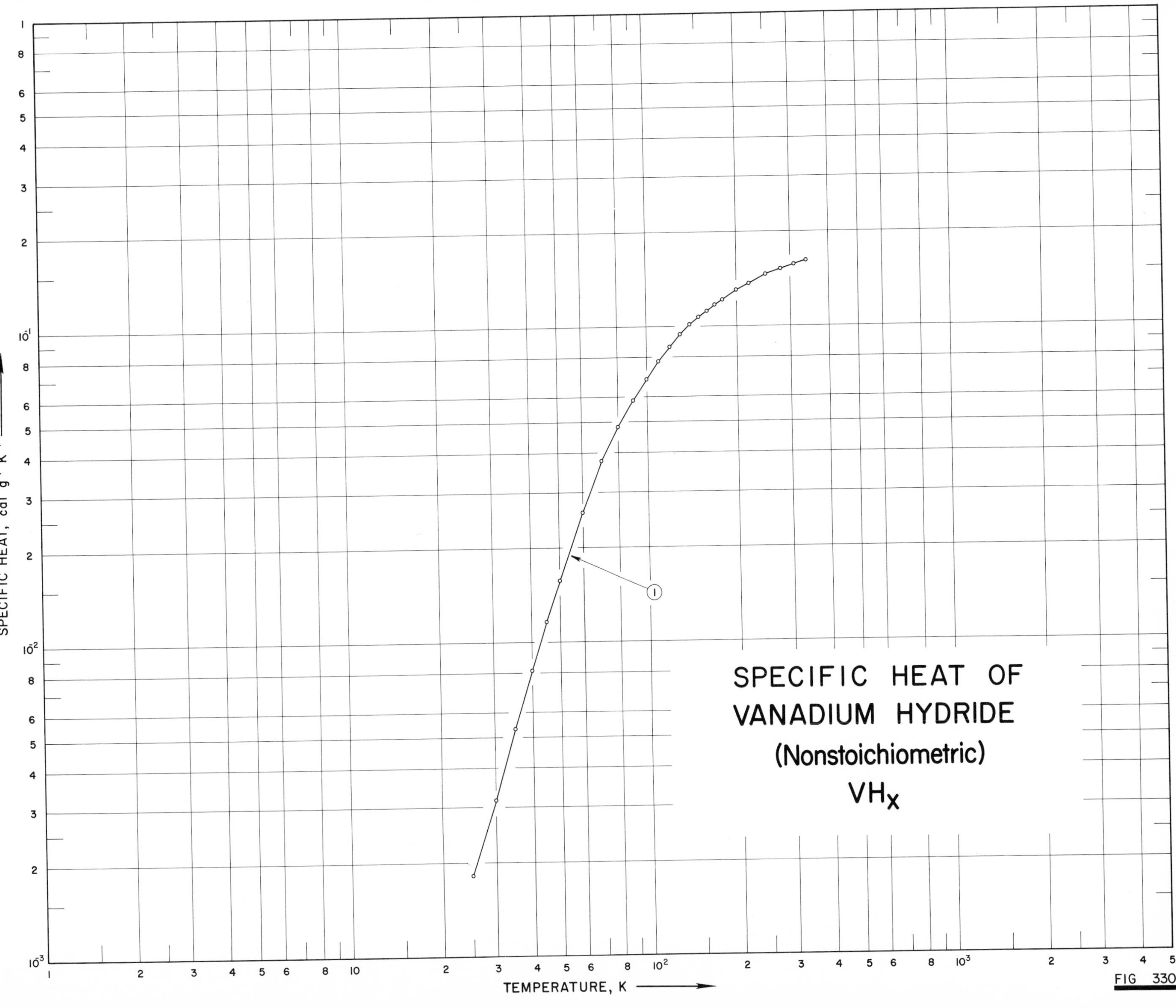
SPECIFIC HEAT, cal g^{-1} K^{-1}
TEMPERATURE, K
1
SPECIFIC HEAT OF
VANADIUM HYDRIDE
(Nonstoichiometric)
VH_x
FIG 330

SPECIFICATION TABLE NO. 330 SPECIFIC HEAT OF VANADIUM HYDRIDE (nonstoichiometric) VH_x

[For Data Reported in Figure and Table No. 330]

Curve No.	Ref. No.	Year	Temp. Range, K	Reported Error, %	Name and Specimen Designation	Composition (weight percent), Specifications and Remarks
1	210	1961	25-340			$VH_{0.734}$; extra pure hydrogen was combined with 99.8 vanadium powder.

DATA TABLE NO. 330 SPECIFIC HEAT OF VANADIUM HYDRIDE VH_x (nonstoichiometric)

[Temperature, T, K; Specific Heat, C_p, Cal $g^{-1}K^{-1}$]

T	C_p
CURVE 1	
25	1.8×10^{-3}
30	3.15
35	5.32
40	8.17
45	1.17×10^{-2}
50	1.58
60	2.597
70	3.773
80	4.837
90	5.882
100	6.869
110	7.817
120	8.669
130	9.462
140	1.018×10^{-1}
150	1.080
160	1.132
170	1.180
180	1.225
190	1.267*
200	1.308
210	1.345*
220	1.378
230	1.413*
240	1.440*
250	1.467
260	1.486*
270	1.507*
280	1.527
290	1.546*
298.16	1.556*
300	1.562*
310	1.579
320	1.592*
330	1.606*
340	1.620

* Not shown on plot

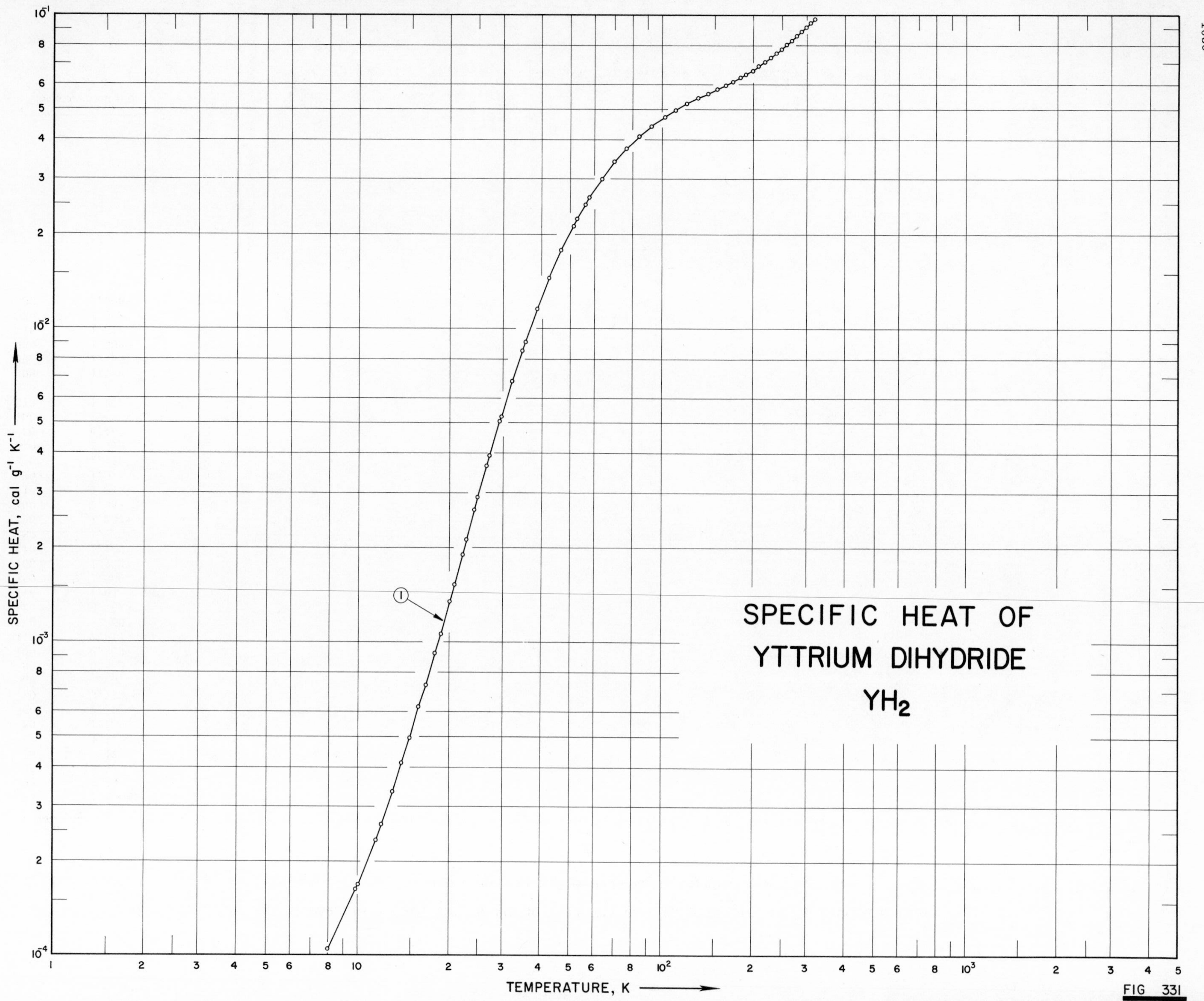

SPECIFIC HEAT OF
YTTRIUM DIHYDRIDE
YH_2
SPECIFIC HEAT, cal g^{-1} K^{-1}
TEMPERATURE, K
1
FIG 331

SPECIFICATION TABLE NO. 331 SPECIFIC HEAT OF YTTRIUM DIHYDRIDE YH_2

[For Data Reported in Figure and Table No. 331]

Curve No.	Ref. No.	Year	Temp. Range, K	Reported Error, %	Name and Specimen Designation	Composition (weight percent), Specifications and Remarks
1	411	1962	6-347			H/Y = 2.003 ±0.004; impurities: 0.0199 O_2, 0.0135 C, 0.0050 F_2, and 0.0017 N_2; prepared by reacting purified Y with H_2 at 400 C; specimen cooled to room temperature in vacuo.

DATA TABLE NO. 331 SPECIFIC HEAT OF YTTRIUM DIHYDRIDE YH_2

[Temperature, T, K; Specific Heat, C_p, Cal $g^{-1}K^{-1}$]

T	C_p	T	C_p
CURVE 1		CURVE 1 (cont.)	
Series I			
		39.06	1.161 x 10^{-2}
281.94	8.619 x 10^{-2}*	42.99	1.457
291.75	8.888*	47.02	1.780
301.83	9.168	51.50	2.128
311.97	9.449	56.46	2.493
322.20	9.733		
332.50	1.002 x 10^{-1}*	Series IV	
342.72	1.030*		
347.32	1.043*	53.07	2.245 x 10^{-2}
		58.23	2.622
Series II		64.03	3.009
		70.28	3.400
6.00	6.4 x 10^{-5}*	77.19	3.757
7.97	1.1 x 10^{-4}	84.81	4.118
9.89	1.63	93.11	4.437
11.51	2.34	102.06	4.723
13.04	3.34	111.72	4.987
14.85	4.96	122.02	5.231
16.79	7.32	132.61	5.448
18.79	1.07 x 10^{-3}	143.18	5.639
20.83	1.537	153.52	5.817
22.84	2.130	163.68	5.987
24.97	2.918	173.83	6.159
27.26	3.941	183.90	6.337
29.76	5.262		
32.32	6.823	Series V	
34.85	8.519		
		181.38	6.291 x 10^{-2}*
Series III		191.00	6.467
		200.85	6.657
5.88	6.2 x 10^{-5}*	210.94	6.862
7.97	1.1 x 10^{-4}*	221.07	7.084
10.03	1.69	231.14	7.313
11.99	2.64	241.22	7.554
13.98	4.12	251.32	7.805
15.98	6.25	261.41	8.068
18.00	9.23	271.47	8.335
20.10	1.354 x 10^{-3}	281.54	8.607
22.14	1.905	291.68	8.888
24.30	2.651		
26.68	3.662		
29.42	5.065		
32.49	6.932*		
35.64	9.076		

* Not shown on plot

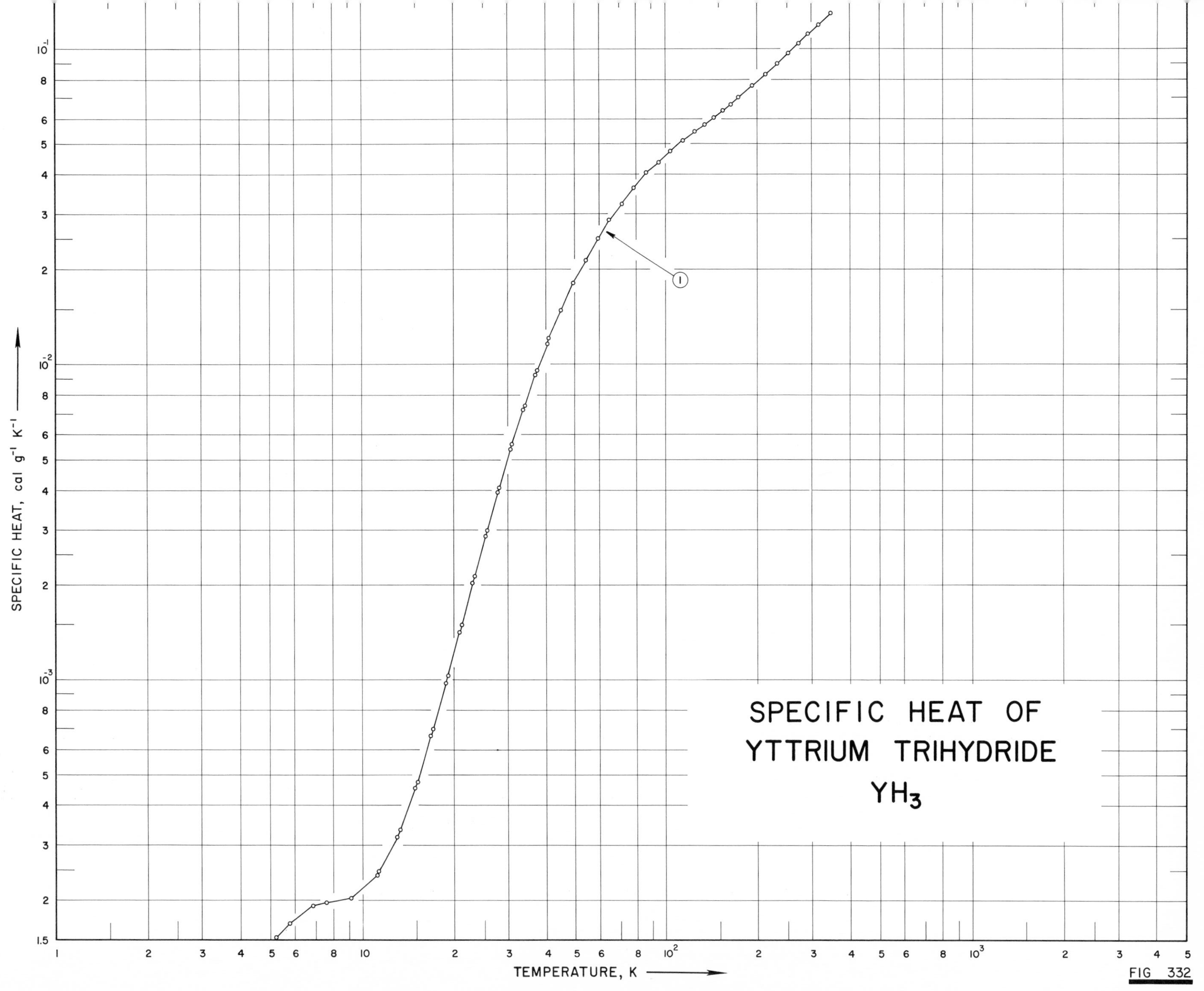

SPECIFIC HEAT OF
YTTRIUM TRIHYDRIDE
YH_3
SPECIFIC HEAT, cal g^{-1} K^{-1}
TEMPERATURE, K
1
FIG 332

SPECIFICATION TABLE NO. 332 SPECIFIC HEAT OF YTTRIUM TRIHYDRIDE YH_3

[For Data Reported in Figure and Table No. 332]

Curve No.	Ref. No.	Year	Temp. Range, K	Reported Error, %	Name and Specimen Designation	Composition (weight percent), Specifications and Remarks
1	211	1963	5-346	<1		0.1 impurities; prepared by the reaction of YH_2 and H_2 at 350 C and 350 mm Hg for 48 hrs; cooled to room temperature over a 4 hr period with hydrogen pressure of 400 mm Hg.

DATA TABLE NO. 332 SPECIFIC HEAT OF YTTRIUM TRIHYDRIDE YH_3

[Temperature, T,K; Specific Heat, C_p, Cal $g^{-1}K^{-1}$]

T	C_p
CURVE 1	
Series I	
5.20	1.52 x 10^{-4}
6.84	1.93
9.11	2.03
11.34	2.48
13.27	3.35
15.09	4.76
17.00	6.97
19.05	1.03 x 10^{-3}
21.12	1.499
23.27	2.130
25.55	2.964
28.07	4.075
30.96	5.583
34.04	7.412
37.36	9.578
40.97	1.209 x 10^{-2}
44.93	1.493
49.30	1.811
54.07	2.149
59.30	2.508
Series II	
5.77	1.70 x 10^{-4}
7.60	1.97
9.40	2.06*
11.16	2.39
12.95	3.19
14.86	4.55
16.80	6.65
18.76	9.72
20.78	1.412 x 10^{-3}
22.93	2.019
25.23	2.840
27.77	3.928
30.60	5.376
33.59	7.133
36.83	9.224
40.41	1.169 x 10^{-2}
44.34	1.451*
48.70	1.766*
53.67	2.121*
CURVE 1 (cont.)	
Series III	
54.05	2.147 x 10^{-2}*
59.00	2.486*
64.79	2.864
71.29	3.243
78.51	3.633
86.47	4.031
95.06	4.392
104.32	4.748
114.55	5.113
125.01	5.467
134.96	5.786
144.74	6.093
154.51	6.395
164.25	6.696
174.08	7.003
Series IV	
173.57	6.988 x 10^{-2}*
183.48	7.303*
193.55	7.626
203.58	7.951*
213.56	8.284
223.64	8.626*
233.75	8.972
243.74	9.327*
253.70	9.677
263.70	1.004 x 10^{-1}*
273.80	1.040
283.87	1.076*
293.86	1.1115
Series V	
286.30	1.085 x 10^{-1}*
296.20	1.1207*
306.37	1.1572*
316.65	1.1943
326.92	1.2310*
337.12	1.2666*
346.18	1.2985
CURVE 1 (cont.)	
Series VI	
160.24	6.572 x 10^{-2}*
170.58	6.893*
180.07	7.224*

* Not shown on plot

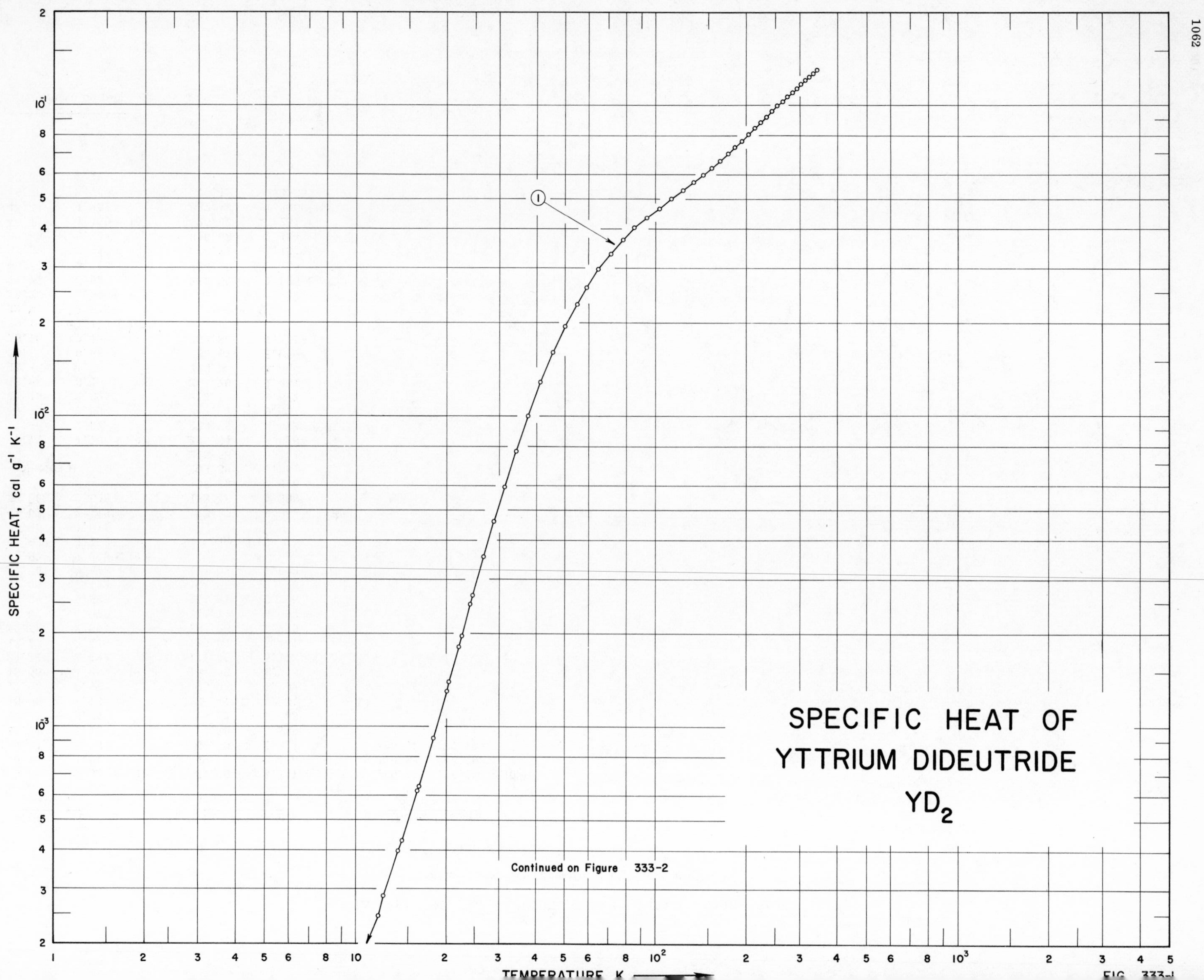
SPECIFIC HEAT OF
YTTRIUM DIDEUTRIDE
YD_2
Continued on Figure 333-2
SPECIFIC HEAT, cal g⁻¹ K⁻¹
TEMPERATURE, K
1

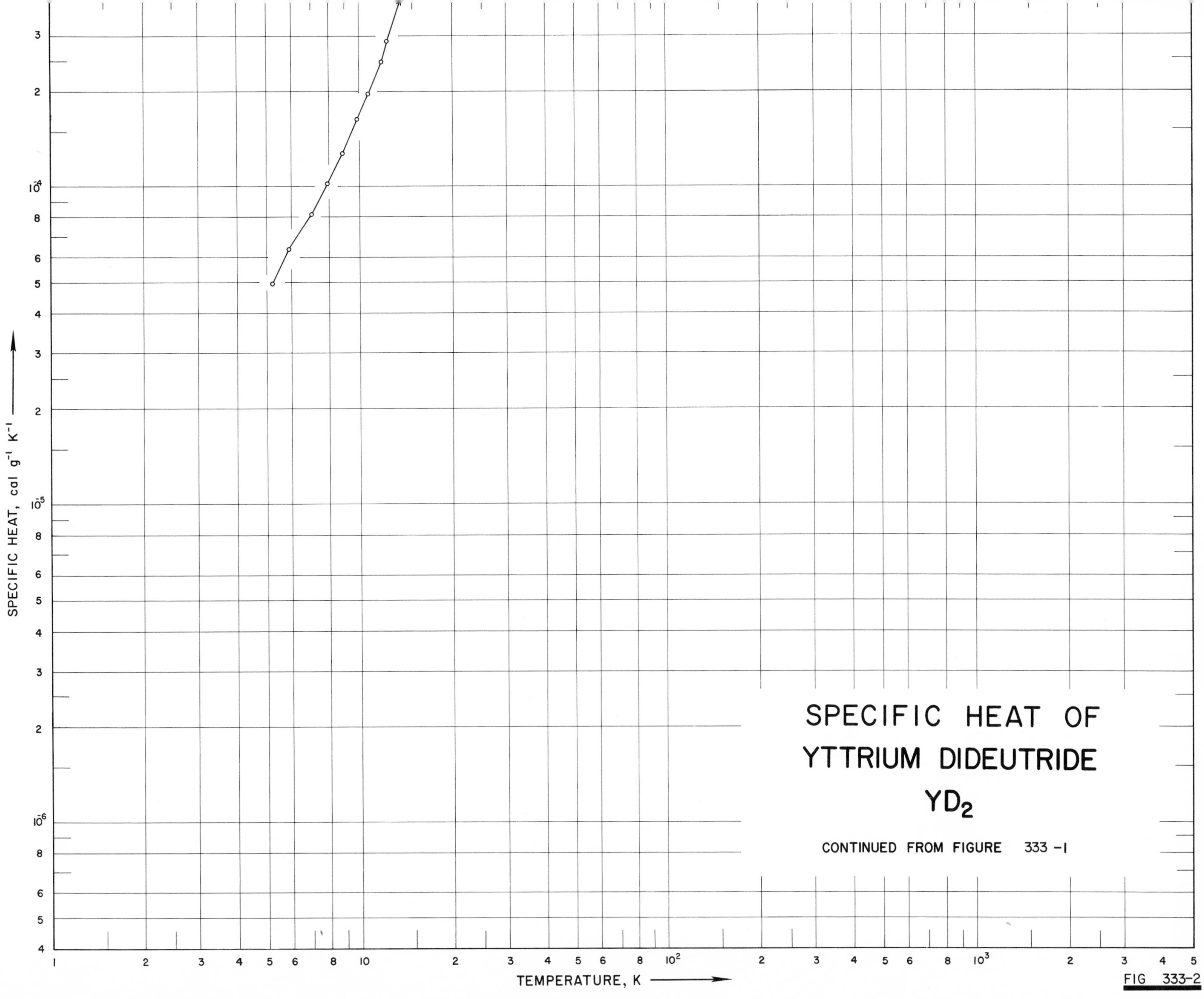
SPECIFIC HEAT, cal g-1 K-1
TEMPERATURE, K
SPECIFIC HEAT OF
YTTRIUM DIDEUTRIDE
YD2
CONTINUED FROM FIGURE 333-I
FIG 333-2

SPECIFICATION TABLE NO. 333 SPECIFIC HEAT OF YTTRIUM DIDEUTRIDE YD_2

[For Data Reported in Figure and Table No. 333]

Curve No.	Ref. No.	Year	Temp. Range, K	Reported Error, %	Name and Specimen Designation	Composition (weight percent), Specifications and Remarks
1	411	1962	5-344			D/Y = 2.003 ±0.004; impurities: 0.0138 F_2, 0.0135 C, 0.0125 O_2, and 0.0022 N_2; prepared by reacting yttrium metal with purified deuterium at 400 C.

DATA TABLE NO. 333 SPECIFIC HEAT OF YTTRIUM DIDEUTRIDE YD_2

[Temperature, T, K; Specific Heat, C_p, Cal $g^{-1}K^{-1}$]

T	C_p
CURVE 1	
Series I	
5.91	6.3 x 10^{-5}
7.94	1.0 x 10^{-4}
9.88	1.63
11.83	2.46
13.87	3.98
16.06	6.21
18.12	9.18
20.06	1.306 x 10^{-3}
22.03	1.817
24.09	2.488
28.63	4.470*
31.57	6.119*
34.67	8.098*
37.84	1.031*
Series II	
5.22	4.9 x 10^{-5}
7.03	8.2
8.89	1.27 x 10^{-4}
10.75	1.96
12.47	2.86
14.27	4.30
16.25	6.43
18.25	9.42*
20.41	1.390 x 10^{-3}
22.53	1.965
24.57	2.664
26.68	3.533
28.85	4.583
31.26	5.933
34.10	7.715
37.45	1.003 x 10^{-2}
41.37	1.291
45.63	1.615
50.16	1.958
55.00	2.311
Series III	
54.18	2.254 x 10^{-2}*
59.28	2.614
64.99	2.985
71.30	3.349
CURVE 1 (cont.)	
78.23	3.708 x 10^{-2}
85.74	4.057
93.96	4.377
103.01	4.687
113.48	5.022
124.47	5.360
134.82	5.680
144.92	5.998
154.81	6.323
164.58	6.657
174.33	7.000
184.11	7.361
193.88	7.726
Series IV	
194.78	7.761 x 10^{-2}*
204.38	8.127
213.91	8.493
223.61	8.866
233.60	9.249
243.72	9.636
253.85	1.002 x 10^{-1}
263.77	1.038
273.59	1.074
283.56	1.109
293.61	1.144
303.70	1.177
313.83	1.210
232.96	1.241
334.18	1.272
344.44	1.302

* Not shown on plot

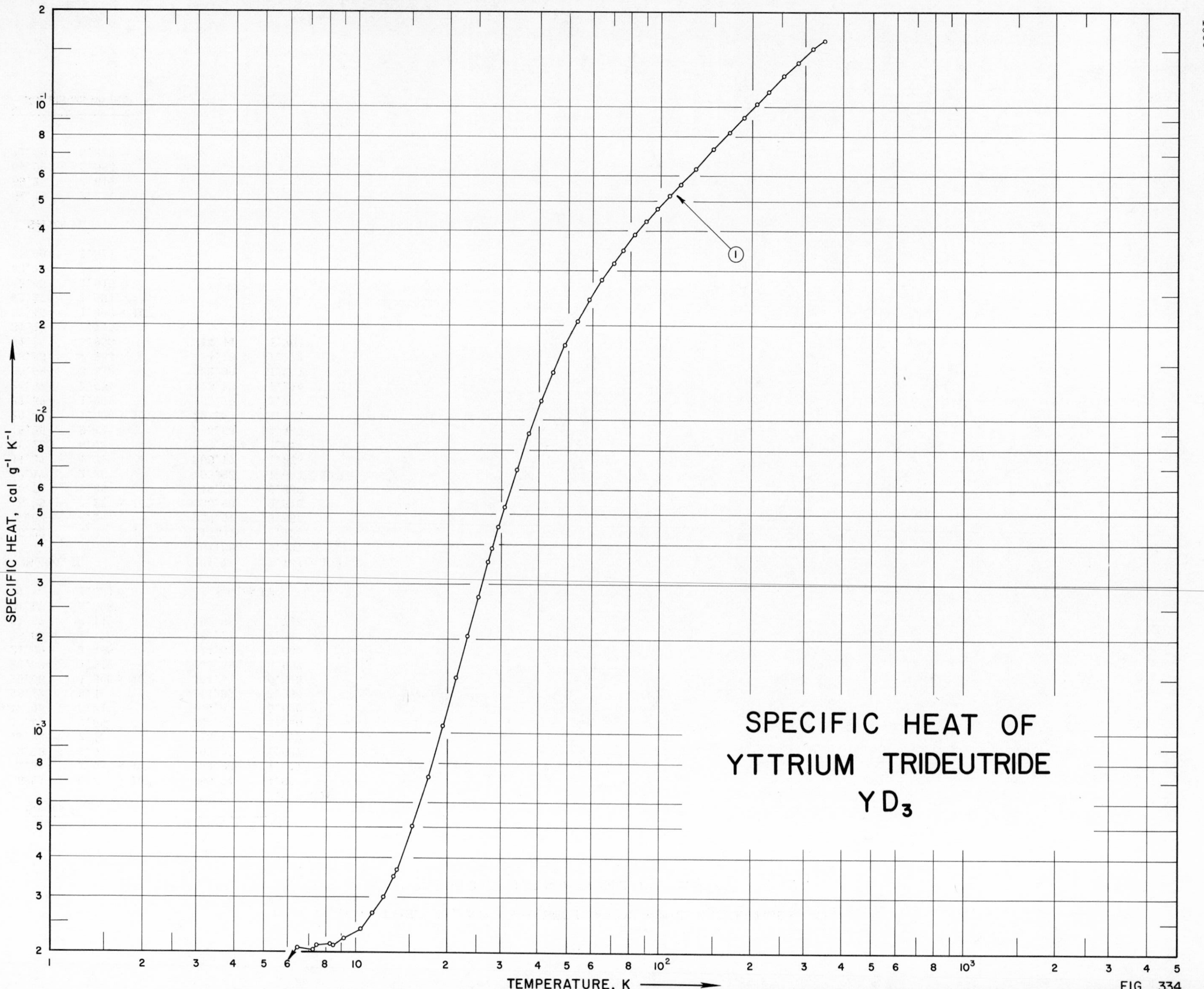
SPECIFIC HEAT OF
YTTRIUM TRIDEUTRIDE
YD$_3$
SPECIFIC HEAT, cal g^{-1} K^{-1}
TEMPERATURE, K
1

FIG 334

SPECIFICATION TABLE NO. 334 SPECIFIC HEAT OF YTTRIUM TRIDEUTRIDE YD_3

[For Data Reported in Figure and Table No. 334]

Curve No.	Ref. No.	Year	Temp. Range, K	Reported Error, %	Name and Specimen Designation	Composition (weight percent), Specifications and Remarks
1	211	1963	5-348	<1		0.1 impurities; prepared by reaction of YD_2 and deuterium gas.

DATA TABLE NO. 334 SPECIFIC HEAT OF YTTRIUM TRIDEUTRIDE YD_3

[Temperature, T, K; Specific Heat, C_p, Cal $g^{-1}K^{-1}$]

T	C_p
CURVE 1	
Series I	
5.26	1.71 x 10^{-4}*
6.42	2.07
7.45	2.12
8.47	2.11
10.45	2.37
11.32	2.66
12.25	2.99
13.59	3.65
15.33	5.05
17.22	7.28
19.22	1.057 x 10^{-3}
21.20	1.503
23.12	2.047
25.04	2.715
27.04	3.531
29.27	4.569
Series II	
5.15	1.60 x 10^{-4}*
6.24	1.96*
7.23	2.03
8.28	2.14
9.18	2.23
10.15	2.34*
11.50	2.67*
13.31	3.48
15.22	4.97*
17.22	7.25*
19.20	1.05 x 10^{-3}*
21.16	1.494*
23.20	2.075*
25.41	2.857*
27.86	3.900
30.55	5.235
33.52	6.930
36.87	9.042
40.52	1.151 x 10^{-2}
44.48	1.428
48.84	1.742
53.56	2.078
58.65	2.433

T	C_p
CURVE 1 (cont.)	
Series III	
54.02	2.109 x 10^{-2}*
58.73	2.440*
64.07	2.800
70.22	3.189
Series IV	
75.21	3.490 x 10^{-2}
82.25	3.907
89.67	4.318
97.95	4.742
107.23	5.207
116.93	5.690
Series V	
121.26	5.905 x 10^{-2}*
130.86	6.380
140.41	6.853*
149.83	7.323
159.30	7.794*
168.94	8.281
178.43	8.756*
188.05	9.252
197.74	9.731*
207.42	1.023 x 10^{-1}
217.11	1.0709*
226.79	1.1187
Series VI	
235.31	1.1607 x 10^{-1}*
245.12	1.2075*
254.85	1.2536
264.65	1.2987*
274.55	1.3440*
284.43	1.3871
Series VII	
289.34	1.4082 x 10^{-1}*
299.26	1.4496*
309.11	1.4902*
318.95	1.5290

T	C_p
CURVE 1 (cont.)	
328.84	1.5663 x 10^{-1}*
338.80	1.6056*
346.30	1.6322
Series VIII*	
51.67	1.944 x 10^{-2}
56.15	2.260
60.61	2.567
66.19	2.938
72.46	3.325
79.02	3.710
86.14	4.127
94.13	4.547
103.12	4.999
112.53	5.471
121.91	5.935
Series IX*	
129.78	6.324 x 10^{-2}
139.32	6.799
148.83	7.273
158.48	7.754
168.16	8.240
177.83	8.733
187.24	1.026 x 10^{-1}
197.18	9.711 x 10^{-2}
207.15	1.021 x 10^{-1}
216.97	1.0700
226.90	1.1192
236.77	1.1674
246.54	1.2144
256.39	1.2604
266.32	1.3066
276.32	1.3516
286.23	1.3937
Series X*	
292.25	1.4209 x 10^{-1}
302.23	1.4625
312.15	1.5033
322.11	1.5421
332.04	1.5800
341.42	1.6142
348.00	1.6376

* Not shown on plot

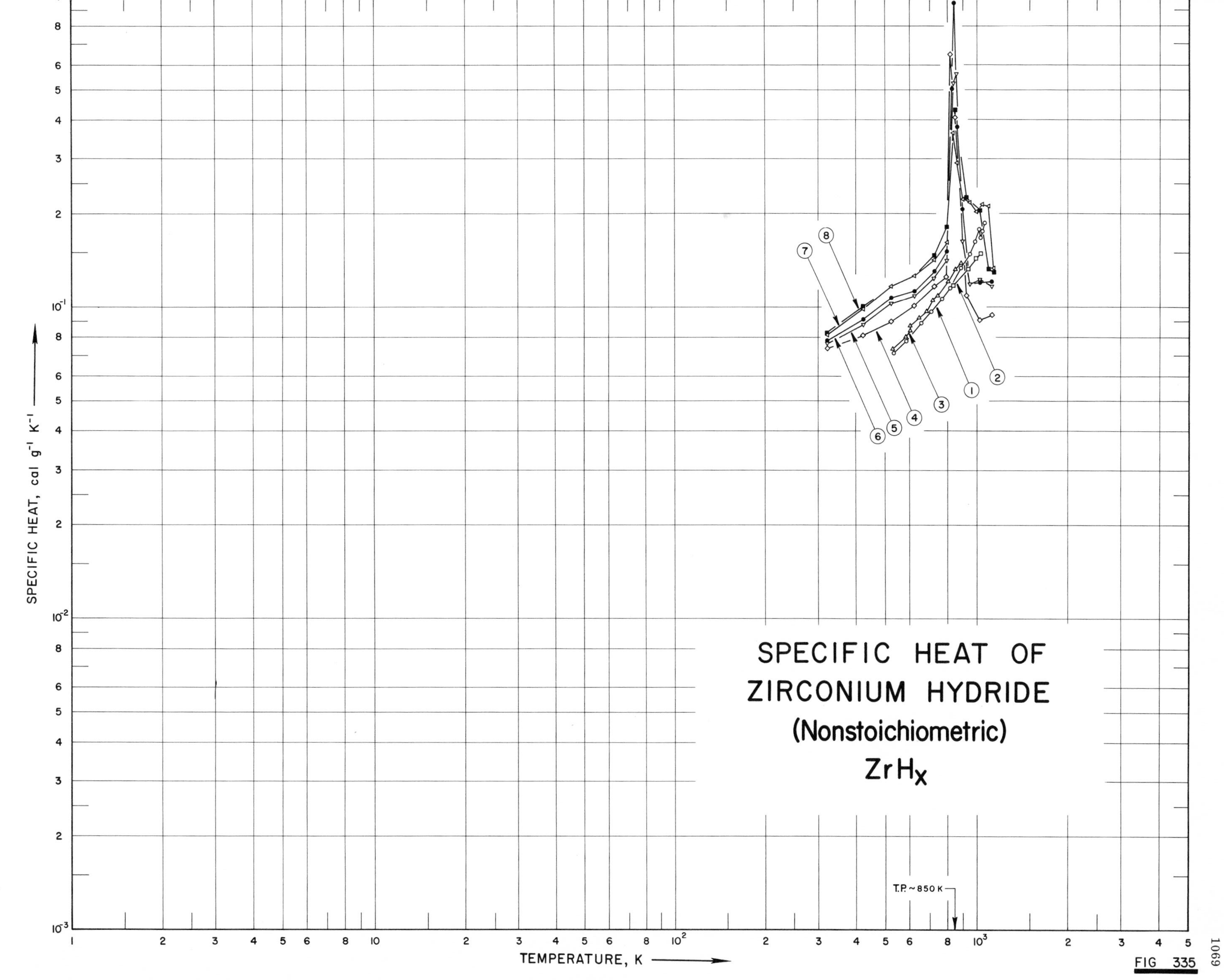

SPECIFIC HEAT OF
ZIRCONIUM HYDRIDE
(Nonstoichiometric)
ZrH$_x$
T.P. ~850 K
SPECIFIC HEAT, cal g^{-1} K^{-1}
TEMPERATURE, K
1
2
3
4
5
6
7
8

FIG 335

SPECIFICATION TABLE NO. 335 SPECIFIC HEAT OF ZIRCONIUM HYDRIDE (nonstoichiometric) ZrH_x

[For Data Reported in Figure and Table No. 335]

Curve No.	Ref. No.	Year	Temp. Range, K	Reported Error, %	Name and Specimen Designation	Composition (weight percent), Specifications and Remarks
1	212	1962	533-1063		50.2 At % H	$ZrH_{1.005}$; 99.10 Zr and 1.90 H.
2	212	1962	543-1038		57.1 At % H	$ZrH_{1.33}$; 98.55 Zr and 1.45 H.
3	212	1962	528-887		64.0 At % H	$ZrH_{1.78}$; 98.08 Zr, 1.92 H.
4	412	1957	323-1123			$ZrH_{0.324}$; 0.356 H, 0.1 Fe, 0.06 C, and 0.01 each Al, Cr, Hf, Mn, N, and O (corrected for impurities).
5	412	1957	323-1123			$ZrH_{0.556}$; 0.88 Fe, 0.611 H, 0.6 Hf, 0.48 Si, 0.32 C, 0.12 Al, and 0.1 Cr (corrected for impurities).
6	412	1957	323-1123			$ZrH_{0.701}$; 0.769 H, 0.53 Fe, 0.44 Hf, 0.35 Si, 0.32 C, 0.12 Al, 0.08 Ni, and 0.06 Cr (corrected for impurities).
7	412	1957	323-1148			$ZrH_{0.999}$; 1.092 H, 0.48 Hf, 0.41 Fe, 0.34 Si, 0.32 C, and 0.17 Al (corrected for impurities).
8	412	1957	323-1148			$ZrH_{1.071}$; 1.17 H, 0.1 Fe, 0.06 C, 0.017 O, and 0.01 each Al, Cr, Hf, Mn, and N.

DATA TABLE NO. 335 SPECIFIC HEAT OF ZIRCONIUM HYDRIDE ZrH_x (nonstoichiometric)

[Temperature, T, K; Specific Heat, C_p, Cal $g^{-1}K^{-1}$]

T	C_p
CURVE 1	
533	7.126 x 10^{-2}
588	7.783
659	8.880
708	9.647
768	1.063 x 10^{-1}
818	1.151
853	1.239*
888	1.348
928	1.447*
948	1.480
978	1.590*
987	1.622
1013	1.644*
1023	1.677
1048	1.754
1018	1.787
1063	1.864
CURVE 2	
543	7.235 x 10^{-2}
648	8.770
708	9.866
763	1.041 x 10^{-1}
838	1.173
938	1.337
993	1.447
1038	1.491
CURVE 3	
528	7.345 x 10^{-2}
583	8.003
603	8.441
603	8.715
648	9.263
683	9.702
713	1.052 x 10^{-1}
743	1.096
803	1.211
848	1.326
887	1.392

T	C_p
CURVE 4	
323	7.38 x 10^{-2}
423	8.12
523	8.99
623	1.013 x 10^{-1}
723	1.174
793	1.255
818	6.48
848	4.08
923	1.096
1023	9.09 x 10^{-2}
1123	9.45
CURVE 5	
323	7.62 x 10^{-2}
423	8.80
523	1.028 x 10^{-1}
623	1.087
723	1.244
798	1.402
835	5.22
860	5.6
898	1.63
948	1.194
1023	1.233
1123	1.165
CURVE 6	
323	7.80 x 10^{-2}
423	9.13
523	1.078 x 10^{-1}
623	1.126
723	1.302
798	1.510
828	5.0
840	9.45
860	3.78
898	2.06
948	1.188
1023	1.203
1123	1.21

T	C_p
CURVE 7	
323	8.15 x 10^{-2}
423	9.85
523	1.169 x 10^{-1}
623	1.261
723	1.419
798	1.614
835	3.62
860	2.9
898	2.228
948	2.176
998	2.022
1048	2.150
1098	2.112
1148	1.34
CURVE 8	
323	8.23 x 10^{-2}
423	1.006 x 10^{-1}
523	1.170
623	1.273
723	1.466
798	1.808
848	4.30
923	2.252
1023	2.050
1098	1.330
1148	1.30

* Not shown on plot

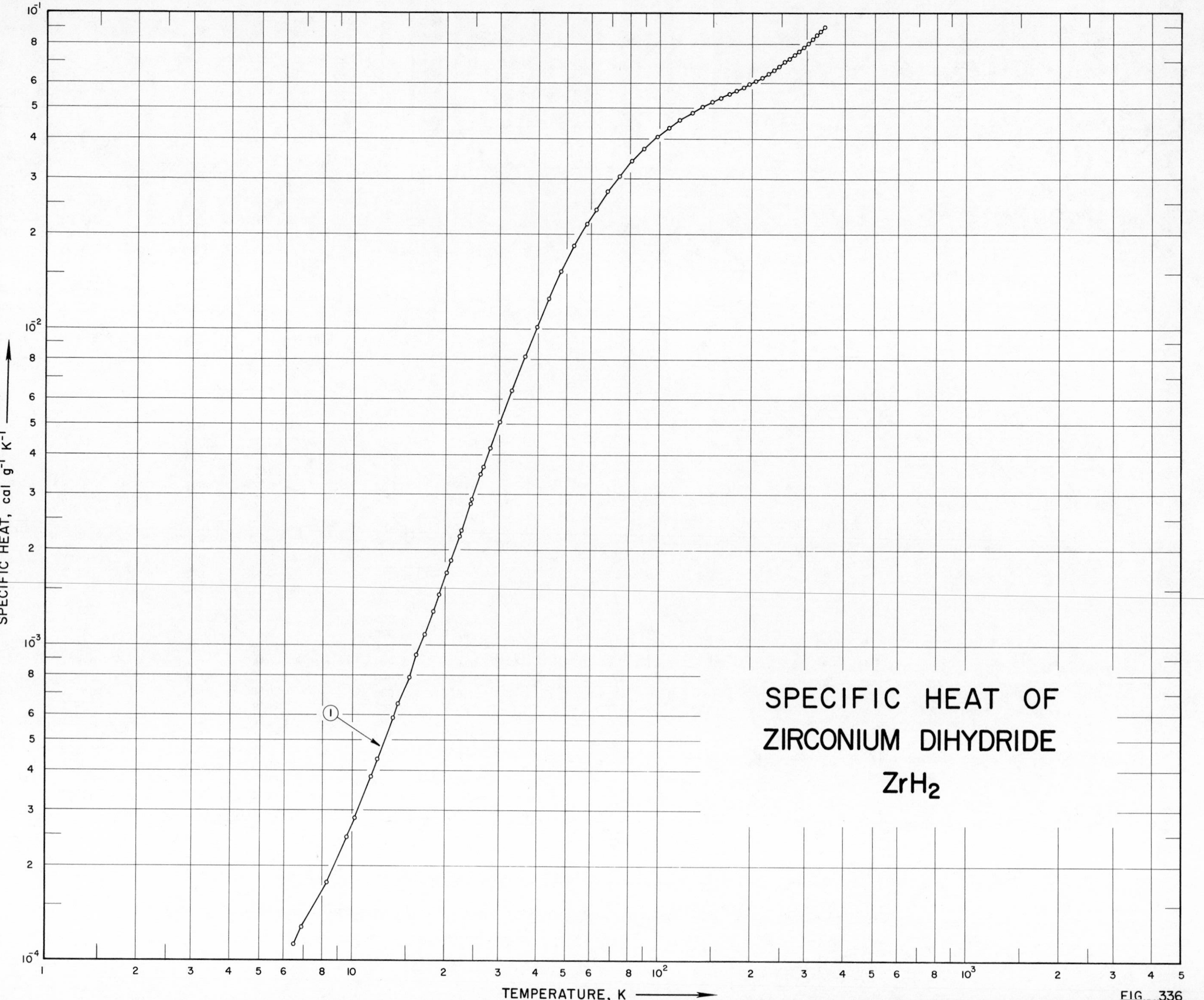
SPECIFIC HEAT OF
ZIRCONIUM DIHYDRIDE
ZrH2
SPECIFIC HEAT, cal g-1 K-1
TEMPERATURE, K
1

FIG. 336

SPECIFICATION TABLE NO. 336 SPECIFIC HEAT OF ZIRCONIUM DIHYDRIDE ZrH_2

[For Data Reported in Figure and Table No. 336]

Curve No.	Ref. No.	Year	Temp. Range, K	Reported Error, %	Name and Specimen Designation	Composition (weight percent), Specifications and Remarks
1	413	1961	7-345			0.02 Fe, 0.01 Cu, 0.0015 Al, 0.0004 each Cr and Ni, 0.0002 Pb, 0.00002 Ag, and 0.00001 B; prepared on a high vacuum line by direct reaction of zirconium metal with hydrogen at about 400 C.

DATA TABLE NO. 336 SPECIFIC HEAT OF ZIRCONIUM DIHYDRIDE ZrH_2

[Temperature, T,K; Specific Heat, C_p, Cal $g^{-1}K^{-1}$]

T	C_p	T	C_p
CURVE 1		**CURVE 1 (cont.)**	
Series I		Series IV	
6.87	1.28 x 10^{-4}	89.50	3.756 x 10^{-2}*
9.60	2.47	98.52	4.078
11.54	3.84	107.49	4.359
13.64	5.88	117.18	4.620
15.30	7.94	127.61	4.867
17.20	1.079 x 10^{-3}	138.07	5.077
19.12	1.446	148.30	5.261
20.96	1.864	158.34	5.421
22.63	2.312	168.19	5.571
24.48	2.898	178.09	5.713
26.65	3.672	188.08	5.858
Series II		Series V	
6.51	1.13 x 10^{-4}	186.63	5.840 x 10^{-2}*
8.34	1.77	196.67	5.988
10.23	2.83	206.59	6.140
12.14	4.35	216.40	6.296
14.17	6.53	226.27	6.459
16.22	9.32	236.14	6.631
18.26	1.276 x 10^{-3}	245.92	6.817
20.24	1.697	255.66	7.009
22.25	2.209	265.42	7.209
24.25	2.816	275.27	7.421
26.17	3.483	285.24	7.638
28.09	4.221	295.39	7.868
30.10	5.078	305.52	8.083
32.90	6.422	315.61	8.346
36.27	8.197	325.79	8.593
39.76	1.020 x 10^{-2}	335.94	8.837
43.63	1.253	345.42	9.084
47.99	1.525		
52.99	1.840		
58.27	2.166		
Series III			
57.30	2.107 x 10^{-2}*		
62.42	2.396		
67.98	2.733		
74.12	3.052		
81.24	3.402		
89.14	3.742		

* Not shown on plot

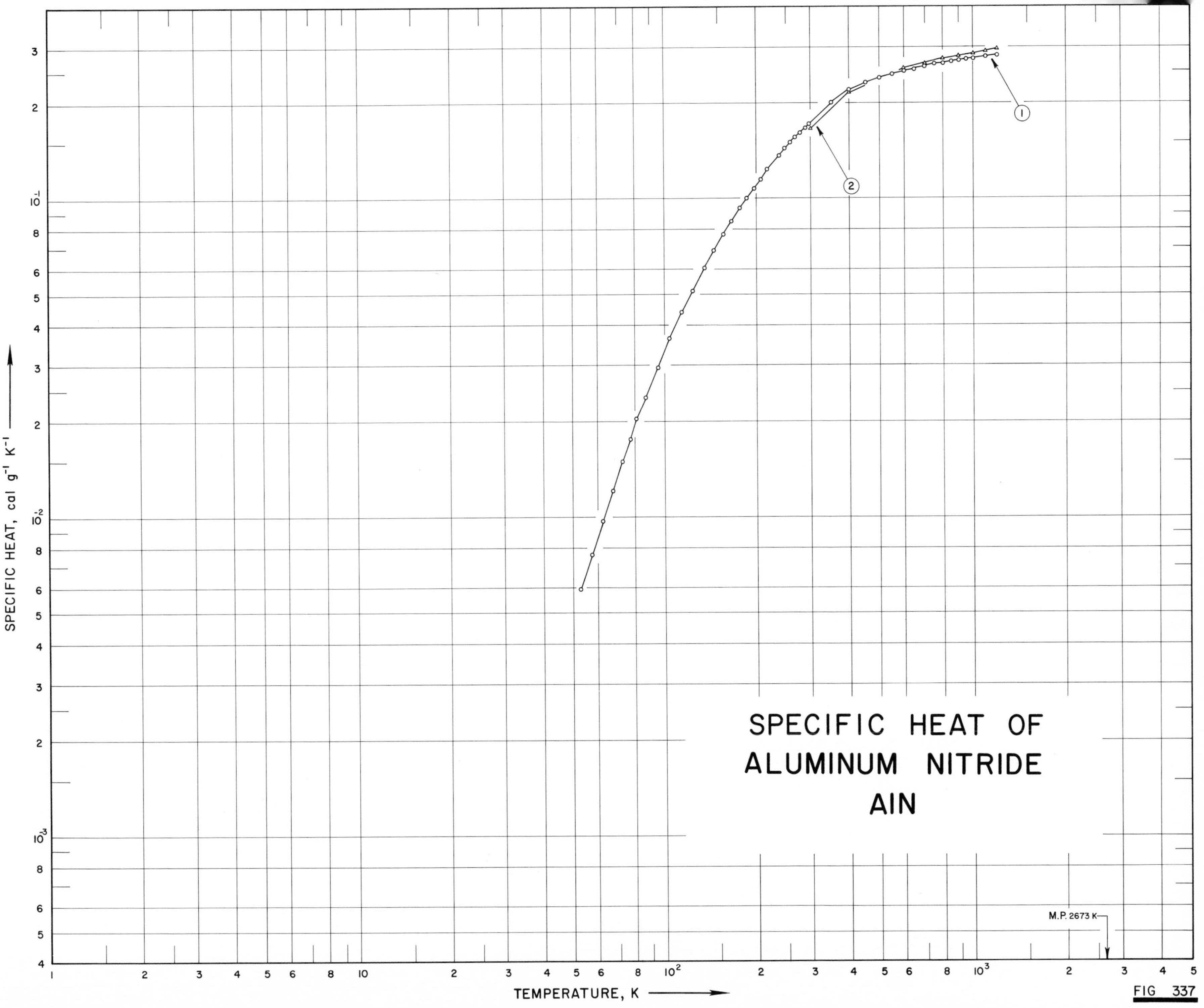
SPECIFIC HEAT OF
ALUMINUM NITRIDE
AlN
1
2
M.P. 2673 K
TEMPERATURE, K
SPECIFIC HEAT, cal g⁻¹ K⁻¹
FIG 337

SPECIFICATION TABLE NO. 337 SPECIFIC HEAT OF ALUMINUM NITRIDE AlN

[For Data Reported in Figure and Table No. 337]

Curve No.	Ref. No.	Year	Temp. Range, K	Reported Error, %	Name and Specimen Designation	Composition (weight percent), Specifications and Remarks
1	213	1961	53-1200	0.3-0.9		98.92 AlN; 1.08 Al_2O_3, 0.01-0.05 Fe, 0.01 Si, <0.001 Ca, Mg, Cu, and <0.0001 Cr.
2	54	1965	300-1200	0.5		99.999 AlN; traces of Cu, Mg and Si; sample supplied by the Aluminum Co. of America.

DATA TABLE NO. 337 SPECIFIC HEAT OF ALUMINUM NITRIDE AlN

[Temperature, T, K; Specific Heat, C_p, Cal $g^{-1}K^{-1}$]

T	C_p	T	C_p
CURVE 1		CURVE 1 (cont.)	
52.91	5.980×10^{-3}	1050.00	2.805×10^{-1}*
57.44	7.683	1100.00	2.821
62.13	9.783	1150.00	2.837
67.15	1.223×10^{-2}	1200.00	2.852
72.49	1.509		
77.19	1.786	CURVE 2	
81.59	2.057		
86.79	2.391	300	1.696×10^{-1}
95.08	2.962	400	2.188
104.95	3.684	500	2.435*
114.74	4.421	600	2.584
124.54	5.194	700	2.689
135.81	6.102	800	2.769
145.91	6.941	900	2.835
156.30	7.802	1000	2.894
165.80	8.573	1100	2.945
176.29	9.417	1200	2.994
185.94	1.019×10^{-1}		
195.99	1.095		
206.46	1.176		
216.42	1.250		
266.17	1.321*		
236.17	1.388		
245.70	1.453		
256.33	1.520		
266.12	1.579		
276.14	1.638		
286.38	1.693		
296.24	1.748		
298.15	1.754*		
300.00	1.767*		
350.00	2.034		
400.00	2.211		
450.00	2.335		
500.00	2.427		
550.00	2.497		
600.00	2.553		
650.00	2.599		
700.00	2.637		
750.00	2.670		
800.00	2.698		
850.00	2.724		
900.00	2.747		
950.00	2.767		
1000.00	2.787		

* Not shown on plot

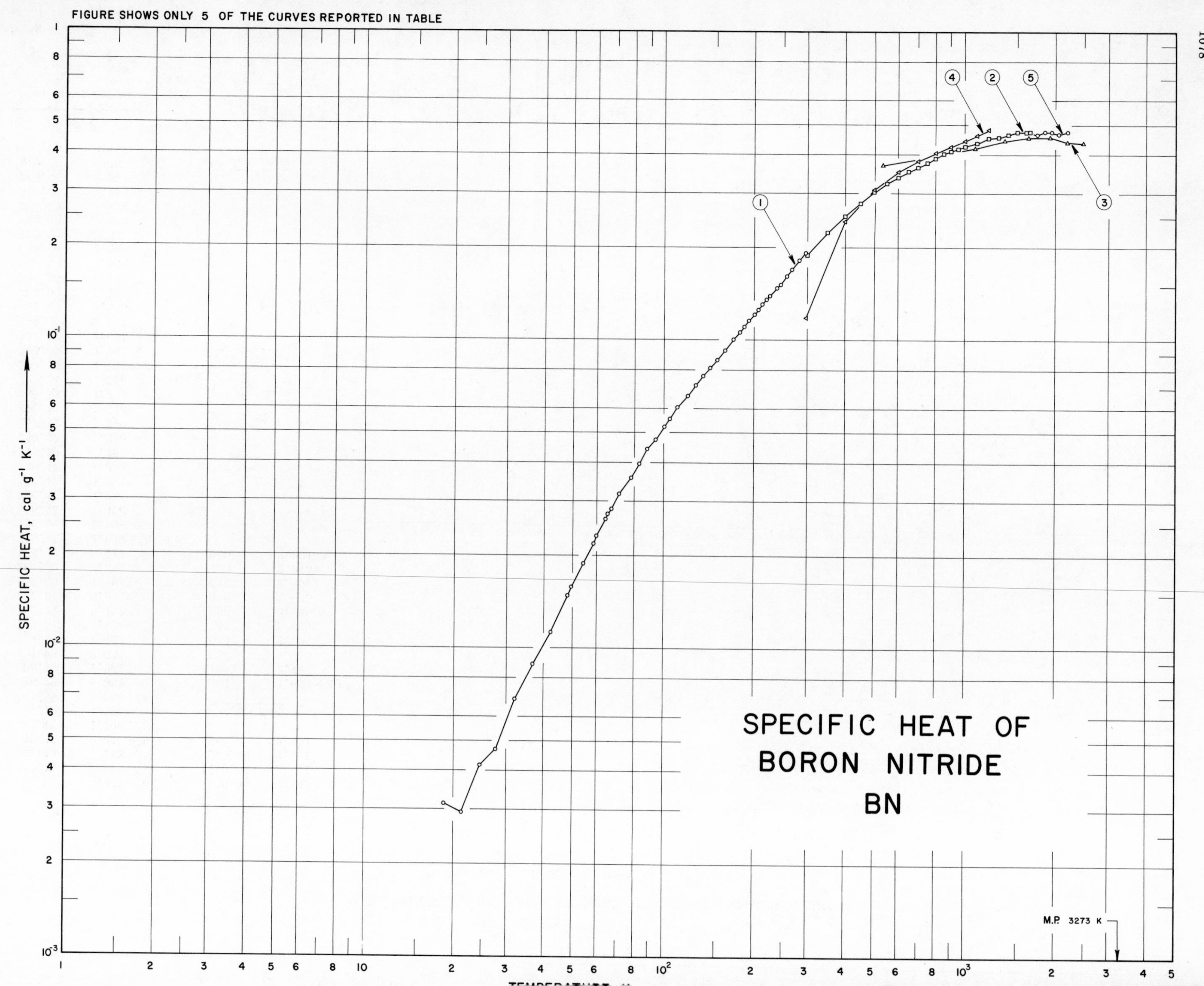
FIGURE SHOWS ONLY 5 OF THE CURVES REPORTED IN TABLE
SPECIFIC HEAT OF
BORON NITRIDE
BN
SPECIFIC HEAT, cal g^{-1} K^{-1}
M.P. 3273 K
1
2
3
4
5

SPECIFICATION TABLE NO. 338 SPECIFIC HEAT OF BORON NITRIDE BN

[For Data Reported in Figure and Table No. 338]

Curve No.	Ref. No.	Year	Temp. Range, K	Reported Error, %	Name and Specimen Designation	Composition (weight percent), Specifications and Remarks
1	214	1954	19-301	1.0		1.5 Fe as Fe_2O_4.
2	215	1961	300-1650			56.85 N and 42.81 B; sample supplied by The Carborundum Co.
3	48	1962	533-2478	≤5.0		Before exposure: 52.4 N, 42.9 B, 0.2 Ca, 0.2 Ti, and 0.1 Si, after exposure: 55.0 N, 41.5 B, and 0.13 C; sample supplied by The Carborundum Co.; hot pressed; crushed in hardened steel mortar to pass 100-mesh screen; density at 25 C, before exposure: apparent density (ASTM method B311-58) 133.5 lb ft^{-3}, true density (by immersion in xylene) 135 lb ft^{-3}, after exposure: apparent density = 122 lb ft^{-3}, true density = 135.5 lb ft^{-3}.
4	163	1962	298-1200	0.5	cubic	Traces of impurities.
5	38	1963	1300-2200	≤5.0		98.0 BN, 1.70 O; measured in an argon atmosphere.
6	414	1926	673-1173			

DATA TABLE NO. 338 SPECIFIC HEAT OF BORON NITRIDE BN

[Temperature, T, K; Specific Heat, C_p, Cal $g^{-1}K^{-1}$]

T	C_p
CURVE 1	
Series I	
65.99	2.732 x 10^{-2}
72.40	3.171
78.09	3.582
83.28	3.981
88.75	4.432
94.71	4.779
100.31	5.226
105.63	5.597
112.57	6.080
121.03	6.604
129.04	7.116
136.68	7.672
144.13	8.176
147.20	8.244*
152.47	8.627
Series II	
54.47	1.886 x 10^{-2}
58.99	2.196
64.92	2.623
71.47	3.095*
77.26	3.530*
82.52	3.937*
87.39	4.295*
92.69	4.682*
99.28	5.121*
106.39	5.605*
113.09	6.072*
119.46	6.483*
125.50	6.798*
132.42	7.281*
139.95	7.797*
147.21	8.296*
154.21	8.788*
161.41	9.280
Series III	
120.86	6.536 x 10^{-2}*
126.92	6.898*
133.70	7.346*
141.18	7.837*
148.98	8.365*

T	C_p
CURVE 1 (cont.)	
157.11	8.949 x 10^{-2}*
164.95	9.513*
172.55	1.007
180.23	1.061
187.14	1.111
194.17	1.168
201.03	1.216
207.75	1.264
214.35	1.308
220.81	1.362
227.16	1.403
233.39	1.451*
Series IV	
49.94	1.575 x 10^{-2}
54.53	1.866*
60.54	2.313
67.75	2.841
200.89	1.214 x 10^{-1}*
214.19	1.303*
220.65	1.357*
226.96	1.402*
233.19	1.447*
239.32	1.488
245.36	1.533
251.31	1.575*
257.17	1.617
262.95	1.659*
268.65	1.704
282.57	1.811
289.82	1.860*
296.97	1.911
Series V	
18.56	3.143 x 10^{-3}
21.42	2.941
24.56	4.191
27.75	4.714
31.91	6.850
36.67	8.865
42.16	1.136 x 10^{-2}
48.27	1.495

T	C_p
CURVE 1 (cont.)	
Series VI	
18.97	3.989 x 10^{-3}
21.99	3.183
25.26	4.311
27.65	4.876
32.59	6.769
39.32	1.011 x 10^{-2}
46.25	1.374
51.43	1.688*
201.65	1.222 x 10^{-1}*
208.33	1.267*
215.92	1.323*
224.44	1.385*
232.77	1.446*
240.91	1.503*
248.90	1.560*
256.73	1.617*
264.41	1.679*
271.97	1.723*
279.25	1.787*
286.54	1.849*
293.72	1.895*
300.80	1.930*
Series VII*	
68.58	2.893 x 10^{-2}
74.64	3.361
80.90	3.804
87.44	4.283
94.39	4.759
101.79	5.291
108.68	5.766
115.23	6.185
121.48	6.528
128.45	6.983
136.11	7.511
143.46	8.010
150.49	8.510
157.42	8.953
164.65	9.465

T	C_p
CURVE 2	
300	1.890 x 10^{-1}
350	2.232
400	2.530
450	2.792
500	3.022
550	3.224
600	3.393
650	3.530
700	3.663
750	3.784
800	3.892
850	4.001
900	4.110
950	4.191
1000	4.271
1050	4.352*
1100	4.392
1150	4.473*
1200	4.513
1250	4.553*
1300	4.593
1350	4.634*
1400	4.674
1450	4.714*
1500	4.714
1550	4.714*
1600	4.714
1650	4.714
CURVE 3	
533	3.700 x 10^{-1}
811	3.850*
1089	4.200
1366	4.450
1644	4.550
1922	4.550
2200	4.400
2478	4.390

T	C_p
CURVE 4	
298.16	1.189 x 10^{-1}
300	1.225*
400	2.446
500	3.087
600	3.501
700	3.804
800	4.054
900	4.267
1000	4.461
1100	4.642
1200	4.811
CURVE 5	
1300	4.525 x 10^{-1}*
1350	4.593*
1400	4.598*
1450	4.694*
1500	4.686*
1550	4.662*
1600	4.634*
1650	4.606*
1700	4.618*
1750	4.662
1800	4.714*
1850	4.743
1900	4.755*
1950	4.730
2000	4.718*
2050	4.674
2100	4.694*
2150	4.706*
2200	4.755
CURVE 6*	
673	1.785 x 10^{-1}
723	1.860
773	1.927
823	1.988
873	2.039
923	2.089
973	2.128
1023	2.162
1073	2.189
1123	2.211
1173	2.224

values should be multiplied by 2

* Not shown on plot

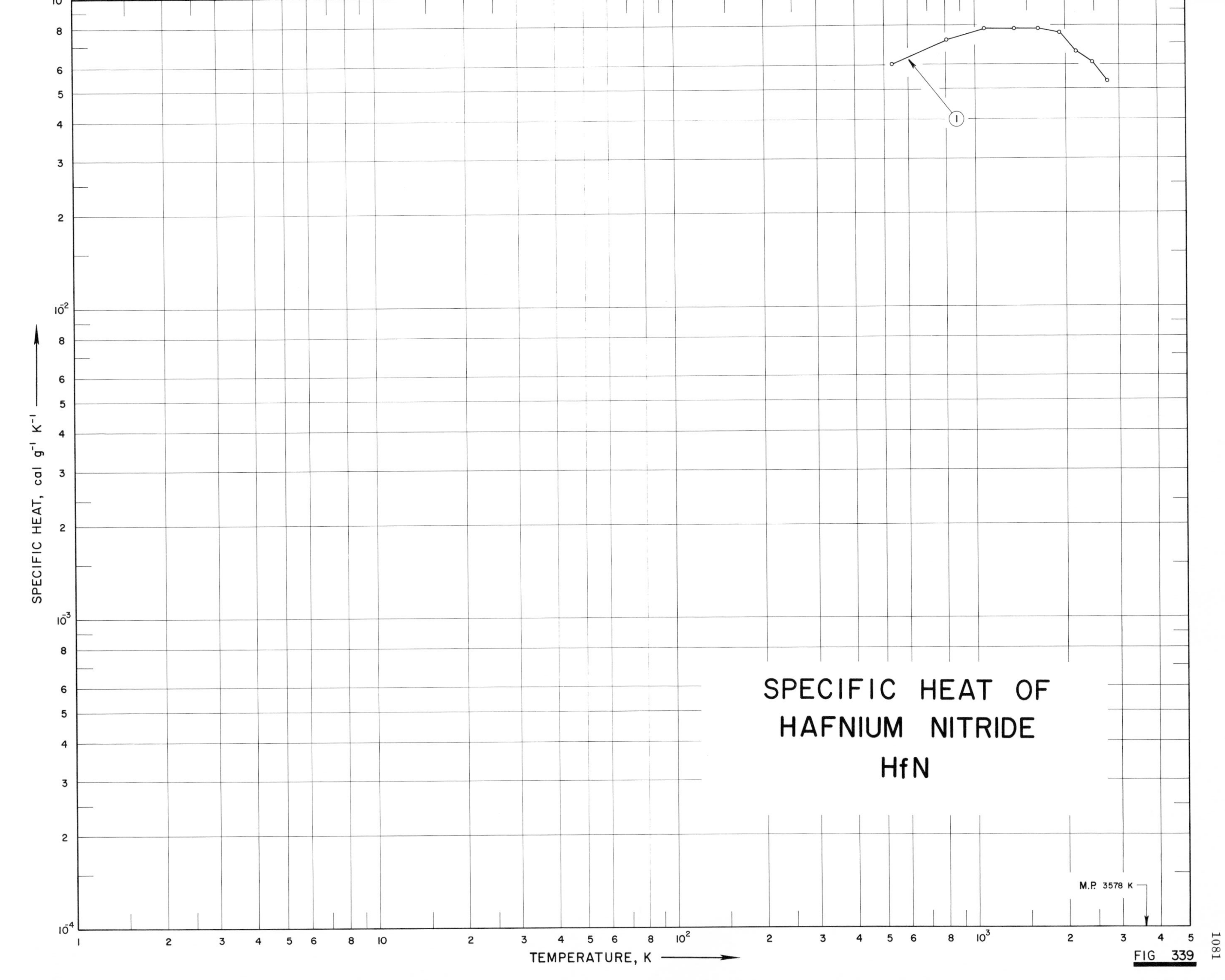
SPECIFIC HEAT OF
HAFNIUM NITRIDE
HfN
SPECIFIC HEAT, cal g⁻¹ K⁻¹
TEMPERATURE, K
M.P. 3578 K
1
FIG 339

SPECIFICATION TABLE NO. 339 SPECIFIC HEAT OF HAFNIUM NITRIDE HfN

[For Data Reported in Figure and Table No. 339]

Curve No.	Ref. No.	Year	Temp. Range, K	Reported Error, %	Name and Specimen Designation	Composition (weight percent), Specifications and Remarks
1	32	1962	533-2755	≤5.0		Wet analysis: 95.4 Hf, 6.61 N_2, and 0.9 O_2; sample supplied by The Carborundum Co.; hot pressed (firing temperature near 6500 F); density = 677 lb ft^{-3}.

DATA TABLE NO. 339 SPECIFIC HEAT OF HAFNIUM NITRIDE HfN

[Temperature, T, K; Specific Heat, C_p, Cal $g^{-1}K^{-1}$]

T	C_p
CURVE 1	
533	6.000×10^{-2}
811	7.200
1089	7.800
1366	7.800
1644	7.800
1922	7.600
2200	6.600
2478	6.100
2755	5.300

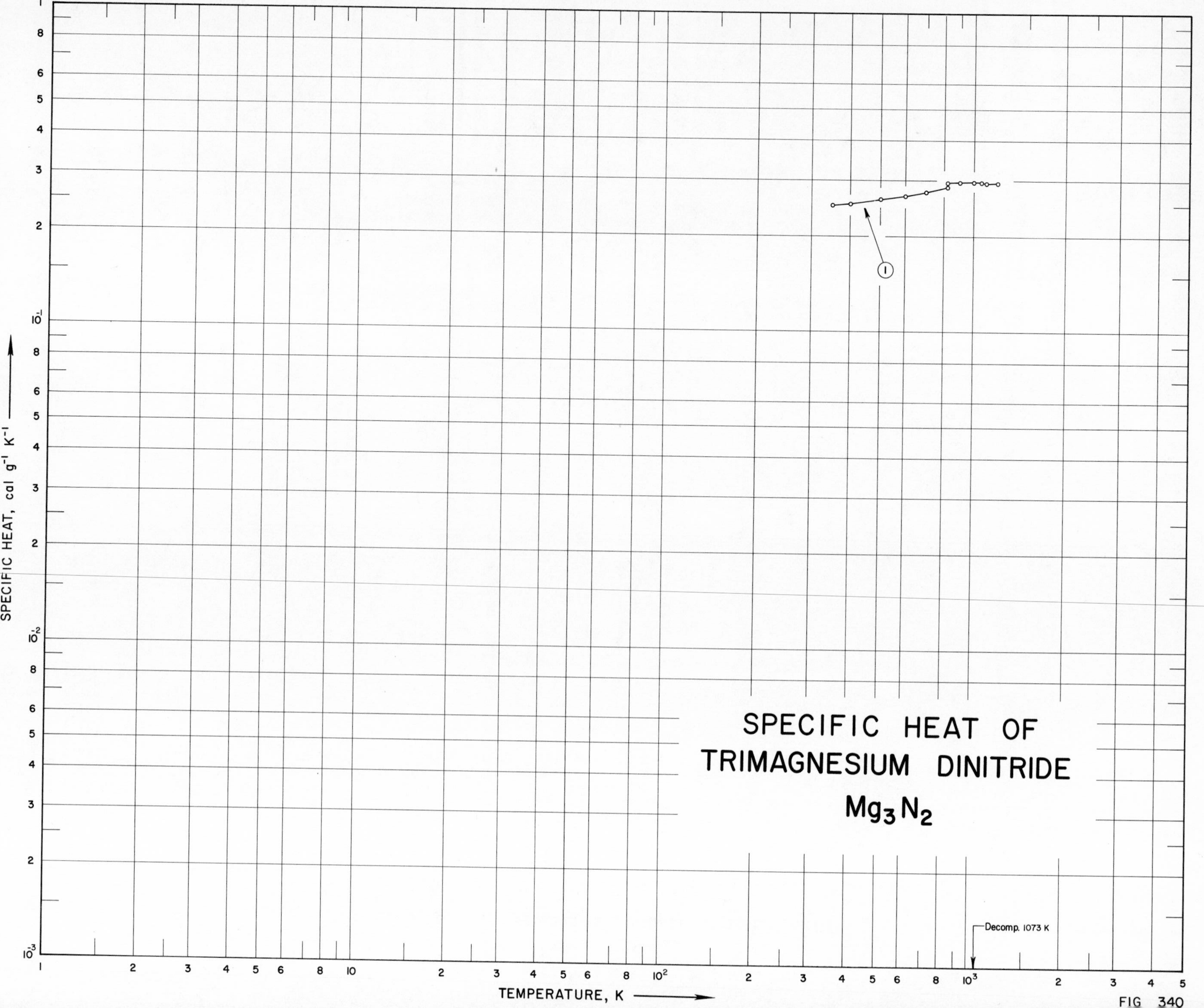
SPECIFIC HEAT OF
TRIMAGNESIUM DINITRIDE
Mg_3N_2
1
Decomp. 1073 K
SPECIFIC HEAT, cal g⁻¹ K⁻¹
TEMPERATURE, K
FIG 340

SPECIFICATION TABLE NO. 340 SPECIFIC HEAT OF TRIMAGNESIUM DINITRIDE Mg_3N_2

[For Data Reported in Figure and Table No. 340]

Curve No.	Ref. No.	Year	Temp. Range, K	Reported Error, %	Name and Specimen Designation	Composition (weight percent), Specifications and Remarks
1	216	1949	350-1200			99.1 Mg_3N_2 and 0.90 MgO; corrected for MgO content.

DATA TABLE NO. 340 SPECIFIC HEAT OF TRIMAGNESIUM DINITRIDE Mg_3N_2

[Temperature, T, K; Specific Heat, C_p, Cal $g^{-1}K^{-1}$]

T	C_p
CURVE 1	
350	2.511 x 10^{-1}
400	2.547
500	2.620
600	2.693
700	2.766
800	2.839
α 823	2.8558*
β 823	2.960*
900	2.960
1000	2.960
β 1061	2.960
γ 1061	2.954*
1100	2.954
1200	2.954

* Not shown on plot

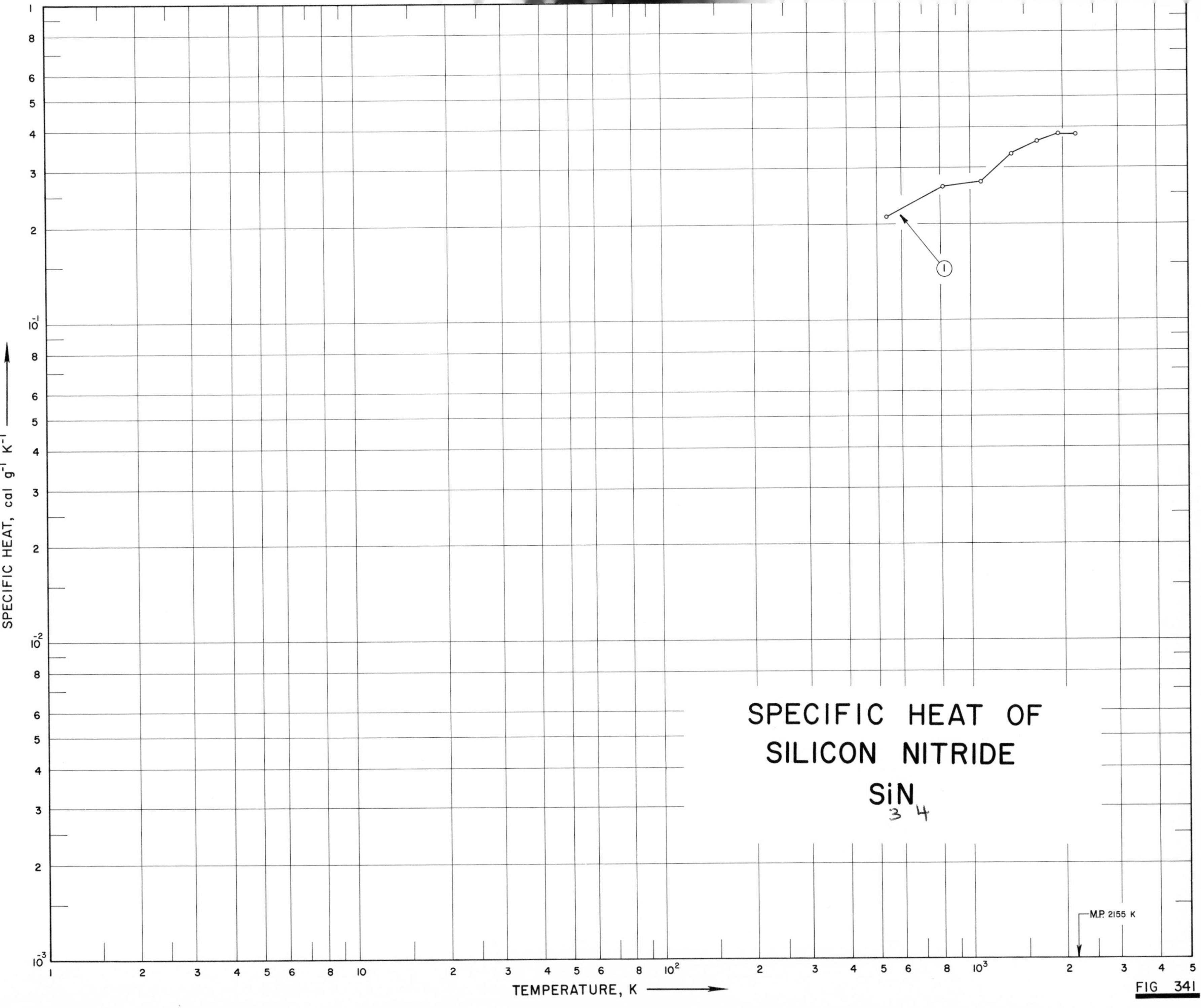
SPECIFIC HEAT OF
SILICON NITRIDE
SiN 3 4
SPECIFIC HEAT, cal g^{-1} K^{-1}
TEMPERATURE, K
M.P. 2155 K
1

FIG 341

SPECIFICATION TABLE NO. 341 SPECIFIC HEAT OF SILICON NITRIDE Si_3N_4

[For Data Reported in Figure and Table No. 341]

Curve No.	Ref. No.	Year	Temp. Range, K	Reported Error, %	Name and Specimen Designation	Composition (weight percent), Specifications and Remarks
1	32	1962	533-2200	≤5.0		98.12 Si_3N_4; 1.5 Fe, 0.3 Al, 0.05 Ca, 0.01 Cu, 0.01 Mg, 0.01 Ti, and traces of Ba, Mn, and Na; sample supplied by The Carborundum Co.; density = 148 lb ft^{-3}.

DATA TABLE NO. 341 SPECIFIC HEAT OF SILICON NITRIDE SiN Si_3N_4

[Temperature, T, K; Specific Heat, C_p, Cal $g^{-1}K^{-1}$]

T	C_p
CURVE 1	
533	2.100×10^{-1}
811	2.600
1089	2.700
1366	3.300
1644	3.600
1922	3.800
2200	3.800

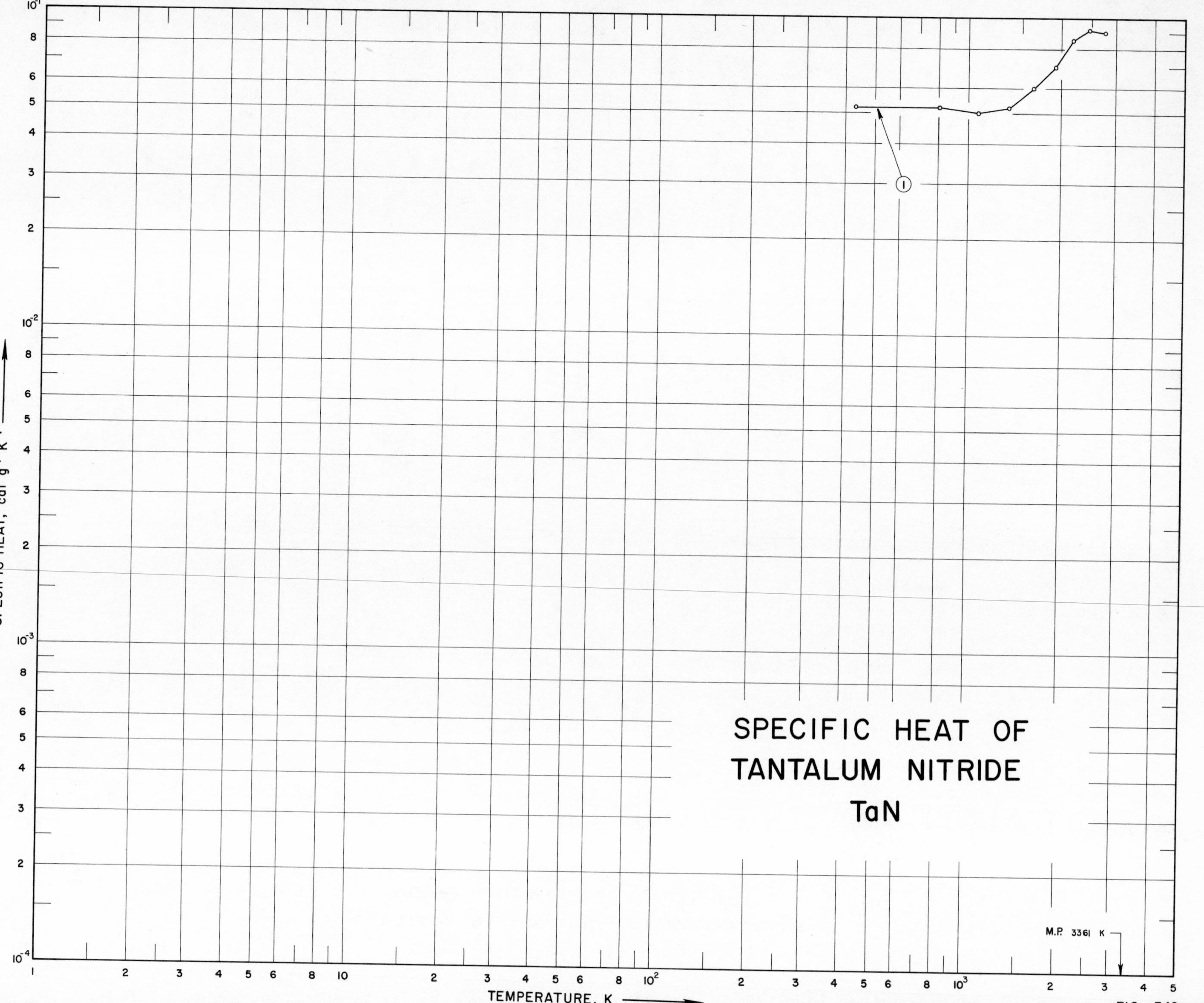
SPECIFIC HEAT OF
TANTALUM NITRIDE
TaN
M.P. 3361 K
1
TEMPERATURE, K
SPECIFIC HEAT, cal g^{-1} K^{-1}
FIG 342

SPECIFICATION TABLE NO. 342 SPECIFIC HEAT OF TANTALUM NITRIDE TaN

[For Data Reported in Figure and Table No. 342]

Curve No.	Ref. No.	Year	Temp. Range, K	Reported Error, %	Name and Specimen Designation	Composition (weight percent), Specifications and Remarks
1	48	1962	533-2755	≤5.0		Before exposure: 95.7 Ta, 3.5 N, 0.3 Fe, 0.2 Si, 0.13 C, and 0.1 Mg, after exposure: 95.5 Ta, 3.1 N, and 0.95 C; sample supplied by The Carborundum Co.; hot pressed; crushed in hardened steel mortar to pass 100-mesh screen; density 25 C, before exposure: apparent density (ASTM method B311-58) 836 lb ft^{-3}, true density (by immersion in xylene) 855 lb ft^{-3}, after exposure: apparent density = 836 lb ft^{-3}, true density = 910 lb ft^{-3}.

DATA TABLE NO. 342 SPECIFIC HEAT OF TANTALUM NITRIDE TaN

[Temperature, T, K; Specific Heat, C_p, Cal $g^{-1}K^{-1}$]

T	C_p
CURVE 1	
533	5.200 x 10^{-2}
811	5.200
1089	5.000
1366	5.200
1644	6.000
1922	7.000
2200	8.500
2478	9.200
2755	9.000

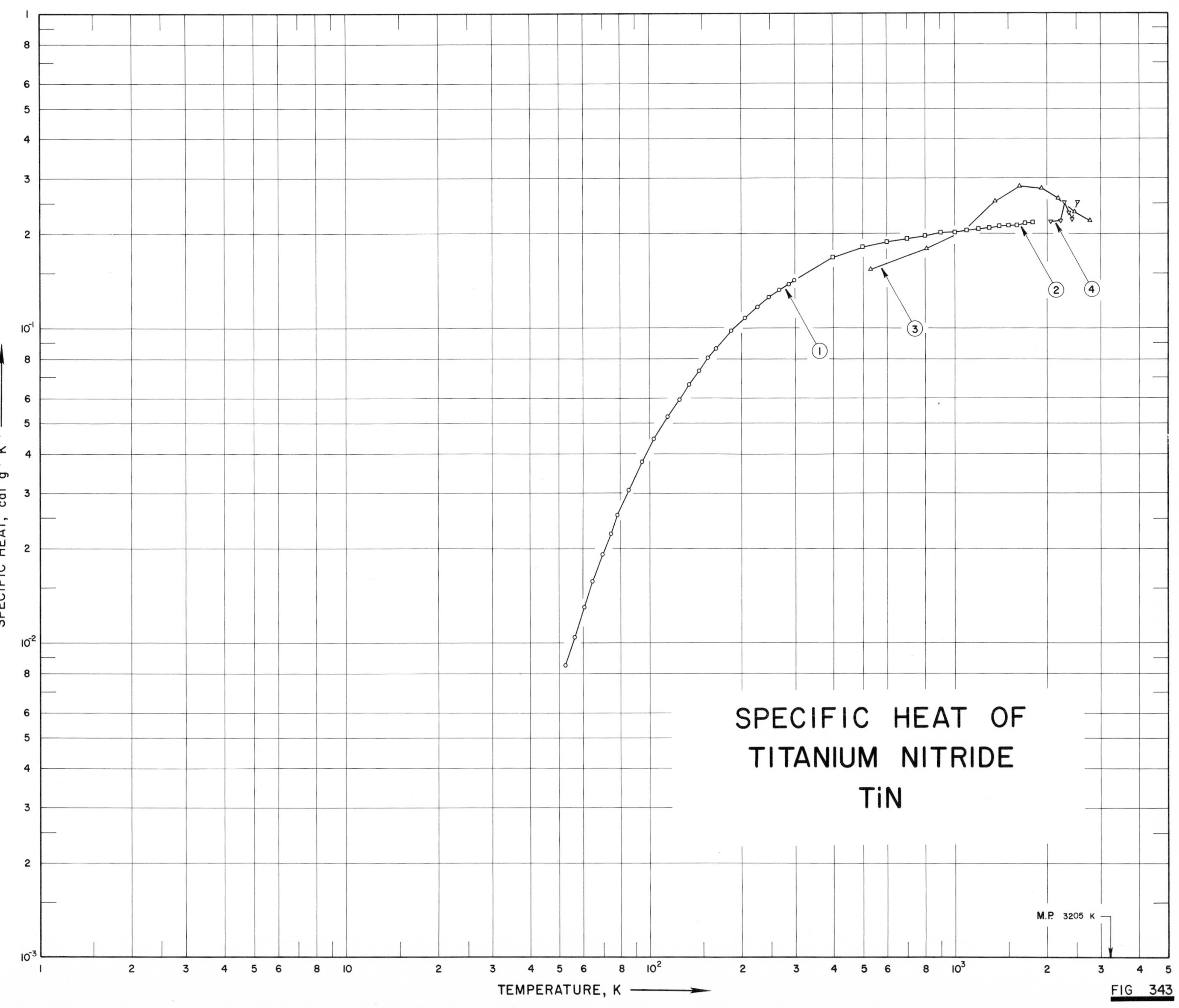
SPECIFIC HEAT OF
TITANIUM NITRIDE
TiN
SPECIFIC HEAT, cal g⁻¹ K⁻¹
TEMPERATURE, K
M.P. 3205 K
1
2
3
4

FIG 343

SPECIFICATION TABLE NO. 343 SPECIFIC HEAT OF TITANIUM NITRIDE TiN

[For Data Reported in Figure and Table No. 343]

Curve No.	Ref. No.	Year	Temp. Range, K	Reported Error, %	Name and Specimen Designation	Composition (weight percent), Specifications and Remarks
1	133	1946	53-298			99.5 TiN, and balance SiN; prepared by passing stream of purified N_2 and H_2 gas over titanium metal at 1400 C.
2	194	1946	298-1800			99.6 Ti; 77.04 Ti and SiN is the major impurity; prepared by heating powdered Ti in purified N_2 and H_2 gas at 1000 C and 10 hrs at 1100 C; density = 327 lb ft^{-3}.
3	32	1962	533-2755	≤5.0		Before exposure: 81.3 Ti, 17.0 N, 0.87 C, and 0.4 Fe, after exposure: 77.6 Ti, 18.9 N, and 1.2 C; sample supplied by the Technical Research Group; hot pressed; density at 25 C, before exposure: apparent density (ASTM method B311-58) 298 lb ft^{-3}, true density (by immersion in xylene) 306 lb ft^{-3}, after exposure: apparent density = 286 lb ft^{-3}, true density = 321 lb ft^{-3}.
4	173	1962	2052-2512	±5.0		

DATA TABLE NO. 343 SPECIFIC HEAT OF TITANIUM NITRIDE TiN

[Temperature, T, K; Specific Heat, C_p, Cal $g^{-1}K^{-1}$]

T	C_p
CURVE 1	
52.5	8.512×10^{-3}
56.4	1.058×10^{-2}
60.5	1.307
64.7	1.573
69.7	1.921
74.1	2.236
78.5	2.555
85.3	3.056
94.9	3.765
104.6	4.492
115.2	5.271
125.2	5.983
135.1	6.689
145.8	7.393
155.6	8.039
165.7	8.654
175.7	9.263*
185.7	9.809
196.1	1.035×10^{-1}*
206.0	1.083
216.3	1.134*
226.3	1.177
235.9	1.215*
246.0	1.255
258.0	1.293*
266.1	1.330
276.4	1.365*
286.2	1.396
296.4	1.426*
298.16	1.431
CURVE 2	
298.15	1.431×10^{-1}*
300	1.438*
400	1.685
500	1.808
600	1.882
700	1.932
800	1.970
900	2.001
1000	2.028
1100	2.051
1200	2.072
1300	2.093

T	C_p
CURVE 2 (cont.)	
1400	2.112×10^{-1}
1500	2.130
1600	2.148
1700	2.165
1800	2.182
CURVE 3	
533	1.550×10^{-1}
811	1.800
1089	2.150*
1366	2.550
1644	2.850
1922	2.800
2200	2.600
2478	2.350
2755	2.200
CURVE 4	
2052	2.190×10^{-1}
2217	2.200
2298	2.530
2360	2.340
2418	2.210
2482	2.230*
2512	2.530

* Not shown on plot

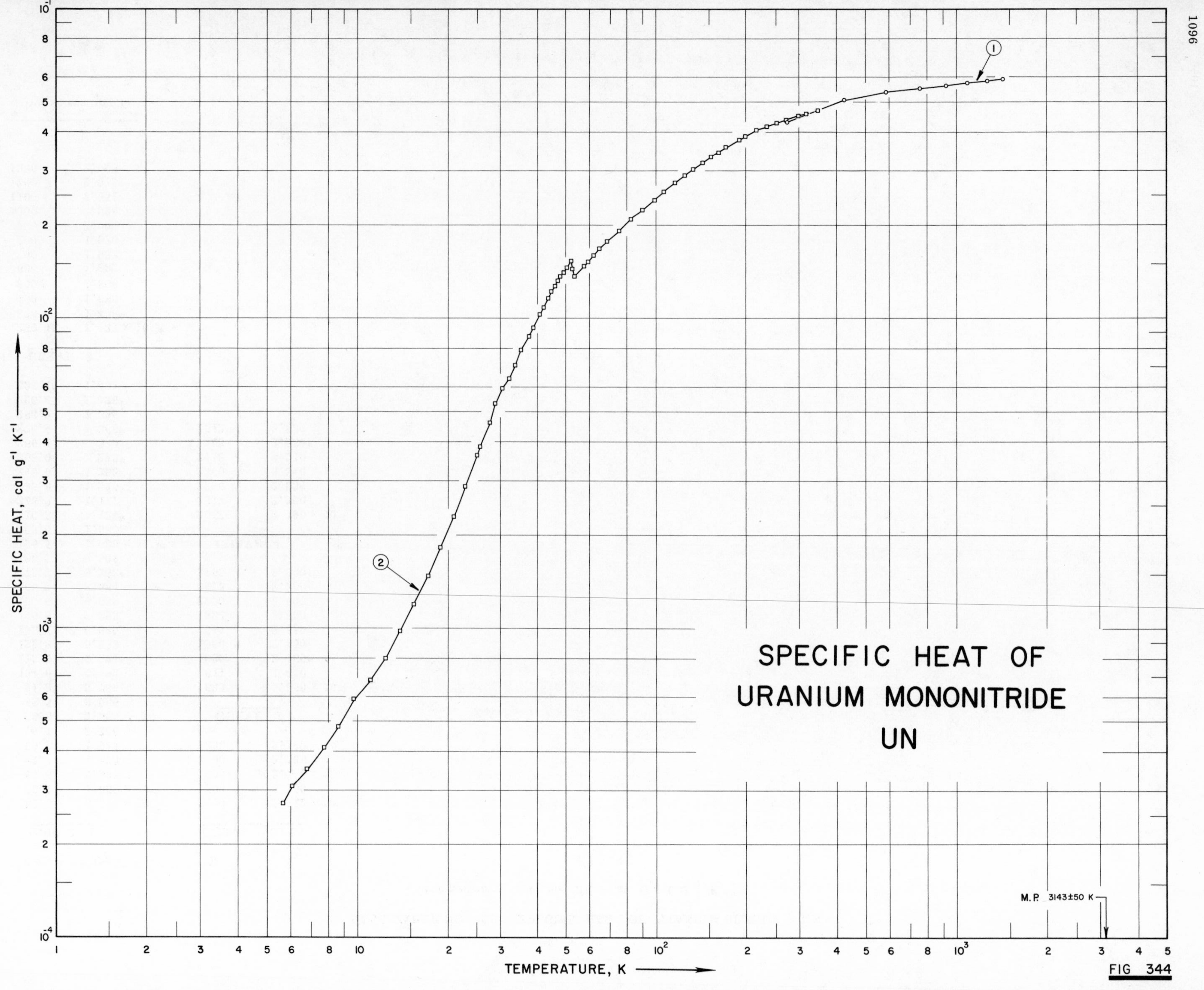
SPECIFIC HEAT OF
URANIUM MONONITRIDE
UN
SPECIFIC HEAT, cal g⁻¹ K⁻¹
TEMPERATURE, K
M.P. 3143±50 K
1
2
FIG 344

SPECIFICATION TABLE NO. 344 SPECIFIC HEAT OF URANIUM NITRIDE UN

[For Data Reported in Figure and Table No. 344]

Curve No.	Ref. No.	Year	Temp. Range, K	Reported Error, %	Name and Specimen Designation	Composition (weight percent), Specifications and Remarks
1	416	1963	273-1422			Impurities: 5.24 N_2, and widely scattered small amount of UO_2; single phase UN; isostatically hot pressed for 2 hrs at 1540 C and 10,000 psi.
2	415	1966	6-346			95.46 UN; 3.36 UO, 1.14 UC and 0.046 Fe (corrected for impurities).

DATA TABLE NO. 344 SPECIFIC HEAT OF URANIUM NITRIDE UN

[Temperature, T, K; Specific Heat, C_p, Cal $g^{-1}K^{-1}$]

T	C_p
CURVE 1	
273	4.30 x 10^{-2}
422	5.02
589	5.33
755	5.50
922	5.62
1089	5.73
1255	5.82
1422	5.92
CURVE 2	
Series I	
118.03	2.785 x 10^{-3}*
124.99	2.908
133.41	3.050
142.59	3.196
151.92	3.332
161.23	3.458
170.55	3.578
179.92	3.684*
189.16	3.783
198.10	3.871
206.88	3.953*
215.56	4.030
224.51	4.101*
233.79	4.170
243.07	4.243*
252.21	4.293
261.31	4.348*
270.37	4.397
279.40	4.443*
288.55	4.491*
297.75	4.535
306.84	4.576*
316.34	4.619
325.97	4.656*
335.82	4.691*
345.88	4.715
Series II	
5.69	2.738 x 10^{-4}
6.09	3.095
6.84	3.531
7.70	4.126
CURVE 2 (cont.)	
8.66	4.841 x 10^{-4}
9.77	5.912
11.04	6.824
12.41	8.015
13.89	9.840
15.46	1.198 x 10^{-3}
17.13	1.480
18.87	1.833
20.78	2.297
22.85	2.888
25.06	3.626
27.64	4.630
30.48	5.955
33.66	7.094
37.36	8.773
38.62	9.356
42.07	1.095 x 10^{-2}
46.07	1.278
61.97	1.607
68.62	1.778
75.38	1.936
82.54	2.102
89.99	2.248
98.98	2.426
106.90	2.581
116.26	2.753
Series III	
35.46	7.943 x 10^{-3}
42.55	1.120 x 10^{-2}*
46.44	1.297*
49.04	1.416
51.65	1.518*
53.67	1.388*
57.53	1.483
59.37	1.536
61.14	1.586*
62.84	1.634*
Series IV	
33.52	7.062 x 10^{-3}*
37.57	8.899*
40.84	1.041 x 10^{-2}
CURVE 2 (cont.)	
43.67	1.171 x 10^{-2}
46.19	1.286*
48.48	1.390*
50.60	1.490*
52.61	1.455
54.62	1.403*
56.59	1.455*
58.47	1.509*
60.27	1.562*
Series V	
25.64	3.872 x 10^{-3}
28.86	5.348
64.81	1.687 x 10^{-2}
70.30	1.819*
Series VI	
32.08	6.424 x 10^{-3}
61.80	1.605 x 10^{-2}*
Series VII	
44.94	1.228 x 10^{-2}
46.05	1.279*
47.10	1.326
48.12	1.376
49.10	1.418*
49.82	1.451*
50.28	1.469
50.74	1.486*
51.18	1.519*
51.62	1.543*
52.06	1.545
52.50	1.456*
52.96	1.380*
53.42	1.378
53.88	1.384*

* Not shown on plot

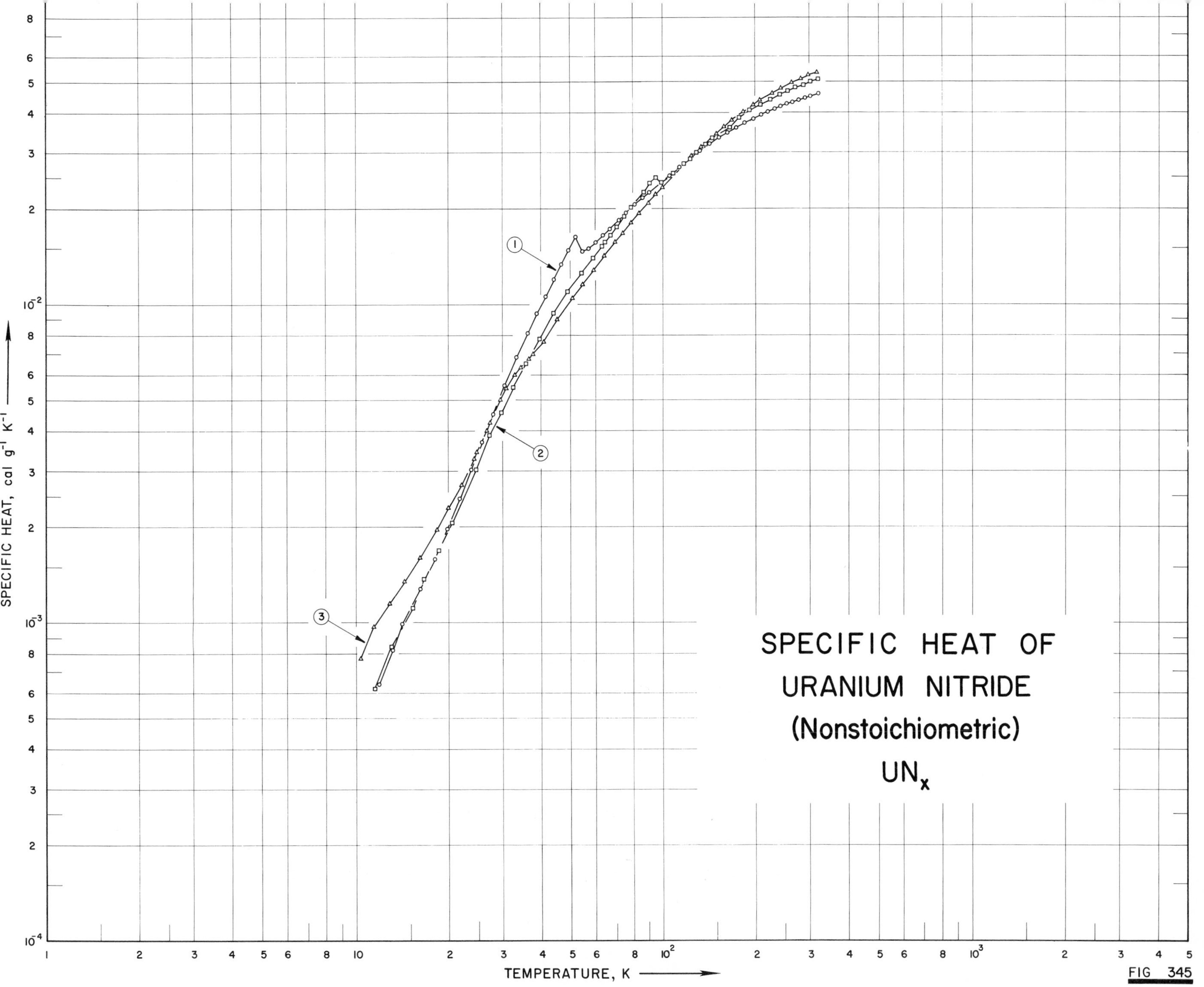
SPECIFIC HEAT OF
URANIUM NITRIDE
(Nonstoichiometric)
UN_x
SPECIFIC HEAT, cal g^{-1} K^{-1}
TEMPERATURE, K
1
2
3
FIG 345

SPECIFICATION TABLE NO. 345 SPECIFIC HEAT OF URANIUM NITRIDE (nonstoichiometric) UN_x

[For Data Reported in Figure and Table No. 345]

Curve No.	Ref. No.	Year	Temp. Range, K	Reported Error, %	Name and Specimen Designation	Composition (weight percent), Specifications and Remarks
1	417	1966	12-324			$UN_{1.01}$; 0.0119 UO_2.
2	417	1966	12-322	±0.1-0.4		$UN_{1.59}$; 0.176 UO_2 and 0.0416 UN.
3	417	1966	10-319	±0.1-0.4		$UN_{1.73}$; 0.0098 UO_2.

DATA TABLE NO. 345 SPECIFIC HEAT OF URANIUM NITRIDE UN_x (nonstoichiometric)

[Temperature, T, K; Specific Heat, C_p, Cal $g^{-1}K^{-1}$]

CURVE 1

T	C_p
11.82	6.40 x 10^{-4}
13.12	8.20
14.61	9.90
16.22	1.27 x 10^{-3}
18.01	1.58
19.81	1.97
21.70	2.46
23.63	3.03
25.72	3.69
28.02	4.53
30.58	5.57
33.43	6.84
36.28	8.14
38.93	9.38
41.54	1.067 x 10^{-2}
44.15	1.201
46.75	1.340
49.36	1.481
52.00	1.631
54.63	1.473
57.30	1.500
60.28	1.572
61.23	1.590*
63.72	1.656
67.04	1.733
72.18	1.857
76.35	1.953
81.59	2.075
86.60	2.180
90.75	2.265
91.64	2.282*
94.22	2.333*
97.95	2.401*
101.85	2.477*
106.00	2.558
110.26	2.639*
114.52	2.713
118.91	2.793*
123.40	2.878*
128.22	2.961*
133.30	3.050
138.41	3.129*
143.48	3.203
148.52	3.283*

CURVE 1 (cont.)

T	C_p
153.61	3.351 x 10^{-2}
158.69	3.420*
158.84	3.420*
164.22	3.480
169.71	3.559*
175.14	3.616
180.59	3.685*
186.05	3.745
191.62	3.812*
192.86	3.813*
198.72	3.862
204.49	3.916*
210.19	3.972
215.79	4.018*
221.31	4.057
226.80	4.109*
232.55	4.145
238.27	4.176*
243.93	4.211
249.58	4.254*
255.16	4.291
257.22	4.285*
266.10	4.333
267.99	4.358*
272.95	4.366*
275.71	4.384*
278.28	4.409
281.12	4.400*
283.57	4.429*
286.28	4.424*
288.83	4.460*
291.64	4.478
294.06	4.495*
297.09	4.478*
299.27	4.502*
300.44	4.503*
302.28	4.508*
304.45	4.533
305.98	4.535*
307.57	4.544*
312.97	4.547*
318.35	4.575*
323.70	4.623

CURVE 2

T	C_p
11.57	6.2 x 10^{-4}
13.07	8.4
15.27	1.16 x 10^{-3}
16.69	1.37
18.62	1.69
20.53	2.06
24.59	3.04
27.17	3.88
29.82	4.57
32.59	5.49
35.70	6.54
39.60	7.83
44.18	9.41
49.03	1.104 x 10^{-2}
54.32	1.261
59.19	1.409
59.47	1.416*
63.22	1.533
64.65	1.574
67.40	1.660
70.59	1.756
71.24	1.778*
75.27	1.906
76.60	1.943*
79.49	2.035
82.06	2.112*
82.10	2.110*
87.13	2.267
87.19	2.267*
87.52	2.278
91.15	2.419
92.43	2.460*
95.33	2.507
97.91	2.447*
99.40	2.421
103.43	2.483*
108.53	2.591
112.56	2.672*
113.72	2.692
117.51	2.773
124.03	2.895
127.83	2.985*
129.16	3.007
131.80	3.062*

CURVE 2 (cont.)

T	C_p
134.35	3.114 x 10^{-2}*
138.35	3.193
139.61	3.215*
144.93	3.316*
146.53	3.345
151.67	3.436*
156.72	3.530*
161.68	3.607
166.85	3.697*
172.50	3.790*
178.04	3.875
183.20	3.946*
188.30	4.021*
193.34	4.088
198.55	4.154*
203.88	4.214*
209.16	4.268
214.58	4.313*
220.29	4.371*
225.93	4.416
231.51	4.467*
237.03	4.517*
242.49	4.572
247.86	4.619*
247.90	4.624*
252.00	4.656*
253.25	4.671*
256.69	4.699
261.38	4.724*
266.11	4.765*
271.28	4.808
276.38	4.817*
281.30	4.866*
286.19	4.896*
289.47	4.917
291.05	4.926*
296.02	4.939*
301.09	4.986*
304.69	5.014
306.21	5.021*
311.36	5.050*
314.54	5.070*
316.50	5.074*
321.59	5.100

CURVE 3

T	C_p
10.38	7.70 x 10^{-4}
11.42	9.7
12.78	1.15 x 10^{-3}
13.15	1.19*
14.39	1.35
16.14	1.60
18.37	1.96
19.99	2.29
20.14	2.31*
22.07	2.72
22.28	2.78*
24.22	3.27
24.70	3.44
26.66	4.00
27.48	4.26
29.56	5.02
30.85	5.45
32.89	6.03
34.34	6.35
36.57	6.76
37.76	6.98
40.73	7.65
45.24	8.99
45.78	9.16*
50.78	1.061 x 10^{-2}
54.54	1.165
54.71	1.169*
55.91	1.198*
59.29	1.288*
59.54	1.292
64.28	1.426
69.68	1.570
74.04	1.687
75.05	1.713*
79.14	1.819
80.07	1.839*
84.16	1.948
90.17	2.095
95.39	2.231
100.07	2.345
100.54	2.356*
105.19	2.466*
110.18	2.583*
115.79	2.718*

CURVE 3 (cont.)

T	C_p
120.56	2.827 x 10^{-2}*
125.78	2.942
130.93	3.054*
134.85	3.140
137.36	3.195*
141.36	3.274*
142.53	3.298*
147.59	3.402*
150.03	3.446
154.88	3.542*
159.64	3.632
164.31	3.719*
168.91	3.802
171.33	3.838*
173.44	3.885*
176.39	3.929*
181.45	4.009*
183.57	4.042
184.48	4.054*
188.54	4.120*
193.44	4.187*
197.88	4.256
206.90	4.368*
208.62	4.403
211.77	4.430*
213.47	4.460*
216.66	4.499*
223.51	4.578*
228.50	4.625
231.61	4.663*
238.36	4.738*
242.83	4.789*
243.29	4.796
248.26	4.849*
253.20	4.907*
258.11	4.960*
264.40	4.997
269.72	5.045*
273.26	5.091*
278.35	5.113*
283.54	5.142
288.57	5.194*
290.33	5.204*
295.32	5.230*
298.55	5.262
300.36	5.262*

* Not shown on plot

DATA TABLE NO. 345 (continued)

T	C_p
CURVE 3 (cont.)	
303.51	5.284×10^{-2}*
305.44	5.300*
308.52	5.330*
313.56	5.347*
318.57	5.386

* Not shown on plot

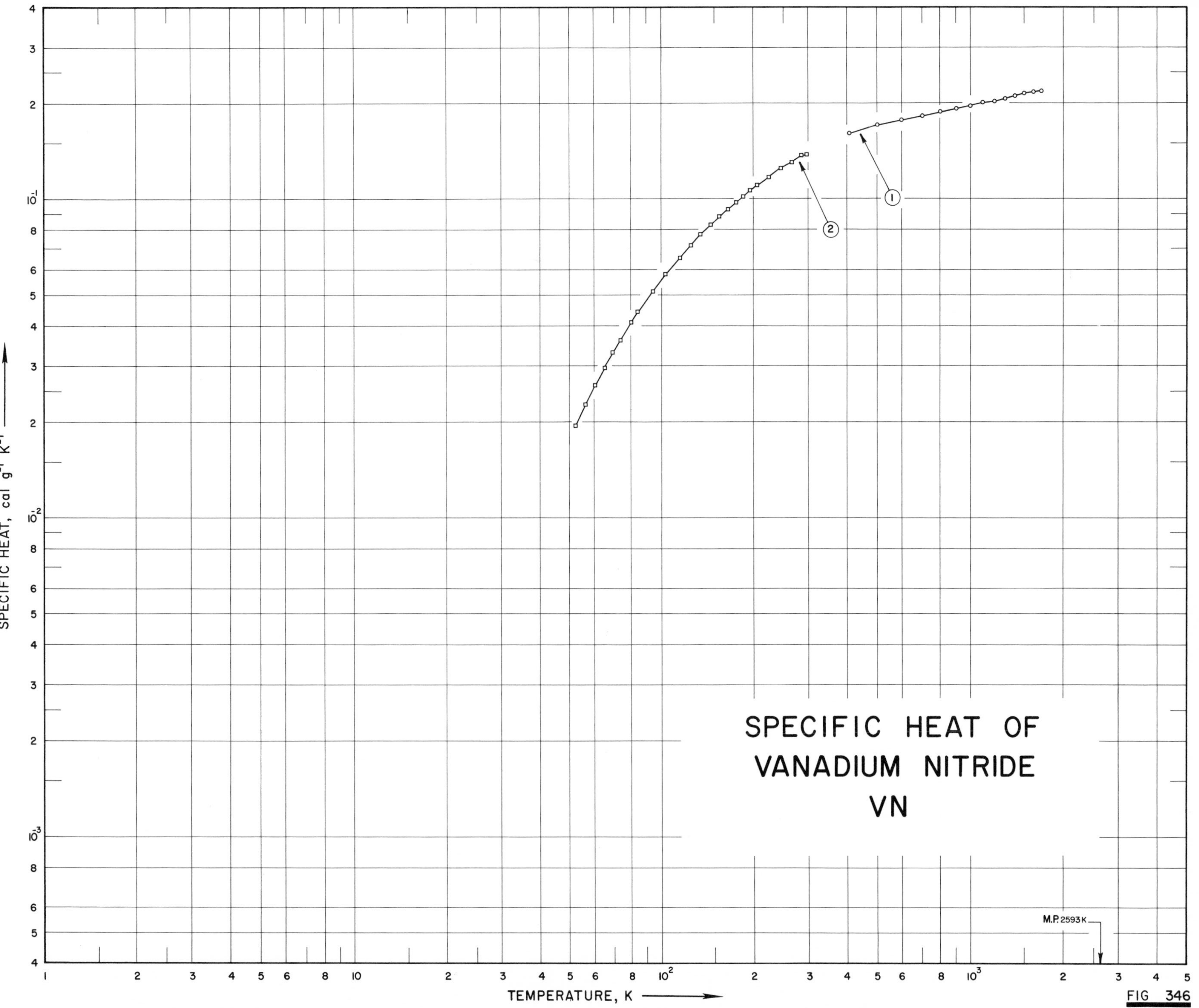

SPECIFIC HEAT OF
VANADIUM NITRIDE
VN
1
2
M.P. 2593 K
TEMPERATURE, K
SPECIFIC HEAT, cal g⁻¹ K⁻¹
FIG 346

SPECIFICATION TABLE NO. 346 SPECIFIC HEAT OF VANADIUM NITRIDE VN

[For Data Reported in Figure and Table No. 346]

Curve No.	Ref. No.	Year	Temp. Range, K	Reported Error, %	Name and Specimen Designation	Composition (weight percent), Specifications and Remarks
1	205	1949	408-1611			78.24 V and 0.05 C (theor. 78.43 V); heated 24 hrs at 1200 C; mixture cooled and ground at 7 hrs interval.
2	204	1949	53-298			78.24 V and 0.05 C (theor. 78.43 V); heated 28 hrs at 1200 C; mixture cooled and reground at 7 hrs interval.

DATA TABLE NO. 346 SPECIFIC HEAT OF VANADIUM NITRIDE VN

[Temperature, T, K; Specific Heat, C_p, Cal $g^{-1}K^{-1}$]

T	C_p
CURVE 1	
408	1.612×10^{-1}
500	1.710
600	1.784
700	1.841
800	1.890
900	1.933
1000	1.973
1100	2.012
1200	2.048
1300	2.084
1400	2.119
1500	2.154
1600	2.188
1611	2.192
CURVE 2	
52.6	1.958×10^{-2}
56.7	2.284
60.9	2.620
65.1	2.963
69.4	3.310
73.5	3.641
80.0	4.138
84.8	4.490
94.7	5.197
104.5	5.864
115.1	6.550
125.1	7.157
135.0	7.752
145.6	8.333
155.5	8.865
165.4	9.367
175.3	9.875
185.6	1.031×10^{-1}
195.9	1.075
205.8	1.115
216.1	1.152*
226.1	1.190
235.8	1.220*
246.1	1.256
256.1	1.290*
266.0	1.316
276.0	1.343*
286.1	1.370
296.3	1.394*
298.16	1.398

* Not shown on plot

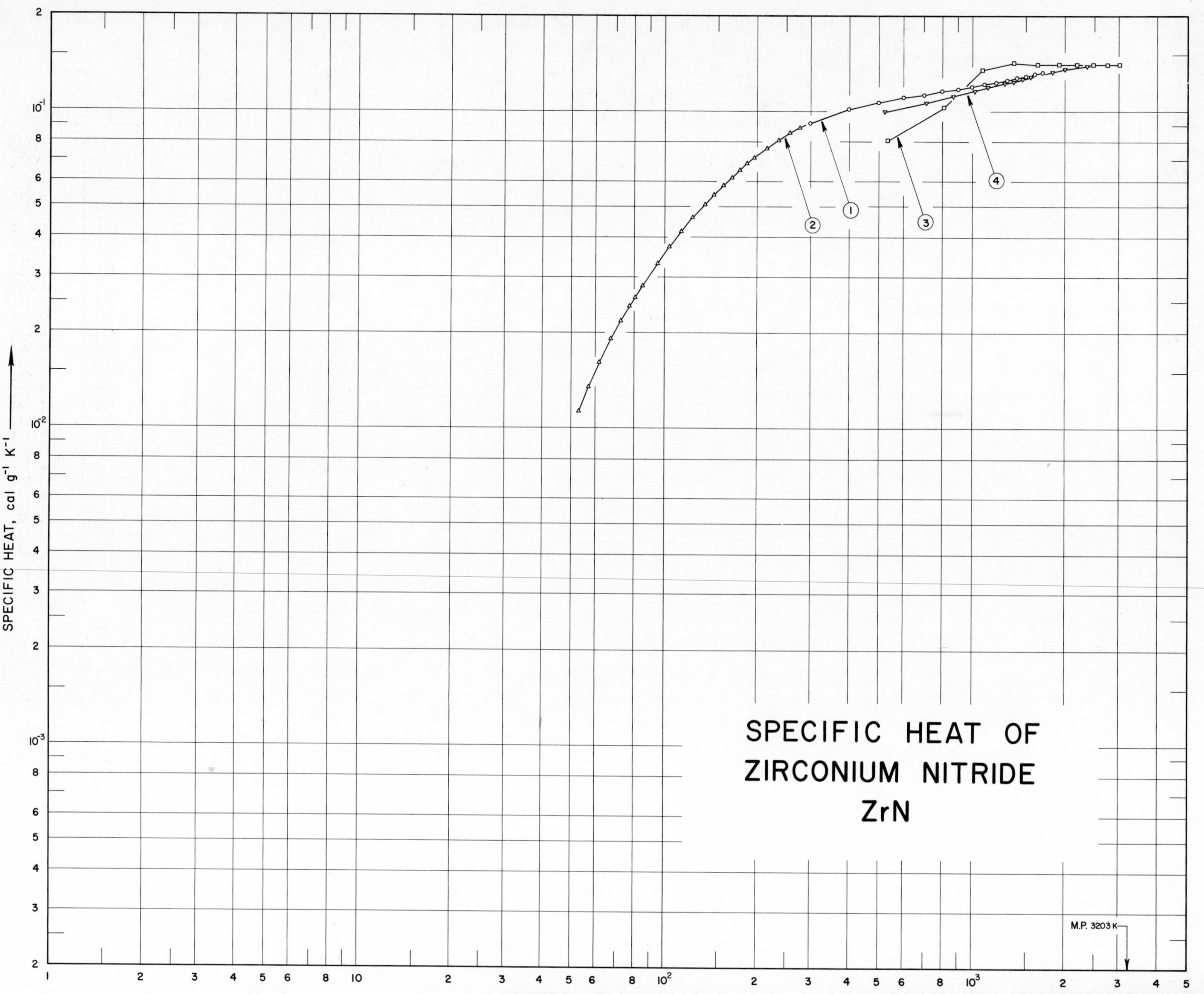
SPECIFIC HEAT OF
ZIRCONIUM NITRIDE
ZrN
M.P. 3203 K
SPECIFIC HEAT, cal g^{-1} K^{-1}
1
2
3
4

SPECIFICATION TABLE NO. 347 SPECIFIC HEAT OF ZIRCONIUM NITRIDE ZrN

[For Data Reported in Figure and Table No. 347]

Curve No.	Ref. No.	Year	Temp. Range, K	Reported Error, %	Name and Specimen Designation	Composition (weight percent), Specifications and Remarks
1	154	1950	298-1700			86.75 Zr including 1.35 Hf (theor. 86.69 Zr).
2	217	1950	53-298			86.75 Zr (theor. 86.69 Zr), 1.35 Hf; corrected for impurities.
3	32	1962	533-3033	≤5.0		Before exposure: 86.9 Zr, 12.8 N, and 0.1 Fe, after exposure: 86.5 Zr, 10.8 N, and 1.19 C; sample supplied by General Electric Co.; pressed and sintered; density at 25 C, before exposure: apparent density (ASTM method B311-58) 450 lb ft^{-3}, true density (by immersion in xylene) 450 lb ft^{-3}, after exposure: apparent density = 425 lb ft^{-3}, true density = 437 lb ft^{-3}.
4	47	1963	522-2770	±5.0		84.6 Zr, 13.5 N, 0.8 H, 0.5 alkali metal oxides, 0.4 Si, and 0.2 Fe; sample supplied by the Norton Co.; hot pressed.

DATA TABLE NO. 347 SPECIFIC HEAT OF ZIRCONIUM NITRIDE ZrN

[Temperature, T, K; Specific Heat, C_p, Cal $g^{-1}K^{-1}$]

T	C_p
CURVE 1	
298	9.183 x 10^{-2}
300	9.211*
400	1.017 x 10^{-1}
500	1.069
600	1.105
700	1.133
800	1.157
900	1.178
1000	1.198
1100	1.217
1200	1.235
1300	1.253
1400	1.270
1500	1.287
1600	1.304
1700	1.321
CURVE 2	
53.1	1.138 x 10^{-2}
57.1	1.355
62.0	1.615
67.5	1.918
72.7	2.193
77.6	2.446
80.5	2.595
85.3	2.830
95.2	3.304
104.8	3.750
114.9	4.197
124.8	4.625
136.2	5.092
145.9	5.461
155.9	5.832
166.2	6.193
176.7	6.534
186.2	6.820
196.4	7.122
206.7	7.397*
216.6	7.648
226.5	7.875*
236.5	8.093
246.0	8.281*
256.4	8.509
266.4	8.673*
CURVE 2 (cont.)	
276.6	8.862 x 10^{-2}
286.8	9.011*
296.7	9.135*
CURVE 3	
533	8.100 x 10^{-2}
811	1.030 x 10^{-1}
1089	1.350
1366	1.420
1644	1.400
1922	1.400
2200	1.400
2478	1.400
2755	1.400
3033	1.400
CURVE 4	
523	9.98 x 10^{-2}
713	1.066 x 10^{-1}
871	1.118
1022	1.163
1133	1.193
1282	1.231
1367	1.240
1457	1.270
1551	1.289
1698	1.318*
1833	1.336
2008	1.359
2108	1.369*
2244	1.380*
2386	1.388
2519	1.392*
2666	1.393*
2772	1.392*

* Not shown on plot

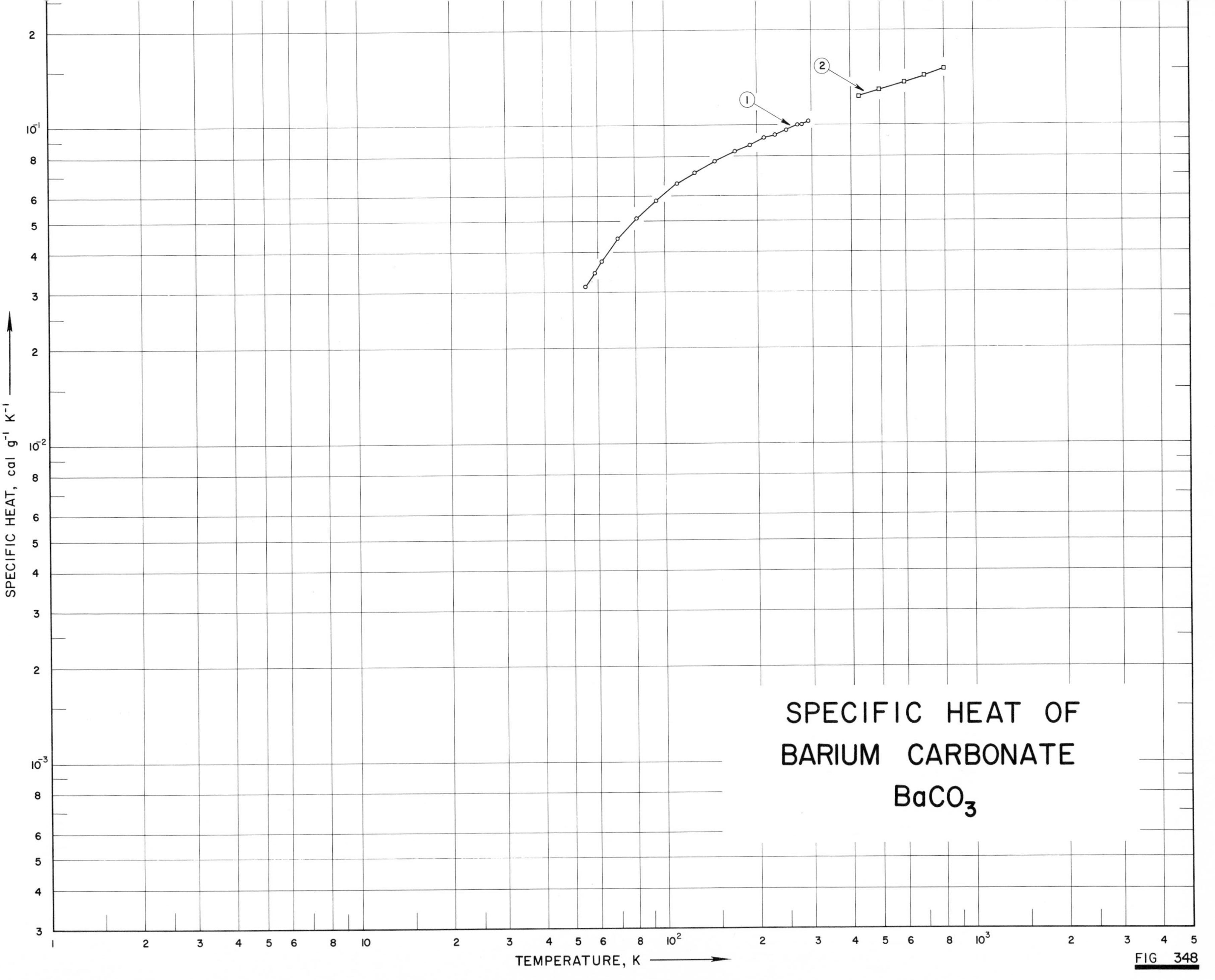
SPECIFIC HEAT OF
BARIUM CARBONATE
$BaCO_3$
SPECIFIC HEAT, cal g^{-1} K^{-1}
TEMPERATURE, K
1
2
FIG 348

SPECIFICATION TABLE NO. 348 SPECIFIC HEAT OF BARIUM CARBONATE, $BaCO_3$

[For Data Reported in Figure and Table No. 348]

Curve No.	Ref. No.	Year	Temp. Range, K	Reported Error, %	Name and Specimen Designation	Composition (weight percent), Specifications and Remarks
1	427	1934	55-296		Witherite	99.9 $BaCO_3$.
2	42	1951	431-806			Exceptionally high purity.

DATA TABLE NO. 348 SPECIFIC HEAT OF BARIUM CARBONATE, $BaCO_3$

[Temperature, T, K; Specific Heat, C_p, Cal $g^{-1} K^{-1}$]

T	C_p
CURVE 1	
54.8	3.140×10^{-2}
58.7	3.457
61.9	3.764
69.9	4.442
81.0	5.118
94.4	5.837
111.8	6.592
126.8	7.129
147.9	7.722
171.1	8.264
191.8	8.675
212.4	9.151
230.9	9.343
250.6	9.668
272.8	1.005×10^{-1}
281.2	1.004
295.9	1.032
CURVE 2	
431	1.230×10^{-1}
506	1.291
600	1.368
706	1.438
800	1.504*
806	1.508

* Not shown on plot

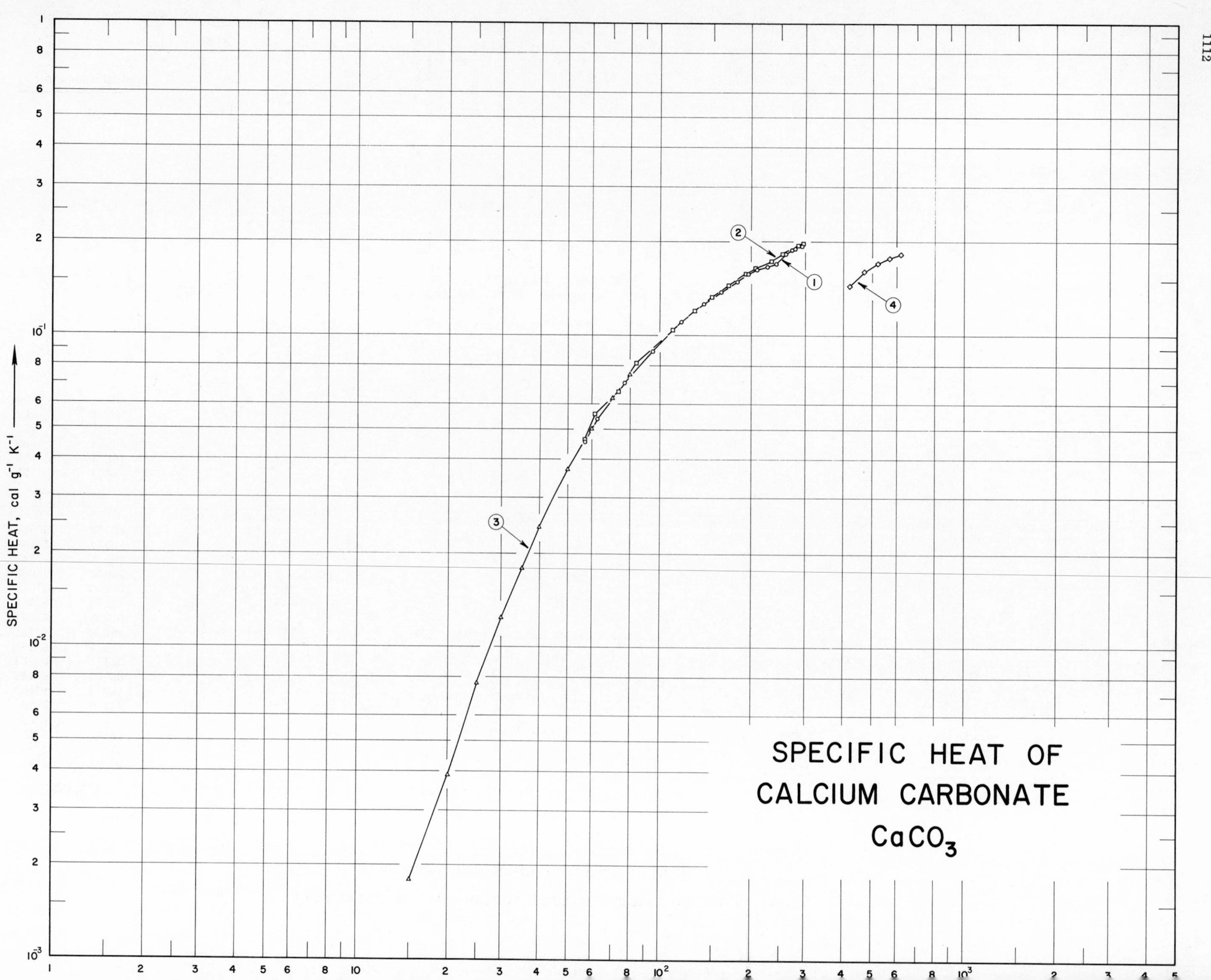
SPECIFIC HEAT OF
CALCIUM CARBONATE
$CaCO_3$
SPECIFIC HEAT, cal g^{-1} K^{-1}
1
2
3
4

SPECIFICATION TABLE NO. 349 SPECIFIC HEAT OF CALCIUM CARBONATE, $CaCO_3$

[For Data Reported in Figure and Table No. 349]

Curve No.	Ref. No.	Year	Temp. Range, K	Reported Error, %	Name and Specimen Designation	Composition (weight percent), Specifications and Remarks
1	427	1934	57-294		Coarse Calcite	99.970 $CaCO_3$.
2	427	1934	57-297		Fine Calcite	98.656 $CaCO_3$, 0.725 CaOH and 0.618 H_2O.
3	314	1935	15-80			
4	428	1961	423-623		Limestone (Madhya Pradesh)	

DATA TABLE NO. 349 SPECIFIC HEAT OF CALCIUM CARBONATE, $CaCO_3$

[Temperature, T, K; Specific Heat, C_p, Cal $g^{-1} K^{-1}$]

T	C_p
CURVE 1	
57.2	4.555×10^{-2}
62.6	5.440
77.5	7.053
95.7	8.938
118.6	1.110×10^{-1}
141.5	1.273
159.7	1.384
180.0	1.492
196.5	1.585
209.2	1.633
227.4	1.668
241.9	1.703
261.0	1.833
272.5	1.887
279.8	1.904
294.3	1.943
CURVE 2	
57.0	4.634×10^{-2}
61.6	5.601
73.5	6.606
84.4	8.158
111.5	1.045×10^{-1}
131.3	1.208
148.5	1.328
168.3	1.453
192.4	1.581
207.7	1.649
233.7	1.731
254.1	1.838
272.1	1.905*
287.5	1.948
296.5	1.978
CURVE 3	
15	1.80×10^{-3}
20	3.90
25	7.69
30	1.25×10^{-2}
35	1.80
40	2.43
50	3.73
CURVE 3 (cont.)	
60	5.03×10^{-2}
70	6.31
80	7.49
CURVE 4	
423	1.445×10^{-1}
473	1.610
523	1.704
573	1.769
623	1.821

* Not shown on plot

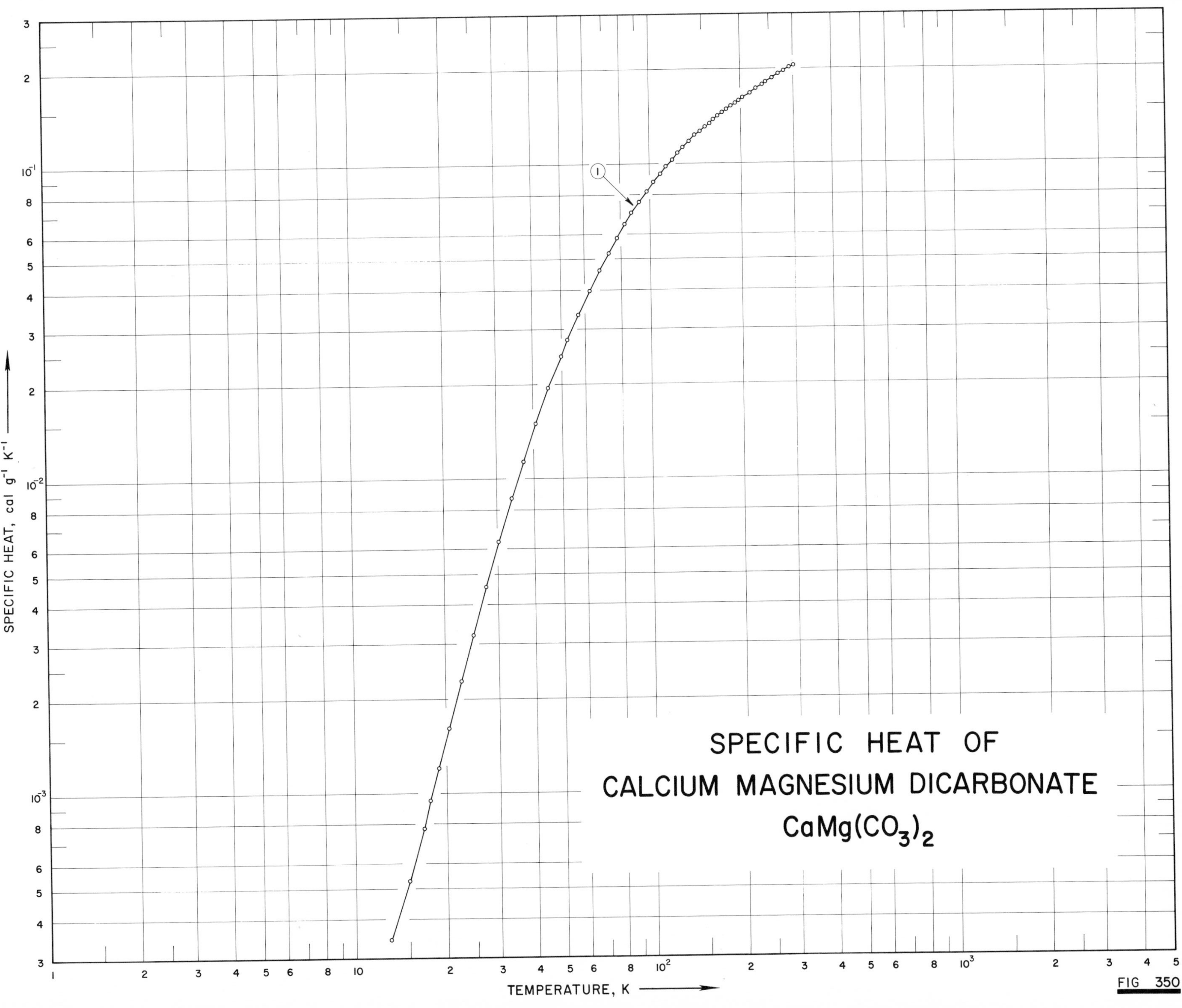
SPECIFIC HEAT OF
CALCIUM MAGNESIUM DICARBONATE
CaMg(CO3)2
SPECIFIC HEAT, cal g-1 K-1
TEMPERATURE, K
1
FIG 350

SPECIFICATION TABLE NO. 350 SPECIFIC HEAT OF CALCIUM MAGNESIUM DICARBONATE, $CaMg(CO_3)_2$

[For Data Reported in Figure and Table No. 350]

Curve No.	Ref. No.	Year	Temp. Range, K	Reported Error, %	Name and Specimen Designation	Composition (weight percent), Specifications and Remarks
1	429	1963	12-301		Dolomite	47.38 CO_2, 30.77 CaO, 21.54 MgO, 0.10 MnO, 0.017 SrO, and 0.008 FeO; crushed in porcelain mortar.

DATA TABLE NO. 350 SPECIFIC HEAT OF CALCIUM MAGNESIUM DICARBONATE, $CaMg(CO_3)_2$

[Temperature, T, K; Specific Heat, C_p, Cal $g^{-1} K^{-1}$]

T	C_p
CURVE 1	
Series I	
52.68	2.771×10^{-2}
57.37	3.326
62.69	3.972
67.95	4.619
73.03	5.236
78.10	5.859
82.99	6.469
87.64	7.017
92.93	7.618
98.50	8.220
103.96	8.801
109.21	9.337
114.52	9.863
120.04	1.038×10^{-1}
125.41	1.088
130.95	1.136
136.56	1.184
142.23	1.240
147.89	1.273
153.49	1.313
159.12	1.353
164.66	1.391
170.20	1.427
175.83	1.462
181.53	1.497
187.19	1.530
193.00	1.564
198.77	1.595
204.37	1.626
210.13	1.656*
216.00	1.687
222.00	1.716*
227.74	1.743
233.60	1.769*
244.85	1.825
250.55	1.850*
256.44	1.876
262.61	1.901*
268.60	1.929
274.46	1.953*
280.38	1.976
286.44	2.000*
292.15	2.021
CURVE 1 (cont.)	
297.89	2.044×10^{-1}*
301.10	2.052
Series II	
11.64	2.408×10^{-4}
13.10	3.459
15.00	5.314
16.74	7.814
17.69	9.565
18.90	1.213×10^{-3}
20.49	1.626
22.51	2.283
24.78	3.205
27.51	4.569
30.47	6.360
33.75	8.714
37.05	1.145×10^{-2}
40.99	1.509
45.34	1.950
50.03	2.467
238.31	1.793×10^{-1}
251.17	1.851*
257.01	1.878*
262.31	1.897*
268.88	1.928*
276.51	1.960*
282.43	1.986*
288.66	2.010*

* Not shown on plot

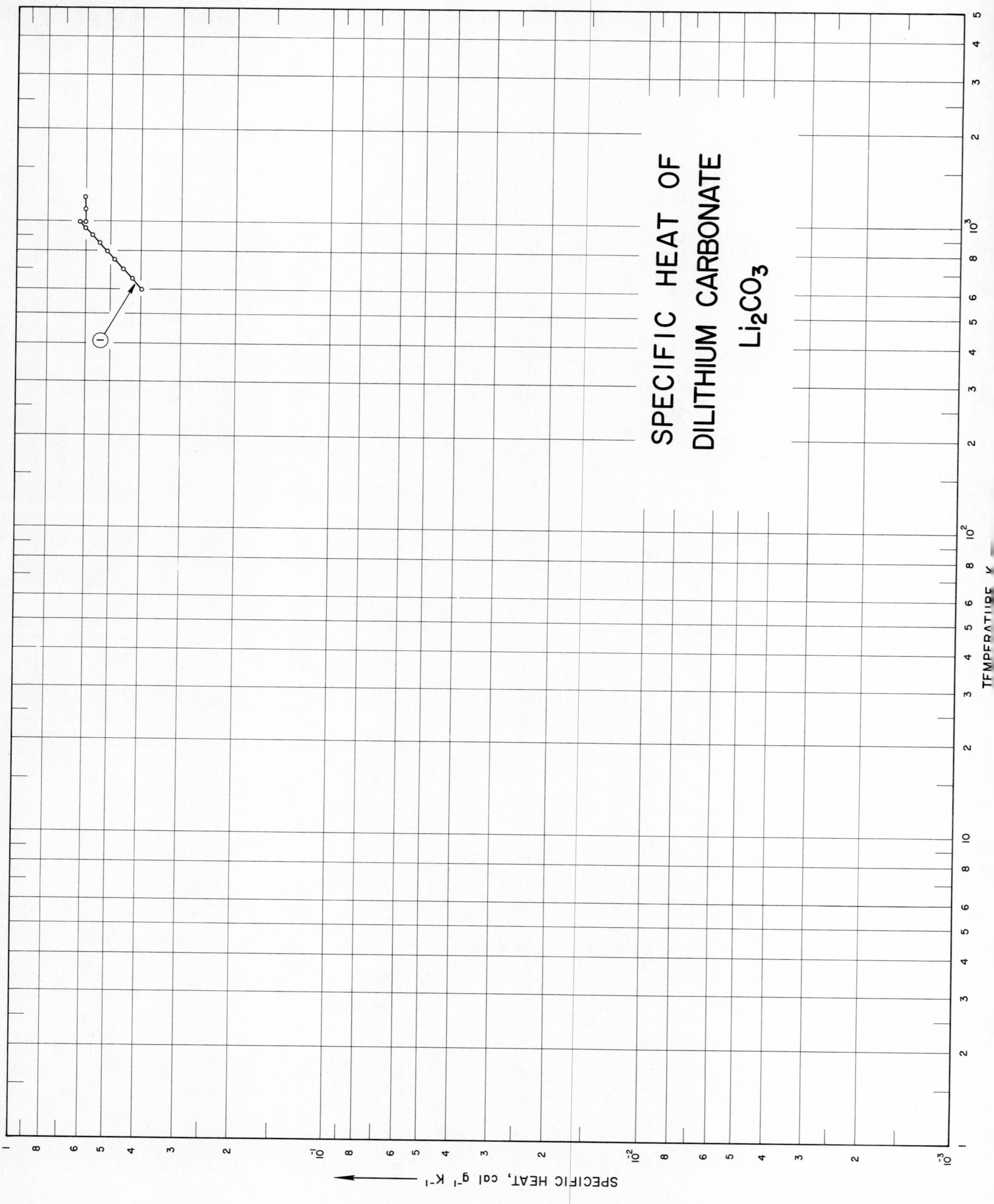

SPECIFIC HEAT OF DILITHIUM CARBONATE

Li_2CO_3

SPECIFIC HEAT, cal g^{-1} K^{-1}

TEMPERATURE, K

1

1 2 3 4 5 6 8 10 2 3 4 5 6 8 10^2 2 3 4 5 6 8 10^3 2 3 4 5

10^{-3} 2 3 4 5 6 8 10^{-2} 2 3 4 5 6 8 10^{-1} 2 3 4 5 6 8 1

SPECIFICATION TABLE NO. 351 SPECIFIC HEAT OF DILITHIUM CARBONATE, Li_2CO_3

[For Data Reported in Figure and Table 351]

Curve No.	Ref. No.	Year	Temp. Range, K	Reported Error, %	Name and Specimen Designation	Composition (weight percent), Specifications and Remarks
1	430	1963	600-1150			Baker analyzed reagents.

DATA TABLE NO. 351 SPECIFIC HEAT OF DILITHIUM CARBONATE, Li_2CO_3

[Temperature, T, K; Specific Heat, C_p, Cal $g^{-1} K^{-1}$]

T	C_p
CURVE 1	
600	3.989×10^{-1}
650	4.278
700	4.567
750	4.856
800	5.145
850	5.435
900	5.724
950	6.013
(s) 996	6.279
(ℓ) 996	6.002
1000	6.009*
1100	6.017
1150	6.027

* Not shown on plot

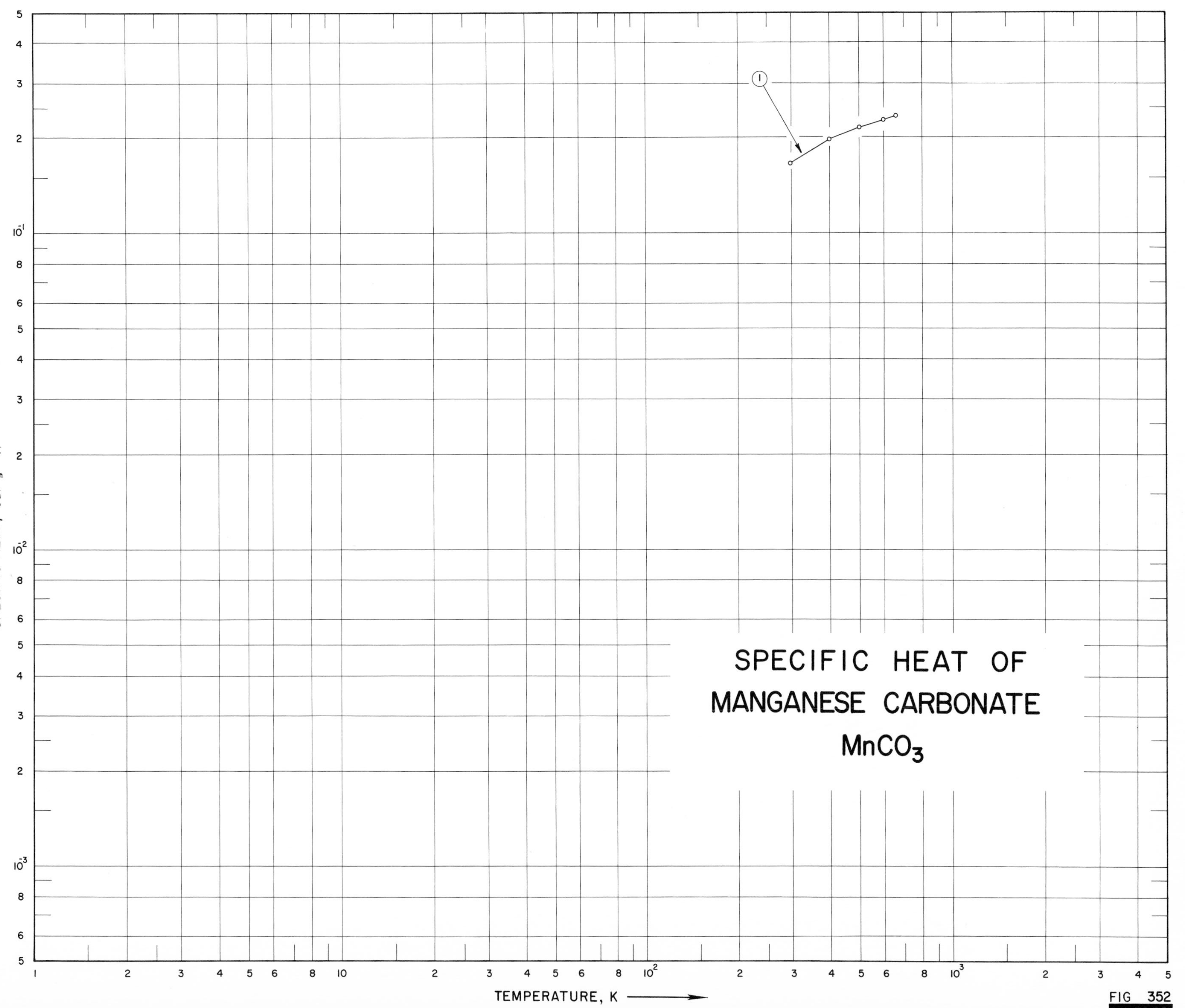
SPECIFIC HEAT OF
MANGANESE CARBONATE
$MnCO_3$
SPECIFIC HEAT, cal g^{-1} K^{-1}
TEMPERATURE, K
1
FIG 352

SPECIFICATION TABLE NO. 352 SPECIFIC HEAT OF MANGANESE CARBONATE, $MnCO_3$

[For Data Reported in Figure and Table No. 352]

Curve No.	Ref. No.	Year	Temp. Range, K	Reported Error, %	Name and Specimen Designation	Composition (weight percent), Specifications and Remarks
1	101	1943	298-660		Rhodoch-rosite	2.0 $CaCO_3$, 0.1 SiO_2, and 0.1 $FeCO_3$; corrected for impurities.

DATA TABLE NO. 352 SPECIFIC HEAT OF MANGANESE CARBONATE, $MnCO_3$

[Temperature, T, K; Specific Heat, C_p, Cal $g^{-1} K^{-1}$]

T	C_p
CURVE 1	
298.15	1.670 x 10^{-1}
300	1.678*
400	1.984
500	2.161
600	2.287
660	2.351

* Not shown on plot

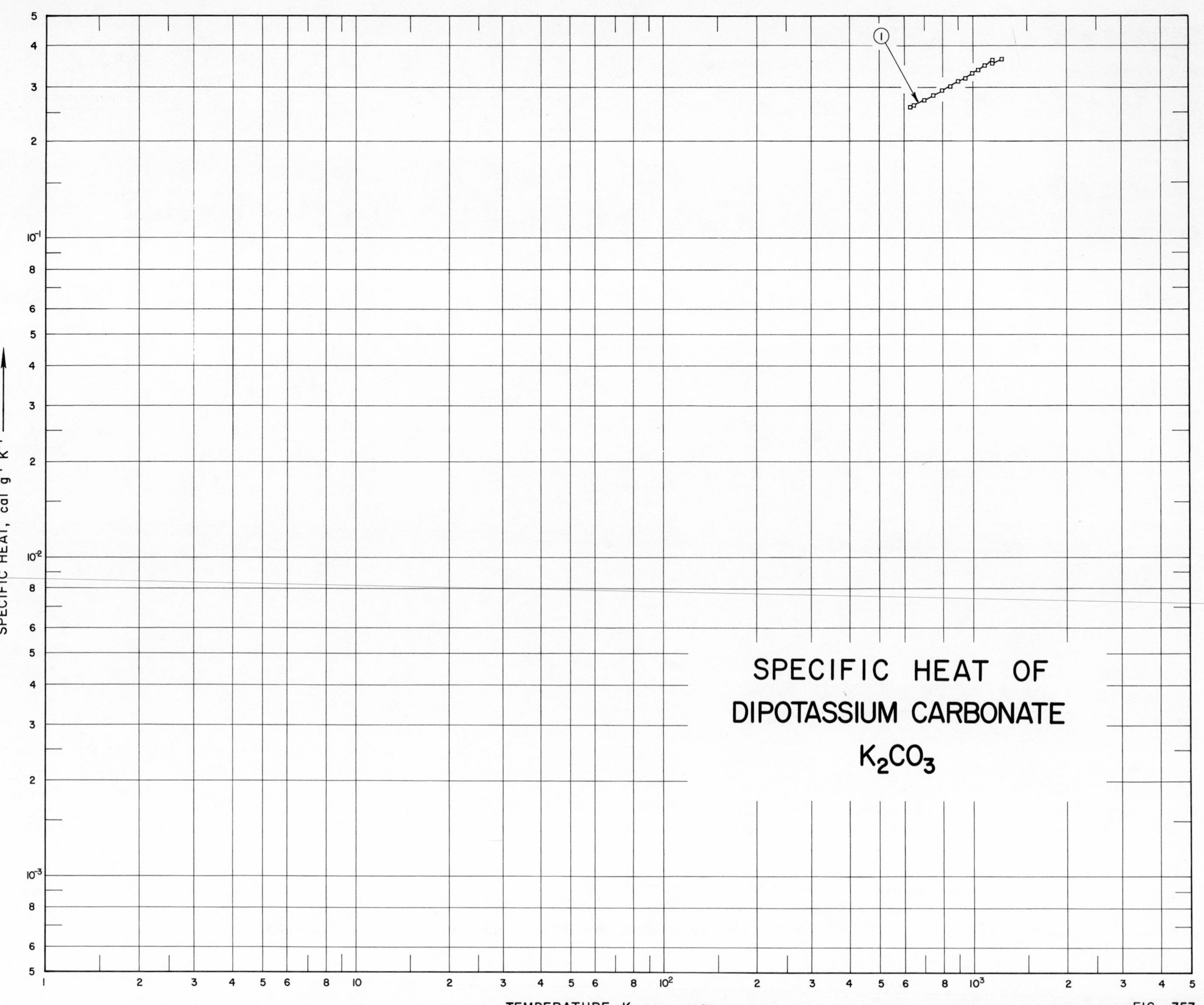
SPECIFIC HEAT OF
DIPOTASSIUM CARBONATE
K_2CO_3
1
SPECIFIC HEAT, cal g⁻¹ K⁻¹
TEMPERATURE, K

FIG. 353

SPECIFICATION TABLE NO. 353 SPECIFIC HEAT OF DIPOTASSIUM CARBONATE, K_2CO_3

[For Data Reported in Figure and Table No. 353]

Curve No.	Ref. No.	Year	Temp. Range, K	Reported Error, %	Name and Specimen Designation	Composition (weight percent), Specifications and Remarks
1	430	1963	630-1250			Baker reagents.

DATA TABLE NO. 353 SPECIFIC HEAT OF DIPOTASSIUM CARBONATE, K_2CO_3

[Temperature, T, K; Specific Heat, C_p, Cal $g^{-1} K^{-1}$]

T	C_p
CURVE 1	
630	2.576 x 10^{-1}
650	2.614
700	2.709
750	2.802
800	2.897
850	2.991
900	3.085
950	3.180
1000	3.274
1050	3.368
1100	3.463
1150	3.557*
(s)1169	3.593
(ℓ)1169	3.574
1200	3.598*
1250	3.636

* Not shown on plot

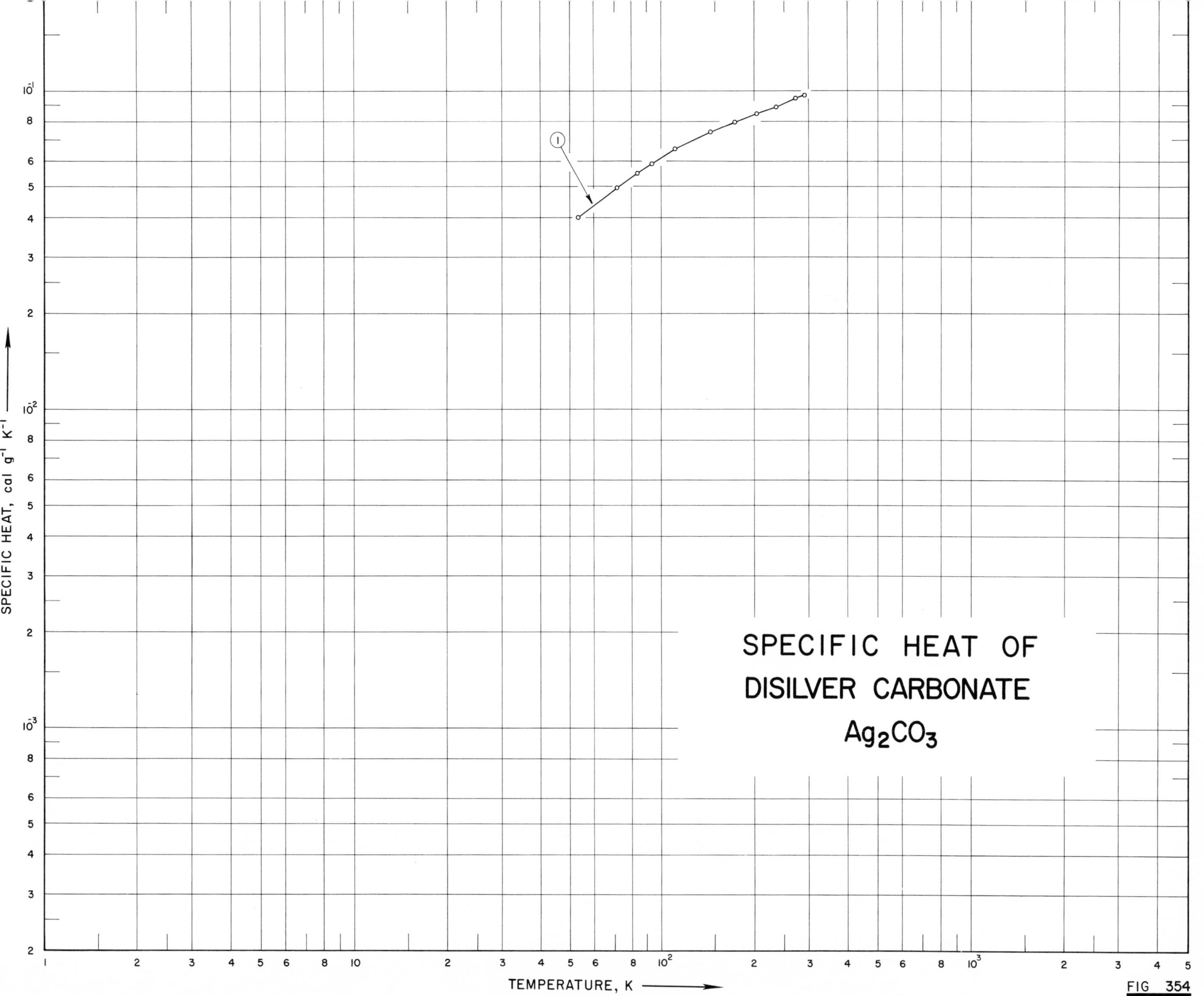
SPECIFIC HEAT OF
DISILVER CARBONATE
Ag_2CO_3
SPECIFIC HEAT, cal g^{-1} K^{-1}
TEMPERATURE, K
1

FIG 354

SPECIFICATION TABLE NO. 354 SPECIFIC HEAT OF DISILVER CARBONATE, Ag_2CO_3

[For Data Reported in Figure and Table No. 354]

Curve No.	Ref. No.	Year	Temp. Range, K	Reported Error, %	Name and Specimen Designation	Composition (weight percent), Specifications and Remarks
1	431	1933	54-290			99.96 Ag_2CO_3; prepared from pure Ag foil and C.P. HNO_3 (recrystallized several times).

DATA TABLE NO. 354 SPECIFIC HEAT OF DISILVER CARBONATE, Ag_2CO_3

[Temperature, T, K; Specific Heat, C_p, Cal $g^{-1} K^{-1}$]

T	C_p
CURVE 1	
53.51	4.011×10^{-2}
54.46	4.025*
71.57	4.972
84.25	5.534
93.53	5.904
93.57	5.904*
112.0	6.571
113.7	6.618
145.2	7.441
173.9	7.989
204.2	8.482
237.4	8.939
272.9	9.487
283.7	9.610
290.4	9.653

* Not shown on plot

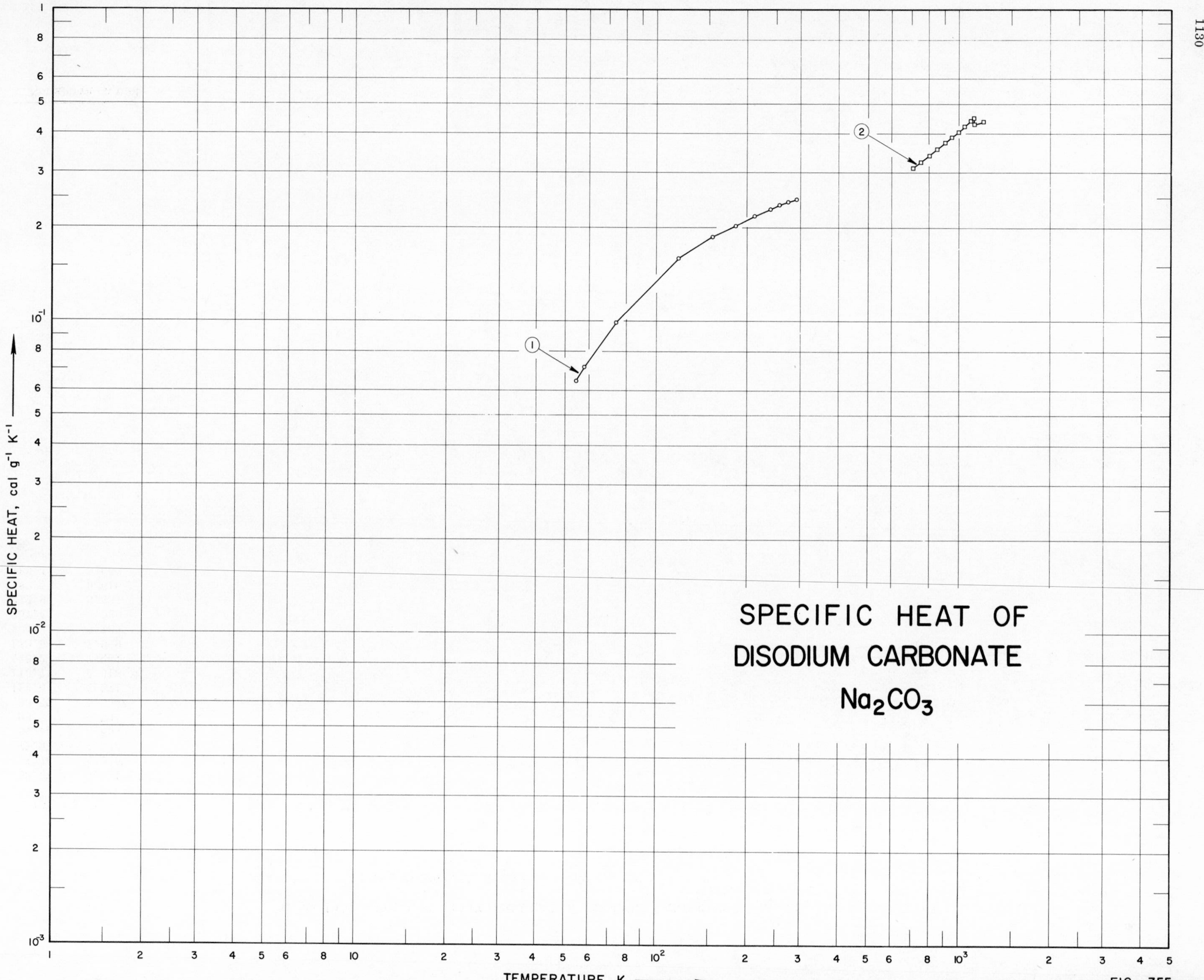

SPECIFIC HEAT OF
DISODIUM CARBONATE
Na2CO3
SPECIFIC HEAT, cal g-1 K-1
TEMPERATURE, K
1
2

FIG 355

SPECIFICATION TABLE NO. 355 SPECIFIC HEAT OF DISODIUM CARBONATE, Na_2CO_3

[For Data Reported in Figure and Table No. 355]

Curve No.	Ref. No.	Year	Temp. Range, K	Reported Error, %	Name and Specimen Designation	Composition (weight percent), Specifications and Remarks
1	431	1933	55-289			Merck anhydrous C.P.; $<$ 0.1 total impurities; hydroden atmosphere.
2	430	1963	707-1210			Baker analyzed reagents.

DATA TABLE NO. 355 SPECIFIC HEAT OF DISODIUM CARBONATE, Na_2CO_3

[Temperature, T, K; Specific Heat, C_p, Cal $g^{-1} K^{-1}$]

T	C_p
CURVE 1	
54.60	6.489 x 10^{-2}
58.43	7.151
74.23	9.935
119.2	1.600
154.6	1.876
183.6	2.034
212.6	2.195
239.7	2.305
256.7	2.370
292.1	2.480
274.1	2.435
280.2	2.454*
289.3	2.476*
CURVE 2	
707	3.123
750	3.266
800	3.432
850	3.599
900	3.766
950	3.932
1000	4.099
1050	4.268
1100	4.432
(s)1127	4.521
(ℓ)1127	4.345
1150	4.369*
1200	4.419*
1210	4.429

* Not shown on plot

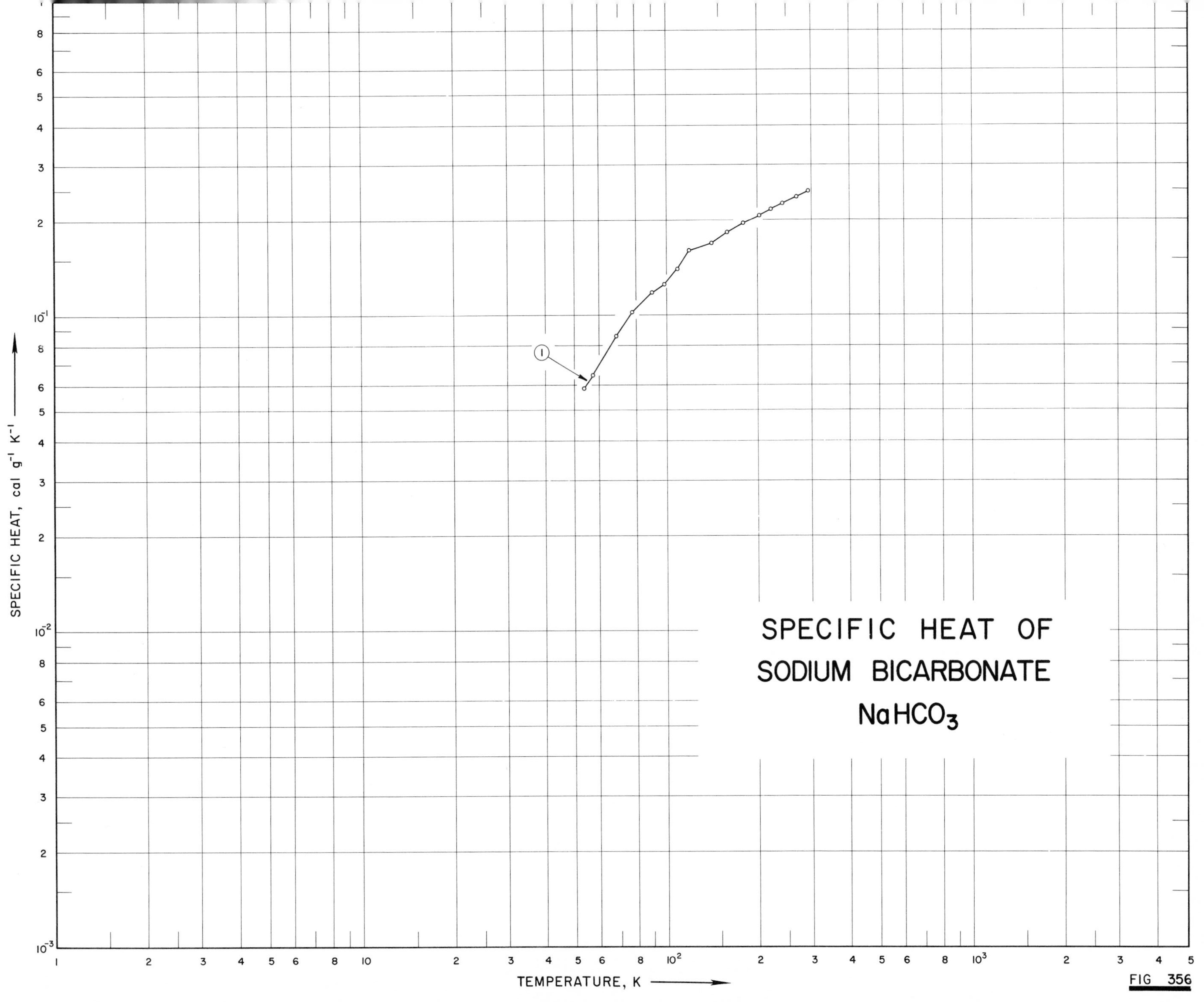
SPECIFIC HEAT OF
SODIUM BICARBONATE
$NaHCO_3$
SPECIFIC HEAT, cal g⁻¹ K⁻¹
TEMPERATURE, K
1

FIG 356

SPECIFICATION TABLE NO. 356 SPECIFIC HEAT OF SODIUM BICARBONATE, $NaHCO_3$

[For Data Reported in Figure and Table No. 356]

Curve No.	Ref. No.	Year	Temp. Range, K	Reported Error, %	Name and Specimen Designation	Composition (weight percent), Specifications and Remarks
1	431	1933	54-295			> 99.8 $NaHCO_3$, 0.1 H_2O, and 0.1 Na_2CO_3; sample supplied by Squibb.

DATA TABLE NO. 356 SPECIFIC HEAT OF SODIUM BICARBONATE, $NaHCO_3$

[Temperature, T, K; Specific Heat, C_p, Cal $g^{-1} K^{-1}$]

T	C_p
CURVE 1	
54.17	5.860 x 10^{-2}
57.54	6.473
68.65	8.601
77.86	1.023 x 10^{-1}
90.07	1.185
99.07	1.258
109.4	1.417
129.8	1.619
142.0	1.700
159.4	1.830
182.3	1.959
203.4	2.077
221.8	2.172
240.6	2.262
270.7	2.388
291.8	2.475*
295.1	2.487

* Not shown on plot

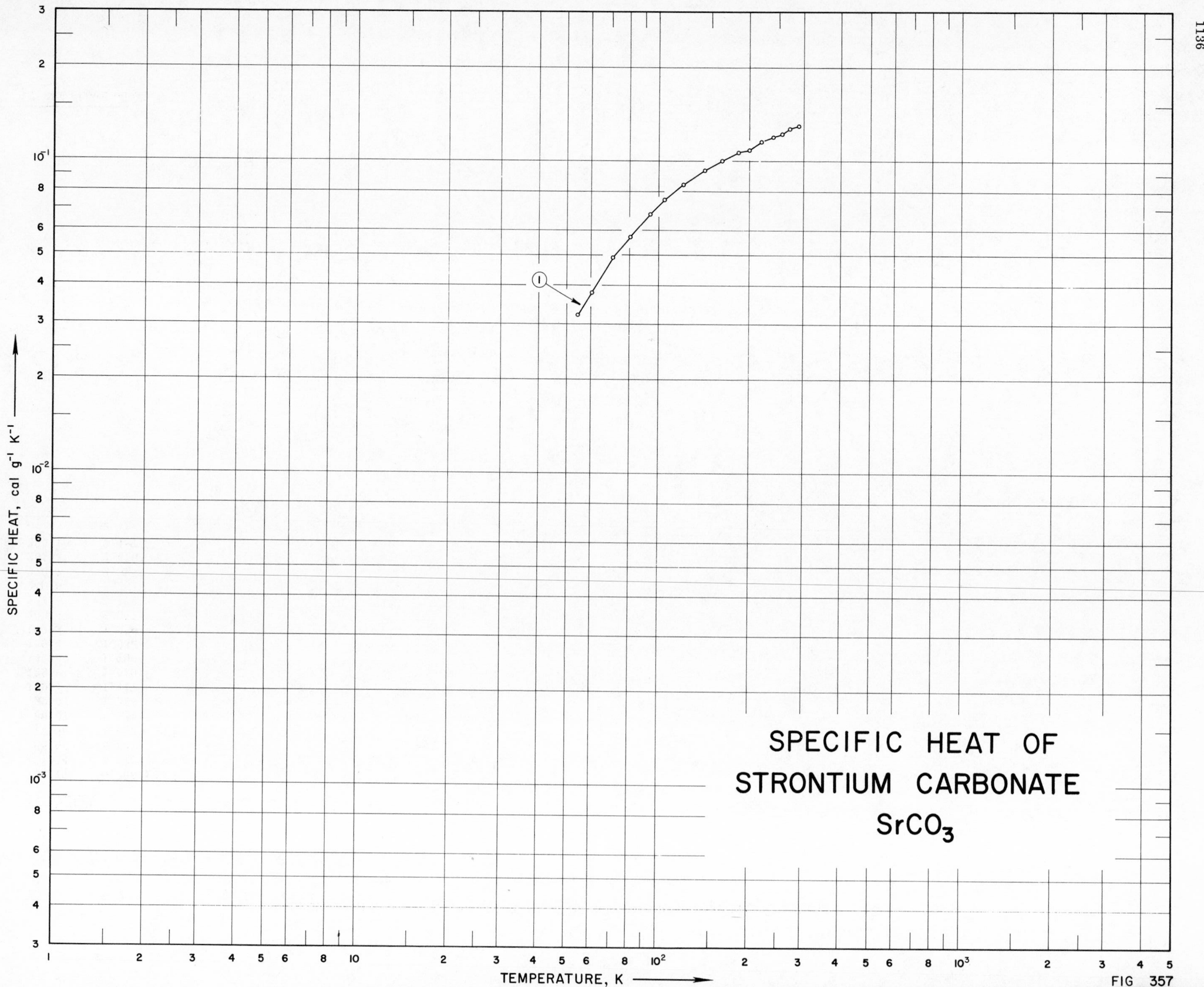
SPECIFIC HEAT OF
STRONTIUM CARBONATE
SrCO3
SPECIFIC HEAT, cal g⁻¹ K⁻¹
TEMPERATURE, K
1
FIG 357

SPECIFICATION TABLE NO. 357 SPECIFIC HEAT OF STRONTIUM CARBONATE, $SrCO_3$

[For Data Reported in Figure and Table No. 357]

Curve No.	Ref. No.	Year	Temp. Range, K	Reported Error, %	Name and Specimen Designation	Composition (weight percent), Specifications and Remarks
1	427	1934	55-291		Strontianite	93.0 $SrCO_3$ and 7.0 $CaCO_3$; corrected for impurities.
2	42	1951	455-1186			Exceptionally high purity.

DATA TABLE NO. 357 SPECIFIC HEAT OF STRONTIUM CARBONATE, $SrCO_3$

[Temperature, T, K; Specific Heat, C_p, Cal $g^{-1} K^{-1}$]

T	C_p
CURVE 1	
54.7	3.221×10^{-2}
60.5	3.791
71.2	4.917
80.8	5.721
94.2	6.774
105.4	7.533
121.3	8.420
142.3	9.369
162.6	1.009×10^{-1}
184.0	1.073
200.4	1.088
218.8	1.162
240.6	1.209
257.6	1.229
272.3	1.280
291.2	1.307
CURVE 2	
455	1.603
500	1.649
600	1.735
700	1.810
800	1.879
900	1.945
1000	2.008
1100	2.070
1186	2.122

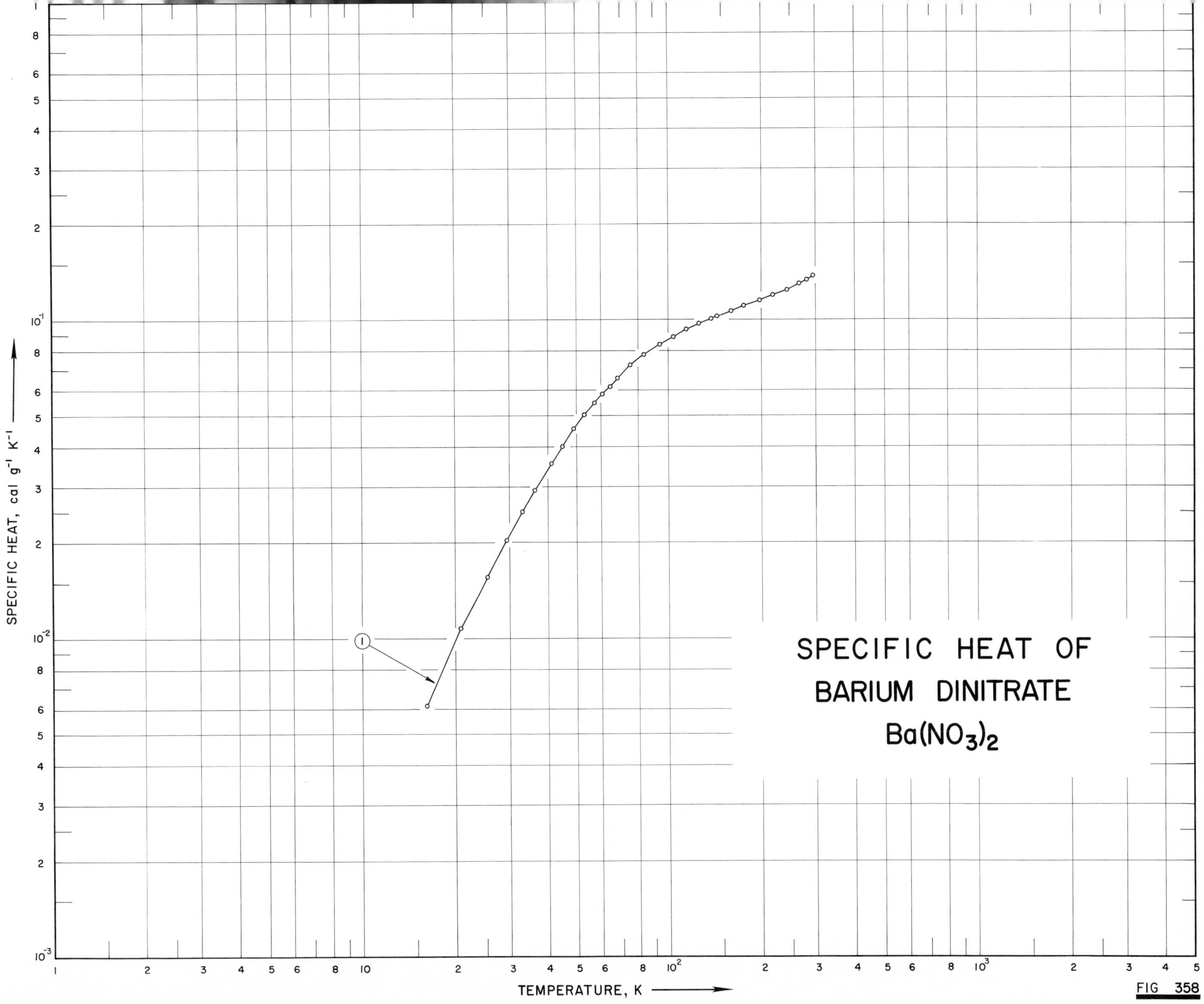

SPECIFIC HEAT OF BARIUM DINITRATE $Ba(NO_3)_2$

SPECIFIC HEAT, cal g^{-1} K^{-1} →

TEMPERATURE, K →

(1)

FIG 358

SPECIFICATION TABLE NO. 358 SPECIFIC HEAT OF BARIUM DINITRATE, $Ba(NO_3)_2$

[For Data Reported in Figure and Table No. 358]

Curve No.	Ref. No.	Year	Temp. Range, K	Reported Error, %	Name and Specimen Designation	Composition (weight percent), Specifications and Remarks
1	432	1930	16-296			

DATA TABLE NO. 358 SPECIFIC HEAT OF BARIUM DINITRATE, $Ba(NO_3)_2$

[Temperature, T, K; Specific Heat, C_p, Cal $g^{-1}K^{-1}$]

T	C_p
CURVE 1	
16.2	6.159×10^{-3}
20.84	1.081×10^{-2}
25.26	1.570
29.39	2.043
33.14	2.517
36.25	2.934
41.22	3.564
44.95	4.036
49.06	4.595
53.00	5.058
57.43	5.536
60.68	5.865
64.45	6.236
68.28	6.622
71.64	6.894*
75.45	7.265
79.15	7.529*
83.75	7.816
89.10	8.145*
94.03	8.424
99.13	8.673*
104.20	8.883
109.75	9.143*
115.25	9.350
121.05	9.622*
127.09	9.832
132.77	9.947*
138.60	1.016×10^{-1}
145.02	1.033
161.43	1.077
168.66	1.094*
176.15	1.111
183.70	1.132*
198.05	1.165
205.48	1.183*
212.24	1.197*
219.00	1.216
226.26	1.224*
233.17	1.239*
242.66	1.259
250.09	1.284*
258.08	1.302*
265.93	1.315
273.87	1.349*
CURVE 1 (cont.)	
281.79	1.354×10^{-1}
289.53	1.377
295.73	1.384

* Not shown on plot

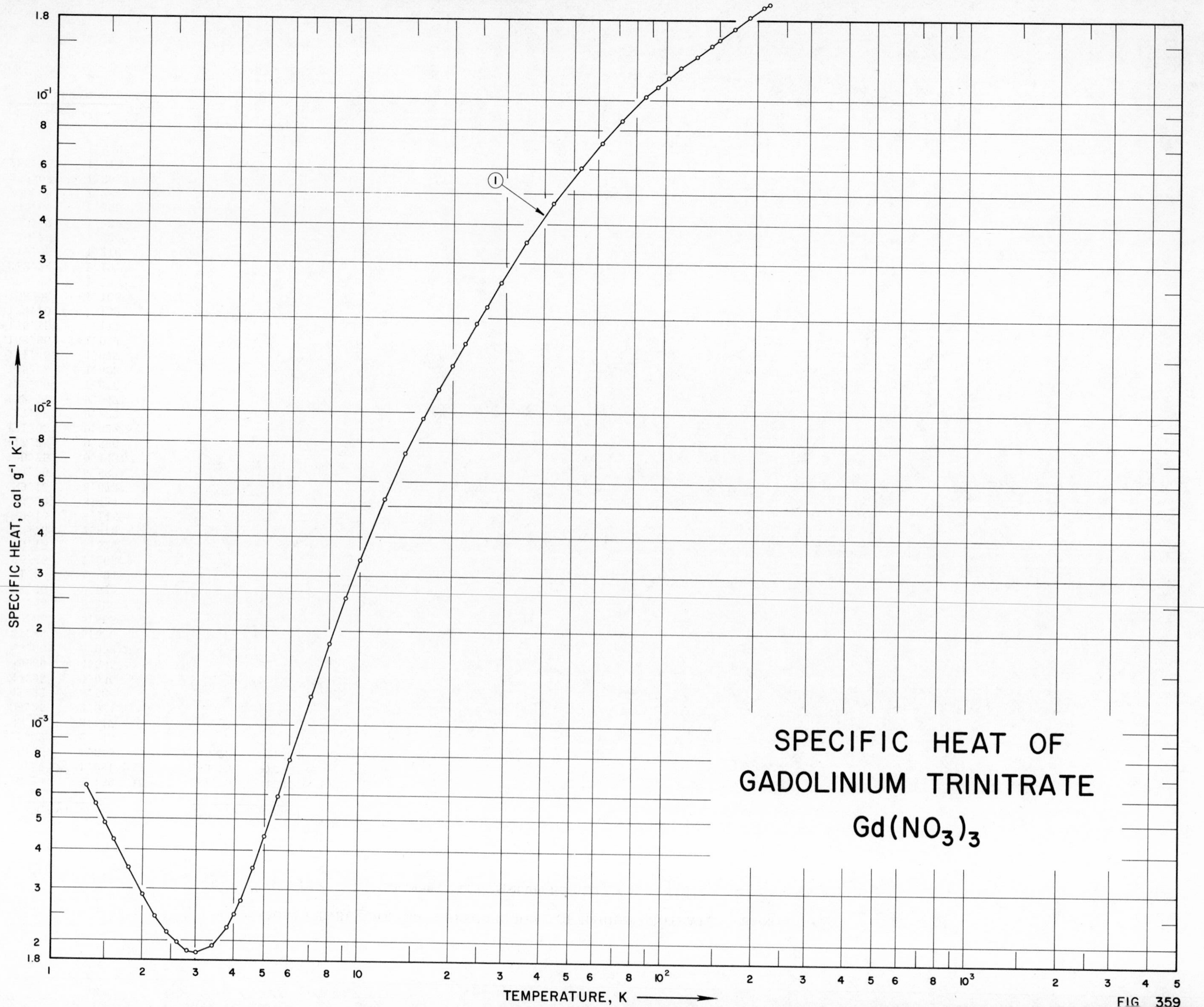
SPECIFIC HEAT OF
GADOLINIUM TRINITRATE
Gd(NO3)3
SPECIFIC HEAT, cal g-1 K-1
TEMPERATURE, K
1
FIG 359

SPECIFICATION TABLE NO. 359 SPECIFIC HEAT OF GADOLINIUM TRINITRATE HEXAHYDRATE, $Gd(NO_3)_3 \cdot 6H_2O$

[For Data Reported in Figure and Table No. 359]

Curve No.	Ref. No.	Year	Temp. Range, K	Reported Error, %	Name and Specimen Designation	Composition (weight percent), Specifications and Remarks
1	433	1962	1.3-220			

DATA TABLE NO. 359 SPECIFIC HEAT OF GADOLINIUM TRINITRITE HEXAHYDRATE, $Gd(NO_3)_3 \cdot 6H_2O$

[Temperature, T, K; Specific Heat, C_p, Cal $g^{-1} K^{-1}$]

T	C_p
CURVE 1	
1.3	6.41 x 10^{-4}
1.4	5.61
1.5	4.86
1.6	4.30
1.8	3.52
2.0	2.87
2.2	2.46
2.4	2.20
2.6	2.03
2.8	1.92
3.0	1.89
3.4	1.99
3.8	2.27
4.0	2.49
4.2	2.76
4.6	3.51
5.0	4.43
5.5	5.947
6.0	7.768
7.0	1.235 x 10^{-3}
8.0	1.823
9.0	2.555
10.0	3.385
12.0	5.301
14.0	7.419
16.0	9.606
18.0	1.181 x 10^{-2}
20.0	1.414
22.0	1.661
24.0	1.922
26.0	2.189
29.0	2.614
35.0	3.516
43.0	4.672
53.0	6.052
62.0	7.297
72.0	8.626
86.0	1.025 x 10^{-1}
94.0	1.105
102.0	1.178
112.0	1.270
127.0	1.389
141.0	1.499
150.0	1.571
CURVE 1 (cont.)	
168.0	1.702 x 10^{-1}
189.0	1.851
210.0	1.985
220.0	2.041

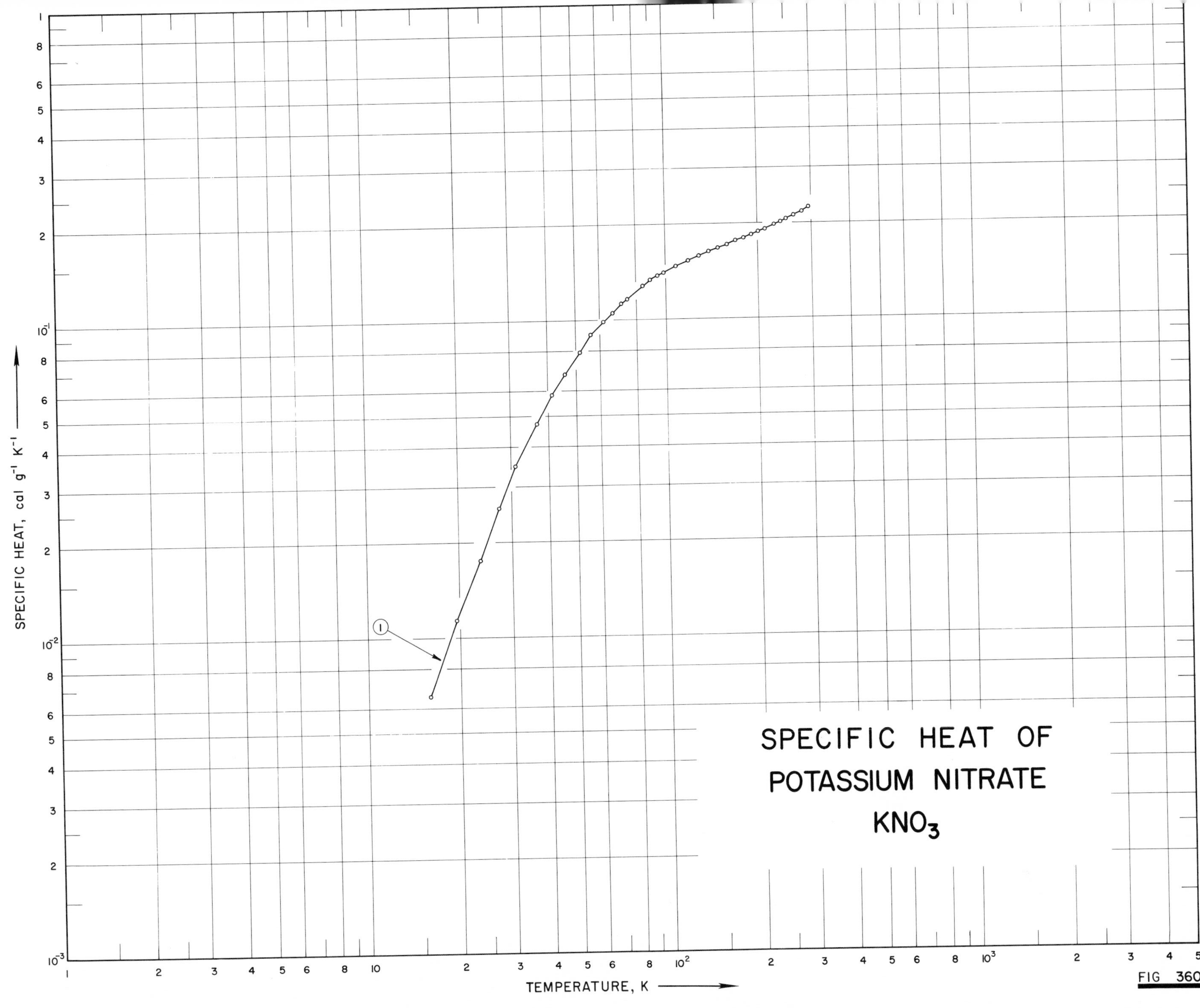

SPECIFIC HEAT OF
POTASSIUM NITRATE
KNO_3
SPECIFIC HEAT, cal g^{-1} K^{-1}
TEMPERATURE, K
1
FIG 360

SPECIFICATION TABLE NO. 360 SPECIFIC HEAT OF POTASSIUM NITRATE, KNO_3

[For Data Reported in Figure and Table No. 360]

Curve No.	Ref. No.	Year	Temp. Range, K	Reported Error, %	Name and Specimen Designation	Composition (weight percent), Specifications and Remarks
1	266	1933	16-296			C.P. quality further purified by 3 or 4 recrystallizations; dried in high vacuum.

DATA TABLE NO. 360 SPECIFIC HEAT OF POTASSIUM NITRATE, KNO_3

[Temperature, T, K; Specific Heat, C_p, Cal $g^{-1} K^{-1}$]

T	C_p	T	C_p
CURVE 1		CURVE 1 (cont.)	
15.96	6.567 x 10^{-3}	196.53	1.872 x 10^{-1}*
19.52	1.146 x 10^{-2}	201.37	1.893
23.45	1.767	201.79	1.891*
27.28	2.572	206.62	1.911*
27.41	2.655*	207.03	1.915*
31.07	3.513	211.87	1.934*
36.78	4.755	212.27	1.934
41.37	5.863	217.11	1.952*
46.20	6.810	222.34	1.977*
51.57	7.981	227.56	1.992
56.86	9.059	233.04	2.017*
61.79	9.940	238.53	2.029*
66.43	1.064 x 10^{-1}	238.90	2.035
71.16	1.129	243.75	2.051*
74.84	1.172	244.12	2.050*
84.24	1.294	248.96	2.074*
89.15	1.340*	249.33	2.074
89.20	1.343	254.53	2.090*
93.95	1.380*	259.73	2.114*
94.01	1.382	264.93	2.129
98.68	1.417	270.11	2.160*
98.74	1.416*	275.29	2.171*
103.38	1.455*	280.46	2.195
108.23	1.487	285.63	2.216*
113.30	1.514*	290.78	2.245*
118.31	1.548	295.92	2.268
123.26	1.574*		
128.17	1.598		
133.31	1.627*		
138.69	1.652		
143.63	1.674*		
144.04	1.678*		
148.98	1.688		
149.36	1.699*		
154.30	1.719*		
154.68	1.714*		
159.62	1.735		
164.93	1.757*		
170.22	1.779		
175.49	1.797*		
180.76	1.813		
186.03	1.834*		
191.28	1.860		
196.12	1.875*		

* Not shown on plot

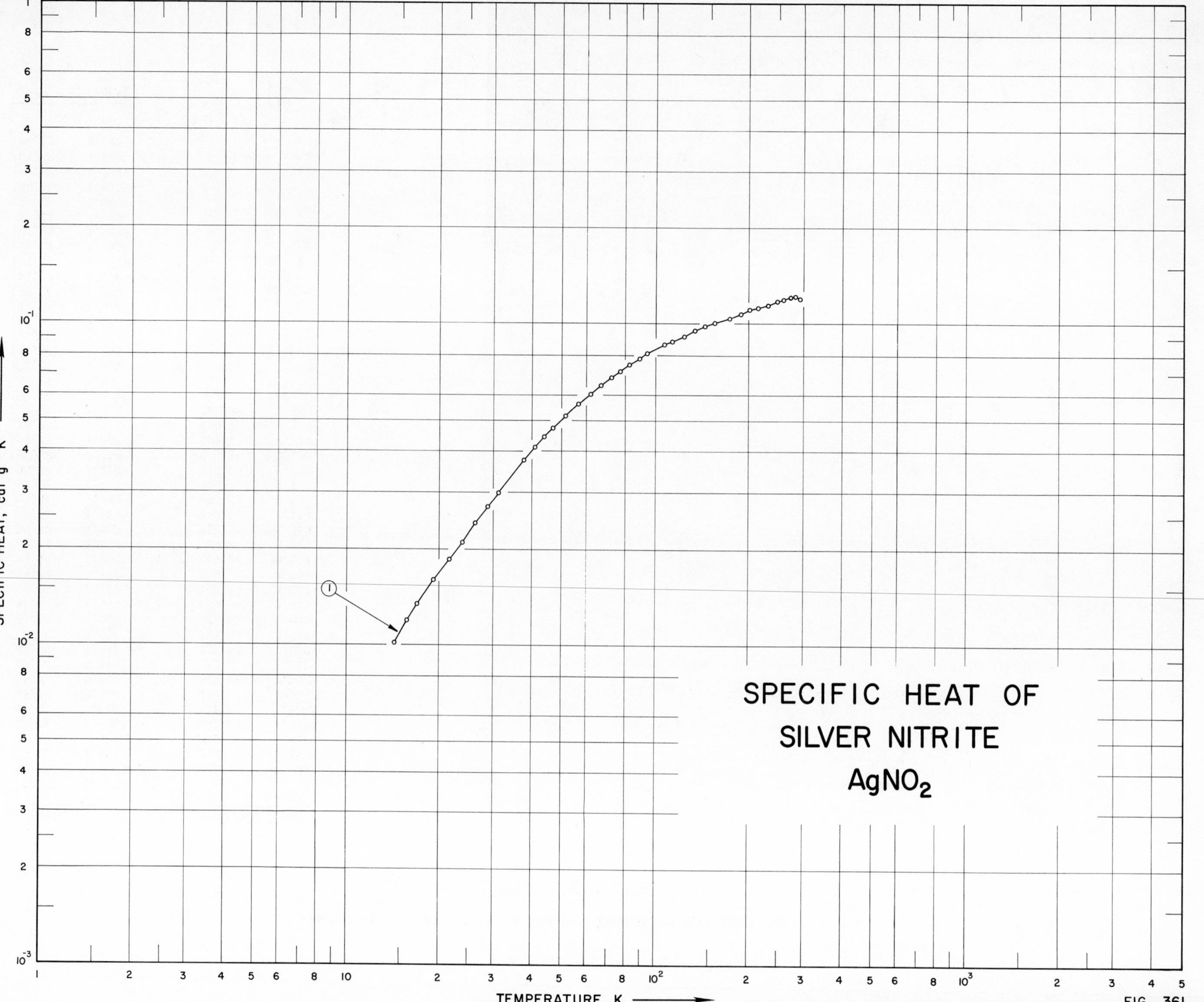
SPECIFIC HEAT OF
SILVER NITRITE
$AgNO_2$
SPECIFIC HEAT, cal g⁻¹ K⁻¹
TEMPERATURE, K
1
FIG 361

SPECIFICATION TABLE NO. 361 SPECIFIC HEAT OF SILVER NITRITE, $AgNO_2$

[For Data Reported in Figure and Table No. 361]

Curve No.	Ref. No.	Year	Temp. Range, K	Reported Error, %	Name and Specimen Designation	Composition (weight percent), Specifications and Remarks
1	434	1937	14-295			69.925 Ag and 29.765 NO_2 (70.10 Ag, 29.96 NO_2 theo); C.P. sample; recrystallized from distilled water and dried in a vacuum desicator for several days.

DATA TABLE NO. 361 SPECIFIC HEAT OF SILVER NITRITE, $AgNO_2$

[Temperature, T, K; Specific Heat, C_p, Cal $g^{-1} K^{-1}$]

T	C_p
CURVE 1	
14.36	1.014 x 10^{-2}
15.77	1.189
17.09	1.345
19.21	1.599
21.65	1.852
23.84	2.093
26.32	2.418
28.82	2.710
31.24	2.989
37.72	3.789
41.04	4.159
44.15	4.471
47.40	4.777
51.67	5.212
56.73	5.693
62.21	6.115
67.37	6.499
72.32	6.876
77.10	7.176
82.86	7.532
89.26	7.863
94.07	8.162
107.35	8.708
113.53	8.916
124.51	9.241
129.71	9.449*
135.02	9.644
140.48	9.833*
145.93	9.950
151.31	1.008 x 10^{-1}*
156.59	1.019
156.60	1.019*
163.45	1.035*
174.86	1.055
189.00	1.090
195.94	1.107*
202.25	1.124
215.27	1.139
223.49	1.148*
232.36	1.164
240.05	1.180*
248.24	1.192
255.45	1.205*
260.44	1.207
CURVE 1 (cont.)	
265.92	1.226 x 10^{-1}*
273.93	1.224*
274.16	1.226
284.13	1.232
284.30	1.224*
294.94	1.213

* Not shown on plot

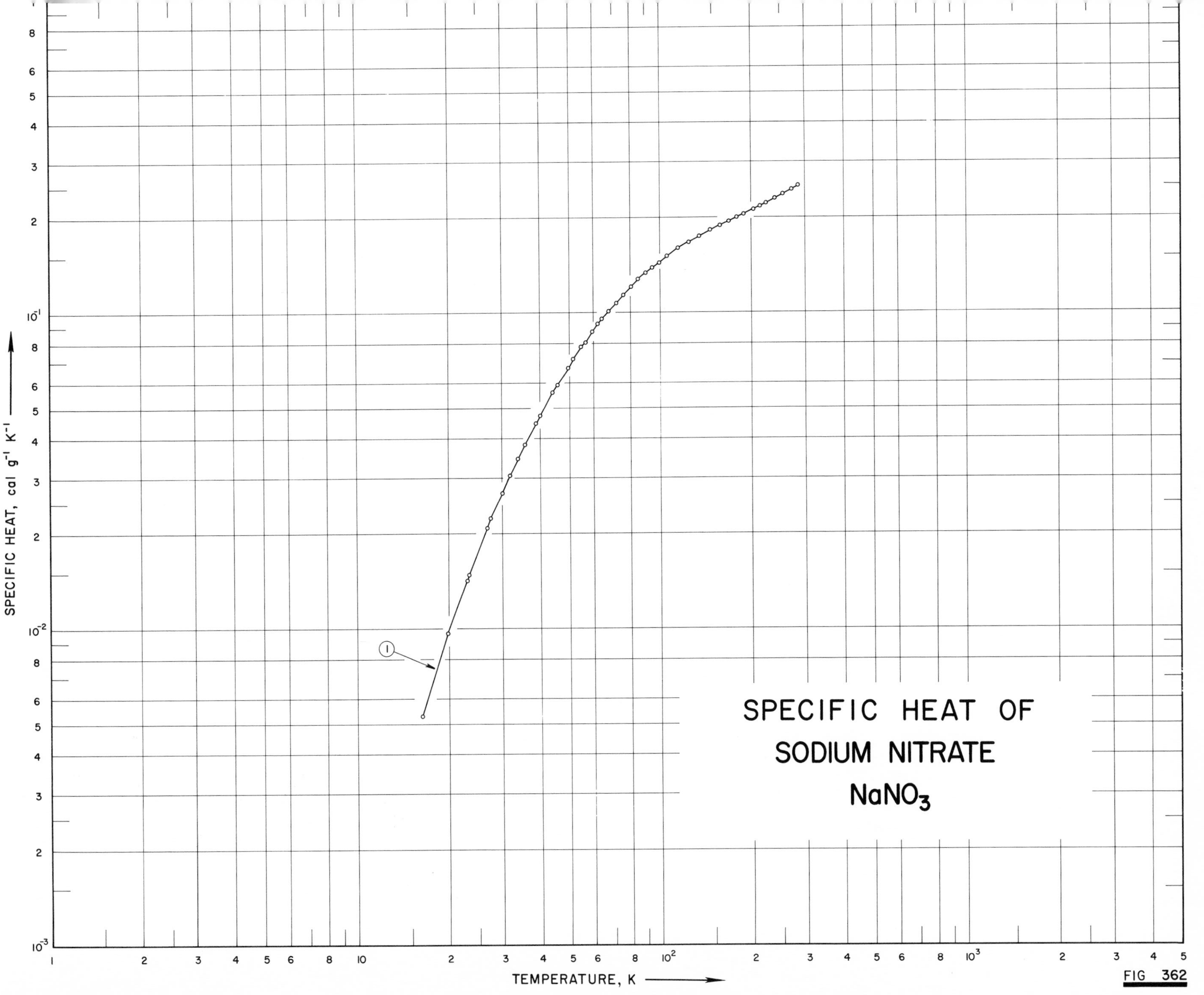
SPECIFIC HEAT OF
SODIUM NITRATE
NaNO3
SPECIFIC HEAT, cal g-1 K-1
TEMPERATURE, K
1

FIG 362

SPECIFICATION TABLE NO. 362 SPECIFIC HEAT OF SODIUM NITRATE, $NaNO_3$

[For Data Reported in Figure and Table No. 362]

Curve No.	Ref. No.	Year	Temp. Range, K	Reported Error, %	Name and Specimen Designation	Composition (weight percent), Specifications and Remarks
1	266	1933	16-287			C.P. quality further purified by 3 or 4 recrystallizations; dried in high vacuum.

DATA TABLE NO. 362 SPECIFIC HEAT OF SODIUM NITRATE, $NaNO_3$

[Temperature, T, K; Specific Heat, C_p, Cal $g^{-1}K^{-1}$]

T	C_p	T	C_p
CURVE 1		CURVE 1 (cont.)	
16.23	5.341×10^{-3}	158.30	1.911×10^{-1}
19.78	9.730	163.59	1.944*
22.92	1.438×10^{-2}	164.48	1.952*
23.29	1.498	168.85	1.967*
26.78	2.100	169.20	1.969
27.56	2.267	174.09	1.995*
30.03	2.720	174.90	1.997*
31.84	3.090	180.09	2.025
33.85	3.499	185.25	2.053*
35.63	3.858	190.39	2.077
38.80	4.526	204.86	2.148
40.08	4.775	207.61	2.164*
44.35	5.653	209.95	2.173*
45.90	5.971	214.99	2.204*
49.73	6.767	215.02	2.205
51.57	7.227	220.06	2.226*
54.91	7.869	220.08	2.225*
56.99	8.143	225.13	2.251
59.64	8.787	230.18	2.274*
62.37	9.280	235.21	2.300*
64.40	9.644	240.24	2.326
67.32	1.020×10^{-1}*	245.25	2.362*
67.67	1.023	247.21	2.358*
71.90	1.088*	250.25	2.374*
71.94	1.087	255.45	2.407
76.22	1.158	258.16	2.414*
80.61	1.224	260.65	2.428*
85.37	1.299	263.04	2.438*
90.20	1.352	267.91	2.471*
91.42	1.360*	272.77	2.488
94.87	1.408	277.23	2.514*
96.60	1.426*	277.62	2.512*
99.42	1.455	282.06	2.539*
101.61	1.477*	282.47	2.537*
106.51	1.525	286.86	2.562
111.30	1.568*		
116.00	1.611		
120.62	1.653*		
125.72	1.689		
131.29	1.734*		
136.79	1.764		
142.24	1.804*		
147.63	1.849		
152.98	1.880*		

* Not shown on plot

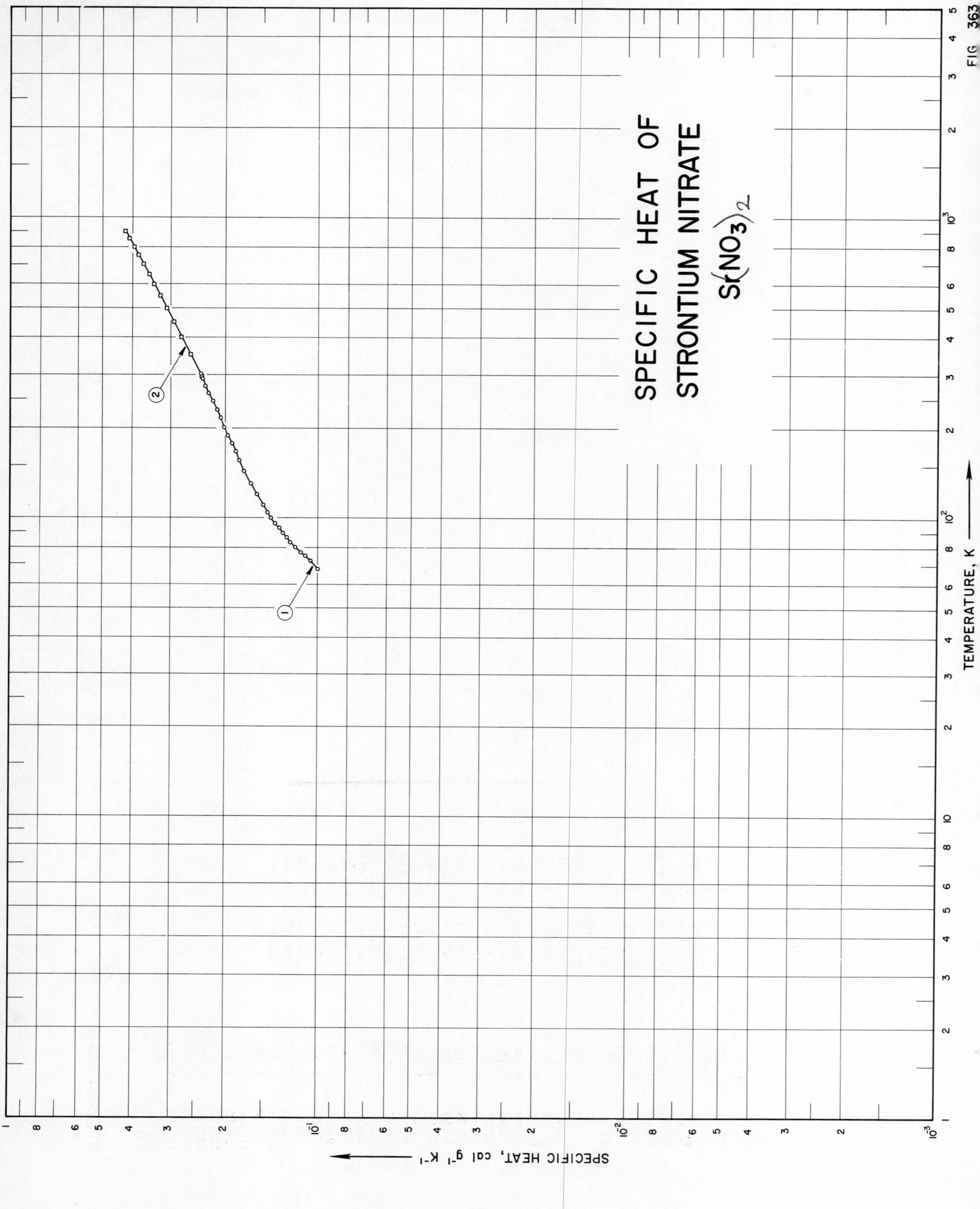

SPECIFIC HEAT OF
STRONTIUM NITRATE
$Sr(NO_3)_2$
SPECIFIC HEAT, cal g^{-1} K^{-1}
TEMPERATURE, K
1
2

FIG 363

SPECIFICATION TABLE NO. 363 SPECIFIC HEAT OF STRONTIUM NITRATE, $Sr(NO_3)_2$

[For Data Reported in Figure and Table No. 363]

Curve No.	Ref. No.	Year	Temp. Range, K	Reported Error, %	Name and Specimen Designation	Composition (weight percent), Specifications and Remarks
1	315	1962	68-304			Reagent grade; dried 4 hrs at 200 C.
2	315	1962	298-900			Same as above.

DATA TABLE NO. 363 SPECIFIC HEAT OF STRONTIUM NITRATE, $Sr(NO_3)_2$

[Temperature, T, K; Specific Heat, C_p, Cal $g^{-1} K^{-1}$]

T	C_p	T	C_p
CURVE 1		CURVE 1 (cont.)	
67.97	1.004 x 10^{-1}	255.77	2.254 x 10^{-1}*
72.34	1.067	259.06	2.264*
75.16	1.101	262.03	2.277
77.85	1.146	264.98	2.287*
80.69	1.192	267.48	2.300*
83.68	1.234	273.65	2.322*
86.57	1.267	276.63	2.324
89.60	1.307	280.59	2.335*
92.81	1.343	288.22	2.361*
96.28	1.381	292.21	2.378
100.59	1.425	296.02	2.391*
102.14	1.447*	299.79	2.398*
105.40	1.464	303.54	2.410
108.91	1.497*		
111.09	1.514	CURVE 2	
115.65	1.557*		
120.20	1.588	298.15	2.392 x 10^{-1}
124.64	1.618*	300	2.400*
131.32	1.663	350	2.605
139.47	1.714*	400	2.784
144.23	1.742	450	2.949
153.53	1.792*	500	3.106
156.22	1.808	550	3.256
162.18	1.837*	600	3.403
167.53	1.857	650	3.546
172.07	1.879*	700	3.688
178.68	1.917	750	3.828
184.82	1.945*	800	3.967
189.60	1.970	850	4.105
193.40	1.984*	900	4.243
198.16	2.006*		
202.37	2.023		
205.89	2.040*		
209.97	2.054*		
213.99	2.067*		
216.71	2.078		
222.30	2.098*		
227.12	2.119*		
230.18	2.134		
234.04	2.153*		
238.42	2.168*		
242.62	2.192*		
246.97	2.212		
250.88	2.230*		

* Not shown on plot

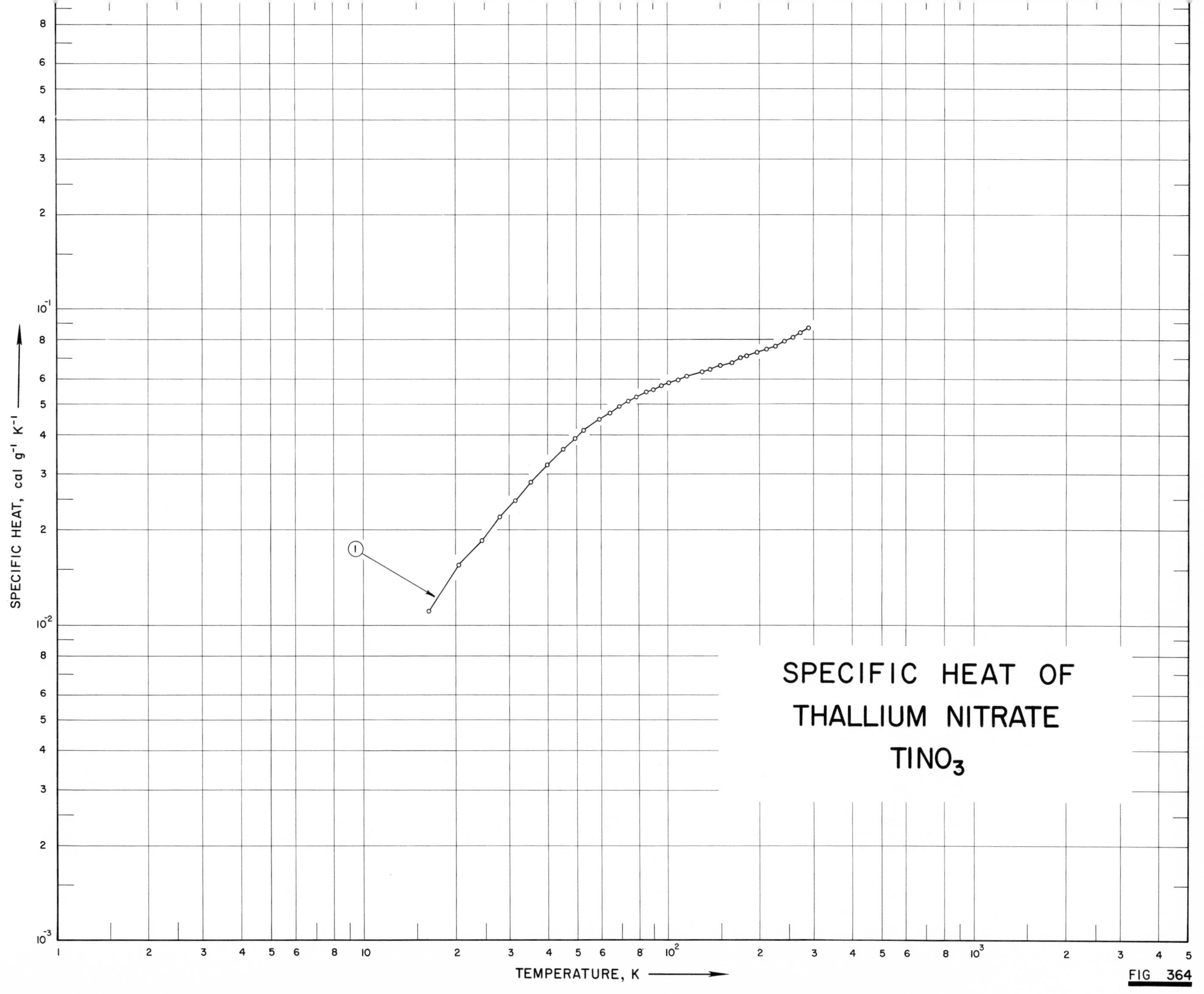
SPECIFIC HEAT OF
THALLIUM NITRATE
$TlNO_3$
SPECIFIC HEAT, cal g^{-1} K^{-1}
TEMPERATURE, K
1

FIG 364

SPECIFICATION TABLE NO. 364 SPECIFIC HEAT OF THALLIUM NITRATE, $TlNO_3$

[For Data Reported in Figure and Table No. 364]

Curve No.	Ref. No.	Year	Temp. Range, K	Reported Error, %	Name and Specimen Designation	Composition (weight percent), Specifications and Remarks
1	436	1932	16–291	1–3		~ 100 $TlNO_3$.

DATA TABLE NO. 364 SPECIFIC HEAT OF THALLIUM NITRATE, $TlNO_3$

[Temperature, T, K; Specific Heat, C_p, Cal g^{-1} K^{-1}]

T	C_p
CURVE 1	
16.45	1.113 x 10^{-2}
20.52	1.564
24.44	1.864
27.95	2.212
31.35	2.495
35.22	2.855
39.94	3.237
45.29	3.622
49.55	3.915
52.91	4.151
59.49	4.504
60.33	4.549*
64.51	4.744
69.19	4.958
74.22	5.157
79.34	5.330
85.54	5.510
90.02	5.619
95.90	5.765
101.71	5.882
108.68	6.032
116.98	6.163
130.25	6.392
137.92	6.497
149.85	6.670
162.78	6.828
174.79	7.064
182.43	7.165
197.32	7.353
211.64	7.541
226.53	7.740
242.59	7.987
250.52	8.119*
258.40	8.224
265.25	8.336*
272.38	8.483
281.69	8.599*
290.60	8.794

* Not shown on plot

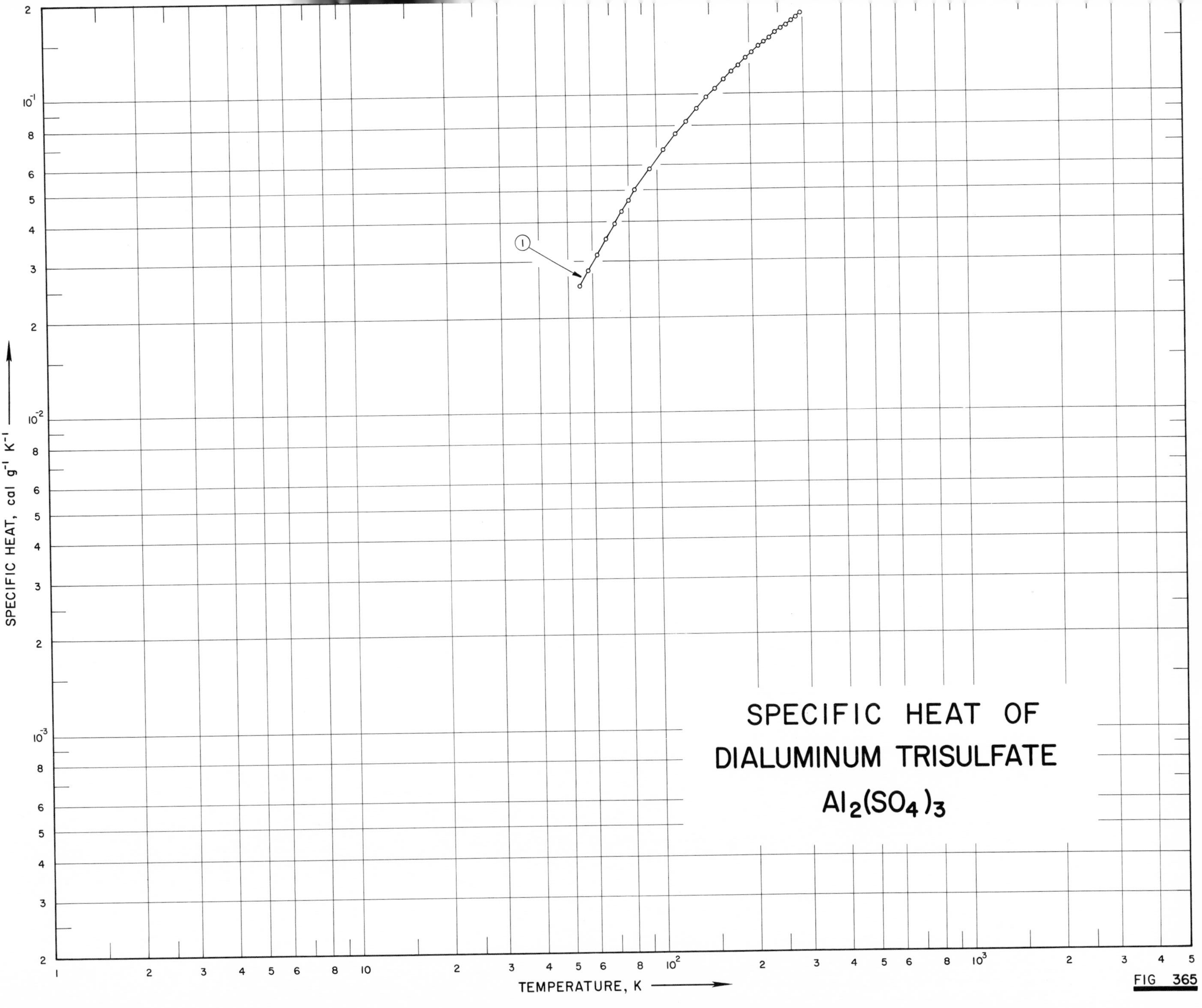
SPECIFIC HEAT OF
DIALUMINUM TRISULFATE
$Al_2(SO_4)_3$
SPECIFIC HEAT, cal g⁻¹ K⁻¹
TEMPERATURE, K
1
FIG 365

SPECIFICATION TABLE NO. 365 SPECIFIC HEAT OF DIALUMINUM TRISULFATE, $Al_2(SO_4)_3$

[For Data Reported in Figure and Table No. 365]

Curve No.	Ref. No.	Year	Temp. Range, K	Reported Error, %	Name and Specimen Designation	Composition (weight percent), Specifications and Remarks
1	359	1946	55-296			99.41 $Al_2(SO_4)_3$, 0.32 H_2O, and 0.26 Al_2O_3; corrected for impurities.

DATA TABLE NO. 365 DIALUMINUM TRISULFATE, $Al_2(SO_4)_3$

[Temperature, T, K; Specific Heat, C_p, Cal $g^{-1}K^{-1}$]

T	C_p
CURVE 1	
54.7	2.534 x 10^{-2}
58.4	2.839
62.4	3.174
66.9	3.566
71.7	3.984
75.9	4.355
80.2	4.721
84.3	5.095
94.4	5.934
104.4	6.752
114.9	7.582
123.9	8.269
134.8	9.099
145.5	9.856
155.2	1.052 x 10^{-1}
165.6	1.123
175.6	1.189
185.4	1.248
196.0	1.310
205.8	1.367
216.7	1.428
226.0	1.474
235.5	1.521
246.5	1.575
256.0	1.625
265.9	1.670
276.4	1.712
286.0	1.761
296.2	1.804

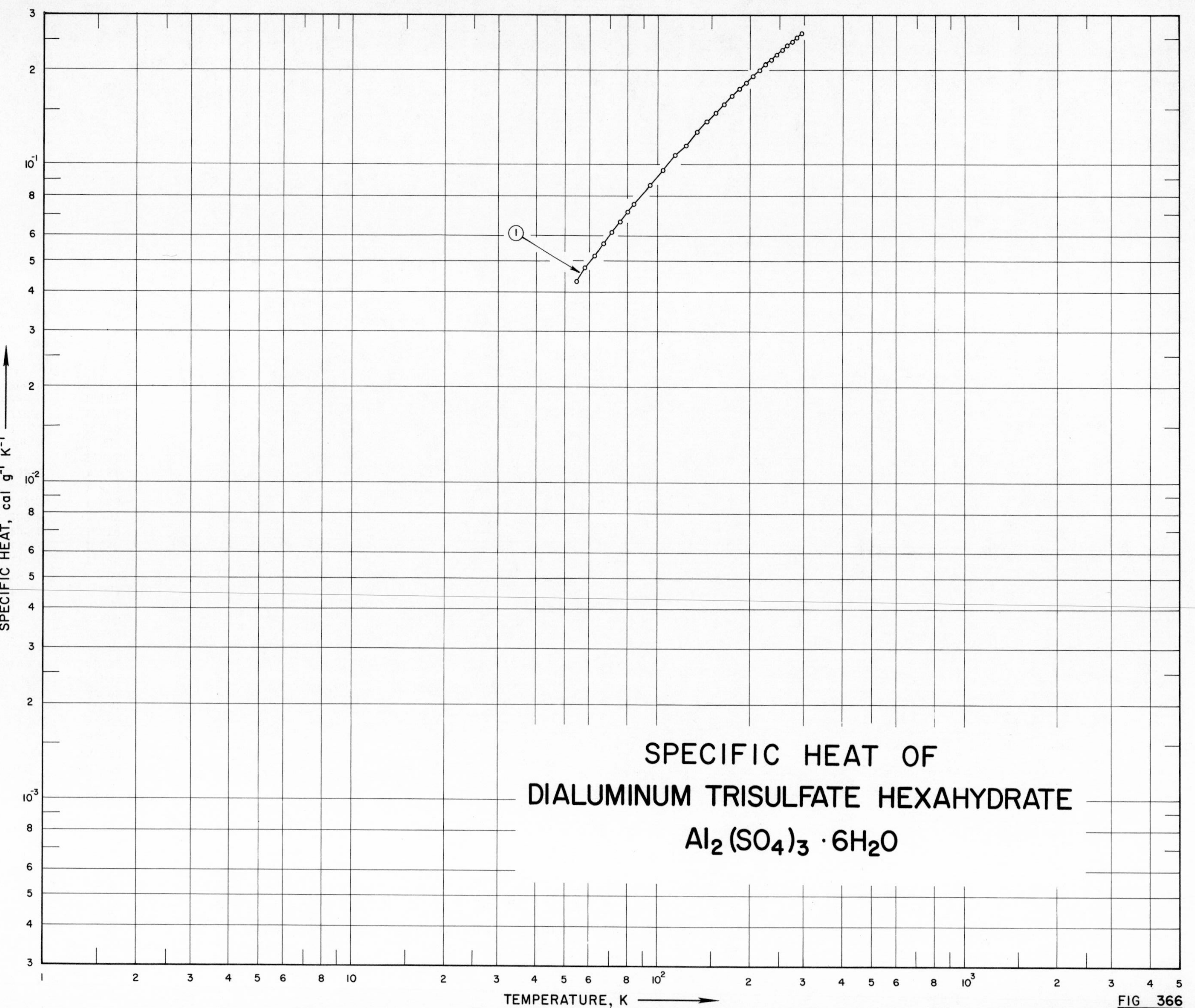
SPECIFIC HEAT, cal g^{-1} K^{-1}
3
2
10^{-1}
8
6
5
4
3
2
10^{-2}
8
6
5
4
3
2
10^{-3}
8
6
5
4
3
1
1
2
3
4
5
6
8
10
2
3
4
5
6
8
10^2
2
3
4
5
6
8
10^3
2
3
4
5
TEMPERATURE, K
SPECIFIC HEAT OF
DIALUMINUM TRISULFATE HEXAHYDRATE
$Al_2(SO_4)_3 \cdot 6H_2O$
FIG 366

SPECIFICATION TABLE NO. 366 SPECIFIC HEAT OF DIALUMINUM TRISULFATE HEXAHYDRATE, $Al_2(SO_4)_3 \cdot 6H_2O$

[For Data Reported in Figure and Table No. 366]

Curve No.	Ref. No.	Year	Temp. Range, K	Reported Error, %	Name and Specimen Designation	Composition (weight percent), Specifications and Remarks
1	359	1946	55-296			22.57 Al_2O_3 (22.64 theo); reagent grade aluminum sulfate octahydrate heated in air for 18 hrs at 140 C; pulverized; aged for 28 hrs at 80 C.

DATA TABLE NO. 366 SPECIFIC HEAT OF DIALUMINUM TRISULFATE HEXAHYDRATE, $Al_2(SO_4)_3 \cdot 6H_2O$

[Temperature, T, K; Specific Heat, C_p, Cal $g^{-1} K^{-1}$]

T	C_p
CURVE 1	
54.5	4.325 x 10^{-2}
58.4	4.755
62.6	5.217
66.9	5.697
71.2	6.177
75.7	6.670
80.4	7.174
84.3	7.572
94.7	8.682
104.3	9.673
114.9	1.076 x 10^{-1}
124.2	1.169
135.1	1.278
145.6	1.379
155.6	1.474
165.4	1.565
175.6	1.658
185.6	1.745
195.9	1.835
206.0	1.920
216.6	2.006
226.3	2.082
235.7	2.155
245.9	2.234
256.2	2.317
265.9	2.390
276.0	2.461
286.1	2.532
296.1	2.610

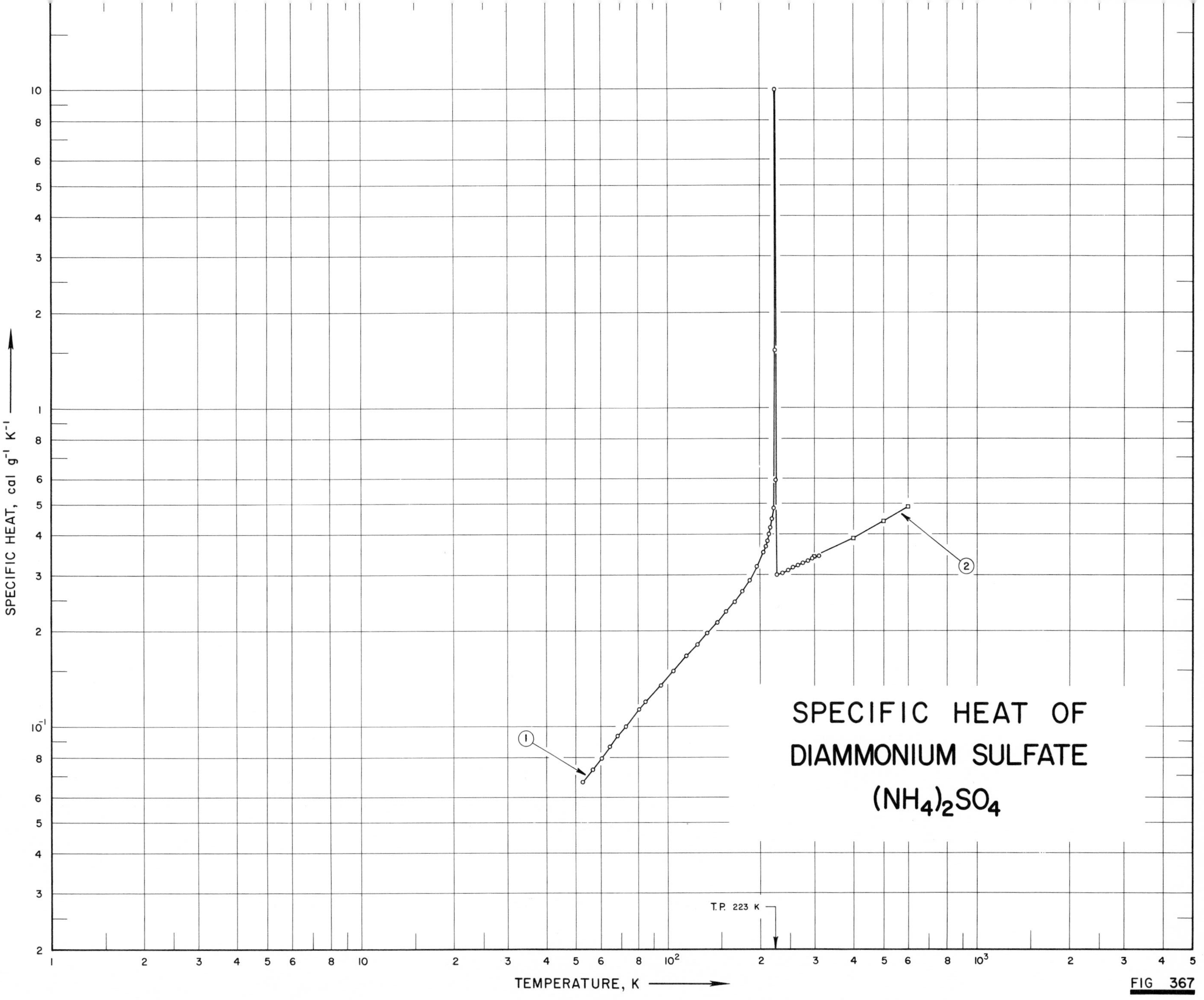
SPECIFIC HEAT OF
DIAMMONIUM SULFATE
$(NH_4)_2SO_4$
SPECIFIC HEAT, cal g^{-1} K^{-1}
TEMPERATURE, K
T.P. 223 K
1
2

FIG 367

SPECIFICATION TABLE NO. 367 SPECIFIC HEAT OF DIAMMONIUM SULFATE, $(NH_4)_2SO_4$

[For Data Reported in Figure and Table No. 367]

Curve No.	Ref. No.	Year	Temp. Range, K	Reported Error, %	Name and Specimen Designation	Composition (weight percent), Specifications and Remarks
1	359	1946	53-311			99.96 $(NH_4)_2SO_4$; reagent grade; dried at 75 C.
2	359	1946	298-600			

DATA TABLE NO. 367 SPECIFIC HEAT OF DIAMMONIUM SULFATE, $(NH_4)_2SO_4$

[Temperature, T, K; Specific Heat, C_p, Cal $g^{-1} K^{-1}$]

T	C_p
CURVE 1	
52.8	6.735×10^{-2}
56.6	7.355
60.4	7.999
64.3	8.665
68.3	9.354
72.5	1.005×10^{-1}
80.6	1.136
84.8	1.202
94.8	1.356
104.4	1.504
115.4	1.674
125.1	1.821
134.8	1.974
145.4	2.136
155.5	2.302
165.3	2.474
175.5	2.677
185.3	2.890
196.0	3.193
205.4	3.540
208.6	3.698
211.5	3.825
214.3	4.024
216.8	4.230
219.3	4.487
221.5	4.869
223.0	1.523×10^{0}
223.4	1.002×10
224.7	5.960×10^{-1}
227.1	3.001
230.2	3.010*
233.3	3.020*
237.0	3.041
246.7	3.103
256.7	3.177
266.0	3.215
276.4	3.273
286.0	3.327
296.1	3.390
305.7	3.432*
310.6	3.449

T	C_p
CURVE 2	
298.15	3.391×10^{-1}*
300	3.400
400	3.909
500	4.417
600	4.926

* Not shown on plot

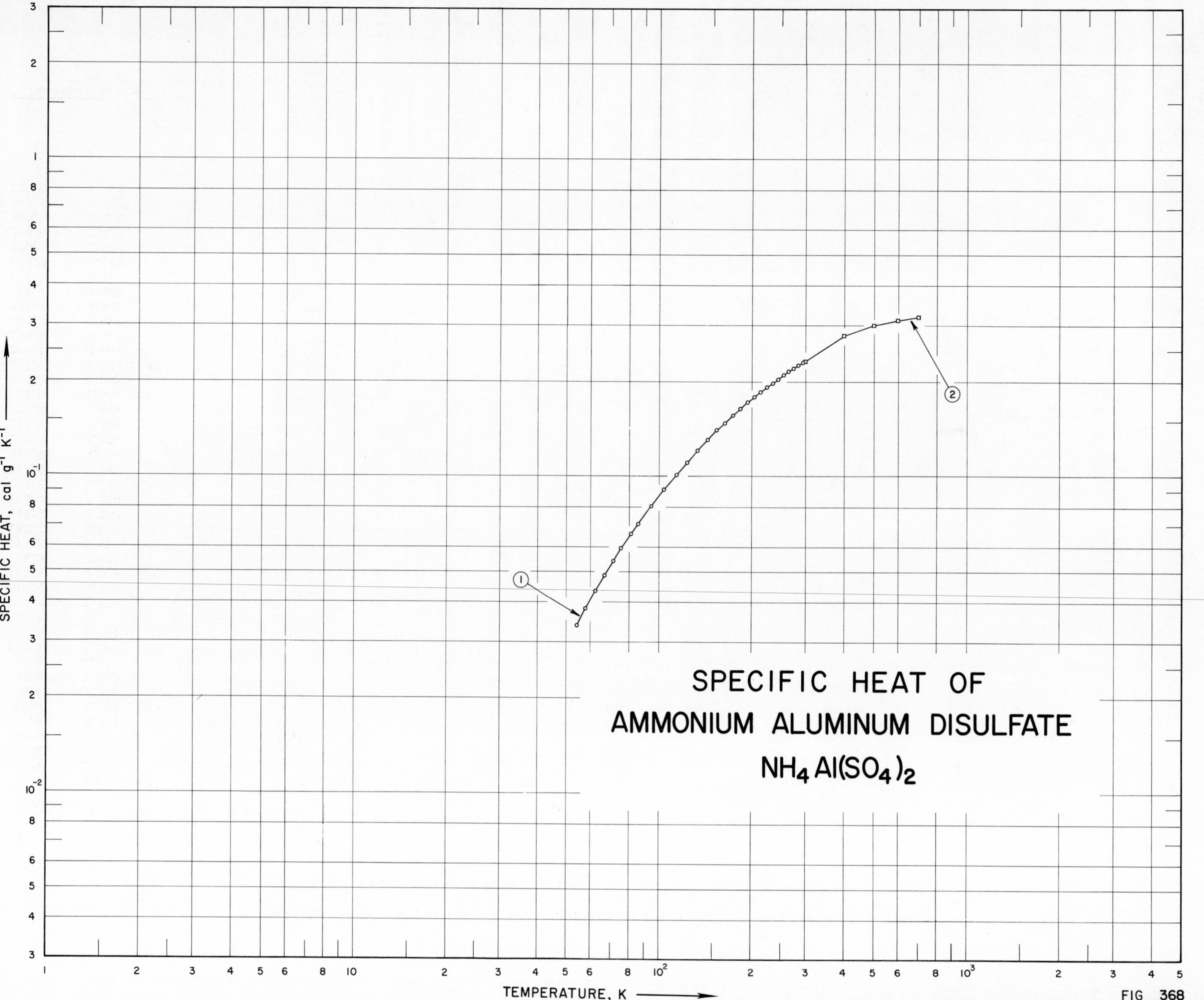

SPECIFIC HEAT OF
AMMONIUM ALUMINUM DISULFATE
$NH_4Al(SO_4)_2$
SPECIFIC HEAT, cal g⁻¹ K⁻¹
TEMPERATURE, K
1
2

FIG 368

SPECIFICATION TABLE NO. 368 SPECIFIC HEAT OF AMMONIUM ALUMINUM DISULFATE, $NH_4Al(SO_4)_2$

[For Data Reported in Figure and Table No. 368]

Curve No.	Ref. No.	Year	Temp. Range, K	Reported Error, %	Name and Specimen Designation	Composition (weight percent), Specifications and Remarks
1	359	1946	55-296			66.37 SO_3, 21.20 Al_2O_3, 10.55 $(HN_4)_2O$ and 0.39 Na_2SO_4.
2	359	1946	298-700			

DATA TABLE NO. 368 SPECIFIC HEAT OF AMMONIUM ALUMINUM DISULFATE, $NH_4Al(SO_4)_2$

[Temperature, T, K; Specific Heat, C_p, Cal $g^{-1} K^{-1}$]

T	C_p
CURVE 1	
54.5	3.367×10^{-2}
58.2	3.796
62.6	4.314
67.1	4.841
71.5	5.364
75.8	5.874
81.5	6.524
86.3	7.043
94.0	7.991
104.5	9.008
115.1	1.008×10^{-1}
124.4	1.099
134.9	1.198
145.8	1.297
155.6	1.384
165.6	1.464
175.8	1.550
185.6	1.623
196.0	1.700
206.0	1.770
215.6	1.836
226.4	1.902
236.0	1.957
246.2	2.014
256.0	2.076
266.3	2.131
276.3	2.178
286.0	2.224
296.2	2.273
CURVE 2	
298.15	2.282×10^{-1}*
300	2.296
400	2.763
500	2.979
600	3.097
700	3.168

* Not shown on plot

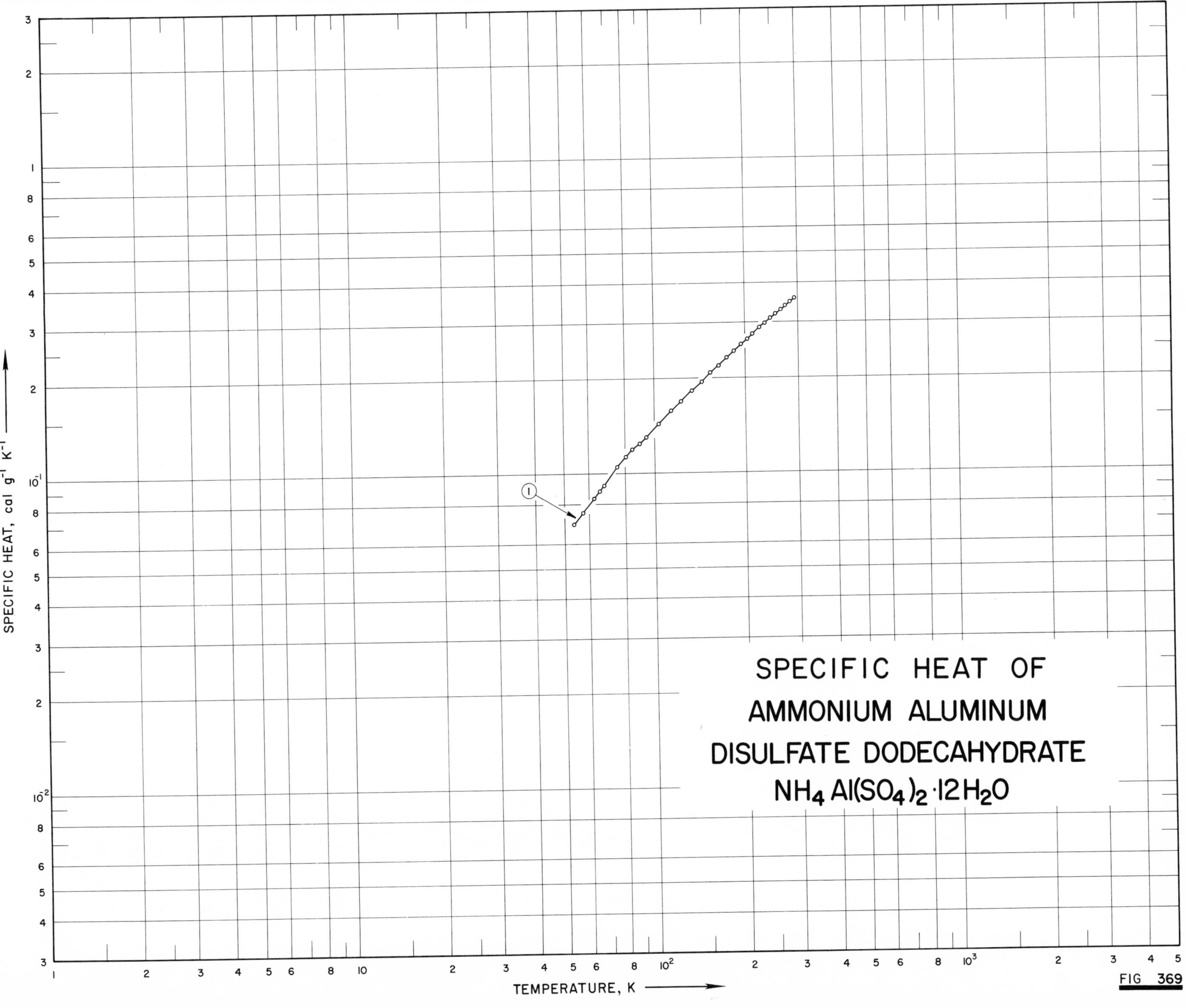
SPECIFIC HEAT OF
AMMONIUM ALUMINUM
DISULFATE DODECAHYDRATE
$NH_4Al(SO_4)_2 \cdot 12H_2O$
1
SPECIFIC HEAT, cal g⁻¹ K⁻¹
TEMPERATURE, K
FIG 369

SPECIFICATION TABLE NO. 369 SPECIFIC HEAT OF AMMONIUM ALUMINUM DISULFATE DODECAHYDRATE, $NH_4Al(SO_4)_2 \cdot 12H_2O$

[For Data Reported in Figure and Table No. 369]

Curve No.	Ref. No.	Year	Temp. Range, K	Reported Error, %	Name and Specimen Designation	Composition (weight percent), Specifications and Remarks
1	359	1946	54-296			11.22 Al_2O_3 (11.24 theo) and 0.20 alkali salts; reagent grade.

DATA TABLE NO. 369 SPECIFIC HEAT OF AMMONIUM ALUMINUM DISULFATE DODECAHYDRATE, $NH_4Al(SO_4)_2 \cdot 12H_2O$

[Temperature, T, K; Specific Heat, C_p, Cal $g^{-1} K^{-1}$]

T	C_p
CURVE 1	
54.0	6.922×10^{-2}
54.9	7.063*
57.8	7.540
59.1	7.741*
63.1	8.383
65.8	8.811
68.1	9.179
75.5	1.050×10^{-1}
75.8	1.060*
81.0	1.126
85.9	1.191
90.1	1.248
94.6	1.309
104.3	1.437
115.2	1.577
124.2	1.692
134.9	1.829
145.3	1.953
155.6	2.078
165.5	2.198
175.7	2.321
185.5	2.431
196.4	2.554
205.8	2.658
210.1	2.711*
214.4	2.757
218.4	2.802*
222.4	2.841*
226.3	2.890
235.7	2.978
246.0	3.084
256.1	3.192
266.0	3.291
276.5	3.393
285.9	3.481
296.1	3.574

* Not shown on plot

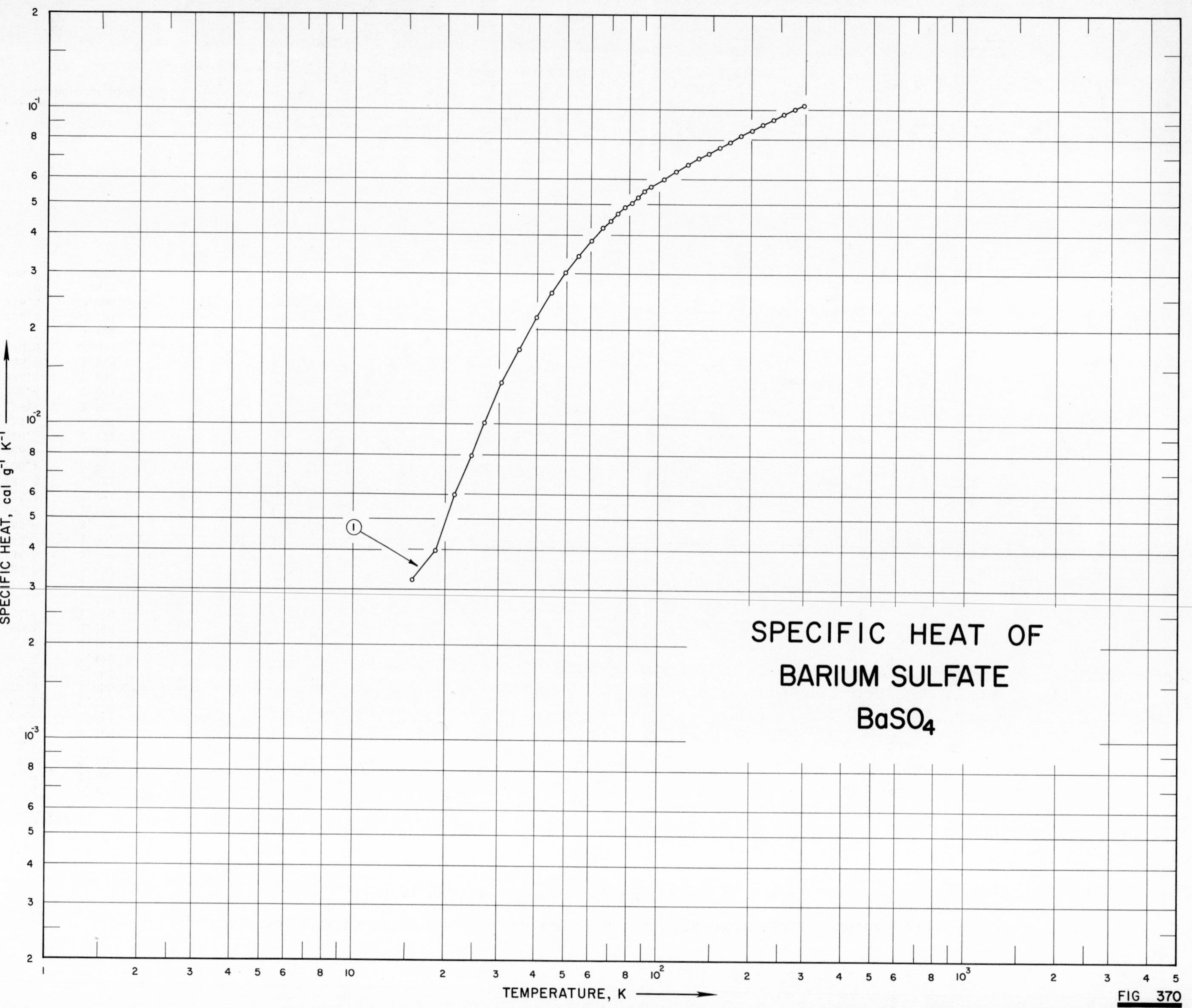
SPECIFIC HEAT OF
BARIUM SULFATE
$BaSO_4$
SPECIFIC HEAT, cal g^{-1} K^{-1}
TEMPERATURE, K
1

FIG 370

SPECIFICATION TABLE NO. 370 SPECIFIC HEAT OF BARIUM SULFATE, $BaSO_4$

[For Data Reported in Figure and Table No. 370]

Curve No.	Ref. No.	Year	Temp. Range, K	Reported Error, %	Name and Specimen Designation	Composition (weight percent), Specifications and Remarks
1	437	1933	16-298			C.P. quality; twice recrystallized.

DATA TABLE NO. 370 SPECIFIC HEAT OF BARIUM SULFATE, $BaSO_4$

[Temperature, T, K; Specific Heat, C_p, Cal $g^{-1} K^{-1}$]

T	C_p
CURVE 1	
15.72	3.213×10^{-3}
18.62	3.985
21.49	5.998
24.42	7.969
26.83	1.011×10^{-2}
30.56	1.367
34.94	1.731
39.83	2.198
44.75	2.626
49.69	3.042
54.95	3.445
60.49	3.852
65.96	4.246
69.61	4.452
73.63	4.687
77.79	4.914
81.73	5.073
85.52	5.287
89.65	5.505
94.22	5.703
104.17	6.045
109.29	6.221*
114.25	6.392
119.44	6.542
124.84	6.718
130.19	6.868*
135.48	7.014
140.83	7.159*
146.18	7.284
151.76	7.459*
158.06	7.613
164.59	7.785*
171.16	7.930
178.13	8.093*
185.78	8.295
193.94	8.487*
201.87	8.620
210.18	8.847*
218.44	8.989
228.03	9.241*
237.28	9.349
246.86	9.391*
256.21	9.683
267.62	1.006×10^{-1}*
CURVE 1 (cont.)	
277.68	1.016×10^{-1}
288.69	1.041*
298.23	1.034

* Not shown on plot

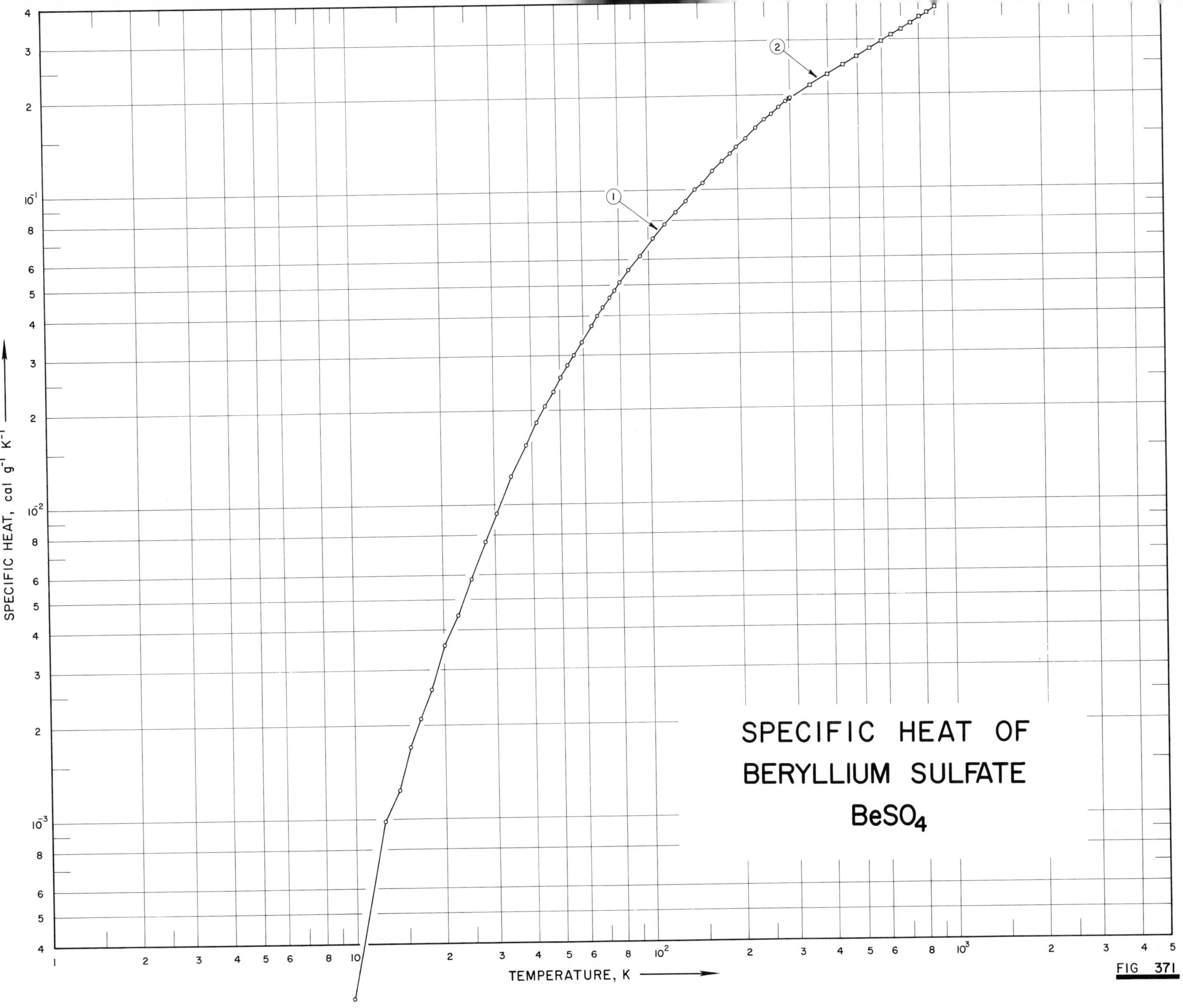
SPECIFIC HEAT OF
BERYLLIUM SULFATE
BeSO4
SPECIFIC HEAT, cal g-1 K-1
TEMPERATURE, K
1
2
FIG 371

SPECIFICATION TABLE NO. 371 SPECIFIC HEAT OF BERYLLIUM SULFATE, $BeSO_4$

[For Data Reported in Figure and Table No. 371]

Curve No.	Ref. No.	Year	Temp. Range, K	Reported Error, %	Name and Specimen Designation	Composition (weight percent), Specifications and Remarks
1	438	1963	10-301			> 99.889 $BeSO_4$, 0.01-0.10 Mg, 0.001-0.01 Al, 0.001-0.01 Fe and 0.0001-0.001 Mn; ground to fine powder and heated overnight to 1000 F to insure complete dehydration.
2	438	1963	298-900			Same as above.

DATA TABLE NO. 371 SPECIFIC HEAT OF BERYLLIUM SULFATE, $BeSO_4$

[Temperature, T, K; Specific Heat, C_p, Cal $g^{-1} K^{-1}$]

T	C_p
CURVE 1	
9.71	2.67 x 10^{-4}
12.56	9.80
14.07	1.24 x 10^{-3}
15.39	1.69
16.73	2.09
18.14	2.58
20.08	3.58
22.30	4.46
24.95	5.84
27.88	7.65
30.31	9.42
34.00	1.235 x 10^{-2}
38.40	1.565
41.88	1.838
44.83	2.073
47.68	2.305
50.46	2.556
53.18	2.787
55.85	3.010
56.29	3.042*
59.42	3.319
60.58	3.426*
64.08	3.729
67.31	4.001
70.62	4.279
74.01	4.564
77.21	4.833
80.48	5.115
83.84	5.397*
86.58	5.617
90.47	5.925*
94.20	6.225
95.96	6.352*
97.78	6.499*
99.98	6.667*
104.54	7.031
107.68	7.264*
111.33	7.535*
114.87	7.800
117.59	8.007*
120.96	8.270*
124.23	8.516
127.43	8.750*
130.82	9.002*
CURVE 1 (cont.)	
134.38	9.254 x 10^{-2}
137.87	9.520*
141.28	9.769*
144.70	1.002 x 10^{-1}
148.35	1.027*
153.23	1.061
158.15	1.096*
162.57	1.125*
166.49	1.154
170.35	1.182*
174.13	1.208*
177.86	1.232
184.17	1.275*
189.15	1.309
194.67	1.347*
198.88	1.373
202.43	1.396*
205.94	1.420*
212.83	1.464
219.70	1.507*
225.79	1.543*
231.12	1.575
236.52	1.606*
241.29	1.635*
247.94	1.673
254.46	1.711*
260.86	1.747
266.57	1.779*
272.02	1.810*
275.34	1.829
281.30	1.864*
286.04	1.889*
290.89	1.913
296.68	1.941*
299.20	1.964*
301.28	1.965
CURVE 2	
298.15	1.949 x 10^{-1}
300	1.959*
350	2.154
400	2.327
450	2.489
CURVE 2 (cont.)	
500	2.642
550	2.791
600	2.937
650	3.080
700	3.221
750	3.361
800	3.500
850	3.639
900	3.777

* Not shown on plot

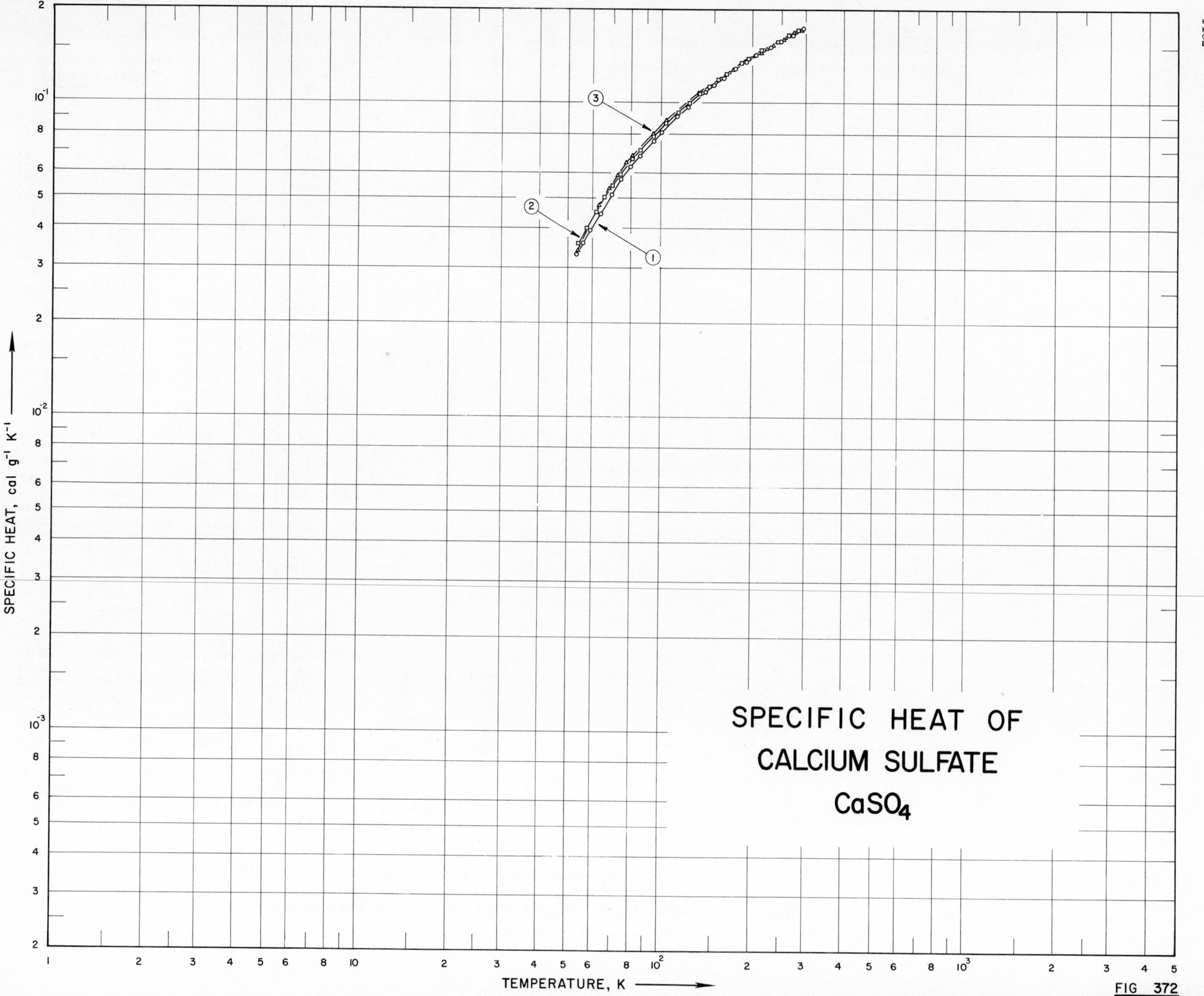
SPECIFIC HEAT OF
CALCIUM SULFATE
CaSO4
SPECIFIC HEAT, cal g⁻¹ K⁻¹
TEMPERATURE, K
1
2
3

FIG 372

SPECIFICATION TABLE NO. 372 SPECIFIC HEAT OF CALCIUM SULFATE, $CaSO_4$

[For Data Reported in Figure and Table No. 372]

Curve No.	Ref. No.	Year	Temp. Range, K	Reported Error, %	Name and Specimen Designation	Composition (weight percent), Specifications and Remarks
1	439	1941	54-296			0.09 $CaSO_4 \cdot 2H_2O$; natural anhydride from Arden, Nevada.
2	439	1941	54-294		α - soluble anhydrite	0.02 H_2O.
3	439	1941	54-295		β - soluble anhydrite	0.25 H_2O.

DATA TABLE NO. 372 SPECIFIC HEAT OF CALCIUM SULFATE, $CaSO_4$

[Temperature, T, K; Specific Heat, C_p, Cal $g^{-1} K^{-1}$]

T	C_p	T	C_p
CURVE 1		CURVE 2 (cont.)	
53.5	3.313 x 10^{-2}	195.5	1.385 x 10^{-1}
56.1	3.599	205.8	1.426*
59.3	3.967	215.6	1.476
63.9	4.459	232.6	1.538*
69.1	5.127	243.3	1.579
74.4	5.759	254.1	1.617*
79.6	6.324	265.5	1.651
85.6	6.817	275.8	1.691*
94.6	7.595	285.3	1.721
101.0	8.146	294.4	1.747*
113.8	9.138		
123.7	9.762	CURVE 3	
140.7	1.094 x 10^{-1}		
149.8	1.149	54.1	3.416 x 10^{-2}
162.0	1.213	58.5	3.989*
177.1	1.295	63.4	4.774
192.2	1.367	68.0	5.362
207.4	1.432	72.5	5.935
230.3	1.523	77.1	6.493
249.8	1.588	80.9	6.824
272.8	1.665	85.4	7.184*
278.7	1.694*	94.6	8.028
295.7	1.744	104.5	8.851
		114.3	9.564*
CURVE 2		123.9	1.021 x 10^{-1}*
		133.8	1.082
54.1	3.585 x 10^{-2}	143.7	1.134*
57.7	4.011	154.1	1.195*
61.8	4.510	163.8	1.241*
65.9	5.017	174.3	1.289
69.8	5.480	183.9	1.337*
73.9	5.942	193.7	1.380*
81.0	6.670	204.0	1.426
85.8	7.125	214.1	1.465
94.5	7.882	223.9	1.506
104.8	8.690	235.0	1.545
114.7	9.424	245.5	1.584*
125.1	1.016 x 10^{-1}	255.6	1.619
135.0	1.079	266.0	1.653*
145.5	1.139	275.7	1.684
155.4	1.191	285.1	1.712*
165.2	1.242	294.9	1.730
175.8	1.299*		
185.6	1.344		

* Not shown on plot

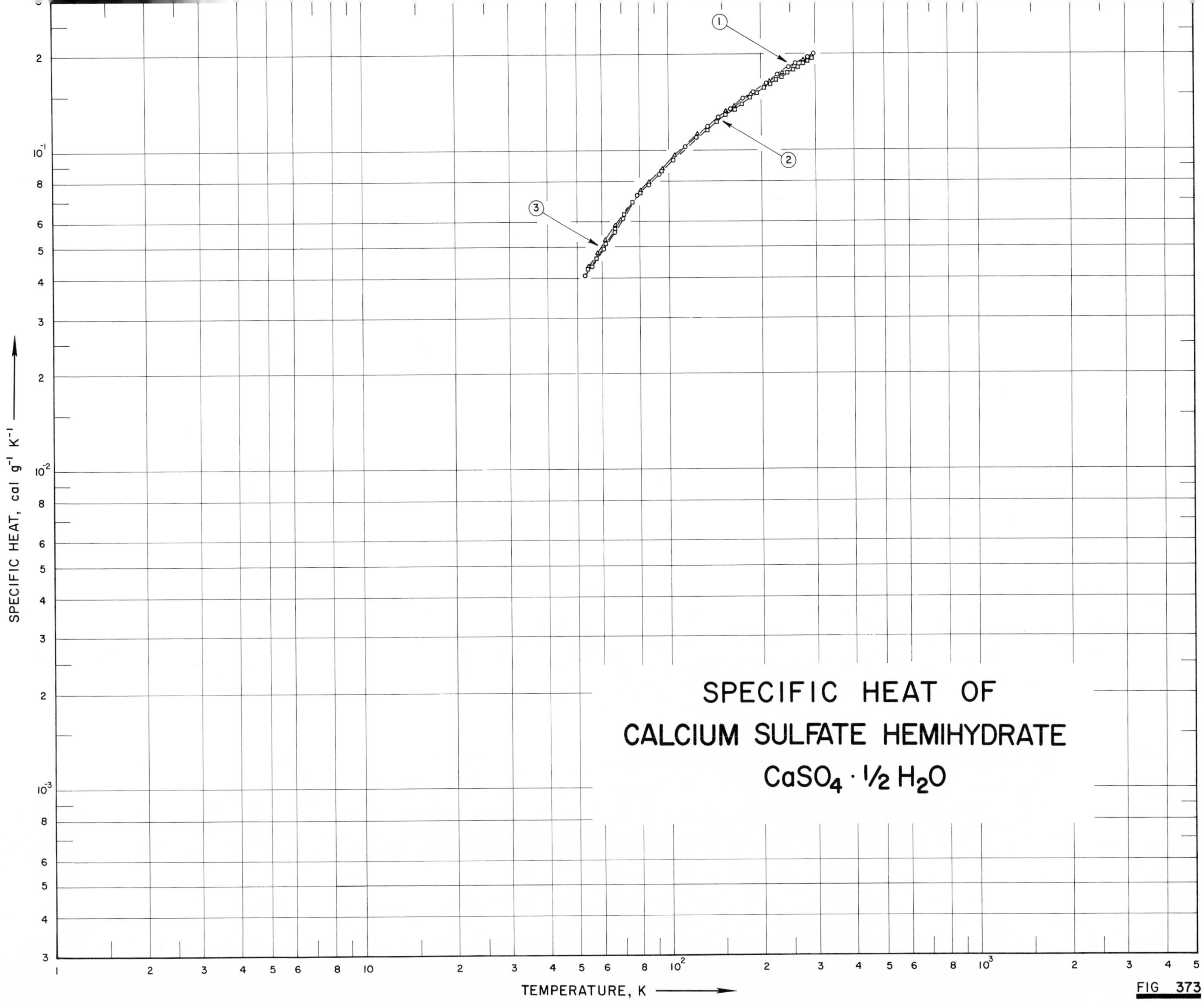

SPECIFIC HEAT OF CALCIUM SULFATE HEMIHYDRATE

$CaSO_4 \cdot 1/2\, H_2O$

SPECIFIC HEAT, cal g^{-1} K^{-1}

TEMPERATURE, K

FIG 373

SPECIFICATION TABLE NO. 373 SPECIFIC HEAT OF CALCIUM SULFATE HEMIHYDRATE, $CaSO_4\cdot\frac{1}{2}H_2O$

[For Data Reported in Figure and Table No. 373]

Curve No.	Ref. No.	Year	Temp. Range, K	Reported Error, %	Name and Specimen Designation	Composition (weight percent), Specifications and Remarks
1	439	1941	53-297		Hemihydrate	99.74 $CaSO_4\cdot\frac{1}{2}H_2O$, 0.15 insoluble and 0.07 $CaCO_3$, trace R_2O_3; 25% α and 75% β - hemihydrate; sample prepared from San Marcos, Mexico gypsum.
2	439	1941	54-295		α-hemihydrate	6.23 H_2O; density = 2.757 g cm^{-3}.
3	439	1941	55-294		metastable β-hemihydrate	6.25 H_2O; annealed 3 days; density = 2.637 g cm^{-3}.

DATA TABLE NO. 373 SPECIFIC HEAT OF CALCIUM SULFATE HEMIHYDRATE, $CaSO_4 \cdot \frac{1}{2}H_2O$

[Temperature, T, K; Specific Heat, C_p, Cal $g^{-1} K^{-1}$]

T	C_p
CURVE 1	
53.3	4.099×10^{-2}
55.8	4.368
60.9	4.960
66.0	5.594
70.3	6.180
78.8	7.344
85.8	7.875
92.7	8.508
102.6	9.425
112.8	1.031×10^{-1}
122.7	1.102
134.9	1.199
146.5	1.286
159.8	1.368
176.4	1.472
188.0	1.535
208.0	1.642
226.7	1.743
247.1	1.830
260.1	1.890
272.0	1.922*
284.8	1.977
297.0	2.025
CURVE 2	
54.4	4.189×10^{-2}
57.9	4.630
61.9	5.126
66.4	5.739
71.0	6.366
75.6	6.938
80.9	7.461
85.7	7.854*
94.7	8.681
104.2	9.507*
114.0	1.031×10^{-1}*
123.3	1.100*
133.6	1.175
143.8	1.242
153.7	1.308
164.0	1.366
173.9	1.417
184.1	1.476
CURVE 2 (cont.)	
194.5	1.527×10^{-1}
205.1	1.580
215.5	1.635
225.3	1.674
234.9	1.718
245.6	1.766
256.1	1.813
265.7	1.848
275.9	1.889
285.4	1.920
294.6	1.957
CURVE 3	
54.5	4.402×10^{-2}
58.0	4.843
61.9	5.346
66.4	5.904
70.8	6.449*
75.0	6.924*
81.3	7.578
86.0	8.033
95.2	8.894
105.0	9.762
114.7	1.055*
124.6	1.131
135.0	1.209*
144.6	1.273*
154.8	1.339
164.6	1.396
175.0	1.456*
184.9	1.513
194.7	1.564*
204.9	1.619*
214.8	1.665
225.3	1.718*
234.9	1.764
245.1	1.802*
255.6	1.858*
265.9	1.898*
275.6	1.941
285.1	1.985*
294.3	2.025*

* Not shown on plot

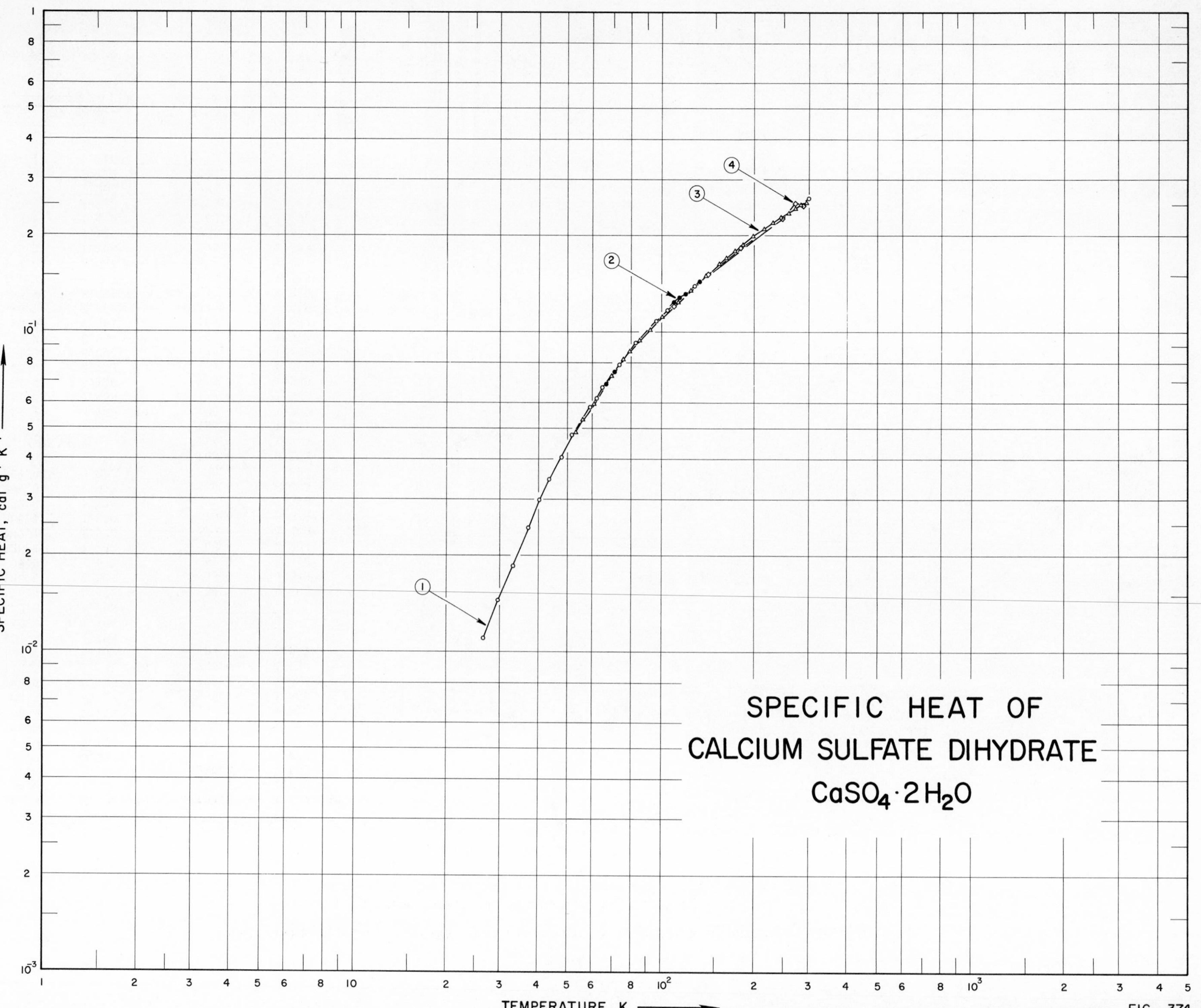

SPECIFIC HEAT OF
CALCIUM SULFATE DIHYDRATE
$CaSO_4 \cdot 2H_2O$
SPECIFIC HEAT, cal g⁻¹ K⁻¹
TEMPERATURE, K
1
2
3
4

FIG 374

SPECIFICATION TABLE NO. 374 SPECIFIC HEAT OF CALCIUM SULFATE DIHYDRATE, $CaSO_4 \cdot 2H_2O$

[For Data Reported in Figure and Table No. 374]

Curve No.	Ref. No.	Year	Temp. Range, K	Reported Error, %	Name and Specimen Designation	Composition (weight percent), Specifications and Remarks
1	437	1933	19-302			20.85 H_2O (20.93 theo); Baker Adamson C.P. $CaSO_4 \cdot 2H_2O$, small crystals; dried and heated to 520 C.
2	437	1933	62-274			20.88 H_2O (20.93 theo); natural gypsum, large single crystal.
3	439	1941	53-295		selenite	0.20 $CaCO_3$; deposit from Ceerlash, Nevada.
4	439	1941	94-298		gypsum	0.37 anhydrite; from deposit near Arden, Nevada.

DATA TABLE NO. 374 SPECIFIC HEAT OF CALCIUM SULFATE DIHYDRATE, $CaSO_4 \cdot 2H_2O$

[Temperature, T, K; Specific Heat, C_p, Cal g^{-1} K^{-1}]

CURVE 1

T	C_p
18.70	3.950×10^{-3}
26.71	1.104×10^{-2}
29.64	1.440
33.18	1.859
37.11	2.445
40.36	2.991
43.77	3.479
47.84	4.054
51.98	4.757
56.56	5.419*
59.47	5.826
62.00	6.232
65.34	6.703
69.29	7.324*
73.72	7.911
78.41	8.544*
83.24	9.171
88.04	9.775*
92.37	1.034×10^{-1}*
96.80	1.079
101.26	1.131*
105.59	1.170
109.84	1.211*
114.04	1.250*
118.22	1.292*
123.62	1.333*
128.39	1.399
134.97	1.460*
140.21	1.499*
145.36	1.550*
150.72	1.598*
155.86	1.650*
160.80	1.680*
165.73	1.720*
170.86	1.763*
179.79	1.840*
185.78	1.878*
191.39	1.931*
197.61	1.971*
203.45	2.011*
210.22	2.071*
217.27	2.110*
224.91	2.161*

CURVE 1 (cont.)

T	C_p
231.90	2.226×10^{-1}*
238.23	2.253*
245.64	2.297*
254.02	2.344*
263.29	2.382*
272.18	2.457*
282.11	2.476*
292.67	2.572*
302.10	2.624

CURVE 2

T	C_p
62.40	6.267×10^{-2}*
66.74	6.883
70.84	7.498
75.20	8.085*
80.05	8.707*
84.70	9.305*
101.38	1.128×10^{-1}*
105.96	1.181*
110.77	1.230
115.89	1.275
120.88	1.321
127.59	1.392
133.12	1.443
138.32	1.493*
143.42	1.538*
148.83	1.589*
153.84	1.635*
158.83	1.675*
162.93	1.713*
168.21	1.760*
173.77	1.817*
179.48	1.852*
185.26	1.900*
190.98	1.933*
196.78	1.978*
203.57	2.023*
216.12	2.111*
231.29	2.203*
239.57	2.229*
247.81	2.287*
256.44	2.337*
265.23	2.401*
274.13	2.454*

CURVE 3

T	C_p
53.0	4.856×10^{-2}
56.4	5.344
60.8	5.953
69.3	7.283
75.5	8.213
79.1	8.718
84.9	9.415
92.3	1.018×10^{-1}
101.2	1.122
113.1	1.246
125.2	1.363
139.5	1.508
153.6	1.633
162.2	1.706
173.1	1.799
184.2	1.882
198.7	1.922
215.9	2.106
229.9	2.206
243.0	2.280
258.7	2.367
271.0	2.439
283.4	2.500
295.1	2.545

CURVE 4

T	C_p
93.6	1.022×10^{-1}*
110.7	1.210
142.2	1.518
181.0	1.843
217.0	2.094*
246.0	2.277
272.1	2.543
289.4	2.496
297.5	2.540*

* Not shown on plot

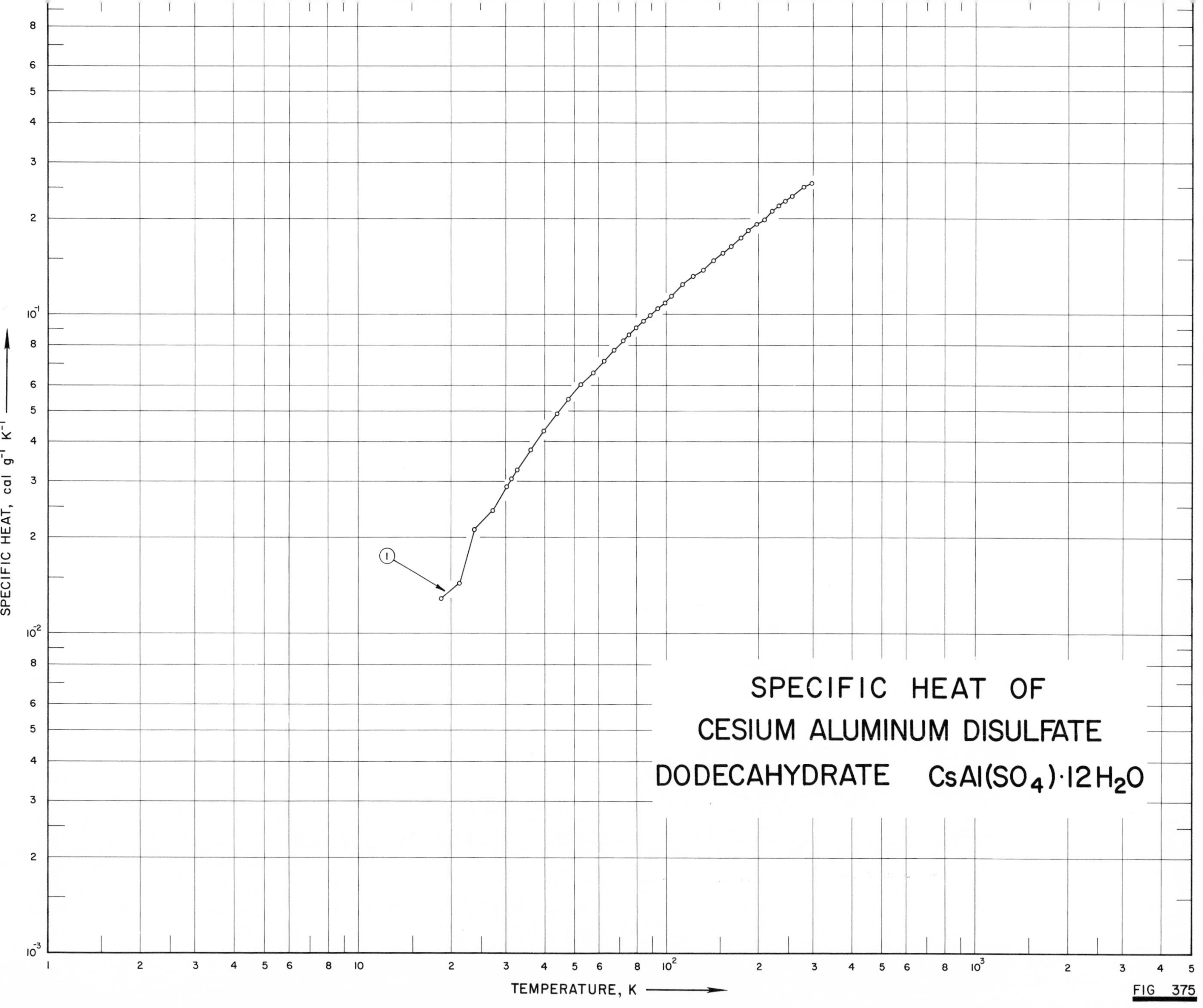
SPECIFIC HEAT OF
CESIUM ALUMINUM DISULFATE
DODECAHYDRATE CsAl(SO4)·12H2O
SPECIFIC HEAT, cal g-1 K-1
TEMPERATURE, K
1
FIG 375

SPECIFICATION TABLE NO. 375 SPECIFIC HEAT OF CESIUM ALUMINUM DISULFATE DODECAHYDRATE, $CsAl(SO_4)_2 \cdot 12H_2O$

[For Data Reported in Figure and Table No. 375]

Curve No.	Ref. No.	Year	Temp. Range, K	Reported Error, %	Name and Specimen Designation	Composition (weight percent), Specifications and Remarks
1	440	1928	19-298	0.25 average deviation	Cesium alum	37.98 H_2O (38.05 theo); density (20C) = 1.978 g cm^{-3}.

DATA TABLE NO. 375 SPECIFIC HEAT OF CESIUM ALUMINUM DISULFATE DODOCAHYDRATE, $CsAl(SO_4)_2 \cdot 12H_2O$

[Temperature, T, K; Specific Heat, C_p, Cal $g^{-1} K^{-1}$]

T	C_p	T	C_p
CURVE 1		CURVE 1 (cont.)	
18.71	1.29 x 10^{-2}	227.88	2.122 x 10^{-1}*
21.26	1.46	233.78	2.208
23.70	2.14	239.29	2.247*
27.23	2.44	245.25	2.275
30.25	2.90	251.62	2.321*
31.31	3.08	257.77	2.356
32.65	3.29	263.77	2.409*
36.13	3.79	281.30	2.521
39.77	4.34	286.73	2.544*
44.17	4.94	292.42	2.578*
48.60	5.46	297.84	2.586
52.87	6.06		
57.71	6.63		
62.68	7.20		
67.57	7.80		
72.63	8.33		
75.88	8.67		
80.07	9.12		
84.84	9.56		
90.26	9.96		
94.20	1.051 x 10^{-1}		
99.62	1.098		
104.61	1.150		
109.24	1.189*		
113.61	1.254		
117.80	1.270*		
122.52	1.322		
127.68	1.363*		
132.57	1.379		
137.90	1.428*		
143.63	1.488		
149.03	1.518*		
154.17	1.569		
159.43	1.603*		
164.62	1.642		
169.46	1.684*		
176.26	1.745		
181.56	1.796*		
187.12	1.840		
192.82	1.882*		
198.19	1.925		
205.72	1.955*		
210.20	1.997		
215.83	2.091*		
221.86	2.124		

* Not shown on plot

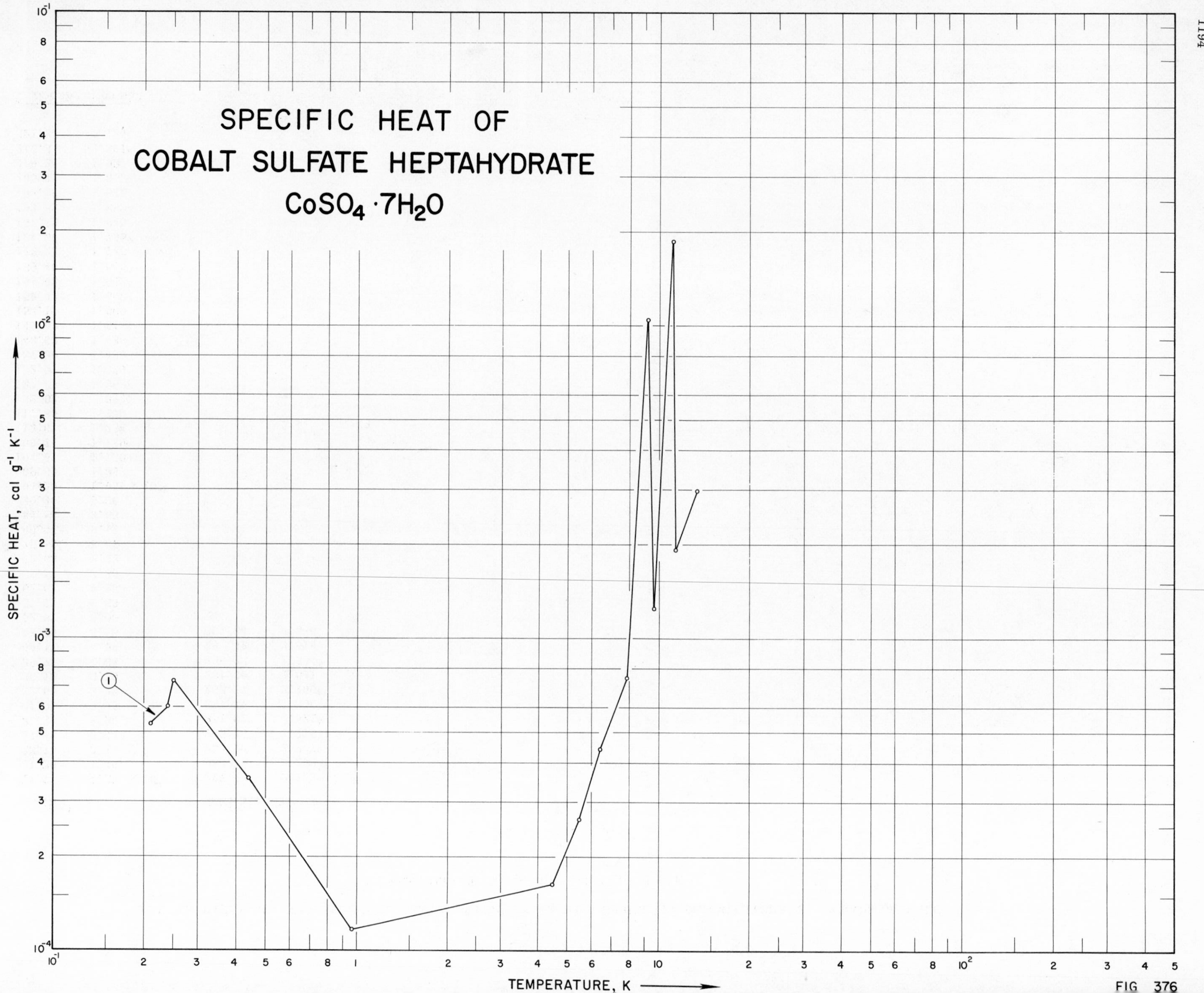
SPECIFIC HEAT OF
COBALT SULFATE HEPTAHYDRATE
CoSO4·7H2O
SPECIFIC HEAT, cal g-1 K-1
TEMPERATURE, K
1

FIG 376

SPECIFICATION TABLE NO. 376 SPECIFIC HEAT OF COBALT SULFATE HEPTAHYDRATE, $CoSO_4 \cdot 7H_2O$

[For Data Reported in Figure and Table No. 376]

Curve No.	Ref. No.	Year	Temp. Range, K	Reported Error, %	Name and Specimen Designation	Composition (weight percent), Specifications and Remarks
1	441	1949	0.2-13			44.76 H_2O (44.86 theo); powdered crystalline.

DATA TABLE NO. 376 SPECIFIC HEAT OF COBALT SULFATE HEPTAHYDRATE, $CoSO_4 \cdot 7H_2O$

[Temperature, T, K; Specific Heat, C_p, Cal $g^{-1} K^{-1}$]

T	C_p
CURVE 1	
Series I	
0.25	7.293×10^{-4}
4.48	1.636
5.46	2.633
6.41	4.447
7.88	7.471
9.17	1.053×10^{-2}
11.19	1.875
Series II	
0.21	5.336×10^{-4}
0.24	6.048
0.44	3.558
0.96	1.174
Series III	
9.66	1.256×10^{-3}
11.37	1.925
13.33	2.974

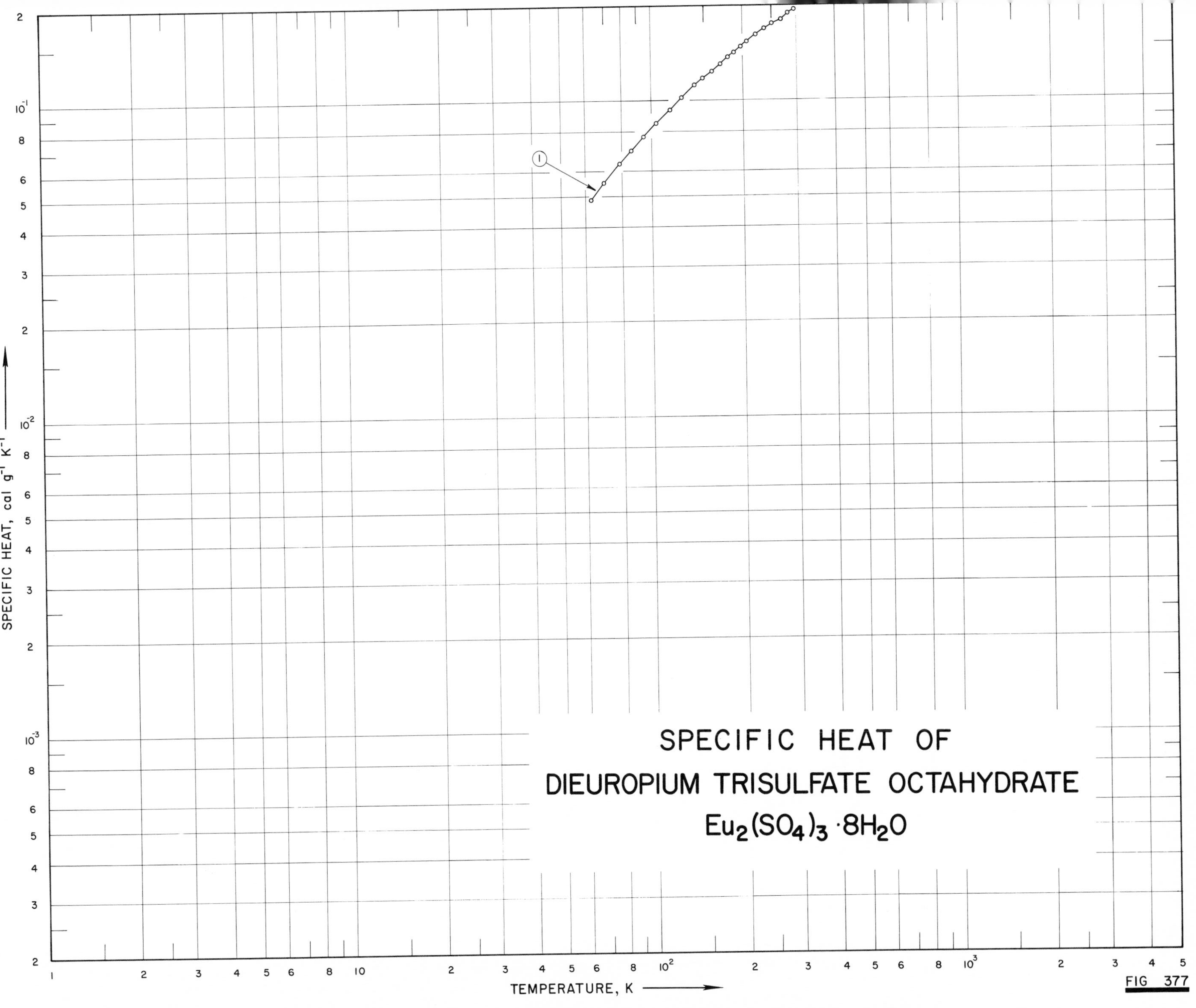

SPECIFIC HEAT OF
DIEUROPIUM TRISULFATE OCTAHYDRATE
$Eu_2(SO_4)_3 \cdot 8H_2O$
SPECIFIC HEAT, cal g^{-1} K^{-1}
TEMPERATURE, K
1

FIG 377

SPECIFICATION TABLE NO. 377 SPECIFIC HEAT OF DIEUROPIUM TRISULFATE OCTAHYDRATE, $Eu_2(SO_4)_3 \cdot 8H_2O$

[For Data Reported in Figure and Table No. 377]

Curve No.	Ref. No.	Year	Temp. Range, K	Reported Error, %	Name and Specimen Designation	Composition (weight percent), Specifications and Remarks
1	442	1942	63-295	0.3-0.5		0.01 Nd; fine crystalline powder.

DATA TABLE NO. 377 SPECIFIC HEAT OF DIEUROPIUM TRISULFATE OCTAHYDRATE, $Eu_2(SO_4)_3 \cdot 8H_2O$

[Temperature, T, K; Specific Heat, C_p, Cal $g^{-1} K^{-1}$]

T	C_p
CURVE 1	
62.83	4.949×10^{-2}
65.67	5.227*
69.49	5.564
74.01	5.949*
78.55	6.372
82.68	6.754*
86.40	7.032
94.51	7.760
99.61	8.213*
104.41	8.585
110.29	9.105*
116.11	9.476
121.16	9.901*
127.37	1.038×10^{-1}
134.48	1.085*
139.90	1.133
144.27	1.168*
148.82	1.191
153.90	1.223*
159.48	1.259
164.87	1.301*
169.93	1.322
174.85	1.364*
179.49	1.383
183.90	1.404*
188.27	1.429
193.03	1.455*
197.96	1.499
202.71	1.524*
207.51	1.551
212.21	1.573*
221.12	1.639
225.35	1.657*
226.91	1.661*
231.65	1.680*
236.70	1.719
242.61	1.726*
246.01	1.753*
250.78	1.774
255.83	1.808*
264.29	1.827*
268.46	1.825
272.84	1.863*
277.28	1.896*
CURVE 1 (cont.)	
281.78	1.911×10^{-1}
286.22	1.923*
290.74	1.960*
295.23	1.967

* Not shown on plot

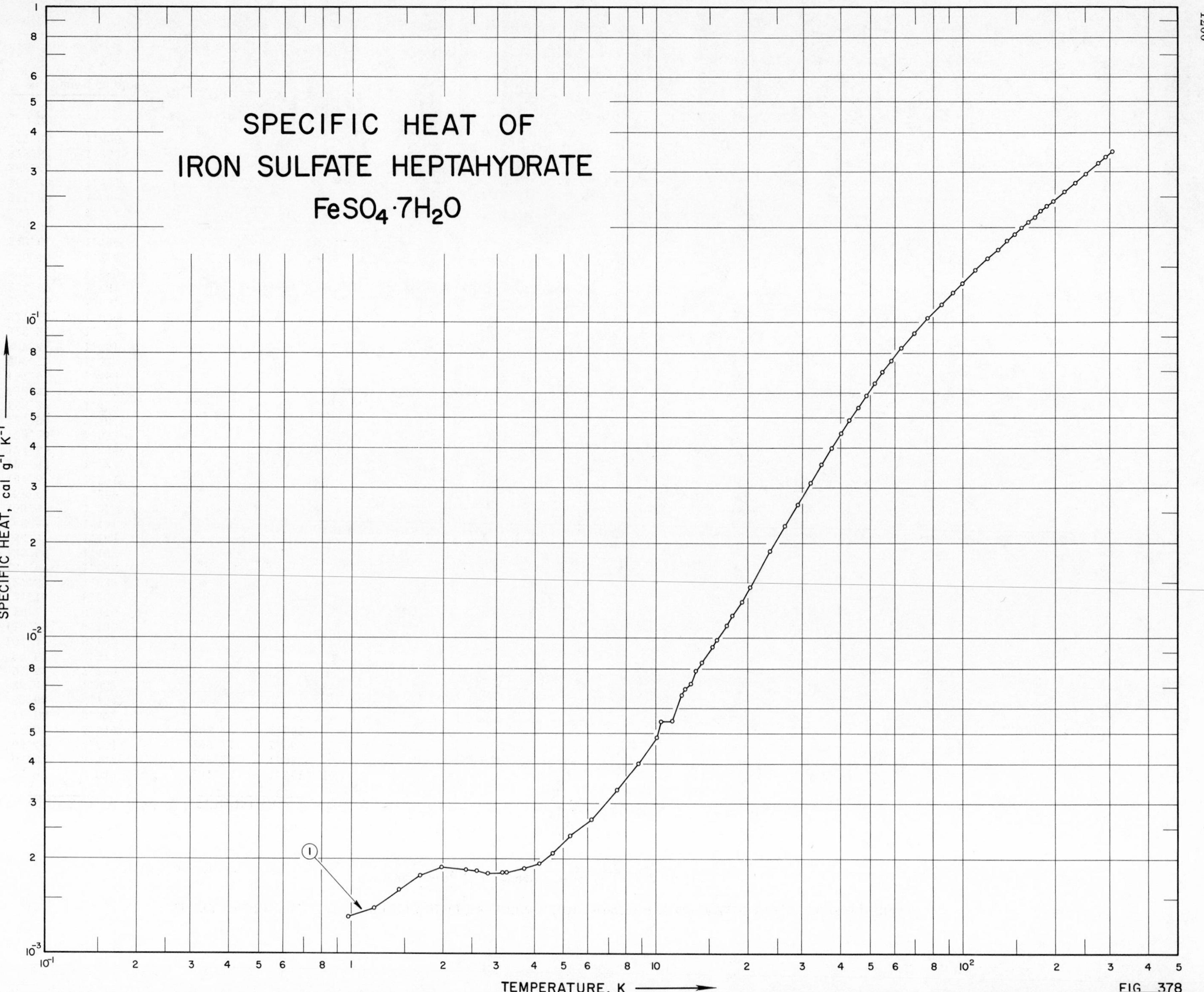
SPECIFIC HEAT OF
IRON SULFATE HEPTAHYDRATE
$FeSO_4 \cdot 7H_2O$
SPECIFIC HEAT, cal g^{-1} K^{-1}
TEMPERATURE, K
1
FIG 378

SPECIFICATION TABLE NO. 378 SPECIFIC HEAT OF IRON SULFATE HEPTAHYDRATE, $FeSO_4 \cdot 7H_2O$

[For Data Reported in Figure and Table No. 378]

Curve No.	Ref. No.	Year	Temp. Range, K	Reported Error, %	Name and Specimen Designation	Composition (weight percent), Specifications and Remarks
1	443	1949	13-308			0.05 alkali and alkaline earth, 0.01 Mn, 0.005 Ni, 0.004 Co and 0.003 Cu; prepared by recrystallizing commercial low manganese and analytical reagent grade $FeSO_4 \cdot 7H_2O$.

DATA TABLE NO. 378 SPECIFIC HEAT OF IRON SULFATE HEPAHYDRATE, $FeSO_4 \cdot 7H_2O$

[Temperature, T, K; Specific Heat, C_p, Cal $g^{-1} K^{-1}$]

T	C_p	T	C_p
CURVE 1 Series I		CURVE 1 (cont.)	
		206.22	2.516×10^{-1}*
92.94	1.247×10^{-1}	215.43	2.604
99.54	1.330	224.12	2.691*
277.55	3.195	232.81	2.778
284.11	3.255*	242.74	2.869*
292.53	3.329	252.01	2.964
		260.80	3.039*
Series II		269.80	3.129*
		279.79	3.233*
13.14	7.193×10^{-3}	289.99	3.331*
14.21	8.416	299.37	3.404*
15.82	9.891	307.67	3.476
17.75	1.180×10^{-2}		
20.32	1.460	Series III	
23.64	1.895		
26.45	2.270	140.13	1.808×10^{-1}
29.16	2.651	147.64	1.896
31.99	3.100		
34.72	3.568		
37.47	4.007		
40.02	4.453		
42.72	4.902		
45.64	5.388		
48.54	5.881		
51.71	6.431		
54.87	6.988		
58.50	7.568		
63.40	8.323		
69.59	9.265		
76.90	1.032×10^{-1}		
85.07	1.144		
93.53	1.244*		
102.38	1.367*		
110.98	1.472		
121.50	1.597		
131.67	1.716		
139.99	1.822		
147.92	1.919		
156.33	1.992		
164.30	2.079		
172.46	2.167		
180.67	2.263		
188.94	2.340		
197.97	2.425		

* Not shown on plot

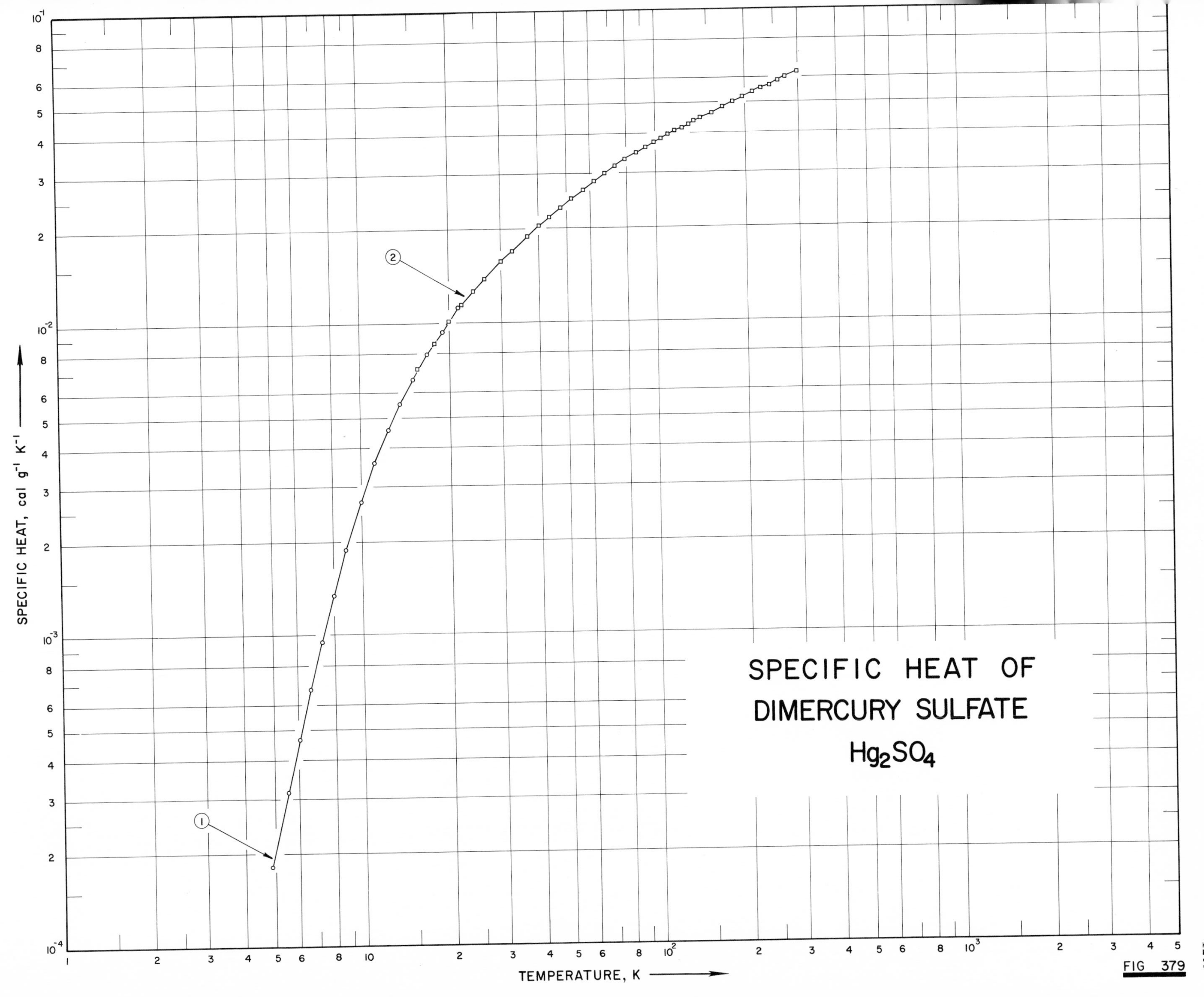
SPECIFIC HEAT OF
DIMERCURY SULFATE
Hg_2SO_4
SPECIFIC HEAT, cal g^{-1} K^{-1}
TEMPERATURE, K
1
2
FIG 379

SPECIFICATION TABLE NO. 379 SPECIFIC HEAT OF DIMERCURY SULFATE, Hg_2SO_4

[For Data Reported in Figure and Table No. 379]

Curve No.	Ref. No.	Year	Temp. Range, K	Reported Error, %	Name and Specimen Designation	Composition (weight percent), Specifications and Remarks
1	444	1960	5-21			Ratio: $SO_4^{-2}/Hg_2SO_4 = 0.19316$ (0.19317 theo).
2	445	1962	16-299			Same as above.

DATA TABLE NO. 379 SPECIFIC HEAT OF DIMERCURY SULFATE, Hg_2SO_4

[Temperature, T, K; Specific Heat, C_p, Cal $g^{-1} K^{-1}$]

T	C_p
CURVE 1	
Series I	
11.156	3.592×10^{-3}
12.412	4.597
13.624	5.578
15.089	6.690
16.869	8.036
19.016	9.441
21.485	1.133×10^{-2}
Series II	
4.851	1.824×10^{-4}
5.523	3.163
6.063	4.673
6.637	6.759
7.291	9.586
8.016	1.362×10^{-3}
8.840	1.892
9.967	2.717

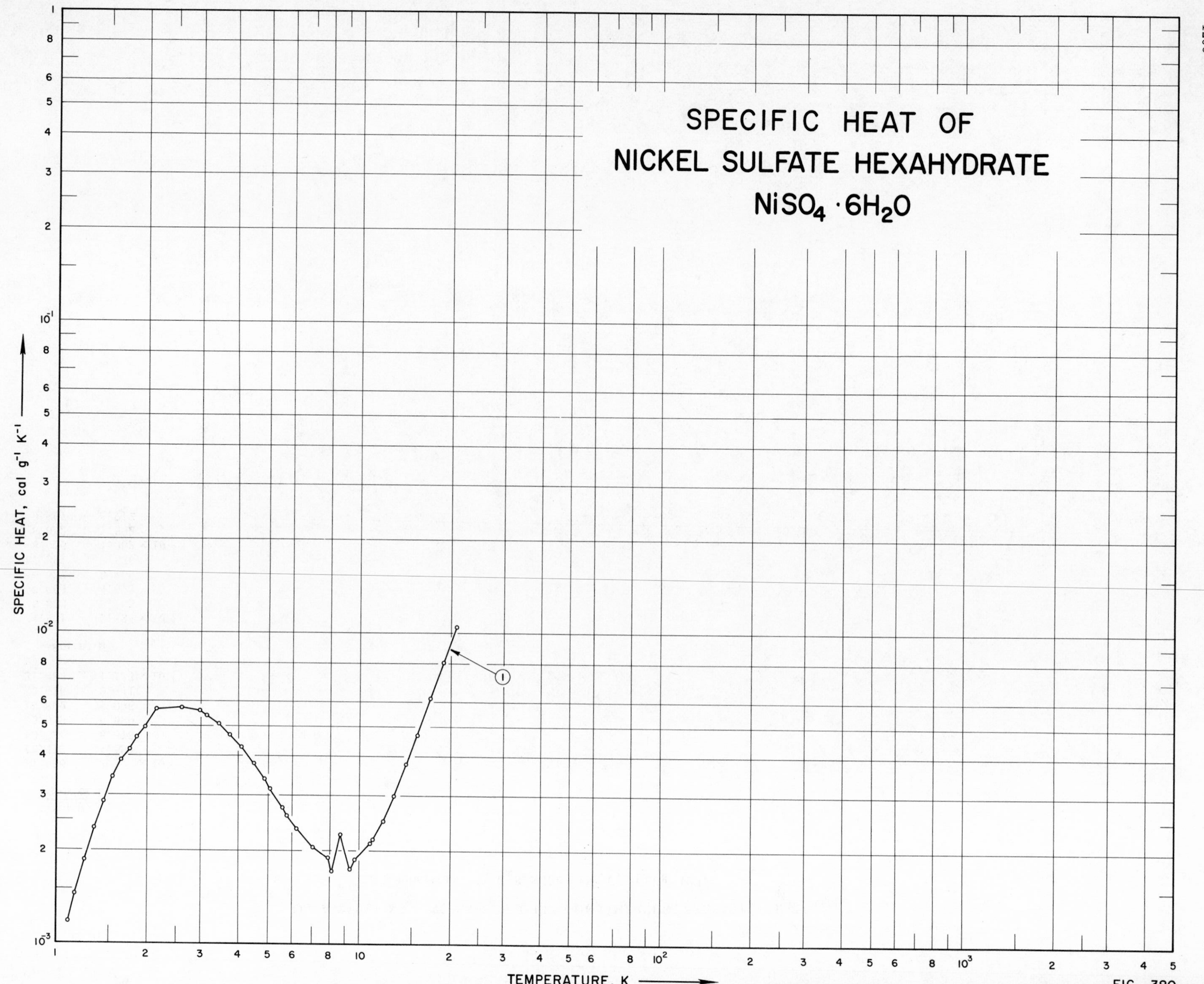
SPECIFIC HEAT OF
NICKEL SULFATE HEXAHYDRATE
$NiSO_4 \cdot 6H_2O$
SPECIFIC HEAT, cal g^{-1} K^{-1}
TEMPERATURE, K
1

FIG 380

SPECIFICATION TABLE NO. 380 SPECIFIC HEAT OF NICKEL SULFATE HEXAHYDRATE, $NiSO_4 \cdot 6H_2O$

[For Data Reported in Figure and Table No. 380]

Curve No.	Ref. No.	Year	Temp. Range, K	Reported Error, %	Name and Specimen Designation	Composition (weight percent), Specifications and Remarks
1	446	1963	5-11			0.04 Mg, 0.02 Co, 0.003 Al, 0.003 Cu, 0.001 Fe and 0.001 Si; tetragonal symmetry.

DATA TABLE NO. 380 SPECIFIC HEAT OF NICKEL SULFATE HEXAHYDRATE, $NiSO_4 \cdot 6H_2O$

[Temperature, T, K; Specific Heat, C_p, Cal $g^{-1} K^{-1}$]

T	C_p
CURVE 1	
Series I	
5.09	3.142×10^{-3}
5.60	2.735
6.25	2.340
7.05	2.035
7.90	1.898
8.68	2.267
9.70	1.868
11.12	2.172
Series II	
10.87	2.115×10^{-3}
12.00	2.488
13.01	2.958
14.26	3.785
15.60	4.721
17.20	6.171
18.95	8.073
20.83	1.050×10^{-2}
Series III	
1.756	4.234×10^{-3}
1.851	4.619
1.978	4.957
2.150	5.402
2.381	5.669
2.630	5.710
2.884	5.634
3.148	5.391
3.445	5.071
3.747	4.676
4.078	4.288
4.461	3.793
4.873	3.375
Series IV	
1.103	1.176×10^{-3}
1.161	1.446
1.246	1.868
1.341	2.359
1.437	2.865

T	C_p
CURVE 1 (cont.)	
1.542	3.432
1.643	3.873
Series V	
2.972	5.634×10^{-3}
5.780	2.576
6.910	2.039
8.120	1.720
9.360	1.742

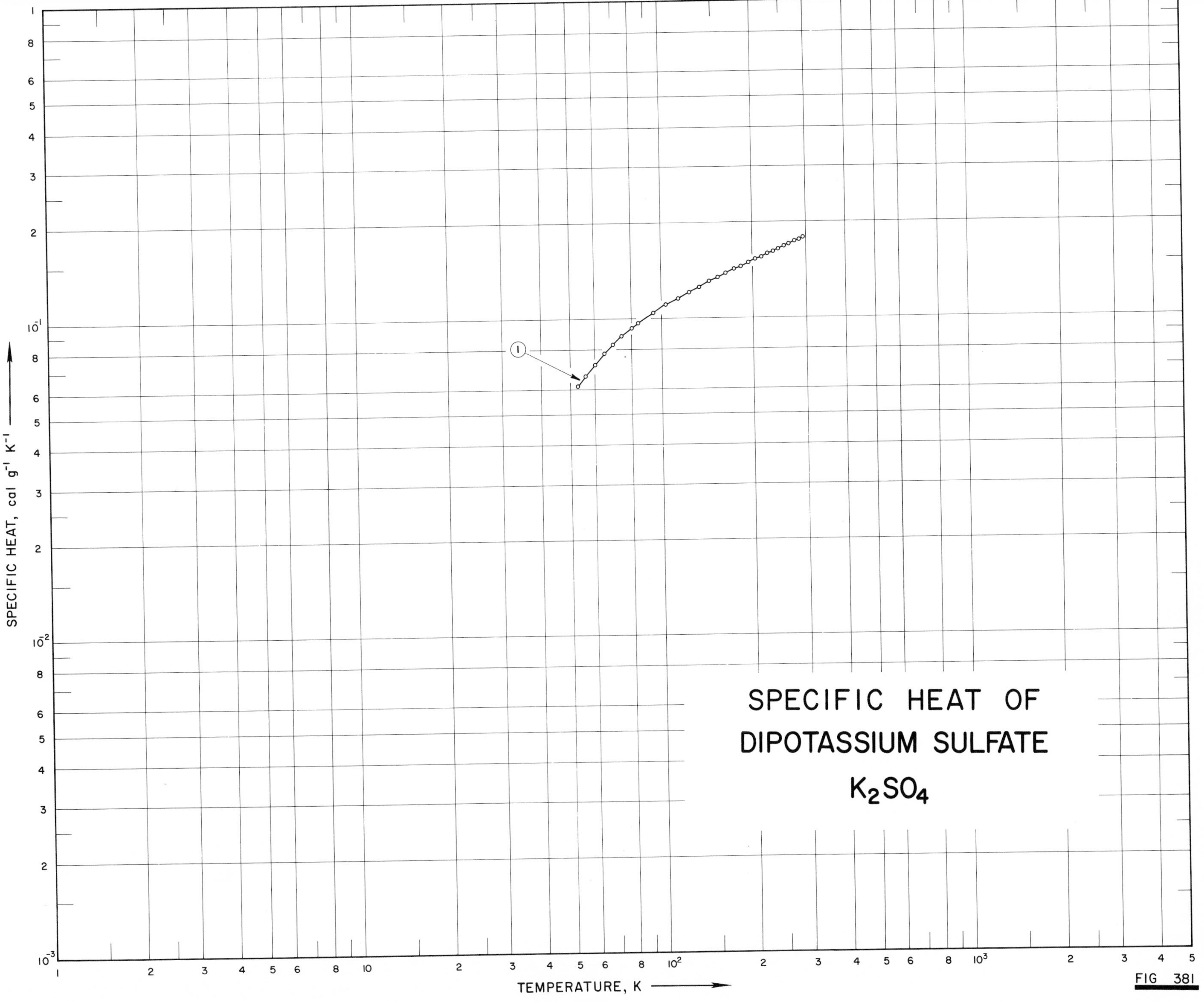

SPECIFIC HEAT OF DIPOTASSIUM SULFATE K_2SO_4

SPECIFIC HEAT, cal g^{-1} K^{-1}

TEMPERATURE, K

1

FIG 381

SPECIFICATION TABLE NO. 381 SPECIFIC HEAT OF DIPOTASSIUM SULFATE, K_2SO_4

[For Data Reported in Figure and Table No. 381]

Curve No.	Ref. No.	Year	Temp. Range, K	Reported Error, %	Name and Specimen Designation	Composition (weight percent), Specifications and Remarks
1	359	1946	53-296			99.7 K_2SO_4; reagent grade; dried several hours at 140 C.

DATA TABLE NO. 381 SPECIFIC HEAT OF DIPOTASSIUM SULFATE, K_2SO_4

[Temperature, T, K; Specific Heat, C_p, Cal $g^{-1} K^{-1}$]

T	C_p
CURVE 1	
52.7	6.158 x 10^{-2}
56.1	6.611
60.3	7.179
64.8	7.770
69.2	8.293
74.2	8.815
80.2	9.349
84.2	9.681
94.5	1.046 x 10^{-1}
104.7	1.110
114.0	1.163
124.2	1.214
134.2	1.264
144.6	1.310
154.7	1.349
164.6	1.389
175.0	1.425
184.8	1.459
195.4	1.498
205.3	1.527
215.1	1.557
225.5	1.592
235.2	1.617
244.8	1.648
255.1	1.677
265.2	1.706
276.0	1.737
286.1	1.756
295.4	1.778

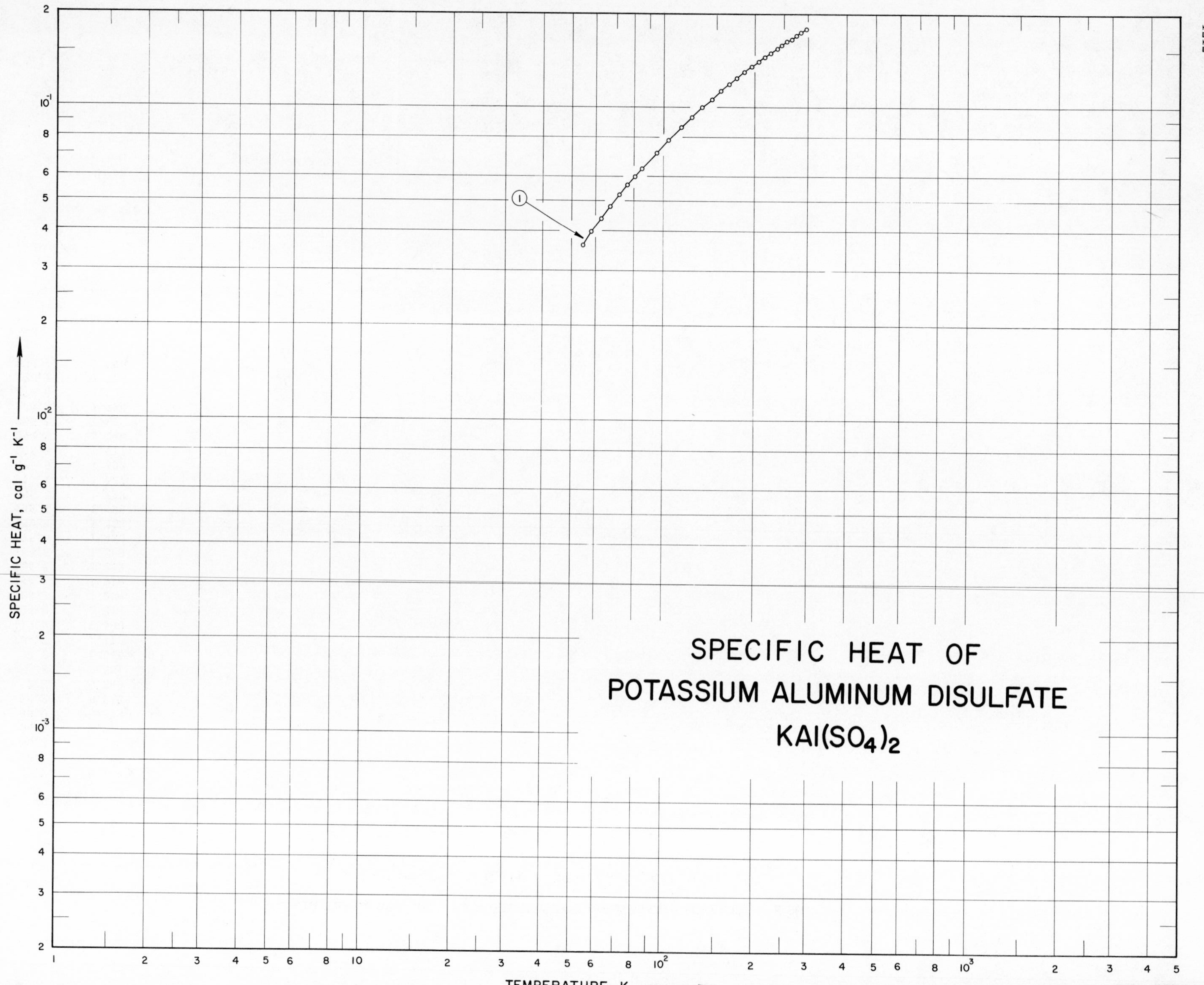
SPECIFIC HEAT OF
POTASSIUM ALUMINUM DISULFATE
$KAl(SO_4)_2$
1
SPECIFIC HEAT, cal g⁻¹ K⁻¹
TEMPERATURE, K
FIG. 382

SPECIFICATION TABLE NO. 382 SPECIFIC HEAT OF POTASSIUM ALUMINUM DISULFATE, $KAl(SO_4)_2$

[For Data Reported in Figure and Table No. 382]

Curve No.	Ref. No.	Year	Temp. Range, K	Reported Error, %	Name and Specimen Designation	Composition (weight percent), Specifications and Remarks
1	359	1946	55-297			19.65 Al_2O_3 (19.74 theo) and 0.3 H_2O.

DATA TABLE NO. 382 SPECIFIC HEAT OF POTASSIUM ALUMINUM DISULFATE, $KAl(SO_4)_2$

[Temperature, T, K; Specific Heat, C_p, Cal $g^{-1} K^{-1}$]

T	C_p
CURVE 1	
54.6	3.626×10^{-2}
58.4	4.009
6.28	4.423
67.3	4.849
71.9	5.271
76.4	5.662
80.6	6.019
85.1	6.371
95.0	7.161
104.6	7.893
115.0	8.660
124.2	9.307
134.9	1.004×10^{-1}
145.7	1.071
155.6	1.132
165.7	1.191
175.6	1.248
186.5	1.306
196.3	1.355
206.2	1.409
216.5	1.454
226.0	1.495
237.9	1.547
246.9	1.586
256.3	1.629
266.3	1.669
276.1	1.705
286.3	1.743
296.5	1.781

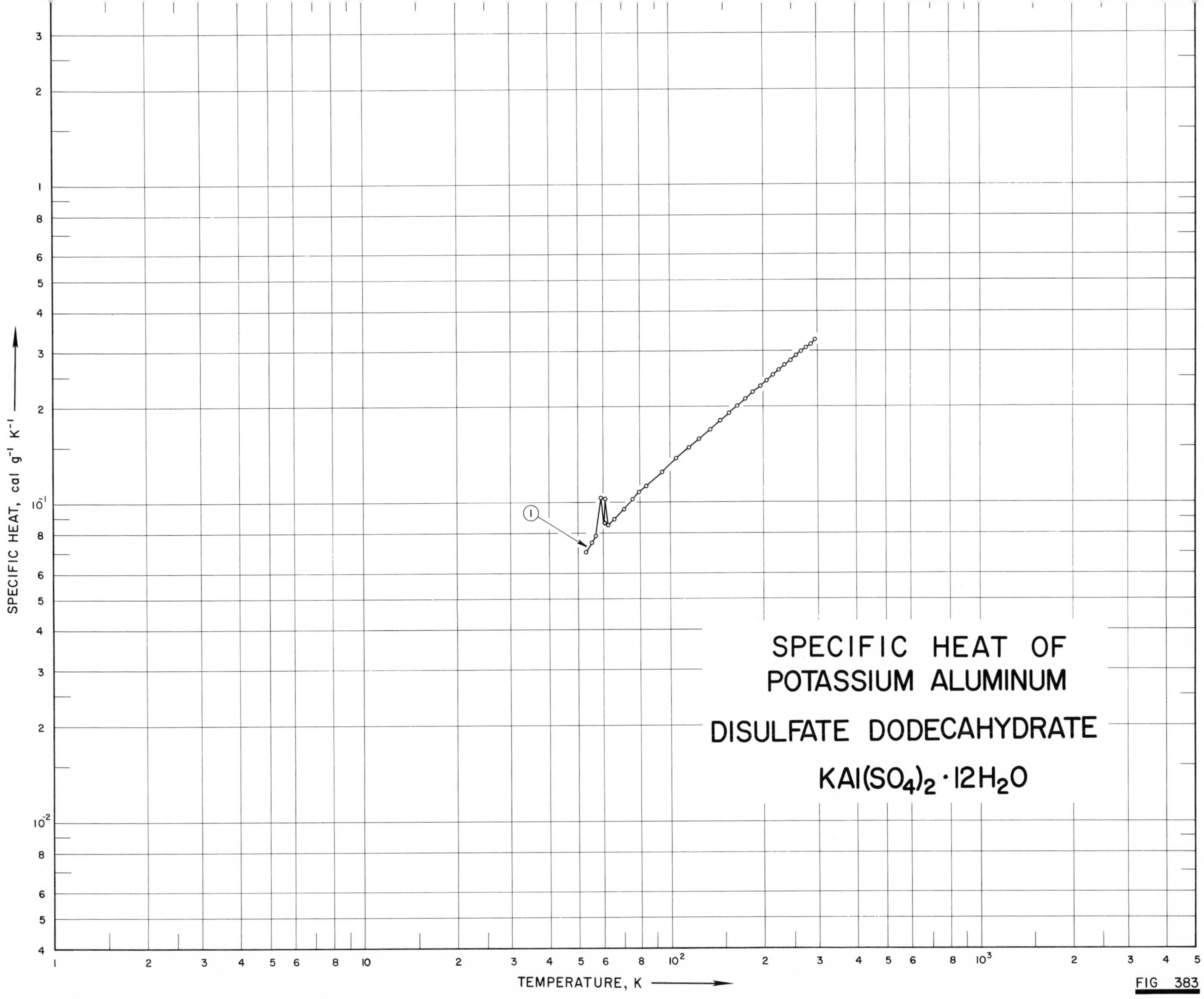

SPECIFIC HEAT OF
POTASSIUM ALUMINUM
DISULFATE DODECAHYDRATE
$KAl(SO_4)_2 \cdot 12H_2O$
SPECIFIC HEAT, cal g⁻¹ K⁻¹
TEMPERATURE, K
1

FIG 383

SPECIFICATION TABLE NO. 383 SPECIFIC HEAT OF POTASSIUM ALUMINUM DISULFATE DODECAHYDRATE, $KAl(SO_4)_2 \cdot 12H_2O$

[For Data Reported in Figure and Table No. 383]

Curve No.	Ref. No.	Year	Temp. Range, K	Reported Error, %	Name and Specimen Designation	Composition (weight percent), Specifications and Remarks
1	359	1946	53-296			10.74 Al_2O_3 and < 0.02 impurities.

DATA TABLE NO. 383 SPECIFIC HEAT OF POTASSIUM ALUMINUM DISULFATE DODECAHYDRATE, $KAl(SO_4)_2 \cdot 12H_2O$

[Temperature, T, K; Specific Heat, C_p, Cal $g^{-1} K^{-1}$]

T	C_p
CURVE 1	
52.7	7.001×10^{-2}
52.9	7.030*
55.3	7.490
56.7	4.920
57.5	8.055*
59.1	1.480×10^{-1}
60.6	8.696×10^{-2}
60.9	1.030×10^{-1}
62.3	8.544×10^{-2}
65.4	8.944
70.0	9.583
75.0	1.026×10^{-1}
79.4	1.003
83.9	1.137
94.6	1.265
104.6	1.380
115.0	1.495
124.2	1.596
135.0	1.713
145.6	1.822
155.5	1.923
165.4	2.023
175.6	2.131
185.5	2.228
196.0	2.334
205.8	2.428
216.4	2.538
226.0	2.624
235.6	2.711
246.1	2.810
256.0	2.911
265.9	3.000
276.1	3.090
286.0	3.168
296.1	3.261

* Not shown on plot

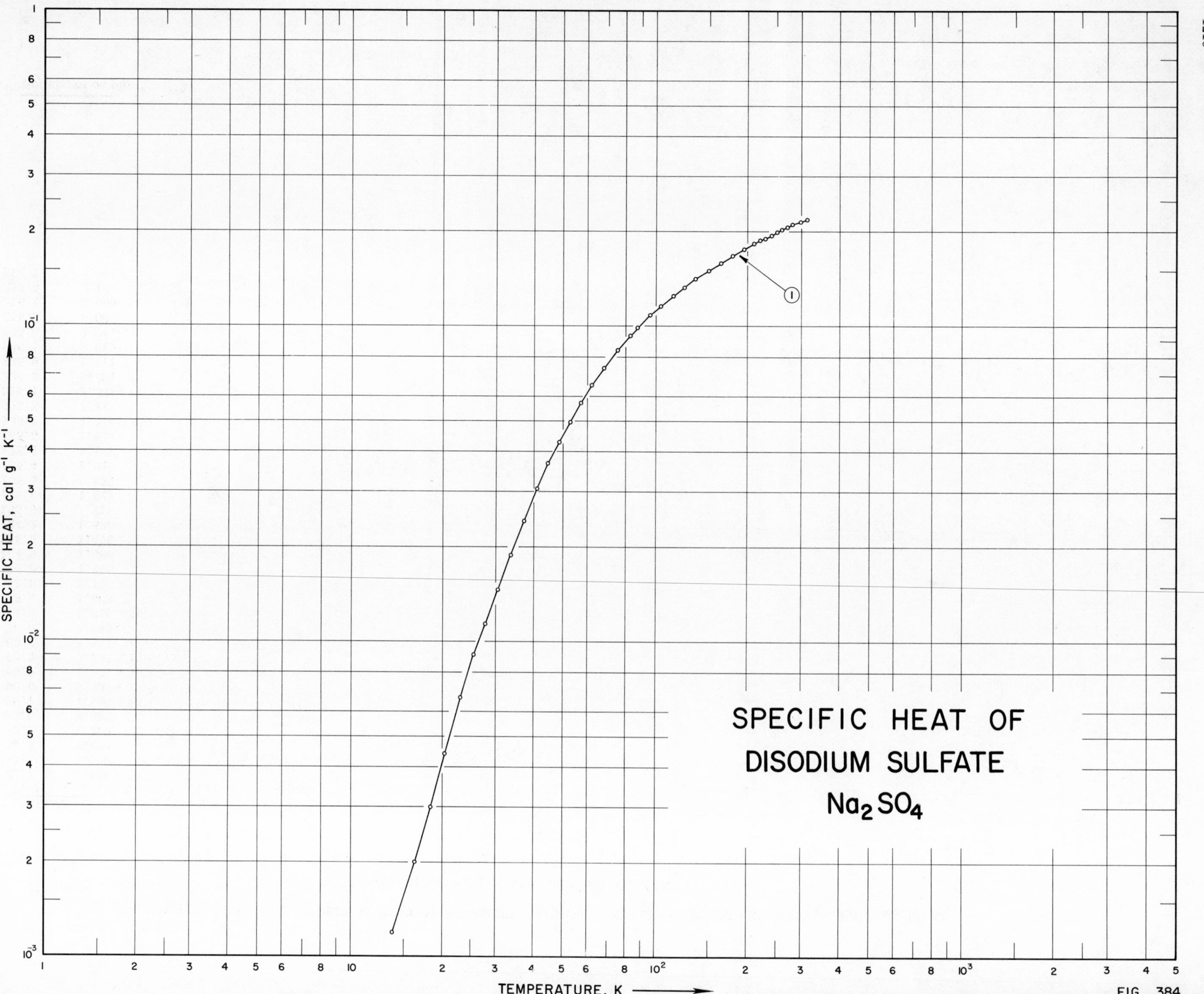
SPECIFIC HEAT OF
DISODIUM SULFATE
Na_2SO_4
SPECIFIC HEAT, cal g^{-1} K^{-1}
TEMPERATURE, K
1

FIG 384

SPECIFICATION TABLE NO. 384 SPECIFIC HEAT OF DISODIUM SULFATE, Na_2SO_4

[For Data Reported in Figure and Table No. 384]

Curve No.	Ref. No.	Year	Temp. Range, K	Reported Error, %	Name and Specimen Designation	Composition (weight percent), Specifications and Remarks
1	447	1938	14-313			< 0.10 impurities; prepared by dehydration of decahydrate under vacuum at temperature below 80 C.

DATA TABLE NO. 384 SPECIFIC HEAT OF DISODIUM SULFATE, Na_2SO_4

[Temperature, T, K; Specific Heat, C_p, Cal $g^{-1} K^{-1}$]

T	C_p
CURVE 1	
13.74	1.204×10^{-3}
16.25	2.014
18.30	2.999
20.43	4.407
23.04	6.611
25.48	9.068
27.73	1.137×10^{-2}
30.52	1.462
33.64	1.885
37.00	2.413
41.11	3.060
44.87	3.657
48.68	4.285
52.72	4.951
57.22	5.708
62.37	6.532
68.15	7.378
75.28	8.371
82.96	9.349
87.55	9.892
95.71	1.079×10^{-1}
104.51	1.166
114.59	1.255
125.07	1.337
136.41	1.420
149.18	1.502
163.43	1.588
179.41	1.675
195.80	1.752
211.35	1.830
220.40	1.875
229.90	1.905
240.09	1.945
250.48	1.993
259.96	2.023
270.37	2.062
281.19	2.102
292.14	2.128*
299.87	2.149
313.44	2.185

* Not shown on plot

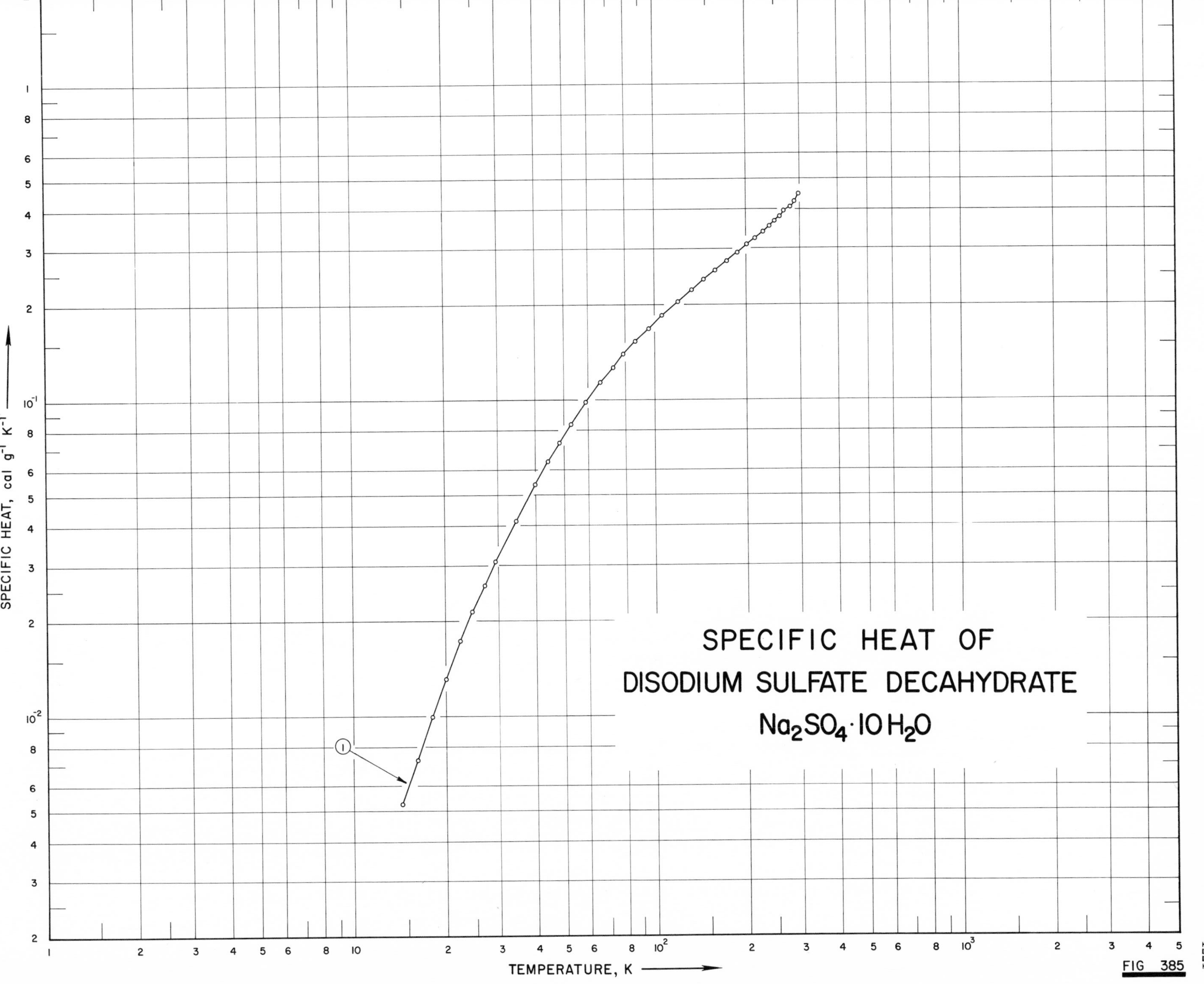

SPECIFIC HEAT OF
DISODIUM SULFATE DECAHYDRATE
$Na_2SO_4 \cdot 10H_2O$
SPECIFIC HEAT, cal g^{-1} K^{-1}
TEMPERATURE, K
1

FIG 385

SPECIFICATION TABLE NO. 385 SPECIFIC HEAT OF DISODIUM SULFATE DECAHYDRATE, $Na_2SO_4 \cdot 10H_2O$

[For Data Reported in Figure and Table No. 385]

Curve No.	Ref. No.	Year	Temp. Range, K	Reported Error, %	Name and Specimen Designation	Composition (weight percent), Specifications and Remarks
1	447	1938	14-298			0.10 impurities; prepared by recrystallization of C.P. sodium sulfate decahydrate.

DATA TABLE NO. 385 SPECIFIC HEAT OF DISODIUM SULFATE DECAHYDRATE, $Na_2SO_4 \cdot 10H_2O$

[Temperature, T, K; Specific Heat, C_p, Cal $g^{-1} K^{-1}$]

T	C_p
CURVE 1	
14.28	5.276×10^{-3}
16.15	7.294
18.11	9.963
20.08	1.322×10^{-2}
22.31	1.744
24.46	2.157
26.79	2.607
29.30	3.104
34.41	4.162
39.79	5.441
44.10	6.422
48.31	7.356
52.95	8.436
59.27	9.876
66.15	1.140×10^{-1}
72.77	1.273
79.42	1.400
86.95	1.534
96.36	1.682
107.29	1.863
120.45	2.052
133.91	2.244
145.99	2.415
158.58	2.578
173.01	2.757
187.67	2.925
201.77	3.119
215.55	3.259
228.56	3.420
239.83	3.557
249.98	3.693
259.33	3.818
266.87	3.982
281.21	4.094
289.83	4.261
298.37	4.528

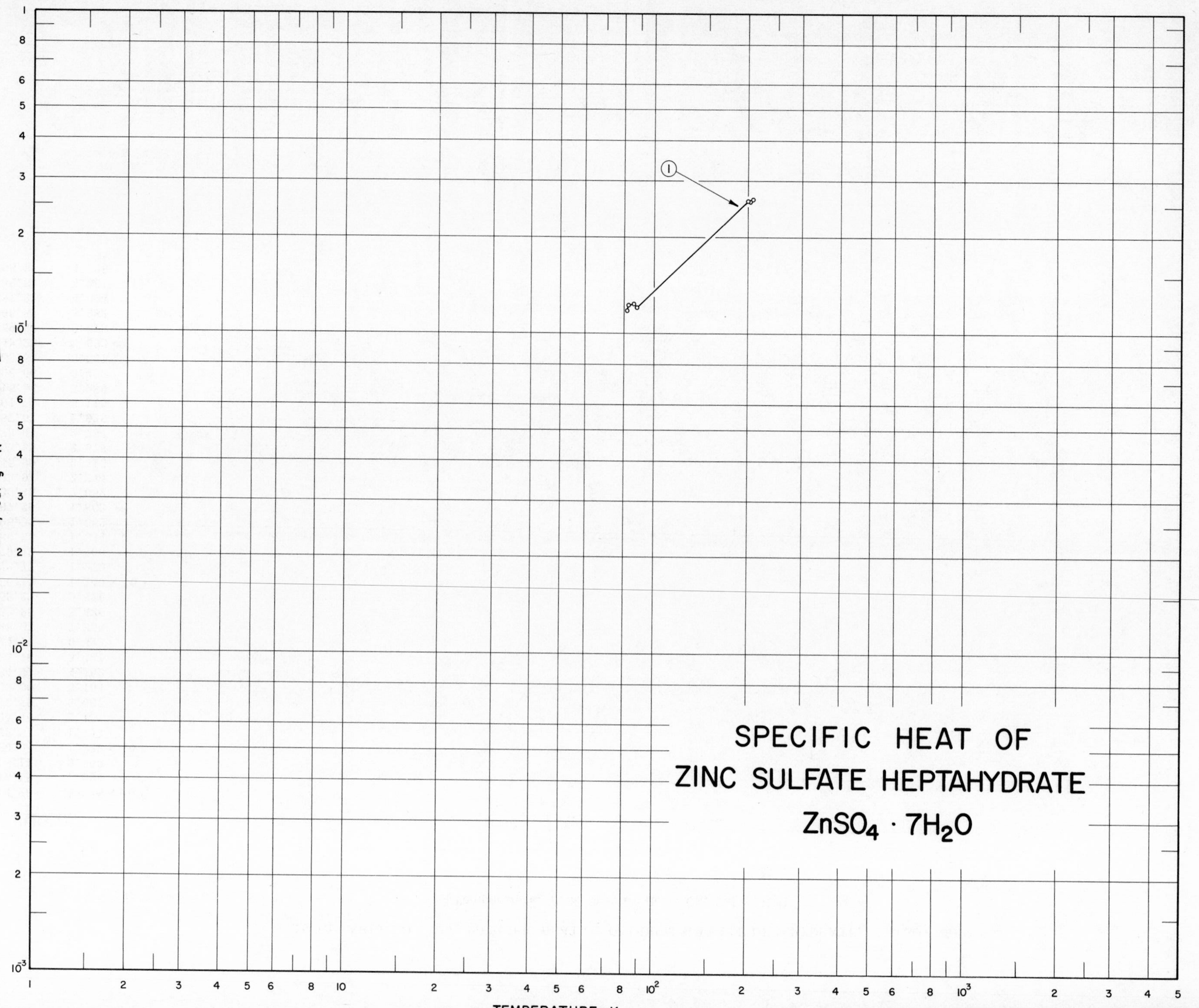
SPECIFIC HEAT OF
ZINC SULFATE HEPTAHYDRATE
$ZnSO_4 \cdot 7H_2O$
SPECIFIC HEAT, cal g^{-1} K^{-1}
1

SPECIFICATION TABLE NO. 386 SPECIFIC HEAT OF ZINC SULFATE HEPTAHYDRATE, $ZnSO_4 \cdot 7H_2O$

[For Data Reported in Figure and Table No. 386]

Curve No.	Ref. No.	Year	Temp. Range, K	Reported Error, %	Name and Specimen Designation	Composition (weight percent), Specifications and Remarks
1	448	1911	82-208			

DATA TABLE NO. 386 SPECIFIC HEAT OF ZINC SULFATE HEPTAHYDRATE, $ZnSO_4 \cdot 7H_2O$

[Temperature, T, K; Specific Heat, C_p, Cal g^{-1} K^{-1}]

T	C_p
CURVE 1	
82.15	1.178×10^{-1}
83.15	1.229
86.15	1.233
88.15	1.214
201.15	2.601
204.15	2.597
208.15	2.636

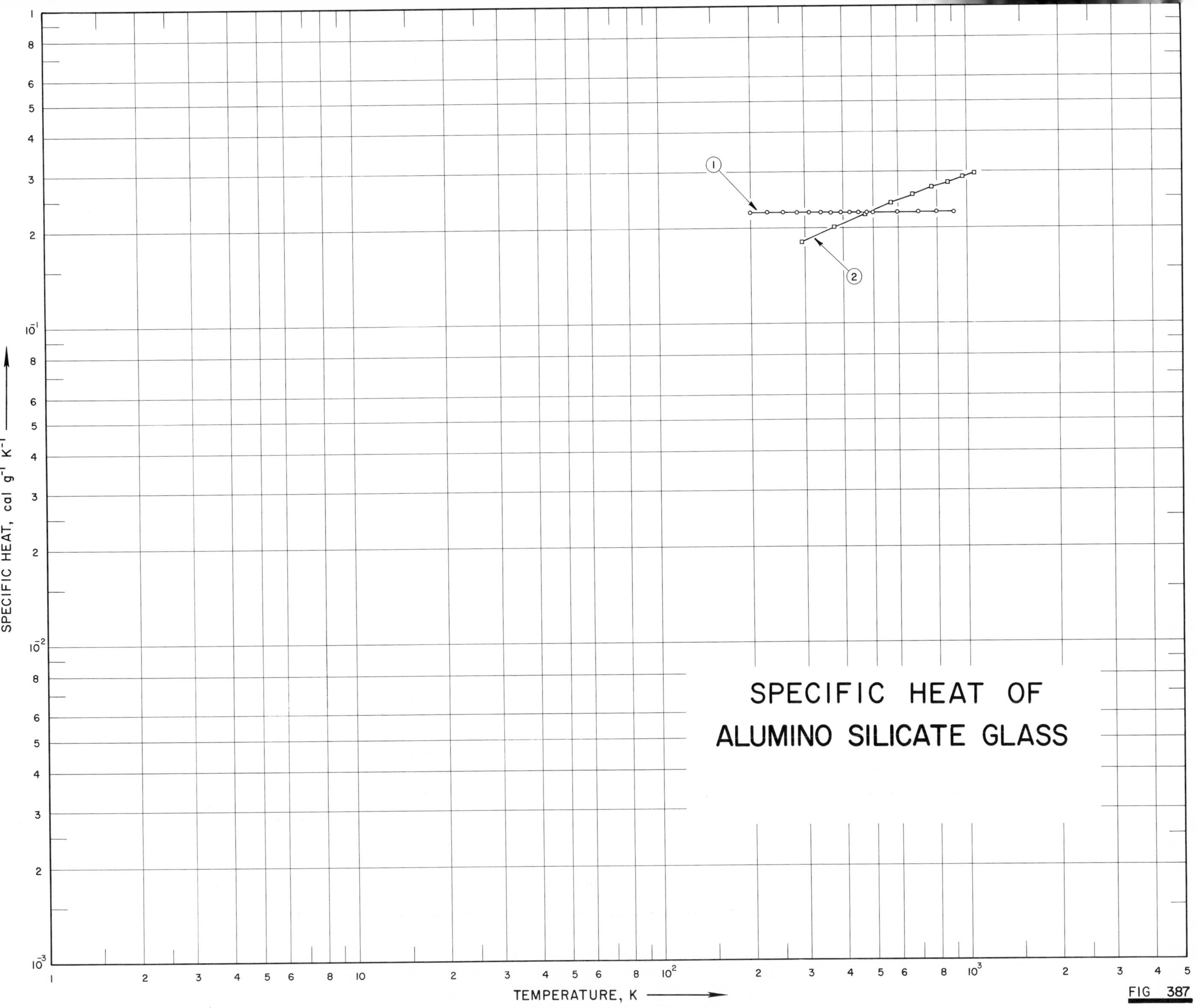
SPECIFIC HEAT OF
ALUMINO SILICATE GLASS
SPECIFIC HEAT, cal g⁻¹ K⁻¹
TEMPERATURE, K
1
2
FIG 387

SPECIFICATION TABLE NO. 387 SPECIFIC HEAT OF ALUMINOSILICATE GLASS

[For Data Reported in Figure and Table No. 387]

Curve No.	Ref. No.	Year	Temp. Range, K	Reported Error, %	Name and Specimen Designation	Composition (weight percent), Specifications and Remarks
1	56	1958	200-912		Corning Glass, no. 1723	
2	344	1958	293-1073		Corning Glass, no. 1723	Density = 164.1 lb ft^{-3}.

DATA TABLE NO. 387 SPECIFIC HEAT OF ALUMINOSILICATE GLASS

[Temperature, T, K; Specific Heat, Cp, Cal $g^{-1}K^{-1}$]

T	Cp
CURVE 1	
199.817	2.3×10^{-1}
227.039	2.3
255.372	2.3
283.16	2.3
310.94	2.3
338.72	2.3
366.49	2.3
394.27	2.3
422.05	2.3
449.83	2.3
477.60	2.3
912.05	2.3
CURVE 2	
293.15	1.80×10^{-1}
373.15	2.02
473.15	2.22
573.15	2.40
673.15	2.54
773.15	2.68
873.15	2.77
973.15	2.88
1073.15	2.97

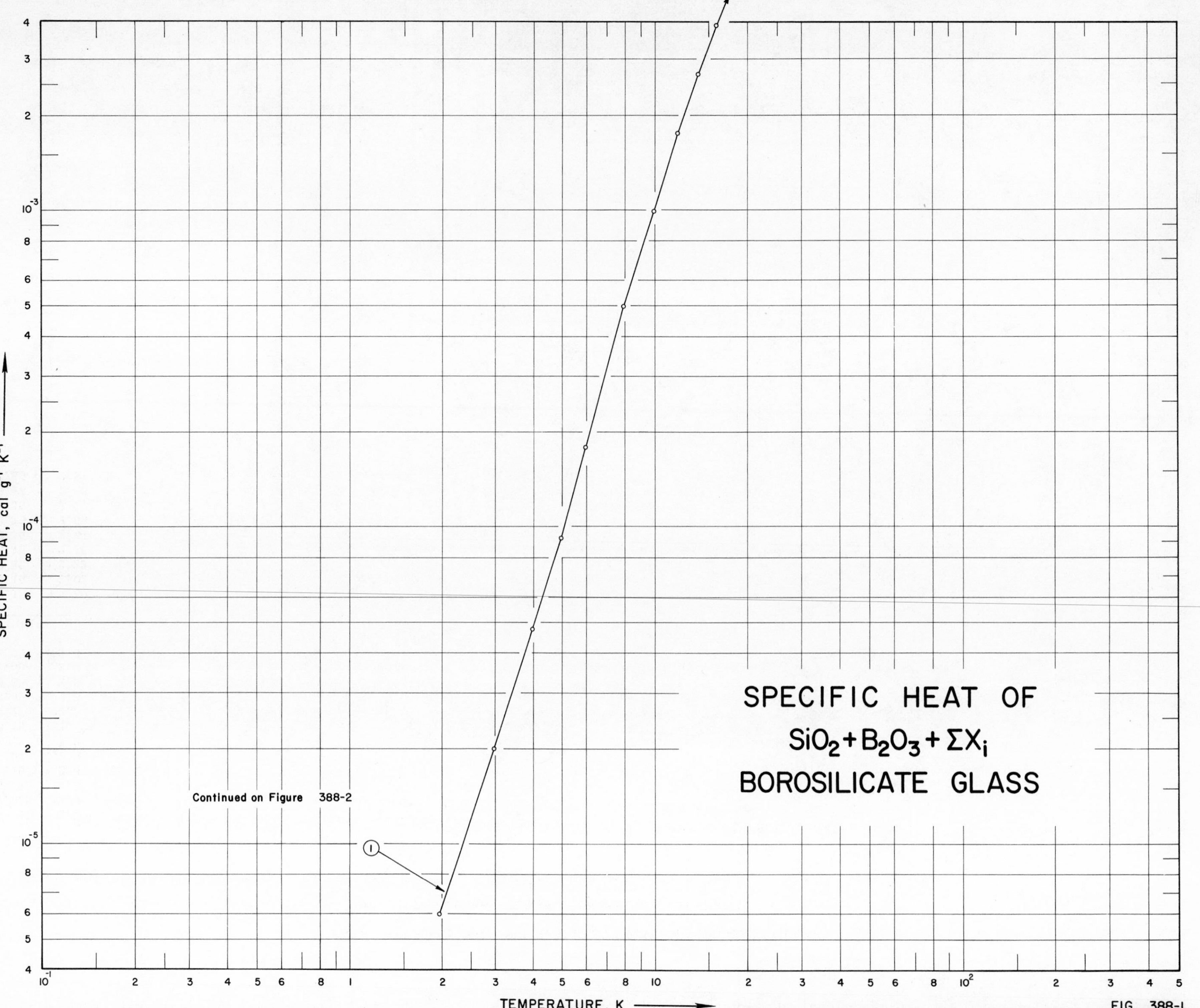

SPECIFIC HEAT OF
$SiO_2 + B_2O_3 + \Sigma X_i$
BOROSILICATE GLASS
TEMPERATURE, K
SPECIFIC HEAT, cal g^{-1} K^{-1}
Continued on Figure 388-2
1

FIG. 388-1

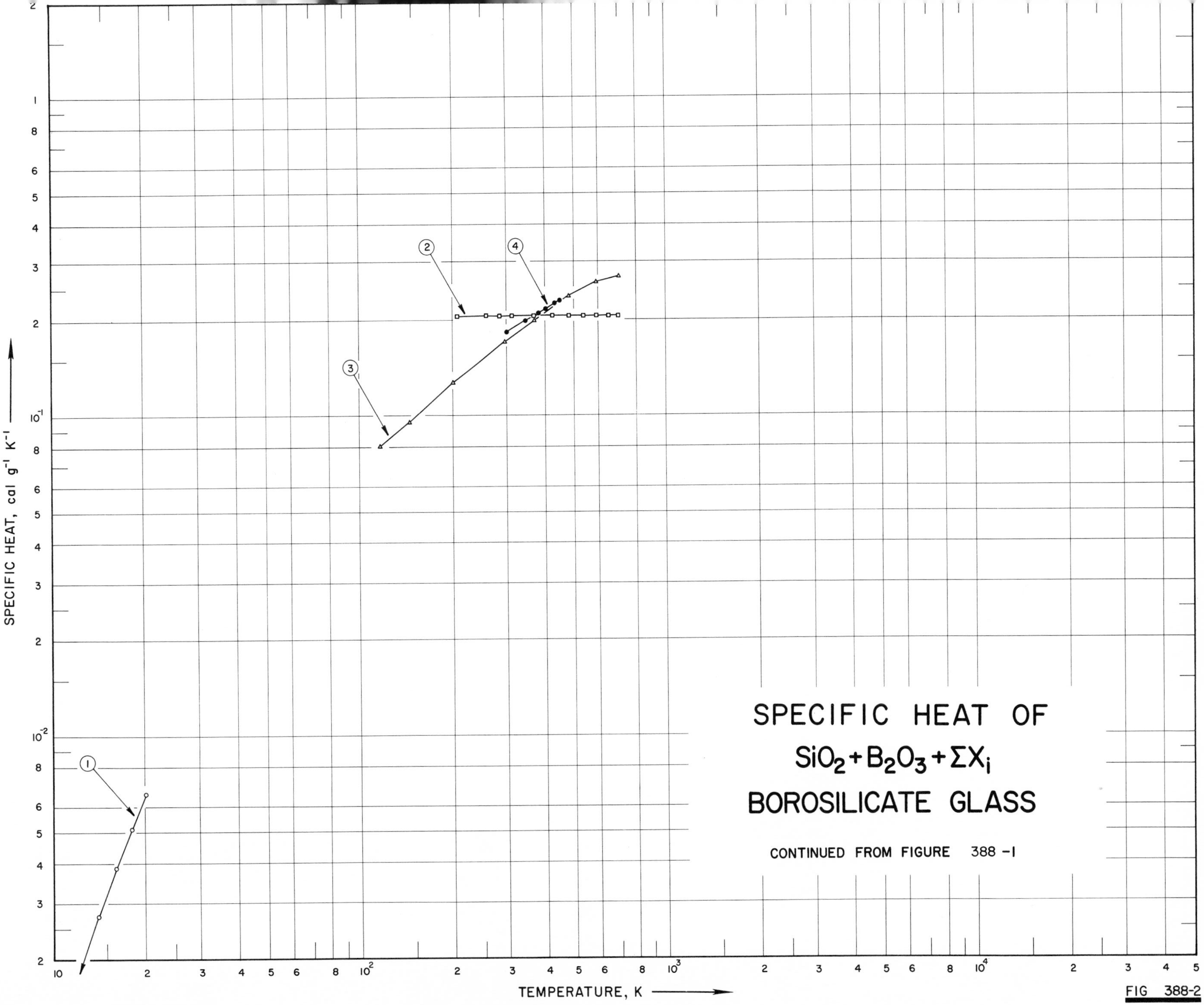

SPECIFIC HEAT OF
$SiO_2 + B_2O_3 + \Sigma X_i$
BOROSILICATE GLASS
CONTINUED FROM FIGURE 388-1
SPECIFIC HEAT, cal g^{-1} K^{-1}
TEMPERATURE, K
1
2
3
4

FIG 388-2

SPECIFICATION TABLE NO. 388 SPECIFIC HEAT OF BOROSILICATE GLASS

[For Data Reported in Figure and Table No. 388]

Curve No.	Ref. No.	Year	Temp. Range, K	Reported Error, %	Name and Specimen Designation	Composition (weight percent), Specifications and Remarks
1	345	1956	2-20			
2	56	1958	206-691		Pittsburgh Plate Glass 3235	80 SiO_2, 14 B_2O_3, 4 Na_2O and 2 Al_2O_3.
3	9	1960	116-700	± 1.0	Pyrex Glass no. 774	80 SiO_2, 14 B_2O_3, 4 Na_2O and 2 Al_2O_3; sample supplied by the Cincinati Gasket and Packing Co; specimen sealed in helium atmosphere in a capsule; density = 138 lb ft^{-3} at 32 F.
4	346	1930	300-446		Pyrex Glass	

DATA TABLE NO. 388 SPECIFIC HEAT OF BOROSILICATE GLASS

[Temperature, T, K; Specific Heat, Cp, Cal $g^{-1}K^{-1}$]

T	Cp
CURVE 1	
2	6.0×10^{-6}
3	2.0×10^{-5}
4	4.8
5	9.3
6	1.80×10^{-4}
8	5.0
10	1.0×10^{-3}
12	1.77
14	2.71
16	3.86
18	5.14
20	6.55
CURVE 2	
206.483	2.07×10^{-1}
255.372	2.07
283.16	2.07
310.94	2.07
366.49	2.07
422.05	2.07
477.60	2.07
533.16	2.07
588.72	2.07
644.27	2.07
691.49	2.07
CURVE 3	
116.483	8.1×10^{-2}
144.261	9.6
199.817	1.28×10^{-1}
293.16	1.72
366.49	2.02
477.60	2.38
588.72	2.66
699.83	2.75
CURVE 4	
299.55	1.859×10^{-1}
344.15	2.009
379.55	2.111
400.55	2.178
428.35	2.261
446.05	2.316

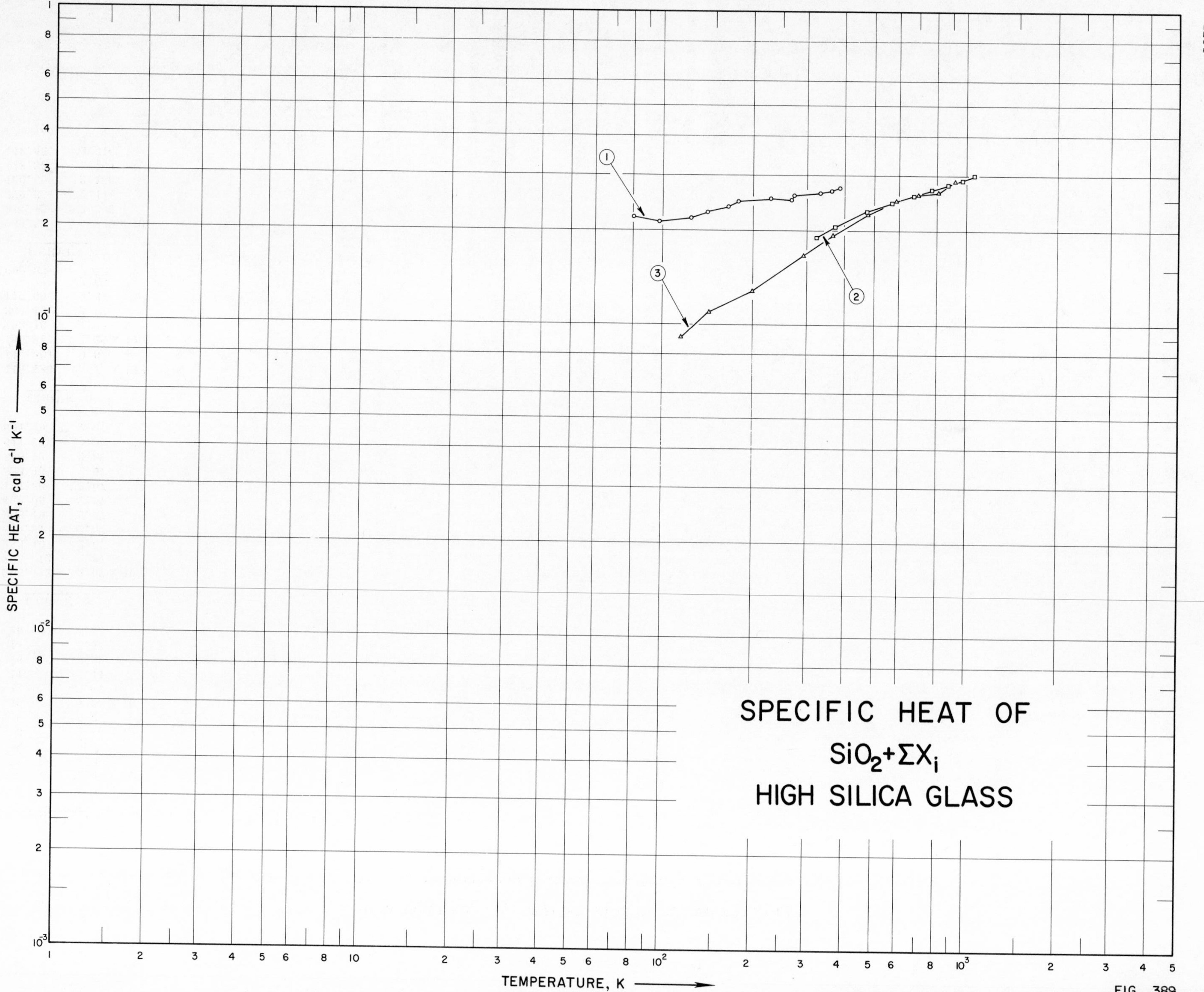
SPECIFIC HEAT OF
$SiO_2 + \Sigma X_i$
HIGH SILICA GLASS
SPECIFIC HEAT, cal g⁻¹ K⁻¹
TEMPERATURE, K
1
2
3

FIG 389

SPECIFICATION TABLE NO. 389 SPECIFIC HEAT OF HIGH SILICA GLASS

[For Data Reported in Figure and Table No. 389]

Curve No.	Ref. No.	Year	Temp. Range, K	Reported Error, %	Name and Specimen Designation	Composition (weight percent), Specifications and Remarks
1	347	1955	81-387	8.0		High purity sample; powdered to pass a 40-mesh screen and retained on a 60-mesh screen.
2	344	1958	323-1073		Vycor brand Clear High Silica 7900	$\geq$96 SiO_2, $\leq$3 B_2O_3 and other oxides; sample supplied by the Corning Glass Works; density = 136 lb ft^{-3}.
3	9	1960	116-922	±1.0	Glass, Vycor	$\geq$96 SiO_2, $\leq$3 B_2O_3 and $\leq$1.0 other oxides; sample supplied by the Corning Glass Works; specimen sealed in helium atmosphere in a capsule; density = 136.5 lb ft^{-3} at 32 F.

DATA TABLE NO. 389 SPECIFIC HEAT OF HIGH SILICA GLASS

[Temperature, T, K; Specific Heat, C_p, Cal g^{-1} K^{-1}]

T	C_p
CURVE 1	
80.5	2.213×10^{-1}
97.9	2.130
125.0	2.196
142.0	2.296
166.0	2.379
179.0	2.479
229.0	2.529
268.0	2.512
274.0	2.596
336.0	2.646
363.0	2.679
387.0	2.745
CURVE 2	
323.15	1.90×10^{-1}
373.15	2.06
473.15	2.30
573.15	2.45
673.15	2.59
773.15	2.71
873.15	2.81
973.15	2.91
1073.15	3.01
CURVE 3	
116.483	9.1×10^{-2}
144.261	1.10×10^{-1}
199.817	1.28
293.16	1.67
366.49	1.93
477.60	2.26
588.72	2.49
699.83	2.60
810.94	2.66
922.05	2.88

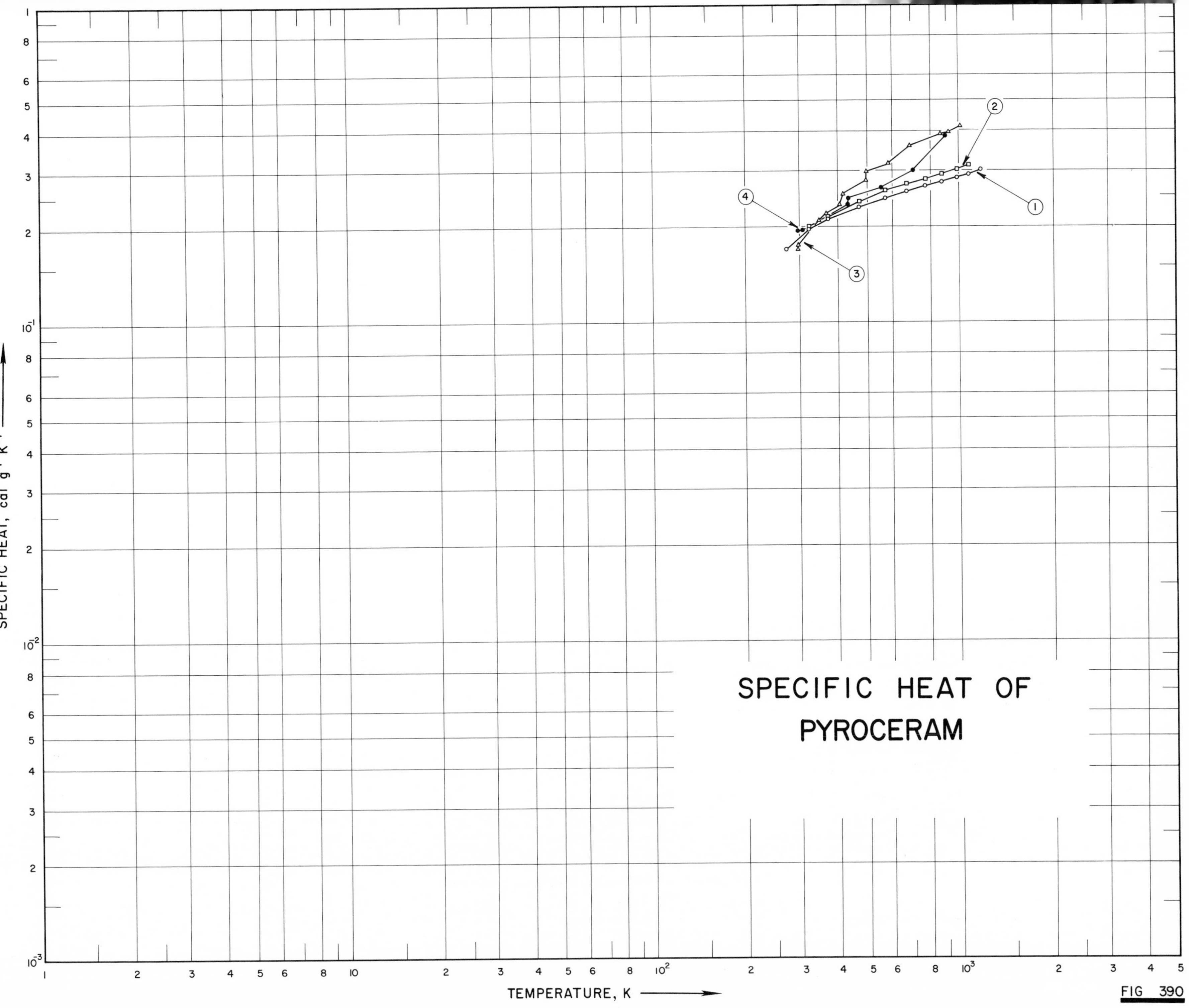
SPECIFIC HEAT OF PYROCERAM
SPECIFIC HEAT, cal g^{-1} K^{-1}
TEMPERATURE, K
1
2
3
4

FIG 390

SPECIFICATION TABLE NO. 390 SPECIFIC HEAT OF PYROCERAM

[For Data Reported in Figure and Table No. 390]

Curve No.	Ref. No.	Year	Temp. Range, K	Reported Error, %	Name and Specimen Designation	Composition (weight percent), Specifications and Remarks
1	344	1958	273-1173		no. 9606	Density = 162.3 lb ft^{-3}.
2	344	1958	323-1073		no. 9608	Density = 156 lb ft^{-3}.
3	348	1963	298-998		sample 1, no. 9608	Sample made by Corning Glass Works; coated with silver paste to make specimen opaque.
4	348	1963	298-893		sample 2, no. 9608	Same as above.

DATA TABLE NO. 390 SPECIFIC HEAT OF PYROCERAM

[Temperature, T, K; Specific Heat, Cp, Cal $g^{-1}K^{-1}$]

T	Cp
CURVE 1	
273.15	1.70 x 10^{-1}
323.15	1.96
373.15	2.12
473.15	2.31
573.15	2.46
673.15	2.58
773.15	2.68
873.15	2.76
973.15	2.84
1073.15	2.92
1173.15	3.01
CURVE 2	
323.15	2.01 x 10^{-1}
373.15	2.15
473.15	2.40
573.15	2.59
673.15	2.71
773.15	2.82
873.15	2.93
973.15	3.02
1073.15	3.12
CURVE 3	
298.15	1.70 x 10^{-1}
298.15	1.75
299.15	1.73*
348.15	2.10
369.15	2.22
408.15	2.34
418.15	2.55
498.15	2.79
498.15	2.98
589.15	3.16
589.15	3.20*
688.15	3.59
863.15	3.90
918.15	3.95
998.15	4.15
CURVE 4	
298.15	1.94 x 10^{-1}
303.15	1.90*
308.15	1.95
435.15	2.35
435.15	2.46
555.15	2.66
704.15	3.00
893.15	3.86

*Not shown on plot

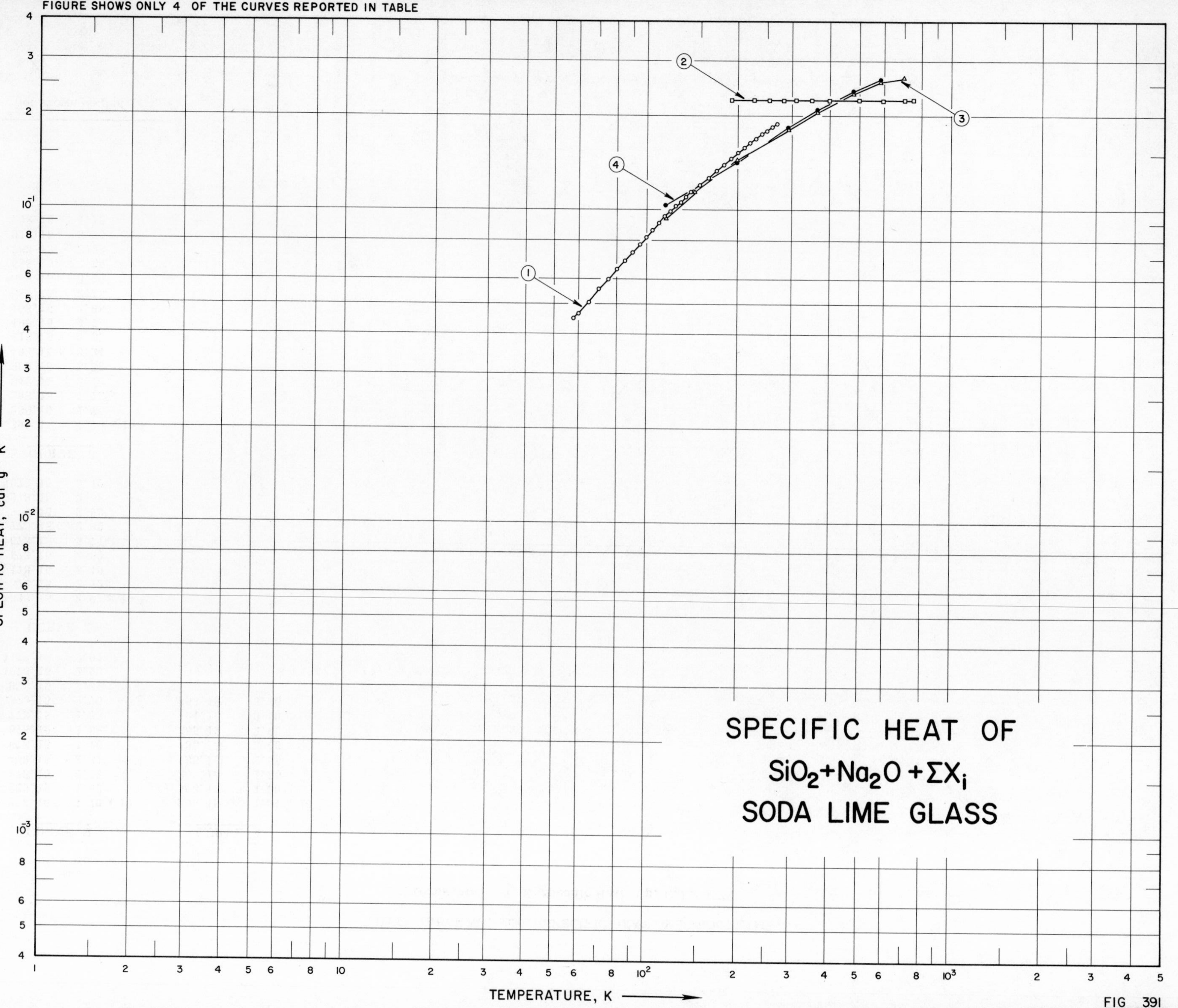

FIGURE SHOWS ONLY 4 OF THE CURVES REPORTED IN TABLE
SPECIFIC HEAT OF
$SiO_2+Na_2O+\Sigma X_i$
SODA LIME GLASS
SPECIFIC HEAT, cal g^{-1} K^{-1}
TEMPERATURE, K
FIG 391

SPECIFICATION TABLE NO. 391 SPECIFIC HEAT OF SODA LIME GLASS

[For Data Reported in Figure and Table No. 391]

Curve No.	Ref. No.	Year	Temp. Range, K	Reported Error, %	Name and Specimen Designation	Composition (weight percent), Specifications and Remarks
1	349	1957	58-270		Glass no. 23	Nominal composition: 72 SiO_2, 15 Na_2O, 9 CaO, 3 MgO and 1 Al_2O_3; commercially pure sample.
2	56	1958	191-753		Libby-Owens-Ford 9330	Nominal composition: 72 SiO_2, 15 Na_2O, 9 CaO, 3 MgO and 1 Al_2O_3; soda-lime silica plate glass.
3	9	1960	116-700	± 1.0	Solex S	Nominal composition: 72 SiO_2, 15 Na_2O, 9 CaO, 3 MgO and 1 Al_2O_3; sample supplied by the Pittsburgh Plate Glass Co; specimen sealed in helium atmosphere in capsule; density = 157 lb ft^{-3} at 32 F.
4	9	1960	116-700	± 1.0	Solex 2808 X	Nominal composition: 72 SiO_2, 15 Na_2O, 9 CaO, 3 MgO and 1 Al_2O_3; sample supplied by the Pittsburgh Plate Glass Co;specimen sealed in helium atmosphere in capsule; density = 158 lb ft^{-3} at 32 F.
5	9	1960	116-700	± 1.0		Nominal composition: 72 SiO_2, 15 Na_2O, 9 CaO, 3 MgO and 1 Al_2O_3; white (clear) plate glass; sample supplied by the Pittsburgh Plate Glass Co; specimen sealed in helium atmosphere in capsule; density = 157 lb ft^{-3} at 32 F.

DATA TABLE NO. 391 SPECIFIC HEAT OF SODA LIME GLASS

[Temperature, T,K; Specific Heat, Cp, Cal $g^{-1}K^{-1}$]

T	Cp
CURVE 1	
58	4.50 x 10^{-2}
60	4.66
65	5.06
70	5.57
75	6.01
80	6.47
85	6.87
90	7.34
95	7.75
100	8.18
105	8.62
110	9.06
115	9.52
120	9.86
125	1.027 x 10^{-1}
130	1.067
135	1.104
140	1.141
145	1.171*
150	1.204
155	1.241*
160	1.271
165	1.302*
170	1.336
175	1.365*
180	1.399
185	1.432*
190	1.466
195	1.496*
200	1.526
205	1.552*
210	1.583
215	1.614*
220	1.640
225	1.661*
230	1.694
235	1.721*
240	1.755
245	1.771*
250	1.798
255	1.822*
260	1.845
265	1.865*
270	1.885

T	Cp
CURVE 2	
191.483	2.24 x 10^{-1}
227.039	2.24
255.372	2.24
283.16	2.24
310.94	2.24
752.60	2.24
CURVE 3	
116.483	9.4 x 10^{-2}
144.261	1.14 x 10^{-1}
199.817	1.44
293.16	1.81
366.49	2.06
477.60	2.36
588.72	2.57
699.83	2.65
CURVE 4	
116.483	1.03 x 10^{-1}
144.261	1.16
199.817	1.42
293.16	1.83
366.49	2.08
477.60	2.38
588.72	2.60
699.83	2.66*
CURVE 5	
116.483	9.7 x 10^{-2}
144.261	1.14 x 10^{-1}
199.817	1.43
293.16	1.82
366.49	2.07
477.60	2.36
588.72	2.57
699.83	2.68

* Not shown on plot

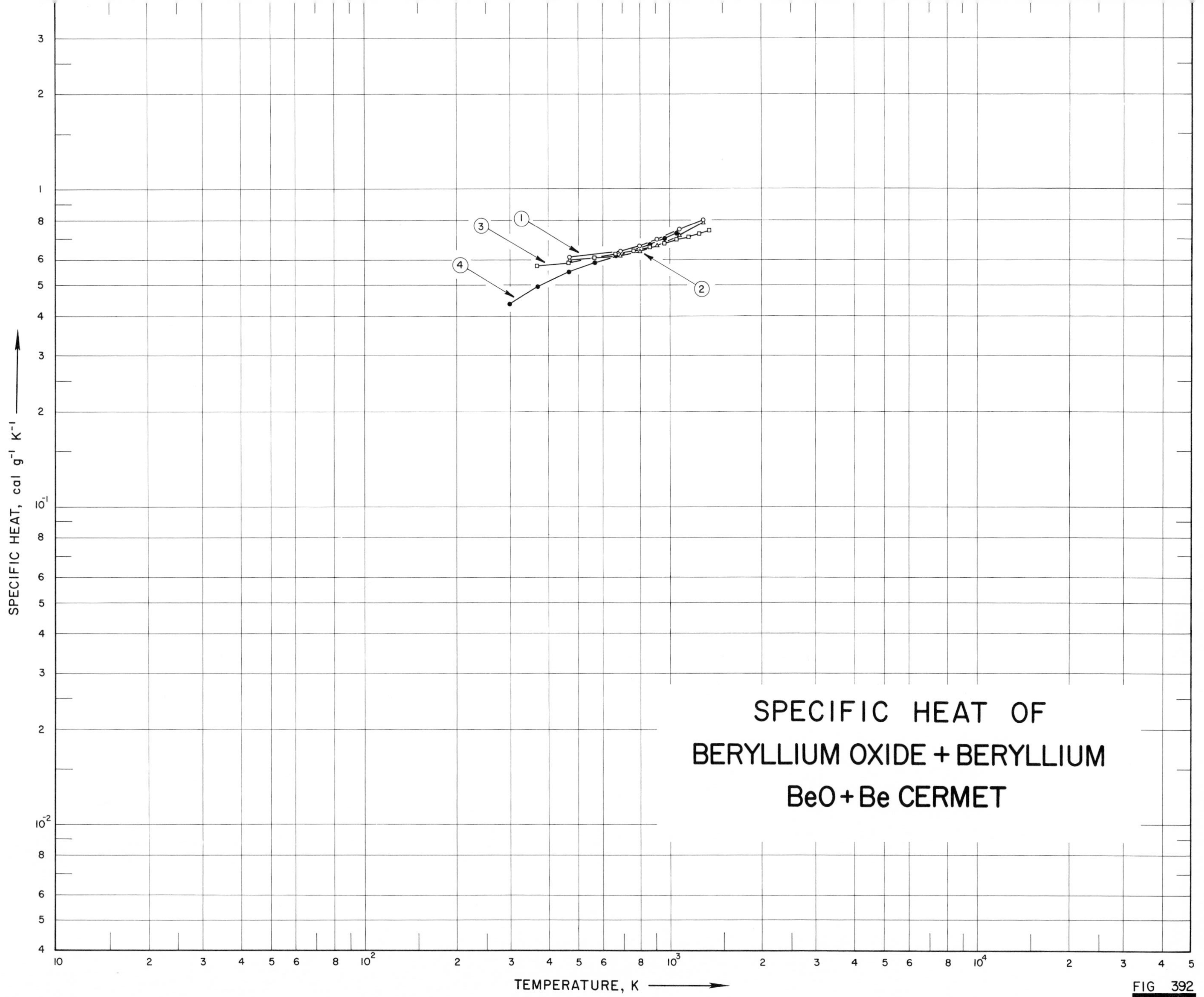
SPECIFIC HEAT OF
BERYLLIUM OXIDE + BERYLLIUM
BeO + Be CERMET
SPECIFIC HEAT, cal g⁻¹ K⁻¹
TEMPERATURE, K
1
2
3
4

FIG 392

SPECIFICATION TABLE NO. 392 SPECIFIC HEAT OF BERYLLIUM + BERYLLIUM OXIDE, Be + BeO CERMET

[For Data Reported in Figure and Table No. 392]

Curve No.	Ref. No.	Year	Temp. Range, K	Reported Error, %	Name and Specimen Designation	Composition (weight percent), Specifications and Remarks
1	189	1958	478-1311	3.0	YB 9052	99.16 Be and 0.84 BeO.
2	189	1958	478-1311	3.0	YB 9054	98.32 Be and 1.68 BeO.
3	350	1960	373-1373	≤2.7	QMV Beryllium	Impurities: 1.16 BeO, 0.15 Al, 0.013 Mg, 0.1 Fe, 0.1 Ni and others; sample manufactured by the Brush Beryllium Corp.
4	45	1962	303-1075	≤3.0		99.3 Be, 0.9 BeO, 0.1 Fe and 0.1 various metals.

DATA TABLE NO. 392 SPECIFIC HEAT OF BERYLLIUM + BERYLLIUM OXIDE, Be + BeO CERMET

[Temperature, T, K; Specific Heat, C_p, Cal $g^{-1} K^{-1}$]

T	C_p
CURVE 1	
478	6.150 x 10^{-1}
700	6.410
811	6.670
922	6.990
1089	7.490
1311	8.040
CURVE 2	
478	6.060 x 10^{-1}
700	6.230
811	6.440
922	6.720
1089	7.220
1311	7.940
CURVE 3	
373	5.780 x 10^{-1}
473	5.940
573	6.110
673	6.280
773	6.440
873	6.610
973	6.780
1073	6.950
1173	7.110
1273	7.280
1373	7.450
CURVE 4	
303	4.37 x 10^{-1}
375	4.98
475	5.53
575	5.91
675	6.20
775	6.47*
875	6.74
975	7.01
1075	7.29

*Not shown on plot

FIGURE SHOWS ONLY 4 OF THE CURVES REPORTED IN TABLE

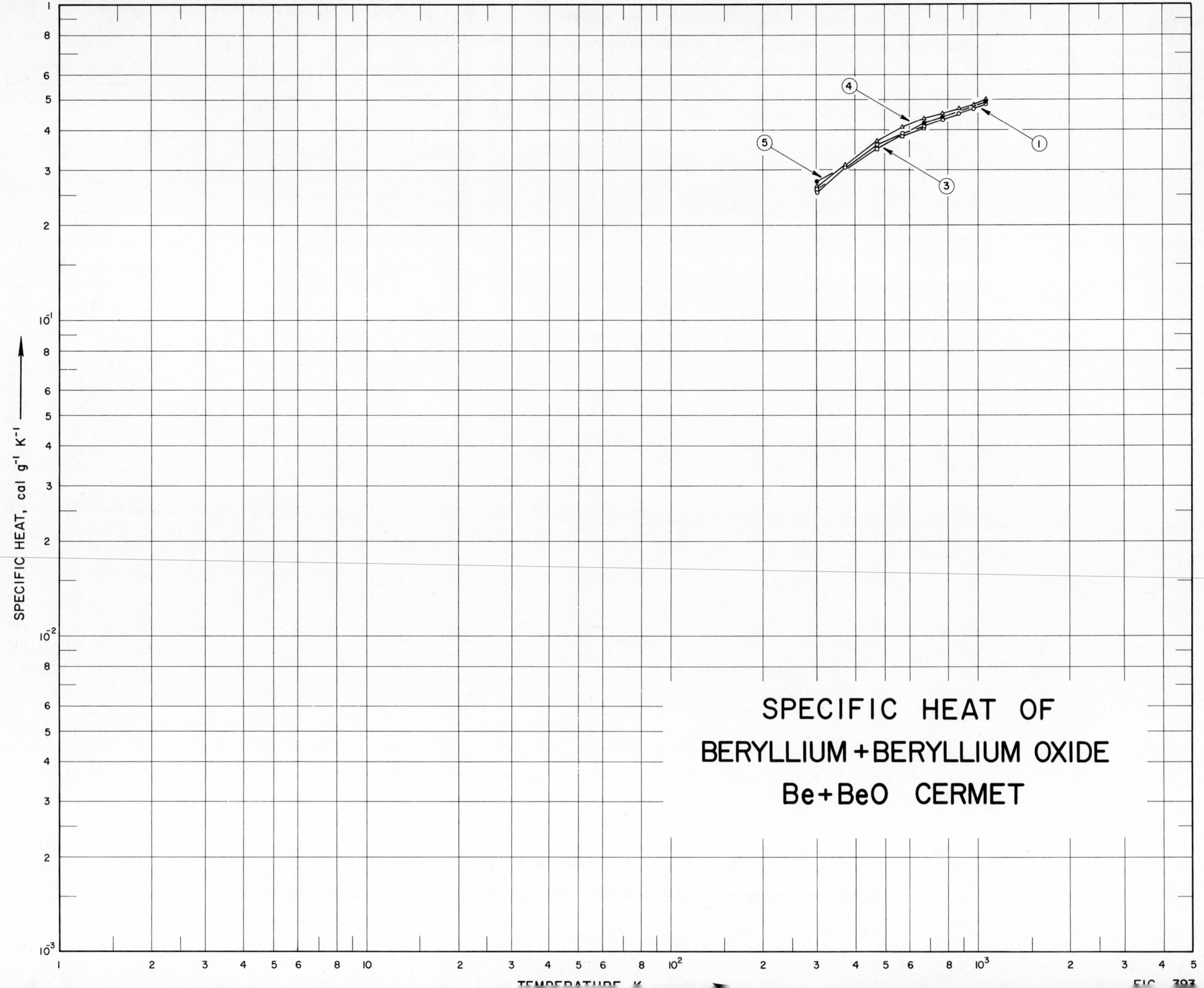
SPECIFIC HEAT OF
BERYLLIUM + BERYLLIUM OXIDE
Be + BeO CERMET
SPECIFIC HEAT, cal g⁻¹ K⁻¹
TEMPERATURE, K
1
3
4
5

FIG. 393

SPECIFICATION TABLE NO. 393 SPECIFIC HEAT OF BERYLLIUM OXIDE + BERYLLIUM. BeO + Be CERMET

[For Data Reported in Figure and Table No. 393]

Curve No.	Ref. No.	Year	Temp. Range, K	Reported Error, %	Name and Specimen Designation	Composition (weight percent), Specifications and Remarks
1	45	1962	303-1075	< 3.0		Nominal composition: 97 BeO and 3 Be.
2	45	1962	303-1073	< 3.0		Nominal composition: 94 BeO and 6 Be.
3	45	1962	303-1073	< 3.0		Nominal composition: 93 BeO and 7 Be.
4	45	1962	303-1073	< 3.0		Nominal composition: 91 BeO and 9 Be.
5	45	1962	303-1073	< 3.0		Nominal composition: 88 BeO and 12 Be.

DATA TABLE NO. 393 SPECIFIC HEAT OF BERYLLIUM OXIDE + BERYLLIUM, BeO + Be CERMET

[Temperature, T, K; Specific Heat, C_p, Cal $g^{-1} K^{-1}$]

T	C_p	T	C_p
CURVE 1		CURVE 5	
303	2.54×10^{-1}	303	2.75×10^{-1}
375	3.05	373	3.13*
475	3.60	473	3.62*
575	3.92	573	3.95*
675	4.14	673	4.20
775	4.34	773	4.41
875	4.51	873	4.57*
975	4.68	973	4.73*
1075	4.85	1073	4.90
CURVE 2			
303	2.62×10^{-1}		
373	3.06		
473	3.58		
573	3.94		
673	4.16		
773	4.36		
873	4.54		
973	4.72		
1073	4.91		
CURVE 3			
303	2.61×10^{-1}		
373	3.03*		
473	3.49		
573	3.84		
673	4.08		
773	4.30*		
873	4.50*		
973	4.68*		
1073	4.86*		
CURVE 4			
303	2.65×10^{-1}		
373	3.11		
473	3.70		
573	4.11		
673	4.36		
773	4.53		
873	4.69		
973	4.83		
1073	5.01		

* Not shown on plot

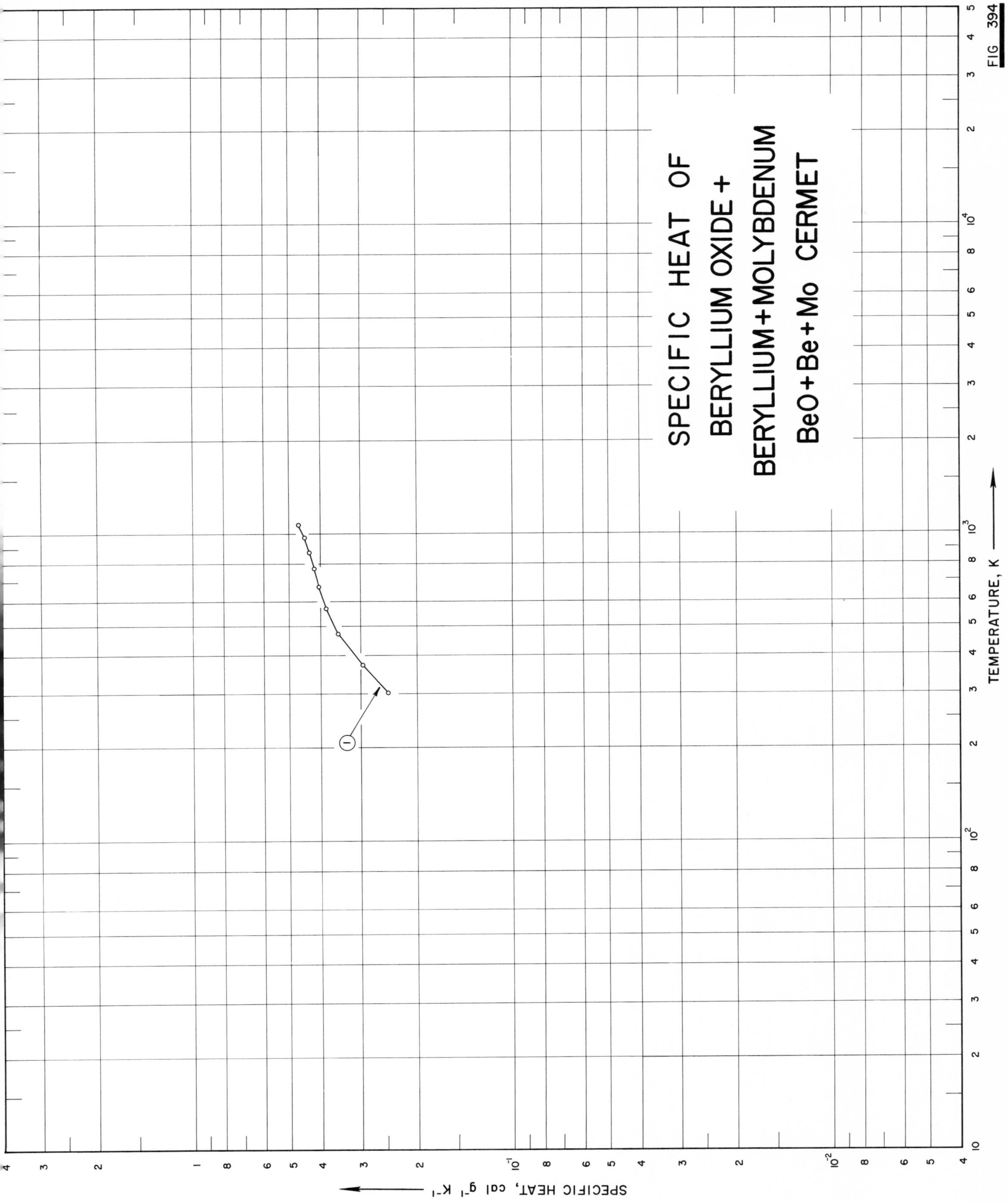

SPECIFIC HEAT OF BERYLLIUM OXIDE + BERYLLIUM + MOLYBDENUM BeO + Be + Mo CERMET

SPECIFIC HEAT, cal g^{-1} K^{-1}

TEMPERATURE, K

1

FIG 394

SPECIFICATION TABLE NO. 394 SPECIFIC HEAT OF BERYLLIUM OXIDE + BERYLLIUM + MOLYBDENUM, BeO + Be + Mo CERMET

[For Data Reported in Figure and Table No. 394]

Curve No.	Ref. No.	Year	Temp. Range, K	Reported Error, %	Name and Specimen Designation	Composition (weight percent), Specifications and Remarks
1	45	1962	303-1073	< 3.0		Nominal composition: 88 BeO, 7 Be and 7 Mo.

DATA TABLE NO. 394 SPECIFIC HEAT OF BERYLLIUM OXIDE + BERYLLIUM + MOLYBDENUM, BeO + Be + Mo CERMET

[Temperature, T, K; Specific Heat, C_p, Cal g^{-1} K^{-1}]

T	C_p
CURVE 1	
303	2.48×10^{-1}
373	2.97
473	3.56
573	3.87
673	4.07
773	4.23
873	4.38
973	4.54
1073	4.70

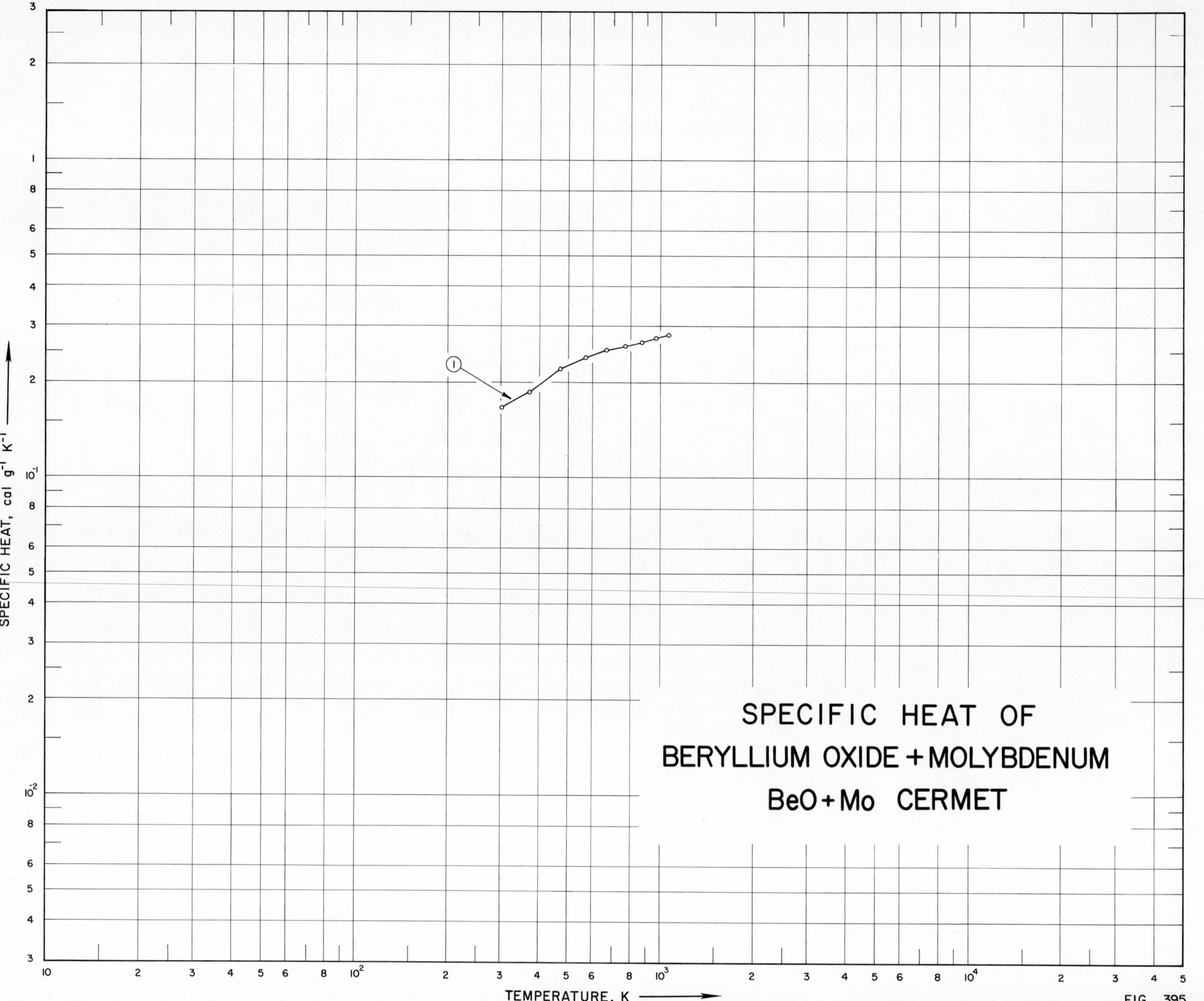
SPECIFIC HEAT OF
BERYLLIUM OXIDE + MOLYBDENUM
BeO + Mo CERMET
SPECIFIC HEAT, cal g⁻¹ K⁻¹
TEMPERATURE, K
1
FIG 395

SPECIFICATION TABLE NO. 395 SPECIFIC HEAT OF BERYLLIUM OXIDE + MOLYBDENUM, BeO + Mo CERMET

[For Data Reported in Figure and Table No. 395]

Curve No.	Ref. No.	Year	Temp. Range, K	Reported Error, %	Name and Specimen Designation	Composition (weight percent), Specifications and Remarks
1	45	1962	303-1073	< 3.0		57.6 BeO and 42.4 Mo.

DATA TABLE NO. 395 SPECIFIC HEAT OF BERYLLIUM OXIDE + MOLYBDENUM, BeO + Mo CERMET

[Temperature, T, K; Specific Heat, C_p, Cal g^{-1} K^{-1}]

T	C_p
CURVE 1	
303	1.68 x 10^{-1}
373	1.88
473	2.23
573	2.42
673	2.55
773	2.63
873	2.70
973	2.78
1073	2.85

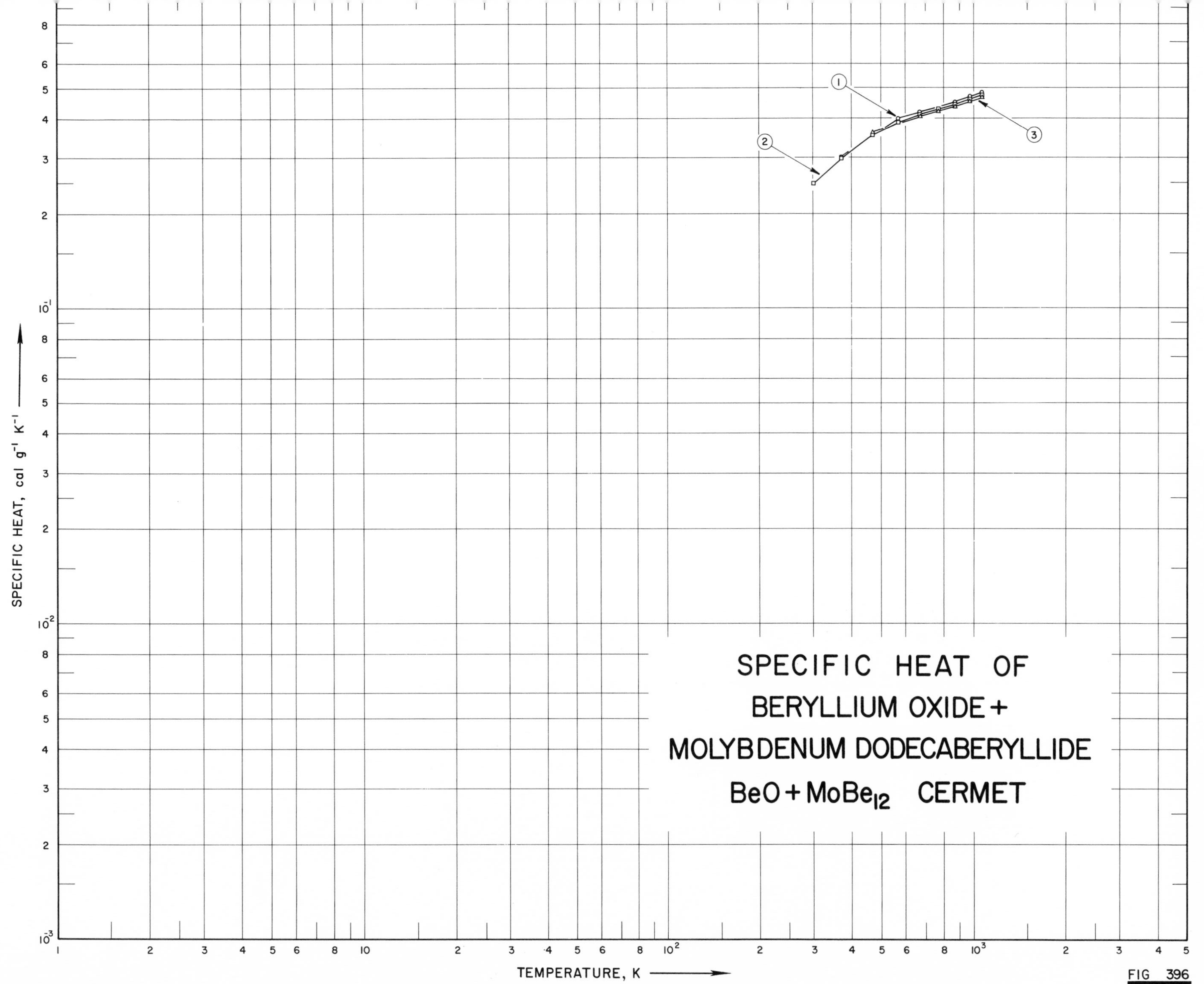
SPECIFIC HEAT OF
BERYLLIUM OXIDE +
MOLYBDENUM DODECABERYLLIDE
BeO + MoBe12 CERMET
SPECIFIC HEAT, cal g-1 K-1
TEMPERATURE, K
1
2
3
FIG 396

SPECIFICATION TABLE NO. 396 SPECIFIC HEAT OF BERYLLIUM OXIDE + MOLYBDENUM DODECABERYLLIDE, $BeO + MoBe_{12}$ CERMET

[For Data Reported in Figure and Table No. 396]

Curve No.	Ref. No.	Year	Temp. Range, K	Reported Error, %	Name and Specimen Designation	Composition (weight percent), Specifications and Remarks
1	45	1962	303-1073	< 3.0		Nominal composition: 94 BeO and 6 $MoBe_{12}$.
2	45	1962	303-1073	< 3.0		Nominal composition: 94 BeO and 9 $MoBe_{12}$.
3	45	1962	303-1073	< 3.0		Nominal composition: 82 BeO and 18 $MoBe_{12}$.

DATA TABLE NO. 396 SPECIFIC HEAT OF BERYLLIUM OXIDE + MOLYBDENUM DODECABERYLLIDE, $BeO + MoBe_{12}$ CERMET

[Temperature, T, K; Specific Heat, C_p, Cal $g^{-1} K^{-1}$]

T	C_p
CURVE 1	
303	2.49 x 10^{-1}*
373	3.00
473	3.60*
573	3.99
673	4.17
773	4.34
873	4.51
973	4.68
1073	4.84
CURVE 2	
303	2.49 x 10^{-1}
373	2.98
473	3.56
573	3.89
673	4.12
773	4.28
873	4.44
973	4.60
1073	4.75
CURVE 3	
303	2.50 x 10^{-1}*
373	2.99*
473	3.62
573	3.92*
673	4.07
773	4.23
873	4.38
973	4.54
1073	4.70

*Not shown on plot

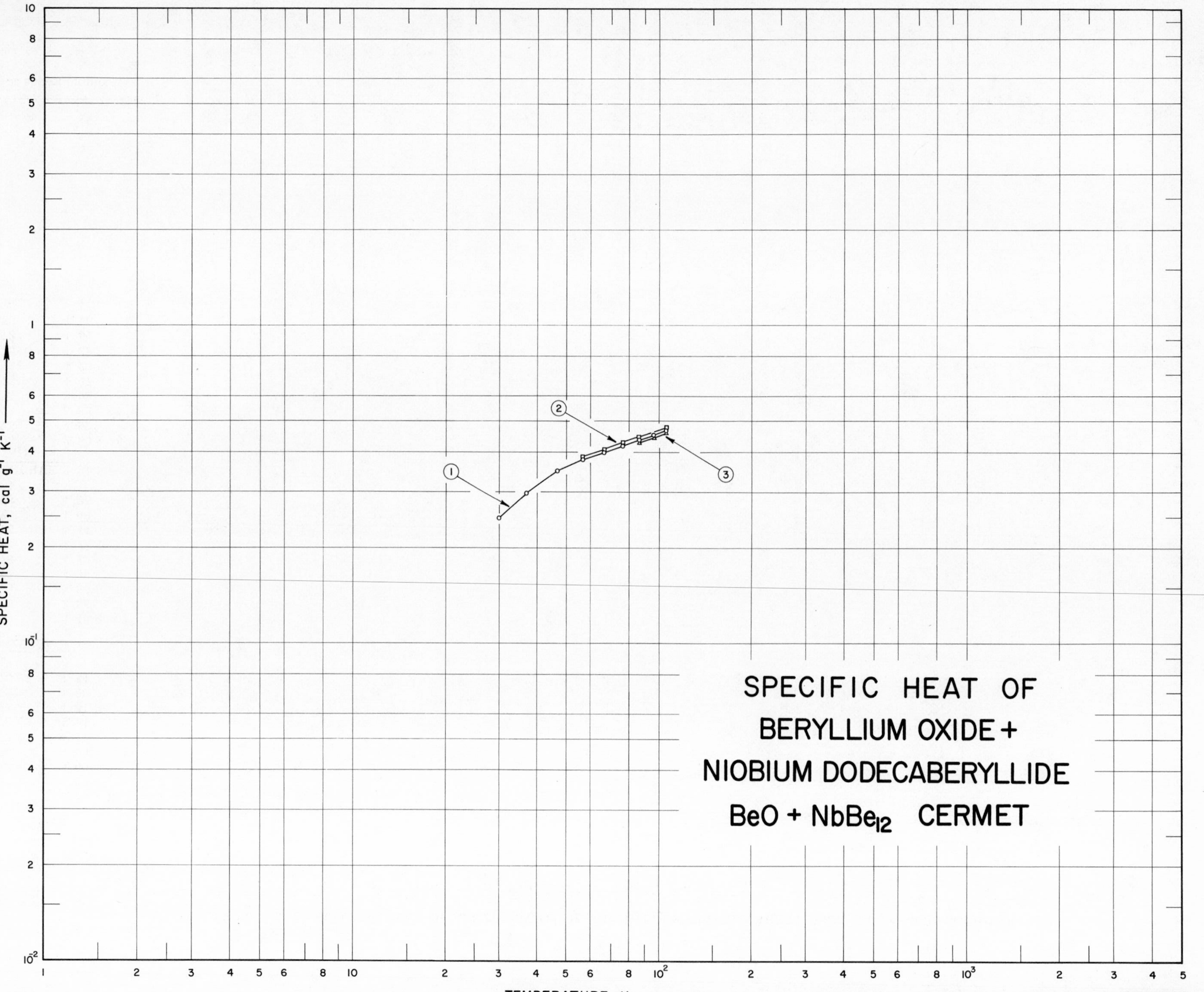
SPECIFIC HEAT OF
BERYLLIUM OXIDE +
NIOBIUM DODECABERYLLIDE
BeO + NbBe12 CERMET
SPECIFIC HEAT, cal g-1 K-1
TEMPERATURE, K
1
2
3

FIG 397

SPECIFICATION TABLE NO. 397 SPECIFIC HEAT OF BERYLLIUM OXIDE + NIOBIUM DODECABERYLLIDE, $BeO + NbBe_{12}$ CERMET

[For Data Reported in Figure and Table No. 397]

Curve No.	Ref. No.	Year	Temp. Range, K	Reported Error, %	Name and Specimen Designation	Composition (weight percent), Specifications and Remarks
1	45	1962	303-1073	< 3.0		Nominal composition: 94 BeO and 6 $NbBe_{12}$.
2	45	1962	303-1073	< 3.0		Nominal composition: 91 BeO and 9 $NbBe_{12}$.
3	45	1962	303-1073	< 3.0		Nominal composition: 82 BeO and 18 $NbBe_{12}$.

DATA TABLE NO. 397 SPECIFIC HEAT OF BERYLLIUM OXIDE + NIOBIUM DODECABERYLLIDE, $BeO + NbBe_{12}$ CERMET

[Temperature, T, K; Specific Heat, C_p, Cal $g^{-1} K^{-1}$]

T	C_p
CURVE 1	
303	2.49×10^{-1}
373	2.99
473	3.52
573	3.83
673	4.04
773	4.24
873	4.41
973	4.56
1073	4.74
CURVE 2	
303	2.49×10^{-1}*
373	2.99*
473	3.53*
573	3.89
673	4.12
773	4.32
873	4.48
973	4.64*
1073	4.80
CURVE 3	
303	2.51×10^{-1}*
373	2.98*
473	3.52*
573	3.87*
673	4.03*
773	4.19*
873	4.34
973	4.49
1073	4.66

* Not shown on plot

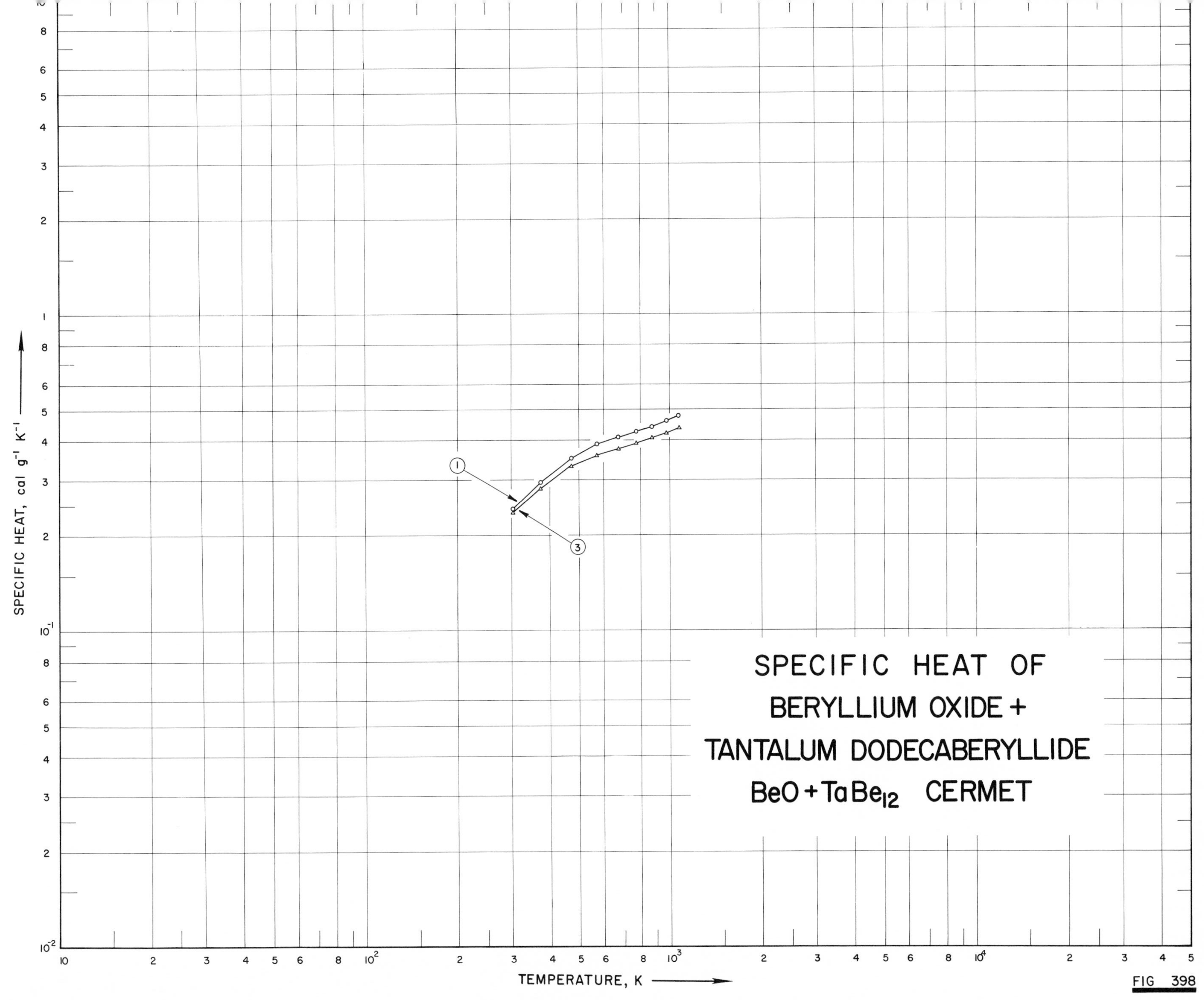
SPECIFIC HEAT OF
BERYLLIUM OXIDE +
TANTALUM DODECABERYLLIDE
BeO + TaBe12 CERMET
SPECIFIC HEAT, cal g-1 K-1
TEMPERATURE, K
1
3

FIG 398

SPECIFICATION TABLE NO. 398 SPECIFIC HEAT OF BERYLLIUM OXIDE + TANTALUM DODECABERYLLIDE, $BeO + TaBe_{12}$ CERMET

[For Data Reported in Figure and Table No. 398]

Curve No.	Ref. No.	Year	Temp. Range, K	Reported Error, %	Name and Specimen Designation	Composition (weight percent), Specifications and Remarks
1	45	1962	303-1073	< 3.0		Nominal composition: 94 BeO and 6 $TaBe_{12}$.
2	45	1962	303-1073	< 3.0		Nominal composition: 91 BeO and 9 $TaBe_{12}$.
3	45	1962	303-1073	< 3.0		Nominal composition: 82 BeO and 18 $TaBe_{12}$.

DATA TABLE NO. 398 SPECIFIC HEAT OF BERYLLIUM OXIDE + TANTALUM DODECABERYLLIDE, $BeO + TaBe_{12}$ CERMET

[Temperature, T, K; Specific Heat, C_p, Cal $g^{-1} K^{-1}$]

T	C_p
CURVE 1	
303	2.44×10^{-1}
373	2.94
473	3.50
573	3.88
673	4.08
773	4.26
873	4.43
973	4.60
1073	4.77
CURVE 2*	
303	2.42×10^{-1}
373	3.02
473	3.55
573	3.88
673	4.10
773	4.28
873	4.44
973	4.59
1073	4.74
CURVE 3	
303	2.37×10^{-1}
373	2.82
473	3.31
573	3.58
673	3.75
773	3.90
873	4.06
973	4.22
1073	4.38

*Not shown on plot

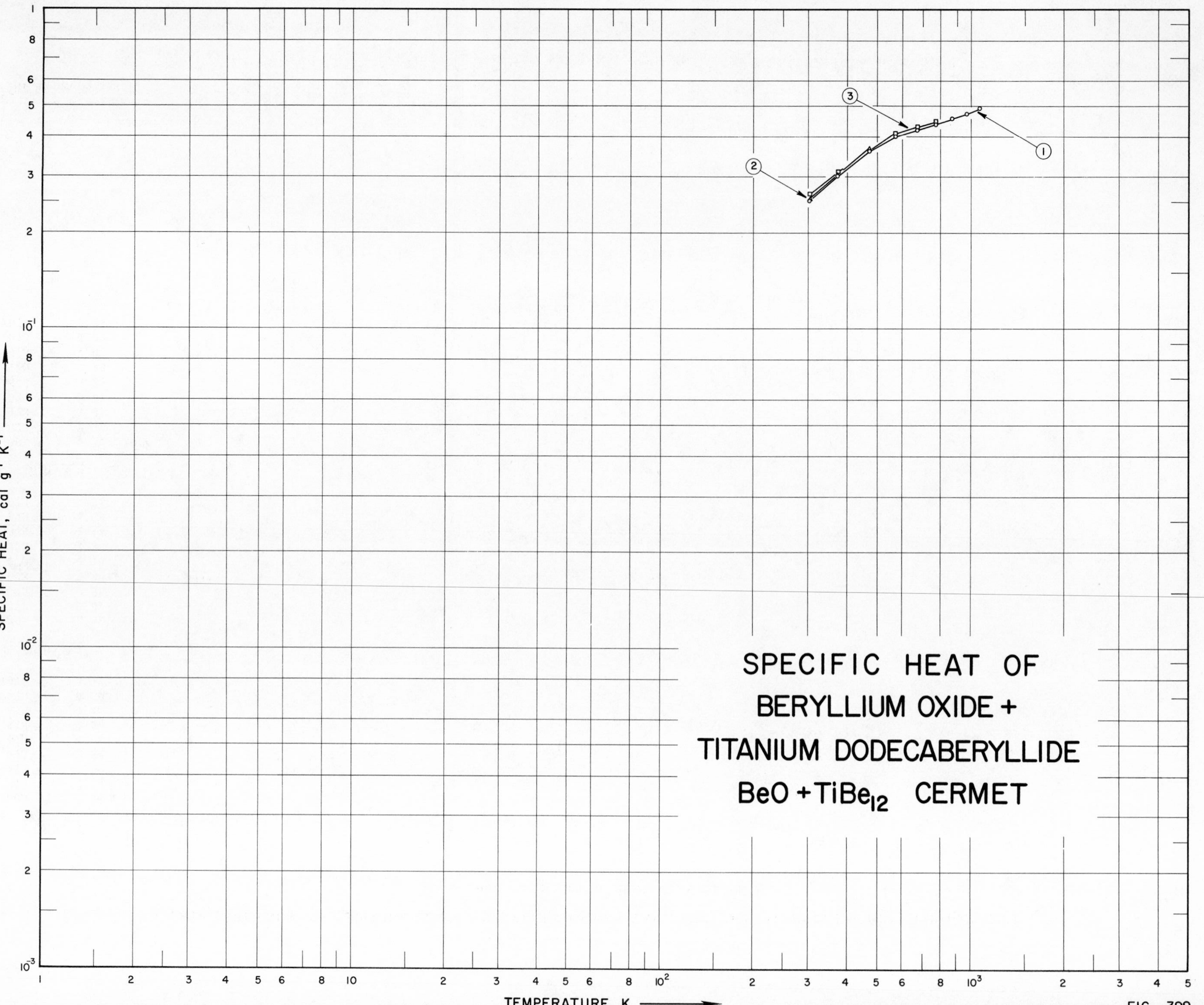

SPECIFIC HEAT OF
BERYLLIUM OXIDE +
TITANIUM DODECABERYLLIDE
BeO + TiBe12 CERMET
SPECIFIC HEAT, cal g-1 K-1
TEMPERATURE, K
1
2
3

FIG. 399

SPECIFICATION TABLE NO. 399 SPECIFIC HEAT OF BERYLLIUM OXIDE + TITANIUM DODECABERYLLIDE, BeO + $TiBe_{12}$ CERMET

[For Data Reported in Figure and Table No. 399]

Curve No.	Ref. No.	Year	Temp. Range, K	Reported Error, %	Name and Specimen Designation	Composition (weight percent), Specifications and Remarks
1	45	1962	303-1073	< 3.0		Nominal composition: 94 BeO and 6 $TiBe_{12}$.
2	45	1962	303-1073	< 3.0		Nominal composition: 91 BeO and 9 $TiBe_{12}$.
3	45	1962	303-1073	< 3.0		Nominal composition: 82 BeO and 18 $TiBe_{12}$.

DATA TABLE NO. 399 SPECIFIC HEAT OF BERYLLIUM OXIDE + TITANIUM DODECABERYLLIDE, $BeO + TiBe_{12}$ CERMET

[Temperature, T, K; Specific Heat, C_p, Cal $g^{-1} K^{-1}$]

T	C_p
CURVE 1	
303	2.53×10^{-1}
373	3.04
473	3.61
573	4.01
673	4.22
773	4.39
873	4.56
973	4.73
1073	4.90
CURVE 2	
303	2.56×10^{-1}
373	3.09
473	3.66
573	4.01*
673	4.23*
773	4.41*
873	4.59*
973	4.75*
1073	4.93*
CURVE 3	
303	2.65×10^{-1}
373	3.10
473	3.65*
573	4.10
673	4.30
773	4.46
873	4.62*
973	4.78*
1073	4.94*

*Not shown on plot

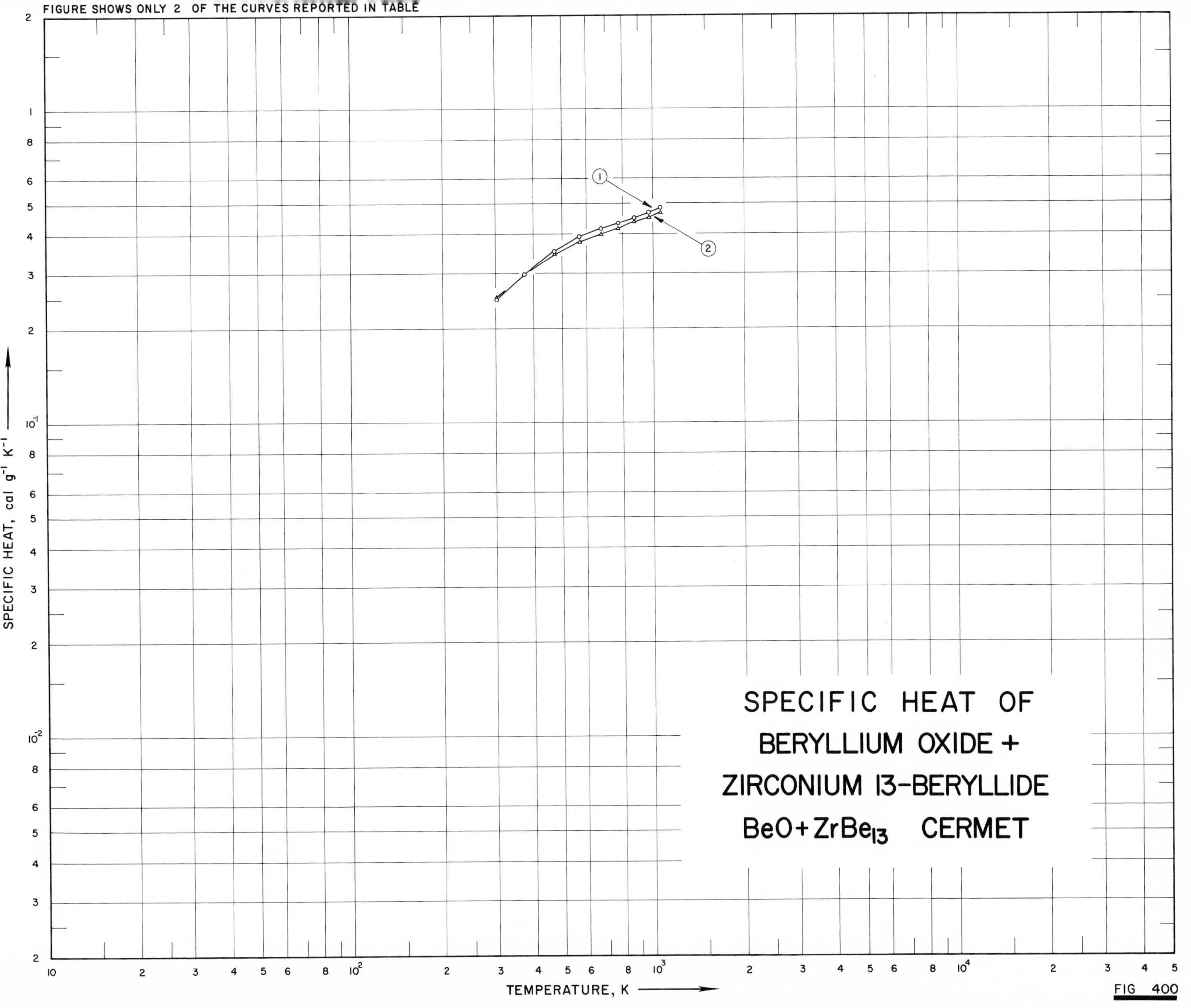
FIGURE SHOWS ONLY 2 OF THE CURVES REPORTED IN TABLE
SPECIFIC HEAT, cal g⁻¹ K⁻¹
TEMPERATURE, K
1
2
SPECIFIC HEAT OF
BERYLLIUM OXIDE +
ZIRCONIUM 13-BERYLLIDE
BeO+ZrBe13 CERMET

FIG 400

SPECIFICATION TABLE NO. 400 SPECIFIC HEAT OF BERYLLIUM OXIDE + ZIRCONIUM 13-BERYLLIDE, $BeO + ZrBe_{13}$ CERMET

[For Data Reported in Figure and Table No. 400]

Curve No.	Ref. No.	Year	Temp. Range, K	Reported Error, %	Name and Specimen Designation	Composition (weight percent), Specifications and Remarks
1	45	1962	303-1073	< 3.0		Nominal composition: 94 BeO and 6 $ZrBe_{13}$.
2	45	1962	303-1073	< 3.0		Nominal composition: 91 BeO and 9 $ZrBe_{13}$.
3	45	1962	303-1073	< 3.0		Nominal composition: 82 BeO and 18 $ZrBe_{13}$.

DATA TABLE NO. 400 SPECIFIC HEAT OF BERYLLIUM OXIDE + ZIRCONIUM 13-BERYLLIDE, BeO + $ZrBe_{13}$ CERMET

[Temperature, T, K; Specific Heat, C_p, Cal g^{-1} K^{-1}]

T	C_p
CURVE 1	
303	2.49×10^{-1}
373	3.00
473	3.54
573	3.95
673	4.19
773	4.36
873	4.54
973	4.71
1073	4.89
CURVE 2*	
303	2.50×10^{-1}
373	3.00
473	3.57
573	3.94
673	4.17
773	4.35
873	4.52
973	4.69
1073	4.86
CURVE 3	
303	2.53×10^{-1}
373	3.00*
473	3.51
573	3.81
673	4.03
773	4.23
873	4.40
973	4.56
1073	4.73

*Not shown on plot

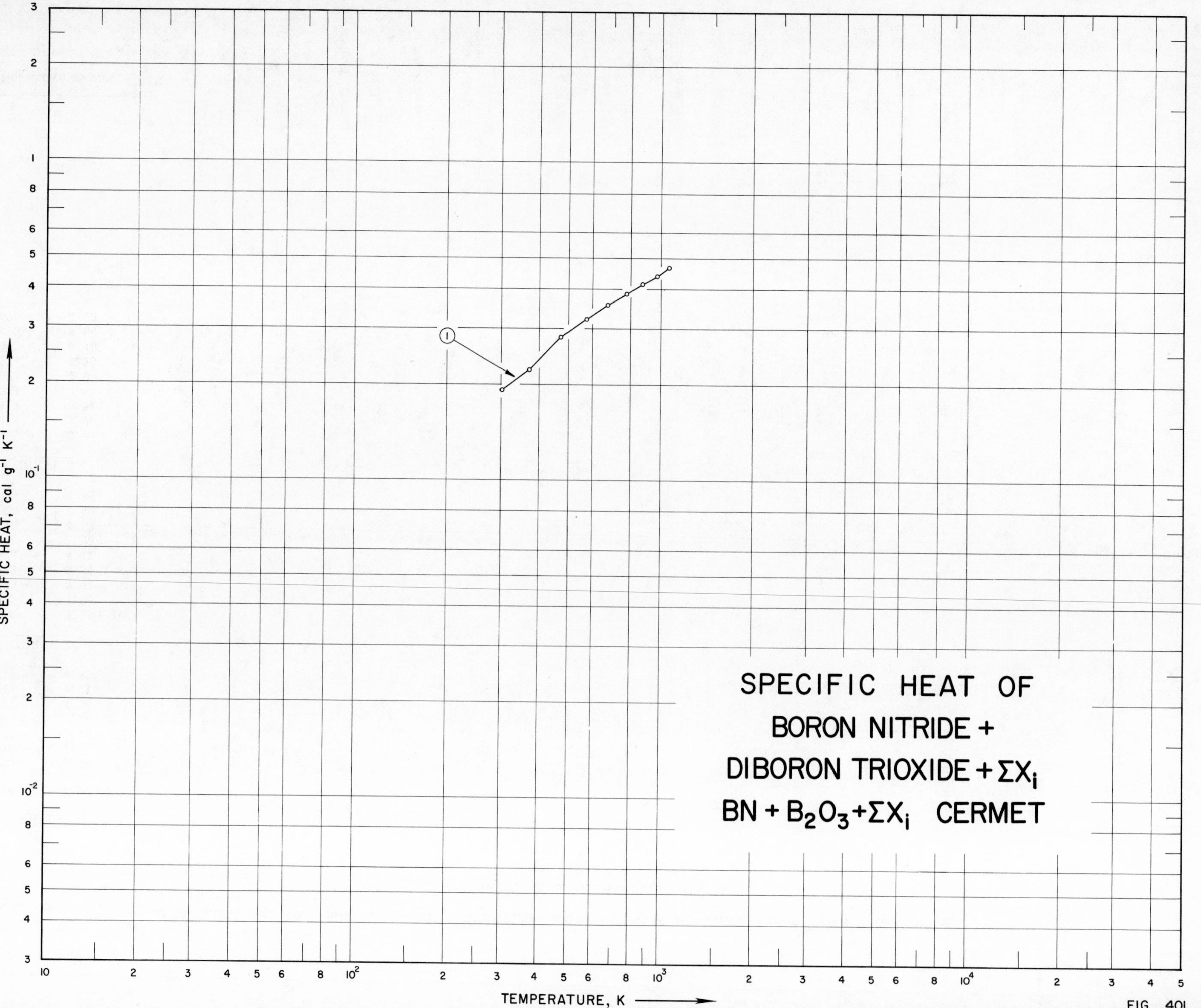
SPECIFIC HEAT OF
BORON NITRIDE +
DIBORON TRIOXIDE + ΣXi
BN + B2O3 + ΣXi CERMET
SPECIFIC HEAT, cal g-1 K-1
TEMPERATURE, K
FIG 401

SPECIFICATION TABLE NO. 401 SPECIFIC HEAT OF BORON NITRIDE + DIBORON TRIOXIDE + ΣXi, $BN + B_2O_3 + \Sigma Xi$ CERMET

[For Data Reported in Figure and Table No. 401]

Curve No.	Ref. No.	Year	Temp. Range, K	Reported Error, %	Name and Specimen Designation	Composition (weight percent), Specifications and Remarks
1	45	1962	303-1073	< 3.0		97.4 BN, 2.4 B_2O_3, 0.2 Al and Si.

DATA TABLE NO. 401 SPECIFIC HEAT OF BORON NITRIDE + DIBORON TRIOXIDE + ΣX_i, $BN + B_2O_3 + \Sigma X_i$ CERMET

[Temperature, T, K; Specific Heat, C_p, Cal $g^{-1} K^{-1}$]

T	C_p
CURVE 1	
303	1.93×10^{-1}
373	2.36
473	2.83
573	3.24
673	3.58
773	3.89
873	4.16
973	4.41
1073	4.68

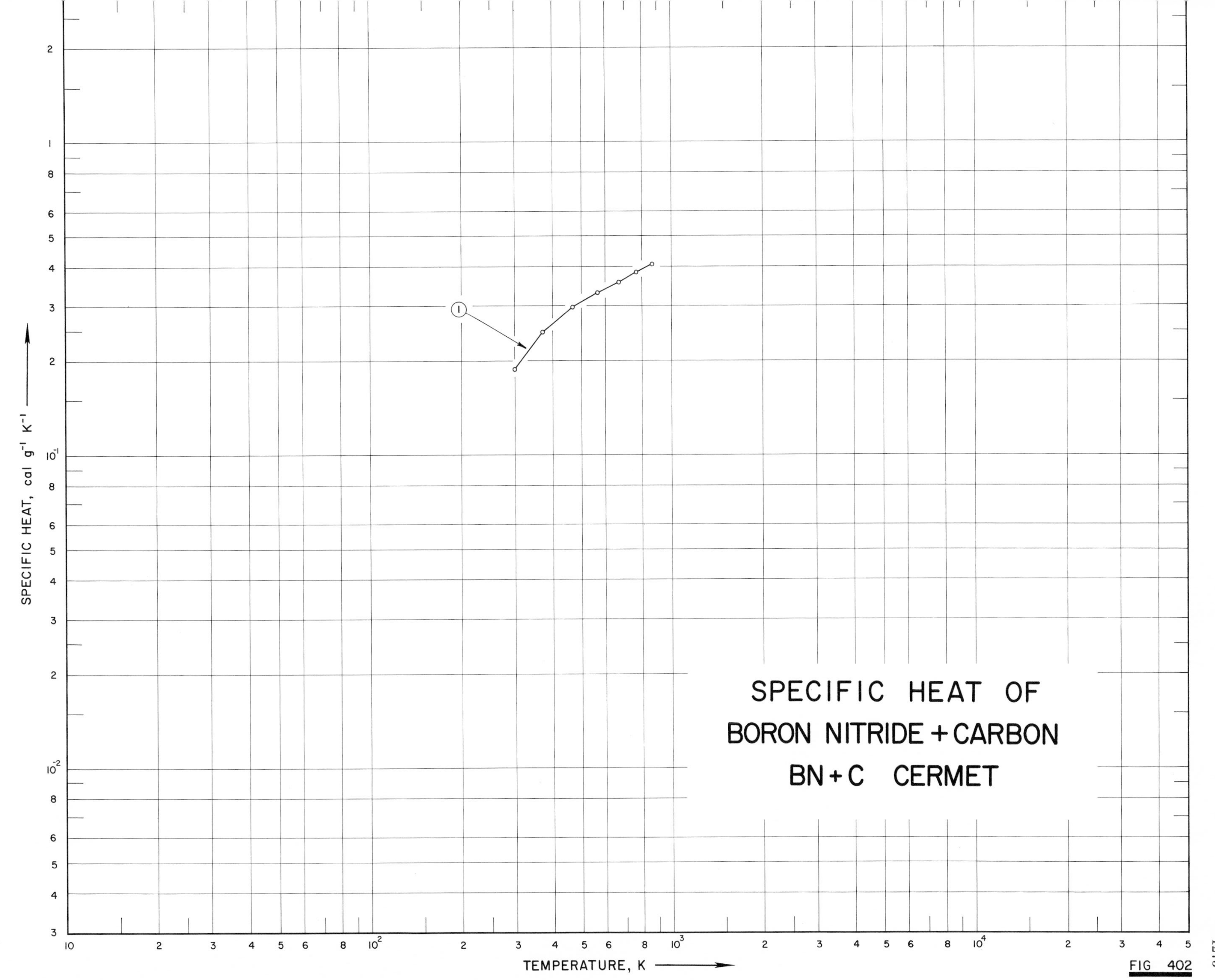
SPECIFIC HEAT OF
BORON NITRIDE + CARBON
BN + C CERMET
TEMPERATURE, K
SPECIFIC HEAT, cal g^{-1} K^{-1}
1
FIG 402

SPECIFICATION TABLE NO. 402 SPECIFIC HEAT OF BORON NITRIDE + CARBON, BN + C CERMET

[For Data Reported in Figure and Table No. 402]

Curve No.	Ref. No.	Year	Temp. Range, K	Reported Error, %	Name and Specimen Designation	Composition (weight percent), Specifications and Remarks
1	45	1962	303-873	< 3.0		Nominal composition: 80 BN and 20 C.

DATA TABLE NO. 402 SPECIFIC HEAT OF BORON NITRIDE + CARBON, BN + C CERMET

[Temperature, T, K; Specific Heat, C_p, Cal $g^{-1} K^{-1}$]

T	C_p
CURVE 1	
303	1.90×10^{-1}
373	2.50
473	3.00
573	3.36
673	3.62
773	3.87
873	4.11

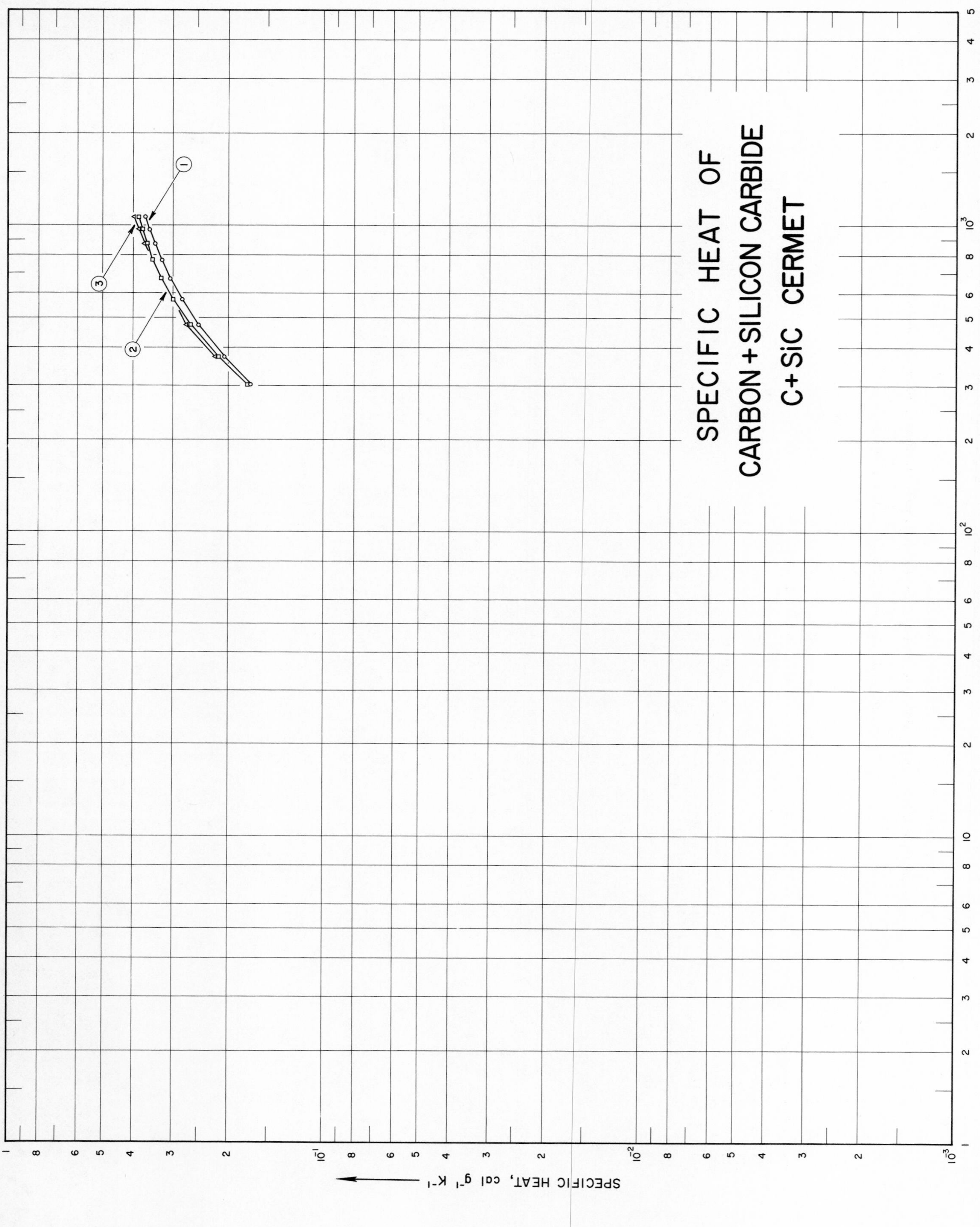

SPECIFIC HEAT OF CARBON + SILICON CARBIDE C+SiC CERMET

FIG 403

SPECIFICATION TABLE NO. 403 SPECIFIC HEAT OF CARBON + SILICON CARBIDE, C + SiC CERMET

[For Data Reported in Figure and Table No. 403]

Curve No.	Ref. No.	Year	Temp. Range, K	Reported Error, %	Name and Specimen Designation	Composition (weight percent), Specifications and Remarks
1	45	1962	303-1073	< 3.0		Nominal composition: 57 C and 43 SiC.
2	45	1962	303-1073	< 3.0		Nominal composition: 72 C and 28 SiC.
3	45	1962	303-1073	< 3.0		Nominal composition: 77 C and 23 SiC.

DATA TABLE NO. 403 SPECIFIC HEAT OF CARBON + SILICON CARBIDE, C + SiC CERMET

[Temperature, T, K; Specific Heat, C_p, Cal $g^{-1} K^{-1}$]

T	C_p
CURVE 1	
303	1.70×10^{-1}
373	2.06
473	2.49
573	2.80
673	3.05
773	3.25
873	3.41
973	3.55
1073	3.65
CURVE 2	
303	1.73×10^{-1}
373	2.15
473	2.63
573	2.99
673	3.26
773	3.47
873	3.60
973	3.73
1073	3.83
CURVE 3	
303	1.73×10^{-1}*
373	2.30
473	2.74
573	3.04*
673	3.28*
773	3.49*
873	3.67
973	3.84
1073	3.98

*Not shown on plot

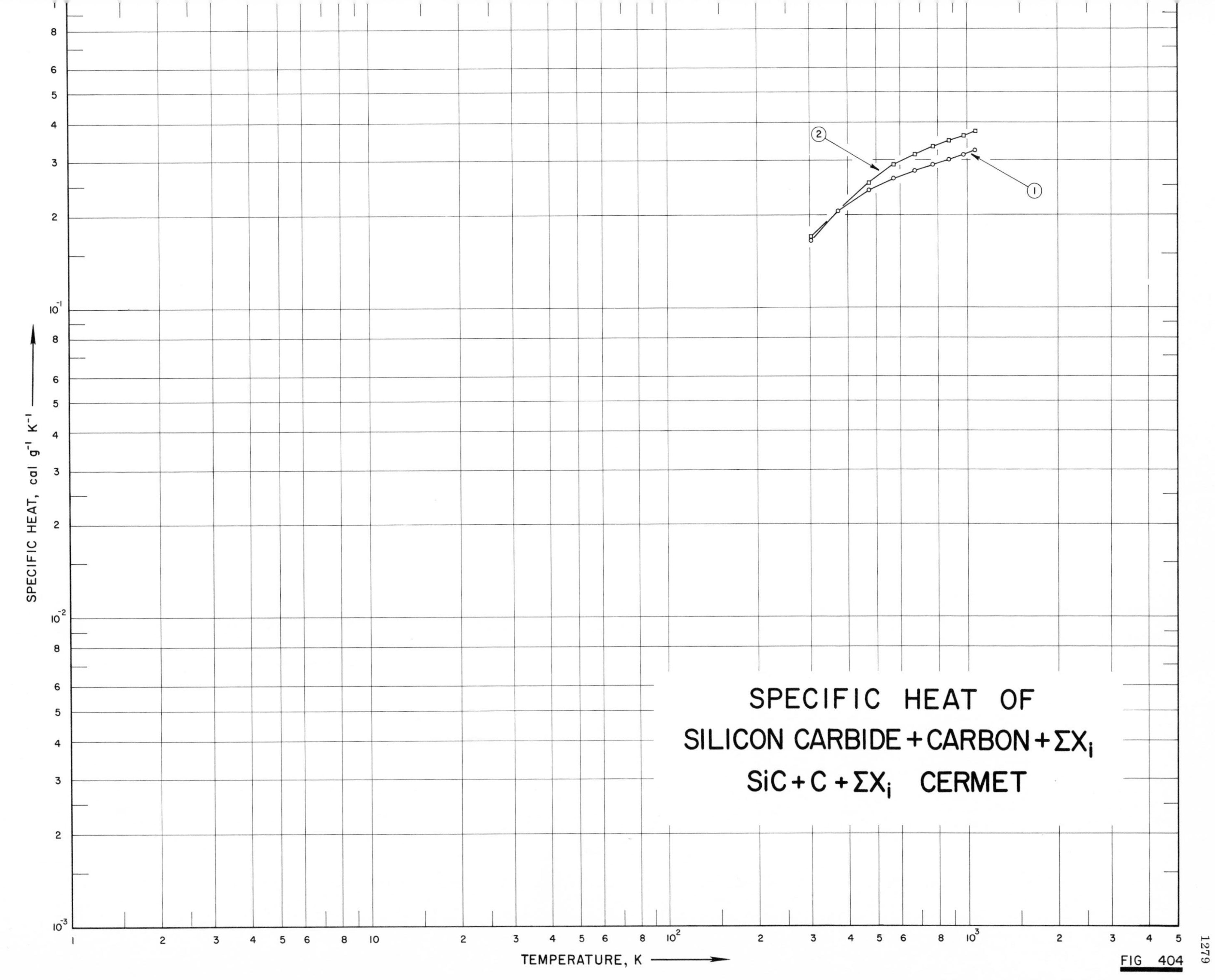
SPECIFIC HEAT OF
SILICON CARBIDE + CARBON + ΣXi
SiC + C + ΣXi CERMET
TEMPERATURE, K
SPECIFIC HEAT, cal g⁻¹ K⁻¹
1
2

FIG 404

SPECIFICATION TABLE NO. 404 SPECIFIC HEAT OF SILICON CARBIDE + CARBON + ΣX_i, SiC + C + ΣX_i CERMET

[For Data Reported in Figure and Table No. 404]

Curve No.	Ref. No.	Year	Temp. Range, K	Reported Error, %	Name and Specimen Designation	Composition (weight percent), Specifications and Remarks
1	45	1962	303-1073	< 3.0	low free carbon	78.0 SiC, 20.0 C, 1.0 Si, and 1.0 various other metals.
2	45	1962	303-1073	< 3.0	high free carbon	50 SiC, 46 C, and 4 Si.

DATA TABLE NO. 404 SPECIFIC HEAT OF SILICON CARBIDE + CARBON + ΣX_i, SiC + C + ΣX_i CERMET

[Temperature, T, K; Specific Heat, C_p, Cal g^{-1} K^{-1}]

T	C_p
CURVE 1	
303	1.66 x 10^{-1}
373	2.07
473	2.41
573	2.63
673	2.78
773	2.92
873	3.03
973	3.14
1073	3.23
CURVE 2	
303	1.71 x 10^{-1}
373	2.07*
473	2.55
573	2.92
673	3.14
773	3.32
873	3.47
973	3.60
1073	3.73

*Not shown on plot

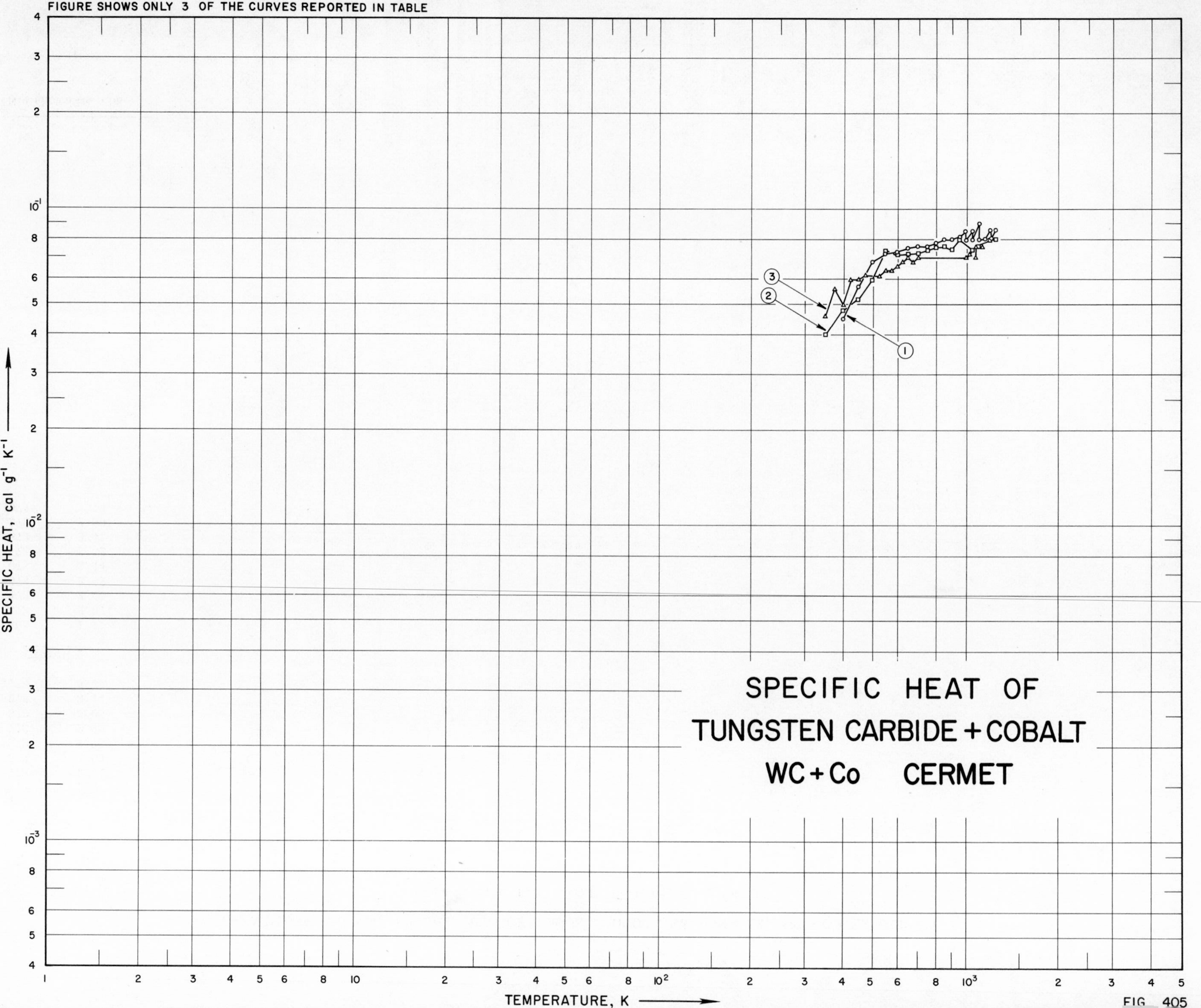
FIGURE SHOWS ONLY 3 OF THE CURVES REPORTED IN TABLE
SPECIFIC HEAT OF
TUNGSTEN CARBIDE + COBALT
WC + Co CERMET
SPECIFIC HEAT, cal g⁻¹ K⁻¹
TEMPERATURE, K
1
2
3
FIG 405

SPECIFICATION TABLE NO. 405 SPECIFIC HEAT OF TUNGSTEN CARBIDE + COBALT, WC + Co CERMET

[For Data Reported in Figure and Table No. 405]

Curve No.	Ref. No.	Year	Temp. Range, K	Reported Error, %	Name and Specimen Designation	Composition (weight percent), Specifications and Remarks
1	351	1959	398-1249	5.0-15.0		7.0 Co; sintered at 1300-1500 C.
2	351	1959	349-1247	5.0-15.0		12.0 Co; sintered at 1300-1500 C.
3	351	1959	348-1200	5.0-15.0		5.0 Co; sintered at 1300-1500 C.
4	351	1959	349-1246	5.0-15.0		3.0 Co; sintered at 1300-1500 C.

DATA TABLE NO. 405 SPECIFIC HEAT OF TUNGSTEN CARBIDE + COBALT, WC + Co CERMET

[Temperature, T, K; Specific Heat, C_p, Cal g^{-1} K^{-1}]

CURVE 1

T	C_p
398	4.51×10^{-2}
446	5.71
497	6.82
549	7.23
596	7.32
647	7.52
696	7.63
749	7.63
797	7.84
847	8.00
897	8.01
947	8.03*
951	8.21
996	8.52
999	7.99
1047	8.53
1050	7.99
1101	9.01
1101	8.00
1150	8.01
1197	8.62
1200	8.13
1249	8.63

CURVE 2

T	C_p
348.9	4.00×10^{-2}
396	4.81
446	5.20
496	5.99
547	7.39
597	7.19
647	7.21
697	7.21
747	7.43
798	7.59
846	7.62
895	7.46
946	8.02
995	8.01*
1045	7.45
1097	7.63
1147	8.02*
1196	8.02
1247	8.02

CURVE 3

T	C_p
347.5	4.60×10^{-2}
373	5.62
376	4.99
422	5.98
446	5.98
472	6.16
497	6.00*
523	6.15
546	6.41
572	6.41
597	6.60
622	6.81
647	7.01
672	6.81
697	7.01
722	7.01*
749	7.01*
774	7.01*
797	7.01*
823	7.01*
849	7.01*
873	7.01*
896	7.01*
923	7.01*
946	7.01*
996	7.01
1022	7.20
1072	7.01
1073	7.61
1097	7.60*
1123	7.60
1200	8.02*

CURVE 4

T	C_p
349	4.60×10^{-2}
396	5.38
447	5.98
496	5.98
546	5.98
596	6.99
649	6.99
696	6.99
747	6.99
797	6.99

CURVE 4 (cont.)

T	C_p
846	6.99×10^{-2}
876	7.58
946	7.39
996	7.40
1046	7.59
1096	7.99
1127	7.39
1196	7.58
1246	7.59

*Not shown on plot

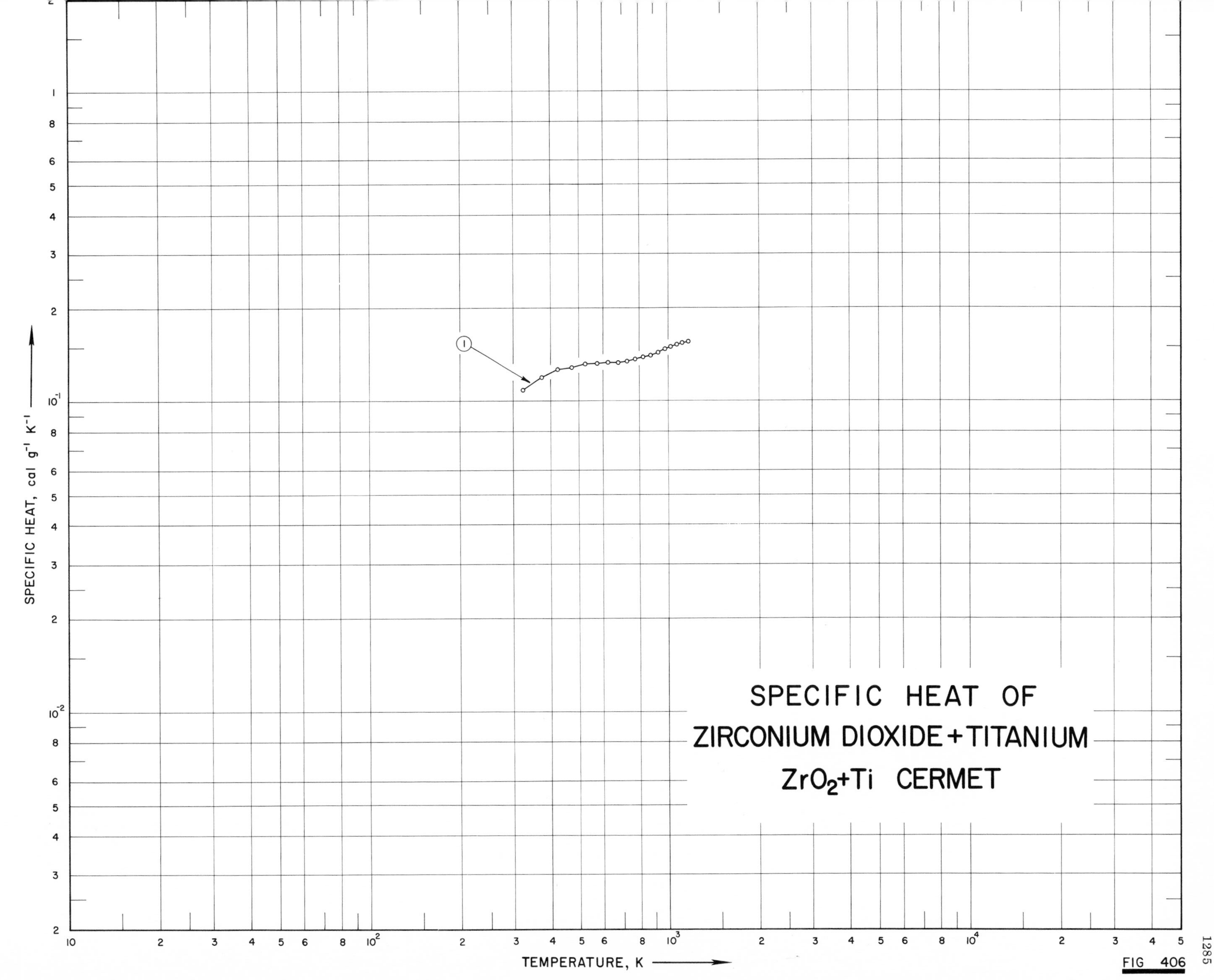

SPECIFIC HEAT OF
ZIRCONIUM DIOXIDE + TITANIUM
ZrO_2+Ti CERMET
SPECIFIC HEAT, cal g^{-1} K^{-1}
TEMPERATURE, K
1

FIG 406

SPECIFICATION TABLE NO. 406 SPECIFIC HEAT OF ZIRCONIUM DIOXIDE + TITANIUM, ZrO_2 + Ti CERMET

[For Data Reported in Figure and Table No. 406]

Curve No.	Ref. No.	Year	Temp. Range, K	Reported Error, %	Name and Specimen Designation	Composition (weight percent), Specifications and Remarks
1	352	1964	323-1173	5.0	Milled Zirconia 15 mole % Ti (ZT-15-M)	93.68 ZrO_2 and 6.32 Ti.

DATA TABLE NO. 406 SPECIFIC HEAT OF ZIRCONIUM DIOXIDE + TITANIUM, ZrO_2 + Ti CERMET

[Temperature, T, K; Specific Heat, C_p, Cal g^{-1} K^{-1}]

T	C_p
CURVE 1	
323	1.10×10^{-1}
373	1.21
423	1.28
473	1.30
529	1.33
573	1.34
623	1.35
673	1.35
723	1.37
773	1.38
823	1.40
873	1.42
923	1.45
973	1.49
1023	1.52
1073	1.54
1123	1.56
1173	1.57

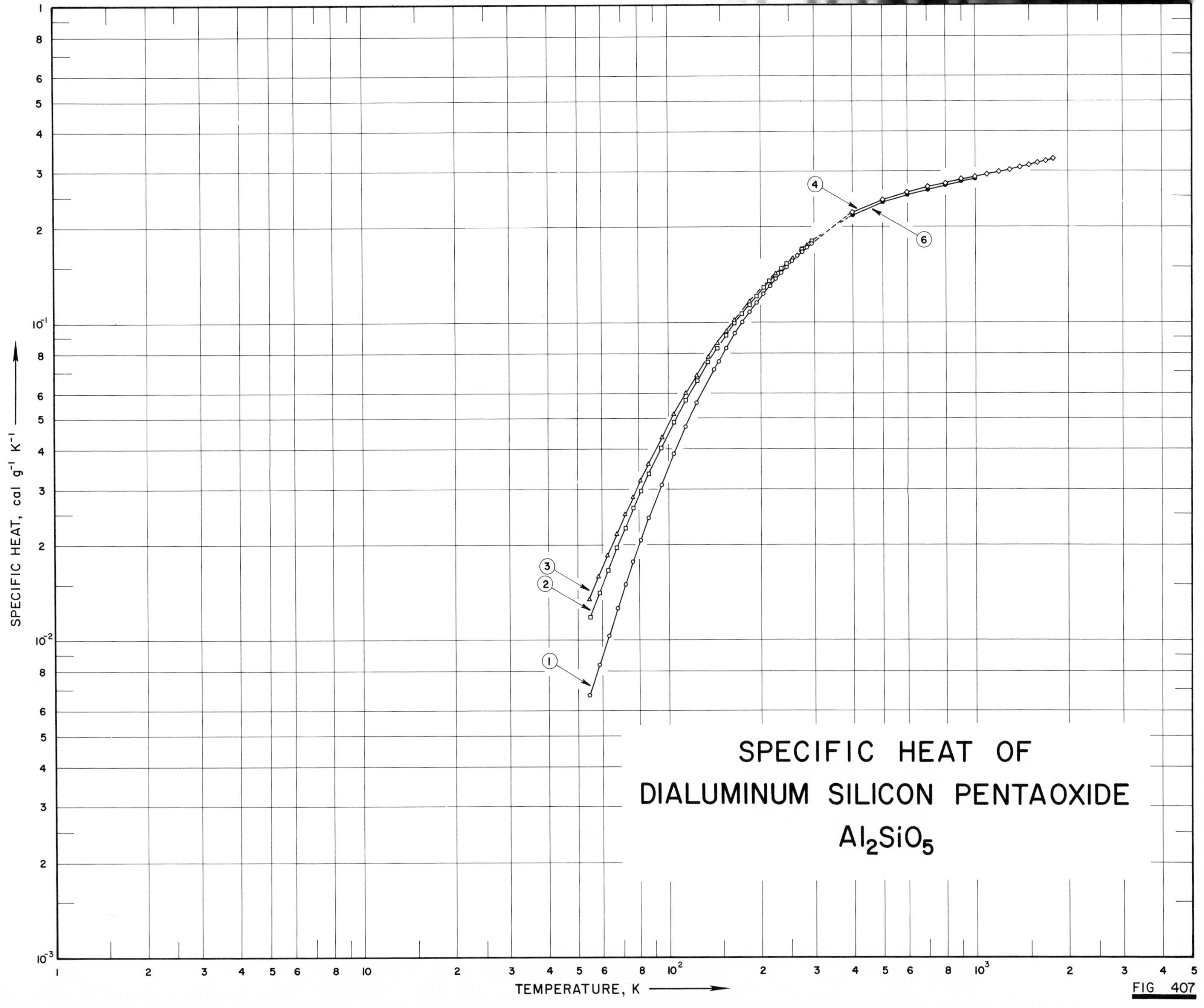
SPECIFIC HEAT OF
DIALUMINUM SILICON PENTAOXIDE
Al_2SiO_5
SPECIFIC HEAT, cal g⁻¹ K⁻¹
TEMPERATURE, K
1
2
3
4
6
FIG 407

SPECIFICATION TABLE NO. 407 SPECIFIC HEAT OF DIALUMINUM SILICON PENTAOXIDE Al_2SiO_5

[For Data Reported in Figure and Table No. 407]

Curve No.	Ref. No.	Year	Temp. Range, K	Reported Error, %	Name and Specimen Designation	Composition (weight percent), Specifications and Remarks
1	353	1950	55-296		Kyanite	62.2 Al_2O_3, 36.9 SiO_2, 0.1 Fe_2O_3, 0.05 CaO.
2	353	1950	55-296		Andalusite	63.15 Al_2O_3, 36.84 SiO_2, 0.11 Fe_2O_3, 0.02 CaO, $<$0.01 TiO_2, $<$0.01 MgO, $<$0.01 MnO.
3	353	1950	54-297		Sillimanite	61.8 Al_2O_3, 36.44 SiO_2, 0.98 Fe_2O_3, 0.28 P_2O_5, 0.24 MgO, 0.14 FeO, 0.07 CaO, 0.04 MnO, 0.04 F, $<N_2O$.
4	354	1964	400-1800		Kyanite	63.20 Al_2O_3, 36.90 SiO_2, 0.10 Fe_2O_3, 0.05 CaO; sample from Burnsville, N. Carolina.
5	354	1964	400-1800		Andalusite	63.15 Al_2O_3, 36.84 SiO_2, 0.11 Fe_2O_3, 0.02 CaO, $<$0.01 MgO, $<$0.01 MnO, $<$0.01 TiO_2; contained a few muscovite [$KAl_3Si_3O_{10}(OH)_2$] inclusions; sample from Standish, Maine.
6	354	1964	400-1800	0.1	Sillimanite	61.8 Al_2O_3, 36.44 SiO_2, 0.98 Fe_2O_3, 0.28 P_2O_5, 0.24 MgO, 0.14 FeO, 0.07 CaO, 0.04 F, 0.04 MnO, $<$0.01 Na_2O; contained $<$0.7 wagnerite ($MgF \cdot MgPO_4$) inclusions; sample from Benson Mines, New York.

DATA TABLE NO. 407 SPECIFIC HEAT OF DIALUMINUM SILICON PENTAOXIDE Al_2SiO_5

[Temperature, T, K; Specific Heat, C_p, Cal g^{-1} K^{-1}]

T	C_p
CURVE 1	
54.76	6.720 x 10^{-3}
58.83	8.362
63.05	1.031 x 10^{-2}
67.56	1.261
71.97	1.510
76.26	1.774
80.72	2.064
86.06	2.429
95.04	3.099
104.61	3.872
114.62	4.721
124.87	5.617
141.9	7.146
146.7	7.553
155.9	8.355
166.4	9.263
175.9	1.007 x 10^{-1}
186.0	1.085
195.8	1.163
206.2	1.240
216.1	1.312
226.0	1.380
236.0	1.445
245.6	1.506
256.0	1.574
266.0	1.630
276.1	1.683
286.3	1.736
296.3	1.786
CURVE 2	
54.9	1.185 x 10^{-2}
58.9	1.417
62.98	1.667
67.23	1.965
71.69	2.256
76.41	2.610
81.00	2.961
86.29	3.351
94.80	4.042
104.8	4.874
114.6	5.706
124.7	6.572
135.9	7.522
145.9	8.343
CURVE 2 (cont.)	
155.7	9.139 x 10^{-2}
165.9	9.954
176.0	1.074 x 10^{-1}
185.9	1.147
195.9	1.219
206.2	1.290
216.2	1.358
226.1	1.418*
236.2	1.479
245.8	1.537
256.2	1.597*
266.2	1.650*
276.1	1.704
286.2	1.752*
296.2	1.801
CURVE 3	
54.40	1.345 x 10^{-2}
58.48	1.594
62.53	1.856
67.00	2.164
71.52	2.491
76.00	2.832
80.86	3.198
86.07	3.610
95.40	4.371
104.8	5.167
114.8	6.017
124.7	6.850
136.0	7.806
146.3	8.645
155.8	9.405
165.9	1.020 x 10^{-1}
175.9	1.096*
186.4	1.170
196.1	1.237*
206.6	1.310*
216.2	1.371*
226.6	1.435
236.3	1.490*
246.2	1.546*
256.2	1.603
266.1	1.656*
276.2	1.705*
286.3	1.752
296.5	1.801*
CURVE 4	
400	2.23 x 10^{-1}
500	2.44
600	2.58
700	2.68
800	2.76
900	2.83
1000	2.89
1100	2.94
1200	3.00
1300	3.05
1400	3.10
1500	3.15
1600	3.19
1700	3.24
1800	3.29
CURVE 5*	
400	2.23 x 10^{-1}
500	2.44
600	2.57
700	2.66
800	2.73
900	2.80
1000	2.85
1100	2.91
1200	2.95
1300	3.00
1400	3.04
1500	3.09
1600	3.13
1700	3.17
1800	3.21
CURVE 6	
400	2.20 x 10^{-1}
500	2.40
600	2.53
700	2.63
800	2.72
900	2.79
1000	2.85
1100	2.92*
1200	2.97*
1300	3.03*
1400	3.09*
CURVE 6 (cont.)	
1500	3.14 x 10^{-1}*
1600	3.19*
1700	3.25*
1800	3.30*

* Not shown on plot

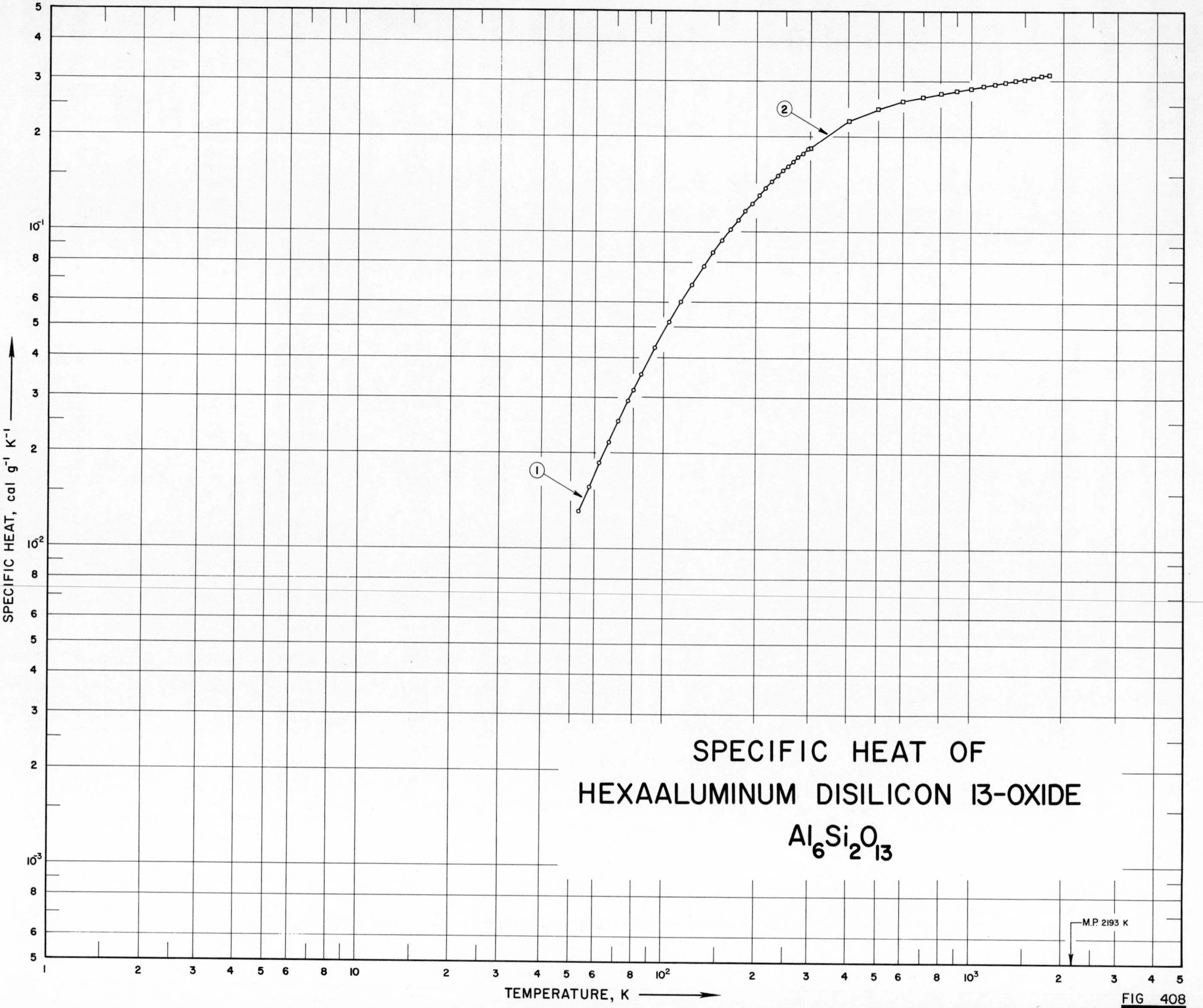
SPECIFIC HEAT OF
HEXAALUMINUM DISILICON 13-OXIDE
$Al_6Si_2O_{13}$
M.P. 2193 K
TEMPERATURE, K
SPECIFIC HEAT, cal g⁻¹ K⁻¹
1
2

FIG 408

SPECIFICATION TABLE NO. 408 SPECIFIC HEAT OF HEXAALUMINUM DISILICON 13-OXIDE $Al_6Si_2O_{13}$

[For Data Reported in Figure and Table No. 408]

Curve No.	Ref. No.	Year	Temp. Range, K	Reported Error, %	Name and Specimen Designation	Composition (weight percent), Specifications and Remarks
1	355	1963	53-296	0.3	Mullite	71.69 Al_2O_3, 28.22 SiO_2; prepared by heating stoichiometric mixture of Al_2O_3 and SiO_2 for 12 days at 1500-1540 C.
2	355	1963	298-1800	0.3	Mullite	Same as above.

DATA TABLE NO. 408 SPECIFIC HEAT OF HEXAALUMINUM DISILICON 13-OXIDE $Al_6Si_2O_{13}$

[Temperature, T, K; Specific Heat, C_p, Cal g^{-1} K^{-1}]

T	C_p
CURVE 1	
53.48	1.31×10^{-2}
57.77	1.57
62.14	1.87
66.74	2.17
71.78	2.523
76.88	2.927
80.05	3.169
84.84	3.556
94.36	4.319
105.04	5.201
114.62	6.004
124.53	6.844
136.00	7.807
145.96	8.614
155.77	9.417
165.95	1.021×10^{-1}
175.92	1.097
186.18	1.171
195.79	1.239
206.27	1.310
216.16	1.377
226.14	1.440
236.09	1.502
246.46	1.563
256.56	1.619
266.45	1.670
276.34	1.723
287.09	1.774
296.35	1.823
CURVE 2	
298.15	1.829×10^{-1}*
300	1.840
400	2.249
500	2.454
600	2.580
700	2.668
800	2.736
900	2.792
1000	2.841
1100	2.885
1200	2.926
1300	2.965
1400	3.002
1500	3.037
CURVE 2 (cont.)	
1600	3.072×10^{-1}
1700	3.106
1800	3.140

* Not shown on plot

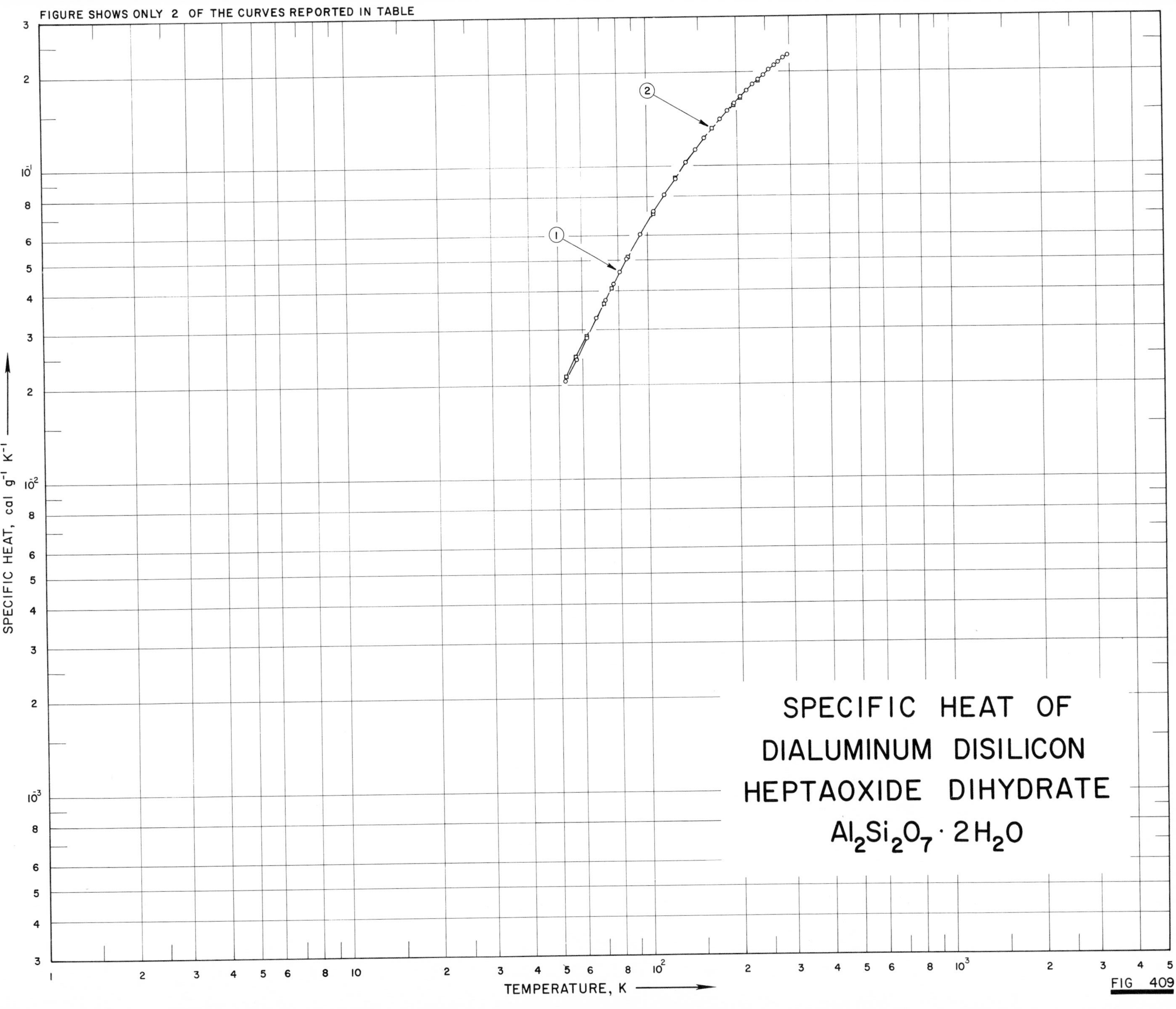
FIGURE SHOWS ONLY 2 OF THE CURVES REPORTED IN TABLE
SPECIFIC HEAT OF
DIALUMINUM DISILICON
HEPTAOXIDE DIHYDRATE
$Al_2Si_2O_7 \cdot 2H_2O$
SPECIFIC HEAT, cal g^{-1} K^{-1}
TEMPERATURE, K
1
2
FIG 409

SPECIFICATION TABLE NO. 409 SPECIFIC HEAT OF DIALUMINUM DISILICON HEPTAOXIDE DIHYDRATE $Al_2Si_2O_7 \cdot 2H_2O$

[For Data Reported in Figure and Table No. 409]

Curve No.	Ref. No.	Year	Temp. Range, K	Reported Error, %	Name and Specimen Designation	Composition (weight percent), Specifications and Remarks
1	356	1961	53-296		Halloysite	45.98 SiO_2, 40.12 Al_2O_3, 13.95 H_2O, 0.20 Na_2O, 0.14 SO_3, 0.03 CaO, 0.02 TiO_2, 0.02 Na_2O, <0.01 Fe_2O_3, nill K_2O, nill MgO; heated to 230 C to remove excess water.
2	356	1961	53-296		Kaolinite	46.59 SiO_2, 39.48 Al_2O_3, 13.74 H_2O; heated to 110 C to remove excess water.
3	356	1961	53-296		Dickite	45.21 SiO_2, 39.92 Al_2O_3, 13.92 H_2O, 0.35 Fe_2O_3, 0.12 TiO_2, 0.08 MgO, 0.04 K_2O and 0.01 CaO; heated to 400 C to remove excess water.

DATA TABLE NO. 409 DIALUMINUM DISILICON HEPTAOXIDE DIHYDRATE $Al_2Si_2O_7 \cdot 2H_2O$

[Temperature, T, K; Specific Heat, C_p, Cal $g^{-1} K^{-1}$]

CURVE 1

T	C_p
52.79	2.077×10^{-2}
57.33	2.437
62.03	2.847
66.88	3.293
71.75	3.754
76.72	4.230
80.80	4.636
85.28	5.086
94.90	6.073
105.62	7.181
114.77	8.115
124.79	9.141
135.82	1.027×10^{-1}
145.70	1.127
155.94	1.227
165.99	1.322*
176.20	1.414
186.38	1.504
196.07	1.582
206.53	1.671
216.29	1.748
226.31	1.822
236.30	1.893
245.76	1.959
256.55	2.032
267.08	2.097
276.54	2.155
286.64	2.212
295.94	2.265

CURVE 2

T	C_p
53.20	2.142×10^{-2}
57.46	2.483
62.05	2.874
66.66	3.283*
71.18	3.692
75.73	4.121
80.97	4.648*
86.10	5.156
94.99	6.077*
104.99	7.127
114.58	8.134*
124.89	9.188
135.90	1.031×10^{-1}*
145.94	1.130*

CURVE 2 (cont.)

T	C_p
156.96	1.236×10^{-1}*
166.43	1.322
176.38	1.413*
186.29	1.495*
196.26	1.576
206.29	1.655
216.55	1.738*
226.07	1.812*
236.05	1.878
245.85	1.947*
256.70	2.017*
266.14	2.088*
276.44	2.145*
286.64	2.205*
296.19	2.261*

CURVE 3*

T	C_p
52.76	2.042×10^{-2}
57.14	2.375
61.55	2.741
66.08	3.135
70.86	3.564
75.63	4.005
81.47	4.563
85.80	4.966
94.95	5.864
105.56	6.910
114.80	7.836
124.89	8.843
136.06	9.947
145.67	1.087×10^{-1}
156.07	1.185
166.10	1.277
176.26	1.367
186.10	1.450
196.36	1.532
206.16	1.614
216.51	1.695
226.32	1.767
236.15	1.836
246.20	1.906
256.44	1.973
267.78	2.037
276.41	2.094
286.64	2.153
296.42	2.209

* Not shown on plot

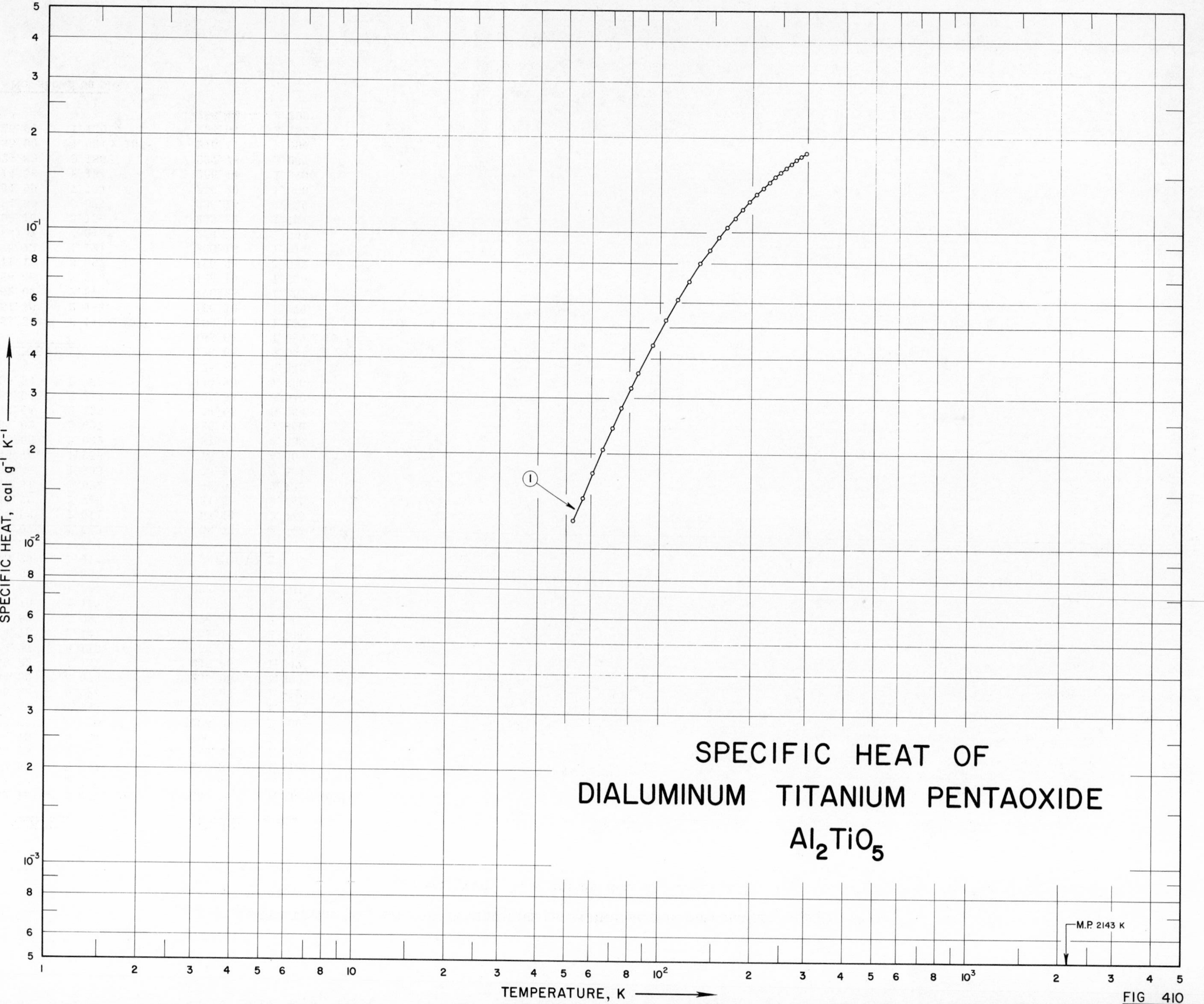
SPECIFIC HEAT OF
DIALUMINUM TITANIUM PENTAOXIDE
Al2TiO5
SPECIFIC HEAT, cal g-1 K-1
TEMPERATURE, K
M.P. 2143 K
1
FIG 410

SPECIFICATION TABLE NO. 410 SPECIFIC HEAT OF DIALUMINUM TITANIUM PENTAOXIDE Al_2TiO_5

[For Data Reported in Figure and Table No. 410]

Curve No.	Ref. No.	Year	Temp. Range, K	Reported Error, %	Name and Specimen Designation	Composition (weight percent), Specifications and Remarks
1	358	1955	53-298			43.95 TiO (43.93 theo.), 0.06 SiO; prepared from pure hydrated alumina and pure titania; pressed into pellets and heated 5 times for 96 hrs between 1400 and 1500 C and 43 hrs between 1500 and 1570 C; quenched to room temperature.

DATA TABLE NO. 410 SPECIFIC HEAT OF DIALUMINUM TITANIUM PENTAOXIDE Al_2TiO_5

[Temperature, T, K; Specific Heat, C_p, Cal g^{-1} K^{-1}]

T	C_p
CURVE 1	
52.74	1.226×10^{-2}
56.55	1.448
60.98	1.739
65.50	2.067
70.22	2.415
75.04	2.782
80.73	3.231
85.01	3.587
94.82	4.417
105.13	5.305
114.75	6.164
124.74	7.038
135.93	8.001
145.50	8.814
155.96	9.656
165.82	1.044×10^{-1}
175.83	1.119
185.70	1.192
195.95	1.261
206.12	1.326
216.20	1.390
226.10	1.449
236.04	1.505
245.67	1.555
256.22	1.608
266.21	1.657
276.27	1.704
286.46	1.748
296.39	1.785*
298.16	1.793

* Not shown on plot

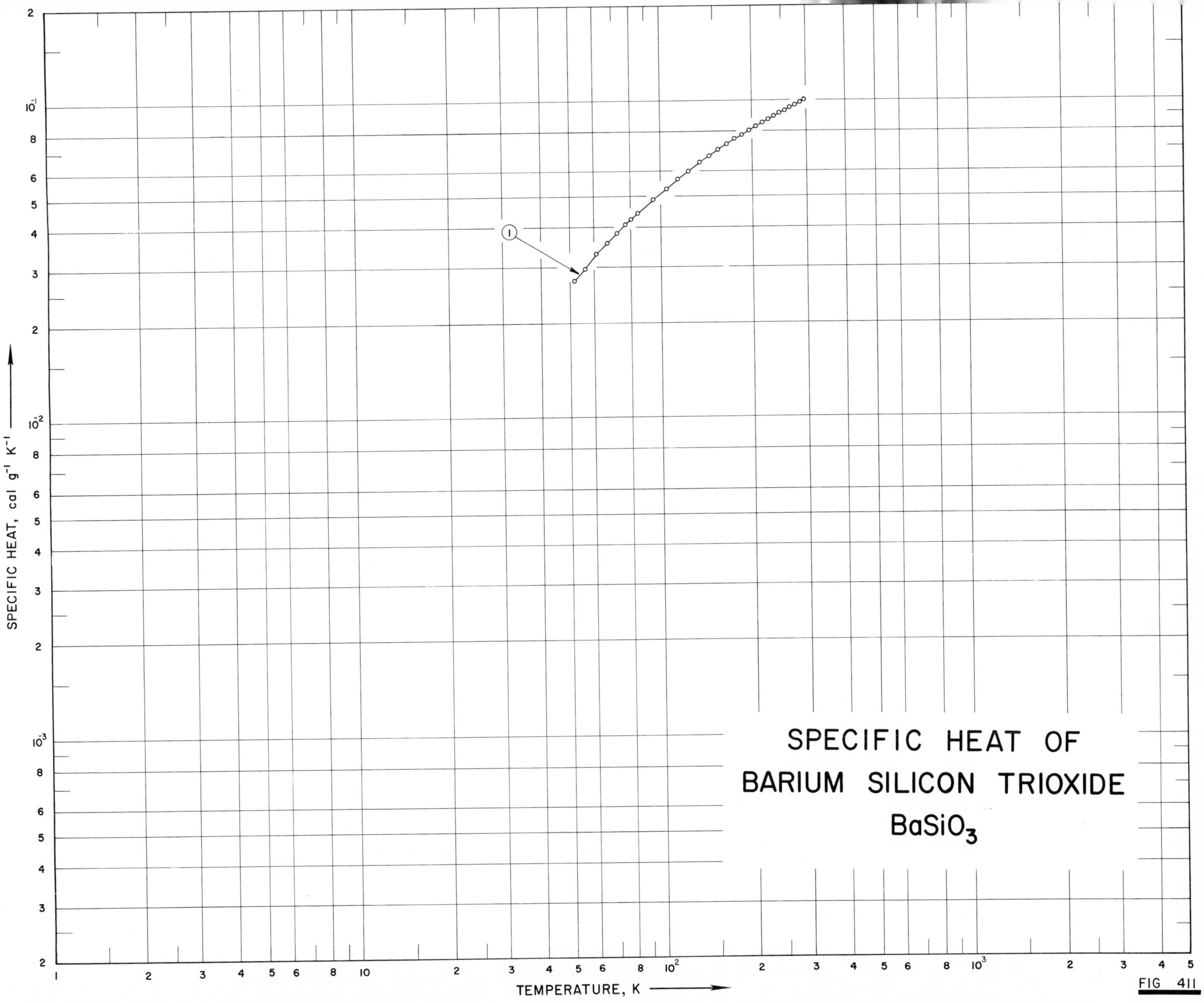

SPECIFIC HEAT OF BARIUM SILICON TRIOXIDE $BaSiO_3$

FIG 411

SPECIFIC HEAT, cal g^{-1} K^{-1}

TEMPERATURE, K

1

SPECIFICATION TABLE NO. 411 SPECIFIC HEAT OF BARIUM SILICON TRIOXIDE $BaSiO_3$

[For Data Reported in Figure and Table No. 411]

Curve No.	Ref. No.	Year	Temp. Range, K	Reported Error, %	Name and Specimen Designation	Composition (weight percent), Specifications and Remarks
1	360	1964	52-296			71.95 Ba, 28.17 SiO_2 (71.85, 28.15 theo.); crystalline; prepared by prolonged heating of stoichiometric mixture of reagent-grade barium carbonate and pure quartz at 950-1350 C.

DATA TABLE NO. 411 SPECIFIC HEAT OF BARIUM SILICON TRIOXIDE $BaSiO_3$

[Temperature, T, K; Specific Heat, C_p, Cal g^{-1} K^{-1}]

T	C_p
CURVE 1	
51.99	2.733×10^{-2}
56.14	2.978
61.32	3.302
66.76	3.594
71.80	3.852
76.60	4.095
80.08	4.259
84.59	4.456
95.26	4.915
105.61	5.328
114.49	5.670
124.62	6.030
135.66	6.401
145.61	6.729
155.79	7.028
165.84	7.310
175.99	7.591
185.94	7.839
196.01	8.073
206.16	8.317
216.15	8.532
226.21	8.753
235.98	8.959
245.85	9.151
256.33	9.343
266.62	9.540
276.46	9.699
287.21	9.896
296.17	1.005×10^{-1}

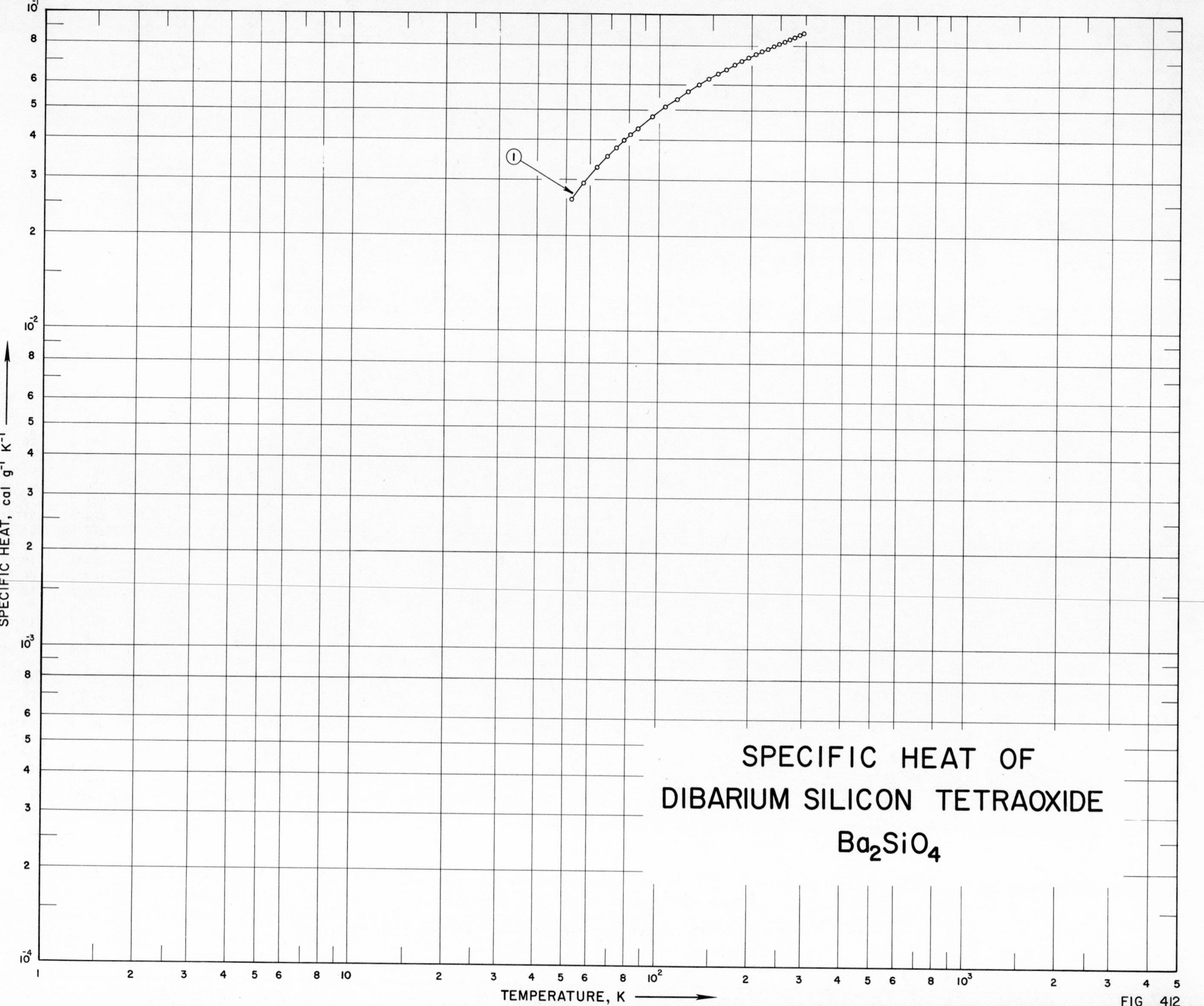
SPECIFIC HEAT OF
DIBARIUM SILICON TETRAOXIDE
Ba_2SiO_4
SPECIFIC HEAT, cal g^{-1} K^{-1}
TEMPERATURE, K
1

FIG 412

SPECIFICATION TABLE NO. 412 SPECIFIC HEAT OF DIBARIUM SILICON TETRAOXIDE Ba_2SiO_4

[For Data Reported in Figure and Table No. 412]

Curve No.	Ref. No.	Year	Temp. Range, K	Reported Error, %	Name and Specimen Designation	Composition (weight percent), Specifications and Remarks
1	360	1964	52-296			83.78 BaO, 16.39 SiO_2 (83.62, 16.38 theo.).

DATA TABLE NO. 412 SPECIFIC HEAT OF DIBARIUM SILICON TETRAOXIDE Ba_2SiO_4

[Temperature, T, K; Specific Heat, C_p, Cal g^{-1} K^{-1}]

T	C_p
CURVE 1	
52.15	2.613×10^{-2}
57.08	2.928
63.00	3.291
68.28	3.566
72.71	3.793
77.07	4.005
80.80	4.177
85.55	4.379
95.16	4.750
105.51	5.118
114.77	5.418
124.51	5.712
135.66	6.020
145.77	6.277
156.23	6.514
166.32	6.740
176.58	6.947
186.36	7.144
196.07	7.315
206.29	7.493
216.11	7.659
226.17	7.823
236.11	7.962
246.00	8.109
256.34	8.256
266.27	8.390
276.50	8.515
286.68	8.654
296.34	8.777

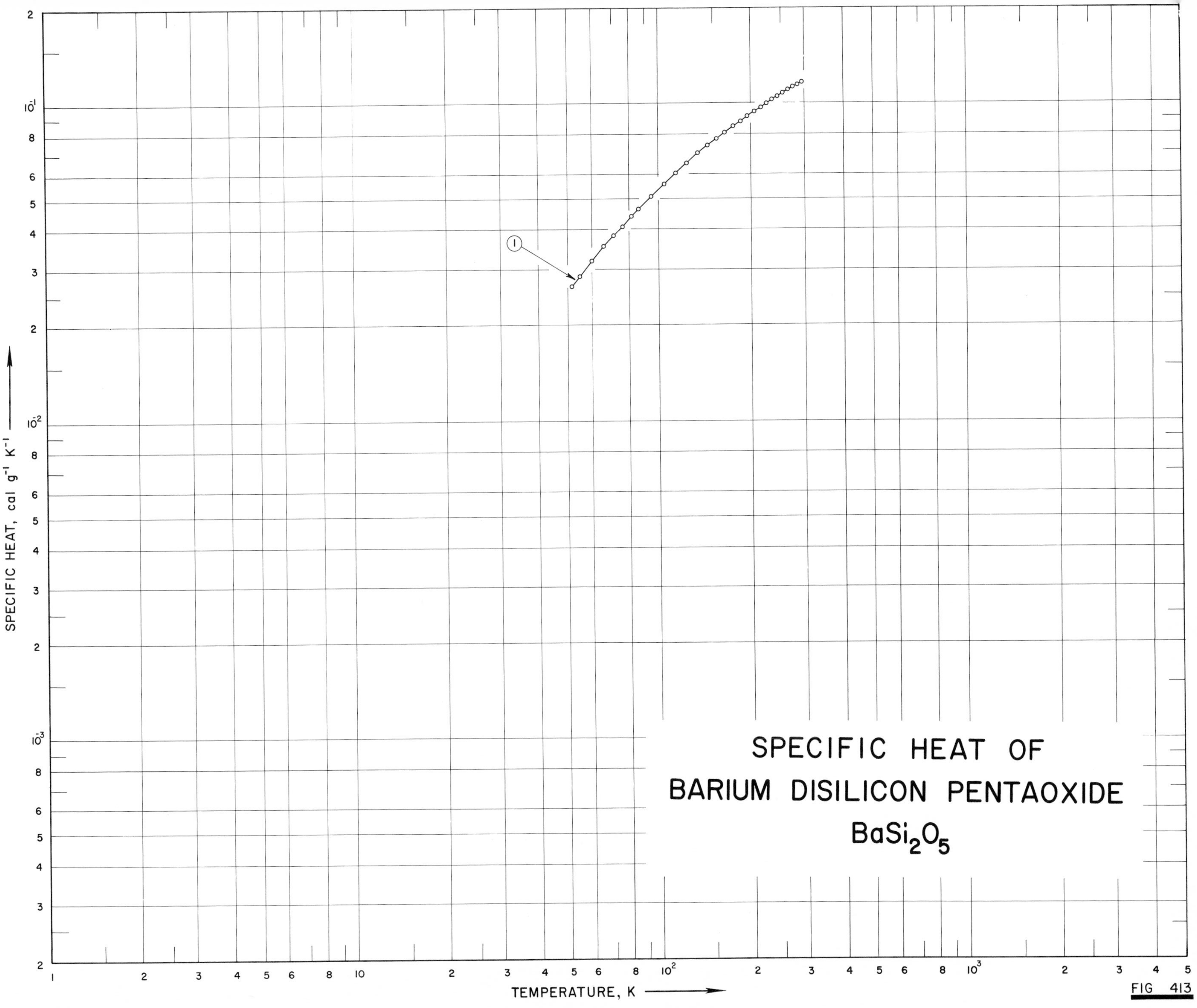
SPECIFIC HEAT OF
BARIUM DISILICON PENTAOXIDE
$BaSi_2O_5$
SPECIFIC HEAT, cal g⁻¹ K⁻¹
TEMPERATURE, K
1
FIG 413

SPECIFICATION TABLE NO. 413 SPECIFIC HEAT OF BARIUM DISILICON PENTAOXIDE $BaSi_2O_5$

[For Data Reported in Figure and Table No. 413]

Curve No.	Ref. No.	Year	Temp. Range, K	Reported Error, %	Name and Specimen Designation	Composition (weight percent), Specifications and Remarks
1	360	1964	52-296			55.85 BaO, 43.77 SiO_2, 0.32 Al_2O_3 (56.06, 43.94 theo.).

DATA TABLE NO. 413 SPECIFIC HEAT OF BARIUM DISILICON PENTAOXIDE $BaSi_2O_5$

[Temperature, T, K; Specific Heat, C_p, Cal g^{-1} K^{-1}]

T	C_p
CURVE 1	
51.63	2.664×10^{-2}
54.92	2.867
60.07	3.195
65.86	3.544
71.17	3.843
76.42	4.131
81.81	4.417
86.41	4.658
94.76	5.071
105.28	5.587
114.77	6.029
124.83	6.479
135.83	6.958
145.71	7.353
155.78	7.725
165.89	8.098
177.05	8.486
185.88	8.771
195.99	9.093
206.03	9.400
216.26	9.703
226.04	9.970
236.16	1.025×10^{-1}
245.80	1.050
256.40	1.077
266.62	1.101
276.47	1.124
286.98	1.148
296.43	1.169

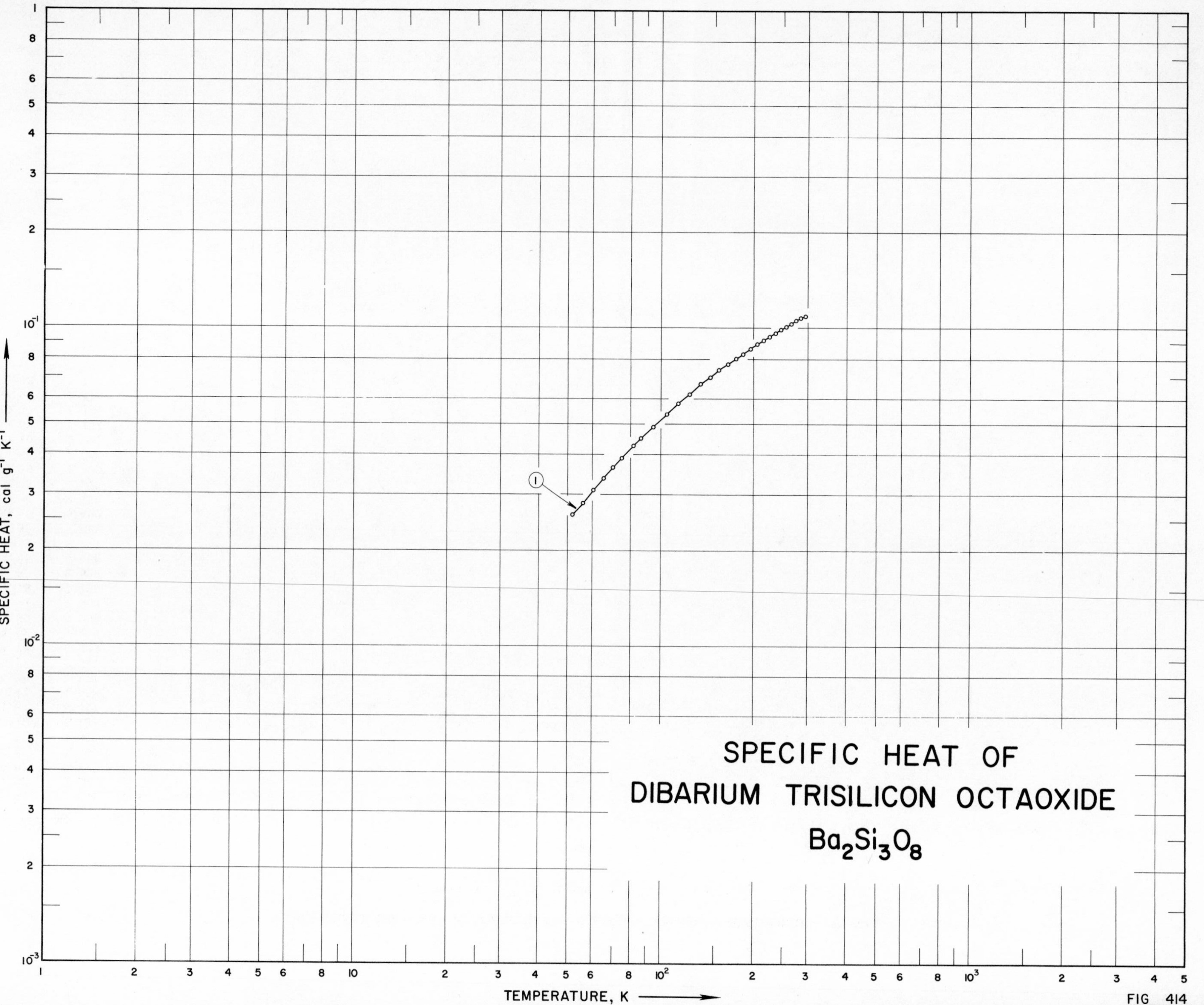
SPECIFIC HEAT OF
DIBARIUM TRISILICON OCTAOXIDE
$Ba_2Si_3O_8$
SPECIFIC HEAT, cal g⁻¹ K⁻¹
TEMPERATURE, K
1

FIG 414

SPECIFICATION TABLE NO. 414 SPECIFIC HEAT OF DIBARIUM TRISILICON OCTAOXIDE $Ba_2Si_3O_8$

[For Data Reported in Figure and Table No. 414]

Curve No.	Ref. No.	Year	Temp. Range, K	Reported Error, %	Name and Specimen Designation	Composition (weight percent), Specifications and Remarks
1	360	1964	52-296			63.00 BaO, 37.05 SiO_2, 0.01 Al_2O_3 (62.98, 37.02 theo.).

DATA TABLE NO. 414 SPECIFIC HEAT OF DIBARIUM TRISILICON OCTAOXIDE $Ba_2Si_3O_8$

[Temperature, T, K; Specific Heat, C_p, Cal g^{-1} K^{-1}]

T	C_p
CURVE 1	
51.94	2.598 x 10^{-2}
55.94	2.828
60.53	3.130
65.56	3.389
70.13	3.656
74.67	3.908
81.65	4.284
86.42	4.518
94.73	4.908
105.09	5.387
114.76	5.816
124.66	6.227
135.86	6.672
145.69	7.032
155.85	7.391
165.84	7.722
175.92	8.044
185.90	8.348
196.01	8.636
206.29	8.921
216.40	9.186
226.11	9.431
236.16	9.687
245.77	9.911
256.45	1.015 x 10^{-1}
266.32	1.037
276.28	1.058
286.79	1.080
296.29	1.098

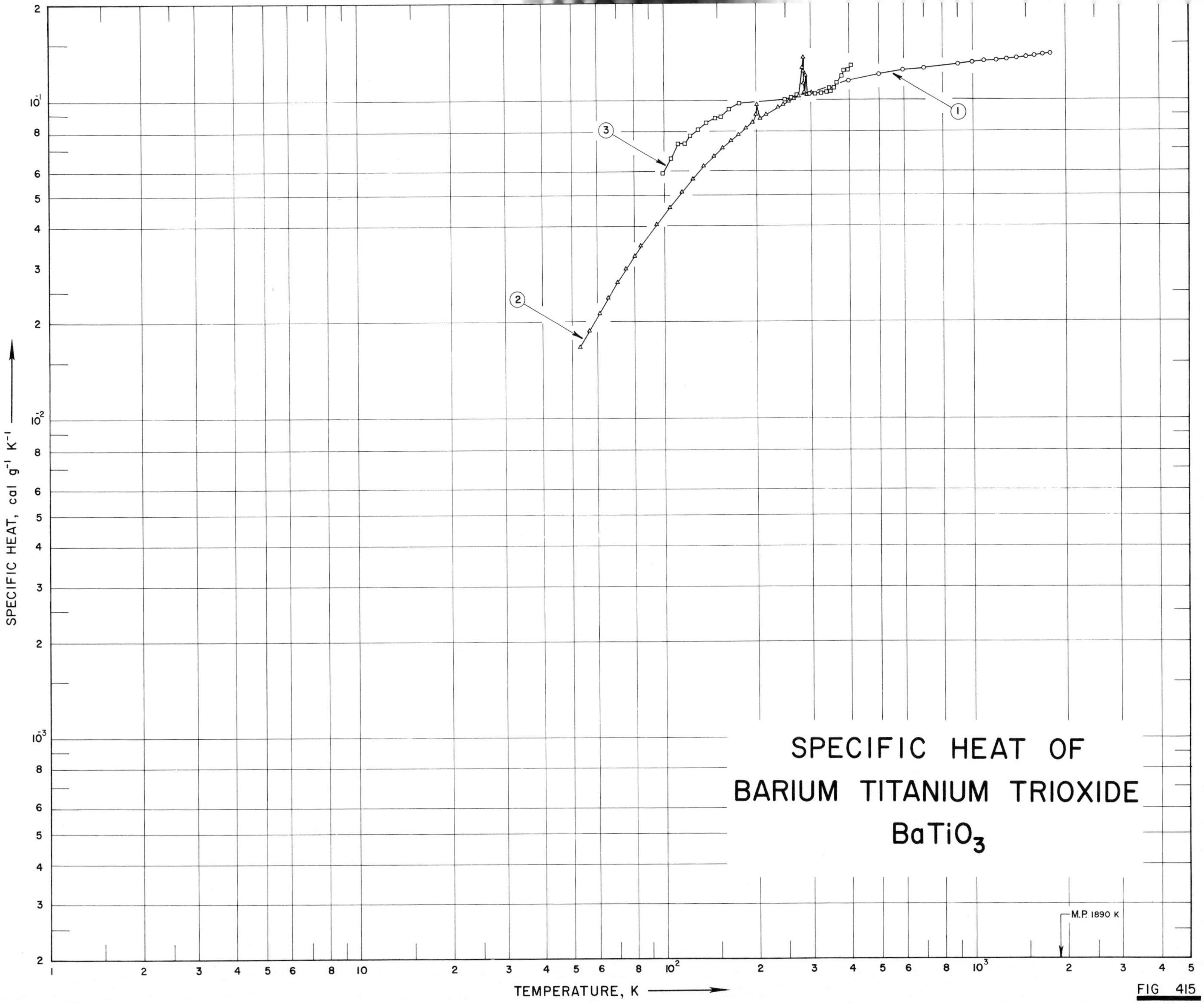

SPECIFIC HEAT OF BARIUM TITANIUM TRIOXIDE $BaTiO_3$

SPECIFIC HEAT, cal g^{-1} K^{-1}

TEMPERATURE, K

M.P. 1890 K

(1) (2) (3)

FIG 415

SPECIFICATION TABLE NO. 415 SPECIFIC HEAT OF BARIUM TITANIUM TRIOXIDE $BaTiO_3$

[For Data Reported in Figure and Table No. 415]

Curve No.	Ref. No.	Year	Temp. Range, K	Reported Error, %	Name and Specimen Designation	Composition (weight percent), Specifications and Remarks
1	361	1953	298-1800			99.7 $BaTiO_3$.
2	362	1952	53-301	±0.1		99.7 $BaTiO_3$; prepared from reagent grade barium hydroxide and titania (99.8 pure after ignition) by prolonged heating at 1350 C.
3	363	1952	99-407			Polycrystalline; sintered.

DATA TABLE NO. 415 SPECIFIC HEAT OF BARIUM TITANIUM TRIOXIDE $BaTiO_3$

[Temperature, T, K; Specific Heat, C_p, Cal g^{-1} K^{-1}]

T	C_p
CURVE 1	
298	1.05 x 10^{-1}
300	1.05*
400	1.16
500	1.21
600	1.24
700	1.27
900	1.30
1000	1.31
1100	1.32
1200	1.34
1300	1.35
1400	1.36
1500	1.37
1600	1.38
1700	1.39
1800	1.40
CURVE 2	
53.05	1.688 x 10^{-2}
56.86	1.895
61.09	2.147
65.59	2.419
70.23	2.691
74.98	2.968
80.02	3.263
84.34	3.516
94.66	4.105
104.44	4.647
114.54	5.179
124.72	5.698
135.99	6.238
146.19	6.696
155.86	7.095
165.96	7.490
175.97	7.850
186.23	8.210
186.89	8.227*
190.61	8.351*
193.28	8.458*
195.32	8.544*
195.83	8.548
197.31	8.651*
199.33	9.033
201.70	9.740
204.84	8.818*
CURVE 2 (cont.)	
206.51	8.793 x 10^{-2}
208.71	8.801*
216.16	9.003
236.44	9.496
245.77	9.697
256.26	9.912
266.08	1.011 x 10^{-1}
CURVE 3	
99	5.89 x 10^{-2}
106	6.6
112	7.3
117	7.3
121.5	7.74
129	8.1
137.5	8.5
147	8.8
153.5	8.86
163.5	9.38
176	9.79
248	1.0 x 10^{-1}
254	1.0*
261	1.0
273	1.030
295	1.040*
302	1.05*
311	1.050
326.5	1.06
338	1.060
346	1.091
349	1.070
358	1.10
365	1.132
373	1.153*
380	1.20
387	1.245
398	1.245
406	1.265*
407	1.286

* Not shown on plot

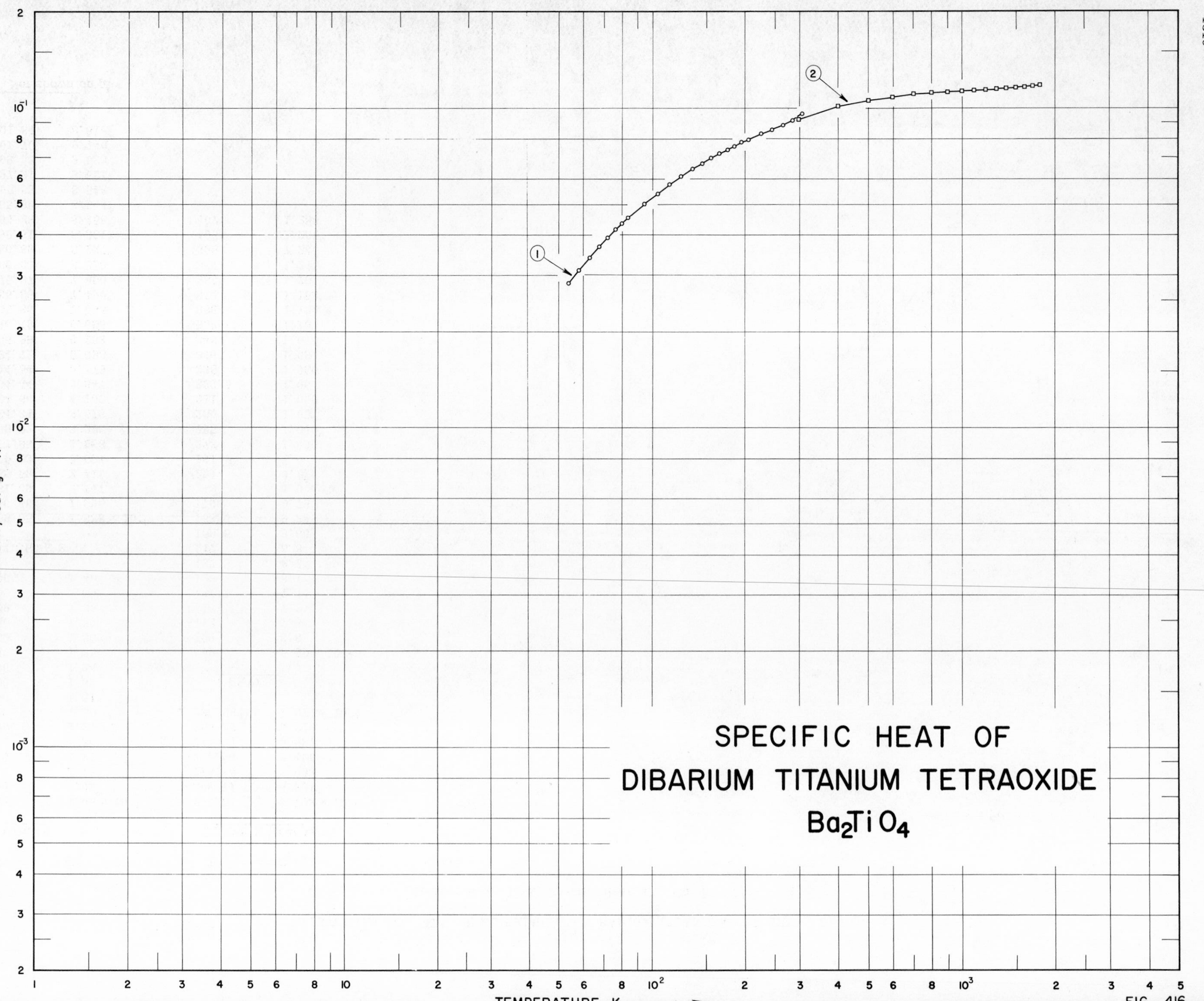
SPECIFIC HEAT OF
DIBARIUM TITANIUM TETRAOXIDE
Ba_2TiO_4
SPECIFIC HEAT, cal g⁻¹ K⁻¹
1
2

SPECIFICATION TABLE NO. 416 SPECIFIC HEAT OF DIBARIUM TITANIUM TETRAOXIDE Ba_2TiO_4

[For Data Reported in Figure and Table No. 416]

Curve No.	Ref. No.	Year	Temp. Range, K	Reported Error, %	Name and Specimen Designation	Composition (weight percent), Specifications and Remarks
1	363	1952	54-306			99.2 Ba_2TiO_4; 20.8 TiO_2 (20.67 theo.), 0.34 CaO, 0.02 SiO_2; no unreacted oxide or metatitanate.
2	361	1953	298-1800			99.2 Ba_2TiO_4; 20.8 TiO_2 (20.67 theo.), 0.34 CaO, 0.02 SiO_2; no unreacted oxide or metatitanate.

DATA TABLE NO. 416 SPECIFIC HEAT OF DIBARIUM TITANIUM TETRAOXIDE Ba_2TiO_4

[Temperature, T, K; Specific Heat, C_p, Cal g^{-1} K^{-1}]

T	C_p
CURVE 1	
54.00	2.848 x 10^{-2}
58.42	3.112
63.14	3.414
67.85	3.694
72.35	3.934
76.82	4.169
80.14	4.340
84.44	4.550
94.90	5.013
104.75	5.406
114.70	5.783
124.79	6.117
136.05	6.464
146.22	6.740
156.08	6.981
166.10	7.211
176.19	7.431
185.98	7.633
196.00	7.809
206.36	7.972
216.50	8.155*
226.39	8.305
236.14	8.448*
245.95	8.598
256.26	8.717*
267.53	8.869
276.46	9.006*
286.52	9.195
296.44	9.203*
296.69	9.407*
298.42	9.224*
302.62	9.508*
306.70	9.622
298.16	9.436*
CURVE 2	
298	9.22 x 10^{-2}
300	9.25*
400	1.02 x 10^{-1}
500	1.06
600	1.09
700	1.10
800	1.12
900	1.13
1000	1.14
CURVE 2 (cont.)	
1100	1.14 x 10^{-1}
1200	1.15
1300	1.16
1400	1.16
1500	1.17
1600	1.17
1700	1.18
1800	1.18

* Not shown on plot

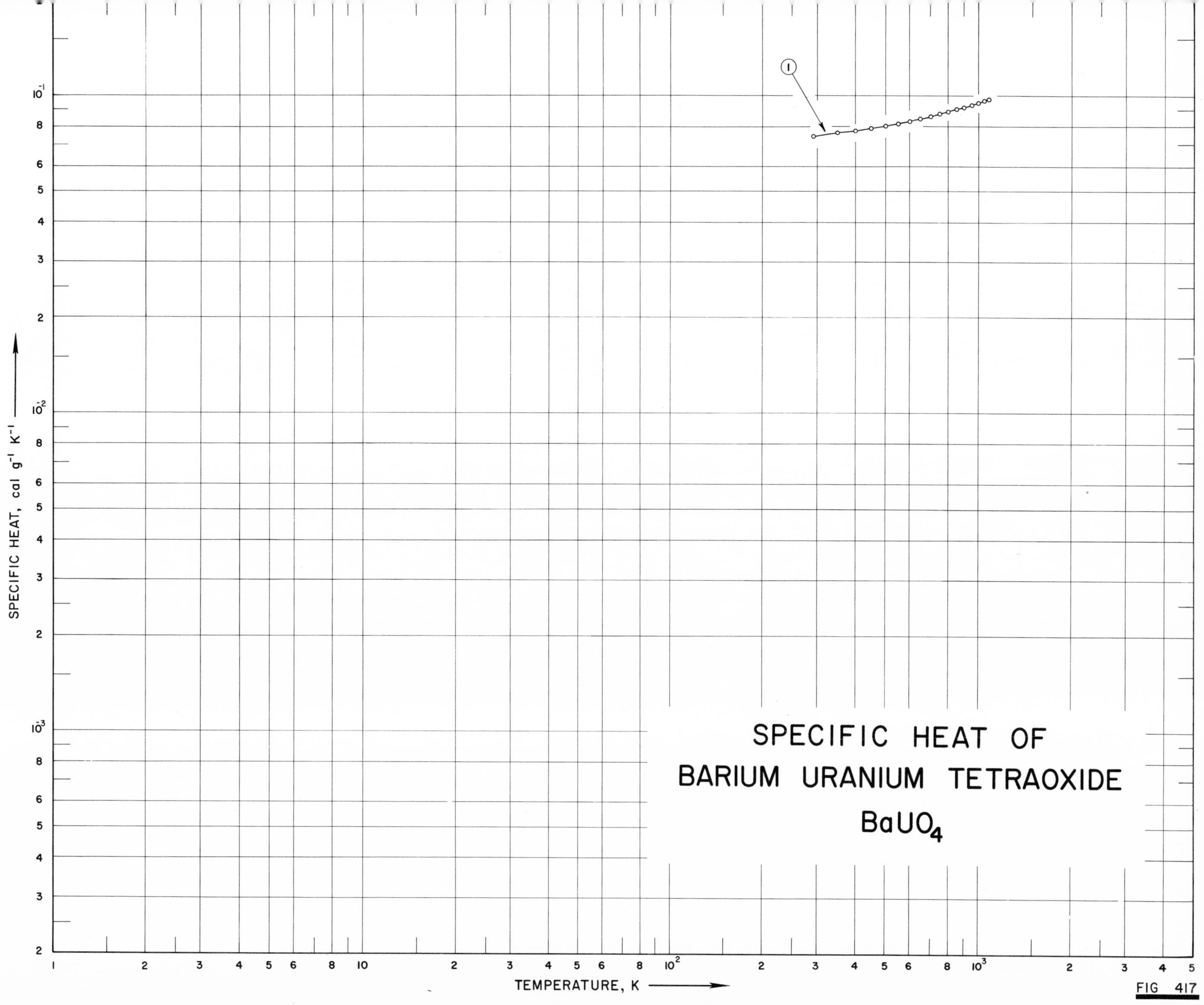
SPECIFIC HEAT OF
BARIUM URANIUM TETRAOXIDE
BaUO4
SPECIFIC HEAT, cal g-1 K-1
TEMPERATURE, K
1

FIG 417

SPECIFICATION TABLE NO. 417 SPECIFIC HEAT OF BARIUM URANIUM TETRAOXIDE $BaUO_4$

[For Data Reported in Figure and Table No. 417]

Curve No.	Ref. No.	Year	Temp. Range, K	Reported Error, %	Name and Specimen Designation	Composition (weight percent), Specifications and Remarks
1	362	1960	293-1084	0.1		53.95 U, 31.10 Ba.

DATA TABLE NO. 417 SPECIFIC HEAT OF BARIUM URANIUM TETRAOXIDE $BaUO_4$

[Temperature, T, K; Specific Heat, C_p, Cal g^{-1} K^{-1}]

T	C_p
CURVE 1	
293	7.46 x 10^{-2}
300	7.48*
350	7.62
400	7.76
450	7.91
500	8.05
550	8.20
600	8.34
650	8.49
700	8.63
750	8.78
800	8.92
850	9.06
900	9.21
950	9.35
1000	9.50
1050	9.64
1084	9.74

* Not shown on plot

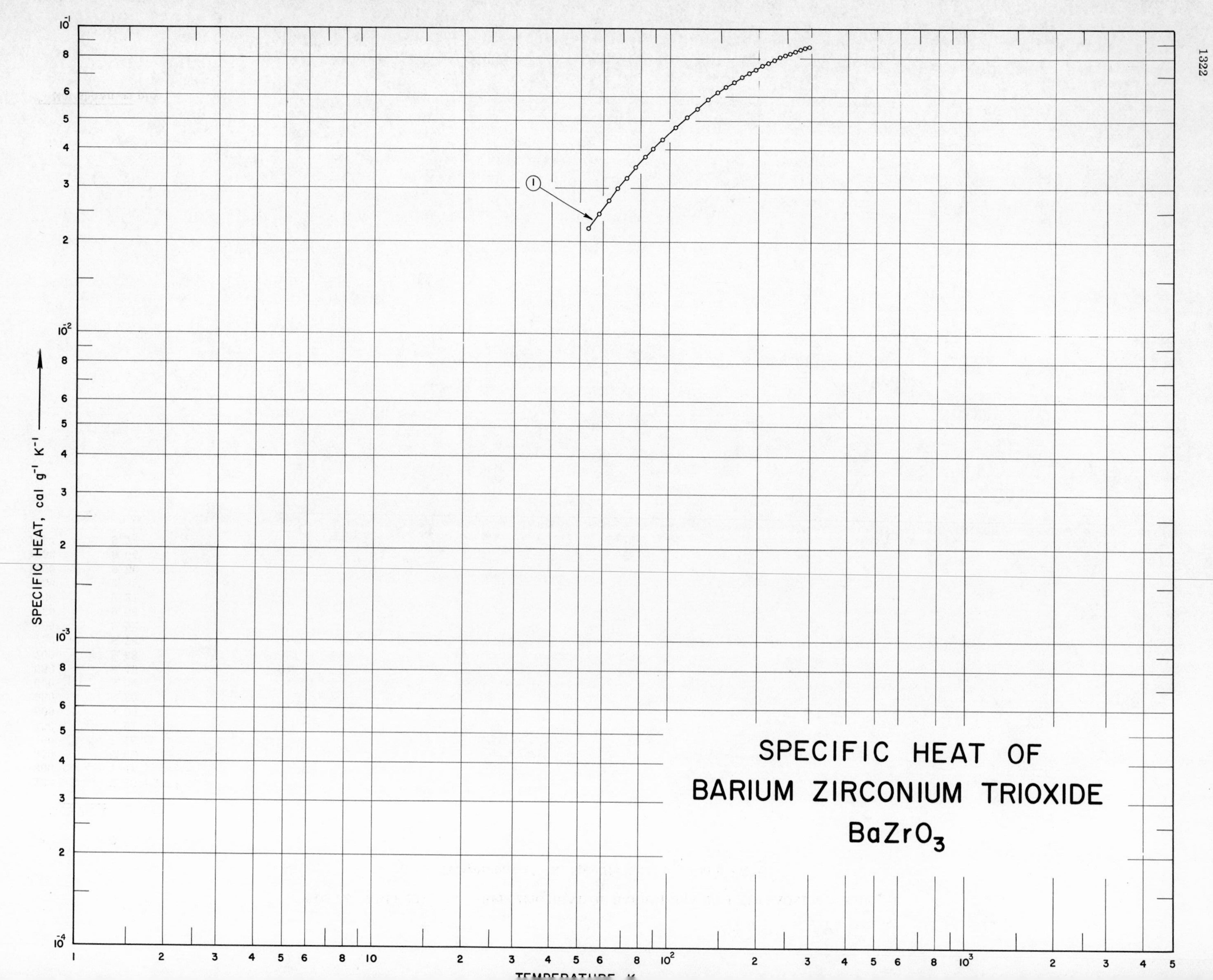

SPECIFIC HEAT OF
BARIUM ZIRCONIUM TRIOXIDE
$BaZrO_3$
SPECIFIC HEAT, cal g^{-1} K^{-1}
1

SPECIFICATION TABLE NO. 418 SPECIFIC HEAT OF BARIUM ZIRCONIUM TRIOXIDE $BaZrO_3$

[For Data Reported in Figure and Table No. 418]

Curve No.	Ref. No.	Year	Temp. Range, K	Reported Error, %	Name and Specimen Designation	Composition (weight percent), Specifications and Remarks
1	364	1960	53-296	0.3		55.40 BaO, 44.63 ZrO_2; prepared by heating reagent-grade barium carbonate and pure zirconia for 24 hrs at 1000 C, 6 hrs at 1350-1400 C, 20 hrs at 1350-1470 C, and 12 hrs at 1300-1350 C.

DATA TABLE NO. 418 SPECIFIC HEAT OF BARIUM ZIRCONIUM TRIOXIDE $BaZrO_3$

[Temperature, T, K; Specific Heat, C_p, Cal $g^{-1} K^{-1}$]

T	C_p
CURVE 1	
53.62	2.236 x 10^{-2}
58.29	2.499
62.70	2.748
67.31	3.003
72.16	3.262
77.30	3.521
83.43	3.822
88.45	4.060
94.97	4.350
105.20	4.783
114.81	5.156
124.56	5.499
136.14	5.890
145.96	6.201
155.99	6.479
165.80	6.729
176.74	6.982
186.00	7.191
196.12	7.394
206.34	7.593
216.04	7.755
226.22	7.933
236.11	8.077
245.55	8.215
256.53	8.348
266.48	8.468
276.36	8.576
286.50	8.674
296.08	8.771

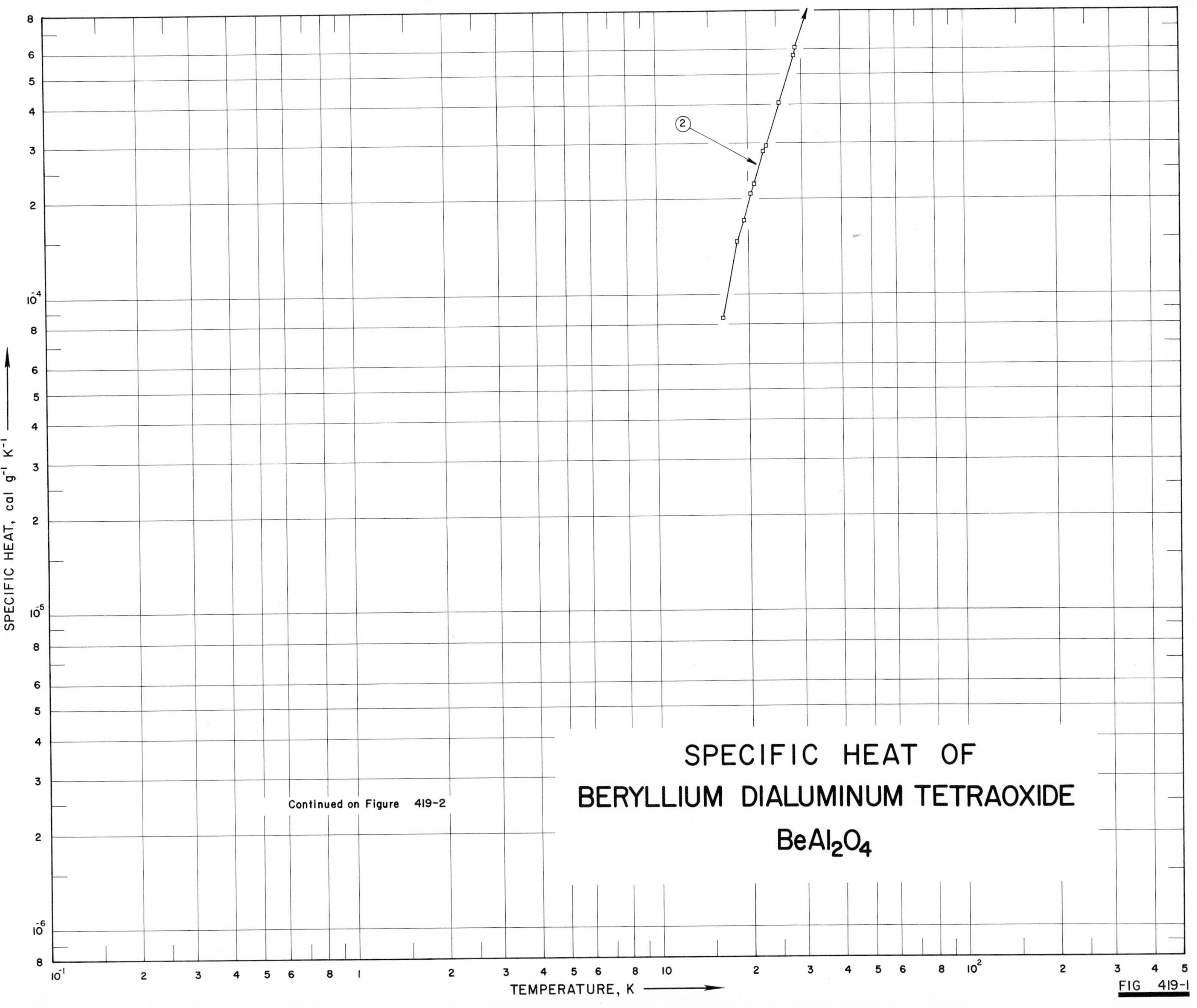
SPECIFIC HEAT OF
BERYLLIUM DIALUMINUM TETRAOXIDE
$BeAl_2O_4$
SPECIFIC HEAT, cal g^{-1} K^{-1}
TEMPERATURE, K
2
Continued on Figure 419-2
FIG 419-1

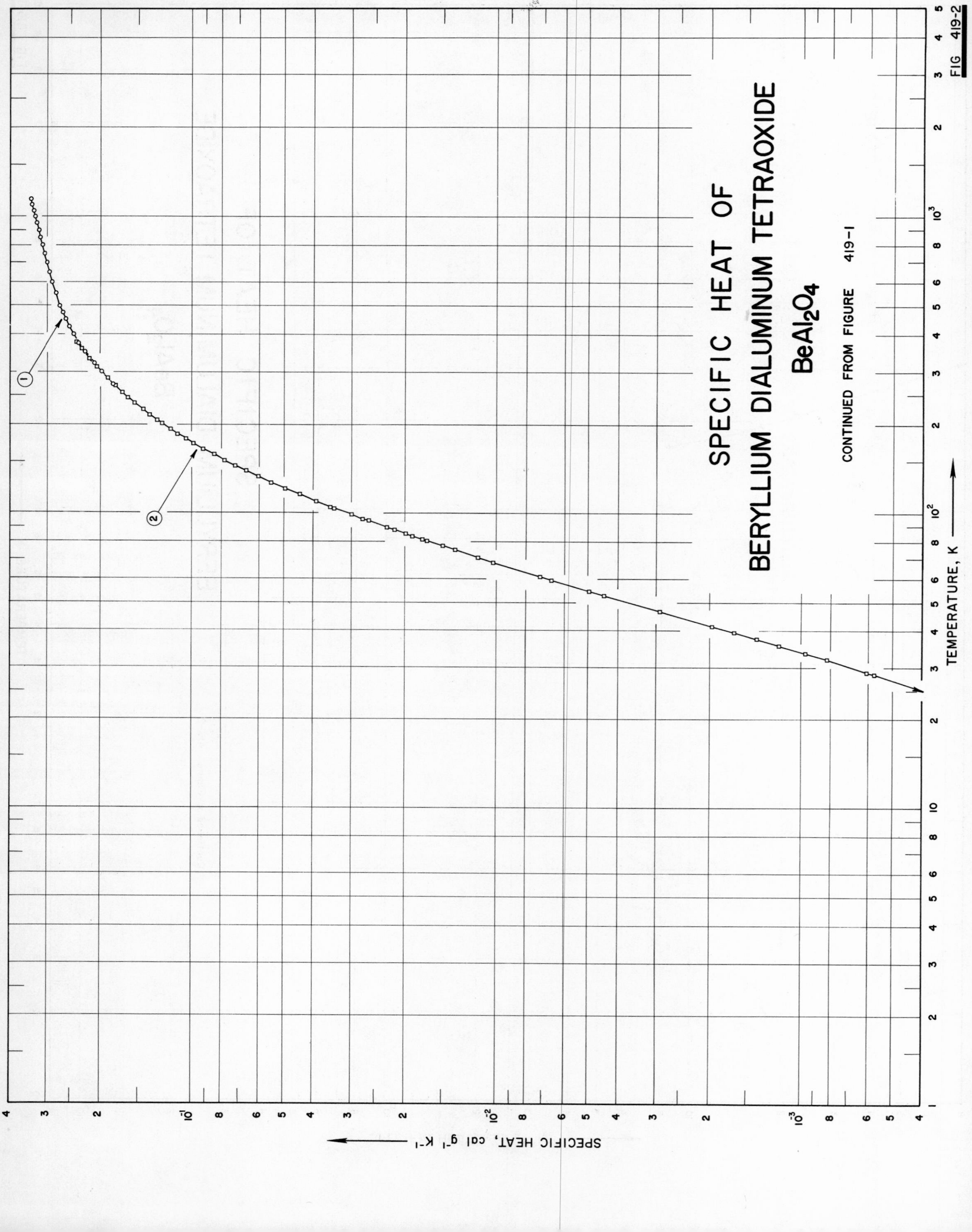

SPECIFIC HEAT OF BERYLLIUM DIALUMINUM TETRAOXIDE $BeAl_2O_4$

CONTINUED FROM FIGURE 419-1

FIG 419-2

SPECIFICATION TABLE NO. 419 SPECIFIC HEAT OF BERYLLIUM DIALUMINUM TETRAOXIDE $BeAl_2O_4$

[For Data Reported in Figure and Table No. 419]

Curve No.	Ref. No.	Year	Temp. Range, K	Reported Error, %	Name and Specimen Designation	Composition (weight percent), Specifications and Remarks
1	170	1964	273-1150			Two samples; average composition: 80.605 Al_2O_3, 19.385 BeO.
2	170	1964	16-376	~0.5		80.33 Al_2O_3, 19.72 BeO, <0.01 Si, 0.01-0.1 Cu, 0.01-0.1 Ni, 0.001-0.01 Fe, 0.001-0.01 Mg, 0.001-0.01 Ca, <0.001 Pb, <0.001 Sn, <0.0001 Ag, 0.0001-0.001 V.

DATA TABLE NO. 419 SPECIFIC HEAT OF BERYLLIUM DIALUMINUM TETRAOXIDE $BeAl_2O_4$

[Temperature, T, K; Specific Heat, C_p, Cal $g^{-1} K^{-1}$]

T	C_p
CURVE 1	
273.15	1.825×10^{-1}
300	1.994
310	2.051*
320	2.107
330	2.159*
340	2.209*
350	2.257
360	2.304*
370	2.349*
373.15	2.363
380	2.393*
390	2.433*
400	2.468
425	2.544
450	2.611
475	2.671
500	2.724
550	2.815
600	2.892
650	2.959
700	3.018
750	3.072
800	3.122
850	3.169
900	3.213
950	3.256
1000	3.296
1050	3.336
1100	3.374
1150	3.411
CURVE 2	
Series 1	
80.55	1.695×10^{-2}
85.15	1.995
89.58	2.299
94.16	2.630
99.08	3.008
104.31	3.432
109.84	3.902
Series 2	
81.62	1.763×10^{-2}
CURVE 2 (cont.)	
Series 2 (cont.)	
87.77	2.175×10^{-2}
93.64	2.593*
99.65	3.054*
105.46	3.529
110.76	3.983*
116.06	4.456
121.65	4.969
127.48	5.518
133.33	6.080
139.15	6.650
145.06	7.230
151.29	7.856
158.10	8.520
165.13	9.211
172.47	9.926
179.45	1.060×10^{-1}
186.11	1.123
193.04	1.187
Series 3	
54.60	4.963×10^{-3}
61.07	7.213
68.19	1.025×10^{-2}
75.12	1.376
81.41	1.750*
87.36	2.147*
Series 4	
70.90	1.154×10^{-2}
77.52	1.512
83.68	1.898
89.79	2.316*
95.99	2.770
Series 5	
28.70	6.080×10^{-4}
33.31	9.694
37.31	1.397×10^{-3}
41.46	1.963
46.63	2.918
52.97	4.446
59.56	6.628
CURVE 2 (cont.)	
Series 6	
19.27	1.732×10^{-4}
20.86	2.259
22.85	2.974
25.21	4.066
Series 7	
16.45	8.471×10^{-5}
18.36	1.487×10^{-4}
20.29	2.108
22.43	2.861
25.25	4.009*
28.29	5.760
31.71	8.245
35.40	1.178×10^{-3}
39.33	1.651
Series 8	
201.80	1.266×10^{-1}
207.60	1.318
215.40	1.385
224.07	1.458
Series 9	
209.69	1.336×10^{-1}*
215.47	1.411*
227.35	1.480*
236.67	1.559
246.31	1.633
255.83	1.703
259.96	1.734*
269.06	1.795
277.90	1.857*
286.53	1.912
294.69	1.964*
303.14	2.013*
311.51	2.061
320.02	2.108*
Series 10	
309.76	2.051×10^{-1}*
317.73	2.095*
CURVE 2 (cont.)	
Series 11	
290.04	1.934×10^{-1}*
298.78	1.989*
307.34	2.037*
315.98	2.086*
324.77	2.133*
333.56	2.179
342.23	2.223*
350.99	2.265*
359.52	2.305
367.87	2.343*
376.16	2.389
Series 12*	
191.34	1.171×10^{-1}
201.13	1.260
210.42	1.343
219.29	1.428

* Not shown on plot

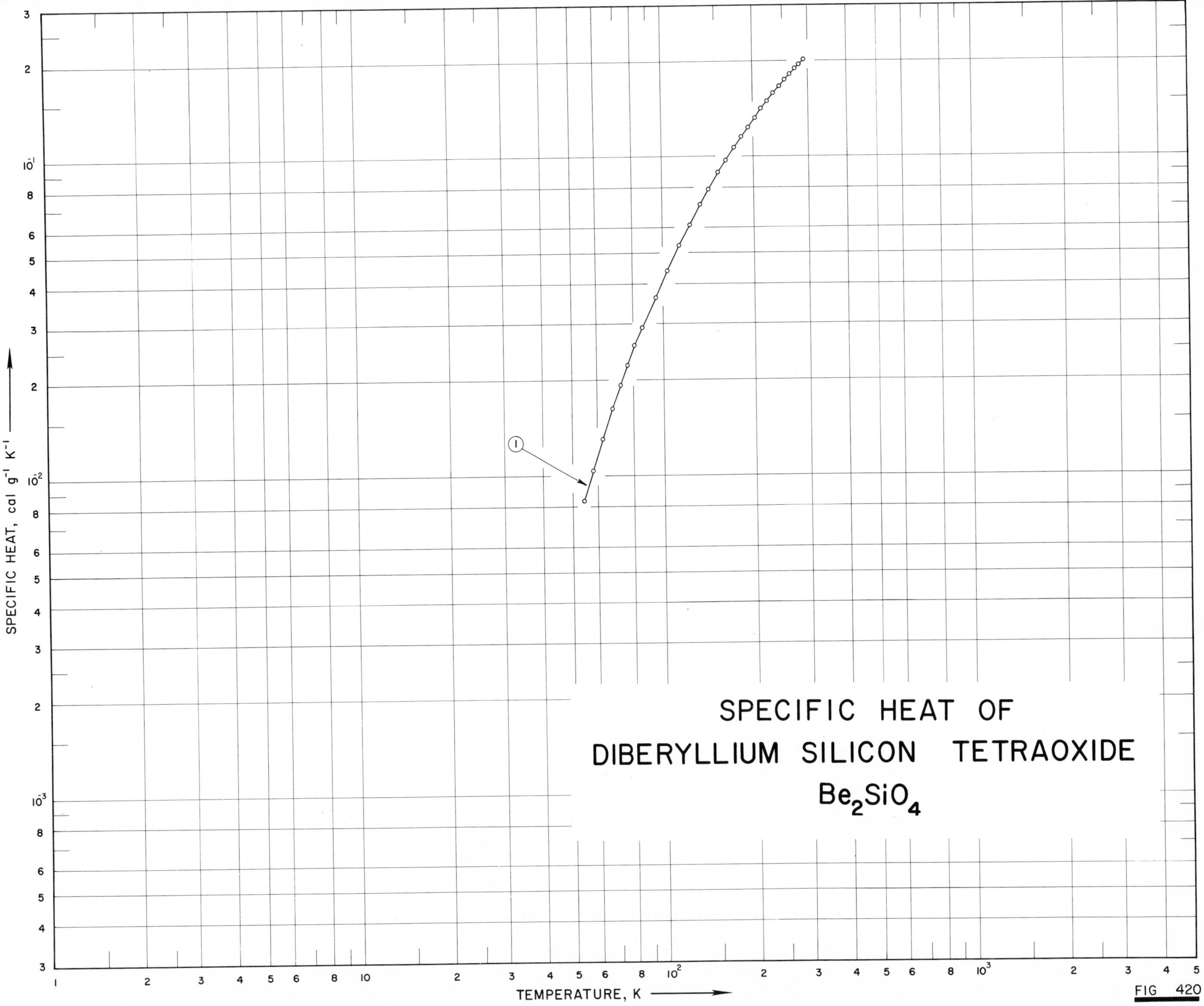

SPECIFIC HEAT OF DIBERYLLIUM SILICON TETRAOXIDE Be_2SiO_4

FIG 420

SPECIFICATION TABLE NO. 420 SPECIFIC HEAT OF DIBERYLLIUM SILICON TETRAOXIDE Be_2SiO_4

[For Data Reported in Figure and Table No. 420]

Curve No.	Ref. No.	Year	Temp. Range, K	Reported Error, %	Name and Specimen Designation	Composition (weight percent), Specifications and Remarks
1	43	1939	55-294			99.9 Be_2SiO_4; well crystallized; sample from Brazil; 99.8 Be_2SiO_4 poorly crystallized; sample from Colorado; measurements made from 126.5 g of Brazil sample and 83.5 g of Colorado sample.

DATA TABLE NO. 420 SPECIFIC HEAT OF DIBERYLLIUM SILICON TETRAOXIDE Be_2SiO_4

[Temperature, T, K; Specific Heat, C_p, Cal g^{-1} K^{-1}]

T	C_p
CURVE 1	
54.7	8.492×10^{-3}
58.6	1.054×10^{-2}
63.3	1.323
68.2	1.653
72.6	1.960
76.6	2.252
81.4	2.607
86.1	2.955
95.0	3.668
104.5	4.461
114.5	5.349
124.3	6.242
134.6	7.178
143.9	8.035
154.3	9.020
163.9	9.873
174.0	1.078×10^{-1}
184.2	1.171
193.6	1.255
203.8	1.345
213.7	1.431
223.8	1.515
234.5	1.604
245.3	1.688
255.5	1.763
265.3	1.836
275.2	1.913
284.4	1.973
294.1	2.042

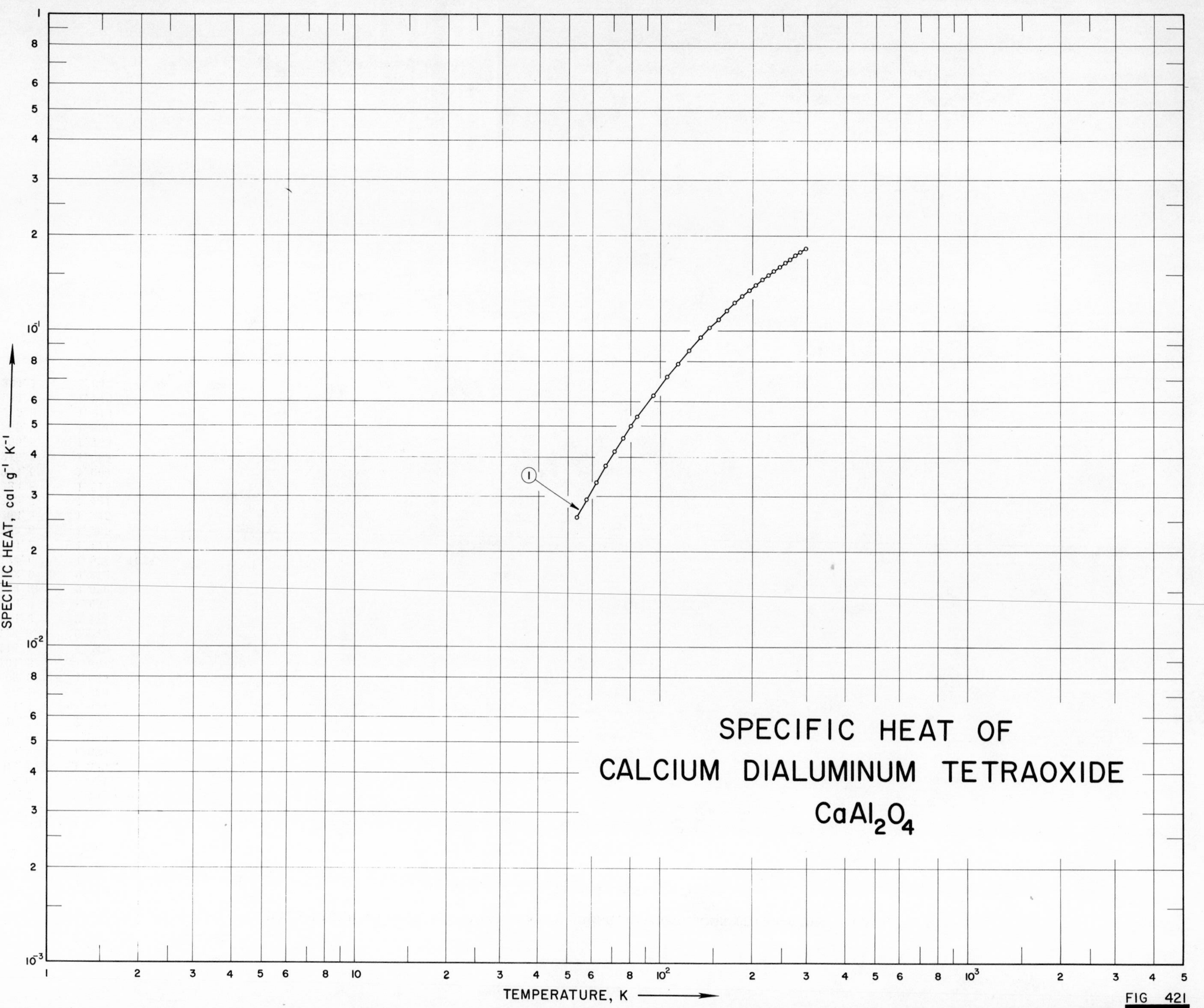
SPECIFIC HEAT OF
CALCIUM DIALUMINUM TETRAOXIDE
$CaAl_2O_4$
1
SPECIFIC HEAT, cal g^{-1} K^{-1}
TEMPERATURE, K
FIG 421

SPECIFICATION TABLE NO. 421 SPECIFIC HEAT OF CALCIUM DIALUMINUM TETRAOXIDE $CaAl_2O_4$

[For Data Reported in Figure and Table No. 421]

Curve No.	Ref. No.	Year	Temp. Range, K	Reported Error, %	Name and Specimen Designation	Composition (weight percent), Specifications and Remarks
1	365	1955	54-298			64.44 Al_2O_3, 35.49 CaO (64.51, 35.49 theo.).

DATA TABLE NO. 421 SPECIFIC HEAT OF CALCIUM DIALUMINUM TETRAOXIDE $CaAl_2O_4$

[Temperature, T, K; Specific Heat, C_p, Cal g^{-1} K^{-1}]

T	C_p
CURVE 1	
53.52	2.577×10^{-2}
57.74	2.937
61.99	3.334
66.52	3.762
71.16	4.186
75.79	4.596
80.40	5.013
84.45	5.382
94.66	6.272
106.15	7.232
114.76	7.941
124.65	8.726
136.22	9.599
145.72	1.029×10^{-1}
155.70	1.097
166.66	1.171
176.04	1.231
185.85	1.291
196.25	1.351
206.07	1.403
216.32	1.462
225.07	1.511
235.98	1.560
246.07	1.609
256.37	1.653
266.38	1.698
276.71	1.741
286.54	1.782
296.30	1.815*
298.16	1.824

* Not shown on plot

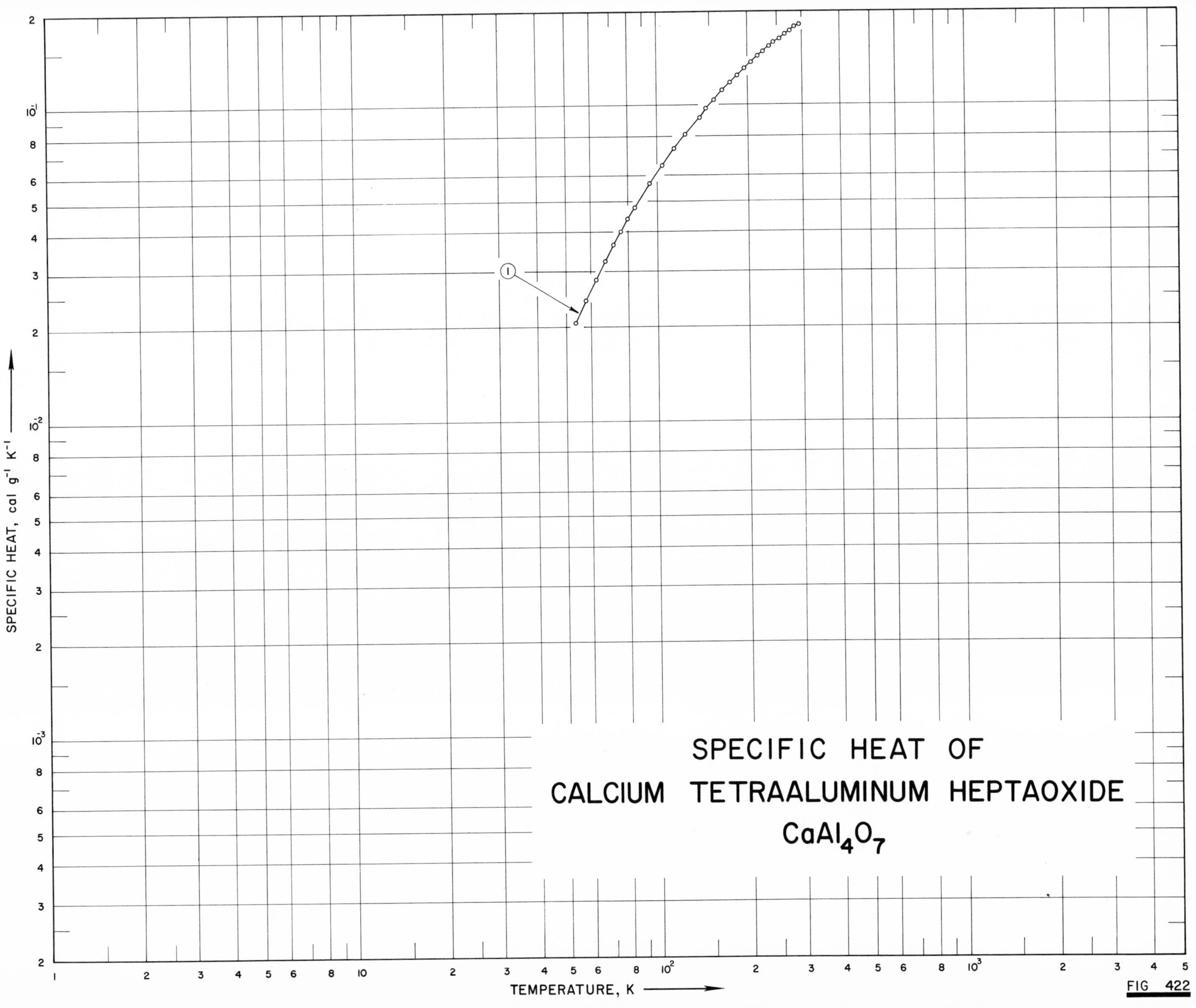
SPECIFIC HEAT OF
CALCIUM TETRAALUMINUM HEPTAOXIDE
$CaAl_4O_7$
1
SPECIFIC HEAT, cal g^{-1} K^{-1}
TEMPERATURE, K
FIG 422

SPECIFICATION TABLE NO. 422 SPECIFIC HEAT OF CALCIUM TETRAALUMINUM HEPTAOXIDE $CaAl_4O_7$

[For Data Reported in Figure and Table No. 422]

Curve No.	Ref. No.	Year	Temp. Range, K	Reported Error, %	Name and Specimen Designation	Composition (weight percent), Specifications and Remarks
1	365	1955	54-296			78.49 Al_2O_3, 21.58 CaO (78.43, 27.58 theo.).

DATA TABLE NO. 422 SPECIFIC HEAT OF CALCIUM TETRAALUMINUM HEPTAOXIDE $CaAl_4O_7$

[Temperature, T, K; Specific Heat, C_p, Cal g^{-1} K^{-1}]

T	C_p
CURVE 1	
53.50	2.061 x 10^{-2}
57.99	2.427
62.54	2.830
67.11	3.233
71.69	3.642
76.24	4.035
80.30	4.404
84.70	4.792
94.76	5.673
104.97	6.538
114.78	7.373
124.71	8.169
138.27	9.242
146.60	9.865
155.94	1.053 x 10^{-1}
165.81	1.124
175.84	1.192
185.76	1.257
195.98	1.321
206.37	1.381
216.13	1.440
226.25	1.495
236.04	1.545
243.88	1.598
256.26	1.649
266.25	1.696
276.42	1.741
286.60	1.790
296.35	1.823

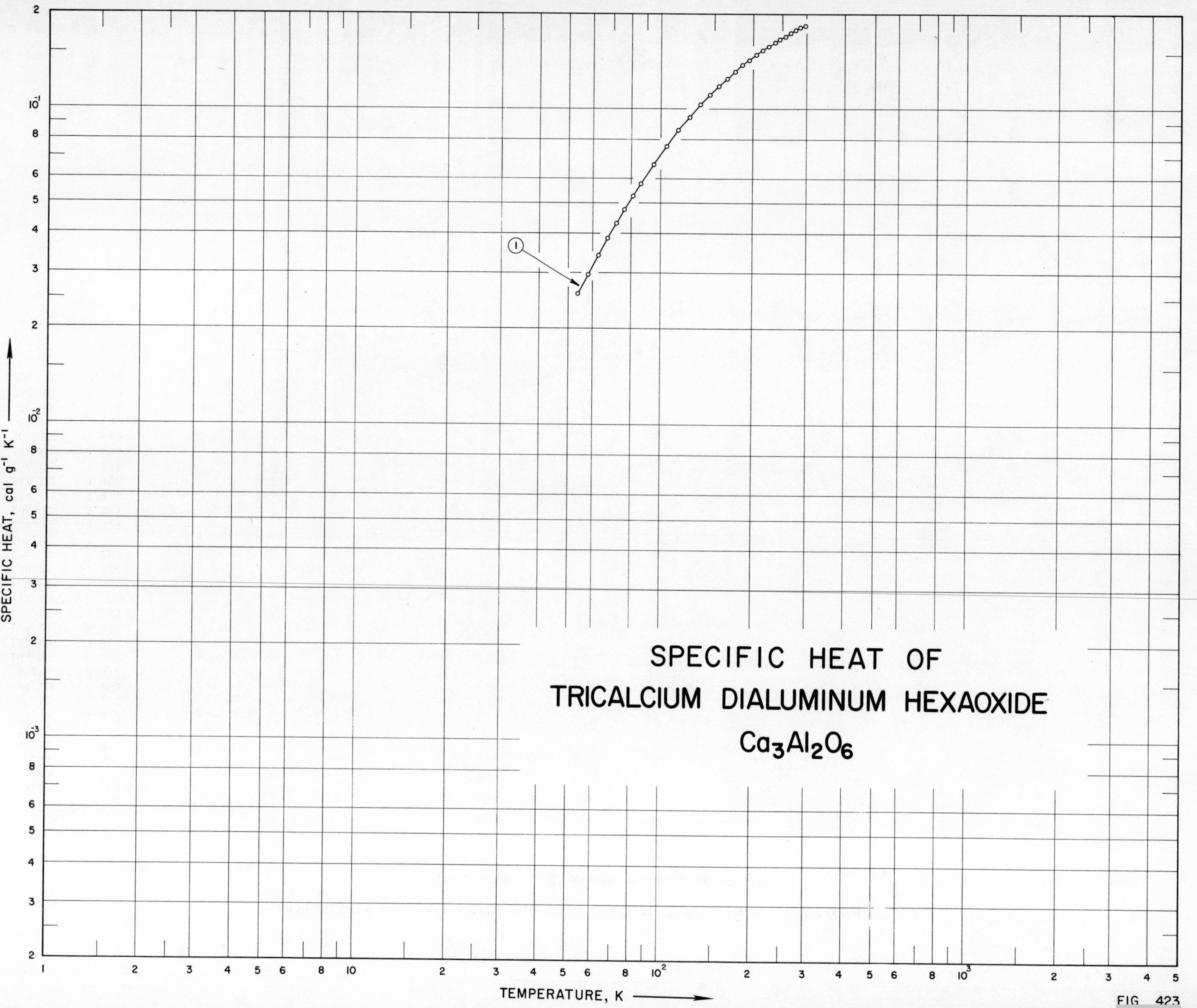

SPECIFIC HEAT OF
TRICALCIUM DIALUMINUM HEXAOXIDE
$Ca_3Al_2O_6$
SPECIFIC HEAT, cal g^{-1} K^{-1}
TEMPERATURE, K
1
FIG 423

SPECIFICATION TABLE NO. 423 SPECIFIC HEAT OF TRICALCIUM DIALUMINUM HEXAOXIDE $Ca_3Al_2O_6$

[For Data Repoted in Figure and Table No. 423]

Curve No.	Ref. No.	Year	Temp. Range, K	Reported Error, %	Name and Specimen Designation	Composition (weight percent), Specifications and Remarks
1	365	1955	54-297			62.25 CaO, 37.84 Al_2O_3 (62.26, 37.74 theo.).

DATA TABLE NO. 423 SPECIFIC HEAT OF TRICALCIUM DIALUMINUM HEXAOXIDE $Ca_3Al_2O_6$

[Temperature, T, K; Specific Heat, C_p, Cal $g^{-1} K^{-1}$]

T	C_p
CURVE 1	
53.86	2.600 x 10^{-2}
58.06	2.988
58.29	3.004*
62.58	3.441
62.80	3.453*
67.09	3.901
71.65	4.356
76.22	4.808
81.20	5.322
86.39	5.840
94.90	6.680
105.00	7.642
114.74	8.571
124.82	9.445
135.88	1.037 x 10^{-1}
145.54	1.111
155.59	1.184
165.92	1.257
176.02	1.321
185.96	1.382
195.92	1.436
206.35	1.491
216.23	1.541
226.03	1.586
236.29	1.633
245.83	1.674
256.54	1.714
266.41	1.753
276.33	1.789
286.51	1.823
296.54	1.849

* Not shown on plot

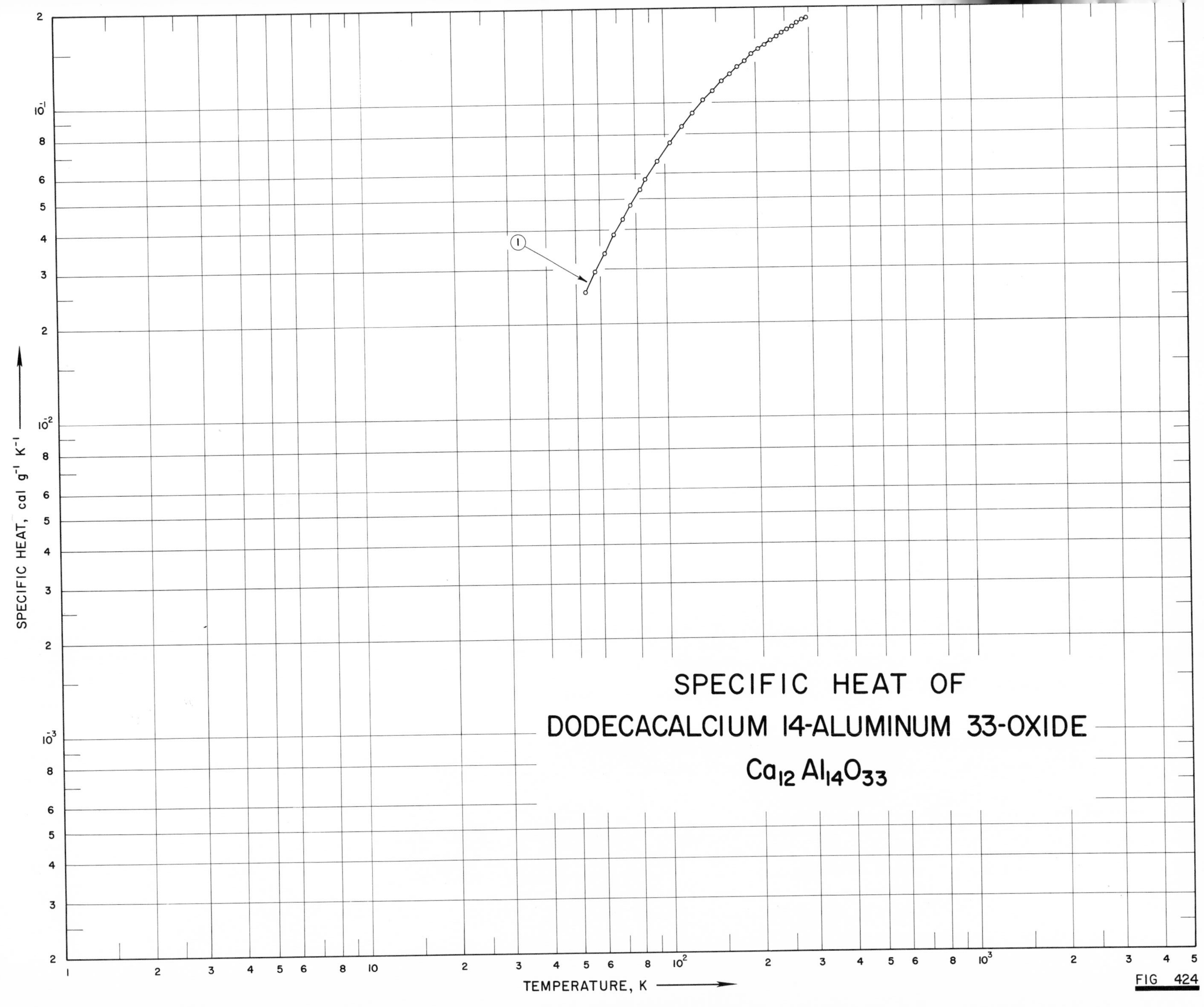
SPECIFIC HEAT OF
DODECACALCIUM 14-ALUMINUM 33-OXIDE
$Ca_{12}Al_{14}O_{33}$
SPECIFIC HEAT, cal g^{-1} K^{-1}
TEMPERATURE, K
1

FIG 424

SPECIFICATION TABLE NO. 424 SPECIFIC HEAT OF DODECACALCIUM 14-ALUMINUM 33-OXIDE $Ca_{12}Al_{14}O_{33}$

[For Data Reported in Figure and Table No. 424]

Curve No.	Ref. No.	Year	Temp. Range, K	Reported Error, %	Name and Specimen Designation	Composition (weight percent), Specifications and Remarks
1	365	1955	54-297			51. 2 Al_2O_3, 48. 32 CaO (51. 47, 48. 32 theo.), 0. 25 MgO + alkali oxides, 0. 10 Fe_2O_3.

DATA TABLE NO. 424 SPECIFIC HEAT OF DODECACALCIUM 14-ALUMINUM 33-OXIDE $Ca_{12}Al_{14}O_{33}$

[Temperature, T, K; Specific Heat, C_p, Cal g^{-1} K^{-1}]

T	C_p
CURVE 1	
53.89	2.529 x 10^{-2}
58.20	2.930
62.50	3.363
67.20	3.842
72.00	4.320
76.63	4.786
83.00	5.331
86.21	5.753
94.55	6.568
104.79	7.536
114.69	8.459
124.87	9.346
135.98	1.025 x 10^{-1}
145.55	1.100
156.38	1.180
165.74	1.245
175.78	1.309
185.56	1.369
195.72	1.429
206.15	1.483
216.04	1.533
225.99	1.581
235.94	1.630
245.78	1.672
256.20	1.714
266.08	1.755
276.25	1.794
286.54	1.831
296.75	1.863

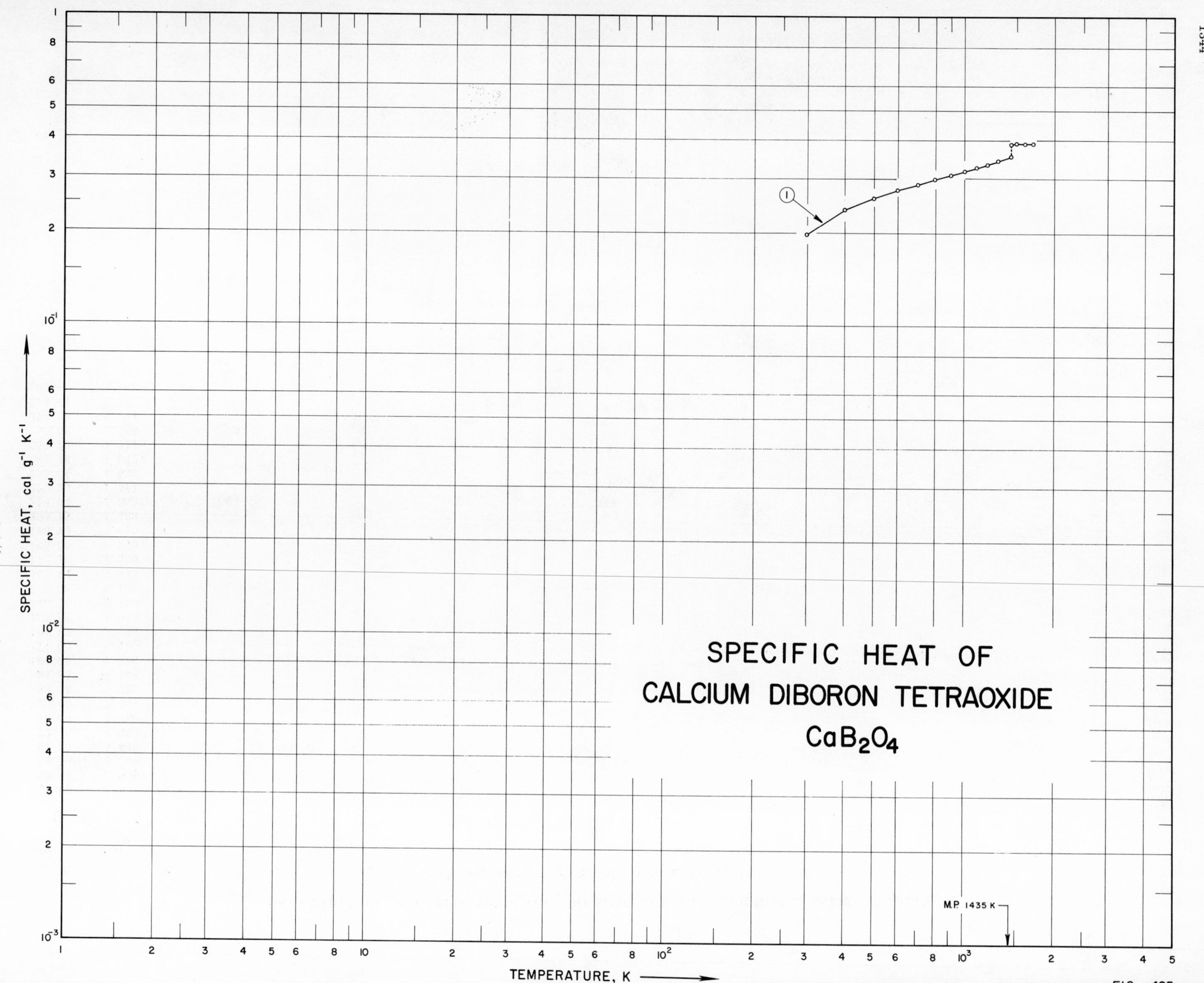
SPECIFIC HEAT OF
CALCIUM DIBORON TETRAOXIDE
CaB_2O_4
M.P. 1435 K
SPECIFIC HEAT, cal g^{-1} K^{-1}
TEMPERATURE, K
1

SPECIFICATION TABLE NO. 425 SPECIFIC HEAT OF CALCIUM DIBORON TETRAOXIDE CaB_2O_4

[For Data Reported in Figure and Table No. 425]

Curve No.	Ref. No.	Year	Temp. Range, K	Reported Error, %	Name and Specimen Designation	Composition (weight percent), Specifications and Remarks
1	366	1948	298-1700			

DATA TABLE NO. 425 SPECIFIC HEAT OF CALCIUM DIBORON TETRAOXIDE CaB_2O_4

[Temperature, T, K; Specific Heat, C_p, Cal g^{-1} K^{-1}]

T	C_p
CURVE 1	
300	1.985 x 10^{-1}
400	2.38
500	2.60
600	2.76
700	2.80
800	2.99
900	3.09
1000	3.18
1100	3.27
1200	3.35
1300	3.44
1400	3.52*
(s) 1435	3.55
(ℓ) 1435	4.91
1500	4.91
1600	4.91
1700	4.91

* Not shown on plot

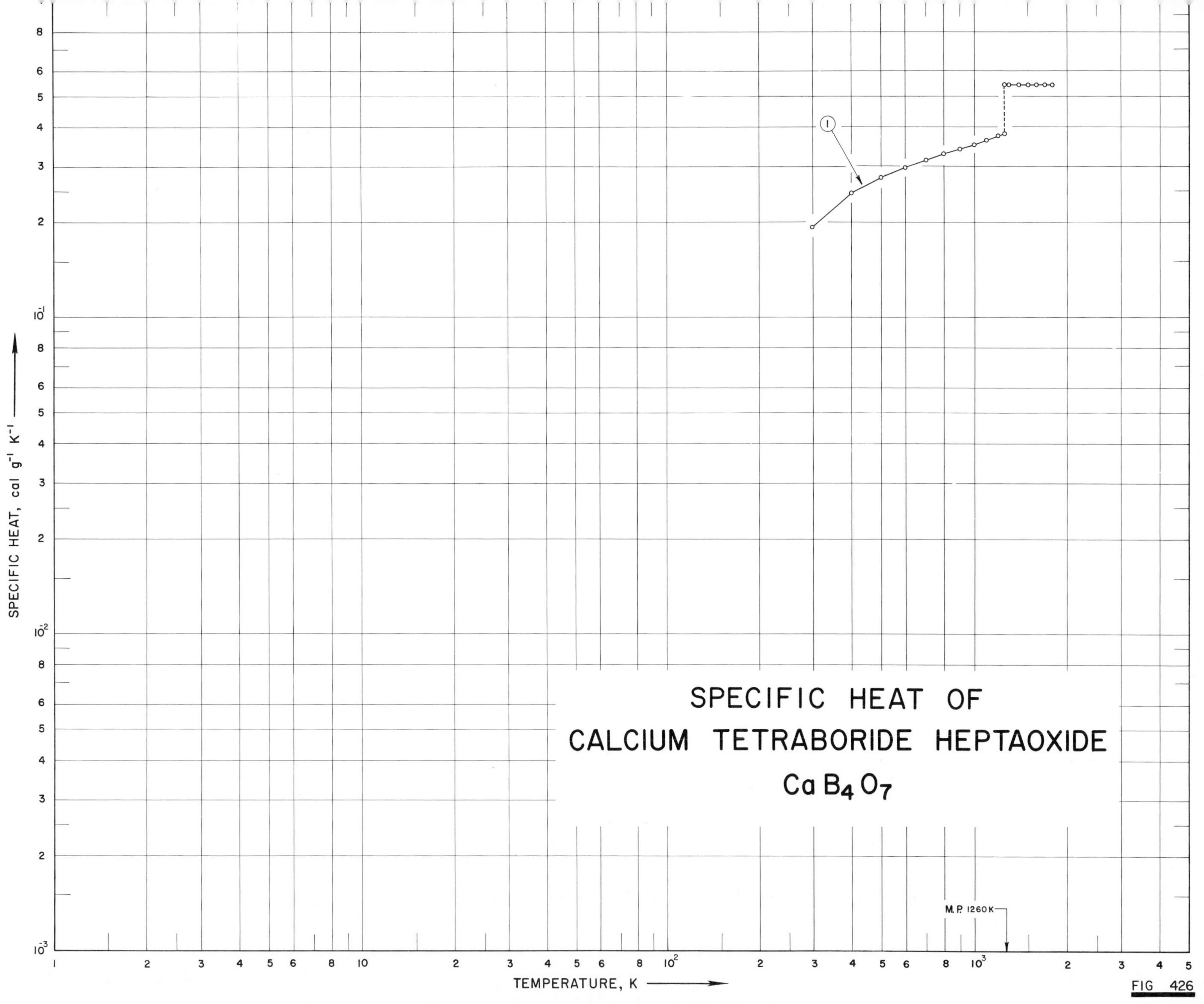
SPECIFIC HEAT OF
CALCIUM TETRABORIDE HEPTAOXIDE
CaB_4O_7
1
M.P. 1260 K
SPECIFIC HEAT, cal g^{-1} K^{-1}
TEMPERATURE, K

FIG 426

SPECIFICATION TABLE NO. 426 SPECIFIC HEAT OF CALCIUM TETRABORON HEPTAOXIDE CaB_4O_7

[For Data Reported in Figure and Table No. 426]

Curve No.	Ref. No.	Year	Temp. Range, K	Reported Error, %	Name and Specimen Designation	Composition (weight percent), Specifications and Remarks
1	366	1948	278-1800			

DATA TABLE NO. 426 SPECIFIC HEAT OF CALCIUM TETRABORON HEPTAOXIDE CaB_4O_7

[Temperature, T, K; Specific Heat, C_p, Cal g^{-1} K^{-1}]

T	C_p
CURVE 1	
298	1.931 x 10^{-1}
300	1.946*
400	2.471
500	2.767
600	2.972
700	3.135
800	3.275
900	3.402
1000	3.521
1100	3.634
1200	3.744
1260	3.808
1260	5.441
1300	5.441
1400	5.441
1500	5.441
1600	5.441
1700	5.441
1800	5.441

* Not shown on plot

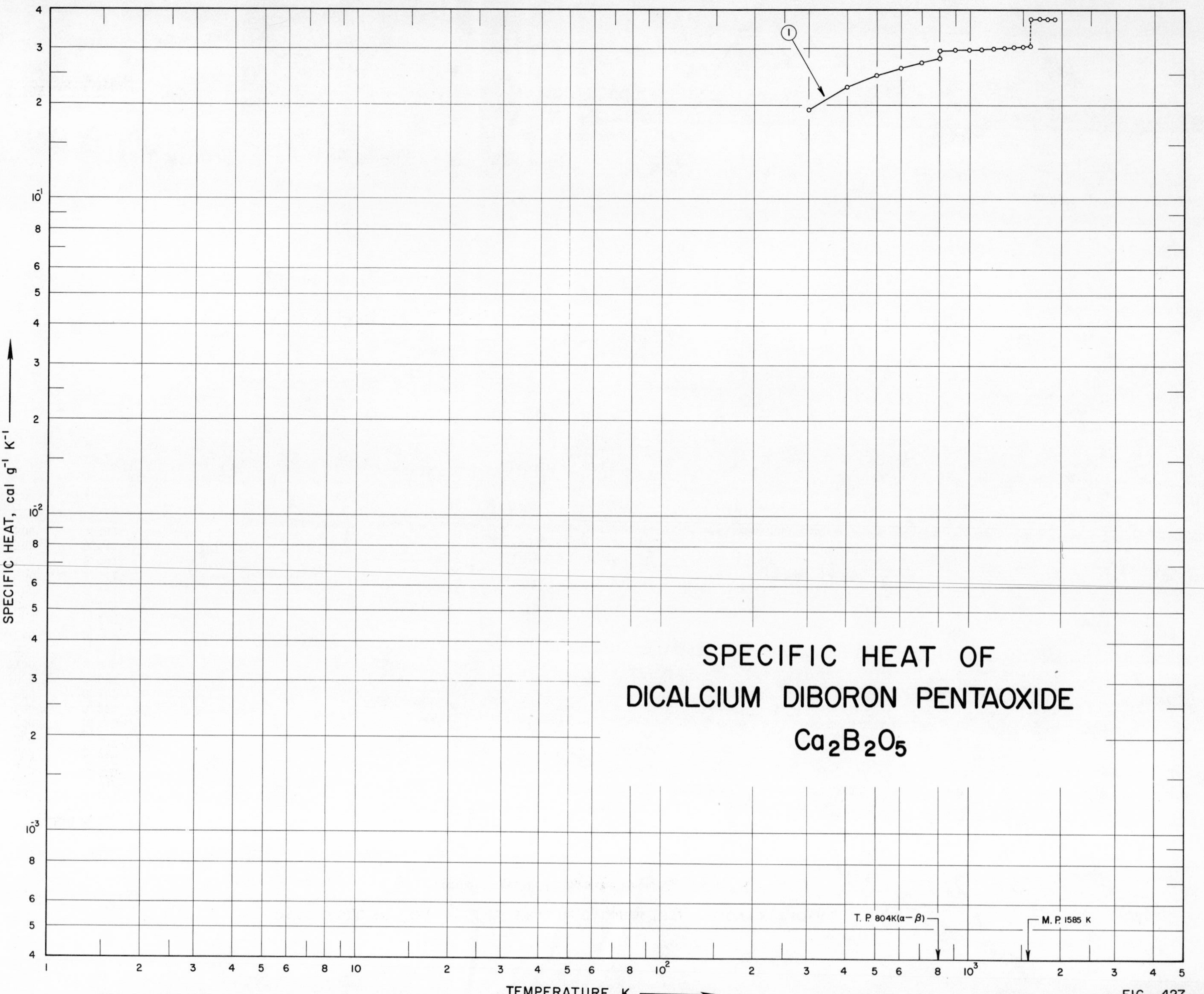
SPECIFIC HEAT OF
DICALCIUM DIBORON PENTAOXIDE
$Ca_2B_2O_5$
T. P. 804K(α − β)
M. P. 1585 K
TEMPERATURE, K
SPECIFIC HEAT, cal g⁻¹ K⁻¹
1

FIG 427

SPECIFICATION TABLE NO. 427 SPECIFIC HEAT OF DICALCIUM DIBORON PENTAOXIDE $Ca_2B_2O_5$

[For Data Reported in Figure and Table No. 427]

Curve No.	Ref. No.	Year	Temp. Range, K	Reported Error, %	Name and Specimen Designation	Composition (weight percent), Specifications and Remarks
1	366	1948	298-1900			α and β crystals.

DATA TABLE NO. 427 SPECIFIC HEAT OF DICALCIUM DIBORON PENTAOXIDE $Ca_2B_2O_5$

[Temperature, T, K; Specific Heat, C_p, Cal g^{-1} K^{-1}]

	T	C_p
	CURVE 1	
	300	1.94 x 10^{-1}
	400	2.29
	500	2.49
	600	2.62
	700	2.73
(α)	804	2.82
(β)	804	2.98
	900	2.99
	1000	3.00
	1100	3.02
	1200	3.03
	1300	3.05
	1400	3.06
	1500	3.07
(β)	1585	3.095
(ℓ)	1585	3.75
	1600	3.75*
	1700	3.75*
	1800	3.75*
	1900	3.75

* Not shown on plot

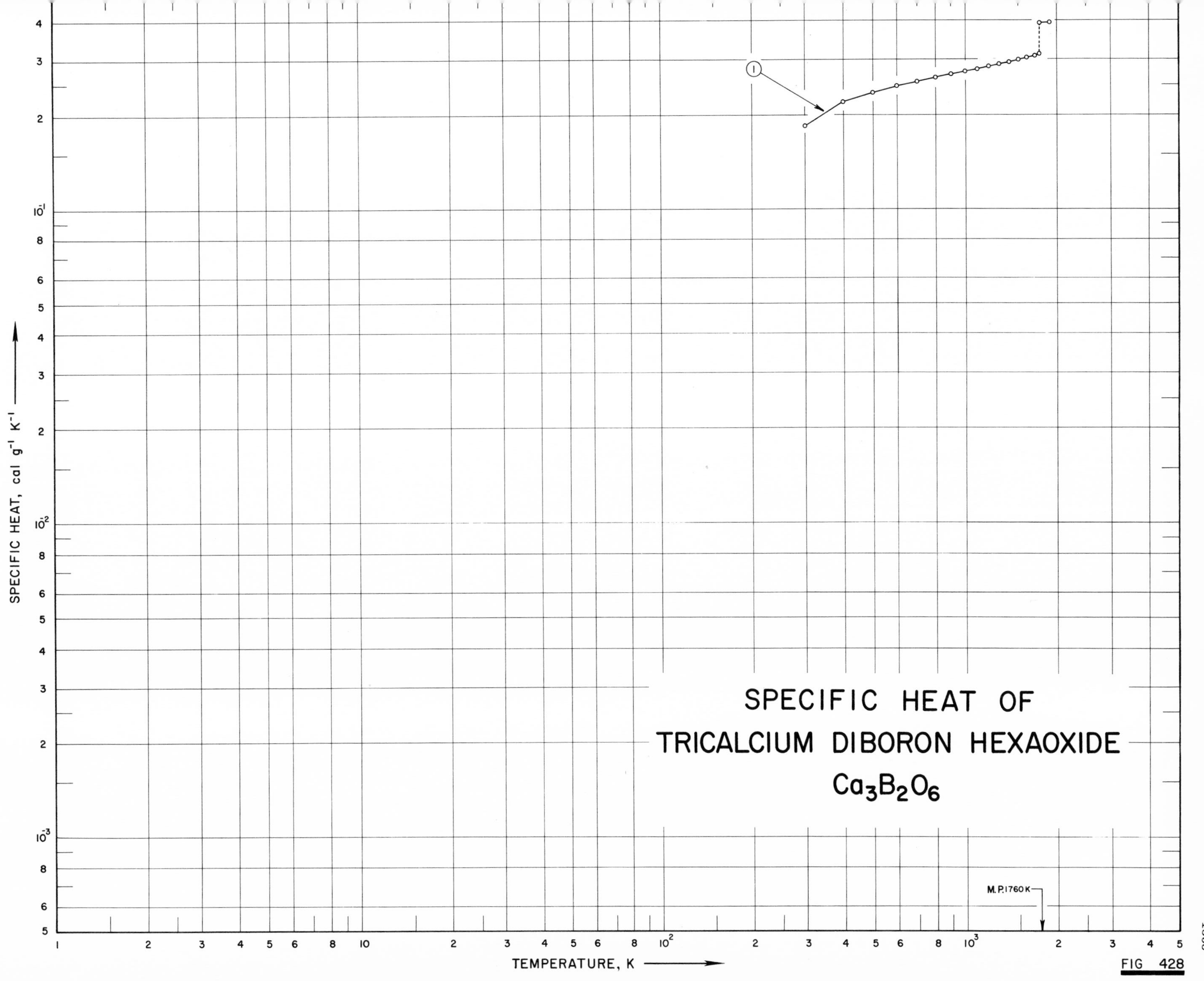

SPECIFIC HEAT OF
TRICALCIUM DIBORON HEXAOXIDE
$Ca_3B_2O_6$
M.P. 1760 K
1
TEMPERATURE, K
SPECIFIC HEAT, cal g^{-1} K^{-1}

FIG 428

SPECIFICATION TABLE NO. 428 SPECIFIC HEAT OF TRICALCIUM DIBORON HEXAOXIDE $Ca_3B_2O_6$

[For Data Reported in Figure and Table No. 428]

Curve No.	Ref. No.	Year	Temp. Range, K	Reported Error, %	Name and Specimen Designation	Composition (weight percent), Specifications and Remarks
1	366	1948	298-1900		Colemanite	

DATA TABLE NO. 428 SPECIFIC HEAT OF TRICALCIUM DIBORON HEXAOXIDE $Ca_3B_2O_6$

[Temperature, T, K; Specific Heat, C_p, Cal g^{-1} K^{-1}]

T	C_p
CURVE 1	
300	1.859×10^{-1}
400	2.20
500	2.37
600	2.48
700	2.57
800	2.64
900	2.70
1000	2.76
1100	2.81
1200	2.87
1300	2.91
1400	2.95
1500	3.00
1600	3.05
1700	3.09
(s) 1760	3.13
(ℓ) 1760	3.95
1800	3.95*
1900	3.95

* Not shown on plot

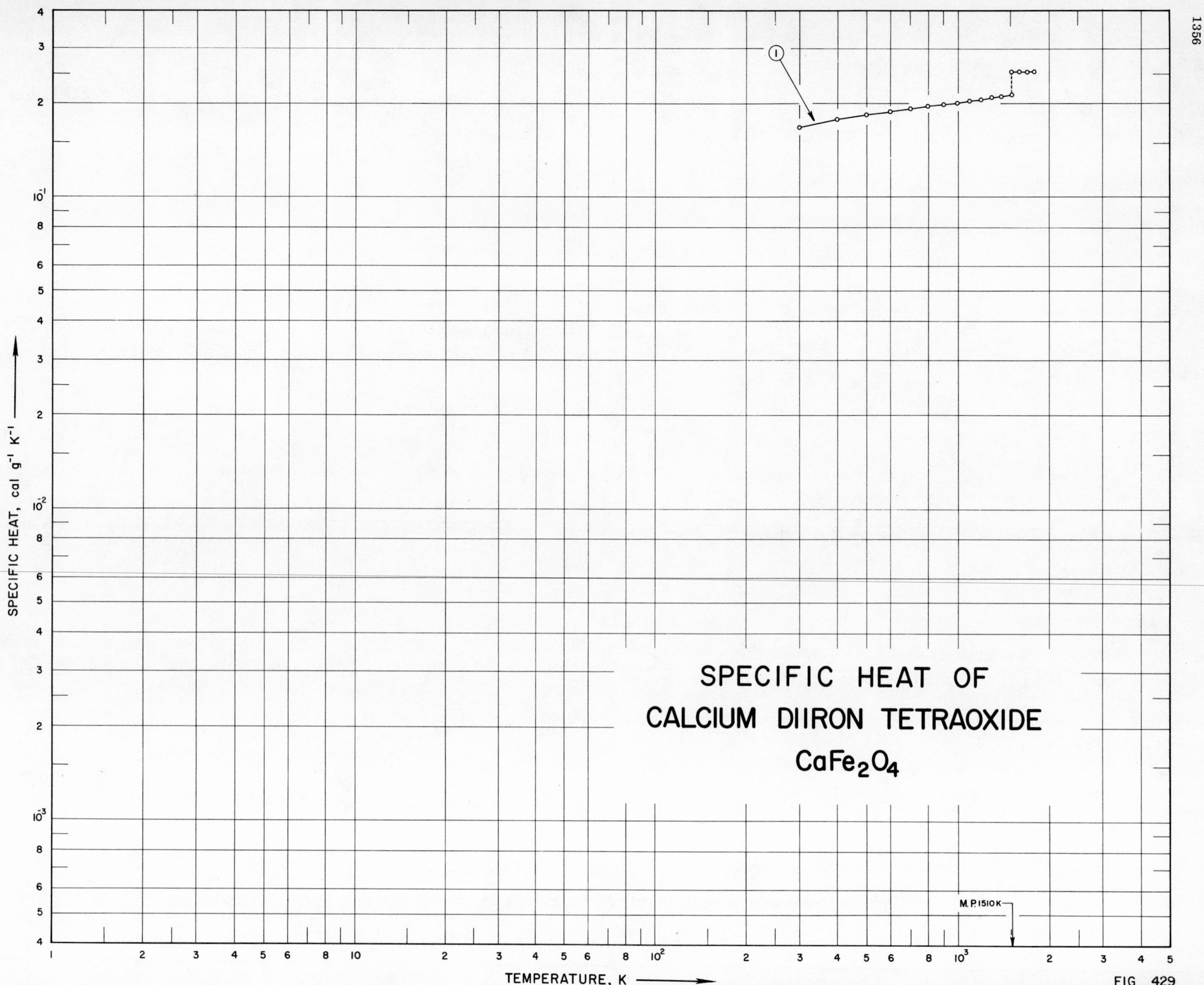
SPECIFIC HEAT OF
CALCIUM DIIRON TETRAOXIDE
$CaFe_2O_4$
1
M.P. 1510 K
SPECIFIC HEAT, cal g^{-1} K^{-1}
TEMPERATURE, K

FIG 429

SPECIFICATION TABLE NO. 429 SPECIFIC HEAT OF CALCIUM DIIRON TETRAOXIDE $CaFe_2O_4$

[For Data Reported in Figure and Table No. 429]

Curve No.	Ref. No.	Year	Temp. Range, K	Reported Error, %	Name and Specimen Designation	Composition (weight percent), Specifications and Remarks
1	367	1954	298-1800			74.05 Fe_2O_3, 26.05 CaO (74.0, 25.99 theo.); prepared from reagent-grade Fe_2O_3 and $CaCO_3$; ground mixed; heated to 1000-1200 C for several hrs; repeated several times adjusting composition between heating cycles.

DATA TABLE NO. 429 SPECIFIC HEAT OF CALCIUM DIIRON TETRAOXIDE $CaFe_2O_4$

[Temperature, T, K; Specific Heat, C_p, Cal $g^{-1} K^{-1}$]

T	C_p
CURVE 1	
300	1.70 x 10^{-1}
400	1.81
500	1.87
600	1.91
700	1.95
800	1.98
900	2.00
1000	2.03
1100	2.06
1200	2.08
1300	2.10
1400	2.13
1500	2.15*
(s) 1510	2.15
(ℓ) 1510	2.54
1600	2.54
1700	2.54
1800	2.54

* Not shown on plot

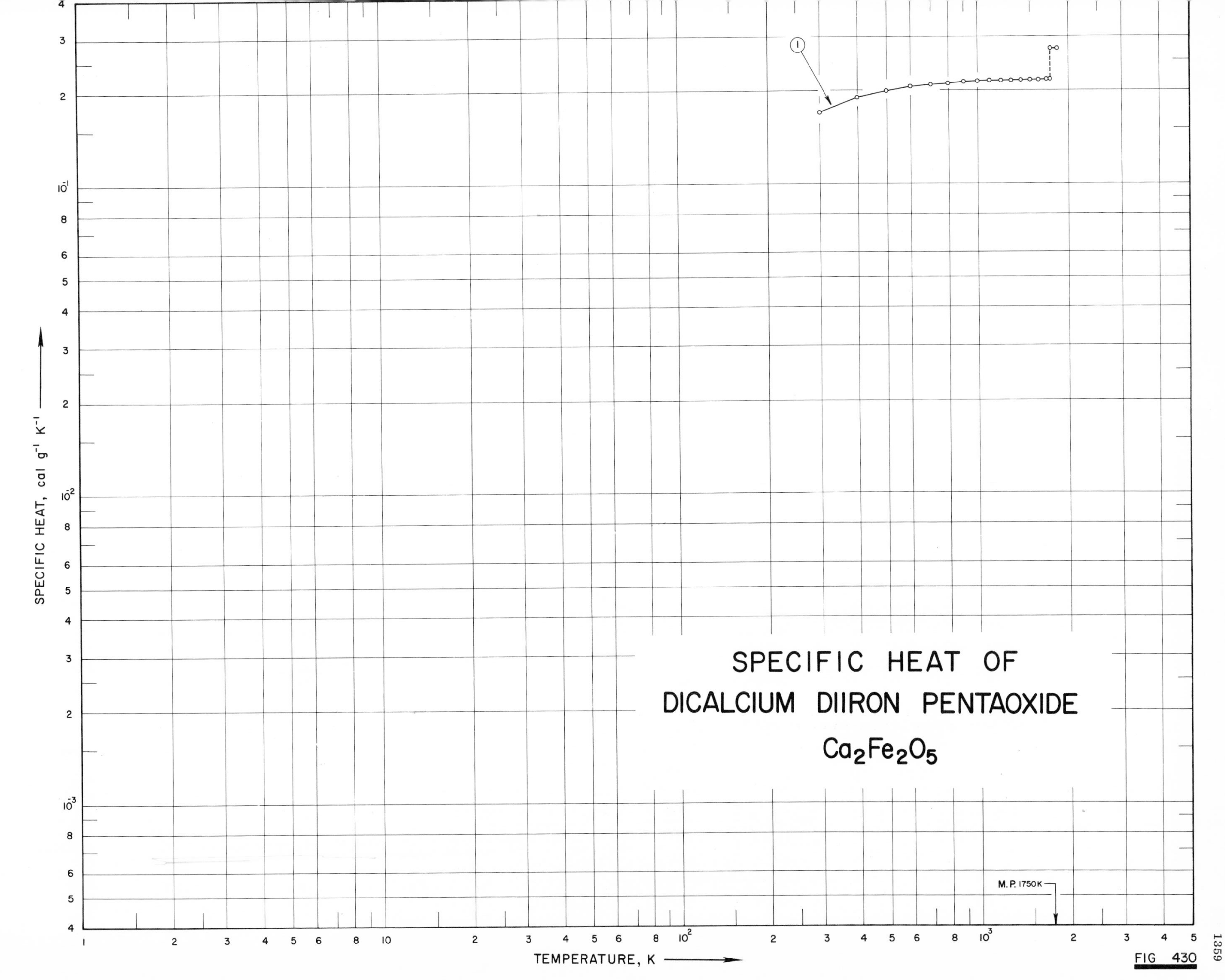
SPECIFIC HEAT OF
DICALCIUM DIIRON PENTAOXIDE
$Ca_2Fe_2O_5$
1
M.P. 1750 K
SPECIFIC HEAT, cal g^{-1} K^{-1}
TEMPERATURE, K
FIG 430

SPECIFICATION TABLE NO. 430 SPECIFIC HEAT OF DICALCIUM DIIRON PENTAOXIDE $Ca_2Fe_2O_5$

[For Data Reported in Figure and Table No. 430]

Curve No.	Ref. No.	Year	Temp. Range, K	Reported Error, %	Name and Specimen Designation	Composition (weight percent), Specifications and Remarks
1	367	1954	298-1850			58.71 Fe_2O_3, 41.27 CaO; prepared from reagent-grade Fe_2O_3 and $CaCO_3$; ground mixed; heated to 850-1230 C for several hrs; repeated several times adjusting composition between heating cycles.

DATA TABLE NO. 430 SPECIFIC HEAT OF DICALCIUM DIIRON PENTAOXIDE $Ca_2Fe_2O_5$

[Temperature, T, K; Specific Heat, C_p, Cal g^{-1} K^{-1}]

T	C_p
CURVE 1	
300	1.70 x 10^{-1}
400	1.91
500	2.01
600	2.06
700	2.09
800	2.11
900	2.13
1000	2.14
1100	2.15
1200	2.15
1300	2.158
1400	2.16
1500	2.165
1600	2.165
1700	2.17
(s) 1750	2.17
(ℓ) 1750	2.73
1800	2.73*
1850	2.73

* Not shown on plot

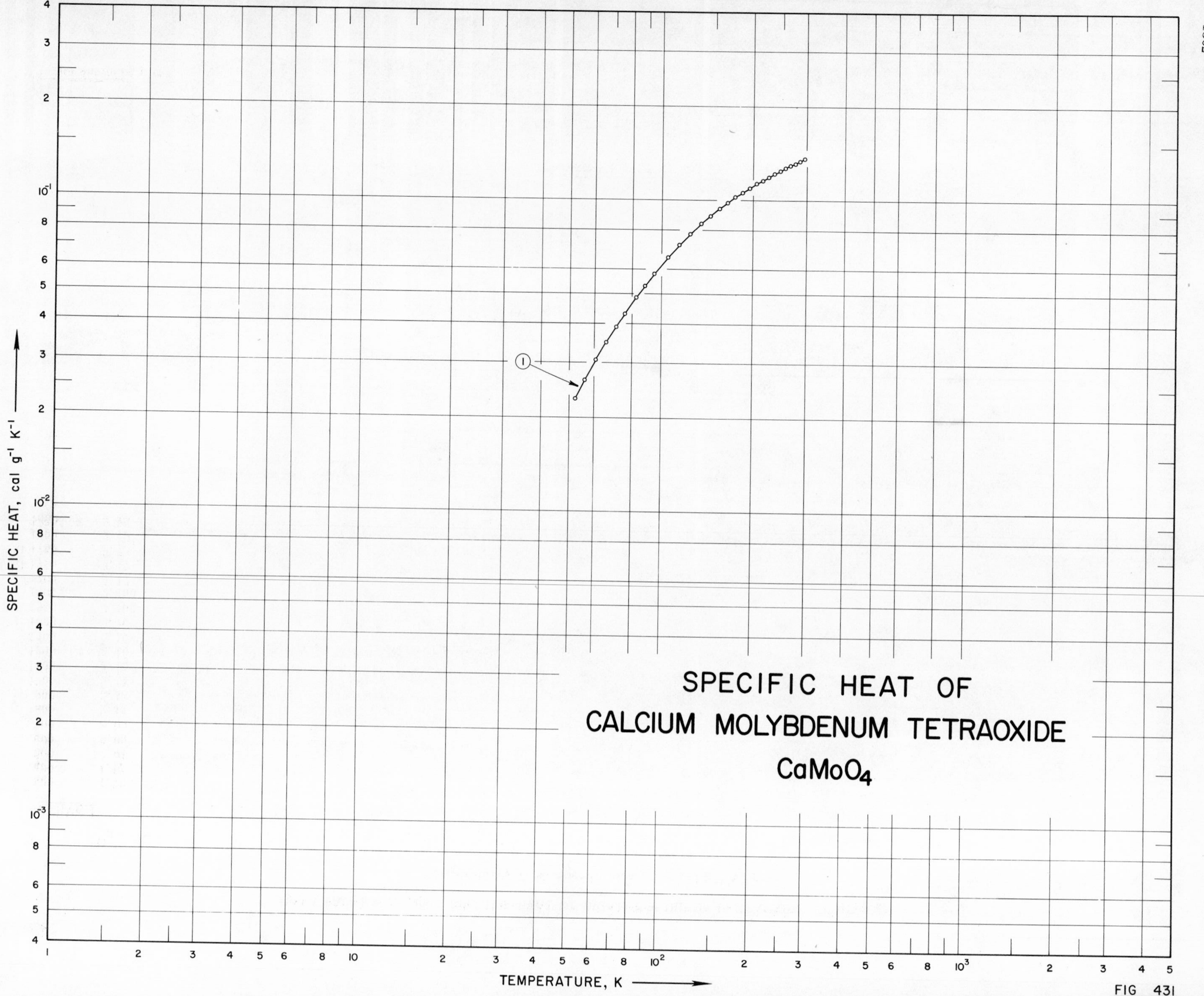

SPECIFIC HEAT OF
CALCIUM MOLYBDENUM TETRAOXIDE
CaMoO4
SPECIFIC HEAT, cal g⁻¹ K⁻¹
TEMPERATURE, K
1

FIG 431

SPECIFICATION TABLE NO. 431 SPECIFIC HEAT OF CALCIUM MOLYBDENUM TETRAOXIDE $CaMoO_4$

[For Data Repoted in Figure and Table No. 431]

Curve No.	Ref. No.	Year	Temp. Range, K	Reported Error, %	Name and Specimen Designation	Composition (weight percent), Specifications and Remarks
1	368	1963	52-296	0.3		71.98 MoO_3, 28.08 CaO; prepared from stoichiometric mixture of reagent-grade $CaCO_3$ and MoO_3 by heating at 500-570 C for 9 days; followed by heating at 600-690 C for 10 days.

DATA TABLE NO. 431 SPECIFIC HEAT OF CALCIUM MOLYBDENUM TETRAOXIDE $CaMoO_4$

[Temperature, T, K; Specific Heat, C_p, Cal $g^{-1} K^{-1}$]

T	C_p
CURVE 1	
52.32	2.300×10^{-2}
56.36	2.641
61.12	3.069
65.99	3.480
71.02	3.900
75.98	4.316
82.71	4.870
88.12	5.294
94.61	5.814
105.56	6.559
114.85	7.174
124.78	7.789
135.84	8.409
145.71	8.924
155.92	9.419
165.95	9.849
175.93	1.025×10^{-1}
186.01	1.063
195.89	1.098
206.36	1.133
216.27	1.165
226.38	1.194
236.29	1.221
246.29	1.248
256.55	1.275
266.45	1.297
276.46	1.319
286.95	1.342
296.43	1.364

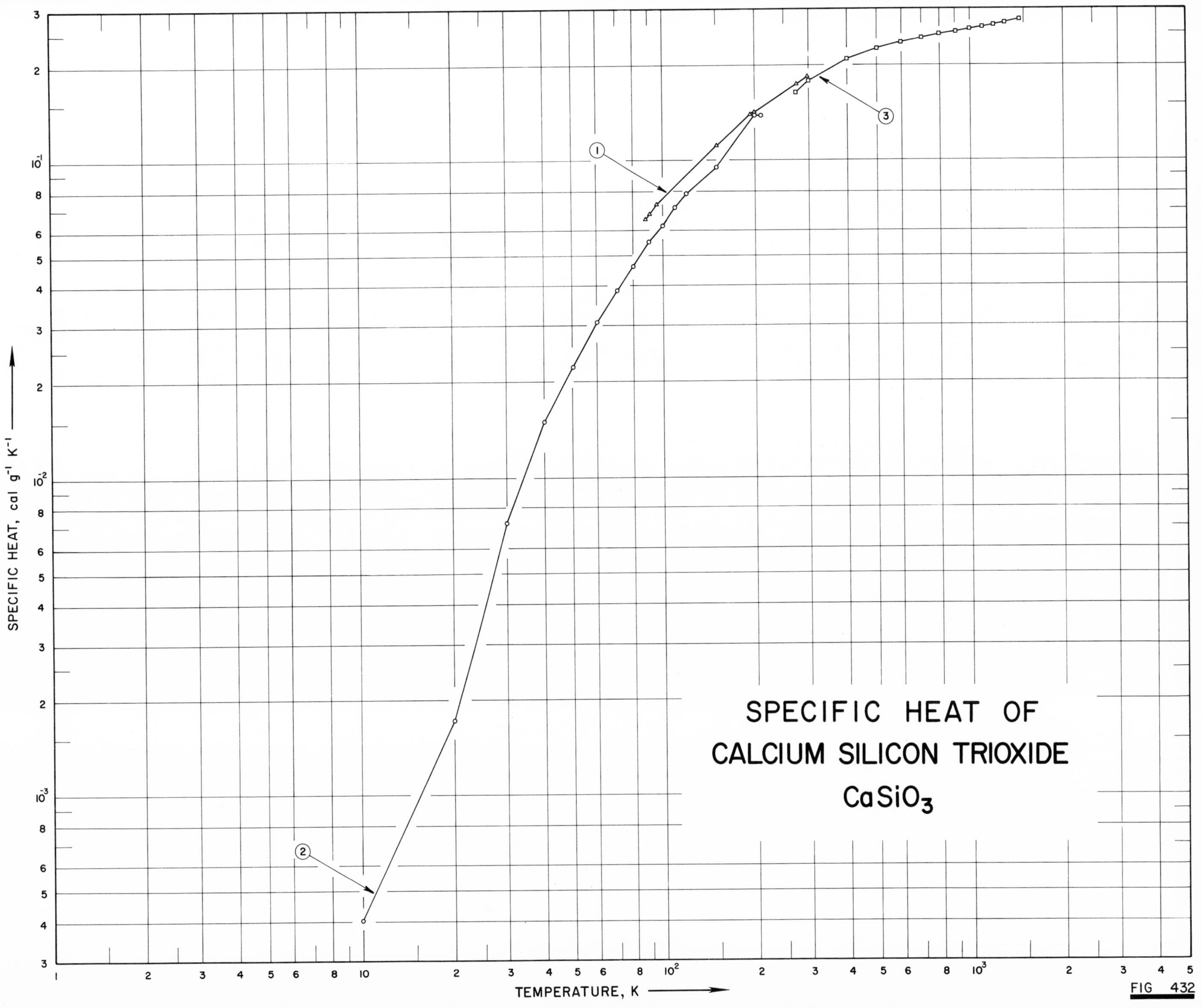
SPECIFIC HEAT OF
CALCIUM SILICON TRIOXIDE
CaSiO$_3$
SPECIFIC HEAT, cal g^{-1} K^{-1}
TEMPERATURE, K
1
2
3
FIG 432

SPECIFICATION TABLE NO. 432 SPECIFIC HEAT OF CALCIUM SILICON TRIOXIDE $CaSiO_3$

[For Data Reported in Figure and Table No. 432]

Curve No.	Ref. No.	Year	Temp. Range, K	Reported Error, %	Name and Specimen Designation	Composition (weight percent), Specifications and Remarks
1	372	1926	88-298	<1	pseudo Wollastonite	Synthetically prepared sample.
2	373	1934	10-210		Wollastonite	β-phase.
3	122	1941	273-1450	0. 25	Wollastonite	51. 52 SiO_2, 47. 85 CaO, 0. 36 R_2O_3; natural mineral from Riverside County, California.

DATA TABLE NO. 432 SPECIFIC HEAT OF CALCIUM SILICON TRIOXIDE, $CaSiO_3$

[Temperature, T, K; Specific Heat, C_p, Cal $g^{-1}K^{-1}$]

T	C_p
CURVE 1	
88.2	6.53×10^{-2}
91.1	6.78
92.3	6.84*
95.7	7.25
151.5	1.133×10^{-1}
194.7	1.396
199.4	1.416
275.4	1.735
278.2	1.754*
295.8	1.829*
298.3	1.832
CURVE 2	
10	4.0×10^{-4}
20	1.7×10^{-3}
30	7.2
40	1.5×10^{-2}
50	2.2
60	3.1
70	3.9
80	4.6
90	5.5
100	6.2
110	7.1
120	7.8
150	9.5
200	1.4×10^{-1}
210	1.4
CURVE 3	
273.15	1.635×10^{-1}
300	1.773
400	2.081
500	2.241
600	2.344
700	2.420
800	2.481
900	2.534
1000	2.581
1100	2.625
1200	2.667
1300	2.707
1400	2.746*
1450	2.765

* Not shown on plot

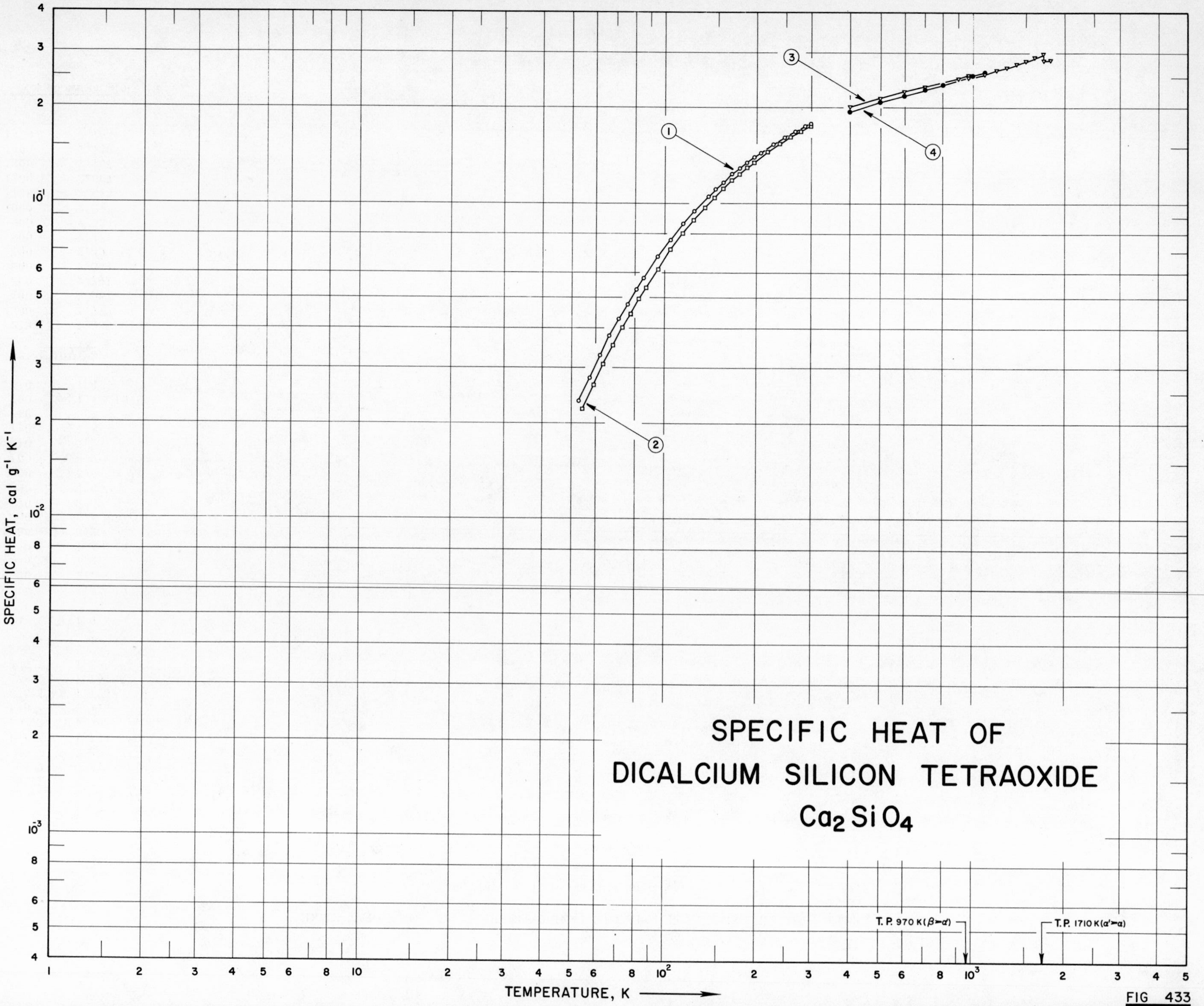

SPECIFIC HEAT OF
DICALCIUM SILICON TETRAOXIDE
Ca2 Si O4
SPECIFIC HEAT, cal g-1 K-1
TEMPERATURE, K
T.P. 970 K (β→α')
T.P. 1710 K (α'→α)
1
2
3
4

FIG 433

SPECIFICATION TABLE NO. 433 SPECIFIC HEAT OF DICALCIUM SILICON TETRAOXIDE Ca_2SiO_4

[For Data Reported in Figure and Table No. 433]

Curve No.	Ref. No.	Year	Temp. Range, K	Reported Error, %	Name and Specimen Designation	Composition (weight percent), Specifications and Remarks
1	369	1951	53-298			64.47 CaO, 34.68 SiO_2 (65.13, 34.57 theo.), 0.32 Fe_2O_3, 0.32 Al_2O_3, 0.14 MgO, 0.02 ignition loss; crystalline, β-phase.
2	370	1957	54-298			γ-phase.
3	371	1957	298-1800			64.47 CaO, 34.68 SiO_2, 0.32 Al_2O_3, 0.32 Fe_2O_3, 0.14 MgO, 0.02 H_2O; β, α, and α' phases.
4	371	1957	298-1100			34.88 SiO_2; γ-phase; prepared from reagent-grade calcium carbonate and silica.

DATA TABLE NO. 433 SPECIFIC HEAT OF DICALCIUM SILICON TETRAOXIDE, Ca_2SiO_4

[Temperature, T, K; Specific Heat, C_p, Cal $g^{-1}K^{-1}$]

T	C_p
CURVE 1	
52.66	2.371 x 10^{-2}
57.02	2.795
61.61	3.280
66.24	3.793
70.92	4.292
75.66	4.790
80.72	5.330
85.20	5.785
94.81	6.753
104.39	7.647
114.59	8.576
124.62	9.412
138.68	1.050 x 10^{-1}
146.69	1.106
155.86	1.166
165.96	1.231
175.94	1.288
185.93	1.344
195.92	1.394
206.22	1.441
216.22	1.490*
226.42	1.532
236.16	1.571*
245.79	1.611
256.19	1.649*
266.31	1.686
276.22	1.717*
286.41	1.752
296.48	1.778*
298.16	1.785
CURVE 2	
54.15	2.228 x 10^{-1}
58.76	2.650
63.28	3.080
67.99	3.544
72.82	4.021
77.35	4.459
82.42	4.963
86.78	5.386
94.74	6.154
104.93	7.106
114.77	7.971
124.82	8.813
CURVE 2 (cont.)	
135.85	9.672 x 10^{-2}
145.64	1.042 x 10^{-1}
155.82	1.113
165.95	1.178
175.76	1.238
185.92	1.296
196.02	1.347
206.24	1.400*
216.34	1.450
226.17	1.496*
236.41	1.538
245 95	1.575*
256.38	1.616
266.21	1.653*
276.10	1.687
286.70	1.722*
296.22	1.757*
298.15	1.760
CURVE 3	
298.15	1.784 x 10^{-1}*
300	1.790*
400	2.023
500	2.162
600	2.263
700	2.346
900	2.488
β 970	2.534
α'970	2.488
1000	2.507*
1100	2.571
1200	2.635
1300	2.699
1400	2.763
1500	2.827
1600	2.891
1700	2.955*
α'1710	2.961
α1710	2.845
1750	2.845*
1800	2.845
CURVE 4	
γ 298.15	1.760 x 10^{-1}*
300	1.765*
400	1.967
500	2.100
600	2.204
700	2.295
800	2.380
900	2.460*
1000	2.538
γ 1100	2.614

* Not shown on plot

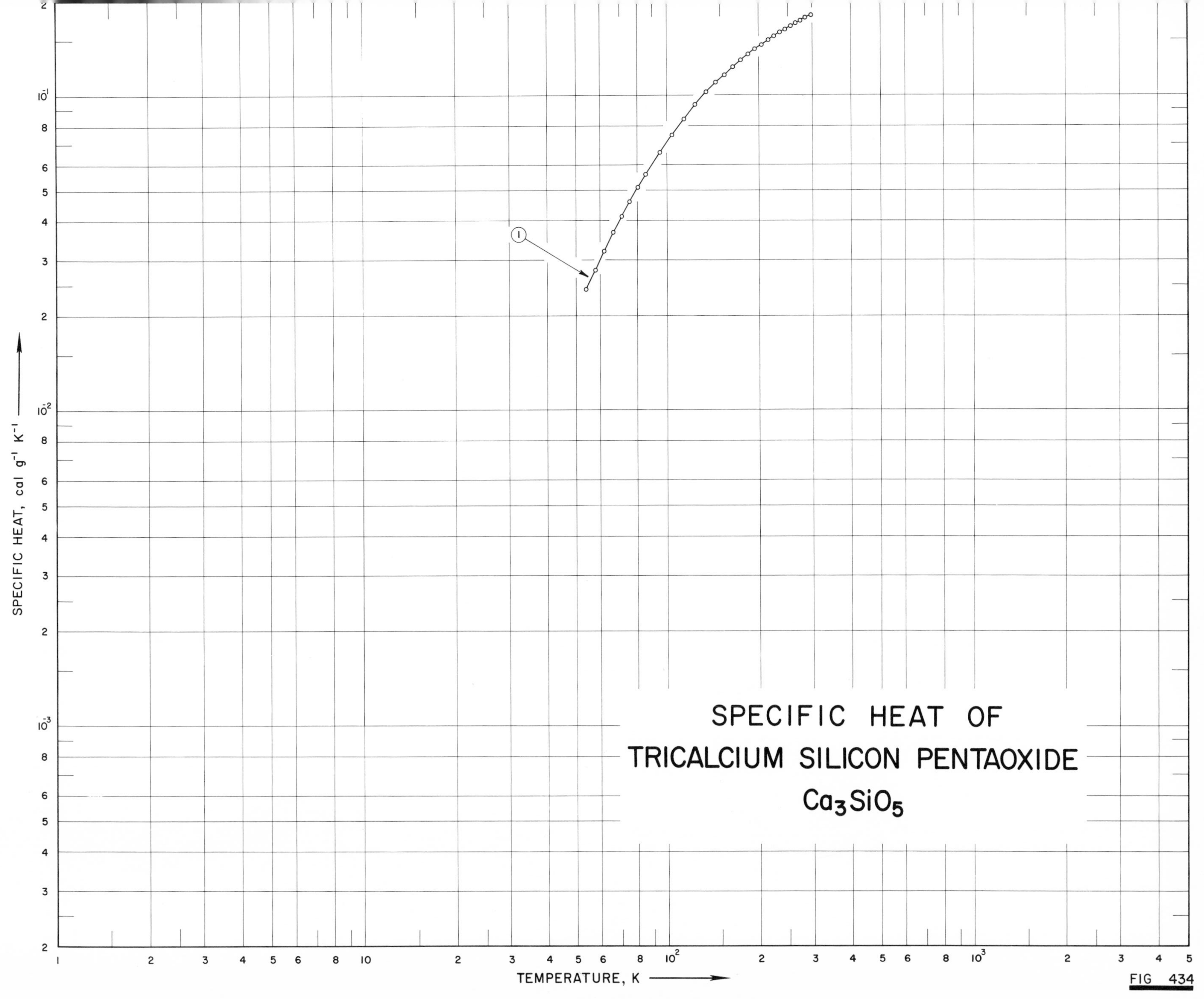
SPECIFIC HEAT OF
TRICALCIUM SILICON PENTAOXIDE
Ca_3SiO_5
1
SPECIFIC HEAT, cal g^{-1} K^{-1}
TEMPERATURE, K
FIG 434

SPECIFICATION TABLE NO. 434 SPECIFIC HEAT OF TRICALCIUM SILICON PENTAOXIDE Ca_3SiO_5

[For Data Reported in Figure and Table No. 434]

Curve No.	Ref. No.	Year	Temp. Range, K	Reported Error, %	Name and Specimen Designation	Composition (weight percent), Specifications and Remarks
1	369	1951	54-298			73.64 CaO, 26.21 SiO_2 (73.69, 26.31 theo.), 0.13 Fe_2O_3, 0.13 Al_2O_3, 0.11 MgO, 0.05 ignition loss.

DATA TABLE NO. 434 SPECIFIC HEAT OF TRICALCIUM SILICON PENTAOXIDE, Ca_3SiO_5

[Temperature, T, K; Specific Heat, C_p, Cal $g^{-1}K^{-1}$]

T	C_p
CURVE 1	
54.26	2.439 x 10^{-2}
58.11	2.792
62.13	3.208
66.58	3.674
71.20	4.148
75.83	4.621
80.73	5.142
85.62	5.642
95.10	6.588
104.67	7.499
114.71	8.445
124.74	9.312
136.27	1.025 x 10^{-1}
146.18	1.102
156.05	1.169
166.07	1.235
176.24	1.297
186.17	1.353
196.00	1.404
206.33	1.453
216.33	1.502
226.01	1.547
236.19	1.587
245.95	1.624
256.33	1.663
266.88	1.698
276.75	1.733
286.50	1.767
296.53	1.791*
298.16	1.799

* Not shown on plot

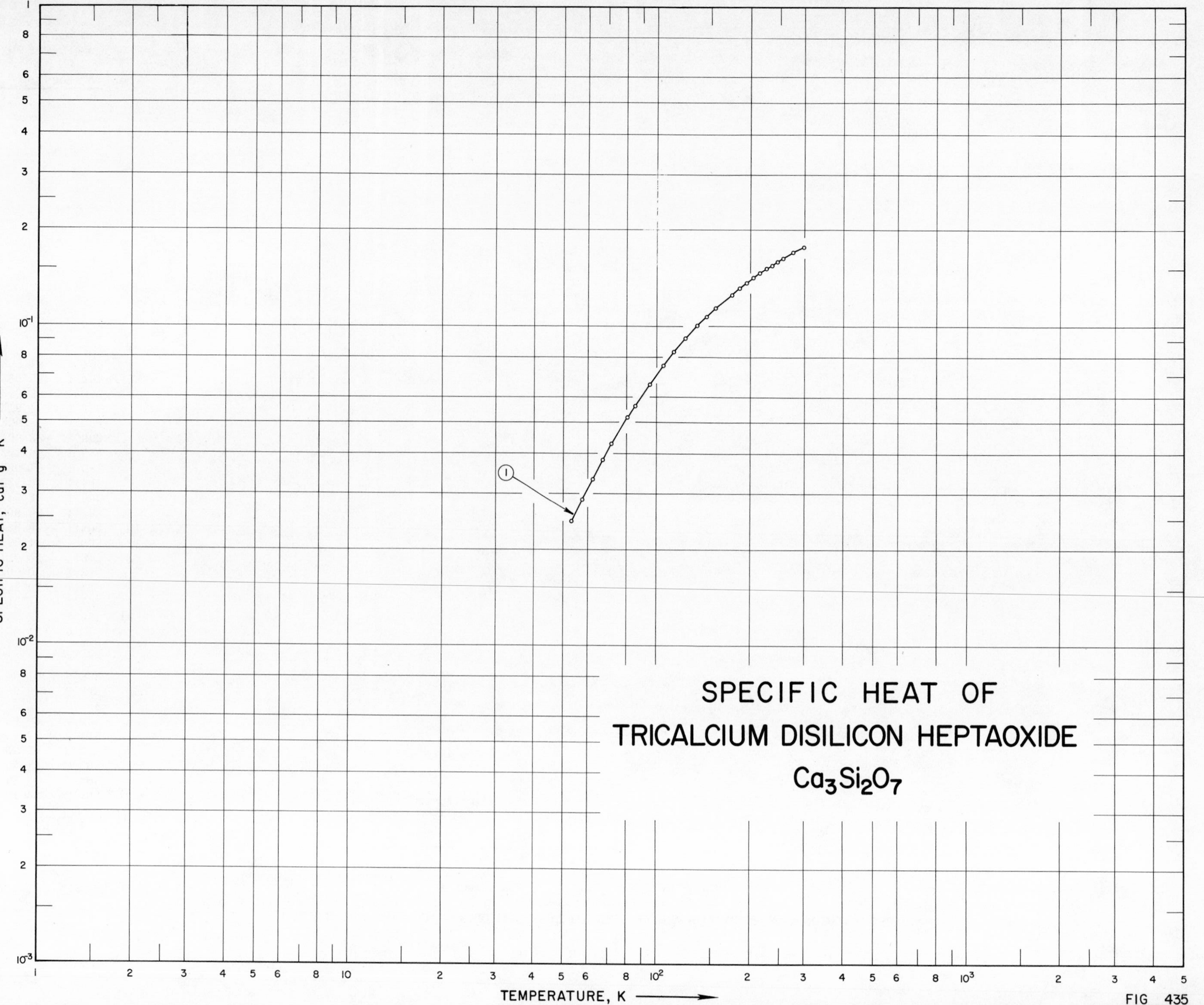

SPECIFIC HEAT OF
TRICALCIUM DISILICON HEPTAOXIDE
$Ca_3Si_2O_7$
SPECIFIC HEAT, cal g^{-1} K^{-1}
TEMPERATURE, K
1

FIG 435

SPECIFICATION TABLE NO. 435 SPECIFIC HEAT OF TRICALCIUM DISILICON HEPTAOXIDE $Ca_3Si_2O_7$

[For Data Reported in Figure and Table No. 435]

Curve No.	Ref. No.	Year	Temp. Range, K	Reported Error, %	Name and Specimen Designation	Composition (weight percent), Specifications and Remarks
1	370	1957	53-298			58.37 lime, 41.62 silica.

DATA TABLE NO. 435 SPECIFIC HEAT OF TRICALCIUM DISILICON HEPTAOXIDE, $Ca_3Si_2O_7$

[Temperature, T, K; Specific Heat, C_p, Cal $g^{-1}K^{-1}$]

T	C_p
CURVE 1	
53.45	2.453 x 10^{-2}
57.86	2.867
62.42	3.326
67.14	3.810
71.75	4.285
80.68	5.183
85.24	5.631
95.04	6.577
105.04	7.531
114.56	8.328
124.51	9.160
136.28	1.006 x 10^{-1}
145.77	1.077
155.90	1.146
175.89	1.267
185.89	1.323
195.82	1.374
206.31	1.426
216.22	1.475
226.31	1.518
236.14	1.559
245.95	1.596
256.41	1.642
276.18	1.708
286.69	1.742*
296.15	1.775*
298.15	1.777

* Not shown on plot

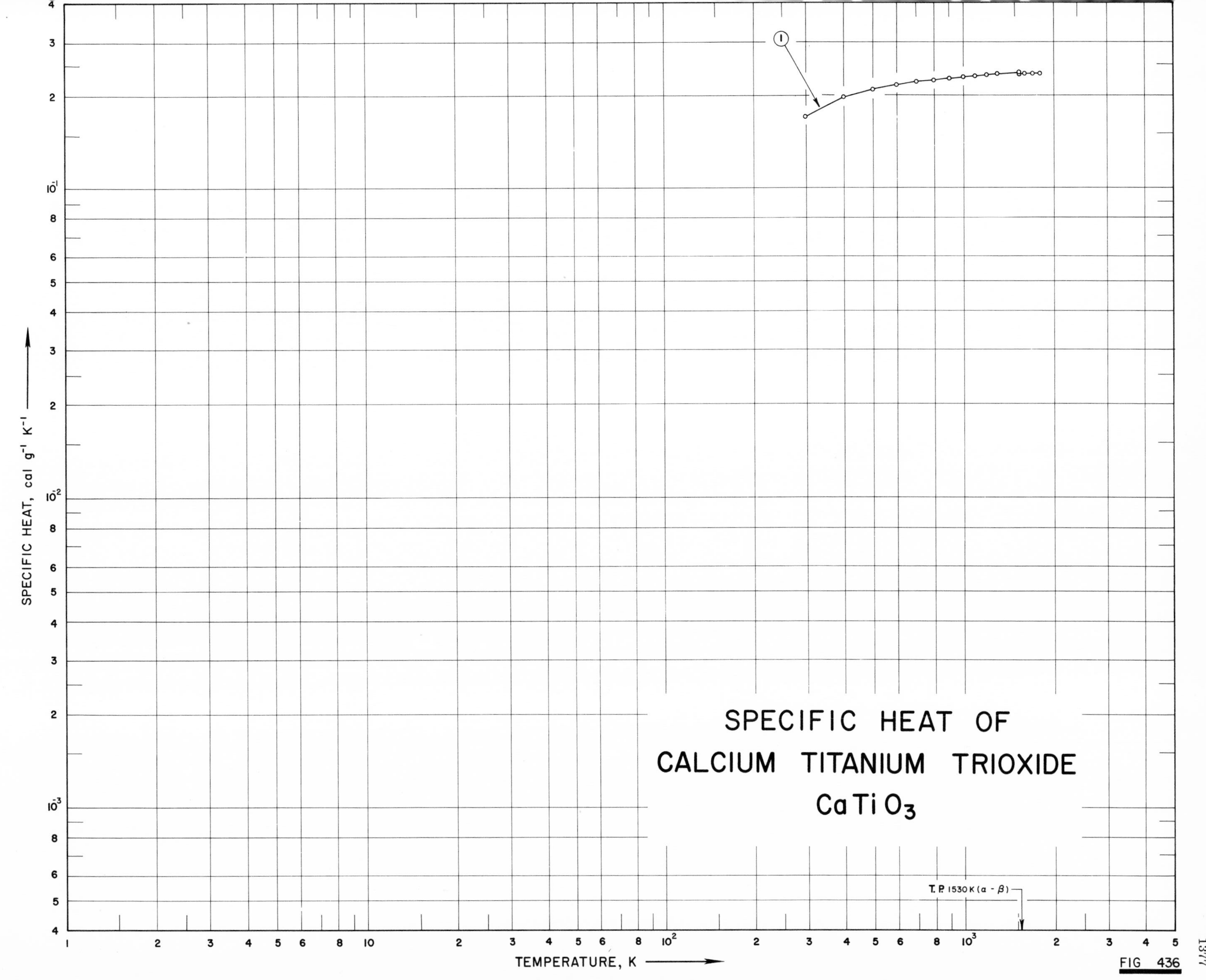

SPECIFIC HEAT OF
CALCIUM TITANIUM TRIOXIDE
CaTiO3
T.P. 1530 K (α - β)
1
SPECIFIC HEAT, cal g-1 K-1
TEMPERATURE, K
FIG 436

SPECIFICATION TABLE NO. 436 SPECIFIC HEAT OF CALCIUM TITANIUM TRIOXIDE $CaTiO_3$

[For Data Reported in Figure and Table No. 436]

Curve No.	Ref. No.	Year	Temp. Range, K	Reported Error, %	Name and Specimen Designation	Composition (weight percent), Specifications and Remarks
1	374	1946	298-1800			0.69 CaO acid soluble, 0.05 Co_2; $\alpha \rightarrow \beta$ transition at 1530 K.

DATA TABLE NO. 436 SPECIFIC HEAT OF CALCIUM TITANIUM TRIOXIDE, $CaTiO_3$

[Temperature, T, K; Specific Heat, C_p, Cal $g^{-1}K^{-1}$]

T	C_p
CURVE 1	
298	1.717×10^{-1}
300	1.724*
400	1.973
500	2.094
600	2.164
700	2.210
800	2.244
900	2.270
1000	2.292
1100	2.310
1200	2.327
1300	2.342
1500	2.369*
1530	2.373
1530	2.355
1600	2.355
1700	2.355
1800	2.355

* Not shown on plot

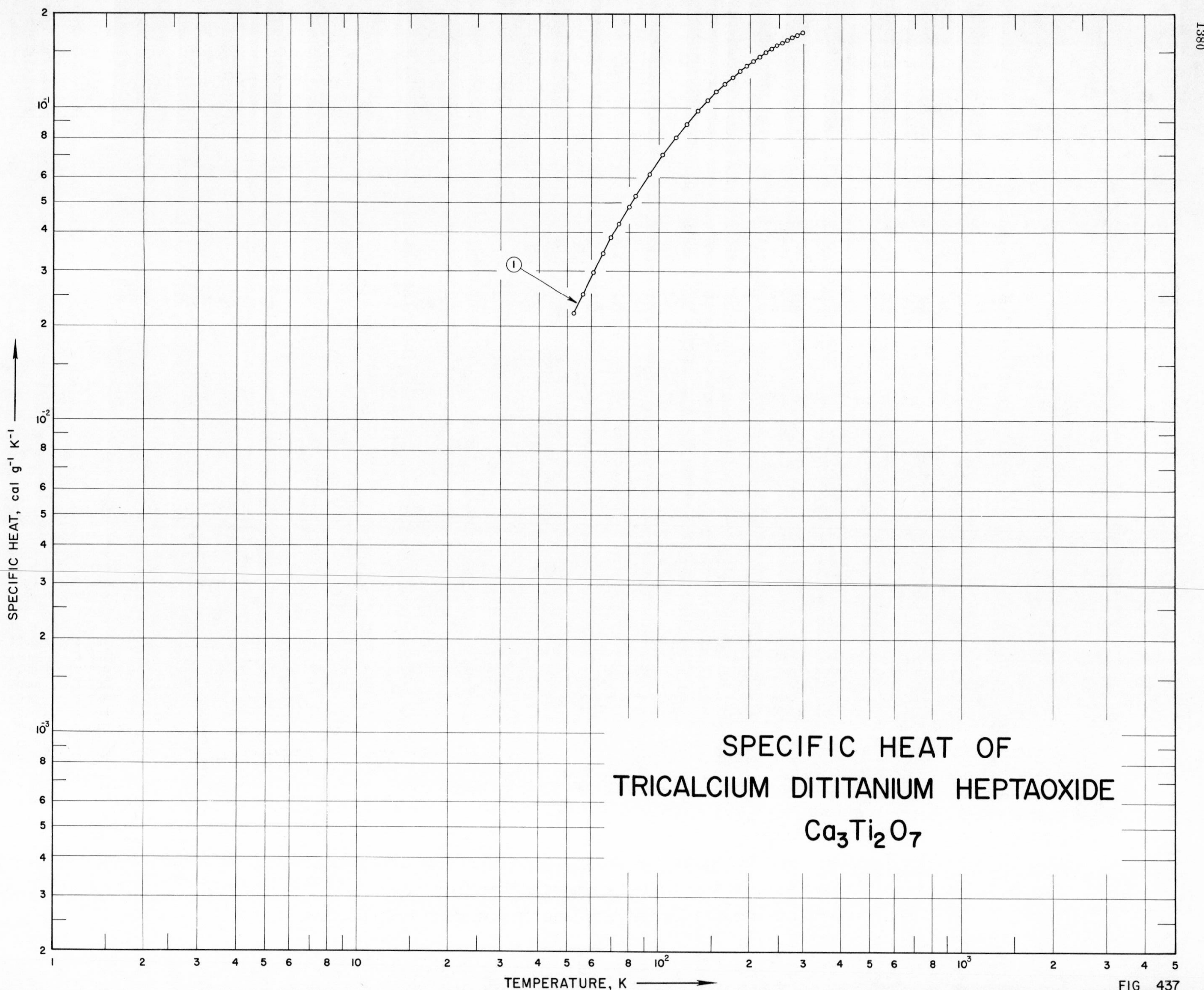
SPECIFIC HEAT, cal g^{-1} K^{-1}
1
SPECIFIC HEAT OF
TRICALCIUM DITITANIUM HEPTAOXIDE
$Ca_3Ti_2O_7$
TEMPERATURE, K

FIG 437

SPECIFICATION TABLE NO. 437 SPECIFIC HEAT OF TRICALCIUM DITITANIUM HEPTAOXIDE $Ca_3Ti_2O_7$

[For Data Reported in Figure and Table No. 437]

Curve No.	Ref. No.	Year	Temp. Range, K	Reported Error, %	Name and Specimen Designation	Composition (weight percent), Specifications and Remarks
1	358	1955	53-298			48.61 titania (48.71 theo.); prepared from reagent-grade calcium carbonate and pure titania; pressed into pellets; heated 5 times for 12 hrs at 1400-1500 C.

DATA TABLE NO. 437 SPECIFIC HEAT OF TRICALCIUM DITITANIUM HEPTAOXIDE, $Ca_3Ti_2O_7$

[Temperature, T, K; Specific Heat, C_p, Cal $g^{-1}K^{-1}$]

T	C_p
CURVE 1	
52.63	2.211 x 10^{-2}
56.50	2.542
61.10	2.976
65.60	3.426
69.86	3.844
74.28	4.262
80.15	4.838
84.36	5.249
93.53	6.130
103.69	7.072
114.31	8.041
124.38	8.898
135.58	9.795
146.01	1.060 x 10^{-1}
155.76	1.129
165.69	1.194
176.52	1.258
186.67	1.319
196.44	1.368
206.18	1.418
216.18	1.466
226.70	1.510
236.29	1.550
245.74	1.584
256.21	1.620
266.17	1.650
276.05	1.683
286.53	1.711
296.87	1.738*
298.16	1.744

* Not shown on plot

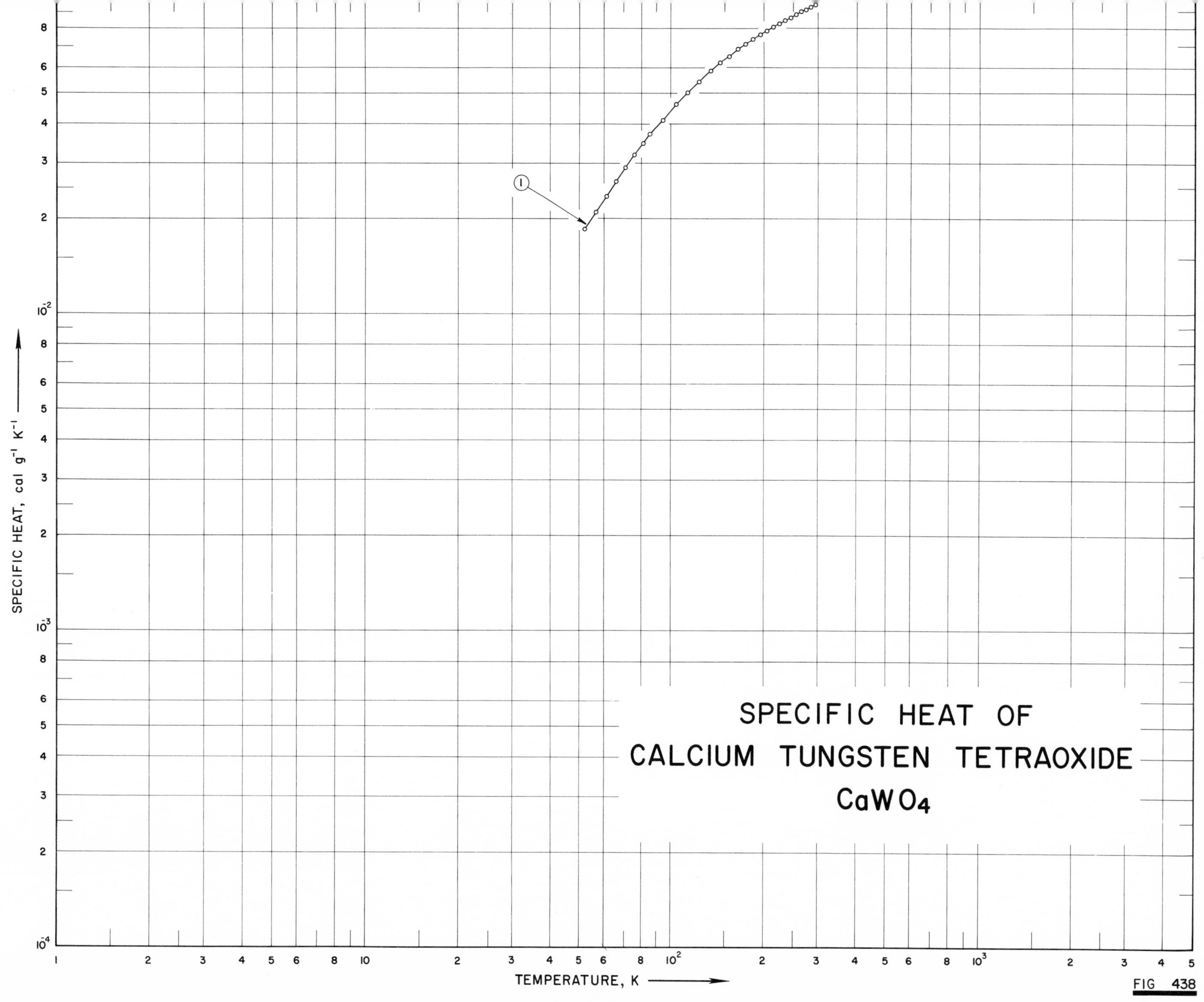
SPECIFIC HEAT OF
CALCIUM TUNGSTEN TETRAOXIDE
CaWO4
SPECIFIC HEAT, cal g-1 K-1
TEMPERATURE, K
1
FIG 438

SPECIFICATION TABLE NO. 438 SPECIFIC HEAT OF CALCIUM TUNGSTEN TETRAOXIDE $CaWO_4$

[For Data Reported in Figure and Table No. 438]

Curve No.	Ref. No.	Year	Temp. Range, K	Reported Error, %	Name and Specimen Designation	Composition (weight percent), Specifications and Remarks
1	375	1961	52-296	0.3		80.59 WO_3, 19.49 CaO; prepared by heating stoichiometric mixture of reagent-grade calcium carbonate and tungstic acid 7 times for total of 2 days at 680 C and 10 days at 800 C.

DATA TABLE NO. 438 SPECIFIC HEAT OF CALCIUM TUNGSTEN TETRAOXIDE, $CaWO_4$

[Temperature, T, K; Specific Heat, C_p, Cal $g^{-1}K^{-1}$]

T	C_p
CURVE 1	
52.28	1.866×10^{-2}
56.82	2.103
61.52	2.366
66.15	2.627
71.26	2.910
76.41	3.186
81.75	3.467
86.45	3.707
94.79	4.106
104.82	4.607
114.15	5.010
124.37	5.447
136.44	5.889
145.88	6.226
155.88	6.549
166.32	6.865
176.19	7.143
186.36	7.396
196.26	7.626
206.26	7.855
216.56	8.077
226.18	8.275
236.06	8.477
246.04	8.647
256.57	8.842
266.46	9.012
276.99	9.158
286.96	9.318
296.40	9.474
CURVE 2	
294	1.031×10^{-1}
300	1.034
400	1.072
500	1.110
600	1.148
700	1.186
800	1.224
900	1.262
1000	1.300
1073	1.328

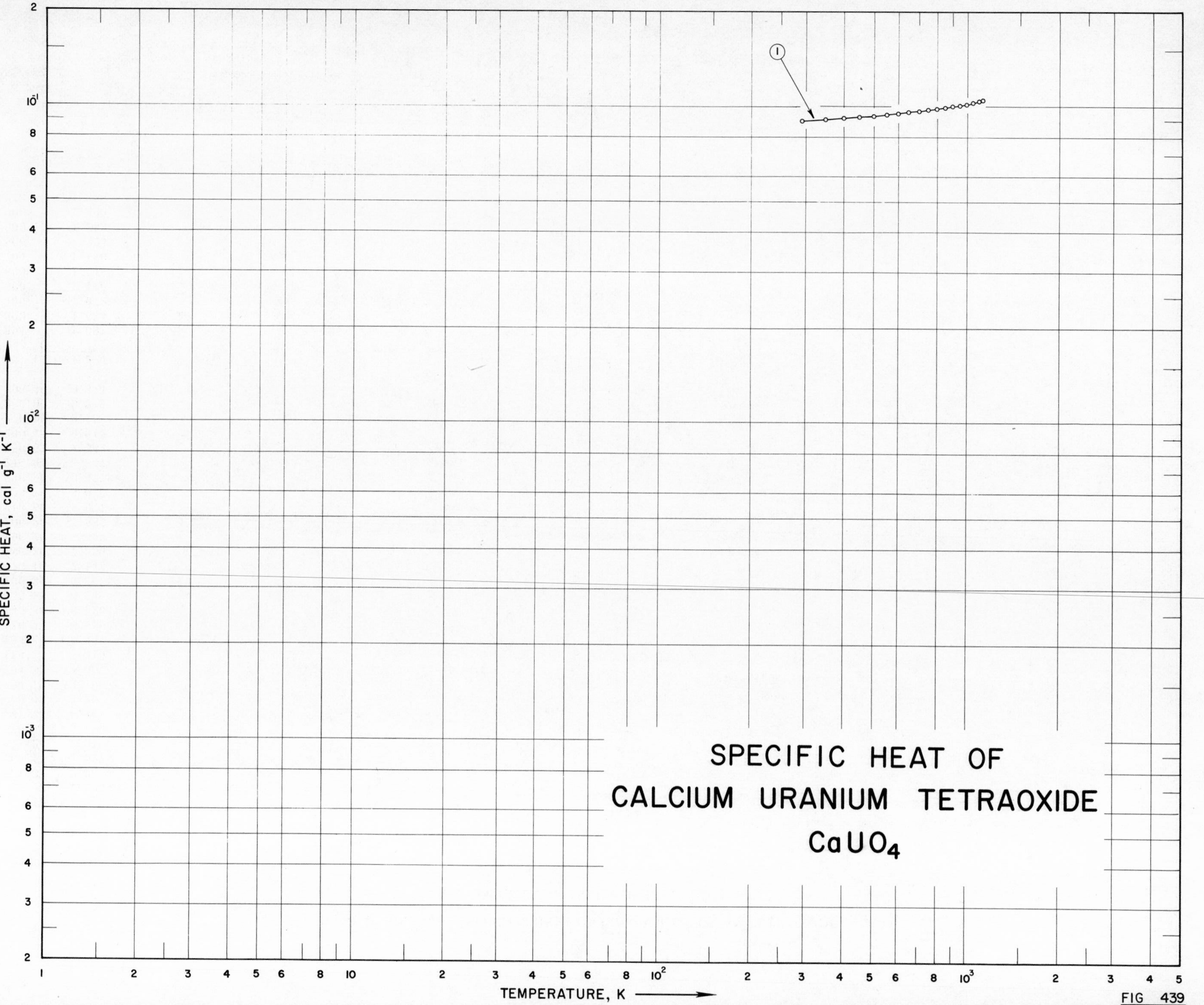

SPECIFIC HEAT OF
CALCIUM URANIUM TETRAOXIDE
$CaUO_4$
SPECIFIC HEAT, cal g^{-1} K^{-1}
TEMPERATURE, K
1

FIG 439

SPECIFICATION TABLE NO. 439 SPECIFIC HEAT OF CALCIUM URANIUM TETRAOXIDE $CaUO_4$

[For Data Reported in Figure and Table No. 439]

Curve No.	Ref. No.	Year	Temp. Range, K	Reported Error, %	Name and Specimen Designation	Composition (weight percent), Specifications and Remarks
1	376	1960	293-1134	0.1		69.40 U; 11.41 Ca.

DATA TABLE NO. 439 SPECIFIC HEAT OF CALCIUM URANIUM TETRAOXIDE, $CaUO_4$

[Temperature, T, K; Specific Heat, C_p, Cal $g^{-1}K^{-1}$]

T	C_p
CURVE 1	
293	9.034×10^{-2}
300	9.046
350	9.128
400	9.209
450	9.291
500	9.373
550	9.455
600	9.537
650	9.618
700	9.700
750	9.782
800	9.864
850	9.946
900	1.003×10^{-1}
950	1.011
1000	1.019
1025	1.023
1025	1.032
1050	1.037
1100	1.046
1134	1.052

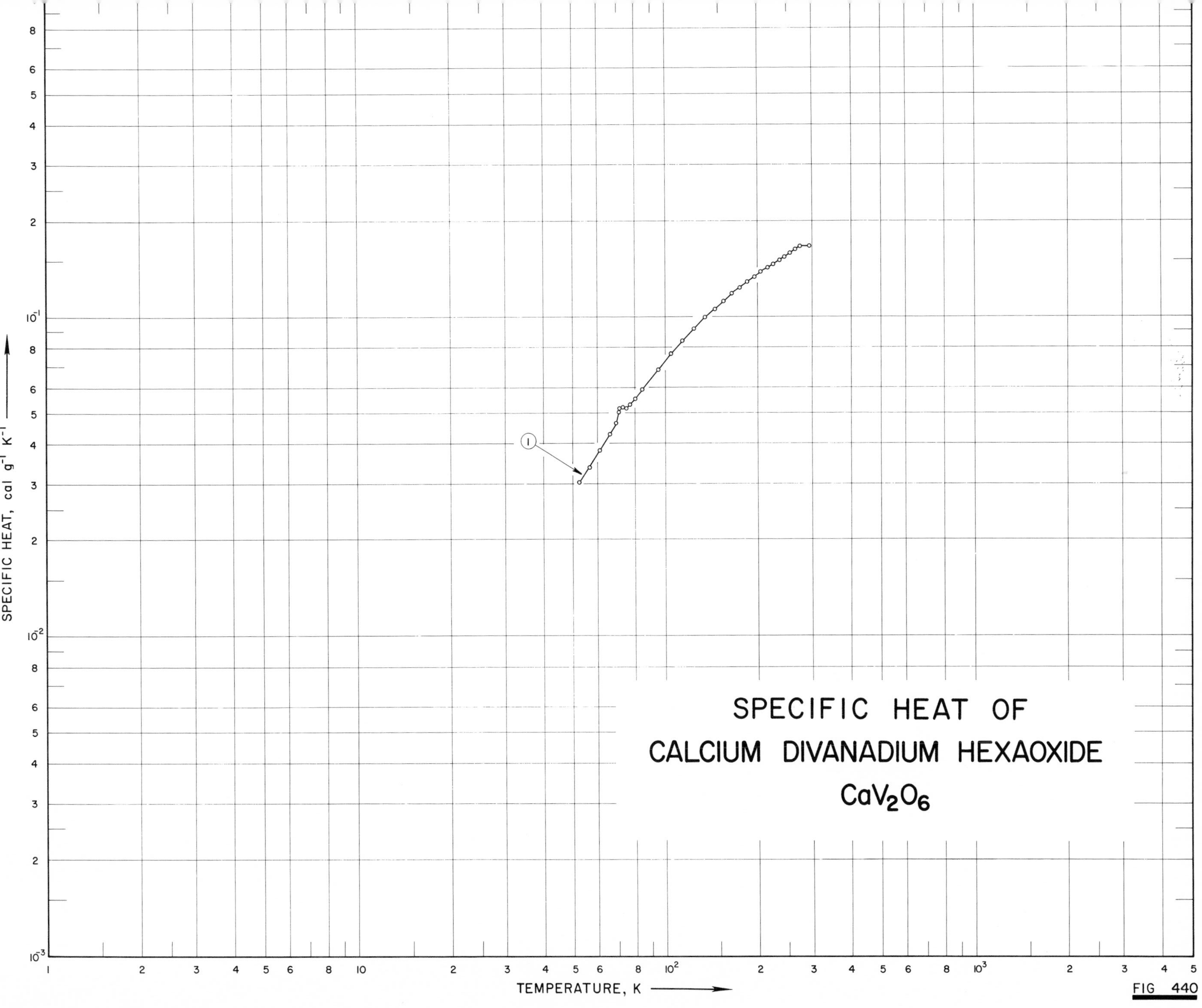
SPECIFIC HEAT OF
CALCIUM DIVANADIUM HEXAOXIDE
CaV_2O_6
SPECIFIC HEAT, cal g⁻¹ K⁻¹
TEMPERATURE, K
1
FIG 440

SPECIFICATION TABLE NO. 440 SPECIFIC HEAT OF CALCIUM DIVANADIUM HEXAOXIDE CaV_2O_6

[For Data Reported in Figure and Table No. 440]

Curve No.	Ref. No.	Year	Temp. Range, K	Reported Error, %	Name and Specimen Designation	Composition (weight percent), Specifications and Remarks
1	378	1961	53-296	0.30		76.32 V_2O_5, 23.48 CaO; prepared by heating stoichiometric mixture of reagent-grade calcium carbonate and vanadium pentaoxide at 610-660 C for 17 days.

DATA TABLE NO. 440 SPECIFIC HEAT OF CALCIUM DIVANADIUM HEXAOXIDE, CaV_2O_6

[Temperature, T, K; Specific Heat, C_p, Cal $g^{-1}K^{-1}$]

T	C_p
CURVE 1	
52.26	3.028×10^{-2}
52.72	3.037*
56.63	3.378
61.31	3.819
61.35	3.818*
66.12	4.286
69.23	4.647
70.64	5.038
71.43	5.194
73.29	5.206
75.15	5.190
75.19	5.240*
77.15	5.337
80.15	5.547
84.59	5.937
95.02	6.841
105.19	7.686
114.60	8.417
124.62	9.165
135.98	9.955
145.73	1.061×10^{-1}
155.83	1.12[illegible]
166.00	1.181
176.23	1.234
186.28	1.288
196.19	1.333
204.70	1.382
216.39	1.425
226.14	1.466
236.17	1.506
245.76	1.543
256.39	1.584
266.20	1.624
266.40	1.624*
270.03	1.638*
273.84	1.650*
275.17	1.658*
276.39	1.667
277.23	1.672*
279.17	1.675*
381.07	1.662*
283.14	1.651*
286.38	1.648*
286.66	1.650*
295.91	1.667*
295.93	1.670

* Not shown on plot

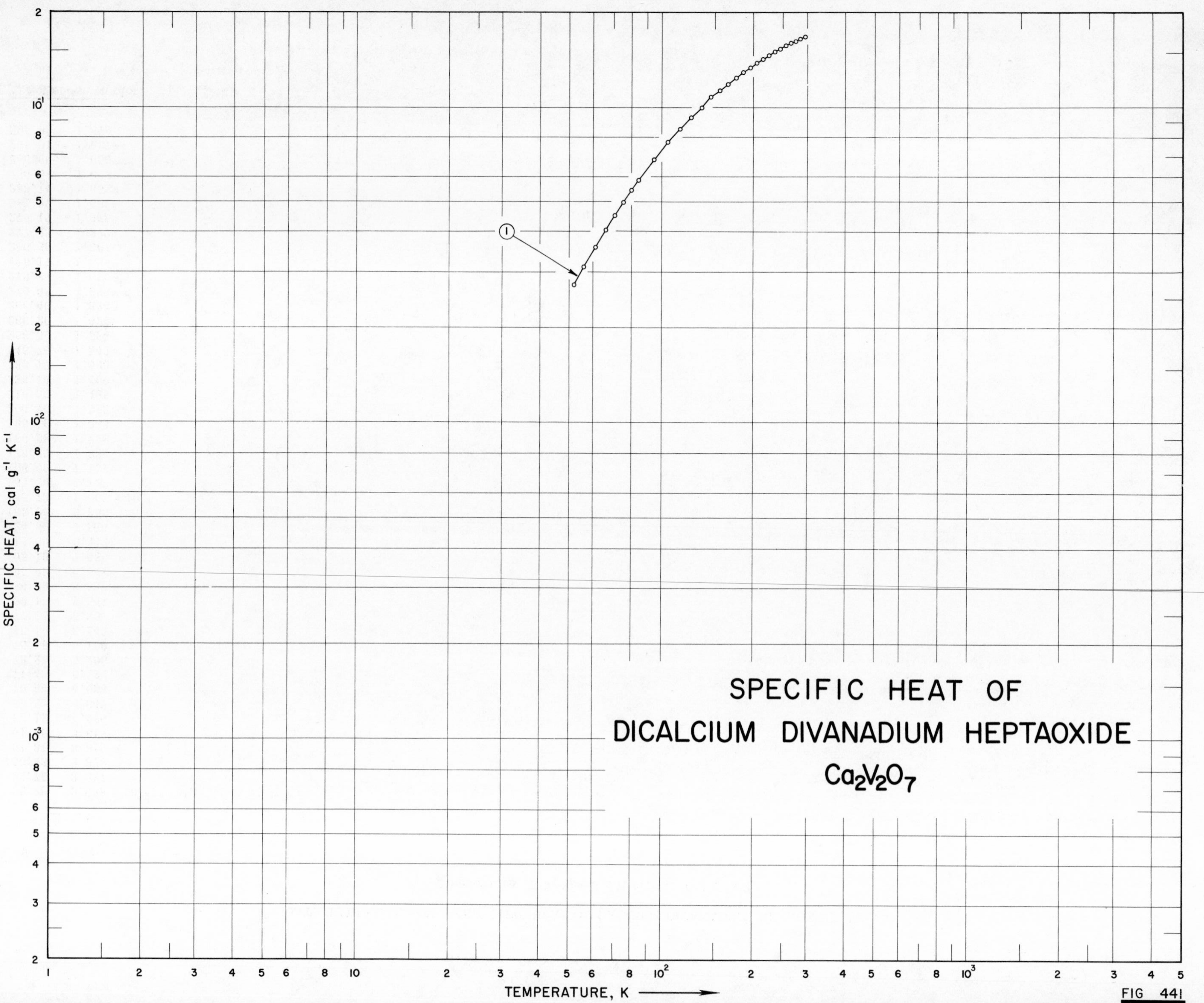
SPECIFIC HEAT OF
DICALCIUM DIVANADIUM HEPTAOXIDE
$Ca_2V_2O_7$
SPECIFIC HEAT, cal g^{-1} K^{-1}
TEMPERATURE, K
1
FIG 441

SPECIFICATION TABLE NO. 441 SPECIFIC HEAT OF DICALCIUM DIVANADIUM HEPTAOXIDE $Ca_2V_2O_7$

[For Data Reported in Figure and Table No. 441]

Curve No.	Ref. No.	Year	Temp. Range, K	Reported Error, %	Name and Specimen Designation	Composition (weight percent), Specifications and Remarks
1	378	1961	52-296	0.3		61.78 V_2O_5, 38.08 CaO; prepared by heating stoichiometric mixture of the oxides at 600-670 C for 28 hrs.

DATA TABLE NO. 441 SPECIFIC HEAT OF DICALCIUM DIVANADIUM HEPTAOXIDE, $Ca_2V_2O_7$

[Temperature, T, K; Specific Heat, C_p, Cal $g^{-1}K^{-1}$]

T	C_p
CURVE 1	
52.14	2.763 x 10^{-2}
56.28	3.150
61.23	3.639
66.12	4.125
70.83	4.594
75.45	5.047
80.22	5.512
84.52	5.921
94.69	6.856
105.59	7.808
115.09	8.576
125.18	9.348
136.31	1.013 x 10^{-1}
145.97	1.080
155.91	1.141
165.87	1.197
176.43	1.254
186.14	1.302
196.26	1.346
206.27	1.391
216.48	1.436
226.20	1.474
236.39	1.512
245.95	1.544
256.62	1.580
266.17	1.612
276.14	1.639
286.71	1.669
296.03	1.697

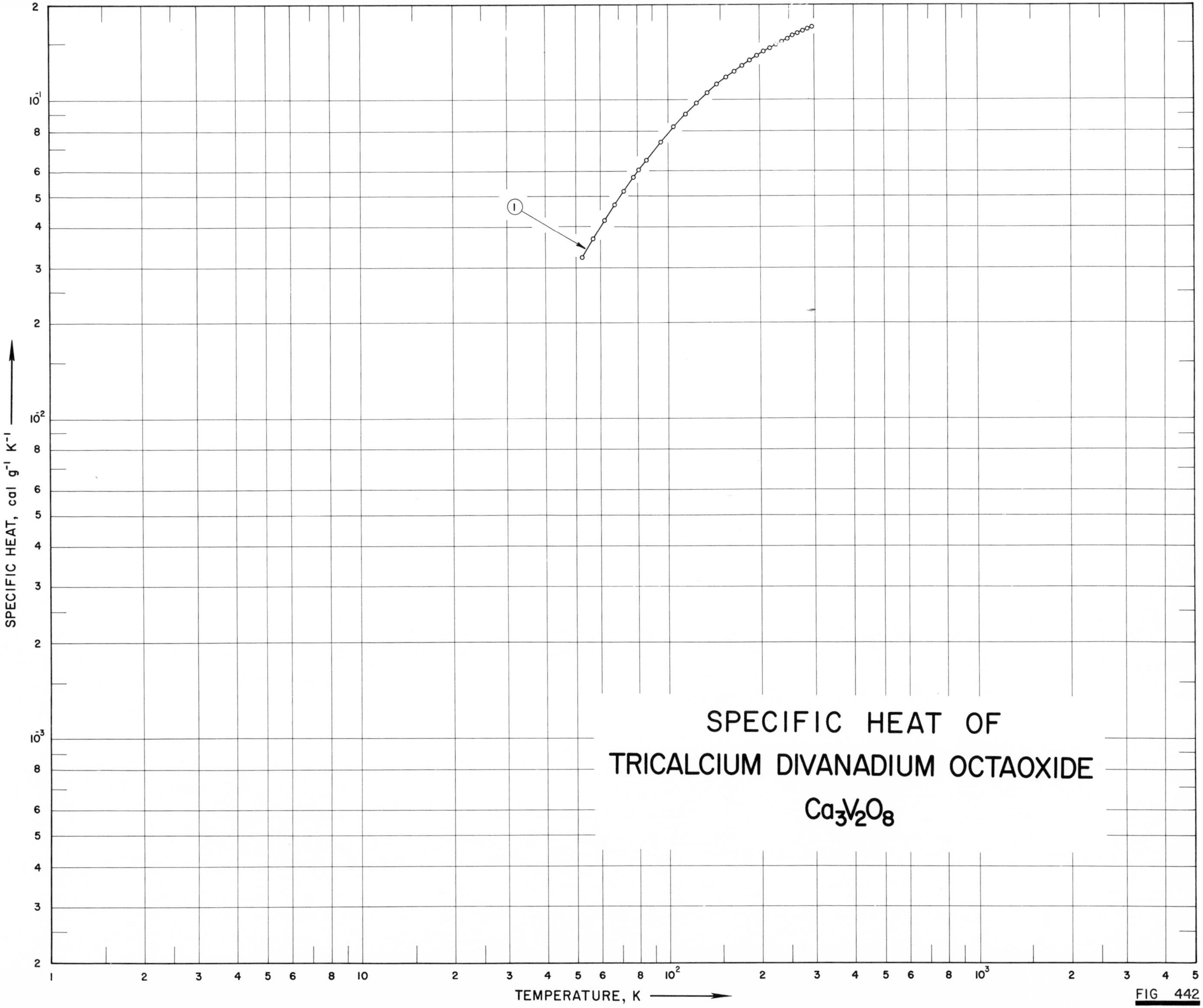

SPECIFIC HEAT OF
TRICALCIUM DIVANADIUM OCTAOXIDE
$Ca_3V_2O_8$

(1)

SPECIFIC HEAT, cal g^{-1} K^{-1}

TEMPERATURE, K

FIG 442

SPECIFICATION TABLE NO. 442 SPECIFIC HEAT OF TRICALCIUM DIVANADIUM OCTAOXIDE $Ca_3V_2O_8$

[For Data Reported in Figure and Table No. 442]

Curve No.	Ref. No.	Year	Temp. Range, K	Reported Error, %	Name and Specimen Designation	Composition (weight percent), Specifications and Remarks
1	378	1961	53-297	0.3		52.00 V_2O_5, 48.01 CaO; prepared by heating stoichiometric oxide mixture at 400 C for 3700 hrs and 1000 C for 16 hrs.

DATA TABLE NO. 442 SPECIFIC HEAT OF TRICALCIUM DIVANADIUM OCTAOXIDE, $Ca_3V_2O_8$

[Temperature, T, K; Specific Heat, C_p, Cal $g^{-1}K^{-1}$]

T	C_p
CURVE 1	
52.73	3.233×10^{-2}
57.43	3.696
62.35	4.210
67.27	4.721
72.20	5.226
77.64	5.766
80.60	6.052
85.60	6.537
95.16	7.411
105.03	8.274
115.11	9.079
124.98	9.830
136.31	1.062×10^{-1}
145.98	1.176
156.06	1.186
166.12	1.240
176.49	1.293
186.70	1.343
196.36	1.384
206.50	1.428
216.40	1.466
226.26	1.503
236.40	1.541
245.91	1.569
256.64	1.605
267.36	1.637
276.93	1.662
287.46	1.695
296.68	1.717

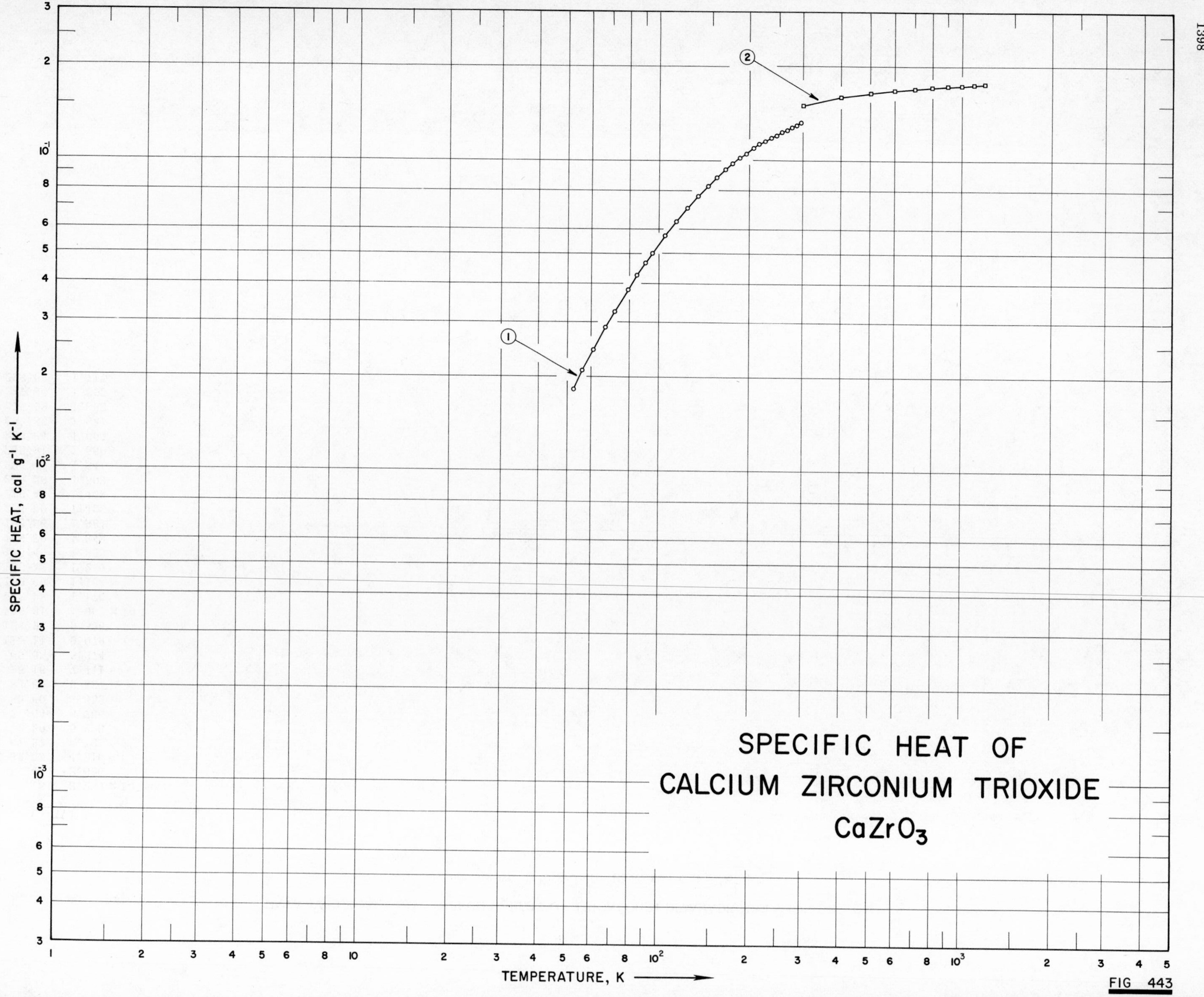
SPECIFIC HEAT OF
CALCIUM ZIRCONIUM TRIOXIDE
CaZrO3
SPECIFIC HEAT, cal g-1 K-1
TEMPERATURE, K
1
2
FIG 443

SPECIFICATION TABLE NO. 443 SPECIFIC HEAT OF CALCIUM ZIRCONIUM TRIOXIDE $CaZrO_3$

[For Data Reported in Figure and Table No. 443]

Curve No.	Ref. No.	Year	Temp. Range, K	Reported Error, %	Name and Specimen Designation	Composition (weight percent), Specifications and Remarks
1	364	1960	53-296	0.3		68.67 ZrO_2, 31.19 CaO; prepared by heating intimate mixture of reagent-grade calcium carbonate and pure zirconia for 30 hrs between 1200 and 1300 C and for 32 hrs between 1400 and 1500 C.
2	54	1965	300-1200	0.8		Sample supplied by Norton Co.

DATA TABLE NO. 443 SPECIFIC HEAT OF CALCIUM ZIRCONIUM TRIOXIDE, $CaZrO_3$

[Temperature, T, K; Specific Heat, C_p, Cal $g^{-1}K^{-1}$]

T	C_p
CURVE 1	
52.81	1.837 x 10^{-2}
56.46	2.100
61.13	2.451
66.91	2.896
71.72	3.263
79.36	3.847
84.90	4.262
90.48	4.670
95.27	5.020
105.15	5.728
114.53	6.352
124.95	7.016
135.78	7.685
145.94	8.277
155.91	8.823
167.20	9.353
176.76	9.777
186.48	1.020 x 10^{-1}
196.11	1.056
207.43	1.098
216.25	1.127
225.94	1.159
236.04	1.189
245.67	1.212
256.68	1.243
266.35	1.264
276.29	1.286
287.01	1.311
296.10	1.327
CURVE 2	
300	1.503 x 10^{-1}
400	1.604
500	1.655
600	1.685
700	1.707
800	1.723
900	1.736
1000	1.747
1100	1.757
1200	1.767

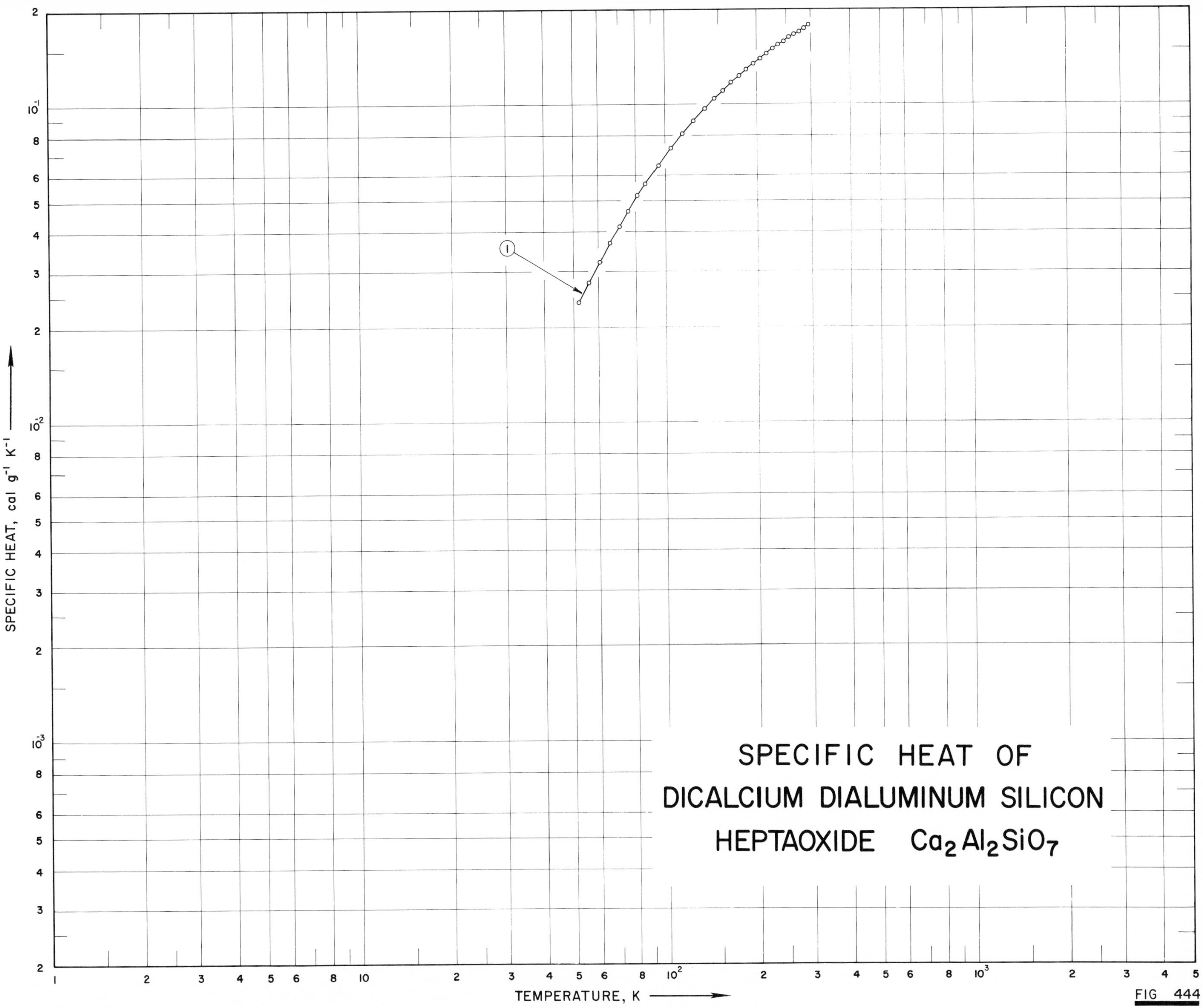

SPECIFIC HEAT OF
DICALCIUM DIALUMINUM SILICON
HEPTAOXIDE $Ca_2Al_2SiO_7$
SPECIFIC HEAT, cal g^{-1} K^{-1}
TEMPERATURE, K
1

FIG 444

SPECIFICATION TABLE NO. 444 SPECIFIC HEAT OF DICALCIUM DIALUMINUM SILICON HEPTAOXIDE $Ca_2Al_2SiO_7$

[For Data Reported in Figure and Table No. 444]

Curve No.	Ref. No.	Year	Temp. Range, K	Reported Error, %	Name and Specimen Designation	Composition (weight percent), Specifications and Remarks
1	379	1963	52-296	0.30	Gehlenite	40.92 CaO, 37.25 Al_2O_3, 21.86 SiO_2; prepared from reagent-grade calcium carbonate, alumina, and silica by heating stoichiometric mixture for 175 hrs at 1050 C, 205 hrs at 1150 C, and 20 hrs at 1300 C.

DATA TABLE NO. 444 SPECIFIC HEAT OF DICALCIUM DIALUMINUM SILICON HEPTAOXIDE, $Ca_2Al_2SiO_7$

[Temperature, T, K; Specific Heat, C_p, Cal $g^{-1}K^{-1}$]

T	C_p
CURVE 1	
52.14	2.403×10^{-2}
56.17	2.773
60.74	3.228
65.64	3.698
70.57	4.161
75.68	4.661
81.48	5.204
86.39	5.660
95.18	6.459
105.00	7.316
114.56	8.115
124.64	8.917
135.81	9.752
146.53	1.052×10^{-1}
155.65	1.112
165.79	1.176
175.90	1.239
185.88	1.295
196.05	1.350
206.17	1.403
216.37	1.454
226.11	1.503
236.69	1.552
245.80	1.589
256.17	1.634
266.41	1.674
276.63	1.713
286.80	1.749
296.23	1.783

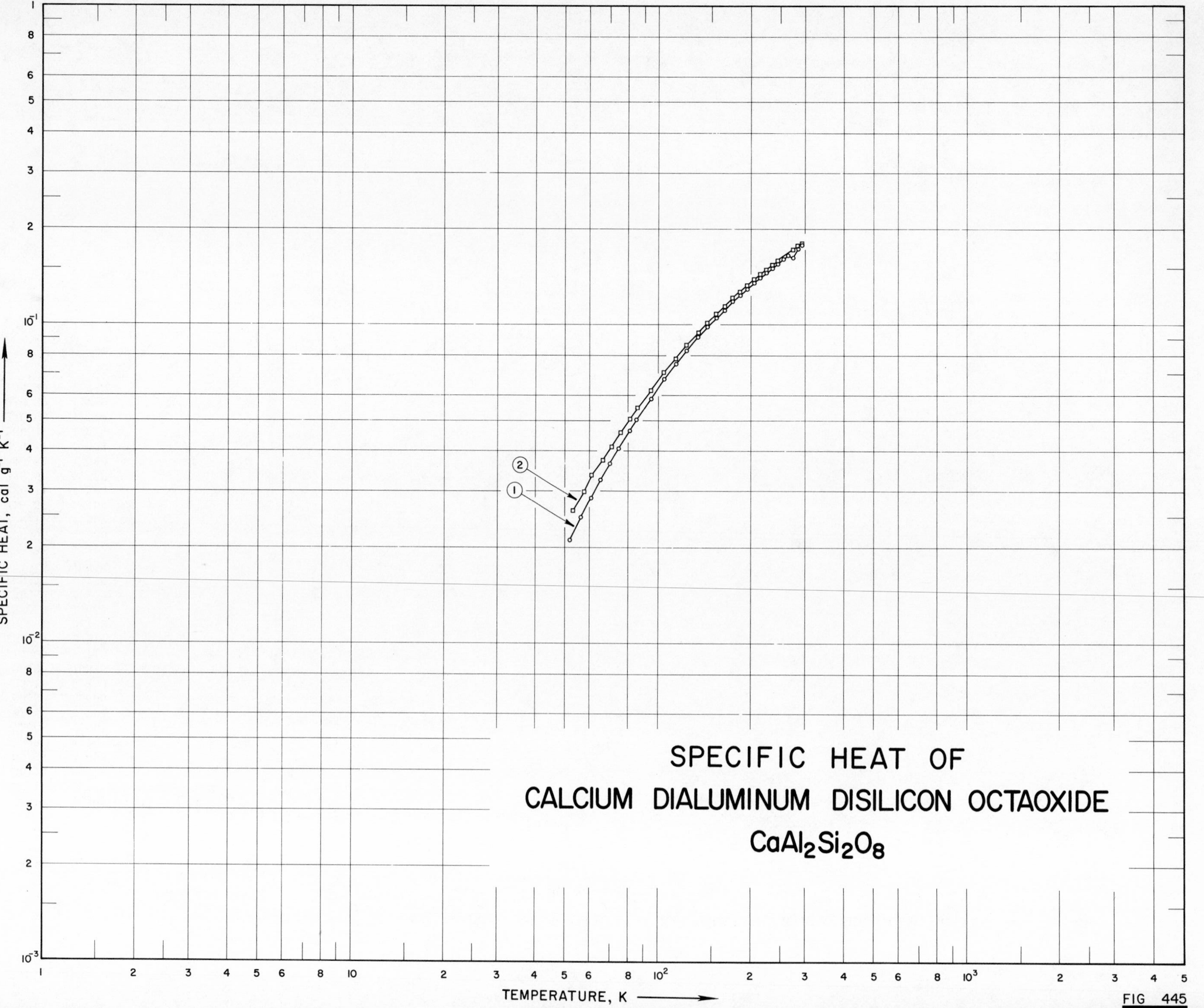
SPECIFIC HEAT OF
CALCIUM DIALUMINUM DISILICON OCTAOXIDE
$CaAl_2Si_2O_8$
SPECIFIC HEAT, cal g⁻¹ K⁻¹
TEMPERATURE, K
1
2

FIG 445

SPECIFICATION TABLE NO. 445 SPECIFIC HEAT OF CALCIUM DIALUMINUM DISILICON OCTAOXIDE $CaAl_2Si_2O_8$

[For Data Reported in Figure and Table No. 445]

Curve No.	Ref. No.	Year	Temp. Range, K	Reported Error, %	Name and Specimen Designation	Composition (weight percent), Specifications and Remarks
1	380	1961	52-296		Hexagonal anorthite	42.99 SiO_2, 34.84 Al_2O_3, 19.77 CaO, 1.76 Fe_2O_3, 0.33 TiO_2, 0.17 H_2O, 0.16 K_2O, 0.15 FeO, 0.07 Na_2O, 0.03 MgO, 0,01 Mg_2O; heated at 600 C for 48 hrs; heated at 830 C for 20 min.
2	370	1957	53-296		Anorthite	43.02 SiO_2, 36.64 Al_2O_3, 20.10 CaO, 0.20 Fe_2O_3.

DATA TABLE NO. 445 SPECIFIC HEAT OF CALCIUM DIALUMINUM DISILICON OCTAOXIDE, $CaAl_2Si_2O_8$

[Temperature, T, K; Specific Heat, C_p, Cal $g^{-1}K^{-1}$]

T	C_p
CURVE 1	
52.28	2.111 x 10^{-2}
56.53	2.481
60.96	2.863
65.44	3.264
69.93	3.666
74.63	4.083
81.05	4.647
85.43	5.028
94.86	5.851
105.69	6.754
115.27	7.530
124.82	8.270
135.88	9.101
145.46	9.816
155.82	1.052 x 10^{-1}
165.15	1.119
176.21	1.182
186.29	1.244
196.16	1.298
206.18	1.358
216.20	1.410
226.30	1.464
236.15	1.513
245.93	1.560
256.73	1.612
266.43	1.655
276.52	1.627
286.56	1.741
296.06	1.782
CURVE 2	
53.43	2.611 x 10^{-2}
57.86	2.998
61.14	3.376
66.41	3.756
70.92	4.144
75.70	4.576
81.13	5.046
85.78	5.470
94.74	6.218
105.23	7.081
114.84	7.832
124.75	8.601
136.11	9.446

T	C_p
CURVE 2 (cont.)	
145.47	1.013 x 10^{-1}
155.60	1.082
165.80	1.149
175.97	1.211
185.90	1.272
195.83	1.329
206.17	1.387
215.93	1.440
226.04	1.493
235.89	1.541
245.73	1.585
256.12	1.633*
266.09	1.679*
276.13	1.725
286.47	1.766
295.93	1.805

* Not shown on plot

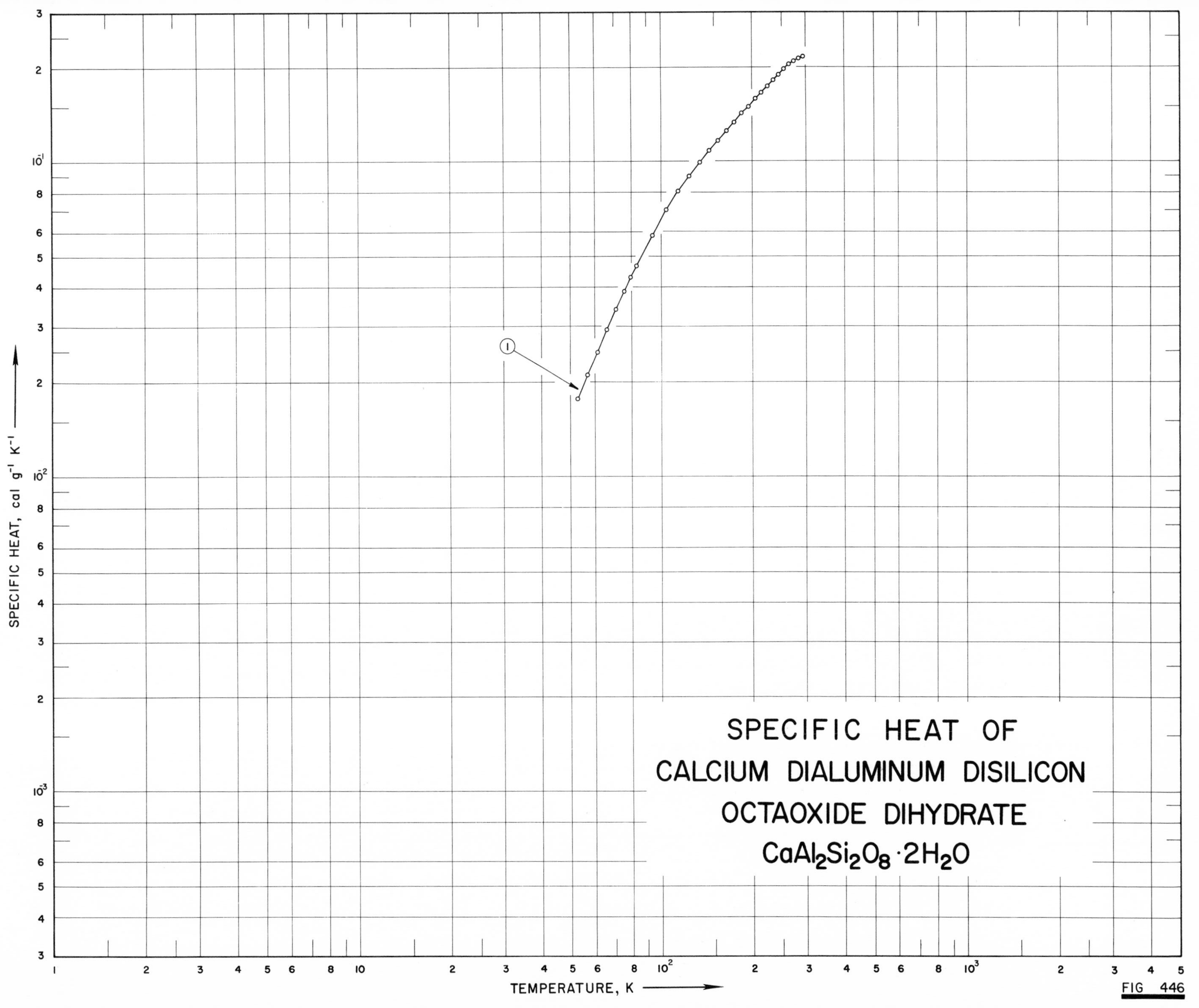
SPECIFIC HEAT OF
CALCIUM DIALUMINUM DISILICON
OCTAOXIDE DIHYDRATE
$CaAl_2Si_2O_8 \cdot 2H_2O$
SPECIFIC HEAT, cal g^{-1} K^{-1}
TEMPERATURE, K
1

FIG 446

SPECIFICATION TABLE NO. 446 SPECIFIC HEAT OF CALCIUM DIALUMINUM DISILICON OCTAOXIDE DIHYDRATE $CaAl_2Si_2O_8 \cdot 2H_2O$

[For Data Reported in Figure and Table No. 446]

Curve No.	Ref. No.	Year	Temp. Range, K	Reported Error, %	Name and Specimen Designation	Composition (weight percent), Specifications and Remarks
1	380	1961	53-296		Lawsonite	38.14 SiO_2, 30.91 Al_2O_3, 17.54 CaO, 11.49 H_2O, 1.56 Fe_2O_3, 0.29 TiO_2, 0.14 K_2O, 0.13 FeO, 0.06 Na_2O, 0.03 MgO, 0.01 Mg_2O.

DATA TABLE NO. 446 SPECIFIC HEAT OF CALCIUM DIALUMINUM DISILICON OCTAOXIDE DIHYDRATE, $CaAl_2Si_2O_8 \cdot 2H_2O$

[Temperature, T, K; Specific Heat, C_p, Cal $g^{-1}K^{-1}$]

T	C_p
CURVE 1	
53.04	1.774×10^{-2}
57.20	2.112
61.50	2.497
66.13	2.935
70.97	3.402
75.67	3.876
79.88	4.309
83.57	4.687
94.38	5.858
105.44	7.080
114.97	8.086
124.78	9.012
135.90	9.995
145.63	1.084×10^{-1}
155.67	1.173
166.04	1.264
175.92	1.349
185.96	1.435
196.15	1.517
206.46	1.597
216.26	1.674
226.34	1.750
236.04	1.827
245.74	1.899
245.82	1.897*
255.96	1.973
256.31	1.979*
265.85	2.051
269.05	2.066*
271.97	2.082*
275.95	2.095
286.19	2.131
286.44	2.126*
296.09	2.163*
296.13	2.162

* Not shown on plot

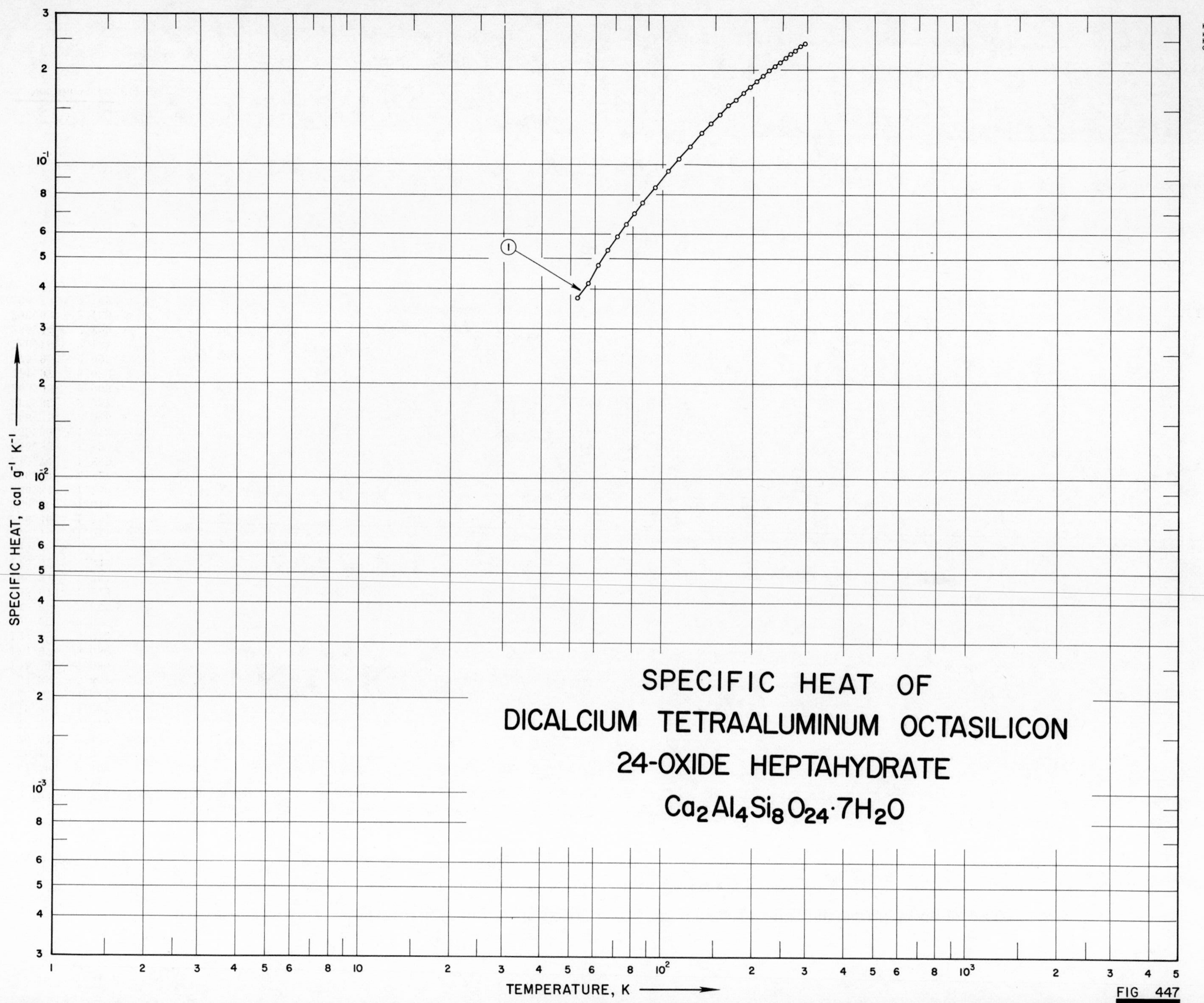
SPECIFIC HEAT OF
DICALCIUM TETRAALUMINUM OCTASILICON
24-OXIDE HEPTAHYDRATE
$Ca_2Al_4Si_8O_{24} \cdot 7H_2O$
1
SPECIFIC HEAT, cal g^{-1} K^{-1}
TEMPERATURE, K

FIG 447

SPECIFICATION TABLE NO. 447 SPECIFIC HEAT OF DICALCIUM TETRAALUMINUM OCTASILICON 24-OXIDE HEPTAHYDRATE $Ca_2Al_4Si_8O_{24} \cdot 7H_2O$

[For Data Reported in Figure and Table No. 447]

Curve No.	Ref. No.	Year	Temp. Range, K	Reported Error, %	Name and Specimen Designation	Composition (weight percent), Specifications and Remarks
1	380	1961	53-296		Leonhardite	51.30 SiO_2, 22.68 Al_2O_3, 13.67 H_2O, 11.43 CaO, 0.18 FeO, 0.15 K_2O, 0.14 Na_2O, 0.09 SrO; air dried to remove 1 mole loosely-bound water.

DATA TABLE NO. 447 SPECIFIC HEAT OF DICALCIUM TETRAALUMINUM OCTASILICON 24-OXIDE HEPTAHYDRATE, $Ca_2Al_4O_{24} \cdot 7H_2O$

[Temperature, T, K; Specific Heat, C_p, Cal $g^{-1}K^{-1}$]

T	C_p
CURVE 1	
52.88	3.779×10^{-2}
57.29	4.206
61.85	4.820
66.46	5.363
71.45	5.944
76.46	6.511
81.29	7.048
86.31	7.601
94.99	8.527
105.21	9.611
114.47	1.053×10^{-1}
124.80	1.158
136.35	1.271
145.79	1.363
155.91	1.454
165.98	1.562
175.86	1.622
185.91	1.702
196.11	1.780
206.57	1.861
216.67	1.937
226.29	2.014
236.13	2.073
245.87	2.136
256.53	2.205
266.47	2.270
276.40	2.330
286.76	2.400
296.42	2.460

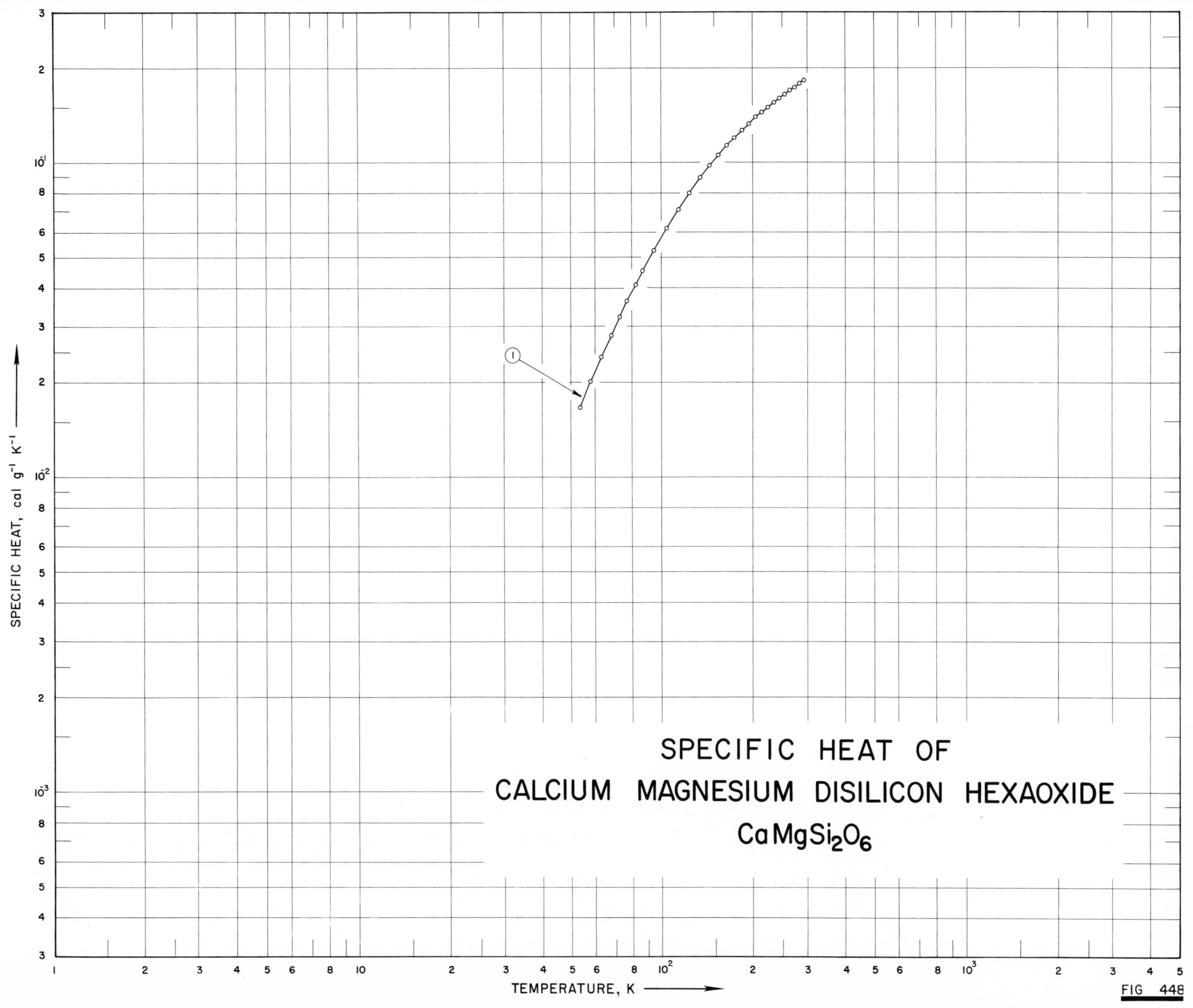
SPECIFIC HEAT OF
CALCIUM MAGNESIUM DISILICON HEXAOXIDE
$CaMgSi_2O_6$
SPECIFIC HEAT, cal g^{-1} K^{-1}
TEMPERATURE, K
1
FIG 448

SPECIFICATION TABLE NO. 448 SPECIFIC HEAT OF CALCIUM MAGNESIUM DISILICON HEXAOXIDE $CaMgSi_2O_6$

[For Data Reported in Figure and Table No. 448]

Curve No.	Ref. No.	Year	Temp. Range, K	Reported Error, %	Name and Specimen Designation	Composition (weight percent), Specifications and Remarks
1	370	1957	54-296			54.78 SiO_2, 25.91 CaO, 18.82 MgO, 0.68 Fe_2O_3, 0.07 FeO.

DATA TABLE NO. 448 SPECIFIC HEAT OF CALCIUM MAGNESIUM DISILICON HEXAOXIDE, $CaMgSi_2O_6$

[Temperature, T, K; Specific Heat, C_p, Cal $g^{-1}K^{-1}$]

T	C_p
CURVE 1	
53.83	1.671×10^{-2}
58.40	2.021
63.33	2.416
68.23	2.835
72.87	3.242
77.47	3.656
82.52	4.119
87.18	4.550
94.82	5.268
105.11	6.238
114.71	7.120
124.66	8.020
135.80	8.985
145.58	9.812
155.87	1.062×10^{-1}
165.78	1.135
175.86	1.206
186.01	1.274
195.86	1.337
206.24	1.401
216.42	1.458
226.05	1.511
235.88	1.562
245.64	1.608
256.29	1.660
266.14	1.704
276.08	1.748
286.63	1.790
295.98	1.831

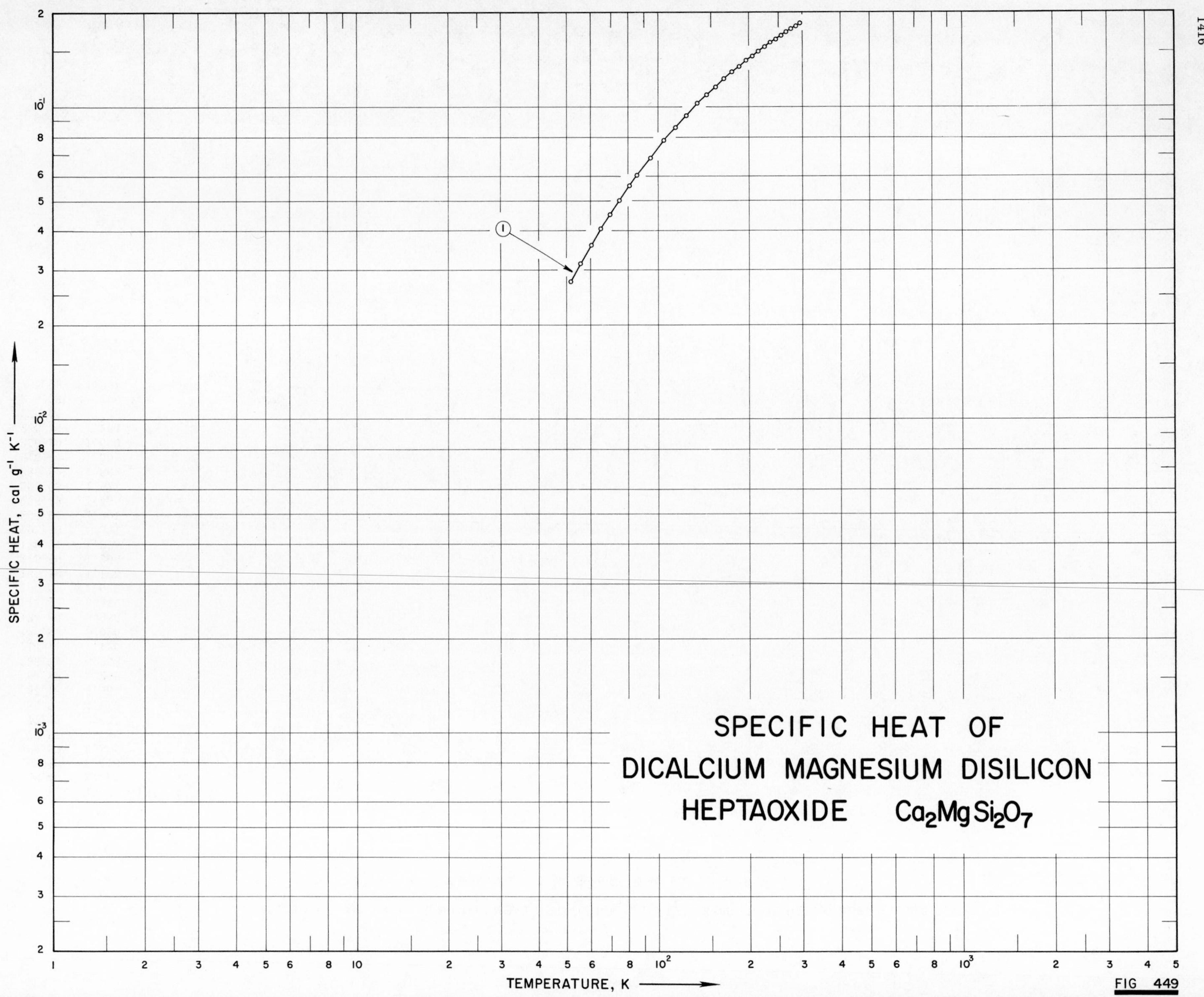
SPECIFIC HEAT OF
DICALCIUM MAGNESIUM DISILICON
HEPTAOXIDE $Ca_2MgSi_2O_7$
SPECIFIC HEAT, cal g⁻¹ K⁻¹
TEMPERATURE, K
1
FIG 449

SPECIFICATION TABLE NO. 449 SPECIFIC HEAT OF DICALCIUM MAGNESIUM DISILICON HEPTAOXIDE $Ca_2MgSi_2O_7$

[For Data Reported in Figure and Table No. 449]

Curve No.	Ref. No.	Year	Temp. Range, K	Reported Error, %	Name and Specimen Designation	Composition (weight percent), Specifications and Remarks
1	379	1963	52-296	0.3	Akermanite	44.14 SiO_2, 41.16 CaO, 14.82 MgO; synthesized by heating stoichiometric mixture of reagent-grade calcium carbonate, magnesium oxide, and silica for 45 hrs at 1050 C, 115 hrs at 1150 C, and 30 hrs at 1250 C.

DATA TABLE NO. 449 SPECIFIC HEAT OF DICALCIUM MAGNESIUM DISILICON HEPTAOXIDE, $Ca_2MgSi_2O_7$

[Temperature, T, K; Specific Heat, C_p, Cal $g^{-1}K^{-1}$]

T	C_p
CURVE 1	
51.93	2.769 x 10^{-2}
55.97	3.154
60.46	3.620
65.06	4.079
69.85	4.541
74.85	5.036
80.65	5.593
85.44	6.048
94.60	6.866
105.51	7.820
114.74	8.590
124.65	9.393
135.89	1.026 x 10^{-1}
145.66	1.095
155.74	1.164
165.73	1.227
176.06	1.290
186.10	1.348
195.97	1.403
206.31	1.458
216.26	1.510
226.29	1.558
236.08	1.606
245.68	1.646
256.37	1.695
266.18	1.734
276.56	1.775
286.65	1.816
296.10	1.851

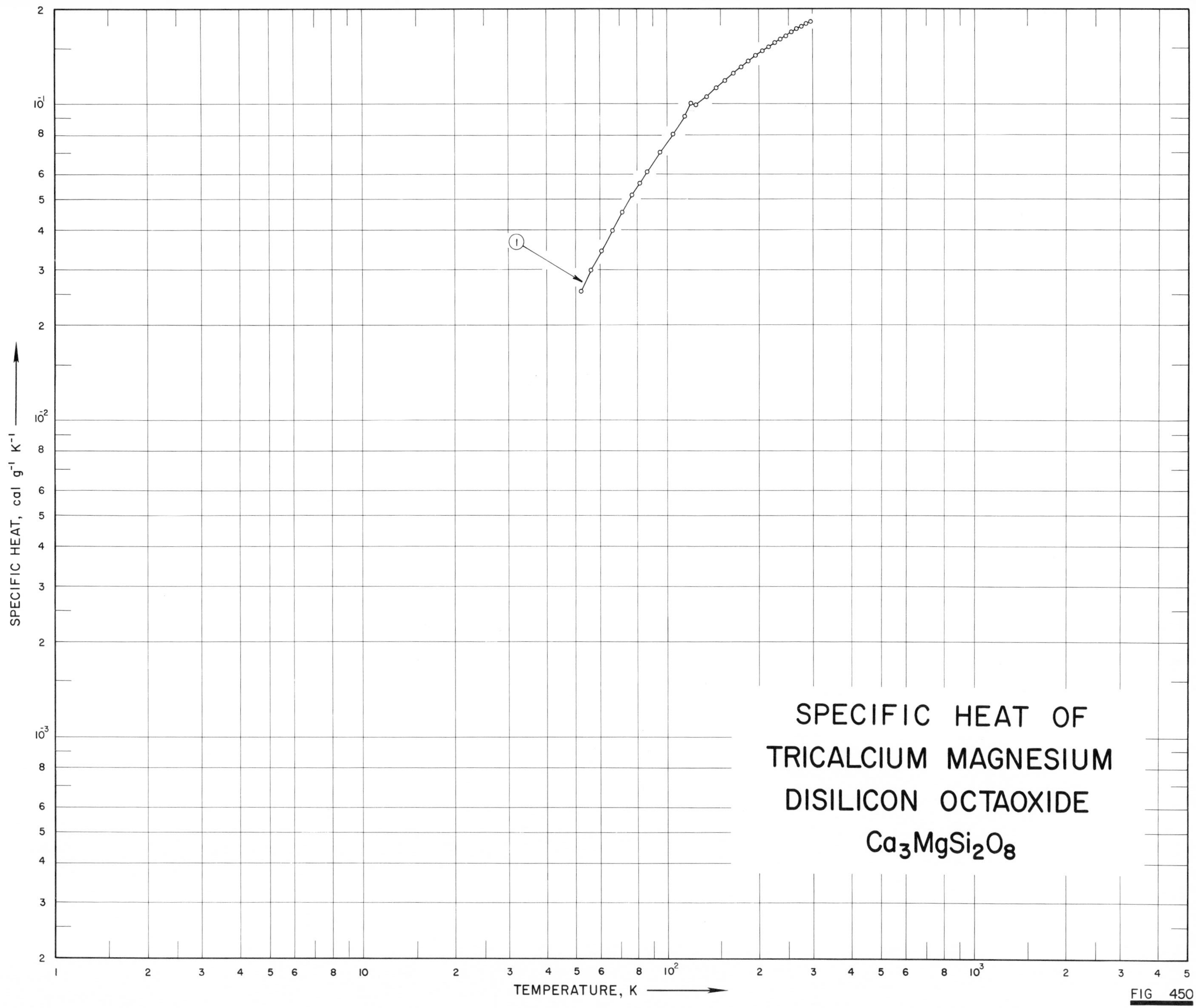

SPECIFIC HEAT OF
TRICALCIUM MAGNESIUM
DISILICON OCTAOXIDE
$Ca_3MgSi_2O_8$

SPECIFIC HEAT, cal g^{-1} K^{-1}

TEMPERATURE, K

FIG 450

SPECIFICATION TABLE NO. 450 SPECIFIC HEAT OF TRICALCIUM MAGNESIUM DISILICON OCTAOXDE $Ca_3MgSi_2O_8$

[For Data Reported in Figure and Table No. 450]

Curve No.	Ref. No.	Year	Temp. Range, K	Reported Error, %	Name and Specimen Designation	Composition (weight percent), Specifications and Remarks
1	379	1963	52-296	0.3	Merwinite	71.14 CaO, 36.57 SiO_2, 12.23 MgO; prepared by heating reagent-grade calcium carbonate, magnesium oxide, and silica in stoichiometric mixture for 175 hrs at 1100 C, 55 hrs at 1200 C, 50 hrs at 1300 C, and 25 hrs at 1150 C.

DATA TABLE NO. 450 SPECIFIC HEAT OF TRICALCIUM MAGNESIUM DISILICON OCTAOXIDE, $Ca_3MgSi_2O_8$

[Temperature, T, K; Specific Heat, C_p, Cal $g^{-1}K^{-1}$]

T	C_p
CURVE 1	
52.05	2.563×10^{-2}
56.18	2.986
60.73	3.444
65.66	3.997
71.20	5.569
76.90	5.178
81.37	5.646
85.98	6.127
94.66	7.006
104.93	8.062
112.50	8.901*
114.75	9.166
115.41	9.251*
118.17	9.647*
120.75	1.008×10^{-1}
123.06	1.009*
124.67	9.963×10^{-2}*
125.38	9.963
136.10	1.061×10^{-1}
146.02	1.128
155.89	1.191
165.79	1.254
176.15	1.315
186.13	1.372
195.88	1.423
206.29	1.474
216.24	1.522
226.32	1.569
236.09	1.611
245.97	1.653
256.32	1.693
266.66	1.731
276.38	1.766
286.92	1.798
296.43	1.829

* Not shown on plot

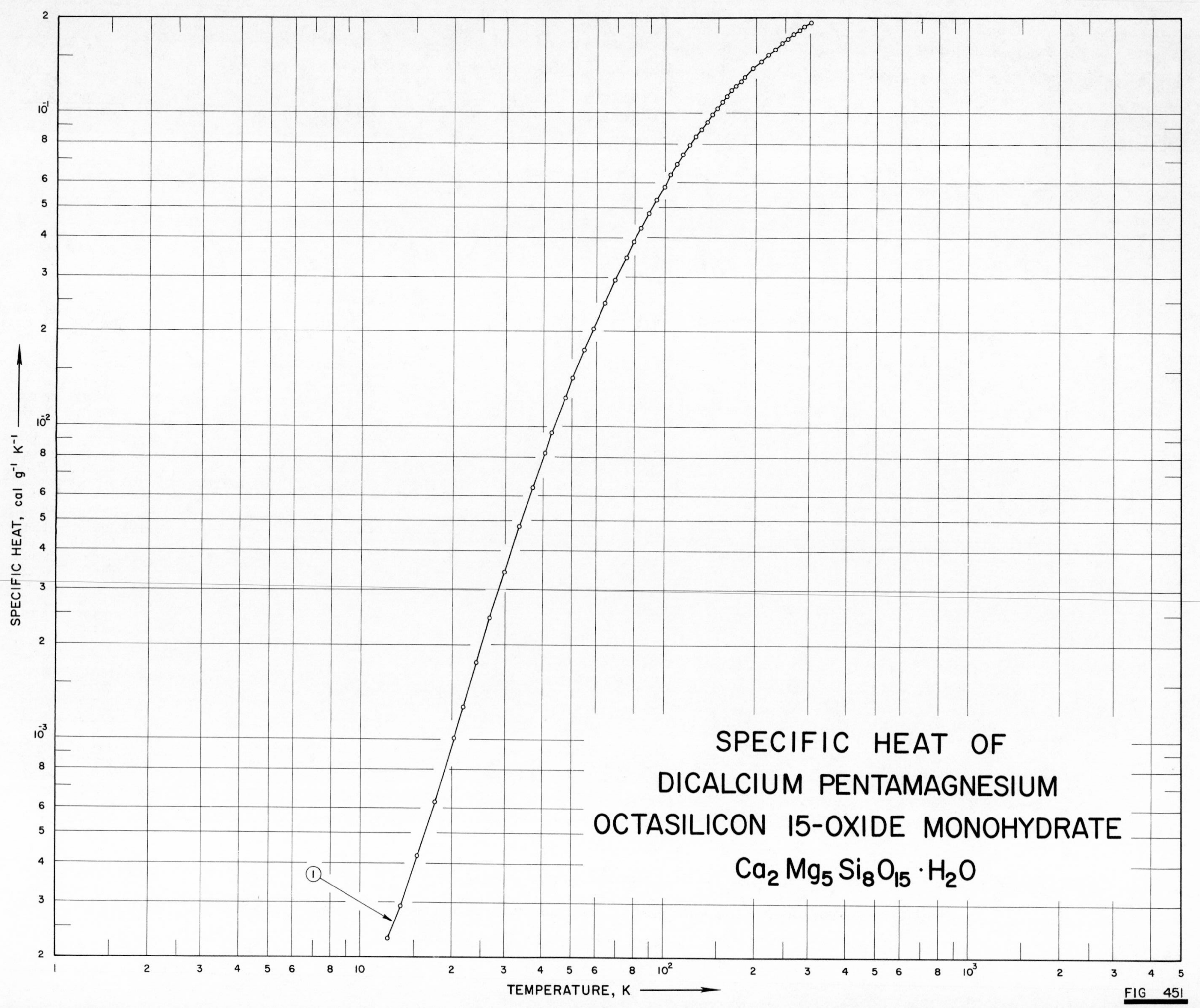
SPECIFIC HEAT OF
DICALCIUM PENTAMAGNESIUM
OCTASILICON 15-OXIDE MONOHYDRATE
$Ca_2Mg_5Si_8O_{15} \cdot H_2O$
SPECIFIC HEAT, cal g^{-1} K^{-1}
TEMPERATURE, K
1
FIG 451

SPECIFICATION TABLE NO. 451 SPECIFIC HEAT OF DICALCIUM PENTAMAGNESIUM OCTASILICON 23-OXIDE MONOHYDRATE $Ca_2Mg_5Si_8O_{23} \cdot H_2O$

[For Data Reported in Figure and Table No. 451]

Curve No.	Ref. No.	Year	Temp. Range, K	Reported Error, %	Name and Specimen Designation	Composition (weight percent), Specifications and Remarks
1	381	1963	12-304	0.3	Tremolite	57.76 SiO_2, 25.21 MgO, 12.96 CaO, 2.13 H_2O^+, 0.51 Al_2O_3, 0.43 Na_2O, 0.31 CO_2, 0.12 K_2O, 0.11 FeO, 0.01 MgO, 0.01 P_2O_5; single phase crystals, dried at 110 for 2 hrs.

DATA TABLE NO. 451 SPECIFIC HEAT OF DICALCIUM PENTAMAGNESIUM OCTASILICON 23-OXIDE MONOHYDRATE $Ca_2Mg_5Si_8O_{23} \cdot H_2O$

[Temperature, T, K; Specific Heat, C_p, Cal $g^{-1}K^{-1}$]

T	C_p
CURVE 1	
Series 1	
54.16	1.696×10^{-2}*
58.91	2.057
64.06	2.486
69.39	2.938
75.26	3.455
79.83	3.888
84.28	4.312
89.33	4.799
94.53	5.292
100.17	5.827
105.79	5.371
110.92	6.860
116.34	7.381
121.95	7.907
127.41	8.415
132.63	8.893
138.26	9.396
144.22	9.913
149.98	1.040×10^{-1}
155.38	1.086
161.13	1.133
167.02	1.178
172.64	1.222
178.88	1.268
184.86	1.311
190.75	1.352*
196.46	1.391
202.20	1.428*
208.24	1.467
214.22	1.504*
220.03	1.539
225.82	1.574*
231.42	1.606
236.91	1.637*
243.04	1.678
249.12	1.706*
254.99	1.737
260.82	1.766*
266.63	1.793
272.62	1.823*
278.33	1.849
284.37	1.875*
289.76	1.898

T	C_p
CURVE 1 (cont.)	
Series 1 (cont.)	
295.48	1.923×10^{-1}*
301.18	1.942*
304.22	1.956
Series 2	
249.17	1.707×10^{-1}*
255.01	1.737*
260.85	1.754*
267.13	1.797*
273.40	1.829*
279.32	1.854*
289.13	1.894*
294.89	1.915*
Series 3	
12.37	2.312×10^{-4}
13.61	2.925
15.49	4.247
17.64	6.264
20.28	1.005×10^{-3}
21.79	1.268
24.08	1.758
26.68	2.429
29.80	3.427
33.38	4.785
36.91	6.365
40.57	8.214
50.31	1.424×10^{-2}
54.93	1.754
42.98	9.520×10^{-3}
47.57	1.235×10^{-2}

* Not shown on plot

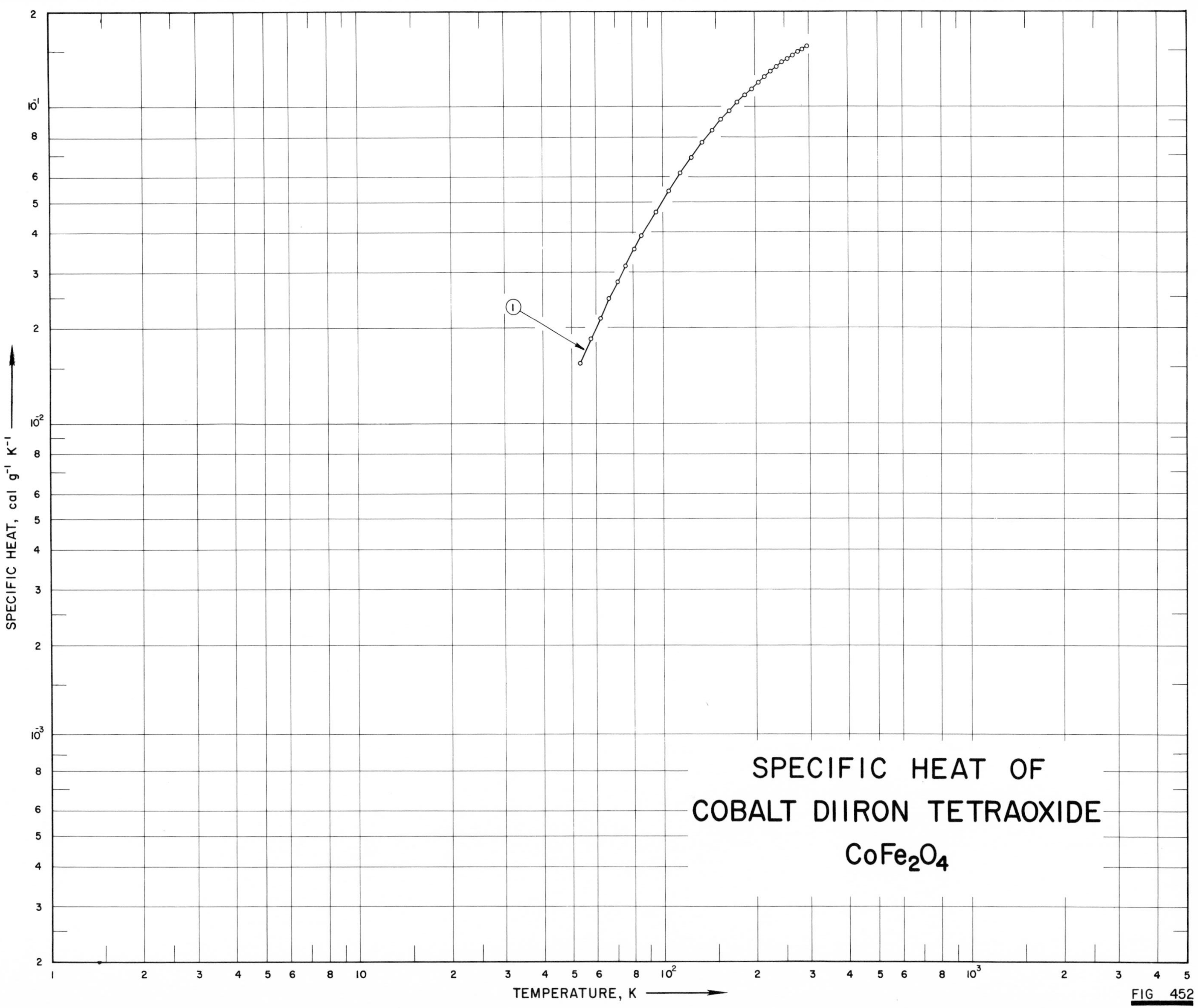
SPECIFIC HEAT OF
COBALT DIIRON TETRAOXIDE
$CoFe_2O_4$
SPECIFIC HEAT, cal g⁻¹ K⁻¹
TEMPERATURE, K
1

FIG 452

SPECIFICATION TABLE NO. 452 SPECIFIC HEAT OF COBALT DIIRON TETRAOXIDE $CoFe_2O_4$

[For Data Reported in Figure and Table No. 452]

Curve No.	Ref. No.	Year	Temp. Range, K	Reported Error, %	Name and Specimen Designation	Composition (weight percent), Specifications and Remarks
1	382	1956	53-296		Cobalt-Iron spinel	68.08 Fe_2O_3, 31.96 CoO (68.06, 31.94 theo.), 0.07 SiO_2; x-ray diffraction agreed with ASTM; no evidence of uncombined oxides; ground mixed; analyzed, composition adjusted; heated 5 times for a total of 9 days at 950-1350 C.

DATA TABLE NO. 452 SPECIFIC HEAT OF COBALT DIIRON TETRAOXIDE, $CoFe_2O_4$

[Temperature, T, K; Specific Heat, C_p, Cal $g^{-1}K^{-1}$]

T	C_p
CURVE 1	
53.29	1.554 x 10^{-2}
57.66	1.856
61.98	2.159
66.32	2.482
70.66	2.810
75.17	3.153
80.58	3.569
85.01	3.903
95.04	4.671
105.02	5.442
114.70	6.188
124.59	6.926
135.90	7.740
145.49	8.421
155.70	9.116
165.81	9.743
175.87	1.034 x 10^{-1}
185.78	1.090
195.98	1.146
206.03	1.199
216.02	1.247
225.80	1.292
236.05	1.336
245.70	1.376
256.42	1.418
266.60	1.454
276.00	1.483
286.35	1.520
296.02	1.550

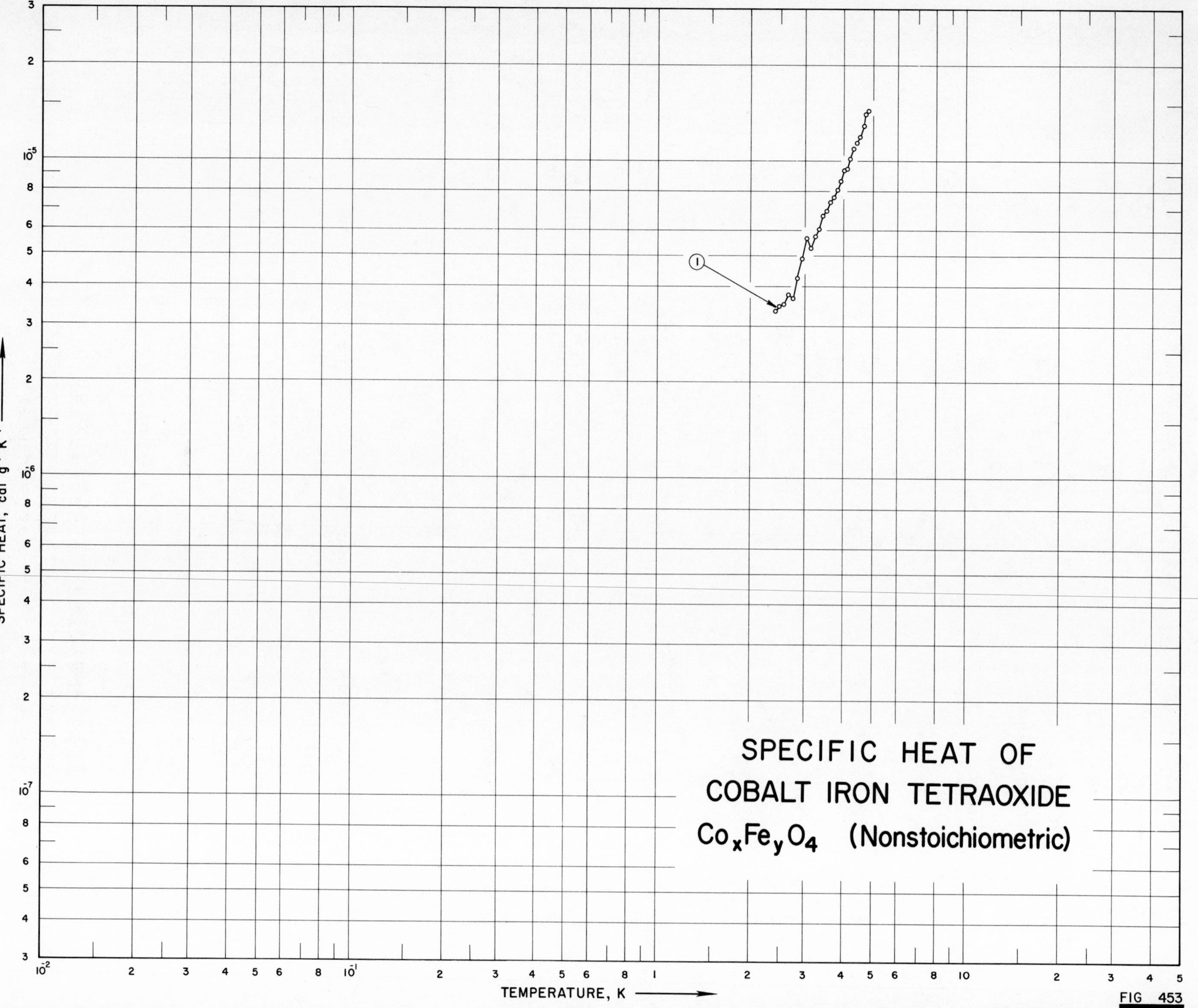

SPECIFIC HEAT OF
COBALT IRON TETRAOXIDE
$Co_xFe_yO_4$ (Nonstoichiometric)
SPECIFIC HEAT, cal g^{-1} K^{-1}
TEMPERATURE, K
1
FIG 453

SPECIFICATION TABLE NO. 453 SPECIFIC HEAT OF COBALT IRON TETRAOXIDE (nonstoichiometric) $Co_xFe_yO_4$

[For Data Reported in Figure and Table No. 453]

Curve No.	Ref. No.	Year	Temp. Range, K	Reported Error, %	Name and Specimen Designation	Composition (weight percent), Specifications and Remarks
1	383	1961	2.4-4.8		$Co_{0.83}Fe_{2.17}O_4$	50.27 Fe, 20.12 Co; x-ray density = 5.29 g cm^{-3}; x-ray lattice parameter = 8.370 Å; Curie temperature = 793 K.

DATA TABLE NO. 453 SPECIFIC HEAT OF COBALT IRON TETRAOXIDE, $Co_xFe_yO_4$ (nonstoichiometric)

[Temperature, T, K; Specific Heat, C_p, Cal $g^{-1}K^{-1}$]

T	C_p
CURVE 1	
2.421	3.36 x 10^{-6}
2.501	3.49
2.583	3.53
2.669	3.79
2.770	3.68
2.854	4.26
2.955	4.916
3.059	5.678
3.151	5.308
3.253	5.765
3.346	6.083
3.437	6.719
3.547	6.945
3.643	7.398
3.741	7.659
3.832	8.073
3.928	8.617
4.024	9.318
4.125	9.392
4.227	1.010 x 10^{-5}
4.321	1.096
4.427	1.142
4.530	1.189
4.568	1.203*
4.662	1.291
4.748	1.407
4.822	1.431

* Not shown on plot

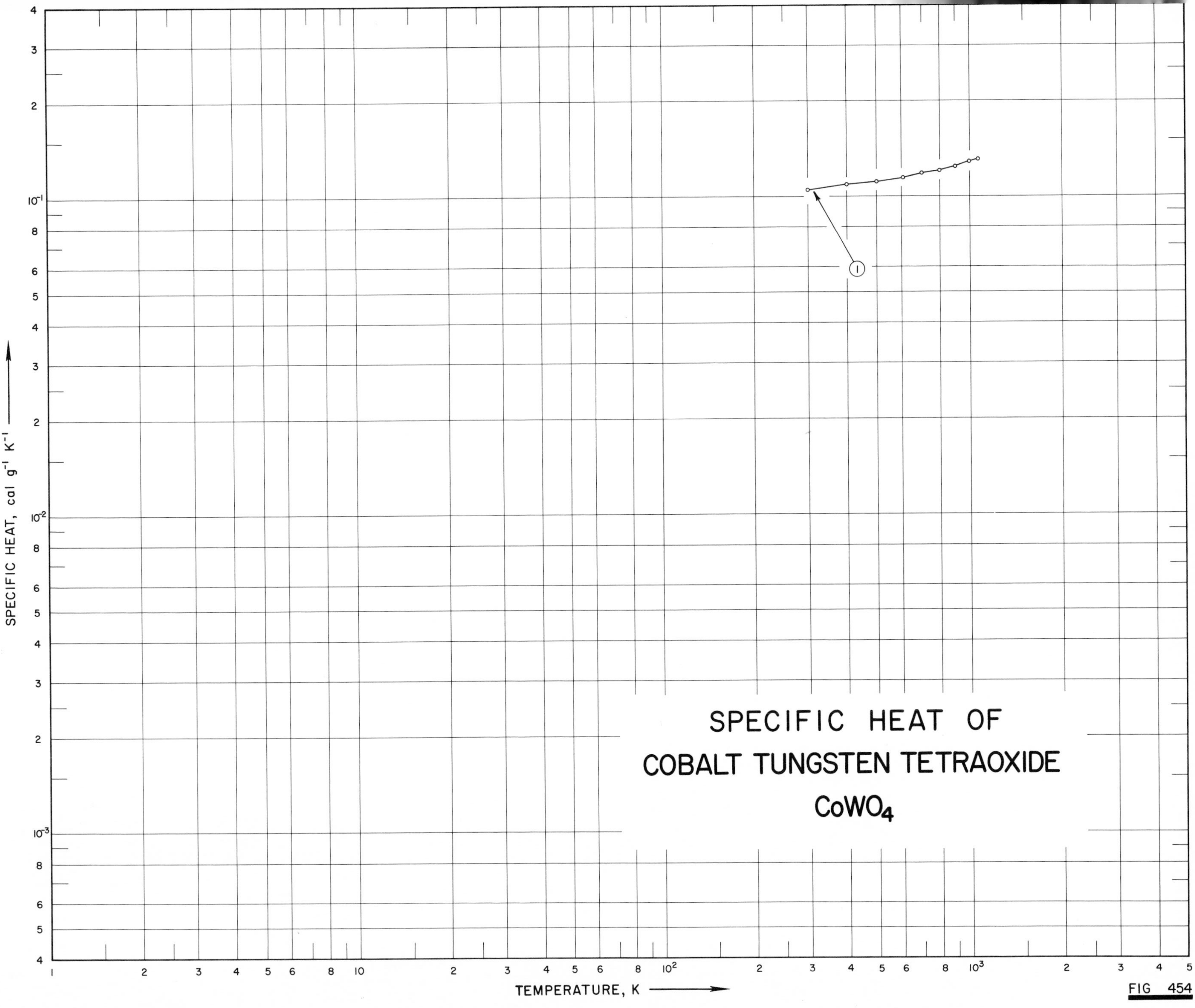
SPECIFIC HEAT OF
COBALT TUNGSTEN TETRAOXIDE
$CoWO_4$
1
SPECIFIC HEAT, cal g^{-1} K^{-1}
TEMPERATURE, K

FIG 454

SPECIFICATION TABLE NO. 454 SPECIFIC HEAT OF COBALT TUNGSTEN TETRAOXIDE $CoWO_4$

[For Data Reported in Figure and Table No. 454]

Curve No.	Ref. No.	Year	Temp. Range, K	Reported Error, %	Name and Specimen Designation	Composition (weight percent), Specifications and Remarks
1	384	1960	300-1073			Prepared by precipitation from $Co(NO_3)_2$ solutions with equivalent quantity of K_2WO_4 solution.

DATA TABLE NO. 454 SPECIFIC HEAT OF COBALT TUNGSTEN TETRAOXIDE $CoWO_4$

[Temperature, T, K; Specific Heat, C_p, Cal $g^{-1} K^{-1}$]

T	C_p
CURVE 1	
300	1.051×10^{-1}
400	1.084
500	1.117
600	1.149
700	1.182
800	1.215
900	1.247
1000	1.280
1073	1.304

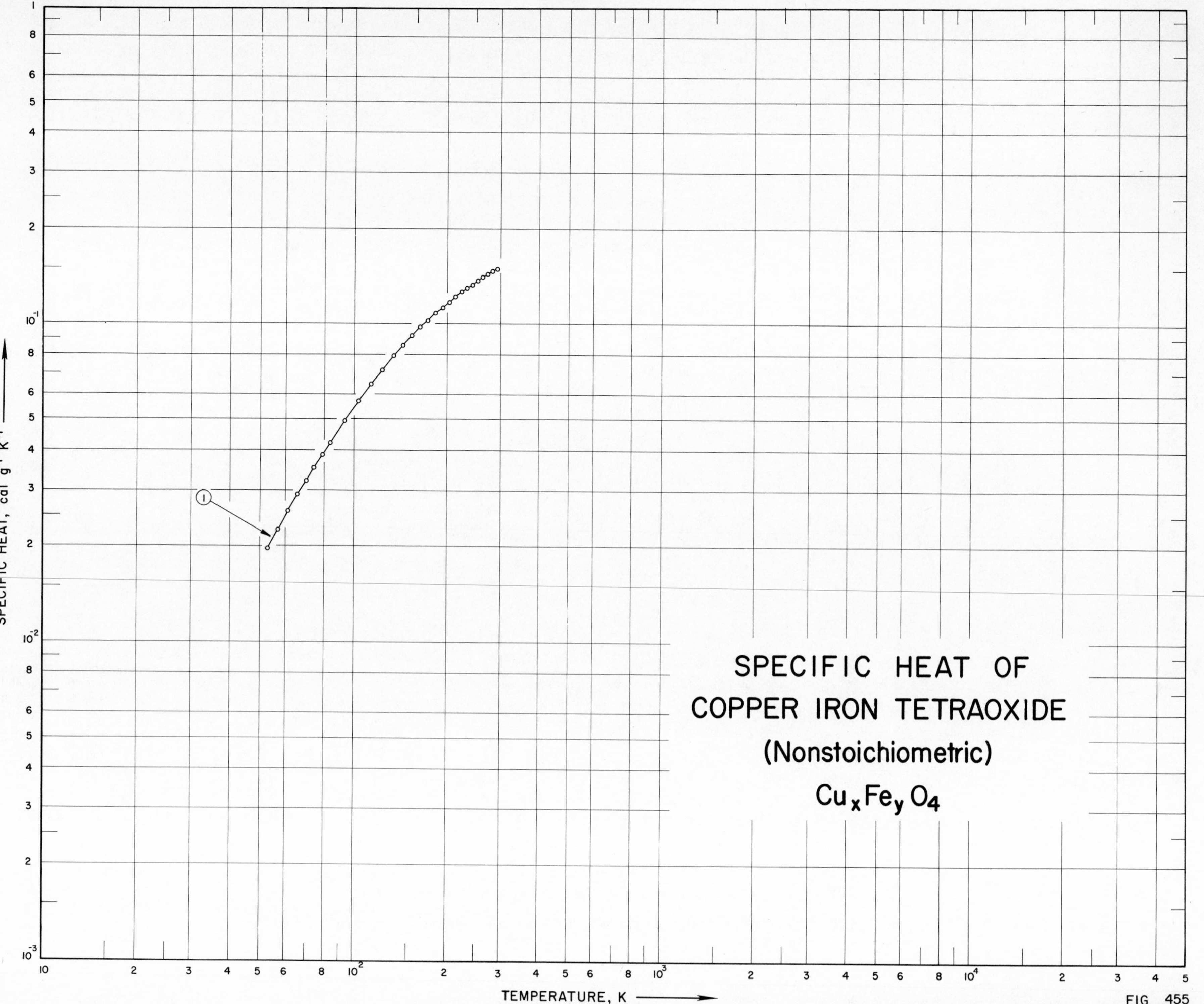

SPECIFIC HEAT OF
COPPER IRON TETRAOXIDE
(Nonstoichiometric)
$Cu_xFe_yO_4$
SPECIFIC HEAT, cal g^{-1} K^{-1}
TEMPERATURE, K
1
FIG 455

SPECIFICATION TABLE NO. 455 SPECIFIC HEAT OF COPPER IRON TETRAOXIDE (nonstoichiometric) $Cu_xFe_yO_4$

[For Data Reported in Figure and Table No. 455]

Curve No.	Ref. No.	Year	Temp. Range, K	Reported Error, %	Name and Specimen Designation	Composition (weight percent), Specifications and Remarks
1	385	1959	53-296			52.99 Fe, 26.91 O, 20.08 Cu, 0,02 Si; prepared by reacting reagent-grade CuO and Fe_2O_3 for 170 hrs at 900-1000 C, 108 hrs at 1000-1100 C, 48 hrs at 1150 C, and 2 hrs at 1250 C.

DATA TABLE NO. 455 SPECIFIC HEAT OF COPPER IRON TETRAOXIDE $Cu_xFe_yO_4$ (nonstoichiometric)

[Temperature, T, K; Specific Heat, C_p, Cal $g^{-1}K^{-1}$]

T	C_p
CURVE 1	
53.10	1.984 x 10^{-2}
57.43	2.286
61.95	2.610
66.47	2.946
70.85	3.271
74.88	3.574
79.69	3.941
84.56	4.290
94.06	5.031
104.73	5.836
114.70	6.561
124.73	7.290
135.93	8.052
145.65	8.706
155.79	9.350
165.95	9.932
175.93	1.048 x 10^{-1}
185.90	1.101
196.09	1.149
206.40	1.199
216.25	1.243
226.01	1.285
236.02	1.324
245.42	1.359
256.11	1.396
266.19	1.430
276.07	1.463
286.47	1.490
295.99	1.517

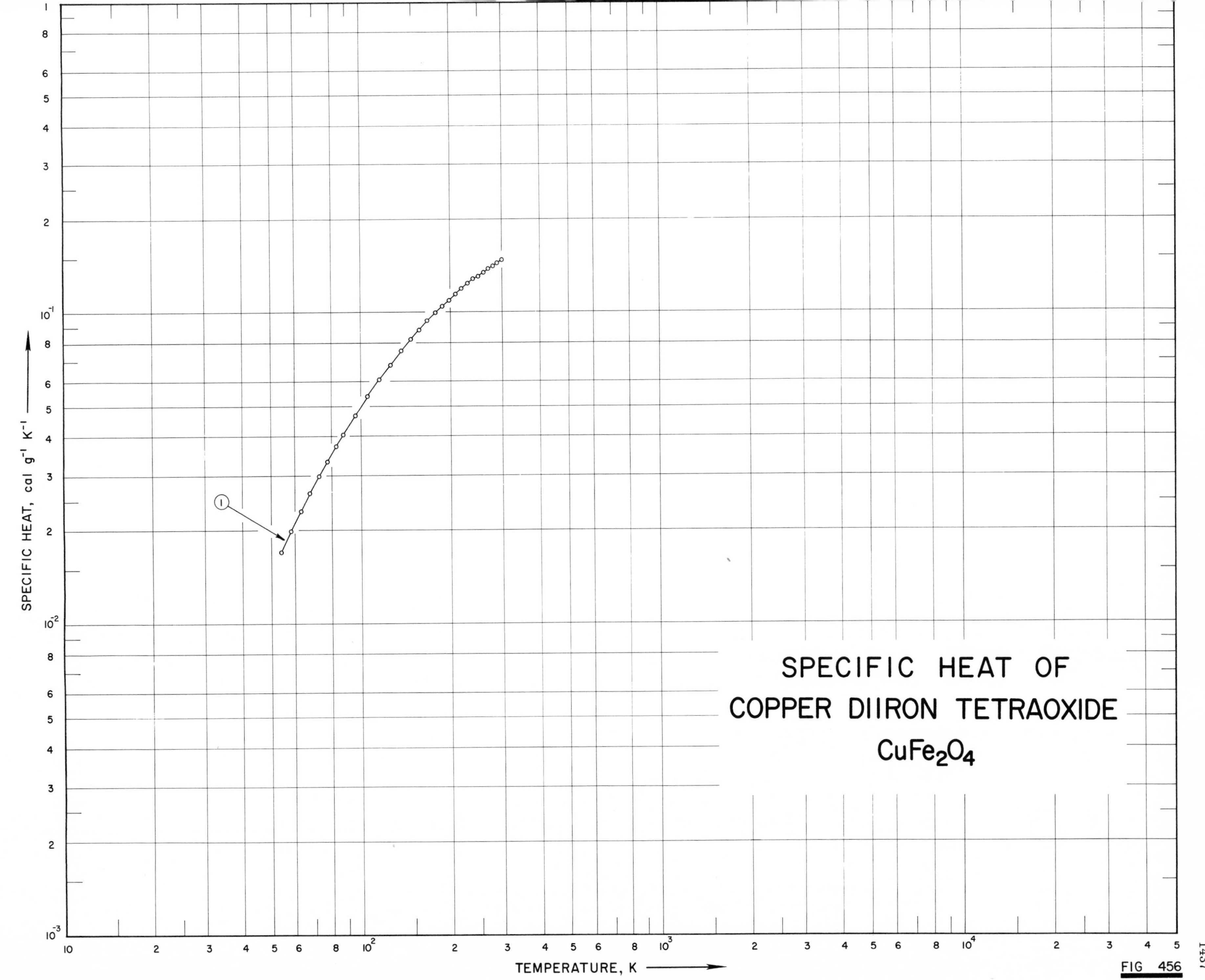
SPECIFIC HEAT OF
COPPER DIIRON TETRAOXIDE
$CuFe_2O_4$
SPECIFIC HEAT, cal g^{-1} K^{-1}
TEMPERATURE, K
1

FIG 456

SPECIFICATION TABLE NO. 456 SPECIFIC HEAT OF COPPER DIIRON TETRAOXIDE $CuFe_2O_4$

[For Data Reported in Figure and Table No. 456]

Curve No.	Ref. No.	Year	Temp. Range, K	Reported Error, %	Name and Specimen Designation	Composition (weight percent), Specifications and Remarks
1	385	1959	54-296		$Cu_{0.75}Fe_{2.25}O_4$	46.67 Fe, 26.69 O, 26.57 Cu, 0.05 Si, 0.02 H_2O; prepared from reagent-grade cupric oxide and ferric oxide by prolonged sintering.

DATA TABLE NO. 456 SPECIFIC HEAT OF COPPER DIIRON TETRAOXIDE $CuFe_2O_4$

[Temperature, T, K; Specific Heat, C_p, Cal $g^{-1}K^{-1}$]

T	C_p
CURVE 1	
53.50	1.704 x 10^{-2}
57.96	1.993
62.23	2.303
66.81	2.627
71.69	2.982
76.47	3.326
81.78	3.725
86.57	4.069
94.76	4.673
104.87	5.417
114.56	6.107
124.55	6.809
135.64	7.561
145.59	8.226
155.71	8.849
165.73	9.451
176.09	1.001 x 10^{-1}
186.03	1.055
196.12	1.105
206.12	1.157
216.14	1.202
227.08	1.250
235.97	1.280
245.92	1.318
256.14	1.354
266.37	1.390
276.19	1.422
286.41	1.450
295.91	1.481

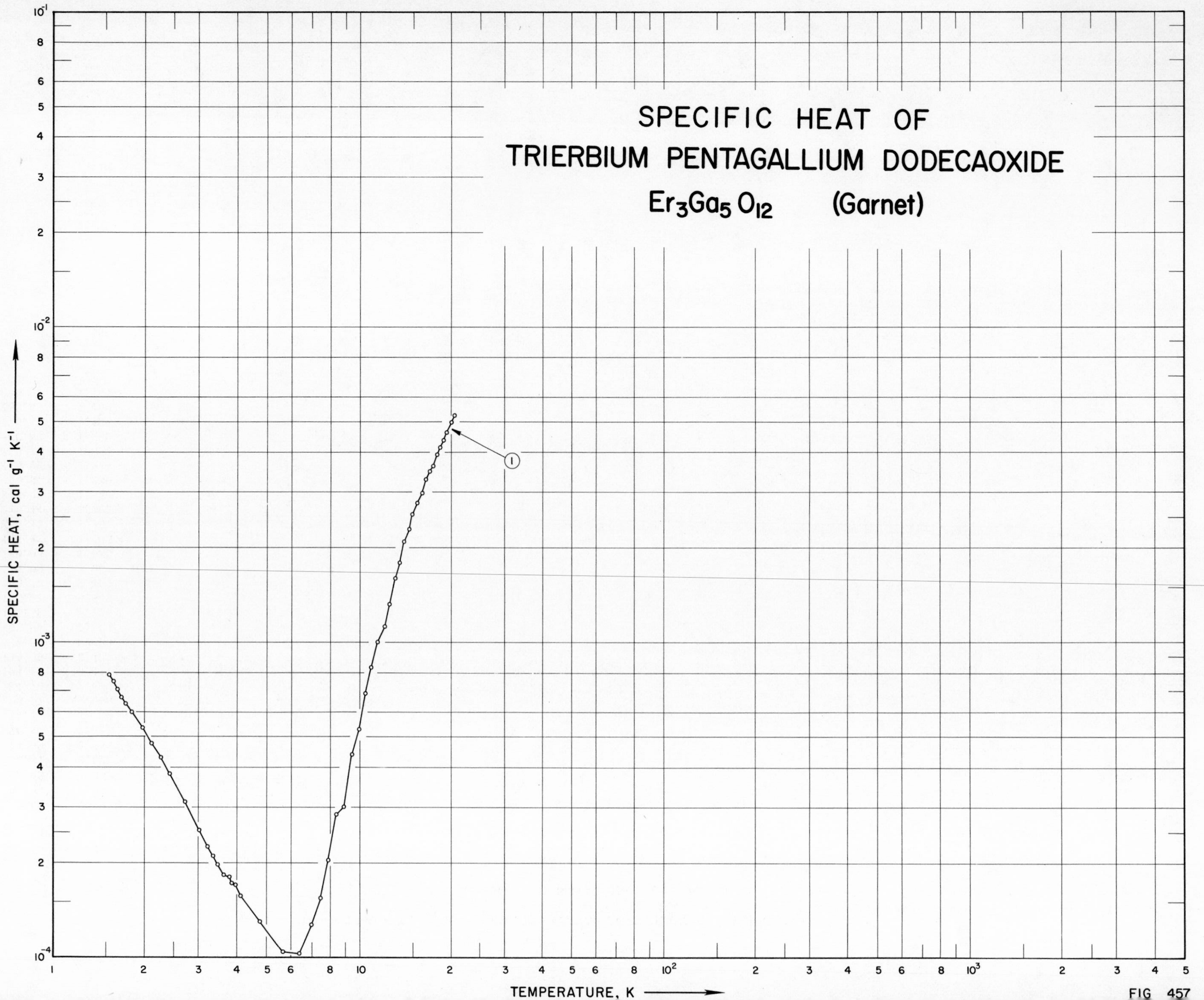
SPECIFIC HEAT OF
TRIERBIUM PENTAGALLIUM DODECAOXIDE
$Er_3Ga_5O_{12}$ (Garnet)
1
SPECIFIC HEAT, cal g^{-1} K^{-1}
TEMPERATURE, K
FIG 457

SPECIFICATION TABLE NO. 457 SPECIFIC HEAT OF TRIERBIUM PENTAGALLIUM DODECAOXIDE, $Er_3Ga_5O_{12}$ (Garnet)

[For Data Reported in Figure and Table No. 457]

Curve No.	Ref. No.	Year	Temp. Range, K	Reported Error, %	Name and Specimen Designation	Composition (weight percent), Specifications and Remarks
1	449	1963	1.5-21		Garnet	99.99 Er_2O_3; supplied by Lindsay Chem. or Johnson, Matthey Co.; 99.9 Ga_2O_3; supplied by the Johnson, Mathey Co.; direct solid state reaction of an intimate mixture of pure rare earth and gallium oxide; sintered block; polycrystalline.

DATA TABLE NO. 457 SPECIFIC HEAT OF TRIERBIUM PENTAGALLIUM DODECAOXIDE, $Er_3Ga_5O_{12}$ (Garnet)

[Temperature, T, K; Specific Heat, C_p, Cal $g^{-1}K^{-1}$]

T	C_p
CURVE 1	
1.543	7.910×10^{-4}
1.585	7.521
1.635	7.131
1.686	6.718
1.745	6.421
1.827	6.007
1.971	5.365
2.114	4.792
2.270	4.311
2.422	3.829
2.710	3.118
3.037	2.545
3.230	2.261
3.365	2.112
3.487	1.972
3.644	1.827
3.801	1.802
3.880	1.720
3.960	1.706
4.133	1.571
4.78	1.300
5.66	1.046
6.38	1.036
6.98	1.275
7.47	1.552
7.92	2.045
8.40	2.866
8.92	3.027
9.45	4.448
9.97	5.319
10.45	6.924
10.94	8.369
11.48	1.007×10^{-3}
12.03	1.126
12.55	1.323
13.02	1.607
13.52	1.798
13.99	2.091
14.54	2.281
14.89	2.568
15.49	2.774
16.05	2.981
16.53	3.302
17.01	3.508
CURVE 1 (cont.)	
17.44	3.646×10^{-3}
17.92	3.967
18.38	4.173
18.81	4.402
19.26	4.654
19.61	4.723*
20.01	5.021
20.51	5.274

* Not shown on plot

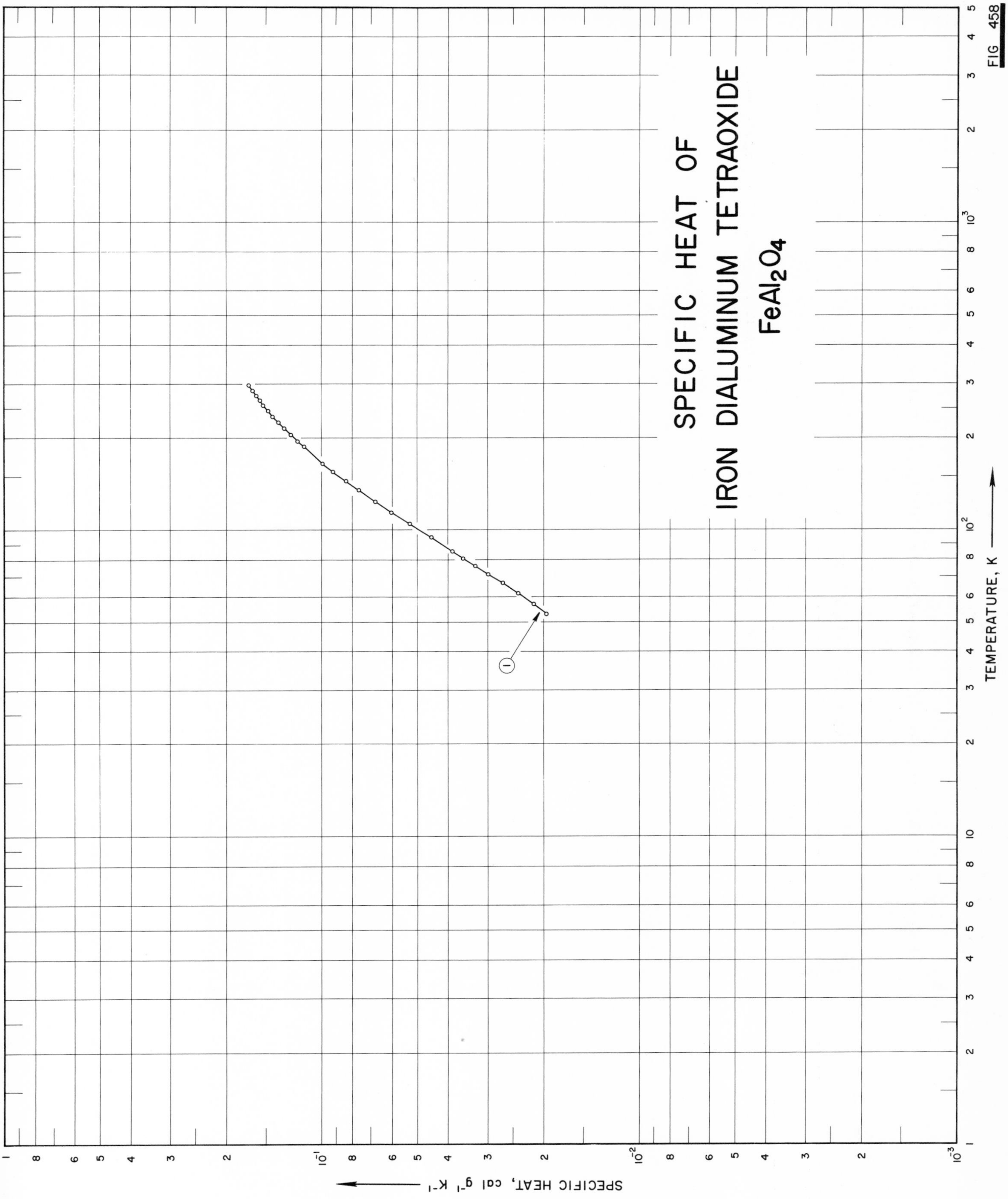

SPECIFIC HEAT OF
IRON DIALUMINUM TETRAOXIDE
$FeAl_2O_4$

SPECIFIC HEAT, cal g^{-1} K^{-1}

TEMPERATURE, K

1

FIG 458

SPECIFICATION TABLE NO. 458 SPECIFIC HEAT OF IRON DIALUMINUM TETRAOXIDE $FeAl_2O_4$

[For Data Reported in Figure and Table No. 458]

Curve No.	Ref. No.	Year	Temp. Range, K	Reported Error, %	Name and Specimen Designation	Composition (weight percent), Specifications and Remarks
1	382	1956	53-298		Iron-aluminum spinel	58.62 Al_2O_3, 41.24 FeO (58.66, 41.34 theo.), 0.12 SiO_2; prepared from reagent-grade powdered iron, ferric oxide, and hydrated alumina; no evidence of uncombined oxides; metallic Fe on magnetic particle found by x-ray diffraction; heated 7 times (40 hrs total) at 1250-1350 C with grinding and mixing in between.

DATA TABLE NO. 458 SPECIFIC HEAT OF IRON DIALUMINUM TETRAOXIDE $FeAl_2O_4$

[Temperature, T, K; Specific Heat, C_p, Cal $g^{-1}K^{-1}$]

T	C_p
CURVE 1	
53.04	1.968×10^{-2}
57.12	2.152
61.94	2.407
66.72	2.689
71.29	2.973
76.14	3.279
80.68	3.580
85.24	3.877
94.77	4.531
105.02	5.285
114.50	6.001
124.51	6.778
135.70	7.641
145.20	8.383
155.71	9.194
165.89	9.919
182.92	1.115×10^{-1}*
187.43	1.140
195.89	1.195
206.08	1.259
216.14	1.317
225.66	1.372
236.04	1.427
245.55	1.475
256.48	1.528
266.61	1.572
276.02	1.609
286.31	1.654
296.03	1.685*
298.16	1.699

* Not shown on plot

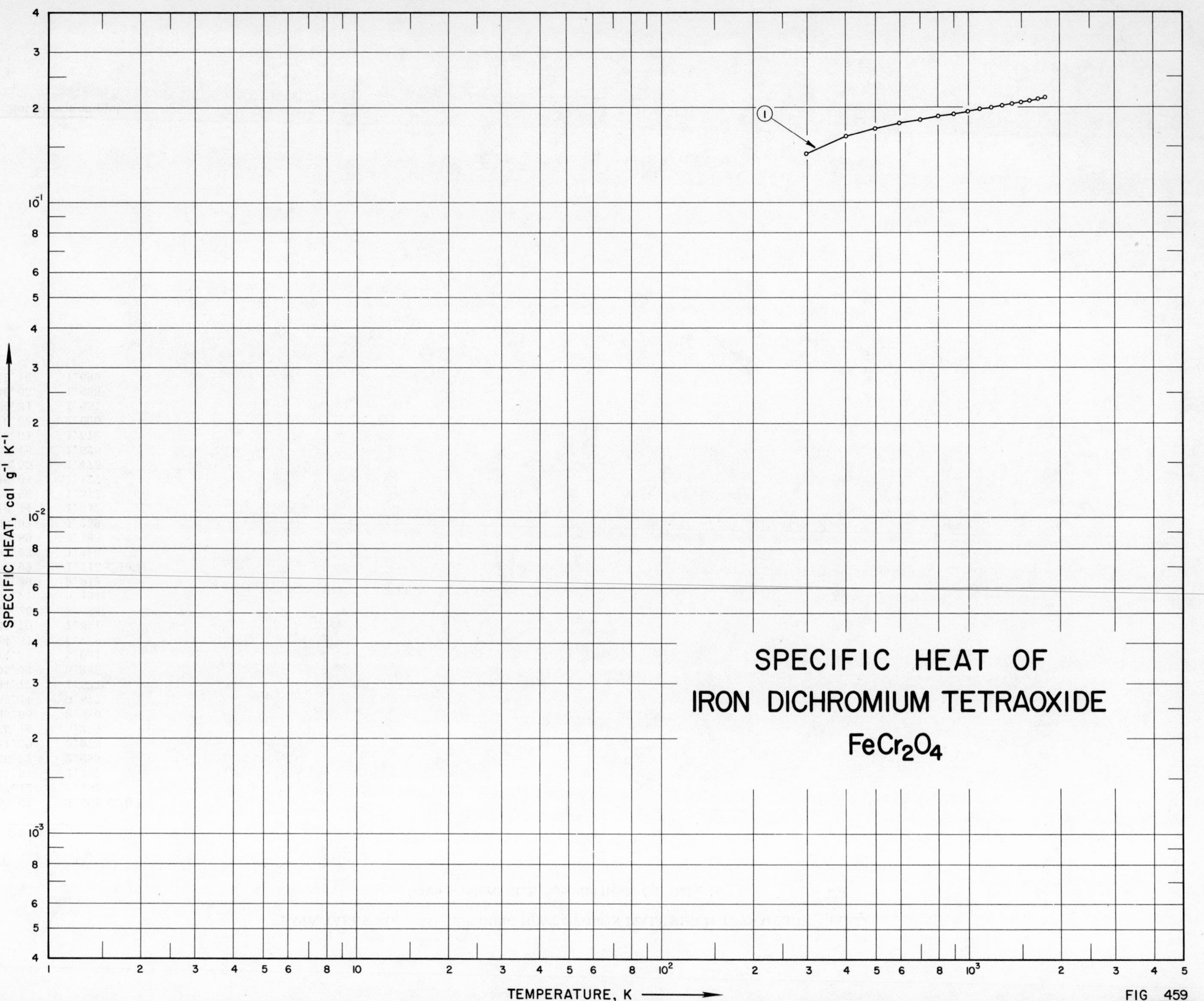
SPECIFIC HEAT OF
IRON DICHROMIUM TETRAOXIDE
FeCr2O4
1
SPECIFIC HEAT, cal g-1 K-1
TEMPERATURE, K

FIG 459

SPECIFICATION TABLE NO. 459 SPECIFIC HEAT OF IRON DICHROMIUM TETRAOXIDE $FeCr_2O_4$

[For Data Reported in Figure and Table No. 459]

Curve No.	Ref. No.	Year	Temp. Range, K	Reported Error, %	Name and Specimen Designation	Composition (weight percent), Specifications and Remarks
1	386	1944	298-1800	0.3		46.09 Cr, 24.77 Fe, 0.75 SiO_2; prepared by heating stoichiometric mixture of high grade sponge iron, reagent-grade ferric oxide, and high-purity chromic oxide at 1300-1350 C for several days in slow stream of He.

DATA TABLE NO. 459 SPECIFIC HEAT OF IRON DICHROMIUM TETRAOXIDE $FeCr_2O_4$

[Temperature, T, K; Specific Heat, C_p, Cal $g^{-1}K^{-1}$]

T	C_p
CURVE 1	
298	1.43×10^{-1}
300	1.435*
400	1.62
500	1.72
600	1.79
700	1.84
800	1.88
900	1.92
1000	1.95
1100	1.98
1200	2.00
1300	2.03
1400	2.06
1500	2.08
1600	2.11
1700	2.13
1800	2.16

* Not shown on plot

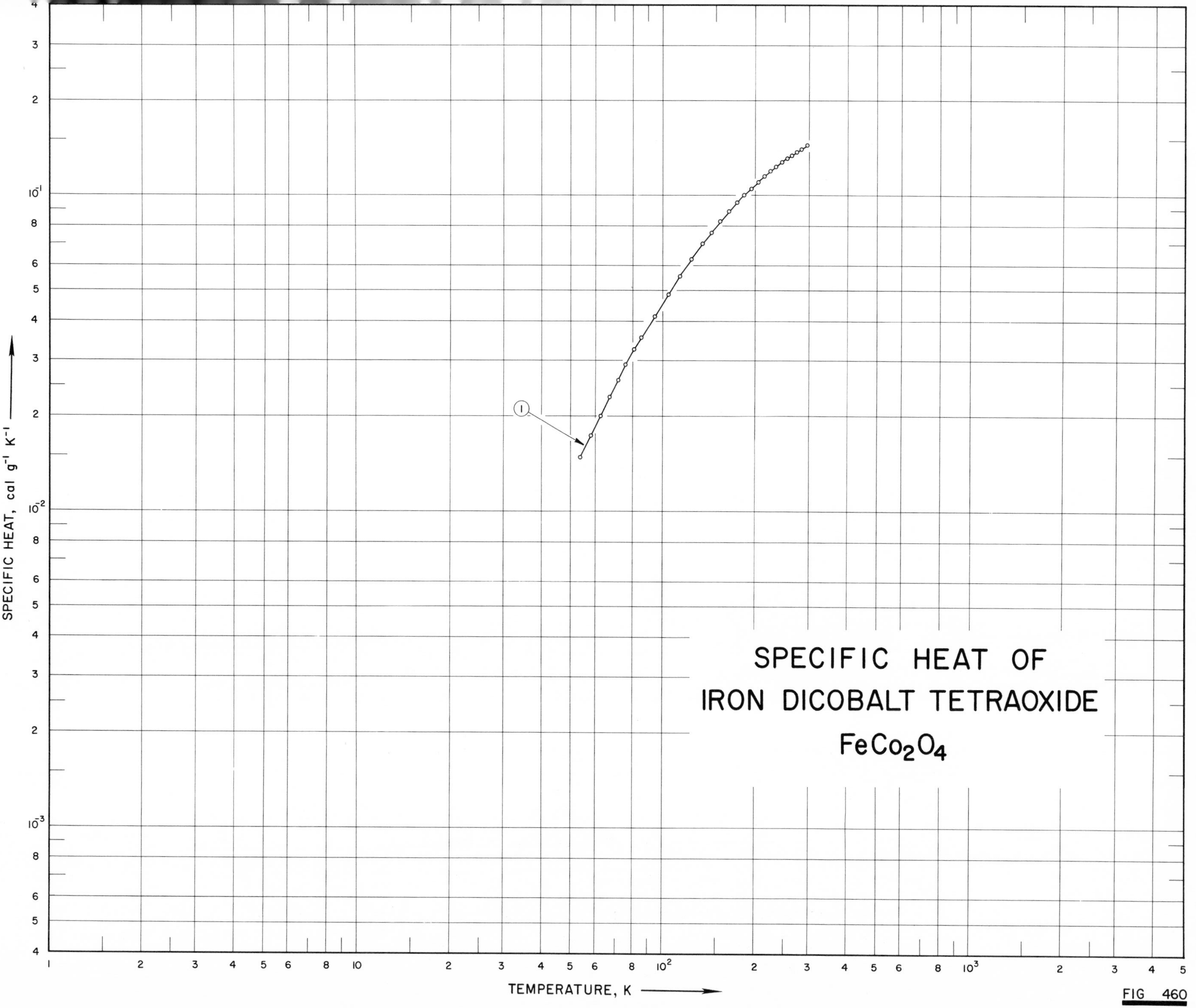
SPECIFIC HEAT OF
IRON DICOBALT TETRAOXIDE
$FeCo_2O_4$
SPECIFIC HEAT, cal g^{-1} K^{-1}
TEMPERATURE, K
1
FIG 460

SPECIFICATION TABLE NO. 460 SPECIFIC HEAT OF IRON DICOBALT TETRAOXIDE $FeCo_2O_4$

[For Data Reported in Figure and Table No. 460]

Curve No.	Ref. No.	Year	Temp. Range, K	Reported Error, %	Name and Specimen Designation	Composition (weight percent), Specifications and Remarks
1	382	1956	54-298		Iron-cobalt	46.6 Co, 26.88 O, 24.37 Fe; no uncombined oxides found by x-ray diffraction; heated 4 times in air for total of 130 hrs at 1050 C with grinding and mixing in between.

DATA TABLE NO. 460 SPECIFIC HEAT OF IRON DICOBALT TETRAOXIDE $FeCo_2O_4$

[Temperature, T, K; Specific Heat, C_p, Cal $g^{-1}K^{-1}$]

T	C_p
CURVE 1	
53.84	1.487×10^{-2}
58.22	1.733
62.68	2.009
67.21	2.303
71.81	2.613
76.38	2.918
81.33	3.257
85.84	3.554
94.55	4.152
105.04	4.884
114.73	5.557
125.12	6.276
136.12	7.021
145.43	7.647
155.56	8.304
165.69	8.909
175.84	9.494
185.63	1.002×10^{-1}
195.84	1.056
206.15	1.108
216.00	1.153
226.03	1.195
236.06	1.237
245.82	1.275
256.26	1.312
266.52	1.346
275.89	1.374
286.46	1.407
295.89	1.437*
298.16	1.442

* Not shown on plot

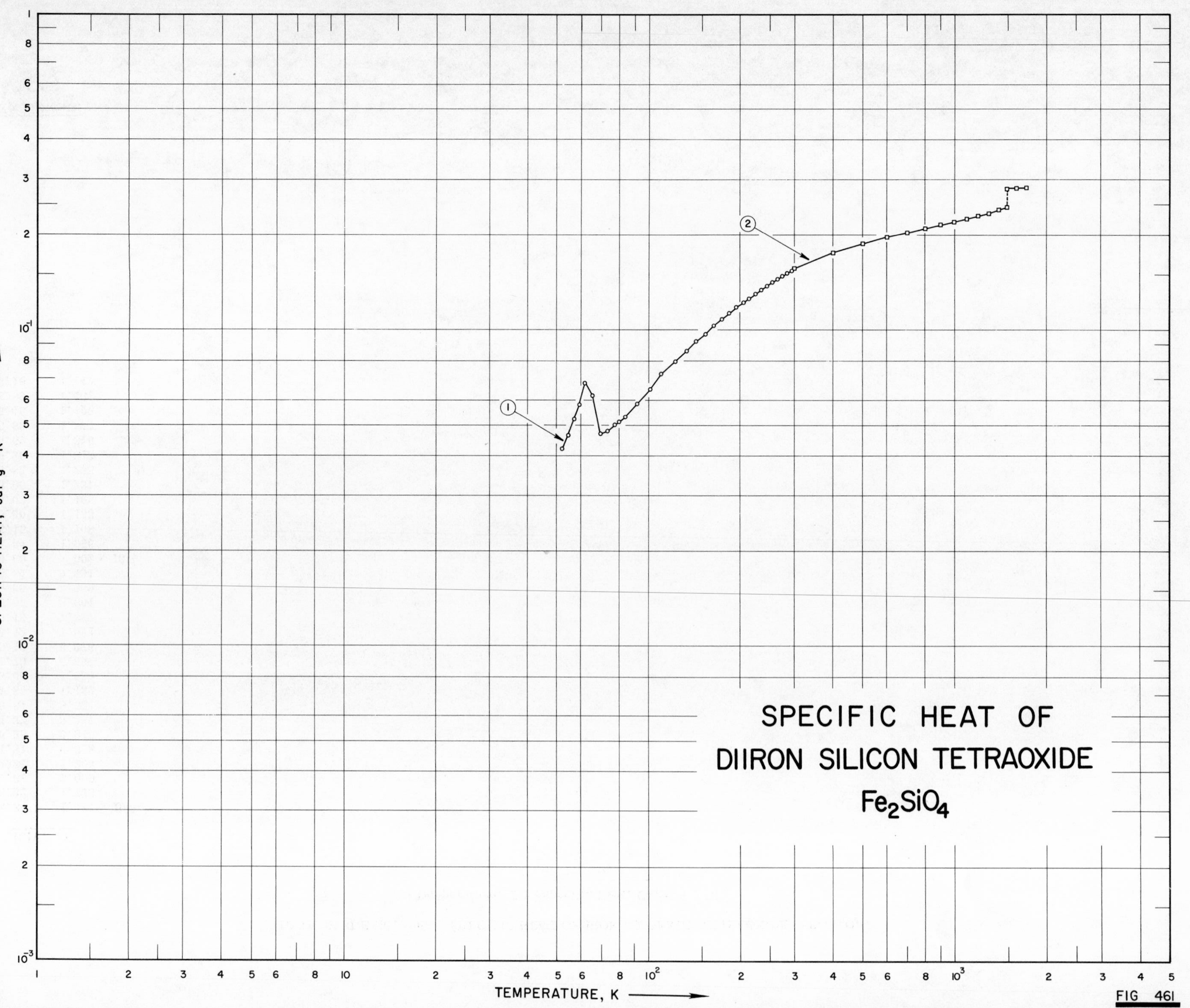
SPECIFIC HEAT OF
DIIRON SILICON TETRAOXIDE
Fe_2SiO_4
SPECIFIC HEAT, cal g^{-1} K^{-1}
TEMPERATURE, K
1
2
FIG 461

SPECIFICATION TABLE NO. 461 SPECIFIC HEAT OF DIIRON SILICON TETRAOXIDE Fe_2SiO_4

[For Data Reported in Figure and Table No. 461]

Curve No.	Ref. No.	Year	Temp. Range, K	Reported Error, %	Name and Specimen Designation	Composition (weight percent), Specifications and Remarks
1	387	1941	52-295			54.5 Fe (54.8 theo.), 29.5 silica; density = 271 lb ft^{-3}.
2	388	1953	298-1724			54.5 Fe (54.8 theo.), 29.5 silica; density = 271 lb ft^{-3}.

DATA TABLE NO. 461 SPECIFIC HEAT OF DIIRON SILICON TETRAOXIDE Fe_2SiO_4

[Temperature, T, K; Specific Heat, C_p, Cal $g^{-1}K^{-1}$]

T	C_p	T	C_p
CURVE 1		CURVE 2 (cont.)	
52.2	4.235 x 10^{-2}	1200	2.32 x 10^{-1}
54.6	4.659	1300	2.37
57.3	5.232	1400	2.42
59.5	5.840	(s) 1490	2.46
62.1	6.812	(l) 1490	2.82
65.5	6.228	1500	2.82*
69.5	4.696	1600	2.82
73.6	4.794	1700	2.82*
77.6	5.006	1724	2.82
80.3	5.114		
84.3	5.345		
92.3	5.855		
102.1	6.547		
113.0	7.278		
123.5	7.951		
133.5	8.579		
144.2	9.212		
154.1	9.737		
164.5	1.036 x 10^{-1}		
174.4	1.083		
184.7	1.135		
194.8	1.180		
205.1	1.226		
214.9	1.263		
224.7	1.305		
235.3	1.347		
245.2	1.383		
255.7	1.422		
265.5	1.457		
275.5	1.485		
285.5	1.522		
295.2	1.549		
CURVE 2			
298	1.56 x 10^{-1}*		
300	1.56		
400	1.77		
500	1.89		
600	1.98		
700	2.05		
800	2.11		
900	2.16		
1000	2.22		
1100	2.27		

* Not shown on plot

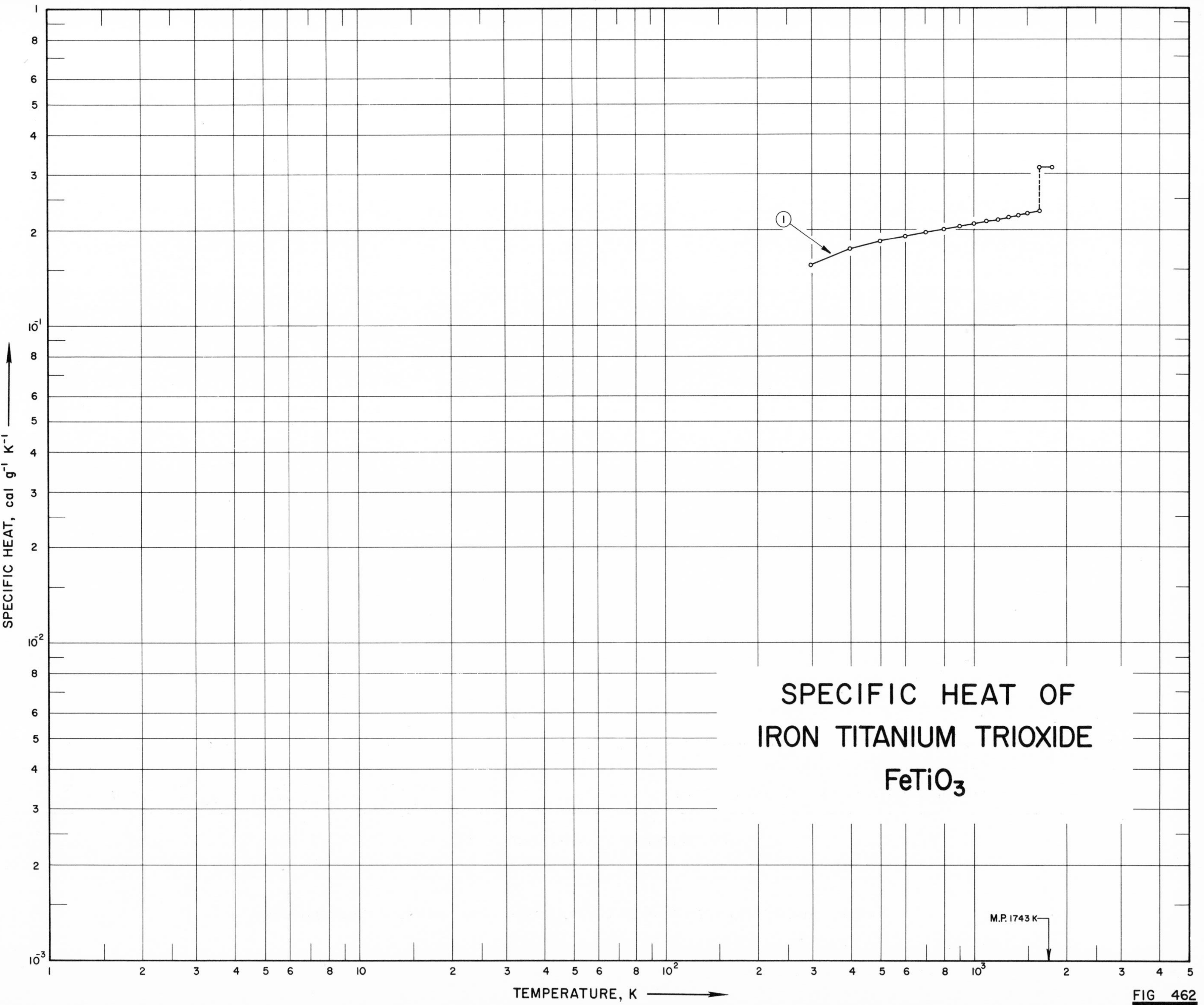

SPECIFIC HEAT OF
IRON TITANIUM TRIOXIDE
FeTiO3
1
M.P. 1743 K
SPECIFIC HEAT, cal g-1 K-1
TEMPERATURE, K

FIG 462

SPECIFICATION TABLE NO. 462 SPECIFIC HEAT OF IRON TITANIUM TRIOXIDE $FeTiO_3$

[For Data Reported in Figure and Table No. 462]

Curve No.	Ref. No.	Year	Temp. Range, K	Reported Error, %	Name and Specimen Designation	Composition (weight percent), Specifications and Remarks
1	374	1946	298-1800		Ilmenite	99.4 $FeTiO_3$; 0.6 silica; powdered raw materials mixed and heated in vacuum for 30 hrs at 1165-1300 C.

DATA TABLE NO. 462 SPECIFIC HEAT OF IRON TITANIUM TRIOXIDE $FeTiO_3$

[Temperature, T, K; Specific Heat, C_p, Cal $g^{-1}K^{-1}$]

T	C_p
CURVE 1	
298	1.55 x 10^{-1}
300	1.56*
400	1.75
500	1.85
600	1.92
700	1.97
800	2.02
900	2.05
1000	2.09
1100	2.13
1200	2.16
1300	2.19
1400	2.22
1500	2.25
1600	2.28*
(s) 1640	2.30
(l) 1640	3.14
1700	3.14*
1800	3.14

* Not shown on plot

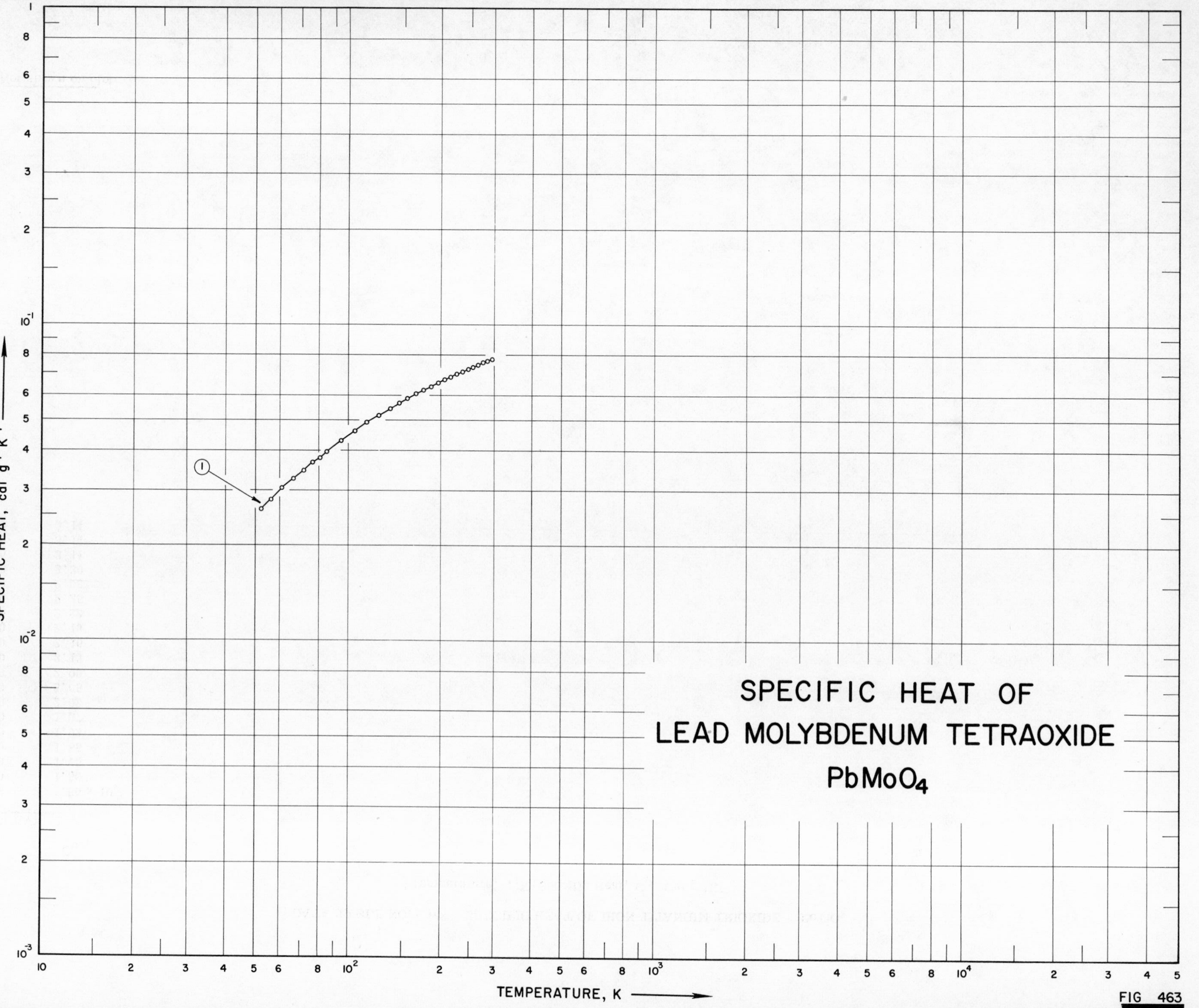
SPECIFIC HEAT OF
LEAD MOLYBDENUM TETRAOXIDE
PbMoO4
SPECIFIC HEAT, cal g-1 K-1
TEMPERATURE, K
1
FIG 463

SPECIFICATION TABLE NO. 463 SPECIFIC HEAT OF LEAD MOLYBDENUM TETRAOXIDE $PbMoO_4$

[For Data Reported in Figure and Table No. 463]

Curve No.	Ref. No.	Year	Temp. Range, K	Reported Error, %	Name and Specimen Designation	Composition (weight percent), Specifications and Remarks
1	389	1964	52-296	0.3		60.70 PbO_2, 39.17 MoO_2; prepared from reagent-grade lead nitrate and ammonium molybdate; dried at 500 C; ground to -80 mesh and dried at 640 C.

DATA TABLE NO. 463 SPECIFIC HEAT OF LEAD MOLYBDENUM TETRAOXIDE $PbMoO_4$

[Temperature, T, K; Specific Heat, C_p, Cal $g^{-1}K^{-1}$]

T	C_p
CURVE 1	
52.41	2.588 x 10^{-2}
56.69	2.795
61.42	3.037
66.54	3.263
71.81	3.465
76.97	3.669
80.61	3.778
85.17	3.969
94.74	4.290
105.23	4.622
114.76	4.900
124.78	5.167
136.03	5.445
145.82	5.663
155.66	5.864
165.64	6.055
175.85	6.235
185.72	6.398
195.80	6.554
206.13	6.709
216.36	6.859
226.42	6.997
236.33	7.120
245.80	7.237
256.22	7.357
266.42	7.471
276.42	7.578
286.86	7.687
296.38	7.774

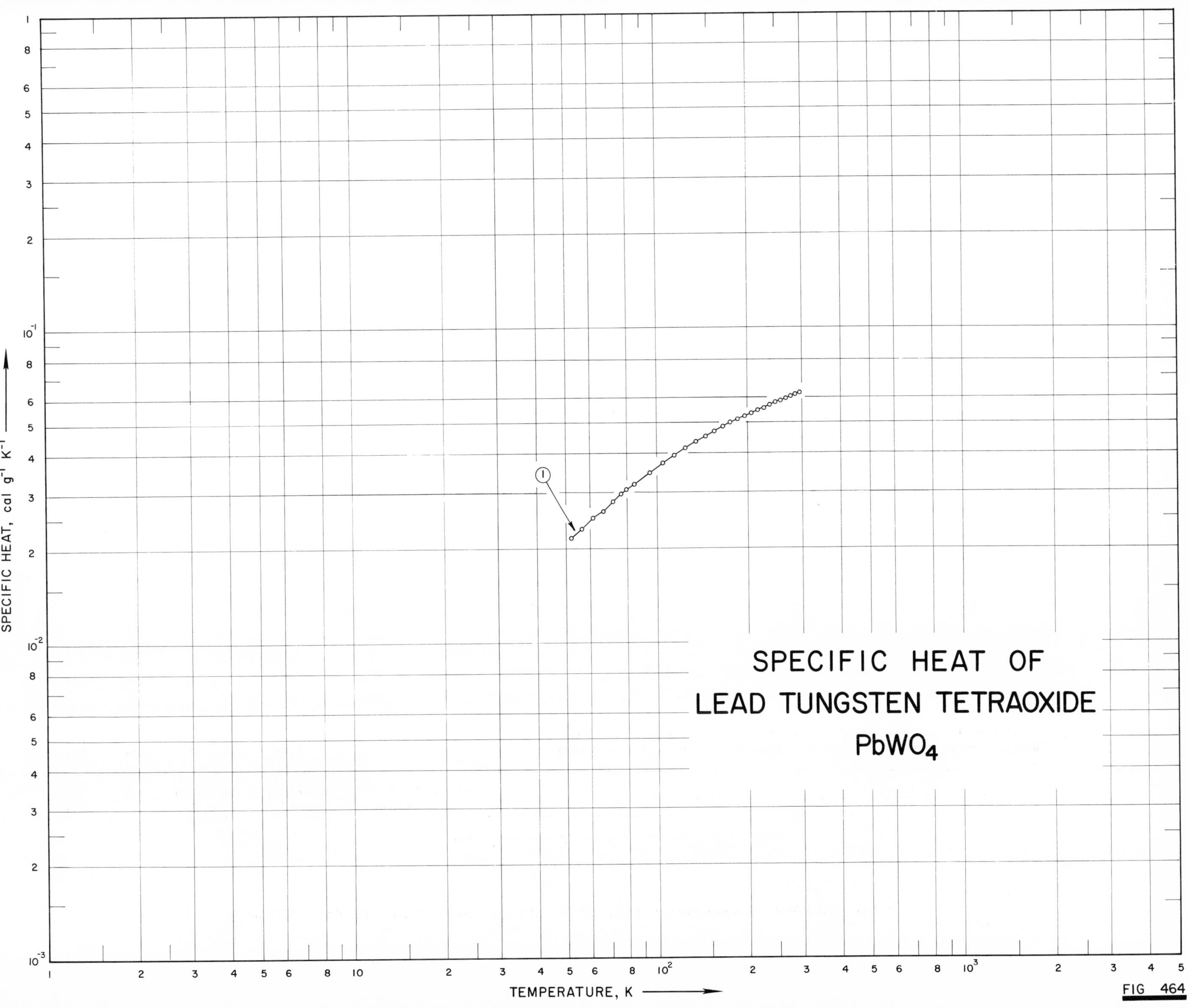
SPECIFIC HEAT OF
LEAD TUNGSTEN TETRAOXIDE
PbWO4
1
SPECIFIC HEAT, cal g⁻¹ K⁻¹
TEMPERATURE, K
FIG 464

SPECIFICATION TABLE NO. 464 SPECIFIC HEAT OF LEAD TUNGSTEN TETRAOXIDE $PbWO_4$

[For Data Reported in Figure and Table No. 464]

Curve No.	Ref. No.	Year	Temp. Range, K	Reported Error, %	Name and Specimen Designation	Composition (weight percent), Specifications and Remarks
1	389	1964	52-296	0.30		50.94 WO_2, 48.94 PbO_2; prepared from reagent-grade lead carbonate and tungstic acid; stoichiometric mixture heated for several days at 600 C and several days at 700 C.

DATA TABLE NO. 464 SPECIFIC HEAT OF LEAD TUNGSTEN TETRAOXIDE $PbWO_4$

[Temperature, T, K; Specific Heat, C_p, Cal $g^{-1}K^{-1}$]

T	C_p
CURVE 1	
51.94	2.178 x 10^{-2}
56.23	2.343
61.11	2.523
66.10	2.650
71.26	2.852
76.31	3.009
79.86	3.116
84.44	3.244
94.89	3.521
105.18	3.771
114.64	3.986
124.66	4.200
135.78	4.408
145.71	4.589
155.72	4.751
166.06	4.912
176.45	5.057
186.08	5.186
195.74	5.303
206.20	5.424
216.40	5.542
226.20	5.643
236.16	5.753
246.04	5.852
256.35	5.945
266.84	6.032
276.54	6.120
286.99	6.202
296.48	6.285

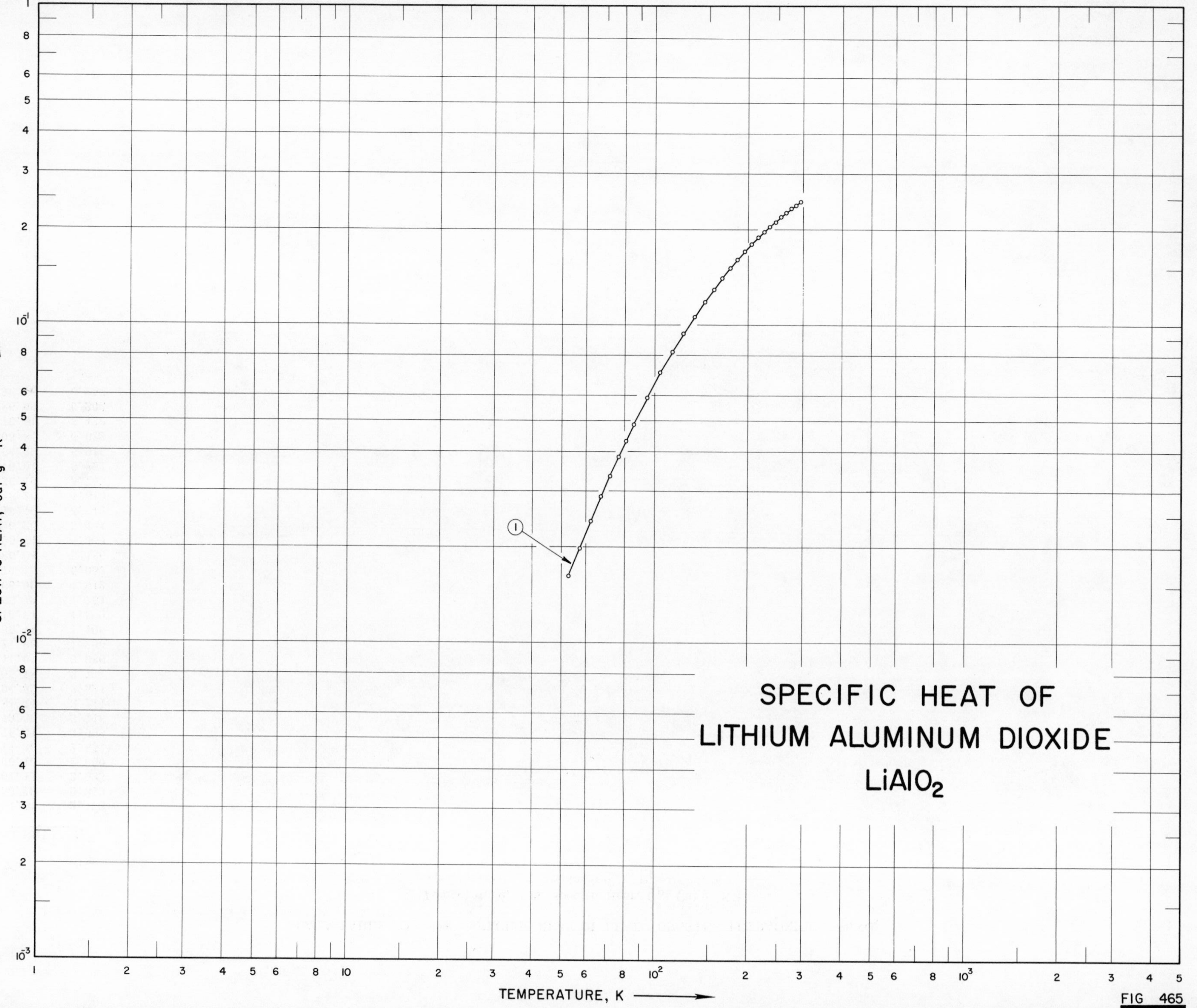
SPECIFIC HEAT OF
LITHIUM ALUMINUM DIOXIDE
LiAlO2
SPECIFIC HEAT, cal g-1 K-1
TEMPERATURE, K
1
FIG 465

SPECIFICATION TABLE NO. 465 SPECIFIC HEAT OF LITHIUM ALUMINUM DIOXIDE $LiAlO_2$

[For Data Reported in Figure and Table No. 465]

Curve No.	Ref. No.	Year	Temp. Range, K	Reported Error, %	Name and Specimen Designation	Composition (weight percent), Specifications and Remarks
1	390	1955	53-296			77.36 Al_2O_3 (77.34 theo.); prepared by prolonged repeated sintering of stoichiometric mixtures of appropriate pure ingredients; heated 50 hrs at 900-1000 C.

DATA TABLE NO. 465 SPECIFIC HEAT OF LITHIUM ALUMINUM DIOXIDE $LiAlO_2$

[Temperature, T, K; Specific Heat, C_p, Cal $g^{-1}K^{-1}$]

T	C_p
CURVE 1	
52.91	1.623 x 10^{-2}
57.24	1.981
62.24	2.418
67.17	2.884
71.78	3.345
76.54	3.849
80.85	4.320
85.69	4.865
94.75	5.900
104.89	7.092
114.71	8.237
124.43	9.398
135.64	1.070
145.98	1.187
155.83	1.296
165.91	1.404
176.18	1.512
186.01	1.608
195.85	1.701
206.24	1.795
216.04	1.884
226.00	1.966
235.96	2.043
245.70	2.115
256.28	2.194
266.35	2.260
276.13	2.323
286.62	2.385
295.98	2.445

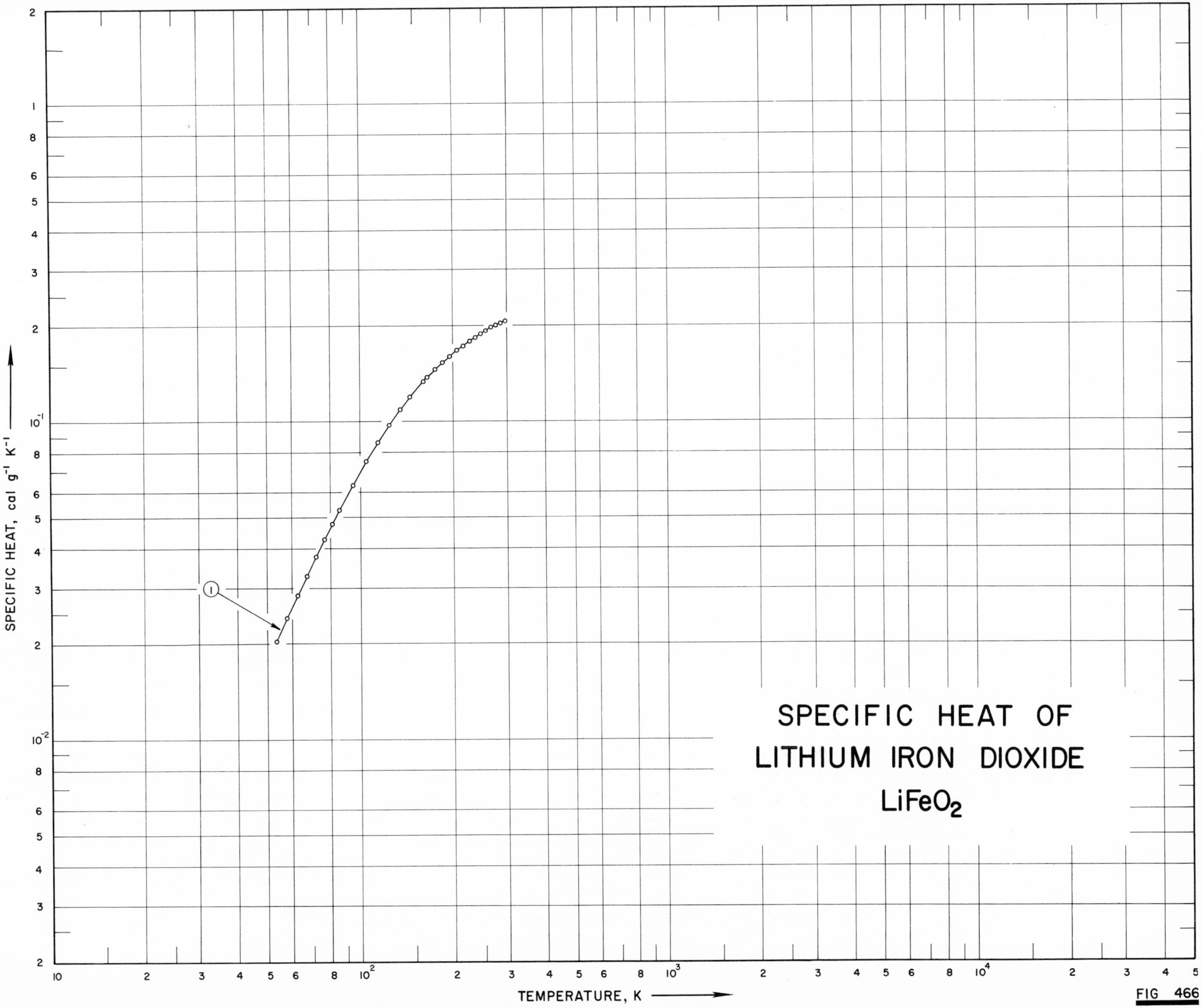

SPECIFIC HEAT OF LITHIUM IRON DIOXIDE $LiFeO_2$

FIG 466

SPECIFICATION TABLE NO. 466 SPECIFIC HEAT OF LITHIUM IRON DIOXIDE $LiFeO_2$

[For Data Reported in Figure and Table No. 466]

Curve No.	Ref. No.	Year	Temp. Range, K	Reported Error, %	Name and Specimen Designation	Composition (weight percent), Specifications and Remarks
1	390	1955	53.4-296			84.15 Fe_2O_3 (84.24 theo.), 0.09 SiO_2; prepared from reagent-grade $LiCO_3$ and pure Fe_2O_3; heated five times for 138 hrs total at 1000-1050 C and 7 hrs at 1150-1200 C.

DATA TABLE NO. 466 SPECIFIC HEAT OF LITHIUM IRON DIOXIDE $LiFeO_2$

[Temperature, T, K; Specific Heat, C_p, Cal $g^{-1}K^{-1}$]

T	C_p
CURVE 1	
53.43	2.049×10^{-2}
57.78	2.423
62.54	2.861
67.12	3.300
71.79	3.773
76.46	4.274
81.09	4.793
85.47	5.297
94.60	6.349
104.75	7.532
114.16	8.620
124.50	9.788
135.78	1.099×10^{-1}
145.48	1.201
160.79	1.346
165.72	1.386
175.93	1.472
185.82	1.548
195.64	1.617
205.96	1.684
215.98	1.745
225.97	1.803
235.98	1.853
245.68	1.897
256.28	1.945
266.33	1.982
276.12	2.016
286.34	2.051
296.01	2.084

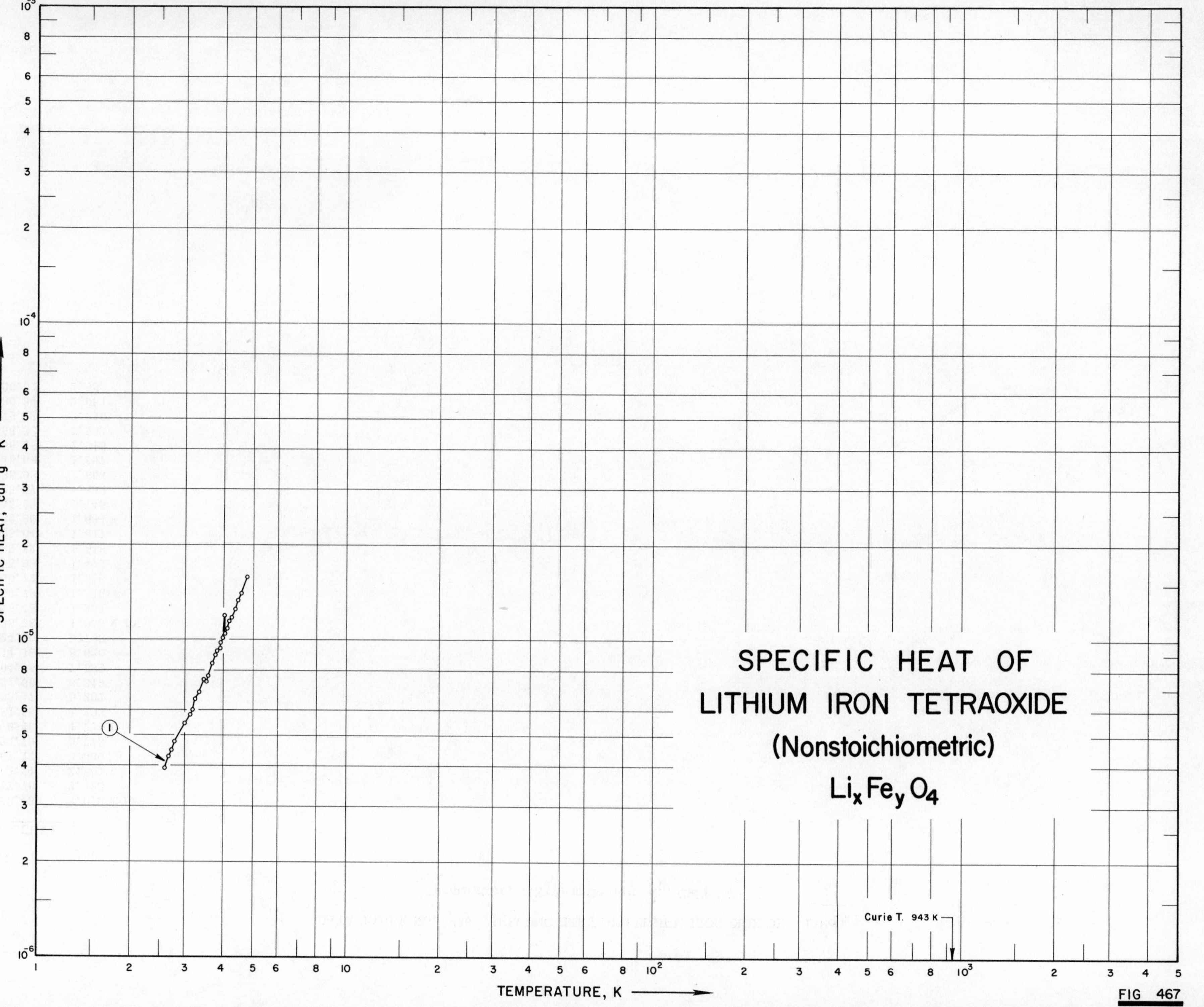
SPECIFIC HEAT OF
LITHIUM IRON TETRAOXIDE
(Nonstoichiometric)
$Li_xFe_yO_4$
Curie T. 943 K
1
SPECIFIC HEAT, cal g⁻¹ K⁻¹
TEMPERATURE, K

FIG 467

SPECIFICATION TABLE NO. 467 SPECIFIC HEAT OF LITHIUM IRON TETRAOXIDE (nonstoichiometric) $Li_xFe_yO_4$

[For Data Reported in Figure and Table No. 467]

Curve No.	Ref. No.	Year	Temp. Range, K	Reported Error, %	Name and Specimen Designation	Composition (weight percent), Specifications and Remarks
1	383	1961	2.6-4.8		$Li_{0.5}Fe_{2.5}O_4$	67.20 Fe, 1.71 Li; x-ray density = 4.75 g cm^{-3}; x-ray lattice parameter = 8.338 Å; ordered structure; Curie temperature = 943 K.

DATA TABLE NO. 467 SPECIFIC HEAT OF LITHIUM IRON TETRAOXIDE $Li_xFe_yO_4$ (nonstoichiometric)

[Temperature, T, K; Specific Heat, C_p, Cal $g^{-1}K^{-1}$]

T	C_p
CURVE 1	
2.599	3.947×10^{-6}
2.669	4.271
2.730	4.480
2.794	4.793
3.006	5.453
3.067	5.481*
3.124	5.819
3.185	6.024
3.242	6.535
3.298	6.535*
3.352	6.855
3.412	6.991*
3.467	7.498
3.527	7.377
3.583	7.735
3.696	8.486
3.778	8.963
3.847	9.251
3.920	9.801
3.920	9.627
3.992	1.019×10^{-5}
4.004	1.016*
4.029	1.214
4.059	1.046
4.132	1.085
4.196	1.158
4.264	1.176
4.395	1.260
4.464	1.341
4.519	1.371*
4.596	1.413
4.653	1.427*
4.720	1.558*
4.779	1.583

* Not shown on plot

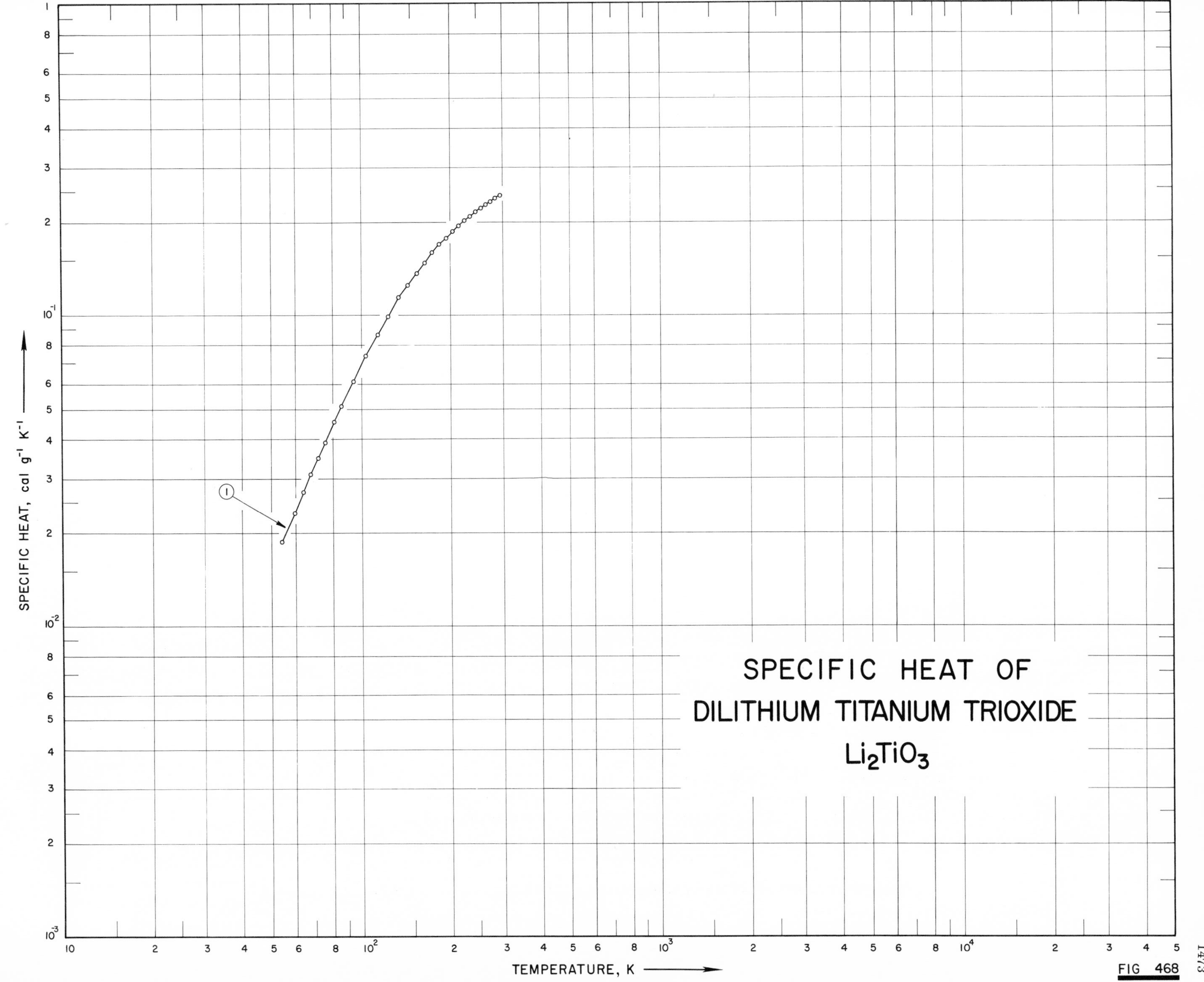
SPECIFIC HEAT OF
DILITHIUM TITANIUM TRIOXIDE
Li_2TiO_3
SPECIFIC HEAT, cal g⁻¹ K⁻¹
TEMPERATURE, K
1
FIG 468

SPECIFICATION TABLE NO. 468 SPECIFIC HEAT OF DILITHIUM TITANIUM TRIOXIDE Li_2TiO_3

[For Data Reported in Figure and Table No. 468]

Curve No.	Ref. No.	Year	Temp. Range, K	Reported Error, %	Name and Specimen Designation	Composition (weight percent), Specifications and Remarks
1	358	1955	55-298			72.70 titania (72.78 theo.), 0.3 silica, 0.06 Pt + Ni; prepared from reagent-grade lithium carbonate and pure titania; heated 6 times for 70 hrs at 1000-1050 C and 30 hrs at 1150 C.

DATA TABLE NO. 468 SPECIFIC HEAT OF DILITHIUM TITANIUM TRIOXIDE Li_2TiO_3

[Temperature, T, K; Specific Heat, C_p, Cal $g^{-1}K^{-1}$]

T	C_p
CURVE 1	
54.58	1.861×10^{-2}
60.12	2.308
64.35	2.694
68.29	3.076
72.14	3.466
76.04	3.889
81.62	4.534
86.23	5.073
94.65	6.099
104.80	7.388
114.33	8.604
124.30	9.893
135.84	1.132×10^{-1}
145.54	1.250
155.80	1.368
165.68	1.475
175.81	1.581
185.77	1.675
196.16	1.769
206.09	1.854
216.33	1.937
226.09	2.010
235.82	2.074
245.67	2.141
256.13	2.205
266.50	2.263
276.20	2.312
286.44	2.364
295.94	2.407*
298.16	2.418

* Not shown on plot

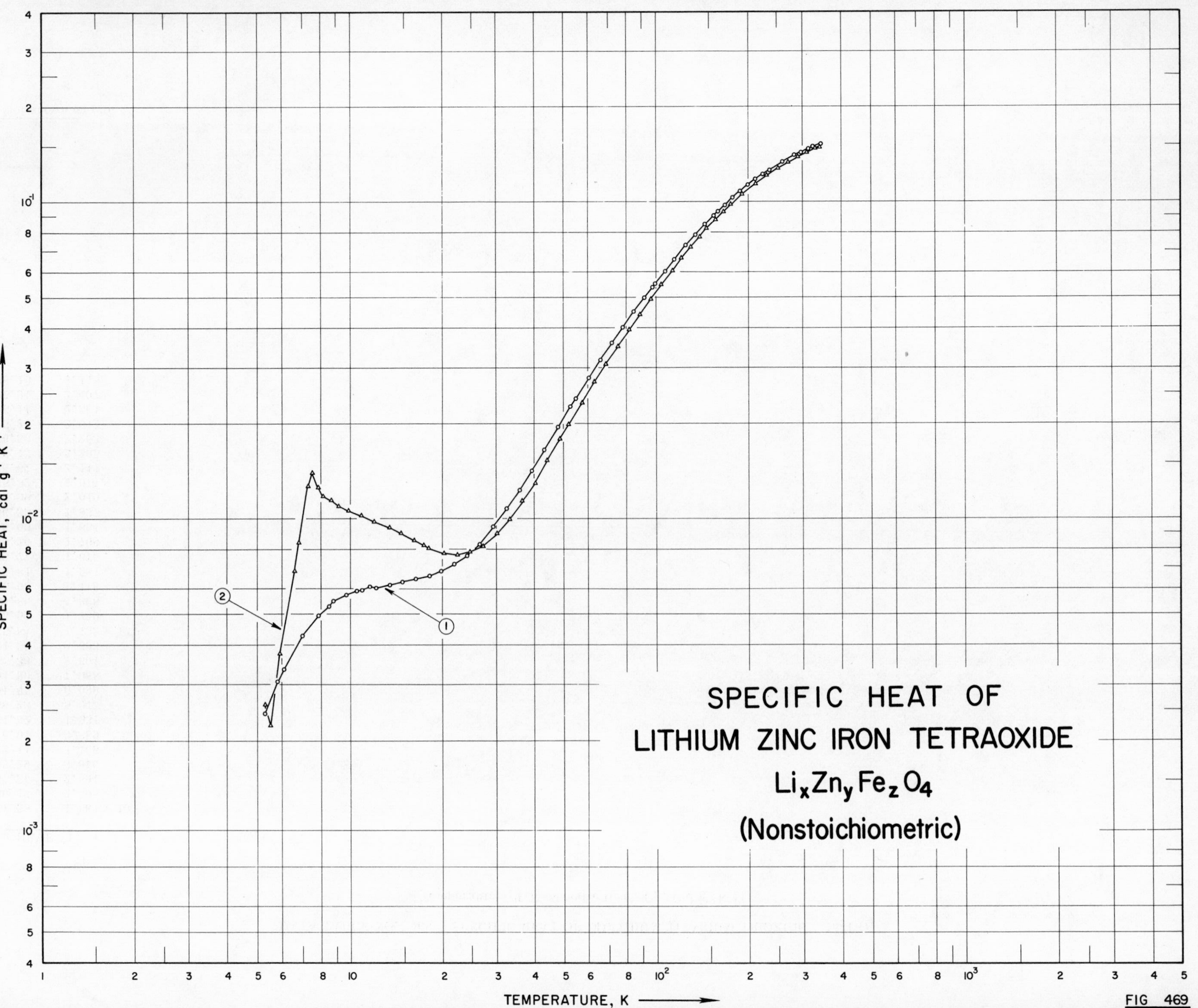
SPECIFIC HEAT OF
LITHIUM ZINC IRON TETRAOXIDE
$Li_xZn_yFe_zO_4$
(Nonstoichiometric)
SPECIFIC HEAT, cal g⁻¹ K⁻¹
TEMPERATURE, K
1
2

FIG 469

SPECIFICATION TABLE NO. 469 SPECIFIC HEAT OF LITHIUM ZINC IRON TETRAOXIDE $Li_xZn_yFe_zO_4$ (nonstoichiometric)

[For Data Reported in Figure and Table No. 469]

Curve No.	Ref. No.	Year	Temp. Range, K	Reported Error, %	Name and Specimen Designation	Composition (weight percent), Specifications and Remarks
1	406	1959	5-346		$Li_{0.05}Zn_{0.90}Fe_{2.05}O_4$	48.0 Fe, 24.4 Zn, 0.12 Li; quenched.
2	406	1959	5-344		$Li_{0.05}Zn_{0.90}Fe_{2.05}O_4$	47.9 Fe, 24.4 Zn, 0.11 Li; annealed.

DATA TABLE NO. 469 SPECIFIC HEAT OF LITHIUM ZINC IRON TETRAOXIDE $Li_xZn_yFe_zO_2$ (nonstoichiometric)

[Temperature, T, K; Specific Heat, C_p, Cal $g^{-1}K^{-1}$]

T	C_p
CURVE 1	
Series I	
97.48	5.370 x 10^{-2}
106.78	6.008*
116.58	6.658*
125.95	7.285
135.26	7.880
145.05	8.482
155.38	9.079
159.74	9.323
167.88	9.765
178.05	1.027 x 10^{-1}
188.81	1.078
200.35	1.130
212.09	1.178
224.23	1.223
236.40	1.266
248.41	1.305*
260.56	1.340
272.60	1.373*
284.59	1.403
296.72	1.430*
Series II	
5.81	3.1 x 10^{-3}
6.91	4.25*
8.51	5.34
8.70	5.47*
9.76	5.76*
10.95	5.97
12.20	6.088
13.43	6.239
14.79	6.353
16.37	6.475
18.11	6.639
19.89	6.866
21.85	7.216
24.12	7.691
26.64	8.301
29.48	9.445
32.54	1.075 x 10^{-2}
35.77	1.231
39.35	1.422
43.41	1.654
48.08	1.950
52.80	2.265
CURVE 1 (cont.)	
Series III	
5.29	2.4 x 10^{-3}
6.08	3.4
6.99	4.29
7.86	4.97
8.76	5.51
9.66	5.76
10.58	5.93
11.61	6.14
Series IV	
54.94	2.407 x 10^{-2}
60.61	2.788
66.28	3.177
72.19	3.591
78.38	4.028
84.89	4.500
91.87	4.984
99.45	5.505
107.54	6.055
115.63	6.593
Series V	
287.47	1.410 x 10^{-1}*
297.28	1.431
307.03	1.451*
316.67	1.472
326.27	1.489
335.96	1.506*
345.79	1.523
CURVE 2	
Series I	
52.10	1.988 x 10^{-2}
57.64	2.336
63.14	2.700
68.96	3.084
75.23	3.505
81.72	3.959
88.65	4.442
96.22	4.955
104.29	5.501
112.84	6.080
CURVE 2 (cont.)	
121.58	6.662 x 10^{-2}
130.20	7.224
138.62	7.751
147.26	8.274
156.38	8.802
165.95	9.328
175.77	9.837
182.75	1.018 x 10^{-1}*
192.27	1.062
201.81	1.104*
211.33	1.143
220.85	1.179*
230.52	1.215
240.20	1.247*
250.07	1.279
260.05	1.309*
270.14	1.336
280.12	1.362*
290.23	1.387
300.48	1.410*
310.90	1.433
321.53	1.453*
332.39	1.474
343.39	1.493
Series II	
5.29	2.6 x 10^{-3}
5.90	3.9
6.58	6.90
7.19	1.28 x 10^{-2}
7.82	1.27
8.63	1.15
9.13	1.11
9.81	1.07
10.76	1.03
11.96	9.85 x 10^{-3}
13.27	9.42
14.64	9.016
16.13	8.579
18.00	8.141
20.14	7.834
22.34	7.754
24.64	7.876
27.16	8.246
CURVE 2 (cont.)	
Series III	
5.05	2.2 x 10^{-3}
5.92	4.0*
6.79	8.46
7.49	1.41 x 10^{-2}
8.14	1.18
9.07	1.12*
10.08	1.06*
11.07	1.02*
27.41	8.293 x 10^{-3}
30.22	8.983
33.21	9.988
36.50	1.133 x 10^{-2}
50.21	1.309
44.35	1.529
48.87	1.791

* Not shown on plot

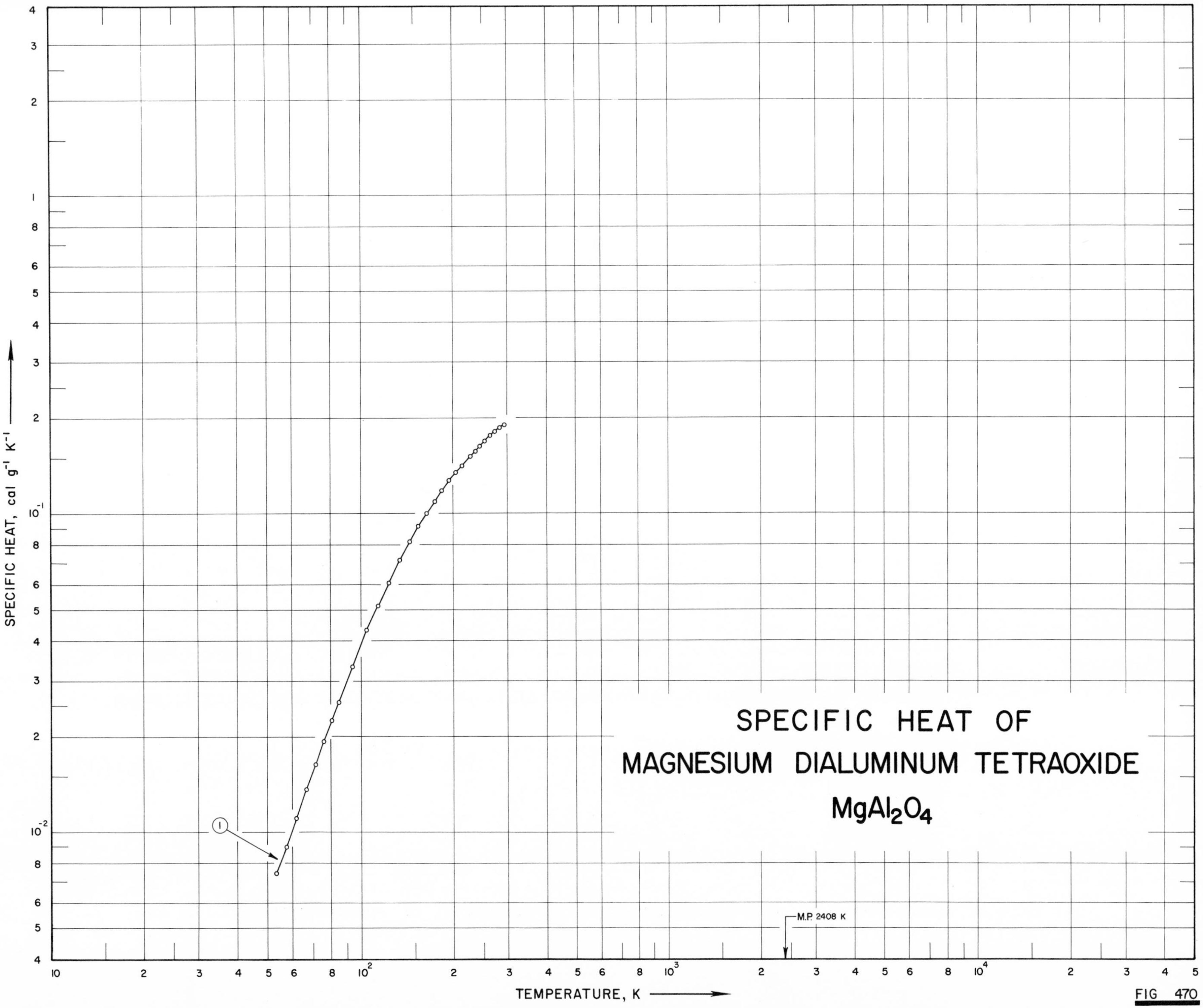

SPECIFIC HEAT OF
MAGNESIUM DIALUMINUM TETRAOXIDE
$MgAl_2O_4$
SPECIFIC HEAT, cal g^{-1} K^{-1}
TEMPERATURE, K
M.P. 2408 K
1

FIG 470

SPECIFICATION TABLE NO. 470 SPECIFIC HEAT OF MAGNESIUM DIALUMINUM TETRAOXIDE $MgAl_2O_4$

[For Data Reported in Figure and Table No. 470]

Curve No.	Ref. No.	Year	Temp. Range, K	Reported Error, %	Name and Specimen Designation	Composition (weight percent), Specifications and Remarks
1	365	1955	54-296		Magnesium-aluminum spinel	71.62 Al_2O_3, 28.33 MgO (71.66, 28.34 theo.).

DATA TABLE NO. 470 SPECIFIC HEAT OF MAGNESIUM DIALUMINUM TETRAOXIDE $MgAl_2O_4$

[Temperature, T, K; Specific Heat, C_p, Cal $g^{-1}K^{-1}$]

T	C_p
CURVE 1	
53.55	7.450×10^{-3}
57.72	9.017
62.19	1.118×10^{-2}
66.84	1.373
71.51	1.645
76.17	1.939
80.80	2.257
85.02	2.572
94.43	3.329
105.62	4.344
114.95	5.190
124.87	6.141
135.89	7.232
146.36	8.216
156.37	9.214
165.90	1.014×10^{-1}
175.91	1.106
185.86	1.193
196.21	1.281
206.36	1.362
216.32	1.427
228.03	1.529
237.36	1.588
245.85	1.650
256.37	1.716
266.66	1.776
276.12	1.829
286.34	1.890
296.31	1.920

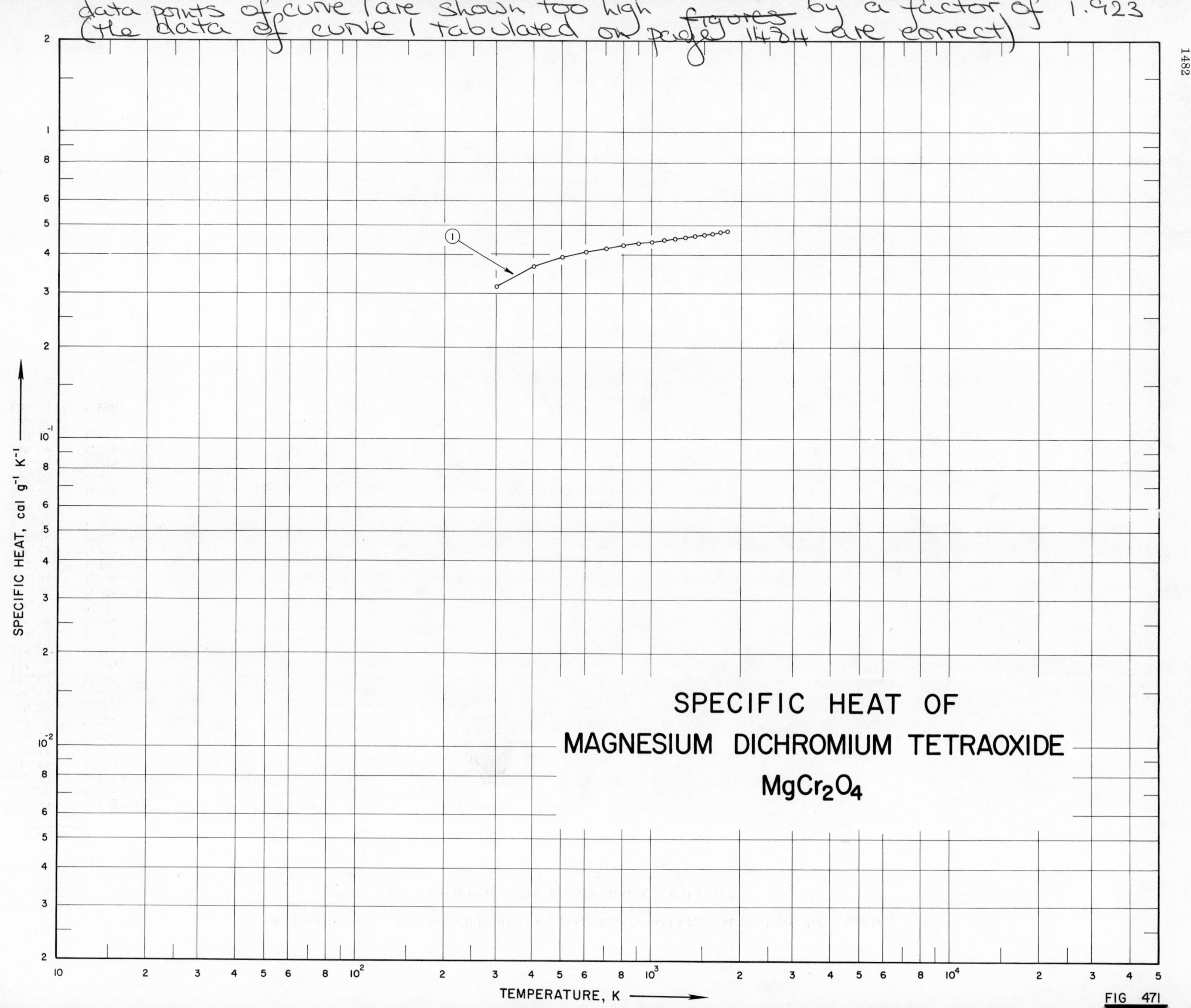

data points of curve 1 are shown too high by a factor of 1.923
(the data of curve 1 tabulated on page 1484 are correct)
1
SPECIFIC HEAT, cal g⁻¹ K⁻¹
TEMPERATURE, K
SPECIFIC HEAT OF
MAGNESIUM DICHROMIUM TETRAOXIDE
$MgCr_2O_4$
FIG 471

SPECIFICATION TABLE NO. 471 SPECIFIC HEAT OF MAGNESIUM DICHROMIUM TETRAOXIDE, $MgCr_2O_4$

[For Data Reported in Figure and Table No. 471]

Curve No.	Ref. No.	Year	Temp. Range, K	Reported Error, %	Name and Specimen Designation	Composition (weight percent), Specifications and Remarks
1	386	1944	298-1800	0.3		54.05 Cr, 12.58 Mg, 0.14 Fe; prepared by reacting reagent-grade MgO and Cr_2O_3 at 1400 C.

DATA TABLE NO. 471 SPECIFIC HEAT OF MAGNESIUM DICHROMIUM TETRAOXIDE $MgCr_2O_4$

[Temperature, T, K; Specific Heat, C_p, Cal $g^{-1}K^{-1}$]

T	C_p
CURVE 1	
298	1.58×10^{-1}
300	1.58*
400	1.84
500	1.97
600	2.05
700	2.11
800	2.15
900	2.19
1000	2.22
1100	2.24
1200	2.27
1300	2.29
1400	2.31
1500	2.34
1600	2.36
1700	2.38
1800	2.40

* Not shown on plot

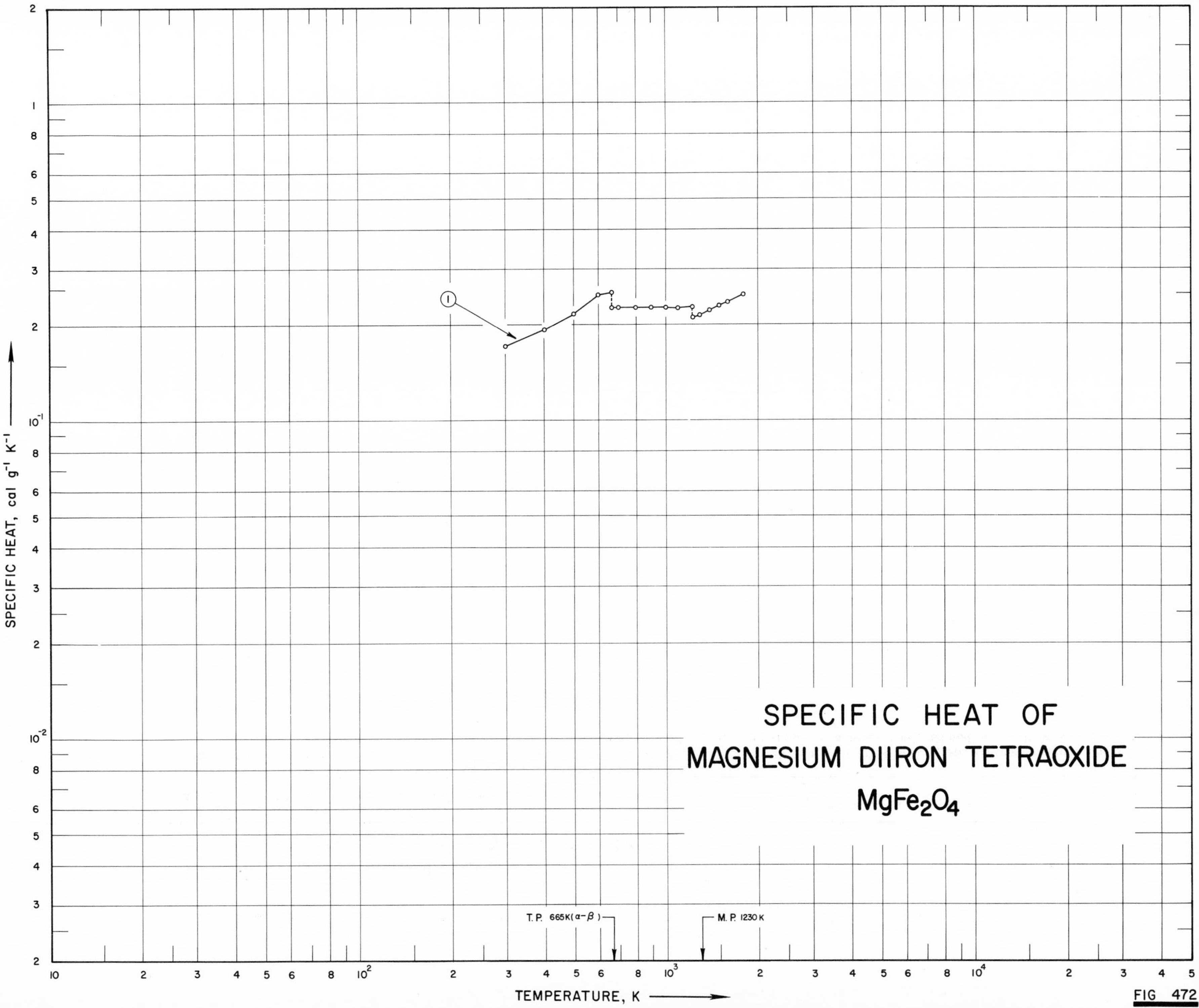
SPECIFIC HEAT OF
MAGNESIUM DIIRON TETRAOXIDE
$MgFe_2O_4$
T. P. 665 K ($\alpha-\beta$)
M. P. 1230 K
TEMPERATURE, K
SPECIFIC HEAT, cal g^{-1} K^{-1}
1
FIG 472

SPECIFICATION TABLE NO. 472 SPECIFIC HEAT OF MAGNESIUM DIIRON TETRAOXIDE $MgFe_2O_4$

[For Data Reported in Figure and Table No. 472]

Curve No.	Ref. No.	Year	Temp. Range, K	Reported Error, %	Name and Specimen Designation	Composition (weight percent), Specifications and Remarks
1	367	1959	298-1800	0.1-1.2		79.74 Fe_2O_3, 20.22 MgO (79.48, 20.16 theo.), 0.14 SiO_2; prepared from reagent-grade Fe_2O_3 and MgO; heated repeated 900-1300 C; material analyzed and composition adjusted between heatings.

DATA TABLE NO. 472 SPECIFIC HEAT OF MAGNESIUM DIIRION TETRAOXIDE $MgFe_2O_4$

[Temperature, T, K; Specific Heat, C_p, Cal $g^{-1}K^{-1}$]

T	C_p
CURVE 1	
298	1.72×10^{-1}
300	1.72
400	1.94
500	2.17
600	2.49
α 665	2.53
β 665	2.27
700	2.27
800	2.27
900	2.27
1000	2.27
1100	2.27
1200	2.27*
β 1230	2.27
γ 1230	2.12
1300	2.16
1400	2.23
1500	2.30
1600	2.37
1800	2.50

* Not shown on plot

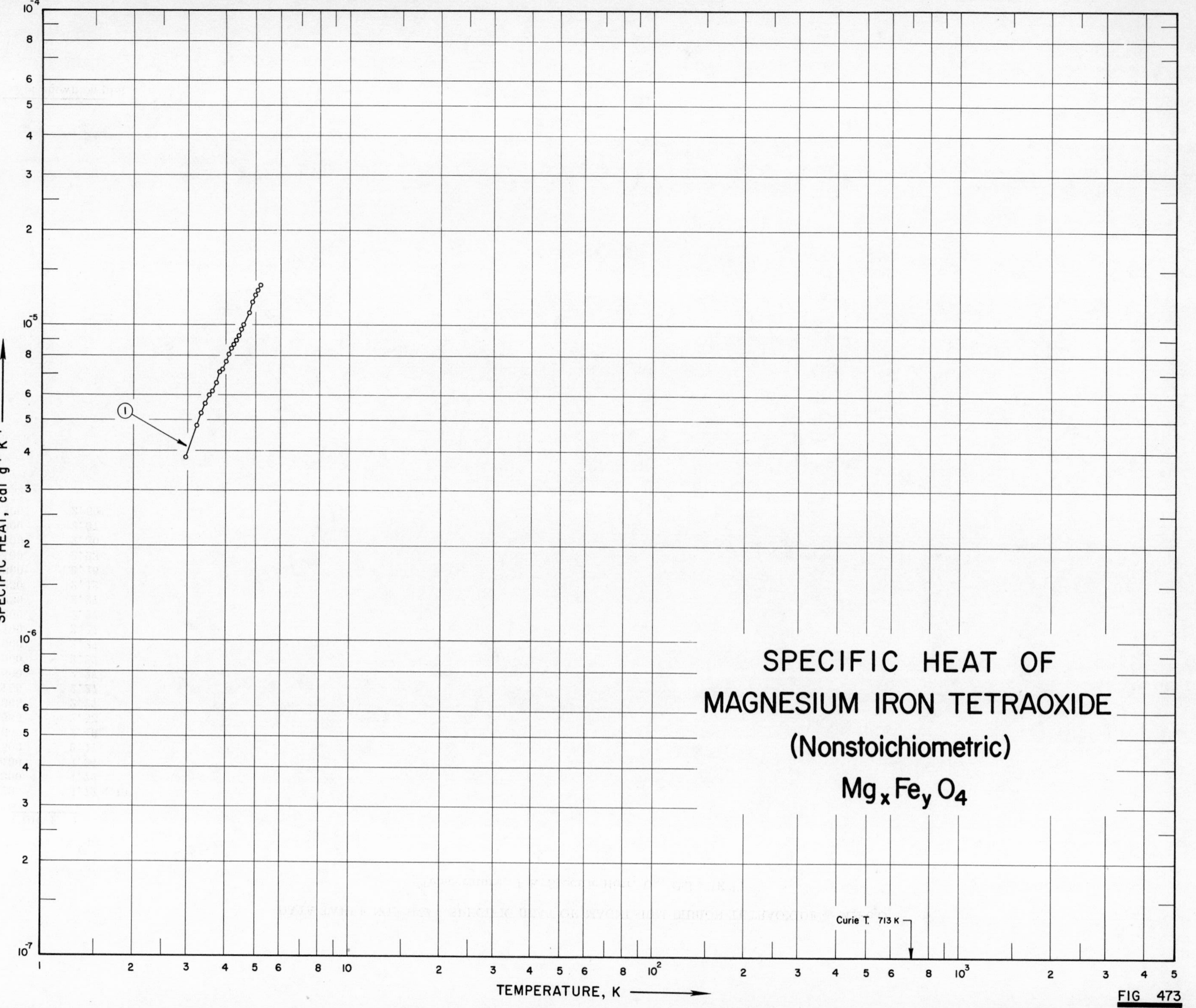
SPECIFIC HEAT OF
MAGNESIUM IRON TETRAOXIDE
(Nonstoichiometric)
$Mg_x Fe_y O_4$
Curie T. 713 K
SPECIFIC HEAT, cal g^{-1} K^{-1}
TEMPERATURE, K
1
FIG 473

SPECIFICATION TABLE NO. 473 SPECIFIC HEAT OF MAGNESIUM IRON TETRAOXIDE (nonstoichiometric) $Mg_xFe_yO_4$

[For Data Reported in Figure and Table No. 473]

Curve No.	Ref. No.	Year	Temp. Range, K	Reported Error, %	Name and Specimen Designation	Composition (weight percent), Specifications and Remarks
1	383	1961	3.0-5.2		$Mg_{0.82}Fe_{2.18}O_4$	58.43 Fe, 9.79 Mg; x-ray density = 4.65 g cm^{-3}; x-ray lattice parameter = 8.36 Å; Curie temperature = 713 K.

DATA TABLE NO. 473 SPECIFIC HEAT OF MAGNESIUM IRON TETRAOXIDE $Mg_xFe_yO_4$ (nonstoichiometric)

[Temperature, T, K; Specific Heat, C_p, Cal $g^{-1}K^{-1}$]

T	C_p
CURVE 1	
2.978	3.828×10^{-6}
3.228	4.848
3.344	5.278
3.448	5.699
3.547	6.044
3.642	6.200
3.734	6.616
3.830	7.164
3.921	7.292
4.019	7.731
4.100	8.100
4.178	8.473
4.257	8.723
4.346	8.979
4.411	9.295
4.487	9.730
4.559	1.019×10^{-5}
4.761	1.111
4.874	1.196
4.985	1.265
5.087	1.309
5.184	1.358

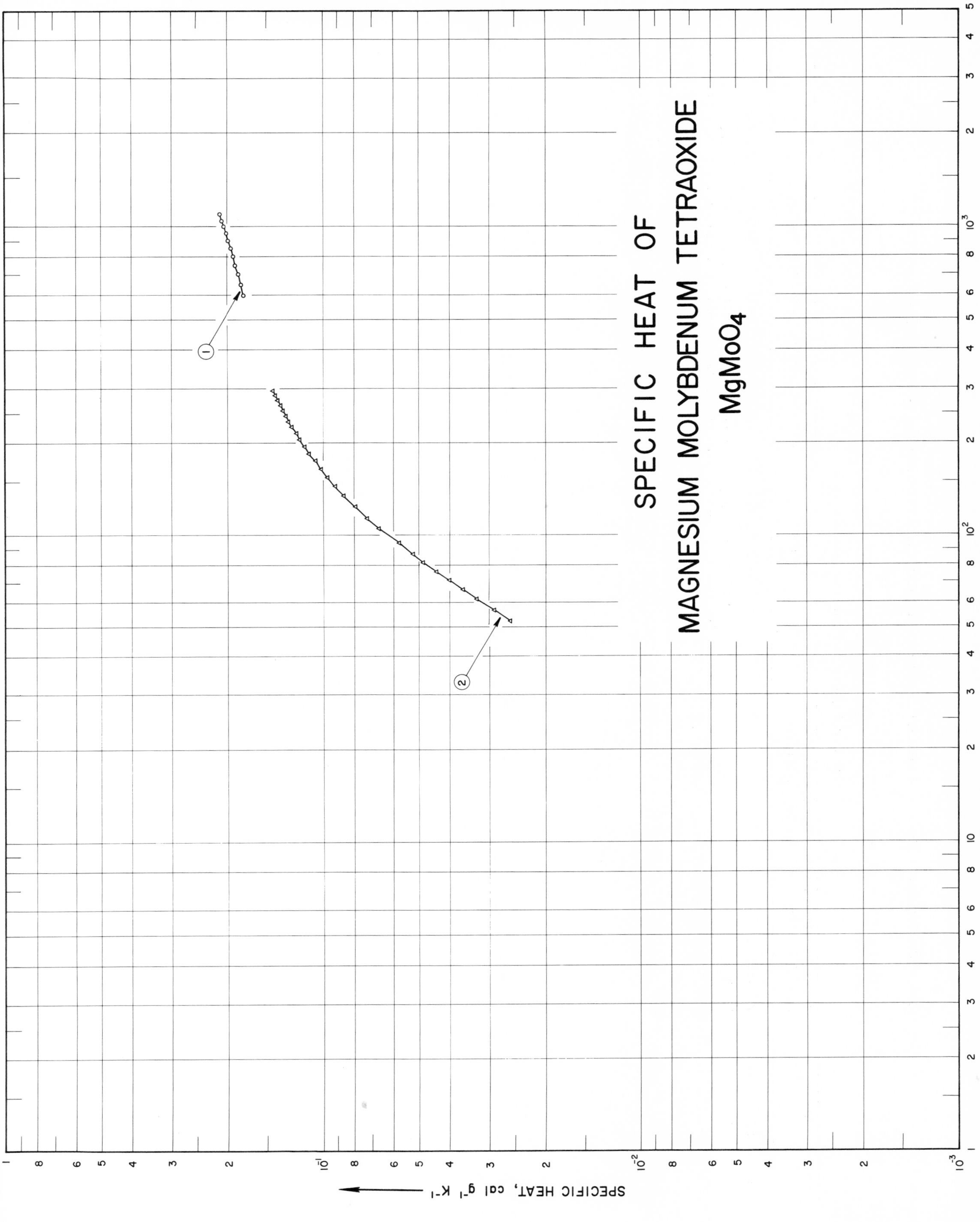

SPECIFIC HEAT OF MAGNESIUM MOLYBDENUM TETRAOXIDE $MgMoO_4$

SPECIFIC HEAT, cal g^{-1} K^{-1}

TEMPERATURE, K

(1)
(2)

FIG 474

SPECIFICATION TABLE NO. 474 SPECIFIC HEAT OF MAGNESIUM MOLYBDENUM TETRAOXIDE $MgMoO_4$

[For Data Reported in Figure and Table No. 474]

Curve No.	Ref. No.	Year	Temp. Range, K	Reported Error, %	Name and Specimen Designation	Composition (weight percent), Specifications and Remarks
1	391	1961	600-1100	±0.1		
2	368	1963	52-296	0.3		78.21 MoO_3, 21.95 MgO; prepared by dissolving stoichiometric amounts of reagent-grade MgO and MoO_3 in boiling water; heated to dryness at 122 C; heated at 870-890 C for 20 hrs.

DATA TABLE NO. 474 SPECIFIC HEAT OF MAGNESIUM MOLYBDENUM TETRAOXIDE $MgMoO_4$

[Temperature, T, K; Specific Heat, C_p, Cal $g^{-1}K^{-1}$]

T	C_p
CURVE 1	
600	1.777×10^{-1}
650	1.812
700	1.846
750	1.880
800	1.914
850	1.948
900	1.983
950	2.017
1000	2.051
1050	2.085
1100	2.119
CURVE 2	
52.43	2.587×10^{-2}
56.74	2.905
61.52	3.287
66.21	3.640
71.11	4.009
76.27	4.414
81.88	4.850
87.03	5.240
94.85	5.770
106.69	6.698
114.94	7.279
124.69	7.924
135.80	8.614
145.69	9.173
155.47	9.694
165.82	1.020×10^{-1}
176.21	1.066
185.94	1.113
195.97	1.147
206.43	1.185
216.44	1.219
226.36	1.252
236.19	1.285
245.70	1.313
256.47	1.342
266.73	1.368
276.61	1.395
286.80	1.417
296.39	1.439

SPECIFIC HEAT OF
MAGNESIUM SILICON TRIOXIDE
MgSiO3
SPECIFIC HEAT, cal g-1 K-1
TEMPERATURE, K
FIG 475

SPECIFICATION TABLE NO. 475 SPECIFIC HEAT OF MAGNESIUM SILICON TRIOXIDE $MgSiO_3$

[For Data Reported in Figure and Table No. 475]

Curve No.	Ref. No.	Year	Temp. Range, K	Reported Error, %	Name and Specimen Designation	Composition (weight percent), Specifications and Remarks
1	392	1943	53-295			92. 0 $MgSiO_3$, 5. 6 Mg_2SiO_4, 2. 4 uncombined silica; corrected for uncombined silica.

DATA TABLE NO. 475 SPECIFIC HEAT OF MAGNESIUM SILICON TRIOXIDE $MgSiO_3$

[Temperature, T, K; Specific Heat, C_p, Cal $g^{-1}K^{-1}$]

T	C_p
CURVE 1	
52.7	1.502×10^{-2}
56.5	1.750
61.1	2.100
65.7	2.484
70.6	2.910
75.5	3.353
80.6	3.827
85.4	4.278
95.1	5.222
104.7	6.140
114.9	7.144
124.5	8.059
135.3	9.075
145.2	9.944
155.4	1.082×10^{-1}
165.5	1.167
176.7	1.251
185.8	1.318
195.9	1.394
206.2	1.461
215.9	1.523
226.2	1.589
236.3	1.645
246.1	1.704
256.4	1.755
266.3	1.803
276.3	1.864
285.9	1.901
295.3	1.938

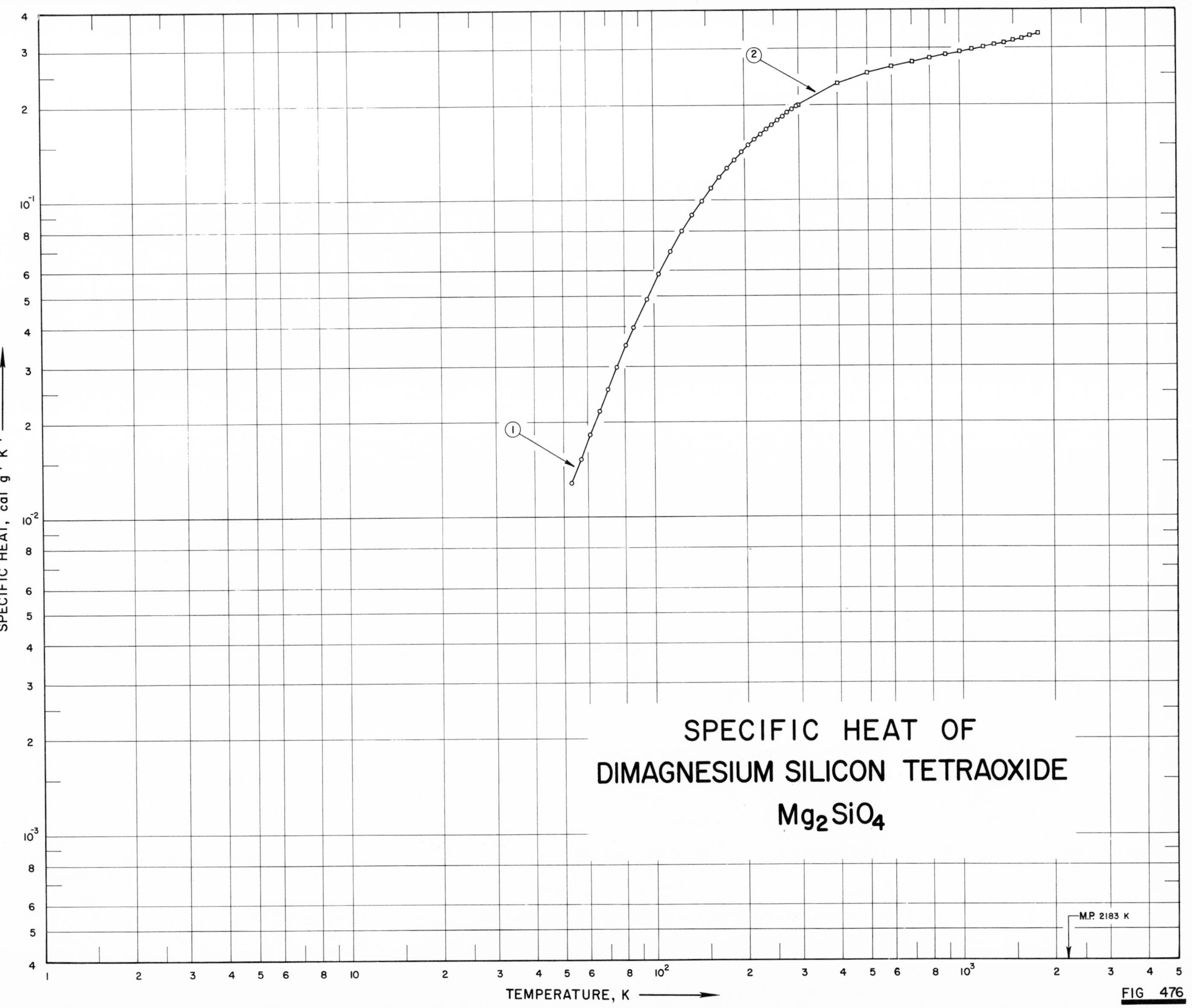
SPECIFIC HEAT OF
DIMAGNESIUM SILICON TETRAOXIDE
Mg_2SiO_4
M.P. 2183 K
1
2
TEMPERATURE, K
SPECIFIC HEAT, cal g^{-1} K^{-1}

FIG 476

SPECIFICATION TABLE NO. 476 SPECIFIC HEAT OF DIMAGNESIUM SILICON TETRAOXIDE Mg_2SiO_4

[For Data Reported in Figure and Table No. 476]

Curve No.	Ref. No.	Year	Temp. Range, K	Reported Error, %	Name and Specimen Designation	Composition (weight percent), Specifications and Remarks
1	392	1943	53.2-295			98.6 Mg_2SiO_4, 0.8 uncombined MgO, no free silica or $MgSiO_3$; corrected for uncombined MgO.
2	388	1953	298-1808			57.51 MgO, 42.60 SiO_2.

DATA TABLE NO. 476 SPECIFIC HEAT OF DIMAGNESIUM SILICON TETRAOXIDE Mg_2SiO_4

[Temperature, T, K; Specific Heat, C_p, Cal $g^{-1}K^{-1}$]

T	C_p	T	C_p
CURVE 1		CURVE 2 (cont.)	
53.2	1.291×10^{-2}	1500	3.21×10^{-1}
57.3	1.544	1600	3.26
61.4	1.846	1700	3.31
65.6	2.182	1800	3.36*
70.0	2.564	1808	3.37
75.0	3.002		
80.6	3.528		
85.5	4.006		
94.7	4.939		
104.3	5.930		
114.4	6.978		
124.8	8.067		
134.7	9.069		
145.3	1.009×10^{-1}		
155.8	1.105		
165.6	1.193		
175.9	1.277		
185.7	1.355		
196.3	1.437		
206.0	1.505		
216.2	1.574		
226.0	1.637		
236.0	1.695		
246.3	1.758		
256.1	1.810		
266.4	1.862		
276.3	1.920		
286.0	1.962		
295.0	1.994		
CURVE 2			
298	2.00×10^{-1}*		
300	2.01		
400	2.35		
500	2.53		
600	2.66		
700	2.75		
800	2.82		
900	2.89		
1000	2.95		
1100	3.01		
1200	3.06		
1300	3.11		
1400	3.16		

* Not shown on plot

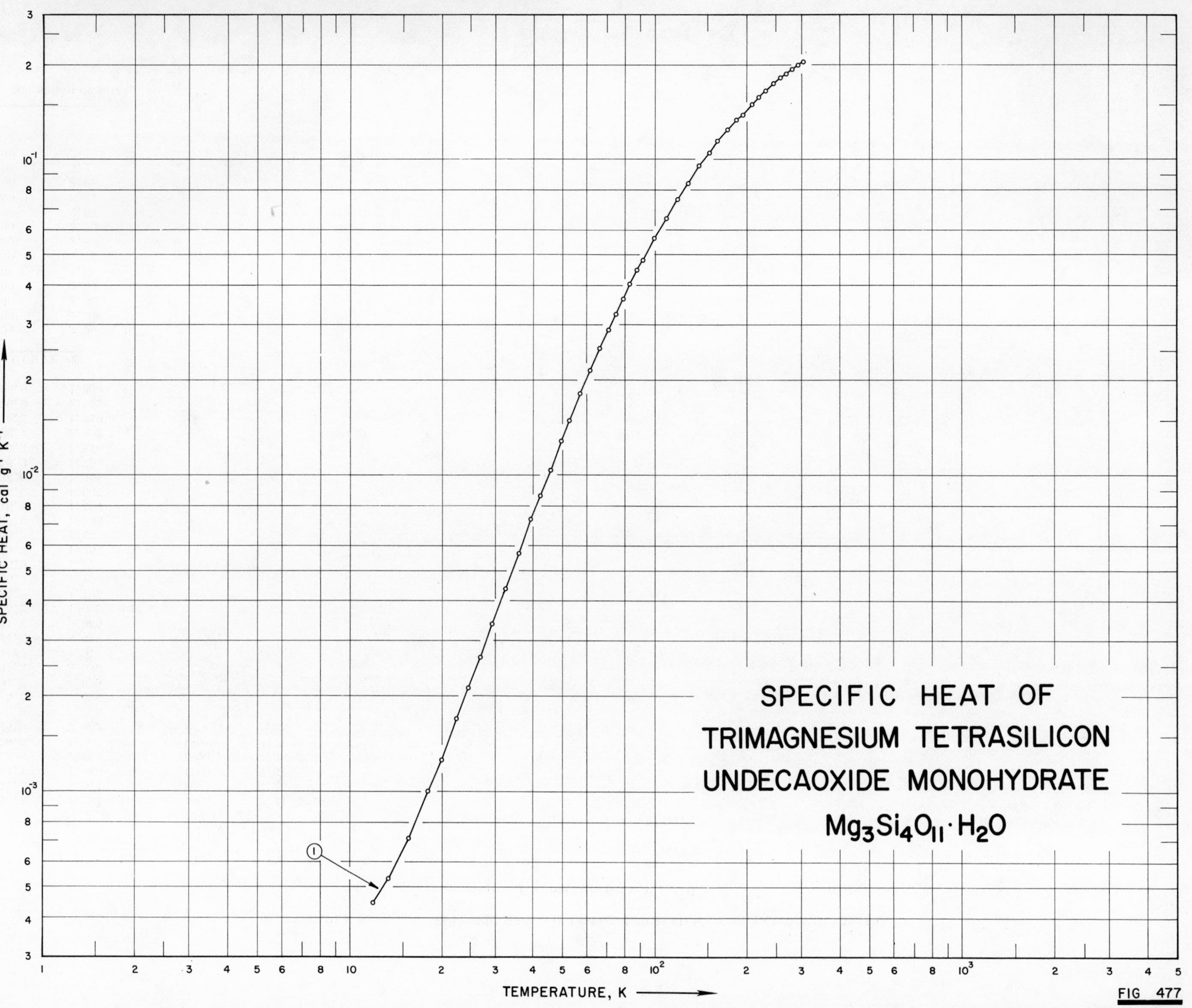

SPECIFIC HEAT OF
TRIMAGNESIUM TETRASILICON
UNDECAOXIDE MONOHYDRATE
$Mg_3Si_4O_{11} \cdot H_2O$
SPECIFIC HEAT, cal g⁻¹ K⁻¹
TEMPERATURE, K
1

FIG 477

SPECIFICATION TABLE NO. 477 SPECIFIC HEAT OF TRIMAGNESIUM TETRASILICON UNDECAOXIDE MONOHYDRATE $Mg_3Si_4O_{11} \cdot H_2O$

[For Data Reported in Figure and Table No. 477]

Curve No.	Ref. No.	Year	Temp. Range, K	Reported Error, %	Name and Specimen Designation	Composition (weight percent), Specifications and Remarks
1	381	1963	53-306	0.3	Talc	62.47 SiO_2, 31.76 MgO, 4.70 H_2O^+, 0.47 Al_2O_3, 0.45 FeO, 0.06 H_2O^-; dried at 115 C for 12 hrs.

DATA TABLE NO. 477 SPECIFIC HEAT OF TRIMAGNESIUM TETRASILICON UNDECAOXIDE MONOHYDRATE $Mg_3Si_4O_{11} \cdot H_2O$

[Temperature, T, K; Specific Heat, C_p, Cal $g^{-1}K^{-1}$]

T	C_p	T	C_p
CURVE 1		**CURVE 1 (cont.)**	
Series I		Series I (cont.)	
52.74	1.502×10^{-2}	288.11	1.982×10^{-1}*
57.48	1.838	294.38	2.010
61.89	2.174	299.39	2.026*
66.37	2.540	299.96	2.033*
70.78	2.911	306.45	2.055
74.92	3.275		
79.12	3.667	Series II	
83.44	4.079		
87.88	4.508	11.85	4.456×10^{-4}
91.53	4.856	13.45	5.318
95.61	5.236*	15.61	7.124
99.95	5.674	18.12	1.005×10^{-3}
104.20	6.085*	20.02	1.272
108.98	6.557	22.45	1.706
113.89	7.042*	24.60	2.141
118.71	7.519	26.79	2.676
123.84	8.026*	29.41	3.404
128.80	8.495	32.43	4.411
134.24	9.022*	35.83	5.726
140.30	9.600	39.47	7.343
146.33	1.015×10^{-1}*	42.34	8.737
151.93	1.065	45.71	1.059×10^{-2}
156.29	1.102*	45.73	1.058*
161.86	1.151	49.71	1.300
168.14	1.205*	53.71	1.567*
174.44	1.258	81.33	3.878*
180.54	1.305*		
186.28	1.350		
190.45	1.382*		
196.15	1.394		
202.23	1.469*		
208.68	1.514		
214.91	1.557*		
219.56	1.587		
225.37	1.625 *		
230.89	1.661		
236.24	1.694*		
238.53	1.710*		
245.27	1.751		
251.56	1.786*		
257.81	1.824		
263.47	1.856*		
269.25	1.887		
274.77	1.917*		
281.54	1.949		

* Not shown on plot

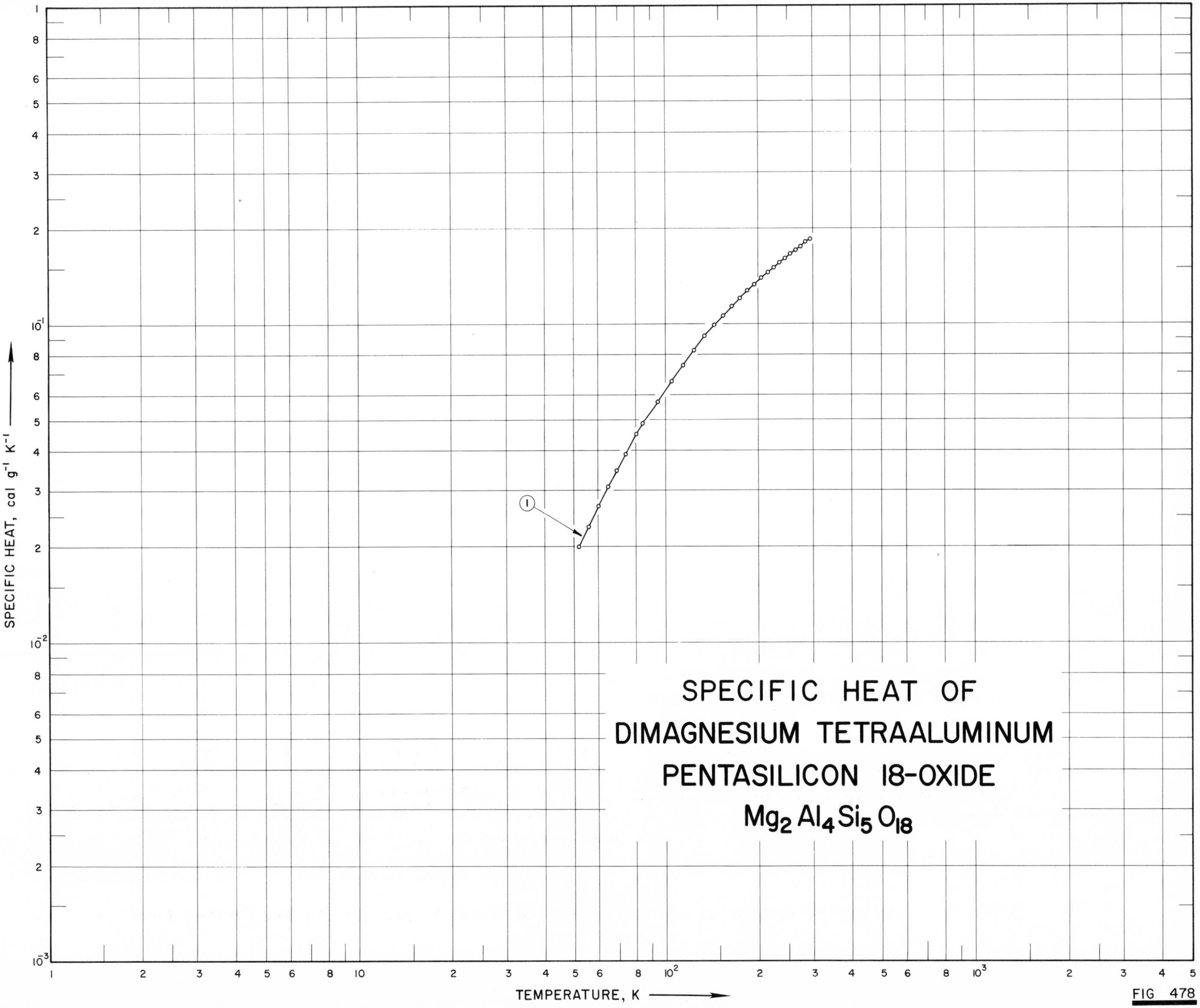

SPECIFIC HEAT OF DIMAGNESIUM TETRAALUMINUM PENTASILICON 18-OXIDE $Mg_2Al_4Si_5O_{18}$

SPECIFIC HEAT, cal g^{-1} K^{-1}

TEMPERATURE, K

FIG 478

SPECIFICATION TABLE NO. 478 SPECIFIC HEAT OF DIMAGNESIUM TETRAALUMINUM PENTASILICON 18-OXIDE $Mg_2Al_4Si_5O_{18}$

[For Data Reported in Figure and Table No. 478]

Curve No.	Ref. No.	Year	Temp. Range, K	Reported Error, %	Name and Specimen Designation	Composition (weight percent), Specifications and Remarks
1	379	1963	52-296	0.3	Cordierite	50.97 SiO_2, 35.45 Al_2O_3, 13.74 MgO.

DATA TABLE NO. 478 SPECIFIC HEAT OF DIMAGNESIUM TETRAALUMINUM PENTASILICON 18-OXIDE $Mg_2Al_4Si_5O_{18}$

[Temperature, T, K; Specific Heat, C_p, Cal $g^{-1}K^{-1}$]

T	C_p
CURVE 1	
51.83	1.997 x 10^{-2}
55.67	2.301
59.93	2.677
64.50	3.078
69.27	3.482
74.06	3.910
80.65	4.506
84.87	4.872
94.58	5.718
105.08	6.636
114.50	7.433
124.46	8.260
135.82	9.173
145.73	9.922
155.99	1.070 x 10^{-1}
165.87	1.139
176.04	1.207
186.05	1.274
195.81	1.333
206.60	1.397
216.16	1.451
225.83	1.506
235.90	1.561
245.71	1.612
256.30	1.663
266.25	1.708
276.34	1.754
286.81	1.802
296.41	1.841

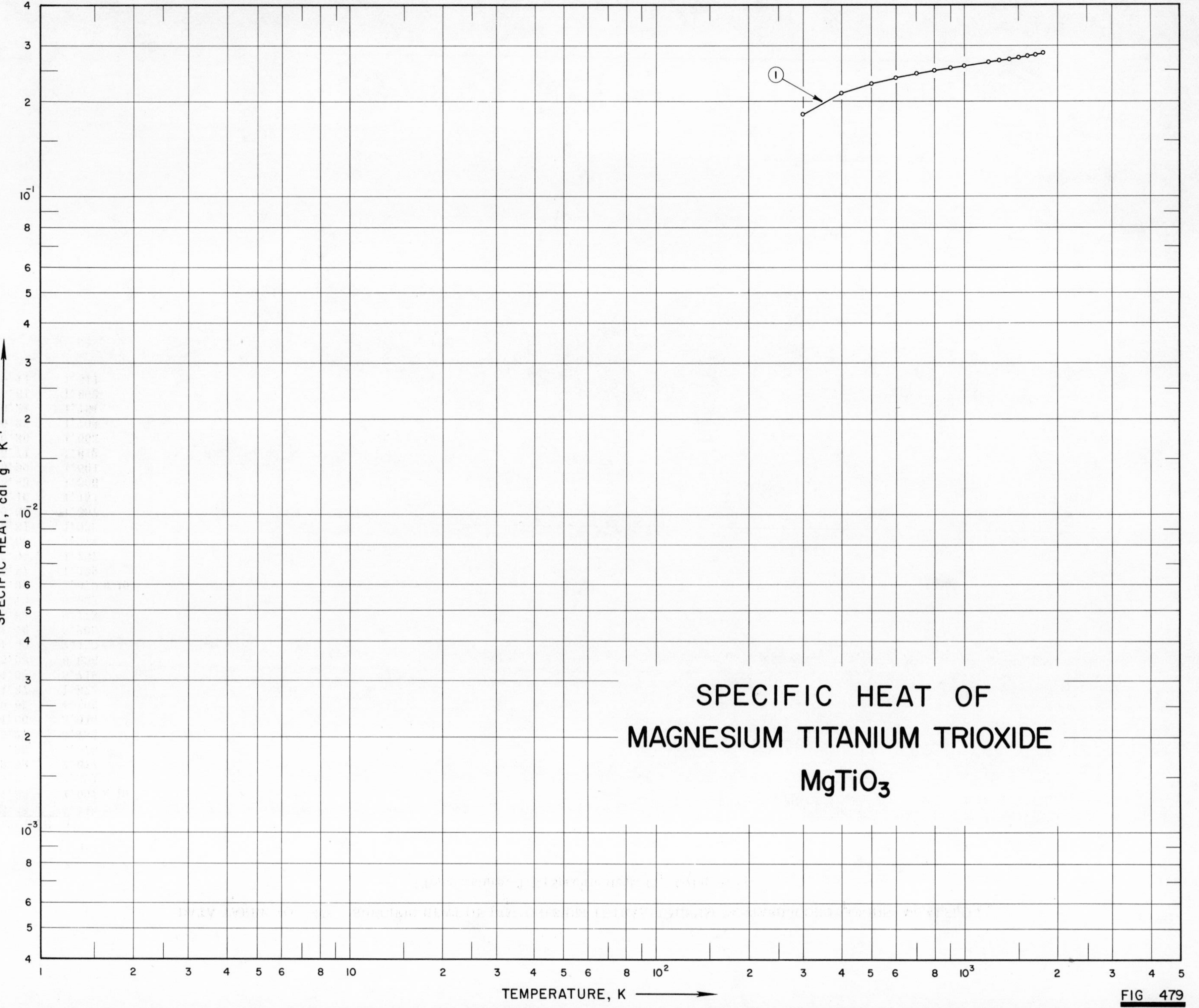
SPECIFIC HEAT OF
MAGNESIUM TITANIUM TRIOXIDE
$MgTiO_3$
1
SPECIFIC HEAT, cal g^{-1} K^{-1}
TEMPERATURE, K

FIG 479

SPECIFICATION TABLE NO. 479 SPECIFIC HEAT OF MAGNESIUM TITANIUM TRIOXIDE $MgTiO_3$

[For Data Reported in Figure and Table No. 479]

Curve No.	Ref. No.	Year	Temp. Range, K	Reported Error, %	Name and Specimen Designation	Composition (weight percent), Specifications and Remarks
1	374	1946	298-1800			99.0 $MgTiO_3$, 0.45 MgO.

DATA TABLE NO. 479 SPECIFIC HEAT OF MAGNESIUM TITANIUM TRIOXIDE $MgTiO_3$

[Temperature, T, K; Specific Heat, C_p, Cal $g^{-1}K^{-1}$]

T	C_p
CURVE 1	
298	1.82×10^{-1}
300	1.83*
400	2.12
500	2.27
600	2.37
700	2.43
800	2.49
900	2.53
1000	2.57
1200	2.64
1300	2.68
1400	2.71
1500	2.74
1600	2.77
1700	2.80
1800	2.83

* Not shown on plot

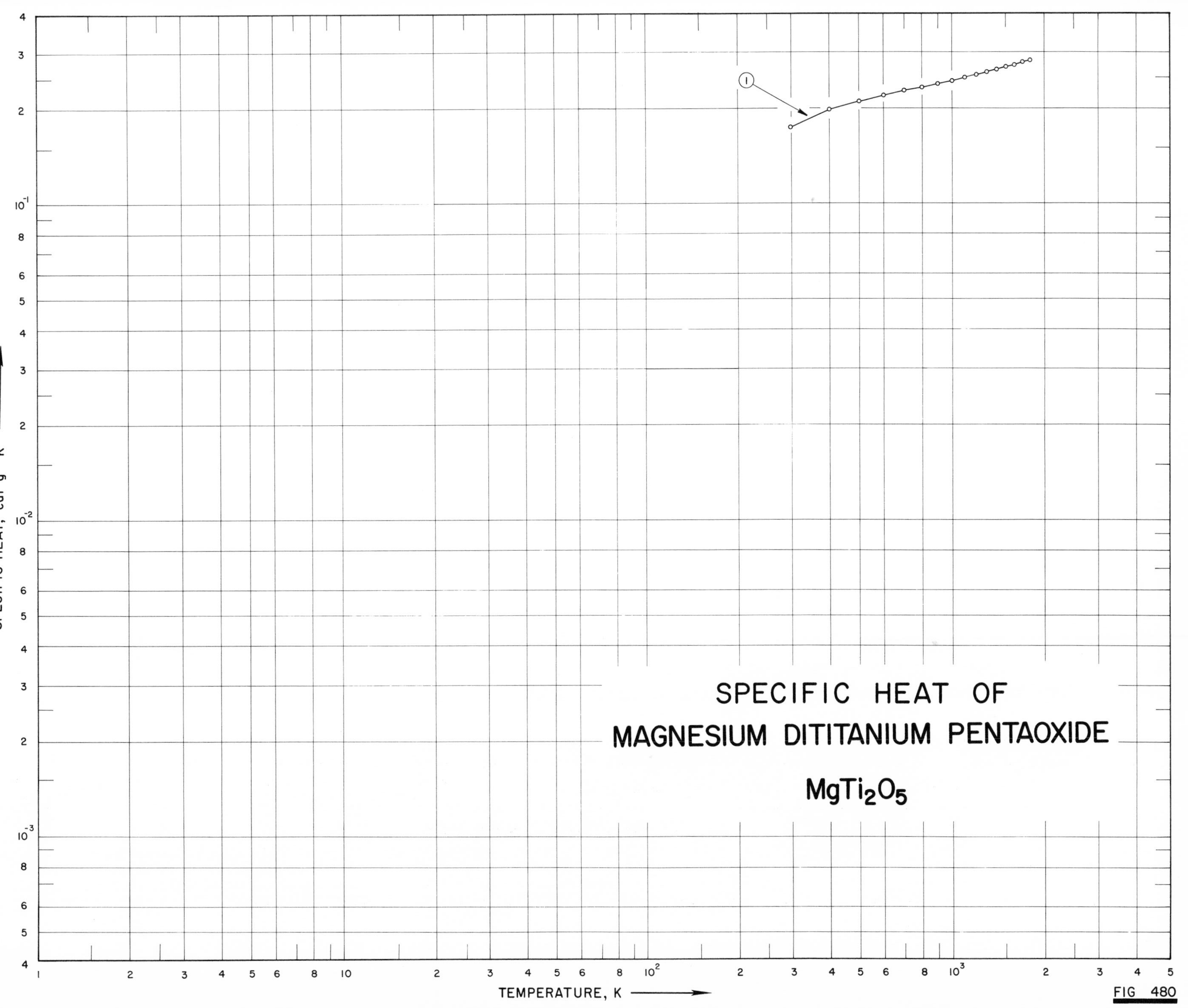
SPECIFIC HEAT OF
MAGNESIUM DITITANIUM PENTAOXIDE
$MgTi_2O_5$
SPECIFIC HEAT, cal g⁻¹ K⁻¹
TEMPERATURE, K
1
FIG 480

SPECIFICATION TABLE NO. 480 SPECIFIC HEAT OF MAGNESIUM DITITANIUM PENTAOXIDE $MgTi_2O_5$

[For Data Reported in Figure and Table No. 480]

Curve No.	Ref. No.	Year	Temp. Range, K	Reported Error, %	Name and Specimen Designation	Composition (weight percent), Specifications and Remarks
1	393	1952	298-1800	0.5 in ΔH		79.63 TiO_2 (79.85 theo.), 0.16 TiO_3; oxides mixed, pressed at 15,000 psi, and heated for long periods at 1300-1500 C.

DATA TABLE NO. 480 SPECIFIC HEAT OF MAGNESIUM DITITANIUM PENTAOXIDE $MgTi_2O_5$

[Temperature, T, K; Specific Heat, C_p, Cal $g^{-1}K^{-1}$]

T	C_p
CURVE 1	
298.15	1.76×10^{-1}
300	1.76*
400	1.99
500	2.12
600	2.21
700	2.28
800	2.34
900	2.40
1000	2.46
1100	2.51
1200	2.56
1300	2.61
1400	2.66
1500	2.71
1600	2.75
1700	2.80
1800	2.85

* Not shown on plot

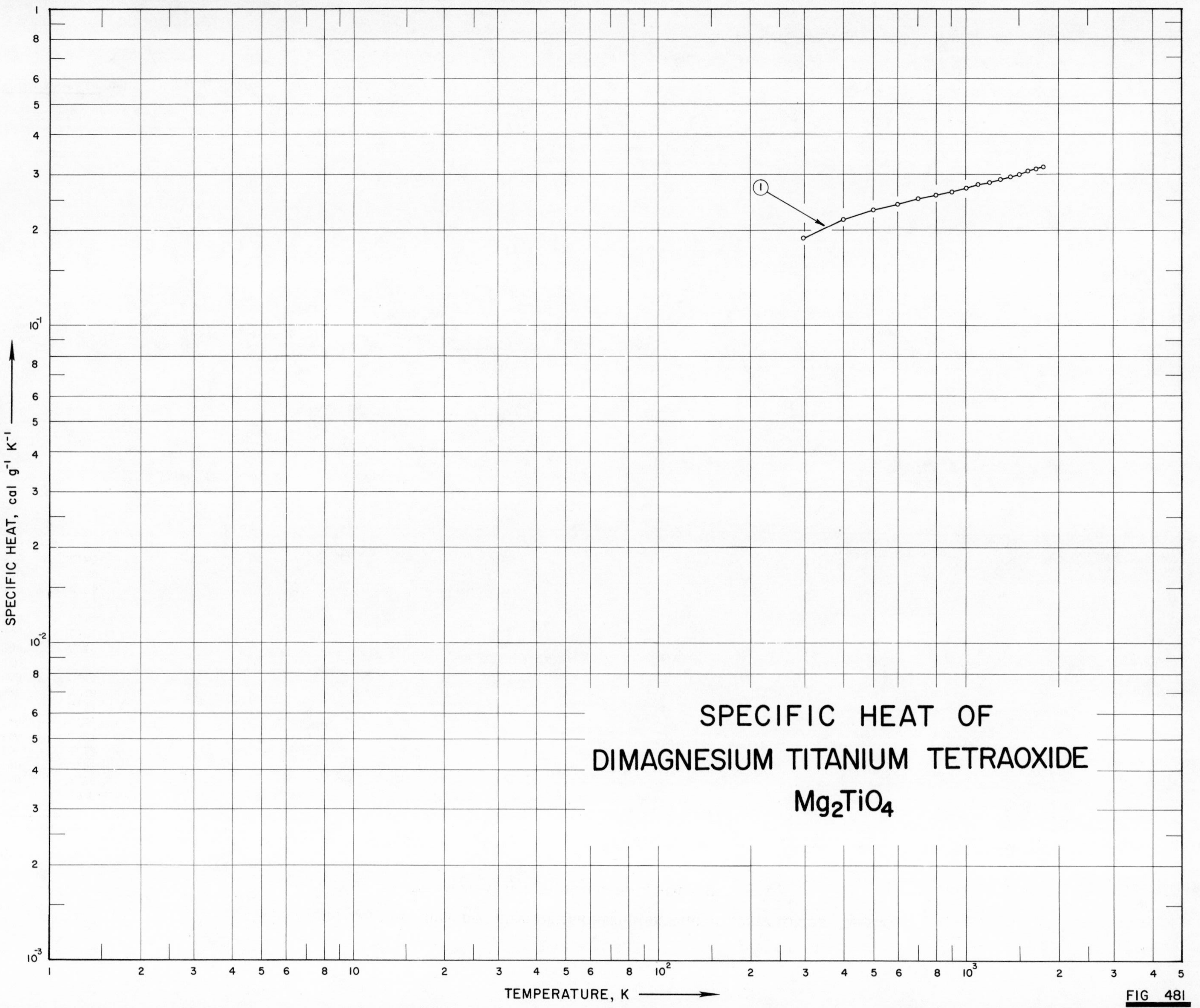
SPECIFIC HEAT OF
DIMAGNESIUM TITANIUM TETRAOXIDE
Mg_2TiO_4
SPECIFIC HEAT, cal g⁻¹ K⁻¹
TEMPERATURE, K
1

FIG 481

SPECIFICATION TABLE NO. 481 SPECIFIC HEAT OF DIMAGNESIUM TITANIUM TETRAOXIDE Mg_2TiO_4

[For Data Reported in Figure and Table No. 481]

Curve No.	Ref. No.	Year	Temp. Range, K	Reported Error, %	Name and Specimen Designation	Composition (weight percent), Specifications and Remarks
1	393	1952	298-1800	0.5 in ΔH		49.53 TiO_2 (49.77 theo.), 0.21 SiO_2; oxides were mixed, pressed at 15,000 psi, and heated for long periods at 1300-1500 C.

DATA TABLE NO. 481 SPECIFIC HEAT OF DIMAGNESIUM TITANIUM TETRAOXIDE Mg_2TiO_4

[Temperature, T, K; Specific Heat, C_p, Cal $g^{-1}K^{-1}$]

T	C_p
CURVE 1	
298	1.92×10^{-1}
300	1.92*
400	2.18
500	2.33
600	2.44
700	2.52
800	2.60
900	2.67
1000	2.73
1100	2.79
1200	2.85
1300	2.91
1400	2.96
1500	3.02
1600	3.07
1700	3.13
1800	3.18

* Not shown on plot

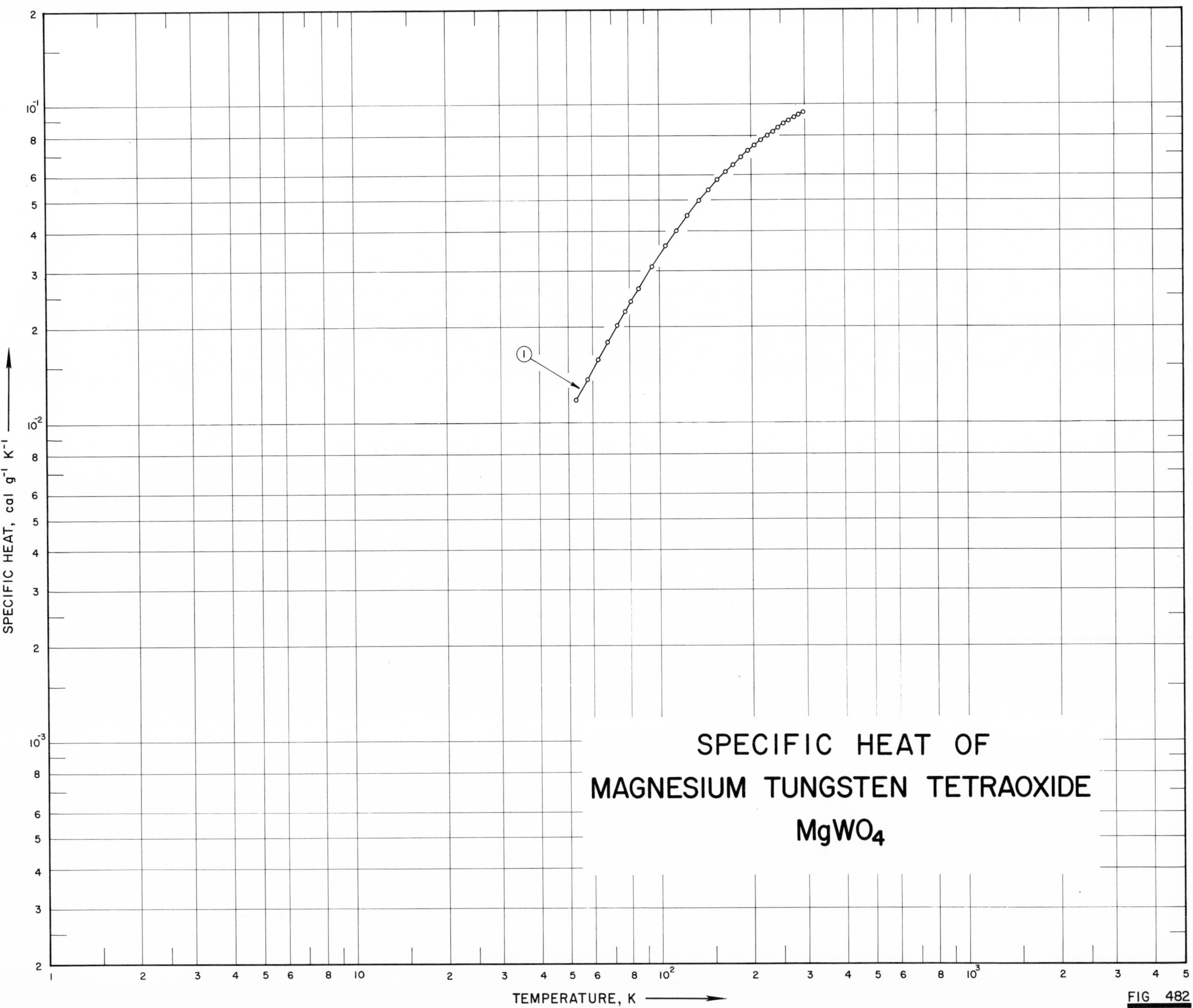
SPECIFIC HEAT OF
MAGNESIUM TUNGSTEN TETRAOXIDE
$MgWO_4$
SPECIFIC HEAT, cal g^{-1} K^{-1}
TEMPERATURE, K
1

FIG 482

SPECIFICATION TABLE NO. 482 SPECIFIC HEAT OF MAGNESIUM TUNGSTEN TETRAOXIDE $MgWO_4$

[For Data Reported in Figure and Table No. 482]

Curve No.	Ref. No.	Year	Temp. Range, K	Reported Error, %	Name and Specimen Designation	Composition (weight percent), Specifications and Remarks
1	383	1961	53-296	0.3		85.24 WO_3, 14.79 MgO; prepared by heating stoichiometric mixture of reagent-grade magnesia and tungstic acid 8 times for total of 5 days at 900 C.

DATA TABLE NO. 482 SPECIFIC HEAT OF MAGNESIUM TUNGSTEN TETRAOXIDE $MgWO_4$

[Temperature, T, K; Specific Heat, C_p, Cal $g^{-1}K^{-1}$]

T	C_p
CURVE 1	
52.92	1.181×10^{-2}
57.77	1.376
62.61	1.584
67.38	1.797
72.41	2.029
77.35	2.260
80.68	2.420
85.67	2.660
94.96	3.114
105.57	3.622
114.75	4.049
124.88	4.512
136.05	5.004
145.86	5.423
155.87	5.812
166.10	6.194
176.15	6.547
185.94	6.874
195.99	7.183
206.12	7.488
216.21	7.774
226.30	8.042
236.08	8.289
245.98	8.509
256.87	8.766
266.48	8.957
276.44	9.163
286.51	9.350
296.01	9.538

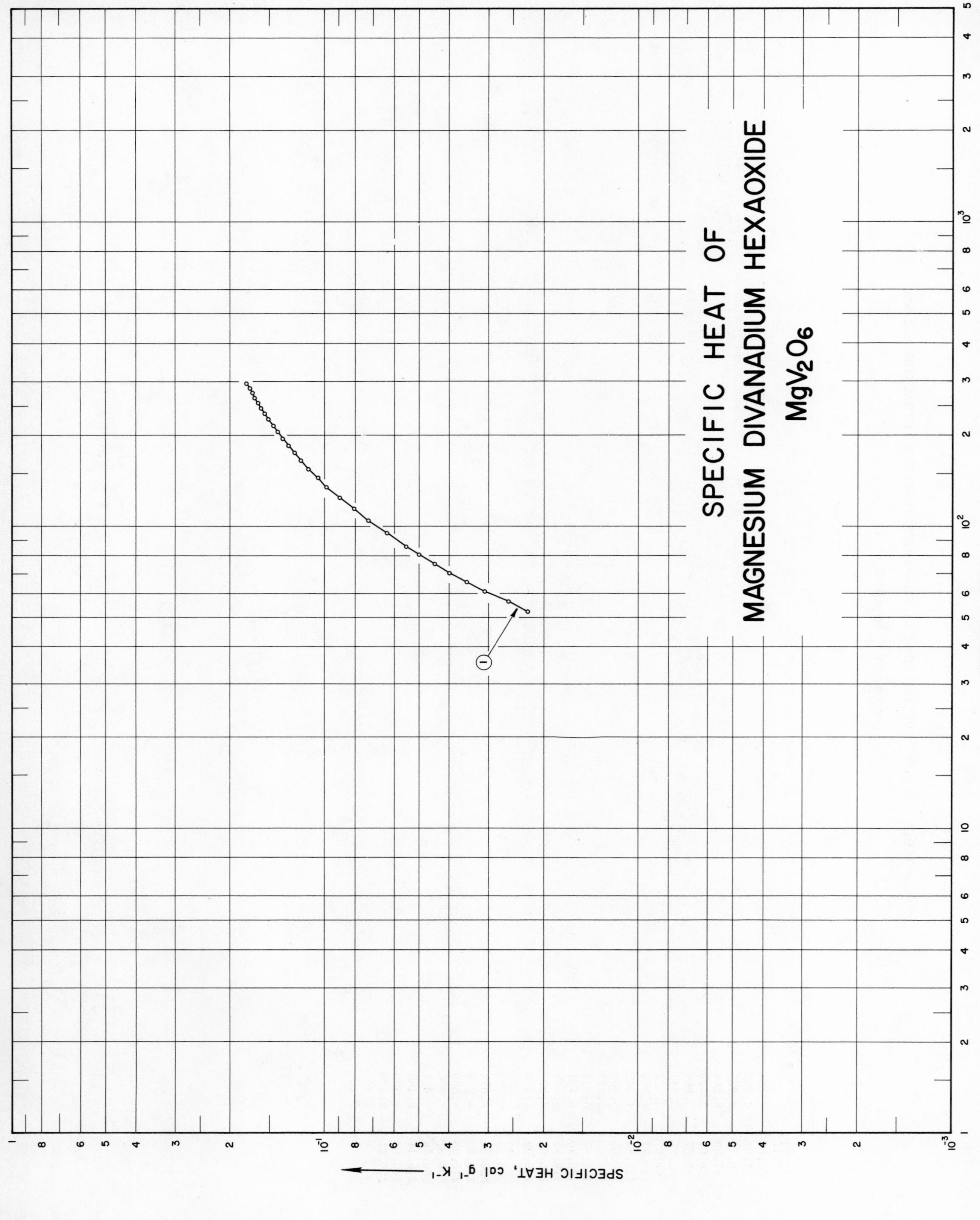

SPECIFIC HEAT OF MAGNESIUM DIVANADIUM HEXAOXIDE MgV_2O_6

FIG 483

SPECIFICATION TABLE NO. 483 SPECIFIC HEAT OF MAGNESIUM DIVANADIUM HEXAOXIDE MgV_2O_6

[For Data Reported in Figure and Table No. 483]

Curve No.	Ref. No.	Year	Temp. Range, K	Reported Error, %	Name and Specimen Designation	Composition (weight percent), Specifications and Remarks
1	394	1962	53-296	0.3		99.9 MgV_2O_6.

DATA TABLE NO. 483 SPECIFIC HEAT OF MAGNESIUM DIVANADIUM HEXAOXIDE MgV_2O_6

[Temperature, T, K; Specific Heat, C_p, Cal $g^{-1}K^{-1}$]

T	C_p
CURVE 1	
52.51	2.261×10^{-2}
56.75	2.596
61.30	3.103
65.79	3.534
70.42	4.016
75.24	4.460
80.92	5.018
86.24	5.500
95.01	6.364
104.97	7.282
114.51	8.070
124.63	8.988
135.89	9.883
145.61	1.059×10^{-1}
155.75	1.129
165.72	1.198
175.84	1.255
186.15	1.316
196.13	1.370
206.21	1.421
216.09	1.468
226.19	1.520
236.63	1.561
246.18	1.599
257.19	1.641
266.70	1.675
276.08	1.705
286.48	1.740
295.95	1.780

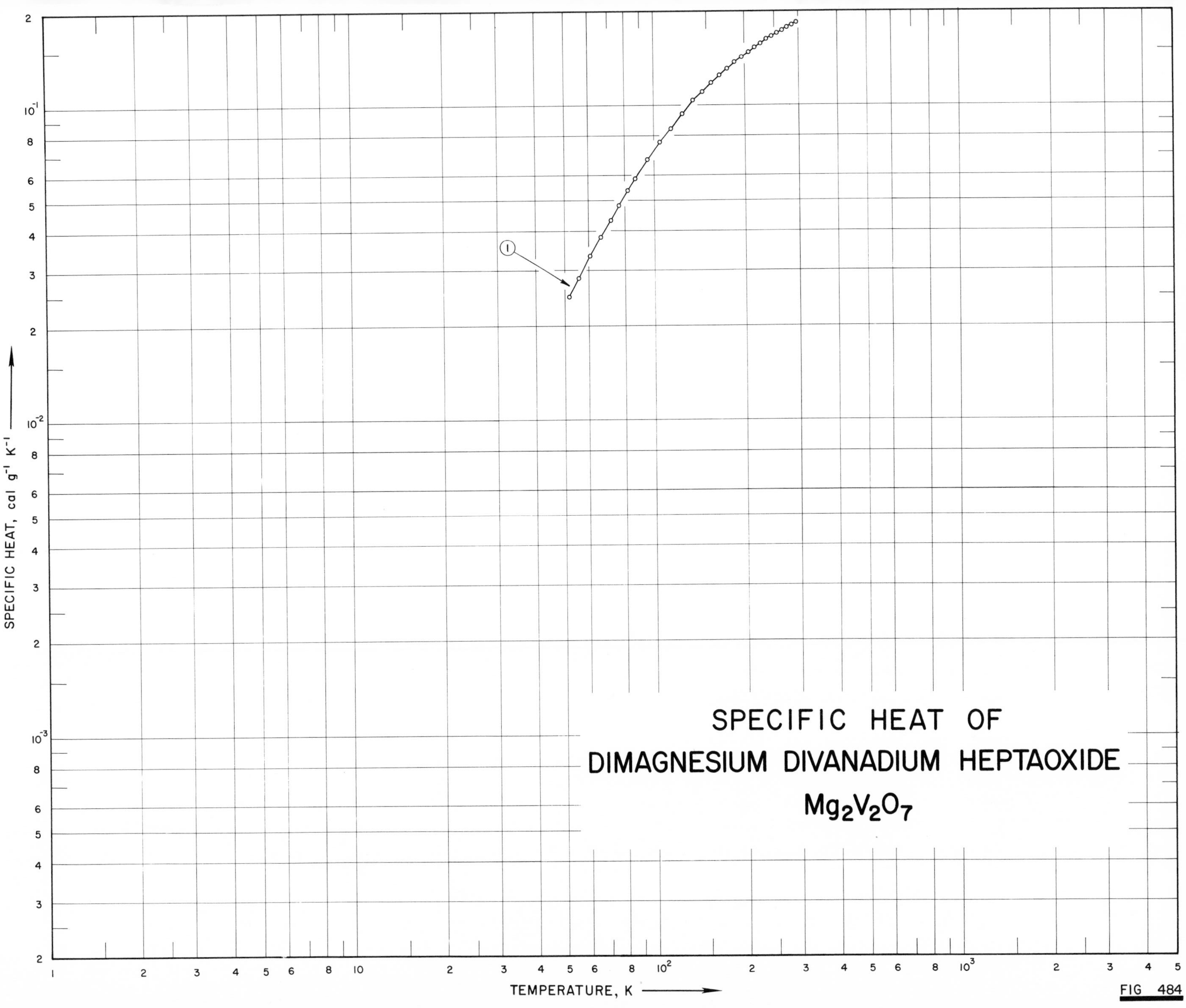

SPECIFIC HEAT OF
DIMAGNESIUM DIVANADIUM HEPTAOXIDE
$Mg_2V_2O_7$
SPECIFIC HEAT, cal g^{-1} K^{-1}
TEMPERATURE, K
1
FIG 484

SPECIFICATION TABLE NO. 484 SPECIFIC HEAT OF DIMAGNESIUM DIVANADIUM HEPTAOXIDE $Mg_2V_2O_7$

[For Data Reported in Figure and Table No. 484]

Curve No.	Ref. No.	Year	Temp. Range, K	Reported Error, %	Name and Specimen Designation	Composition (weight percent), Specifications and Remarks
1	394	1962	52-296	0.3		99.9 $Mg_2V_2O_7$.

DATA TABLE NO. 484 SPECIFIC HEAT OF DIMAGNESIUM DIVANDIUM HEPTAOXIDE $Mg_2V_2O_7$

[Temperature, T, K; Specific Heat, C_p, Cal $g^{-1}K^{-1}$]

T	C_p
CURVE 1	
51.90	2.489×10^{-2}
55.83	2.851
60.84	3.356
66.00	3.855
71.21	4.354
76.24	4.865
81.62	5.406
86.52	5.882
94.87	6.777
105.08	7.672
114.77	8.476
124.58	9.448
136.61	1.046×10^{-1}
145.82	1.116
156.06	1.189
165.86	1.256
176.16	1.320
185.98	1.378
195.94	1.433
206.33	1.486
216.83	1.540
226.15	1.587
236.32	1.629
245.96	1.669
256.84	1.710
266.87	1.746
276.71	1.781
281.86	1.816
296.32	1.848

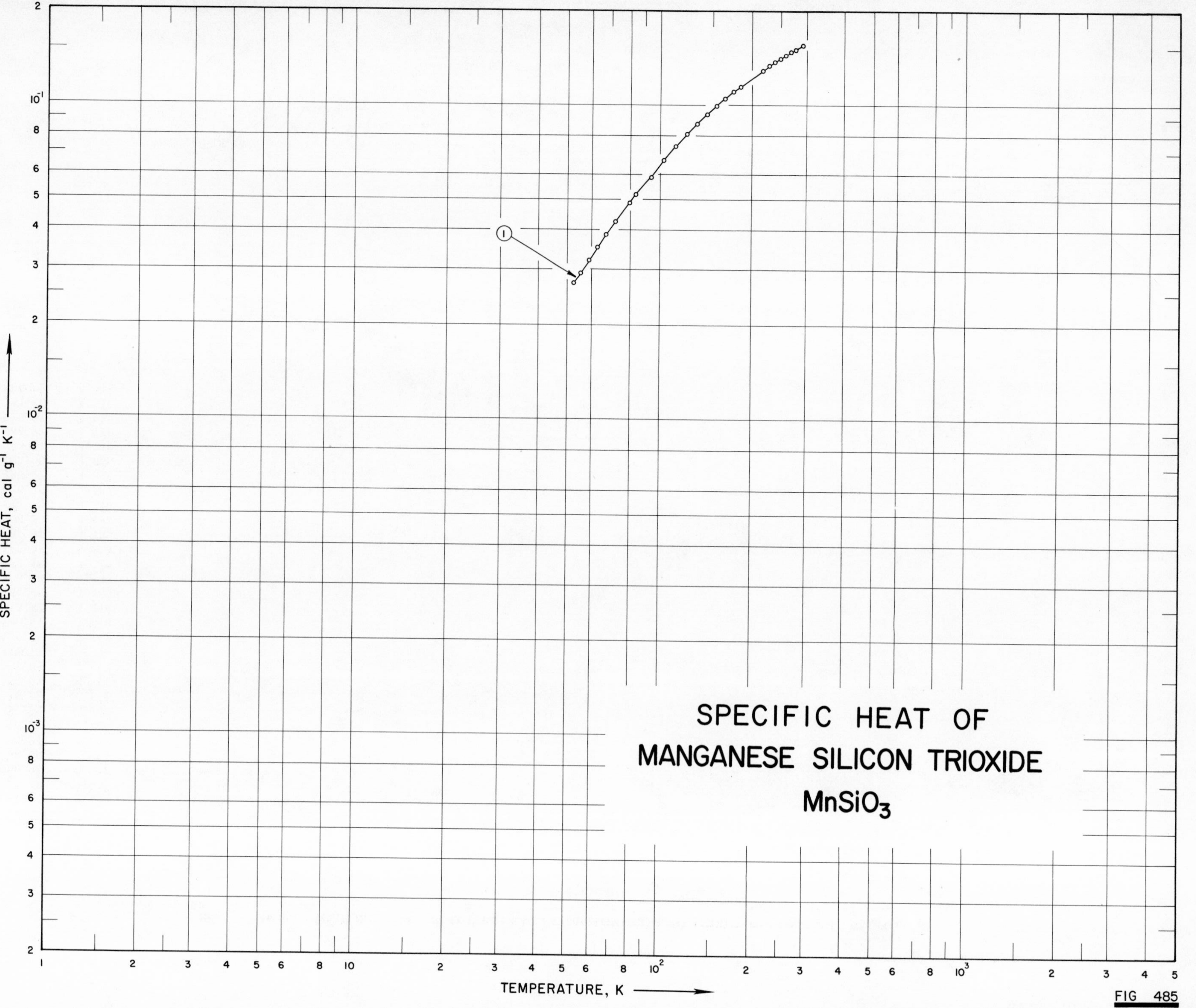
SPECIFIC HEAT OF
MANGANESE SILICON TRIOXIDE
MnSiO3
SPECIFIC HEAT, cal g-1 K-1
TEMPERATURE, K
1
FIG 485

SPECIFICATION TABLE NO. 485 SPECIFIC HEAT OF MANGANESE SILICON TRIOXIDE $MnSiO_3$

[For Data Reported in Figure and Table No. 485]

Curve No.	Ref. No.	Year	Temp. Range, K	Reported Error, %	Name and Specimen Designation	Composition (weight percent), Specifications and Remarks
1	387	1941	53-295			54.18 MnO (54.15 theo.); density = 230 lb ft^{-3}.

DATA TABLE NO. 485 SPECIFIC HEAT OF MANGANESE SILICON TRIOXIDE $MnSiO_3$

[Temperature, T, K; Specific Heat, C_p, Cal $g^{-1}K^{-1}$]

T	C_p
CURVE 1	
52.60	2.741 x 10^{-2}
55.50	2.937
59.20	3.227
63.10	3.545
67.40	3.897
72.20	4.294
80.30	4.926
84.30	5.235
93.60	5.930
103.30	6.683
113.60	7.427
123.70	8.107
134.20	8.787
144.60	9.428
154.90	9.978
164.80	1.059 x 10^{-1}
175.20	1.111
185.10	1.158
218.10	1.299
228.30	1.344
238.30	1.378
248.10	1.418
257.90	1.448
268.20	1.486
278.10	1.514
286.80	1.545*
294.80	1.567

* Not shown on plot

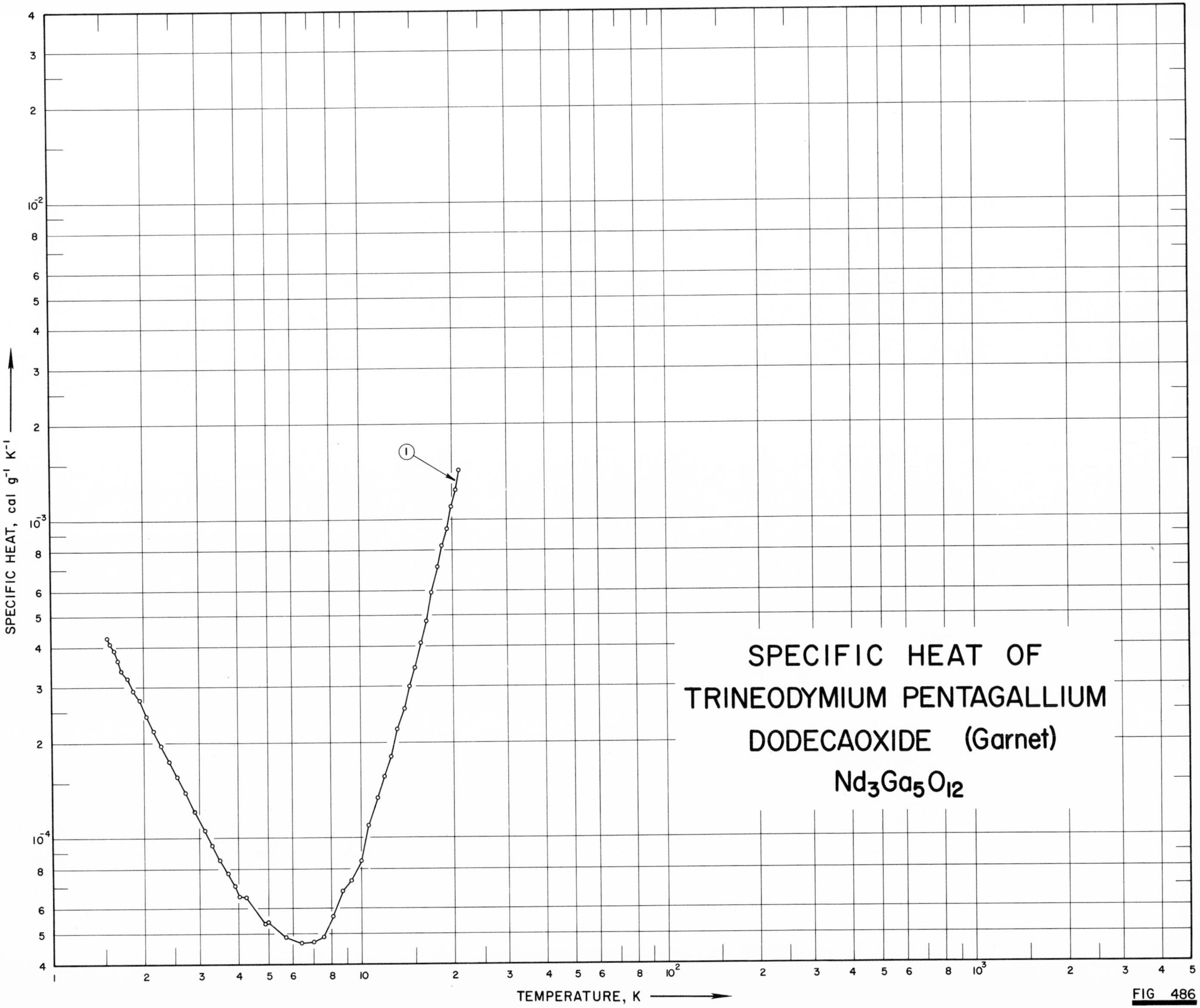

SPECIFIC HEAT OF
TRINEODYMIUM PENTAGALLIUM
DODECAOXIDE (Garnet)
$Nd_3Ga_5O_{12}$
SPECIFIC HEAT, cal g^{-1} K^{-1}
TEMPERATURE, K
1

FIG 486

SPECIFICATION TABLE NO. 486 SPECIFIC HEAT OF TRINEODYMIUM PENTAGALLIUM DODECAOXIDE, $Nd_3Ga_5O_{12}$ (Garnet)

[For Data Reported in Figure and Table No. 486]

Curve No.	Ref. No.	Year	Temp. Range, K	Reported Error, %	Name and Specimen Designation	Composition (weight percent), Specifications and Remarks
1	449	1963	1.5-21		Garnet	99.99 Nd_2O_3; supplied by the Lindsay Chem. or Johnson, Matthey and Co.; 99.99 Ga_2O_3; supplied by Johnson, Matthey and Co.; direct solid state reaction of an intimate mixture of pure rare earth and gallium oxide; sintered block; polycrystalline.

DATA TABLE NO. 486 SPECIFIC HEAT OF TRINEODYMIUM PENTAGALLIUM DODECAOXIDE, $Nd_3Ga_5O_{12}$ (Garnet)

[Temperature, T, K; Specific Heat, C_p, Cal $g^{-1}K^{-1}$]

T	C_p	T	C_p
CURVE 1		CURVE 1 (cont.)	
1.535	4.295×10^{-4}	19.29	9.350×10^{-4}
1.569	4.123	19.90	1.104×10^{-3}
1.612	3.902	20.56	1.247
1.655	3.657	21.14	1.433
1.699	3.387		
1.780	3.190		
1.853	2.920		
1.946	2.724		
2.037	2.425		
2.158	2.189		
2.273	1.963		
2.405	1.757		
2.556	1.571		
2.721	1.391		
2.905	1.215		
3.127	1.065		
3.313	9.473×10^{-5}		
3.501	8.540		
3.707	7.755		
3.903	7.068		
4.045	6.552		
4.25	6.503		
4.90	5.374		
5.00	5.424		
5.72	4.884		
6.42	4.687		
7.01	4.712		
7.57	4.884		
8.13	5.669		
8.70	6.798		
9.34	7.362		
10.01	8.491		
10.68	1.099×10^{-4}		
11.34	1.337		
11.99	1.563		
12.61	1.804		
13.22	2.201		
13.87	2.552		
14.48	3.019		
15.09	3.436		
15.74	4.123		
16.49	4.835		
17.16	5.914		
17.92	7.117		
18.55	8.270		

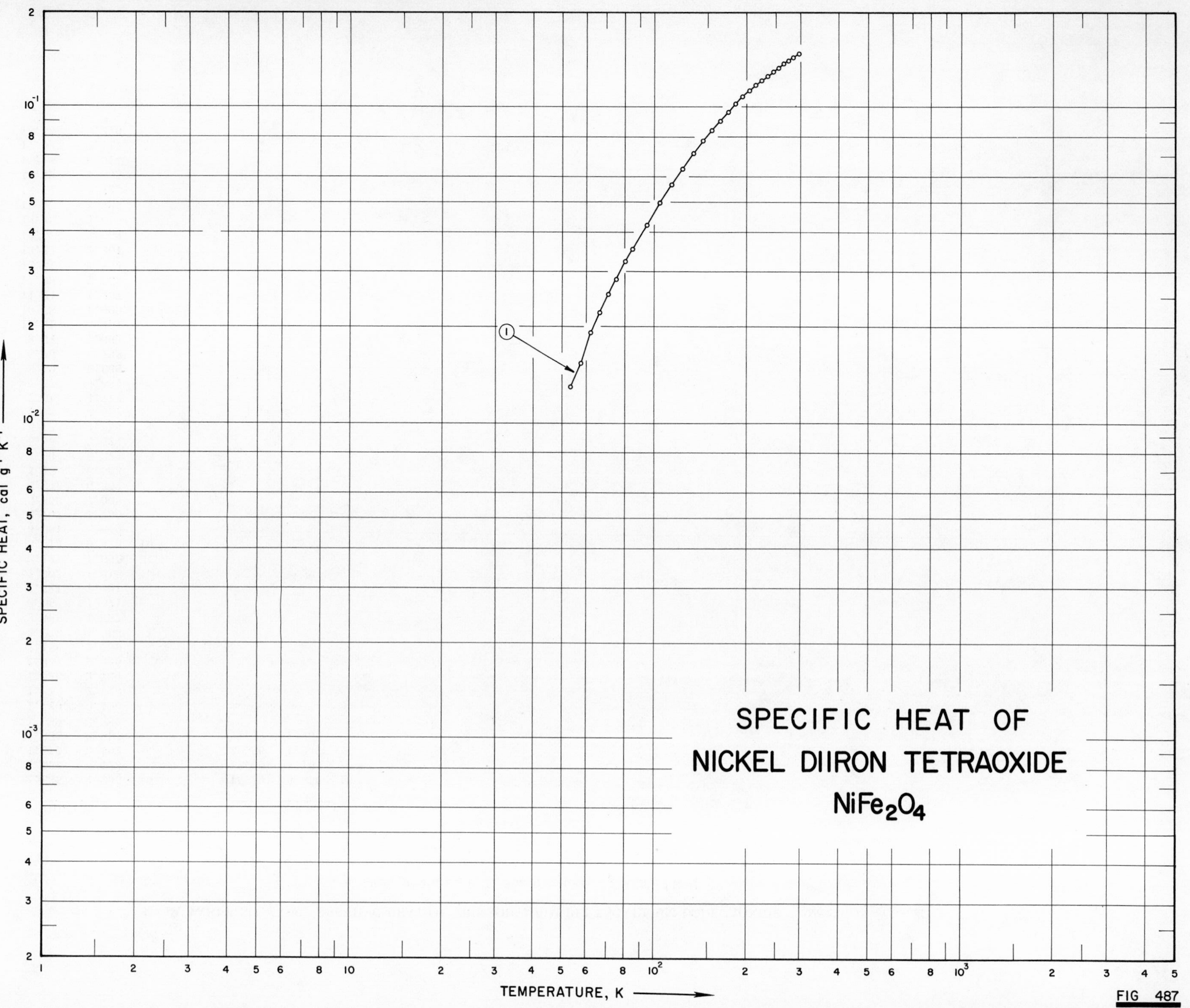
SPECIFIC HEAT OF
NICKEL DIIRON TETRAOXIDE
NiFe2O4
SPECIFIC HEAT, cal g-1 K-1
TEMPERATURE, K
1
FIG 487

SPECIFICATION TABLE NO. 487 SPECIFIC HEAT OF NICKEL DIIRON TETRAOXIDE $NiFe_2O_4$

[For Data Reported in Figure and Table No. 487]

Curve No.	Ref. No.	Year	Temp. Range, K	Reported Error, %	Name and Specimen Designation	Composition (weight percent), Specifications and Remarks
1	382	1956	54-298		Nickel-iron spinel	68.11 Fe_2O_3, 31.86 NiO (68.13, 31.87 theo.), 0.03 SiO_2, 27.22 O (27.33 theo.); x-ray diffraction agreed with ASTM, no impurity line detected; heated to 990-1270 C for prolonged periods with grinding and mixing in between heatings.

DATA TABLE NO. 487 SPECIFIC HEAT OF NICKEL DIIRON TETRAOXIDE $NiFe_2O_4$

[Temperature, T, K; Specific Heat, C_p, Cal $g^{-1}K^{-1}$]

T	C_p
CURVE 1	
53.58	1.403×10^{-2}
57.89	1.658
62.32	1.942
66.85	2.256
71.25	2.568
75.53	2.873
80.85	3.256
85.44	3.577
94.75	4.256
104.95	5.009
114.74	5.717
124.57	6.421
135.63	7.189
145.44	7.863
155.78	8.537
165.70	9.139
175.81	9.732
185.61	1.026×10^{-1}
195.67	1.080
206.03	1.132
215.96	1.179
225.85	1.222
236.38	1.268
245.78	1.306
256.20	1.345
266.39	1.381
276.26	1.414
286.48	1.449
295.93	1.479*
298.16	1.485

* Not shown on plot

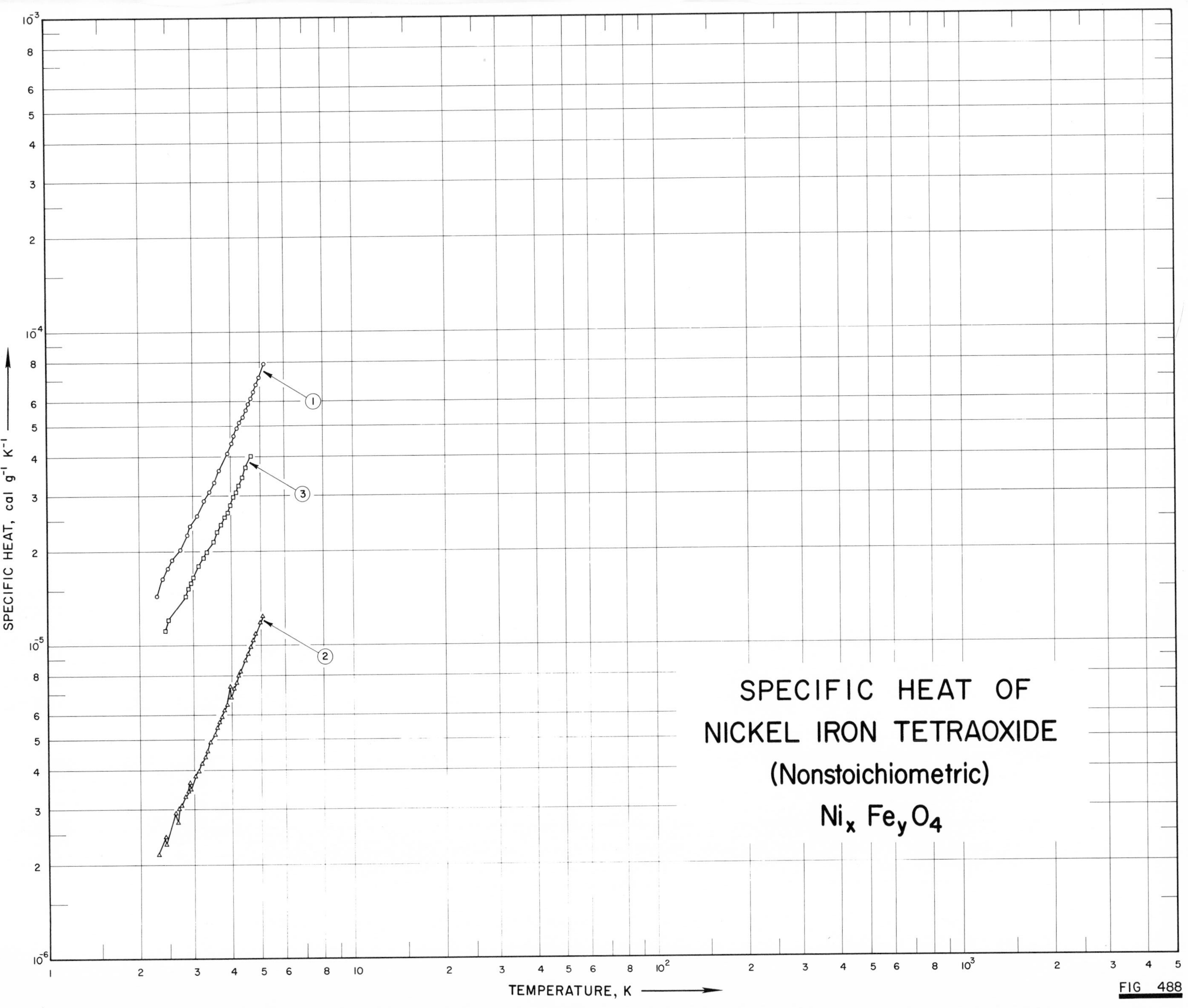
SPECIFIC HEAT OF
NICKEL IRON TETRAOXIDE
(Nonstoichiometric)
Nix Fey O4
TEMPERATURE, K
SPECIFIC HEAT, cal g-1 K-1
1
2
3
FIG 488

SPECIFICATION TABLE NO. 488 SPECIFIC HEAT OF NICKEL IRON TETRAOXIDE (nonstoichiometric) $Ni_xFe_yO_4$

[For Data Reported in Figure and Table No. 488]

Curve No.	Ref. No.	Year	Temp. Range, K	Reported Error, %	Name and Specimen Designation	Composition (weight percent), Specifications and Remarks
1	383	1961	2.3-5.2		$Ni_{0.88}Fe_{2.12}O_4$	50.14 Fe, 22.02 Ni; x-ray lattice parameter = 8.343 Å.
2	383	1961	2.3-5.1		$Ni_{0.94}Fe_{2.06}O_4$	48.80 Fe, 23.53 Ni; x-ray lattice parameter = 8.339 Å.
3	383	1961	2.4-4.7		$Ni_{0.58}Fe_{2.42}O_4$	60.14 Fe, 12.34 Ni; x-ray lattice parameter = 8.362 Å.

DATA TABLE NO. 488 SPECIFIC HEAT OF NICKEL IRON TETRAOXIDE $Ni_xFe_yO_4$ (nonstoichiometric)

[Temperature, T, K; Specific Heat, C_p, Cal g^{-1} K^{-1}]

T	C_p
CURVE 1	
2.294	2.339 x 10^{-6}
2.387	2.663
2.488	6.863
2.562	3.065
2.646	3.075*
2.723	3.300
2.804	3.390*
2.877	3.654
2.946	3.926
3.020	3.994*
3.104	4.236
3.182	4.456*
3.263	4.741
3.337	4.901*
3.409	5.041
3.476	5.249*
3.538	5.393
3.610	5.705*
3.666	5.886
3.794	6.182*
3.909	6.670
4.030	7.187
4.109	7.579
4.201	8.028
4.298	8.368
4.409	8.708
4.505	9.170
4.590	9.584
4.681	1.001 x 10^{-5}
4.777	1.052
4.871	1.105
4.975	1.167
5.073	1.201*
5.160	1.283
CURVE 2	
Series 1	
2.296	2.165 x 10^{-6}
2.415	2.452
2.422	2.339
2.630	2.892*
2.725	3.083
2.801	3.304
2.882	3.437
CURVE 2 (cont.)	
Series 1 (cont.)	
2.949	3.580 x 10^{-6}*
3.020	3.836
3.100	3.993
3.178	4.232
3.260	4.413
3.320	4.621
3.393	4.941
3.455	5.047*
3.525	5.204
3.533	5.350*
3.582	5.480*
3.643	5.715
3.654	5.739*
3.713	5.903
3.772	6.090*
3.776	6.233
3.836	6.448*
3.869	6.486
3.890	6.724
3.954	7.410
3.988	6.963*
4.088	7.321
4.150	7.632
4.188	7.645*
4.283	8.276
4.344	8.429
4.438	8.978
4.530	9.418
4.623	9.903
4.714	1.042 x 10^{-5}
4.803	1.090
4.888	1.136*
4.973	1.184
5.056	1.243
Series 2	
2.602	2.929 x 10^{-6}
2.651	2.735
2.671	3.025
2.705	3.124*
2.758	3.202*
2.793	3.154*
2.802	3.328*
CURVE 2 (cont.)	
Series 2 (cont.)	
2.852	3.325 x 10^{-6}*
2.913	3.669
2.932	3.471
2.943	3.563*
3.014	3.826*
3.035	3.795*
3.064	3.812*
3.114	4.007*
3.123	4.126*
3.177	4.126*
3.191	4.259*
3.233	4.474
3.297	4.443*
3.323	4.743*
3.423	4.771*
3.424	4.955*
3.530	4.989*
3.588	5.473
3.647	5.551*
3.715	5.848*
3.759	5.937*
3.852	6.380*
3.879	6.366*
3.970	7.062*
3.979	6.837
4.088	7.175*
4.105	7.512*
4.214	8.044
CURVE 3	
2.417	4.404 x 10^{-6}
2.481	4.785
2.840	5.702
2.897	6.030
2.954	6.232
3.005	6.512
3.060	6.700*
3.133	7.072
3.194	7.255*
3.259	7.535
3.333	7.848
3.405	8.133*
3.510	8.475
CURVE 3 (cont.)	
3.615	9.093 x 10^{-6}
3.712	9.570
3.817	1.015 x 10^{-5}
3.902	1.049
3.993	1.108
4.080	1.182
4.166	1.220
4.251	1.283
4.337	1.360
4.412	1.383
4.492	1.456
4.573	1.497
4.650	1.587

* Not shown on plot

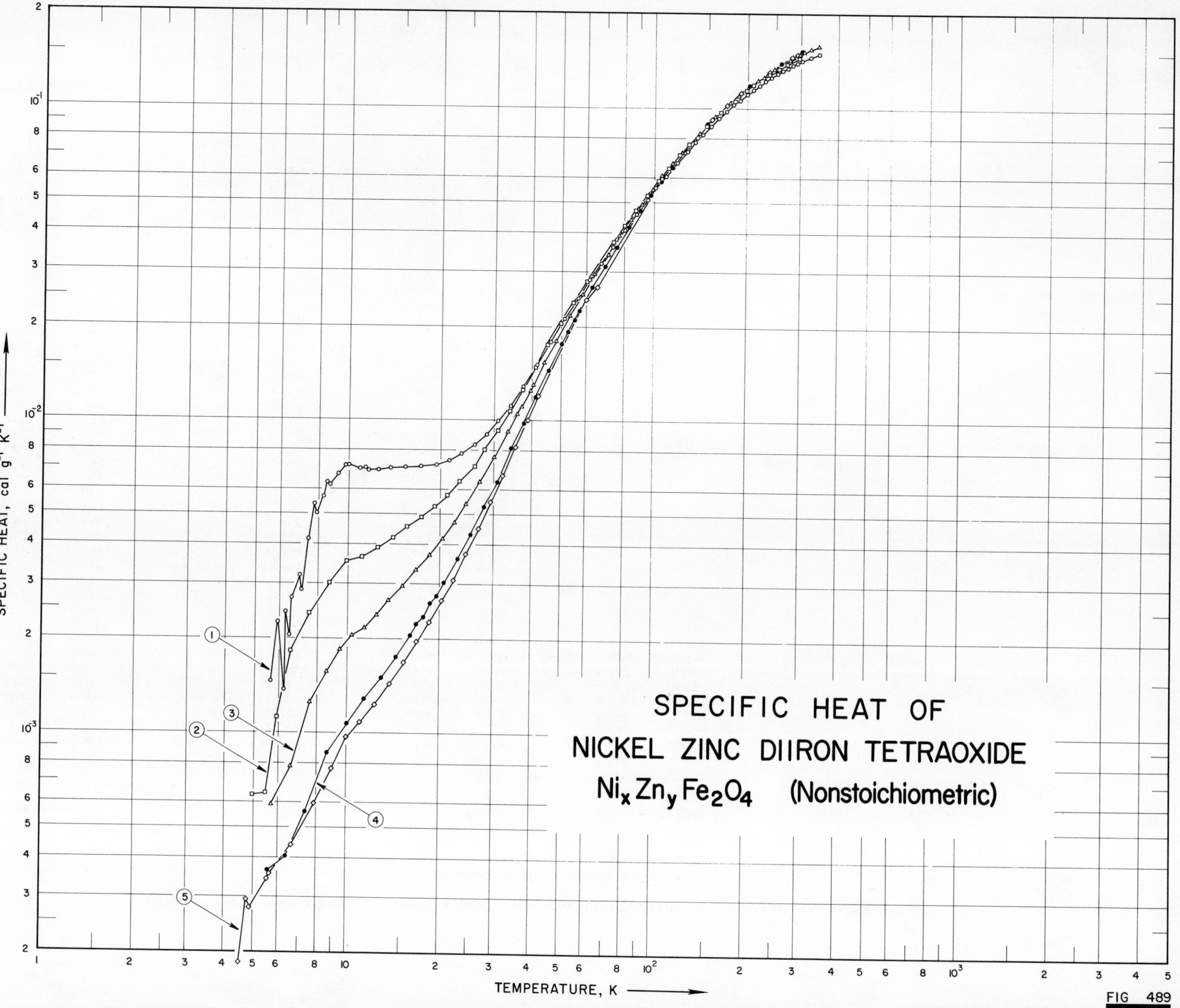
SPECIFIC HEAT OF
NICKEL ZINC DIIRON TETRAOXIDE
$Ni_x Zn_y Fe_2O_4$ (Nonstoichiometric)
SPECIFIC HEAT, cal g⁻¹ K⁻¹
TEMPERATURE, K
1
2
3
4
5

FIG 489

SPECIFICATION TABLE NO. 489 SPECIFIC HEAT OF NICKEL ZINC DIIRON TETRAOXIDE (nonstoichiometric) $Ni_xZn_yFe_2O_4$

[For Data Reported in Figure and Table No. 489]

Curve No.	Ref. No.	Year	Temp. Range, K	Reported Error, %	Name and Specimen Designation	Composition (weight percent), Specifications and Remarks
1	395	1957	6-345		$Ni_{0.1}Zn_{0.9}Fe_2O_4$	Very pure and carefully prepared; annealed at 1200 C.
2	395	1957	5-298		$Ni_{0.2}Zn_{0.8}Fe_2O_4$	Same as above.
3	395	1957	6-346		$Ni_{0.3}Zn_{0.7}Fe_2O_4$	Same as above.
4	395	1957	6-304		$Ni_{0.4}Zn_{0.6}Fe_2O_4$	Commercial grade material.
5	395	1957	5-300		$Ni_{0.4}Zn_{0.6}Fe_2O_4$	46.8 ±0.1 Fe (46.84 theo.); very pure and carefully prepared; annealed at 900 C.

DATA TABLE NO. 489 SPECIFIC HEAT OF NICKEL ZINC DIIRON TETRAOXIDE $Ni_xZn_yFe_2O_4$ (nonstoichiometric)

[Temperature, T, K; Specific Heat, C_p, Cal g^{-1} K^{-1}]

T	C_p
CURVE 1	
Series 1	
37.40	1.294×10^{-2}
41.55	1.518
46.25	1.797
51.44	2.125
56.87	2.478
61.98	2.828
67.43	3.197
73.39	3.603
80.18	4.080
87.50	4.596
95.15	5.115
Series 2	
80.54	4.104×10^{-2}*
87.37	4.585*
94.52	5.071*
101.80	5.562
109.70	6.092
118.99	6.706
128.10	7.298
136.30	7.807
144.47	8.298
153.07	8.797
162.33	9.305
172.04	9.808
181.55	1.027×10^{-1}
190.90	1.070
200.43	1.111
210.27	1.151
220.43	1.189
230.85	1.226
241.44	1.261
252.14	1.294
262.75	1.324
273.26	1.351
Series 3	
6.22	1.373×10^{-3}
6.48	2.055
7.08	2.862
7.76	5.366
8.54	6.364
CURVE 1 (cont.)	
Series 3 (cont.)	
9.29	6.738×10^{-3}
10.00	7.189
11.32	7.016
Series 4	
5.66	1.468×10^{-3}
5.93	2.259
6.27	2.429
6.59	2.691
6.98	3.178
7.43	4.160
7.89	5.033
8.28	5.699
8.73	6.239
9.23	6.780*
9.75	7.154
10.33	7.206*
10.90	6.975
11.61	6.926
12.56	6.932
13.71	7.004
15.33	7.055
17.28	7.114
19.34	7.227
21.37	7.455
23.49	7.812
25.88	8.352
28.31	9.047
30.90	9.983
33.91	1.123
Series 5	
283.77	1.377×10^{-1}
294.16	1.401
304.19	1.423
314.78	1.445*
324.97	1.462
335.16	1.480*
345.44	1.497
CURVE 2	
4.95	6.340×10^{-4}
5.45	6.424
5.94	1.122×10^{-3}
6.55	1.827
7.49	2.411
8.70	3.007
9.91	3.529
11.17	3.659
12.50	3.915
14.02	4.213
15.66	4.551
17.35	4.897
19.15	5.281
21.05	5.769
23.15	6.378
25.46	7.112
27.98	8.059
30.76	9.252
33.87	1.077×10^{-2}
37.27	1.261
41.13	1.485
45.42	1.750
49.92	2.049
54.87	2.389
60.66	2.792
67.19	3.257*
73.80	3.727
80.35	4.207
87.37	4.724
95.11	5.265
103.38	5.858
112.09	6.439
121.39	7.073
130.54	7.672
139.40	8.232*
148.06	8.764
156.53	9.247*
165.21	9.723
174.01	1.020×10^{-1}*
182.75	1.063
191.53	1.104*
200.38	1.144
209.27	1.180*
210.47	1.185
219.36	1.219*
CURVE 2 (cont.)	
228.17	1.251×10^{-1}*
237.02	1.282
245.89	1.312*
254.64	1.338
263.23	1.364*
271.68	1.386
280.14	1.408*
288.75	1.429*
297.66	1.448
CURVE 3	
Series 1	
35.79	1.055×10^{-2}
39.51	1.254
48.00	1.789*
53.47	2.176
58.77	2.549
64.43	2.961
70.56	3.402
77.46	3.903
84.48	4.437
91.47	4.947
99.32	5.505
107.50	6.093
115.56	6.658
123.76	7.225
132.32	7.799*
140.88	8.350
149.47	8.888*
158.52	9.420
168.10	9.963*
177.84	1.049×10^{-1}
187.66	1.098
197.62	1.145*
207.21	1.187*
207.31	1.187*
217.42	1.229
227.23	1.268
237.48	1.305*
247.58	1.341
257.62	1.374*
267.72	1.405*
277.94	1.434
288.21	1.461*
CURVE 3 (cont.)	
Series 1 (cont.)	
298.73	1.487×10^{-1}
309.38	1.513
319.94	1.538*
Series 2	
173.08	1.023×10^{-1}*
182.60	1.072*
192.25	1.119*
201.97	1.163*
211.82	1.206*
221.95	1.247*
232.37	1.287*
242.87	1.325*
253.32	1.360*
263.78	1.392*
274.26	1.423*
284.62	1.451*
294.90	1.478*
305.22	1.503*
315.47	1.526*
325.60	1.545
335.70	1.566*
345.83	1.583
Series 3	
5.71	5.940×10^{-4}
6.58	7.864
7.59	1.255×10^{-3}
8.56	1.573
9.47	1.857
10.33	2.059
11.30	2.169
12.41	2.390
13.65	2.657
15.10	2.951
16.75	3.324
18.55	3.713
20.42	4.189
22.32	4.729
24.43	5.401
27.06	6.362
30.05	7.603
33.43	9.236
CURVE 3 (cont.)	
Series 3 (cont.)	
36.94	1.113×10^{-2}
40.31	1.310
44.10	1.543
48.11	1.809
CURVE 4	
Series 1	
5.59	3.640×10^{-4}
6.36	4.040
7.38	5.640
8.65	8.160
10.00	1.074×10^{-3}
11.45	1.292
12.94	1.501
14.46	1.755
16.04	2.051
17.72	2.350
19.54	2.739
Series 2	
16.78	2.226×10^{-3}
18.64	2.599
20.61	3.038
22.80	3.617
25.27	4.320
27.97	5.277
30.95	6.351
34.11	8.154
37.47	9.773
41.26	1.193×10^{-2}
45.54	1.457
50.25	1.768
52.74	1.937
55.23	2.109
57.45	2.267
63.00	2.673
69.23	3.125
75.89	3.621
83.03	4.164
90.35	4.723
97.86	5.281
105.87	5.868

* Not shown on plot

DATA TABLE NO. 489 (continued)

T	C_p
CURVE 4 (cont.)	
Series 2 (cont.)	
114.12	6.472×10^{-2}
122.47	7.068*
131.20	7.676*
140.13	8.276*
149.01	8.851
158.03	9.404*
167.25	9.945*
168.47	1.001×10^{-1}*
177.45	1.051*
186.57	1.098*
195.75	1.143*
204.81	1.186
213.75	1.225*
222.67	1.262*
231.62	1.297*
240.59	1.331*
249.57	1.364*
258.62	1.393
267.81	1.422*
276.99	1.451*
286.10	1.476*
295.17	1.502*
304.24	1.525
CURVE 5	
Series 1	
63.23	2.653×10^{-2}*
69.11	3.085*
75.83	3.589*
83.16	4.153*
91.49	4.7894*
90.54	4.7210*
98.30	5.2906*
106.78	5.9143
115.28	6.5389*
123.12	7.1027*
130.73	7.6295*
138.60	8.1609
147.06	8.712*
155.74	9.253
164.52	9.773*
173.47	1.028×10^{-1}
182.58	1.076*
191.87	1.123

T	C_p
CURVE 5 (cont.)	
Series 1 (cont.)	
201.18	1.167×10^{-1}*
210.40	1.209*
219.59	1.247*
228.67	1.284*
237.58	1.318
246.56	1.350*
255.62	1.380*
264.65	1.410*
273.70	1.439*
282.71	1.465
291.70	1.490
300.83	1.514*
Series 2	
4.50	2.9×10^{-4}
4.87	2.8
5.65	3.6
4.75	2.9
5.58	3.4
5.49	3.4*
6.66	4.40
7.87	5.96
8.96	7.72
10.00	9.719
11.11	1.078×10^{-3}
12.38	1.237
13.77	1.432
15.27	1.685
16.86	1.952
18.52	2.262
20.28	2.632
22.17	3.088
24.37	3.717
26.80	4.501
29.37	5.470
32.14	6.674
35.29	8.205
38.67	1.004×10^{-2}
42.16	1.208
46.10	1.455*
50.74	1.763*
55.57	2.099*
60.34	2.440
65.57	2.674

* Not shown on plot

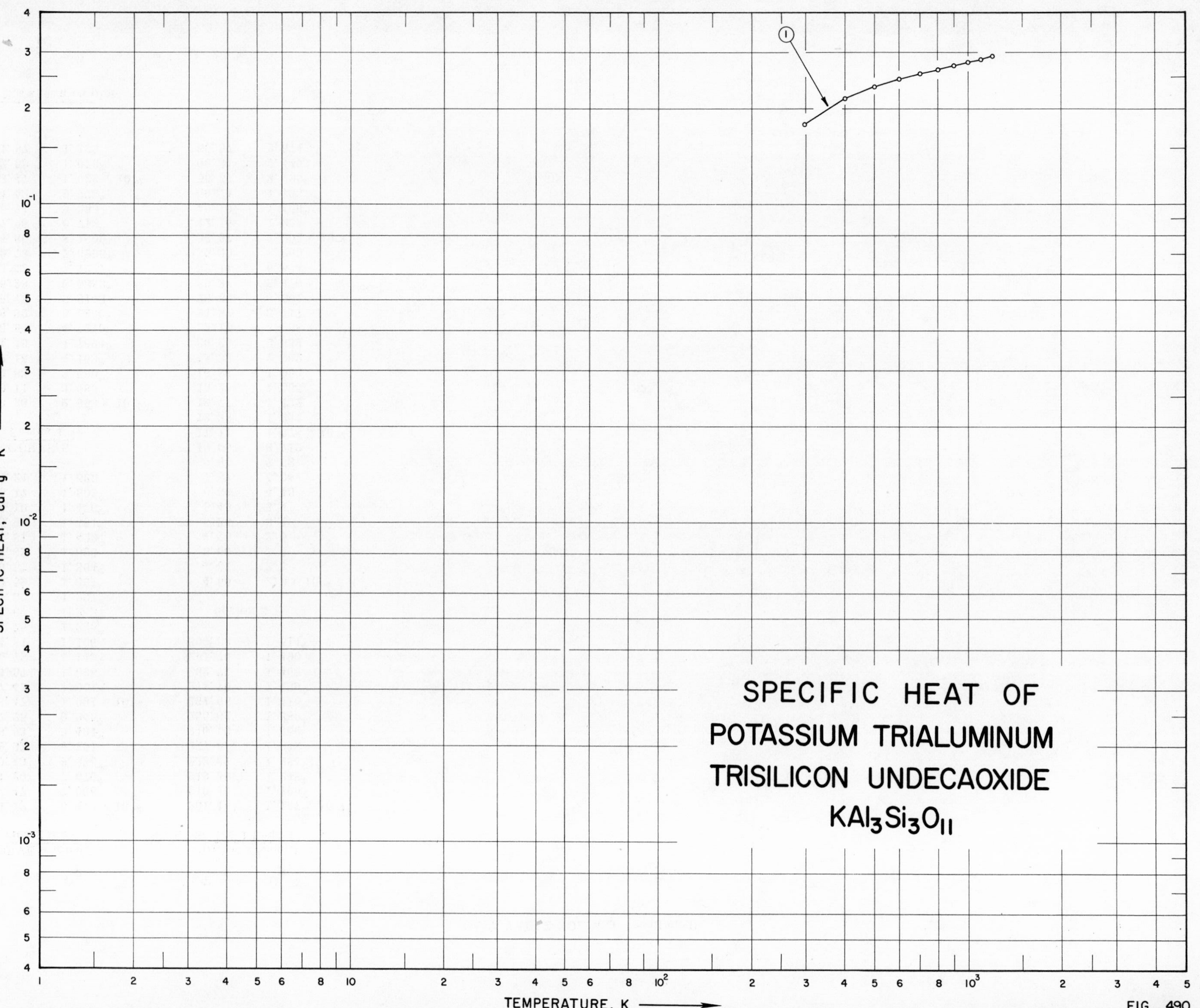

SPECIFIC HEAT OF
POTASSIUM TRIALUMINUM
TRISILICON UNDECAOXIDE
$KAl_3Si_3O_{11}$
SPECIFIC HEAT, cal g^{-1} K^{-1}
TEMPERATURE, K
1
FIG 490

SPECIFICATION TABLE NO. 490 SPECIFIC HEAT OF POTASSIUM TRIALUMINUM TRISILICON UNDECAOXIDE $KAl_3Si_3O_{11}$

[For Data Reported in Figure and Table No. 490]

Curve No.	Ref. No.	Year	Temp. Range, K	Reported Error, %	Name and Specimen Designation	Composition (weight percent), Specifications and Remarks
1	397	1964	298-1200	0.1	Dehydrated Muscovite	47.97 SiO_2, 35.00 Al_2O_3, 11.22 K_2O, 3.37 Fe_2O_3, 1.04 Na_2O, 0.83 MgO, 0.49 FeO, 0.08 TiO_2; prepared by heating muscovite for 35 hrs at 900-1000 K.

DATA TABLE NO. 490 SPECIFIC HEAT OF POTASSIUM TRIALUMINUM TRISILICON UNDECAOXIDE $KAl_3Si_3O_{11}$

[Temperature, T, K; Specific Heat, C_p, Cal $g^{-1}K^{-1}$]

T	C_p
CURVE 1	
298.15	1.783 x 10^{-1}
300	1.792*
400	2.153
500	2.348
600	2.478
700	2.576
800	2.659
900	2.732
1000	2.799
1100	2.861
1200	2.922

* Not shown on plot

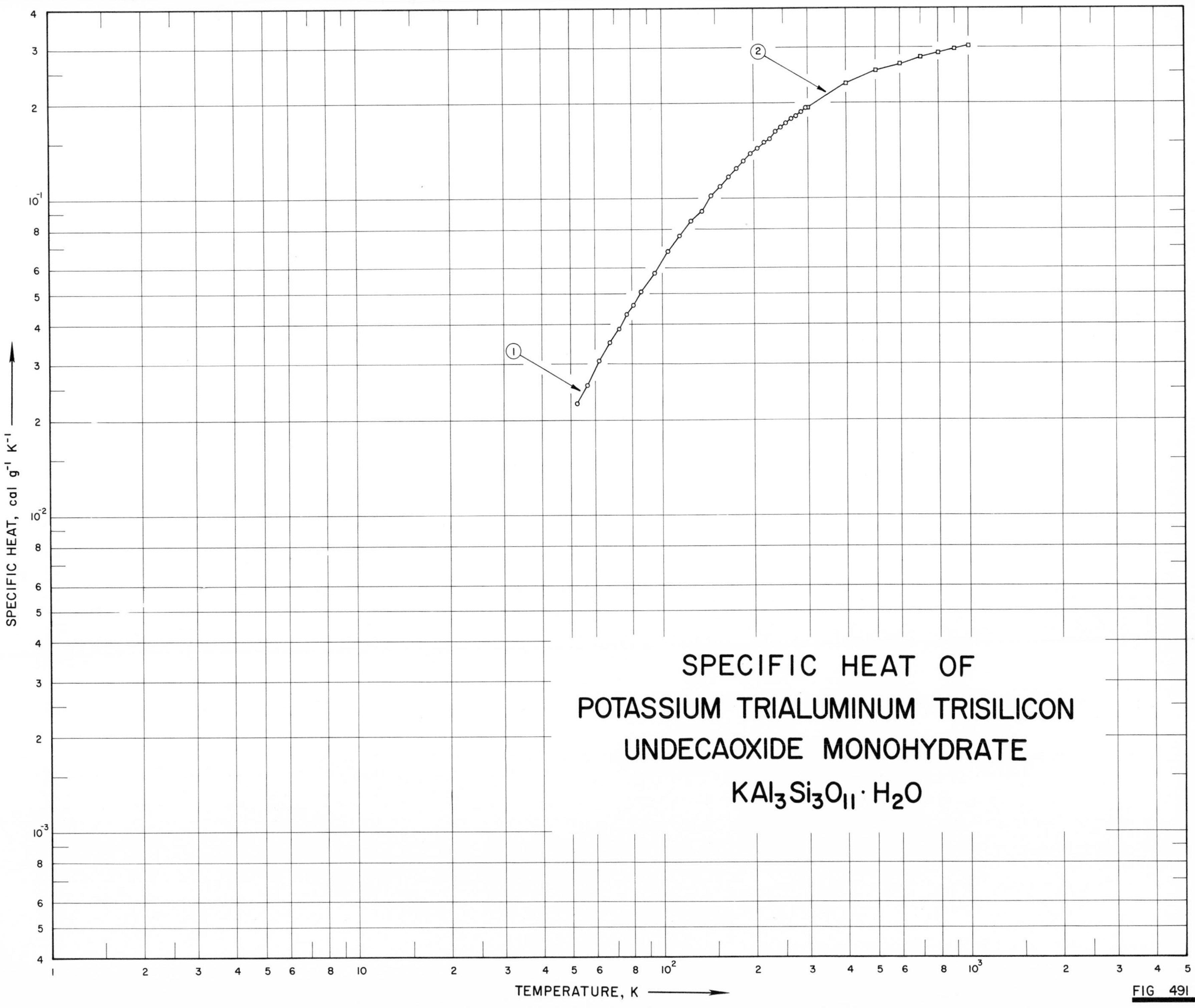
SPECIFIC HEAT OF
POTASSIUM TRIALUMINUM TRISILICON
UNDECAOXIDE MONOHYDRATE
$KAl_3Si_3O_{11} \cdot H_2O$
SPECIFIC HEAT, cal g⁻¹ K⁻¹
TEMPERATURE, K
1
2

FIG 491

SPECIFICATION TABLE NO. 491 SPECIFIC HEAT OF POTASSIUM TRIALUMINUM TRISILICON UNDECAOXIDE MONOHYDRATE $KAl_3Si_3O_{11} \cdot H_2O$

[For Data Reported in Figure and Table No. 491]

Curve No.	Ref. No.	Year	Temp. Range, K	Reported Error, %	Name and Specimen Designation	Composition (weight percent), Specifications and Remarks
1	396	1963	53-297	0.3	Muscovite	45.79 SiO_2, 33.47 Al_2O_3, 10.73 K_2O, 4.47 H_2O, 3.22 Fe_2O_3, 0.99 Na_2O, 0.79 MgO, 0.47 FeO, 0.08 TiO_2, nil CaO; ground to -20 mesh; dried at 110 C.
2	397	1964	298-1000	0.2	Muscovite	45.75 SiO_2, 33.47 Al_2O_3, 10.73 K_2O, 4.46 H_2O, 3.22 Fe_2O_3, 0.99 Na_2O, 0.79 MgO, 0.47 FeO, 0.08 TiO_2; natural sample of thin plates about 1 in. size; ground to -20 mesh; dried at 110 C.

DATA TABLE NO. 491 SPECIFIC HEAT OF POTASSIUM TRIALUMINUM TRISILICON UNDECAOXIDE MONOHYDRATE $KAl_3Si_3O_{11} \cdot H_2O$

[Temperature, T, K; Specific Heat, C_p, Cal g^{-1} K^{-1}]

T	C_p
CURVE 1	
52.59	2.252 x 10^{-2}
56.81	2.588
61.89	3.073
67.13	3.502
72.05	3.886
76.81	4.321
80.61	4.625
85.58	5.099
94.73	5.825
105.09	6.804
114.96	7.642
124.84	8.478
136.09	9.139
146.15	1.021 x 10^{-1}
156.09	1.095
166.20	1.171
176.27	1.245
186.18	1.312
196.30	1.381
206.40	1.439
216.90	1.505
226.74	1.539
236.42	1.622
246.10	1.674
256.75	1.727
266.98	1.780
276.64	1.820
287.02	1.876
296.61	1.928
CURVE 2	
298.15	1.928 x 10^{-1}*
300	1.938
400	2.315
500	2.525
600	2.669
700	2.783
800	2.879
900	2.967
1000	3.048

* Not shown on plot

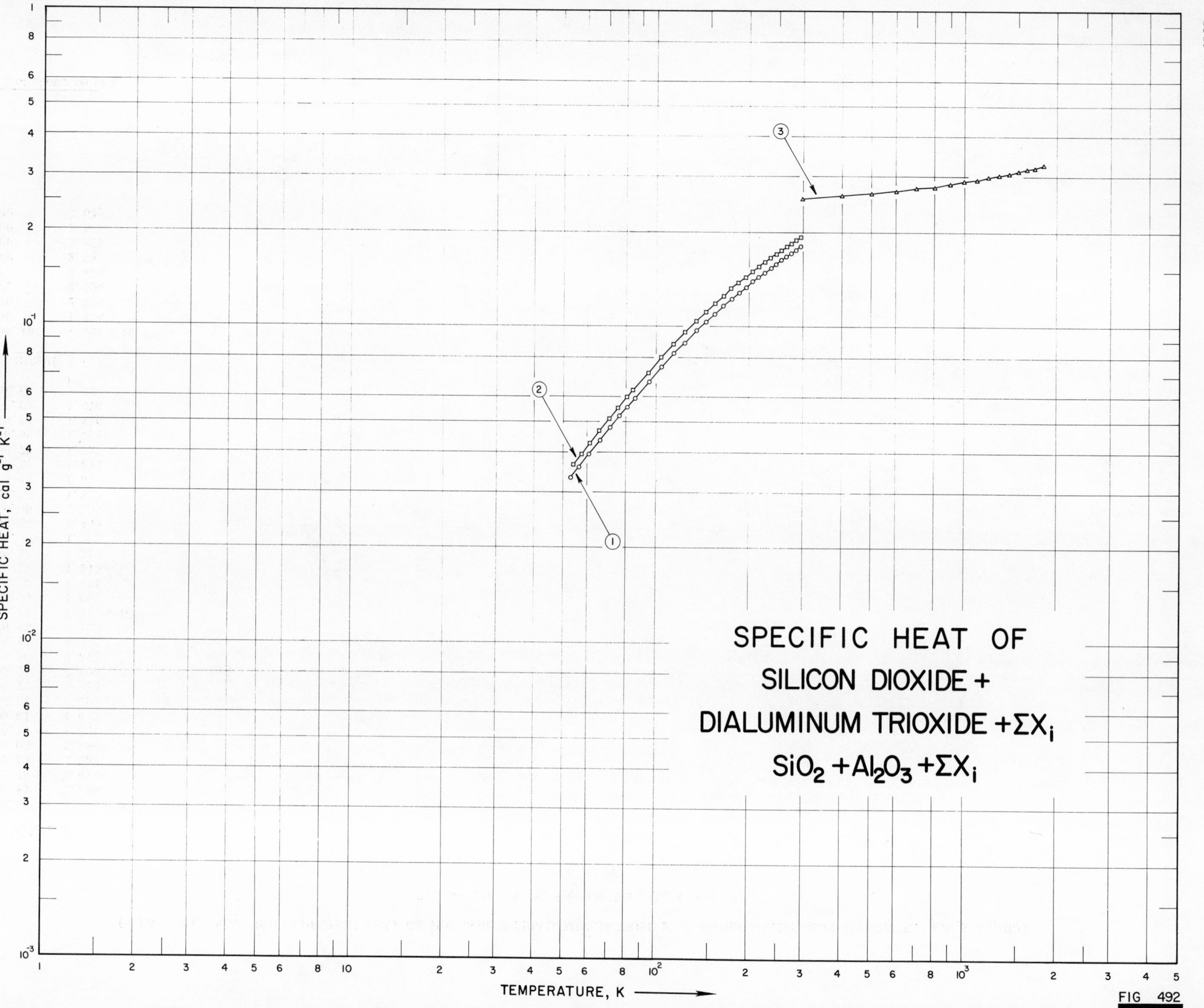
SPECIFIC HEAT OF
SILICON DIOXIDE +
DIALUMINUM TRIOXIDE $+\Sigma X_i$
$SiO_2 + Al_2O_3 + \Sigma X_i$
SPECIFIC HEAT, cal g^{-1} K^{-1}
TEMPERATURE, K
1
2
3
FIG 492

SPECIFICATION TABLE NO. 492 SPECIFIC HEAT OF SILICON DIOXIDE + DIALUMINUM TRIOXIDE + ΣX_i $SiO_2 + Al_2O_3 + \Sigma X_i$

[For Data Reported in Figure and Table No. 492]

Curve No.	Ref. No.	Year	Temp. Range, K	Reported Error, %	Name and Specimen Designation	Composition (weight percent), Specifications and Remarks
1	357	1948	53-297		Perlite, dehydrated Sample B	Completely dehydrated by igniting at constant weight at 1000-1100 C.
2	357	1948	54.3-297		Perlite, hydrated Sample A	73.61 SiO_2, 12.17 Al_2O_3, 5.08 K_2O, 3.34 H_2O, 2.97 Na_2O, 1.51 Fe_2O_3, 0.84 CaO; sample from Superior, Arizona; crushed in diamond mortar and screened; dried at 120 C.
3	357	1948	298-1816		Perlite, dehydrated Sample B	Completely dehydrated by igniting at constant weight at 1000-1100 C; high temperature test.

DATA TABLE NO. 492 SPECIFIC HEAT OF SILICON DIOXIDE + DIALUMINUM TRIOXIDE + ΣX_i $SiO_2 + Al_2O_3 + \Sigma X_i$

[Temperature, T, K; Specific Heat, C_p, Cal g^{-1} K^{-1}]

T	C_p
CURVE 1	
53.0	3.335 x 10^{-2}
56.5	3.608
60.8	3.962
66.0	4.385
71.2	4.814
76.2	5.225
80.8	5.578
85.6	5.958
95.3	6.718
104.9	7.470
114.8	8.263
124.4	8.929
136.1	9.773
146.3	1.046 x 10^{-1}
155.8	1.109
166.2	1.175
176.3	1.236
186.2	1.291
196.3	1.347
206.5	1.401
216.5	1.452
225.6	1.494
227.5	1.503*
236.4	1.546
236.4	1.547*
246.2	1.588
246.2	1.594*
256.4	1.645
256.5	1.642*
266.4	1.685
266.8	1.686*
276.6	1.726
276.9	1.730*
286.7	1.771
286.7	1.771*
296.8	1.813
CURVE 2	
54.3	3.655 x 10^{-2}
57.5	3.933
61.3	4.275
65.9	4.680
70.6	5.100
75.4	5.534
80.7	5.991

T	C_p
CURVE 2 (cont.)	
84.5	6.319 x 10^{-2}
94.5	7.162
104.5	8.000
114.7	8.840
124.4	9.612
135.7	1.049 x 10^{-1}
146.1	1.123
156.1	1.196
166.0	1.262
175.9	1.329
186.1	1.387
196.3	1.446
206.[illegible]	1.[illegible]03
216.6	1.561
226.8	1.613
235.9	1.662
246.1	1.709
256.2	1.759
266.1	1.803
276.5	1.845
286.6	1.8918
296.2	1.9334
CURVE 3	
298.15	2.56 x 10^{-1}
300	2.57*
400	2.61
500	2.66
600	2.71
700	2.75
800	2.80
900	2.84
1000	2.89
1100	2.94
1200	2.98
1300	3.03
1400	3.08
1500	3.12
1600	3.17
1700	3.22
1810	3.27*
1816	3.27

* Not shown on plot

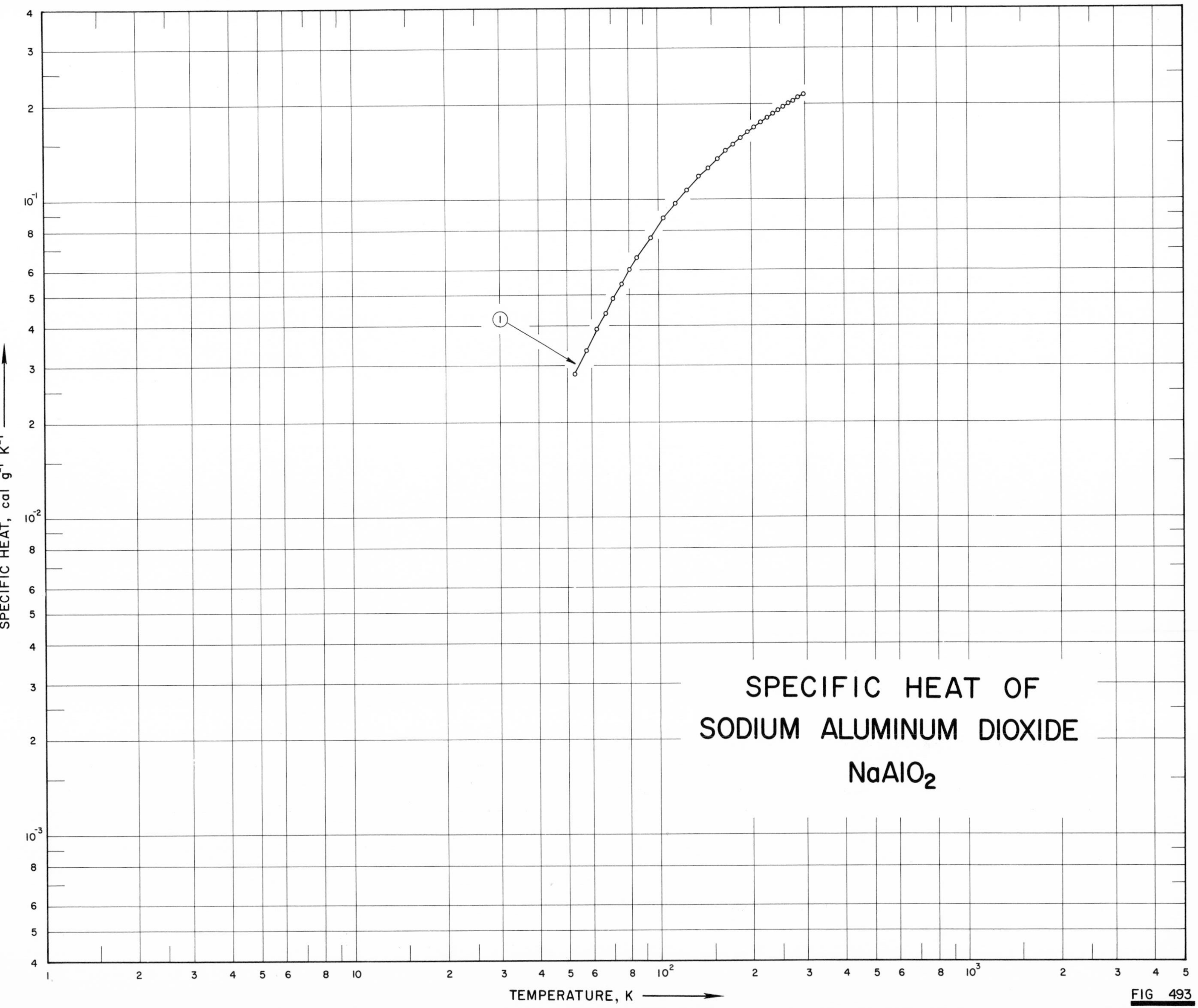
SPECIFIC HEAT OF
SODIUM ALUMINUM DIOXIDE
$NaAlO_2$
SPECIFIC HEAT, cal g^{-1} K^{-1}
TEMPERATURE, K
1
FIG 493

SPECIFICATION TABLE NO. 493 SPECIFIC HEAT OF SODIUM ALUMINUM DIOXIDE $NaAlO_2$

[For Data Reported in Figure and Table No. 493]

Curve No.	Ref. No.	Year	Temp. Range, K	Reported Error, %	Name and Specimen Designation	Composition (weight percent), Specifications and Remarks
1	390	1955	53-298			62.08 Al_2O_3 (62.19 theo.); prepared from reagent-grade Na_2CO_3 and pure hydrated Al_2O_3; heated six times for total 89 hrs at 1000-1050 C.

DATA TABLE NO. 493 SPECIFIC HEAT OF SODIUM ALUMINUM DIOXIDE $NaAlO_2$

[Temperature, T, K; Specific Heat, C_p, Cal $g^{-1}K^{-1}$]

T	C_p
CURVE 1	
53.10	2.829 x 10^{-2}
57.90	3.359
62.61	3.909
66.71	4.386
70.91	4.879
75.58	5.434
80.57	6.029
85.14	6.565
94.57	7.635
104.83	8.754
114.46	9.743
124.64	1.074 x 10^{-1}
136.21	1.181
145.78	1.264
155.84	1.347
165.90	1.424
175.94	1.499
185.79	1.565
195.94	1.630
206.09	1.693
216.38	1.757
226.04	1.813
236.05	1.863
245.70	1.913
256.15	1.965
266.66	2.009
276.23	2.047
286.50	2.096
296.10	2.130*
298.16	2.137

* Not shown on plot

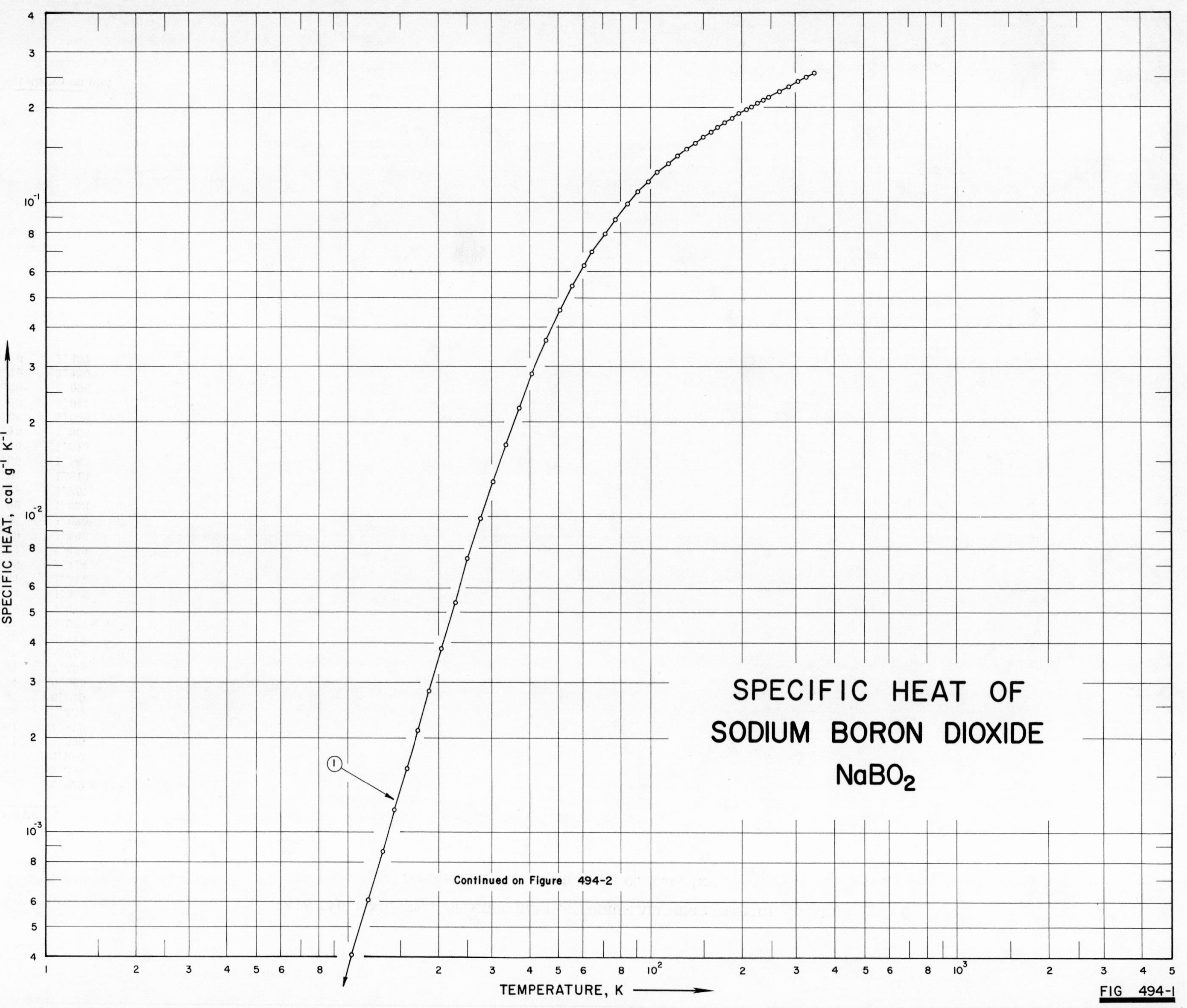
SPECIFIC HEAT OF
SODIUM BORON DIOXIDE
$NaBO_2$
Continued on Figure 494-2
TEMPERATURE, K
SPECIFIC HEAT, cal g^{-1} K^{-1}
FIG 494-1

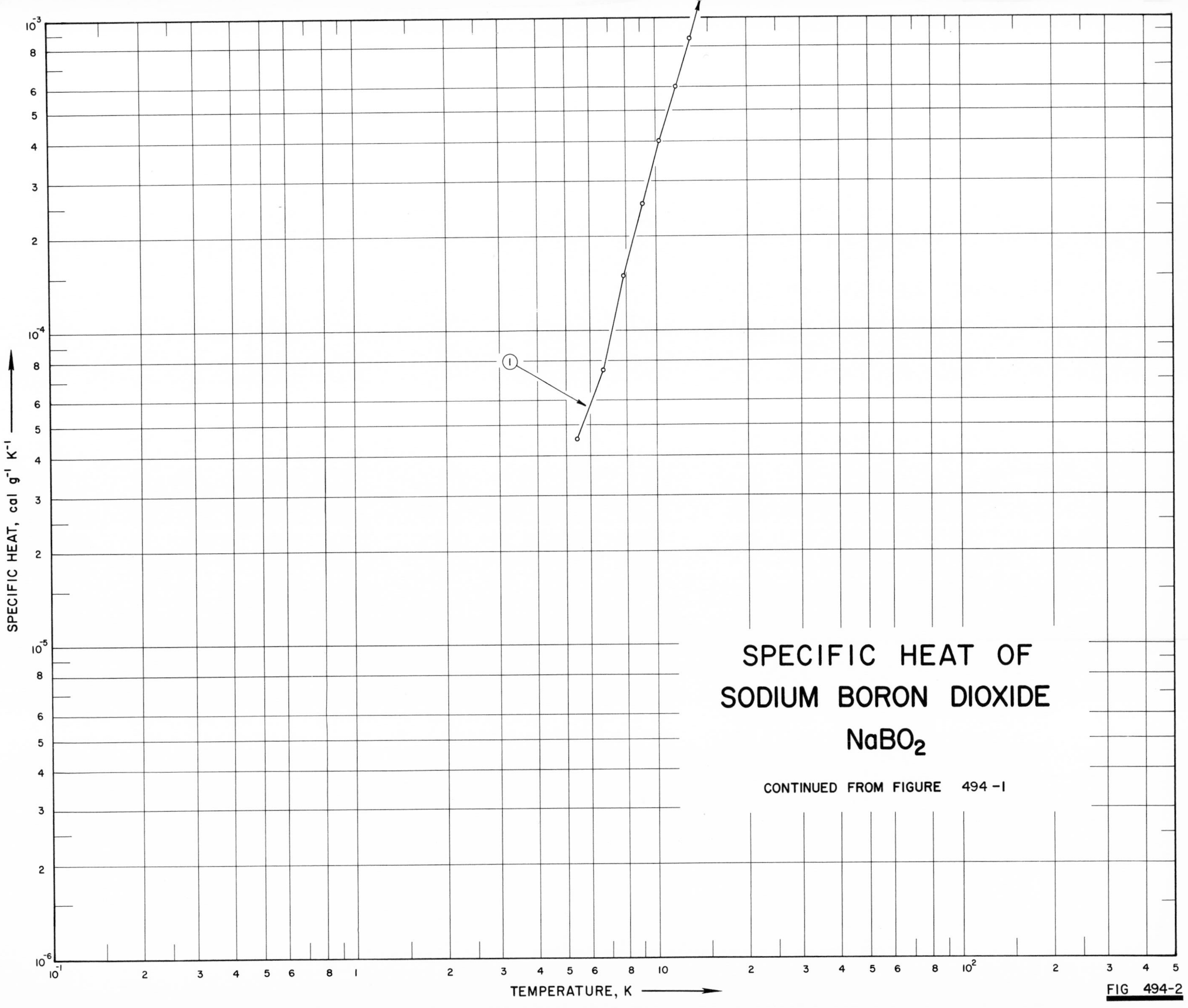
SPECIFIC HEAT OF
SODIUM BORON DIOXIDE
$NaBO_2$
CONTINUED FROM FIGURE 494-1
SPECIFIC HEAT, cal g⁻¹ K⁻¹
TEMPERATURE, K
1
FIG 494-2

SPECIFICATION TABLE NO. 494 SPECIFIC HEAT OF SODIUM BORON DIOXIDE $NaBO_2$

[For Data Reported in Figure and Table No. 494]

Curve No.	Ref. No.	Year	Temp. Range, K	Reported Error, %	Name and Specimen Designation	Composition (weight percent), Specifications and Remarks
1	399	1960	5-350	0.10		52.91 B_2O_3, 47.11 Na_2O; recrystallized evacuated at 25 C for 3 days.

DATA TABLE NO. 494 SPECIFIC HEAT OF SODIUM BORON DIOXIDE $NaBO_2$

[Temperature, T, K; Specific Heat, C_p, Cal $g^{-1}K^{-1}$]

T	C_p
CURVE 1	
5.48	4.558 x 10^{-5}
6.68	7.597
7.83	1.520 x 10^{-4}
9.08	2.583
10.38	4.103
11.72	6.078
13.06	8.661
14.40	1.185 x 10^{-3}
15.74	1.595
17.14	2.112
18.69	2.811
20.53	3.844
22.74	5.379
25.15	7.400
27.64	9.907
30.36	1.305 x 10^{-2}
33.41	1.709
36.89	2.225
40.89	2.861
45.57	3.654
50.72	4.553
55.96	5.467
60.88	6.309
64.98	6.996
71.20	7.968
77.18	8.872
77.94	8.977*
84.46	9.900
91.34	1.079 x 10^{-1}
99.00	1.166
107.24	1.255
115.59	1.337
123.94	1.415
132.48	1.489
140.98	1.558
149.44	1.620
158.07	1.681
167.33	1.744
177.16	1.805
187.21	1.866
197.36	1.924
207.75	1.978
215.06	2.018
225.14	2.070
235.23	2.118
CURVE 1 (cont.)	
245.47	2.168 x 10^{-1}
255.80	2.215*
265.98	2.261
276.10	2.304*
286.16	2.348
296.22	2.387*
306.34	2.430
316.45	2.468*
326.55	2.504
336.56	2.542*
345.49	2.577

* Not shown on plot

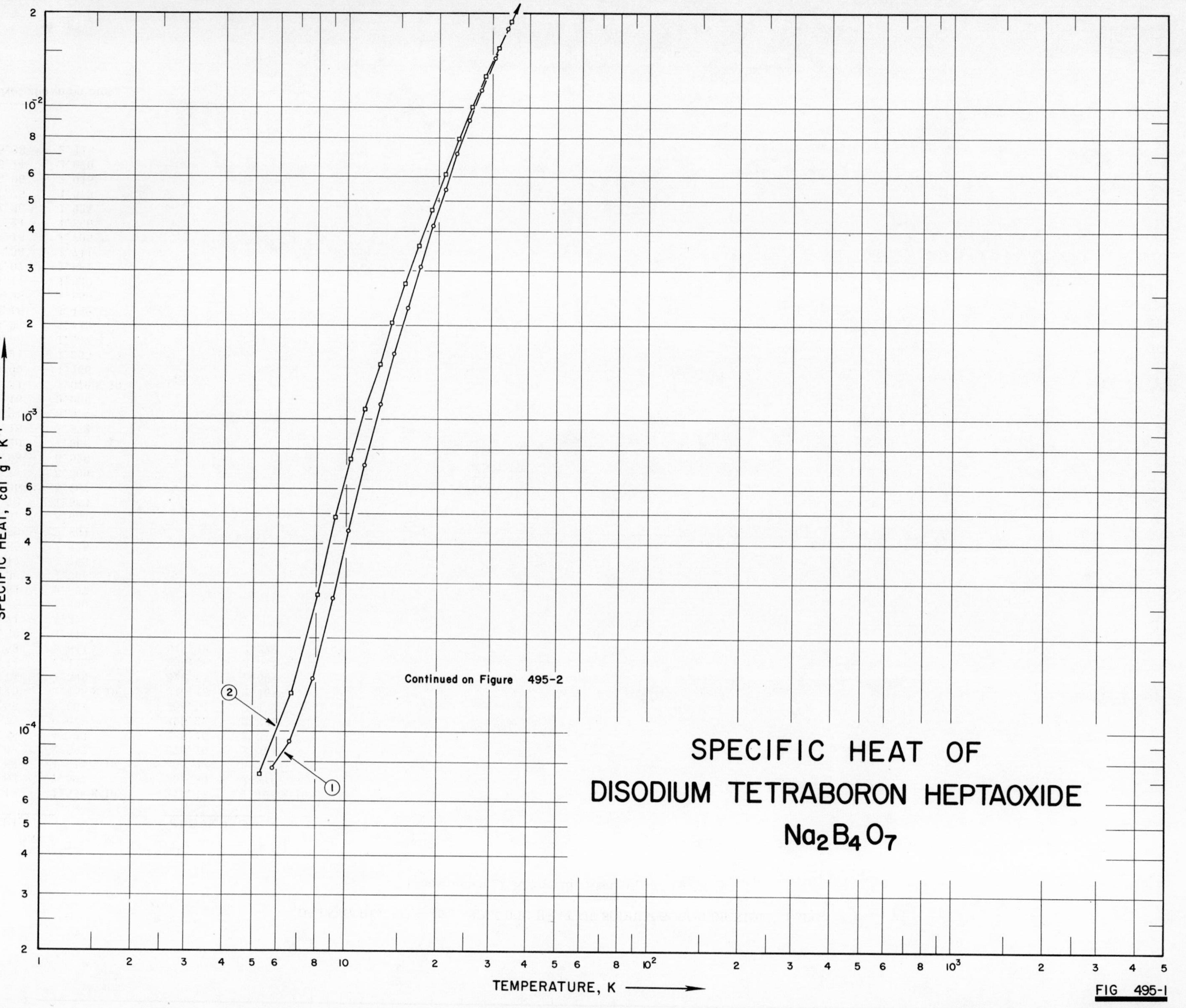

SPECIFIC HEAT, cal g^{-1} K^{-1}
TEMPERATURE, K
Continued on Figure 495-2
SPECIFIC HEAT OF
DISODIUM TETRABORON HEPTAOXIDE
$Na_2B_4O_7$
FIG 495-1

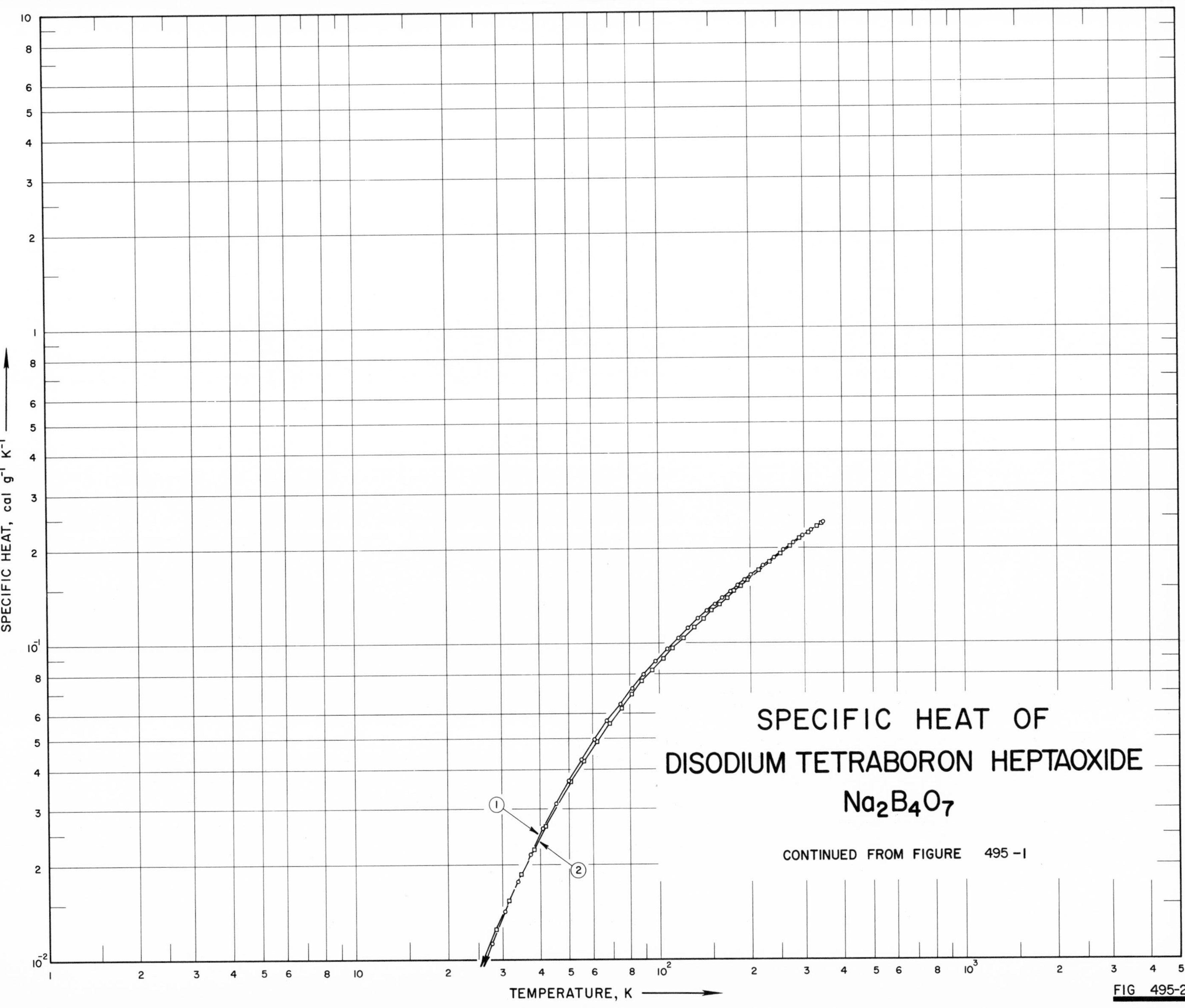
SPECIFIC HEAT OF
DISODIUM TETRABORON HEPTAOXIDE
$Na_2B_4O_7$
CONTINUED FROM FIGURE 495-1
SPECIFIC HEAT, cal g⁻¹ K⁻¹
TEMPERATURE, K
1
2

FIG 495-2

SPECIFICATION TABLE NO. 495 SPECIFIC HEAT OF DISODIUM TETRABORON HEPTAOXIDE $Na_2B_4O_7$

[For Data Reported in Figure and Table No. 495]

Curve No.	Ref. No.	Year	Temp. Range, K	Reported Error, %	Name and Specimen Designation	Composition (weight percent), Specifications and Remarks
1	399	1960	5-350	≤5	Crystalline	99. 87 ±0. 04 $Na_2B_4O_7$; prepared by crystallizing dehydrated sample of analytic reagent-grade sodium, tetraborate decahydrate from molten state under carefully controlled conditions; helium atmosphere.
2	399	1960	5-343	≤5	Vitreous	100. 00 ± 0. 04 $Na_2B_4O_7$; reagent-grade sodium tetraborate decahydrate was heated to 820 for 30 min; annealed for 15 min at 420 C; cooled in anhydrous atmosphere; helium atmosphere.

DATA TABLE NO. 495 SPECIFIC HEAT OF DISODIUM TETRABORON HEPTAOXIDE $Na_2B_4O_7$

[Temperature, T, K; Specific Heat, C_p, Cal $g^{-1}K^{-1}$]

T	C_p
CURVE 1	
Series I	
5.82	7.652×10^{-5}
6.60	9.341
7.81	1.481×10^{-4}
9.07	2.673
10.21	4.422
11.50	7.155
12.92	1.123×10^{-3}
14.37	1.630
15.92	2.286
17.52	3.095
19.21	4.189
21.03	5.451
23.02	7.105
25.23	9.063
27.73	1.138×10^{-2}
30.61	1.432
33.80	1.784
37.25	2.176
41.09	2.631
45.45	3.153
50.18	3.729
55.32	4.346
61.01	5.023
67.36	5.759
74.48	6.534
81.61	7.304
89.18	8.084
97.84	8.889
107.24	9.738
117.03	1.057×10^{-1}
126.62	1.138
135.89	1.211
144.71	1.278
153.70	1.345
162.66	1.410
172.08	1.476
181.88	1.541
191.80	1.606
201.60	1.669
211.13	1.730*
220.53	1.789
229.84	1.844*
238.94	1.897
248.07	1.949*

T	C_p
CURVE 1 (cont.)	
Series I (cont.)	
257.40	2.003×10^{-1}
266.90	2.0651*
276.50	2.110
286.39	2.153*
296.65	2.251
306.88	2.260*
317.02	2.310
327.09	2.360*
337.30	2.40*
347.69	2.45
Series II*	
253.26	1.977×10^{-1}
262.55	2.027
271.91	2.081
281.41	2.132
291.36	2.182
301.61	2.236
311.91	2.287
322.22	2.335
333.05	2.386
343.97	2.438
CURVE 2	
5.31	7.304×10^{-5}
6.68	1.337×10^{-4}
8.09	2.758
9.22	4.894
10.32	7.453
11.54	1.078×10^{-3}
12.82	1.515
14.13	2.042
15.59	2.728
17.21	3.597
19.01	4.705
21.08	6.111
23.41	7.940
25.95	1.009×10^{-2}
28.71	1.263
31.69	1.559
34.85	1.884
38.23	2.256

T	C_p
CURVE 2 (cont.)	
41.92	2.663×10^{-2}
46.13	3.143*
51.00	3.699
56.38	4.319
62.26	4.969
68.55	5.654
75.01	6.320
81.43	6.996
88.06	7.716
95.20	8.313
103.03	9.013
111.83	9.773
121.62	1.059×10^{-1}
131.54	1.140
140.88	1.213
149.95	1.282
159.04	1.348
168.25	1.415
177.49	1.480
186.79	1.542
196.08	1.605
205.13	1.664*
214.14	1.722
223.13	1.779*
232.21	1.834
241.51	1.891*
250.99	1.947
260.78	2.005*
270.61	2.058
280.35	2.112*
290.09	2.164
300.00	2.216*
310.30	2.270
320.84	2.324*
331.13	2.370
343.63	2.431

* Not shown on plot

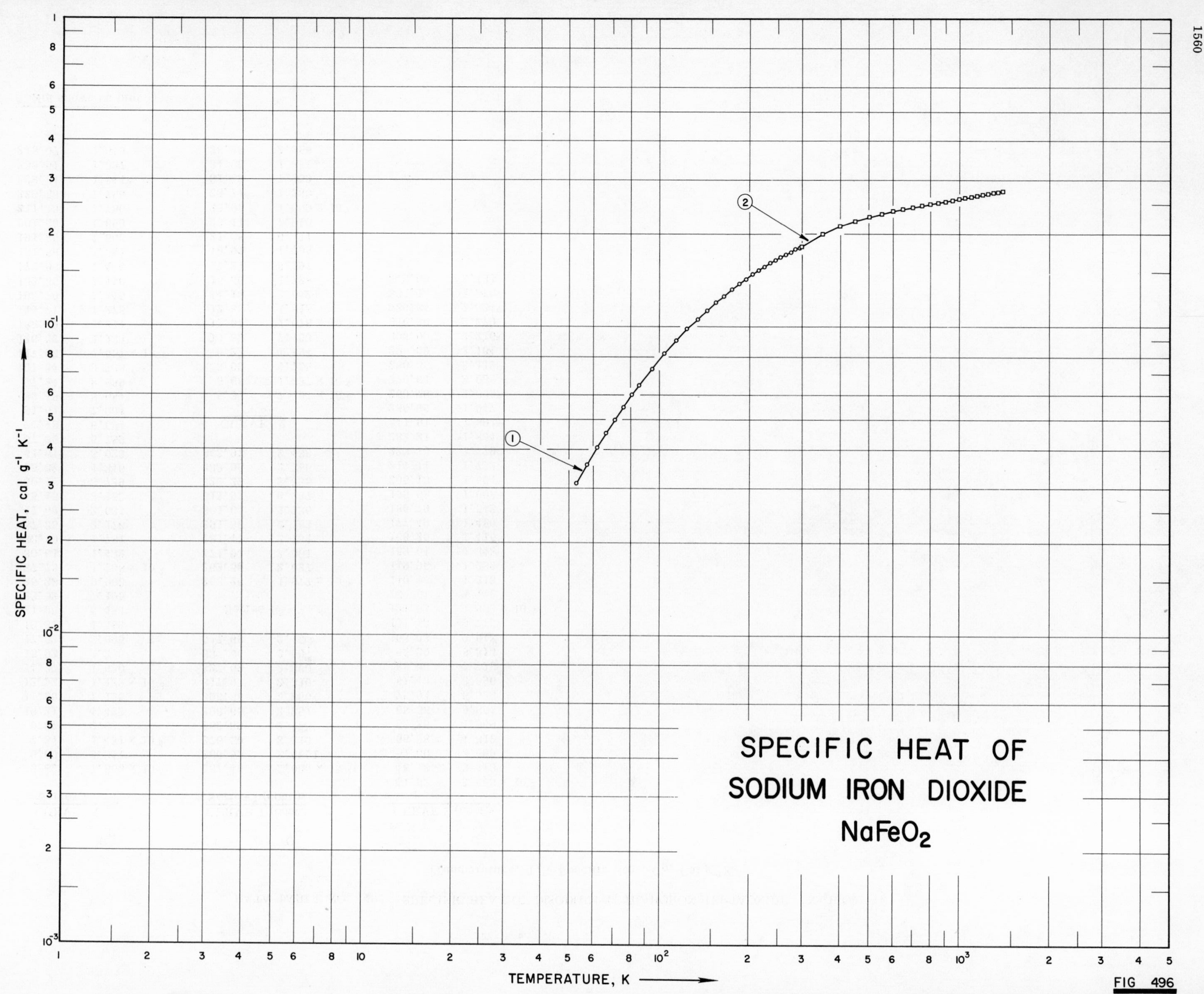

SPECIFIC HEAT OF
SODIUM IRON DIOXIDE
$NaFeO_2$
SPECIFIC HEAT, cal g^{-1} K^{-1}
TEMPERATURE, K
1
2

FIG 496

SPECIFICATION TABLE NO. 496 SPECIFIC HEAT OF SODIUM IRON DIOXIDE $NaFeO_2$

[For Data Reported in Figure and Table No. 496]

Curve No.	Ref. No.	Year	Temp. Range, K	Reported Error, %	Name and Specimen Designation	Composition (weight percent), Specifications and Remarks
1	400	1958	298-1400	0.3		98.6 $NaFeO_2$; argon atmosphere.
2	390	1955	53-296			72.05 Fe_2O_3 (72.04 theo.), 0.06 SiO_2; prepared from reagent-grade Na_2CO_3 and pure ferric oxide; heated 12 times for total 60 hrs at 1000-1050 C and 10 hrs at 1100-1150 C.

DATA TABLE NO. 496 SPECIFIC HEAT OF SODIUM IRON DIOXIDE $NaFeO_2$

[Temperature, T, K; Specific Heat, C_p, Cal $g^{-1}K^{-1}$]

T	C_p	T	C_p
CURVE 1		CURVE 2 (cont.)	
298.15	1.823×10^{-1}	196.09	1.434×10^{-1}
300	1.831	206.14	1.481
350	2.007	216.22	1.529
400	2.127	225.93	1.573
450	2.215	236.18	1.613
500	2.277	245.65	1.652
550	2.336	256.29	1.691
600	2.381	266.34	1.724
650	2.420	276.20	1.753
700	2.454	286.67	1.791
750	2.485	296.11	1.817
800	2.513		
850	2.539		
900	2.563		
950	2.587		
1000	2.609		
1050	2.631		
1100	2.652		
1150	2.672		
1200	2.691		
1250	2.711		
1300	2.729		
1350	2.748		
1400	2.767		
CURVE 2			
53.16	3.119×10^{-2}		
57.70	3.572		
62.23	4.054		
66.82	4.529		
71.34	4.995		
75.95	5.475		
81.17	6.019		
85.68	6.465		
94.69	7.320		
104.84	8.242		
114.65	9.049		
124.32	9.870		
135.73	1.068×10^{-1}		
145.63	1.139		
155.71	1.206		
166.10	1.269		
176.17	1.326		
186.17	1.383		

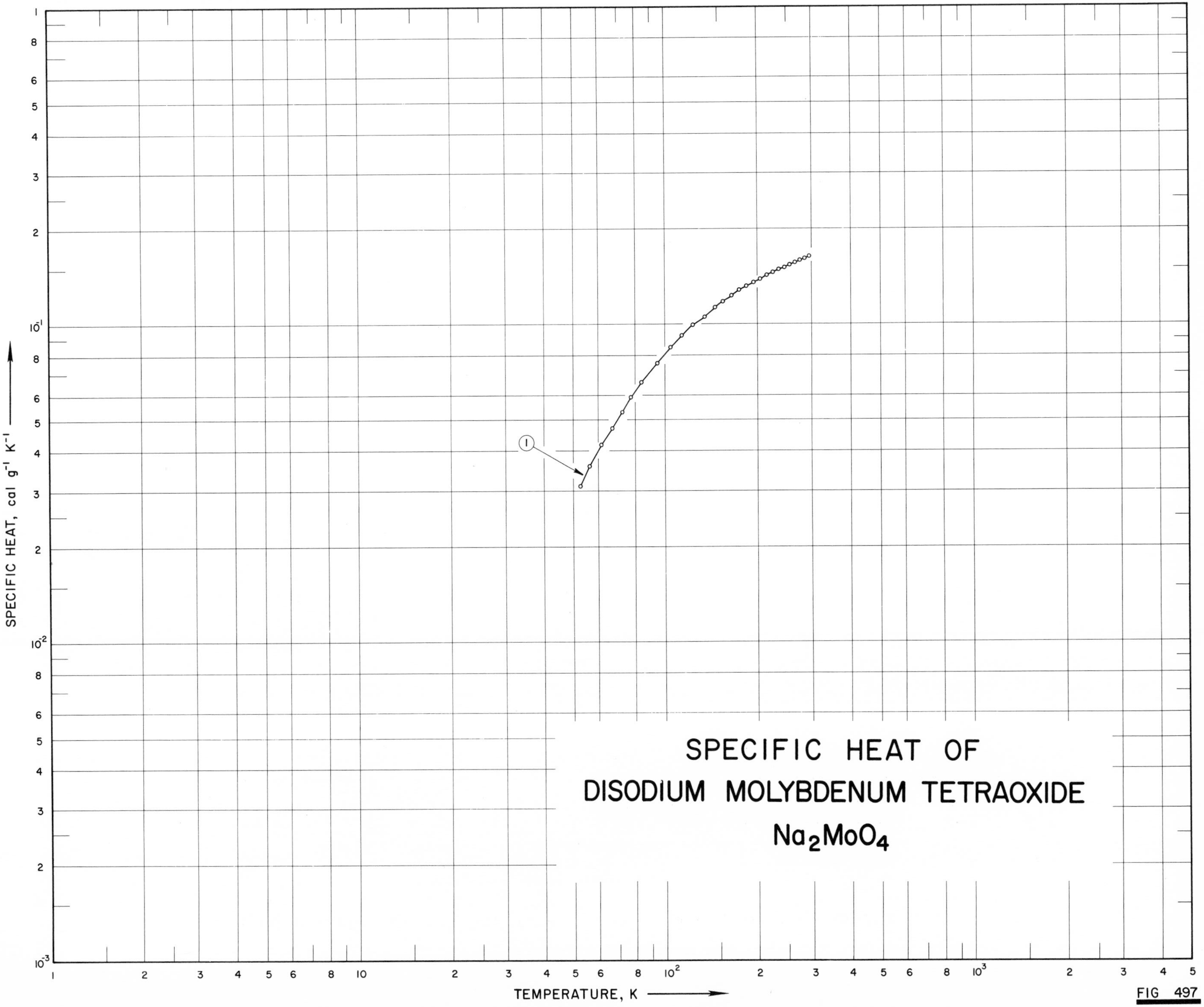

SPECIFIC HEAT OF DISODIUM MOLYBDENUM TETRAOXIDE

Na_2MoO_4

SPECIFIC HEAT, cal g^{-1} K^{-1}

TEMPERATURE, K

FIG 497

SPECIFICATION TABLE NO. 497 SPECIFIC HEAT OF DISODIUM MOLYBDENUM TETRAOXIDE Na_2MoO_4

[For Data Reported in Figure and Table No. 497]

Curve No.	Ref. No.	Year	Temp. Range, K	Reported Error, %	Name and Specimen Designation	Composition (weight percent), Specifications and Remarks
1	368	1963	53-300	<0.3		69.89 MoO_3.

DATA TABLE NO. 497 SPECIFIC HEAT OF DISODIUM MOLYBDENUM TETRAOXIDE Na_2MoO_4

[Temperature, T, K; Specific Heat, C_p, Cal $g^{-1}K^{-1}$]

T	C_p
CURVE 1	
52.67	3.123×10^{-2}
56.82	3.588
61.98	4.190
67.04	4.746
72.54	5.337
77.91	5.934
79.86	6.231*
84.17	6.570
94.84	7.566
105.03	8.494
114.27	9.232
124.36	9.975
136.55	1.059×10^{-1}
145.78	1.135
155.80	1.187
166.29	1.238
176.18	1.282
186.13	1.322
196.27	1.360
206.13	1.396
216.39	1.431
226.11	1.462
236.36	1.491
245.83	1.519
256.52	1.548
266.43	1.572
276.69	1.595
286.76	1.618
296.35	1.642

* Not shown on plot

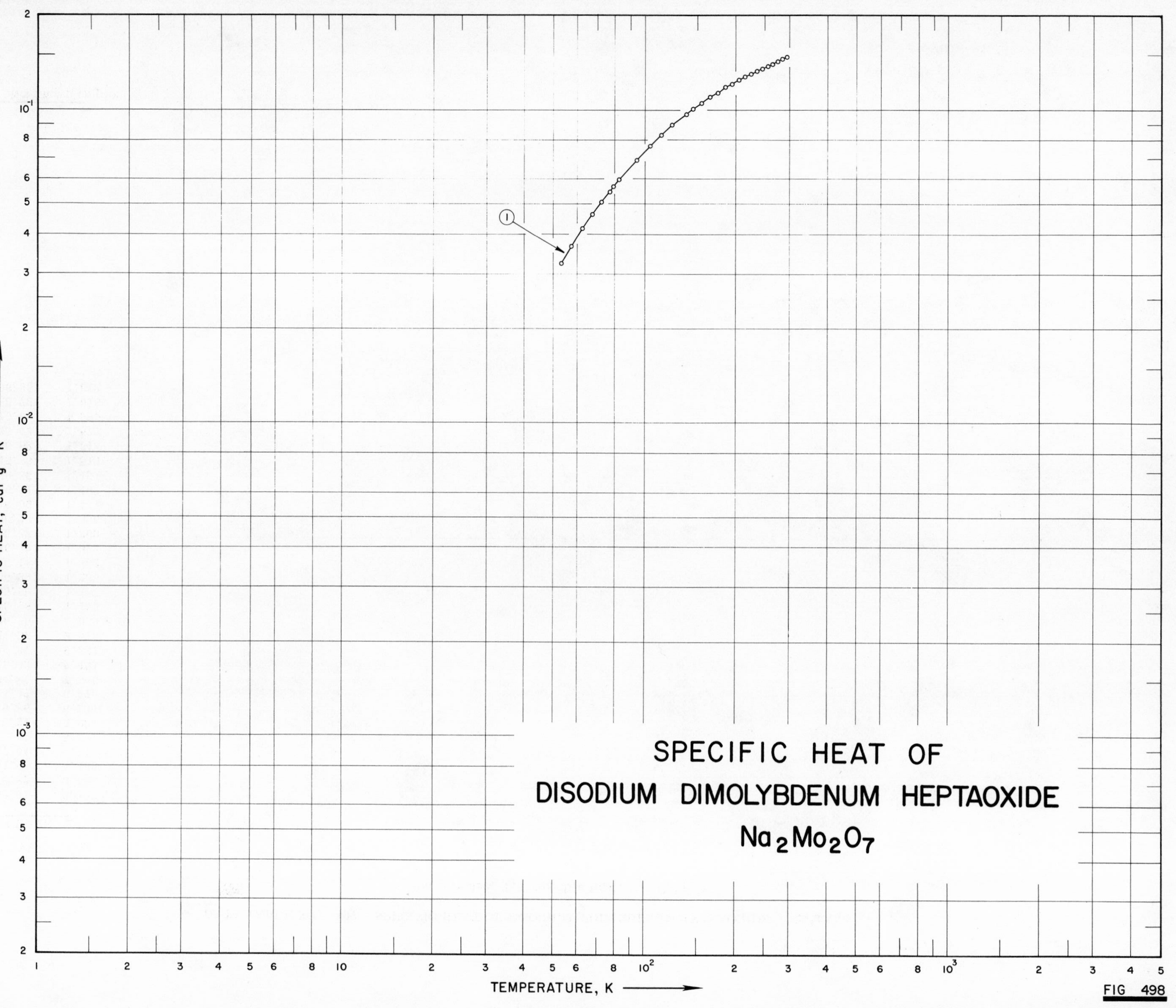
SPECIFIC HEAT OF
DISODIUM DIMOLYBDENUM HEPTAOXIDE
$Na_2Mo_2O_7$
SPECIFIC HEAT, cal g⁻¹ K⁻¹
TEMPERATURE, K
1

FIG 498

SPECIFICATION TABLE NO. 498 SPECIFIC HEAT OF DISODIUM DIMOLYBDENUM HEPTAOXIDE $Na_2Mo_2O_7$

[For Data Reported in Figure and Table No. 498]

Curve No.	Ref. No.	Year	Temp. Range, K	Reported Error, %	Name and Specimen Designation	Composition (weight percent), Specifications and Remarks
1	401	1963	53-296	0.3		82.29 MoO_3.

DATA TABLE NO. 498 SPECIFIC HEAT OF DISODIUM DIMOLYBDENUM HEPTAOXIDE $Na_2Mo_2O_7$

[Temperature, T,K; Specific Heat, C_p, Cal $g^{-1}K^{-1}$]

T	C_p
CURVE 1	
53.24	3.250 x 10^{-2}
57.58	3.684
62.62	4.196
67.65	4.662
72.43	5.099
77.01	5.516
79.41	5.717
82.93	6.025
94.71	6.966
105.02	7.717
114.52	8.366
124.49	8.989
137.84	9.744
145.80	1.015 x 10^{-1}
155.78	1.061
166.00	1.106
176.00	1.143
186.64	1.186
196.30	1.219
206.29	1.252
216.17	1.282
226.25	1.310
235.99	1.339
245.97	1.365
256.35	1.385
266.58	1.414
276.47	1.438
286.75	1.459
296.41	1.480

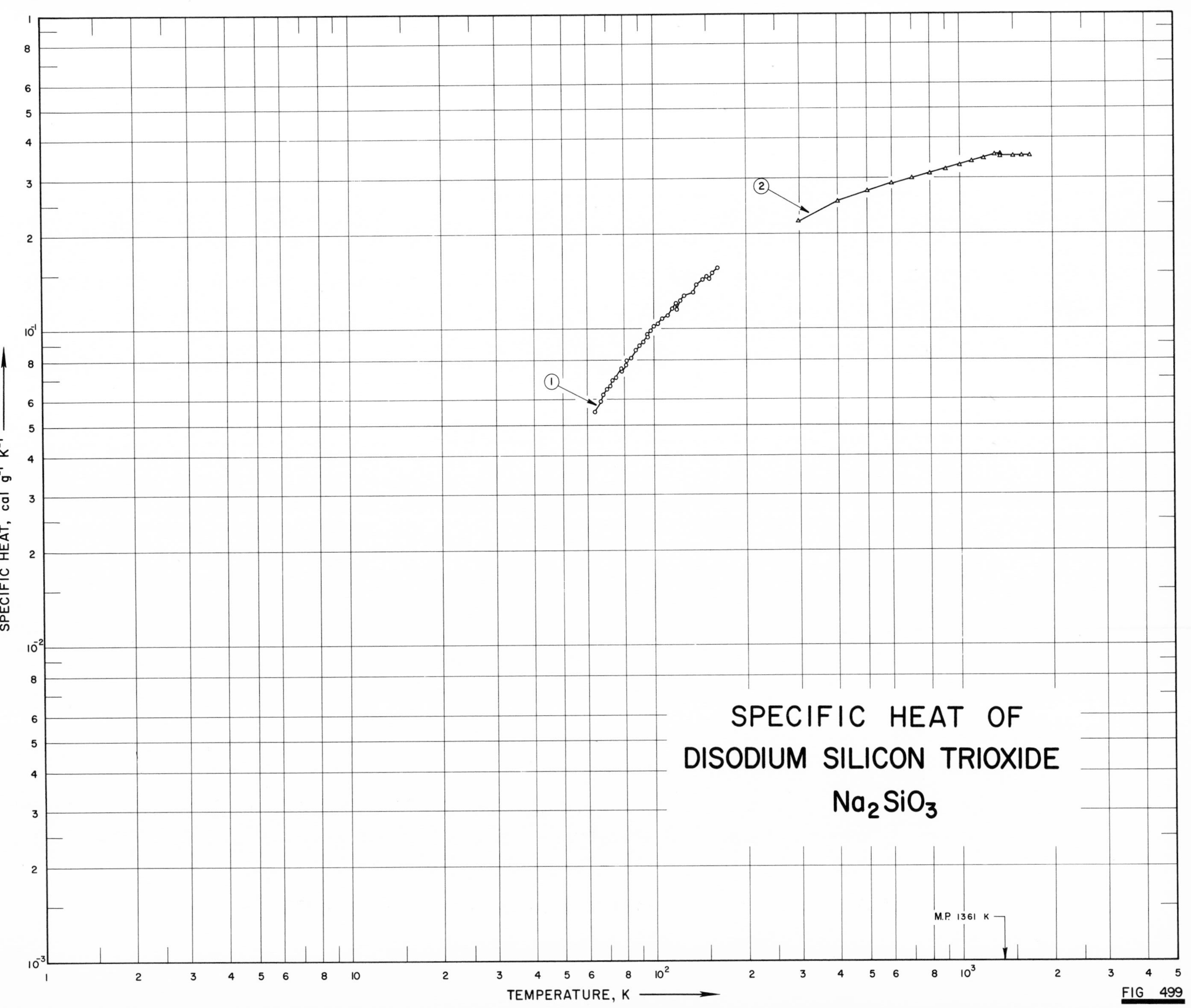

SPECIFIC HEAT OF
DISODIUM SILICON TRIOXIDE
Na_2SiO_3
M.P. 1361 K
TEMPERATURE, K
SPECIFIC HEAT, cal g⁻¹ K⁻¹
1
2

FIG 499

SPECIFICATION TABLE NO. 499 SPECIFIC HEAT OF DISODIUM SILICON TRIOXIDE Na_2SiO_3

[For Data Reported in Figure and Table No. 499]

Curve No.	Ref. No.	Year	Temp. Range, K	Reported Error, %	Name and Specimen Designation	Composition (weight percent), Specifications and Remarks
1	402	1953	64-162			
2	403	1945	298-1800	0.1-1		99.5 Na_2SiO_3.

DATA TABLE NO. 499 SPECIFIC HEAT OF DISODIUM SILICON TRIOXIDE Na_2SiO_3

[Temperature, T, K; Specific Heat, C_p, Cal $g^{-1}K^{-1}$]

T	C_p	T	C_p
CURVE 1		CURVE 2	
63.740	5.484 x 10^{-2}	298.15	2.189 x 10^{-1}
66.752	5.936	300	2.198*
67.682	6.228	400	2.535
69.886	6.474	500	2.733
71.635	6.630	600	2.876
72.754	6.902	700	2.994
74.876	7.019	800	3.098
75.480	7.238*	900	3.194
78.086	7.506	1000	3.285
78.226	7.393	1100	3.373
81.052	7.715	1200	3.458
81.494	7.987	1300	2.542
84.425	8.101	(s) 1361	3.593
84.524	8.141*	(l) 1361	3.507
87.328	8.595	1400	3.507*
89.854	8.857	1500	3.507
92.546	9.103	1600	3.507
95.550	9.414	1700	3.507
95.344	9.635		
97.924	9.889		
97.976	9.881*		
100.71	1.022 x 10^{1}		
101.16	1.016*		
103.60	1.029*		
104.27	1.045		
106.45	1.063*		
107.33	1.079		
109.23	1.090*		
110.33	1.113*		
112.08	1.108		
113.29	1.149*		
115.59	1.162		
118.99	1.209		
119.41	1.163		
122.95	1.230		
126.44	1.275		
129.83	1.296*		
135.58	1.314		
138.86	1.385		
145.30	1.434		
149.10	1.469		
152.47	1.447		
155.87	1.520		
159.23	1.538*		
162.48	1.571		

* Not shown on plot

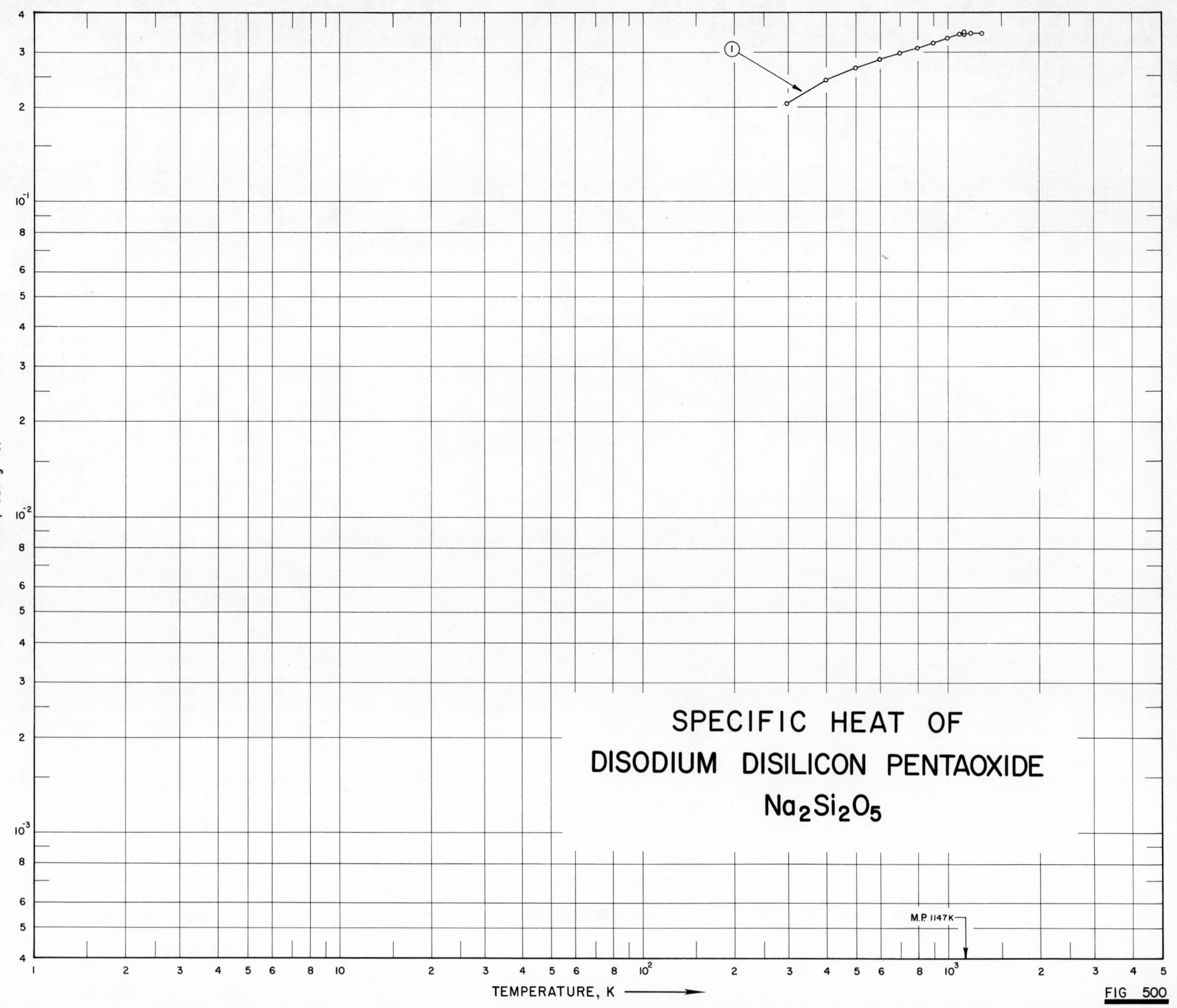
SPECIFIC HEAT OF
DISODIUM DISILICON PENTAOXIDE
$Na_2Si_2O_5$
M.P. 1147 K
TEMPERATURE, K
SPECIFIC HEAT, cal g^{-1} K^{-1}
FIG 500

SPECIFICATION TABLE NO. 500 SPECIFIC HEAT OF DISODIUM DISILICON PENTAOXIDE $Na_2Si_2O_5$

[For Data Reported in Figure and Table No. 500]

Curve No.	Ref. No.	Year	Temp. Range, K	Reported Error, %	Name and Specimen Designation	Composition (weight percent), Specifications and Remarks
1	403	1945	298-1300	1		99.0 $Na_2Si_2O_5$.

DATA TABLE NO. 500 SPECIFIC HEAT OF DISODIUM DISILICON PENTAOXIDE $Na_2Si_2O_5$

[Temperature, T, K; Specific Heat, C_p, Cal $g^{-1}K^{-1}$]

T	C_p
CURVE 1	
298.15	2.054 x 10^{-1}
300	2.064*
400	2.441
500	2.665
600	2.830
700	2.965
800	3.086
900	3.198
1000	3.304
1100	3.407
(s) 1147	3.454
(ℓ) 1147	3.422
1200	3.422
1300	3.422

* Not shown on plot

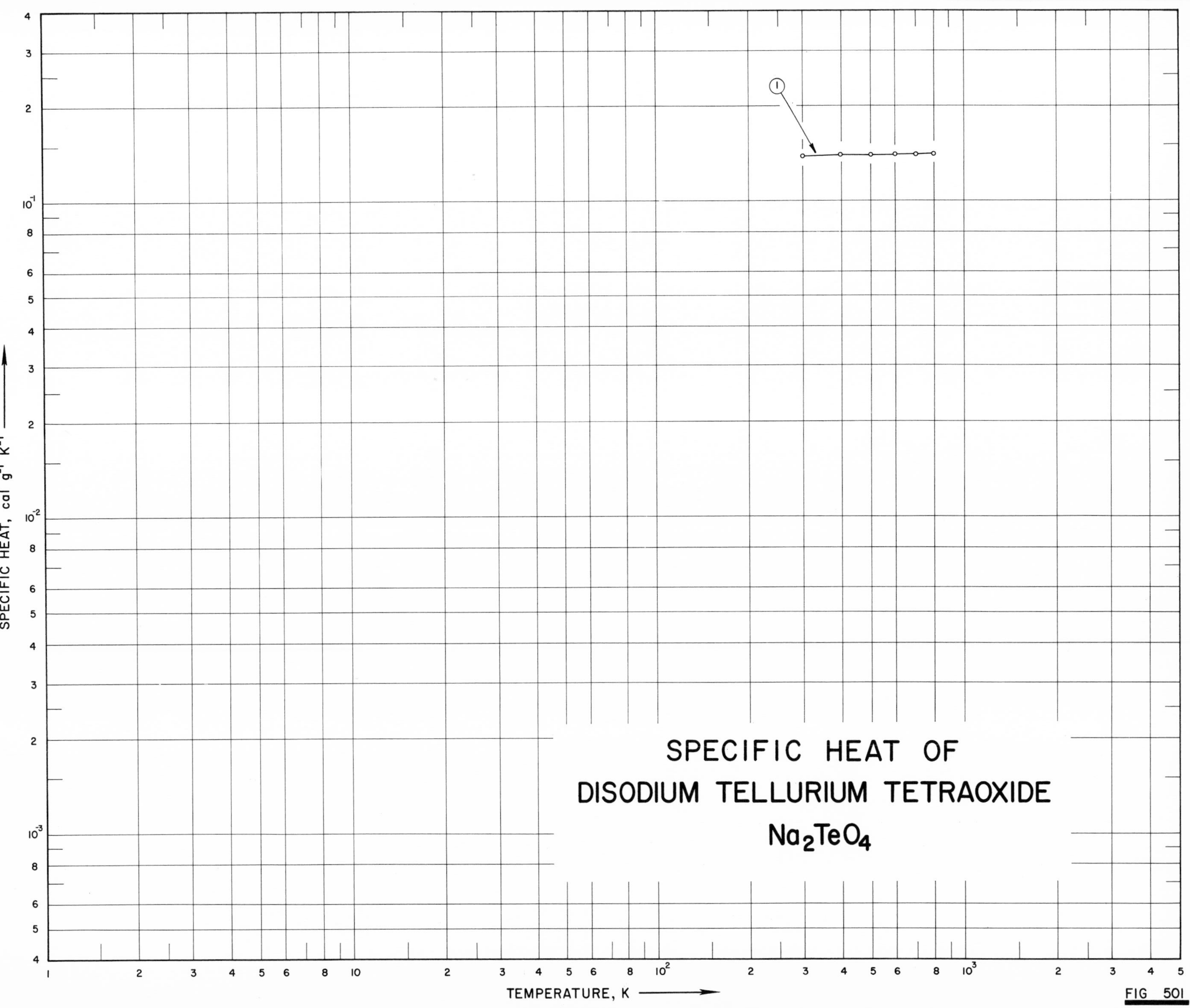
SPECIFIC HEAT OF
DISODIUM TELLURIUM TETRAOXIDE
Na_2TeO_4
SPECIFIC HEAT, cal g⁻¹ K⁻¹
TEMPERATURE, K
1

FIG 501

SPECIFICATION TABLE NO. 501 SPECIFIC HEAT OF DISODIUM TELLURIUM TETRAOXIDE Na_2TeO_4

[For Data Reported in Figure and Table No. 501]

Curve No.	Ref. No.	Year	Temp. Range, K	Reported Error, %	Name and Specimen Designation	Composition (weight percent), Specifications and Remarks
1	128	1962	300-800	0.5		$>$99.99 Na_2TeO_4, $<$0.1 Ca, traces of Ag and Al; sample supplied by E. H. Sargent and Co. Inc.; sealed in argon atmosphere.

DATA TABLE NO. 501 SPECIFIC HEAT OF DISODIUM TELLURIUM TETRAOXIDE Na_2TeO_4

[Temperature, T, K; Specific Heat, C_p, Cal $g^{-1}K^{-1}$]

T	C_p
CURVE 1	
300	1.397×10^{-1}
400	1.398
500	1.400
600	1.402
700	1.403
800	1.405

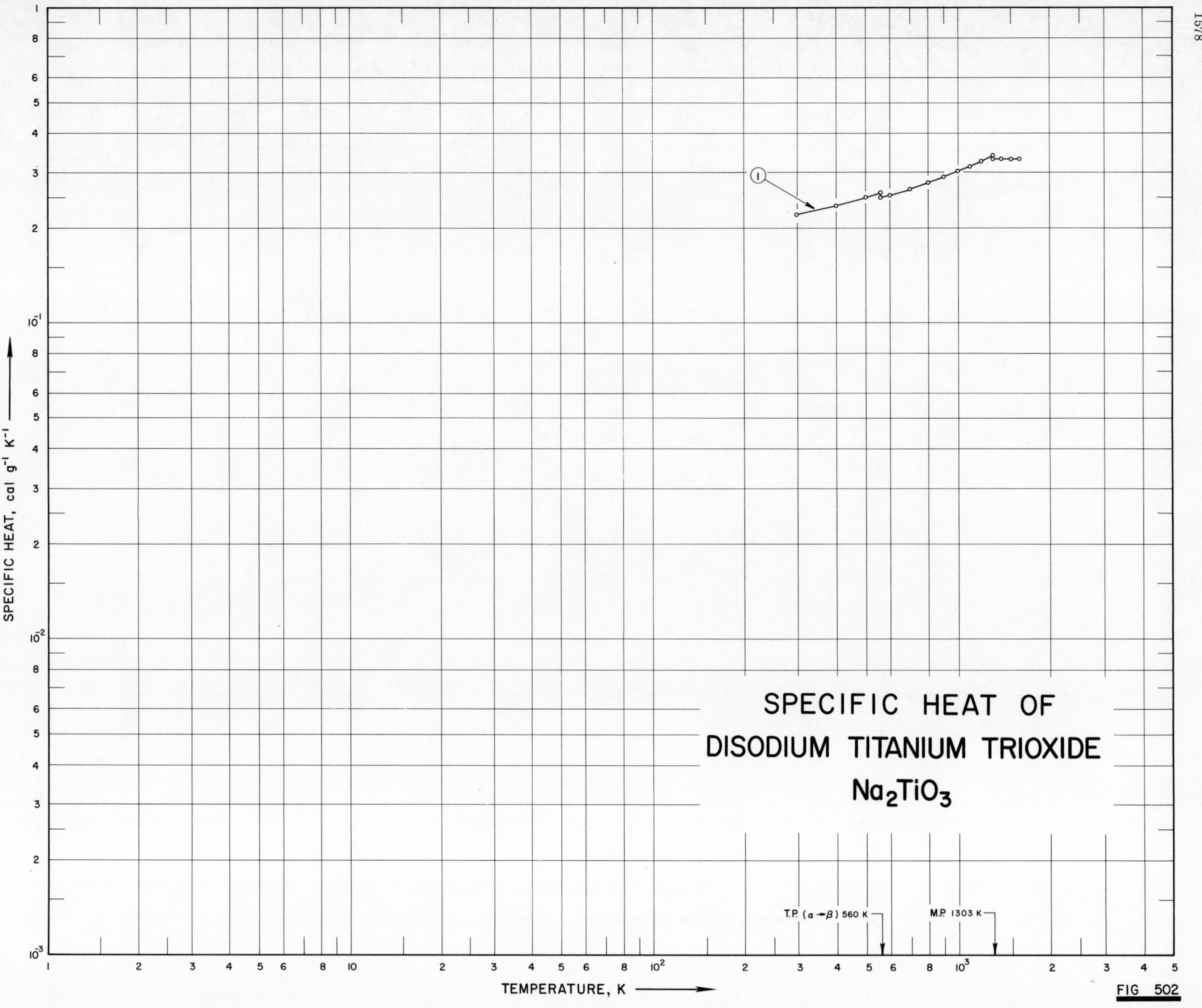

SPECIFIC HEAT OF
DISODIUM TITANIUM TRIOXIDE
Na_2TiO_3
1
T.P. (α → β) 560 K
M.P. 1303 K
SPECIFIC HEAT, cal g⁻¹ K⁻¹
TEMPERATURE, K

FIG 502

SPECIFICATION TABLE NO. 502 SPECIFIC HEAT OF DISODIUM TITANIUM TRIOXIDE Na_2TiO_3

[For Data Reported in Figure and Table No. 502]

Curve No.	Ref. No.	Year	Temp. Range, K	Reported Error, %	Name and Specimen Designation	Composition (weight percent), Specifications and Remarks
1	404	1945	298-1600			98.4 Na_2TiO_3; prepared by treating stoichiometric weights of Na_2CO_3 (prepared from reagent-grade $NaHCO_3$) with TiO_2; heated for several hrs with constant pumping to remove CO_2.

DATA TABLE NO. 502 SPECIFIC HEAT OF DISODIUM TITANIUM TRIOXIDE Na_2TiO_3

[Temperature, T,K; Specific Heat, C_p, Cal $g^{-1}K^{-1}$]

T	C_p
CURVE 1	
298.15	2.210 x 10^{-1}
300	2.213*
400	2.359
500	2.505
(α) 560	2.592
(β) 560	2.500
600	2.548
700	2.667
800	2.787
900	2.907
1000	3.027
1100	3.147
1200	3.266
1300	3.386*
(β) 1303	3.390
(ℓ) 1303	3.305
1400	3.305
1500	3.305
1600	3.305

* Not shown on plot

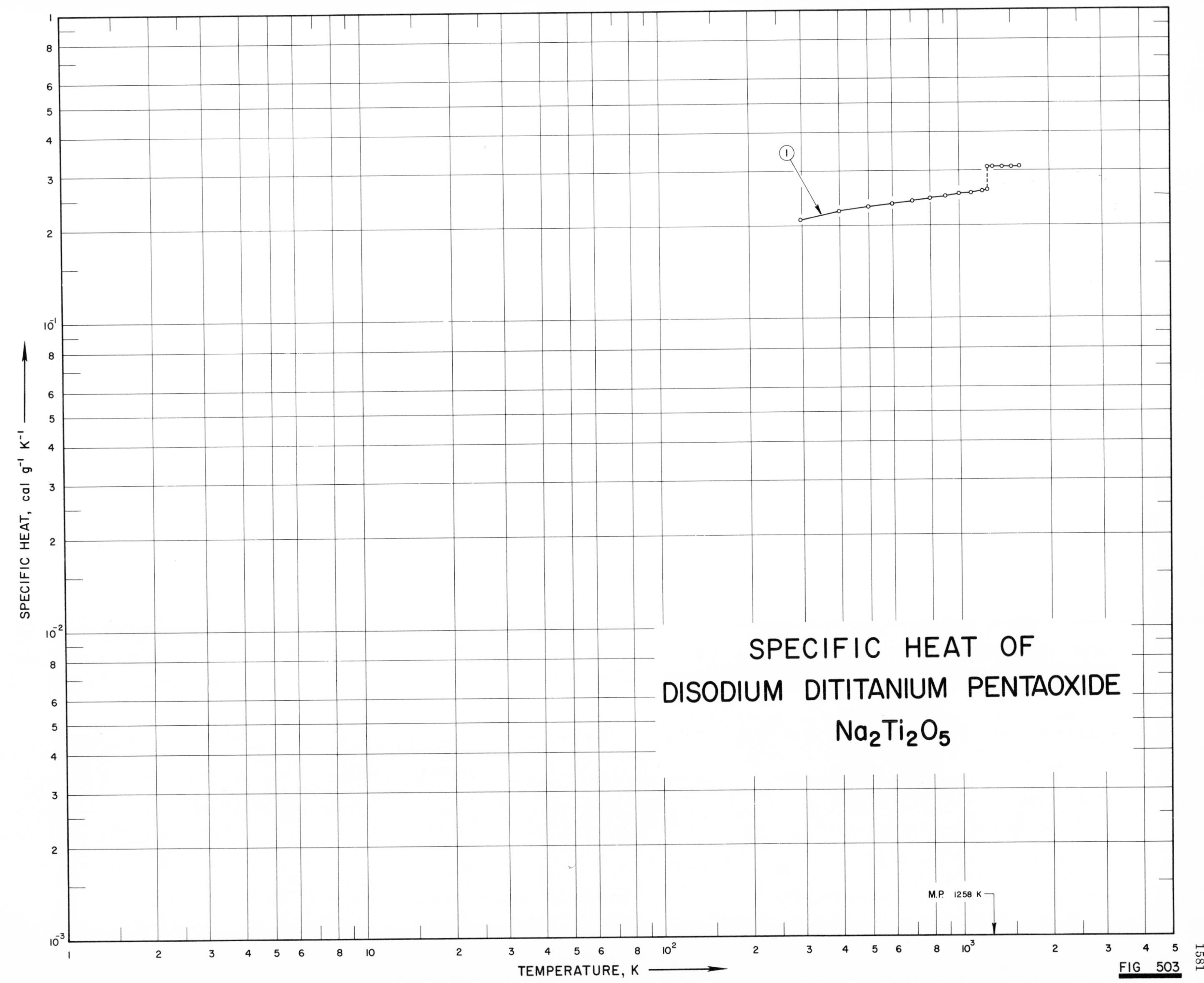
SPECIFIC HEAT OF
DISODIUM DITITANIUM PENTAOXIDE
$Na_2Ti_2O_5$
1
M.P. 1258 K
SPECIFIC HEAT, cal g^{-1} K^{-1}
TEMPERATURE, K
FIG 503

SPECIFICATION TABLE NO. 503 SPECIFIC HEAT OF DISODIUM DITITANIUM PENTAOXIDE $Na_2Ti_2O_5$

[For Data Reported in Figure and Table No. 503]

Curve No.	Ref. No.	Year	Temp. Range, K	Reported Error, %	Name and Specimen Designation	Composition (weight percent), Specifications and Remarks
1	404	1945	298-1600			98. 9 $Na_2Ti_2O_5$; prepared by treating stoichiometric weights of Na_2CO_3 (prepared from reagent-grade $NaHCO_3$) with 98. 6 pure TiO_2; heated several hrs at 900-1100 C with constant pumping to remove CO_2.

DATA TABLE NO. 503 SPECIFIC HEAT OF DISODIUM DITITANIUM PENTAOXIDE $Na_2Ti_2O_5$

[Temperature, T, K; Specific Heat, C_p, Cal $g^{-1}K^{-1}$]

T	C_p
CURVE 1	
298.15	2.085×10^{-1}
300	2.089*
400	2.221
500	2.300
600	2.357
700	2.404
800	2.446
900	2.485
1000	2.521
1100	2.557
1200	2.591
(s) 1258	2.611
(ℓ) 1258	3.088
1300	3.088
1400	3.088
1500	3.088
1600	3.088

* Not shown on plot

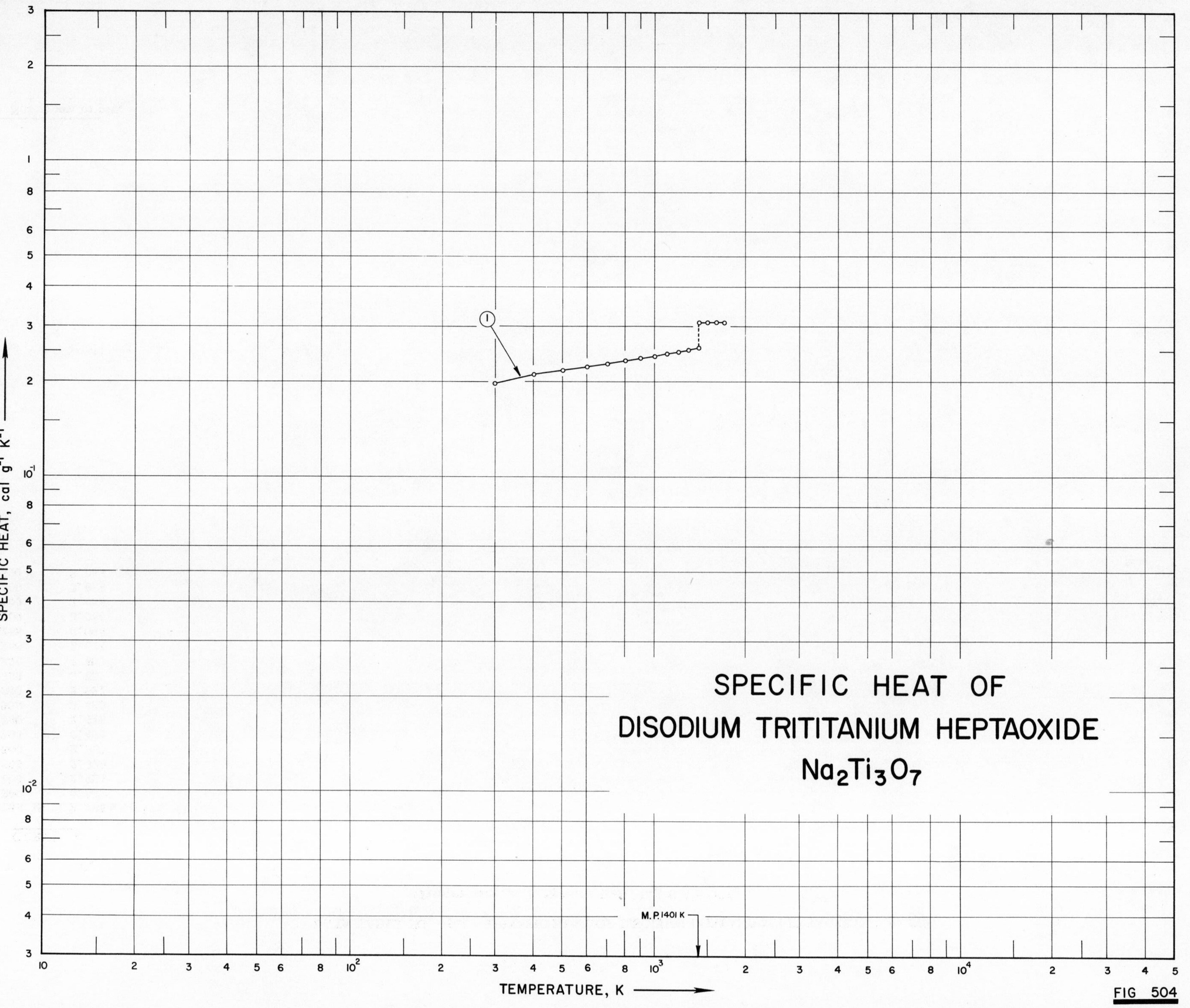
SPECIFIC HEAT OF
DISODIUM TRITITANIUM HEPTAOXIDE
$Na_2Ti_3O_7$
1
M.P. 1401 K
SPECIFIC HEAT, cal g⁻¹ K⁻¹
TEMPERATURE, K
FIG 504

SPECIFICATION TABLE NO. 504 SPECIFIC HEAT OF DISODIUM TRITITANIUM HEPTAOXIDE $Na_2Ti_3O_7$

[For Data Reported in Figure and Table No. 504]

Curve No.	Ref. No.	Year	Temp. Range, K	Reported Error, %	Name and Specimen Designation	Composition (weight percent), Specifications and Remarks
1	404	1945	298-1700			98.6 $Na_2Ti_3O_7$; prepared by treating stoichiometric weights Na_2CO_3 (prepared from reagent-grade $NaHCO_3$) with 98.6 pure TiO_2; heated several hrs with constant pumping to remove CO_2.

DATA TABLE NO. 504 SPECIFIC HEAT OF DISODIUM TRITITANIUM HEPTAOXIDE $Na_2Ti_3O_7$

[Temperature, T, K; Specific Heat, C_p, Cal $g^{-1}K^{-1}$]

T	C_p
CURVE 1	
298	1.998×10^{-1}
300	2.001*
400	2.128
500	2.205
600	2.263
700	2.312
800	2.356
900	2.398
1000	2.437
1100	2.476
1200	2.514
1300	2.551
(s) 1401	2.588
(ℓ) 1401	3.121
1500	3.121
1600	3.121
1700	3.121

* Not shown on plot

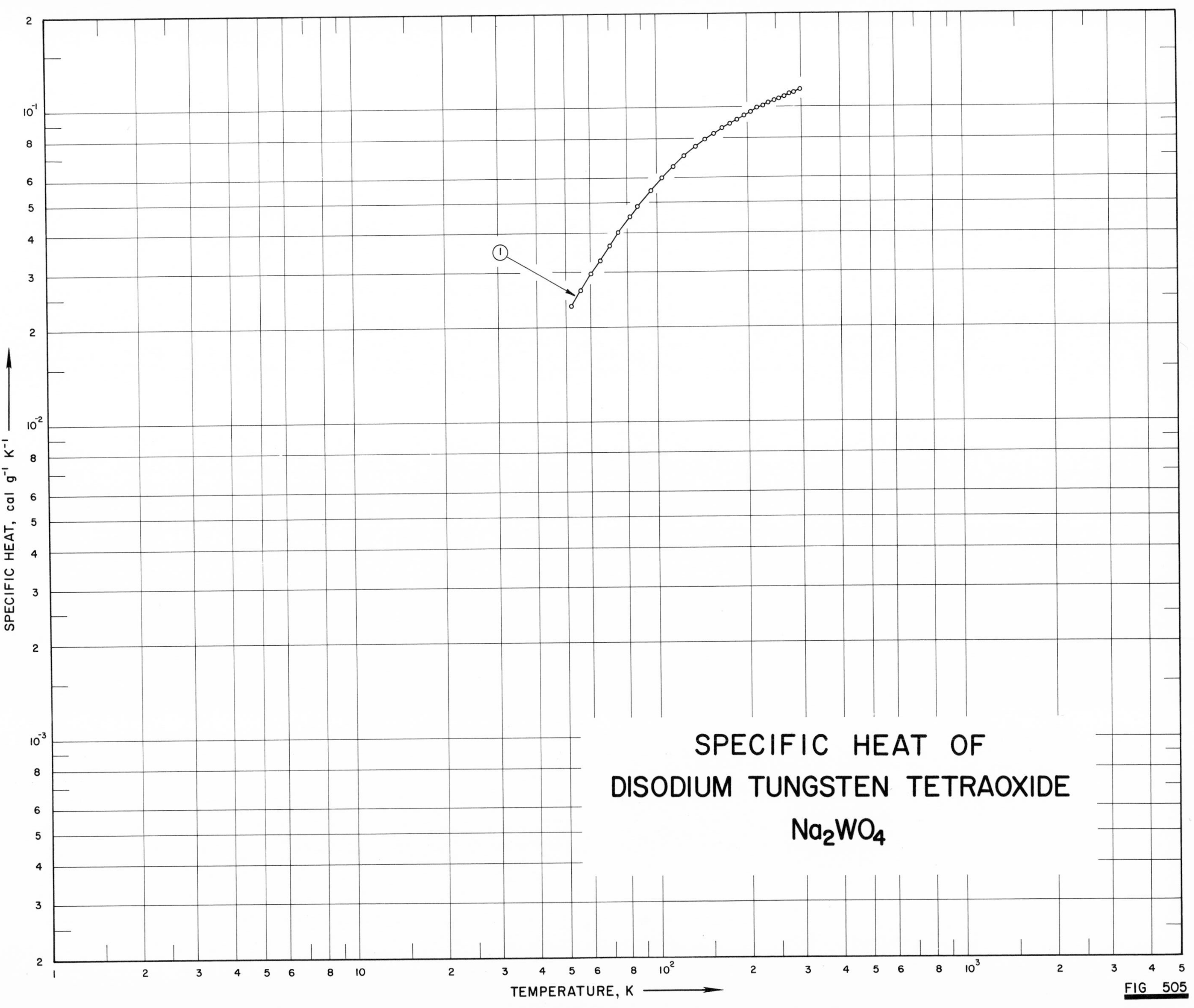
SPECIFIC HEAT OF
DISODIUM TUNGSTEN TETRAOXIDE
Na_2WO_4
SPECIFIC HEAT, cal g^{-1} K^{-1}
TEMPERATURE, K
1

FIG 505

SPECIFICATION TABLE NO. 505 SPECIFIC HEAT OF DISODIUM TUNGSTEN TETRAOXIDE Na_2WO_4

[For Data Reported in Figure and Table No. 505]

Curve No.	Ref. No.	Year	Temp. Range, K	Reported Error, %	Name and Specimen Designation	Composition (weight percent), Specifications and Remarks
1	375	1961	52-300	0.3		78.95 WO_3; prepared by melting twice stoichiometric mixture of reagent-grade sodium carbonate and tungstic acid; heated 4 times for total of 4.5 days at 460-600 C.

DATA TABLE NO. 505 SPECIFIC HEAT OF DISODIUM TUNGSTEN TETRAOXIDE Na_2WO_4

[Temperature, T, K; Specific Heat, C_p, Cal $g^{-1}K^{-1}$]

T	C_p
CURVE 1	
52.01	2.358×10^{-2}
55.97	2.644
60.42	2.974
64.83	3.298
69.61	3.658
74.62	4.026
82.13	4.506
86.81	4.860
95.18	5.397
105.12	5.983
114.65	6.497
124.74	6.990
136.04	7.501
146.02	7.909
155.91	8.277
166.07	8.600
176.07	8.899
186.09	9.185
196.09	9.430
206.18	9.682
216.29	9.954
226.34	1.013×10^{-1}
236.20	1.032
246.06	1.051
256.61	1.070
266.80	1.087
276.49	1.103
286.66	1.118
296.07	1.133*
299.87	1.140

* Not shown on plot

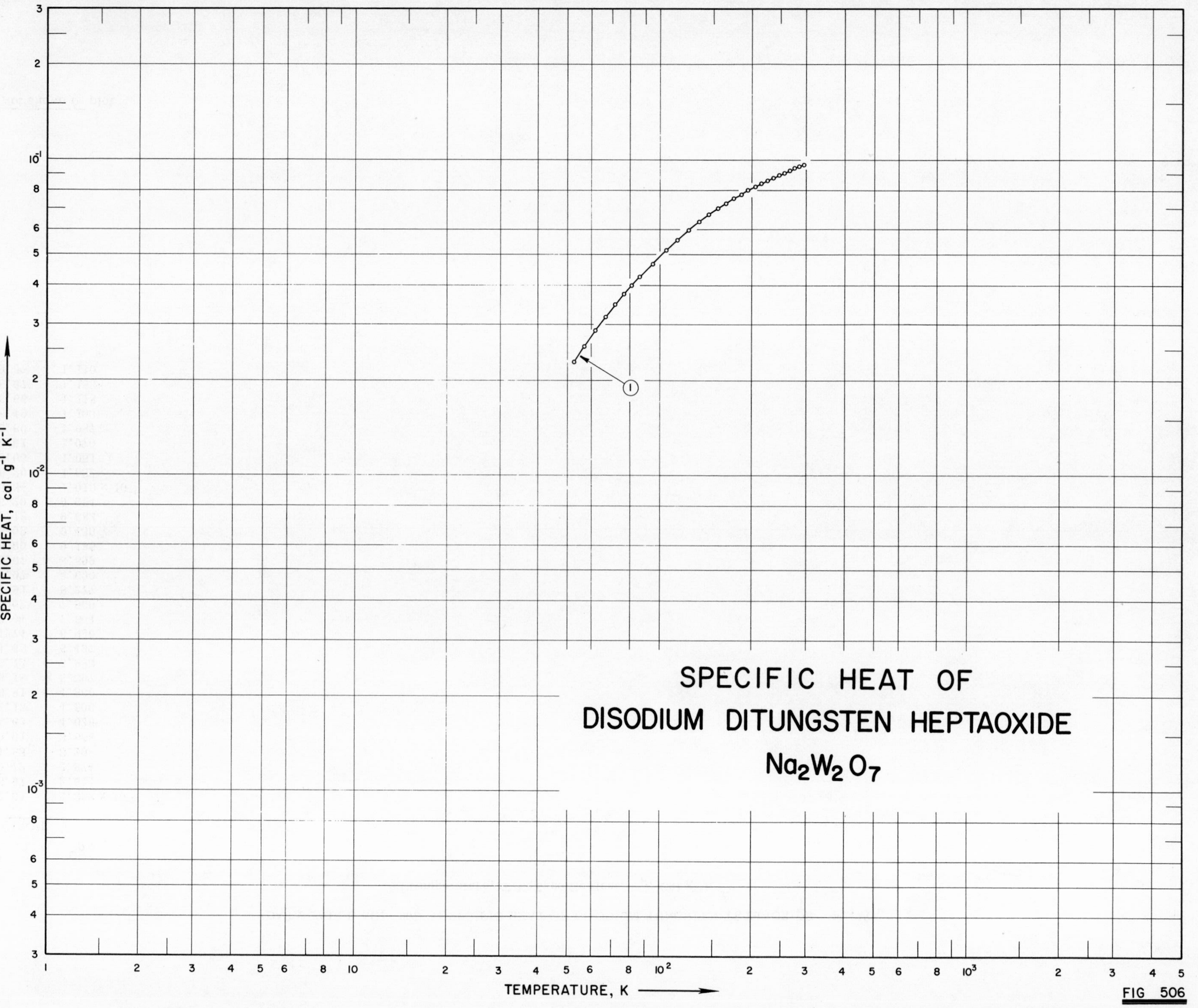

SPECIFIC HEAT OF
DISODIUM DITUNGSTEN HEPTAOXIDE
$Na_2W_2O_7$
SPECIFIC HEAT, cal g^{-1} K^{-1}
TEMPERATURE, K
1

FIG 506

SPECIFICATION TABLE NO. 506 SPECIFIC HEAT OF DISODIUM DITUNGSTEN HEPTAOXIDE $Na_2W_2O_7$

[For Data Reported in Figure and Table No. 506]

Curve No.	Ref. No.	Year	Temp. Range, K	Reported Error, %	Name and Specimen Designation	Composition (weight percent), Specifications and Remarks
1	401	1963	53-296	0.3		87.78 WO_3, 0.24 Al_2O_3, 0.24 SiO_2, 11.79 Na_2O.

DATA TABLE NO. 506 SPECIFIC HEAT OF DISODIUM DITUNGSTEN HEPTAOXIDE $Na_2W_2O_7$

[Temperature, T, K; Specific Heat, C_p, Cal $g^{-1}K^{-1}$]

T	C_p
CURVE 1	
52.75	2.311 x 10^{-2}
56.90	2.574
61.80	2.897
66.79	3.198
71.66	3.487
76.59	3.772
81.02	4.018
86.14	4.292
94.64	4.697
104.99	5.205
114.68	5.625
124.69	6.023
135.84	6.430
145.61	6.753
155.71	7.056
165.64	7.330
176.25	7.602
185.99	7.832
195.89	8.056
206.21	8.265
216.05	8.465
226.27	8.657
236.05	8.834
245.79	9.000
256.20	9.169
266.47	9.319
276.31	9.468
286.84	9.609
296.26	9.744

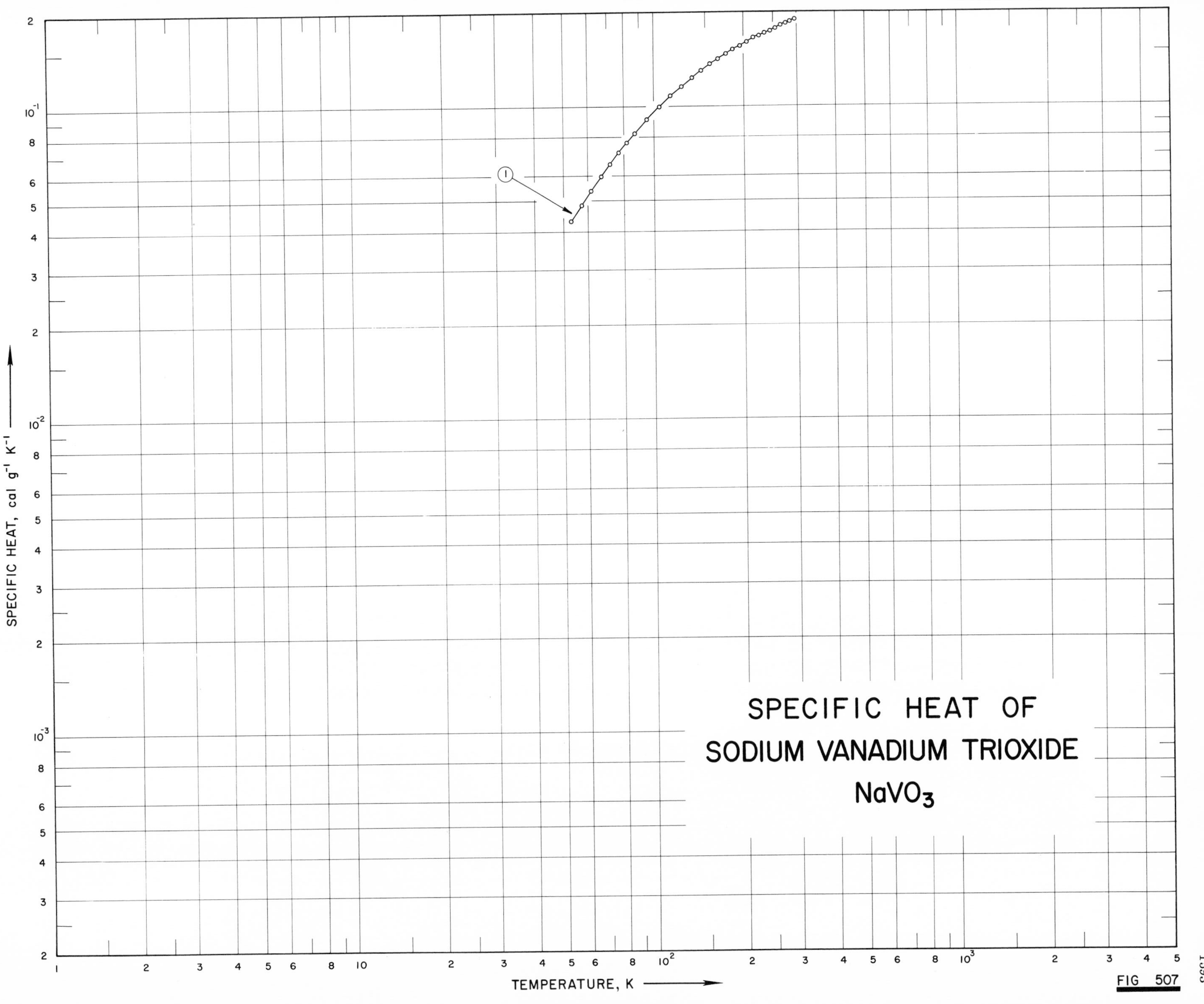
SPECIFIC HEAT OF
SODIUM VANADIUM TRIOXIDE
NaVO3
SPECIFIC HEAT, cal g-1 K-1
TEMPERATURE, K
1
FIG 507

SPECIFICATION TABLE NO. 507 SPECIFIC HEAT OF SODIUM VANADIUM TRIOXIDE $NaVO_3$

[For Data Reported in Figure and Table No. 507]

Curve No.	Ref. No.	Year	Temp. Range, K	Reported Error, %	Name and Specimen Designation	Composition (weight percent), Specifications and Remarks
1	405	1961	53-296	0.3		99.8 $NaVO_3$; heated at 500 C for 3 hrs.

DATA TABLE NO. 507 SPECIFIC HEAT OF SODIUM VANADIUM TRIOXIDE $NaVo_3$

[Temperature, T, K; Specific Heat, C_p, Cal $g^{-1}K^{-1}$]

T	C_p
CURVE 1	
52.79	4.345×10^{-2}
57.22	4.876
61.86	5.433
66.61	6.006
71.57	6.593
76.62	7.168
81.88	7.745
86.80	8.258
94.20	9.119
104.92	1.002×10^{-1}
114.95	1.087
124.87	1.166
136.07	1.246
146.32	1.315
155.97	1.375
165.95	1.427
176.02	1.479
186.02	1.527
196.00	1.570
206.17	1.614
217.57	1.661
226.10	1.692
236.30	1.728
246.28	1.758
256.65	1.794
266.75	1.825
276.68	1.853
286.85	1.880
296.36	1.908

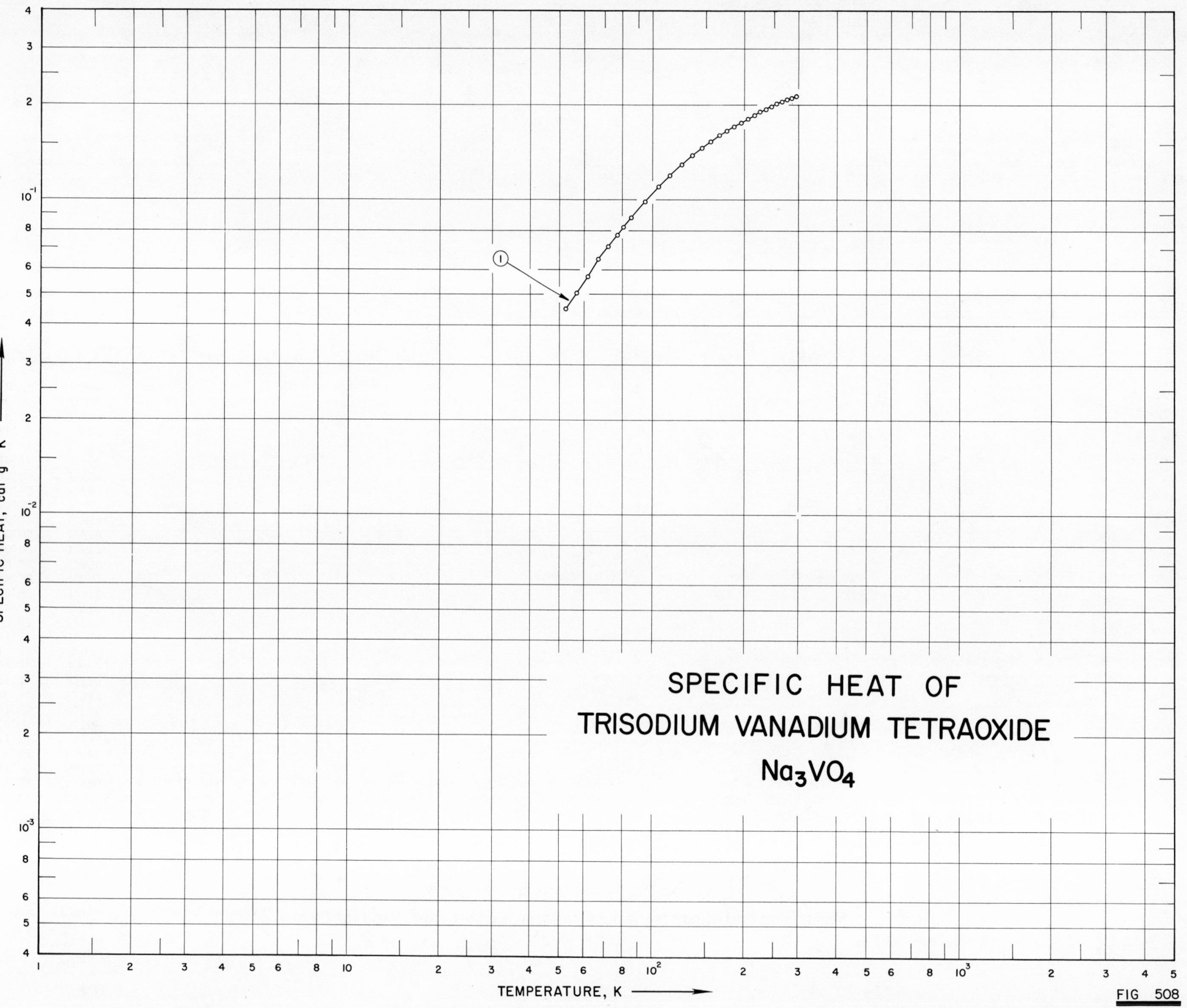
SPECIFIC HEAT OF
TRISODIUM VANADIUM TETRAOXIDE
Na_3VO_4
SPECIFIC HEAT, cal g^{-1} K^{-1}
TEMPERATURE, K
1

FIG 508

SPECIFICATION TABLE NO. 508 SPECIFIC HEAT OF TRISODIUM VANADIUM TETRAOXIDE Na_3VO_4

[For Data Reported in Figure and Table No. 508]

Curve No.	Ref. No.	Year	Temp. Range, K	Reported Error, %	Name and Specimen Designation	Composition (weight percent), Specifications and Remarks
1	405	1961	53-296	0.3		99.8 Na_3VO_4; heated 5 hrs at 725 C.

DATA TABLE NO. 508 SPECIFIC HEAT OF TRISODIUM VANADIUM TETRAOXIDE Na_3VO_4

[Temperature, T,K; Specific Heat, C_p, Cal $g^{-1}K^{-1}$]

T	C_p
CURVE 1	
52.59	4.526×10^{-2}
56.96	5.085
61.73	5.725
66.78	6.465
71.99	7.095
77.11	7.742
80.52	8.194
85.48	8.803
94.84	9.896
105.64	1.108×10^{-1}
114.60	1.196
125.33	1.296
236.18	1.389
145.64	1.463
155.79	1.535
165.90	1.601
175.96	1.660
186.11	1.718
196.17	1.766
206.26	1.808
216.04	1.862
226.37	1.903
236.32	1.944
245.84	1.977
256.42	2.021
266.26	2.051
276.28	2.081
286.40	2.107
295.99	2.140

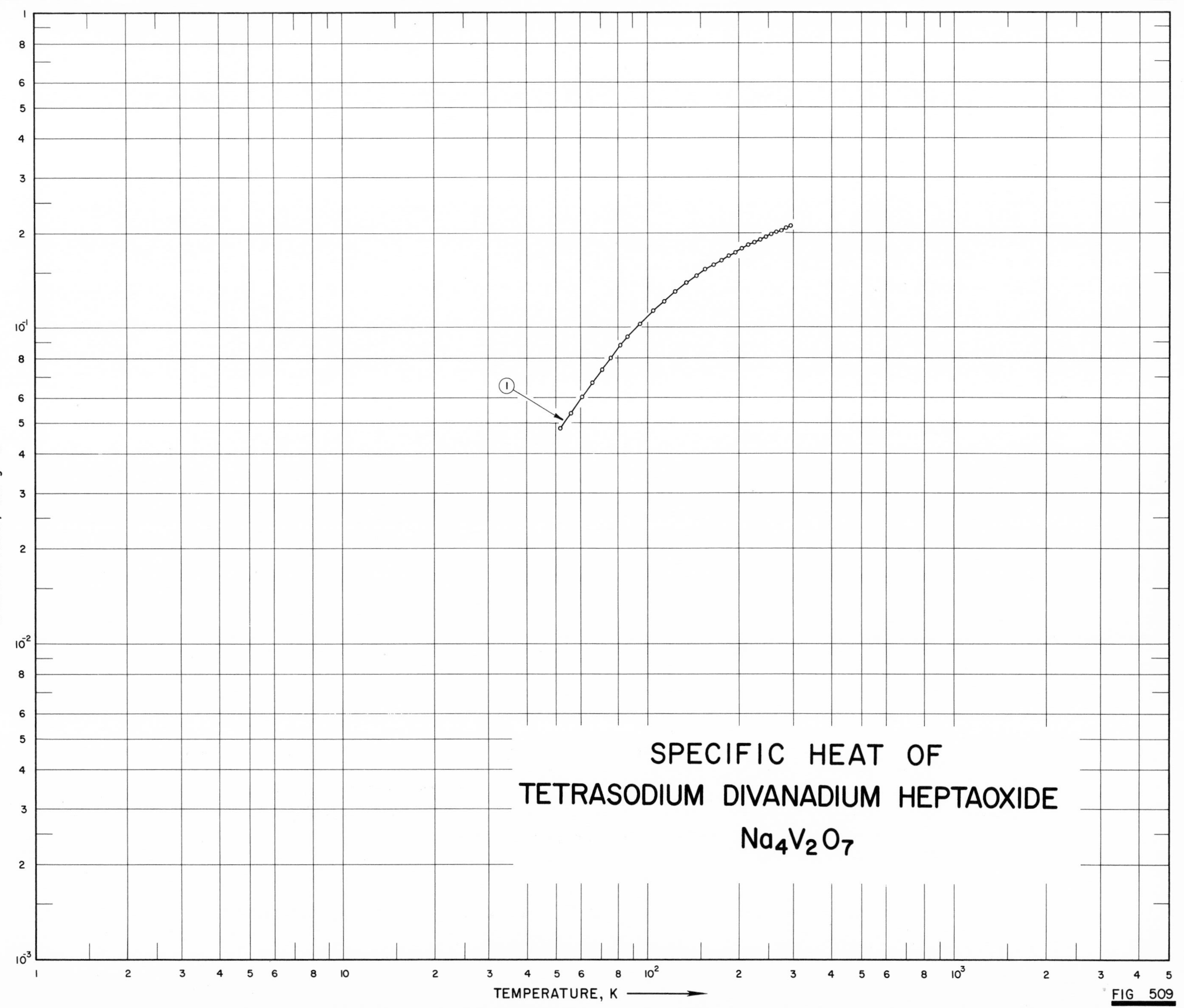
SPECIFIC HEAT OF
TETRASODIUM DIVANADIUM HEPTAOXIDE
$Na_4V_2O_7$
SPECIFIC HEAT, cal g⁻¹ K⁻¹
TEMPERATURE, K
1

FIG 509

SPECIFICATION TABLE NO. 509 SPECIFIC HEAT OF TETRASODIUM DIVANADIUM HEPTAOXIDE $Na_4V_2O_7$

[For Data Reported in Figure and Table No. 509]

Curve No.	Ref. No.	Year	Temp. Range, K	Reported Error, %	Name and Specimen Designation	Composition (weight percent), Specifications and Remarks
1	405	1961	52-296	0.3		99.8 $Na_4V_2O_7$; heated for 5 hrs at 500 C.

DATA TABLE NO. 509 SPECIFIC HEAT OF TETRASODIUM DIVANADIUM HEPTAOXIDE $Na_4V_2O_7$

[Temperature, T, K; Specific Heat, C_p, Cal $g^{-1}K^{-1}$]

T	C_p
CURVE 1	
51.93	4.816×10^{-2}
56.04	5.372
60.88	6.035
65.77	6.712
70.77	7.389
75.83	8.030
81.99	8.792
86.65	9.341
94.83	1.025×10^{-1}
105.03	1.130
114.56	1.218
124.75	1.306
135.88	1.392
145.79	1.464
155.95	1.530
166.00	1.589
176.16	1.644
186.13	1.695
195.97	1.738
206.33	1.787
216.28	1.830
226.28	1.869
236.40	1.907
245.90	1.940
256.47	1.979
266.46	2.011
276.14	2.038
286.76	2.074
295.94	2.104

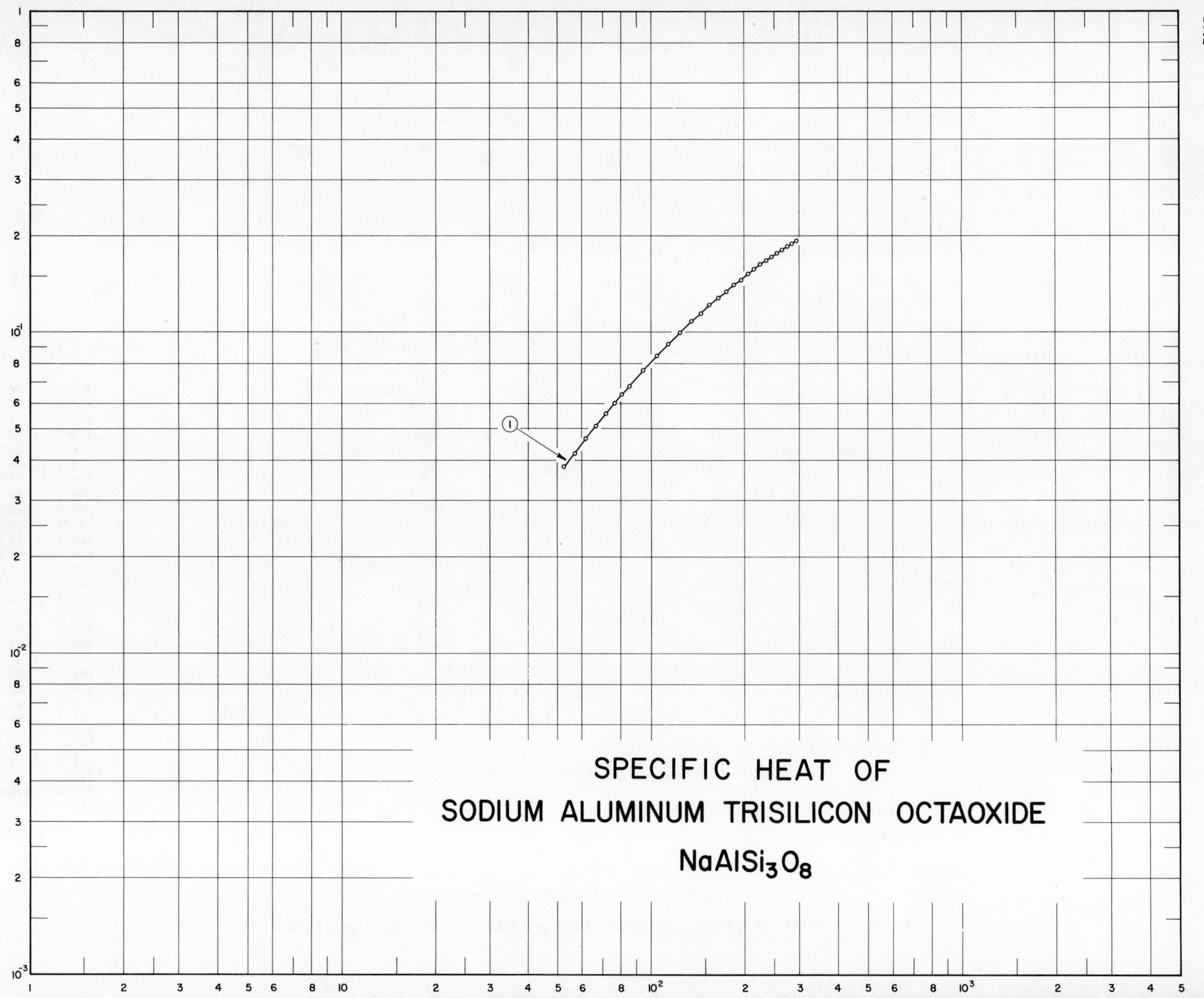
SPECIFIC HEAT, cal g^{-1} K^{-1}
1 8 6 5 4 3 2 10^{-1} 8 6 5 4 3 2 10^{-2} 8 6 5 4 3 2 10^{-3}
1 2 3 4 5 6 8 10 2 3 4 5 6 8 10^2 2 3 4 5 6 8 10^3 2 3 4 5
1
SPECIFIC HEAT OF
SODIUM ALUMINUM TRISILICON OCTAOXIDE
$NaAlSi_3O_8$

SPECIFICATION TABLE NO. 510 SPECIFIC HEAT OF SODIUM ALUMINUM TRISILICON OCTAOXIDE $NaAlSi_3O_8$

[For Data Reported in Figure and Table No. 510]

Curve No.	Ref. No.	Year	Temp. Range, K	Reported Error, %	Name and Specimen Designation	Composition (weight percent), Specifications and Remarks
1	380	1961	53-297		Dehydrated analcite	56.05 SiO_2, 22.36 Al_2O_3, 13.44 Na_2O, 8.13 combined H_2O, 0.10 K_2O, 0.02 MgO, 0.01 TiO_2, 0.03 Fe_2O_3, 0.01 absorbed H_2O, $<$0.005 CaO; water free basis: 61.01 SiO_2, 24.34 Al_2O_3, 14.63 Na_2O. 0.11 K_2O, 0.03 Fe_2O_3, 0.02 MgO, 0.01 TiO; prepared from analcite; heated at 500 C for 20 hrs.

DATA TABLE NO. 510 SPECIFIC HEAT OF SODIUM ALUMINUM TRISILICON OCTAOXIDE $NaAlSi_3O_8$

[Temperature, T, K; Specific Heat, C_p, Cal $g^{-1}K^{-1}$]

T	C_p
CURVE 1	
52.67	3.805 x 10^{-2}
57.33	4.226
61.98	4.670
66.76	5.120
71.63	5.575
76.61	6.020
80.97	6.411
85.51	6.812
94.76	7.603
105.02	8.464
114.52	9.211
124.81	9.998
136.04	1.083 x 10^{-1}
145.74	1.152
155.94	1.223
165.90	1.285
176.07	1.349
186.30	1.412
196.11	1.466
206.26	1.525
216.39	1.577
226.27	1.629
236.11	1.676
245.69	1.720
256.75	1.770
266.60	1.813
276.81	1.855
286.77	1.896
296.59	1.939

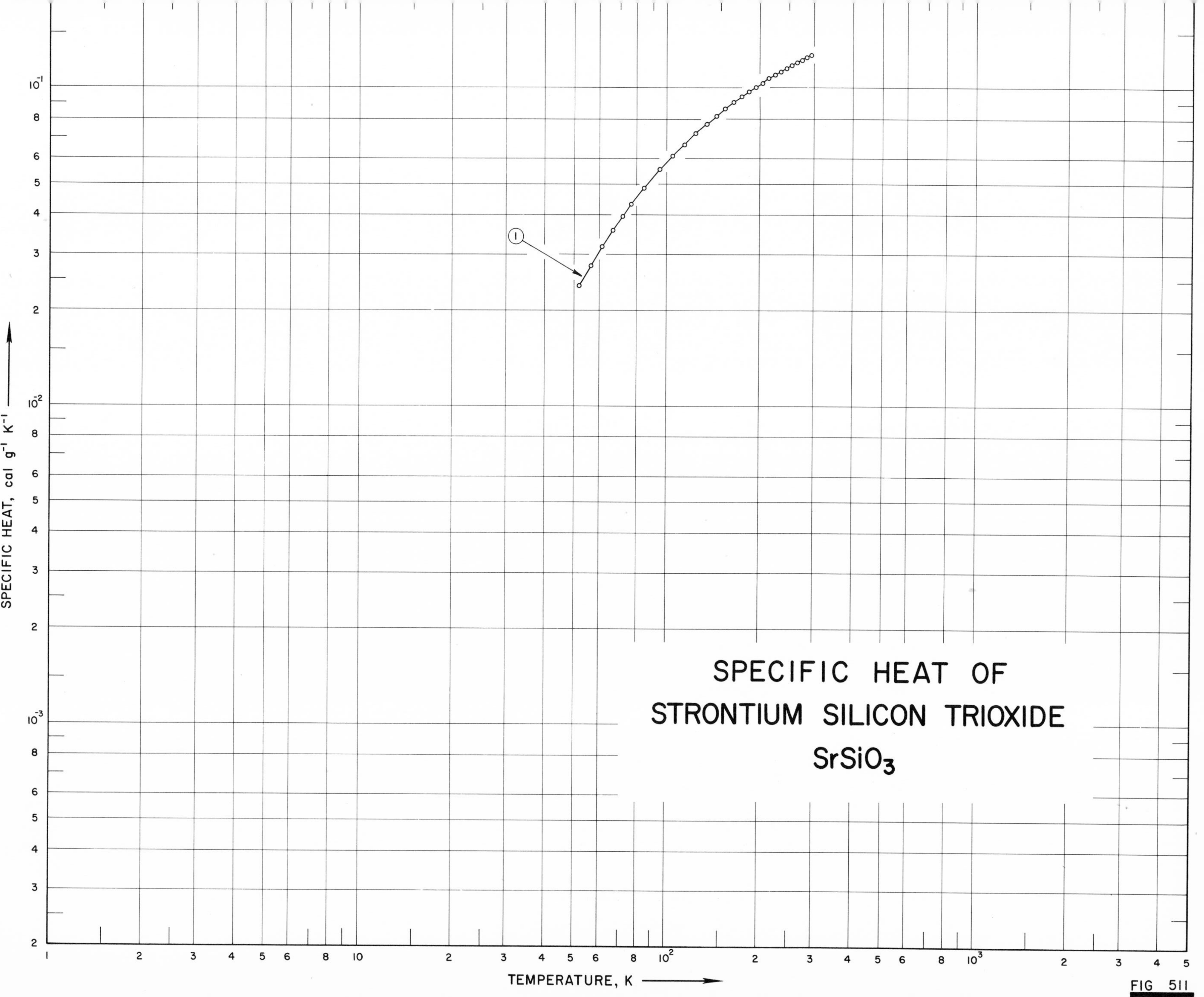
SPECIFIC HEAT OF
STRONTIUM SILICON TRIOXIDE
$SrSiO_3$
SPECIFIC HEAT, cal g^{-1} K^{-1}
TEMPERATURE, K
1
FIG 511

SPECIFICATION TABLE NO. 511 SPECIFIC HEAT OF STRONTIUM SILICON TRIOXIDE $SrSiO_3$

[For Data Reported in Figure and Table No. 511]

Curve No.	Ref. No.	Year	Temp. Range, K	Reported Error, %	Name and Specimen Designation	Composition (weight percent), Specifications and Remarks
1	360	1964	52-296			63.26 SrO, 36.66 SiO_2 (63.30, 36.70 theo.), 0.12 Al_2O_3.

DATA TABLE NO. 511 SPECIFIC HEAT OF STRONTIUM SILICON TRIOXIDE $SrSiO_3$

[Temperature, T, K; Specific Heat, C_p, Cal $g^{-1}K^{-1}$]

T	C_p
CURVE 1	
52.11	2.422×10^{-2}
56.58	2.786
61.50	3.202
66.84	3.613
71.95	3.999
76.71	4.354
80.07	4.594
84.75	4.912
95.48	5.600
105.23	6.194
114.96	6.744
124.81	7.269
135.83	7.825
145.64	8.277
155.71	8.717
166.07	9.133
176.11	9.524
186.13	9.878
196.14	1.021×10^{-1}
206.24	1.054
216.52	1.086
226.28	1.116
236.34	1.145
245.94	1.171
256.35	1.196
266.58	1.222
276.55	1.246
286.92	1.269
296.39	1.289

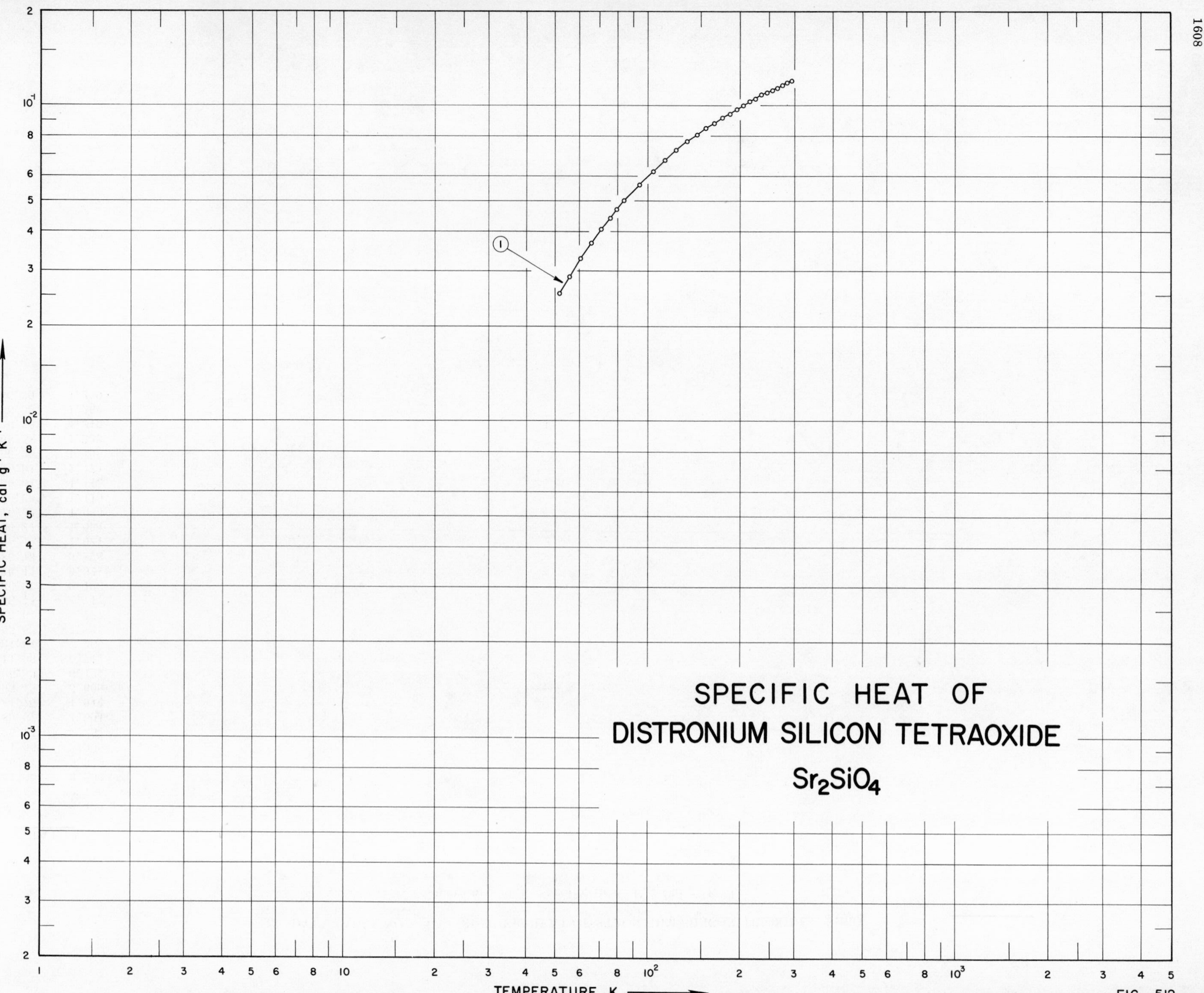
SPECIFIC HEAT OF
DISTRONIUM SILICON TETRAOXIDE
Sr_2SiO_4
SPECIFIC HEAT, cal g^{-1} K^{-1}
TEMPERATURE, K
1
FIG 512

SPECIFICATION TABLE NO. 512 SPECIFIC HEAT OF DISTRONTIUM SILICON TETRAOXIDE Sr_2SiO_4

[For Data Reported in Figure and Table No. 512]

Curve No.	Ref. No.	Year	Temp. Range, K	Reported Error, %	Name and Specimen Designation	Composition (weight percent), Specifications and Remarks
1	360	1964	52-296			77.51 SrO, 22.47 SiO_2 (77.53, 22.47 theo.).

DATA TABLE NO. 512 SPECIFIC HEAT OF DISTRONTIUM SILICON TETRAOXIDE Sr_2SiO_4

[Temperature, T,K; Specific Heat, C_p, Cal $g^{-1}K^{-1}$]

T	C_p
CURVE 1	
51.99	2.56 x 10^{-2}
55.84	2.88
60.56	3.28
65.62	3.68
70.64	4.06
75.67	4.42
79.56	4.71
84.08	5.01
94.41	5.64
105.13	6.23
114.68	6.73
124.70	7.22
135.78	7.70
145.63	8.09
155.69	8.47
166.25	8.83
176.10	9.15
185.94	9.44
195.97	9.74
206.15	1.00 x 10^{-1}
216.35	1.03
226.09	1.05
236.24	1.08
245.78	1.10
256.45	1.12
266.39	1.14
276.53	1.16
286.74	1.18
296.27	1.20

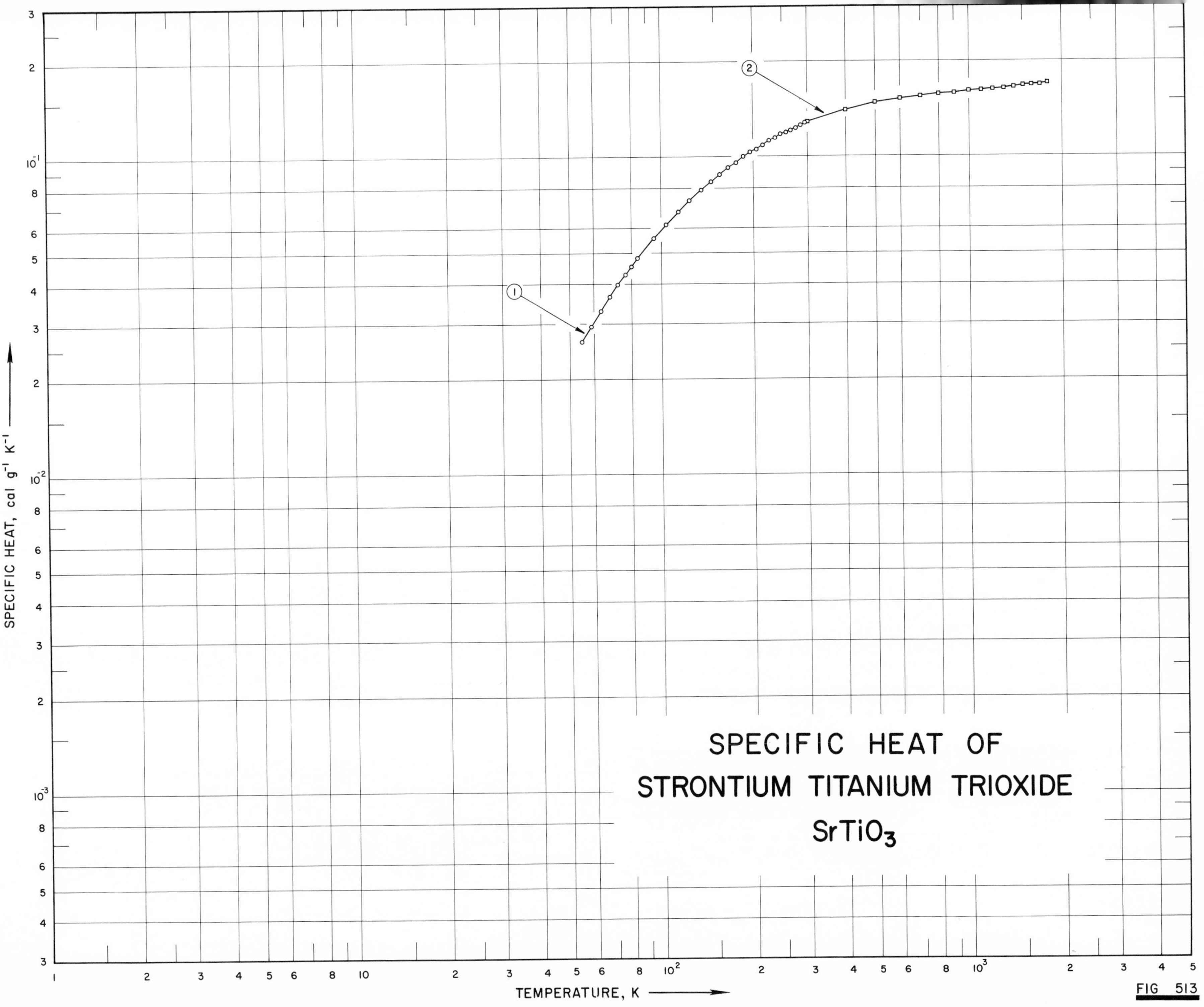
SPECIFIC HEAT OF
STRONTIUM TITANIUM TRIOXIDE
SrTiO$_3$
SPECIFIC HEAT, cal g^{-1} K^{-1}
TEMPERATURE, K
1
2
FIG 513

SPECIFICATION TABLE NO. 513 SPECIFIC HEAT OF STRONTIUM TITANIUM TRIOXIDE $SrTiO_3$

[For Data Reported in Figure and Table No. 513]

Curve No.	Ref. No.	Year	Temp. Range, K	Reported Error, %	Name and Specimen Designation	Composition (weight percent), Specifications and Remarks
1	362	1952	55-298			99.5 $SrTiO_3$; prepared from reagent-grade strontium carbonate and titania by prolonged heating at 1350 C.
2	361	1953	298-1800			99.5 $SrTiO_3$; prepared from reagent-grade strontium carbonate and titania by prolonged heating at 1350 C.

DATA TABLE NO. 513 SPECIFIC HEAT OF STRONTIUM TITANIUM TRIOXIDE $SrTiO_3$

[Temperature, T, K; Specific Heat, C_p, Cal $g^{-1}K^{-1}$]

T	C_p
CURVE 1	
54.84	2.639×10^{-2}
58.79	2.947
63.10	3.301
67.58	3.651
71.96	3.974
76.44	4.297
80.04	4.559
84.13	4.848
94.62	5.552
104.37	6.162
114.75	6.773
124.78	7.307
136.25	7.912
146.27	8.396
156.01	8.821
166.02	9.252
176.15	9.644
186.02	1.001×10^{-1}
195.95	1.036
206.34	1.067
216.32	1.098
226.26	1.127
236.39	1.153
247.32	1.177
256.49	1.198
266.22	1.219
276.19	1.240
286.38	1.261
296.47	1.276*
298.16	1.281
CURVE 2	
298	1.28×10^{-1}*
300	1.28
400	1.42
500	1.48
600	1.53
700	1.55
800	1.58
900	1.59
1000	1.61
1100	1.62
1200	1.64
CURVE 2 (cont.)	
1300	1.65×10^{-1}
1400	1.66
1500	1.67
1600	1.68
1700	1.69
1800	1.70

* Not shown on plot

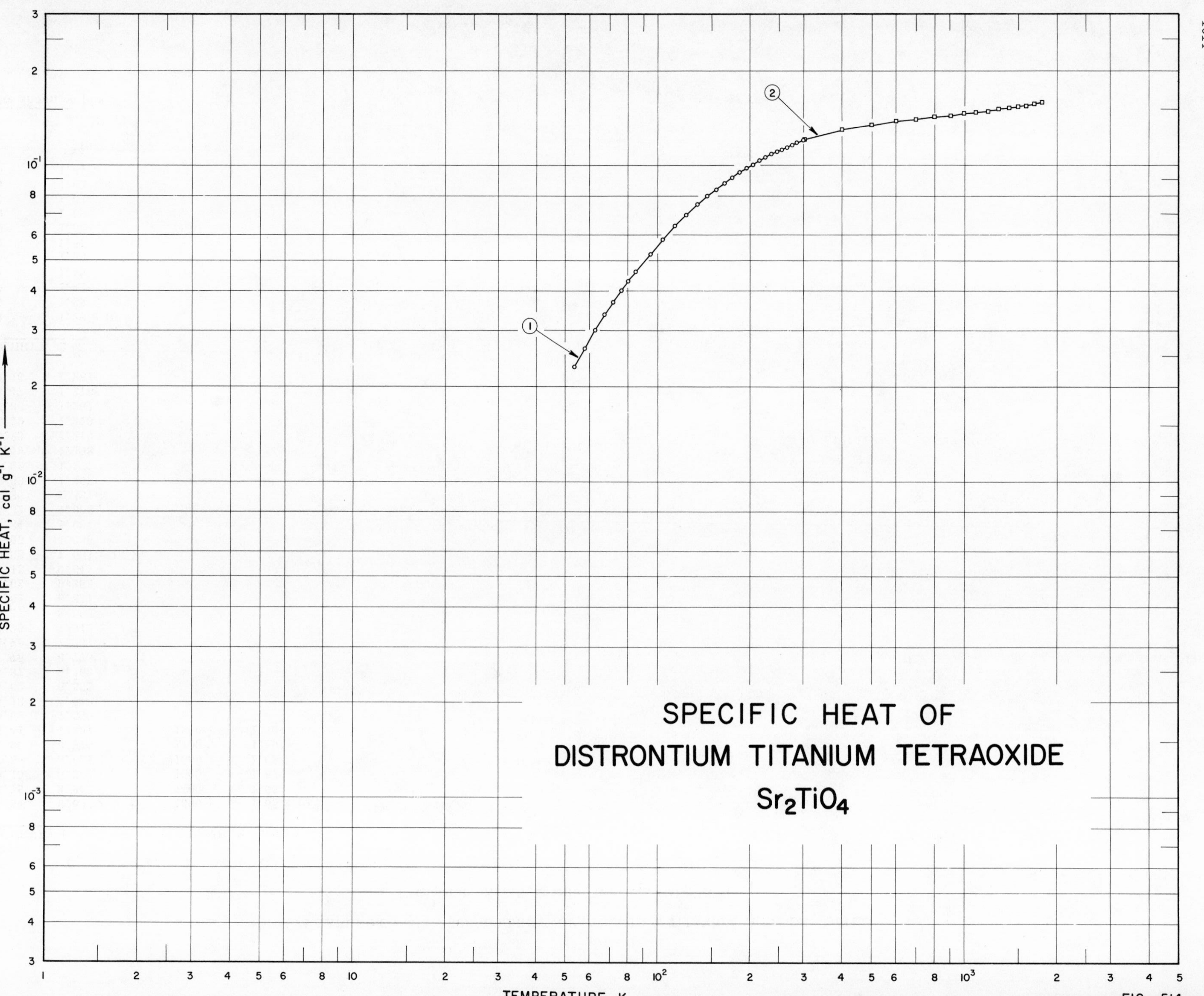
SPECIFIC HEAT OF
DISTRONTIUM TITANIUM TETRAOXIDE
Sr_2TiO_4
SPECIFIC HEAT, cal g^{-1} K^{-1}
1
2

SPECIFICATION TABLE NO. 514 SPECIFIC HEAT OF DISTRONTIUM TITANIUM TETRAOXIDE Sr_2TiO_4

[For Data Reported in Figure and Table No. 514]

Curve No.	Ref. No.	Year	Temp. Range, K	Reported Error, %	Name and Specimen Designation	Composition (weight percent), Specifications and Remarks
1	363	1952	54-298			99.5 Sr_2TiO_4; 27.85 TiO_2 (27.82 theo.), 0.17 CaO, 0.03 SiO_2.
2	361	1953	298-1800			99.5 Sr_2TiO_4; 27.85 TiO_2 (27.82 theo.), 0.17 CaO, 0.03 SiO_2.

DATA TABLE NO. 514 SPECIFIC HEAT OF DISTRONTIUM TITANIUM TETRAOXIDE Sr_2TiO_4

[Temperature, T, K; Specific Heat, C_p, Cal $g^{-1}K^{-1}$]

T	C_p
CURVE 1	
53.64	2.310 x 10^{-2}
58.13	2.643
62.83	3.012
67.41	3.366
71.98	3.695
76.50	4.015
80.47	4.301
85.02	4.611
94.86	5.248
104.69	5.836
114.70	6.425
124.64	6.951
136.01	7.522
146.00	7.982
155.66	8.382
165.88	8.793
175.86	9.152
185.92	9.486
195.97	9.789
206.27	1.008 x 10^{-1}
216.73	1.037
226.24	1.060
236.33	1.082
246.02	1.103
256.22	1.122
266.67	1.142
276.38	1.160
286.49	1.177
296.66	1.193*
298.16	1.196
CURVE 2	
298	1.20 x 10^{-1}*
300	1.20
400	1.29
500	1.34
600	1.37
700	1.40
800	1.42
900	1.44
1000	1.46
1100	1.47
1200	1.49
CURVE 2 (cont.)	
1300	1.50 x 10^{-1}
1400	1.52
1500	1.53
1600	1.55
1700	1.56
1800	1.57

* Not shown on plot

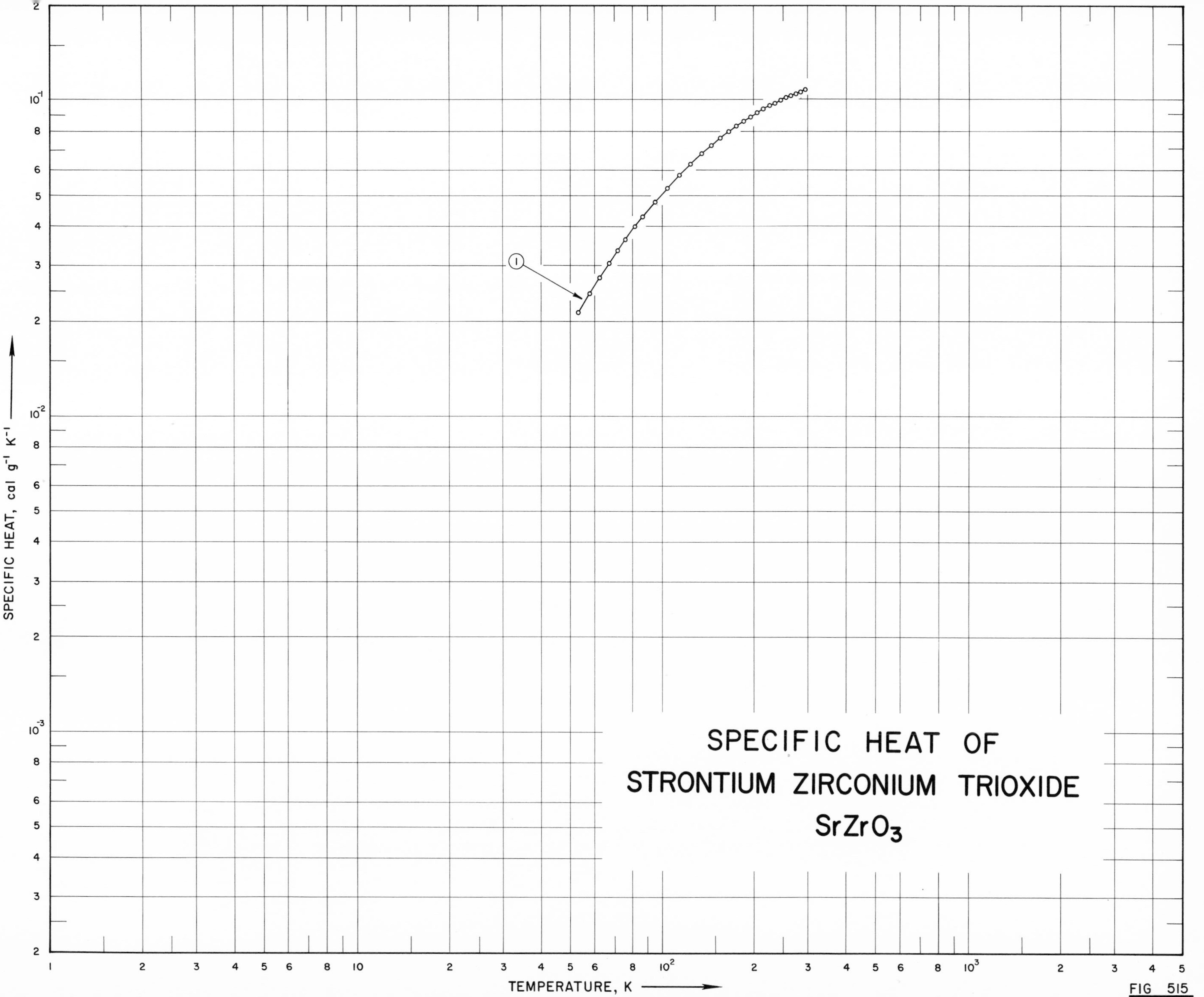
SPECIFIC HEAT OF
STRONTIUM ZIRCONIUM TRIOXIDE
$SrZrO_3$
TEMPERATURE, K
SPECIFIC HEAT, cal g^{-1} K^{-1}
1
FIG 515

SPECIFICATION TABLE NO. 515 SPECIFIC HEAT OF STRONTIUM ZIRCONIUM TRIOXIDE $SrZrO_3$

[For Data Reported in Figure and Table No. 515]

Curve No.	Ref. No.	Year	Temp. Range, K	Reported Error, %	Name and Specimen Designation	Composition (weight percent), Specifications and Remarks
1	364	1960	53-296	0.3		54.32 ZrO_2, 45.56 SrO (54.32, 45.68 theo.); prepared by heating reagent-grade strontium carbonate and pure zirconia for 24 hrs at 1000 C, 6 hrs at 1350-1400 C, 8 hrs at 1350-1470 C, and 12 hrs at 1300-1350 C.

DATA TABLE NO. 515 SPECIFIC HEAT OF STRONTIUM ZIRCONIUM TRIOXIDE $SrZrO_3$

[Temperature, T, K; Specific Heat, C_p, Cal $g^{-1}K^{-1}$]

T	C_p
CURVE 1	
53.44	2.145×10^{-2}
57.97	2.454
62.44	2.758
67.07	3.065
71.51	3.353
76.19	3.648
81.72	3.995
86.63	4.286
94.94	4.770
104.94	5.325
114.82	5.823
124.49	6.290
136.28	6.833
146.12	7.247
156.12	7.631
166.15	7.988
176.50	8.314
185.91	8.614
195.88	8.883
206.07	9.165
215.81	9.407
225.91	9.650
235.82	9.843
245.65	1.003×10^{-1}
256.16	1.023
265.99	1.040
275.94	1.055
286.17	1.071
295.79	1.086

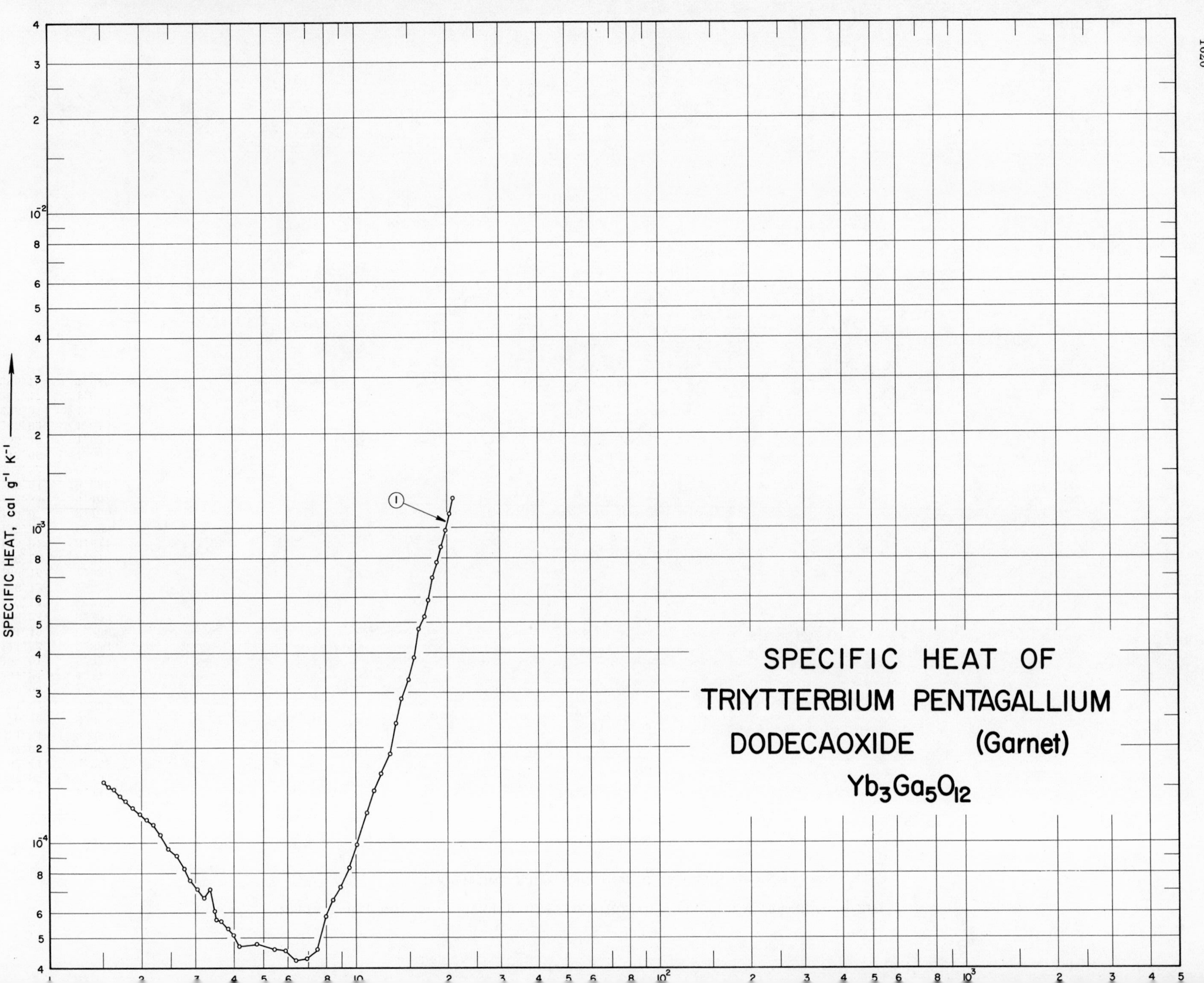
SPECIFIC HEAT OF
TRIYTTERBIUM PENTAGALLIUM
DODECAOXIDE (Garnet)
$Yb_3Ga_5O_{12}$
SPECIFIC HEAT, cal g^{-1} K^{-1}
1

SPECIFICATION TABLE NO. 516 SPECIFIC HEAT OF TRIYTTERBIUM PENTAGALLIUM DODECAOXIDE, $Yb_3Ga_5O_{12}$ (Garnet)

[For Data Reported in Figure and Table No. 516]

Curve No.	Ref. No.	Year	Temp. Range, K	Reported Error, %	Name and Specimen Designation	Composition (weight percent), Specifications and Remarks
1	449	1963	1.5-21		Garnet	99.99 Yb_2O_3; supplied by Lindsay Chem.; 99.99 Ga_2O_3; supplied by Johnson, Matthey and Co.; direct solid state reaction of an intimate mixture of pure rare earth and gallium oxide; sintered block; polycrystalline.

DATA TABLE NO. 516 SPECIFIC HEAT OF TRIYTTERBIUM PENTAGALLIUM DODECAOXIDE, $Yb_3Ga_5O_{12}$ (Garnet)

[Temperature, T, K; Specific Heat, C_p, Cal $g^{-1}K^{-1}$]

T	C_p
CURVE 1	
1.518	1.565×10^{-4}
1.568	1.502
1.630	1.486
1.705	1.416
1.778	1.360
1.870	1.290
1.974	1.231
2.088	1.186
2.195	1.148
2.310	1.062
2.448	9.540×10^{-5}
2.612	9.067
2.766	8.255
2.892	7.578
3.053	7.127
3.202	6.676
3.350	7.127
3.462	6.022
3.505	5.684
3.645	5.661
3.675	5.593*
3.828	5.345
3.835	5.322
3.995	5.052
4.011	5.030*
4.015	5.142*
4.166	4.736*
4.195	4.691
4.78	4.759
5.45	4.578
5.93	4.533
6.38	4.218
6.94	4.285
7.49	4.555
7.96	5.841
8.43	6.563
8.93	7.240
9.50	8.300
10.15	9.834
10.88	1.247×10^{-4}
11.55	1.468
12.19	1.655
12.88	1.910
13.55	2.391
CURVE 1 (cont.)	
14.19	2.864×10^{-4}
14.85	3.293
15.54	3.857
16.21	4.759
16.82	5.210
17.40	5.887
17.99	6.947
18.58	7.759
19.21	8.683
19.83	9.788
20.42	1.101×10^{-3}
20.94	1.238

* Not shown on plot

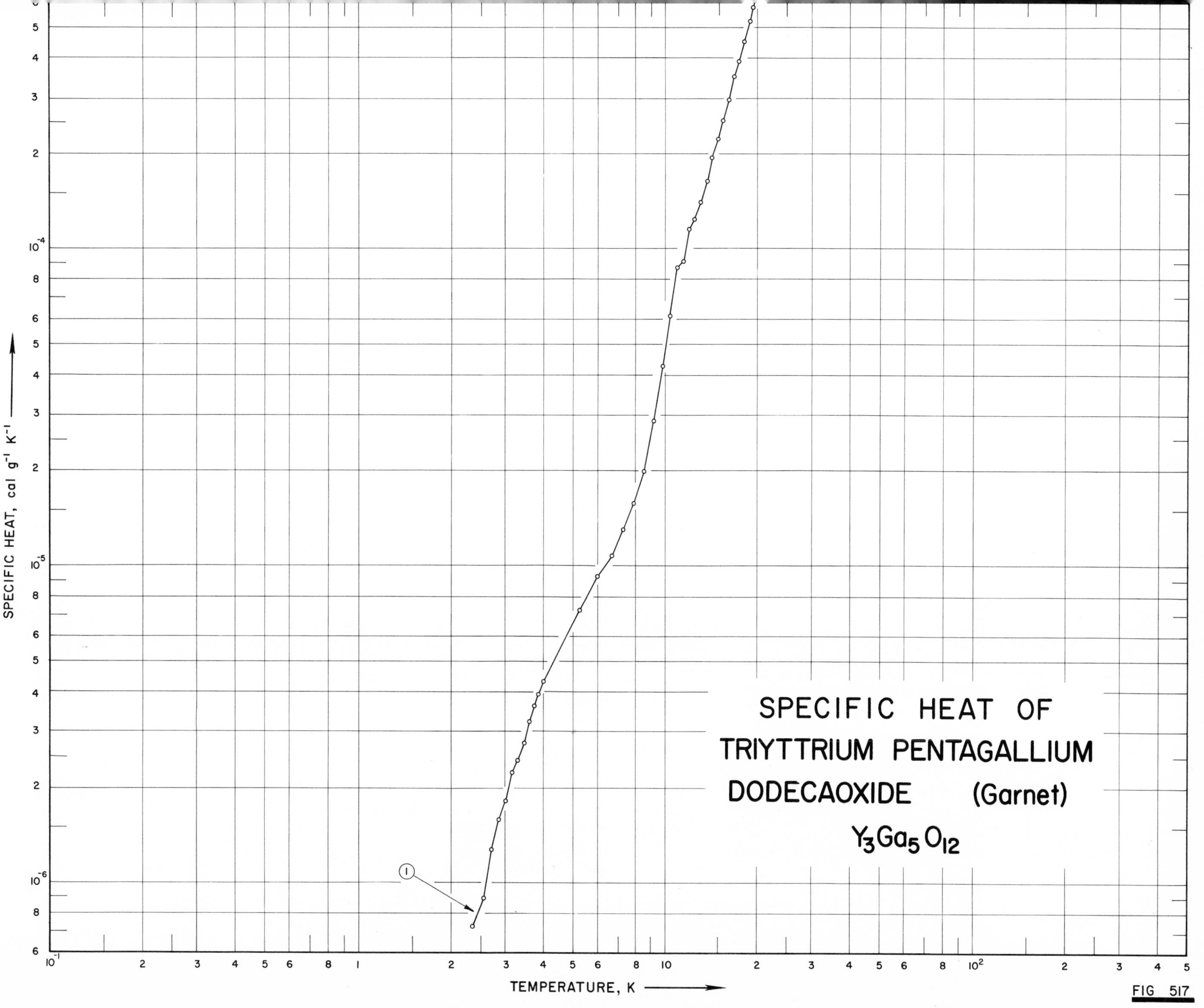
SPECIFIC HEAT OF
TRIYTTRIUM PENTAGALLIUM
DODECAOXIDE (Garnet)
$Y_3Ga_5O_{12}$
SPECIFIC HEAT, cal g⁻¹ K⁻¹
TEMPERATURE, K
1

FIG 517

SPECIFICATION TABLE NO. 517 SPECIFIC HEAT OF TRIYTTRIUM PENTAGALLIUM DODECAOXIDE, $Y_3Ga_5O_{12}$ (Garnet)

[For Data Reported in Figure and Table No. 517]

Curve No.	Ref. No.	Year	Temp. Range, K	Reported Error, %	Name and Specimen Designation	Composition (weight percent), Specifications and Remarks
1	449	1963	2.3-20		Garnet	99.99 Y_2O_3; supplied by Lindsay Chemical Division; 99.99 Ga_2O_3; supplied by Johnson, Matthey and Co.; direct solid state reaction of an intimate mixture of rare earth and gallium oxide; sintered block; polycrystalline.

DATA TABLE NO. 517 SPECIFIC HEAT OF TRIYTTRIUM PENTAGALLIUM DODECAOXIDE, $Y_3Ga_5O_{12}$ (Garnet)

[Temperature, T, K; Specific Heat, C_p, Cal $g^{-1} K^{-1}$]

T	C_p
CURVE 1	
2.368	7.312×10^{-7}
2.560	8.941
2.725	1.275×10^{-6}
2.885	1.587
3.042	1.818
3.193	2.223
3.338	2.439
3.480	2.765
3.625	3.227
3.766	3.611
3.893	3.937
4.020	4.322
5.32	7.224
6.03	9.207
6.73	1.075×10^{-5}
7.34	1.309
7.95	1.584
8.60	1.989
9.25	2.889
9.92	4.263
10.53	6.158
11.12	8.733
11.64	9.148
12.12	1.158×10^{-4}
12.67	1.243
13.23	1.418
13.80	1.640
14.40	1.927
15.00	2.220
15.64	2.537
16.32	2.952
17.02	3.493
17.70	3.908
18.37	4.500
19.02	5.240
19.66	5.773
20.30	6.602

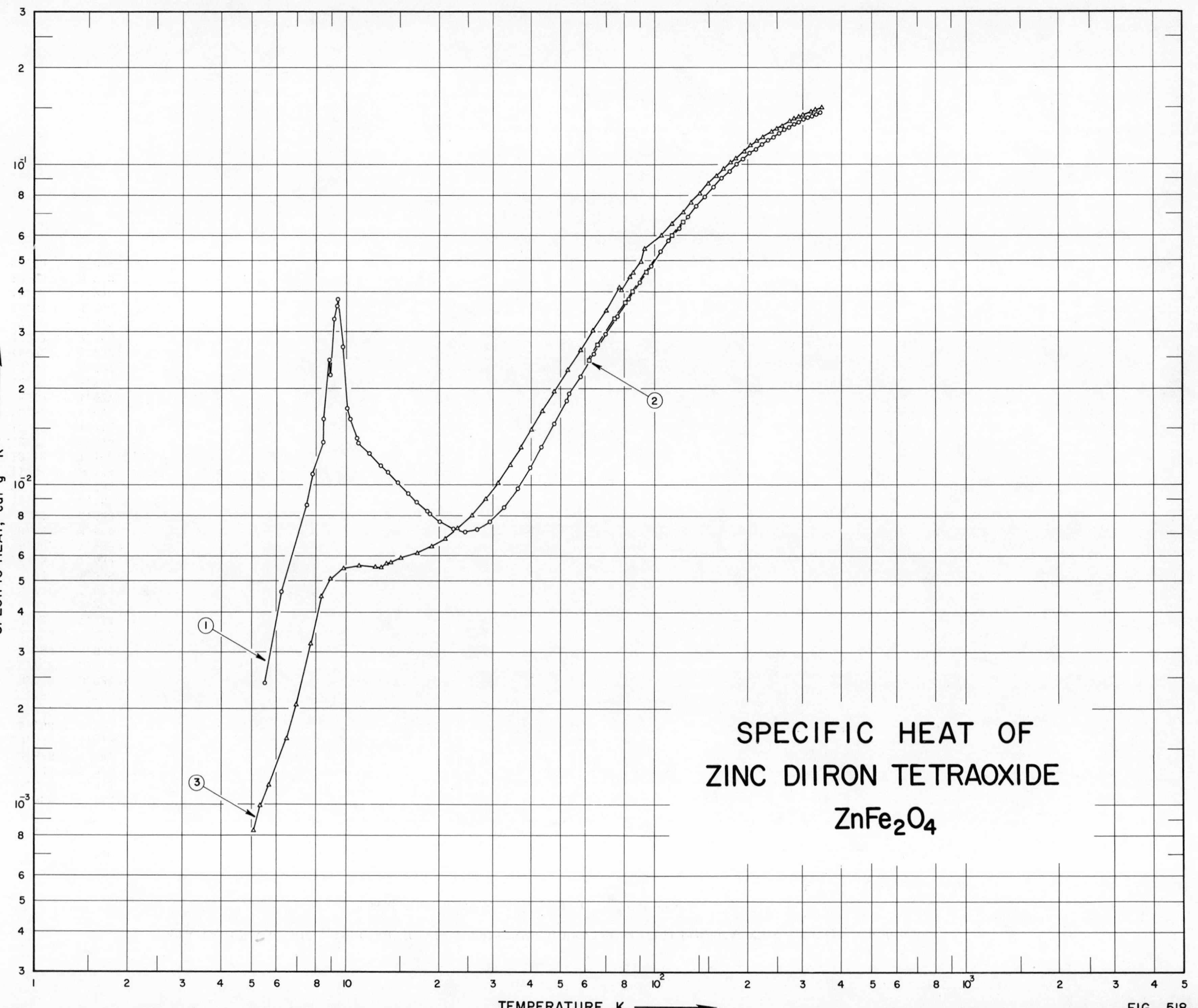
SPECIFIC HEAT OF
ZINC DIIRON TETRAOXIDE
$ZnFe_2O_4$
SPECIFIC HEAT, cal g^{-1} K^{-1}
TEMPERATURE, K
1
2
3

SPECIFICATION TABLE NO. 518 SPECIFIC HEAT OF ZINC DIIRON TETRAOXIDE $ZnFe_2O_4$

[For Data Reported in Figure and Table No. 518]

Curve No.	Ref. No.	Year	Temp. Range, K	Reported Error, %	Name and Specimen Designation	Composition (weight percent), Specifications and Remarks
1	395	1957	6-344			46.24 ±0.1 Fe, 27.2 ±0.1 Zn (46.33, 27.12 theo.), <0.1 ferrous Fe, 0.01-0.1 Al, 0.01-0.1 Mn, 0.001-0.01 Co, 0.001-0.01 Cu, 0.001-0.01 Mg, 0.001-0.01 Ni, 0.001-0.01 S; pressed; fired 14 hrs at 1100 C in air; fragmented to pass 30-mesh screen; reformed into slugs; fired 12 hrs at 1100 C and furnace cooled in 16 hrs.
2	382	1956	53-298		Zinc-iron Spinel	66.36 Fe_2O_3, 33.89 ZnO; heated to 940-1280 C several times (total 18 days) with grinding and mixing between heatings.
3	406	1959	5-348			46.0 Fe, 27.0 Zn.

DATA TABLE NO. 518 SPECIFIC HEAT OF ZINC DIIRON TETRAOXIDE $ZnFe_2O_4$

[Temperature, T,K; Specific Heat, C_p, Cal $g^{-1}K^{-1}$]

T	C_p
CURVE 1	
Series I	
184.44	9.955 x 10^{-2}*
193.52	1.037 x 10^{-1}
203.27	1.078
212.90	1.117
222.68	1.153
232.53	1.188
242.39	1.221
252.28	1.251
262.21	1.280
272.28	1.304
282.53	1.332
292.86	1.357
303.14	1.379
313.31	1.401
323.41	1.420
333.65	1.438
343.93	1.456
Series II	
7.81	1.1 x 10^{-2}
8.47	1.6
8.87	2.4
9.18	3.3
9.41	3.8
9.76	2.7
10.18	1.73
10.91	1.40
Series III	
5.50	2.4 x 10^{-3}
6.24	4.60
6.82	6.22
7.48	8.23
8.46	1.36 x 10^{-2}
8.89	2.2
9.39	3.6
10.38	1.60
11.10	1.357
11.82	1.259
12.95	1.156
13.66	1.093
14.70	1.019
CURVE 1 (cont.)	
15.85	9.420 x 10^{-3}
16.97	8.823
18.31	8.250
20.14	7.666
22.28	7.280
24.45	7.147
26.74	7.255
29.39	7.666
32.55	8.483
36.01	9.710
39.61	1.127 x 10^{-2}
43.47	1.317
47.72	1.551
52.45	1.831
Series IV	
17.04	8.806 x 10^{-3}*
18.69	8.118
52.37	1.783 x 10^{-2}*
58.05	2.176
64.00	2.561
69.98	2.951
76.16	3.359
82.61	3.809
89.79	4.2870
97.61	4.8150
105.53	5.3509
111.99	5.7728
120.25	6.3182
128.58	6.8550
137.08	7.388
146.07	7.927
155.56	8.470
165.25	9.001
175.00	9.499
184.94	9.980
CURVE 2	
53.26	1.933 x 10^{-2}*
57.63	2.186*
61.92	2.451
66.19	2.735
70.51	3.033*
CURVE 2 (cont.)	
74.85	3.319 x 10^{-2}
80.73	3.711
85.36	4.013
94.92	4.637
104.74	5.314*
114.74	5.977
124.74	6.629
136.03	7.330*
145.55	7.919*
155.99	8.524*
166.01	9.051*
175.93	9.553*
185.96	1.002 x 10^{-1}*
195.93	1.047*
206.18	1.091*
216.14	1.129*
225.56	1.162*
236.04	1.199*
245.76	1.229*
256.44	1.263*
266.82	1.290*
276.21	1.313*
286.51	1.340*
295.96	1.365*
298.16	1.368*
CURVE 3	
Series I	
77.38	4.130 x 10^{-2}
83.97	4.455
90.94	4.951
93.28	5.448
106.20	5.981
114.41	6.531
122.86	7.084
131.53	7.636
140.24	8.079
148.94	8.673
157.90	9.180
167.26	9.686
176.68	1.015 x 10^{-1}
183.62	1.048
195.05	1.099
204.89	1.140
CURVE 3 (cont.)	
214.49	1.178 x 10^{-1}
224.03	1.213
229.61	1.233*
238.91	1.263
248.35	1.292
258.02	1.320
Series II	
253.41	1.307 x 10^{-1}*
263.22	1.334*
273.00	1.360
282.63	1.383
292.23	1.405
301.82	1.424
311.29	1.444*
320.81	1.461
330.45	1.478
339.30	1.493*
347.80	1.507
Series III	
5.08	8.3 x 10^{-4}
5.12	9.1*
5.33	2.0
5.68	1.2 x 10^{-3}
6.48	1.6
6.93	2.0
7.71	3.2
8.31	4.48
8.93	5.19
9.82	5.48
11.03	5.571
12.39	5.533
13.82	5.724
15.35	5.923*
Series IV	
12.18	5.517 x 10^{-3}
13.44	5.691
15.07	5.927
17.03	6.110
18.94	6.413
CURVE 3 (cont.)	
20.97	6.786 x 10^{-3}
23.16	7.300
25.64	8.014
28.42	9.005
28.15	8.893*
31.12	1.012 x 10^{-2}
34.14	1.155
37.12	1.312
40.19	1.485
43.71	1.697
47.84	1.962
52.66	2.285
58.00	2.633
63.60	3.029
70.41	3.497
78.50	4.059
85.67	4.578
Series V	
7.36	2.7 x 10^{-3}
8.02	3.8
8.77	5.02
9.74	5.43

* Not shown on plot

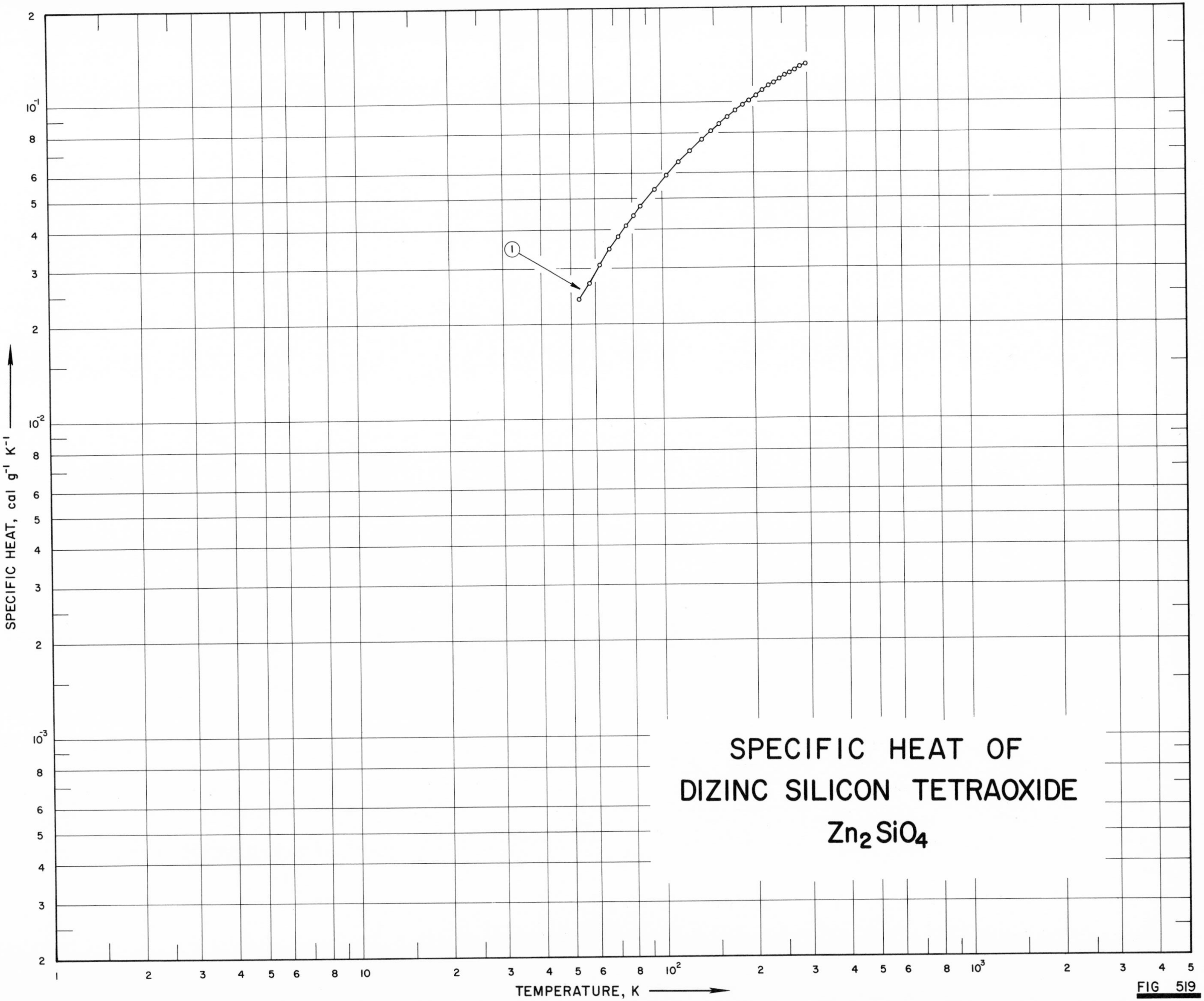

SPECIFIC HEAT OF DIZINC SILICON TETRAOXIDE Zn_2SiO_4

FIG 519

SPECIFICATION TABLE NO. 519 SPECIFIC HEAT OF DIZINC SILICON TETRAOXIDE Zn_2SiO_4

[For Data Reported in Figure and Table No. 519]

Curve No.	Ref. No.	Year	Temp. Range, K	Reported Error, %	Name and Specimen Designation	Composition (weight percent), Specifications and Remarks
1	369	1951	53-298		Willemite	72.95 ZnO, 26.92 SiO_2 (73.05, 26.95 theo.), remainder impurities probably Fe_2O_3, Al_2O_3, and MgO.

DATA TABLE NO. 519 SPECIFIC HEAT OF DIZINC SILICON TETRAOXIDE Zn_2SiO_4

[Temperature, T, K; Specific Heat, C_p, Cal $g^{-1}K^{-1}$]

T	C_p
CURVE 1	
53.01	2.406×10^{-2}
57.34	2.714
62.00	3.078
66.68	3.441
71.55	3.790
76.37	4.117
80.63	4.417
85.08	4.712
94.71	5.327
104.42	5.902
114.59	6.490
124.69	7.037
136.01	7.621
146.09	8.114
155.90	8.554
165.83	8.994
175.95	9.407
185.88	9.811
195.89	1.018×10^{-1}
206.18	1.054
216.21	1.091
225.86	1.125
236.15	1.156
245.68	1.185
256.06	1.214
266.03	1.244
276.00	1.270
286.51	1.296
296.24	1.318*
298.16	1.321

* Not shown on plot

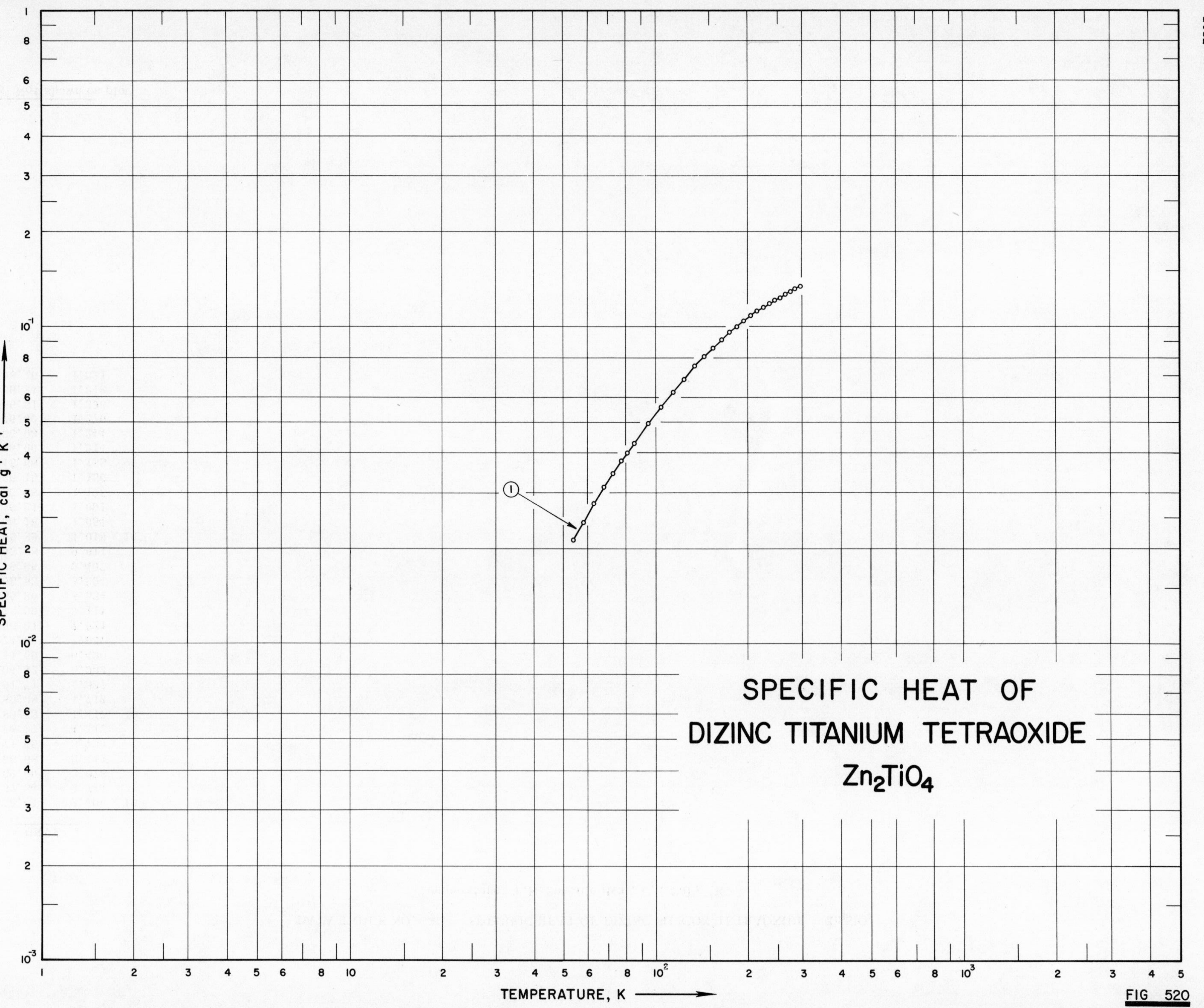
SPECIFIC HEAT OF
DIZINC TITANIUM TETRAOXIDE
Zn_2TiO_4
SPECIFIC HEAT, cal g^{-1} K^{-1}
TEMPERATURE, K
1
FIG 520

SPECIFICATION TABLE NO. 520 SPECIFIC HEAT OF DIZINC TITANIUM TETRAOXIDE Zn_2TiO_4

[For Data Reported in Figure and Table No. 520]

Curve No.	Ref. No.	Year	Temp. Range, K	Reported Error, %	Name and Specimen Designation	Composition (weight percent), Specifications and Remarks
1	358	1955	54-298			67.08 ZnO, 32.86 TiO (67.07, 32.93 theo.), 0.05 insoluble in HCl; prepared from reagent-grade zinc oxide and pure titania; pressed into pellets; heated 4 times for total of 65 hrs above 1000 C; quenched to room temperature.

DATA TABLE NO. 520 SPECIFIC HEAT OF DIZINC TITANIUM TETRAOXIDE Zn_2TiO_4

[Temperature, T, K; Specific Heat, C_p, Cal $g^{-1}K^{-1}$]

T	C_p
CURVE 1	
53.88	2.137 x 10^{-2}
58.15	2.429
62.94	2.781
67.80	3.130
72.68	3.465
77.30	3.780
80.83	4.038
85.11	4.315
94.80	4.974
104.72	5.625
114.28	6.256
124.63	6.894
135.83	7.562
145.47	8.110
155.71	8.650
165.52	9.153
175.98	9.655
185.75	1.009 x 10^{-1}
195.99	1.051
206.62	1.091
216.08	1.125
226.04	1.161
235.80	1.192
245.68	1.223
256.13	1.250
266.07	1.278
276.25	1.303
286.39	1.327
296.51	1.348*
298.16	1.353

* Not shown on plot

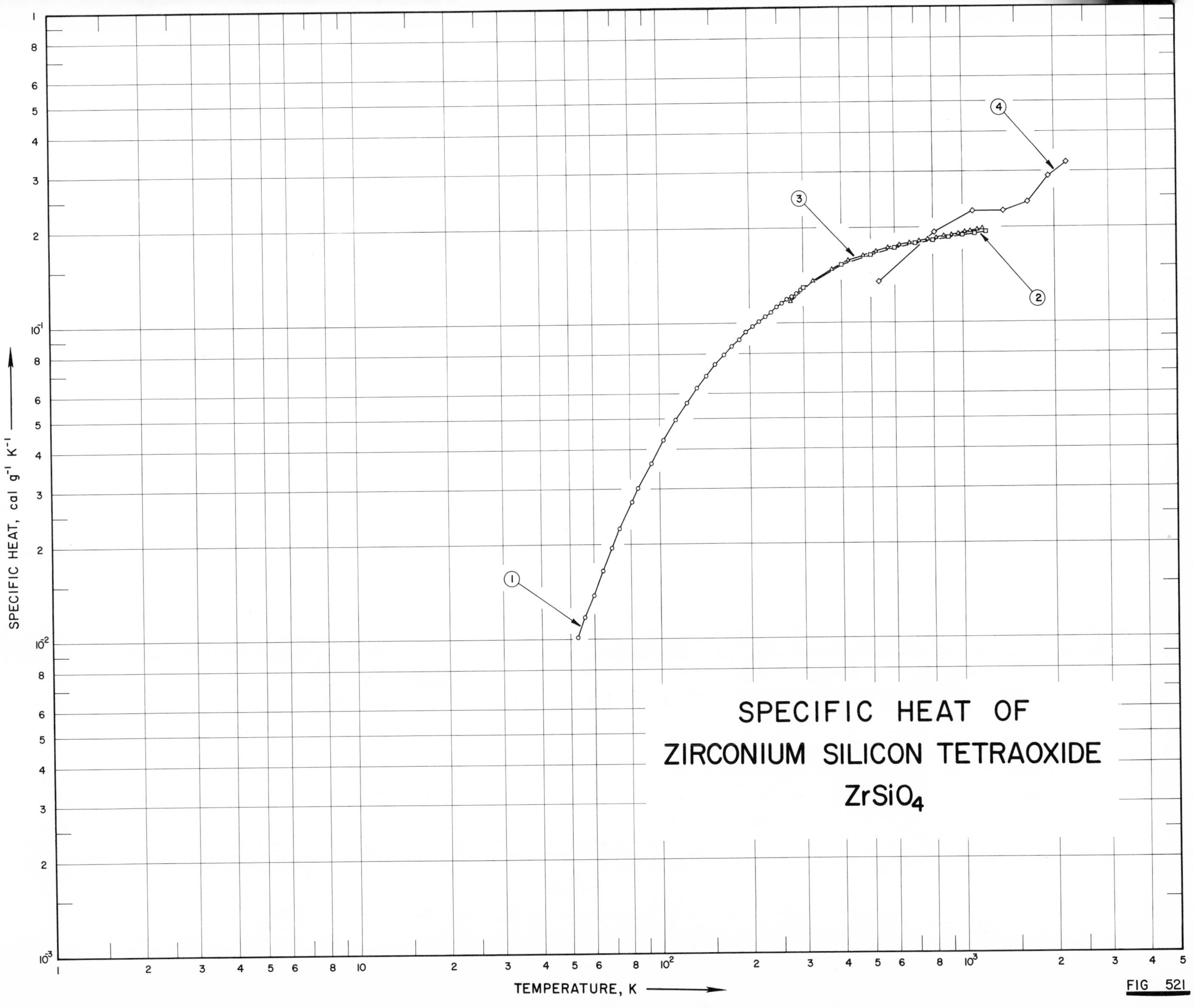
SPECIFIC HEAT OF
ZIRCONIUM SILICON TETRAOXIDE
$ZrSiO_4$
SPECIFIC HEAT, cal g^{-1} K^{-1}
TEMPERATURE, K
1
2
3
4
FIG 521

SPECIFICATION TABLE NO. 521 SPECIFIC HEAT OF ZIRCONIUM SILICON TETRAOXIDE $ZrSiO_4$

[For Data Reported in Figure and Table No. 521]

Curve No.	Ref. No.	Year	Temp. Range, K	Reported Error, %	Name and Specimen Designation	Composition (weight percent), Specifications and Remarks
1	387	1941	53-295			98.6 $ZrSiO_4$, 1.3 SiO_2, 0.04 Fe_2O_3; corrected for Fe_2O_3 and excess SiO_2.
2	154	1950	298-1800			66.30 ZrO_2 including HfO_2 with 1.15 Hf, 33.6 SiO_2, 0.4 Fe_2O_3.
3	94	1960	273-1173	0.25		65.4 ZrO_2 + HfO_2, 33.2 SiO_2, 0.1-1 Al, 0.1-1 Ti, 0.01-0.1 Fe, <0.01 Ag, <0.01 Ca, <0.01 Cu, <0.0001 Mn; fired and sintered to form of small cylinders.
4	32	1960	533-2200	≤ 5	Taylor Zircon CZ-5	65-66 ZrO_2, 33-34 SiO_2, 1 max Al_2O_3, 0.3 max TiO_2, 0.1 max Fe_2O_3, 0.2 max others; sample supplied by Charles Taylor and Sons Co.; slip-cast and sintered; density = 252 lb ft^{-3}.

DATA TABLE NO. 521 SPECIFIC HEAT OF ZIRCONIUM SILICON TETRAOXIDE $ZrSiO_4$

[Temperature, T, K; Specific Heat, C_p, Cal $g^{-1}K^{-1}$]

T	C_p
CURVE 1	
52.7	1.013×10^{-2}
55.8	1.177
59.7	1.384
64.1	1.654
68.7	1.949
73.3	2.250
80.6	2.728
84.9	3.012
93.9	3.631
103.4	4.275
113.9	4.959
124.1	5.631
134.3	6.258
144.8	6.820
154.7	7.404
165.0	7.982
175.2	8.457
185.0	8.899
195.5	9.368
205.6	9.783
215.4	1.014×10^{-1}
225.2	1.052
235.7	1.090
245.6	1.127
255.4	1.161
265.8	1.192
275.8	1.219
285.5	1.244
294.8	1.275
CURVE 2	
298	1.28×10^{-1}
300	1.29
400	1.53
500	1.65
600	1.72
700	1.78
800	1.82
900	1.86
1000	1.89
1100	1.91
1200	1.94
1300	1.97
1400	1.99
CURVE 2 (cont.)	
1500	2.02×10^{-1}
1600	2.04
1700	2.07
1800	2.09
CURVE 3	
273.15	1.178×10^{-1}
323.15	1.366
373.15	1.488
423.15	1.574
473.15	1.637
523.15	1.687
573.15	1.727
623.15	1.760
673.15	1.789
723.15	1.815
773.15	1.838
823.15	1.859
873.15	1.879
923.15	1.897
973.15	1.915
1023.15	1.931
1073.15	1.947
1123.15	1.963
1173.15	1.978
CURVE 4	
533.16	1.35×10^{-1}
810.94	1.93
1088.72	2.25
1366.49	2.25
1644.27	2.40
1922.05	2.90
2199.83	3.20

REFERENCES TO DATA SOURCES

Ref. No.	TPRC No.	
1	9820	Desnoyers, J.E. and Morrison, J.A., Phil. Mag., 3 (25), 42-8, 1958.
2	23953	Burk, D.L. and Friedberg, S.A., Phys. Rev., 111 (5), 1275-82, 1958.
3	27469	Victor, A.C., J. Am. Chem. Soc., 36, 1903-11, 1962.
4	882	DeSorbo, W., J. Chem. Phys., 21, 876-80, 1953.
5	20786	Rasor, N.S. and McClelland, J.D., J. Phys. Chem. Solids, 15, 17-26, 1960.
6	26335	Barriault, R.J., Bender, S.L., Dreikorn, R.E., Einwohner, T.H., Feber, R.C., Gannon, R.E., Hanst, P.L., Ihnat, M.E., Phaneuf, J.P., Schick, H.L., and Ward, C.H., ASD TR 61-260, 1-404, 1962. [AD 278 633]
7	37904	West, E.D. and Ishihara, S., Advances Thermophys. Properties Extrem. Temp. Pres., 3rd ASME Symp., Lafayette, Indiana. 146-51, 1965.
8	28005	Svikis, V.D., Can. Dept. Mines Tech. Surv., Mines Branch, Res. Rept. R96, 1-42, 1962.
9	10947	Lucks, C.F., Deem, H.W., and Wood, W.D., Am. Ceram. Soc. Bull., 39 (6), 313-9, 1960.
10	26074	Neel, D.S., Pears, C.D., and Oglesby, S., Jr., WADD TR 60-924, 1-201, 1962. [AD 275 536]
11	16590	Fieldhouse, I.B., Lang, J.I., and Blau, H.H., WADC TR 59-744, 4, 1-78, 1960. [AD 249 166]
12	869	De Sorbo, W., J. Am. Chem. Soc., 77, 4713-5, 1955.
13	12245	De Sorbo, W. and Nichols, G.E., J. Phys. Chem. Solids, 6 (4), 352-66, 1958.
14	23306	VanderHoeven, B.J.C., Jr. and Keesom, P.H., Phys. Rev., 130 (4), 1318-21, 1963.
15	32615	De Sorbo, W. and Tyler, W.W., J. Chem. Phys., 21 (10), 1660-3, 1953.
16	6560	Lucks, C.F. and Deem, H.W., WADC TR 55-496, 1-65, 1956. [AD 97 185]
17	428	De Sorbo, W. and Tyler, W.W., J. Chem. Phys., 26 (2), 244-7, 1957.
18	21864	Parks, G.S. and Kelley, K.K., J. Phys. Chem., 30, 47-55, 1926.
19	22766	Roth, W.A. and Bertram, W.W., Z. Electrochem., 35, 297-308, 1929.
20	1895	Shomate, C.H. and Naylor, B.F., J. Am. Chem. Soc., 67 (1), 72-5, 1945.
21	26376	Ginnings, D.C. and Corruccini, R.J., RP-1797, J. Res. Natl. Bur. Std., 38 (6), 593-600, 1947.
22	20017	Kerr, E.C., Johnston, H.L., and Hallet, N.C., J. Am. Chem. Soc., 72, 4740-2, 1950.
23	1699	Ginnings, D.C. and Furukawa, G.T., J. Am. Chem. Soc., 75, 522-7, 1953.
24	452	Furukawa, G.T., Douglas, T.B., McCoskey, R.E., and Ginnings, D.C., J. Res. Natl. Bur. Std., RP 2964, 57, 67-82, 1956.
25	28902	Walker, B.E., Grand, J.A., and Ewing, C.T., WADC TR 54-185, Pt. 3, NRLRD 657, 1-15, 1956.
26	18576	West, E.D. and Ginnings, D.C., J. Res. Natl. Bur. Std., 60, 309-16, 1958.
27	7689	Fieldhouse, I.B., Hedge, J.C., and Lang, J.I., WADC TR 58-274, 1-79, 1958. [AD 206 892], [PB 151 583]
28	26504	Romanovskii, V.A. and Tarasov, V.V., Soviet Phys.-Solid State, 2 (6), 1170-5, 1960.
29	29885	Wiebelt, J.A., Ph.D. Thesis, Oklahoma State Univ., 1-91, 1960.
30	20084	Schmidt, N.E. and Sokolov, V.A., Russ. J. Inorg. Chem., 5, 797-802, 1960.
31	16590	Fieldhouse, I.B., Lang, J.I., and Blau, H.H., WADC TR 59-744, 4, 1-78, 1960. [AD 249 166], [PB 171 390]
32	26074	Neel, D.S., Pears, C.D., and Oglesby, S., Jr., WADD TR 60-924, 1-216, 1962. [AD 275 536]
33	21364	Oetting, F.L., Univ. Microfilms, Mic. 60-5976, 1-74, 1961.
34	23817	Hoch, M. and Johnston, H.L., J. Phys. Chem., 65, 1184-5, 1961.
35	26393	Edwards, J.W. and Kington, G.L., Trans. Faraday Soc. (G.B.), 58, Pt. 7, 1313-22, 1962.
36	28756	Martin, D.L., Can. J. Phys., 40, 1166-73, 1962.

Ref. No.	TPRC No.	
37	25229	Ewing, C.T., Stone, J.P., Spann, J.R., Kovacina, T.A., and Miller, R.R., NRL-5964, 1-13, 1963. [AD 405 673]
38	29980	Prophet, H. and Stull, D.R., J. Chem. Eng. Data, 8 (1), 78-81, 1963.
39	25497	Sterett, K.F., Blackburn, D.H., Bestul, A.B., Chang, S.S., and Horman, J., Natl. Bur. Std. J. Res., C69 (1), 19-26, 1965.
40	21846	Anderson, C.T., J. Am. Chem. Soc., 52, 2712-20, 1930.
41	10350	Anderson, C.T., J. Am. Chem. Soc., 57, 429-31, 1935.
42	26388	Lander, J.J., J. Am. Chem. Soc., 73, 5794-7, 1951.
43	11861	Kelley, K.K., J. Am. Chem. Soc., 61, 1217-8, 1939.
44	15021	Kandyba, V.V., Kantor, P.B., Krasovitskaya, R.M., and Formichev, E.N., AEC-TR-4310, 1-3, 1960.
45	29978	Walker, B.E., Jr., Ewign, C.T., and Miller, R.R., J. Chem. Eng. Data, 7, 595-7, 1962.
46	30541	Victor, A.C. and Douglas, T.B., J. Res. Natl. Bur. Std., 67A (4), 325-9, 1963.
47	25961	Hedge, J.C., Kostenko, C., and Lang, J.I., ASD-TDR 63-597, 1-128, 1963. [AD 424 375]
48	26008	Pears, C.D., ASD-TDR 62-765, 1-420, 1963. [AD 298 061]
49	21834	Anderson, C.T., J. Am. Chem. Soc., 52, 2720-3, 1930.
50	28368	Kelley, K.K., J. Am. Chem. Soc., 63, 1137-9, 1941.
51	30685	Southard, J.C., J. Am. Chem. Soc., 63, 3147-50, 1941.
52	30663	Kerr, E.C., Hersh, H.N., and Johnston, H.L., J. Am. Chem. Soc., 72, 4738-40, 1950.
53	21864	Parks, G.S. and Kelley, K.K., J. Phys. Chem., 30, 47-55, 1926.
54	37903	Mezaki, R., Tilleux, E.W., Jambois, T.F., and Margrave, J.L., Advances Thermophys. Properties Extreme Temp. Pres., 3rd ASME Symp., Lafayette, Indiana, 138-45, 1965.
55	31661	Kuznetsov, F.A. and Rezukhina, T.N., Russ. J. Phys. Chem., 34 (11), 1164, 1960.
56	7695	Melonas, J.V., Covington, P.C., and Pears, C.D., WADC TR 58-129, 1-15, 1958. [AD 155 816]
57	29312	King. E.G. and Christensen, A.U., U.S. Bur. Mines, Rept. Invest., RI-5789, 1-6, 1961.
58	31603	Weller, W.W. and King, E.G., U.S. Bur. Mines, Rept. Invest., 6245, 1-6, 1963.
59	31604	Pankratz, L.B. and Kelley, K.K., U.S. Bur. Mines, Rept. Invest., 6248, 1-11, 1963.
60	12111	Anderson, C.T., J. Am. Chem. Soc., 59, 488-91, 1937.
61	1871	Kelley, K.K., Boericke, F.S., Moore, G.E., Huffman, E.H., and Bangert, W.M., U.S. Bur. Mines, Tech. Paper, 662, 1-43, 1944.
62	521	Volger, J., Nature (England), 170, 1027, 1952.
63	1351	King, E.G., J. Am. Chem. Soc., 79, 2399-400, 1957.
64	9764	King, E.G. and Christensen, A.U., Jr., J. Am. Chem. Soc., 80, 1800-1, 1958.
65	21814	Millar, R.W., J. Am. Chem. Soc., 51, 215-22, 1929.
66	7768	Hu, J.H. and Johnston, H.L., J. Am. Chem. Soc., 73, 4550-1, 1951.
67	27444	Gregor, L.V., J. Phys. Chem., 66, 1645-7, 1962.
68	7025	Clusius, K. and Harteck, P., Z. Physik. Chem., 134, 243-63, 1928.
69	20978	Wohler, L. and Jochum, N., Z. Physik. Chem., 167A (3), 169-79, 1933.
70	932	Hu, J.H. and Johnston, H.L., J. Am. Chem. Soc., 75, 2471-3, 1953.
71	1213	Assayag, G. and Bizette, H., Compt. Rend. (France), 239, 238-40, 1954.
72	27075	Justice, B.H., U.S. At. Energy Comm., TID-12722, 1-175, 1961.
73	26759	Pankratz, L.B., King, E.G., and Kelley, K.K., U.S. Bur. Mines, Rept. Invest., 6033, 1-18, 1962.
74	499	Adams, G.B., Jr. and Johnston, H.L., J. Am. Chem. Soc., 74, 4788-9, 1952.
75	18112	King, E.G., J. Am. Chem. Soc., 80 (8), 1799-1800, 1958.
76	21359	Levinstein, M.A., WADD TR 60-654, 1-91, 1961. [AD 264 223]
77	1295	Todd, S.S. and Bonnickson, K.R., J. Am. Chem. Soc., 73, 3894-5, 1951.
78	1296	Coughlin, J.P., King, E.G. and Bonnickson, K.R., J. Am. Chem. Soc., 73, 3891-3, 1951.
79	19866	Brown, G.G. and Furnas, C.C., Trans. Am. Inst. Chem. Eng., 18, 309-46, 1926.
80	18139	Gronvold, F. and Westrum, E.F., Jr., J. Am. Chem. Soc., 81, 1780-3, 1959.
81	1000	Kouvel, J.S., Phys. Rev., 102, 1489-90, 1956.

Ref. No.	TPRC No.	
82	878	Blomeke, J.O. and Ziegler, W.T., J. Am. Chem. Soc., 73, 5099-5102, 1951.
83	10033	Goldstein, H.W., Neilson, E.F., Walsh, P.N., and White, D., J. Phys. Chem., 63, 1445-9, 1959.
84	27999	King, E.G., Weller, W.W., and Pankratz, L.B., U.S. Bur. Mines, Rept. Invest., 5857, 1-6, 1961.
85	28374	Spencer, H.M. and Spicer, W.M., J. Am. Chem. Soc., 64, 617-21, 1942.
86	9824	King, E.G., J. Am. Chem. Soc., 80, 2400-1, 1958.
87	28169	Kostryukov, V.N. and Morozova, G.K., Russ. J. Phys. Chem., 34 (8), 873-4, 1960.
88	25509	Rodigina, E.N., Gomel'skii, K.Z., and Luginina, V.F., Zhur. Fiz. Khim., 35, 1799-1802, 1961.
89	9	Johnston, H.L. and Bauer, T.W., J. Am. Chem. Soc., 73, 1119-22, 1951.
90	706	Shomate, H. and Cohen, A.J., J. Am. Chem. Soc., 77, 285-6, 1955.
91	12112	Giauque, W.F. and Archibald, R.C., J. Am. Chem. Soc., 59, 561-9, 1937.
92	1550	Arthur, J.S., J. Appl. Phys., 21, 732-3, 1950.
93	19407	Barron, T.H.K., Berg, W.T., and Morrison, J.A., Proc. Roy. Soc. (London), A250, 70-83, 1959.
94	18701	Victor, A.C. and Douglas, T.B., WADC TR 57-374, Pt. VI, 1-16, 1960. [AD 258 383]
95	24812	Fieldhouse, I.B. and Lang, J.I., WADD TR 60-904, 1-119, 1961. [AD 268 304]
96	19870	Lien, W.H., UCRL-9880, 1-78, 1962.
97	25901	Makarounis, O. and Jenkins, R.J., NRDL TR 599, 1-24, 1962. [AD 295 887]
98	31606	Pankratz, L.B. and Kelley, K.K., U.S. Bur. Mines, Rept. Invest., 6295, 1-7, 1963.
99	21807	Millar, R.W., J. Am. Chem. Soc., 50, 1875-83, 1928.
100	28381	Kelley, K.K. and Moore, G.E., J. Am. Chem. Soc., 65, 782-5, 1943.
101	23113	Moore, G.E., J. Am. Chem. Soc., 65, 1398-9, 1943.
102	30770	Orr, R.L., J. Am. Chem. Soc., 76, 857-8, 1954.
103	386	King, E.G., J. Am. Chem. Soc., 76, 3289-91, 1954.
104	21830	Anderson, C.T., J. Am. Chem. Soc., 52, 2296-300, 1930.
105	534	Bauer, T.W. and Johnston, H.L., J. Am. Chem. Soc., 75, 2217-19, 1953.
106	5123	Seltz, H., Dunkerley, F.J., and DeWitt, B.J., J. Am. Chem. Soc., 65, 600-2, 1943.
107	2515	Cosgrove, L.A. and Snyder, P.E., J. Am. Chem. Soc., 75, 1227-8, 1953.
108	1581	Smith, D.F., Brown, D., Dworkin, A.S., Sasmor, D.J., and Van Artsdalen, E.R., J. Am. Chem. Soc., 78, 1533-6, 1956.
109	517	Westrum, E.F., Jr., Hatcher, J.B., and Osborne, D.W., J. Chem. Phys., 21, 419-23, 1953.
110	20445	Seltz, H., DeWitt, B.J., and McDonald, H.J., J. Am. Chem. Soc., 62, 88-9, 1940.
111	633	Tomlinson, J.R., Domash, L., Hay, R.G., and Montgomery, C.W., J. Am. Chem. Soc., 77, 909-10, 1955.
112	25326	Gel'd, P.V. and Kusenko, F.G., Russ. Met. Fuels, 2 (2), 43-5, 1960.
113	20317	Kusenko, F.G. and Gel'd, P.V., Izv. Vysshikh Uchebn. Zavedenii. Tsvetn. Met., 3 (4), 102-6, 1960.
114	529	Orr, R.L., J. Am. Chem. Soc., 75, 2808-9, 1953.
115	22924	Sandenaw, T.A., TID 15183, LADC 5260, 1-23, 1962.
116	26364	Sandenaw, T.A., LADC 5555, 1-24, 1963.
117	27231	Gerkin, R.E. and Pitzer, K.S., J. Am. Chem. Soc., 84 (14), 2662-4, 1962.
118	29642	Nernst, W., Ann. Physik, 36 (4), 395-439, 1911.
119	30241	Wietzel, R., Z. Anorg. Chem., 116, 71-95, 1921.
120	26210	Westrum, E.F., Jr., IVe Congress International du Verre, Paris, 396-9, 1956.
121	6314	Moser, H., Physik Z., 37 (21), 737-53, 1936.
122	25447	Southard, J.C., J. Am. Chem. Soc., 63, 3142-6, 1941.
123	23932	Flubacher, P., Leadbetter, A.J., Morrison, J.A., and Stoicheff, B.P., J. Phys. Chem. Solids, 12 (1), 53-65, 1959.
124	26662	Clark, A.E. and Strakna, R.E., Phys. Chem. Glasses, 3 (4), 121-6, 1962.
125	15366	Anderson, C.T., J. Am. Chem. Soc., 58, 568-70, 1936.
126	21319	Grimley, R.T. and Margrave, J.L., J. Phys. Chem., 64, 1763-4, 1960.
127	4311	Kelley, K.K., J. Am. Chem. Soc., 62, 818-9, 1940.

Ref. No.	TPRC No.	
128	32377	Mezaki, R. and Margrave, J.L., J. Phys. Chem., 66, 1713-4, 1962.
129	7048	Osborne, D.W. and Westrum, E.F., Jr., ANL 5085, 1-12, 1953. [AD 15189]
130	916	Osborne, D.W. and Westrum, E.F., Jr., J. Chem. Phys., 21, 1884-7, 1953.
131	20571	Victor, A.C. and Douglas, T.B., J. Res. Natl. Bur. Std., 65 (2), 105-11, 1961.
132	21815	Millar, R.W., J. Am. Chem. Soc., 51, 207-14, 1929.
133	7830	Shomate, C.H., J. Am. Chem. Soc., 68, 310-2, 1946.
134	5622	Naylor, B.F., J. Am. Chem. Soc., 68, 1077-80, 1946.
135	11873	McDonald, H.J. and Seltz, H., J. Am. Chem. Soc., 61, 2405-7, 1939.
136	32767	Shomate, C.H., J. Am. Chem. Soc., 69, 218-9, 1947.
137	887	Lietz, J., Hamburger Beitr. Angew. Mineral. Kristallphysik (Germany), 1, 229-38, 1956.
138	19225	Keesom, P.H. and Pearlman, N., Phys. Rev., 112, 800-4, 1958.
139	1741	Moore, G.E. and Kelley, K.K., J. Am. Chem. Soc., 69, 2105-7, 1947.
140	28951	Cabbage, A., Welch, F., and Trice, J.B., NEPA 935-SCR-35 (USAEC), 1-11, 1962. [PB 162012]
141	2109	Jones, W.M., Gordon, J., and Long, E.A., J. Chem. Phys., 20, 695-9, 1952.
142	33559	Long, E.A., Jones, W.M., and Gordon, J., USAEC Rept. A-329, 1-13, 1942.
143	20197	Popov, M.M., Gal'chenko, G.L., and Senin, M.D., Russ. J. Inorgan. Chem., 3 (8), 18-21, 1958.
144	28950	Powers, H., Welch, F., and Trice, J.B., NEPA 934-SCR-34, (USAEC), 1-16, 1962. [PB 162011]
145	17358	Westrum, E.F., Jr. and Gronvold, F., J. Am. Chem. Soc., 81 (8), 1777-80, 1959.
146	27554	Khomyakov, K.G., Spitsyn, V.I., and Zhvanko, S.A., Issled. v Obl. Khin. Urana, Sb. Statei, 141-4, 1961.
147	930	Osborne, D.W., Westrum, E.F., Jr., and Lohr, H.R., J. Am. Chem. Soc., 79, 529-30, 1957.
148	38590	Gotoo, K. and Naito, K., J. Phys. Chem. Solids, 26 (11), 1673-7, 1957.
149	23116	Cook, O.A., J. Am. Chem. Soc., 69, 331-3, 1947.
150	877	Jaffray, J. and Lyand, R., Compt. Rend., 233, 133-5, 1951.
151	15368	Anderson, C.T., J. Am. Chem. Soc., 58, 564-6, 1936.
152	21806	Millar, R.W., J. Am. Chem. Soc., 50, 2653-6, 1928.
153	4542	Kelley, K.K., Ind. Eng. Chem., 36, 377, 1944.
154	1363	Coughlin, J.P. and King, E.G., J. Am. Chem. Soc., 72, 2262-5, 1950.
155	15248	Cutler, M., Snodgrass, H.R., Cheney, G.T., Appel, J., Mallon, C.E., and Meyer, C.H., Jr., USAEC GA-1939, 1-99, 1961.
156	32919	Piesbergen, Von. U., Z. Naturforsch., A18 (2), 141-47, 1963.
157	33434	Lundin, C.E., Pool, M.J., and Sullivan, R.W., AF CRL-63-156, DRI No. 2115, 1-55, 1963. [AD 420015]
158	26264	Kochetkova, N.M. and Rezukhina, T.N., Proc. 4th Simiconductor Materials Conf., Moscow, 26-8, 1961.
159	9117	Nachtrieb, N.H. and Clement, N., J. Phys. Chem., 62 (7), 876-7, 1958.
160	17357	Gul'tyaev, P.V. and Petrov, A.V., Soviet Phys.-Solid State, 1, 330-4, 1959.
161	30501	Wulff, C.A. and Westrum, E.F., Jr., J. Phys. Chem., 67 (11), 2376-8, 1963.
162	24810	Booker, J., Paine, R.M., and Stonehouse, A.J., WADD TR 60-889, 1-133, 1961. [AD 265625]
163	29086	Mezaki, R., Tilleux, E.W., Barnes, D.W., and Margrave, J.L., Proc. Intern. Symp. Thermodyn. Nuclear Materials (Vienna), Paper S-M 26/48, 775-88, 1962.
164	28935	Arthur D. Little, Inc., ASD-TDR 62-204, 1-83, 1962. [AD 277500]
165	37348	Westrum, E.F., Jr. (author), McClaine, L.A. (editor), ASD-TDR-62-204, Pt. III, 1-22? 1964. [AD 601424]
166	6530	White, D. and Swift, R.M., 1-37, 1956. [AD 103561]
167	1346	Swift, R.M. and White, D., J. Am. Chem. Soc., 79, 3641-4, 1957.
168	30503	Westrum, E.F., Jr. and Clay, G.A., J. Phys. Chem., 67 (11), 2385-7, 1963.
169	26774	Kaufman, L. and Clougherty, E.V., RTD-TDR 63-4096, 1-375, 1963. [AD 428006]
170	33446	Furukawa, G.T., Douglas, T.B., Saba, W.G., and Victor, A.C., Natl. Bur. Std., Rept. No. 8186, 1-178, 1964. [AD 437168]

Ref. No.	TPRC No.	
171	24224	Kresfovnikov, A. N. and Vendrikh, M. S., Izv. Vysshikh Uchebn. Zavedenii, Tsvet. Met., (2), 54-7, 1959.
172	36433	Kirillin, V. A., Sheindlin, A. E., Chekhovskoi, V. Ya., and Tyukaev, V. I., High Temperature, 2 (5), 640-4, 1964.
173	26335	Barriault, R. J., et al., ASD-TR 61-260, 1-404, 1962. [AD 278 633]
174	33091	Valentine, R. H., Jambois, T. F., and Margrave, J. L., J. Chem. Eng. Data, 9 (2), 182-4, 1964.
175	6381	Trice, J. B., Neely, J. J., and Teeter, C. E., Jr., AECU-19, NEPA 816, 1-14, 1948.
176	9423	Trice, J. B., Neely, J. J., and Teeter, C. E., Jr., USAEC, NEPA 821, 1-9, 1948.
177	189	Neely, J. J., Teeter, C. E., Jr., and Trice, J. B., J. Am. Ceram. Soc., 33, 363-4, 1950.
178	4298	Kelley, K. K., Ind. Eng. Chem., 33, 1314-15, 1941.
179	883	De Sorbo, W., J. Am. Chem. Soc., 75, 1825-7, 1953.
180	925	Oriani, R. A. and Murphy, W. K., J. Am. Chem. Soc., 76, 343-5, 1954.
181	31301	Seltz, H., McDonald, H. J., and Wells, C., Am. Inst. Mining Met. Engrs. Tech. Publ. 1137, 1-11, 1939.
182	27756	Swanson, M. L., Can. J. Phys., 40, 719-24, 1962.
183	28381	Kelley, K. K. and Moore, G. E., J. Am. Chem. Soc., 65, 782-5, 1943.
184	30425	Levinson, L. S., J. Chem. Phys., 42 (9), 3342, 1965.
185	29100	Levinstien, M. A., ASD-TDR 62-201, 1-93, 1962. [AD 283 967]
186	37099	Pankratz, L. B., Weller, W. W., and Kelley, K. K., U. S. Bur. Mines, RI-6446, 1-9, 1964.
187	22004	Kruger, O. L. and Savage, H., Argonne Natl. Lab. Annual Rept. 1963, Met. Div., ANL-6868, 133-8, 1963.
188	29311	Humprey, G. L., Todd, S. S., Coughlin, J. P., and King, E. G., U. S. Bur. Mines, BM-RI-4888, 1-23, 1952.
189	6970	Fieldhouse, I. B., Hedge, J. C., Lang, J. I., and Waterman, T. I., WADC-TR 57-487, 1-78, 1957. [AD 150 954] [PB 131 718]
190	37451	Kirillin, V. A., Sheindlin, A. E., and Chekhovskoi, V. Ya., High Temperature, 2 (1), 6-10, 1964.
191	36205	Westrum, E. F., Jr., Takahashi, Y., and Stout, N. D., J. Phys. Chem., 69 (5), 1520-4, 1965.
192	36207	Takahashi, Y., Westrum, E. F., Jr., and Kent, R. A., J. Chem. Eng. Data, 10 (2), 128-9, 1965.
193	1877	Kelley, K. K., Ind. Eng. Chem., 36, 865-6, 1944.
194	4768	Naylor, B. F., J. Am. Chem. Soc., 68, 370-1, 1946.
195	33540	Taylor, R. E. and Nakata, M. M., WADD-TR 60-581, Pt. 4, 1-109, 1963. [AD 428 669] [AD 441 079]
196	36206	Levinson, L. S., J. Chem. Phys., 42 (8), 2891-2, 1965.
197	19423	Boettcher, A. B. and Schneider, G., Proc. U. N. Intern. Conf. Peaceful Uses At. Energy, 2nd Geneva, 6, 561-3, 1958.
198	26961	Mukaibo, T., Naito, K., Sato, K., and Uchijima, T., Thermodyn. Nucl. Mater. Proc. Symp., Vienna, 645-51, 1962.
199	29037	Westrum, E. F., Jr., USAEC, TID-16987, 1-15, 1962.
200	32764	Harrington, L. C. and Rowe, G. H., TID-4500, PWAC-426, 1-13, 1964.
201	37906	Westrum, E. F., Jr., Suits, E., and Lonsdale, H. F., Advances Thermophys. Properties Extreme Temp. Pres., 3rd ASME Symp., Lafayette, Indiana, 156-61, 1965.
202	31300	Levinson, L. S., J. Chem. Phys., 38 (9), 2105-6, 1963.
203	37907	Farr, J. D., Witteman, W. G., Stone, P. L., and Westrum, E. F., Jr., Advances Thermophys. Properties Extreme Temp. Pres., 3rd ASME Symp., Lafayette, Indiana, 162-6, 1965.
204	7462	Shomate, C. H. and Kelley, K. K., J. Am. Chem. Soc., 71, 314-5, 1949.
205	6563	King, E. G., J. Am. Chem. Soc., 71, 316-7, 1949.
206	24389	Arthur D. Little, Inc., ASD-TDR 62-204, Pt. 2, AF 33-616-7472, 1-125, 1963.
207	24203	Kostryukov, V. N., Russian J. Phys. Chem., 35 (8), 865-7, 1961.
208	25747	Stalinski, B. and Bieganski, Z., Bull. Acad. Polon. Sci., Ser. Sci. Chim., 8 (5), 243-8, 1960.
209	18131	Flowtow, H. E., Lohr, H. R., Abraham, B. M., and Osborne, D. W., J. Am. Chem. Soc., 81, 3529-33, 1959.
210	28734	Bieganski, Z. and Stalinski, B., Bull. Acad. Polon. Sci., Ser. Sci. Chim., 9 (5), 367-72, 1961.

Ref. No.	TPRC No.	
211	29939	Flotow, H.E., Osborne, D.W., Otto, K., and Abraham, B.M., J. Chem. Phys., 38, 2620-6, 1963.
212	26006	Beck, R.L., ASM Trans., Quart., 55 (3), 556-64, 1962.
213	27996	Mah, A.D., King, E.G., Weller, W.W., and Christenson, A.U., U.S. Bur. Mines, RI-5716, 1-8, 1961.
214	27512	Gronvold, F. and Westrum, E.F., Jr., Inorg. Chem., 1 (1), 36-48, 1962.
215	29941	McDonald, R.A. and Stull, D.R., J. Phys. Chem., 65, 1918, 1961.
216	29490	Mitchell, D.W., Ind. Eng. Chem., 41, 2027-31, 1949.
217	1478	Todd, S.S., J. Am. Chem. Soc., 72 (7), 2914-5, 1950.
218	29404	Itskevich, E.S., Russ. J. Phys. Chem., 35 (8), 891-2, 1961.
219	14603	Bolling, G.F., J. Chem. Phys., 33 (1), 305-6, 1960, Addendum, 36 (4), 1085-6, 1962.
220	18271	Westrum, E.F., Jr., Chou, C., and Gronvold, F., J. Chem. Phys., 30 (3), 761-4, 1959.
221	31499	Kelley, K.K., J. Am. Chem. Soc., 61, 203-7, 1939.
222	8737	Westrum, E.F., Jr., Chou, C., Machol, R.E., and Gronvold, F., J. Chem. Phys., 28 (3), 497-503, 1958.
223	24522	Gronvold, F., Thurmann-Moe, T., Westrum, E.F., Jr., and Chang, E., J. Chem. Phys., 35 (5), 1665-9, 1961.
224	25641	Westrum, E.F., Jr., Carlson, H.G., Gronvold, F., and Kjekshus, A., J. Chem. Phys., 35 (5), 1670-6, 1961.
225	20836	Harris, P.M., Macwood, G.E., and White, D., WADD-TR 60-771, 1-76, 1962. [AD 275 508]
226	13174	Gronvold, F. and Westrum, E.F., Jr., Acta Chem. Scand., 13, 241-8, 1959.
227	20004	Gronvold, F., Thurmann-Moe, T., Westrum, E.F., Jr., and Levitin, N.E., Acta Chem. Scand., 14 (3), 634-40, 1960.
228	22207	Baer, Y., Busch, G., Frohlich, C., and Steigmeier, E., Z. Naturforsch (Germany), 17A (10), 886-9, 1962.
229	28690	Golutvin, Yu.M., and Liang, C., Russ. J. Phys. Chem., 35 (1), 62-7, 1961.
230	37449	Kalishevich, G.I., Gel'd, P.V., and Krentsis, R.P., High Temperature, 2 (1), 11-4, 1964.
231	37419	Dismukes, J.P., Ekstrom, L., Hockings, E.F., Kudman, I., and Lindenbald, N.E., Contr. No. NObs-88595, Proj. Ser-NO-SR 007 1201, Task 802, 1-100, 1964. [AD 441 794]
232	31709	Golutvin, Yu.M., Kozlovskaya, T.M., and Maslennikova, E.G., Russ. J. Phys. Chem., 37 (6), 723-7, 1963.
233	7088	Douglas, T.B. and Logan, W.M., Natl. Bur. Std., WADC TR 53-201, Pt. 3, 1-13, 1953. [AD 24 019]
234	7132	Ewing, C.T. and Walker, B.E., WADC TR 54-185, 1-27, 1954. [AD 50 565]
235	276	Walker, B.E., Grand, J.A., and Miller, R.R., J. Phys. Chem., 60, 231-3, 1956.
236	23764	Golutvin, Yu.M., Russ. J. Phys. Chem., 33 (8), 164-8, 1959.
237	16770	Snyder, M.J., Duckworth, W.H., BMI-1223, 1-34, 1957. [AD 145 106]
238	31675	Golutvin, Yu.M. and Kozlovskaya, T.M., Russ. J. Phys. Chem., 36 (2), 183-4, 1962.
239	31602	Pankratz, L.B. and Kelley, K.K., U.S. Bur. Mines Dept., Rept. Invest. No. 6241, 1-5, 1963.
240	1206	Todd, S.S., J. Am. Chem. Soc., 75, 1229-31, 1953.
241	18483	Chandrasekharaiah, M.S., Grimley, R.T., and Margrave, J.L., J. Phys. Chem., 63, 1505-6, 1959.
242	29243	Topol, L.E. and Ransom, L.D., NAA-SR-6518, 1-2, 1961.
243	956	Clusius, K., Goldmann, J., and Perlick, A., Z. Naturforch, 4A, 424-32, 1949.
244	6647	Cooper, C.B., Univ. of Maryland, Tech. Rept. No. 11, Pt. I, 1-8, 1955. [AD 69 453]
245	6361	Berg, W.T. and Morrison, J.A., Proc. Roy. Soc. (London), 242A, 467-92, 1957.
246	37909	Gardner, E.T. and Taylor, A.R., Jr., U.S. Bur. Mines, RI-6435, 1-8, 1964.
247	20196	King, E.G., Weller, W.W., Christensen, A.U., and Kelley, K.K., U.S. Bur. Mines, RI-5799, 1-20, 1961.
248	22239	Itskevich, E.S. and Strelkov, P.G., Russ. J. Phys. Chem., 33 (7), 60-2, 1959.
249	16023	Topol, L.E. and Ransom, L.D., USAEC, NAN-SR-5111, 1-3, 1960.
250	28384	Kelley, K.K. and Moore, G.E., J. Am. Chem. Soc., 65, 1264-7, 1943.
251	19933	Kaylor, C.E., Ph.D. Thesis, Univ. Alabama, Univ. Microfilms Publ. 59-983, 1-103, 1959.

Ref. No.	TPRC No.	
252	37911	Smith, D. F., Kaylor, C. E., Walden, G. E., Taylor, A. R., Jr., and Gayle, J. B., U. S. Bur. Mines, BM-RI-5832, 1-20, 1961.
253	29699	Taylor, A. R., Gardner, T. E., and Smith, D. F., U. S. Bur. Mines, RI-6157, 1-7, 1963.
254	16233	Stout, J. W. and Chisholm, R. C., Inst. Study of Metals, Univ. of Chicago Tech. Rept. 7, 1-50, 1961. [AD 262 535]
255	16225	Chisholm, R. C. and Stout, J. W., Inst. Study of Metals, Univ. of Chicago Tech. Rept. 6, 1-30, 1961. [AD 262 537]
256	9968	Robinson, W. K. and Friedberg, S. A., 1-15, 1959. [AD 219 870]
257	647	Friedberg, S. A., Physica, 18, 714-22, 1952.
258	23813	Oetting, F. L. and Gregory, N. W., J. Phys. Chem., 65, 138-40, 1961.
259	27203	Friedberg, S. A., Cohen, A. F., and Schelleng, J. H., J. Phys. Soc. (Japan), 17, Suppl. B-1, 515-17, 1962.
260	26118	Pfeffer, W., Z. Physik, 168 (3), 305-15, 1962.
261	22945	Murray, R. B., Phys. Rev., 128 (4), 1570-4, 1962.
262	330	Friedberg, S. A. and Wasscher, J. D., Physica, 19, 1072-8, 1953.
263	23918	Voorhoeve, W. H. M. and Dokoupil, Z., Physica, 27 (8), 777-82, 1961.
264	535	Busey, R. H. and Giaque, W. F., J. Am. Chem. Soc., 74, 4443-6, 1952.
265	19850	Furukawa, G. T., NBS Rept. 6848, 1-17, 1960. [AD 238 686]
266	15333	Southard, J. C. and Nelson, R. A., J. Am. Chem. Soc., 55 (11), 4865-9, 1933.
267	11907	Keesom, W. H. and Clark, C. W., Physica, 2, 698-706, 1935.
268	1792	Mustajoki, A., Ann. Acad. Sci. Fennicae, 98, Series A, I, 7-45, 1951.
269	469	Keesom, P. H. and Pearlman, N., Phys. Rev., 91, 1354-5, 1953.
270	9746	Strelkov, P. G., Itskevich, E. S., Kostryukov, V. N., and Mirskaya, G. G., Zhur. Fiz. Khim., 28, 645-9, 1954.
271	1190	Webb, F. J. and Wilks, J., Proc. Roy. Soc. (London), 230A, 549-59, 1955.
272	23209	Yagfarov, M. Sh., Russ. J. Inorg. Chem., 6 (11), 1236-8, 1961.
273	31679	Kolesov, V. P., Paukov, I. E., and Skuratov, S. M., Russ. J. Phys. Chem., 36 (4), 400-5, 1962.
274	34377	Bevan, R. B., Jr., Gilbert, R. A., and Busey, R. H., J. Phys. Chem., 70 (1), 147-50, 1966.
275	591	Cerney, C. and Erdoes, E., Chem. Listy, 47, 1742-4, 1953.
276	32685	Smith, D. F., Gardner, T. E., Letson, B. B., and Taylor, A. R., Jr., U. S. Bur. Mines, RI-3616, 1-8, 1963.
277	807	Ginnings, D. C. and Corruccini, R. J., Natl. Bur. Std. U. S. J. Res., 39, 309-16, 1947.
278	1363	Coughlin, J. P. and King, E. G., J. Am. Chem. Soc., 72, 2262-5, 1950.
279	1818	King, E. G., J. Am. Chem. Soc., 79, 2056-7, 1957.
280	10354	Pitzer, K. S., Smith, W. V., and Latimer, W. M., J. Am. Chem. Soc., 60, 1826-8, 1938.
281	34816	Taylor, A. R., Jr. and Gardner, T. E., U. S. Bur. Mines, RI 6664, 1-19, 1965.
282	1893	Naylor, B. F., J. Am. Chem. Soc., 67, 150-2, 1945.
283	955	Todd, S. S., J. Am. Chem. Soc., 71, 4115-6, 1949.
284	16843	King, E. G. and Christensen, A. U., U. S. Bur. Mines, RI-5510, 1-7, 1959.
285	24526	Westrum, E. F., Jr. and Beale, A. F., Jr., J. Phys. Chem., 65, 353-5, 1961.
286	24586	Burney, G. A. and Westrum, E. F., Jr., J. Phys. Chem., 65, 349-52, 1961.
287	903	Catalano, E. and Stout, J. W., J. Chem. Phys., 23, 1803-8, 1955.
288	7078	Douglas, T. B. and Dever, J. L., WADC TR 53-201, Pt. 1, 1-24, 1953. [AD 22 234]
289	1704	Martin, D. L., Phil. Mag., 46, 751-8, 1955.
290	24585	Westrum, E. F., Jr. and Burney, G. A., J. Phys. Chem., 65, 344-8, 1961.
291	28380	Stout, J. W. and Adams, H. E., J. Am. Chem. Soc., 64, 1535-8, 1942.
292	10538	Brady, A. P., Clauss, J. K., and Myers, O. E., WADC TR 56-4, Pt. 1, 1-49, 1957. [AD 131 096] [PB 163 987]
293	1304	Catalano, E. and Stout, J. W., J. Chem. Phys., 23 (7), 1284-9, 1955.
294	30964	Rao, R. V. G., Das, C. D., Keer, H. V., and Biswas, A. B., Proc. Phys. Soc. (GB), 81, Pt. 1 (519), 191-2, 1963.

Ref. No.	TPRC No.	
295	25508	Pace, E. L. and Mosser, J. S., J. Chem. Phys., 39 (1), 154-8, 1963.
296	321	Lohr, H.R., Osborn, D.W., and Westrum, E.F., Jr., J. Am. Chem. Soc., 76, 3837-9, 1954.
297	23812	Euler, R.D. and Westrum, E.F., Jr., J. Phys. Chem., 65, 132-4, 1961.
298	1563	Brickwedde, F.G., Hoge, H.J., and Scott, R.B., J. Chem. Phys., 16 (5), 429-36, 1948.
299	928	Osborne, D.W., Westrum, E.F., Jr., and Lohr, H.R., J. Am. Chem. Soc., 77, 2737-9, 1955.
300	16678	Burns, J.H., Osborne, D.W., and Westrum, E.F., Jr., J. Chem. Phys., 33 (2), 387-94, 1960.
301	5322	Llewellyn, D.R., J. Am. Chem. Soc., 28-36, 1953.
302	870	Stout, J.W. and Catalano, E., J. Chem. Phys., 23 (11), 2013-22, 1955.
303	713	Dworkin, A.S., Sasmore, D.J., and VanArtsdalen, E.R., J. Am. Chem. Soc., 77, 1304-6, 1955.
304	310	Davies, T. and Staveley, L.A.K., Trans. Faraday Soc., 53, 19-30, 1957.
305	17982	King, E.G. and Weller, W.W., U.S. Bur. Mines, RI-5590, 1-5, 1960.
306	17120	King, E.G. and Weller, W.W., U.S. Bur. Mines, RI-5485, 1-5, 1959.
307	18257	Gronvold, F., Westrum, E.F., Jr., and Chou, C., J. Chem. Phys., 30 (2), 528-31, 1959.
308	20920	Eastman, E.D. and Rodebush, W.H., J. Am. Chem. Soc., 40, 489-500, 1918.
309	21830	Anderson, C.T., J. Am. Chem. Soc., 52, 2296-300, 1930.
310	31158	Walker, B.E., Ewing, C.F., and Miller, R.R., J. Phys. Chem., 61, 1682-3, 1957.
311	21815	Millar, R.W., J. Am. Chem. Soc., 51, 207-14, 1929.
312	27909	Itskevich, E.S. and Strelkov, P.G., Russ. J. Phys. Chem., 34 (6), 627-9, 1960.
313	5948	Hu, J.-H. and Johnston, H.L., J. Am. Chem. Soc., 74, 4771-2, 1952.
314	28986	O'Neal, H.R., Univ. California, UCRL-10426, 1-66, 1963.
315	27995	Taylor, A.R., Jr. and Smith, D.F., U.S. Bur. Mines, RI-5967, 1-12, 1962.
316	39227	Martin, D.L. and Snowdon, R.L., Can. J. Phys., 44 (7), 1449-65, 1966.
317	33170	Goodman, R.M. and Westrum, E.F., Jr., J. Chem. Eng. Data, 11 (3), 294-5, 1966.
318	10352	Brown, O.L.I., Smith, W.V., and Latimer, W.M., J. Am. Chem. Soc., 58, 1758-9, 1936.
319	23983	Pfeffer, W., Z. Physik, 164, 295-302, 1961.
320	24314	Pfeffer, W., Z. Physik, 162, 413-20, 1961.
321	21222	Hellwege, K.H., Kuch, F., Niemann, K., and Pfeffer, W., Z. Physik, 162, 358-62, 1961.
322	23911	Shirley, D.A., J. Am. Chem. Soc., 82 (15), 3841-3, 1960.
323	1006	Kelley, K.K. and Moore, G.E., J. Am. Chem. Soc., 65, 2340-2, 1943.
324	25650	Chisholm, R.C. and Stout, J.W., J. Chem. Phys., 36 (4), 972-9, 1962.
325	22472	Popov, M.M., Gal'chenko, G.L., and Senin, M.D., Russ. J. Inorg. Chem., 4 (6), 560-2, 1959.
326	28398	Shomate, C.H., J. Am. Chem. Soc., 69, 220-1, 1947.
327	26883	McBride, J.J., Ph.D. Thesis, Univ. Michigan, Univ. Microfilms Publ. 58-950, 1-131, 1958.
328	1274	Voskresenskaya, N.K., Sokolov, V.A., Banashek, E.I., and Schmidt, N.E., Izv. Sektora Fiz.-Khim, Anal., Inst. Obshchei. Neorg. Khim., Akad. Nauk, SSSR, 27, 233-8, 1956.
329	33921	Osborne, D.W., Schreiner, F., Malm, J.G., Selig, H., and Rochester, L., J. Chem. Phys., 44 (7), 2802-9, 1966.
330	39468	Hornung, E.W., Brodale, G.E., Fisher, R.A., and Giauque, W.F., J. Chem. Phys., 45 (2), 614-24, 1966.
331	40814	Hornung, E.W., Fisher, R.A., Brodale, G.E., and Giauque, W.F., J. Chem. Phys., 46 (1), 67-72, 1967.
332	18220	Myers, O.E., WADC TR 56-4 (Pt. II), 1-47, 1957. [AD 131 097]
333	5095	Westrum, E.F., Jr. and Pitzer, K.S., J. Am. Chem. Soc., 71, 1940-9, 1949.
334	32418	Johnston, W.V., Pilipovich, D., and Sheenan, D.E., Chicago Univ. Press, 139-43, 1963.
335	27248	McDonald, R.A., Sinke, G.C., and Stull, D.R., J. Chem. Eng. Data, 7 (1), 83, 1962.
336	36294	Mori, T., Tamura, H., and Sawaguchi, E., J. Phys. Soc. Japan, 20 (2), 281, 1965.
337	20296	Claytor, R.N. and Marshall, B.J., Phys. Rev., 120, 332-4, 1960.
338	27992	King, E.G. and Weller, W.W., U.S. Bur. Mines, RI-6040, 1-5, 1962.
339	37754	Weller, W.W. and Kelley, K.K., U.S. Bur. Mines, RI-6511, 1-11, 1964.

Ref. No.	TPRC No.	
340	21847	Anderson, C. T., J. Am. Chem. Soc., 53, 476-83, 1931.
341	21851	Anderson, C. T., J. Am. Chem. Soc., 54, 107-11, 1932.
342	39410	Huffman, D. R. and Wild, R. L., Phys. Rev., 148 (2), 526-7, 1966.
343	28000	King, E. G. and Weller, W. W., U. S. Bur. Mines, RI-6001, 1-4, 1962.
344	25075	Corning Materials Handbook, Corning Glass Works Tech. Products Div., 1961.
345	307	Smith, P. L. and Wolcott, N. M., Phil. Mag., 1 (8), 854-65, 1956.
346	21724	DeVries, T., Ind. Eng. Chem., 22, 617-8, 1930.
347	901	Skogan, H. S., Ph. D. Thesis, Rutgers Univ., Univ. Microfilms Publ. 14058, 1-147, 1955.
348	24384	Rudkin, R. L., ASD-TDR-62-24 (Pt. 2), 1-16, 1963. [AD 413 005]
349	17360	Sochava, I. V. and Trapeznikova, O. N., Soviet Phys. Doklady, 2, 164-6, 1957.
350	15793	Lockheed Aircraft Corp., Missiles and Space Div., LMSD-288140, 1-28, 1960. [AD 241 410]
351	19708	Booss, H. J., Trans. Met. Soc. AIME, 215, 395-7, 1959.
352	33351	Arias, A., Lewis Research Center (Cleveland, Ohio), NASA-TN-D-2464, 1-72, 1964.
353	408	Todd, S. S., J. Am. Chem. Soc., 72, 4742-3, 1950.
354	32979	Pankratz, L. B. and Kelley, K. K., U. S. Bur. Mines, RI-6370, 1-7, 1964.
355	31605	Pankratz, L. B., Weller, W. W., and Kelley, K. K., U. S. Bur. Mines, RI-6287, 1-7, 1963.
356	26326	King, E. G. and Weller, W. W., U. S. Bur. Mines, RI-5810, 1-6, 1961.
357	30254	King, E. G., Todd, S. S., and Kelley, K. K., U. S. Bur. Mines, RI-4394, 1-15, 1948.
358	362	King, E. G., J. Am. Chem. Soc., 77, 2150-2, 1955.
359	28624	Kelley, K. K., Shomate, C. H., Young, F. E., Naylor, B. F., Salo, A. E., and Huffman, E. H., U. S. Bur. Mines, Tech. Papers 688, 1-104, 1946.
360	36238	Weller, W. W. and Kelley, K. K., U. S. Bur. Mines, RI-6556, 1-8, 1964.
361	26614	Coughlin, J. P. and Orr, R. L., J. Am. Chem. Soc., 75 (3), 530-1, 1953.
362	1793	Todd, S. S. and Lorenson, R. E., J. Am. Chem. Soc., 74, 2043-5, 1952.
363	2104	Todd, S. S. and Lorenson, R. E., J. Am. Chem. Soc., 74, 3764-5, 1952.
364	17198	King, E. G. and Weller, W. W., U. S. Bur. Mines, RI-5571, 1-3, 1960.
365	672	King, E. G., J. Phys. Chem., 59, 218-9, 1955.
366	2132	King, E. G., Torgeson, D. R., and Cook, O. A., J. Am. Chem. Soc., 70, 2160-3, 1948.
367	1040	Bonnickson, K. R., J. Am. Chem. Soc., 76, 1480-2, 1954.
368	29697	Weller, W. W. and King, E. G., U. S. Bur. Mines, RI-6147, 1-6, 1963.
369	1297	Todd, S. S., J. Am. Chem. Soc., 73, 3277-8, 1951.
370	8941	King, E. G., J. Am. Chem. Soc., 79, 5437-8, 1957.
371	29553	Coughlin, J. P. and O'Brien, C. J., J. Phys. Chem., 61, 767-9, 1957.
372	21862	Parks, G. S. and Kelley, K. K., J. Phys. Chem., 30, 1175-8, 1926.
373	10409	Cristescu, S. and Simon, F., Z. Physik Chem., B25, 273-82, 1934.
374	5624	Naylor, B. F. and Cook, O. A., J. Am. Chem. Soc., 68, 1003-5, 1946.
375	23762	King, E. G. and Weller, W. W., U. S. Bur. Mines, RI-5791, 1-6, 1961.
376	28689	Leonidov, V. Ya., Rezukhina, T. N., and Bereznikova, I. A., Russ. J. Phys. Chem., 34 (8), 885-6, 1960.
377	34594	Sandenaw, T. A. and Storms, E. K., U. S. At. Energy Comm., LA-331, UC-34, 1-10, 1965.
378	29660	King, E. G. and Weller, W. W., U. S. Bur. Mines, RI-5954, 1-6, 1961.
379	32686	Weller, W. W. and Kelley, K. K., U. S. Bur. Mines, RI-6343, 1-7, 1963.
380	26327	King, E. G. and Weller, W. W., U. S. Bur. Mines, RI-5855, 1-8, 1961.
381	30497	Robie, R. A. and Stout, J. W., J. Phys. Chem., 67 (11), 2252-6, 1963.
382	1579	King, E. G., J. Phys. Chem., 60, 410-2, 1956.
383	20869	Atkins, K. R. and Pollack, S. R., Univ. Pennsylvania, 1-140, 1961. [AD 256 661]
384	27471	Yakovleva, R. A. and Rezukhina, T. N., Russ. J. Phys. Chem., 34 (4), 390-2, 1960.
385	16966	King, E. G. and Kelley, K. K., U. S. Bur. Mines, RI-5502, 1-6, 1959.
386	1868	Naylor, B. F., Ind. Eng. Chem., 36, 933-41, 1944.

Ref. No.	TPRC No.	
387	3791	Kelley, K.K., J. Am. Chem. Soc., 63, 2750-2, 1941.
388	26392	Orr, R.L., J. Am. Chem. Soc., 75 (3), 528-9, 1953.
389	32936	Weller, W.W. and Kelley, K.K., U.S. Bur. Mines, RI-6357, 1-5, 1964.
390	1191	King, E.G., J. Am. Chem. Soc., 77, 3189-90, 1955.
391	31672	Rezukhina, T.N., Levitskii, V.A., and Kazimirova, N.M., Russ. J. Phys. Chem., 35 (11), 1305-6, 1961.
392	1387	Kelley, K.K., J. Am. Chem. Soc., 65, 339-41, 1943.
393	26389	Orr, R.L. and Coughlin, J.P., J. Am. Chem. Soc., 74 (12) 3186-7, 1952.
394	29696	Weller, W.W. and King, E.G., U.S. Bur. Mines, RI-6130, 1-5, 1962.
395	10587	Grimes, D.M., Westrum, E.F., Jr., and Legvold, S., 1-37, 1957. [AD 135 955] [PB 140 340]
396	27914	Weller, W.W. and King, E.G., U.S. Bur. Mines, RI-6281, 1-4, 1963.
397	32980	Pankratz, L.B., U.S. Bur. Mines, RI-6371, 1-6, 1964.
398	17122	Kelley, K.K., Barany, R., King, E.G., and Chisenson, A.U., U.S. Bur. Mines, RI-5436, 1-16, 1959.
399	16773	Westrum, E.F., Jr., U.S. At. Energy Comm. Rept. No. AECU-3925, 1-27, 1958.
400	33605	Margrave, J.L. and Grimley, R.T., J. Phys. Chem., 62, 1436-8, 1958.
401	29769	Weller, W.W. and Kelley, K.K., U.S. Bur. Mines, RI-6191, 1-5, 1963.
402	322	Tarasov, V.V. and Savitskaya, Ya.S., Doklady Akad. Nauk. SSSR, 88, 1019-22, 1953.
403	1909	Naylor, B.F., J. Am. Chem. Soc., 67, 466-7, 1945.
404	2679	Naylor, B.F., J. Am. Chem. Soc., 67, 2120-2, 1945.
405	24291	King, E.G. and Weller, W.W., U.S. Bur. Mines, RI-5715, 1-5, 1961.
406	7678	Westrum, E.F., Jr. and Grimes, D.M., J. Phys. Chem. Solids, 6 (2), 280-6, 1958.
407	33793	Gerstein, B.C., Chung, P.L., and Danielson, G.C., J. Phys. Chem. Solids, 27 (6/7), 1161-5, 1966.
408	28643	Clusius, K. and Faber, G., Z. Physik Chem., 51B, 352-70, 1942.
409	23858	Saba, W.G., Ph.D. Thesis, Univ. Pittsburgh, Univ. Microfilms Publ. 61-1885, 1-99, 1961.
410	27006	Stalinski, B. and Bieganski, Z., Bull. Akad. Polon. Sci., Ser. Sci. Chim., 10, 247-51, 1962.
411	25597	Flotow, H.E., Osborne, D.W., and Otto, K., J. Chem. Phys., 36, 866-72, 1962.
412	7000	Douglas, T.B. and Victor, A.C., WADC TR57-374, Pt. II, 1-75, 1957. [AD 150 128]
413	20428	Flotow, H.E. and Osborne, D.W., J. Chem. Phys., 34 (4), 1418-25, 1961.
414	21405	Magnus, A. and Danz, H., Ann. Physik, 87, 407-27, 1926.
415	39469	Westrum, E.F., Jr. and Barber, C.M., J. Chem. Phys., 45 (2), 635-9, 1966.
416	25118	Speidel, E.O. and Keller, D.L., BMI-1633, EURAEC-706, 1-68, 1963.
417	39379	Counsell, J.F., Dell, R.M., and Martin, J.F., Trans. Faraday Soc., 62 (7), 1736-47, 1966.
418	29729	Kalishevich, G.I., Gel'd, P.V., and Krentsis, R.P., Russ. J. Phys. Chem., 39 (12), 1602-3, 1965.
419	32213	Krentsis, R.P., Gel'd, P.V., and Kalishevich, G.I., Izv., Vysshikh Uchebn., Zavedenii Chernaya Met., 6 (9), 161-8, 1963.
420	29920	Krentsis, R.P. and Gel'd, P.V., Izv. Vuz-Chernaya Met., 11, 12-41, 1962.
421	39630	Letun, S.M., Gel'd, P.V., and Serebrennikov, N.N., Russ. Metallurgy (Metally), 6, 97-103, 1965.
422	9116	King, E.G. and Christensen, A.U., J. Phys. Chem., 62 (4), 499-500, 1958.
423	31318	Nadzhafov, Yu.B. and Sharifov, K.A., Tr. Inst. Fiz. Akad. Nauk. Azerb. SSR, 11, 31-5, 1963.
424	38934	Raub, C.J., Compton, V.B., Geballe, T.H., Matthias, B.T., Maita, J.P., and Hull, G.W., Jr., J. Phys. Chem. Solids, 26 (12), 2051-7, 1965.
425	329	Dworkin, A.S., Sasmor, D.J., and Van Artsdalen, E.R., J. Chem. Phys., 22, 837-42, 1954.
426	20619	Lyubimov, A.P. and Belashchenko, D.K., Sbor. Moskovskogo Inst. Stali., 33, 3-11, 1955.
427	10348	Anderson, C.T., J. Am. Chem. Soc., 56, 340-2, 1934.
428	23287	Dhar, P.R., Gupta, I.N., and Mahapatra, U.P., Res. and Ind. (New Delhi), 6, 6-7, 1961.
429	30496	Stout, J.W. and Robie, R.A., J. Phys. Chem., 67 (11), 2248-52, 1963.
430	31267	Janz, G.J., Neuenschwander, E., and Kelley, F.J., Trans. Faraday Soc., 59, 841-5, 1963.
431	12462	Anderson, C.T., J. Am. Chem. Soc., 55, 3621-3, 1933.

Ref. No.	TPRC No.	
432	10447	Latimer, W.M. and Ahlberg, J.E., Z. Physik Chem., A148, 464-70, 1930.
433	26119	Hellwege, K.H., Pfeffer, W., and Thiel, H.J., Z. Physik, 168, 474-7, 1962.
434	10353	Brown, O.L.I. Smith, W.V., and Latimer, W.M., J. Am. Chem. Soc., 59 921-4 1937.
435	21893	Thomas, S.B. and Parks, G.S., J. Phys. Chem., 35, 2091-102, 1931.
436	21850	Latimer, W.M. and Ahlberg, J.E., J. Am. Chem. Soc., 54, 1900-4, 1932.
437	10359	Latimer, W.M., Hicks, J.F.G., Jr., and Shutz, P.W., J. Chem. Phys., 1, 620-4, 1933.
438	31601	Taylor, A.R., Jr., Gardner, T.E., and Smith, D.F., U.S. Bur. Mines, RI-6240, 1-8, 1963.
439	28623	Kelley, K.K., Southard, J.C., and Anderson, C.T., U.S. Bur. Mines, Tech. Paper No. 625, 1-73, 1941.
440	21809	Latimer W.M. and Greensfelder, B.S., J. Am. Chem. Soc., 50, 2202-13, 1928.
441	31012	Fritz, J.J. and Giauque, W.F., J. Am. Chem. Soc., 71, 2168-76, 1949.
442	28379	Long, E.A. and Degraff, R.A., J. Am. Chem. Soc., 64, 1346-9, 1942.
443	20046	Lyon, D.N. and Giauque, W.F., J. Am. Chem. Soc., 71, 1647-57, 1959.
444	27660	Brackett, T.E., Hornung, E.W., and Hopkins, T.E., J. Am. Chem. Soc., 82, 4155-7, 1960.
445	27452	Papadopoulos, M.N. and Giauque, W.F., J. Phys. Chem., 66, 2049-50, 1962.
446	33432	Stout, J.W. and Hadley, W.B., Univ. of Chicago, 1-35, 1963. [AD 420 803]
447	11848	Pitzer, K.S. and Coulter, L.V., J. Am. Chem. Soc., 60, 1310-3, 1938.
448	30698	Pollitzer, F., Z. Elektrochem., 17, 5-14, 1911.
449	33489	Bleaney, B. (Editor), AFCRL-63-192, 1-205, 1963. [AD 411 557]
450	11843	Eastman, E.D. and McGavock, W.C., J. Am. Chem. Soc., 59, 145-51, 1937.
451	323	Braune, H. and Möller, O., Z. Naturforsch, 9A, 210-7, 1954.
452	8410	West, E.D., J. Am. Chem. Soc., 81, 29-37, 1959.
453	27515	Gunther, P., Ann. Physik, 51, 828-46, 1916.
454	13522	Carpenter, L.G. and Harle, T.F., Phil. Mag., 23, 193-208, 1937.
455	11849	Frederick, K.J. and Hildebrand, J.H., J. Am. Chem. Soc., 60, 1437-9, 1938.
456	3356	Young, F.E. and Hildebrand, J.H., J. Am. Chem. Soc., 64, 839-40, 1942.
457	38605	Paukov, I.E., Strelkov, P.G., Nogreva, V.V., and Belyi, V.I., Soviet Phys. Doklady, 10 (5), 460-2, 1965.
458	29984	Westrum, E.F., Jr. and Feick, G. J. Chem. Eng. Data, 8, 176-7, 1963.

Material Index

MATERIAL INDEX TO SPECIFIC HEAT COMPANION VOLUMES 4, 5, AND 6